Full-Color Art in Commen...

20-A small-appliance branch circuits

15-A general-purpose branch circuit

15-A individual branch circuit

15-A general-purpose branch circuit

20-A small-appliance branch circuits

G = GFCI protection required

Spare raceways or pull lines from electrical service panel and communications center to accessible attic space

Junction boxes

Raceway or pull lines

Unfinished utility area

Finished basement area

Service panel

Communications center

Tunnel Lighting

18 ft

Parking Lot Lighting

22ft

Clearance from building openings (600 V or less)

3 ft

3 ft

Clearance from ground (600 V or less)

National Electrical Code®
Handbook

Tenth Edition
International Electrical Code® Series

Mark W. Earley, P.E.
Editor-in-Chief

Jeffrey S. Sargent
Senior Editor

Joseph V. Sheehan, P.E.
Editor

John M. Caloggero
Editor

With the complete text of the 2005 edition of the *National Electrical Code*®

NFPA
National Fire Protection Association, Quincy, Massachusetts

Product Manager: Charles Durang
Editorial Manager: Sylvia Dovner
Project Editor: Joyce Grandy
Copy Editor: Nancy Wirtes
Editorial Assistant: Carol A. Henderson
Text Processing: Lynn Lupo, Maureen White
Composition: Modern Graphics
Art Coordinator: Cheryl Langway
Illustrations: Rollin Graphics, George Nichols, J. Philip Simmons
Interior Design: Cheryl Langway and The Davis Group
Chapter Opening Photo: © Christopher Lucas/IPNSTOCK
Cover Design: Cameron, Inc.
Manufacturing Manager: Ellen Glisker
Printer: R.R. Donnelley/Willard

Copyright © 2005
National Fire Protection Association, Inc.
One Batterymarch Park
Quincy, Massachusetts 02169-7471

Notice Concerning Liability: Publication of this handbook is for the purpose of circulating information and opinion among those concerned for fire and electrical safety and related subjects. While every effort has been made to achieve a work of high quality, neither the NFPA nor the contributors to this handbook guarantee the accuracy or completeness of or assume any liability in connection with the information and opinions contained in this handbook. The NFPA and the contributors shall in no event be liable for any personal injury, property, or other damages of any nature whatsoever, whether special, indirect, consequential, or compensatory, directly or indirectly resulting from the publication, use of, or reliance upon this handbook.

This handbook is published with the understanding that the NFPA and the contributors to this handbook are supplying information and opinion but are not attempting to render engineering or other professional services. If such services are required, the assistance of an appropriate professional should be sought.

Notice Concerning Code Interpretations: This tenth edition of the *National Electrical Code® Handbook* is based on the 2005 edition of NFPA 70, *National Electrical Code*. All NFPA codes, standards, recommended practices, and guides, of which the document contained herein is one, are developed through a consensus standards development process approved by the American National Standards Institute. This process brings together volunteers representing varied viewpoints and interests to achieve consensus on fire and other safety issues. The handbook contains the complete text of NFPA 70 and any applicable Formal Interpretations issued by the Association. These documents are accompanied by explanatory commentary and other supplementary materials.

The commentary and supplementary materials in this handbook are not a part of the *Code* and do not constitute Formal Interpretations of the NFPA (which can be obtained only through requests processed by the responsible technical committees in accordance with the published procedures of the NFPA). The commentary and supplementary materials, therefore, solely reflect the personal opinions of the editor or other contributors and do not necessarily represent the official position of the NFPA or its technical committees.

®Registered Trademark National Fire Protection Association, Inc.

NFPA No.: 70HB05
ISBN (book): 0-87765-625-8
ISBN (CD): 0-87765-627-4
Library of Congress Control No.: 2004116336
Printed in the United States of America
05   06   07   08   09   5   4   3   2

Dedication

The 2005 *NEC Handbook* is dedicated to Philip H. Cox, who recently retired as Executive Director of the International Association of Electrical Inspectors (IAEI). Phil has had a long and distinguished career in the electrical industry. He has been a member of the *NEC* code-making panels in three different capacities. He represented IAEI as a member, NEMA as a field representative, and IAEI as CEO and Executive Director. He served on CMP 6 and as chairman of CMP 1. He also served on the *NEC* Technical Correlating Committee, where his activities included a number of task groups. The membership group, which he chaired, greatly streamlined the membership process into a model of efficiency.

Phil is a tireless advocate for electrical safety, which he believes depends on *Code* education. Toward this end, Phil has continued to improve and expand the IAEI product line through products such as the *IAEI Analysis of Changes to the National Electrical Code.* He is a warm person who cares deeply about his work and the people he works with. Phil is a graduate of the University of Arkansas, where he earned bachelor's and master's degrees.

Contents

Preface

This handbook contains the 50th edition of the *National Electrical Code*. Nearly 109 years have passed since those cold days of March 18–19, 1896, when a group of 23 persons representing a wide variety of organizations met at the headquarters of the American Society of Mechanical Engineers in New York City. Their purpose was to develop a national code of rules for electrical construction and operation. (It is interesting to note that this meeting took place a mere 17 years after the invention of the incandescent light bulb.) This attempt was not the first to establish consistent rules for electrical installations, but it was the first national effort. The number of electrical fires was increasing, and the need for standardization was becoming urgent. By 1881, one insurer had reported electrical fires in 65 textile mills in New England.

The major problem was the lack of an authoritative, nationwide electrical installation standard. As one of the early participants noted, "We were without standards and inspectors, while manufacturers were without experience and knowledge of real installation needs. The workmen frequently created the standards as they worked, and rarely did two men think and work alike."

By 1895, five electrical installation codes had come into use in the United States, causing considerable controversy and confusion. The manufacture of products that met the requirements of all five codes was difficult, so something had to be done to develop a single, national code. The committee that met in 1896 recognized that the five existing codes should be used collectively as the basis for the new code. In the first known instance of international harmonization, the group also referred to the German code, the code of the British Board of Trade, and the Phoenix Rules of England. The importance of industry consensus was immediately recognized; before the committee met again in 1897, the new code was reviewed by 1200 individuals in the United States and Europe. Shortly thereafter, the first standardized U.S. electrical code, the *National Electrical Code®*, was published.

The *National Electrical Code* has become the most widely adopted code in the United States. It is the installation code used in all 50 states and all U.S. territories. Moreover, it has grown well beyond the borders of the United States and is now used in numerous other countries. Because the *Code* is a living document, constantly changing to reflect changes in technology, its use continues to grow.

Some things have not changed. The *National Electrical Code* continues to offer an open-consensus process. Anyone can submit a proposal for change or a public comment, and all proposals and comments are subject to a rigorous public review process. The *NEC* still provides the best technical information, ensuring the practical safeguarding of persons and property from the hazards arising from the use of electricity.

Throughout its history, the National Electrical Code Committee has been guided by giants in the electrical industry. The names are too numerous to mention. Certainly the first chairman, William J. Hammer, should be applauded for providing the leadership necessary to get the *Code* started. More recently, the *Code* has been chaired by outstanding leaders such as Richard L. Loyd, Richard W. Osborne, Richard G. Biermann, D. Harold Ware, and James W. Carpenter. Each of these men has devoted many years to the National Electrical Code Committee.

The editors wish to note the passing of some long-term committee members who made numerous contributions to the *National Electrical Code*: Anthony Montourri, CMP 9; Leland J. Hall, former chair of CMP 14; and James N. Pearse, CMP 20 and CMP 17.

The editors have conferred closely with members of the National Electrical Code Committee in developing the revisions incorporated into the 2005 edition of the *Code*. The assistance and cooperation of code-making panel chairs and various committee members are herein gratefully acknowledged.

This edition of the *NEC Handbook* would not have been possible without the invaluable technical assistance of Kenneth G. Mastrullo, Senior Electrical Specialist; Lee F. Richardson, Senior Electrical Engineer; Richard J. Roux, Senior Electrical Specialist; and Donald W. Shields, Senior Electrical Specialist. Their contributions are greatly appreciated.

The editors acknowledge with thanks the manufacturers and their representatives who generously supplied photographs, drawings, and data upon request. Special thanks also to the editors of and contributors to past editions. Their work provided an excellent foundation on which to build.

The editors express special thanks to Joyce Grandy for her long hours and extraordinary effort in attending to all of the editorial details that we technical types often overlook. Special thanks are also due to Sylvia Dovner, an outstanding manager who kept this project on track. Without the efforts of Joyce and Sylvia, this new and improved edition of the *NEC Handbook* would not have been possible.

We also wish to thank the electrical support staff: Carol Henderson, Mary Warren-Pilson, and Kathleen Stevens, along with their leader, Jean O'Connor, for their support on this project.

The editors express their sincere appreciation to Richard Berman, Philip H. Cox, Allan Manche, Brian Phelan, David Kendall, Lori Tennant, Ray C. Mullin, James Pauley, Vincent Saporita, Peter J. Schram, and John C. Wiles for special help on specific articles. We also wish to thank Mr. Schram for his work on developing the summaries of *Code* changes. Finally, we also thank the following for contributing photos and graphics for this edition:

3M Co., Electrical Markets Division
AFC Cable Systems, Inc.
Agfa Corporation
Ajax Tocco Magnethermic, Park Ohio Industries
Alcoa Inc.
Allied Tube & Conduit, a Tyco International Co.
American Society of Mechanical Engineers
Appleton Electric Co., EGS Electrical Group
Bose Corp.
Bussmann Division, Cooper Industries
Cable Tray Institute
Carlon®, Lamson & Sessions
Caterpillar Inc.
Colortran, Inc.
Cooper Crouse-Hinds
Daniel Woodhead Co.
Dranetz-BMI
Dual-Lite, Inc.
Electronic Theatre Controls, Inc.
Fire Control Instruments
Fluke Corp.
Ford Motor Co.
General Electric Co.

H. H. Robertson Floor Systems
Hubbell, Inc.
Hubbell Inc., Kellems Division
Hubbell RACO
International Association of Electrical Inspectors/ Michael Johnston, Director of Education
Kieffer and Company, Inc./ Stephen Kieffer, Chairman & CEO
Kliegl Bros.
L. E. Mason Co.
Lithonia Lighting, Reloc Wiring Systems
MPHusky Corp.
NAPCO Security Systems, Inc.
National Electrical Manufacturers Association
O-Z./Gedney, a division of EGS Electrical Group
Pass & Seymour/Legrand®
Production Arts Lighting, Inc. and Production Resource Group, L.L.C.
Pyrotenax Cables, Ltd.
Radyne Corp.

Reading Municipal Light Department
RKL Lighting Company
Rockbestos-Suprenant Cable Corp.
S&C Electric Co.
Schneider Electric SA
Smart House
Solar Design Associates, Inc.
Southwest Technology Development Institute/ John Wiles, Program Manager
Square D Co.
State of New Hampshire Electricians Board/Chief Inspector Mark Hilbert
Strayfield Ltd.
Texas Instruments
Thermatool Corp.
Thomas & Betts Corp.
Tyco Electronics Corp.
Underwriters Laboratories Inc.
Uniloy Milacron USA Inc.
Union Connector Co., Inc.
Walker Systems, a Wiremold Co.
The Wiremold Co.

Article 90

Introduction

Contents

90.1 Purpose

(A) Practical Safeguarding The purpose of this *Code* is the practical safeguarding of persons and property from hazards arising from the use of electricity.

The *National Electrical Code (NEC)* is prepared by the National Electrical Code Committee, which consists of a Technical Correlating Committee and 19 code-making panels. The code-making panels have specific subject responsibility within the *Code*. The scope of the National Electrical Code Committee is as follows:

> This committee shall have primary responsibility for documents on minimizing the risk of electricity as a source of electric shock and as a potential ignition source of fires and explosions. It shall also be responsible for text to minimize the propagation of fire and explosions due to electrical installations.

In addition to its overall responsibility for the *National Electrical Code*, the Technical Correlating Committee is responsible for NFPA 70A, *National Electrical Code Requirements for One- and Two-Family Dwellings,* and for correlation of the following:

1. NFPA 70B, *Recommended Practice for Electrical Equipment Maintenance*
2. NFPA 70E, *Standard for Electrical Safety in the Workplace*
3. NFPA 73, *Electrical Inspection Code for Existing Dwellings*
4. NFPA 79, *Electrical Standard for Industrial Machinery*
5. NFPA 110, *Standard for Emergency and Standby Power Systems*
6. NFPA 111, *Standard on Stored Electrical Energy Emergency and Standby Power Systems*

1

(B) Adequacy This *Code* contains provisions that are considered necessary for safety. Compliance therewith and proper maintenance results in an installation that is essentially free from hazard but not necessarily efficient, convenient, or adequate for good service or future expansion of electrical use.

> FPN: Hazards often occur because of overloading of wiring systems by methods or usage not in conformity with this *Code*. This occurs because initial wiring did not provide for increases in the use of electricity. An initial adequate installation and reasonable provisions for system changes provide for future increases in the use of electricity.

Consideration should always be given to future expansion of the electrical system. Future expansion might be unlikely in some occupancies, but for others it is wise to plan an initial installation comprised of service-entrance conductors and equipment, feeder conductors, and panelboards that allows for future additions, alterations, designs, and so on.

(C) Intention This *Code* is not intended as a design specification or an instruction manual for untrained persons.

The *NEC* is intended for use by capable engineers and electrical contractors in the design and/or installation of electrical equipment; by inspection authorities exercising legal jurisdiction over electrical installations; by property insurance inspectors; by qualified industrial, commercial, and residential electricians; and by instructors of electrical apprentices or students.

(D) Relation to Other International Standards The requirements in this *Code* address the fundamental principles of protection for safety contained in Section 131 of International Electrotechnical Commission Standard 60364-1, *Electrical Installations of Buildings*.

> FPN: IEC 60364-1, Section 131, contains fundamental principles of protection for safety that encompass protection against electric shock, protection against thermal effects, protection against overcurrent, protection against fault currents, and protection against overvoltage. All of these potential hazards are addressed by the requirements in this *Code*.

In addition to being the most widely adopted code for the built environment in the United States, the *NEC* is also adopted and used extensively in many foreign countries. Section 90.1(D) makes it clear that the *NEC* is compatible with international safety principles. Added as a Tentative Interim Amendment (TIA) to the 1999 *Code*, this section calls attention to the fact that installations meeting the requirements of the *NEC* are also in compliance with the fundamental principles outlined in IEC 60364-1, *Electrical Installations of Buildings*, Section 131. That TIA allowed countries that do not have formalized rules for electrical installations to adopt the *NEC* and by so doing to be fully compatible with the safety principles of IEC 60364-1, Section 131. The addition of 90.1(D) will promote acceptance and adoption of the *NEC* internationally.

The *NEC* is an essential part of the safety system of the Americas, and its future will be enhanced by increased international acceptance.

90.2 Scope

(A) Covered This *Code* covers the installation of electrical conductors, equipment, and raceways; signaling and communications conductors, equipment, and raceways; and optical fiber cables and raceways for the following:

(1) Public and private premises, including buildings, structures, mobile homes, recreational vehicles, and floating buildings
(2) Yards, lots, parking lots, carnivals, and industrial substations

> FPN to (2): For additional information concerning such installations in an industrial or multibuilding complex, see ANSI C2-2002, *National Electrical Safety Code*.

Requirements for locations such as these are found throughout the *Code*. Specific items such as outside feeders and branch circuits can be found in Article 225, grounding in Article 250, surge arresters in Article 280, switches in Article 404, outside lighting in Article 410, transformers in Article 450, and carnivals in Article 525.

(3) Installations of conductors and equipment that connect to the supply of electricity

Often, but not always, the source of supply of electricity is the serving electric utility. The point of connection from a premises wiring system to a serving electric utility system is, by definition, referred to as the *service point*. The conductors on the premises side of the service point are, by definition, referred to as *service conductors*. (These definitions are found in Article 100.) The requirements for service conductors as well as for service-related equipment are found in Article 230. Article 230 applies only where the source of supply of electricity is from a utility.

Where the source of supply of electricity is not the serving electric utility, the source may be a generator, a battery system, a solar photovoltaic system, a fuel cell, or a combination of those sources. Requirements for such sources of supply are found in Article 445 and Articles 700 through 702 for generators, Article 480 for storage batteries, Article 690 for solar photovoltaic systems, and Article 692 for fuel cells. The associated delivery wiring requirements are found in Chapters 2 and 3 (except Article 230) and in Articles 700 through 702 for emergency, legally required, and optional standby power system circuits.

(4) Installations used by the electric utility, such as office buildings, warehouses, garages, machine shops, and recreational buildings, that are not an integral part of a generating plant, substation, or control center.

Section 90.2(A), which was rewritten for the 2002 *Code*, provides order and clarity concerning the portions of electric utility facilities covered by the *NEC*. [See 90.2(B) and the related commentary for information on facilities and specific lighting not covered by the *NEC*.] Exhibit 90.1 illustrates the distinction between electric utility facilities to which the *NEC* does and does not apply.

Industrial and multibuilding complexes and campus-style wiring often include substations and other installations that employ construction and wiring similar to those of electric utility installations. Although such nonutility installations are within the scope of the *NEC*, the *NEC* requirements may not always be all-inclusive, for example, in clearances of conductors or in clearances from buildings or structures for nominal voltages over 600 volts. In such cases, the user can find additional information in the *National Electrical Safety Code (NESC)*, published by the Institute of Electrical and Electronics Engineers, Inc., P.O. Box 1331, 445 Hoes Lane, Piscataway, NJ 08855-1331.

(B) Not Covered This *Code* does not cover the following:

(1) Installations in ships, watercraft other than floating buildings, railway rolling stock, aircraft, or automotive vehicles other than mobile homes and recreational vehicles

> FPN: Although the scope of this *Code* indicates that the *Code* does not cover installations in ships, portions of this *Code* are incorporated by reference into Title 46, *Code of Federal Regulations*, Parts 110–113.

The *NEC* does not specifically cover shipboard wiring. Title 46 of the *Code of Federal Regulations*, Parts 110–113, however, does contain many specific *NEC*-referenced requirements. These requirements, which originated in the *NEC*, are enforced by the U.S. Coast Guard. Installation require-

Exhibit 90.1 Typical electric utility complexes showing examples of facilities covered and not covered by the provisions of the *NEC*.

ments for floating buildings are covered in the *NEC* and are found in Article 553.

(2) Installations underground in mines and self-propelled mobile surface mining machinery and its attendant electrical trailing cable
(3) Installations of railways for generation, transformation, transmission, or distribution of power used exclusively for operation of rolling stock or installations used exclusively for signaling and communications purposes
(4) Installations of communications equipment under the exclusive control of communications utilities located outdoors or in building spaces used exclusively for such installations
(5) Installations under the exclusive control of an electric utility where such installations

a. Consist of service drops or service laterals, and associated metering, or

b. Are located in legally established easements, rights-of-way, or by other agreements either designated by or recognized by public service commissions, utility commissions, or other regulatory agencies having jurisdiction for such installations, or

c. Are on property owned or leased by the electric utility for the purpose of communications, metering, generation, control, transformation, transmission, or distribution of electric energy.

FPN to (4) and (5): Examples of utilities may include those entities that are typically designated or recognized by governmental law or regulation by public service/utility commissions and that install, operate, and maintain electric supply (such as generation, transmission, or distribution systems) or communication systems (such as telephone, CATV, Internet, satellite, or data services). Utilities may be subject to compliance with codes and standards covering their regulated activities as adopted under governmental law or regulation. Additional information can be found through consultation with the appropriate governmental bodies, such as state regulatory commissions, Federal Energy Regulatory Commission, and Federal Communications Commission.

An FPN was added to the 2005 *Code* to clarify the use of the word *utility* as used in 90.2(B)(4) and 90.2(B)(5). This explanatory information now provides the authority having jurisdiction a basis for judgment concerning this issue.

It is not the intent of 90.2(B)(5) to exclude the *NEC* as an installation regulatory document. After all, the *NEC* is fully capable of being utilized for electrical installations in most cases, and 90.2(B)(5) does not pertain to areas where portions of the *NEC* could not be used. Rather, 90.2(B)(5) lists specific areas where the nature of the installation requires specialized rules or where other installation rules, standards, and guidelines have been developed for specific uses and industries. For example, the electrical utility industry uses the *NESC* as its primary requirement in the generation, transmission, distribution, and metering of electrical energy. See Exhibit 90.1 for examples of electric utility facilities that may or may not be covered by the *Code*.

(C) Special Permission The authority having jurisdiction for enforcing this *Code* may grant exception for the installation of conductors and equipment that are not under the exclusive control of the electric utilities and are used to connect the electric utility supply system to the service-entrance conductors of the premises served, provided such installations are outside a building or terminate immediately inside a building wall.

90.3 Code Arrangement

This *Code* is divided into the introduction and nine chapters, as shown in Figure 90.3. Chapters 1, 2, 3, and 4 apply

generally; Chapters 5, 6, and 7 apply to special occupancies, special equipment, or other special conditions. These latter chapters supplement or modify the general rules. Chapters 1 through 4 apply except as amended by Chapters 5, 6, and 7 for the particular conditions.

Chapter 8 covers communications systems and is not subject to the requirements of Chapters 1 through 7 except where the requirements are specifically referenced in Chapter 8.

Chapter 9 consists of tables.

Annexes are not part of the requirements of this *Code* but are included for informational purposes only.

The reference to "the introduction" is intended to include Article 90 in the application of the *Code*. Chapters 1 through 4 apply generally, except as amended or specifically referenced in Chapters 5, 6, and 7 (Articles 500 through 780). For example, 300.22 (Chapter 3) is modified by 725.3(C) and 760.3(B) and is specifically referenced in 800.133(D) and 830.3(B). A graphic explanation of the *NEC* arrangement, Figure 90.3, was added to the 2002 *Code*.

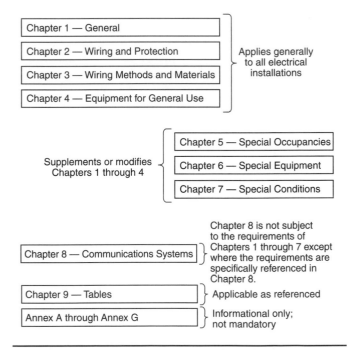

Figure 90.3 Code Arrangement.

90.4 Enforcement

This *Code* is intended to be suitable for mandatory application by governmental bodies that exercise legal jurisdiction over electrical installations, including signaling and communications systems, and for use by insurance inspectors. The authority having jurisdiction for enforcement of the *Code* has the responsibility for making interpretations of the rules, for deciding on the approval of equipment and materials, and for granting the special permission contemplated in a number of the rules.

Some localities do not adopt the *NEC*, but even in those localities, installations that comply with the current *Code* are prima facie evidence that the electrical installation is safe.

Section 90.4 advises that all materials and equipment used under the requirements of the *Code* are subject to the approval of the authority having jurisdiction. The text of 90.7, 110.2, and 110.3, along with the definitions of the terms *approved, identified (as applied to equipment), labeled,* and *listed,* is intended to provide a basis for the authority having jurisdiction to make the judgments that fall within that particular area of responsibility.

The phrase "including signaling and communication systems" was added to the 2002 *Code* to emphasize that, indeed, these systems are also subject to enforcement.

By special permission, the authority having jurisdiction may waive specific requirements in this *Code* or permit alternative methods where it is assured that equivalent objectives can be achieved by establishing and maintaining effective safety.

It is the responsibility of the authority having jurisdiction to interpret the specific rules of the *Code*. This paragraph empowers the authority having jurisdiction, using special permission (written consent), to permit alternative methods where specific rules are not established in the *Code*. For example, the authority having jurisdiction may waive specific requirements in industrial occupancies, research and testing laboratories, and other occupancies where the specific type of installation is not covered in the *Code*.

This *Code* may require new products, constructions, or materials that may not yet be available at the time the *Code* is adopted. In such event, the authority having jurisdiction may permit the use of the products, constructions, or materials that comply with the most recent previous edition of this *Code* adopted by the jurisdiction.

This paragraph of 90.4 permits the authority having jurisdiction to waive a new *Code* requirement during the interim period between acceptance of a new edition of the *NEC* and the availability of a new product, construction, or material redesigned to comply with the increased safety required by the latest edition. It is difficult to establish a viable future effective date in each section of the *NEC* because the time needed to change existing products and standards, as well as to develop new materials and test methods, usually is not known at the time the latest edition of the *Code* is adopted.

90.5 Mandatory Rules, Permissive Rules, and Explanatory Material

(A) Mandatory Rules Mandatory rules of this *Code* are those that identify actions that are specifically required or prohibited and are characterized by the use of the terms *shall* or *shall not.*

Section 90.5, which was revised and reorganized for the 1999 *Code,* clarifies that two distinctive types of rules are stated in the *Code.* Mandatory rules, characterized by the terms *shall* and *shall not,* are covered in 90.5(A).

(B) Permissive Rules Permissive rules of this *Code* are those that identify actions that are allowed but not required, are normally used to describe options or alternative methods, and are characterized by the use of the terms *shall be permitted* or *shall not be required.*

Permissive rules are simply options or alternative methods of achieving equivalent safety — they are not requirements. A close reading of permissive terms is important, because permissive rules are often misinterpreted. For example, the frequently used permissive term *shall be permitted* can be mistaken for a requirement. Substituting "the inspector must allow [item A or method A]" for "[item A or method A] shall be permitted" generally clarifies the interpretation.

(C) Explanatory Material Explanatory material, such as references to other standards, references to related sections of this *Code,* or information related to a *Code* rule, is included in this *Code* in the form of fine print notes (FPNs). Fine print notes are informational only and are not enforceable as requirements of this *Code.*

Brackets containing section references to another NFPA document are for informational purposes only and are provided as a guide to indicate the source of the extracted text. These bracketed references immediately follow the extracted text.

A number of requirements in the *NEC* have been extracted from other NFPA codes and standards. Therefore, a second paragraph was added for the 2005 *Code* to prevent any misunderstanding about the purpose of bracketed references to other NFPA codes and standards — they are provided only to indicate the section of the NFPA document from which the material in the *NEC* was extracted. Although *NEC* requirements based on extracted material are under the jurisdiction of the technical committee responsible for the particular document in which the extracted material resides, this revision to 90.5(C) makes it clear that the *NEC* requirements stand on their own and that extracted material with bracketed references does not indicate that other NFPA documents are adopted through reference.

Fine print notes (FPNs) do not contain statements of intent or recommendations. They present additional supplementary material that aids in the application of the requirement. In addition to explanatory material being in fine print (small type), the material is further identified in the *Code* by the abbreviation *FPN* preceding the paragraph. Fine print notes are not requirements of the *NEC* and are not enforceable.

Footnotes to tables, although also in fine print, are not explanatory material unless they are identified by the abbreviation *FPN*. Table footnotes are part of the tables and are necessary for proper use of the tables. For example, the footnotes at the end of Table 310.13 are necessary for the use of the table and therefore are mandatory and enforceable *Code* text.

Additional explanatory material is also found in the annexes at the back of this handbook. Annex A is a reference list of product safety standards used for product listing where that listing is required by the *Code*. Annex B provides guidance on the use of the general formula for ampacity found in 310.15(C). Annex C consists of wire fill tables for conduit and tubing. Annex D contains example calculations. Annex E presents various tables showing fire resistance ratings for Types I-V construction to correlate with the uses of Type NM cable, Annex F contains cross-reference tables for Chapter 3 realignment with the 2002 edition, and Annex G contains model administration and enforcement legislation.

FPN: The format and language used in this *Code* follows guidelines established by NFPA and published in the *NEC Style Manual*. Copies of this manual can be obtained from NFPA.

This fine print note informs the user that a style manual is available for the *NEC*. A style manual is basically a "how-to" pamphlet for editors. The *NEC Style Manual* contains a list of rules and regulations used by the panels and editors who prepare the *NEC*. The *NEC Style Manual*, which was revised for the 2002 edition of the *Code*, is available from NFPA.

90.6 Formal Interpretations

To promote uniformity of interpretation and application of the provisions of this *Code,* formal interpretation procedures have been established and are found in the NFPA Regulations Governing Committee Projects.

The procedures for implementing Formal Interpretations of the provisions of the *NEC* are outlined in the NFPA Regulations Governing Committee Projects. These regulations are included in the *NFPA Directory*, which is published annually and can be obtained from the Secretary of the NFPA Standards Council. The Formal Interpretations procedure can be found in Section 6 of the Regulations.

The National Electrical Code Committee cannot be responsible for subsequent actions of authorities enforcing the *NEC* that accept or reject its findings. The authority having jurisdiction is responsible for interpreting *Code* rules and should attempt to resolve all disagreements at the local level. Two general forms of Formal Interpretations are recognized: (1) those that are interpretations of the literal text and (2) those that are interpretations of the intent of the Committee at the time the particular text was issued.

Interpretations of the *NEC* not subject to processing are those that involve (1) a determination of compliance of a design, installation, product, or equivalency of protection; (2) a review of plans or specifications or judgment or knowledge that can be acquired only as a result of on-site inspection; (3) text that clearly and decisively provides the requested information; or (4) subjects not previously considered by the Technical Committee or not addressed in the document. Formal Interpretations of *Code* rules are published in several venues, including *necdigest*™, the NFPA Electrical Section News segment found in the *NFPA Journal*, in *NFPA News*, and in the *National Fire Codes* subscription service and are sent to interested trade publications.

Most interpretations of the *NEC* are rendered as the personal opinions of NFPA electrical engineering staff or of an involved member of the National Electrical Code Committee because the request for interpretation does not qualify for processing as a Formal Interpretation in accordance with NFPA Regulations Governing Committee Projects. Such opinions are rendered in writing only in response to written requests. The correspondence contains a disclaimer indicating that it is not a Formal Interpretation issued pursuant to NFPA Regulations and that any opinion expressed is the personal opinion of the author and does not necessarily represent the official position of NFPA or the National Electrical Code Committee.

90.7 Examination of Equipment for Safety

For specific items of equipment and materials referred to in this *Code*, examinations for safety made under standard conditions provide a basis for approval where the record is made generally available through promulgation by organizations properly equipped and qualified for experimental testing, inspections of the run of goods at factories, and service-value determination through field inspections. This avoids the necessity for repetition of examinations by different examiners, frequently with inadequate facilities for such work, and the confusion that would result from conflicting reports on the suitability of devices and materials examined for a given purpose.

It is the intent of this *Code* that factory-installed internal wiring or the construction of equipment need not be inspected at the time of installation of the equipment, except to detect alterations or damage, if the equipment has been listed by a qualified electrical testing laboratory that is recognized as having the facilities described in the preceding paragraph and that requires suitability for installation in accordance with this *Code*.

FPN No. 1: See requirements in 110.3.

FPN No. 2: *Listed* is defined in Article 100.

FPN No. 3: Annex A contains an informative list of product safety standards for electrical equipment.

Testing laboratories, inspection agencies, and other organizations concerned with product evaluation publish lists of equipment and materials that have been tested and meet nationally recognized standards or that have been found suitable for use in a specified manner. The *Code* does not contain detailed information on equipment or materials but refers to products as "listed," "labeled," or "identified." See Article 100 for definitions of these terms.

NFPA does not approve, inspect, or certify any installations, procedures, equipment, or materials, nor does it approve or evaluate testing laboratories. In determining the acceptability of installations or procedures, equipment, or materials, the authority having jurisdiction may base acceptance on compliance with NFPA or other appropriate standards. In the absence of such standards, the authority may require evidence of proper installation, procedures, or use. The authority having jurisdiction may also refer to the listing or labeling practices of an organization concerned with product evaluations that is able to determine compliance with appropriate standards for the current production of listed items.

Annex A contains a list of product safety standards used for product listing. The list includes only product safety standards for which a listing is required by the *Code*. For example, 344.6 requires that rigid metal conduit, Type RMC, be listed. By using Annex A, the user finds that the listing standard for rigid metal conduit is UL 6, *Rigid Metal Conduit*. Because associated conduit fittings are required to be listed, UL 514B, *Fittings for Cable and Conduit*, is found in Annex A also.

90.8 Wiring Planning

(A) Future Expansion and Convenience Plans and specifications that provide ample space in raceways, spare race-

ways, and additional spaces allow for future increases in electric power and communication circuits. Distribution centers located in readily accessible locations provide convenience and safety of operation.

The requirement for providing the exclusively dedicated equipment space mandated by 110.26(F) supports the intent of 90.8(A) regarding future increases in the use of electricity. The phrase "and communications circuits" was added for the 2005 *Code* to point out the importance of considering communications circuits when planning future needs. Electrical and communications distribution centers should contain additional space and capacity for future additions and should be conveniently located for easy accessibility.

Where electrical and communications distribution equipment is installed so that easy access cannot be achieved, a spare raceway(s) or pull line(s) should be run at the initial installation, as illustrated in Exhibit 90.2.

(B) Number of Circuits in Enclosures It is elsewhere provided in this *Code* that the number of wires and circuits confined in a single enclosure be varyingly restricted. Limiting the number of circuits in a single enclosure minimizes the effects from a short circuit or ground fault in one circuit.

These limitations minimize the heating effects inherently present wherever current-carrying conductors are grouped together. See 408.35 for restrictions on the number of overcurrent devices on one panelboard.

90.9 Units of Measurement

(A) Measurement System of Preference For the purpose of this *Code*, metric units of measurement are in accordance with the modernized metric system known as the International System of Units (SI).

According to a recent report titled "A Metric for Success" by the National Institute of Standards and Technology (NIST), most U.S. industries that do business abroad are predominantly metric already because of global sourcing of parts, service, components, and production. However, quite a few domestic industries still use U.S. Customary units. The NIST report warns that domestic industries that ignore global realities and continue to design and manufacture with nonmetric measures will find that they risk increasing their costs. Nonmetric modular products (the building construction industry uses great quantities of modular parts) and those that interface with outside industry products are especially

Exhibit 90.2 A residential distribution system showing spare raceways or pull lines that allow for future circuits and loads for both electrical and communications systems.

Spare raceways or pull lines from electrical service panel and communications center to accessible attic space

Junction boxes

Raceway or pull lines

Unfinished utility area

Finished basement area

Service panel

Communications center

vulnerable to the added costs of adapting to a metric environment.

Metric standards are beginning to appear in the domestic building construction industry because our national standards are being harmonized with international standards. The *National Electrical Code* is an important building construction standard and moves another step in the metric direction.

(B) Dual System of Units SI units shall appear first, and inch-pound units shall immediately follow in parentheses. Conversion from inch-pound units to SI units shall be based on hard conversion except as provided in 90.9(C).

Hard conversion is explained in FPN No. 1 following 90.9(D). Calculations to convert measurements from inch-pound units to metric units must be made using hard conversion. The hard-conversion method is mandatory except for trade sizes [e.g., raceway sizes in Table 300.1(C)], extracted material (e.g., class and zone measurements from other

NFPA documents), and safety calculations (e.g., minimum distances taken from Table 110.31).

Example

Using the hard-conversion method, determine the equivalent metric conversion for 24 in., generally the minimum cover requirements for direct burial cables and conductors in nonspecific locations taken from row 1 of Table 300.5.

Solution

Step 1.

$$24 \text{ in.} \times \frac{25.4 \text{ mm}}{1 \text{ in.}} = 609.6 \text{ mm}$$

Step 2. Because the calculation is being performed as a hard conversion, the 609.6 mm dimension may be changed, and the selected equivalent cover requirement is 600 mm.

For the 2005 *Code* as well as the 2002 *Code*, the measurements of 600 mm and 24 in. appear in Table 300.5 for the minimum cover requirements for direct burial cables and conductors in nonspecific locations. For the 1999 *NEC*, the selected SI unit of measure was required to be 609.6 mm.

The 2005 *Code* (as well as the 2002 *Code*) permits much more latitude for the final selected dimension, and so the equivalent minimum cover requirement of 600 mm is a more practical solution. Basically, a hard conversion permits a change in a dimension or allows rounding up or down to better fit the physical constraints of the installation.

(C) Permitted Uses of Soft Conversion The cases given in 90.9(C)(1) through (C)(4) shall not be required to use hard conversion and shall be permitted to use soft conversion.

(1) Trade Sizes Where the actual measured size of a product is not the same as the nominal size, trade size designators shall be used rather than dimensions. Trade practices shall be followed in all cases.

Metric trade sizes (metric designators) of conduits were added in the 1996 *Code* as fine print notes in each raceway article. Since the 2002 *Code*, these metric designators appear in the *Code* text, preceding the trade size equivalents, in the raceway articles.

For example, in 350.20(A) of this *Code*, the size requirement is stated as follows: "LFMC smaller than metric designator 16 (trade size 1/2) shall not be used." In 351-5(a) of the 1999 *NEC*, the size requirement was stated as follows: "Liquidtight flexible metallic conduit smaller than 1/2-in. electrical trade size shall not be used."

This change does not reflect a technical change but rather provides acceptable language to both domestic and international users of the *NEC*. For ease of use, in Table 4 of Chapter 9, metric designators are separate columns.

(2) Extracted Material Where material is extracted from another standard, the context of the original material shall not be compromised or violated. Any editing of the extracted text shall be confined to making the style consistent with that of the *NEC*.

(3) Industry Practice Where industry practice is to express units in inch-pound units, the inclusion of SI units shall not be required.

The following examples illustrate conversions from U.S. Customary units to SI units. Example 1 shows the process of converting a dimension from feet to meters, where safety is a concern. Table 110.31 contains minimum permitted distances from a fence to a live part for voltages 601 and greater. Example 1 calculates the equivalent metric conversion for 10 ft using the minimum distance of 10 ft in Table 110.31 where the measurement is from a fence to a live part from 601 volts to 13,799 volts.

Example 1

Determine the equivalent metric conversion for 10 ft where the calculation could have a negative impact on safety, such as the minimum distance of 10 ft given in Table 110.31, and where the measurement is from a fence to a live part from 601 volts to 13,799 volts.

Solution

Step 1.
$$10 \text{ ft} \times \frac{0.3048 \text{ m}}{1 \text{ ft}} = 3.048 \text{ m}$$

Step 2. Round up the calculation to 3.05 m, because a distance less than 3.048 could have a negative impact on safety. The answer, 3.05 m, matches the minimum distance in Table 110.31 from a fence to a live part from 601 volts to 13,799 volts.

Because safety is a concern for this conversion calculation, the original *Code* distance (the U.S. Customary units for this example) remains the shortest permitted distance. The final metric equivalent ends up slightly larger. The exact difference is of no practical concern, however, because 0.2 mm is less than 1/32 in. From a practical point of view, a variance of 1/32 in. in a length of 10 ft is insignificant.

Example 2

Using the soft-conversion method, determine the equivalent metric conversion for 30 in. where the calculation could have a negative impact on safety, such as a 30 in. minimum horizontal working space requirement in the rear of equipment that requires access to nonelectrical parts according to 110.26(A)(1)(a).

Solution

Step 1.
$$30 \text{ in.} \times \frac{25.4 \text{ mm}}{1 \text{ in.}} = 762 \text{ mm}$$

Step 2. Do not round off the calculation, because even a slight reduction in the original distance could have a negative impact on safety. The answer is 762 mm, which matches the minimum distance of 110.26(A)(1)(a) for a minimum horizontal working space.

(4) Safety Where a negative impact on safety would result, soft conversion shall be used.

(D) Compliance Conversion from inch-pound units to SI units shall be permitted to be an approximate conversion. Compliance with the numbers shown in either the SI system or the inch-pound system shall constitute compliance with this *Code*.

> FPN No. 1: Hard conversion is considered a change in dimensions or properties of an item into new sizes that might or might not be interchangeable with the sizes used in the original measurement. Soft conversion is considered a direct mathematical conversion and involves

a change in the description of an existing measurement but not in the actual dimension.

FPN No. 2: SI conversions are based on IEEE/ASTM SI 10-1997, *Standard for the Use of the International System of Units (SI): The Modern Metric System.*

Commentary Table 90.1 offers some examples of the hard-conversion process. U.S. Customary units were used in the 1993, 1996, and 1999 *Code* and were still valid for the 2002 *Code*. Soft-conversion SI units were used in the 1996 and 1999 *Code*. The hard-conversion SI units, which were added to the 2002 *Code*, were listed with their equivalent U.S. Customary units. The equivalent U.S. units are given only to show the small variance between customary units and the hard-conversion units.

Warning signs that state specific clearances, such as required in 513.10(B), permit distance measurements in either inch-pound units or metric units.

Commentary Table 90.1 Conversions Using the Hard-Conversion Method

U.S. Customary Units	Soft Conversions, SI Units	Hard Conversions, SI Units	Equivalent U. S. Customary Units
½ in.	12.7 mm	13 mm	0.51 in.
¾ in.	19 mm	19 mm	0.75 in.
1 in.	25.4 mm	25 mm	0.98 in.
4 in.	102 mm	100 mm	3.94 in.
12 in.	305 mm	300 mm	11.81 ft
2 ft	610 mm	600 mm	1.97 ft
3 ft	914 mm	900 mm	2.95 ft
6 ft	1.83 m	1.8 m	5.91 ft
15 ft	4.57 m	4.5 m	14.76 ft

General

ARTICLE 100
Definitions

Summary of Changes

- **Bonding Jumper, System:** Added definition for use in place of *bonding jumper* or the often misused term *main bonding jumper*.

- **Coordination (Selective):** Relocated definition from 240.2 and expanded it.

- **Device:** Changed *carry* to *carry or control*.

- **Dwelling Unit:** Revised to coordinate with definition in NFPA 1, NFPA *101®*, and *NFPA 5000®*. Does not impact usage within the *NEC*.

- **Grounded, Solidly:** Added definition to define *solidly grounded* as the term is used in other *NEC* articles. Replaces definition in 230.95.

- **Grounding Electrode:** Added definition that defines the function of a grounding electrode.

- **Grounding Electrode Conductor:** Revised to include applications where feeders or branch circuits supply a building or structure.

- **Guest Room:** Added definition to coordinate with revised definition of *dwelling unit*.

- **Guest Suite:** Added definition to coordinate with revised definition of *dwelling unit*. Ensures that a suite with more than one room is covered by *NEC* requirements.

- **Handhole Enclosure:** Added definition in Article 100 because term is used in both Articles 300 and 314.

- **Outline Lighting:** Revised to include other electrically powered light sources.

- **Qualified Person:** Added FPN to reference NFPA 70E.

- **Separately Derived System:** Revised to cover any premises wiring system whose power is derived from other than a service.

- **Supplementary Overcurrent Protective Device:** Added definition to distinguish between general use devices such as branch circuit overcurrent protective where such devices are extremely application oriented and where, prior to applying the devices, the differences and limitations for these devices must be investigated and found acceptable.

Contents

Scope. This article contains only those definitions essential to the proper application of this *Code*. It is not intended to include commonly defined general terms or commonly defined technical terms from related codes and standards. In general, only those terms that are used in two or more articles are defined in Article 100. Other definitions are included in the article in which they are used but may be referenced in Article 100.

Part I of this article contains definitions intended to apply wherever the terms are used throughout this *Code*. Part II contains definitions applicable only to the parts of articles specifically covering installations and equipment operating at over 600 volts, nominal.

Commonly defined general terms include those terms defined in general English language dictionaries and terms that are not used in a unique or restricted manner in the *NEC*. Commonly defined technical terms such as *volt* (abbreviated V) and *ampere* (abbreviated A) are found in the *IEEE Standard Dictionary of Electrical and Electronic Terms*.

Definitions that are not listed in Article 100 are included in their appropriate article. For articles that follow the common format according to the *NEC Style Manual*, the section number is generally XXX.2 Definition(s). For example, the definition of *nonmetallic-sheathed cable* is found in 334.2 Definitions. The 2005 edition of the *Code* does contain some isolated exceptions to this general rule because the *NEC* has not been entirely converted to a common numbering system.

I. General

Accessible (as applied to equipment). Admitting close approach; not guarded by locked doors, elevation, or other effective means.

Exhibit 100.1 illustrates examples of equipment considered accessible (as applied to equipment). The main rule for switches and circuit breakers used as switches is shown in (a) and is according to 404.8(A). In (b), the busway installation is according to 368.17(C). The exceptions to the main rule are illustrated in (c), the installation of busway switches installed according to 404.8(A), Exception No. 1; (d), a switch installed adjacent to a motor according to 404.8(A), Exception No. 2; and (e), a hookstick-operated isolating switch installed according to 404.8(A), Exception No. 3.

Accessible (as applied to wiring methods). Capable of being removed or exposed without damaging the building structure or finish or not permanently closed in by the structure or finish of the building.

Wiring methods located behind removable panels designed to allow access are not considered permanently enclosed and are considered exposed as applied to wiring methods.

Exhibit 100.1 Example of busway and of switches considered accessible even if located above 6 ft 7 in.

(c) Installation of busway switches

(b) Busway installation

Chain- or rope-operated switch

(d) Switch installed adjacent to motor

Hookstick-operated switch

(e) Hookstick-operated isolating switch

(a) Manually operated disconnect switch

6 ft 7 in. max.

See 300.4(C) regarding cables located in spaces behind accessible panels.

Exhibit 100.2 illustrates examples of wiring methods and equipment that are considered accessible.

Accessible, Readily (Readily Accessible). Capable of being reached quickly for operation, renewal, or inspections without requiring those to whom ready access is requisite to climb over or remove obstacles or to resort to portable ladders, and so forth.

The definition of *readily accessible* does not preclude the use of a locked door for service equipment or rooms containing service equipment, provided those for whom ready access is necessary have a key (or lock combination) available. For example, 230.70(A)(1) and 230.205(A) require service-disconnecting means to be readily accessible. Section 225.32 requires that feeder disconnecting means for separate buildings be readily accessible. A commonly used, permitted practice is to locate the disconnecting means in the electrical equipment room of an office building or large apartment building and to keep the door to that room locked to

Busway

Junction box

Lay-in luminaire

Suspended ceiling with lift-out panels

Exhibit 100.2 Examples of busways and junction boxes considered accessible even if located behind hung ceilings having lift-out panels.

prevent access by unauthorized persons. Section 240.24(A) requires that overcurrent devices be so located as to be readily accessible.

Ampacity. The current, in amperes, that a conductor can carry continuously under the conditions of use without exceeding its temperature rating.

The definition of the term *ampacity* states that the maximum current a conductor carries continuously varies with the conditions of use as well as with the temperature rating of the conductor insulation. For example, ambient temperature is a condition of use. A conductor with insulation rated at 60°C and installed near a furnace where the ambient temperature is continuously maintained at 60°C has no current-carrying capacity. Any current flowing through the conductor will raise its temperature above the 60°C insulation rating. Therefore, the ampacity of this conductor, regardless of its size, is zero. See the ampacity correction factors for temperature at the bottom of Table 310.16 through Table 310.20, or see Annex B. The temperature limitations on conductors is further explained and examples given in 310.10 and in the commentary following that section.

Another condition of use is the number of conductors in a raceway or cable. [See 310.15(B)(2).]

Appliance. Utilization equipment, generally other than industrial, that is normally built in standardized sizes or types and is installed or connected as a unit to perform one or more functions such as clothes washing, air conditioning, food mixing, deep frying, and so forth.

Approved. Acceptable to the authority having jurisdiction.

See the definition of *authority having jurisdiction* and 110.2 for a better understanding of the approval process. Understanding *NEC* terms such as *listed*, *labeled*, and *identified* (as applied to equipment) will also assist the user in understanding the approval process.

Askarel. A generic term for a group of nonflammable synthetic chlorinated hydrocarbons used as electrical insulating media. Askarels of various compositional types are used. Under arcing conditions, the gases produced, while consisting predominantly of noncombustible hydrogen chloride, can include varying amounts of combustible gases, depending on the askarel type.

Attachment Plug (Plug Cap) (Plug). A device that, by insertion in a receptacle, establishes a connection between the conductors of the attached flexible cord and the conductors connected permanently to the receptacle.

Standard attachment caps are available with built-in options, such as switching, fuses, or even ground-fault circuit-interrupter (GFCI) protection.

Attachment plug contact blades have specific shapes, sizes, and configurations so that a receptacle or cord connector will not accept an attachment plug of a voltage or current rating different from that for which the device is intended. Configuration charts from NEMA WD 6, *Wiring Devices — Dimensional Requirements*, for general-purpose nonlocking and specific-purpose locking plugs and receptacles are shown in Exhibit 406.3 and Exhibit 406.4, respectively.

Authority Having Jurisdiction (AHJ). The organization, office, or individual responsible for approving equipment, materials, an installation, or a procedure.

FPN: The phrase "authority having jurisdiction" is used in NFPA documents in a broad manner, since jurisdictions and approval agencies vary, as do their responsibilities. Where public safety is primary, the AHJ may be a federal, state, local, or other regional department or individual such as a fire chief; fire marshal; chief of a fire prevention bureau, labor department, or health department; building official; electrical inspector; or others having statutory authority. For insurance purposes, an insurance inspection department, rating bureau, or other insurance company representative may be the AHJ. In many circumstances, the property owner or his or her designated agent assumes the role of the AHJ; at government installations, the commanding officer or departmental official may be the AHJ.

The important role of the authority having jurisdiction (AHJ) cannot be overstated in the current North American safety system. The basic role of the AHJ is to verify that an installation complies with the *Code*. The definition of *authority having jurisdiction* and the accompanying explanation (the FPN) bring a sense of uniformity to the *Code*, since this exact definition has appeared in many other NFPA documents for quite some time. This definition is very helpful in understanding *Code* enforcement, the inspection process, the definition of *approved*, and 90.7 and 110.2.

Automatic. Self-acting, operating by its own mechanism when actuated by some impersonal influence, as, for example, a change in current, pressure, temperature, or mechanical configuration.

Bathroom. An area including a basin with one or more of the following: a toilet, a tub, or a shower.

Bonding (Bonded). The permanent joining of metallic parts to form an electrically conductive path that ensures electrical continuity and the capacity to conduct safely any current likely to be imposed.

The purpose of bonding is to establish an effective path for fault current that, in turn, facilitates the operation of the overcurrent protective device. This is explained in 250.4(A)(3) and (4) and 250.4(B)(3) and (4). Specific bonding requirements are found in Part V of Article 250 and in other sections of the *Code* as referenced in 250.3.

Bonding Jumper. A reliable conductor to ensure the required electrical conductivity between metal parts required to be electrically connected.

Both concentric- and eccentric-type knockouts can impair the electrical conductivity between metal parts and may actually introduce unnecessary impedance into the grounding path. Installing bonding jumper(s) is one method often used between metal raceways and metal parts to ensure electrical conductivity. Bonding jumpers may be found at service equipment [250.92(B)], bonding for over 250 volts (250.97), and expansion fittings in metal raceways (250.98). Exhibit 100.3 shows the difference between concentric- and eccentric-type knockouts. Exhibit 100.3 also illustrates one method of applying bonding jumpers at these types of knockouts.

Exhibit 100.3 Bonding jumpers installed around concentric or eccentric knockouts.

Bonding Jumper, Equipment. The connection between two or more portions of the equipment grounding conductor.

Bonding Jumper, Main. The connection between the grounded circuit conductor and the equipment grounding conductor at the service.

Exhibit 100.4 shows a main bonding jumper used to provide the connection between the grounded service conductor and the equipment grounding conductor at the service. Bonding jumpers may be located throughout the electrical system,

Exhibit 100.4 A main bonding jumper installed at the service between the grounded service conductor and the equipment grounding conductor.

but a main bonding jumper is located only at the service. Main bonding jumper requirements are found in 250.28.

Bonding Jumper, System. The connection between the grounded circuit conductor and the equipment grounding conductor at a separately derived system.

Exhibit 100.5 shows a system bonding jumper used to provide the connection between the grounded conductor and the equipment grounding conductor(s) of a transformer used as a separately derived system.

System bonding jumpers are located near the source of the separately derived system. A system bonding jumper is used at the derived system if the derived system contains a grounded conductor. Like the main bonding jumper at the service equipment, the system bonding jumper provides the necessary link between the equipment grounding conductors and the system grounded conductor in order to establish an effective path for ground-fault current. The requirements for system bonding jumper(s) are found in 250.30(A)(1).

Branch Circuit. The circuit conductors between the final overcurrent device protecting the circuit and the outlet(s).

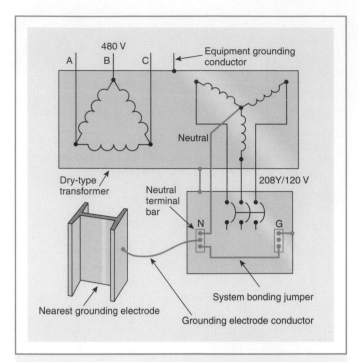

Exhibit 100.5 A system bonding jumper installed near the source of a separately derived system between the system grounded conductor and the equipment grounding conductor(s).

Exhibit 100.6 shows the difference between branch circuits and feeders. Conductors between the overcurrent devices in the panelboards and the duplex receptacles are branch-circuit conductors. Conductors between the service equipment or source of separately derived systems and the panelboards are feeders.

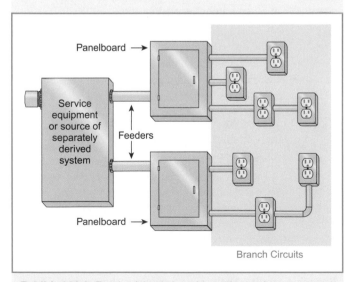

Exhibit 100.6 Feeder (circuits) and branch circuits.

Branch Circuit, Appliance. A branch circuit that supplies energy to one or more outlets to which appliances are to be connected and that has no permanently connected luminaires (lighting fixtures) that are not a part of an appliance.

Two or more 20-ampere small-appliance branch circuits are required by 210.11(C)(1) for dwelling units. Section 210.52(B)(1) requires that these circuits supply receptacle outlets located in such rooms as the kitchen, pantry, and so on. These small-appliance branch circuits are not permitted to supply other outlets or permanently connected lighting fixtures. (See 210.52 for exact details.)

Branch Circuit, General-Purpose. A branch circuit that supplies two or more receptacles or outlets for lighting and appliances.

Branch Circuit, Individual. A branch circuit that supplies only one utilization equipment.

An individual branch circuit is a circuit that supplies only one piece of utilization equipment (e.g., one range, one space heater, one motor). See 210.23 regarding permissible loads for branch circuits.

An individual branch circuit supplies only one single receptacle for the connection of a single attachment plug. This single receptacle is required to have an ampere rating not less than that of the branch circuit, as stated in 210.21(B)(1).

Exhibit 100.7 illustrates an individual branch circuit with a single receptacle intended for the connection of one piece of utilization equipment. A branch circuit that supplies one duplex receptacle that can accommodate two cord-and-

Exhibit 100.7 An individual branch circuit, which supplies only one utilization equipment via a single receptacle.

plug-connected appliances or similar equipment is not an individual branch circuit.

Branch Circuit, Multiwire. A branch circuit that consists of two or more ungrounded conductors that have a voltage between them, and a grounded conductor that has equal voltage between it and each ungrounded conductor of the circuit and that is connected to the neutral or grounded conductor of the system.

For the 2002 edition, this definition was editorially modified by substituting the word *voltage* for the term *potential difference*. See 210.4, 240.20(B)(1), and 300.13(B) for specific information about multiwire branch circuits.

Building. A structure that stands alone or that is cut off from adjoining structures by fire walls with all openings therein protected by approved fire doors.

A building is generally considered to be a roofed or walled structure that may be used or intended for supporting or sheltering any use or occupancy. However, it may also be a separate structure such as a pole, billboard sign, or water tower.

Definitions of the terms *fire walls* and *fire doors* are the responsibility of building codes. Generically, a fire wall may be defined as a wall that separates buildings or subdivides a building to prevent the spread of fire and that has a fire resistance rating and structural stability. Fire doors (and fire windows) are used to protect openings in walls, floors, and ceilings against the spread of fire and smoke within, into, or out of buildings.

Cabinet. An enclosure that is designed for either surface mounting or flush mounting and is provided with a frame, mat, or trim in which a swinging door or doors are or can be hung.

Both cabinets and cutout boxes are covered in Article 312. Cabinets are designed for surface or flush mounting with a trim to which a swinging door(s) is hung. Cutout boxes are designed for surface mounting with a swinging door(s) secured directly to the box. Panelboards are electrical assemblies designed to be placed in a cabinet or cutout box. (See the definitions of *cutout box* and *panelboard*.)

Circuit Breaker. A device designed to open and close a circuit by nonautomatic means and to open the circuit automatically on a predetermined overcurrent without damage to itself when properly applied within its rating.

FPN: The automatic opening means can be integral, direct acting with the circuit breaker, or remote from the circuit breaker.

Adjustable (as applied to circuit breakers). A qualifying term indicating that the circuit breaker can be set to trip at various values of current, time, or both, within a predetermined range.

Instantaneous Trip (as applied to circuit breakers). A qualifying term indicating that no delay is purposely introduced in the tripping action of the circuit breaker.

Inverse Time (as applied to circuit breakers). A qualifying term indicating that there is purposely introduced a delay in the tripping action of the circuit breaker, which delay decreases as the magnitude of the current increases.

Nonadjustable (as applied to circuit breakers). A qualifying term indicating that the circuit breaker does not have any adjustment to alter the value of current at which it will trip or the time required for its operation.

Setting (of circuit breakers). The value of current, time, or both, at which an adjustable circuit breaker is set to trip.

Concealed. Rendered inaccessible by the structure or finish of the building. Wires in concealed raceways are considered concealed, even though they may become accessible by withdrawing them.

Raceways and cables supported or located within hollow frames or permanently closed in by the finish of buildings are considered concealed. Open-type work — such as raceways and cables in exposed areas, in unfinished basements, in accessible underfloor areas or attics, attached to the surface of finished areas, or behind, above, or below panels designed to allow access and that may be removed without damage to the building structure or finish — is not considered concealed. [See definition of *exposed (as applied to wiring methods)*.]

Conductor, Bare. A conductor having no covering or electrical insulation whatsoever.

Conductor, Covered. A conductor encased within material of composition or thickness that is not recognized by this *Code* as electrical insulation.

Typical covered conductors are the green-covered equipment grounding conductors contained within a nonmetallic-sheathed cable or the uninsulated grounded system conductors within the overall exterior jacket of a Type SE cable. Covered conductors should always be treated as bare conductors for working clearances, because they are really uninsulated conductors.

Conductor, Insulated. A conductor encased within material of composition and thickness that is recognized by this *Code* as electrical insulation.

For the covering on a conductor to be considered insulation, the conductor with the covering material generally is required to pass minimum testing required by a product standard. One such product standard is UL 83, *Thermoplastic-Insulated Wires and Cables*. To meet the requirements of UL 83, specimens of finished single-conductor wires must pass specified tests that measure (1) resistance to flame propagation, (2) dielectric strength, even while immersed, and (3) resistance to abrasion, cracking, crushing, and impact. Only wires and cables that meet the minimum fire, electrical, and physical properties required by the applicable standards are permitted to be marked with the letter designations found in Table 310.13 and Table 310.61. See 310.13 for the exact requirements of insulated conductor construction and applications.

Conduit Body. A separate portion of a conduit or tubing system that provides access through a removable cover(s) to the interior of the system at a junction of two or more sections of the system or at a terminal point of the system.

Boxes such as FS and FD or larger cast or sheet metal boxes are not classified as conduit bodies.

Conduit bodies are a portion of a raceway system with removable covers to allow access to the interior of the system. They include the short-radius type as well as capped elbows and service-entrance elbows.

Some conduit bodies are referred to in the trade as "condulets" and include the LB, LL, LR, C, T, and X designs. (See 300.15 and Article 314 for rules on the usage of conduit bodies.)

Type FS and Type FD boxes are not classified as conduit bodies; they are listed with boxes in Table 314.16(A).

Connector, Pressure (Solderless). A device that establishes a connection between two or more conductors or between one or more conductors and a terminal by means of mechanical pressure and without the use of solder.

Continuous Load. A load where the maximum current is expected to continue for 3 hours or more.

Controller. A device or group of devices that serves to govern, in some predetermined manner, the electric power delivered to the apparatus to which it is connected.

A controller may be a remote-controlled magnetic contactor, switch, circuit breaker, or device that is normally used to start and stop motors and other apparatus and, in the case of motors, is required to be capable of interrupting the stalled-rotor current of the motor. Stop-and-start stations and similar control circuit components that do not open the power conductors to the motor are not considered controllers.

Cooking Unit, Counter-Mounted. A cooking appliance designed for mounting in or on a counter and consisting of one or more heating elements, internal wiring, and built-in or mountable controls.

Coordination (Selective). Localization of an overcurrent condition to restrict outages to the circuit or equipment affected, accomplished by the choice of overcurrent protective devices and their ratings or settings.

Moved from 240.2 to Article 100 and slightly revised for the 2005 *Code*, this definition is no longer limited to just Article 240. For the 2005 *Code*, selective coordination requirements have been expanded to include emergency and legally required systems of 700.27 and 701.18. The past *Code* requirements regarding selective coordination remain for elevator feeders in 620.62.

The main goal of selective coordination is to isolate the faulted portion of the electrical circuit quickly while at the same time maintaining power to the remainder of the electrical system. The electrical system overcurrent protection must guard against short circuits and ground faults to ensure that the resulting damage is minimized while other parts of the system not directly involved with the fault are kept on until other protective devices clear the fault.

Overcurrent protective devices, such as fuses and circuit breakers, have time/current characteristics that determine the time it takes to clear the fault for a given value of fault current. Selectivity occurs when the device closest to the fault opens before the next device upstream operates. For example, any fault on a branch circuit should open the branch circuit breaker rather than the feeder overcurrent protection. All faults on a feeder should open the feeder overcurrent protection rather than the service overcurrent protection. When selectivity occurs, the electrical system is considered to be coordinated.

With coordinated overcurrent protection, the faulted or overloaded circuit is isolated by the selective operation of only the overcurrent protective device closest to the overcurrent condition. This isolation prevents power loss to unaffected loads.

Copper-Clad Aluminum Conductors. Conductors drawn from a copper-clad aluminum rod with the copper metallurgically bonded to an aluminum core. The copper forms a minimum of 10 percent of the cross-sectional area of a solid conductor or each strand of a stranded conductor.

Cutout Box. An enclosure designed for surface mounting that has swinging doors or covers secured directly to and telescoping with the walls of the box proper.

Dead Front. Without live parts exposed to a person on the operating side of the equipment.

Demand Factor. The ratio of the maximum demand of a system, or part of a system, to the total connected load of a system or the part of the system under consideration.

Device. A unit of an electrical system that is intended to carry or control but not utilize electric energy.

The definition of *device* was revised and made a bit broader for the 2005 *Code*. Components (such as switches, circuit breakers, fuseholders, receptacles, attachment plugs, and lampholders) that distribute or control but do not consume electrical energy are considered devices. Devices that consume incidental amounts of electrical energy in the performance of carrying or controlling electricity are now also considered devices. Some examples of these components include a switch with an internal pilot light, a GFCI receptacle, and even a magnetic contactor.

Disconnecting Means. A device, or group of devices, or other means by which the conductors of a circuit can be disconnected from their source of supply.

For disconnecting means for service equipment, see Part VI of Article 230; for fuses, see Part IV of Article 240; for circuit breakers, see Part VII of Article 240; for appliances, see Part III of Article 422; for space-heating equipment, see Part III of Article 424; for motors and controllers, see Part IX of Article 430; and for air-conditioning and refrigerating equipment, see Part II of Article 440. (See also references for *disconnecting means* in the index.)

Dusttight. Constructed so that dust will not enter the enclosing case under specified test conditions.

Table 430.91, Motor Controller Enclosure Selection, provides a basis for selecting enclosure types that are dusttight. (See also the commentary following the definition of *enclosure*.)

The term *dustproof* was removed from the *Code* for the 2002 edition because it was no longer applicable or used in the *Code*.

Duty, Continuous. Operation at a substantially constant load for an indefinitely long time.

Duty, Intermittent. Operation for alternate intervals of (1) load and no load; or (2) load and rest; or (3) load, no load, and rest.

Duty, Periodic. Intermittent operation in which the load conditions are regularly recurrent.

Duty, Short-Time. Operation at a substantially constant load for a short and definite, specified time.

Duty, Varying. Operation at loads, and for intervals of time, both of which may be subject to wide variation.

Information on the protection of intermittent, periodic, short-time, and varying-duty motors against overload can be found in 430.33.

Dwelling Unit. A single unit, providing complete and independent living facilities for one or more persons, including permanent provisions for living, sleeping, cooking, and sanitation.

A mobile home may be considered to be a dwelling unit. Where dwelling units are referenced throughout the *Code*, it is important to note that rooms in motels, hotels, and similar occupancies could be classified as dwelling units if they satisfy the requirements of the definition. For example, the motel or hotel room illustrated in Exhibit 100.8 clearly meets the definition because it has permanent provisions for living, sleeping, cooking, and sanitation.

Exhibit 100.8 Example of motel or hotel room considered to be a dwelling unit.

Dwelling, One-Family. A building that consists solely of one dwelling unit.

Dwelling, Two-Family. A building that consists solely of two dwelling units.

Dwelling, Multifamily. A building that contains three or more dwelling units.

Electric Sign. A fixed, stationary, or portable self-contained, electrically illuminated utilization equipment with words or symbols designed to convey information or attract attention.

Enclosed. Surrounded by a case, housing, fence, or wall(s) that prevents persons from accidentally contacting energized parts.

Enclosure. The case or housing of apparatus, or the fence or walls surrounding an installation to prevent personnel from accidentally contacting energized parts or to protect the equipment from physical damage.

FPN: See Table 430.91 for examples of enclosure types.

The information in Commentary Table 1.1 is taken from the 2003 UL *General Information Directory* (White Book), category AALZ, "Electrical Equipment for Use in Ordinary Locations." The table summarizes the intended uses of the various types of enclosures for nonhazardous locations.

Enclosures that comply with the requirements for more than one type of enclosure may be marked with multiple designations. Enclosures marked with a type may also be marked as follows:

A Type 1 may be marked "Indoor Use Only."

A Type 3, 3S, 4, 4X, 6, or 6P may be marked "Raintight."

A Type 3R may be marked "Rainproof."

A Type 4, 4X, 6, or 6P may be marked "Watertight."

A Type 4X or 6P may be marked "Corrosion Resistant."

A Type 2, 5, 12, 12K, or 13 may be marked "Driptight."

A Type 3, 3S, 5, 12K, or 13 may be marked "Dusttight."

For equipment designated "Raintight," testing designed to simulate exposure to a beating rain will not result in entrance of water. For equipment designated "Rainproof," testing designed to simulate exposure to a beating rain will not interfere with the operation of the apparatus or result in wetting of live parts and wiring within the enclosure. "Watertight" equipment is so constructed that water does not enter the enclosure when subjected to a stream of water. "Corrosion resistant" equipment is constructed so that it provides a degree of protection against exposure to corrosive agents such as salt spray. "Driptight" equipment is con-

Commentary Table 1.1 Environmental Protections for Nonhazardous Locations, by Type of Enclosure

Enclosure Type Number	Provides a Degree of Protection Against the Following Environmental Conditions*
1	Indoor use
2	Indoor use, limited amounts of falling water
3R	Outdoor use, undamaged by the formation of ice on the enclosure†
3	Same as 3R plus windblown dust
3S	Same as 3R plus windblown dust; external mechanisms remain operable while ice laden
4	Outdoor use, splashing water, windblown dust, hose-directed water, undamaged by the formation of ice on the enclosure†
4X	Same as 4 plus resists corrosion
5	Indoor use to provide a degree of protection against settling airborne dust, falling dirt, and dripping noncorrosive liquids
6	Same as 3R plus entry of water during temporary submersion at a limited depth
6P	Same as 3R plus entry of water during prolonged submersion at a limited depth
12, 12K	Indoor use, dust, dripping noncorrosive liquids
13	Indoor use, dust, spraying water, oil, and noncorrosive coolants

*All enclosure types provide a degree of protection against ordinary corrosion and against accidental contact with the enclosed equipment when doors or covers are closed and in place. All types of enclosures provide protection against a limited amount of falling dirt.
†All outdoor-type enclosures provide a degree of protection against rain, snow, and sleet. Outdoor enclosures are also suitable for use indoors if they meet the environmental conditions present.
Source: Underwriters Laboratories, *General Information Directory*, 2003 edition.

structed so that falling moisture or dirt does not enter the enclosure. "Dusttight" equipment is constructed so that circulating or airborne dust does not enter the enclosure.

Energized. Electrically connected to, or is, a source of voltage.

The definition of *energized* was broadened for the 2005 *Code* to point out that equipment such as batteries, capacitors, and conductors with induced voltages must also be considered energized. This term is no longer limited to just "connected to a source of voltage." For a more thorough understanding of *energized*, also see the definitions of *exposed (as applied to live parts)* and *live parts*.

Equipment. A general term including material, fittings, devices, appliances, luminaires (fixtures), apparatus, and the like used as a part of, or in connection with, an electrical installation.

Explosionproof Apparatus. Apparatus enclosed in a case that is capable of withstanding an explosion of a specified gas or vapor that may occur within it and of preventing the ignition of a specified gas or vapor surrounding the enclosure by sparks, flashes, or explosion of the gas or vapor within, and that operates at such an external temperature that a surrounding flammable atmosphere will not be ignited thereby.

> FPN: For further information, see ANSI/UL 1203-1999, *Explosion-Proof and Dust-Ignition-Proof Electrical Equipment for Use in Hazardous (Classified) Locations.*

Exposed (as applied to live parts). Capable of being inadvertently touched or approached nearer than a safe distance by a person. It is applied to parts that are not suitably guarded, isolated, or insulated.

For a more thorough understanding of *exposed (as applied to live parts)*, also see the definitions of *energized* and *live parts*. Requirements for guarding of live parts may be found in 110.27.

Exposed (as applied to wiring methods). On or attached to the surface or behind panels designed to allow access.

See Exhibit 100.2, where wiring methods located behind a suspended ceiling with lift-out panels are considered exposed (as applied to wiring methods).

Externally Operable. Capable of being operated without exposing the operator to contact with live parts.

Feeder. All circuit conductors between the service equipment, the source of a separately derived system, or other power supply source and the final branch-circuit overcurrent device.

See the commentary following the definition of *branch circuit*, including Exhibit 100.6, which illustrates the difference between branch circuits and feeders.

Festoon Lighting. A string of outdoor lights that is suspended between two points.

The general requirements for festoon lighting are located in 225.6(B). Use the index to find specific requirements.

Fitting. An accessory such as a locknut, bushing, or other part of a wiring system that is intended primarily to perform a mechanical rather than an electrical function.

Items such as condulets, conduit couplings, EMT connectors and couplings, and threadless connectors are considered fittings.

Garage. A building or portion of a building in which one or more self-propelled vehicles can be kept for use, sale, storage, rental, repair, exhibition, or demonstration purposes.

Revised for the 2002 *Code*, the definition of *garage* was simplified and includes the garages for electric vehicles covered in Article 625.

> FPN: For commercial garages, repair and storage, see Article 511.

Ground. A conducting connection, whether intentional or accidental, between an electrical circuit or equipment and the earth or to some conducting body that serves in place of the earth.

Grounded. Connected to earth or to some conducting body that serves in place of the earth.

Grounded, Effectively. Intentionally connected to earth through a ground connection or connections of sufficiently low impedance and having sufficient current-carrying capacity to prevent the buildup of voltages that may result in undue hazards to connected equipment or to persons.

Grounded, Solidly. Connected to ground without inserting any resistor or impedance device.

Grounded Conductor. A system or circuit conductor that is intentionally grounded.

Ground-Fault Circuit Interrupter (GFCI). A device intended for the protection of personnel that functions to de-energize a circuit or portion thereof within an established period of time when a current to ground exceeds the values established for a Class A device.

> FPN: Class A ground-fault circuit interrupters trip when the current to ground has a value in the range of 4 mA to 6 mA. For further information, see UL 943, *Standard for Ground-Fault Circuit Interrupters.*

Ground-Fault Protection of Equipment. A system intended to provide protection of equipment from damaging line-to-ground fault currents by operating to cause a disconnecting means to open all ungrounded conductors of the faulted circuit. This protection is provided at current levels less than those required to protect conductors from damage through the operation of a supply circuit overcurrent device.

Grounding Conductor. A conductor used to connect equipment or the grounded circuit of a wiring system to a grounding electrode or electrodes.

Grounding Conductor, Equipment. The conductor used to connect the non–current-carrying metal parts of equipment, raceways, and other enclosures to the system grounded conductor, the grounding electrode conductor, or both, at the service equipment or at the source of a separately derived system.

Grounding Electrode. A device that establishes an electrical connection to the earth.

Grounding Electrode Conductor. The conductor used to connect the grounding electrode(s) to the equipment grounding conductor, to the grounded conductor, or to both, at the service, at each building or structure where supplied by a feeder(s) or branch circuit(s), or at the source of a separately derived system.

Guarded. Covered, shielded, fenced, enclosed, or otherwise protected by means of suitable covers, casings, barriers, rails, screens, mats, or platforms to remove the likelihood of approach or contact by persons or objects to a point of danger.

Guest Room. An accommodation combining living, sleeping, sanitary, and storage facilities within a compartment.

Guest Suite. An accommodation with two or more contiguous rooms comprising a compartment, with or without doors between such rooms, that provides living, sleeping, sanitary, and storage facilities.

Handhole Enclosure. An enclosure identified for use in underground systems, provided with an open or closed bottom, and sized to allow personnel to reach into, but not enter, for the purpose of installing, operating, or maintaining equipment or wiring or both.

The term *handhole enclosure* is a much needed addition to Article 100 and is new in the 2005 *Code*. Requirements for handhole enclosures are found in 314.30. Exhibit 100.9 shows the installation of one type of handhole enclosure.

Exhibit 100.9 Example of handhole enclosure installation. (Courtesy of Strongwell)

Hoistway. Any shaftway, hatchway, well hole, or other vertical opening or space in which an elevator or dumbwaiter is designed to operate.

See Article 620 for the installation of electrical equipment and wiring methods in hoistways.

Identified (as applied to equipment). Recognizable as suitable for the specific purpose, function, use, environment, application, and so forth, where described in a particular *Code* requirement.

> FPN: Some examples of ways to determine suitability of equipment for a specific purpose, environment, or application include investigations by a qualified testing laboratory (listing and labeling), an inspection agency, or other organizations concerned with product evaluation.

In Sight From (Within Sight From, Within Sight). Where this *Code* specifies that one equipment shall be "in sight from," "within sight from," or "within sight," and so forth, of another equipment, the specified equipment is to be visible and not more than 15 m (50 ft) distant from the other.

Exhibit 430.20 depicts requirements for the placement of a disconnecting means that is not in sight.

Interrupting Rating. The highest current at rated voltage that a device is intended to interrupt under standard test conditions.

Interrupting ratings are essential in the coordination of electrical systems so that available fault currents can be properly controlled. Other sections specifically dealing with interrupting ratings are 110.9, 240.60(C), 240.83(C), and 240.86.

> FPN: Equipment intended to interrupt current at other than fault levels may have its interrupting rating implied in other ratings, such as horsepower or locked rotor current.

Isolated (as applied to location). Not readily accessible to persons unless special means for access are used.

See the definition of *accessible, readily.*

Labeled. Equipment or materials to which has been attached a label, symbol, or other identifying mark of an organization that is acceptable to the authority having jurisdiction and concerned with product evaluation, that maintains periodic inspection of production of labeled equipment or materials, and by whose labeling the manufacturer indicates compliance with appropriate standards or performance in a specified manner.

Equipment and conductors required or permitted by this *Code* are acceptable only if they have been approved for a specific environment or application by the authority having jurisdiction, as stated in 110.2. See 90.7 regarding the examination of equipment for safety. Listing or labeling by a qualified testing laboratory provides a basis for approval.

Lighting Outlet. An outlet intended for the direct connection of a lampholder, a luminaire (lighting fixture), or a pendant cord terminating in a lampholder.

Listed. Equipment, materials, or services included in a list published by an organization that is acceptable to the authority having jurisdiction and concerned with evaluation of products or services, that maintains periodic inspection of production of listed equipment or materials or periodic evaluation of services, and whose listing states that the equipment, material, or services either meets appropriate designated standards or has been tested and found suitable for a specified purpose.

FPN: The means for identifying listed equipment may vary for each organization concerned with product evaluation, some of which do not recognize equipment as listed unless it is also labeled. Use of the system employed by the listing organization allows the authority having jurisdiction to identify a listed product.

The *NEC* definition of *listed* matches the definition of *listed* found in the NFPA Regulations Governing Committee Projects. Reviewing other *NEC*-defined terms such as *approved, authority having jurisdiction (AHJ), identified (as applied to equipment),* and *labeled* will help the user understand the approval process.

Live Parts. Energized conductive components.

The definition of *live parts* is associated with all voltage levels, not just voltage levels that present a shock hazard.

Location, Damp. Locations protected from weather and not subject to saturation with water or other liquids but subject to moderate degrees of moisture. Examples of such locations include partially protected locations under canopies, marquees, roofed open porches, and like locations, and interior locations subject to moderate degrees of moisture, such as some basements, some barns, and some cold-storage warehouses.

Location, Dry. A location not normally subject to dampness or wetness. A location classified as dry may be temporarily subject to dampness or wetness, as in the case of a building under construction.

Location, Wet. Installations under ground or in concrete slabs or masonry in direct contact with the earth; in locations subject to saturation with water or other liquids, such as vehicle washing areas; and in unprotected locations exposed to weather.

It is intended that the inside of a raceway in a wet location or a raceway installed underground be considered a wet location. Therefore, any conductors contained therein would be required to be suitable for wet locations.

See 300.6(D) for some examples of wet locations and 410.4(A) for information on luminaires installed in wet locations. See *patient care area* in 517.2 for a definition of wet locations in a patient care area.

Luminaire. A complete lighting unit consisting of a lamp or lamps together with the parts designed to distribute the light, to position and protect the lamps and ballast (where applicable), and to connect the lamps to the power supply.

The term *luminaire* replaced the generic term *lighting fixture* throughout the *Code* for the 2002 edition. Although new lighting techniques such as light pipes and glass fiber optics are sometimes referred to as "lighting systems," the definition of *luminaire* does not necessarily preclude such systems, because light pipes and fiber optics are actually "parts designed to distribute the light."

Metal-Enclosed Power Switchgear. A switchgear assembly completely enclosed on all sides and top with sheet metal (except for ventilating openings and inspection windows) containing primary power circuit switching, interrupting devices, or both, with buses and connections. The assembly may include control and auxiliary devices. Access to the interior of the enclosure is provided by doors, removable covers, or both.

Motor Control Center. An assembly of one or more enclosed sections having a common power bus and principally containing motor control units.

Multioutlet Assembly. A type of surface, flush, or freestanding raceway designed to hold conductors and receptacles, assembled in the field or at the factory.

The definition of *multioutlet assembly* now includes a reference to a freestanding assembly with multiple outlets, commonly called a power pole. In dry locations, metallic and nonmetallic multioutlet assemblies are permitted; however, they are not permitted to be installed if concealed. See Article 380 for details on recessing multioutlet assemblies. Exhibit 100.10 shows a multioutlet assembly used for countertop appliances.

Nonautomatic. Action requiring personal intervention for its control. As applied to an electric controller, nonautomatic control does not necessarily imply a manual controller, but only that personal intervention is necessary.

Nonlinear Load. A load where the wave shape of the steady-state current does not follow the wave shape of the applied voltage.

Nonlinear loads are a major cause of harmonic currents in modern circuits. Additional conductor heating is just one of the undesirable operational effects often associated with harmonic currents. FPN No. 1 following 310.10 points out that harmonic current, as well as fundamental current, should be used in determining the heat generated internally in a conductor.

Actual circuit measurements of current for nonlinear

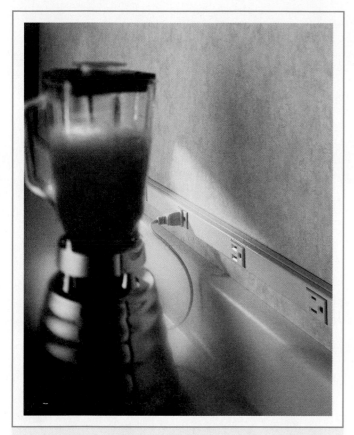

Exhibit 100.10 Multioutlet assembly installed to serve countertop appliances. (Courtesy of The Wiremold Co.)

loads should be made using only true rms-measuring ammeter instruments. Averaging ammeters produces inaccurate values if used to measure nonlinear loads. [See the associated commentary in 310.15(B)(4)(c).]

> FPN: Electronic equipment, electronic/electric-discharge lighting, adjustable-speed drive systems, and similar equipment may be nonlinear loads.

Outlet. A point on the wiring system at which current is taken to supply utilization equipment.

An example is a lighting outlet or a receptacle outlet.

Outline Lighting. An arrangement of incandescent lamps, electric discharge lighting, or other electrically powered light sources to outline or call attention to certain features such as the shape of a building or the decoration of a window.

Revised for the 2005 *Code*, the definition of *outline lighting* now clearly includes low-voltage light-emitting diodes as well as other luminaires installed to form various shapes. See Article 600 for details on outline lighting.

Overcurrent. Any current in excess of the rated current of equipment or the ampacity of a conductor. It may result from overload, short circuit, or ground fault.

> FPN: A current in excess of rating may be accommodated by certain equipment and conductors for a given set of conditions. Therefore, the rules for overcurrent protection are specific for particular situations.

Overload. Operation of equipment in excess of normal, full-load rating, or of a conductor in excess of rated ampacity that, when it persists for a sufficient length of time, would cause damage or dangerous overheating. A fault, such as a short circuit or ground fault, is not an overload.

Panelboard. A single panel or group of panel units designed for assembly in the form of a single panel, including buses and automatic overcurrent devices, and equipped with or without switches for the control of light, heat, or power circuits; designed to be placed in a cabinet or cutout box placed in or against a wall, partition, or other support; and accessible only from the front.

See Article 408, Parts I and III, for detailed requirements concerning panelboards.

Plenum. A compartment or chamber to which one or more air ducts are connected and that forms part of the air distribution system.

The definition of *plenum* in the *Code* is essentially the same as the definition of *plenum* in NFPA 90A, *Standard for the Installation of Air-Conditioning and Ventilating Systems*. For information on wiring methods permitted within plenums, see 300.22(B).

The definition of *plenum* is not intended to apply to the space above a suspended ceiling that is used for environmental air as referred to in 300.22(C).

The air-handling space under a computer room floor has special requirements. See Article 645.

Power Outlet. An enclosed assembly that may include receptacles, circuit breakers, fuseholders, fused switches, buses, and watt-hour meter mounting means; intended to supply and control power to mobile homes, recreational vehicles, park trailers, or boats or to serve as a means for distributing power required to operate mobile or temporarily installed equipment.

Premises Wiring (System). That interior and exterior wiring, including power, lighting, control, and signal circuit wiring together with all their associated hardware, fittings, and wiring devices, both permanently and temporarily installed, that extends from the service point or source of power, such as a battery, a solar photovoltaic system, or a generator, transformer, or converter windings, to the outlet(s). Such wiring does not include wiring internal to appliances, luminaires (fixtures), motors, controllers, motor control centers, and similar equipment.

Qualified Person. One who has skills and knowledge related to the construction and operation of the electrical equipment and installations and has received safety training on the hazards involved.

> FPN: Refer to NFPA 70E-2004, *Standard for Electrical Safety in the Workplace*, for electrical safety training requirements.

The following excerpt on training requirements is taken from 110.6 in the 2004 edition of NFPA 70E, *Standard for Electrical Safety in the Workplace*. These training requirements are presented here only as an aid to understanding the requisite minimum training requirements specified in NFPA 70E, a recognized and widely used workplace safety standard. It is imporrtant to understand that this commentary, like the fine print note following the definition of *qualified person*, is informational only, and mandatory application of these safety training provisions is dependent on whether NFPA 70E has been specifically adopted by the enforcing jurisdiction.

Excerpt from NFPA 70E-2004, *Standard for Electrical Safety in the Workplace*

110.6 Training Requirements.

(A) **Safety Training.** The training requirements contained in this section shall apply to employees who face a risk of electrical hazard that is not reduced to a safe level by the electrical installation requirements of Chapter 4 [of NFPA 70E]. Such employees shall be trained to understand the specific hazards associated with electrical energy. They shall be trained in safety-related work practices and procedural requirements as necessary to provide protection from the electrical hazards associated with their respective job or task assignments. Employees shall be trained to identify and understand the relationship between electrical hazards and possible injury.

(B) **Type of Training.** The training required by this section shall be classroom or on-the-job type, or a combination of the two. The degree of training provided shall be determined by the risk to the employee.

(C) **Emergency Procedures.** Employees working on or near exposed energized electrical conductors or circuit parts shall be trained in methods of release of victims from contact with exposed energized conductors or circuit parts. Employees shall be regularly instructed in methods of first aid and emergency procedures, such as approved methods of resuscitation, if their duties warrant such training.

(D) **Employee Training.**

(1) **Qualified Person.** A qualified person shall be trained and knowledgeable of the construction and operation of equipment or a specific work method and be trained to recognize and avoid the electrical hazards that might be present with respect to that equipment or work method.

(a) Such persons shall also be familiar with the proper use of the special precautionary techniques, personal protective equipment, including arc-flash, insulating and shielding materials, and insulated tools and test equipment. A person can be considered qualified with respect to certain equipment and methods but still be unqualified for others.

(b) An employee who is undergoing on-the-job training and who, in the course of such training, has demonstrated an ability to perform duties safely at his or her level of training and who is under the direct supervision of a qualified person shall be considered to be a qualified person for the performance of those duties.

(c) Such persons permitted to work within the Limited Approach Boundary of exposed live parts operating at 50 volts or more shall, at a minimum, be additionally trained in all of the following:

(1) The skills and techniques necessary to distinguish exposed energized parts from other parts of electrical equipment

(2) The skills and techniques necessary to determine the nominal voltage of exposed live parts

(3) The approach distances specified in Table 130.2(C) [of NFPA 70E] and the corresponding voltages to which the qualified person will be exposed

(4) The decision-making process necessary to determine the degree and extent of the hazard and the personal protective equipment and job planning necessary to perform the task safely

(2) **Unqualified Persons.** Unqualified persons shall be trained in and be familiar with any of the electrical safety-related practices that might not be addressed specifically by Chapter 1 [of NFPA 70E] but are necessary for their safety.

Raceway. An enclosed channel of metal or nonmetallic materials designed expressly for holding wires, cables, or busbars, with additional functions as permitted in this *Code*. Raceways include, but are not limited to, rigid metal conduit, rigid nonmetallic conduit, intermediate metal conduit, liquidtight flexible conduit, flexible metallic tubing, flexible metal conduit, electrical nonmetallic tubing, electrical metallic tubing, underfloor raceways, cellular concrete floor raceways, cellular metal floor raceways, surface raceways, wireways, and busways.

Raceways are covered generally within Article 300 and specifically throughout Chapter 3. Cable trays (see Article

392) are support systems for wiring methods and are not considered to be raceways.

Rainproof. Constructed, protected, or treated so as to prevent rain from interfering with the successful operation of the apparatus under specified test conditions.

See the commentary following the definition of *enclosure*.

Raintight. Constructed or protected so that exposure to a beating rain will not result in the entrance of water under specified test conditions.

Raceways on exterior surfaces of buildings are required to be made raintight (see 225.22 and 230.53). For boxes and cabinets, see 300.6. Also see the commentary following the definitions of *location*, *wet*, and *enclosure*.

Receptacle. A receptacle is a contact device installed at the outlet for the connection of an attachment plug. A single receptacle is a single contact device with no other contact device on the same yoke. A multiple receptacle is two or more contact devices on the same yoke.

Exhibit 100.11 shows one single and two multiple receptacles.

| Single Receptacle | Multiple Receptacle (Duplex) | Multiple Receptacle |

Exhibit 100.11 Receptacles.

Receptacle Outlet. An outlet where one or more receptacles are installed.

See Exhibit 100.11 and the commentary following 220.3(B)(9).

Remote-Control Circuit. Any electric circuit that controls any other circuit through a relay or an equivalent device.

Exhibit 100.12 illustrates a remote-control circuit that starts and stops an electric motor.

Remote-control circuit

Stop

Start

M

Exhibit 100.12 Remote-control circuit for starting and stopping an electric motor.

Sealable Equipment. Equipment enclosed in a case or cabinet that is provided with a means of sealing or locking so that live parts cannot be made accessible without opening the enclosure. The equipment may or may not be operable without opening the enclosure.

Separately Derived System. A premises wiring system whose power is derived from a source of electric energy or equipment other than a service. Such systems have no direct electrical connection, including a solidly connected grounded circuit conductor, to supply conductors originating in another system.

Revised for the 2005 *Code*, the definition of *separately derived system* more accurately describes the term, but the examples of such systems have been deleted from the definition. Some examples of a separately derived system may include a generator, a battery, converter windings, a transformer, and a solar photovoltaic system provided they "have no direct electrical connection" to another source.

Service. The conductors and equipment for delivering electric energy from the serving utility to the wiring system of the premises served.

The definition of *service* includes the statement that electric energy to a service can be supplied only by the serving utility. If electric energy is supplied by other than the serving utility, the supplied conductors and equipment are considered feeders, not a service.

Service Cable. Service conductors made up in the form of a cable.

Service Conductors. The conductors from the service point to the service disconnecting means.

Service conductors is a broad term and may include service drops, service laterals, and service-entrance conductors. This term specifically excludes, however, any wiring on the supply side (serving utility side) of the service point. Simply put, the service conductors originate at the service point (where the serving utility ends) and end at the service disconnect. These service conductors may originate only from the serving utility.

If the utility has specified that the service point is at the utility pole, then the service conductors from an overhead distribution system originate at the utility pole and terminate at the service disconnecting means.

If the utility has specified that the service point is at the utility manhole, then the service conductors from an underground distribution system originate at the utility manhole and terminate at the service disconnecting means. Where utility-owned primary conductors are extended to outdoor pad-mounted transformers on private property, the service conductors originate at the secondary connections of the transformers only if the utility has specified that the service point is at the secondary connections.

See Article 230, Part VIII, and the commentary following 230.200 for service conductors exceeding 600 volts, nominal.

Service Drop. The overhead service conductors from the last pole or other aerial support to and including the splices, if any, connecting to the service-entrance conductors at the building or other structure.

In Exhibit 100.13, the overhead service-drop conductors run from the utility pole and connect to the service-entrance conductors at the service point. Conductors on the utility side of the service point are not covered by the *NEC*. The utility specifies the location of the service point. Exact locations of the service point may vary from utility to utility, as well as from occupancy to occupancy.

Exhibit 100.13 Overhead system showing a service drop from a utility pole to attachment on a house and service-entrance conductors from point of attachment (spliced to service-drop conductors), down the side of the house, through the meter socket, and terminating in the service equipment.

Service-Entrance Conductors, Overhead System. The service conductors between the terminals of the service equipment and a point usually outside the building, clear of building walls, where joined by tap or splice to the service drop.

See Exhibit 100.13 for an illustration of service-entrance conductors in an overhead system.

Service-Entrance Conductors, Underground System. The service conductors between the terminals of the service equipment and the point of connection to the service lateral.

FPN: Where service equipment is located outside the building walls, there may be no service-entrance conductors or they may be entirely outside the building.

See Exhibit 100.14 for an illustration of service-entrance conductors in an underground system.

Exhibit 100.14 Underground systems showing service laterals run from a pole and from a transformer.

Service Equipment. The necessary equipment, usually consisting of a circuit breaker(s) or switch(es) and fuse(s) and their accessories, connected to the load end of service conductors to a building or other structure, or an otherwise designated area, and intended to constitute the main control and cutoff of the supply.

Service equipment may consist of circuit breakers or fused switches that are provided to disconnect all ungrounded conductors in a building or other structure from the service-entrance conductors. It is important to understand that individual meter socket enclosures are not considered service equipment according to 230.66. A case could be made that potential and current transformer cabinets associated with utility meter enclosures are also excluded from the definition of *service equipment.*

The disconnecting means at any one location is not allowed to consist of more than six circuit breakers or six

switches and is required to be readily accessible either outside the building or structure or inside nearest the point of entrance of the service-entrance conductors. See 230.6 for service conductors outside the building and Article 230, Part VI, for service disconnecting means.

Service Lateral. The underground service conductors between the street main, including any risers at a pole or other structure or from transformers, and the first point of connection to the service-entrance conductors in a terminal box or meter or other enclosure, inside or outside the building wall. Where there is no terminal box, meter, or other enclosure, the point of connection is considered to be the point of entrance of the service conductors into the building.

As Exhibit 100.14 shows, the underground service laterals may be run from poles or from transformers and with or without terminal boxes, provided they begin at the service point. Conductors on the utility side of the service point are not covered by the *NEC.* The utility specifies the location of the service point. Exact locations of the service point may vary from utility to utility, as well as from occupancy to occupancy.

Service Point. The point of connection between the facilities of the serving utility and the premises wiring.

The service point is the point of demarcation between the serving utility and the premises wiring. The service point is the point on the wiring system where the serving utility ends and the premises wiring begins. The serving utility generally specifies the location of the service point.

Because the location of the service point is generally determined by the utility, the service-drop conductors and the service-lateral conductors may or may not be part of the service covered by the *NEC.* For these types of conductors to be covered, they must be physically located on the premises wiring side of the service point. If the conductors are located on the utility side of the service point, they are not covered by the definition of *service conductors* and are therefore not covered by the *NEC.*

Based on the definitions of the terms *service point* and *service conductors,* any conductor on the serving utility side of the service point generally is not covered by the *NEC.* For example, a typical suburban residence has an overhead service drop from the utility pole to the house. If the utility specifies that the service point is at the point of attachment of the service drop to the house, then the service-drop conductors are not considered service conductors because the service drop is not on the premises wiring side of the service point. Alternatively, if the utility specifies that the service point is "at the pole," then the service-drop conductors are considered service conductors, and the *NEC* would apply to the service drop.

Exact locations for a service point may vary from utility to utility, as well as from occupancy to occupancy.

Show Window. Any window used or designed to be used for the display of goods or advertising material, whether it is fully or partly enclosed or entirely open at the rear and whether or not it has a platform raised higher than the street floor level.

See 220.14(G), 220.43(A), and Exhibit 220.1 for show-window lighting load requirements.

Signaling Circuit. Any electric circuit that energizes signaling equipment.

Solar Photovoltaic System. The total components and subsystems that, in combination, convert solar energy into electrical energy suitable for connection to a utilization load.

See Article 690 for solar photovoltaic system requirements.

Special Permission. The written consent of the authority having jurisdiction.

The authority having jurisdiction for enforcement of the *Code* is responsible for making interpretations and granting special permission contemplated in a number of the rules, as stated in 90.4. For specific examples of special permission, see 110.26(A)(1)(b), 230.2(B), and 426.14.

Structure. That which is built or constructed.

Added for the 2002 *Code*, this definition of *structure* allows architects, electrical engineers, general contractors, electrical contractors, and all building officials to use the same definition.

Supplementary Overcurrent Protective Device. A device intended to provide limited overcurrent protection for specific applications and utilization equipment such as luminaires (lighting fixtures) and appliances. This limited protection is in addition to the protection provided in the required branch circuit by the branch circuit overcurrent protective device.

There are two levels of overcurrent protection within branch circuits: branch circuit overcurrent protection and supplementary overcurrent protection. The devices used to provide

overcurrent protection are different, and the differences are found in the product standards UL 489, *Molded-Case Circuit Breakers, Molded-Case Switches and Circuit-Breaker Enclosures,* and UL 1077, *Supplementary Protectors for Use in Electrical Equipment.*

Provided as a generalization for understanding, the *NEC* requires that all branch circuits use only branch circuit "rated" overcurrent protective devices to protect branch circuits, but it permits supplementary overcurrent protection devices for limited use downstream of the branch circuit "rated" overcurrent protective device.

Added for the 2005 *Code*, the definition of *supplementary overcurrent protection device* contains two important distinctions between supplementary overcurrent protection devices and branch circuit overcurrent protective devices. First, the use of a supplementary device is specifically limited to only a few applications. Second, where it is used, the supplementary device must be in addition to and be protected by the more robust branch circuit overcurrent protective device.

Switch, Bypass Isolation. A manually operated device used in conjunction with a transfer switch to provide a means of directly connecting load conductors to a power source and of disconnecting the transfer switch.

See 700.6(B) and 701.7(B) for further information on bypass isolation transfer switches.

Switch, General-Use. A switch intended for use in general distribution and branch circuits. It is rated in amperes, and it is capable of interrupting its rated current at its rated voltage.

Switch, General-Use Snap. A form of general-use switch constructed so that it can be installed in device boxes or on box covers, or otherwise used in conjunction with wiring systems recognized by this *Code*.

Switch, Isolating. A switch intended for isolating an electric circuit from the source of power. It has no interrupting rating, and it is intended to be operated only after the circuit has been opened by some other means.

Switch, Motor-Circuit. A switch rated in horsepower that is capable of interrupting the maximum operating overload current of a motor of the same horsepower rating as the switch at the rated voltage.

Switch, Transfer. An automatic or nonautomatic device for transferring one or more load conductor connections from one power source to another.

Switchboard. A large single panel, frame, or assembly of panels on which are mounted on the face, back, or both, switches, overcurrent and other protective devices, buses, and usually instruments. Switchboards are generally accessible from the rear as well as from the front and are not intended to be installed in cabinets.

Busbars are required to be arranged to avoid inductive overheating. Service busbars are required to be isolated by barriers from the remainder of the switchboard. Most modern switchboards are totally enclosed to minimize the probability of spreading fire to adjacent combustible materials and to guard live parts. See Article 408 for more information regarding switchboards.

Thermally Protected (as applied to motors). The words *Thermally Protected* appearing on the nameplate of a motor or motor-compressor indicate that the motor is provided with a thermal protector.

Thermal Protector (as applied to motors). A protective device for assembly as an integral part of a motor or motor-compressor that, when properly applied, protects the motor against dangerous overheating due to overload and failure to start.

FPN: The thermal protector may consist of one or more sensing elements integral with the motor or motor-compressor and an external control device.

Utilization Equipment. Equipment that utilizes electric energy for electronic, electromechanical, chemical, heating, lighting, or similar purposes.

Ventilated. Provided with a means to permit circulation of air sufficient to remove an excess of heat, fumes, or vapors.

See the commentary following 110.13(B).

Volatile Flammable Liquid. A flammable liquid having a flash point below 38°C (100°F), or a flammable liquid whose temperature is above its flash point, or a Class II combustible liquid that has a vapor pressure not exceeding 276 kPa (40 psia) at 38°C (100°F) and whose temperature is above its flash point.

The flash point of a liquid is defined as the minimum temperature at which it gives off sufficient vapor to form an ignitible mixture, with the air near the surface of the liquid or within the vessel used to contain the liquid. An ignitible mixture is defined as a mixture within the explosive or flammable range (between upper and lower limits) that is capable of the propagation of flame away from the source of ignition when ignited. Some emission of vapors takes place below the flash point but not in sufficient quantities to form an ignitible mixture.

Voltage (of a circuit). The greatest root-mean-square (rms) (effective) difference of potential between any two conductors of the circuit concerned.

Common 3-phase, 4-wire wye systems are 480/277 volts and 208/120 volts. The voltage of the circuit is the higher voltage between any two conductors (i.e., 480 volts or 208 volts). The voltage of the circuit of a 2-wire feeder or branch circuit (single phase and the grounded conductor) derived from these systems would be the lower voltage between two conductors (i.e., 277 volts or 120 volts). The same applies to dc or single-phase, 3-wire systems where there are two voltages.

FPN: Some systems, such as 3-phase 4-wire, single-phase 3-wire, and 3-wire direct current, may have various circuits of various voltages.

Voltage, Nominal. A nominal value assigned to a circuit or system for the purpose of conveniently designating its voltage class (e.g., 120/240 volts, 480Y/277 volts, 600 volts). The actual voltage at which a circuit operates can vary from the nominal within a range that permits satisfactory operation of equipment.

See 220.5(A) for a list of nominal voltages used in computing branch-circuit and feeder loads.

FPN: See ANSI C84.1-1995, *Voltage Ratings for Electric Power Systems and Equipment (60 Hz).*

Voltage to Ground. For grounded circuits, the voltage between the given conductor and that point or conductor of the circuit that is grounded; for ungrounded circuits, the greatest voltage between the given conductor and any other conductor of the circuit.

The voltage to ground of a 277/480-volt wye system would be 277 volts; of a 120/208-volt wye system, 120 volts; and of a 3-phase, 3-wire ungrounded 480-volt system, 480 volts.

For a 3-phase, 4-wire delta system with the center of one leg grounded, there are two voltages to ground. For example, on a 240-volt system, two legs would each have 120 volts to ground, and the third, or "high," leg would have 208 volts to ground. See 110.15, 230.56, and 408.3(E) for special marking and arrangements on such circuit conductors.

Watertight. Constructed so that moisture will not enter the enclosure under specified test conditions.

Unless an enclosure is hermetically sealed, it is possible for moisture to enter the enclosure. See the commentary following the definition of *enclosure* and following Table 430.91.

Weatherproof. Constructed or protected so that exposure to the weather will not interfere with successful operation.

> FPN: Rainproof, raintight, or watertight equipment can fulfill the requirements for weatherproof where varying weather conditions other than wetness, such as snow, ice, dust, or temperature extremes, are not a factor.

See the commentary following the definition of *enclosure*. Industry standards for enclosures are found in the commentary following 430.91.

II. Over 600 Volts, Nominal

Whereas the preceding definitions are intended to apply wherever the terms are used throughout this *Code*, the following definitions are applicable only to parts of the article specifically covering installations and equipment operating at over 600 volts, nominal.

Electronically Actuated Fuse. An overcurrent protective device that generally consists of a control module that provides current sensing, electronically derived time–current characteristics, energy to initiate tripping, and an interrupting module that interrupts current when an overcurrent occurs. Electronically actuated fuses may or may not operate in a current-limiting fashion, depending on the type of control selected.

Although they are called fuses because they interrupt current by melting a fusible element, electronically actuated fuses respond to a signal from an electronic control rather than from the heat generated by actual current passing through a fusible element. Electronically actuated fuses have controls similar to those of electronic circuit breakers.

Fuse. An overcurrent protective device with a circuit-opening fusible part that is heated and severed by the passage of overcurrent through it.

> FPN: A fuse comprises all the parts that form a unit capable of performing the prescribed functions. It may or may not be the complete device necessary to connect it into an electrical circuit.

Controlled Vented Power Fuse. A fuse with provision for controlling discharge circuit interruption such that no solid material may be exhausted into the surrounding atmosphere.

> FPN: The fuse is designed so that discharged gases will not ignite or damage insulation in the path of the dis-

charge or propagate a flashover to or between grounded members or conduction members in the path of the discharge where the distance between the vent and such insulation or conduction members conforms to manufacturer's recommendations.

Expulsion Fuse Unit (Expulsion Fuse). A vented fuse unit in which the expulsion effect of gases produced by the arc and lining of the fuseholder, either alone or aided by a spring, extinguishes the arc.

Nonvented Power Fuse. A fuse without intentional provision for the escape of arc gases, liquids, or solid particles to the atmosphere during circuit interruption.

Power Fuse Unit. A vented, nonvented, or controlled vented fuse unit in which the arc is extinguished by being drawn through solid material, granular material, or liquid, either alone or aided by a spring.

Vented Power Fuse. A fuse with provision for the escape of arc gases, liquids, or solid particles to the surrounding atmosphere during circuit interruption.

Multiple Fuse. An assembly of two or more single-pole fuses.

Switching Device. A device designed to close, open, or both, one or more electric circuits.

Circuit Breaker. A switching device capable of making, carrying, and interrupting currents under normal circuit conditions, and also of making, carrying for a specified time, and interrupting currents under specified abnormal circuit conditions, such as those of short circuit.

Cutout. An assembly of a fuse support with either a fuseholder, fuse carrier, or disconnecting blade. The fuseholder or fuse carrier may include a conducting element (fuse link) or may act as the disconnecting blade by the inclusion of a nonfusible member.

Disconnecting (or Isolating) Switch (Disconnector, Isolator). A mechanical switching device used for isolating a circuit or equipment from a source of power.

Disconnecting Means. A device, group of devices, or other means whereby the conductors of a circuit can be disconnected from their source of supply.

Interrupter Switch. A switch capable of making, carrying, and interrupting specified currents.

Oil Cutout (Oil-Filled Cutout). A cutout in which all or part of the fuse support and its fuse link or disconnecting blade is mounted in oil with complete immersion of the contacts and the fusible portion of the conducting element (fuse link) so that arc interruption by severing of the fuse link or by opening of the contacts will occur under oil.

Oil Switch. A switch having contacts that operate under oil (or askarel or other suitable liquid).

Regulator Bypass Switch. A specific device or combination of devices designed to bypass a regulator.

ARTICLE 110
Requirements for Electrical Installations

Summary of Changes

- **110.1:** Revised paragraph to include enclosures intended for personnel entry.

- **110.12:** Added FPN referencing ANSI-approved standards.

- **110.15:** Revised paragraph to clarify application of special identification to the high leg only.

- **110.16:** Revised paragraph to include meter socket enclosures.

- **110.26(C)(2):** Deleted six-ft width limitation so that requirement applies to all equipment rated 1200 amperes and greater and containing overcurrent devices, switching devices, and control devices.

- **Part V, 110.70–110.79:** Moved Article 314, Part IV to Article 110.

Contents

I. General

110.1 Scope

This article covers general requirements for the examination and approval, installation and use, access to and spaces about electrical conductors and equipment; enclosures intended for personnel entry; and tunnel installations.

110.2 Approval

The conductors and equipment required or permitted by this *Code* shall be acceptable only if approved.

> FPN: See 90.7, Examination of Equipment for Safety, and 110.3, Examination, Identification, Installation, and Use of Equipment. See definitions of *Approved, Identified, Labeled,* and *Listed.*

All electrical equipment is required to be approved as defined in Article 100 and, as such, to be acceptable to the authority having jurisdiction (also defined in Article 100). Section 110.3 provides guidance for the evaluation of equipment and recognizes listing or labeling as a means of establishing suitability.

Approval of equipment is the responsibility of the electrical inspection authority, and many such approvals are based on tests and listings of testing laboratories.

110.3 Examination, Identification, Installation, and Use of Equipment

(A) Examination In judging equipment, considerations such as the following shall be evaluated:

(1) Suitability for installation and use in conformity with the provisions of this *Code*

> FPN: Suitability of equipment use may be identified by a description marked on or provided with a product to identify the suitability of the product for a specific purpose, environment, or application. Suitability of equipment may be evidenced by listing or labeling.

(2) Mechanical strength and durability, including, for parts designed to enclose and protect other equipment, the adequacy of the protection thus provided
(3) Wire-bending and connection space
(4) Electrical insulation
(5) Heating effects under normal conditions of use and also under abnormal conditions likely to arise in service
(6) Arcing effects
(7) Classification by type, size, voltage, current capacity, and specific use
(8) Other factors that contribute to the practical safeguarding of persons using or likely to come in contact with the equipment

For wire-bending and connection space in cabinets and cutout boxes, see 312.6, Table 312.6(A), Table 312.6(B), 312.7, 312.9, and 312.11. For wire-bending and connection space in other equipment, see the appropriate *NEC* article and section. For example, see 314.16 and 314.28 for outlet, device, pull, and junction boxes, as well as conduit bodies; 404.3 and 404.18 for switches; 408.3(F) for switchboards and panelboards; and 430.10 for motors and motor controllers.

(B) Installation and Use Listed or labeled equipment shall be installed and used in accordance with any instructions included in the listing or labeling.

Manufacturers usually supply installation instructions with equipment for use by general contractors, erectors, electrical contractors, electrical inspectors, and others concerned with an installation. It is important to follow the listing or labeling installation instructions. For example, 210.52, second paragraph, permits permanently installed electric baseboard heaters to be equipped with receptacle outlets that meet the requirements for the wall space utilized by such heaters. The installation instructions for such permanent baseboard heaters indicate that the heaters should not be mounted beneath a receptacle. In dwelling units, it is common to use low-density heating units that measure in excess of 12 ft in length. Therefore, to meet the provisions of 210.52(A) and also the installation instructions, a receptacle must either be part of the heating unit or be installed in the floor close to the wall but not above the heating unit. (See 210.52, FPN, and Exhibit 210.23 for more specific details.)

In itself, 110.3 does not require listing or labeling of equipment. It does, however, require considerable evaluation of equipment. Section 110.2 requires that equipment be acceptable only if approved. The term *approved* is defined in Article 100 as acceptable to the authority having jurisdiction (AHJ). Before issuing approval, the authority having jurisdiction may require evidence of compliance with 110.3(A). The most common form of evidence considered acceptable by authorities having jurisdiction is a listing or labeling by a third party.

Some sections in the *Code* require listed or labeled equipment. For example, 250.8 includes the phrase "listed pressure connectors, listed clamps, or other listed means."

110.4 Voltages

Throughout this *Code*, the voltage considered shall be that at which the circuit operates. The voltage rating of electrical equipment shall not be less than the nominal voltage of a circuit to which it is connected.

Voltages used for computing branch-circuit and feeder loads are nominal voltages as listed in 220.5. See the definitions of *voltage (of a circuit)*; *voltage, nominal*; and *voltage to ground* in Article 100. See also 300.2 and 300.3(C), which specify the voltage limitations of conductors of circuits rated 600 volts, nominal, or less, and over 600 volts, nominal.

110.5 Conductors

Conductors normally used to carry current shall be of copper unless otherwise provided in this *Code*. Where the conductor material is not specified, the material and the sizes given in this *Code* shall apply to copper conductors. Where other materials are used, the size shall be changed accordingly.

FPN: For aluminum and copper-clad aluminum conductors, see 310.15.

See 310.14 for aluminum conductor material.

110.6 Conductor Sizes

Conductor sizes are expressed in American Wire Gage (AWG) or in circular mils.

For copper, aluminum, or copper-clad aluminum conductors up to size 4/0 AWG, this *Code* uses the American Wire Gage (AWG) for size identification, which is the same as the Brown and Sharpe (BS) Gage. Changed for the 2002 *Code*, wire sizes up to size 4/0 AWG are now expressed as XX AWG, XX being the size wire. For example, a wire size expressed as *No. 12* in prior editions of the *Code* is now

expressed as *12 AWG*. The resulting expression would therefore appear as *six 12 AWG conductors* instead of *6 No. 12 conductors*.

Conductors larger than 4/0 AWG are sized in circular mils, beginning with 250,000 circular mils. Prior to the 1990 edition, a 250,000-circular-mil conductor was labeled *250 MCM*. The term *MCM* was defined as 1000 circular mils (the first *M* being the Roman numeral designation for 1000). Beginning in the 1990 edition, the notation was changed to 250 *kcmil* to recognize the accepted convention that *k* indicates 1000. UL standards and IEEE standards also use the notation *kcmil* rather than *MCM*.

The circular mil area of a conductor is equal to its diameter in mils squared (1 in. = 1000 mils). For example, the circular mil area of an 8 AWG solid conductor that has a 0.1285-in. diameter is calculated as follows:

$$0.1285 \text{ in.} \times 1000 = 128.5 \text{ mils}$$

$$128.5 \times 128.5 = 16{,}512.25 \text{ circular mils}$$

or 16,510 circular mils (rounded off)

According to Table 8 in Chapter 9, this rounded value represents the circular mil area for one conductor. Where stranded conductors are used, the circular mil area of each strand must be multiplied by the number of strands to determine the circular mil area of the conductor.

110.7 Insulation Integrity

Completed wiring installations shall be free from short circuits and from grounds other than as required or permitted in Article 250.

Insulation is the material that prevents the flow of electricity between points of different potential in an electrical system. Failure of the insulation system is one of the most common causes of problems in electrical installations, in both high-voltage and low-voltage systems.

Insulation tests are performed on new or existing installations to determine the quality or condition of the insulation of conductors and equipment. The principal causes of insulation failures are heat, moisture, dirt, and physical damage (abrasion or nicks) occurring during and after installation. Insulation can also fail due to chemical attack, sunlight, and excessive voltage stresses.

Insulation integrity must be maintained during overcurrent conditions. Overcurrent protective devices must be selected and coordinated using tables of insulation thermal-withstand ability to ensure that the damage point of an insulated conductor is never reached. These tables, entitled "Allowable Short-Circuit Currents for Insulated Copper (or Aluminum) Conductors," are contained in the Insulated Cable Engineers Association's publication ICEA P-32-382. See 110.10 for other circuit components.

In an insulation resistance test, a voltage ranging from

100 to 5000 (usually 500 to 1000 volts for systems of 600 volts or less), supplied from a source of constant potential, is applied across the insulation. A megohmmeter is usually the potential source, and it indicates the insulation resistance directly on a scale calibrated in megohms (MÙ). The quality of the insulation is evaluated based on the level of the insulation resistance.

The insulation resistance of many types of insulation varies with temperature, so the field data obtained should be corrected to the standard temperature for the class of equipment being tested. The megohm value of insulation resistance obtained is inversely proportional to the volume of insulation tested. For example, a cable 1000 ft long would be expected to have one-tenth the insulation resistance of a cable 100 ft long, if all other conditions are identical.

The insulation resistance test is relatively easy to perform and is useful on all types and classes of electrical equipment. Its main value lies in the charting of data from periodic tests, corrected for temperature, over a long period so that deteriorative trends can be detected.

Manuals on this subject are available from instrument manufacturers. Thorough knowledge in the use of insulation testers is essential if the test results are to be meaningful. Exhibit 110.1 shows a typical megohmmeter insulation tester.

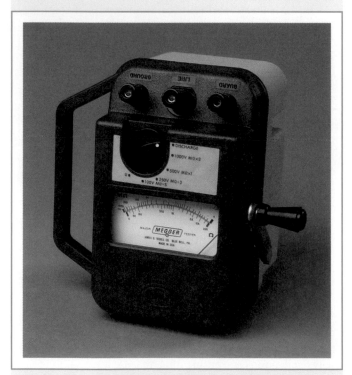

Exhibit 110.1 A manual multivoltage, multirange insulation tester.

110.8 Wiring Methods

Only wiring methods recognized as suitable are included in this *Code*. The recognized methods of wiring shall be permitted to be installed in any type of building or occupancy, except as otherwise provided in this *Code*.

The scope of Article 300 applies generally to all wiring methods, except as amended, modified, or supplemented by other *NEC* chapters. The application statement is found in 90.3, Code Arrangement.

110.9 Interrupting Rating

Equipment intended to interrupt current at fault levels shall have an interrupting rating sufficient for the nominal circuit voltage and the current that is available at the line terminals of the equipment.

Equipment intended to interrupt current at other than fault levels shall have an interrupting rating at nominal circuit voltage sufficient for the current that must be interrupted.

The interrupting rating of overcurrent protective devices is determined under standard test conditions. It is important that the test conditions match the actual installation needs. Section 110.9 states that all fuses and circuit breakers intended to interrupt the circuit at fault levels must have an adequate interrupting rating wherever they are used in the electrical system. Fuses or circuit breakers that do not have adequate interrupting ratings could rupture while attempting to clear a short circuit.

Interrupting ratings should not be confused with short-circuit current ratings. Short-circuit current ratings are further explained in the commentary following 110.10.

110.10 Circuit Impedance and Other Characteristics

The overcurrent protective devices, the total impedance, the component short-circuit current ratings, and other characteristics of the circuit to be protected shall be selected and coordinated to permit the circuit-protective devices used to clear a fault to do so without extensive damage to the electrical components of the circuit. This fault shall be assumed to be either between two or more of the circuit conductors or between any circuit conductor and the grounding conductor or enclosing metal raceway. Listed products applied in accordance with their listing shall be considered to meet the requirements of this section.

In the 1999 *Code*, the word *current* was substituted for the obsolete word *withstand*. That change correlated the *Code* language with the standard marking language used on equipment. Withstand ratings are not marked on equipment, but short-circuit current ratings are. This marking appears on many pieces of equipment, such as panelboards, switchboards, busways, contactors, and starters. Additionally, the last sentence of 110.10 is meant to address concerns of what exactly constitutes "extensive damage." Because, under product safety requirements, electrical equipment is evaluated for indications of extensive damage, listed products used within their ratings are considered to have met the requirements of 110.10.

The basic purpose of overcurrent protection is to open the circuit before conductors or conductor insulation is damaged when an overcurrent condition occurs. An overcurrent condition can be the result of an overload, a ground fault, or a short circuit and must be eliminated before the conductor insulation damage point is reached.

Overcurrent protective devices (such as fuses and circuit breakers) should be selected to ensure that the short-circuit current rating of the system components is not exceeded should a short circuit or high-level ground fault occur.

System components include wire, bus structures, switching, protection and disconnect devices, and distribution equipment, all of which have limited short-circuit ratings and would be damaged or destroyed if those short-circuit ratings were exceeded. Merely providing overcurrent protective devices with sufficient interrupting rating would not ensure adequate short-circuit protection for the system components. When the available short-circuit current exceeds the short-circuit current rating of an electrical component, the overcurrent protective device must limit the let-through energy to within the rating of that electrical component.

Utility companies usually determine and provide information on available short-circuit current levels at the service equipment. Literature on how to calculate short-circuit currents at each point in any distribution generally can be obtained by contacting the manufacturers of overcurrent protective devices or by referring to IEEE 141-1993, *IEEE Recommended Practice for Electric Power Distribution for Industrial Plants* (Red Book).

For a typical one-family dwelling with a 100-ampere service using 2 AWG aluminum supplied by a 37½ kVA transformer with 1.72 percent impedance located at a distance of 25 ft, the available short-circuit current would be approximately 6000 amperes.

Available short-circuit current to multifamily structures, where pad-mounted transformers are located close to the multimetering location, can be relatively high. For example, the line-to-line fault current values close to a low-impedance transformer could exceed 22,000 amperes. At the secondary of a single-phase, center-tapped transformer, the line-to-neutral fault current is approximately one and one-half times that of the line-to-line fault current. The short-circuit current rating of utilization equipment located and connected near the service equipment should be known. For example, HVAC equipment is tested at 3500 amperes through a 40-ampere load rating and at 5000 amperes for loads rated more than 40 amperes.

Adequate short-circuit protection can be provided by

fuses, molded-case circuit breakers, and low-voltage power circuit breakers, depending on specific circuit and installation requirements.

110.11 Deteriorating Agents

Unless identified for use in the operating environment, no conductors or equipment shall be located in damp or wet locations; where exposed to gases, fumes, vapors, liquids, or other agents that have a deteriorating effect on the conductors or equipment; or where exposed to excessive temperatures.

> FPN No. 1: See 300.6 for protection against corrosion.
> FPN No. 2: Some cleaning and lubricating compounds can cause severe deterioration of many plastic materials used for insulating and structural applications in equipment.

Equipment identified only as "dry locations," "Type 1," or "indoor use only" shall be protected against permanent damage from the weather during building construction.

110.12 Mechanical Execution of Work

Electrical equipment shall be installed in a neat and workmanlike manner.

> FPN: Accepted industry practices are described in ANSI/NECA 1-2000, *Standard Practices for Good Workmanship in Electrical Contracting,* and other ANSI-approved installation standards.

The regulation in 110.12 calling for "neat and workmanlike" installations has appeared in the *NEC* as currently worded for more than a half-century. It stands as a basis for pride in one's work and has been emphasized by persons involved in the training of apprentice electricians for many years.

Many *Code* conflicts or violations have been cited by the authority having jurisdiction based on the authority's interpretation of "neat and workmanlike manner." Many electrical inspection authorities use their own experience or precedents in their local areas as the basis for their judgments.

Examples of installations that do not qualify as "neat and workmanlike" include exposed runs of cables or raceways that are improperly supported (e.g., sagging between supports or use of improper support methods); field-bent and kinked, flattened, or poorly measured raceways; or cabinets, cutout boxes, and enclosures that are not plumb or not properly secured.

The FPN, new for the 2005 *Code,* directs the user to an industry accepted ANSI standard that clearly describes and illustrates "neat and workmanlike" electrical installations. See Exhibit 110.2.

(A) Unused Openings Unused cable or raceway openings in boxes, raceways, auxiliary gutters, cabinets, cutout boxes, meter socket enclosures, equipment cases, or housings shall

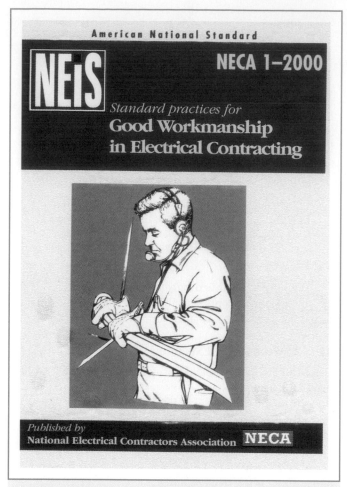

American National Standard

NEiS

NECA 1–2000

Standard practices for

Good Workmanship in Electrical Contracting

Published by
National Electrical Contractors Association **NECA**

Exhibit 110.2 Exhibit 110.2. ANSI/NECA 1-2000, *Standard Practice for Good Workmanship in Electrical Contracting*, one example of the many ANSI standards that describe "neat and workmanlike" installations.

be effectively closed to afford protection substantially equivalent to the wall of the equipment. Where metallic plugs or plates are used with nonmetallic enclosures, they shall be recessed at least 6 mm (¼ in.) from the outer surface of the enclosure.

The phrase *unused cable or raceway openings* clarifies that openings used for normal operation, such as weep holes, are not required to be closed up.

See 408.7 for requirements on unused openings in switchboard and panelboard enclosures.

(B) Subsurface Enclosures Conductors shall be racked to provide ready and safe access in underground and subsurface enclosures into which persons enter for installation and maintenance.

(C) Integrity of Electrical Equipment and Connections Internal parts of electrical equipment, including busbars,

wiring terminals, insulators, and other surfaces, shall not be damaged or contaminated by foreign materials such as paint, plaster, cleaners, abrasives, or corrosive residues. There shall be no damaged parts that may adversely affect safe operation or mechanical strength of the equipment such as parts that are broken; bent; cut; or deteriorated by corrosion, chemical action, or overheating.

110.13 Mounting and Cooling of Equipment

(A) Mounting Electrical equipment shall be firmly secured to the surface on which it is mounted. Wooden plugs driven into holes in masonry, concrete, plaster, or similar materials shall not be used.

(B) Cooling Electrical equipment that depends on the natural circulation of air and convection principles for cooling of exposed surfaces shall be installed so that room airflow over such surfaces is not prevented by walls or by adjacent installed equipment. For equipment designed for floor mounting, clearance between top surfaces and adjacent surfaces shall be provided to dissipate rising warm air.

Electrical equipment provided with ventilating openings shall be installed so that walls or other obstructions do not prevent the free circulation of air through the equipment.

Ventilated is defined in Article 100. Panelboards, transformers, and other types of equipment are adversely affected if enclosure surfaces normally exposed to room air are covered or tightly enclosed. Ventilating openings in equipment are provided to allow the circulation of room air around internal components of the equipment; the blocking of such openings can cause dangerous overheating. For example, a ventilated busway must be located where there are no walls or other objects that might interfere with the natural circulation of air and convection principles for cooling. Ventilation for motor locations is covered in 430.14(A) and 430.16. Ventilation for transformer locations is covered in 450.9 and 450.45. In addition to 110.13, proper placement of equipment requiring ventilation becomes enforceable using the requirements of 110.3(B).

110.14 Electrical Connections

Because of different characteristics of dissimilar metals, devices such as pressure terminal or pressure splicing connectors and soldering lugs shall be identified for the material of the conductor and shall be properly installed and used. Conductors of dissimilar metals shall not be intermixed in a terminal or splicing connector where physical contact occurs between dissimilar conductors (such as copper and aluminum, copper and copper-clad aluminum, or aluminum and copper-clad aluminum), unless the device is identified for the purpose and conditions of use. Materials such as solder, fluxes, inhibitors, and compounds, where employed, shall be suitable for the use and shall be of a type that will not adversely affect the conductors, installation, or equipment.

FPN: Many terminations and equipment are marked with a tightening torque.

Section 110.3(B) applies where terminations and equipment are marked with tightening torques.

For the testing of wire connectors for which the manufacturer has not assigned another value appropriate for the design, Commentary Tables 1.2 through 1.5 provide data on the tightening torques that Underwriters Laboratories uses. These tables should be used for guidance only if no tightening information on a specific wire connector is available. They should not be used to replace the manufacturer's instructions, which should always be followed.

The information in the tables was taken from UL 486B, *Wire Connections for Use with Aluminum Conductors*. Similar information can be found in UL 486A, *Wire Connections and Solder Lugs for Use with Copper Conductors*.

(A) Terminals Connection of conductors to terminal parts shall ensure a thoroughly good connection without damaging the conductors and shall be made by means of pressure connectors (including set-screw type), solder lugs, or splices to flexible leads. Connection by means of wire-binding screws or studs and nuts that have upturned lugs or the equivalent shall be permitted for 10 AWG or smaller conductors.

Terminals for more than one conductor and terminals used to connect aluminum shall be so identified.

(B) Splices Conductors shall be spliced or joined with splicing devices identified for the use or by brazing, welding,

Commentary Table 1.2 Tightening Torques for Screws,* in Pound-Inches

| Wire Size (AWG or kcmil) | Slotted Head No. 10 and Larger | | Hexagonal Head-External Drive Socket Wrench | |
	Slot Width to 3/64 in. or Slot Length to 1/4 in.†	Slot Width Over 3/64 in. or Slot Length Over 1/4 in.†	Split-Bolt Connectors	Other Connectors
30-10	20	35	80	75
8	25	40	80	75
6	35	45	165	110
4	35	45	165	110
3	35	50	275	150
2	40	50	275	150
1	—	50	275	150
1/0	—	50	385	180
2/0	—	50	385	180
3/0	—	50	500	250
4/0	—	50	500	250
250	—	50	650	325
300	—	50	650	325
350	—	50	650	325
400	—	50	825	325
500	—	50	825	375
600	—	50	1000	375
700	—	50	1000	375
750	—	50	1000	375
800	—	50	1100	500
900	—	50	1100	500
1000	—	50	1100	500
1250	—	—	1100	600
1500	—	—	1100	600
1750	—	—	1100	600
2000	—	—	1100	600

*Clamping screws with multiple tightening means. For example, for a slotted hexagonal head screw, use the torque value associated with the tool used in the installation. UL uses both values when testing.
†For values of slot width or length other than those specified, select the largest torque value associated with conductor size.

Commentary Table 1.3 Torques in Pound-Inches for Slotted Head Screws* Smaller Than No. 10, for Use with 8 AWG and Smaller Conductors

Screw-Slot Length (in.)†	Screw-Slot Width Less Than 3/64 in.	Screw-Slot Width 3/64 in. and Larger
To 5/32	7	9
5/32	7	12
3/16	7	12
7/32	7	12
1/4	9	12
9/32	—	15
Above 9/32	—	20

*Clamping screws with multiple tightening means. For example, for a slotted hexagonal head screw, use the torque value associated with the tool used in the installation. UL uses both values when testing.
†For slot lengths of intermediate values, select torques pertaining to next-shorter slot length.

Commentary Table 1.4 Torques for Recessed Allen Head Screws

Socket Size Across Flats (in.)	Torque (lb-in.)
1/8	45
5/32	100
3/16	120
7/32	150
1/4	200
5/16	275
3/8	375
1/2	500
9/16	600

Commentary Table 1.5 Lug-Bolting Torques for Connection of Wire Connectors to Busbars

Bolt Diameter	Tightening Torque (lb-ft)
No. 8 or smaller	1.5
No. 10	2
1/4 in. or less	6
5/16 in.	11
3/8 in.	19
7/16 in.	30
1/2 in.	40
9/16 in. or larger	55

or soldering with a fusible metal or alloy. Soldered splices shall first be spliced or joined so as to be mechanically and electrically secure without solder and then be soldered. All splices and joints and the free ends of conductors shall be covered with an insulation equivalent to that of the conductors or with an insulating device identified for the purpose.

Wire connectors or splicing means installed on conductors for direct burial shall be listed for such use.

Field observations and trade magazine articles indicate that electrical connection failures have been determined to be the cause of many equipment burnouts and fires. Many of these failures are attributable to improper terminations, poor workmanship, the differing characteristics of dissimilar metals, and improper binding screws or splicing devices.

UL's requirements for listing solid aluminum conductors in 12 AWG and 10 AWG and for listing snap switches and receptacles for use on 15- and 20-ampere branch circuits incorporate stringent tests that take into account the factors listed in the preceding paragraph. For further information regarding receptacles and switches using CO/ALR-rated terminals, refer to 404.14(C) and 406.2(C).

Screwless pressure terminal connectors of the conductor push-in type are for use with solid copper and copper-clad aluminum conductors only.

Instructions that describe proper installation techniques and emphasize the need to follow those techniques and practice good workmanship are required to be included with each coil of 12 AWG and 10 AWG insulated aluminum wire or cable. See also the commentary on tightening torque that follows 110.14, FPN.

New product and material designs that provide increased levels of safety of aluminum wire terminations have been developed by the electrical industry. To assist all concerned parties in the proper and safe use of solid aluminum wire in making connections to wiring devices used on 15- and 20-ampere branch circuits, the following information is presented. Understanding and using this information is essential for proper application of materials and devices now available.

For New Installations

The following commentary is based on a report prepared by the Ad Hoc Committee on Aluminum Terminations prior to publication of the 1975 *Code*. This information is still pertinent today and is necessary for compliance with 110.14(A) when aluminum wire is used in new installations.

New Materials and Devices. For direct connection, only 15- and 20-ampere receptacles and switches marked "CO/ALR" and connected as follows under Installation Method should be used.

The "CO/ALR" marking is on the device mounting yoke or strap. The "CO/ALR" marking means the devices have been tested to stringent heat-cycling requirements to determine their suitability for use with UL-labeled aluminum, copper, or copper-clad aluminum wire.

Listed solid aluminum wire, 12 AWG or 10 AWG, marked with the aluminum insulated wire label should be used. The installation instructions that are packaged with the wire should be used.

Installation Method. Exhibit 110.3 illustrates the following correct method of connection:

1. The freshly stripped end of the wire is wrapped two-thirds to three-quarters of the distance around the wire-binding screw post, as shown in Step A of Exhibit 110.3. The loop is made so that rotation of the screw during tightening will tend to wrap the wire around the post rather than unwrap it.
2. The screw is tightened until the wire is snugly in contact with the underside of the screw head and with the contact plate on the wiring device, as shown in Step B of Exhibit 110.3.
3. The screw is tightened an additional half-turn, thereby providing a firm connection, as shown in Step C of Exhibit 110.3. If a torque screwdriver is used, the screw is tightened to 12 lb-in.

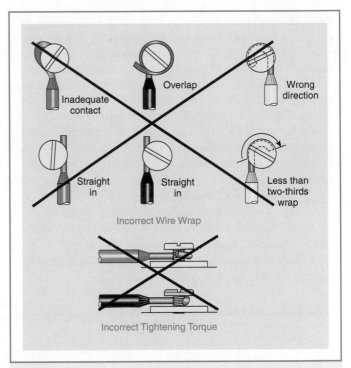

Exhibit 110.4 Incorrect methods of terminating aluminum wire at wire-binding screw terminals of receptacles and snap switches. (Redrawn courtesy of Underwriters Laboratories Inc.)

label should be used with wiring devices marked "CO/ALR" and connected as described under Installation Method. This is the preferred and recommended method for using such wire.

For the following types of devices, the terminals should not be directly connected to aluminum conductors but may be used with labeled copper or copper-clad conductors:

1. Receptacles and snap switches marked "AL-CU"
2. Receptacles and snap switches having no conductor marking
3. Receptacles and snap switches that have back-wired terminals or screwless terminals of the push-in type

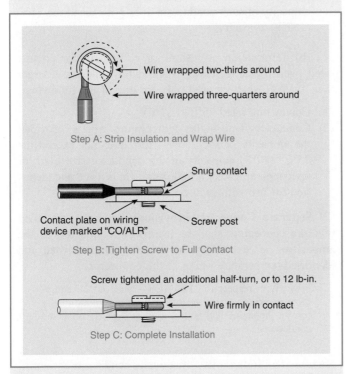

Exhibit 110.3 Correct method of terminating aluminum wire at wire-binding screw terminals of receptacles and snap switches. (Redrawn courtesy of Underwriters Laboratories Inc.)

4. The wires should be positioned behind the wiring device to decrease the likelihood of the terminal screws loosening when the device is positioned into the outlet box.

Exhibit 110.4 illustrates incorrect methods of connection. These methods should *not* be used.

Existing Inventory. Labeled 12 AWG or 10 AWG solid aluminum wire that does not bear the new aluminum wire

For Existing Installations

If examination discloses overheating or loose connections, the recommendations described under Existing Inventory should be followed.

Twist-On Wire Connectors

Because 110.14(B) requires conductors to be spliced with "splicing devices identified for the use," wire connectors are required to be marked for conductor suitability. Twist-on wire connectors are not suitable for splicing aluminum conductors or copper-clad aluminum to copper conductors unless it is so stated and marked as such on the shipping carton. The marking is typically "AL-CU (dry locations)."

Presently, one style of wire nut and one style of crimp-type connector have been listed as having met these requirements.

On February 2, 1995, Underwriters Laboratories announced the listing of a twist-on wire connector suitable for use with aluminum-to-copper conductors, in accordance with UL 486C, *Splicing Wire Connectors*. That was the first listing of a twist-on type connector for aluminum-to-copper conductors since 1987. The UL listing does *not* cover aluminum-to-aluminum combinations. However, more than one aluminum or copper conductor is allowed when used in combination.

These listed wire-connecting devices are available for pigtailing short lengths of copper conductors to the original aluminum branch-circuit conductors, as shown in Exhibit 110.5. Primarily, these pigtailed conductors supply 15- and 20-ampere wiring devices. Pigtailing is permitted, provided there is suitable space within the enclosure.

Exhibit 110.5 Pigtailing copper to aluminum conductors using two listed devices.

(C) Temperature Limitations The temperature rating associated with the ampacity of a conductor shall be selected and coordinated so as not to exceed the lowest temperature rating of any connected termination, conductor, or device. Conductors with temperature ratings higher than specified for terminations shall be permitted to be used for ampacity adjustment, correction, or both.

(1) Equipment Provisions The determination of termination provisions of equipment shall be based on 110.14(C)(1)(a) or (C)(1)(b). Unless the equipment is listed

and marked otherwise, conductor ampacities used in determining equipment termination provisions shall be based on Table 310.16 as appropriately modified by 310.15(B)(6).

(a) Termination provisions of equipment for circuits rated 100 amperes or less, or marked for 14 AWG through 1 AWG conductors, shall be used only for one of the following:

(1) Conductors rated 60°C (140°F).
(2) Conductors with higher temperature ratings, provided the ampacity of such conductors is determined based on the 60°C (140°F) ampacity of the conductor size used.
(3) Conductors with higher temperature ratings if the equipment is listed and identified for use with such conductors.
(4) For motors marked with design letters B, C, or D, conductors having an insulation rating of 75°C (167°F) or higher shall be permitted to be used, provided the ampacity of such conductors does not exceed the 75°C (167°F) ampacity.

(b) Termination provisions of equipment for circuits rated over 100 amperes, or marked for conductors larger than 1 AWG, shall be used only for one of the following:

(1) Conductors rated 75°C (167°F)
(2) Conductors with higher temperature ratings, provided the ampacity of such conductors does not exceed the 75°C (167°F) ampacity of the conductor size used, or up to their ampacity if the equipment is listed and identified for use with such conductors

(2) Separate Connector Provisions Separately installed pressure connectors shall be used with conductors at the ampacities not exceeding the ampacity at the listed and identified temperature rating of the connector.

> FPN: With respect to 110.14(C)(1) and (C)(2), equipment markings or listing information may additionally restrict the sizing and temperature ratings of connected conductors.

Section 110.14(C)(1) states that where conductors are terminated in equipment, the selected conductor ampacities must be based on Table 310.16, unless the equipment is specifically listed and marked otherwise. The intent of this requirement is to clarify which ampacities are used to determine the proper conductor size at equipment terminations.

When equipment of 600 volts or less is evaluated relative to the appropriate temperature characteristics of the terminations, conductors sized according to Table 310.16 are required to be used. The UL *General Information Directory* (White Book, page 3) clearly indicates that the 60°C and 75°C provisions for equipment have been determined using conductors from Table 310.16. However, installers or designers unaware of the UL guide card information might attempt to select conductors based on a table other than Table 310.16, especially if a wiring method that allows the use of ampacities such as those in Table 310.17 is used. That use can result in overheated terminations at the equipment. Clearly, the ampacities shown in other tables (such as Table 310.17) could be used for various conditions to which the wiring method is subject (ambient, ampacity correction, etc.), but the conductor size at the termination must be based on ampacities from Table 310.16. This change does not introduce any new impact on the equipment or the wiring methods; it simply adds a rule from the listing information into the *Code* because it is an installation and equipment selection issue.

Section 110.14(C)(1)(a) requires that conductor terminations, as well as conductors, be rated for the operating temperature of the circuit. For example, the load on an 8 AWG THHN, 90°C copper wire is limited to 40 amperes where connected to a disconnect switch with terminals rated at 60°C. The same 8 AWG THHN, 90°C wire is limited to 50 amperes where connected to a fusible switch with terminals rated at 75°C. The conductor ampacities were selected from Table 310.16. Not only does this requirement apply to conductor terminations of breakers and fusible switches, but the equipment enclosure must also permit terminations above 60°C. Exhibit 110.6 shows an example of termination temperature markings.

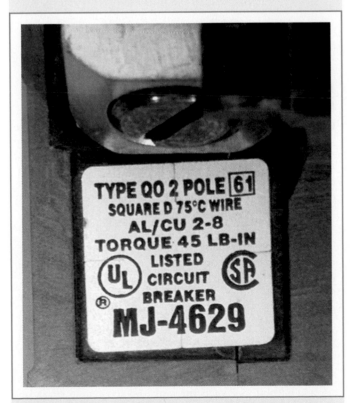

Exhibit 110.6 An example of termination temperature markings on a main circuit breaker. (Courtesy of Square D Co.)

110.15 High-Leg Marking

On a 4-wire, delta-connected system where the midpoint of one phase winding is grounded, only the conductor or busbar having the higher phase voltage to ground shall be durably and permanently marked by an outer finish that is orange in color or by other effective means. Such identification shall be placed at each point on the system where a connection is made if the grounded conductor is also present.

The high leg is common on a 240/120-volt 3-phase, 4-wire delta system. It is typically designated as "B phase." The high-leg marking, which is required to be the color orange or other similar effective means, is intended to prevent problems due to the lack of complete standardization where metered and nonmetered equipment are installed in the same installation. Electricians should always test each phase relative to ground with suitable equipment to determine exactly where the high leg is located in the system. The requirement in 110.15 previously appeared in 384-3(e) of the 1999 *NEC*. It was moved to Article 110 in 2002, when the application became a more general requirement. For the 2005 *Code*, 110.15 was editorially modified for clarity.

110.16 Flash Protection

Switchboards, panelboards, industrial control panels, meter socket enclosures, and motor control centers that are in other than dwelling occupancies and are likely to require examination, adjustment, servicing, or maintenance while energized shall be field marked to warn qualified persons of potential electric arc flash hazards. The marking shall be located so as to be clearly visible to qualified persons before examination, adjustment, servicing, or maintenance of the equipment.

This requirement was added in the 2002 *Code*. Field marking that warns electrical workers of potential electrical arc flash hazards is now required because significant numbers of electricians have been seriously burned or killed by accidental electrical arc flash while working on "hot" (energized) equipment. Most of those accidents could have been prevented or their severity significantly reduced if electricians had been wearing the proper type of protective clothing. Requiring switchboards, panelboards, and motor control centers to be individually field marked with proper warning labels will raise the level of awareness of electrical arc flash hazards and thereby decrease the number of accidents.

Exhibit 110.7 shows an electrical employee working inside the flash protection boundary and in front of a large-capacity service-type switchboard that has not been de-energized and that is not under the lockout/tagout procedure. The worker is wearing personal protective equipment (PPE) considered appropriate flash protection clothing for the flash

Exhibit 110.7 Electrical worker clothed in personal protective equipment (PPE) appropriate for the hazard involved.

hazard involved. Suitable PPE appropriate to a particular hazard is described in NFPA 70E, *Standard for Electrical Safety in the Workplace*.

Exhibit 110.8 displays one example of a warning sign required by 110.16.

> # ⚠WARNING
> ## Potential Arc Flash Hazard
> ## Appropriate PPE and Tools Required when working on this equipment

Exhibit 110.8 One example of an arc flash warning sign required by 110.16.

Accident reports continue to confirm the fact that workers responsible for the installation or maintenance of electrical equipment often do not turn off the power source before working on the equipment. Working electrical equipment energized is a major safety concern in the electrical industry. The real purpose of this additional code requirement is to alert electrical contractors, electricians, facility owners and managers, and other interested parties to some of the hazards of working on or near energized equipment and to emphasize the importance of turning off the power before working on electrical circuits.

The information in fine print notes is not mandatory. Employers can be assured that they are providing a safe workplace for their employees if safety-related work prac-tices required by NFPA 70E have been implemented and are being followed. (See also the commentary following the definition of *qualified person* in Article 100.)

In addition to the standards referenced in the fine print notes and their individual bibliographies, additional information on this subject can be found in the 1997 report "Hazards of Working Electrical Equipment Hot," published by the National Electrical Manufacturers Association.

> FPN No. 1: NFPA 70E-2004, *Standard for Electrical Safety in the Workplace*, provides assistance in determining severity of potential exposure, planning safe work practices, and selecting personal protective equipment.
> FPN No. 2: ANSI Z535.4-1998, *Product Safety Signs and Labels*, provides guidelines for the design of safety signs and labels for application to products.

110.18 Arcing Parts

Parts of electric equipment that in ordinary operation produce arcs, sparks, flames, or molten metal shall be enclosed or separated and isolated from all combustible material.

Examples of electrical equipment that may produce sparks during ordinary operation include open motors having a centrifugal starting switch, open motors with commutators, and collector rings. Adequate separation from combustible material is essential if open motors with those features are used.

> FPN: For hazardous (classified) locations, see Articles 500 through 517. For motors, see 430.14.

110.19 Light and Power from Railway Conductors

Circuits for lighting and power shall not be connected to any system that contains trolley wires with a ground return.

Exception: Such circuit connections shall be permitted in car houses, power houses, or passenger and freight stations operated in connection with electric railways.

110.21 Marking

The manufacturer's name, trademark, or other descriptive marking by which the organization responsible for the product can be identified shall be placed on all electric equipment. Other markings that indicate voltage, current, wattage, or other ratings shall be provided as specified elsewhere in this *Code*. The marking shall be of sufficient durability to withstand the environment involved.

The *Code* requires that equipment ratings be marked on the equipment and that such markings be located so as to be visible or easily accessible during or after installation.

110.22 Identification of Disconnecting Means

Each disconnecting means shall be legibly marked to indicate its purpose unless located and arranged so the purpose is evident. The marking shall be of sufficient durability to withstand the environment involved.

Where circuit breakers or fuses are applied in compliance with the series combination ratings marked on the equipment by the manufacturer, the equipment enclosure(s) shall be legibly marked in the field to indicate the equipment has been applied with a series combination rating. The marking shall be readily visible and state the following:

> CAUTION — SERIES COMBINATION SYSTEM
> RATED _____ AMPERES. IDENTIFIED
> REPLACEMENT COMPONENTS REQUIRED.

FPN: See 240.86(B) for interrupting rating marking for end-use equipment.

Proper identification needs to be specific. For example, the marking should indicate not simply "motor" but rather "motor, water pump"; not simply "lights" but rather "lights, front lobby." Consideration also should be given to the form of identification. Marking often fades or is covered by paint after installation. See 408.4 and its associated commentary for further information on circuit directories for switchboards and panelboards. See 408.4 and its associated commentary for further information on circuit directories for switchboards and panelboards.

The second paragraph of 110.22 requires series-rated overcurrent devices to be legibly marked. The equipment manufacturer can mark the equipment to be used with series combination ratings. If the equipment is installed in the field at its marked series combination rating, the equipment must have an additional label, as specified in 110.22, to indicate that the series combination rating has been used.

110.23 Current Transformers

Unused current transformers associated with potentially energized circuits shall be short-circuited.

Because Article 450 specifically exempts current transformers, the practical solution to prevent damage to current transformers not connected to a load or for unused current transformers has been placed in 110.23.

II. 600 Volts, Nominal, or Less

110.26 Spaces About Electrical Equipment

Sufficient access and working space shall be provided and maintained about all electric equipment to permit ready and safe operation and maintenance of such equipment. Enclosures housing electrical apparatus that are controlled by a lock(s) shall be considered accessible to qualified persons.

Key to understanding 110.26 is the division of requirements for spaces about electrical equipment in two separate and distinct categories: working space and dedicated equipment space. The term *working space* generally applies to the protection of the worker, and *dedicated equipment space* applies to the space reserved for future access to electrical equipment and to protection of the equipment from intrusion by non-electrical equipment. The performance requirements for all spaces about electrical equipment are set forth in the first sentence. Storage of materials that blocks access or prevents safe work practices must be avoided at all times.

(A) **Working Space** Working space for equipment operating at 600 volts, nominal, or less to ground and likely to require examination, adjustment, servicing, or maintenance while energized shall comply with the dimensions of 110.26(A)(1), (A)(2), and (A)(3) or as required or permitted elsewhere in this *Code.*

The intent of 110.26(A) is to provide enough space for personnel to perform any of the operations listed without jeopardizing worker safety. These operations include examination, adjustment, servicing, and maintenance of equipment. Examples of such equipment include panelboards, switches, circuit breakers, controllers, and controls on heating and air-conditioning equipment. It is important to understand that the word *examination*, as used in 110.26(A), includes such tasks as checking for the presence of voltage using a portable voltmeter.

Minimum working clearances are not required if the equipment is such that it is not likely to require examination, adjustment, servicing, or maintenance while energized. However, "sufficient" access and working space are still required by the opening paragraph of 110.26.

(1) **Depth of Working Space** The depth of the working space in the direction of live parts shall not be less than that specified in Table 110.26(A)(1) unless the requirements of 110.26(A)(1)(a), (A)(1)(b), or (A)(1)(c) are met. Distances shall be measured from the exposed live parts or from the enclosure or opening if the live parts are enclosed.

For the 2005 *Code*, some of the text associated with Conditions 1 and 2 was edited for clarity and enforceability. Also, the Condition 2 metric clearance for 151 to 600 volts was revised from 1 m to 1.1 m to reflect an accurate metric conversion.

Included in these clearance requirements is the step-back distance from the face of the equipment. Table 110.26(A)(1) provides requirements for clearances away from the equipment, based on the circuit voltage to ground

Table 110.26(A)(1) Working Spaces

Nominal Voltage to Ground	Minimum Clear Distance		
	Condition 1	Condition 2	Condition 3
0–150	900 mm (3 ft)	900 mm (3 ft)	900 mm (3 ft)
151–600	900 mm (3 ft)	1.1 m (3½ ft)	1.2 m (4 ft)

Note: Where the conditions are as follows:
Condition 1 — Exposed live parts on one side of the working space and no live or grounded parts on the other side of the working space, or exposed live parts on both sides of the working space that are effectively guarded by insulating materials.
Condition 2 — Exposed live parts on one side of the working space and grounded parts on the other side of the working space. Concrete, brick, or tile walls shall be considered as grounded.
Condition 3 — Exposed live parts on both sides of the working space.

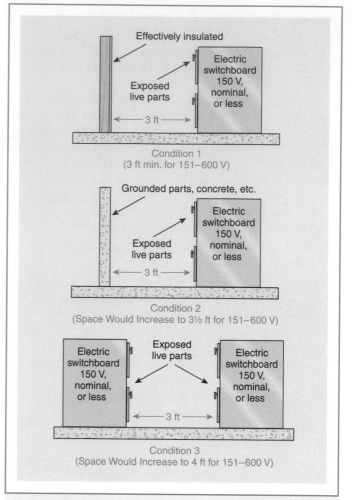

Condition 1
(3 ft min. for 151–600 V)

Condition 2
(Space Would Increase to 3½ ft for 151–600 V)

Condition 3
(Space Would Increase to 4 ft for 151–600 V)

Exhibit 110.9 Distances measured from the live parts if the live parts are exposed or from the enclosure front if the live parts are enclosed. If any assemblies, such as switchboards or motor-control centers, are accessible from the back and expose live parts, the working clearance dimensions would be required at the rear of the equipment, as illustrated. Note that for Condition 3, where there is an enclosure on opposite sides of the working space, the clearance for only one working space is required.

and whether there are grounded or ungrounded objects in the step-back space or exposed live parts across from each other. The voltages to ground consist of two groups: 0 to 150, inclusive, and 151 to 600, inclusive. Examples of common electrical supply systems covered in the 0 to 150 volts to ground group include 120/240-volt, single-phase, 3-wire and 208Y/120-volt, 3-phase, 4-wire. Examples of common electrical supply systems covered in the 151 to 600 volts to ground group include 240-volt, 3-phase, 3-wire; 480Y/277-volt, 3-phase, 4-wire; and 480-volt, 3-phase, 3-wire (ungrounded and corner grounded). Remember, where an ungrounded system is utilized, the voltage to ground (by definition) is the greatest voltage between the given conductor and any other conductor of the circuit. For example, the voltage to ground for a 480-volt ungrounded delta system is 480 volts. See Exhibit 110.9 for the general working clearance requirements for each of the three conditions listed in Table 110.26(A)(1).

(a) **Dead-Front Assemblies.** Working space shall not be required in the back or sides of assemblies, such as dead-front switchboards or motor control centers, where all connections and all renewable or adjustable parts, such as fuses or switches, are accessible from locations other than the back or sides. Where rear access is required to work on nonelectrical parts on the back of enclosed equipment, a minimum horizontal working space of 762 mm (30 in.) shall be provided.

The intent of this section is to point out that work space is required only from the side(s) of the enclosure that requires access. The general rule still applies: Equipment that requires front, rear, or side access for electrical activities described in 110.26(A) must meet the requirements of Table 110.26(A)(1). In many cases, equipment of "dead-front" assemblies requires only front access. For equipment that requires rear access for nonelectrical activity, however, a reduced working space of at least 30 in. must be provided. Exhibit 110.10 shows a reduced working space of 30 in. at the rear of equipment to allow work on nonelectrical parts.

(b) **Low Voltage.** By special permission, smaller working spaces shall be permitted where all exposed live parts operate at not greater than 30 volts rms, 42 volts peak, or 60 volts dc.

(c) **Existing Buildings.** In existing buildings where electrical equipment is being replaced, Condition 2 working clearance shall be permitted between dead-front switchboards, panelboards, or motor control centers located across

Exhibit 110.10 Example of the 30 in. minimum working space at the rear of equipment to allow work on nonelectrical parts, such as the replacement of an air filter.

Exhibit 110.11 Permitted reduction from a Condition 3 to a Condition 2 clearance according to 110.26(A)(1)(c).

the aisle from each other where conditions of maintenance and supervision ensure that written procedures have been adopted to prohibit equipment on both sides of the aisle from being open at the same time and qualified persons who are authorized will service the installation.

This section permits some relief for installations that are being upgraded. When assemblies such as dead-front switchboards, panelboards, or motor-control centers are replaced in an existing building, the working clearance allowed is that required by Table 110.26(A)(1), Condition 2. The reduction from a Condition 3 to a Condition 2 clearance is allowed only where a written procedure prohibits facing doors of equipment from being open at the same time and where only authorized and qualified persons service the installation. Exhibit 110.11 illustrates this relief for existing buildings.

(2) Width of Working Space The width of the working space in front of the electric equipment shall be the width of the equipment or 750 mm (30 in.), whichever is greater. In all cases, the work space shall permit at least a 90 degree opening of equipment doors or hinged panels.

Regardless of the width of the electrical equipment, the working space cannot be less than 30 in. wide. This space allows an individual to have at least shoulder-width space in front of the equipment. The 30 in. measurement can be made from either the left or the right edge of the equipment and can overlap other electrical equipment, provided the other equipment does not extend beyond the clearance re-

quired by Table 110.26(A)(1). If the equipment is wider than 30 in., the left-to-right space must be equal to the width of the equipment. See Exhibit 110.12 for an explanation of the 30 in. width requirement.

Exhibit 110.12 The 30 in. wide front working space, which is not required to be directly centered on the electrical equipment if space is sufficient for safe operation and maintenance of such equipment.

Sufficient depth in the working space also must be provided to allow a panel or a door to open at least 90 degrees. If doors or hinged panels are wider than 3 ft, more than a 3 ft deep working space must be provided to allow a full 90-degree opening. (See Exhibit 110.13.)

(3) Height of Working Space The work space shall be clear and extend from the grade, floor, or platform to the height required by 110.26(E). Within the height requirements

90° Opening

of this section, other equipment that is associated with the electrical installation and is located above or below the electrical equipment shall be permitted to extend not more than 150 mm (6 in.) beyond the front of the electrical equipment.

In addition to requiring a working space to be clear from the floor to a height of 61/2 ft or to the height of the equipment, whichever is greater, 110.26(A)(3) permits electrical equipment located above or below other electrical equipment to extend into the working space not more than 6 in. This requirement allows the placement of a 12 in. × 12 in. wireway on the wall directly above or below a 6 in. deep panelboard without impinging on the working space or compromising practical working clearances. The requirement continues to prohibit large differences in depth of equipment below or above other equipment that specifically requires working space. In order to minimize the amount of space required for electrical equipment, it was not uncommon to find installations of large free-standing, dry-type transformers within the required work space for a wall-mounted panelboard. Clear access to the panelboard is compromised by the location of the transformer with its grounded enclosure and this type of installation and is clearly not permitted by this section. Electrical equipment that produces heat or that otherwise requires ventilation also must comply with 110.3(B) and 110.13.

(B) Clear Spaces Working space required by this section shall not be used for storage. When normally enclosed live parts are exposed for inspection or servicing, the working space, if in a passageway or general open space, shall be suitably guarded.

Section 110.26(B), as well as the rest of 110.26, does not prohibit the placement of panelboards in corridors or passageways. For that reason, when the covers of corridor-mounted panelboards are removed for servicing or other work, access to the area around the panelboard should be guarded or limited to protect unqualified persons using the corridor.

(C) Entrance to Working Space
(1) Minimum Required At least one entrance of sufficient area shall be provided to give access to working space about electrical equipment.

(2) Large Equipment For equipment rated 1200 amperes or more that contains overcurrent devices, switching devices, or control devices, there shall be one entrance to the required working space not less than 610 mm (24 in.) wide and 2.0 m (6½ ft) high at each end of the working space. Where the entrance has a personnel door(s), the door(s) shall open in the direction of egress and be equipped with panic bars, pressure plates, or other devices that are normally latched but open under simple pressure.

A single entrance to the required working space shall be permitted where either of the conditions in 110.26(C)(2)(a) or (C)(2)(b) is met.

The stipulation that large equipment must be at least 6 ft wide was deleted for the 2005 *Code*. Now, for the purposes of this section, large equipment is simply equipment rated 1200 amperes or more. The removal of the 6 ft condition has the effect of broadening the scope of this requirement to now include all spaces containing "equipment rated 1200 amperes or more that contains overcurrent devices, switching devices, or control devices." The effect of this revision is that the required working space for one 1200-ampere safety switch with a width of approximately 3 ft is now required to be provided with two entrances/exits unless one of the provisions permitting a single entrance can be applied to that space. For equipment of this type, it is not unusual that the provision calling for a continuous and unobstructed way of exit travel from the working space can be applied.

Where the entrance(s) to the working space is through a door, each door must comply with the requirements for swinging open in the direction of egress and have door opening hardware that does not require turning of a door knob or similar action that may preclude quick exit from the area in the event of an emergency.

This requirement affords safety for workers exposed to energized conductors by allowing an injured worker to safely and quickly exit an electrical room without having to turn knobs or pull doors open.

For a graphical explanation of access and entrance requirements to a working space, see Exhibits 110.14 and 110.15. Notice the unacceptable and hazardous situation shown in Exhibit 110.16.

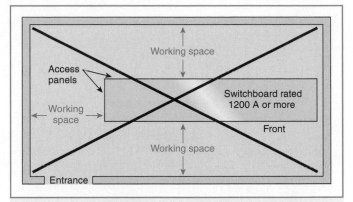

Exhibit 110.16 Unacceptable arrangement of a large switchboard. A person could be trapped behind arcing electrical equipment.

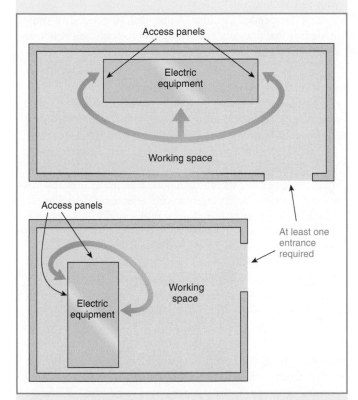

Exhibit 110.14 Basic Rule, first paragraph. At least one entrance is required to provide access to the working space around electrical equipment [110.26(C)(1)]. The lower installation would not be acceptable for a switchboard rated 1200 amperes or more.

Exhibit 110.15 Basic Rule, second paragraph. For equipment rated 1200 amperes or more, one entrance not less than 24 in. wide and 6½ ft high is required at each end [110.26(C)(2)].

(a) Unobstructed Exit. Where the location permits a continuous and unobstructed way of exit travel, a single entrance to the working space shall be permitted.

(b) Extra Working Space. Where the depth of the working space is twice that required by 110.26(A)(1), a single entrance shall be permitted. It shall be located so that the distance from the equipment to the nearest edge of the entrance is not less than the minimum clear distance specified in Table 110.26(A)(1) for equipment operating at that voltage and in that condition.

For an explanation of paragraphs 110.26(C)(2)(a) and 110.26(C)(2)(b), see Exhibits 110.17 and 110.18.

(D) Illumination Illumination shall be provided for all working spaces about service equipment, switchboards, panelboards, or motor control centers installed indoors. Additional lighting outlets shall not be required where the work space is illuminated by an adjacent light source or as permitted by 210.70(A)(1), Exception No. 1, for switched receptacles. In electrical equipment rooms, the illumination shall not be controlled by automatic means only.

(E) Headroom The minimum headroom of working spaces about service equipment, switchboards, panelboards, or motor control centers shall be 2.0 m (6½ ft). Where the electrical equipment exceeds 2.0 m (6½ ft) in height, the minimum headroom shall not be less than the height of the equipment.

Exception: In existing dwelling units, service equipment or panelboards that do not exceed 200 amperes shall be permitted in spaces where the headroom is less than 2.0 m (6½ ft).

(F) Dedicated Equipment Space All switchboards, panelboards, distribution boards, and motor control centers shall be located in dedicated spaces and protected from damage.

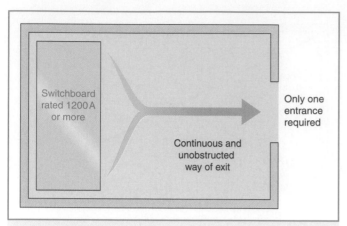

Exhibit 110.17 Equipment location that allows a continuous and unobstructed way of exit travel.

X = minimum allowable distance

Exhibit 110.18 Working space with one entrance. Only one entrance is required if the working space required by 110.26(A) is doubled. See Table 110.26(A)(1) for permitted dimensions of X.

Exception: Control equipment that by its very nature or because of other rules of the Code must be adjacent to or within sight of its operating machinery shall be permitted in those locations.

(1) Indoor Indoor installations shall comply with 110.26(F)(1)(a) through (F)(1)(d).

(a) Dedicated Electrical Space. The space equal to the width and depth of the equipment and extending from the floor to a height of 1.8 m (6 ft) above the equipment or to the structural ceiling, whichever is lower, shall be dedicated to the electrical installation. No piping, ducts, leak protection apparatus, or other equipment foreign to the electrical installation shall be located in this zone.

Exception: Suspended ceilings with removable panels shall be permitted within the 1.8-m (6-ft) zone.

(b) Foreign Systems. The area above the dedicated space required by 110.26(F)(1)(a) shall be permitted to contain foreign systems, provided protection is installed to avoid damage to the electrical equipment from condensation, leaks, or breaks in such foreign systems.

(c) Sprinkler Protection. Sprinkler protection shall be permitted for the dedicated space where the piping complies with this section.

(d) Suspended Ceilings. A dropped, suspended, or similar ceiling that does not add strength to the building structure shall not be considered a structural ceiling.

The dedicated electrical space includes the space defined by extending the footprint of the switchboard or panelboard from the floor to a height of 6 ft above the height of the equipment or to the structural ceiling, whichever is lower. This reserved space permits busways, conduits, raceways, and cables to enter the equipment. The dedicated electrical space must be clear of piping, ducts, leak protection apparatus, or equipment foreign to the electrical installation. Plumbing, heating, ventilation, and air-conditioning piping, ducts, and equipment must be installed outside the width and depth zone.

Foreign systems installed directly above the dedicated space reserved for electrical equipment must include protective equipment that ensures that occurrences such as leaks, condensation, and even breaks do not damage the electrical equipment located below.

Sprinkler protection is permitted for the dedicated spaces as long as the sprinkler or other suppression system piping complies with 110.26(F)(1)(d). A dropped, suspended, or similar ceiling is permitted to be located directly in the dedicated space, as are building structural members.

The electrical equipment also must be protected from physical damage. Damage can be caused by activities performed near the equipment, such as material handling by personnel or the operation of a forklift or other mobile equipment. See 110.27(B) for other provisions relating to the protection of electrical equipment.

Exhibits 110.19, 110.20, and 110.21 illustrate the two distinct indoor installation spaces required by 110.26(A) and 110.26(F), that is, the working space and the dedicated electrical space.

In Exhibit 110.19, the dedicated electrical space required by 110.26(F) is the space outlined by the width and the depth of the equipment (the footprint) and extending from the floor to 6 ft above the equipment or to the structural ceiling (whichever is lower). The dedicated electrical space is reserved for the installation of electrical equipment and for the installation of conduits, cable trays, and so on, entering or exiting that equipment. The outlined area in front of the electrical equipment in Exhibit 110.19 is the working space required by 110.26(A). Note that sprinkler protection is afforded the entire dedicated electrical space and working

Exhibit 110.19 The two distinct indoor installation spaces required by 110.26(A) and 110.26(F): the working space and the dedicated electrical space.

space without actually entering either space. Also note that the exhaust duct is not located in or directly above the dedicated electrical space. Although not specifically required to be located here, this duct location may be a cost-effective solution that avoids the substantial physical protection requirements of 110.26(F)(1)(b).

Exhibit 110.20 illustrates the working space required in front of the panelboard by 110.26(A). No equipment, electrical or otherwise, is allowed in the working space.

Exhibit 110.21 illustrates the dedicated electrical space above and below the panelboard required by 110.26(F)(1). This space is for the cables, raceways, and so on, that run to and from the panelboard.

(2) Outdoor Outdoor electrical equipment shall be installed in suitable enclosures and shall be protected from accidental contact by unauthorized personnel, or by vehicular traffic, or by accidental spillage or leakage from piping systems. The working clearance space shall include the zone described in 110.26(A). No architectural appurtenance or other equipment shall be located in this zone.

Extreme care should be taken where protection from unauthorized personnel or vehicular traffic is added to existing installations in order to comply with 110.26(F)(2). Any excavation or driving of steel into the ground for the placement of fencing, vehicle stops, or bollards should be done only after a thorough investigation of the belowgrade wiring.

110.27 Guarding of Live Parts

(A) Live Parts Guarded Against Accidental Contact Except as elsewhere required or permitted by this *Code,* live parts of electrical equipment operating at 50 volts or more shall be guarded against accidental contact by approved enclosures or by any of the following means:

Exhibit 110.20 The working space in front of a panelboard required by 110.26(A). This illustration supplements the dedicated electrical space shown in Exhibit 110.19.

Exhibit 110.21 The dedicated electrical space above and below a panelboard required by 110.26(F)(1).

Structural ceiling

Suspended ceiling

Permitted

Permitted

6 ft or less

Permitted

Luminaire permitted

Dedicated electrical space

6½ ft min.

Side View

Front View

(1) By location in a room, vault, or similar enclosure that is accessible only to qualified persons.

(2) By suitable permanent, substantial partitions or screens arranged so that only qualified persons have access to the space within reach of the live parts. Any openings in such partitions or screens shall be sized and located so that persons are not likely to come into accidental contact with the live parts or to bring conducting objects into contact with them.

(3) By location on a suitable balcony, gallery, or platform elevated and arranged so as to exclude unqualified persons.

(4) By elevation of 2.5 m (8 ft) or more above the floor or other working surface.

Contact conductors used for traveling cranes are permitted to be bare by 610.13(B) and 610.21(A). Although contact conductors obviously have to be bare for contact shoes on the moving member to make contact with the conductor, it is possible to place guards near the conductor to prevent its accidental contact with persons and still have slots or spaces through which the moving contacts can operate. The *Code* also recognizes the guarding of live parts by elevation.

(B) Prevent Physical Damage In locations where electric equipment is likely to be exposed to physical damage, enclosures or guards shall be so arranged and of such strength as to prevent such damage.

(C) Warning Signs Entrances to rooms and other guarded locations that contain exposed live parts shall be marked with conspicuous warning signs forbidding unqualified persons to enter.

> FPN: For motors, see 430.232 and 430.233. For over 600 volts, see 110.34.

Live parts of electrical equipment should be covered, shielded, enclosed, or otherwise protected by covers, barriers, mats, or platforms to prevent the likelihood of contact by persons or objects. See the definitions of *dead front* and *isolated (as applied to location)* in Article 100.

III. Over 600 Volts, Nominal

110.30 General

Conductors and equipment used on circuits over 600 volts, nominal, shall comply with Part I of this article and with the following sections, which supplement or modify Part I. In no case shall the provisions of this part apply to equipment on the supply side of the service point.

See "Over 600 volts" in the index to this *Handbook* for articles, parts, and sections that include requirements for installations over 600 volts.

Equipment on the supply side of the service point is outside the scope of the *NEC*. Such equipment is covered by ANSI C2, *National Electrical Safety Code*, published by the Institute of Electrical and Electronics Engineers (IEEE).

110.31 Enclosure for Electrical Installations

Electrical installations in a vault, room, or closet or in an area surrounded by a wall, screen, or fence, access to which is controlled by a lock(s) or other approved means, shall be considered to be accessible to qualified persons only. The type of enclosure used in a given case shall be designed and constructed according to the nature and degree of the hazard(s) associated with the installation.

For installations other than equipment as described in 110.31(D), a wall, screen, or fence shall be used to enclose an outdoor electrical installation to deter access by persons who are not qualified. A fence shall not be less than 2.1 m (7 ft) in height or a combination of 1.8 m (6 ft) or more of fence fabric and a 300-mm (1-ft) or more extension utilizing three or more strands of barbed wire or equivalent. The distance from the fence to live parts shall be not less than given in Table 110.31.

Table 110.31 Minimum Distance from Fence to Live Parts

Nominal Voltage	Minimum Distance to Live Parts	
	m	ft
601 – 13,799	3.05	10
13,800 – 230,000	4.57	15
Over 230,000	5.49	18

Note: For clearances of conductors for specific system voltages and typical BIL ratings, see ANSI C2-2002, *National Electrical Safety Code*.

FPN: See Article 450 for construction requirements for transformer vaults.

(A) Fire Resistivity of Electrical Vaults The walls, roof, floors, and doorways of vaults containing conductors and equipment over 600 volts, nominal, shall be constructed of materials that have adequate structural strength for the conditions, with a minimum fire rating of 3 hours. The floors of vaults in contact with the earth shall be of concrete that is not less than 4 in. (102 mm) thick, but where the vault is constructed with a vacant space or other stories below it, the floor shall have adequate structural strength for the load imposed on it and a minimum fire resistance of 3 hours. For the purpose of this section, studs and wallboards shall not be considered acceptable.

(B) Indoor Installations

(1) In Places Accessible to Unqualified Persons Indoor electrical installations that are accessible to unqualified per-

sons shall be made with metal-enclosed equipment. Metal-enclosed switchgear, unit substations, transformers, pull boxes, connection boxes, and other similar associated equipment shall be marked with appropriate caution signs. Openings in ventilated dry-type transformers or similar openings in other equipment shall be designed so that foreign objects inserted through these openings are deflected from energized parts.

(2) In Places Accessible to Qualified Persons Only Indoor electrical installations considered accessible only to qualified persons in accordance with this section shall comply with 110.34, 110.36, and 490.24.

(C) Outdoor Installations

(1) In Places Accessible to Unqualified Persons Outdoor electrical installations that are open to unqualified persons shall comply with Parts I, II, and III of Article 225.

FPN: For clearances of conductors for system voltages over 600 volts, nominal, see ANSI C2-2002, *National Electrical Safety Code.*

(2) In Places Accessible to Qualified Persons Only Outdoor electrical installations that have exposed live parts shall be accessible to qualified persons only in accordance with the first paragraph of this section and shall comply with 110.34, 110.36, and 490.24.

(D) Enclosed Equipment Accessible to Unqualified Persons Ventilating or similar openings in equipment shall be designed such that foreign objects inserted through these openings are deflected from energized parts. Where exposed to physical damage from vehicular traffic, suitable guards shall be provided. Nonmetallic or metal-enclosed equipment located outdoors and accessible to the general public shall be designed such that exposed nuts or bolts cannot be readily removed, permitting access to live parts. Where nonmetallic or metal-enclosed equipment is accessible to the general public and the bottom of the enclosure is less than 2.5 m (8 ft) above the floor or grade level, the enclosure door or hinged cover shall be kept locked. Doors and covers of enclosures used solely as pull boxes, splice boxes, or junction boxes shall be locked, bolted, or screwed on. Underground box covers that weigh over 45.4 kg (100 lb) shall be considered as meeting this requirement.

110.32 Work Space About Equipment

Sufficient space shall be provided and maintained about electric equipment to permit ready and safe operation and maintenance of such equipment. Where energized parts are exposed, the minimum clear work space shall not be less than 2.0 m (6½ ft) high (measured vertically from the floor or platform) or less than 900 mm (3 ft) wide (measured parallel to the equipment). The depth shall be as required in 110.34(A). In all cases, the work space shall permit at least a 90 degree opening of doors or hinged panels.

110.33 Entrance and Access to Work Space

(A) Entrance At least one entrance not less than 610 mm (24 in.) wide and 2.0 m (6½ ft) high shall be provided to give access to the working space about electric equipment. Where the entrance has a personnel door(s), the door(s) shall open in the direction of egress and be equipped with panic bars, pressure plates, or other devices that are normally latched but open under simple pressure.

(1) Large Equipment On switchboard and control panels exceeding 1.8 m (6 ft) in width, there shall be one entrance at each end of the equipment. A single entrance to the required working space shall be permitted where either of the conditions in 110.33(A)(1)(a) or (A)(1)(b) is met.

(a) Unobstructed Exit. Where the location permits a continuous and unobstructed way of exit travel, a single entrance to the working space shall be permitted.

(b) Extra Working Space. Where the depth of the working space is twice that required by 110.34(A), a single entrance shall be permitted. It shall be located so that the distance from the equipment to the nearest edge of the entrance is not less than the minimum clear distance specified in Table110.34(A) for equipment operating at that voltage and in that condition.

(2) Guarding Where bare energized parts at any voltage or insulated energized parts above 600 volts, nominal, to ground are located adjacent to such entrance, they shall be suitably guarded.

Section 110.33(A) contains requirements very similar to those of 110.26(C). For further information, see the commentary following 110.26(C)(2), most of which also is valid for over-600-volt installations.

(B) Access Permanent ladders or stairways shall be provided to give safe access to the working space around electric equipment installed on platforms, balconies, or mezzanine floors or in attic or roof rooms or spaces.

110.34 Work Space and Guarding

(A) Working Space Except as elsewhere required or permitted in this *Code,* the minimum clear working space in the direction of access to live parts of electrical equipment shall not be less than specified in Table 110.34(A). Distances shall be measured from the live parts, if such are exposed, or from the enclosure front or opening if such are enclosed.

Exception: Working space shall not be required in back of equipment such as dead-front switchboards or control

Table 110.34(A) Minimum Depth of Clear Working Space at Electrical Equipment

Nominal Voltage to Ground	Minimum Clear Distance		
	Condition 1	Condition 2	Condition 3
601–2500 V	900 mm (3 ft)	1.2 m (4 ft)	1.5 m (5 ft)
2501–9000 V	1.2 m (4 ft)	1.5 m (5 ft)	1.8 m (6 ft)
9001–25,000 V	1.5 m (5 ft)	1.8 m (6 ft)	2.8 m (9 ft)
25,001V–75 kV	1.8 m (6 ft)	2.5 m (8 ft)	3.0 m (10 ft)
Above 75 kV	2.5 m (8 ft)	3.0 m (10 ft)	3.7 m (12 ft)

Note: Where the conditions are as follows:
Condition 1 — Exposed live parts on one side of the working space and no live or grounded parts on the other side of the working space, or exposed live parts on both sides of the working space that are effectively guarded by insulating materials.
Condition 2 — Exposed live parts on one side of the working space and grounded parts on the other side of the working space. Concrete, brick, or tile walls shall be considered as grounded.
Condition 3 — Exposed live parts on both sides of the working space.

assemblies where there are no renewable or adjustable parts (such as fuses or switches) on the back and where all connections are accessible from locations other than the back. Where rear access is required to work on de-energized parts on the back of enclosed equipment, a minimum working space of 750 mm (30 in.) horizontally shall be provided.

(B) Separation from Low-Voltage Equipment Where switches, cutouts, or other equipment operating at 600 volts, nominal, or less are installed in a vault, room, or enclosure where there are exposed live parts or exposed wiring operating at over 600 volts, nominal, the high-voltage equipment shall be effectively separated from the space occupied by the low-voltage equipment by a suitable partition, fence, or screen.

Exception: Switches or other equipment operating at 600 volts, nominal, or less and serving only equipment within the high-voltage vault, room, or enclosure shall be permitted to be installed in the high-voltage vault, room or enclosure without a partition, fence, or screen if accessible to qualified persons only.

(C) Locked Rooms or Enclosures The entrance to all buildings, vaults, rooms, or enclosures containing exposed live parts or exposed conductors operating at over 600 volts, nominal, shall be kept locked unless such entrances are under the observation of a qualified person at all times.

Where the voltage exceeds 600 volts, nominal, permanent and conspicuous warning signs shall be provided, reading as follows:

DANGER — HIGH VOLTAGE — KEEP OUT

Equipment used on circuits over 600 volts, nominal, and containing exposed live parts or exposed conductors is required to be located in a locked room or in an enclosure. The provisions for locking are not required if the room or enclosure is under observation at all times, as is the case with some engine rooms.

(D) Illumination Illumination shall be provided for all working spaces about electrical equipment. The lighting outlets shall be arranged so that persons changing lamps or making repairs on the lighting system are not endangered by live parts or other equipment.

The points of control shall be located so that persons are not likely to come in contact with any live part or moving part of the equipment while turning on the lights.

(E) Elevation of Unguarded Live Parts Unguarded live parts above working space shall be maintained at elevations not less than required by Table 110.34(E).

Table 110.34(E) Elevation of Unguarded Live Parts Above Working Space

Nominal Voltage Between Phases	Elevation	
	m	ft
601–7500 V	2.8	9
7501–35,000 V	2.9	9½
Over 35 kV	2.9 m + 9.5 mm/kV above 35	9½ ft + 0.37 in./kV above 35

(F) Protection of Service Equipment, Metal-Enclosed Power Switchgear, and Industrial Control Assemblies Pipes or ducts foreign to the electrical installation and requiring periodic maintenance or whose malfunction would endanger the operation of the electrical system shall not be located in the vicinity of the service equipment, metal-enclosed power switchgear, or industrial control assemblies. Protection shall be provided where necessary to avoid damage from condensation leaks and breaks in such foreign systems. Piping and other facilities shall not be considered foreign if provided for fire protection of the electrical installation.

110.36 Circuit Conductors

Circuit conductors shall be permitted to be installed in raceways; in cable trays; as metal-clad cable, as bare wire, cable, and busbars; or as Type MV cables or conductors as provided in 300.37, 300.39, 300.40, and 300.50. Bare live conductors shall conform with 490.24.

Insulators, together with their mounting and conductor attachments, where used as supports for wires, single-conductor cables, or busbars, shall be capable of safely withstanding the maximum magnetic forces that would prevail

when two or more conductors of a circuit were subjected to short-circuit current.

Exposed runs of insulated wires and cables that have a bare lead sheath or a braided outer covering shall be supported in a manner designed to prevent physical damage to the braid or sheath. Supports for lead-covered cables shall be designed to prevent electrolysis of the sheath.

110.40 Temperature Limitations at Terminations

Conductors shall be permitted to be terminated based on the 90°C (194°F) temperature rating and ampacity as given in Tables 310.67 through 310.86, unless otherwise identified.

IV. Tunnel Installations over 600 Volts, Nominal

110.51 General

(A) Covered The provisions of this part shall apply to the installation and use of high-voltage power distribution and utilization equipment that is portable, mobile, or both, such as substations, trailers, cars, mobile shovels, draglines, hoists, drills, dredges, compressors, pumps, conveyors, underground excavators, and the like.

(B) Other Articles The requirements of this part shall be additional to, or amendatory of, those prescribed in Articles 100 through 490 of this *Code*. Special attention shall be paid to Article 250.

(C) Protection Against Physical Damage Conductors and cables in tunnels shall be located above the tunnel floor and so placed or guarded to protect them from physical damage.

110.52 Overcurrent Protection

Motor-operated equipment shall be protected from overcurrent in accordance with Parts III, IV, and V of Article 430. Transformers shall be protected from overcurrent in accordance with 450.3.

110.53 Conductors

High-voltage conductors in tunnels shall be installed in metal conduit or other metal raceway, Type MC cable, or other approved multiconductor cable. Multiconductor portable cable shall be permitted to supply mobile equipment.

110.54 Bonding and Equipment Grounding Conductors

(A) Grounded and Bonded All non–current-carrying metal parts of electric equipment and all metal raceways and cable sheaths shall be effectively grounded and bonded to all metal pipes and rails at the portal and at intervals not exceeding 300 m (1000 ft) throughout the tunnel.

(B) Equipment Grounding Conductors An equipment grounding conductor shall be run with circuit conductors inside the metal raceway or inside the multiconductor cable jacket. The equipment grounding conductor shall be permitted to be insulated or bare.

110.55 Transformers, Switches, and Electrical Equipment

All transformers, switches, motor controllers, motors, rectifiers, and other equipment installed below ground shall be protected from physical damage by location or guarding.

110.56 Energized Parts

Bare terminals of transformers, switches, motor controllers, and other equipment shall be enclosed to prevent accidental contact with energized parts.

110.57 Ventilation System Controls

Electrical controls for the ventilation system shall be arranged so that the airflow can be reversed.

110.58 Disconnecting Means

A switch or circuit breaker that simultaneously opens all ungrounded conductors of the circuit shall be installed within sight of each transformer or motor location for disconnecting the transformer or motor. The switch or circuit breaker for a transformer shall have an ampere rating not less than the ampacity of the transformer supply conductors. The switch or circuit breaker for a motor shall comply with the applicable requirements of Article 430.

110.59 Enclosures

Enclosures for use in tunnels shall be dripproof, weatherproof, or submersible as required by the environmental conditions. Switch or contactor enclosures shall not be used as junction boxes or as raceways for conductors feeding through or tapping off to other switches, unless the enclosures comply with 312.8.

V. Manholes and Other Electric Enclosures Intended for Personnel Entry, All Voltages

Prior to the 2005 *Code*, the requirements for manholes were found in Part IV of Article 314. For the 2005 edition, manhole requirements were moved to Article 110 and placed there as the new Part V. Placing the manhole requirements in Article 110 makes sense because manhole working space issues for cabling and other equipment here parallel those same working space issues elsewhere in Article 110. For handhole installations, see Article 314.

110.70 General

Electric enclosures intended for personnel entry and specifically fabricated for this purpose shall be of sufficient size to provide safe work space about electric equipment with live parts that is likely to require examination, adjustment, servicing, or maintenance while energized. Such enclosures shall have sufficient size to permit ready installation or withdrawal of the conductors employed without damage to the conductors or to their insulation. They shall comply with the provisions of this part.

Exception: Where electric enclosures covered by Part V of this article are part of an industrial wiring system operating under conditions of maintenance and supervision that ensure that only qualified persons monitor and supervise the system, they shall be permitted to be designed and installed in accordance with appropriate engineering practice. If required by the authority having jurisdiction, design documentation shall be provided.

The provisions of Part V are conditional, just like the requirements in 110.26, that is, some of the requirements are applicable only where the equipment "is likely to require examination, adjustment, servicing, or maintenance while energized."

110.71 Strength

Manholes, vaults, and their means of access shall be designed under qualified engineering supervision and shall withstand all loads likely to be imposed on the structures.

> FPN: See ANSI C2-2002, *National Electrical Safety Code,* for additional information on the loading that can be expected to bear on underground enclosures.

110.72 Cabling Work Space

A clear work space not less than 900 mm (3 ft) wide shall be provided where cables are located on both sides, and not less than 750 mm (2½ ft) where cables are only on one side. The vertical headroom shall not be less than 1.8 m (6 ft) unless the opening is within 300 mm (1 ft), measured horizontally, of the adjacent interior side wall of the enclosure.

Exception: A manhole containing only one or more of the following shall be permitted to have one of the horizontal work space dimensions reduced to 600 mm (2 ft) where the other horizontal clear work space is increased so the sum of the two dimensions is not less than 1.8 m (6 ft):

(1) Optical fiber cables as covered in Article 770
(2) Power-limited fire alarm circuits supplied in accordance with 760.41(A)

(3) Class 2 or Class 3 remote-control and signaling circuits, or both, supplied in accordance with 725.41

110.73 Equipment Work Space

Where electric equipment with live parts that is likely to require examination, adjustment, servicing, or maintenance while energized is installed in a manhole, vault, or other enclosure designed for personnel access, the work space and associated requirements in 110.26 shall be met for installations operating at 600 volts or less. Where the installation is over 600 volts, the work space and associated requirements in 110.34 shall be met. A manhole access cover that weighs over 45 kg (100 lb) shall be considered as meeting the requirements of 110.34(C).

110.74 Bending Space for Conductors

Bending space for conductors operating at 600 volts or below shall be provided in accordance with the requirements of 314.28. Conductors operating over 600 volts shall be provided with bending space in accordance with 314.71(A) and 314.71(B), as applicable. All conductors shall be cabled, racked up, or arranged in an approved manner that provides ready and safe access for persons to enter for installation and maintenance.

Exception: Where 314.71(B) applies, each row or column of ducts on one wall of the enclosure shall be calculated individually, and the single row or column that provides the maximum distance shall be used.

110.75 Access to Manholes

(A) Dimensions Rectangular access openings shall not be less than 650 mm × 550 mm (26 in. × 22 in.). Round access openings in a manhole shall not be less than 650 mm (26 in.) in diameter.

Exception: A manhole that has a fixed ladder that does not obstruct the opening or that contains only one or more of the following shall be permitted to reduce the minimum cover diameter to 600 mm (2 ft):

(1) Optical fiber cables as covered in Article 770
(2) Power-limited fire alarm circuits supplied in accordance with 760.41
(3) Class 2 or Class 3 remote-control and signaling circuits, or both, supplied in accordance with 725.41

(B) Obstructions Manhole openings shall be free of protrusions that could injure personnel or prevent ready egress.

(C) Location Manhole openings for personnel shall be located where they are not directly above electric equipment or conductors in the enclosure. Where this is not practicable, either a protective barrier or a fixed ladder shall be provided.

(D) Covers Covers shall be over 45 kg (100 lb) or otherwise designed to require the use of tools to open. They shall be designed or restrained so they cannot fall into the manhole or protrude sufficiently to contact electrical conductors or equipment within the manhole.

(E) Marking Manhole covers shall have an identifying mark or logo that prominently indicates their function, such as "electric."

110.76 Access to Vaults and Tunnels

(A) Location Access openings for personnel shall be located where they are not directly above electric equipment or conductors in the enclosure. Other openings shall be permitted over equipment to facilitate installation, maintenance, or replacement of equipment.

(B) Locks In addition to compliance with the requirements of 110.34(C), if applicable, access openings for personnel shall be arranged such that a person on the inside can exit when the access door is locked from the outside, or in the case of normally locking by padlock, the locking arrangement shall be such that the padlock can be closed on the locking system to prevent locking from the outside.

110.77 Ventilation

Where manholes, tunnels, and vaults have communicating openings into enclosed areas used by the public, ventilation to open air shall be provided wherever practicable.

110.78 Guarding

Where conductors or equipment, or both, could be contacted by objects falling or being pushed through a ventilating grating, both conductors and live parts shall be protected in accordance with the requirements of 110.27(A)(2) or 110.31(B)(1), depending on the voltage.

110.79 Fixed Ladders

Fixed ladders shall be corrosion resistant.

Wiring and Protection

ARTICLE 200
Use and Identification of Grounded Conductors

Summary of Changes

- **200.6(D):** Revised to reorganize requirements on identification of a grounded conductor where there is more than one nominal voltage system and to require that the means of identification be permanently posted at each branch-circuit panelboard.

- **200.7(C)(1):** Revised to require that the identification method used to reidentify a white or gray conductor in a cable assembly as an ungrounded conductor must encircle the insulation and be a color other than white, gray, or green.

Contents

200.1 Scope

This article provides requirements for the following:

(1) Identification of terminals
(2) Grounded conductors in premises wiring systems
(3) Identification of grounded conductors

> FPN: See Article 100 for definitions of *Grounded Conductor* and *Grounding Conductor*.

The requirements of Article 200 cover the grounded conductor's use in premises wiring systems and the acceptable methods for identifying grounded conductors and the terminals to which they are connected. Identification of grounded conductors is a long-standing, fundamental safety concept that helps ensure proper connection of the conductor throughout an electrical system. Proper connection and maintaining correct polarity are essential to ensuring safe interface with wiring devices, appliances, and portable and permanently installed luminaires.

The grounded circuit conductor is referred to throughout the *Code* as the *grounded conductor*. In accordance with the Article 100 definition of *grounded conductor*, it is a conductor that is intentionally connected to earth or some conducting body that serves as earth. A common example of being connected to a conducting body that serves as earth is one in which the grounded conductor of a transformer-supplied, separately derived system is connected to effectively grounded building steel. The building steel is not earth but serves in its place for the purposes of grounding the separately derived system.

The grounded conductor is often, but not always, the neutral conductor. For example, in a single-phase 2-wire or in a 3-phase corner-grounded delta system, the intentionally grounded conductor is not a neutral conductor. Through its very nature of being connected to the same grounding electrode system as the non–current-carrying metal parts of electrical equipment, there is generally no potential difference between the grounded conductor and those grounded metal parts. However, unlike an equipment grounding conductor, the grounded conductor is a circuit conductor and as such is a current-carrying conductor.

Electric shock injuries and electrocutions have occurred as a result of working on the grounded conductor while the circuit is energized. Extreme caution must be exercised where the grounded (neutral) conductor is part of a multiwire branch circuit, and it should be noted that 300.13 does not permit the wiring terminals of a device, such as a receptacle, to be the means of maintaining the continuity of the grounded conductor in that type of branch circuit.

In addition to the requirements in this article, use and installation of the grounded conductor are covered extensively by the requirements in Article 250.

200.2 General

All premises wiring systems, other than circuits and systems exempted or prohibited by 210.10, 215.7, 250.21, 250.22, 250.162, 503.155, 517.63, 668.11, 668.21, and 690.41 Exception, shall have a grounded conductor that is identified in accordance with 200.6.

The grounded conductor, where insulated, shall have insulation that is (1) suitable, other than color, for any ungrounded conductor of the same circuit on circuits of less than 1000 volts or impedance grounded neutral systems of 1 kV and over, or (2) rated not less than 600 volts for solidly grounded neutral systems of 1 kV and over as described in 250.184(A).

200.3 Connection to Grounded System

Premises wiring shall not be electrically connected to a supply system unless the latter contains, for any grounded conductor of the interior system, a corresponding conductor that is grounded. For the purpose of this section, *electrically connected* shall mean connected so as to be capable of carrying current, as distinguished from connection through electromagnetic induction.

Grounded conductors of premises wiring (other than separately derived systems) must be connected to the supply system grounded conductor to ensure a common, continuous, grounded system.

200.6 Means of Identifying Grounded Conductors

(A) Sizes 6 AWG or Smaller An insulated grounded conductor of 6 AWG or smaller shall be identified by a continuous white or gray outer finish or by three continuous white stripes on other than green insulation along its entire length. Wires that have their outer covering finished to show a white or gray color but have colored tracer threads in the braid identifying the source of manufacture shall be considered as meeting the provisions of this section. Insulated grounded conductors shall also be permitted to be identified as follows:

(1) The grounded conductor of a mineral-insulated, metal-sheathed cable shall be identified at the time of installation by distinctive marking at its terminations.
(2) A single-conductor, sunlight-resistant, outdoor-rated cable used as a grounded conductor in photovoltaic power systems as permitted by 690.31 shall be identified at the time of installation by distinctive white marking at all terminations.
(3) Fixture wire shall comply with the requirements for grounded conductor identification as specified in 402.8.
(4) For aerial cable, the identification shall be as above, or by means of a ridge located on the exterior of the cable so as to identify it.

The use of white insulation or white marking is the most common method of identifying the grounded conductor. However, Article 200 provides a number of alternative identification means, including the use of gray insulation or markings, the use of three continuous white stripes along the conductor insulation, surface markings, and colored braids

or separators. The required identification of the grounded conductor is performed either by the wire or cable manufacturer or by the installer at the time of installation.

The general rule of 200.6(A) requires insulated conductors 6 AWG or smaller to be white or gray for their entire length where they are used as grounded conductors. Beginning with the 1999 edition, the *Code* also permits three continuous white stripes along the entire length of conductor insulation that is colored other than green as a means to identify a conductor as the grounded conductor. The three white stripes method of identification is permitted for all conductor sizes and is the method that most typically would be employed by a wire or cable manufacturer.

Other methods of identification are also permitted in 200.6(A). For example, the grounded conductor of mineral-insulated (MI) cable, due to its unique construction, is permitted to be identified at the time of installation. Aerial cable may have its grounded conductor identified by a ridge along its insulated surface, and fixture wires are permitted to have the grounded conductor identified by various methods, including colored insulation, stripes on the insulation, colored braid, colored separator, and tinned conductors. These identification methods are found in 402.8 and explained in detail in 400.22(A) through 400.22(E).

For 6 AWG or smaller, identification of the grounded conductor solely by distinctive white marking or gray at the time of installation is not permitted except as described for flexible cords and multiconductor cables in 200.6(C) and 200.6(E) and for single conductors in outdoor photovoltaic power installations in accordance with 200.6(A)(2).

(B) Sizes Larger Than 6 AWG An insulated grounded conductor larger than 6 AWG shall be identified by one of the following means:

The general rule of 200.6(B) requires that insulated grounded conductors larger than 6 AWG be identified using one of three acceptable methods. As is allowed by 200.6(A) for 6 AWG and smaller insulated conductors, 200.6(B) permits the use of a continuous white or gray color along the entire length of the conductor insulation or the use of three continuous white stripes on the entire length of the insulated (other than green-colored insulation) conductor. The most common identification method used by installers to identify a single conductor as a grounded conductor is application of a white or gray marking to the insulation at all termination points at the time of installation. To be clearly visible, this field-applied white or gray marking must completely encircle the conductor insulation. This coloring can be applied by using marking tape or by painting the insulation. This method of identification is shown in Exhibit 200.1.

Exhibit 200.1 Field-applied identification, as permitted by 200.6(B), of a 4 AWG conductor to identify it as the grounded conductor.

(1) By a continuous white or gray outer finish.

(2) By three continuous white stripes along its entire length on other than green insulation.

(3) At the time of installation, by a distinctive white or gray marking at its terminations. This marking shall encircle the conductor or insulation.

(C) Flexible Cords An insulated conductor that is intended for use as a grounded conductor, where contained within a flexible cord, shall be identified by a white or gray outer finish or by methods permitted by 400.22.

(D) Grounded Conductors of Different Systems Where grounded conductors of different systems are installed in the same raceway, cable, box, auxiliary gutter, or other type of enclosure, each grounded conductor shall be identified by system. Identification that distinguishes each system grounded conductor shall be permitted by one of the following means:

(1) One system grounded conductor shall have an outer covering conforming to 200.6(A) or 200.6(B).

(2) The grounded conductor(s) of other systems shall have a different outer covering conforming to 200.6(A) or 200.6(B) or by an outer covering of white or gray with a readily distinguishable colored stripe other than green running along the insulation.

(3) Other and different means of identification as allowed by 200.6(A) or 200.6(B) that will distinguish each system grounded conductor.

This means of identification shall be permanently posted at each branch-circuit panelboard.

The requirements found in 200.6(D) have remained essentially the same since the 1987 edition of the *NEC*. However, these requirements are often misapplied. As Exhibit 200.2 shows, where grounded conductors of different systems are present in the same enclosure, these grounded conductors must be distinguished from each other, which can be accomplished through the use of different colors or marking schemes. The use of colored stripes (other than green) on white insulation for the entire conductor insulation length is an acceptable method to distinguish one system grounded conductor from one with white insulation or white marking. It is important to note that this requirement applies only where grounded conductors of different systems are installed in a common enclosure, such as a junction or pull box or a wireway.

First introduced in the 2002 *Code*, it is now permitted to identify one system grounded conductor with white insulation or field-installed white marking and the other system grounded conductor with gray insulation or field-installed gray marking. Gray and white are considered to be different means of identification.

The 2005 *Code* requires the identification method or scheme used to distinguish the grounded conductors of different systems to be posted at all panelboards that supply branch circuits. In addition, industry practice of using white for lower voltage systems and gray for higher voltage systems is permitted but not mandated by the *Code*.

Exhibit 200.2 Grounded conductors of different systems in the same enclosure. The grounded conductors of the different systems are identified by color through the use of white and gray colored insulation, one of the methods specified by 200.6(D).

(E) Grounded Conductors of Multiconductor Cables
The insulated grounded conductors in a multiconductor cable

shall be identified by a continuous white or gray outer finish or by three continuous white stripes on other than green insulation along its entire length. Multiconductor flat cable 4 AWG or larger shall be permitted to employ an external ridge on the grounded conductor.

Exception No. 1: Where the conditions of maintenance and supervision ensure that only qualified persons service the installation, grounded conductors in multiconductor cables shall be permitted to be permanently identified at their terminations at the time of installation by a distinctive white marking or other equally effective means.

Exception No. 1 to 200.6(E) introduces the concept of identifying grounded conductors of multiconductor cables at termination locations. This exception allows identification of a conductor that is part of a multiconductor cable as the grounded conductor at the time of installation by use of a distinctive white marking or other equally effective means, such as numbering, lettering, or tagging, as shown in Exhibit 200.3. Exception No. 1 to 200.6(E) is intended to apply to installations in facilities that have a regulated system of maintenance and supervision that ensures that only qualified persons service the installation. Permission to reidentify a conductor within a cable assembly is not predicated on the conductor size.

Exhibit 200.3 Field-applied identification to the conductor of a multiconductor armored cable that will be used as the grounded conductor as permitted by 200.6(E), Exception No. 1.

Exception No. 2: The grounded conductor of a multiconductor varnished-cloth-insulated cable shall be permitted to be identified at its terminations at the time of installation by a distinctive white marking or other equally effective means.

> FPN: The color gray may have been used in the past as an ungrounded conductor. Care should be taken when working on existing systems.

The term *natural gray* was changed to *gray* in the 2002 *Code* because the phrase "natural gray outer finish" was deemed obsolete. This change reserves all shades of gray insulation and marking for grounded conductors. The FPN following 200.6 warns the user to exercise caution when working on existing systems because gray may have been used on those existing systems.

200.7 Use of Insulation of a White or Gray Color or with Three Continuous White Stripes

(A) General The following shall be used only for the grounded circuit conductor, unless otherwise permitted in 200.7(B) and 200.7(C):

(1) A conductor with continuous white or gray covering
(2) A conductor with three continuous white stripes on other than green insulation
(3) A marking of white or gray color at the termination

(B) Circuits of Less Than 50 Volts A conductor with white or gray color insulation or three continuous white stripes or having a marking of white or gray at the termination for circuits of less than 50 volts shall be required to be grounded only as required by 250.20(A).

(C) Circuits of 50 Volts or More The use of insulation that is white or gray or that has three continuous white stripes for other than a grounded conductor for circuits of 50 volts or more shall be permitted only as in (1) through (3).

(1) If part of a cable assembly and where the insulation is permanently reidentified to indicate its use as an ungrounded conductor, by painting or other effective means at its termination, and at each location where the conductor is visible and accessible. Identification shall encircle the insulation and shall be a color other than white, gray, or green.
(2) Where a cable assembly contains an insulated conductor for single-pole, 3-way or 4-way switch loops and the conductor with white or gray insulation or a marking of three continuous white stripes is used for the supply to the switch but not as a return conductor from the switch to the switched outlet. In these applications, the conductor with white or gray insulation or with three continuous white stripes shall be permanently reidentified to indicate its use by painting or other effective means at its terminations and at each location where the conductor is visible and accessible.

Previous editions of the *Code* permitted switch loops using a white insulated conductor to supply the switch but not as the return conductor to supply the lighting outlet. Prior to the 1999 *NEC*, re-identification of this particular ungrounded conductor was not required. However, many electronic automation devices requiring a grounded conductor are now available for installation into switch outlets. Therefore, re-identification of all ungrounded conductors that are white or otherwise identified by one of the methods permitted for grounded conductors is now required at every termination point to avoid confusion and improper wiring at the time a switching device is installed or replaced. The required re-identification must be effective, permanent, and suitable for the environment, to clearly identify the insulated conductor as an ungrounded conductor.

(3) Where a flexible cord, having one conductor identified by a white or gray outer finish or three continuous white stripes or by any other means permitted by 400.22, is used for connecting an appliance or equipment permitted by 400.7. This shall apply to flexible cords connected to outlets whether or not the outlet is supplied by a circuit that has a grounded conductor.

The term *natural gray* was changed to *gray* in the 2002 *Code* because the phrase "natural gray outer finish" was deemed obsolete. This change reserves all shades of gray insulation and marking for identification of grounded conductors. The FPN following 200.6 warns the user to exercise caution when working on existing systems because gray may have been used on those existing systems.

FPN: The color gray may have been used in the past as an ungrounded conductor. Care should be taken when working on existing systems.

200.9   Means of Identification of Terminals

The identification of terminals to which a grounded conductor is to be connected shall be substantially white in color. The identification of other terminals shall be of a readily distinguishable different color.

Exception: Where the conditions of maintenance and supervision ensure that only qualified persons service the installations, terminals for grounded conductors shall be permitted to be permanently identified at the time of installation by a distinctive white marking or other equally effective means.

200.10   Identification of Terminals

(A) Device Terminals All devices, excluding panelboards, provided with terminals for the attachment of conductors and intended for connection to more than one side of the circuit shall have terminals properly marked for identification, unless the electrical connection of the terminal intended to be connected to the grounded conductor is clearly evident.

Exception: Terminal identification shall not be required for devices that have a normal current rating of over 30 amperes, other than polarized attachment plugs and polarized receptacles for attachment plugs as required in 200.10(B).

(B) Receptacles, Plugs, and Connectors Receptacles, polarized attachment plugs, and cord connectors for plugs and polarized plugs shall have the terminal intended for connection to the grounded conductor identified as follows:

(1) Identification shall be by a metal or metal coating that is substantially white in color or by the word *white* or the letter *W* located adjacent to the identified terminal.

(2) If the terminal is not visible, the conductor entrance hole for the connection shall be colored white or marked with the word *white* or the letter *W.*

Section 200.10(B) requires that terminals of receptacles, plugs, and connectors intended for the connection of the grounded conductor be marked by one of several methods, including the word *white*, the letter *W*, or a distinctive white color. The variety of these methods allows the plating of all screws and terminals to meet other requirements of specific applications, such as corrosion-resistant devices.

FPN: See 250.126 for identification of wiring device equipment grounding conductor terminals.

(C) Screw Shells For devices with screw shells, the terminal for the grounded conductor shall be the one connected to the screw shell.

(D) Screw Shell Devices with Leads For screw shell devices with attached leads, the conductor attached to the screw shell shall have a white or gray finish. The outer finish of the other conductor shall be of a solid color that will not be confused with the white or gray finish used to identify the grounded conductor.

FPN: The color gray may have been used in the past as an ungrounded conductor. Care should be taken when working on existing systems.

The term *natural gray* was changed to *gray* in the 2002 *Code* because the phrase "natural gray outer finish" was deemed obsolete. This change reserves all shades of gray insulation and marking for identification of grounded conductors. The FPN following 200.6 warns the user to exercise caution when working on existing systems because gray may have been used on those existing systems.

(E) Appliances Appliances that have a single-pole switch or a single-pole overcurrent device in the line or any line-connected screw shell lampholders, and that are to be connected by (1) a permanent wiring method or (2) field-installed attachment plugs and cords with three or more wires (including the equipment grounding conductor), shall have means to identify the terminal for the grounded circuit conductor (if any).

200.11 Polarity of Connections

No grounded conductor shall be attached to any terminal or lead so as to reverse the designated polarity.

ARTICLE 210
Branch Circuits

Summary of Changes

- **210.4(B):** Revised the requirement on disconnecting all ungrounded conductors of a multiwire branch circuit supplying devices or equipment on the same strap or yoke to apply to all occupancies.

- **210.5(C):** Added new requirement that each ungrounded branch-circuit conductor be identified by where there is more than one nominal voltage system on the premises and that the means of identification be posted at each branch-circuit panelboard or similar distribution equipment.

- **210.6(D)(2):** Revised to clarify that for the purposes of this requirement, luminaires are not included as utilization equipment.

- **210.7(B):** Revised requirement to apply where devices or equipment are installed on the same mounting yoke.

- **210.8(A)(7):** Revised to add laundry and utility sinks to the GFCI requirement.

- **210.8(B)(2):** Revised to describe what constitutes commercial and institutional kitchens.

- **210.8(B)(4):** Added new requirement for GFCI protection of 125-volt, 15- and 20-ampere receptacles installed outdoors in public spaces.

- **210.8(C):** Added new requirement for GFCI protection of 125-volt, 15- and 20-ampere circuits supplying boat hoists.

- **210.12(B):** Revised to require listed combination-type arc-fault circuit-interrupters to protect 120-volt, 15- and 20-ampere branch circuits that supply bedrooms in dwelling units. Branch/feeder AFCIs are permitted to meet this requirement until January 1, 2008. An exception permits the AFCI to be at other than the origination of the branch circuit under specified conditions.

- **210.18:** Added new requirement that guest rooms and guest suites that are provided with permanent provisions for cooking have branch circuits and outlets installed to meet the rules for dwelling units.

- **210.19(A)(3):** Revised Exception No.1 to include the leads supplied with the appliance as branch circuit tap conductors.

- **210.23(A)(1):** Revised to limit the 80 percent load requirement to cord-and-plug-connected equipment that is not fastened in place.

- **210.52(C)(1), Exception:** Added exception exempting certain wall spaces directly behind a rangetop or sink from receptacle outlet requirement, with a new Figure 210.52 illustrating the exempted area.

- **210.52(D):** Added exception to allow bathroom basins to have GFCI-protected receptacles on the side or face of the basin cabinet not more than 12 in. below the countertop.

- **210.52(E):** Revised to require that the grade-level dwelling units (with direct entrance/exit to grade) of multifamily dwellings be provided with an easily accessible outdoor GFCI-protected receptacle.

- **210.60(A):** Revised to require that guest suites with permanent provisions for cooking comply with all applicable rules of 210.52.

- **210.63:** Added exception exempting evaporative coolers installed in one- and two-family dwellings from service receptacle requirement.

- **210.70(B):** Revised requirements for lighting outlets in guest rooms and guest suites to parallel the requirements of 210.70(A)(1) for dwelling units.

Contents

I. General Provisions

210.1 Scope

This article covers branch circuits except for branch circuits that supply only motor loads, which are covered in Article 430. Provisions of this article and Article 430 apply to branch circuits with combination loads.

According to 668.3(C)(1), electrolytic cell line conductors, cells, cell line attachments, and the wiring of auxiliary equipment and devices within the cell line working zone are not required to comply with the provisions of Article 210.

210.2 Other Articles for Specific-Purpose Branch Circuits

Branch circuits shall comply with this article and also with the applicable provisions of other articles of this *Code*. The provisions for branch circuits supplying equipment listed in Table 210.2 amend or supplement the provisions in this article and shall apply to branch circuits referred to therein.

210.3 Rating

Branch circuits recognized by this article shall be rated in accordance with the maximum permitted ampere rating or setting of the overcurrent device. The rating for other than individual branch circuits shall be 15, 20, 30, 40, and 50 amperes. Where conductors of higher ampacity are used for any reason, the ampere rating or setting of the specified overcurrent device shall determine the circuit rating.

Where the length of the branch circuit conductors is determined to cause an unacceptable voltage drop, larger conductors with a higher ampacity commonly are used. For example, a branch circuit wired with 10 AWG copper conductors has an allowable ampacity of at least 30 amperes per Table 310.16. However, if the branch circuit overcurrent protective device is a 20-ampere circuit breaker or fuse, the rating of this branch circuit is 20 amperes, based on the size or rating of the overcurrent protective device.

Exception: Multioutlet branch circuits greater than 50 amperes shall be permitted to supply nonlighting outlet loads on industrial premises where conditions of maintenance and

Table 210.2 Specific-Purpose Branch Circuits

Equipment	Article	Section
Air-conditioning and refrigerating equipment		440.6, 440.31, 440.32
Audio signal processing, amplification, and reproduction equipment		640.8
Busways		368.17
Circuits and equipment operating at less than 50 volts	720	
Central heating equipment other than fixed electric space-heating equipment		422.12
Class 1, Class 2, and Class 3 remote-control, signaling, and power-limited circuits	725	
Closed-loop and programmed power distribution	780	
Cranes and hoists		610.42
Electric signs and outline lighting		600.6
Electric welders	630	
Elevators, dumbwaiters, escalators, moving walks, wheelchair lifts, and stairway chair lifts		620.61
Fire alarm systems	760	
Fixed electric heating equipment for pipelines and vessels		427.4
Fixed electric space-heating equipment		424.3
Fixed outdoor electric deicing and snow-melting equipment		426.4
Information technology equipment		645.5
Infrared lamp industrial heating equipment		422.48, 424.3
Induction and dielectric heating equipment	665	
Marinas and boatyards		555.19
Mobile homes, manufactured homes, and mobile home parks	550	
Motion picture and television studios and similar locations	530	
Motors, motor circuits, and controllers	430	
Pipe organs		650.7
Recreational vehicles and recreational vehicle parks	551	
Switchboards and panelboards		408.52
Theaters, audience areas of motion picture and television studios, and similar locations		520.41, 520.52, 520.62
X-ray equipment		660.2, 517.73

supervision ensure that only qualified persons service the equipment.

It is common in industrial establishments to provide several single receptacles with ratings of 50 amperes or higher on a single branch circuit to allow quick relocation of equipment for production or maintenance use, such as in the case of electric welders. Generally, only one piece of equipment at a time is supplied from this type of receptacle circuit. The type of receptacle used in this situation is generally a configuration known as a pin-and-sleeve receptacle, although the *Code* does not preclude the use of other configurations and designs. Pin-and-sleeve receptacles may or may not be horsepower rated.

210.4 Multiwire Branch Circuits

(A) General Branch circuits recognized by this article shall be permitted as multiwire circuits. A multiwire circuit shall be permitted to be considered as multiple circuits. All conductors shall originate from the same panelboard or similar distribution equipment.

> FPN: A 3-phase, 4-wire, wye-connected power system used to supply power to nonlinear loads may necessitate that the power system design allow for the possibility of high harmonic neutral currents.

The power supplies for equipment such as computers, printers, and adjustable-speed motor drives can introduce harmonic currents in the system neutral conductor. The resulting total harmonic distortion current could exceed the load current of the device itself. See the commentary following 310.15(B)(4)(c) for a discussion of neutral conductor ampacity.

(B) Devices or Equipment Where a multiwire branch circuit supplies more than one device or equipment on the same yoke, a means shall be provided to disconnect simultaneously all ungrounded conductors supplying those devices or equipment at the point where the branch circuit originates.

Where a multiwire branch circuit supplies multiple devices or pieces of equipment supported on the same mounting strap or yoke, 210.4(B) specifically requires simultaneous disconnection of all ungrounded conductors and requires that it take place at the panelboard or other distribution equipment where the multiwire circuit originates. In previous editions of the *Code*, this requirement covered installations in dwelling units only. For the 2005 edition, the requirement has been expanded in scope and applies to all occupancies.

Multiwire branch circuits can be dangerous when not

all the ungrounded circuit conductors are de-energized and equipment supplied from a multiwire circuit is being serviced. Equipment and devices on a common mounting yoke or strap pose a significant risk because of the close proximity of their wiring terminals or connections. For that reason, all ungrounded conductors supplying the devices or equipment on that strap must be simultaneously disconnected to reduce the risk of shock to personnel working on equipment supplied by the multiwire branch circuit. The simultaneous disconnecting means requirement takes the guesswork out of ensuring safe conditions for maintenance.

Most commonly, duplex or other multiple receptacle configurations supported on common mounting hardware are the focus of this requirement. However, equipment mounted on a yoke can include devices such as receptacles, switches, and lampholders, as well as other items such as dimmers, pilot lights, and home automation controls.

Many 125-volt, 15- and 20-ampere duplex receptacles have a break-off tab that permits each of the two receptacles to be supplied from different circuits or a 3-wire (multiwire) branch circuit. This arrangement is commonly called a *split-wired receptacle* (i.e., one circuit supplies half the duplex receptacle, and another circuit supplies the other half). The simultaneous opening of both "hot" conductors at the panelboard effectively protects personnel from inadvertent contact during servicing with an energized conductor or device terminal. The simultaneous disconnection can be achieved by a 2-pole circuit breaker, as shown in Exhibit 210.1 (top), or by two single-pole circuit breakers with an identified handle tie, as shown in Exhibit 210.1 (bottom). Where fuses

are used for the branch circuit overcurrent protection, a 2-pole disconnect switch is required.

(C) Line-to-Neutral Loads Multiwire branch circuits shall supply only line-to-neutral loads.

Exception No. 1: A multiwire branch circuit that supplies only one utilization equipment.

Exception No. 2: Where all ungrounded conductors of the multiwire branch circuit are opened simultaneously by the branch-circuit overcurrent device.

> FPN: See 300.13(B) for continuity of grounded conductor on multiwire circuits.

The term *multiwire branch circuit* is defined in Article 100 as "a branch circuit that consists of two or more ungrounded conductors that have a voltage between them and a grounded conductor that has equal voltage between it and each ungrounded conductor of the circuit and that is connected to the neutral or grounded conductor of the system." Although defined as "a" branch circuit, 210.4(A) permits a multiwire branch circuit to be considered as multiple circuits and could be used, for instance, to satisfy the requirement for providing two small appliance branch circuits for countertop receptacle outlets in a dwelling-unit kitchen.

The circuit most commonly used as a multiwire branch circuit consists of two ungrounded conductors and one grounded conductor supplied from a 120/240-volt, single-phase, 3-wire system. Such multiwire circuits supply appliances that have both line-to-line and line-to-neutral connected loads, such as electric ranges and clothes dryers, and also supply loads that are line-to-neutral connected only, such as the split-wired receptacle shown in Exhibit 210.1. A multiwire branch circuit is also permitted to supply a device with a 250-volt receptacle and a 125-volt receptacle, as shown in Exhibit 210.2, provided the branch circuit overcurrent device simultaneously opens both of the ungrounded conductors.

Multiwire branch circuits have many advantages, in-

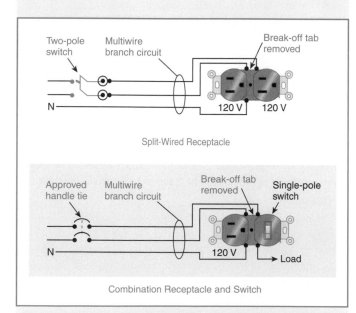

Split-Wired Receptacle

Combination Receptacle and Switch

Exhibit 210.1 Examples where 210.4(B) requires the simultaneous disconnection of all ungrounded conductors to multiwire branch circuits supplying more than one device or equipment on the same yoke.

Exhibit 210.2 An example of 210.4(C), Exception No. 2, which permits a multiwire branch circuit to supply line-to-neutral and line-to-line connected loads provided the ungrounded conductors are opened simultaneously by the branch-circuit overcurrent device.

cluding using three wires to do the work of four (in place of two 2-wire circuits), less raceway fill, easier balancing and phasing of a system, and less voltage drop. See the commentary following 215.2(A)(3), FPN No. 3, for further information on voltage drop for branch circuits.

Multiwire branch circuits may be derived from a 120/ 240-volt, single-phase; a 208Y/120-volt and 480Y/277-volt, 3-phase, 4-wire; or a 240/120-volt, 3-phase, 4-wire delta system. Section 210.11(B) requires multiwire branch circuits to be properly balanced. If two ungrounded conductors and a common neutral are used as a multiwire branch circuit supplied from a 208Y/120-volt, 3-phase, 4-wire system, the neutral carries the same current as the phase conductor with the highest current and, therefore, should be the same size. The neutral for a 2-phase, 3-wire or a 2-phase, 5-wire circuit must be sized to carry 140 percent of the ampere rating of the circuit, as required by 220.61(A) Exception. See the commentary following 210.4(A), FPN, for further information on 3-phase, 4-wire system neutral conductors.

If loads are connected line-to-line (i.e., utilization equipment connected between 2 or 3 phases), 2-pole or 3-pole circuit breakers are required to disconnect all ungrounded conductors simultaneously. In testing 240-volt equipment, it is quite possible not to realize that the circuit is still energized with 120 volts if one pole of the overcurrent device is open. See 210.10 and 240.20(B) for further information on circuit breaker overcurrent protection of ungrounded conductors. Other precautions concerning device removal on multiwire branch circuits are found in the commentary following 300.13(B).

210.5 Identification for Branch Circuits

(A) Grounded Conductor The grounded conductor of a branch circuit shall be identified in accordance with 200.6.

(B) Equipment Grounding Conductor The equipment grounding conductor shall be identified in accordance with 250.119.

(C) Ungrounded Conductors Where the premises wiring system has branch circuits supplied from more than one nominal voltage system, each ungrounded conductor of a branch circuit, where accessible, shall be identified by system. The means of identification shall be permitted to be by separate color coding, marking tape, tagging, or other approved means and shall be permanently posted at each branch-circuit panelboard or similar branch-circuit distribution equipment.

The requirement to identify ungrounded branch circuit conductors has been expanded in the 2005 *Code* to cover all branch circuit configurations and is not applicable to only multiwire circuits. As was the case in the 2002 edition, the

identification requirement applies only to those premises that have more than one nominal voltage system supplying branch circuits (e.g., a 208Y/120-volt system and a 480Y/277-volt system). Unlike the requirement of 200.6(D) for identifying the grounded conductors supplied from different voltage systems, application of this revised rule for identification of the ungrounded conductors does not depend on the different system conductors being installed in the same raceway, cabinet, or enclosure.

The method of identification can be unique to the premises, and although color coding is a popular method, other types of marking or tagging are acceptable alternatives. It is intended that whatever method of identification is used it be consistent throughout the premises. To that end, the identification legend is required to be posted at each branch circuit panelboard or other equipment from which branch circuits are supplied. The expansion of this requirement is based on the need to provide a higher level of safety for personnel working on premises electrical systems with multiple supply voltages.

Exhibit 210.3 shows an example of two different nominal voltage systems in a building. Each ungrounded system conductor is identified by color-coded marking tape. A notice indicating the means of the identification is permanently located at each panelboard. It should be noted that this requirement now applies to all ungrounded branch circuit conductors.

Exhibit 210.3 Examples of accessible (ungrounded) phase conductors identified by marking tape.

210.6 Branch-Circuit Voltage Limitations

The nominal voltage of branch circuits shall not exceed the values permitted by 210.6(A) through 210.6(E).

(A) Occupancy Limitation In dwelling units and guest rooms or guest suites of hotels, motels, and similar occupancies, the voltage shall not exceed 120 volts, nominal, between conductors that supply the terminals of the following:

(1) Luminaires (lighting fixtures)
(2) Cord-and-plug-connected loads 1440 volt-amperes, nominal, or lcss or less than ¼ hp

The term *similar occupancies* in 210.6(A) refers to sleeping rooms in dormitories, fraternities, sororities, nursing homes, and other such facilities. This requirement is intended to reduce the exposure of residents in dwellings and similar occupancies from electric shock hazards when using or servicing permanently installed luminaires and cord-and-plug-connected portable lamps and appliances. For the 2005 *Code*, 210.6(A) has been revised to specifically identify a hotel or motel guest suite as an area where this voltage limitation is also mandatory.

Small loads, such as those of 1440 volt-amperes or less and motors of less than ¼ horsepower, are limited to 120-volt circuits. High-wattage cord-and-plug-connected loads, such as electric ranges, clothes dryers, and some window air conditioners, may be connected to a 208-volt or 240-volt circuit.

(B) 120 Volts Between Conductors Circuits not exceeding 120 volts, nominal, between conductors shall be permitted to supply the following:

(1) The terminals of lampholders applied within their voltage ratings

Section 210.6(B)(1) allows lampholders to be used only within their voltage ratings. See the commentary following 210.6(C)(2) for details on voltage limitations for listed incandescent luminaires.

(2) Auxiliary equipment of electric-discharge lamps

Auxiliary equipment includes ballasts and starting devices for fluorescent and high-intensity-discharge (e.g., mercury vapor, metal halide, and sodium) lamps.

(3) Cord-and-plug-connected or permanently connected utilization equipment

(C) 277 Volts to Ground Circuits exceeding 120 volts, nominal, between conductors and not exceeding 277 volts, nominal, to ground shall be permitted to supply the following:

(1) Listed electric-discharge luminaires (lighting fixtures)

Section 210.6(C)(1) allows listed electric-discharge luminaires to be used only within their ratings. See 225.7(C) and 225.7(D) for additional restrictions on the installation of outdoor luminaires.

(2) Listed incandescent luminaires (lighting fixtures), where supplied at 120 volts or less from the output of a stepdown autotransformer that is an integral component of the luminaire (fixture) and the outer shell terminal is electrically connected to a grounded conductor of the branch circuit

Section 210.6(C)(2) permits an incandescent luminaire on a 277-volt circuit only if it is a listed luminaire with an integral autotransformer and an output to the lampholder that does not exceed 120 volts. In this application, the autotransformer supplies 120 volts to the lampholder, and the grounded conductor is connected to the screw shell of the lampholder. This application is similar to a branch circuit derived from an autotransformer, except that the 120-volt circuit is the internal wiring of the luminaire.

(3) Luminaires (lighting fixtures) equipped with mogul-base screw shell lampholders
(4) Lampholders, other than the screw shell type, applied within their voltage ratings
(5) Auxiliary equipment of electric-discharge lamps
(6) Cord-and-plug-connected or permanently connected utilization equipment

Exhibit 210.4 shows some examples of luminaires permitted to be connected to branch circuits. Medium-base screw shell lampholders cannot be directly connected to 277-volt branch circuits. Other types of lampholders may be connected to 277-volt circuits but only if the lampholders have a 277-volt rating. A 277-volt branch circuit may be connected to a listed electric-discharge fixture or to a listed autotransformer-type incandescent fixture with a medium-base screw shell lampholder.

Typical examples of the cord-and-plug-connected equipment listed under 210.6(C)(6) are through-the-wall heating and air-conditioning units and restaurant deep fat fryers that operate at 480 volts, 3 phase, from a grounded wye system.

The requirement in 210.6 is often misapplied because 210.6(C) describes the voltage as "volts to ground," whereas 210.6(A), 210.6(B), 210.6(D), and 210.6(E) describe voltage as "volts between conductors." Luminaires listed for and connected to a 480-volt source may be used in applications permitted by 210.6(C) provided the 480-volt system is in fact a grounded wye system that contains a grounded conduc-

Exhibit 210.4 Examples of luminaires permitted by 210.6(B) and 210.6(C) to be connected to branch circuits.

tor (thus limiting the system "voltage to ground" to the 277-volt level).

(D) 600 Volts Between Conductors Circuits exceeding 277 volts, nominal, to ground and not exceeding 600 volts, nominal, between conductors shall be permitted to supply the following:

(1) The auxiliary equipment of electric-discharge lamps mounted in permanently installed luminaires (fixtures) where the luminaires (fixtures) are mounted in accordance with one of the following:

   a. Not less than a height of 6.7 m (22 ft) on poles or similar structures for the illumination of outdoor areas such as highways, roads, bridges, athletic fields, or parking lots

   b. Not less than a height of 5.5 m (18 ft) on other structures such as tunnels

The minimum mounting heights required by 210.6(D)(1) are for circuits that exceed 277 volts to ground and do not exceed 600 volts phase to phase. These circuits supply the auxiliary equipment of electric-discharge lamps. Exhibit 210.5 (left) shows the minimum mounting height of 18 ft for luminaires installed in tunnels and similar structures. Exhibit 210.5 (right) illustrates the minimum mounting height of 22 ft for luminaires in outdoor areas such as parking lots.

(2) Cord-and-plug-connected or permanently connected utilization equipment other than luminaires (fixtures)

The addition of the words "other than luminaires" clarifies that for the purposes of this requirement, utilization equip-

ment does not include luminaires except those covered in 200.6(D)(1). For luminaire installations that are not on poles or in a tunnel, the branch circuit voltage is limited to 277 volts to ground.

FPN: See 410.78 for auxiliary equipment limitations.

Exception No. 1 to (B), (C), and (D): For lampholders of infrared industrial heating appliances as provided in 422.14.

Exception No. 2 to (B), (C), and (D): For railway properties as described in 110.19.

(E) Over 600 Volts Between Conductors Circuits exceeding 600 volts, nominal, between conductors shall be permitted to supply utilization equipment in installations where conditions of maintenance and supervision ensure that only qualified persons service the installation.

210.7 Branch Circuit Receptacle Requirements

(A) Receptacle Outlet Location Receptacle outlets shall be located in branch circuits in accordance with Part III of Article 210.

(B) Multiple Branch Circuits Where two or more branch circuits supply devices or equipment on the same yoke, a means to simultaneously disconnect the ungrounded conductors supplying those devices shall be provided at the point at which the branch circuits originate.

The requirements for replacement of receptacles formerly contained in this section (1999 and previous editions) are now located in 406.3(D).

In 210.7(B), specifying a means to simultaneously disconnect the ungrounded conductors is a safety issue that

Tunnel Lighting

Parking Lot Lighting

Exhibit 210.5 Minimum mounting heights for tunnel and parking lot lighting as required by 210.6(D)(1) for circuits exceeding 277 volts to ground and not exceeding 600 volts between conductors supplying auxiliary equipment of electric-discharge lampholders.

applies to devices (actually, the single yoke) where more than one branch circuit is involved. Note that this requirement applies to devices or equipment on the same yoke that are supplied by multiple branch circuits. For installations where multiwire branch circuits supply devices or equipment on a common yoke, see 210.4(B).

210.8 Ground-Fault Circuit-Interrupter Protection for Personnel

Section 210.8 is the main rule for the application of ground-fault circuit interrupters (GFCIs). Since the introduction of the GFCI in the 1971 *Code*, these devices have proved to their users and to the electrical community that they are worth the added cost during construction or remodeling. Published data from the Consumer Product Safety Commission show a decreasing trend in the number of electrocutions in the United States since the introduction of GFCI devices. Unfortunately, no statistics are available for the actual number of lives saved by GFCI devices or the actual number of injuries prevented by GFCI devices. However, most experts in the field would agree that the number of saved lives and prevented injuries is substantial.

Exhibit 210.6 shows a typical circuit arrangement of a GFCI. The line conductors are passed through a sensor and are connected to a shunt-trip device. As long as the current in the conductors is equal, the device remains in a closed position. If one of the conductors comes in contact with a grounded object, either directly or through a person's body,

Exhibit 210.6 The circuitry and components of a typical GFCI.

some of the current returns by an alternative path, resulting in an unbalanced current. The toroidal coil senses the unbalanced current, and a circuit is established to the shunt-trip mechanism that reacts and opens the circuit. Note that the circuit design does not require the presence of an equipment grounding conductor, which is the reason 406.3(D)(3)(b) permits the use of GFCIs as replacements for receptacles where a grounding means does not exist.

GFCIs operate on currents of 5 mA. Listing standards permit a differential of 4 to 6 mA. At trip levels of 5 mA (the instantaneous current could be much higher), a shock can be felt during the time of the fault. The shock can lead to involuntary reactions that may cause secondary accidents such as falls. GFCIs do not protect persons from shock hazards where contact is between phase and neutral or between phase-to-phase conductors.

A variety of GFCIs are available, including portable and plug-in types and circuit-breaker types, types built into attachment plug caps, and receptacle types. Each type has a test switch so that units can be checked periodically to ensure proper operation. See Exhibits 210.7 and 210.8.

Exhibit 210.8 A 15-ampere duplex receptacle with integral GFCI that also protects downstream loads. (Courtesy of Pass & Seymour/Legrand®)

Note that there are no exceptions to the bathroom GFCI requirement. For example, if a washing machine is located in the bathroom, the 15- or 20- ampere, 125 volt receptacle that is required to be supplied from the laundry branch circuit must be GFCI protected.

A *bathroom* is defined in Article 100 as "an area including a basin with one or more of the following: a toilet, a tub, or a shower." The term applies to the entire area, whether or not a separating door, as illustrated in Exhibit 210.9, is present. Note that 210.52(D) requires that a receptacle be located on the wall or partition adjacent to each basin location or in the side or face of the basin cabinet. However, if the basins are adjacent and in close proximity, then one receptacle outlet may satisfy the requirement, as shown in Exhibit 210.9 (top).

Exhibit 210.7 A portable plug-in type of GFCI. (Courtesy of Pass & Seymour/Legrand®)

Although 210.8 is the main rule for GFCIs, other specific applications require the use of GFCIs. These additional specific applications are listed in Commentary Table 210.1.

> FPN: See 215.9 for ground-fault circuit-interrupter protection for personnel on feeders.

(A) Dwelling Units All 125-volt, single-phase, 15- and 20-ampere receptacles installed in the locations specified in (1) through (8) shall have ground-fault circuit-interrupter protection for personnel.

(1) Bathrooms

GFCI receptacles in bathrooms prevent accidents. Therefore, 210.8(A)(1) requires that all 125-volt, single-phase, 15- and 20-ampere receptacles in bathrooms have GFCI protection, including receptacles that are integral with luminaires and, of course, wall-mounted receptacles adjacent to the basin.

(2) Garages, and also accessory buildings that have a floor located at or below grade level not intended as habitable rooms and limited to storage areas, work areas, and areas of similar use

Exception No. 1 to (2): Receptacles that are not readily accessible.

Exception No. 2 to (2): A single receptacle or a duplex receptacle for two appliances located within dedicated space for each appliance that, in normal use, is not easily moved from one place to another and that is cord-and-plug connected in accordance with 400.7(A)(6), (A)(7), or (A)(8).

Commentary Table 210.1 Additional Requirements for the Application of GFCI Protection

Location	Applicable Section(s)
Aircraft Hangars	513.12
Audio system equipment	640.10(A)
Boathouses	555.19(B)(1)
Carnivals, circuses, fairs, and similar events	525.23
Commercial garages	511.12
Electric vehicle charging systems	625.22
Electronic equipment, sensitive	647.7(A)
Elevators, escalators, and moving walkways	620.85
Feeders	215.9
Fountains	680.51(A)
Health care facilities	517.20(A), 517.21
High-pressure spray washers	422.49
Hydromassage bathtubs	680.71
Marinas	555.19(B)(1)
Mobile and manufactured homes	550.13(B), 550.13(E), 550.32(E)
Natural and artificially made bodies of water	682.15
Park trailers	552.41(C)
Pools, permanently installed	680.22(A)(1), 680.22(A)(5), 680.22(B)(4), 680.23(A)(3)
Pools, storable	680.32
Sensitive electronic equipment	647.7(A)
Signs with fountains	680.57(B)
Signs, mobile or portable	600.10(C)(2)
Recreational vehicles	551.40(C), 551.41(C)
Recreational vehicle parks	551.71
Replacement receptacles	406.3(D)(2)
Temporary installations	590.6

Exhibit 210.9 GFCI-protected receptacles in bathrooms in accordance with 210.8(A)(1).

Exhibit 210.10 Examples of receptacles in a garage that are required by 210.8(A)(2) to have GFCI protection. Some receptacles are exempt because they are not readily accessible or are for an appliance that occupies dedicated space.

Receptacles installed under the exceptions to 210.8(A)(2) shall not be considered as meeting the requirements of 210.52(G).

The requirement for GFCI receptacles in garages and sheds, as illustrated in Exhibit 210.10, improves safety for persons using portable hand-held tools, gardening appliances, lawn mowers, string trimmers, snow blowers, and so on, that might be connected to these receptacles, which are often the closest ones available. GFCI protection is also required in garage areas where auto repair work and general workshop electrical tools are used.

Exception No. 1 to 210.8(A)(2) permits a ceiling-mounted receptacle that is installed for connection of a garage door opener to be exempt from the GFCI requirement. Exception No. 2 to 210.8(A)(2) allows a duplex receptacle

located where two cord-and-plug-connected appliances occupy a dedicated space to be exempt from the GFCI requirement. If only a single cord-and-plug-connected appliance, such as a food freezer, occupies the dedicated space, then a single receptacle must be used.

(3) Outdoors

Exception to (3): Receptacles that are not readily accessible and are supplied by a dedicated branch circuit for electric snow-melting or deicing equipment shall be permitted to be installed in accordance with 426.28.

The dwelling unit shown in Exhibit 210.11 has four outdoor receptacles. Three of the receptacles are considered to be at direct grade-level access and must have GFCI protection for personnel. The fourth receptacle located adjacent to the gutter for the roof-mounted snow-melting cable is not readily accessible and, therefore, is exempt from the GFCI requirements of 210.8(A)(3). However as indicated in the exception, this receptacle is covered by the equipment protection requirements of 426.28. See the commentary following 210.52(E) and 406.8(B)(1) regarding the installation of outdoor receptacles in wet and damp locations.

Exhibit 210.11 A dwelling unit with three receptacles that are required by 210.8(A)(3) to have GFCI protection and one that is exempt because it supplies a roof heating tape and is covered by the requirement of 426.28.

(4) Crawl spaces — at or below grade level
(5) Unfinished basements — for purposes of this section, unfinished basements are defined as portions or areas of the basement not intended as habitable rooms and limited to storage areas, work areas, and the like

Exception No. 1 to (5): Receptacles that are not readily accessible.

Exception No. 2 to (5): A single receptacle or a duplex receptacle for two appliances located within dedicated space for each appliance that, in normal use, is not easily moved

from one place to another and that is cord-and-plug connected in accordance with 400.7(A)(6), (A)(7), or (A)(8).

Exception No. 3 to (5): A receptacle supplying only a permanently installed fire alarm or burglar alarm system shall not be required to have ground-fault circuit-interrupter protection.

Receptacles installed under the exceptions to 210.8(A)(5) shall not be considered as meeting the requirements of 210.52(G).

An unfinished portion of a basement is limited to storage areas, work areas, and the like. The receptacles in the work area of the basement shown in Exhibit 210.12 must have GFCI protection. Section 210.8(A)(5) does not apply to finished areas in basements, such as sleeping rooms or family rooms, and GFCI protection of receptacles in those areas is not required. In addition, freezer and laundry receptacles do not require GFCI protection, in accordance with 210.8(A)(5), Exception No. 2.

Exception No. 3 was added for the 2002 *Code* to permit the omission of GFCI protection for outlets that serve burglar and fire alarm systems, thus adding a degree of reliability to those important systems.

G = GFCI protection required

Exhibit 210.12 A basement floor plan with GFCI-protected receptacles in the work area, in accordance with 210.8(A)(5), and non-GFCI receptacles elsewhere.

(6) Kitchens — where the receptacles are installed to serve the countertop surfaces

Many countertop kitchen appliances are ungrounded, and the presence of water and grounded surfaces contributes to a hazardous environment, leading to the requirement in 210.8(A)(6) for GFCI protection around a kitchen sink. See Exhibit 210.13 and Exhibit 210.26. The requirement is intended for receptacles serving the countertop. Receptacles installed for disposals, dishwashers, and trash compactors are not required to be protected by GFCIs. According to 406.4(E), receptacles installed to serve countertops cannot be installed in the countertop in the face-up position because liquid, dirt, and other foreign material can enter the receptacle.

G = GFCI protection required

Exhibit 210.13 GFCI-protected receptacles shown in accordance with 210.8(A)(6) to serve countertop surfaces in dwelling unit kitchens.

(7) Laundry, utility, and wet bar sinks — where the receptacles are installed within 1.8 m (6 ft) of the outside edge of the sink

Recognizing that sinks at wet bars are not the only location where a ground-fault shock hazard exists, this requirement now also covers sinks in laundry and utility areas. With this change, GFCI protection requirements are now in place for all areas in a dwelling unit in which a sink is installed. The revised text of this requirement does not limit the GFCI requirement to only receptacles serving countertop surfaces; rather, it covers all 125-volt, 15- and 20-ampere receptacles that are within 6 ft of any point along the outside edge of the sink. Many appliances used in these locations are ungrounded, and the presence of water and grounded surfaces contributes to a hazardous environment, leading to the revision of this requirement for GFCI protection around

sinks. Unlike the GFCI requirements for garages and unfinished basements, there are no exceptions to GFCI protection for receptacles installed within 6 ft of laundry, utility, and wet bar sinks. As illustrated in Exhibit 210.14, any 125-volt, 15- and 20-ampere receptacles installed within 6 ft of a wet bar, laundry, or utility sink is required to be GFCI protected.

Exhibit 210.14 GFCI protection of receptacles located within 6 ft of a wet bar sink in accordance with 210.8(A)(7).

(8) Boathouses

(B) Other Than Dwelling Units All 125-volt, single-phase, 15- and 20-ampere receptacles installed in the locations specified in (1) through (5) shall have ground-fault circuit-interrupter protection for personnel:

(1) Bathrooms

If receptacles are provided in bathroom areas of hotels and motels, GFCI-protected receptacles are required. Lavatories in airports, commercial buildings, industrial facilities, and other nondwelling occupancies are required to have *all* their receptacles GFCI protected. The only exception to this requirement is found in 517.21, which permits receptacles in hospital critical care areas to be non-GFCI if the toilet and basin are installed in the patient room rather than in a separate bathroom. Some motel and hotel bathrooms, like the one shown in Exhibit 210.15, have the basin located outside the door to the room containing the tub, toilet, or another basin. The definition of *bathroom* as found in Article 100 applies to motel and hotel bathrooms, as does the GFCI requirement of 210.8(B)(1).

(2) Commercial and institutional kitchens — for the purposes of this section, a kitchen is an area with a sink and permanent facilities for food preparation and cooking

General lighting circuit

20-A bathroom receptacle circuit

S   G

G   S

G = GFCI protection required

Exhibit 210.15 GFCI protection of receptacles in a motel/hotel bathroom where one basin is located outside the door to the rest of the bathroom area, in accordance with 210.8(B)(1).

Section 210.8(B)(2), which was new for the 2002 *Code*, requires all 15- and 20-ampere, 125-volt receptacles in non-dwelling-type kitchens to be GFCI protected. This requirement applies to all 15- and 20-ampere, 125-volt kitchen receptacles, whether or not the receptacle serves countertop areas.

Accident data related to electrical incidents in nondwelling kitchens reveal the presence of many hazards, including poorly maintained electrical apparatus, damaged electrical cords, wet floors, and employees without proper electrical safety training. Mandating some limited form of GFCI protection for high-hazard areas such as nondwelling kitchens should help prevent electrical accidents. This requirement now provides specific information on what is considered to be a commercial or institutional kitchen. A location with a sink and a portable cooking appliance (e.g., cord-and-plug-connected microwave oven) is not considered a commercial or institutional kitchen for the purposes of applying this requirement. Kitchens in restaurants, hotels, schools, churches, dining halls, and similar facilities are examples of the types of kitchens covered by this requirement.

(3) Rooftops

Section 210.8(B)(3) requires all rooftop 15- and 20-ampere receptacles in nondwelling occupancies to be GFCI protected. For rooftops that also have heating, air-conditioning, and refrigeration equipment, see 210.63.

(4) Outdoors in public spaces—for the purpose of this section a public space is defined as any space that is for use by, or is accessible to, the public

Electrocution and electrical shock accident data provided by the U.S. Consumer Product Safety Commission indicate that such accidents are occurring at locations other than dwelling units and construction sites. The accident data indicate that a number of electrical accidents have occurred at outdoor locations where there is access to the general public and implicate faulty equipment supplied from outdoor receptacles as the cause. This requirement specifies GFCI protection for all 125-volt, 15- and 20-ampere receptacles installed outdoors where these receptacles are accessible to the general public. In other words, unless it can be determined that the location of outdoor receptacles restricts access to only authorized personnel (such as employees or maintenance personnel of a particular facility), GFCI protection of all 125-volt, 15- and 20- ampere receptacle(s) installed outdoors is required if they can be accessed by the general public.

Exception to (3) and (4): Receptacles that are not readily accessible and are supplied from a dedicated branch circuit for electric snow-melting or deicing equipment shall be permitted to be installed in accordance with the applicable provisions of Article 426.

(5) Outdoors, where installed to comply with 210.63

Section 210.63, which requires the installation of a 125-volt receptacle within 25 ft of heating, air-conditioning, and refrigeration (HACR) equipment for use by service personnel, has been expanded since its first appearance in the *Code*, from applying to only equipment installed on rooftop to now applying to any location where HACR equipment is installed, including all outdoor locations. This new GFCI requirement correlates with the expanded coverage of 210.63 and affords service personnel a permanently installed, GFCI-protected receptacle for servicing outdoor HACR equipment for all occupancies not covered by the dwelling unit requirements in 210.8(A)(3).

(C) Boat Hoists Ground-fault circuit-interrupter protection for personnel shall be provided for outlets that supply boat hoists installed in dwelling unit locations and supplied by 125-volt, 15- and 20-ampere branch circuits.

The proximity of this type of equipment to water and the wet or damp environment inherent to the location in which boat hoists are used is the reason for this new GFCI requirement. Documented cases of electrocutions associated with the use of boat hoists have been compiled by the U.S. Consumer Product Safety Commission. This requirement applies only to dwelling unit locations, and GFCI protection must be provided for boat hoists supplied by 15- or 20-ampere,

120-volt branch circuits. It is important to note that in contrast to the requirements in 210.8(A) and 210.8(B), 210.8(C) applies to all outlets supplied from 15- and 20-ampere, 120-volt branch circuits, not just to receptacle outlets. Therefore, cord-and-plug-connected and hard-wired boat hoists are covered by this requirement.

210.9 Circuits Derived from Autotransformers

Branch circuits shall not be derived from autotransformers unless the circuit supplied has a grounded conductor that is electrically connected to a grounded conductor of the system supplying the autotransformer.

Exhibits 210.16 through 210.19 illustrate typical applications of autotransformers. In Exhibit 210.16, a 120-volt supply is derived from a 240-volt system. The grounded conductor of the primary system is electrically connected to the grounded conductor of the secondary system.

Exhibit 210.16 Circuitry for an autotransformer used to derive a 2-wire, 120-volt system for lighting or convenience receptacles from a 240-volt corner-grounded delta system.

A buck-boost transformer is classified as an autotransformer. A buck-boost transformer provides a means of raising (boosting) or lowering (bucking) a supply line voltage by a small amount (usually no more than 20 percent). A buck-boost is a transformer with two primary windings (H1-H2 and H3-H4) and two secondary windings (X1-X2 and X3-X4). Its primary and secondary windings are connected so that the electrical characteristics are changed from a transformer that has its primary and secondary windings insulated from each other to one that has primary and secondary windings connected to buck or boost the voltage as an autotransformer, correcting voltage by up to 20 percent.

A single unit is used to boost or buck single-phase voltage, but two or three units are used to boost or buck 3-phase voltage. An autotransformer requires little physical space, is economical, and, above all, is efficient.

One common application of a boost transformer is to derive a single-phase, 240-volt supply system for ranges, air conditioners, heating elements, and motors from a 3-phase, 208Y/120-volt source system. The boosted leg should not be used to supply line-to-neutral loads because the boosted line-to-neutral voltage will be higher than 120 volts.

Another common boost transformer application is to increase a single-phase, 240-volt source to a single-phase, 277-volt supply for lighting systems. One common 3-phase application is to boost 440 volts to 550 volts for power equipment.

Other common applications of a buck transformer include transforming 240 volts to 208 volts for use with 208-volt appliances and converting a 480Y/277-volt source to a 416Y/240-volt supply system.

Literature containing diagrams for connection and application of autotransformers is available from manufacturers.

Exception No. 1: An autotransformer shall be permitted without the connection to a grounded conductor where transforming from a nominal 208 volts to a nominal 240-volt supply or similarly from 240 volts to 208 volts.

Exception No. 1 to 210.9 allows an autotransformer (without an electrical connection to a grounded conductor) to extend or add an individual branch circuit in an existing installation where transforming (boosting) 208 volts to 240 volts, as shown in Exhibit 210.17. Exhibits 210.18 and 210.19 illustrate typical single-phase and 3-phase buck and boost transformers connected as autotransformers to change 240 volts to 208 volts and vice versa.

Exhibit 210.17 Circuitry for an autotransformer used to derive a 240-volt system for appliances from a 208Y/120-volt source, in accordance with 210.9, Exception No.1.

Exception No. 2: In industrial occupancies, where conditions of maintenance and supervision ensure that only qualified persons service the installation, autotransformers shall be permitted to supply nominal 600-volt loads from nominal 480-volt systems, and 480-volt loads from nominal 600-

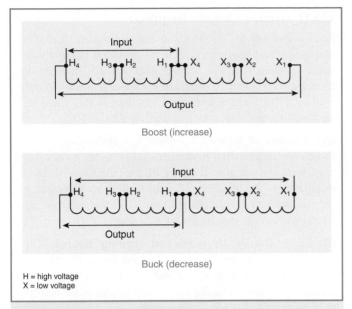

Boost (increase)

Buck (decrease)

H = high voltage
X = low voltage

Exhibit 210.18 Typical single-phase connection diagrams for buck or boost transformers connected as autotransformers to change 240 volts single-phase to 208 volts and vice versa.

volt systems, without the connection to a similar grounded conductor.

In industrial locations, Exception No. 2 to 210.9 allows the use of an autotransformer to supply a 600-volt load from 480-volt systems, provided there are qualified personnel to service the installation. It also allows 480-volt loads to be supplied through an autotransformer supplied by a 600-volt system.

210.10 Ungrounded Conductors Tapped from Grounded Systems

Two-wire dc circuits and ac circuits of two or more ungrounded conductors shall be permitted to be tapped from the ungrounded conductors of circuits that have a grounded neutral conductor. Switching devices in each tapped circuit shall have a pole in each ungrounded conductor. All poles of multipole switching devices shall manually switch together where such switching devices also serve as a disconnecting means as required by the following:

(1) 410.48 for double-pole switched lampholders
(2) 410.54(B) for electric-discharge lamp auxiliary equipment switching devices

Exhibit 210.19 Typical connection diagrams for buck or boost transformers connected in 3-phase open delta as autotransformers to change 240 volts to 208 volts and vice versa.

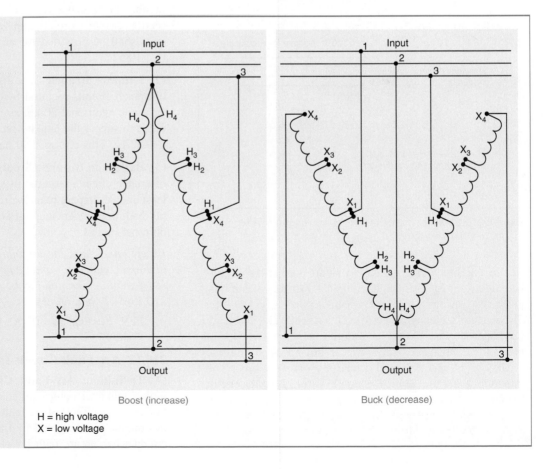

Boost (increase)

Buck (decrease)

H = high voltage
X = low voltage

(3) 422.31(B) for an appliance
(4) 424.20 for a fixed electric space-heating unit
(5) 426.51 for electric deicing and snow-melting equipment
(6) 430.85 for a motor controller
(7) 430.103 for a motor

Two-wire ungrounded branch circuits may be tapped from ac or dc circuits of two or more ungrounded conductors that have a grounded neutral conductor. Exhibit 210.20 (top) illustrates an ungrounded 2-wire branch circuit tapped from the ungrounded conductors of a dc or single-phase system to supply a small motor. Exhibit 210.20 (bottom) illustrates a 3-phase, 4-wire wye system.

Exhibit 210.20 Branch circuits tapped from ungrounded conductors of multiwire systems.

Circuit breakers or switches that are used as the disconnecting means for a branch circuit must open all poles simultaneously using only the manual operation of the disconnecting means. Therefore, if switches and fuses are used and one fuse blows, or if circuit breakers (two single-pole circuit breakers with a handle tie) are used and one breaker trips, one pole could possibly remain closed. The intention is not to require a common trip of fuses or circuit breakers but rather to disconnect (manually) the ungrounded conductors of the branch circuit with one manual operation. See 240.20(B) for information on handle ties.

210.11 Branch Circuits Required

Branch circuits for lighting and for appliances, including motor-operated appliances, shall be provided to supply the loads calculated in accordance with 220.10. In addition, branch circuits shall be provided for specific loads not covered by 220.10 where required elsewhere in this *Code* and for dwelling unit loads as specified in 210.11(C).

(A) Number of Branch Circuits The minimum number of branch circuits shall be determined from the total calculated load and the size or rating of the circuits used. In all installations, the number of circuits shall be sufficient to supply the load served. In no case shall the load on any circuit exceed the maximum specified by 220.18.

(B) Load Evenly Proportioned Among Branch Circuits Where the load is calculated on the basis of volt-amperes per square meter or per square foot, the wiring system up to and including the branch-circuit panelboard(s) shall be provided to serve not less than the calculated load. This load shall be evenly proportioned among multioutlet branch circuits within the panelboard(s). Branch-circuit overcurrent devices and circuits shall only be required to be installed to serve the connected load.

(C) Dwelling Units

(1) Small-Appliance Branch Circuits In addition to the number of branch circuits required by other parts of this section, two or more 20-ampere small-appliance branch circuits shall be provided for all receptacle outlets specified by 210.52(B).

(2) Laundry Branch Circuits In addition to the number of branch circuits required by other parts of this section, at least one additional 20-ampere branch circuit shall be provided to supply the laundry receptacle outlet(s) required by 210.52(F). This circuit shall have no other outlets.

(3) Bathroom Branch Circuits In addition to the number of branch circuits required by other parts of this section, at least one 20-ampere branch circuit shall be provided to supply bathroom receptacle outlet(s). Such circuits shall have no other outlets.

Exception: Where the 20-ampere circuit supplies a single bathroom, outlets for other equipment within the same bathroom shall be permitted to be supplied in accordance with 210.23(A)(1) and (A)(2).

> FPN: See Examples D1(A), D1(B), D2(B), and D4(A) in Annex D.

210.12 Arc-Fault Circuit-Interrupter Protection

(A) Definition: Arc-Fault Circuit Interrupter An arc-fault circuit interrupter is a device intended to provide protection from the effects of arc faults by recognizing characteristics unique to arcing and by functioning to de-energize the circuit when an arc fault is detected.

(B) Dwelling Unit Bedrooms All 120-volt, single phase, 15- and 20-ampere branch circuits supplying outlets installed in dwelling unit bedrooms shall be protected by a listed arc-fault circuit interrupter, combination type installed to provide protection of the branch circuit.

Branch/feeder AFCIs shall be permitted to be used to meet the requirements of 210.12(B) until January 1, 2008.

FPN: For information on types of arc-fault circuit interrupters, see UL 1699-1999, *Standard for Arc-Fault Circuit Interrupters.*

Exception: The location of the arc-fault circuit interrupter shall be permitted to be at other than the origination of the branch circuit in compliance with (a) and (b):

(a) The arc-fault circuit interrupter installed within 1.8 m (6 ft) of the branch circuit overcurrent device as measured along the branch circuit conductors.

(b) The circuit conductors between the branch circuit overcurrent device and the arc-fault circuit interrupter shall be installed in a metal raceway or a cable with a metallic sheath.

The definition of *arc-fault circuit interrupter* given in 210.12(A) explains its function. The basic objective is to de-energize the branch circuit when an arc fault is detected.

Arc-fault circuit interrupters are evaluated in UL 1699, *Standard for Arc-Fault Circuit-Interrupters*, using testing methods that create or simulate arcing conditions to determine the product's ability to detect and interrupt arcing faults. These devices are also tested to verify that arc detection is not unduly inhibited by the presence of loads and circuit characteristics that may mask the hazardous arcing condition. In addition, these devices are evaluated to determine resistance to unwanted tripping due to the presence of arcing that occurs in control and utilization equipment under normal operating conditions or to a loading condition that closely mimics an arcing fault, such as a solid-state electronic ballast or a dimmed load.

UL 1699 is the standard covering arc-fault devices that have a maximum rating of 20 amperes intended for use in 120-volt ac, 60-Hz circuits. These devices may also have the capability to perform other functions such as overcurrent protection, ground-fault circuit interruption, and surge suppression. UL 1699 currently recognizes five types of arc-fault circuit interrupters: branch/feeder AFCI, combination AFCI, cord AFCI, outlet AFCI, and portable AFCI.

Placement of the device in the circuit and a review of the UL guide information must be considered when complying with 210.12. The *NEC* is clear that the objective is to provide protection of the entire branch circuit. (See Article 100 for the definition of *branch circuit*.) For instance, a cord AFCI cannot be used to comply with the requirement of 210.12 to protect the entire branch circuit.

The type of AFCI required to comply with 210.12(B)

is the subject of a revision in the 2005 *Code*. To expand the level of AFCI protection for cord sets that are plugged into receptacles supplied by AFCI-protected branch circuits, the use of combination-type AFCI devices is now required. However, mandatory use of only combination-type AFCI devices to comply with 210.12(B) becomes effective January 1, 2008. Until that effective date, the use of either a combination-type or a branch/feeder-type AFCI device meets the requirement of 210.12(B). In addition to the revised type of AFCI protection required, the location of where the AFCI device is to be located in the circuit now provides a new option. Because the protection requirement is for the entire branch circuit, location of the device at the point the branch circuit originates (service or feeder panelboard or similar distribution equipment) has been and continues to be the main requirement. However, the new exception permits the AFCI device to be located in close vicinity to the point of origin as long as the branch-circuit conductors that are not AFCI protected do not exceed 6 ft in length and the portion of the circuit between the point of origin and the AFCI location is installed in a metal raceway or a metallic-sheathed cable.

Section 210.12(B) requires that AFCI protection be provided for all 15- and 20-ampere 120-volt branch circuits that supply outlets (including receptacle, lighting, and other outlets; see definition of *outlet* in Article 100) in dwelling unit bedrooms regardless of whether the circuit supplies only outlets in the bedroom(s) or supplies outlets in the bedroom and other areas of the dwelling. Because circuits are often shared between a bedroom and other areas such as closets and hallways, providing AFCI protection on the complete circuit would comply with 210.12. There is no prohibition against using AFCI protection on other circuits or in locations other than bedrooms.

210.18 Guest Rooms and Guest Suites

Guest rooms and guest suites that are provided with permanent provisions for cooking shall have branch circuits and outlets installed to meet the rules for dwelling units.

This new requirement ensures that guest rooms and guest suites equipped with permanent provisions for cooking are treated the same as dwelling units in regard to the branch circuit requirements contained in Parts I, II, and III of Article 210.

II. Branch-Circuit Ratings

210.19 Conductors — Minimum Ampacity and Size

(A) Branch Circuits Not More Than 600 Volts

(1) General Branch-circuit conductors shall have an ampacity not less than the maximum load to be served. Where a branch circuit supplies continuous loads or any combination of continuous and noncontinuous loads, the minimum branch-circuit conductor size, before the application of any adjustment or correction factors, shall have an allowable ampacity not less than the noncontinuous load plus 125 percent of the continuous load.

Exception: Where the assembly, including the overcurrent devices protecting the branch circuit(s), is listed for operation at 100 percent of its rating, the allowable ampacity of the branch circuit conductors shall be permitted to be not less than the sum of the continuous load plus the noncontinuous load.

Conductors of branch circuits rated not more than 600 volts must be able to supply power to loads without overheating. The requirements in 210.19(A)(1) establish minimum size and ampacity requirements to allow that to happen. The requirements for the minimum size of overcurrent protection devices are found in 210.20. An example showing these minimum-size calculations is found in the commentary following 210.20(A), Exception.

> FPN No. 1: See 310.15 for ampacity ratings of conductors.
> FPN No. 2: See Part II of Article 430 for minimum rating of motor branch-circuit conductors.
> FPN No. 3: See 310.10 for temperature limitation of conductors.
> FPN No. 4: Conductors for branch circuits as defined in Article 100, sized to prevent a voltage drop exceeding 3 percent at the farthest outlet of power, heating, and lighting loads, or combinations of such loads, and where the maximum total voltage drop on both feeders and branch circuits to the farthest outlet does not exceed 5 percent, provide reasonable efficiency of operation. See FPN No. 2 of 215.2(A)(3) for voltage drop on feeder conductors.

FPN No. 4 expresses a warning about improper voltage due to a voltage drop in supply conductors, a major source of trouble and inefficient operation in electrical equipment. Undervoltage conditions reduce the capability and reliability of motors, lighting sources, heaters, and solid-state equipment. Sample voltage-drop calculations are found in the commentary following 215.2(A)(3), FPN No. 3, and following Table 9 in Chapter 9.

(2) Multioutlet Branch Circuits Conductors of branch circuits supplying more than one receptacle for cord-and-plug-connected portable loads shall have an ampacity of not less than the rating of the branch circuit.

Because the loading of branch-circuit conductors that supply receptacles for cord-and-plug-connected portable loads is unpredictable, it is safest simply to require such circuits to have an ampacity that is not less than the rating of the branch circuit. According to 210.3, the rating of the branch circuit is actually the rating of the overcurrent device.

(3) Household Ranges and Cooking Appliances Branch-circuit conductors supplying household ranges, wall-mounted ovens, counter-mounted cooking units, and other household cooking appliances shall have an ampacity not less than the rating of the branch circuit and not less than the maximum load to be served. For ranges of 8¾ kW or more rating, the minimum branch-circuit rating shall be 40 amperes.

Based on the basic requirement of 110.14(C)(1)(a), the minimum 40-ampere rated branch-circuit would require the use of 8 AWG, Type THW copper or 6 AWG, Type XHHW aluminum conductors. See Table 310.16 for other applications.

Exception No. 1: Tap conductors supplying electric ranges, wall-mounted electric ovens, and counter-mounted electric cooking units from a 50-ampere branch circuit shall have an ampacity of not less than 20 and shall be sufficient for the load to be served. These tap conductors include any conductors that are a part of the leads supplied with the appliance that are smaller than the branch circuit conductors. The taps shall not be longer than necessary for servicing the appliance.

Exception No. 1 to 210.19(A)(3) covers factory-installed and field-installed tap conductors. A revision to the 2005 *Code* clarifies that the supply conductors included in a factory-installed pigtail are considered to be tap conductors in applying this exception. As illustrated in Exhibit 210.21, this exception permits a 20-ampere tap conductor from a range, oven, or cooking unit to be connected to a 50-ampere branch circuit if the following four conditions are met:

1. The taps are not longer than necessary to service or permit access to the junction box.
2. The taps to each unit are properly spliced.
3. The junction box is adjacent to each unit.
4. The taps are of sufficient size for the load to be served.

Exception No. 2: The neutral conductor of a 3-wire branch circuit supplying a household electric range, a wall-mounted oven, or a counter-mounted cooking unit shall be permitted to be smaller than the ungrounded conductors where the maximum demand of a range of 8¾ kW or more rating has been calculated according to Column C of Table 220.55, but such conductor shall have an ampacity of not less than 70 percent of the branch-circuit rating and shall not be smaller than 10 AWG.

Exhibit 210.21 Tap conductors permitted by 210.19(A)(3), Exception No. 1, sized smaller than the branch-circuit conductors and not to be longer than necessary for servicing the appliances.

Column C of Table 220.55 indicates that the maximum demand for one range (not over 12 kW rating) is 8 kW (8 kW = 8000 volt-amperes; 8000 volt-amperes ÷ 240 volts = 33.3 amperes). In accordance with the fundamental termination rule of 110.14(C)(1)(a), the allowable ampacity of an 8 AWG, copper conductor from the 60°C column of Table 310.16 is 40 amperes, and it may be used for the range branch circuit. According to this computation, the neutral of this 3-wire circuit can be smaller than 8 AWG but not smaller than 10 AWG, which has an allowable ampacity of 30 amperes (30 amperes is more than 70 percent of 40 amperes, per Exception No. 2). The maximum demand for the neutral of an 8-kW range circuit seldom exceeds 25 amperes, because the only line-to-neutral connected loads are lights, clocks, timers, and the heating elements of some ranges when the control is adjusted to the low-heat setting.

(4) Other Loads Branch-circuit conductors that supply loads other than those specified in 210.2 and other than cooking appliances as covered in 210.19(A)(3) shall have an ampacity sufficient for the loads served and shall not be smaller than 14 AWG.

Exception No. 1: Tap conductors shall have an ampacity sufficient for the load served. In addition, they shall have an ampacity of not less than 15 for circuits rated less than

40 amperes and not less than 20 for circuits rated at 40 or 50 amperes and only where these tap conductors supply any of the following loads:

(a) Individual lampholders or luminaires (fixtures) with taps extending not longer than 450 mm (18 in.) beyond any portion of the lampholder or luminaire (fixture).

(b) A fixture having tap conductors as provided in 410.67.

(c) Individual outlets, other than receptacle outlets, with taps not over 450 mm (18 in.) long.

(d) Infrared lamp industrial heating appliances.

(e) Nonheating leads of deicing and snow-melting cables and mats.

Tap conductors are generally required to have the same ampacity as the branch-circuit overcurrent device. Exception No. 1 to 210.19(A)(4) lists specific applications in items (a) through (e) where the tap conductors are permitted with reduced ampacities. These tap conductors are required to have an ampacity of 15 amperes or more (14 AWG copper conductors) for circuits rated less than 40 amperes. The tap conductors must have an ampacity of 20 amperes or more (12 AWG copper conductors) for circuits rated 40 or 50 amperes.

Exception No. 2: Fixture wires and flexible cords shall be permitted to be smaller than 14 AWG as permitted by 240.5.

(B) Branch Circuits Over 600 Volts The ampacity of conductors shall be in accordance with 310.15 and 310.60, as applicable. Branch-circuit conductors over 600 volts shall be sized in accordance with 210.19(B)(1) or (B)(2).

(1) General The ampacity of branch-circuit conductors shall not be less than 125 percent of the designed potential load of utilization equipment that will be operated simultaneously.

(2) Supervised Installations For supervised installations, branch-circuit conductor sizing shall be permitted to be determined by qualified persons under engineering supervision. Supervised installations are defined as those portions of a facility where both of the following conditions are met:

(1) Conditions of design and installation are provided under engineering supervision.
(2) Qualified persons with documented training and experience in over 600-volt systems provide maintenance, monitoring, and servicing of the system.

Part II of Article 210 was revised for the 2002 *Code* to include requirements for the branch circuits over 600 volts in 210.19(B). Basically, branch circuits over 600 volts must be sized at 125 percent of the combined simultaneous load, unless the branch circuits over 600 volts are located at facilities that qualify as supervised installations.

210.20 Overcurrent Protection

Branch-circuit conductors and equipment shall be protected by overcurrent protective devices that have a rating or setting that complies with 210.20(A) through (D).

(A) Continuous and Noncontinuous Loads Where a branch circuit supplies continuous loads or any combination of continuous and noncontinuous loads, the rating of the overcurrent device shall not be less than the noncontinuous load plus 125 percent of the continuous load.

An example calculation for a continuous load only is illustrated in Exhibit 210.22.

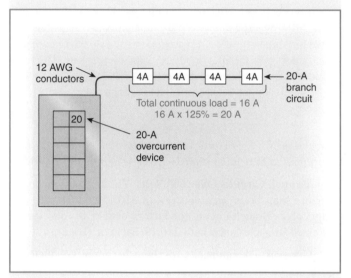

Exhibit 210.22 A continuous load (store lighting) calculated at 125 percent to determine the ampacity of the conductor and the branch-circuit size.

Exception: Where the assembly, including the overcurrent devices protecting the branch circuit(s), is listed for operation at 100 percent of its rating, the ampere rating of the overcurrent device shall be permitted to be not less than the sum of the continuous load plus the noncontinuous load.

According to 210.20, an overcurrent device that supplies continuous and noncontinuous loads must have a rating that is not less than the sum of 100 percent of the noncontinuous load plus 125 percent of the continuous load, calculated in accordance with Article 210.

In addition, 210.19(A)(1) requires that the circuit conductors, chosen from the ampacity tables, must have an

initial ampacity of not less than the sum of 100 percent of the noncontinuous load plus 125 percent of the continuous load, the same as calculated for the overcurrent device.

The rating of the overcurrent device cannot exceed the final ampacity of the circuit conductors after all the derating or correction factors have been applied, such as for temperature or number of conductors.

Example

Determine the minimum-size overcurrent protective device and the minimum conductor size for the following circuit:

- 25 amperes of continuous load
- 60°C overcurrent device terminal rating
- Type THWN conductors
- Four current-carrying copper conductors in a raceway

Solution

STEP 1. Determine the size of the overcurrent protective device (OCPD). Referring to 210.20(A), 125 percent of 25 amperes is 31.25 amperes. Thus, the minimum standard-size overcurrent device, according to 240.6(A), is 35 amperes.

STEP 2. Determine the minimum conductor size. The ampacity of the conductor must not be less than 125 percent of the 25-ampere continuous load, which results in 31.25 amperes. The conductor must have an allowable ampacity of not less than 31.25 amperes before any adjustment or correction factors are applied. Because there are four current-carrying conductors in the raceway, Table 310.15(B)(2)(a) applies. First, calculate the ampacity of the conductor using the ampacity value calculated above:

$$\text{Conductor ampacity} = \frac{\text{Computed load}}{\text{Percent adjustment factor from Table } 310/15(B)(2)(a)}$$

$$\frac{31.25 \text{ amperes}}{0.80} = 39.06 \text{ amperes}$$

Because of the 60°C rating of the overcurrent device terminal, it is necessary to choose a conductor based on the ampacities in the 60°C column of Table 310.16. The calculated load must not exceed the conductor ampacity. Therefore, an 8 AWG conductor with a 60°C allowable ampacity of 40 amperes is the minimum size permitted. Conductors with a higher allowable ampacity based on their insulation temperature rating may be used, but only at a 60°C allowable ampacity.

(B) Conductor Protection Conductors shall be protected in accordance with 240.4. Flexible cords and fixture wires shall be protected in accordance with 240.5.

(C) Equipment The rating or setting of the overcurrent protective device shall not exceed that specified in the applicable articles referenced in Table 240.3 for equipment.

(D) Outlet Devices The rating or setting shall not exceed that specified in 210.21 for outlet devices.

210.21 Outlet Devices

Outlet devices shall have an ampere rating that is not less than the load to be served and shall comply with 210.21(A) and (B).

(A) Lampholders Where connected to a branch circuit having a rating in excess of 20 amperes, lampholders shall be of the heavy-duty type. A heavy-duty lampholder shall have a rating of not less than 660 watts if of the admedium type, or not less than 750 watts if of any other type.

The intent of 210.21(A) is to restrict a fluorescent lighting branch-circuit rating to not more than 20 amperes because most lampholders manufactured for use with fluorescent lights have a rating less than that required for heavy-duty lampholders (660 for admedium type or 750 watts for all other types).

Branch-circuit conductors for fluorescent electric-discharge lighting are usually connected to ballasts rather than to lampholders, and, by specifying a wattage rating for these lampholders, a limit of 20 amperes is applied to ballast circuits.

Only the admedium-base lampholder is recognized as heavy duty at the rating of 660 watts. Other lampholders are required to have a rating of not less than 750 watts to be recognized as heavy duty. The requirement of 210.21(A) prohibits the use of medium-base screw shell lampholders on branch circuits that are in excess of 20 amperes.

(B) Receptacles
(1) Single Receptacle on an Individual Branch Circuit A single receptacle installed on an individual branch circuit shall have an ampere rating not less than that of the branch circuit.

Exception No. 1: A receptacle installed in accordance with 430.81(B).

Exception No. 2: A receptacle installed exclusively for the use of a cord-and-plug-connected arc welder shall be permitted to have an ampere rating not less than the minimum branch-circuit conductor ampacity determined by 630.11(A) for arc welders.

FPN: See the definition of *receptacle* in Article 100.

(2) Total Cord-and-Plug-Connected Load Where connected to a branch circuit supplying two or more receptacles or outlets, a receptacle shall not supply a total cord-and-plug-connected load in excess of the maximum specified in Table 210.21(B)(2).

Table 210.21(B)(2) Maximum Cord-and-Plug-Connected Load to Receptacle

Circuit Rating (Amperes)	Receptacle Rating (Amperes)	Maximum Load (Amperes)
15 or 20	15	12
20	20	16
30	30	24

(3) Receptacle Ratings Where connected to a branch circuit supplying two or more receptacles or outlets, receptacle ratings shall conform to the values listed in Table 210.21(B)(3), or where larger than 50 amperes, the receptacle rating shall not be less than the branch-circuit rating.

Table 210.21(B)(3) Receptacle Ratings for Various Size Circuits

Circuit Rating (Amperes)	Receptacle Rating (Amperes)
15	Not over 15
20	15 or 20
30	30
40	40 or 50
50	50

Exception No. 1: Receptacles for one or more cord-and-plug-connected arc welders shall be permitted to have ampere ratings not less than the minimum branch-circuit conductor ampacity permitted by 630.11(A) or (B) as applicable for arc welders.

Exception No. 2: The ampere rating of a receptacle installed for electric discharge lighting shall be permitted to be based on 410.30(C).

A single receptacle installed on an individual branch circuit must have an ampere rating not less than that of the branch circuit. For example, a single receptacle on a 20-ampere individual branch circuit must be rated at 20 amperes; however, two or more 15-ampere receptacles or duplex receptacles are permitted on a 20-ampere general-purpose branch circuit. This requirement does not apply to specific types of cord-and-plug-connected arc welders.

(4) Range Receptacle Rating The ampere rating of a range receptacle shall be permitted to be based on a single range demand load as specified in Table 220.55.

210.23 Permissible Loads

In no case shall the load exceed the branch-circuit ampere rating. An individual branch circuit shall be permitted to supply any load for which it is rated. A branch circuit supplying two or more outlets or receptacles shall supply only the loads specified according to its size as specified in 210.23(A) through (D) and as summarized in 210.24 and Table 210.24.

The requirements of 210.23 are often misunderstood. An individual (single-outlet) branch circuit can supply any load within its rating. On the other side, the load, of course, cannot be greater than the branch-circuit rating.

(A) 15- and 20-Ampere Branch Circuits A 15- or 20-ampere branch circuit shall be permitted to supply lighting units or other utilization equipment, or a combination of both, and shall comply with 210.23(A)(1) and (A)(2).

Section 210.23(A) permits a 15- or 20-ampere branch circuit for lighting to also supply utilization equipment fastened in place, such as an air conditioner. The equipment load must not exceed 50 percent of the branch-circuit ampere rating (7.5 amperes on a 15-ampere circuit and 10 amperes on a 20-ampere circuit). However, according to 210.52(B), such fastened-in-place equipment is not permitted on the small-appliance branch circuits required in a kitchen, dining room, and so on. A revision to 210.23(A)(1) clarifies that only cord-and-plug-connected utilization equipment that is not fastened in place can have a rating of up to 80 percent of the branch circuit rating where the circuit also supplies other loads. Equipment that is fastened in place, whether direct wired or cord and plug connected (waste disposers and dishwashers for example), is covered by the 50-percent requirement in 210.23(A)(2).

Exception: The small appliance branch circuits, laundry branch circuits, and bathroom branch circuits required in a dwelling unit(s) by 210.11(C)(1), (C)(2), and (C)(3) shall supply only the receptacle outlets specified in that section.

(1) Cord-and-Plug-Connected Equipment Not Fastened in Place The rating of any one cord-and-plug-connected utilization equipment not fastened in place shall not exceed 80 percent of the branch-circuit ampere rating.

(2) Utilization Equipment Fastened in Place The total rating of utilization equipment fastened in place, other than luminaires (lighting fixtures), shall not exceed 50 percent of the branch-circuit ampere rating where lighting units, cord-and-plug-connected utilization equipment not fastened in place, or both, are also supplied.

(B) 30-Ampere Branch Circuits A 30-ampere branch circuit shall be permitted to supply fixed lighting units with heavy-duty lampholders in other than a dwelling unit(s) or utilization equipment in any occupancy. A rating of any

one cord-and-plug-connected utilization equipment shall not exceed 80 percent of the branch-circuit ampere rating.

(C) 40- and 50-Ampere Branch Circuits A 40- or 50-ampere branch circuit shall be permitted to supply cooking appliances that are fastened in place in any occupancy. In other than dwelling units, such circuits shall be permitted to supply fixed lighting units with heavy-duty lampholders, infrared heating units, or other utilization equipment.

A branch circuit that supplies two or more outlets is permitted to supply only the loads specified according to its size, in accordance with 210.23(A) through 210.23(C) and as summarized in 210.24 and Table 210.24. Other circuits are not permitted to have more than one outlet and are considered individual branch circuits. However, 517.71 and 660.4(B) do not require individual branch circuits for portable, mobile, and transportable medical X-ray equipment requiring a capacity of not over 60 amperes.

(D) Branch Circuits Larger Than 50 Amperes Branch circuits larger than 50 amperes shall supply only nonlighting outlet loads.

See the commentary following 210.3, Exception, regarding multioutlet branch circuits greater than 50 amperes that are permitted to supply nonlighting outlet loads in industrial establishments.

210.24 Branch-Circuit Requirements — Summary

The requirements for circuits that have two or more outlets or receptacles, other than the receptacle circuits of 210.11(C)(1) and (C)(2), are summarized in Table 210.24. This table provides only a summary of minimum requirements. See 210.19, 210.20, and 210.21 for the specific requirements applying to branch circuits.

Table 210.24 summarizes the branch-circuit requirements of conductors, overcurrent protection, outlet devices, maximum load, and permissible load where two or more outlets are supplied.

    If the branch circuit serves a fixture load and supplies two or more fixture outlets, 210.23 requires the branch circuit to have a specific ampere rating that is based on the rating of the overcurrent device, as stated in 210.3. Thus, if the circuit breaker that protects the branch circuit is rated 20 amperes, the conductors supplying the circuit must have an ampacity not less than 20 amperes.

    Note that in accordance with the Article 100 definition of *ampacity*, the ampacity is determined after all derating (adjustment and correction) factors, such as those in

Table 210.24 Summary of Branch-Circuit Requirements

Circuit Rating	15 A	20 A	30 A	40 A	50 A
Conductors (min. size):					
Circuit wires[1]	14	12	10	8	6
Taps	14	14	14	12	12
Fixture wires and cords — see 240.5					
Overcurrent Protection	**15 A**	**20 A**	**30 A**	**40 A**	**50 A**
Outlet devices:					
Lampholders permitted	Any type	Any type	Heavy duty	Heavy duty	Heavy duty
Receptacle rating[2]	15 max. A	15 or 20 A	30 A	40 or 50 A	50 A
Maximum Load	**15 A**	**20 A**	**30 A**	**40 A**	**50 A**
Permissible load	See 210.23(A)	See 210.23(A)	See 210.23(B)	See 210.23(C)	See 210.23(C)

[1]These gauges are for copper conductors.
[2]For receptacle rating of cord-connected electric-discharge luminaires (lighting fixtures), see 410.30(C).

310.15(B)(2)(a), have been applied. If seven to nine such conductors are in one conduit, a 12 AWG, Type THHN copper conductor (30 amperes, per Table 310.16) adjusted to 70 percent, per Table 310.15(B)(2)(a), would have an allowable ampacity of 21 amperes and would be suitable for a load of 20 amperes. Thus, this conductor would be acceptable for use on the 20-ampere multioutlet branch circuit.

210.25 Common Area Branch Circuits

Branch circuits in dwelling units shall supply only loads within that dwelling unit or loads associated only with that dwelling unit. Branch circuits required for the purpose of lighting, central alarm, signal, communications, or other needs for public or common areas of a two-family or multifamily dwelling shall not be supplied from equipment that supplies an individual dwelling unit.

Not only does 210.25 prohibit branch circuits from feeding more than one dwelling unit, it also prohibits the sharing of systems, equipment, or common lighting if that equipment is fed from any of the dwelling units. The systems, equipment, or lighting for public or common areas is required to be supplied from a separate "house load" panelboard. This requirement permits access to the branch-circuit disconnecting means without the need to enter the space of any tenants. The requirement also prevents a tenant from turning off important circuits that may affect other tenants.

III. Required Outlets

210.50 General

Receptacle outlets shall be installed as specified in 210.52 through 210.63.

(A) Cord Pendants A cord connector that is supplied by a permanently connected cord pendant shall be considered a receptacle outlet.

(B) Cord Connections A receptacle outlet shall be installed wherever flexible cords with attachment plugs are used. Where flexible cords are permitted to be permanently connected, receptacles shall be permitted to be omitted for such cords.

Flexible cords are permitted to be permanently connected to boxes or fittings where specifically permitted by the *Code*. However, plugging a cord into a lampholder by inserting a screw-plug adapter is not permitted, because 410.47 requires lampholders of the screw shell type to be installed for use as lampholders only.

(C) Appliance Outlets Appliance receptacle outlets installed in a dwelling unit for specific appliances, such as laundry equipment, shall be installed within 1.8 m (6 ft) of the intended location of the appliance.

See 210.52(F) and 210.11(C)(2) for requirements regarding laundry receptacle outlets and branch circuits.

210.52 Dwelling Unit Receptacle Outlets

This section provides requirements for 125-volt, 15- and 20-ampere receptacle outlets. Receptacle outlets required by this section shall be in addition to any receptacle that is part of a luminaire (lighting fixture) or appliance, located within cabinets or cupboards, or located more than 1.7 m (5½ ft) above the floor.

Permanently installed electric baseboard heaters

equipped with factory-installed receptacle outlets or outlets provided as a separate assembly by the manufacturer shall be permitted as the required outlet or outlets for the wall space utilized by such permanently installed heaters. Such receptacle outlets shall not be connected to the heater circuits.

> FPN: Listed baseboard heaters include instructions that may not permit their installation below receptacle outlets.

The requirements of 210.52 apply to dwelling unit receptacles that are rated 125 volts and 15 or 20 amperes and that are not part of a luminaire or an appliance. These receptacles are normally used to supply lighting and general-purpose electrical equipment and are in addition to the ones that are 5 ½ ft above the floor and within cupboards and cabinets.

According to listing requirements [see 110.3(B)], permanent electric baseboard heaters may not be located beneath wall receptacles. If the receptacle is part of the heater, appliance or lamp cords are less apt to be exposed to the heating elements, as might occur should the cords fall into convector slots. Many electric baseboard heaters are of the low-density type and are longer than 12 ft. To meet the spacing requirements of 210.52(A)(1), the required receptacle may be located as a part of the heater unit, as shown as Exhibit 210.23.

Exhibit 210.23 Permanent electric baseboard heater equipped with a receptacle outlet to meet the spacing requirements of 210.52(A).

(A) General Provisions In every kitchen, family room, dining room, living room, parlor, library, den, sunroom, bedroom, recreation room, or similar room or area of dwelling units, receptacle outlets shall be installed in accordance with the general provisions specified in 210.52(A)(1) through (A)(3).

(1) Spacing Receptacles shall be installed so that no point measured horizontally along the floor line in any wall space is more than 1.8 m (6 ft) from a receptacle outlet.

Receptacles are required to be located so that no point in any wall space is more than 6 ft from a receptacle. This rule intends that an appliance or lamp with a flexible cord attached may be placed anywhere in the room near a wall and be within 6 ft of a receptacle, thus eliminating the need for extension cords. Although not an enforceable requirement, receptacles may be placed equal distances apart where there is no specific room layout for the general use of electrical equipment. Section 210.52(A)(1) does not prohibit a receptacle layout designed for intended utilization equipment or practical room use. For example, receptacles in a living room, family room, or den that are intended to serve home entertainment equipment or home office equipment may be placed in corners, may be grouped, or may be placed in a convenient location. Receptacles that are intended for window-type holiday lighting may be placed under windows. In any event, even if more receptacles than the minimum are installed in a room, no point in any wall space is permitted to be more than 6 ft from a receptacle.

(2) Wall Space As used in this section, a wall space shall include the following:

(1) Any space 600 mm (2 ft) or more in width (including space measured around corners) and unbroken along the floor line by doorways, fireplaces, and similar openings
(2) The space occupied by fixed panels in exterior walls, excluding sliding panels
(3) The space afforded by fixed room dividers such as freestanding bar-type counters or railings

A *wall space* is a wall unbroken along the floor line by doorways, fireplaces, archways, and similar openings and may include two or more walls of a room (around corners), as illustrated in Exhibit 210.24.

Fixed room dividers, such as bar-type counters and railings, are to be included in the 6-ft measurement. Fixed panels in exterior walls are counted as regular wall space, and a floor-type receptacle close to the wall can be used to meet the required spacing. Isolated, individual wall spaces 2 ft or more in width are often used for small pieces of furniture on which a lamp or an appliance may be placed, and to preclude the use of an extension cord to supply equipment in such an isolated space, a receptacle outlet is required.

The word *usable* does not appear at all in 210.52 as a condition for determining compliance with the receptacle-spacing requirements. As an example, to correctly determine the dimension of the wall line in a room, the wall space

Exhibit 210.24 Typical room plan view of the location of dwelling unit receptacles meeting the requirements of 210.52(A).

behind the swing of a door is included in the measurement. This does not mean that the receptacle outlet has to be located in that space, only that the space has been included in the wall-line measurement.

(3) Floor Receptacles Receptacle outlets in floors shall not be counted as part of the required number of receptacle outlets unless located within 450 mm (18 in.) of the wall.

(B) Small Appliances

(1) Receptacle Outlets Served In the kitchen, pantry, breakfast room, dining room, or similar area of a dwelling unit, the two or more 20-ampere small-appliance branch circuits required by 210.11(C)(1) shall serve all wall and floor receptacle outlets covered by 210.52(A), all countertop outlets covered by 210.52(C), and receptacle outlets for refrigeration equipment.

Section 210.52(B) requires a minimum of two 20-ampere circuits for all receptacle outlets for the small-appliance loads, including refrigeration equipment, in the kitchen, dining room, pantry, and breakfast room of a dwelling unit. The limited exceptions to what can be connected to these receptacle circuits allows the full capacity of the small-appliance circuits to be dedicated to the kitchen/dining area wall and countertop receptacles for the purposes of supplying cord-and-plug-connected portable appliance loads.

Connecting fastened-in-place appliances such as waste disposers or dishwashers to these circuits would reduce the capacity to supply the typical higher wattage portable loads

used in these areas, such as toasters, coffee makers, skillets, mixers, and the like. The *Code* can control the outlets that these circuits supply but cannot control the number of portable appliances that occupants use in these areas.

No restriction is placed on the number of outlets connected to a general-lighting or small-appliance branch circuit. The minimum number of receptacle outlets in a room is determined by 210.52(A) based on the room perimeter and 210.52(C) for counter spaces. It may be desirable to provide more than the minimum number of receptacle outlets required, thereby further reducing the need for extension cords and cords lying across counters.

Exhibit 210.25 illustrates the application of the requirements of 210.52(B)(1), 210.52(B)(2), and 210.52(B)(3). The small-appliance branch circuits illustrated in Exhibit 210.25 are not permitted to serve any other outlets, such as might be connected to exhaust hoods or fans, disposals, or dishwashers. The countertop receptacles are also required to be supplied by these two circuits if only the minimum of two circuits is provided for that dwelling. Note that only the counter area is required to be supplied by both of the small-appliance branch circuits. The wall receptacle outlets in the kitchen and dining room are permitted to be supplied by one or both of the circuits, as shown in the two diagrams in Exhibit 210.25.

The dining room switched receptacle on a 15-ampere general-purpose branch circuit is permitted according to 210.52(B)(1), Exception No. 1. The refrigerator receptacle supplied by a 15-ampere individual branch circuit (Exhibit 210.25, bottom) is permitted by 210.52(B)(1), Exception No. 2.

Exception No. 1: In addition to the required receptacles specified by 210.52, switched receptacles supplied from a general-purpose branch circuit as defined in 210.70(A)(1), Exception No. 1, shall be permitted.

Exception No. 1 to 210.52(B)(1) permits switched receptacles supplied from general-purpose 15-ampere branch circuits to be located in kitchens, pantries, breakfast rooms, and similar areas. See 210.70(A) and Exhibit 210.25 for details.

Exception No. 2: The receptacle outlet for refrigeration equipment shall be permitted to be supplied from an individual branch circuit rated 15 amperes or greater.

Exception No. 2 to 210.52(B)(1) allows a choice for refrigeration equipment receptacle outlets located in a kitchen or similar area. An individual 15-ampere or larger branch cir-

20-A small-appliance branch circuits

15-A general-purpose branch circuit

15-A individual branch circuit

15-A general-purpose branch circuit

20-A small-appliance branch circuits

G = GFCI protection required

Exhibit 210.25 Small-appliance branch circuits as required by 210.52(B)(1), 210.52(B)(2), and 210.52(B)(3) for all receptacle outlets in the kitchen (including refrigerator), pantry, and dining room.

cuit may serve this equipment, or it may be included in the 20-ampere small-appliance branch circuit. Refrigeration equipment is also exempt from the GFCI requirements of 210.8 where the receptacle outlet for the refrigerator is located as shown in Exhibit 210.25.

(2) No Other Outlets The two or more small-appliance branch circuits specified in 210.52(B)(1) shall have no other outlets.

Exception No. 1: A receptacle installed solely for the electrical supply to and support of an electric clock in any of the rooms specified in 210.52(B)(1).

Exception No. 2: Receptacles installed to provide power for supplemental equipment and lighting on gas-fired ranges, ovens, or counter-mounted cooking units.

Exception No. 2 to 210.52(B)(2) allows the small electrical loads associated with gas-fired appliances to be connected to small-appliance branch circuits. See Exhibit 210.25 for an illustration.

(3) Kitchen Receptacle Requirements Receptacles installed in a kitchen to serve countertop surfaces shall be supplied by not fewer than two small-appliance branch circuits, either or both of which shall also be permitted to supply receptacle outlets in the same kitchen and in other rooms specified in 210.52(B)(1). Additional small-appliance branch circuits shall be permitted to supply receptacle outlets in the kitchen and other rooms specified in 210.52(B)(1). No small-appliance branch circuit shall serve more than one kitchen.

Because the countertop receptacle outlets generally supply more of the portable cooking appliances than the wall receptacles in the kitchen and dining areas, the counter areas must be supplied by no fewer than two small-appliance branch circuits. The *Code* does not specify that both circuits be installed to serve the receptacle outlet(s) at each separate counter area in a kitchen, but rather that the total counter area of a kitchen must be supplied by no fewer than two circuits, and the arrangement of these circuits is determined by the designer or installer.

For example, a single receptacle outlet on a kitchen island is not required to be supplied by both of the small-appliance circuits serving the counter area. To provide efficient distribution of the small-appliance load, the number of receptacles connected to each small-appliance circuit should be carefully analyzed. The concept of evenly proportioning the load as specified in 210.11(A) (for loads calculated on the basis of volt-amperes per square foot) can be used as a best practice in distributing the number of receptacle outlets to be supplied by each of the small-appliance branch circuits. Where additional small-appliance branch circuits are installed, they are subject to all the requirements that apply to the minimum two required circuits.

The two circuits that supply the countertop receptacle outlets may also supply receptacle outlets in the pantry, dining room, and breakfast room, as well as an electric clock receptacle and electric loads associated with gas-fired appliances, but these circuits are to supply no other outlets. See 210.8(A)(6) for GFCI requirements applicable to receptacles serving kitchen counters.

(C) Countertops In kitchens and dining rooms of dwelling units, receptacle outlets for counter spaces shall be installed in accordance with 210.52(C)(1) through (C)(5).

(1) Wall Counter Spaces A receptacle outlet shall be installed at each wall counter space that is 300 mm (12 in.) or wider. Receptacle outlets shall be installed so that no point along the wall line is more than 600 mm (24 in.) measured horizontally from a receptacle outlet in that space.

Exception: Receptacle outlets shall not be required on a wall directly behind a range or sink in the installation described in Figure 210.52.

Outlet within 600 mm (24 in.) | Outlet not required if *X* <300 mm (12 in.) | Outlet within 600 mm (24 in.)

X

Sink or range extending from face of counter

Outlets not required if *X* <450 mm (18 in.) | Outlet within 600 mm (24 in.)

X

Outlet within 600 mm (24 in.)

Sink or range mounted in corner

Figure 210.52 Determination of Area Behind Sink or Range

This exception and the associated figure (Figure 210.52) were added to the 2005 *Code* to define the wall space behind a sink or range that is not required to be provided with a receptacle outlet. Figure 210.52 also shows where the edge of a sink or range is considered to be on the wall behind or to the side of the sink or range. Using these benchmarks, compliance with the wall counter receptacle outlet spacing required by 210.52(C)(1) can be determined. Note that where the space behind a sink or range is 12 in. or more or 18 in. or more (depending on the counter configuration), the space behind the sink or range must be included in measuring the wall counter space.

(2) Island Counter Spaces At least one receptacle shall be installed at each island counter space with a long dimension of 600 mm (24 in.) or greater and a short dimension of 300 mm (12 in.) or greater. Where a rangetop or sink is installed in an island counter and the width of the counter behind the rangetop or sink is less than 300 mm (12 in.), the rangetop or sink is considered to divide the island into two separate countertop spaces as defined in 210.52(C)(4).

(3) Peninsular Counter Spaces At least one receptacle outlet shall be installed at each peninsular counter space with a long dimension of 600 mm (24 in.) or greater and a short dimension of 300 mm (12 in.) or greater. A peninsular countertop is measured from the connecting edge.

(4) Separate Spaces Countertop spaces separated by rangetops, refrigerators, or sinks shall be considered as separate countertop spaces in applying the requirements of 210.52(C)(1), (C)(2), and (C)(3).

(5) Receptacle Outlet Location Receptacle outlets shall be located above, but not more than 500 mm (20 in.) above, the countertop. Receptacle outlets rendered not readily accessible by appliances fastened in place, appliance garages, sinks, or rangetops as covered in 210.52(C)(1), Exception, or appliances occupying dedicated space shall not be considered as these required outlets.

Exception to (5): To comply with the conditions specified in (1) or (2), receptacle outlets shall be permitted to be mounted not more than 300 mm (12 in.) below the countertop. Receptacles mounted below a countertop in accordance with this exception shall not be located where the countertop extends more than 150 mm (6 in.) beyond its support base.

(1) Construction for the physically impaired
(2) On island and peninsular countertops where the countertop is flat across its entire surface (no backsplashes, dividers, etc.) and there are no means to mount

a receptacle within 500 mm (20 in.) above the countertop, such as an overhead cabinet

Dwelling unit receptacles that serve countertop spaces in kitchens, dining areas, and similar rooms, as illustrated in Exhibit 210.26, are required to be installed as follows:

1. In each wall space wider than 12 in. and spaced so that no point along the wall line is more than 24 in. from a receptacle
2. Not more than 20 in. above the countertop [According to 406.4(E), receptacles cannot be installed in a face-up position. Receptacles installed in a face-up position in a countertop could collect crumbs, liquids, and other debris, resulting in a potential fire or shock hazard.]
3. At each countertop island and peninsular countertop with a short dimension of at least 12 in. and a long dimension of at least 24 in. (The measurement of a peninsular-type countertop is from the edge connecting to the nonpeninsular counter.)
4. Accessible for use and not blocked by appliances occupying dedicated space or fastened in place
5. Fed from two or more of the required 20-ampere small-appliance branch circuits and GFCI protected according to 210.8(A)(6)

G = GFCI protection required

Exhibit 210.26 Dwelling unit receptacles serving countertop spaces in a kitchen and installed in accordance with 210.52(C).

The maximum permitted height of a receptacle serving a countertop specified in 210.52(C)(5) was revised in the 2002 *Code* from 18 in. to 20 in. as a practical consideration based on standard cabinet and counter dimensions.

(D) Bathrooms In dwelling units, at least one receptacle outlet shall be installed in bathrooms within 900 mm (3 ft) of the outside edge of each basin. The receptacle outlet shall be located on a wall or partition that is adjacent to the basin or basin countertop.

Section 210.52(D) requires one wall receptacle in each bathroom of a dwelling unit to be installed adjacent to (within 36 in. of) the basin. Added to the 2005 *Code,* the exception permits an alternative to locating the required outlet in the wall adjacent to the basin. Different in application from the exception to 210.52(C)(5), the permission to install a receptacle outlet in the side or face of the basin cabinet is not contingent on the adjacent wall location being unfeasible or inaccessible to a handicapped person. Like the kitchen counter rule, the outlet must be located so that the receptacle(s) is not more than 12 in. below the basin countertop.

This receptacle is required in addition to any receptacle that may be part of any luminaire or medicine cabinet. If there is more than one basin, a receptacle outlet is required adjacent to each basin location. If the basins are in close proximity, one receptacle outlet installed between the two basins can be used to satisfy this requirement. See 406.8(C), which prohibits installation of a receptacle over a bathtub or inside a shower stall. See Exhibit 210.9 for a sample electrical layout of a bathroom.

Section 210.11(C)(3) requires the receptacle outlets to be supplied from a 20-ampere branch circuit with no other outlets. However, this circuit is permitted to supply the required receptacles in more than one bathroom. If the circuit supplies the required receptacle outlet in only one bathroom, it is allowed to also supply lighting and an exhaust fan in that bathroom provided the lighting and fan load does not exceed that permitted by 210.23(A)(2). This receptacle is also required to be GFCI protected in accordance with 210.8(A)(1).

Exception: The receptacle shall not be required to be mounted in the wall or partition where it is installed on the side or face of the basin cabinet not more than 300 mm (12 in.) below the countertop.

(E) Outdoor Outlets For a one-family dwelling and each unit of a two-family dwelling that is at grade level, at least one receptacle outlet accessible at grade level and not more than 2.0 m (6½ ft) above grade shall be installed at the front and back of the dwelling.

For each dwelling unit of a multifamily dwelling where the dwelling unit is located at grade level and provided with individual exterior entrance/egress, at least one receptacle outlet accessible from grade level and not more than 2.0 m (6½ ft) above grade shall be installed. See 210.8(A)(3).

The rule for one- and two-family dwellings requires two outdoor receptacle outlets for each dwelling unit. One receptacle outlet is required at the front of the dwelling, and one is required at the back of the dwelling, as shown in Exhibit 210.27. For one- and two-family dwellings, the phrase *accessible at grade level* clearly requires that the two required receptacle outlets are to be available to a person standing on the ground (at grade level). Where outdoor heating, air-conditioning, or refrigeration (HACR) equipment is located at grade level, the receptacle outlets required by this section can be used to comply with the receptacle outlet requirement of 210.63, provided that at least one of the outlets is located within 25 ft of the HACR equipment. The outlets required by 210.52(E) comply with the "located on the same level" requirement of 210.63. Outdoor receptacle outlets on decks, porches, and similar structures can be used to meet 210.52(E) as long as the receptacle outlet is not more than 6 ½ ft above grade and can be accessed by a person standing at grade.

Multifamily dwellings (those with three or more dwelling units) are now required to be provided with at least one outdoor receptacle outlet accessible from grade level. This 2005 *Code* change applies to those dwelling units in a multifamily structure that are located at grade level and have entrance/egress doorways that lead directly to the exterior of the structure. Concerns over the unauthorized use of outdoor receptacle outlets installed at multifamily dwellings by other than the dwelling occupant(s) can be allayed by using a switch inside the dwelling unit to control the outlet. Due to the spatial constraints often associated with the construction

of multifamily units, the required receptacle outlet is permitted to be accessible "from grade," and an outdoor receptacle outlet located 6 ½ ft or less above grade and accessible by walking up a set of deck or porch steps can be used to meet this requirement.

Outdoor receptacles must be installed so that the receptacle faceplate rests securely on the supporting surface to prevent moisture from entering the enclosure. On uneven surfaces, such as brick, stone, or stucco, it may be necessary to close openings with caulking compound or mastic. See 406.8 for further information on receptacles installed in damp or wet locations.

(F) Laundry Areas In dwelling units, at least one receptacle outlet shall be installed for the laundry.

Exception No. 1: In a dwelling unit that is an apartment or living area in a multifamily building where laundry facilities are provided on the premises and are available to all building occupants, a laundry receptacle shall not be required.

Exception No. 2: In other than one-family dwellings where laundry facilities are not to be installed or permitted, a laundry receptacle shall not be required.

A laundry receptacle outlet(s) is supplied by a 20-ampere branch circuit. This circuit can have no other outlets. See 210.11(C)(2) for further information.

(G) Basements and Garages For a one-family dwelling, at least one receptacle outlet, in addition to any provided for laundry equipment, shall be installed in each basement and in each attached garage, and in each detached garage with electric power. See 210.8(A)(2) and (A)(5). Where a portion of the basement is finished into one or more habitable rooms, each separate unfinished portion shall have a receptacle outlet installed in accordance with this section.

In a one-family dwelling, a receptacle must be installed in the basement (in addition to the laundry receptacle), in each attached garage, and in each detached garage with electric power.

Section 210.8(A)(5) requires receptacles in unfinished basements to be protected by GFCIs. Section 210.8(A)(2) requires receptacles installed in garages to be protected by GFCIs. Where detached garages are not supplied with electricity, receptacle outlets do not have to be installed.

(H) Hallways In dwelling units, hallways of 3.0 m (10 ft) or more in length shall have at least one receptacle outlet.

As used in this subsection, the hall length shall be considered the length along the centerline of the hall without passing through a doorway.

Exhibit 210.27 Row housing with GFCI-protected receptacles located at the front and the back of each one-family dwelling, as required by 210.52(E).

The requirement in 210.52(H) is intended to minimize strain or damage to cords and receptacles for dwelling unit receptacles. The requirement does not apply to common hallways of hotels, motels, apartment buildings, condominiums, and so on.

210.60 Guest Rooms or Guest Suites

(A) General Guest rooms or guest suites in hotels, motels, and similar occupancies shall have receptacle outlets installed in accordance with 210.52(A) and 210.52(D). Guest rooms or guest suites provided with permanent provisions for cooking shall have receptacle outlets installed in accordance with all of the applicable rules in 210.52.

(B) Receptacle Placement In applying the provisions of 210.52(A), the total number of receptacle outlets shall not be less than the minimum number that would comply with the provisions of that section. These receptacle outlets shall be permitted to be located conveniently for permanent furniture layout. At least two receptacle outlets shall be readily accessible. Where receptacles are installed behind the bed, the receptacle shall be located to prevent the bed from contacting any attachment plug that may be installed or the receptacle shall be provided with a suitable guard.

Section 210.60(B) permits the receptacles in guest rooms and guest suites of hotels and motels to be placed in accessible locations that are compatible with permanent furniture. However, the minimum number of receptacles required by 210.52 is not permitted to be reduced. The minimum number of receptacle outlets should be determined by assuming there is no furniture in the room. The practical locations of that minimum number of receptacles are then determined based on the permanent furniture layout.

Hotel and motel rooms and suites are commonly used as remote offices for businesspeople who use laptop computers and other plug-in devices. The *Code* requires two receptacle outlets to be available without moving furniture to access those receptacles. To reduce the risk of fire to bedding material, receptacles located behind beds must include guards if attachment plugs might contact the bed.

Bathroom areas for guest rooms and suites in hotels and motels are required to be provided with a receptacle outlet adjacent to the basin location, in accordance with 210.8(B)(1).

Extended-stay hotels and motels are often equipped with permanent provisions for cooking and countertop areas. All applicable receptacle spacing and supply requirements in 210.52 apply to guest rooms or suites that contain such provisions. A portable microwave oven is not considered to be a permanently installed cooking appliance. See 210.18 and its associated commentary for more information regard-

ing hotel and motel guest rooms and guest suites that are equipped with permanent provisions for cooking.

Exhibit 210.28 shows receptacle outlets in a hotel guest room located conveniently with respect to the permanent furniture layout. Some spaces that are 2 ft or more in width have no receptacle outlets because 210.60(B) permits the required number of outlets to be placed in convenient locations that are compatible with the permanent furniture layout. In Exhibit 210.28, the receptacle outlet adjacent to the permanent dresser is needed because 210.60(B) applies only to the location of receptacle outlets, not to the minimum number of receptacle outlets.

Exhibit 210.28 Floor plan of a hotel guest room with receptacles located as permitted by 210.60(B) with respect to permanent furniture.

210.62 Show Windows

At least one receptacle outlet shall be installed directly above a show window for each 3.7 linear m (12 linear ft) or major fraction thereof of show window area measured horizontally at its maximum width.

Show windows usually extend from floor to ceiling for maximum display. To discourage floor receptacles and unsightly extension cords likely to cause physical injury, receptacles must be installed directly above a show window, and one receptacle is required for every 12 linear ft or "major fraction thereof" (6 ft or more). See 220.14(G) and 220.43(A) for information regarding load computations for show windows.

210.63 Heating, Air-Conditioning, and Refrigeration Equipment Outlet

A 125-volt, single-phase, 15- or 20-ampere-rated receptacle outlet shall be installed at an accessible location for the servicing of heating, air-conditioning, and refrigeration equipment. The receptacle shall be located on the same level and within 7.5 m (25 ft) of the heating, air-conditioning, and refrigeration equipment. The receptacle outlet shall not be connected to the load side of the equipment disconnecting means.

Section 210.63 is intended to prevent makeshift methods of obtaining 125-volt power for servicing and troubleshooting heating, air-conditioning, and refrigeration (HACR) equipment. The reference to 210.8 in the fine print note to 210.63 reminds the Code user of the GFCI requirements for these receptacle outlets. The requirements in 210.52(E) for outdoor dwelling unit receptacles located within 25 ft of HACR equipment meet the requirements of 210.63.

The requirements of 210.63 were expanded in the 2002 Code to improve worker safety. As a result, a receptacle outlet is now required for troubleshooting HACR equipment at grade-accessible outdoor equipment and at rooftop units associated with one- and two-family dwelling units. A new exception added in the 2005 Code exempts evaporative coolers (commonly referred to as "swamp coolers") from the receptacle requirement where the cooler is installed at a one- or two-family dwelling. It should be noted that although this type of cooling equipment is exempt from 210.63, one- and two-family dwellings are required to have outdoor receptacle outlets at the front and the back of the structure in accordance with 210.52(E).

Exception: A receptacle outlet shall not be required at one- and two-family dwellings for the service of evaporative coolers.

> FPN: See 210.8 for ground-fault circuit-interrupter requirements.

210.70 Lighting Outlets Required

Lighting outlets shall be installed where specified in 210.70(A), (B), and (C).

(A) Dwelling Units In dwelling units, lighting outlets shall be installed in accordance with 210.70(A)(1), (A)(2), and (A)(3).

(1) Habitable Rooms At least one wall switch-controlled lighting outlet shall be installed in every habitable room and bathroom.

Exception No. 1: In other than kitchens and bathrooms, one or more receptacles controlled by a wall switch shall be permitted in lieu of lighting outlets.

A receptacle is not permitted to be switched as a lighting outlet on a small-appliance branch circuit. A receptacle can be switched as a lighting outlet (e.g., in the dining room)

supplied by a branch circuit other than a small-appliance branch circuit. See Exhibit 210.25, which shows a dining room switched receptacle on a 15-ampere general-purpose branch circuit.

Exception No. 2: Lighting outlets shall be permitted to be controlled by occupancy sensors that are (1) in addition to wall switches or (2) located at a customary wall switch location and equipped with a manual override that will allow the sensor to function as a wall switch.

(2) Additional Locations Additional lighting outlets shall be installed in accordance with (A)(2)(a), (A)(2)(b), and (A)(2)(c).

(a) At least one wall switch-controlled lighting outlet shall be installed in hallways, stairways, attached garages, and detached garages with electric power.

(b) For dwelling units, attached garages, and detached garages with electric power, at least one wall switch–controlled lighting outlet shall be installed to provide illumination on the exterior side of outdoor entrances or exits with grade level access. A vehicle door in a garage shall not be considered as an outdoor entrance or exit.

(c) Where one or more lighting outlet(s) are installed for interior stairways, there shall be a wall switch at each floor level, and landing level that includes an entryway, to control the lighting outlet(s) where the stairway between floor levels has six risers or more.

Exception to (A)(2)(a), (A)(2)(b), and (A)(2)(c): In hallways, stairways, and at outdoor entrances, remote, central, or automatic control of lighting shall be permitted.

Section 210.70 points out that adequate lighting and proper control and location of switching are as essential to the safety of occupants of dwelling units, hotels, motels, and so on, as are proper wiring requirements. Proper illumination ensures safe movement for persons of all ages, thus preventing many accidents.

Although the requirement in 210.70(A)(2)(b) calls for a switched lighting outlet at outdoor entrances and exits, it does not prohibit a single lighting outlet, if suitably located, from serving more than one door.

A wall switch–controlled lighting outlet is required in the kitchen and in the bathroom. A receptacle outlet controlled by a wall switch is not permitted to serve as a lighting outlet in these rooms. Occupancy sensors are permitted to be used for switching these lighting outlets, provided they are equipped with a manual override or are used in addition to regular switches.

(3) Storage or Equipment Spaces For attics, underfloor spaces, utility rooms, and basements, at least one lighting outlet containing a switch or controlled by a wall switch shall be installed where these spaces are used for storage or contain equipment requiring servicing. At least one point of control shall be at the usual point of entry to these spaces. The lighting outlet shall be provided at or near the equipment requiring servicing.

Installation of lighting outlets in attics, underfloor spaces or crawl areas, utility rooms, and basements is required when these spaces are used for storage (e.g., holiday decorations or luggage).

If such spaces contain equipment that requires servicing (e.g., air-handling units, cooling and heating equipment, water pumps, or sump pumps), 210.70(C) requires that a lighting outlet be installed in these spaces.

(B) Guest Rooms or Guest Suites In hotels, motels, or similar occupancies, guest rooms or guest suites shall have at least one wall switch–controlled lighting outlet installed in every habitable room and bathroom.

In addition to adding guest suites to the lighting outlet requirement, 210.70(B) has been revised to mirror the language of 210.70(A) and requires a switched lighting outlet in every habitable room (the hotel room or rooms in a suite) and in bathrooms. The bathroom for each guest room or suite and, where provided, the kitchen are required to have a least one lighting outlet controlled by a wall switch. All other rooms can employ a switched receptacle to meet this lighting requirement. Exception No. 2 permits the use of occupancy sensors to control the lighting outlet provided that it is in the typical switch location and can be manually controlled.

Exception No. 1: In other than bathrooms and kitchens where provided, one or more receptacles controlled by a wall switch shall be permitted in lieu of lighting outlets.

Exception No. 2: Lighting outlets shall be permitted to be controlled by occupancy sensors that are (1) in addition to wall switches or (2) located at a customary wall switch location and equipped with a manual override that will allow the sensor to function as a wall switch.

(C) Other Than Dwelling Units For attics and underfloor spaces containing equipment requiring servicing, such as heating, air-conditioning, and refrigeration equipment, at least one lighting outlet containing a switch or controlled by a wall switch shall be installed in such spaces. At least one point of control shall be at the usual point of entry to these spaces. The lighting outlet shall be provided at or near the equipment requiring servicing.

ARTICLE 215
Feeders

Summary of Changes

- **215.2(A):** Revised to delete exception and requirements for specific circuit arrangements and to add new requirement establishing the minimum size of a feeder grounded conductor.

- **215.12:** Added requirement that each ungrounded feeder-circuit conductor be identified by some means where there is more than one nominal voltage system on the premises and that the means of identification be posted at each feeder panelboard or similar distribution equipment.

Contents

215.1 Scope

This article covers the installation requirements, overcurrent protection requirements, minimum size, and ampacity of conductors for feeders supplying branch-circuit loads.

Exception: Feeders for electrolytic cells as covered in 668.3(C)(1) and (C)(4).

The scope section for Article 215 includes a reference to overcurrent protection requirements for feeders, which encompasses both the fact that Article 215 references Article 240 for overcurrent protection of feeder circuits and the fact that 215.3 includes the 125-percent sizing rule for feeder overcurrent devices supplying continuous loads.

The total connected load to be supplied by the feeder must be calculated to accurately determine feeder conductor ampacity. The sum of the computed and connected loads supplied by a feeder is multiplied by the demand factor to determine the load that the feeder conductors must be sized to serve. See Article 100 for the definition of *demand factor*.

When the total connected load is operated simultaneously, the demand factor is 100 percent; that is, the maximum demand is equal to the total connected load. Due to diversity, the maximum operating load carried at any time may be only three-quarters of the total connected load; thus, the demand factor is 75 percent.

On a new installation, a minimum value for the demand factor can be determined by applying the requirements and tables of Article 220, Branch-Circuit, Feeder, and Service Calculations.

Feeder conductor sizes are determined by calculating the total volt-amperes (VA) of the feeder load at the nominal voltage of the feeder circuit. See 220.5 for the nominal system voltages used in computing branch-circuit and feeder loads. See 310.15 for allowable ampacities and sizes of insulated conductors.

Feeder circuits must have sufficient ampacity to safely supply the calculated load. Wiring systems that do not provide for increases in the use of electricity often create hazards. It is good practice to allow for future expansion and convenience increases, as stated in 90.8(A).

215.2 Minimum Rating and Size

(A) Feeders Not More Than 600 Volts

The 2005 *Code* contains a new requirement on the minimum size for the grounded conductor of a feeder circuit. In addition to sizing the conductor per 220.61 (formerly 220.22), the absolute minimum is now based on 250.122 and Table 250.122, the section and table covering equipment grounding conductor sizing.

In past editions of the *Code*, the minimum size of the grounded (neutral) feeder conductor was determined using the requirement of 220.61 where the grounded conductor was used solely as a normal circuit conductor and was not used as the conductor to create the effective ground-fault current path. Therefore, where a feeder circuit supplied primarily line-to-line connected loads and only a small line-to-neutral load, the grounded conductor of the feeder circuit could be significantly smaller than the ungrounded feeder conductors and the equipment grounding conductor because it only had to be sized to carry the maximum unbalanced load. In such cases, a line-to-neutral fault could result in significant damage to the comparatively small grounded conductor and failure to open the overcurrent device under this particular short-circuit condition.

Using 250.122 provides a minimum size for the grounded conductor that has a direct sizing relationship to the ungrounded conductors for the purposes of facilitating operation of the feeder overcurrent device under short-circuit conditions. For feeder circuits installed in parallel in separate raceways or cables, the requirements of 220.61 and 310.4 are to be used to determine the minimum grounded conductor size.

The requirement for sizing the grounded feeder conductor has also been added to 215.2(B) for feeder circuits over 600 volts.

(1) General Feeder conductors shall have an ampacity not less than required to supply the load as calculated in Parts III, IV, and V of Article 220. The minimum feeder-circuit conductor size, before the application of any adjustment or correction factors, shall have an allowable ampacity not less than the noncontinuous load plus 125 percent of the continuous load.

Exception: Where the assembly, including the overcurrent devices protecting the feeder(s), is listed for operation at 100 percent of its rating, the allowable ampacity of the feeder conductors shall be permitted to be not less than the sum of the continuous load plus the noncontinuous load.

The size of the feeder circuit grounded conductor shall not be smaller than that required by 250.122, except that 250.122(F) shall not apply where grounded conductors are run in parallel.

Additional minimum sizes shall be as specified in 215.2(A)(2) and (A)(3) under the conditions stipulated.

(2) Ampacity Relative to Service Conductors The feeder conductor ampacity shall not be less than that of the service conductors where the feeder conductors carry the total load supplied by service conductors with an ampacity of 55 amperes or less.

(3) Individual Dwelling Unit or Mobile Home Conductors Feeder conductors for individual dwelling units or mobile homes need not be larger than service conductors. Paragraph 310.15(B)(6) shall be permitted to be used for conductor size.

For example, according to Table 310.16, a 3/0 AWG, Type THW copper wire has an ampacity of 200 amperes. However, for a 3-wire, single-phase dwelling service, as shown in Exhibit 215.1, Table 310.15(B)(6) permits 2/0 AWG, Type THW copper conductors or 4/0 AWG, Type THW aluminum conductors for services or feeders rated at 200 amperes. Feeder conductors carrying the total load supplied by the service are not required to be sized larger than the service-entrance conductors.

Service-entrance conductors

Feeder conductors need not be larger than service-entrance conductors

200-A service disconnect

200-A panelboard

Exhibit 215.1 A 3-wire, single-phase dwelling service with an ampacity of 200 amperes for 2/0 AWG copper or 4/0 AWG aluminum conductors used as service-entrance conductors and feeder conductors, according to 215.2(A)(3).

FPN No. 1: See Examples D1 through D11 in Annex D.

FPN No. 2: Conductors for feeders as defined in Article 100, sized to prevent a voltage drop exceeding 3 percent at the farthest outlet of power, heating, and lighting loads, or combinations of such loads, and where the maximum total voltage drop on both feeders and branch circuits to the farthest outlet does not exceed 5 percent, will provide reasonable efficiency of operation.

FPN No. 3: See 210.19(A), FPN No. 4, for voltage drop for branch circuits.

Reasonable operating efficiency is achieved if the voltage drop of a feeder or the voltage drop of a branch circuit is limited to 3 percent. However, the total voltage drop of a branch circuit plus a feeder can reach 5 percent and still achieve reasonable operating efficiency. See Article 100 for the definitions of *feeder* and *branch circuit*.

The 5-percent voltage-drop value is explanatory material and, as such, appears as a fine print note. Fine print notes are not mandatory (see 90.5). However, where circuit conductors are increased due to voltage drop, 250.122(B) requires an increase in circular mil area for the associated equipment grounding conductors.

The resistance or impedance of conductors may cause a substantial difference between voltage at service equipment and voltage at the point-of-utilization equipment. Excessive voltage drop impairs the starting and the operation of electrical equipment. Undervoltage can result in inefficient operation of heating, lighting, and motor loads. An applied voltage of 10 percent below rating can result in a decrease in efficiency of substantially more than 10 percent. For example,

fluorescent light output would be reduced by 15 percent, and incandescent light output would be reduced by 30 percent. Induction motors would run hotter and produce less torque. With an applied voltage of 10 percent below rating, the running current would increase 11 percent, and the operating temperature would increase by 12 percent. At the same time, torque would be reduced by 19 percent.

In addition to resistance or impedance, the type of raceway or cable enclosure, the type of circuit (ac, dc, single-phase, 3-phase), and the power factor should be considered to determine voltage drop.

The following basic formula can be used to determine the voltage drop in a 2-wire dc circuit, a 2-wire ac circuit, or a 3-wire ac single-phase circuit, all with a balanced load at 100 percent power factor and where reactance can be neglected.

$$VD = \frac{2 \times L \times R \times I}{1000}$$

where:

VD = voltage drop (based on conductor temperature of 75°C)

L = one-way length of circuit (ft)

R = conductor resistance in ohms (Ω) per 1000 ft (from Chapter 9, Table 8)

I = load current (amperes)

For 3-phase circuits (at 100 percent power factor), the voltage drop between any two phase conductors is 0.866 times the voltage drop calculated by the preceding formula.

Example

Determine the voltage drop in a 240-volt, 2-wire heating circuit with a load of 50 amperes. The circuit size is 6 AWG, Type THHN copper, and the one-way circuit length is 100 ft.

Solution

STEP 1. Find the conductor resistance in Chapter 9, Table 8.

STEP 2. Substitute values into the voltage-drop formula:

$$VD = \frac{2 \times L \times R \times I}{1000}$$
$$= \frac{2 \times 100 \times 0.491 \times 50}{1000} = 4.91 \text{ volts}$$

STEP 3. Determine the percentage of the voltage drop:

$$\% \, VD = \frac{4.91 \text{ V}}{240 \text{ V}} = 0.02 \text{ or } 2\%$$

A 12-volt drop on a 240-volt circuit is a 5-percent drop. A 4.91-volt drop falls within this percentage. If the total voltage drop exceeds 5 percent, or 12 volts, larger-size conductors should be used, the circuit length should be shortened, or the circuit load should be reduced.

See the commentary following Chapter 9, Table 9, for an example of voltage-drop calculation using ac reactance and resistance. Voltage-drop tables and calculations are also available from various manufacturers.

(B) Feeders Over 600 Volts The ampacity of conductors shall be in accordance with 310.15 and 310.60 as applicable. Where installed, the size of the feeder circuit grounded conductor shall not be smaller than that required by 250.122, except that 250.122(F) shall not apply where grounded conductors are run in parallel. Feeder conductors over 600 volts shall be sized in accordance with 215.2(B)(1), (B)(2), or (B)(3).

(1) Feeders Supplying Transformers The ampacity of feeder conductors shall not be less than the sum of the nameplate ratings of the transformers supplied when only transformers are supplied.

(2) Feeders Supplying Transformers and Utilization Equipment The ampacity of feeders supplying a combination of transformers and utilization equipment shall not be less than the sum of the nameplate ratings of the transformers and 125 percent of the designed potential load of the utilization equipment that will be operated simultaneously.

(3) Supervised Installations For supervised installations, feeder conductor sizing shall be permitted to be determined by qualified persons under engineering supervision. Supervised installations are defined as those portions of a facility where all of the following conditions are met:

(1) Conditions of design and installation are provided under engineering supervision.
(2) Qualified persons with documented training and experience in over 600-volt systems provide maintenance, monitoring, and servicing of the system.

Section 215.2(B) sets the minimum requirements for feeders over 600 volts. Unless the circuit is part of a supervised installation [defined in 215.2(B)(3)], the minimum ampacity for feeder circuit conductors over 600 volts can be no less than 100 percent of the transformer nameplate load plus 125 percent of any additional utilization equipment. The overcurrent protection requirements for feeders over 600 volts must be in accordance with Article 240, Part IX.

215.3 Overcurrent Protection

Feeders shall be protected against overcurrent in accordance with the provisions of Part I of Article 240. Where a feeder

supplies continuous loads or any combination of continuous and noncontinuous loads, the rating of the overcurrent device shall not be less than the noncontinuous load plus 125 percent of the continuous load.

Exception: Where the assembly, including the overcurrent devices protecting the feeder(s), is listed for operation at 100 percent of its rating, the ampere rating of the overcurrent device shall be permitted to be not less than the sum of the continuous load plus the noncontinuous load.

The feeder overcurrent protection requirements in 215.3 are somewhat similar to the branch-circuit overcurrent protection requirements in 210.20(A).

Exception: Overcurrent protection for feeders over 600 volts, nominal, shall comply with Part XI of Article 240.

215.4 Feeders with Common Neutral

(A) Feeders with Common Neutral Two or three sets of 3-wire feeders or two sets of 4-wire or 5-wire feeders shall be permitted to utilize a common neutral.

(B) In Metal Raceway or Enclosure Where installed in a metal raceway or other metal enclosure, all conductors of all feeders using a common neutral shall be enclosed within the same raceway or other enclosure as required in 300.20.

If feeder conductors carrying ac current, including the neutral, are installed in metal raceways, the conductors are required to be grouped together to avoid induction heating of the surrounding metal. If it is necessary to run parallel conductors through multiple metal raceways, conductors from each phase plus the neutral must be run in each raceway. See 250.102(E), 250.134(B), 300.3, 300.5(I), and 300.20 for requirements associated with conductor grouping of feeder circuits.

A 3-phase, 4-wire (208Y/120-volt, 480Y/277-volt) system is often used to supply both lighting and motor loads. The 3-phase motor loads are typically not connected to the neutral and thus will not cause current in the neutral conductor. The maximum current on the neutral, therefore, is due to lighting loads or circuits where the neutral is used. On this type of system (3-phase, 4-wire), a demand factor of 70 percent is permitted by 220.61 for that portion of the neutral load in excess of 200 amperes.

For example, if the maximum possible unbalanced load is 500 amperes, the neutral would have to be large enough to carry 410 amperes (200 amperes plus 70 percent of 300 amperes, or 410 amperes). No reduction of the neutral capacity for that portion of the load consisting of electric-discharge lighting is permitted.

Section 310.15(B)(4)(c) points out that a neutral conductor must be counted as a current-carrying conductor if the load it serves consists of harmonic currents. See 220.61 for other systems in which the 70 percent demand factor may be applied. The maximum unbalanced load for feeders supplying clothes dryers, household ranges, wall-mounted ovens, and counter-mounted cooking units is required to be considered 70 percent of the load on the ungrounded conductors. See Examples D1(a) through D5(b) of Annex D.

215.5 Diagrams of Feeders

If required by the authority having jurisdiction, a diagram showing feeder details shall be provided prior to the installation of the feeders. Such a diagram shall show the area in square feet of the building or other structure supplied by each feeder, the total calculated load before applying demand factors, the demand factors used, the calculated load after applying demand factors, and the size and type of conductors to be used.

215.6 Feeder Conductor Grounding Means

Where a feeder supplies branch circuits in which equipment grounding conductors are required, the feeder shall include or provide a grounding means, in accordance with the provisions of 250.134, to which the equipment grounding conductors of the branch circuits shall be connected.

215.7 Ungrounded Conductors Tapped from Grounded Systems

Two-wire dc circuits and ac circuits of two or more ungrounded conductors shall be permitted to be tapped from the ungrounded conductors of circuits having a grounded neutral conductor. Switching devices in each tapped circuit shall have a pole in each ungrounded conductor.

Section 215.7 does not require a common trip or simultaneous opening of circuit breakers or fuses but rather requires a switching device to manually disconnect the ungrounded conductors of the feeder. See 210.10 for similar requirements related to the ungrounded conductors of the branch circuit.

215.9 Ground-Fault Circuit-Interrupter Protection for Personnel

Feeders supplying 15- and 20-ampere receptacle branch circuits shall be permitted to be protected by a ground-fault circuit interrupter in lieu of the provisions for such interrupters as specified in 210.8 and 590.6(A).

Several manufacturers offer double-pole 120/240-volt circuit-breaker-type ground-fault circuit interrupters (GFCIs) for application to a feeder, thereby protecting all branch circuits supplied by that feeder. This type of GFCI installation is in lieu of the provisions of 210.8 for outdoor, bathroom, garage, kitchen, basement, and boathouse receptacles. GFCI protection in the feeder can also be used to protect construction-site receptacles, as covered in 590.6(A), provided the feeder supplies no lighting branch circuits.

It may be more economical or convenient to install GFCIs for feeders. However, consideration should be given to the fact that a GFCI may be monitoring several branch circuits and will de-energize all branch circuits in response to a line-to-ground fault from one branch circuit. As stated in 90.1(B), the installation may be "free from hazard but not necessarily efficient, convenient, or adequate for good service" where GFCI protection of a feeder is used in lieu of GFCI protection for each branch circuit or at the individual receptacles.

215.10 Ground-Fault Protection of Equipment

Each feeder disconnect rated 1000 amperes or more and installed on solidly grounded wye electrical systems of more than 150 volts to ground, but not exceeding 600 volts phase-to-phase, shall be provided with ground-fault protection of equipment in accordance with the provisions of 230.95.

> FPN: For buildings that contain healthcare occupancies, see the requirements of 517.17.

Exception No. 1: The provisions of this section shall not apply to a disconnecting means for a continuous industrial process where a nonorderly shutdown will introduce additional or increased hazards.

Exception No. 2: The provisions of this section shall not apply to fire pumps.

Exception No. 3: The provisions of this section shall not apply if ground-fault protection of equipment is provided on the supply side of the feeder.

The intent of 215.10 is to require ground-fault protection of equipment for feeder disconnects that are rated 1000 amperes or more at 480Y/277 volts. A similar requirement for services is found in 230.95. The reason for the requirement is the unusually high number of burndowns reported on feeders and services in this voltage range. Prior to being put into service, each ground-fault protection system must be performance tested and documented according to the requirements of 230.95(C).

It should be noted that ground-fault protection of feeder equipment is not required if protection is provided on an upstream feeder or at the service. However, additional levels of ground-fault protection on feeders may be desired so that a single ground fault does not de-energize the whole electrical system. See 230.95 for further commentary on ground-fault protection of services. Also according to the new fine print note, see 517.17, which requires an additional level of ground-fault protection for health care facilities.

For emergency feeders, according to Article 700, the ground-fault protection requirements are different. See 700.26 for further details.

215.11 Circuits Derived from Autotransformers

Feeders shall not be derived from autotransformers unless the system supplied has a grounded conductor that is electrically connected to a grounded conductor of the system supplying the autotransformer.

Exception No. 1: An autotransformer shall be permitted without the connection to a grounded conductor where transforming from a nominal 208 volts to a nominal 240-volt supply or similarly from 240 volts to 208 volts.

Exception No. 2: In industrial occupancies, where conditions of maintenance and supervision ensure that only qualified persons service the installation, autotransformers shall be permitted to supply nominal 600-volt loads from nominal 480-volt systems, and 480-volt loads from nominal 600-volt systems, without the connection to a similar grounded conductor.

Section 215.11 addresses autotransformers for feeders and is similar to the requirements in 210.9 for branch circuits. See the commentary following 210.9 for further information on autotransformers that supply branch circuits.

215.12 Identification for Feeders

(A) Grounded Conductor The grounded conductor of a feeder shall be identified in accordance with 200.6.

(B) Equipment Grounding Conductor The equipment grounding conductor shall be identified in accordance with 250.119.

(C) Ungrounded Conductors Where the premises wiring system has feeders supplied from more than one nominal voltage system, each ungrounded conductor of a feeder, where accessible, shall be identified by system. The means of identification shall be permitted to be by separate color coding, marking tape, tagging, or other approved means and shall be permanently posted at each feeder panelboard or similar feeder distribution equipment.

Parallel with the new requirement for ungrounded branch circuit conductors in 210.5(C), 215.12(C) requires identification of ungrounded feeder conductors where there is more

than one nominal voltage supply system to a building, structure, or other premises. The identification scheme is not specified, but whatever is used is required to be consistent throughout the premises. A permanent legend or directory indicating the feeder identification system for the premises is required to be posted at each point in the distribution system from which feeder circuits are supplied.

ARTICLE 220
Branch-Circuit, Feeder, and Service Calculations

Summary of Changes

- **General:** Changed the word *compute* and its derivatives to *calculate* or *equal* in Article 220 and throughout the *Code*.
- **Figure 220.1:** Added diagram showing the revised arrangement of Article 220.
- **220.3:** Completely reorganized. New Table 220.3 identifies requirements in other articles that are in addition to, or modifications of, the requirements in Article 220.
- **220.14(D):** Revised to apply to all luminaire loads, not only to recessed luminaires.
- **220.14(K):** Added requirement for calculating receptacle outlet loads in banks and office buildings.
- **220.82(C)(4):** Revised requirements for calculating dwelling unit heat pump compressor and supplemental heating loads.

Contents

I. General

220.1 Scope

This article provides requirements for calculating branch-circuit, feeder, and service loads. Part I provides for general requirements for calculation methods. Part II provides calculation methods for branch circuit loads. Parts III and IV provide calculation methods for feeders and services. Part V provides calculation methods for farms.

Article 220 contains requirements for calculating branch-circuit, feeder, and service loads. Revised for the 2005 *Code* to provide better organization of the calculation rules, Article 220 now contains a new Part I with general requirements, and Parts II through V contain the calculation requirements for branch circuits, feeders, services, and farm loads. The organization chart of the revised Article 220 is shown in Figure 220.1. Although Article 220 does not contain the requirements for determining the minimum number of branch circuits, the loads calculated in accordance with this article are used in conjunction with the rules of 210.11 to determine how many branch circuits are needed at a premises. A global change in the 2005 *Code* is use of the word *calculate* (or a derivative thereof) consistently in all load calculation requirements. Table 220.3, which is new in the 2005 *Code*, identifies other articles and sections with load calculation requirements.

FPN: See Figure 220.1 for information on the organization of Article 220.

220.3 Application of Other Articles

In other articles applying to the calculation of loads in specialized applications, there are requirements provided in Table 220.3 that are in addition to, or modifications of, those within this article.

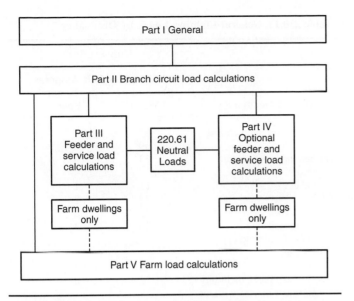

Figure 220.1 Branch-Circuit, Feeder, and Service Calculation Methods

220.5 Calculations

(A) Voltages Unless other voltages are specified, for purposes of calculating branch-circuit and feeder loads, nominal system voltages of 120, 120/240, 208Y/120, 240, 347, 480Y/277, 480, 600Y/347, and 600 volts shall be used.

(B) Fractions of an Ampere Where calculations result in a fraction of an ampere that is less than 0.5, such fractions shall be permitted to be dropped.

For uniform calculation of load, nominal voltages, as listed in 220.5(A), are required to be used in computing the ampere load on the conductors. To select conductor sizes, refer to 310.15(A) and 310.15(B).

Loads are computed on the basis of volt-amperes (VA) or kilovolt-amperes (kVA), rather than watts or kilowatts (kW), to calculate the true ampere values. However, the rating of equipment is given in watts or kilowatts for noninductive loads. Such ratings are considered to be the equivalent of the same rating in volt-amperes or kilovolt-amperes.

Table 220.3 Additional Load Calculation References

Calculation	Article	Section (or Part)
Air-Conditioning and Refrigerating Equipment, Branch-Circuit Conductor Sizing	440	Part IV
Cranes and Hoists, Rating and Size of Conductors	610	610.14
Electric Welders, ampacity calculations	630	630.11, 630.31
Electrically Driven or Controlled Irrigation Machines	675	675.7(A), 675.22(A)
Electrolytic Cell Lines	668	668.3(C)
Electroplating, Branch-Circuit Conductor Sizing	669	669.5
Elevator Feeder Demand Factors	620	620.14
Fire Pumps, Voltage Drop (mandatory calculation)	695	695.7
Fixed Electric Heating Equipment for Pipelines and Vessels, Branch-Circuit Sizing	427	427.4
Fixed Electric Space Heating Equipment, Branch-Circuit Sizing	424	424.3
Fixed Outdoor Electric Deicing and Snow-Melting Equipment, Branch-Circuit Sizing	426	426.4
Industrial Machinery, Supply Conductor Sizing	670	670.4(A)
Marinas and Boatyards, Feeder and Service Load Calculations	555	555.12
Mobile Homes, Manufactured Homes, and Mobile Home Parks, Total Load for Determining Power Supply	550	550.18(B)
Mobile Homes, Manufactured Homes, and Mobile Home Parks, Allowable Demand Factors for Park Electrical Wiring Systems	550	550.31
Motion Picture and Television Studios and Similar Locations – Sizing of Feeder Conductors for Television Studio Sets	530	530.19
Motors, Feeder Demand Factor	430	430.26
Motors, Multimotor and Combination-Load Equipment	430	430.25
Motors, Several Motors or a Motor(s) and Other Load(s)	430	430.24
Over 600 Volt Branch Circuit Calculations	210	210.19(B)
Over 600 Volt Feeder Calculations	215	215.2(B)
Phase Converters, Conductors	455	455.6
Recreational Vehicle Parks, Basis of Calculations	551	551.73(A)
Sensitive Electrical Equipment, Voltage Drop (mandatory calculation)	647	647.4(D)
Solar Photovoltaic Systems, Circuit Sizing and Current	690	690.8
Storage-Type Water Heaters	422	422.11(E)
Theaters, Stage Switchboard Feeders	520	520.27

See, for example, 220.55. This concept recognizes that load calculations determine conductor and circuit sizes, that the power factor of the load is often unknown, and that the conductor "sees" the circuit volt-amperes only, not the circuit power (watts).

See Examples D1(a) through D5(b) in Annex D. The results of these examples are generally expressed in amperes. Unless the computations result in a major fraction of an ampere (0.5 or larger), such fractions (less than 0.5) may be dropped, in accordance with 220.5(B).

II. Branch Circuit Load Calculations

220.10 General

Branch-circuit loads shall be calculated as shown in 220.12, 220.14, and 220.16.

220.12 Lighting Load for Specified Occupancies

A unit load of not less than that specified in Table 220.12 for occupancies specified therein shall constitute the minimum lighting load. The floor area for each floor shall be calculated from the outside dimensions of the building, dwelling unit, or other area involved. For dwelling units, the calculated floor area shall not include open porches, garages, or unused or unfinished spaces not adaptable for future use.

> FPN: The unit values herein are based on minimum load conditions and 100 percent power factor and may not provide sufficient capacity for the installation contemplated.

General lighting loads determined by 220.12 are in fact minimum lighting loads, and there are no exceptions to these requirements. Therefore, energy saving–type calculations are not permitted to be used to determine the minimum calculated lighting load if they produce loads less than the load calculated according to 220.12. On the other hand, energy saving–type calculations can be a useful tool to reduce the connected lighting load and actual power consumption.

Examples of unused or unfinished spaces for dwelling units are some attics, cellars, and crawl spaces.

220.14 Other Loads — All Occupancies

In all occupancies, the minimum load for each outlet for general-use receptacles and outlets not used for general illumination shall not be less than that calculated in 220.14(A) through (L), the loads shown being based on nominal branch-circuit voltages.

Exception: The loads of outlets serving switchboards and switching frames in telephone exchanges shall be waived from the calculations.

Table 220.12 General Lighting Loads by Occupancy

Type of Occupancy	Unit Load	
	Volt-Amperes per Square Meter	Volt-Amperes per Square Foot
Armories and auditoriums	11	1
Banks	39[b]	3½[b]
Barber shops and beauty parlors	33	3
Churches	11	1
Clubs	22	2
Court rooms	22	2
Dwelling units[a]	33	3
Garages — commercial (storage)	6	½
Hospitals	22	2
Hotels and motels, including apartment houses without provision for cooking by tenants[a]	22	2
Industrial commercial (loft) buildings	22	2
Lodge rooms	17	1½
Office buildings	39[b]	3½[b]
Restaurants	22	2
Schools	33	3
Stores	33	3
Warehouses (storage)	3	¼
In any of the preceding occupancies except one-family dwellings and individual dwelling units of two-family and multifamily dwellings:		
Assembly halls and auditoriums	11	1
Halls, corridors, closets, stairways	6	½
Storage spaces	3	¼

[a]See 220.14(J).
[b]See 220.14(K).

(A) Specific Appliances or Loads An outlet for a specific appliance or other load not covered in 220.14(B) through (L) shall be calculated based on the ampere rating of the appliance or load served.

(B) Electric Dryers and Household Electric Cooking Appliances Load calculations shall be permitted as specified in 220.54 for electric dryers and in 220.55 for electric ranges and other cooking appliances.

(C) Motor Loads Outlets for motor loads shall be calculated in accordance with the requirements in 430.22, 430.24, and 440.6.

(D) Luminaires (Lighting Fixtures) An outlet supplying luminaire(s) [lighting fixture(s)] shall be calculated based on the maximum volt-ampere rating of the equipment and lamps for which the luminaire(s) [fixture(s)] is rated.

In general, no additional calculation is required for luminaires (recessed and surface mounted) installed in or on a dwelling unit, because the load of such luminaires is covered in the 3 volt-amperes per square foot calculation specified by Table 220.12. Where the rating of the luminaires installed for general lighting exceeds the minimum load provided for in Table 220.12, the minimum general lighting load for that premises is to be based on the installed luminaires. Distinguishing between the luminaires installed for general lighting versus those installed for accent, specialty or display lighting is much easier to delineate in commercial (particularly mercantile) occupancies.

(E) Heavy-Duty Lampholders Outlets for heavy-duty lampholders shall be calculated at a minimum of 600 volt-amperes.

(F) Sign and Outline Lighting Sign and outline lighting outlets shall be calculated at a minimum of 1200 volt-amperes for each required branch circuit specified in 600.5(A).

Section 220.14(F) assigns 1200 volt-amperes as a minimum circuit load for the signs and outline lighting outlets required by 600.5(A). If the specific load is known to be larger, then, according to 220.14, the actual load is used for calculation purposes.

(G) Show Windows Show windows shall be calculated in accordance with either of the following:

(1) The unit load per outlet as required in other provisions of this section
(2) At 200 volt-amperes per 300 mm (1 ft) of show window

The following two options are permitted for the load calculations for branch circuits serving show windows:

1. 180 volt-amperes per receptacle according to 210.62, which requires one receptacle per 12 linear ft
2. 200 volt-amperes per linear foot of show-window space

As shown in Exhibit 220.1, the linear-foot calculation method is permitted in lieu of the specified unit load per outlet for branch circuits serving show windows.

(H) Fixed Multioutlet Assemblies Fixed multioutlet assemblies used in other than dwelling units or the guest rooms or guest suites of hotels or motels shall be calculated in accordance with (H)(1) or (H)(2). For the purposes of this

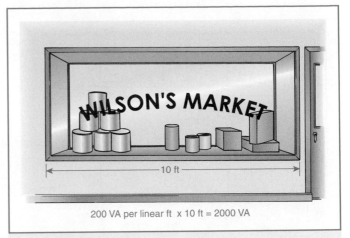

200 VA per linear ft x 10 ft = 2000 VA

Exhibit 220.1 An example of the linear-foot load calculation for branch circuits serving a show window.

section, the calculation shall be permitted to be based on the portion that contains receptacle outlets.

(1) Where appliances are unlikely to be used simultaneously, each 1.5 m (5 ft) or fraction thereof of each separate and continuous length shall be considered as one outlet of not less than 180 volt-amperes.
(2) Where appliances are likely to be used simultaneously, each 300 mm (1 ft) or fraction thereof shall be considered as an outlet of not less than 180 volt-amperes.

Fixed multioutlet assemblies are commonly used in commercial and industrial locations. The use of multioutlet assemblies is divided into two broad areas. The first area of use is light use, which means that not all the cord-connected equipment is expected to be used at the same time, as noted in 220.14(H)(1). An example of light use is a workbench area where one worker uses one electrical tool at a time. The second area of use is heavy use, which is characterized by all the cord-connected equipment generally operating at the same time, as noted in 220.14(H)(2). An example of heavy use is a retail outlet displaying television sets, where most, if not all, sets are operating simultaneously.

As shown in Exhibit 220.2, the requirement of 220.14(H)(1) states that each 5 ft of a fixed multioutlet assembly must be considered as one outlet rated 180 volt-amperes. The requirement of 220.14(H)(2) states that where appliances are likely to be used simultaneously, each foot of multioutlet assembly is to be considered as one outlet rated 180 volt-amperes.

(I) Receptacle Outlets Except as covered in 220.14(J) and (K), receptacle outlets shall be calculated at not less than 180 volt-amperes for each single or for each multiple receptacle on one yoke. A single piece of equipment consisting of

Exhibit 220.2 The requirements of 220.14(H)(1) and (H)(2) as applied to fixed multioutlet assemblies.

Exhibit 220.3 The load requirement of 180 volt-amperes per 220.14(I) as applied to single- and multiple-receptacle outlets on single straps and the load of 360 volt-amperes applied to each receptacle that consists of four receptacles.

a multiple receptacle comprised of four or more receptacles shall be calculated at not less than 90 volt-amperes per receptacle. This provision shall not be applicable to the receptacle outlets specified in 210.11(C)(1) and (C)(2).

As illustrated in Exhibit 220.3, the load of 180 volt-amperes is applied to single and multiple receptacles mounted on a single yoke or strap, and a load of 360 volt-amperes is applied to each receptacle that consists of four receptacles. These are considered receptacle outlets, in accordance with 220.14(I). The receptacle outlets are not the lighting outlets installed for general illumination or the small-appliance branch circuits, as indicated in 220.14(J). The receptacle load for outlets for general illumination in one- and two-family and multifamily dwellings and in guest rooms of hotels and motels is included in the general lighting load value assigned by Table 220.12. The load requirement for the small-appliance branch circuits is 1500 volt-amperes per circuit, as described in 220.52(A).

Note in Exhibit 220.3 that the last outlet of the top circuit consists of two duplex receptacles on separate straps. That outlet is calculated at 360 volt-amperes because each duplex receptacle is on one yoke. The multiple receptacle supplied from the bottom circuit in the exhibit, which comprises four receptacles, is calculated at 90 volt-amperes per receptacle (4 × 90 VA = 360 VA). For example, single-strap and multiple-receptacle devices are calculated as follows:

Device	Computed Load
Duplex receptacle	180 VA
Triplex receptacle	180 VA
Double duplex receptacle	360 VA (180 × 2)
Quad or four-plex-type receptacle	360 VA (90 × 4)

A load of 180 volt-amperes is not required to be considered for outlets supplying recessed lighting fixtures, lighting outlets for general illumination, and small-appliance branch circuits. To apply the requirement of 180 volt-amperes in those cases would be unrealistic, because it would unnecessarily restrict the number of lighting or receptacle outlets on branch circuits in dwelling units. See the note below Table 220.12 that references 220.14(J). This note indicates that the requirement of 180 volt-amperes does not apply to most receptacle outlets in dwellings.

In Exhibit 220.4, the maximum number of outlets permitted on 15- and 20-ampere branch circuits is 10 and 13

Exhibit 220.4 Maximum number of outlets permitted on 15- and 20-ampere branch circuits.

outlets, respectively. This restriction does not apply to outlets connected to general lighting or small-appliance branch circuits in dwelling units.

(J) Dwelling Occupancies In one-family, two-family, and multifamily dwellings and in guest rooms or guest suites of hotels and motels, the outlets specified in (J)(1), (J)(2), and (J)(3) are included in the general lighting load calculations of 220.12. No additional load calculations shall be required for such outlets.

(1) All general-use receptacle outlets of 20-ampere rating or less, including receptacles connected to the circuits in 210.11(C)(3)
(2) The receptacle outlets specified in 210.52(E) and (G)
(3) The lighting outlets specified in 210.70(A) and (B)

(K) Banks and Office Buildings In banks or office buildings, the receptacle loads shall be calculated to be the larger of (1) or (2):

(1) The computed load from 220.14(I)
(2) 11 volt-amperes/m^2 or 1 volt-ampere/ft^2

(L) Other Outlets Other outlets not covered in 220.14(A) through (K) shall be calculated based on 180 volt-amperes per outlet.

220.16 Loads for Additions to Existing Installations

(A) Dwelling Units Loads added to an existing dwelling unit(s) shall comply with the following as applicable:

(1) Loads for structural additions to an existing dwelling unit or for a previously unwired portion of an existing dwelling unit, either of which exceeds 46.5 m^2 (500 ft^2), shall be calculated in accordance with 220.12 and 220.14.
(2) Loads for new circuits or extended circuits in previously wired dwelling units shall be calculated in accordance with either 220.12 or 220.14, as applicable.

(B) Other Than Dwelling Units Loads for new circuits or extended circuits in other than dwelling units shall be calculated in accordance with either 220.12 or 220.14, as applicable.

220.18 Maximum Loads

The total load shall not exceed the rating of the branch circuit, and it shall not exceed the maximum loads specified in 220.18(A) through (C) under the conditions specified therein.

(A) Motor-Operated and Combination Loads Where a circuit supplies only motor-operated loads, Article 430 shall

apply. Where a circuit supplies only air-conditioning equipment, refrigerating equipment, or both, Article 440 shall apply. For circuits supplying loads consisting of motor-operated utilization equipment that is fastened in place and has a motor larger than 1/8 hp in combination with other loads, the total calculated load shall be based on 125 percent of the largest motor load plus the sum of the other loads.

(B) Inductive Lighting Loads For circuits supplying lighting units that have ballasts, transformers, or autotransformers, the calculated load shall be based on the total ampere ratings of such units and not on the total watts of the lamps.

(C) Range Loads It shall be permissible to apply demand factors for range loads in accordance with Table 220.55, including Note 4.

III. Feeder and Service Load Calculations

220.40 General

The calculated load of a feeder or service shall not be less than the sum of the loads on the branch circuits supplied, as determined by Part II of this article, after any applicable demand factors permitted by Parts III or IV or required by Part V have been applied.

> FPN: See Examples D1(A) through D10 in Annex D. See 220.18(B) for the maximum load in amperes permitted for lighting units operating at less than 100 percent power factor.

In the example shown in Exhibit 220.5, each panel serves a computed load of 80 amperes. The main feeder is sized to carry the total computed load of 240 amperes (3 × 80

Exhibit 220.5 Feeder conductors sized in accordance with 220.40.

amperes). The feeder tap conductors from the meter enclosure to the panelboards are sized to supply a computed load of 80 amperes. The main feeder is not intended to be sized to carry 300 amperes based on the sum of the panelboards.

See Exhibit 230.13 for a similar example for service conductors. The ungrounded service conductors are no longer required to be sized for the sum of the main overcurrent device rating of 300 amperes. Service conductors are required to have sufficient ampacity to carry the loads calculated in accordance with Article 220, with the appropriate demand factors applied. See 230.23, 230.31, and 230.42 for specifics on size and rating of service conductors.

Part III of Article 220 contains the requirements for calculating feeder and service loads. Part IV provides optional methods for calculating feeder and service loads in dwelling units and multifamily dwellings.

Except as permitted in 240.4 and 240.6, the rating of the overcurrent device cannot exceed the final ampacity of the circuit conductors after all the correction and adjustment factors have been applied, such as where the ambient temperature, the number of current-carrying conductors, or both exceed the parameters on which the allowable ampacity table values are based.

Example

Determine the minimum-size overcurrent protective device (OCPD) and the minimum conductor size for a feeder circuit with the following characteristics:

- 3-phase, 4-wire feeder (full-size neutral)
- 125-ampere noncontinuous load
- 200-ampere continuous load
- 75°C overcurrent device terminal rating
- Type THWN insulated conductors
- Four current-carrying conductors in a raceway
- A load the majority of which is nonlinear

Solution

STEP 1. Select the feeder OCPD rating by first totaling the continuous and noncontinuous loads according to 215.3:

$$\text{OCPD rating} = 125\% \text{ of continuous load} + \text{noncontinuous load}$$
$$= (200\text{ A} \times 1.25) + 125\text{ A}$$
$$= 250\text{ A} + 125\text{ A}$$
$$= 375\text{ A}$$

Using 240.4(B) and 240.6(A), adjust the minimum standard-size OCPD to 400 amperes.

STEP 2. Select the feeder conductor size before derating by first summing the continuous and noncontinuous loads according to 215.2(A)(1).

$$\text{Feeder size} \atop \text{(before derating)} = {125\% \text{ of continuous load} \atop + \text{ noncontinuous load}}$$
$$= (200\text{ A} \times 1.25) + 125\text{ A}$$
$$= 250\text{ A} + 125\text{ A}$$
$$= 375\text{ A}$$

Using Table 310.16 and using the 75°C column (because of the overcurrent device terminal), the minimum-size Type THWN copper conductor that can supply a calculated load of 375 amperes is 500-kcmil copper, which has an ampacity of 380 amperes.

STEP 3. Apply the derating factors to the feeder conductor size. Section 310.15(B)(4)(c) requires that the neutral conductor be counted as a current-carrying conductor because a major portion of the load consists of fluorescent and high-intensity discharge (HID) luminaires. Therefore, this feeder circuit consists of four current-carrying conductors in the same raceway. Section 310.15(B)(2) requires an 80-percent adjustment factor for four current-carrying conductors in the same raceway. According to Table 310.16, 500-kcmil, Type THWN conductors have an ampacity of 380 amperes. The adjustment factors are applied to this ampacity as follows:

$$\text{Adjusted ampacity} = \text{Table ampacity} \times \text{adjustment factor}$$
$$= 380\text{ A} \times 0.80$$
$$= 304\text{ A}$$

According to 240.4(B) and 240.6(A), a conductor with a calculated ampacity of 304 amperes is not permitted to be protected by a 400-ampere OCPD. Therefore, the 500-kcmil, Type THWN copper conductor cannot be used.

STEP 4. Revise the feeder conductor selection and perform a check. The next standard-size conductor listed in Table 310.16 is a 600-kcmil copper conductor in the 90°C column. If higher-temperature insulations are used, adjustment factors can be applied to the higher ampacity. Because Type THWN is a 75°C insulation, a 90°C Type THHN is selected. If a 600-kcmil Type THHN copper conductor is used, perform the check as follows. According to Table 310.16, a 600-kcmil conductor has an ampacity of 475 amperes in the Type THHN 90°C column:

$$\text{Adjusted ampacity} = \text{table ampacity} \times \text{adjustment factor}$$
$$= 475\text{ A} \times 0.80$$
$$= 380\text{ A}$$

A conductor with a calculated ampacity of 380 amperes is allowed to be protected by a 400-ampere OCPD, in accordance with 240.4(B). However, because the OCPD terminations are rated 75°C, the load current cannot exceed the ampacity of a 600-kcmil conductor in the 75°C column of Table 310.16, which has a value of 420 amperes.

STEP 5. Evaluate the circuit. The calculation in Step 4 results in four 600-kcmil Type THHN copper conductors in one raceway, each with an ampacity of 380 amperes, supplying a 375-ampere continuous load and protected by a 400-ampere OCPD. It is important to note here that a 90°C, 600-kcmil copper conductor with a final calculated ampacity of 380 amperes is permitted to terminate on a 75°C-rated terminal, according to 110.14(C)(1)(b)(2).

220.42 General Lighting

The demand factors specified in Table 220.42 shall apply to that portion of the total branch-circuit load calculated for general illumination. They shall not be applied in determining the number of branch circuits for general illumination.

Table 220.42 Lighting Load Demand Factors

Type of Occupancy	Portion of Lighting Load to Which Demand Factor Applies (Volt-Amperes)	Demand Factor (Percent)
Dwelling units	First 3000 or less at	100
	From 3001 to 120,000 at	35
	Remainder over 120,000 at	25
Hospitals*	First 50,000 or less at	40
	Remainder over 50,000 at	20
Hotels and motels, including apartment houses without provision for cooking by tenants*	First 20,000 or less at	50
	From 20,001 to 100,000 at	40
	Remainder over 100,000 at	30
Warehouses (storage)	First 12,500 or less at	100
	Remainder over 12,500 at	50
All others	Total volt-amperes	100

*The demand factors of this table shall not apply to the calculated load of feeders or services supplying areas in hospitals, hotels, and motels where the entire lighting is likely to be used at one time, as in operating rooms, ballrooms, or dining rooms.

220.43 Show-Window and Track Lighting

(A) Show Windows For show-window lighting, a load of not less than 660 volt-amperes/linear meter or 200 volt-amperes/linear foot shall be included for a show window, measured horizontally along its base.

> FPN: See 220.14(G) for branch circuits supplying show windows.

The calculation of 200 volt-amperes for each linear foot of a show window is required to determine the feeder load. See the commentary following 220.14(G) for load calculations for branch circuits in show windows.

(B) Track Lighting For track lighting in other than dwelling units or guest rooms or guest suites of hotels or motels, an additional load of 150 volt-amperes shall be included for every 600 mm (2 ft) of lighting track or fraction thereof. Where multicircuit track is installed, the load shall be considered to be divided equally between the track circuits.

Example

A lighting plan shows 62.5 linear ft of single-circuit track lighting for a small department store featuring clothing. Because the actual track lighting fixtures are owner supplied, neither the quantity of track lighting fixtures nor the lamp size is specified. What is the minimum calculated load associated with the track lighting that must be added to the service or feeder supplying this store?

Solution

According to 220.43(B), the minimum calculated load to be added to the service or feeder supplying this track light installation is calculated as follows:

$$\frac{62.5 \text{ ft}}{2 \text{ ft}} = 31.25, \text{ rounded up to } 32$$

$$32 \times 150 \text{ VA} = 4800 \text{ VA}$$

Thus, the minimum load that must be added to the service or feeder calculation is 4800 volt-amperes.

It is important to note that the branch circuits supplying this installation are covered in 410.101(B). For the track lighting branch-circuit load, the maximum load on the track cannot exceed the rating of the branch circuit supplying the track. Also, the track must be supplied by a branch circuit that has a rating not exceeding the rating of the track. The track length does not enter into the branch-circuit calculation.

Section 220.43(B) is not intended to limit the number of feet of track on a single branch circuit, nor is it intended to limit the number of fixtures on an individual track. Rather, 220.43(B) is meant to be used solely for load calculations of feeders and services.

220.44 Receptacle Loads — Other Than Dwelling Units

Receptacle loads calculated in accordance with 220.14(H) and (I) shall be permitted to be made subject to the demand factors given in Table 220.42 or Table 220.44.

Table 220.44 Demand Factors for Non-dwelling Receptacle Loads

Portion of Receptacle Load to Which Demand Factor Applies (Volt-Amperes)	Demand Factor (Percent)
First 10 kVA or less at	100
Remainder over 10 kVA at	50

Section 220.44 permits receptacle loads, calculated at not more than 180 volt-amperes per strap, to be computed by either of the following methods:

1. The receptacle loads are added to the lighting load. The demand factors (if applicable) in Table 220.12 are then applied to the combined load.
2. The receptacle loads are calculated (without the lighting load) with demand factors from Table 220.44 applied.

220.50 Motors

Motor loads shall be calculated in accordance with 430.24, 430.25, and 430.26 and with 440.6 for hermetic refrigerant motor compressors.

220.51 Fixed Electric Space Heating

Fixed electric space heating loads shall be calculated at 100 percent of the total connected load. However, in no case shall a feeder or service load current rating be less than the rating of the largest branch circuit supplied.

Exception: Where reduced loading of the conductors results from units operating on duty-cycle, intermittently, or from all units not operating at the same time, the authority having jurisdiction may grant permission for feeder and service conductors to have an ampacity less than 100 percent, provided the conductors have an ampacity for the load so determined.

220.52 Small Appliance and Laundry Loads — Dwelling Unit

(A) Small Appliance Circuit Load In each dwelling unit, the load shall be calculated at 1500 volt-amperes for each 2-wire small-appliance branch circuit required by 210.11(C)(1). Where the load is subdivided through two or more feeders, the calculated load for each shall include not less than 1500 volt-amperes for each 2-wire small-appliance branch circuit. These loads shall be permitted to be included with the general lighting load and subjected to the demand factors provided in Table 220.42.

Exception: The individual branch circuit permitted by 210.52(B)(1), Exception No. 2, shall be permitted to be excluded from the calculation required by 220.52.

See the commentary following 210.52(B) regarding required receptacle outlets for small-appliance branch circuits.

(B) Laundry Circuit Load A load of not less than 1500 volt-amperes shall be included for each 2-wire laundry branch circuit installed as required by 210.11(C)(2). This load shall be permitted to be included with the general lighting load and subjected to the demand factors provided in Table 220.42.

In each dwelling unit, the feeder load is required to be calculated at 1500 volt-amperes for each of the two or more (2-wire) small-appliance branch circuits and at 1500 volt-amperes for each (2-wire) laundry branch circuit. Where additional small-appliance and laundry branch circuits are provided, they also are calculated at 1500 volt-amperes per circuit. These loads are permitted to be totaled and then added to the general lighting load. The demand factors in Table 220.42 can then be applied to the combined total load of the small-appliance branch circuits, the laundry branch circuit, and the general lighting from Table 220.12.

220.53 Appliance Load — Dwelling Unit(s)

It shall be permissible to apply a demand factor of 75 percent to the nameplate rating load of four or more appliances fastened in place, other than electric ranges, clothes dryers, space-heating equipment, or air-conditioning equipment, that are served by the same feeder or service in a one-family, two-family, or multifamily dwelling.

For appliances fastened in place (other than ranges, clothes dryers, and space-heating and air-conditioning equipment), feeder capacity must be provided for the sum of these loads; for a total load of four or more such appliances, a demand factor of 75 percent may be applied. See Table 430.248 for the full-load current, in amperes, for single-phase ac motors, in accordance with 220.50.

Example

Determine the feeder capacity needed for a 120/240-volt fastened-in-place appliance load in a dwelling unit for the following:

Appliance	Rating	Load
Water heater	4000 W, 240 V	4000 VA
Kitchen disposal	1/2 hp, 120 V	1176 VA
Dishwasher	1200 W, 120 V	1200 VA
Furnace motor	1/4 hp, 120 V	696 VA
Attic fan	1/4 hp, 120 V	696 VA
Water pump	1/2 hp, 240 V	1176 VA

Solution

STEP 1. Calculate the total of the six fastened-in-place appliances:

$$\text{Total load} = 4000 \text{ VA} + 1176 \text{ VA} + 1200 \text{ VA}$$
$$+ 697 \text{ VA} + 696 \text{ VA} + 1176 \text{ VA}$$
$$= 8944 \text{ VA}$$

STEP 2. Because the load is for more than four appliances, apply a demand factor of 75 percent:

$$8944 \text{ VA} \times 0.75 = 6708 \text{ VA}$$

Thus, 6708 volt-amperes is the load to be added to the other determined loads for calculating the size of service and feeder conductors.

220.54 Electric Clothes Dryers — Dwelling Unit(s)

The load for household electric clothes dryers in a dwelling unit(s) shall be either 5000 watts (volt-amperes) or the nameplate rating, whichever is larger, for each dryer served. The use of the demand factors in Table 220.54 shall be permitted. Where two or more single-phase dryers are supplied by a

Table 220.54 Demand Factors for Household Electric Clothes Dryers

Number of Dryers	Demand Factor(Percent)
1–4	100%
5	85%
6	75%
7	65%
8	60%
9	55%
10	50%
11	47%
12–22	% = 47 − (number of dryers − 11)
23	35%
24–42	% = 35 − [0.5 × (number of dryers − 23)]
43 and over	25%

The exact method of calculation presented in Table 220.54 was revised in the 2002 *Code* to produce a more accurate load for all quantities of dryers. To calculate the load of household electric dryers, 220.54 specifies a minimum demand of 5 kVA for the calculation of feeder conductors. If the nameplate rating is known and exceeds 5 kW, the larger rating is applied.

3-phase, 4-wire feeder or service, the total load shall be calculated on the basis of twice the maximum number connected between any two phases.

220.55 Electric Ranges and Other Cooking Appliances — Dwelling Unit(s)

The load for household electric ranges, wall-mounted ovens, counter-mounted cooking units, and other household cooking appliances individually rated in excess of 1¾ kW shall be permitted to be calculated in accordance with Table 220.55. Kilovolt-amperes (kVA) shall be considered equivalent to kilowatts (kW) for loads calculated under this section.

Where two or more single-phase ranges are supplied by a 3-phase, 4-wire feeder or service, the total load shall be calculated on the basis of twice the maximum number connected between any two phases.

> FPN No. 1: See Example D5(A) in Annex D.
> FPN No. 2: See Table 220.56 for commercial cooking equipment.
> FPN No. 3: See the examples in Annex D.

220.56 Kitchen Equipment — Other Than Dwelling Unit(s)

It shall be permissible to calculate the load for commercial electric cooking equipment, dishwasher booster heaters, water heaters, and other kitchen equipment in accordance with Table 220.56. These demand factors shall be applied to all equipment that has either thermostatic control or intermittent use as kitchen equipment. These demand factors shall not apply to space-heating, ventilating, or air-conditioning equipment.

However, in no case shall the feeder or service calculated load be less than the sum of the largest two kitchen equipment loads.

220.60 Noncoincident Loads

Where it is unlikely that two or more noncoincident loads will be in use simultaneously, it shall be permissible to use only the largest load(s) that will be used at one time for calculating the total load of a feeder or service.

220.61 Feeder or Service Neutral Load

(A) Basic Calculation The feeder or service neutral load shall be the maximum unbalance of the load determined by this article. The maximum unbalanced load shall be the maximum net calculated load between the neutral and any one ungrounded conductor.

Exception: For 3-wire, 2-phase or 5-wire, 2-phase systems, the maximum unbalanced load shall be the maximum net calculated load between the neutral and any one ungrounded conductor multiplied by 140 percent.

(B) Permitted Reductions A service or feeder supplying the following loads shall be permitted to have an additional

Table 220.55 Demand Factors and Loads for Household Electric Ranges, Wall-Mounted Ovens, Counter-Mounted Cooking Units, and Other Household Cooking Appliances over 1¾ kW Rating (Column C to be used in all cases except as otherwise permitted in Note 3.)

Number of Appliances	Demand Factor (Percent) (See Notes)		Column C Maximum Demand (kW) (See Notes) (Not over 12 kW Rating)
	Column A (Less than 3½ kW Rating)	Column B (3½ kW to 8¾ kW Rating)	
1	80	80	8
2	75	65	11
3	70	55	14
4	66	50	17
5	62	45	20
6	59	43	21
7	56	40	22
8	53	36	23
9	51	35	24
10	49	34	25
11	47	32	26
12	45	32	27
13	43	32	28
14	41	32	29
15	40	32	30
16	39	28	31
17	38	28	32
18	37	28	33
19	36	28	34
20	35	28	35
21	34	26	36
22	33	26	37
23	32	26	38
24	31	26	39
25	30	26	40
26–30	30	24	15 kW + 1 kW for each range
31–40	30	22	
41–50	30	20	25 kW + ¾ kW for each range
51–60	30	18	
61 and over	30	16	

1. Over 12 kW through 27 kW ranges all of same rating. For ranges individually rated more than 12 kW but not more than 27 kW, the maximum demand in Column C shall be increased 5 percent for each additional kilowatt of rating or major fraction thereof by which the rating of individual ranges exceeds 12 kW.

For household electric ranges and other cooking appliances, the size of the conductors must be determined by the rating of the range. According to Table 220.55, for one range rated 12 kW or less, the maximum demand load is 8 kW (8 kVA per 220.55), and 8 AWG copper conductors with 60°C insulation would suffice. Note that 210.19(A)(3) does not permit the branch-circuit rating of a circuit supplying household ranges with a nameplate rating of 8¾ kW to be less than 40 amperes.

2. Over 8¾ kW through 27 kW ranges of unequal ratings. For ranges individually rated more than 8¾ kW and of different ratings, but none exceeding 27 kW, an average value of rating shall be calculated by adding together the ratings of all ranges to obtain the total connected load (using 12 kW for any range rated less than 12 kW) and dividing by the total number of ranges. Then the maximum demand in Column C shall be increased 5 percent for each kilowatt or major fraction thereof by which this average value exceeds 12 kW.

Table 220.55 *Continued*

Note 2 to Table 220.55 provides for ranges larger than 8¾ kW. Note 4 covers installations where the circuit supplies multiple cooking components, which are combined and treated as a single range.

3. Over 1¾ kW through 8¾ kW. In lieu of the method provided in Column C, it shall be permissible to add the nameplate ratings of all household cooking appliances rated more than 1¾ kW but not more than 8¾ kW and multiply the sum by the demand factors specified in Column A or B for the given number of appliances. Where the rating of cooking appliances falls under both Column A and Column B, the demand factors for each column shall be applied to the appliances for that column, and the results added together.

The branch-circuit load for one range is permitted to be computed by using either the nameplate rating of the appliance or Table 220.55. If a single branch circuit supplies a counter-mounted cooking unit and not more than two wall-mounted ovens, all of which are located in the same room, the nameplate ratings of these appliances can be added and the total treated as the equivalent of one range, according to Note 4 of Table 220.55.

Example

Calculate the load for a single branch circuit that supplies the following cooking units:

- One counter-mounted cooking unit with rating of 8 kW
- One wall-mounted oven with rating of 7 kW
- A second wall-mounted oven with rating of 6 kW

Solution

STEP 1. The combined cooking appliances can be treated as one range, according to Note 4 of Table 220.55. In Table 220.55, find the maximum demand for one range not over 12 kW, which is 8 kW (from Column C).

STEP 2. According to Note 1 in Table 220.55, for ranges that are over 12 kW but not more than 27 kW, the maximum demand in Column C (8 kW) is increased 5 percent for each kW that exceeds 12 kW. Determine the additional kilowatts:

$$\text{Combined unit rating} = 8\text{ kW} + 7\text{ kW} + 6\text{ kW}$$
$$= 21\text{ kW}$$
$$\text{Additional kW} = 21\text{ kW} - 12\text{ kW} = 9\text{ kW}$$

STEP 3. Calculate by how much the maximum load in Column C in Table 220.55 must be increased for the combined appliances:

$$\text{Increase} = 5\%\text{ per kW} \times 9\text{ kW} = 45\%$$
$$= 0.45 \times 8\text{ kW} = 3.6\text{ kW}$$

STEP 4. Calculate the total load in amperes, as follows:

$$\text{Total load} = 8\text{ kW} + 3.6\text{ kW} = 11.6\text{ kW}$$
$$= 11.6\text{ kW} = 11,600\text{ W} = 11,600\text{ VA}$$
$$= \frac{11,600\text{ VA}}{240\text{V}}$$
$$= 48.3\text{ A}$$

4. Branch-Circuit Load. It shall be permissible to calculate the branch-circuit load for one range in accordance with Table 220.55. The branch-circuit load for one wall-mounted oven or one counter-mounted cooking unit shall be the nameplate rating of the appliance. The branch-circuit load for a counter-mounted cooking unit and not more than two wall-mounted ovens, all supplied from a single branch circuit and located in the same room, shall be calculated by adding the nameplate rating of the individual appliances and treating this total as equivalent to one range.

5. This table also applies to household cooking appliances rated over 1¾ kW and used in instructional programs.

The nameplate ratings of all household cooking appliances rated more than 1¾ kW but not more than 8¾ kW may be added and the sum multiplied by the demand factor specified in Column A or B of Table 220.55 for the given number of appliances. For feeder demand factors for other than dwelling units—that is, commercial electric cooking equipment, dishwasher booster heaters, water heaters, and so on—see Table 220.56.

The demand factors in the *Code* are based on the diversified use of household appliances, because it is unlikely that all appliances will be used simultaneously or that all cooking units and the oven of a range will be at maximum heat for any length of time.

Table 220.56 Demand Factors for Kitchen Equipment — Other Than Dwelling Unit(s)

Number of Units of Equipment	Demand Factor (Percent)
1	100
2	100
3	90
4	80
5	70
6 and over	65

demand factor of 70 percent applied to the amount in 220.61(B)(1) or portion of the amount in 220.61(B)(2) determined by the basic calculation:

(1) A feeder or service supplying household electric ranges, wall-mounted ovens, counter-mounted cooking units, and electric dryers, where the maximum unbalanced load has been determined in accordance with Table 220.55 for ranges and Table 220.54 for dryers

(2) That portion of the unbalanced load in excess of 200 amperes where the feeder or service is supplied from a 3-wire dc or single-phase ac system, or a 4-wire, 3-phase; 3-wire, 2-phase system, or a 5-wire, 2-phase system

(C) Prohibited Reductions There shall be no reduction of the neutral or grounded conductor capacity applied to the amount in 220.61(C)(1), or portion of the amount in (C)(2), from that determined by the basic calculation:

(1) Any portion of a 3-wire circuit consisting of 2-phase wires and the neutral of a 4-wire, 3-phase, wye-connected system

(2) That portion consisting of nonlinear loads supplied from a 4-wire, wye-connected, 3-phase system

> FPN No. 1: See Examples D1(A), D1(B), D2(B), D4(A), and D5(A) in Annex D.
> FPN No. 2: A 3-phase, 4-wire, wye-connected power system used to supply power to nonlinear loads may necessitate that the power system design allow for the possibility of high harmonic neutral currents.

Section 220.61, which describes the basis for calculating the neutral load of feeders or services as the maximum unbalanced load that can occur between the neutral and any other ungrounded conductor, has been revised for the 2005 *Code* to provide a better organized approach to the calculation requirements.

For a household electric range or clothes dryer, the maximum unbalanced load may be assumed to be 70 percent, so the neutral may be sized on that basis. Section 220.61(B) permits the reduction of the feeder neutral conductor size

under specific conditions of use, and 220.61(C)(1) and 220.61(C)(2) cite a circuit arrangement and a load characteristic as applications where it is not permitted to reduce the capacity of a neutral or grounded conductor of a feeder or service.

If the system also supplies nonlinear loads such as electric-discharge lighting, including fluorescent and HID, or data-processing or similar equipment, the neutral is considered a current-carrying conductor if the load of the electric-discharge lighting, data-processing, or similar equipment on the feeder neutral consists of more than half the total load, in accordance with 310.15(B)(4)(c). Electric-discharge lighting and data-processing equipment may have harmonic currents in the neutral that may exceed the load current in the ungrounded conductors. It would be appropriate to require a full-size or larger feeder neutral conductor, depending on the total harmonic distortion contributed by the equipment to be supplied (see 220.61, FPN No. 2).

In some instances, the neutral current may exceed the current in the phase conductors. See the commentary following 310.15(B)(4)(c) regarding neutral conductor ampacity.

IV. Optional Feeder and Service Load Calculations

220.80 General

Optional feeder and service load calculations shall be permitted in accordance with Part IV.

220.82 Dwelling Unit

(A) Feeder and Service Load This section applies to a dwelling unit having the total connected load served by a single 120/240-volt or 208Y/120-volt set of 3-wire service or feeder conductors with an ampacity of 100 or greater. It shall be permissible to calculate the feeder and service loads in accordance with this section instead of the method specified in Part III of this article. The calculated load shall be the result of adding the loads from 220.82(B) and (C). Feeder and service-entrance conductors whose calculated load is determined by this optional calculation shall be permitted to have the neutral load determined by 220.61.

The optional method given in 220.82 applies to a single dwelling unit, whether it is a separate building or located in a multifamily dwelling. The optional calculation permitted by 220.82 may be used only if the service-entrance or feeder conductors have an ampacity of at least 100 amperes. See Article 100 for the definition of *dwelling unit*.

Examples of the optional calculation for a dwelling unit are given in Examples D2(a), D2(b), D2(c), and D4(b) of Annex D.

(B) General Loads The general calculated load shall be not less than 100 percent of the first 10 kVA plus 40 percent of the remainder of the following loads:

(1) 33 volt-amperes/m^2 or 3 volt-amperes/ft^2 for general lighting and general-use receptacles. The floor area for each floor shall be calculated from the outside dimensions of the dwelling unit. The calculated floor area shall not include open porches, garages, or unused or unfinished spaces not adaptable for future use.

(2) 1500 volt-amperes for each 2-wire, 20-ampere small-appliance branch circuit and each laundry branch circuit specified in 220.52.

(3) The nameplate rating of all appliances that are fastened in place, permanently connected, or located to be on a specific circuit, ranges, wall-mounted ovens, counter-mounted cooking units, clothes dryers, and water heaters.

Section 220.82(B)(3) includes appliances that may not be fastened in place but that may be permanently connected or on a specific circuit, such as clothes dryers, dishwashers, and freezers.

(4) The nameplate ampere or kVA rating of all motors and of all low-power-factor loads.

(C) Heating and Air-Conditioning Load The largest of the following six selections (load in kVA) shall be included:

(1) 100 percent of the nameplate rating(s) of the air conditioning and cooling.

(2) 100 percent of the nameplate rating(s) of the heating when a heat pump is used without any supplemental electric heating.

(3) 100 percent of the nameplate ratings of electric thermal storage and other heating systems where the usual load is expected to be continuous at the full nameplate value. Systems qualifying under this selection shall not be calculated under any other selection in 220.82(C).

(4) 100 percent of the nameplate rating(s) of the heat pump compressor and 65 percent of the supplemental electric heating for central electric space heating systems. If the heat pump compressor is prevented from operating at the same time as the supplementary heat, it does not need to be added to the supplementary heat for the total central space heating load.

Where the heat pump compressor and supplemental heating will operate at the same time, 100 percent of the compressor load plus 65 percent of the supplemental heating load are considered to be the central space heating load. If the equipment operates such that the compressor cannot operate concurrently with the supplemental heating, the central space

heating load is based on only 65 percent of the supplemental heating load.

(5) 65 percent of the nameplate rating(s) of electric space heating if less than four separately controlled units.

(6) 40 percent of the nameplate rating(s) of electric space heating if four or more separately controlled units.

Section 220.60 states that for loads that do not operate simultaneously, the largest load being considered is used. In concert with 220.60, 220.82(C) requires that only the largest of the six choices needs to be included in the feeder or service calculation. Examples of calculations using air conditioning and heating are found in Annex D, Examples D2(a), (b), and (c).

220.83 Existing Dwelling Unit

This section shall be permitted to be used to determine if the existing service or feeder is of sufficient capacity to serve additional loads. Where the dwelling unit is served by a 120/240-volt or 208Y/120-volt, 3-wire service, it shall be permissible to calculate the total load in accordance with 220.83(A) or (B).

(A) Where Additional Air-Conditioning Equipment or Electric Space-Heating Equipment Is Not to Be Installed The following formula shall be used for existing and additional new loads.

Load (kVA)	Percent of Load
First 8 kVA of load at	100
Remainder of load at	40

Load calculations shall include the following:

(1) General lighting and general-use receptacles at 33 volt-amperes/m^2 or 3 volt-amperes/ft^2 as determined by 220.12

(2) 1500 volt-amperes for each 2-wire, 20-ampere small-appliance branch circuit and each laundry branch circuit specified in 220.52

(3) Household range(s), wall-mounted oven(s), and counter-mounted cooking unit(s)

(4) All other appliances that are permanently connected, fastened in place, or connected to a dedicated circuit, at nameplate rating

(B) Where Additional Air-Conditioning Equipment or Electric Space-Heating Equipment Is to Be Installed The following formula shall be used for existing and additional new loads. The larger connected load of air-conditioning or space-heating, but not both, shall be used.

Load	Percent of Load
Air-conditioning equipment	100
Central electric space heating	100
Less than four separately controlled space-heating units	100
First 8 kVA of all other loads	100
Remainder of all other loads	40

Other loads shall include the following:

(1) General lighting and general-use receptacles at 33 volt-amperes/m^2 or 3 volt-amperes/ft^2 as determined by 220.12
(2) 1500 volt-amperes for each 2-wire, 20-ampere small-appliance branch circuit and each laundry branch circuit specified in 220.52
(3) Household range(s), wall-mounted oven(s), and counter-mounted cooking unit(s)
(4) All other appliances that are permanently connected, fastened in place, or connected to a dedicated circuit, including four or more separately controlled space-heating units, at nameplate rating

The optional methods described in Section 220.83(A) and 220.83(B) allow an additional load to be supplied by an existing service.

Example

An existing dwelling unit is served by a 100-ampere service. An additional load of a single 5-kVA, 240-volt air-conditioning unit is to be installed. Because the existing load does not contain heating or air-conditioning equipment, the existing load is calculated according to 220.83(A). The load of the existing dwelling unit consists of the following:

General lighting, 24 ft × 40 ft = 960 ft^2 × 3 VA per ft^2	2,880 VA
Small-appliance circuits (3 × 1500 VA)	4,500 VA
Laundry circuit at 1500 VA	1,500 VA
Electric range rated 10.5 kW	10,500 VA
Electric water heater rated 3.0 kW	3,000 VA
Total existing load	22,380 VA

STEP 1. Following the requirements of 220.83(A), calculate the existing dwelling unit load before adding any equipment:

First 8 kVA of load at 100%	8,000 VA
Remainder of load at 40% (22,380 − 8,000) = 14,380 × 40%	5,752 VA
Total load (without air-conditioning equipment)	13,752 VA
13,752 VA ÷ 240 V	57.3 amperes

STEP 2. Prepare a list of the existing and new loads of the dwelling unit.

General lighting, 24 ft × 40 ft = 960 ft^2 × 3 VA per ft^2	2,880 VA
Small-appliance circuits (3 × 1500 VA)	4,500 VA
Laundry circuit at 1500 VA	1,500 VA
Electric range rated 10.5 kW	10,500 VA
Electric water heater rated 3.0 kW	3,000 VA
Added air-conditioning equipment	5,000 VA
Total new load	27,380 VA

STEP 3. Following the requirements in 220.83(B), calculate the dwelling unit total load after adding any new equipment.

First 8 kVA of other load at 100%	8,000 VA
Remainder of other load at 40% (22,380 − 8,000) = 14,380 × 40%	5,752 VA
100% of air-conditioning equipment	5,000 VA
Total load (with added air-conditioning equipment)	18,752 VA
18,752 VA ÷ 240 V	78.13 amperes

The additional load contributed by the added 5-kVA air conditioning does not exceed the allowable load permitted on a 100-ampere service.

220.84 Multifamily Dwelling

(A) Feeder or Service Load It shall be permissible to calculate the load of a feeder or service that supplies three or more dwelling units of a multifamily dwelling in accordance with Table 220.84 instead of Part III of this article if all the following conditions are met:

(1) No dwelling unit is supplied by more than one feeder.
(2) Each dwelling unit is equipped with electric cooking equipment.

Exception: When the calculated load for multifamily dwellings without electric cooking in Part III of this article exceeds that calculated under Part IV for the identical load plus electric cooking (based on 8 kW per unit), the lesser of the two loads shall be permitted to be used.

According to 220.84(A)(2), each dwelling unit must be equipped with electric cooking equipment in order for the load calculation method found in 220.84(A) to be used. The exception to 220.84(A)(2) permits load calculation for dwelling units that do not have electric cooking equipment, by using a simulated electric cooking equipment load of 8 kW per unit and comparing this calculated load to the load for the dwellings without electric cooking equipment calculated in accordance with Part III of Article 220 (commonly referred to as the "standard calculation"). Whichever of these calculations yields the smallest load is permitted to be used.

(3) Each dwelling unit is equipped with either electric space heating or air conditioning, or both. Feeders and service conductors whose calculated load is determined by this optional calculation shall be permitted to have the neutral load determined by 220.61.

(B) House Loads House loads shall be calculated in accordance with Part III of this article and shall be in addition to the dwelling unit loads calculated in accordance with Table 220.84.

(C) Connected Loads The calculated load to which the demand factors of Table 220.84 apply shall include the following:

(1) 33 volt-amperes/m^2 or 3 volt-amperes/ft^2 for general lighting and general-use receptacles.

(2) 1500 volt-amperes for each 2-wire, 20-ampere small-appliance branch circuit and each laundry branch circuit specified in 220.52.

(3) The nameplate rating of all appliances that are fastened in place, permanently connected or located to be on a specific circuit, ranges, wall-mounted ovens, counter-mounted cooking units, clothes dryers, water heaters, and space heaters. If water heater elements are interlocked so that all elements cannot be used at the same time, the maximum possible load shall be considered the nameplate load.

(4) The nameplate ampere or kilovolt-ampere rating of all motors and of all low-power-factor loads.

(5) The larger of the air-conditioning load or the space-heating load.

220.85 Two Dwelling Units

Where two dwelling units are supplied by a single feeder and the calculated load under Part III of this article exceeds that for three identical units calculated under 220.84, the lesser of the two loads shall be permitted to be used.

220.86 Schools

The calculation of a feeder or service load for schools shall be permitted in accordance with Table 220.86 in lieu of Part III of this article where equipped with electric space heating, air conditioning, or both. The connected load to which the demand factors of Table 220.86 apply shall include all of the interior and exterior lighting, power, water heating, cooking, other loads, and the larger of the air-conditioning load or space-heating load within the building or structure.

Feeders and service-entrance conductors whose calculated load is determined by this optional calculation shall be permitted to have the neutral load determined by 220.61. Where the building or structure load is calculated by this optional method, feeders within the building or structure shall have ampacity as permitted in Part III of this article; however, the ampacity of an individual feeder shall not be required to be larger than the ampacity for the entire building.

Table 220.84 Optional Calculations — Demand Factors for Three or More Multifamily Dwelling Units

Number of Dwelling Units	Demand Factor (Percent)
3–5	45
6–7	44
8–10	43
11	42
12–13	41
14–15	40
16–17	39
18–20	38
21	37
22–23	36
24–25	35
26–27	34
28–30	33
31	32
32–33	31
34–36	30
37–38	29
39–42	28
43–45	27
46–50	26
51–55	25
56–61	24
62 and over	23

The method of load calculation under 220.84 is optional and applies only where one service or feeder supplies the entire load of a dwelling unit. If all the stated conditions prevail, the optional calculations in 220.84 may be used instead of those in Part III of Article 220.

This section shall not apply to portable classroom buildings.

Table 220.86 Optional Method — Demand Factors for Feeders and Service-Entrance Conductors for Schools

Connected Load		Demand Factor (Percent)
First 33 VA/m^2 Plus,	(3 VA/ft^2) at	100
Over 33 to 220 VA/m^2 Plus,	(3 to 20 VA/ft^2) at	75
Remainder over 220 VA/m^2	(20 VA/ft^2) at	25

Many schools add small, portable classroom buildings. The air-conditioning load in these portable classrooms must comply with Article 440, and the lighting load must be considered continuous. The demand factors in Table 220.86 do not apply to portable classrooms, because those demand factors would decrease the feeder or service size to below that required for the connected continuous load.

220.87 Determining Existing Loads

The calculation of a feeder or service load for existing installations shall be permitted to use actual maximum demand to determine the existing load under all of the following conditions:

(1) The maximum demand data is available for a 1-year period.

Exception: If the maximum demand data for a 1-year period is not available, the calculated load shall be permitted to be based on the maximum demand (measure of average power demand over a 15-minute period) continuously recorded over a minimum 30-day period using a recording ammeter or power meter connected to the highest loaded phase of the feeder or service, based on the initial loading at the start of the recording. The recording shall reflect the maximum demand of the feeder or service by being taken when the building or space is occupied and shall include by measurement or calculation the larger of the heating or cooling equipment load, and other loads that may be periodic in nature due to seasonal or similar conditions.

(2) The maximum demand at 125 percent plus the new load does not exceed the ampacity of the feeder or rating of the service.
(3) The feeder has overcurrent protection in accordance with 240.4, and the service has overload protection in accordance with 230.90.

Additional loads may be connected to existing services and feeders under the following conditions:

1. The maximum demand kVA data for a minimum 1-year period (or the 30-day alternative method from the exception) is available.
2. The installation complies with 220.87(2) and 220.87(3).

220.88 New Restaurants

Calculation of a service or feeder load, where the feeder serves the total load, for a new restaurant shall be permitted in accordance with Table 220.88 in lieu of Part III of this article.

The overload protection of the service conductors shall be in accordance with 230.90 and 240.4.

Feeder conductors shall not be required to be of greater ampacity than the service conductors.

Section 220.88 recognizes the effects of load diversity that are typical of restaurant occupancies. It also recognizes the amount of continuous loads as a percentage of the total connected load. The exact method of calculation presented in Table 220.88 was revised in the 2002 *Code* to more accurately reflect the original load study data.

The National Restaurant Association, the Edison Electric Institute, and the Electric Power Research Institute based the data for the change in 220.88 on load studies of 262 restaurants. These studies show that the demand factors were lower for restaurants with larger connected loads. Based on this information, it was determined that demand factors for restaurant loads are appropriate.

When using the optional method found in 220.88, it is important to notice that, first, all loads are added, even heating and air conditioning, and then the appropriate demand load is calculated from Table 220.88. The service or feeder size is calculated after application of the demand load factor.

Example 1

A new, all-electric restaurant has a total connected load of 348 kVA at 208Y/120 volts. Using Table 220.88, calculate the demand load and determine the size of the service-entrance conductors and the maximum-size overcurrent device for the service.

Solution

STEP 1. Use the value in Table 220.88 for a connected load of 348 kVA (Row 3, Column 2) to calculate the demand load for an all-electric restaurant.

$$\begin{aligned} \text{Demand load} &= 50\% \text{ of amount over 325 kVA} + 172.5 \text{ kVA} \\ &= 0.50 \times (348 \text{ kVA} - 325 \text{ kVA}) + 172.5 \text{ kVA} \\ &= (0.50 \times 23) + 172.5 \\ &= 11.5 + 172.5 \\ &= 184 \text{ kVA} \end{aligned}$$

STEP 2. Calculate the service size using the calculated demand load in STEP 1.

$$\begin{aligned} \text{Service size} &= \frac{\text{kVA}_{\text{Demand load}} \times 1000}{\text{voltage} \times \sqrt{3}} \\ &= \frac{184 \text{ kVA} \times 1000}{203 \text{ V} \times \sqrt{3}} \\ &= 510.7 \text{ or } 511 \text{ A} \end{aligned}$$

STEP 3. Determine the size of the overcurrent device. The next larger standard-size overcurrent device is 600 amperes. The minimum size of the conductors must be adequate to handle the load, but 240.4(B) permits the next larger standard-rated overcurrent device to be used.

Table 220.88 Optional Method — Permitted Load Calculations for Service and Feeder Conductors for New Restaurants

Total Connected Load (kVA)	All Electric Restaurant Calculated Loads (kVA)	Not All Electric Restaurant Calculated Loads (kVA)
0–200	80%	100%
201–325	10% (amount over 200) + 160.0	50% (amount over 200) + 200.0
326–800	50% (amount over 325) + 172.5	45% (amount over 325) + 262.5
Over 800	50% (amount over 800) + 410.0	20% (amount over 800) + 476.3

Note: Add all electrical loads, including both heating and cooling loads, to calculate the total connected load. Select the one demand factor that applies from the table, then multiply the total connected load by this single demand factor.

Example 2

A new restaurant has gas cooking appliances plus a total connected electrical load of 348 kVA at 208Y/120 volts. Calculate the demand load using Table 220.88 and the service size. Then determine the maximum-size overcurrent device for the service.

Solution

STEP 1. Calculate the demand load for a new restaurant using the value in Table 220.88 for a connected load of 348 kVA (Column 3, Row 3) as follows:

$$\text{Demand load} = 45\% \text{ of amount over } 325 \text{ kVA} + 262.5 \text{ kVA}$$
$$= 0.45 \times (348 - 325) + 262.5$$
$$= (0.45 \times 23) + 262.5$$
$$= 10.35 + 262.5$$
$$= 272.85 \text{ kVA}$$

STEP 2. Calculate the service size using the calculated demand load in Step 1.

$$\text{Service size} = \frac{\text{kVA}_{\text{Demand load}} \times 1000}{\text{voltage} \times \sqrt{3}}$$
$$= \frac{272.85 \text{ kVA} \times 1000}{208 \text{ V} \times \sqrt{3}}$$
$$= 757.36 \quad \text{or} \quad 757 \text{ A}$$

STEP 3. Determine the size of the overcurrent device. The next higher standard rating for an overcurrent device, according to 240.6, is 800 amperes. Section 230.79 requires that the service disconnecting means have a rating that is not less than the calculated load (757 amperes). The minimum size of the conductors must be adequate to handle the load, but 240.4(B) permits the next larger standard-rated overcurrent device to be used.

Service or feeder conductors whose calculated load is determined by this optional calculation shall be permitted to have the neutral load determined by 220.61.

V. Farm Load Calculation

220.100 General

Farm loads shall be calculated in accordance with Part V.

220.102 Farm Loads — Buildings and Other Loads

(A) Dwelling Unit The feeder or service load of a farm dwelling unit shall be calculated in accordance with the provisions for dwellings in Part III or IV of this article. Where the dwelling has electric heat and the farm has electric grain-drying systems, Part IV of this article shall not be used to calculate the dwelling load where the dwelling and farm load are supplied by a common service.

(B) Other Than Dwelling Unit Where a feeder or service supplies a farm building or other load having two or more separate branch circuits, the load for feeders, service conductors, and service equipment shall be calculated in accordance with demand factors not less than indicated in Table 220.102.

Table 220.102 Method for Calculating Farm Loads for Other Than Dwelling Unit

Ampere Load at 240 Volts Maximum	Demand Factor (Percent)
Loads expected to operate simultaneously, but not less than 125 percent full-load current of the largest motor and not less than the first 60 amperes of load	100
Next 60 amperes of all other loads	50
Remainder of other load	25

220.103 Farm Loads — Total

Where supplied by a common service, the total load of the farm for service conductors and service equipment shall be calculated in accordance with the farm dwelling unit load and demand factors specified in Table 220.103. Where there is equipment in two or more farm equipment buildings or for loads having the same function, such loads shall be calculated in accordance with Table 220.102 and shall be permitted to be combined as a single load in Table 220.103 for calculating the total load.

Table 220.103 Method for Calculating Total Farm Load

Individual Loads Calculated in Accordance with Table 220.102	Demand Factor (Percent)
Largest load	100
Second largest load	75
Third largest load	65
Remaining loads	50

Note: To this total load, add the load of the farm dwelling unit calculated in accordance with Part III or IV of this article. Where the dwelling has electric heat and the farm has electric grain-drying systems, Part IV of this article shall not be used to calculate the dwelling load.

ARTICLE 225
Outside Branch Circuits and Feeders

Summary of Changes

- **225.17:** Added new requirements for masts that support overhead branch circuits and feeders.

- **225.22:** Revised to specify that raintight raceways are required only in wet locations.

- **225.30(A)(6):** Added condition to have multiple supplies for the purposes of enhancing reliability.

Contents

Table 225.2 Other Articles

Equipment/Conductors	Article
Branch circuits	210
Class 1, Class 2, and Class 3 remote-control, signaling, and power-limited circuits	725
Communications circuits	800
Community antenna television and radio distribution systems	820
Conductors for general wiring	310
Electrically driven or controlled irrigation machines	675
Electric signs and outline lighting	600
Feeders	215
Fire alarm systems	760
Fixed outdoor electric deicing and snow-melting equipment	426
Floating buildings	553
Grounding	250
Hazardous (classified) locations	500
Hazardous (classified) locations — specific	510
Marinas and boatyards	555
Messenger supported wiring	396
Mobile homes, manufactured homes, and mobile home parks	550
Open wiring on insulators	398
Over 600 volts, general	490
Overcurrent protection	240
Radio and television equipment	810
Services	230
Solar photovoltaic systems	690
Swimming pools, fountains, and similar installations	680
Use and identification of grounded conductors	200

225.1 Scope

This article covers requirements for outside branch circuits and feeders run on or between buildings, structures, or poles on the premises; and electric equipment and wiring for the supply of utilization equipment that is located on or attached to the outside of buildings, structures, or poles.

> FPN: For additional information on wiring over 600 volts, see ANSI C2-2002, *National Electrical Safety Code.*

225.2 Other Articles

Application of other articles, including additional requirements to specific cases of equipment and conductors, is shown in Table 225.2.

I. General

225.3 Calculation of Loads 600 Volts, Nominal, or Less

(A) Branch Circuits The load on outdoor branch circuits shall be as determined by 220.10.

(B) Feeders The load on outdoor feeders shall be as determined by Part III of Article 220.

225.4 Conductor Covering

Where within 3.0 m (10 ft) of any building or structure other than supporting poles or towers, open individual (aerial) overhead conductors shall be insulated or covered. Conductors in cables or raceways, except Type MI cable, shall be of the rubber-covered type or thermoplastic type and, in wet locations, shall comply with 310.8. Conductors for festoon lighting shall be of the rubber-covered or thermoplastic type.

Exception: Equipment grounding conductors and grounded circuit conductors shall be permitted to be bare or covered as specifically permitted elsewhere in this Code.

The exception to 225.4 and Exception No. 2 to 250.184(A)(1) correlate to permit the use of the bare messenger wire of an overhead cable assembly as the grounded (neutral) conductor of an outdoor feeder circuit. In accordance with 250.32(B)(2), a grounding electrode conductor connection to the grounded conductor is permitted at the load end of such outdoor overhead feeder circuits.

225.5 Size of Conductors 600 Volts, Nominal, or Less

The ampacity of outdoor branch-circuit and feeder conductors shall be in accordance with 310.15 based on loads as determined under 220.10 and Part III of Article 220.

225.6 Conductor Size and Support

(A) Overhead Spans Open individual conductors shall not be smaller than the following:

(1) For 600 volts, nominal, or less, 10 AWG copper or 8 AWG aluminum for spans up to 15 m (50 ft) in length, and 8 AWG copper or 6 AWG aluminum for a longer span unless supported by a messenger wire

(2) For over 600 volts, nominal, 6 AWG copper or 4 AWG aluminum where open individual conductors, and 8 AWG copper or 6 AWG aluminum where in cable

The size limitation of copper and aluminum conductors for overhead spans is based on the need for adequate mechanical strength to support the weight of the conductors and to withstand wind, ice, and other similar conditions. Exhibit 225.1 illustrates overhead spans that are not messenger supported; that are run between buildings, structures, or poles; and that are 600 volts or less. If the conductors are supported on a messenger, the messenger cable provides the necessary mechanical strength. See 396.10 for wiring methods permitted to be messenger supported.

Exhibit 225.1 Minimum sizes of conductors in overhead spans as specified by 225.6(A)(1) for 600 volts, nominal, or less.

(B) Festoon Lighting Overhead conductors for festoon lighting shall not be smaller than 12 AWG unless the conductors are supported by messenger wires. In all spans exceeding 12 m (40 ft), the conductors shall be supported by messenger wire. The messenger wire shall be supported by strain insulators. Conductors or messenger wires shall not be attached to any fire escape, downspout, or plumbing equipment.

Article 100 defines *festoon lighting* as "a string of outdoor lights that is suspended between two points." The conductors for festoon lighting must be larger than 12 AWG unless a messenger wire supports them. On all spans of festoon lighting exceeding 40 ft, messenger wire is required and must be supported by strain insulators. See Exhibit 225.2. If no

Exhibit 225.2 Messenger wire required by 225.6(B) for festoon lighting conductors in a span exceeding 40 ft.

messenger wire is required, the 12 AWG or larger conductors are required to be supported by strain insulators.

Attachment of festoon lighting to fire escapes, plumbing equipment, or metal drain spouts is prohibited because the attachment could provide a path to ground. Moreover, such methods of attachment could not be relied on for a permanent or secure means of support.

225.7 Lighting Equipment Installed Outdoors

(A) General For the supply of lighting equipment installed outdoors, the branch circuits shall comply with Article 210 and 225.7(B) through (D).

(B) Common Neutral The ampacity of the neutral conductor shall not be less than the maximum net computed load current between the neutral and all ungrounded conductors connected to any one phase of the circuit.

Multiwire branch circuits consisting of a neutral and two or more ungrounded conductors are permitted, provided the neutral capacity is not less than the total load of all ungrounded conductors connected to any one phase of the circuit.

Exhibit 225.3 illustrates a 120/240-volt, single-phase, 3-wire system, and Exhibit 225.4 illustrates a 208Y/120-volt, 3-phase, 4-wire system. In Exhibit 225.3, all branch circuits are rated at 20 amperes. The maximum unbalanced current that can occur is four times 20 amperes, or 80 amperes. In Exhibit 225.4, all branch circuits also are rated at 20 amperes. The maximum unbalanced current that can occur on a 3-phase system with the load connected as shown is 80 amperes, due to the load on phase A.

(C) 277 Volts to Ground Circuits exceeding 120 volts, nominal, between conductors and not exceeding 277 volts,

Exhibit 225.3 A 120/240-volt, single-phase, 3-wire system (branch circuits rated at 20 amperes; maximum unbalanced current of 80 amperes).

Exhibit 225.4 A 208Y/120-volt, 3-phase, 4-wire system (branch circuits rated at 20 amperes; maximum unbalanced current of 80 amperes).

nominal, to ground shall be permitted to supply luminaires (lighting fixtures) for illumination of outdoor areas of industrial establishments, office buildings, schools, stores, and other commercial or public buildings where the luminaires (fixtures) are not less than 900 mm (3 ft) from windows, platforms, fire escapes, and the like.

Branch circuits for the outdoor illumination of industrial establishments, office buildings, schools, stores, and other commercial or public buildings are permitted to operate with a maximum voltage to ground of 277 volts. See 210.6(D) for tunnel and pole-mounted luminaires with voltages greater than 277 volts to ground. The restrictions outlined in 225.7(C) are in addition to those in 210.6(C).

(D) 600 Volts Between Conductors Circuits exceeding 277 volts, nominal, to ground and not exceeding 600 volts, nominal, between conductors shall be permitted to supply the auxiliary equipment of electric-discharge lamps in accordance with 210.6(D)(1).

Section 210.6(D)(1) contains the minimum height requirements for circuits exceeding 277 volts, nominal, to ground but not exceeding 600 volts, nominal, between conductors for circuits that supply the auxiliary equipment of electric-discharge lamps.

225.10 Wiring on Buildings

The installation of outside wiring on surfaces of buildings shall be permitted for circuits of not over 600 volts, nominal, as open wiring on insulators, as multiconductor cable, as Type MC cable, as Type MI cable, as messenger supported wiring, in rigid metal conduit, in intermediate metal conduit, in rigid nonmetallic conduit, in cable trays, as cablebus, in wireways, in auxiliary gutters, in electrical metallic tubing, in flexible metal conduit, in liquidtight flexible metal conduit, in liquidtight flexible nonmetallic conduit, and in busways. Circuits of over 600 volts, nominal, shall be installed as provided in 300.37.

225.11 Circuit Exits and Entrances

Where outside branch and feeder circuits leave or enter a building, the requirements of 230.52 and 230.54 shall apply.

Section 225.11 references 230.52, Individual Conductors Entering Buildings or Other Structures, and 230.54, Overhead Service Locations. Section 225.18 covers the requirements for clearances from ground (not over 600 volts), and 225.19 covers the requirements for clearances from buildings for conductors not over 600 volts. See 225.19(D) for final span clearances from windows, doors, fire escapes, and so on. Exhibit 225.5 shows an example where these requirements apply.

225.12 Open-Conductor Supports

Open conductors shall be supported on glass or porcelain knobs, racks, brackets, or strain insulators.

225.14 Open-Conductor Spacings

(A) 600 Volts, Nominal, or Less Conductors of 600 volts, nominal, or less, shall comply with the spacings provided in Table 230.51(C).

(B) Over 600 Volts, Nominal Conductors of over 600 volts, nominal, shall comply with the spacings provided in 110.36 and 490.24.

Clearance from
building openings
(600 V or less)

3 ft

3 ft

Clearance
from ground
(600 V or less)

Exhibit 225.5 Examples of the requirements in 225.11, which references 230.52 and 230.54; the requirements in 225.18 for clearances from ground for conductors not over 600 volts; the requirements in 225.19 for clearances from buildings for conductors not over 600 volts; and the requirements in 225.19(D) for clearances from windows, doors, fire escapes, and so on.

(C) Separation from Other Circuits Open conductors shall be separated from open conductors of other circuits or systems by not less than 100 mm (4 in.).

(D) Conductors on Poles Conductors on poles shall have a separation of not less than 300 mm (1 ft) where not placed on racks or brackets. Conductors supported on poles shall provide a horizontal climbing space not less than the following:

(1) Power conductors below communications conductors — 750 mm (30 in.)
(2) Power conductors alone or above communications conductors:
    a. 300 volts or less — 600 mm (24 in.)
    b. Over 300 volts — 750 mm (30 in.)
(3) Communications conductors below power conductors — same as power conductors
(4) Communications conductors alone — no requirement

Sufficient space is required for linemen to climb over or through conductors to safely work with conductors on the pole.

225.15 Supports over Buildings

Supports over a building shall be in accordance with 230.29.

225.16 Attachment to Buildings

(A) Point of Attachment The point of attachment to a building shall be in accordance with 230.26.

(B) Means of Attachment The means of attachment to a building shall be in accordance with 230.27.

225.17 Masts as Supports

Where a mast is used for the support of final spans of feeders or branch circuits, it shall be of adequate strength or be supported by braces or guys to withstand safely the strain imposed by the overhead drop. Where raceway-type masts are used, all raceway fittings shall be identified for use with masts. Only the feeder or branch circuit conductors specified within this section shall be permitted to be attached to the feeder and/or branch circuit mast.

Section 225.17 is new in the 2005 *Code* and provides the same rules for masts associated with and supporting overhead branch circuits and feeders as are required for masts associated with and supporting service drops in 230.28. A mast supporting an overhead branch circuit or feeder span is not permitted to support conductors of other systems, such as overhead conductor spans for signaling, communications, or CATV systems.

225.18 Clearance from Ground

Overhead spans of open conductors and open multiconductor cables of not over 600 volts, nominal, shall have a clearance of not less than the following:

(1) 3.0 m (10 ft) — above finished grade, sidewalks, or from any platform or projection from which they might be reached where the voltage does not exceed 150 volts to ground and accessible to pedestrians only
(2) 3.7 m (12 ft) — over residential property and driveways, and those commercial areas not subject to truck traffic where the voltage does not exceed 300 volts to ground
(3) 4.5 m (15 ft) — for those areas listed in the 3.7-m (12-ft) classification where the voltage exceeds 300 volts to ground
(4) 5.5 m (18 ft) — over public streets, alleys, roads, parking areas subject to truck traffic, driveways on other than residential property, and other land traversed by vehicles, such as cultivated, grazing, forest, and orchard

225.19 Clearances from Buildings for Conductors of Not Over 600 Volts, Nominal

(A) Above Roofs Overhead spans of open conductors and open multiconductor cables shall have a vertical clearance of not less than 2.5 m (8 ft) above the roof surface. The vertical clearance above the roof level shall be maintained for a distance not less than 900 mm (3 ft) in all directions from the edge of the roof.

Exception No. 1: The area above a roof surface subject to pedestrian or vehicular traffic shall have a vertical clearance from the roof surface in accordance with the clearance requirements of 225.18.

Exception No. 2: Where the voltage between conductors does not exceed 300, and the roof has a slope of 100 mm in 300 mm (4 in. in 12 in.) or greater, a reduction in clearance to 900 mm (3 ft) shall be permitted.

Exception No. 3: Where the voltage between conductors does not exceed 300, a reduction in clearance above only the overhanging portion of the roof to not less than 450 mm (18 in.) shall be permitted if (1) not more than 1.8 m (6 ft) of the conductors, 1.2 m (4 ft) horizontally, pass above the roof overhang and (2) they are terminated at a through-the-roof raceway or approved support.

Exception No. 4: The requirement for maintaining the vertical clearance 900 mm (3 ft) from the edge of the roof shall not apply to the final conductor span where the conductors are attached to the side of a building.

(B) From Nonbuilding or Nonbridge Structures From signs, chimneys, radio and television antennas, tanks, and other nonbuilding or nonbridge structures, clearances — vertical, diagonal, and horizontal — shall not be less than 900 mm (3 ft).

(C) Horizontal Clearances Clearances shall not be less than 900 mm (3 ft).

(D) Final Spans Final spans of feeders or branch circuits shall comply with 225.19(D)(1), (D)(2), and (D)(3).

(1) Clearance from Windows Final spans to the building they supply, or from which they are fed, shall be permitted to be attached to the building, but they shall be kept not less than 900 mm (3 ft) from windows that are designed to be opened, and from doors, porches, balconies, ladders, stairs, fire escapes, or similar locations.

Exception: Conductors run above the top level of a window shall be permitted to be less than the 900-mm (3-ft) requirement.

(2) Vertical Clearance The vertical clearance of final spans above, or within 900 mm (3 ft) measured horizontally of, platforms, projections, or surfaces from which they might be reached shall be maintained in accordance with 225.18.

(3) Building Openings The overhead branch-circuit and feeder conductors shall not be installed beneath openings through which materials may be moved, such as openings

in farm and commercial buildings, and shall not be installed where they obstruct entrance to these buildings' openings.

(E) Zone for Fire Ladders Where buildings exceed three stories or 15 m (50 ft) in height, overhead lines shall be arranged, where practicable, so that a clear space (or zone) at least 1.8 m (6 ft) wide will be left either adjacent to the buildings or beginning not over 2.5 m (8 ft) from them to facilitate the raising of ladders when necessary for fire fighting.

225.20 Mechanical Protection of Conductors

Mechanical protection of conductors on buildings, structures, or poles shall be as provided for services in 230.50.

225.21 Multiconductor Cables on Exterior Surfaces of Buildings

Supports for multiconductor cables on exterior surfaces of buildings shall be as provided in 230.51.

225.22 Raceways on Exterior Surfaces of Buildings or Other Structures

Raceways on exteriors of buildings or other structures shall be arranged to drain and shall be raintight in wet locations.

Exception: Flexible metal conduit, where permitted in 348.12(1), shall not be required to be raintight.

Raintight is defined in Article 100 as "constructed or protected so that exposure to a beating rain will not result in the entrance of water under specified test conditions." To ensure this, all conduit bodies, fittings, and boxes used in wet locations are required to be provided with threaded hubs or other approved means. Threadless couplings and connectors used with metal conduit or electrical metallic tubing installed on the exterior of a building must be of the raintight type [see 342.42(A), 344.42(A), and 348.42].

If raceways are exposed to weather or rain through weatherhead openings, condensation is likely to occur, causing moisture to accumulate within raceways at low points of the installation and in junction boxes. Therefore, raceways should be installed to permit drainage through drain holes at appropriate locations. This section has been revised so that the raintight requirement applies only to raceways installed in wet locations.

225.24 Outdoor Lampholders

Where outdoor lampholders are attached as pendants, the connections to the circuit wires shall be staggered. Where such lampholders have terminals of a type that puncture the insulation and make contact with the conductors, they shall be attached only to conductors of the stranded type.

Splices to branch-circuit conductors for outdoor lampholders of the Edison-base type or "pigtail" sockets are required to be staggered so that splices will not be in close proximity to each other. Pin-type terminal sockets must be attached to stranded conductors only and are intended for installations for temporary lighting or decorations, signs, or specifically approved applications.

225.25 Location of Outdoor Lamps

Locations of lamps for outdoor lighting shall be below all energized conductors, transformers, or other electric utilization equipment, unless either of the following apply:

(1) Clearances or other safeguards are provided for relamping operations.
(2) Equipment is controlled by a disconnecting means that can be locked in the open position.

Because 225.18 requires a minimum clearance for open conductors of 10 ft above grade or platforms, it may be difficult to keep all electrical equipment above the lamps. Section 225.25(1) allows other clearances or safeguards to permit safe relamping, and 225.25(2) permits the use of a disconnecting means to de-energize the circuit.

225.26 Vegetation as Support

Vegetation such as trees shall not be used for support of overhead conductor spans.

Where overhead conductor spans are attached to a tree, normal tree growth around the attachment device causes the mounting insulators to break and the conductor insulation to be degraded. The requirement in 225.26 reduces the likelihood of chafing of the conductor insulation and the danger of shock to tree trimmers and tree climbers. The exception to 225.26 permitting trees as a support method for overhead conductor spans on a temporary basis was deleted in the 2002 *Code*. However, outdoor luminaires and associated equipment are permitted by 410.16(H) to be supported by trees. To prevent the chafing damage, conductors are run up the tree from an underground wiring method. See 300.5(D) for requirements on the protection of direct-buried conductors emerging from below grade.

II. More Than One Building or Other Structure

Part II covers outside branch circuits and feeders on single managed properties where outside branch circuits and feeders are the source of electrical supply for buildings and structures. Important in the application of the Part II requirements are the Article 100 definitions of *service point, service, service equipment, feeder,* and *branch circuit.* Determining what constitutes a set of feeder or branch-circuit conductors versus a set of service conductors depends on a clear understanding of where the service point is located and where the service and service equipment for a premises are located. In some cases, particularly with medium- and high-voltage distribution, the service location of a campus or multibuilding facility is a switchyard or substation. With the location of the service point and service equipment established, the requirements for outside branch circuits and feeders from Part II (and Part III if over 600 volts) can be properly applied.

Included in Part II of Article 225 are the requirements for overhead and underground feeders that supply buildings or structures on college and other institutional campuses, multibuilding industrial facilities, multibuilding commercial facilities, and other facilities where the electrical supply is an outdoor feeder or branch circuit. Such distribution is permitted under the condition that the entire premises is under a single management. Many of the requirements in Part II covering the number of feeders or branch circuits and the location and type of disconnecting means are similar to the requirements for services in Article 230.

225.30 Number of Supplies

Where more than one building or other structure is on the same property and under single management, each additional building or other structure that is served by a branch circuit or feeder on the load side of the service disconnecting means shall be supplied by only one feeder or branch circuit unless permitted in 225.30(A) through (E). For the purpose of this section, a multiwire branch circuit shall be considered a single circuit.

(A) Special Conditions Additional feeders or branch circuits shall be permitted to supply the following:

(1) Fire pumps
(2) Emergency systems
(3) Legally required standby systems
(4) Optional standby systems
(5) Parallel power production systems
(6) Systems designed for connection to multiple sources of supply for the purpose of enhanced reliability

The fundamental requirement of 225.30 is that a building or structure be supplied by a single branch circuit or feeder, similar to the rule for a single service in 230.2. Sections 225.30(A) through 225.30(E) identify conditions under which a building or structure is permitted to be supplied by multiple sources. To address the need for increased reliability of the power source to a premises where critical operational loads (not classed as emergency or legally required standby) are supplied, 225.30(A)(6), which is a new item, has been included as a condition where multiple feeder or branch circuit sources are permitted. Double-ended (main-tie-main) switchgear supplied by two feeders is an example of the type of source covered by this new provision.

(B) Special Occupancies By special permission, additional feeders or branch circuits shall be permitted for either of the following:

(1) Multiple-occupancy buildings where there is no space available for supply equipment accessible to all occupants
(2) A single building or other structure sufficiently large to make two or more supplies necessary

(C) Capacity Requirements Additional feeders or branch circuits shall be permitted where the capacity requirements are in excess of 2000 amperes at a supply voltage of 600 volts or less.

(D) Different Characteristics Additional feeders or branch circuits shall be permitted for different voltages, frequencies, or phases or for different uses, such as control of outside lighting from multiple locations.

(E) Documented Switching Procedures Additional feeders or branch circuits shall be permitted to supply installations under single management where documented safe switching procedures are established and maintained for disconnection.

Buildings on college campuses, multibuilding industrial facilities, and multibuilding commercial facilities are permitted to be supplied by secondary loop supply (secondary selective) networks, provided that documented switching procedures are established. These switching procedures must establish a method to safely operate switches for the facility during maintenance and during alternative supply and emergency supply conditions. Keyed interlock systems are often used to reduce the likelihood of inappropriate switching procedures that could result in hazardous conditions.

225.31 Disconnecting Means

Means shall be provided for disconnecting all ungrounded conductors that supply or pass through the building or structure.

225.32 Location

The disconnecting means shall be installed either inside or outside of the building or structure served or where the conductors pass through the building or structure. The dis-

connecting means shall be at a readily accessible location nearest the point of entrance of the conductors. For the purposes of this section, the requirements in 230.6 shall be utilized.

Exception No. 1: For installations under single management, where documented safe switching procedures are established and maintained for disconnection, and where the installation is monitored by qualified individuals, the disconnecting means shall be permitted to be located elsewhere on the premises.

Exception No. 2: For buildings or other structures qualifying under the provisions of Article 685, the disconnecting means shall be permitted to be located elsewhere on the premises.

Exception No. 3: For towers or poles used as lighting standards, the disconnecting means shall be permitted to be located elsewhere on the premises.

Exception No. 4: For poles or similar structures used only for support of signs installed in accordance with Article 600, the disconnecting means shall be permitted to be located elsewhere on the premises.

The basic requirement on locating the disconnecting means for a feeder or branch circuit supplying a structure is essentially the same as that specified for services in 230.70(A), but an important difference applies to feeder and branch-circuit sources. Unlike a premises supplied by a service, where a building or structure is supplied by a feeder or branch circuit, there always has to be a feeder or branch circuit disconnecting means at the building or structure supplied unless one of the conditions in Exception No. 1 through Exception No. 4 can be applied. Particularly for campus-style facilities supplied by a single utility service, the service disconnecting means may be remote from the buildings or structures supplied. In such installations, the supply conductors to the buildings or structures are feeders or branch circuits, and the main requirement of this section is that the feeder or branch-circuit disconnecting means is to be located inside or outside the building or structure supplied, at the point nearest to where the supply conductors enter the building or structure. This requirement applies to conductors that supply a building and to conductors that pass through a building.

This requirement ensures that, where the service disconnecting means is remote from a building or structure, there is a disconnecting means for the feeder or branch circuit located at the building or structure to facilitate ready disconnection of the power. Outside disconnecting means are not required to be physically attached to the building or structure supplied, as would be the case with feeder-supplied outside free-standing switchgear located at the building or structure. In addition, this disconnecting means requirement is modified by the rules in 700.12(B)(6), 701.11(B)(5), and 702.11 for emergency, legally required standby, and optional standby feeders supplied by an outdoor generator set.

225.33 Maximum Number of Disconnects

(A) General The disconnecting means for each supply permitted by 225.30 shall consist of not more than six switches or six circuit breakers mounted in a single enclosure, in a group of separate enclosures, or in or on a switchboard. There shall be no more than six disconnects per supply grouped in any one location.

Exception: For the purposes of this section, disconnecting means used solely for the control circuit of the ground-fault protection system, or the control circuit of the power-operated supply disconnecting means, installed as part of the listed equipment, shall not be considered a supply disconnecting means.

(B) Single-Pole Units Two or three single-pole switches or breakers capable of individual operation shall be permitted on multiwire circuits, one pole for each ungrounded conductor, as one multipole disconnect, provided they are equipped with handle ties or a master handle to disconnect all ungrounded conductors with no more than six operations of the hand.

225.34 Grouping of Disconnects

(A) General The two to six disconnects as permitted in 225.33 shall be grouped. Each disconnect shall be marked to indicate the load served.

Exception: One of the two to six disconnecting means permitted in 225.33, where used only for a water pump also intended to provide fire protection, shall be permitted to be located remote from the other disconnecting means.

(B) Additional Disconnecting Means The one or more additional disconnecting means for fire pumps or for emergency, legally required standby or optional standby system permitted by 225.30 shall be installed sufficiently remote from the one to six disconnecting means for normal supply to minimize the possibility of simultaneous interruption of supply.

225.35 Access to Occupants

In a multiple-occupancy building, each occupant shall have access to the occupant's supply disconnecting means.

Exception: In a multiple-occupancy building where electric supply and electrical maintenance are provided by the building management and where these are under continuous building management supervision, the supply disconnecting means supplying more than one occupancy shall be permit-ted to be accessible to authorized management personnel only.

225.36 Suitable for Service Equipment

The disconnecting means specified in 225.31 shall be suitable for use as service equipment.

Exception: For garages and outbuildings on residential property, a snap switch or a set of 3-way or 4-way snap switches shall be permitted as the disconnecting means.

225.37 Identification

Where a building or structure has any combination of feeders, branch circuits, or services passing through it or supplying it, a permanent plaque or directory shall be installed at each feeder and branch-circuit disconnect location denoting all other services, feeders, or branch circuits supplying that building or structure or passing through that building or structure and the area served by each.

The requirement of 225.37 correlates with 230.2(E) in that where a building has multiple sources of supply, permanent identification at each supply (service, feeder, and branch circuit) disconnecting means is required. The term *permanent* indicates that this required identification has to have the same permanency as the disconnecting means itself. This identification is an important safety feature in an emergency condition, because in many cases first responders are not familiar with the electrical distribution system of a facility. In addition to 225.37 and 230.2(E), identification of power sources is covered by the requirements of 700.8(A) for emergency sources, 701.9(A) for legally required standby sources, 702.8(A) for optional standby sources, and 705.10 for parallel power production sources.

Exception No. 1: A plaque or directory shall not be required for large-capacity multibuilding industrial installations under single management, where it is ensured that disconnection can be accomplished by establishing and maintaining safe switching procedures.

Exception No. 2: This identification shall not be required for branch circuits installed from a dwelling unit to a second building or structure.

225.38 Disconnect Construction

Disconnecting means shall meet the requirements of 225.38(A) through (D).

Exception: For garages and outbuildings on residential property, snap switches or sets of 3-way or 4-way snap switches shall be permitted as the disconnecting means.

(A) Manually or Power Operable The disconnecting means shall consist of either (1) a manually operable switch or a circuit breaker equipped with a handle or other suitable operating means or (2) a power-operable switch or circuit breaker, provided the switch or circuit breaker can be opened by hand in the event of a power failure.

(B) Simultaneous Opening of Poles Each building or structure disconnecting means shall simultaneously disconnect all ungrounded supply conductors that it controls from the building or structure wiring system.

(C) Disconnection of Grounded Conductor Where the building or structure disconnecting means does not disconnect the grounded conductor from the grounded conductors in the building or structure wiring, other means shall be provided for this purpose at the location of disconnecting means. A terminal or bus to which all grounded conductors can be attached by means of pressure connectors shall be permitted for this purpose.

In a multisection switchboard, disconnects for the grounded conductor shall be permitted to be in any of the switchboard, provided any such switchboard is marked.

(D) Indicating The building or structure disconnecting means shall plainly indicate whether it is in the open or closed position.

225.39 Rating of Disconnect

The feeder or branch-circuit disconnecting means shall have a rating of not less than the load to be supplied, determined in accordance with Parts I and II of Article 220 for branch circuits, Parts III or IV of Article 220 for feeders, or Part V of Article 220 for farm loads. In no case shall the rating be lower than specified in 225.39(A), (B), (C), or (D).

(A) One-Circuit Installation For installations to supply only limited loads of a single branch circuit, the branch circuit disconnecting means shall have a rating of not less than 15 amperes.

(B) Two-Circuit Installations For installations consisting of not more than two 2-wire branch circuits, the feeder or branch-circuit disconnecting means shall have a rating of not less than 30 amperes.

(C) One-Family Dwelling For a one-family dwelling, the feeder disconnecting means shall have a rating of not less than 100 amperes, 3-wire.

(D) All Others For all other installations, the feeder or branch-circuit disconnecting means shall have a rating of not less than 60 amperes.

225.40 Access to Overcurrent Protective Devices

Where a feeder overcurrent device is not readily accessible, branch-circuit overcurrent devices shall be installed on the load side, shall be mounted in a readily accessible location, and shall be of a lower ampere rating than the feeder overcurrent device.

III. Over 600 Volts

225.50 Sizing of Conductors

The sizing of conductors over 600 volts shall be in accordance with 210.19(B) for branch circuits and 215.2(B) for feeders.

225.51 Isolating Switches

Where oil switches or air, oil, vacuum, or sulfur hexafluoride circuit breakers constitute a building disconnecting means, an isolating switch with visible break contacts and meeting the requirements of 230.204(B), (C), and (D) shall be installed on the supply side of the disconnecting means and all associated equipment.

Exception: The isolating switch shall not be required where the disconnecting means is mounted on removable truck panels or metal-enclosed switchgear units that cannot be opened unless the circuit is disconnected and that, when removed from the normal operating position, automatically disconnect the circuit breaker or switch from all energized parts.

225.52 Location

A building or structure disconnecting means shall be located in accordance with 225.32, or it shall be electrically operated by a similarly located remote-control device.

225.53 Type

Each building or structure disconnect shall simultaneously disconnect all ungrounded supply conductors it controls and shall have a fault-closing rating not less than the maximum available short-circuit current available at its supply terminals.

Where fused switches or separately mounted fuses are installed, the fuse characteristics shall be permitted to contribute to the fault closing rating of the disconnecting means.

The requirement for a disconnecting means for over-600-volt feeders to buildings or structures is similar to the requirements found in 230.205(B). This disconnect can be an air, oil, SF_6, or vacuum breaker. It also can be an air, oil, SF_6, or vacuum switch. Where a switch is used, fuses are permitted to help with the fault-closing capability of the switch. The common practices of using fused load-break cutouts to switch sections of overhead lines and using load-break elbows to switch sections of underground lines are permitted. However, the building disconnecting means must be gang-operated to simultaneously open and close all ungrounded supply conductors. Load-break elbows and fused cutouts cannot be used as the building disconnection means.

225.60 Clearances over Roadways, Walkways, Rail, Water, and Open Land

(A) 22 kV Nominal to Ground or Less The clearances over roadways, walkways, rail, water, and open land for conductors and live parts up to 22 kV nominal to ground or less shall be not less than the values shown in Table 225.60.

Table 225.60 Clearances over Roadways, Walkways, Rail, Water, and Open Land

	Clearance	
Location	m	ft
Open land subject to vehicles, cultivation, or grazing	5.6	18.5
Roadways, driveways, parking lots, and alleys	5.6	18.5
Walkways	4.1	13.5
Rails	8.1	26.5
Spaces and ways for pedestrians and restricted traffic	4.4	14.5
Water areas not suitable for boating	5.2	17

(B) Over 22 kV Nominal to Ground Clearances for the categories shown in Table 225.60 shall be increased by 10 mm (0.4 in.) per kV above 22,000 volts.

(C) Special Cases For special cases, such as where crossings will be made over lakes, rivers, or areas using large vehicles such as mining operations, specific designs shall be engineered considering the special circumstances and shall be approved by the authority having jurisdiction.

> FPN: For additional information, see ANSI C2-2002, *National Electrical Safety Code.*

Section 225.60 and its associated table add clearance requirements and specific distances that correlate with requirements in the *National Electrical Safety Code®* (*NESC®*).

225.61 Clearances over Buildings and Other Structures

(A) 22 kV Nominal to Ground or Less The clearances over buildings and other structures for conductors and live parts up to 22 kV, nominal, to ground or less shall be not less than the values shown in Table 225.61.

Table 225.61 Clearances over Buildings and Other Structures

Clearance from Conductors or Live Parts from:	Horizontal		Vertical	
	m	ft	m	ft
Building walls, projections, and windows	2.3	7.5	—	—
Balconies, catwalks, and similar areas accessible to people	2.3	7.5	4.1	13.5
Over or under roofs or projections not readily accessible to people	—	—	3.8	12.5
Over roofs accessible to vehicles but not trucks	—	—	4.1	13.5
Over roofs accessible to trucks	—	—	5.6	18.5
Other structures	2.3	7.5	—	—

(B) Over 22 kV Nominal to Ground Clearances for the categories shown in Table 225.61 shall be increased by 10 mm (0.4 in.) per kV above 22,000 volts.

> FPN: For additional information, see ANSI C2-2002, *National Electrical Safety Code.*

Section 225.61 and its associated table add clearance requirements and specific distances over buildings and structures that correlate with requirements in the *National Electrical Safety Code (NESC).*

ARTICLE 230
Services

Summary of Changes

- **230.2(A)(6):** Added condition to have multiple services for the purposes of enhancing reliability.
- **230.40, Exception No.1:** Revised to clarify that each occupancy is permitted to be supplied by a set of service-entrance conductors from each service described in 230.2.
- **230.44:** Revised to permit other than service conductors if conductors are separated by a barrier.
- **230.71(A):** Revised to permit additional disconnecting means for transient voltage surge suppressors that are installed in listed equipment.
- **230.72(B):** Revised to include emergency systems in the permission to have additional, remote disconnecting means.

- **230.82(3):** Revised to require meter disconnect switches to have a short-circuit current rating not less than the available short-circuit current.

- **230.82(8):** Revised to permit transient voltage surge suppressors installed in listed equipment to be connected on the supply side of the service disconnecting means.

- **230.95:** Deleted definition of *solidly grounded* and incorporated the concept into the requirement. Added new fine print note referencing 517.17(A).

Contents

230.1 Scope

This article covers service conductors and equipment for control and protection of services and their installation requirements.

FPN: See Figure 230.1.

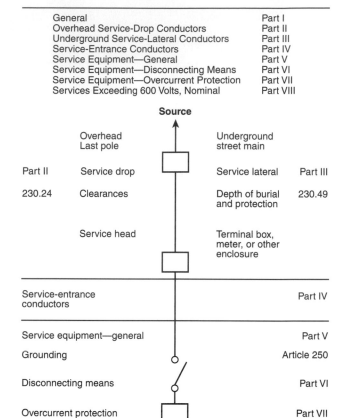

General	Part I
Overhead Service-Drop Conductors	Part II
Underground Service-Lateral Conductors	Part III
Service-Entrance Conductors	Part IV
Service Equipment—General	Part V
Service Equipment—Disconnecting Means	Part VI
Service Equipment—Overcurrent Protection	Part VII
Services Exceeding 600 Volts, Nominal	Part VIII

Figure 230.1 Services

The requirements for the subjects covered in Parts I through VIII of Article 230 are arranged as follows:

Subject	Part	Section
General	I	230.2–230.10
Overhead Service-Drop Conductors	II	230.22–230.29
Underground Service-Lateral Conductors	III	230.30–230.33
Service-Entrance Conductors	IV	230.40–230.56
Service-Equipment—General	V	230.62–230.66

Subject	Part	Section
Service-Equipment— Disconnecting Means	VI	230.70–230.82
Service-Equipment—Overcurrent Protection	VII	230.90–230.95
Services Exceeding 600 Volts, Nominal	VIII	230.200–230.212

I. General

230.2 Number of Services

A building or other structure served shall be supplied by only one service unless permitted in 230.2(A) through (D). For the purpose of 230.40, Exception No. 2 only, underground sets of conductors, 1/0 AWG and larger, running to the same location and connected together at their supply end but not connected together at their load end shall be considered to be supplying one service.

The basic requirement of 230.2 is that a building or other structure can be supplied by only one service. However, under certain conditions, where a single service may not be adequate to supply a building or structure, 230.2 permits the installation of additional services. Sections 230.2(A) through 230.2(D) describe those conditions under which more than one service is permitted. Where more than one service (or combination of service, feeder, and branch circuit) is installed, 230.2(E) requires that a permanent plaque or directory with the pertinent information on the multiple sources of supply be located at each supply source disconnecting means.

(A) Special Conditions Additional services shall be permitted to supply the following:

(1) Fire pumps
(2) Emergency systems
(3) Legally required standby systems
(4) Optional standby systems
(5) Parallel power production systems

The intent of 230.2(A) is that a disruption of the main building service should not disconnect fire pump equipment, emergency systems, legally required standby systems, optional standby systems, or parallel power production systems.

(6) Systems designed for connection to multiple sources of supply for the purpose of enhanced reliability

(B) Special Occupancies By special permission, additional services shall be permitted for either of the following:

(1) Multiple-occupancy buildings where there is no available space for service equipment accessible to all occupants
(2) A single building or other structure sufficiently large to make two or more services necessary

Section 230.2(B) permits additional services for certain occupancies by special permission (written consent of the authority having jurisdiction; see the definition of *authority having jurisdiction* in Article 100). A separate service in multiple-occupancy buildings may be necessary if no space is available for service equipment accessible to all occupants. Also, if a building is large, more than one service is permitted. The expansion of buildings, shopping centers, and industrial plants often necessitates the addition of one or more services. It may, for example, be impractical or impossible to install one service for an industrial plant with sufficient capacity for any and all future loads. It is also impractical to run extremely long feeders. The serving utility and the authority having jurisdiction should be consulted before the use of this special permission is contemplated.

(C) Capacity Requirements Additional services shall be permitted under any of the following:

(1) Where the capacity requirements are in excess of 2000 amperes at a supply voltage of 600 volts or less
(2) Where the load requirements of a single-phase installation are greater than the serving agency normally supplies through one service
(3) By special permission

Section 230.2(C) permits a building or structure to be served by two or more services if capacity requirements are in excess of 2000 amperes at a supply voltage of 600 or less. In such cases, it is not required that each service be rated 2000 amperes or that there be one service rated 2000 amperes and the additional service(s) be rated for the calculated load in excess of 2000 amperes. For example, this provision permits a building with a calculated load of 2300 amperes to have two 1200-ampere services. Additional services for lesser loads are also allowed by special permission.

Many electric power companies have specifications for and have adopted special regulations covering certain types of electrical loads and service equipment that may be energized from their lines. Consultation with the serving utility is advised to determine line and transformer capacities before electrical services for large buildings and facilities are designed.

(D) Different Characteristics Additional services shall be permitted for different voltages, frequencies, or phases, or for different uses, such as for different rate schedules.

Section 230.2(D) permits the installation of more than one service for different characteristics, such as different voltages, frequencies, single-phase or 3-phase services, or different utility rate schedules. For example, different service characteristics exist between a 3-wire, 120/240-volt, single-phase service and a 3-phase, 4-wire, 480Y/277-volt service. For different applications, such as different rate schedules, this requirement allows a second service for supplying a second meter on a different rate. Curtailable loads, interruptible loads, electric heating, and electric water heating are examples of loads that may be on a different rate.

(E) Identification Where a building or structure is supplied by more than one service, or any combination of branch circuits, feeders, and services, a permanent plaque or directory shall be installed at each service disconnect location denoting all other services, feeders, and branch circuits supplying that building or structure and the area served by each. See 225.37.

Section 230.2(E) states that where any combination of branch circuits, feeders, and services supplies power to a building or structure, a permanent plaque or directory must be installed at each service disconnect location to indicate where the other disconnects that feed the building are located, as illustrated in Exhibit 230.1. All the other services on or in the building or structure and the area served by

each must be noted on the plaques or directories. The plaques or directories should be of sufficient durability to withstand the ambient environment. See the commentary following 225.37 for further information on identification of multiple supply sources to a building or structure.

Exhibits 230.2 through 230.13 illustrate examples of permitted service configurations in accordance with 230.2(B), 230.2(C), 230.40 Exceptions No.1 and No. 2, 230.71, and 230.72. The figures are intended to clarify some of the *Code* rules that affect services that are often misunderstood. No attempt has been made to include every type of service arrangement. It should be understood that the

One building with more than one occupancy

Exhibit 230.2 Two services. Two service drops supplying two services installed at separate locations for a building where there is no available space for service equipment accessible to all occupants. Maximum of six service disconnecting means for each service.

Service no. 1 - Suite 10
Service no. 2 located at northeast corner of building

Service no. 2 - Suite 20
Service no. 1 located at southeast corner of building

Exhibit 230.1 An example of more than one service installed for one building and permanent plaques or directories denoting all other services and the area served by each.

A single-occupancy building with more than one service

Exhibit 230.3 Two services. Two service drops supplying two services installed at separate locations for a building with capacity requirements exceeding 2000 amperes. Maximum of six service disconnecting means for each service.

One building with two groups of occupancies

Exhibit 230.4 One service. One service drop supplying two service equipment enclosures installed at separate locations, each with maximum of six service disconnecting means.

One building with more than one occupancy with separate loads

Exhibit 230.5 One service. One service drop supplying maximum of six service equipment enclosures grouped at one location.

term *one location*, as applied to services, is determined by the authority having jurisdiction.

230.3 One Building or Other Structure Not to Be Supplied Through Another

Service conductors supplying a building or other structure shall not pass through the interior of another building or other structure.

Service conductors are permitted to be installed along the exterior of one building to supply another building. However, service conductors supplying a building are not permitted to pass through the interior of a building. Each building

One building with more than one occupancy with separate loads

Exhibit 230.6 One service. One service drop supplying a single service equipment enclosure with maximum of six service disconnecting means grouped at one location. Optional arrangement to that in Exhibit 230.5.

One building with more than one occupancy with separate loads

Exhibit 230.7 One service. One service lateral consisting of six sets of conductors 1/0 AWG or larger, terminating in a single service equipment enclosure with maximum of six service disconnecting means.

served in this manner is required to be provided with a disconnecting means for all ungrounded conductors, in accordance with Part VI, Service Equipment—Disconnecting Means.

For example, in Exhibit 230.14, Building No. 2 service is *not* to be supplied through Building No. 1 service. The service disconnecting means shown for Building No. 1 and Building No. 2 are located on the exterior walls. A disconnecting means suitable for use as service equipment is required to be provided for each building. It is important to note that this requirement applies only to service conductors

One building with more than one occupancy with separate loads

Exhibit 230.8 Two services. Two service laterals, terminating in two service equipment enclosures installed at separate locations, with each enclosure permitted to have maximum of six service disconnecting means.

One building with more than one occupancy with separate loads

Exhibit 230.10 One service. Two service laterals, each consisting of conductors 1/0 AWG or larger, supplying two sets of service-entrance conductors terminating in two service equipment enclosures grouped in one location. Not more than six service disconnecting means permitted at this service location.

One building with more than one occupancy with separate loads

Exhibit 230.9 One service. One service lateral supplying two sets of service-entrance conductors terminating in two service equipment enclosures grouped in one location. Not more than six service disconnecting means permitted at this service location.

Four services to four separate buildings

Exhibit 230.11 Four services. Four service laterals supplying four service equipment enclosures installed at separate locations on a contiguous structure, each enclosure with a maximum of six service disconnecting means. Note presence of firewalls. See definition of *building* in Article 100.

and does not apply to feeder or branch-circuit conductors that originate in one building, pass through that building, and exit the building of origin en route to supplying a separate building or structure. Feeder and branch-circuit conductors are provided with overcurrent protection at the point they receive their supply unless otherwise permitted by 240.21.

230.6 Conductors Considered Outside the Building

Conductors shall be considered outside of a building or other structure under any of the following conditions:

(1) Where installed under not less than 50 mm (2 in.) of concrete beneath a building or other structure
(2) Where installed within a building or other structure in a raceway that is encased in concrete or brick not less than 50 mm (2 in.) thick

Service-entrance conductors are considered to be "outside" a building if they are installed beneath the building under not less than 2 in. of concrete or concealed in a raceway

One building with more than one occupancy with separate loads

Exhibit 230.12 Two services. Two service laterals supplying two service equipment enclosures installed at separate locations, each enclosure with a maximum of six service disconnecting means.

One building with more than one occupancy with separate loads

Exhibit 230.13 One service. One service lateral supplying four service equipment enclosures installed at different locations, with each enclosure permitted to have a maximum of six service disconnecting means.

within the building and encased by not less than 2 in. of concrete or brick, as illustrated in Exhibit 230.15. Service-entrance conductors are also considered outside the building if they are installed in a vault. See 450.41 through 450.48 for the construction requirements of transformer vaults. Service conductors installed under 18 in. of earth beneath the building are also considered outside the building according to 230.6(4).

Exhibit 230.14 Service conductors installed in accordance with 230.3 so as not to pass through the interior of Building No. 1 to supply Building No. 2.

Exhibit 230.15 Service conductors considered outside a building where installed under not less than 2 in. of concrete beneath the building or in a raceway encased by not less than 2 in. of concrete or brick within the building.

(3) Where installed in any vault that meets the construction requirements of Article 450, Part III
(4) Where installed in conduit and under not less than 450 mm (18 in.) of earth beneath a building or other structure

230.7 Other Conductors in Raceway or Cable

Conductors other than service conductors shall not be installed in the same service raceway or service cable.

All feeder and branch-circuit conductors must be separated from service conductors. Service conductors are not provided with overcurrent protection where they receive their supply; they are protected against overload conditions at their load end by the service disconnect fuses or circuit breakers. The amount of current that could be imposed on feeder or branch-circuit conductors, should they be in the same raceway and a fault occur, would be much higher than the ampacity of the feeder or branch-circuit conductors.

Exception No. 1: Grounding conductors and bonding jumpers.

Exception No. 2: Load management control conductors having overcurrent protection.

Because load management control conductors, control circuits, and switch leg conductors for use with special rate meters are usually short, they are allowed in the service raceway or cable. (See Exhibit 230.24 for an example.)

230.8 Raceway Seal

Where a service raceway enters a building or structure from an underground distribution system, it shall be sealed in accordance with 300.5(G). Spare or unused raceways shall also be sealed. Sealants shall be identified for use with the cable insulation, shield, or other components.

Sealant, such as duct seal or a bushing incorporating the physical characteristics of a seal, must be used to seal the ends of service raceways. The intent of this requirement is to prevent water, usually the result of condensation due to temperature differences, from entering the service equipment via the raceway. The sealant material should be compatible with the conductor insulation and should not cause deterioration of the insulation over time. For underground services over 600 volts, nominal, refer to 300.50(E) for raceway seal requirements. (See Exhibit 300.10 for an example.)

230.9 Clearances on Buildings

Service conductors and final spans shall comply with 230.9(A), (B), and (C).

(A) Clearances Service conductors installed as open conductors or multiconductor cable without an overall outer jacket shall have a clearance of not less than 900 mm (3 ft) from windows that are designed to be opened, doors, porches, balconies, ladders, stairs, fire escapes, or similar locations.

Exception: Conductors run above the top level of a window shall be permitted to be less than the 900-mm (3-ft) requirement.

As illustrated in Exhibit 230.16, the clearance of 3 ft applies to open conductors, not to a raceway or to a cable assembly

that has an overall outer jacket such as Types SE, MC, and MI cables. The intent is to protect the conductors from physical damage and to protect personnel from accidental contact with the conductors. The exception permits service conductors, including drip loops and service-drop conductors, to be located just above window openings, because they are considered out of reach.

Exhibit 230.16 Required dimensions for service conductors located alongside a window (left) and service conductors above the top level of a window designed to be opened (right).

(B) Vertical Clearance The vertical clearance of final spans above, or within 900 mm (3 ft) measured horizontally of, platforms, projections, or surfaces from which they might be reached shall be maintained in accordance with 230.24(B).

Where service conductors are located within 3 ft measured horizontally of a balcony, stair landing, or other platform, clearance to the platform must be at least 10 ft, as shown in Exhibit 230.17. See 230.24(B) for vertical clearances from ground.

(C) Building Openings Overhead service conductors shall not be installed beneath openings through which materials may be moved, such as openings in farm and commercial buildings, and shall not be installed where they obstruct entrance to these building openings.

Elevated openings in buildings through which materials may be moved, such as barns and storage buildings, are often high enough for the service conductors to be installed below the opening. However, 230.9(C) prohibits such placement in order to reduce the likelihood of damage to the service conductors and the potential for electric shock to persons using the openings.

Exhibit 230.17 Required dimensions for service conductors located above a stair landing, according to 230.9(B) and 230.24(B).

230.10 Vegetation as Support

Vegetation such as trees shall not be used for support of overhead service conductors.

II. Overhead Service-Drop Conductors

230.22 Insulation or Covering

Individual conductors shall be insulated or covered.

Exception: The grounded conductor of a multiconductor cable shall be permitted to be bare.

The intent of 230.22 is to prevent problems created by weather and abrasion and other deleterious effects that reduce the insulating quality of the covering or insulation. The grounded conductor (neutral) of triplex or quadraplex service-drop conductors is often bare and used to mechanically support the ungrounded conductors.

230.23 Size and Rating

(A) General Conductors shall have sufficient ampacity to carry the current for the load as calculated in accordance with Article 220 and shall have adequate mechanical strength.

When a load is added to any service, the installer must be aware of all existing loads. The potential for overloading the service conductors must be governed by installer responsibility and inspector awareness; they should not rely solely on the usual method of overcurrent protection. The serving electric utility should be notified whenever load is added, to ensure that adequate power is available.

(B) Minimum Size The conductors shall not be smaller than 8 AWG copper or 6 AWG aluminum or copper-clad aluminum.

Exception: Conductors supplying only limited loads of a single branch circuit — such as small polyphase power, controlled water heaters, and similar loads — shall not be smaller than 12 AWG hard-drawn copper or equivalent.

(C) Grounded Conductors The grounded conductor shall not be less than the minimum size as required by 250.24(C).

230.24 Clearances

Service-drop conductors shall not be readily accessible and shall comply with 230.24(A) through (D) for services not over 600 volts, nominal.

(A) Above Roofs Conductors shall have a vertical clearance of not less than 2.5 m (8 ft) above the roof surface. The vertical clearance above the roof level shall be maintained for a distance of not less than 900 mm (3 ft) in all directions from the edge of the roof.

Service-drop conductors are not permitted to be readily accessible. This main rule applies to services rated 600 volts and less, grounded or ungrounded. An 8-ft vertical clearance is required over the roof surface, extending 3 ft in all directions from the edge. Note that Exception No. 4 to 230.24(A) allows the final span of the service drop to enter this space in order to attach to the building or service mast.

Exception No. 1: The area above a roof surface subject to pedestrian or vehicular traffic shall have a vertical clearance from the roof surface in accordance with the clearance requirements of 230.24(B).

Exception No. 1 to 230.24(A) requires service-drop conductor clearance above a roof surface subject to vehicular or pedestrian traffic, such as the rooftop parking area shown in Exhibit 230.18, to meet the clearance requirements of 230.24(B).

Exception No. 2: Where the voltage between conductors does not exceed 300 and the roof has a slope of 100 mm in

Exhibit 230.18 Service-drop conductor clearance required by 230.24(A), Exception No. 1.

300 mm (4 in. in 12 in.) or greater, a reduction in clearance to 900 mm (3 ft) shall be permitted.

Exception No. 2 to 230.24(A) permits a reduction in service-drop conductor clearance above the roof from 8 ft to 3 ft, as illustrated in Exhibit 230.19, where the voltage between conductors does not exceed 300 volts (e.g., 120/240, 208Y/ 120 services) and the roof is sloped not less than 4 in. vertically in 12 in. horizontally. Steeply sloped roofs are less likely to be walked on by other than those who have to work on the roof. There are no restrictions on the length of the conductors over the roof.

Exhibit 230.19 Reduction in clearance above a roof as permitted by 230.24(A), Exception No. 2.

Exception No. 3: Where the voltage between conductors does not exceed 300, a reduction in clearance above only the overhanging portion of the roof to not less than 450 mm

(18 in.) shall be permitted if (1) not more than 1.8 m (6 ft) of service-drop conductors, 1.2 m (4 ft) horizontally, pass above the roof overhang, and (2) they are terminated at a through-the-roof raceway or approved support.

FPN: See 230.28 for mast supports.

Exception No. 3 to 230.24(A) permits a reduction of service-drop conductor clearances to 18 in. above the roof, as illustrated in Exhibit 230.20. This reduction is for service-mast (through-the-roof) installations where the voltage between conductors does not exceed 300 volts (e.g., 120/240, 208Y/ 120 services) and the mast is located within 4 ft of the edge of the roof, measured horizontally. Exception No. 3 applies to sloped or flat roofs that are easily walked on. Not more than 6 ft of conductors is permitted to pass over the roof.

Exhibit 230.20 Reduction in clearance above a roof as permitted by 230.24(A), Exception No. 3.

Exception No. 4: The requirement for maintaining the vertical clearance 900 mm (3 ft) from the edge of the roof shall not apply to the final conductor span where the service drop is attached to the side of a building.

Section 230.24(A) applies to the vertical clearance above roofs for service-drop conductors up to 600 volts. This main rule requires a vertical clearance of 8 ft above the roof, including those areas 3 ft in all directions beyond the edge of the roof.

Exception No. 4 to 230.24(A) exempts the final span of a service drop attached to the side of a building from the 8-ft and 3-ft requirements, to allow the service conductors to be attached to the building, as illustrated in Exhibit 230.21. Exception No. 2 and Exception No. 3 permit lesser clearances for service drops of 300 volts or less, as illustrated in Exhibits 230.19 and 230.20.

Exhibit 230.21 Clearance of the final span of a service drop, as permitted by 230.24(A), Exception No. 4.

If the roof is subject to pedestrian or vehicular traffic, the vertical clearance of the service drop must be the same as the vertical clearance from the ground, in accordance with 230.24(B).

(B) Vertical Clearance from Ground Service-drop conductors, where not in excess of 600 volts, nominal, shall have the following minimum clearance from final grade:

(1) 3.0 m (10 ft) — at the electric service entrance to buildings, also at the lowest point of the drip loop of the building electric entrance, and above areas or sidewalks accessible only to pedestrians, measured from final grade or other accessible surface only for service-drop cables supported on and cabled together with a grounded bare messenger where the voltage does not exceed 150 volts to ground

(2) 3.7 m (12 ft) — over residential property and driveways, and those commercial areas not subject to truck traffic where the voltage does not exceed 300 volts to ground

(3) 4.5 m (15 ft) — for those areas listed in the 3.7-m (12-ft) classification where the voltage exceeds 300 volts to ground

(4) 5.5 m (18 ft) — over public streets, alleys, roads, parking areas subject to truck traffic, driveways on other than residential property, and other land such as cultivated, grazing, forest, and orchard

Exhibit 230.22 illustrates the 10-ft, 12-ft, 15-ft, and 18-ft vertical clearances from ground for service-drop conductors up to 600 volts, as specified by 230.24(B). The voltages given in 230.24(B)(1), 230.24(B)(2), and 230.24(B)(3) are nominal voltages to ground, not the nominal voltage between circuit conductors as specified in 230.24(A), Exceptions No. 2 and No. 3. This is important to note, because a 480Y/277-volt system (277 volts to ground) is covered by the 12-ft clearance requirement in 230.24(B)(2), but an ungrounded 480-volt system (considered to be 480 volts to ground) is required to have a 15-ft clearance over commercial areas not subject to truck traffic, in accordance with (B)(3).

Exhibit 230.22 Clearances in accordance with 230.24(B).

(C) Clearance from Building Openings See 230.9.

(D) Clearance from Swimming Pools See 680.8.

230.26 Point of Attachment

The point of attachment of the service-drop conductors to a building or other structure shall provide the minimum clearances as specified in 230.9 and 230.24. In no case shall this point of attachment be less than 3.0 m (10 ft) above finished grade.

230.27 Means of Attachment

Multiconductor cables used for service drops shall be attached to buildings or other structures by fittings identified for use with service conductors. Open conductors shall be attached to fittings identified for use with service conductors or to noncombustible, nonabsorbent insulators securely attached to the building or other structure.

See 230.51 for mounting and supporting of service cables and individual open service conductors. See 230.54 for connections at service heads.

230.28 Service Masts as Supports

Where a service mast is used for the support of service-drop conductors, it shall be of adequate strength or be supported by braces or guys to withstand safely the strain imposed by the service drop. Where raceway-type service masts are used, all raceway fittings shall be identified for use with service masts. Only power service-drop conductors shall be permitted to be attached to a service mast.

Where the service drop is secured to the mast, a guy wire may be installed to support the mast and provide adequate mechanical strength to support the service drop. Communications conductors such as those for cable TV or telephone service are not permitted to be attached to the service mast.

230.29 Supports over Buildings

Service-drop conductors passing over a roof shall be securely supported by substantial structures. Where practicable, such supports shall be independent of the building.

III. Underground Service-Lateral Conductors

230.30 Insulation

Service-lateral conductors shall be insulated for the applied voltage.

Exception: A grounded conductor shall be permitted to be uninsulated as follows:

(1) Bare copper used in a raceway.

(2) Bare copper for direct burial where bare copper is judged to be suitable for the soil conditions.

(3) Bare copper for direct burial without regard to soil conditions where part of a cable assembly identified for underground use.

(4) Aluminum or copper-clad aluminum without individual insulation or covering where part of a cable assembly identified for underground use in a raceway or for direct burial.

Exhibit 230.23 shows various applications of bare grounded service-lateral and service-entrance conductors for underground locations. Aluminum or copper-clad aluminum conductors must be insulated if they are run in a raceway or direct buried, unless they are part of a cable assembly identified for the use.

Exhibit 230.23 Bare grounded service-lateral and service-entrance conductors for underground locations.

230.31 Size and Rating

(A) General Service-lateral conductors shall have sufficient ampacity to carry the current for the load as calculated in accordance with Article 220 and shall have adequate mechanical strength.

(B) Minimum Size The conductors shall not be smaller than 8 AWG copper or 6 AWG aluminum or copper-clad aluminum.

Exception: Conductors supplying only limited loads of a single branch circuit — such as small polyphase power, controlled water heaters, and similar loads — shall not be smaller than 12 AWG copper or 10 AWG aluminum or copper-clad aluminum.

(C) Grounded Conductors The grounded conductor shall not be less than the minimum size required by 250.24(C).

See 310.15(B)(3) for requirements on the allowable ampacity of bare and covered conductors. For further information on sizing the grounded conductor for dwelling services, refer to 310.15(B)(6); for more information on sizing the grounded service conductor for any occupancy, see the commentary following 230.42(C).

230.32 Protection Against Damage

Underground service-lateral conductors shall be protected against damage in accordance with 300.5. Service-lateral conductors entering a building shall be installed in accordance with 230.6 or protected by a raceway wiring method identified in 230.43.

230.33 Spliced Conductors

Service-lateral conductors shall be permitted to be spliced or tapped in accordance with 110.14, 300.5(E), 300.13, and 300.15.

IV. Service-Entrance Conductors

230.40 Number of Service-Entrance Conductor Sets

Each service drop or lateral shall supply only one set of service-entrance conductors.

Exception No. 1: A building shall be permitted to have one set of service-entrance conductors for each service, as defined in 230.2, run to each occupancy or group of occupancies.

Exception No. 2: Where two to six service disconnecting means in separate enclosures are grouped at one location and supply separate loads from one service drop or lateral, one set of service-entrance conductors shall be permitted to supply each or several such service equipment enclosures.

See Exhibit 230.2 through Exhibit 230.13 for examples of permitted service configurations. Note that in Exhibit 230.7 and Exhibit 230.10 the multiple sets of service lateral conductors are considered to be a single service lateral in accordance with 230.2.

Exception No. 3: A single-family dwelling unit and a separate structure shall be permitted to have one set of service-entrance conductors run to each from a single service drop or lateral.

Each set of service-drop or service-lateral conductors is allowed to supply only one set of service-entrance conductors. However, if a service drop or a service lateral supplies a building with more than one occupancy, such as multifamily dwellings, strip malls, and office buildings, each service drop or service lateral is allowed to supply more than one set of service-entrance conductors, provided they are run to each occupancy or group of occupancies.

Exception No. 3 to 230.40 allows a second set of service-entrance conductors supplied by a single service drop or lateral at a single-family dwelling unit to also supply another building on the premises, such as a garage or storage shed.

Exception No. 4: A two-family dwelling or a multifamily dwelling shall be permitted to have one set of service-entrance conductors installed to supply the circuits covered in 210.25.

Exception No. 5: One set of service-entrance conductors connected to the supply side of the normal service disconnecting means shall be permitted to supply each or several systems covered by 230.82(4) or 230.82(5).

230.41 Insulation of Service-Entrance Conductors

Service-entrance conductors entering or on the exterior of buildings or other structures shall be insulated.

Exception: A grounded conductor shall be permitted to be uninsulated as follows:

(1) *Bare copper used in a raceway or part of a service cable assembly.*
(2) *Bare copper for direct burial where bare copper is judged to be suitable for the soil conditions.*
(3) *Bare copper for direct burial without regard to soil conditions where part of a cable assembly identified for underground use.*
(4) *Aluminum or copper-clad aluminum without individual insulation or covering where part of a cable assembly or identified for underground use in a raceway, or for direct burial.*
(5) *Bare conductors used in an auxiliary gutter.*

Service-entrance conductors must be insulated; however, bare grounded conductors are permitted under the same conditions as underground service laterals. Bare grounded conductors are permitted in an auxiliary gutter that is used to supplement the space of a metering enclosure, service panelboard, or service switchboard. See Exhibit 230.23, which shows examples of bare grounded conductors for underground locations.

230.42 Minimum Size and Rating

(A) General The ampacity of the service-entrance conductors before the application of any adjustment or correction factors shall not be less than either (A)(1) or (A)(2). Loads shall be determined in accordance with Article 220. Ampacity shall be determined from 310.15. The maximum allowable current of busways shall be that value for which the busway has been listed or labeled.

(1) The sum of the noncontinuous loads plus 125 percent of continuous loads
(2) The sum of the noncontinuous load plus the continuous load if the service-entrance conductors terminate in an overcurrent device where both the overcurrent device and its assembly are listed for operation at 100 percent of their rating

(B) Specific Installations In addition to the requirements of 230.42(A), the minimum ampacity for ungrounded conductors for specific installations shall not be less than the rating of the service disconnecting means specified in 230.79(A) through (D).

The basic rule in 230.42(B) requires ungrounded service conductors to be sized large enough to carry the calculated load and to have an ampacity not less than the minimum rating of the service disconnecting means required by 230.79(A) through 230.79(D).

(C) Grounded Conductors The grounded conductor shall not be less than the minimum size as required by 250.24(C).

The minimum conductor needed to carry the maximum unbalanced load calculated in accordance with 220.61 and the minimum size conductor necessary to complete the effective ground fault current path required by 250.24(C) are the two sections of the *Code* that directly affect the minimum size required for the grounded service conductor. The largest conductor determined in accordance with these two sections is to be used as the minimum size for the grounded service conductor. Section 220.61 does have demand factors that can be applied to certain portions of the unbalanced load.

The additional heating effect of harmonic currents due to nonlinear loads should be considered in the sizing of the neutral conductor of a 3-phase, 4-wire wye system. If the service to a building is a single-phase, 3-wire, 120/240-volt system with no 240-volt loads, the maximum current in the neutral would be the same as the maximum current in the ungrounded conductor. If all loads connected to one leg are "on" and all the other loads on the other leg are "off," the neutral will carry the maximum current. In such cases, the size of the service neutral cannot be reduced and would be sized the same as the ungrounded conductors. See 310.15(B)(4) for more information on when the grounded conductor has to be considered a current-carrying conductor for the purposes of ampacity adjustment.

230.43 Wiring Methods for 600 Volts, Nominal, or Less

Service-entrance conductors shall be installed in accordance with the applicable requirements of this *Code* covering the type of wiring method used and shall be limited to the following methods:

(1) Open wiring on insulators
(2) Type IGS cable
(3) Rigid metal conduit
(4) Intermediate metal conduit
(5) Electrical metallic tubing
(6) Electrical nonmetallic tubing (ENT)
(7) Service-entrance cables
(8) Wireways
(9) Busways
(10) Auxiliary gutters
(11) Rigid nonmetallic conduit

(12) Cablebus
(13) Type MC cable
(14) Mineral-insulated, metal-sheathed cable
(15) Flexible metal conduit not over 1.8 m (6 ft) long or liquidtight flexible metal conduit not over 1.8 m (6 ft) long between raceways, or between raceway and service equipment, with equipment bonding jumper routed with the flexible metal conduit or the liquidtight flexible metal conduit according to the provisions of 250.102(A), (B), (C), and (E)
(16) Liquidtight flexible nonmetallic conduit

Where flexible metal conduit or liquidtight flexible metal conduit is installed for services, a bonding jumper must be installed between both ends within the raceway. The bonding jumper is allowed to be installed outside the raceway, but it must follow the path of the raceway and cannot exceed 6 ft in length. The bonding jumper must not be wrapped or spiraled around the flexible conduit.

230.44 Cable Trays

Cable tray systems shall be permitted to support service-entrance conductors. Cable trays used to support service-entrance conductors shall contain only service-entrance conductors.

Exception: Conductors other than service-entrance conductors shall be permitted to be installed in a cable tray with service-entrance conductors, provided a solid fixed barrier of a material compatible with the cable tray is installed to separate the service-entrance conductors from other conductors installed in the cable tray.

230.46 Spliced Conductors

Service-entrance conductors shall be permitted to be spliced or tapped in accordance with 110.14, 300.5(E), 300.13, and 300.15.

Splices are permitted in service-entrance conductors if the splice meets the requirements of 230.46. Splices must be in an enclosure or be direct buried using a listed underground splice kit. It is common to have an underground service lateral terminate at a terminal box either inside or outside the building. At that point, service conductors may be spliced or run directly to the service equipment.

Splices are permitted where, for example, the cable enters a terminal box and a different wiring method, such as conduit, continues to the service equipment. Splices are most common where metering equipment is located on the line side of service equipment, service busways, and taps for supplying up to six disconnecting means. See Exhibit 230.24 for splice/flash connections permitted in metering equipment.

Exhibit 230.24 Time clock and control switch integral to a meter for use, generally, with water heaters.

230.49 Protection Against Physical Damage — Underground

Underground service-entrance conductors shall be protected against physical damage in accordance with 300.5.

230.50 Protection of Open Conductors and Cables Against Damage — Above Ground

Service-entrance conductors installed above ground shall be protected against physical damage as specified in 230.50(A) or (B).

(A) Service Cables Service cables, where subject to physical damage, shall be protected by any of the following:

(1) Rigid metal conduit
(2) Intermediate metal conduit
(3) Schedule 80 rigid nonmetallic conduit
(4) Electrical metallic tubing
(5) Other approved means

(B) Other Than Service Cable Individual open conductors and cables other than service cables shall not be installed within 3.0 m (10 ft) of grade level or where exposed to physical damage.

Exception: Type MI and Type MC cable shall be permitted within 3.0 m (10 ft) of grade level where not exposed to physical damage or where protected in accordance with 300.5(D).

230.51 Mounting Supports

Cables or individual open service conductors shall be supported as specified in 230.51(A), (B), or (C).

(A) Service Cables Service cables shall be supported by straps or other approved means within 300 mm (12 in.) of every service head, gooseneck, or connection to a raceway or enclosure and at intervals not exceeding 750 mm (30 in.).

(B) Other Cables Cables that are not approved for mounting in contact with a building or other structure shall be mounted on insulating supports installed at intervals not exceeding 4.5 m (15 ft) and in a manner that maintains a clearance of not less than 50 mm (2 in.) from the surface over which they pass.

(C) Individual Open Conductors Individual open conductors shall be installed in accordance with Table 230.51(C). Where exposed to the weather, the conductors shall be mounted on insulators or on insulating supports attached to racks, brackets, or other approved means. Where not exposed to the weather, the conductors shall be mounted on glass or porcelain knobs.

230.52 Individual Conductors Entering Buildings or Other Structures

Where individual open conductors enter a building or other structure, they shall enter through roof bushings or through

Table 230.51(C) Supports

Maximum Volts	Maximum Distance Between Supports		Minimum Clearance			
			Between Conductors		From Surface	
	m	ft	mm	in.	mm	in.
600	2.7	9	150	6	50	2
600	4.5	15	300	12	50	2
300	1.4	4½	75	3	50	2
600*	1.4*	4½*	65*	2½*	25*	1*

*Where not exposed to weather.

the wall in an upward slant through individual, noncombustible, nonabsorbent insulating tubes. Drip loops shall be formed on the conductors before they enter the tubes.

230.53 Raceways to Drain

Where exposed to the weather, raceways enclosing service-entrance conductors shall be raintight and arranged to drain. Where embedded in masonry, raceways shall be arranged to drain.

Exception: As permitted in 348.12(1).

The goal of 230.53 is to prevent water from entering internal electrical equipment through the raceway system. Service raceways exposed to the weather must have raintight fittings and drain holes. During the installation of raceways in masonry, provisions to drain and divert water should be made to prevent the entrance of surface water, rain, or water from poured concrete.

230.54 Overhead Service Locations

(A) Raintight Service Head Service raceways shall be equipped with a raintight service head at the point of connection to service-drop conductors.

(B) Service Cable Equipped with Raintight Service Head or Gooseneck Service cables shall be equipped with a raintight service head.

Exception: Type SE cable shall be permitted to be formed in a gooseneck and taped with a self-sealing weather-resistant thermoplastic.

(C) Service Heads and Goosenecks Above Service-Drop Attachment Service heads and goosenecks in service-entrance cables shall be located above the point of attachment of the service-drop conductors to the building or other structure.

Exception: Where it is impracticable to locate the service head or gooseneck above the point of attachment, the service head or gooseneck location shall be permitted not farther than 600 mm (24 in.) from the point of attachment.

(D) Secured Service cables shall be held securely in place.

(E) Separately Bushed Openings Service heads shall have conductors of different potential brought out through separately bushed openings.

Exception: For jacketed multiconductor service cable without splice.

(F) Drip Loops Drip loops shall be formed on individual conductors. To prevent the entrance of moisture, service-entrance conductors shall be connected to the service-drop conductors either (1) below the level of the service head or

(2) below the level of the termination of the service-entrance cable sheath.

(G) Arranged That Water Will Not Enter Service Raceway or Equipment Service-drop conductors and service-entrance conductors shall be arranged so that water will not enter service raceway or equipment.

Service raceways and service cables are required to be equipped with a raintight service head (weatherhead). Type SE service-entrance cables may be installed without a service head if they are run continuously from a utility pole to metering or service equipment or if they are shaped in a downward direction (forming a "gooseneck") and sealed by taping and painting, as shown in Exhibit 230.25.

Service heads and goosenecks are required to be located above the service-drop point of attachment to the building or structure unless such location is not feasible, in which case the service head or gooseneck is permitted to be located not farther than 24 in. from the point of attachment, in accordance with the exception to 230.54(C). Individual conductors should extend in a downward direction, as shown in Exhibit 230.25, or drip loops should be formed so that, where splices are made, they are at the lowest point of the drip loop.

Exhibit 230.25 A service-entrance cable that terminates in a gooseneck without a raintight service head (weatherhead).

230.56 Service Conductor with the Higher Voltage to Ground

On a 4-wire, delta-connected service where the midpoint of one phase winding is grounded, the service conductor having the higher phase voltage to ground shall be durably and permanently marked by an outer finish that is orange in color, or by other effective means, at each termination or junction point.

Proper service connections require the service conductors having the higher voltage to ground to be durably marked by an outer finish of orange, such as by painting, colored adhesive tagging, or taping. Marking should be at both the point of connection to the service-drop conductors and the point of connection to the service disconnecting means. See 110.15 and 408.3(E) for high-leg marking and phase arrangement requirements.

V. Service Equipment — General

230.62 Service Equipment — Enclosed or Guarded

Energized parts of service equipment shall be enclosed as specified in 230.62(A) or guarded as specified in 230.62(B).

(A) Enclosed Energized parts shall be enclosed so that they will not be exposed to accidental contact or shall be guarded as in 230.62(B).

(B) Guarded Energized parts that are not enclosed shall be installed on a switchboard, panelboard, or control board and guarded in accordance with 110.18 and 110.27. Where energized parts are guarded as provided in 110.27(A)(1) and (A)(2), a means for locking or sealing doors providing access to energized parts shall be provided.

230.66 Marking

Service equipment rated at 600 volts or less shall be marked to identify it as being suitable for use as service equipment. Individual meter socket enclosures shall not be considered service equipment.

According to the listing information, panelboards with the neutral factory-bonded to the enclosure are to be marked "Suitable Only for Use as Service Equipment." Other types of equipment intended for optional use, either as service equipment or as subdistribution panelboards for feeders on the load side of the service disconnect, are required by 230.66 to be marked as suitable for use as service equipment. Section 225.36 requires feeder and branch circuit disconnecting means to be suitable for use as service equipment.

VI. Service Equipment — Disconnecting Means

230.70 General

Means shall be provided to disconnect all conductors in a building or other structure from the service-entrance conductors.

(A) Location The service disconnecting means shall be installed in accordance with 230.70(A)(1), (A)(2), and (A)(3).

No maximum distance is specified from the point of entrance of service conductors to a readily accessible location for the installation of a service disconnecting means. The authority enforcing this *Code* has the responsibility for, and is charged with, making the decision as to how far inside the building the service-entrance conductors are allowed to travel to the service disconnecting means. The length of service-entrance conductors should be kept to a minimum inside buildings, because power utilities provide limited overcurrent protection. In the event of a fault, the service conductors could ignite nearby combustible materials.

Some local jurisdictions have ordinances that allow service-entrance conductors to run within the building up to a specified length to terminate at the disconnecting means. The authority having jurisdiction may permit service conductors to bypass fuel storage tanks or gas meters and the like, permitting the service disconnecting means to be located in a readily accessible location.

However, if the authority judges the distance as being excessive, the disconnecting means may be required to be located on the outside of the building or near the building at a readily accessible location that is not necessarily nearest the point of entrance of the conductors. See also 230.6 and Exhibit 230.15 for conductors considered to be outside a building.

See 404.8(A) for mounting-height restrictions for switches and for circuit breakers used as switches.

(1) Readily Accessible Location The service disconnecting means shall be installed at a readily accessible location either outside of a building or structure or inside nearest the point of entrance of the service conductors.

(2) Bathrooms Service disconnecting means shall not be installed in bathrooms.

(3) Remote Control Where a remote control device(s) is used to actuate the service disconnecting means, the service disconnecting means shall be located in accordance with 230.70(A)(1).

(B) Marking Each service disconnect shall be permanently marked to identify it as a service disconnect.

(C) Suitable for Use Each service disconnecting means shall be suitable for the prevailing conditions. Service equipment installed in hazardous (classified) locations shall comply with the requirements of Articles 500 through 517.

230.71 Maximum Number of Disconnects

(A) General The service disconnecting means for each service permitted by 230.2, or for each set of service-entrance conductors permitted by 230.40, Exception Nos. 1, 3, 4, or

5, shall consist of not more than six switches or sets of circuit breakers, or a combination of not more than six switches and sets of circuit breakers, mounted in a single enclosure, in a group of separate enclosures, or in or on a switchboard. There shall be not more than six sets of disconnects per service grouped in any one location. For the purpose of this section, disconnecting means used solely for power monitoring equipment, transient voltage surge suppressors, or the control circuit of the ground-fault protection system or power-operable service disconnecting means, installed as part of the listed equipment, shall not be considered a service disconnecting means.

Section 230.71(A) covers the maximum number of disconnects permitted as the disconnecting means for the service conductors that supply the building or structure. One set of service-entrance conductors, either overhead or underground, is permitted to supply two to six service disconnecting means in lieu of a single main disconnect. A single-occupancy building can have up to six disconnects for each set of service-entrance conductors. Multiple-occupancy buildings (residential or other than residential) can be provided with one main service disconnect or up to six main disconnects for each set of service-entrance conductors.

Multiple-occupancy buildings may have service-entrance conductors run to each occupancy, and each such set of service-entrance conductors may have from one to six disconnects (see 230.40, Exception No. 1).

Where service-entrance conductors are routed outside the building (see 230.6 and Exhibit 230.15), each set of service-entrance conductors is permitted to supply not more than six disconnecting means at each occupancy of a multiple-occupancy building. See Exhibit 230.2 through Exhibit 230.13 for examples of permitted service configurations.

Exhibit 230.26 shows a single enclosure for grouping service equipment that consists of six circuit breakers or six fused switches. This arrangement does not require a main switch. Six separate enclosures also would be permitted as the service equipment. Revised in the 2005 *Code* to also include switches that disconnect power to transient voltage surge suppressors and power monitoring equipment, the last sentence of 230.71(A) specifies that the disconnect switch for such equipment installed as part of the listed equipment does *not* count as one of the six service disconnecting means permitted by 230.71(A). The disconnecting means for the control circuit of ground-fault protection equipment or for a power-operable service disconnecting means are also not considered to be service disconnecting means where such disconnecting means are installed as a component of listed equipment.

(B) Single-Pole Units Two or three single-pole switches or breakers, capable of individual operation, shall be permit-

Exhibit 230.26 An enclosure for grouping service equipment consisting of six circuit breakers or six fused switches.

ted on multiwire circuits, one pole for each ungrounded conductor, as one multipole disconnect, provided they are equipped with handle ties or a master handle to disconnect all conductors of the service with no more than six operations of the hand.

> FPN: See 408.36(A) for service equipment in panelboards, and see 430.95 for service equipment in motor control centers.

230.72 Grouping of Disconnects

(A) General The two to six disconnects as permitted in 230.71 shall be grouped. Each disconnect shall be marked to indicate the load served.

Exception: One of the two to six service disconnecting means permitted in 230.71, where used only for a water pump also intended to provide fire protection, shall be permitted to be located remote from the other disconnecting means.

The water pump in 230.72(A), Exception, is not the fire pump covered by the requirements of Article 695; rather, it is a water pump used for normal water supply and also for fire protection. This application is common in agricultural settings and permits separation of the water pump disconnect so it can remain operational in the event of a problem at the location of the other service disconnecting means.

(B) Additional Service Disconnecting Means The one or more additional service disconnecting means for fire pumps, emergency systems, legally required standby, or optional standby services permitted by 230.2 shall be installed remote from the one to six service disconnecting means for normal service to minimize the possibility of simultaneous interruption of supply.

The intent of 230.2(A) is to permit separate services, where necessary, for fire pumps (with one to six disconnects) or for emergency, legally required standby, or optional standby systems (with one to six disconnects), in addition to the one to six disconnects for the normal building service. Article 230 recognizes that a disruption of the normal building service should not disconnect the fire pump, emergency system, or other exempted systems. Because these services are in addition to the normal services, the one to six disconnects allowed for them are not included as one of the six disconnects for the normal supply. These separate services are permitted by 230.2 and are required to be installed in accordance with all the applicable requirements of Article 230.

(C) Access to Occupants In a multiple-occupancy building, each occupant shall have access to the occupant's service disconnecting means.

A multiple-occupancy building may have any number of dwelling units, offices, and the like that are independent of each other. Unless electric service and maintenance are provided by and under continuous supervision of the building management, the occupants of a multiple-occupancy building must have ready access to their disconnecting means, as required by 240.24(B).

Exception: In a multiple-occupancy building where electric service and electrical maintenance are provided by the building management and where these are under continuous building management supervision, the service disconnecting means supplying more than one occupancy shall be permitted to be accessible to authorized management personnel only.

230.74 Simultaneous Opening of Poles

Each service disconnect shall simultaneously disconnect all ungrounded service conductors that it controls from the premises wiring system.

230.75 Disconnection of Grounded Conductor

Where the service disconnecting means does not disconnect the grounded conductor from the premises wiring, other means shall be provided for this purpose in the service equipment. A terminal or bus to which all grounded conductors can be attached by means of pressure connectors shall be permitted for this purpose. In a multisection switchboard,

disconnects for the grounded conductor shall be permitted to be in any section of the switchboard, provided any such switchboard section is marked.

Provisions are required at the service equipment for disconnecting the grounded conductor from the premises wiring. This disconnection does not have to be by operation of the service disconnecting means. Disconnection can be, and most commonly is, accomplished by manually removing the grounded conductor from the bus or terminal bar to which it is lugged or bolted. This location is often referred to as the *neutral disconnect link*.

Manufacturers design neutral terminal bars for service equipment so that grounded conductors must be cut to be attached; that is, the grounded conductor cannot be run straight through the service equipment without means of disconnection from the premises wiring.

230.76 Manually or Power Operable

The service disconnecting means for ungrounded service conductors shall consist of one of the following:

(1) A manually operable switch or circuit breaker equipped with a handle or other suitable operating means
(2) A power-operated switch or circuit breaker, provided the switch or circuit breaker can be opened by hand in the event of a power supply failure

230.77 Indicating

The service disconnecting means shall plainly indicate whether it is in the open or closed position.

230.79 Rating of Service Disconnecting Means

The service disconnecting means shall have a rating not less than the load to be carried, determined in accordance with Article 220. In no case shall the rating be lower than specified in 230.79(A), (B), (C), or (D).

Three-wire services that supply one-family dwellings are required to be installed using wire with the capacity to supply a 100-ampere service for all single-family dwellings.

A conductor ampacity of 60 amperes is permitted for other loads. Smaller sizes are permitted down to 14 AWG copper (12 AWG aluminum) for installations with one circuit. Two-circuit installations must have a rating of at least 30 amperes. Exhibit 230.27 illustrates the conductor sizing requirements of 230.79 for ungrounded service-entrance conductors. A single service disconnecting means is required to have a rating of not less than the load to be carried.

(A) One-Circuit Installation For installations to supply only limited loads of a single branch circuit, the service disconnecting means shall have a rating of not less than 15 amperes.

(B) Two-Circuit Installations For installations consisting of not more than two 2-wire branch circuits, the service disconnecting means shall have a rating of not less than 30 amperes.

(C) One-Family Dwelling For a one-family dwelling, the service disconnecting means shall have a rating of not less than 100 amperes, 3-wire.

(D) All Others For all other installations, the service disconnecting means shall have a rating of not less than 60 amperes.

230.80 Combined Rating of Disconnects

Where the service disconnecting means consists of more than one switch or circuit breaker, as permitted by 230.71, the combined ratings of all the switches or circuit breakers used shall not be less than the rating required by 230.79.

Section 230.71(A) permits up to six individual switches or circuit breakers, mounted in a single enclosure, in a group of separate enclosures, or in or on a switchboard or several switchboards, to serve as the required service disconnecting means at any one location. Section 230.80 refers to situations in which more than one switch or circuit breaker is used as the disconnecting means and indicates that the combined rating of all the switches or circuit breakers used cannot be less than the rating required for a single switch or circuit breaker.

Section 230.90 requires an overcurrent device to provide overload protection in each ungrounded service conductor. A single overcurrent device must have a rating or setting that is not higher than the allowable ampacity of the service conductors. However, Exception No. 3 to 230.90(A) allows not more than six circuit breakers or six sets of fuses to be considered the overcurrent device. None of these individual overcurrent devices can have a rating or setting higher than the ampacity of the service conductors.

In complying with these rules, it is possible for the total of the six overcurrent devices to be greater than the rating of the service-entrance conductors. However, the size of the service-entrance conductors is required to be adequate for the computed load only, and each individual service disconnecting means is required to be large enough for the individual load it supplies. See the commentary following 230.90(A), Exception No. 3.

230.81 Connection to Terminals

The service conductors shall be connected to the service disconnecting means by pressure connectors, clamps, or other approved means. Connections that depend on solder shall not be used.

230.82 Equipment Connected to the Supply Side of Service Disconnect

Only the following equipment shall be permitted to be connected to the supply side of the service disconnecting means:

(1) Cable limiters or other current-limiting devices

Cable limiters or other current-limiting devices are applied ahead of the service disconnecting means for the following reasons:

1. Individually isolates faulted cable(s).
2. Continuity of service is maximized even though one or more cables are faulted.
3. The possibility of severe equipment damage or burn-down as a result of a fault on the service conductors is reduced.
4. The current-limiting feature of cable limiters can be used to provide protection against high short-circuit currents for services and to provide compliance with 110.10.

(2) Meters and meter sockets nominally rated not in excess of 600 volts, provided all metal housings and service enclosures are grounded
(3) Meter disconnect switches nominally rated not in excess of 600 volts that have a short-circuit current rating equal to or greater than the available short circuit current, provided all metal housings and service enclosures are grounded

Meter sockets and meter disconnect switches are permitted to be connected on the supply side of the service disconnecting means. Although the *NEC* recognizes that meter sockets have commonly been used on the supply side of the service disconnect, meter sockets have been omitted from 230.82. The meter disconnect is a load-break disconnect switch designed to interrupt the service load on 480Y/277-volt services with self-contained meter sockets. The meter disconnect is not the service disconnecting means. The purpose of the meter disconnect switch is to facilitate meter change, maintenance, or disconnecting of the service. A revision to the 2005 *Code* requires meter disconnect switches to have a short-circuit current rating that is not less than the available short-circuit current at the line terminals of the meter disconnect switch.

Self-contained meters do not have external potential transformers or current transformers. The load current of the service travels through the meter itself. Neither the self-contained meter nor the meter bypass switch in the meter socket is designed to break the load current on a 480Y/277-volt system.

Self-contained meters or internal meter bypass switches should not be used to break the load current of a service having a voltage of over 150 volts to ground, because a hazardous arc could be generated. Arcs generated at voltages greater than 150 volts are considered self-sustaining and can transfer from the energized portions of the equipment to the grounded portions of the equipment.

An arc created while breaking load current on a 480Y/277-volt system (277 volts to ground) could transfer to the grounded equipment enclosure, creating a high-energy arcing ground fault and arc flash that could develop into a 3-phase short circuit. This hazardous arcing could burn down the meter socket and injure the person performing the work.

(4) Instrument transformers (current and voltage), impedance shunts, load management devices, and arresters

(5) Taps used only to supply load management devices, circuits for standby power systems, fire pump equipment, and fire and sprinkler alarms, if provided with service equipment and installed in accordance with requirements for service-entrance conductors

Systems such as emergency lighting, fire alarms, fire pumps, standby power, and sprinkler alarms are permitted to be connected ahead of the normal service disconnecting means only if such systems are provided with a separate disconnecting means and overcurrent protection.

(6) Solar photovoltaic systems, fuel cell systems, or interconnected electric power production sources

(7) Control circuits for power-operable service disconnecting means, if suitable overcurrent protection and disconnecting means are provided

(8) Ground-fault protection systems or transient voltage surge suppressors, where installed as part of listed equipment, if suitable overcurrent protection and disconnecting means are provided

Transient voltage surge suppressors (TVSSs) have been added to the types of equipment permitted to be connected on the line side of the service disconnecting means. Such applications are permitted only in listed equipment, and the conductors connected to the TVSS device must have a disconnecting means and overcurrent protection. Field connection of TVSS devices to the line terminals of the service equipment is not permitted.

VII. Service Equipment — Overcurrent Protection

230.90 Where Required

Each ungrounded service conductor shall have overload protection.

Service-entrance conductors, overhead or underground, are the supply conductors between the point of connection to the service-drop or service-lateral conductors and the service equipment. Service equipment is intended to constitute the main control and means of cutoff of the electrical supply to the premises wiring system. At this point, an overcurrent device, usually a circuit breaker or a fuse, must be installed in series with each ungrounded service conductor to provide overload protection only.

The service overcurrent device will not protect the service conductors under short-circuit or ground-fault conditions on the line side of the disconnect. Protection against ground faults and short circuits is provided by the special requirements for service conductor protection and the location of the conductors.

On multiwire circuits, two or three single-pole switches or circuit breakers that are capable of individual operation are permitted as one protective device. This allowance is acceptable, provided the switches or circuit breakers are equipped with handle ties or a master handle, so that all ungrounded conductors of a service can be disconnected with not more than six operations of the hand, per 230.71(B).

(A) Ungrounded Conductor Such protection shall be provided by an overcurrent device in series with each ungrounded service conductor that has a rating or setting not higher than the allowable ampacity of the conductor. A set of fuses shall be considered all the fuses required to protect all the ungrounded conductors of a circuit. Single-pole circuit breakers, grouped in accordance with 230.71(B), shall be considered as one protective device.

Exception No. 1: For motor-starting currents, ratings that conform with 430.52, 430.62, and 430.63 shall be permitted.

If a service supplies a motor load as well as lighting or a lighting and appliance load, then the overcurrent protective device is required to have a rating that is sufficient for the lighting and/or appliance load, in accordance with Articles 210 and 220. For an individual motor, the rating is specified by 430.52; for two or more motors, the rating is specified by 430.62.

Example

Determine the minimum-size service conductors to supply a 100-ampere lighting and appliance load plus three squirrel-cage induction motors rated 460 volts, 3 phase, code letter F, service factor 1.15, 40°C, full-voltage starting one 100-hp and two 25-hp motors on a 480-volt, 3-phase system.

Solution

STEP 1. Calculate the conductor loads. The full-load current of the 100-hp motor is 124 amperes (from Table 430.250). The full-load current of each 25-hp motor is 34 amperes (from Table 430.250). The service-entrance conductors are calculated at 125 percent of 124 amperes (155 amperes) plus two motors at 34 amperes each (68 amperes), for a total of 223 amperes (see 430.24). The motor load of 223 amperes plus the lighting and appliance load of 100 amperes equals a total load of 323 amperes. Based on this calculation, the service-entrance conductors cannot be smaller than 400-kcmil copper or 600-kcmil aluminum (see Table 310.16).

STEP 2. Determine the overcurrent protection. The maximum rating of the service overcurrent protective device is based on the lighting and appliance loads calculated in accordance with Article 220 plus the largest motor branch-circuit overcurrent device plus the sum of the other full-load motor currents.

Using an inverse-time circuit breaker (see Table 430.52), 250 percent of 124 amperes (100-hp motor) is 310 amperes. The next standard size allowed is 350 amperes plus 2 times 34 amperes, for a total of 418 amperes, plus the lighting and appliance load (100 amperes), for a total of 518 amperes. Because going up to the next standard-size overcurrent device is not permitted by 430.62, the next lower standard size is 500 amperes. See 240.6 and 430.63 for feeder overcurrent device rating.

Exception No. 2: Fuses and circuit breakers with a rating or setting that conforms with 240.4(B) or (C) and 240.6 shall be permitted.

Where the conductor rating does not correspond to the standard ampere rating of a circuit breaker or fuse, the next-larger-size circuit breaker or fuse may be used, provided its rating does not exceed 800 amperes, as permitted in 240.4(B)(3). See 240.6 for standard ampere ratings of fuses and circuit breakers.

Exception No. 3: Two to six circuit breakers or sets of fuses shall be permitted as the overcurrent device to provide the overload protection. The sum of the ratings of the circuit breakers or fuses shall be permitted to exceed the ampacity of the service conductors, provided the calculated load does not exceed the ampacity of the service conductors.

Circuit breaker or fuse ampere ratings are permitted to be greater than the ampacity of the service conductors. If multi-

ple disconnects are used as the disconnecting means, the ampacity of the service conductors must be equal to or greater than the load calculated in accordance with Article 220; however, they are not required to be sized equal to or greater than the sum of the multiple disconnects.

For example, the computed load for a service is 350 amperes. The ampacity of a 500-kcmil, Type XHHW copper conductor is 380 amperes (from Table 310.16), and the conductor is allowed to be protected by a 400-ampere fuse or circuit breaker in accordance with 240.4(B). The rating of the fuse or circuit breaker is based on the ampacity of the service conductor, not on the rating of the service disconnect switch.

In this example, a 400-ampere fuse or circuit breaker may be considered properly sized for the protection of 500-kcmil, Type XHHW copper service conductors. If the service disconnecting means [see Exception No. 3 to 230.90(A)] consists of six circuit breakers or six sets of fuses, the combined ratings must not be less than the rating required for a single switch or circuit breaker, in accordance with 230.80. See the commentary following 230.80.

As Exhibit 230.27 shows, the combined ratings of the overcurrent devices equal the size of the overcurrent device required by 240.4 and 230.90(A), Exception No. 3. However, the combined ratings of the overcurrent devices are not required to be equal to or less than that value. For example, all disconnects could be rated 100 amperes. See 230.23(A) and the associated commentary.

Exhibit 230.27 An example in which the combined ratings of the overcurrent devices are permitted to exceed the ampacity of the service conductors.

Exception No. 4: Overload protection for fire pump supply conductors shall conform with 695.4(B)(1).

Exception No. 5: Overload protection for 120/240-volt, 3-wire, single-phase dwelling services shall be permitted in accordance with the requirements of 310.15(B)(6).

(B) Not in Grounded Conductor No overcurrent device shall be inserted in a grounded service conductor except a circuit breaker that simultaneously opens all conductors of the circuit.

230.91 Location

The service overcurrent device shall be an integral part of the service disconnecting means or shall be located immediately adjacent thereto.

230.92 Locked Service Overcurrent Devices

Where the service overcurrent devices are locked or sealed or are not readily accessible to the occupant, branch-circuit overcurrent devices shall be installed on the load side, shall be mounted in a readily accessible location, and shall be of lower ampere rating than the service overcurrent device.

230.93 Protection of Specific Circuits

Where necessary to prevent tampering, an automatic overcurrent device that protects service conductors supplying only a specific load, such as a water heater, shall be permitted to be locked or sealed where located so as to be accessible.

230.94 Relative Location of Overcurrent Device and Other Service Equipment

The overcurrent device shall protect all circuits and devices.

Exception No. 1: The service switch shall be permitted on the supply side.

Exception No. 2: High-impedance shunt circuits, surge arresters, surge-protective capacitors, and instrument transformers (current and voltage) shall be permitted to be connected and installed on the supply side of the service disconnecting means as permitted in 230.82.

Exception No. 3: Circuits for load management devices shall be permitted to be connected on the supply side of the service overcurrent device where separately provided with overcurrent protection.

Exception No. 4: Circuits used only for the operation of fire alarm, other protective signaling systems, or the supply to fire pump equipment shall be permitted to be connected on the supply side of the service overcurrent device where separately provided with overcurrent protection.

Exception No. 5: Meters nominally rated not in excess of 600 volts shall be permitted, provided all metal housings and service enclosures are grounded.

Exception No. 6: Where service equipment is power operable, the control circuit shall be permitted to be connected ahead of the service equipment if suitable overcurrent protection and disconnecting means are provided.

230.95 Ground-Fault Protection of Equipment

Ground-fault protection of equipment shall be provided for solidly grounded wye electrical services of more than 150 volts to ground but not exceeding 600 volts phase-to-phase for each service disconnect rated 1000 amperes or more. The grounded conductor for the solidly grounded wye system shall be connected directly to ground without inserting any resistor or impedance device.

The rating of the service disconnect shall be considered to be the rating of the largest fuse that can be installed or the highest continuous current trip setting for which the actual overcurrent device installed in a circuit breaker is rated or can be adjusted.

See the definition of *ground-fault protection of equipment* in Article 100. Ground-fault protection of equipment on services rated 1000 amperes or more operating at 480Y/277 volts was first required in the 1971 *Code* because of the unusually high number of burndowns reported on those types of service. This requirement does not apply to systems where the grounded conductor is not solidly grounded, as is the case with high-impedance grounded neutral systems covered in 250.36.

Ground-fault protection of services does not protect the conductors on the supply side of the service disconnecting means, but it is designed to provide protection from line-to-ground faults that occur on the load side of the service disconnecting means. An alternative to installing ground-fault protection may be to provide multiple disconnects rated less than 1000 amperes. For instance, up to six 800-ampere disconnecting means may be used, and in that case ground-fault protection would not be required. FPN No. 2 to 230.95(C) recognizes that ground-fault protection may be desirable at lesser amperages on solidly grounded systems for voltages exceeding 150 volts to ground but not exceeding 600 volts phase to phase.

In addition to providing ground-fault protection, engineering studies are recommended to determine the circuit impedance and short-circuit currents that would be available at the supply terminals, so that equipment and overcurrent protection of the proper interrupting rating are used. See 110.9 and 110.10 for details on interrupting rating and circuit impedance.

The two basic types of ground-fault equipment protectors are illustrated in Exhibits 230.28 and 230.29. In Exhibit 230.28, the ground-fault sensor is installed around all the circuit conductors, and a stray current on a line-to-ground fault sets up an unbalance of the currents flowing in individ-

Exhibit 230.28 A ground-fault sensor encircling all circuit conductors, including the neutral.

Exhibit 230.29 A ground-fault sensor encircling only the bonding jumper conductor.

ual conductors installed through the ground-fault sensor. When this current exceeds the setting of the ground-fault sensor, the shunt trip operates and opens the circuit breakers.

The ground-fault sensor illustrated in Exhibit 230.29 is installed around the bonding jumper only. When an unbalanced current from a line-to-ground fault occurs, the current flows through the bonding jumper and the shunt trip causes

the circuit breaker to operate, removing the load from the line. See also 250.24(A)(4), which permits a grounding electrode conductor connection to the equipment grounding terminal bar or bus.

Exception No. 1: The ground-fault protection provisions of this section shall not apply to a service disconnect for a continuous industrial process where a nonorderly shutdown will introduce additional or increased hazards.

Exception No. 2: The ground-fault protection provisions of this section shall not apply to fire pumps.

Most fire pumps rated 100 hp and over would require a disconnecting means rated at 1000 amperes or more. However, due to the emergency nature of their use, fire pumps are exempt from the provisions of 230.95.

(A) Setting The ground-fault protection system shall operate to cause the service disconnect to open all ungrounded conductors of the faulted circuit. The maximum setting of the ground-fault protection shall be 1200 amperes, and the maximum time delay shall be one second for ground-fault currents equal to or greater than 3000 amperes.

The maximum setting for ground-fault sensors is 1200 amperes. There is no minimum, but it should be noted that settings at low levels increase the likelihood of unwanted shutdowns. The requirements of 230.95 place a restriction on fault currents greater than 3000 amperes and limit the duration of the fault to not more than 1 second. This restriction minimizes the amount of damage done by an arcing fault, which is directly proportional to the time the arcing fault is allowed to burn.

Care should be taken to ensure that interconnecting multiple supply systems does not negate proper sensing by the ground-fault protection equipment. A careful engineering study must be made to ensure that fault currents do not take parallel paths to the supply system, thereby bypassing the ground-fault detection device. See 215.10, 240.13, 517.17, and 705.32 for further information on ground-fault protection of equipment.

(B) Fuses If a switch and fuse combination is used, the fuses employed shall be capable of interrupting any current higher than the interrupting capacity of the switch during a time that the ground-fault protective system will not cause the switch to open.

(C) Performance Testing The ground-fault protection system shall be performance tested when first installed on site. The test shall be conducted in accordance with instructions that shall be provided with the equipment. A written record of this test shall be made and shall be available to the authority having jurisdiction.

The requirement for ground-fault protection system performance testing is a result of numerous reports of ground-fault protection systems that were improperly wired and could not or did not perform the function for which they were intended. This *Code* and qualified testing laboratories require a set of performance testing instructions to be supplied with the equipment. Evaluation and listing of the instructions fall under the jurisdiction of those best qualified to make such judgments, the qualified electrical testing laboratory (see 90.7). If listed equipment is not installed in accordance with the instructions provided, the installation does not comply with 110.3(B).

FPN No. 1: Ground-fault protection that functions to open the service disconnect affords no protection from faults on the line side of the protective element. It serves only to limit damage to conductors and equipment on the load side in the event of an arcing ground fault on the load side of the protective element.

FPN No. 2: This added protective equipment at the service equipment may make it necessary to review the overall wiring system for proper selective overcurrent protection coordination. Additional installations of ground-fault protective equipment may be needed on feeders and branch circuits where maximum continuity of electrical service is necessary.

FPN No. 3: Where ground-fault protection is provided for the service disconnect and interconnection is made with another supply system by a transfer device, means or devices may be needed to ensure proper ground-fault sensing by the ground-fault protection equipment.

FPN No. 4: See 517.17(A) for information on where an additional step of ground fault protection is required for hospitals and other buildings with critical areas or life support equipment.

VIII. Services Exceeding 600 Volts, Nominal

230.200 General

Service conductors and equipment used on circuits exceeding 600 volts, nominal, shall comply with all the applicable provisions of the preceding sections of this article and with the following sections that supplement or modify the preceding sections. In no case shall the provisions of Part VIII apply to equipment on the supply side of the service point.

Where services rated over 600 volts supply utility-owned and utility-maintained transformers, the conductors on the line side and load side of the service point are service-lateral conductors; however, only those conductors on the load side

of the service point come under the requirements of the *NEC*. The service point is a specific location where the supply conductors of the electric utility and the customer-owned (premises wiring) conductors connect.

Exhibit 230.30 depicts an installation where the transformer and service-lateral conductors to the service point are owned by the electric utility. The transformer secondary conductors between the service point and the service disconnecting means at the building are service-entrance conductors. In the installation depicted in Exhibit 230.31, the main

Exhibit 230.30 Service rated over 600 volts supplying a utility-owned transformer.

Exhibit 230.31 Service rated over 600 volts supplying a customer-owned transformer.

service disconnecting means is located at the customer-owned transformer primary. The conductors between the transformer secondary and the line side of the building disconnecting means are feeders. The connection point may be in a belowground or aboveground junction box, where a change in the wiring method might occur. Conductors on the load side of the building service disconnecting means are feeders. Each building or structure is required to have a disconnecting means, in accordance with 225.31.

Where services rated over 600 volts supply customer-owned and customer-maintained transformers, the conductors from the service point to the transformer service disconnect are service-lateral conductors, as illustrated in Exhibit 230.31. The conductors between the transformer secondary and the building disconnecting means are feeders, as defined in Article 100.

> FPN: For clearances of conductors of over 600 volts, nominal, see ANSI C2-2002, *National Electrical Safety Code.*

230.202 Service-Entrance Conductors

Service-entrance conductors to buildings or enclosures shall be installed to conform to 230.202(A) and (B).

(A) Conductor Size Service-entrance conductors shall not be smaller than 6 AWG unless in multiconductor cable. Multiconductor cable shall not be smaller than 8 AWG.

(B) Wiring Methods Service-entrance conductors shall be installed by one of the wiring methods covered in 300.37 and 300.50.

230.204 Isolating Switches

(A) Where Required Where oil switches or air, oil, vacuum, or sulfur hexafluoride circuit breakers constitute the service disconnecting means, an isolating switch with visible break contacts shall be installed on the supply side of the disconnecting means and all associated service equipment.

Exception: An isolating switch shall not be required where the circuit breaker or switch is mounted on removable truck panels or metal-enclosed switchgear units where both of the following conditions apply:

(1) Cannot be opened unless the circuit is disconnected.
(2) Where all energized parts are automatically disconnected when the circuit breaker or switch is removed from the normal operating position.

(B) Fuses as Isolating Switch Where fuses are of the type that can be operated as a disconnecting switch, a set of such fuses shall be permitted as the isolating switch.

(C) Accessible to Qualified Persons Only The isolating switch shall be accessible to qualified persons only.

(D) Grounding Connection Isolating switches shall be provided with a means for readily connecting the load side conductors to ground when disconnected from the source of supply.

A means for grounding the load side conductors shall not be required for any duplicate isolating switch installed and maintained by the electric supply company.

Exhibit 230.32 illustrates a two-position isolating switch for grounding a load-side conductor when it is disconnected from high-voltage line buses.

Exhibit 230.32 Two-position isolating switch for grounding a load-side conductor disconnected from high-voltage line buses.

230.205 Disconnecting Means

(A) Location The service disconnecting means shall be located in accordance with 230.70.

(B) Type Each service disconnect shall simultaneously disconnect all ungrounded service conductors that it controls and shall have a fault-closing rating that is not less than the maximum short-circuit current available at its supply terminals.

Where fused switches or separately mounted fuses are installed, the fuse characteristics shall be permitted to contribute to the fault-closing rating of the disconnecting means.

(C) Remote Control For multibuilding, industrial installations under single management, the service disconnecting means shall be permitted to be located at a separate building or structure. In such cases, the service disconnecting means shall be permitted to be electrically operated by a readily accessible, remote-control device.

230.206 Overcurrent Devices as Disconnecting Means

Where the circuit breaker or alternative for it, as specified in 230.208 for service overcurrent devices, meets the requirements specified in 230.205, they shall constitute the service disconnecting means.

230.208 Protection Requirements

A short-circuit protective device shall be provided on the load side of, or as an integral part of, the service disconnect, and shall protect all ungrounded conductors that it supplies. The protective device shall be capable of detecting and interrupting all values of current, in excess of its trip setting or melting point, that can occur at its location. A fuse rated in continuous amperes not to exceed three times the ampacity of the conductor, or a circuit breaker with a trip setting of not more than six times the ampacity of the conductors, shall be considered as providing the required short-circuit protection.

FPN: See Table 310.67 through Table 310.86 for ampacities of conductors rated 2001 volts and above.

Overcurrent devices shall conform to 230.208(A) and (B).

(A) Equipment Type Equipment used to protect service-entrance conductors shall meet the requirements of Article 490, Part II.

(B) Enclosed Overcurrent Devices The restriction to 80 percent of the rating for an enclosed overcurrent device for continuous loads shall not apply to overcurrent devices installed in systems operating at over 600 volts.

230.209 Surge Arresters (Lightning Arresters)

Surge arresters installed in accordance with the requirements of Article 280 shall be permitted on each ungrounded overhead service conductor.

230.210 Service Equipment — General Provisions

Service equipment, including instrument transformers, shall conform to Article 490, Part I.

230.211 Metal-Enclosed Switchgear

Metal-enclosed switchgear shall consist of a substantial metal structure and a sheet metal enclosure. Where installed over a combustible floor, suitable protection thereto shall be provided.

Exhibit 230.33 shows an assembly of metal-enclosed switchgear. Metal-enclosed switchgear can be used in lieu of a vault in accordance with 230.212.

230.212 Over 35,000 Volts

Where the voltage exceeds 35,000 volts between conductors that enter a building, they shall terminate in a metal-enclosed switchgear compartment or a vault conforming to the requirements of 450.41 through 450.48.

Exhibit 230.33 Metal-enclosed switchgear. (Courtesy of Square D Co.)

ARTICLE 240
Overcurrent Protection

Summary of Changes

- **240.5(B)(1):** Revised to indicate that the supply cords of specific listed appliances and portable lamps are considered protected when used in accordance with the listing requirements for the appliance or portable lamp.
- **240.5(B)(3):** Revised to indicate that listed extension cord sets are considered protected when used in accordance with the listing requirements for the extension cord.
- **240.5(B)(4):** Relocated the requirements for field-assembled extension cord sets.
- **240.20(B)(1) through (3):** Revised to require *identified* handle ties where they are used with individual single-pole circuit breakers.
- **240.21(B):** Revised to clarify that "tap conductors" must have an ampacity not less than the rating of the overcurrent protective device in which they terminate and that the use of the "next higher standard rating" is not permitted.
- **240.21(C):** Revised to clarify that transformer secondary conductors must have an ampacity not less than the rating of the overcurrent protective device in which they terminate and that the use of the "next higher standard rating" is not permitted.
- **240.21(C)(2)(1):** Added new requirement for determining the minimum ampacity of transformer secondary conductors not over 10 ft long.

- **240.24(A):** Revised to provide specific maximum height for operating handle of a switch or circuit breaker.
- **240.60(D):** Added requirement limiting Class H renewable cartridge fuses for use only in existing installations if there is no evidence of overfusing or tampering.
- **240.86:** Revised to allow engineering of series rated systems for existing installations only, when that engineering is completed by a licensed professional engineer and the stamped, documented design is made available to the authority having jurisdiction. Field marking to identify upstream protective devices is required on end-use equipment.

Contents

I. General

240.1 Scope

Parts I through VII of this article provide the general requirements for overcurrent protection and overcurrent protective devices not more than 600 volts, nominal. Part VIII covers overcurrent protection for those portions of supervised industrial installations operating at voltages of not more than 600 volts, nominal. Part IX covers overcurrent protection over 600 volts, nominal.

> FPN: Overcurrent protection for conductors and equipment is provided to open the circuit if the current reaches a value that will cause an excessive or dangerous temperature in conductors or conductor insulation. See also 110.9 for requirements for interrupting ratings and 110.10 for requirements for protection against fault currents.

240.2 Definitions

Current-Limiting Overcurrent Protective Device. A device that, when interrupting currents in its current-limiting range, reduces the current flowing in the faulted circuit to a magnitude substantially less than that obtainable in the same circuit if the device were replaced with a solid conductor having comparable impedance.

Most electrical distribution systems can deliver high ground-fault or short-circuit currents to components such as conductors and service equipment. These components may not be able to handle short-circuit currents; they may be damaged or destroyed, and serious burndowns and fires could result. Properly selected current-limiting overcurrent protective devices, such as the ones shown in Exhibit 240.1, limit the let-through energy to an amount that does not exceed the rating of the components, in spite of high available short-circuit currents. A current-limiting protective device is one that cuts off a fault current in less than one-half cycle. It thus prevents short-circuit currents from building up to their full available values.

Exhibit 240.1 Class R current-limiting fuses with rejection feature to prohibit the installation of non–current-limiting fuses. (Courtesy of Bussmann Division, Cooper Industries)

Proper selection of current-limiting devices may depend on the type of device selected. For example, a Class RK5 fuse is not as current limiting as a Class RK1 fuse. Furthermore, a Class RK1 fuse is not as current limiting as a high-speed semiconductor fuse. See 110.9 and 110.10 for details on interrupting ratings, circuit impedance, and other characteristics.

Supervised Industrial Installation. For the purposes of Part VIII, the industrial portions of a facility where all of the following conditions are met:

(1) Conditions of maintenance and engineering supervision ensure that only qualified persons monitor and service the system.
(2) The premises wiring system has 2500 kVA or greater of load used in industrial process(es), manufacturing activities, or both, as calculated in accordance with Article 220.
(3) The premises has at least one service or feeder that is more than 150 volts to ground and more than 300 volts phase-to-phase.

This definition excludes installations in buildings used by the industrial facility for offices, warehouses, garages, machine shops, and recreational facilities that are not an integral part of the industrial plant, substation, or control center.

For a facility to be recognized as an industrial establishment, it must have the following: a combined process and manufacturing load greater than 2500 kVA; at least one service, 480Y/277V, nominal, or greater; and maintenance and engineering supervision that ensures that only qualified persons monitor and service the electrical system. All process or manufacturing loads from each low-, medium-, and high-voltage system can be added together to satisfy the load requirements of Part VIII. Loads are calculated in accordance with Article 220. However, loads not associated with manufacturing or processing cannot be used to meet the 2500-kVA minimum requirement.

The provisions of Part VIII apply only to low-voltage electrical systems (600 volts, nominal, or less) used for process or manufacturing. Part VIII does not apply to electrical systems operating at over 600 volts, nominal, or to electrical systems that serve separate facilities, such as offices, warehouses, garages, machine shops, or recreational buildings, that are not a part of the manufacturing or industrial process. However, if a part of the industrial process or manufacturing electrical system is used to serve an office, warehouse, garage, machine shop, or recreational facility that is an integral part of the industrial plant, control center, or substation, Part VIII can still apply to the total process or manufacturing electrical system.

Tap Conductors. As used in this article, a tap conductor is defined as a conductor, other than a service conductor, that has overcurrent protection ahead of its point of supply

that exceeds the value permitted for similar conductors that are protected as described elsewhere in 240.4.

240.3 Other Articles
Equipment shall be protected against overcurrent in accordance with the article in this *Code* that covers the type of equipment specified in Table 240.3.

240.4 Protection of Conductors
Conductors, other than flexible cords, flexible cables, and fixture wires, shall be protected against overcurrent in accordance with their ampacities specified in 310.15, unless otherwise permitted or required in 240.4(A) through (G).

(A) Power Loss Hazard Conductor overload protection shall not be required where the interruption of the circuit would create a hazard, such as in a material-handling magnet circuit or fire pump circuit. Short-circuit protection shall be provided.

FPN: See NFPA 20-2003, *Standard for the Installation of Stationary Pumps for Fire Protection.*

(B) Devices Rated 800 Amperes or Less The next higher standard overcurrent device rating (above the ampacity of the conductors being protected) shall be permitted to be used, provided all of the following conditions are met:

(1) The conductors being protected are not part of a multioutlet branch circuit supplying receptacles for cord-and-plug-connected portable loads.
(2) The ampacity of the conductors does not correspond with the standard ampere rating of a fuse or a circuit breaker without overload trip adjustments above its rating (but that shall be permitted to have other trip or rating adjustments).
(3) The next higher standard rating selected does not exceed 800 amperes.

Table 210.24 summarizes the requirements for the size of conductors and the size of the overcurrent protection for branch circuits where two or more outlets are required. The first footnote indicates that the wire sizes are for copper conductors. Section 210.3 indicates that branch-circuit conductors rated 15, 20, 30, 40, and 50 amperes must be protected at their ratings. Section 210.19(A) requires that branch-circuit conductors have an ampacity not less than the rating of the branch circuit and not less than the maximum load to be served. These specific requirements take precedence over 240.4(B), which applies generally.

Table 310.16 through Table 310.86 list the ampacities of conductors. Section 240.6 lists the standard ratings of overcurrent devices. Where the ampacity of the conductor specified in these tables does not match the rating of the

Table 240.3 Other Articles

Equipment	Article
Air-conditioning and refrigerating equipment	440
Appliances	422
Assembly occupancies	518
Audio signal processing, amplification, and reproduction equipment	640
Branch circuits	210
Busways	368
Capacitors	460
Class 1, Class 2, and Class 3 remote-control, signaling, and power-limited circuits	725
Closed-loop and programmed power distribution	780
Cranes and hoists	610
Electric signs and outline lighting	600
Electric welders	630
Electrolytic cells	668
Elevators, dumbwaiters, escalators, moving walks, wheelchair lifts, and stairway chair lifts	620
Emergency systems	700
Fire alarm systems	760
Fire pumps	695
Fixed electric heating equipment for pipelines and vessels	427
Fixed electric space-heating equipment	424
Fixed outdoor electric deicing and snow-melting equipment	426
Generators	445
Health care facilities	517
Induction and dielectric heating equipment	665
Industrial machinery	670
Luminaires (lighting fixtures), lampholders, and lamps	410
Motion picture and television studios and similar locations	530
Motors, motor circuits, and controllers	430
Phase converters	455
Pipe organs	650
Receptacles	406
Services	230
Solar photovoltaic systems	690
Switchboards and panelboards	408
Theaters, audience areas of motion picture and television studios, and similar locations	520
Transformers and transformer vaults	450
X-ray equipment	660

standard overcurrent device, 240.4 permits the use of the next larger standard overcurrent device. All three conditions in 240.4(B) must be met in order for this permission to apply. However, if the ampacity of a conductor matches the

standard rating of 240.6, that conductor must be protected at the standard size device. For example, in Table 310.16, 3 AWG, 75°C copper, Type THWN, the ampacity is listed as 100 amperes. That conductor would be protected by a 100-ampere overcurrent device.

The provisions of 240.4(B) do not modify or change the allowable ampacity of the conductor — they only serve to provide a reasonable increase in the permitted overcurrent protective device rating where the allowable ampacity and the standard overcurrent protective device ratings do not correspond. For circuits rated 600 volts and under, the allowable ampacity of branch circuit, feeder, or service conductors always has to be capable of supplying the calculated load in accordance with the requirements of 210.19(A)(1), 215.2(A)(1), and 230.42(A). For example, a 500-kcmil THWN copper conductor has an allowable ampacity of 380 amperes from Table 310.16. This conductor can supply a load not exceeding 380 amperes and, in accordance with 240.4(B), can be protected by a 400-ampere overcurrent protective device.

In contrast to 240.4(B), 310.15(B)(6) does permit the conductor types and sizes specified in Table 310.15(B)(6) to supply calculated loads based on their ratings from that table that exceed their allowable ampacities specified in Table 310.16. The overcurrent protection for these residential supply conductors is also permitted to be based on the increased rating allowed by this Article 310 table. Application of 310.15(B)(6) and its table is permitted only for single-phase, 120/240-volt, residential services and main power feeders. The increased ratings given in Table 310.15(B)(6) are based on the significant diversity inherent to most dwelling unit loads and the fact that only the two ungrounded service or feeder conductors are considered to be current carrying.

(C) Devices Rated Over 800 Amperes Where the overcurrent device is rated over 800 amperes, the ampacity of the conductors it protects shall be equal to or greater than the rating of the overcurrent device defined in 240.6.

(D) Small Conductors Unless specifically permitted in 240.4(E) or 240.4(G), the overcurrent protection shall not exceed 15 amperes for 14 AWG, 20 amperes for 12 AWG, and 30 amperes for 10 AWG copper; or 15 amperes for 12 AWG and 25 amperes for 10 AWG aluminum and copper-clad aluminum after any correction factors for ambient temperature and number of conductors have been applied.

(E) Tap Conductors Tap conductors shall be permitted to be protected against overcurrent in accordance with the following:

(1) 210.19(A)(3) and (A)(4) Household Ranges and Cooking Appliances and Other Loads

(2) 240.5(B)(2) Fixture Wire
(3) 240.21 Location in Circuit
(4) 368.17(B) Reduction in Ampacity Size of Busway
(5) 368.17(C) Feeder or Branch Circuits (busway taps)
(6) 430.53(D) Single Motor Taps

(F) Transformer Secondary Conductors Single-phase (other than 2-wire) and multiphase (other than delta-delta, 3-wire) transformer secondary conductors shall not be considered to be protected by the primary overcurrent protective device. Conductors supplied by the secondary side of a single-phase transformer having a 2-wire (single-voltage) secondary, or a three-phase, delta-delta connected transformer having a 3-wire (single-voltage) secondary, shall be permitted to be protected by overcurrent protection provided on the primary (supply) side of the transformer, provided this protection is in accordance with 450.3 and does not exceed the value determined by multiplying the secondary conductor ampacity by the secondary to primary transformer voltage ratio.

The fundamental requirement of 240.4 specifies that conductors are to be protected against overcurrent in accordance with their ampacity, and 240.21 requires that the protection be provided at the point the conductor receives its supply. Section 240.4(F) permits the secondary circuit conductors from a transformer to be protected by overcurrent devices in the primary circuit conductors of the transformer only in the following two special cases:

1. A transformer with a 2-wire primary and a 2-wire secondary, provided the transformer primary is protected in accordance with 450.3
2. A 3-phase, delta-delta-connected transformer having a 3-wire, single-voltage secondary, provided its primary is protected in accordance with 450.3

Except for those two special cases, transformer secondary conductors must be protected by the use of overcurrent devices, because the primary overcurrent devices do not provide such protection. As an example, consider a single-phase transformer with a 2-wire secondary that is provided with primary overcurrent protection rated at 50 amperes. The transformer is rated 480/240 volts. Conductors supplied by the secondary have an ampacity of 100 amperes. Is the 50-ampere overcurrent protection allowed to protect the conductors that are connected to the secondary?

The secondary-to-primary voltage ratio in this example is 240 ÷ 480, a ratio of 0.5. Multiplying the secondary conductor ampacity of 100 amperes by 0.5 yields 50 amperes. Thus, the maximum rating of the overcurrent device allowed on the primary of the transformer that will also provide overcurrent protection for the secondary conductors is 50 amperes. These secondary conductors are not tap conductors, are not limited in length, and do not require overcurrent protection where they receive their supply, which is at the transformer secondary terminals.

However, if the secondary consisted of a 3-wire, 240/120-volt system, a 120-volt line-to-neutral load could draw up to 200 amperes before the overcurrent device in the primary actuated. That would be the result of the 1:4 secondary-to-primary voltage ratio of the 120-volt winding of the transformer secondary, which can cause dangerous overloading of the secondary conductors.

(G) Overcurrent Protection for Specific Conductor Applications Overcurrent protection for the specific conductors shall be permitted to be provided as referenced in Table 240.4(G).

Table 240.4(G) Specific Conductor Applications

Conductor	Article	Section
Air-conditioning and refrigeration equipment circuit conductors	440, Parts III, VI	
Capacitor circuit conductors	460	460.8(B) and 460.25(A)–(D)
Control and instrumentation circuit conductors (Type ITC)	727	727.9
Electric welder circuit conductors	630	630.12 and 630.32
Fire alarm system circuit conductors	760	760.23, 760.24, 760.41, and Chapter 9, Tables 12(A) and 12(B)
Motor-operated appliance circuit conductors	422, Part II	
Motor and motor-control circuit conductors	430, Parts III, IV, V, VI, VII	
Phase converter supply conductors	455	455.7
Remote-control, signaling, and power- limited circuit conductors	725	725.23, 725.24, 725.41, and Chapter 9, Tables 11(A) and 11(B)
Secondary tie conductors	450	450.6

240.5 Protection of Flexible Cords, Flexible Cables, and Fixture Wires

Flexible cord and flexible cable, including tinsel cord and extension cords, and fixture wires shall be protected against overcurrent by either 240.5(A) or (B).

(A) Ampacities Flexible cord and flexible cable shall be protected by an overcurrent device in accordance with their ampacity as specified in Tables 400.5(A) and 400.5(B). Fixture wire shall be protected against overcurrent in accordance with its ampacity as specified in Table 402.5. Supplementary overcurrent protection, as in 240.10, shall be permitted to be an acceptable means for providing this protection.

(B) Branch Circuit Overcurrent Device Flexible cord shall be protected where supplied by a branch circuit in accordance with one of the methods described in 240.5(B)(1), (B)(2), (B)(3), or (B)(4).

(1) Supply Cord of Listed Appliance or Portable Lamps Where flexible cord or tinsel cord is approved for and used with a specific listed appliance or portable lamp, it shall be considered to be protected when applied within the appliance or portable lamp listing requirements.

(2) Fixture Wire Fixture wire shall be permitted to be tapped to the branch circuit conductor of a branch circuit in accordance with the following:

(1) 20-ampere circuits — 18 AWG, up to 15 m (50 ft) of run length
(2) 20-ampere circuits — 16 AWG, up to 30 m (100 ft) of run length
(3) 20-ampere circuits — 14 AWG and larger
(4) 30-ampere circuits — 14 AWG and larger

Section 240.5(A) references Tables 400.5(A) and 400.5(B) for flexible cords and flexible cables and Table 402.5 for fixture wire ampacity. Supplementary protection, as described in 240.10, is also acceptable as an alternative for protection of either flexible cord or fixture wire.

Sections 240.5(B)(1) through 240.5(B)(4) permit smaller conductors to be connected to branch circuits of a greater rating. For flexible cords, 240.5(B)(1) and 240.5(B)(3) now specify that flexible cord connected to a listed appliance or portable lamp or used in a listed extension cord set is considered to be protected as long as the appliance, lamp, or extension cord is used in accordance with its listing requirements. These listing requirements are developed by the third-party testing and listing organizations with technical input from cord, appliance, and lamp manufacturers. For other than field-assembled extension cords, the *Code* no longer contains specific provisions for the overcurrent protection of flexible cord based on cord conductor size. For fixture wire, 240.5(B)(2) establishes a maximum protective device rating based on a minimum conductor size and a maximum conductor length.

(5) 40-ampere circuits — 12 AWG and larger
(6) 50-ampere circuits — 12 AWG and larger

(3) Extension Cord Sets Flexible cord used in listed extension cord sets shall be considered to be protected when applied within the extension cord listing requirements.

(4) Field Assembled Extension Cord Sets Flexible cord used in extension cords made with separately listed and installed components shall be permitted to be supplied by a branch circuit in accordance with the following:
    20-ampere circuits — 16 AWG and larger

Field-assembled extension cords are permitted provided the conductors are 16 AWG or larger and the overcurrent protection for the branch circuit to which the cord is connected does not exceed 20 amperes. The cord and the cord caps and connectors used for this type of assembly are required to be listed.

240.6 Standard Ampere Ratings

(A) Fuses and Fixed-Trip Circuit Breakers The standard ampere ratings for fuses and inverse time circuit breakers shall be considered 15, 20, 25, 30, 35, 40, 45, 50, 60, 70, 80, 90, 100, 110, 125, 150, 175, 200, 225, 250, 300, 350, 400, 450, 500, 600, 700, 800, 1000, 1200, 1600, 2000, 2500, 3000, 4000, 5000, and 6000 amperes. Additional standard ampere ratings for fuses shall be 1, 3, 6, 10, and 601. The use of fuses and inverse time circuit breakers with nonstandard ampere ratings shall be permitted.

(B) Adjustable-Trip Circuit Breakers The rating of adjustable-trip circuit breakers having external means for adjusting the current setting (long-time pickup setting), not meeting the requirements of 240.6(C), shall be the maximum setting possible.

(C) Restricted Access Adjustable-Trip Circuit Breakers A circuit breaker(s) that has restricted access to the adjusting means shall be permitted to have an ampere rating(s) that is equal to the adjusted current setting (long-time pickup setting). Restricted access shall be defined as located behind one of the following:

(1) Removable and sealable covers over the adjusting means
(2) Bolted equipment enclosure doors
(3) Locked doors accessible only to qualified personnel

The set long-time pickup rating (as opposed to the instantaneous trip rating) of an adjustable-trip circuit breaker can be considered the circuit breaker rating where access to the adjustment means is limited. This access limitation can be provided by locating the adjustment means behind sealable covers, as shown in Exhibit 240.2, behind bolted equipment enclosures, or behind locked equipment room doors with access available only to qualified personnel. The purpose of limiting access to the adjustment prevents tampering or readjustment by unqualified personnel.

Exhibit 240.2 An adjustable-trip circuit breaker with a transparent, removable, and sealable cover. (Courtesy of Square D Co.)

Manufacturers of both fuses and inverse time circuit breakers have, or can make available, products with ampere ratings other than those listed in 240.6(A). Selection of such nonstandard ratings is not required by the *Code* but may permit better protection for conductors.

240.8 Fuses or Circuit Breakers in Parallel

Fuses and circuit breakers shall be permitted to be connected in parallel where they are factory assembled in parallel and listed as a unit. Individual fuses, circuit breakers, or combinations thereof shall not otherwise be connected in parallel.

Section 240.8 prohibits the use of fuses or circuit breakers in parallel unless they are factory assembled in parallel and listed as a unit. Section 404.17 prohibits the use of fuses in parallel in fused switches except as permitted by 240.8 for listed assemblies.

It is not the intent of 240.8 to restore the use of standard fuses in parallel in disconnect switches. However, 240.8 recognizes parallel low-voltage circuit breakers or fuses and parallel high-voltage circuit breakers or fuses if they are tested and factory assembled in parallel and listed as a unit.

High-voltage fuses have long been recognized in parallel when they are assembled in an identified common mounting.

240.9 Thermal Devices

Thermal relays and other devices not designed to open short circuits or ground faults shall not be used for the protection

of conductors against overcurrent due to short circuits or ground faults, but the use of such devices shall be permitted to protect motor branch-circuit conductors from overload if protected in accordance with 430.40.

240.10 Supplementary Overcurrent Protection

Where supplementary overcurrent protection is used for luminaires (lighting fixtures), appliances, and other equipment or for internal circuits and components of equipment, it shall not be used as a substitute for required branch-circuit overcurrent devices or in place of the required branch-circuit protection. Supplementary overcurrent devices shall not be required to be readily accessible.

240.12 Electrical System Coordination

Where an orderly shutdown is required to minimize the hazard(s) to personnel and equipment, a system of coordination based on the following two conditions shall be permitted:

(1) Coordinated short-circuit protection

With coordinated overcurrent protection, the faulted or overloaded circuit is isolated by the selective operation of only the overcurrent protective device closest to the overcurrent condition. This prevents power loss to unaffected loads. Examples of overcurrent protection without coordination and coordinated protection are illustrated in Exhibit 240.3.

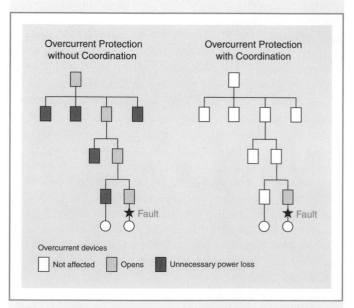

Exhibit 240.3 Overcurrent protection schemes without system coordination and with system coordination.

(2) Overload indication based on monitoring systems or devices

FPN: The monitoring system may cause the condition to go to alarm, allowing corrective action or an orderly shutdown, thereby minimizing personnel hazard and equipment damage.

240.13 Ground-Fault Protection of Equipment

Ground-fault protection of equipment shall be provided in accordance with the provisions of 230.95 for solidly grounded wye electrical systems of more than 150 volts to ground but not exceeding 600 volts phase-to-phase for each individual device used as a building or structure main disconnecting means rated 1000 amperes or more.

The provisions of this section shall not apply to the disconnecting means for the following:

(1) Continuous industrial processes where a nonorderly shutdown will introduce additional or increased hazards
(2) Installations where ground-fault protection is provided by other requirements for services or feeders
(3) Fire pumps

Section 240.13 extends the requirement of 230.95 to building disconnects, regardless of how the disconnects are classified (service disconnects or building disconnects for feeders or even branch circuits). See 215.10 and Article 225, Part II, for the requirements for building disconnects not on the utility service.

Section 240.13 requires each building or structure disconnect that is rated 1000 amperes or more, on a solidly grounded system of more than 150 volts to ground (e.g., a 480Y/277-volt system), to be provided with ground-fault protection for equipment. Provisions are included for fire pumps and for continuous industrial processes in which nonorderly shutdowns would introduce additional hazards.

Where ground-fault protection for equipment is installed at the service equipment and other buildings or structures are supplied by feeders or branch circuits, 250.24 requires regrounding of the grounded conductor if an equipment grounding conductor is not included in the run. However, 240.13(2) exempts the grounded conductor from being regrounded downstream from the ground-fault protected service. Regrounding of the neutral at the second building may nullify the ground-fault protection of the second building that would otherwise be provided by ground-fault protection at the main service.

II. Location

240.20 Ungrounded Conductors

(A) Overcurrent Device Required A fuse or an overcurrent trip unit of a circuit breaker shall be connected in series with each ungrounded conductor. A combination of a current transformer and overcurrent relay shall be considered equivalent to an overcurrent trip unit.

FPN: For motor circuits, see Parts III, IV, V, and XI of Article 430.

(B) Circuit Breaker as Overcurrent Device Circuit breakers shall open all ungrounded conductors of the circuit both manually and automatically unless otherwise permitted in 240.20(B)(1), (B)(2), and (B)(3).

(1) Multiwire Branch Circuit Except where limited by 210.4(B), individual single-pole circuit breakers, with or without identified handle ties, shall be permitted as the protection for each ungrounded conductor of multiwire branch circuits that serve only single-phase line-to-neutral loads.

(2) Grounded Single-Phase and 3-Wire dc Circuits In grounded systems, individual single-pole circuit breakers with identified handle ties shall be permitted as the protection for each ungrounded conductor for line-to-line connected loads for single-phase circuits or 3-wire, direct-current circuits.

(3) 3-Phase and 2-Phase Systems For line-to-line loads in 4-wire, 3-phase systems or 5-wire, 2-phase systems having a grounded neutral and no conductor operating at a voltage greater than permitted in 210.6, individual single-pole circuit breakers with identified handle ties shall be permitted as the protection for each ungrounded conductor.

Before discussing handle ties, it is important to understand the Article 100 definition of the term *multiwire branch circuit*, as well as 210.4(C) and its two exceptions. Multiwire branch circuits are permitted to supply line-to-line connected loads where the loads are associated with a single piece of utilization equipment or where all of the ungrounded conductors are opened simultaneously by the branch-circuit overcurrent device (automatic opening in response to overcurrent). See the commentary following 210.4(C) for additional information. In addition, a revision to 210.4(B) in the 2005 *Code* expands the requirement for a means to simultaneously disconnect all ungrounded conductors of a multiwire circuit supplying devices or equipment on the same mounting strap or yoke to apply to all occupancies.

The basic rule in 240.20(B) requires circuit breakers to open all ungrounded conductors of the circuit when it trips (automatic operation in response to overcurrent) or is manually operated as a disconnecting means. For 2-wire circuits with one conductor grounded, this rule is simple and needs no further explanation. For multiwire branch circuits of 600 volts or less, however, there are three acceptable methods of complying with this rule.

The first, and most widely used, method is to install a multipole circuit breaker with an internal common trip mechanism. The use of such multipole devices ensures compliance with all of the *Code* requirements for overcurrent protection and disconnecting of multiwire branch circuits.

This breaker is operated by an external single lever internally attached to the two or three poles of the circuit breaker, or the external lever may be attached to multiple handles operated as one, provided the breaker is a factory-assembled unit in accordance with 240.8. Underwriters Laboratories refers to these devices as *multipole common trip circuit breakers.* This type of circuit breaker is required to be used for branch circuits that comprise multiple ungrounded conductors supplied by ungrounded three-phase and single-phase systems. Where circuit breakers are used on ungrounded systems, it is important to verify compliance with the application requirements in 240.85. Of course, multipole common trip circuit breakers are permitted to be installed on any branch circuit supplied from a grounded system where used within their ratings.

The second option permitted for multiwire branch circuits is to use two or three single-pole circuit breakers and add an identified handle tie to function as a common operating handle. This multipole circuit breaker is field assembled by externally attaching an identified common lever (handle tie) onto the two or three individual circuit breakers. It is important to understand that handle ties do not cause the circuit breaker to function as a common trip device; rather, it only allows common operation as a disconnecting means. Handle tie mechanism circuit breakers are permitted as a substitute for internal common trip mechanism circuit breakers only for limited applications. Unless specifically prohibited elsewhere, circuit breakers with identified handle ties are permitted for multiwire branch circuits only where the circuit is supplied from grounded 3-phase or grounded single-phase systems. The single-pole circuit breakers used together in this fashion must be rated for the dual voltage encountered, such as 120/240 volts. It is important to note that the term *approved* has been revised to *identified* for the 2005 *Code* in order to require the use of hardware that has been designed specifically to perform this common disconnecting means function. The use of approved, homemade hardware to perform this function is no longer permitted.

The third method is to use individual single-pole circuit breakers without common trip mechanisms or without handle ties for multiwire branch circuits. Unless limited by other sections of the *Code,* such as 210.4(B), this method is permitted for multiwire circuits, provided the multiwire branch circuit supplies only single-phase line-to-neutral loads.

Exhibit 240.4 through Exhibit 240.6 illustrate some examples of how the requirements in 240.20(B) are applied. In Exhibit 240.4, where multipole common trip circuit breakers are required, handle ties are not permitted because the circuits are supplied from ungrounded systems. In Exhibit 240.5, where the supply systems are grounded, single-pole circuit breakers are permitted and handle ties or common trip operation are not required because the circuits supply

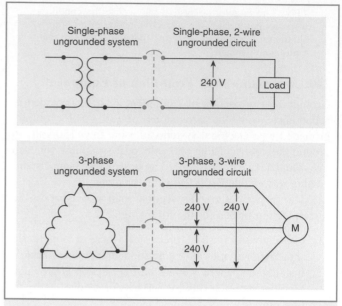

Exhibit 240.4 Examples of circuits that require multipole common trip–type circuit breakers, in accordance with 240.20(B).

Exhibit 240.5 Examples of circuits in which single-pole circuit breakers are permitted, in accordance with 240.20(B)(1), because they open the ungrounded conductor of the circuit.

line-to-neutral loads. In Exhibit 240.6, in which line-to-line loads are supplied from single-phase or 4-wire, 3-phase systems, identified handle ties or multipole common trip circuit breakers are permitted.

Exhibit 240.6 Examples of circuits in which identified handle ties are permitted to provide the simultaneous disconnecting function in accordance with 240.20(B)(2) or 240.20(B)(3).

(C) Closed-Loop Power Distribution Systems Listed devices that provide equivalent overcurrent protection in closed-loop power distribution systems shall be permitted as a substitute for fuses or circuit breakers.

240.21 Location in Circuit

Overcurrent protection shall be provided in each ungrounded circuit conductor and shall be located at the point where the conductors receive their supply except as specified in 240.21(A) through (G). No conductor supplied under the provisions of 240.21(A) through (G) shall supply another conductor under those provisions, except through an overcurrent protective device meeting the requirements of 240.4.

(A) Branch-Circuit Conductors Branch-circuit tap conductors meeting the requirements specified in 210.19 shall be permitted to have overcurrent protection located as specified in that section.

(B) Feeder Taps Conductors shall be permitted to be tapped, without overcurrent protection at the tap, to a feeder as specified in 240.21(B)(1) through (B)(5). The provisions of 240.4(B) shall not be permitted for tap conductors.

An important addition in the 2005 *Code* is the last sentence in 240.21(B). The use of the next standard higher standard size provision of 240.4(B) is not permitted for feeder tap conductor applications. For instance, the use of a 500-kcmil THWN copper conductor (380 amperes, per Table 310.16) as a tap conductor to supply a 400-ampere rated device is clearly not permitted by this revision.

Exhibit 240.7 illustrates how a smaller 1/0 AWG, Type THW copper conductor (150 amperes, from Table 310.16) is supplied from a larger 3/0 AWG, Type THW copper feeder conductor with an ampacity of 200 amperes (sized to compensate for voltage drop) that is protected by a 150-ampere overcurrent protective device. Because the ampacity of the 1/0 AWG conductor is not exceeded by the rating of the overcurrent device, the 1/0 AWG conductor is not considered to be a tap conductor based on the definition of tap conductor in 240.2. The overcurrent device protects both sets of conductors in accordance with the basic rule of 240.4, and additional overcurrent protection is not required at the point where the 1/0 AWG conductor is supplied.

Exhibit 240.7 An example in which the circuit breaker protecting the feeder conductors is permitted by 240.21(A) to protect the smaller conductors supplying the panelboard.

(1) Taps Not Over 3 m (10 ft) Long Where the length of the tap conductors does not exceed 3 m (10 ft) and the tap conductors comply with all of the following:

(1) The ampacity of the tap conductors is
   a. Not less than the combined calculated loads on the circuits supplied by the tap conductors, and
   b. Not less than the rating of the device supplied by the tap conductors or not less than the rating of the overcurrent-protective device at the termination of the tap conductors.

(2) The tap conductors do not extend beyond the switchboard, panelboard, disconnecting means, or control devices they supply.

(3) Except at the point of connection to the feeder, the tap conductors are enclosed in a raceway, which shall extend from the tap to the enclosure of an enclosed switchboard, panelboard, or control devices, or to the back of an open switchboard.

(4) For field installations where the tap conductors leave the enclosure or vault in which the tap is made, the rating of the overcurrent device on the line side of the tap conductors shall not exceed 10 times the ampacity of the tap conductor.

FPN: For overcurrent protection requirements for lighting and appliance branch-circuit panelboards and certain power panelboards, see 408.36(A), (B), and (E).

(2) Taps Not Over 7.5 m (25 ft) Long Where the length of the tap conductors does not exceed 7.5 m (25 ft) and the tap conductors comply with all the following:

(1) The ampacity of the tap conductors is not less than one-third of the rating of the overcurrent device protecting the feeder conductors.

(2) The tap conductors terminate in a single circuit breaker or a single set of fuses that will limit the load to the ampacity of the tap conductors. This device shall be permitted to supply any number of additional overcurrent devices on its load side.

(3) The tap conductors are protected from physical damage by being enclosed in an approved raceway or by other approved means.

Exhibit 240.8 illustrates the conditions of 240.21(B)(2), in which three 3/0 AWG, Type THW copper tap conductors are protected from physical damage in a raceway. The tap conductors are not more than 25 ft in length between terminations, and the conductors are tapped from 500-kcmil, Type THW copper feeders and terminate in a circuit breaker. It is important to note that the lengths specified in 240.21(B) and 240.21(C) apply to the conductors and not to a raceway enclosing the conductors or to the distance between the enclosures in which the tap conductors originate and terminate.

Note that a 3/0 AWG, Type THW copper conductor (200 amperes) is more than one-third the rating of the overcurrent

Exhibit 240.8 An example in which the feeder taps terminate in a single circuit breaker, per 240.21(B)(2)(2).

device (400 amperes) protecting the feeder circuit. See Table 310.16 for the ampacity of copper conductors in conduit.

(3) Taps Supplying a Transformer [Primary Plus Secondary Not Over 7.5 m (25 ft) Long] Where the tap conductors supply a transformer and comply with all the following conditions:

(1) The conductors supplying the primary of a transformer have an ampacity at least one-third the rating of the overcurrent device protecting the feeder conductors.

(2) The conductors supplied by the secondary of the transformer shall have an ampacity that is not less than the value of the primary-to-secondary voltage ratio multiplied by one-third of the rating of the overcurrent device protecting the feeder conductors.

(3) The total length of one primary plus one secondary conductor, excluding any portion of the primary conductor that is protected at its ampacity, is not over 7.5 m (25 ft).

(4) The primary and secondary conductors are protected from physical damage by being enclosed in an approved raceway or by other approved means.

(5) The secondary conductors terminate in a single circuit breaker or set of fuses that limit the load current to not more than the conductor ampacity that is permitted by 310.15.

Exhibit 240.9 illustrates the conditions of 240.21(B)(3). The overcurrent protection requirements of 408.36 for panelboards and 450.3(B) for transformers also apply.

(4) Taps Over 7.5 m (25 ft) Long Where the feeder is in a high bay manufacturing building over 11 m (35 ft) high at walls and the installation complies with all the following conditions:

(1) Conditions of maintenance and supervision ensure that only qualified persons service the systems.

Secondary-to-primary voltage ratio = 208/480 = 0.433

Exhibit 240.9 An example in which the transformer feeder taps (primary plus secondary) are not over 25 ft long, per 240.21(B)(3) and 240.21(C)(5).

(2) The tap conductors are not over 7.5 m (25 ft) long horizontally and not over 30 m (100 ft) total length.

(3) The ampacity of the tap conductors is not less than one-third the rating of the overcurrent device protecting the feeder conductors.

(4) The tap conductors terminate at a single circuit breaker or a single set of fuses that limit the load to the ampacity of the tap conductors. This single overcurrent device shall be permitted to supply any number of additional overcurrent devices on its load side.

(5) The tap conductors are protected from physical damage by being enclosed in an approved raceway or by other approved means.

(6) The tap conductors are continuous from end-to-end and contain no splices.

(7) The tap conductors are sized 6 AWG copper or 4 AWG aluminum or larger.

(8) The tap conductors do not penetrate walls, floors, or ceilings.

(9) The tap is made no less than 9 m (30 ft) from the floor.

Exhibit 240.10 illustrates the requirements of 240.21(B)(4). It permits a tap of 100 ft for manufacturing buildings with walls that are over 35 ft high where the tap connection is made not less than 30 ft from the floor and conditions of maintenance and supervision ensure that only qualified persons service these systems.

(5) Outside Taps of Unlimited Length Where the conductors are located outdoors of a building or structure, except at the point of load termination, and comply with all of the following conditions:

(1) The conductors are protected from physical damage in an approved manner.

Exhibit 240.10 Application of 240.21(B)(4) in which the horizontal length of the feeder tap conductors does not exceed 25 ft, the total vertical plus horizontal length of the tap conductors does not exceed 100 ft, and the tap connection is made at a point not less than 30 ft from the floor.

(2) The conductors terminate at a single circuit breaker or a single set of fuses that limit the load to the ampacity of the conductors. This single overcurrent device shall be permitted to supply any number of additional overcurrent devices on its load side.

(3) The overcurrent device for the conductors is an integral part of a disconnecting means or shall be located immediately adjacent thereto.

(4) The disconnecting means for the conductors is installed at a readily accessible location complying with one of the following:

   a. Outside of a building or structure

   b. Inside, nearest the point of entrance of the conductors

   c. Where installed in accordance with 230.6, nearest the point of entrance of the conductors

Section 240.21(B)(5) permits outside conductors to be tapped from a feeder without any limitations on the length of the tap conductors. The tap conductors must be protected against physical damage and must terminate in a single, fused disconnect or a single circuit breaker with a rating that does not exceed the ampacity of the tap conductors. Also, this fused disconnect or circuit breaker must be installed at a readily accessible location either inside or outside a building or structure. Furthermore, if the fused disconnect or circuit breaker is installed inside a building or structure, it must be located nearest the point of entrance of the tap conductors.

(C) Transformer Secondary Conductors Each set of conductors feeding separate loads shall be permitted to be connected to a transformer secondary, without overcurrent protection at the secondary, as specified in 240.21(C)(1) through (C)(6). The provisions of 240.4(B) shall not be permitted for transformer secondary conductors.

> FPN: For overcurrent protection requirements for transformers, see 450.3.

Transformer secondary conductors are permitted without an overcurrent protective device at the point the secondary conductors receive their supply under any of the following five conditions:

1. The primary overcurrent protective device, as described in 240.21(C)(1), can protect single-phase (2-wire) and 3-phase (delta-delta) transformer secondary conductors.
2. The transformer secondary conductors do not exceed 10 ft.
3. The transformer secondary conductors do not exceed 25 ft (two applications).
4. The transformer primary plus the secondary conductors do not exceed 25 ft.
5. The transformer secondary conductors are located outdoors.

Like the change in 240.21(B) for the 2005 *Code*, 240.21(C) has been revised to specifically prohibit application of 240.4(B) with transformer secondary conductors covered by the requirements of 240.21(C)(1) through 240.21(C)(6). See the commentary for 240.21(B).

The question of whether it is permissible to connect more than one set of secondary conductors to the secondary terminals of a transformer has been clearly answered in the 2005 *Code* by revising the first sentence of 240.21(C) to specify that the requirements apply to "each set of conductors feeding separate loads." First, it should be noted that it has never been the intent of these secondary conductor rules to limit their application to one set per transformer, and in fact there was no prohibition on installing multiple sets of secondary conductors and applying the 25-ft secondary conductor rule of 240.21(C)(6) to one set of conductors and the 10-ft secondary conductor rule to another set of conductors. Each set is treated individually in applying these requirements. For example, two panelboards could be supplied by two separate sets of transformer secondary conductors, with each set of conductors complying with one of the sets of rules from 240.21(C)(1) through (C)(6). Important

to remember, though, is the coordination between the requirements of 240.21(C) and, where used, the transformer secondary protection requirements in 450.3(A) and 450.3(B).

(1) Protection by Primary Overcurrent Device Conductors supplied by the secondary side of a single-phase transformer having a 2-wire (single-voltage) secondary, or a three-phase, delta-delta connected transformer having a 3-wire (single-voltage) secondary, shall be permitted to be protected by overcurrent protection provided on the primary (supply) side of the transformer, provided this protection is in accordance with 450.3 and does not exceed the value determined by multiplying the secondary conductor ampacity by the secondary to primary transformer voltage ratio.

Single-phase (other than 2-wire) and multiphase (other than delta-delta, 3-wire) transformer secondary conductors are not considered to be protected by the primary overcurrent protective device.

(2) Transformer Secondary Conductors Not Over 3 m (10 ft) Long Where the length of secondary conductor does not exceed 3 m (10 ft) and complies with all of the following:

(1) The ampacity of the secondary conductors is
   a. Not less than the combined calculated loads on the circuits supplied by the secondary conductors, and
   b. Not less than the rating of the device supplied by the secondary conductors or not less than the rating of the overcurrent-protective device at the termination of the secondary conductors, and
   c. Not less than one-tenth of the rating of the overcurrent device protecting the primary of the transformer, multiplied by the primary to secondary transformer voltage ratio
(2) The secondary conductors do not extend beyond the switchboard, panelboard, disconnecting means, or control devices they supply.
(3) The secondary conductors are enclosed in a raceway, which shall extend from the transformer to the enclosure of an enclosed switchboard, panelboard, or control devices or to the back of an open switchboard.

> FPN: For overcurrent protection requirements for lighting and appliance branch-circuit panelboards and certain power panelboards, see 408.36(A), (B), and (E).

For the 2005 *Code*, the minimum size requirement for 10-ft transformer secondary conductors has been revised to establish a relationship between the size of the ungrounded secondary conductors and the rating of the transformer primary overcurrent protective device, in a vein similar to that required for 10-ft feeder tap conductors. This size–rating relationship is necessary because the transformer primary device also provides short-circuit ground-fault protection for the transformer secondary conductors. It is important to understand that this revision is only one piece to establishing the minimum size for secondary conductors, and it is necessary to also ensure that the ampacity of the conductors is adequate for the calculated load and is not less than the rating of the device or overcurrent protective device in which the conductors terminate. The following example illustrates the application of this new requirement.

Example

Background Information

   75 kVA, 3-phase, 480-volt primary to 208Y/120-volt secondary

   Transformer primary overcurrent protective device is rated 125 amperes

Calculation

   1/10 of the primary OCPD rating:

$$125 \text{ amperes} \div 10 = 12.5 \text{ amperes}$$

   Line-to-line primary to secondary voltage ratio (480/208) is 2.31

$$12.5 \text{ amperes} \times 2.31 = 29 \text{ amperes}$$

Minimum Ampacity for Ungrounded Transformer Secondary Conductor

   29 amperes – 10 AWG copper THWN from Table 310.16 (30 amperes from 60°C column)

Conclusion

A 10 AWG copper conductor is permitted to be tapped from the secondary of this transformer with primary overcurrent protection rated 125 amperes. The load supplied by this secondary conductor cannot exceed the conductor's allowable ampacity from Table 310.16 coordinated with the temperature rating of the conductor terminations in accordance with 110.14(C)(1)(a).

(3) Industrial Installation Secondary Conductors Not Over 7.5 m (25 ft) Long For industrial installations only, where the length of the secondary conductors does not exceed 7.5 m (25 ft) and complies with all of the following:

(1) The ampacity of the secondary conductors is not less than the secondary current rating of the transformer, and the sum of the ratings of the overcurrent devices does not exceed the ampacity of the secondary conductors.
(2) All overcurrent devices are grouped.
(3) The secondary conductors are protected from physical damage by being enclosed in an approved raceway or by other approved means.

(4) Outside Secondary Conductors Where the conductors are located outdoors of a building or structure, except at the point of load termination, and comply with all of the following conditions:

(1) The conductors are protected from physical damage in an approved manner.
(2) The conductors terminate at a single circuit breaker or a single set of fuses that limit the load to the ampacity of the conductors. This single overcurrent device shall be permitted to supply any number of additional overcurrent devices on its load side.
(3) The overcurrent device for the conductors is an integral part of a disconnecting means or shall be located immediately adjacent thereto.
(4) The disconnecting means for the conductors is installed at a readily accessible location complying with one of the following:
   a. Outside of a building or structure
   b. Inside, nearest the point of entrance of the conductors
   c. Where installed in accordance with 230.6, nearest the point of entrance of the conductors

(5) Secondary Conductors from a Feeder Tapped Transformer Transformer secondary conductors installed in accordance with 240.21(B)(3) shall be permitted to have overcurrent protection as specified in that section.

(6) Secondary Conductors Not Over 7.5 m (25 ft) Long Where the length of secondary conductor does not exceed 7.5 m (25 ft) and complies with all of the following:

(1) The secondary conductors shall have an ampacity that is not less than the value of the primary-to-secondary voltage ratio multiplied by one-third of the rating of the overcurrent device protecting the primary of the transformer.
(2) The secondary conductors terminate in a single circuit breaker or set of fuses that limit the load current to not more than the conductor ampacity that is permitted by 310.15.
(3) The secondary conductors are protected from physical damage by being enclosed in an approved raceway or by other approved means.

(D) Service Conductors Service-entrance conductors shall be permitted to be protected by overcurrent devices in accordance with 230.91.

(E) Busway Taps Busways and busway taps shall be permitted to be protected against overcurrent in accordance with 368.17.

(F) Motor Circuit Taps Motor-feeder and branch-circuit conductors shall be permitted to be protected against overcurrent in accordance with 430.28 and 430.53, respectively.

(G) Conductors from Generator Terminals Conductors from generator terminals that meet the size requirement in 445.13 shall be permitted to be protected against overload by the generator overload protective device(s) required by 445.12.

240.22 Grounded Conductor

No overcurrent device shall be connected in series with any conductor that is intentionally grounded, unless one of the following two conditions is met:

(1) The overcurrent device opens all conductors of the circuit, including the grounded conductor, and is designed so that no pole can operate independently.
(2) Where required by 430.36 or 430.37 for motor overload protection.

240.23 Change in Size of Grounded Conductor

Where a change occurs in the size of the ungrounded conductor, a similar change shall be permitted to be made in the size of the grounded conductor.

Section 240.23 acknowledges that the size of the grounded circuit conductor may be increased (e.g., because of voltage-drop problems) or reduced to correspond to a reduction made in the size of the ungrounded circuit conductor(s), as in the case of feeder tap conductors, provided of course that the grounded and ungrounded conductors comprise the same circuit.

240.24 Location in or on Premises

(A) Accessibility Overcurrent devices shall be readily accessible and shall be installed so that the center of the grip of the operating handle of the switch or circuit breaker, when in its highest position, is not more than 2.0 m (6 ft 7 in.) above the floor or working platform unless one of the following applies:

(1) For busways, as provided in 368.12.
(2) For supplementary overcurrent protection, as described in 240.10.
(3) For overcurrent devices, as described in 225.40 and 230.92.
(4) For overcurrent devices adjacent to utilization equipment that they supply, access shall be permitted to be by portable means.

Section 240.24(A)(4) recognizes the need for overcurrent protection in locations that are not readily accessible, such as above suspended ceilings. It permits overcurrent devices to be located so that they are not readily accessible, as long as they are located next to the appliance, motor, or other equipment they supply and can be reached by using a ladder. For the purposes of this requirement, ready access to the operating handle of a fusible switch or circuit breaker is considered to be not more than 6 ft 7 in. above the finished floor or working platform.

The measurement is made from the center of the device operating handle when the handle is at its highest position. This text, added for the 2005 *Code*, parallels the requirement of 404.8(A), which applies to all switches and to circuit breakers used as switches. For information regarding the accessibility of supplementary overcurrent devices, refer to 240.10.

(B) Occupancy Each occupant shall have ready access to all overcurrent devices protecting the conductors supplying that occupancy.

Exception No. 1: Where electric service and electrical maintenance are provided by the building management and where these are under continuous building management supervision, the service overcurrent devices and feeder overcurrent devices supplying more than one occupancy shall be permitted to be accessible to only authorized management personnel in the following:

(1) Multiple-occupancy buildings
(2) Guest rooms or guest suites of hotels and motels that are intended for transient occupancy

Exception No. 2: Where electric service and electrical maintenance are provided by the building management and where these are under continuous building management supervision, the branch circuit overcurrent devices supplying any guest rooms or guest suites shall be permitted to be accessible to only authorized management personnel for guest rooms of hotels and motels that are intended for transient occupancy.

(C) Not Exposed to Physical Damage Overcurrent devices shall be located where they will not be exposed to physical damage.

    FPN: See 110.11, Deteriorating Agents.

(D) Not in Vicinity of Easily Ignitible Material Overcurrent devices shall not be located in the vicinity of easily ignitible material, such as in clothes closets.

Examples of locations where combustible materials may be stored are linen closets, paper storage closets, and clothes closets.

(E) Not Located in Bathrooms In dwelling units and guest rooms or guest suites of hotels and motels, overcurrent devices, other than supplementary overcurrent protection, shall not be located in bathrooms.

III. Enclosures

240.30 General

(A) Protection from Physical Damage Overcurrent devices shall be protected from physical damage by one of the following:

(1) Installation in enclosures, cabinets, cutout boxes, or equipment assemblies
(2) Mounting on open-type switchboards, panelboards, or control boards that are in rooms or enclosures free from dampness and easily ignitible material and are accessible only to qualified personnel

Properly selected overcurrent protective devices are designed to open a circuit before an overcurrent condition can seriously damage conductor insulation. Requirements that overcurrent devices be enclosed in cabinets or cutout boxes ensure that hot metal particles will not be ejected in the vicinity of combustible materials. Also, use of an enclosure prevents contact with live parts by personnel.

Overcurrent devices mounted on open-type switchboards, panelboards, or control boards and having exposed energized parts are to be located where accessible only to qualified persons.

(B) Operating Handle The operating handle of a circuit breaker shall be permitted to be accessible without opening a door or cover.

240.32 Damp or Wet Locations

Enclosures for overcurrent devices in damp or wet locations shall comply with 312.2(A).

240.33 Vertical Position

Enclosures for overcurrent devices shall be mounted in a vertical position unless that is shown to be impracticable. Circuit breaker enclosures shall be permitted to be installed horizontally where the circuit breaker is installed in accordance with 240.81. Listed busway plug-in units shall be permitted to be mounted in orientations corresponding to the busway mounting position.

The general rule of 240.33 requires enclosures for overcurrent devices to be installed in a vertical position. A wall-mounted vertical position for enclosures for overcurrent devices is desirable to afford easier access, natural hand operation, normal swinging or closing of doors or covers, and legibility of the manufacturer's markings. In addition, this section does not permit a panelboard or fusible switch enclosure to be installed in a horizontal position such that the back of the enclosure is mounted on the ceiling or the floor. Compliance with the up position of the handle being on or closed, and the down position of the handle being off or open, in accordance with 240.81 limits the number of pole spaces available on a panelboard where its cabinet is mounted in a horizontal position on a wall.

IV. Disconnecting and Guarding

240.40 Disconnecting Means for Fuses

A disconnecting means shall be provided on the supply side of all fuses in circuits over 150 volts to ground and cartridge fuses in circuits of any voltage where accessible to other than qualified persons, so that each circuit containing fuses can be independently disconnected from the source of power. A current-limiting device without a disconnecting means shall be permitted on the supply side of the service disconnecting means as permitted by 230.82. A single disconnecting means shall be permitted on the supply side of more than one set of fuses as permitted by 430.112, Exception, for group operation of motors and 424.22(C) for fixed electric space-heating equipment.

A single disconnect switch is allowed to serve more than one set of fuses, such as in multimotor installations or for electric space-heating equipment where the heating element load is required to be subdivided, each element with its own set of fuses. A revision in the 2005 *Code* recognizes the installation of cable limiters or similar current-limiting devices on the supply side of the service disconnecting means, as permitted by 230.82(1). No disconnecting means is required on the supply side of such devices.

240.41 Arcing or Suddenly Moving Parts

Arcing or suddenly moving parts shall comply with 240.41(A) and (B).

(A) Location Fuses and circuit breakers shall be located or shielded so that persons will not be burned or otherwise injured by their operation.

(B) Suddenly Moving Parts Handles or levers of circuit breakers, and similar parts that may move suddenly in such a way that persons in the vicinity are likely to be injured by being struck by them, shall be guarded or isolated.

Arcing or sudden-moving parts are usually associated with switchboards or control boards that may be of the open type. Switchboards and control boards should be under competent supervision and accessible only to qualified persons. Fuses or circuit breakers must be located or shielded so that, under an abnormal condition, the subsequent arc across the opening device will not injure persons in the vicinity.

Guardrails may be provided in the vicinity of disconnecting means because sudden-moving handles may be capable of causing injury. Modern switchboards, for example, are equipped with removable handles. See Article 100 for the definition of *guarded*. See also 110.27 for the guarding of live parts (600 volts, nominal, or less).

V. Plug Fuses, Fuseholders, and Adapters

240.50 General

(A) Maximum Voltage Plug fuses shall be permitted to be used in the following circuits:

(1) Circuits not exceeding 125 volts between conductors
(2) Circuits supplied by a system having a grounded neutral where the line-to-neutral voltage does not exceed 150 volts

Plug fuses can be installed in circuits supplied by 120/240-volt, single-phase, 3-wire systems and by 208Y/120-volt, 3-phase, 4-wire systems.

(B) Marking Each fuse, fuseholder, and adapter shall be marked with its ampere rating.

(C) Hexagonal Configuration Plug fuses of 15-ampere and lower rating shall be identified by a hexagonal configuration of the window, cap, or other prominent part to distinguish them from fuses of higher ampere ratings.

Exhibit 240.11 shows two examples of Edison-base plug fuses and a Type S plug fuse. Note the hexagonal feature on the 15-ampere fuse in accordance with 240.50(C).

(D) No Energized Parts Plug fuses, fuseholders, and adapters shall have no exposed energized parts after fuses or fuses and adapters have been installed.

(E) Screw Shell The screw shell of a plug-type fuseholder shall be connected to the load side of the circuit.

Exhibit 240.12 shows a Type S nonrenewable plug fuse and its corresponding adapter.

240.51 Edison-Base Fuses

(A) Classification Plug fuses of the Edison-base type shall be classified at not over 125 volts and 30 amperes and below.

(B) Replacement Only Plug fuses of the Edison-base type shall be used only for replacements in existing installations where there is no evidence of overfusing or tampering.

Exhibit 240.11 Two plug fuses and a Type S fuse. (Courtesy of Bussmann Division, Cooper Industries)

Exhibit 240.12 Type S nonrenewable plug fuse and adapter. (Redrawn from Bussmann Division, Cooper Industries)

240.52 Edison-Base Fuseholders

Fuseholders of the Edison-base type shall be installed only where they are made to accept Type S fuses by the use of adapters.

240.53 Type S Fuses

Type S fuses shall be of the plug type and shall comply with 240.53(A) and (B).

(A) Classification Type S fuses shall be classified at not over 125 volts and 0 to 15 amperes, 16 to 20 amperes, and 21 to 30 amperes.

(B) Noninterchangeable Type S fuses of an ampere classification as specified in 240.53(A) shall not be interchangeable with a lower ampere classification. They shall be designed so that they cannot be used in any fuseholder other than a Type S fuseholder or a fuseholder with a Type S adapter inserted.

240.54 Type S Fuses, Adapters, and Fuseholders

(A) To Fit Edison-Base Fuseholders Type S adapters shall fit Edison-base fuseholders.

(B) To Fit Type S Fuses Only Type S fuseholders and adapters shall be designed so that either the fuseholder itself or the fuseholder with a Type S adapter inserted cannot be used for any fuse other than a Type S fuse.

(C) Nonremovable Type S adapters shall be designed so that once inserted in a fuseholder, they cannot be removed.

(D) Nontamperable Type S fuses, fuseholders, and adapters shall be designed so that tampering or shunting (bridging) would be difficult.

(E) Interchangeability Dimensions of Type S fuses, fuseholders, and adapters shall be standardized to permit interchangeability regardless of the manufacturer.

VI. Cartridge Fuses and Fuseholders

240.60 General

(A) Maximum Voltage — 300-Volt Type Cartridge fuses and fuseholders of the 300-volt type shall be permitted to be used in the following circuits:

(1) Circuits not exceeding 300 volts between conductors
(2) Single-phase line-to-neutral circuits supplied from a 3-phase, 4-wire, solidly grounded neutral source where the line-to-neutral voltage does not exceed 300 volts

(B) Noninterchangeable — 0–6000-Ampere Cartridge Fuseholders Fuseholders shall be designed so that it will be difficult to put a fuse of any given class into a fuseholder that is designed for a current lower, or voltage higher, than that of the class to which the fuse belongs. Fuseholders for current-limiting fuses shall not permit insertion of fuses that are not current-limiting.

(C) Marking Fuses shall be plainly marked, either by printing on the fuse barrel or by a label attached to the barrel showing the following:

(1) Ampere rating
(2) Voltage rating
(3) Interrupting rating where other than 10,000 amperes
(4) Current limiting where applicable
(5) The name or trademark of the manufacturer

The interrupting rating shall not be required to be marked on fuses used for supplementary protection.

Exhibit 240.13 shows two examples of Class G fuses rated 300 volts. Note the plainly marked barrels. Class H-type cartridge fuses have an interrupting capacity (IC) rating of 10,000 amperes, which need not be marked on the fuse. However, Class CC, G, J, K, L, R, and T cartridge fuses exceed the 10,000-ampere IC rating and must be marked with the IC rating. Section 240.83(C) requires that the IC rating for circuit breakers for other than 5000 amperes be indicated on the circuit breaker. Fuses or circuit breakers used for supplementary overcurrent protection of fluorescent fixtures, semiconductor rectifiers, motor-operated appliances, and so on, need not be marked for IC. See also the commentary on circuit breakers following 110.10.

Exhibit 240.13 Two Class G fuses rated 300 volts. (Courtesy of Bussmann Division, Cooper Industries)

(D) Renewable Fuses Class H cartridge fuses of the renewable type shall only be permitted to be used for replacement in existing installations where there is no evidence of overfusing or tampering.

Class H renewable fuses are now permitted only as a replacement in existing installations. Where overfusing and/or tampering are detected with an existing installation, the use of nonrenewable fuses as a replacement is mandatory. An important consideration in the use of the traditional Class H renewable fuse is its 10,000-ampere interrupting rating.

In addition, caution must be exercised where the use of renewable fuses is contemplated because the manufacturer's directions provided in some modern fusible switches do not permit the use of renewable fuses or strongly recommend against their use.

240.61 Classification

Cartridge fuses and fuseholders shall be classified according to voltage and amperage ranges. Fuses rated 600 volts, nominal, or less shall be permitted to be used for voltages at or below their ratings.

See 490.21(B) for application of high-voltage fuses.

VII. Circuit Breakers

240.80 Method of Operation

Circuit breakers shall be trip free and capable of being closed and opened by manual operation. Their normal method of operation by other than manual means, such as electrical or pneumatic, shall be permitted if means for manual operation are also provided.

240.81 Indicating

Circuit breakers shall clearly indicate whether they are in the open "off" or closed "on" position.

Where circuit breaker handles are operated vertically rather than rotationally or horizontally, the "up" position of the handle shall be the "on" position.

See 240.83(D), 404.11, and 410.81(A) for requirements for circuit breakers used as switches. To ensure that operating the handle in a downward motion turns the device off, 240.81 prohibits a circuit breaker from being inverted.

240.82 Nontamperable

A circuit breaker shall be of such design that any alteration of its trip point (calibration) or the time required for its operation requires dismantling of the device or breaking of a seal for other than intended adjustments.

240.83 Marking

(A) Durable and Visible Circuit breakers shall be marked with their ampere rating in a manner that will be durable and visible after installation. Such marking shall be permitted to be made visible by removal of a trim or cover.

(B) Location Circuit breakers rated at 100 amperes or less and 600 volts or less shall have the ampere rating molded, stamped, etched, or similarly marked into their handles or escutcheon areas.

(C) Interrupting Rating Every circuit breaker having an interrupting rating other than 5000 amperes shall have its interrupting rating shown on the circuit breaker. The interrupting rating shall not be required to be marked on circuit breakers used for supplementary protection.

Section 240.83(C) recognizes series-rated circuit breakers and requires that the end-use equipment be marked with the series combination rating. For example, a circuit breaker with an interrupting rating of 10,000 amperes may perform safely on a circuit with an available fault current that is greater than 10,000 amperes, under the following conditions:

1. It is protected on its line side by a circuit breaker with a suitable interrupting rating.
2. The series combination has been tested and demonstrated to safely open a short-circuit current higher than the 10,000 amperes on the load side of the downstream breaker.

The UL *General Information Directory* ("White Book"), under the category "Switchboards, Dead Front Type (WEVZ)," provides the following information:

Short Circuit Ratings: Dead-front switchboard sections or interiors are marked with their DC or RMS symmetrical short-circuit current rating in amperes. The marking states that short-circuit ratings are limited to the lowest short-circuit rating of (1) any switchboard section connected in series or (2) the lowest short-circuit rating of any device installed or intended to be installed therein. However, for combination series-connected devices, the short-circuit current rating marked on the switchboard may be higher than the short-circuit current rating of a specific circuit breaker installed or to be installed in the switchboard. This higher rating is valid only if the specific overcurrent devices identified in the marking are used within or ahead of the switchboard in accordance with the marked instructions. In many cases, the short-circuit ratings are associated with instructions for securing supply wiring within the switchboard. The reference to "securing supply wiring within the switchboard" alludes to the necessity to prevent the supply wiring from violently moving due to magnetic forces under short-circuit or ground-fault conditions.

(D) Used as Switches Circuit breakers used as switches in 120-volt and 277-volt fluorescent lighting circuits shall be listed and shall be marked SWD or HID. Circuit breakers used as switches in high-intensity discharge lighting circuits shall be listed and shall be marked as HID.

Circuit breakers marked "SWD" are 15- or 20-ampere breakers that have been subjected to additional endurance and temperature testing to assess their ability to be used as the regular control device for fluorescent lighting circuits. A change in the 2002 *Code* instituted a requirement that circuit breakers marked "HID" are also acceptable for switching applications, and that this marking must be on circuit breakers used as the regular switching device to control high-intensity discharge (HID) lighting such as mercury vapor, high-pressure or low-pressure sodium, or metal halide lighting. Circuit breakers marked "HID" can be used for switching both high-intensity discharge and fluorescent lighting loads; however, a circuit breaker marked "SWD" can be used only as a switching device for fluorescent lighting loads.

(E) Voltage Marking Circuit breakers shall be marked with a voltage rating not less than the nominal system voltage that is indicative of their capability to interrupt fault currents between phases or phase to ground.

240.85 Applications

A circuit breaker with a straight voltage rating, such as 240V or 480V, shall be permitted to be applied in a circuit in which the nominal voltage between any two conductors does not exceed the circuit breaker's voltage rating. A two-pole circuit breaker shall not be used for protecting a 3-phase, corner-grounded delta circuit unless the circuit breaker is marked 1ф–3ф to indicate such suitability.

A circuit breaker with a slash rating, such as 120/240V or 480Y/277V, shall be permitted to be applied in a solidly grounded circuit where the nominal voltage of any conductor to ground does not exceed the lower of the two values of the circuit breaker's voltage rating and the nominal voltage between any two conductors does not exceed the higher value of the circuit breaker's voltage rating.

> FPN: Proper application of molded case circuit breakers on 3-phase systems, other than solidly grounded wye, particularly on corner grounded delta systems, considers the circuit breakers' individual pole-interrupting capability.

A circuit breaker marked 480Y/277V is not intended for use on a 480-volt system with up to 480 volts to ground, such as a 480-volt circuit derived from a corner-grounded, delta-connected system. A circuit breaker marked either 480V or 600V should be used on such a system. In like manner, a circuit breaker marked 120/240V is not intended for use on a delta-connected 240-volt circuit. A 240-volt, 480-volt, or 600-volt circuit breaker should be used on such a circuit. The slash (/) between the lower and higher voltage ratings in the marking indicates that the circuit breaker has been tested for use on a circuit with the higher voltage between phases and with the lower voltage to ground.

240.86 Series Ratings

Where a circuit breaker is used on a circuit having an available fault current higher than the marked interrupting rating by

being connected on the load side of an acceptable overcurrent protective device having a higher rating, the circuit breaker shall meet the requirements specified in (A) or (B), and (C).

A series rated system is a combination of circuit breakers or a combination of fuses and circuit breakers that can be applied at available short-circuit levels above the interrupting rating of the load-side circuit breakers but not above that of the main or line-side device. Series rated systems can consist of fuses that protect circuit breakers or of circuit breakers that protect circuit breakers. The arrangement of protective components in a series rated system can be as specified in 240.86(A) for engineered systems applied to existing installations or in 240.86(B) for tested combinations that can be applied in any new or existing installation.

(A) Selected Under Engineering Supervision in Existing Installations The series rated combination devices shall be selected by a licensed professional engineer engaged primarily in the design or maintenance of electrical installations. The selection shall be documented and stamped by the professional engineer. This documentation shall be available to those authorized to design, install, inspect, maintain, and operate the system. This series combination rating, including identification of the upstream device, shall be field marked on the end use equipment.

This new provision allows for an engineering solution at existing facilities where an increase in the available fault current (due to factors such as increases in transformer size, lowering of transformer impedances, and changes in utility distribution systems) puts the existing circuit overcurrent protection equipment at peril in regard to interrupting fault currents as required by 110.9. The objective of this "engineered system" is to maintain compliance with 110.9 by redesigning the overcurrent protection scheme to accommodate the increase in available fault current and not having to undertake a wholesale replacement of electrical distribution equipment. Where the increase in fault current causes existing equipment to be "underrated," the engineering approach is to provide upstream protection that functions in concert with the existing protective devices to safely open the circuit under fault conditions. The requirement specifies that the design of such systems is to be performed only by licensed professional engineers whose credentials substantiate their ability to perform this type of engineering. Documentation in the form of stamped drawings and field marking of end-use equipment to indicate it is a component of a series rated system is required.

Designing a series rated system requires careful consideration of the fault-clearing characteristics of the existing protective devices and their ability to interact with the newly installed upstream protective device(s) when subjected to fault conditions. This new provision does not ensure that an engineered series rated system can be applied to all existing installations. The operating parameters of the existing overcurrent protection equipment dictate what can be done in a field-engineered protection scheme.

Compatibility with series rated systems will in all likelihood be limited to circuit breakers that (1) remain closed during the first ½ cycle of a fault and (2) have an interrupting rating that is not less than the let-through current of an upstream protective device (such as a current-limiting fuse). In those cases where the opening of a circuit breaker, under any level of fault current, begins in less than ½ cycle, the use of a field engineered series rated system will in all likelihood be contrary to acceptable application practices specified by the circuit breaker manufacturer. Where there is any doubt over the proper application of existing downstream circuit breakers with new upstream overcurrent protective devices, the manufacturers of the existing circuit breakers and the new upstream overcurrent protective devices must be consulted.

The safety objective of any overcurrent protection scheme is to ensure compliance with 110.9.

(B) Tested Combinations The combination of line-side overcurrent device and load-side circuit breaker(s) is tested and marked on the end use equipment, such as switchboards and panelboards.

Section 240.86(B) requires that, when a series rating is used, the switchboards, panelboards, and load centers be marked for use with the series rated combinations that may be used. Therefore, the enclosures must have a label affixed by the equipment manufacturer that provides the series rating of the combination(s). Because there is often not enough room in the equipment to show all the legitimate series rated combinations, UL 67, *Standard for Panelboards*, allows a bulletin to be referenced and supplied with the panelboard. These bulletins typically provide all the acceptable combinations. Note that the installer of a series rated system also has to provide the additional labeling on equipment enclosures required by 110.22, indicating that the equipment has been applied in a series-rated system.

(C) Motor Contribution Series ratings shall not be used where

(1) Motors are connected on the load side of the higher-rated overcurrent device and on the line side of the lower-rated overcurrent device, and
(2) The sum of the motor full-load currents exceeds 1 percent of the interrupting rating of the lower-rated circuit breaker.

One critical requirement limits the use of series rated systems in which motors are connected between the line-side (protecting) device and the load-side (protected) circuit breaker. Section 240.86(C) requires that series ratings developed under the parameters of either 240.86(A) or 240.86(B) are not to be used where the sum of motor full-load currents exceeds 1 percent of the interrupting rating of the load-side (protected) circuit breaker, as illustrated in Exhibit 240.14.

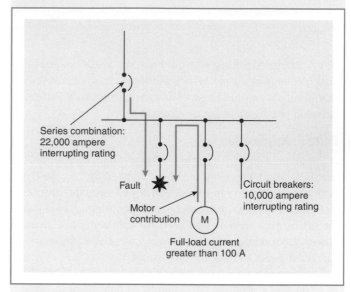

Exhibit 240.14 Example of installation where level of motor contribution exceeds 1 percent of interrupting rating for the lowest-rated circuit breaker in this series rated system.

VIII. Supervised Industrial Installations
240.90 General
Overcurrent protection in areas of supervised industrial installations shall comply with all of the other applicable provisions of this article, except as provided in Part VIII. The provisions of Part VIII shall be permitted only to apply to those portions of the electrical system in the supervised industrial installation used exclusively for manufacturing or process control activities.

Part VIII provides alternative approaches to overcurrent protection for low-voltage distribution systems (600 volts, nominal, and under) that supply large manufacturing plants or industrial processes. For further elaboration on the definition of "supervised industrial installation," see the commentary following 240.2. Section 240.21 contains the requirements that specify the point in the circuit at which overcurrent protection for conductors is to be located. The general rule of 240.21 is that conductors are to be protected at the point they receive their supply. However, 240.21(B) and 240.21(C) contain provisions that allow the overcurrent protection for feeders and transformer secondary conductors to be located at other than the point of supply. The rules in Part VIII modify the 240.21(B) and 240.21(C) requirements based on the condition that only qualified personnel monitor and maintain the installation. Included in the requirements of 240.92(A) through 240.92(D) are provisions for longer conductor lengths, for use of differential relays as the means for providing short-circuit ground-fault protection, and for using up to six circuit breakers or fuses as the overload protection for outside feeder taps and outside transformer secondary conductors.

240.92 Location in Circuit

An overcurrent device shall be connected in each ungrounded circuit conductor as required in 240.92(A) through (D).

(A) Feeder and Branch-Circuit Conductors Feeder and branch-circuit conductors shall be protected at the point the conductors receive their supply as permitted in 240.21 or as otherwise permitted in 240.92(B), (C), or (D).

(B) Transformer Secondary Conductors of Separately Derived Systems Conductors shall be permitted to be connected to a transformer secondary of a separately derived system, without overcurrent protection at the connection, where the conditions of 240.92(B)(1), (B)(2), and (B)(3) are met.

(1) Short-Circuit and Ground-Fault Protection The conductors shall be protected from short-circuit and ground-fault conditions by complying with one of the following conditions:

(1) The length of the secondary conductors does not exceed 30 m (100 ft) and the transformer primary overcurrent device has a rating or setting that does not exceed 150 percent of the value determined by multiplying the secondary conductor ampacity by the secondary-to-primary transformer voltage ratio.

(2) The conductors are protected by a differential relay with a trip setting equal to or less than the conductor ampacity.

FPN: A differential relay is connected to be sensitive only to short-circuit or fault currents within the protected zone and is normally set much lower than the conductor ampacity. The differential relay is connected to trip protective devices that will de-energize the protected conductors if a short-circuit condition occurs.

(3) The conductors shall be considered to be protected if calculations, made under engineering supervision, determine that the system overcurrent devices will protect the conductors within recognized time vs. current limits for all short-circuit and ground-fault conditions.

(2) Overload Protection The conductors shall be protected against overload conditions by complying with one of the following:

(1) The conductors terminate in a single overcurrent device that will limit the load to the conductor ampacity.

(2) The sum of the overcurrent devices at the conductor termination limits the load to the conductor ampacity. The overcurrent devices shall consist of not more than six circuit breakers or sets of fuses, mounted in a single enclosure, in a group of separate enclosures, or in or on a switchboard. There shall be no more than six overcurrent devices grouped in any one location.

(3) Overcurrent relaying is connected [with a current transformer(s), if needed] to sense all of the secondary conductor current and limit the load to the conductor ampacity by opening upstream or downstream devices.

(4) Conductors shall be considered to be protected if calculations, made under engineering supervision, determine that the system overcurrent devices will protect the conductors from overload conditions.

(3) Physical Protection The secondary conductors are protected from physical damage by being enclosed in an approved raceway or by other approved means.

(C) Outside Feeder Taps Outside conductors shall be permitted to be tapped to a feeder or to be connected at a transformer secondary, without overcurrent protection at the tap or connection, where all the following conditions are met:

(1) The conductors are protected from physical damage in an approved manner.

(2) The sum of the overcurrent devices at the conductor termination limits the load to the conductor ampacity. The overcurrent devices shall consist of not more than six circuit breakers or sets of fuses mounted in a single enclosure, in a group of separate enclosures, or in or on a switchboard. There shall be no more than six overcurrent devices grouped in any one location.

(3) The tap conductors are installed outdoors of a building or structure except at the point of load termination.

(4) The overcurrent device for the conductors is an integral part of a disconnecting means or shall be located immediately adjacent thereto.

(5) The disconnecting means for the conductors are installed at a readily accessible location complying with one of the following:

 a. Outside of a building or structure
 b. Inside, nearest the point of entrance of the conductors
 c. Where installed in accordance with 230.6, nearest the point of entrance of the conductors

(D) Protection by Primary Overcurrent Device Conductors supplied by the secondary side of a transformer shall be permitted to be protected by overcurrent protection provided on the primary (supply) side of the transformer, provided the primary device time–current protection characteristic, multiplied by the maximum effective primary-to-secondary transformer voltage ratio, effectively protects the secondary conductors.

IX. Overcurrent Protection Over 600 Volts, Nominal

240.100 Feeders and Branch Circuits

(A) Location and Type of Protection Feeder and branch-circuit conductors shall have overcurrent protection in each ungrounded conductor located at the point where the conductor receives its supply or at an alternative location in the circuit when designed under engineering supervision that includes but is not limited to considering the appropriate fault studies and time–current coordination analysis of the protective devices and the conductor damage curves. The overcurrent protection shall be permitted to be provided by either 240.100(A)(1) or (A)(2).

(1) Overcurrent Relays and Current Transformers Circuit breakers used for overcurrent protection of 3-phase circuits shall have a minimum of three overcurrent relay elements operated from three current transformers. The separate overcurrent relay elements (or protective functions) shall be permitted to be part of a single electronic protective relay unit.

On 3-phase, 3-wire circuits, an overcurrent relay element in the residual circuit of the current transformers shall be permitted to replace one of the phase relay elements.

An overcurrent relay element, operated from a current transformer that links all phases of a 3-phase, 3-wire circuit, shall be permitted to replace the residual relay element and one of the phase-conductor current transformers. Where the neutral is not regrounded on the load side of the circuit as permitted in 250.184(B), the current transformer shall be permitted to link all 3-phase conductors and the grounded circuit conductor (neutral).

(2) Fuses A fuse shall be connected in series with each ungrounded conductor.

(B) Protective Devices The protective device(s) shall be capable of detecting and interrupting all values of current that can occur at their location in excess of their trip-setting or melting point.

(C) Conductor Protection The operating time of the protective device, the available short-circuit current, and the conductor used shall be coordinated to prevent damaging or dangerous temperatures in conductors or conductor insulation under short-circuit conditions.

240.101 Additional Requirements for Feeders

(A) Rating or Setting of Overcurrent Protective Devices The continuous ampere rating of a fuse shall not exceed three times the ampacity of the conductors. The long-time trip element setting of a breaker or the minimum trip setting of an electronically actuated fuse shall not exceed six times the ampacity of the conductor. For fire pumps, conductors shall be permitted to be protected for overcurrent in accordance with 695.4(B).

(B) Feeder Taps Conductors tapped to a feeder shall be permitted to be protected by the feeder overcurrent device where that overcurrent device also protects the tap conductor.

ARTICLE 250
Grounding and Bonding

Summary of Changes

- **250.2:** Revised definition of *effective ground-fault current path* to include its function of facilitating the operation of overcurrent devices or ground-fault detectors.
- **Table 250.3:** Added reference to Article 392 grounding requirements for cable trays.
- **250.4(A)(5):** Revised to include facilitating the operation of overcurrent devices or ground-fault detectors as part of the performance requirements for the *effective ground-fault current path.*
- **250.8:** Revised to prohibit use of sheet metal screws as a means to attach connection devices for grounding conductors.
- **250.20(E):** Added new requirement to correlate 250.20 with 250.36 and 250.186.
- **250.21:** Revised to require ground detectors on ungrounded ac systems unless the voltage to ground is less than 120 volts.
- **250.24(B):** Added requirement from 250.28 covering purpose of the main bonding jumper at service equipment.
- **250.28:** Added the term *system bonding jumper* throughout this section for application with separately derived systems.
- **250.30:** Reorganized this requirement to improve usability and integrated the new term *system bonding jumper* where applicable. Revised requirement for sizing the *common grounding electrode conductor* to require a minimum 3/0 AWG copper or 250 kcmil aluminum conductor.
- **250.32:** Revised the title of the requirement to better convey the type of supply to a building or structure that is covered by these rules. Added new provision in the exception to 250.32(A) permitting multiwire circuits to be considered as a single branch circuit.
- **250.50:** Revised to require the use of a concrete-encased electrode if a building has a footing or a foundation. Exception added to exempt existing buildings or structures where access to concrete-encased electrode requires damaging the concrete.

- **250.52(A)(2):** Deleted the phrase *effectively grounded* and provided a list of conditions under which the metal frame of a building can be used as a grounding electrode.

- **250.64(B):** Revised to delete the word *severe* from the protection requirement for 4 AWG or larger grounding electrode conductors that are subject to physical damage.

- **250.64(C):** Revised to permit the use of a copper or aluminum busbar as a connection point for grounding electrode conductors or bonding jumpers.

- **250.64(D):** Revised to clarify requirements for sizing the grounding electrode conductor and the grounding electrode "taps" used in multiple service disconnecting means arrangements.

- **250.64(E):** Revised to limit the bonding requirement to ferrous metal enclosures and to indicate that nonferrous metal enclosures are not required to be electrically continuous.

- **250.68(A):** Added an exception exempting grounding electrode connections to structural metal encapsulated with fire-proofing material from having to be accessible.

- **250.84(B):** Revised requirement to apply only to metal raceways that contain metal sheathed or armored cable.

- **250.92(B)(4):** Revised to require that bonding fitting used at services be listed.

- **250.100:** Revised to require specific bonding methods for raceways, enclosures, and equipment installed in hazardous (classified) locations regardless of the presence of a supplementary equipment grounding conductor in the raceways or enclosures.

- **250.104(D):** Relocated requirements for bonding water piping and structural metal to separately derived systems from 250.104(A)(4) of 2002 *Code.*

- **250.118(5)d and 250.118(6)e:** Revised to indicate that only where the flexible metal or liquidtight flexible metal conduit requires the ability to flex or move after the initial installation is a "wire type" equipment grounding conductor required.

- **250.118(14):** Added surface metal raceways listed for grounding as a permitted type of equipment grounding conductor.

- **250.122(E):** Revised to require sizing equipment grounding conductor per Table 250.122 for cords and fixture wire with circuit conductors larger than 10 AWG.

- **250.122(G):** Added new requirement for sizing wire-type equipment grounding conductors run with feeder tap conductors.

- **250.126:** Revised to permit other grounding symbols.

- **250.146(A):** Revised to permit a listed self-grounding contact yoke or device that complies with 250.146(B).

Added rule to require that at least one of the insulating mounting screw retention washers be removed from receptacles that are not the listed self-grounding type.

- **250.184:** Revised section to contain specific rules for single-point grounded neutral systems. Clarified that both single-point or multigrounded neutral systems are permitted by this requirement. Added specific installation requirements for each grounding option.

Contents

I. General

250.1 Scope

This article covers general requirements for grounding and bonding of electrical installations, and specific requirements in (1) through (6).

The complete revision of Article 250 is one of the most significant changes to occur in the recent history of the *Code.* Undertaken during the 1999 *Code* revision cycle, the task of reorganizing the large amount of subject matter contained in this article for the purpose of creating a more logical approach to the subject of grounding and bonding was a collective effort of the NEC Usability Task Group, Code-Making Panel 5, and *NEC* users who submitted pro-

posals and comments. To better organize the existing requirements, similar rules that previously appeared in different parts of Article 250 were relocated and grouped in the same part of the article. In addition, many of the exceptions were converted into positive code language. The overall new approach to the layout provides a more user-friendly Article 250. As an aid to the users of the *Code,* the commentary for Annex F provides two cross-reference lists. Exhibit F.1 references the 1996, 1999, 2002, and 2005 sections to the 1996 Article 250 topics, and Exhibit F.2 references the 2002, 1999, and 1996 sections to the 2005 Article 250 topics. For the 2005 *Code,* the title of Article 250 has been changed to *Grounding and Bonding* to reinforce that grounding and bonding are two separate concepts but are not mutually exclusive and in fact are directly interrelated through the requirements of Article 250.

(1) Systems, circuits, and equipment required, permitted, or not permitted to be grounded
(2) Circuit conductor to be grounded on grounded systems
(3) Location of grounding connections
(4) Types and sizes of grounding and bonding conductors and electrodes
(5) Methods of grounding and bonding
(6) Conditions under which guards, isolation, or insulation may be substituted for grounding

250.2 Definitions

Effective Ground-Fault Current Path. An intentionally constructed, permanent, low-impedance electrically conductive path designed and intended to carry current under ground-fault conditions from the point of a ground fault on a wiring system to the electrical supply source and that facilitates the operation of the overcurrent protective device or ground fault detectors on high-impedance grounded systems.

Ground Fault. An unintentional, electrically conducting connection between an ungrounded conductor of an electrical circuit and the normally non–current-carrying conductors, metallic enclosures, metallic raceways, metallic equipment, or earth.

Ground-Fault Current Path. An electrically conductive path from the point of a ground fault on a wiring system through normally non–current-carrying conductors, equipment, or the earth to the electrical supply source.

> FPN: Examples of ground-fault current paths could consist of any combination of equipment grounding conductors, metallic raceways, metallic cable sheaths, electrical equipment, and any other electrically conductive material such as metal water and gas piping, steel framing members, stucco mesh, metal ducting, reinforcing steel, shields of communications cables, and the earth itself.

Section 250.2 was new for the 2002 *Code.* Following a common numbering sequence throughout the *NEC,* definitions that are specific to an article and not generally used elsewhere now appear in X.2 of their respective articles. For examples of article-related definitions, see 240.2, 280.2, 285.2, 517.2, and 680.2.

One of the keys to proper application of the Article 250 requirements is understanding the definitions of terms used throughout the *Code* that relate to bonding and grounding. Some of the most basic and widely used terms are *bonding, grounded, grounded conductor, equipment grounding conductor,* and *grounding electrode conductor.* These terms are defined in Article 100.

250.3 Application of Other Articles

In other articles applying to particular cases of installation of conductors and equipment, requirements are identified in Table 250.3 that are in addition to, or modifications of, those of this article.

250.4 General Requirements for Grounding and Bonding

The following general requirements identify what grounding and bonding of electrical systems are required to accomplish. The prescriptive methods contained in Article 250 shall be followed to comply with the performance requirements of this section.

Section 250.4 provides the performance requirements for grounding and bonding of electrical systems and equipment. Performance-based requirements provide an overall objective without stating the specifics for accomplishing that objective. The first paragraph of 250.4 indicates that the performance objectives stated in 250.4(A) for grounded systems and in 250.4(B) for ungrounded systems are accomplished by complying with the prescriptive requirements found in the rest of Article 250.

The requirements of 250.4 do not provide a specific rule for the sizing or connection of grounding conductors; rather, it states overall performance considerations for grounding conductors and applies to both grounded and ungrounded systems. Sections 250.4(A)(5) for grounded systems and 250.4(B)(4) for ungrounded systems contain fault current path objectives that were stated in 250.51 of the 1996 and earlier editions of the *Code.*

(A) Grounded Systems

(1) Electrical System Grounding Electrical systems that are grounded shall be connected to earth in a manner that will limit the voltage imposed by lightning, line surges, or unintentional contact with higher-voltage lines and that will stabilize the voltage to earth during normal operation.

(2) Grounding of Electrical Equipment Non–current-carrying conductive materials enclosing electrical conductors or equipment, or forming part of such equipment, shall be connected to earth so as to limit the voltage to ground on these materials.

(3) Bonding of Electrical Equipment Non–current-carrying conductive materials enclosing electrical conductors or equipment, or forming part of such equipment, shall be connected together and to the electrical supply source in a manner that establishes an effective ground-fault current path.

(4) Bonding of Electrically Conductive Materials and Other Equipment Electrically conductive materials that are likely to become energized shall be connected together and to the electrical supply source in a manner that establishes an effective ground-fault current path.

(5) Effective Ground-Fault Current Path Electrical equipment and wiring and other electrically conductive material likely to become energized shall be installed in a manner that creates a permanent, low-impedance circuit facilitating the operation of the overcurrent device or ground detector for high-impedance grounded systems. It shall be capable of safely carrying the maximum ground-fault current likely to be imposed on it from any point on the wiring system where a ground fault may occur to the electrical supply source. The earth shall not be considered as an effective ground-fault current path.

This section was revised for the 2005 *Code* to recognize that the performance objective for the effective ground-fault current path is not always to facilitate operation of an overcurrent protective device. In the case of a high-impedance grounded system installed in accordance with 250.36, the performance objective is to ensure operation of the required ground detector to provide annunciation of a ground-fault condition.

(B) Ungrounded Systems

(1) Grounding Electrical Equipment Non–current-carrying conductive materials enclosing electrical conductors or equipment, or forming part of such equipment, shall be connected to earth in a manner that will limit the voltage imposed by lightning or unintentional contact with higher-voltage lines and limit the voltage to ground on these materials.

(2) Bonding of Electrical Equipment Non–current-carrying conductive materials enclosing electrical conductors or equipment, or forming part of such equipment, shall be connected together and to the supply system grounded equipment in a manner that creates a permanent, low-imped-

Table 250.3 Additional Grounding Requirements

Conductor/Equipment	Article	Section
Agricultural buildings		547.9 and 547.10
Audio signal processing, amplification, and reproduction equipment		640.7
Branch circuits		210.5, 210.6, 406.3
Cablebus		370.9
Cable trays	392	392.3(C), 392.7
Capacitors		460.10, 460.27
Circuits and equipment operating at less than 50 volts	720	
Closed-loop and programmed power distribution		780.3
Communications circuits	800	
Community antenna television and radio distribution systems		820.93, 820.100, 820.103
Conductors for general wiring	310	
Cranes and hoists	610	
Electrically driven or controlled irrigation machines		675.11(C), 675.12, 675.13, 675.14, 675.15
Electric signs and outline lighting	600	
Electrolytic cells	668	
Elevators, dumbwaiters, escalators, moving walks, wheelchair lifts, and stairway chair lifts	620	
Fire alarm systems		760.9
Fixed electric heating equipment for pipelines and vessels		427.29, 427.48
Fixed outdoor electric deicing and snow-melting equipment		426.27
Flexible cords and cables		400.22, 400.23
Floating buildings		553.8, 553.10, 553.11
Grounding-type receptacles, adapters, cord connectors, and attachment plugs		406.9
Hazardous (classified) locations	500–517	
Health care facilities	517	
Induction and dielectric heating equipment	665	
Industrial machinery	670	
Information technology equipment		645.15
Intrinsically safe systems		504.50
Luminaires (lighting fixtures) and lighting equipment		410.17, 410.18, 410.20, 410.21, 410.105(B)
Luminaires (fixtures), lampholders, and lamps	410	
Marinas and boatyards		555.15
Mobile homes and mobile home park	550	
Motion picture and television studios and similar locations		530.20, 530.64(B)
Motors, motor circuits, and controllers	430	
Outlet, device, pull, and junction boxes; conduit bodies; and fittings		314.4, 314.25
Over 600 volts, nominal, underground wiring methods		300.50(B)
Panelboards		408.40
Pipe organs	650	
Radio and television equipment	810	
Receptacles and cord connectors		406.3
Recreational vehicles and recreational vehicle parks	551	
Services	230	
Solar photovoltaic systems		690.41, 690.42, 690.43, 690.45, 690.47
Swimming pools, fountains, and similar installations	680	
Switchboards and panelboards		408.3(D)
Switches		404.12
Theaters, audience areas of motion picture and television studios, and similar locations		520.81
Transformers and transformer vaults		450.10
Use and identification of grounded conductors	200	
X-ray equipment	660	517.78

ance path for ground-fault current that is capable of carrying the maximum fault current likely to be imposed on it.

(3) Bonding of Electrically Conductive Materials and Other Equipment Electrically conductive materials that are likely to become energized shall be connected together and to the supply system grounded equipment in a manner that creates a permanent, low-impedance path for ground-fault current that is capable of carrying the maximum fault current likely to be imposed on it.

(4) Path for Fault Current Electrical equipment, wiring, and other electrically conductive material likely to become energized shall be installed in a manner that creates a permanent, low-impedance circuit from any point on the wiring system to the electrical supply source to facilitate the operation of overcurrent devices should a second fault occur on the wiring system. The earth shall not be considered as an effective fault-current path.

> FPN No. 1: A second fault that occurs through the equipment enclosures and bonding is considered a ground fault.
> FPN No. 2: See Figure 250.4 for information on the organization of Article 250.

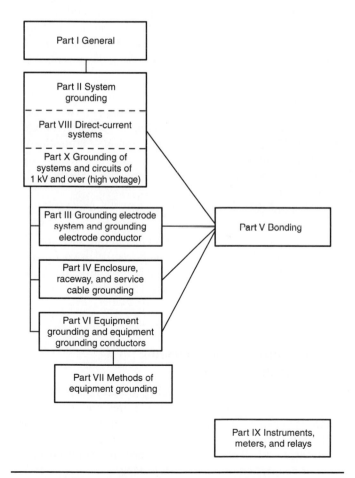

Figure 250.4 Grounding

Grounding can be divided into two areas: system grounding and equipment grounding. These two areas are kept separate except at the point where they receive their source of power, such as at the service equipment or at a separately derived system.

Grounding is the intentional connection of a current-carrying conductor to ground or something that serves in place of ground. In most instances, this connection is made at the supply source, such as a transformer, and at the main service disconnecting means of the premises using the energy.

There are three basic reasons for grounding:

1. To limit the voltages caused by lightning or by accidental contact of the supply conductors with conductors of higher voltage
2. To stabilize the voltage under normal operating conditions (which maintains the voltage at one level relative to ground, so that any equipment connected to the system will be subject only to that potential difference)
3. To facilitate the operation of overcurrent devices, such as fuses, circuit breakers, or relays, under ground-fault conditions

Exhibit 250.1 shows a typical grounding system for a single-phase, 3-wire service supplied from a utility transformer. Inside the service disconnecting means, the grounded conductor of the system is intentionally connected to a grounding electrode via the grounding electrode conductor. Bonding the equipment grounding bus to the grounded or neutral bus via the main bonding jumper within the service disconnecting means provides a ground reference for exposed, non–current-carrying parts of the electrical system and a circuit through the grounded service conductor back to the utility transformer (source of supply) for ground-fault cur-

Exhibit 250.1 A typical grounding system for a single-phase, 3-wire service.

rent. At the utility transformer, oftentimes, an additional connection from the grounded conductor to a separate grounding electrode is made.

250.6 Objectionable Current over Grounding Conductors

(A) Arrangement to Prevent Objectionable Current The grounding of electrical systems, circuit conductors, surge arresters, and conductive non–current-carrying materials and equipment shall be installed and arranged in a manner that will prevent objectionable current over the grounding conductors or grounding paths.

(B) Alterations to Stop Objectionable Current If the use of multiple grounding connections results in objectionable current, one or more of the following alterations shall be permitted to be made, provided that the requirements of 250.4(A)(5) or (B)(4) are met:

(1) Discontinue one or more but not all of such grounding connections.
(2) Change the locations of the grounding connections.
(3) Interrupt the continuity of the conductor or conductive path interconnecting the grounding connections.
(4) Take other suitable remedial and approved action.

An increase in the use of electronic controls and computer equipment, which are sensitive to stray currents, has caused installation designers to look for ways to isolate electronic equipment from the effects of such stray circulating currents. Circulating currents on equipment grounding conductors, metal raceways, and building steel develop potential differences between ground and the neutral of electronic equipment.

A solution often recommended by inexperienced individuals is to isolate the electronic equipment from all other power equipment by disconnecting it from the power equipment ground. In this ill-conceived corrective action, the equipment grounding means is removed or nonmetallic spacers are installed in the metallic raceway system. The electronic equipment is then grounded to an earth ground isolated from the common power system ground. Isolating equipment in this manner creates a potential difference that is a shock hazard. The error is compounded because such isolation does not establish a low-impedance ground-fault return path to the power source, which is necessary to actuate the overcurrent protection device. Section 250.6(B) is not intended to allow disconnection of all power grounding connections to the electronic equipment. See also the commentary following 250.6(D).

(C) Temporary Currents Not Classified as Objectionable Currents Temporary currents resulting from accidental con-

ditions, such as ground-fault currents, that occur only while the grounding conductors are performing their intended protective functions shall not be classified as objectionable current for the purposes specified in 250.6(A) and (B).

(D) Limitations to Permissible Alterations The provisions of this section shall not be considered as permitting electronic equipment from being operated on ac systems or branch circuits that are not grounded as required by this article. Currents that introduce noise or data errors in electronic equipment shall not be considered the objectionable currents addressed in this section.

Section 250.6(D) indicates that currents that result in noise or data errors in electronic equipment are not considered to be the objectionable currents referred to in 250.6, which limits the alterations permitted by 250.6(C). See 250.96(B) and 250.146(D) for requirements that provide safe bonding and grounding methods to minimize noise and data errors.

(E) Isolation of Objectionable Direct-Current Ground Currents Where isolation of objectionable dc ground currents from cathodic protection systems is required, a listed ac coupling/dc isolating device shall be permitted in the equipment grounding path to provide an effective return path for ac ground-fault current while blocking dc current.

The dc ground current on grounding conductors as a result of a cathodic protection system may be considered objectionable. Because of the required grounding and bonding connections associated with metal piping systems, it is inevitable that where cathodic protection for the piping system is provided, dc current will be present on grounding and bonding conductors.

Section 250.6(E) allows the use of a listed ac coupling/dc isolating device. This device prevents the dc current on grounding and bonding conductors and allows the ground-fault return path to function properly. To be listed for this function, these devices are evaluated by the product testing organizations for proper performance under ground-fault conditions.

250.8 Connection of Grounding and Bonding Equipment

Grounding conductors and bonding jumpers shall be connected by exothermic welding, listed pressure connectors, listed clamps, or other listed means. Connection devices or fittings that depend solely on solder shall not be used. Sheet metal screws shall not be used to connect grounding conductors or connection devices to enclosures.

Section 250.8 prohibits the use of sheet metal screws as a means for directly attaching equipment grounding conductors to equipment or as a means for attaching connection devices for equipment grounding conductors to equipment. Connection means that are listed, that are part of listed equipment, or that are exothermically welded are required to ensure a permanent and low-resistance connection. Exhibit 250.2 and Exhibit 250.3 illustrate two methods of attaching an equipment bonding jumper to a grounded metal box.

Exhibit 250.2 Use of a grounding screw to attach equipment bonding jumper to a metal box.

Exhibit 250.3 Use of a listed grounding clip to attach a grounding conductor to a metal box.

250.10 Protection of Ground Clamps and Fittings

Ground clamps or other fittings shall be approved for general use without protection or shall be protected from physical damage as indicated in (1) or (2) as follows:

(1) In installations where they are not likely to be damaged
(2) Where enclosed in metal, wood, or equivalent protective covering

250.12 Clean Surfaces

Nonconductive coatings (such as paint, lacquer, and enamel) on equipment to be grounded shall be removed from threads and other contact surfaces to ensure good electrical continuity or be connected by means of fittings designed so as to make such removal unnecessary.

II. System Grounding

250.20 Alternating-Current Systems to Be Grounded

Alternating-current systems shall be grounded as provided for in 250.20(A), (B), (C), or (D). Other systems shall be permitted to be grounded. If such systems are grounded, they shall comply with the applicable provisions of this article.

> FPN: An example of a system permitted to be grounded is a corner-grounded delta transformer connection. See 250.26(4) for conductor to be grounded.

(A) Alternating-Current Systems of Less Than 50 Volts
Alternating-current systems of less than 50 volts shall be grounded under any of the following conditions:

(1) Where supplied by transformers, if the transformer supply system exceeds 150 volts to ground
(2) Where supplied by transformers, if the transformer supply system is ungrounded
(3) Where installed as overhead conductors outside of buildings

(B) Alternating-Current Systems of 50 Volts to 1000 Volts
Alternating-current systems of 50 volts to 1000 volts that supply premises wiring and premises wiring systems shall be grounded under any of the following conditions:

(1) Where the system can be grounded so that the maximum voltage to ground on the ungrounded conductors does not exceed 150 volts

Exhibit 250.4 illustrates the grounding requirements of 250.20(B)(1) as applied to a 120-volt, single-phase, 2-wire system and to a 120/240-volt, single-phase, 3-wire system.

120-V, single-phase, 2-wire system

120/240-V, single-phase, 3-wire system

Exhibit 250.4 Typical systems required to be grounded in accordance with 250.20(B)(1). The conductor to be grounded is in accordance with 250.26.

The selection of which conductor is to be grounded is covered by 250.26.

(2) Where the system is 3-phase, 4-wire, wye connected in which the neutral is used as a circuit conductor
(3) Where the system is 3-phase, 4-wire, delta connected in which the midpoint of one phase winding is used as a circuit conductor

Exhibit 250.5 illustrates which conductor is required to be grounded for all wye systems if the neutral is used as a circuit conductor. Where the midpoint of one phase of a 3-phase, 4-wire delta system is used as a circuit conductor, it must be grounded and the high-leg conductor must be identified. See 250.20(B)(2) and 250.20(B)(3), as well as 250.26.

(C) Alternating-Current Systems of 1 kV and Over Alternating-current systems supplying mobile or portable equipment shall be grounded as specified in 250.188. Where supplying other than mobile or portable equipment, such systems shall be permitted to be grounded.

(D) Separately Derived Systems Separately derived systems, as covered in 250.20(A) or (B), shall be grounded as specified in 250.30.

208Y/120-V, 3-phase, 4-wire wye system

120/240-V, 3-phase, 4-wire delta system

Exhibit 250.5 Typical systems required to be grounded by 250.20(B)(2) and 250.20(B)(3). The conductor to be grounded is in accordance with 250.26.

Two of the most common sources of separately derived systems in premises wiring are transformers and generators. An autotransformer or step-down transformer that is part of electrical equipment and that does not supply premises wiring is not the source of a separately derived system. See the definition of *premises wiring* in Article 100.

FPN No. 1: An alternate ac power source such as an on-site generator is not a separately derived system if the neutral is solidly interconnected to a service-supplied system neutral.

Exhibit 250.6 and Exhibit 250.7 depict a 208Y/120-volt, 3-phase, 4-wire electrical service supplying a service disconnecting means to a building. The system is fed through a transfer switch connected to a generator intended to provide power for an emergency or standby system.

In Exhibit 250.6, the neutral conductor from the generator to the load is not disconnected by the transfer switch. There is a direct electrical connection between the normal grounded system conductor (neutral) and the generator neutral through the neutral bus in the transfer switch, thereby grounding the generator neutral. Because the generator is grounded by connection to the normal system ground, it is

Exhibit 250.6 A 208Y/120-volt, 3-phase, 4-wire system that has a direct electrical connection of the grounded circuit conductor (neutral) to the generator and is therefore not considered a separately derived system.

Exhibit 250.7 A 208Y/120-volt, 3-phase, 4-wire system that does not have a direct electrical connection of the grounded circuit conductor (neutral) to the generator and is therefore considered a separately derived system.

not a separately derived system, and there are no requirements for grounding the neutral at the generator. Under these conditions, it is necessary to run an equipment grounding conductor from the service equipment to the 3-pole transfer switch and from the 3-pole transfer switch to the generator. This can be in the form of any of the items listed in 250.118.

In Exhibit 250.7, the grounded conductor (neutral) is connected to the switching contacts of a 4-pole transfer switch. Therefore, the generator system does not have a direct electrical connection to the other supply system grounded conductor (neutral), and the system supplied by the generator is considered separately derived. This separately derived system (3-phase, 4-wire, wye-connected system that supplies line-to-neutral loads) is required to be grounded in accordance with 250.20(B) and 250.20(D). The methods for grounding the system are specified in 250.30(A).

Section 250.30(A)(1) requires separately derived systems to have a system bonding jumper connected between the generator frame and the grounded circuit conductor (neutral). The grounding electrode conductor from the generator is required to be connected to a grounding electrode. This conductor should be located as close to the generator as practicable, according to 250.30(A)(4). If the generator is in a building, the preferred grounding electrode is required to be one of the following, depending on which grounding electrode is closest to the generator location: (1) effectively grounded structural metal member or (2) the first 5 ft of water pipe into a building where the piping is effectively grounded. (The exception to 250.52(A)(1) permits the grounding connection to the water piping beyond the first 5 ft.) For buildings or structures in which the preferred electrodes are not available, the choice can be made from any of the grounding electrodes specified in 250.52(A)(3) through 250.52(A)(7).

FPN No. 2: For systems that are not separately derived and are not required to be grounded as specified in 250.30, see 445.13 for minimum size of conductors that must carry fault current.

(E) Impedance Grounded Neutral Systems Impedance grounded neutral systems shall be grounded in accordance with 250.36 or 250.186.

250.21 Alternating-Current Systems of 50 Volts to 1000 Volts Not Required to Be Grounded

The following ac systems of 50 volts to 1000 volts shall be permitted to be grounded but shall not be required to be grounded:

(1) Electric systems used exclusively to supply industrial electric furnaces for melting, refining, tempering, and the like
(2) Separately derived systems used exclusively for rectifiers that supply only adjustable-speed industrial drives
(3) Separately derived systems supplied by transformers that have a primary voltage rating less than 1000 volts, provided that all the following conditions are met:
  a. The system is used exclusively for control circuits.
  b. The conditions of maintenance and supervision ensure that only qualified persons service the installation.

    c. Continuity of control power is required.

    d. Ground detectors are installed on the control system.

(4) Other systems that are not required to be grounded in accordance with the requirements of 250.20(B).

Where an alternating-current system is not grounded as permitted in 250.21(1) through (4), ground detectors shall be installed on the system.

Exception: Systems of less than 120 volts to ground as permitted by this Code shall not be required to have ground detectors.

New for the 2005 *Code,* ungrounded electrical systems as permitted in 250.21 are required to be provided with ground detectors. In the 2002 and previous editions, the installation of ground detectors was required only for some very specific applications of ungrounded systems (and in impedance grounded neutral systems), but there was only a recommendation that they be installed on all ungrounded electrical systems. The exception to this requirement permits the operation of specific-purpose ungrounded systems without ground detectors where the voltage to ground is less than 120 volts. For further information on what is considered to be the voltage to ground in an ungrounded system, see the definition of *voltage to ground* in Article 100.

Ungrounded electrical systems are permitted by the *NEC* for the specific functions described in 250.21(1), (2), and (3) and for general power distribution systems in accordance with 250.21(4). Delta-connected, 3-phase, 3-wire, 240-volt and 480-volt systems are examples of common electrical distribution systems that are permitted but are not required to have a circuit conductor that is intentionally grounded. The operational advantage in using an ungrounded electrical system is continuity of operation, which in some processes might create a safer condition than would be achieved by the automatic and unplanned opening of the supply circuit.

Unlike solidly grounded systems, in which the first line-to-ground fault causes the overcurrent protective device to automatically open the circuit, the same line-to-ground fault in an ungrounded system does not result in the operation of the overcurrent device — it simply results in the faulted circuit conductor becoming a grounded conductor until a repair of the damaged conductor insulation can be performed. However, this latent ground-fault condition will remain undetected unless ground detectors are installed in the ungrounded system or until another insulation failure on a different ungrounded conductor results in a line-to-line-to-ground fault, with the potential for more extensive damage to electrical equipment.

Ground detectors are used to provide a visual indication, an audible signal, or both, to alert system operators and maintainers of a ground-fault condition in the electrical sys-

tem. With notification of the ground-fault condition, rather than automatic interruption of the circuit, the operators of the process supplied by the ungrounded system can then take the necessary steps to effect an orderly shutdown, determine where the ground fault is located in the system, and safely perform the necessary repair.

It should be noted that ungrounded systems are simply systems without an intentionally grounded circuit conductor that is part of normal circuit operation, as is the case in 120/240-volt, single-phase, 3-wire; 208Y/120-volt, 3-phase, 4-wire; and 480Y/277-volt, 3-phase, 4-wire systems in which there is a grounded conductor that is used as a circuit conductor. The fact that a system operates without a grounded conductor does not exempt that system from complying with all of the applicable requirements in Article 250 for establishing a grounding electrode system and for equipment grounding. These protective features are required for grounded and ungrounded electrical distribution systems.

250.22 Circuits Not to Be Grounded

The following circuits shall not be grounded:

(1) Circuits for electric cranes operating over combustible fibers in Class III locations, as provided in 503.155
(2) Circuits in health care facilities as provided in 517.61 and 517.160
(3) Circuits for equipment within electrolytic cell working zone as provided in Article 668
(4) Secondary circuits of lighting systems as provided in 411.5(A)

250.24 Grounding Service-Supplied Alternating-Current Systems

(A) System Grounding Connections A premises wiring system supplied by a grounded ac service shall have a grounding electrode conductor connected to the grounded service conductor, at each service, in accordance with 250.24(A)(1) through (A)(5).

(1) General The connection shall be made at any accessible point from the load end of the service drop or service lateral to and including the terminal or bus to which the grounded service conductor is connected at the service disconnecting means.

The grounded conductor of an ac service is connected to a grounding electrode system to limit the voltage to ground imposed on the system by lightning, line surges, and (unintentional) high-voltage crossovers. Another reason for requiring this connection is to stabilize the voltage to ground during normal operation, including short circuits. These performance requirements are stated in 250.4(A) and 250.4(B).

The actual connection of the grounded service conductor

to the grounded electrode conductor is permitted to be made at various locations, according to 250.24(A)(1). Allowing various locations for the connection to be made continues to meet the overall objectives for grounding while allowing the installer a variety of practical solutions. Exhibit 250.8 illustrates three possible connection point solutions to where the grounded conductor of the service could be connected to the grounding electrode conductor.

At load end of service drop

In accessible meter enclosure

In service disconnecting means

Exhibit 250.8 An ac service supplied from an overhead distribution system illustrating three accessible connection points where the grounded service conductor is connected to the grounding electrode conductor according to 250.24(A)(1).

FPN: See definitions of *Service Drop* and *Service Lateral* in Article 100.

(2) Outdoor Transformer Where the transformer supplying the service is located outside the building, at least one additional grounding connection shall be made from the grounded service conductor to a grounding electrode, either at the transformer or elsewhere outside the building.

See Exhibit 250.9 for an illustration of an outdoor distribution system transformer connected to an additional grounding electrode.

Exception: The additional grounding connection shall not be made on high-impedance grounded neutral systems. The system shall meet the requirements of 250.36.

(3) Dual Fed Services For services that are dual fed (double ended) in a common enclosure or grouped together in separate enclosures and employing a secondary tie, a single

Grounding electrode connection at service

One additional connection to a grounding electrode where transformer is located outside the building

Exhibit 250.9 A 3-wire, 120/240-volt ac, single-phase, secondary distribution system in which grounding connections are required on the secondary side of the transformer according to 250.24(A)(2) and the supply side of the service disconnecting means according to 250.24(A)(1).

grounding electrode connection to the tie point of the grounded conductor(s) from each power source shall be permitted.

(4) Main Bonding Jumper as Wire or Busbar Where the main bonding jumper specified in 250.28 is a wire or busbar and is installed from the grounded conductor terminal bar or bus to the equipment grounding terminal bar or bus in the service equipment, the grounding electrode conductor shall be permitted to be connected to the equipment grounding terminal, bar, or bus to which the main bonding jumper is connected.

(5) Load-Side Grounding Connections A grounding connection shall not be made to any grounded conductor on the load side of the service disconnecting means except as otherwise permitted in this article.

FPN: See 250.30(A) for separately derived systems, 250.32 for connections at separate buildings or structures, and 250.142 for use of the grounded circuit conductor for grounding equipment.

The power for ac premises wiring systems is either separately derived, in accordance with 250.20(D), or supplied by the service. See the definition of *service* in Article 100. Section 250.30 covers grounding requirements for separately derived ac systems. Section 250.24(A) covers system grounding requirements for service-supplied ac systems.

According to 250.24, a premises wiring system supplied by an ac service that is required to be grounded must have a grounding electrode conductor at each service connected to the grounding electrodes that meets the requirements in Part III of Article 250. Note that the grounding electrode requirements for a separately derived system are specified in 250.30(A)(3) and 250.30(A)(4).

The grounding electrode conductor connection to the grounded conductor is specific. The *Code* requires that the connection be made to the grounded service conductor and describes where this connection is permitted. Where the transformer supplying a service is located outside of a building or structure, a grounding connection must be made at the transformer secondary or at another outdoor location under the conditions specified in 250.24(A)(2). In addition, the conductor that is grounded at the transformer is required to be grounded again at the building or structure, according to 250.24(A)(1).

Section 250.24(A)(5) prohibits regrounding of the grounded conductor on the load side of the service disconnecting means. This requirement is also in concert with 250.142(B).

(B) Main Bonding Jumper For a grounded system, an unspliced main bonding jumper shall be used to connect the equipment grounding conductor(s) and the service-disconnect enclosure to the grounded conductor within the enclosure for each service disconnect in accordance with 250.28.

Where the service equipment of a grounded system consists of multiple disconnecting means, a main bonding jumper for each separate service disconnecting means is required to connect the grounded service conductor, the equipment grounding conductor, and the service equipment enclosure. See Exhibit 250.10 and Exhibit 250.11, which accompany the commentary following 250.24(C), Exception.

Exception No. 1: Where more than one service disconnecting means is located in an assembly listed for use as service equipment, an unspliced main bonding jumper shall bond the grounded conductor(s) to the assembly enclosure.

Where multiple service disconnecting means are part of an assembly listed as service equipment, all grounded service conductors are required to be run to and bonded to the assembly. However, only one section of the assembly is required to have the main bonding jumper connection. See Exhibit 250.12, which accompanies the commentary following 250.28(D).

Exception No. 2: Impedance grounded neutral systems shall be permitted to be connected as provided in 250.36 and 250.186.

(C) Grounded Conductor Brought to Service Equipment Where an ac system operating at less than 1000 volts is grounded at any point, the grounded conductor(s) shall be run to each service disconnecting means and shall be bonded to each disconnecting means enclosure. The grounded conductor(s) shall be installed in accordance with 250.24(C)(1) through (C)(3).

Exception: Where more than one service disconnecting means are located in an assembly listed for use as service equipment, it shall be permitted to run the grounded conductor(s) to the assembly, and the conductor(s) shall be bonded to the assembly enclosure.

If the utility service that supplies premises wiring is grounded, the grounded conductor, whether or not it is used to supply a load, must be run to the service equipment, be bonded to the equipment, and be connected to a grounding electrode system. Exhibit 250.10 shows an example of the main rule in 250.24(C), which requires the grounded service conductor to be brought in and bonded to each service disconnecting means enclosure. This requirement is based on the grounded conductor being used to complete the ground-fault current path between the service equipment and the utility source. The grounded service conductor's other func-

Exhibit 250.10 A grounded system in which the grounded service conductor is brought into a 3-phase, 4-wire service equipment enclosure and to the 3-phase, 3-wire service equipment enclosure, where it is bonded to each service disconnecting means.

tion, as a circuit conductor for normal loads, is covered in 200.3 and 220.61.

The exception to 250.24(C) permits a single connection of the grounded service conductor to a listed service assembly (such as a switchboard) that contains more than one service disconnecting means, as shown in Exhibit 250.11.

φ N

Assembly listed for service equipment

Main bonding jumper

Main tenant A

Main tenant B

Main tenant C

Exhibit 250.11 One connection of the grounded service conductor to a listed service assembly containing multiple service disconnecting means, in accordance with 250.24(C), Exception.

(1) Routing and Sizing This conductor shall be routed with the phase conductors and shall not be smaller than the required grounding electrode conductor specified in Table 250.66 but shall not be required to be larger than the largest ungrounded service-entrance phase conductor. In addition, for service-entrance phase conductors larger than 1100 kcmil copper or 1750 kcmil aluminum, the grounded conductor shall not be smaller than 12½ percent of the area of the largest service-entrance phase conductor. The grounded conductor of a 3-phase, 3-wire delta service shall have an ampacity not less than that of the ungrounded conductors.

(2) Parallel Conductors Where the service-entrance phase conductors are installed in parallel, the size of the grounded conductor shall be based on the total circular mil area of the parallel conductors as indicated in this section. Where installed in two or more raceways, the size of the grounded conductor in each raceway shall be based on the size of the ungrounded service-entrance conductor in the raceway but not smaller than 1/0 AWG.

> FPN: See 310.4 for grounded conductors connected in parallel.

For a multiple raceway or cable service installation, the minimum size for the grounded conductor in each raceway or cable where conductors are in parallel cannot be less than 1/0 AWG. Although the cumulative size of the parallel grounded conductors may be larger than is required by 250.24(C)(1), the minimum 1/0 AWG per raceway or cable

correlates with the requirements for parallel conductors contained in 310.4.

(3) High Impedance The grounded conductor on a high-impedance grounded neutral system shall be grounded in accordance with 250.36.

(D) Grounding Electrode Conductor A grounding electrode conductor shall be used to connect the equipment grounding conductors, the service-equipment enclosures, and, where the system is grounded, the grounded service conductor to the grounding electrode(s) required by Part III of this article.

High-impedance grounded neutral system connections shall be made as covered in 250.36.

> FPN: See 250.24(A) for ac system grounding connections.

(E) Ungrounded System Grounding Connections A premises wiring system that is supplied by an ac service that is ungrounded shall have, at each service, a grounding electrode conductor connected to the grounding electrode(s) required by Part III of this article. The grounding electrode conductor shall be connected to a metal enclosure of the service conductors at any accessible point from the load end of the service drop or service lateral to the service disconnecting means.

250.26 Conductor to Be Grounded — Alternating-Current Systems

For ac premises wiring systems, the conductor to be grounded shall be as specified in the following:

(1) Single-phase, 2-wire — one conductor
(2) Single-phase, 3-wire — the neutral conductor
(3) Multiphase systems having one wire common to all phases — the common conductor
(4) Multiphase systems where one phase is grounded — one phase conductor
(5) Multiphase systems in which one phase is used as in (2) — the neutral conductor

Section 250.26 works in conjunction with 250.20(B). Once the requirements of 250.20(B) establish that a system is required to be grounded, the requirements of 250.26 identify the conductor in the system that is required to be grounded. In addition to covering systems where it is mandatory to ground the system, this requirement also identifies which conductor is to be grounded in systems that are permitted to be grounded, such as a corner-grounded delta system.

250.28 Main Bonding Jumper and System Bonding Jumper

For a grounded system, main bonding jumpers and system bonding jumpers shall be installed as follows:

The term *system bonding jumper* is new for the 2005 *Code*. This new term distinguishes the system bonding jumper installed in the disconnecting means enclosure supplied from a separately derived system from the main bonding jumper installed only in a service disconnecting means enclosure. See the commentary following the Article 100 definition of *bonding jumper, system.*

(A) Material Main bonding jumpers and system bonding jumpers shall be of copper or other corrosion-resistant material. A main bonding jumper and a system bonding jumper shall be a wire, bus, screw, or similar suitable conductor.

(B) Construction Where a main bonding jumper or a system bonding jumper is a screw only, the screw shall be identified with a green finish that shall be visible with the screw installed.

The requirement in 250.28(B) specifies that where a screw is used for the main or system bonding jumper, the screw must have a green color that is visible when it is installed. This identification requirement makes it possible to readily distinguish the bonding jumper screw from other screws in the grounded conductor terminal bar, to ensure that the required bonding connection has been made.

(C) Attachment Main bonding jumpers and system bonding jumpers shall be attached in the manner specified by the applicable provisions of 250.8.

(D) Size Main bonding jumpers and system bonding jumpers shall not be smaller than the sizes shown in Table 250.66. Where the supply conductors are larger than 1100 kcmil copper or 1750 kcmil aluminum, the bonding jumper shall have an area that is not less than 12½ percent of the area of the largest phase conductor except that, where the phase conductors and the bonding jumper are of different materials (copper or aluminum), the minimum size of the bonding jumper shall be based on the assumed use of phase conductors of the same material as the bonding jumper and with an ampacity equivalent to that of the installed phase conductors.

The minimum size for the main bonding jumper on the supply side of a service and for the system bonding jumper on the supply side of a separately derived system is determined through the use of a table and a calculation that establishes a proportional relationship between the size of the largest ungrounded supply conductor and the minimum conductor size (cross-sectional area) necessary to create an effective ground-fault current return path for short-time high-current conditions.

In a grounded system, the primary function of the main bonding jumper and of the system bonding jumpers is to create the link for ground-fault current between the equipment grounding conductors and the grounded conductor. Table 250.66, Grounding Electrode Conductor for Alternating-Current Systems, has a number of functions in Article 250 in addition to its use for sizing the grounding electrode conductor. It is used in several sections of Article 250 for sizing various supply-side conductors of the grounding and bonding system. Section 250.28(D) refers to Table 250.66 for directly sizing main and system bonding jumpers where the ungrounded conductors do not exceed 1100-kcmil copper or 1750-kcmil aluminum.

Unlike the function of the grounding electrode conductor, which carries current to the ground or to the conducting body that serves as ground (via connection to a grounding electrode), the main and system bonding jumpers are placed directly in the supply side ground-fault current return path. Therefore, where the largest ungrounded supply conductor exceeds the parameters of Table 250.66, it is necessary to maintain a proportional relationship between the ungrounded conductor and the main or system bonding jumper. Grounding electrode conductors are not required to be larger than 3/0 AWG copper or 250-kcmil aluminum conductors, but to establish the minimum size for the main or system bonding jumper for ungrounded conductors exceeding 1100-kcmil copper or 1750-kcmil aluminum, its circular mil area cannot be less than 12½ percent of the circular mil area of the largest ungrounded conductor (for conductors in parallel, the total area of the largest phase set). It should be noted that where a main or system bonding jumper is provided as part of listed equipment, such as is the case with many panelboards and switchboards listed for use as service equipment, it is not necessary to replicate this bonding jumper with another one sized in accordance with 250.28(D).

To apply the bonding jumper requirements, each line-side service equipment enclosure is treated separately, as depicted in Exhibit 250.12. The main bonding jumper in the left enclosure is a 4 AWG copper conductor. Based on the 3/0 AWG ungrounded service conductors supplying the 200-ampere circuit breaker and Table 250.66, the minimum-size main bonding jumper for this service equipment enclosure is 4 AWG copper. Similarly, the 1/0 AWG main bonding jumper for the enclosure on the right is derived from Table 250.66 using the 500-kcmil ungrounded service conductors.

In addition to the main bonding jumpers for the two disconnecting means enclosures, other conductors shown in Exhibit 250.12 are sized using Table 250.66. First, the grounding electrode conductor at 2/0 AWG is full-sized based on Table 250.66 using the 750-kcmil ungrounded service conductor as the basis for selection. It should be noted that there are conditions where the grounding electrode conductor is permitted to sized smaller than what is required in the table; see 250.66.

Next, the grounded conductor run to each enclosure in

750 kcmil → Supply conductors
← Minimum 2/0 AWG grounded conductor
2/0 AWG supply-side equipment bonding jumper

Neutral terminal block

2/0 AWG grounding electrode conductor

4 AWG supply-side equipment bonding jumper

3/0 AWG

200-A circuit breaker

1/0 AWG supply-side equipment bonding jumper

500 kcmil

400-A circuit breaker

1/0 AWG grounded conductor

4 AWG grounded conductor

4 AWG main bonding jumper

1/0 AWG main bonding jumper

Exhibit 250.12 An example of the bonding requirements for service equipment.

accordance with 250.24(B) is sized using Table 250.66 as the minimum size permitted. For the grounded conductor, the reference to Table 250.66 is found in 250.24(C)(1). For each enclosure, the minimum size grounded conductor is established based on the largest ungrounded conductor serving that enclosure. The grounded conductor is also subject to the requirements of 220.61 covering the conductor's capacity for unbalanced load, which could result in having to increase the size to larger than what was determined from Table 250.66.

Finally, supply-side equipment bonding jumpers are used for the three bonding metal conduits containing service conductors and the metal wireway located above the two service equipment enclosures. These bonding jumpers are also sized from Table 250.66 via a reference from 250.102(C). The bonding jumpers for the raceways are sized based on the ungrounded conductors contained in each metal service raceway.

For the metal conduit entering the top of the wireway and the wireway itself, the bonding jumper is sized from Table 250.66 based on the 750-kcmil main service-entrance conductors and cannot be less than a 2/0 AWG copper conductor. The service-entrance conductors to the enclosures are 3/0 AWG and 500-kcmil copper, based on the loads supplied from each enclosure. The bonding jumpers for the short nipples are sized based on the size of the phase conductors supplying each disconnecting means. In this case, the metal raceway nipples containing the 3/0 AWG and 500-kcmil ungrounded service conductors require minimum 4 AWG and 1/0 AWG copper supply-side equipment bonding jumpers, respectively.

There are instances, particularly with large capacity services or separately derived systems, where the main or system bonding jumper is required to be larger than the grounding electrode conductor. Section 250.28(D) requires that where the service-entrance conductors are larger than 1100-kcmil copper or 1750-kcmil aluminum, the bonding jumper is to have a cross-sectional area of not less than 12 ½ percent of the cross-sectional area of the largest phase conductor. For example, if a service is supplied by four 500-kcmil conductors in parallel for each phase, the minimum cross-sectional area of the bonding jumper is calculated as follows: 4 × 500 kcmil = 2000 kcmil. Therefore, the main or system bonding jumper cannot be less than 12½ percent of 2000 kcmil, which results in a 250-kcmil copper conductor. The copper grounding electrode conductor for this set of conductors, based on Table 250.66, is not required to be larger than 3/0 AWG.

250.30 Grounding Separately Derived Alternating-Current Systems

(A) Grounded Systems A separately derived ac system that is grounded shall comply with 250.30(A)(1) through (A)(8). A grounding connection shall not be made to any grounded circuit conductor on the load side of the point of grounding of the separately derived system except as otherwise permitted in this article.

> FPN: See 250.32 for connections at separate buildings or structures, and 250.142 for use of the grounded circuit conductor for grounding equipment.

Exception: Impedance grounded neutral system grounding connections shall be made as specified in 250.36 or 250.186.

Section 250.30(A) provides the requirements for bonding and grounding the separately derived systems described in 250.20(D). A *separately derived system* is defined in Article 100 as a premises wiring system in which power is derived from a battery, a solar photovoltaic system, a generator, a transformer, or converter windings. It has no direct electrical connection, including a solidly connected grounded circuit conductor, to supply conductors originating in another system.

The requirements of 250.30 are commonly applied to 480-volt transformers that transform a 480-volt supply to a 208Y/120-volt system for lighting and appliance loads. These requirements provide for a low-impedance path to ground so that line-to-ground faults on circuits supplied by the transformer result in a sufficient amount of current to operate the overcurrent devices. These requirements also apply to generators or systems that are derived from converter windings, although these systems do not have the same wide use as separately derived systems that are derived from transformers.

(1) System Bonding Jumper An unspliced system bonding jumper in compliance with 250.28(A) through (D) that is sized based on the derived phase conductors shall be used to connect the equipment grounding conductors of the separately derived system to the grounded conductor. This connection shall be made at any single point on the separately derived system from the source to the first system disconnecting means or overcurrent device, or it shall be made at the source of a separately derived system that has no disconnecting means or overcurrent devices.

Where a separately derived system provides a grounded conductor, a system bonding jumper must be installed to connect the equipment grounding conductors to the grounded conductor. Equipment grounding conductors are connected to the grounding electrode system by the grounding electrode conductor. The system bonding jumper is sized according to 250.28(D) and may be located at any point between the source terminals (transformer, generator, etc.) and the first disconnecting means or overcurrent device. See the commentary following 250.28(D) for further information on sizing the system bonding jumper.

Exception No. 1: For separately derived systems that are dual fed (double ended) in a common enclosure or grouped together in separate enclosures and employing a secondary tie, a single system bonding jumper connection to the tie point of the grounded circuit conductors from each power source shall be permitted.

Exception No. 2: A system bonding jumper at both the source and the first disconnecting means shall be permitted where doing so does not establish a parallel path for the grounded conductor. Where a grounded conductor is used in this manner, it shall not be smaller than the size specified for the system bonding jumper but shall not be required to be larger than the ungrounded conductor(s). For the purposes of this exception, connection through the earth shall not be considered as providing a parallel path.

Exception No. 3: The size of the system bonding jumper for a system that supplies a Class 1, Class 2, or Class 3 circuit, and is derived from a transformer rated not more than 1000 volt-amperes, shall not be smaller than the derived phase conductors and shall not be smaller than 14 AWG copper or 12 AWG aluminum.

Section 250.30(A)(1) requires the system bonding jumper to be not smaller than the sizes given in Table 250.66, that is, not smaller than 8 AWG copper. Exception No. 3 to 250.30(A)(1) permits a system bonding jumper for a Class 1, Class 2, or Class 3 circuit to be not smaller than 14 AWG copper or 12 AWG aluminum.

(2) Equipment Bonding Jumper Size Where a bonding jumper of the wire type is run with the derived phase conductors from the source of a separately derived system to the first disconnecting means, it shall be sized in accordance with 250.102(C), based on the size of the derived phase conductors.

(3) Grounding Electrode Conductor, Single Separately Derived System A grounding electrode conductor for a single separately derived system shall be sized in accordance with 250.66 for the derived phase conductors and shall be used to connect the grounded conductor of the derived system to the grounding electrode as specified in 250.30(A)(7). This connection shall be made at the same point on the separately derived system where the system bonding jumper is installed.

Exception No. 1: Where the system bonding jumper specified in 250.30(A)(1) is a wire or busbar, it shall be permitted to connect the grounding electrode conductor to the equipment grounding terminal, bar, or bus, provided the equipment grounding terminal, bar, or bus is of sufficient size for the separately derived system.

Exception No. 2: Where a separately derived system originates in listed equipment suitable as service equipment, the grounding electrode conductor from the service or feeder equipment to the grounding electrode shall be permitted as the grounding electrode conductor for the separately derived system, provided the grounding electrode conductor is of sufficient size for the separately derived system. Where the equipment ground bus internal to the equipment is not smaller than the required grounding electrode conductor for the separately derived system, the grounding electrode connection for the separately derived system shall be permitted to be made to the bus.

Exception No. 3: A grounding electrode conductor shall not be required for a system that supplies a Class 1, Class 2, or Class 3 circuit and is derived from a transformer rated not more than 1000 volt-amperes, provided the grounded conductor is bonded to the transformer frame or enclosure by a jumper sized in accordance with 250.30(A)(1), Exception No. 3, and the transformer frame or enclosure is grounded by one of the means specified in 250.134.

If a separately derived system is required to be grounded, the conductor to be grounded is allowed to be connected to the grounding electrode system at any location between the source terminals (transformer, generator, etc.) and the first disconnecting means or overcurrent device. The location of the grounding electrode conductor connection to the grounded conductor must be at the same point as where the bonding jumper is connected to the grounded conductor. By establishing a common point of connection, normal neutral current will be carried only on the system grounded conductor. Metal raceways, piping systems, and structural steel must not provide a parallel circuit for neutral current. Exhibits 250.13 and 250.14 illustrate examples of grounding electrode connections for separately derived systems.

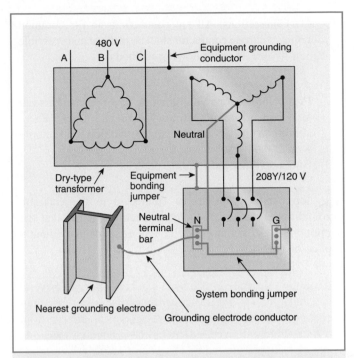

Exhibit 250.14 A grounding arrangement for a separately derived system in which the grounding electrode conductor connection is made at the first disconnecting means.

Exhibit 250.13 A grounding arrangement for a separately derived system in which the grounding electrode conductor connection is made at the transformer.

(4) Grounding Electrode Conductor, Multiple Separately Derived Systems Where more than one separately derived system is installed, it shall be permissible to connect a tap from each separately derived system to a common grounding electrode conductor. Each tap conductor shall connect the grounded conductor of the separately derived system to the common grounding electrode conductor. The grounding electrode conductors and taps shall comply with 250.30(A)(4)(a) through (A)(4)(c).

Exception No. 1: Where the system bonding jumper specified in 250.30(A)(1) is a wire or busbar, it shall be permitted to connect the grounding electrode conductor to the equipment grounding terminal, bar, or bus, provided the equipment grounding terminal, bar, or bus is of sufficient size for the separately derived system.

Exception No. 2: A grounding electrode conductor shall not be required for a system that supplies a Class 1, Class 2, or Class 3 circuit and is derived from a transformer rated not more than 1000 volt-amperes, provided the system grounded conductor is bonded to the transformer frame or enclosure by a jumper sized in accordance with 250.30(A)(1), Exception No. 3 and the transformer frame or enclosure is grounded by one of the means specified in 250.134.

(a) Common Grounding Electrode Conductor Size. The common grounding electrode conductor shall not be smaller than 3/0 AWG copper or 250 kcmil aluminum.

(b) Tap Conductor Size. Each tap conductor shall be sized in accordance with 250.66 based on the derived phase conductors of the separately derived system it serves.

Exception: Where a separately derived system originates in listed equipment suitable as service equipment, the grounding electrode conductor from the service or feeder equipment to the grounding electrode shall be permitted as the grounding electrode conductor for the separately derived system, provided the grounding electrode conductor is of sufficient size for the separately derived system. Where the equipment ground bus internal to the equipment is not smaller than the required grounding electrode conductor for the separately derived system, the grounding electrode connection for the separately derived system shall be permitted to be made to the bus.

(3) Connections. All tap connections to the common grounding electrode conductor shall be made at an accessible location by one of the following methods:

(1) A listed connector.
(2) Listed connections to aluminum or copper busbars not less than 6 mm × 50 mm (¼ in. × 2 in.). Where aluminum busbars are used, the installation shall comply with 250.64(A).
(3) By the exothermic welding process.

Tap conductors shall be connected to the common grounding electrode conductor in such a manner that the common grounding electrode conductor remains without a splice or joint.

A common grounding electrode conductor serving several separately derived systems is permitted instead of installing separate individual grounding electrode conductors from each separately derived system to the grounding electrode system. A tapped grounding electrode conductor is installed from the common grounding electrode conductor to the point of connection to the individual separately derived system grounded conductor. This tap is sized from Table 250.66 based on the size of the ungrounded conductors for that individual separately derived system.

The sizing requirement for the common grounding electrode conductor was revised for the 2005 *Code*. So that the grounding electrode conductor always has sufficient size to accommodate the multiple separately derived systems that it serves, the minimum size for this conductor is now 3/0 AWG copper or 250-kcmil aluminum. Note that this new minimum size for the common grounding electrode conductor correlates with the maximum size grounding electrode conductor required by Table 250.66; therefore, the 3/0 AWG copper or 250-kcmil aluminum becomes the maximum size required for the common grounding electrode conductor. The sizing requirement for the common grounding electrode conductor is specified in 250.30(A)(4)(a), and the sizing requirement for the individual taps to the common grounding electrode conductor is specified in 250.30(A)(4)(b). The rules covering the method of connection of the tap conductor to the common grounding electrode conductor are specified in 250.30(A)(4)(c). The following example, together with Exhibit 250.15, illustrates this new permitted installation method.

Example

A large post-and-beam loft-type building is being renovated for use as an office building. The building is being furnished with four 45-kVA, 480 to 120/208-volt, 3-phase, 4-wire, wye-connected transformers. Each transformer secondary supplies an adjacent 150-ampere main circuit breaker panelboard using 1/0 AWG, Type THHN copper conductors. The transformers are strategically placed throughout the building to facilitate efficient distribution. Because the building contains no effectively grounded structural steel, each transformer secondary must be grounded to the water service electrode within the first 5 ft of entry into the building. A common grounding electrode conductor has been selected as the method to connect all the transformers to the grounding electrode system.

What is the minimum-size common grounding electrode conductor that must be used to connect the four transformers to the grounding electrode system? What is the minimum-size grounding electrode conductor to connect each of the four transformers to the common grounding electrode conductor?

Solution

STEP 1. Determine the minimum size for the common grounding electrode conductor. In accordance with 250.30(A)(4)(a), the minimum size required is 3/0 copper or 250-kcmil aluminum. No calculation is necessary, and the common grounding electrode conductor does not have to be sized larger than specified by this requirement. Additional transformers installed in the building can be connected to this common grounding electrode conductor, and no increase in its size is required.

STEP 2. Determine the size of each individual grounding electrode tap conductor for each of the separately derived systems. According to Table 250.66, a 1/0 AWG copper derived phase conductor requires a conductor not smaller than 6 AWG copper for each transformer grounding electrode tap conductor. This individual grounding electrode conductor will be used as the permitted tap conductor and will run from the conductor to be grounded of each separately derived system to a connection point located on the common grounding electrode conductor. This conductor is labeled "Conductor B" in Exhibit 250.15.

(5) Installation The installation of all grounding electrode conductors shall comply with 250.64(A), (B), (C), and (E).

(6) Bonding Structural steel and metal piping shall be bonded in accordance with 250.104(D).

(7) Grounding Electrode The grounding electrode shall be as near as practicable to and preferably in the same area as the grounding electrode conductor connection to the system. The grounding electrode shall be the nearest one of the following:

(1) Metal water pipe grounding electrode as specified in 250.52(A)(1)

Exhibit 250.15 The grounding arrangement for multiple separately derived systems using taps from a common grounding electrode conductor, according to 250.30(A)(4)(a) and 250.30(A)(4)(b).

Labels in exhibit:
- 150-A main circuit breaker panelboard (typical for 4)
- Floor-mounted 45-kVA transformer (typical for 4)
- Conductor A
- Individual grounding electrode conductor 6 AWG copper typical to the 4 transformers
- Common grounding electrode conductor, 3/0 AWG copper
- Conductor B
- Bonding jumper
- Metal underground water pipe
- Water meter
- Supplementary ground rod electrode (10-ft ground rod)

(2) Structural metal grounding electrode as specified in 250.52(A)(2)

Exception No. 1: Any of the other electrodes identified in 250.52(A) shall be used where the electrodes specified by 250.30(A)(7) are not available.

Exception No. 2 to (1) and (2): Where a separately derived system originates in listed equipment suitable for use as service equipment, the grounding electrode used for the service or feeder equipment shall be permitted as the grounding electrode for the separately derived system.

> FPN: See 250.104(D) for bonding requirements of interior metal water piping in the area served by separately derived systems.

Section 250.30(A)(7) requires that the grounding electrode be as near as is practicable to the grounding conductor connection to the system to minimize the impedance to ground. If an effectively grounded structural metal member of the building structure or an effectively grounded metal water pipe is available nearby, 250.30(A)(7) requires that it be used as the grounding electrode. For example, where a transformer is installed on the fiftieth floor, the grounding electrode conductor is not required to be run to the service

grounding electrode system. However, where an effectively grounded metal water pipe is used as an electrode for a separately derived system, 250.52(A) specifies that only the first 5 ft of water piping entering the building can be used as a grounding electrode. Therefore, the grounding electrode conductor connection to the metal water piping must be made at some point on this first 5 ft of piping.

Concern over the use of nonmetallic piping or fittings is the basis for the "within 5 ft" requirement. Where the piping system is located in an industrial or commercial building and is serviced only by qualified persons and the entire length that will be used as an electrode is exposed, the connection may be made at any point on the piping system.

The practice of grounding the secondary of an isolating transformer to a ground rod or running the grounding electrode conductor back to the service ground (usually to reduce electrical noise on data processing systems) is not permitted where either of the electrodes covered in item (1) or item (2) of 250.30(A)(7) is available. However, an isolation transformer that is part of a listed power supply for a data processing room is not required to be grounded in accordance with 250.30(A)(7), but it must be grounded in accordance with the manufacturer's instructions.

Exhibit 250.13 and Exhibit 250.14 are typical wiring

diagrams for dry-type transformers supplied from a 480-volt, 3-phase feeder to derive a 208Y/120-volt or 480Y/277-volt secondary. As indicated in 250.30(A)(1), the bonding jumper connection is required to be sized according to 250.28(D). In Exhibit 250.13, this connection is made at the source of the separately derived system, in the transformer enclosure. In Exhibit 250.14, the bonding jumper connection is made at the first disconnecting means. With the grounding electrode conductor, the bonding jumper, and the bonding of the grounded circuit conductor (neutral) connected as shown, line-to-ground fault currents are able to return to the supply source through a short, low-impedance path. A path of lower impedance is provided that facilitates the operation of overcurrent devices, in accordance with 250.4(A)(5). The grounding electrode conductor from the secondary grounded circuit conductor is sized according to Table 250.66.

(8) Grounded Conductor Where a grounded conductor is installed and the system bonding jumper is not located at the source of the separately derived system, 250.30(A)(8)(a), (A)(8)(b), and (A)(8)(c) shall apply.

(a) Routing and Sizing. This conductor shall be routed with the derived phase conductors and shall not be smaller than the required grounding electrode conductor specified in Table 250.66 but shall not be required to be larger than the largest ungrounded derived phase conductor. In addition, for phase conductors larger than 1100 kcmil copper or 1750 kcmil aluminum, the grounded conductor shall not be smaller than 12½ percent of the area of the largest derived phase conductor. The grounded conductor of a 3-phase, 3-wire delta system shall have an ampacity not less than that of the ungrounded conductors.

(b) Parallel Conductors. Where the derived phase conductors are installed in parallel, the size of the grounded conductor shall be based on the total circular mil area of the parallel conductors, as indicated in this section. Where installed in two or more raceways, the size of the grounded conductor in each raceway shall be based on the size of the ungrounded conductors in the raceway but not smaller than 1/0 AWG.

FPN: See 310.4 for grounded conductors connected in parallel.

(c) Impedance Grounded System. The grounded conductor of an impedance grounded neutral system shall be installed in accordance with 250.36 or 250.186.

(B) Ungrounded Systems The equipment of an ungrounded separately derived system shall be grounded as specified in 250.30(B)(1) and (B)(2).

(1) Grounding Electrode Conductor A grounding electrode conductor, sized in accordance with 250.66 for the derived phase conductors, shall be used to connect the metal enclosures of the derived system to the grounding electrode as specified in 250.30(B)(2). This connection shall be made at any point on the separately derived system from the source to the first system disconnecting means.

For ungrounded separately derived systems, a grounding electrode conductor is required to be connected to the metal enclosure of the system disconnecting means. The grounding electrode conductor is sized from Table 250.66 based on the largest ungrounded supply conductor. This connection establishes a reference to ground for all exposed non–current-carrying metal equipment supplied from the ungrounded system. The equipment grounding conductors of circuits supplied from the ungrounded system are connected to ground via this grounding electrode conductor connection.

(2) Grounding Electrode Except as permitted by 250.34 for portable and vehicle-mounted generators, the grounding electrode shall comply with 250.30(A)(7).

250.32 Buildings or Structures Supplied by Feeder(s) or Branch Circuit(s)

(A) Grounding Electrode Building(s) or structure(s) supplied by feeder(s) or branch circuit(s) shall have a grounding electrode or grounding electrode system installed in accordance with 250.50. The grounding electrode conductor(s) shall be connected in accordance with 250.32(B) or (C). Where there is no existing grounding electrode, the grounding electrode(s) required in 250.50 shall be installed.

Where a building or structure is supplied by a feeder, 250.32(A) requires that a grounding electrode system be established at each building or structure supplied, unless one already exists. The equipment grounding bus must be bonded to the grounding electrode system, and the disconnecting means enclosure, building steel, and interior metal water piping are also required to be bonded to the grounding electrode system. All exposed non–current-carrying metal parts of electrical equipment are required to be grounded through equipment grounding conductor connections to the equipment grounding bus at the building disconnecting means. The connection of the grounded (neutral) conductor to the grounding electrode system, as shown in Exhibit 250.16, is permitted only where it can be ensured that such a connection does not establish a parallel circuit path for normal neutral current on equipment grounding conductors, metal shields of cables not intended to be used as a current-carrying conductor, metal piping systems, or other metal structures that are continuous between buildings.

Service conductors — Feeder — Feeder

N G — N G — N G

Building 1 — Building 2 — Building 3

No continuous metallic paths between buildings

Exhibit 250.16 Example of grounding electrode systems required at feeder-supplied Building 2 and Building 3, in accordance with 250.32(A).

Exception: A grounding electrode shall not be required where only a single branch circuit supplies the building or structure and the branch circuit includes an equipment grounding conductor for grounding the conductive non–current-carrying parts of equipment. For the purpose of this section, a multiwire branch circuit shall be considered as a single branch circuit.

Where a building is supplied by a single branch circuit (2-wire or multiwire) and is installed in or has a wire-type equipment grounding conductor, as covered in 250.118, it is not required to establish a grounding electrode system or connect to an existing one. Where the installation occurs at other than a dwelling unit, the disconnecting means at the remote building is required to be suitable for service equipment in accordance with 225.36. See Exhibit 250.17 for an example of this provision.

(B) Grounded Systems For a grounded system at the separate building or structure, the connection to the grounding electrode and grounding or bonding of equipment, structures, or frames required to be grounded or bonded shall comply with either 250.32(B)(1) or (B)(2).

(1) Equipment Grounding Conductor An equipment grounding conductor as described in 250.118 shall be run with the supply conductors and connected to the building or structure disconnecting means and to the grounding electrode(s). The equipment grounding conductor shall be used for grounding or bonding of equipment, structures, or frames required to be grounded or bonded. The equipment grounding conductor shall be sized in accordance with 250.122. Any installed grounded conductor shall not be connected to the equipment grounding conductor or to the grounding electrode(s).

Main building service equipment

Remote building

Connection to grounding electrode not required for single branch-circuit disconnecting means

One branch circuit with equipment grounding conductor

PVC conduit

Exhibit 250.17 An installation where a connection from the single branch-circuit disconnecting means enclosure to a grounding electrode system is not required at the remote building because an equipment grounding conductor is installed with the circuit conductors.

Where a feeder supplies a building and an equipment grounding conductor is run with or encloses the feeder, the grounded conductor (neutral) is not permitted to be connected to the equipment grounding conductor or to the grounding electrode system, as illustrated in Exhibit 250.18.

(2) Grounded Conductor Where (1) an equipment grounding conductor is not run with the supply to the building or structure, (2) there are no continuous metallic paths bonded to the grounding system in each building or structure involved, and (3) ground-fault protection of equipment has not been installed on the supply side of the feeder(s), the grounded conductor run with the supply to the building or structure shall be connected to the building or structure disconnecting means and to the grounding electrode(s) and shall be used for grounding or bonding of equipment, structures, or frames required to be grounded or bonded. The size of the grounded conductor shall not be smaller than the larger of either of the following:

(1) That required by 220.61
(2) That required by 250.122

Similar to the provisions of 250.30(A)(3), the requirement in 250.32(B)(2) eliminates the creation of parallel paths for normal neutral current on grounding conductors, metal raceways, metal piping, and other metal structures. In the

Main building
service equipment

Remote building

Isolated
neutral

Feeder with equipment
grounding conductor to
additional building

Grounding electrode required

Exhibit 250.18 An installation in which connection between the grounded conductor (neutral) and equipment grounding terminal bar is not allowed. A connection from the equipment grounding terminal bus to the grounding electrode is required.

1999 and previous editions of the *Code*, the grounding electrode conductor and equipment grounding conductors were permitted to be connected to the grounded conductor at a separate building or structure. This multiple-location grounding arrangement could provide parallel paths for neutral current along the electrical system and along other continuous metallic piping and mechanical systems as well. Connection of the grounded conductor to a grounding electrode system at a separate building or structure is permitted only if these parallel paths are not created and if there is no common ground-fault protection of equipment provided at the service where the feeder or branch circuit originates.

Where the grounded conductor is used as part of the ground-fault current return circuit, it is required to be sized no less than that required by 250.122 for equipment grounding conductors, but it also has to be sized to carry the maximum unbalanced load, as specified in 220.61.

Like the grounded service conductor, a branch-circuit or feeder grounded conductor used in the application permitted by 250.32(B)(2) is a circuit conductor for normal neutral current and is also the circuit conductor used to create an effective ground-fault current return path. Therefore, it is necessary to size the grounded conductor in this application based on which of those two functions requires the larger conductor. Of course there is no prohibition on installing a full-size grounded (neutral) conductor, thus ensuring compliance with both 250.122 and 220.61.

(C) Ungrounded Systems The grounding electrode(s) shall be connected to the building or structure disconnecting means.

(D) Disconnecting Means Located in Separate Building or Structure on the Same Premises Where one or more disconnecting means supply one or more additional buildings or structures under single management, and where these disconnecting means are located remote from those buildings or structures in accordance with the provisions of 225.32, Exception Nos. 1 and 2, all of the following conditions shall be met:

(1) The connection of the grounded conductor to the grounding electrode at a separate building or structure shall not be made.

(2) An equipment grounding conductor for grounding any non–current-carrying equipment, interior metal piping systems, and building or structural metal frames is run with the circuit conductors to a separate building or structure and bonded to existing grounding electrode(s) required in Part III of this article, or, where there are no existing electrodes, the grounding electrode(s) required in Part III of this article shall be installed where a separate building or structure is supplied by more than one branch circuit.

(3) Bonding the equipment grounding conductor to the grounding electrode at a separate building or structure shall be made in a junction box, panelboard, or similar enclosure located immediately inside or outside the separate building or structure.

Exhibit 250.19 illustrates an installation in which the disconnect for Building 2 is located in Building 1. Section 250.32(D) applies to separate buildings or structures that do not have a disconnect, as permitted by Exception No. 1 and Exception No. 2 to 225.32. The feeder conductors must terminate in a panelboard, junction box, or similar enclosure that is located immediate to the point the supply conductors enter the building or structure inside or outside the building.

An equipment grounding conductor must be run with the feeder conductors, the grounded conductor must not be bonded to the enclosure or equipment grounding bus, and the equipment grounding bus must be connected to a new or existing grounding electrode system at the second building. All non–current-carrying metal parts of equipment, building steel, and interior metal piping systems must be connected to the grounding electrode system.

(E) Grounding Electrode Conductor The size of the grounding electrode conductor to the grounding electrode(s) shall not be smaller than given in 250.66, based on the

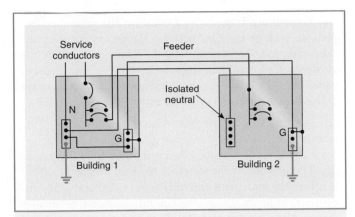

Exhibit 250.19 Grounding and bonding requirements for a separate building under single management with the disconnect remotely located from the building.

largest ungrounded supply conductor. The installation shall comply with Part III of this article.

A grounding electrode system is connected to the grounded conductor and/or the equipment enclosures by the grounding electrode conductor (see definition in Article 100) according to 250.24 for services, to 250.30 for separately derived systems, and to 250.32 for two or more buildings supplied from a common service. Each of these sections directs the user to the same general requirements; that is, the grounding electrode conductor must comply with Part III of Article 250. A revision to the 2002 *Code* clarified that where a feeder or branch circuit supplies a building or structure, the conductor used to connect the equipment grounding conductor and, as permitted by 250.32(B)(2), the grounded conductor to the grounding electrode system is a grounding electrode conductor and must be sized in accordance with 250.66.

250.34 Portable and Vehicle-Mounted Generators

(A) Portable Generators The frame of a portable generator shall not be required to be connected to a grounding electrode as defined in 250.52 for a system supplied by the generator under the following conditions:

(1) The generator supplies only equipment mounted on the generator, cord-and-plug-connected equipment through receptacles mounted on the generator, or both, and

(2) The non–current-carrying metal parts of equipment and the equipment grounding conductor terminals of the receptacles are bonded to the generator frame.

Portable describes equipment that is easily carried by personnel from one location to another. *Mobile* describes equip-

ment, such as vehicle-mounted generators, that is capable of being moved on wheels or rollers.

The frame of a portable generator is not required to be connected to earth (ground rod, water pipe, etc.) if the generator has receptacles mounted on the generator panel and the receptacles have equipment grounding terminals bonded to the generator frame.

(B) Vehicle-Mounted Generators The frame of a vehicle shall not be required to be connected to a grounding electrode as defined in 250.52 for a system supplied by a generator located on this vehicle under the following conditions:

(1) The frame of the generator is bonded to the vehicle frame, and

(2) The generator supplies only equipment located on the vehicle or cord-and-plug-connected equipment through receptacles mounted on the vehicle, or both equipment located on the vehicle and cord-and-plug-connected equipment through receptacles mounted on the vehicle or on the generator, and

(3) The non–current-carrying metal parts of equipment and the equipment grounding conductor terminals of the receptacles are bonded to the generator frame.

Vehicle-mounted generators that provide a neutral conductor and are installed as separately derived systems supplying equipment and receptacles on the vehicle are required to have the neutral conductor bonded to the generator frame and to the vehicle frame. The non–current-carrying parts of the equipment must be bonded to the generator frame.

(C) Grounded Conductor Bonding A system conductor that is required to be grounded by 250.26 shall be bonded to the generator frame where the generator is a component of a separately derived system.

> FPN: For grounding portable generators supplying fixed wiring systems, see 250.20(D).

Portable and vehicle-mounted generators that are installed as separately derived systems and that provide a neutral conductor (such as 3-phase, 4-wire wye connected; single-phase 240/120 volt; or 3-phase, 4-wire delta connected) are required to have the neutral conductor bonded to the generator frame.

250.36 High-Impedance Grounded Neutral Systems

High-impedance grounded neutral systems in which a grounding impedance, usually a resistor, limits the ground-fault current to a low value shall be permitted for 3-phase ac systems of 480 volts to 1000 volts where all the following conditions are met:

(1) The conditions of maintenance and supervision ensure that only qualified persons service the installation.
(2) Continuity of power is required.
(3) Ground detectors are installed on the system.
(4) Line-to-neutral loads are not served.

Section 250.36 covers high-impedance grounded neutral systems of 480 to 1000 volts. Systems rated over 1000 volts are covered in 250.186. For information on the differences between solidly grounded systems and high-impedance grounded neutral systems, see "Grounding for Emergency and Standby Power Systems," by Robert B. West, *IEEE Transactions on Industry Applications*, Vol. IA-15, No. 2, March/April 1979.

As the schematic diagram in Exhibit 250.20 shows, a high-impedance grounded neutral system is designed to minimize the amount of fault current during a ground fault. The grounding impedance is usually selected to limit fault current to a value that is slightly greater than or equal to the capacitive charging current. This system is used where continuity of power is required. Therefore, a ground fault results in an alarm condition rather than in the tripping of a circuit breaker, which allows a safe and orderly shutdown of a process in which a non-orderly shutdown can introduce additional or increased hazards.

480 V to 1000 V

Neutral point

Insulated neutral conductor

Grounding impedance

Unspliced equipment bonding jumper

No line-to-neutral loads

Exhibit 250.20 Schematic diagram of a high-impedance grounded neutral system.

High-impedance grounded neutral systems shall comply with the provisions of 250.36(A) through (G).

(A) Grounding Impedance Location The grounding impedance shall be installed between the grounding electrode conductor and the system neutral. Where a neutral is not available, the grounding impedance shall be installed between the grounding electrode conductor and the neutral derived from a grounding transformer.

(B) Neutral Conductor The neutral conductor from the neutral point of the transformer or generator to its connection point to the grounding impedance shall be fully insulated.

The neutral conductor shall have an ampacity of not less than the maximum current rating of the grounding impedance. In no case shall the neutral conductor be smaller than 8 AWG copper or 6 AWG aluminum or copper-clad aluminum.

The current through the neutral conductor is limited by the grounding impedance. Therefore, the neutral conductor is not required to be sized to carry high-fault current. The neutral conductor cannot be smaller than 8 AWG copper or 6 AWG aluminum.

(C) System Neutral Connection The system neutral conductor shall not be connected to ground except through the grounding impedance.

> FPN: The impedance is normally selected to limit the ground-fault current to a value slightly greater than or equal to the capacitive charging current of the system. This value of impedance will also limit transient overvoltages to safe values. For guidance, refer to criteria for limiting transient overvoltages in ANSI/IEEE 142-1991, *Recommended Practice for Grounding of Industrial and Commercial Power Systems*.

Additional information can be found in "Charging Current Data for Guesswork-Free Design of High-Resistance Grounded Systems," by D. S. Baker, *IEEE Transactions on Industry Applications*, Vol. IA-15, No. 2, March/April 1979; and "High-Resistance Grounding," by Baldwin Bridger, Jr., *IEEE Transactions on Industry Applications*, Vol. IA-19, No. 1, January/February 1983.

(D) Neutral Conductor Routing The conductor connecting the neutral point of the transformer or generator to the grounding impedance shall be permitted to be installed in a separate raceway. It shall not be required to run this conductor with the phase conductors to the first system disconnecting means or overcurrent device.

(E) Equipment Bonding Jumper The equipment bonding jumper (the connection between the equipment grounding conductors and the grounding impedance) shall be an unspliced conductor run from the first system disconnecting means or overcurrent device to the grounded side of the grounding impedance.

(F) Grounding Electrode Conductor Location The grounding electrode conductor shall be attached at any point from the grounded side of the grounding impedance to the equipment grounding connection at the service equipment or first system disconnecting means.

(G) Equipment Bonding Jumper Size The equipment bonding jumper shall be sized in accordance with (1) or (2) as follows:

(1) Where the grounding electrode conductor connection is made at the grounding impedance, the equipment bonding jumper shall be sized in accordance with 250.66, based on the size of the service entrance conductors for a service or the derived phase conductors for a separately derived system.

(2) Where the grounding electrode conductor is connected at the first system disconnecting means or overcurrent device, the equipment bonding jumper shall be sized the same as the neutral conductor in 250.36(B).

III. Grounding Electrode System and Grounding Electrode Conductor

250.50 Grounding Electrode System

All grounding electrodes as described in 250.52(A)(1) through (A)(6) that are present at each building or structure served shall be bonded together to form the grounding electrode system. Where none of these grounding electrodes exist, one or more of the grounding electrodes specified in 250.52(A)(4) through (A)(7) shall be installed and used.

Section 250.50 introduces the important concept of a "grounding electrode system," in which all electrodes are bonded together, as illustrated in Exhibit 250.21. Rather than total reliance on a single grounding electrode to perform its function over the life of the electrical installation, the *NEC* encourages the formation of a system of electrodes "that are present at each building or structure served." There is no doubt that building a system of electrodes adds a level of reliability and helps ensure system performance over a long period of time.

This section was revised for the 2005 *Code* to clearly require the inclusion of a concrete-encased electrode, described in 250.52(A)(3), in the grounding electrode system for buildings or structures having a concrete footing or foundation with not less than 20 ft of surface area in direct contact with the earth. This requirement applies to all buildings and structures with a foundation and/or footing having 20 ft or more of ½ in. or greater electrically conductive reinforcing steel or 20 ft or more of bare copper not smaller than 4 AWG. However, an exception does exempt existing buildings and structures where access to the concrete-encased electrode

Exhibit 250.21 A grounding electrode system that uses the metal frame of a building, a ground ring, a concrete-encased electrode, a metal underground water pipe, and a ground rod.

would involve some type of demolition or similar activity that would disturb the existing construction. Because the installation of the footings and foundation is one of the first elements of a construction project and in most cases has long been completed by the time the electric service is installed, this revised text necessitates an awareness and coordinated effort on the part of designers and the construction trades in making sure that the concrete-encased electrode is incorporated into the grounding electrode system.

Exception: Concrete-encased electrodes of existing buildings or structures shall not be required to be part of the grounding electrode system where the steel reinforcing bars or rods are not accessible for use without disturbing the concrete.

250.52 Grounding Electrodes

(A) Electrodes Permitted for Grounding

(1) Metal Underground Water Pipe A metal underground water pipe in direct contact with the earth for 3.0 m (10 ft) or more (including any metal well casing effectively bonded to the pipe) and electrically continuous (or made electrically continuous by bonding around insulating joints or insulating pipe) to the points of connection of the grounding electrode conductor and the bonding conductors. Interior metal water piping located more than 1.52 m (5 ft) from the point of entrance to the building shall not be used as a part of the grounding electrode system or as a conductor to interconnect electrodes that are part of the grounding electrode system.

Exception: In industrial and commercial buildings or structures where conditions of maintenance and supervision ensure that only qualified persons service the installation, interior metal water piping located more than 1.52 m (5 ft) from the point of entrance to the building shall be permitted as a part of the grounding electrode system or as a conductor to interconnect electrodes that are part of the grounding electrode system, provided that the entire length, other than short sections passing perpendicular through walls, floors, or ceilings, of the interior metal water pipe that is being used for the conductor is exposed.

The effectiveness of underground water piping as a grounding electrode for electrical systems has long been recognized, but in the early years of the *NEC* concerns over the effect of electric current on metal water piping created some uncertainty as to whether metal water piping systems should be used as grounding electrodes. To address those concerns, the electrical industry and the waterworks industry formed a committee to evaluate the use of metal underground water piping systems as grounding electrodes. Based on its findings, the committee issued an authoritative report on the subject. The International Association of Electrical Inspectors published the report, *Interim Report of the American Research Committee on Grounding*, in January 1944 (reprinted March 1949).

The National Institute of Standards and Technology (NIST) has monitored the electrolysis of metal systems, because current at a grounding electrode on dc systems can cause displacement of metal. The results of this monitoring have shown that problems are minimal.

The last sentence of 250.52(A)(1) prohibits the use of that portion of the interior metal water piping system that extends more than 5 ft beyond the point of entrance into the building to interconnect grounding electrodes and the grounding electrode conductor, because there are concerns over the use of nonmetallic piping or fittings causing an interruption in the interior electrical continuity of the metal water piping. The exception to 250.52(A)(1), however, permits this practice, provided there is qualified maintenance and the entire length of the water piping used as an electrode is exposed. This 5-ft limit also applies to the replacement of nongrounding receptacles with grounding-type or branch-circuit extensions in accordance with 250.130(C). See the commentary following 250.130(C) and the illustration that accompanies that commentary, Exhibit 250.49.

(2) Metal Frame of the Building or Structure The metal frame of the building or structure, where any of the following methods are used to make an earth connection:

(1) 3.0 m (10 ft) or more of a single structural metal member in direct contact with the earth or encased in concrete that is in direct contact with the earth

(2) The structural metal frame is bonded to one or more of the grounding electrodes as defined in 250.52(A)(1), (A)(3), or (A)(4)

(3) The structural metal frame is bonded to one or more of the grounding electrodes as defined in 250.52(A)(5) or (A)(6) that comply with 250.56, or

(4) Other approved means of establishing a connection to earth.

The 2005 *NEC* revision to 250.52(A)(2) provides four means by which the metal frame of a building or structure can be judged suitable for use as a grounding electrode. This revision defines what is considered to be effectively grounded as applied to the metal frame of a building. The metal frame of the building can be considered an electrode through 10 ft of direct contact with the earth or through connection to one of the electrode types described in 250.52(A)(1), 250.52(A)(3), or 250.52(A)(4) or through connection to rod or plate type electrodes that comply with the requirement of 250.56. If building steel is grounded through a connection to an underground metal water pipe, replacement of the water pipe with nonmetallic piping will result in the building steel no longer being "effectively grounded."

(3) Concrete-Encased Electrode An electrode encased by at least 50 mm (2 in.) of concrete, located within and near the bottom of a concrete foundation or footing that is in direct contact with the earth, consisting of at least 6.0 m (20 ft) of one or more bare or zinc galvanized or other electrically conductive coated steel reinforcing bars or rods of not less than 13 mm (½ in.) in diameter, or consisting of at least 6.0 m (20 ft) of bare copper conductor not smaller than 4 AWG. Reinforcing bars shall be permitted to be bonded together by the usual steel tie wires or other effective means.

Exhibit 250.22 shows an example of a concrete-encased electrode.

(4) Ground Ring A ground ring encircling the building or structure, in direct contact with the earth, consisting of at least 6.0 m (20 ft) of bare copper conductor not smaller than 2 AWG.

(5) Rod and Pipe Electrodes Rod and pipe electrodes shall not be less than 2.5 m (8 ft) in length and shall consist of the following materials.

(a) Electrodes of pipe or conduit shall not be smaller than metric designator 21 (trade size ¾) and, where of iron or steel, shall have the outer surface galvanized or otherwise metal-coated for corrosion protection.

4 AWG or larger bare copper conductor or steel reinforcing bar or rod not less than ½-in. diameter at least 20 ft long

Grounding electrode conductor

Nonmetallic protective sleeve

Connection listed for the purpose

Foundation in direct contact with earth

2 in. min.

Exhibit 250.22 A concrete-encased electrode.

(b) Electrodes of rods of iron or steel shall be at least 15.87 mm (⅝ in.) in diameter. Stainless steel rods less than 16 mm (⅝ in.) in diameter, nonferrous rods, or their equivalent shall be listed and shall not be less than 13 mm (½ in.) in diameter.

(6) Plate Electrodes Each plate electrode shall expose not less than 0.186 m² (2 ft²) of surface to exterior soil. Electrodes of iron or steel plates shall be at least 6.4 mm (¼ in.) in thickness. Electrodes of nonferrous metal shall be at least 1.5 mm (0.06 in.) in thickness.

(7) Other Local Metal Underground Systems or Structures Other local metal underground systems or structures such as piping systems, underground tanks, and underground metal well casings that are not effectively bonded to a metal water pipe.

(B) Electrodes Not Permitted for Grounding The following shall not be used as grounding electrodes:

(1) Metal underground gas piping system
(2) Aluminum electrodes

FPN: See 250.104(B) for bonding requirements of gas piping.

250.53 Grounding Electrode System Installation

FPN: See 547.9 and 547.10 for special grounding and bonding requirements for agricultural buildings.

(A) Rod, Pipe, and Plate Electrodes Where practicable, rod, pipe, and plate electrodes shall be embedded below permanent moisture level. Rod, pipe, and plate electrodes shall be free from nonconductive coatings such as paint or enamel.

(B) Electrode Spacing Where more than one of the electrodes of the type specified in 250.52(A)(5) or (A)(6) are

used, each electrode of one grounding system (including that used for air terminals) shall not be less than 1.83 m (6 ft) from any other electrode of another grounding system. Two or more grounding electrodes that are effectively bonded together shall be considered a single grounding electrode system.

(C) Bonding Jumper The bonding jumper(s) used to connect the grounding electrodes together to form the grounding electrode system shall be installed in accordance with 250.64(A), (B), and (E), shall be sized in accordance with 250.66, and shall be connected in the manner specified in 250.70.

(D) Metal Underground Water Pipe Where used as a grounding electrode, metal underground water pipe shall meet the requirements of 250.53(D)(1) and (D)(2).

(1) Continuity Continuity of the grounding path or the bonding connection to interior piping shall not rely on water meters or filtering devices and similar equipment.

(2) Supplemental Electrode Required A metal underground water pipe shall be supplemented by an additional electrode of a type specified in 250.52(A)(2) through (A)(7). Where the supplemental electrode is a rod, pipe, or plate type, it shall comply with 250.56. The supplemental electrode shall be permitted to be bonded to the grounding electrode conductor, the grounded service-entrance conductor, the nonflexible grounded service raceway, or any grounded service enclosure.

Exception: The supplemental electrode shall be permitted to be bonded to the interior metal water piping at any convenient point as covered in 250.52(A)(1), Exception.

Section 250.53(D)(2) specifically requires that rod, pipe, or plate electrodes used to supplement metal water piping be installed in accordance with 250.56. This requirement clarifies that the supplemental electrode system must be installed as if it were the sole grounding electrode for the system. If 25 ohms or less of earth resistance cannot be achieved with one rod, pipe, or plate, another electrode (other than the metal piping that is being supplemented) must be provided. One of the permitted methods of bonding a supplemental grounding electrode conductor to the primary electrode system is to connect it to the service enclosure.

The requirement to supplement the metal water pipe is based on the practice of using a plastic pipe for replacement when the original metal water pipe fails. This type of replacement leaves the system without a grounding electrode unless a supplemental electrode is provided.

(E) Supplemental Electrode Bonding Connection Size Where the supplemental electrode is a rod, pipe, or plate electrode, that portion of the bonding jumper that is the sole connection to the supplemental grounding electrode shall not be required to be larger than 6 AWG copper wire or 4 AWG aluminum wire.

Section 250.53(E) correlates with 250.52(A)(5) or 250.52(A)(6) and with 250.66(A). For example, if a metal underground water pipe or the metal frame of the building or structure is used as the grounding electrode or as part of the grounding electrode system, Table 250.66 must be used for sizing the grounding electrode conductor. The size of the grounding electrode conductor or bonding jumper for ground rod or pipe or for plate electrodes between the service equipment and the electrodes is not required to be larger than 6 AWG copper or 4 AWG aluminum.

(F) Ground Ring The ground ring shall be buried at a depth below the earth's surface of not less than 750 mm (30 in.).

(G) Rod and Pipe Electrodes The electrode shall be installed such that at least 2.44 m (8 ft) of length is in contact with the soil. It shall be driven to a depth of not less than 2.44 m (8 ft) except that, where rock bottom is encountered, the electrode shall be driven at an oblique angle not to exceed 45 degrees from the vertical or, where rock bottom is encountered at an angle up to 45 degrees, the electrode shall be permitted to be buried in a trench that is at least 750 mm (30 in.) deep. The upper end of the electrode shall be flush with or below ground level unless the aboveground end and the grounding electrode conductor attachment are protected against physical damage as specified in 250.10.

All rod and pipe electrodes must have at least 8 ft of length in contact with the soil, regardless of rock bottom. Where rock bottom is encountered, the electrodes must either be driven at not more than a 45-degree angle or buried in a 2½-ft-deep trench. It should be noted that driving the rod at an angle is permitted only if it is not possible to drive the rod vertically to obtain at least 8 ft of earth contact. Burying the ground rod is permitted only if it is not possible to drive the rod vertically or at an angle.

Ground clamps used on buried electrodes must be listed for direct earth burial. Ground clamps installed aboveground must be protected where subject to physical damage. Exhibit 250.23 illustrates these requirements.

(H) Plate Electrode Plate electrodes shall be installed not less than 750 mm (30 in.) below the surface of the earth.

250.54 Supplementary Grounding Electrodes

Supplementary grounding electrodes shall be permitted to be connected to the equipment grounding conductors specified in 250.118 and shall not be required to comply with

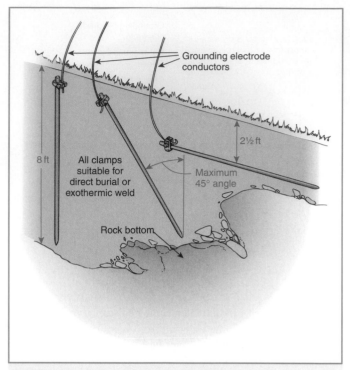

Exhibit 250.23 Installation requirements for rod and pipe electrodes as specified by 250.53(G).

the electrode bonding requirements of 250.50 or 250.53(C) or the resistance requirements of 250.56, but the earth shall not be used as an effective ground-fault current path as specified in 250.4(A)(5) and 250.4(B)(4).

Grounding electrodes, such as ground rods, that are connected to equipment are not permitted to be used in lieu of the equipment grounding conductor, but they may be used for supplementary protection at electrical equipment locations. For example, grounding electrodes may be used for lightning protection or to establish a reference to ground in the area of electrically operated equipment. Sections 250.4(A)(5) and 250.4(B)(4) also specify that the earth not be used as the sole equipment grounding conductor or effective (ground) fault current path. Supplementary grounding electrodes are not required to be incorporated into the grounding electrode system for the service or other source of electrical supply.

250.56 Resistance of Rod, Pipe, and Plate Electrodes

A single electrode consisting of a rod, pipe, or plate that does not have a resistance to ground of 25 ohms or less shall be augmented by one additional electrode of any of the types specified by 250.52(A)(2) through (A)(7). Where multiple rod, pipe, or plate electrodes are installed to meet the requirements of this section, they shall not be less than 1.8 m (6 ft) apart.

> FPN: The paralleling efficiency of rods longer than 2.5 m (8 ft) is improved by spacing greater than 1.8 m (6 ft).

A supplemental rod, pipe, or plate electrode must be spaced at least 6 ft from any other rod, pipe, and plate electrode. See Exhibit 250.24.

The resistance to ground of a driven grounding electrode can be measured by a ground tester used in the manner shown in Exhibit 250.25.

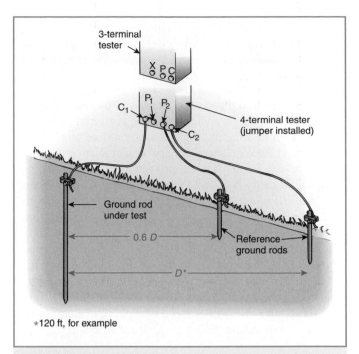

Exhibit 250.25 The resistance to ground of a ground rod being measured by a ground tester.

Exhibit 250.24 The 6-ft spacing between electrodes required by 250.53(B) and 250.56.

250.58 Common Grounding Electrode

Where an ac system is connected to a grounding electrode in or at a building or structure, the same electrode shall be used to ground conductor enclosures and equipment in or on that building or structure. Where separate services, feeders, or branch circuits supply a building and are required to be connected to a grounding electrode(s), the same grounding electrode(s) shall be used.

Two or more grounding electrodes that are effectively bonded together shall be considered as a single grounding electrode system in this sense.

250.60 Use of Air Terminals

Air terminal conductors and driven pipes, rods, or plate electrodes used for grounding air terminals shall not be used

in lieu of the grounding electrodes required by 250.50 for grounding wiring systems and equipment. This provision shall not prohibit the required bonding together of grounding electrodes of different systems.

> FPN No. 1: See 250.106 for spacing from air terminals. See 800.100(D), 810.21(J), and 820.100(D) for bonding of electrodes.
>
> FPN No. 2: Bonding together of all separate grounding electrodes will limit potential differences between them and between their associated wiring systems.

250.62 Grounding Electrode Conductor Material

The grounding electrode conductor shall be of copper, aluminum, or copper-clad aluminum. The material selected shall be resistant to any corrosive condition existing at the installation or shall be suitably protected against corrosion. The conductor shall be solid or stranded, insulated, covered, or bare.

250.64 Grounding Electrode Conductor Installation

Grounding electrode conductors shall be installed as specified in 250.64(A) through (F).

(A) Aluminum or Copper-Clad Aluminum Conductors

Bare aluminum or copper-clad aluminum grounding conductors shall not be used where in direct contact with masonry or the earth or where subject to corrosive conditions. Where used outside, aluminum or copper-clad aluminum grounding conductors shall not be terminated within 450 mm (18 in.) of the earth.

(B) Securing and Protection Against Physical Damage
Where exposed, a grounding electrode conductor or its enclosure shall be securely fastened to the surface on which it is carried. A 4 AWG or larger copper or aluminum grounding electrode conductor shall be protected where exposed to physical damage. A 6 AWG grounding electrode conductor that is free from exposure to physical damage shall be permitted to be run along the surface of the building construction without metal covering or protection where it is securely fastened to the construction; otherwise, it shall be in rigid metal conduit, intermediate metal conduit, rigid nonmetallic conduit, electrical metallic tubing, or cable armor. Grounding electrode conductors smaller than 6 AWG shall be in rigid metal conduit, intermediate metal conduit, rigid nonmetallic conduit, electrical metallic tubing, or cable armor.

See 250.64(E) for additional information on situations in which raceways enclose the grounding electrode conductor. Also see the commentary following 250.92(A)(3) and the illustration that accompanies that commentary, Exhibit 250.32, for installation requirements for metal raceways used to install and physically protect the grounding electrode conductor(s).

(C) Continuous Grounding electrode conductor(s) shall be installed in one continuous length without a splice or joint except as permitted in (1) through (4):

(1) Splicing shall be permitted only by irreversible compression-type connectors listed as grounding and bonding equipment or by the exothermic welding process.
(2) Sections of busbars shall be permitted to be connected together to form a grounding electrode conductor.
(3) Bonding jumper(s) from grounding electrode(s) and grounding electrode conductor(s) shall be permitted to be connected to an aluminum or copper busbar not less than 6 mm × 50 mm (¼ in. × 2 in.). The busbar shall be securely fastened and shall be installed in an accessible location. Connections shall be made by a listed connector or by the exothermic welding process.
(4) Where aluminum busbars are used, the installation shall comply with 250.64(A).

Although an infrequent occurrence, there are conditions under which it may be necessary to splice the grounding electrode conductor, such as in the case of a remodeling project within a building or the replacement of existing

electrical equipment. Section 250.64(C) permits splicing a wire-type grounding electrode conductor with irreversible compression-type fittings specifically listed as grounding equipment or by exothermic welding. These methods create connections that are equated to have the same permanency as an unspliced conductor. This section also recognizes the normal bolted connections between sections of busbar that are joined to form the grounding electrode conductor.

A new method for connecting sections of the grounding electrode conductor is recognized in the 2005 Code. A securely fastened section of copper or aluminum busbar, not less than ¼ in. thick by 2 in. wide (the length can be whatever is necessary to make the connections), is permitted as a connection point for multiple grounding electrode conductors or for bonding jumpers that are used to bond multiple grounding electrodes together. The connection of the wire to the busbar must be via an exothermic weld or by a listed connector that is attached to the busbar using the typical bolted connection.

(D) Grounding Electrode Conductor Taps Where a service consists of more than a single enclosure as permitted in 230.71(A), it shall be permitted to connect taps to the common grounding electrode conductor. Each such tap conductor shall extend to the inside of each such enclosure. The common grounding electrode conductor shall be sized in accordance with 250.66, based on the sum of the circular mil area of the largest ungrounded service entrance conductors. Where more than one set of service entrance conductors as permitted by 230.40, Exception No. 2 connect directly to a service drop or lateral, the common grounding electrode conductor shall be sized in accordance with Table 250.66 Note 1. The tap conductors shall be permitted to be sized in accordance with the grounding electrode conductors specified in 250.66 for the largest conductor serving the respective enclosures. The tap conductors shall be connected to the common grounding electrode conductor in such a manner that the common grounding electrode conductor remains without a splice or joint.

Grounding electrode (tap) conductors must be sized using Table 250.66 and are based on the size of the largest phase conductor serving each service disconnecting means enclosure. The main grounding electrode conductor from which the taps are made is sized from Table 250.66 based on the sum of the cross-sectional areas of the largest ungrounded service-entrance conductors or equivalent cross-sectional area for parallel conductors that supply the multiple service disconnecting means. As illustrated in Exhibit 250.26, the tap method eliminates the difficulties found in looping grounding electrode conductors from one enclosure to another. The 2 AWG grounding electrode conductor (based

Exhibit 250.26 The tap method of connecting grounding electrode conductors from one enclosure to another.

on the 350-kcmil ungrounded conductor) shown in Exhibit 250.26 is required to be installed without a splice or joint, except as permitted in 250.64(C), and the 8 AWG and 4 AWG taps are sized from Table 250.66 based on the size of the ungrounded conductor serving the respective service disconnecting means.

(E) Enclosures for Grounding Electrode Conductors Ferrous metal enclosures for grounding electrode conductors shall be electrically continuous from the point of attachment to cabinets or equipment to the grounding electrode and shall be securely fastened to the ground clamp or fitting. Nonferrous metal enclosures shall not be required to be electrically continuous. Ferrous metal enclosures that are not physically continuous from cabinets or equipment to the grounding electrode shall be made electrically continuous by bonding each end of the raceway or enclosure to the grounding electrode conductor. Bonding shall apply at each end and to all intervening ferrous raceways, boxes, and enclosures between the service equipment and the grounding electrode. The bonding jumper for a grounding electrode conductor raceway or cable armor shall be the same size as, or larger than, the required enclosed grounding electrode conductor. Where a raceway is used as protection for a grounding electrode conductor, the installation shall comply with the requirements of the appropriate raceway article.

Bonding jumpers installed to ensure the electrical continuity of ferrous metal enclosures must be sized in accordance with 250.102(C). Exhibit 250.32, which appears in the commentary following 250.92(A)(3), shows the bonding of a ferrous metal raceway to a grounding electrode conductor at both ends to ensure that the raceway and conductor are in parallel.

(F) To Electrode(s) A grounding electrode conductor shall be permitted to be run to any convenient grounding electrode available in the grounding electrode system, or to one or more grounding electrode(s) individually, or to the aluminum or copper busbar as permitted in 250.64(C). The grounding electrode conductor shall be sized for the largest grounding electrode conductor required among all the electrodes connected to it.

Exhibit 250.27 shows an example of a grounding electrode system. The single grounding electrode conductor is permitted to run "to any convenient grounding electrode available," and the other electrodes are connected together using bonding jumpers sized in accordance with 250.66. For the 2005 *Code*, a permitted alternative to running the grounding

Exhibit 250.27 An example of running the grounding electrode conductor to any convenient electrode available as well as bonding electrodes together to form the grounding electrode system required by 250.50.

electrode conductor to an electrode is to run it to a busbar used as a connection point for bonding jumpers from multiple electrodes that form the grounding electrode system. See the commentary on 250.64(C) for more information on this method.

250.66 Size of Alternating-Current Grounding Electrode Conductor

The size of the grounding electrode conductor of a grounded or ungrounded ac system shall not be less than given in Table 250.66, except as permitted in 250.66(A) through (C).

> FPN: See 250.24(C) for size of ac system conductor brought to service equipment.

Table 250.66 Grounding Electrode Conductor for Alternating-Current Systems

Size of Largest Ungrounded Service-Entrance Conductor or Equivalent Area for Parallel Conductors[a] (AWG/kcmil)		Size of Grounding Electrode Conductor (AWG/kcmil)	
Copper	Aluminum or Copper-Clad Aluminum	Copper	Aluminum or Copper-Clad Aluminum[b]
2 or smaller	1/0 or smaller	8	6
1 or 1/0	2/0 or 3/0	6	4
2/0 or 3/0	4/0 or 250	4	2
Over 3/0 through 350	Over 250 through 500	2	1/0
Over 350 through 600	Over 500 through 900	1/0	3/0
Over 600 through 1100	Over 900 through 1750	2/0	4/0
Over 1100	Over 1750	3/0	250

Notes:
1. Where multiple sets of service-entrance conductors are used as permitted in 230.40, Exception No. 2, the equivalent size of the largest service-entrance conductor shall be determined by the largest sum of the areas of the corresponding conductors of each set.
2. Where there are no service-entrance conductors, the grounding electrode conductor size shall be determined by the equivalent size of the largest service-entrance conductor required for the load to be served.
[a]This table also applies to the derived conductors of separately derived ac systems.
[b]See installation restrictions in 250.64(A).

Example

Apply the sizing requirements in Table 250.66 to Exhibit 250.28 to determine the size of the grounding electrode conductor.

Solution

STEP 1. Using Table 8 in Chapter 9, calculate the total circular mil area of both grounded service conductors.

$$3 \text{ AWG} = 52,620 \text{ circular mils}$$
$$3/0 \text{ AWG} = 167,800 \text{ mills}$$
$$\text{Total area} = 220,420 \text{ circular mils}$$

From Table 8, the next larger standard size is 250 kcmil.

STEP 2. Use Table 250.66 to size the grounding electrode conductor. According to the fourth row, "Over 3/0 through 350," the size should be 2 AWG copper or 1/0 AWG aluminum.

Note that the taps to the grounding electrode conductor from each service disconnecting means enclosure in Exhibit 250.28 are sized from Table 250.66 based on the size of the service-entrance conductors supplying the enclosures.

Exhibit 250.28 A grounding electrode conductor with multiple sets of service conductors, sized according to Table 250.66, Note 1.

(A) Connections to Rod, Pipe, or Plate Electrodes Where the grounding electrode conductor is connected to rod, pipe, or plate electrodes as permitted in 250.52(A)(5) or (A)(6), that portion of the conductor that is the sole connection to the grounding electrode shall not be required to be larger than 6 AWG copper wire or 4 AWG aluminum wire.

(B) Connections to Concrete-Encased Electrodes Where the grounding electrode conductor is connected to a concrete-encased electrode as permitted in 250.52(A)(3), that portion of the conductor that is the sole connection to the grounding electrode shall not be required to be larger than 4 AWG copper wire.

(C) Connections to Ground Rings Where the grounding electrode conductor is connected to a ground ring as permitted in 250.52(A)(4), that portion of the conductor that is the sole connection to the grounding electrode shall not be required to be larger than the conductor used for the ground ring.

As illustrated in Exhibit 250.29, where a grounding electrode conductor is run from the service equipment or separately derived system to a water pipe or structural metal building member and from that point to one of the electrodes mentioned in 250.66(A), that portion of the grounding electrode between the service equipment or separately derived system and the water pipe or structural metal building member must be a full-size conductor, per Table 250.66. If the grounding electrode conductor from the service equipment was run, for example, to the ground rod first and then to the water pipe, the conductor to the ground rod would also have to be full size, per Table 250.66. Note that Exhibit 250.29 is not intended to show the physical routing and connection of the bonding jumpers. The sizes for the bonding jumpers to the ground rod and the concrete-encased electrode shown in Exhibit 250.29 are the maximum sizes required by the *Code*. The use of bonding jumpers or grounding electrode

conductors larger than required by 250.66 is certainly not prohibited.

250.68 Grounding Electrode Conductor and Bonding Jumper Connection to Grounding Electrodes

(A) Accessibility The connection of a grounding electrode conductor or bonding jumper to a grounding electrode shall be accessible.

Exception No. 1: An encased or buried connection to a concrete-encased, driven, or buried grounding electrode shall not be required to be accessible.

Exception No. 2: An exothermic or irreversible compression connection to fire-proofed structural metal shall not be required to be accessible.

Where the exposed portion of an encased, driven, or buried electrode is used for the termination of a grounding electrode conductor, the terminations must be accessible. However, if the connection is buried or encased, terminations are not required to be accessible. Ground clamps and other connectors suitable for use where buried in earth or embedded in concrete must be listed for such use, either by a marking on the connector or by a tag attached to the connector. See Exhibit 250.22 and Exhibit 250.24 for illustrations of encased and buried electrodes. For the 2005 *Code*, an exception has been added to permit connections to fireproofed structural steel to be encapsulated by the fireproofing material. Because these connections are not required to be accessible for inspection, the method of connection to the structural member must be either an exothermic weld or an irreversible compression connector. This new exception recognizes the importance of maintaining the integrity of the structural fireproofing.

(B) Effective Grounding Path The connection of a grounding electrode conductor or bonding jumper to a grounding electrode shall be made in a manner that will ensure a permanent and effective grounding path. Where necessary to ensure the grounding path for a metal piping system used as a grounding electrode, effective bonding shall be provided around insulated joints and around any equipment likely to be disconnected for repairs or replacement. Bonding conductors shall be of sufficient length to permit removal of such equipment while retaining the integrity of the bond.

Examples of equipment likely to be disconnected for repairs or replacement are water meters and water filter systems.

Exhibit 250.29 Grounding electrode conductor and bonding jumpers sized in accordance with 250.66 for a service supplied by 3/0 AWG ungrounded conductors.

250.70 Methods of Grounding and Bonding Conductor Connection to Electrodes

The grounding or bonding conductor shall be connected to the grounding electrode by exothermic welding, listed lugs, listed pressure connectors, listed clamps, or other listed means. Connections depending on solder shall not be used. Ground clamps shall be listed for the materials of the grounding electrode and the grounding electrode conductor and, where used on pipe, rod, or other buried electrodes, shall also be listed for direct soil burial or concrete encasement. Not more than one conductor shall be connected to the grounding electrode by a single clamp or fitting unless the clamp or fitting is listed for multiple conductors. One of the following methods shall be used:

(1) A pipe fitting, pipe plug, or other approved device screwed into a pipe or pipe fitting

(2) A listed bolted clamp of cast bronze or brass, or plain or malleable iron

(3) For indoor telecommunications purposes only, a listed sheet metal strap-type ground clamp having a rigid metal base that seats on the electrode and having a strap of such material and dimensions that it is not likely to stretch during or after installation

(4) An equally substantial approved means

Where a ground clamp is used and terminates, for example, on a galvanized water pipe, the clamp must be of a material that is compatible with steel, to prevent galvanic corrosion. The same type of compatibility requirement applies to ground clamps on copper water pipe.

Exhibit 250.30 shows a listed ground clamp generally used with 8 AWG through 4 AWG grounding electrode conductors. Exothermic weld kits acceptable for this purpose are commercially available.

Exhibit 250.31 shows a listed U-bolt ground clamp. These clamps are available for all pipe sizes and all grounding

Exhibit 250.31 An application of a listed U-bolt ground clamp.

electrode conductor sizes. Where grounding electrode conductors are run in conduit, conduit hubs may be bolted to the threaded portion of the U-bolt.

IV. Enclosure, Raceway, and Service Cable Grounding

250.80 Service Raceways and Enclosures

Metal enclosures and raceways for service conductors and equipment shall be grounded.

Exception: A metal elbow that is installed in an underground installation of rigid nonmetallic conduit and is isolated from possible contact by a minimum cover of 450 mm (18 in.) to any part of the elbow shall not be required to be grounded.

The exception to 250.80 recognizes that metal sweep elbows are often installed in underground installations of rigid nonmetallic conduit. The metal elbows are installed because nonmetallic elbows can be damaged by friction from the pulling ropes used during conductor installation. The elbows are isolated from physical contact by burying the entire elbow at a depth not less than 18 in. below grade.

250.84 Underground Service Cable or Raceway

(A) Underground Service Cable The sheath or armor of a continuous underground metal-sheathed or armored service cable system that is bonded to the grounded underground system shall not be required to be grounded at the building or structure. The sheath or armor shall be permitted to be insulated from the interior metal raceway conduit or piping.

(B) Underground Service Raceway Containing Cable An underground metal service raceway that contains a metal-

Exhibit 250.30 An application of a listed ground clamp.

sheathed or armored cable bonded to the grounded underground system shall not be required to be grounded at the building or structure. The sheath or armor shall be permitted to be insulated from the interior metal raceway or piping.

250.86 Other Conductor Enclosures and Raceways

Except as permitted by 250.112(I), metal enclosures and raceways for other than service conductors shall be grounded.

Section 250.86 requires grounding, bonding, and ensured electrical continuity of all enclosures and metal raceways. Connectors, couplings, or other similar fittings that perform mechanical and electrical functions must ensure bonding and grounding continuity between the fitting, the metal raceway, and the enclosure. Metal enclosures must be grounded so that when a fault occurs between an ungrounded (hot) conductor and ground, the potential difference between the non–current-carrying parts of the electrical installation is minimized, thereby reducing the risk of shock.

Exception No. 1: Metal enclosures and raceways for conductors added to existing installations of open wire, knob and tube wiring, and nonmetallic-sheathed cable shall not be required to be grounded where these enclosures or wiring methods comply with (1) through (4) as follows:

(1) Do not provide an equipment ground

(2) Are in runs of less than 7.5 m (25 ft)

(3) Are free from probable contact with ground, grounded metal, metal lath, or other conductive material

(4) Are guarded against contact by persons

Exception No. 2: Short sections of metal enclosures or raceways used to provide support or protection of cable assemblies from physical damage shall not be required to be grounded.

Exception No. 3: A metal elbow shall not be required to be grounded where it is installed in a nonmetallic raceway and is isolated from possible contact by a minimum cover of 450 mm (18 in.) to any part of the elbow or is encased in not less than 50 mm (2 in.) of concrete.

V. Bonding

250.90 General

Bonding shall be provided where necessary to ensure electrical continuity and the capacity to conduct safely any fault current likely to be imposed.

250.92 Services

(A) Bonding of Services The non–current-carrying metal parts of equipment indicated in 250.92(A)(1), (A)(2), and (A)(3) shall be effectively bonded together.

(1) The service raceways, cable trays, cablebus framework, auxiliary gutters, or service cable armor or sheath except as permitted in 250.84.

(2) All service enclosures containing service conductors, including meter fittings, boxes, or the like, interposed in the service raceway or armor.

(3) Any metallic raceway or armor enclosing a grounding electrode conductor as specified in 250.64(B). Bonding shall apply at each end and to all intervening raceways, boxes, and enclosures between the service equipment and the grounding electrode.

Section 250.92(A)(3) is intended to clarify that where metal raceways, boxes, or enclosures contain a grounding electrode conductor, both ends of the raceway, box, or enclosure must be bonded to the grounding electrode conductor, as illustrated in Exhibit 250.32. Bonding the raceway to the conductor reduces the impedance and minimizes the potential difference between the electrical equipment and ground. It should be noted that a change in 250.64(E) for the 2005 *Code* requires bonding of only ferrous metal enclosures that contain a grounding electrode conductor. See also 250.64(E) and 250.102(A) for requirements covering the installation of protective enclosures for grounding electrode conductors and for materials permitted as bonding jumpers.

Exhibit 250.32 Bonding of a metal raceway that contains a grounding electrode conductor to the conductor at both ends, as required by 250.64(E).

(B) Method of Bonding at the Service Electrical continuity at service equipment, service raceways, and service conductor enclosures shall be ensured by one of the following methods:

(1) Bonding equipment to the grounded service conductor in a manner provided in 250.8

Exhibit 250.33 illustrates grounding and bonding at an individual service. Exhibit 250.34 illustrates a grounding and bonding arrangement for up to six switches (three are shown) that serve as the service disconnecting means for an individual service. Section 250.24(C) clarifies that the grounded service conductor must be run to each service disconnecting means and be bonded to the disconnecting means enclosure. Section 250.92(B)(1) permits the bonding of service equipment enclosures to be accomplished by bonding the grounded service conductor to the enclosure.

Exhibit 250.34 A grounding and bonding arrangement for multiple switches that serve as the service disconnecting means for an individual service.

Note that method (4) in 250.92(B) requires other similar devices, such as listed bonding-type locknuts or bushings. Standard locknuts or sealing locknuts are not acceptable as the "sole means" for bonding on the line side of service equipment.

Grounding and bonding bushings for use with rigid or intermediate metal conduit are provided with means (usually one or more set screws that make positive contact with the conduit) for reliably bonding the bushing and the conduit on which it is threaded to the metal equipment enclosure or box.

Grounding bushings used with rigid or intermediate metal conduit or with tubing (EMT) fittings, such as those shown in Exhibits 250.35 and 250.36, have provisions for connecting a bonding jumper or have means provided by the manufacturer for use in mounting a wire connector. This type of bushing may also have means (usually one or more set screws) to reliably bond the bushing to the conduit. Exhibit 250.37 shows a bonding-type wedge lug used to connect a conduit to a box.

Exhibit 250.33 Grounding and bonding for a service with one disconnecting means.

(2) Connections utilizing threaded couplings or threaded bosses on enclosures where made up wrenchtight
(3) Threadless couplings and connectors where made up tight for metal raceways and metal-clad cables
(4) Other listed devices, such as bonding-type locknuts, bushings, or bushings with bonding jumpers

Bonding jumpers meeting the other requirements of this article shall be used around concentric or eccentric knockouts that are punched or otherwise formed so as to impair the electrical connection to ground. Standard locknuts or bushings shall not be the sole means for the bonding required by this section.

For an example of concentric and eccentric knockouts, see the commentary following the definition of *bonding jumper* in Article 100 and Exhibit 100.3.

Exhibit 250.35 Grounding bushings used to connect a copper bonding or grounding wire to conduits. (Courtesy of Thomas & Betts Corp.)

Exhibit 250.37 A grounding wedge lug used to provide an electrical connection between a conduit and a box. (Courtesy of Thomas & Betts Corp.)

FPN No. 1: A 6 AWG copper conductor with one end bonded to the grounded nonflexible metallic raceway or equipment and with 150 mm (6 in.) or more of the other end made accessible on the outside wall is an example of the approved means covered in 250.94(3).

Other accessible external means for intersystem bonding that comply with 250.94, FPN No. 1, are illustrated in Exhibit 250.38. On the left is an illustration of accessible means for the connection. The illustration on the right shows a method of providing the required bonding means when the panelboard is a flush type.

FPN No. 2: See 800.100, 810.21, and 820.100 for bonding and grounding requirements for communications circuits, radio and television equipment, and CATV circuits.

An external accessible bonding means is equally important for separate buildings and mobile homes. In these occupancies, the disconnecting means enclosure on the load side of the service can be considered the equivalent of the service equipment for the purpose of intersystem bonding.

The *Code* requires that separate systems be bonded together to reduce the differences of potential between them due to lightning or accidental contact with power lines. Lightning protection systems, communications, radio and TV, and CATV systems must be bonded together to minimize the potential differences between the systems.

Lack of interconnection can result in a severe shock

Exhibit 250.36 A threaded grounding bushing with set screws used to ensure electrical and mechanical connection and a terminal for connection of a grounding conductor or bonding jumper. (Courtesy of Thomas & Betts Corp.)

250.94 Bonding for Other Systems

An accessible means external to enclosures for connecting intersystem bonding and grounding electrode conductors shall be provided at the service equipment and at the disconnecting means for any additional buildings or structures by at least one of the following means:

(1) Exposed nonflexible metallic raceways
(2) Exposed grounding electrode conductor
(3) Approved means for the external connection of a copper or other corrosion-resistant bonding or grounding conductor to the grounded raceway or equipment

Exhibit 250.38 Examples of accessible external means for intersystem bonding, as required by 250.94 for service equipment and building or structure disconnecting means.

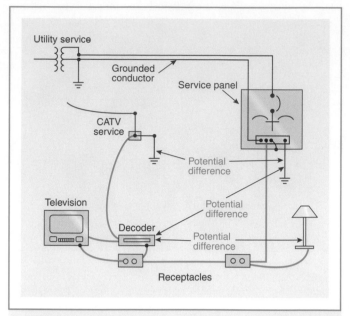

Exhibit 250.39 A CATV installation that does not comply with the *Code*, illustrating why bonding between different systems is necessary.

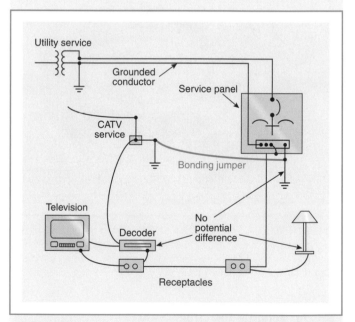

Exhibit 250.40 A cable TV installation that complies with 250.94.

and fire hazard. The reason for this potential hazard is illustrated in Exhibit 250.39, which shows a CATV cable with its jacket grounded to a separate ground rod and not bonded to the power ground. The cable is connected to the cable decoder and the tuner of a television set. Also connected to the decoder and the television is the 120-volt supply, with one conductor grounded at the service (the power ground). In each case, resistance to ground is present at the grounding electrode. This resistance to ground varies widely, depending on soil conditions and the type of grounding electrode. The resistance at the CATV ground is likely to be higher than the power ground resistance, because the power ground is often an underground metal water piping system or concrete-encased electrode, whereas the CATV ground is commonly a ground rod.

For example, for the CATV installation shown in Exhibit 250.39, assume that a current is induced in the power line by a switching surge or a nearby lightning strike, so that a momentary current of 1000 amperes occurs over the power line to the power line ground. This amount of current is not unusual under such circumstances — the amount could be, and often is, considerably higher. Also assume that the power ground has a resistance of 10 ohms, a very low value in most circumstances (a single ground rod in average soil has a resistance to ground in the neighborhood of 40 ohms).

According to Ohm's law, the current through the equipment connected to the electrical system will be raised momentarily to a potential of 10,000 volts (1000 amperes × 10 ohms). This potential of 10,000 volts would exist between the CATV system and the electrical system and between the grounded conductor within the CATV cable and the grounded surfaces in the walls of the home, such as water pipes (which are connected to the power ground), over which the cable runs. This potential could also appear across a person with one hand on the CATV cable and the other hand on a metal surface connected to the power ground (e.g., a radiator or a refrigerator).

Actual voltage is likely to be many times the 10,000 volts calculated, because extremely low (below normal) values were assumed for both resistance to ground and current. Most insulation systems, however, are not designed to withstand even 10,000 volts. Even if the insulation system does withstand a 10,000-volt surge, it is likely to be damaged, and breakdown of the insulation system will result in sparking.

The same situation would exist if the current surge were on the CATV cable or a telephone line. The only difference would be the voltage involved, which would depend on the individual resistance to ground of the grounding electrodes.

The solution is to bond the two grounding electrode systems together, as shown in Exhibit 250.40, or to connect the CATV cable jacket to the power ground, which is exactly what the *Code* requires. When one system is raised above ground potential, the second system rises to the same potential, and no voltage exists between the two grounding systems.

These bonding rules are provided to address the difficulties that communications and CATV installers encounter in complying with *Code* grounding and bonding requirements. These difficulties arise from the increasing use of plastic for water pipe, fittings, water meters, and service conduit. In the past, bonding between communications, CATV, and power systems was usually achieved by connecting the communications protector grounds or cable shield to an interior metallic water pipe, because the pipe was often used as the power grounding electrode. Thus, the requirement that the power, communications, CATV cable shield, and metallic water piping systems be bonded together was easily satisfied. If the power was grounded to one of the other electrodes permitted by the *Code*, usually by a made electrode such as a ground rod, the bond was connected to the power grounding electrode conductor or to a metallic service raceway, since at least one of these was usually accessible.

With the proliferation of plastic water pipe and the increasing tendency for service equipment (often flush-mounted) to be installed in finished areas, where the grounding electrode conductor is often concealed, as well as the increased use of plastic service-entrance conduit, communications and CATV installers no longer have access to a point for connecting bonding jumpers or grounding conductors. See Exhibit 250.39 and also the commentary following 820.100(D), FPN No. 2.

250.96 Bonding Other Enclosures

(A) General Metal raceways, cable trays, cable armor, cable sheath, enclosures, frames, fittings, and other metal non–current-carrying parts that are to serve as grounding conductors, with or without the use of supplementary equipment grounding conductors, shall be effectively bonded where necessary to ensure electrical continuity and the capacity to conduct safely any fault current likely to be imposed on them. Any nonconductive paint, enamel, or similar coating shall be removed at threads, contact points, and contact surfaces or be connected by means of fittings designed so as to make such removal unnecessary.

(B) Isolated Grounding Circuits Where required for the reduction of electrical noise (electromagnetic interference) on the grounding circuit, an equipment enclosure supplied by a branch circuit shall be permitted to be isolated from a raceway containing circuits supplying only that equipment by one or more listed nonmetallic raceway fittings located at the point of attachment of the raceway to the equipment enclosure. The metal raceway shall comply with provisions of this article and shall be supplemented by an internal insulated equipment grounding conductor installed in accordance with 250.146(D) to ground the equipment enclosure.

> FPN: Use of an isolated equipment grounding conductor does not relieve the requirement for grounding the raceway system.

To reduce electromagnetic interference, 250.96(B) permits electronic equipment to be isolated from the raceway in a manner similar to that for cord-and-plug-connected equipment. Section 250.96(B) specifies that a metal equipment enclosure supplied by a branch circuit is the subject of the requirement and that subsequent wiring, raceways, or other equipment beyond the insulating fitting is not permitted.

Exhibits 250.41 and 250.42 show examples of installations. In Exhibit 250.41, note that the metal raceway is grounded in the usual manner, by attachment to the grounded service enclosure, satisfying the concern mentioned in the FPN to 250.96(B). In Exhibit 250.42, note that 408.40, Exception, permits, but does not require, the isolated equipment grounding conductor (which is required to be insulated) to pass through the subpanel and run back to the service equipment. The key to this method of grounding electronic equipment is to always ensure that the insulated equipment grounding conductor, regardless of where it terminates in the distribution system, is connected in a manner that creates an effective path for ground-fault current, as required by 250.4(A)(5).

250.97 Bonding for Over 250 Volts

For circuits of over 250 volts to ground, the electrical continuity of metal raceways and cables with metal sheaths that contain any conductor other than service conductors shall be ensured by one or more of the methods specified for services in 250.92(B), except for (B)(1).

Exhibit 250.41 An installation in which the electronic equipment is grounded through the isolated equipment grounding conductor.

Exhibit 250.42 An installation in which the isolated equipment grounding conductor is allowed to pass through the subpanel without connecting to the grounding bus to terminate at the service grounding bus.

Exception: Where oversized, concentric, or eccentric knockouts are not encountered, or where a box or enclosure with concentric or eccentric knockouts is listed to provide a per-

manent, reliable electrical bond, the following methods shall be permitted:

(1) Threadless couplings and connectors for cables with metal sheaths
(2) Two locknuts, on rigid metal conduit or intermediate metal conduit, one inside and one outside of boxes and cabinets
(3) Fittings with shoulders that seat firmly against the box or cabinet, such as electrical metallic tubing connectors, flexible metal conduit connectors, and cable connectors, with one locknut on the inside of boxes and cabinets
(4) Listed fittings

Bonding around prepunched concentric or eccentric knock-outs is not required if the enclosure containing the knockouts has been tested and is listed as suitable for bonding. Guide card information from the UL *General Information for Electrical Equipment Directory* (the UL "White Book") indicates that concentric and eccentric knockouts of all metallic outlet boxes evaluated in accordance with UL 514A, *Metallic Outlet Boxes,* are suitable for bonding in circuits of above or below 250 volts to ground without the use of additional bonding equipment. Metallic outlet boxes are permitted, but not required, to be marked to indicate this condition of use.

The methods in items (1), (2), (3), and (4) in the exception to 250.97 are permitted for circuits over 250 volts to ground only where there are no oversize, concentric, or eccentric knockouts. Note that method (3) permits fittings, such as EMT connectors, cable connectors, and similar fittings with shoulders that seat firmly against the metal of a box or cabinet, to be installed with only one locknut on the inside of the box.

250.98 Bonding Loosely Jointed Metal Raceways

Expansion fittings and telescoping sections of metal raceways shall be made electrically continuous by equipment bonding jumpers or other means.

250.100 Bonding in Hazardous (Classified) Locations

Regardless of the voltage of the electrical system, the electrical continuity of non–current-carrying metal parts of equipment, raceways, and other enclosures in any hazardous (classified) location as defined in Article 500 shall be ensured by any of the methods specified in 250.92(B)(2) through (B)(4) that are approved for the wiring method used. One or more of these bonding methods shall be used whether or not supplementary equipment grounding conductors are installed.

250.102 Equipment Bonding Jumpers

(A) Material Equipment bonding jumpers shall be of copper or other corrosion-resistant material. A bonding jumper shall be a wire, bus, screw, or similar suitable conductor.

(B) Attachment Equipment bonding jumpers shall be attached in the manner specified by the applicable provisions of 250.8 for circuits and equipment and by 250.70 for grounding electrodes.

(C) Size — Equipment Bonding Jumper on Supply Side of Service The bonding jumper shall not be smaller than the sizes shown in Table 250.66 for grounding electrode conductors. Where the service-entrance phase conductors are larger than 1100 kcmil copper or 1750 kcmil aluminum, the bonding jumper shall have an area not less than 12½ percent of the area of the largest phase conductor except that, where the phase conductors and the bonding jumper are of different materials (copper or aluminum), the minimum size of the bonding jumper shall be based on the assumed use of phase conductors of the same material as the bonding jumper and with an ampacity equivalent to that of the installed phase conductors. Where the service-entrance conductors are paralleled in two or more raceways or cables, the equipment bonding jumper, where routed with the raceways or cables, shall be run in parallel. The size of the bonding jumper for each raceway or cable shall be based on the size of the service-entrance conductors in each raceway or cable.

(D) Size — Equipment Bonding Jumper on Load Side of Service The equipment bonding jumper on the load side of the service overcurrent devices shall be sized, as a minimum, in accordance with the sizes listed in Table 250.122, but shall not be required to be larger than the largest ungrounded circuit conductors supplying the equipment and shall not be smaller than 14 AWG.

A single common continuous equipment bonding jumper shall be permitted to bond two or more raceways or cables where the bonding jumper is sized in accordance with Table 250.122 for the largest overcurrent device supplying circuits therein.

(E) Installation The equipment bonding jumper shall be permitted to be installed inside or outside of a raceway or enclosure. Where installed on the outside, the length of the equipment bonding jumper shall not exceed 1.8 m (6 ft) and shall be routed with the raceway or enclosure. Where installed inside of a raceway, the equipment bonding jumper shall comply with the requirements of 250.119 and 250.148.

Exception: An equipment bonding jumper longer than 1.8 m (6 ft) shall be permitted at outside pole locations for the purpose of bonding or grounding isolated sections of metal raceways or elbows installed in exposed risers of metal conduit or other metal raceway.

In many applications, equipment bonding jumpers must be installed on the outside of metal raceways and enclosures. For example, it would be impractical to install the bonding jumper for a conduit expansion joint on the inside of the conduit. For some metal raceway and rigid conduit systems and conduit systems in hazardous (classified) locations, installing the bonding jumper where it is visible and accessible for inspection and maintenance is desirable. An external bonding jumper has a higher impedance than an internal bonding jumper, but by limiting the length of the bonding jumper to 6 ft and routing it with the raceway, the increase in the impedance of the equipment grounding circuit is insignificant. Exhibit 250.43 illustrates a bonding jumper run outside a length of flexible metal conduit. Because the function of a bonding jumper is readily apparent, color identification is permitted, but not required.

Exhibit 250.43 A bonding jumper around the outside of a flexible metal conduit.

250.104 Bonding of Piping Systems and Exposed Structural Steel

(A) Metal Water Piping The metal water piping system shall be bonded as required in (A)(1), (A)(2), or (A)(3) of this section. The bonding jumper(s) shall be installed in accordance with 250.64(A), (B), and (E). The points of attachment of the bonding jumper(s) shall be accessible.

(1) General Metal water piping system(s) installed in or attached to a building or structure shall be bonded to the service equipment enclosure, the grounded conductor at the service, the grounding electrode conductor where of sufficient size, or to the one or more grounding electrodes used. The bonding jumper(s) shall be sized in accordance with Table 250.66 except as permitted in 250.104(A)(2) and (A)(3).

Bonding the metal water piping system of a building or structure is not the same as using the metal water piping system as a grounding electrode. Bonding to the grounding electrode system places the bonded components at the same voltage level. For example, a current of 2000 amperes across 25 ft of 6 AWG copper conductor produces a voltage differential of approximately 26 volts. Sections 250.104(A)(1) and 250.104(A)(3) require the metal water piping system of a building or structure to be bonded to the service equipment or grounding electrode conductor or, where supplied by a feeder or branch circuit, to the building or structure disconnecting means or grounding electrode conductor. Information concerning bonding provisions for buildings with multiple occupancies and isolated metal water piping systems is contained in the commentary for 250.104(A)(2).

In those cases where it cannot be reasonably concluded that the hot and cold water pipes are reliably bonded through mechanical connections, an electrical bonding jumper is required to ensure that this connection is made. Some judgment must be exercised for each installation. The special installation requirements provided in 250.64(A), 250.64(B), and 250.64(E) also apply to the water piping bonding jumper.

(2) Buildings of Multiple Occupancy In buildings of multiple occupancy where the metal water piping system(s) installed in or attached to a building or structure for the individual occupancies is metallically isolated from all other occupancies by use of nonmetallic water piping, the metal water piping system(s) for each occupancy shall be permitted to be bonded to the equipment grounding terminal of the panelboard or switchboard enclosure (other than service equipment) supplying that occupancy. The bonding jumper shall be sized in accordance with Table 250.122.

Section 250.104(A)(2) recognizes that the increased use of nonmetallic water piping mains can result in the interior metal piping system of a multiple-occupancy building to be isolated from ground and from the other occupancies. Therefore, the water pipe is permitted to be bonded to the panelboard or switchboard that serves only that particular occupancy. The bonding jumper, in this case, is permitted to be sized according to Table 250.122, based on the size of the main overcurrent device supplying the occupancy.

(3) Multiple Buildings or Structures Supplied by a Feeder(s) or Branch Circuit(s) The metal water piping sys-

tem(s) installed in or attached to a building or structure shall be bonded to the building or structure disconnecting means enclosure where located at the building or structure, to the equipment grounding conductor run with the supply conductors, or to the one or more grounding electrodes used. The bonding jumper(s) shall be sized in accordance with 250.66, based on the size of the feeder or branch circuit conductors that supply the building. The bonding jumper shall not be required to be larger than the largest ungrounded feeder or branch circuit conductor supplying the building.

(B) Other Metal Piping Where installed in or attached to a building or structure, metal piping system(s), including gas piping, that is likely to become energized shall be bonded to the service equipment enclosure, the grounded conductor at the service, the grounding electrode conductor where of sufficient size, or to the one or more grounding electrodes used. The bonding jumper(s) shall be sized in accordance with 250.122, using the rating of the circuit that is likely to energize the piping system(s). The equipment grounding conductor for the circuit that is likely to energize the piping shall be permitted to serve as the bonding means. The points of attachment of the bonding jumper(s) shall be accessible.

> FPN: Bonding all piping and metal air ducts within the premises will provide additional safety.

Unlike the metal piping systems covered in 250.104(A), this requirement applies only to metal piping systems that are likely to become energized. What this means is that where metal piping systems and electrical circuits interface through mechanical and electrical connections within equipment, a failure of electrical insulation can result in the connected piping system(s) becoming energized. Gas appliances are a common example of metal gas piping and electrical circuits being connected to a common piece of equipment, and in this case the 250.104(B) requirements apply. The required bonding of these other piping systems can occur at the same locations specified in 250.104(A), or an additional provision within this paragraph permits the equipment grounding conductor of the circuit that is likely to energize the piping as the means for bonding the piping. Typically, the use of an additional bonding jumper is not necessary to comply with this requirement because the equipment grounding connection to the non–current-carrying metal parts of the appliance also provides a bonding connection to the metal piping attached to the appliance. This is a bonding requirement, and the other piping is not being used as an electrode. Therefore, this requirement does not conflict with 250.52(B)(1), which prohibits the use of metal underground gas piping as a grounding electrode for electrical services or other sources of supply.

(C) Structural Metal Exposed structural metal that is interconnected to form a metal building frame and is not intentionally grounded and is likely to become energized shall be bonded to the service equipment enclosure, the grounded conductor at the service, the grounding electrode conductor where of sufficient size, or the one or more grounding electrodes used. The bonding jumper(s) shall be sized in accordance with Table 250.66 and installed in accordance with 250.64(A), (B), and (E). The points of attachment of the bonding jumper(s) shall be accessible.

Section 250.104(C) requires exposed metal building framework that is not intentionally or inherently grounded to be bonded to the service equipment or grounding electrode system. Revised for the 2005 *Code*, this requirement applies to all metal framework, not only to steel framework.

(D) Separately Derived Systems Metal water piping systems and structural metal that is interconnected to form a building frame shall be bonded to separately derived systems in accordance with (D)(1) through (D)(3).

(1) Metal Water Piping System(s) The grounded conductor of each separately derived system shall be bonded to the nearest available point of the metal water piping system(s) in the area served by each separately derived system. This connection shall be made at the same point on the separately derived system where the grounding electrode conductor is connected. Each bonding jumper shall be sized in accordance with Table 250.66 based on the largest ungrounded conductor of the separately derived system.

Exception No. 1: A separate bonding jumper to the metal water piping system shall not be required where the metal water piping system is used as the grounding electrode for the separately derived system.

Exception No. 2: A separate water piping bonding jumper shall not be required where the metal frame of a building or structure is used as the grounding electrode for a separately derived system and is bonded to the metal water piping in the area served by the separately derived system.

(2) Structural Metal Where exposed structural metal that is interconnected to form the building frame exists in the area served by the separately derived system, it shall be bonded to the grounded conductor of each separately derived system. This connection shall be made at the same point on the separately derived system where the grounding electrode conductor is connected. Each bonding jumper shall be sized in accordance with Table 250.66 based on the largest ungrounded conductor of the separately derived system.

Exception No. 1: A separate bonding jumper to the building structural metal shall not be required where the metal frame of a building or structure is used as the grounding electrode for the separately derived system.

Exception No. 2: A separate bonding jumper to the building structural metal shall not be required where the water piping of a building or structure is used as the grounding electrode for a separately derived system and is bonded to the building structural metal in the area served by the separately derived system.

Section 250.104(D) requires that where a separately derived system supplies the power, the metal piping system and the exposed structural metal in the area supplied by the separately derived system must be bonded to the grounded conductor at the point nearest the derived system and that this connection must be accessible. Where either of these two building elements is used as the grounding electrode for the separately derived system, it is not necessary to provide an additional bonding jumper.

In addition, two new exceptions for the 2005 *Code* permit the following approaches for bonding of metal piping or metal structures to separately derived systems. Where the building metal structure is used as the grounding electrode for a separately derived system, it is permitted to install a bonding jumper between the metal structure and the water piping, thus eliminating the need to run a separate bonding jumper from the separately derived system source or distribution equipment to the water piping. The same approach can be taken for the structural metal where metal water piping is serving as the electrode for the separately derived system and a bonding jumper is installed from the piping to the metal framework in the area served by that system. Any bonding jumper used for this application is sized from 250.66 based on the largest ungrounded supply conductor of the separately derived system.

(3) Common Grounding Electrode Conductor Where a common grounding electrode conductor is installed for multiple separately derived systems as permitted by 250.30(A)(4), and exposed structural metal that is interconnected to form the building frame or interior metal piping exists in the area served by the separately derived system, the metal piping and the structural metal member shall be bonded to the common grounding electrode conductor.

Exception: A separate bonding jumper from each derived system to metal water piping and to structural metal members shall not be required where the metal water piping and the structural metal members in the area served by the separately derived system are bonded to the common grounding electrode conductor.

250.106 Lightning Protection Systems
The lightning protection system ground terminals shall be bonded to the building or structure grounding electrode system.

FPN No. 1: See 250.60 for use of air terminals. For further information, see NFPA 780-2004, *Standard for the Installation of Lightning Protection Systems*, which contains detailed information on grounding, bonding, and spacing from lightning protection systems.

FPN No. 2: Metal raceways, enclosures, frames, and other non–current-carrying metal parts of electric equipment installed on a building equipped with a lightning protection system may require bonding or spacing from the lightning protection conductors in accordance with NFPA 780-2004, *Standard for the Installation of Lightning Protection Systems*. Separation from lightning protection conductors is typically 1.8 m (6 ft) through air or 900 mm (3 ft) through dense materials such as concrete, brick, or wood.

Section 250.106 specifies that the grounding electrode system of the lightning protection system be bonded to the electrical service grounding electrode system, as shown in Exhibit 250.44. A similar requirement is found in 4.14 of NFPA 780, *Standard for the Installation of Lightning Protection Systems*. Additional bonding between the lightning protection system and the electrical system may be necessary based on proximity and whether separation between the systems is through air or building materials.

FPN No. 2 to 250.106 references NFPA 780 for guidance on determining the need for additional bonding connections. Section 4.21.2 of NFPA 780 includes a method for calculating flashover distances.

Exposed, non–current-carrying metal parts of fixed equipment that are not likely to become energized are not required to be grounded. These parts include some metal nameplates on nonmetallic enclosures and small parts, such

Exhibit 250.44 Bonding between the lightning system ground terminals and the electrical service grounding electrode system, in accordance with 250.106.

as bolts and screws, if they are located so that they are not likely to become energized.

VI. Equipment Grounding and Equipment Grounding Conductors

250.110 Equipment Fastened in Place or Connected by Permanent Wiring Methods (Fixed)

Exposed non–current-carrying metal parts of fixed equipment likely to become energized shall be grounded under any of the following conditions:

(1) Where within 2.5 m (8 ft) vertically or 1.5 m (5 ft) horizontally of ground or grounded metal objects and subject to contact by persons
(2) Where located in a wet or damp location and not isolated
(3) Where in electrical contact with metal
(4) Where in a hazardous (classified) location as covered by Articles 500 through 517
(5) Where supplied by a metal-clad, metal-sheathed, metal-raceway, or other wiring method that provides an equipment ground, except as permitted by 250.86, Exception No. 2, for short sections of metal enclosures
(6) Where equipment operates with any terminal at over 150 volts to ground

Exception No. 1: Metal frames of electrically heated appliances, exempted by special permission, in which case the frames shall be permanently and effectively insulated from ground.

Exception No. 2: Distribution apparatus, such as transformer and capacitor cases, mounted on wooden poles, at a height exceeding 2.5 m (8 ft) above ground or grade level.

Exception No. 3: Listed equipment protected by a system of double insulation, or its equivalent, shall not be required to be grounded. Where such a system is employed, the equipment shall be distinctively marked.

250.112 Fastened in Place or Connected by Permanent Wiring Methods (Fixed) — Specific

Exposed, non–current-carrying metal parts of the kinds of equipment described in 250.112(A) through (K), and non–current-carrying metal parts of equipment and enclosures described in 250.112(L) and (M), shall be grounded regardless of voltage.

(A) Switchboard Frames and Structures Switchboard frames and structures supporting switching equipment, except frames of 2-wire dc switchboards where effectively insulated from ground.

Section 250.112(A) clarifies that dc switchboards insulated from ground are not required to be grounded.

(B) Pipe Organs Generator and motor frames in an electrically operated pipe organ, unless effectively insulated from ground and the motor driving it.

(C) Motor Frames Motor frames, as provided by 430.242.

(D) Enclosures for Motor Controllers Enclosures for motor controllers unless attached to ungrounded portable equipment.

(E) Elevators and Cranes Electric equipment for elevators and cranes.

(F) Garages, Theaters, and Motion Picture Studios Electric equipment in commercial garages, theaters, and motion picture studios, except pendant lampholders supplied by circuits not over 150 volts to ground.

(G) Electric Signs Electric signs, outline lighting, and associated equipment as provided in Article 600.

(H) Motion Picture Projection Equipment Motion picture projection equipment.

(I) Power-Limited Remote-Control, Signaling, and Fire Alarm Circuits Equipment supplied by Class 1 power-limited circuits and Class 1, Class 2, and Class 3 remote-control and signaling circuits, and by fire alarm circuits, shall be grounded where system grounding is required by Part II or Part VIII of this article.

(J) Luminaires (Lighting Fixtures) Luminaires (lighting fixtures) as provided in Part V of Article 410.

(K) Skid Mounted Equipment Permanently mounted electrical equipment and skids shall be grounded with an equipment bonding jumper sized as required by 250.122.

(L) Motor-Operated Water Pumps Motor-operated water pumps, including the submersible type.

The requirement of 250.112(L) is intended to reduce stray voltages and minimize shock hazard during maintenance, when the pump is hauled out of the well casing and might be tested in a barrel of water.

(M) Metal Well Casings Where a submersible pump is used in a metal well casing, the well casing shall be bonded to the pump circuit equipment grounding conductor.

Section 250.112(M) is intended to prevent a shock hazard that could exist due to a potential difference between the pump, which is grounded to the system ground, and the metal well casing.

250.114 Equipment Connected by Cord and Plug

Under any of the conditions described in 250.114(1) through (4), exposed non–current-carrying metal parts of cord-and-plug-connected equipment likely to become energized shall be grounded.

Exception: Listed tools, listed appliances, and listed equipment covered in 250.114(2) through (4) shall not be required to be grounded where protected by a system of double insulation or its equivalent. Double insulated equipment shall be distinctively marked.

The exception to 250.114 recognizes listed double-insulated appliances, motor-operated hand-held tools, stationary and fixed motor-operated tools, and light industrial motor-operated tools as not requiring equipment grounding connections.

(1) In hazardous (classified) locations (see Articles 500 through 517)
(2) Where operated at over 150 volts to ground

Exception No. 1: Motors, where guarded, shall not be required to be grounded.

Exception No. 2: Metal frames of electrically heated appliances, exempted by special permission, shall not be required to be grounded, in which case the frames shall be permanently and effectively insulated from ground.

(3) In residential occupancies:
  a. Refrigerators, freezers, and air conditioners
  b. Clothes-washing, clothes-drying, dish-washing machines; kitchen waste disposers; information technology equipment; sump pumps and electrical aquarium equipment
  c. Hand-held motor-operated tools, stationary and fixed motor-operated tools, light industrial motor-operated tools
  d. Motor-operated appliances of the following types: hedge clippers, lawn mowers, snow blowers, and wet scrubbers
  e. Portable handlamps
(4) In other than residential occupancies:
  a. Refrigerators, freezers, and air conditioners
  b. Clothes-washing, clothes-drying, dish-washing machines; information technology equipment; sump pumps and electrical aquarium equipment
  c. Hand-held motor-operated tools, stationary and fixed motor-operated tools, light industrial motor-operated tools
  d. Motor-operated appliances of the following types: hedge clippers, lawn mowers, snow blowers, and wet scrubbers

e. Portable handlamps
f. Cord-and-plug-connected appliances used in damp or wet locations or by persons standing on the ground or on metal floors or working inside of metal tanks or boilers
g. Tools likely to be used in wet or conductive locations

Exception: Tools and portable handlamps likely to be used in wet or conductive locations shall not be required to be grounded where supplied through an isolating transformer with an ungrounded secondary of not over 50 volts.

Tools must be grounded by an equipment grounding conductor within the cord or cable supplying the tool, except where the tool is supplied by an isolating transformer, as permitted by the exception following 250.114(4). Portable tools and appliances protected by an approved system of double insulation must be listed by a qualified electrical testing laboratory as being suitable for the purpose, and the equipment must be distinctively marked as double insulated.

Cord-connected portable tools or appliances are not intended to be used in damp, wet, or conductive locations unless they are grounded, supplied by an isolation transformer with a secondary of not more than 50 volts, or protected by an approved system of double insulation.

Exhibit 250.45 shows an example of lighting equipment supplied through an isolating transformer operating at 6 or 12 volts that provides safe illumination for work inside boilers, tanks, and similar locations that may be metal or wet.

Exhibit 250.45 Lighting equipment supplied through an isolating transformer operating at 6 or 12 volts and therefore not required to be grounded. (Courtesy of Daniel Woodhead Co.)

250.116 Nonelectric Equipment

The metal parts of nonelectric equipment described in this section shall be grounded.

(1) Frames and tracks of electrically operated cranes and hoists
(2) Frames of nonelectrically driven elevator cars to which electric conductors are attached
(3) Hand-operated metal shifting ropes or cables of electric elevators

> FPN: Where extensive metal in or on buildings may become energized and is subject to personal contact, adequate bonding and grounding will provide additional safety.

Because metal siding on buildings is not electrical equipment, it is outside the scope of the *Code* [see 90.2(A)]. Therefore, the *Code* cannot require that it be grounded. Quite often, however, luminaires, signs, or receptacles are installed on buildings with metal siding that could become energized. Grounding of metal siding reduces the risk of shock to persons who may come in contact with siding that has become energized.

250.118 Types of Equipment Grounding Conductors

The equipment grounding conductor run with or enclosing the circuit conductors shall be one or more or a combination of the following:

(1) A copper, aluminum, or copper-clad aluminum conductor. This conductor shall be solid or stranded; insulated, covered, or bare; and in the form of a wire or a busbar of any shape.
(2) Rigid metal conduit.
(3) Intermediate metal conduit.
(4) Electrical metallic tubing.
(5) Listed flexible metal conduit meeting all the following conditions:

   a. The conduit is terminated in fittings listed for grounding.
   b. The circuit conductors contained in the conduit are protected by overcurrent devices rated at 20 amperes or less.
   c. The combined length of flexible metal conduit and flexible metallic tubing and liquidtight flexible metal conduit in the same ground return path does not exceed 1.8 m (6 ft).
   d. Where used to connect equipment where flexibility is necessary after installation, an equipment grounding conductor shall be installed.

(6) Listed liquidtight flexible metal conduit meeting all the following conditions:

    a. The conduit is terminated in fittings listed for grounding.

    b. For metric designators 12 through 16 (trade sizes ⅜ through ½), the circuit conductors contained in the conduit are protected by overcurrent devices rated at 20 amperes or less.

    c. For metric designators 21 through 35 (trade sizes ¾ through 1¼), the circuit conductors contained in the conduit are protected by overcurrent devices rated not more than 60 amperes and there is no flexible metal conduit, flexible metallic tubing, or liquidtight flexible metal conduit in trade sizes metric designators 12 through 16 (trade sizes ⅜ through ½) in the grounding path.

    d. The combined length of flexible metal conduit and flexible metallic tubing and liquidtight flexible metal conduit in the same ground return path does not exceed 1.8 m (6 ft).

    e. Where used to connect equipment where flexibility is necessary after installation, an equipment grounding conductor shall be installed.

In those cases where liquidtight flexible metal conduit (LFMC) or flexible metal conduit (FMC) is used to connect to equipment that requires a degree of movement or flexibility as part of its anticipated operating conditions, it is required that an equipment grounding conductor be installed in the LFMC or FMC. Flexible raceways that are used to facilitate connection to equipment but that remain stationary after the connection is made are not covered by the provision of 250.118(5)(d) or 250.118(6)(e).

(7) Flexible metallic tubing where the tubing is terminated in fittings listed for grounding and meeting the following conditions:

    a. The circuit conductors contained in the tubing are protected by overcurrent devices rated at 20 amperes or less.

    b. The combined length of flexible metal conduit and flexible metallic tubing and liquidtight flexible metal conduit in the same ground return path does not exceed 1.8 m (6 ft).

(8) Armor of Type AC cable as provided in 320.108.

(9) The copper sheath of mineral-insulated, metal-sheathed cable.

(10) Type MC cable where listed and identified for grounding in accordance with the following:

    a. The combined metallic sheath and grounding conductor of interlocked metal tape–type MC cable

    b. The metallic sheath or the combined metallic sheath and grounding conductors of the smooth or corrugated tube type MC cable

(11) Cable trays as permitted in 392.3(C) and 392.7.

(12) Cablebus framework as permitted in 370.3.

(13) Other listed electrically continuous metal raceways and listed auxiliary gutters.

(14) Surface metal raceways listed for grounding.

Exhibit 250.46 illustrates the various sizes of rigid metal conduit that enclose the feeder circuit conductors and are equipment grounding conductors with or without the installation of a wire-type equipment grounding conductor in the conduits.

250.119 Identification of Equipment Grounding Conductors

Unless required elsewhere in this *Code,* equipment grounding conductors shall be permitted to be bare, covered, or insulated. Individually covered or insulated equipment grounding conductors shall have a continuous outer finish that is either green or green with one or more yellow stripes except as permitted in this section. Conductors with insulation or individual covering that is green, green with one or more yellow stripes, or otherwise identified as permitted by this section shall not be used for ungrounded or grounded circuit conductors.

A new condition in this section of the 2005 *Code* precludes re-identification (such as marking tape to the insulation) of any conductor having green- or green-with-yellow-stripe-colored insulation or covering for use as an ungrounded or grounded conductor.

(A) Conductors Larger Than 6 AWG Equipment grounding conductors larger than 6 AWG shall comply with 250.119(A)(1) and (A)(2).

(1) An insulated or covered conductor larger than 6 AWG shall be permitted, at the time of installation, to be permanently identified as an equipment grounding conductor at each end and at every point where the conductor is accessible.

Exception: Conductors larger than 6 AWG shall not be required to be marked in conduit bodies that contain no splices or unused hubs.

(2) Identification shall encircle the conductor and shall be accomplished by one of the following:

    a. Stripping the insulation or covering from the entire exposed length

Exhibit 250.46 Various sizes of enclosing metal conduits used as equipment grounding conductors, as they apply to a service and feeder system.

Rigid metal conduits suitable for equipment grounding purposes (typical)

Feeders

Three 2/0 AWG, 1½ in.
Three 1 AWG, 1¼ in.
Three 1 AWG, 1¼ in.

Three 6 AWG, 1 in.
Three 3 AWG, 1 in.

Service disconnecting means with 400-A fuses

175 A   125 A   125 A
Feeder distribution panelboard No. 1

100 A   60 A
Feeder distribution panelboard No. 2

2½ in. RMC   2½ in. RMC

Wireway

Main feeder 500-kcmil Cu, 3-in. rigid metal conduit (RMC)

b. Coloring the exposed insulation or covering green
c. Marking the exposed insulation or covering with green tape or green adhesive labels

(B) Multiconductor Cable Where the conditions of maintenance and supervision ensure that only qualified persons service the installation, one or more insulated conductors in a multiconductor cable, at the time of installation, shall be permitted to be permanently identified as equipment grounding conductors at each end and at every point where the conductors are accessible by one of the following means:

(1) Stripping the insulation from the entire exposed length
(2) Coloring the exposed insulation green
(3) Marking the exposed insulation with green tape or green adhesive labels

(C) Flexible Cord An uninsulated equipment grounding conductor shall be permitted, but, if individually covered, the covering shall have a continuous outer finish that is either green or green with one or more yellow stripes.

250.120 Equipment Grounding Conductor Installation

An equipment grounding conductor shall be installed in accordance with 250.120(A), (B), and (C).

(A) Raceway, Cable Trays, Cable Armor, Cablebus, or Cable Sheaths Where it consists of a raceway, cable tray, cable armor, cablebus framework, or cable sheath or where it is a wire within a raceway or cable, it shall be installed

in accordance with the applicable provisions in this *Code* using fittings for joints and terminations approved for use with the type raceway or cable used. All connections, joints, and fittings shall be made tight using suitable tools.

(B) Aluminum and Copper-Clad Aluminum Conductors Equipment grounding conductors of bare or insulated aluminum or copper-clad aluminum shall be permitted. Bare conductors shall not come in direct contact with masonry or the earth or where subject to corrosive conditions. Aluminum or copper-clad aluminum conductors shall not be terminated within 450 mm (18 in.) of the earth.

(C) Equipment Grounding Conductors Smaller Than 6 AWG Equipment grounding conductors smaller than 6 AWG shall be protected from physical damage by a raceway or cable armor except where run in hollow spaces of walls or partitions, where not subject to physical damage, or where protected from physical damage.

250.122 Size of Equipment Grounding Conductors

(A) General Copper, aluminum, or copper-clad aluminum equipment grounding conductors of the wire type shall not be smaller than shown in Table 250.122 but shall not be required to be larger than the circuit conductors supplying the equipment. Where a raceway or a cable armor or sheath is used as the equipment grounding conductor, as provided in 250.118 and 250.134(A), it shall comply with 250.4(A)(5) or (B)(4).

Table 250.122 Minimum Size Equipment Grounding Conductors for Grounding Raceway and Equipment

Rating or Setting of Automatic Overcurrent Device in Circuit Ahead of Equipment, Conduit, etc., Not Exceeding (Amperes)	Size (AWG or kcmil)	
	Copper	Aluminum or Copper-Clad Aluminum*
15	14	12
20	12	10
30	10	8
40	10	8
60	10	8
100	8	6
200	6	4
300	4	2
400	3	1
500	2	1/0
600	1	2/0
800	1/0	3/0
1000	2/0	4/0
1200	3/0	250
1600	4/0	350
2000	250	400
2500	350	600
3000	400	600
4000	500	800
5000	700	1200
6000	800	1200

Note: Where necessary to comply with 250.4(A)(5) or (B)(4), the equipment grounding conductor shall be sized larger than given in this table.
*See installation restrictions in 250.120.

The last sentence of 250.122(A) alerts users that, if a long distance exists between a power source and utilization equipment, some of the wiring methods permitted by 250.118 for equipment grounding purposes must be evaluated to ensure that they can provide an effective path for ground-fault current.

(B) Increased in Size Where ungrounded conductors are increased in size, equipment grounding conductors, where installed, shall be increased in size proportionally according to circular mil area of the ungrounded conductors.

Equipment grounding conductors on the load side of the service disconnecting means and overcurrent devices are sized based on the size of the feeder or branch-circuit overcurrent devices ahead of them. Where the ungrounded circuit

conductors are increased in size to compensate for voltage drop or for any other reason related to proper circuit operation, the equipment grounding conductors must be increased proportionally.

Example

A 240-volt, single-phase, 250-ampere load is supplied from a 300-ampere breaker located in a panelboard 500 ft away. The conductors are 250-kcmil copper, installed in rigid nonmetallic conduit, with a 4 AWG copper equipment grounding conductor. If the conductors are increased to 350 kcmil, what is the minimum size for the equipment grounding conductor based on the proportional-increase requirement?

Solution

STEP 1. Calculate the size ratio of the new conductors to the existing conductors:

$$\text{Size ratio} = \frac{350,000 \text{ circular mils}}{250,000 \text{ circular mils}} = 1.4$$

STEP 2. Calculate the cross-sectional area of the new equipment grounding conductor. According to Chapter 9, Table 8, 4 AWG, the size of the existing grounding conductor, has a cross-sectional area of 41,740 circular mils.

STEP 3. Determine the size of the new equipment grounding conductor. Again, referring to Chapter 9, Table 8, we find that 58,436 circular mils is larger than 3 AWG. The next larger size is 66,360 circular mils, which converts to a 2 AWG copper equipment grounding conductor.

(C) Multiple Circuits Where a single equipment grounding conductor is run with multiple circuits in the same raceway or cable, it shall be sized for the largest overcurrent device protecting conductors in the raceway or cable.

A single equipment grounding conductor must be sized for the largest overcurrent device. It is not required to be sized for the composite of all the circuits in the raceway because it is not anticipated that all circuits will develop faults at the same time. For example, three 3-phase circuits in the same raceway, protected by overcurrent devices rated 30, 60, and 100 amperes, would require only one equipment grounding conductor, sized according to the largest overcurrent device (in this case, 100 amperes). Therefore, an 8 AWG copper or 6 AWG aluminum conductor or copper-clad aluminum conductor is required, according to Table 250.122.

(D) Motor Circuits Where the overcurrent device consists of an instantaneous trip circuit breaker or a motor short-circuit protector, as allowed in 430.52, the equipment grounding conductor size shall be permitted to be based on the rating of the motor overload protective device but shall not be less than the size shown in Table 250.122.

(E) Flexible Cord and Fixture Wire The equipment grounding conductor in a flexible cord with the largest circuit conductor 10 AWG or smaller, and the equipment grounding conductor used with fixture wires of any size in accordance with 240.5, shall not be smaller than 18 AWG copper and shall not be smaller than the circuit conductors. The equipment grounding conductor in a flexible cord with a circuit conductor larger than 10 AWG shall be sized in accordance with Table 250.122.

(F) Conductors in Parallel Where conductors are run in parallel in multiple raceways or cables as permitted in 310.4, the equipment grounding conductors, where used, shall be run in parallel in each raceway or cable. One of the methods in 250.122(F)(1) or (F)(2) shall be used to ensure the equipment grounding conductors are protected.

(1) Based on Rating of Overcurrent Protective Device Each parallel equipment grounding conductor shall be sized on the basis of the ampere rating of the overcurrent device protecting the circuit conductors in the raceway or cable in accordance with Table 250.122.

Where wire-type equipment grounding conductors are installed in multiple raceways or cables used to enclose conductors in parallel, a full-sized equipment grounding conductor selected from Table 250.122 based on the size of the overcurrent device protecting the paralleled circuit is required in each raceway or cable.

The full-sized equipment grounding conductor is required to prevent overloading and possible burnout of the conductor should a ground fault occur along one of the parallel branches. The installation conditions for paralleled conductors prescribed in 310.4 result in proportional distribution of the current-time duty among the several paralleled grounding conductors only for overcurrent conditions downstream of the paralleled set of circuit conductors.

Exhibit 250.47 shows a parallel arrangement with two nonmetallic conduits installed underground. For clarity, a one-line diagram with equipment grounding conductors is shown. A ground fault at the enclosure will cause the equipment grounding conductor in the top conduit to carry more than its proportionate share of fault current. Note that the fault is fed by two different conductors of the same phase, one from the left and one from the right. The shortest and lowest-impedance path to ground from the fault to the supply panelboard is through the equipment grounding conductor in the top conduit. The grounding path from the fault through the bottom conduit is longer and of higher impedance. Therefore, the equipment grounding conductor in each raceway

Exhibit 250.47 Grounding paths for ground fault at the load supplied by parallel conductors in two nonmetallic raceways, illustrating the reason for the requirement of 250.122(F)(1).

must be capable of carrying a major portion of the fault current without burning open.

(2) Ground-Fault Protection of Equipment Installed Where ground-fault protection of equipment is installed, each parallel equipment grounding conductor in a multiconductor cable shall be permitted to be sized in accordance with Table 250.122 on the basis of the trip rating of the ground-fault protection where the following conditions are met:

(1) Conditions of maintenance and supervision ensure that only qualified persons will service the installation.
(2) The ground-fault protection equipment is set to trip at not more than the ampacity of a single ungrounded conductor of one of the cables in parallel.
(3) The ground-fault protection is listed for the purpose of protecting the equipment grounding conductor.

Section 250.122(F)(2) applies to cables that are installed in parallel. Because cable assemblies are manufactured in standard conductor size configurations, the equipment grounding conductor in a cable is properly sized for some circuit arrangements but not necessarily for all parallel circuit arrangements. Where the cable is used in large-capacity parallel circuits, the equipment grounding conductor in each cable may not be large enough to comply with Table 250.122, depending on the size of the overcurrent device protecting the circuit.

To address this problem, 250.122(F)(2) permits the sizing of the equipment grounding conductor within a multiconductor cable to be based on the trip rating of an equipment ground-fault protection device. This method of protection is permitted only where the installation is serviced by qualified personnel and the ground-fault device is specifically listed for protecting the equipment grounding conductor. The adjustable trip setting of the ground-fault protection device cannot be set higher than the ampacity of a single ungrounded conductor installed as part of the parallel circuit arrangement. This provision is intended to permit the use of standard cable assembly configurations in large-capacity parallel circuits without having to custom-manufacture the cable to include an equipment grounding conductor sized for a specific parallel circuit arrangement.

(G) Feeder Taps Equipment grounding conductors run with feeder taps shall not be smaller than shown in Table 250.122 based on the rating of the overcurrent device ahead of the feeder but shall not be required to be larger than the tap conductors.

This paragraph, which is new for the 2005 *Code,* clarifies how to size a wire-type equipment grounding conductor for feeder tap installations covered in 240.21(B). This requirement specifies that it is the rating of the overcurrent device on the line or supply side of the feeders that is the basis for selection from Table 250.122 rather than the rating of the overcurrent or other device at the load end of the tap conductors. This stands to reason because it is the device on the supply side of the tap conductors that needs to be opened under a ground-fault condition between the point at which they are supplied and the point at which they terminate. In accordance with this paragraph and 250.122(A), the equipment grounding conductor is not required to be larger than the ungrounded conductors under any circumstance.

For example, a 600-kcmil copper conductor is tapped to a 1200-ampere feeder and supplies a fusible switch with 400-ampere fuses. Where the 400-ampere overcurrent protection is installed at the point the 600-kcmil conductors receive their supply, the equipment grounding conductor from Table 250.122 is a 3 AWG copper or 1 AWG aluminum conductor. However, in this tap conductor application, it is the 1200-ampere device that is on the line side of the 600-kcmil tap conductors, and the equipment grounding conductor from Table 250.122 is based on the 1200-ampere device. In this case, the equipment grounding conductor is required to be a 3/0 AWG copper or 250-kcmil aluminum conductor. This provision applies only where a wire-type equipment grounding conductor is run with the feeder tap conductors. Other equipment grounding conductors permitted in 250.118 can also be used where they meet the requirements for tap conductor wiring methods specified in 240.21(B)(1) through 240.21(B)(5).

250.124 Equipment Grounding Conductor Continuity

(A) Separable Connections Separable connections such as those provided in drawout equipment or attachment plugs

and mating connectors and receptacles shall provide for first-make, last-break of the equipment grounding conductor. First-make, last-break shall not be required where interlocked equipment, plugs, receptacles, and connectors preclude energization without grounding continuity.

(B) Switches No automatic cutout or switch shall be placed in the equipment grounding conductor of a premises wiring system unless the opening of the cutout or switch disconnects all sources of energy.

250.126 Identification of Wiring Device Terminals

The terminal for the connection of the equipment grounding conductor shall be identified by one of the following:

(1) A green, not readily removable terminal screw with a hexagonal head.
(2) A green, hexagonal, not readily removable terminal nut.
(3) A green pressure wire connector. If the terminal for the grounding conductor is not visible, the conductor entrance hole shall be marked with the word *green* or *ground*, the letters *G* or *GR*, a grounding symbol, or otherwise identified by a distinctive green color. If the terminal for the equipment grounding conductor is readily removable, the area adjacent to the terminal shall be similarly marked.

FPN: See FPN Figure 250.126.

FPN Figure 250.126 One Example of a Symbol Used to Identify the Grounding Termination Point for an Equipment Grounding Conductor.

VII. Methods of Equipment Grounding

250.130 Equipment Grounding Conductor Connections

Equipment grounding conductor connections at the source of separately derived systems shall be made in accordance with 250.30(A)(1). Equipment grounding conductor connections at service equipment shall be made as indicated in 250.130(A) or (B). For replacement of non–grounding-type receptacles with grounding-type receptacles and for branch-circuit extensions only in existing installations that do not have an equipment grounding conductor in the branch circuit, connections shall be permitted as indicated in 250.130(C).

(A) For Grounded Systems The connection shall be made by bonding the equipment grounding conductor to the grounded service conductor and the grounding electrode conductor.

The grounding arrangement for a grounded system is illustrated in Exhibit 250.48.

Exhibit 250.48 Grounding arrangement for grounded systems, per 250.130(A), illustrating connection of the equipment grounding conductor (bus) to the enclosures and the grounded service conductor.

(B) For Ungrounded Systems The connection shall be made by bonding the equipment grounding conductor to the grounding electrode conductor.

(C) Nongrounding Receptacle Replacement or Branch Circuit Extensions The equipment grounding conductor of a grounding-type receptacle or a branch-circuit extension shall be permitted to be connected to any of the following:

(1) Any accessible point on the grounding electrode system as described in 250.50
(2) Any accessible point on the grounding electrode conductor
(3) The equipment grounding terminal bar within the enclosure where the branch circuit for the receptacle or branch circuit originates
(4) For grounded systems, the grounded service conductor within the service equipment enclosure
(5) For ungrounded systems, the grounding terminal bar within the service equipment enclosure

FPN: See 406.3(D) for the use of a ground-fault circuit-interrupting type of receptacle.

Section 250.130(C) applies to both ungrounded and grounded systems. It permits a nongrounding-type receptacle to be replaced with a grounding-type receptacle under the following conditions.

1. The branch circuit does not contain an equipment ground.
2. An existing branch circuit is being extended for additional receptacle outlets.

3. An equipment grounding conductor is connected between the receptacle grounding terminal to any accessible point on the grounding electrode system, to any accessible point on the grounding electrode conductor, to the grounded service conductor within the service equipment enclosure, or to the equipment grounding terminal bar in the enclosure from which the circuit is supplied.

The requirement in 250.52(A)(1) does not permit this separate equipment grounding conductor to be connected to the metal water piping of a building or structure beyond the first 5 ft of where the piping enters the building or structure unless the conditions of the exception to 250.52(A)(1) can be met.

Exhibit 250.49 shows a branch-circuit extension made from an existing installation. This method is also permitted to ground a replacement 3-wire receptacle in the existing ungrounded box on the left, where no grounding conductor is available.

Exhibit 250.49 Branch-circuit extension to an existing installation, per 250.130(C), illustrating a separate equipment grounding conductor connected to the grounding electrode system.

250.132 Short Sections of Raceway

Isolated sections of metal raceway or cable armor, where required to be grounded, shall be grounded in accordance with 250.134.

250.134 Equipment Fastened in Place or Connected by Permanent Wiring Methods (Fixed) — Grounding

Unless grounded by connection to the grounded circuit conductor as permitted by 250.32, 250.140, and 250.142, non–current-carrying metal parts of equipment, raceways, and other enclosures, if grounded, shall be grounded by one of the following methods.

Section 250.134 eliminates any conflict between 250.134(A), which requires an equipment grounding conductor to be used for equipment grounding, and 250.32, 250.140, and 250.142, which permit the grounded circuit conductor to be used for equipment grounding if certain specified conditions are met.

(A) Equipment Grounding Conductor Types By any of the equipment grounding conductors permitted by 250.118.

(B) With Circuit Conductors By an equipment grounding conductor contained within the same raceway, cable, or otherwise run with the circuit conductors.

One of the functions of an equipment grounding conductor is to provide a low-impedance ground-fault path between a ground fault and the electrical source. This path allows the overcurrent protective device to actuate, interrupting the current. To keep the impedance at a minimum, it is necessary to run the equipment grounding conductor in the same raceway or cable as the circuit conductor(s). This practice allows the magnetic field developed by the circuit conductor and the equipment grounding conductor to cancel, reducing their impedance.

Magnetic flux strength is inversely proportional to the square of the distance between the two conductors. By placing an equipment grounding conductor away from the conductor delivering the fault current, the magnetic flux cancellation decreases. This increases the impedance of the fault path and delays operation of the protective device.

Exception No. 1: As provided in 250.130(C), the equipment grounding conductor shall be permitted to be run separately from the circuit conductors.

Exception No. 1 to 250.134(B) permits an equipment grounding conductor to be run to the grounding electrode separately from the other conductors of an ac circuit. This practice applies only where a grounding-type receptacle is used on a circuit that does not include an equipment grounding conductor. See the commentary following 250.130(C) for further explanation.

Exception No. 2: For dc circuits, the equipment grounding conductor shall be permitted to be run separately from the circuit conductors.

> FPN No. 1: See 250.102 and 250.168 for equipment bonding jumper requirements.
> FPN No. 2: See 400.7 for use of cords for fixed equipment.

250.136 Equipment Considered Effectively Grounded

Under the conditions specified in 250.136(A) and (B), the non–current-carrying metal parts of the equipment shall be considered effectively grounded.

(A) Equipment Secured to Grounded Metal Supports Electrical equipment secured to and in electrical contact with a metal rack or structure provided for its support and grounded by one of the means indicated in 250.134. The structural metal frame of a building shall not be used as the required equipment grounding conductor for ac equipment.

Exhibit 250.50 shows an example of electrical equipment secured to and in electrical contact with a metal rack that is effectively grounded in accordance with 250.136(A).

All equipment in good electrical contact with metal rack

Metal raceway

Nonmetallic raceway (typical)

Metal rack effectively grounded

Exhibit 250.50 An example of electrical equipment considered to be effectively grounded through mechanical connections to a grounded metal rack.

(B) Metal Car Frames Metal car frames supported by metal hoisting cables attached to or running over metal sheaves or drums of elevator machines that are grounded by one of the methods indicated in 250.134.

250.138 Cord-and-Plug-Connected Equipment

Non–current-carrying metal parts of cord-and-plug-connected equipment, if grounded, shall be grounded by one of the methods in 250.138(A) or (B).

(A) By Means of an Equipment Grounding Conductor By means of an equipment grounding conductor run with the power supply conductors in a cable assembly or flexible cord properly terminated in a grounding-type attachment plug with one fixed grounding contact.

Exception: The grounding contacting pole of grounding-type plug-in ground-fault circuit interrupters shall be permitted to be of the movable, self-restoring type on circuits operating at not over 150 volts between any two conductors or over 150 volts between any conductor and ground.

(B) By Means of a Separate Flexible Wire or Strap By means of a separate flexible wire or strap, insulated or bare, protected as well as practicable against physical damage, where part of equipment.

250.140 Frames of Ranges and Clothes Dryers

Frames of electric ranges, wall-mounted ovens, counter-mounted cooking units, clothes dryers, and outlet or junction boxes that are part of the circuit for these appliances shall be grounded in the manner specified by 250.134 or 250.138.

Exception: For existing branch circuit installations only where an equipment grounding conductor is not present in the outlet or junction box, the frames of electric ranges, wall-mounted ovens, counter-mounted cooking units, clothes dryers, and outlet or junction boxes that are part of the circuit for these appliances shall be permitted to be grounded to the grounded circuit conductor if all the following conditions are met.

(1) The supply circuit is 120/240-volt, single-phase, 3-wire; or 208Y/120-volt derived from a 3-phase, 4-wire, wye-connected system.

(2) The grounded conductor is not smaller than 10 AWG copper or 8 AWG aluminum.

(3) The grounded conductor is insulated, or the grounded conductor is uninsulated and part of a Type SE service-entrance cable and the branch circuit originates at the service equipment.

(4) Grounding contacts of receptacles furnished as part of the equipment are bonded to the equipment.

The exception to 250.140 applies only to existing branch circuits supplying the appliances specified in 250.140. The grounded conductor (neutral) of newly installed branch circuits supplying ranges and clothes dryers is no longer permitted to be used for grounding the non–current-carrying metal parts of the appliances. Branch circuits installed for new appliance installations are required to provide an equipment grounding conductor sized in accordance with 250.122 for grounding the non–current-carrying metal parts.

Caution should be exercised to ensure that new appliances connected to an existing branch circuit are properly grounded. An older appliance connected to a new branch circuit must have its 3-wire cord and plug replaced with a 4-conductor cord, with one of those conductors being an equipment grounding conductor. The bonding jumper between the neutral and the frame of the appliance must be removed. Where a new range or clothes dryer is connected to an existing branch circuit without an equipment grounding conductor, in which the neutral conductor is used for grounding the appliance frame, it must be ensured that a bonding jumper is in place between the neutral terminal of the appliance and the frame of the appliance.

The grounded circuit conductor of an existing branch circuit is still permitted to be used to ground the frame of an electric range, wall-mounted oven, or counter-mounted cooking unit, provided all four conditions of 250.140, Exception, are met. In addition, a revision in this provision for the 2005 *Code* permits application of the exception only where the existing branch-circuit wiring method does not provide an equipment grounding conductor. There are many existing branch circuits in which nonmetallic sheath cable with three insulated circuit conductors and a bare equipment grounding conductor was used to supply a range or clothes dryer. The bare equipment grounding conductor was simply not used because it was permitted to ground the equipment with the insulated neutral conductor of the NM cable. This "extra" conductor was on account of the fact that the bare conductor in a Type NM cable is to be used only as an equipment grounding conductor and cannot be used as a grounded (neutral) conductor in the same manner as is permitted for an uninsulated conductor in the service entrance.

In addition to grounding the frame of the range or clothes dryer, the grounded circuit conductor of these existing branch circuits is also permitted to be used to ground any junction boxes in the circuit supplying the appliance, and a 3-wire pigtail and range receptacle are permitted to be used.

Prior to the 1996 *Code,* use of the grounded circuit conductor as a grounding conductor was permitted for all installations. In many instances, the wiring method was service-entrance cable with an uninsulated neutral conductor covered by the cable jacket. Where Type SE cable was used to supply ranges and dryers, the branch circuit was required to originate at the service equipment to avoid neutral current from downstream panelboards on metal objects, such as pipes or ducts.

Exhibit 250.51 shows an existing installation in which Type SE service-entrance cable was used for ranges, dryers, wall-mounted ovens, and counter-mounted cooking units. Junction boxes in the supply circuit were also permitted to be grounded from the grounded neutral conductor.

250.142 Use of Grounded Circuit Conductor for Grounding Equipment

(A) Supply-Side Equipment A grounded circuit conductor shall be permitted to ground non–current-carrying metal parts of equipment, raceways, and other enclosures at any of the following locations:

(1) On the supply side or within the enclosure of the ac service-disconnecting means

(2) On the supply side or within the enclosure of the main disconnecting means for separate buildings as provided in 250.32(B)

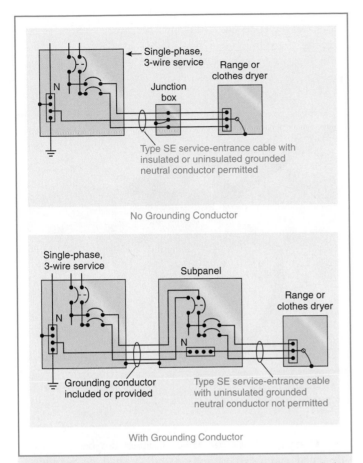

No Grounding Conductor

With Grounding Conductor

Exhibit 250.51 An existing installation in which the grounded conductor in Type SE service-entrance cable was used for grounding the frames of ranges and clothes dryers, plus associated metal junction boxes, in accordance with 250.140.

(3) On the supply side or within the enclosure of the main disconnecting means or overcurrent devices of a separately derived system where permitted by 250.30(A)(1)

In separately derived systems, the grounded circuit conductor is permitted to ground non–current-carrying metal parts of equipment, raceways, and other enclosures only on the supply side of the main disconnecting means.

(B) Load-Side Equipment Except as permitted in 250.30(A)(1) and 250.32(B), a grounded circuit conductor shall not be used for grounding non–current-carrying metal parts of equipment on the load side of the service disconnecting means or on the load side of a separately derived system disconnecting means or the overcurrent devices for a separately derived system not having a main disconnecting means.

Exception No. 1: The frames of ranges, wall-mounted ovens, counter-mounted cooking units, and clothes dryers under

the conditions permitted for existing installations by 250.140 shall be permitted to be grounded by a grounded circuit conductor.

Exception No. 2: It shall be permissible to ground meter enclosures by connection to the grounded circuit conductor on the load side of the service disconnect where all of the following conditions apply:

(1) No service ground-fault protection is installed.
(2) All meter socket enclosures are located immediately adjacent to the service disconnecting means.
(3) The size of the grounded circuit conductor is not smaller than the size specified in Table 250.122 for equipment grounding conductors.

Exception No. 3: Direct-current systems shall be permitted to be grounded on the load side of the disconnecting means or overcurrent device in accordance with 250.164.

Exception No. 4: Electrode-type boilers operating at over 600 volts shall be grounded as required in 490.72(E)(1) and 490.74.

One major reason the grounded circuit conductor is not permitted to be grounded on the load side of the service [except as permitted in 250.30, 250.32(B)(2), and the four exceptions to 250.142(B)] is that, should the grounded service conductor become disconnected at any point on the line side of the ground, the equipment grounding conductor and all conductive parts connected to it would carry the neutral current, raising the potential to ground of exposed metal parts not normally intended to carry current. This could result in arcing in concealed spaces and could pose a severe shock hazard, particularly if the path is inadvertently opened by a person servicing or repairing piping or ductwork. Even without an open grounded conductor (usually referred to as an open neutral), the equipment grounding conductor path would become a parallel path with the grounded conductor, and there would be some potential drop on exposed and concealed dead metal parts. The magnitude of this potential difference would be determined by the relative impedances of the equipment grounding path and the grounded conductor circuits. Not only would the equipment grounding conductor path be affected, but all parallel paths not intended as equipment grounding conductors would be affected as well. This could involve current through metal building structures, piping, and ducts. The requirements of 250.30 and 250.32(B) have been revised in recent editions of the *Code* to prohibit the creation of parallel paths for normal neutral current.

250.144 Multiple Circuit Connections

Where equipment is required to be grounded and is supplied by separate connection to more than one circuit or grounded premises wiring system, a means for grounding shall be provided for each such connection as specified in 250.134 and 250.138.

250.146 Connecting Receptacle Grounding Terminal to Box

An equipment bonding jumper shall be used to connect the grounding terminal of a grounding-type receptacle to a grounded box unless grounded as in 250.146(A) through (D).

(A) Surface Mounted Box Where the box is mounted on the surface, direct metal-to-metal contact between the device yoke and the box or a contact yoke or device that complies with 250.146(B) shall be permitted to ground the receptacle to the box. At least one of the insulating washers shall be removed from receptacles that do not have a contact yoke or device that complies with 250.146(B) to ensure direct metal-to-metal contact. This provision shall not apply to cover-mounted receptacles unless the box and cover combination are listed as providing satisfactory ground continuity between the box and the receptacle.

The main rule of 250.146 requires an equipment bonding jumper to be installed between the device box and the receptacle grounding terminal. However, 250.146(A) permits the equipment bonding jumper to be omitted where the metal yoke of the device is in direct metal-to-metal contact with the metal device box and at least one of the fiber retention washers for the receptacle mounting screws is removed, as illustrated in Exhibit 250.52.

Cover-mounted wiring devices, such as on 4-in. square covers, are not considered grounded. Section 250.146(A)

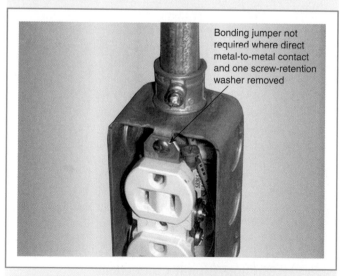

Bonding jumper not required where direct metal-to-metal contact and one screw-retention washer removed

Exhibit 250.52 An example of a box-mounted receptacle attached to a surface box where a bonding jumper is not required provided at least one of the insulating washers is removed.

does not apply to cover-mounted receptacles, such as the one illustrated in Exhibit 250.53. Box-cover and device combinations listed as providing grounding continuity are permitted.

Bonding jumper required from grounded metal box to cover-mounted receptacles

Exhibit 250.53 An example of a cover-mounted receptacle attached to a surface box where a bonding jumper is required.

(B) Contact Devices or Yokes Contact devices or yokes designed and listed as self-grounding shall be permitted in conjunction with the supporting screws to establish the grounding circuit between the device yoke and flush-type boxes.

Section 250.146(B) is illustrated by Exhibit 250.54, which shows a receptacle designed with a spring-type grounding strap for holding the mounting screw and establishing the grounding circuit so that an equipment bonding jumper is not required. Such devices are listed as "self-grounding."

(C) Floor Boxes Floor boxes designed for and listed as providing satisfactory ground continuity between the box and the device shall be permitted.

(D) Isolated Receptacles Where required for the reduction of electrical noise (electromagnetic interference) on the grounding circuit, a receptacle in which the grounding terminal is purposely insulated from the receptacle mounting means shall be permitted. The receptacle grounding terminal shall be grounded by an insulated equipment grounding conductor run with the circuit conductors. This grounding conductor shall be permitted to pass through one or more panelboards without connection to the panelboard grounding terminal as permitted in 408.40, Exception, so as to terminate within the same building or structure directly at an equipment

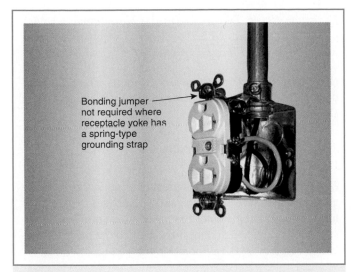

Exhibit 250.54 A receptacle designed with a listed spring-type grounding strap. The strap that holds the mounting screw captive establishes a grounding circuit and eliminates the need to provide a wire-type equipment bonding jumper to the box, in accordance with 250.146(B).

Exhibit 250.55 An isolated-ground–type receptacle with an insulated equipment grounding conductor and with the device box grounded through the metal raceway.

grounding conductor terminal of the applicable derived system or service.

> FPN: Use of an isolated equipment grounding conductor does not relieve the requirement for grounding the raceway system and outlet box.

Section 250.146(D) allows an isolated-ground–type receptacle to be installed without a bonding jumper between the metal device box and the receptacle grounding terminal. An insulated equipment grounding conductor, as shown in Exhibit 250.55, is installed with the branch-circuit conductors. This conductor may originate in the service panel, pass through any number of subpanels without being connected to the equipment grounding bus, and terminate at the isolated-ground–type receptacle ground terminal. However, this does not exempt the metal device box from being grounded. The metal device box must be grounded either by an equipment grounding conductor run with the circuit conductors or by a wiring method that serves as an equipment grounding conductor. See 250.118 for types of equipment grounding conductors.

According to 250.146(D), where isolated-ground–type receptacles are used, the isolated equipment grounding conductor can terminate at an equipment grounding terminal of the applicable service or derived system in the same building as the receptacle. If the isolated equipment grounding conductor terminates at a separate building, a large voltage difference may exist between buildings during lightning transients. Such transients could cause damage to equipment connected to an isolated-ground–type receptacle and present

a shock hazard between the isolated equipment frame and other grounded surfaces.

The fine print note to 250.146(D) is a reminder that metallic raceways and boxes are still required to be grounded by one of the usual required methods. This could require a separate grounding conductor, for example, to ground a metal box in a nonmetallic raceway system or to ground a metal box supplied by flexible metal conduit. Where an ordinary grounding-type receptacle is being replaced with an isolated-ground–type receptacle, use of an existing insulated equipment grounding conductor as the isolated receptacle grounding conductor could effectively defeat or seriously compromise the required box or raceway equipment ground.

250.148 Continuity and Attachment of Equipment Grounding Conductors to Boxes

Where circuit conductors are spliced within a box, or terminated on equipment within or supported by a box, any equipment grounding conductor(s) associated with those circuit conductors shall be spliced or joined within the box or to the box with devices suitable for the use in accordance with 250.148(A) through (E).

Where a metal box is used in a metal raceway system and there is a wire-type equipment grounding conductor installed in the raceway, it is not required that the wire-type equipment grounding conductor be connected to the pull box provided the box is effectively grounded by the metal raceway and the circuit conductors are not spliced or terminated to equipment in the metal box. An example of this provision would be where conductors are run unbroken through a pull box.

Exception: The equipment grounding conductor permitted in 250.146(D) shall not be required to be connected to the other equipment grounding conductors or to the box.

(A) Connections Connections and splices shall be made in accordance with 110.14(B) except that insulation shall not be required.

(B) Grounding Continuity The arrangement of grounding connections shall be such that the disconnection or the removal of a receptacle, luminaire (fixture), or other device fed from the box does not interfere with or interrupt the grounding continuity.

(C) Metal Boxes A connection shall be made between the one or more equipment grounding conductors and a metal box by means of a grounding screw that shall be used for no other purpose or a listed grounding device.

(D) Nonmetallic Boxes One or more equipment grounding conductors brought into a nonmetallic outlet box shall be arranged such that a connection can be made to any fitting or device in that box requiring grounding.

(E) Solder Connections depending solely on solder shall not be used.

VIII. Direct-Current Systems

250.160 General

Direct-current systems shall comply with Part VIII and other sections of Article 250 not specifically intended for ac systems.

250.162 Direct-Current Circuits and Systems to Be Grounded

Direct-current circuits and systems shall be grounded as provided for in 250.162(A) and (B).

(A) Two-Wire, Direct-Current Systems A 2-wire, dc system supplying premises wiring and operating at greater than 50 volts but not greater than 300 volts shall be grounded.

Exception No. 1: A system equipped with a ground detector and supplying only industrial equipment in limited areas shall not be required to be grounded.

Exception No. 2: A rectifier-derived dc system supplied from an ac system complying with 250.20 shall not be required to be grounded.

Exception No. 3: Direct-current fire alarm circuits having a maximum current of 0.030 amperes as specified in Article 760, Part III, shall not be required to be grounded.

(B) Three-Wire, Direct-Current Systems The neutral onductor of all 3-wire, dc systems supplying premises wiring shall be grounded.

250.164 Point of Connection for Direct-Current Systems

(A) Off-Premises Source Direct-current systems to be grounded and supplied from an off-premises source shall have the grounding connection made at one or more supply stations. A grounding connection shall not be made at individual services or at any point on the premises wiring.

As shown in the 3-wire dc distribution system in Exhibit 250.56, the neutral is grounded at the off-premises generator site. Grounding of a 2-wire dc system would be accomplished in the same manner. For an on-premises generator, a grounding connection is required and is to be located at the source of the first system disconnecting means or overcurrent device. Other equivalent means that use equipment listed and identified for such use are permitted.

3-Wire, 120/240-V dc Distribution System

Exhibit 250.56 A 3-wire, 120/240-volt dc distribution system with the neutral grounded at the off-premises generator site.

(B) On-Premises Source Where the dc system source is located on the premises, a grounding connection shall be made at one of the following:

(1) The source
(2) The first system disconnection means or overcurrent device
(3) By other means that accomplish equivalent system protection and that utilize equipment listed and identified for the use

250.166 Size of Direct-Current Grounding Electrode Conductor

The size of the grounding electrode conductor for a dc system shall be as specified in 250.166(A) through (E).

(A) Not Smaller Than the Neutral Conductor Where the dc system consists of a 3-wire balancer set or a balancer winding with overcurrent protection as provided in 445.12(D), the grounding electrode conductor shall not be smaller than the neutral conductor and not smaller than 8 AWG copper or 6 AWG aluminum.

(B) Not Smaller Than the Largest Conductor Where the dc system is other than as in 250.166(A), the grounding electrode conductor shall not be smaller than the largest conductor supplied by the system, and not smaller than 8 AWG copper or 6 AWG aluminum.

(C) Connected to Rod, Pipe, or Plate Electrodes Where connected to rod, pipe, or plate electrodes as in 250.52(A)(5) or 250.52(A)(6), that portion of the grounding electrode conductor that is the sole connection to the grounding electrode shall not be required to be larger than 6 AWG copper wire or 4 AWG aluminum wire.

(D) Connected to a Concrete-Encased Electrode Where connected to a concrete-encased electrode as in 250.52(A)(3), that portion of the grounding electrode conductor that is the sole connection to the grounding electrode shall not be required to be larger than 4 AWG copper wire.

(E) Connected to a Ground Ring Where connected to a ground ring as in 250.52(A)(4), that portion of the grounding electrode conductor that is the sole connection to the grounding electrode shall not be required to be larger than the conductor used for the ground ring.

250.168 Direct-Current Bonding Jumper

For dc systems, the size of the bonding jumper shall not be smaller than the system grounding electrode conductor specified in 250.166.

250.169 Ungrounded Direct-Current Separately Derived Systems

Except as otherwise permitted in 250.34 for portable and vehicle-mounted generators, an ungrounded dc separately derived system supplied from a stand-alone power source (such as an engine–generator set) shall have a grounding electrode conductor connected to an electrode that complies with Part III to provide for grounding of metal enclosures, raceways, cables, and exposed non–current-carrying metal parts of equipment. The grounding electrode conductor connection shall be to the metal enclosure at any point on the separately derived system from the source to the first system disconnecting means or overcurrent device, or it shall be made at the source of a separately derived system that has no disconnecting means or overcurrent devices.

The size of the grounding electrode conductor shall be in accordance with 250.166.

IX. Instruments, Meters, and Relays

250.170 Instrument Transformer Circuits

Secondary circuits of current and potential instrument transformers shall be grounded where the primary windings are connected to circuits of 300 volts or more to ground and, where on switchboards, shall be grounded irrespective of voltage.

Exception: Circuits where the primary windings are connected to circuits of less than 1000 volts with no live parts or wiring exposed or accessible to other than qualified persons.

250.172 Instrument Transformer Cases

Cases or frames of instrument transformers shall be grounded where accessible to other than qualified persons.

Exception: Cases or frames of current transformers, the primaries of which are not over 150 volts to ground and that are used exclusively to supply current to meters.

250.174 Cases of Instruments, Meters, and Relays Operating at Less Than 1000 Volts

Instruments, meters, and relays operating with windings or working parts at less than 1000 volts shall be grounded as specified in 250.174(A), (B), or (C).

(A) Not on Switchboards Instruments, meters, and relays not located on switchboards, operating with windings or working parts at 300 volts or more to ground, and accessible to other than qualified persons, shall have the cases and other exposed metal parts grounded.

(B) On Dead-Front Switchboards Instruments, meters, and relays (whether operated from current and potential transformers or connected directly in the circuit) on switchboards having no live parts on the front of the panels shall have the cases grounded.

(C) On Live-Front Switchboards Instruments, meters, and relays (whether operated from current and potential transformers or connected directly in the circuit) on switchboards having exposed live parts on the front of panels shall not have their cases grounded. Mats of insulating rubber or other suitable floor insulation shall be provided for the operator where the voltage to ground exceeds 150.

250.176 Cases of Instruments, Meters, and Relays — Operating Voltage 1 kV and Over

Where instruments, meters, and relays have current-carrying parts of 1 kV and over to ground, they shall be isolated by elevation or protected by suitable barriers, grounded metal, or insulating covers or guards. Their cases shall not be grounded.

Exception: Cases of electrostatic ground detectors where the internal ground segments of the instrument are connected to the instrument case and grounded and the ground detector is isolated by elevation.

250.178 Instrument Grounding Conductor

The grounding conductor for secondary circuits of instrument transformers and for instrument cases shall not be smaller than 12 AWG copper or 10 AWG aluminum. Cases of instrument transformers, instruments, meters, and relays that are mounted directly on grounded metal surfaces of enclosures or grounded metal switchboard panels shall be considered to be grounded, and no additional grounding conductor shall be required.

X. Grounding of Systems and Circuits of 1 kV and Over (High Voltage)

250.180 General

Where high-voltage systems are grounded, they shall comply with all applicable provisions of the preceding sections of this article and with 250.182 through 250.190, which supplement and modify the preceding sections.

250.182 Derived Neutral Systems

A system neutral derived from a grounding transformer shall be permitted to be used for grounding high-voltage systems.

250.184 Solidly Grounded Neutral Systems

Solidly grounded neutral systems shall be permitted to be either single point grounded or multigrounded neutral.

For systems over 1000 volts, the *Code* permits solidly grounded neutral systems that are either single-point grounded or multigrounded systems. For the 2005 *Code*, 250.184 was reorganized, and new requirements for the installation of single-point grounded systems were added. Circuits supplied from a single-point grounded system are required to have an equipment grounding conductor run with the circuit conductors, and this conductor is not to be used as a conductor for continuous line-to-neutral load.

(A) Neutral Conductor
(1) Insulation Level The minimum insulation level for neutral conductors of solidly grounded systems shall be 600 volts.

Exception No. 1: Bare copper conductors shall be permitted to be used for the neutral of service entrances and the neutral of direct-buried portions of feeders.

Exception No. 2: Bare conductors shall be permitted for the neutral of overhead portions installed outdoors.

Exception No. 3: The neutral grounded conductor shall be permitted to be a bare conductor if isolated from phase conductors and protected from physical damage.

FPN: See 225.4 for conductor covering where within 3.0 m (10 ft) of any building or other structure.

(2) Ampacity The neutral conductor shall be of sufficient ampacity for the load imposed on the conductor but not less than 33⅓ percent of the ampacity of the phase conductors.

Exception: In industrial and commercial premises under engineering supervision, it shall be permissible to size the ampacity of the neutral conductor to not less than 20 percent of the ampacity of the phase conductor.

(B) Single Point Grounded System Where a single point grounded neutral system is used, the following shall apply:

(1) A single point grounded system shall be permitted to be supplied from (a) or (b):
  a. A separately derived system
  b. A multigrounded neutral system with an equipment grounding conductor connected to the multigrounded neutral at the source of the single point grounded system
(2) A grounding electrode shall be provided for the system.
(3) A grounding electrode conductor shall connect the grounding electrode to the system neutral.
(4) A bonding jumper shall connect the equipment grounding conductor to the grounding electrode conductor.
(5) An equipment bonding conductor shall be provided to each building, structure, and equipment enclosure.
(6) A neutral shall only be required where phase to neutral loads are supplied.
(7) The neutral, where provided, shall be insulated and isolated from earth except at one location.
(8) An equipment grounding conductor shall be run with the phase conductors and shall comply with (a), (b), and (c):
  a. Shall not carry continuous load
  b. May be bare or insulated
  c. Shall have sufficient ampacity for fault current duty

(C) Multigrounded Neutral Systems Where a multigrounded neutral system is used, the following shall apply:

(1) The neutral of a solidly grounded neutral system shall be permitted to be grounded at more than one point. Grounding shall be permitted at one or more of the following locations:

a. Transformers supplying conductors to a building or other structure

b. Underground circuits where the neutral is exposed

c. Overhead circuits installed outdoors

(2) The multigrounded neutral conductor shall be grounded at each transformer and at other additional locations by connection to a made or existing electrode.

(3) At least one grounding electrode shall be installed and connected to the multigrounded neutral circuit conductor every 400 m (1300 ft).

(4) The maximum distance between any two adjacent electrodes shall not be more than 400 m (1300 ft).

(5) In a multigrounded shielded cable system, the shielding shall be grounded at each cable joint that is exposed to personnel contact.

250.186 Impedance Grounded Neutral Systems

Impedance grounded neutral systems in which a grounding impedance, usually a resistor, limits the ground-fault current, shall be permitted where all of the following conditions are met:

(1) The conditions of maintenance and supervision ensure that only qualified persons will service the installation.

(2) Ground detectors are installed on the system.

(3) Line-to-neutral loads are not served.

Impedance grounded neutral systems shall comply with the provisions of 250.186(A) through (D).

(A) Location The grounding impedance shall be inserted in the grounding conductor between the grounding electrode of the supply system and the neutral point of the supply transformer or generator.

(B) Identified and Insulated The neutral conductor of an impedance grounded neutral system shall be identified, as well as fully insulated with the same insulation as the phase conductors.

(C) System Neutral Connection The system neutral shall not be connected to ground, except through the neutral grounding impedance.

(D) Equipment Grounding Conductors Equipment grounding conductors shall be permitted to be bare and shall be electrically connected to the ground bus and grounding electrode conductor.

250.188 Grounding of Systems Supplying Portable or Mobile Equipment

Systems supplying portable or mobile high-voltage equipment, other than substations installed on a temporary basis, shall comply with 250.188(A) through (F).

Portable describes equipment that is easily carried from one location to another. Mobile describes equipment that is easily moved on wheels, treads, and so on.

(A) Portable or Mobile Equipment Portable or mobile high-voltage equipment shall be supplied from a system having its neutral grounded through an impedance. Where a delta-connected high-voltage system is used to supply portable or mobile equipment, a system neutral shall be derived.

(B) Exposed Non–Current-Carrying Metal Parts Exposed non–current-carrying metal parts of portable or mobile equipment shall be connected by an equipment grounding conductor to the point at which the system neutral impedance is grounded.

(C) Ground-Fault Current The voltage developed between the portable or mobile equipment frame and ground by the flow of maximum ground-fault current shall not exceed 100 volts.

(D) Ground-Fault Detection and Relaying Ground-fault detection and relaying shall be provided to automatically de-energize any high-voltage system component that has developed a ground fault. The continuity of the equipment grounding conductor shall be continuously monitored so as to de-energize automatically the high-voltage circuit to the portable or mobile equipment upon loss of continuity of the equipment grounding conductor.

(E) Isolation The grounding electrode to which the portable or mobile equipment system neutral impedance is connected shall be isolated from and separated in the ground by at least 6.0 m (20 ft) from any other system or equipment grounding electrode, and there shall be no direct connection between the grounding electrodes, such as buried pipe and fence, and so forth.

(F) Trailing Cable and Couplers High-voltage trailing cable and couplers for interconnection of portable or mobile equipment shall meet the requirements of Part III of Article 400 for cables and 490.55 for couplers.

250.190 Grounding of Equipment

All non–current-carrying metal parts of fixed, portable, and mobile equipment and associated fences, housings, enclosures, and supporting structures shall be grounded.

Exception: Where isolated from ground and located so as to prevent any person who can make contact with ground from contacting such metal parts when the equipment is energized.

Grounding conductors not an integral part of a cable assembly shall not be smaller than 6 AWG copper or 4 AWG aluminum.

FPN: See 250.110, Exception No. 2, for pole-mounted distribution apparatus.

ARTICLE 280
Surge Arresters

Summary of Changes

- **280.4(3):** Added new requirement for short-circuit current rating marking limiting surge arrester use to applications where this rating is not exceeded.

- **280.4(4):** Added new requirement for specific application listing of surge arresters used on ungrounded systems, corner grounded systems, and impedance grounded systems.

- **280.24(A)(2):** Added reference to *static wires* being used as the grounded conductor to the requirement covering the number of grounding connections for each mile of line.

Contents

I. General

280.1 Scope

This article covers general requirements, installation requirements, and connection requirements for surge arresters installed on premises wiring systems.

Voltage surges with peaks of several thousand volts, even on 120-volt circuits, are not uncommon. These surges occur because of induced voltages in power and transmission lines resulting from lightning strikes in the vicinity of the line. Surges also occur as a result of switching inductive circuits on the premises. Surge arresters for installation as part of an electric service are commercially available. The basic standards on surge arresters are ANSI/IEEE C62.1, *Standard for Gapped Silicon-Carbide Surge Arresters for AC Power Circuits*, and ANSI/IEEE C62.11, *Standard for Metal-Oxide Surge Arresters for AC Power Circuits*.

280.2 Definition

Surge Arrester. A protective device for limiting surge voltages by discharging or bypassing surge current, and it also prevents continued flow of follow current while remaining capable of repeating these functions.

280.3 Number Required

Where used at a point on a circuit, a surge arrester shall be connected to each ungrounded conductor. A single installation of such surge arresters shall be permitted to protect a number of interconnected circuits, provided that no circuit is exposed to surges while disconnected from the surge arresters.

Means must be provided for protection of circuits that may be disconnected from the generating station bus. A switch with double-throw action used to disconnect the outside circuits from the station generator and alternatively connect those circuits to ground would satisfy the condition of a single set of arresters protecting more than one circuit.

Surge arresters are required to be installed on circuits in buildings that house explosives. For details, see Chapter 6 of NFPA 495, *Explosive Materials Code*.

280.4 Surge Arrester Selection

(A) Circuits of Less Than 1000 Volts Surge arresters installed on a circuit of less than 1000 volts shall comply with all of the following:

(1) The rating of the surge arrester shall be equal to or greater than the maximum continuous phase-to-ground power frequency voltage available at the point of application.

(2) Surge arresters installed on circuits of less than 1000 volts shall be listed.

(3) Surge arresters shall be marked with a short circuit current rating and shall not be installed at a point on the system where the available fault current is in excess of that rating.

(4) Surge arresters shall not be installed on ungrounded systems, impedance grounded systems, or corner grounded delta systems unless listed specifically for use on these systems.

(B) Circuits of 1 kV and Over — Silicon Carbide Types
The rating of a silicon carbide-type surge arrester shall be not less than 125 percent of the maximum continuous phase-to-ground voltage available at the point of application.

> FPN No. 1: For further information on surge arresters, see ANSI/IEEE C62.1-1989, *Standard for Gapped Silicon-Carbide Surge Arresters for AC Power Circuits*; ANSI/IEEE C62.2-1987, *Guide for the Application of Gapped Silicon-Carbide Surge Arresters for Alternating-Current Systems*; ANSI/IEEE C62.11-1993, *Standard for Metal-Oxide Surge Arresters for Alternating-Current Power Circuits*; and ANSI/IEEE C62.22-1991, *Guide for the Application of Metal-Oxide Surge Arresters for Alternating-Current Systems*.

> FPN No. 2: The selection of a properly rated metal oxide arrester is based on considerations of maximum continuous operating voltage and the magnitude and duration of overvoltages at the arrester location as affected by phase-to-ground faults, system grounding techniques, switching surges, and other causes. See the manufacturer's application rules for selection of the specific arrester to be used at a particular location.

II. Installation

280.11 Location

Surge arresters shall be permitted to be located indoors or outdoors. Surge arresters shall be made inaccessible to unqualified persons, unless listed for installation in accessible locations.

Maximum protection is achieved where the surge protective device is located as close as practicable to the equipment to be protected. When a surge passes through an arrester, a wave is reflected in both directions on the conductors connected to the surge arrester. The magnitude of the reflected wave increases as the distance from the arrester increases. If the length of the conductor between the protected equipment and the surge arrester is short, the magnitude of the wave reflected through the equipment is minimized.

280.12 Routing of Surge Arrester Connections

The conductor used to connect the surge arrester to line or bus and to ground shall not be any longer than necessary and shall avoid unnecessary bends.

Arrester conductors should be as short and be run as straight as practicable, avoiding any sharp bends and turns, which increase the impedance.

III. Connecting Surge Arresters

280.21 Installed at Services of Less Than 1000 Volts

Line and ground connecting conductors shall not be smaller than 14 AWG copper or 12 AWG aluminum. The arrester grounding conductor shall be connected to one of the following:

(1) Grounded service conductor
(2) Grounding electrode conductor
(3) Grounding electrode for the service
(4) Equipment grounding terminal in the service equipment

High-frequency currents, such as those common to lightning discharges, tend to reduce the effectiveness of a grounding conductor. Single-phase or 3-phase grounded or ungrounded services are permitted to have the surge arrester grounded to the equipment grounding terminal in the service equipment. Exhibit 280.1 shows three methods of grounding the ground terminals of surge arresters at service entrances. In the upper left diagram, an arrester is connected to a neutral service conductor. In the upper right diagram, an arrester is connected to a grounding electrode conductor. In the lower diagram, an arrester is connected to a grounding electrode conductor of an ungrounded system.

280.22 Installed on the Load Side Services of Less Than 1000 Volts

Line and ground connecting conductors shall not be smaller than 14 AWG copper or 12 AWG aluminum. A surge arrester shall be permitted to be connected between any two conductors — ungrounded conductor(s), grounded conductor, grounding conductor. The grounded conductor and the grounding conductor shall be interconnected only by the normal operation of the surge arrester during a surge.

280.23 Circuits of 1 kV and Over — Surge-Arrester Conductors

The conductor between the surge arrester and the line and the surge arrester and the grounding connection shall not be smaller than 6 AWG copper or aluminum.

280.24 Circuits of 1 kV and Over — Interconnections

The grounding conductor of a surge arrester protecting a transformer that supplies a secondary distribution system shall be interconnected as specified in 280.24(A), (B), or (C).

(A) Metallic Interconnections A metallic interconnection shall be made to the secondary grounded circuit conductor or the secondary circuit grounding conductor provided that,

Single-Phase Grounded System

3-Phase Ungrounded System

Exhibit 280.1 Three methods of grounding surge arresters at service entrances.

in addition to the direct grounding connection at the surge arrester, the following occurs:

(1) The grounded conductor of the secondary has elsewhere a grounding connection to a continuous metal underground water piping system. However, in urban water-pipe areas where there are at least four water-pipe connections on the neutral and not fewer than four such connections in each mile of neutral, the metallic interconnection shall be permitted to be made to the secondary neutral with omission of the direct grounding connection at the surge arrester.

(2) The grounded conductor of the secondary system is a part of a multiground neutral system or static wire of which the primary neutral or static wire has at least four ground connections in each mile of line in addition to a ground at each service.

(B) Through Spark Gap or Device Where the surge arrester grounding conductor is not connected as in 280.24(A) or where the secondary is not grounded as in 280.24(A) but

is otherwise grounded as in 250.52, an interconnection shall be made through a spark gap or listed device as follows:

(1) For ungrounded or unigrounded primary systems, the spark gap or listed device shall have a 60-Hz breakdown voltage of at least twice the primary circuit voltage but not necessarily more than 10 kV, and there shall be at least one other ground on the grounded conductor of the secondary that is not less than 6.0 m (20 ft) distant from the surge arrester grounding electrode.

(2) For multigrounded neutral primary systems, the spark gap or listed device shall have a 60-Hz breakdown of not more than 3 kV, and there shall be at least one other ground on the grounded conductor of the secondary that is not less than 6.0 m (20 ft) distant from the surge arrester grounding electrode.

(C) By Special Permission An interconnection of the surge arrester ground and the secondary neutral, other than as provided in 280.24(A) or (B), shall be permitted to be made only by special permission.

280.25 Grounding

Except as indicated in this article, surge arrester grounding connections shall be made as specified in Article 250. Grounding conductors shall not be run in metal enclosures unless bonded to both ends of such enclosure.

ARTICLE 285
Transient Voltage Surge
Suppressors: TVSSs

Summary of Changes

- **285.3(2):** Added new requirement for specific application listing of surge arresters used on ungrounded systems, corner grounded systems, and impedance grounded systems.
- **285.21(A)(1):** Added reference to 230.82(8) for applications permitting TVSS devices to be installed on the line side of the service disconnecting means.

Contents

I. General

285.1 Scope

This article covers general requirements, installation requirements, and connection requirements for transient voltage surge suppressors (TVSSs) permanently installed on premises wiring systems.

Article 285, which was added in the 2002 *Code,* was created to provide installation requirements for new technology for the protection of persons and electronic equipment.

A transient voltage surge suppressor (TVSS) is a common component of an electrical system that provides a protection function similar to that of a surge arrester (see Article 280). A TVSS is generally installed to protect sensitive electronic equipment such as computers, telecommunications equipment, security systems, and electronic appliances. Compared to a surge arrester, a TVSS should begin to divert or limit the surge current from a transient (or surge) event much closer to the operating voltage. Two examples of transient voltage surge suppressors are shown in Exhibit 285.1. This article covers only permanently installed TVSS devices; the portable cord-and-plug-connected types are intended to be used in accordance with their listings and the manufacturers' instructions.

Exhibit 285.1 Two TVSS devices suitable for service-entrance installation, one for direct connection to panelboard busbars and one for mounting in an cabinet or enclosure knockout. (Courtesy of General Electric Co.)

285.2 Definition

Transient Voltage Surge Suppressor (TVSS). A protective device for limiting transient voltages by diverting or limiting surge current; it also prevents continued flow of follow current while remaining capable of repeating these functions.

285.3 Uses Not Permitted

A TVSS device shall not be installed in the following:

(1) Circuits exceeding 600 volts
(2) On ungrounded systems, impedance grounded systems, or corner grounded delta systems unless listed specifically for use on these systems.
(3) Where the rating of the TVSS is less than the maximum continuous phase-to-ground power frequency voltage available at the point of application

> FPN: For further information on TVSSs, see NEMA LS 1-1992, *Standard for Low Voltage Surge Suppression Devices.* The selection of a properly rated TVSS is based on criteria such as maximum continuous operating voltage, the magnitude and duration of overvoltages at the suppressor location as affected by phase-to-ground faults, system grounding techniques, and switching surges.

UL 1449, *Safety Standard for Transient Voltage Surge Suppressors,* limits TVSS applications to 600 volts and less. TVSS devices are permitted to be installed on ungrounded systems, impedance grounded systems, and corner grounded systems where the TVSS is specifically listed for such systems.

285.4 Number Required

Where used at a point on a circuit, the TVSS shall be connected to each ungrounded conductor.

285.5 Listing

A TVSS shall be a listed device.

285.6 Short Circuit Current Rating

The TVSS shall be marked with a short circuit current rating and shall not be installed at a point on the system where the available fault current is in excess of that rating. This marking requirement shall not apply to receptacles.

The first TVSS device is commonly installed in the electrical system either as an integral component of or near to the service entrance equipment in residential and commercial structures. It is imperative that the available fault current at the point of installation not exceed the short-circuit current rating of the TVSS. The installed TVSS device must match or exceed the system's available fault current at its point of installation on a system. Of course, as an alternative to the TVSS at the service, an industrial or large commercial facility may elect to install arresters (Article 280) at the service equipment and at intermediate points in the distribution system and then install TVSS devices downstream at panelboards that serve loads susceptible to transients. See Exhibit 285.2.

Exhibit 285.2 A TVSS as an integral component of a receptacle, providing local point-of-use protection of equipment when transient events occur within the facility. (Courtesy of Pass & Seymour/ Legrand®)

Another type of TVSS is the point-of-use TVSS. These devices (e.g., receptacles and permanently installed power strips) may be installed at the equipment (computers, equipment with electronic controls, etc.). The function of a point-of-use TVSS is to remove small transients that pass through the more robust surge devices located at the service. Point-of-use TVSS devices are also useful in removing small transients that have been generated within the building.

II. Installation

285.11 Location

TVSSs shall be permitted to be located indoors or outdoors and shall be made inaccessible to unqualified persons, unless listed for installation in accessible locations.

285.12 Routing of Connections

The conductors used to connect the TVSS to the line or bus and to ground shall not be any longer than necessary and shall avoid unnecessary bends.

The conductor length used to connect the TVSS plays an important role in protection performance. As the length of the conductor increases, so does the impedance in the conduction path. This drives the clamping voltage higher and reduces the protection provided by the TVSS unit. Maximum protection is achieved where the TVSS is located as close as practicable to the equipment being protected, as shown in Exhibit 285.3.

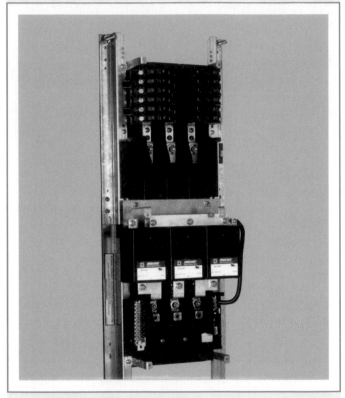

Exhibit 285.3 A TVSS mounted as an integral component of a panelboard, which minimizes conductor length between the electrical system and the TVSS. (Courtesy of Square D Co.)

III. Connecting Transient Voltage Surge Suppressors

285.21 Connection

Where a TVSS is installed, it shall comply with 285.21(A) through (C).

UL 1449, *Safety Standard for Transient Voltage Surge Suppressors*, is used to investigate the safety of a TVSS. In accordance with the scope of UL 1449, a TVSS must be installed on the load side of the service disconnect overcurrent protection. The requirement for connection on the load side of the first overcurrent protection device in a feeder-supplied structure is necessary due to the exposure of external feeder conductors to a more hostile surge environment such as lightning.

Two requirements in Article 230 have been revised for the 2005 *Code* to permit the installation of TVSS equipment on the line side of the service disconnecting means. First, a revision to 230.71(A) permits an additional disconnecting means at the service equipment for TVSS equipment installed as part of listed equipment. The disconnecting means for the TVSS device does not count as one of the six permitted by 230.71(A) where the TVSS and its disconnecting means are provided in the listed equipment by the manufacturer. The second revision is in 230.82(8), in which TVSS equipment installed in listed equipment is permitted to be connected on the line side of the service disconnecting means where the TVSS equipment is provided with a disconnecting means and overcurrent protection.

(A) Location

(1) Service Supplied Building or Structure The transient voltage surge suppressor shall be connected on the load side of a service disconnect overcurrent device required in 230.91, unless installed in accordance with 230.82(8).

(2) Feeder Supplied Building or Structure The transient voltage surge suppressor shall be connected on the load side of the first overcurrent device at the building or structure.

Exception to (1) and (2): Where the TVSS is also listed as a surge arrester, the connection shall be as permitted by Article 280.

(3) Separately Derived System The TVSS shall be connected on the load side of the first overcurrent device in a separately derived system.

(B) Conductor Size Line and ground connecting conductors shall not be smaller than 14 AWG copper or 12 AWG aluminum.

(C) Connection Between Conductors A TVSS shall be permitted to be connected between any two conductors — ungrounded conductor(s), grounded conductor, grounding conductor. The grounded conductor and the grounding conductor shall be interconnected only by the normal operation of the TVSS during a surge.

285.25 Grounding

Grounding conductors shall not be run in metal enclosures unless bonded to both ends of such enclosure.

See 250.64(E) and 250.92(A)(3) and the associated commentary, including Exhibit 250.32, for requirements to bond both ends of a metal raceway that encloses a grounding electrode conductor.

Wiring Methods and Materials

There were two significant editorial changes in Chapter 3 for the 2002 *Code*. Both changes resulted from the collective efforts of the NEC Usability Task Group, the code-making panels (CMPs) having specific subject responsibility within Chapter 3, and the users who submitted proposals and comments.

First and foremost, many common requirements for wiring methods in Chapter 3 were aligned using a common numbering system. All the circular raceway articles and many of the cable articles now have a common numbering format, which assists users in locating common requirements within an article. Additionally, students of the *Code* will find it much easier to understand, learn, and compare the many common requirements of various wiring methods.

The second significant editorial change was the renumbering of many articles in Chapter 3. Achieving a true common numbering format in each article necessitated splitting a few of them into two separate articles. These additional articles forced the NEC Usability Task Group to consider and then adopt a new article renumbering scheme for Chapter 3. This new format allows for a grouping of like articles and leaves space for new wiring method articles should they be added in the future.

Annex F contains three cross-reference tables to guide the user through the Chapter 3 reorganization.

ARTICLE 300
Wiring Methods

Summary of Changes

- **300.3(A):** Added exception to permit single overhead conductors.
- **300.4(A)(1):** Added Exception No. 2 to permit a listed and marked steel plate less than ¹⁄₁₆ in. thick that provides equal or better protection against nail or screw penetration.
- **300.4(B)(2):** Added Exception No. 2 to permit a listed and marked steel plate less than ¹⁄₁₆ in. thick that provides equal or better protection against nail or screw penetration.
- **300.4(D):** Revised to add a reference to furring strips, clarifying that even though the strips may not be framing members, the 1¼ in. clearance from any edge subject to nail or screw penetration is required to be maintained. Exception No. 3 added to permit a listed and marked steel plate less than ¹⁄₁₆ in. thick that provides equal or better protection against nail or screw penetration.
- **300.4(E):** Added Exception No. 2 to permit a listed and marked steel plate less than ¹⁄₁₆ in. thick that provides equal or better protection against nail or screw penetration.

- **300.5(B):** Added new requirement for listing for use in wet locations applies to cables and insulated conductors in underground raceways and enclosures.
- **300.6:** Revised title to clarify that this section addresses corrosion and other deterioration of metal and nonmetallic electrical equipment. Also revised to provide separate protection requirements for ferrous metal equipment and nonferrous metal equipment.
- **300.18(A):** Added exception for short lengths used to protect conductors or cable assemblies from physical damage.
- **300.22(B):** Deleted permission to use limited lengths of liquidtight flexible metal conduit (LFMC) as a wiring method in ducts and plenums used for environmental air.
- **Table 300.50:** Expanded to include special conditions formerly contained in six exceptions.

Contents

I. General Requirements

300.1 Scope

(A) All Wiring Installations This article covers wiring methods for all wiring installations unless modified by other articles.

(B) Integral Parts of Equipment The provisions of this article are not intended to apply to the conductors that form an integral part of equipment, such as motors, controllers, motor control centers, or factory assembled control equipment or listed utilization equipment.

Generally, wiring within equipment is covered by product standards. As an example, integral wiring of motors is covered by NEMA MG 1, *Motors and Generators*; integral wiring of industrial control equipment by UL 508, *Standard for Industrial Control Equipment*; and wiring that forms an integral part of industrial machinery by NFPA 79, *Electrical Standard for Industrial Machinery*.

(C) Metric Designators and Trade Sizes Metric designators and trade sizes for conduit, tubing, and associated fittings and accessories shall be as designated in Table 300.1(C).

Table 300.1(C) Metric Designator and Trade Sizes

Metric Designator	Trade Size
12	⅜
16	½
21	¾
27	1
35	1¼
41	1½
53	2
63	2½
78	3
91	3½
103	4
129	5
155	6

Note: The metric designators and trade sizes are for identification purposes only and are not actual dimensions.

Using metric designators to describe circular raceways is one more step in the metrication of the *NEC*, as stated in both 90.9 and its associated commentary. Metric designators for conduits first appeared in 1989 in IEC 981, *Extra-Heavy Duty Rigid Steel Conduits for Electrical Installations*. Since then, both NEMA and the *NEC* have recognized metric designators. The *NEC* did so in the 1996 edition, allowing metric designators to appear as fine print notes following the mention of trade sizes of circular raceways. Assigning metric designators to traditional trade size threaded conduit does not change the physical dimensions or the traditional "NPT type" threads of the conduit. Using metric designators is simply another method of identifying the size of a circular raceway.

Table 300.1(C) identifies a distinct metric designator for each circular raceway trade size. Dimensions or descriptions of circular raceways have traditionally included an inch size or unit of measure. The unit of measure associated with a circular raceway has not been included in Table 300.1(C) or throughout the *NEC* because it reflects not a true measure but rather a "modular" or "relative" measure. Many examples of modular or relative measures can be found in the building trades. For example, in North America, items such as a 2 ft × 4 ft drop-in luminaire, an 8 ft fluorescent lamp, and a 2 in. × 4 in. piece of lumber do not reflect true dimensions but rather are loosely associated dimensions common in modular construction. As stated in the footnote to Table 300.1(C), the metric designators and trade sizes are not actual dimensions.

According to Table 4 of Chapter 9, each metric designator sized circular raceway is identical in internal and external dimensions (including manufacturing tolerances) to its trade size counterpart. Therefore, Annex C wire fill tables are applicable to both metric designator and trade size circular raceways, and so for installation practices, introducing an associated metric designator for traditional circular raceways trade sizes should not be a concern.

What is a concern for installation practices, however, is the use of a circular raceway with threaded joints, where the threaded joints are not according to the product standard. For example, rigid metal conduit (RMC) is required to be listed, according to 344.6, and the appropriate product standard for this listing is UL 6, *Electrical Rigid Metal Conduit — Steel*. Intermediate metal conduit (IMC) is required to be listed, according to 342.6, and the appropriate product standard for this listing is UL 1242, *Standard for Electrical Intermediate Metal Conduit — Steel*. Both listed conduits must be threaded in accordance with ANSI/ASME B.1.20.1-1993, *Pipe Threads, General Purpose (Inch)*. Therefore, only conduits threaded to the traditional dimension of ³/₄-in. taper per foot are acceptable. Simply stated, an installation using metric threaded conduit is not permitted by the *NEC*. However, an installation using threads according to ANSI/ASME B1.20.1-1993 is in compliance with the *NEC*.

300.2 Limitations

(A) Voltage Wiring methods specified in Chapter 3 shall be used for 600 volts, nominal, or less where not specifically limited in some section of Chapter 3. They shall be permitted for over 600 volts, nominal, where specifically permitted elsewhere in this *Code*.

(B) Temperature Temperature limitation of conductors shall be in accordance with 310.10.

See 110.14(C) and its commentary for information on temperature limitations of conductor terminations.

300.3 Conductors

(A) Single Conductors Single conductors specified in Table 310.13 shall only be installed where part of a recognized wiring method of Chapter 3.

Exception: Individual conductors shall be permitted where installed as separate overhead conductors in accordance with 225.6.

Section 300.3(A) clearly states that building wire, such as individual insulated conductors identified as THHN, is prohibited from use outside a recognized wiring method. This exception, added for the 2005 *Code*, points out two long-time permissions: first, allowing individual conductors as festoon lighting and, second, allowing individual conductors as overhead spans.

(B) Conductors of the Same Circuit All conductors of the same circuit and, where used, the grounded conductor and all equipment grounding conductors and bonding conductors shall be contained within the same raceway, auxiliary gutter, cable tray, cablebus assembly, trench, cable, or cord, unless otherwise permitted in accordance with 300.3(B)(1) through (B)(4).

This general rule remains consistent with electrical theory; that is, to reduce inductive heating and to avoid increases in overall circuit impedance, all circuit conductors of an individual circuit must be grouped. Similar requirements are found in 300.5(I).

(1) Paralleled Installations Conductors shall be permitted to be run in parallel in accordance with the provisions of 310.4. The requirement to run all circuit conductors within the same raceway, auxiliary gutter, cable tray, trench, cable, or cord shall apply separately to each portion of the paralleled installation, and the equipment grounding conductors shall comply with the provisions of 250.122. Parallel runs in cable tray shall comply with the provisions of 392.8(D).

Exception: Conductors installed in nonmetallic raceways run underground shall be permitted to be arranged as isolated phase installations. The raceways shall be installed in close proximity, and the conductors shall comply with the provisions of 300.20(B).

(2) Grounding and Bonding Conductors Equipment grounding conductors shall be permitted to be installed outside a raceway or cable assembly where in accordance with the provisions of 250.130(C) for certain existing installations or in accordance with 250.134(B), Exception No. 2, for dc circuits. Equipment bonding conductors shall be permitted to be installed on the outside of raceways in accordance with 250.102(E).

Section 300.3(B)(2) recognizes that some types of grounding and bonding conductors can be run as single conductors on the exterior of the raceway or outside a cable assembly.

(3) Nonferrous Wiring Methods Conductors in wiring methods with a nonmetallic or other nonmagnetic sheath, where run in different raceways, auxiliary gutters, cable trays, trenches, cables, or cords, shall comply with the provisions of 300.20(B). Conductors in single-conductor Type MI cable with a nonmagnetic sheath shall comply with the provisions of 332.31. Conductors of single-conductor Type MC cable with a nonmagnetic sheath shall comply with the provisions of 330.31, 330.116, and 300.20(B).

Section 300.3(B)(3) was revised for the 2002 *Code* to address the installation of single-conductor-Type MC cable using a nonferrous (nonmagnetic) sheath.

(4) Enclosures Where an auxiliary gutter runs between a column-width panelboard and a pull box, and the pull box includes neutral terminations, the neutral conductors of circuits supplied from the panelboard shall be permitted to originate in the pull box.

Section 300.3(B)(4) recognizes the practice of supplying narrow, column-type panelboard through an auxiliary gutter from an overhead pull box and running only the ungrounded conductors down from the pull box to the panelboard. As shown in Exhibit 300.1, the feeder and branch-circuit neutral

Exhibit 300.1 An installation where an auxiliary gutter extends from the panelboard up to a pull box used as a termination point for the feeder and branch-circuit grounded conductors (neutrals).

conductors are terminated in the overhead pull box and are not carried with the ungrounded conductors. Inductive heating does not occur, because the load-carrying conductors extend down and back up within the same enclosure.

(C) Conductors of Different Systems

(1) 600 Volts, Nominal, or Less Conductors of circuits rated 600 volts, nominal, or less, ac circuits, and dc circuits shall be permitted to occupy the same equipment wiring enclosure, cable, or raceway. All conductors shall have an insulation rating equal to at least the maximum circuit voltage applied to any conductor within the enclosure, cable, or raceway.

Exception: For solar photovoltaic systems in accordance with 690.4(B).

> FPN: See 725.55(A) for Class 2 and Class 3 circuit conductors.

Section 300.3(C)(1) makes it clear that it is the maximum circuit voltage in the raceway, not the maximum insulation voltage rating of the conductors in the raceway, that determines the minimum voltage rating required for the insulation of conductors for systems of 600 volts or less.

The conductors of a 3-phase, 4-wire, 208Y/120-volt ac circuit; a 3-phase, 4-wire, 480Y/277-volt ac circuit; and a 3-wire, 120/240-volt dc circuit may occupy the same equipment wiring enclosure, cable, or raceway if all the conductors are insulated for the maximum circuit voltage of any conductor. In that case, the maximum circuit voltage would be 480 volts, and 600-volt insulation would be suitable for all the conductors.

If a 2-wire, 120-volt circuit is included in the same raceway with a 3-wire, 120/240-volt circuit having 600-volt conductors, then the 2-wire, 120-volt circuit conductors could use 300-volt insulation because the maximum circuit voltage is only 240 volts.

Section 690.4(B) prohibits the location of solar photovoltaic circuits within the same enclosure as conductors of other systems unless the conductors are separated by a partition or are connected together.

Section 725.55(A) prohibits Class 2 and Class 3 circuit conductors from occupying the same enclosure, cable, or raceway as Class 1, electric light, and power conductors, unless specifically permitted in 725.55(B) through 725.55(J).

(2) Over 600 Volts, Nominal Conductors of circuits rated over 600 volts, nominal, shall not occupy the same equipment wiring enclosure, cable, or raceway with conductors of circuits rated 600 volts, nominal, or less unless otherwise permitted in (C)(2)(a) through (C)(2)(e).

(a) Secondary wiring to electric-discharge lamps of 1000 volts or less, if insulated for the secondary voltage involved, shall be permitted to occupy the same luminaire (fixture), sign, or outline lighting enclosure as the branch-circuit conductors.

(b) Primary leads of electric-discharge lamp ballasts insulated for the primary voltage of the ballast, where contained within the individual wiring enclosure, shall be permitted to occupy the same luminaire (fixture), sign, or outline lighting enclosure as the branch-circuit conductors.

(c) Excitation, control, relay, and ammeter conductors used in connection with any individual motor or starter shall be permitted to occupy the same enclosure as the motor-circuit conductors.

(d) In motors, switchgear and control assemblies, and similar equipment, conductors of different voltage ratings shall be permitted.

(e) In manholes, if the conductors of each system are permanently and effectively separated from the conductors of the other systems and securely fastened to racks, insulators, or other approved supports, conductors of different voltage ratings shall be permitted.

Conductors having nonshielded insulation and operating at different voltage levels shall not occupy the same enclosure, cable, or raceway.

300.4 Protection Against Physical Damage

Where subject to physical damage, conductors shall be protected.

(A) Cables and Raceways Through Wood Members

(1) Bored Holes In both exposed and concealed locations, where a cable- or raceway-type wiring method is installed through bored holes in joists, rafters, or wood members, holes shall be bored so that the edge of the hole is not less than 32 mm (1¼ in.) from the nearest edge of the wood member. Where this distance cannot be maintained, the cable or raceway shall be protected from penetration by screws or nails by a steel plate or bushing, at least 1.6 mm (¹⁄₁₆ in.) thick, and of appropriate length and width installed to cover the area of the wiring.

Exception No. 1: Steel plates shall not be required to protect rigid metal conduit, intermediate metal conduit, rigid nonmetallic conduit, or electrical metallic tubing.

Exception No. 2: A listed and marked steel plate less than 1.6 mm (¹⁄₁₆ in.) thick that provides equal or better protection against nail or screw penetration shall be permitted.

(2) Notches in Wood Where there is no objection because of weakening the building structure, in both exposed and concealed locations, cables or raceways shall be permitted to be laid in notches in wood studs, joists, rafters, or other wood members where the cable or raceway at those points is protected against nails or screws by a steel plate at least 1.6 mm (¹⁄₁₆ in.) thick, and of appropriate length and width, installed to cover the area of the wiring. The steel plate shall be installed before the building finish is applied.

Exception No. 1: Steel plates shall not be required to protect rigid metal conduit, intermediate metal conduit, rigid non-metallic conduit, or electrical metallic tubing.

Exception No. 2: A listed and marked steel plate less than 1.6 mm (¹⁄₁₆ in.) thick that provides equal or better protection against nail or screw penetration shall be permitted.

The intent of 300.4(A)(1) is to prevent nails and screws from being driven into cables and raceways. Keeping the edge of a drilled hole 1¼ in. from the nearest edge of a stud, as shown in Exhibit 300.2, should prevent nails from penetrating the stud far enough to injure a cable. Building codes limit the maximum size of bored or notched holes in studs, and 300.4(A)(2) indicates that consideration should be given to the size of notches in studs, so as not to affect the strength of the structure.

Exhibit 300.2 A steel plate used to protect a nonmetallic-sheathed cable within 1¼ in. of the edge of a wood stud. (Courtesy of Hubbell RACO)

Exception No. 1 to 300.4(A)(1) and Exception No. 1 to 300.4(A)(2) permit intermediate metal conduit, rigid metal conduit, rigid nonmetallic conduit, and electrical metallic

tubing to be installed through bored holes or laid in notches less than 1¼ in. from the nearest edge of the stud, without a steel plate or bushing. Exception No. 2 to 300.4(A)(1) and Exception No. 2 to 300.4(A)(2), new for the 2005 *Code*, permit steel plates thinner than ¹⁄₁₆ in. to protect cables and raceways, but only if the plates are specifically listed and marked.

(B) Nonmetallic-Sheathed Cables and Electrical Non-metallic Tubing Through Metal Framing Members

(1) Nonmetallic-Sheathed Cable In both exposed and concealed locations where nonmetallic-sheathed cables pass through either factory or field punched, cut, or drilled slots or holes in metal members, the cable shall be protected by listed bushings or listed grommets covering all metal edges that are securely fastened in the opening prior to installation of the cable.

The phrase "listed bushing or listed grommet covering all metal edges" was new to the 2002 *Code*. This change requires the use of listed grommets or listed bushings that completely encircle Type NM cables as they pass through holes in metal studs. These listed devices must also securely seat in the stud opening and meet pull-out requirements of the product standard. This requirement affords physical protection for nonmetallic-sheathed cables as the cables are pulled through the openings in metal studs. Notice, too, that this requirement mandates all metal studs to be positioned in place before cable is pulled through protected openings. Fastening the listed grommet or listed bushing in place prior to installing cable is also mandatory. Should additional metal studs become necessary after installation of a cable, the cable must be removed before the stud is added. Field notching of metal studs and then installation of the stud around a nonmetallic sheathed cable already installed or in place leads to cable damage and can result in insulation failure.

(2) Nonmetallic-Sheathed Cable and Electrical Nonmetallic Tubing Where nails or screws are likely to penetrate nonmetallic-sheathed cable or electrical nonmetallic tubing, a steel sleeve, steel plate, or steel clip not less than 1.6 mm (¹⁄₁₆ in.) in thickness shall be used to protect the cable or tubing.

Exception: A listed and marked steel plate less than 1.6 mm (¹⁄₁₆ in.) thick that provides equal or better protection against nail or screw penetration shall be permitted.

This new exception for the 2005 *Code* permits the use of steel plates thinner than ¹⁄₁₆ in. to protect cables and raceways, but only if the plates are specifically listed and marked.

(C) Cables Through Spaces Behind Panels Designed to Allow Access Cables or raceway-type wiring methods, installed behind panels designed to allow access, shall be supported according to their applicable articles.

Cable- or raceway-type wiring methods installed above suspended ceilings with lift-up panels must not be laid on the suspended ceiling so as to inhibit access. They are required to be supported according to 300.11(A), 300.23, and the requirements of the article applicable to the wiring method involved. Similarly, low-voltage, optical fiber, broadband, and communications cables are not permitted to block access to equipment above the suspended ceiling. Examples of this requirement are also found in 725.7, 760.7, 770.21, 800.21, 820.21, and 830.21. For support of low-voltage cables, optical fiber, broadband, and communications cables, see 720.11, 725.8, 760.8, 770.24, 800.24, 820.24, and 830.24.

(D) Cables and Raceways Parallel to Framing Members and Furring Strips In both exposed and concealed locations, where a cable- or raceway-type wiring method is installed parallel to framing members, such as joists, rafters, or studs, or is installed parallel to furring strips, the cable or raceway shall be installed and supported so that the nearest outside surface of the cable or raceway is not less than 32 mm (1¼ in.) from the nearest edge of the framing member or furring strips where nails or screws are likely to penetrate. Where this distance cannot be maintained, the cable or raceway shall be protected from penetration by nails or screws by a steel plate, sleeve, or equivalent at least 1.6 mm (¹⁄₁₆ in.) thick.

As shown in Exhibit 300.3, NM cables are positioned so as to equal or exceed the minimum clear distance of 1¼ in. that is required between the furring strip (wood strapping in this case) and the nearest edge of the NM cable as required by 300.4(D).

Exception No. 1: Steel plates, sleeves, or the equivalent shall not be required to protect rigid metal conduit, intermediate metal conduit, rigid nonmetallic conduit, or electrical metallic tubing.

Exception No. 2: For concealed work in finished buildings, or finished panels for prefabricated buildings where such supporting is impracticable, it shall be permissible to fish the cables between access points.

Exception No. 3: A listed and marked steel plate less than 1.6 mm (¹⁄₁₆ in.) thick that provides equal or better protection against nail or screw penetration shall be permitted.

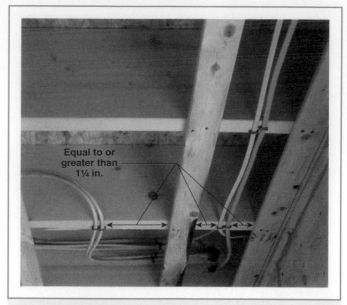

Equal to or greater than 1¼ in.

Exhibit 300.3 Nonmetallic sheathed cables adjacent to furring strips in a wood frame structure.

The intent of 300.4(D) is to prevent mechanical damage to cables and raceways from nails and screws. Revised for the 2005 *Code*, this section recognizes that cables and raceways need the same level of physical protection at furring strips as they do at framing members where nails and screws are likely to penetrate. The *Code* offers two means of protection. The first method is to fasten the cable or raceway so that it is at least 1¼ in. from the edge of the framing member, as illustrated in Exhibit 300.4. This requirement generally applies to exposed and concealed work. The second method permits the cable or raceway to be installed closer than 1¼ in. from the edge of the framing member if physical protection, such as a steel plate, its equivalent, or a sleeve, is provided. (A steel plate is illustrated in Exhibit 300.2.) New for the 2005 *Code*, Exception No. 3 to 300.4(D) permits the use of steel plates thinner than ¹⁄₁₆ in. to protect cables and raceways, but only if the plates are specifically listed and marked. As stated in Exception No. 1, the steel plate requirement does not apply to rigid metal conduit, rigid nonmetallic conduit, intermediate metal conduit, or electrical metallic tubing wiring methods because those methods provide physical protection for the conductors.

(E) Cables and Raceways Installed in Shallow Grooves Cable- or raceway-type wiring methods installed in a groove, to be covered by wallboard, siding, paneling, carpeting, or similar finish, shall be protected by 1.6 mm (¹⁄₁₆ in.) thick steel plate, sleeve, or equivalent or by not less than 32-mm (1¼-in.) free space for the full length of the groove in which the cable or raceway is installed.

Exhibit 300.4 Cables and raceways installed parallel to framing members in accordance with 300.4(D).

12-2 W/G Type NM-B

1¼ in. min.

1¼ in. min.

Exception No. 1: Steel plates, sleeves, or the equivalent shall not be required to protect rigid metal conduit, intermediate metal conduit, rigid nonmetallic conduit, or electrical metallic tubing.

Exception No. 2: A listed and marked steel plate less than 1.6 mm (¹/₁₆ in.) thick that provides equal or better protection against nail or screw penetration shall be permitted.

(F) Insulated Fittings Where raceways containing ungrounded conductors 4 AWG or larger enter a cabinet, box enclosure, or raceway, the conductors shall be protected by a substantial fitting providing a smoothly rounded insulating surface, unless the conductors are separated from the fitting or raceway by substantial insulating material that is securely fastened in place.

Exception: Where threaded hubs or bosses that are an integral part of a cabinet, box enclosure, or raceway provide a smoothly rounded or flared entry for conductors.

Conduit bushings constructed wholly of insulating material shall not be used to secure a fitting or raceway. The insulating fitting or insulating material shall have a temperature rating not less than the insulation temperature rating of the installed conductors.

Heavy conductors and cables tend to stress the conductor insulation at raceway terminating points. Providing insulated bushing or smooth rounded entries at raceway and cable terminations reduces the risk of insulation failure at conductor insulation stress points. The temperature ratings of insulating bushing must coordinate with the insulation of the conductor to ensure that the protection remains intact over the life cycle of the insulated conductor.

Because this requirement is located in 300.4(F), it applies generally to all wiring methods and all enclosures. See also 342.46, 344.46, and 352.46 for information relating to bushings.

Listed insulating bushings provided separately or as part of a fitting are colored black or brown if they are suitable for a temperature of 150°C and any other color for 90°C, unless specifically marked for a higher temperature. Exhibit 300.5 shows an insulated thermoplastic or fiber bushing that is used to protect the conductors from chafing against a metal conduit fitting. Note the use of a double locknut.

Insulated thermoplastic or fiber bushing (Note: double locknut)

Exhibit 300.5 An insulating bushing used to protect conductors from chafing against a metal conduit fitting.

300.5 Underground Installations

(A) Minimum Cover Requirements Direct-buried cable or conduit or other raceways shall be installed to meet the minimum cover requirements of Table 300.5.

Conductors under residential driveways must be at least 18 in. below grade. However, if the conductors are protected by an overcurrent device rated at not more than 20 amperes

Table 300.5 Minimum Cover Requirements, 0 to 600 Volts, Nominal, Burial in Millimeters (Inches)

	Type of Wiring Method or Circuit									
Location of Wiring Method or Circuit	**Column 1 Direct Burial Cables or Conductors**		**Column 2 Rigid Metal Conduit or Intermediate Metal Conduit**		**Column 3 Nonmetallic Raceways Listed for Direct Burial Without Concrete Encasement or Other Approved Raceways**		**Column 4 Residential Branch Circuits Rated 120 Volts or Less with GFCI Protection and Maximum Overcurrent Protection of 20 Amperes**		**Column 5 Circuits for Control of Irrigation and Landscape Lighting Limited to Not More Than 30 Volts and Installed with Type UF or in Other Identified Cable or Raceway**	
	mm	in.	mm	in.	mm	in.	mm	in.	mm	in.
All locations not specified below	600	24	150	6	450	18	300	12	150	6
In trench below 50-mm (2-in.) thick concrete or equivalent	450	18	150	6	300	12	150	6	150	6
Under a building	0 (in raceway only)	0	0	0	0	0	0 (in raceway only)	0	0 (in raceway only)	0
Under minimum of 102-mm (4-in.) thick concrete exterior slab with no vehicular traffic and the slab extending not less than 152 mm (6 in.) beyond the underground installation	450	18	100	4	100	4	150 (direct burial) / 100 (in raceway)	6 / 4	150	6
Under streets, highways, roads, alleys, driveways, and parking lots	600	24	600	24	600	24	600	24	600	24
One- and two-family dwelling driveways and outdoor parking areas, and used only for dwelling-related purposes	450	18	450	18	450	18	300	12	450	18
In or under airport runways, including adjacent areas where trespassing prohibited	450	18	450	18	450	18	450	18	450	18

Notes:

1. Cover is defined as the shortest distance in millimeters (inches) measured between a point on the top surface of any direct-buried conductor, cable, conduit, or other raceway and the top surface of finished grade, concrete, or similar cover.

2. Raceways approved for burial only where concrete encased shall require concrete envelope not less than 50 mm (2 in.) thick.

3. Lesser depths shall be permitted where cables and conductors rise for terminations or splices or where access is otherwise required.

4. Where one of the wiring method types listed in Columns 1–3 is used for one of the circuit types in Columns 4 and 5, the shallowest depth of burial shall be permitted.

5. Where solid rock prevents compliance with the cover depths specified in this table, the wiring shall be installed in metal or nonmetallic raceway permitted for direct burial. The raceways shall be covered by a minimum of 50 mm (2 in.) of concrete extending down to rock.

and provided with ground-fault circuit-interrupter (GFCI) protection for personnel, the burial depth may be reduced to 12 in. Exhibits 300.6 and 300.7 show examples of underground installations of 18 in. and 12 in., respectively. See 300.50 where circuits exceed 600 volts.

(B) Listing Cables and insulated conductors installed in enclosures or raceways in underground installations shall be listed for use in wet locations.

Section 310.8(C) and Table 310.13 are used to determine which general wiring conductor types are permitted to be installed in wet locations. For the 2005 *Code*, this listing requirement was relocated from 300.5(D)(5) to 300.5(B) so that it would apply to all underground installations covered by 300.5.

Rigid nonmetallic conduit elbows installed as part of a long run of conduit are often damaged in the process of pulling the conductors, due to friction at the bend. For service raceways, 250.80, Exception, permits a metal elbow to be installed without being grounded, provided it is isolated from possible contact by at least 18 in. of cover to any part of the elbow, as shown in Exhibit 300.8. For other than service raceways, Exception No. 3 to 250.86 applies.

Exhibit 300.6 PVC rigid nonmetallic conduit buried in compliance with Table 300.5 and installed in accordance with 300.5(A).

Exhibit 300.8 An application of 250.80, Exception, which permits the metal elbow to be ungrounded, provided it is isolated from contact by a minimum cover of 18 in. to any part of the elbow.

(C) Underground Cables Under Buildings Underground cable installed under a building shall be in a raceway that is extended beyond the outside walls of the building.

(D) Protection from Damage Direct-buried conductors and cables shall be protected from damage in accordance with 300.5(D)(1) through (D)(4).

(1) Emerging from Grade Direct-buried conductors and enclosures emerging from grade shall be protected by enclosures or raceways extending from the minimum cover distance below grade required by 300.5(A) to a point at least

Exhibit 300.7 A 20-ampere, GFCI-protected residential branch circuit installed with a minimum burial depth of 12 in. beneath a residential driveway.

2.5 m (8 ft) above finished grade. In no case shall the protection be required to exceed 450 mm (18 in.) below finished grade.

(2) Conductors Entering Buildings Conductors entering a building shall be protected to the point of entrance.

(3) Service Conductors Underground service conductors that are not encased in concrete and that are buried 450 mm (18 in.) or more below grade shall have their location identified by a warning ribbon that is placed in the trench at least 300 mm (12 in.) above the underground installation.

Providing a warning ribbon reduces the risk of an accident, electrocution, or an arc-flash incident during excavation near underground service conductors that are not encased in concrete. This provision requiring a warning ribbon does not extend to feeders and branch circuits because these circuits contain short-circuit and overload protection.

(4) Enclosure or Raceway Damage Where the enclosure or raceway is subject to physical damage, the conductors shall be installed in rigid metal conduit, intermediate metal conduit, Schedule 80 rigid nonmetallic conduit, or equivalent.

(E) Splices and Taps Direct-buried conductors or cables shall be permitted to be spliced or tapped without the use of splice boxes. The splices or taps shall be made in accordance with 110.14(B).

There is a difference between multiconductor cables labeled for direct burial and single conductors labeled for direct burial. Because direct-burial multiconductor cables may or may not contain individual conductors labeled for direct burial, the overall cable jacket may be the only underground protection technique for the contained conductors. Although the direct-burial splicing techniques used on multiconductor cables can differ widely from the techniques used on direct-burial single-conductor cables, the *Code* requirements are generally the same. The splicing technique should be listed for the cable type and listed for direct burial, due to the identified requirements and the listing requirements of 110.14(B) and 250.8. An example of an underground splicing method used with single-conductor direct-burial cables is shown in Exhibit 300.9.

(F) Backfill Backfill that contains large rocks, paving materials, cinders, large or sharply angular substances, or corrosive material shall not be placed in an excavation where materials may damage raceways, cables, or other substruc-

Exhibit 300.9 An underground splicing method. (Courtesy of 3M Co., Electrical Markets Division)

tures or prevent adequate compaction of fill or contribute to corrosion of raceways, cables, or other substructures.

Where necessary to prevent physical damage to the raceway or cable, protection shall be provided in the form of granular or selected material, suitable running boards, suitable sleeves, or other approved means.

(G) Raceway Seals Conduits or raceways through which moisture may contact live parts shall be sealed or plugged at either or both ends.

> FPN: Presence of hazardous gases or vapors may also necessitate sealing of underground conduits or raceways entering buildings.

Section 300.5(G) was editorially revised for the 2005 *Code* to remove the term *energized* before the phrase *live part*. Exhibit 300.10 shows a conduit sealing bushing used to prevent the entrance of gas or moisture. See 230.8 for sealing service raceways.

(H) Bushing A bushing, or terminal fitting, with an integral bushed opening shall be used at the end of a conduit or other raceway that terminates underground where the conductors or cables emerge as a direct burial wiring method. A seal incorporating the physical protection characteristics of a bushing shall be permitted to be used in lieu of a bushing.

Exhibit 300.11 shows a Type UF cable buried in compliance with Table 300.5. Note the protective bushing where the cable is used with metal conduit. The commentary following 300.4(F) applies to this requirement also.

Exhibit 300.10 A conduit sealing bushing used to prevent the entrance of gas or moisture. (Redrawn courtesy of O-Z/Gedney, a division of EGS Electrical Group)

(I) Conductors of the Same Circuit All conductors of the same circuit and, where used, the grounded conductor and all equipment grounding conductors shall be installed in the same raceway or cable or shall be installed in close proximity in the same trench.

Exception No. 1: Conductors in parallel in raceways or cables shall be permitted, but each raceway or cable shall contain all conductors of the same circuit including grounding conductors.

Conductors of the same circuit are also addressed in 300.3(B). Section 300.5(I), Exception No. 1, permits the installation of paralleled conductors in different raceways. See 310.4 for conductors in parallel.

Exception No. 2: Isolated phase, polarity, grounded conductor, and equipment grounding and bonding conductor installations shall be permitted in nonmetallic raceways or cables with a nonmetallic covering or nonmagnetic sheath in close proximity where conductors are paralleled as permitted in 310.4, and where the conditions of 300.20(B) are met.

Isolated phase installations contain only one phase per raceway or cable. The spacing between isolated phase raceways and cables should be as small as possible and the length of the run limited to avoid increased circuit impedance and the resulting increase in voltage drop inherent in an installation involving ac circuits. Isolated phase installations may be used in ac circuits to limit available fault current at downstream equipment.

Isolated phase installations present an inherent hazard of overheating, a risk that must be understood and carefully controlled. This hazard results from induced currents in metal surrounding a raceway that contains only one phase conductor. [See 300.20(A) and 300.20(B) for more information on induced currents in raceways.] The surrounding metal acts as a shorted transformer turn. In underground installations, a single conductor is unlikely to be installed in a metal raceway or, if it were, is unlikely to present a fire hazard. This is not true, however, for aboveground raceways, and

Exhibit 300.11 A Type UF cable buried in compliance with Table 300.5.

it is the reason isolated phase installations have limited application for aboveground installations.

See 300.3(B)(3) together with 330.31 and 332.31, which recognize single-conductor Type MI cable and single-conductor Type MC cable.

(J) Ground Movement Where direct-buried conductors, raceways, or cables are subject to movement by settlement or frost, direct-buried conductors, raceways, or cables shall be arranged so as to prevent damage to the enclosed conductors or to equipment connected to the raceways.

> FPN: This section recognizes "S" loops in underground direct burial to raceway transitions, expansion fittings in raceway risers to fixed equipment, and, generally, the provision of flexible connections to equipment subject to settlement or frost heaves.

Section 300.5(J) points out the practical need for installers to allow for movement of direct-buried equipment, cables, and raceways. Slack must be allowed in cables or expansion joints, or other measures must be taken if ground movement due to frost or settlement is anticipated.

(K) Directional Boring Cables or raceways installed using directional boring equipment shall be approved for the purpose.

Manufacturers of both metal and nonmetallic raceways suitable for underground use offer products that can endure the rigors of boring-type installation methods. One wiring method introduced in the 2005 *Code*, high density polyethylene conduit: Type HDPE conduit (Article 353), also can be used with boring-type installation methods.

300.6 Protection Against Corrosion and Deterioration

Raceways, cable trays, cablebus, auxiliary gutters, cable armor, boxes, cable sheathing, cabinets, elbows, couplings, fittings, supports, and support hardware shall be of materials suitable for the environment in which they are to be installed.

Section 300.6 was reorganized and expanded for the 2005 *Code* to include protection from deterioration. Information on chemical exposure as well as exposure to sunlight for nonmetallic equipment was added for this edition of the *Code*. Section 300.6 applies generally. For specific applications, it is necessary to review manufacturers' information as well as the particular *Code* article covering the wiring method and equipment under consideration.

(A) Ferrous Metal Equipment Ferrous metal raceways, cable trays, cablebus, auxiliary gutters, cable armor, boxes, cable sheathing, cabinets, metal elbows, couplings, nipples, fittings, supports, and support hardware shall be suitably protected against corrosion inside and outside (except threads at joints) by a coating of approved corrosion-resistant material. Where corrosion protection is necessary and the conduit is threaded in the field, the threads shall be coated with an approved electrically conductive, corrosion-resistant compound.

Exhibit 300.12 shows one example of an anti-corrosion paste-type compound that could be approved for use on field-cut conduit threads.

Exhibit 300.12 KOPR-Shield® (a registered trademark of Jet Lube), a conductive anti-corrosion surface compound suitable for application on field-cut conduit threads where protection from corrosion is necessary. (Courtesy of Thomas & Betts)

Exception: Stainless steel shall not be required to have protective coatings.

(1) Protected from Corrosion Solely by Enamel Where protected from corrosion solely by enamel, ferrous metal raceways, cable trays, cablebus, auxiliary gutters, cable armor, boxes, cable sheathing, cabinets, metal elbows, couplings, nipples, fittings, supports, and support hardware shall not be used outdoors or in wet locations as described in 300.6(D).

(2) Organic Coatings on Boxes or Cabinets Where boxes or cabinets have an approved system of organic coatings and are marked "Raintight," "Rainproof," or "Outdoor Type," they shall be permitted outdoors.

(3) In Concrete or in Direct Contact with the Earth Ferrous metal raceways, cable armor, boxes, cable sheathing, cabinets, elbows, couplings, nipples, fittings, supports, and support hardware shall be permitted to be installed in concrete or in direct contact with the earth, or in areas subject to severe corrosive influences where made of material approved for the condition, or where provided with corrosion protection approved for the condition.

Where ferrous or nonferrous metal conduit has corrosion protection and is judged suitable for the condition, it may be installed in concrete, in contact with the earth, or in areas exposed to severe corrosive influence. Special precautions are normally necessary for installing aluminum conduits in concrete, and specific approval by the authority having jurisdiction may be necessary.

Metal raceways installed in the earth can be coated with an asphalt compound, plastic sheath, or other equivalent protection to help prevent deterioration. Also, metallic raceways are available with a bonded PVC coating.

Galvanized rigid steel conduit and steel intermediate metal conduit do not generally require supplementary corrosion protection.

(B) Non-Ferrous Metal Equipment Non-ferrous raceways, cable trays, cablebus, auxiliary gutters, cable armor, boxes, cable sheathing, cabinets, elbows, couplings, nipples, fittings, supports, and support hardware embedded or encased in concrete or in direct contact with the earth shall be provided with supplementary corrosion protection.

(C) Nonmetallic Equipment Nonmetallic raceways, cable trays, cablebus, auxiliary gutters, boxes, cables with a non-metallic outer jacket and internal metal armor or jacket, cable sheathing, cabinets, elbows, couplings, nipples, fittings, supports, and support hardware shall be made of material approved for the condition and shall comply with (C)(1) and (C)(2) as applicable to the specific installation.

(1) Exposed to Sunlight Where exposed to sunlight, the materials shall be listed as sunlight resistant or shall be identified as sunlight resistant.

(2) Chemical Exposure Where subject to exposure to chemical solvents, vapors, splashing, or immersion, materials or coatings shall either be inherently resistant to chemicals based on its listing or be identified for the specific chemical reagent.

(D) Indoor Wet Locations In portions of dairy processing facilities, laundries, canneries, and other indoor wet locations, and in locations where walls are frequently washed or where there are surfaces of absorbent materials, such as damp paper or wood, the entire wiring system, where in-

stalled exposed, including all boxes, fittings, raceways, and cable used therewith, shall be mounted so that there is at least a 6-mm (¼-in.) airspace between it and the wall or supporting surface.

Exception: Nonmetallic raceways, boxes, and fittings shall be permitted to be installed without the airspace on a concrete, masonry, tile, or similar surface.

> FPN: In general, areas where acids and alkali chemicals are handled and stored may present such corrosive conditions, particularly when wet or damp. Severe corrosive conditions may also be present in portions of meatpacking plants, tanneries, glue houses, and some stables; in installations immediately adjacent to a seashore and swimming pool areas; in areas where chemical deicers are used; and in storage cellars or rooms for hides, casings, fertilizer, salt, and bulk chemicals.

The exception to 300.6(D) is in harmony with 547.5(B), permitting nonmetallic boxes, fittings, conduit, and cables to be installed without the airspace in corrosive locations of agricultural buildings.

300.7 Raceways Exposed to Different Temperatures

(A) Sealing Where portions of a cable raceway or sleeve are known to be subjected to different temperatures and where condensation is known to be a problem, as in cold storage areas of buildings or where passing from the interior to the exterior of a building, the raceway or sleeve shall be filled with an approved material to prevent the circulation of warm air to a colder section of the raceway or sleeve. An explosionproof seal shall not be required for this purpose.

Where a raceway is used to enclose the lighting and refrigeration branch-circuit conductors within a walk-in chest, the circulation of air through the raceway from a warmer to a colder section could cause condensation within the raceway. Circulation of air can be prevented by sealing the raceway with a suitable pliable compound at a conduit body or junction box, usually installed in the raceway before it enters the colder section. Special sealing fittings, such as those used in hazardous (classified) locations, are not necessary.

(B) Expansion Fittings Raceways shall be provided with expansion fittings where necessary to compensate for thermal expansion and contraction.

> FPN: Table 352.44(A) provides the expansion information for polyvinyl chloride (PVC). A nominal number for steel conduit can be determined by multiplying the expansion length in this table by 0.20. The coefficient of expansion for steel electrical metallic tubing, intermediate metal conduit, and rigid conduit is 11.70×10^{-6}

(0.0000117 mm per mm of conduit for each °C in temperature change) [6.50×10^{-6} (0.0000065 in. per inch of conduit for each °F in temperature change)].

This fine print note provides the relationship of linear expansion of PVC rigid nonmetallic conduit to steel conduit. For example, if a calculation indicated a linear expansion of 1¼ in. for PVC conduit, the steel conduit equivalent of expansion would be only ¼ in.

300.8 Installation of Conductors with Other Systems

Raceways or cable trays containing electric conductors shall not contain any pipe, tube, or equal for steam, water, air, gas, drainage, or any service other than electrical.

Section 300.8 specifically prohibits installation of an electrical conductor in a raceway or cable tray that includes a drain, water, oil, air, or similar pipe.

300.10 Electrical Continuity of Metal Raceways and Enclosures

Metal raceways, cable armor, and other metal enclosures for conductors shall be metallically joined together into a continuous electric conductor and shall be connected to all boxes, fittings, and cabinets so as to provide effective electrical continuity. Unless specifically permitted elsewhere in this *Code*, raceways and cable assemblies shall be mechanically secured to boxes, fittings, cabinets, and other enclosures.

Sections 250.4(A) and 250.4(B) set forth in detail what must be accomplished by the grounding and bonding of metal parts of the electrical system. The metal parts must form an effective low-impedance path to ground in order to safely conduct any fault current and facilitate the operation of overcurrent devices protecting the enclosed circuit conductors.

Exception No. 1: Short sections of raceways used to provide support or protection of cable assemblies from physical damage shall not be required to be made electrically continuous.

Exception No. 2: Equipment enclosures to be isolated, as permitted by 250.96(B), shall not be required to be metallically joined to the metal raceway.

300.11 Securing and Supporting

(A) Secured in Place Raceways, cable assemblies, boxes, cabinets, and fittings shall be securely fastened in place. Support wires that do not provide secure support shall not be permitted as the sole support. Support wires and associated fittings that provide secure support and that are installed in addition to the ceiling grid support wires shall be permitted as the sole support. Where independent support wires are used, they shall be secured at both ends. Cables and raceways shall not be supported by ceiling grids.

(1) Fire-Rated Assemblies Wiring located within the cavity of a fire-rated floor–ceiling or roof–ceiling assembly shall not be secured to, or supported by, the ceiling assembly, including the ceiling support wires. An independent means of secure support shall be provided and shall be permitted to be attached to the assembly. Where independent support wires are used, they shall be distinguishable by color, tagging, or other effective means from those that are part of the fire-rated design.

Exception: The ceiling support system shall be permitted to support wiring and equipment that have been tested as part of the fire-rated assembly.

Wiring methods of any type and all luminaires are not allowed to be supported or secured to the support wires or T bars of a fire-rated ceiling assembly unless the assembly has been tested and listed for that use. If support wires are selected as the supporting means for the electrical system within the fire-rated ceiling cavity, they must be distinguishable from the ceiling support wires and must be secured at both ends.

Generally, the rule for supporting electrical equipment is "securely fastened in place." This phrase means not only that vertical support for the weight of the equipment must be provided but also that the equipment must be secured to prevent horizontal movement or sway. The intention is to prevent the loss of grounding continuity provided by the raceway that could result from horizontal movement.

Sections 300.11(A)(1) and 300.11(A)(2) are quite similar. Unless the exceptions apply, these sections clearly prohibit all types of wiring from being attached in any way to the support wires of a ceiling assembly. Unless ceiling grids are part of the building structure, they, too, are prohibited from furnishing support for cables and raceways. However, if wiring and equipment are located within the ceiling cavity and rigidly supported independent of the ceiling, without the use of ceiling-type hanger wire, then the requirements of this section are met.

Refer to the appropriate wiring method article in Chapter 3 of the *Code* for cable and raceway supporting requirements. See 410.15(A) and 410.16 for the proper support of luminaires; 314.23 for the support of outlet boxes; and 725.8, 760.7, and 770.24 for various low-voltage fire alarm and optical fiber cable supports. See Chapter 8 for communications cable supports.

FPN: One method of determining fire rating is testing in accordance with NFPA 251-1999, *Standard Methods of Tests of Fire Endurance of Building Construction and Materials.*

(2) Non–Fire-Rated Assemblies Wiring located within the cavity of a non–fire-rated floor–ceiling or roof–ceiling assembly shall not be secured to, or supported by, the ceiling assembly, including the ceiling support wires. An independent means of secure support shall be provided.

Exception: The ceiling support system shall be permitted to support branch-circuit wiring and associated equipment where installed in accordance with the ceiling system manufacturer's instructions.

(B) Raceways Used as Means of Support Raceways shall be used only as a means of support for other raceways, cables, or nonelectric equipment under any of the following conditions:

(1) Where the raceway or means of support is identified for the purpose
(2) Where the raceway contains power supply conductors for electrically controlled equipment and is used to support Class 2 circuit conductors or cables that are solely for the purpose of connection to the equipment control circuits
(3) Where the raceway is used to support boxes or conduit bodies in accordance with 314.23 or to support luminaires (fixtures) in accordance with 410.16(F)

The purpose of 300.11(B)(3) is to prevent cables from being attached to the exterior of a raceway. Electrical, telephone, and computer cables wrapped around a raceway can prevent dissipation of heat from the raceway and affect the temperature of the conductors therein. This section also prohibits the use of a raceway as a means of support for nonelectric equipment, such as suspended ceilings, water pipes, nonelectric signs, and the like, which could cause a mechanical failure of the raceway.

However, 300.11(B)(2) does allow the installation of Class 2 thermostat conductors for a boiler or air conditioner unit to be supported by the conduit supplying power to the unit, as shown in Exhibit 300.13. These Class 2 circuits are functionally associated with the branch-circuit wiring method.

(C) Cables Not Used as Means of Support Cable wiring methods shall not be used as a means of support for other cables, raceways, or nonelectrical equipment.

This section prohibits cables from being used as a means of support for other cables, raceways, or nonelectric equipment. Taking the requirements of 300.11(B) and 300.11(C) together, the indiscriminate practice of using one supported

Exhibit 300.13 Raceways used to support Class 2 thermostat cables.

cable or raceway to support other raceways and cables is properly limited.

300.12 Mechanical Continuity — Raceways and Cables

Metal or nonmetallic raceways, cable armors, and cable sheaths shall be continuous between cabinets, boxes, fittings, or other enclosures or outlets.

Exception: Short sections of raceways used to provide support or protection of cable assemblies from physical damage shall not be required to be mechanically continuous.

300.13 Mechanical and Electrical Continuity — Conductors

(A) General Conductors in raceways shall be continuous between outlets, boxes, devices, and so forth. There shall be no splice or tap within a raceway unless permitted by 300.15; 368.56(A); 376.56; 378.56; 384.56; 386.56; 388.56; or 390.6.

Splices or taps are prohibited within raceways unless the raceways are equipped with hinged or removable covers. Busway conductors are exempt from this requirement. Splices and taps must be accessible according to 300.15.

(B) Device Removal In multiwire branch circuits, the continuity of a grounded conductor shall not depend on device connections such as lampholders, receptacles, and so forth, where the removal of such devices would interrupt the continuity.

Grounded conductors (neutrals) of multiwire branch circuits supplying receptacles, lampholders, or other such devices are not permitted to depend on terminal connections for continuity between devices. For such installations (3- or 4-wire circuits), a splice is made and a jumper is connected to the terminal, unless the neutral is looped; that is, a receptacle or lampholder could be replaced without interrupting the continuity of energized downstream line-to-neutral loads (see commentary to 300.14). Opening the neutral could cause unbalanced voltages, and a considerably higher voltage would be impressed on one part of a multiwire branch circuit, especially if the downstream line-to-neutral loads were appreciably unbalanced. This requirement does not apply to individual 2-wire circuits or other circuits that do not contain a grounded (neutral) conductor.

Exhibit 300.14 Two different boxes with free conductor lengths.

300.14 Length of Free Conductors at Outlets, Junctions, and Switch Points

At least 150 mm (6 in.) of free conductor, measured from the point in the box where it emerges from its raceway or cable sheath, shall be left at each outlet, junction, and switch point for splices or the connection of luminaires (fixtures) or devices. Where the opening to an outlet, junction, or switch point is less than 200 mm (8 in.) in any dimension, each conductor shall be long enough to extend at least 75 mm (3 in.) outside the opening.

Exception: Conductors that are not spliced or terminated at the outlet, junction, or switch point shall not be required to comply with 300.14.

A conductor looping through an outlet box and intended for connection to receptacles, switches, lampholders, or other such devices requires slack so that terminal connections can be made easily. See 314.16(B)(1) for more details about looped conductors. Conductors running through a box should have sufficient slack to prevent physical damage from the insertion of devices or from the use of fixture studs, hickeys, or other fixture supports within the box.

Since the 1999 *Code*, 300.14 is more specific about the measurements of free conductor length required at each splice point or device outlet. For these free conductor length measurements, see Exhibit 300.14.

300.15 Boxes, Conduit Bodies, or Fittings — Where Required

A box shall be installed at each outlet and switch point for concealed knob-and-tube wiring.

Fittings and connectors shall be used only with the specific wiring methods for which they are designed and listed.

Where the wiring method is conduit, tubing, Type AC cable, Type MC cable, Type MI cable, nonmetallic-sheathed cable, or other cables, a box or conduit body shall be installed at each conductor splice point, outlet point, switch point, junction point, termination point, or pull point, unless otherwise permitted in 300.15(A) through (M).

(A) Wiring Methods with Interior Access A box or conduit body shall not be required for each splice, junction, switch, pull, termination, or outlet points in wiring methods with removable covers, such as wireways, multioutlet assemblies, auxiliary gutters, and surface raceways. The covers shall be accessible after installation.

(B) Equipment An integral junction box or wiring compartment as part of approved equipment shall be permitted in lieu of a box.

(C) Protection A box or conduit body shall not be required where cables enter or exit from conduit or tubing that is used to provide cable support or protection against physical damage. A fitting shall be provided on the end(s) of the conduit or tubing to protect the cable from abrasion.

Section 300.15(C) permits conduit or tubing to be used as support and protection against physical damage without terminating in a box. It also permits conduit or tubing to be used as physical protection for underground cables that exit from buildings or that are located outdoors on poles, without a box being required on the end of the conduit. A fitting to protect the wires or cables against physical damage is required on the ends of the conduit or tubing.

(D) Type MI Cable A box or conduit body shall not be required where accessible fittings are used for straight-through splices in mineral-insulated metal-sheathed cable.

(E) Integral Enclosure A wiring device with integral enclosure identified for the use, having brackets that securely fasten the device to walls or ceilings of conventional on-site frame construction, for use with nonmetallic-sheathed cable, shall be permitted in lieu of a box or conduit body.

FPN: See 334.30(C); 545.10; 550.15(I); 551.47(E), Exception No. 1; and 552.48(E), Exception No. 1.

Section 300.15(E) applies to a device with an integral enclosure (boxless device) such as the one shown in Exhibit 300.15.

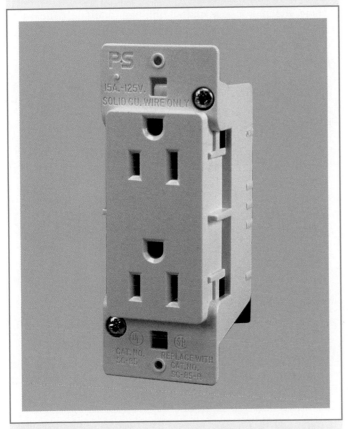

Exhibit 300.15 A self-contained device (SCD) receptacle. (Courtesy of Pass & Seymour/Legrand®)

(F) Fitting A fitting identified for the use shall be permitted in lieu of a box or conduit body where conductors are not spliced or terminated within the fitting. The fitting shall be accessible after installation.

Where a cable system makes a transition to a raceway to provide mechanical protection against damage, 300.15(F) permits the use of a fitting instead of a box. For example, where nonmetallic-sheathed cable that runs overhead on floor joists and drops down on a masonry wall to supply a receptacle needs to be protected from physical damage, a short length of raceway is installed to the outlet device box. The cable sheath is removed for the length of the raceway. The cable is then inserted in the raceway and secured by a combination fitting that is fastened to the end of the raceway.

(G) Direct-Buried Conductors As permitted in 300.5(E), a box or conduit body shall not be required for splices and taps in direct-buried conductors and cables.

(H) Insulated Devices As permitted in 334.40(B), a box or conduit body shall not be required for insulated devices supplied by nonmetallic-sheathed cable.

(I) Enclosures A box or conduit body shall not be required where a splice, switch, terminal, or pull point is in a cabinet or cutout box, in an enclosure for a switch or overcurrent device as permitted in 312.8, in a motor controller as permitted in 430.10(A), or in a motor control center.

(J) Luminaires (Fixtures) A box or conduit body shall not be required where a luminaire (fixture) is used as a raceway as permitted in 410.31 and 410.32.

(K) Embedded A box or conduit body shall not be required for splices where conductors are embedded as permitted in 424.40, 424.41(D), 426.22(B), 426.24(A), and 427.19(A).

(L) Manholes and Handhole Enclosures Where accessible only to qualified persons, a box or conduit body shall not be required for conductors in manholes or handhole enclosures, except where connecting to electrical equipment. The installation shall comply with the provisions of Part V of Article 110 for manholes, and 314.30 for handhole enclosures.

For the 2005 *Code*, handhole enclosures are also included in this section. See the definition of *handhole enclosure* in Article 100.

(M) Closed Loop A box shall not be required with a closed-loop power distribution system where a device identified and listed as suitable for installation without a box is used.

See Article 780, Closed-Loop and Programmed Power Distribution.

300.16 Raceway or Cable to Open or Concealed Wiring

(A) Box or Fitting A box or terminal fitting having a separately bushed hole for each conductor shall be used wherever a change is made from conduit, electrical metallic tubing, electrical nonmetallic tubing, nonmetallic-sheathed cable, Type AC cable, Type MC cable, or mineral-insulated, metal-sheathed cable and surface raceway wiring to open wiring or to concealed knob-and-tube wiring. A fitting used for this purpose shall contain no taps or splices and shall not be used at luminaire (fixture) outlets.

(B) Bushing A bushing shall be permitted in lieu of a box or terminal where the conductors emerge from a raceway and enter or terminate at equipment, such as open switchboards, unenclosed control equipment, or similar equipment. The bushing shall be of the insulating type for other than lead-sheathed conductors.

300.17 Number and Size of Conductors in Raceway

The number and size of conductors in any raceway shall not be more than will permit dissipation of the heat and ready installation or withdrawal of the conductors without damage to the conductors or to their insulation.

> FPN: See the following sections of this *Code*: intermediate metal conduit, 342.22; rigid metal conduit, 344.22; flexible metal conduit, 348.22; liquidtight flexible metal conduit, 350.22; rigid nonmetallic conduit, 352.22; liquidtight nonmetallic flexible conduit, 356.22; electrical metallic tubing, 358.22; flexible metallic tubing, 360.22; electrical nonmetallic tubing, 362.22; cellular concrete floor raceways, 372.11; cellular metal floor raceways, 374.5; metal wireways, 376.22; nonmetallic wireways, 378.22; surface metal raceways, 386.22; surface nonmetallic raceways, 388.22; underfloor raceways, 390.5; fixture wire, 402.7; theaters, 520.6; signs, 600.31(C); elevators, 620.33; audio signal processing, amplification, and reproduction equipment, 640.23(A) and 640.24; Class 1, Class 2, and Class 3 circuits, Article 725; fire alarm circuits, Article 760; and optical fiber cables and raceways, Article 770.

Listed wire-pulling compounds are available to assist in the process of pulling wires into raceways. As pointed out in 310.9, wire-pulling compounds should not be used if they have a harmful effect on either the conductor or the conductor insulation.

300.18 Raceway Installations

(A) Complete Runs Raceways, other than busways or exposed raceways having hinged or removable covers, shall be installed complete between outlet, junction, or splicing points prior to the installation of conductors. Where required to facilitate the installation of utilization equipment, the raceway shall be permitted to be initially installed without a terminating connection at the equipment. Prewired raceway assemblies shall be permitted only where specifically permitted in this *Code* for the applicable wiring method.

Exception: Short sections of raceways used to contain conductors or cable assemblies for protection from physical damage shall not be required to be installed complete between outlet, junction, or splicing points.

One of the primary functions of a raceway is to provide physical protection for conductors. If raceways are incomplete at the time of conductor installation, a greater possibility exists for damage to the conductors.

Section 300.18(A) does, however, permit the installation of conductors in a raceway prior to the complete installation of the raceway, up to the point of utilization. The motor installation shown in Exhibit 300.16 is a typical application, where the motor is supplied through liquidtight flexible metal conduit that terminates in the motor terminal box through a 90 degree angle connector. Wiring a fixture whip prior to connecting a luminaire is also permitted by this section.

The exception, new for the 2005 *Code*, points out that these requirements do not apply to certain short sections of raceways used for physical protection of cables.

Exhibit 300.16 An application of 300.18(A), which permits the conductors supplying a motor through liquidtight flexible metal conduit to be installed prior to the connection of the raceway to the motor terminal box.

(B) Welding Metal raceways shall not be supported, terminated, or connected by welding to the raceway unless specifically designed to be or otherwise specifically permitted to be in this *Code*.

300.19 Supporting Conductors in Vertical Raceways

(A) Spacing Intervals — Maximum Conductors in vertical raceways shall be supported if the vertical rise exceeds

Table 300.19(A) Spacings for Conductor Supports

		Conductors			
		Aluminum or Copper-Clad Aluminum		Copper	
Size of Wire	Support of Conductors in Vertical Raceways	m	ft	m	ft
18 AWG through 8 AWG	Not greater than	30	100	30	100
6 AWG through 1/0 AWG	Not greater than	60	200	30	100
2/0 AWG through 4/0 AWG	Not greater than	55	180	25	80
Over 4/0 AWG through 350 kcmil	Not greater than	41	135	18	60
Over 350 kcmil through 500 kcmil	Not greater than	36	120	15	50
Over 500 kcmil through 750 kcmil	Not greater than	28	95	12	40
Over 750 kcmil	Not greater than	26	85	11	35

the values in Table 300.19(A). One cable support shall be provided at the top of the vertical raceway or as close to the top as practical. Intermediate supports shall be provided as necessary to limit supported conductor lengths to not greater than those values specified in Table 300.19(A).

Exception: Steel wire armor cable shall be supported at the top of the riser with a cable support that clamps the steel wire armor. A safety device shall be permitted at the lower end of the riser to hold the cable in the event there is slippage of the cable in the wire-armored cable support. Additional wedge-type supports shall be permitted to relieve the strain on the equipment terminals caused by expansion of the cable under load.

(B) Support Methods One of the following methods of support shall be used.

(1) By clamping devices constructed of or employing insulating wedges inserted in the ends of the raceways. Where clamping of insulation does not adequately support the cable, the conductor also shall be clamped.

(2) By inserting boxes at the required intervals in which insulating supports are installed and secured in a satisfactory manner to withstand the weight of the conductors attached thereto, the boxes being provided with covers.

(3) In junction boxes, by deflecting the cables not less than 90 degrees and carrying them horizontally to a distance not less than twice the diameter of the cable, the cables being carried on two or more insulating supports and additionally secured thereto by tie wires if desired. Where this method is used, cables shall be supported at intervals not greater than 20 percent of those mentioned in the preceding tabulation.

(4) By a method of equal effectiveness.

Conductors in long vertical runs must be supported if the vertical rise exceeds the values in Table 300.19(A). This requirement prevents the weight of the conductors from damaging the insulation where they leave the conduit and prevents the conductors from being pulled out of the terminals. Support bushings or cleats such as those shown in Exhibits 300.17 and 300.18 may be used, in addition to many other types of grips manufactured for this purpose.

Example

A vertical raceway contains 1/0 AWG copper conductors. One cable support near the top of the run would be required if the vertical run exceeds 100 ft. Intermediate supports may be required to limit the supported length to the table values. If the vertical run is less than 100 ft, cable supports would not be required.

Support bushing with properly sized support wedge

Exhibit 300.17 A support bushing, located at the top of a vertical conduit at a cabinet or pull box, used to prevent the weight of the conductors from damaging insulation or placing strain on termination points. (Redrawn courtesy of O-Z/Gedney, a division of EGS Electrical Group)

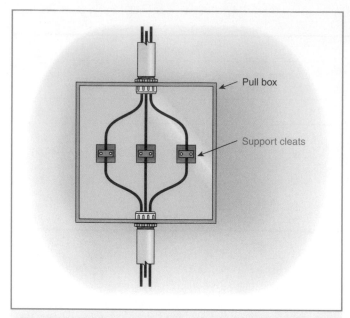

Exhibit 300.18 Support cleats used to prevent the weight of vertical conductors from damaging insulation or placing strain on termination points.

300.20 Induced Currents in Metal Enclosures or Metal Raceways

(A) Conductors Grouped Together Where conductors carrying alternating current are installed in metal enclosures or metal raceways, they shall be arranged so as to avoid heating the surrounding metal by induction. To accomplish this, all phase conductors and, where used, the grounded conductor and all equipment grounding conductors shall be grouped together.

Exception No. 1: Equipment grounding conductors for certain existing installations shall be permitted to be installed separate from their associated circuit conductors where run in accordance with the provisions of 250.130(C).

Exception No. 2: A single conductor shall be permitted to be installed in a ferromagnetic enclosure and used for skin-effect heating in accordance with the provisions of 426.42 and 427.47.

(B) Individual Conductors Where a single conductor carrying alternating current passes through metal with magnetic properties, the inductive effect shall be minimized by (1) cutting slots in the metal between the individual holes through which the individual conductors pass or (2) passing all the conductors in the circuit through an insulating wall sufficiently large for all of the conductors of the circuit.

Exception: In the case of circuits supplying vacuum or electric-discharge lighting systems or signs or X-ray apparatus, the currents carried by the conductors are so small that the

inductive heating effect can be ignored where these conductors are placed in metal enclosures or pass through metal.

> FPN: Because aluminum is not a magnetic metal, there will be no heating due to hysteresis; however, induced currents will be present. They will not be of sufficient magnitude to require grouping of conductors or special treatment in passing conductors through aluminum wall sections.

Section 300.3(B)(3) permits single-conductor Type MI cable as well as single-conductor Type MC cable. In addition to requirements contained in their respective articles (Article 332 for Type MI and Article 330 for Type MC), the installation must conform to 300.20 regarding inductive effects.

300.21 Spread of Fire or Products of Combustion

Electrical installations in hollow spaces, vertical shafts, and ventilation or air-handling ducts shall be made so that the possible spread of fire or products of combustion will not be substantially increased. Openings around electrical penetrations through fire-resistant–rated walls, partitions, floors, or ceilings shall be firestopped using approved methods to maintain the fire resistance rating.

> FPN: Directories of electrical construction materials published by qualified testing laboratories contain many listing installation restrictions necessary to maintain the fire-resistive rating of assemblies where penetrations or openings are made. Building codes also contain restrictions on membrane penetrations on opposite sides of a fire-resistance–rated wall assembly. An example is the 600-mm (24-in.) minimum horizontal separation that usually applies between boxes installed on opposite sides of the wall. Assistance in complying with 300.21 can be found in building codes, fire resistance directories, and product listings.

The intent of 300.21 is that cables, cable trays, and raceways be installed through rated wall, floor, and ceiling assemblies in such a manner that they do not contribute to the spread of fire or the products of combustion. NFPA 221, *Standard for Fire Walls and Fire Barrier Walls*, defines *fire resistance rating* as "the time, in minutes or hours, that materials or assemblies have withstood a fire test exposure" that "should be established in accordance with the test procedures of NFPA 251, *Standard Methods of Tests of Fire Endurance of Building Construction and Materials*." (ASTM E 119, *Standard Test Methods for Fire Tests of Building Construction and Materials*, and ANSI/UL 263, *Standard for Fire Tests of Building Construction and Materials*, are similar to NFPA 251.)

Further, NFPA 221, Section 4.2, Penetration Seals, states the following:

All through-penetration protection systems shall be tested and rated in accordance with ASTM E 814, *Standard Test Method for Fire Tests of Through-Penetration Fire Stops*, or ANSI/UL 1479, *Fire Test of Through-Penetration Fire Stops*. The positive pressure difference between the exposed and unexposed surfaces of the test assembly shall not be less than 0.01 in. (2.5 Pa) water gauge. A through-penetration protection system shall have an F rating not less than the required fire resistance rating of the fire wall or fire barrier wall.

Exception: Concrete, mortar, or grout shall be permitted with maximum 6-in. (153-mm) nominal diameter steel or copper pipe or steel conduit. Concrete, mortar, or grout shall be the thickness required to maintain the required fire resistance rating of the wall being penetrated. The maximum opening size shall be 144 in.^{2} (0.094 m^2).

According to the 2004 UL *Fire Resistance Directory — Volume 2A, Category XHEX, Through-Penetration Firestop Systems*, a firestop system is a specific construction consisting of a wall or floor assembly, a penetrating item passing through an opening in the wall or floor assembly, and the materials designed to prevent the spread of fire through the openings. The specifications for materials in a firestop system and the assembly of the materials are details that directly relate to the established ratings. Information concerning these details is described in the individual systems. The hourly ratings apply only to the complete systems. Individual components are designated for use in a specific system to achieve specified ratings. The individual components are not assigned ratings and are not intended to be interchanged between systems. Additionally, the substitution or elimination of components required in a system should not be made unless specifically permitted in the individual system or in the general guidelines.

The basic standard used to investigate products in this category is ANSI/UL 1479 (ASTM E 814-02), *Standard for Fire Tests of Through-Penetration Firestops*. This document defines the criteria for hourly F, T, and L ratings for firestop systems. The F rating criteria prohibit flame passage through the system and require acceptable hose stream test performance. The T rating criteria prohibit flame passage through the system and require the maximum temperature rise on the unexposed surface of the wall or floor assembly, on the penetrating item, and on the fill material not to exceed 325°F (181°C) above ambient and require acceptable hose stream test performance. The L rating criteria determine the amount of air leakage, in cubic feet per minute per square foot of opening (CFM/sq ft), through the firestop system at ambient and/or 400°F air temperatures at an air pressure differential of 0.30 in. W.C. The L ratings are intended to assist authorities having jurisdiction and others in determining the suitability of firestop systems for the protection of penetrations and miscellaneous openings in floors and smoke barriers for the purpose of restricting the movement of smoke in accordance with the NFPA *101, Life Safety Code*.

Materials used in firestop systems are to be installed in accordance with the manufacturers' instructions provided with the materials. The structural integrity of the floor or wall assembly needs to be evaluated when providing openings for the penetrating items.

A firestop system, the seals for which are shown in Exhibit 300.19, may be used to meet the requirements of 300.21.

Exhibit 300.19 Fire seals used in a through-penetration firestop system to maintain the fire resistance rating of the wall, as required by 300.21. (Courtesy of O-Z/Gedney, a division of EGS Electrical Group)

300.22 Wiring in Ducts, Plenums, and Other Air-Handling Spaces

The provisions of this section apply to the installation and uses of electric wiring and equipment in ducts, plenums, and other air-handling spaces.

FPN: See Article 424, Part VI, for duct heaters.

(A) Ducts for Dust, Loose Stock, or Vapor Removal No wiring systems of any type shall be installed in ducts used to transport dust, loose stock, or flammable vapors. No wiring system of any type shall be installed in any duct, or shaft containing only such ducts, used for vapor removal or for ventilation of commercial-type cooking equipment.

(B) Ducts or Plenums Used for Environmental Air Only wiring methods consisting of Type MI cable, Type MC cable employing a smooth or corrugated impervious metal sheath without an overall nonmetallic covering, electrical metallic tubing, flexible metallic tubing, intermediate metal conduit, or rigid metal conduit without an overall nonmetallic covering shall be installed in ducts or plenums specifically fabricated to transport environmental air. Flexible metal conduit shall be permitted, in lengths not to exceed 1.2 m

(4 ft), to connect physically adjustable equipment and devices permitted to be in these ducts and plenum chambers. The connectors used with flexible metal conduit shall effectively close any openings in the connection. Equipment and devices shall be permitted within such ducts or plenum chambers only if necessary for their direct action upon, or sensing of, the contained air. Where equipment or devices are installed and illumination is necessary to facilitate maintenance and repair, enclosed gasketed-type luminaires (fixtures) shall be permitted.

Section 300.22 limits the use of materials that would contribute smoke and products of combustion during a fire in an area that handles environmental air, and 300.22(B) provides for an effective barrier against the spread of products of combustion into the ducts or plenums.

Section 300.22(B) applies to sheet metal ducts and other ducts and plenums specifically constructed to transport environmental air. Because equipment and devices such as luminaires and motors are not normally permitted in ducts or plenums, the wiring methods in 300.22(B) differ from those permitted in 300.22(C). For the 2005 Code, the limited list of wiring methods permitted to be used in the spaces given in 300.22(B) became further restricted by removing the permission to use liquidtight flexible metallic conduit. Prior to the 2005 edition, liquidtight flexible metallic conduit was permitted to be used in lengths not exceeding 4 ft within ducts or plenums used for environmental air. However, the permission to use liquidtight flexible metallic conduit was removed because the overall nonmetallic outer covering is not evaluated for the proper fire retardancy and smoke index rating where used within ducts or plenums used for environmental air.

(C) Other Space Used for Environmental Air This section applies to space used for environmental air-handling purposes other than ducts and plenums as specified in 300.22(A) and (B). It does not include habitable rooms or areas of buildings, the prime purpose of which is not air handling.

> FPN: The space over a hung ceiling used for environmental air-handling purposes is an example of the type of other space to which this section applies.

Section 300.22(C) applies to other spaces that are used to transport environmental air and that are not specifically manufactured as ducts or plenums, such as the space or cavity between a structural floor or roof and a suspended (hung) ceiling. Many spaces above suspended ceilings are intended to transport return air. Some spaces are also used for supply air, but they are far less common than those used for return

air. This section does not apply to habitable rooms and other areas whose prime purpose is other than air handling. Such an area is shown in Exhibit 300.20. If the prime purpose of the room or space is air handling as depicted in Exhibit 300.20, then the restrictions in 300.22(C) apply, whether or not electrical equipment is located in the room.

Exhibit 300.20 An application of 300.22(C). Ordinary wiring methods are permitted in habitable areas that are not used primarily for air handling. Only wiring methods described in 300.22(C)(1) are permitted in rooms or spaces that are primarily used for air handling.

Exception: This section shall not apply to the joist or stud spaces of dwelling units where the wiring passes through such spaces perpendicular to the long dimension of such spaces.

The exception to 300.22(C) permits cable to pass through joist or stud spaces of a dwelling unit, as illustrated in Exhibit 300.21. The joist space is covered with sheet metal and used as a cold-air return for a forced warm-air central heating system. Equipment such as junction boxes or device enclosures is not permitted in this location.

(1) Wiring Methods The wiring methods for such other space shall be limited to totally enclosed, nonventilated, insulated busway having no provisions for plug-in connections, Type MI cable, Type MC cable without an overall nonmetallic covering, Type AC cable, or other factory-assembled multiconductor control or power cable that is specifically listed for the use, or listed prefabricated cable assemblies of metallic manufactured wiring systems without nonmetallic sheath. Other types of cables and conductors shall be installed in electrical metallic tubing, flexible metal-

Exhibit 300.21 A cable passing through joist spaces of a dwelling unit, as permitted by 300.22(C), Exception.

lic tubing, intermediate metal conduit, rigid metal conduit without an overall nonmetallic covering, flexible metal conduit, or, where accessible, surface metal raceway or metal wireway with metal covers or solid bottom metal cable tray with solid metal covers.

Revised for the 2002 *Code*, 300.22(C)(1) no longer permits liquidtight flexible metal conduit as covered in Article 350 to be installed within "other spaces used for environmental air." The previous exception permitting this application for single lengths up to 6 ft was removed.

(2) Equipment Electrical equipment with a metal enclosure, or with a nonmetallic enclosure listed for the use and having adequate fire-resistant and low-smoke-producing characteristics, and associated wiring material suitable for the ambient temperature shall be permitted to be installed in such other space unless prohibited elsewhere in this *Code*.

Electrical equipment with metal enclosures is allowed within spaces used for environmental air. However, nonmetallic enclosures must be listed for this use.

Exception: Integral fan systems shall be permitted where specifically identified for such use.

It is not intended that the requirements of 300.22(B) or 300.22(C) apply to air-handling areas beneath raised floors in information technology rooms. See Article 645, Information Technology Equipment.

(D) Information Technology Equipment Electric wiring in air-handling areas beneath raised floors for information technology equipment shall be permitted in accordance with Article 645.

300.23 Panels Designed to Allow Access

Cables, raceways, and equipment installed behind panels designed to allow access, including suspended ceiling panels, shall be arranged and secured so as to allow the removal of panels and access to the equipment.

Section 300.23 is intended to prevent the excess accumulation of wires and cables that could limit access to electrical equipment by preventing the removal of access panels.

II. Requirements for Over 600 Volts, Nominal

Part II of Article 300 was changed for the 1999 *Code*. Some of the requirements previously located in Article 710, Over 600 Volts, Nominal, General, in the 1996 *Code* are now in Article 300, Part II. The flow chart shown in Exhibit 300.22 shows the relocated and reorganized sections.

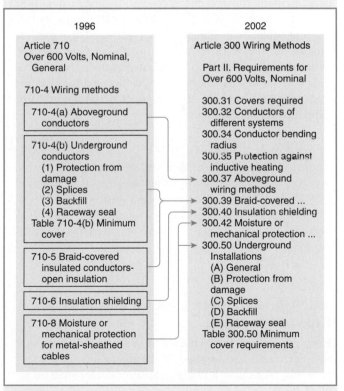

Exhibit 300.22 The relocation of Article 710, Over 600 Volts, Nominal, General (1996 *Code*) into Article 300, Part II (2002 *Code*).

300.31 Covers Required

Suitable covers shall be installed on all boxes, fittings, and similar enclosures to prevent accidental contact with energized parts or physical damage to parts or insulation.

300.32 Conductors of Different Systems

See 300.3(C)(2).

300.34 Conductor Bending Radius

The conductor shall not be bent to a radius less than 8 times the overall diameter for nonshielded conductors or 12 times the overall diameter for shielded or lead-covered conductors during or after installation. For multiconductor or multiplexed single conductor cables having individually shielded conductors, the minimum bending radius is 12 times the diameter of the individually shielded conductors or 7 times the overall diameter, whichever is greater.

300.35 Protection Against Induction Heating

Metallic raceways and associated conductors shall be arranged so as to avoid heating of the raceway in accordance with the applicable provisions of 300.20.

300.37 Aboveground Wiring Methods

Aboveground conductors shall be installed in rigid metal conduit, in intermediate metal conduit, in electrical metallic tubing, in rigid nonmetallic conduit, in cable trays, as busways, as cablebus, in other identified raceways, or as exposed runs of metal-clad cable suitable for the use and purpose. In locations accessible to qualified persons only, exposed runs of Type MV cables, bare conductors, and bare busbars shall also be permitted. Busbars shall be permitted to be either copper or aluminum.

In transformer vaults, switch rooms, and similar areas restricted to qualified personnel, any suitable wiring method may be used. Exposed wiring using bare or insulated conductors on insulators is commonly employed, as is rigid metal conduit and rigid nonmetallic conduit. Throughout 300.37 as well as 300.39, the term *exposed* was substituted for the term *open*. Exposed is preferred since it is a defined term in Article 100.

300.39 Braid-Covered Insulated Conductors — Exposed Installation

Exposed runs of braid-covered insulated conductors shall have a flame-retardant braid. If the conductors used do not have this protection, a flame-retardant saturant shall be applied to the braid covering after installation. This treated braid covering shall be stripped back a safe distance at conductor terminals, according to the operating voltage. Where practicable, this distance shall not be less than 25

mm (1 in.) for each kilovolt of the conductor-to-ground voltage of the circuit.

300.40 Insulation Shielding

Metallic and semiconducting insulation shielding components of shielded cables shall be removed for a distance dependent on the circuit voltage and insulation. Stress reduction means shall be provided at all terminations of factory-applied shielding.

Metallic shielding components such as tapes, wires, or braids, or combinations thereof, and their associated conducting or semiconducting components shall be grounded.

300.42 Moisture or Mechanical Protection for Metal-Sheathed Cables

Where cable conductors emerge from a metal sheath and where protection against moisture or physical damage is necessary, the insulation of the conductors shall be protected by a cable sheath terminating device.

300.50 Underground Installations

(A) General Underground conductors shall be identified for the voltage and conditions under which they are installed. Direct burial cables shall comply with the provisions of 310.7. Underground cables shall be installed in accordance with 300.50(A)(1) or (A)(2), and the installation shall meet the depth requirements of Table 300.50.

(1) Shielded Cables and Nonshielded Cables in Metal-Sheathed Cable Assemblies Underground cables, including nonshielded, Type MC and moisture-impervious metal sheath cables, shall have those sheaths grounded through an effective grounding path meeting the requirements of 250.4(A)(5) or (B)(4). They shall be direct buried or installed in raceways identified for the use.

(2) Other Nonshielded Cables Other nonshielded cables not covered in 300.50(A)(1) shall be installed in rigid metal conduit, intermediate metal conduit, or rigid nonmetallic conduit encased in not less than 75 mm (3 in.) of concrete.

(B) Protection from Damage Conductors emerging from the ground shall be enclosed in listed raceways. Raceways installed on poles shall be of rigid metal conduit, intermediate metal conduit, PVC Schedule 80, or equivalent, extending from the minimum cover depth specified in Table 300.50 to a point 2.5 m (8 ft) above finished grade. Conductors entering a building shall be protected by an approved enclosure or raceway from the minimum cover depth to the point of entrance. Where direct-buried conductors, raceways, or cables are subject to movement by settlement or frost, they shall be installed to prevent damage to the enclosed conductors or to the equipment connected to the raceways. Metallic enclosures shall be grounded.

Table 300.50 Minimum Cover[1] Requirements

	General Conditions (not otherwise specified)						Special Conditions (use if applicable)					
	(1) Direct-Buried Cables		(2) Rigid Nonmetallic Conduit[2]		(3) Rigid Metal Conduit and Intermediate Metal Conduit		(4) Raceways under buildings or exterior concrete slabs, 100 mm (4 in.) minimum thickness[3]		(5) Cables in airport runways or adjacent areas where trespass is prohibited		(6) Areas subject to vehicular traffic, such as thoroughfares and commercial parking areas	
Circuit Voltage	mm	in.	mm	in.	mm	in.	mm	in.	mm	in.	mm	in.
Over 600 V through 22 kV	750	30	450	18	150	6	100	4	450	18	600	24
Over 22 kV through 40 kV	900	36	600	24	150	6	100	4	450	18	600	24
Over 40 kV	1000	42	750	30	150	6	100	4	450	18	600	24

Notes:

1. Lesser depths shall be permitted where cables and conductors rise for terminations or splices or where access is otherwise required.

2. Where solid rock prevents compliance with the cover depths specified in this table, the wiring shall be installed in a metal or nonmetallic raceway permitted for direct burial. The raceways shall be covered by a minimum of 50 mm (2 in.) of concrete extending down to rock.

[1] Cover is defined as the shortest distance in millimeters (inches) measured between a point on the top surface of any direct-buried conductor, cable, conduit, or other raceway and the top surface of finished grade, concrete, or similar cover.

[2] Listed by a qualified testing agency as suitable for direct burial without encasement. All other nonmetallic systems shall require 50 mm (2 in.) of concrete or equivalent above conduit in addition to the table depth.

[3] The slab shall extend a minimum of 150 mm (6 in.) beyond the underground installation, and a warning ribbon or other effective means suitable for the conditions shall be placed above the underground installation.

(C) Splices Direct burial cables shall be permitted to be spliced or tapped without the use of splice boxes, provided they are installed using materials suitable for the application. The taps and splices shall be watertight and protected from mechanical damage. Where cables are shielded, the shielding shall be continuous across the splice or tap.

Exception: At splices of an engineered cabling system, metallic shields of direct-buried single-conductor cables with maintained spacing between phases shall be permitted to be interrupted and overlapped. Where shields are interrupted and overlapped, each shield section shall be grounded at one point.

(D) Backfill Backfill containing large rocks, paving materials, cinders, large or sharply angular substances, or corrosive materials shall not be placed in an excavation where materials can damage or contribute to the corrosion of raceways, cables, or other substructures or where it may prevent adequate compaction of fill.

Protection in the form of granular or selected material or suitable sleeves shall be provided to prevent physical damage to the raceway or cable.

(E) Raceway Seal Where a raceway enters from an underground system, the end within the building shall be sealed with an identified compound so as to prevent the entrance

of moisture or gases, or it shall be so arranged to prevent moisture from contacting live parts.

ARTICLE 310
Conductors for General Wiring

Summary of Changes

- **310.4:** Revised to require that where parallel conductors are in separate raceways or cables, each separate cable or raceway must contain the same number of conductors.

- **310.6, Exception:** Revised to permit the use of non-shielded insulated conductors to limit such applications to 2400 volts.

- **310.8(D):** Added item No. 3 to permit listed or listed and marked taping or sleeving as a means to make conductors sunlight resistant.

- **310.10, FPN No. 2:** Added FPN warning that conductors in close proximity to rooftops may experience a temperature rise of 17°C (30°F) above ambient.

- **310.15(B)(2):** Revised to clarify that each current-carrying conductor of a parallel set of conductors is to be counted as a current-carrying conductor.

• **Figure 310.60:** Deleted detail 4 (9 electrical ducts) because it did not have a corresponding cross-reference in Table 310.77 through Table 310.86.

Contents

310.1 Scope

This article covers general requirements for conductors and their type designations, insulations, markings, mechanical strengths, ampacity ratings, and uses. These requirements do not apply to conductors that form an integral part of equipment, such as motors, motor controllers, and similar equipment, or to conductors specifically provided for elsewhere in this *Code*.

FPN: For flexible cords and cables, see Article 400. For fixture wires, see Article 402.

310.2 Conductors

As the electrical industry moves toward global suppliers of insulated conductors, it is important for the user to understand that all "insulated" conductors are not equal. Only wires and cables that meet the minimum fire, electrical, and physical properties required by the applicable standards are permitted to be marked with the letter designations found in Table 310.13, Table 310.61, and Table 310.62. See 310.13 for the requirements of insulated conductor construction and applications.

(A) Insulated Conductors shall be insulated.

Exception: Where covered or bare conductors are specifically permitted elsewhere in this Code.

FPN: See 250.184 for insulation of neutral conductors of a solidly grounded high-voltage system.

(B) Conductor Material Conductors in this article shall be of aluminum, copper-clad aluminum, or copper unless otherwise specified.

310.3 Stranded Conductors

Where installed in raceways, conductors of size 8 AWG and larger shall be stranded.

Exception: As permitted or required elsewhere in this Code.

Larger-size conductors are required to be stranded to provide greater flexibility. This requirement does not apply to bus-bars and the conductors of Type MI mineral-insulated metal-sheathed cable. In addition, the bonding conductors of a common bonding grid of a permanently installed swimming pool are required to be solid, nonstranded conductors of 8 AWG or larger, according to 680.26(C).

310.4 Conductors in Parallel

Aluminum, copper-clad aluminum, or copper conductors of size 1/0 AWG and larger, comprising each phase, polarity, neutral, or grounded circuit conductor, shall be permitted to be connected in parallel (electrically joined at both ends).

For the 2005 *Code*, one change to 310.4 is the addition of the word *polarity* throughout the section, specifically allowing the inclusion of dc circuits. Conductors connected in parallel, in accordance with 310.4, are considered a single conductor with a total cross-sectional area of all conductors in parallel. Therefore, if individual conductors are tapped from conductors in parallel, the tap connection must include all the conductors in parallel for that particular phase. Tapping into only one of the parallel conductors results in unbalanced distribution of tap load current between parallel conductors. This unbalance results in one of the conductors carrying more than its share of the load, which can cause overheating and conductor insulation failure. For example, if a 250-kcmil conductor is tapped from a set of two 500-kcmil conductors in parallel, the splicing device must include both 500-kcmil conductors and the single 250-kcmil tap conductor.

Exception No. 1: As permitted in 620.12(A)(1).

Exception No. 2: Conductors in sizes smaller than 1/0 AWG shall be permitted to be run in parallel to supply control power to indicating instruments, contactors, relays, solenoids, and similar control devices, provided all of the following apply:

(a) They are contained within the same raceway or cable.

(b) The ampacity of each individual conductor is sufficient to carry the entire load current shared by the parallel conductors.

(c) The overcurrent protection is such that the ampacity of each individual conductor will not be exceeded if one or more of the parallel conductors become inadvertently disconnected.

Exception No. 3: Conductors in sizes smaller than 1/0 AWG shall be permitted to be run in parallel for frequencies of 360 Hz and higher where conditions (a), (b), and (c) of Exception No. 2 are met.

For example, in control wiring and circuits that operate at frequencies greater than 360 Hz, it may be necessary to reduce cable capacitance or voltage drop in long lengths of wire. A 14 AWG conductor might have more than sufficient capacity to carry the load, but by installing two conductors in parallel, the voltage drop can be reduced to acceptable limits. This method is permissible, provided the safeguards listed in Exception No. 2 are followed.

Exception No. 4: Under engineering supervision, grounded neutral conductors in sizes 2 AWG and larger shall be permitted to be run in parallel for existing installations.

> FPN: Exception No. 4 can be used to alleviate overheating of neutral conductors in existing installations due to high content of triplen harmonic currents.

The word *triplen* refers to a third-order harmonic current, such as the third, sixth, ninth, and so on. Our concern is limited to odd-number triplen harmonic currents, such as

the third, ninth, and fifteenth, since these are additive currents in the neutral conductor and do not cancel. See Chapter 27 of NFPA 70B, *Recommended Practice for Electrical Equipment Maintenance*, for additional information on power quality and harmonics.

The paralleled conductors in each phase, polarity, neutral, or grounded circuit conductor shall comply with all of the following:

(1) Be the same length
(2) Have the same conductor material
(3) Be the same size in circular mil area
(4) Have the same insulation type
(5) Be terminated in the same manner

Where run in separate raceways or cables, the raceways or cables shall have the same physical characteristics. Where conductors are in separate raceways or cables, the same number of conductors shall be used in each raceway or cable. Conductors of one phase, polarity, neutral, or grounded circuit conductor shall not be required to have the same physical characteristics as those of another phase, polarity, neutral, or grounded circuit conductor to achieve balance.

> FPN: Differences in inductive reactance and unequal division of current can be minimized by choice of materials, methods of construction, and orientation of conductors.

For example, the conductors in phases A and B may be copper, and those in phase C may be aluminum. However, all raceways or cables must be of the same size, material, and length. *Cables*, in this case, means wiring method type cables such as Type MC. A new sentence was added to this paragraph for the 2005 *Code* making it clear that all conduits or cables enclosing parallel conductors must contain the same number of conductors.

Where equipment grounding conductors are used with conductors in parallel, they shall comply with the requirements of this section except that they shall be sized in accordance with 250.122.

Conductors installed in parallel shall comply with the provisions of 310.15(B)(2)(a).

Section 310.4 permits a practical means for installing large-capacity conductors for feeders or services. The paralleling of two or more conductors in place of using one large conductor to ensure equal division of current depends on a number of factors. Therefore, several conditions must be satisfied so as not to overload any of the individual paralleled conductors. Other than as permitted in 250.122 and the exceptions to 310.4, there does not appear to be any practical need to parallel conductors smaller than 1/0 AWG.

To avoid excessive voltage drop and also to ensure equal division of current, it is essential that different phase conductors be located close together and that each phase conductor, grounded conductor, and the grounding conductor (if used) be grouped together in each raceway or cable. However, isolated phase installations are permitted underground where the phase conductors are run in nonmetallic raceways that are in close proximity.

The impedance of a circuit in an aluminum raceway or aluminum-sheathed cable will differ from the impedance of the same circuit in a steel raceway or steel-sheathed cable; therefore, separate raceways and cables must have the same physical characteristics. Also, the same number of conductors must be used in each raceway or cable. See 300.20 regarding induced currents in metal enclosures or raceways.

Note that all conductors of the same phase or neutral are required to be of the same conductor material. For example, if 12 conductors are paralleled for a 3-phase, 4-wire, 480Y/277-volt ac circuit, four conductors could be installed in each of three raceways. The *Code* does not intend that all 12 conductors be copper or aluminum but does intend that the individual conductors in parallel for each phase, grounded conductor, and neutral be the same material, insulation type, length, and so on. Also, it is intended that the three raceways have the same physical characteristics (e.g., three rigid aluminum conduits, three steel IMC conduits, three EMTs, or three nonmetallic conduits), not a mixture (e.g., two rigid aluminum conduits and one rigid steel conduit).

It is neither economical nor practical to use conductors larger than 1000 kcmil in raceways unless the conductor size is governed by voltage drop. The ampacity of larger sizes would increase very little in proportion to the increase in the size of the conductor. Where the cross-sectional area of a conductor increases 50 percent (e.g., from 1000 to 1500 kcmil), a Type THW conductor ampacity increases only 80 amperes (less than 15 percent). A 100 percent increase (from 1000 to 2000 kcmil) causes an increase of only 120 amperes (approximately 2 percent). Generally, where cost is a factor, installation of two (or more) paralleled conductors per phase may be beneficial.

310.5 Minimum Size of Conductors

The minimum size of conductors shall be as shown in Table 310.5, except as permitted elsewhere in this *Code*.

For the 2005 *Code*, 10 exceptions were deleted from 310.5 and the phrase "except as permitted elsewhere in this *Code*" was added. As a reminder, the 10 exceptions from the 2002 *Code* dealt with the following subjects. These subjects may

Table 310.5 Minimum Size of Conductors

Conductor Voltage Rating (Volts)	Minimum Conductor Size (AWG)	
	Copper	Aluminum or Copper-Clad Aluminum
0–2000	14	12
2001–8000	8	8
8001–15,000	2	2
15,001–28,000	1	1
28,001–35,000	1/0	1/0

permit different minimum conductor sizes than the sizes stated in Table 310.5. Although this list may not be totally inclusive, the 10 subjects and their applicable sections are as follows:

1. Flexible cords as permitted by 400.12
2. Fixture wire as permitted by 402.6
3. Motors rated 1 hp or less as permitted by 430.22(F)
4. Cranes and hoists as permitted by 610.14
5. Elevator control and signaling circuits as permitted by 620.12
6. Class 1, Class 2, and Class 3 circuits as permitted by 725.27(A) and 725.51, Exception
7. Fire alarm circuits as permitted by 760.27(A); 760.51, Exception; and 760.82(B)
8. Motor-control circuits as permitted by 430.72
9. Control and instrumentation circuits as permitted by 727.6
10. Electric signs and outline lighting as permitted in 600.31(B) and 600.32(B)

310.6 Shielding

Solid dielectric insulated conductors operated above 2000 volts in permanent installations shall have ozone-resistant insulation and shall be shielded. All metallic insulation shields shall be grounded through an effective grounding path meeting the requirements of 250.4(A)(5) or 250.4(B)(4). Shielding shall be for the purpose of confining the voltage stresses to the insulation.

Exception: Nonshielded insulated conductors listed by a qualified testing laboratory shall be permitted for use up to 2400 volts under the following conditions:

(a) Conductors shall have insulation resistant to electric discharge and surface tracking, or the insulated conductor(s) shall be covered with a material resistant to ozone, electric discharge, and surface tracking.

(b) Where used in wet locations, the insulated conductor(s) shall have an overall nonmetallic jacket or a continuous metallic sheath.

(c) Insulation and jacket thicknesses shall be in accordance with Table 310.63.

Solid dielectric insulated conductors that are permanently installed and that operate at greater than 2000 volts must have ozone-resistant insulation and must be shielded with a grounded metallic shield (note exception). Shielding is accomplished by applying a metal tape or nonmetallic semi-conducting tape around the conductor surface, to prevent corona from forming and to reduce high-voltage stresses.

Corona is a faint glow adjacent to the surface of the electrical conductor at high voltage. If high-voltage stresses and a charging current are flowing between the conductor and ground (usually due to moisture), the surrounding atmosphere is ionized, and ozone—generated by an electric discharge in ordinary oxygen or air—is formed, which will attack the conductor jacket and insulation and may eventually break them down. The shield is at ground potential; therefore, no voltage above ground is present on the jacket outside the shield, thus preventing a discharge from the jacket and the subsequent formation of ozone.

For the 2005 *Code*, a significant change occurred to the exception of 310.6. No longer does the exception permit unshielded insulated conductors up to an 8 kV level under certain conditions. The revised exception limits the omission of shielding up to the 2.4 kV level only.

Specialized training and close adherence to manufacturers' instructions are absolutely essential for high-voltage cable installations.

Exhibits 310.1, 310.2, and 310.3 show some examples of shielded cable installations: a three-conductor cable of the shielded type, a stress-relief cone for an indoor cable terminator, and a stress cone on a single-conductor shielded cable terminating inside a pothead. In Exhibit 310.3, a clamping ring provides a grounding connection between the copper shielding tape and the shield to the metallic base of the pothead.

Exhibit 310.2 A one-piece, premolded stress-relief cone for indoor cable terminations of up to 35 kV phase-to-phase.

Exhibit 310.3 A stress cone on a single-conductor shielded cable terminating inside a pothead.

Exhibit 310.1 A three-conductor cable of the shielded type.

310.7 Direct Burial Conductors

Conductors used for direct burial applications shall be of a type identified for such use.

Cables rated above 2000 volts shall be shielded.

Exception: Nonshielded multiconductor cables rated 2001–5000 volts shall be permitted if the cable has an overall metallic sheath or armor.

The metallic shield, sheath, or armor shall be grounded through an effective grounding path meeting the requirements of 250.4(A)(5) or (B)(4).

FPN No. 1: See 300.5 for installation requirements for conductors rated 600 volts or less.

FPN No. 2: See 300.50 for installation requirements for conductors rated over 600 volts.

310.8 Locations

(A) Dry Locations Insulated conductors and cables used in dry locations shall be any of the types identified in this *Code*.

(B) Dry and Damp Locations Insulated conductors and cables used in dry and damp locations shall be Types FEP, FEPB, MTW, PFA, RHH, RHW, RHW-2, SA, THHN, THW, THW-2, THHW, THHW-2, THWN, THWN-2, TW, XHH, XHHW, XHHW-2, Z, or ZW.

(C) Wet Locations Insulated conductors and cables used in wet locations shall be

(1) Moisture-impervious metal-sheathed;
(2) Types MTW, RHW, RHW-2, TW, THW, THW-2, THHW, THHW-2, THWN, THWN-2, XHHW, XHHW-2, ZW; or
(3) Of a type listed for use in wet locations.

(D) Locations Exposed to Direct Sunlight Insulated conductors or cables used where exposed to direct rays of the sun shall comply with one of the following:

(1) Cables listed, or listed and marked, as being sunlight resistant
(2) Conductors listed, or listed and marked, as being sunlight resistant
(3) Covered with insulating material, such as tape or sleeving, that is listed, or listed and marked, as being sunlight resistant

310.9 Corrosive Conditions

Conductors exposed to oils, greases, vapors, gases, fumes, liquids, or other substances having a deleterious effect on the conductor or insulation shall be of a type suitable for the application.

See the commentary following 501.20 regarding gasoline-resistant conductors. Before being used, wire pulling compounds should first be investigated to determine compliance with 310.9.

310.10 Temperature Limitation of Conductors

No conductor shall be used in such a manner that its operating temperature exceeds that designated for the type of insulated conductor involved. In no case shall conductors be associated together in such a way, with respect to type of circuit, the wiring method employed, or the number of conductors, that the limiting temperature of any conductor is exceeded.

Most terminations are normally designed for 60°C or 75°C maximum temperatures, although some are now being designed for 90°C. Therefore, the higher-rated ampacities for conductors of 90°C, 110°C, and so on, cannot be used unless the terminals at which the conductors terminate have comparable ratings.

Table 310.16 through Table 310.20 have ampacity correction factors for ambient temperatures greater or less than the ambient temperature identified in the table heading. To assign the proper ampacity to a conductor in an ambient above 30°C (86°F), the appropriate temperature correction factor must be used. This correction factor is applied in addition to any adjustment factor, such as in 310.15(B)(2)(a).

Example

Determine the adequacy of 2 AWG THHN copper conductors to be installed in a raceway in an ambient temperature of 50°C (122°F).

Solution

Table 310.16 shows that the allowable ampacity of the conductor at 30°C is 130 amperes, which is multiplied by 0.82 (taken from the correction factors at the bottom of the table). Thus, the allowable ampacity of the 2 AWG conductor at 50°C is reduced to 106.6 amperes (130 amperes × 0.82 = 106.6 amperes).

If six of these conductors were run in a raceway, 310.15(B)(2)(a) would require the allowable ampacity to be further reduced to 80 percent, which, in this case, would be 106.6 amperes × 0.8 = 85.28 amperes. Under these conditions, the 2 AWG conductors would be suitable for an 80-ampere circuit.

The basis for determining the ampacities of conductors for Table 310.16 and Table 310.17 was the NEMA "Report of Determination of Maximum Permissible Current-Carrying Capacity of Code Insulated Wires and Cables for Building Purposes," dated June 27, 1938. The basis for determining the ampacities of conductors for Table 310.18 and Table 310.19 and the ampacity tables in Annex B was the Neher-McGrath method. See the commentary following 310.15(C) for further explanation.

Conductors should be chosen that have a rating above the anticipated maximum ambient temperature. The operating temperature of conductors should be controlled at or below the conductor rating by coordinating conductor size, number of associated conductors, and ampacity for the particular conductor rating and ambient temperature. All tabulations should be corrected for the anticipated ambient temperature, using the correction factors at the bottom of the ampacity tables. If more than three conductors are associated together, the additional adjustment shown in 310.15(B)(2)(a) must be applied.

> FPN No. 1: The temperature rating of a conductor (see Tables 310.13 and 310.61) is the maximum temperature, at any location along its length, that the conductor can withstand over a prolonged time period without serious degradation. The allowable ampacity tables, the ampacity tables of Article 310 and the ampacity tables of Annex B, the correction factors at the bottom of these tables, and the notes to the tables provide guidance for coordinating conductor sizes, types, allowable ampacities, ampacities, ambient temperatures, and number of associated conductors.
>
> The principal determinants of operating temperature are as follows:
>
> (1) Ambient temperature — ambient temperature may vary along the conductor length as well as from time to time.
> (2) Heat generated internally in the conductor as the result of load current flow, including fundamental and harmonic currents.
> (3) The rate at which generated heat dissipates into the ambient medium. Thermal insulation that covers or surrounds conductors affects the rate of heat dissipation.
> (4) Adjacent load-carrying conductors — adjacent conductors have the dual effect of raising the ambient temperature and impeding heat dissipation.

FPN No. 1 focuses attention on the necessity for derating conductors where high ambient temperatures are encountered and provides users with helpful information in coordinating ampacities, ambient temperatures, conductor size and number, and so on, to ensure operation at or below rating.

The second principal determinant of FPN No. 1 explains that heating due to harmonic current should also be considered. In certain cases, this may require using larger-sized conductors. For existing installations, see 310.4, Exception No. 4, as well as the commentary associated with 310.4.

> FPN No. 2: Conductors installed in conduit exposed to direct sunlight in close proximity to rooftops have been shown, under certain conditions, to experience a temperature rise of 17°C (30°F) above ambient temperature on which the ampacity is based.

FPN No. 2, new for the 2005 Code, points out the possibility of additional heat rise for some roof top conductors in conduit placed near the roof in direct sunlight. During the 2005 Code cycle, the panel received information related to this subject and prudently decided to add this material not as a requirement to derate the conductors but as a fine print note. As described in 90.5(C), fine print notes are not mandatory requirements in the NEC but rather explanatory information to assist the Code user.

310.11 Marking

(A) Required Information All conductors and cables shall be marked to indicate the following information, using the applicable method described in 310.11(B):

(1) The maximum rated voltage
(2) The proper type letter or letters for the type of wire or cable as specified elsewhere in this Code
(3) The manufacturer's name, trademark, or other distinctive marking by which the organization responsible for the product can be readily identified
(4) The AWG size or circular mil area

> FPN: See Conductor Properties, Table 8 of Chapter 9, for conductor area expressed in SI units for conductor sizes specified in AWG or circular mil area.

(5) Cable assemblies where the neutral conductor is smaller than the ungrounded conductors shall be so marked

(B) Method of Marking

(1) Surface Marking The following conductors and cables shall be durably marked on the surface. The AWG size or circular mil area shall be repeated at intervals not exceeding 610 mm (24 in.). All other markings shall be repeated at intervals not exceeding 1.0 m (40 in.).

(1) Single-conductor and multiconductor rubber- and thermoplastic-insulated wire and cable
(2) Nonmetallic-sheathed cable
(3) Service-entrance cable
(4) Underground feeder and branch-circuit cable
(5) Tray cable
(6) Irrigation cable
(7) Power-limited tray cable
(8) Instrumentation tray cable

(2) Marker Tape Metal-covered multiconductor cables shall employ a marker tape located within the cable and running for its complete length.

Exception No. 1: Mineral-insulated, metal-sheathed cable.

Exception No. 2: Type AC cable.

Exception No. 3: The information required in 310.11(A) shall be permitted to be durably marked on the outer nonmetallic covering of Type MC, Type ITC, or Type PLTC cables at intervals not exceeding 1.0 m (40 in.).

Exception No. 4: The information required in 310.11(A) shall be permitted to be durably marked on a nonmetallic covering under the metallic sheath of Type ITC or Type PLTC cable at intervals not exceeding 1.0 m (40 in.).

Type PLTC cable is permitted to have a metallic sheath or armor over a nonmetallic jacketed cable. A second nonmetallic jacket covering the metallic sheath is optional. Exception No. 3 and Exception No. 4 define the marking requirements for either case.

FPN: Included in the group of metal-covered cables are Type AC cable (Article 320), Type MC cable (Article 330), and lead-sheathed cable.

(3) Tag Marking The following conductors and cables shall be marked by means of a printed tag attached to the coil, reel, or carton:

(1) Mineral-insulated, metal-sheathed cable
(2) Switchboard wires
(3) Metal-covered, single-conductor cables
(4) Type AC cable

(4) Optional Marking of Wire Size The information required in 310.11(A)(4) shall be permitted to be marked on the surface of the individual insulated conductors for the following multiconductor cables:

(1) Type MC cable
(2) Tray cable
(3) Irrigation cable
(4) Power-limited tray cable
(5) Power-limited fire alarm cable
(6) Instrumentation tray cable

(C) Suffixes to Designate Number of Conductors A type letter or letters used alone shall indicate a single insulated conductor. The letter suffixes shall be indicated as follows:

(1) D — For two insulated conductors laid parallel within an outer nonmetallic covering
(2) M — For an assembly of two or more insulated conductors twisted spirally within an outer nonmetallic covering

(D) Optional Markings All conductors and cables contained in Chapter 3 shall be permitted to be surface marked to indicate special characteristics of the cable materials. These markings include, but are not limited to, markings for limited smoke, sunlight resistant, and so forth.

New cable insulations that have special characteristics are frequently developed. An example is the family of limited-smoke cables, which are permitted to be marked "LS." Other materials that have other characteristics such as sunlight resistance and low corrosivity have been developed or are in development. Section 310.11(D) allows these developments without each characteristic being identified in the *Code*.

310.12 Conductor Identification

(A) Grounded Conductors Insulated or covered grounded conductors shall be identified in accordance with 200.6.

Section 200.6 permits a recently developed method of identification described as "three continuous white stripes on other than green insulation along the conductor's entire length."

(B) Equipment Grounding Conductors Equipment grounding conductors shall be in accordance with 250.119.

(C) Ungrounded Conductors Conductors that are intended for use as ungrounded conductors, whether used as a single conductor or in multiconductor cables, shall be finished to be clearly distinguishable from grounded and grounding conductors. Distinguishing markings shall not conflict in any manner with the surface markings required by 310.11(B)(1). Branch-circuit ungrounded conductors shall be identified in accordance with 210.5(C). Feeders shall be identified in accordance with 215.12.

Two new sentences were added to 310.12(C) for the 2005 *Code* that point to revised identification requirements for ungrounded conductors of branch circuits and feeders. The revised identification requirements apply only where a premises wiring system has circuits from more than one nominal voltage system. See the commentary following 210.5(C) and 215.12 for additional information on these 2005 *Code* changes.

Exception: Conductor identification shall be permitted in accordance with 200.7.

Ungrounded conductors with white or gray insulation are permitted if the conductors are permanently reidentified at termination points and if the conductor is visible and accessible. The normal methods of reidentification include colored tape, tagging, or paint. Other applications where white conductors are permitted include flexible cords and circuits less than 50 volts. A white conductor used in single-pole, 3-way and 4-way switch loops also requires reidentification (a color other than white, gray, or green) if it is used as an ungrounded conductor. See 200.7(C)(2) for exact details about this required reidentification.

310.13 Conductor Constructions and Applications

Insulated conductors shall comply with the applicable provisions of one or more of the following: Tables 310.13, 310.61, 310.62, 310.63, and 310.64.

Table 310.13 Conductor Application and Insulations

Trade Name	Type Letter	Maximum Operating Temperature	Application Provisions	Insulation	Thickness of Insulation				Outer Covering[1]
					AWG or kcmil	mm		mils	
Fluorinated ethylene propylene	FEP or FEPB	90°C 194°F	Dry and damp locations	Fluorinated ethylene propylene	14–10 8–2	0.51 0.76		20 30	None
		200°C 392°F	Dry locations — special applications[2]	Fluorinated ethylene propylene	14–8	0.36		14	Glass braid
					6–2	0.36		14	Glass or other suitable braid material
Mineral insulation (metal sheathed)	MI	90°C 194°F	Dry and wet locations	Magnesium oxide	18–16[3] 16–10	0.58 0.91		23 36	Copper or alloy steel
		250°C 482°F	For special applications[2]		9–4 3–500	1.27 1.40		50 55	
Moisture-, heat-, and oil-resistant thermoplastic	MTW	60°C 140°F 90°C 194°F	Machine tool wiring in wet locations Machine tool wiring in dry locations. FPN: See NFPA 79.	Flame-retardant moisture-, heat-, and oil-resistant thermoplastic	(A) (B) 22–12 10 8 6 4–2 1–4/0 213–500 501–1000	(A) 0.76 0.76 1.14 1.52 1.52 2.03 2.41 2.79	(B) 0.38 0.51 0.76 0.76 1.02 1.27 1.52 1.78	(A) (B) 30 15 30 20 45 30 60 30 60 40 80 50 95 60 110 70	(A) None (B) Nylon jacket or equivalent
Paper		85°C 185°F	For underground service conductors, or by special permission	Paper					Lead sheath
Perfluoro-alkoxy	PFA	90°C 194°F 200°C 392°F	Dry and damp locations Dry locations — special applications[2]	Perfluoro-alkoxy	14–10 8–2 1–4/0	0.51 0.76 1.14		20 30 45	None
Perfluoro-alkoxy	PFAH	250°C 482°F	Dry locations only. Only for leads within apparatus or within raceways connected to apparatus (nickel or nickel-coated copper only)	Perfluoro-alkoxy	14–10 8–2 1–4/0	0.51 0.76 1.14		20 30 45	None
Thermoset	RHH	90°C 194°F	Dry and damp locations		14–10 8–2 1–4/0 213–500 501–1000 1001–2000 For 601–2000 see Table 310.62.	1.14 1.52 2.03 2.41 2.79 3.18		45 60 80 95 110 125	Moisture-resistant, flame-retardant, nonmetallic covering[1]

(continues)

Table 310.13 *Continued*

Trade Name	Type Letter	Maximum Operating Temperature	Application Provisions	Insulation	Thickness of Insulation			Outer Covering[1]
					AWG or kcmil	mm	mils	
Moisture-resistant thermoset	RHW[4]	75°C 167°F	Dry and wet locations	Flame-retardant, moisture-resistant thermoset	14–10 8–2 1–4/0 213–500 501–1000 1001–2000 For 601–2000, see Table 310.62.	1.14 1.52 2.03 2.41 2.79 3.18	45 60 80 95 110 125	Moisture-resistant, flame-retardant, nonmetallic covering[5]
Moisture-resistant thermoset	RHW-2	90°C 194°F	Dry and wet locations	Flame-retardant moisture-resistant thermoset	14–10 8–2 1–4/0 213–500 501–1000 1001–2000 For 601–2000, see Table 310.62.	1.14 1.52 2.03 2.41 2.79 3.18	45 60 80 95 110 125	Moisture-resistant, flame-retardant, nonmetallic covering[5]
Silicone	SA	90°C 194°F 200°C 392°F	Dry and damp locations For special application[2]	Silicone rubber	14–10 8–2 1–4/0 213–500 501–1000 1001–2000	1.14 1.52 2.03 2.41 2.79 3.18	45 60 80 95 110 125	Glass or other suitable braid material
Thermoset	SIS	90°C 194°F	Switchboard wiring only	Flame-retardant thermoset	14–10 8–2 1–4/0	0.76 1.14 2.41	30 45 95	None
Thermoplastic and fibrous outer braid	TBS	90°C 194°F	Switchboard wiring only	Thermoplastic	14–10 8 6–2 1–4/0	0.76 1.14 1.52 2.03	30 45 60 80	Flame-retardant, nonmetallic covering
Extended polytetra-fluoro-ethylene	TFE	250°C 482°F	Dry locations only. Only for leads within apparatus or within raceways connected to apparatus, or as open wiring (nickel or nickel-coated copper only)	Extruded polytetra-fluoro-ethylene	14–10 8–2 1–4/0	0.51 0.76 1.14	20 30 45	None
Heat-resistant thermoplastic	THHN	90°C 194°F	Dry and damp locations	Flame-retardant, heat-resistant thermoplastic	14–12 10 8–6 4–2 1–4/0 250–500 501–1000	0.38 0.51 0.76 1.02 1.27 1.52 1.78	15 20 30 40 50 60 70	Nylon jacket or equivalent
Moisture- and heat-resistant thermoplastic	THHW	75°C 167°F 90°C 194°F	Wet location Dry location	Flame-retardant, moisture- and heat-resistant thermoplastic	14–10 8 6–2 1–4/0 213–500 501–1000	0.76 1.14 1.52 2.03 2.41 2.79	30 45 60 80 95 110	None

Table 310.13 *Continued*

Trade Name	Type Letter	Maximum Operating Temperature	Application Provisions	Insulation	Thickness of Insulation			Outer Covering[1]
					AWG or kcmil	mm	mils	
Moisture- and heat-resistant thermoplastic	THW[4]	75°C 167°F 90°C 194°F	Dry and wet locations Special applications within electric discharge lighting equipment. Limited to 1000 open-circuit volts or less. (size 14-8 only as permitted in 410.33)	Flame-retardant, moisture- and heat-resistant thermoplastic	14–10 8 6–2 1–4/0 213–500 501–1000 1001–2000	0.76 1.14 1.52 2.03 2.41 2.79 3.18	30 45 60 80 95 110 125	None
Moisture- and heat-resistant thermoplastic	THWN[4]	75°C 167°F	Dry and wet locations	Flame-retardant, moisture- and heat-resistant thermoplastic	14–12 10 8–6 4–2 1–4/0 250–500 501–1000	0.38 0.51 0.76 1.02 1.27 1.52 1.78	15 20 30 40 50 60 70	Nylon jacket or equivalent
Moisture-resistant thermoplastic	TW	60°C 140°F	Dry and wet locations	Flame-retardant, moisture-resistant thermoplastic	14–10 8 6–2 1–4/0 213–500 501–1000 1001–2000	0.76 1.14 1.52 2.03 2.41 2.79 3.18	30 45 60 80 95 110 125	None
Underground feeder and branch-circuit cable — single conductor (for Type UF cable employing more than one conductor, see Article 340.)	UF	60°C 140°F 75°C 167°F[7]	See Article 340.	Moisture-resistant Moisture- and heat-resistant	14–10 8–2 1–4/0	1.52 2.03 2.41	60[6] 80[6] 95[6]	Integral with insulation
Underground service-entrance cable — single conductor (for Type USE cable employing more than one conductor, see Article 338.)	USE[4]	75°C 167°F	See Article 338.	Heat- and moisture-resistant	14–10 8–2 1–4/0 213–500 501–1000 1001–2000	1.14 1.52 2.03 2.41 2.79 3.18	45 60 80 95[8] 110 125	Moisture resistant nonmetallic covering (See 338.2.)
Thermoset	XHH	90°C 194°F	Dry and damp locations	Flame-retardant thermoset	14–10 8–2 1–4/0 213–500 501–1000 1001–2000	0.76 1.14 1.40 1.65 2.03 2.41	30 45 55 65 80 95	None
Moisture-resistant thermoset	XHHW[4]	90°C 194°F 75°C 167°F	Dry and damp locations Wet locations	Flame-retardant, moisture-resistant thermoset	14–10 8–2 1–4/0 213–500 501–1000 1001–2000	0.76 1.14 1.40 1.65 2.03 2.41	30 45 55 65 80 95	None

(continues)

Table 310.13 *Continued*

Trade Name	Type Letter	Maximum Operating Temperature	Application Provisions	Insulation	Thickness of Insulation			Outer Covering[1]
					AWG or kcmil	mm	mils	
Moisture-resistant thermoset	XHHW-2	90°C 194°F	Dry and wet locations	Flame-retardant, moisture-resistant thermoset	14–10	0.76	30	None
					8–2	1.14	45	
					1–4/0	1.40	55	
					213–500	1.65	65	
					501–1000	2.03	80	
					1001–2000	2.41	95	
Modified ethylene tetrafluoro-ethylene	Z	90°C 194°F 150°C 302°F	Dry and damp locations Dry locations — special applications[2]	Modified ethylene tetrafluoro-ethylene	14–12	0.38	15	None
					10	0.51	20	
					8–4	0.64	25	
					3–1	0.89	35	
					1/0–4/0	1.14	45	
Modified ethylene tetrafluoro-ethylene	ZW[4]	75°C 167°F 90°C 194°F 150°C 302°F	Wet locations Dry and damp locations Dry locations — special applications[2]	Modified ethylene tetrafluoro-ethylene	14–10	0.76	30	None
					8–2	1.14	45	

[1] Some insulations do not require an outer covering.
[2] Where design conditions require maximum conductor operating temperatures above 90°C (194°F).
[3] For signaling circuits permitting 300-volt insulation.
[4] Listed wire types designated with the suffix "2," such as RHW-2, shall be permitted to be used at a continuous 90°C (194°F) operating temperature, wet or dry.
[5] Some rubber insulations do not require an outer covering.
[6] Includes integral jacket.
[7] For ampacity limitation, see 340.80.
[8] Insulation thickness shall be permitted to be 2.03 mm (80 mils) for listed Type USE conductors that have been subjected to special investigations. The nonmetallic covering over individual rubber-covered conductors of aluminum-sheathed cable and of lead-sheathed or multiconductor cable shall not be required to be flame retardant. For Type MC cable, see 330.104. For nonmetallic-sheathed cable, see Article 334, Part III. For Type UF cable, see Article 340, Part III.

These conductors shall be permitted for use in any of the wiring methods recognized in Chapter 3 and as specified in their respective tables or as permitted elsewhere in this *Code*.

> FPN: Thermoplastic insulation may stiffen at temperatures lower than −10°C (+14°F). Thermoplastic insulation may also be deformed at normal temperatures where subjected to pressure, such as at points of support. Thermoplastic insulation, where used on dc circuits in wet locations, may result in electroendosmosis between conductor and insulation.

Table 310.13 lists the various types of insulated conductors covered by the requirements of this *Code*. More detailed wire classification information from sizes 14 AWG through 2000 kcmil are available in standards and directories such as those published by Underwriters Laboratories Inc.

Table 310.13 also includes conductor applications and maximum operating temperatures. Some conductors have dual ratings. For example, Type XHHW is rated 90°C for dry and damp locations and 75°C for wet locations; Type

THW is rated 75°C for dry and wet locations and 90°C for special applications within electric-discharge lighting equipment.

Types RHW-2, XHHW-2, and other types identified by the suffix "2" are rated 90°C for wet locations as well as dry and damp locations. Conductors permitted to be identified by the suffix "2," other than Types RHW-2 and XHHW-2, are identified in the table by footnote 4.

The maximum continuous ampacities for copper, copper-clad aluminum, and aluminum conductors, rated 0 through 2000 volts, are listed in Table 310.16 through Table 310.20 and accompanying adjustment factors of 310.15(B)(1) through 310.15(B)(6), or they can be calculated in accordance with 310.15(C).

Copper-clad aluminum conductors are drawn from a copper-clad aluminum rod with the copper metallurgically bonded to an aluminum core. The copper forms a minimum of 10 percent of the cross-sectional area of a solid conductor or of each strand of a stranded conductor. See the commentary following 110.14 for making electrical connections with different types of conductor material. A comparison of the characteristics of copper, copper-clad aluminum, and aluminum conductors can be made from Commentary Table 3.1.

Commentary Table 3.1 Conductor Characteristics

Characteristic	Copper	Copper-Clad Aluminum	Aluminum
Density (lb/in.³)	0.323	0.121	0.098
Density (g/cm³)	8.91	3.34	2.71
Resistivity (ohms/CMF)	10.37	16.08	16.78
Resistivity microhm — CM	1.724	2.673	2.790
Conductivity (IACS %)	100	61–63	61.0
Weight % copper	100	26.8	—
Tensile K psi — hard	65.0	30.0	27.0
Tensile kg/mm² — hard	45.7	21.1	19.0
Tensile K psi — annealed	35.0	17.0	17.0*
Tensile kg/mm² — annealed	24.6	12.0	12.0
Specific gravity	8.91	3.34	2.71

*Semiannealed

310.14 Aluminum Conductor Material

Solid aluminum conductors 8, 10, and 12 AWG shall be made of an AA-8000 series electrical grade aluminum alloy conductor material. Stranded aluminum conductors 8 AWG through 1000 kcmil marked as Type RHH, RHW, XHHW, THW, THHW, THWN, THHN, service-entrance Type SE Style U and SE Style R shall be made of an AA-8000 series electrical grade aluminum alloy conductor material.

Section 310.14 provides proper recognition of approved AA-8000 series electrical-grade aluminum alloy conductor materials for wire and cable products. It also provides coordination with the UL listing requirements for testing terminations, such as CO/ALR devices and other connectors, suitable for use with aluminum conductors. The electrical industry has developed AA-8000 series aluminum alloy materials and the connectors suitable for use with aluminum conductors to provide for safe and stable connections. Connections suitable for use with aluminum conductors are also generally listed as suitable for use with copper conductors and are marked accordingly, such as AL7CU or AL9CU. Numbers 7 and 9 identify the temperature ratings of 75°C and 90°C, respectively, for these connectors.

310.15 Ampacities for Conductors Rated 0–2000 Volts

(A) General

(1) Tables or Engineering Supervision Ampacities for conductors shall be permitted to be determined by tables as provided in 310.15(B) or under engineering supervision, as provided in 310.15(C).

FPN No. 1: Ampacities provided by this section do not take voltage drop into consideration. See 210.19(A), FPN No. 4, for branch circuits and 215.2(A), FPN No. 2, for feeders.

FPN No. 2: For the allowable ampacities of Type MTW wire, see Table 13.5.1 in NFPA 79-2002, *Electrical Standard for Industrial Machinery.*

Editorially revised for the 2005 *Code,* 310.15(A)(1) permits two methods of determining conductor ampacity for conductors rated 0 through 2000 volts: selecting the ampacity from a table, using correction factors in the table or notes where necessary, or calculating the ampacity. The latter method can be complex and time consuming and requires engineering supervision. It can, however, result in lower installation costs, in some cases, and if calculated properly, it provides a mathematically exact ampacity. See the commentary following 310.15(C) and Annex B for further explanation.

(2) Selection of Ampacity Where more than one calculated or tabulated ampacity could apply for a given circuit length, the lowest value shall be used.

Exception: Where two different ampacities apply to adjacent portions of a circuit, the higher ampacity shall be permitted to be used beyond the point of transition, a distance equal to 3.0 m (10 ft) or 10 percent of the circuit length figured at the higher ampacity, whichever is less.

Example

Three 500-kcmil THW conductors in a rigid conduit are run from a motor control center for 12 ft past a heat-treating furnace to a pump motor located 150 ft from the motor control center. Where run in a 78°F to 86°F ambient temperature, the conductors have an ampacity of 380 amperes, per Table 310.16. The ambient temperature near the furnace, where the conduit is run, is found to be 113°F, and the length of this particular part of the run is greater than 10 ft and more than 10 percent of the total length of the run at the 78°F to 86°F ambient. Determine the ampacity of total run in accordance with 310.15(A)(2).

Solution

In accordance with the correction factors for temperature at the bottom of Table 310.16, the ampacity is 0.82 × 380 amperes, or 311.6 amperes. This, therefore, is the ampacity of the total run, in accordance with 310.15(A)(2).

Had the run near the furnace at the 113°F ambient been 10 ft or less in length, the ampacity of the entire run would have been 380 amperes, in accordance with the exception to 310.15(A)(2). The heat-sinking effect of the run at the lower ambient temperature would have been sufficient to reduce the temperature of the conductor near the furnace.

FPN: See 110.14(C) for conductor temperature limitations due to termination provisions.

(B) Tables Ampacities for conductors rated 0 to 2000 volts shall be as specified in the Allowable Ampacity Table 310.16 through Table 310.19, and Ampacity Table 310.20 and Table 310.21 as modified by (B)(1) through (B)(6).

FPN: Table 310.16 through Table 310.19 are application tables for use in determining conductor sizes on loads calculated in accordance with Article 220. Allowable ampacities result from consideration of one or more of the following:

(1) Temperature compatibility with connected equipment, especially the connection points.
(2) Coordination with circuit and system overcurrent protection.
(3) Compliance with the requirements of product listings or certifications. See 110.3(B).
(4) Preservation of the safety benefits of established industry practices and standardized procedures.

Ampacity tables, particularly Table 310.16, do not take into account all the many factors affecting ampacity. See the commentary following 310.15(C) and Annex B for further explanation. However, experience over many years has proved the table values to be adequate for loads calculated in accordance with Article 220, because not all diversity factors and load factors found in most actual installations are specifically provided for in Article 220. If loads are not calculated in accordance with the requirements of Article 220, the table ampacities, even when corrected in accordance with ambient correction factors and the notes to the tables, might be too high. This result can be particularly true where many cables or raceways are routed close to one another underground. However, load diversity and thermal conductance fill around buried cable could result in increased ampacity. See Annex B for further information.

(1) General For explanation of type letters used in tables and for recognized sizes of conductors for the various conductor insulations, see 310.13. For installation requirements, see 310.1 through 310.10 and the various articles of this *Code*. For flexible cords, see Tables 400.4, 400.5(A), and 400.5(B).

(2) Adjustment Factors

(a) More Than Three Current-Carrying Conductors in a Raceway or Cable. Where the number of current-carrying conductors in a raceway or cable exceeds three, or where single conductors or multiconductor cables are stacked or bundled longer than 600 mm (24 in.) without maintaining spacing and are not installed in raceways, the allowable ampacity of each conductor shall be reduced as shown in Table 310.15(B)(2)(a). Each current-carrying conductor of

a paralleled set of conductors shall be counted as a current-carrying conductor.

Table 310.15(B)(2)(a) Adjustment Factors for More Than Three Current-Carrying Conductors in a Raceway or Cable

Number of Current-Carrying Conductors	Percent of Values in Tables 310.16 through 310.19 as Adjusted for Ambient Temperature if Necessary
4–6	80
7–9	70
10–20	50
21–30	45
31–40	40
41 and above	35

FPN No. 1: See Annex B, Table B.310.11, for adjustment factors for more than three current-carrying conductors in a raceway or cable with load diversity.

FPN No. 2: See 366.23(A) for correction factors for conductors in sheet metal auxiliary gutters and 376.22 for correction factors for conductors in metal wireways.

The factors in the second column of Table 310.15(B)(2)(a) are based on no diversity, meaning that all conductors in the raceway or cable are loaded to their maximum rated load. For load diversity, the user is directed to Annex B.

Specific cross references for raceway fill and adjustment factors of 310.15(B)(2) can be found in the fine print note following 300.17. For Class 1 conductors, see 725.28(A); for fire alarm systems, 760.28 and 760.52; for optical fiber cables and raceways, 770.12; and for communications wires and cables within buildings, 800.110.

The last sentence was added for the 2005 *Code*, to clarify that for parallel conductors, each separate conductor that is considered a current-carrying conductor must be counted when applying the adjustment factors of Table 310.15(B)(2)(a). FPN No. 2, new for the 2005 *Code*, points to different adjustment factors for the number of conductors contained in metal auxiliary gutters and metal wireways.

Exception No. 1: Where conductors of different systems, as provided in 300.3, are installed in a common raceway or cable, the derating factors shown in Table 310.15(B)(2)(a) shall apply only to the number of power and lighting conductors (Articles 210, 215, 220, and 230).

Exception No. 1 to 310.15(B)(2)(a) assumes that the watt loss (heating) from any control and signal conductors in the same raceway or cable will not be enough to significantly increase the temperature of the power and lighting conductors. See 725.26 and 725.54 for limitations on the installation of control and signal conductors in the same raceway or cable as power and lighting conductors.

Exception No. 2: For conductors installed in cable trays, the provisions of 392.11 shall apply.

Exception No. 3: Derating factors shall not apply to conductors in nipples having a length not exceeding 600 mm (24 in.).

Exception No. 4: Derating factors shall not apply to underground conductors entering or leaving an outdoor trench if those conductors have physical protection in the form of rigid metal conduit, intermediate metal conduit, or rigid nonmetallic conduit having a length not exceeding 3.05 m (10 ft) and if the number of conductors does not exceed four.

Exhibit 310.4 illustrates Exception No. 4 to 310.15(B)(2)(a), in that derating factors do not apply to the conductors because they have physical protection (conduit) that does not exceed 10 ft in length and the number of conductors does not exceed four.

Exhibit 310.4 An application of 310.15(B)(2)(a), Exception No. 4.

Exception No. 5: Adjustment factors shall not apply to Type AC cable or to Type MC cable without an overall outer jacket under the following conditions:

(1) Each cable has not more than three current-carrying conductors.

(2) The conductors are 12 AWG copper.

(3) Not more than 20 current-carrying conductors are bundled, stacked, or supported on "bridle rings."

A 60 percent adjustment factor shall be applied where the current-carrying conductors in these cables that are stacked or bundled longer than 600 mm (24 in.) without maintaining spacing exceeds 20.

Example

A commercial office space requires fourteen 277-volt fluorescent lighting circuits to serve a single open office area. The office area lighting is assumed to be a continuous load, and the office ambient temperature will not exceed 30°C (86°F). Each circuit will be arranged so that it has a calculated load not exceeding 16 amperes. The selected wiring method is Type MC cable, 3-conductor (with an additional equipment grounding conductor), 12 AWG THHN copper. Each individual MC cable will contain a 3-wire multiwire branch circuit. To serve the entire area, this arrangement requires a total of seven Type MC cables bundled for a distance of about 25 ft, without maintaining spacing between them where they leave the electrical room and enter the office area.

Determine the ampacity of each circuit conductor in accordance with 310.15, applying Exception No. 5 to 310.15(B)(2)(a) to account for the bundled cables. Then determine the maximum branch-circuit overcurrent protection permitted for these bundled MC cables.

Solution

STEP 1. To apply Exception No. 5, first determine the quantity of current-carrying conductors. According to 310.15(B)(5), equipment grounding conductors are not counted as current-carrying conductors. According to 310.15(B)(4)(c), fluorescent lighting is considered a nonlinear load, so the grounded conductor of each Type MC cable must be counted as a current-carrying conductor.

7 cables × 3 conductors each = 21 current-carrying
conductors

Because the quantity of current-carrying conductors exceeds 20, a 60 percent adjustment factor is required by 310.15(B)(2)(a), Exception No. 5.

STEP 2. Determine the ampacity of each current-carrying conductor due to these MC cables with more than 20 current-carrying conductors being bundled.

From Table 310.16, 12 AWG THHN = 30 amps

30 amps × 0.60 = 18 amps

Because the actual calculated load is 16 amperes of continuous load, 210.19(A)(1) is applicable. The conductors must have an ampacity equal to or greater than the load before the adjustment factor is applied. Because the ampacity of the conductors after the adjustment factor is applied is 18 amperes, no further adjustment is necessary and the conductors are suitable for this installation.

STEP 3. Finally, determine the maximum size of overcurrent protection device permitted for these bundled MC cable branch circuits. Section 240.4(B) permits the use of the next higher standard rating of overcurrent protection device. Therefore, although the conductors have a calculated ampacity of 18 amperes, a 20-ampere overcurrent protective device is permitted. In addition, and of significance, the 20-ampere overcurrent protective device is in compliance with 210.20(A), given that the actual 16-ampere continuous load would require a 20-ampere overcurrent protective device, based on the listing of the overcurrent device.

(b) More Than One Conduit, Tube, or Raceway. Spacing between conduits, tubing, or raceways shall be maintained.

Spacing is normally maintained between individual conduits in groups of conduit runs from junction box to junction box because of the need to separate the conduits where they enter the junction box, to allow room for locknuts and bushings. Field experience has indicated that this degree of spacing between runs has not caused any problems.

(3) **Bare or Covered Conductors** Where bare or covered conductors are used with insulated conductors, their allowable ampacities shall be limited to those permitted for the adjacent insulated conductors.

(4) **Neutral Conductor**

(a) A neutral conductor that carries only the unbalanced current from other conductors of the same circuit shall not be required to be counted when applying the provisions of 310.15(B)(2)(a).

(b) In a 3-wire circuit consisting of two phase wires and the neutral of a 4-wire, 3-phase, wye-connected system, a common conductor carries approximately the same current as the line-to-neutral load currents of the other conductors and shall be counted when applying the provisions of 310.15(B)(2)(a).

(c) On a 4-wire, 3-phase wye circuit where the major portion of the load consists of nonlinear loads, harmonic currents are present in the neutral conductor; the neutral shall therefore be considered a current-carrying conductor.

During the 1996 *NEC* cycle, a task group composed of interested parties was created to recommend to the National Electrical Code Committee the direction its standards should take to improve the safeguarding of persons and property from conditions that can be introduced by nonlinear loads. This group was designated the NEC Correlating Committee Ad Hoc Subcommittee on Nonlinear Loads. The scope of the subcommittee was as follows:

1. To study the effects of electrical loads producing substantial current distortion upon electrical system distribution components including but not limited to
   a. Distribution transformers, current transformers, and others
   b. Switchboards and panelboards
   c. Phase and neutral feeder conductors
   d. Phase and neutral branch-circuit conductors
   e. Proximate data and communications conductors
2. To study harmful effects, if any, to the system components from overheating resulting from these load characteristics
3. To make recommendations for methods to minimize the harmful effects of nonlinear loads considering all means, including compensating methods at load sources
4. To prepare proposals, if necessary, to amend the 1996 *National Electrical Code*, where amelioration to fire safety may be achieved

The subcommittee reviewed technical literature and electrical theory on the fundamental nature of harmonic distortion, as well as the requirements in and proposals for the 1993 *NEC* regarding nonlinear loads. The subcommittee concluded that, while nonlinear loads can cause undesirable operational effects, including additional heating, no significant threat to persons and property had been substantiated.

The subcommittee agreed with the existing *Code* text regarding nonlinear loads. However, the subcommittee submitted many proposals for the 1996 *NEC*, including a definition of *nonlinear load*, revised text reflecting that definition, fine print notes calling attention to the effects of nonlinear loads, and proposals permitting the paralleling of neutral conductors in existing installations under engineering supervision.

As part of the subcommittee's final report, nine proposals for changes to the 1993 *NEC* were submitted. All were accepted without modification as changes in the 1996 *NEC*. Also included in this report and now pertinent to 310.15(B)(4)(c) in the 2002 *NEC* is the following discussion.

Should Neutral Conductors Be Oversized?

There is concern that, because the theoretical maximum neutral current is 1.73 times the balanced phase conductor current, a potential exists for neutral conductor overheating in 3-phase, 4-wire, wye-connected power systems. The subcommittee acknowledged this theoretical basis, although a review of documented information could not identify fires attributed to the use of nonlinear loads.

The subcommittee reviewed all available data regarding measurements of circuits that contain nonlinear loads. The data were obtained from consultants, equipment manufacturers, and testing laboratories, and included hundreds of feeder and branch circuits involving 3-phase, 4-wire, wye-connected systems with nonlinear loads. The data revealed that many circuits had neutral conductor current greater than the phase conductor current, and approximately 5 percent of all circuits reported had neutral conductor current exceeding 125 percent of the highest phase conductor current. One documented survey with data collected in 1988 from 146 three-phase computer power system sites determined that 3.4 percent of the sites had neutral current in excess of the rated system full-load current.

According to 384-16(C) of the 1993 *NEC* [for the 2005 *NEC*, refer to 210.19(A)(1) and 215.2(A)(1)], the total continuous load on any overcurrent device located in a panelboard should not exceed 80 percent of its rating (the exception being assemblies listed for continuous operation at 100 percent of its rating). Because the neutral conductor is usually not connected to an overcurrent device, derating for continuous operation is not necessary. Therefore, neutral conductor ampacity is usually 125 percent of the maximum continuous current allowed by the overcurrent device.

Also important for gathering electrically measured data from existing installations is the measurement of nonsinusoidal voltages and currents.

Measurement of Nonsinusoidal Voltages and Currents

The measurement of nonsinusoidal voltages and currents may require instruments different from the conventional meters used to measure sinusoidal waveforms. Many voltage and current meters respond only to the peak value of a waveform and indicate a value that is equivalent to the rms value of a sinusoidal waveform. For a sinusoidal waveform, the rms value will be 70.7 percent of the peak value. Meters of this type are known as "average responding meters" and will give a true indication only if the waveform being measured is sinusoidal. Both analog and digital meters may be average responding instruments. Voltages and currents that are nonsinusoidal, such as those with harmonic frequencies, cannot be accurately measured using an average responding meter. Only a meter that measures "true rms" can be used to correctly measure the rms value of a nonsinusoidal waveform.

Exhibit 310.5 shows an example of a clamp-on ammeter that uses true rms measurements. Exhibit 310.6 shows an example of a portable diagnostic analyzer used for more sophisticated power measurements, including measuring harmonic distortion.

Exhibit 310.5 A clamp-on ammeter that uses true rms measurements. (Courtesy of Fluke Corp.)

Exhibit 310.6 A portable tool for such tasks as diagnostic power analysis, including harmonic distortion. (Courtesy of Dranetz-BMI)

(5) Grounding or Bonding Conductor A grounding or bonding conductor shall not be counted when applying the provisions of 310.15(B)(2)(a).

(6) 120/240-Volt, 3-Wire, Single-Phase Dwelling Services and Feeders For individual dwelling units of one family, two-family, and multifamily dwellings, conductors, as listed in Table 310.15(B)(6), shall be permitted as 120/240-volt, 3-wire, single-phase service-entrance conductors, service lateral conductors, and feeder conductors that serve as the main power feeder to each dwelling unit and are installed in raceway or cable with or without an equipment grounding conductor. For application of this section, the main power feeder shall be the feeder(s) between the main disconnect and the lighting and appliance branch-circuit panelboards(s). The feeder conductors to a dwelling unit shall not be required to have an allowable ampacity rating greater than their service-entrance conductors. The grounded conductor shall be permitted to be smaller than the ungrounded conductors, provided the requirements of 215.2, 220.61, and 230.42 are met.

Table 310.15(B)(6) Conductor Types and Sizes for 120/240-Volt, 3-Wire, Single-Phase Dwelling Services and Feeders. Conductor Types RHH, RHW, RHW-2, THHN, THHW, THW, THW-2, THWN, THWN-2, XHHW, XHHW-2, SE, USE, USE-2

Conductor (AWG or kcmil)		Service or
Copper	Aluminum or Copper-Clad Aluminum	Feeder Rating (Amperes)
4	2	100
3	1	110
2	1/0	125
1	2/0	150
1/0	3/0	175
2/0	4/0	200
3/0	250	225
4/0	300	250
250	350	300
350	500	350
400	600	400

For the 2005 *Code*, only editorial changes were made to 310.15(B)(6). If a single set of 3-wire, single-phase, service-entrance conductors in raceway or cable supplies a one-family, two-family, or multifamily dwelling, the reduced conductor size permitted by 310.15(B)(6) is applicable to the service-entrance conductors, service-lateral conductors, or any feeder conductors that supply the main power feeder to a dwelling unit.

Section 310.15(B)(6) permits the main feeder to a dwelling unit to be sized according to the conductor sizes in Table 310.15(B)(6) even if other loads, such as ac units and pool loads, are fed from the same service. The feeder conductors to a dwelling unit are not required to be larger than its service-entrance conductors.

Exhibits 310.7 and 310.8 illustrate the application of 310.15(B)(6). In Exhibit 310.7, the reduced conductor size permitted is applicable to the service-entrance conductors run to each apartment from the meters. In Exhibit 310.8, the reduced conductor size permitted is also applicable to the feeder conductors run to each apartment from the service disconnecting means, because these feeders carry the entire load to each apartment.

Previously limited to a two-wire size reduction but

Exhibit 310.7 An application of 310.15(B)(6).

Exhibit 310.8 Another application of 310.15(B)(6).

changed in the 1996 *Code*, the grounded conductor is now permitted to be reduced more than two sizes. The stipulation is that the requirements from the other applicable *Code* sections are observed, including 250.24(B).

Other sections of the *Code* must also be applied for determining the size of service or feeder conductors. Section 230.42 requires service conductors to be of sufficient size to carry the load calculated in accordance with Article 220. It should not be taken for granted that the grounded (neutral) conductor can be automatically reduced. In addition to the load, the grounded (neutral) conductor must also provide a low-impedance fault path capable of conducting fault current back to the transformer. Section 250.24(B) provides the requirements for determining the minimum size grounded (neutral) conductor.

Section 250.24(B) is not applicable to the grounded (neutral) conductor of feeders; therefore, the minimum size is governed by 220.61. Section 220.61 requires the grounded (neutral) conductor to be large enough to carry the maximum unbalance of the net calculated load connected to the neutral and any one ungrounded conductor. In the event that there are no 240-volt loads, the neutral, under severe unbalanced conditions, carries the same current as the ungrounded conductor supplying the load.

(C) Engineering Supervision Under engineering supervision, conductor ampacities shall be permitted to be calculated by means of the following general formula:

$$I = \sqrt{\frac{TC - (TA + \Delta TD)}{RDC(1 + YC)RCA}}$$

where:

TC = conductor temperature in degrees Celsius (°C)

TA = ambient temperature in degrees Celsius (°C)

ΔTD = dielectric loss temperature rise

RDC = dc resistance of conductor at temperature TC

YC = component ac resistance resulting from skin effect and proximity effect

RCA = effective thermal resistance between conductor and surrounding ambient

FPN: See Annex B for examples of formula applications.

The formula in 310.15(C) was developed by J. H. Neher and M. H. McGrath to determine conductor ampacity. It is actually a composite of a number of separate formulas. A description of this method of calculation was given in AIEE paper No. 5-660, "The Calculation of the Temperature Rise and Load Capability of Cable Systems," by J. H. Neher and M. H. McGrath. This paper was presented to the AIEE general meeting in Montreal, Quebec, on June 24–28, 1956, and was published in *AIEE Transactions, Part III (Power Apparatus and Systems)*, Vol. 76, October 1957, pp. 752–772. The AIEE (American Institute of Electrical Engineers) is now the Institute of Electrical and Electronic Engineers (IEEE).

The Neher-McGrath formula in 310.15(C) is a heat-transfer formula, comprising a series of heat-transfer calculations, that takes into account all heat sources and the thermal resistances between the heat sources and free air. The most common use for the Neher-McGrath formula is to calculate the ampacity of conductors in underground electrical ducts

(raceways), although the formula is applicable to all conductor installations.

It is not the intent of the following discussion to provide instruction on the use of the Neher-McGrath method of calculation. The intent is to identify the many factors affecting the calculations. It is because of the many variables and the complexities of the many formulas involved that the *Code* requires the calculation to be made under engineering supervision.

Current passing through a conductor produces I^2R losses in the form of heat, which results from conductor losses and appears as a temperature rise in the conductor. This heat must pass through the cable insulation, the air in the raceway, and the raceway itself to the surrounding medium, usually earth or concrete, where it is dissipated into the air by radiation and convection. Unless the heat is dissipated, the temperature in the conductor will exceed the rating of the conductor insulation.

The conductor's ampacity is based on the rate of heat dissipation through the thermal resistances surrounding the conductor. Current traveling through a material with a specific resistance at a specified temperature generates this heat. Additional heat is caused by skin and proximity effects, because usually the current is ac and there are other conductors in the same duct.

For conductors in underground electrical ducts, there are several heat sources, as follows, and as illustrated in Exhibit 310.9:

1. *Conductor losses due to the load current I^2R.* These losses vary with the load current, conductor material, and conductor cross-sectional area (conductor size).
2. *Skin-effect heating if the current is alternating current.* The heat developed by the skin effect is due to the shape of the conductor and is based on the configuration of the conductors (i.e., solid, stranded, or compact).
3. *Hysteresis losses if the duct is steel or other magnetic material.* These losses are dependent on the magnetic properties of the electrical duct and the shape of the duct.
4. *Heating from other conductors in the duct.* This heating is based on the number, location, and proximity of other conductors as well as the losses in the other conductors. The more conductors in the raceway, the greater the heating effect from these conductors is likely to be. This factor replaces the adjustment factors in 310.15(B)(2)(a) to the ampacity tables.
5. *Mutual heating from other ducts, cables, and so forth, in the vicinity.* The closer the other heat sources and the more they surround the duct for which calculations are being made, the greater the heating effect. For example, in the case of a symmetrical nine-duct bank, three ducts high and three ducts wide, the center duct will receive the most heat as a result of mutual heating.

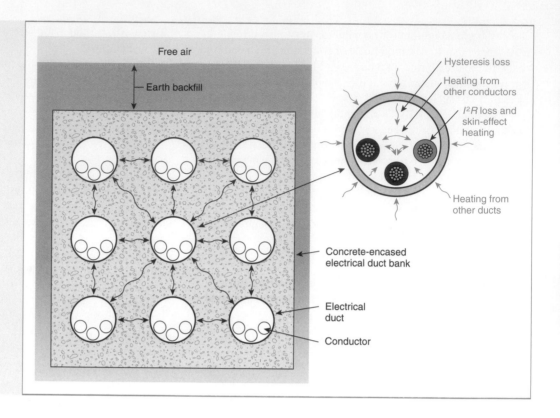

Exhibit 310.9 Heat sources.

Heat generated by the following various types of losses is conducted through the different thermal barriers or resistances, as illustrated in Exhibit 310.10.

Conductor Insulation. The conductor Insulation, which is designed to perform as a good electrical insulator, also serves as a good thermal insulator. It presents a thermal resistance

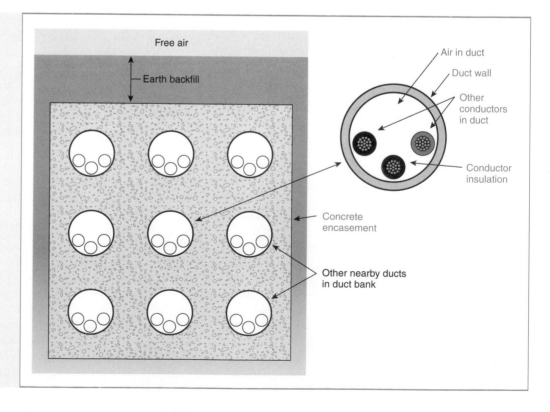

Exhibit 310.10 Thermal barriers (resistances).

Table 310.16 Allowable Ampacities of Insulated Conductors Rated 0 Through 2000 Volts, 60°C Through 90°C (140°F Through 194°F), Not More Than Three Current-Carrying Conductors in Raceway, Cable, or Earth (Directly Buried), Based on Ambient Temperature of 30°C (86°F)

Size AWG or kcmil	Temperature Rating of Conductor (See Table 310.13.)						Size AWG or kcmil
	60°C (140°F)	75°C (167°F)	90°C (194°F)	60°C (140°F)	75°C (167°F)	90°C (194°F)	
	Types TW, UF	Types RHW, THHW, THW, THWN, XHHW, USE, ZW	Types TBS, SA, SIS, FEP, FEPB, MI, RHH, RHW-2, THHN, THHW, THW-2, THWN-2, USE-2, XHH, XHHW, XHHW-2, ZW-2	Types TW, UF	Types RHW, THHW, THW, THWN, XIIHW, USE	Types TBS, SA, SIS, THHN, THHW, THW-2, THWN-2, RHH, RHW-2, USE-2, XHH, XHHW, XHHW-2, ZW-2	
	COPPER			ALUMINUM OR COPPER-CLAD ALUMINUM			
18	—	—	14	—	—	—	—
16	—	—	18	—	—	—	—
14*	20	20	25	—	—	—	—
12*	25	25	30	20	20	25	12*
10*	30	35	40	25	30	35	10*
8	40	50	55	30	40	45	8
6	55	65	75	40	50	60	6
4	70	85	95	55	65	75	4
3	85	100	110	65	75	85	3
2	95	115	130	75	90	100	2
1	110	130	150	85	100	115	1
1/0	125	150	170	100	120	135	1/0
2/0	145	175	195	115	135	150	2/0
3/0	165	200	225	130	155	175	3/0
4/0	195	230	260	150	180	205	4/0
250	215	255	290	170	205	230	250
300	240	285	320	190	230	255	300
350	260	310	350	210	250	280	350
400	280	335	380	225	270	305	400
500	320	380	430	260	310	350	500
600	355	420	475	285	340	385	600
700	385	460	520	310	375	420	700
750	400	475	535	320	385	435	750
800	410	490	555	330	395	450	800
900	435	520	585	355	425	480	900
1000	455	545	615	375	445	500	1000
1250	495	590	665	405	485	545	1250
1500	520	625	705	435	520	585	1500
1750	545	650	735	455	545	615	1750
2000	560	665	750	470	560	630	2000

CORRECTION FACTORS

Ambient Temp. (°C)	For ambient temperatures other than 30°C (86°F), multiply the allowable ampacities shown above by the appropriate factor shown below.						Ambient Temp. (°F)
21–25	1.08	1.05	1.04	1.08	1.05	1.04	70–77
26–30	1.00	1.00	1.00	1.00	1.00	1.00	78–86
31–35	0.91	0.94	0.96	0.91	0.94	0.96	87–95
36–40	0.82	0.88	0.91	0.82	0.88	0.91	96–104
41–45	0.71	0.82	0.87	0.71	0.82	0.87	105–113
46–50	0.58	0.75	0.82	0.58	0.75	0.82	114–122
51–55	0.41	0.67	0.76	0.41	0.67	0.76	123–131
56–60	—	0.58	0.71	—	0.58	0.71	132–140
61–70	—	0.33	0.58	—	0.33	0.58	141–158
71–80	—	—	0.41	—	—	0.41	159–176

* See 240.4(D).

Table 310.17 Allowable Ampacities of Single-Insulated Conductors Rated 0 Through 2000 Volts in Free Air, Based on Ambient Air Temperature of 30°C (86°F)

	Temperature Rating of Conductor (See Table 310.13.)						
	60°C (140°F)	75°C (167°F)	90°C (194°F)	60°C (140°F)	75°C (167°F)	90°C (194°F)	
Size AWG or kcmil	Types TW, UF	Types RHW, THHW, THW, THWN, XHHW, ZW	Types TBS, SA, SIS, FEP, FEPB, MI, RHH, RHW-2, THHN, THHW, THW-2, THWN-2, USE-2, XHH, XHHW, XHHW-2, ZW-2	Types TW, UF	Types RHW, THHW, THW, THWN, XHHW	Types TBS, SA, SIS, THHN, THHW, THW-2, THWN-2, RHH, RHW-2, USE-2, XHH, XHHW, XHHW-2, ZW-2	Size AWG or kcmil
	COPPER			ALUMINUM OR COPPER-CLAD ALUMINUM			
18	—	—	18	—	—	—	—
16	—	—	24	—	—	—	—
14*	25	30	35	—	—	—	—
12*	30	35	40	25	30	35	12*
10*	40	50	55	35	40	40	10*
8	60	70	80	45	55	60	8
6	80	95	105	60	75	80	6
4	105	125	140	80	100	110	4
3	120	145	165	95	115	130	3
2	140	170	190	110	135	150	2
1	165	195	220	130	155	175	1
1/0	195	230	260	150	180	205	1/0
2/0	225	265	300	175	210	235	2/0
3/0	260	310	350	200	240	275	3/0
4/0	300	360	405	235	280	315	4/0
250	340	405	455	265	315	355	250
300	375	445	505	290	350	395	300
350	420	505	570	330	395	445	350
400	455	545	615	355	425	480	400
500	515	620	700	405	485	545	500
600	575	690	780	455	540	615	600
700	630	755	855	500	595	675	700
750	655	785	885	515	620	700	750
800	680	815	920	535	645	725	800
900	730	870	985	580	700	785	900
1000	780	935	1055	625	750	845	1000
1250	890	1065	1200	710	855	960	1250
1500	980	1175	1325	795	950	1075	1500
1750	1070	1280	1445	875	1050	1185	1750
2000	1155	1385	1560	960	1150	1335	2000

CORRECTION FACTORS

Ambient Temp. (°C)	For ambient temperatures other than 30°C (86°F), multiply the allowable ampacities shown above by the appropriate factor shown below.						Ambient Temp. (°F)
21–25	1.08	1.05	1.04	1.08	1.05	1.04	70–77
26–30	1.00	1.00	1.00	1.00	1.00	1.00	78–86
31–35	0.91	0.94	0.96	0.91	0.94	0.96	87–95
36–40	0.82	0.88	0.91	0.82	0.88	0.91	96–104
41–45	0.71	0.82	0.87	0.71	0.82	0.87	105–113
46–50	0.58	0.75	0.82	0.58	0.75	0.82	114–122
51–55	0.41	0.67	0.76	0.41	0.67	0.76	123–131
56–60	—	0.58	0.71	—	0.58	0.71	132–140
61–70	—	0.33	0.58	—	0.33	0.58	141–158
71–80	—	—	0.41	—	—	0.41	159–176

* See 240.4(D).

Table 310.18 Allowable Ampacities of Insulated Conductors Rated 0 Through 2000 Volts, 150°C Through 250°C (302°F Through 482°F). Not More Than Three Current-Carrying Conductors in Raceway or Cable, Based on Ambient Air Temperature of 40°C (104°F)

Size AWG or kcmil	Temperature Rating of Conductor (See Table 310.13.)				Size AWG or kcmil
	150°C (302°F)	200°C (392°F)	250°C (482°F)	150°C (302°F)	
	Type Z	Types FEP, FEPB, PFA, SA	Types PFAH, TFE	Type Z	
	COPPER		NICKEL OR NICKEL-COATED COPPER	ALUMINUM OR COPPER-CLAD ALUMINUM	
14	34	36	39	—	14
12	43	45	54	30	12
10	55	60	73	44	10
8	76	83	93	57	8
6	96	110	117	75	6
4	120	125	148	94	4
3	143	152	166	109	3
2	160	171	191	124	2
1	186	197	215	145	1
1/0	215	229	244	169	1/0
2/0	251	260	273	198	2/0
3/0	288	297	308	227	3/0
4/0	332	346	361	260	4/0

CORRECTION FACTORS

Ambient Temp. (°C)	For ambient temperatures other than 40°C (104°F), multiply the allowable ampacities shown above by the appropriate factor shown below.				Ambient Temp. (°F)
41–50	0.95	0.97	0.98	0.95	105–122
51–60	0.90	0.94	0.95	0.90	123–140
61–70	0.85	0.90	0.93	0.85	141–158
71–80	0.80	0.87	0.90	0.80	159–176
81–90	0.74	0.83	0.87	0.74	177–194
91–100	0.67	0.79	0.85	0.67	195–212
101–120	0.52	0.71	0.79	0.52	213–248
121–140	0.30	0.61	0.72	0.30	249–284
141–160	—	0.50	0.65	—	285–320
161–180	—	0.35	0.58	—	321–356
181–200	—	—	0.49	—	357–392
201–225	—	—	0.35	—	393–437

to heat generated by the conductor due to the I^2R losses, including any dielectric losses. This thermal resistance value depends on the thickness of the insulation and the type of insulating material used. Materials such as polyvinyl chloride, used in Type THW and other conductors; cross-linked polyethylene, used in Type XHHW and other conductors; and rubber, used in Type RHW and other conductors, have different thermal resistivities. In addition, the thickness of the conductor insulation varies from one type of insulation to another, even for the same size conductor.

Airspace. The next thermal barrier encountered by the heat flow generated in the conductor is the airspace between the conductor insulation and the surrounding wall or raceway. The thermal resistance of this airspace is based on the number of conductors in the duct, the assumed mean value of the temperature of the air in the duct, and the constants provided in the Neher-McGrath paper, which were determined from experimental data.

Duct Wall. After it passes through the airspace around the conductors, the heat encounters the thermal resistance of the duct wall. This thermal resistance is based on the thermal resistivity of the type of material used and the thickness of the duct wall. Metallic materials have less thermal resistance than nonmetallic materials. The thicker the wall, the greater the thermal resistance.

Earth Backfill. The thermal resistance that must be considered next is that offered by the earth or other backfill material above the duct. This incorporates not only the thermal resistivity and ambient temperature of the earth but also the number of current-carrying conductors within the duct, the outside diameter of the duct, the burial depth, a loss factor, and the mutual heating factor caused by other nearby ducts. The deeper the duct is buried, the greater the thermal resistance.

To prevent the temperature of the conductors from exceeding the rated temperature of the insulation, heat dissipation through thermal resistances must be equal to or greater than the heat developed. Thus, the thermal resistances of all the components of a conductor must be determined, and the allowable temperature differential above the ambient temperature and between the conductors and the surface of the earth must be known.

In addition to the Neher-McGrath paper itself, as described in the first paragraph of this commentary, the following references provide more detailed information on the use of this method of calculation:

IPCEA P-46-426 (IEEE S-135-1), *Power Cable Ampacity Tables*, Insulated Power Cable Engineers Association.

IEEE 835-1994, *IEEE Standard Power Cable Ampacity Tables*, Institute of Electrical and Electronic Engineers.

P. Pollak, "Neher-McGrath Calculations for Insulated Power Cables," IEEE Paper No. CH2040-4/84/0000-0172, presented at the IEEE Industrial and Commercial Power Systems Technical Conference, Atlanta, GA, May 9, 1984.

M. T. Brown et al., "Cable Ampacities, the NEC and Computerized Applications," IEEE Paper No. CH2207-9/85/00000-0323.

J. M. Caloggero, "How to Use the Neher-McGrath Method to Calculate Ampacity of Underground Conductors," *NFPA Fire Journal*, May/June 1988, pp. 17-18.

310.60 Conductors Rated 2001 to 35,000 Volts

(A) Definitions

Electrical Ducts. As used in Article 310, electrical ducts shall include any of the electrical conduits recognized in Chapter 3 as suitable for use underground; other raceways round in cross section, listed for underground use, and embedded in earth or concrete.

The term *electrical ducts* is used to differentiate them from other ducts, for example, those used for air handling. It is intended to include nonmetallic electrical ducts commonly used for underground wiring, as well as other raceways (e.g., rigid metal conduit, intermediate metal conduit, rigid nonmetallic conduit) listed for use underground in earth or concrete.

Thermal Resistivity. As used in this *Code*, the heat transfer capability through a substance by conduction. It is the reciprocal of thermal conductivity and is designated Rho and expressed in the units °C-cm/watt.

(B) Ampacities of Conductors Rated 2001 to 35,000 Volts Ampacities for solid dielectric-insulated conductors shall be permitted to be determined by tables or under engineering supervision, as provided in 310.60(C) and (D).

(1) Selection of Ampacity Where more than one calculated or tabulated ampacity could apply for a given circuit length, the lowest value shall be used.

Exception: Where two different ampacities apply to adjacent portions of a circuit, the higher ampacity shall be permitted to be used beyond the point of transition, a distance equal to 3.0 m (10 ft) or 10 percent of the circuit length figured at the higher ampacity, whichever is less.

FPN: See 110.40 for conductor temperature limitations due to termination provisions.

Table 310.19 Allowable Ampacities of Single-Insulated Conductors, Rated 0 Through 2000 Volts, 150°C Through 250°C (302°F Through 482°F), in Free Air, Based on Ambient Air Temperature of 40°C (104°F)

Size AWG or kcmil	Temperature Rating of Conductor (See Table 310.13.)				Size AWG or kcmil
	150°C (302°F)	200°C (392°F)	250°C (482°F)	150°C (302°F)	
	Type Z	Types FEP, FEPB, PFA, SA	Types PFAH, TFE	Type Z	
	COPPER		NICKEL, OR NICKEL-COATED COPPER	ALUMINUM OR COPPER-CLAD ALUMINUM	
14	46	54	59	—	14
12	60	68	78	47	12
10	80	90	107	63	10
8	106	124	142	83	8
6	155	165	205	112	6
4	190	220	278	148	4
3	214	252	327	170	3
2	255	293	381	198	2
1	293	344	440	228	1
1/0	339	399	532	263	1/0
2/0	390	467	591	305	2/0
3/0	451	546	708	351	3/0
4/0	529	629	830	411	4/0

CORRECTION FACTORS

Ambient Temp. (°C)	For ambient temperatures other than 40°C (104°F), multiply the allowable ampacities shown above by the appropriate factor shown below.				Ambient Temp. (°F)
41–50	0.95	0.97	0.98	0.95	105–122
51–60	0.90	0.94	0.95	0.90	123–140
61–70	0.85	0.90	0.93	0.85	141–158
71–80	0.80	0.87	0.90	0.80	159–176
81–90	0.74	0.83	0.87	0.74	177–194
91–100	0.67	0.79	0.85	0.67	195–212
101–120	0.52	0.71	0.79	0.52	213–248
121–140	0.30	0.61	0.72	0.30	249–284
141–160	—	0.50	0.65	—	285–320
161–180	—	0.35	0.58	—	321–356
181–200	—	—	0.49	—	357–392
201–225	—	—	0.35	—	393–437

Table 310.20 Ampacities of Not More Than Three Single Insulated Conductors, Rated 0 Through 2000 Volts, Supported on a Messenger, Based on Ambient Air Temperature of 40°C (104°F)

Size AWG or kcmil	Temperature Rating of Conductor (See Table 310.13.)				Size AWG or kcmil
	75°C (167°F)	90°C (194°F)	75°C (167°F)	90°C (194°F)	
	Types RHW, THHW, THW, THWN, XHHW, ZW	Types MI, THHN, THHW, THW-2, THWN-2, RHH, RHW-2, USE-2, XHHW, XHHW-2, ZW-2	Types RHW, THW, THWN, THHW, XHHW	Types THHN, THHW, RHH, XHHW, RHW-2, XHHW-2, THW-2, THWN-2, USE-2, ZW-2	
	COPPER		ALUMINUM OR COPPER-CLAD ALUMINUM		
8	57	66	44	51	8
6	76	89	59	69	6
4	101	117	78	91	4
3	118	138	92	107	3
2	135	158	106	123	2
1	158	185	123	144	1
1/0	183	214	143	167	1/0
2/0	212	247	165	193	2/0
3/0	245	287	192	224	3/0
4/0	287	335	224	262	4/0
250	320	374	251	292	250
300	359	419	282	328	300
350	397	464	312	364	350
400	430	503	339	395	400
500	496	580	392	458	500
600	553	647	440	514	600
700	610	714	488	570	700
750	638	747	512	598	750
800	660	773	532	622	800
900	704	826	572	669	900
1000	748	879	612	716	1000

CORRECTION FACTORS

Ambient Temp. (°C)	For ambient temperatures other than 40°C (104°F), multiply the allowable ampacities shown above by the appropriate factor shown below.				Ambient Temp. (°F)
21–25	1.20	1.14	1.20	1.14	70–77
26–30	1.13	1.10	1.13	1.10	79–86
31–35	1.07	1.05	1.07	1.05	88–95
36–40	1.00	1.00	1.00	1.00	97–104
41–45	0.93	0.95	0.93	0.95	106–113
46–50	0.85	0.89	0.85	0.89	115–122
51–55	0.76	0.84	0.76	0.84	124–131
56–60	0.65	0.77	0.65	0.77	133–140
61–70	0.38	0.63	0.38	0.63	142–158
71–80	—	0.45	—	0.45	160–176

Table 310.21 Ampacities of Bare or Covered Conductors in Free Air, Based on 40°C (104°F) Ambient, 80°C (176°F) Total Conductor Temperature, 610 mm/sec (2 ft/sec) Wind Velocity

| Copper Conductors | | | | AAC Aluminum Conductors | | | |
| Bare | | Covered | | Bare | | Covered | |
AWG or kcmil	Amperes	AWG or kcmil	Amperes	AWG or kcmil	Amperes	AWG or kcmil	Amperes
8	98	8	103	8	76	8	80
6	124	6	130	6	96	6	101
4	155	4	163	4	121	4	127
2	209	2	219	2	163	2	171
1/0	282	1/0	297	1/0	220	1/0	231
2/0	329	2/0	344	2/0	255	2/0	268
3/0	382	3/0	401	3/0	297	3/0	312
4/0	444	4/0	466	4/0	346	4/0	364
250	494	250	519	266.8	403	266.8	423
300	556	300	584	336.4	468	336.4	492
500	773	500	812	397.5	522	397.5	548
750	1000	750	1050	477.0	588	477.0	617
1000	1193	1000	1253	556.5	650	556.5	682
—	—	—	—	636.0	709	636.0	744
—	—	—	—	795.0	819	795.0	860
—	—	—	—	954.0	920	—	—
—	—	—	—	1033.5	968	1033.5	1017
—	—	—	—	1272	1103	1272	1201
—	—	—	—	1590	1267	1590	1381
—	—	—	—	2000	1454	2000	1527

Table 310.61 Conductor Application and Insulation

Trade Name	Type Letter	Maximum Operating Tempera- ture	Application Provision	Insulation	Outer Covering
Medium voltage solid dielectric	MV-90 MV-105*	90°C 105°C	Dry or wet locations rated 2001 volts and higher	Thermo- plastic or thermo- setting	Jacket, sheath, or armor

*Where design conditions require maximum conductor temperatures above 90°C.

(C) Tables Ampacities for conductors rated 2001 to 35,000 volts shall be as specified in the Ampacity Tables 310.67 through 310.86. Ampacities at ambient temperatures other than those shown in the tables shall be determined by the formula in 310.60(C)(4).

FPN No. 1: For ampacities calculated in accordance with 310.60(B), reference IEEE 835-1994 (IPCEA Pub. No. P-46-426), *Standard Power Cable Ampacity Tables*, and

Table 310.62 Thickness of Insulation for 601- to 2000-Volt Nonshielded Types RHH and RHW

| Conductor Size (AWG or kcmil) | Column A[1] | | Column B[2] | |
	mm	mils	mm	mils
14–10	2.03	80	1.52	60
8	2.03	80	1.78	70
6–2	2.41	95	1.78	70
1–2/0	2.79	110	2.29	90
3/0–4/0	2.79	110	2.29	90
213–500	3.18	125	2.67	105
501–1000	3.56	140	3.05	120
1001–2000	3.56	140	3.56	140

[1]Column A insulations are limited to natural, SBR, and butyl rubbers.
[2]Column B insulations are materials such as cross-linked polyethylene, ethylene propylene rubber, and composites thereof.

the references therein for availability of all factors and constants.

FPN No. 2: Ampacities provided by this section do not take voltage drop into consideration. See 210.19(A), FPN No. 4, for branch circuits and 215.2(A), FPN No. 2, for feeders.

(1) Grounded Shields Ampacities shown in Tables 310.69, 310.70, 310.81, and 310.82 are for cable with shields grounded at one point only. Where shields are grounded at more than one point, ampacities shall be adjusted to take into consideration the heating due to shield currents.

(2) Burial Depth of Underground Circuits Where the burial depth of direct burial or electrical duct bank circuits is modified from the values shown in a figure or table, ampacities shall be permitted to be modified as indicated in (C)(a) and (C)(b).

(a) Where burial depths are increased in part(s) of an electrical duct run, no decrease in ampacity of the conductors

is needed, provided the total length of parts of the duct run increased in depth is less than 25 percent of the total run length.

(b) Where burial depths are deeper than shown in a specific underground ampacity table or figure, an ampacity derating factor of 6 percent per 300-mm (1-ft) increase in depth for all values of rho shall be permitted.

No rating change is needed where the burial depth is decreased.

(3) Electrical Ducts in Figure 310.60 At locations where electrical ducts enter equipment enclosures from under

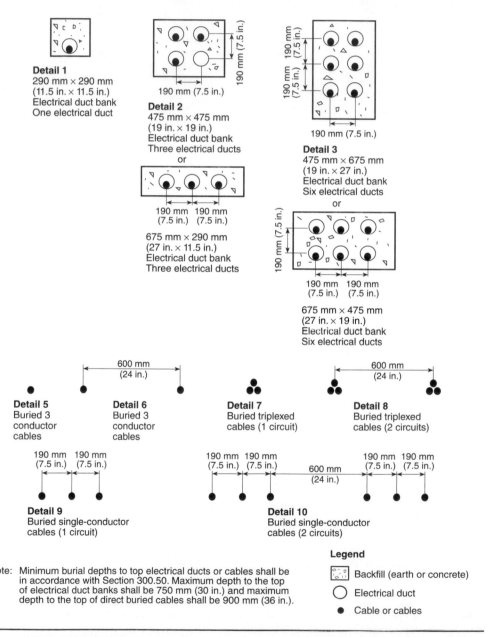

Figure 310.60 Cable Installation Dimensions for Use with Table 310.77 Through Table 310.86.

Table 310.63 Thickness of Insulated Conductors Rated 2400 Volts and Jacket for Nonshielded Solid Dielectric Insulation

Conductor Size (AWG or kcmil)	Dry Locations, Single Conductor						Wet or Dry Locations					
	Without Jacket Insulation		With Jacket				Single Conductor				Multiconductor Insulation*	
			Insulation		Jacket		Insulation		Jacket			
	mm	mils	mm	mils	mm	mils	mm	mils	mm	mils	mm	mils
8	2.79	110	2.29	90	0.76	30	3.18	125	2.03	80	2.29	90
6	2.79	110	2.29	90	0.76	30	3.18	125	2.03	80	2.29	90
4–2	2.79	110	2.29	90	1.14	45	3.18	125	2.03	80	2.29	90
1–2/0	2.79	110	2.29	90	1.14	45	3.18	125	2.03	80	2.29	90
3/0–4/0	2.79	110	2.29	90	1.65	65	3.18	125	2.41	95	2.29	90
213–500	3.05	120	2.29	90	1.65	65	3.56	140	2.79	110	2.29	90
501–750	3.30	130	2.29	90	1.65	65	3.94	155	3.18	125	2.29	90
751–1000	3.30	130	2.29	90	1.65	65	3.94	155	3.18	125	2.29	90

*Under a common overall covering such as a jacket, sheath, or armor.

Table 310.64 Thickness of Insulation for Shielded Solid Dielectric Insulated Conductors Rated 2001 to 35,000 Volts

Conductor Size (AWG or kcmil)	2001–5000 Volts		5001–8000 volts						8001–15,000 volts						15,001–25,000 volts					
			100 Percent Insulation Level[1]		133 Percent Insulation Level[2]		173 Percent Insulation Level[3]		100 Percent Insulation Level[1]		133 Percent Insulation Level[2]		173 Percent Insulation Level[3]		100 Percent Insulation Level[1]		133 Percent Insulation Level[2]		173 Percent Insulation Level[3]	
	mm	mils	mm	mils	mm	mils	mm	mils	mm	mils	mm	mils	mm	mils	mm	mils	mm	mils	mm	mils
8	2.29	90	—	—	—	—	—	—	—	—	—	—	—	—	—	—	—	—	—	—
6–4	2.29	90	2.92	115	3.56	140	4.45	175	—	—	—	—	—	—	—	—	—	—	—	—
2	2.29	90	2.92	115	3.56	140	4.45	175	4.45	175	5.59	220	6.60	260	—	—	—	—	—	—
1	2.29	90	2.92	115	3.56	140	4.45	175	4.45	175	5.59	220	6.60	260	6.60	260	8.13	320	10.67	420
1/0–2000	2.29	90	2.92	115	3.56	140	4.45	175	4.45	175	5.59	220	6.60	260	6.60	260	8.13	320	10.67	420

Conductor Size (AWG or kcmil)	25,001–28000 volts						28,001–35,000 volts					
	100 Percent Insulation Level[1]		133 Percent Insulation Level[2]		173 Percent Insulation Level[3]		100 Percent Insulation Level[1]		133 Percent Insulation Level[2]		173 Percent Insulation Level[3]	
	mm	mils	mm	mils	mm	mils	mm	mils	mm	mils	mm	mils
1	7.11	280	8.76	345	11.30	445	—	—	—	—	—	—
1/0–2000	7.11	280	8.76	345	11.30	445	8.76	345	10.67	420	14.73	580

Notes:

[1]**100 Percent Insulation Level.** Cables in this category shall be permitted to be applied where the system is provided with relay protection such that ground faults will be cleared as rapidly as possible but, in any case, within 1 minute. While these cables are applicable to the great majority of cable installations that are on grounded systems, they shall be permitted to be used also on other systems for which the application of cables is acceptable, provided the above clearing requirements are met in completely de-energizing the faulted section.

[2]**133 Percent Insulation Level.** This insulation level corresponds to that formerly designated for ungrounded systems. Cables in this category shall be permitted to be applied in situations where the clearing time requirements of the 100 percent level category cannot be met and yet there is adequate assurance that the faulted section will be de-energized in a time not exceeding 1 hour. Also, they shall be permitted to be used in 100 percent insulation level applications where additional insulation is desirable.

[3]**173 Percent Insulation Level.** Cables in this category shall be permitted to be applied under all of the following conditions:
(1) In industrial establishments where the conditions of maintenance and supervision ensure that only qualified persons service the installation
(2) Where the fault clearing time requirements of the 133 percent level category cannot be met
(3) Where an orderly shutdown is essential to protect equipment and personnel
(4) There is adequate assurance that the faulted section will be de-energized in an orderly shutdown
Also, cables with this insulation thickness shall be permitted to be used in 100 or 133 percent insulation level applications where additional insulation strength is desirable.

ground, spacing between such ducts, as shown in Figure 310.60, shall be permitted to be reduced without requiring the ampacity of conductors therein to be reduced.

(4) Ambients Not in Tables Ampacities at ambient temperatures other than those shown in the tables shall be determined by means of the following formula:

$$I_2 = I_1 \sqrt{\frac{TC - TA_2 - \Delta TD}{TC - TA_1 - \Delta TD}}$$

where:

I_1 = ampacity from tables at ambient TA_1

I_2 = ampacity at desired ambient TA_2

TC = conductor temperature in degrees Celsius (°C)

TA_1 = surrounding ambient from tables in degrees Celsius (°C)

TA_2 = desired ambient in degrees Celsius (°C)

ΔTD = dielectric loss temperature rise

(D) Engineering Supervision Under engineering supervision, conductor ampacities shall be permitted to be calculated by means of the following general formula:

$$I = \sqrt{\frac{TC - (TA + \Delta TD)}{RDC(1 + YC)RCA}}$$

where:

TC = conductor temperature in °C

TA = ambient temperature in °C

ΔTD = dielectric loss temperature rise

RDC = dc resistance of conductor at temperature TC

YC = component ac resistance resulting from skin effect and proximity effect

RCA = effective thermal resistance between conductor and surrounding ambient

FPN: See Annex B for examples of formula applications.

Table 310.67 Ampacities of Insulated Single Copper Conductor Cables Triplexed in Air Based on Conductor Temperatures of 90°C (194°F) and 105°C (221°F) and Ambient Air Temperature of 40°C (104°F)

Conductor Size (AWG or kcmil)	Temperature Rating of Conductor (See Table 310.61.)			
	2001–5000 Volts Ampacity		5001–35,000 Volts Ampacity	
	90°C (194°F) Type MV-90	105°C (221°F) Type MV-105	90°C (194°F) Type MV-90	105°C (221°F) Type MV-105
8	65	74	—	—
6	90	99	100	110
4	120	130	130	140
2	160	175	170	195
1	185	205	195	225
1/0	215	240	225	255
2/0	250	275	260	295
3/0	290	320	300	340
4/0	335	375	345	390
250	375	415	380	430
350	465	515	470	525
500	580	645	580	650
750	750	835	730	820
1000	880	980	850	950

Table 310.68 Ampacities of Insulated Single Aluminum Conductor Cables Triplexed in Air Based on Conductor Temperatures of 90°C (194°F) and 105°C (221°F) and Ambient Air Temperature of 40°C (104°F)

Conductor Size (AWG or kcmil)	Temperature Rating of Conductor (See Table 310.61.)			
	2001–5000 Volts Ampacity		5001–35,000 Volts Ampacity	
	90°C (194°F) Type MV-90	105°C (221°F) Type MV-105	90°C (194°F) Type MV-90	105°C (221°F) Type MV-105
8	50	57	—	—
6	70	77	75	84
4	90	100	100	110
2	125	135	130	150
1	145	160	150	175
1/0	170	185	175	200
2/0	195	215	200	230
3/0	225	250	230	265
4/0	265	290	270	305
250	295	325	300	335
350	365	405	370	415
500	460	510	460	515
750	600	665	590	660
1000	715	800	700	780

Table 310.69 Ampacities of Insulated Single Copper Conductor Isolated in Air Based on Conductor Temperatures of 90°C (194°F) and 105°C (221°F) and Ambient Air Temperature of 40°C (104°F)

	Temperature Rating of Conductor (See Table 310.61.)					
	2001–5000 Volts Ampacity		5001–15,000 Volts Ampacity		15,001–35,000 Volts Ampacity	
Conductor Size (AWG or kcmil)	90°C (194°F) Type MV-90	105°C (221°F) Type MV-105	90°C (194°F) Type MV-90	105°C (221°F) Type MV-105	90°C (194°F) Type MV-90	105°C (221°F) Type MV-105
8	83	93	—	—	—	—
6	110	120	110	125	—	—
4	145	160	150	165	—	—
2	190	215	195	215	—	—
1	225	250	225	250	225	250
1/0	260	290	260	290	260	290
2/0	300	330	300	335	300	330
3/0	345	385	345	385	345	380
4/0	400	445	400	445	395	445
250	445	495	445	495	440	490
350	550	615	550	610	545	605
500	695	775	685	765	680	755
750	900	1000	885	990	870	970
1000	1075	1200	1060	1185	1040	1160
1250	1230	1370	1210	1350	1185	1320
1500	1365	1525	1345	1500	1315	1465
1750	1495	1665	1470	1640	1430	1595
2000	1605	1790	1575	1755	1535	1710

Table 310.71 Ampacities of an Insulated Three-Conductor Copper Cable Isolated in Air Based on Conductor Temperatures of 90°C (194°F) and 105°C (221°F) and Ambient Air Temperature of 40°C (104°F)

	Temperature Rating of Conductor (See Table 310.61.)			
	2001–5000 Volts Ampacity		5001–35,000 Volts Ampacity	
Conductor Size (AWG or kcmil)	90°C (194°F) Type MV-90	105°C (221°F) Type MV-105	90°C (194°F) Type MV-90	105°C (221°F) Type MV-105
8	59	66	—	—
6	79	88	93	105
4	105	115	120	135
2	140	154	165	185
1	160	180	185	210
1/0	185	205	215	240
2/0	215	240	245	275
3/0	250	280	285	315
4/0	285	320	325	360
250	320	355	360	400
350	395	440	435	490
500	485	545	535	600
750	615	685	670	745
1000	705	790	770	860

Table 310.70 Ampacities of Insulated Single Aluminum Conductor Isolated in Air Based on Conductor Temperatures of 90°C (194°F) and 105°C (221°F) and Ambient Air Temperature of 40°C (104°F)

	Temperature Rating of Conductor (See Table 310.61.)					
	2001–5000 Volts Ampacity		5001–15,000 Volts Ampacity		15,001–35,000 Volts Ampacity	
Conductor Size (AWG or kcmil)	90°C (194°F) Type MV-90	105°C (221°F) Type MV-105	90°C (194°F) Type MV-90	105°C (221°F) Type MV-105	90°C (194°F) Type MV-90	105°C (221°F) Type MV-105
8	64	71	—	—	—	—
6	85	95	87	97	—	—
4	115	125	115	130	—	—
2	150	165	150	170	—	—
1	175	195	175	195	175	195
1/0	200	225	200	225	200	225
2/0	230	260	235	260	230	260
3/0	270	300	270	300	270	300
4/0	310	350	310	350	310	345
250	345	385	345	385	345	380
350	430	480	430	480	430	475
500	545	605	535	600	530	590
750	710	790	700	780	685	765
1000	855	950	840	940	825	920
1250	980	1095	970	1080	950	1055
1500	1105	1230	1085	1215	1060	1180
1750	1215	1355	1195	1335	1165	1300
2000	1320	1475	1295	1445	1265	1410

Table 310.72 Ampacities of an Insulated Three-Conductor Aluminum Cable Isolated in Air Based on Conductor Temperatures of 90°C (194°F) and 105°C (221°F) and Ambient Air Temperature of 40°C (104°F)

	Temperature Rating of Conductor (See Table 310.61.)			
	2001–5000 Volts Ampacity		5001–35,000 Volts Ampacity	
Conductor Size (AWG or kcmil)	90°C (194°F) Type MV-90	105°C (221°F) Type MV-105	90°C (194°F) Type MV-90	105°C (221°F) Type MV-105
8	46	51	—	—
6	61	68	72	80
4	81	90	95	105
2	110	120	125	145
1	125	140	145	165
1/0	145	160	170	185
2/0	170	185	190	215
3/0	195	215	220	245
4/0	225	250	255	285
250	250	280	280	315
350	310	345	345	385
500	385	430	425	475
750	495	550	540	600
1000	585	650	635	705

Table 310.73 Ampacities of an Insulated Triplexed or Three Single-Conductor Copper Cables in Isolated Conduit in Air Based on Conductor Temperatures of 90°C (194°F) and 105°C (221°F) and Ambient Air Temperature of 40°C (104°F)

| Conductor Size (AWG or kcmil) | Temperature Rating of Conductor (See Table 310.61.) | | | |
| | 2001–5000 Volts Ampacity | | 5001–35,000 Volts Ampacity | |
	90°C (194°F) Type MV-90	105°C (221°F) Type MV-105	90°C (194°F) Type MV-90	105°C (221°F) Type MV-105
8	55	61	—	—
6	75	84	83	93
4	97	110	110	120
2	130	145	150	165
1	155	175	170	190
1/0	180	200	195	215
2/0	205	225	225	255
3/0	240	270	260	290
4/0	280	305	295	330
250	315	355	330	365
350	385	430	395	440
500	475	530	480	535
750	600	665	585	655
1000	690	770	675	755

Table 310.75 Ampacities of an Insulated Three-Conductor Copper Cable in Isolated Conduit in Air Based on Conductor Temperatures of 90°C (194°F) and 105°C (221°F) and Ambient Air Temperature of 40°C (104°F)

| Conductor Size (AWG or kcmil) | Temperature Rating of Conductor (See Table 310.61.) | | | |
| | 2001–5000 Volts Ampacity | | 5001–35,000 Volts Ampacity | |
	90°C (194°F) Type MV-90	105°C (221°F) Type MV-105	90°C (194°F) Type MV-90	105°C (221°F) Type MV-105
8	52	58	—	—
6	69	77	83	92
4	91	100	105	120
2	125	135	145	165
1	140	155	165	185
1/0	165	185	195	215
2/0	190	210	220	245
3/0	220	245	250	280
4/0	255	285	290	320
250	280	315	315	350
350	350	390	385	430
500	425	475	470	525
750	525	585	570	635
1000	590	660	650	725

Table 310.74 Ampacities of an Insulated Triplexed or Three Single-Conductor Aluminum Cables in Isolated Conduit in Air Based on Conductor Temperatures of 90°C (194°F) and 105°C (221°F) and Ambient Air Temperature of 40°C (104°F)

| Conductor Size (AWG or kcmil) | Temperature Rating of Conductor (See Table 310.61.) | | | |
| | 2001–5000 Volts Ampacity | | 5001–35,000 Volts Ampacity | |
	90°C (194°F) Type MV-90	105°C (221°F) Type MV-105	90°C (194°F) Type MV-90	105°C (221°F) Type MV-105
8	43	48	—	—
6	58	65	65	72
4	76	85	84	94
2	100	115	115	130
1	120	135	130	150
1/0	140	155	150	170
2/0	160	175	175	200
3/0	190	210	200	225
4/0	215	240	230	260
250	250	280	255	290
350	305	340	310	350
500	380	425	385	430
750	490	545	485	540
1000	580	645	565	640

Table 310.76 Ampacities of an Insulated Three-Conductor Aluminum Cable in Isolated Conduit in Air Based on Conductor Temperatures of 90°C (194°F) and 105°C (221°F) and Ambient Air Temperature of 40°C (104°F)

| Conductor Size (AWG or kcmil) | Temperature Rating of Conductor (See Table 310.61.) | | | |
| | 2001–5000 Volts Ampacity | | 5001–35,000 Volts Ampacity | |
	90°C (194°F) Type MV-90	105°C (221°F) Type MV-105	90°C (194°F) Type MV-90	105°C (221°F) Type MV-105
8	41	46	—	—
6	53	59	64	71
4	71	79	84	94
2	96	105	115	125
1	110	125	130	145
1/0	130	145	150	170
2/0	150	165	170	190
3/0	170	190	195	220
4/0	200	225	225	255
250	220	245	250	280
350	275	305	305	340
500	340	380	380	425
750	430	480	470	520
1000	505	560	550	615

Table 310.77 Ampacities of Three Single-Insulated Copper Conductors in Underground Electrical Ducts (Three Conductors per Electrical Duct) Based on Ambient Earth Temperature of 20°C (68°F), Electrical Duct Arrangement per Figure 310.60, 100 Percent Load Factor, Thermal Resistance (RHO) of 90, Conductor Temperatures of 90°C (194°F) and 105°C (221°F)

	Temperature Rating of Conductor (See Table 310.61.)			
	2001–5000 Volts Ampacity		5001–35,000 Volts Ampacity	
Conductor Size (AWG or kcmil)	90°C (194°F) Type MV-90	105°C (221°F) Type MV-105	90°C (194°F) Type MV-90	105°C (221°F) Type MV-105
One Circuit (See Figure 310.60, Detail 1.)				
8	64	69	—	—
6	85	92	90	97
4	110	120	115	125
2	145	155	155	165
1	170	180	175	185
1/0	195	210	200	215
2/0	220	235	230	245
3/0	250	270	260	275
4/0	290	310	295	315
250	320	345	325	345
350	385	415	390	415
500	470	505	465	500
750	585	630	565	610
1000	670	720	640	690
Three Circuits (See Figure 310.60, Detail 2.)				
8	56	60	—	—
6	73	79	77	83
4	95	100	99	105
2	125	130	130	135
1	140	150	145	155
1/0	160	175	165	175
2/0	185	195	185	200
3/0	210	225	210	225
4/0	235	255	240	255
250	260	280	260	280
350	315	335	310	330
500	375	405	370	395
750	460	495	440	475
1000	525	565	495	535
Six Circuits (See Figure 310.60, Detail 3.)				
8	48	52	—	—
6	62	67	64	68
4	80	86	82	88
2	105	110	105	115
1	115	125	120	125
1/0	135	145	135	145
2/0	150	160	150	165
3/0	170	185	170	185
4/0	195	210	190	205
250	210	225	210	225
350	250	270	245	265
500	300	325	290	310
750	365	395	350	375
1000	410	445	390	415

Table 310.78 Ampacities of Three Single-Insulated Aluminum Conductors in Underground Electrical Ducts (Three Conductors per Electrical Duct) Based on Ambient Earth Temperature of 20°C (68°F), Electrical Duct Arrangement per Figure 310.60, 100 Percent Load Factor, Thermal Resistance (RHO) of 90, Conductor Temperatures of 90°C (194°F) and 105°C (221°F)

	Temperature Rating of Conductor (See Table 310.61.)			
	2001–5000 Volts Ampacity		5001–35,000 Volts Ampacity	
Conductor Size (AWG or kcmil)	90°C (194°F) Type MV-90	105°C (221°F) Type MV-105	90°C (194°F) Type MV-90	105°C (221°F) Type MV-105
One Circuit (See Figure 310.60, Detail 1.)				
8	50	54	—	—
6	66	71	70	75
4	86	93	91	98
2	115	125	120	130
1	130	140	135	145
1/0	150	160	155	165
2/0	170	185	175	190
3/0	195	210	200	215
4/0	225	245	230	245
250	250	270	250	270
350	305	325	305	330
500	370	400	370	400
750	470	505	455	490
1000	545	590	525	565
Three Circuits (See Figure 310.60, Detail 2.)				
8	44	47	—	—
6	57	61	60	65
4	74	80	77	83
2	96	105	100	105
1	110	120	110	120
1/0	125	135	125	140
2/0	145	155	145	155
3/0	160	175	165	175
4/0	185	200	185	200
250	205	220	200	220
350	245	265	245	260
500	295	320	290	315
750	370	395	355	385
1000	425	460	405	440
Six Circuits (See Figure 310.60, Detail 3.)				
8	38	41	—	—
6	48	52	50	54
4	62	67	64	69
2	80	86	80	88
1	91	98	90	99
1/0	105	110	105	110
2/0	115	125	115	125
3/0	135	145	130	145
4/0	150	165	150	160
250	165	180	165	175
350	195	210	195	210
500	240	255	230	250
750	290	315	280	305
1000	335	360	320	345

Table 310.79 Ampacities of Three Insulated Copper Conductors Cabled Within an Overall Covering (Three-Conductor Cable) in Underground Electrical Ducts (One Cable per Electrical Duct) Based on Ambient Earth Temperature of 20°C (68°F), Electrical Duct Arrangement per Figure 310.60, 100 Percent Load Factor, Thermal Resistance (RHO) of 90, Conductor Temperatures of 90°C (194°F) and 105°C (221°C)

| Conductor Size (AWG or kcmil) | Temperature Rating of Conductor (See Table 310.61.) | | | |
| | 2001–5000 Volts Ampacity | | 5001–35,000 Volts Ampacity | |
	90°C (194°F) Type MV-90	105°C (221°F) Type MV-105	90°C (194°F) Type MV-90	105°C (221°F) Type MV-105
One Circuit (See Figure 310.60, Detail 1.)				
8	59	64	—	—
6	78	84	88	95
4	100	110	115	125
2	135	145	150	160
1	155	165	170	185
1/0	175	190	195	210
2/0	200	220	220	235
3/0	230	250	250	270
4/0	265	285	285	305
250	290	315	310	335
350	355	380	375	400
500	430	460	450	485
750	530	570	545	585
1000	600	645	615	660
Three Circuits (See Figure 310.60, Detail 2.)				
8	53	57	—	—
6	69	74	75	81
4	89	96	97	105
2	115	125	125	135
1	135	145	140	155
1/0	150	165	160	175
2/0	170	185	185	195
3/0	195	210	205	220
4/0	225	240	230	250
250	245	265	255	270
350	295	315	305	325
500	355	380	360	385
750	430	465	430	465
1000	485	520	485	515
Six Circuits (See Figure 310.60, Detail 3.)				
8	46	50	—	—
6	60	65	63	68
4	77	83	81	87
2	98	105	105	110
1	110	120	115	125
1/0	125	135	130	145
2/0	145	155	150	160
3/0	165	175	170	180
4/0	185	200	190	200
250	200	220	205	220
350	240	270	245	275
500	290	310	290	305
750	350	375	340	365
1000	390	420	380	405

Table 310.80 Ampacities of Three Insulated Aluminum Conductors Cabled Within an Overall Covering (Three-Conductor Cable) in Underground Electrical Ducts (One Cable per Electrical Duct) Based on Ambient Earth Temperature of 20°C (68°F), Electrical Duct Arrangement per Figure 310.60, 100 Percent Load Factor, Thermal Resistance (RHO) of 90, Conductor Temperatures of 90°C (194°F) and 105°C (221°C)

| Conductor Size (AWG or kcmil) | Temperature Rating of Conductor (See Table 310.61.) | | | |
| | 2001–5000 Volts Ampacity | | 5001–35,000 Volts Ampacity | |
	90°C (194°F) Type MV-90	105°C (221°F) Type MV-105	90°C (194°F) Type MV-90	105°C (221°F) Type MV-105
One Circuit (See Figure 310.60, Detail 1.)				
8	46	50	—	—
6	61	66	69	74
4	80	86	89	96
2	105	110	115	125
1	120	130	135	145
1/0	140	150	150	165
2/0	160	170	170	185
3/0	180	195	195	210
4/0	205	220	220	240
250	230	245	245	265
350	280	310	295	315
500	340	365	355	385
750	425	460	440	475
1000	495	535	510	545
Three Circuits (See Figure 310.60, Detail 2.)				
8	41	44	—	—
6	54	58	59	64
4	70	75	75	81
2	90	97	100	105
1	105	110	110	120
1/0	120	125	125	135
2/0	135	145	140	155
3/0	155	165	160	175
4/0	175	185	180	195
250	190	205	200	215
350	230	250	240	255
500	280	300	285	305
750	345	375	350	375
1000	400	430	400	430
Six Circuits (See Figure 310.60, Detail 3.)				
8	36	39	—	—
6	46	50	49	53
4	60	65	63	68
2	77	83	80	86
1	87	94	90	98
1/0	99	105	105	110
2/0	110	120	115	125
3/0	130	140	130	140
4/0	145	155	150	160
250	160	170	160	170
350	190	205	190	205
500	230	245	230	245
750	280	305	275	295
1000	320	345	315	335

Table 310.81 Ampacities of Single Insulated Copper Conductors Directly Buried in Earth Based on Ambient Earth Temperature of 20°C (68°F), Arrangement per Figure 310.60, 100 Percent Load Factor, Thermal Resistance (RHO) of 90, Conductor Temperatures of 90°C (194°F) and 105°C (221°C)

| Conductor Size (AWG or kcmil) | Temperature Rating of Conductor (See Table 310.61.) | | | |
| | 2001–5000 Volts Ampacity | | 5001–35,000 Volts Ampacity | |
	90°C (194°F) Type MV-90	105°C (221°F) Type MV-105	90°C (194°F) Type MV-90	105°C (221°F) Type MV-105
One Circuit, Three Conductors (See Figure 310.60, Detail 9.)				
8	110	115	—	—
6	140	150	130	140
4	180	195	170	180
2	230	250	210	225
1	260	280	240	260
1/0	295	320	275	295
2/0	335	365	310	335
3/0	385	415	355	380
4/0	435	465	405	435
250	470	510	440	475
350	570	615	535	575
500	690	745	650	700
750	845	910	805	865
1000	980	1055	930	1005
Two Circuits, Six Conductors (See Figure 310.60, Detail 10.)				
8	100	110	—	—
6	130	140	120	130
4	165	180	160	170
2	215	230	195	210
1	240	260	225	240
1/0	275	295	255	275
2/0	310	335	290	315
3/0	355	380	330	355
4/0	400	430	375	405
250	435	470	410	440
350	520	560	495	530
500	630	680	600	645
750	775	835	740	795
1000	890	960	855	920

Table 310.82 Ampacities of Single Insulated Aluminum Conductors Directly Buried in Earth Based on Ambient Earth Temperature of 20°C (68°F), Arrangement per Figure 310.60, 100 Percent Load Factor, Thermal Resistance (RHO) of 90, Conductor Temperatures of 90°C (194°F) and 105°C (221°F)

| Conductor Size (AWG or kcmil) | Temperature Rating of Conductor (See Table 310.61) | | | |
| | 2001–5000 Volts Ampacity | | 5001–35,000 Volts Ampacity | |
	90°C (194°F) Type MV-90	105°C (221°F) Type MV-105	90°C (194°F) Type MV-90	105°C (221°F) Type MV-105
One Circuit, Three Conductors (See Figure 310.60, Detail 9.)				
8	85	90	—	—
6	110	115	100	110
4	140	150	130	140
2	180	195	165	175
1	205	220	185	200
1/0	230	250	215	230
2/0	265	285	245	260
3/0	300	320	275	295
4/0	340	365	315	340
250	370	395	345	370
350	445	480	415	450
500	540	580	510	545
750	665	720	635	680
1000	780	840	740	795
Two Circuits, Six Conductors (See Figure 310.60, Detail 10.)				
8	80	85	—	—
6	100	110	95	100
4	130	140	125	130
2	165	180	155	165
1	190	200	175	190
1/0	215	230	200	215
2/0	245	260	225	245
3/0	275	295	255	275
4/0	310	335	290	315
250	340	365	320	345
350	410	440	385	415
500	495	530	470	505
750	610	655	580	625
1000	710	765	680	730

Table 310.83 Ampacities of Three Insulated Copper Conductors Cabled Within an Overall Covering (Three-Conductor Cable), Directly Buried in Earth Based on Ambient Earth Temperature of 20°C (68°F), Arrangement per Figure 310.60, 100 Percent Load Factor, Thermal Resistance (RHO) of 90, Conductor Temperatures of 90°C (194°F) and 105°C (221°F)

Conductor Size (AWG or kcmil)	Temperature Rating of Conductor (See Table 310.61.)			
	2001–5000 Volts Ampacity		5001–35,000 Volts Ampacity	
	90°C (194°F) Type MV-90	105°C (221°F) Type MV-105	90°C (194°F) Type MV-90	105°C (221°F) Type MV-105
One Circuit (See Figure 310.60, Detail 5.)				
8	85	89	—	—
6	105	115	115	120
4	135	150	145	155
2	180	190	185	200
1	200	215	210	225
1/0	230	245	240	255
2/0	260	280	270	290
3/0	295	320	305	330
4/0	335	360	350	375
250	365	395	380	410
350	440	475	460	495
500	530	570	550	590
750	650	700	665	720
1000	730	785	750	810
Two Circuits (See Figure 310.60, Detail 6.)				
8	80	84	—	—
6	100	105	105	115
4	130	140	135	145
2	165	180	170	185
1	185	200	195	210
1/0	215	230	220	235
2/0	240	260	250	270
3/0	275	295	280	305
4/0	310	335	320	345
250	340	365	350	375
350	410	440	420	450
500	490	525	500	535
750	595	640	605	650
1000	665	715	675	730

Table 310.84 Ampacities of Three Insulated Aluminum Conductors Cabled Within an Overall Covering (Three-Conductor Cable), Directly Buried in Earth Based on Ambient Earth Temperature of 20°C (68°F), Arrangement per Figure 310.60, 100 Percent Load Factor, Thermal Resistance (RHO) of 90, Conductor Temperatures of 90°C (194°F) and 105°C (221°F)

Conductor Size (AWG or kcmil)	Temperature Rating of Conductor (See Table 310.61.)			
	2001–5000 Volts Ampacity		5001–35,000 Volts Ampacity	
	90°C (194°F) Type MV-90	105°C (221°F) Type MV-105	90°C (194°F) Type MV-90	105°C (221°F) Type MV-105
One Circuit (See Figure 310.60, Detail 5.)				
8	65	70	—	—
6	80	88	90	95
4	105	115	115	125
2	140	150	145	155
1	155	170	165	175
1/0	180	190	185	200
2/0	205	220	210	225
3/0	230	250	240	260
4/0	260	280	270	295
250	285	310	300	320
350	345	375	360	390
500	420	450	435	470
750	520	560	540	580
1000	600	650	620	665
Two Circuits (See Figure 310.60, Detail 6.)				
8	60	66	—	—
6	75	83	80	95
4	100	110	105	115
2	130	140	135	145
1	145	155	150	165
1/0	165	180	170	185
2/0	190	205	195	210
3/0	215	230	220	240
4/0	245	260	250	270
250	265	285	275	295
350	320	345	330	355
500	385	415	395	425
750	480	515	485	525
1000	550	590	560	600

Table 310.85 Ampacities of Three Triplexed Single Insulated Copper Conductors Directly Buried in Earth Based on Ambient Earth Temperature of 20°C (68°F), Arrangement per Figure 310.60, 100 Percent Load Factor, Thermal Resistance (RHO) of 90, Conductor Temperatures 90°C (194°F) and 105°C (221°F)

| Conductor Size (AWG or kcmil) | Temperature Rating of Conductor (See Table 310.61.) | | | |
| | 2001–5000 Volts Ampacity | | 5001–35,000 Volts Ampacity | |
	90°C (194°F) Type MV-90	105°C (221°F) Type MV-105	90°C (194°F) Type MV-90	105°C (221°F) Type MV-105
One Circuit, Three Conductors (See Figure 310.60, Detail 7.)				
8	90	95	—	—
6	120	130	115	120
4	150	165	150	160
2	195	205	190	205
1	225	240	215	230
1/0	255	270	245	260
2/0	290	310	275	295
3/0	330	360	315	340
4/0	375	405	360	385
250	410	445	390	410
350	490	580	470	505
500	590	635	565	605
750	725	780	685	740
1000	825	885	770	830
Two Circuits, Six Conductors (See Figure 310.60, Detail 8.)				
8	85	90	—	—
6	110	115	105	115
4	140	150	140	150
2	180	195	175	190
1	205	220	200	215
1/0	235	250	225	240
2/0	265	285	255	275
3/0	300	320	290	315
4/0	340	365	325	350
250	370	395	355	380
350	445	480	425	455
500	535	575	510	545
750	650	700	615	660
1000	740	795	690	745

Table 310.86 Ampacities of Three Triplexed Single Insulated Aluminum Conductors Directly Buried in Earth Based on Ambient Earth Temperature of 20°C (68°F), Arrangement per Figure 310.60, 100 Percent Load Factor, Thermal Resistance (RHO) of 90, Conductor Temperatures 90°C (194°F) and 105°C (221°F)

| Conductor Size (AWG or kcmil) | Temperature Rating of Conductor (See Table 310.61.) | | | |
| | 2001–5000 Volts Ampacity | | 5001–35,000 Volts Ampacity | |
	90°C (194°F) Type MV-90	105°C (221°F) Type MV-105	90°C (194°F) Type MV-90	105°C (221°F) Type MV-105
One Circuit, Three Conductors (See Figure 310.60, Detail 7.)				
8	70	75	—	—
6	90	100	90	95
4	120	130	115	125
2	155	165	145	155
1	175	190	165	175
1/0	200	210	190	205
2/0	225	240	215	230
3/0	255	275	245	265
4/0	290	310	280	305
250	320	350	305	325
350	385	420	370	400
500	465	500	445	480
750	580	625	550	590
1000	670	725	635	680
Two Circuits, Six Conductors (See Figure 310.60, Detail 8.)				
8	65	70	—	—
6	85	95	85	90
4	110	120	105	115
2	140	150	135	145
1	160	170	155	170
1/0	180	195	175	190
2/0	205	220	200	215
3/0	235	250	225	245
4/0	265	285	255	275
250	290	310	280	300
350	350	375	335	360
500	420	455	405	435
750	520	560	485	525
1000	600	645	565	605

ARTICLE 312
Cabinets, Cutout Boxes, and Meter Socket Enclosures

Summary of Changes

- **312.2(A):** For enclosures installed in wet locations, revised to require any cable or raceway that enters on the top or on the side above the level of uninsulated live parts in the enclosure be provided with fittings listed for wet locations.

- **312.4:** Added requirement that repairs be made to broken or incomplete openings in plaster, drywall, or plasterboard surfaces of flush-mounted cabinets, including panelboard cabinets

- **312.5(B):** Added messenger supported wiring as a wiring method covered by this requirement.

Contents

312.1 Scope

This article covers the installation and construction specifications of cabinets, cutout boxes, and meter socket enclosures.

See the definitions of *cabinet* and *cutout box* in Article 100. Cabinets and cutout boxes are designed with a swinging door(s) to enclose potential transformers, current transformers, switches, overcurrent devices, meters, or control equipment. Some cabinets for circuit breaker panelboards or load centers may not have doors, as permitted by 240.30(A)(2). Cabinets and cutout boxes are required to be of sufficient size to accommodate all devices and conductors without overcrowding or jamming. This condition can be prevented by the use of auxiliary gutters (Article 366).

The serving electric utility often has equipment specifications or service requirements beyond the *Code* for meter sockets, metering cabinets, and metering compartments within switchgear, switchboards, and panelboards. Consulting with the local electric utility on these requirements will help identify suitable equipment for an installation. One organization, the Electric Utility Service Equipment Requirements Committee (EUSERC), promotes uniform electric utility metering service requirements for these enclosures that meet the requirements of the *Code* and the serving utility. Many electric equipment manufacturers identify their equipment as meeting the EUSERC metering space requirements.

I. Installation

312.2 Damp, Wet, or Hazardous (Classified) Locations

(A) Damp and Wet Locations In damp or wet locations, surface-type enclosures within the scope of this article shall be placed or equipped so as to prevent moisture or water from entering and accumulating within the cabinet or cutout box, and shall be mounted so there is at least 6-mm (¼-in.) airspace between the enclosure and the wall or other supporting surface. Enclosures installed in wet locations shall be weatherproof. For enclosures in wet locations, raceways or cables entering above the level of uninsulated live parts shall use fittings listed for wet locations.

The last sentence was added for the 2005 *Code* to point out the need for watertight connections in areas where the wiring could lead water directly into uninsulated live parts.

Exception: Nonmetallic enclosures shall be permitted to be installed without the airspace on a concrete, masonry, tile, or similar surface.

FPN: For protection against corrosion, see 300.6.

(B) Hazardous (Classified) Locations Installations in hazardous (classified) locations shall conform to Articles 500 through 517.

312.3 Position in Wall

In walls of concrete, tile, or other noncombustible material, cabinets shall be installed so that the front edge of the cabinet is not set back of the finished surface more than 6 mm (¼ in.). In walls constructed of wood or other combustible material, cabinets shall be flush with the finished surface or project therefrom.

312.4 Repairing Plaster and Drywall or Plasterboard

Plaster, drywall, or plasterboard surfaces that are broken or incomplete shall be repaired so there will be no gaps or open spaces greater than 3 mm (⅛ in.) at the edge of the cabinet or cutout box employing a flush-type cover.

Section 312.4 was added for the 2005 *Code* out of the need for requirements for cabinets and cutout boxes similar to the requirements found in 314.21.

312.5 Cabinets, Cutout Boxes, and Meter Socket Enclosures

Conductors entering enclosures within the scope of this article shall be protected from abrasion and shall comply with 312.5(A) through (C).

(A) Openings to Be Closed Openings through which conductors enter shall be adequately closed.

(B) Metal Cabinets, Cutout Boxes, and Meter Socket Enclosures Where metal enclosures within the scope of this article are installed with messenger supported wiring, open wiring on insulators, or concealed knob-and-tube wiring, conductors shall enter through insulating bushings or, in dry locations, through flexible tubing extending from the last insulating support and firmly secured to the enclosure.

Editorial corrections were made for the 2005 *Code* to specifically address the terms *messenger-supported wiring* and *open wiring on insulators* replacing the general term *open wiring*.

(C) Cables Where cable is used, each cable shall be secured to the cabinet, cutout box, or meter socket enclosure.

The main rule of 312.5(C) prohibits the installation of several cables bunched together and run through a knockout or chase nipple. Individual cable clamps or connectors are required

to be used with only one cable per clamp or connector, unless the clamp or connector is identified for more than a single cable.

Exception: Cables with entirely nonmetallic sheaths shall be permitted to enter the top of a surface-mounted enclosure through one or more nonflexible raceways not less than 450 mm (18 in.) and not more than 3.0 m (10 ft) in length, provided all of the following conditions are met:

(a) Each cable is fastened within 300 mm (12 in.), measured along the sheath, of the outer end of the raceway.

(b) The raceway extends directly above the enclosure and does not penetrate a structural ceiling.

(c) A fitting is provided on each end of the raceway to protect the cable(s) from abrasion and the fittings remain accessible after installation.

(d) The raceway is sealed or plugged at the outer end using approved means so as to prevent access to the enclosure through the raceway.

(e) The cable sheath is continuous through the raceway and extends into the enclosure beyond the fitting not less than 6 mm (¼ in.).

(f) The raceway is fastened at its outer end and at other points in accordance with the applicable article.

(g) Where installed as conduit or tubing, the allowable cable fill does not exceed that permitted for complete conduit or tubing systems by Table 1 of Chapter 9 of this Code and all applicable notes thereto.

This exception allows multiple nonmetallic cables such as Type NM, NMC, NMS, UF, SE, and USE to enter the top of a surface-mounted enclosure through a nonflexible raceway sleeve or nipple. These sleeves or nipples are permitted to be between 18 in. and 10 ft in length. However, if the nipple length exceeds 24 in., the ampacity adjustment factors of 310.15(B)(2) apply.

FPN: See Table 1 in Chapter 9, including Note 9, for allowable cable fill in circular raceways. See 310.15(B)(2)(a) for required ampacity reductions for multiple cables installed in a common raceway.

312.6 Deflection of Conductors

Conductors at terminals or conductors entering or leaving cabinets or cutout boxes and the like shall comply with 312.6(A) through (C).

Exception: Wire-bending space in enclosures for motor controllers with provisions for one or two wires per terminal shall comply with 430.10(B).

(A) Width of Wiring Gutters Conductors shall not be deflected within a cabinet or cutout box unless a gutter having a width in accordance with Table 312.6(A) is provided. Conductors in parallel in accordance with 310.4 shall be judged on the basis of the number of conductors in parallel.

(B) Wire-Bending Space at Terminals Wire-bending space at each terminal shall be provided in accordance with 312.6(B)(1) or (B)(2).

(1) Conductors Not Entering or Leaving Opposite Wall Table 312.6(A) shall apply where the conductor does not enter or leave the enclosure through the wall opposite its terminal.

(2) Conductors Entering or Leaving Opposite Wall Table 312.6(B) shall apply where the conductor does enter or leave the enclosure through the wall opposite its terminal.

Exception No. 1: Where the distance between the wall and its terminal is in accordance with Table 312.6(A), a conductor shall be permitted to enter or leave an enclosure through the wall opposite its terminal, provided the conductor enters or leaves the enclosure where the gutter joins an adjacent gutter that has a width that conforms to Table 312.6(B) for the conductor.

Exception No. 2: A conductor not larger than 350 kcmil shall be permitted to enter or leave an enclosure containing only a meter socket(s) through the wall opposite its terminal, provided the distance between the terminal and the opposite wall is not less than that specified in Table 312.6(A) and the terminal is a lay-in type, where the terminal is either of the following:

(a) Directed toward the opening in the enclosure and within a 45 degree angle of directly facing the enclosure wall

(b) Directly facing the enclosure wall and offset not greater than 50 percent of the bending space specified in Table 312.6(A)

FPN: *Offset* is the distance measured along the enclosure wall from the axis of the centerline of the terminal to a line passing through the center of the opening in the enclosure.

Section 312.6(B)(2) and Table 312.6(B) provide the requirements for wire-bending space where straight-in wiring or offset (double bends) is employed at terminals. Section 312.6(B)(1) applies only to 90 degree bends.

The notes to Table 312.6(B) permit a reduction in required bending space for removable and lay-in wire terminals. The removable terminal wire connectors can be either compression type or setscrew type. However, connectors are required to be of the type intended for a single conductor (single barrel). Removable connectors designed for multiple wires are not permitted to have a reduction in bending space.

To facilitate wiring, a terminal may be placed on the stripped end of the conductor, which has been cut to the proper length. The terminal is crimped or lightly torqued on the wire as intended. The wire is bent and routed to facilitate mounting onto the stud or landing pad for proper connection. All mechanical screws, bolts, and nuts involved should then be torqued to the proper value. See the commentary following the fine print note to 110.14 regarding tightening torques.

Table 312.6(A) Minimum Wire-Bending Space at Terminals and Minimum Width of Wiring Gutters

Wire Size (AWG or kcmil)	Wires per Terminal									
	1		2		3		4		5	
	mm	in.	mm	in.	mm	in.	mm	in.	mm	in.
14–10	Not specified		—	—	—	—	—	—	—	—
8–6	38.1	1½	—	—	—	—	—	—	—	—
4–3	50.8	2	—	—	—	—	—	—	—	—
2	63.5	2½	—	—	—	—	—	—	—	—
1	76.2	3	—	—	—	—	—	—	—	—
1/0–2/0	88.9	3½	127	5	178	7	—	—	—	—
3/0–4/0	102	4	152	6	203	8	—	—	—	—
250	114	4½	152	6	203	8	254	10	—	—
300–350	127	5	203	8	254	10	305	12	—	—
400–500	152	6	203	8	254	10	305	12	356	14
600–700	203	8	254	10	305	12	356	14	406	16
750–900	203	8	305	12	356	14	406	16	457	18
1000–1250	254	10	—	—	—	—	—	—	—	—
1500–2000	305	12	—	—	—	—	—	—	—	—

Note: Bending space at terminals shall be measured in a straight line from the end of the lug or wire connector (in the direction that the wire leaves the terminal) to the wall, barrier, or obstruction.

Table 312.6(B) Minimum Wire-Bending Space at Terminals

Wire Size (AWG or kcmil)		Wires per Terminal							
		1		2		3		4 or More	
All Other Conductors	Compact Stranded AA-8000 Aluminum Alloy Conductors (See Note 3.)	mm	in.	mm	in.	mm	in.	mm	in.
14–10	12–8	Not specified		—	—	—	—	—	—
8	6	38.1	1½	—	—	—	—	—	—
6	4	50.8	2	—	—	—	—	—	—
4	2	76.2	3	—	—	—	—	—	—
3	1	76.2	3	—	—	—	—	—	—
2	1/0	88.9	3½	—	—	—	—	—	—
1	2/0	114	4½	—	—	—	—	—	—
1/0	3/0	140	5½	140	5½	178	7	—	—
2/0	4/0	152	6	152	6	190	7½	—	—
3/0	250	165[a]	6½[a]	165[a]	6½[a]	203	8	—	—
4/0	300	178[b]	7[b]	190[c]	7½[c]	216[a]	8½[a]	—	—
250	350	216[d]	8½[d]	229[d]	8½[d]	254[b]	9[b]	254	10
300	400	254[e]	10[e]	254[d]	10[d]	279[b]	11[b]	305	12
350	500	305[e]	12[e]	305[e]	12[e]	330[e]	13[e]	356[d]	14[d]
400	600	330[e]	13[e]	330[e]	13[e]	356[e]	14[e]	381[e]	15[e]
500	700–750	356[e]	14[e]	356[e]	14[e]	381[e]	15[e]	406[e]	16[e]
600	800–900	381[e]	15[e]	406[e]	16[e]	457[e]	18[e]	483[e]	19[e]
700	1000	406[e]	16[e]	457[e]	18[e]	508[e]	20[e]	559[e]	22[e]
750	—	432[e]	17[e]	483[e]	19[e]	559[e]	22[e]	610[e]	24[e]
800	—	457	18	508	20	559	22	610	24
900	—	483	19	559	22	610	24	610	24
1000	—	508	20	—	—	—	—	—	—
1250	—	559	22	—	—	—	—	—	—
1500	—	610	24	—	—	—	—	—	—
1750	—	610	24	—	—	—	—	—	—
2000	—	610	24	—	—	—	—	—	—

1. Bending space at terminals shall be measured in a straight line from the end of the lug or wire connector in a direction perpendicular to the enclosure wall.

2. For removable and lay-in wire terminals intended for only one wire, bending space shall be permitted to be reduced by the following number of millimeters (inches):

[a] 12.7 mm (½ in.)
[b] 25.4 mm (1 in.)
[c] 38.1 mm (1½ in.)
[d] 50.8 mm (2 in.)
[e] 76.2 mm (3 in.)

3. This column shall be permitted to determine the required wire-bending space for compact stranded aluminum conductors in sizes up to 1000 kcmil and manufactured using AA-8000 series electrical grade aluminum alloy conductor material in accordance with 310.14.

Note that in accordance with the notes to Table 312.6(A) and Table 312.6(B), when Table 312.6(A) is used, bending space is measured in the direction in which the wire leaves the terminal, and when Table 312.6(B) is used, it is measured in a direction perpendicular to the enclosure wall.

A lay-in-type terminal is a pressure wire connector in which part of the connector is removable or swings away so that the stripped end of the conductor can be laid into the fixed portion of the connector. The removable or swing-away portion is then put back in place and the connector tightened down on the conductor.

In conjunction with the following list, Exhibit 312.1

shows the application of the rules of 312.6(B)(1), 312.6(B)(2), Table 312.6(A), and Table 312.6(B) to the wiring for a lay-in-type terminal.

T_1, 312.6(B)(2): Table 312.6(B) applies for conductors M.

T_2, 312.6(B)(2): Table 312.6(B) applies for conductors BR_2 unless Exception No. 2 of 312.6(B)(2) applies. This exception allows Table 312.6(A) to apply to T_2 as long as BR_2 enters or leaves the enclosure where gutter G_2 joins gutter G_1, and gutter G_1 has a width conforming to Table 312.6(B) for BR_2.

T_3, 312.6(B)(2): Table 312.6(B) applies for conductors BR_3.

T_4, 312.6(B)(1): Table 312.6(A) applies for conductor N.

G_1, 312.6(A): Table 312.6(A) applies for conductors M. Table 312.6(B) applies for conductors BR_2 where T_2 does not comply with Table 312.6(B).

G_2, 312.6(A): Table 312.6(A) applies for conductors BR_2.

G_3, 312.6(A): Table 312.6(A) applies for conductors BR_3.

G_4, 312.6(A): Table 312.6(A) applies for conductor N.

Exhibit 312.1 Wiring for a lay-in-type terminal to which the list of rules in the commentary following 312.6(B)(2), Exception No. 2 applies.

Exhibit 312.2 illustrates the conditions under which 312.6(B)(2), Exception No. 2, is applicable. The terminal on the left has an offset not greater than 50 percent of bending space, per condition (b) of Exception No. 2. The terminal on the right is within a 45 degree angle of the enclosure, per condition (a) of Exception No. 2. See also Exhibit 430.1, which shows an example of wire-bending space in enclosures for motor controllers.

Exhibit 312.2 Conditions under which 312.6(B)(2), Exception No. 2, is applicable.

(C) Conductors 4 AWG or Larger Installation shall comply with 300.4(F).

312.7 Space in Enclosures

Cabinets and cutout boxes shall have sufficient space to accommodate all conductors installed in them without crowding.

312.8 Enclosures for Switches or Overcurrent Devices

Enclosures for switches or overcurrent devices shall not be used as junction boxes, auxiliary gutters, or raceways for conductors feeding through or tapping off to other switches or overcurrent devices, unless adequate space for this purpose is provided. The conductors shall not fill the wiring space at any cross section to more than 40 percent of the cross-sectional area of the space, and the conductors, splices, and taps shall not fill the wiring space at any cross section to more than 75 percent of the cross-sectional area of that space.

Most enclosures are intended to accommodate only those conductors that will be connected to terminals for switches or overcurrent devices within the enclosures themselves. Where adequate space is provided for additional conductors, such as control circuits, the total conductor fill in the enclosure may not exceed 40 percent of the cross section of the wiring space in the enclosure and no more than 75 percent if splices or taps are necessary.

Example

If an enclosure has a wiring space of 4 in. wide by 3 in. deep, the cross-sectional area is 12 in.^{2} Thus, the total conductor fill (see Chapter 9, Table 5 for dimensions of conductors) at any cross section cannot exceed 4.8 in.^{2} (40 percent of 12 in.^{2}), and the maximum space for conductors and splices or taps at any cross section cannot exceed 9 in.^{2} (75 percent of 12 in.^{2}).

In general, the best way to avoid overcrowding enclosures is to use properly sized auxiliary gutters (366.22, 366.56, and 366.58) or junction boxes (314.16 and 314.28). See 430.10 and commentary for wiring space in enclosures for motor controllers and disconnecting means. See also 110.59 for tunnel installations over 600 volts.

312.9 Side or Back Wiring Spaces or Gutters

Cabinets and cutout boxes shall be provided with back-wiring spaces, gutters, or wiring compartments as required by 312.11(C) and (D).

II. Construction Specifications

312.10 Material

Cabinets, cutout boxes, and meter socket enclosures shall comply with 312.10(A) through (C).

(A) Metal Cabinets and Cutout Boxes Metal enclosures within the scope of this article shall be protected both inside and outside against corrosion.

> FPN: For information on protection against corrosion, see 300.6.

(B) Strength The design and construction of enclosures within the scope of this article shall be such as to secure ample strength and rigidity. If constructed of sheet steel, the metal thickness shall not be less than 1.35 mm (0.053 in.) uncoated.

(C) Nonmetallic Cabinets Nonmetallic cabinets shall be listed, or they shall be submitted for approval prior to installation.

312.11 Spacing

The spacing within cabinets and cutout boxes shall comply with 312.11(A) through (D).

(A) General Spacing within cabinets and cutout boxes shall be sufficient to provide ample room for the distribution of wires and cables placed in them and for a separation between metal parts of devices and apparatus mounted within them as follows.

(1) Base Other than at points of support, there shall be an airspace of at least 1.59 mm (0.0625 in.) between the base of the device and the wall of any metal cabinet or cutout box in which the device is mounted.

(2) Doors There shall be an airspace of at least 25.4 mm (1.00 in.) between any live metal part, including live metal parts of enclosed fuses, and the door.

Exception: Where the door is lined with an approved insulating material or is of a thickness of metal not less than 2.36 mm (0.093 in.) uncoated, the airspace shall not be less than 12.7 mm (0.500 in.).

(3) Live Parts There shall be an airspace of at least 12.7 mm (0.500 in.) between the walls, back, gutter partition, if of metal, or door of any cabinet or cutout box and the nearest exposed current-carrying part of devices mounted within the cabinet where the voltage does not exceed 250. This spacing shall be increased to at least 25.4 mm (1.00 in.) for voltages of 251 to 600, nominal.

Exception: Where the conditions in 312.11(A)(2), Exception, are met, the airspace for nominal voltages from 251 to 600 shall be permitted to be not less than 12.7 mm (0.500 in.).

(B) Switch Clearance Cabinets and cutout boxes shall be deep enough to allow the closing of the doors when 30-ampere branch-circuit panelboard switches are in any position, when combination cutout switches are in any position, or when other single-throw switches are opened as far as their construction permits.

(C) Wiring Space Cabinets and cutout boxes that contain devices or apparatus connected within the cabinet or box to more than eight conductors, including those of branch circuits, meter loops, feeder circuits, power circuits, and similar circuits, but not including the supply circuit or a continuation thereof, shall have back-wiring spaces or one or more side-wiring spaces, side gutters, or wiring compartments.

(D) Wiring Space — Enclosure Side-wiring spaces, side gutters, or side-wiring compartments of cabinets and cutout boxes shall be made tight enclosures by means of covers, barriers, or partitions extending from the bases of the devices contained in the cabinet, to the door, frame, or sides of the cabinet.

Exception: Side-wiring spaces, side gutters, and side-wiring compartments of cabinets shall not be required to be made tight enclosures where those side spaces contain only conductors that enter the cabinet directly opposite to the devices where they terminate.

Partially enclosed back-wiring spaces shall be provided with covers to complete the enclosure. Wiring spaces that are required by 312.11(C) and are exposed when doors are open shall be provided with covers to complete the enclosure. Where adequate space is provided for feed-through conductors and for splices as required in 312.8, additional barriers shall not be required.

ARTICLE 314
Outlet, Device, Pull, and Junction Boxes; Conduit Bodies; Fittings; and Handhole Enclosures

Summary of Changes

- **Article 314 title and 314.1:** Deleted manholes and added handhole enclosures.

- **314.16(B)(1):** Revised to require a two-conductor addition to the box fill calculation where a looped, unbroken conductor has a total length that is not less than twice the minimum length required for free conductors in 300.14.

- **314.17(B) and (C):** Added messenger supported wiring to requirements covering conductors entering metal and nonmetallic boxes.

- **314.20:** Revised to permit measurement to be made from plaster rings, extension rings, and listed extenders.

- **314.23(B)(1):** Revised to prohibit unprotected screw threads from passing through the box.

- **314.23(F):** Revised to clarify that the requirement does not apply to raceway-supported enclosures containing splicing devices.

- **314.27(D):** Added section covering outlet boxes and box-support systems for support of ceiling paddle fans; moved from Article 422.

- **314.30:** Added text to provide the sizing and installation requirements for using handhole enclosures in electrical systems.

- **Part IV, Manholes:** Moved entire article to Article 110, Part V.

Contents

I. Scope and General

314.1 Scope

This article covers the installation and use of all boxes and conduit bodies used as outlet, device, junction, or pull boxes, depending on their use, and handhole enclosures. Cast, sheet metal, nonmetallic, and other boxes such as FS, FD, and larger boxes are not classified as conduit bodies. This article also includes installation requirements for fittings used to join raceways and to connect raceways and cables to boxes and conduit bodies.

Revised for the 2005 *Code*, Article 314 now includes handhole enclosures. See the definition of *handhole enclosure* in Article 100. Manholes were previously covered by Article 314, but now those requirements are found in Article 110. Manholes were moved to Article 110 because of the close association with working space requirements, which are also found in Article 110.

314.2 Round Boxes

Round boxes shall not be used where conduits or connectors requiring the use of locknuts or bushings are to be connected to the side of the box.

Section 314.2 requires the use of rectangular or octagonal boxes having a flat bearing surface at each knockout for locknuts and bushings to ensure effective grounding continuity. Round boxes, however, can be used if the conduit or cable is secured by clamps within the box or if the cable does not need attachment to the box, as permitted by 314.17(C), Exception.

314.3 Nonmetallic Boxes

Nonmetallic boxes shall be permitted only with open wiring on insulators, concealed knob-and-tube wiring, cabled wiring methods with entirely nonmetallic sheaths, flexible cords, and nonmetallic raceways.

Exception No. 1: Where internal bonding means are provided between all entries, nonmetallic boxes shall be permitted to be used with metal raceways or metal-armored cables.

Exception No. 2: Where integral bonding means with a provision for attaching an equipment bonding jumper inside the box are provided between all threaded entries in nonmetallic boxes listed for the purpose, nonmetallic boxes shall be permitted to be used with metal raceways or metal-armored cables.

The main rule of 314.3 was revised for the 2002 *Code* to permit the use of nonmetallic boxes with flexible cords. Exception No. 1 applies to nonmetallic boxes without threaded entries and permits the use of metal raceways and metal-armored cables with nonmetallic boxes. Internal bonding means must be installed to ensure ground continuity between the metal raceways or metal-armored cables. For the purposes of Exception No. 1, the term *metal-armored cable* includes cables with a metal covering such as mineral-insulated, metal-sheathed cable (Type MI), metal-clad cable (Type MC), and armored cable (Type AC).

Exception No. 2 permits the use of metal raceways and metal-armored cables with listed nonmetallic boxes equipped with threaded entries. An integral bonding means is required to ensure ground continuity between the threaded entries. The requirement for means to attach a bonding jumper accommodates devices or equipment attached to the box. For the purposes of Exception No. 2, the term *metal-armored cable* includes cables with a metal covering such as mineral-insulated, metal-sheathed cable (Type MI), metal-clad cable (Type MC), and armored cable (Type AC).

314.4 Metal Boxes

All metal boxes shall be grounded in accordance with the provisions of Article 250.

314.5 Short-Radius Conduit Bodies

Conduit bodies such as capped elbows and service-entrance elbows that enclose conductors 6 AWG or smaller, and are only intended to enable the installation of the raceway and the contained conductors, shall not contain splices, taps, or devices and shall be of sufficient size to provide free space for all conductors enclosed in the conduit body.

Short-radius conduit bodies are not permitted to contain splices.

II. Installation

314.15 Damp, Wet, or Hazardous (Classified) Locations

(A) Damp or Wet Locations In damp or wet locations, boxes, conduit bodies, and fittings shall be placed or equipped so as to prevent moisture from entering or accumulating within the box, conduit body, or fitting. Boxes, conduit bodies, and fittings installed in wet locations shall be listed for use in wet locations.

FPN No. 1: For boxes in floors, see 314.27(C).
FPN No. 2: For protection against corrosion, see 300.6.

Article 100 defines the term *weatherproof* as "constructed or protected so that exposure to the weather will not interfere with successful operation." Rainproof, raintight, or watertight equipment can fulfill the requirements for this definition where varying weather conditions other than wetness, such as snow, ice, dust, or temperature extremes, are not a factor. A weatherhead fitting is considered to be weatherproof because the openings for the conductors are placed in a downward position so that rain or snow cannot enter the fitting.

See the definitions of *damp location* and *wet location* under *location* in Article 100, as well as the commentary following the definition of *enclosure*, for further explanation.

(B) Hazardous (Classified) Locations Installations in hazardous (classified) locations shall conform to Articles 500 through 517.

314.16 Number of Conductors in Outlet, Device, and Junction Boxes, and Conduit Bodies

Boxes and conduit bodies shall be of sufficient size to provide free space for all enclosed conductors. In no case shall the volume of the box, as calculated in 314.16(A), be less than the fill calculation as calculated in 314.16(B). The minimum volume for conduit bodies shall be as calculated in 314.16(C).

The provisions of this section shall not apply to terminal housings supplied with motors.

FPN: For volume requirements of motor terminal housings, see 430.12.

Boxes and conduit bodies enclosing conductors 4 AWG or larger shall also comply with the provisions of 314.28.

(A) Box Volume Calculations The volume of a wiring enclosure (box) shall be the total volume of the assembled sections and, where used, the space provided by plaster rings, domed covers, extension rings, and so forth, that are marked with their volume or are made from boxes the dimensions of which are listed in Table 314.16(A).

(1) Standard Boxes The volumes of standard boxes that are not marked with their volume shall be as given in Table 314.16(A).

(2) Other Boxes Boxes 1650 cm³ (100 in.³) or less, other than those described in Table 314.16(A), and nonmetallic boxes shall be durably and legibly marked by the manufacturer with their volume. Boxes described in Table 314.16(A) that have a volume larger than is designated in the table shall be permitted to have their volume marked as required by this section.

(B) Box Fill Calculations The volumes in paragraphs 314.16(B)(1) through (B)(5), as applicable, shall be added together. No allowance shall be required for small fittings such as locknuts and bushings.

(1) Conductor Fill Each conductor that originates outside the box and terminates or is spliced within the box shall be counted once, and each conductor that passes through the box without splice or termination shall be counted once. A looped, unbroken conductor not less than twice the minimum length required for free conductors in 300.14 shall be counted twice. The conductor fill shall be calculated using Table 314.16(B). A conductor, no part of which leaves the box, shall not be counted.

Exception: An equipment grounding conductor or conductors or not over four fixture wires smaller than 14 AWG, or both, shall be permitted to be omitted from the calculations where they enter a box from a domed luminaire (fixture) or similar canopy and terminate within that box.

(2) Clamp Fill Where one or more internal cable clamps, whether factory or field supplied, are present in the box, a single volume allowance in accordance with Table 314.16(B) shall be made based on the largest conductor present in the

Table 314.16(A) Metal Boxes

Box Trade Size			Minimum Volume		Maximum Number of Conductors*						
mm	in.		cm³	in.³	18	16	14	12	10	8	6
100 × 32	(4 × 1¼)	round/octagonal	205	12.5	8	7	6	5	5	5	2
100 × 38	(4 × 1½)	round/octagonal	254	15.5	10	8	7	6	6	5	3
100 × 54	(4 × 2⅛)	round/octagonal	353	21.5	14	12	10	9	8	7	4
100 × 32	(4 × 1¼)	square	295	18.0	12	10	9	8	7	6	3
100 × 38	(4 × 1½)	square	344	21.0	14	12	10	9	8	7	4
100 × 54	(4 × 2⅛)	square	497	30.3	20	17	15	13	12	10	6
120 × 32	(4¹¹⁄₁₆× 1¼)	square	418	25.5	17	14	12	11	10	8	5
120 × 38	(4¹¹⁄₁₆× 1½)	square	484	29.5	19	16	14	13	11	9	5
120 × 54	(4¹¹⁄₁₆× 2⅛)	square	689	42.0	28	24	21	18	16	14	8
75 × 50 × 38	(3 × 2 × 1½)	device	123	7.5	5	4	3	3	3	2	1
75 × 50 × 50	(3 × 2 × 2)	device	164	10.0	6	5	5	4	4	3	2
75 × 50 × 57	(3 × 2 × 2¼)	device	172	10.5	7	6	5	4	4	3	2
75 × 50 × 65	(3 × 2 × 2½)	device	205	12.5	8	7	6	5	5	4	2
75 × 50 × 70	(3 × 2 × 2¾)	device	230	14.0	9	8	7	6	5	4	2
75 × 50 × 90	(3 × 2 × 3½)	device	295	18.0	12	10	9	8	7	6	3
100 × 54 × 38	(4 × 2⅛× 1½)	device	169	10.3	6	5	5	4	4	3	2
100 × 54 × 48	(4 × 2⅛× 1⅞)	device	213	13.0	8	7	6	5	5	4	2
100 × 54 × 54	(4 × 2⅛× 2⅛)	device	238	14.5	9	8	7	6	5	4	2
95 × 50 × 65	(3¾× 2 × 2½)	masonry box/gang	230	14.0	9	8	7	6	5	4	2
95 × 50 × 90	(3¾× 2 × 3½)	masonry box/gang	344	21.0	14	12	10	9	8	7	4
min. 44.5 depth	FS — single cover/gang (1¾)		221	13.5	9	7	6	6	5	4	2
min. 60.3 depth	FD — single cover/gang (2⅜)		295	18.0	12	10	9	8	7	6	3
min. 44.5 depth	FS — multiple cover/gang (1¾)		295	18.0	12	10	9	8	7	3	
min. 60.3 depth	FD — multiple cover/gang (2⅜)		395	24.0	16	13	12	10	9	8	4

*Where no volume allowances are required by 314.16(B)(2) through (B)(5).

box. No allowance shall be required for a cable connector with its clamping mechanism outside the box.

(3) Support Fittings Fill Where one or more luminaire (fixture) studs or hickeys are present in the box, a single volume allowance in accordance with Table 314.16(B) shall be made for each type of fitting based on the largest conductor present in the box.

(4) Device or Equipment Fill For each yoke or strap containing one or more devices or equipment, a double volume allowance in accordance with Table 314.16(B) shall be made for each yoke or strap based on the largest conductor connected to a device(s) or equipment supported by that yoke or strap.

Section 314.16 provides the requirements and identifies the allowances for the number of conductors permitted to be enclosed within a box. This section requires that the *total box volume* be equal to or greater than the *total box fill*.

Table 314.16(B) Volume Allowance Required per Conductor

Size of Conductor (AWG)	Free Space Within Box for Each Conductor	
	cm³	in.³
18	24.6	1.50
16	28.7	1.75
14	32.8	2.00
12	36.9	2.25
10	41.0	2.50
8	49.2	3.00
6	81.9	5.00

General Requirements

The total box volume is determined by adding the individual volumes of the box components. The components include the box itself plus any attachments to it, such as a plaster ring, an extension ring, or a dome cover. The volume of each box component is determined either from the volume marking on the component itself or from the standard volumes listed in Table 314.16(A). If a box is marked with a larger volume than listed in Table 314.16(A), the larger volume can be used instead of the table value.

Adding all the volume allowances for all items contributing to box fill determines the total box fill. The volume allowance for each fill item is based on the volume listed in Table 314.16(B) for the conductor size indicated. Commentary Table 314.1 summarizes the components contributing to box fill.

Specific Examples

The following three examples illustrate the applicable requirements of 314.16 and the accompanying tables.

Example 1

Using the simple method [according to the footnote of Table 314.16(A)], select a standard-sized box for use where all the conductors are the same size and the box does not contain any cable clamps, support fittings, devices, or equipment grounding conductors. Refer to Exhibit 314.1 for an example.

Solution

To determine the number of conductors permitted in a standard 4 in. × 1½ in. square box (21.0 in.^{3}), count the conductors in the box and compare the total to the maximum number of conductors permitted by Table 314.16(A). Each unspliced conductor running through the box is counted as one conductor, and each other conductor is counted as one conductor. Therefore, the total conductor count for this box is nine conductors. Table 314.16(A) indicates that the maximum fill for this box is nine 12 AWG conductors, so the box is adequately sized.

Example 2

The standard method for determining adequate box size calculates the total box volume and then subtracts the total box fill to ensure compliance. Using this method, refer to Exhibit 314.2 and determine whether the box is adequately sized.

Solution

For a standard 3 in. × 2 in. × 3½ in. device box (18 in.^{3}), Table 314.16(A) allows up to a maximum of nine 14 AWG

Commentary Table 314.1 Summary of Items Contributing to Box Fill

Items Contained Within Box	Volume Allowance	Based on [see Table 314.16(B)]
Conductors that originate outside box	One for each conductor	Actual conductor size
Conductors that pass through box without splice or connection (*less than 12 in. in total length)	One for each conductor	Actual conductor size
*Conductors 12 in. or greater that are looped and unbroken (see 300.15 for exact measurement)	Two for a single (entire) unbroken conductor	Actual conductor size
Conductors that originate within box and do not leave box	None (these conductors not counted)	n.a.
Fixture conductors [per 314.16(B)(1), Exception]	None (these conductors not counted)	n.a.
Internal cable clamps (one or more)	One only	Largest-sized conductor present
Support fittings (such as fixture studs, hickeys)	One for each type of support fitting	Largest-sized conductor present
Devices (such as receptacles, switches)	Two for each yoke or mounting strap	Largest-sized conductor connected to device or equipment
Equipment grounding conductor (one or more)	One only	Largest equipment grounding conductor present
Isolated equipment grounding conductor (one or more) [see 250.146(D)]	One only	Largest isolated and insulated equipment grounding conductor present

*2005 *Code* change.

4 in. × 1½ in. square box (21.0 in.³)

Standard 3 in. × 2 in. × 3½ in. device box (18 in.³)

Exhibit 314.1 Example 1: A standard-sized 4 in. × 1½ in. square box (21.0 in.³) containing no fittings or devices, such as fixture studs, cable clamps, switches, receptacles, or equipment grounding conductors.

conductors. The box fill for this situation as given in Commentary Table 314.2 is 16 in.³. Therefore, because the total box fill of 16 in.³ is less than the 18 in.³ total box volume permitted, the box is adequately sized.

Example 3

Using the standard method, determine the adequacy of the device illustrated in Exhibit 314.3. The exhibit illustrates two 3 in. × 2 in. × 3½ in. device boxes assembled to configure a single box.

Solution

The total box volume, referring to Table 314.16(A), is 36 in.³ (2 × 18 in.³). The total box fill, referring to Table

Exhibit 314.2 Example 2: A device box that contains components and conductors requiring deductions in accordance with 314.16.

314.16(B), is determined as given in Commentary Table 314.3. With only 26 in.³ of the 36 in.³ filled, the box is adequately sized.

(5) Equipment Grounding Conductor Fill Where one or more equipment grounding conductors or equipment bonding jumpers enter a box, a single volume allowance in accordance with Table 314.16(B) shall be made based on the largest equipment grounding conductor or equipment bonding jumper present in the box. Where an additional set of equipment grounding conductors, as permitted by 250.146(D), is present in the box, an additional volume allowance shall be made based on the largest equipment grounding conductor in the additional set.

Commentary Table 314.2 Total Box Fill for Example No. 2

Items Contained Within Box	Volume Allowance	Unit Volume Based on Table 314.16(B) (in.³)	Total Box Fill (in.³)
4 conductors	4 volume allowances for 14 AWG conductors	2.00	8.00
1 clamp	1 volume allowance (based on 14 AWG conductors)	2.00	2.00
1 device	2 volume allowances (based on 14 AWG conductors)	2.00	4.00
Equipment grounding conductors (all)	1 volume allowance (based on 14 AWG conductors)	2.00	2.00
Total			16.00

Exhibit 314.3 Example 3: Two standard gangable device boxes containing conductors of different sizes.

(C) Conduit Bodies

(1) General Conduit bodies enclosing 6 AWG conductors or smaller, other than short-radius conduit bodies as described in 314.5, shall have a cross-sectional area not less than twice the cross-sectional area of the largest conduit or tubing to which it is attached. The maximum number of conductors permitted shall be the maximum number permitted by Table 1 of Chapter 9 for the conduit or tubing to which it is attached.

(2) With Splices, Taps, or Devices Only those conduit bodies that are durably and legibly marked by the manufacturer with their volume shall be permitted to contain splices,

taps, or devices. The maximum number of conductors shall be calculated in accordance with 314.16(B). Conduit bodies shall be supported in a rigid and secure manner.

As illustrated in Exhibit 314.4, conduit bodies other than the short-radius type are permitted to contain splices or taps, provided the conduit bodies are marked with their cubic-inch capacity. Such conduit bodies are required to have a cross-sectional area not less than twice that of the conduit to which they are attached and are not permitted to contain more conductors than the attached raceway. The volume requirements for splicing or tapping are provided in 314.16(C).

Conduit bodies must be rigidly supported. See 314.23(D)(2), the exception to 314.23(E), and Exception No. 1 of 314.23(F), which permit the raceway to support the conduit body, provided the conduit body is not larger than the attached raceway.

See 314.28 for requirements that apply to conduit bodies used as pull and junction boxes.

Exhibit 314.4 An example of splices in a raceway-supported conduit body.

Commentary Table 314.3 Total Box Fill for Example No. 3

Items Contained Within Box	Volume Allowance	Unit Volume Based on Table 314.16(B) (in.³)	Total Box Fill (in.³)
6 conductors	2 volume allowances for 14 AWG conductors	2.00	4.00
	4 volume allowances for 12 AWG conductors	2.25	9.00
2 clamps	1 volume allowance (based on 12 AWG conductors)	2.25	2.25
2 devices	2 volume allowances (based on 14 AWG conductors)	2.00	4.00
	2 volume allowances (based on 12 AWG conductors)	2.25	4.50
Equipment grounding conductors (all)	1 volume allowance (based on 12 AWG conductors)	2.25	2.25
Total			26.00

314.17 Conductors Entering Boxes, Conduit Bodies, or Fittings

Conductors entering boxes, conduit bodies, or fittings shall be protected from abrasion and shall comply with 314.17(A) through (D).

(A) Openings to Be Closed Openings through which conductors enter shall be adequately closed.

(B) Metal Boxes and Conduit Bodies Where metal boxes or conduit bodies are installed with messenger supported wiring, open wiring on insulators, or concealed knob-and-tube wiring, conductors shall enter through insulating bushings or, in dry locations, through flexible tubing extending from the last insulating support to not less than 6 mm (¼ in.) inside the box and beyond any cable clamps. Except as provided in 300.15(C), the wiring shall be firmly secured to the box or conduit body. Where raceway or cable is installed with metal boxes or conduit bodies, the raceway or cable shall be secured to such boxes and conduit bodies.

Editorial corrections were made for the 2005 *Code* in both 314.17(B) and 314.17(C) to specifically address *messenger supported wiring* and *open wiring on insulators*, which replace the general term *open wiring*.

(C) Nonmetallic Boxes and Conduit Bodies Nonmetallic boxes and conduit bodies shall be suitable for the lowest temperature-rated conductor entering the box. Where nonmetallic boxes and conduit bodies are used with messenger supported wiring, open wiring on insulators, or concealed knob-and-tube wiring, the conductors shall enter the box through individual holes. Where flexible tubing is used to enclose the conductors, the tubing shall extend from the last insulating support to not less than 6 mm (¼ in.) inside the box and beyond any cable clamp. Where nonmetallic-sheathed cable or multiconductor Type UF cable is used, the sheath shall extend not less than 6 mm (¼ in.) inside the box and beyond any cable clamp. In all instances, all permitted wiring methods shall be secured to the boxes.

Standard nonmetallic boxes are permitted for use with 90°C insulated conductors. A nonmetallic box used for splicing a conductor of a higher temperature rating to a conductor of a lower temperature rating is required to be identified as suitable for the temperature rating of the lower-rated conductor. The intent is to avoid the necessity of giving a high temperature rating to boxes in a normal temperature location simply because high-temperature conductors enter the box from or exit to a high-temperature location. However, where insulated conductors rated at higher temperatures are necessary in a high-temperature environment, the box is required to be suitably identified by a marking on the box or in the listing of the box to comply with 110.3(B).

Exception: Where nonmetallic-sheathed cable or multiconductor Type UF cable is used with single gang boxes not larger than a nominal size 57 mm × 100 mm (2¼ in. × 4 in.) mounted in walls or ceilings, and where the cable is fastened within 200 mm (8 in.) of the box measured along the sheath and where the sheath extends through a cable knockout not less than 6 mm (¼ in.), securing the cable to the box shall not be required. Multiple cable entries shall be permitted in a single cable knockout opening.

Some cable-securing means is required in all single gang boxes larger than 2¼ in. × 4 in. The requirement is based on the width of the box and the likelihood that the cable will be pushed back out of the box when the conductors and device, if any, are folded back into the box during installation of receptacles, switches, dimmers, and so on.

(D) Conductors 4 AWG or Larger Installation shall comply with 300.4(F).

FPN: See 110.12(A) for requirements on closing unused cable and raceway knockout openings.

314.19 Boxes Enclosing Flush Devices

Boxes used to enclose flush devices shall be of such design that the devices will be completely enclosed on back and sides and substantial support for the devices will be provided. Screws for supporting the box shall not be used in attachment of the device contained therein.

314.20 In Wall or Ceiling

In walls or ceilings with a surface of concrete, tile, gypsum, plaster, or other noncombustible material, boxes employing a flush-type cover or faceplate shall be installed so that the front edge of the box, plaster ring, extension ring, or listed extender will not be set back of the finished surface more than 6 mm (¼ in.).

In walls and ceilings constructed of wood or other combustible surface material, boxes, plaster rings, extension rings, or listed extenders shall be flush with the finished surface or project therefrom.

For the 2005 *Code*, "plaster ring, extension ring, or listed extender" was added to make a list of the front faces in addition to just a box. This list of options of possible front surfaces clarifies from where the maximum clearance dimension could be measured.

The terms *surface* and *with a surface of* make it clear that the requirements of this section apply only to the construction of the surface of the wall or ceiling, not to the structure or subsurface of the wall or ceiling. Therefore, a wall constructed of wood but sheathed with an outer layer of gypsum board is permitted to contain boxes set back or recessed not more than ¼ in. Using an opposite example, a wall constructed of metal studs but finished with wood panels requires that contained outlet boxes be mounted flush with the combustible finish.

314.21 Repairing Plaster and Drywall or Plasterboard

Plaster, drywall, or plasterboard surfaces that are broken or incomplete around boxes employing a flush-type cover or faceplate shall be repaired so there will be no gaps or open spaces greater than 3 mm (⅛ in.) at the edge of the box.

For the 2005 *Code*, a clarifying phrase was added to point out that repair is necessary only for boxes employing a flush-type cover or faceplate.

314.22 Exposed Surface Extensions

Surface extensions from a flush-mounted box shall be made by mounting and mechanically securing an extension ring over the flush box. Equipment grounding and bonding shall be in accordance with Article 250.

Exception: A surface extension shall be permitted to be made from the cover of a flush-mounted box where the cover is designed so it is unlikely to fall off or be removed if its securing means becomes loose. The wiring method shall be flexible for a length sufficient to permit removal of the cover and provide access to the box interior, and arranged so that any bonding or grounding continuity is independent of the connection between the box and cover.

Exhibit 314.5 shows an example of a flexible surface extension from a flush-mounted outlet box. The exception to 314.22, which permits this technique, requires the use of a cover that will not fall off if the securing screws become loose.

314.23 Supports

Enclosures within the scope of this article shall be supported in accordance with one or more of the provisions in 314.23(A) through (H).

(A) Surface Mounting An enclosure mounted on a building or other surface shall be rigidly and securely fastened in place. If the surface does not provide rigid and secure

Exhibit 314.5 An example of a flexible surface extension from a flush-mounted outlet box.

support, additional support in accordance with other provisions of this section shall be provided.

(B) Structural Mounting An enclosure supported from a structural member of a building or from grade shall be rigidly supported either directly or by using a metal, polymeric, or wood brace.

(1) Nails and Screws Nails and screws, where used as a fastening means, shall be attached by using brackets on the outside of the enclosure, or they shall pass through the interior within 6 mm (¼ in.) of the back or ends of the enclosure. Screws shall not be permitted to pass through the box unless exposed threads in the box are protected using approved means to avoid abrasion of conductor insulation.

This requirement prevents nails from interfering with the installation of devices. Permitting nails inside the box within ¼ in. of the ends reduces splitting of the smaller wooden studs used in some frame-type construction. However, splitting sometimes occurs where nails are within ¼ in. of the back of the box. The last sentence was added for the 2005 *Code* to point out that the coarse screw threads found on most screws used for mounting boxes present a severe abrasion hazard to conductor insulation when the threads are left exposed inside a box.

(2) Braces Metal braces shall be protected against corrosion and formed from metal that is not less than 0.51 mm (0.020 in.) thick uncoated. Wood braces shall have a cross section not less than nominal 25 mm × 50 mm (1 in. × 2 in.). Wood braces in wet locations shall be treated for the conditions. Polymeric braces shall be identified as being suitable for the use.

(C) Mounting in Finished Surfaces An enclosure mounted in a finished surface shall be rigidly secured thereto by clamps, anchors, or fittings identified for the application.

Where structural members are lacking or where boxes are cut into existing walls, boxes are permitted to be secured by clamps or anchors. Exhibit 314.6 shows one example of an acceptable mounting method. Shown in the right portion of the exhibit is a device box mounted in an opening in an existing wall by means of a securing bracket that is part of the box.

Securing bracket

Exhibit 314.6 One type of device box used for old work.

(D) Suspended Ceilings An enclosure mounted to structural or supporting elements of a suspended ceiling shall be not more than 1650 cm³ (100 in.³) in size and shall be securely fastened in place in accordance with either (D)(1) or (D)(2).

(1) Framing Members An enclosure shall be fastened to the framing members by mechanical means such as bolts, screws, or rivets, or by the use of clips or other securing means identified for use with the type of ceiling framing member(s) and enclosure(s) employed. The framing members shall be adequately supported and securely fastened to each other and to the building structure.

(2) Support Wires The installation shall comply with the provisions of 300.11(A). The enclosure shall be secured, using methods identified for the purpose, to ceiling support wire(s), including any additional support wire(s) installed for that purpose. Support wire(s) used for enclosure support shall be fastened at each end so as to be taut within the ceiling cavity.

(E) Raceway Supported Enclosure, Without Devices, Luminaires (Fixtures), or Lampholders An enclosure that

does not contain a device(s) other than splicing devices or support a luminaire(s) [fixture(s)], lampholder, or other equipment and is supported by entering raceways shall not exceed 1650 cm³ (100 in.³) in size. It shall have threaded entries or have hubs identified for the purpose. It shall be supported by two or more conduits threaded wrenchtight into the enclosure or hubs. Each conduit shall be secured within 900 mm (3 ft) of the enclosure, or within 450 mm (18 in.) of the enclosure if all conduit entries are on the same side.

Boxes are not permitted to be supported by rigid raceways using locknuts and bushings. Enclosures without devices or luminaires, however, are considered to be adequately supported, provided the conduit is connected to the enclosure by threaded hubs and the threaded conduits enter the box on two or more sides and are supported within 3 ft of the enclosure. A box is not permitted to be supported by a single raceway.

Exception: Rigid metal, intermediate metal, or rigid nonmetallic conduit or electrical metallic tubing shall be permitted to support a conduit body of any size, including a conduit body constructed with only one conduit entry, provided the trade size of the conduit body is not larger than the largest trade size of the conduit or electrical metallic tubing.

(F) Raceway Supported Enclosures, with Devices, Luminaires (Fixtures), or Lampholders An enclosure that contains a device(s), other than splicing devices, or supports a luminaire(s) [fixture(s)], lampholder, or other equipment and is supported by entering raceways shall not exceed 1650 cm³ (100 in.³) in size. It shall have threaded entries or have hubs identified for the purpose. It shall be supported by two or more conduits threaded wrenchtight into the enclosure or hubs. Each conduit shall be secured within 450 mm (18 in.) of the enclosure.

The conduit is required to be supported within 18 in. if the enclosure contains devices (other than splicing devices) or luminaires.

Exception No. 1: Rigid metal or intermediate metal conduit shall be permitted to support a conduit body of any size, including a conduit body constructed with only one conduit entry, provided the trade size of the conduit body is not larger than the largest trade size of the conduit.

Exception No. 2: An unbroken length(s) of rigid or intermediate metal conduit shall be permitted to support a box used for luminaire (fixture) or lampholder support, or to support a wiring enclosure that is an integral part of a luminaire (fixture) and used in lieu of a box in accordance with 300.15(B), where all of the following conditions are met:

(a) The conduit is securely fastened at a point so that the length of conduit beyond the last point of conduit support does not exceed 900 mm (3 ft).

(b) The unbroken conduit length before the last point of conduit support is 300 mm (12 in.) or greater, and that portion of the conduit is securely fastened at some point not less than 300 mm (12 in.) from its last point of support.

(c) Where accessible to unqualified persons, the luminaire (fixture) or lampholder, measured to its lowest point, is at least 2.5 m (8 ft) above grade or standing area and at least 900 mm (3 ft) measured horizontally to the 2.5 m (8 ft) elevation from windows, doors, porches, fire escapes, or similar locations.

(d) A luminaire (fixture) supported by a single conduit does not exceed 300 mm (12 in.) in any direction from the point of conduit entry.

(e) The weight supported by any single conduit does not exceed 9 kg (20 lb).

(f) At the luminaire (fixture) or lampholder end, the conduit(s) is threaded wrenchtight into the box, conduit body, or integral wiring enclosure, or into hubs identified for the purpose. Where a box or conduit body is used for support, the luminaire (fixture) shall be secured directly to the box or conduit body, or through a threaded conduit nipple not over 75 mm (3 in.) long.

(G) Enclosures in Concrete or Masonry An enclosure supported by embedment shall be identified as suitably protected from corrosion and securely embedded in concrete or masonry.

Boxes are permitted to be embedded in masonry or concrete, provided they are rigid and secure. Exhibit 314.7 shows a mud box installed in a concrete ceiling. Additional support is not required.

Concrete ceiling

Exhibit 314.7 A mud box installed in a concrete ceiling.

(H) Pendant Boxes An enclosure supported by a pendant shall comply with 314.23(H)(1) or (H)(2).

(1) Flexible Cord A box shall be supported from a multiconductor cord or cable in an approved manner that protects the conductors against strain, such as a strain-relief connector threaded into a box with a hub.

(2) Conduit A box supporting lampholders or luminaires (lighting fixtures), or wiring enclosures within luminaires (fixtures) used in lieu of boxes in accordance with 300.15(B), shall be supported by rigid or intermediate metal conduit stems. For stems longer than 450 mm (18 in.), the stems shall be connected to the wiring system with flexible fittings suitable for the location. At the luminaire (fixture) end, the conduit(s) shall be threaded wrenchtight into the box or wiring enclosure, or into hubs identified for the purpose.

Where supported by only a single conduit, the threaded joints shall be prevented from loosening by the use of setscrews or other effective means, or the luminaire (fixture), at any point, shall be at least 2.5 m (8 ft) above grade or standing area and at least 900 mm (3 ft) measured horizontally to the 2.5 m (8 ft) elevation from windows, doors, porches, fire escapes, or similar locations. A luminaire (fixture) supported by a single conduit shall not exceed 300 mm (12 in.) in any horizontal direction from the point of conduit entry.

314.24 Depth of Outlet Boxes

No box shall have an internal depth of less than 12.7 mm (½ in.). Boxes intended to enclose flush devices shall have an internal depth of not less than 23.8 mm (15/16 in.).

The use of a shallow box might become necessary because of old work or existing construction where, for example, very shallow partitions, plumbing pipes, or ductwork is encountered within the partition. The selection of a box used in this type of situation is required to be based on its having sufficient cubic-inch capacity.

314.25 Covers and Canopies

In completed installations, each box shall have a cover, faceplate, lampholder, or luminaire (fixture) canopy, except where the installation complies with 410.14(B).

(A) Nonmetallic or Metal Covers and Plates Nonmetallic or metal covers and plates shall be permitted. Where metal covers or plates are used, they shall comply with the grounding requirements of 250.110.

> FPN: For additional grounding requirements, see 410.18(A) for metal luminaire (fixture) canopies, and 404.12 and 406.5(B) for metal faceplates.

(B) Exposed Combustible Wall or Ceiling Finish Where a luminaire (fixture) canopy or pan is used, any combustible wall or ceiling finish exposed between the edge of the canopy or pan and the outlet box shall be covered with noncombustible material.

Because heat from a short circuit, from a ground fault, or due to overlamping could create a fire hazard within a fixture canopy or pan, any exposed combustible wall or ceiling space between the edge of the outlet box and the perimeter of the luminaire is required to be covered with noncombustible material. The noncombustible material need not be metal. Glass fiber pads commonly provided as thermal barriers within the ceiling pan of luminaires can be used to meet this requirement. Where the wall or ceiling finish is concrete, tile, gypsum, plaster, or other noncombustible material, the requirements of this section do not apply. Section 314.20 contains requirements for flush-mounted boxes and boxes "set back" from finish surfaces.

(C) Flexible Cord Pendants Covers of outlet boxes and conduit bodies having holes through which flexible cord pendants pass shall be provided with bushings designed for the purpose or shall have smooth, well-rounded surfaces on which the cords may bear. So-called hard rubber or composition bushings shall not be used.

314.27 Outlet Boxes

(A) Boxes at Luminaire (Lighting Fixture) Outlets Boxes used at luminaire (lighting fixture) or lampholder outlets shall be designed for the purpose. At every outlet used exclusively for lighting, the box shall be designed or installed so that a luminaire (lighting fixture) may be attached.

Exception: A wall-mounted luminaire (fixture) weighing not more than 3 kg (6 lb) shall be permitted to be supported on other boxes or plaster rings that are secured to other boxes, provided the luminaire (fixture) or its supporting yoke is secured to the box with no fewer than two No. 6 or larger screws.

Device boxes are designed for the mounting of snap switches, receptacles, and other devices, usually with 6-32 screws (No. 6 screws with 32 threads per inch). Generally, device boxes are not suitable for supporting other than lightweight wall-mounted luminaires. The exception to 314.27(A) permits luminaires such as wall-bracket types or sconces weighing less than 6 lb to be supported by a device box using No. 6 or larger screws. For heavier or ceiling-mounted lighting luminaires, see 410.16, Means of Support. The outlet box is required to provide "adequate" support.

(B) Maximum Luminaire (Fixture) Weight Outlet boxes or fittings installed as required by 314.23 shall be permitted

to support luminaires (lighting fixtures) weighing 23 kg (50 lb) or less. A luminaire (lighting fixture) that weighs more than 23 kg (50 lb) shall be supported independently of the outlet box unless the outlet box is listed for the weight to be supported.

Moved to Chapter 3 in the 2002 *Code*, this requirement previously appeared in 410.16. Regardless of whether a luminaire is attached to an outlet box or is supported independently of the outlet box, care should be taken to securely fasten the supporting means of the luminaire. Exhibit 314.8 illustrates one method of supporting a fixture in accordance with 314.27(B).

Exhibit 314.8 A wooden brace used to support a heavy luminaire according to 314.27(B).

(C) Floor Boxes Boxes listed specifically for this application shall be used for receptacles located in the floor.

Exception: Where the authority having jurisdiction judges them free from likely exposure to physical damage, moisture, and dirt, boxes located in elevated floors of show windows and similar locations shall be permitted to be other than those listed for floor applications. Receptacles and covers shall be listed as an assembly for this type of location.

(D) Boxes at Ceiling-Suspended (Paddle) Fan Outlets Outlet boxes or outlet box systems used as the sole support of a ceiling-suspended (paddle) fan shall be listed, shall be marked by their manufacturer as suitable for this purpose, and shall not support ceiling-suspended (paddle) fans that weigh more than 32 kg (70 lb). For outlet boxes or outlet box systems designed to support ceiling-suspended (paddle) fans that weigh more than 16 kg (35 lb), the required marking shall include the maximum weight to be supported.

Outlet boxes specifically listed to adequately support ceiling-mounted paddle fans are available, as are several alternative and retrofit methods that can provide suitable support for a paddle fan. For the 2005 *Code*, 314.27(D) was rewritten to specifically address listed fan boxes, listed fan box assemblies, as well as fan weights up to 70 pounds.

One method of supporting a ceiling fan so that the box does not serve as the sole support is shown in Exhibit 314.9. For a detailed view of an outlet box specifically listed to support a paddle fan, see 422.18, the associated commentary, and Exhibit 422.2.

Exhibit 314.9 A ceiling fan supported without depending on the box for sole support.

314.28 Pull and Junction Boxes and Conduit Bodies

Boxes and conduit bodies used as pull or junction boxes shall comply with 314.28(A) through (D).

Exception: Terminal housings supplied with motors shall comply with the provisions of 430.12.

(A) Minimum Size For raceways containing conductors of 4 AWG or larger, and for cables containing conductors of 4

AWG or larger, the minimum dimensions of pull or junction boxes installed in a raceway or cable run shall comply with (A)(1) through (A)(3). Where an enclosure dimension is to be calculated based on the diameter of entering raceways, the diameter shall be the metric designator (trade size) expressed in the units of measurement employed.

(1) Straight Pulls In straight pulls, the length of the box shall not be less than eight times the metric designator (trade size) of the largest raceway.

Section 314.28(A)(1) applies to minimum dimensions of pull and junction boxes or conduit bodies used with raceways or cables containing conductors 4 AWG or larger. For straight pulls, for example, trade size 2 conduit containing four 4/0 AWG, Type THHW conductors (see Annex C, Table C8) requires a 16-in.-long pull box (8 × 2 in. = 16 in.). It should be understood that although 16 in. is the required minimum length, a longer pull box may be desired for maximum ease in handling this size conductor.

(2) Angle or U Pulls Where splices or where angle or U pulls are made, the distance between each raceway entry inside the box and the opposite wall of the box shall not be less than six times the metric designator (trade size) of the largest raceway in a row. This distance shall be increased for additional entries by the amount of the sum of the diameters of all other raceway entries in the same row on the same wall of the box. Each row shall be calculated individually, and the single row that provides the maximum distance shall be used.

Exception: Where a raceway or cable entry is in the wall of a box or conduit body opposite a removable cover, the distance from that wall to the cover shall be permitted to comply with the distance required for one wire per terminal in Table 312.6(A).

The distance between raceway entries enclosing the same conductor shall not be less than six times the metric designator (trade size) of the larger raceway.

When transposing cable size into raceway size in 314.28(A)(1) and (A)(2), the minimum metric designator (trade size) raceway required for the number and size of conductors in the cable shall be used.

Exhibits 314.10 and 314.11 provide examples of calculations required by 314.28(A)(2). As the example in Exhibit 314.10 illustrates, where splices, angle pulls, or U pulls are made, the distance between each raceway entry inside the box and the opposite wall of the box must not be less than six times the trade diameter of the largest raceway, plus the distance for additional raceway entries. This additional distance is

6 × 2 in. (trade diameter of largest raceway) = 12 in.
12 in. + 5 in. (sum of diameters of other entries, row 2 only)
  = 17 in. (min. required from each entry to opposite wall)

Exhibit 314.10 An example showing calculations required by 314.28(A)(2) for splices, angle pulls, or U pulls.

$A = (6 \times 4 \text{ in.}) + 2 \text{ in.} + 2 \text{ in.} = 28 \text{ in. min.}$
$B = (6 \times 4 \text{ in.}) + 2 \text{ in.} + 2 \text{ in.} = 28 \text{ in. min.}$
$C = 6 \times 2 \text{ in.} = 12 \text{ in. min.}$ required between raceways enclosing the same conductor
$D = 6 \times 2 \text{ in.} = 12 \text{ in. min.}$ required between raceways enclosing the same conductor
$E = 6 \times 4 \text{ in.} = 24 \text{ in. min.}$ required between raceways enclosing the same conductor

Exhibit 314.11 An example showing calculations required by 314.28(A)(2) for raceways enclosing the same conductor.

calculated by adding the diameters of the other raceway entries in one row on the same side of the box. The example in Exhibit 314.11 shows that raceway entries enclosing the same conductor are required to have a minimum separation between them. The intent is to provide adequate space for the conductor to make the bend.

(3) Smaller Dimensions Boxes or conduit bodies of dimensions less than those required in 314.28(A)(1) and (A)(2) shall be permitted for installations of combinations of conductors that are less than the maximum conduit or tubing fill (of conduits or tubing being used) permitted by Table 1 of Chapter 9, provided the box or conduit body has been listed for, and is permanently marked with, the maximum number and maximum size of conductors permitted.

(B) Conductors in Pull or Junction Boxes In pull boxes or junction boxes having any dimension over 1.8 m (6 ft), all conductors shall be cabled or racked up in an approved manner.

(C) Covers All pull boxes, junction boxes, and conduit bodies shall be provided with covers compatible with the box or conduit body construction and suitable for the conditions of use. Where used, metal covers shall comply with the grounding requirements of 250.110. An extension from the cover of an exposed box shall comply with 314.22, Exception.

(D) Permanent Barriers Where permanent barriers are installed in a box, each section shall be considered as a separate box.

314.29 Boxes, Conduit Bodies, and Handhole Enclosures to Be Accessible

Boxes, conduit bodies, and handhole enclosures shall be installed so that the wiring contained in them can be rendered accessible without removing any part of the building or, in underground circuits, without excavating sidewalks, paving, earth, or other substance that is to be used to establish the finished grade.

Exception: Listed boxes and handhole enclosures shall be permitted where covered by gravel, light aggregate, or non-cohesive granulated soil if their location is effectively identified and accessible for excavation.

Consideration should be given to the accessibility of junction boxes installed on a structural ceiling above a suspended ceiling. A box is permitted to be used at any point for the connection of conduit, tubing, or cable, provided it is not rendered inaccessible. See Article 100 for the definition of *accessible (as applied to wiring methods)*. See 300.15 for other requirements for boxes, conduit bodies, or fittings. For the 2005 *Code*, this section covers handhole enclosures as well.

314.30 Handhole Enclosures

Handhole enclosures shall be designed and installed to withstand all loads likely to be imposed.

> FPN: See ANSI/SCTE 77-2002, *Specification for Underground Enclosure Integrity*, for additional information on deliberate and nondeliberate traffic loading that can be expected to bear on underground enclosures.

(A) Size Handhole enclosures shall be sized in accordance with 314.28(A) for conductors operating at 600 volts or below, and in accordance with 314.71 for conductors operating at over 600 volts. For handhole enclosures without bottoms where the provisions of 314.28(A)(2), Exception, or 314.71(B)(1), Exception No. 1, apply, the measurement to the removable cover shall be taken from the end of the conduit or cable assembly.

(B) Wiring Entries Underground raceways and cable assemblies entering a handhole enclosure shall extend into the enclosure, but they shall not be required to be mechanically connected to the enclosure.

(C) Handhole Enclosures Without Bottoms Where handhole enclosures without bottoms are installed, all enclosed conductors and any splices or terminations, if present, shall be listed as suitable for wet locations.

Due to the possibility of flooding within a subterranean bottomless handhole enclosure, it is required that the conductors and splices be listed for use in a wet location. Although not specifically addressed in 314.30(C), handholes with a bottom (complete enclosure) are also subject to the potential of water infiltration and are not required to comply with 314.15 covering boxes, conduit bodies, and fittings installed in wet locations. Since the water that enters the handhole with a bottom may accumulate, it is necessary to treat the inside of the enclosure as a wet location.

(D) Covers Handhole enclosure covers shall have an identifying mark or logo that prominently identifies the function of the enclosure, such as "electric." Handhole enclosure covers shall require the use of tools to open, or they shall weigh over 45 kg (100 lb). Metal covers and other exposed conductive surfaces shall be bonded in accordance with 250.96(A).

This section is new for the 2005 *Code*. The definition of *handhole enclosure*, found in Article 100, states it is "an enclosure identified for use in underground systems, provided with an open or closed bottom, and sized to allow personnel to reach into, but not enter, for the purpose of

installing, operating, or maintaining equipment or wiring or both. Also see the commentary following 314.1.

III. Construction Specifications

314.40 Metal Boxes, Conduit Bodies, and Fittings

(A) Corrosion Resistant Metal boxes, conduit bodies, and fittings shall be corrosion resistant or shall be well-galvanized, enameled, or otherwise properly coated inside and out to prevent corrosion.

> FPN: See 300.6 for limitation in the use of boxes and fittings protected from corrosion solely by enamel.

(B) Thickness of Metal Sheet steel boxes not over 1650 cm^3 (100 in.^{3}) in size shall be made from steel not less than 1.59 mm (0.0625 in.) thick. The wall of a malleable iron box or conduit body and a die-cast or permanent-mold cast aluminum, brass, bronze, or zinc box or conduit body shall not be less than 2.38 mm ($\frac{3}{32}$ in.) thick. Other cast metal boxes or conduit bodies shall have a wall thickness not less than 3.17 mm ($\frac{1}{8}$ in.).

Exception No. 1: Listed boxes and conduit bodies shown to have equivalent strength and characteristics shall be permitted to be made of thinner or other metals.

Exception No. 2: The walls of listed short radius conduit bodies, as covered in 314.5, shall be permitted to be made of thinner metal.

(C) Metal Boxes Over 1650 cm^3 (100 in.^{3}) Metal boxes over 1650 cm^3 (100 in.^{3}) in size shall be constructed so as to be of ample strength and rigidity. If of sheet steel, the metal thickness shall not be less than 1.35 mm (0.053 in.) uncoated.

(D) Grounding Provisions A means shall be provided in each metal box for the connection of an equipment grounding conductor. The means shall be permitted to be a tapped hole or equivalent.

For device boxes and other standard outlet boxes, the means provided by the box manufacturer is usually in the form of a 10-32 tapped hole marked "GR" or "GRD," or the equivalent, next to the hole. It should be noted, however, that the means provided may not necessarily be used.

314.41 Covers

Metal covers shall be of the same material as the box or conduit body with which they are used, or they shall be lined with firmly attached insulating material that is not less than 0.79 mm ($\frac{1}{32}$ in.) thick, or they shall be listed for the purpose. Metal covers shall be the same thickness as the boxes or conduit body for which they are used, or they shall be listed for the purpose. Covers of porcelain or other approved insulating materials shall be permitted if of such form and thickness as to afford the required protection and strength.

314.42 Bushings

Covers of outlet boxes and conduit bodies having holes through which flexible cord pendants may pass shall be provided with approved bushings or shall have smooth, well-rounded surfaces on which the cord may bear. Where individual conductors pass through a metal cover, a separate hole equipped with a bushing of suitable insulating material shall be provided for each conductor. Such separate holes shall be connected by a slot as required by 300.20.

314.43 Nonmetallic Boxes

Provisions for supports or other mounting means for nonmetallic boxes shall be outside of the box, or the box shall be constructed so as to prevent contact between the conductors in the box and the supporting screws.

314.44 Marking

All boxes and conduit bodies, covers, extension rings, plaster rings, and the like shall be durably and legibly marked with the manufacturer's name or trademark.

IV. Pull and Junction Boxes for Use on Systems Over 600 Volts, Nominal

314.70 General

Where pull and junction boxes are used on systems over 600 volts, the installation shall comply with the provisions of Part IV and also with the following general provisions of this article:

(1) Part I, 314.2, 314.3, and 314.4
(2) Part II, 314.15; 314.17; 314.20; 314.23(A), (B), or (G); 314.28(B); and 314.29
(3) Part III, 314.40(A) and (C) and 314.41

314.71 Size of Pull and Junction Boxes

Pull and junction boxes shall provide adequate space and dimensions for the installation of conductors, and they shall comply with the specific requirements of this section.

Exception: Terminal housings supplied with motors shall comply with the provisions of 430.12.

(A) For Straight Pulls The length of the box shall not be less than 48 times the outside diameter, over sheath, of the largest shielded or lead-covered conductor or cable entering the box. The length shall not be less than 32 times the outside diameter of the largest nonshielded conductor or cable.

(B) For Angle or U Pulls

(1) Distance to Opposite Wall The distance between each cable or conductor entry inside the box and the opposite wall of the box shall not be less than 36 times the outside diameter, over sheath, of the largest cable or conductor.

This distance shall be increased for additional entries by the amount of the sum of the outside diameters, over sheath, of all other cables or conductor entries through the same wall of the box.

Exception No. 1: Where a conductor or cable entry is in the wall of a box opposite a removable cover, the distance from that wall to the cover shall be permitted to be not less than the bending radius for the conductors as provided in 300.34.

Exception No. 2: Where cables are nonshielded and not lead covered, the distance of 36 times the outside diameter shall be permitted to be reduced to 24 times the outside diameter.

(2) Distance Between Entry and Exit The distance between a cable or conductor entry and its exit from the box shall not be less than 36 times the outside diameter, over sheath, of that cable or conductor.

Exception: Where cables are nonshielded and not lead covered, the distance of 36 times the outside diameter shall be permitted to be reduced to 24 times the outside diameter.

(C) Removable Sides One or more sides of any pull box shall be removable.

314.72 Construction and Installation Requirements

(A) Corrosion Protection Boxes shall be made of material inherently resistant to corrosion or shall be suitably protected, both internally and externally, by enameling, galvanizing, plating, or other means.

(B) Passing Through Partitions Suitable bushings, shields, or fittings having smooth, rounded edges shall be provided where conductors or cables pass through partitions and at other locations where necessary.

(C) Complete Enclosure Boxes shall provide a complete enclosure for the contained conductors or cables.

(D) Wiring Is Accessible Boxes shall be installed so that the wiring is accessible without removing any part of the building. Working space shall be provided in accordance with 110.34.

(E) Suitable Covers Boxes shall be closed by suitable covers securely fastened in place. Underground box covers that weigh over 45 kg (100 lb) shall be considered meeting this requirement. Covers for boxes shall be permanently marked "DANGER — HIGH VOLTAGE — KEEP OUT." The marking shall be on the outside of the box cover and shall be readily visible. Letters shall be block type and at least 13 mm (½ in.) in height.

(F) Suitable for Expected Handling Boxes and their covers shall be capable of withstanding the handling to which they are likely to be subjected.

ARTICLE 320
Armored Cable: Type AC

Summary of Changes

• **320.10:** Added FPN to clarify that the uses permitted list is not intended to include all of the permitted uses of Type AC cable.

• **320.12:** Revised to clarify where AC cable is not permitted to be used, specifically identifying locations where the cable is subject to physical damage, where the cable is subject to excessive moisture or dampness, or where the location is determined to be damp or wet.

• **320.30:** Revised to reorganize supporting and securing requirements. Also clarified permitted installation of cable in up to a 6-ft unsupported length from the last point of support to the connection to a luminaire(s) or other equipment in an accessible ceiling. Added text to permit Type AC cable fittings to be considered as the last means of support.

• **320.80:** Revised to permit 90°C conductor ampacity to be used for ampacity adjustment purposes.

Contents

I. General

320.1 Scope

This article covers the use, installation, and construction specifications for armored cable, Type AC.

Type AC (armored) cable is listed by Underwriters Laboratories in sizes 14 AWG through 1 AWG copper and 12 AWG through 1 AWG aluminum or copper-clad aluminum and is rated at 600 volts or less. However, the *NEC* does not specifically require that Type AC cable be listed. Exhibit 320.1 shows an example of Type AC cable.

Exhibit 320.1 An example of Type AC cable. (Courtesy of AFC Cable Systems, Inc.)

320.2 Definition

Armored Cable, Type AC. A fabricated assembly of insulated conductors in a flexible metallic enclosure. See 320.100.

II. Installation

320.10 Uses Permitted

Type AC cable shall be permitted as follows:

(1) In both exposed and concealed work
(2) In cable trays
(3) In dry locations
(4) Embedded in plaster finish on brick or other masonry, except in damp or wet locations

(5) To be run or fished in the air voids of masonry block or tile walls where such walls are not exposed or subject to excessive moisture or dampness

FPN: The "Uses Permitted" is not an all-inclusive list.

An FPN was added for the 2005 *Code* to clarify that the list of *uses permitted* is not necessarily a complete list.

320.12 Uses Not Permitted

Type AC cable shall not be used as follows:

(1) Where subject to physical damage
(2) In damp or wet locations
(3) In air voids of masonry block or tile walls where such walls are exposed or subject to excessive moisture or dampness
(4) Where exposed to corrosive fumes or vapors
(5) Embedded in plaster finish on brick or other masonry in damp or wet locations

The list of uses not permitted has been changed for the 2005 *Code*, clarifying that Type AC cable is not permitted to be used where it is subject to physical damage and not permitted to be used in damp or wet locations. There are no exceptions to this list.

320.15 Exposed Work

Exposed runs of cable, except as provided in 300.11(A), shall closely follow the surface of the building finish or of running boards. Exposed runs shall also be permitted to be installed on the underside of joists where supported at each joist and located so as not to be subject to physical damage.

320.17 Through or Parallel to Framing Members

Type AC cable shall be protected in accordance with 300.4(A), (C), and (D) where installed through or parallel to framing members.

320.23 In Accessible Attics

Type AC cables in accessible attics or roof spaces shall be installed as specified in 320.23(A) and (B).

(A) Where Run Across the Top of Floor Joists Where run across the top of floor joists, or within 2.1 m (7 ft) of floor or floor joists across the face of rafters or studding, in attics and roof spaces that are accessible, the cable shall be protected by substantial guard strips that are at least as high as the cable. Where this space is not accessible by permanent stairs or ladders, protection shall only be required within

1.8 m (6 ft) of the nearest edge of the scuttle hole or attic entrance.

In accessible attics, Type AC cable installed across the top of floor joists or within 7 ft of the floor or floor joists across the face of rafters or studs must be protected by guard strips. Where the attic is not accessible by a permanent ladder or stairs, guard strips are required only within 6 ft of the scuttle hole or opening.

(B) Cable Installed Parallel to Framing Members Where the cable is installed parallel to the sides of rafters, studs, or floor joists, neither guard strips nor running boards shall be required, and the installation shall also comply with 300.4(D).

320.24 Bending Radius

Bends in Type AC cable shall be made such that the cable is not damaged. The radius of the curve of the inner edge of any bend shall not be less than five times the diameter of the Type AC cable.

320.30 Securing and Supporting

(A) General Type AC cable shall be supported and secured by staples, cable ties, straps, hangers, or similar fittings, designed and installed so as not to damage the cable.

Section 320.30 requires that Type AC cable be secured. Simply draping the cable over air ducts or lower members of bar joists, pipes, and ceiling grid members is not permitted.

(B) Securing Unless otherwise provided, Type AC cable shall be secured within 300 mm (12 in.) of every outlet box, junction box, cabinet, or fitting and at intervals not exceeding 1.4 m (4½ ft) where installed on or across framing members.

(C) Supporting Unless otherwise provided, Type AC cable shall be supported at intervals not exceeding 1.4 m (4½ ft).

Horizontal runs of Type AC cable installed in wooden or metal framing members or similar supporting means shall be considered supported where such support does not exceed 1.4-m (4½-ft) intervals.

Section 320.30(C) permits Type AC cable, where run horizontally through framing members, to be passed through bored or punched holes in framing members without additional securing, provided the cable is secured within 12 in. of the outlet and the framing members are less than 54 in. apart.

(D) Unsupported Cables Type AC cable shall be permitted to be unsupported where the cable complies with any of the following:

(1) Is fished between access points through concealed spaces in finished buildings or structures and supporting is impracticable

(2) Is not more than 600 mm (2 ft) in length at terminals where flexibility is necessary

(3) Is not more than 1.8 m (6 ft) in length from the last point of cable support to the point of connection to a luminaire(s) [lighting fixture(s)] or other electrical equipment and the cable and point of connection are within an accessible ceiling. For the purposes of this section, Type AC cable fittings shall be permitted as a means of cable support.

Section 320.30(D)(3) has been editorially revised for the 2005 *Code*. This section now points out that these permitted runs of Type AC cable (in lengths up to 6 ft) must be in an accessible ceiling.

320.40 Boxes and Fittings

At all points where the armor of AC cable terminates, a fitting shall be provided to protect wires from abrasion, unless the design of the outlet boxes or fittings is such as to afford equivalent protection, and, in addition, an insulating bushing or its equivalent protection shall be provided between the conductors and the armor. The connector or clamp by which the Type AC cable is fastened to boxes or cabinets shall be of such design that the insulating bushing or its equivalent will be visible for inspection. Where change is made from Type AC cable to other cable or raceway wiring methods, a box, fitting, or conduit body shall be installed at junction points as required in 300.15.

Armored cable connectors are considered suitable for equipment grounding if installed in accordance with 300.10.

320.80 Ampacity

The ampacity shall be determined by 310.15.

(A) Thermal Insulation Armored cable installed in thermal insulation shall have conductors rated at 90°C (194°F). The ampacity of cable installed in these applications shall be that of 60°C (140°F) conductors. The 90°C (194°F) rating shall be permitted to be used for ampacity derating purposes, provided the final derated ampacity does not exceed that for a 60°C (140°F) rated conductor.

The requirements for armored cable installed in thermal insulation recognize the decrease in heat dissipation capability of cables. New for the 2005 *Code*, the last sentence brings this section into harmony with 334.80 for Type NM cable and clarifies that the 90°C degree insulation may be used for derating purposes.

Cable marked "ACTH" indicates an armored cable rated 75°C and employing conductors having thermoplastic insulation. Cable marked "ACTHH" indicates an armored cable rated 90°C and employing conductors having thermoplastic insulation. Cable marked "ACHH" indicates armored cable rated 90°C and employing conductors having thermosetting insulation.

(B) Cable Tray The ampacity of Type AC cable installed in cable tray shall be determined in accordance with 392.11.

III. Construction Specifications

320.100 Construction

Type AC cable shall have an armor of flexible metal tape and shall have an internal bonding strip of copper or aluminum in intimate contact with the armor for its entire length.

The armor of Type AC cable is recognized as an equipment grounding conductor by 250.118. The required internal bonding strip can be simply cut off at the termination of the armored cable, or it can be bent back on the armor. It is not necessary to connect it to an equipment grounding terminal. It reduces the inductive reactance of the spiral armor and increases the armor's effectiveness as an equipment ground. Many installers use this strip to help prevent the insulating bushing required by 320.40 (the "red head") from falling out during rough wiring.

320.104 Conductors

Insulated conductors shall be of a type listed in Table 310.13 or those identified for use in this cable. In addition, the conductors shall have an overall moisture-resistant and fire-retardant fibrous covering. For Type ACT, a moisture-resistant fibrous covering shall be required only on the individual conductors.

320.108 Equipment Grounding

Type AC cable shall provide an adequate path for equipment grounding as required by 250.4(A)(5) or 250.4(B)(4).

320.120 Marking

The cable shall be marked in accordance with 310.11, except that Type AC shall have ready identification of the manufacturer by distinctive external markings on the cable sheath throughout its entire length.

ARTICLE 322
Flat Cable Assemblies: Type FC

Contents

Exhibit 322.1 Basic components and accessories used for an installation of Type FC cable assembly. (Redrawn from The Wiremold Co.)

I. General

322.1 Scope

This article covers the use, installation, and construction specifications for flat cable assemblies, Type FC.

Type FC (flat) cable is an assembly of three or four parallel 10 AWG special stranded copper wires formed integrally with an insulating material web. The cable is marked with the size of the maximum branch circuit to which it may be connected, the cable type designation, manufacturer's identification, maximum working voltage, conductor size, and temperature rating. A marking accompanying the cable on a tag or reel indicates the special metal raceways and specific FC cable fittings with which the cable is intended to be used. Exhibits 322.1 and 322.2 show the basic components of this wiring method.

322.2 Definition

Flat Cable Assembly, Type FC. An assembly of parallel conductors formed integrally with an insulating material web

specifically designed for field installation in surface metal raceway.

II. Installation

322.10 Uses Permitted

Flat cable assemblies shall be permitted only as follows:

(1) As branch circuits to supply suitable tap devices for lighting, small appliances, or small power loads. The rating of the branch circuit shall not exceed 30 amperes.
(2) Where installed for exposed work.
(3) In locations where they will not be subjected to physical damage. Where a flat cable assembly is installed less than 2.5 m (8 ft) above the floor or fixed working platform, it shall be protected by a cover identified for the use.

Exhibit 322.2 A fixture hanger used with Type FC cable assembly. (Courtesy of The Wiremold Co.)

(4) In surface metal raceways identified for the use. The channel portion of the surface metal raceway systems shall be installed as complete systems before the flat cable assemblies are pulled into the raceways.

322.12 Uses Not Permitted

Flat cable assemblies shall not be used as follows:

(1) Where subject to corrosive vapors unless suitable for the application
(2) In hoistways or on elevators or escalators
(3) In any hazardous (classified) location
(4) Outdoors or in wet or damp locations unless identified for the use

322.30 Securing and Supporting

The flat cable assemblies shall be supported by means of their special design features, within the surface metal raceways.

The surface metal raceways shall be supported as required for the specific raceway to be installed.

322.40 Boxes and Fittings

(A) Dead Ends Each flat cable assembly dead end shall be terminated in an end-cap device identified for the use.

The dead-end fitting for the enclosing surface metal raceway shall be identified for the use.

(B) Luminaire (Fixture) Hangers Luminaire (fixture) hangers installed with the flat cable assemblies shall be identified for the use.

(C) Fittings Fittings to be installed with flat cable assemblies shall be designed and installed to prevent physical damage to the cable assemblies.

(D) Extensions All extensions from flat cable assemblies shall be made by approved wiring methods, within the junction boxes, installed at either end of the flat cable assembly runs.

322.56 Splices and Taps

(A) Splices Splices shall be made in listed junction boxes.

(B) Taps Taps shall be made between any phase conductor and the grounded conductor or any other phase conductor by means of devices and fittings identified for the use. Tap devices shall be rated at not less than 15 amperes, or more than 300 volts to ground, and shall be color-coded in accordance with the requirements of 322.120(C).

III. Construction

322.100 Construction

Flat cable assemblies shall consist of two, three, four, or five conductors.

322.104 Conductors

Flat cable assemblies shall have conductors of 10 AWG special stranded copper wires.

322.112 Insulation

The entire flat cable assembly shall be formed to provide a suitable insulation covering all the conductors and using one of the materials recognized in Table 310.13 for general branch-circuit wiring.

322.120 Marking

(A) Temperature Rating In addition to the provisions of 310.11, Type FC cable shall have the temperature rating durably marked on the surface at intervals not exceeding 600 mm (24 in.).

(B) Identification of Grounded Conductor The grounded conductor shall be identified throughout its length by means of a distinctive and durable white or gray marking.

> FPN: The color gray may have been used in the past as an ungrounded conductor. Care should be taken when working on existing systems.

(C) Terminal Block Identification Terminal blocks identified for the use shall have distinctive and durable markings for color or word coding. The grounded conductor section shall have a white marking or other suitable designation.

The next adjacent section of the terminal block shall have a black marking or other suitable designation. The next section shall have a red marking or other suitable designation. The final or outer section, opposite the grounded conductor section of the terminal block, shall have a blue marking or other suitable designation.

ARTICLE 324
Flat Conductor Cable: Type FCC

Summary of Changes

• **324.6:** New requirement that Type FCC cable be listed.

• **324.101:** New requirement for corrosion resistance of metal components.

Contents

I. General

324.1 Scope

This article covers a field-installed wiring system for branch circuits incorporating Type FCC cable and associated accessories as defined by the article. The wiring system is designed for installation under carpet squares.

The FCC (flat conductor cable) system is designed to provide a completely accessible, flexible power system. As shown in Exhibit 324.1, it also provides an easy method for reworking obsolete wiring systems currently in use in many office facilities. The carpet squares are not permitted to be larger than 36 in. × 36 in. to comply with 324.10(H). This limitation is judged necessary to provide ready access to the cable by lifting a carpet square. It also reduces the likelihood of damage to the cable by an individual cutting through the carpet above the cable with a knife or razor blade and possibly penetrating the top shield of the cable in the process.

Exhibit 324.1 Type FCC cable installed beneath carpet squares. (Courtesy of Tyco Electronics Corp.)

324.2 Definitions

Bottom Shield. A protective layer that is installed between the floor and Type FCC flat conductor cable to protect the cable from physical damage and may or may not be incorporated as an integral part of the cable.

Cable Connector. A connector designed to join Type FCC cables without using a junction box.

FCC System. A complete wiring system for branch circuits that is designed for installation under carpet squares. The FCC system includes Type FCC cable and associated shielding, connectors, terminators, adapters, boxes, and receptacles.

Insulating End. An insulator designed to electrically insulate the end of a Type FCC cable.

Metal Shield Connections. Means of connection designed to electrically and mechanically connect a metal shield to another metal shield, to a receptacle housing or self-contained device, or to a transition assembly.

Top Shield. A grounded metal shield covering under-carpet components of the FCC system for the purposes of providing protection against physical damage.

Transition Assembly. An assembly to facilitate connection of the FCC system to other wiring systems, incorporating (1) a means of electrical interconnection and (2) a suitable box or covering for providing electrical safety and protection against physical damage.

Type FCC Cable. Three or more flat copper conductors placed edge-to-edge and separated and enclosed within an insulating assembly.

324.6 Listing Requirements

Type FCC cable and associated fittings shall be listed.

This listing requirement for Type FCC cable and fittings is new for the 2005 *Code*.

II. Installation

324.10 Uses Permitted

(A) Branch Circuits Use of FCC systems shall be permitted both for general-purpose and appliance branch circuits and for individual branch circuits.

(B) Branch-Circuit Ratings

(1) Voltage Voltage between ungrounded conductors shall not exceed 300 volts. Voltage between ungrounded conductors and the grounded conductor shall not exceed 150 volts.

(2) Current General-purpose and appliance branch circuits shall have ratings not exceeding 20 amperes. Individual branch circuits shall have ratings not exceeding 30 amperes.

(C) Floors Use of FCC systems shall be permitted on hard, sound, smooth, continuous floor surfaces made of concrete, ceramic, or composition flooring, wood, and similar materials.

(D) Walls Use of FCC systems shall be permitted on wall surfaces in surface metal raceways.

(E) Damp Locations Use of FCC systems in damp locations shall be permitted.

(F) Heated Floors Materials used for floors heated in excess of 30°C (86°F) shall be identified as suitable for use at these temperatures.

(G) System Height Any portion of an FCC system with a height above floor level exceeding 2.3 mm (0.090 in.) shall be tapered or feathered at the edges to floor level.

(H) Coverings Floor-mounted Type FCC cable, cable connectors, and insulating ends shall be covered with carpet squares not larger than 914 mm (36 in.) square. Carpet squares that are adhered to the floor shall be attached with release-type adhesives.

(I) Corrosion Resistance Metal components of the system shall be either corrosion resistant, coated with corrosion-resistant materials, or insulated from contact with corrosive substances.

(J) Metal-Shield Connectors Metal shields shall be connected to each other and to boxes, receptacle housings, self-contained devices, and transition assemblies using metal-shield connectors.

324.12 Uses Not Permitted

FCC systems shall not be used in the following locations:

(1) Outdoors or in wet locations
(2) Where subject to corrosive vapors
(3) In any hazardous (classified) location
(4) In residential, school, and hospital buildings

Section 324.12(4) prohibits Type FCC wiring systems throughout school and hospital buildings, even though parts of these buildings may be office or administrative spaces.

324.18 Crossings

Crossings of more than two Type FCC cable runs shall not be permitted at any one point. Crossings of a Type FCC cable over or under a flat communications or signal cable shall be permitted. In each case, a grounded layer of metal shielding shall separate the two cables, and crossings of more than two flat cables shall not be permitted at any one point.

324.30 Securing and Supporting

All FCC system components shall be firmly anchored to the floor or wall using an adhesive or mechanical anchoring system identified for this use. Floors shall be prepared to ensure adherence of the FCC system to the floor until the carpet squares are placed.

324.40 Boxes and Fittings

(A) Cable Connections and Insulating Ends All Type FCC cable connections shall use connectors identified for their use, installed such that electrical continuity, insulation, and sealing against dampness and liquid spillage are provided. All bare cable ends shall be insulated and sealed against dampness and liquid spillage using listed insulating ends.

(B) Polarization of Connections All receptacles and connections shall be constructed and installed so as to maintain proper polarization of the system.

(C) Shields

(1) Top Shield A metal top shield shall be installed over all floor-mounted Type FCC cable, connectors, and insulating ends. The top shield shall completely cover all cable runs, corners, connectors, and ends.

(2) Bottom Shield A bottom shield shall be installed beneath all Type FCC cable, connectors, and insulating ends.

(D) Connection to Other Systems Power feed, grounding connection, and shield system connection between the FCC system and other wiring systems shall be accomplished in a transition assembly identified for this use.

(E) Metal-Shield Connectors Metal shields shall be connected to each other and to boxes, receptacle housings, self-contained devices, and transition assemblies using metal-shield connectors.

324.41 Floor Coverings

Floor-mounted Type FCC cable, cable connectors, and insulating ends shall be covered with carpet squares not larger than 914 mm (36 in.) square. Carpet squares that are adhered to the floor shall be attached with release-type adhesives.

324.42 Devices

(A) Receptacles All receptacles, receptacle housings, and self-contained devices used with the FCC system shall be identified for this use and shall be connected to the Type FCC cable and metal shields. Connection from any grounding conductor of the Type FCC cable shall be made to the shield system at each receptacle.

(B) Receptacles and Housings Receptacle housings and self-contained devices designed either for floor mounting or

for in-wall or on-wall mounting shall be permitted for use with the FCC system. Receptacle housings and self-contained devices shall incorporate means for facilitating entry and termination of Type FCC cable and for electrically connecting the housing or device with the metal shield. Receptacles and self-contained devices shall comply with 406.3. Power and communications outlets installed together in common housing shall be permitted in accordance with 800.133(A)(1)(c), Exception No. 2.

324.56 Splices and Taps

(A) FCC Systems Alterations Alterations to FCC systems shall be permitted. New cable connectors shall be used at new connection points to make alterations. It shall be permitted to leave unused cable runs and associated cable connectors in place and energized. All cable ends shall be covered with insulating ends.

(B) Transition Assemblies All transition assemblies shall be identified for their use. Each assembly shall incorporate means for facilitating entry of the Type FCC cable into the assembly, for connecting the Type FCC cable to grounded conductors, and for electrically connecting the assembly to the metal cable shields and to equipment grounding conductors.

324.60 Grounding

All metal shields, boxes, receptacle housings, and self-contained devices shall be electrically continuous to the equipment grounding conductor of the supplying branch circuit. All such electrical connections shall be made with connectors identified for this use. The electrical resistivity of such shield system shall not be more than that of one conductor of the Type FCC cable used in the installation.

III. Construction

324.100 Construction

(A) Type FCC Cable Type FCC cable shall be listed for use with the FCC system and shall consist of three, four, or five flat copper conductors, one of which shall be an equipment grounding conductor.

Section 324.100 requires all FCC cables to be listed. There was no listing requirement for FCC cable prior to the 1996 *Code*.

(B) Shields

(1) Materials and Dimensions All top and bottom shields shall be of designs and materials identified for their use. Top shields shall be metal. Both metallic and nonmetallic materials shall be permitted for bottom shields.

(2) Resistivity Metal shields shall have cross-sectional areas that provide for electrical resistivity of not more than that of one conductor of the Type FCC cable used in the installation.

324.101 Corrosion Resistance

Metal components of the system shall be either corrosion resistant, coated with corrosion-resistant materials, or insulated from contact with corrosive substances.

324.112 Insulation

The insulating material of the cable shall be moisture resistant and flame retardant. All insulating materials in the FCC systems shall be identified for their use.

324.120 Markings

(A) Cable Marking Type FCC cable shall be clearly and durably marked on both sides at intervals of not more than 610 mm (24 in.) with the information required by 310.11(A) and with the following additional information:

(1) Material of conductors
(2) Maximum temperature rating
(3) Ampacity

(B) Conductor Identification Conductors shall be clearly and durably identified on both sides throughout their length as specified in 310.12.

ARTICLE 326
Integrated Gas Spacer Cable: Type IGS

Contents

I. General

326.1 Scope

This article covers the use, installation, and construction specifications for integrated gas spacer cable, Type IGS.

As illustrated in Exhibit 326.1, Type IGS (integrated gas spacer) cable consists of solid aluminum rod conductors, 250-kcmil minimum size. These conductors are insulated with dry kraft paper and are factory installed in a medium-density polyethylene gas pipe, minimum trade size 2, which is then filled with sulfur hexafluoride (SF_6) gas at a pressure of approximately 20 psi.

Exhibit 326.1 Cross section of single-conductor, 4750-kcmil Type IGS cable.

326.2 Definition

Integrated Gas Spacer Cable, Type IGS. A factory assembly of one or more conductors, each individually insulated and enclosed in a loose fit, nonmetallic flexible conduit as an integrated gas spacer cable rated 0 through 600 volts.

II. Installation

326.10 Uses Permitted

Type IGS cable shall be permitted for use under ground, including direct burial in the earth, as the following:

(1) Service-entrance conductors
(2) Feeder or branch-circuit conductors

326.12 Uses Not Permitted

Type IGS cable shall not be used as interior wiring or be exposed in contact with buildings.

326.24 Bending Radius

Where the coilable nonmetallic conduit and cable is bent for installation purposes or is flexed or bent during shipment or installation, the radii of bends measured to the inside of the bend shall not be less than specified in Table 326.24.

Table 326.24 Minimum Radii of Bends

Conduit Size		Minimum Radii	
Metric Designator	Trade Size	mm	in.
53	2	600	24
78	3	900	35
103	4	1150	45

326.26 Bends

A run of Type IGS cable between pull boxes or terminations shall not contain more than the equivalent of four quarter bends (360 degrees total), including those bends located immediately at the pull box or terminations.

326.40 Fittings

Terminations and splices for Type IGS cable shall be identified as a type that is suitable for maintaining the gas pressure within the conduit. A valve and cap shall be provided for each length of the cable and conduit to check the gas pressure or to inject gas into the conduit.

326.80 Ampacity

The ampacity of Type IGS cable shall not exceed the values shown in Table 326.80.

Table 326.80 Ampacity of Type IGS Cable

Size (kcmil)	Amperes	Size (kcmil)	Amperes
250	119	2500	376
500	168	3000	412
750	206	3250	429
1000	238	3500	445
1250	266	3750	461
1500	292	4000	476
1750	315	4250	491
2000	336	4500	505
2250	357	4750	519

III. Construction Specifications

326.104 Conductors

The conductors shall be solid aluminum rods, laid parallel, consisting of one to nineteen 12.7 mm (½ in.) diameter rods. The minimum conductor size shall be 250 kcmil, and the maximum size shall be 4750 kcmil.

326.112 Insulation

The insulation shall be dry kraft paper tapes and a pressurized sulfur hexafluoride gas (SF_6), both approved for electrical use. The nominal gas pressure shall be 138 kPa gauge (20 pounds per square inch gauge). The thickness of the paper spacer shall be as specified in Table 326.112.

Table 326.112 Paper Spacer Thickness

Size (kcmil)	Thickness	
	mm	in.
250–1000	1.02	0.040
1250–4750	1.52	0.060

326.116 Conduit

The conduit shall be a medium density polyethylene identified as suitable for use with natural gas rated pipe in metric designator 53, 78, or 103 (trade size 2, 3, or 4). The percent fill dimensions for the conduit are shown in Table 326.116.

Table 326.116 Conduit Dimensions

Conduit Size		Actual Outside Diameter		Actual Inside Diameter	
Metric Designator	Trade Size	mm	in.	mm	in.
53	2	60	2.375	49.46	1.947
78	3	89	3.500	73.30	2.886
103	4	114	4.500	94.23	3.710

The size of the conduit permitted for each conductor size shall be calculated for a percent fill not to exceed those found in Table 1, Chapter 9.

326.120 Marking

The cable shall be marked in accordance with 310.11(A), 310.11(B)(1), and 310.11(D).

ARTICLE 328
Medium Voltage Cable: Type MV

Summary of Changes

- **328.10:** Added FPN to clarify that the uses permitted list is not intended to include all of the permitted uses of Type MV cable.

- **328.12:** Revised to permit cable tray use if specified in 392(B)(2), and to add direct burial unless in accordance with 300.50.

Contents

I. General

328.1 Scope

This article covers the use, installation, and construction specifications for medium voltage cable, Type MV.

Type MV (medium voltage) cables are rated 2001 to 35,000 volts. Cables rated 2001 to 8000 volts may be shielded or nonshielded. All insulated conductors 8001 volts and higher have electrostatic shielding. When non-shielded cables are used, they must comply with the exception to 310.6. If these cables are installed in underground conduits, they must be encased in 3 in. of concrete to comply with 300.50(A)(2).

328.2 Definition

Medium Voltage Cable, Type MV. A single or multiconductor solid dielectric insulated cable rated 2001 volts or higher.

II. Installation

328.10 Uses Permitted

Type MV cable shall be permitted for use on power systems rated up to 35,000 volts nominal as follows:

(1) In wet or dry locations
(2) In raceways
(3) In cable trays as specified in 392.3(B)(2)
(4) Direct buried in accordance with 300.50
(5) In messenger-supported wiring

    FPN: The "Uses Permitted" is not an all-inclusive list.

328.12 Uses Not Permitted

Unless identified for the use, Type MV cable shall not be used as follows:

(1) Where exposed to direct sunlight
(2) In cable trays, unless specified in 392.3(B)(2)
(3) Direct buried, unless in accordance with 300.50

Type MV cables intended for installation in cable trays in accordance with Article 392 are marked "For CT Use" or "For Use in Cable Trays."

328.80 Ampacity

The ampacity of Type MV cable shall be determined in accordance with 310.60. The ampacity of Type MV cable installed in cable tray shall be determined in accordance with 392.13.

III. Construction Specifications

328.100 Construction

Type MV cables shall have copper, aluminum, or copper-clad aluminum conductors and shall comply with Table 310.61 and Tables 310.63 or 310.64.

Cables with aluminum conductors are marked with the word *Aluminum* or the letters *AL*.

328.120 Marking

Medium voltage cable shall be marked as required by 310.11.

Cables are marked with their conductor size, voltage rating, and insulation level (100 percent or 133 percent).

ARTICLE 330
Metal-Clad Cable: Type MC

Summary of Changes

- **330.10:** Added new FPN to clarify that the uses permitted list is not intended to include all of the permitted uses of Type MC cable.

- **330.12:** Revised to add *where subject to physical damage* as a use that is not permitted. New fine print note added to indicate that Type MC cable identified for direct burial is suitable for installation in concrete.

- **330.30:** Revised to reorganize supporting and securing requirements. Also clarified permitted installation of cable in up to a 6-ft unsupported length from the last point of support to the connection to a luminaire(s) or other equipment in an accessible ceiling. Added text indicating that Type MC cable fittings can be considered as the last means of support.

Contents

I. General

330.1 Scope

This article covers the use, installation, and construction specifications of metal-clad cable, Type MC.

The basic standard to investigate products in this category is UL 1569, *Standard for Metal-Clad Cables*. Summary information regarding listed metal-clad cable may be found in the UL *General Information for Electrical Equipment Directory* under category PJAZ.

330.2 Definition

Metal Clad Cable, Type MC. A factory assembly of one or more insulated circuit conductors with or without optical fiber members enclosed in an armor of interlocking metal tape, or a smooth or corrugated metallic sheath.

Type MC (metal-clad) cable is rated for use up to 2000 volts and is listed in sizes 18 AWG and larger for copper and 12 AWG and larger for aluminum or copper-clad aluminum, and it employs thermoset or thermoplastic insulated conductors. Composite electrical MC and optical fiber cables are permitted by 770.9(C) and are marked "MC-OF." See Exhibit 770.2 for an example.

Type MC cable must be marked with the maximum rated voltage, the proper insulation type letter or letters, and the AWG size or circular mil area. This marking may be on a marker tape located within the cable and running its complete length, or, if the metallic covering is of smooth construction that permits surface marking, the MC cable may be durably marked on the outer covering at intervals not exceeding 24 in. For MC cable with an outer nonmetallic covering, the marking is permitted on the surface of the nonmetallic jacket. See 310.11 for marking requirements.

Type MC cable is available in three designs: interlocked metal tape, corrugated metal tube, and smooth metal tube. A nonmetallic jacket may be provided over the metal sheath. Cables suitable for use in cable trays, direct sunlight, or direct burial applications are so marked. Type MC cable includes Type CS (copper sheath) and Type ALS (aluminum sheath). Exhibit 330.1 shows some examples of Type MC cable.

Exhibit 330.1 Examples of Type MC cable. (Courtesy of AFC Cable Systems, Inc.)

II. Installation

330.10 Uses Permitted

(A) General Uses Type MC cable shall be permitted as follows:

(1) For services, feeders, and branch circuits
(2) For power, lighting, control, and signal circuits
(3) Indoors or outdoors
(4) Exposed or concealed
(5) To be direct buried where identified for such use
(6) In cable tray where identified for such use
(7) In any raceway
(8) As aerial cable on a messenger
(9) In hazardous (classified) locations as permitted
(10) In dry locations and embedded in plaster finish on brick or other masonry except in damp or wet locations
(11) In wet locations where any of the following conditions are met:
    a. The metallic covering is impervious to moisture.
    b. A lead sheath or moisture-impervious jacket is provided under the metal covering.
    c. The insulated conductors under the metallic covering are listed for use in wet locations.
(12) Where single-conductor cables are used, all phase conductors and, where used, the neutral conductor shall be grouped together to minimize induced voltage on the sheath.

Single-conductor Type MC cable is permitted to be used according to 330.10(A)(12). This single-conductor wiring method was introduced in the 2002 *NEC*. Installation practice for single-conductor wiring methods dictates close circuit conductor spacing not only to minimize induced voltage on the sheath, but also to minimize overall circuit impedance.

(B) Specific Uses Type MC cable shall be permitted to be installed in compliance with Parts II and III of Article 725 and 770.133 as applicable and in accordance with 330.10(B)(1) through (B)(4).

(1) Cable Tray Type MC cable installed in cable tray shall comply with 392.3, 392.4, 392.6, and 392.8 through 392.13.

(2) Direct Buried Direct-buried cable shall comply with 300.5 or 300.50, as appropriate.

(3) Installed as Service-Entrance Cable Type MC cable installed as service-entrance cable shall be permitted in accordance with 230.43.

(4) Installed Outside of Buildings or as Aerial Cable Type MC cable installed outside of buildings or as aerial cable shall comply with 225.10, 396.10, and 396.12.

    FPN: The "Uses Permitted" is not an all-inclusive list.

An FPN was added for the 2005 *Code* clarifying that the list of *uses permitted* is not necessarily a complete list. Editorial changes for the 2005 *Code* stating more specific cross references were also made in 330.10(B).

330.12 Uses Not Permitted

Type MC cable shall not be used where exposed to the following destructive corrosive conditions, unless the metallic sheath is suitable for the conditions or is protected by material suitable for the conditions:

(1) Where subject to physical damage
(2) Direct burial in the earth
(3) In concrete
    FPN to (3): MC cable that is identified for direct burial applications is suitable for installation in concrete.
(4) Where subject to cinder fills, strong chlorides, caustic alkalis, or vapors of chlorine or of hydrochloric acids

The list of uses not permitted was changed for the 2005 *Code*. One such change clarifies that Type MC cable is not permitted to be used where it is subject to physical damage.

330.17 Through or Parallel to Framing Members

Type MC cable shall be protected in accordance with 300.4(A), (C), and (D) where installed through or parallel to framing members.

330.23 In Accessible Attics

The installation of Type MC cable in accessible attics or roof spaces shall also comply with 320.23.

In accessible attics, Type MC cable installed across the top of floor joists or within 7 ft of the floor or floor joists across the face of rafters or studs must be protected by guard strips. Where the attic is not accessible by a permanent ladder or stairs, guard strips are required only within 6 ft of the scuttle hole or opening.

330.24 Bending Radius

Bends in Type MC cable shall be so made that the cable will not be damaged. The radius of the curve of the inner edge of any bend shall not be less than required in 330.24(A) through (C).

The minimum bending radius of 12 times the outside diameter (OD) of a single shielded conductor [330.24(A)(2)] is consistent with Insulated Cable Engineers Association (ICEA) [formerly Insulated Power Cable Engineers Association (IPCEA)] requirements and good engineering practice. The same minimum on a shielded multiconductor cable, however, would be excessive.

(A) Smooth Sheath
(1) Ten times the external diameter of the metallic sheath for cable not more than 19 mm (¾ in.) in external diameter

(2) Twelve times the external diameter of the metallic sheath for cable more than 19 mm (¾ in.) but not more than 38 mm (1½ in.) in external diameter

(3) Fifteen times the external diameter of the metallic sheath for cable more than 38 mm (1½ in.) in external diameter

(B) Interlocked-Type Armor or Corrugated Sheath Seven times the external diameter of the metallic sheath.

(C) Shielded Conductors Twelve times the overall diameter of one of the individual conductors or seven times the overall diameter of the multiconductor cable, whichever is greater.

330.30 Securing and Supporting

(A) General Type MC cable shall be supported and secured by staples, cable ties, straps, hangers, or similar fittings or other approved means designed and installed so as not to damage the cable.

Type MC cable that runs horizontally through or on framing members or racks (spaced less than 6 ft apart) without additional securing is considered supported. Cable ties are not required as the cable passes through or on these members. However, the MC cable must be secured (fastened in place) within 12 in. of the outlet box.

Section 330.30(A) contains the general support requirements for Type MC cable, that is, supported and secured at least every 6 ft.

(B) Securing Unless otherwise provided, cables shall be secured at intervals not exceeding 1.8 m (6 ft). Cables containing four or fewer conductors sized no larger than 10 AWG shall be secured within 300 mm (12 in.) of every box, cabinet, fitting, or other cable termination.

According to 330.30(B), MC cable containing four or fewer conductors of size 10 AWG or less is required to be secured within 12 in. from every box, cabinet, or fitting. Both requirements are illustrated in Exhibit 330.2.

(C) Supporting Unless otherwise provided, cables shall be supported at intervals not exceeding 1.8-m (6-ft).

Horizontal runs of Type MC cable installed in wooden or metal framing members or similar supporting means shall be considered supported and secured where such support does not exceed 1.8-m (6-ft) intervals.

(D) Unsupported Cables Type MC cable shall be permitted to be unsupported where the cable:

(1) Is fished between access points through concealed spaces in finished buildings or structures and supporting is impractical; or

Exhibit 330.2 An application of 330.30(A) and 330.30(B), showing Type MC cable supported and secured at intervals not exceeding 6 ft and within 12 in. of the box.

(2) Is not more than 1.8 m (6 ft) in length from the last point of cable support to the point of connection to a luminaire (lighting fixture) or other piece of electrical equipment and the cable and point of connection are within an accessible ceiling. For the purpose of this section, Type MC cable fittings shall be permitted as a means of cable support.

Section 330.30(D) permits Type MC cable to be fished, as shown in Exhibit 330.3.

330.31 Single Conductors

Where single-conductor cables with a nonferrous armor or sheath are used, the installation shall comply with 300.20.

Exhibit 330.3 An application of 330.30(D), which permits Type MC cable to be fished in walls, floors, or ceilings.

330.40 Boxes and Fitting

Fittings used for connecting Type MC cable to boxes, cabinets, or other equipment shall be listed and identified for such use.

Connectors should be selected in accordance with the size and type of cable for which they are designated. Bronze connectors are intended for use only with cable employing corrugated copper armor. Some Type AC cable connectors are also acceptable for use with Type MC cable when specifically indicated on the fitting or the shipping carton.

330.80 Ampacity

The ampacity of Type MC cable shall be determined in accordance with 310.15 or 310.60 for 14 AWG and larger

conductors and in accordance with Table 402.5 for 18 AWG and 16 AWG conductors. The installation shall not exceed the temperature ratings of terminations and equipment.

(A) Type MC Cable Installed in Cable Tray The ampacities for Type MC cable installed in cable tray shall be determined in accordance with 392.11 and 392.13.

(B) Single Type MC Conductors Grouped Together Where single Type MC conductors are grouped together in a triangular or square configuration and installed on a messenger or exposed with a maintained free airspace of not less than 2.15 times one conductor diameter (2.15 × O.D.) of the largest conductor contained within the configuration and adjacent conductor configurations or cables, the ampacity of the conductors shall not exceed the allowable ampacities in the following tables:

(1) Table 310.20 for conductors rated 0 through 2000 volts
(2) Tables 310.67 and 310.68 for conductors rated over 2000 volts

III. Construction Specifications

330.104 Conductors

The conductors shall be of copper, aluminum, or copper-clad aluminum, solid or stranded. The minimum conductor size shall be 18 AWG copper and 12 AWG aluminum or copper-clad aluminum.

330.108 Equipment Grounding

Where Type MC cable is used for equipment grounding, it shall comply with 250.118(10) and 250.122.

The armor of interlocking Type MC cable is not usually recognized by UL as the sole means of providing an equipment grounding circuit, but it may be used to supplement the internal grounding conductor. A new construction where an aluminum equipment grounding conductor within a multiconductor cable is in continuous contact with the aluminum interlocked armor has been approved by UL to serve as the equipment grounding conductor.

If an equipment grounding conductor, either single or segmented, is included in a multiconductor Type MC cable, the equipment grounding conductor must be an integral part of the multiconductor Type MC cable. The size of the equipment grounding conductor is determined according to 250.122.

The following explanations apply only to installations of single-conductor Type MC cable. If single Type MC conductors are installed as open runs or as messenger-supported wiring, the smooth or corrugated metallic sheath on each conductor must have sufficient cross-sectional area to comply with Table 250.122, or concentric conductors may be provided over the conductor under the metallic sheath. Either the metallic sheath or the combination of the concentric conductors in parallel with the metallic sheath may be used to provide the required equipment grounding path. For single Type MC conductors with an interlocking metal tape armor installed as open runs or as messenger-supported wiring, concentric conductors must be provided over the conductor under the metallic sheath. The total cross-sectional area of the concentric conductors must comply with Table 250.122.

If single Type MC conductors are installed in cable tray, the cable tray, an equipment grounding conductor(s) within the cable tray, and/or the equipment grounding provided with the single conductor may be used individually or in any parallel combination to provide the equipment grounding path required by 392.3(C) and 250.118.

If single Type MC conductors are installed underground in a trench, a separate equipment grounding conductor may be installed in close proximity to the circuit conductors within the same trench. The separate equipment grounding conductor may be used alone or in parallel with the equipment grounding provided with the single conductor cables to provide the equipment grounding path required by 300.5(I) and 250.118.

If single Type MC conductors are installed in parallel, the equipment grounding path provided with each conductor must be sized on the basis of the ampere rating of the overcurrent device protecting the circuit conductors according to 250.122(F). A smaller equipment grounding conductor may be permitted only if ground-fault protection of equipment is installed in accordance with 250.122(F)(2).

330.112 Insulation

Insulated conductors shall comply with 330.112(A) or (B).

(A) 600 Volts Insulated conductors in sizes 18 AWG and 16 AWG shall be of a type listed in Table 402.3, with a maximum operating temperature not less than 90°C (194°F) and as permitted by 725.27. Conductors larger than 16 AWG shall be of a type listed in Table 310.13 or of a type identified for use in Type MC cable.

(B) Over 600 Volts Insulated conductors shall be of a type listed in Tables 310.61 through 310.64.

330.116 Sheath

Metallic covering shall be one of the following types: smooth metallic sheath, corrugated metallic sheath, interlocking metal tape armor. The metallic sheath shall be continuous and close fitting. A nonmagnetic sheath or armor shall be used on single conductor Type MC. Supplemental protection of an outer covering of corrosion-resistant material shall be permitted and shall be required where such protection is

needed. The sheath shall not be used as a current-carrying conductor.

FPN: See 300.6 for protection against corrosion.

ARTICLE 332
Mineral-Insulated, Metal-Sheathed Cable: Type MI

Summary of Changes

- **332.10:** New FPN added to clarify that the uses permitted list is not intended to include all of the permitted uses of Type MI cable.

- **332.12:** Underground runs without necessary physical protection added to uses not permitted list.

Contents

I. General

332.1 Scope

This article covers the use, installation, and construction specifications for mineral-insulated, metal-sheathed cable, Type MI.

332.2 Definition

Mineral-Insulated, Metal-Sheathed Cable, Type MI. A factory assembly of one or more conductors insulated with a highly compressed refractory mineral insulation and enclosed in a liquidtight and gastight continuous copper or alloy steel sheath.

Type MI mineral-insulated, metal-sheathed cable consists of one or more solid copper conductors insulated with highly compressed magnesium oxide and enclosed in a continuous copper or alloy steel (e.g., stainless steel) sheath with or without a nonmetallic jacket. It is manufactured in size 16 AWG to 500 kcmil, single conductor; 16 AWG to 4 AWG, two and three conductor; 16 AWG to 6 AWG, four conductor; and 16 AWG to 10 AWG, seven conductor. The cable is rated 600 volts.

II. Installation

332.10 Uses Permitted

Type MI cable shall be permitted as follows:

(1) For services, feeders, and branch circuits
(2) For power, lighting, control, and signal circuits

Type MI cable is also constructed with a 300-volt rating for signal circuit applications.

(3) In dry, wet, or continuously moist locations
(4) Indoors or outdoors
(5) Where exposed or concealed
(6) Where embedded in plaster, concrete, fill, or other masonry, whether above or below grade
(7) In any hazardous (classified) location
(8) Where exposed to oil and gasoline
(9) Where exposed to corrosive conditions not deteriorating to its sheath
(10) In underground runs where suitably protected against physical damage and corrosive conditions
(11) In or attached to cable tray

In lieu of an overall continuous copper sheath, Type MI cable is available with an overall continuous stainless steel sheath. Type MI cable is also available with an optional overall nonmetallic jacket.

FPN: The "Uses Permitted" is not an all-inclusive list.

An FPN was added for the 2005 *Code* clarifying that the list of *uses permitted* is not necessarily a complete list.

332.12 Uses Not Permitted

Type MI cable shall not be used under the following conditions or in the following locations:

(1) In underground runs unless protected from physical damage, where necessary
(2) Where exposed to conditions that are destructive and corrosive to the metallic sheath, unless additional protection is provided

In lieu of an overall continuous copper sheath, Type MI cable is available with an overall continuous stainless steel sheath. Type MI cable is also available with an optional overall nonmetallic jacket.

332.17 Through or Parallel to Framing Members

Type MI cable shall be protected in accordance with 300.4 where installed through or parallel to framing members.

332.24 Bending Radius

Bends in Type MI cable shall be so made that the cable will not be damaged. The radius of the inner edge of any bend shall not be less than required as follows:

(1) Five times the external diameter of the metallic sheath for cable not more than 19 mm (¾ in.) in external diameter
(2) Ten times the external diameter of the metallic sheath for cable greater than 19 mm (¾ in.) but not more than 25 mm (1 in.) in external diameter

The minimum bending radius is intended to prevent mechanical damage to the conductor insulation or the sheath that could result in cracking, a hot spot at the point of damage, or both.

As illustrated in Exhibit 332.1, for cables with an external diameter (OD) not greater than ¾ in., the minimum bending radius *(R)* is 5 times the cable OD. For cables greater than ¾ in. but not greater than 1 in. in diameter, the minimum bending radius is 10 times the cable OD.

332.30 Securing and Supporting

Type MI cable shall be supported and secured by staples, straps, hangers, or similar fittings, designed and installed so as not to damage the cable, at intervals not exceeding 1.8 m (6 ft).

(A) Horizontal Runs Through Holes and Notches In other than vertical runs, cables installed in accordance with 300.4 shall be considered supported and secured where such support does not exceed 1.8 m (6 ft) intervals.

External diameter (OD)

Bending radius ($R_{min.}$)

Cable ¾-in. and smaller OD:
$R_{min.} = 5 \times OD$

Cable ¾-in. to 1-in. OD:
$R_{min.} = 10 \times OD$

Exhibit 332.1 An illustration of 332.24, for bends in Type MI cable.

(B) Unsupported Cable Type MI cable shall be permitted to be unsupported where the cable is fished between access points through concealed spaces in finished buildings or structures and supporting is impracticable.

(C) Cable Trays All MI cable installed in cable trays shall comply with 392.8(B).

332.31 Single Conductors

Where single-conductor cables are used, all phase conductors and, where used, the neutral conductor shall be grouped together to minimize induced voltage on the sheath.

Section 332.31 permits Type MI cables to be used as single conductors. The larger sizes of Type MI cable are available only as single-conductor cables. Because single conductors in a metal sheath can result in induced voltage on the sheath, this section requires all conductors of the circuit to be grouped together to minimize the voltage on the sheath.

Where single conductors enter a ferrous metal enclosure, inductive heating can occur due to hysteresis loss caused by the magnetic flux occurring in ferrous metals and I^2R losses from the currents induced by the conductor. To minimize this magnetic heating of enclosures, 300.20 requires additional measures, including cutting slots in the metal between the individual holes for each conductor connector. Cable manufacturers offer nonferrous connecting plates that accept individual threaded connections of all circuit conductors, thereby eliminating circulating currents and fully complying with 300.20.

332.40 Boxes and Fittings

(A) Fittings Fittings used for connecting Type MI cable to boxes, cabinets, or other equipment shall be identified for such use.

(B) Terminal Seals Where Type MI cable terminates, an end seal fitting shall be installed immediately after stripping to prevent the entrance of moisture into the insulation. The conductors extending beyond the sheath shall be individually provided with an insulating material.

Terminations specifically investigated for use with this cable are listed as "mineral-insulated cable fittings." Fittings for use on Type MI cable are suitable for use at a maximum operating temperature of 90°C in dry locations and 60°C in wet locations. As shown in Exhibit 332.2, a complete box connector consists of a connector body and a screw-on potting fitting that may be used separately as an end fitting for change to open wiring. The screw-on potting fitting is to be assembled with a special tool and consists of a screw-on pot, insulating cap, insulating sleeving, anchoring bead, and sealing compound.

332.80 Ampacity

The ampacity of Type MI cable shall be determined in accordance with 310.15. The conductor temperature at the end seal fitting shall not exceed the temperature rating of the listed end seal fitting, and the installation shall not exceed the temperature ratings of terminations or equipment.

(A) Type MI Cable Installed in Cable Tray The ampacities for Type MI cable installed in cable tray shall be determined in accordance with 392.11.

(B) Single Type MI Conductors Grouped Together Where single Type MI conductors are grouped together in a triangular or square configuration, as required by 332.31, and installed on a messenger or exposed with a maintained free air space of not less than 2.15 times one conductor diameter (2.15 × O.D.) of the largest conductor contained within the configuration and adjacent conductor configurations or cables, the ampacity of the conductors shall not exceed the allowable ampacities of Table 310.17.

Determining Type MI cable ampacities according to Table 310.16 or Table 310.17 is by far the most common approach. However, where MI cables are terminated at electrical equipment such as circuit breakers, distribution switchgear, transfer switches, and the like, it is important to understand the temperature limitations of electrical equipment terminals and to coordinate those temperature limitations with the ampacity of the MI cables. As stated in both the UL *General Information Directory* and in 110.14(C)(1) of the 2005 *NEC*,

Exhibit 332.2 A Type MI cable fitting used for terminating cable to an enclosure, to a box, or directly to equipment. (Courtesy of Pyrotenax Cables Ltd.)

unless equipment is listed and marked otherwise, conductor ampacities used in determining equipment terminations must be based on Table 310.16 as modified by 310.15(B)(1) through 310.15(B)(6).

III. Construction Specifications

332.104 Conductors

Type MI cable conductors shall be of solid copper, nickel, or nickel-coated copper with a resistance corresponding to standard AWG and kcmil sizes.

332.108 Equipment Grounding

Where the outer sheath is made of copper, it shall provide an adequate path for equipment grounding purposes. Where made of steel, an equipment grounding conductor shall be provided.

The copper sheath of Type MI cable is permitted as an equipment grounding conductor according to 250.118(10), but an alloy steel outer sheath of Type MI cable is not. According to the product standard, for Type MI cables with an alloy steel outer sheath, one of the conductors (within the cable assembly) is to be used for equipment grounding. This product standard statement is in concert with the requirement of 300.3 prohibiting the installations of single conductors unless they are part of a wiring method of Chapter 3 of the *Code*. The size of an equipment grounding conductor is determined according to 250.122. If an individual equipment grounding conductor is installed with a multiconductor Type MI cable installation, the separate equipment grounding conductor must be an integral part of the multiconductor Type MI cable.

The following explanation applies only to installations of single-conductor Type MI cable. If single Type MI conductors are installed as open runs or as messenger-supported wiring, the copper sheath on each conductor must have sufficient cross-sectional area to comply with Table 250.122.

If single Type MI conductors are installed in cable tray, the cable tray, an equipment grounding conductor(s) within the cable tray, and/or the equipment grounding provided with the single conductor may be used individually or in any parallel combination to provide the equipment grounding path according to the requirements of 392.3(C) and 250.118.

If single Type MI conductors are installed underground in a trench, a separate equipment grounding conductor may be installed in close proximity to the circuit conductors within the same trench. The separate equipment grounding conductor may be used alone or in parallel with the equipment grounding provided with the single conductor to provide the equipment grounding path required by 300.5(I) and 250.118.

If single Type MI conductors are installed in parallel, the equipment grounding path provided with each conductor must be sized on the basis of the ampere rating of the overcurrent device protecting the circuit conductors in accordance with 250.122(F). A smaller equipment grounding conductor may be permitted only if ground-fault protection of equipment is installed in accordance with 250.122(F)(2).

332.112 Insulation

The conductor insulation in Type MI cable shall be a highly compressed refractory mineral that provides proper spacing for all conductors.

332.116 Sheath

The outer sheath shall be of a continuous construction to provide mechanical protection and moisture seal.

ARTICLE 334
Nonmetallic-Sheathed Cable: Types NM, NMC, and NMS

Summary of Changes

- **334.10(4):** Revised to limit Types NM, NMC, and NMS cable use in cable trays to structures permitted to be of Types III, IV, or V construction.

- **334.10(C):** Revised to delete reference to Article 780 as permitted use for Type NMS cable.

- **334.12:** Revised to clarify that Types NM, NMC, and NMS cables are not permitted in dwellings or other structures unless specifically permitted in 334.10(1), (2), and (3).

- **334.15(B):** Revised to limit the types of Chapter 3 raceways permitted to be used as a means to provide physical protection for Types NM, NMC and NMS cables. Alternatives to the methods identified in this section must be approved by the authority having jurisdiction.

- **334.15(C):** Added provision covering the use of conduit or tubing to provide physical protection and support for Types NM, NMC, and NMS cables installed on the walls of an unfinished basement. Requires abrasion protection at the point the cable enters the conduit or tubing.

- **334.80:** Added second paragraph covering ampacity adjustment for bundled cables that pass through fire- or draft-stopped holes in wood framing.

Contents

I. General

334.1 Scope

This article covers the use, installation, and construction specifications of nonmetallic-sheathed cable.

334.2 Definitions

Nonmetallic-Sheathed Cable. A factory assembly of two or more insulated conductors enclosed within an overall nonmetallic jacket.

Type NM. Insulated conductors enclosed within an overall nonmetallic jacket.

Type NMC. Insulated conductors enclosed within an overall, corrosion resistant, nonmetallic jacket.

Type NMS. Insulated power or control conductors with signaling, data, and communications conductors within an overall nonmetallic jacket.

Types NM, NMC, and NMS nonmetallic-sheathed cable may be used for either exposed or concealed wiring. Where exposed, the cable should not be subject to physical damage. Nonmetallic-sheathed cable was first recognized in the 1928 *NEC* as a substitute for concealed knob-and-tube wiring (Article 394) and open wiring on insulators (Article 398). The original advantages of nonmetallic-sheathed cable over knob-and-tube wiring were that the outer sheath provides continuous protection in addition to the insulation applied to the conductors; the cable is easily fished in partitions of finished buildings; no insulating supports are required; and only one hole need be bored that can accommodate more than one cable passing through a wood cross member.

334.6 Listed

Type NM, Type NMC, and Type NMS cables shall be listed.

Nonmetallic-sheathed cables must be listed. UL 719, *Standard for Nonmetallic-Sheathed Cables*, requires a construction and performance evaluation, including testing relating to flammability, dielectric voltage-withstand, unwinding at low temperatures, pulling through joists, conductor pullout, crushing, and abrasion.

II. Installation

334.10 Uses Permitted

Type NM, Type NMC, and Type NMS cables shall be permitted to be used in the following:

(1) One- and two-family dwellings.
(2) Multifamily dwellings permitted to be of Types III, IV, and V construction except as prohibited in 334.12.
(3) Other structures permitted to be of Types III, IV, and V construction except as prohibited in 334.12. Cables shall be concealed within walls, floors, or ceilings that provide a thermal barrier of material that has at least a 15-minute finish rating as identified in listings of fire-rated assemblies.

> FPN No. 1: Types of building construction and occupancy classifications are defined in NFPA 220-1999, *Standard on Types of Building Construction*, or the applicable building code, or both.
>
> FPN No. 2: See Annex E for determination of building types [NFPA 220, Table 3-1].

A well-established means of codifying fire protection and fire safety requirements is to classify buildings by types of construction, based on materials used for the structural elements and the degree of fire resistance afforded by each element. The five fundamental construction types used by the model building codes are Type I (fire resistive), Type II (noncombustible), Type III (combination of combustible and noncombustible), Type IV (heavy timber), and Type V (wood frame). Types I and II basically require all structural elements to be noncombustible, whereas Types III, IV, and V allow some or all of the structural elements to be combustible (wood).

The selection of building construction types is regulated by the building code, based on the occupancy, height, and area of the building. The local code official or the architect for a building project can be consulted to determine the minimum allowable (permitted) construction type for the building under consideration. When a building of a selected height (in feet or stories above grade) and area is permitted to be built of combustible construction (i.e., Types III, IV, or V), the installation of nonmetallic-sheathed cable is permitted. The common areas (corridors) and incidental and

subordinate uses (laundry rooms, lounge rooms, etc.) that serve a multifamily dwelling occupancy are also considered part of the multifamily occupancy, thereby allowing the use of nonmetallic-sheathed cable in those areas.

If a building is to be of noncombustible construction (i.e., Type I or II) by the owner's choice, even though the building code would permit combustible construction, the building is allowed to be wired with nonmetallic-sheathed cable. In such an instance, nonmetallic-sheathed cable may be installed in the noncombustible building because the *Code* would have permitted the building to be of combustible construction.

Annex E provides charts and other explanatory information to assist the user in understanding and categorizing the exact types of construction under consideration. A table to cross reference building types to the various building code types of construction is also provided in Annex E.

(4) Cable trays in structures permitted to be Types III, IV, or V where the cables are identified for the use.

> FPN: See 310.10 for temperature limitation of conductors.

(A) Type NM Type NM cable shall be permitted as follows:

(1) For both exposed and concealed work in normally dry locations except as prohibited in 334.10(3)
(2) To be installed or fished in air voids in masonry block or tile walls

For concealed work, nonmetallic-sheathed cable should be installed where it is protected from physical damage often caused by nails or screws. Where practical, care should be taken to avoid areas where trim, door and window casings, baseboards, moldings, and so on, are likely to be nailed. See 300.4 for details on protection against physical damage.

(B) Type NMC Type NMC cable shall be permitted as follows:

(1) For both exposed and concealed work in dry, moist, damp, or corrosive locations, except as prohibited by 334.10(3)
(2) In outside and inside walls of masonry block or tile
(3) In a shallow chase in masonry, concrete, or adobe protected against nails or screws by a steel plate at least 1.59 mm (¹⁄₁₆ in.) thick and covered with plaster, adobe, or similar finish

Type NMC (corrosion-resistant) cable is required for installation in dairy barns and similar farm buildings (see Article 547), where cable will be exposed to fumes, vapors, or liquids such as ammonia and barnyard acids. Under such circumstances, ordinary types of nonmetallic-sheathed cable have in some cases deteriorated rapidly due to ammonia fumes or the growth of fungus or mold.

In addition to insulated conductors, nonmetallic-sheathed cable must have an insulated or bare conductor for equipment grounding purposes only, in order to comply with 334.108. See 250.119 and Table 250.122 for identification requirements and minimum conductor sizes.

(C) Type NMS Type NMS cable shall be permitted as follows:

(1) For both exposed and concealed work in normally dry locations except as prohibited by 334.10(3)
(2) To be installed or fished in air voids in masonry block or tile walls

The construction of hybrid cable, Type NMS, is fully described in 334.116(C), and the permitted conductors are described in 334.100.

334.12 Uses Not Permitted

Users should be aware that the list of uses not permitted is not a complete list. For example, Type NM, nonmetallic-sheathed cables are not permitted to be installed in ducts, plenums, and other air-handling spaces. See 300.22, which limits the use of materials in ducts, plenums, and other air-handling spaces that may contribute smoke and products of combustion during a fire.

(A) Types NM, NMC, and NMS Types NM, NMC, and NMS cables shall not be permitted as follows:

(1) In any dwelling or structure not specifically permitted in 334.10(1), (2), and (3)
(2) Exposed in dropped or suspended ceilings in other than one- and two-family and multifamily dwellings

Revised for the 2005 *Code*, 334.12(A)(1) clearly limits the use of the cables covered by Article 334.

Revised for the 2005 *Code*, 334.12(A)(2) prohibits any nonmetallic sheathed cables installed as exposed in the space above accessible hung ceilings. This change does not affect dwelling-type occupancies. The term *exposed*, as used in this requirement, closely follows the definition of *exposed (as applied to wiring methods)* found in Article 100, which states "on or attached to the surface or behind panels designed to allow access."

For example, cables installed above a dropped sheet rock ceiling or dropped sheet rock soffit would not be considered exposed cable, provided the area above the ceiling is not accessible (does not have removable tiles or does not contain an access panel). Very often, hung or dropped ceil-

ings are accessible; therefore, cables installed above these types of ceilings would be considered exposed cables if the cables do not have additional protection.

Additionally, a simple change to an architectural finish schedule during construction could change what is an acceptable wiring method. For example, if a corridor ceiling in an occupancy (other than a dwelling type) called for a painted gypsum board ceiling and the finish schedule changed the ceiling construction to a 2 ft × 2 ft accessible tile ceiling, the wiring method would no longer be permitted to be nonmetallic-sheathed cable unless the nonmetallic-sheathed cable was installed using additional protection. Examples of additional protection are found in 334.15(B).

(3) As service-entrance cable
(4) In commercial garages having hazardous (classified) locations as defined in 511.3
(5) In theaters and similar locations, except where permitted in 518.4(B)
(6) In motion picture studios
(7) In storage battery rooms
(8) In hoistways or on elevators or escalators
(9) Embedded in poured cement, concrete, or aggregate
(10) In hazardous (classified) locations, except where permitted by the following:
    a. 501.10(B)(3)
    b. 502.10(B)(3)
    c. 504.20

(B) Types NM and NMS Types NM and NMS cables shall not be used under the following conditions or in the following locations:

(1) Where exposed to corrosive fumes or vapors
(2) Where embedded in masonry, concrete, adobe, fill, or plaster
(3) In a shallow chase in masonry, concrete, or adobe and covered with plaster, adobe, or similar finish
(4) Where exposed or subject to excessive moisture or dampness

334.15 Exposed Work

In exposed work, except as provided in 300.11(A), cable shall be installed as specified in 334.15(A) through (C).

(A) To Follow Surface Cable shall closely follow the surface of the building finish or of running boards.

(B) Protection from Physical Damage Cable shall be protected from physical damage where necessary by rigid metal conduit, intermediate metal conduit, electrical metallic tubing, Schedule 80 PVC rigid nonmetallic conduit, or other approved means. Where passing through a floor, the cable shall be enclosed in rigid metal conduit, intermediate metal conduit, electrical metallic tubing, Schedule 80 PVC rigid nonmetallic conduit, or other approved means extending at least 150 mm (6 in.) above the floor.

Where Type NMC cable is installed in shallow chases in masonry, concrete, or adobe, the cable shall be protected against nails or screws by a steel plate at least 1.59 mm (¹⁄₁₆ in.) thick and covered with plaster, adobe, or similar finish.

(C) In Unfinished Basements Where cable is run at angles with joists in unfinished basements, it shall be permissible to secure cables not smaller than two 6 AWG or three 8 AWG conductors directly to the lower edges of the joists. Smaller cables shall be run either through bored holes in joists or on running boards. NM cable used on a wall of an unfinished basement shall be permitted to be installed in a listed conduit or tubing. Conduit or tubing shall utilize a nonmetallic bushing or adapter at the point the cable enters the raceway. Metal conduit and tubings and metal outlet boxes shall be grounded.

Section 334.15(C) was changed for the 2005 *Code*. The means of providing physical protection in 334.15(B) is now a more specific list of the required protection techniques. Notice that rigid nonmetallic conduit, Type RNC, Schedule 40 is omitted from this list unless it is judged as an "approved" means by the authority having jurisdiction (AHJ). Also, where NMC is installed close to the surface in masonry, concrete, or adobe-type construction, physical protection must be afforded to the cable by using steel plate–type protectors.

For exposed work in unfinished basements, as described in 334.15(C), physical protection for nonmetallic-sheathed cables run on unfinished walls can take the form of any listed conduit or tubing (including rigid nonmetallic conduit, Type RNC, Schedule 40).

Exhibit 334.1 Nonmetallic-sheathed cables installed in an unfinished basement.

As illustrated in Exhibit 334.1, nonmetallic-sheathed cables installed in an unfinished basement can be run through joists and attached to the side of joists or beams and running boards. Section 300.4(D) requires cables that are run parallel to framing members be installed at least 1¼ in. from the nearest edge of studs, joists, or rafters.

334.17 Through or Parallel to Framing Members

Types NM, NMC, or NMS cable shall be protected in accordance with 300.4 where installed through or parallel to framing members. Grommets used as required in 300.4(B)(1) shall remain in place and be listed for the purpose of cable protection.

In accordance with 300.4(B)(1), where cable passes through factory- or field-punched holes in metal studs or similar members, it is required to be protected by listed bushings or listed grommets covering all metal edges and securely fastened in the opening before being installed. See the commentary following 300.4(B)(1) for further information regarding physical protection of NM cables.

334.23 In Accessible Attics

The installation of cable in accessible attics or roof spaces shall also comply with 320.23.

334.24 Bending Radius

Bends in Types NM, NMC, and NMS cable shall be so made that the cable will not be damaged. The radius of the curve of the inner edge of any bend during or after installation shall not be less than five times the diameter of the cable.

334.30 Securing and Supporting

Nonmetallic-sheathed cable shall be supported and secured by staples, cable ties, straps, hangers, or similar fittings designed and installed so as not to damage the cable, at intervals not exceeding 1.4 m (4½ ft) and within 300 mm (12 in.) of every outlet box, junction box, cabinet, or fitting. Flat cables shall not be stapled on edge.

Sections of cable protected from physical damage by raceway shall not be required to be secured within the raceway.

For the 2005 *Code*, the changes to 334.30 were for the most part editorial. The general requirement of 334.30 requires that the cable be secured. Simply draping the cable over air ducts, rafters, timbers, joists, pipes, and ceiling grid members is not permitted, except where fished as allowed in 334.30(B)(1).

(A) Horizontal Runs Through Holes and Notches In other than vertical runs, cables installed in accordance with

300.4 shall be considered to be supported and secured where such support does not exceed 1.4-m (4½-ft) intervals and the nonmetallic-sheathed cable is securely fastened in place by an approved means within 300 mm (12 in.) of each box, cabinet, conduit body, or other nonmetallic-sheathed cable termination.

> FPN: See 314.17(C) for support where nonmetallic boxes are used.

Nonmetallic-sheathed cable that runs horizontally through framing members (spaced less than 54 in. apart) and passes through bored or punched holes in framing members without additional securing is considered supported by the framing members. Cable ties are not required as the cable passes through these members. However, the nonmetallic-sheathed cable must be secured (fastened in place) within 12 in. of the outlet box. Where the cable terminates at a nonmetallic outlet box that does not contain a cable clamping device, the cable may be secured (fastened in place) within 8 in. of the outlet box, according to 314.17(C), Exception.

Section 334.30 also prohibits two-conductor nonmetallic-sheathed cable (or other flat configurations) from being stapled on edge. The intent is to prohibit the cable from being installed with its short dimension against a wood joist. When stapled in this manner, two cables are usually placed side by side under the staple. If the staple is driven too far into the stud, damage to the insulation and conductors can occur. See 300.4(C) for support requirements of cables through spaces behind panels designed to allow access.

(B) Unsupported Cables Nonmetallic-sheathed cable shall be permitted to be unsupported where the cable:

(1) Is fished between access points through concealed spaces in finished buildings or structures and supporting is impracticable.

(2) Is not more than 1.4 m (4½ ft) from the last point of cable support to the point of connection to a luminaire (lighting fixture) or other piece of electrical equipment and the cable and point of connection are within an accessible ceiling.

Section 334.30(B)(2) permits short, unsupported lengths of nonmetallic-sheathed cable for luminaire and equipment connections.

(C) Wiring Device Without a Separate Outlet Box A wiring device identified for the use, without a separate outlet box, and incorporating an integral cable clamp shall be permitted where the cable is secured in place at intervals not exceeding 1.4 m (4½ ft) and within 300 mm (12 in.) from the wiring device wall opening, and there shall be at least a 300 mm (12 in.) loop of unbroken cable or 150 mm (6 in.) of a cable end available on the interior side of the finished wall to permit replacement.

334.40 Boxes and Fittings

(A) Boxes of Insulating Material Nonmetallic outlet boxes shall be permitted as provided by 314.3.

Nonmetallic boxes and nonmetallic wiring systems are recommended for use in some corrosive atmospheres. See 314.3, 314.17(C), and Article 547 for details.

(B) Devices of Insulating Material Switch, outlet, and tap devices of insulating material shall be permitted to be used without boxes in exposed cable wiring and for rewiring in existing buildings where the cable is concealed and fished. Openings in such devices shall form a close fit around the outer covering of the cable, and the device shall fully enclose the part of the cable from which any part of the covering has been removed. Where connections to conductors are by binding-screw terminals, there shall be available as many terminals as conductors.

(C) Devices with Integral Enclosures Wiring devices with integral enclosures identified for such use shall be permitted as provided by 300.15(E).

334.80 Ampacity

The ampacity of Types NM, NMC, and NMS cable shall be determined in accordance with 310.15. The ampacity shall be in accordance with the 60°C (140°F) conductor temperature rating. The 90°C (194°F) rating shall be permitted to be used for ampacity derating purposes, provided the final derated ampacity does not exceed that for a 60°C (140°F) rated conductor. The ampacity of Types NM, NMC, and NMS cable installed in cable tray shall be determined in accordance with 392.11.

Where more than two NM cables containing two or more current-carrying conductors are bundled together and pass through wood framing that is to be fire- or draft-stopped using thermal insulation or sealing foam, the allowable ampacity of each conductor shall be adjusted in accordance with Table 310.15(B)(2)(a).

Section 310.15(B)(2)(a) states in part: "or where single conductors or multiconductor cables are stacked or bundled longer than 600 mm (24 in.) without maintaining spacing and are not installed in raceways, the allowable ampacity of each conductor shall be reduced as shown in Table 310.15(B)(2)(a)." Failure to comply with the appropriate adjustment ampacity derating called for by this table, where nonmetallic-sheathed cables may be stacked or bundled, can lead to overheating of conductors. For the 2005 *Code*, a new derating requirement was added to prevent overheating of Type NM conductors where passing though draft- and fire-stopping material.

III. Construction Specifications

334.100 Construction

The outer cable sheath of nonmetallic-sheathed cable shall be a nonmetallic material.

334.104 Conductors

The 600 volt insulated conductors shall be sizes 14 AWG through 2 AWG copper conductors or sizes 12 AWG through 2 AWG aluminum or copper-clad aluminum conductors. The signaling conductors shall comply with 780.5. The communication conductors shall comply with Part V of Article 800.

334.108 Equipment Grounding

In addition to the insulated conductors, the cable shall have an insulated or bare conductor for equipment grounding purposes only.

Changed for the 2005 *Code*, this section no longer *permits* an equipment grounding conductor; it *requires* an equipment grounding conductor. The required equipment grounding conductor must be sized in accordance with 250.122 and comply with UL 719, *Standard for Nonmetallic-Sheathed Cables*.

334.112 Insulation

The insulated power conductors shall be one of the types listed in Table 310.13 that are suitable for branch circuit wiring or one that is identified for use in these cables. Conductor insulation shall be rated at 90°C (194°F).

> FPN: Types NM, NMC, and NMS cable identified by the markings NM-B, NMC-B, and NMS-B meet this requirement.

334.116 Sheath

The outer sheath of nonmetallic-sheathed cable shall comply with 334.116(A), (B), and (C).

(A) Type NM The overall covering shall be flame retardant and moisture resistant.

(B) Type NMC The overall covering shall be flame retardant, moisture resistant, fungus resistant, and corrosion resistant.

(C) Type NMS The overall covering shall be flame retardant and moisture resistant. The sheath shall be applied so as to separate the power conductors from the communications and signaling conductors. The signaling conductors shall be permitted to be shielded. An optional outer jacket shall be permitted.

> FPN: For composite optical cable, see 770.9 and 770.133.

ARTICLE 336
Power and Control Tray Cable: Type TC

Summary of Changes

- **336.10:** Deleted text limiting use of TC cable in hazardous (classified) locations only in industrial establishments under the conditions of supervision and maintenance.
- **336.10(4):** Revised to clarify permitted use of Type TC cable supported by a messenger wire in outdoor locations.
- **336.10(7):** Revised to clarify installation requirements for equipment grounding conductors in cables 6 AWG or smaller. Revised the marking requirement for identifying Type TC cable that is suitable to be installed outside of a cable tray or raceway where the cable is supported on struts, angles, or channels.
- **336.12(2):** Revised to prohibit TC cable outside a raceway or cable tray system except as permitted in 336.10(7).

Contents

I. General

336.1 Scope

This article covers the use, installation, and construction specifications for power and control tray cable, Type TC.

The basic standard to investigate products in this category is UL 1277, *Electrical Power and Control Tray Cables with Optional Optical-Fiber Members.* Summary information regarding listed power and control tray cable may be found in the UL *General Information for Electrical Equipment Directory* under category QPOR.

336.2 Definition

Power and Control Tray Cable, Type TC. A factory assembly of two or more insulated conductors, with or without associated bare or covered grounding conductors, under a nonmetallic jacket.

II. Installation

336.10 Uses Permitted

Type TC cable shall be permitted to be used as follows:

(1) For power, lighting, control, and signal circuits.
(2) In cable trays.
(3) In raceways.
(4) In outdoor locations supported by a messenger wire.
(5) For Class 1 circuits as permitted in Parts II and III of Article 725.
(6) For non–power-limited fire alarm circuits if conductors comply with the requirements of 760.27.
(7) In industrial establishments where the conditions of maintenance and supervision ensure that only qualified persons service the installation, and where the cable is continuously supported and protected against physical damage using mechanical protection, such as struts, angles, or channels, Type TC tray cable that complies with the crush and impact requirements of Type MC cable and is identified for such use with the marking Type TC–ER shall be permitted between a cable tray and the utilization equipment or device. The cable shall be secured at intervals not exceeding 1.8 m (6 ft). Equipment grounding for the utilization equipment shall be provided by an equipment grounding conductor within the cable. In cables containing conductors sized 6 AWG or smaller, the equipment grounding conductor shall be provided within the cable or, at the time of installation, one or more insulated conductors shall be permanently identified as an equipment grounding conductor in accordance with 250.119(B).
(8) Where installed in wet locations, Type TC cable shall also be resistant to moisture and corrosive agents.

FPN: See 310.10 for temperature limitation of conductors.

According to 336.10(3), Type TC cable is permitted to be installed in a hazardous location only if that location is in an industrial establishment where conditions of maintenance and supervision ensure that only qualified persons service the installation. The overall jacket on Type TC cable is a "gas/vaportight continuous sheath" in the sense discussed in 501.15(D) and 501.15(E). However, Type TC cable is not investigated for transmission of gases or vapors through the core; thus, when these cables are used in hazardous (classified) locations, they may need to be sealed according to 501.15(D) and 501.15(E).

Section 336.10(6) permits Type TC tray cable to be used for non–power-limited fire alarm circuits. According to 760.27, the cable must be listed and the conductor material must be copper. Aluminum and copper-clad aluminum conductors are not permitted for fire alarm circuits.

According to 336.10(7), specific types of TC tray cable used in qualifying occupancies are permitted to extend from a cable tray to a piece of equipment without the use of conduit. Prior to the 2002 *Code*, a restriction of 50 ft applied to this type of installation. For later editions of the *Code*, the 50-ft restriction was removed, allowing virtually any length of cable to be used for the circuit extension from the cable tray to the equipment served. According to UL 1277, *Electrical Power and Control Tray Cables with Optional Optical-Fiber Members*, cables suitable for use as exposed wiring between cable tray and the utilization equipment according to 336.10(7) are surface marked "Type TC-ER" (tray cable for exposed runs).

336.12 Uses Not Permitted

Type TC tray cable shall not be installed or used as follows:

(1) Installed where it will be exposed to physical damage
(2) Installed outside a raceway or cable tray system, except as permitted in 336.10(7)
(3) Used where exposed to direct rays of the sun, unless identified as sunlight resistant
(4) Direct buried, unless identified for such use

336.24 Bending Radius

Bends in Type TC cable shall be made so as not to damage the cable. For Type TC cable without metal shielding, the minimum bending radius shall be as follows:

(1) Four times the overall diameter for cables 25 mm (1 in.) or less in diameter
(2) Five times the overall diameter for cables larger than 25 mm (1 in.) but not more than 50 mm (2 in.) in diameter
(3) Six times the overall diameter for cables larger than 50 mm (2 in.) in diameter

Type TC cables with metallic shielding shall have a minimum bending radius of not less than 12 times the cable overall diameter.

336.80 Ampacity

The ampacity of Type TC tray cable shall be determined in accordance with 392.11 for 14 AWG and larger conductors, in accordance with 402.5 for 18 AWG through 16 AWG conductors where installed in cable tray, and in accordance with 310.15 where installed in a raceway or as messenger supported wiring.

III. Construction Specifications

336.100 Construction

A metallic sheath or armor as defined in 330.116 shall not be permitted either under or over the nonmetallic jacket. Metallic shield(s) shall be permitted over groups of conductors, under the outer jacket, or both.

Type TC cables may contain one or more metal shields, but they do not have a metal sheath or armor. Electrical cables with a metal sheath or armor are covered in either Article 320 as armored cable, Type AC, or Article 330 as metal-clad cable, Type MC.

336.104 Conductors

The insulated conductors of Type TC tray cable shall be in sizes 18 AWG through 1000 kcmil copper and sizes 12 AWG through 1000 kcmil aluminum or copper-clad aluminum. Insulated conductors of sizes 14 AWG and larger copper and sizes 12 AWG and larger aluminum or copper-clad aluminum shall be one of the types listed in Table 310.13 or Table 310.62 that is suitable for branch circuit and feeder circuits or one that is identified for such use.

(A) Fire Alarm Systems Where used for fire alarm systems, conductors shall also be in accordance with 760.27.

(B) Thermocouple Circuits Conductors in Type TC cable used for thermocouple circuits in accordance with Part III of Article 725 shall also be permitted to be any of the materials used for thermocouple extension wire.

(C) Class 1 Circuit Conductors Insulated conductors of 18 AWG and 16 AWG copper shall also be in accordance with 725.27.

336.116 Jacket

The outer jacket shall be a flame-retardant, nonmetallic material.

336.120 Marking

There shall be no voltage marking on a Type TC cable employing thermocouple extension wire.

ARTICLE 338
Service-Entrance Cable: Types SE and USE

Summary of Changes

- **338.10(A):** Revised to reference specific sections and parts of Article 230 that pertain to the use of Types SE and USE cables for services.

Contents

I. General

338.1 Scope

This article covers the use, installation, and construction specifications of service-entrance cable.

According to the 2004 UL *Electrical Construction Materials Directory*, category TXKT (service cable) and category TYLZ (service-entrance cable rated 600 volts) are listed in sizes 14 AWG and larger for copper and 12 AWG and larger for aluminum or copper-clad aluminum. Type SE cable contains Types RHW, RHW–2, XHHW, XHHW–2, THWN, or THWN–2 conductors. Type USE cable contains conductors with insulation equivalent to RHW or XHHW. Type USE-2 contains insulation equivalent to RHW-2 or XHHW-2 and is rated 90°C wet or dry.

The type designation of the conductors may be marked on the surface of the cable. When used, this marking indicates the temperature rating for the cable corresponding to the temperature rating of the conductors. When this marking does not appear, the temperature rating of the cable is 75°C. The cables are designated as Type SE, Type USE or USE-2, and submersible water pump cable.

Type SE—Indicates cable for aboveground installation. Both the individual insulated conductors and the outer jacket or finish of Type SE are suitable for use where exposed to sun.

Type USE or USE-2—Indicates cable for underground installation, including burial directly in the earth. Cable in sizes 4/0 AWG and smaller and having all conductors insulated is suitable for all of the underground uses for which Type UF cable is permitted by the *Code*. Types USE and USE-2 are not suitable for use in premises or above ground except to terminate at the service equipment or metering equipment. Both the insulation and the outer covering, when used on single and multiconductor Types USE and USE-2, are suitable for use where exposed to sun.

Submersible water pump cable—Indicates a multiconductor cable in which two, three, or four single-conductor,

Type USE or USE-2 cables are provided in a flat or twisted assembly. The cable is listed in sizes 14 AWG to 4/0 AWG inclusive for copper and 12 AWG to 4/0 AWG inclusive for aluminum or copper-clad aluminum. The cable is tag-marked "For Use Within the Well Casing for Wiring Deep-Well Water Pumps Where the Cable Is Not Subject to Repetitive Handling Caused by Frequent Servicing of the Pump Units." The insulation may also be surface-marked "Pump Cable." The cable may be directly buried in the earth in conjunction with this use.

338.2 Definitions

Service-Entrance Cable. A single conductor or multiconductor assembly provided with or without an overall covering, primarily used for services, and of the following types:

Type SE. Service-entrance cable having a flame-retardant, moisture-resistant covering.

Type USE. Service-entrance cable, identified for underground use, having a moisture-resistant covering, but not required to have a flame-retardant covering.

II. Installation

338.10 Uses Permitted

(A) Service-Entrance Conductors Service-entrance cable shall be permitted to be used as service-entrance conductors and shall be installed in accordance with 230.6, 230.7, and Parts II, III, and IV of Article 230.

Type USE used for service laterals shall be permitted to emerge from the ground outside at terminations in meter bases or other enclosures where protected in accordance with 300.5(D).

(B) Branch Circuits or Feeders

(1) Grounded Conductor Insulated Type SE service-entrance cables shall be permitted in wiring systems where all of the circuit conductors of the cable are of the rubber-covered or thermoplastic type.

Branch circuits using service-entrance cable as a wiring method are permitted only if all circuit conductors within the cable are fully insulated according to 310.13. The equipment grounding conductor is the only conductor permitted to be bare within service-entrance cable used for branch circuits.

(2) Grounded Conductor Not Insulated Type SE service-entrance cable shall be permitted for use where the insulated conductors are used for circuit wiring and the uninsulated conductor is used only for equipment grounding purposes.

Exception: Uninsulated conductors shall be permitted as a grounded conductor in accordance with 250.140, 250.32, and 225.30 through 225.40.

Service-entrance cable containing a bare grounded (neutral) conductor is not permitted for new installations where it is used as a branch circuit to supply appliances such as ranges, wall-mounted ovens, counter-mounted cooking units, or clothes dryers. This exception permits a bare neutral service-entrance cable for existing installations only and is coordinated with the sections listed.

(3) Temperature Limitations Type SE service-entrance cable used to supply appliances shall not be subject to conductor temperatures in excess of the temperature specified for the type of insulation involved.

(4) Installation Methods for Branch Circuits and Feeders

(a) *Interior Installations.* In addition to the provisions of this article, Type SE service-entrance cable used for interior wiring shall comply with the installation requirements of Parts I and II of Article 334, excluding 334.80.

FPN: See 310.10 for temperature limitation of conductors.

(b) *Exterior Installations.* In addition to the provisions of this article, service-entrance cable used for feeders or branch circuits, where installed as exterior wiring, shall be installed in accordance with Part I of Article 225. The cable shall be supported in accordance with 334.30, unless used as messenger-supported wiring as permitted in Part II of Article 396. Type USE cable installed as underground feeder and branch circuit cable shall comply with Part II of Article 340. Where Type USE cable emerges from the ground at terminations, it shall be protected in accordance with 300.5(D). Multiconductor service-entrance cable shall be permitted to be installed as messenger-supported wiring in accordance with 225.10 and Part II of Article 396.

In accordance with 338.2, cables marked only as "Type USE service-entrance cable" are not required to have a flame-retardant covering.

338.24 Bending Radius

Bends in Types USE and SE cable shall be so made that the cable will not be damaged. The radius of the curve of the inner edge of any bend, during or after installation, shall not be less than five times the diameter of the cable.

III. Construction

338.100 Construction

Cabled, single-conductor, Type USE constructions recognized for underground use shall be permitted to have a bare copper conductor cabled with the assembly. Type USE single, parallel, or cabled conductor assemblies recognized for underground use shall be permitted to have a bare copper concentric conductor applied. These constructions shall not require an outer overall covering.

> FPN: See 230.41, Exception, item (2), for directly buried, uninsulated service-entrance conductors.

Type SE or USE cable containing two or more conductors shall be permitted to have one conductor uninsulated.

338.120 Marking

Service-entrance cable shall be marked as required in 310.11. Cable with the neutral conductor smaller than the ungrounded conductors shall be so marked.

ARTICLE 340
Underground Feeder and Branch-Circuit Cable: Type UF

Summary of Changes

- **340.6:** Added requirement that Type UF cable be listed.
- **340.10(4):** Revised to reference Parts II and III of Article 334.
- **340.12:** Revised to reference Part II of Article 396.
- **340.112:** Revised to require a 90°C rating where used as a substitute cable for Type NM cable.

Contents

I. General

340.1 Scope

This article covers the use, installation, and construction specifications for underground feeder and branch-circuit cable, Type UF.

340.2 Definition

Underground Feeder and Branch-Circuit Cable, Type UF. A factory assembly of one or more insulated conductors with an integral or an overall covering of nonmetallic material suitable for direct burial in the earth.

340.6 Listing Requirements

Type UF cable shall be listed.

New for the 2005 *Code*, Type UF cable is now required to be listed.

II. Installation

340.10 Uses Permitted

Type UF cable shall be permitted as follows:
(1) For use underground, including direct burial in the earth. For underground requirements, see 300.5.
(2) As single-conductor cables. Where installed as single-conductor cables, all conductors of the feeder grounded conductor or branch circuit, including the grounded conductor and equipment grounding conductor, if any, shall be installed in accordance with 300.3.
(3) For wiring in wet, dry, or corrosive locations under the recognized wiring methods of this *Code*.
(4) Installed as nonmetallic-sheathed cable. Where so installed, the installation and conductor requirements shall comply with Parts II and III of Article 334 and shall be of the multiconductor type.

Where UF cable is installed as nonmetallic-sheathed cable, the ampacity of Type UF cable is determined according to 334.80. For Type UF cable used for interior wiring, see the installation requirements and the associated commentary in Parts I and II of Article 334.

(5) For solar photovoltaic systems in accordance with 690.31.
(6) As single-conductor cables as the nonheating leads for heating cables as provided in 424.43.
(7) Supported by cable trays. Type UF cable supported by cable trays shall be of the multiconductor type.

> FPN: See 310.10 for temperature limitation of conductors.

340.12 Uses Not Permitted

Type UF cable shall not be used as follows:

(1) As service-entrance cable
(2) In commercial garages
(3) In theaters and similar locations

(4) In motion picture studios

(5) In storage battery rooms

(6) In hoistways or on elevators or escalators

(7) In hazardous (classified) locations

(8) Embedded in poured cement, concrete, or aggregate, except where embedded in plaster as nonheating leads where permitted in 424.43

(9) Where exposed to direct rays of the sun, unless identified as sunlight resistant

(10) Where subject to physical damage

(11) As overhead cable, except where installed as messenger-supported wiring in accordance with Part II of Article 396

Type UF cable suitable for exposure to the direct rays of the sun is indicated by tag marking and marking on the cable surface with the designation "Sunlight Resistant."

This physical protection requirement ensures that Type UF cable, as it emerges from underground, is protected from physical damage.

340.24 Bending Radius

Bends in Type UF cable shall be so made that the cable is not damaged. The radius of the curve of the inner edge of any bend shall not be less than five times the diameter of the cable.

340.80 Ampacity

The ampacity of Type UF cable shall be that of 60°C (140°F) conductors in accordance with 310.15.

III. Construction Specifications

340.104 Conductors

The conductors shall be sizes 14 AWG copper or 12 AWG aluminum or copper-clad aluminum through 4/0 AWG.

340.108 Equipment Grounding

In addition to the insulated conductors, the cable shall be permitted to have an insulated or bare conductor for equipment grounding purposes only.

340.112 Insulation

The conductors of Type UF shall be one of the moisture-resistant types listed in Table 310.13 that is suitable for branch-circuit wiring or one that is identified for such use. Where installed as a substitute wiring method for NM cable, the conductor insulation shall be rated 90°C (194°F).

The last sentence was added for the 2005 *Code* to require UF to meet the same conductor temperature requirements where UF is used as a substitute for Type NMC or NM.

340.116 Sheath

The overall covering shall be flame retardant; moisture, fungus, and corrosion resistant; and suitable for direct burial in the earth.

ARTICLE 342
Intermediate Metal Conduit: Type IMC

Summary of Changes

- **342.22:** Revised to permit cables in IMC if the cable article does not specifically prohibit such use. The number of cables is limited to the percentage fill of Table 1, Chapter 9.

- **342.42:** Revised to specifically reference 314.15(A) where threadless couplings and connectors are installed in wet locations.

Contents

I. General

342.1 Scope

This article covers the use, installation, and construction specifications for intermediate metal conduit (IMC) and associated fittings.

342.2 Definition

Intermediate Metal Conduit (IMC). A steel threadable raceway of circular cross section designed for the physical protection and routing of conductors and cables and for use as an equipment grounding conductor when installed with its integral or associated coupling and appropriate fittings.

Intermediate metal conduit (IMC) is a thinner-walled rigid metal conduit that is satisfactory for uses in all locations where rigid metal conduit (RMC) is permitted to be used. Threaded fittings, couplings, connectors, and so on, are interchangeable between IMC and RMC. Threadless fittings for IMC are suitable only for the type of conduit indicated by the marking on the carton.

342.6 Listing Requirements

IMC, factory elbows and couplings, and associated fittings shall be listed.

II. Installation

342.10 Uses Permitted

(A) All Atmospheric Conditions and Occupancies Use of IMC shall be permitted under all atmospheric conditions and occupancies.

(B) Corrosion Environments IMC, elbows, couplings, and fittings shall be permitted to be installed in concrete, in direct contact with the earth, or in areas subject to severe corrosive influences where protected by corrosion protection and judged suitable for the condition.

(C) Cinder Fill IMC shall be permitted to be installed in or under cinder fill where subject to permanent moisture where protected on all sides by a layer of noncinder concrete not less than 50 mm (2 in.) thick; where the conduit is not less than 450 mm (18 in.) under the fill; or where protected by corrosion protection and judged suitable for the condition.

(D) Wet Locations All supports, bolts, straps, screws, and so forth, shall be of corrosion-resistant materials or protected against corrosion by corrosion-resistant materials.

FPN: See 300.6 for protection against corrosion.

Galvanized IMC installed in concrete does not require supplementary corrosion protection. Similarly, galvanized IMC installed in contact with soil does not generally require supplementary corrosion protection. As a guide in the absence of experience with the corrosive effects of soil in a specific location, soils producing severe corrosive effects are generally characterized by low resistivity, less than 2000 ohm-cm. Wherever ferrous metal conduit runs directly from concrete encasement to soil burial, the metal in contact with the soil can be severely corroded.

342.14 Dissimilar Metals

Where practicable, dissimilar metals in contact anywhere in the system shall be avoided to eliminate the possibility of galvanic action.

Aluminum fittings and enclosures shall be permitted to be used with IMC.

342.20 Size

(A) Minimum IMC smaller than metric designator 16 (trade size ½) shall not be used.

(B) Maximum IMC larger than metric designator 103 (trade size 4) shall not be used.

FPN: See 300.1(C) for the metric designators and trade sizes. These are for identification purposes only and do not relate to actual dimensions.

Table 300.1(C) identifies a distinct metric designator for each circular raceway trade size. For further explanation of metric designators, see 90.9 and the commentary following Table 300.1(C).

342.22 Number of Conductors

The number of conductors shall not exceed that permitted by the percentage fill specified in Table 1, Chapter 9.

Cables shall be permitted to be installed where such use is not prohibited by the respective cable articles. The number of cables shall not exceed the allowable percentage fill specified in Table 1, Chapter 9.

Table 1 of Chapter 9 specifies the maximum fill percentage of a conduit or tubing. Table 4 provides the usable area within the selected conduit or tubing, and Table 5 provides the required area for each conductor. Examples using these tables to calculate a conduit or tubing size are provided in the commentary following Chapter 9, Table 1, Notes to Tables, Note 6.

If the conductors are of the same wire size, the tables in Annex C can be used instead of doing the calculations. Annex C, which contains 12 sets of tables, accurately indicates the maximum number of conductors permitted in a conduit or tubing. Examples using Annex C to select a conduit or tubing size are provided in the commentary following the introduction in the annex.

To select the proper trade size intermediate metal conduit, the section entitled "Article 342—Intermediate Metal Conduit (IMC)" in Table 4 of Chapter 9 should be followed. Tables C.4 and C.4(A) for intermediate metal conduit in Annex C are also permissible.

Editorial changes were made to the last paragraph of this section for the 2005 *Code*.

342.24 Bends — How Made

Bends of IMC shall be so made that the conduit will not be damaged and the internal diameter of the conduit will not be effectively reduced. The radius of the curve of any field bend to the centerline of the conduit shall not be less than indicated in Table 2, Chapter 9.

The term *field bend* means any bend or offset made by installers, using proper tools and equipment, during the installation of conduit systems. Table 2 of Chapter 9 was added for the 2005 *Code* to replace the 2002 *Code* Table 344.24, Radius of Conduit Bends. Table 2 of Chapter 9 is now the common table for raceway field bend measurements.

342.26 Bends — Number in One Run

There shall not be more than the equivalent of four quarter bends (360 degrees total) between pull points, for example, conduit bodies and boxes.

See the commentary following 344.26 for the rationale behind limiting the number of bends.

342.28 Reaming and Threading

All cut ends shall be reamed or otherwise finished to remove rough edges. Where conduit is threaded in the field, a standard cutting die with a taper of 1 in 16 (¾ in. taper per foot) shall be used.

FPN: See ANSI/ASME B.1.20.1-1983, *Standard for Pipe Threads, General Purpose (Inch)*.

Conduit is cut using a saw or a roll cutter (pipe cutter). Care should be taken to ensure a straight cut, given that crooked threads will result from a die not started on the pipe squarely. After the cut is made, the conduit must be reamed. Proper reaming removes burrs from the interior of the cut conduit so that, as wires and cables are pulled through the conduit, no chafing of the insulation or cable jacket occurs. Finally, the conduit is threaded. The number of threads is important, because cutting too many threads prevents a conduit from being made up properly. If a threaded ring gauge is not

available, cut the same number of threads on the conduit as are present on the factory (threaded) end of the conduit.

342.30 Securing and Supporting

IMC shall be installed as a complete system in accordance with 300.18 and shall be securely fastened in place and supported in accordance with 342.30(A) and (B).

(A) Securely Fastened Each IMC shall be securely fastened within 900 mm (3 ft) of each outlet box, junction box, device box, cabinet, conduit body, or other conduit termination. Fastening shall be permitted to be increased to a distance of 1.5 m (5 ft) where structural members do not readily permit fastening within 900 mm (3 ft). Where approved, conduit shall not be required to be securely fastened within 900 mm (3 ft) of the service head for above-the-roof termination of a mast.

As illustrated in Exhibit 342.1, intermediate metal conduit (IMC) is required to be securely fastened at least every 10 ft. Fastening is also required within 3 ft of outlet boxes, junction boxes, cabinets, and conduit bodies. However, where structural support members do not permit fastening within 3 ft, the support may be located up to 5 ft away.

(B) Supports IMC shall be supported in accordance with one of the following:

(1) Conduit shall be supported at intervals not exceeding 3 m (10 ft).
(2) The distance between supports for straight runs of conduit shall be permitted in accordance with Table 344.30(B)(2), provided the conduit is made up with threaded couplings and such supports prevent transmission of stresses to termination where conduit is deflected between supports.
(3) Exposed vertical risers from industrial machinery or fixed equipment shall be permitted to be supported at intervals not exceeding 6 m (20 ft) if the conduit is made up with threaded couplings, the conduit is supported and securely fastened at the top and bottom of the riser, and no other means of intermediate support is readily available.
(4) Horizontal runs of IMC supported by openings through framing members at intervals not exceeding 3 m (10 ft) and securely fastened within 900 mm (3 ft) of termination points shall be permitted.

Section 342.30(B)(4) permits lengths of IMC to be supported (but not necessarily secured) by framing members at 10-ft intervals, provided the IMC is secured and supported

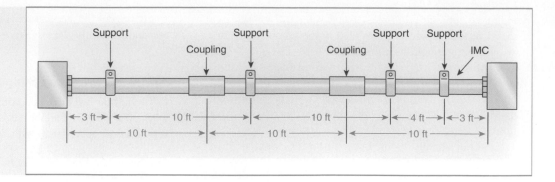

Exhibit 342.1 Minimum fastening requirements for intermediate metal conduit according to 342.30(A).

at least 3 ft from the box or enclosure. Installations where the IMC is installed through the bar joists is just one example and is shown in Exhibit 342.2.

Exhibit 342.2 An example of intermediate metal conduit supported by framing members and securely fastened at the 3-ft distance from the box, as required by 342.30(B)(4).

342.42 Couplings and Connectors

(A) Threadless Threadless couplings and connectors used with conduit shall be made tight. Where buried in masonry or concrete, they shall be the concretetight type. Where installed in wet locations, they shall comply with 314.15(A). Threadless couplings and connectors shall not be used on threaded conduit ends unless listed for the purpose.

(B) Running Threads Running threads shall not be used on conduit for connection at couplings.

See the commentary following 344.42(B) for examples of threadless fittings.

342.46 Bushings

Where a conduit enters a box, fitting, or other enclosure, a bushing shall be provided to protect the wire from abrasion unless the design of the box, fitting, or enclosure is such as to afford equivalent protection.

> FPN: See 300.4(F) for the protection of conductors 4 AWG and larger at bushings.

342.56 Splices and Taps

Splices and taps shall be made in accordance with 300.15.

342.60 Grounding

IMC shall be permitted as an equipment grounding conductor.

III. Construction Specifications

342.120 Marking

Each length shall be clearly and durably marked at least every 1.5 m (5 ft) with the letters IMC. Each length shall be marked as required in 110.21.

342.130 Standard Lengths

The standard length of IMC shall be 3.05 m (10 ft), including an attached coupling, and each end shall be threaded. Longer or shorter lengths with or without coupling and threaded or unthreaded shall be permitted.

ARTICLE 344
Rigid Metal Conduit: Type RMC

Summary of Changes

* **344.22:** Revised to permit cables in RMC if the cable article does not specifically prohibit such use. The number of cables is limited to the percentage fill of Table 1, Chapter 9.

• **Table 344.24:** Relocated to Chapter 9, Table 2.
• **344.42:** Revised to specifically reference 314.15(A) where threadless couplings and connectors are installed in wet locations.

Contents

I. General

344.1 Scope

This article covers the use, installation, and construction specifications for rigid metal conduit (RMC) and associated fittings.

344.2 Definition

Rigid Metal Conduit (RMC). A threadable raceway of circular cross section designed for the physical protection and routing of conductors and cables and for use as an equipment grounding conductor when installed with its integral or associated coupling and appropriate fittings. RMC is generally made of steel (ferrous) with protective coatings

or aluminum (nonferrous). Special use types are silicon bronze and stainless steel.

344.6 Listing Requirements

RMC, factory elbows and couplings, and associated fittings shall be listed.

II. Installation

344.10 Uses Permitted

(A) All Atmospheric Conditions and Occupancies Use of RMC shall be permitted under all atmospheric conditions and occupancies. Ferrous raceways and fittings protected from corrosion solely by enamel shall be permitted only indoors and in occupancies not subject to severe corrosive influences.

Section 344.10(A) makes it clear that aluminum rigid conduit can be used with steel fittings and enclosures, as can aluminum fittings and enclosures with steel rigid conduit. Tests have shown that the galvanic corrosion at steel and aluminum interfaces is minor compared to the natural corrosion on the combination of steel and steel or of aluminum and aluminum.

(B) Corrosion Environments RMC, elbows, couplings, and fittings shall be permitted to be installed in concrete, in direct contact with the earth, or in areas subject to severe corrosive influences where protected by corrosion protection and judged suitable for the condition.

Section 344.10(B) indicates the permitted uses for listed ferrous and nonferrous conduit, including their use in concrete, in direct contact with the earth, and in corrosive areas. The fine print note to 344.10(D) references 300.6 for additional information on protection against corrosion and specific types of corrosion-resistant materials.

The authority having jurisdiction for enforcing this *Code* should be consulted for approval of corrosion-resistant materials or for requirements prior to the installation of nonferrous metal (aluminum) conduit in concrete, since chloride additives in the concrete mix may cause corrosion.

(C) Cinder Fill RMC shall be permitted to be installed in or under cinder fill where subject to permanent moisture where protected on all sides by a layer of noncinder concrete not less than 50 mm (2 in.) thick; where the conduit is not less than 450 mm (18 in.) under the fill; or where protected by corrosion protection and judged suitable for the condition.

Although cinder fill is not commonly used in modern construction, it is still encountered at older building sites. Care should be taken to ensure the proper installation of rigid metal conduit as permitted by 344.10. Where cinders have been used as fill, they may contain sulfur, and where they have combined with moisture, sulfuric acid is formed, which can corrode metal raceways.

(D) Wet Locations All supports, bolts, straps, screws, and so forth, shall be of corrosion-resistant materials or protected against corrosion by corrosion-resistant materials.

> FPN: See 300.6 for protection against corrosion.

344.14 Dissimilar Metals

Where practicable, dissimilar metals in contact anywhere in the system shall be avoided to eliminate the possibility of galvanic action. Aluminum fittings and enclosures shall be permitted to be used with steel RMC, and steel fittings and enclosures shall be permitted to be used with aluminum RMC where not subject to severe corrosive influences.

344.20 Size

(A) Minimum RMC smaller than metric designator 16 (trade size ½) shall not be used.

Exception: For enclosing the leads of motors as permitted in 430.245(B).

(B) Maximum RMC larger than metric designator 155 (trade size 6) shall not be used.

> FPN: See 300.1(C) for the metric designators and trade sizes. These are for identification purposes only and do not relate to actual dimensions.

Table 300.1(C) identifies a distinct metric designator for each circular raceway trade size. For further explanation of metric designators, see 90.9 and the commentary following Table 300.1(C).

344.22 Number of Conductors

The number of conductors shall not exceed that permitted by the percentage fill specified in Table 1, Chapter 9.

Cables shall be permitted to be installed where such use is not prohibited by the respective cable articles. The number of cables shall not exceed the allowable percentage fill specified in Table 1, Chapter 9.

Table 1 of Chapter 9 specifies the maximum fill percentage of a conduit or tubing. Table 4 provides the usable area within the selected conduit or tubing, and Table 5 provides the required area for each conductor. Examples using these tables to calculate a conduit or tubing size are provided in the commentary following Chapter 9, Table 1, Notes to Tables, Note 6.

If the conductors are of the same wire size, Annex C can be used instead of doing the calculations. Annex C,

through 12 sets of tables, accurately indicates the maximum number of conductors permitted in a conduit or tubing. Examples using Annex C to select a conduit or tubing size are provided in the commentary following the introduction in the annex.

To select the proper trade size of rigid metal conduit, the section entitled "Article 344—Rigid Metal Conduit (RMC)" in Table 4 of Chapter 9 should be followed. Annex C, Tables C.8 and C.8(A) for rigid metal conduit are also permissible.

344.24 Bends — How Made

Bends of RMC shall be so made that the conduit will not be damaged and so that the internal diameter of the conduit will not be effectively reduced. The radius of the curve of any field bend to the centerline of the conduit shall not be less than indicated in Table 2, Chapter 9.

The term *field bend* means any bend or offset made by installers, using proper tools and equipment, during the installation of conduit systems. Table 2 of Chapter 9 was added for the 2005 *Code* to replace the 2002 *Code* Table 344.24, Radius of Conduit Bends. Table 2 of Chapter 9 is now the common table for raceway field bend measurements.

344.26 Bends — Number in One Run

There shall not be more than the equivalent of four quarter bends (360 degrees total) between pull points, for example, conduit bodies and boxes.

Limiting the number of bends in a conduit run reduces pulling tension on conductors and helps ensure easy insertion or removal of conductors during later phases of construction, when the conduit may be permanently enclosed by the finish of the building. Adjustments during that time are often impossible. The *Code* does not limit the pull points to conduit bodies and boxes; these are only examples of pull points.

344.28 Reaming and Threading

All cut ends shall be reamed or otherwise finished to remove rough edges. Where conduit is threaded in the field, a standard cutting die with a 1 in 16 taper (¾-in. taper per foot) shall be used.

> FPN: See ANSI/ASME B.1.20.1-1983, *Standard for Pipe Threads, General Purpose (Inch).*

Conduit is cut using a saw or a roll cutter (pipe cutter). Care should be taken to ensure a straight cut because crooked threads will result from a die not started on the pipe squarely. After the cut is made, the conduit must be reamed. Proper reaming removes burrs from the interior of the cut conduit so that as wires and cables are pulled through the conduit, no chafing of the insulation or cable jacket occurs. Finally, the conduit is threaded. The number of threads is important. To determine the correct number of threads for a conduit end, cut the same number of threads on the conduit as are present on the factory (threaded) end of the conduit. Where excessive threads are cut on the conduit and threaded couplings are installed, the conduit will butt within the coupling, resulting in a weak mechanical joint and poor grounding continuity.

344.30 Securing and Supporting

RMC shall be installed as a complete system in accordance with 300.18 and shall be securely fastened in place and supported in accordance with 344.30(A) and (B).

(A) Securely Fastened RMC shall be securely fastened within 900 mm (3 ft) of each outlet box, junction box, device box, cabinet, conduit body, or other conduit termination. Fastening shall be permitted to be increased to a distance of 1.5 m (5 ft) where structural members do not readily permit fastening within 900 mm (3 ft). Where approved, conduit shall not be required to be securely fastened within 900 mm (3 ft) of the service head for above-the-roof termination of a mast.

As illustrated in Exhibit 344.1, rigid metal conduit is required to be securely fastened at least every 10 ft. Secure fastening is also required within 3 ft of outlet boxes, junction boxes, cabinets, and conduit bodies. However, where structural support members do not permit fastening within 3 ft, secure fastening may be located up to 5 ft away.

(B) Supports RMC shall be supported in accordance with one of the following:

(1) Conduit shall be supported at intervals not exceeding 3 m (10 ft).
(2) The distance between supports for straight runs of conduit shall be permitted in accordance with Table

Table 344.30(B)(2) Supports for Rigid Metal Conduit

Conduit Size		Maximum Distance Between Rigid Metal Conduit Supports	
Metric Designator	Trade Size	m	ft
16–21	½–¾	3.0	10
27	1	3.7	12
35–41	1¼–1½	4.3	14
53–63	2–2½	4.9	16
78 and larger	3 and larger	6.1	20

344.30(B)(2), provided the conduit is made up with threaded couplings, and such supports prevent transmission of stresses to termination where conduit is deflected between supports.
(3) Exposed vertical risers from industrial machinery or fixed equipment shall be permitted to be supported at intervals not exceeding 6 m (20 ft) if the conduit is made up with threaded couplings, the conduit is supported and securely fastened at the top and bottom of the riser, and no other means of intermediate support is readily available.
(4) Horizontal runs of RMC supported by openings through framing members at intervals not exceeding 3 m (10 ft) and securely fastened within 900 mm (3 ft) of termination points shall be permitted.

Section 344.30(B)(4) permits lengths of rigid metal conduit to be supported (but not necessarily secured) by framing members at 10-ft intervals, provided the rigid metal conduit is secured and supported at least 3 ft from the box or enclosure. Installations where the rigid metal conduit is installed through the bar joists is just one example and is shown in Exhibit 344.2.

344.42 Couplings and Connectors

(A) Threadless Threadless couplings and connectors used with conduit shall be made tight. Where buried in masonry

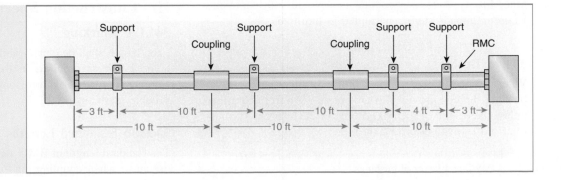

Exhibit 344.1 Minimum support required for rigid metal conduit according to 344.30(A).

Exhibit 344.2 An example of rigid metal conduit supported by framing members and securely fastened at the 3-ft distance from the box, as required by 344.30(B)(4).

Exhibit 344.3 Conduit body with a threadless connector. (Courtesy of Appleton Electric Co., EGS Electrical Group)

Exhibit 344.4 A three-piece-type (union-type) coupling. (Courtesy of Appleton Electric Co., EGS Electrical Group)

or concrete, they shall be the concretetight type. Where installed in wet locations, they shall comply with 314.15(A). Threadless couplings and connectors shall not be used on threaded conduit ends unless listed for the purpose.

(B) Running Threads Running threads shall not be used on conduit for connection at couplings.

Exhibit 344.3 illustrates a threadless connection integral to a conduit body, FS box, and so on. This type of connection may be separate from the conduit body or box as an individual fitting of the compression type (raintight) suitable for wet locations, or it may be of the set-screw type.

Threadless fittings are not intended for use over threads because the fitting will not seat properly. The threaded end of the conduit should be cut off and reamed before installation.

Exhibit 344.4 illustrates a three-piece coupling (the electrical equivalent of a pipe union), which is used to join two lengths of conduit where it is impossible to turn either length, such as in underground or concrete slab construction. Another fitting for joining conduit is a bolted split coupling. Running threads are not permitted to join two conduits.

344.46 Bushings

Where a conduit enters a box, fitting, or other enclosure, a bushing shall be provided to protect the wire from abrasion unless the design of the box, fitting, or enclosure is such as to afford equivalent protection.

    FPN: See 300.4(F) for the protection of conductors sizes 4 AWG and larger at bushings.

344.56 Splices and Taps

Splices and taps shall be made in accordance with 300.15.

344.60 Grounding

RMC shall be permitted as an equipment grounding conductor.

III. Construction Specifications

344.120 Marking

Each length shall be clearly and durably identified in every 3 m (10 ft) as required in the first sentence of 110.21. Nonferrous conduit of corrosion-resistant material shall have suitable markings.

344.130 Standard Lengths

The standard length of RMC shall be 3.05 m (10 ft), including an attached coupling, and each end shall be threaded.

Longer or shorter lengths with or without coupling and threaded or unthreaded shall be permitted.

ARTICLE 348
Flexible Metal Conduit: Type FMC

Summary of Changes

- **348.22:** Revised to permit cables in FMC if the cable article does not specifically prohibit such use. The number of cables is limited to the percentage fill of Table 1, Chapter 9.

- **348.30: Exception No. 2:** Revised to permit 4-ft and 5-ft unsupported lengths at terminals where flexibility is required, depending on the trade size of the conduit.

Contents

I. General

348.1 Scope

This article covers the use, installation, and construction specifications for flexible metal conduit (FMC) and associated fittings.

348.2 Definition

Flexible Metal Conduit (FMC). A raceway of circular cross section made of helically wound, formed, interlocked metal strip.

348.6 Listing Requirements

FMC and associated fittings shall be listed.

All flexible metal conduit as well as all flexible metal conduit fittings are required to be listed.

II. Installation

348.10 Uses Permitted

FMC shall be permitted to be used in exposed and concealed locations.

348.12 Uses Not Permitted

FMC shall not be used in the following:

(1) In wet locations unless the conductors are approved for the specific conditions and the installation is such that liquid is not likely to enter raceways or enclosures to which the conduit is connected

Listed flexible metal conduit (FMC) is permitted for use in wet locations, provided the completed installation prevents water from entering enclosures or other raceways to which the conduit is connected. Also, for this application, the conductors must be suitable for wet locations. Listed FMC ½ in. and larger may be installed in unlimited lengths, provided an equipment grounding conductor is installed with the circuit conductors. See 250.118(5) as well as 348.60 for specific requirements related to the use of FMC as an equipment grounding conductor.

(2) In hoistways, other than as permitted in 620.21(A)(1)
(3) In storage battery rooms
(4) In any hazardous (classified) location other than as permitted in 501.10(B) and 504.20
(5) Where exposed to materials having a deteriorating effect on the installed conductors, such as oil or gasoline
(6) Underground or embedded in poured concrete or aggregate
(7) Where subject to physical damage

348.20 Size

(A) Minimum FMC less than metric designator 16 (trade size ½) shall not be used unless permitted in 348.20(A)(1) through (A)(5) for metric designator 12 (trade size ⅜).

(1) For enclosing the leads of motors as permitted in 430.245(B)
(2) In lengths not in excess of 1.8 m (6 ft) for any of the following uses:

a. For utilization equipment
b. As part of a listed assembly
c. For tap connections to luminaires (lighting fixtures) as permitted in 410.67(C)

Section 348.20(A)(2) makes it clear that ⅜-in. flexible metal conduit is permitted to be used as the manufactured or field-installed metal raceway (1½ ft to 6 ft in length) to enclose tap conductors between the outlet box and the terminal housing of recessed luminaires. Flexible metal conduit is also permitted to be used as a 6-ft fixture whip from an outlet box to a luminaire.

Section 604.6(A) permits a smaller minimum size for manufactured wiring systems because the conductors are not as prone to physical damage when assembled under factory-controlled conditions.

(3) For manufactured wiring systems as permitted in 604.6(A)
(4) In hoistways as permitted in 620.21(A)(1)
(5) As part of a listed assembly to connect wired luminaire (fixture) sections as permitted in 410.77(C)

(B) Maximum FMC larger than metric designator 103 (trade size 4) shall not be used.

> FPN: See 300.1(C) for the metric designators and trade sizes. These are for identification purposes only and do not relate to actual dimensions.

348.22 Number of Conductors

The number of conductors shall not exceed that permitted by the percentage fill specified in Table 1, Chapter 9, or as permitted in Table 348.22, or for metric designator 12 (trade size ⅜).

Cables shall be permitted to be installed where such use is not prohibited by the respective cable articles. The number of cables shall not exceed the allowable percentage fill specified in Table 1, Chapter 9.

Table 1 of Chapter 9 specifies the maximum fill percentage of a conduit or tubing. Table 4 provides the usable area within the selected conduit or tubing, and Table 5 provides the required area for each of the conductors. Examples using these tables to calculate a conduit or tubing size are provided in the commentary following Chapter 9, Table 1, Notes to Tables, Note 6.

If the conductors are of the same wire size, the tables in Annex C may be used instead of doing the calculations. Annex C, with 12 sets of tables, accurately indicates the maximum number of conductors permitted in a conduit or tubing. Examples using Annex C to select a conduit or tubing size are provided in the commentary following the introduction in the annex.

To select the proper trade size of flexible metal conduit, the section entitled "Article 348 — Flexible Metal Conduit (FMC)" in Table 4 of Chapter 9 should be followed. Annex C, Tables C.3 and C.3(A) for flexible metal conduit are also permissible.

348.24 Bends — How Made

Bends in conduit shall be made so that the conduit is not damaged and the internal diameter of the conduit is not effectively reduced. Bends shall be permitted to be made manually without auxiliary equipment. The radius of the curve to the centerline of any bend shall not be less than shown in Table 2, Chapter 9 using the column "Other Bends."

Table 2 of Chapter 9 was added for the 2005 *Code* to replace the 2002 *Code* Table 344.24, Radius of Conduit Bends. Table 2 of Chapter 9 is now the common table for raceway field bend measurements.

Table 348.22 Maximum Number of Insulated Conductors in Metric Designator 12 (Trade Size ⅜) Flexible Metal Conduit*

Size (AWG)	Types RFH-2, SF-2		Types TF, XHHW, TW		Types TFN, THHN, THWN		Types FEP, FEBP, PF, PGF	
	Fittings Inside Conduit	Fittings Outside Conduit	Fittings Inside Conduit	Fittings Outside Conduit	Fittings Inside Conduit	Fittings Outside Conduit	Fittings Inside Conduit	Fittings Outside Conduit
18	2	3	3	5	5	8	5	8
16	1	2	3	4	4	6	4	6
14	1	2	2	3	3	4	3	4
12	—	—	1	2	2	3	2	3
10	—	—	1	1	1	1	1	2

*In addition, one covered or bare equipment grounding conductor of the same size shall be permitted.

348.26 Bends — Number in One Run

There shall not be more than the equivalent of four quarter bends (360 degrees total) between pull points, for example, conduit bodies and boxes.

As with other raceways, a run of flexible metal conduit installed between boxes, conduit bodies, and so on, is not permitted to contain more than the equivalent of four quarter bends (360 degrees total). Proper shaping and support of this flexible wiring method ensures that conductors can be easily installed or withdrawn at any time.

348.28 Trimming

All cut ends shall be trimmed or otherwise finished to remove rough edges, except where fittings that thread into the convolutions are used.

348.30 Securing and Supporting

FMC shall be securely fastened in place and supported in accordance with 348.30(A) and (B).

(A) Securely Fastened FMC shall be securely fastened in place by an approved means within 300 mm (12 in.) of each box, cabinet, conduit body, or other conduit termination and shall be supported and secured at intervals not to exceed 1.4 m (4½ ft).

Exception No. 1: Where FMC is fished.

Exception No. 2: At terminals where flexibility is required, lengths shall not exceed the following:

(1) 900 mm (3 ft) for metric designators 16 through 35 (trade sizes ½ through 1¼)

(2) 1200 mm (4 ft) for metric designators 41 through 53 (trade sizes 1½ through 2)

(3) 1500 mm (5 ft) for metric designators 63 (trade size 2½) and larger

Exception No. 3: Lengths not exceeding 1.8 m (6 ft) from a luminaire (fixture) terminal connection for tap connections to luminaires (light fixtures) as permitted in 410.67(C).

Exception No. 4: Lengths not exceeding 1.8 m (6 ft) from the last point where the raceway is securely fastened for connections within an accessible ceiling to luminaire(s) [lighting fixture(s)] or other equipment.

Exception No. 2 was revised for the 2005 *Code* to make the simple 3 ft rule more practical and usable for larger flex sizes. Exception No. 4 was revised for the 2005 *Code* to correlate this permission of unsupported flexible metal conduit with other wiring methods permitting the same exception throughout Chapter 3. One such example is 320.30(D)(3) for Type AC cable.

(B) Supports Horizontal runs of FMC supported by openings through framing members at intervals not greater than 1.4 m (4½ ft) and securely fastened within 300 mm (12 in.) of termination points shall be permitted.

348.42 Couplings and Connectors

Angle connectors shall not be used for concealed raceway installations.

348.56 Splices and Taps

Splices and taps shall be made in accordance with 300.15.

348.60 Grounding and Bonding

Where used to connect equipment where flexibility is required, an equipment grounding conductor shall be installed.

Where flexibility is not required, FMC shall be permitted to be used as an equipment grounding conductor when installed in accordance with 250.118(5).

Where required or installed, equipment grounding conductors shall be installed in accordance with 250.134(B).

Where required or installed, equipment bonding jumpers shall be installed in accordance with 250.102.

According to the product standard UL 1, *Standard for Flexible Metal Conduit*, FMC longer than 6 ft has not been judged to be suitable for grounding purposes. The general rules for permitting or not permitting FMC for grounding purposes are found in 250.118(5). The requirements in 250.118 for flexible metal conduit were changed for the 2005 *Code*. According to 348.60, an additional equipment grounding conductor is always required where FMC is used for flexibility. Examples of such installations include using FMC to minimize the transmission of vibration from equipment such as motors or to provide flexibility for floodlights, spotlights, or other equipment that requires adjustment after installation.

Another specific exception is the requirement for a bonding jumper where FMC is used in hazardous (classified) locations. See 501.30(B), 502.30(B), and 503.30(B) for details on types of equipment grounding conductors. In addition, 250.102(E) permits the routing of equipment bonding jumpers on the outside of the raceway in lengths that are no longer than 6 ft and bonded at each end.

According to 250.118(5), where the length of the total ground-fault return path exceeds 6 ft or the circuit overcurrent protection exceeds 20 amperes, a separate equipment grounding conductor must be installed with the circuit conductors. The upper sketch in Exhibit 348.1 shows an acceptable application of flexible metal conduit where the total length of any ground return path is limited to 6 ft. The lower sketch shows an application that is unacceptable because the

Exhibit 348.1 An example of acceptable and unacceptable applications of flexible metal conduit without separate equipment grounding conductors used as a fixture whip, in accordance with 250.118(5)(c).

grounding return path for luminaire 2 exceeds the permitted maximum of 6 ft to the box.

ARTICLE 350
Liquidtight Flexible Metal Conduit: Type LFMC

Summary of Changes

- **350.22:** Revised to permit cables in LFMC if the cable article does not specifically prohibit such use. The number of cables is limited to the percentage fill of Table 1, Chapter 9.

- **350.30:** Added Exception No. 4 to permit 6-ft unsupported length within an accessible ceiling to connect to luminaires or other equipment.

- **350.60:** Revised to permit use for equipment grounding where flexibility of the raceway is not required.

Contents

I. General

350.1 Scope

This article covers the use, installation, and construction specifications for liquidtight flexible metal conduit (LFMC) and associated fittings.

350.2 Definition

Liquidtight Flexible Metal Conduit (LFMC). A raceway of circular cross section having an outer liquidtight, nonmetallic, sunlight-resistant jacket over an inner flexible metal core with associated couplings, connectors, and fittings for the installation of electric conductors.

350.6 Listing Requirements

LFMC and associated fittings shall be listed.

II. Installation

350.10 Uses Permitted

LFMC shall be permitted to be used in exposed or concealed locations as follows:

(1) Where conditions of installation, operation, or maintenance require flexibility or protection from liquids, vapors, or solids

(2) As permitted by 501.10(B), 502.10, 503.10, and 504.20 and in other hazardous (classified) locations where specifically approved, and by 553.7(B)

(3) For direct burial where listed and marked for the purpose

Liquidtight flexible metal conduit (LFMC) is intended for use in wet locations or where exposed to mineral oil, both at a maximum temperature of 140°F. LFMC is not intended for use where exposed to gasoline or similar light petroleum solvents unless so marked on the product. LFMC is required to be a listed product. If properly marked for the application, LFMC is permitted for direct burial in the earth. Note that the requirements of 300.5 are applicable to LFMC if installed underground. LFMC is on the permitted list of wiring methods allowed for services (230.43), provided the length does not exceed 6 ft and an equipment bonding jumper is installed in accordance with 250.102. LFMC may be installed in unlimited lengths, provided it meets the other requirements of Article 350 and a separate equipment grounding conductor is installed with the circuit conductors.

Liquidtight flexible nonmetallic conduit (LFNC) is also used extensively in the machine tool and related industries. See 14.5.5 in NFPA 79, *Electrical Standard for Industrial Machinery*, for the uses permitted on an industrial machine.

350.12 Uses Not Permitted

LFMC shall not be used as follows:

(1) Where subject to physical damage
(2) Where any combination of ambient and conductor temperature produces an operating temperature in excess of that for which the material is approved

350.20 Size

(A) Minimum LFMC smaller than metric designator 16 (trade size ½) shall not be used.

Exception: LFMC of metric designator 12 (trade size ⅜) shall be permitted as covered in 348.20(A).

(B) Maximum The maximum size of LFMC shall be metric designator 103 (trade size 4).

> FPN: See 300.1(C) for the metric designators and trade sizes. These are for identification purposes only and do not relate to actual dimensions.

350.22 Number of Conductors or Cables

(A) Metric Designators 16 through 103 (Trade Sizes ½ through 4) The number of conductors shall not exceed that permitted by the percentage fill specified in Table 1, Chapter 9.

Cables shall be permitted to be installed where such use is not prohibited by the respective cable articles. The number of cables shall not exceed the allowable percentage fill specified in Table 1, Chapter 9.

(B) Metric Designator 12 (Trade Size ⅜) The number of conductors shall not exceed that permitted in Table 348.22, "Fittings Outside Conduit" columns.

Table 1 of Chapter 9 specifies the maximum percent fill of a conduit or tubing. Table 4 provides the usable area within the selected conduit or tubing, and Table 5 provides the required area for each of the conductors. Examples using these tables to calculate a conduit or tubing size are provided in the commentary following Chapter 9, Table 1, Notes to Tables, Note 6.

If the conductors are of the same wire size, the tables in Annex C may be used instead of doing the calculations. Annex C, with 12 sets of tables, accurately indicates the maximum number of conductors permitted in a conduit or tubing. Examples using Annex C to select a conduit or tubing size are provided in the commentary following the introduction in the annex.

To select the proper trade size of liquidtight flexible metal conduit, the section entitled "Article 350—Liquidtight Flexible Metal Conduit (LFMC)" in Table 4 of Chapter 9 should be followed. Annex C, Tables C.7 and C.7(A) for liquidtight flexible metal conduit are also permissible.

350.24 Bends — How Made

Bends in conduit shall be so made that the conduit will not be damaged and the internal diameter of the conduit will not be effectively reduced. Bends shall be permitted to be made manually without auxiliary equipment. The radius of the curve to the centerline of any bend shall not be less than required in Table 2, Chapter 9 using the column "Other Bends."

Table 2 of Chapter 9 was added for the 2005 *Code* to replace the 2002 *Code* Table 344.24, Radius of Conduit Bends. Table 2 of Chapter 9 is now the common table for raceway field bend measurements.

350.26 Bends — Number in One Run

There shall not be more than the equivalent of four quarter bends (360 degrees total) between pull points, for example, conduit bodies and boxes.

350.30 Securing and Supporting

LFMC shall be securely fastened in place and supported in accordance with 350.30(A) and (B).

(A) Securely Fastened LFMC shall be securely fastened in place by an approved means within 300 mm (12 in.) of each box, cabinet, conduit body, or other conduit termination and shall be supported and secured at intervals not to exceed 1.4 m (4½ ft).

Exception No. 1: Where LFMC is fished.

Exception No. 2: Lengths not exceeding 900 mm (3 ft) at terminals where flexibility is necessary.

Exception No. 3: Lengths not exceeding 1.8 m (6 ft) from a luminaire (fixture) terminal connection for tap conductors to luminaires (lighting fixtures), as permitted in 410.67(C).

Exception No. 4: Lengths not exceeding 1.8 m (6 ft) from the last point where the raceway is securely fastened for connections within an accessible ceiling to luminaire(s) [lighting fixture(s)] or other equipment.

Exception No. 4 was revised for the 2005 *Code* to correlate this permission of unsupported liquidtight flexible metal conduit with other wiring methods permitting the same exception throughout Chapter 3. One such example is 320.30(D)(3) for Type AC cable.

(B) Supports Horizontal runs of LFMC supported by openings through framing members at intervals not greater than 1.4 m (4½ ft) and securely fastened within 300 mm (12 in.) of termination points shall be permitted.

350.42 Couplings and Connectors

Angle connectors shall not be used for concealed raceway installations.

350.56 Splices and Taps

Splices and taps shall be made in accordance with 300.15.

350.60 Grounding and Bonding

Where used to connect equipment where flexibility is required, an equipment grounding conductor shall be installed.

Where flexibility is not required, LFMC shall be permitted to be used as an equipment grounding conductor when installed in accordance with 250.118(6).

Where required or installed, equipment grounding conductors shall be installed in accordance with 250.134(B).

Where required or installed, equipment bonding jumpers shall be installed in accordance with 250.102.

FPN: See 501.30(B), 502.30(B), and 503.30(B) for types of equipment grounding conductors.

III. Construction Specifications

350.120 Marking

LFMC shall be marked according to 110.21. The trade size and other information required by the listing shall also be marked on the conduit. Conduit suitable for direct burial shall be so marked.

ARTICLE 352
Rigid Nonmetallic Conduit: Type RNC

Summary of Changes

- **352.10(H):** Revised to prohibit nonmetallic conduit bodies that are supported by nonmetallic conduit from supporting equipment or containing devices other than splicing devices.
- **352.12(E):** Added exception permitting the installation of conductors or cables with temperature ratings higher than the temperature rating of the rigid nonmetallic conduit, provided the conductors or cables are operated at a temperature that does not exceed the temperature rating of the RNC.
- **352.22:** Revised to permit cables in RNC if the cable article does not specifically prohibit such use. The number of cables is limited to the percentage fill of Table 1, Chapter 9.

Contents

I. General

352.1 Scope

This article covers the use, installation, and construction specifications for rigid nonmetallic conduit (RNC) and associated fittings.

352.2 Definition

Rigid Nonmetallic Conduit (RNC). A nonmetallic raceway of circular cross section, with integral or associated couplings, connectors, and fittings for the installation of electrical conductors and cables.

The 2004 UL *General Information for Electrical Equipment Directory* (White Book) describes two types of rigid nonmetallic conduit recognized for use in accordance with Article 352 as described in the following extract.

1. Rigid nonmetallic Schedule 40 and Schedule 80 PVC conduit (DZYR)
2. Reinforced thermosetting resin conduit (DZKT)

Rigid Nonmetallic Schedule 40 and Schedule 80 PVC Conduit (DZYR)

Rigid nonmetallic Schedule 40 PVC conduit is suitable for underground use by direct burial or encasement in concrete. Unless marked "Underground Use Only" or equivalent wording, Schedule 40 conduit is also suitable for above ground use indoors or outdoors exposed to sunlight and weather where not subject to physical damage.

Schedule 80 conduit has a reduced cross-sectional area available for wiring space and is suitable for use wherever Schedule 40 conduit may be used. The marking "Schedule 80" identifies the conduit as suitable for use where exposed to physical damage and for installations on poles to comply with Section 352.10(F) and 352.12(C).

Unless marked for higher temperature, rigid nonmetallic conduit is intended for use with wires rated 75°C or less, including where it is encased in concrete within buildings and where ambient temperature is 50°C or less. Where encased in concrete in trenches outside of buildings, it is suitable for use with wires rated 90°C or less.

Listed PVC conduit is inherently resistant to atmosphere containing common industrial corrosive agents and will also withstand vapors or mist of caustic, pickling acids, plating bath, and hydrofluoric and chromic acids.

PVC conduit, elbows, and bends (including couplings) that have been investigated for direct exposure to other reagents may be identified by the designation "Reagent Resistant" printed on the surface of the product. Further information on reagent resistance may be found in the 2004 UL *Electrical Construction Materials Directory* and in UL 651, "Schedule 40 and 80 Rigid PVC Conduit."

PVC conduit is designed for connection to couplings, fittings, and boxes by the use of a suitable solvent-type cement. Instructions supplied by the solvent-type cement manufacturer describe the method of assembly and precautions to be followed.

Reinforced Thermosetting Resin Conduit (DZKT)

[formerly referred to as rigid nonmetallic fiberglass conduit and sometimes fiberglass reinforced epoxy conduit (FRE) conduit.]

Reinforced thermosetting resin conduit (RTRC) marked "Below Ground" or ("BG") has been investigated for underground use only—for direct burial, with or without encasement in concrete.

RTRC conduit marked "Above Ground" (or "AG") has been investigated for use above ground, under ground, and for direct burial with or without encasement in concrete. This conduit has been evaluated for concealed or exposed work where not subject to physical damage.

Reinforced thermosetting resin conduit has been investigated for use with wires rated 90°C or less.

Reinforced thermosetting resin conduit is listed in trade sizes ½ to 6 in. inclusive in IPS, ID, RTRC 40 and RTRC 80 dimensions, as marked on the product. Listing includes straight conduit, elbows, bends, and other fittings, unless otherwise noted.

Reinforced thermosetting resin conduit, elbows, bends, and other fittings, which have been investigated for direct exposure to reagents, are identified by the designation "Reagent Resistant" and are marked to indicate the specific reagents.

Reinforced thermosetting resin conduit is designed for connection to couplings, fittings, and boxes by use of a suitable epoxy-type cement or drive-on bell and spigot. Instructions supplied by the epoxy-type cement manufacturer describe the method of assembly and precautions to be followed.

For use of Schedule 80, see 300.5(D), 551.80(B), and 300.50(B).

352.6 Listing Requirements

RNC, factory elbows, and associated fittings shall be listed.

II. Installation

352.10 Uses Permitted

The use of RNC shall be permitted in accordance with 352.10(A) through (H).

> FPN: Extreme cold may cause some nonmetallic conduits to become brittle and therefore more susceptible to damage from physical contact.

(A) Concealed RNC shall be permitted in walls, floors, and ceilings.

(B) Corrosive Influences RNC shall be permitted in locations subject to severe corrosive influences as covered in 300.6 and where subject to chemicals for which the materials are specifically approved.

(C) Cinders RNC shall be permitted in cinder fill.

(D) Wet Locations RNC shall be permitted in portions of dairies, laundries, canneries, or other wet locations and in locations where walls are frequently washed, the entire conduit system including boxes and fittings used therewith shall be installed and equipped so as to prevent water from entering the conduit. All supports, bolts, straps, screws, and so forth, shall be of corrosion-resistant materials or be protected against corrosion by approved corrosion-resistant materials.

(E) Dry and Damp Locations RNC shall be permitted for use in dry and damp locations not prohibited by 352.12.

(F) Exposed RNC shall be permitted for exposed work where not subject to physical damage if identified for such use.

(G) Underground Installations For underground installations, see 300.5 and 300.50.

> FPN: Refer to Article 353 for High Density Polyethylene Conduit: Type HDPE.

(H) Support of Conduit Bodies Rigid nonmetallic conduit shall be permitted to support nonmetallic conduit bodies not larger than the largest trade size of an entering raceway. These conduit bodies shall not support luminaires (fixtures) or other equipment and shall not contain devices other than splicing devices as permitted by 110.14(B) and 314.16(C)(2).

352.12 Uses Not Permitted

RNC shall not be used under the following conditions.

(A) Hazardous (Classified) Locations

(1) In hazardous (classified) locations, except as permitted in 503.10(A), 504.20, 514.8 Exception No. 2, and 515.8

(2) In Class I, Division 2 locations, except as permitted in 501.10(B)(3)

(B) Support of Luminaires (Fixtures) For the support of luminaires (fixtures) or other equipment not described in 352.10(H).

(C) Physical Damage Where subject to physical damage unless identified for such use.

(D) Ambient Temperatures Where subject to ambient temperatures in excess of 50°C (122°F) unless listed otherwise.

(E) Insulation Temperature Limitations For conductors or cables operating at a temperature higher than the RNC listed operating temperature rating.

Exception: Conductors or cables rated at a temperature higher than the RNC listed temperature rating shall be permitted to be installed in RNC, provided they are not operated at a temperature higher than the RNC listed temperature rating.

An exception was added to 352.12(E) for the 2005 *Code*. One example of an application for this exception is the use of 105°C rated medium voltage cables, Type MV, where the cable ampacity at the 105°C rating is reduced to the cable ampacity at 75°C or 90°C and, thus, matches the listed operating temperature rating of the nonmetallic conduit (75°C or 90°C.)

(F) Theaters and Similar Locations In theaters and similar locations, except as provided in 518.4 and 520.5.

Nonmetallic conduits are not permitted to be installed in ducts, plenums, and other air-handling spaces. See 300.22, which limits the use of materials in ducts, plenums, and other air-handling spaces that may contribute smoke and products of combustion during a fire.

Additionally, rigid nonmetallic conduit is not permitted in places of assembly or theaters except as permitted in 518.4 and 520.5. See those sections for specific details.

352.20 Size

(A) Minimum RNC smaller than metric designator 16 (trade size ½) shall not be used.

(B) Maximum RNC larger than metric designator 155 (trade size 6) shall not be used.

> FPN: The trade sizes and metric designators are for identification purposes only and do not relate to actual dimensions. See 300.1(C).

352.22 Number of Conductors

The number of conductors shall not exceed that permitted by the percentage fill specified in Table 1, Chapter 9.

Cables shall be permitted to be installed where such use is not prohibited by the respective cable articles. The number of cables shall not exceed the allowable percentage fill specified in Table 1, Chapter 9.

Table 1 of Chapter 9 specifies the maximum percent fill of a conduit or tubing. Table 4 provides the usable area within the selected conduit or tubing, and Table 5 provides the required area for each of the conductors. Examples using these tables to calculate a conduit or tubing size are provided in the commentary following Chapter 9, Table 1, Notes to Tables, Note 6.

If the conductors are of the same wire size, the tables in Annex C may be used instead of doing the calculations. Annex C, which contains 12 sets of tables, accurately indicates the maximum number of conductors permitted in a conduit or tubing. Examples using Annex C to select a conduit or tubing size are provided in the commentary following the introduction in the annex.

To permit selection of the proper trade size rigid nonmetallic conduit, Table 4 of Chapter 9 contains four separate subtables, one for each type of rigid nonmetallic conduit. The appropriate table for the given type of rigid nonmetallic conduit should be followed. Annex C Tables C.9 and C.9(A) through C.12 and C.12(A) are also permissible, provided the appropriate table for the given type of rigid nonmetallic conduit is used.

Where Schedule 80 RNC is used, notice that the cross-sectional area available for wires is considerably less than that of other raceways of the same trade size due to the extra-thick wall of a Schedule 80 conduit.

352.24 Bends — How Made

Bends shall be so made that the conduit will not be damaged and the internal diameter of the conduit will not be effectively reduced. Field bends shall be made only with bending equipment identified for the purpose. The radius of the curve to the centerline of such bends shall not be less than shown in Table 2, Chapter 9.

The installation of rigid metal conduit in runs of PVC conduit installed underground is covered in 300.5(D) and the associated commentary. The term *field bend* means any bend or offset made by installers, using proper tools and equipment, during the installation of conduit systems. Table 2 of Chapter 9 was added for the 2005 *Code* to replace the 2002 *Code* Table 344.24, Radius of Conduit Bends. Table 2 of Chapter 9 is now the common table for raceway field bend measurements.

352.26 Bends — Number in One Run

There shall not be more than the equivalent of four quarter bends (360 degrees total) between pull points, for example, conduit bodies and boxes.

Limiting the number of bends in a conduit run reduces pulling tension on conductors and helps ensure easy insertion or removal of conductors during later phases of construction, when the conduit may be permanently enclosed by the finish of the building. Adjustments during that time are often impossible. The *Code* does not limit the pull points to conduit bodies and boxes; these are only examples of pull points.

352.28 Trimming

All cut ends shall be trimmed inside and outside to remove rough edges.

352.30 Securing and Supporting

RNC shall be installed as a complete system as provided in 300.18 and shall be fastened so that movement from thermal expansion or contraction is permitted. RNC shall be securely fastened and supported in accordance with 352.30(A) and (B).

The requirements of 352.30 are fairly stringent because they are based on ambient temperatures higher than normally encountered and use horizontal-support tests only. Expansion can cause damage to the raceway or its supports. Expansion fittings, therefore, should be used, and the supports must allow expansion/contraction cycles without damage. See the commentary on expansion fittings following 352.44 for details.

(A) Securely Fastened RNC shall be securely fastened within 900 mm (3 ft) of each outlet box, junction box, device box, conduit body, or other conduit termination. Conduit listed for securing at other than 900 mm (3 ft) shall be permitted to be installed in accordance with the listing.

(B) Supports RNC shall be supported as required in Table 352.30(B). Conduit listed for support at spacings other than as shown in Table 352.30(B) shall be permitted to be installed in accordance with the listing. Horizontal runs of RNC supported by openings through framing members at intervals not exceeding those in Table 352.30(B) and securely fastened within 900 mm (3 ft) of termination points shall be permitted.

Table 352.30(B) Support of Rigid Nonmetallic Conduit (RNC)

Conduit Size		Maximum Spacing Between Supports	
Metric Designator	Trade Size	mm or m	ft
16–27	½–1	900 mm	3
35–53	1¼–2	1.5 m	5
63–78	2½–3	1.8 m	6
91–129	3½–5	2.1 m	7
155	6	2.5 m	8

352.44 Expansion Fittings

Expansion fittings for RNC shall be provided to compensate for thermal expansion and contraction where the length change, in accordance with Table 352.44(A) or Table 352.44(B), is expected to be 6 mm (¼ in.) or greater in a straight run between securely mounted items such as boxes, cabinets, elbows, or other conduit terminations.

Expansion fittings are generally provided in exposed runs of rigid nonmetallic conduit where (1) the run is long, (2) the run is subject to large temperature variations during or after installation, or (3) expansion and contraction measures are provided for the building or other structures. Rigid nonmetallic conduit exhibits a considerably greater change in length per degree change in temperature than do metal raceway systems.

In some parts of the United States and other countries, outdoor temperature variations of over 100°F are common. According to Table 352.44(A), a 100-ft run of PVC rigid nonmetallic conduit will change 4.06 in. in length if the temperature change is 100°F.

The normal expansion range of most larger sizes of rigid nonmetallic conduit expansion couplings is generally 6 in. Information concerning installation and application of this type of coupling may be obtained from manufacturers' instructions.

Expansion fittings are seldom used underground, where temperatures are relatively constant. If rigid nonmetallic conduit is buried or covered immediately, expansion and contraction are not considered a problem.

352.46 Bushings

Where a conduit enters a box, fitting, or other enclosure, a bushing or adapter shall be provided to protect the wire from abrasion unless the box, fitting, or enclosure design provides equivalent protection.

Table 352.44(A) Expansion Characteristics of PVC Rigid Nonmetallic Conduit Coefficient of Thermal Expansion = 6.084 × 10^{-5} mm/mm/°C (3.38 × 10^{-5} in./in./°F)

Temperature Change (°C)	Length Change of PVC Conduit (mm/m)	Temperature Change (°F)	Length Change of PVC Conduit (in./100 ft)	Temperature Change (°F)	Length Change of PVC Conduit (in./100 ft)
5	0.30	5	0.20	105	4.26
10	0.61	10	0.41	110	4.46
15	0.91	15	0.61	115	4.66
20	1.22	20	0.81	120	4.87
25	1.52	25	1.01	125	5.07
30	1.83	30	1.22	130	5.27
35	2.13	35	1.42	135	5.48
40	2.43	40	1.62	140	5.68
45	2.74	45	1.83	145	5.88
50	3.04	50	2.03	150	6.08
55	3.35	55	2.23	155	6.29
60	3.65	60	2.43	160	6.49
65	3.95	65	2.64	165	6.69
70	4.26	70	2.84	170	6.90
75	4.56	75	3.04	175	7.10
80	4.87	80	3.24	180	7.30
85	5.17	85	3.45	185	7.50
90	5.48	90	3.65	190	7.71
95	5.78	95	3.85	195	7.91
100	6.08	100	4.06	200	8.11

Table 352.44(B) Expansion Characteristics of Reinforced Thermosetting Resin Conduit (RTRC)
Coefficient of Thermal Expansion = 2.7×10^{-5} mm/mm/°C (1.5×10^{-5} in./in./°F)

Temperature Change (°C)	Length Change of RTRC Conduit (mm/m)	Temperature Change (°F)	Length Change of RTRC Conduit (in./100 ft)	Temperature Change (°F)	Length Change of RTRC Conduit (in./100 ft)
5	0.14	5	0.09	105	1.89
10	0.27	10	0.18	110	1.98
15	0.41	15	0.27	115	2.07
20	0.54	20	0.36	120	2.16
25	0.68	25	0.45	125	2.25
30	0.81	30	0.54	130	2.34
35	0.95	35	0.63	135	2.43
40	1.08	40	0.72	140	2.52
45	1.22	45	0.81	145	2.61
50	1.35	50	0.90	150	2.70
55	1.49	55	0.99	155	2.79
60	1.62	60	1.08	160	2.88
65	1.76	65	1.17	165	2.97
70	1.89	70	1.26	170	3.06
75	2.03	75	1.35	175	3.15
80	2.16	80	1.44	180	3.24
85	2.30	85	1.53	185	3.33
90	2.43	90	1.62	190	3.42
95	2.57	95	1.71	195	3.51
100	2.70	100	1.80	200	3.60

FPN: See 300.4(F) for the protection of conductors 4 AWG and larger at bushings.

352.48 Joints

All joints between lengths of conduit, and between conduit and couplings, fittings, and boxes, shall be made by an approved method.

352.56 Splices and Taps

Splices and taps shall be made in accordance with 300.15.

352.60 Grounding

Where equipment grounding is required, a separate equipment grounding conductor shall be installed in the conduit.

Exception No. 1: As permitted in 250.134(B), Exception No. 2, for dc circuits and 250.134(B), Exception No. 1, for separately run equipment grounding conductors.

Exception No. 2: Where the grounded conductor is used to ground equipment as permitted in 250.142.

III. Construction Specifications

352.100 Construction

RNC and fittings shall be composed of suitable nonmetallic material that is resistant to moisture and chemical atmospheres. For use above ground, it shall also be flame retardant, resistant to impact and crushing, resistant to distortion from heat under conditions likely to be encountered in service, and resistant to low temperature and sunlight effects. For use underground, the material shall be acceptably resistant to moisture and corrosive agents and shall be of sufficient strength to withstand abuse, such as by impact and crushing, in handling and during installation. Where intended for direct burial, without encasement in concrete, the material shall also be capable of withstanding continued loading that is likely to be encountered after installation.

352.120 Marking

Each length of RNC shall be clearly and durably marked at least every 3 m (10 ft) as required in the first sentence of 110.21. The type of material shall also be included in the marking unless it is visually identifiable. For conduit recognized for use above ground, these markings shall be permanent. For conduit limited to underground use only, these markings shall be sufficiently durable to remain legible until the material is installed. Conduit shall be permitted to be surface marked to indicate special characteristics of the material.

FPN: Examples of these markings include but are not limited to "limited smoke" and "sunlight resistant."

ARTICLE 353
High Density Polyethylene Conduit: Type HDPE Conduit

Summary of Changes

- Added article covering installation, use, and construction of Type HDPE conduit.

Contents

I. General

353.1 Scope

This article covers the use, installation, and construction specifications for high density polyethylene (HDPE) conduit and associated fittings.

353.2 Definition

High Density Polyethylene (HDPE) Conduit. A nonmetallic raceway of circular cross section, with associated couplings, connectors, and fittings for the installation of electrical conductors.

Article 353, new for the 2005 *Code*, deals only with high density polyethylene conduit, Type HDPE. The following

information is taken from the UL Guide Card (EAZX) in the 2004 UL *General Information Directory* (White Book).

Rigid Nonmetallic Underground Conduit, Plastic (EAZX)

One type of conduit under this guide card is High-density Polyethylene Conduit: Type HDPE Conduit, Schedule 40. High-density polyethylene (HDPE) Schedule 40 conduit is a plastic type of rigid nonmetallic conduit, for use only when installed underground. The HDPE Schedule 40 conduit is intended for direct burial with or without being encased in concrete. The conduit is intended for use in ambient temperatures of 50°C or less. Unless marked otherwise, HDPE Schedule 40 conduit is intended for use with wires rated 75°C or less. HDPE Schedule 40 conduit, when directly buried or encased in concrete, may be used with wires rated 90°C or less. Where conduit emerges from underground installation, the wiring method shall be of a type recognized for the purpose. HDPE conduit is designed for joining by threaded couplings, drive-on couplings, or a butt-fusing process. Instructions supplied by the solvent-type, HDPE specific, cement manufacturer describe the method of assembly and precautions to be followed. Samples of High Density Polyethylene Conduit: Type HDPE Conduit, are shown in Exhibit 353.1.

Exhibit 353.1 Listed high density polyethylene conduit: Type HDPE conduit. (Courtesy of Carlon®, Lamson & Sessions)

353.6 Listing Requirements

HDPE conduit and associated fittings shall be listed.

II. Installation

353.10 Uses Permitted

The use of HDPE conduit shall be permitted under the following conditions:

(1) In discrete lengths or in continuous lengths from a reel
(2) In locations subject to severe corrosive influences as covered in 300.6 and where subject to chemicals for which the conduit is listed
(3) In cinder fill
(4) In direct burial installations in earth or concrete

FPN to (4): Refer to 300.5 and 300.50 for underground installations.

353.12 Uses Not Permitted

HDPE conduit shall not be used under the following conditions:

(1) Where exposed
(2) Within a building
(3) In hazardous (classified) locations, except as permitted in 504.20
(4) Where subject to ambient temperatures in excess of 50°C (122°F) unless listed otherwise
(5) For conductors or cables operating at a temperature higher than the HDPE conduit listed operating temperature rating

Exception: Conductors or cables rated at a temperature higher than the HDPE conduit listed temperature rating shall be permitted to be installed in HDPE conduit, provided they are not operated at a temperature higher than the HDPE conduit listed temperature rating.

353.20 Size

(A) Minimum HDPE conduit smaller than metric designator 16 (trade size ½) shall not be used.

(B) Maximum HDPE conduit larger than metric designator 103 (trade size 4) shall not be used.

FPN: The trade sizes and metric designators are for identification purposes only and do not relate to actual dimensions. See 300.1(C).

353.22 Number of Conductors

The number of conductors shall not exceed that permitted by the percentage fill specified in Table 1, Chapter 9.

Cables shall be permitted to be installed where such use is not prohibited by the respective cable articles. The number of cables shall not exceed the allowable percentage fill specified in Table 1, Chapter 9.

353.24 Bends — How Made

Bends shall be so made that the conduit will not be damaged and the internal diameter of the conduit will not be effectively reduced. Bends shall be permitted to be made manually without auxiliary equipment, and the radius of the curve to the centerline of such bends shall not be less than shown in Table 354.24.

Bending Type HDPE conduit does not follow the common bending radius in Chapter 9 Table 2 but rather must conform to Table 354.24.

353.26 Bends — Number in One Run

There shall not be more than the equivalent of four quarter bends (360 degrees total) between pull points, for example, conduit bodies and boxes.

353.28 Trimming

All cut ends shall be trimmed inside and outside to remove rough edges.

353.46 Bushings

Where a conduit enters a box, fitting, or other enclosure, a bushing or adapter shall be provided to protect the wire from abrasion unless the box, fitting, or enclosure design provides equivalent protection.

FPN: See 300.4(F) for the protection of conductors 4 AWG and larger at bushings.

353.48 Joints

All joints between lengths of conduit, and between conduit and couplings, fittings, and boxes, shall be made by an approved method.

353.56 Splices and Taps

Splices and taps shall be made in accordance with 300.15.

353.60 Grounding

Where equipment grounding is required, a separate equipment grounding conductor shall be installed in the conduit.

Exception No. 1: The equipment grounding conductor shall be permitted to be run separately from the conduit where used for grounding dc circuits as permitted in 250.134, Exception No. 2.

Exception No. 2: The equipment grounding conductor shall not be required where the grounded conductor is used to ground equipment as permitted in 250.142.

III. Construction Specifications

353.100 Construction

HDPE conduit shall be composed of high density polyethylene that is resistant to moisture and chemical atmospheres. The material shall be resistant to moisture and corrosive agents and shall be of sufficient strength to withstand abuse, such as by impact and crushing, in handling and during installation. Where intended for direct burial, without encasement in concrete, the material shall also be capable of withstanding continued loading that is likely to be encountered after installation.

353.120 Marking

Each length of HDPE shall be clearly and durably marked at least every 3 m (10 ft) as required in 110.21. The type of material shall also be included in the marking.

ARTICLE 354
Nonmetallic Underground Conduit with Conductors: Type NUCC

Contents

I. General

354.1 Scope

This article covers the use, installation, and construction specifications for nonmetallic underground conduit with conductors (NUCC).

354.2 Definition

Nonmetallic Underground Conduit with Conductors (NUCC). A factory assembly of conductors or cables inside a nonmetallic, smooth wall conduit with a circular cross section.

Nonmetallic underground conduit with conductors (preassembled conductors in conduit) has been used by electric utilities for outdoor lighting for several years. It is supplied in continuous lengths on coils or reels or in cartons. It consists of nonmetallic conduit with the conductors preinstalled by the manufacturer. The product is designed to allow conductors to be removed and reinserted; therefore, maintenance is an issue.

354.6 Listing Requirements

NUCC and associated fittings shall be listed.

II. Installation

354.10 Uses Permitted

The use of NUCC and fittings shall be permitted in the following:

(1) For direct burial underground installation (For minimum cover requirements, see Tables 300.5 and 300.50 under Rigid Nonmetallic Conduit.)
(2) Encased or embedded in concrete
(3) In cinder fill
(4) In underground locations subject to severe corrosive influences as covered in 300.6 and where subject to chemicals for which the assembly is specifically approved

354.12 Uses Not Permitted

NUCC shall not be used in the following:

(1) In exposed locations
(2) Inside buildings

Exception: The conductor or the cable portion of the assembly, where suitable, shall be permitted to extend within the building for termination purposes in accordance with 300.3.

(3) In hazardous (classified) locations except as permitted by 503.10(A), 504.20, 514.8, and 515.8, and in Class I, Division 2 locations as permitted in 501.10(B)(3)

354.20 Size

(A) Minimum NUCC smaller than metric designator 16 (trade size ½) shall not be used.

(B) Maximum NUCC larger than metric designator 103 (trade size 4) shall not be used.

FPN: See 300.1(C) for the metric designators and trade sizes. These are for identification purposes only and do not relate to actual dimensions.

354.22 Number of Conductors

The number of conductors or cables shall not exceed that permitted by the percentage fill in Table 1, Chapter 9.

354.24 Bends — How Made

Bends shall be manually made so that the conduit will not be damaged and the internal diameter of the conduit will not be effectively reduced. The radius of the curve of the centerline of such bends shall not be less than shown in Table 354.24.

Table 354.24 Minimum Bending Radius for Nonmetallic Underground Conduit with Conductors (NUCC)

Conduit Size		Minimum Bending Radius	
Metric Designator	Trade Size	mm	in.
16	½	250	10
21	¾	300	12
27	1	350	14
35	1¼	450	18
41	1½	500	20
53	2	650	26
63	2½	900	36
78	3	1200	48
103	4	1500	60

Bending NUCC does not follow Chapter 9 Table 2 but rather must conform to Table 354.24.

354.26 Bends — Number in One Run

There shall not be more than the equivalent of four quarter bends (360 degrees total) between termination points.

354.28 Trimming

For termination, the conduit shall be trimmed away from the conductors or cables using an approved method that will not damage the conductor or cable insulation or jacket. All conduit ends shall be trimmed inside and out to remove rough edges.

354.46 Bushings

Where the NUCC enters a box, fitting, or other enclosure, a bushing or adapter shall be provided to protect the conductor or cable from abrasion unless the design of the box, fitting, or enclosure provides equivalent protection.

> FPN: See 300.4(F) for the protection of conductors size 4 AWG or larger.

354.48 Joints

All joints between conduit, fittings, and boxes shall be made by an approved method.

354.50 Conductor Terminations

All terminations between the conductors or cables and equipment shall be made by an approved method for that type of conductor or cable.

354.56 Splices and Taps

Splices and taps shall be made in junction boxes or other enclosures.

354.60 Grounding

Where equipment grounding is required, an assembly containing a separate equipment grounding conductor shall be used.

III. Construction Specifications

354.100 Construction

(A) General NUCC is an assembly that is provided in continuous lengths shipped in a coil, reel, or carton.

(B) Nonmetallic Underground Conduit The nonmetallic underground conduit shall be listed and composed of a material that is resistant to moisture and corrosive agents. It shall also be capable of being supplied on reels without damage or distortion and shall be of sufficient strength to withstand abuse, such as impact or crushing, in handling and during installation without damage to conduit or conductors.

(C) Conductors and Cables Conductors and cables used in NUCC shall be listed and shall comply with 310.8(C). Conductors of different systems shall be installed in accordance with 300.3(C).

(D) Conductor Fill The maximum number of conductors or cables in NUCC shall not exceed that permitted by the percentage fill in Table 1, Chapter 9.

354.120 Marking

NUCC shall be clearly and durably marked at least every 3.05 m (10 ft) as required by 110.21. The type of conduit material shall also be included in the marking.

Identification of conductors or cables used in the assembly shall be provided on a tag attached to each end of the assembly or to the side of a reel. Enclosed conductors or cables shall be marked in accordance with 310.11.

ARTICLE 356
Liquidtight Flexible Nonmetallic Conduit: Type LFNC

Summary of Changes

- **356.22:** Revised to permit cables in LFNC if the cable article does not specifically prohibit such use. The number of cables is limited to the percentage fill of Table 1, Chapter 9.

- **356.30(1):** Revised securing requirements for LFNC-B where installed in lengths exceeding 6 ft.

- **356.30(4):** Added provision permitting unsupported lengths of LFNC-B not to exceed 6 ft for connection of luminaires or other equipment located in accessible ceilings.
- **356.42:** Revised to require that fittings used with LFNC be listed and to permit straight fittings only where directly buried or encased in concrete.

Contents

I. General

356.1 Scope

This article covers the use, installation, and construction specifications for liquidtight flexible nonmetallic conduit (LFNC) and associated fittings.

356.2 Definition

Liquidtight Flexible Nonmetallic Conduit (LFNC). A raceway of circular cross section of various types as follows:

(1) A smooth seamless inner core and cover bonded together and having one or more reinforcement layers between the core and covers, designated as Type LFNC-A

(2) A smooth inner surface with integral reinforcement within the conduit wall, designated as Type LFNC-B

(3) A corrugated internal and external surface without integral reinforcement within the conduit wall, designated as LFNC-C

LFNC is flame resistant and with fittings and is approved for the installation of electrical conductors.

FPN: FNMC is an alternative designation for LFNC.

356.6 Listing Requirements

LFNC and associated fittings shall be listed.

II. Installation

356.10 Uses Permitted

LFNC shall be permitted to be used in exposed or concealed locations for the following purposes:

FPN: Extreme cold may cause some types of nonmetallic conduits to become brittle and therefore more susceptible to damage from physical contact.

(1) Where flexibility is required for installation, operation, or maintenance

(2) Where protection of the contained conductors is required from vapors, liquids, or solids

(3) For outdoor locations where listed and marked as suitable for the purpose

(4) For direct burial where listed and marked for the purpose

(5) Type LFNC-B shall be permitted to be installed in lengths longer than 1.8 m (6 ft) where secured in accordance with 356.30

(6) Type LFNC-B as a listed manufactured prewired assembly, metric designator 16 through 27 (trade size ½ through 1) conduit

Prewired Type LFNC-B is a listed assembly where the conductors are required to be installed at the manufacturing facility, where controlled conditions prevent damage to the conductor insulation. Special cutting tools are required to be used when cutting prewired Type LFNC-B to prevent nicking the conductor installation. This prewired assembly is shown in Exhibit 356.1.

Exhibit 356.1 Listed manufactured prewired assembly of liquidtight flexible nonmetallic conduit, Type B. (Courtesy of Carlon®, Lamson & Sessions)

356.12 Uses Not Permitted

LFNC shall not be used as follows:

(1) Where subject to physical damage
(2) Where any combination of ambient and conductor temperatures is in excess of that for which the LFNC is approved
(3) In lengths longer than 1.8 m (6 ft), except as permitted by 356.10(5) or where a longer length is approved as essential for a required degree of flexibility
(4) Where the operating voltage of the contained conductors is in excess of 600 volts, nominal, except as permitted in 600.32(A)
(5) In any hazardous (classified) location other than as permitted in 501.10(B), 502.10(A) and (B), 503.10(A), and 504.20

356.20 Size

(A) Minimum LFNC smaller than metric designator 16 (trade size ½) shall not be used unless permitted in 356.20(A)(1) through (A)(3) for metric designator 12 (trade size ⅜).

(1) For enclosing the leads of motors as permitted in 430.245(B)
(2) In lengths not exceeding 1.8 m (6 ft) as part of a listed assembly for tap connections to luminaires (lighting fixtures) as required in 410.67(C), or for utilization equipment
(3) For electric sign conductors in accordance with 600.32(A)

(B) Maximum LFNC larger than metric designator 103 (trade size 4) shall not be used.

> FPN: See 300.1(C) for the metric designators and trade sizes. These are for identification purposes only and do not relate to actual dimensions.

356.22 Number of Conductors

The number of conductors shall not exceed that permitted by the percentage fill specified in Table 1, Chapter 9.

Cables shall be permitted to be installed where such use is not prohibited by the respective cable articles. The number of cables shall not exceed the allowable percentage fill specified in Table 1, Chapter 9.

Table 1 of Chapter 9 specifies the maximum fill percentage of a conduit or tubing. Table 4 provides the usable area within the selected conduit or tubing, and Table 5 provides the required area for each of the conductors. Examples using these tables to calculate a conduit or tubing size are provided in the commentary following Chapter 9, Table 1, Notes to Tables, Note 6.

If the conductors are of the same wire size, the tables in Annex C may be used instead of doing the calculations. Annex C, with 12 sets of tables, accurately indicates the maximum number of conductors permitted in a conduit or tubing. Examples using Annex C to select a conduit or tubing size are provided in the commentary following the introduction in the annex.

To permit selection of the proper trade size of liquidtight flexible nonmetallic conduit, Table 4 of Chapter 9 contains two sections entitled "Article 356—Liquidtight Flexible Nonmetallic Conduit (LFNC-B)" and "Article 356—Liquidtight Flexible Nonmetallic Conduit (LFNC-A)." The construction differences between these two raceway types are explained in 356.2. Annex C, Tables C.5 and C.5(A) or C.6 and C.6(A) for liquidtight flexible nonmetallic conduit (LFNC-A or LFNC-B) are also permissible.

356.24 Bends — How Made

Bends in conduit shall be so made that the conduit is not damaged and the internal diameter of the conduit is not effectively reduced. Bends shall be permitted to be made manually without auxiliary equipment. The radius of the curve to the centerline of any bend shall not be less than shown in Table 2, Chapter 9 using the column "Other Bends."

Table 2 of Chapter 9 was added for the 2005 *Code* to replace the 2002 *Code* Table 344.24, Radius of Conduit Bends. Table 2 of Chapter 9 is now the common table for raceway field bend measurements.

356.26 Bends — Number in One Run

There shall not be more than the equivalent of four quarter bends (360 degrees total) between pull points, for example, conduit bodies and boxes.

356.28 Trimming

All cut ends of conduit shall be trimmed inside and outside to remove rough edges.

356.30 Securing and Supporting

Type LFNC-B shall be securely fastened and supported in accordance with one of the following:

(1) Where installed in lengths exceeding 1.8 m (6 ft), the conduit shall be securely fastened at intervals not exceeding 900 mm (3 ft) and within 300 mm (12 in.) on each side of every outlet box, junction box, cabinet, or fitting.
(2) Securing or supporting of the conduit shall not be required where it is fished, installed in lengths not

exceeding 900 mm (3 ft) at terminals where flexibility is required, or installed in lengths not exceeding 1.8 m (6 ft) from a luminaire (fixture) terminal connection for tap conductors to luminaires (lighting fixtures) permitted in 410.67(C).

(3) Horizontal runs of LFNC supported by openings through framing members at intervals not exceeding 900 mm (3 ft) and securely fastened within 300 mm (12 in.) of termination points shall be permitted.

(4) Securing or supporting of LFNC-B shall not be required where installed in lengths not exceeding 1.8 m (6 ft) from the last point where the raceway is securely fastened for connections within an accessible ceiling to luminaire(s) [lighting fixture(s)] or other equipment.

Item (4) was added for the 2005 *Code* to correlate this permission of unsupported liquidtight flexible nonmetallic conduit with other wiring methods permitting the same exception throughout Chapter 3. One such example is 320.30(D)(3) for Type AC cable.

356.42 Couplings and Connectors

Only fittings listed for use with LFNC shall be used. Angle connectors shall not be used for concealed raceway installations. Straight LFNC fittings are permitted for direct burial or encasement in concrete.

356.56 Splices and Taps

Splices and taps shall be made in accordance with 300.15.

356.60 Grounding and Bonding

Where used to connect equipment where flexibility is required, an equipment grounding conductor shall be installed.

Where required or installed, equipment grounding conductors shall be installed in accordance with 250.134(B).

Where required or installed, equipment bonding jumpers shall be installed in accordance with 250.102.

III. Construction Specifications

356.100 Construction

LFNC-B as a prewired manufactured assembly shall be provided in continuous lengths capable of being shipped in a coil, reel, or carton without damage.

356.120 Marking

LFNC shall be marked at least every 600 mm (2 ft) in accordance with 110.21. The marking shall include a type designation in accordance with 356.2 and the trade size. Conduit that is intended for outdoor use or direct burial shall be marked.

The type, size, and quantity of conductors used in pre-

wired manufactured assemblies shall be identified by means of a printed tag or label attached to each end of the manufactured assembly and either the carton, coil, or reel. The enclosed conductors shall be marked in accordance with 310.11.

ARTICLE 358
Electrical Metallic Tubing: Type EMT

Contents

I. General

358.1 Scope

This article covers the use, installation, and construction specifications for electrical metallic tubing (EMT) and associated fittings.

358.2 Definition

Electrical Metallic Tubing (EMT). An unthreaded thin-wall raceway of circular cross section designed for the physical protection and routing of conductors and cables and for use as an equipment grounding conductor when installed utilizing appropriate fittings. EMT is generally made of steel (ferrous) with protective coatings or aluminum (nonferrous).

358.6 Listing Requirements

EMT, factory elbows, and associated fittings shall be listed.

II. Installation

358.10 Uses Permitted

(A) **Exposed and Concealed** The use of EMT shall be permitted for both exposed and concealed work.

(B) **Corrosion Protection** Ferrous or nonferrous EMT, elbows, couplings, and fittings shall be permitted to be installed in concrete, in direct contact with the earth, or in areas subject to severe corrosive influences where protected by corrosion protection and judged suitable for the condition.

(C) **Wet Locations** All supports, bolts, straps, screws, and so forth shall be of corrosion-resistant materials or protected against corrosion by corrosion-resistant materials.

> FPN: See 300.6 for protection against corrosion.

According to the 2004 UL *General Information for Electrical Equipment Directory* (White Book), category FJMX, galvanized steel electrical metallic tubing (EMT) installed in concrete, on grade or above, generally requires no supplementary corrosion protection. Galvanized steel EMT in concrete slab below grade level may require supplementary corrosion protection. In general, galvanized steel EMT in contact with soil requires supplementary corrosion protection. Where galvanized steel EMT without supplementary corrosion protection extends directly from concrete encasement to soil burial, severe corrosive effects are likely to occur on the metal in contact with the soil.

358.12 Uses Not Permitted

EMT shall not be used under the following conditions:

(1) Where, during installation or afterward, it will be subject to severe physical damage
(2) Where protected from corrosion solely by enamel
(3) In cinder concrete or cinder fill where subject to permanent moisture unless protected on all sides by a layer of noncinder concrete at least 50 mm (2 in.) thick or unless the tubing is at least 450 mm (18 in.) under the fill
(4) In any hazardous (classified) location except as permitted by 502.10, 503.10, and 504.20
(5) For the support of luminaires (fixtures) or other equipment except conduit bodies no larger than the largest trade size of the tubing

(6) Where practicable, dissimilar metals in contact anywhere in the system shall be avoided to eliminate the possibility of galvanic action

Exception: Aluminum fittings and enclosures shall be permitted to be used with steel EMT where not subject to severe corrosive influences.

358.20 Size

(A) **Minimum** EMT smaller than metric designator 16 (trade size ½) shall not be used.

Exception: For enclosing the leads of motors as permitted in 430.245(B).

(B) **Maximum** The maximum size of EMT shall be metric designator 103 (trade size 4).

> FPN: See 300.1(C) for the metric designators and trade sizes. These are for identification purposes only and do not relate to actual dimensions.

358.22 Number of Conductors

The number of conductors shall not exceed that permitted by the percentage fill specified in Table 1, Chapter 9.

Cables shall be permitted to be installed where such use is not prohibited by the respective cable articles. The number of cables shall not exceed the allowable percentage fill specified in Table 1, Chapter 9.

Table 1 of Chapter 9 specifies the maximum fill percentage of a conduit or tubing. Table 4 provides the usable area within the selected conduit or tubing, and Table 5 provides the required area for each conductor. Examples using these tables to calculate a conduit or tubing size are provided in the commentary following Chapter 9, Table 1, Notes to Tables, Note 6.

If the conductors are of the same wire size, instead of doing the calculations, the tables in Annex C may be used. Annex C, with 12 sets of tables, accurately indicates the maximum number of conductors permitted in a conduit or tubing. Examples using Annex C to select a conduit or tubing size are provided in the commentary following the introduction in the annex.

To select the proper trade size of electrical metallic tubing, the section entitled "Article 358—Electrical Metallic Tubing (EMT)" in Table 4 of Chapter 9 should be followed. Annex C, Tables C.1 and C.1(A) for electrical metallic tubing are also permissible.

358.24 Bends — How Made

Bends shall be made so that the tubing is not damaged and the internal diameter of the tubing is not effectively reduced.

The radius of the curve of any field bend to the centerline of the tubing shall not be less than shown in Table 2, Chapter 9 for one-shot and full shoe benders.

Table 2 of Chapter 9 was added for the 2005 *Code* to replace the 2002 *Code* Table 344.24, Radius of Conduit Bends. Table 2 of Chapter 9 is now the common table for raceway field bend measurements.

358.26   Bends — Number in One Run

There shall not be more than the equivalent of four quarter bends (360 degrees total) between pull points, for example, conduit bodies and boxes.

358.28   Reaming and Threading

(A) Reaming All cut ends of EMT shall be reamed or otherwise finished to remove rough edges.

In addition to a reamer, a half-round file has proved practical for removing rough edges. The steel handle of a pair of pump pliers, the nose of side-cutting pliers, or an electrician's knife can be effective on the smaller sizes of EMT as well.

(B) Threading EMT shall not be threaded.

Exception: EMT with factory threaded integral couplings complying with 358.100.

358.30   Securing and Supporting

EMT shall be installed as a complete system in accordance with 300.18 and shall be securely fastened in place and supported in accordance with 358.30(A) and (B).

(A) Securely Fastened EMT shall be securely fastened in place at least every 3 m (10 ft). In addition, each EMT run between termination points shall be securely fastened within 900 mm (3 ft) of each outlet box, junction box, device box, cabinet, conduit body, or other tubing termination.

"Securely fastened in place" means the EMT must be supported and secured at the prescribed intervals, as illustrated in Exhibit 358.1.

Exception No. 1: Fastening of unbroken lengths shall be permitted to be increased to a distance of 1.5 m (5 ft) where structural members do not readily permit fastening within 900 mm (3 ft).

As illustrated in Exhibit 358.2, Exception No. 1 to 358.30(A) permits boxes secured to ceiling or roof support structural members that are spaced not more than 5 ft apart to serve as support for runs of EMT perpendicular to the axis of the ceiling or roof support members.

Exception No. 2: For concealed work in finished buildings or prefinished wall panels where such securing is impracticable, unbroken lengths (without coupling) of EMT shall be permitted to be fished.

(B) Supports Horizontal runs of EMT supported by openings through framing members at intervals not greater than 3 m (10 ft) and securely fastened within 900 mm (3 ft) of termination points shall be permitted.

See the commentary following 342.30(B)(4), which applies to horizontal runs of intermediate metal conduit.

358.42   Couplings and Connectors

Couplings and connectors used with EMT shall be made up tight. Where buried in masonry or concrete, they shall be concretetight type. Where installed in wet locations, they shall comply with 314.15(A).

Fittings have been tested for use only with steel EMT unless there is specific marking on the device or carton to indicate the fittings are suitable for use with aluminum or other material.

According to 358.6, only listed fittings are permitted to

Exhibit 358.1 Minimum requirements for securely fastening electrical metallic tubing (EMT) unless an exception applies.

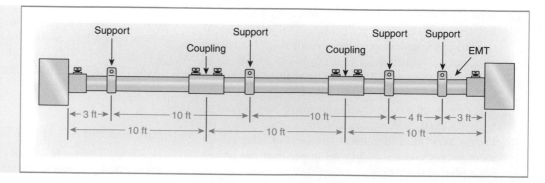

Exhibit 358.2 An example of an EMT installation complying with 358.30(A), Exception No. 1.

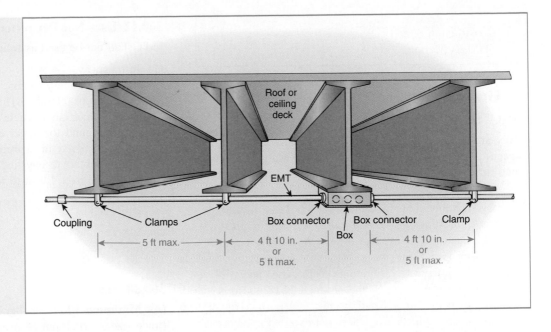

Roof or ceiling deck

EMT

Coupling — Clamps — Box connector — Box connector — Clamp

Box

5 ft max. — 4 ft 10 in. or 5 ft max. — 4 ft 10 in. or 5 ft max.

be used with EMT. According to UL 797, *Electrical Metallic Tubing — Steel*, listed fittings that are suitable for use in poured concrete or where exposed to rain are so indicated on the fitting or carton. The term *raintight* or the equivalent on the carton indicates suitability for use where directly exposed to rain. The term *concretetight* or equivalent on the carton indicates suitability for use in poured concrete. See 225.22 and 230.54(A) for raintight requirements as applied to raceways on exterior surfaces of buildings and to service raceways. Fittings for use in wet locations are listed for use in wet locations.

Indentor-type fittings are for use with metallic-coated EMT only and require a special tool supplied by the manufacturer for proper installation. Diametrically opposed indentor-type tools require two sets of indentations nominally 90 degrees apart. Triple-indent tools require one set of indentations.

358.56 Splices and Taps

Splices and taps shall be made in accordance with 300.15.

358.60 Grounding

EMT shall be permitted as an equipment grounding conductor.

III. Construction Specifications

358.100 Construction

Factory-threaded integral couplings shall be permitted. Where EMT with a threaded integral coupling is used, threads for both the tubing and coupling shall be factory-made. The coupling and EMT threads shall be designed so as to prevent bending of the tubing at any part of the thread.

358.120 Marking

EMT shall be clearly and durably marked at least every 3 m (10 ft) as required in the first sentence of 110.21.

ARTICLE 360
Flexible Metallic Tubing:
Type FMT

Contents

III. Construction Specifications
360.120 Marking

I. General

360.1 Scope

This article covers the use, installation, and construction specifications for flexible metallic tubing (FMT) and associated fittings.

Flexible metallic tubing is a type of raceway used for certain specific applications, particularly under the requirements of 300.22(B) and (C) for wiring in ducts, plenums, and other air-handling spaces. Flexible metallic tubing is very flexible and is rarely affected by vibration or other movement. It is an effective barrier to gases and products of combustion if installed with matching listed fittings and is of adequate mechanical strength for use where not exposed to physical damage.

Flexible metallic tubing not greater than 6 ft in length is suitable for use as a raceway for branch-circuit tap conductors that supply luminaires where outlet boxes are located not less than 1 ft from the luminaire.

360.2 Definition

Flexible Metallic Tubing (FMT). A raceway that is circular in cross section, flexible, metallic, and liquidtight without a nonmetallic jacket.

360.6 Listing Requirements

FMT and associated fittings shall be listed.

II. Installation

360.10 Uses Permitted

FMT shall be permitted to be used for branch circuits as follows:

(1) In dry locations
(2) Where concealed
(3) In accessible locations
(4) For system voltages of 1000 volts maximum

A common application of flexible metallic tubing is as a branch-circuit wiring method for equipment or luminaires mounted on or above suspended ceilings. The 1000-volt limitation prohibits the use of flexible metallic tubing for the secondary circuits of sign ballasts, sign transformers, electronic sign power supplies, or oil burner ignition transformers unless these circuits are less than 1000 volts.

360.12 Uses Not Permitted

FMT shall not be used as follows:

(1) In hoistways
(2) In storage battery rooms
(3) In hazardous (classified) locations unless otherwise permitted under other articles in this *Code*
(4) Under ground for direct earth burial, or embedded in poured concrete or aggregate
(5) Where subject to physical damage
(6) In lengths over 1.8 m (6 ft)

Unlike flexible metal conduit or liquidtight flexible metal conduit, flexible metallic tubing is limited in use to 6-ft lengths.

360.20 Size

(A) Minimum FMT smaller than metric designator 16 (trade size ½) shall not be used.

Exception No. 1: FMT of metric designator 12 (trade size ⅜) shall be permitted to be installed in accordance with 300.22(B) and (C).

Exception No. 2: FMT of metric designator 12 (trade size ⅜) shall be permitted in lengths not in excess of 1.8 m (6 ft) as part of an approved assembly or for luminaires (lighting fixtures). See 410.67(C).

(B) Maximum The maximum size of FMT shall be metric designator 21 (trade size ¾).

> FPN: See 300.1(C) for the metric designators and trade sizes. These are for identification purposes only and do not relate to actual dimensions.

360.22 Number of Conductors

(A) FMT — Metric Designators 16 and 21 (Trade Sizes ½ and ¾) The number of conductors in metric designators 16 (trade size ½) and 21 (trade size ¾) shall not exceed that permitted by the percentage fill specified in Table 1, Chapter 9.

Cables shall be permitted to be installed where such use is not prohibited by the respective cable articles. The number of cables shall not exceed the allowable percentage fill specified in Table 1, Chapter 9.

Table 1 of Chapter 9 specifies the maximum fill percentage of a conduit or tubing. Table 4 provides the usable area within the selected conduit or tubing, and Table 5 provides the required area for each of the conductors. Examples using these tables to calculate a conduit or tubing size are provided in the commentary following Chapter 9, Table 1, Notes to Tables, Note 6.

If the conductors are of the same wire size, the tables

in Annex C may be used instead of doing the calculations. Annex C, with 12 sets of tables, accurately indicates the maximum number of conductors permitted in a conduit or tubing. Examples using Annex C to select a conduit or tubing size are provided in the commentary following the introduction in the annex.

To select the proper trade size of flexible metallic tubing, the section entitled "348—Flexible Metal Conduit (FMC)" in Table 4 of Chapter 9 or the manufacturer's instruction should be followed. Annex C, Tables C.3 and C.3(A) for flexible metal conduit for sizes ½ in. and ¾ in. are also permissible.

(B) FMT — Metric Designator 12 (Trade Size ⅜) The number of conductors in metric designator 12 (trade size ⅜) shall not exceed that permitted in Table 348.22.

360.24 Bends

(A) Infrequent Flexing Use Where FMT may be infrequently flexed in service after installation, the radii of bends measured to the inside of the bend shall not be less than specified in Table 360.24(A).

Table 360.24(A) Minimum Radii for Flexing Use

Metric Designator	Trade Size	Minimum Radii for Flexing Use	
		mm	in.
12	⅜	25.4	10
16	½	317.5	12½
21	¾	444.5	17½

(B) Fixed Bends Where FMT is bent for installation purposes and is not flexed or bent as required by use after installation, the radii of bends measured to the inside of the bend shall not be less than specified in Table 360.24(B).

Table 360.24(B) Minimum Radii for Fixed Bends

Metric Designator	Trade Size	Minimum Radii for Fixed Bends	
		mm	in.
12	⅜	88.9	3½
16	½	101.6	4
21	¾	127.0	5

360.40 Boxes and Fittings

Fittings shall effectively close any openings in the connection.

360.56 Splices and Taps

Splices and taps shall be made in accordance with 300.15.

360.60 Grounding

FMT shall be permitted as an equipment grounding conductor where installed in accordance with 250.118(7).

III. Construction Specifications

360.120 Marking

FMT shall be marked according to 110.21.

ARTICLE 362
Electrical Nonmetallic Tubing: Type ENT

Summary of Changes

- **362.12(4):** Added exception permitting the installation of conductors or cables with temperature ratings higher than the temperature rating of the rigid nonmetallic conduit, provided the conductors or cables are operated at a temperature that does not exceed the temperature rating of the ENT.

- **362.22:** Revised to permit cables in ENT if the cable article does not specifically prohibit such use. The number of cables is limited to the percentage fill of Table 1, Chapter 9.

- **362.30:** Added Exception No. 2 permitting unsupported lengths of ENT not to exceed 6 ft for connection of luminaires or other equipment located in accessible ceilings.

Contents

I. General

362.1 Scope

This article covers the use, installation, and construction specifications for electrical nonmetallic tubing (ENT) and associated fittings.

362.2 Definition

Electrical Nonmetallic Tubing (ENT). A nonmetallic pliable corrugated raceway of circular cross section with integral or associated couplings, connectors, and fittings for the installation of electric conductors. ENT is composed of a material that is resistant to moisture and chemical atmospheres and is flame retardant.

A pliable raceway is a raceway that can be bent by hand with a reasonable force but without other assistance.

Electrical nonmetallic tubing (ENT) is made of the same material (PVC) used for rigid nonmetallic conduit (Article 362) suitable for aboveground use. The outside diameters of ENT (½-in. through 2-in. trade sizes only) are such that standard couplings and other fittings for rigid PVC conduit can be used.

Because of the corrugations, the raceway can be bent by hand and has some degree of flexibility. ENT is not intended for use where flexibility is necessary, as at motor terminations to prevent transmission of noise and vibration, or for connection of adjustable luminaires or moving parts. ENT is suitable for the installation of conductors having a temperature rating as indicated on the product. The maximum allowable ambient temperature is 122°F. Exhibit 362.1 shows an example of electrical nonmetallic tubing.

362.6 Listing Requirements

ENT and associated fittings shall be listed.

II. Installation

362.10 Uses Permitted

For the purpose of this article, the first floor of a building shall be that floor that has 50 percent or more of the exterior

Exhibit 362.1 An example of electrical nonmetallic tubing (ENT). (Courtesy of Carlon®, Lamson & Sessions)

wall surface area level with or above finished grade. One additional level that is the first level and not designed for human habitation and used only for vehicle parking, storage, or similar use shall be permitted. The use of ENT and fittings shall be permitted in the following:

(1) In any building not exceeding three floors above grade as follows:
   a. For exposed work, where not prohibited by 362.12
   b. Concealed within walls, floors, and ceilings

Electrical nonmetallic tubing (ENT) is permitted to be installed either concealed or exposed. Where exposed and subject to physical damage, ENT is required to be protected and is limited to use in buildings not exceeding three floors above grade, as explained in the first paragraph of 362.10. Where concealed or above a suspended ceiling (exposed), ENT is permitted to be installed within walls, floors, or ceilings in buildings of three floors or less without the need for fire-rated construction. The three-floor limitation is based on the likelihood that only a small quantity of ENT will be exposed to fire and that the occupants will have adequate time to exit the building before the products of combustion make the building untenable. Exhibit 362.2 illustrates permitted uses of ENT in a building of three floors or less.

(2) In any building exceeding three floors above grade, ENT shall be concealed within walls, floors, and ceilings where the walls, floors, and ceilings provide a thermal barrier of material that has at least a 15-minute finish rating as identified in listings of fire-rated assemblies.

Exhibit 362.2 Examples of permitted uses of ENT in a building not exceeding three floors.

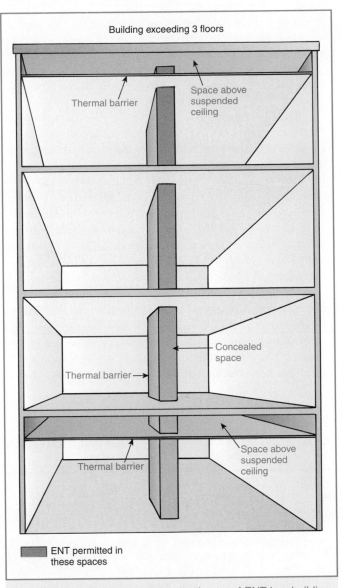

Exhibit 362.3 Examples of permitted uses of ENT in a building exceeding three floors.

The 15-minute-finish-rated thermal barrier shall be permitted to be used for combustible or noncombustible walls, floors, and ceilings.

Exception to (2): Where a fire sprinkler system(s) is installed in accordance with NFPA 13-2002, Standard for the Installation of Sprinkler Systems, on all floors, ENT shall be permitted to be used within walls, floors, and ceilings, exposed or concealed, in buildings exceeding three floors above grade.

ENT is permitted to be installed within the walls, floors, or ceilings of a building of any height where the walls, floors, or ceilings provide a thermal barrier of material that has at least a 15-minute finish rating. It is not permitted or intended that ENT be used exposed in the first three floors of a building that exceeds three floors except as permitted in 362.10(5). Where installed in a building over three floors, ENT must be installed behind the 15-minute thermal barrier on all floors. Exhibit 362.3 illustrates permitted uses of ENT in a building exceeding three floors. An addition to the 2002 *Code*, the exception to (5) concerning a fire sprinkler system, negates the requirement for a 15-minute finish rating.

FPN: A finish rating is established for assemblies containing combustible (wood) supports. The finish rating is defined as the time at which the wood stud or wood joist reaches an average temperature rise of 121°C (250°F) or an individual temperature of 163°C (325°F) as measured on the plane of the wood nearest the fire. A finish rating is not intended to represent a rating for a membrane ceiling.

Interior finish is generally considered to consist of those materials or combinations of materials that form the exposed interior surface of walls and ceilings in a building. Common interior finish materials include plaster, gypsum wallboard, wood, plywood paneling, fibrous ceiling tiles, and a variety of wall coverings. Ordinary paint, wallpaper, or other similar wall coverings not exceeding 1/28 in. in thickness are generally considered incidental to interior finish, except where the authority having jurisdiction deems them a hazard. For more information regarding classification of interior finish material, refer to 10.2.1 of NFPA *101, Life Safety Code.*

The finish rating of a wall or ceiling finish material is the time required for the unexposed surface of the finish membrane to reach an average temperature rise of 250°F above ambient or an individual temperature rise at any one point not exceeding 325°F when the assembly is tested in accordance with NFPA 251, *Standard Methods of Tests of Fire Endurance of Building Construction and Materials* (also known as ANSI/UL 263, *Standard for Fire Tests of Building Construction and Materials,* or ASTM E119, *Standard Test Methods for Fire Tests of Building Construction and Materials*).

The finish rating of wall and ceiling finish materials tested and rated by UL as part of wall and ceiling assemblies can be found in the UL *Fire Resistance Directory,* immediately following the assembly rating and just below the design number. Only assemblies containing combustible support members, however, have published finish ratings. Obviously,

it is not the intent to limit ENT to constructions consisting of combustible support members. This section is intended to provide a 15-minute thermal barrier as a minimum threshold of acceptability.

Commentary Table 362.1, reproduced from the NFPA *Fire Protection Handbook,* 19th edition (Section 12, Chapter 4, Table 12.4.15), provides ratings for common finish materials. If the finish rating concealing the ENT is unknown or is less than 15 minutes, the ENT can still be used if the installation meets the criteria in 362.10, including the three-floor limitation, where required, and the installation is not prohibited by 362.12. For finish materials not tested and rated in the UL *Fire Resistance Directory,* use Commentary Table 362.1.

(3) In locations subject to severe corrosive influences as covered in 300.6 and where subject to chemicals for which the materials are specifically approved.

(4) In concealed, dry, and damp locations not prohibited by 362.12.

(5) Above suspended ceilings where the suspended ceilings provide a thermal barrier of material that has at least a

Commentary Table 362.1 Various Finishes over Wood Framing, One Side (Combustible) with Exposure on Finish Side

Material	Fire Resistance Rating[a] (min)
Fiberboard, 1/2 in. thick	5
Fiberboard, flameproofed, 1/2 in. thick	10
Fiberboard, 1/2 in. thick, with 1/2 in. 1:2, 1:2 gypsum-sand plaster	15
Gypsum wallboard, 3/8 in. thick	10
Gypsum wallboard, 1/2 in. thick	15
Gypsum wallboard, 5/8 in. thick	20
Gypsum wallboard, laminated, two 3/8 in.	28
Gypsum wallboard, laminated, one 3/8 in. plus one 1/2 in. thick	37
Gypsum wallboard, laminated, two 1/2 in. thick	47
Gypsum wallboard, laminated, two 5/8 in. thick	60
Gypsum lath, plain or indented, 3/8 in. thick, with 1/2 in. 1:2, 1:2 gypsum-sand plaster	20
Gypsum lath, perforated, 3/8 in. thick, with 1/2 in. 1:2, 1:2 gypsum-sand plaster	30
Gypsum-sand plaster, 1:2, 1:3, 1/2 in. thick, on wood lath	15
Lime-sand plaster, 1:5, 1:7.5, 1/2 in. thick, on wood lath	15
Gypsum-sand plaster, 1:2, 1:2, 3/4 in. thick, on metal lath (no paper backing)	15
Neat gypsum plaster, 3/4 in. thick on metal lath (no paper backing)[b]	30
Neat gypsum plaster, 1 in. thick, on metal lath (no paper backing)[b]	35
Lime-sand plaster, 1:5, 1:7.5, 3/4 in. thick, on metal lath (no paper backing)	10
Portland cement plaster, 3/4 in. thick, on metal lath (no paper backing)	10
Gypsum-sand plaster, 1:2, 1:3, 3/4 in. thick, on paper-backed metal lath	20

Note: For SI units, 1 in. = 25.4 mm.
[a]From National Bureau of Standards [now known as the National Institute for Standards and Technology], BMS-92.
[b]Unsanded wood-fiber plaster.

15-minute finish rating as identified in listings of fire-rated assemblies, except as permitted in 362.10(1)(a).

Exception to (5): ENT shall be permitted to be used above suspended ceilings in buildings exceeding three floors above grade where the building is protected throughout by a fire sprinkler system installed in accordance with NFPA 13-2002, Standard for the Installation of Sprinkler Systems.

(6) Encased in poured concrete, or embedded in a concrete slab on grade where ENT is placed on sand or approved screenings, provided fittings identified for this purpose are used for connections.

(7) For wet locations indoors as permitted in this section or in a concrete slab on or below grade, with fittings listed for the purpose.

(8) Metric designator 16 through 27 (trade size ½ through 1) as listed manufactured prewired assembly.

> FPN: Extreme cold may cause some types of nonmetallic conduits to become brittle and therefore more susceptible to damage from physical contact.

Prewired ENT is a listed assembly whose conductors must be installed at the manufacturing facility, where controlled conditions prevent damage to the conductor insulation. Special tools are required when cutting prewired ENT to prevent nicking of the conductor installation. A prewired assembly is shown in Exhibit 362.4.

Exhibit 362.4 Listed manufactured prewired ENT assembly. (Courtesy of Carlon®, Lamson & Sessions)

362.12 Uses Not Permitted

ENT shall not be used in the following:

(1) In hazardous (classified) locations, except as permitted by 504.20 and 505.15(A)(1)

(2) For the support of luminaires (fixtures) and other equipment

(3) Where subject to ambient temperatures in excess of 50°C (122°F) unless listed otherwise

(4) For conductors or cables operating at a temperature higher than the ENT listed temperature rating

Exception to (4): Conductors or cables rated at a temperature higher than the ENT listed temperature rating shall be permitted to be installed in ENT, provided they are not operated at a temperature higher than the ENT listed temperature rating.

(5) For direct earth burial

(6) Where the voltage is over 600 volts

(7) In exposed locations, except as permitted by 362.10(1), 362.10(5), and 362.10(7)

(8) In theaters and similar locations, except as provided in 518.4 and 520.5

See 518.4 and 520.5 for details of permitted wiring methods.

(9) Where exposed to the direct rays of the sun, unless identified as sunlight resistant

(10) Where subject to physical damage

362.20 Size

(A) Minimum ENT smaller than metric designator 16 (trade size ½) shall not be used.

(B) Maximum ENT larger than metric designator 53 (trade size 2) shall not be used.

> FPN: See 300.1(C) for the metric designators and trade sizes. These are for identification purposes only and do not relate to actual dimensions.

362.22 Number of Conductors

The number of conductors shall not exceed that permitted by the percentage fill in Table 1, Chapter 9.

Cables shall be permitted to be installed where such use is not prohibited by the respective cable articles. The number of cables shall not exceed the allowable percentage fill specified in Table 1, Chapter 9.

Table 1 of Chapter 9 specifies the maximum fill percentage of a conduit or tubing. Table 4 provides the usable area within the selected conduit or tubing, and Table 5 provides the required area for each conductor. Examples using these tables to calculate a conduit or tubing size are provided in the commentary following Chapter 9, Table 1, Notes to Tables, Note 6.

If the conductors are of the same wire size, Annex C may be consulted. Annex C, which contains 12 sets of tables, accurately indicates the maximum number of conductors permitted in a conduit or tubing. Examples using Annex C to select a conduit or tubing size are provided in the commentary following the introduction in the annex.

To select the proper trade size electrical nonmetallic tubing, the section entitled "Article 362 — Electrical Nonmetallic Tubing (ENT)" in Table 4 of Chapter 9 should be followed. Annex C, Tables C.2 and C.2(A) for electrical nonmetallic tubing are also permissible.

362.24 Bends — How Made

Bends shall be so made that the tubing will not be damaged and the internal diameter of the tubing will not be effectively reduced. Bends shall be permitted to be made manually without auxiliary equipment, and the radius of the curve to the centerline of such bends shall not be less than shown in Table 2, Chapter 9 using the column "Other Bends."

362.26 Bends — Number in One Run

There shall not be more than the equivalent of four quarter bends (360 degrees total) between pull points, for example, conduit bodies and boxes.

362.28 Trimming

All cut ends shall be trimmed inside and outside to remove rough edges.

362.30 Securing and Supporting

ENT shall be installed as a complete system in accordance with 300.18 and shall be securely fastened in place and supported in accordance with 362.30(A) and (B).

(A) Securely Fastened ENT shall be securely fastened at intervals not exceeding 900 mm (3 ft). In addition, ENT shall be securely fastened in place within 900 mm (3 ft) of each outlet box, device box, junction box, cabinet, or fitting where it terminates.

Exception No. 1: Lengths not exceeding a distance of 1.8 m (6 ft) from a luminaire (fixture) terminal connection for tap connections to lighting luminaires (fixtures) shall be permitted without being secured.

Exception No. 2: Lengths not exceeding 1.8 m (6 ft) from the last point where the raceway is securely fastened for connections within an accessible ceiling to luminaire(s) [lighting fixture(s)] or other equipment.

As illustrated in Exhibit 362.5, where ENT is run on the surface of framing members, it is required to be fastened to the framing member every 3 ft and within 3 ft of every box. See 300.4(D) for provisions on protection against physical damage.

As illustrated in Exhibit 362.6, ENT is permitted to be used as fixture whip without support for lengths not exceeding 6 ft. See 410.67(C) for details on tap conductor wiring.

Exhibit 362.5 An application of 362.30(A), showing ENT supported every 3 ft and within 3 ft of the box.

Exhibit 362.6 An application of 362.30(A), Exception No. 1, showing ENT unsupported in lengths not exceeding 6 ft.

(B) Supports Horizontal runs of ENT supported by openings in framing members at intervals not exceeding 900 mm (3 ft) and securely fastened within 900 mm (3 ft) of termination points shall be permitted.

362.46 Bushings

Where a tubing enters a box, fitting, or other enclosure, a bushing or adapter shall be provided to protect the wire from abrasion unless the box, fitting, or enclosure design provides equivalent protection.

> FPN: See 300.4(F) for the protection of conductors size 4 AWG or larger.

362.48 Joints

All joints between lengths of tubing and between tubing and couplings, fittings, and boxes shall be by an approved method.

362.56 Splices and Taps

Splices and taps shall be made only in accordance with 300.15.

> FPN: See Article 314 for rules on the installation and use of boxes and conduit bodies.

362.60 Grounding

Where equipment grounding is required, a separate equipment grounding conductor shall be installed in the raceway.

III. Construction Specifications

362.100 Construction

ENT shall be made of material that does not exceed the ignitibility, flammability, smoke generation, and toxicity characteristics of rigid (nonplasticized) polyvinyl chloride.

ENT, as a prewired manufactured assembly, shall be provided in continuous lengths capable of being shipped in a coil, reel, or carton without damage.

362.120 Marking

ENT shall be clearly and durably marked at least every 3 m (10 ft) as required in the first sentence of 110.21. The type of material shall also be included in the marking. Marking for limited smoke shall be permitted on the tubing that has limited smoke-producing characteristics.

The type, size, and quantity of conductors used in prewired manufactured assemblies shall be identified by means of a printed tag or label attached to each end of the manufactured assembly and either the carton, coil, or reel. The enclosed conductors shall be marked in accordance with 310.11.

ARTICLE 366
Auxiliary Gutters

Summary of Changes

- **General:** Rearranged and renumbered to coordinate with format of other Chapter 3 raceway articles.
- **366.2:** Added section with definitions of *metallic auxiliary gutters* and *nonmetallic auxiliary gutters*.
- **366.10 and 366.12:** Added new sections detailing uses permitted and uses not permitted of auxiliary gutters.
- **366.58(A):** Revised to specify that the minimum dimension for conductor deflection space is based on the one wire per terminal dimension from Table 312.6(A).
- **366.58(B):** Added requirement for sizing an auxiliary gutter used as a pull box.

Contents

III. Construction Specifications
    366.100 Construction
        (A) Electrical and Mechanical Continuity
        (B) Substantial Construction
        (C) Smooth Rounded Edges
        (D) Covers
        (E) Clearance of Bare Live Parts
    366.120 Marking
        (A) Outdoors
        (B) Indoors

I. General

366.1 Scope

This article covers the use, installation, and construction requirements of metal auxiliary gutters and nonmetallic auxiliary gutters and associated fittings.

366.2 Definitions

These definitions are new for the 2005 *Code*.

Metallic Auxiliary Gutters. Sheet metal enclosures with hinged or removable covers for housing and protecting electric wires, cable, and busbars in which conductors are laid in place after the wireway has been installed as a complete system.

Nonmetallic Auxiliary Gutters. Flame retardant, nonmetallic enclosures with removable covers for housing and protecting electric wires, cable, and busbars in which conductors are laid in place after the wireway has been installed as a complete system.

366.6 Listing Requirements

(A) Outdoors Nonmetallic auxiliary gutters installed outdoors shall comply with the following:

(1) Be listed as suitable for exposure to sunlight
(2) Be listed as suitable for use in wet locations
(3) Be listed for maximum ambient temperature of the installation

(B) Indoors Nonmetallic auxiliary gutters installed indoors shall be listed for the maximum ambient temperature of the installation.

Only nonmetallic auxiliary gutters are required to be listed. Auxiliary gutter sections and associated fittings are identical to those of wireways and if listed by UL bear the UL listing mark "Listed Wireway" or "Auxiliary Gutter." They differ only in their intended use. See the commentary following 376.2 for a comparative discussion. Gutters (and wireways)

are required to be constructed and installed to ensure adequate electrical and mechanical continuity of the complete system per 250.118(14).

II. Installation

366.10 Uses Permitted

Auxiliary gutters shall be permitted to supplement wiring spaces at meter centers, distribution centers, switchboards, and similar points of wiring systems and may enclose conductors or busbars.

(A) Sheet Metal Auxiliary Gutters

(1) Indoor and Outdoor Use Sheet metal auxiliary gutters shall be permitted for indoor and outdoor use.

(2) Wet Locations Sheet metal auxiliary gutters installed in wet locations shall be suitable for such locations.

(B) Nonmetallic Auxiliary Gutters Nonmetallic auxiliary gutters shall be listed for the maximum ambient temperature of the installation and marked for the installed conductor insulation temperature rating.

(1) Outdoors Nonmetallic auxiliary gutters shall be permitted to be installed outdoors where listed and marked as suitable for the purpose.

> FPN: Extreme cold may cause nonmetallic auxiliary gutter to become brittle and therefore more susceptible to damage from physical contact.

(2) Indoors Nonmetallic auxiliary gutters shall be permitted to be installed indoors.

This section provides requirements for both indoor and outdoor installations. Both metal and nonmetallic gutters must have expansion fittings where temperature changes are expected to change gutter length more than ¼ in. according to 366.44. See the fine print note following 378.44 regarding expansion characteristics of PVC rigid nonmetallic conduit and PVC nonmetallic wireway.

366.12 Uses Not Permitted

Auxiliary gutters shall not be used under the following conditions:

(1) To enclose switches, overcurrent devices, appliances, or other similar equipment.
(2) To extend a greater distance than 9 m (30 ft) beyond the equipment that it supplements.

Exception: As permitted in 620.35 for elevators, an auxiliary gutter shall be permitted to extend a distance greater than 9 m (30 ft) beyond the equipment it supplements.

FPN: For wireways, see Articles 376 and 378. For busways, see Article 368.

366.22 Number of Conductors

(A) Sheet Metal Auxiliary Gutters The sum of the cross-sectional areas of all contained conductors at any cross section of a sheet metal auxiliary gutter shall not exceed 20 percent of the interior cross-sectional area of the sheet metal auxiliary gutter. The derating factors in 310.15(B)(2)(a) shall be applied only where the number of current-carrying conductors, including neutral conductors classified as current-carrying under the provisions of 310.15(B)(4), exceeds 30. Conductors for signaling circuits or controller conductors between a motor and its starter and used only for starting duty shall not be considered as current-carrying conductors.

(B) Nonmetallic Auxiliary Gutters The sum of cross-sectional areas of all contained conductors at any cross section of the nonmetallic auxiliary gutter shall not exceed 20 percent of the interior cross-sectional area of the nonmetallic auxiliary gutter.

Section 366.22 calls out the requirements for metal and nonmetallic auxiliary gutters. A common requirement for both types of auxiliary gutters is that all the contained conductors must not exceed 20 percent fill of the interior cross-sectional area of the gutter. The dimensions of insulated conductors, found in Tables 5 and 5A of Chapter 9, may be used to calculate the size of auxiliary gutters.

Where sheet metal auxiliary gutters contain 30 or fewer current-carrying conductors, the correction factors in 310.15(B)(2) do not apply. However, if more than 30 conductors are installed in a sheet metal auxiliary gutter, the ampacity adjustment factors of 310.15(B)(2) apply, and there is no limit on the number of current-carrying conductors up to the 20 percent fill.

The requirements for nonmetallic auxiliary gutters limit the cross-sectional area of all conductors to 20 percent. There is no 30-conductor allowance. The derating factors specified in 310.15(B)(2) must be applied as stated in 366.23(B).

See the example for calculating the size of a wireway in the commentary following 376.22. This calculation method is also applicable to auxiliary gutters.

No limit is placed on the size of conductors that may be installed in an auxiliary gutter; however, see 366.23(A) for ampacity limitations of bare copper or aluminum busbars enclosed in gutters.

366.23 Ampacity of Conductors

(A) Sheet Metal Auxiliary Gutters Where the number of current-carrying conductors contained in the sheet metal auxiliary gutter is 30 or less, the correction factors specified in 310.15(B)(2)(a) shall not apply. The current carried continuously in bare copper bars in sheet metal auxiliary gutters shall not exceed 1.55 amperes/mm^2 (1000 amperes/in.^{2}) of cross section of the conductor. For aluminum bars, the current carried continuously shall not exceed 1.09 amperes/mm^2 (700 amperes/in.^{2}) of cross section of the conductor.

(B) Nonmetallic Auxiliary Gutters The derating factors specified in 310.15(B)(2)(a) shall be applicable to the current-carrying conductors in the nonmetallic auxiliary gutter.

366.30 Securing and Supporting

(A) Sheet Metal Auxiliary Gutters Sheet metal auxiliary gutters shall be supported throughout their entire length at intervals not exceeding 1.5 m (5 ft).

(B) Nonmetallic Auxiliary Gutters Nonmetallic auxiliary gutters shall be supported at intervals not to exceed 900 mm (3 ft) and at each end or joint, unless listed for other support intervals. In no case shall the distance between supports exceed 3 m (10 ft).

366.44 Expansion Fittings

Expansion fittings shall be installed where expected length change, due to expansion and contraction due to temperature change, is more than 6 mm (0.25 in.).

366.56 Splices and Taps

Splices and taps shall comply with 366.56(A) through (D).

(A) Within Gutters Splices or taps shall be permitted within gutters where they are accessible by means of removable covers or doors. The conductors, including splices and taps, shall not fill the gutter to more than 75 percent of its area.

(B) Bare Conductors Taps from bare conductors shall leave the gutter opposite their terminal connections, and conductors shall not be brought in contact with uninsulated current-carrying parts of different potential.

(C) Suitably Identified All taps shall be suitably identified at the gutter as to the circuit or equipment that they supply.

(D) Overcurrent Protection Tap connections from conductors in auxiliary gutters shall be provided with overcurrent protection as required in 240.21.

To prevent abrasion of the conductor insulation, suitable bushings, shields, and so on, must be provided where conductors pass around bends or between gutters and cabinets and other locations. Sections 366.56(C) and 366.56(D) require all taps from gutters to be identified (as to circuits or equipment) and protected with overcurrent devices per 240.21.

366.58 Insulated Conductors

(A) Deflected Insulated Conductors Where insulated conductors are deflected within an auxiliary gutter, either at the ends or where conduits, fittings, or other raceways or cables enter or leave the gutter, or where the direction of the gutter is deflected greater than 30 degrees, dimensions corresponding to one wire per terminal in Table 312.6(A) shall apply.

Also, conductors are required to be shaped or formed in a permanent manner so that they are not in contact with bare busbars within the gutter.

(B) Auxiliary Gutters Used as Pullboxes Where insulated conductors 4 AWG or larger are pulled through an auxiliary gutter, the distance between raceway and cable entries enclosing the same conductor shall not be less than that required in 314.28(A)(1) for straight pulls and 314.28(A)(2) for angle pulls.

366.60 Grounding

Metal auxiliary gutters shall be grounded.

III. Construction Specifications

366.100 Construction

(A) Electrical and Mechanical Continuity Gutters shall be constructed and installed so that adequate electrical and mechanical continuity of the complete system is secured.

(B) Substantial Construction Gutters shall be of substantial construction and shall provide a complete enclosure for the contained conductors. All surfaces, both interior and exterior, shall be suitably protected from corrosion. Corner joints shall be made tight, and where the assembly is held together by rivets, bolts, or screws, such fasteners shall be spaced not more than 300 mm (12 in.) apart.

(C) Smooth Rounded Edges Suitable bushings, shields, or fittings having smooth, rounded edges shall be provided where conductors pass between gutters, through partitions, around bends, between gutters and cabinets or junction boxes, and at other locations where necessary to prevent abrasion of the insulation of the conductors.

(D) Covers Covers shall be securely fastened to the gutter.

(E) Clearance of Bare Live Parts Bare conductors shall be securely and rigidly supported so that the minimum clearance between bare current-carrying metal parts of different potential mounted on the same surface will not be less than 50 mm (2 in.), nor less than 25 mm (1 in.) for parts that are held free in the air. A clearance not less than 25 mm (1 in.) shall be secured between bare current-carrying metal parts and any metal surface. Adequate provisions shall be made for the expansion and contraction of busbars.

366.120 Marking

(A) Outdoors Nonmetallic auxiliary gutters installed outdoors shall have the following markings:

(1) Suitable for exposure to sunlight
(2) Suitable for use in wet locations
(3) Installed conductor insulation temperature rating

(B) Indoors Nonmetallic auxiliary gutters installed indoors shall have the following markings:

(1) Installed conductor insulation temperature rating

ARTICLE 368
Busways

Summary of Changes

• **General:** Rearranged and renumbered to coordinate with format of other Chapter 3 raceway articles.

Contents

I. General Requirements

368.1 Scope

This article covers service-entrance, feeder, and branch-circuit busways and associated fittings.

368.2 Definition

Busway. A grounded metal enclosure containing factory-mounted, bare or insulated conductors, which are usually copper or aluminum bars, rods, or tubes.

> FPN: For cablebus, refer to Article 370.

Exhibit 368.1 shows a 10-ft section of feeder busway. See Exhibit 100.1 for other example illustrations.

Exhibit 368.1 A 10-ft section of feeder busway.

According to the 2004 UL *Electrical Construction Equipment Directory*, category CWFT, busways and short-run busways are provided with metal enclosures. These enclosures and in some cases an additional ground bus are

intended for use as equipment grounding conductors. Some busways are not intended for use ahead of service equipment and are marked with the maximum rating of overcurrent protection to be used on the supply side of the busway. Busways that have been investigated to determine suitability for installation in a specified position, for use in vertical runs, for support at intervals greater than 5 ft, or for outdoor use are so marked. This marking is on or contiguous to the nameplate incorporating the manufacturer's name and electrical rating. A busway or fitting containing a vapor seal is so marked, but unless marked otherwise, the busway or fitting has not been investigated for passage through a fire wall.

Short-run busways are marked to limit the run to 30 ft or less and no more than 10 ft vertically. They are intended primarily to feed switchboards. Except for transformer stubs, short-run busways are not intended to have intermediate taps. Short-run busways are not ventilated and may be marked for outdoor use.

Busways and associated fittings marked "Short Circuit Current Rating(s) Maximum rms Symmetrical Amps _____ Volts _____" have been investigated for the rating indicated.

Busways that are intended to supply and support industrial and commercial luminaires are classified as "Lighting Busway" and are so marked. Trolley busway is marked "Trolley Busway" and is additionally marked "Lighting Busway" if intended to supply and support industrial and commercial luminaires. Busway with provision for insertion of plug-in devices at any point along its length and intended for general use is classified as "Continuous Plug-in Busway" and is so marked. The marking is contiguous to the marking of the manufacturer's name and the electrical rating.

Busway marked "Lighting Busway" and protected by overcurrent devices rated in excess of 20 amperes is intended for use only with luminaires employing heavy-duty lampholders unless additional overcurrent protection is provided for the luminaire in accordance with this *Code*.

Trolley busway should be installed out of the reach of people, or it should be otherwise installed to prevent accidental contact with exposed conductors.

II. Installation

368.10 Uses Permitted

Busways shall be permitted to be installed where they are located in accordance with 368.10(A) through (C).

(A) Exposed Busways shall be permitted to be located in the open where visible, except as permitted in 368.10(C).

(B) Concealed Busways shall be permitted to be installed behind access panels, provided the busways are totally enclosed, of nonventilating-type construction, and installed so that the joints between sections and at fittings are accessible for maintenance purposes. Where installed behind access panels, means of access shall be provided, and either of the following conditions shall be met:

(1) The space behind the access panels shall not be used for air-handling purposes.
(2) Where the space behind the access panels is used for environmental air, other than ducts and plenums, there shall be no provisions for plug-in connections, and the conductors shall be insulated.

Unless busways are mounted in the open and are visible, the installation must comply with 368.10(B). Busways are commonly used as feeders and are mounted horizontally in industrial buildings or mounted vertically in high-rise buildings. See Exhibit 100.2 for an example of a busway mounted above a hung ceiling.

(C) Through Walls and Floors Busways shall be permitted to be installed through walls or floors in accordance with (C)(1) and (C)(2).

(1) Walls Unbroken lengths of busway shall be permitted to be extended through dry walls.

(2) Floors Floor penetrations shall comply with (a) and (b):

(a) Busways shall be permitted to be extended vertically through dry floors if totally enclosed (unventilated) where passing through and for a minimum distance of 1.8 m (6 ft) above the floor to provide adequate protection from physical damage.

(b) In other than industrial establishments, where a vertical riser penetrates two or more dry floors, a minimum 100-mm (4-in.) high curb shall be installed around all floor openings for riser busways to prevent liquids from entering the opening. The curb shall be installed within 300 mm (12 in.) of the floor opening. Electrical equipment shall be located so that it will not be damaged by liquids that are retained by the curb.

FPN: See 300.21 for information concerning the spread of fire or products of combustion.

A busway or fitting containing a vapor seal is so marked, but, unless marked otherwise, the busway or fitting has not been investigated for passage through a fire-rated wall. The requirements of 300.21 are extremely important in order to confine a fire and the products of combustion at their origin.

Section 368.10(C)(2)(b) requires that a curb be placed around a busway if the busway penetrates two or more dry floors. Experience has shown that if liquid spills occur on upper floors of normally dry buildings, the spilled liquid often flows to the busway floor penetration and down the

vertical rise of the busway. Spills can cause extensive damage to the busway and the building's electrical system. The addition of a 4-in. curb encircling the busway can help eliminate this potentially dangerous situation.

368.12 Uses Not Permitted

(A) Physical Damage Busways shall not be installed where subject to severe physical damage or corrosive vapors.

(B) Hoistways Busways shall not be installed in hoistways.

(C) Hazardous Locations Busways shall not be installed in any hazardous (classified) location, unless specifically approved for such use.

FPN: See 501.10(B).

(D) Wet Locations Busways shall not be installed outdoors or in wet or damp locations unless identified for such use.

(E) Working Platform Lighting busway and trolley busway shall not be installed less than 2.5 m (8 ft) above the floor or working platform unless provided with a cover identified for the purpose.

368.17 Overcurrent Protection

Overcurrent protection shall be provided in accordance with 368.17(A) through (D).

The rated ampacity of a busway is based on the allowable temperature rise of the conductors and can be determined in the field only by reference to the nameplate data. The requirements of 240.4(B) and 240.4(C) are applicable to busways.

(A) Rating of Overcurrent Protection — Feeders A busway shall be protected against overcurrent in accordance with the allowable current rating of the busway.

Exception No. 1: The applicable provisions of 240.4 shall be permitted.

Exception No. 2: Where used as transformer secondary ties, the provisions of 450.6(A)(3) shall be permitted.

(B) Reduction in Ampacity Size of Busway Overcurrent protection shall be required where busways are reduced in ampacity.

Exception: For industrial establishments only, omission of overcurrent protection shall be permitted at points where busways are reduced in ampacity, provided that the length of the busway having the smaller ampacity does not exceed 15 m (50 ft) and has an ampacity at least equal to one-third the rating or setting of the overcurrent device next back on the line, and provided that such busway is free from contact with combustible material.

In industrial establishments, where the size of a smaller busway is kept within the specified limits, the additional cost of providing overcurrent protection at the point where the size is changed is not warranted. For example, a busway protected by a 1200-ampere overcurrent device may be reduced in size, provided the smaller busway has a current rating of 400 amperes (⅓ of 1200 amperes) and does not extend more than 50 ft. In this case, overcurrent protection would be required if the smaller busway were rated less than 400 amperes (e.g., 200 amperes, 300 amperes).

(C) Feeder or Branch Circuits Where a busway is used as a feeder, devices or plug-in connections for tapping off feeder or branch circuits from the busway shall contain the overcurrent devices required for the protection of the feeder or branch circuits. The plug-in device shall consist of an externally operable circuit breaker or an externally operable fusible switch. Where such devices are mounted out of reach and contain disconnecting means, suitable means such as ropes, chains, or sticks shall be provided for operating the disconnecting means from the floor.

Externally operated fused switches and circuit breakers that are plugged into busways and mounted out of reach are to be considered accessible if operated by means such as ropes, chains, or hooksticks.

Exception No. 1: As permitted in 240.21.

Exception No. 2: For fixed or semifixed luminaires (lighting fixtures), where the branch-circuit overcurrent device is part of the luminaire (fixture) cord plug on cord-connected luminaires (fixtures).

Exception No. 3: Where luminaires (fixtures) without cords are plugged directly into the busway and the overcurrent device is mounted on the luminaire (fixture).

(D) Rating of Overcurrent Protection — Branch Circuits A busway used as a branch circuit shall be protected against overcurrent in accordance with 210.20.

368.30 Support

Busways shall be securely supported at intervals not exceeding 1.5 m (5 ft) unless otherwise designed and marked.

368.56 Branches from Busways

Branches from busways shall be permitted to be made in accordance with 368.56(A), (B), and (C).

Exhibit 368.2 shows an example of a cable or cord branch from a busway installed according to the provisions of 368.56(B)(2).

Exhibit 368.2 An example of a cable or cord branch from a busway installed according to 368.56(B).

Section 400.9 specifically prohibits the installation of spliced cords. Exhibit 368.3 shows an example of a cable or cord branch from a busway installed according to 368.56(B)(2), Exception.

Exhibit 368.3 An example of an installation permitted only in industrial occupancies with other restrictions, according to 368.56(B)(2), Exception.

(A) General Branches from busways shall be permitted to use any of the following wiring methods:

(1) Type AC armored cable
(2) Type MC metal-clad cable
(3) Type MI mineral-insulated, metal-sheathed cable
(4) Type IMC intermediate metal conduit
(5) Type RMC rigid metal conduit
(6) Type FMC flexible metal conduit
(7) Type LFMC liquidtight flexible metal conduit
(8) Type RNC rigid nonmetallic conduit
(9) Type LFNC liquidtight flexible nonmetal conduit
(10) Type EMT electrical metallic tubing
(11) Type ENT electrical nonmetallic tubing
(12) Busways
(13) Strut-type channel raceway
(14) Surface metal raceways
(15) Surface nonmetallic raceways

Where a separate equipment grounding conductor is used, connection of the equipment grounding conductor to the busway shall comply with 250.8 and 250.12.

(B) Cord and Cable Assemblies Suitable cord and cable assemblies approved for extra-hard usage or hard usage, and listed bus drop cable shall be permitted as branches from busways for the connection of portable equipment or the connection of stationary equipment to facilitate their interchange in accordance with 400.7 and 400.8 and the following conditions:

(1) The cord or cable shall be attached to the building by an approved means.
(2) The length of the cord or cable from a busway plug-in device to a suitable tension take-up support device shall not exceed 1.8 m (6 ft).
(3) The cord and cable shall be installed as a vertical riser from the tension take-up support device to the equipment served.
(4) Strain relief cable grips shall be provided for the cord or cable at the busway plug-in device and equipment terminations.

Exception to (B)(2): In industrial establishments only, where the conditions of maintenance and supervision ensure that only qualified persons service the installation, lengths exceeding 1.8 m (6 ft) shall be permitted between the busway plug-in device and the tension take-up support device where the cord or cable is supported at intervals not exceeding 2.5 m (8 ft).

(C) Branches from Trolley-Type Busways Suitable cord and cable assemblies approved for extra-hard usage or hard usage and listed bus drop cable shall be permitted as branches from trolley-type busways for the connection of movable equipment in accordance with 400.7 and 400.8.

368.58 Dead Ends

A dead end of a busway shall be closed.

368.60 Grounding

Busway shall be grounded.

III. Construction
368.120 Marking

Busways shall be marked with the voltage and current rating for which they are designed, and with the manufacturer's name or trademark in such a manner as to be visible after installation.

IV. Requirements for Over 600 Volts, Nominal
368.214 Adjacent and Supporting Structures

Metal-enclosed busways shall be installed so that temperature rise from induced circulating currents in any adjacent metallic parts will not be hazardous to personnel or constitute a fire hazard.

368.234 Barriers and Seals

(A) Vapor Seals Busway runs that have sections located both inside and outside of buildings shall have a vapor seal at the building wall to prevent interchange of air between indoor and outdoor sections.

Exception: Vapor seals shall not be required in forced-cooled bus.

(B) Fire Barriers Fire barriers shall be provided where fire walls, floors, or ceilings are penetrated.

FPN: See 300.21 for information concerning the spread of fire or products of combustion.

368.236 Drain Facilities

Drain plugs, filter drains, or similar methods shall be provided to remove condensed moisture from low points in busway run.

368.237 Ventilated Bus Enclosures

Ventilated busway enclosures shall be installed in accordance with Article 110, Part III, and 490.24.

368.238 Terminations and Connections

Where bus enclosures terminate at machines cooled by flammable gas, seal-off bushings, baffles, or other means shall be provided to prevent accumulation of flammable gas in the busway enclosures.

All conductor termination and connection hardware

shall be accessible for installation, connection, and maintenance.

368.239 Switches

Switching devices or disconnecting links provided in the busway run shall have the same momentary rating as the busway. Disconnecting links shall be plainly marked to be removable only when bus is de-energized. Switching devices that are not load-break shall be interlocked to prevent operation under load, and disconnecting link enclosures shall be interlocked to prevent access to energized parts.

368.240 Wiring 600 Volts or Less, Nominal

Secondary control devices and wiring that are provided as part of the metal-enclosed bus run shall be insulated by fire-retardant barriers from all primary circuit elements with the exception of short lengths of wire, such as at instrument transformer terminals.

368.244 Expansion Fittings

Flexible or expansion connections shall be provided in long, straight runs of bus to allow for temperature expansion or contraction, or where the busway run crosses building vibration insulation joints.

368.258 Neutral

Neutral bus, where required, shall be sized to carry all neutral load current, including harmonic currents, and shall have adequate momentary and short-circuit rating consistent with system requirements.

368.260 Grounding

Metal-enclosed busway shall be grounded.

368.320 Marking

Each busway run shall be provided with a permanent nameplate on which the following information shall be provided:

(1) Rated voltage.
(2) Rated continuous current; if bus is forced-cooled, both the normal forced-cooled rating and the self-cooled (not forced-cooled) rating for the same temperature rise shall be given.
(3) Rated frequency.
(4) Rated impulse withstand voltage.
(5) Rated 60-Hz withstand voltage (dry).
(6) Rated momentary current.
(7) Manufacturer's name or trademark.

> FPN: See ANSI C37.23-1987 (R1991), *Guide for Metal-Enclosed Bus and Calculating Losses in Isolated-Phase Bus,* for construction and testing requirements for metal-enclosed buses.

ARTICLE 370
Cablebus

Contents

370.1 Scope

This article covers the use and installation requirements of cablebus and associated fittings.

370.2 Definition

Cablebus. An assembly of insulated conductors with fittings and conductor terminations in a completely enclosed, ventilated protective metal housing. Cablebus is ordinarily assembled at the point of installation from the components furnished or specified by the manufacturer in accordance with instructions for the specific job. This assembly is designed to carry fault current and to withstand the magnetic forces of such current.

As shown in Exhibit 370.1, cablebus consists of a metal structure or framework installed in a manner similar to that for a cable tray support system. Insulated conductors, 1/0 AWG or larger, are field installed within the framework on special insulating blocks at specified intervals to provide controlled spacing between conductors. To completely enclose the conductors, a ventilated top cover is attached to the framework.

370.3 Use

Approved cablebus shall be permitted at any voltage or current for which spaced conductors are rated and shall be installed only for exposed work, except as permitted in 370.6.

Exhibit 370.1 A section of cablebus with conductors in place and the ventilated top cover ready to be attached to the busway frame. (Courtesy of MPHusky Corp.)

Cablebus installed outdoors or in corrosive, wet, or damp locations shall be identified for such use. Cablebus shall not be installed in hoistways or hazardous (classified) locations unless specifically approved for such use. Cablebus shall be permitted to be used for branch circuits, feeders, and services.

Cablebus framework, where bonded, shall be permitted to be used as the equipment grounding conductor for branch circuits and feeders.

370.4 Conductors

(A) Types of Conductors The current-carrying conductors in cablebus shall have an insulation rating of 75°C (167°F) or higher and be an approved type suitable for the application.

(B) Ampacity of Conductors The ampacity of conductors in cablebus shall be in accordance with Tables 310.17 and 310.19, or with Tables 310.69 and 310.70 for installations over 600 volts.

(C) Size and Number of Conductors The size and number of conductors shall be that for which the cablebus is designed, and in no case smaller than 1/0 AWG.

(D) Conductor Supports The insulated conductors shall be supported on blocks or other mounting means designed for the purpose.

The individual conductors in a cablebus shall be sup-

ported at intervals not greater than 900 mm (3 ft) for horizontal runs and 450 mm (1½ ft) for vertical runs. Vertical and horizontal spacing between supported conductors shall not be less than one conductor diameter at the points of support.

370.5 Overcurrent Protection

Cablebus shall be protected against overcurrent in accordance with the allowable ampacity of the cablebus conductors in accordance with 240.4.

Exception: Overcurrent protection shall be permitted in accordance with 240.100 and 240.101 for over 600 volts, nominal.

370.6 Support and Extension Through Walls and Floors

(A) Support Cablebus shall be securely supported at intervals not exceeding 3.7 m (12 ft).

Exception: Where spans longer than 3.7 m (12 ft) are required, the structure shall be specifically designed for the required span length.

(B) Transversely Routed Cablebus shall be permitted to extend transversely through partitions or walls, other than fire walls, provided the section within the wall is continuous, protected against physical damage, and unventilated.

(C) Through Dry Floors and Platforms Except where firestops are required, cablebus shall be permitted to extend vertically through dry floors and platforms, provided the cablebus is totally enclosed at the point where it passes through the floor or platform and for a distance of 1.8 m (6 ft) above the floor or platform.

(D) Through Floors and Platforms in Wet Locations Except where firestops are required, cablebus shall be permitted to extend vertically through floors and platforms in wet locations where (1) there are curbs or other suitable means to prevent waterflow through the floor or platform opening, and (2) where the cablebus is totally enclosed at the point where it passes through the floor or platform and for a distance of 1.8 m (6 ft) above the floor or platform.

370.7 Fittings

A cablebus system shall include approved fittings for the following:

(1) Changes in horizontal or vertical direction of the run
(2) Dead ends
(3) Terminations in or on connected apparatus or equipment or the enclosures for such equipment
(4) Additional physical protection where required, such as guards where subject to severe physical damage

370.8 Conductor Terminations

Approved terminating means shall be used for connections to cablebus conductors.

370.9 Grounding

A cablebus installation shall be grounded and bonded in accordance with Article 250, excluding 250.86, Exception No. 2.

370.10 Marking

Each section of cablebus shall be marked with the manufacturer's name or trade designation and the maximum diameter, number, voltage rating, and ampacity of the conductors to be installed. Markings shall be located so as to be visible after installation.

ARTICLE 372
Cellular Concrete Floor Raceways

Summary of Changes

• **372.17:** Added requirement that the ampacity adjustment factors of 310.15(B)(2) apply to the installed conductors.

Contents

372.1 Scope

This article covers cellular concrete floor raceways, the hollow spaces in floors constructed of precast cellular concrete slabs, together with suitable metal fittings designed to provide access to the floor cells.

Cellular concrete floor raceways are a form of floor deck construction commonly used in high-rise office buildings.

This construction method is very similar in design, application, and adaptation to cellular metal floor raceways. Basically, this wiring method consists of floor cells (that are part of the structural floor system), header ducts laid at right angles to the cells and used to carry conductors from cabinets to cells, and junction boxes.

372.2 Definitions

Cell. A single, enclosed tubular space in a floor made of precast cellular concrete slabs, the direction of the cell being parallel to the direction of the floor member.

Header. Transverse metal raceways for electric conductors, providing access to predetermined cells of a precast cellular concrete floor, thereby permitting the installation of electric conductors from a distribution center to the floor cells.

372.4 Uses Not Permitted

Conductors shall not be installed in precast cellular concrete floor raceways as follows:

(1) Where subject to corrosive vapor
(2) In any hazardous (classified) locations except as permitted by 504.20, and in Class I, Division 2 locations as permitted in 501.10(B)(3)
(3) In commercial garages, other than for supplying ceiling outlets or extensions to the area below the floor but not above

FPN: See 300.8 for installation of conductors with other systems.

Section 300.8 prohibits the installation of electric conductors in raceways or cable trays containing any pipes, tubes, or other means for carrying steam, water, air, gas, or drainage or for any service other than electrical.

372.5 Header

The header shall be installed in a straight line at right angles to the cells. The header shall be mechanically secured to the top of the precast cellular concrete floor. The end joints shall be closed by a metal closure fitting and sealed against the entrance of concrete. The header shall be electrically continuous throughout its entire length and shall be electrically bonded to the enclosure of the distribution center.

372.6 Connection to Cabinets and Other Enclosures

Connections from headers to cabinets and other enclosures shall be made by means of listed metal raceways and listed fittings.

372.7 Junction Boxes

Junction boxes shall be leveled to the floor grade and sealed against the free entrance of water or concrete. Junction boxes shall be of metal and shall be mechanically and electrically continuous with the header.

372.8 Markers

A suitable number of markers shall be installed for the future location of cells.

372.9 Inserts

Inserts shall be leveled and sealed against the entrance of concrete. Inserts shall be of metal and shall be fitted with grounded-type receptacles. A grounding conductor shall connect the insert receptacles to a positive ground connection provided on the header. Where cutting through the cell wall for setting inserts or other purposes (such as providing access openings between header and cells), chips and other dirt shall not be allowed to remain in the raceway, and the tool used shall be designed so as to prevent the tool from entering the cell and damaging the conductors.

372.10 Size of Conductors

No conductor larger than 1/0 AWG shall be installed, except by special permission.

372.11 Maximum Number of Conductors

The combined cross-sectional area of all conductors or cables shall not exceed 40 percent of the cross-sectional area of the cell or header.

372.12 Splices and Taps

Splices and taps shall be made only in header access units or junction boxes.

For the purposes of this section, so-called loop wiring (continuous unbroken conductor connecting the individual outlets) shall not be considered to be a splice or tap.

372.13 Discontinued Outlets

When an outlet is abandoned, discontinued, or removed, the sections of circuit conductors supplying the outlet shall be removed from the raceway. No splices or reinsulated conductors, such as would be the case of abandoned outlets on loop wiring, shall be allowed in raceways.

372.17 Ampacity of Conductors

The ampacity adjustment factors, provided in 310.15(B)(2), shall apply to conductors installed in cellular concrete floor raceways.

ARTICLE 374
Cellular Metal Floor Raceways

Summary of Changes

- **374.17:** New requirement that the ampacity adjustment factors of 310.15(B)(2) apply to the installed conductors.

Contents

374.1 Scope

This article covers the use and installation requirements for cellular metal floor raceways.

Cellular metal floor raceways, as shown in Exhibit 374.1, are a form of metal floor deck construction designed for use in steel-frame buildings and consisting of sheet metal formed into shapes that are combined to form cells or raceways. The cells extend across the building and, depending on the structural strength required, can have various shapes and sizes.

374.2 Definitions

Cellular Metal Floor Raceway. The hollow spaces of cellular metal floors, together with suitable fittings, that may be approved as enclosures for electric conductors.

Cell. A single enclosed tubular space in a cellular metal floor member, the axis of the cell being parallel to the axis of the metal floor member.

Header. A transverse raceway for electric conductors, providing access to predetermined cells of a cellular metal floor, thereby permitting the installation of electric conductors from a distribution center to the cells.

Exhibit 374.1 An artistic rendition of a cross section of cellular metal floor raceway system. (Photo courtesy of H.H. Robertson Floor Systems)

374.3 Uses Not Permitted

Conductors shall not be installed in cellular metal floor raceways as follows:

(1) Where subject to corrosive vapor
(2) In any hazardous (classified) location except as permitted by 504.20, and in Class I, Division 2 locations as permitted by 501.10(B)(3)
(3) In commercial garages, other than for supplying ceiling outlets or extensions to the area below the floor but not above

FPN: See 300.8 for installation of conductors with other systems.

Section 300.8 prohibits the installation of electric conductors in raceways or cable trays containing any pipes, tubes, or other carriers of steam, water, air, gas, or drainage or for any service other than electrical.

I. Installation

374.4 Size of Conductors

No conductor larger than 1/0 AWG shall be installed, except by special permission.

374.5 Maximum Number of Conductors in Raceway

The combined cross-sectional area of all conductors or cables shall not exceed 40 percent of the interior cross-sectional area of the cell or header.

Connections to the cells are made by means of headers extending across the cells and connecting only to those cells that are to be used as raceways for the conductors. Two or three separate headers, connecting to different sets of cells, may be used for different systems, such as light and power, signaling, and communications systems.

374.6 Splices and Taps

Splices and taps shall be made only in header access units or junction boxes.

For the purposes of this section, so-called loop wiring (continuous unbroken conductor connecting the individual outlets) shall not be considered to be a splice or tap.

374.7 Discontinued Outlets

When an outlet is abandoned, discontinued, or removed, the sections of circuit conductors supplying the outlet shall be removed from the raceway. No splices or reinsulated conductors, such as would be the case with abandoned outlets on loop wiring, shall be allowed in raceways.

374.8 Markers

A suitable number of markers shall be installed for locating cells in the future.

Markers are brass flat-head screws set into the top side of the cells and adjusted so that their heads are flush with the floor finish and are exposed in order to aid in the location of cells for future installations.

374.9 Junction Boxes

Junction boxes shall be leveled to the floor grade and sealed against the free entrance of water or concrete. Junction boxes used with these raceways shall be of metal and shall be electrically continuous with the raceway.

Connections to wall outlets are to be made with metal raceways unless there are provisions for equipment grounding termination, as required by 374.11. Installation instructions are supplied by the manufacturer for use by the general contractor, erector, electrical contractor, inspector, and others concerned with the installation.

374.10 Inserts

Inserts shall be leveled to the floor grade and sealed against the entrance of concrete. Inserts shall be of metal and shall be electrically continuous with the raceway. In cutting through the cell wall and setting inserts, chips and other dirt shall not be allowed to remain in the raceway, and tools shall be used that are designed to prevent the tool from entering the cell and damaging the conductors.

374.11 Connection to Cabinets and Extensions from Cells

Connections between raceways and distribution centers and wall outlets shall be made by means of liquidtight flexible metal conduit, flexible metal conduit where not installed in concrete, rigid metal conduit, intermediate metal conduit, electrical metallic tubing, or approved fittings. Where there are provisions for the termination of an equipment grounding conductor, nonmetallic conduit, electrical nonmetallic tubing, or liquidtight flexible nonmetallic conduit shall be permitted. Where installed in concrete, liquidtight flexible nonmetallic conduit shall be listed and marked for direct burial.

> FPN: Liquidtight flexible metal conduit and liquidtight flexible nonmetallic conduit that is suitable for installation in concrete is listed and marked for direct burial.

374.17 Ampacity of Conductors

The ampacity adjustment factors in 310.15(B)(2) shall apply to conductors installed in cellular metal floor raceways.

Added for the 2005 *Code* and placed here as a reminder, ampacity adjustment factors apply to cellular metal floor raceways.

II. Construction Specifications

374.100 General

Cellular metal floor raceways shall be constructed so that adequate electrical and mechanical continuity of the complete system will be secured. They shall provide a complete enclosure for the conductors. The interior surfaces shall be free from burrs and sharp edges, and surfaces over which conductors are drawn shall be smooth. Suitable bushings or fittings having smooth rounded edges shall be provided where conductors pass.

ARTICLE 376
Metal Wireways

Summary of Changes

- **376.23(A):** Revised to specify that the minimum dimension for conductor deflection space is based on the one wire per terminal dimension from Table 312.6(A).

- **376.23(B):** Added text on transposing cable size into the minimum raceway size for the purposes of sizing wireways used as pull boxes.

- **376.56(B):** Added requirements covering the installation of power distribution blocks in metal wireways.

Contents

I. General

376.1 Scope

This article covers the use, installation, and construction specifications for metal wireways and associated fittings.

376.2 Definition

Metal Wireways. Sheet metal troughs with hinged or removable covers for housing and protecting electric wires and cable and in which conductors are laid in place after the wireway has been installed as a complete system.

Wireways are sheet-metal enclosures equipped with hinged or removable covers and are manufactured in 1-ft to 10-ft lengths and various widths and depths. Couplings, elbows, end plates, and accessories such as T and X fittings are available. Unlike auxiliary gutters, which are not permitted to extend more than 30 ft from the equipment they supplement, wireways may be run throughout an entire area.

II. Installation

376.10 Uses Permitted

The use of metal wireways shall be permitted in the following:

(1) For exposed work
(2) In concealed spaces as permitted in 376.10(4)
(3) In hazardous (classified) locations as permitted by 501.10(B) for Class I, Division 2 locations; 502.10(B) for Class II, Division 2 locations; and 504.20 for intrinsically safe wiring. Where installed in wet locations, wireways shall be listed for the purpose.
(4) As extensions to pass transversely through walls if the length passing through the wall is unbroken. Access to the conductors shall be maintained on both sides of the wall.

376.12 Uses Not Permitted

Metal wireways shall not be used in the following:

(1) Where subject to severe physical damage
(2) Where subject to severe corrosive environments

376.21 Size of Conductors

No conductor larger than that for which the wireway is designed shall be installed in any wireway.

376.22 Number of Conductors

The sum of the cross-sectional areas of all contained conductors at any cross section of a wireway shall not exceed 20 percent of the interior cross-sectional area of the wireway. The derating factors in 310.15(B)(2)(a) shall be applied only where the number of current-carrying conductors, including neutral conductors classified as current-carrying under the provisions of 310.15(B)(4), exceeds 30. Conductors for signaling circuits or controller conductors between a motor and its starter and used only for starting duty shall not be considered as current-carrying conductors.

The main requirement of 376.22 is that the total of the cross-sectional areas of all conductors must not exceed 20 percent of the interior cross-sectional area of the wireway. If the quantity of conductors does not exceed 30, the adjustment factors of 310.15(B)(2) do not apply. Where the quantity of conductors does exceed 30, however, the adjustment factors of 310.15(B)(2) do apply. The following example uses a wireway with only 26 current-carrying conductors; therefore, the adjustment factors of 310.15(B)(2) do not apply.

Example

A wireway contains 26 conductors: ten 3/0 AWG XHHW-2, three 6 AWG THWN, three 8 AWG THHN, and ten 12 AWG THHN. Find the minimum standard size wireway required by Article 376.

Solution

Conductor Type and Size	Quantity	Individual Area* (in.^{2})	Total Area (in.^{2})
3/0 AWG, XHHW-2	10 ×	0.2642	= 2.6420
6 AWG, THWN	3 ×	0.0507	= 0.1521
8 AWG, THHN	3 ×	0.0366	= 0.1098
12 AWG, THHN	10 ×	0.0133	= 0.1330
			3.0369

Total area occupied by conductors

Minimum wireway area required:

$$3.0369 \text{ in.}^2 \div 20\% \text{ fill} = 15.1845 \text{ in.}^2$$

Minimum size square wireway required:

$$15.1845 \text{ in.}^2 = 3.9 \text{ in.} \times 3.9 \text{ in.} \quad \text{or} \quad 4 \text{ in.} \times 4 \text{ in.}$$
$$\text{wireway}$$

*Individual area dimensions of conductors are from Chapter 9, Table 5.

376.23 Insulated Conductors

Insulated conductors installed in a metallic wireway shall comply with 376.23(A) and (B).

(A) **Deflected Insulated Conductors** Where insulated conductors are deflected within a metallic wireway, either at the ends or where conduits, fittings, or other raceways or cables enter or leave the metallic wireway, or where the direction of the metallic wireway is deflected greater than 30 degrees, dimensions corresponding to one wire per terminal in Table 312.6(A) shall apply.

The intent of 376.23 is to provide adequate space for installing and bending conductors without damaging the conductor insulation.

(B) **Metallic Wireways Used as Pull Boxes** Where insulated conductors 4 AWG or larger are pulled through a wireway, the distance between raceway and cable entries enclosing the same conductor shall not be less than that required by 314.28(A)(1) for straight pulls and 314.28(A)(2) for angle pulls. When transposing cable size into raceway size, the minimum metric designator (trade size) raceway required for the number and size of conductors in the cable shall be used.

Section 376.23(A) was revised for the 2005 *Code* to ensure that the column of one wire per terminal was used in Table 312.6(A). Section 376.23(B) was revised for the 2005 *Code* to make sure that the same adequate space requirements that apply to conduits also apply to cables as they enter a wireway.

The requirements in 376.23(B) ensure that where wireways are used as pull boxes, the same minimum dimension requirements associated with raceway entries of pull boxes apply.

376.30 Securing and Supporting

Metal wireways shall be supported in accordance with 376.30(A) and (B).

(A) Horizontal Support Wireways shall be supported where run horizontally at each end and at intervals not to exceed 1.5 m (5 ft) or for individual lengths longer than 1.5 m (5 ft) at each end or joint, unless listed for other support intervals. The distance between supports shall not exceed 3 m (10 ft).

(B) Vertical Support Vertical runs of wireways shall be securely supported at intervals not exceeding 4.5 m (15 ft) and shall not have more than one joint between supports. Adjoining wireway sections shall be securely fastened together to provide a rigid joint.

376.56 Splices, Taps, and Power Distribution Blocks

(A) Splices and Taps Splices and taps shall be permitted within a wireway, provided they are accessible. The conductors, including splices and taps, shall not fill the wireway to more than 75 percent of its area at that point.

Conductors in wireways are accessible through hinged or removable covers. Circuits, taps, or splices may be added or altered if necessary. See 376.22 and the associated example in the commentary regarding the number of conductors permitted. For the 2005 *Code,* 376.56(B) was added to ensure that power distribution blocks are safely placed within metal wireways.

(B) Power Distribution Blocks

(1) Installation Power distribution blocks installed in metal wireways shall be listed.

(2) Size of Enclosure In addition to the wiring space requirement in 376.56(A), the power distribution block shall be installed in a wireway with dimensions not smaller than specified in the installation instructions of the power distribution block.

(3) Wire Bending Space Wire bending space at the terminals of power distribution blocks shall comply with 312.6(B).

(4) Live Parts Power distribution blocks shall not have exposed live parts in the wireway after installation.

376.58 Dead Ends

Dead ends of metal wireways shall be closed.

376.70 Extensions from Metal Wireways

Extensions from wireways shall be made with cord pendants installed in accordance with 400.10 or with any wiring method in Chapter 3 that includes a means for equipment grounding. Where a separate equipment grounding conductor is employed, connection of the equipment grounding conductors in the wiring method to the wireway shall comply with 250.8 and 250.12.

Extensions from wireways using metal raceways, metal-sheathed cables, and nonmetallic-sheathed cables are made through knockouts provided on the wireway or field punched. Rigid nonmetallic conduit, electrical nonmetallic tubing, and liquidtight flexible nonmetallic conduit may also be used. Cables and nonmetallic raceways as well as the wireway must include a means for ensuring an effective continuation of the equipment grounding conductor. Sections of wireways, including accessory fittings (elbows, endplates, flanges, etc.), are bolted together, ensuring a rigid mechanical and electrical connection.

III. Construction Specifications

376.120 Marking

Metal wireways shall be so marked that their manufacturer's name or trademark will be visible after installation.

ARTICLE 378
Nonmetallic Wireways

Summary of Changes

- **378.23(A):** Revised to specify that the minimum dimension for conductor deflection space is based on the one wire per terminal dimension from Table 312.6(A).
- **378.23(B):** Added text on transposing cable size into the minimum raceway size for the purposes of sizing wireways used as pull boxes.

Contents

I. General

378.1 Scope

This article covers the use, installation, and construction specifications for nonmetallic wireways and associated fittings.

378.2 Definition

Nonmetallic Wireways. Flame retardant, nonmetallic troughs with removable covers for housing and protecting electric wires and cables in which conductors are laid in place after the wireway has been installed as a complete system.

Nonmetallic wireways are troughs with removable covers in which conductors are laid after the wireway has been installed as a complete system. The wireway must be installed on the surface of the structure, not concealed. It is allowed to pass through walls, provided that an unbroken length passes through the wall.

378.6 Listing Requirements

Nonmetallic wireways and associated fittings shall be listed.

II. Installation

378.10 Uses Permitted

The use of nonmetallic wireways shall be permitted in the following:

(1) Only for exposed work, except as permitted in 378.10(4).
(2) Where subject to corrosive environments where identified for the use.
(3) In wet locations where listed for the purpose.

FPN: Extreme cold may cause nonmetallic wireways to become brittle and therefore more susceptible to damage from physical contact.

(4) As extensions to pass transversely through walls if the length passing through the wall is unbroken. Access to the conductors shall be maintained on both sides of the wall.

378.12 Uses Not Permitted

Nonmetallic wireways shall not be used in the following:

(1) Where subject to physical damage
(2) In any hazardous (classified) location, except as permitted in 504.20
(3) Where exposed to sunlight unless listed and marked as suitable for the purpose
(4) Where subject to ambient temperatures other than those for which nonmetallic wireway is listed
(5) For conductors whose insulation temperature limitations would exceed those for which the nonmetallic wireway is listed

378.21 Size of Conductors

No conductor larger than that for which the nonmetallic wireway is designed shall be installed in any nonmetallic wireway.

378.22 Number of Conductors

The sum of cross-sectional areas of all contained conductors at any cross section of the nonmetallic wireway shall not exceed 20 percent of the interior cross-sectional area of the nonmetallic wireway. Conductors for signaling circuits or controller conductors between a motor and its starter and used only for starting duty shall not be considered as current-carrying conductors.

The derating factors specified in 310.15(B)(2)(a) shall be applicable to the current-carrying conductors up to and including the 20 percent fill specified above.

378.23 Insulated Conductors

Insulated conductors installed in a nonmetallic wireway shall comply with 378.23(A) and (B).

The intent of 378.23 is to provide adequate space for installing and bending conductors without damaging the conductor insulation.

Section 378.23(A) was revised for the 2005 *Code* to ensure that the column of one wire per terminal was used in Table 312.6(A). Section 378.23(B) was revised for the 2005 *Code* to make sure that the same adequate space requirements that apply to conduits also apply to cables as they enter a wireway. The requirements in 378.23(B) ensure that where wireways are used as pull boxes, the same minimum dimension requirements associated with raceway entries of pull boxes apply.

(A) Deflected Insulated Conductors Where insulated conductors are deflected within a nonmetallic wireway, either at the ends or where conduits, fittings, or other raceways or cables enter or leave the nonmetallic wireway, or where the direction of the nonmetallic wireway is deflected greater than 30 degrees, dimensions corresponding to one wire per terminal in Table 312.6(A) shall apply.

(B) Nonmetallic Wireways Used as Pull Boxes Where insulated conductors 4 AWG or larger are pulled through a wireway, the distance between raceway and cable entries enclosing the same conductor shall not be less than that required in 314.28(A)(1) for straight pulls and in 314.28(A)(2) for angle pulls. When transposing cable size into raceway size, the minimum metric designator (trade size) raceway required for the number and size of conductors in the cable shall be used.

378.30 Securing and Supporting

Nonmetallic wireway shall be supported in accordance with 378.30(A) and (B).

(A) Horizontal Support Nonmetallic wireways shall be supported where run horizontally at intervals not to exceed 900 mm (3 ft), and at each end or joint, unless listed for other support intervals. In no case shall the distance between supports exceed 3 m (10 ft).

(B) Vertical Support Vertical runs of nonmetallic wireway shall be securely supported at intervals not exceeding 1.2 m (4 ft), unless listed for other support intervals, and shall not have more than one joint between supports. Adjoining nonmetallic wireway sections shall be securely fastened together to provide a rigid joint.

378.44 Expansion Fittings

Expansion fittings for nonmetallic wireway shall be provided to compensate for thermal expansion and contraction where the length change is expected to be 6 mm (0.25 in.) or greater in a straight run.

> FPN: See Table 352.44(A) for expansion characteristics of PVC rigid nonmetallic conduit. The expansion characteristics of PVC nonmetallic wireway are identical.

378.56 Splices and Taps

Splices and taps shall be permitted within a nonmetallic wireway, provided they are accessible. The conductors, including splices and taps, shall not fill the nonmetallic wireway to more than 75 percent of its area at that point.

378.58 Dead Ends

Dead ends of nonmetallic wireway shall be closed using listed fittings.

378.60 Grounding

Where equipment grounding is required, a separate equipment grounding conductor shall be installed in the nonmetallic wireway. A separate equipment grounding conductor shall not be required where the grounded conductor is used to ground equipment as permitted in 250.142.

378.70 Extensions from Nonmetallic Wireways

Extensions from nonmetallic wireway shall be made with cord pendants or any wiring method of Chapter 3. A separate equipment grounding conductor shall be installed in, or an equipment grounding connection shall be made to, any of the wiring methods used for the extension.

III. Construction Specifications

378.120 Marking

Nonmetallic wireways shall be marked so that the manufacturer's name or trademark and interior cross-sectional area in square inches shall be visible after installation. Marking for limited smoke shall be permitted on the nonmetallic wireways that have limited smoke-producing characteristics.

ARTICLE 380
Multioutlet Assembly

Contents

380.1 Scope

This article covers the use and installation requirements for multioutlet assemblies.

Multioutlet assemblies are metal and nonmetallic raceways that are usually surface mounted and designed to contain branch-circuit conductors and receptacles. Exhibit 380.1 provides an illustration of a multioutlet assembly. Receptacles may be spaced at desired intervals and may be assembled at the factory or in the field. See 220.14(H) and Exhibit 220.4 for load calculations.

Exhibit 380.1 A typical multioutlet assembly shown in assembly form.

380.2 Use

(A) Permitted The use of a multioutlet assembly shall be permitted in dry locations.

(B) Not Permitted A multioutlet assembly shall not be installed as follows:

(1) Where concealed, except that it shall be permissible to surround the back and sides of a metal multioutlet assembly by the building finish or recess a nonmetallic multioutlet assembly in a baseboard
(2) Where subject to severe physical damage
(3) Where the voltage is 300 volts or more between conductors unless the assembly is of metal having a thickness of not less than 1.02 mm (0.040 in.)
(4) Where subject to corrosive vapors
(5) In hoistways
(6) In any hazardous (classified) locations except Class I, Division 2 locations as permitted in 501.10(B)(3)

380.3 Metal Multioutlet Assembly Through Dry Partitions

It shall be permissible to extend a metal multioutlet assembly through (not run within) dry partitions if arrangements are made for removing the cap or cover on all exposed portions and no outlet is located within the partitions.

ARTICLE 382
Nonmetallic Extensions

Contents

I. General
   382.1 Scope
   382.2 Definition

II. Installation
   382.10 Uses Permitted
      (A) From an Existing Outlet
      (B) Exposed and in a Dry Location
      (C) Residential or Offices
   382.12 Uses Not Permitted
   382.15 Exposed
   382.26 Bends
   382.30 Securing and Supporting
   382.40 Boxes and Fittings
   382.56 Splices and Taps

I. General

382.1 Scope

This article covers the use, installation, and construction specifications for nonmetallic extensions.

382.2 Definition

Nonmetallic Extension. An assembly of two insulated conductors within a nonmetallic jacket or an extruded thermoplastic covering. The classification includes surface extensions intended for mounting directly on the surface of walls or ceilings.

II. Installation

382.10 Uses Permitted

Nonmetallic extensions shall be permitted only in accordance with 382.10(A), (B), and (C).

(A) From an Existing Outlet The extension shall be from an existing outlet on a 15- or 20-ampere branch circuit.

(B) Exposed and in a Dry Location The extension shall be run exposed and in a dry location.

(C) Residential or Offices For nonmetallic surface extensions mounted directly on the surface of walls or ceilings, the building shall be occupied for residential or office purposes and shall not exceed three floors above grade.

> FPN No. 1: See 310.10 for temperature limitation of conductors.
> FPN No. 2: See 362.10 for definition of *first floor.*

382.12 Uses Not Permitted

Nonmetallic extensions shall not be used as follows:

(1) In unfinished basements, attics, or roof spaces
(2) Where the voltage between conductors exceeds 150 volts for nonmetallic surface extension and 300 volts for aerial cable

(3) Where subject to corrosive vapors
(4) Where run through a floor or partition, or outside the room in which it originates

382.15 Exposed

One or more extensions shall be permitted to be run in any direction from an existing outlet, but not on the floor or within 50 mm (2 in.) from the floor.

382.26 Bends

A bend that reduces the normal spacing between the conductors shall be covered with a cap to protect the assembly from physical damage.

382.30 Securing and Supporting

Nonmetallic surface extensions shall be secured in place by approved means at intervals not exceeding 200 mm (8 in.), with an allowance for 300 mm (12 in.) to the first fastening where the connection to the supplying outlet is by means of an attachment plug. There shall be at least one fastening between each two adjacent outlets supplied. An extension shall be attached to only woodwork or plaster finish and shall not be in contact with any metal work or other conductive material other than with metal plates on receptacles.

382.40 Boxes and Fittings

Each run shall terminate in a fitting that covers the end of the assembly. All fittings and devices shall be of a type identified for the use.

382.56 Splices and Taps

Extensions shall consist of a continuous unbroken length of the assembly, without splices, and without exposed conductors between fittings. Taps shall be permitted where approved fittings completely covering the tap connections are used. Aerial cable and its tap connectors shall be provided with an approved means for polarization. Receptacle-type tap connectors shall be of the locking type.

ARTICLE 384
Strut-Type Channel Raceway

Contents

I. General

384.1 Scope

This article covers the use, installation, and construction specifications of strut-type channel raceway.

384.2 Definition

Strut-Type Channel Raceway. A metallic raceway that is intended to be mounted to the surface of or suspended from a structure, with associated accessories for the installation of electrical conductors and cables.

384.6 Listing Requirements

Strut-type channel raceways, closure strips, and accessories shall be listed and identified for such use.

II. Installation

384.10 Uses Permitted

The use of strut-type channel raceways shall be permitted in the following:

(1) Where exposed.
(2) In dry locations.
(3) In locations subject to corrosive vapors where protected by finishes judged suitable for the condition.
(4) Where the voltage is 600 volts or less.
(5) As power poles.
(6) In Class I, Division 2 hazardous (classified) locations as permitted in 501.10(B)(3).
(7) As extensions of unbroken lengths through walls, partitions, and floors where closure strips are removable from either side and the portion within the wall, partition, or floor remains covered.

(8) Ferrous channel raceways and fittings protected from corrosion solely by enamel shall be permitted only indoors.

The installation shown in Exhibit 384.1 is typical of how a strut-type channel raceway can be used.

Exhibit 384.1 An example of a strut-type channel raceway using accessories to support and supply power to luminaires. (Redrawn courtesy of Allied Tube & Conduit, a Tyco International Co.)

384.12 Uses Not Permitted

Strut type channel raceways shall not be used as follows:

(1) Where concealed.
(2) Ferrous channel raceways and fittings protected from corrosion solely by enamel shall not be permitted where subject to severe corrosive influences.

384.21 Size of Conductors

No conductor larger than that for which the raceway is listed shall be installed in strut-type channel raceways.

384.22 Number of Conductors

The number of conductors permitted in strut-type channel raceways shall not exceed the percentage fill using Table 384.22 and applicable outside diameter (O.D.) dimensions of specific types and sizes of wire given in the tables in Chapter 9.

The derating factors of 310.15(B)(2)(a) shall not apply to conductors installed in strut-type channel raceways where all of the following conditions are met:

(1) The cross-sectional area of the raceway exceeds 2500 mm² (4 in.²).
(2) The current-carrying conductors do not exceed 30 in number.
(3) The sum of the cross-sectional areas of all contained conductors does not exceed 20 percent of the interior cross-sectional area of the strut-type channel raceways,

Table 384.22 Channel Size and Inside Diameter Area

Size Channel	Area in.²	mm²	40% Area* in.²	mm²	25% Area** in.²	mm²
1⅝ × ¹³/₁₆	0.887	572	0.355	229	0.222	143
1⅝ × 1	1.151	743	0.460	297	0.288	186
1⅝ × 1⅜	1.677	1076	0.671	433	0.419	270
1⅝ × 1⅝	2.028	1308	0.811	523	0.507	327
1⅝ × 2⁷/₁₆	3.169	2045	1.267	817	0.792	511
1⅝ × 3¼	4.308	2780	1.723	1112	1.077	695
1½ × ¾	0.849	548	0.340	219	0.212	137
1½ × 1½	1.828	1179	0.731	472	0.457	295
1½ × 1⅞	2.301	1485	0.920	594	0.575	371
1½ × 3	3.854	2487	1.542	995	0.964	622

*Raceways with external joiners shall use a 40 percent wire fill calculation to determine the number of conductors permitted.
**Raceways with internal joiners shall use a 25 percent wire fill calculation to determine the number of conductors permitted.

calculated in accordance with the following formula for wire fill:

$$n = \frac{ca}{wa}$$

where:

n = number of wires
ca = channel area in square inches
wa = wire area

The adjustment factors of Table 310.15(B)(2) are applicable to strut-type channel raceways because Table 384.22 does not contain any raceways with cross-sectional areas greater than 4 in.²

Example

Calculate the maximum quantity of 10 AWG Type THWN-2 copper conductors permitted in a normal 1½ in. × 1½ in. strut-type channel raceway whose joiners are mounted internally. (Generally, ordinary-duty strut-type channel raceway couplings or joiners are of the internal type, and heavy-duty strut raceway couplings are of the external type.)

Solution

STEP 1. Because the strut-type channel raceway joiners are mounted on the internal surface of the raceway, the second note in Table 384.22 requires the use of the "25% Area" column for the maximum usable internal area of the raceway. According to Table 384.22 and using the "25% ½ in. strut-type channel is 0.457 in.² According to Chapter 9, Table 5, a 10 AWG, Type THWN-2 copper conductor has an area of 0.0211 in.²

STEP 2. Using the formula for wire fill,

$$n = \frac{ca}{wa}$$

where:

n = number of wires
ca = channel area (in.^{2})
wa = wire area (in.^{2})

and substituting the table values,

$$n = \frac{0.457 \text{ in.}^2}{0.0211 \text{ in.}^2} = 21.66$$

or not more than twenty-one 10 AWG Type THWN-2 copper conductors. However, the adjustment factors of Table 310.15(B)(2)(a) are applicable where a raceway contains more than three current-carrying conductors.

384.30 Securing and Supporting

(A) Surface Mount A surface mount strut-type channel raceway shall be secured to the mounting surface with retention straps external to the channel at intervals not exceeding 3 m (10 ft) and within 900 mm (3 ft) of each outlet box, cabinet, junction box, or other channel raceway termination.

(B) Suspension Mount Strut-type channel raceways shall be permitted to be suspension mounted in air with approved appropriate methods designed for the purpose at intervals not to exceed 3 m (10 ft) and within 900 mm (3 ft) of channel raceway terminations and ends.

384.56 Splices and Taps

Splices and taps shall be permitted in raceways that are accessible after installation by having a removable cover. The conductors, including splices and taps, shall not fill the raceway to more than 75 percent of its area at that point. All splices and taps shall be made by approved methods.

384.60 Grounding

Strut-type channel raceway enclosures providing a transition to or from other wiring methods shall have a means for connecting an equipment grounding conductor. Strut-type channel raceways shall be permitted as an equipment grounding conductor in accordance with 250.118(14). Where a snap-fit metal cover for strut-type channel raceways is used to achieve electrical continuity in accordance with the listing, this cover shall not be permitted as the means for providing electrical continuity for a receptacle mounted in the cover.

III. Construction Specifications

384.100 Construction

Strut-type channel raceways and their accessories shall be of a construction that distinguishes them from other raceways.

Raceways and their elbows, couplings, and other fittings shall be designed such that the sections can be electrically and mechanically coupled together and installed without subjecting the wires to abrasion. They shall comply with 384.100(A), (B), and (C).

(A) Material Raceways and accessories shall be formed of steel, stainless steel, or aluminum.

(B) Corrosion Protection Steel raceways and accessories shall be protected against corrosion by galvanizing or by an organic coating.

> FPN: Enamel and PVC coatings are examples of organic coatings that provide corrosion protection.

(C) Cover Covers of strut-type channel raceways shall be either metallic or nonmetallic.

384.120 Marking

Each length of strut-type channel raceways shall be clearly and durably identified as required in the first sentence of 110.21.

ARTICLE 386
Surface Metal Raceways

Summary of Changes
- **386.30:** Added requirement for securing and supporting of surface metal raceways.
- **386.70:** Revised to permit stamping or imprinting for the purposes of identifying separate compartments.

Contents

I. General

386.1 Scope

This article covers the use, installation, and construction specifications for surface metal raceways and associated fittings.

386.2 Definition

Surface Metal Raceway. A metallic raceway that is intended to be mounted to the surface of a structure, with associated couplings, connectors, boxes, and fittings for the installation of electrical conductors.

386.6 Listing Requirements

Surface metal raceway and associated fittings shall be listed.

II. Installation

386.10 Uses Permitted

The use of surface metal raceways shall be permitted in the following:

(1) In dry locations.
(2) In Class I, Division 2 hazardous (classified) locations as permitted in 501.10(B)(3).
(3) Under raised floors, as permitted in 645.5(D)(2).
(4) Extension through walls and floors. Surface metal raceway shall be permitted to pass transversely through dry walls, dry partitions, and dry floors if the length passing through is unbroken. Access to the conductors shall be maintained on both sides of the wall, partition, or floor.

The installation shown in Exhibit 386.1 is just one of many ways a surface metal raceway can be used.

386.12 Uses Not Permitted

Surface metal raceways shall not be used in the following:

(1) Where subject to severe physical damage, unless otherwise approved
(2) Where the voltage is 300 volts or more between conductors, unless the metal has a thickness of not less than 1.02 mm (0.040 in.) nominal
(3) Where subject to corrosive vapors
(4) In hoistways
(5) Where concealed, except as permitted in 386.10

386.21 Size of Conductors

No conductor larger than that for which the raceway is designed shall be installed in surface metal raceway.

386.22 Number of Conductors or Cables

The number of conductors or cables installed in surface metal raceway shall not be greater than the number for which

Exhibit 386.1 An example of a surface metal raceway extending from an existing receptacle outlet.

the raceway is designed. Cables shall be permitted to be installed where such use is not prohibited by the respective cable articles.

The derating factors of 310.15(B)(2)(a) shall not apply to conductors installed in surface metal raceways where all of the following conditions are met:

(1) The cross-sectional area of the raceway exceeds 2500 mm^2 (4 in.^{2})
(2) The current-carrying conductors do not exceed 30 in number
(3) The sum of the cross-sectional areas of all contained conductors does not exceed 20 percent of the interior cross-sectional area of the surface metal raceway

The number, type, and sizes of conductors permitted to be installed in a listed surface metal raceway are marked on the raceway or on the package in which it is shipped. Typically, this information is available in detail in the manufacturer's catalog.

386.30 Securing and Supporting

Surface metal raceways shall be supported at intervals in accordance with the manufacturer's installation instructions.

386.56 Splices and Taps

Splices and taps shall be permitted in surface metal raceways having a removable cover that is accessible after installation.

The conductors, including splices and taps, shall not fill the raceway to more than 75 percent of its area at that point. Splices and taps in surface metal raceways without removable covers shall be made only in boxes. All splices and taps shall be made by approved methods.

Taps of Type FC cable installed in surface metal raceway shall be made in accordance with 322.56(B).

386.60 Grounding

Surface metal raceway enclosures providing a transition from other wiring methods shall have a means for connecting an equipment grounding conductor.

As the example in Exhibit 386.2 shows, where a surface metal raceway is supplied by Type MC or NM cable, a means (e.g., grounding terminal screw or lug) must be available at the surface metal raceway for terminating the equipment grounding conductor provided within the cable.

Exhibit 386.2 An example of providing a means for terminating an equipment grounding conductor at a surface metal raceway.

386.70 Combination Raceways

When combination surface metallic raceways are used for both signaling and for lighting and power circuits, the different systems shall be run in separate compartments identified by stamping, imprinting, or color coding of the interior finish.

III. Construction Specifications

386.100 Construction

Surface metal raceways shall be of such construction as will distinguish them from other raceways. Surface metal

raceways and their elbows, couplings, and similar fittings shall be designed so that the sections can be electrically and mechanically coupled together and installed without subjecting the wires to abrasion.

Where covers and accessories of nonmetallic materials are used on surface metal raceways, they shall be identified for such use.

ARTICLE 388
Surface Nonmetallic Raceways

Summary of Changes
- **388.70:** Revised to permit stamping or imprinting for the purposes of identifying separate compartments.

Contents

I. General

388.1 Scope

This article covers the use, installation, and construction specifications for surface nonmetallic raceways and associated fittings.

388.2 Definition

Surface Nonmetallic Raceway. A nonmetallic raceway that is intended to be mounted to the surface of a structure, with associated couplings, connectors, boxes, and fittings for the installation of electrical conductors.

388.6 Listing Requirements

Surface nonmetallic raceway and associated fittings shall be listed.

II. Installation

388.10 Uses Permitted

Surface nonmetallic raceway shall be permitted as follows:

(1) The use of surface nonmetallic raceways shall be permitted in dry locations.

(2) Extension through walls and floors shall be permitted. Surface nonmetallic raceway shall be permitted to pass transversely through dry walls, dry partitions, and dry floors if the length passing through is unbroken. Access to the conductors shall be maintained on both sides of the wall, partition, or floor.

The installations shown in Exhibits 388.1 and 388.2 are typical of how a surface nonmetallic raceway can be used.

Exhibit 388.2 An example of a surface nonmetallic raceway supplying a speed control switch and a paddle fan outlet. (Courtesy of The Wiremold Co.)

Exhibit 388.1 An example of a surface nonmetallic raceway extending from an existing receptacle outlet. (Courtesy of The Wiremold Co.)

388.12 Uses Not Permitted

Surface nonmetallic raceways shall not be used in the following:

(1) Where concealed, except as permitted in 388.10(2)

(2) Where subject to severe physical damage

(3) Where the voltage is 300 volts or more between conductors, unless listed for higher voltage

(4) In hoistways

(5) In any hazardous (classified) location except Class I, Division 2 locations as permitted in 501.10(B)(3)

(6) Where subject to ambient temperatures exceeding those for which the nonmetallic raceway is listed

(7) For conductors whose insulation temperature limitations would exceed those for which the nonmetallic raceway is listed

388.21 Size of Conductors

No conductor larger than that for which the raceway is designed shall be installed in surface nonmetallic raceway.

388.22 Number of Conductors or Cables

The number of conductors or cables installed in surface nonmetallic raceway shall not be greater than the number for which the raceway is designed. Cables shall be permitted to be installed where such use is not prohibited by the respective cable articles.

388.56 Splices and Taps

Splices and taps shall be permitted in surface nonmetallic raceways having a removable cover that is accessible after installation. The conductors, including splices and taps, shall not fill the raceway to more than 75 percent of its area at that point. Splices and taps in surface nonmetallic raceways without removable covers shall be made only in boxes. All splices and taps shall be made by approved methods.

388.60 Grounding

Where equipment grounding is required, a separate equipment grounding conductor shall be installed in the raceway.

388.70 Combination Raceways

When combination surface nonmetallic raceways are used both for signaling and for lighting and power circuits, the different systems shall be run in separate compartments identified by stamping, imprinting, or color coding of the interior finish.

III. Construction Specifications

388.100 Construction

Surface nonmetallic raceways shall be of such construction as will distinguish them from other raceways. Surface non-metallic raceways and their elbows, couplings, and similar fittings shall be designed so that the sections can be mechanically coupled together and installed without subjecting the wires to abrasion.

Surface nonmetallic raceways and fittings are made of suitable nonmetallic material that is resistant to moisture and chemical atmospheres. It shall also be flame retardant, resistant to impact and crushing, resistant to distortion from heat under conditions likely to be encountered in service, and resistant to low-temperature effects.

388.120 Marking

Surface nonmetallic raceways that have limited smoke-producing characteristics shall be permitted to be so identified.

ARTICLE 390
Underfloor Raceways

Summary of Changes

- **390.17:** Added requirement that the ampacity adjustment factors of 310.15(B)(2) apply to the installed conductors.

Contents

390.1 Scope

This article covers the use and installation requirements for underfloor raceways.

390.2 Use

An underfloor raceway is a practical means of bringing light, power, and signal and communications systems to desks, work benches, or tables that are not located adjacent to wall space. This wiring method offers flexibility in layout where used with movable partitions and is commonly used in large retail stores and office buildings to supply power at any desired location.

Underfloor raceways are permitted beneath the surface of concrete, wood, or other flooring material. The wiring method between cabinets, raceway junction boxes, and outlet boxes may be rigid metal conduit, intermediate metal conduit, rigid nonmetallic conduit, liquidtight flexible nonmetallic conduit, electrical nonmetallic tubing, or electrical metallic tubing. Flexible metal conduit may be used where not installed in concrete.

(A) Permitted The installation of underfloor raceways shall be permitted beneath the surface of concrete or other flooring material or in office occupancies where laid flush with the concrete floor and covered with linoleum or equivalent floor covering.

(B) Not Permitted Underfloor raceways shall not be installed (1) where subject to corrosive vapors or (2) in any hazardous (classified) locations, except as permitted by 504.20 and in Class I, Division 2 locations as permitted in 501.10(B)(3). Unless made of a material judged suitable for the condition or unless corrosion protection approved for the condition is provided, ferrous or nonferrous metal underfloor raceways, junction boxes, and fittings shall not be installed in concrete or in areas subject to severe corrosive influences.

390.3 Covering

Raceway coverings shall comply with 390.3(A) through (D).

(A) Raceways Not Over 100 mm (4 in.) Wide Half-round and flat-top raceways not over 100 mm (4 in.) in width shall have not less than 20 mm (¾ in.) of concrete or wood above the raceway.

Exception: As permitted in 390.3(C) and (D) for flat-top raceways.

As Exhibit 390.1 illustrates, a ¾-in. concrete or wood covering is required over underfloor raceways not over 4 in. wide, except for trench-type and other raceways that are flush with concrete.

As Exhibit 390.2 illustrates, flat-top underfloor raceways over 4 in. wide and spaced less than 1 in. apart must be covered with at least 1½ in. of concrete.

Approved flush-type underfloor raceways may be installed flush with the floor surface, provided they have covers at least equal to those of junction box covers.

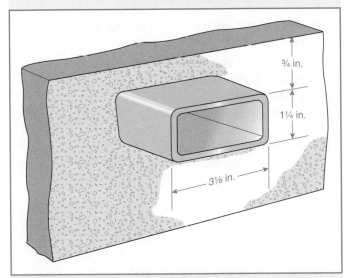

Exhibit 390.1 An underfloor raceway not over 4 in. installed in compliance with 390.3(A). (Redrawn from Walker Systems, a Wiremold Co.)

Exhibit 390.2 Two side-by-side underfloor raceways over 4 in. installed in compliance with 390.3(B). (Redrawn from Walker Systems, a Wiremold Co.)

(B) Raceways Over 100 mm (4 in.) Wide But Not Over 200 mm (8 in.) Wide Flat-top raceways over 100 mm (4 in.) but not over 200 mm (8 in.) wide with a minimum of 25 mm (1 in.) spacing between raceways shall be covered with concrete to a depth of not less than 25 mm (1 in.). Raceways spaced less than 25 mm (1 in.) apart shall be covered with concrete to a depth of 38 mm (1½ in.).

(C) Trench-Type Raceways Flush with Concrete Trench-type flush raceways with removable covers shall be permitted to be laid flush with the floor surface. Such approved raceways shall be designed so that the cover plates provide adequate mechanical protection and rigidity equivalent to junction box covers.

(D) Other Raceways Flush with Concrete In office occupancies, approved metal flat-top raceways, if not over 100 mm (4 in.) in width, shall be permitted to be laid flush with the concrete floor surface, provided they are covered with substantial linoleum that is not less than 1.6 mm (¹⁄₁₆ in.) thick or with equivalent floor covering. Where more than one and not more than three single raceways are each installed flush with the concrete, they shall be contiguous with each other and joined to form a rigid assembly.

390.4 Size of Conductors

No conductor larger than that for which the raceway is designed shall be installed in underfloor raceways.

390.5 Maximum Number of Conductors in Raceway

The combined cross-sectional area of all conductors or cables shall not exceed 40 percent of the interior cross-sectional area of the raceway.

390.6 Splices and Taps

Splices and taps shall be made only in junction boxes.

For the purposes of this section, so-called loop wiring (continuous, unbroken conductor connecting the individual outlets) shall not be considered to be a splice or tap.

Exception: Splices and taps shall be permitted in trench-type flush raceway having a removable cover that is accessible after installation. The conductors, including splices and taps, shall not fill more than 75 percent of the raceway area at that point.

390.7 Discontinued Outlets

When an outlet is abandoned, discontinued, or removed, the sections of circuit conductors supplying the outlet shall be removed from the raceway. No splices or reinsulated conductors, such as would be the case with abandoned outlets on loop wiring, shall be allowed in raceways.

Loop wiring (continuous, unbroken conductors) is recognized where it runs from the underfloor raceway up to the terminals of attached receptacles, back into the raceway, and then on to the next device, as illustrated in Exhibit 390.3. When an outlet is removed, the sections of conductors supplying the outlet must be removed from the raceway as well. As is the case with abandoned outlets on loop wiring, reinsulated or spliced conductors are not allowed in raceways, except in trench-type raceways as covered in the exception to 390.6.

Exhibit 390.3 A receptacle outlet supplied from an underfloor raceway by the loop method of wiring. (Courtesy of Walker Systems, a Wiremold Co.)

390.8 Laid in Straight Lines

Underfloor raceways shall be laid so that a straight line from the center of one junction box to the center of the next junction box coincides with the centerline of the raceway system. Raceways shall be firmly held in place to prevent disturbing this alignment during construction.

390.9 Markers at Ends

A suitable marker shall be installed at or near each end of each straight run of raceways to locate the last insert.

390.10 Dead Ends

Dead ends of raceways shall be closed.

390.13 Junction Boxes

Junction boxes shall be leveled to the floor grade and sealed to prevent the free entrance of water or concrete. Junction boxes used with metal raceways shall be metal and shall be electrically continuous with the raceways.

390.14 Inserts

Inserts shall be leveled and sealed to prevent the entrance of concrete. Inserts used with metal raceways shall be metal and shall be electrically continuous with the raceway. Inserts set in or on fiber raceways before the floor is laid shall be mechanically secured to the raceway. Inserts set in fiber raceways after the floor is laid shall be screwed into the raceway. When cutting through the raceway wall and setting inserts, chips and other dirt shall not be allowed to remain in the raceway, and tools shall be used that are designed so as to prevent the tool from entering the raceway and damaging conductors that may be in place.

390.15 Connections to Cabinets and Wall Outlets

Connections from underfloor raceways to distribution centers and wall outlets shall be made by approved fittings or by any of the wiring methods in Chapter 3, where installed in accordance with the provisions of the respective articles.

390.17 Ampacity of Conductors

The ampacity adjustment factors, in 310.15(B)(2), shall apply to conductors installed in underfloor raceways.

Added for the 2005 *Code* and placed here as a reminder, ampacity adjustment factors apply to underfloor raceways.

ARTICLE 392
Cable Trays

Summary of Changes

- **392.10(A)(1):** Revised to require that cables sized 1000 kcmil or larger be installed in a single layer except where conductors installed in parallel are bound together to form a single circuit group.

- **392.11(B)(3):** Added exception for solid bottom cable trays containing single conductor cables that require the conductor ampacity to be determined in accordance with the "Engineering Supervision" provisions of 310.15(C).

Contents

392.1 Scope

This article covers cable tray systems, including ladder, ventilated trough, ventilated channel, solid bottom, and other similar structures.

Cable trays are mechanical support systems. Cable trays are not raceways. See the definition of *raceway* in Article 100.

FPN: For further information on cable trays, see ANSI/NEMA–VE 1-1998, *Metal Cable Tray Systems*; NEMA–VE 2-1996, *Metal Cable Tray Installation Guidelines*; and NEMA–FG-1998, *Nonmetallic Cable Tray Systems*.

392.2 Definition

Cable Tray System. A unit or assembly of units or sections and associated fittings forming a structural system used to securely fasten or support cables and raceways.

392.3 Uses Permitted

Cable tray shall be permitted to be used as a support system for service conductors, feeders, branch circuits, communications circuits, control circuits, and signaling circuits. Cable tray installations shall not be limited to industrial establishments. Where exposed to direct rays of the sun, insulated conductors and jacketed cables shall be identified as being sunlight resistant. Cable trays and their associated fittings shall be identified for the intended use.

Cable tray installations are typically an industrial-type wiring method. However, cable tray installations have never been restricted to industrial installations. Cable tray installations are being applied more in commercial installations than ever before, especially as a wire-and-cable management system for telecommunications/data installations.

(A) Wiring Methods The wiring methods in Table 392.3(A) shall be permitted to be installed in cable tray systems under the conditions described in their respective articles and sections.

Section 392.3(A) identifies the raceways and many of the cable types that may be supported in commercial and industrial cable tray installations. Cable tray is rarely used as a major raceway support system. For raceway support systems, the versatility of strut systems exceeds that of cable tray support systems.

Section 392.3(A) does not identify all the specific cable types that may be installed in commercial and industrial cable tray systems. According to Table 392.3(A), other factory-assembled, multiconductor control, signal, and power

Table 392.3(A) Wiring Methods

Wiring Method	Article	Section
Armored cable	320	
Communication raceways	800	
Electrical metallic tubing	358	
Electrical nonmetallic tubing	362	
Fire alarm cables	760	
Flexible metal conduit	348	
Flexible metallic tubing	360	
Instrumentation tray cable	727	
Intermediate metal conduit	342	
Liquidtight flexible metal conduit	350	
Liquidtight flexible nonmetallic conduit	356	
Metal-clad cable	330	
Mineral-insulated, metal-sheathed cable	332	
Multiconductor service-entrance cable	338	
Multiconductor underground feeder and branch-circuit cable	340	
Multipurpose and communications cables	800	
Nonmetallic-sheathed cable	334	
Power and control tray cable	336	
Power-limited tray cable		725.61(C) and 725.82(E)
Optical fiber cables	770	
Optical fiber raceways	770	
Other factory-assembled, multiconductor control, signal, or power cables that are specifically approved for installation in cable trays		
Rigid metal conduit	344	
Rigid nonmetallic conduit	352	

cables that are specifically approved for installation in cable trays are permitted as well.

(B) In Industrial Establishments The wiring methods in Table 392.3(A) shall be permitted to be used in any industrial establishment under the conditions described in their respective articles. In industrial establishments only, where conditions of maintenance and supervision ensure that only qualified persons service the installed cable tray system, any of the cables in 392.3(B)(1) and (B)(2) shall be permitted to be installed in ladder, ventilated trough, solid bottom, or ventilated channel cable trays.

Section 392.3(B) permits single-conductor cables (rated 0 to 2000 volts) and Type MV cables to be installed in ladder, ventilated-trough, or ventilated-channel cable trays, provided the installation is located in a qualifying industrial facility.

(1) Single Conductors Single-conductor cables shall be permitted to be installed in accordance with (B)(1)(a) through (B)(1)(c).

(a) Single-conductor cable shall be 1/0 AWG or larger and shall be of a type listed and marked on the surface for use in cable trays. Where 1/0 AWG through 4/0 AWG single-conductor cables are installed in ladder cable tray, the maximum allowable rung spacing for the ladder cable tray shall be 225 mm (9 in.).

(b) Welding cables shall comply with the provisions of Article 630, Part IV.

Cable trays used to support welding cables are required to be dedicated for welding cable installation. See 630.42 for installation details.

(c) Single conductors used as equipment grounding conductors shall be insulated, covered, or bare, and they shall be 4 AWG or larger.

(2) Medium Voltage Single- and multiconductor medium voltage cables shall be Type MV cable. Single conductors shall be installed in accordance with 392.3(B)(1).

(C) Equipment Grounding Conductors Metallic cable trays shall be permitted to be used as equipment grounding conductors where continuous maintenance and supervision ensure that qualified persons service the installed cable tray system and the cable tray complies with provisions of 392.7.

Section 392.3(C) expanded the optional use of the cable tray as the equipment grounding conductor. No longer is this practice limited to qualifying industrial installations. To qualify as an equipment grounding conductor, the cable tray system must meet all four requirements of 392.7(B).

(D) Hazardous (Classified) Locations Cable trays in hazardous (classified) locations shall contain only the cable types permitted in 501.10, 502.10, 503.10, 504.20, and 505.15.

(E) Nonmetallic Cable Tray In addition to the uses permitted elsewhere in 392.3, nonmetallic cable tray shall be permitted in corrosive areas and in areas requiring voltage isolation.

Fiberglass cable trays are often used to support cables in corrosive environments or in electrolytic cell rooms where voltage isolation is required. See Article 668, Electrolytic Cells.

392.4 Uses Not Permitted

Cable tray systems shall not be used in hoistways or where subject to severe physical damage. Cable tray systems shall not be used in ducts, plenums, and other air-handling spaces, except as permitted in 300.22, to support wiring methods recognized for use in such spaces.

Section 300.22(C) specifically limits the types of wiring methods that may be used within other spaces used for environmental air. Metallic cable trays may be used within these spaces to support only the recognized wiring methods permitted in these spaces. The cable tray types may be ladder, ventilated trough, ventilated channel, or solid bottom. Metal cable trays are not the limiting factor; rather, the cable or wiring method is the limiting factor.

392.5 Construction Specifications

(A) Strength and Rigidity Cable trays shall have suitable strength and rigidity to provide adequate support for all contained wiring.

(B) Smooth Edges Cable trays shall not have sharp edges, burrs, or projections that could damage the insulation or jackets of the wiring.

(C) Corrosion Protection Cable tray systems shall be corrosion resistant. If made of ferrous material, the system shall be protected from corrosion as required by 300.6.

(D) Side Rails Cable trays shall have side rails or equivalent structural members.

(E) Fittings Cable trays shall include fittings or other suitable means for changes in direction and elevation of runs.

(F) Nonmetallic Cable Tray Nonmetallic cable trays shall be made of flame-retardant material.

392.6 Installation

(A) Complete System Cable trays shall be installed as a complete system. Field bends or modifications shall be so made that the electrical continuity of the cable tray system and support for the cables is maintained. Cable tray systems shall be permitted to have mechanically discontinuous segments between cable tray runs or between cable tray runs and equipment. The system shall provide for the support of the cables in accordance with their corresponding articles.

Where cable trays support individual conductors and where the conductors pass from one cable tray to another, or from a cable tray to raceway(s) or from a cable tray to equipment where the conductors are terminated, the distance between cable trays or between the cable tray and the raceway(s) or the equipment shall not exceed 1.8 m (6 ft). The conductors shall be secured to the cable tray(s) at the transition, and they shall be protected, by guarding or by location, from physical damage.

A bonding jumper sized in accordance with 250.102 shall connect the two sections of cable tray, or the cable tray and the raceway or equipment. Bonding shall be in accordance with 250.96.

Runs of cable tray are not required to be totally mechanically continuous from the equipment source to the equipment termination. Breaks in the mechanical continuity of cable tray systems are permitted and often occur at tees, crossovers, elevation changes, fire stops, or for thermal contraction and expansion. Also, cable tray systems are not required to be mechanically connected to the equipment they serve.

The 6-ft distance limit applies to mechanically discontinuous cable tray segments for individual conductors but not to trays containing multiconductor cables. For further information regarding multiconductor Type TC tray cable used with discontinuous cable tray, refer to 336.10(6).

Most important, especially for discontinuous cable tray segments, is the bonding of the entire cable tray system. According to 250.96, properly sized and installed bonding conductors must be installed across any mechanical discontinuities in the cable tray system and across any space between the cable tray and the conductor termination equipment enclosure or its equipment ground bus.

Of course, the cables installed within cable tray systems must always be supported to the minimum requirements of the applicable article. This requirement either limits the gap distance in cable tray runs and between the cable tray and the equipment enclosures or requires intermediate cable supports at the appropriate distances in place of the cable tray.

(B) Completed Before Installation Each run of cable tray shall be completed before the installation of cables.

(C) Supports Supports shall be provided to prevent stress on cables where they enter raceways or other enclosures from cable tray systems.

Cable trays shall be supported at intervals in accordance with the installation instructions.

(D) Covers In portions of runs where additional protection is required, covers or enclosures providing the required protection shall be of a material that is compatible with the cable tray.

(E) Multiconductor Cables Rated 600 Volts or Less Multiconductor cables rated 600 volts or less shall be permitted to be installed in the same cable tray.

(F) Cables Rated Over 600 Volts Cables rated over 600 volts and those rated 600 volts or less installed in the same cable tray shall comply with either of the following:

(1) The cables rated over 600 volts are Type MC.
(2) The cables rated over 600 volts are separated from the cables rated 600 volts or less by a solid fixed barrier of a material compatible with the cable tray.

(G) Through Partitions and Walls Cable trays shall be permitted to extend transversely through partitions and walls or vertically through platforms and floors in wet or dry locations where the installations, complete with installed cables, are made in accordance with the requirements of 300.21.

(H) Exposed and Accessible Cable trays shall be exposed and accessible except as permitted by 392.6(G).

(I) Adequate Access Sufficient space shall be provided and maintained about cable trays to permit adequate access for installing and maintaining the cables.

(J) Raceways, Cables, Boxes, and Conduit Bodies Supported from Cable Tray Systems In industrial facilities where conditions of maintenance and supervision ensure that only qualified persons service the installation and where the cable tray systems are designed and installed to support the load, such systems shall be permitted to support raceways and cables, and boxes and conduit bodies covered in 314.1. For raceways terminating at the tray, a listed cable tray clamp or adapter shall be used to securely fasten the raceway to the cable tray system. Additional supporting and securing of the raceway shall be in accordance with the requirements of the appropriate raceway article.

For raceways or cables running parallel to and attached to the bottom or side of a cable tray system, fastening and supporting shall be in accordance with the requirements of the appropriate raceway or cable article.

For boxes and conduit bodies attached to the bottom or side of a cable tray system, fastening and supporting shall be in accordance with the requirements of 314.23.

Section 392.6(J) permits conduit and cable termination supports as well as outlet boxes supported solely by the cable tray in qualifying industrial facilities only. These items are not permitted to be supported solely by the cable tray in commercial installations.

For commercial installations (and nonqualifying industrial facilities), conduits must be supported within 3 ft of the cable tray or within 5 ft if structural members do not permit fastening within 3 ft of the cable tray. Cables connect-

ing to equipment outside the cable tray system must be supported according to their respective article. For example, Type MC cable in the larger sizes is required to be supported outside a cable tray system at intervals not exceeding 6 ft, according to 330.30.

392.7 Grounding

(A) Metallic Cable Trays Metallic cable trays that support electrical conductors shall be grounded as required for conductor enclosures in accordance with 250.96.

Section 392.7(A), together with 250.96, requires all cable tray systems that support electrical conductors (whether mechanically continuous or with isolated segments) to be electrically continuous and effectively bonded and grounded. This requirement applies whether or not the cable tray is used as an equipment grounding conductor.

(B) Steel or Aluminum Cable Tray Systems Steel or aluminum cable tray systems shall be permitted to be used as equipment grounding conductors, provided that all the following requirements are met:

(1) The cable tray sections and fittings shall be identified for grounding purposes.
(2) The minimum cross-sectional area of cable trays shall conform to the requirements in Table 392.7(B).
(3) All cable tray sections and fittings shall be legibly and durably marked to show the cross-sectional area of metal in channel cable trays, or cable trays of one-piece construction, and the total cross-sectional area of both side rails for ladder or trough cable trays.
(4) Cable tray sections, fittings, and connected raceways shall be bonded in accordance with 250.96, using bolted mechanical connectors or bonding jumpers sized and installed in accordance with 250.102.

For cable tray systems in commercial occupancies, designers are afforded the option to specify multiconductor cables without equipment grounding conductors (EGCs) and to use the cable tray system as the required EGC, provided the cable tray system meets the requirements of 392.7(A) and 392.7(B). Exhibit 392.1 shows an example of the grounding and bonding of multiconductor cables in cable trays with conduit runs to power equipment.

For cable tray systems in industrial establishments that qualify, the designer is also afforded the option to specify single-conductor cables without a cable EGC and to use the cable tray as the required EGC. Again, this option is available only if the cable tray system meets the requirements of 392.7(A) and 392.7(B).

Table 392.7(B) Metal Area Requirements for Cable Trays Used as Equipment Grounding Conductor

Maximum Fuse Ampere Rating, Circuit Breaker Ampere Trip Setting, or Circuit Breaker Protective Relay Ampere Trip Setting for Ground-Fault Protection of Any Cable Circuit in the Cable Tray System	Minimum Cross-Sectional Area of Metal[a]			
	Steel Cable Trays		Aluminum Cable Trays	
	mm^2	in.^{2}	mm^2	in.^{2}
60	129	0.20	129	0.20
100	258	0.40	129	0.20
200	451.5	0.70	129	0.20
400	645	1.00	258	0.40
600	967.5	1.50[b]	258	0.40
1000	—	—	387	0.60
1200	—	—	645	1.00
1600	—	—	967.5	1.50
2000	—	—	1290	2.00[b]

[a]Total cross-sectional area of both side rails for ladder or trough cable trays; or the minimum cross-sectional area of metal in channel cable trays or cable trays of one-piece construction.
[b]Steel cable trays shall not be used as equipment grounding conductors for circuits with ground-fault protection above 600 amperes. Aluminum cable trays shall not be used as equipment grounding conductors for circuits with ground-fault protection above 2000 amperes.

392.8 Cable Installation

(A) Cable Splices Cable splices made and insulated by approved methods shall be permitted to be located within a cable tray, provided they are accessible and do not project above the side rails.

(B) Fastened Securely In other than horizontal runs, the cables shall be fastened securely to transverse members of the cable trays.

Other fastening requirements are found in 392.6(A) and 392.6(C).

(C) Bushed Conduit and Tubing A box shall not be required where cables or conductors are installed in bushed conduit and tubing used for support or for protection against physical damage.

(D) Connected in Parallel Where single conductor cables comprising each phase, neutral, or grounded conductor of an alternating-current circuit are connected in parallel as permitted in 310.4, the conductors shall be installed in groups consisting of not more than one conductor per phase, neutral, or grounded conductor to prevent current imbalance in the paralleled conductors due to inductive reactance.

Single conductors shall be securely bound in circuit groups to prevent excessive movement due to fault-current magnetic forces unless single conductors are cabled together, such as triplexed assemblies.

The binding or otherwise grouping of 3-phase circuits is good engineering practice. If properly done, it results in the phase reactances of the conductors being balanced, which reduces the voltage unbalance between the phases of the 3-phase circuit.

(E) Single Conductors Where any of the single conductors installed in ladder or ventilated trough cable trays are 1/0 through 4/0 AWG, all single conductors shall be installed in a single layer. Conductors that are bound together to comprise each circuit group shall be permitted to be installed in other than a single layer.

392.9 Number of Multiconductor Cables, Rated 2000 Volts or Less, in Cable Trays

The number of multiconductor cables, rated 2000 volts or less, permitted in a single cable tray shall not exceed the requirements of this section. The conductor sizes herein apply to both aluminum and copper conductors.

(A) Any Mixture of Cables Where ladder or ventilated trough cable trays contain multiconductor power or lighting cables, or any mixture of multiconductor power, lighting, control, and signal cables, the maximum number of cables shall conform to the following:

(1) Where all of the cables are 4/0 AWG or larger, the sum of the diameters of all cables shall not exceed the cable tray width, and the cables shall be installed in a single layer.

(2) Where all of the cables are smaller than 4/0 AWG, the sum of the cross-sectional areas of all cables shall not exceed the maximum allowable cable fill area in Column 1 of Table 392.9 for the appropriate cable tray width.

(3) Where 4/0 AWG or larger cables are installed in the same cable tray with cables smaller than 4/0 AWG, the sum of the cross-sectional areas of all cables smaller than 4/0 AWG shall not exceed the maximum allowable fill area resulting from the calculation in Column 2 of Table 392.9 for the appropriate cable tray width. The 4/0 AWG and larger cables shall be installed in a single layer, and no other cables shall be placed on them.

(B) Multiconductor Control and/or Signal Cables Only Where a ladder or ventilated trough cable tray having a usable inside depth of 150 mm (6 in.) or less contains multiconductor control and/or signal cables only, the sum of the cross-sectional areas of all cables at any cross section shall not exceed 50 percent of the interior cross-sectional area of the cable tray. A depth of 150 mm (6 in.) shall be used to

Exhibit 392.1 An example of multiconductor cables in cable trays with conduit runs to power equipment where bonding is provided in accordance with 392.7(B)(4). (Redrawn courtesy of Cable Tray Institute)

calculate the allowable interior cross-sectional area of any cable tray that has a usable inside depth of more than 150 mm (6 in.).

(C) Solid Bottom Cable Trays Containing Any Mixture Where solid bottom cable trays contain multiconductor power or lighting cables, or any mixture of multiconductor power, lighting, control, and signal cables, the maximum number of cables shall conform to the following:

(1) Where all of the cables are 4/0 AWG or larger, the sum of the diameters of all cables shall not exceed 90 percent of the cable tray width, and the cables shall be installed in a single layer.

(2) Where all of the cables are smaller than 4/0 AWG, the sum of the cross-sectional areas of all cables shall not

exceed the maximum allowable cable fill area in Column 3 of Table 392.9 for the appropriate cable tray width.

(3) Where 4/0 AWG or larger cables are installed in the same cable tray with cables smaller than 4/0 AWG, the sum of the cross-sectional areas of all cables smaller than 4/0 AWG shall not exceed the maximum allowable fill area resulting from the computation in Column 4 of Table 392.9 for the appropriate cable tray width. The 4/0 AWG and larger cables shall be installed in a single layer, and no other cables shall be placed on them.

(D) Solid Bottom Cable Tray — Multiconductor Control and/or Signal Cables Only Where a solid bottom cable tray having a usable inside depth of 150 mm (6 in.) or less

Table 392.9 Allowable Cable Fill Area for Multiconductor Cables in Ladder, Ventilated Trough, or Solid Bottom Cable Trays for Cables Rated 2000 Volts or Less

Inside Width of Cable Tray		Maximum Allowable Fill Area for Multiconductor Cables							
		Ladder or Ventilated Trough Cable Trays, 392.9(A)				Solid Bottom Cable Trays, 392.9(C)			
		Column 1 Applicable for 392.9(A)(2) Only		Column 2[a] Applicable for 392.9(A)(3) Only		Column 3 Applicable for 392.9(C)(2) Only		Column 4[a] Applicable for 392.9(C)(3) Only	
mm	in.	mm^2	$in.^2$	mm^2	$in.^2$	mm^2	$in.^2$	mm^2	$in.^2$
150	6.0	4,500	7.0	$4,500 - (30\ Sd)$[b]	$7 - (1.2\ Sd)$[b]	3,500	5.5	$3,500 - (25\ Sd)$[b]	$5.5 - Sd$[b]
225	9.0	6,800	10.5	$6,800 - (30\ Sd)$	$10.5 - (1.2\ Sd)$	5,100	8.0	$5,100 - (25\ Sd)$	$8.0 - Sd$
300	12.0	9,000	14.0	$9,000 - (30\ Sd)$	$14 - (1.2\ Sd)$	7,100	11.0	$7,100 - (25\ Sd)$	$11.0 - Sd$
450	18.0	13,500	21.0	$13,500 - (30\ Sd)$	$21 - (1.2\ Sd)$	10,600	16.5	$10,600 - (25\ Sd)$	$16.5 - Sd$
600	24.0	18,000	28.0	$18,000 - (30\ Sd)$	$28 - (1.2\ Sd)$	14,200	22.0	$14,200 - (25\ Sd)$	$22.0 - Sd$
750	30.0	22,500	35.0	$22,500 - (30\ Sd)$	$35 - (1.2\ Sd)$	17,700	27.5	$17,700 - (25\ Sd)$	$27.5 - Sd$
900	36.0	27,000	42.0	$27,000 - (30\ Sd)$	$42 - (1.2\ Sd)$	21,300	33.0	$21,300 - (25\ Sd)$	$33.0 - Sd$

[a]The maximum allowable fill areas in Columns 2 and 4 shall be calculated. For example, the maximum allowable fill in mm^2 for a 150-mm wide cable tray in Column 2 shall be 4500 minus (30 multiplied by Sd) [the maximum allowable fill, in square inches, for a 6-in. wide cable tray in Column 2 shall be 7 minus (1.2 multiplied by Sd)].

[b]The term *Sd* in Columns 2 and 4 is equal to the sum of the diameters, in mm, of all cables 107.2 mm (in inches, of all 4/0 AWG) and larger multiconductor cables in the same cable tray with smaller cables.

contains multiconductor control and/or signal cables only, the sum of the cross-sectional areas of all cables at any cross section shall not exceed 40 percent of the interior cross-sectional area of the cable tray. A depth of 150 mm (6 in.) shall be used to calculate the allowable interior cross-sectional area of any cable tray that has a usable inside depth of more than 150 mm (6 in.).

(E) Ventilated Channel Cable Trays Where ventilated channel cable trays contain multiconductor cables of any type, the following shall apply:

(1) Where only one multiconductor cable is installed, the cross-sectional area shall not exceed the value specified in Column 1 of Table 392.9(E).

(2) Where more than one multiconductor cable is installed, the sum of the cross-sectional area of all cables shall not exceed the value specified in Column 2 of Table 392.9(E).

Cable trays may be ladder, ventilated trough, ventilated channel, or solid bottom of various widths. The depth of a cable tray is a structural consideration. The depth (up to 6 in.) is related to fill only where the cable tray contains signal and control cables or if the cable tray contains large cable splices. Heat dissipation is generally not a problem for signal and control cables. Inspectors or contractors, therefore, are not expected to calculate the various combinations of cable tray

Table 392.9(E) Allowable Cable Fill Area for Multiconductor Cables in Ventilated Channel Cable Trays for Cables Rated 2000 Volts or Less

Inside Width of Cable Tray		Maximum Allowable Fill Area for Multiconductor Cables			
		Column 1 One Cable		Column 2 More Than One Cable	
mm	in.	mm^2	$in.^2$	mm^2	$in.^2$
75	3	1500	2.3	850	1.3
100	4	2900	4.5	1600	2.5
150	6	4500	7.0	2450	3.8

fill in the field. Installation handbooks for cable tray applications are available from various cable tray manufacturers.

(F) Solid Channel Cable Trays Where solid channel cable trays contain multiconductor cables of any type, the following shall apply:

(1) Where only one multiconductor cable is installed, the cross-sectional area of the cable shall not exceed the value specified in Column 1 of Table 392.9(F).

(2) Where more than one multiconductor cable is installed, the sum of the cross-sectional area of all cable shall

not exceed the value specified in Column 2 of Table 392.9(F).

Table 392.9(F) Allowable Cable Fill Area for Multiconductor Cables in Solid Channel Cable Trays for Cables Rated 2000 Volts or Less

Inside Width of Cable Tray		Column 1 One Cable		Column 2 More Than One Cable	
mm	in.	mm²	in.²	mm²	in.²
50	2	850	1.3	500	0.8
75	3	1300	2.0	700	1.1
100	4	2400	3.7	1400	2.1
150	6	3600	5.5	2100	3.2

392.10 Number of Single-Conductor Cables, Rated 2000 Volts or Less, in Cable Trays

The number of single-conductor cables, rated 2000 volts or less, permitted in a single cable tray section shall not exceed the requirements of this section. The single conductors, or conductor assemblies, shall be evenly distributed across the cable tray. The conductor sizes herein apply to both aluminum and copper conductors.

(A) Ladder or Ventilated Trough Cable Trays Where ladder or ventilated trough cable trays contain single-conductor cables, the maximum number of single conductors shall conform to the following:

(1) Where all of the cables are 1000 kcmil or larger, the sum of the diameters of all single conductor cables shall not exceed cable tray width, and the cables shall be installed in a single layer. Conductors that are bound together to comprise each circuit group shall be permitted to be installed in other than a single layer.

Section 392.10(A)(1) was expanded for the 2005 *Code* to clarify that cables 1000 kcmil and larger should be installed in a single layer except where these large cables are bundled according to circuit.

(2) Where all of the cables are from 250 kcmil up to 1000 kcmil, the sum of the cross-sectional areas of all single-conductor cables shall not exceed the maximum allowable cable fill area in Column 1 of Table 392.10(A) for the appropriate cable tray width.

(3) Where 1000 kcmil or larger single-conductor cables are installed in the same cable tray with single-conductor cables smaller than 1000 kcmil, the sum of the cross-sectional areas of all cables smaller than 1000 kcmil shall not exceed the maximum allowable fill area re-

sulting from the computation in Column 2 of Table 392.10(A) for the appropriate cable tray width.

(4) Where any of the single conductor cables are 1/0 through 4/0 AWG, the sum of the diameters of all single conductor cables shall not exceed the cable tray width.

(B) Ventilated Channel Cable Trays Where 50 mm (2 in.), 75 mm (3 in.), 100 mm (4 in.), or 150 mm (6 in.) wide ventilated channel cable trays contain single-conductor cables, the sum of the diameters of all single conductors shall not exceed the inside width of the channel.

392.11 Ampacity of Cables, Rated 2000 Volts or Less, in Cable Trays

(A) Multiconductor Cables The allowable ampacity of multiconductor cables, nominally rated 2000 volts or less, installed according to the requirements of 392.9 shall be as given in Tables 310.16 and 310.18, subject to the provisions of (1), (2), (3), and 310.15(A)(2).

(1) The derating factors of 310.15(B)(2)(a) shall apply only to multiconductor cables with more than three current-carrying conductors. Derating shall be limited to the number of current-carrying conductors in the cable and not to the number of conductors in the cable tray.

(2) Where cable trays are continuously covered for more than 1.8 m (6 ft) with solid unventilated covers, not over 95 percent of the allowable ampacities of Tables 310.16 and 310.18 shall be permitted for multiconductor cables.

(3) Where multiconductor cables are installed in a single layer in uncovered trays, with a maintained spacing of not less than one cable diameter between cables, the ampacity shall not exceed the allowable ambient temperature-corrected ampacities of multiconductor cables, with not more than three insulated conductors rated 0 through 2000 volts in free air, in accordance with 310.15(C).

FPN: See Table B.310.3.

Exhibit 392.2 illustrates the requirement of 392.11(A)(3). Note that the cables, rated 2000 volts or less, are installed in a single layer in an uncovered tray, with not less than one cable diameter between cables and not more than three conductors per cable. Refer to Table B.310.3 in Annex B for the ampacity of the conductors in this configuration.

(B) Single-Conductor Cables The allowable ampacity of single-conductor cables shall be as permitted by 310.15(A)(2). The derating factors of 310.15(B)(2)(a) shall not apply to the ampacity of cables in cable trays. The ampacity of single-conductor cables, or single conductors

Table 392.10(A) Allowable Cable Fill Area for Single-Conductor Cables in Ladder or Ventilated Trough Cable Trays for Cables Rated 2000 Volts or Less

Inside Width of Cable Tray		Maximum Allowable Fill Area for Single-Conductor Cables in Ladder or Ventilated Trough Cable Trays			
		Column 1 Applicable for 392.10(A)(2) Only		Column 2[a] Applicable for 392.10(A)(3) Only	
mm	in.	mm²	in.²	mm²	in.²
150	6	4,200	6.5	4,200 – (28 Sd)[b]	6.5 – (1.1 Sd)[b]
225	9	6,100	9.5	6,100 – (28 Sd)	9.5 – (1.1 Sd)
300	12	8,400	13.0	8,400 – (28 Sd)	13.0 – (1.1 Sd)
450	18	12,600	19.5	12,600 – (28 Sd)	19.5 – (1.1 Sd)
600	24	16,800	26.0	16,800 – (28 Sd)	26.0 – (1.1 Sd)
750	30	21,000	32.5	21,000 – (28 Sd)	32.5 – (1.1 Sd)
900	36	25,200	39.0	25,200 – (28 Sd)	39.0 – (1.1 Sd)

[a]The maximum allowable fill areas in Column 2 shall be calculated. For example, the maximum allowable fill, in mm², for a 150 mm wide cable tray in Column 2 shall be 4200 minus (28 multiplied by Sd) [the maximum allowable fill, in square inches, for a 6-in. wide cable tray in Column 2 shall be 6.5 minus (1.1 multiplied by Sd)].

[b]The term *Sd* in Column 2 is equal to the sum of the diameters, in mm, of all cables 507 mm² (in inches, of all 1000 kcmil) and larger single-conductor cables in the same ladder or ventilated trough cable tray with small cables.

Not less than one cable-diameter spacing or separation between cables

Exhibit 392.2 An illustration of 392.11(A)(3) for multiconductor cables, 2000 volts or less, with not more than three conductors per cable (ampacity to be determined from Table B.310.3 in Annex B).

cabled together (triplexed, quadruplexed, etc.), nominally rated 2000 volts or less, shall comply with the following:

(1) Where installed according to the requirements of 392.10, the ampacities for 600 kcmil and larger single-conductor cables in uncovered cable trays shall not exceed 75 percent of the allowable ampacities in Tables 310.17 and 310.19. Where cable trays are continuously covered for more than 1.8 m (6 ft) with solid unventilated covers, the ampacities for 600 kcmil and larger cables shall not exceed 70 percent of the allowable ampacities in Tables 310.17 and 310.19.

(2) Where installed according to the requirements of 392.10, the ampacities for 1/0 AWG through 500 kcmil single-conductor cables in uncovered cable trays shall not exceed 65 percent of the allowable ampacities in Tables 310.17 and 310.19. Where cable trays are continuously covered for more than 1.8 m (6 ft) with solid unventilated covers, the ampacities for 1/0 AWG through 500 kcmil cables shall not exceed 60 percent of the allowable ampacities in Tables 310.17 and 310.19.

(3) Where single conductors are installed in a single layer in uncovered cable trays, with a maintained space of not less than one cable diameter between individual conductors, the ampacity of 1/0 AWG and larger cables shall not exceed the allowable ampacities in Tables 310.17 and 310.19.

Exception to (B)(3): For solid bottom cable trays the ampacity of single conductor cables shall be determined by 310.15(C).

(4) Where single conductors are installed in a triangular or square configuration in uncovered cable trays, with a maintained free airspace of not less than 2.15 times one conductor diameter (2.15 × O.D.) of the largest conductor contained within the configuration and adjacent conductor configurations or cables, the ampacity of 1/0 AWG and larger cables shall not exceed the allowable ampacities of two or three single insulated conductors rated 0 through 2000 volts supported on a messenger in accordance with 310.15(B).

FPN: See Table 310.20.

Section 392.11(B)(4) recognizes single conductors in a triangular configuration installed in a cable tray with maintained spacing as having the same ampacity as three single insulated conductors on a messenger. The maintained spacing allows air to circulate around the cable.

Where three single conductors, nominally rated 2000 volts or less, are cabled together in a triangular configuration, with not less than 2.15 times the conductor diameter (2.15 × OD) between groups, as illustrated in Exhibit 392.3, the ampacity of the conductors is determined in accordance with Table 310.20.

Exhibit 392.3 An illustration of 392.11(B)(4), for three single conductors installed in a triangular configuration with spacing between groups of not less than 2.15 times the conductor diameter (ampacities to be determined from Table 310.20).

Where single conductors are installed in cable trays, their ampacities are permitted to be calculated using many variations of Table 310.16, Table 310.17, and Table 310.19. Where these single conductor cables emerge from a cable tray installation and are terminated at circuit breakers, distribution switchgear, and similar electrical equipment, it is important to understand the temperature limitations of the electrical equipment terminals and to coordinate those temperature limitations with the ampacity of the single conductor cables. As stated in both the UL *General Information for Electrical Equipment Directory* and in 110.14(C)(1), unless the equipment is listed and marked otherwise, conductor ampacities used in determining equipment terminations must be based on Table 310.16 as modified by 310.15(B)(1) through 310.15(B)(6).

392.12 Number of Type MV and Type MC Cables (2001 Volts or Over) in Cable Trays

The number of cables rated 2001 volts or over permitted in a single cable tray shall not exceed the requirements of this section.

The sum of the diameters of single-conductor and multiconductor cables shall not exceed the cable tray width, and the cables shall be installed in a single layer. Where single conductor cables are triplexed, quadruplexed, or bound together in circuit groups, the sum of the diameters of the single conductors shall not exceed the cable tray width, and these groups shall be installed in single layer arrangement.

392.13 Ampacity of Type MV and Type MC Cables (2001 Volts or Over) in Cable Trays

The ampacity of cables, rated 2001 volts, nominal, or over, installed according to 392.12 shall not exceed the requirements of this section.

(A) Multiconductor Cables (2001 Volts or Over) The allowable ampacity of multiconductor cables shall be as given in Tables 310.75 and 310.76, subject to the following provisions:

(1) Where cable trays are continuously covered for more than 1.8 m (6 ft) with solid unventilated covers, not more than 95 percent of the allowable ampacities of Tables 310.75 and 310.76 shall be permitted for multiconductor cables.

(2) Where multiconductor cables are installed in a single layer in uncovered cable trays, with maintained spacing of not less than one cable diameter between cables, the ampacity shall not exceed the allowable ampacities of Tables 310.71 and 310.72.

(B) Single-Conductor Cables (2001 Volts or Over) The ampacity of single-conductor cables, or single conductors cabled together (triplexed, quadruplexed, etc.), shall comply with the following:

(1) The ampacities for 1/0 AWG and larger single-conductor cables in uncovered cable trays shall not exceed 75 percent of the allowable ampacities in Tables 310.69 and 310.70. Where the cable trays are covered for more than 1.8 m (6 ft) with solid unventilated covers, the ampacities for 1/0 AWG and larger single-conductor cables shall not exceed 70 percent of the allowable ampacities in Tables 310.69 and 310.70.

(2) Where single-conductor cables are installed in a single layer in uncovered cable trays, with a maintained space of not less than one cable diameter between individual conductors, the ampacity of 1/0 AWG and larger cables shall not exceed the allowable ampacities in Tables 310.69 and 310.70.

(3) Where single conductors are installed in a triangular or square configuration in uncovered cable trays, with a maintained free air space of not less than 2.15 times the diameter (2.15 × O.D.) of the largest conductor contained within the configuration and adjacent conductor configurations or cables, the ampacity of 1/0 AWG and larger cables shall not exceed the allowable ampacities in Tables 310.67 and 310.68.

ARTICLE 394
Concealed Knob-and-Tube Wiring

Contents

I. General

394.1 Scope

This article covers the use, installation, and construction specifications of concealed knob-and-tube wiring.

394.2 Definition

Concealed Knob-and-Tube Wiring. A wiring method using knobs, tubes, and flexible nonmetallic tubing for the protection and support of single insulated conductors.

Open wiring on insulators (Article 398) is required to be exposed, whereas knob-and-tube wiring is allowed to be concealed. Conductors used for knob-and-tube work may be of any general-use type specified by Article 310.

II. Installation

394.10 Uses Permitted

Concealed knob-and-tube wiring shall be permitted to be installed in the hollow spaces of walls and ceilings or in unfinished attics and roof spaces as provided by 394.23 only as follows:

(1) For extensions of existing installations
(2) Elsewhere by special permission

Concealed knob-and-tube wiring is permitted to be installed only for extensions of existing installations or where special permission is granted by the authority having jurisdiction of enforcement of the *Code*. See definition of *special permission* in Article 100.

394.12 Uses Not Permitted

Concealed knob-and-tube wiring shall not be used in the following:

(1) Commercial garages
(2) Theaters and similar locations
(3) Motion picture studios
(4) Hazardous (classified) locations
(5) Hollow spaces of walls, ceilings, and attics where such spaces are insulated by loose, rolled, or foamed-in-place insulating material that envelops the conductors

Concealed knob-and-tube wiring is designed for use in hollow spaces of walls, ceilings, and attics and utilizes the free air in such spaces for heat dissipation. Weatherization of hollow spaces by blown-in, foamed-in, or rolled insulation prevents the dissipation of heat into the free air space, resulting in higher conductor temperature, which could cause insulation breakdown and possible ignition of the insulation.

394.17 Through or Parallel to Framing Members

Conductors shall comply with 398.17 where passing through holes in structural members. Where passing through wood cross members in plastered partitions, conductors shall be protected by noncombustible, nonabsorbent, insulating tubes extending not less than 75 mm (3 in.) beyond the wood member.

The provision for insulated tubes where knob-and-tube wiring passes through wood cross members in plastered partitions is intended to protect the wire from contact with plaster that is likely to accumulate on horizontal wood members.

394.19 Clearances

(A) General A clearance of not less than 75 mm (3 in.) shall be maintained between conductors and a clearance of not less than 25 mm (1 in.) between the conductor and the surface over which it passes.

(B) Limited Conductor Space Where space is too limited to provide these minimum clearances, such as at meters, panelboards, outlets, and switch points, the individual conductors shall be enclosed in flexible nonmetallic tubing, which shall be continuous in length between the last support and the enclosure or terminal point.

(C) Clearance from Piping, Exposed Conductors, and So Forth Conductors shall comply with 398.19 for clearances from other exposed conductors, piping, and so forth.

394.23 In Accessible Attics

Conductors in unfinished attics and roof spaces shall comply with 394.23(A) or (B).

> FPN: See 310.10 for temperature limitation of conductors.

(A) Accessible by Stairway or Permanent Ladder Conductors shall be installed along the side of or through bored holes in floor joists, studs, or rafters. Where run through bored holes, conductors in the joists and in studs or rafters to a height of not less than 2.1 m (7 ft) above the floor or floor joists shall be protected by substantial running boards extending not less than 25 mm (1 in.) on each side of the conductors. Running boards shall be securely fastened in place. Running boards and guard strips shall not be required where conductors are installed along the sides of joists, studs, or rafters.

Exhibit 394.1 illustrates the "running board" method of protecting open-type conductors where they are installed at a height less than 7 ft above the floor or floor joists in an accessible attic. This method is applied in attics that are accessible by stairways or permanent ladders and where such spaces are generally used for storage.

(B) Not Accessible by Stairway or Permanent Ladder Conductors shall be installed along the sides of or through bored holes in floor joists, studs, or rafters.

Exception: In buildings completed before the wiring is installed, attic and roof spaces that are not accessible by stairway or permanent ladder and have headroom at all points less than 900 mm (3 ft), the wiring shall be permitted to be installed on the edges of rafters or joists facing the attic or roof space.

394.30 Securing and Supporting

(A) Supporting Conductors shall be rigidly supported on noncombustible, nonabsorbent insulating materials and shall

Exhibit 394.1 Open wiring in an accessible attic protected by running boards, as required by 394.23(A).

not contact any other objects. Supports shall be installed as follows:

(1) Within 150 mm (6 in.) of each side of each tap or splice, and
(2) At intervals not exceeding 1.4 m (4½ ft).

Where it is impracticable to provide supports, conductors shall be permitted to be fished through hollow spaces in dry locations, provided each conductor is individually enclosed in flexible nonmetallic tubing that is in continuous lengths between supports, between boxes, or between a support and a box.

(B) Securing Where solid knobs are used, conductors shall be securely tied thereto by tie wires having insulation equivalent to that of the conductor.

394.42 Devices

Switches shall comply with 404.4 and 404.10(B).

394.56 Splices and Taps

Splices shall be soldered unless approved splicing devices are used. In-line or strain splices shall not be used.

III. Construction Specifications

394.104 Conductors

Conductors shall be of a type specified by Article 310.

ARTICLE 396
Messenger Supported Wiring

Contents

I. General

396.1 Scope

This article covers the use, installation, and construction specifications for messenger supported wiring.

Messenger supported wiring systems have been manufactured and successfully used in industrial installations for many years. They have also been used for many years as service drops by utilities for commercial and residential installations. See references to messenger supported wiring in 225.6(A)(1) and 225.6(B).

396.2 Definition

Messenger Supported Wiring. An exposed wiring support system using a messenger wire to support insulated conductors by any one of the following:

(1) A messenger with rings and saddles for conductor support
(2) A messenger with a field-installed lashing material for conductor support
(3) Factory-assembled aerial cable
(4) Multiplex cables utilizing a bare conductor, factory assembled and twisted with one or more insulated conductors, such as duplex, triplex, or quadruplex type of construction

II. Installation

396.10 Uses Permitted

(A) Cable Types The cable types in Table 396.10(A) shall be permitted to be installed in messenger supported wiring

under the conditions described in the article or section referenced for each.

Table 396.10(A) Cable Types

Cable Type	Section	Article
Metal-clad cable		330
Mineral-insulated, metal-sheathed cable		332
Multiconductor service-entrance cable		338
Multiconductor underground feeder and branch-circuit cable		340
Other factory-assembled, multiconductor control, signal, or power cables that are identified for the use		
Power and control tray cable		336
Power-limited tray cable	725.61(C) and 725.82(E)	

(B) In Industrial Establishments In industrial establishments only, where conditions of maintenance and supervision ensure that only qualified persons service the installed messenger supported wiring, the following shall be permitted:

(1) Any of the conductor types shown in Table 310.13 or Table 310.62
(2) MV cable

Where exposed to weather, conductors shall be listed for use in wet locations. Where exposed to direct rays of the sun, conductors or cables shall be sunlight resistant.

Some of the triplex and quadruplex cable used by utilities as service-drop cable does not use conductors recognized in Table 310.13 and does not meet the requirements of Article 310. Such triplex and quadruplex cable would, therefore, be acceptable only where approved by the authority having jurisdiction.

See 310.15(B) and Table 310.20 for two or three single-insulated conductors supported on a messenger. See 310.15(C) and Annex B, Table B.310.3, for ampacities of conductors for other cable types.

(C) Hazardous (Classified) Locations Messenger supported wiring shall be permitted to be used in hazardous (classified) locations where the contained cables are permitted for such use in 501.10, 502.10, 503.10, and 504.20.

396.12 Uses Not Permitted

Messenger supported wiring shall not be used in hoistways or where subject to physical damage.

396.30 Messenger Support

The messenger shall be supported at dead ends and at intermediate locations so as to eliminate tension on the conductors. The conductors shall not be permitted to come into contact with the messenger supports or any structural members, walls, or pipes.

396.56 Conductor Splices and Taps

Conductor splices and taps made and insulated by approved methods shall be permitted in messenger supported wiring.

396.60 Grounding

The messenger shall be grounded as required by 250.80 and 250.86 for enclosure grounding.

ARTICLE 398
Open Wiring on Insulators

Contents

I. General

398.1 Scope

This article covers the use, installation, and construction specifications of open wiring on insulators.

398.2 Definition

Open Wiring on Insulators. An exposed wiring method using cleats, knobs, tubes, and flexible tubing for the protection and support of single insulated conductors run in or on buildings.

Open wiring on insulators is an exposed wiring method that is not permitted to be concealed by the structure or finish of the building. It is permitted indoors or outdoors, in dry or wet locations, and where subject to corrosive vapors, provided the insulation choice from Table 310.13 is suitable for use in a corrosive environment.

This method of wiring is no longer permitted for temporary lighting and power circuits on construction sites but is permitted for lighting and power circuits in agricultural buildings [see 547.5(A)]. It may also be used for services (see 230.43).

II. Installation

398.10 Uses Permitted

Open wiring on insulators shall be permitted only for industrial or agricultural establishments on systems of 600 volts, nominal, or less, as follows:

(1) Indoors or outdoors
(2) In wet or dry locations
(3) Where subject to corrosive vapors
(4) For services

See Tables 310.17 and 310.19 for ampacities of conductors.

398.12 Uses Not Permitted

Open wiring on insulators shall not be installed where concealed by the building structure.

398.15 Exposed Work

(A) Dry Locations In dry locations, where not exposed to physical damage, conductors shall be permitted to be separately enclosed in flexible nonmetallic tubing. The tubing shall be in continuous lengths not exceeding 4.5 m (15 ft) and secured to the surface by straps at intervals not exceeding 1.4 m (4½ ft).

(B) Entering Spaces Subject to Dampness, Wetness, or Corrosive Vapors Conductors entering or leaving locations subject to dampness, wetness, or corrosive vapors shall have drip loops formed on them and shall then pass upward and inward from the outside of the buildings, or from the damp, wet, or corrosive location, through noncombustible, nonabsorbent insulating tubes.

FPN: See 230.52 for individual conductors entering buildings or other structures.

(C) Exposed to Physical Damage Conductors within 2.1 m (7 ft) from the floor shall be considered exposed to physical damage. Where open conductors cross ceiling joists and wall studs and are exposed to physical damage, they shall be protected by one of the following methods:

(1) Guard strips not less than 25 mm (1 in.) nominal in thickness and at least as high as the insulating supports, placed on each side of and close to the wiring.

(2) A substantial running board at least 13 mm (½ in.) thick in back of the conductors with side protections. Running boards shall extend at least 25 mm (1 in.) outside the conductors, but not more than 50 mm (2 in.), and the protecting sides shall be at least 50 mm (2 in.) high and at least 25 mm (1 in.) nominal in thickness.

(3) Boxing made in accordance with 398.15(C)(1) or (C)(2) and furnished with a cover kept at least 25 mm (1 in.) away from the conductors within. Where protecting vertical conductors on side walls, the boxing shall be closed at the top and the holes through which the conductors pass shall be bushed.

(4) Rigid metal conduit, intermediate metal conduit, rigid nonmetallic conduit, or electrical metallic tubing. When installed in metal piping, the conductors shall be encased in continuous lengths of approved flexible tubing.

398.17 Through or Parallel to Framing Members

Open conductors shall be separated from contact with walls, floors, wood cross members, or partitions through which they pass by tubes or bushings of noncombustible, nonabsorbent insulating material. Where the bushing is shorter than the hole, a waterproof sleeve of noninductive material shall be inserted in the hole and an insulating bushing slipped into the sleeve at each end in such a manner as to keep the conductors absolutely out of contact with the sleeve. Each conductor shall be carried through a separate tube or sleeve.

> FPN: See 310.10 for temperature limitation of conductors.

398.19 Clearances

Open conductors shall be separated at least 50 mm (2 in.) from metal raceways, piping, or other conducting material, and from any exposed lighting, power, or signaling conductor, or shall be separated therefrom by a continuous and firmly fixed nonconductor in addition to the insulation of the conductor. Where any insulating tube is used, it shall be secured at the ends. Where practicable, conductors shall pass over rather than under any piping subject to leakage or accumulations of moisture.

The provision for additional protective insulation on open wiring is to prevent contact with metal piping, metal objects, or exposed conductors of other circuits.

398.23 In Accessible Attics

Conductors in unfinished attics and roof spaces shall comply with 398.23(A) or (B).

(A) Accessible by Stairway or Permanent Ladder Conductors shall be installed along the side of or through bored holes in floor joists, studs, or rafters. Where run through bored holes, conductors in the joists and in studs or rafters to a height of not less than 2.1 m (7 ft) above the floor or floor joists shall be protected by substantial running boards extending not less than 25 mm (1 in.) on each side of the conductors. Running boards shall be securely fastened in place. Running boards and guard strips shall not be required for conductors installed along the sides of joists, studs, or rafters.

(B) Not Accessible by Stairway or Permanent Ladder Conductors shall be installed along the sides of or through bored holes in floor joists, studs, or rafters.

Exception: In buildings completed before the wiring is installed, in attic and roof spaces that are not accessible by stairway or permanent ladder and have headroom at all points less than 900 mm (3 ft), the wiring shall be permitted to be installed on the edges of rafters or joists facing the attic or roof space.

398.30 Securing and Supporting

(A) Conductor Sizes Smaller Than 8 AWG Conductors smaller than 8 AWG shall be rigidly supported on noncombustible, nonabsorbent insulating materials and shall not contact any other objects. Supports shall be installed as follows:

(1) Within 150 mm (6 in.) from a tap or splice
(2) Within 300 mm (12 in.) of a dead-end connection to a lampholder or receptacle
(3) At intervals not exceeding 1.4 m (4½ ft) and at closer intervals sufficient to provide adequate support where likely to be disturbed

(B) Conductor Sizes 8 AWG and Larger Supports for conductors 8 AWG or larger installed across open spaces shall be permitted up to 4.5 m (15 ft) apart if noncombustible, nonabsorbent insulating spacers are used at least every 1.4 m (4½ ft) to maintain at least 65 mm (2½ in.) between conductors.

Where not likely to be disturbed in buildings of mill construction, 8 AWG and larger conductors shall be permitted to be run across open spaces if supported from each wood cross member on approved insulators maintaining 150 mm (6 in.) between conductors.

Mill construction is generally considered to be a building where the floors and ceilings are supported by wood timbers or beams or wood cross members spaced approximately 15 ft apart. This type of construction is sometimes referred to as plank-on-timber construction. Section 398.30(B) permits 8 AWG and larger conductors to span this distance where the ceilings are high and free of obstructions and the conductors are unlikely to come into contact with other objects.

(C) Industrial Establishments In industrial establishments only, where conditions of maintenance and supervision ensure that only qualified persons service the system, conductors of sizes 250 kcmil and larger shall be permitted to be run across open spaces where supported at intervals up to 9.0 m (30 ft) apart.

It is common practice in industrial buildings to install open feeders on insulators, which are mounted on the bottom of roof trusses at every bay location. Many bays are more than 15 ft wide. Therefore, 398.30(C) permits size 250 kcmil and larger conductors to be supported at 30-ft intervals where it is ensured that qualified persons will service the system.

In addition to the ease and economy of installation or alteration of open wiring, it is to be noted that, by close spacing of conductors, the reactance of a circuit is reduced; hence, the voltage drop is reduced.

(D) Mounting of Conductor Supports Where nails are used to mount knobs, they shall not be smaller than tenpenny. Where screws are used to mount knobs, or where nails or screws are used to mount cleats, they shall be of a length sufficient to penetrate the wood to a depth equal to at least one-half the height of the knob and the full thickness of the cleat. Cushion washers shall be used with nails.

(E) Tie Wires 8 AWG or larger conductors supported on solid knobs shall be securely tied thereto by tie wires having an insulation equivalent to that of the conductor.

398.42 Devices

Surface-type snap switches shall be mounted in accordance with 404.10(A), and boxes shall not be required. Other type switches shall be installed in accordance with 404.4.

III. Construction Specifications

398.104 Conductors

Conductors shall be of a type specified by Article 310.

4

Equipment for General Use

ARTICLE 400
Flexible Cords and Cables

Summary of Changes

• **400.5:** Revised to require that where cords are used in ambient temperatures exceeding 30°C (86°F), ampacity correction factors are to be applied.

• **400.8:** Added new item (7) prohibiting use where subject to physical damage.

• **400.14:** Revised to permit cords and cables to be installed in aboveground raceways in industrial establishments, with restrictions.

Contents

I. General

400.1 Scope

This article covers general requirements, applications, and construction specifications for flexible cords and flexible cables.

Flexible cords and cables, because of the nature of their use, are not considered to be wiring methods. Wiring methods are covered in Chapter 3 of the *Code*. Careful study of 400.7, Uses Permitted, and 400.8, Uses Not Permitted, is required before choosing flexible cords or cables for a specific application.

400.2 Other Articles

Flexible cords and flexible cables shall comply with this article and with the applicable provisions of other articles of this *Code*.

400.3 Suitability

Flexible cords and cables and their associated fittings shall be suitable for the conditions of use and location.

400.4 Types

Flexible cords and flexible cables shall conform to the description in Table 400.4. Types of flexible cords and flexible cables other than those listed in the table shall be the subject of special investigation.

400.5 Ampacities for Flexible Cords and Cables

(A) Ampacity Tables Table 400.5(A) provides the allowable ampacities, and Table 400.5(B) provides the ampacities for flexible cords and cables with not more than three current-carrying conductors. These tables shall be used in conjunction with applicable end-use product standards to ensure selection of the proper size and type. Where cords are used in ambient temperatures exceeding 30°C (86°F),

Table 400.4 Flexible Cords and Cables (See 400.4.)

Trade Name	Type Letter	Voltage	AWG or kcmil	Number of Conductors	Insulation	Nominal Insulation Thickness[1]			Braid on Each Conductor	Outer Covering	Use		
						AWG or kcmil	mm	mils					
Lamp cord	C	300 600	18–16 14–10	2 or more	Thermoset or thermo-plastic	18–16 14–10	0.76 1.14	30 45	Cotton	None	Pendant or portable	Dry locations	Not hard usage
Elevator cable	E See Note 5. See Note 9. See Note 10.	300 or 600	20–2	2 or more	Thermoset	20–16 14–12 12–10 8–2	0.51 0.76 1.14 1.52	20 30 45 60	Cotton	Three cotton, Outer one flame-retardant & moisture-resistant. See Note 3.	Elevator lighting and control	Unclassified locations	
						20–16 14–12 12–10 8–2	0.51 0.76 1.14 1.52	20 30 45 60	Flexible nylon jacket				
Elevator cable	EO See Note 5. See Note 10.	300 or 600	20–2	2 or more	Thermoset	20–16 14–12 12–10 8–2	0.51 0.76 1.14 1.52	20 30 45 60	Cotton	Outer one Three cotton, flame-retardant & moisture-resistant. See Note 3.	Elevator lighting and control	Unclassified locations	
										One cotton and a neoprene jacket. See Note 3.		Hazardous (classified) locations	
Elevator cable	ETP See Note 5. See Note 10.	300 or 600							Rayon	Thermoplastic	Hazardous (classified) locations		
	ETT See Note 5. See Note 10.	300 or 600							None	One cotton or equivalent and a thermoplastic jacket			
Electric vehicle cable	EV	600	18–500 See Note 11.	2 or more plus grounding conductor(s), plus optional hybrid data, signal communications, and optical fiber cables	Thermoset with optional nylon See Note 12.	18–16 14–10 8–2 1–4/0 250–500	0.76 (0.51) 1.14 (0.76) 1.52 (1.14) 2.03 (1.52) 2.41 (1.90)	30 (20) 45 (30) 60 (45) 80 (60) 95 (75) See Note 12.	Optional	Thermoset	Electric vehicle charging	Wet locations	Extra hard usage
	EVJ	300	18–12 See Note 11.			18–12	0.76 (0.51)	30 (20) See Note 12.					Hard usage
	EVE	600	18–500 See Note 11.	2 or more plus grounding conductor(s), plus optional hybrid data, signal communications, and optical fiber cables	Thermo-plastic elastomer with optional nylon See Note 12.	18–16 14–10 8–2 1–4/0 250–500	0.76 (0.51) 1.14 (0.76) 1.52 (1.14) 2.03 (1.52) 2.41 (1.90)	30 (20) 45 (30) 60 (45) 80 (60) 95 (75) See Note 12.		Thermoplastic elastomer			Extra hard usage
	EVJE	300	18–12 See Note 11.			18–12	0.76 (0.51)	30 (20) See Note 12.					Hard usage

(continues)

Table 400.4 *Continued*

Trade Name	Type Letter	Voltage	AWG or kcmil	Number of Conductors	Insulation	Nominal Insulation Thickness[1]			Braid on Each Conductor	Outer Covering	Use		
						AWG or kcmil	mm	mils					
	EVT	600	18–500 See Note 11.	2 or more plus grounding conductor(s), plus optional hybrid data, signal communications, and optical fiber cables	Thermoplastic with optional nylon See Note 12.	18–16 14–10 8–2 1–4/0 250–500	0.76 (0.51) 1.14 (0.76) 1.52 (1.14) 2.03 (1.52) 2.41 (1.90)	30 (20) 45 (30) 60 (45) 80 (60) 95 (75) See Note 12.	Optional	Thermoplastic	Electric vehicle charging	Wet Locations	Extra hard usage
	EVJT	300	18–12 See Note 11.			18–12	0.76 (0.51)	30 (20) See Note 12.					Hard usage
Portable power cable	G	2000	12–500	2–6 plus grounding conductor(s)	Thermoset	12–2 1–4/0 250–500	1.52 2.03 2.41	60 80 95		Oil-resistant thermoset	Portable and extra hard usage		
	G-GC	2000	12–500	3–6 plus grounding conductors and 1 ground check conductor	Thermoset	12–2 1–4/0 250–500	1.52 2.03 2.41	60 80 95		Oil-resistant thermoset			
Heater cord	HPD	300	18–12	2, 3, or 4	Thermoset	18–16 14–12	0.38 0.76	15 30	None	Cotton or rayon	Portable heaters	Dry locations	Not hard usage
Parallel heater cord	HPN See Note 6.	300	18–12	2 or 3	Oil-resistant thermoset	18–16 14–12	1.14 1.52 2.41	45 60 95	None	Oil-resistant thermoset	Portable	Damp locations	Not hard usage
Thermoset jacketed heater cords	HSJ	300	18–12	2, 3, or 4	Thermoset	18–16	0.76	30	None	Cotton and Thermoset	Portable or portable heater	Damp locations	Hard usage
	HSJO	300	18–12		Oil-resistant thermoset	14–12	1.14	45		Cotton and oil-resistant thermoset			
	HSJOO	300	18–12										
Non-integral parallel cords	NISP-1 See Note 6.	300	20–18	2 or 3	Thermoset	20–18	0.38	15	None	Thermoset	Pendant or portable	Damp locations	Not hard usage
	NISP-2 See Note 6.	300	18–16			18–16	0.76	30					
	NISPE-1 See Note 6.	300	20–18		Thermoplastic elastomer	20–18	0.38	15		Thermoplastic elastomer			
	NISPE-2 See Note 6.	300	18–16			18–16	0.76	30					
	NISPT-1 See Note 6.	300	20–18		Thermoplastic	20–18	0.38	15		Thermoplastic			
	NISPT-2 See Note 6.	300	18–16			18–16	0.76	30					
Twisted portable cord	PD	300 600	18–16 14–10	2 or more	Thermoset or thermoplastic	18–16 14–10	0.76 1.14	30 45	Cotton	Cotton or rayon	Pendant or portable	Dry locations	Not hard usage
Portable power cable	PPE	2000	12–500	1–6 plus optional grounding conductor(s)	Thermoplastic elastomer	12–2 1–4/0 250–500	1.52 2.03 2.41	60 80 95		Oil-resistant thermoplastic elastomer	Portable, extra hard usage		
Hard service cord	S See Note 4.	600	18–12	2 or more	Thermoset	18–16 14–10 8–2	0.76 1.14 1.52	30 45 60	None	Thermoset	Pendant or portable	Damp locations	Extra hard usage

Table 400.4 *Continued*

Trade Name	Type Letter	Voltage	AWG or kcmil	Number of Conductors	Insulation	Nominal Insulation Thickness[1] AWG or kcmil	mm	mils	Braid on Each Conductor	Outer Covering	Use		
Flexible stage and lighting power cable	SC	600	8–250	1 or more		8–2 1–4/0 250	1.52 2.03 2.41	60 80 95		Thermoset[2]	Portable, extra hard usage		
	SCE	600			Thermoplastic elastomer					Thermoplastic elastomer[2]			
	SCT	600			Thermoplastic					Thermoplastic[2]			
Hard service cord	SE See Note 4.	600	18–2	2 or more	Thermoplastic elastomer	18–16 14–10 8–2	0.76 1.14 1.52	30 45 60	None	Thermoplastic elastomer	Pendant or portable	Damp locations	Extra hard usage
	SEW See Note 4. See Note 13.	600										Damp and wet locations	
	SEO See Note 4.	600								Oil-resistant thermoplastic elastomer		Damp locations	
	SEOW See Note 4. See Note 13.	600										Damp and wet locations	
	SEOO See Note 4.	600			Oil-resistant thermoplastic elastomer							Damp locations	
	SEOOW See Note 4. See Note 13.	600										Damp and wet locations	
Junior hard service cord	SJ	300	18–10	2–6	Thermoset	18–12	0.76	30	None	Thermoset	Pendant or portable	Damp locations	Hard usage
	SJE	300			Thermoplastic elastomer					Thermoplastic elastomer			
	SJEW See Note 13.	300										Damp and wet locations	
	SJEO	300								Oil-resistant thermoplastic elastomer		Damp locations	
	SJEOW See Note 13.	300										Damp and wet locations	
	SJEOO	300			Oil-resistant thermoplastic elastomer							Damp locations	
	SJEOOW See Note 13.	300										Damp and wet locations	
	SJO	300			Thermoset					Oil-resistant thermoset		Damp locations	
	SJOW See Note 13.	300										Damp and wet locations	
	SJOO	300			Oil-resistant thermoset							Damp locations	

(continues)

Table 400.4 *Continued*

Trade Name	Type Letter	Voltage	AWG or kcmil	Number of Conductors	Insulation	Nominal Insulation Thickness[1] AWG or kcmil	mm	mils	Braid on Each Conductor	Outer Covering	Use		
	SJOOW See Note 13.	300										Damp and wet locations	
	SJT	300			Thermo-plastic	10	1.14	45		Thermoplastic		Damp locations	
	SJTW See Note 13.	300										Damp and wet locations	
	SJTO	300				18–12	0.76	30		Oil-resistant thermoplastic		Damp locations	
	SJTOW See Note 13.	300										Damp and wet locations	
	SJTOO	300			Oil-resistant thermo-plastic							Damp locations	
	SJTOOW See Note 13.	300										Damp and wet locations	
Hard service cord	SO See Note 4.	600	18–2	2 or more	Thermoset	18–16	0.76	30	None	Oil-resistant thermoset	Pendant or portable	Damp locations	Extra hard usage
	SOW See Note 4. See Note 13.	600										Damp and wet locations	
	SOO See Note 4.	600			Oil-resistant thermoset	14–10 8–2	1.14 1.52	45 60				Damp locations	
	SOOW See Note 4. See Note 13.	600										Damp and wet locations	
All thermoset parallel cord	SP-1 See Note 6.	300	20–18	2 or 3	Thermoset	20–18	0.76	30	None	None	Pendant or portable	Damp locations	Not hard usage
	SP-2 See Note 6.	300	18–16			18–16	1.14	45					
	SP-3 See Note 6.	300	18–10			18–16 14 12 10	1.52 2.03 2.41 2.80	60 80 95 110			Refrigerators, room air conditioners, and as permitted in 422.16(B)		
All elastomer (thermo-plastic) parallel cord	SPE-1 See Note 6.	300	20–18	2 or 3	Thermo-plastic elastomer	20–18	0.76	30	None	None	Pendant or portable	Damp locations	Not hard usage
	SPE-2 See Note 6.	300	18–16			18–16	1.14	45					
	SPE-3 See Note 6.	300	18–10			18–16 14 12 10	1.52 2.03 2.41 2.80	60 80 95 110			Refrigerators, room air conditioners, and as permitted in 422.16(B)		
All plastic parallel cord	SPT 1 See Note 6.	300	20 18	2 or 3	Thermo-plastic	20–18	0.76	30	None	None	Pendant or portable	Damp locations	Not hard usage

Table 400.4 *Continued*

Trade Name	Type Letter	Voltage	AWG or kcmil	Number of Conductors	Insulation	Nominal Insulation Thickness[1]			Braid on Each Conductor	Outer Covering	Use		
						AWG or kcmil	mm	mils					
	SPT-1W See Note 6. See Note 13.	300										Damp and wet locations	
	SPT-2 See Note 6.	300	18–16			18–16	1.14	45				Damp locations	
	SPT-2W See Note 6. See Note 13.	300										Damp and wet locations	
	SPT-3 See Note 6.	300	18–10			18 16 14 12 10	1.52 2.03 2.41 2.80	60 80 95 110			Refrigerators, room air conditioners, and as permitted in 422.16(B)	Damp locations	Not hard usage
Range, dryer cable	SRD	300	10–4	3 or 4	Thermoset	10–4	1.14	45	None	Thermoset	Portable	Damp locations	Ranges, dryers
	SRDE	300	10–4	3 or 4	Thermoplastic elastomer				None	Thermoplastic elastomer			
	SRDT	300	10–4	3 or 4	Thermoplastic				None	Thermoplastic			
Hard service cord	ST See Note 4.	600	18–2	2 or more	Thermoplastic	18–16 14–10 8–2	0.76 1.14 1.52	30 45 60	None	Thermoplastic	Pendant or portable	Damp locations	Extra hard usage
	STW See Note 4. See Note 13.	600										Damp and wet locations	
	STO See Note 4.	600								Oil-resistant thermoplastic		Damp locations	
	STOW See Note 4. See Note 13.	600										Damp and wet locations	
	STOO See Note 4.	600				Oil-resistant thermoplastic						Damp locations	
	STOOW See Note 4.	600										Damp and wet locations	
Vacuum cleaner cord	SV See Note 6.	300	18–16	2 or 3	Thermoset	18–16	0.38	15	None	Thermoset	Pendant or portable	Damp locations	Not hard usage
	SVE See Note 6.	300			Thermoplastic elastomer					Thermoplastic elastomer			
	SVEO See Note 6.	300								Oil-resistant thermoplastic elastomer			
	SVEOO See Note 6.	300			Oil-resistant thermoplastic elastomer								
	SVO	300			Thermoset					Oil-resistant thermoset			
	SVOO	300			Oil-resistant thermoset					Oil-resistant thermoset			

(continues)

Table 400.4 *Continued*

Trade Name	Type Letter	Voltage	AWG or kcmil	Number of Conductors	Insulation	Nominal Insulation Thickness[1]			Braid on Each Conductor	Outer Covering	Use		
						AWG or kcmil	mm	mils					
	SVT See Note 6.	300			Thermo-plastic					Thermoplastic			
	SVTO See Note 6.	300			Thermo-plastic Oil-resistant thermo-plastic					Oil-resistant thermoplastic			
	SVTOO	300											
Parallel tinsel cord	TPT See Note 2.	300	27	2	Thermo-plastic	27	0.76	30	None	Thermoplastic	Attached to an appliance	Damp locations	Not hard usage
Jacketed tinsel cord	TST See Note 2.	300	27	2	Thermo-plastic	27	0.38	15	None	Thermoplastic	Attached to an appliance	Damp locations	Not hard usage
Portable power-cable	W	2000	12–500 501–1000	1–6 1	Thermoset	12–2 1–4/0 250–500 501–1000	1.52 2.03 2.41 2.80	60 80 95 110		Oil-resistant thermoset	Portable, extra hard usage		

*See Note 8.

**The required outer covering on some single conductor cables may be integral with the insulation.

Notes:

1. All types listed in Table 400.4 shall have individual conductors twisted together except for Types HPN, SP-1, SP-2, SP-3, SPE-1, SPE-2, SPE-3, SPT-1, SPT-2, SPT-3, TPT, NISP-1, NISP-2, NISPT-1, NISPT-2, NISPE-1, NISPE-2, and three-conductor parallel versions of SRD, SRDE, and SRDT.

2. Types TPT and TST shall be permitted in lengths not exceeding 2.5 m (8 ft) where attached directly, or by means of a special type of plug, to a portable appliance rated at 50 watts or less and of such nature that extreme flexibility of the cord is essential.

3. Rubber-filled or varnished cambric tapes shall be permitted as a substitute for the inner braids.

4. Types G, G-GC, S, SC, SCE, SCT, SE, SEO, SEOO, SO, SOO, ST, STO, STOO, PPE, and W shall be permitted for use on theater stages, in garages, and elsewhere where flexible cords are permitted by this *Code*.

5. Elevator traveling cables for operating control and signal circuits shall contain nonmetallic fillers as necessary to maintain concentricity. Cables shall have steel supporting members as required for suspension by 620.41. In locations subject to excessive moisture or corrosive vapors or gases, supporting members of other materials shall be permitted. Where steel supporting members are used, they shall run straight through the center of the cable assembly and shall not be cabled with the copper strands of any conductor.

    In addition to conductors used for control and signaling circuits, Types E, EO, ETP, and ETT elevator cables shall be permitted to incorporate in the construction one or more 20 AWG telephone conductor pairs, one or more coaxial cables, or one or more optical fibers. The 20 AWG conductor pairs shall be permitted to be covered with suitable shielding for telephone, audio, or higher frequency communications circuits; the coaxial cables consist of a center conductor, insulation, and shield for use in video or other radio frequency communications circuits. The optical fiber shall be suitably covered with flame-retardant thermoplastic. The insulation of the conductors shall be rubber or thermoplastic of thickness not less than specified for the other conductors of the particular type of cable. Metallic shields shall have their own protective covering. Where used, these components shall be permitted to be incorporated in any layer of the cable assembly but shall not run straight through the center.

6. The third conductor in these cables shall be used for equipment grounding purpose only. The insulation of the grounding conductor for Types SPE-1, SPE-2, SPE-3, SPT-1, SPT-2, SPT-3, NISPT-1, NISPT-2, NISPE-1, and NISPE-2 shall be permitted to be thermoset polymer.

7. The individual conductors of all cords, except those of heat-resistant cords, shall have a thermoset or thermoplastic insulation, except that the equipment grounding conductor where used shall be in accordance with 400.23(B).

8. Where the voltage between any two conductors exceeds 300, but does not exceed 600, flexible cord of 10 AWG and smaller shall have thermoset or thermoplastic insulation on the individual conductors at least 1.14 mm (45 mils) in thickness, unless Type S, SE, SEO, SEOO, SO, SOO, ST, STO, or STOO cord is used.

9. Insulations and outer coverings that meet the requirements as flame retardant, limited smoke, and are so listed, shall be permitted to be marked for limited smoke after the code type designation.

10. Elevator cables in sizes 20 AWG through 14 AWG are rated 300 volts, and sizes 10 through 2 are rated 600 volts. 12 AWG is rated 300 volts with a 0.76-mm (30-mil) insulation thickness and 600 volts with a 1.14-mm (45-mil) insulation thickness.

11. Conductor size for Types EV, EVJ, EVE, EVJE, EVT, and EVJT cables apply to nonpower-limited circuits only. Conductors of power-limited (data, signal, or communications) circuits may extend beyond the stated AWG size range. All conductors shall be insulated for the same cable voltage rating.

12. Insulation thickness for Types EV, EVJ, EVEJE, EVT, and EVJT cables of nylon construction is indicated in parentheses.

13. Cords that comply with the requirements for outdoor cords and are so listed shall be permitted to be designated as weather and water resistant with the suffix "W" after the code type designation. Cords with the "W" suffix are suitable for use in wet locations.

Type G-GC cable was added to Table 400.4 in the 1999 *Code*. This cable is similar to Type G cable, except that it incorporates an insulated ground-check (GC) conductor. The ground-check conductor is used as part of a low-voltage circuit that monitors the grounding conductor continuity. Notes 11 and 12 to Table 400.4 coordinate these cable types with Article 625, Electric Vehicle Charging System.

Table 400.5(A) Allowable Ampacity for Flexible Cords and Cables [Based on Ambient Temperature of 30°C (86°F). See 400.13 and Table 400.4.]

Size (AWG)	Thermoplastic Types TPT, TST	Thermoset Types C, E, EO, PD, S, SJ, SJO, SJOW, SJOO, SJOOW, SO, SOW, SOO, SOOW, SP-1, SP-2, SP-3, SRD, SV, SVO, SVOO / Thermoplastic Types ET, ETLB, ETP, ETT, SE, SEW, SEO, SEOW, SEOOW, SJE, SJEW, SJEO, SJEOW, SJEOOW, SJT, SJTW, SJTO, SJTOW, SJTOO, SJTOOW, SPE-1, SPE-2, SPE-3, SPT-1, SPT-1W, SPT-2, SPT-2W, SPT-3, ST, SRDE, SRDT, STO, STOW, STOO, STOOW, SVE, SVEO, SVT, SVTO, SVTOO		Types HPD, HPN, HSJ, HSJO, HSJOO
		A+	B+	
27*	0.5	—	—	—
20	—	5**	***	—
18	—	7	10	10
17	—	—	12	13
16	—	10	13	15
15	—	—	—	17
14	—	15	18	20
12	—	20	25	30
10	—	25	30	35
8	—	35	40	—
6	—	45	55	—
4	—	60	70	—
2	—	80	95	—

*Tinsel cord.
**Elevator cables only.
***7 amperes for elevator cables only; 2 amperes for other types.
+The allowable currents under Column A apply to 3-conductor cords and other multiconductor cords connected to utilization equipment so that only 3 conductors are current-carrying. The allowable currents under Column B apply to 2-conductor cords and other multiconductor cords connected to utilization equipment so that only 2 conductors are current carrying.

the temperature correction factors from Table 310.16 that correspond to the temperature rating of the cord shall be applied to the ampacity from Table 400.5(B). Where the number of current-carrying conductors exceeds three, the allowable ampacity or the ampacity of each conductor shall be reduced from the 3-conductor rating as shown in Table 400.5.

Where power cables are used in an ambient temperature exceeding 30°C (86°F), a new requirement in the 2005 *Code* specifies that "correction factors" are to be applied to the Table 400.5(B) cable ampacities. This new provision parallels the Article 310 requirements for ampacity correction of single conductors used in elevated ambient temperatures. In fact, the ambient correction factors that are to be used for power cables are those specified in Table 310.16. The spe-

cific correction factor to be applied is predicated on the temperature rating of the power cable.

Example

Section 555.13 permits the use of portable power cable in applications where flexibility is necessary on floating piers, and in this application a 200-ampere feeder using Type W portable power cable is to be installed to distribution equipment on a floating pier. The specifications for the project indicate that the design ambient temperature is 95°F. The feeder is rated 208Y/120 volts, three-phase, four-wire and, due to the load characteristics, the grounded (neutral) conductor is to be counted as a current-carrying conductor per 400.5(B). The type W portable power cable has copper conductors and a 90°C temperature rating. With this information, the minimum ampacity for the feeder is determined as follows.

Table 400.5(B) Ampacity of Cable Types SC, SCE, SCT, PPE, G, G-GC, and W. [Based on Ambient Temperature of 30°C (86°F). See Table 400.4.]

Size (AWG or kcmil)	Temperature Rating of Cable								
	60°C (140°F)			75°C (167°F)			90°C (194°F)		
	D[1]	E[2]	F[3]	D[1]	E[2]	F[3]	D[1]	E[2]	F[3]
12	—	31	26	—	37	31	—	42	35
10	—	44	37	—	52	43	—	59	49
8	60	55	48	70	65	57	80	74	65
6	80	72	63	95	88	77	105	99	87
4	105	96	84	125	115	101	140	130	114
3	120	113	99	145	135	118	165	152	133
2	140	128	112	170	152	133	190	174	152
1	165	150	131	195	178	156	220	202	177
1/0	195	173	151	230	207	181	260	234	205
2/0	225	199	174	265	238	208	300	271	237
3/0	260	230	201	310	275	241	350	313	274
4/0	300	265	232	360	317	277	405	361	316
250	340	296	259	405	354	310	455	402	352
300	375	330	289	445	395	346	505	449	393
350	420	363	318	505	435	381	570	495	433
400	455	392	343	545	469	410	615	535	468
500	515	448	392	620	537	470	700	613	536
600	575	—	—	690	—	—	780	—	—
700	630	—	—	755	—	—	855	—	—
750	655	—	—	785	—	—	885	—	—
800	680	—	—	815	—	—	920	—	—
900	730	—	—	870	—	—	985	—	—
1000	780	—	—	935	—	—	1055	—	—

[1]The ampacities under subheading D shall be permitted for single-conductor Types SC, SCE, SCT, PPE, and W cable only where the individual conductors are not installed in raceways and are not in physical contact with each other except in lengths not to exceed 600 mm (24 in.) where passing through the wall of an enclosure.

[2]The ampacities under subheading E apply to two-conductor cables and other multiconductor cables connected to utilization equipment so that only two conductors are current carrying.

[3]The ampacities under subheading F apply to three-conductor cables and other multiconductor cables connected to utilization equipment so that only three conductors are current carrying.

Table 400.5 Adjustment Factors for More Than Three Current-Carrying Conductors in a Flexible Cord or Cable

Number of Conductors	Percent of Value in Tables 400.5(A) and 400.5(B)
4 – 6	80
7 – 9	70
10 – 20	50
21 – 30	45
31 – 40	40
41 and above	35

STEP 1. Determine minimum ampacity for conductors based on application of derating factors.

Calculated minimum feeder ampacity [includes 215.2(A)(1) and 215.3 increase in size for continuous load] = 200 amperes

Ampacity adjustment factor for 4 current-carrying conductors = 80% of "3 current-carrying conductor" column in Table 400.5.

Temperature correction factor for 95°F ambient = 0.96 from Table 310.16, 90°C column for copper conductors.

200/0.8/0.96 = 260.4 amperes   or 260 amperes

STEP 2. Select cable. From Table 400.5(B), a 3/0 AWG copper Type W portable power cable has an ampacity of 274 amperes from the "3 current-carrying conductor" column. In accordance with 110.14, the 90°C cable ampacity can be used where it is required to "derate" the cable ampacity. For this particular installation, both an ambient temperature correction factor and an ampacity adjustment factor for more than 3 current-carrying conductors in the cable have to be applied.

STEP 3. Termination analysis. Compliance with the termination temperature requirement of 110.14(C)(1)(b) is met with this 3/0 AWG cable, since its 75°C ampacity is 275 amperes. It should be noted that compliance with 110.14(C) is based on Table 310.16 conductor ampacities, and the ampacity from that table for a 75°C, 3/0 AWG copper conductor is 200 amperes. Therefore, this cable with 3/0 conductors operating at 200 amperes (noncontinuous plus continuous loads) is of sufficient physical size to dissipate the heat occurring at equipment terminations.

The important factor to remember is that the load (200 amperes minus the 25 percent increase for those loads considered to be continuous) supplied by this cable does not cause the cable to operate at an ampacity greater than that specified in Table 310.16 for a 75°C, 3/0 AWG copper conductor.

(B) Ultimate Insulation Temperature In no case shall conductors be associated together in such a way with respect to the kind of circuit, the wiring method used, or the number of conductors such that the limiting temperature of the conductors is exceeded.

A neutral conductor that carries only the unbalanced current from other conductors of the same circuit shall not be required to meet the requirements of a current-carrying conductor.

In a 3-wire circuit consisting of two phase wires and the neutral of a 4-wire, 3-phase, wye-connected system, a common conductor carries approximately the same current as the line-to-neutral currents of the other conductors and shall be considered to be a current-carrying conductor.

On a 4-wire, 3-phase, wye circuit where the major portion of the load consists of nonlinear loads, there are harmonic currents present in the neutral conductor and the neutral shall be considered to be a current-carrying conductor.

An equipment grounding conductor shall not be considered a current-carrying conductor.

Where a single conductor is used for both equipment grounding and to carry unbalanced current from other conductors, as provided for in 250.140 for electric ranges and electric clothes dryers, it shall not be considered as a current-carrying conductor.

Exception: For other loading conditions, adjustment factors shall be permitted to be calculated under 310.15(C).

FPN: See Annex B, Table B.310.11, for adjustment factors for more than three current-carrying conductors in a raceway or cable with load diversity.

400.6 Markings

(A) Standard Markings Flexible cords and cables shall be marked by means of a printed tag attached to the coil reel or carton. The tag shall contain the information required in 310.11(A). Types S, SC, SCE, SCT, SE, SEO, SEOO, SJ, SJE, SJEO, SJEOO, SJO, SJT, SJTO, SJTOO, SO, SOO, ST, STO, STOO, SEW, SEOW, SEOOW, SJEW, SJEOW, SJEOOW, SJOW, SJTW, SJTOW, SJTOOW, SOW, SOOW, STW, STOW, and STOOW flexible cords and G, G-GC,

PPE, and W flexible cables shall be durably marked on the surface at intervals not exceeding 610 mm (24 in.) with the type designation, size, and number of conductors.

(B) Optional Markings Flexible cords and cable types listed in Table 400.4 shall be permitted to be surface marked to indicate special characteristics of the cable materials. These markings include, but are not limited to, markings for limited smoke, sunlight resistance, and so forth.

The UL *Electrical Construction Materials Directory*, under the category Flexible Cord (ZJCZ), lists the following additional markings:

"Water Resistant" indicates the cord is suitable for immersion in water.

"For Mobile Home Use," "For Recreational Vehicle Use," or "For Mobile Home and Recreational Vehicle Use" followed by the current rating in amperes indicates suitability for use in mobile homes or recreational vehicles.

"Outdoor" or "W-A" indicates suitability for use outdoors. The minimum temperature rating for these cords is −40°C, unless otherwise marked on the cord.

"W" indicates suitability for use outdoors and for immersion in water. The low temperature rating for these cords is −40°C, unless otherwise marked on the cord with optional ratings of −50°C, −60°C, or −70°C. The low temperature ratings are determined by means of a bend test (not a suppleness test) at the given temperature.

"VW-1" indicates that the cord complies with a vertical flame test.

Cords that have been evaluated for leakage currents between the circuit conductor and the grounding conductor and between the circuit conductor and the outer surface of the jacket may have the leakage current values marked on the cable jacket.

400.7 Uses Permitted

(A) Uses Flexible cords and cables shall be used only for the following:

A reference to the provisions of 645.5 that allow the use of cords as data processing cables has been deleted from 400.7 in the 2005 *NEC*. In accordance with 90.3, the structure of the *Code* permits requirements in Chapters 5, 6, and 7 to amend or modify requirements in Chapters 1 through 4. In this particular case, 645.5 modifies 400.7 with specific provisions for the use of flexible cord in the space beneath the raised floor of an information technology equipment room.

(1) Pendants
(2) Wiring of luminaires (fixtures)
(3) Connection of portable lamps, portable and mobile signs, or appliances
(4) Elevator cables
(5) Wiring of cranes and hoists
(6) Connection of utilization equipment to facilitate frequent interchange
(7) Prevention of the transmission of noise or vibration
(8) Appliances where the fastening means and mechanical connections are specifically designed to permit ready removal for maintenance and repair, and the appliance is intended or identified for flexible cord connection
(9) Connection of moving parts
(10) Where specifically permitted elsewhere in this *Code*

(B) Attachment Plugs Where used as permitted in 400.7(A)(3), (A)(6), and (A)(8), each flexible cord shall be equipped with an attachment plug and shall be energized from a receptacle outlet.

Exception: As permitted in 368.56.

Flexible cords are permitted to be hard-wired into a junction box if the cord is used for the following:

1. Luminaires and fixtures mentioned in 400.7(A)
2. Supplies to pendant pushbutton stations for cranes
3. Portable lamp (droplight) connections

400.8 Uses Not Permitted

Unless specifically permitted in 400.7, flexible cords and cables shall not be used for the following:

(1) As a substitute for the fixed wiring of a structure
(2) Where run through holes in walls, structural ceilings, suspended ceilings, dropped ceilings, or floors
(3) Where run through doorways, windows, or similar openings
(4) Where attached to building surfaces

Exception to (4): Flexible cord and cable shall be permitted to be attached to building surfaces in accordance with the provisions of 368.56(B)

Section 368.56(B) provides the requirements for the installation of flexible cords installed as branches from busways.

(5) Where concealed by walls, floors, or ceilings or located above suspended or dropped ceilings

(6) Where installed in raceways, except as otherwise permitted in this *Code*

The flexible cords and cables referred to in Article 400 are not limited to use with portable equipment. They may not be used, however, as a substitute for the fixed wiring of a structure or where concealed behind building walls, floors, or ceilings (including structural, suspended, or dropped-type ceilings). See 240.5, 590.4(B), and 590.4(C) for the uses of multiconductor flexible cords for feeder and branch-circuit installations and for overcurrent protection requirements for flexible cord. See 410.30 for cord-connected luminaires.

(7) Where subject to physical damage

400.9 Splices

Flexible cord shall be used only in continuous lengths without splice or tap where initially installed in applications permitted by 400.7(A). The repair of hard-service cord and junior hard-service cord (see Trade Name column in Table 400.4) 14 AWG and larger shall be permitted if conductors are spliced in accordance with 110.14(B) and the completed splice retains the insulation, outer sheath properties, and usage characteristics of the cord being spliced.

The requirements of 400.9 are intended to ensure that flexible cords and cables first installed under any of the uses permitted in 400.7(A)(1) through 400.7(A)(10) are in their original or near-original condition. Damage to a cord can occur under the sometimes extreme conditions of use to which the cord is subjected. The provisions of this section permit repair of a cord in such a manner that the cord will retain its original operating and use integrity. However, if the repaired cord is reused or reinstalled at a new location, the in-line repair is no longer permitted, and the cord can be used only in lengths that do not contain a splice.

400.10 Pull at Joints and Terminals

Flexible cords and cables shall be connected to devices and to fittings so that tension is not transmitted to joints or terminals.

Exception: Listed portable single-pole devices that are intended to accommodate such tension at their terminals shall be permitted to be used with single-conductor flexible cable.

> FPN: Some methods of preventing pull on a cord from being transmitted to joints or terminals are knotting the cord, winding with tape, and fittings designed for the purpose.

400.11 In Show Windows and Show Cases

Flexible cords used in show windows and show cases shall be Type S, SE, SEO, SEOO, SJ, SJE, SJEO, SJEOO, SJO,

SJOO, SJT, SJTO, SJTOO, SO, SOO, ST, STO, STOO, SEW, SEOW, SEOOW, SJEW, SJEOW, SJEOOW, SJOW, SJOOW, SJTW, SJTOW, SJTOOW, SOW, SOOW, STW, STOW, or STOOW.

Exception No. 1: For the wiring of chain-supported luminaires (lighting fixtures).

Exception No. 2: As supply cords for portable lamps and other merchandise being displayed or exhibited.

Flexible cords listed for hard usage or extra-hard usage should be used in show windows and show cases because such cords may come in contact with combustible materials, such as fabrics or paper products, usually present at these locations and because they are exposed to wear and tear from continual housekeeping and display changes. Flexible cords used in show windows and show cases should be maintained in good condition.

400.12 Minimum Size

The individual conductors of a flexible cord or cable shall not be smaller than the sizes in Table 400.4.

Added in the 1999 *Code*, the exception to 400.12 correlates with cable Type G-GC, which was added to Table 400.4 in 1999. This cable is similar to Type G, except that it incorporates an insulated ground-check (GC) conductor. The ground-check conductor is used as part of a low-voltage circuit that monitors the grounding conductor continuity.

Exception: The size of the insulated ground-check conductor of Type G-GC cables shall be not smaller than 10 AWG.

400.13 Overcurrent Protection

Flexible cords not smaller than 18 AWG, and tinsel cords or cords having equivalent characteristics of smaller size approved for use with specific appliances, shall be considered as protected against overcurrent by the overcurrent devices described in 240.5.

400.14 Protection from Damage

Flexible cords and cables shall be protected by bushings or fittings where passing through holes in covers, outlet boxes, or similar enclosures.

A variety of bushings and fittings are available for protecting flexible cords and cables, both insulated and noninsulated. Some bushings or fittings include pull-relief means, as required in 400.10. Many insulating bushings are listed by Underwriters Laboratories Inc. in the following product categories:

1. Conduit fittings (bushings and fittings for use on the ends of conduit in boxes and gutters)
2. Insulating devices and materials
3. Bushings (for the protection of cords where they pass through walls or barriers of metal)
4. Outlet bushings and fittings (for use on the ends of conduit, electrical metallic tubing, or armored cable where a change to open wiring is made)

Section 400.14 allows flexible cords and cables installed in an industrial establishment. A second paragraph was added to the 2005 *Code* to permit flexible cords and cables to be installed in aboveground raceways where the length of the raceway does not exceed 50 ft and the conductors are protected at their ampacity after application of the adjustment factors for more than three current-carrying conductors as specified in Table 400.5.

In industrial establishments where the conditions of maintenance and supervision ensure that only qualified persons service the installation, flexible cords and cables shall be permitted to be installed in aboveground raceways that are no longer than 15 m (50 ft) to protect the flexible cord or cable from physical damage. Where more than three current-carrying conductors are installed within the raceway, the allowable ampacity shall be reduced in accordance with Table 400.5.

II. Construction Specifications

400.20 Labels

Flexible cords shall be examined and tested at the factory and labeled before shipment.

See the definition of *labeled* in Article 100.

400.21 Nominal Insulation Thickness

The nominal thickness of insulation for conductors of flexible cords and cables shall not be less than specified in Table 400.4.

Exception: The nominal insulation thickness for the ground-check conductors of Type G-GC cables shall not be less than 1.14 mm (45 mils) for 8 AWG and not less than 0.76 mm (30 mils) for 10 AWG.

400.22 Grounded-Conductor Identification

One conductor of flexible cords that is intended to be used as a grounded circuit conductor shall have a continuous marker that readily distinguishes it from the other conductor or conductors. The identification shall consist of one of the methods indicated in 400.22(A) through 400.22(F).

(A) Colored Braid A braid finished to show a white or gray color and the braid on the other conductor or conductors finished to show a readily distinguishable solid color or colors.

(B) Tracer in Braid A tracer in a braid of any color contrasting with that of the braid and no tracer in the braid of the other conductor or conductors. No tracer shall be used in the braid of any conductor of a flexible cord that contains a conductor having a braid finished to show white or gray.

Exception: In the case of Types C and PD and cords having the braids on the individual conductors finished to show white or gray. In such cords, the identifying marker shall be permitted to consist of the solid white or gray finish on one conductor, provided there is a colored tracer in the braid of each other conductor.

(C) Colored Insulation A white or gray insulation on one conductor and insulation of a readily distinguishable color or colors on the other conductor or conductors for cords having no braids on the individual conductors.

For jacketed cords furnished with appliances, one conductor having its insulation colored light blue, with the other conductors having their insulation of a readily distinguishable color other than white or gray.

Exception: Cords that have insulation on the individual conductors integral with the jacket.

The insulation shall be permitted to be covered with an outer finish to provide the desired color.

(D) Colored Separator A white or gray separator on one conductor and a separator of a readily distinguishable solid color on the other conductor or conductors of cords having insulation on the individual conductors integral with the jacket.

Since the 2002 edition of the *Code*, identification of grounded conductors in flexible cords and cables can be accomplished through the use of a white- or gray-colored braid, a white- or gray-colored tracer in the braid, white- or gray-colored insulation, or a white- or gray-colored separator. In existing installations where a gray-colored braid, tracer, or conductor insulation is encountered, caution should be exercised because gray may have been used as means to identify ungrounded conductors.

(E) Tinned Conductors One conductor having the individual strands tinned and the other conductor or conductors having the individual strands untinned for cords having insulation on the individual conductors integral with the jacket.

(F) Surface Marking One or more ridges, grooves, or white stripes located on the exterior of the cord so as to

identify one conductor for cords having insulation on the individual conductors integral with the jacket.

An additional method is now permitted to identify the grounded conductor in a cord where the insulation of the conductor is part of the molded jacket and is not separable. A white stripe on the exterior of the cord serves to identify the segment of the cord that contains the grounded conductor. An example of a type cord with these characteristics is zip cord, which is commonly used for floor lamps and table lamps. It is important that the grounded conductor be easily identified where zip cord is used for wiring lampholders, because the *Code* requires the grounded conductor to be connected to the device terminal that connects to the screw-shell of the lampholder.

400.23 Equipment Grounding Conductor Identification

A conductor intended to be used as an equipment grounding conductor shall have a continuous identifying marker readily distinguishing it from the other conductor or conductors. Conductors having a continuous green color or a continuous green color with one or more yellow stripes shall not be used for other than equipment grounding purposes. The identifying marker shall consist of one of the methods in 400.23(A) or 400.23(B).

(A) Colored Braid A braid finished to show a continuous green color or a continuous green color with one or more yellow stripes.

(B) Colored Insulation or Covering For cords having no braids on the individual conductors, an insulation of a continuous green color or a continuous green color with one or more yellow stripes.

400.24 Attachment Plugs

Where a flexible cord is provided with an equipment grounding conductor and equipped with an attachment plug, the attachment plug shall comply with 250.138(A) and 250.138(B).

III. Portable Cables Over 600 Volts, Nominal

400.30 Scope

This part applies to multiconductor portable cables used to connect mobile equipment and machinery.

400.31 Construction

(A) Conductors The conductors shall be 8 AWG copper or larger and shall employ flexible stranding.

Exception: The size of the insulated ground-check conductor of Type G-GC cables shall be not smaller than 10 AWG.

(B) Shields Cables operated at over 2000 volts shall be shielded. Shielding shall be for the purpose of confining the voltage stresses to the insulation.

(C) Equipment Grounding Conductor(s) An equipment grounding conductor(s) shall be provided. The total area shall not be less than that of the size of the equipment grounding conductor required in 250.122.

400.32 Shielding

All shields shall be grounded.

400.33 Grounding

Grounding conductors shall be connected in accordance with Part V of Article 250.

400.34 Minimum Bending Radii

The minimum bending radii for portable cables during installation and handling in service shall be adequate to prevent damage to the cable.

400.35 Fittings

Connectors used to connect lengths of cable in a run shall be of a type that lock firmly together. Provisions shall be made to prevent opening or closing these connectors while energized. Suitable means shall be used to eliminate tension at connectors and terminations.

400.36 Splices and Terminations

Portable cables shall not contain splices unless the splices are of the permanent molded, vulcanized types in accordance with 110.14(B). Terminations on portable cables rated over 600 volts, nominal, shall be accessible only to authorized and qualified personnel.

ARTICLE 402
Fixture Wires

Summary of Changes

• **402.8:** Revised to clarify use of continuous white stripes to identify the grounded conductor.

Contents

402.1 Scope

This article covers general requirements and construction specifications for fixture wires.

402.2 Other Articles

Fixture wires shall comply with this article and also with the applicable provisions of other articles of this *Code*.

> FPN: For application in luminaires (lighting fixtures), see Article 410.

402.3 Types

Fixture wires shall be of a type listed in Table 402.3, and they shall comply with all requirements of that table. The fixture wires listed in Table 402.3 are all suitable for service at 600 volts, nominal, unless otherwise specified.

> FPN: Thermoplastic insulation may stiffen at temperatures colder than −10°C (+14°F), requiring that care be exercised during installation at such temperatures. Thermoplastic insulation may also be deformed at normal temperatures where subjected to pressure, requiring that care be exercised during installation and at points of support.

402.5 Allowable Ampacities for Fixture Wires

The allowable ampacity of fixture wire shall be as specified in Table 402.5.

No conductor shall be used under such conditions that its operating temperature exceeds the temperature specified in Table 402.3 for the type of insulation involved.

> FPN: See 310.10 for temperature limitation of conductors.

402.6 Minimum Size

Fixture wires shall not be smaller than 18 AWG.

402.7 Number of Conductors in Conduit or Tubing

The number of fixture wires permitted in a single conduit or tubing shall not exceed the percentage fill specified in Table 1, Chapter 9.

Table 402.3 Fixture Wires

Name	Type Letter	Insulation	Thickness of Insulation			Outer Covering	Maximum Operating Temperature	Application Provisions
			AWG	mm	mils			
Heat-resistant rubber-covered fixture wire — flexible stranding	FFH-2	Heat-resistant rubber	18–16	0.76	30	Nonmetallic covering	75°C 167°F	Fixture wiring
		Cross-linked synthetic polymer	18–16	0.76	30			
ECTFE — solid or 7-strand	HF	Ethylene chlorotrifluoroethylene	18–14	0.38	15	None	150°C 302°F	Fixture wiring
ECTFE — flexible stranding	HFF	Ethylene chlorotrifluoroethylene	18–14	0.38	15	None	150°C 302°F	Fixture wiring
Tape insulated fixture wire — solid or 7-strand	KF-1	Aromatic polyimide tape	18–10	0.14	5.5	None	200°C 392°F	Fixture wiring — limited to 300 volts
	KF-2	Aromatic polyimide tape	18–10	0.21	8.4	None	200°C 392°F	Fixture wiring
Tape insulated fixture wire — flexible stranding	KFF-1	Aromatic polyimide tape	18–10	0.14	5.5	None	200°C 392°F	Fixture wiring — limited to 300 volts
	KFF-2	Aromatic polyimide tape	18–10	0.21	8.4	None	200°C 392°F	Fixture wiring
Perfluoro-alkoxy — solid or 7-strand (nickel or nickel-coated copper)	PAF	Perfluoro-alkoxy	18–14	0.51	20	None	250°C 482°F	Fixture wiring (nickel or nickel-coated copper)
Perfluoro-alkoxy — flexible stranding	PAFF	Perfluoro-alkoxy	18–14	0.51	20	None	150°C 302°F	Fixture wiring
Fluorinated ethylene propylene fixture wire — solid or 7-strand	PF	Fluorinated ethylene propylene	18–14	0.51	20	None	200°C 392°F	Fixture wiring
Fluorinated ethylene propylene fixture wire — flexible stranding	PFF	Fluorinated ethylene propylene	18–14	0.51	20	None	150°C 302°F	Fixture wiring
Fluorinated ethylene propylene fixture wire — solid or 7-strand	PGF	Fluorinated ethylene propylene	18–14	0.36	14	Glass braid	200°C 392°F	Fixture wiring
Fluorinated ethylene propylene fixture wire — flexible stranding	PGFF	Fluorinated ethylene propylene	18–14	0.36	14	Glass braid	150°C 302°F	Fixture wiring
Extruded polytetrafluoroethylene — solid or 7-strand (nickel or nickel-coated copper)	PTF	Extruded polytetrafluoroethylene	18–14	0.51	20	None	250°C 482°F	Fixture wiring (nickel or nickel-coated copper)

Table 402.3 *Continued*

Name	Type Letter	Insulation	Thickness of Insulation			Outer Covering	Maximum Operating Temperature	Application Provisions
			AWG	mm	mils			
Extruded polytetrafluoroethylene — flexible stranding 26-36 (AWG silver or nickel-coated copper)	PTFF	Extruded polytetrafluoroethylene	18–14	0.51	20	None	150°C 302°F	Fixture wiring (silver or nickel-coated copper)
Heat-resistant rubber-covered fixture wire — solid or 7-strand	RFH-1	Heat-resistant rubber	18	0.38	15	Nonmetallic covering	75°C 167°F	Fixture wiring — limited to 300 volts
	RFH-2	Heat-resistant rubber Cross-linked synthetic polymer	18–16	0.76	30	None or non-metallic covering	75°C 167°F	Fixture wiring
Heat-resistant cross-linked synthetic polymer-insulated fixture wire — solid or 7-strand	RFHH-2*	Cross-linked synthetic polymer	18–16	0.76	30	None or non-metallic covering	90°C 194°F	Fixture wiring
	RFHH-3*		18–16	1.14	45			
Silicone insulated fixture wire — solid or 7-strand	SF-1	Silicone rubber	18	0.38	15	Nonmetallic covering	200°C 392°F	Fixture wiring — limited to 300 volts
	SF-2	Silicone rubber	18–12 10	0.76 1.14	30 45	Nonmetallic covering	200°C 392°F	Fixture wiring
Silicone insulated fixture wire — flexible stranding	SFF-1	Silicone rubber	18	0.38	15	Nonmetallic covering	150°C 302°F	Fixture wiring — limited to 300 volts
	SFF-2	Silicone rubber	18–12 10	0.76 1.14	30 45	Nonmetallic covering	150°C 302°F	Fixture wiring
Thermoplastic covered fixture wire — solid or 7-strand	TF*	Thermoplastic	18–16	0.76	30	None	60°C 140°F	Fixture wiring
Thermoplastic covered fixture wire — flexible stranding	TFF*	Thermoplastic	18–16	0.76	30	None	60°C 140°F	Fixture wiring
Heat-resistant thermoplastic covered fixture wire — solid or 7-strand	TFN*	Thermoplastic	18–16	0.38	15	Nylon-jacketed or equivalent	90°C 194°F	Fixture wiring
Heat-resistant thermoplastic covered fixture wire — flexible stranded	TFFN*	Thermoplastic	18–16	0.38	15	Nylon-jacketed or equivalent	90°C 194°F	Fixture wiring
Cross-linked polyolefin insulated fixture wire — solid or 7-strand	XF*	Cross-linked polyolefin	18–14 12-10	0.76 1.14	30 45	None	150°C 302°F	Fixture wiring — limited to 300 volts

(continues)

Table 402.3 *Continued*

Name	Type Letter	Insulation	Thickness of Insulation			Outer Covering	Maximum Operating Temperature	Application Provisions
			AWG	mm	mils			
Cross-linked polyolefin insulated fixture wire — flexible stranded	XFF*	Cross-linked polyolefin	18–14 12–10	0.76 1.14	30 45	None	150°C 302°F	Fixture wiring — limited to 300 volts
Modified ETFE — solid or 7-strand	ZF	Modified ethylene tetrafluoro-ethylene	18–14	0.38	15	None	150°C 302°F	Fixture wiring
Flexible stranding	ZFF	Modified ethylene tetrafluoro-ethylene	18–14	0.38	15	None	150°C 302°F	Fixture wiring
High temp. modified ETFE — solid or 7-strand	ZHF	Modified ethylene tetrafluoro-ethylene	18–14	0.38	15	None	200°C 392°F	Fixture wiring

*Insulations and outer coverings that meet the requirements of flame retardant, limited smoke, and are so listed shall be permitted to be marked for limited smoke after the *Code* type designation.

Table 402.5 Allowable Ampacity for Fixture Wires

Size (AWG)	Allowable Ampacity
18	6
16	8
14	17
12	23
10	28

Table 1 of Chapter 9 specifies the maximum percent fill of a conduit or tubing. Table 4 of Chapter 9 provides the usable area within the selected conduit or tubing, and Table 5 of Chapter 9 provides the required area for each of the conductors. Examples using these tables to calculate a conduit or tubing size are provided according to Chapter 9, Table 1, Note 6. The following examples show how to determine the minimum size conduit where the conductors are different sizes and how to select the minimum size conduit directly from the tables in Annex C where the conductors are all the same size.

Example 1

A remote ballast installation requires a single flexible metal conduit to contain fourteen 16 AWG TFFN fixture wires and three 12 AWG THHN conductors. What size flexible metal conduit will be required?

Solution

The solution is found by using Table 1 of Chapter 9 and the accompanying Note 6 following Table 1. Table 1 sets the maximum percentage of conduit and tubing fill based on the internal cross-sectional area of the raceway in question. Note 6 states, "For [calculating] combinations of conductors of different sizes, use Table 5 and Table 5A for dimensions of conductors and Table 4 for the applicable conduit or tubing dimensions."

STEP 1. Using Table 1, look up the maximum percent of cross section of conduit permitted for conductors. Table 1 sets the limit of conductor fill for over two conductors at 40 percent of the total cross-sectional area of the raceway.

STEP 2. Look up the individual conductor cross-sectional areas in Chapter 9, Table 5.

$$16 \text{ AWG TFFN} = 0.0072 \text{ in.}^2$$

$$12 \text{ AWG THHN} = 0.0133 \text{ in.}^2$$

STEP 3. Calculate the total area occupied by the wires as follows:

$$\text{Fourteen 16 AWG TFFN} \times 0.0072 = 0.1008 \text{ in.}^2$$

$$\text{Three 12 AWG THHN} \times 0.0133 = 0.0399 \text{ in.}^2$$

$$\text{Total area} = \overline{0.1407 \text{ in.}^2}$$

STEP 4. Using the 40 percent column of Table 4, Article 348 – Flexible Metal Conduit, look up the appropriate flexible metal conduit size based on 40 percent fill and a total conductor area fill of 0.1407 in.2 Because 0.1407 in.2 is greater than 0.127 and less than 0.213, select trade size ¾ flexible metal conduit.

If the conductors in a raceway or tubing are all of the same wire size, the Annex C tables may be used instead of doing the calculations. The following example uses Annex C tables to determine electrical metallic tubing size.

Example 2

A fire alarm system requires a riser to contain thirty-six 16 AWG TFFN conductors. What size electrical metallic tubing will be required?

Solution

First, in Annex C, Table C1, find TFFN insulation in the first column. Next, find 16 AWG in the second column. Proceed across the table until the desired number of conductors is equal to or less than the number shown in the table for the respective conduit and tubing sizes. Using this method, a 1-in. EMT is required.

402.8 Grounded Conductor Identification

Fixture wires that are intended to be used as grounded conductors shall be identified by one or more continuous white stripes on other than green insulation or by the means described in 400.22(A) through 400.22(E).

A new rule similar to that required for flexible cords and cable was added to the 2005 *Code* to ensure that the grounded conductor in fixture wires is easily recognized. Because it is necessary that the grounded conductor be connected to the screwshell of lampholders, it must be easily recognized.

402.9 Marking

(A) Method of Marking Thermoplastic insulated fixture wire shall be durably marked on the surface at intervals not exceeding 610 mm (24 in.). All other fixture wire shall be marked by means of a printed tag attached to the coil, reel, or carton.

(B) Optional Marking Fixture wire types listed in Table 402.3 shall be permitted to be surface marked to indicate special characteristics of the cable materials. These markings include, but are not limited to, markings for limited smoke, sunlight resistance, and so forth.

402.10 Uses Permitted

Fixture wires shall be permitted (1) for installation in luminaires (lighting fixtures) and in similar equipment where enclosed or protected and not subject to bending or twisting in use, or (2) for connecting luminaires (lighting fixtures) to the branch-circuit conductors supplying the luminaires (fixtures).

Fixture wire is permitted to be used as a tap conductor to connect a luminaire(s) to the branch circuit conductors. There must be a transition from the branch circuit wiring

method to the fixture wire tap conductors. This can be done in a junction box or other fitting that is allowed to contain splices.

402.11 Uses Not Permitted

Fixture wires shall not be used as branch-circuit conductors.

402.12 Overcurrent Protection

Overcurrent protection for fixture wires shall be as specified in 240.5.

ARTICLE 404
Switches

Summary of Changes

- **404.6(A) and (B):** Changed terminology to reduce confusion from a lock-out device to a means built in the switch to hold the blades in the open position when set.
- **404.8(B):** Revised to require the separating barriers to be identified.
- **404.9(B):** Added the option of use of GFCI in lieu of grounding of a nonconductive faceplate.

Contents

I. Installation

404.1 Scope

The provisions of this article shall apply to all switches, switching devices, and circuit breakers where used as switches.

As part of the Chapter 3 reorganization in the 2002 *Code*, several articles were relocated to new chapters. Article 404 was formerly Article 380. Chapter 3 contains articles with requirements on wiring methods, and Chapter 4 contains requirements on equipment for general use. For a complete matrix on the reorganization of Chapter 3, see Annex F.

404.2 Switch Connections

(A) Three-Way and Four-Way Switches Three-way and four-way switches shall be wired so that all switching is done only in the ungrounded circuit conductor. Where in metal raceways or metal-armored cables, wiring between switches and outlets shall be in accordance with 300.20(A).

Exception: Switch loops shall not require a grounded conductor.

The *NEC* does not specifically prohibit the use of two 2-conductor nonmetallic-sheathed cables instead of a single

3-conductor cable for wiring three-way and four-way switches. However, using two 2-conductor cables could easily result in a violation of 300.20 if metal boxes are used and the cables enter the box through separate knockouts. Also, use of the same clamp or section of a clamp for both cables would, in most cases, be in violation of 110.3(B), because clamps have been tested for only one cable per clamp or section of clamp.

The grounded conductor is not needed in a switch loop [see 300.20(A)] because the ungrounded conductor both enters and leaves the enclosure in the same cable or raceway, thus avoiding inductive heating.

(B) Grounded Conductors Switches or circuit breakers shall not disconnect the grounded conductor of a circuit.

Exception: A switch or circuit breaker shall be permitted to disconnect a grounded circuit conductor where all circuit conductors are disconnected simultaneously, or where the device is arranged so that the grounded conductor cannot be disconnected until all the ungrounded conductors of the circuit have been disconnected.

404.3 Enclosure

(A) General Switches and circuit breakers shall be of the externally operable type mounted in an enclosure listed for the intended use. The minimum wire-bending space at terminals and minimum gutter space provided in switch enclosures shall be as required in 312.6.

Exception No. 1: Pendant- and surface-type snap switches and knife switches mounted on an open-face switchboard or panelboard shall be permitted without enclosures.

Exception No. 2: Switches and circuit breakers installed in accordance with 110.27(A)(1), (A)(2), (A)(3), or (A)(4) shall be permitted without enclosures.

Exception No. 2 to 404.3 recognizes the variety of means allowed by 110.27(A) for the guarding of live parts. Switches and circuit breakers guarded by these means are permitted without enclosures.

(B) Used as a Raceway Enclosures shall not be used as junction boxes, auxiliary gutters, or raceways for conductors feeding through or tapping off to other switches or overcurrent devices, unless the enclosure complies with 312.8.

404.4 Wet Locations

A switch or circuit breaker in a wet location or outside of a building shall be enclosed in a weatherproof enclosure or cabinet that shall comply with 312.2(A). Switches shall not be installed within wet locations in tub or shower spaces unless installed as part of a listed tub or shower assembly.

404.5 Time Switches, Flashers, and Similar Devices

Time switches, flashers, and similar devices shall be of the enclosed type or shall be mounted in cabinets or boxes or equipment enclosures. Energized parts shall be barriered to prevent operator exposure when making manual adjustments or switching.

Exception: Devices mounted so they are accessible only to qualified persons shall be permitted without barriers, provided they are located within an enclosure such that any energized parts within 152 mm (6.0 in.) of the manual adjustment or switch are covered by suitable barriers.

404.6 Position and Connection of Switches

(A) Single-Throw Knife Switches Single-throw knife switches shall be placed so that gravity will not tend to close them. Single-throw knife switches, approved for use in the inverted position, shall be provided with an integral mechanical means that ensures that the blades remain in the open position when so set.

(B) Double-Throw Knife Switches Double-throw knife switches shall be permitted to be mounted so that the throw is either vertical or horizontal. Where the throw is vertical, integral mechanical means shall be provided to hold the blades in the open position when so set.

A revision in the 2005 *Code* to 404.6(A) and 404.6(B) clarifies that an integral "mechanical means" that does not necessarily have to be a "locking device" is required for single-throw and double-throw switches to ensure the switch blades remain disengaged regardless of their orientation when the switch is in the "off" (open) position. New switch designs incorporate mechanical means other than a catch or a latch to ensure that the blades cannot accidentally close from the "off" position.

(C) Connection of Switches Single-throw knife switches and switches with butt contacts shall be connected such that their blades are de-energized when the switch is in the open position. Bolted pressure contact switches shall have barriers that prevent inadvertent contact with energized blades. Single-throw knife switches, bolted pressure contact switches, molded case switches, switches with butt contacts, and circuit breakers used as switches shall be connected so that the terminals supplying the load are de-energized when the switch is in the open position.

New for the 2002 *Code*, bolted pressure switches that have energized blades when open, such as bottom feed designs,

must be provided with barriers or a means to guard against inadvertent contact with the energized blades. This requirement is intended to provide protection against accidental contact with live parts in those cases where personnel are working on energized equipment.

Exception: The blades and terminals supplying the load of a switch shall be permitted to be energized when the switch is in the open position where the switch is connected to circuits or equipment inherently capable of providing a back-feed source of power. For such installations, a permanent sign shall be installed on the switch enclosure or immediately adjacent to open switches with the following words or equivalent: WARNING — LOAD SIDE TERMINALS MAY BE ENERGIZED BY BACKFEED.

Batteries, generators, and double-ended switchboard ties are typical backfeed sources. These sources may cause the load side of the switch or circuit breaker to be energized when it is in the open position, a condition inherent to the circuitry.

404.7 Indicating

General-use and motor-circuit switches, circuit breakers, and molded case switches, where mounted in an enclosure as described in 404.3, shall clearly indicate whether they are in the open (off) or closed (on) position.

Where these switch or circuit breaker handles are operated vertically rather than rotationally or horizontally, the up position of the handle shall be the (on) position.

Exception No. 1: Vertically operated double-throw switches shall be permitted to be in the closed (on) position with the handle in either the up or down position.

Exception No. 2: On busway installations, tap switches employing a center-pivoting handle shall be permitted to be open or closed with either end of the handle in the up or down position. The switch position shall be clearly indicating and shall be visible from the floor or from the usual point of operation.

Exception No. 2 was added to the 2005 *Code* to clarify the off and on positions and the operation to turn the switch off. Some busway switches are designed with a center pivot, so that at any time, one end of the switch handle is in the up position and the other side is down. This allows the switch to be pulled down to turn it off and also pulled down to turn it on, a configuration not in accordance with the requirement in the main rule. The added exception allows this time-proven method of operating busway switches.

404.8 Accessibility and Grouping

(A) **Location** All switches and circuit breakers used as switches shall be located so that they may be operated from a readily accessible place. They shall be installed such that the center of the grip of the operating handle of the switch or circuit breaker, when in its highest position, is not more than 2.0 m (6 ft 7 in.) above the floor or working platform.

Exception No. 1: On busway installations, fused switches and circuit breakers shall be permitted to be located at the same level as the busway. Suitable means shall be provided to operate the handle of the device from the floor.

Exception No. 2: Switches and circuit breakers installed adjacent to motors, appliances, or other equipment that they supply shall be permitted to be located higher than 2.0 m (6 ft 7 in.) and to be accessible by portable means.

Exception No. 3: Hookstick operable isolating switches shall be permitted at greater heights.

(B) **Voltage Between Adjacent Devices** A snap switch shall not be grouped or ganged in enclosures with other snap switches, receptacles, or similar devices, unless they are arranged so that the voltage between adjacent devices does not exceed 300 volts, or unless they are installed in enclosures equipped with identified, securely installed barriers between adjacent devices.

Barriers are required between switches that are ganged in a box and used to control 277-volt lighting on 480Y/277-volt systems where two or more phase conductors enter the box. Permanent barriers would be required between devices fed from two different phases of this system because the voltage between the phase conductors would be 480 volts, nominal, and would exceed the 300-volt limit. Barriers are required even if one device space is left empty because the two remaining devices fed from different phase conductors still would be adjacent to each other. This requirement now applies to switches ganged together with any wiring device where the voltage between adjacent conductors exceeds 300 volts.

404.9 Provisions for General-Use Snap Switches

(A) **Faceplates** Faceplates provided for snap switches mounted in boxes and other enclosures shall be installed so as to completely cover the opening and, where the switch is flush mounted, seat against the finished surface.

(B) **Grounding** Snap switches, including dimmer and similar control switches, shall be effectively grounded and shall provide a means to ground metal faceplates, whether or not a metal faceplate is installed. Snap switches shall be considered effectively grounded if either of the following conditions is met:

(1) The switch is mounted with metal screws to a metal box or to a nonmetallic box with integral means for grounding devices.
(2) An equipment grounding conductor or equipment bonding jumper is connected to an equipment grounding termination of the snap switch.

Exception to (B): Where no grounding means exists within the snap-switch enclosure or where the wiring method does not include or provide an equipment ground, a snap switch without a grounding connection shall be permitted for replacement purposes only. A snap switch wired under the provisions of this exception and located within reach of earth, grade, conducting floors, or other conducting surfaces shall be provided with a faceplate of nonconducting, noncombustible material or shall be protected by a ground-fault circuit interrupter.

The exception to 404.9(B) was expanded in the 2005 *Code* to require GFCI protection for the circuits terminating in an ungrounded device box where a metal faceplate is installed and it can be touched by persons in contact with a grounded surface. This exception provides additional safety to persons where older electrical installations did not provide an equipment grounding conductor in the wiring method.

The provisions of 404.9(B) specify that switching devices, including snap switches, dimmers, and similar control devices, must be grounded. Although the non–current-carrying metal parts of these devices typically are not subject to contact by personnel, there is concern about the use of metal faceplates, which do pose a shock hazard if they become energized. Therefore, the switch must provide a means for connection of an equipment grounding conductor to ground the metal faceplate if one is installed.

The requirements in 404.9(B)(1) or 404.9(B)(2) describe the provisions to satisfy the main requirement. Switch plates in existing installations attached to switches in boxes without an equipment grounding conductor must be made of insulating material. See Exhibit 404.1, following the commentary in 404.12, for an example of the typical method by which a metal faceplate is grounded. Section 404.9(B)(2) was revised to clarify that the metal yoke of the device, where in direct contact with a grounded metal device box, serves to adequately ground the yoke, thereby grounding the faceplate.

(C) **Construction** Metal faceplates shall be of ferrous metal not less than 0.76 mm (0.030 in.) in thickness or of nonferrous metal not less than 1.02 mm (0.040 in.) in thickness. Faceplates of insulating material shall be noncombustible and not less than 2.54 mm (0.010 in.) in thickness, but they shall be permitted to be less than 2.54 mm (0.010 in.) in thickness if formed or reinforced to provide adequate mechanical strength.

NM cable

Nonmetallic box

Off

Equipment grounding conductor

Metal screws

Metal faceplate

Exhibit 404.1 Grounding of a metal faceplate through attachment to the grounded yoke of a snap switch.

404.10 Mounting of Snap Switches

(A) Surface-Type Snap switches used with open wiring on insulators shall be mounted on insulating material that separates the conductors at least 13 mm (½ in.) from the surface wired over.

(B) Box Mounted Flush-type snap switches mounted in boxes that are set back of the finished surface as permitted in 314.20 shall be installed so that the extension plaster ears are seated against the surface. Flush-type snap switches mounted in boxes that are flush with the finished surface or project from it shall be installed so that the mounting yoke or strap of the switch is seated against the box.

Section 404.10(B) was revised in the 2005 *Code* to make it clear that regardless of where a box for a flush-type switch is installed (in a wall or in a ceiling), the switch yoke or strap must be seated against the box, or where the box is set back, the yoke or strap must be against the finished surface of the wall, ceiling, or other location in which the box is installed.

Cooperation is necessary among the building trades (carpenters, drywall installers, plasterers, and so on) in order

for electricians to properly set device boxes flush with the finish surface, thereby ensuring a secure seating of the switch yoke and permitting the maximum projection of switch handles through the installed switch plate.

404.11 Circuit Breakers as Switches

A hand-operable circuit breaker equipped with a lever or handle, or a power-operated circuit breaker capable of being opened by hand in the event of a power failure, shall be permitted to serve as a switch if it has the required number of poles.

FPN: See the provisions contained in 240.81 and 240.83.

Circuit breakers that are capable of being hand operated must clearly indicate whether they are in the open (off) or closed (on) position. See 404.7 for details on handle positions. See 240.83(D) for SWD and HID marking for circuit breakers used as switches for 120-volt and 277-volt fluorescent and high-intensity discharge lighting circuits.

404.12 Grounding of Enclosures

Metal enclosures for switches or circuit breakers shall be grounded as specified in Article 250. Where nonmetallic enclosures are used with metal raceways or metal-armored cables, provision shall be made for grounding continuity.

Except as covered in 404.9(B), Exception, nonmetallic boxes for switches shall be installed with a wiring method that provides or includes an equipment ground.

404.13 Knife Switches

(A) Isolating Switches Knife switches rated at over 1200 amperes at 250 volts or less, and at over 600 amperes at 251 to 600 volts, shall be used only as isolating switches and shall not be opened under load.

(B) To Interrupt Currents To interrupt currents over 1200 amperes at 250 volts, nominal, or less, or over 600 amperes at 251 to 600 volts, nominal, a circuit breaker or a switch of special design listed for such purpose shall be used.

(C) General-Use Switches Knife switches of ratings less than specified in 404.13(A) and 404.13(B) shall be considered general-use switches.

FPN: See the definition of *General-Use Switch* in Article 100.

(D) Motor-Circuit Switches Motor-circuit switches shall be permitted to be of the knife-switch type.

FPN: See the definition of a *Motor-Circuit Switch* in Article 100.

404.14 Rating and Use of Snap Switches

Snap switches shall be used within their ratings and as indicated in 404.14(A) through 404.14(D).

> FPN No. 1: For switches on signs and outline lighting, see 600.6.
>
> FPN No. 2: For switches controlling motors, see 430.83, 430.109, and 430.110.

(A) Alternating Current General-Use Snap Switch A form of general-use snap switch suitable only for use on ac circuits for controlling the following:

(1) Resistive and inductive loads, including electric-discharge lamps, not exceeding the ampere rating of the switch at the voltage involved

(2) Tungsten-filament lamp loads not exceeding the ampere rating of the switch at 120 volts

(3) Motor loads not exceeding 80 percent of the ampere rating of the switch at its rated voltage

(B) Alternating-Current or Direct-Current General-Use Snap Switch A form of general-use snap switch suitable for use on either ac or dc circuits for controlling the following:

(1) Resistive loads not exceeding the ampere rating of the switch at the voltage applied.

(2) Inductive loads not exceeding 50 percent of the ampere rating of the switch at the applied voltage. Switches rated in horsepower are suitable for controlling motor loads within their rating at the voltage applied.

(3) Tungsten-filament lamp loads not exceeding the ampere rating of the switch at the applied voltage if T-rated.

(C) CO/ALR Snap Switches Snap switches rated 20 amperes or less directly connected to aluminum conductors shall be listed and marked CO/ALR.

(D) Alternating-Current Specific-Use Snap Switches Rated for 347 Volts Snap switches rated 347 volts ac shall be listed and shall be used only for controlling the loads permitted by (D)(1) and (D)(2).

(1) Noninductive Loads Noninductive loads other than tungsten-filament lamps not exceeding the ampere and voltage ratings of the switch.

(2) Inductive Loads Inductive loads not exceeding the ampere and voltage ratings of the switch. Where particular load characteristics or limitations are specified as a condition of the listing, those restrictions shall be observed regardless of the ampere rating of the load.

The ampere rating of the switch shall not be less than 15 amperes at a voltage rating of 347 volts ac. Flush-type snap switches rated 347 volts ac shall not be readily interchangeable in box mounting with switches identified in 404.14(A) and 404.14(B).

Although they are not commonly used in the United States, 600Y/347-volt systems are permitted by the *Code*. In accordance with 210.6 and 225.7(D), these systems can be used to supply installations of outdoor lighting. For the purposes of controlling lighting circuits on these systems, 404.14(D)(2) permits a relatively new type of ac specific-use snap switch that is 347-volt rated. Such switches, unless specifically restricted, are permitted to be used on circuits of a lower voltage, such as 277- and 120-volt circuits.

(E) Dimmer Switches General-use dimmer switches shall be used only to control permanently installed incandescent luminaires (lighting fixtures) unless listed for the control of other loads and installed accordingly.

General-use dimmers are not permitted to control receptacles or cord-and-plug-connected table and floor lamps. Section 404.14(E) does not apply to commercial dimmers or theater dimmers that can be used for fluorescent lighting and portable lighting. If a dimmer that has been evaluated only for the control of incandescent luminaires is used, the potential for connecting incompatible equipment such as a cord-and-plug-connected motor-operated appliance or a portable fluorescent lamp is increased by using the dimmer to control a receptacle(s).

II. Construction Specifications

404.15 Marking

(A) Ratings Switches shall be marked with the current voltage, and, if horsepower rated, the maximum rating for which they are designed.

(B) Off Indication Where in the off position, a switching device with a marked OFF position shall completely disconnect all ungrounded conductors to the load it controls.

404.16 600-Volt Knife Switches

Auxiliary contacts of a renewable or quick-break type or the equivalent shall be provided on all knife switches rated 600 volts and designed for use in breaking current over 200 amperes.

404.17 Fused Switches

A fused switch shall not have fuses in parallel except as permitted in 240.8.

404.18 Wire-Bending Space

The wire-bending space required by 404.3 shall meet Table 312.6(B) spacings to the enclosure wall opposite the line and load terminals.

ARTICLE 406
Receptacles, Cord Connectors, and Attachment Plugs (Caps)

Summary of Changes

- **406.4(D):** Added two new exceptions covering listed kits or assemblies and listed nonmetallic face plates.

- **406.6 and 406.6(D):** Added requirement covering the use of flanged surface inlets.

- **406.8(B)(1):** Revised to include all wet locations, not just those outdoors.

- **406.8(C):** Revised to prohibit receptacles directly over a bathtub or shower stall.

- **406.9(B)(4):** Revised to permit other grounding symbols.

Contents

406.1 Scope

This article covers the rating, type, and installation of receptacles, cord connectors, and attachment plugs (cord caps).

To consolidate the requirements for receptacles, cord connectors, and attachment caps into one location dedicated to these devices, Article 406 was added to the 2002 *NEC*. Article 406 comprises information that was taken from Article 410, Part L, 410-56, 410-57, and 410-58, and 210-7(d) of the 1999 *NEC* and that was reorganized and editorially rewritten for the 2002 *NEC*.

406.2 Receptacle Rating and Type

(A) Receptacles Receptacles shall be listed and marked with the manufacturer's name or identification and voltage and ampere ratings.

Section 406.2(A) was revised for the 2005 *Code* to clarify what was meant by "listed for the purpose." It was often interpreted that the receptacle had to be identified for a specific use. The intent is to ensure that the receptacles are suitable for the branch-circuit voltage and the load current that the receptacle must conduct.

(B) Rating Receptacles and cord connectors shall be rated not less than 15 amperes, 125 volts, or 15 amperes, 250 volts, and shall be of a type not suitable for use as lampholders.

  FPN: See 210.21(B) for receptacle ratings where installed on branch circuits.

(C) Receptacles for Aluminum Conductors Receptacles rated 20 amperes or less and designed for the direct connection of aluminum conductors shall be marked CO/ALR.

Section 406.2(C) requires that 15- and 20-ampere receptacles directly connected to aluminum conductors be suitable for such use. If the receptacle is not of the CO/ALR type, it can be connected with a copper pigtail to an aluminum branch-circuit conductor only if the wire connector is suitable for such a connection and is marked with the letters AL and CU. The commentary following 110.14(B) further explains the suitability of wire connectors used to join copper and aluminum conductors.

(D) Isolated Ground Receptacles Receptacles incorporating an isolated grounding connection intended for the reduction of electrical noise (electromagnetic interference) as permitted in 250.146(D) shall be identified by an orange triangle located on the face of the receptacle.

(1) Isolated Equipment Grounding Conductor Required Receptacles so identified shall be used only with grounding conductors that are isolated in accordance with 250.146(D).

(2) Installation in Nonmetallic Boxes Isolated ground receptacles installed in nonmetallic boxes shall be covered with a nonmetallic faceplate.

Exception: Where an isolated ground receptacle is installed in a nonmetallic box, a metal faceplate shall be permitted if the box contains a feature or accessory that permits the effective grounding of the faceplate.

406.3 General Installation Requirements

Receptacle outlets shall be located in branch circuits in accordance with Part III of Article 210. General installation requirements shall be in accordance with 406.3(A) through 406.3(F).

(A) Grounding Type Receptacles installed on 15- and 20-ampere branch circuits shall be of the grounding type. Grounding-type receptacles shall be installed only on circuits of the voltage class and current for which they are rated, except as provided in Table 210.21(B)(2) and Table 210.21(B)(3).

Exception: Nongrounding-type receptacles installed in accordance with 406.3(D).

(B) To Be Grounded Receptacles and cord connectors that have grounding contacts shall have those contacts effectively grounded.

Exception No. 1: Receptacles mounted on portable and vehicle-mounted generators in accordance with 250.34.

Exception No. 2: Replacement receptacles as permitted by 406.3(D).

(C) Methods of Grounding The grounding contacts of receptacles and cord connectors shall be grounded by connection to the equipment grounding conductor of the circuit supplying the receptacle or cord connector.

FPN: For installation requirements for the reduction of electrical noise, see 250.146(D).

The branch-circuit wiring method shall include or provide an equipment-grounding conductor to which the grounding contacts of the receptacle or cord connector are connected.

FPN No. 1: See 250.118 for acceptable grounding means.
FPN No. 2: For extensions of existing branch circuits, see 250.130.

(D) Replacements Replacement of receptacles shall comply with 406.3(D)(1), (D)(2), and (D)(3) as applicable.

(1) Grounding-Type Receptacles Where a grounding means exists in the receptacle enclosure or a grounding conductor is installed in accordance with 250.130(C), grounding-type receptacles shall be used and shall be connected to the grounding conductor in accordance with 406.3(C) or 250.130(C).

(2) Ground-Fault Circuit Interrupters Ground-fault circuit-interrupter protected receptacles shall be provided where replacements are made at receptacle outlets that are required to be so protected elsewhere in this *Code*.

(3) Non–grounding-Type Receptacles Where grounding means does not exist in the receptacle enclosure, the installation shall comply with (D)(3)(a), (D)(3)(b), or (D)(3)(c).

(a) A non–grounding-type receptacle(s) shall be permitted to be replaced with another non–grounding-type receptacle(s).

(b) A non–grounding-type receptacle(s) shall be permitted to be replaced with a ground-fault circuit interrupter-type of receptacle(s). These receptacles shall be marked "No Equipment Ground." An equipment grounding conductor shall not be connected from the ground-fault circuit-interrupter-type receptacle to any outlet supplied from the ground-fault circuit-interrupter receptacle.

(c) A non–grounding-type receptacle(s) shall be permitted to be replaced with a grounding-type receptacle(s) where supplied through a ground-fault circuit interrupter. Grounding-type receptacles supplied through the ground-fault circuit interrupter shall be marked "GFCI Protected" and "No Equipment Ground." An equipment grounding conductor shall not be connected between the grounding-type receptacles.

(E) Cord-and-Plug-Connected Equipment The installation of grounding-type receptacles shall not be used as a requirement that all cord-and-plug-connected equipment be of the grounded type.

> FPN: See 250.114 for types of cord-and-plug-connected equipment to be grounded.

(F) Noninterchangeable Types Receptacles connected to circuits that have different voltages, frequencies, or types of current (ac or dc) on the same premises shall be of such design that the attachment plugs used on these circuits are not interchangeable.

406.4 Receptacle Mounting

Receptacles shall be mounted in boxes or assemblies designed for the purpose, and such boxes or assemblies shall be securely fastened in place unless otherwise permitted elsewhere in this *Code*.

This section has been revised to clarify that boxes in which receptacles are installed are not always securely fastened in place. Receptacles in pendant boxes are permitted, provided the box is supported from the flexible cord in accordance with 314.23(H)(1). A pendant box that is properly suspended is not required to be securely fastened in place.

(A) Boxes That Are Set Back Receptacles mounted in boxes that are set back from the finished surface as permitted in 314.20 shall be installed such that the mounting yoke or strap of the receptacle is held rigidly at the finished surface.

A revision to 406.4(A) in the 2005 *Code* clarifies that this rule applies to all finished surfaces where the device box is set back from the finished surface of a wall, a ceiling, or other location within a building.

(B) Boxes That Are Flush Receptacles mounted in boxes that are flush with the finished surface or project therefrom shall be installed such that the mounting yoke or strap of the receptacle is held rigidly against the box or box cover.

To comply with 406.4(B), the outlet box used to enclose a receptacle must be rigidly and securely supported according to 314.23(B) or 314.23(C). In addition, mounting outlet boxes with the proper setback, according to 314.20, requires the cooperation of other construction trades (drywall installers, plasterers, and carpenters) and the building designers.

The intent of 406.4(A) through 406.4(C) is to allow attachment plugs to be inserted or removed without moving the receptacle. Additionally, by restricting movement of the receptacle, effective grounding continuity can be maintained for contact devices or receptacle yokes where the box is

installed flush with the wall surface or where it projects therefrom. The proper installation of receptacles helps ensure that attachment plugs can be fully inserted, thus providing a better contact.

(C) Receptacles Mounted on Covers Receptacles mounted to and supported by a cover shall be held rigidly against the cover by more than one screw or shall be a device assembly or box cover listed and identified for securing by a single screw.

Receptacles mounted on raised covers, such as the receptacle illustrated in Exhibit 406.1, are not permitted to be secured by a single screw unless listed and identified for the use.

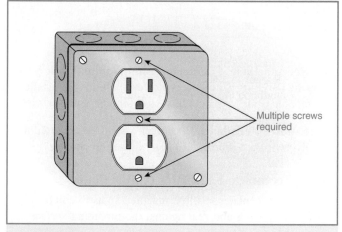

Exhibit 406.1 A receptacle mounted on a raised cover.

(D) Position of Receptacle Faces After installation, receptacle faces shall be flush with or project from faceplates of insulating material and shall project a minimum of 0.4 mm (0.015 in.) from metal faceplates.

The reason for requiring receptacles to project from metal faceplates is to prevent faults between the blades of attachment plugs and metal faceplates. The proper mounting of faceplates ensures that attachment plugs can be fully inserted, thus providing a better contact. The *NEC* does not specify the position (blades up or blades down) of a common vertically mounted 15- or 20-ampere duplex receptacle. Although many drawings in this handbook, such as Exhibit 406.1, show the slots for blades up, the receptacle may be installed with the slots for blades down. Receptacles can also be installed horizontally as well as vertically. Refer to 406.8(B) for information on receptacles installed in wet locations.

Exception No. 1: Listed kits or assemblies encompassing receptacles and nonmetallic faceplates that cover the receptacle face, where the plate cannot be installed on any other receptacle, shall be permitted.

Exception No. 2: Listed nonmetallic faceplates that cover the receptacle face to a maximum thickness of 1 mm (0.040 in.) shall be permitted.

Two exceptions have been added to 406.4(D) in the 2005 *Code*. The first allows the use of listed kits that include the receptacle and a nonmetallic faceplate and that have been evaluated by a recognized testing laboratory to ensure that sufficient blade contact is achieved by the attachment plug when inserted in the receptacle. In addition, the faceplate would not fit the standard style receptacle. The second new exception permits the use of a faceplate with an integral receptacle to cover the face of the box-mounted receptacle. These listed faceplates can be installed with any properly installed, flush-mounted device.

(E) Receptacles in Countertops and Similar Work Surfaces in Dwelling Units Receptacles shall not be installed in a face-up position in countertops or similar work surfaces.

(F) Exposed Terminals Receptacles shall be enclosed so that live wiring terminals are not exposed to contact.

406.5 Receptacle Faceplates (Cover Plates)

Receptacle faceplates shall be installed so as to completely cover the opening and seat against the mounting surface.

(A) Thickness of Metal Faceplates Metal faceplates shall be of ferrous metal not less than 0.76 mm (0.030 in.) in thickness or of nonferrous metal not less than 1.02 mm (0.040 in.) in thickness.

(B) Grounding Metal faceplates shall be grounded.

Section 406.5(B) requires that metal receptacle faceplates be grounded. Generally, this requirement is easily met by grounding the metal box. However, isolated ground receptacles installed in nonmetallic boxes are problematic because grounding the receptacle in this case does not ground the faceplate. Section 406.2(D)(2) contains two solutions concerning the receptacle faceplate. First, the general solution is to use only nonmetallic faceplates. Second, the exception to 406.2(D)(2) allows a nonmetallic box manufacturer to add a feature or accessory to accomplish effective grounding of a metal faceplate.

(C) Faceplates of Insulating Material Faceplates of insulating material shall be noncombustible and not less than

2.54 mm (0.10 in.) in thickness but shall be permitted to be less than 2.54 mm (0.10 in.) in thickness if formed or reinforced to provide adequate mechanical strength.

406.6 Attachment Plugs, Cord Connectors, and Flanged Surface Devices

All attachment plugs, cord connectors, and flanged surface devices (inlets and outlets) shall be listed and marked with the manufacturer's name or identification and voltage and ampere ratings.

Section 406.6 has been revised to include requirements governing the use of flanged surface inlet devices or motor base inlet plugs, often called motor plugs.

An energized cord cap is often improperly used to supply power to a building from a portable generator when a power failure occurs. The revision to 406.6 in the 2005 *Code* is aimed at prohibiting the improper use of a cord cap, where the blades are exposed and energized, to supply power to a cord body or plug into a receptacle to backfeed it. Prongs or blades that are exposed to contact by persons must not be energized unless an energized cord connector is installed in a flanged inlet device. Exhibit 406.2 illustrates a flanged inlet device.

Exhibit 406.2 Flanged inlet device. (Courtesy of Pass & Seymour/Legrand®)

(A) Construction of Attachment Plugs and Cord Connectors Attachment plugs and cord connectors shall be constructed so that there are no exposed current-carrying parts except the prongs, blades, or pins. The cover for wire terminations shall be a part that is essential for the operation of an attachment plug or connector (dead-front construction).

(B) Connection of Attachment Plugs Attachment plugs shall be installed so that their prongs, blades, or pins are not energized unless inserted into an energized receptacle. No receptacle shall be installed so as to require the insertion of an energized attachment plug as its source of supply.

The design requirements found in 406.6(B) (referred to as dead-front construction) minimize the occurrence of electrical faults between metal plates and attachment plugs with terminal screws exposed on the face of the plug.

The requirements in 406.6(B) were originally found in product information only. However, as an aid to the inspection community, these requirements are now clearly stated in the *NEC*. A live attachment plug cap can be a dangerous situation. Attachment plug caps should never be installed so as to allow the blades to be energized without being plugged into a device.

(C) Attachment Plug Ejector Mechanisms Attachment plug ejector mechanisms shall not adversely affect engagement of the blades of the attachment plug with the contacts of the receptacle.

Section 406.6(C) permits a device that reduces the likelihood of damage to the cord when the cord is pulled to remove the plug. This device is designed for use by persons with mobility or visual impairment.

(D) Flanged Surface Inlet A flanged surface inlet shall be installed such that the prongs, blades, or pins are not energized unless an energized cord connector is inserted into it.

Section 406.6(D) was added to cover flanged inlet devices and prohibit the prongs, blades, or pins of motor base inlet plugs from being energized before a cord body is inserted in it.

406.7 Noninterchangeability

Receptacles, cord connectors, and attachment plugs shall be constructed such that receptacle or cord connectors do not accept an attachment plug with a different voltage or current rating from that for which the device is intended. However, a 20-ampere T-slot receptacle or cord connector shall be permitted to accept a 15-ampere attachment plug of the same voltage rating. Non–grounding-type receptacles and connectors shall not accept grounding-type attachment plugs.

For information on receptacle and attachment cap configurations, see Exhibits 406.3 and 406.4 of ANSI C73 standard configuration charts.

406.8 Receptacles in Damp or Wet Locations

The requirements of 406.8(A) and (B) as they apply to the covers that are typically used with lower rated receptacles (15 through 60 amperes) are summarized in Exhibit 406.5.

(A) Damp Locations A receptacle installed outdoors in a location protected from the weather or in other damp locations shall have an enclosure for the receptacle that is weatherproof when the receptacle is covered (attachment plug cap not inserted and receptacle covers closed).

An installation suitable for wet locations shall also be considered suitable for damp locations.

A receptacle shall be considered to be in a location protected from the weather where located under roofed open porches, canopies, marquees, and the like, and will not be subjected to a beating rain or water runoff.

(B) Wet Locations

(1) 15- and 20-Ampere Receptacles in a Wet Location 15- and 20-ampere, 125- and 250-volt receptacles installed in a wet location shall have an enclosure that is weatherproof whether or not the attachment plug cap is inserted.

To ensure the weatherproof integrity of the cord-and-plug connection to receptacles located in a wet location, 406.8(B)(1) requires receptacle covers that provide a weatherproof enclosure at all times regardless of whether the plug is inserted or not. The requirement for this type of cover is not contingent on the anticipated use of the receptacle. This requirement applies to all 15- and 20-ampere, 125- and 250-volt receptacles that are installed in wet locations, including those receptacle outlets at dwelling units specified by 210.52(E). Exhibits 406.6 and 406.7 are examples of the type of receptacle enclosure required by 406.8(B)(1).

(2) Other Receptacles All other receptacles installed in a wet location shall comply with (B)(2)(a) or (B)(2)(b).

Examples of wet locations that were listed in 406.8(B)(2)(a) have been deleted from the 2005 *Code* to eliminate the possibility of missing some locations and to emphasize that receptacles placed on any surface, horizontal or vertical, in a wet location require a cover of some type to prevent water from entering the receptacle whether it is in use or not.

DESCRIPTION		NEMA NUMBER	15 AMPERE		20 AMPERE		30 AMPERE		50 AMPERE		60 AMPERE	
			RECEPTACLE	PLUG	RECEPTACLE	PLUG	RECEPTACLE	PLUG	RECEPTACLE	PLUG	RECEPTACLE	PLUG
2-POLE 2-WIRE	125V	1	1-15R	1-15P (POLARIZED) / (NON POLAR.)		2-20P MATES WITH 5-20R		1-30P MATES WITH 5-30R				
	250V	2		2-15P MATES WITH 6-15R	2-20R	2-20P	2-30R	2-30P				
	277V AC	3										
	600V	4										
2-POLE 3-WIRE GROUNDING	125V	5	5-15R	5-15P	5-20R	5-20P	5-30R	5-30P	5-50R	5-50P		
	125V	5ALT			5ALT-20R							
	250V	6	6-15R	6-15P	6-20R	6-20P	6-30R	6-30P	6-50R	6-50P		
	250V	6ALT			6ALT-20R							
	277V AC	7	7-15R	7-15P	7-20R	7-20P	7-30R	7-30P	7-50R	7-50P		
	347V AC	24	24-15R	24-15P	24-20R	24-20P	24-30R	24-30P	24-50R	24-50P		
	480V AC	8										
	600V AC	9										
3-POLE 3-WIRE	125 / 250V	10			10-20R	10-20P	10-30R	10-30P	10-50R	10-50P		
	3 Ø 250V	11	11-15R	11-15P	11-20R	11-20P	11-30R	11-30P	11-50R	11-50P		
	3 Ø 480V	12										
	3 Ø 600V	13										
3-POLE 4-WIRE GROUNDING	125 / 250V	14	14-15R	14-15P	14-20R	14-20P	14-30R	14-30P	14-50R	14-50P	14-60R	14-60P
	3 Ø 250V	15	15-15R	15-15P	15-20R	15-20P	15-30R	15-30P	15-50R	15-50P	15-60R	15-60P
	3 Ø 480V	16										
	3 Ø 600V	17										
4-POLE 4-WIRE	3 Ø Y 120 / 280V	18	18-15R	18-15P	18-20R	18-20P	18-30R	18-30P	18-50R	18-50P	18-60R	18-60P
	3 Ø Y 277 / 480V	19										
	3 Ø Y 347 / 600V	20										
4-POLE 5-WIRE GROUNDING	3 Ø Y 120 / 208V	21										
	3 Ø Y 277 / 480V	22										
	3 Ø Y 347 / 600V	23										

Note: Blank spaces reserved for future configurations.

Exhibit 406.3 Configuration chart for general-purpose locking plugs and receptacles. (Reprinted courtesy of NEMA WD 6-1997, *Wiring, Devices—Dimensional Requirements,* by permission of the National Electrical Manufacturers Association, Copyright 1997)

DESCRIPTION		NEMA NUMBER	15 AMPERE		20 AMPERE		30 AMPERE		50 AMPERE		60 AMPERE	
			RECEPTACLE	PLUG	RECEPTACLE	PLUG	RECEPTACLE	PLUG	RECEPTACLE	PLUG	RECEPTACLE	PLUG
2-POLE 2-WIRE	125V	1	L1-15R	L1-15P								
	250V	2			L2-20R	L2-20P						
	277V AC	3										
	600V	4										
2-POLE 3-WIRE GROUNDING	125V	5	L5-15R	L5-15P	L5-20R	L5-20P	L5-30R	L5-30P	L5-50R	L5-50P	L5-60R	L5-60P
	250V	6	L6-15R	L6-15P	L6-20R	L6-20P	L6-30R	L6-30P	L6-50R	L6-50P	L6-60R	L6-60P
	277V AC	7	L7-15R	L7-15P	L7-20R	L7-20P	L7-30R	L7-30P	L7-50R	L7-50P	L7-60R	L7-60P
	347V AC	24			L24-20R	L24-20P						
	480V AC	8			L8-20R	L8-20P	L8-30R	L8-30P	L8-50R	L8-50P	L8-60R	L8-60P
	600V AC	9			L9-20R	L9-20P	L9-30R	L9-30P	L9-50R	L9-50P	L9-60R	L9-60P
3-POLE 3-WIRE	125 / 250V	10			L10-20R	L10-20P	L10-30R	L10-30P				
	3 Ø 250V	11	11-15R	11-15P	L11-20R	L11-20P	L11-30R	L11-30P				
	3 Ø 480V	12			L12-20R	L12-20P	L12-30R	L12-30P				
	3 Ø 600V	13					L13-30R	L13-30P				
3-POLE 4-WIRE GROUNDING	125 / 250V	14			L14-20R	L14-20P	L14-30R	L14-30P	L14-50R	L14-50P	L14-60R	L14-60P
	3 Ø 250V	15			L15-20R	L15-20P	L15-30R	L15-30P	L15-50R	L15-50P	L15-60R	L15-60P
	3 Ø 480V	16			L16-20R	L16-20P	L16-30R	L16-30P	L16-50R	L16-50P	L16-60R	L16-60P
	3 Ø 600V	17					L17-30R	L17-30P	L17-50R	L17-50P	L17-60R	L17-60P
4-POLE 4-WIRE	3 Ø Y 120 / 208V	18			L18-20R	L18-20P	L18-30R	L18-30P				
	3 Ø Y 277 / 480V	19			L19-20R	L19-20P	L19-30R	L19-30P				
	3 Ø Y 347 / 600V	20			L20-20R	L20-20P	L20-30R	L20-30P				
4-POLE 5-WIRE GROUNDING	3 Ø Y 120 / 208V	21			L21-20R	L21-20P	L21-30R	L21-30P	L21-50R	L21-50P	L21-60R	L21-60P
	3 Ø Y 277 / 480V	22			L22-20R	L22-20P	L22-30R	L22-30P	L22-50R	L22-50P	L22-60R	L22-60P
	3 Ø Y 347 / 600V	23			L23-20R	L23-20P	L23-30R	L23-30P	L23-50R	L23-50P	L23-60R	L23-60P

Note: Blank spaces reserved for future configurations.

Exhibit 406.4 Configuration chart for specific-purpose locking plugs and receptacles. (Reprinted courtesy of NEMA WD 6-1997, *Wiring Devices—Dimensional Requirements,* by permission of the National Electrical Manufacturers Association, Copyright 1997)

Exhibit 406.5 Requirements for receptacle cover (enclosure) types.

Damp and Wet Receptacle Locations	Receptacle Cover (Enclosure) Type Requirements	
	Cover that *is not* weatherproof, with attachment plug cap inserted into receptacle	Cover that *is* weatherproof, with attachment plug cap inserted into receptacle ("in-use" type)
406.8(A): Outdoor damp locations	Minimum type required Note: "In-use" type covers permitted	Permitted
406.8(A): Indoor damp locations	Minimum type required Note: "In-use" type covers permitted	Permitted
406.8(B)(1)&(2): Outdoor wet locations	Required for receptacle types other than those rated 15- and 20-amperes, 125 and 250 volts, where the tool, appliance, or other utilization equipment plugged into the receptacle *is* attended while in use. Note: "In-use" type covers permitted	(a) Required for receptacles rated 15- and 20-amperes, 125 and 250 volts (b) Required for receptacles other than those rated 15- and 20-amperes, 125 and 250 volts, where the tool, appliance, or other utilization equipment plugged into the receptacle *is not* attended while in use.
406.8(B)(2): Indoor wet locations	Required for receptacle types other than those rated 15- and 20-amperes, 125 and 250 volts, where the tool, appliance, or other utilization equipment plugged into the receptacle *is* attended while in use. Note: "In-use" type covers permitted	(a) Required for receptacles rated 15- and 20-amperes, 125 and 250 volts (b) Required for receptacles other than those rated 15- and 20-amperes, 125 and 250 volts, where the tool, appliance, or other utilization equipment plugged into the receptacle *is not* attended while in use.

Section 406.8(B)(2)(a) applies to receptacles other than those rated 15 and 20 amperes, 125 and 250 volts, that supply cord-and-plug-connected equipment likely to be used outdoors or in a wet location for long periods of time. A portable pump motor is an example of such equipment. Receptacles for this application should remain weatherproof while they are in use.

(a) A receptacle installed in a wet location, where the product intended to be plugged into it is not attended while in use, shall have an enclosure that is weatherproof with the attachment plug cap inserted or removed.

(b) A receptacle installed in a wet location where the product intended to be plugged into it will be attended while in use (e.g., portable tools) shall have an enclosure that is weatherproof when the attachment plug is removed.

Section 406.8(B)(2)(b) applies to receptacles other than those rated 15 and 20 amperes, 125 and 250 volts, that supply cord-and-plug-connected portable tools or other portable equipment likely to be used outdoors for a specific purpose and then removed.

(C) Bathtub and Shower Space Receptacles shall not be installed within or directly over a bathtub or shower stall.

Section 406.8(C) prohibits the installation of receptacles inside bathtub and shower spaces or above their footprint, even if the receptacles are installed in a weatherproof enclosure. Prohibiting such installation helps minimize the use of shavers, radios, hair dryers, and so on, in these areas.

Exhibit 406.6 A single-gang weatherproof cover suitable for use in wet locations. (Courtesy of Thomas & Betts Corp.)

Exhibit 406.7 A two-gang weatherproof cover suitable for use in wet locations. (Courtesy of Carlon®, Lamson & Sessions)

The unprotected-line side of GFCI-protected receptacles installed in bathtub and shower spaces could possibly become wet and therefore create a shock hazard by energizing surrounding wet surfaces.

(D) Protection for Floor Receptacles Standpipes of floor receptacles shall allow floor-cleaning equipment to be operated without damage to receptacles.

(E) Flush Mounting with Faceplate The enclosure for a receptacle installed in an outlet box flush-mounted in a finished surface shall be made weatherproof by means of a weatherproof faceplate assembly that provides a watertight connection between the plate and the finished surface.

The term *wall surface* was deleted from this section of the 2005 *Code* to make it clear that flush-mounted receptacle outlets installed in walls, ceilings, or other building appurtenances or structures must have watertight integrity between the faceplate and the finished surface.

406.9 Grounding-Type Receptacles, Adapters, Cord Connectors, and Attachment Plugs

(A) Grounding Poles Grounding-type receptacles, cord connectors, and attachment plugs shall be provided with one fixed grounding pole in addition to the circuit poles. The grounding contacting pole of grounding-type plug-in ground-fault circuit interrupters shall be permitted to be of the movable, self-restoring type on circuits operating at not over 150 volts between any two conductors or any conductor and ground.

(B) Grounding-Pole Identification Grounding-type receptacles, adapters, cord connections, and attachment plugs shall have a means for connection of a grounding conductor to the grounding pole.

A terminal for connection to the grounding pole shall be designated by one of the following:

(1) A green-colored hexagonal-headed or -shaped terminal screw or nut, not readily removable.
(2) A green-colored pressure wire connector body (a wire barrel).
(3) A similar green-colored connection device, in the case of adapters. The grounding terminal of a grounding adapter shall be a green-colored rigid ear, lug, or similar device. The grounding connection shall be so designed that it cannot make contact with current-carrying parts of the receptacle, adapter, or attachment plug. The adapter shall be polarized.

Section 406.9(B)(3) requires the grounding terminal of an adapter to be a green-colored ear, lug, or similar device, thereby prohibiting use of an adapter with an attached pigtail grounding wire, which had been used for many years.

(4) If the terminal for the grounding conductor is not visible, the conductor entrance hole shall be marked with the

word *green* or *ground*, the letters *G* or *GR*, a grounding symbol, or otherwise identified by a distinctive green color. If the terminal for the equipment grounding conductor is readily removable, the area adjacent to the terminal shall be similarly marked.

FPN: See FPN Figure 406.9(B)(4).

FPN Figure 406.9(B)(4) One Example of a Symbol Used to Identify the Termination Point for an Equipment Grounding Conductor.

A change to 406.9(B)(4) makes the grounding symbol shown in FPN Figure 406.9(B)(4) an example of a symbol that can be used to identify the grounding pole of electrical equipment. As a "fine print note," the use of this symbol is not mandatory and other recognized symbols can be used to identify the terminal for connection of the equipment grounding conductors. A similar revision has been made in 250.126.

(C) Grounding Terminal Use A grounding terminal or grounding-type device shall not be used for purposes other than grounding.

(D) Grounding-Pole Requirements Grounding-type attachment plugs and mating cord connectors and receptacles shall be designed such that the grounding connection is made before the current-carrying connections. Grounding-type devices shall be so designed that grounding poles of attachment plugs cannot be brought into contact with current-carrying parts of receptacles or cord connectors.

The grounding blade of the attachment plug cap of most grounding-type combinations is longer than the circuit conductor blades and is used to ensure a "make-first, break-last" grounding connection. In some non-ANSI-approved pin-and-sleeve-type connections, the grounding contact of the receptacle is closer to the face of the receptacle than it is to other contacts, serving the same purpose.

(E) Use Grounding-type attachment plugs shall be used only with a cord having an equipment grounding conductor.

FPN: See 200.10(B) for identification of grounded conductor terminals.

406.10 Connecting Receptacle Grounding Terminal to Box

The connection of the receptacle grounding terminal shall comply with 250.146.

ARTICLE 408
Switchboards and Panelboards

Summary of Changes

- Much of Article 408 has been reorganized and renumbered.

- **408.4:** Revised wording to emphasize the importance of circuit directory for the safe operation of an electrical system and to require explicit identification of all circuits originating in switchboards and panelboards.

- **408.7:** Added requirement that all unused openings in switchboards and panelboards be effectively closed.

Contents

I. General

408.1 Scope

This article covers the following:

(1) All switchboards, panelboards, and distribution boards installed for the control of light and power circuits

(2) Battery-charging panels supplied from light or power circuits

See the definitions of *panelboard* and *switchboard* in Article 100.

408.2 Other Articles

Switches, circuit breakers, and overcurrent devices used on switchboards, panelboards, and distribution boards, and their enclosures shall comply with this article and also with the requirements of Articles 240, 250, 312, 314, 404, and other articles that apply. Switchboards and panelboards in hazardous (classified) locations shall comply with the requirements of Articles 500 through 517.

408.3 Support and Arrangement of Busbars and Conductors

(A) Conductors and Busbars on a Switchboard or Panelboard Conductors and busbars on a switchboard or panelboard shall comply with 408.3(A)(1), (A)(2), and (A)(3) as applicable.

(1) Location Conductors and busbars shall be located so as to be free from physical damage and shall be held firmly in place.

(2) Service Switchboards Barriers shall be placed in all service switchboards such that no uninsulated, ungrounded service busbar or service terminal is exposed to inadvertent contact by persons or maintenance equipment while servicing load terminations.

Where it can be demonstrated that it is unfeasible to disconnect or de-energize the service conductors supplying a service switchboard, qualified electricians may be required to work on these switchboards with the load terminals de-energized but with the service bus energized. Barriers are required in service switchboards to provide physical separation (adequate distance or obstacle) between load terminals and the service busbars and terminals, thus providing some measure of safety against inadvertent contact with line-energized parts during maintenance and installation of new feeders or branch circuits. In most multisection switchboards, barriers are not required, because the line-side conductors and busbars are not in the same switchboard sections that contain the load terminals.

It must be clearly understood that de-energizing the load side of a switchboard by operation of the disconnecting means does not de-energize the ungrounded service conductors. Every effort should be made to completely disconnect power from the equipment before performing any work inside. If complete disconnection is not feasible, the installer should become familiar with NFPA 70E, *Standard for Electrical Safety in the Workplace*. This industry-recognized document provides guidance for protective equipment and appropriate work rules that must be followed for working on energized equipment.

(3) Same Vertical Section Other than the required interconnections and control wiring, only those conductors that are intended for termination in a vertical section of a switchboard shall be located in that section.

Exception: Conductors shall be permitted to travel horizontally through vertical sections of switchboards where such conductors are isolated from busbars by a barrier.

The exception to 408.3(A)(3) permits conductors to travel horizontally through vertical sections of a switchboard where barriers are provided to isolate the conductors from the busbars. Horizontal travel of conductors through more than one section of a multisection switchboard is necessary where a raceway or cable entry is made into a switchboard section other than the one at which the conductors are terminated.

(B) Overheating and Inductive Effects The arrangement of busbars and conductors shall be such as to avoid overheating due to inductive effects.

(C) Used as Service Equipment Each switchboard or panelboard, if used as service equipment, shall be provided with a main bonding jumper sized in accordance with 250.28(D) or the equivalent placed within the panelboard or one of the sections of the switchboard for connecting the grounded service conductor on its supply side to the switchboard or panelboard frame. All sections of a switchboard shall be bonded together using an equipment bonding conductor sized in accordance with Table 250.122 or Table 250.66 as appropriate.

Exception: Switchboards and panelboards used as service equipment on high-impedance grounded-neutral systems in accordance with 250.36 shall not be required to be provided with a main bonding jumper.

(D) Terminals In switchboards and panelboards, load terminals for field wiring, including grounded circuit conductor load terminals and connections to the ground bus for load equipment grounding conductors, shall be so located that it is not necessary to reach across or beyond an uninsulated ungrounded line bus in order to make connections.

(E) Phase Arrangement The phase arrangement on 3-phase buses shall be A, B, C from front to back, top to bottom, or left to right, as viewed from the front of the switchboard or panelboard. The B phase shall be that phase having the higher voltage to ground on 3-phase, 4-wire, delta-connected systems. Other busbar arrangements shall be permitted for additions to existing installations and shall be marked.

The high leg is common on a 240/120-volt, 3-phase, 4-wire delta system. It is typically designated as "B phase." Section 110.15 requires the high-leg marking to be the color orange or other similar effective means of identification. Electricians should always test each phase to ground with suitable equipment in order to know exactly where this high leg is located in the system.

The exception to 408.3(E) permits the phase leg having the higher voltage to ground to be located at the right-hand position (C phase), making it unnecessary to transpose the panelboard or switchboard busbar arrangement ahead of and beyond a metering compartment. The exception recognizes the fact that metering compartments have been standardized with the high leg at the right position (C phase) rather than in the center on B phase.

See also 110.15 and 230.56 for further information on identifying conductors with the higher voltage to ground. Other busbar arrangements for making additions to existing installations are permitted by 408.3(E).

Exception: Equipment within the same single section or multisection switchboard or panelboard as the meter on 3-phase, 4-wire, delta-connected systems shall be permitted

to have the same phase configuration as the metering equipment.

> FPN: See 110.15 for requirements on marking the busbar or phase conductor having the higher voltage to ground where supplied from a 4-wire, delta-connected system.

(F) Minimum Wire-Bending Space The minimum wire-bending space at terminals and minimum gutter space provided in panelboards and switchboards shall be as required in 312.6.

Section 408.3(F) requires that installations in the field comply with 312.6. See also the commentary following 408.35, which covers the size of the enclosure.

408.4 Circuit Directory or Circuit Identification

Every circuit and circuit modification shall be legibly identified as to its clear, evident, and specific purpose or use. The identification shall include sufficient detail to allow each circuit to be distinguished from all others. The identification shall be included in a circuit directory that is located on the face or inside of the panel door in the case of a panelboard, and located at each switch on a switchboard.

The requirement to provide an up-to-date, accurate, and legible circuit directory applies to panelboards and switchboards covered in Article 408. The circuit directory is an important feature for the safe operation of an electrical system under normal and emergency conditions. The purpose of an accurate and legible circuit directory in these types of equipment is to provide clear identification of circuit breakers and switches that may need to be operated by service personnel or those who need to operate a switch or circuit breaker in an emergency. This requirement is specific to switchboards and panelboards; however, the identification requirements of 110.22 apply to all disconnecting means.

Section 408.4 was revised for the 2005 *Code* to require that the identification for every circuit supplied by a panelboard or switchboard be legible and clearly state the specific purpose for which the circuit is used. Circuits used for the same purpose must be identified as to their location. For example, small appliance branch circuits can supply outlets in the kitchen, dining room, and kitchen countertops. Identifying the circuits as small appliance branch circuits is not acceptable; instead, they should be identified as "kitchen wall receptacles," "dining room floor receptacle," or "kitchen countertop receptacles left of sink." Circuit directories containing multiple entries with only "lights" or "outlets" do not provide the sufficient detail required by this section.

408.5 Clearance for Conductor Entering Bus Enclosures

Where conduits or other raceways enter a switchboard, floor-standing panelboard, or similar enclosure at the bottom, sufficient space shall be provided to permit installation of conductors in the enclosure. The wiring space shall not be less than shown in Table 408.5 where the conduit or raceways enter or leave the enclosure below the busbars, their supports, or other obstructions. The conduit or raceways, including their end fittings, shall not rise more than 75 mm (3 in.) above the bottom of the enclosure.

Table 408.5 Clearance for Conductors Entering Bus Enclosures

Conductor	Minimum Spacing Between Bottom of Enclosure and Busbars, Their Supports, or Other Obstructions	
	mm	in.
Insulated busbars, their supports, or other obstructions	200	8
Noninsulated busbars	250	10

In the 2002 edition of the *Code*, the clearance requirement was located in Part II, Switchboards, and was applicable only to switchboards. For the 2005 *Code*, the requirements for clearances for conductors entering open-bottom switchboards was relocated to Part I, General, and now applies to floor-standing, open-bottom panelboards as well as switchboards.

Section 408.5 should be carefully considered in the installation of underground conduit or raceways that terminate in the open bottom of any enclosure containing busbars. For example, larger sizes of conduit used for service laterals or feeders and extending more than 3 in. above the bottom of the enclosure are difficult to shorten. On the other hand, conduits or raceways should not be installed flush with the finished floor under switchboards that are located on the outside of buildings or in other locations where water could enter the raceways.

408.7 Unused Openings

Unused openings for circuit breakers and switches shall be closed using identified closures, or other approved means that provide protection substantially equivalent to the wall of the enclosure.

This new requirement for closing unused switch and circuit breaker openings is specific to the equipment covered within the scope of Article 408. In addition, the requirement of 110.12(A) for closing unused cable and conduit openings applies to all electrical enclosures including panelboard cabinets and switchboard enclosures. Unused openings often occur during renovation and alteration of existing electrical

systems and equipment. These two requirements are necessary to restore the electrical equipment enclosure integrity to a condition that minimizes the possibility of an escaping arc, spark, or molten metal igniting surrounding combustible material and also minimizes the potential for accidental contact with live parts.

II. Switchboards

408.16 Switchboards in Damp or Wet Locations

Switchboards in damp or wet locations shall be installed in accordance with 312.2(A).

408.17 Location Relative to Easily Ignitible Material

Switchboards shall be placed so as to reduce to a minimum the probability of communicating fire to adjacent combustible materials. Where installed over a combustible floor, suitable protection thereto shall be provided.

One way to comply with the requirement of 408.17 is to form and attach a piece of sheet steel or other suitable noncombustible material to the floor under the electrical equipment.

408.18 Clearances

(A) From Ceiling For other than a totally enclosed switchboard, a space not less than 900 mm (3 ft) shall be provided between the top of the switchboard and any combustible ceiling, unless a noncombustible shield is provided between the switchboard and the ceiling.

(B) Around Switchboards Clearances around switchboards shall comply with the provisions of 110.26.

Sufficient access and working space are required to permit safe operation and maintenance of switchboards. Table 110.26(A)(1) indicates minimum working clearances from 0 to 600 volts, and Table 110.34(A) is used for voltages over 600 volts.

408.19 Conductor Insulation

An insulated conductor used within a switchboard shall be listed, shall be flame retardant, and shall be rated not less than the voltage applied to it and not less than the voltage applied to other conductors or busbars with which it may come in contact.

408.20 Location of Switchboards

Switchboards that have any exposed live parts shall be located in permanently dry locations and then only where under competent supervision and accessible only to qualified persons. Switchboards shall be located such that the probability of damage from equipment or processes is reduced to a minimum.

408.22 Grounding of Instruments, Relays, Meters, and Instrument Transformers on Switchboards

Instruments, relays, meters, and instrument transformers located on switchboards shall be grounded as specified in 250.170 through 250.178.

III. Panelboards

408.30 General

All panelboards shall have a rating not less than the minimum feeder capacity required for the load calculated in accordance with Article 220. Panelboards shall be durably marked by the manufacturer with the voltage and the current rating and the number of phases for which they are designed and with the manufacturer's name or trademark in such a manner so as to be visible after installation, without disturbing the interior parts or wiring.

> FPN: See 110.22 for additional requirements.

Many panelboards are suitable for use as service equipment and are so marked by the manufacturer.

Listed panelboards are used with copper conductors, unless they are marked to indicate which terminals are suitable for use with aluminum conductors. Such marking must be independent of any marking on terminal connectors and must appear on a wiring diagram or other readily visible location. If all terminals are suitable for use with aluminum conductors as well as with copper conductors, the panelboard is marked "Use Copper or Aluminum Wire." A panelboard using terminals or main or branch-circuit units individually marked AL-CU is marked "Use Copper or Aluminum Wire" or "Use Copper Wire Only." The latter marking indicates that wiring space or other factors make the panelboard unsuitable for aluminum conductors. [See 110.14(C).]

Panelboards to which units (circuit breakers, switches, etc.) may be added in the field are marked with the name or trademark of the manufacturer and the catalog number or equivalent of those units intended for installation in the field.

Unless the panelboard is marked to indicate otherwise, the termination provisions are based on the use of 60°C ampacities for wire sizes 14 AWG through 1 AWG and 75°C ampacities for wire sizes 1/0 AWG and larger.

408.34 Classification of Panelboards

Panelboards shall be classified for the purposes of this article as either lighting and appliance branch-circuit panelboards

or power panelboards, based on their content. A lighting and appliance branch circuit is a branch circuit that has a connection to the neutral of the panelboard and that has overcurrent protection of 30 amperes or less in one or more conductors.

(A) Lighting and Appliance Branch-Circuit Panelboard A lighting and appliance branch-circuit panelboard is one having more than 10 percent of its overcurrent devices protecting lighting and appliance branch circuits.

A lighting and appliance branch-circuit panelboard is one in which more than 10 percent of the installed overcurrent devices are rated 15, 20, or 30 amperes and supply circuits with a grounded (neutral) conductor. For example, a 24-position (space for 24 full-size circuit breakers), 120/240-volt, residential panelboard contains 21 overcurrent devices. If three or more (10 percent of 21 = 2.1) of those overcurrent devices supply 15-, 20-, or 30-ampere circuits with a neutral conductor, the panelboard is considered to be a lighting and appliance branch-circuit panelboard. However, if 12 two-pole, 30 ampere circuit breakers for electric heat are installed in the panelboard, there would be no branch circuits with neutral connections, and this panelboard therefore would be considered a power panelboard and subject to the overcurrent protection requirements of 408.36(B). If the supply circuit to this panelboard includes a neutral conductor, overcurrent protection not exceeding the rating of the panelboard is required on its supply side. This overcurrent protection can be located in the panelboard or, where supplied by a feeder, can be protected by the feeder overcurrent protective device.

(B) Power Panelboard A power panelboard is one having 10 percent or fewer of its overcurrent devices protecting lighting and appliance branch circuits.

A power panelboard is a panelboard that has 10 percent or less of the installed overcurrent devices supplying lighting and appliance branch circuits. Any panelboard that is not classified as a lighting and appliance branch-circuit panelboard is a power panelboard. A typical power panelboard could be located near the service and be designed to supply facility feeder circuits. The feeders from a power panelboard supply other utilization equipment or other panelboards that supply either other branch circuits or other feeders.

408.35 Number of Overcurrent Devices on One Panelboard

Not more than 42 overcurrent devices (other than those provided for in the mains) of a lighting and appliance branch-circuit panelboard shall be installed in any one cabinet or cutout box.

A lighting and appliance branch-circuit panelboard shall be provided with physical means to prevent the installation of more overcurrent devices than that number for which the panelboard was designed, rated, and approved.

For the purposes of this article, a 2-pole circuit breaker shall be considered two overcurrent devices; a 3-pole circuit breaker shall be considered three overcurrent devices.

Class CTL is the Underwriters Laboratories Inc. designation for the *Code* requirement for circuit limitation within a lighting and appliance branch-circuit panelboard and means "circuit limiting."

Class CTL panelboards incorporate physical features that, in conjunction with the physical size, configuration, or other means provided in Class CTL circuit breakers, fuseholders, or fusible switches, are designed to prevent the installation of more overcurrent protective poles than the number for which the panelboard is designed and rated.

It should be noted that switchboards, unlike panelboards, are not limited to 42 overcurrent devices or 42 switches or devices.

408.36 Overcurrent Protection

(A) Lighting and Appliance Branch-Circuit Panelboard Individually Protected Each lighting and appliance branch-circuit panelboard shall be individually protected on the supply side by not more than two main circuit breakers or two sets of fuses having a combined rating not greater than that of the panelboard.

Exception No. 1: Individual protection for a lighting and appliance panelboard shall not be required if the panelboard feeder has overcurrent protection not greater than the rating of the panelboard.

Main overcurrent protection may be an integral part of a panelboard or may be located remote from the panelboard. See also the commentary following 408.35. Exhibit 408.1 shows a panelboard with a 200-ampere main circuit breaker. Exhibit 408.2 illustrates overcurrent protection for the panelboard feeders having a rating not greater than the rating of the panelboard.

Exception No. 2: For existing installations, individual protection for lighting and appliance branch-circuit panelboards shall not be required where such panelboards are used as service equipment in supplying an individual residential occupancy.

The phrase "for existing installations" in Exception No. 2 refers to the existing panelboard. It is not intended that a

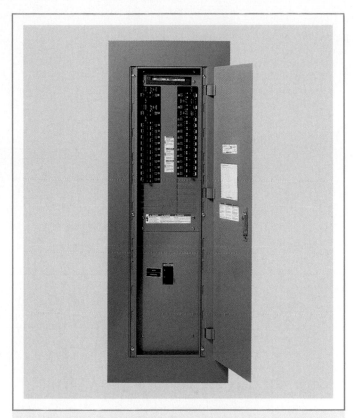

Exhibit 408.1 A panelboard with main circuit breaker disconnect, suitable for use as service equipment. (Courtesy of Square D Co.)

split-bus panelboard used in an individual residential occupancy be replaced if a circuit is added to the existing panelboard. It does mean, however, that for the installation of new panelboards in new or existing residential occupancies, a split-bus six-disconnect panelboard (with more than two circuit breakers or sets of fuses protecting the panelboard) is not permitted for the service equipment.

An individual residential occupancy could be a dwelling unit in a multifamily dwelling where the panelboard is used as service equipment. See the definition of *dwelling unit* in Article 100.

Exhibit 408.3 shows the split-bus circuitry for a 200-ampere (*left*) and a 150-ampere (*right*) panelboard. The left panel has two 100-ampere main breakers installed as disconnecting means and 200-ampere main lugs. The right is a split-bus panel with 150-ampere main lugs and six main breaker disconnecting means. The 150-ampere panelboard is suitable for use as service equipment only if it is not a lighting and appliance panelboard or if it presently exists in an individual residential occupancy.

(B) Power Panelboard Protection In addition to the requirements of 408.30, a power panelboard with supply conductors that include a neutral, and having more than 10

Exhibit 408.2 An arrangement of three individual lighting and appliance branch-circuit panelboards with main overcurrent protection remote from the panelboards.

Wireway

MLO = Main lug only

Exhibit 408.3 Circuitry for a 200-ampere (left) and a 150-ampere (right) split-bus panelboard.

percent of its overcurrent devices protecting branch circuits rated 30 amperes or less, shall be protected by an overcurrent protective device having a rating not greater than that of the panelboard. This overcurrent protective device shall be located within or at any point on the supply side of the panelboard.

When a power panelboard is required to have overcurrent protection, such protection can be provided by an overcurrent

device in the panelboard or by an overcurrent device protecting the conductors that supply the panelboard. In either case, the rating of the overcurrent device is not permitted to exceed the rating of the panelboard. For example, a feeder protected by a 450-ampere overcurrent device supplies a panelboard with a 600-ampere rating. Because the panelboard is obviously large enough to supply the calculated load and the overcurrent device protecting the feeder does not exceed the rating of the panelboard, an individual overcurrent device in the panelboard is not required. Note that where the panelboard is used as service equipment, the exception to 408.36(B) allows the panelboard overcurrent protection to be multiple devices in accordance with the "six-disconnect rule" in 230.71.

Exception: This individual protection shall not be required for a power panelboard used as service equipment with multiple disconnecting means in accordance with 230.71.

(C) Snap Switches Rated at 30 Amperes or Less Panelboards equipped with snap switches rated at 30 amperes or less shall have overcurrent protection of 200 amperes or less.

The requirement of 408.36(C) is limited to snap switches; it does not apply to panelboards equipped with circuit breakers.

(D) Supplied Through a Transformer Where a panelboard is supplied through a transformer, the overcurrent protection required by 408.36(A), (B), and (C) shall be located on the secondary side of the transformer.

Exception: A panelboard supplied by the secondary side of a transformer shall be considered as protected by the overcurrent protection provided on the primary side of the transformer where that protection is in accordance with 240.21(C)(1).

(E) Delta Breakers A 3-phase disconnect or overcurrent device shall not be connected to the bus of any panelboard that has less than 3-phase buses. Delta breakers shall not be installed in panelboards.

(F) Back-Fed Devices Plug-in-type overcurrent protection devices or plug-in type main lug assemblies that are backfed and used to terminate field-installed ungrounded supply conductors shall be secured in place by an additional fastener that requires other than a pull to release the device from the mounting means on the panel.

408.37 Panelboards in Damp or Wet Locations

Panelboards in damp or wet locations shall be installed to comply with 312.2(A).

408.38 Enclosure

Panelboards shall be mounted in cabinets, cutout boxes, or enclosures designed for the purpose and shall be dead-front.

Exception: Panelboards other than of the dead-front, externally operable type shall be permitted where accessible only to qualified persons.

408.39 Relative Arrangement of Switches and Fuses

In panelboards, fuses of any type shall be installed on the load side of any switches.

Exception: Fuses installed as part of service equipment in accordance with the provisions of 230.94 shall be permitted on the line side of the service switch.

Sections 230.82 and 230.94 permit the service switch on either the supply side or the load side of fuses such as cable limiters and other current-limiting devices. Where fuses of panelboards are accessible to other than qualified persons, such as occupants of a multifamily dwelling, 240.40 requires that disconnecting means be located on the supply side of all fuses in circuits of over 150 volts to ground and in cartridge-type fuses in circuits of any voltage. Thus, when the disconnect switch is opened, the fuses are de-energized, and danger from shock is reduced.

408.40 Grounding of Panelboards

Panelboard cabinets and panelboard frames, if of metal, shall be in physical contact with each other and shall be grounded. Where the panelboard is used with nonmetallic raceway or cable or where separate grounding conductors are provided, a terminal bar for the grounding conductors shall be secured inside the cabinet. The terminal bar shall be bonded to the cabinet and panelboard frame, if of metal; otherwise it shall be connected to the grounding conductor that is run with the conductors feeding the panelboard.

A separate equipment grounding conductor terminal bar must be installed and bonded to the panelboard for the termination of feeder and branch-circuit equipment grounding conductors. Where installed within service equipment, this terminal is bonded to the neutral terminal bar. Any other connection between the equipment grounding terminal bar and the neutral bar, other than allowed in 250.32, is not permitted. If this downstream connection occurs, current in the neutral or grounded conductor would take parallel paths through the equipment grounding conductors (the raceway, the building structure, or earth, for example) back to the service equipment. Normal load currents on the equipment grounding conductors could create a shock hazard. Exposed metal parts of equipment could have a potential difference of several volts created by the load current on the grounding conductors. Another safety hazard created by this effect, where subpanels are used, is arcing or loose connections at connectors and raceway fittings, creating a potential fire hazard. Exhibit 408.4 illustrates the connection of the grounded conductor (neutral bar) to the metallic service equipment enclosure via the main bonding jumper.

Exhibit 408.4 A split-bus lighting and appliance branch-circuit panelboard supplying an individual residential occupancy.

Exception: Where an isolated equipment grounding conductor is provided as permitted by 250.146(D), the insulated equipment grounding conductor that is run with the circuit conductors shall be permitted to pass through the panelboard without being connected to the panelboard's equipment grounding terminal bar.

Grounding conductors shall not be connected to a terminal bar provided for grounded conductors (may be a neutral) unless the bar is identified for the purpose and is located where interconnection between equipment grounding conductors and grounded circuit conductors is permitted or required by Article 250.

The grounding of electronic equipment as well as overall power quality is of concern to the electrical industry. Sensitive electronic equipment used in industrial and commercial power systems may fail to perform properly if electrical noise is present in the equipment grounding conductor.

An isolated equipment grounding terminal is permitted if it is necessary for the reduction of electrical noise on the grounding circuit. This equipment grounding terminal must be grounded by an insulated equipment grounding conductor that is run with the circuit conductors. The isolated equipment grounding conductor is also permitted to pass through one or more panelboards (without connection to the panelboard grounding terminal), but it is important that the equipment grounding conductor terminate directly at the applicable separately derived system or service grounding terminal. If the isolated equipment grounding conductor is run in a separate building, however, 250.146(D) requires the isolated equipment grounding conductor to terminate at a panelboard within the same building.

A connection to only a separate grounding electrode that places the earth in the fault return path may prevent sufficient current for opening overcurrent protection when a ground fault occurs. See the commentary following 250.146(D) FPN and 250.54.

408.41 Grounded Conductor Terminations

Each grounded conductor shall terminate within the panelboard in an individual terminal that is not also used for another conductor.

In accordance with their listing and the requirement of 110.14(A), conductor terminations are suitable for a single conductor unless the terminal is marked or otherwise identified as suitable for more than one conductor. This requirement applies to the termination of grounded conductors in panelboards. The use of a single termination point within a panelboard to connect more than one grounded conductor or to connect a grounded conductor and an equipment grounding conductor can be problematic when it is necessary to isolate a particular grounded conductor for testing purposes. For example, if the grounded conductors of two branch circuits were terminated at a single connection point and it was necessary to isolate one branch circuit for the purposes of troubleshooting, the fact that the circuit not being tested remains energized can create an unsafe working condition for service personnel disconnecting the grounded conductor of the circuit that is being tested. In some cases there are panelboard instructions that permit the use of a single conductor termination for more than one equipment grounding conductor. See 408.40 for the requirements on panelboard terminations for grounded and equipment grounding conductors.

Exception: Grounded conductors of circuits with parallel conductors shall be permitted to terminate in a single terminal if the terminal is identified for connection of more than one conductor.

IV. Construction Specifications

408.50 Panels

The panels of switchboards shall be made of moisture-resistant, noncombustible material.

408.51 Busbars

Insulated or bare busbars shall be rigidly mounted.

408.52 Protection of Instrument Circuits

Instruments, pilot lights, potential transformers, and other switchboard devices with potential coils shall be supplied by a circuit that is protected by standard overcurrent devices rated 15 amperes or less.

Exception No. 1: Overcurrent devices rated more than 15 amperes shall be permitted where the interruption of the circuit could create a hazard. Short-circuit protection shall be provided.

Exception No. 2: For ratings of 2 amperes or less, special types of enclosed fuses shall be permitted.

408.53 Component Parts

Switches, fuses, and fuseholders used on panelboards shall comply with the applicable requirements of Articles 240 and 404.

408.55 Wire-Bending Space in Panelboards

The enclosure for a panelboard shall have the top and bottom wire-bending space sized in accordance with Table 312.6(B) for the largest conductor entering or leaving the enclosure. Side wire-bending space shall be in accordance with Table 312.6(A) for the largest conductor to be terminated in that space.

Exception No. 1: Either the top or bottom wire-bending space shall be permitted to be sized in accordance with Table 312.6(A) for a lighting and appliance branch-circuit panelboard rated 225 amperes or less.

Exception No. 2: Either the top or bottom wire-bending space for any panelboard shall be permitted to be sized in accordance with Table 312.6(A) where at least one side wire-bending space is sized in accordance with Table 312.6(B) for the largest conductor to be terminated in any side wire-bending space.

Exception No. 3: The top and bottom wire-bending space shall be permitted to be sized in accordance with Table 312.6(A) spacings if the panelboard is designed and constructed for wiring using only one single 90 degree bend for each conductor, including the grounded circuit conductor, and the wiring diagram shows and specifies the method of wiring that shall be used.

Exception No. 4: Either the top or the bottom wire-bending space, but not both, shall be permitted to be sized in accordance with Table 312.6(A) where there are no conductors terminated in that space.

Section 408.55 covers the size of the enclosure for a panelboard. Using Exhibit 408.5 as a reference (see 312.6), the general rule calls for wire-bending spaces T_1 and T_4 to be in accordance with Table 312.6(B) for size M conductors (assuming these are the largest conductors entering the enclosure). Side wire-bending space T_2 must be in accordance with Table 312.6(A) for the wire size to be used with the largest-rated unit facing that side space, and T_3 must be

similarly sized for the largest-rated unit facing the right side of the enclosure.

Exception No. 1 to 408.55 permits either T_1 or T_4 (not both) to be reduced to the space required by Table 312.6(A) for size M conductors for a panelboard rated 225 amperes or less.

Exception No. 2 to 408.55 permits either T_1 or T_4 (not both) to be reduced to the space required by Table 312.6(A) for size M conductors for any panelboard. Exception No. 2 is valid where either T_2 or T_3 (or both) is sized in accordance with Table 312.6(B) for the largest conductor to be terminated in either the left or the right side space. Under the construction rules of 408.55, a panelboard enclosure might not be of adequate size for all manner of wiring; therefore 312.6 must be considered when wiring is planned.

Exception No. 3 to 408.55 permits both the top and the bottom wire-bending space to be reduced as noted. A single 90-degree bend, meaning one and only one 90-degree bend, must be present for the ungrounded conductors. A grounded conductor is permitted to be wired straight in if spacing is provided per Table 312.6(B) for the grounded conductor.

Exception No. 4 to 408.55 permits a reduction to the Table 312.6(A) spacing for the top or bottom space where no terminals face that space. In this case, the space is a gutter space, and measurement is on a line perpendicular to the wall of the enclosure and to the closest barrier post or side of a switch, fuse, or circuit breaker unit that is, or may be, installed. Exhibit 408.5 illustrates that exception.

408.56 Minimum Spacings

The distance between bare metal parts, busbars, and so forth shall not be less than specified in Table 408.56.

Where close proximity does not cause excessive heating, parts of the same polarity at switches, enclosed fuses,

Table 408.56 Minimum Spacings Between Bare Metal Parts

Voltage	Opposite Polarity Where Mounted on the Same Surface		Opposite Polarity Where Held Free in Air		Live Parts to Ground*	
	mm	in.	mm	in.	mm	in.
Not over 125 volts, nominal	19.1	¾	12.7	½	12.7	½
Not over 250 volts, nominal	31.8	1¼	19.1	¾	12.7	½
Not over 600 volts, nominal	50.8	2	25.4	1	25.4	1

*For spacing between live parts and doors of cabinets, see 312.11(A)(1), (2), and (3).

Exhibit 408.5 Panelboard wire-bending space per 312.6 and 408.55.

and so forth shall be permitted to be placed as close together as convenience in handling will allow.

Exception: The distance shall be permitted to be less than that specified in Table 408.56 at circuit breakers and switches and in listed components installed in switchboards and panelboards.

ARTICLE 409
Industrial Control Panels

Summary of Changes

- This new article sets installation requirements to incorporate listed control components into a listed enclosure in a manner that produces a safe control panel for various industrial applications such as lighting controls, conveyer controls, air-conditioning controls, etc.

Contents

I. General

409.1 Scope

This article covers industrial control panels intended for general use and operating at 600 volts or less.

> FPN: UL 508A is a safety standard for industrial control panels.

Article 409 is new for the 2005 *NEC* and provides installation and construction requirements for industrial control panels. These field- and factory-assembled control panels are used for the control and operation of a multitude of processes, from a sewerage pump station to an industrial process line. Similar in function to motor control centers in some regards, control panels also contain control, overcurrent protection, and power distribution equipment for operation of industrial heating processes, robotics, spray painting and powder coating lines, and countless other processes.

409.2 Definition

Section 409.2 provides a definition for *industrial control panel* that is compatible with industry product standards such as UL 508A, *Industrial Control Panels*, and specifies that a control panel consists of two or more devices such as motor controllers installed with related control devices, including control relays and timers.

Industrial Control Panel. An assembly of a systematic and standard arrangement of two or more components such as motor controllers, overload relays, fused disconnect switches, and circuit breakers and related control devices such as pushbutton stations, selector switches, timers, switches, control relays, and the like with associated wiring, terminal blocks, pilot lights, and similar components. The industrial control panel does not include the controlled equipment.

409.3 Other Articles

In addition to the requirements of Article 409, industrial control panels that contain branch circuits for specific loads or components, or are for control of specific types of equipment addressed in other articles of this *Code*, shall be constructed and installed in accordance with the applicable requirements from the specific articles in Table 409.3.

Table 409.3 Other Articles

Equipment/Occupancy	Article	Section
Branch circuits	210	
Luminaires	410	
Motors, motor circuits, and controllers	430	
Air-conditioning and refrigerating equipment	440	
Capacitors		460.8, 460.9
Hazardous (classified) locations	500, 501, 502, 503, 504, 505	
Commercial garages; aircraft hangars; motor fuel dispensing facilities; bulk storage plants; spray application, dipping, and coating processes; and inhalation anesthetizing locations	511, 513, 514, 515, 516, and 517 Part IV	
Cranes and hoists	610	
Electrically driven or controlled irrigation machines	675	
Elevators, dumbwaiters, escalators, moving walks, wheelchair lifts, and stairway chair lifts	620	
Industrial machinery	670	
Resistors and reactors	470	
Transformers	450	
Class 1, Class 2, and Class 3 remote-control, signaling, and power-limited circuits	725	

II. Installation

409.20 Conductor — Minimum Size and Ampacity

The size of the industrial control panel supply conductor shall have an ampacity not less than 125 percent of the full-

load current rating of all resistance heating loads plus 125 percent of the full-load current rating of the highest rated motor plus the sum of the full-load current ratings of all other connected motors and apparatus based on their duty cycle that may be in operation at the same time.

409.21 Overcurrent Protection

(A) General Industrial control panels shall be provided with overcurrent protection in accordance with Parts I, II, and IX of Article 240.

(B) Location This protection shall be provided by either of the following:

(1) An overcurrent protective device located ahead of the industrial control panel.
(2) A single main overcurrent protective device located within the industrial control panel. Where overcurrent protection is provided as part of the industrial control panel, the supply conductors shall be considered as either feeders or taps as covered by 240.21.

(C) Rating The rating or setting of the overcurrent protective device for the circuit supplying the industrial control panel shall not be greater than the sum of the largest rating or setting of the branch-circuit short-circuit and ground-fault protective device provided with the industrial control panel, plus 125 percent of the full-load current rating of all resistance heating loads, plus the sum of the full-load currents of all other motors and apparatus that could be in operation at the same time.

The rating or setting of the overcurrent protective device provided for the circuit supplying the industrial control panel is calculated by the sum of the following:

1. The largest rating or setting of the branch-circuit and ground-fault protective device installed in the industrial control panel
2. The sum of full-load currents of all other motors and apparatus that could operate simultaneously
3. 125 percent of the full-load current rating of all the resistance heating loads

Where the control panel does not provide short-circuit and ground-fault protection for the motor(s) that it controls, the overcurrent device installed on the line side of the control panel is sized in accordance with 430.52 for a single motor or 430.53 for a group of motors.

Exception: Where one or more instantaneous trip circuit breakers or motor short-circuit protectors are used for motor branch-circuit short-circuit and ground-fault protection as permitted by 430.52(C), the procedure specified above for determining the maximum rating of the protective device for the circuit supplying the industrial control panel shall apply with the following provision: For the purpose of the calculation, each instantaneous trip circuit breaker or motor short-circuit protector shall be assumed to have a rating not exceeding the maximum percentage of motor full-load current permitted by Table 430.52 for the type of control panel supply circuit protective device employed.

Where no branch-circuit short-circuit and ground-fault protective device is provided with the industrial control panel for motor or combination of motor and non-motor loads, the rating or setting of the overcurrent protective device shall be based on 430.52 and 430.53, as applicable.

409.30 Disconnecting Means

Disconnecting means that supply motor loads shall comply with Part IX of Article 430.

409.60 Grounding

Multisection industrial control panels shall be bonded together with an equipment grounding conductor or an equivalent grounding bus sized in accordance with Table 250.122. Equipment grounding conductors shall terminate on this grounding bus or to a grounding termination point provided in a single-section industrial control panel.

III. Construction Specifications

In addition to the field supply wiring requirements for all control panels contained in Part II, construction requirements for control panels are covered in Part III of Article 409. The *Code* does not require that control panels be listed, and Part III provides the authority having jurisdiction with a set of requirements that can be used as a benchmark for approval of a field-constructed control panel.

409.100 Enclosures

Table 430.91 shall be used as the basis for selecting industrial control panel enclosures for use in specific locations other than hazardous (classified) locations. The enclosures are not intended to protect against conditions such as condensation, icing, corrosion, or contamination that may occur within the enclosure or enter via the conduit or unsealed openings.

409.102 Busbars and Conductors

Industrial control panels utilizing busbars shall comply with 409.102(A) and 409.102(B).

(A) Support and Arrangement Busbars shall be protected from physical damage and be held firmly in place.

(B) Phase Arrangement The phase arrangement on 3-phase horizontal common power and vertical buses shall be A, B, C from front to back, top to bottom, or left to right, as viewed from the front of the industrial control panel. The B phase shall be that phase having the higher voltage to ground on 3-phase, 4-wire, delta-connected systems. Other busbar arrangements shall be permitted for additions to existing installations and shall be marked.

409.104 Wiring Space in Industrial Control Panels

(A) General Industrial control panel enclosures shall not be used as junction boxes, auxiliary gutters, or raceways for conductors feeding through or tapping off to other switches or overcurrent devices, unless adequate space for this purpose is provided. The conductors shall not fill the wiring space at any cross section to more than 40 percent of the cross-sectional area of the space, and the conductors, splices, and taps shall not fill the wiring space at any cross section to more than 75 percent of the cross-sectional area of that space.

(B) Wire Bending Space Wire bending space for the main supply terminals shall be in accordance with the requirements in 312.6. Wire bending space for other terminals shall be in accordance with the requirements in 430.10(B). The gutter space shall comply with 312.8.

409.108 Service-Entrance Equipment

Where used as service equipment, each industrial control panel shall be of the type that is suitable for use as service equipment.

Where a grounded conductor is provided, the industrial control panel shall be provided with a main bonding jumper, sized in accordance with 250.28(D), for connecting the grounded conductor, on its supply side, to the industrial control panel equipment ground bus or terminal.

409.110 Marking

An industrial control panel shall be marked with the following information that is plainly visible after installation:

Not unlike other types of electrical distribution and control equipment, one of the primary concerns associated with a group of components assembled in a common enclosure for the purposes of operation, control, and overcurrent protection is the ability of this equipment to limit and contain the effects of an internal fault such as a short-circuit or ground fault in such a manner that at the very least the internal fault does not pose an external ignition threat. To that end, 409.110(3) requires markings on all industrial control panels indicating their short circuit current rating.

In many control panel installations, the available fault energy at the line terminals of components within the control panel is significant, and the absence of a short-circuit current rating and compliance therewith is a recipe for the possibility of a failure that extends beyond the control panel enclosure. Not only are there high levels of short-circuit current available at the line terminals of many industrial control panels, there is also an interaction of the protective and control components under fault conditions that is assessed as part of the evaluation of control panels by conformity testing organizations. Assembling interrelated and interactive control and protective components in an enclosure presents a dynamic that in many cases can be evaluated for safety only under strict conformity assessment guidelines.

However, in the case of some simple assemblies consisting of a control component(s) installed in an enclosure, the authority having jurisdiction may not consider the assembly to be an industrial control panel and require a short-circuit current marking where it can be determined that the enclosure and component(s) have been installed in accordance with their listing and the assembly complies with 110.10. For each assembly that the AHJ must approve, it is necessary to be keenly aware of the short-circuit current considerations and whether there is any interaction between the assembled listed components that may impact compliance with 110.10.

(1) Manufacturer's name, trademark, or other descriptive marking by which the organization responsible for the product can be identified.
(2) Supply voltage, phase, frequency, and full-load current.
(3) Short-circuit current rating of the industrial control panel based on one of the following:
   a. Short-circuit current rating of a listed and labeled assembly
   b. Short-circuit current rating established utilizing an approved method

   FPN: UL 508A-2001, Supplement SB, is an example of an approved method.

(4) If the industrial control panel is intended as service equipment, it shall be marked to identify it as being suitable for use as service equipment.
(5) Electrical wiring diagram or the number of the index to the electrical drawings showing the electrical wiring diagram.
(6) An enclosure type number shall be marked on the industrial control panel enclosure.

ARTICLE 410
Luminaires (Lighting Fixtures), Lampholders, and Lamps

Summary of Changes

- **410.1:** Added to the scope decorative lighting products, lighting accessories for temporary seasonal and holiday use, and portable flexible lighting products.
- **410.4(D):** Revised to require listing for damp locations or, where subject to shower spray, wet locations.
- **410.4(E):** Added section to require lamp protection for mercury vapor or metal halide lamps subject to physical damage in venues such as sports arenas.
- **410.15(B):** Revised to allow nonmetallic poles for support of luminaires.
- **410.18, Exception No 2:** Added exception to permit GFCI protection in lieu of grounding where no equipment grounding conductor exists at the lighting outlet.
- **410.73(F)(5):** Added paragraph to require a lamp containment barrier for certain types of metal halide lamps.
- **410.73(G):** Added requirement, with exceptions, for a local disconnecting means, for use when servicing indoor fluorescent luminaires that contain ballasts, use double-ended lamps, and can be serviced in place. January 1, 2008, effective date.

Contents

I. General

410.1 Scope

This article covers luminaires (lighting fixtures), lampholders, pendants, incandescent filament lamps, arc lamps, electric-discharge lamps, decorative lighting products, lighting accessories for temporary seasonal and holiday use, portable flexible lighting products, and the wiring and equipment forming part of such products and lighting installations.

The identification and installation requirements for receptacles and receptacle covers that were located in Article 410, Part L in previous editions of the *Code* were relocated in the 2002 *Code* to Article 406, Receptacles, Cord Connectors, and Attachment Plugs (Caps). This article consolidated receptacle requirements that were formerly located in Articles 210 and 410 into an article specifically on receptacles, cord connectors, and attachment caps. See the commentary following 406.1.

410.2 Application of Other Articles

Equipment for use in hazardous (classified) locations shall conform to Articles 500 through 517. Lighting systems operating at 30 volts or less shall conform to Article 411. Arc lamps used in theaters shall comply with 520.61, and arc lamps used in projection machines shall comply with 540.20. Arc lamps used on constant-current systems shall comply with the general requirements of Article 490.

410.3 Live Parts

Luminaires (fixtures), lampholders, and lamps shall have no live parts normally exposed to contact. Exposed accessible terminals in lampholders and switches shall not be installed in metal luminaire (fixture) canopies or in open bases of portable table or floor lamps.

Exception: Cleat-type lampholders located at least 2.5 m (8 ft) above the floor shall be permitted to have exposed terminals.

II. Luminaire (Fixture) Locations

410.4 Luminaires (Fixtures) in Specific Locations

A pamphlet entitled *Luminaires Marking Guide*, available from Underwriters Laboratories Inc., was developed to help the authority having jurisdiction quickly determine whether common types of UL-listed fluorescent, high-intensity discharge, and incandescent fixtures are installed correctly.

(A) Wet and Damp Locations Luminaires (fixtures) installed in wet or damp locations shall be installed so that water cannot enter or accumulate in wiring compartments, lampholders, or other electrical parts. All luminaires (fixtures) installed in wet locations shall be marked, "Suitable for Wet Locations." All luminaires (fixtures) installed in damp locations shall be marked, "Suitable for Wet Locations" or "Suitable for Damp Locations."

Where luminaires are exposed to the weather or subject to water saturation, they must be of a type marked "Suitable for Wet Locations." Construction, design, and installation of these luminaires prevents the entrance of rain, snow, ice, and dust. Outdoor parks and parking lots, outdoor recreational areas (tennis, golf, baseball, etc.), car wash areas, and building exteriors are examples of wet locations.

Locations protected from the weather and not subject to water saturation but still exposed to moisture, such as the following, may be considered damp locations:

1. The underside of store or gasoline station canopies or theater marquees
2. Some cold-storage warehouses
3. Some agricultural buildings
4. Some basements
5. Roofed open porches and carports

Luminaires used in these locations must be marked "Suitable for Damp Locations." See the definitions of *location, damp*; *location, dry*; and *location, wet* in Article 100.

(B) Corrosive Locations Luminaires (fixtures) installed in corrosive locations shall be of a type suitable for such locations.

(C) In Ducts or Hoods Luminaires (fixtures) shall be permitted to be installed in commercial cooking hoods where all of the following conditions are met:

(1) The luminaire (fixture) shall be identified for use within commercial cooking hoods and installed such that the temperature limits of the materials used are not exceeded.
(2) The luminaire (fixture) shall be constructed so that all exhaust vapors, grease, oil, or cooking vapors are excluded from the lamp and wiring compartment. Diffusers shall be resistant to thermal shock.
(3) Parts of the luminaire (fixture) exposed within the hood shall be corrosion resistant or protected against corrosion, and the surface shall be smooth so as not to collect deposits and to facilitate cleaning.
(4) Wiring methods and materials supplying the luminaire(s) [fixture(s)] shall not be exposed within the cooking hood.

FPN: See 110.11 for conductors and equipment exposed to deteriorating agents.

The requirement in 410.4(C)(4) was initially taken from NFPA 96, *Standard for Ventilation Control and Fire Protection of Commercial Cooking Operations.* NFPA 96 provides the minimum fire safety requirements (preventive and operative) related to the design, installation, operation, inspection, and maintenance of all public and private cooking operations, except single-family residential dwellings. This coverage includes, but is not limited to, all cooking equipment, exhaust hoods, grease removal devices, exhaust ductwork, exhaust fans, dampers, fire-extinguishing equipment, and all other auxiliary or ancillary components or systems that are involved in the capture, containment, and control of grease-laden cooking effluent.

NFPA 96 is intended to include residential cooking equipment where used for purposes other than residential family use, such as employee kitchens or break areas and church and meeting hall kitchens, regardless of frequency of use.

Grease may cause the deterioration of conductor insulation, resulting in short circuits or ground faults in wiring, hence the requirement prohibiting wiring methods and materials (raceways, cables, lampholders) within ducts or hoods. Conventional enclosed and gasketed-type luminaires located in the path of travel of exhaust products are not permitted because a fire could result from the high temperatures on grease-coated glass bowls or globes enclosing the lamps.

Recessed or surface gasketed-type luminaires intended for location within hoods must be identified as suitable for the specific purpose and should be installed with the required clearances maintained. Note that wiring systems, including rigid metal conduit, are not permitted to be run exposed within the cooking hood.

For further information, refer to UL 710, *Standard for Safety for Exhaust Hoods for Commercial Cooking Equipment.*

(D) Bathtub and Shower Areas No parts of cord-connected luminaires (fixtures), chain-, cable-, or cord-suspended-luminaires (fixtures), lighting track, pendants, or ceiling-suspended (paddle) fans shall be located within a zone measured 900 mm (3 ft) horizontally and 2.5 m (8 ft) vertically from the top of the bathtub rim or shower stall threshold. This zone is all encompassing and includes the zone directly over the tub or shower stall. Luminaires (lighting fixtures) located in this zone shall be listed for damp locations, or listed for wet locations where subject to shower spray.

A revision of 410.4(D) clarifies that securely fastened luminaires installed in or on the ceiling or wall are permitted to be located in the bathtub or shower area. Where they are subject to shower spray, the luminaires must be listed for a wet location. Luminaires installed in the tub or shower zone and not subject to shower spray are required to be listed for use in a damp location. GFCI protection is required only where specified in the installation instructions for the luminaire.

The intent of 410.4(D) is to keep cord-connected, chain-hanging, or pendant luminaires and suspended fans out of the reach of an individual standing on a bathtub rim. The list of prohibited items recognizes that the same risk of electric shock is present for each one.

Exhibit 410.1 illustrates the restricted zone in which the specified luminaires, lighting track, and paddle fans are prohibited. This requirement applies to hydromassage bathtubs, as defined in 680.2, as well as other bathtub types and shower areas. See 680.43 for installation requirements for spas and hot tubs (as defined in 680.2) installed indoors.

(E) Luminaires (Fixtures) in Indoor Sports, Mixed-Use, and All-Purpose Facilities Luminaires (fixtures) subject to physical damage, using a mercury vapor or metal halide lamp, installed in playing and spectator seating areas of indoor sports, mixed-use, or all-purpose facilities shall be of the type that protects the lamp with a glass or plastic lens. Such luminaires (fixtures) shall be permitted to have an additional guard.

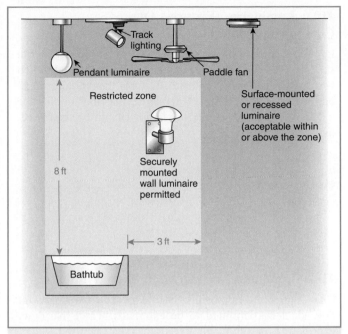

Exhibit 410.1 Luminaires, lighting track, and suspended (paddle) fan located near a bathtub.

Section 410.4(E) was added to the 2005 edition of the *NEC*. There have been instances of accidental breakage of mercury or metal halide lamp outer jackets when such lamps are used in open luminaires in sports facilities and other similar locations. If the lamp is damaged, glass shards can fall on players or spectators. If the envelope is damaged, the arc tube continues to operate even though the outer jacket may be cracked or missing. This new section requires luminaires to have their lamps protected by a glass or plastic lens, and it also permits an additional protective guard over the lens cover.

410.5 Luminaires (Fixtures) Near Combustible Material

Luminaires (fixtures) shall be constructed, installed, or equipped with shades or guards so that combustible material is not subjected to temperatures in excess of 90°C (194°F).

Nearly every fire requires an initial heat source, an initial fuel source, and an action that brings them together. The requirements of 410.5, 410.6, 410.7, and 410.8 regulate only the placement of the heat source. It is important to remember that successful fire prevention is most likely to come about if the initial heat source and initial fuel source are treated with due care. Tests have shown that hot particles from broken incandescent lamps can ignite combustibles below the lamps.

410.6 Luminaires (Fixtures) Over Combustible Material

Lampholders installed over highly combustible material shall be of the unswitched type. Unless an individual switch is provided for each luminaire (fixture), lampholders shall be located at least 2.5 m (8 ft) above the floor or shall be located or guarded so that the lamps cannot be readily removed or damaged.

Section 410.6 refers to pendants and fixed lighting equipment installed above highly combustible material. If the lamp cannot be located out of reach, the requirement can be met by equipping the lamp with a suitable guard. Section 410.6 does not apply to portable lamps.

410.7 Luminaires (Fixtures) in Show Windows

Chain-supported luminaires (fixtures) used in a show window shall be permitted to be externally wired. No other externally wired luminaires (fixtures) shall be used.

410.8 Luminaires (Fixtures) in Clothes Closets

(A) Definition

Storage Space The volume bounded by the sides and back closet walls and planes extending from the closet floor vertically to a height of 1.8 m (6 ft) or to the highest clothes-hanging rod and parallel to the walls at a horizontal distance of 600 mm (24 in.) from the sides and back of the closet walls, respectively, and continuing vertically to the closet ceiling parallel to the walls at a horizontal distance of 300 mm (12 in.) or the width of the shelf, whichever is greater; for a closet that permits access to both sides of a hanging rod, this space includes the volume below the highest rod extending 300 mm (12 in.) on either side of the rod on a plane horizontal to the floor extending the entire length of the rod.

    FPN: See Figure 410.8.

The 24-in. rule is intended to cover the clothes-hanging space, even if no clothes-hanging rod is installed. If a clothes-hanging rod is installed, the space extends from the floor to the top of the highest rod. If no clothes-hanging rod is installed, the space extends from the floor to a height of 6 ft.

    In addition to the space in which clothing is hung from the closet pole or rod, this requirement also establishes a 12-in.-wide shelf space to cover those installations where shelving is not in place at the time of fixture installation. If shelving is installed and the shelves are wider than 12 in., the greater width must be applied in establishing this space.

    The storage space for closets that permit access to both sides of the clothes-hanging rod is based on a horizontal plane extending 12 in. from both sides of the rod, from the

Figure 410.8 Closet Storage Space

rod down to the floor. This equates to the 24-in. space required for the closet rod where there is only one direction of access.

(B) Luminaire (Fixture) Types Permitted
Listed luminaires (fixtures) of the following types shall be permitted to be installed in a closet:

(1) A surface-mounted or recessed incandescent luminaire (fixture) with a completely enclosed lamp
(2) A surface-mounted or recessed fluorescent luminaire (fixture)

(C) Luminaire (Fixture) Types Not Permitted
Incandescent luminaires (fixtures) with open or partially enclosed lamps and pendant luminaires (fixtures) or lampholders shall not be permitted.

See the commentary following 410.8(D)(3).

(D) Location
Luminaires (fixtures) in clothes closets shall be permitted to be installed as follows:

(1) Surface-mounted incandescent luminaires (fixtures) installed on the wall above the door or on the ceiling, provided there is a minimum clearance of 300 mm (12 in.) between the luminaire (fixture) and the nearest point of a storage space
(2) Surface-mounted fluorescent luminaires (fixtures) installed on the wall above the door or on the ceiling, provided there is a minimum clearance of 150 mm

(6 in.) between the luminaire (fixture) and the nearest point of a storage space

(3) Recessed incandescent luminaires (fixtures) with a completely enclosed lamp installed in the wall or the ceiling, provided there is a minimum clearance of 150 mm (6 in.) between the luminaire (fixture) and the nearest point of a storage space

The requirement in 410.8(D)(3) results from tests that have shown that a hot filament falling from a broken incandescent lamp can ignite combustible material below the luminaire in which the lamp is installed.

(4) Recessed fluorescent luminaires (fixtures) installed in the wall or the ceiling, provided there is a minimum clearance of 150 mm (6 in.) between the luminaire (fixture) and the nearest point of a storage space

Note that the clearance measurement for each requirement in 410.8(D) is to the luminaire and not to the lamp itself.

It is not mandatory to install a luminaire in a clothes closet; if one is installed, however, the conditions for installation are as required by 410.8(D).

The requirements of 410.8(D) apply to incandescent and fluorescent lighting in clothes closets of various kinds of occupancies. The requirement is intended to prevent hot lamps or parts of broken lamps from coming in contact with boxes, cartons, blankets, and the like, stored on shelves and with clothing hung in closets.

410.9 Space for Cove Lighting

Coves shall have adequate space and shall be located so that lamps and equipment can be properly installed and maintained.

Adequate space is necessary for easy access for relamping luminaires or replacing lampholders, ballasts, and so on; adequate space also improves ventilation.

III. Provisions at Luminaire (Fixture) Outlet Boxes, Canopies, and Pans

410.10 Space for Conductors

Canopies and outlet boxes taken together shall provide adequate space so that luminaire (fixture) conductors and their connecting devices can be properly installed.

410.11 Temperature Limit of Conductors in Outlet Boxes

Luminaires (fixtures) shall be of such construction or installed so that the conductors in outlet boxes shall not be

subjected to temperatures greater than that for which the conductors are rated.

Branch-circuit wiring, other than 2-wire or multiwire branch circuits supplying power to luminaires (fixtures) connected together, shall not be passed through an outlet box that is an integral part of a luminaire (fixture) unless the luminaire (fixture) is identified for through-wiring.

FPN: See 410.32 for wiring supplying power to fixtures connected together.

Branch-circuit conductors run to a lighting outlet box are not permitted to be subjected to higher temperatures than those temperatures for which they are rated. Take, for example, conductors that are rated 75°C and that are to supply a ceiling outlet box for the connection of a surface-mounted luminaire or attached outlet box of a recessed luminaire. The design and installation of the luminaire should be such that the heat of the lamps does not subject the conductors to a greater temperature than 75°C. These types of luminaire are listed by Underwriters Laboratories Inc. based on a heat-contributing factor of the supply conductors of not more than the maximum permitted lamp wattage of the luminaire.

Exhibit 410.2 illustrates recessed luminaires listed by Underwriters Laboratories Inc. for one set of supply conductors, and Exhibit 410.3 illustrates luminaires listed for a feed-through installation.

Exhibit 410.2 Recessed luminaires designed for branch-circuit conductors terminating at each luminaire (no feed-through).

410.12 Outlet Boxes to Be Covered

In a completed installation, each outlet box shall be provided with a cover unless covered by means of a luminaire (fixture) canopy, lampholder, receptacle, or similar device.

Exhibit 410.3 Recessed luminaires designed for feed-through branch-circuit conductors.

410.13 Covering of Combustible Material at Outlet Boxes

Any combustible wall or ceiling finish exposed between the edge of a luminaire (fixture) canopy or pan and an outlet box shall be covered with noncombustible material.

Luminaires must be designed and installed not only to prevent overheating of conductors but also to prevent overheating of adjacent combustible wall or ceiling finishes. Hence, it is required that any combustible finish between the edge of a luminaire canopy and an outlet box be covered with a noncombustible material or luminaire accessory. See 314.20 for the requirements covering combustible finishes.

Where luminaires are not directly mounted on outlet boxes, suitable outlet box covers are required.

410.14 Connection of Electric-Discharge Luminaires (Lighting Fixtures)

(A) Independent of the Outlet Box Electric-discharge luminaires (lighting fixtures) supported independently of the outlet box shall be connected to the branch circuit through metal raceway, nonmetallic raceway, Type MC cable, Type AC cable, Type MI cable, nonmetallic sheathed cable, or by flexible cord as permitted in 410.30(B) or 410.30(C).

(B) Access to Boxes Electric-discharge luminaires (fixtures) surface mounted over concealed outlet, pull, or junction boxes and designed not to be supported solely by the outlet box shall be provided with suitable openings in the back of the luminaire (fixture) to provide access to the wiring in the box.

IV. Luminaire (Fixture) Supports

410.15 Supports

(A) General Luminaires (fixtures) and lampholders shall be securely supported. A luminaire (fixture) that weighs more than 3 kg (6 lb) or exceeds 400 mm (16 in.) in any dimension shall not be supported by the screw shell of a lampholder.

(B) Metal or Nonmetallic Poles Supporting Luminaires (Lighting Fixtures) Metal or nonmetallic poles shall be permitted to be used to support luminaires (lighting fixtures) and as a raceway to enclose supply conductors, provided the following conditions are met:

(1) A pole shall have a handhole not less than 50 mm × 100 mm (2 in. × 4 in.) with a raintight cover to provide access to the supply terminations within the pole or pole base.

Exception No. 1: No handhole shall be required in a pole 2.5 m (8 ft) or less in height above grade where the supply wiring method continues without splice or pull point, and where the interior of the pole and any splices are accessible by removing the luminaire (fixture).

Exception No. 1 to 410.15(B)(1) typically applies to both landscape (bollard-type) lighting and pole lights at residential dwellings.

Exception No. 2: No handhole shall be required in a pole 6.0 m (20 ft) or less in height above grade that is provided with a hinged base.

Exception No. 2 to 410.15(B)(1) recognizes metal light poles that do not have a handhole but instead use a hinged-base pole to permit access to splices made in the pole base. The height of the pole is limited to 20 ft. Both parts of the pole must be bonded in accordance with 250.96 (systems operating at 250 volts or less) or 250.97 (circuits operating at over 250 volts), depending on the voltage of the system.

Exhibit 410.4 illustrates a metal light pole with a hinged baseplate that meets the requirements of Exception No. 2 to 410.15(B)(1). A handhole is not necessary because the pole can be tilted to allow access to terminations in the base.

(2) Where raceway risers or cable is not installed within the pole, a threaded fitting or nipple shall be brazed, welded, or attached to the pole opposite the handhole for the supply connection.

(3) A metal pole shall be provided with a grounding terminal as follows:

Exhibit 410.4 A hinged-base metal pole supporting a luminaire.

a. A pole with a handhole shall have the grounding terminal accessible from the handhole.
b. A pole with a hinged base shall have the grounding terminal accessible within the base.

Exception to (3): No grounding terminal shall be required in a pole 2.5 m (8 ft) or less in height above grade where the supply wiring method continues without splice or pull, and where the interior of the pole and any splices are accessible by removing the luminaire (fixture).

(4) A metal pole with a hinged base shall have the hinged base and pole bonded together.
(5) Metal raceways or other equipment grounding conductors shall be bonded to the metal pole with an equipment grounding conductor recognized by 250.118 and sized in accordance with 250.122.
(6) Conductors in vertical poles used as raceway shall be supported as provided in 300.19.

In 410.15(B) metal poles are permitted to be used as a raceway. If the pole is used as a raceway, the user is reminded to maintain the required separation between power wiring

and any communications, signaling, and power-limited circuits that may also be installed within the pole.

For example, where a light pole supports a luminaire and a security camera, the security camera signaling and power-limited wiring may be installed within the pole cavity. Because the pole contains open circuit conductors (e.g., THW or XHHW conductors) to supply the luminaire already, the separation requirements can be fulfilled by enclosing the camera conductors within a flexible raceway and installing that raceway within the pole. Of course, the use of a cable or cord assembly for the lighting circuit conductors remains an optional choice for the user. The provisions of 410.15(B)(6) remind the user that the conductors are to be supported in accordance with 300.19, the same as a vertical raceway.

Section 410.15(B) is not intended to necessarily require the placement of a raceway for communications cables on the exterior of a lighting pole.

410.16 Means of Support

(A) Outlet Boxes Outlet boxes or fittings installed as required by 314.23 and complying with the provisions of 314.27(A) and 314.27(B) shall be permitted to support luminaires (fixtures).

Regardless of whether a luminaire is attached to an outlet box or is supported independently of the outlet box, care should be taken to securely fasten the outlet box or support the independent rod or hanger to ensure that the luminaire is securely mounted in place. The luminaire may be securely mounted to the box; however, if the box is not secured, it becomes the weak link in the luminaire support.

(B) Inspection Luminaires (fixtures) shall be installed such that the connections between the luminaire (fixture) conductors and the circuit conductors can be inspected without requiring the disconnection of any part of the wiring unless the luminaires (fixtures) are connected by attachment plugs and receptacles.

(C) Suspended Ceilings Framing members of suspended ceiling systems used to support luminaires (fixtures) shall be securely fastened to each other and shall be securely attached to the building structure at appropriate intervals. Luminaires (fixtures) shall be securely fastened to the ceiling framing member by mechanical means such as bolts, screws, or rivets. Listed clips identified for use with the type of ceiling framing member(s) and luminaire(s) [fixture(s)] shall also be permitted.

Section 410.16(C) provides requirements for luminaire installations where the luminaires are supported by a suspended, or hung, ceiling. Where luminaires are supported independent of a suspended ceiling, 410.16(C) does not apply.

Section 410.16(C) requires that if clips are used to support luminaires to the framing members of a suspended ceiling, those clips must be of a type listed for the application. However, the use of listed clips for luminaire support does not complete the requirements of this section. Additionally, the ceiling framing members must be securely attached to each other and to the building structure. For the support of wiring that is located in the cavity of floor-ceiling assemblies, see 300.11(A). This requirement applies to all luminaires supported by a suspended ceiling assembly, including lay-in and surface-mounted types.

(D) Luminaire (Fixture) Studs Luminaire (fixture) studs that are not a part of outlet boxes, hickeys, tripods, and crowfeet shall be made of steel, malleable iron, or other material suitable for the application.

(E) Insulating Joints Insulating joints that are not designed to be mounted with screws or bolts shall have an exterior metal casing, insulated from both screw connections.

(F) Raceway Fittings Raceway fittings used to support a luminaire(s) [lighting fixture(s)] shall be capable of supporting the weight of the complete fixture assembly and lamp(s).

(G) Busways Luminaires (fixtures) shall be permitted to be connected to busways in accordance with 368.17(C).

(H) Trees Outdoor luminaires (lighting fixtures) and associated equipment shall be permitted to be supported by trees.

FPN No. 1: See 225.26 for restrictions for support of overhead conductors.

The support of overhead conductor spans on trees is prohibited by 225.26.

FPN No. 2: See 300.5(D) for protection of conductors.

Section 300.5(D) requires buried conductors and cables to be protected from physical damage by the use of raceways from a specified point belowgrade to a point at least 8 ft above finish grade.

V. Grounding

410.17 General

Luminaires (fixtures) and lighting equipment shall be grounded as required in Article 250 and Part V of this article.

410.18 Exposed Luminaire (Fixture) Parts

(A) Exposed Conductive Parts Exposed metal parts shall be grounded or insulated from ground and other conducting surfaces or be inaccessible to unqualified personnel. Lamp tie wires, mounting screws, clips, and decorative bands on glass spaced at least 38 mm (1½ in.) from lamp terminals shall not be required to be grounded.

(B) Made of Insulating Material Luminaires (fixtures) directly wired or attached to outlets supplied by a wiring method that does not provide a ready means for grounding shall be made of insulating material and shall have no exposed conductive parts.

Exception No. 1: Replacement luminaires (fixtures) shall be permitted to connect an equipment grounding conductor from the outlet in compliance with 250.130(C). The luminaire (fixture) shall then be grounded in accordance with 410.18(A).

The exception to 410.18(B) provides a method by which a luminaire with exposed conductive parts can be installed at an outlet where the wiring method is not an equipment grounding conductor per 250.118 or does not provide an equipment grounding conductor. In older installations where luminaires are replaced, the requirement to ground exposed metal parts of the luminaire is not negated simply because there is no means of grounding provided by the existing wiring system. The means allowed by the exception is the same as is permitted for receptacles installed at outlets where no grounding means exists. A single grounding conductor can be run independently of the circuit conductors, from the outlet to a point on the wiring system where an effective grounding connection can be made. The acceptable termination points for this separate grounding conductor are specified by 250.130(C).

Exception No. 2: Where no equipment grounding conductor exists at the outlet, replacement luminaires (fixtures) that are GFCI protected shall not be required to be connected to an equipment grounding conductor.

Exception No. 2 to 410.18(B) was added to the 2005 *Code* to permit a GFCI to provide protection for personnel where luminaires are supplied by a circuit that does not have an equipment grounding conductor. This exception provides added protection similar to that provided for receptacles supplied from older circuits without an equipment grounding conductor. However, it does not allow the installation of a new circuit without an EGC to supply luminaires.

410.20 Equipment Grounding Conductor Attachment

Luminaires (fixtures) with exposed metal parts shall be provided with a means for connecting an equipment grounding conductor for such luminaires (fixtures).

410.21 Methods of Grounding

Luminaires (fixtures) and equipment shall be considered grounded where mechanically connected to an equipment grounding conductor as specified in 250.118 and sized in accordance with 250.122.

VI. Wiring of Luminaires (Fixtures)

410.22 Luminaire (Fixture) Wiring — General

Wiring on or within fixtures shall be neatly arranged and shall not be exposed to physical damage. Excess wiring shall be avoided. Conductors shall be arranged so that they are not subjected to temperatures above those for which they are rated.

410.23 Polarization of Luminaires (Fixtures)

Luminaires (fixtures) shall be wired so that the screw shells of lampholders are connected to the same luminaire (fixture) or circuit conductor or terminal. The grounded conductor, where connected to a screw-shell lampholder, shall be connected to the screw shell.

410.24 Conductor Insulation

Luminaires (fixtures) shall be wired with conductors having insulation suitable for the environmental conditions, current, voltage, and temperature to which the conductors will be subjected.

> FPN: For ampacity of luminaire (fixture) wire, maximum operating temperature, voltage limitations, minimum wire size, and so forth, see Article 402.

410.27 Pendant Conductors for Incandescent Filament Lamps

(A) **Support** Pendant lampholders with permanently attached leads, where used for other than festoon wiring, shall be hung from separate stranded rubber-covered conductors that are soldered directly to the circuit conductors but supported independently thereof.

(B) **Size** Unless part of listed decorative lighting assemblies, pendant conductors shall not be smaller than 14 AWG for mogul-base or medium-base screw-shell lampholders or smaller than 18 AWG for intermediate or candelabra-base lampholders.

(C) **Twisted or Cabled** Pendant conductors longer than 900 mm (3 ft) shall be twisted together where not cabled in a listed assembly.

410.28 Protection of Conductors and Insulation

(A) **Properly Secured** Conductors shall be secured in a manner that does not tend to cut or abrade the insulation.

(B) **Protection Through Metal** Conductor insulation shall be protected from abrasion where it passes through metal.

(C) **Luminaire (Fixture) Stems** Splices and taps shall not be located within luminaire (fixture) arms or stems.

(D) **Splices and Taps** No unnecessary splices or taps shall be made within or on a luminaire (fixture).

> FPN: For approved means of making connections, see 110.14.

(E) **Stranding** Stranded conductors shall be used for wiring on luminaire (fixture) chains and on other movable or flexible parts.

(F) **Tension** Conductors shall be arranged so that the weight of the luminaire (fixture) or movable parts does not put tension on the conductors.

410.29 Cord-Connected Showcases

Individual showcases, other than fixed, shall be permitted to be connected by flexible cord to permanently installed receptacles, and groups of not more than six such showcases shall be permitted to be coupled together by flexible cord and separable locking-type connectors with one of the group connected by flexible cord to a permanently installed receptacle.

The installation shall comply with 410.29(A) through 410.29(E).

(A) **Cord Requirements** Flexible cord shall be of the hard-service type, having conductors not smaller than the branch-circuit conductors, having ampacity at least equal to the branch-circuit overcurrent device, and having an equipment grounding conductor.

> FPN: See Table 250.122 for size of equipment grounding conductor.

(B) **Receptacles, Connectors, and Attachment Plugs** Receptacles, connectors, and attachment plugs shall be of a listed grounding type rated 15 or 20 amperes.

(C) **Support** Flexible cords shall be secured to the undersides of showcases such that all of the following conditions are ensured:

(1) The wiring is not exposed to mechanical damage.
(2) The separation between cases is not in excess of 50 mm (2 in.), or more than 300 mm (12 in.) between the first case and the supply receptacle.
(3) The free lead at the end of a group of showcases has a female fitting not extending beyond the case.

(D) No Other Equipment Equipment other than show-cases shall not be electrically connected to showcases.

(E) Secondary Circuit(s) Where showcases are cord-connected, the secondary circuit(s) of each electric-discharge lighting ballast shall be limited to one showcase.

410.30 Cord-Connected Lampholders and Luminaires (Fixtures)

(A) Lampholders Where a metal lampholder is attached to a flexible cord, the inlet shall be equipped with an insulating bushing that, if threaded, is not smaller than metric designator 12 (trade size ⅜) pipe size. The cord hole shall be of a size appropriate for the cord, and all burrs and fins shall be removed in order to provide a smooth bearing surface for the cord.

Bushing having holes 7 mm (9⁄32 in.) in diameter shall be permitted for use with plain pendant cord and holes 11 mm (13⁄32 in.) in diameter with reinforced cord.

Metal lampholders (brass- and aluminum-shell type) used with flexible-cord pendants are required to be equipped with smooth and permanently secured insulating bushings. Non-metallic-type lampholders do not require a bushing because the material and design afford equivalent protection.

(B) Adjustable Luminaires (Fixtures) Luminaires (fixtures) that require adjusting or aiming after installation shall not be required to be equipped with an attachment plug or cord connector, provided the exposed cord is of the hard-usage or extra-hard-usage type and is not longer than that required for maximum adjustment. The cord shall not be subject to strain or physical damage.

(C) Electric-Discharge Luminaires (Fixtures)

(1) Cord-Connected Installation A listed luminaire (fixture) or a listed assembly shall be permitted to be cord connected if the following conditions apply:

(1) The luminaire (fixture) is located directly below the outlet or busway.
(2) The flexible cord meets all the following:
  a. Is visible for its entire length outside the luminaire (fixture)
  b. Is not subject to strain or physical damage
  c. Is terminated in a grounding-type attachment plug cap or busway plug, or is a part of a listed assembly incorporating a manufactured wiring system connector in accordance with 604.6(C), or has a luminaire (fixture) assembly with a strain relief and canopy

Section 410.30(C)(1)(2)(c) was expanded for the 2005 *Code* to include a listed manufactured wiring system connector

as part of the fabricated assembly. This revision helps to clarify that a manufactured wiring system connector is permitted to be used to supply electric-discharge luminaires in place of a grounding type attachment plug.

Section 410.30(C)(1) applies to listed cord-and-plug-connected electric-discharge luminaires, such as the luminaire illustrated in Exhibit 410.5, and electric-discharge luminaire assemblies. The supply cord is not permitted to penetrate a suspended ceiling, because the cord is required to be continuously visible along its entire length.

Exhibit 410.5 A listed cord-and-plug-connected electric-discharge luminaire.

Supply cords cannot be used as a supporting means, and the luminaires are suspended directly below the outlet boxes supplying each luminaire. Electric-discharge luminaires are not permitted to be supplied by cord if they are installed in lift-out-type suspended ceilings. If the electric-discharge luminaire is suspended below the lift-out-type ceiling, the cord is not permitted to penetrate the ceiling. Section 400.8 further explains the uses not permitted for cords. According to 368.17(C), Exceptions No. 2 and No. 3, electric-discharge luminaires are permitted to be connected to busways by cords plugged directly into the busway or suspended from the busway.

(2) Provided with Mogul-Base, Screw-Shell Lampholders Electric-discharge luminaires (lighting fixtures) provided with mogul-base, screw-shell lampholders shall be permitted to be connected to branch circuits of 50 amperes or less by cords complying with 240.5. Receptacles and attachment plugs shall be permitted to be of a lower ampere rating than the branch circuit but not less than 125 percent of the luminaire (fixture) full-load current.

(3) Equipped with Flanged Surface Inlet Electric-discharge luminaires (lighting fixtures) equipped with a flanged surface inlet shall be permitted to be supplied by cord pendants equipped with cord connectors. Inlets and connectors shall be permitted to be of a lower ampere rating than the branch circuit but not less than 125 percent of the luminaire (fixture) load current.

410.31 Luminaires (Fixtures) as Raceways

Luminaires (fixtures) shall not be used as a raceway for circuit conductors unless listed and marked for use as a raceway.

According to the UL *Luminaire Marking Guide*, luminaires listed for use as raceways are marked "Suitable for Use as a Raceway" and also "Maximum of ____°C permitted in raceway." Without these markings, a row of luminaires connected end to end cannot be used as a raceway for circuit conductors other than the 2-wire or multiwire circuit supplying the luminaires. Luminaires identified for use as a raceway have been evaluated for the heat contribution caused by additional current-carrying conductors.

410.32 Wiring Supplying Luminaires (Fixtures) Connected Together

Luminaires (fixtures) designed for end-to-end connection to form a continuous assembly, or luminaires (fixtures) connected together by recognized wiring methods, shall be permitted to contain the conductors of a 2-wire branch circuit, or one multiwire branch circuit, supplying the connected luminaires (fixtures) and need not be listed as a raceway. One additional 2-wire branch circuit separately supplying one or more of the connected luminaires (fixtures) shall also be permitted.

The provisions of 410.32 facilitate convenient switching and supply circuit arrangements for a physically continuous row of luminaires or a row that is made continuous via the wiring method. A single 2-wire or a single multiwire branch circuit supplying the luminaires is permitted to be run through the continuous row(s), and the luminaires are not required to be listed for use as a raceway. An additional 2-wire branch circuit is permitted to be run through these luminaires. This circuit may supply only luminaires in the connected row(s) and is commonly employed to switch night lighting as an energy conservation method.

> FPN: See Article 100 for the definition of *Multiwire Branch Circuit.*

410.33 Branch Circuit Conductors and Ballasts

Branch-circuit conductors within 75 mm (3 in.) of a ballast shall have an insulation temperature rating not lower than

90°C (194°F) unless supplying a luminaire (fixture) listed and marked as suitable for a different insulation temperature.

Temperature ratings, along with other insulated conductor specifications, are found in Table 310.13. Note the 90°C rating that may be applied to Type THW conductors for special application in electric-discharge lighting equipment.

VII. Construction of Luminaires (Fixtures)

410.34 Combustible Shades and Enclosures

Adequate airspace shall be provided between lamps and shades or other enclosures of combustible material.

410.35 Luminaire (Fixture) Rating

(A) Marking All luminaires (fixtures) shall be marked with the maximum lamp wattage or electrical rating, manufacturer's name, trademark, or other suitable means of identification. A luminaire (fixture) requiring supply wire rated higher than 60°C (140°F) shall be marked in letters not smaller than 6 mm (¼ in.) high, prominently displayed on the luminaire (fixture) and shipping carton or equivalent.

(B) Electrical Rating The electrical rating shall include the voltage and frequency and shall indicate the current rating of the unit, including the ballast, transformer, or autotransformer.

410.36 Design and Material

Luminaires (fixtures) shall be constructed of metal, wood, or other material suitable for the application and shall be designed and assembled so as to secure requisite mechanical strength and rigidity. Wiring compartments, including their entrances, shall be designed and constructed to permit conductors to be drawn in and withdrawn without physical damage.

410.37 Nonmetallic Luminaires (Fixtures)

When luminaire (fixture) wiring compartments are constructed from combustible material, armored or lead-covered conductors with suitable fittings shall be used or the wiring compartment shall be lined with metal.

410.38 Mechanical Strength

(A) Tubing for Arms Tubing used for arms and stems where provided with cut threads shall not be less than 1.02 mm (0.040 in.) in thickness and, where provided with rolled (pressed) threads, shall not be less than 0.64 mm (0.025 in.) in thickness. Arms and other parts shall be fastened to prevent turning.

(B) Metal Canopies Metal canopies supporting lampholders, shades, and so forth exceeding 4 kg (8 lb), or incorporating attachment-plug receptacles, shall not be less than 0.51 mm (0.020 in.) in thickness. Other canopies shall not be less than 0.41 mm (0.016 in.) if made of steel and not less than 0.51 mm (0.020 in.) if of other metals.

(C) Canopy Switches Pull-type canopy switches shall not be inserted in the rims of metal canopies that are less than 0.64 mm (0.025 in.) in thickness, unless the rims are reinforced by the turning of a bead or the equivalent. Pull-type canopy switches, whether mounted in the rims or elsewhere in sheet metal canopies, shall not be located more than 90 mm (3½ in.) from the center of the canopy. Double set-screws, double canopy rings, a screw ring, or equal method shall be used where the canopy supports a pull-type switch or pendant receptacle.

The thickness requirements in the preceding paragraph shall apply to measurements made on finished (formed) canopies.

410.39 Wiring Space

Bodies of luminaires (fixtures), including portable lamps, shall provide ample space for splices and taps and for the installation of devices, if any. Splice compartments shall be of nonabsorbent, noncombustible material.

410.42 Portable Lamps

(A) General Portable lamps shall be wired with flexible cord recognized by 400.4 and an attachment plug of the polarized or grounding type. Where used with Edison-base lampholders, the grounded conductor shall be identified and attached to the screw shell and the identified blade of the attachment plug.

(B) Portable Handlamps In addition to the provisions of 410.42(A), portable handlamps shall comply with the following:

(1) Metal shell, paper-lined lampholders shall not be used.
(2) Handlamps shall be equipped with a handle of molded composition or other insulating material.
(3) Handlamps shall be equipped with a substantial guard attached to the lampholder or handle.
(4) Metallic guards shall be grounded by means of an equipment grounding conductor run with circuit conductors within the power-supply cord.
(5) Portable handlamps shall not be required to be grounded where supplied through an isolating transformer with an ungrounded secondary of not over 50 volts.

See Exhibit 410.6 for an example of a portable handlamp (drop light) meeting the requirements of 410.42.

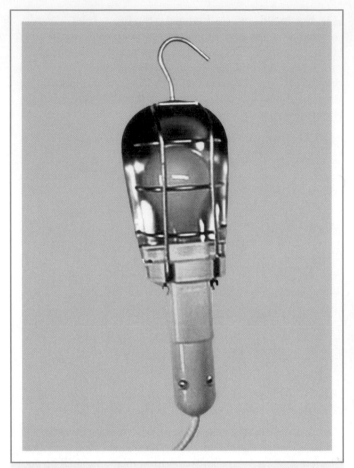

Exhibit 410.6 A portable handlamp with a grounded metal guard and reflector and a swivel-type hook that permits the lamp to be adjusted in different positions. (Courtesy of Daniel Woodhead Co.)

410.44 Cord Bushings

A bushing or the equivalent shall be provided where flexible cord enters the base or stem of a portable lamp. The bushing shall be of insulating material unless a jacketed type of cord is used.

410.45 Tests

All wiring shall be free from short circuits and grounds and shall be tested for these defects prior to being connected to the circuit.

410.46 Live Parts

Exposed live parts within porcelain luminaires (fixtures) shall be suitably recessed and located so as to make it improbable that wires come in contact with them. There shall be a spacing of at least 13 mm (½ in.) between live parts and the mounting plane of the luminaire (fixture).

VIII. Installation of Lampholders

410.47 Screw-Shell Type

Lampholders of the screw-shell type shall be installed for use as lampholders only. Where supplied by a circuit having a grounded conductor, the grounded conductor shall be connected to the screw shell.

Many years ago it was common practice to install screw shell lampholders with screw shell adapters in baseboards and walls to connect cord-connected appliances and lighting equipment. This now-prohibited practice permitted exposed live parts to be contacted by persons when the adapters were removed. See 406.2(B) for permitted uses of receptacles.

410.48 Double-Pole Switched Lampholders

Where supplied by the ungrounded conductors of a circuit, the switching device of lampholders of the switched type shall simultaneously disconnect both conductors of the circuit.

Single-pole switching may be used to interrupt the ungrounded conductor of a 2-wire circuit having one conductor grounded. The grounded conductor must be connected to the screw shell of the lampholder.

Where a 2-wire circuit is derived from the two ungrounded conductors of a multiwire circuit (3- or 4-wire system) or from the two ungrounded conductors of a 2-wire circuit (3-wire system) and is used with switched lampholders, the switching device is required to be double-pole and to simultaneously disconnect both ungrounded conductors of the circuit. See 410.52 for information on the construction of switched lampholders.

410.49 Lampholders in Wet or Damp Locations

Lampholders installed in wet or damp locations shall be of the weatherproof type.

IX. Construction of Lampholders

410.50 Insulation

The outer metal shell and the cap shall be lined with insulating material that prevents the shell and cap from becoming a part of the circuit. The lining shall not extend beyond the metal shell more than 3 mm (⅛ in.) but shall prevent any current-carrying part of the lamp base from being exposed when a lamp is in the lampholding device.

410.52 Switched Lampholders

Switched lampholders shall be of such construction that the switching mechanism interrupts the electrical connection to the center contact. The switching mechanism shall also be permitted to interrupt the electrical connection to the screw shell if the connection to the center contact is simultaneously interrupted.

X. Lamps and Auxiliary Equipment

410.53 Bases, Incandescent Lamps

An incandescent lamp for general use on lighting branch circuits shall not be equipped with a medium base if rated over 300 watts, or with a mogul base if rated over 1500 watts. Special bases or other devices shall be used for over 1500 watts.

410.54 Electric-Discharge Lamp Auxiliary Equipment

(A) Enclosures Auxiliary equipment for electric-discharge lamps shall be enclosed in noncombustible cases and treated as sources of heat.

The UL *General Information Electrical Equipment Directory* contains two categories for ballasts under Electric Discharge Lamp Control Equipment (FKOT): fluorescent ballasts (FKVS) and HID (high intensity discharge) ballasts (FLCR).

Fluorescent ballast enclosures are categorized by UL as open, indoor, outdoor (both Type 1 and Type 2), and weatherproof. HID ballasts are categorized the same, except there is no open-type ballast for HID ballasts.

Open Type. Open core and coil constructions (i.e., ballasts without complete metal enclosures) are intended for use within suitable enclosures.

Indoor Ballasts. Indoor ballasts are suitable for use in an indoor, dry location only.

Outdoor Ballasts. Type 1 outdoor ballasts are suitable for use in (1) outdoor equipment, (2) fixtures intended for wet or damp locations, or (3) an outdoor sign if the ballasts are within an overall electrical enclosure. Ballasts of this type are marked "Type 1 Outdoor" or "Type 1." Type 2 outdoor ballasts are suitable for use in (1) outdoor equipment, (2) fixtures intended for wet or damp locations, or (3) an outdoor sign if the ballasts, in addition to their own enclosure, are within an overall enclosure. Ballasts of this type are marked "Type 2 Outdoor" or "Type 2."

Weatherproof Ballasts. Weatherproof ballasts are suitable for use where completely exposed to the weather without an additional enclosure and are marked "Weatherproof" or "WP."

(B) Switching Where supplied by the ungrounded conductors of a circuit, the switching device of auxiliary equipment shall simultaneously disconnect all conductors.

XI. Special Provisions for Flush and Recessed Luminaires (Fixtures)

410.64 General

Luminaires (fixtures) installed in recessed cavities in walls or ceilings shall comply with 410.65 through 410.72.

410.65 Temperature

(A) Combustible Material Luminaires (fixtures) shall be installed so that adjacent combustible material will not be subjected to temperatures in excess of 90°C (194°F).

(B) Fire-Resistant Construction Where a luminaire (fixture) is recessed in fire-resistant material in a building of fire-resistant construction, a temperature higher than 90°C (194°F) but not higher than 150°C (302°F) shall be considered acceptable if the luminaire (fixture) is plainly marked that it is listed for that service.

(C) Recessed Incandescent Luminaires (Fixtures) Incandescent luminaires (fixtures) shall have thermal protection and shall be identified as thermally protected.

Because many recessed incandescent luminaires are suitable for a wide variety of lamp sizes and types and finish trims, the temperature close to the lamp can vary widely. Therefore, many manufacturers have chosen to locate thermal protectors away from the source of heat, such as in the outlet box, and to design the protector so that it detects a change in temperature resulting from the addition of thermal insulation around the luminaire. This design prevents nuisance tripping of the protector (as a result of changing lamp wattage, for example) but still provides protection against overheating arising from thermal insulation around a recessed luminaire not designed for such use.

Exception No. 1: Thermal protection shall not be required in a recessed luminaire (fixture) identified for use and installed in poured concrete.

Exception No. 2: Thermal protection shall not be required in a recessed luminaire (fixture) whose design, construction, and thermal performance characteristics are equivalent to a thermally protected luminaire (fixture) and are identified as inherently protected.

Recessed incandescent luminaires without thermal protection are not permitted unless they are listed and identified as providing equivalent temperature protection by construction design.

Exhibit 410.7 illustrates a listed Type IC recessed luminaire installed in direct contact with thermal insulation. Thermal protection is provided to deactivate the lamp should the luminaire be mislamped so that it overheats.

Exhibit 410.7 A listed Type IC recessed luminaire suitable for use in insulated ceilings installed in direct contact with thermal insulation. (Redrawn courtesy of Thomas Industries Inc.)

410.66 Clearance and Installation

(A) Clearance

(1) Non-Type IC A recessed luminaire (fixture) that is not identified for contact with insulation shall have all recessed parts spaced not less than 13 mm (½ in.) from combustible materials. The points of support and the trim finishing off the opening in the ceiling or wall surface shall be permitted to be in contact with combustible materials.

(2) Type IC A recessed luminaire (fixture) that is identified for contact with insulation, Type IC, shall be permitted to be in contact with combustible materials at recessed parts, points of support, and portions passing through or finishing off the opening in the building structure.

(B) Installation Thermal insulation shall not be installed above a recessed luminaire (fixture) or with 75 mm (3 in.) of the recessed luminaire's (fixture's) enclosure, wiring compartment, or ballast unless it is identified for contact with insulation, Type IC.

410.67 Wiring

(A) General Conductors that have insulation suitable for the temperature encountered shall be used.

(B) Circuit Conductors Branch-circuit conductors that have an insulation suitable for the temperature encountered shall be permitted to terminate in the luminaire (fixture).

(C) Tap Conductors Tap conductors of a type suitable for the temperature encountered shall be permitted to run from the luminaire (fixture) terminal connection to an outlet box placed at least 300 mm (1 ft) from the luminaire (fixture). Such tap conductors shall be in suitable raceway or Type AC or MC cable of at least 450 mm (18 in.) but not more than 1.8 m (6 ft) in length.

XII. Construction of Flush and Recessed Luminaires (Fixtures)

410.68 Temperature

Luminaires (fixtures) shall be constructed such that adjacent combustible material is not subject to temperatures in excess of 90°C (194°F).

410.70 Lamp Wattage Marking

Incandescent lamp luminaires (fixtures) shall be marked to indicate the maximum allowable wattage of lamps. The markings shall be permanently installed, in letters at least 6 mm (¼ in.) high, and shall be located where visible during relamping.

410.71 Solder Prohibited

No solder shall be used in the construction of a luminaire (fixture) box.

410.72 Lampholders

Lampholders of the screw-shell type shall be of porcelain or other suitable insulating materials. Where used, cements shall be of the high-heat type.

XIII. Special Provisions for Electric-Discharge Lighting Systems of 1000 Volts or Less

410.73 General

(A) Open-Circuit Voltage of 1000 Volts or Less Equipment for use with electric-discharge lighting systems and

designed for an open-circuit voltage of 1000 volts or less shall be of a type intended for such service.

(B) Considered as Energized The terminals of an electric-discharge lamp shall be considered as energized where any lamp terminal is connected to a circuit of over 300 volts.

(C) Transformers of the Oil-Filled Type Transformers of the oil-filled type shall not be used.

(D) Additional Requirements In addition to complying with the general requirements for luminaires (lighting fixtures), such equipment shall comply with Part XIII of this article.

(E) Thermal Protection — Fluorescent Luminaires (Fixtures)

(1) Integral Thermal Protection The ballast of a fluorescent luminaire (fixture) installed indoors shall have integral thermal protection. Replacement ballasts shall also have thermal protection integral with the ballast.

(2) Simple Reactance Ballasts A simple reactance ballast in a fluorescent luminaire (fixture) with straight tubular lamps shall not be required to be thermally protected.

(3) Exit Fixtures A ballast in a fluorescent exit luminaire (fixture) shall not have thermal protection.

(4) Egress Luminaires (Fixtures) A ballast in a fluorescent luminaire (fixture) that is used for egress lighting and energized only during a failure of the normal supply shall not have thermal protection.

Thermal protection that is integral with the ballast is required for fluorescent luminaires installed indoors. Thermally protected ballasts are required as replacements for nonthermally protected ballasts in older fixtures. Thermally protected fluorescent lamp ballasts intended for use in accordance with 410.73(E) are marked "Class P."

Because different Class P ballasts have different heating characteristics, the heating characteristics should be considered when selecting replacements for nonthermally protected ballasts. This type of ballast protection is set to open the circuit at a predetermined temperature, to prevent abnormal ballast heat buildup caused by a fault in one or more of the ballast components or by some lampholder or wiring fault.

Section 410.73(E)(3) exempts exit sign fixtures from the thermal protection requirement because overheating during high ambient conditions could cause the thermal protection to operate. This action could impair evacuation during a fire.

Section 410.73(E)(4) exempts egress lighting from the thermal protection requirement for the same reason that exit signs are exempt. However, this exemption applies to egress lighting that is energized only during the emergency condition.

Exhibit 410.8 illustrates a reactance-type ballast used in series with a preheat-type fluorescent lamp 30 watts or less. This type of ballast does not require thermal protection, and the luminaire may be equipped with automatic-type starters (such as used with medicine cabinet luminaires) or a manual momentary contact starter (such as used with desk lamps and some small under-cabinet luminaires).

Exhibit 410.8 The circuitry for a simple reactance-type ballast for fluorescent lighting.

(F) High-Intensity Discharge Luminaires (Fixtures)

(1) Recessed Recessed high-intensity luminaires (fixtures) designed to be installed in wall or ceiling cavities shall have thermal protection and be identified as thermally protected.

(2) Inherently Protected Thermal protection shall not be required in a recessed high-intensity luminaire (fixture) whose design, construction, and thermal performance characteristics are equivalent to a thermally protected luminaire (fixture) and are identified as inherently protected.

(3) Installed in Poured Concrete Thermal protection shall not be required in a recessed high-intensity discharge luminaire (fixture) identified for use and installed in poured concrete.

(4) Recessed Remote Ballasts A recessed remote ballast for a high-intensity discharge luminaire (fixture) shall have thermal protection that is integral with the ballast and be identified as thermally protected.

(5) Metal Halide Lamp Containment Luminaires (fixtures) that use a metal halide lamp other than a thick-glass parabolic reflector lamp (PAR) shall be provided with a containment barrier that encloses the lamp, or shall be pro-

vided with a physical means that only allows the use of a lamp that is Type O.

Section 410.73(F)(5) was added to the 2005 *Code* because of fire reports from different agencies. Metal halide lamps have been identified as the likely cause of ignition in several recent major fires by insurers of large industrial facilities. HID lighting fixtures can create an ignition source when physical damage occurs to the arc tube in open luminaires. Lamp types that are suitable for use in open luminaires (those that do not require lamp enclosures) are classified by the lamp manufacturers as Type O or Type S. Type-O lamps are provided with a shroud around the arc tube and are containment-tested in accordance with ANSI C78.387, *Electric Lamps — Metal-Halide Lamps — Methods of Measuring Characteristics*, and are rated for use in open luminaires. The *Code* now requires luminaires that use a metal halide lamp to have either a containment barrier that encloses the lamp or some means that allows only a Type-O lamp to be installed in the luminaire.

FPN: See ANSI Standard C78.387, *American National Standard for Electric Lamps — Metal Halide Lamps, Methods of Measuring Characteristics*.

(G) Disconnecting Means In indoor locations, other than dwellings and associated accessory structures, fluorescent luminaires (fixtures) that utilize double-ended lamps and contain ballast(s) that can be serviced in place or ballasted luminaires that are supplied from multiwire branch circuits and contain ballast(s) that can be serviced in place shall have a disconnecting means either internal or external to each luminaire (fixture), to disconnect simultaneously from the source of supply all conductors of the ballast, including the grounded conductor if any. The line side terminals of the disconnecting means shall be guarded. The disconnecting means shall be located so as to be accessible to qualified persons before servicing or maintaining the ballast. This requirement shall become effective January 1, 2008.

Section 410.73(G) was added to the 2005 *Code* with an effective date of January 1, 2008. It requires a disconnecting means for the following types of fluorescent luminaires that can be serviced in place:

1. Luminaires that utilize double ended lamps
2. Luminaires containing a ballast(s) and supplied from multiwire branch circuits

The disconnect can be either inside or outside the luminaire and must disconnect all supply conductors simultaneously, including the grounded conductor if one is provided. Exceptions are provided for; hazardous (classified) locations, emergency illumination, cord-and-plug connected luminaires, industrial facilities, and luminaires not supplied by a multiwire branch circuit and in which disconnection does not leave the illuminated space in total darkness.

Exception No. 1: A disconnecting means shall not be required for luminaires (fixtures) installed in hazardous (classified) location(s).

Exception No. 2: A disconnecting means shall not be required for emergency illumination required in 700.16.

Exception No. 3: For cord-and-plug-connected luminaires, an accessible separable connector or an accessible plug and receptacle shall be permitted to serve as the disconnecting means.

Exception No. 4: A disconnecting means shall not be required in industrial establishments with restricted public access where conditions of maintenance and supervision ensure that only qualified persons service the installation by written procedures.

Exception No. 5: Where more than one luminaire is installed and supplied by other than a multiwire branch circuit, a disconnecting means shall not be required for every luminaire when the design of the installation includes locally accessible disconnects, such that the illuminated space cannot be left in total darkness.

410.74 Direct-Current Equipment

Luminaires (fixtures) installed on dc circuits shall be equipped with auxiliary equipment and resistors designed for dc operation. The luminaires (fixtures) shall be marked for dc operation.

410.75 Open-Circuit Voltage Exceeding 300 Volts

Equipment having an open-circuit voltage exceeding 300 volts shall not be installed in dwelling occupancies unless such equipment is designed so that there will be no exposed live parts when lamps are being inserted, are in place, or are being removed.

Luminaires intended for use in nondwelling occupancies are so marked. This marking usually indicates that the luminaire has maintenance features that are considered beyond the capabilities of the ordinary homeowner or that the luminaire involves voltages in excess of those permitted by this *Code* for dwelling occupancies. For other references to voltage limitations within dwelling units, see 210.6(A), 210.6(B), and 410.80(B).

410.76 Luminaire (Fixture) Mounting

(A) Exposed Ballasts Luminaires (fixtures) that have exposed ballasts or transformers shall be installed so that such ballasts or transformers will not be in contact with combustible material.

(B) Combustible Low-Density Cellulose Fiberboard Where a surface-mounted luminaire (fixture) containing a ballast is to be installed on combustible low-density cellulose fiberboard, it shall be listed for this condition or shall be spaced not less than 38 mm (1½ in.) from the surface of the fiberboard. Where such luminaires (fixtures) are partially or wholly recessed, the provisions of 410.64 through 410.72 shall apply.

FPN: Combustible low-density cellulose fiberboard includes sheets, panels, and tiles that have a density of 320 kg/m³ (20 lb/ft³) or less and that are formed of bonded plant fiber material but does not include solid or laminated wood or fiberboard that has a density in excess of 320 kg/m³ (20 lb/ft³) or is a material that has been integrally treated with fire-retarding chemicals to the degree that the flame spread in any plane of the material will not exceed 25, determined in accordance with tests for surface burning characteristics of building materials. See ANSI/ASTM E84-1997, *Test Method for Surface Burning Characteristics of Building Materials.*

Fluorescent lamp luminaires intended for mounting on combustible low-density cellulose fiberboard ceilings have been evaluated with thermal insulation above the ceiling in the vicinity of the luminaire and are marked "Suitable for Surface Mounting on Combustible Low-Density Cellulose Fiberboard."

Fluorescent lamp luminaires not so marked may be directly mounted against a ceiling surface constructed of a material other than combustible low-density fiberboard or may be spaced not less than 1½ in. from the surface of the low-density fiberboard.

410.77 Equipment Not Integral with Luminaire (Fixture)

(A) Metal Cabinets Auxiliary equipment, including reactors, capacitors, resistors, and similar equipment, where not installed as part of a luminaire (lighting fixture) assembly, shall be enclosed in accessible, permanently installed metal cabinets.

(B) Separate Mounting Separately mounted ballasts that are intended for direct connection to a wiring system shall not be required to be separately enclosed.

(C) Wired Luminaire (Fixture) Sections Wired luminaire (fixture) sections are paired, with a ballast(s) supplying a lamp or lamps in both. For interconnection between paired units, it shall be permissible to use metric designator 12 (trade size ⅜) flexible metal conduit in lengths not exceeding 7.5 m (25 ft), in conformance with Article 348. Luminaire (fixture) wire operating at line voltage, supplying only the ballast(s) of one of the paired luminaires (fixtures), shall be permitted in the same raceway as the lamp supply wires of the paired luminaires (fixtures).

Wired luminaire sections are shipped in pairs and marked for use in pairs. Each individual unit includes lamps in odd-numbered quantities (one or three is most common), with the odd lamp in each luminaire supplied by a two-lamp ballast located in one luminaire of the pair. Two-lamp ballasts are more energy efficient than single-lamp or three-lamp ballasts.

410.78 Autotransformers

An autotransformer that is used to raise the voltage to more than 300 volts, as part of a ballast for supplying lighting units, shall be supplied only by a grounded system.

410.79 Switches

Snap switches shall comply with 404.14.

XIV. Special Provisions for Electric-Discharge Lighting Systems of More Than 1000 Volts

Sections 410.80 through 410.87 apply to interior electric-discharge neon tube–type lighting that contains neon, helium, or argon gas, with or without mercury, at low vapor pressure; long-length fluorescent tube lighting requiring more than 1000 volts; and cold-cathode fluorescent lamp installations arranged to operate with several tubes in series.

410.80 General

(A) **Listing** Electric-discharge lighting systems with an open-circuit voltage exceeding 1000 volts shall be listed and installed in conformance with that listing.

(B) **Dwelling Occupancies** Equipment that has an open-circuit voltage exceeding 1000 volts shall not be installed in or on dwelling occupancies.

Section 410.80(B) specifically prohibits electric-discharge lighting systems, such as high-intensity-discharge and fluorescent lighting, that have an open-circuit voltage greater than 1000 volts to be installed in dwelling occupancies. These lighting systems are often used as decorative lighting as well as for outline lighting and signs. Installation of neon tubing as an art form is covered by Article 600.

(C) **Live Parts** The terminal of an electric-discharge lamp shall be considered as a live part.

(D) **Additional Requirements** In addition to complying with the general requirements for luminaires (lighting fixtures), such equipment shall comply with Part XIV of this article.

FPN: For signs and outline lighting, see Article 600.

410.81 Control

(A) **Disconnection** Luminaires (fixtures) or lamp installation shall be controlled either singly or in groups by an externally operable switch or circuit breaker that opens all ungrounded primary conductors.

(B) **Within Sight or Locked Type** The switch or circuit breaker shall be located within sight from the luminaires (fixtures) or lamps, or it shall be permitted elsewhere if it is provided with a means for locking in the open position.

The requirement in 410.81(B) is intended to help protect the service person from the disconnecting means being turned on or closed while the equipment is being serviced.

410.82 Lamp Terminals and Lampholders

Parts that must be removed for lamp replacement shall be hinged or held captive. Lamps or lampholders shall be designed so that there are no exposed live parts when lamps are being inserted or removed.

410.83 Transformers

(A) **Type** Transformers shall be enclosed, identified for the use, and listed.

(B) **Voltage** The secondary-circuit voltage shall not exceed 15,000 volts, nominal, under any load condition. The voltage to ground of any output terminals of the secondary circuit shall not exceed 7500 volts, under any load conditions.

(C) **Rating** Transformers shall have a secondary short-circuit current rating of not more than 150 mA if the open-circuit voltage is over 7500 volts, and not more than 300 mA if the open-circuit voltage rating is 7500 volts or less.

(D) **Secondary Connections** Secondary circuit outputs shall not be connected in parallel or in series.

410.84 Transformer Locations

(A) **Accessible** Transformers shall be accessible after installation.

(B) Secondary Conductors Transformers shall be installed as near to the lamps as practicable to keep the secondary conductors as short as possible.

(C) Adjacent to Combustible Materials Transformers shall be located so that adjacent combustible materials are not subjected to temperatures in excess of 90°C (194°F).

410.85 Exposure to Damage

Lamps shall not be located where normally exposed to physical damage.

410.86 Marking

Each luminaire (fixture) or each secondary circuit of tubing having an open-circuit voltage of over 1000 volts shall have a clearly legible marking in letters not less than 6 mm (¼ in.) high reading "Caution ____ volts." The voltage indicated shall be the rated open-circuit voltage.

410.87 Switches

Snap switches shall comply with 404.4.

Refer to 600.32(A) for wiring methods for electric-discharge neon tube–type lighting.

XV. Lighting Track

410.100 Definition

Lighting Track. A manufactured assembly designed to support and energize luminaires (lighting fixtures) that are capable of being readily repositioned on the track. Its length can be altered by the addition or subtraction of sections of track.

410.101 Installation

(A) Lighting Track Lighting track shall be permanently installed and permanently connected to a branch circuit. Only lighting track fittings shall be installed on lighting track. Lighting track fittings shall not be equipped with general-purpose receptacles.

A lighting track fitting differs from a fitting as defined in Article 100 in that it usually performs both an electrical and a mechanical function. Such assemblies are not intended to be used for locating convenience receptacles or as an alternative for required receptacle outlets such as those required in 210.62 for show windows. Lighting track can be removed and relocated and therefore is not a substitute for required outlets.

(B) Connected Load The connected load on lighting track shall not exceed the rating of the track. Lighting track shall be supplied by a branch circuit having a rating not more than that of the track.

Section 220.43(B) addresses track lighting loads.

The volt-ampere (VA) load for 2 ft of track is 150 VA because a value of 150 VA is more consistent with standard lamp values for 2 ft of track. It should be understood that 220.43(B) is not intended to limit the number of feet of track on a single branch circuit, nor is it intended to limit the number of fixtures on an individual track. Rather, 220.43(B) is intended to be used for load calculations of feeders and services.

Example

Suppose a lighting plan shows 62.5 ft of single circuit lighting track for a small department store featuring clothing. Because the actual track luminaires are owner supplied, neither the quantity of track luminaires nor the lamp size is specified. What is the minimum calculated load associated with the lighting track that must be added to the service or feeder supplying this store?

Solution

According to 220.43(B), the minimum calculated load to be added to the service or feeder supplying this track light installation is determined as follows:

$$62.5 \text{ ft} \div 2 \text{ ft} = 31.25 \text{ ft} \quad (\text{round up to 32})$$

$$32 \times 150 \text{ VA} = 4800 \text{ VA}$$

The minimum load that must be added to the service or feeder is 4800 VA.

It is important to note that the branch circuits supplying this installation are covered in 410.101(B). For the lighting track branch-circuit load, the maximum load on the track cannot exceed the rating of the branch circuit supplying the track. Also, the track must be supplied by a branch circuit having a rating not exceeding the rating of the track. The track length does not enter into the branch-circuit calculation.

(C) Locations Not Permitted Lighting track shall not be installed in the following locations:

(1) Where likely to be subjected to physical damage
(2) In wet or damp locations
(3) Where subject to corrosive vapors
(4) In storage battery rooms
(5) In hazardous (classified) locations
(6) Where concealed
(7) Where extended through walls or partitions

(8) Less than 1.5 m (5 ft) above the finished floor except where protected from physical damage or track operating at less than 30 volts rms open-circuit voltage

(9) Where prohibited by 410.4(D)

Low-voltage lighting track operating at less than 30 volts is permitted to be installed less than 5 ft above the floor.

(D) Support Fittings identified for use on lighting track shall be designed specifically for the track on which they are to be installed. They shall be securely fastened to the track, shall maintain polarization and grounding, and shall be designed to be suspended directly from the track.

410.103 Heavy-Duty Lighting Track

Heavy-duty lighting track is lighting track identified for use exceeding 20 amperes. Each fitting attached to a heavy-duty lighting track shall have individual overcurrent protection.

410.104 Fastening

Lighting track shall be securely mounted so that each fastening is suitable for supporting the maximum weight of luminaires (fixtures) that can be installed. Unless identified for supports at greater intervals, a single section 1.2 m (4 ft) or shorter in length shall have two supports, and, where installed in a continuous row, each individual section of not more than 1.2 m (4 ft) in length shall have one additional support.

410.105 Construction Requirements

(A) Construction The housing for the lighting track system shall be of substantial construction to maintain rigidity. The conductors shall be installed within the track housing, permitting insertion of a luminaire (fixture), and designed to prevent tampering and accidental contact with live parts. Components of lighting track systems of different voltages shall not be interchangeable. The track conductors shall be a minimum 12 AWG or equal and shall be copper. The track system ends shall be insulated and capped.

(B) Grounding Lighting track shall be grounded in accordance with Article 250, and the track sections shall be securely coupled to maintain continuity of the circuitry, polarization, and grounding throughout.

XVI. Decorative Lighting and Similar Accessories

410.110 Listing of Decorative Lighting

Decorative lighting and similar accessories used for holiday lighting and similar purposes, in accordance with 590.3(B), shall be listed.

Section 410.110 was added to the 2005 *Code* to require decorative lighting and their accessories to be listed.

ARTICLE 411
Lighting Systems Operating at 30 Volts or Less

Summary of Changes

- **411.4(A)(2):** Added new provision permitting concealed wiring where supplied by a listed Class 2 power source.
- **411.5(C):** Revised to limit the use of bare conductors and current-carrying parts to indoor use only.

Contents

411.1 Scope
411.2 Definition
411.3 Listing Required
411.4 Locations Not Permitted
411.5 Secondary Circuits
  (A) Grounding
  (B) Isolation
  (C) Bare Conductors
411.6 Branch Circuit
411.7 Hazardous (Classified) Locations

411.1 Scope

This article covers lighting systems operating at 30 volts or less and their associated components.

Article 411 addresses low-voltage lighting systems. Article 411 is intended to cover low-voltage interior lighting and low-voltage exterior (landscape) lighting installations. It covers systems operating at 30 volts rms or 42.4 volts peak or less, having a maximum output of 600 VA.

411.2 Definition

Lighting Systems Operating at 30 Volts or Less. A lighting system consisting of an isolating power supply operating at 30 volts (42.4 volts peak) or less under any load condition, with one or more secondary circuits, each limited to 25 amperes maximum, supplying luminaires (lighting fixtures) and associated equipment identified for the use.

411.3 Listing Required

Lighting systems operating at 30 volts or less shall be listed.

411.4 Locations Not Permitted

Lighting systems operating at 30 volts or less shall not be installed in the locations described in 411.4(A) and 411.4(B).

The installation requirements of 411.4 recognize that shock and fire hazards still exist, even with low-voltage systems, and this section was expanded in the 2005 *Code* to be more specific. Low-voltage illumination systems are allowed to be installed concealed or extending through a wall, provided the system is installed in accordance either with one of the wiring methods in Chapter 3 or with Section 725.52 and is supplied from a listed Class 2 power supply. These systems are not allowed to be installed within 10 ft of pools, spas, fountains, or similar locations, unless allowed by Article 680.

(A) Where concealed or extended through a building wall unless permitted in (1) or (2):

(1) Installed using any of the wiring methods specified in Chapter 3
(2) Installed using wiring supplied by a listed Class 2 power source and installed in accordance with 725.52

(B) Where installed within 3.0 m (10 ft) of pools, spas, fountains, or similar locations, unless permitted by Article 680.

411.5 Secondary Circuits

(A) Grounding Secondary circuits shall not be grounded.

(B) Isolation The secondary circuit shall be insulated from the branch circuit by an isolating transformer.

(C) Bare Conductors Exposed bare conductors and current-carrying parts shall be permitted for indoor installations only. Bare conductors shall not be installed less than 2.1 m (7 ft) above the finished floor, unless specifically listed for a lower installation height.

The permission to use bare conductors for low-voltage illumination is now limited to indoor installations only.

411.6 Branch Circuit

Lighting systems operating at 30 volts or less shall be supplied from a maximum 20-ampere branch circuit.

411.7 Hazardous (Classified) Locations

Where installed in hazardous (classified) locations, these systems shall conform with Articles 500 through 517 in addition to this article.

ARTICLE 422
Appliances

Summary of Changes

- **422.12:** Added Exception No. 2 permitting permanently connected air-conditioning equipment on the same branch circuit with central heating equipment.

- **422.13:** Added requirement that storage-type water heaters with a capacity of 120 gal or less be considered a continuous load.

- **422.16(B)(4):** Added new section permitting range hoods to be cord-and-plug connected under specified conditions.

- **422.18:** Revised to defer to 314.27(D) for weight restrictions on outlet boxes supporting ceiling paddle fans.

- **422.31(B):** Added new last sentence requiring the provision for locking be permanently installed on or at the circuit breaker or switch used as the disconnecting means and remain in place with or without the lock installed.

- **422.51:** Added requirement for GFCI protection of cord-and-plug-connected vending machines, effective for machines manufactured or re-manufactured on or after January 1, 2005.

Contents

I. General

422.1 Scope

This article covers electric appliances used in any occupancy.

Article 422 covers electric appliances that may be used in a dwelling unit or in commercial and industrial locations. It also covers appliances that may be fastened in place or be cord-and-plug-connected, such as air-conditioning units, dishwashers, heating appliances, water heaters, infrared heating lamps, and so on. See 422.3 for the requirements of other articles. Also see Article 100 for the definition of *appliance*.

422.3 Other Articles

The requirements of Article 430 shall apply to the installation of motor-operated appliances, and the requirements of Article 440 shall apply to the installation of appliances containing a hermetic refrigerant motor-compressor(s), except as specifically amended in this article.

422.4 Live Parts

Appliances shall have no live parts normally exposed to contact other than those parts functioning as open-resistance heating elements, such as the heating element of a toaster, which are necessarily exposed.

II. Installation

422.10 Branch-Circuit Rating

This section specifies the ratings of branch circuits capable of carrying appliance current without overheating under the conditions specified.

Conductors that form integral parts of appliances are tested as part of the listing or labeling process.

(A) Individual Circuits The rating of an individual branch circuit shall not be less than the marked rating of the appliance or the marked rating of an appliance having combined loads as provided in 422.62.

The rating of an individual branch circuit for motor-operated appliances not having a marked rating shall be in accordance with Part II of Article 430.

The branch-circuit rating for an appliance that is continuously loaded, other than a motor-operated appliance, shall not be less than 125 percent of the marked rating, or not less than 100 percent of the marked rating if the branch-circuit device and its assembly are listed for continuous loading at 100 percent of its rating.

Branch circuits for household cooking appliances shall be permitted to be in accordance with Table 220.55.

(B) Circuits Supplying Two or More Loads For branch circuits supplying appliance and other loads, the rating shall be determined in accordance with 210.23.

422.11 Overcurrent Protection

Appliances shall be protected against overcurrent in accordance with 422.11(A) through 422.11(G) and 422.10.

(A) Branch-Circuit Overcurrent Protection Branch circuits shall be protected in accordance with 240.4.

If a protective device rating is marked on an appliance, the branch-circuit overcurrent device rating shall not exceed the protective device rating marked on the appliance.

If a labeled or listed appliance is provided with installation instructions from the manufacturer, the branch-circuit size is not permitted to be less than the minimum size stated in the installation instructions. See 110.3(B) and its related commentary regarding the installation and use of listed or labeled equipment.

(B) Household-Type Appliances with Surface Heating Elements Household-type appliances with surface heating elements having a maximum demand of more than 60 amperes calculated in accordance with Table 220.19 shall have its power supply subdivided into two or more circuits, each of which shall be provided with overcurrent protection rated at not over 50 amperes.

(C) Infrared Lamp Commercial and Industrial Heating Appliances Infrared lamp commercial and industrial heating appliances shall have overcurrent protection not exceeding 50 amperes.

(D) Open-Coil or Exposed Sheathed-Coil Types of Surface Heating Elements in Commercial-Type Heating Appliances Open-coil or exposed sheathed-coil types of surface heating elements in commercial-type heating appliances shall be protected by overcurrent protective devices rated at not over 50 amperes.

(E) Single Non–Motor-Operated Appliance If the branch circuit supplies a single non–motor-operated appliance, the rating of overcurrent protection shall:

(1) Not exceed that marked on the appliance;
(2) Not exceed 20 amperes if the overcurrent protection rating is not marked and the appliance is rated 13.3 amperes or less; or
(3) Not exceed 150 percent of the appliance rated current if the overcurrent protection rating is not marked and the appliance is rated over 13.3 amperes. Where 150 percent of the appliance rating does not correspond to a standard overcurrent device ampere rating, the next higher standard rating shall be permitted.

(F) Electric Heating Appliances Employing Resistance-Type Heating Elements Rated More Than 48 Amperes

(1) Electric Heating Appliances Electric heating appliances employing resistance-type heating elements rated more than 48 amperes, other than household appliances with surface heating elements covered by 422.11(B), and commercial-type heating appliances covered by 422.11(D), shall have the heating elements subdivided. Each subdivided load shall not exceed 48 amperes and shall be protected at not more than 60 amperes.

These supplementary overcurrent protective devices shall be (1) factory-installed within or on the heater enclosure or provided as a separate assembly by the heater manufacturer; (2) accessible; and (3) suitable for branch-circuit protection.

The main conductors supplying these overcurrent protective devices shall be considered branch-circuit conductors.

(2) Commercial Kitchen and Cooking Appliances Commercial kitchen and cooking appliances using sheathed-type heating elements not covered in 422.11(D) shall be permitted to be subdivided into circuits not exceeding 120 amperes and protected at not more than 150 amperes where one of the following is met:

(1) Elements are integral with and enclosed within a cooking surface.
(2) Elements are completely contained within an enclosure identified as suitable for this use.
(3) Elements are contained within an ASME-rated and stamped vessel.

(3) Water Heaters and Steam Boilers Water heaters and steam boilers employing resistance-type immersion electric heating elements contained in an ASME-rated and stamped vessel or listed instantaneous water heaters shall be permitted to be subdivided into circuits not exceeding 120 amperes and protected at not more than 150 amperes.

(G) Motor-Operated Appliances Motors of motor-operated appliances shall be provided with overload protection in accordance with Part III of Article 430. Hermetic refrigerant motor-compressors in air-conditioning or refrigerating equipment shall be provided with overload protection in accordance with Part VI of Article 440. Where appliance overcurrent protective devices that are separate from the appliance are required, data for selection of these devices shall be marked on the appliance. The minimum marking shall be that specified in 430.7 and 440.4.

422.12 Central Heating Equipment

Central heating equipment other than fixed electric space-heating equipment shall be supplied by an individual branch circuit.

Exception No. 1 to 422.12 permits the electric motors, ignition systems, controls, and so on, of fossil-fuel-fired central heating equipment to be connected to the same individual branch circuit, as defined in Article 100 under *branch circuit, individual.*

Exception No. 2 was added to the 2005 *Code* to allow a permanently connected air-conditioning unit to be supplied from a branch circuit that supplies central heating equipment other than fixed electric space heating equipment, because central heating equipment and air-conditioning equipment are considered to be noncoincident loads, meaning that the two loads are unlikely to operate at the same time.

Exception No. 1: Auxiliary equipment, such as a pump, valve, humidifier, or electrostatic air cleaner directly associated with the heating equipment, shall be permitted to be connected to the same branch circuit.

Exception No. 2: Permanently connected air-conditioning equipment shall be permitted to be connected to the same branch circuit.

422.13 Storage-Type Water Heaters

A fixed storage-type water heater that has a capacity of 450 L (120 gal) or less shall be considered a continuous load.

The revision to 422.13 specifically identifies an electric water heater with a storage capacity of 120 gal or less as a continuous load. This revision requires the branch circuit overcurrent device and conductors to be sized based on 125 percent of the water heater nameplate rating unless the overcurrent device and the assembly it is installed in are listed to be used at 100 percent of its continuous current rating. In addition, feeders and services that supply water heater branch circuits are also impacted by the fact that this type of equipment is considered to be a continuous load.

FPN: For branch-circuit rating, see 422.10.

422.14 Infrared Lamp Industrial Heating Appliances

In industrial occupancies, infrared heating appliance lampholders shall be permitted to be operated in series on circuits of over 150 volts to ground, provided the voltage rating of the lampholders is not less than the circuit voltage.

Each section, panel, or strip carrying a number of infrared lampholders (including the internal wiring of such section, panel, or strip) shall be considered an appliance. The terminal connection block of each such assembly shall be considered an individual outlet.

422.15 Central Vacuum Outlet Assemblies

(A) Listed central vacuum outlet assemblies shall be permitted to be connected to a branch circuit in accordance with 210.23(A).

(B) The ampacity of the connecting conductors shall not be less than the ampacity of the branch circuit conductors to which they are connected.

(C) An equipment grounding conductor shall be used where the central vacuum outlet assembly has accessible non–current-carrying metal parts.

Section 422.15 permits listed central vacuum outlet devices to be connected to the ordinary 15- or 20-ampere general-purpose branch circuits that may be located in the same area in which the vacuum outlet is installed. Starting and stopping of the central vacuum system is achieved by a Class 2 control circuit that originates at the main unit of the central vacuum system. The circuit is switched at each outlet by the insertion or removal of the matching vacuum hose in the outlet.

422.16 Flexible Cords

(A) General Flexible cord shall be permitted (1) for the connection of appliances to facilitate their frequent interchange or to prevent the transmission of noise or vibration or (2) to facilitate the removal or disconnection of appliances that are fastened in place, where the fastening means and mechanical connections are specifically designed to permit ready removal for maintenance or repair and the appliance is intended or identified for flexible cord connection.

It should be understood that a cord-connected appliance is required to be specifically designed, mechanically and electrically, to be readily removable for maintenance and repair.

(B) Specific Appliances

(1) Electrically Operated Kitchen Waste Disposers Electrically operated kitchen waste disposers shall be permitted to be cord-and-plug connected with a flexible cord identified as suitable for the purpose in the installation instructions of the appliance manufacturer, where all of the following conditions are met:

(1) The flexible cord shall be terminated with a grounding-type attachment plug.

Exception: A listed kitchen waste disposer distinctly marked to identify it as protected by a system of double insulation, or its equivalent, shall not be required to be terminated with a grounding-type attachment plug.

(2) The length of the cord shall not be less than 450 mm (18 in.) and not over 900 mm (36 in.).
(3) Receptacles shall be located to avoid physical damage to the flexible cord.
(4) The receptacle shall be accessible.

The kitchen waste disposer illustrated in Exhibit 422.1 is an example of a cord-and-plug-connected appliance with mechanical connections designed to permit removal. The cord and receptacle are designed and installed in accordance with 422.16(B)(1). To facilitate control of the waste disposer from a wall location, this receptacle is permitted to be switched by a general-use snap switch as long as the load does not exceed the requirements of 404.14(A)(3).

Cord-and-plug-connected appliance

Exhibit 422.1 A cord-and-plug-connected kitchen waste disposer.

(2) Built-in Dishwashers and Trash Compactors Built-in dishwashers and trash compactors shall be permitted to be cord-and-plug connected with a flexible cord identified as suitable for the purpose in the installation instructions

of the appliance manufacturer where all of the following conditions are met:

(1) The flexible cord shall be terminated with a grounding-type attachment plug.

Exception: A listed dishwasher or trash compactor distinctly marked to identify it as protected by a system of double insulation, or its equivalent, shall not be required to be terminated with a grounding-type attachment plug.

(2) The length of the cord shall be 0.9 m to 1.2 m (3 ft to 4 ft) measured from the face of the attachment plug to the plane of the rear of the appliance.
(3) Receptacles shall be located to avoid physical damage to the flexible cord.
(4) The receptacle shall be located in the space occupied by the appliance or adjacent thereto.
(5) The receptacle shall be accessible.

(3) Wall-Mounted Ovens and Counter-Mounted Cooking Units Wall-mounted ovens and counter-mounted cooking units complete with provisions for mounting and for making electrical connections shall be permitted to be permanently connected or, only for ease in servicing or for installation, cord-and-plug connected.

A separable connector or a plug and receptacle combination in the supply line to an oven or cooking unit shall be approved for the temperature of the space in which it is located.

(4) Range Hoods Range hoods shall be permitted to be cord-and-plug connected with a flexible cord identified as suitable for use on range hoods in the installation instructions of the appliance manufacturer, where all of the following conditions are met:

(1) The flexible cord is terminated with a grounding-type attachment plug.

Exception: A listed range hood distinctly marked to identify it as protected by a system of double insulation, or its equivalent, shall not be required to be terminated with a grounding-type attachment plug.

(2) The length of the cord is not less than 450 mm (18 in.) and not over 900 mm (36 in.).
(3) Receptacles are located to avoid physical damage to the flexible cord.
(4) The receptacle is accessible.
(5) The receptacle is supplied by an individual branch circuit.

Section 422.16(B)(4) was added to the 2005 *Code* to allow range hoods to be cord-and-plug-connected under five specific prerequisite conditions. A grounding-type attachment plug is not required where the range hood is identified as protected by a system of double insulation. One method of verifying such protection is to look for the mark of a testing laboratory.

422.17 Protection of Combustible Material

Each electrically heated appliance that is intended by size, weight, and service to be located in a fixed position shall be placed so as to provide ample protection between the appliance and adjacent combustible material.

422.18 Support of Ceiling-Suspended (Paddle) Fans

Ceiling-suspended (paddle) fans shall be supported independently of an outlet box or by listed outlet box or outlet box systems identified for the use and installed in accordance with 314.27(D).

A revision to this section in the 2005 *Code* makes it clear that paddle fans must be supported independently of the outlet box or be supported by a listed box identified for fan support (see Exhibit 422.2).

Ceiling-suspended paddle fans must be supported by an outlet box that is listed or identified for support of a paddle fan. Where the outlet box is not listed or identified to support a paddle fan, the fan must be supported from the building structure, not from the outlet box. The maximum paddle fan weight allowed to be supported by an outlet box identified for support of a paddle fan is 70 lb. When the weight of a paddle fan exceeds 35 lb, the listed outlet box must be marked with the amount of weight that it is allowed to support. See 314.27(D) for the requirements for the support of boxes that supply power to paddle fans. Exhibit 314.9 provides an example of a paddle fan supported independent of the outlet box.

422.20 Other Installation Methods

Appliances employing methods of installation other than covered by this article shall be permitted to be used only by special permission.

III. Disconnecting Means

422.30 General

A means shall be provided to disconnect each appliance from all ungrounded conductors in accordance with the following sections of Part III. If an appliance is supplied by more than one source, the disconnecting means shall be grouped and identified.

Exhibit 422.2 Supporting a ceiling-suspended (paddle) fan (35 lb or less) with a box identified for such use. (Courtesy of Hubbell RACO)

422.31 Disconnection of Permanently Connected Appliances

(A) Rated at Not Over 300 Volt-Amperes or ⅛ Horse-power For permanently connected appliances rated at not over 300 volt-amperes or ⅛ hp, the branch-circuit overcurrent device shall be permitted to serve as the disconnecting means.

(B) Appliances Rated Over 300 Volt-Amperes or ⅛ Horsepower For permanently connected appliances rated over 300 volt-amperes or ⅛ hp, the branch-circuit switch or circuit breaker shall be permitted to serve as the disconnecting means where the switch or circuit breaker is within sight from the appliance or is capable of being locked in the open position. The provision for locking or adding a lock to the disconnecting means shall be installed on or at the switch or circuit breaker used as the disconnecting means and shall remain in place with or without the lock installed.

Section 422.31(B) has been expanded to increase safety to maintenance and service personnel working on electrical appliances. The requirement now provides detailed information pertaining to the method of providing the disconnection required by Article 422, Part III. A device that is attached to the circuit breaker handle by a set screw is not an acceptable means to serve as a safe method of locking the device in the off position. The device must have provisions for placement of a lock on it to secure the device in the off position. The lock-out device must be part of the disconnect assembly and must remain in place after the padlock is removed, whether it is a fused disconnect switch, a single circuit breaker, or a circuit breaker in a panelboard.

FPN: For appliances employing unit switches, see 422.34.

422.32 Disconnecting Means for Motor-Driven Appliance

If a switch or circuit breaker serves as the disconnecting means for a permanently connected motor-driven appliance of more than ⅛ hp, it shall be located within sight from the motor controller and shall comply with Part IX of Article 430.

Exception: If a motor-driven appliance of more than ⅛ hp is provided with a unit switch that complies with 422.34(A), (B), (C), or (D), the switch or circuit breaker serving as the other disconnecting means shall be permitted to be out of sight from the motor controller.

422.33 Disconnection of Cord-and-Plug-Connected Appliances

(A) Separable Connector or an Attachment Plug and Receptacle For cord-and-plug-connected appliances, an accessible separable connector or an accessible plug and receptacle shall be permitted to serve as the disconnecting means. Where the separable connector or plug and receptacle are not accessible, cord-and-plug-connected appliances shall be provided with disconnecting means in accordance with 422.31.

(B) Connection at the Rear Base of a Range For cord-and-plug-connected household electric ranges, an attachment plug and receptacle connection at the rear base of a range, if it is accessible from the front by removal of a drawer, shall be considered as meeting the intent of 422.33(A).

(C) Rating The rating of a receptacle or of a separable connector shall not be less than the rating of any appliance connected thereto.

Exception: Demand factors authorized elsewhere in this Code shall be permitted to be applied to the rating of a receptacle or of a separable connector.

422.34 Unit Switch(es) as Disconnecting Means

A unit switch(es) with a marked-off position that is a part of an appliance and disconnects all ungrounded conductors shall be permitted as the disconnecting means required by this article where other means for disconnection are provided in occupancies specified in 422.34(A) through 422.34(D).

(A) Multifamily Dwellings In multifamily dwellings, the other disconnecting means shall be within the dwelling unit, or on the same floor as the dwelling unit in which the appliance is installed, and shall be permitted to control lamps and other appliances.

(B) Two-Family Dwellings In two-family dwellings, the other disconnecting means shall be permitted either inside or outside of the dwelling unit in which the appliance is installed. In this case, an individual switch or circuit breaker for the dwelling unit shall be permitted and shall also be permitted to control lamps and other appliances.

(C) One-Family Dwellings In one-family dwellings, the service disconnecting means shall be permitted to be the other disconnecting means.

(D) Other Occupancies In other occupancies, the branch-circuit switch or circuit breaker, where readily accessible for servicing of the appliance, shall be permitted as the other disconnecting means.

422.35 Switch and Circuit Breaker to Be Indicating

Switches and circuit breakers used as disconnecting means shall be of the indicating type.

IV. Construction

422.40 Polarity in Cord-and-Plug-Connected Appliances

If the appliance is provided with a manually operated, line-connected, single-pole switch for appliance on–off operation, an Edison-base lampholder, or a 15- or 20-ampere receptacle, the attachment plug shall be of the polarized or grounding type.

A 2-wire, nonpolarized attachment plug shall be permitted to be used on a listed double-insulated shaver.

FPN: For polarity of Edison-base lampholders, see 410.42(A).

422.41 Cord-and-Plug-Connected Appliances Subject to Immersion

Cord-and-plug-connected portable, freestanding hydromassage units and hand-held hair dryers shall be constructed to provide protection for personnel against electrocution when immersed while in the "on" or "off" position.

Although receptacles in bathrooms of dwelling units have been required to be protected by ground-fault circuit interrupters since the 1975 edition of the *Code*, many receptacles in existing bathrooms are not so protected. Cord-and-plug-connected appliances such as hand-held hair dryers, curling irons, and so on, which can and have accidentally fallen into bathtubs, causing fatalities, are required to be provided with some form of protective device that is part of the appliance. Three types of protectors comply with this requirement:

1. Appliance-leakage circuit interrupters (ALCIs)
2. Immersion-detector circuit interrupters (IDCIs)
3. Ground-fault circuit interrupters (GFCIs)

ALCIs de-energize the supply to the appliance when leakage current exceeds a predetermined value. IDCIs de-energize the supply when a liquid causes a conductive path between a live part and a sensor, and GFCIs de-energize the supply when the current to ground exceeds a predetermined value.

422.42 Signals for Heated Appliances

In other than dwelling-type occupancies, each electrically heated appliance or group of appliances intended to be applied to combustible material shall be provided with a signal or an integral temperature-limiting device.

A common way to provide a signal light for electrically heated appliances in commercial or industrial locations is to use a red light connected to and within sight of the appliance that indicates that the appliance is energized and operating.

No signal is required for an electrically heated appliance provided with an integral high-temperature limiting device, such as a thermostat, that limits the temperature to which the appliance can heat.

422.43 Flexible Cords

(A) Heater Cords All cord-and-plug-connected smoothing irons and electrically heated appliances that are rated at more than 50 watts and produce temperatures in excess of 121°C (250°F) on surfaces with which the cord is likely to be in contact shall be provided with one of the types of approved heater cords listed in Table 400.4.

(B) Other Heating Appliances All other cord-and-plug-connected electrically heated appliances shall be connected with one of the approved types of cord listed in Table 400.4, selected in accordance with the usage specified in that table.

422.44 Cord-and-Plug-Connected Immersion Heaters

Electric heaters of the cord-and-plug-connected immersion type shall be constructed and installed so that current-carrying parts are effectively insulated from electrical contact with the substance in which they are immersed.

422.45 Stands for Cord-and-Plug-Connected Appliances

Each smoothing iron and other cord-and-plug-connected electrically heated appliance intended to be applied to combustible material shall be equipped with an approved stand, which shall be permitted to be a separate piece of equipment or a part of the appliance.

422.46 Flatirons

Electrically heated smoothing irons shall be equipped with an identified temperature-limiting means.

422.47 Water Heater Controls

All storage or instantaneous-type water heaters shall be equipped with a temperature-limiting means in addition to its control thermostat to disconnect all ungrounded conductors. Such means shall comply with both of the following:

(1) Installed to sense maximum water temperature.
(2) Be either a trip-free, manually reset type or a type having a replacement element. Such water heaters shall be marked to require the installation of a temperature and pressure relief valve.

Exception No. 1: Storage water heaters that are identified as being suitable for use with supply water temperature of 82°C (180°F) or above and a capacity of 60 kW or above.

Exception No. 2: Instantaneous-type water heaters that are identified as being suitable for such use, with a capacity of 4 L (1 gal) or less.

FPN: See ANSI Z21.22-1999/CSA 4.4-M99, *Relief Valves for Hot Water Supply Systems.*

422.48 Infrared Lamp Industrial Heating Appliances

(A) 300 Watts or Less Infrared heating lamps rated at 300 watts or less shall be permitted with lampholders of the medium-base, unswitched porcelain type or other types identified as suitable for use with infrared heating lamps rated 300 watts or less.

(B) Over 300 Watts Screw-shell lampholders shall not be used with infrared lamps rated over 300 watts, unless the lampholders are identified as being suitable for use with infrared heating lamps rated over 300 watts.

Infrared (heat) radiation lamps are tungsten-filament incandescent lamps similar in appearance to lighting lamps. However, they are designed to operate at a lower temperature, thus transferring more heat radiation and less light intensity.

Infrared lamps are used for a variety of heating and drying purposes in industrial locations.

422.49 High-Pressure Spray Washers

All single-phase cord-and-plug-connected high-pressure spray washing machines rated at 250 volts or less shall be provided with factory-installed ground-fault circuit-interrupter protection for personnel. The ground-fault circuit interrupter shall be an integral part of the attachment plug or shall be located in the supply cord within 300 mm (12 in.) of the attachment plug.

High-pressure spray washers may be used without a GFCI as part of the supply cord if the washers are rated 3-phase or are over 250 volts.

422.50 Cord-and-Plug-Connected Pipe Heating Assemblies

Cord-and-plug-connected pipe heating assemblies intended to prevent freezing of piping shall be listed.

The listing requirement was added as a result of data that substantiated numerous fires initiated by heat tapes. Additional requirements for ground-fault protection equipment are found in 427.22.

422.51 Cord-and-Plug-Connected Vending Machines

Cord-and-plug-connected vending machines manufactured or re-manufactured on or after January 1, 2005, shall include a ground-fault circuit-interrupter as an integral part of the attachment plug or located in the power supply cord within 300 mm (12 in.) of the attachment plug. Cord-and-plug-connected vending machines not incorporating integral GFCI protection shall be connected to a GFCI protected outlet.

Section 422.51 was added to the 2005 *Code* to provide added safety to consumers who purchase items from cord-and-plug-connected vending machines. The U.S. Consumer Product Safety Commission (CPSC) has investigated four electrocutions in four separate incidents since 1995 and three incidents involving vending machines that were nonfatal shocks. Those investigations spurred this new requirement, with an enforcement date of January 1, 2005, that new and remanufactured vending machines be provided with a GFCI as an integral part of the power supply cord within 12 in. of the attachment plug. However, a machine plugged into a GFCI-protected outlet is permitted.

V. Marking

422.60 Nameplate

(A) Nameplate Marking Each electric appliance shall be provided with a nameplate giving the identifying name and the rating in volts and amperes, or in volts and watts. If the appliance is to be used on a specific frequency or frequencies, it shall be so marked.

Where motor overload protection external to the appliance is required, the appliance shall be so marked.

> FPN: See 422.11 for overcurrent protection requirements.

(B) To Be Visible Marking shall be located so as to be visible or easily accessible after installation.

422.61 Marking of Heating Elements

All heating elements that are rated over one ampere, replaceable in the field, and a part of an appliance shall be legibly marked with the ratings in volts and amperes, or in volts and watts, or with the manufacturer's part number.

422.62 Appliances Consisting of Motors and Other Loads

(A) Nameplate Horsepower Markings Where a motor-operated appliance nameplate includes a horsepower rating, that rating shall not be less than the horsepower rating on the motor nameplate. Where an appliance consists of multiple motors, or one or more motors and other loads, the nameplate value shall not be less than the equivalent horsepower of the combined loads, calculated in accordance with 430.110(C)(1).

(B) Additional Nameplate Markings Appliances, other than those factory-equipped with cords and attachment plugs and with nameplates in compliance with 422.60, shall be marked in accordance with 422.62(B)(1) or (B)(2).

(1) Marking In addition to the marking required in 422.60, the marking on an appliance consisting of a motor with other load(s) or motors with or without other load(s) shall specify the minimum supply circuit conductor ampacity and the maximum rating of the circuit overcurrent protective device. This requirement shall not apply to an appliance with a nameplate in compliance with 422.60 where both the minimum supply circuit conductor ampacity and maximum rating of the circuit overcurrent protective device are not more than 15 amperes.

(2) Alternate Marking Method An alternative marking method shall be permitted to specify the rating of the largest motor in volts and amperes, and the additional load(s) in volts and amperes, or volts and watts in addition to the marking required in 422.60. The ampere rating of a motor

⅛ horsepower or less or a nonmotor load 1 ampere or less shall be permitted to be omitted unless such loads constitute the principal load.

ARTICLE 424
Fixed Electric Space-Heating Equipment

Summary of Changes

- **424.3(B):** Added requirement that electric space heating equipment be considered a continuous load.

- **424.6:** Added requirement that electric baseboard heaters, heating cables, duct heaters, and radiant heating systems be listed and labeled.

- **424.44(G):** Deleted spa locations from electrically heated floor GFCI requirement. See 680.27(C)(3).

Contents

I. General

424.1 Scope

This article covers fixed electric equipment used for space heating. For the purpose of this article, heating equipment shall include heating cable, unit heaters, boilers, central systems, or other approved fixed electric space-heating equipment. This article shall not apply to process heating and room air conditioning.

424.2 Other Articles

Fixed electric space-heating equipment incorporating a hermetic refrigerant motor-compressor shall also comply with Article 440.

424.3 Branch Circuits

(A) Branch-Circuit Requirements Individual branch circuits shall be permitted to supply any size fixed electric space-heating equipment.

Branch circuits supplying two or more outlets for fixed electric space-heating equipment shall be rated 15, 20, 25, or 30 amperes. In nondwelling occupancies, fixed infrared heating equipment shall be permitted to be supplied from branch circuits rated not over 50 amperes.

(B) Branch-Circuit Sizing Fixed electric space heating equipment shall be considered continuous load.

Section 424.3(B) has been simplified by stating that the branch circuits sized for fixed electric space heating equipment must be considered a continuous load.

The sizing of branch-circuit conductors and overcurrent devices supplying fixed electric space-heating equipment at

125 percent of the total load of the heaters (and motors) is predicated on the need to protect overcurrent devices, particularly in panelboards, and conductors from overheating during periods of prolonged operation.

The requirement in 424.3(B) is phrased to eliminate any question about whether the heating equipment is a continuous load, as defined in Article 100. Note that this revision to consider fixed electric space heating as a continuous load also impacts feeders and services that supply fixed electric space-heating equipment branch circuits.

424.6 Listed Equipment

Electric baseboard heaters, heating cables, duct heaters, and radiant heating systems shall be listed and labeled.

II. Installation

424.9 General

All fixed electric space-heating equipment shall be installed in an approved manner.

Permanently installed electric baseboard heaters equipped with factory-installed receptacle outlets, or outlets provided as a separate listed assembly, shall be permitted in lieu of a receptacle outlet(s) that is required by 210.50(B). Such receptacle outlets shall not be connected to the heater circuits.

FPN: Listed baseboard heaters include instructions that may not permit their installation below receptacle outlets.

The second paragraph of 424.9 restates the permission granted in the second paragraph of 210.52, that is, it allows factory-installed receptacle outlets in electric baseboard heaters to satisfy the spacing requirements for receptacle outlets in dwelling units according to 210.52(A).

Heating equipment and systems often have special installation instructions for spacings, types of supply wires, or special control equipment, which must be considered in determining the suitability of the installation.

424.10 Special Permission

Fixed electric space-heating equipment and systems installed by methods other than covered by this article shall be permitted only by special permission.

424.11 Supply Conductors

Fixed electric space-heating equipment requiring supply conductors with over 60°C insulation shall be clearly and permanently marked. This marking shall be plainly visible after installation and shall be permitted to be adjacent to the field connection box.

Fixed electric space-heating equipment may require supply conductors with a temperature rating greater than 60°C, due to their proximity to the heating elements and the installation instructions provided with a listed product.

424.12 Locations

(A) Exposed to Physical Damage Where subject to physical damage, fixed electric space-heating equipment shall be protected in an approved manner.

(B) Damp or Wet Locations Heaters and related equipment installed in damp or wet locations shall be listed for such locations and shall be constructed and installed so that water or other liquids cannot enter or accumulate in or on wired sections, electrical components, or ductwork.

FPN No. 1: See 110.11 for equipment exposed to deteriorating agents.
FPN No. 2: See 680.27(C) for pool deck areas.

424.13 Spacing from Combustible Materials

Fixed electric space-heating equipment shall be installed to provide the required spacing between the equipment and adjacent combustible material, unless it is listed to be installed in direct contact with combustible material.

Section 424.13 was made more restrictive in the 2005 *Code*. Where electric space heating equipment is to be installed in direct contact with combustible material, the equipment is now required to be listed for installation in direct contact with combustible material.

III. Control and Protection of Fixed Electric Space-Heating Equipment

424.19 Disconnecting Means

Means shall be provided to disconnect the heater, motor controller(s), and supplementary overcurrent protective device(s) of all fixed electric space-heating equipment from all ungrounded conductors. Where heating equipment is supplied by more than one source, the disconnecting means shall be grouped and marked.

Section 424.19(C) permits a unit switch to serve as the disconnecting means, provided that it has a marked "off" position and disconnects all ungrounded conductors and that other means are also provided in accordance with paragraphs 424.19(C)(1) through 424.19(C)(4). Such other means are not required to be capable of being locked in the open position as required by 424.19(B)(1).

See 424.20 for thermostatically controlled switching devices.

(A) Heating Equipment with Supplementary Overcurrent Protection The disconnecting means for fixed electric space-heating equipment with supplementary overcurrent protection shall be within sight from the supplementary overcurrent protective device(s), on the supply side of these devices, if fuses, and, in addition, shall comply with either 424.19(A)(1) or (A)(2).

(1) Heater Containing No Motor Rated Over ⅛ Horsepower The above disconnecting means or unit switches complying with 424.19(C) shall be permitted to serve as the required disconnecting means for both the motor controller(s) and heater under either of the following conditions:

(1) The disconnecting means provided is also within sight from the motor controller(s) and the heater.
(2) The disconnecting means provided is capable of being locked in the open position.

(2) Heater Containing a Motor(s) Rated Over ⅛ Horsepower The above disconnecting means shall be permitted to serve as the required disconnecting means for both the motor controller(s) and heater by one of the following means:

(1) Where the disconnecting means is also in sight from the motor controller(s) and the heater.
(2) Where the disconnecting means is not within sight from the heater, a separate disconnecting means shall be installed, or the disconnecting means shall be capable of being locked in the open position, or unit switches complying with 424.19(C) shall be permitted.
(3) Where the disconnecting means is not within sight from the motor controller location, a disconnecting means complying with 430.102 shall be provided.
(4) Where the motor is not in sight from the motor controller location, 430.102(B) shall apply.

(B) Heating Equipment Without Supplementary Overcurrent Protection

(1) Without Motor or with Motor Not Over ⅛ Horsepower For fixed electric space-heating equipment without a motor rated over ⅛ hp, the branch-circuit switch or circuit breaker shall be permitted to serve as the disconnecting means where the switch or circuit breaker is within sight from the heater or is capable of being locked in the open position.

(2) Over ⅛ Horsepower For motor-driven electric space-heating equipment with a motor rated over ⅛ hp, a disconnecting means shall be located within sight from the motor controller or shall be permitted to comply with the requirements in 424.19(A)(2).

(C) Unit Switch(es) as Disconnecting Means A unit switch(es) with a marked "off" position that is part of a fixed heater and disconnects all ungrounded conductors shall be permitted as the disconnecting means required by this article where other means for disconnection are provided in the types of occupancies in 424.19(C)(1) through (C)(4).

(1) Multifamily Dwellings In multifamily dwellings, the other disconnecting means shall be within the dwelling unit, or on the same floor as the dwelling unit in which the fixed heater is installed, and shall also be permitted to control lamps and appliances.

(2) Two-Family Dwellings In two-family dwellings, the other disconnecting means shall be permitted either inside or outside of the dwelling unit in which the fixed heater is installed. In this case, an individual switch or circuit breaker for the dwelling unit shall be permitted and shall also be permitted to control lamps and appliances.

(3) One-Family Dwellings In one-family dwellings, the service disconnecting means shall be permitted to be the other disconnecting means.

(4) Other Occupancies In other occupancies, the branch-circuit switch or circuit breaker, where readily accessible for servicing of the fixed heater, shall be permitted as the other disconnecting means.

424.20 Thermostatically Controlled Switching Devices

(A) Serving as Both Controllers and Disconnecting Means Thermostatically controlled switching devices and combination thermostats and manually controlled switches shall be permitted to serve as both controllers and disconnecting means, provided all of the following conditions are met:

(1) Provided with a marked "off" position
(2) Directly open all ungrounded conductors when manually placed in the "off" position
(3) Designed so that the circuit cannot be energized automatically after the device has been manually placed in the "off" position
(4) Located as specified in 424.19

(B) Thermostats That Do Not Directly Interrupt All Ungrounded Conductors Thermostats that do not directly interrupt all ungrounded conductors and thermostats that operate remote-control circuits shall not be required to meet the requirements of 424.20(A). These devices shall not be permitted as the disconnecting means.

424.21 Switch and Circuit Breaker to Be Indicating

Switches and circuit breakers used as disconnecting means shall be of the indicating type.

424.22 Overcurrent Protection

(A) Branch-Circuit Devices Electric space-heating equipment, other than such motor-operated equipment as required by Article 430 and Article 440 to have additional overcurrent protection, shall be permitted to be protected against overcurrent where supplied by one of the branch circuits in Article 210.

(B) Resistance Elements Resistance-type heating elements in electric space-heating equipment shall be protected at not more than 60 amperes. Equipment rated more than 48 amperes and employing such elements shall have the heating elements subdivided, and each subdivided load shall not exceed 48 amperes. Where a subdivided load is less than 48 amperes, the rating of the supplementary overcurrent protective device shall comply with 424.3(B). A boiler employing resistance-type immersion heating elements contained in an ASME rated and stamped vessel shall be permitted to comply with 424.72(A).

The reason for subdividing the overcurrent protection is to minimize the amount of damaging energy released into the heating elements during a short circuit, thereby reducing the risk of fire. In addition to safety, a second benefit may be partial continuity of service.

When a short circuit occurs, large amounts of damaging energy are released. This damage comes in the form of both heat and magnetic energy. By limiting the size of the overcurrent device protecting the individual heating elements, the damaging short-circuit energy released at the element is greatly reduced, thereby greatly reducing the risk of fire.

Historically, it has been stated that the subdivision size of 60 amperes was originally selected to use the maximum fuseholder size of 60 amperes while maintaining up to a 48-ampere heating element size (48 amperes $\times$ 125% = 60 amperes).

The following example makes a strong case for the 60-ampere subdivision requirement when a short circuit occurs at the element level. Although this example uses fuses, the same case can be made using circuit breakers.

Example

A 200-kW, 3-phase, 480-volt resistance-type unit heater has a phase current of 240.6 amperes. If the current *Code* rule for subdivision [424.22(B)] was not observed and this load were to be protected by just one device, the selected overcurrent device could be sized by multiplying 240.6 amperes times 125 percent and selecting a maximum overcurrent protective device sized 350 amperes. If the subdivision requirement were followed, the heater would probably contain six separately protected internal circuits limited in size to 60 amperes.

Solution

By using the UL *Electrical Construction Equipment Directory* (Green Book), the energy let-through of a 350-ampere fuse can be compared to the energy let-through of a 60-ampere fuse. In the fuse section (JCQR), the let-through energy, approximated by the current squared and then multiplied by the time, or I^2t, is provided for various fuse classes (UL).

For this example, a 600-volt, 60-ampere Class T fuse could have a let-through, I^2t, as high as 30,000 ampere squared seconds. But a 600-volt, 350-ampere Class T fuse could have a let-through, I^2t, as high as 1,100,000 ampere squared seconds. That means the 350-ampere fuse could let through 36.67 times as much damaging energy as the 60-ampere fuse during a short circuit. The difference in energy let-through between these two overcurrent devices (the single 350-ampere device and the group of 60-ampere devices) is significant enough to make the difference between replacing a single element or replacing a good portion of the entire system.

This example illustrates that subdivision of a circuit greatly reduces the risk of fire.

(C) Overcurrent Protective Devices The supplementary overcurrent protective devices for the subdivided loads specified in 424.22(B) shall be (1) factory-installed within or on the heater enclosure or supplied for use with the heater as a separate assembly by the heater manufacturer; (2) accessible, but shall not be required to be readily accessible; and (3) suitable for branch-circuit protection.

> FPN: See 240.10.

Where cartridge fuses are used to provide this overcurrent protection, a single disconnecting means shall be permitted to be used for the several subdivided loads.

> FPN No. 1: For supplementary overcurrent protection, see 240.10.
> FPN No. 2: For disconnecting means for cartridge fuses in circuits of any voltage, see 240.40.

Where supplementary overcurrent protection is required, the heating equipment manufacturer is required to furnish the necessary overcurrent protective devices.

(D) Branch-Circuit Conductors The conductors supplying the supplementary overcurrent protective devices shall be considered branch-circuit conductors.

Where the heaters are rated 50 kW or more, the conductors supplying the supplementary overcurrent protective devices specified in 424.22(C) shall be permitted to be sized at not less than 100 percent of the nameplate rating of the heater, provided all of the following conditions are met:

(1) The heater is marked with a minimum conductor size.

(2) The conductors are not smaller than the marked minimum size.

(3) A temperature-actuated device controls the cyclic operation of the equipment.

(E) Conductors for Subdivided Loads Field-wired conductors between the heater and the supplementary overcurrent protective devices shall be sized at not less than 125 percent of the load served. The supplementary overcurrent protective devices specified in 424.22(C) shall protect these conductors in accordance with 240.4.

Where the heaters are rated 50 kW or more, the ampacity of field-wired conductors between the heater and the supplementary overcurrent protective devices shall be permitted to be not less than 100 percent of the load of their respective subdivided circuits, provided all of the following conditions are met:

(1) The heater is marked with a minimum conductor size.

(2) The conductors are not smaller than the marked minimum size.

(3) A temperature-activated device controls the cyclic operation of the equipment.

IV. Marking of Heating Equipment

424.28 Nameplate

(A) Marking Required Each unit of fixed electric space-heating equipment shall be provided with a nameplate giving the identifying name and the normal rating in volts and watts or in volts and amperes.

Electric space-heating equipment intended for use on alternating current only or direct current only shall be marked to so indicate. The marking of equipment consisting of motors over ⅛ hp and other loads shall specify the rating of the motor in volts, amperes, and frequency, and the heating load in volts and watts or in volts and amperes.

(B) Location This nameplate shall be located so as to be visible or easily accessible after installation.

424.29 Marking of Heating Elements

All heating elements that are replaceable in the field and are part of an electric heater shall be legibly marked with the ratings in volts and watts or in volts and amperes.

V. Electric Space-Heating Cables

424.34 Heating Cable Construction

Heating cables shall be furnished complete with factory-assembled nonheating leads at least 2.1 m (7 ft) in length.

424.35 Marking of Heating Cables

Each unit shall be marked with the identifying name or identification symbol, catalog number, and ratings in volts and watts or in volts and amperes.

Each unit length of heating cable shall have a permanent legible marking on each nonheating lead located within 75 mm (3 in.) of the terminal end. The lead wire shall have the following color identification to indicate the circuit voltage on which it is to be used:

(1) 120 volt, nominal — yellow

(2) 208 volt, nominal — blue

(3) 240 volt, nominal — red

(4) 277 volt, nominal — brown

(5) 480 volt, nominal — orange

424.36 Clearances of Wiring in Ceilings

Wiring located above heated ceilings shall be spaced not less than 50 mm (2 in.) above the heated ceiling and shall be considered as operating at an ambient temperature of 50°C (122°F). The ampacity of conductors shall be calculated on the basis of the correction factors shown in the 0–2000 volt ampacity tables of Article 310. If this wiring is located above thermal insulation having a minimum thickness of 50 mm (2 in.), the wiring shall not require correction for temperature.

424.38 Area Restrictions

(A) Shall Not Extend Beyond the Room or Area Heating cables shall not extend beyond the room or area in which they originate.

(B) Uses Prohibited Heating cables shall not be installed in the following:

(1) In closets

(2) Over walls

(3) Over partitions that extend to the ceiling, unless they are isolated single runs of embedded cable

(4) Over cabinets whose clearance from the ceiling is less than the minimum horizontal dimension of the cabinet to the nearest cabinet edge that is open to the room or area

(C) In Closet Ceilings as Low-Temperature Heat Sources to Control Relative Humidity The provisions of 424.38(B) shall not prevent the use of cable in closet ceilings as low-temperature heat sources to control relative humidity, provided they are used only in those portions of the ceiling that are unobstructed to the floor by shelves or other permanent luminaires (fixtures).

424.39 Clearance from Other Objects and Openings

Heating elements of cables shall be separated at least 200 mm (8 in.) from the edge of outlet boxes and junction boxes that are to be used for mounting surface luminaires (lighting fixtures). A clearance of not less than 50 mm (2 in.) shall be provided from recessed luminaires (fixtures) and their trims, ventilating openings, and other such openings in room surfaces. Sufficient area shall be provided to ensure that no heating cable is covered by any surface-mounted units.

424.40 Splices

Embedded cables shall be spliced only where necessary and only by approved means, and in no case shall the length of the heating cable be altered.

424.41 Installation of Heating Cables on Dry Board, in Plaster, and on Concrete Ceilings

(A) In Walls Cables shall not be installed in walls unless it is necessary for an isolated single run of cable to be installed down a vertical surface to reach a dropped ceiling.

(B) Adjacent Runs Adjacent runs of cable not exceeding 9 watts/m (2¾ watts/ft) shall not be installed less than 38 mm (1½ in.) on centers.

(C) Surfaces to Be Applied Heating cables shall be applied only to gypsum board, plaster lath, or other fire-resistant material. With metal lath or other electrically conductive surfaces, a coat of plaster shall be applied to completely separate the metal lath or conductive surface from the cable.

> FPN: See also 424.41(F).

(D) Splices All heating cables, the splice between the heating cable and nonheating leads, and 75-mm (3-in.) minimum of the nonheating lead at the splice shall be embedded in plaster or dry board in the same manner as the heating cable.

(E) Ceiling Surface The entire ceiling surface shall have a finish of thermally noninsulating sand plaster that has a nominal thickness of 13 mm (½ in.), or other noninsulating material identified as suitable for this use and applied according to specified thickness and directions.

(F) Secured Cables shall be secured by means of approved stapling, tape, plaster, nonmetallic spreaders, or other approved means either at intervals not exceeding 400 mm (16 in.) or at intervals not exceeding 1.8 m (6 ft) for cables identified for such use. Staples or metal fasteners that straddle the cable shall not be used with metal lath or other electrically conductive surfaces.

(G) Dry Board Installations In dry board installations, the entire ceiling below the heating cable shall be covered with gypsum board not exceeding 13 mm (½ in.) thickness. The void between the upper layer of gypsum board, plaster lath, or other fire-resistant material and the surface layer of gypsum board shall be completely filled with thermally conductive, nonshrinking plaster or other approved material or equivalent thermal conductivity.

(H) Free from Contact with Conductive Surfaces Cables shall be kept free from contact with metal or other electrically conductive surfaces.

(I) Joists In dry board applications, cable shall be installed parallel to the joist, leaving a clear space centered under the joist of 65 mm (2½ in.) (width) between centers of adjacent runs of cable. A surface layer of gypsum board shall be mounted so that the nails or other fasteners do not pierce the heating cable.

(J) Crossing Joists Cables shall cross joists only at the ends of the room unless the cable is required to cross joists elsewhere in order to satisfy the manufacturer's instructions that the installer avoid placing the cable too close to ceiling penetrations and luminaires (lighting fixtures).

424.42 Finished Ceilings

Finished ceilings shall not be covered with decorative panels or beams constructed of materials that have thermal insulating properties, such as wood, fiber, or plastic. Finished ceilings shall be permitted to be covered with paint, wallpaper, or other approved surface finishes.

424.43 Installation of Nonheating Leads of Cables

(A) Free Nonheating Leads Free nonheating leads of cables shall be installed in accordance with approved wiring methods from the junction box to a location within the ceiling. Such installations shall be permitted to be single conductors in approved raceways, single or multiconductor Type UF, Type NMC, Type MI, or other approved conductors.

(B) Leads in Junction Box Not less than 150 mm (6 in.) of free nonheating lead shall be within the junction box. The marking of the leads shall be visible in the junction box.

(C) Excess Leads Excess leads of heating cables shall not be cut but shall be secured to the underside of the ceiling and embedded in plaster or other approved material, leaving only a length sufficient to reach the junction box with not less than 150 mm (6 in.) of free lead within the box.

424.44 Installation of Cables in Concrete or Poured Masonry Floors

(A) Watts per Linear Foot Constant wattage heating cables shall not exceed 54 watts/linear meter (16½ watts/linear foot) of cable.

(B) Spacing Between Adjacent Runs The spacing between adjacent runs of cable shall not be less than 25 mm (1 in.) on centers.

(C) Secured in Place Cables shall be secured in place by nonmetallic frames or spreaders or other approved means while the concrete or other finish is applied.

Cables shall not be installed where they bridge expansion joints unless protected from expansion and contraction.

(D) Spacings Between Heating Cable and Metal Embedded in the Floor Spacings shall be maintained between the heating cable and metal embedded in the floor, unless the cable is a grounded metal-clad cable.

(E) Leads Protected Leads shall be protected where they leave the floor by rigid metal conduit, intermediate metal conduit, rigid nonmetallic conduit, electrical metallic tubing, or by other approved means.

(F) Bushings or Approved Fittings Bushings or approved fittings shall be used where the leads emerge within the floor slab.

(G) Ground-Fault Circuit-Interrupter Protection Ground-fault circuit-interrupter protection for personnel shall be provided for cables installed in electrically heated floors of bathrooms and in hydromassage bathtub locations.

Section 424.44(G) requires the use of GFCI protection where cables are installed in concrete or poured masonry floors, thereby reducing shock hazards to persons with bare feet in these areas. In the 2002 *Code*, the reference to conductive floor coverings was removed to clarify the requirement for GFCI protection in all the areas identified in 424.44(G), regardless of the type of floor covering over the concrete or poured masonry. This section was revised for the 2005 *NEC* by deleting the reference to spa and hot tub locations. In accordance with 680.27(C)(3), electric radiant heating cables are not permitted in these locations.

424.45 Inspection and Tests

Cable installations shall be made with due care to prevent damage to the cable assembly and shall be inspected and approved before cables are covered or concealed.

VI. Duct Heaters

424.57 General

Part VI shall apply to any heater mounted in the airstream of a forced-air system where the air-moving unit is not provided as an integral part of the equipment.

424.58 Identification

Heaters installed in an air duct shall be identified as suitable for the installation.

424.59 Airflow

Means shall be provided to ensure uniform and adequate airflow over the face of the heater in accordance with the manufacturer's instructions.

> FPN: Heaters installed within 1.2 m (4 ft) of the outlet of an air-moving device, heat pump, air conditioner, elbows, baffle plates, or other obstructions in ductwork may require turning vanes, pressure plates, or other devices on the inlet side of the duct heater to ensure an even distribution of air over the face of the heater.

424.60 Elevated Inlet Temperature

Duct heaters intended for use with elevated inlet air temperature shall be identified as suitable for use at the elevated temperatures.

424.61 Installation of Duct Heaters with Heat Pumps and Air Conditioners

Heat pumps and air conditioners having duct heaters closer than 1.2 m (4 ft) to the heat pump or air conditioner shall have both the duct heater and heat pump or air conditioner identified as suitable for such installation and so marked.

424.62 Condensation

Duct heaters used with air conditioners or other air-cooling equipment that could result in condensation of moisture shall be identified as suitable for use with air conditioners.

424.63 Fan Circuit Interlock

Means shall be provided to ensure that the fan circuit is energized when any heater circuit is energized. However, time- or temperature-controlled delay in energizing the fan motor shall be permitted.

424.64 Limit Controls

Each duct heater shall be provided with an approved, integral, automatic-reset temperature-limiting control or controllers to de-energize the circuit or circuits.

In addition, an integral independent supplementary control or controllers shall be provided in each duct heater that disconnects a sufficient number of conductors to interrupt current flow. This device shall be manually resettable or replaceable.

424.65 Location of Disconnecting Means

Duct heater controller equipment shall be either accessible with the disconnecting means installed at or within sight from the controller or as permitted by 424.19(A).

424.66 Installation

Duct heaters shall be installed in accordance with the manufacturer's instructions in such a manner that operation does not create a hazard to persons or property. Furthermore, duct heaters shall be located with respect to building construction and other equipment so as to permit access to the heater. Sufficient clearance shall be maintained to permit replacement of controls and heating elements and for adjusting and cleaning of controls and other parts requiring such attention. See 110.26.

> FPN: For additional installation information, see NFPA 90A-2002, *Standard for the Installation of Air-Conditioning and Ventilating Systems*, and NFPA 90B-2002, *Standard for the Installation of Warm Air Heating and Air-Conditioning Systems*.

VII. Resistance-Type Boilers

424.70 Scope

The provisions in Part VII of this article shall apply to boilers employing resistance-type heating elements. Electrode-type boilers shall not be considered as employing resistance-type heating elements. See Part VIII of this article.

424.71 Identification

Resistance-type boilers shall be identified as suitable for the installation.

424.72 Overcurrent Protection

(A) Boiler Employing Resistance-Type Immersion Heating Elements in an ASME Rated and Stamped Vessel A boiler employing resistance-type immersion heating elements contained in an ASME rated and stamped vessel shall have the heating elements protected at not more than 150 amperes. Such a boiler rated more than 120 amperes shall have the heating elements subdivided into loads not exceeding 120 amperes.

Where a subdivided load is less than 120 amperes, the rating of the overcurrent protective device shall comply with 424.3(B).

(B) Boiler Employing Resistance-Type Heating Elements Rated More Than 48 Amperes and Not Contained in an ASME Rated and Stamped Vessel A boiler employing resistance-type heating elements not contained in an ASME rated and stamped vessel shall have the heating elements protected at not more than 60 amperes. Such a boiler rated more than 48 amperes shall have the heating elements subdivided into loads not exceeding 48 amperes.

Where a subdivided load is less than 48 amperes, the rating of the overcurrent protective device shall comply with 424.3(B).

See the commentary following 424.22(B) for an explanation of the subdivision requirement.

(C) Supplementary Overcurrent Protective Devices The supplementary overcurrent protective devices for the subdi-

vided loads as required by 424.72(A) and 424.72(B) shall be as follows:

(1) Factory-installed within or on the boiler enclosure or provided as a separate assembly by the boiler manufacturer
(2) Accessible, but need not be readily accessible
(3) Suitable for branch-circuit protection

Where cartridge fuses are used to provide this overcurrent protection, a single disconnecting means shall be permitted for the several subdivided circuits. See 240.40.

(D) Conductors Supplying Supplementary Overcurrent Protective Devices The conductors supplying these supplementary overcurrent protective devices shall be considered branch-circuit conductors.

Where the heaters are rated 50 kW or more, the conductors supplying the overcurrent protective device specified in 424.72(C) shall be permitted to be sized at not less than 100 percent of the nameplate rating of the heater, provided all of the following conditions are met:

(1) The heater is marked with a minimum conductor size.
(2) The conductors are not smaller than the marked minimum size.
(3) A temperature- or pressure-actuated device controls the cyclic operation of the equipment.

(E) Conductors for Subdivided Loads Field-wired conductors between the heater and the supplementary overcurrent protective devices shall be sized at not less than 125 percent of the load served. The supplementary overcurrent protective devices specified in 424.72(C) shall protect these conductors in accordance with 240.4.

Where the heaters are rated 50 kW or more, the ampacity of field-wired conductors between the heater and the supplementary overcurrent protective devices shall be permitted to be not less than 100 percent of the load of their respective subdivided circuits, provided all of the following conditions are met:

(1) The heater is marked with a minimum conductor size.
(2) The conductors are not smaller than the marked minimum size.
(3) A temperature-activated device controls the cyclic operation of the equipment.

424.73 Overtemperature Limit Control

Each boiler designed so that in normal operation there is no change in state of the heat transfer medium shall be equipped with a temperature-sensitive limiting means. It shall be installed to limit maximum liquid temperature and shall directly or indirectly disconnect all ungrounded conductors to the heating elements. Such means shall be in addition to a temperature regulating system and other devices protecting the tank against excessive pressure.

424.74 Overpressure Limit Control

Each boiler designed so that in normal operation there is a change in state of the heat transfer medium from liquid to vapor shall be equipped with a pressure-sensitive limiting means. It shall be installed to limit maximum pressure and shall directly or indirectly disconnect all ungrounded conductors to the heating elements. Such means shall be in addition to a pressure regulating system and other devices protecting the tank against excessive pressure.

VIII. Electrode-Type Boilers

424.80 Scope

The provisions in Part VIII of this article shall apply to boilers for operation at 600 volts, nominal, or less, in which heat is generated by the passage of current between electrodes through the liquid being heated.

> FPN: For over 600 volts, see Part V of Article 490.

424.81 Identification

Electrode-type boilers shall be identified as suitable for the installation.

424.82 Branch-Circuit Requirements

The size of branch-circuit conductors and overcurrent protective devices shall be calculated on the basis of 125 percent of the total load (motors not included). A contactor, relay, or other device, approved for continuous operation at 100 percent of its rating, shall be permitted to supply its full-rated load. See 210.19(A), Exception. The provisions of this section shall not apply to conductors that form an integral part of an approved boiler.

Where an electrode boiler is rated 50 kW or more, the conductors supplying the boiler electrode(s) shall be permitted to be sized at not less than 100 percent of the nameplate rating of the electrode boiler, provided all the following conditions are met:

(1) The electrode boiler is marked with a minimum conductor size.
(2) The conductors are not smaller than the marked minimum size.
(3) A temperature- or pressure-actuated device controls the cyclic operation of the equipment.

424.83 Overtemperature Limit Control

Each boiler, designed so that in normal operation there is no change in state of the heat transfer medium, shall be equipped with a temperature-sensitive limiting means. It shall be installed to limit maximum liquid temperature and shall directly or indirectly interrupt all current flow through the electrodes. Such means shall be in addition to the temperature regulating system and other devices protecting the tank against excessive pressure.

424.84 Overpressure Limit Control

Each boiler, designed so that in normal operation there is a change in state of the heat transfer medium from liquid to vapor, shall be equipped with a pressure-sensitive limiting means. It shall be installed to limit maximum pressure and shall directly or indirectly interrupt all current flow through the electrodes. Such means shall be in addition to a pressure regulating system and other devices protecting the tank against excessive pressure.

424.85 Grounding

For those boilers designed such that fault currents do not pass through the pressure vessel, and the pressure vessel is electrically isolated from the electrodes, all exposed non–current-carrying metal parts, including the pressure vessel, supply, and return connecting piping, shall be grounded in accordance with Article 250.

For all other designs, the pressure vessel containing the electrodes shall be isolated and electrically insulated from ground.

424.86 Markings

All electrode-type boilers shall be marked to show the following:

(1) The manufacturer's name
(2) The normal rating in volts, amperes, and kilowatts
(3) The electrical supply required specifying frequency, number of phases, and number of wires
(4) The marking "Electrode-Type Boiler"
(5) A warning marking, "All Power Supplies Shall Be Disconnected Before Servicing, Including Servicing the Pressure Vessel"

The nameplate shall be located so as to be visible after installation.

IX. Electric Radiant Heating Panels and Heating Panel Sets

424.90 Scope

The provisions of Part IX of this article shall apply to radiant heating panels and heating panel sets.

424.91 Definitions

Heating Panel. A complete assembly provided with a junction box or a length of flexible conduit for connection to a branch circuit.

Heating Panel Set. A rigid or nonrigid assembly provided with nonheating leads or a terminal junction assembly identified as being suitable for connection to a wiring system.

424.92 Markings

(A) Location Markings shall be permanent and in a location that is visible prior to application of panel finish.

(B) Identified as Suitable Each unit shall be identified as suitable for the installation.

(C) Required Markings Each unit shall be marked with the identifying name or identification symbol, catalog number, and rating in volts and watts or in volts and amperes.

(D) Labels Provided by Manufacturer The manufacturers of heating panels or heating panel sets shall provide marking labels that indicate that the space-heating installation incorporates heating panels or heating panel sets and instructions that the labels shall be affixed to the panelboards to identify which branch circuits supply the circuits to those space-heating installations. If the heating panels and heating panel set installations are visible and distinguishable after installation, the labels shall not be required to be provided and affixed to the panelboards.

424.93 Installation

(A) General

(1) Manufacturer's Instructions Heating panels and heating panel sets shall be installed in accordance with the manufacturer's instructions.

(2) Locations Not Permitted The heating portion shall not be installed as follows:

(1) In or behind surfaces where subject to physical damage
(2) Run through or above walls, partitions, cupboards, or similar portions of structures that extend to the ceiling.
(3) Run in or through thermal insulation, but shall be permitted to be in contact with the surface of thermal insulation.

(3) Separation from Outlets for Luminaires (Lighting Fixtures) Edges of panels and panel sets shall be separated by not less than 200 mm (8 in.) from the edges of any outlet boxes and junction boxes that are to be used for mounting surface luminaires (lighting fixtures). A clearance of not less than 50 mm (2 in.) shall be provided from recessed luminaires (fixtures) and their trims, ventilating openings, and other such openings in room surfaces, unless the heating panels and panel sets are listed and marked for lesser clearances, in which case they shall be permitted to be installed at the marked clearances. Sufficient area shall be provided

to ensure that no heating panel or heating panel set is to be covered by any surface-mounted units.

(4) Surfaces Covering Heating Panels After the heating panels or heating panel sets are installed and inspected, it shall be permitted to install a surface that has been identified by the manufacturer's instructions as being suitable for the installation. The surface shall be secured so that the nails or other fastenings do not pierce the heating panels or heating panel sets.

(5) Surface Coverings Surfaces permitted by 424.93(A)(4) shall be permitted to be covered with paint, wallpaper, or other approved surfaces identified in the manufacturer's instructions as being suitable.

(B) Heating Panel Sets

(1) Mounting Location Heating panel sets shall be permitted to be secured to the lower face of joists or mounted in between joists, headers, or nailing strips.

(2) Parallel to Joists or Nailing Strips Heating panel sets shall be installed parallel to joists or nailing strips.

(3) Installation of Nails, Staples, or Other Fasteners Nailing or stapling of heating panel sets shall be done only through the unheated portions provided for this purpose. Heating panel sets shall not be cut through or nailed through any point closer than 6 mm (¼ in.) to the element. Nails, staples, or other fasteners shall not be used where they penetrate current-carrying parts.

(4) Installed as Complete Unit Heating panel sets shall be installed as complete units unless identified as suitable for field cutting in an approved manner.

424.94 Clearances of Wiring in Ceilings

Wiring located above heated ceilings shall be spaced not less than 50 mm (2 in.) above the heated ceiling and shall be considered as operating at an ambient of 50°C (122°F). The ampacity shall be calculated on the basis of the correction factors given in the 0–2000 volt ampacity tables of Article 310. If this wiring is located above thermal insulations having a minimum thickness of 50 mm (2 in.), the wiring shall not require correction for temperature.

424.95 Location of Branch-Circuit and Feeder Wiring in Walls

(A) Exterior Walls Wiring methods shall comply with Article 300 and 310.10.

(B) Interior Walls Any wiring behind heating panels or heating panel sets located in interior walls or partitions shall be considered as operating at an ambient temperature of 40°C (104°F), and the ampacity shall be calculated on the basis of the correction factors given in the 0–2000 volt ampacity tables of Article 310.

424.96 Connection to Branch-Circuit Conductors

(A) General Heating panels or heating panel sets assembled together in the field to form a heating installation in one room or area shall be connected in accordance with the manufacturer's instructions.

(B) Heating Panels Heating panels shall be connected to branch-circuit wiring by an approved wiring method.

(C) Heating Panel Sets

(1) Connection to Branch Circuit Wiring Heating panel sets shall be connected to branch-circuit wiring by a method identified as being suitable for the purpose.

(2) Panel Sets with Terminal Junction Assembly A heating panel set provided with terminal junction assembly shall be permitted to have the nonheating leads attached at the time of installation in accordance with the manufacturer's instructions.

424.97 Nonheating Leads

Excess nonheating leads of heating panels or heating panel sets shall be permitted to be cut to the required length. They shall meet the installation requirements of the wiring method employed in accordance with 424.96. Nonheating leads shall be an integral part of a heating panel and a heating panel set and shall not be subjected to the ampacity requirements of 424.3(B) for branch circuits.

424.98 Installation in Concrete or Poured Masonry

(A) Maximum Heated Area Heating panels or heating panel sets shall not exceed 355 watts/m^2 (33 watts/ft^2) of heated area.

(B) Secured in Place and Identified as Suitable Heating panels or heating panel sets shall be secured in place by means specified in the manufacturer's instructions and identified as suitable for the installation.

(C) Expansion Joints Heating panels or heating panel sets shall not be installed where they bridge expansion joints unless provision is made for expansion and contraction.

(D) Spacings Spacings shall be maintained between heating panels or heating panel sets and metal embedded in the floor. Grounded metal-clad heating panels shall be permitted to be in contact with metal embedded in the floor.

(E) Protection of Leads Leads shall be protected where they leave the floor by rigid metal conduit, intermediate metal conduit, rigid nonmetallic conduit, or electrical metallic tubing, or by other approved means.

(F) Bushings or Fittings Required Bushings or approved fittings shall be used where the leads emerge within the floor slabs.

424.99 Installation Under Floor Covering

(A) Identification Heating panels or heating panel sets for installation under floor covering shall be identified as suitable for installation under floor covering.

(B) Maximum Heated Area Heating panels or panel sets installed under floor covering shall not exceed 160 watts/m^2 (15 watts/ft^2) of heated area.

(C) Installation Listed heating panels or panel sets, if installed under floor covering, shall be installed on floor surfaces that are smooth and flat in accordance with the manufacturer's instructions and shall also comply with 424.99(C)(1) through (C)(5).

(1) Expansion Joints Heating panels or heating panel sets shall not be installed where they bridge expansion joints unless protected from expansion and contraction.

(2) Connection to Conductors Heating panels and heating panel sets shall be connected to branch-circuit and supply wiring by wiring methods recognized in Chapter 3.

(3) Anchoring Heating panels and heating panel sets shall be firmly anchored to the floor using an adhesive or anchoring system identified for this use.

(4) Coverings After heating panels or heating panel sets are installed and inspected, they shall be permitted to be covered by a floor covering that has been identified by the manufacturer as being suitable for the installation. The covering shall be secured to the heating panel or heating panel sets with release-type adhesives or by means identified for this use.

(5) Fault Protection A device to open all ungrounded conductors supplying the heating panels or heating panel sets, provided by the manufacturer, shall function when a low- or high-resistance line-to-line, line-to-grounded conductor, or line-to-ground fault occurs, such as the result of a penetration of the element or element assembly.

> FPN: An integral grounding shield may be required to provide this protection.

A system that uses conductive-film heating elements is an example of a heating system that could be installed under a floor covering.

ARTICLE 426
Fixed Outdoor Electric Deicing and Snow-Melting Equipment

Summary of Changes

• **426.4:** Added requirement that deicing and snow-melting equipment be considered a continuous load.

Contents

I. General

426.1 Scope

The requirements of this article shall apply to electrically energized heating systems and the installation of these systems.

(A) Embedded Embedded in driveways, walks, steps, and other areas.

(B) Exposed Exposed on drainage systems, bridge structures, roofs, and other structures.

Article 426 includes requirements for resistance heating elements, impedance heating systems, or skin-effect heating systems used for deicing and snow melting. These systems are defined in 426.2.

426.2 Definitions

Heating System. A complete system consisting of components such as heating elements, fastening devices, nonheating circuit wiring, leads, temperature controllers, safety signs, junction boxes, raceways, and fittings.

Impedance Heating System. A system in which heat is generated in a pipe or rod, or combination of pipes and rods, by causing current to flow through the pipe or rod by direct connection to an ac voltage source from a dual-winding transformer. The pipe or rod shall be permitted to be embedded in the surface to be heated, or constitute the exposed components to be heated.

Resistance Heating Element. A specific separate element to generate heat that is embedded in or fastened to the surface to be heated.

> FPN: Tubular heaters, strip heaters, heating cable, heating tape, and heating panels are examples of resistance heaters.

Skin-Effect Heating System. A system in which heat is generated on the inner surface of a ferromagnetic envelope embedded in or fastened to the surface to be heated.

> FPN: Typically, an electrically insulated conductor is routed through and connected to the envelope at the other end. The envelope and the electrically insulated conductor are connected to an ac voltage source from a dual-winding transformer.

426.3 Application of Other Articles

Cord-and-plug-connected fixed outdoor electric deicing and snow-melting equipment intended for specific use and identified as suitable for this use shall be installed according to Article 422.

426.4 Continuous Load

Fixed outdoor electric deicing and snow-melting equipment shall be considered as a continuous load.

Section 426.4 was completely revised for the 2005 *Code* to clarify that fixed outdoor electric deicing and snow-melting equipment must be considered a continuous load when sizing the branch circuits, feeders, and service conductors over overcurrent protective devices.

II. Installation

426.10 General

Equipment for outdoor electric deicing and snow melting shall be identified as being suitable for the following:

(1) The chemical, thermal, and physical environment
(2) Installation in accordance with the manufacturer's drawings and instructions

426.11 Use

Electrical heating equipment shall be installed in such a manner as to be afforded protection from physical damage.

Underwriters Laboratories Inc. requires that manufacturers of UL-listed mat or cable deicing and snow-melting equipment provide specific installation instructions for their products. The instructions supplement the requirements contained in Article 426. For example, if the equipment can be installed only in concrete that is double poured (poured in two parts), the installation instructions are to specifically require that installation technique. Where the instructions do not specifically require that a double-pour installation process be used, it is acceptable to use either a single- or double-pour method of installation. See 110.3(B) regarding the installation and use of listed or labeled equipment.

426.12 Thermal Protection

External surfaces of outdoor electric deicing and snow-melting equipment that operate at temperatures exceeding 60°C (140°F) shall be physically guarded, isolated, or thermally insulated to protect against contact by personnel in the area.

426.13 Identification

The presence of outdoor electric deicing and snow-melting equipment shall be evident by the posting of appropriate caution signs or markings where clearly visible.

426.14 Special Permission

Fixed outdoor deicing and snow-melting equipment employing methods of construction or installation other than covered by this article shall be permitted only by special permission.

See the definition of *special permission* in Article 100.

III. Resistance Heating Elements

426.20 Embedded Deicing and Snow-Melting Equipment

(A) Watt Density Panels or units shall not exceed 1300 watts/m^2 (120 watts/ft^2) of heated area.

(B) Spacing The spacing between adjacent cable runs is dependent upon the rating of the cable and shall be not less than 25 mm (1 in.) on centers.

(C) Cover Units, panels, or cables shall be installed as follows:

(1) On a substantial asphalt or masonry base at least 50 mm (2 in.) thick and have at least 38 mm (1½ in.) of asphalt or masonry applied over the units, panels, or cables; or
(2) They shall be permitted to be installed over other approved bases and embedded within 90 mm (3½ in.) of

masonry or asphalt but not less than 38 mm (1½ in.) from the top surface; or

(3) Equipment that has been specially investigated for other forms of installation shall be installed only in the manner for which it has been investigated.

(D) Secured Cables, units, and panels shall be secured in place by frames or spreaders or other approved means while the masonry or asphalt finish is applied.

(E) Expansion and Contraction Cables, units, and panels shall not be installed where they bridge expansion joints unless provision is made for expansion and contraction.

426.21 Exposed Deicing and Snow-Melting Equipment

(A) Secured Heating element assemblies shall be secured to the surface being heated by approved means.

(B) Overtemperature Where the heating element is not in direct contact with the surface being heated, the design of the heater assembly shall be such that its temperature limitations shall not be exceeded.

(C) Expansion and Contraction Heating elements and assemblies shall not be installed where they bridge expansion joints unless provision is made for expansion and contraction.

(D) Flexural Capability Where installed on flexible structures, the heating elements and assemblies shall have a flexural capability that is compatible with the structure.

426.22 Installation of Nonheating Leads for Embedded Equipment

(A) Grounding Sheath or Braid Nonheating leads having a grounding sheath or braid shall be permitted to be embedded in the masonry or asphalt in the same manner as the heating cable without additional physical protection.

(B) Raceways All but 25 mm to 150 mm (1 in. to 6 in.) of nonheating leads of Type TW and other approved types not having a grounding sheath shall be enclosed in a rigid conduit, electrical metallic tubing, intermediate metal conduit, or other raceways within asphalt or masonry; and the distance from the factory splice to raceway shall not be less than 25 mm (1 in.) or more than 150 mm (6 in.).

(C) Bushings Insulating bushings shall be used in the asphalt or masonry where leads enter conduit or tubing.

See 300.4(F) and the associated commentary for further information regarding insulating bushings.

(D) Expansion and Contraction Leads shall be protected in expansion joints and where they emerge from masonry

or asphalt by rigid conduit, electrical metallic tubing, intermediate metal conduit, other raceways, or other approved means.

(E) Leads in Junction Boxes Not less than 150 mm (6 in.) of free nonheating lead shall be within the junction box.

426.23 Installation of Nonheating Leads for Exposed Equipment

(A) Nonheating Leads Power supply nonheating leads (cold leads) for resistance elements shall be suitable for the temperature encountered. Not less than 150 mm (6 in.) of nonheating leads shall be provided within the junction box. Preassembled factory supplied and field assembled nonheating leads on approved heaters shall be permitted to be shortened if the markings specified in 426.25 are retained.

(B) Protection Nonheating power supply leads shall be enclosed in a rigid conduit, intermediate metal conduit, electrical metallic tubing, or other approved means.

426.24 Electrical Connection

(A) Heating Element Connections Electrical connections, other than factory connections of heating elements to nonheating elements embedded in masonry or asphalt or on exposed surfaces, shall be made with insulated connectors identified for the use.

(B) Circuit Connections Splices and terminations at the end of the nonheating leads, other than the heating element end, shall be installed in a box or fitting in accordance with 110.14 and 300.15.

426.25 Marking

Each factory-assembled heating unit shall be legibly marked within 75 mm (3 in.) of each end of the nonheating leads with the permanent identification symbol, catalog number, and ratings in volts and watts or in volts and amperes.

426.26 Corrosion Protection

Ferrous and nonferrous metal raceways, cable armor, cable sheaths, boxes, fittings, supports, and support hardware shall be permitted to be installed in concrete or in direct contact with the earth, or in areas subject to severe corrosive influences, where made of material suitable for the condition, or where provided with corrosion protection identified as suitable for the condition.

426.27 Grounding Braid or Sheath

Grounding means, such as copper braid, metal sheath, or other approved means, shall be provided as part of the heated section of the cable, panel, or unit.

426.28 Equipment Protection

Ground-fault protection of equipment shall be provided for fixed outdoor electric deicing and snow-melting equipment, except for equipment that employs mineral-insulated, metal-sheathed cable embedded in a noncombustible medium.

Section 426.28 was revised for the 1999 *Code* to require ground-fault protection of equipment for fixed outdoor electric deicing and snow-melting equipment. Rather than protecting the entire branch circuit, the ground-fault protection requirement is focused on protecting just the equipment itself. This affords the manufacturer and the user an option of providing both circuit and equipment protection, or just the required equipment protection. This required protection for fixed outdoor deicing and snow-melting equipment may be accomplished by using circuit breakers equipped with ground-fault equipment protection (GFEP) or an integral device supplied as part of the deicing or snow-melting equipment that is sensitive to leakage currents in the magnitude of 6 milliamperes to 50 milliamperes [referred to by UL as ground-fault equipment protection circuit interrupters (GFEPCIs)]. These protection devices, if applied properly, will substantially reduce the risk of a fire being started by low-level electrical arcing.

It is important to understand that this required equipment protection is not the same as a GFCI used for personal protection that trips at 5 milliamperes (± 1 milliampere).

For further information regarding ground-fault equipment protection used to comply with 426.28 and 427.22, refer to the UL *General Information for Electrical Equipment Directory* (White Book), category KCZI.

IV. Impedance Heating

426.30 Personnel Protection

Exposed elements of impedance heating systems shall be physically guarded, isolated, or thermally insulated with a weatherproof jacket to protect against contact by personnel in the area.

426.31 Isolation Transformer

A dual-winding transformer with a grounded shield between the primary and secondary windings shall be used to isolate the distribution system from the heating system.

426.32 Voltage Limitations

Unless protected by ground-fault circuit-interrupter protection for personnel, the secondary winding of the isolation transformer connected to the impedance heating elements shall not have an output voltage greater than 30 volts ac.

Where ground-fault circuit-interrupter protection for personnel is provided, the voltage shall be permitted to be greater than 30 but not more than 80 volts.

426.33 Induced Currents

All current-carrying components shall be installed in accordance with 300.20.

426.34 Grounding

An impedance heating system that is operating at a voltage greater than 30 but not more than 80 shall be grounded at a designated point(s).

V. Skin-Effect Heating

426.40 Conductor Ampacity

The current through the electrically insulated conductor inside the ferromagnetic envelope shall be permitted to exceed the ampacity values shown in Article 310, provided it is identified as suitable for this use.

426.41 Pull Boxes

Where pull boxes are used, they shall be accessible without excavation by location in suitable vaults or above grade. Outdoor pull boxes shall be of watertight construction.

426.42 Single Conductor in Enclosure

The provisions of 300.20 shall not apply to the installation of a single conductor in a ferromagnetic envelope (metal enclosure).

426.43 Corrosion Protection

Ferromagnetic envelopes, ferrous or nonferrous metal raceways, boxes, fittings, supports, and support hardware shall be permitted to be installed in concrete or in direct contact with the earth, or in areas subjected to severe corrosive influences, where made of material suitable for the condition, or where provided with corrosion protection identified as suitable for the condition. Corrosion protection shall maintain the original wall thickness of the ferromagnetic envelope.

426.44 Grounding

The ferromagnetic envelope shall be grounded at both ends; and, in addition, it shall be permitted to be grounded at intermediate points as required by its design.

The provisions of 250.30 shall not apply to the installation of skin-effect heating systems.

FPN: For grounding methods, see Article 250.

VI. Control and Protection

426.50 Disconnecting Means

(A) Disconnection All fixed outdoor deicing and snow-melting equipment shall be provided with a means for disconnection from all ungrounded conductors. Where readily accessible to the user of the equipment, the branch-circuit switch or circuit breaker shall be permitted to serve as the disconnecting means. The disconnecting means shall be of the indicating type and be provided with a positive lockout in the "off" position.

> The requirements for the disconnecting means have been made more stringent in the 2005 *Code*. The disconnect must indicate when it is in the on or off position. It must also be provided with a means for locking the disconnect in the off position. The disconnecting means is allowed to be the branch circuit switch or circuit breaker where it is "readily accessible"; in that case, it must have a means for a positive lockout in the off position.

(B) Cord-and-Plug-Connected Equipment The factory-installed attachment plug of cord-and-plug-connected equipment rated 20 amperes or less and 150 volts or less to ground shall be permitted to be the disconnecting means.

426.51 Controllers

(A) Temperature Controller with "Off" Position Temperature controlled switching devices that indicate an "off" position and that interrupt line current shall open all ungrounded conductors when the control device is in the "off" position. These devices shall not be permitted to serve as the disconnecting means unless provided with a positive lockout in the "off" position.

(B) Temperature Controller Without "Off" Position Temperature controlled switching devices that do not have an "off" position shall not be required to open all ungrounded conductors and shall not be permitted to serve as the disconnecting means.

(C) Remote Temperature Controller Remote controlled temperature-actuated devices shall not be required to meet the requirements of 426.51(A). These devices shall not be permitted to serve as the disconnecting means.

(D) Combined Switching Devices Switching devices consisting of combined temperature-actuated devices and manually controlled switches that serve both as the controller and the disconnecting means shall comply with all of the following conditions:

(1) Open all ungrounded conductors when manually placed in the "off" position
(2) Be so designed that the circuit cannot be energized automatically if the device has been manually placed in the "off" position
(3) Be provided with a positive lockout in the "off" position

426.52 Overcurrent Protection

Fixed outdoor electric deicing and snow-melting equipment shall be permitted to be protected against overcurrent where supplied by a branch circuit as specified in 426.4.

426.54 Cord-and-Plug-Connected Deicing and Snow-Melting Equipment

Cord-and-plug-connected deicing and snow-melting equipment shall be listed.

> According to the UL *General Information for Electrical Equipment Directory* (White Book), category KOBQ, UL listed deicing and snow-melting equipment is provided with means for permanent wiring connection, except the equipment rated 20 amperes or less and 150 volts or less to ground may be of cord-and-plug-connected construction. See the definition of *listed* in Article 100.

ARTICLE 427
Fixed Electric Heating Equipment for Pipelines and Vessels

Summary of Changes

- **427.4:** Added requirement that fixed electric heating equipment for pipelines and vessels is considered to be a continuous load.
- **427.27, Exception:** Added exception to allow an increase in the supply voltage, from 80 volts to 132 volts to ground, in industrial establishments where GFCI protection is provided and the four conditions are met.

Contents

I. General

427.1 Scope

The requirements of this article shall apply to electrically energized heating systems and the installation of these systems used with pipelines or vessels or both.

Article 427 includes requirements for impedance heating, induction heating, and skin-effect heating, in addition to resistance heating elements. Definitions of the various systems are provided in 427.2.

FPN: For further information, see ANSI/IEEE Std. 515-1997, *Standard for the Testing, Design, Installation and*

Maintenance of Electrical Resistance Heat Tracing for Industrial Applications, and ANSI/IEEE Std. 844-2000, *Recommended Practice for Electrical Impedance, Induction, and Skin Effect Heating of Pipelines and Vessels.*

427.2 Definitions

Impedance Heating System. A system in which heat is generated in a pipeline or vessel wall by causing current to flow through the pipeline or vessel wall by direct connection to an ac voltage source from a dual-winding transformer.

Induction Heating System. A system in which heat is generated in a pipeline or vessel wall by inducing current and hysteresis effect in the pipeline or vessel wall from an external isolated ac field source.

Integrated Heating System. A complete system consisting of components such as pipelines, vessels, heating elements, heat transfer medium, thermal insulation, moisture barrier, nonheating leads, temperature controllers, safety signs, junction boxes, raceways, and fittings.

Pipeline. A length of pipe including pumps, valves, flanges, control devices, strainers, and/or similar equipment for conveying fluids.

Resistance Heating Element. A specific separate element to generate heat that is applied to the pipeline or vessel externally or internally.

FPN: Tubular heaters, strip heaters, heating cable, heating tape, heating blankets, and immersion heaters are examples of resistance heaters.

Skin-Effect Heating System. A system in which heat is generated on the inner surface of a ferromagnetic envelope attached to a pipeline or vessel, or both.

FPN: Typically, an electrically insulated conductor is routed through and connected to the envelope at the other end. The envelope and the electrically insulated conductor are connected to an ac voltage source from a dual-winding transformer.

Vessel. A container such as a barrel, drum, or tank for holding fluids or other material.

427.3 Application of Other Articles

Cord-connected pipe heating assemblies intended for specific use and identified as suitable for this use shall be installed according to Article 422.

427.4 Continuous Load

Fixed electric heating equipment for pipelines and vessels shall be considered continuous load.

The requirement in 427.4 has been amended and is phrased to eliminate any question about whether the heating equipment is a continuous load, as defined in Article 100.

II. Installation

427.10 General

Equipment for pipeline and vessel electrical heating shall be identified as being suitable for (1) the chemical, thermal, and physical environment and (2) installation in accordance with the manufacturer's drawings and instructions.

427.11 Use

Electrical heating equipment shall be installed in such a manner as to be afforded protection from physical damage.

427.12 Thermal Protection

External surfaces of pipeline and vessel heating equipment that operate at temperatures exceeding 60°C (140°F) shall be physically guarded, isolated, or thermally insulated to protect against contact by personnel in the area.

427.13 Identification

The presence of electrically heated pipelines, vessels, or both, shall be evident by the posting of appropriate caution signs or markings at frequent intervals along the pipeline or vessel.

III. Resistance Heating Elements

427.14 Secured

Heating element assemblies shall be secured to the surface being heated by means other than the thermal insulation.

427.15 Not in Direct Contact

Where the heating element is not in direct contact with the pipeline or vessel being heated, means shall be provided to prevent overtemperature of the heating element unless the design of the heater assembly is such that its temperature limitations will not be exceeded.

427.16 Expansion and Contraction

Heating elements and assemblies shall not be installed where they bridge expansion joints unless provisions are made for expansion and contraction.

427.17 Flexural Capability

Where installed on flexible pipelines, the heating elements and assemblies shall have a flexural capability that is compatible with the pipeline.

427.18 Power Supply Leads

(A) Nonheating Leads Power supply nonheating leads (cold leads) for resistance elements shall be suitable for the temperature encountered. Not less than 150 mm (6 in.) of nonheating leads shall be provided within the junction box. Preassembled factory supplied and field assembled nonheating leads on approved heaters shall be permitted to be shortened if the markings specified in 427.20 are retained.

(B) Power Supply Leads Protection Nonheating power supply leads shall be protected where they emerge from electrically heated pipeline or vessel heating units by rigid metal conduit, intermediate metal conduit, electrical metallic tubing, or other raceways identified as suitable for the application.

(C) Interconnecting Leads Interconnecting nonheating leads connecting portions of the heating system shall be permitted to be covered by thermal insulation in the same manner as the heaters.

427.19 Electrical Connections

(A) Nonheating Interconnections Nonheating interconnections, where required under thermal insulation, shall be made with insulated connectors identified as suitable for this use.

(B) Circuit Connections Splices and terminations outside the thermal insulation shall be installed in a box or fitting in accordance with 110.14 and 300.15.

427.20 Marking

Each factory-assembled heating unit shall be legibly marked within 75 mm (3 in.) of each end of the nonheating leads with the permanent identification symbol, catalog number, and ratings in volts and watts or in volts and amperes.

427.22 Equipment Protection

Ground-fault protection of equipment shall be provided for electric heat tracing and heating panels. This requirement shall not apply in industrial establishments where there is alarm indication of ground faults and the following conditions apply:

(1) Conditions of maintenance and supervision ensure that only qualified persons service the installed systems.
(2) Continued circuit operation is necessary for safe operation of equipment or processes.

Section 427.22 states that electric heat-tracing and heating panels must have ground-fault protection for equipment except in industrial establishments where an alarm indication of a ground fault is provided and the two conditions of 427.22 are met.

Rather than protecting the entire branch circuit, the ground-fault protection requirement is focused on protecting just the equipment itself. This affords the manufacturer and the user an option of providing both circuit and equipment protection or just the required equipment protection. This required protection may be accomplished by using circuit breakers equipped with ground-fault equipment protection (GFEP) or an integral device supplied as part of the pipeline or vessel heating equipment that is sensitive to leakage currents in the magnitude of 6 milliamperes to 50 milliamperes [referred to by UL as ground-fault equipment protection circuit-interrupters (GFEPCIs)]. These protection devices, if applied properly, substantially reduce the risk of a fire being started by low-level electrical arcing.

It is important to understand that this required equipment protection is not the same as a GFCI used for personal protection that trips at 5 milliamperes (± 1 milliampere).

For further information regarding ground-fault equipment protection used to comply with 426.28 and 427.22, refer to the UL *General Information for Electrical Equipment Directory* (White Book), category KCZI.

427.23 Grounded Conductive Covering

Electric heating equipment shall be listed and have a grounded conductive covering in accordance with 427.23(A) or 427.23(B). The conductive covering shall provide an effective ground path for equipment protection.

Section 427.23 requires a listed grounded conductive covering on all heaters. This grounded conductive covering is intended to provide a ground fault current path in order to trip circuit or ground-fault protective devices, thus reducing the potential for fire and electric shock. It also provides added mechanical protection of the heating cable or panel.

(A) Heating Wires or Cables Heating wires or cables shall have a grounded conductive covering that surrounds the heating element and bus wires, if any, and their electrical insulation.

(B) Heating Panels Heating panels shall have a grounded conductive covering over the heating element and its electrical insulation on the side opposite the side attached to the surface to be heated.

IV. Impedance Heating

427.25 Personnel Protection

All accessible external surfaces of the pipeline, vessel, or both, being heated shall be physically guarded, isolated, or thermally insulated (with a weatherproof jacket for outside installations) to protect against contact by personnel in the area.

427.26 Isolation Transformer

A dual-winding transformer with a grounded shield between the primary and secondary windings shall be used to isolate the distribution system from the heating system.

427.27 Voltage Limitations

Unless protected by ground-fault circuit-interrupter protection for personnel, the secondary winding of the isolation transformer connected to the pipeline or vessel being heated shall not have an output voltage greater than 30 volts ac.

Where ground-fault circuit-interrupter protection for personnel is provided, the voltage shall be permitted to be greater than 30 but not more than 80 volts.

Exception: In industrial establishments, the isolation transformer connected to the pipeline or vessel being heated shall be permitted to have an output voltage not greater than 132 volts ac to ground where all of the following conditions apply:

(1) Conditions of maintenance and supervision ensure that only qualified persons service the installed systems.
(2) Ground fault protection of equipment is provided.
(3) The pipeline or vessel being heated is completely enclosed in a grounded metal enclosure.
(4) The transformer secondary connections to the pipeline or vessel being heated are completely enclosed in a grounded metal mesh or metal enclosure.

An exception to 427.27 was added to the 2005 *Code* to allow a higher voltage of 132 volts for impedance heating where installed in industrial establishments, provided the four prerequisites are followed. For other than industrial establishments, the maximum operating voltage allowed is 30 volts ac; where GFCI protection is used, a maximum of 80 volts ac is permitted.

427.28 Induced Currents

All current-carrying components shall be installed in accordance with 300.20.

427.29 Grounding

The pipeline, vessel, or both, that is being heated and operating at a voltage greater than 30 but not more than 80 shall be grounded at designated points.

427.30 Secondary Conductor Sizing

The ampacity of the conductors connected to the secondary of the transformer shall be at least 100 percent of the total load of the heater.

V. Induction Heating

427.35 Scope

This part covers the installation of line frequency induction heating equipment and accessories for pipelines and vessels.

> FPN: See Article 665 for other applications.

427.36 Personnel Protection

Induction coils that operate or may operate at a voltage greater than 30 volts ac shall be enclosed in a nonmetallic or split metallic enclosure, isolated, or made inaccessible by location to protect personnel in the area.

427.37 Induced Current

Induction coils shall be prevented from inducing circulating currents in surrounding metallic equipment, supports, or structures by shielding, isolation, or insulation of the current paths. Stray current paths shall be bonded to prevent arcing.

VI. Skin-Effect Heating

427.45 Conductor Ampacity

The ampacity of the electrically insulated conductor inside the ferromagnetic envelope shall be permitted to exceed the values given in Article 310, provided it is identified as suitable for this use.

427.46 Pull Boxes

Pull boxes for pulling the electrically insulated conductor in the ferromagnetic envelope shall be permitted to be buried under the thermal insulation, provided their locations are indicated by permanent markings on the insulation jacket surface and on drawings. For outdoor installations, pull boxes shall be of watertight construction.

427.47 Single Conductor in Enclosure

The provisions of 300.20 shall not apply to the installation of a single conductor in a ferromagnetic envelope (metal enclosure).

427.48 Grounding

The ferromagnetic envelope shall be grounded at both ends, and, in addition, it shall be permitted to be grounded at intermediate points as required by its design. The ferromagnetic envelope shall be bonded at all joints to ensure electrical continuity.

The provisions of 250.30 shall not apply to the installation of skin-effect heating systems.

> FPN: See Article 250 for grounding methods.

VII. Control and Protection

427.55 Disconnecting Means

(A) Switch or Circuit Breaker Means shall be provided to disconnect all fixed electric pipeline or vessel heating equipment from all ungrounded conductors. The branch-circuit switch or circuit breaker, where readily accessible to the user of the equipment, shall be permitted to serve as the disconnecting means. The disconnecting means shall be of the indicating type and shall be provided with a positive lockout in the "off" position.

(B) Cord-and-Plug-Connected Equipment The factory-installed attachment plug of cord-and-plug-connected equipment rated 20 amperes or less and 150 volts or less to ground shall be permitted to be the disconnecting means.

427.56 Controls

(A) Temperature Control with "Off" Position Temperature controlled switching devices that indicate an "off" position and that interrupt line current shall open all ungrounded conductors when the control device is in this "off" position. These devices shall not be permitted to serve as the disconnecting means unless provided with a positive lockout in the "off" position.

(B) Temperature Control Without "Off" Position Temperature controlled switching devices that do not have an "off" position shall not be required to open all ungrounded conductors and shall not be permitted to serve as the disconnecting means.

(C) Remote Temperature Controller Remote controlled temperature-actuated devices shall not be required to meet the requirements of 427.56(A) and 427.56(B). These devices shall not be permitted to serve as the disconnecting means.

(D) Combined Switching Devices Switching devices consisting of combined temperature-actuated devices and manually controlled switches that serve both as the controllers and the disconnecting means shall comply with all the following conditions:

(1) Open all ungrounded conductors when manually placed in the "off" position
(2) Be designed so that the circuit cannot be energized automatically if the device has been manually placed in the "off" position
(3) Be provided with a positive lockout in the "off" position

427.57 Overcurrent Protection

Heating equipment shall be considered as protected against overcurrent where supplied by a branch circuit as specified in 210.3 and 210.23.

ARTICLE 430
Motors, Motor Circuits, and Controllers

Summary of Changes

- **Design E Motors:** Deleted all references to Design E motors from Article 430.

- **430.2:** Added new section with definitions applicable to Article 430.

- **430.6(A)(1):** Moved information on ratings of motors built for low speeds or high torques from table headings of Tables 430.248, 430.249, and 430.250 to new last sentence.

- **430.8:** Added requirement to include short-circuit current rating on motor controllers, with four exceptions.

- **430.53(C)(3):** Revised to no longer require the circuit breaker be listed for group installation, recognizing that all listed molded-case circuit breakers are acceptable on group motor installations.

- **430.102(B), Exception:** Revised to clarify intent that a locking means be available at all times (i.e., to prohibit a portable locking means that is removed when the lock is removed).

- **430.109(A)(6):** Revised to permit listed manual motor controllers additionally marked "Suitable as Motor Disconnect" as disconnecting means on the line side of the fuses permitted in 430.52 (C)(5).

- **430.109(A)(7):** Added requirement to recognize the use of system isolation equipment as a disconnecting means. This equipment is required to be specifically listed for disconnection purposes.

- **Part X:** Revised to consolidate all requirements for adjustable speed drive systems, formerly scattered throughout Article 430, in a new Part X in a central location in Article 430.

- **Tables 430.248, 430.249, 430.250:** Moved sentence referencing low-speed or high-torque motors from table headers to 430.6(A)(1).

Contents

I. General

Most electrical equipment is rated in volt-amperes (VA) or watt input. Basic to the understanding of Article 430 is the fact that motors traditionally have been rated in horsepower output. Circuits supplying motors are sized according to the input to the motor (input equals output plus losses of the motor). The losses are not the type of information found on the nameplate of a motor. Tables 430.249 and 430.250 contain accurate industry-wide input ampere ratings for motors.

However, some motors are available with their output ratings expressed in watts and kilowatts. (One horsepower equals approximately 746 watts.) It is important to understand that circuits that supply motors not rated in horsepower still must be sized according to the input of the motor, rated in amperes. Sizing circuits based solely on kilowatt output results in seriously undersized conductors and the improper application of overcurrent devices. See 430.6 for ampacity and motor rating determination.

430.1 Scope

This article covers motors, motor branch-circuit and feeder conductors and their protection, motor overload protection, motor control circuits, motor controllers, and motor control centers.

> FPN No. 1: Installation requirements for motor control centers are covered in 110.26(F). Air conditioning and refrigerating equipment are covered in Article 440.
>
> FPN No. 2: Figure 430.1 is for information only.

430.2 Definitions

In accordance with the *NEC Style Manual*, the __.2 section of each article is used for definitions unique to that article. Two definitions (*controller* and *motor control circuits*) have been relocated from previous editions of the *Code*. In addition, three new definitions (*adjustable speed drive*, *adjustable-speed drive system*, and *system isolation equipment*) are included in Article 430 for the 2005 *Code*.

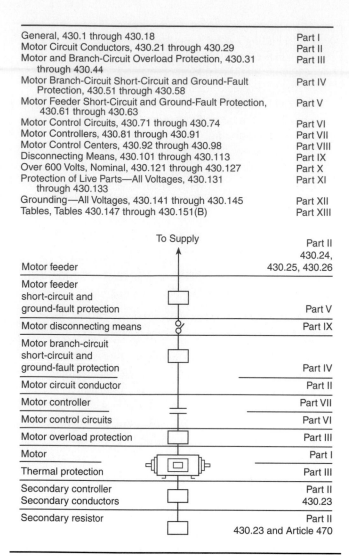

Figure 430.1 Article 430 Contents

Article 430 was rearranged for the 2005 *Code* by consolidating all material pertinent to adjustable speed drive systems into Part X and moving Parts X, XI, XII, and XIII forward.

Figure 430.1 is intended to assist the user in following the provisions of Article 430. The requirements for motors in Article 430 are broken down into 14 individual parts. Many of these parts are pictorially identified in Figure 430.1 as they pertain to the installation of motors.

Adjustable Speed Drive. A combination of the power converter, motor, and motor mounted auxiliary devices such as encoders, tachometers, thermal switches and detectors, air blowers, heaters, and vibration sensors.

Adjustable-Speed Drive System. An interconnected combination of equipment that provides a means of adjusting the speed of a mechanical load coupled to a motor. A drive

system typically consists of an adjustable speed drive and auxiliary electrical apparatus.

Controller. For the purpose of this article, a controller is any switch or device that is normally used to start and stop a motor by making and breaking the motor circuit current.

Motor Control Circuit. The circuit of a control apparatus or system that carries the electric signals directing the performance of the controller but does not carry the main power current.

System Isolation Equipment. A redundantly monitored, remotely operated contactor-isolating system, packaged to provide the disconnection/isolation function, capable of verifiable operation from multiple remote locations by means of lockout switches, each having the capability of being padlocked in the "off" (open) position.

430.4 Part-Winding Motors

A part-winding start induction or synchronous motor is one that is arranged for starting by first energizing part of its primary (armature) winding and, subsequently, energizing the remainder of this winding in one or more steps. A standard part-winding start induction motor is arranged so that one-half of its primary winding can be energized initially, and, subsequently, the remaining half can be energized, both halves then carrying equal current. A hermetic refrigerant compressor motor shall not be considered a standard part-winding start induction motor.

Where separate overload devices are used with a standard part-winding start induction motor, each half of the motor winding shall be individually protected in accordance with 430.32 and 430.37 with a trip current one-half that specified.

Each motor-winding connection shall have branch-circuit short-circuit and ground-fault protection rated at not more than one-half that specified by 430.52.

Exception: A short-circuit and ground-fault protective device shall be permitted for both windings if the device will allow the motor to start. Where time-delay (dual-element) fuses are used, they shall be permitted to have a rating not exceeding 150 percent of the motor full-load current.

430.5 Other Articles

Motors and controllers shall also comply with the applicable provisions of Table 430.5.

430.6 Ampacity and Motor Rating Determination

The size of conductors supplying equipment covered by Article 430 shall be selected from the allowable ampacity tables in accordance with 310.15(B) or shall be calculated in accordance with 310.15(C). Where flexible cord is used,

Table 430.5 Other Articles

Equipment/Occupancy	Article	Section
Air-conditioning and refrigerating equipment	440	
Capacitors		460.8, 460.9
Commercial garages; aircraft hangars; motor fuel dispensing facilities; bulk storage plants; spray application, dipping, and coating processes; and inhalation anesthetizing locations	511, 513, 514, 515, 516, and 517 Part IV	
Cranes and hoists	610	
Electrically driven or controlled irrigation machines	675	
Elevators, dumbwaiters, escalators, moving walks, wheelchair lifts, and stairway chair lifts	620	
Fire pumps	695	
Hazardous (classified) locations	500–503 and 505	
Industrial machinery	670	
Motion picture projectors		540.11 and 540.20
Motion picture and television studios and similar locations	530	
Resistors and reactors	470	
Theaters, audience areas of motion picture and television studios, and similar locations		520.48
Transformers and transformer vaults	450	

the size of the conductor shall be selected in accordance with 400.5. The required ampacity and motor ratings shall be determined as specified in 430.6(A), (B), and (C).

(A) General Motor Applications For general motor applications, current ratings shall be determined based on (A)(1) and (A)(2).

(1) Table Values Other than for motors built for low speeds (less than 1200 RPM) or high torques, and for multispeed motors, the values given in Tables 430.247, 430.248, 430.249, and 430.250 shall be used to determine the ampacity of conductors or ampere ratings of switches, branch-circuit short-circuit and ground-fault protection, instead of the actual current rating marked on the motor nameplate. Where a motor is marked in amperes, but not horsepower, the horsepower rating shall be assumed to be that corresponding to the value given in Tables 430.247, 430.248,

430.249, and 430.250, interpolated if necessary. Motors built for low speeds (less than 1200 RPM) or high torques may have higher full-load currents, and multispeed motors will have full-load current varying with speed, in which case the nameplate current ratings shall be used.

Exception No. 1: Multispeed motors shall be in accordance with 430.22(A) and 430.52.

Exception No. 2: For equipment that employs a shaded-pole or permanent-split capacitor-type fan or blower motor that is marked with the motor type, the full load current for such motor marked on the nameplate of the equipment in which the fan or blower motor is employed shall be used instead of the horsepower rating to determine the ampacity or rating of the disconnecting means, the branch-circuit conductors, the controller, the branch-circuit short-circuit and ground-fault protection, and the separate overload protection. This marking on the equipment nameplate shall not be less than the current marked on the fan or blower motor nameplate.

Exception No. 3: For a listed motor-operated appliance that is marked with both motor horsepower and full-load current, the motor full-load current marked on the nameplate of the appliance shall be used instead of the horsepower rating on the appliance nameplate to determine the ampacity or rating of the disconnecting means, the branch-circuit conductors, the controller, the branch-circuit short-circuit and ground-fault protection, and any separate overload protection.

Exception No. 3 to 430.6(A)(1) is intended to resolve confusion that may result when motor-operated appliances are labeled with both horsepower and ampere ratings. The nameplate current rating in amperes is closely evaluated by the testing laboratories and is considered more accurate than the marked horsepower rating for motor-operated appliances.

(2) Nameplate Values Separate motor overload protection shall be based on the motor nameplate current rating.

For general motor applications other than motors built for low speeds (less than 1200 rpm), multispeed motors, or high-torque motors, the ampacity of motor branch-circuit conductors, branch-circuit and ground-fault protection, and ampere rating of the motor disconnecting means are determined by the ampere values listed in Table 430.247 through Table 430.250, not the ampere values marked on the motor nameplate. The ampere values are based on the horsepower rating and nominal voltage listed on the motor nameplate.

The ampere rating provided on the motor nameplate is used to size the overload protective devices intended to protect the motor, motor control apparatus, and motor branch-circuit conductors. Where the motor is marked with

(B) Torque Motors For torque motors, the rated current shall be locked-rotor current, and this nameplate current shall be used to determine the ampacity of the branch-circuit conductors covered in 430.22 and 430.24, the ampere rating of the motor overload protection, and the ampere rating of motor branch-circuit short-circuit and ground-fault protection in accordance with 430.52(B).

> FPN: For motor controllers and disconnecting means, see 430.83(D) and 430.110.

(C) Alternating-Current Adjustable Voltage Motors For motors used in alternating-current, adjustable voltage, variable torque drive systems, the ampacity of conductors, or ampere ratings of switches, branch-circuit short-circuit and ground-fault protection, and so forth, shall be based on the maximum operating current marked on the motor or control nameplate, or both. If the maximum operating current does not appear on the nameplate, the ampacity determination shall be based on 150 percent of the values given in Tables 430.249 and 430.250.

430.7 Marking on Motors and Multimotor Equipment

(A) Usual Motor Applications A motor shall be marked with the following information.

(1) Manufacturer's name.
(2) Rated volts and full-load current. For a multispeed motor, full-load current for each speed, except shaded-pole and permanent-split capacitor motors where amperes are required only for maximum speed.
(3) Rated frequency and number of phases if an ac motor.
(4) Rated full-load speed.
(5) Rated temperature rise or the insulation system class and rated ambient temperature.
(6) Time rating. The time rating shall be 5, 15, 30, or 60 minutes, or continuous.
(7) Rated horsepower if ⅛ hp or more. For a multispeed motor ⅛ hp or more, rated horsepower for each speed, except shaded-pole and permanent-split capacitor motors ⅛ hp or more where rated horsepower is required only for maximum speed. Motors of arc welders are not required to be marked with the horsepower rating.
(8) Code letter or locked-rotor amperes if an alternating-current motor rated ½ hp or more. On polyphase wound-rotor motors, the code letter shall be omitted.

> FPN: See 430.7(B).

(9) Design letter for design B, C, or D motors.

(10) Secondary volts and full-load current if a wound-rotor induction motor.
(11) Field current and voltage for dc excited synchronous motors.
(12) Winding — straight shunt, stabilized shunt, compound, or series, if a dc motor. Fractional horsepower dc motors 175 mm (7 in.) or less in diameter shall not be required to be marked.
(13) A motor provided with a thermal protector complying with 430.32(A)(2) or (B)(2) shall be marked "Thermally Protected." Thermally protected motors rated 100 watts or less and complying with 430.32(B)(2) shall be permitted to use the abbreviated marking "T.P."
(14) A motor complying with 430.32(B)(4) shall be marked "Impedance Protected." Impedance protected motors rated 100 watts or less and complying with 430.32(B)(4) shall be permitted to use the abbreviated marking "Z.P."
(15) Motors equipped with electrically powered condensation prevention heaters shall be marked with the rated heater voltage, number of phases, and the rated power in watts.

(B) Locked-Rotor Indicating Code Letters Code letters marked on motor nameplates to show motor input with locked rotor shall be in accordance with Table 430.7(B).

The code letter indicating motor input with locked rotor shall be in an individual block on the nameplate, properly designated.

(1) Multispeed Motors Multispeed motors shall be marked with the code letter designating the locked-rotor kilovolt-ampere (kVA) per horsepower for the highest speed at which the motor can be started.

Table 430.7(B) Locked-Rotor Indicating Code Letters

Code Letter	Kilovolt-Amperes per Horsepower with Locked Rotor
A	0–3.14
B	3.15–3.54
C	3.55–3.99
D	4.0–4.49
E	4.5–4.99
F	5.0–5.59
G	5.6–6.29
H	6.3–7.09
J	7.1–7.99
K	8.0–8.99
L	9.0–9.99
M	10.0–11.19
N	11.2–12.49
P	12.5–13.99
R	14.0–15.99
S	16.0–17.99
T	18.0–19.99
U	20.0–22.39
V	22.4 and up

Exception: Constant horsepower multispeed motors shall be marked with the code letter giving the highest locked-rotor kilovolt-ampere (kVA) per horsepower.

(2) Single-Speed Motors Single-speed motors starting on wye connection and running on delta connections shall be marked with a code letter corresponding to the locked-rotor kilovolt-ampere (kVA) per horsepower for the wye connection.

(3) Dual-Voltage Motors Dual-voltage motors that have a different locked-rotor kilovolt-ampere (kVA) per horsepower on the two voltages shall be marked with the code letter for the voltage giving the highest locked-rotor kilovolt-ampere (kVA) per horsepower.

(4) 50/60 Hz Motors Motors with 50- and 60-Hz ratings shall be marked with a code letter designating the locked-rotor kilovolt-ampere (kVA) per horsepower on 60 Hz.

(5) Part-Winding Motors Part-winding start motors shall be marked with a code letter designating the locked-rotor kilovolt-ampere (kVA) per horsepower that is based on the locked-rotor current for the full winding of the motor.

The following example shows how to determine the locked-rotor current for a specific motor using the values from Table 430.7(B).

Example

Using Table 430.7(B), find the maximum locked-rotor current for a 20-hp, 460-volt, 3-phase motor with the nameplate kilovolt-ampere code letter "G."

Solution

Table 430.7(B) lists a range of values for "G." The maximum value in the range is 6.29, or 6.29 kilovolt-amperes per horsepower. Use the following formula to find the maximum locked-rotor current.

$$\text{Locked-rotor amperes} = \text{motor hp} \times \text{maximum code letter value} \frac{kVA}{hp} \times \frac{1000}{\text{volts} \times 1.73}$$

Substitute as follows:

$$\text{Locked-rotor amperes} = 20\text{ hp} \times 6.29 \frac{kVA}{hp} \times \frac{1000}{460\text{ volts} \times 1.73}$$

$$= 158\text{ amperes}$$

Therefore, the maximum locked-rotor current for a 20-hp, 460-volt motor with code letter "G" is 158 amperes.

(C) Torque Motors Torque motors are rated for operation at standstill and shall be marked in accordance with 430.7(A), except that locked-rotor torque shall replace horsepower.

(D) Multimotor and Combination-Load Equipment

(1) Factory-Wired Multimotor and combination-load equipment shall be provided with a visible nameplate marked with the manufacturer's name, the rating in volts, frequency, number of phases, minimum supply circuit conductor ampacity, and the maximum ampere rating of the circuit short-circuit and ground-fault protective device. The conductor ampacity shall be calculated in accordance with 430.24 and counting all of the motors and other loads that will be operated at the same time. The short-circuit and ground-fault protective device rating shall not exceed the value calculated in accordance with 430.53. Multimotor equipment for use on two or more circuits shall be marked with the preceding information for each circuit.

Section 110.3(B) requires listed and labeled equipment to be used and installed in accordance with the manufacturer's instructions accompanying the equipment or marked on the nameplate.

The nameplate marking for the maximum ampere rating of the branch-circuit short-circuit and ground-fault protective device may limit the type of protective device to a fuse by stipulating "fuse" without reference to a circuit breaker.

That means the circuit to the equipment must be protected by fuses, such as by a fused disconnect switch. The fused switch may be supplied from a circuit breaker in a panelboard.

(2) Not Factory-Wired Where the equipment is not factory-wired and the individual nameplates of motors and other loads are visible after assembly of the equipment, the individual nameplates shall be permitted to serve as the required marking.

430.8 Marking on Controllers

A controller shall be marked with the manufacturer's name or identification, the voltage, the current or horsepower rating, the short-circuit current rating, and such other necessary data to properly indicate the applications for which it is suitable.

Section 430.8 now requires controllers to be marked with their short-circuit current rating. Four exceptions have been added to the 2005 *Code* to allow the short-circuit rating to be excluded from the controller nameplate under specified conditions.

Exception No. 1: The short-circuit current rating is not required for controllers applied in accordance with 430.81(A) or (B).

Exception No. 2: The short-circuit rating is not required to be marked on the controller when the short-circuit current rating of the controller is marked elsewhere on the assembly.

Exception No. 3: The short-circuit rating is not required to be marked on the controller when the assembly into which it is installed has a marked short-circuit current rating.

Exception No. 4: Short-circuit ratings are not required for controllers rated less than 2 hp at 300 V or less and listed exclusively for general-purpose branch circuits.

A controller that includes motor overload protection suitable for group motor application shall be marked with the motor overload protection and the maximum branch-circuit short-circuit and ground-fault protection for such applications.

Combination controllers that employ adjustable instantaneous trip circuit breakers shall be clearly marked to indicate the ampere settings of the adjustable trip element.

Where a controller is built-in as an integral part of a motor or of a motor-generator set, individual marking of the controller shall not be required if the necessary data are on the nameplate. For controllers that are an integral part of equipment approved as a unit, the above marking shall be permitted on the equipment nameplate.

FPN: See 110.10 for information on circuit impedance and other characteristics.

430.9 Terminals

(A) Markings Terminals of motors and controllers shall be suitably marked or colored where necessary to indicate the proper connections.

(B) Conductors Motor controllers and terminals of control circuit devices shall be connected with copper conductors unless identified for use with a different conductor.

Terminals for motor controllers are tested, designed, and listed using copper conductors, unless marked for other conductors. Section 430.9(B) highlights this limitation while permitting other conductors to be used if they are determined to be suitable and are identified as such.

(C) Torque Requirements Control circuit devices with screw-type pressure terminals used with 14 AWG or smaller copper conductors shall be torqued to a minimum of 0.8 N·m (7 lb-in.) unless identified for a different torque value.

Proper torque is essential for safe and reliable connections. Section 430.9(C) enhances safety by providing a minimum torque value for screw-type pressure terminals. See the commentary following 110.14(B) for more information on electrical connections.

430.10 Wiring Space in Enclosures

(A) General Enclosures for motor controllers and disconnecting means shall not be used as junction boxes, auxiliary gutters, or raceways for conductors feeding through or tapping off to the other apparatus unless designs are employed that provide adequate space for this purpose.

During the planning stages of a motor(s) installation, location and proper working spaces (as required by 110.26 and 110.34) for motor controllers and disconnects should be considered, including provisions for the use of auxiliary gutters or junction boxes, to ensure space for conductors feeding through or tapping off to other apparatus. For switch and overcurrent device enclosures, see 312.8 and the commentary that follows.

FPN: See 312.8 for switch and overcurrent-device enclosures.

(B) Wire-Bending Space in Enclosures Minimum wire-bending space within the enclosures for motor controllers shall be in accordance with Table 430.10(B) where measured in a straight line from the end of the lug or wire connector (in the direction the wire leaves the terminal) to the wall or barrier. Where alternate wire termination means are substi-

Table 430.10(B) Minimum Wire-Bending Space at the Terminals of Enclosed Motor Controllers

| Size of Wire (AWG or kcmil) | Wires per Terminal* | | | |
| | 1 | | 2 | |
	mm	in.	mm	in.
14–10	Not specified		—	—
8–6	38	1½	—	—
4–3	50	2	—	—
2	65	2½	—	—
1	75	3	—	—
1/0	125	5	125	5
2/0	150	6	150	6
3/0–4/0	175	7	175	7
250	200	8	200	8
300	250	10	250	10
350–500	300	12	300	12
600–700	350	14	400	16
750–900	450	18	475	19

*Where provision for three or more wires per terminal exists, the minimum wire-bending space shall be in accordance with the requirements of Article 312.

Exhibit 430.1 Wire-bending space in enclosures for motor controllers.

tuted for that supplied by the manufacturer of the controller, they shall be of a type identified by the manufacturer for use with the controller and shall not reduce the minimum wire-bending space.

Exhibit 430.1 illustrates application of the wiring-bending space requirements of either 430.10(B) or 312.6(B) within an enclosure for a motor controller.

430.11 Protection Against Liquids

Suitable guards or enclosures shall be provided to protect exposed current-carrying parts of motors and the insulation of motor leads where installed directly under equipment, or in other locations where dripping or spraying oil, water, or other injurious liquid may occur, unless the motor is designed for the existing conditions.

Exposed current-carrying parts and insulated leads of motors should be suitably protected from liquids (dripping or spraying oil or other lubricants, water, or excessive moisture). The presence of liquids may cause deterioration and insulation breakdown.

430.12 Motor Terminal Housings

(A) Material Where motors are provided with terminal housings, the housings shall be of metal and of substantial construction.

Exception: In other than hazardous (classified) locations, substantial, nonmetallic, nonburning housings shall be permitted, provided an internal grounding means between the motor frame and the equipment grounding connection is incorporated within the housing.

Nonmetallic terminal housings are permitted on motors of any size, provided the housing material has been determined to be nonburning, with, for example, a flammability rating of at least 94-5V in accordance with UL 746C, *Polymeric Materials — Use in Electrical Equipment Evaluations.*

(B) Dimensions and Space — Wire-to-Wire Connections Where these terminal housings enclose wire-to-wire connections, they shall have minimum dimensions and usable volumes in accordance with Table 430.12(B).

(C) Dimensions and Space — Fixed Terminal Connections Where these terminal housings enclose rigidly mounted motor terminals, the terminal housing shall be of sufficient size to provide minimum terminal spacings and usable volumes in accordance with Table 430.12(C)(1) and Table 430.12(C)(2).

(D) Large Wire or Factory Connections For motors with larger ratings, greater number of leads, or larger wire sizes,

Table 430.12(B) Terminal Housings — Wire-to-Wire Connections

Motors 275 mm (11 in.) in Diameter or Less

Horsepower	Cover Opening Minimum Dimension		Usable Volume Minimum	
	mm	in.	cm³	in.³
1 and smaller[a]	41	1⅝	170	10.5
1½, 2, and 3[b]	45	1¾	275	16.8
5 and 7½	50	2	365	22.4
10 and 15	65	2½	595	36.4

Motors Over 275 mm (11 in.) in Diameter — Alternating-Current Motors

Maximum Full Load Current for 3-Phase Motors with Maximum of 12 Leads (Amperes)	Terminal Box Cover Opening Minimum Dimension		Usable Volume Minimum		Typical Maximum Horsepower 3-Phase	
	mm	in.	cm³	in.³	230 Volt	460 Volt
45	65	2.5	595	36.4	15	30
70	84	3.3	1,265	77	25	50
110	100	4.0	2,295	140	40	75
160	125	5.0	4,135	252	60	125
250	150	6.0	7,380	450	100	200
400	175	7.0	13,775	840	150	300
600	200	8.0	25,255	1540	250	500

Direct-Current Motors

Maximum Full-Load Current for Motors with Maximum of 6 Leads (Amperes)	Terminal Box Minimum Dimensions		Usable Volume Minimum	
	mm	in.	cm³	in.³
68	65	2.5	425	26
105	84	3.3	900	55
165	100	4.0	1,640	100
240	125	5.0	2,950	180
375	150	6.0	5,410	330
600	175	7.0	9,840	600
900	200	8.0	18,040	1,100

Note: Auxiliary leads for such items as brakes, thermostats, space heaters, and exciting fields shall be permitted to be neglected if their current-carrying area does not exceed 25 percent of the current-carrying area of the machine power leads.

[a]For motors rated 1 hp and smaller and with the terminal housing partially or wholly integral with the frame or end shield, the volume of the terminal housing shall not be less than 18.0 cm³ (1.1 in.³) per wire-to-wire connection. The minimum cover opening dimension is not specified.

[b]For motors rated 1½, 2, and 3 hp and with the terminal housing partially or wholly integral with the frame or end shield, the volume of the terminal housing shall not be less than 23.0 cm³ (1.4 in.³) per wire-to-wire connection. The minimum cover opening dimension is not specified.

Table 430.12(C)(1) Terminal Spacings — Fixed Terminals

Nominal Volts	Minimum Spacing			
	Between Line Terminals		Between Line Terminals and Other Uninsulated Metal Parts	
	mm	in.	mm	in.
240 or less	6	¼	6	¼
Over 250 – 600	10	⅜	10	⅜

Table 430.12(C)(2) Usable Volumes — Fixed Terminals

Power-Supply Conductor Size (AWG)	Minimum Usable Volume per Power-Supply Conductor	
	cm³	in.³
14	16	1
12 and 10	20	1¼
8 and 6	37	2¼

or where motors are installed as a part of factory-wired equipment, without additional connection being required at the motor terminal housing during equipment installation, the terminal housing shall be of ample size to make connections, but the foregoing provisions for the volumes of terminal housings shall not be considered applicable.

(E) Equipment Grounding Connections A means for attachment of an equipment grounding conductor termination in accordance with 250.8 shall be provided at motor terminal housings for wire-to-wire connections or fixed terminal connections. The means for such connections shall be permitted to be located either inside or outside the motor terminal housing.

Exception: Where a motor is installed as a part of factory-wired equipment that is required to be grounded and without additional connection being required at the motor terminal housing during equipment installation, a separate means for motor grounding at the motor terminal housing shall not be required.

430.13 Bushing

Where wires pass through an opening in an enclosure, conduit box, or barrier, a bushing shall be used to protect the conductors from the edges of openings having sharp edges. The bushing shall have smooth, well-rounded surfaces where it may be in contact with the conductors. If used where oils, greases, or other contaminants may be present, the bushing shall be made of material not deleteriously affected.

FPN: For conductors exposed to deteriorating agents, see 310.9.

430.14 Location of Motors

(A) Ventilation and Maintenance Motors shall be located so that adequate ventilation is provided and so that maintenance, such as lubrication of bearings and replacing of brushes, can be readily accomplished.

Exception: Ventilation shall not be required for submersible types of motors.

(B) Open Motors Open motors that have commutators or collector rings shall be located or protected so that sparks cannot reach adjacent combustible material.

Exception: Installation of these motors on wooden floors or supports shall be permitted.

430.16 Exposure to Dust Accumulations

In locations where dust or flying material collects on or in motors in such quantities as to seriously interfere with the ventilation or cooling of motors and thereby cause dangerous temperatures, suitable types of enclosed motors that do not overheat under the prevailing conditions shall be used.

FPN: Especially severe conditions may require the use of enclosed pipe-ventilated motors, or enclosure in separate dusttight rooms, properly ventilated from a source of clean air.

For motors exposed to combustible dust or readily ignitible flying material, see the requirements of 502.125 (Class II, Divisions 1 and 2) and 503.125 (Class III, Divisions 1 and 2). For classification of locations, see 500.5(C) (Class II locations) and 500.5(D) (Class III locations).

430.17 Highest Rated or Smallest Rated Motor

In determining compliance with 430.24, 430.53(B), and 430.53(C), the highest rated or smallest rated motor shall be based on the rated full-load current as selected from Tables 430.247, 430.248, 430.249, and 430.250.

430.18 Nominal Voltage of Rectifier Systems

The nominal value of the ac voltage being rectified shall be used to determine the voltage of a rectifier derived system.

Exception: The nominal dc voltage of the rectifier shall be used if it exceeds the peak value of the ac voltage being rectified.

II. Motor Circuit Conductors

430.21 General

Part II specifies ampacities of conductors that are capable of carrying the motor current without overheating under the conditions specified.

The provisions of Part II shall not apply to motor circuits rated over 600 volts, nominal.

The provisions of Articles 250, 300, and 310 shall not apply to conductors that form an integral part of equipment, such as motors, motor controllers, motor control centers, or other factory-assembled control equipment.

FPN No. 1: See 300.1(B) and 310.1 for similar requirements.
FPN No. 2: See 110.14(C) and 430.9(B) for equipment device terminal requirements.
FPN No. 3: For over 600 volts, nominal, see Part XI.

430.22 Single Motor

(A) General Conductors that supply a single motor used in a continuous duty application shall have an ampacity of not less than 125 percent of the motor's full-load current rating as determined by 430.6(A)(1).

Section 430.22 describes the branch-circuit requirements for single motor installations. Generally, the branch circuit that serves a continuous-duty motor must be sized at 125 percent of the motor full-load current or greater. The provision for a conductor with an ampacity of at least 125 percent of the motor full-load current rating does not constitute a conductor derating; rather, it is based on the need to provide for a sustained running current that is greater than the rated full-load current and for protection of the conductors by the motor overload protective device set above the motor full-load current rating.

The ampacity of the motor branch-circuit conductors is based on the full-load current rating values provided in Table 430.248 through Table 430.250. Motor nameplate full-load current is not to be used to size branch-circuit conductors.

Exhibit 430.2 illustrates each motor on an individual branch circuit with branch-circuit short-circuit and ground-fault protective devices and disconnecting means in one location, and controllers and overload protection at the motor locations.

Exception No. 2 to 430.22 was moved in the 2005 *Code* to the new Part X that addresses adjustable-speed drive systems.

Exhibit 430.3 also illustrates each motor on an individual branch circuit, but, unlike Exhibit 430.2, the branch circuits are tapped from a feeder at a convenient location, such as a junction box or wireway, or from open wiring. The tap conductors are required to terminate in a branch-circuit protective device located not more than 25 ft from where the taps are connected to the feeder, in accordance with 430.28. Also see 430.28, Exception, which permits a 100-ft tap under some conditions in high-bay manufacturing facilities.

Where motors are connected to a 15- or 20-ampere

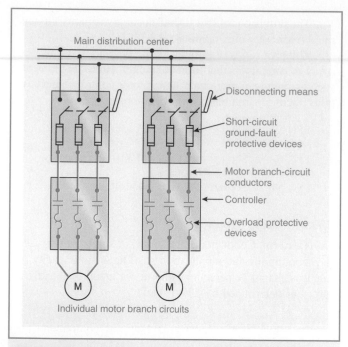

Exhibit 430.2 A main distribution center supplying individual branch circuits to each motor (branch-circuit short-circuit and ground-fault protective devices).

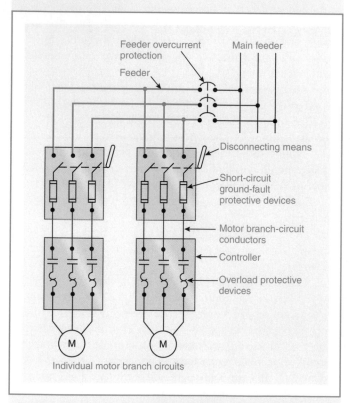

Exhibit 430.3 A feeder supplying individual branch circuits to each motor.

branch circuit that also supplies lighting or other appliance loads, as illustrated in Exhibit 430.4, the provisions of Article 210 and Article 430 apply. Motors rated less than 1 hp may be connected to these circuits, and they must be provided with overload protective devices unless the motors are not permanently installed, are started manually, and are within sight from the controller location. For additional information on the installation of motors (1 hp or less), see 430.32(B), 430.32(C), and 430.53(A).

Exhibit 430.4 A 20-ampere branch circuit supplying lighting, small motors, and appliances.

Exhibit 430.5 illustrates the following essential parts of a motor branch circuit:

1. Branch-circuit conductors
2. Disconnecting means
3. Branch-circuit short-circuit and ground-fault protective devices
4. Motor-controller
5. Motor overload protective devices

The branch-circuit short-circuit and ground-fault protective device may be a fuse or a circuit breaker and must be capable of carrying the starting current of the motor without opening the circuit. See Table 430.52.

In general, every motor must be provided with overload protective devices intended to protect the motor windings, motor-control apparatus, and motor branch-circuit conduc-

Feeder conductors

Tap conductors

Disconnecting means

Motor controller

Motor overload protective devices

M

Motor branch-circuit short-circuit and ground-fault protective devices

Motor branch-circuit conductor ampacity 125% of motor full-load current rating

Exhibit 430.5 A motor branch circuit showing the essential parts.

tors against excessive heating due to motor overloads and failure to start. Overload in equipment is defined as operation in excess of normal full-load rating, which, when it persists for a sufficient length of time, will cause damage or dangerous overheating. Overload in a motor includes a stalled rotor but does not include fault currents due to short circuits or ground faults. See 430.44 for conditions where providing automatic opening of a motor circuit due to overload may be objectionable.

Exception: For dc motors operating from a rectified single-phase power supply, the conductors between the field wiring terminals of the rectifier and the motor shall have an ampacity of not less than the following percent of the motor full-load current rating:

(a) Where a rectifier bridge of the single-phase half-wave type is used, 190 percent.

(b) Where a rectifier bridge of the single-phase full-wave type is used, 150 percent.

(B) Multispeed Motor For a multispeed motor, the selection of branch-circuit conductors on the line side of the controller shall be based on the highest of the full-load current ratings shown on the motor nameplate. The selection of branch-circuit conductors between the controller and the motor shall be based on the current rating of the winding(s) that the conductors energize.

(C) Wye-Start, Delta-Run Motor For a wye-start, delta-run connected motor, the selection of branch-circuit conductors on the line side of the controller shall be based on the motor full-load current. The selection of conductors between the controller and the motor shall be based on 58 percent of the motor full-load current.

A wye-start, delta-run winding configuration is a method of providing reduced-voltage starting for a polyphase induction motor. This method requires a specific type of motor controller and a delta-wired motor with all leads brought out to the terminal box. This method of starting finds wide application in certain compressors used for air conditioning and where the driven machinery is allowed to start unloaded. During starting, the windings are configured in a wye configuration. The wye-start configuration results in a reduced starting voltage of a mathematical ratio of $1/\sqrt{3} = 0.5774$, or 58 percent of the full line voltage, which results in approximately 58 percent starting current and about one-third of the normal starting torque. Once the motor attains speed, the windings are reconfigured to run as delta, giving full line voltage to the individual windings, which allows the motor to have full torque capability.

In Exhibit 430.6, conductors from terminals T_1, T_2, and T_3 to the motor, as well as the conductors from terminals T_4, T_5, and T_6 to the motor, are all sized at 58 percent of the full-load current used to size the conductors that supply L_1, L_2, and L_3.

Exhibit 430.6 An elementary wiring diagram of a typical wye-start, delta-run motor and controller. During START, contacts 1M and S are closed and contacts 2M are open. During RUN, contacts 1M and 2M are closed and contacts S are open. (Redrawn courtesy of Square D Co.)

(D) Part-Winding Motor For a part-winding connected motor, the selection of branch-circuit conductors on the line side of the controller shall be based on the motor full-load current. The selection of conductors between the controller and the motor shall be based on 50 percent of the motor full-load current.

(E) Other Than Continuous Duty Conductors for a motor used in a short-time, intermittent, periodic, or varying duty application shall have an ampacity of not less than the percentage of the motor nameplate current rating shown in Table 430.22(E), unless the authority having jurisdiction grants special permission for conductors of lower ampacity.

Table 430.22(E) Duty-Cycle Service

Classification of Service	Nameplate Current Rating Percentages			
	5-Minute Rated Motor	15-Minute Rated Motor	30- & 60-Minute Rated Motor	Continuous Rated Motor
Short-time duty operating valves, raising or lowering rolls, etc.	110	120	150	—
Intermittent duty freight and passenger elevators, tool heads, pumps, drawbridges, turntables, etc. (for arc welders, see 630.11)	85	85	90	140
Periodic duty rolls, ore- and coal-handling machines, etc.	85	90	95	140
Varying duty	110	120	150	200

Note: Any motor application shall be considered as continuous duty unless the nature of the apparatus it drives is such that the motor will not operate continuously with load under any condition of use.

Most motor applications are continuous duty, meaning they operate at a constant load for an indefinitely long time. For motors that are not continuous duty, the motor nameplate currents and Table 430.22(E) are used to determine the branch circuit ampacity. A motor is considered to be for continuous duty unless the nature of the apparatus it drives is such that the motor cannot operate continuously with load under any condition of use. Conductors for a motor used for short-time, intermittent, periodic, or varying duty are required to have an ampacity in accordance with Table

430.22(E). Branch-circuit conductors for a motor with a rated horsepower used for 5-minute short-time duty service are permitted to be sized smaller than for the same motor with a 60-minute rating, due to the cooling intervals between operating periods. The terms *continuous duty, intermittent duty, periodic duty, short-time duty,* and *varying duty* are defined in Article 100.

(F) Separate Terminal Enclosure The conductors between a stationary motor rated 1 hp or less and the separate terminal enclosure permitted in 430.245(B) shall be permitted to be smaller than 14 AWG but not smaller than 18 AWG, provided they have an ampacity as specified in 430.22(A).

430.23 Wound-Rotor Secondary

Before the advent of adjustable-speed drives, wound-rotor ac motors were generally used where speed control was desired; where high starting torque for a rapid, smooth acceleration to full load was required; for frequent starting; and for low starting current. Wound-rotor ac motors are also known as slip-ring motors because three slip rings are mounted on the shaft, and brushes in contact with the slip rings are connected to field-installed external resistance units and a controller, as shown in Exhibit 430.7. The resistors are a part of the rotor circuit, and the full value of the external resistance is in the circuit when the motor starts. This value is gradually reduced until the motor attains the desired speed.

Exhibit 430.7 A branch circuit to a wound-rotor induction motor showing a controller and separate bank of resistors for motor starting and speed regulation.

Controllers used for speed regulation, usually a dial-type or a drum-type switch, are basically for two types of loads: constant torque (machine loads) and variable torque (fan loads).

The ampacities of the conductors between the controller and the resistance units are the allowable percentages of Table 430.23(C) for the resistor duty classification.

(A) Continuous Duty For continuous duty, the conductors connecting the secondary of a wound-rotor ac motor to its controller shall have an ampacity not less than 125 percent of the full-load secondary current of the motor.

(B) Other Than Continuous Duty For other than continuous duty, these conductors shall have an ampacity, in percent of full-load secondary current, not less than that specified in Table 430.22(E).

(C) Resistor Separate from Controller Where the secondary resistor is separate from the controller, the ampacity of the conductors between controller and resistor shall not be less than that shown in Table 430.23(C).

Table 430.23(C) Secondary Conductor

Resistor Duty Classification	Ampacity of Conductor in Percent of Full-Load Secondary Current
Light starting duty	35
Heavy starting duty	45
Extra-heavy starting duty	55
Light intermittent duty	65
Medium intermittent duty	75
Heavy intermittent duty	85
Continuous duty	110

430.24 Several Motors or a Motor(s) and Other Load(s)

Conductors supplying several motors, or a motor(s) and other load(s), shall have an ampacity not less than 125 percent of the full-load current rating of the highest rated motor plus the sum of the full-load current ratings of all the other motors in the group, as determined by 430.6(A), plus the ampacity required for the other loads.

FPN: See Annex D, Example No. D8.

Where feeders serve motors and/or other electrical loads, the highest rating or setting of the feeder short-circuit and ground-fault protective devices for the minimum-size feeder conductor permitted by 430.24 is specified in 430.62.

Where the selection of a feeder protective device of higher rating or setting is based on the simultaneous starting of two or more motors, the size of the feeder conductors is required to be increased accordingly.

These requirements and those of 430.62 for the short-circuit and ground-fault protection of power feeders are based on the principle that the power feeder conductors should be sized to have an ampacity equal to 125 percent of the full-load current of the largest motor plus the full-load currents of all other motors and all other loads supplied by the feeder.

Except where two or more motors may be started simultaneously, the heaviest load that a power feeder will ever be required to carry occurs when the largest motor is started and all the other motors supplied by the same feeder are running and delivering their full-rated horsepower.

Where the conductors are branch-circuit conductors to multimotor equipment, 430.53 specifies the maximum rating of the branch-circuit short-circuit and ground-fault protective device, and 430.7(D)(1) requires the maximum ampere rating of the short-circuit and ground-fault protective device to be marked on multimotor equipment.

See 430.62(B) and the associated commentary if the size of the feeder conductors is larger than the minimum size.

Exception No. 1: Where one or more of the motors of the group are used for short-time, intermittent, periodic, or varying duty, the ampere rating of such motors to be used in the summation shall be determined in accordance with 430.22(E). For the highest rated motor, the greater of either the ampere rating from 430.22(E) or the largest continuous duty motor full-load current multiplied by 1.25 shall be used in the summation.

Exception No. 2: The ampacity of conductors supplying motor-operated fixed electric space-heating equipment shall conform with 424.3(B).

Exception No. 3: Where the circuitry is interlocked so as to prevent operation of selected motors or other loads at the same time, the conductor ampacity shall be permitted to be based on the summation of the currents of the motors and other loads to be operated at the same time that results in the highest total current.

430.25 Multimotor and Combination-Load Equipment

The ampacity of the conductors supplying multimotor and combination-load equipment shall not be less than the minimum circuit ampacity marked on the equipment in accordance with 430.7(D). Where the equipment is not factory-wired and the individual nameplates are visible in accordance with 430.7(D)(2), the conductor ampacity shall be determined in accordance with 430.24.

When computing the load for the minimum allowable conductor size for a combination lighting (or lighting and appliance) load and motor load, the capacity for the lighting load is determined in accordance with Article 220 (and other applicable articles and sections, for example, Article 422, for appliances), and the motor load is determined in accordance with 430.22 (single motor) or 430.24 (two or more motors). The lighting load and the motor load are added together to determine the minimum conductor ampacity.

430.26 Feeder Demand Factor

Where reduced heating of the conductors results from motors operating on duty-cycle, intermittently, or from all motors not operating at one time, the authority having jurisdiction may grant permission for feeder conductors to have an ampacity less than specified in 430.24, provided the conductors have sufficient ampacity for the maximum load determined in accordance with the sizes and number of motors supplied and the character of their loads and duties.

The authority having jurisdiction may grant permission to allow a demand factor of less than 100 percent if operational procedures, production demands, or the nature of the work is such that not all the motors are running at one time. Engineering study or evaluation of motor operation may provide information that will allow a demand factor of less than 100 percent. A fine print note was added to 430.26 in the 2005 *Code* to allow demand factors to be applied to the design of new installations, based on historical experience and data for similar installations. Application of demand factors is subject to the approval of the AHJ.

FPN: Demand factors determined in the design of new facilities can often be validated against actual historical experience from similar installations. Refer to ANSI/IEEE Std. 141, *IEEE Recommended Practice for Electric Power Distribution for Industrial Plants,* and ANSI/IEEE Std. 241, *Recommended Practice for Electric Power Systems in Commercial Buildings,* for information on the calculation of loads and demand factor.

430.27 Capacitors with Motors

Where capacitors are installed in motor circuits, conductors shall comply with 460.8 and 460.9.

430.28 Feeder Taps

Feeder tap conductors shall have an ampacity not less than that required by Part II, shall terminate in a branch-circuit protective device, and, in addition, shall meet one of the following requirements:

Section 430.28 contains three basic requirements for feeder taps that supply motor circuits.

First, the tap conductor from the feeder to the motor overcurrent device must be sized according to Part II of Article 430. For a single motor load, the tap conductors are sized the same as the motor branch-circuit conductors, that is, according to 430.22. Section 430.22 requires that motor branch-circuit conductors be sized at least 125 percent of the full-load current value for the motor given in Table 430.248 through Table 430.250. The table value, rather than the nameplate value, is the full-load current used for conductor sizing according to 430.6(A).

Second, the tap conductors must terminate in a set of fuses or a circuit breaker, thus limiting the load on the tap conductors. It is important to point out that reduced-size tap conductors are protected from overload by the terminal overcurrent device but protected from short-circuit (and ground-fault) only from the feeder overcurrent device.

Third, where the tap conductor ampacity is less than the ampacity of the feeder, the tap conductor installation must meet the additional requirements associated with their tap conductor distance limits, that is, 10 ft, 25 ft, or, by exception, 100 ft.

The requirements for tap conductors that supply motor loads are somewhat similar to the basic tap requirements found in 240.21. For example, where tap conductors supply a motor load and do not exceed 10 ft, the tap conductors must be sized for the load, terminate in a set of fuses or a circuit breaker, be enclosed by a controller or a raceway, and be protected by a feeder overcurrent device not exceeding 10 times the tap conductor ampacity.

Where the tap conductors supply a motor load and do not exceed 25 ft, the tap conductors must be sized for the load, terminate in a set of fuses or a circuit breaker, be protected from physical damage or be enclosed in a raceway, and have an ampacity at least one-third that of the feeder conductor.

In a high-bay manufacturing building, feeder taps up to 100 ft long are conditionally permitted under 430.28, Exception.

Example

A 15-hp, 230-volt, 3-phase, NEMA Design B, squirrel-cage induction motor with a service factor of 1.15 is to be supplied by a tap from a 250-kcmil feeder. Assuming three conductors in an individual raceway, all Type THWN copper, and no ambient correction factor, the feeder has an ampacity of 255 amperes (from Table 310.16, 75°C column). Where the tap conductors are not over 25 ft long (see Exhibit 430.8), 4 AWG conductors with an ampacity of 85 amperes are permitted ($\frac{1}{3} \times 255$ amperes = 85 amperes).

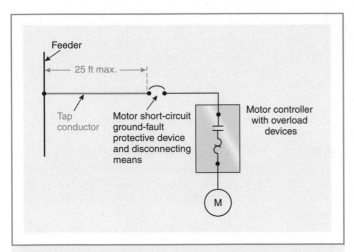

Exhibit 430.8 Protective devices (branch-circuit short-circuit and ground-fault) for a motor branch circuit located not more than 25 ft from the point where the conductors are tapped to a feeder.

Solution

Based on 430.6(A) and Table 430.250, the full-load current of the 15-hp motor is 42 amperes. According to 430.52(C)(1), the motor branch-circuit short-circuit and ground-fault protective device cannot exceed the values given in Table 430.52. The maximum time-delay fuse value is $42 \times 1.75 = 73.5$ amperes. The maximum inverse time circuit breaker value is $42 \times 2.50 = 105$ amperes.

Section 430.52(C)(1), Exception No. 1, allows the next higher standard size — 80 and 110 amperes, respectively. A higher size, based on Exception No. 2, is allowed if the 80- or 110-ampere size is not adequate to start the motor.

Based on 430.32, the motor overload protective devices (heaters) are required to be set at a value not greater than 125 percent of the full-load ampere rating marked on the motor nameplate. A higher-sized motor overload protective device setting of up to 140 percent may be used according to the permissive rules set forth in 430.32(C). Regardless of the exact setting, with the motor overload protection set at approximately 50 amperes, the 4 AWG THWN copper motor branch-circuit tap conductors are well protected from overload.

For additional information concerning taps supplying motor circuits for group installations, see 430.53(D) and the associated commentary.

(1) Be enclosed either by an enclosed controller or by a raceway, be not more than 3.0 m (10 ft) in length, and, for field installation, be protected by an overcurrent device on the line side of the tap conductor, the rating or setting of which shall not exceed 1000 percent of the tap conductor ampacity

(2) Have an ampacity of at least one-third that of the feeder conductors, be suitably protected from physical damage

or enclosed in a raceway, and be not more than 7.5 m (25 ft) in length

(3) Have an ampacity not less than the feeder conductors

Exception: Feeder taps over 7.5 m (25 ft) long. In high-bay manufacturing buildings [over 11 m (35 ft) high at walls], where conditions of maintenance and supervision ensure that only qualified persons service the systems, conductors tapped to a feeder shall be permitted to be not over 7.5 m (25 ft) long horizontally and not over 30.0 m (100 ft) in total length where all of the following conditions are met:

(a) The ampacity of the tap conductors is not less than one-third that of the feeder conductors.

(b) The tap conductors terminate with a single circuit breaker or a single set of fuses conforming with (1) Part IV, where the load-side conductors are a branch circuit, or (2) Part V, where the load-side conductors are a feeder.

(c) The tap conductors are suitably protected from physical damage and are installed in raceways.

(d) The tap conductors are continuous from end-to-end and contain no splices.

(e) The tap conductors shall be 6 AWG copper or 4 AWG aluminum or larger.

(f) The tap conductors shall not penetrate walls, floors, or ceilings.

(g) The tap shall not be made less than 9.0 m (30 ft) from the floor.

430.29 Constant Voltage Direct-Current Motors — Power Resistors

Conductors connecting the motor controller to separately mounted power accelerating and dynamic braking resistors in the armature circuit shall have an ampacity not less than the value calculated from Table 430.29 using motor full-load current. If an armature shunt resistor is used, the power accelerating resistor conductor ampacity shall be calculated using the total of motor full-load current and armature shunt resistor current.

Table 430.29 Conductor Rating Factors for Power Resistors

Time in Seconds		Ampacity of Conductor in Percent of Full-Load Current
On	Off	
5	75	35
10	70	45
15	75	55
15	45	65
15	30	75
15	15	85
Continuous Duty		110

Armature shunt resistor conductors shall have an ampacity of not less than that calculated from Table 430.29 using rated shunt resistor current as full-load current.

III. Motor and Branch-Circuit Overload Protection

430.31 General

Part III specifies overload devices intended to protect motors, motor-control apparatus, and motor branch-circuit conductors against excessive heating due to motor overloads and failure to start.

Overload in electrical apparatus is an operating overcurrent that, when it persists for a sufficient length of time, would cause damage or dangerous overheating of the apparatus. It does not include short circuits or ground faults.

These provisions shall not be interpreted as requiring overload protection where it might introduce additional or increased hazards, as in the case of fire pumps.

> FPN: For protection of fire pump supply conductors, see 695.6.

The provisions of Part III shall not apply to motor circuits rated over 600 volts, nominal.

> FPN No. 1: For over 600 volts, nominal, see Part XI.
> FPN No. 2: See Annex D, Example No. D8.

Section 430.31 sets the general guidelines for motor overload protection. The purpose of this protection is to guard the motor against abnormal operating conditions such as failure to start from a locked rotor, single-phase condition or too much friction on the driven load. The overload protection also guards against excessive heating in the motor caused by an overload condition or from a loss of phase condition.

Adequately applied overload protection should protect the motor from any overload condition prior to damage occurring in the motor. Overload protection is not designed or may not be capable of breaking short-circuit current or ground-fault current. Overload protection may not be installed where it could cause increased hazards as identified for fire pumps.

430.32 Continuous-Duty Motors

(A) More Than 1 Horsepower Each motor used in a continuous duty application and rated more than 1 hp shall be protected against overload by one of the means in 430.32(A)(1) through (A)(4).

The basic premise behind 430.32(A) through 430.32(E) is that the operation of a motor in excess of its normal full-load rating for a prolonged period of time causes damage

or dangerous overheating that may start a fire. Overload protection is intended to protect the motor and the system components from damaging overload currents.

A continuous-duty motor with a marked service factor of 1.15 or greater or with a marked temperature rise of 40°C or less can carry a 25 percent overload for an extended period without damage to the motor. Other similar types of motors are those with a service factor of less than 1.15 or those with a marked temperature rise of greater than 40°C that are incapable of withstanding a prolonged overload, where the motor overload protective device opens the circuit if the motor continues to draw 115 percent of its rated full-load current.

Section 430.32(A) was revised for the 2005 *Code* to make it clear that *continuous-duty motor* is not the same as *continuous load*. The duty of a motor is determined by the application of the motor as defined in Article 100 under *duty*.

(1) Separate Overload Device A separate overload device that is responsive to motor current. This device shall be selected to trip or shall be rated at no more than the following percent of the motor nameplate full-load current rating:

Motors with a marked service factor 1.15 or greater	125%
Motors with a marked temperature rise 40°C or less	125%
All other motors	115%

Modification of this value shall be permitted as provided in 430.32(C). For a multispeed motor, each winding connection shall be considered separately.

Where a separate motor overload device is connected so that it does not carry the total current designated on the motor nameplate, such as for wye-delta starting, the proper percentage of nameplate current applying to the selection or setting of the overload device shall be clearly designated on the equipment, or the manufacturer's selection table shall take this into account.

> FPN: Where power factor correction capacitors are installed on the load side of the motor overload device, see 460.9.

Motors are required to be protected from overloads. To protect a motor from an overload, the motor nameplate full-load current is used to select the overload protection rather than the full-load current values from Table 430.248 through Table 430.250, which are used to design the feeder and branch circuit wiring.

(2) Thermal Protector A thermal protector integral with the motor, approved for use with the motor it protects on the basis that it will prevent dangerous overheating of the motor due to overload and failure to start. The ultimate trip current of a thermally protected motor shall not exceed the following percentage of motor full-load current given in Table 430.248, Table 430.249, and Table 430.250:

Motor full-load current 9 amperes or less	170%
Motor full-load current from 9.1 to, and including, 20 amperes	156%
Motor full-load current greater than 20 amperes	140%

If the motor current-interrupting device is separate from the motor and its control circuit is operated by a protective device integral with the motor, it shall be arranged so that the opening of the control circuit will result in interruption of current to the motor.

(3) Integral with Motor A protective device integral with a motor that will protect the motor against damage due to failure to start shall be permitted if the motor is part of an approved assembly that does not normally subject the motor to overloads.

(4) Larger Than 1500 Horsepower For motors larger than 1500 hp, a protective device having embedded temperature detectors that cause current to the motor to be interrupted when the motor attains a temperature rise greater than marked on the nameplate in an ambient temperature of 40°C.

Continuous-duty-rated motors of more than 1 hp can be protected against overload conditions by any one of the following four methods, in accordance with 430.32(A):

1. An overload device located in the motor controller, such as a bimetallic element or eutectic material
2. A thermal protector located in the motor that senses excessive current or temperature
3. A protective device in the motor, if the motor is part of an assembly that does not normally subject the motor to overloads
4. For motors larger than 1500 hp, a temperature-sensitive device embedded in the motor windings that will de-energize the motor

(B) One Horsepower or Less, Automatically Started Any motor of 1 hp or less that is started automatically shall be protected against overload by one of the following means.

(1) Separate Overload Device By a separate overload device following the requirements of 430.32(A)(1).

For a multispeed motor, each winding connection shall be considered separately. Modification of this value shall be permitted as provided in 430.32(C).

(2) Thermal Protector A thermal protector integral with the motor, approved for use with the motor that it protects on the basis that it will prevent dangerous overheating of the motor due to overload and failure to start. Where the motor current-interrupting device is separate from the motor

and its control circuit is operated by a protective device integral with the motor, it shall be arranged so that the opening of the control circuit results in interruption of current to the motor.

A thermal protector located inside the motor housing, as shown in Exhibit 430.9, is connected in series with the motor winding. This protective device commonly consists of a set of normally closed contacts attached to a bimetallic disk through which the circuit is normally closed. The thermal protector heating coil (in series with the motor winding) causes the disk to heat rapidly. The heat-actuated disk snaps the contacts open to protect the motor windings from overheating due to failure to start, a sudden heavy overload, or a prolonged overload.

After the circuit opens and the motor has cooled to a normal temperature, the contacts automatically close and restart the motor. In some cases, this may not be desirable. For such applications, the protective device is designed so that it must be returned to the closed position by a manually controlled reset, as required by 430.43. For larger motors (usually over 1 hp), a similar device is used. This device, upon abnormal overload, acts as a control-circuit switch and operates the control circuit of a motor current-interrupting device, usually a motor contactor or starter, located separately from the motor. A thermal protector and circuit-interrupting device should be approved for use with the motor it protects and is required to open the circuit on an overcurrent, as specified in 430.32(A)(2).

(3) Integral with Motor A protective device integral with a motor that protects the motor against damage due to failure to start shall be permitted (1) if the motor is part of an approved assembly that does not subject the motor to overloads, or (2) if the assembly is also equipped with other safety controls (such as the safety combustion controls on a domestic oil burner) that protect the motor against damage due to failure to start. Where the assembly has safety controls that protect the motor, it shall be so indicated on the nameplate of the assembly where it will be visible after installation.

(4) Impedance-Protected If the impedance of the motor windings is sufficient to prevent overheating due to failure to start, the motor shall be permitted to be protected as specified in 430.32(D)(2)(a) for manually started motors if the motor is part of an approved assembly in which the motor will limit itself so that it will not be dangerously overheated.

FPN: Many ac motors of less than 1/20 hp, such as clock motors, series motors, and so forth, and also some larger motors such as torque motors, come within this classification. It does not include split-phase motors having automatic switches that disconnect the starting windings.

Exhibit 430.9 A thermal protector for a motor, in which a heat-sensitive snap-action disk opens contacts and protects the motor against dangerous overheating. This device is integrally mounted within the motor. (Redrawn courtesy of Texas Instruments, Inc.)

(C) Selection of Overload Relay Where the sensing element or setting of the overload relay selected in accordance with 430.32(A)(1) and 430.32(B)(1) is not sufficient to start the motor or to carry the load, higher size sensing elements

or incremental settings shall be permitted to be used, provided the trip current of the overload relay does not exceed the following percentage of motor nameplate full-load current rating:

Motors with marked service factor 1.15 or greater	140%
Motors with a marked temperature rise 40°C or less	140%
All other motors	130%

If not shunted during the starting period of the motor as provided in 430.35, the overload device shall have sufficient time delay to permit the motor to start and accelerate its load.

> FPN: A Class 20 or Class 30 overload relay will provide a longer motor acceleration time than a Class 10 or Class 20, respectively. Use of a higher class overload relay may preclude the need for selection of a higher trip current.

(D) One Horsepower or Less, Nonautomatically Started

(1) Permanently Installed Overload protection shall be in accordance with 430.32(B).

(2) Not Permanently Installed

Motors rated 1 hp or less that are not permanently installed and not automatically started, such as for bench grinders, drill presses, and portable electric tools, are not required to have overload protection and may be protected by the branch-circuit short-circuit fuse or circuit breaker. This type of equipment is usually attended by the operator, who can immediately shut off power to the motor should it overheat and start smoking.

   (a) Within Sight from Controller. Overload protection shall be permitted to be furnished by the branch circuit short-circuit and ground-fault protective device; such device, however, shall not be larger than that specified in Part IV of Article 430.

Exception: Any such motor shall be permitted on a nominal 120 volt branch circuit protected at not over 20 amperes.

   (b) Not Within Sight from Controller. Overload protection shall be in accordance with 430.32(B).

(E) Wound-Rotor Secondaries The secondary circuits of wound-rotor ac motors, including conductors, controllers, resistors, and so forth, shall be permitted to be protected against overload by the motor-overload device.

430.33 Intermittent and Similar Duty

A motor used for a condition of service that is inherently short-time, intermittent, periodic, or varying duty, as illustrated by Table 430.22(E), shall be permitted to be protected against overload by the branch-circuit short-circuit and ground-fault protective device, provided the protective device rating or setting does not exceed that specified in Table 430.52.

Any motor application shall be considered to be for continuous duty unless the nature of the apparatus it drives is such that the motor cannot operate continuously with load under any condition of use.

If a motor is selected for duty-cycle service (short-time, intermittent, periodic, or varying), it can be assumed that the motor will not operate continuously, due to the nature of the apparatus or machinery it drives. Therefore, prolonged overloads are rare unless mechanical failure in the driven apparatus stalls the motor; in that case, the branch-circuit protective device would open the circuit. The omission of overload protective devices for such motors is based on the type of duty and not on the time rating of the motor.

430.35 Shunting During Starting Period

(A) Nonautomatically Started For a nonautomatically started motor, the overload protection shall be permitted to be shunted or cut out of the circuit during the starting period of the motor if the device by which the overload protection is shunted or cut out cannot be left in the starting position and if fuses or inverse time circuit breakers rated or set at not over 400 percent of the full-load current of the motor are located in the circuit so as to be operative during the starting period of the motor.

(B) Automatically Started The motor overload protection shall not be shunted or cut out during the starting period if the motor is automatically started.

Exception: The motor overload protection shall be permitted to be shunted or cut out during the starting period on an automatically started motor where the following apply:

(a) The motor starting period exceeds the time delay of available motor overload protective devices, and

(b) Listed means are provided to perform the following:

(1) Sense motor rotation and automatically prevent the shunting or cutout in the event that the motor fails to start, and

(2) Limit the time of overload protection shunting or cutout to less than the locked rotor time rating of the protected motor, and

(3) Provide for shutdown and manual restart if motor running condition is not reached.

If not shunted during the starting period of the motor, the overload device must have sufficient time delay to start and

accelerate its load. If shunting is employed, the overload protection is permitted to be bypassed only during the starting period of the motor. See Exhibit 430.10, which illustrates this type of bypass during motor starting. When the switch is thrown momentarily in one direction (starting position), the overload protective fuses are shunted or cut out of the circuit. The switch is then thrown in the opposite direction (running position) and must be designed so that it cannot be left in the starting position.

Exhibit 430.10 Arrangement for across-the-line, or full-voltage, starting of a motor.

If fuses are used as overload protection, they may be shunted or cut out of the circuit during the starting period by a device (in this case a double-throw switch) designed so that it cannot be left in the starting position. Therefore, during the starting period, the motor is protected only by the branch-circuit fuses that are always rated within the limits of 430.35(A). If there are no branch-circuit fuses, as permitted by 430.53, then a starter (shunting) device is not allowed during the starting period unless the feeder protection is within the limits of 430.35(A) (not over 400 percent of the full-load motor current).

430.36 Fuses — In Which Conductor

Where fuses are used for motor overload protection, a fuse shall be inserted in each ungrounded conductor and also in the grounded conductor if the supply system is 3-wire, 3-phase ac with one conductor grounded.

430.37 Devices Other Than Fuses — In Which Conductor

Where devices other than fuses are used for motor overload protection, Table 430.37 shall govern the minimum allowable number and location of overload units such as trip coils or relays.

Table 430.37 Overload Units

Kind of Motor	Supply System	Number and Location of Overload Units, Such as Trip Coils or Relays
1-phase ac or dc	2-wire, 1-phase ac or dc ungrounded	1 in either conductor
1-phase ac or dc	2-wire, 1-phase ac or dc, one conductor grounded	1 in ungrounded conductor
1-phase ac or dc	3-wire, 1-phase ac or dc, grounded neutral	1 in either ungrounded conductor
1-phase ac	Any 3-phase	1 in ungrounded conductor
2-phase ac	3-wire, 2-phase ac, ungrounded	2, one in each phase
2-phase ac	3-wire, 2-phase ac, one conductor grounded	2 in ungrounded conductors
2-phase ac	4-wire, 2-phase ac, grounded or ungrounded	2, one per phase in ungrounded conductors
2-phase ac	Grounded neutral or 5-wire, 2-phase ac, ungrounded	2, one per phase in any ungrounded phase wire
3-phase ac	Any 3-phase	3, one in each phase*

Exception: An overload unit in each phase shall not be required where overload protection is provided by other approved means.

All 3-phase motors, except those protected by other approved means, must be provided with three overload units, one in each phase. Examples of those motors protected by other means include specially designed or integral-type detectors, with or without supplementary external protective devices. See 430.36 for instances in which fuses used as overloads are required even in the grounded conductor.

430.38 Number of Conductors Opened by Overload Device

Motor overload devices, other than fuses or thermal protectors, shall simultaneously open a sufficient number of ungrounded conductors to interrupt current flow to the motor.

430.39 Motor Controller as Overload Protection

A motor controller shall also be permitted to serve as an overload device if the number of overload units complies with Table 430.37 and if these units are operative in both the starting and running position in the case of a dc motor, and in the running position in the case of an ac motor.

For the purpose of Article 430, a controller may be a switch, a circuit breaker, a contactor, or any other device that starts and stops a motor by making and breaking the motor circuit current. The controller must be capable of interrupting the stalled-rotor current of the motor and must have a horsepower rating that is not lower than the horsepower rating of the motor. Motor controllers are covered in Part VII of Article 430.

Dual-element fuses can be sized to provide motor overload protection (see 430.36). Automatically operated contactors or circuit breakers (with trip units) are governed by the requirements of 430.37 and 430.38 where these devices are used to provide overload protection.

430.40 Overload Relays

Overload relays and other devices for motor overload protection that are not capable of opening short circuits or ground faults shall be protected by fuses or circuit breakers with ratings or settings in accordance with 430.52 or by a motor short-circuit protector in accordance with 430.52.

Some overload devices are marked with a maximum short-circuit and ground-fault protective device rating or setting. This rating sets the limit on the maximum fuse or breaker size that may be upstream from the overload device. The rating also notifies the user that coordination between the overload device and the short-circuit and ground-fault device is required, which is most often the case for group motor installation.

Exception: Where approved for group installation and marked to indicate the maximum size of fuse or inverse time circuit breaker by which they must be protected, the overload devices shall be protected in accordance with this marking.

FPN: For instantaneous trip circuit breakers or motor short-circuit protectors, see 430.52.

430.42 Motors on General-Purpose Branch Circuits

Overload protection for motors used on general-purpose branch circuits as permitted in Article 210 shall be provided as specified in 430.42(A), (B), (C), or (D).

(A) Not Over 1 Horsepower One or more motors without individual overload protection shall be permitted to be connected to a general-purpose branch circuit only where the installation complies with the limiting conditions specified in 430.32(B) and 430.32(D) and 430.53(A)(1) and (A)(2).

(B) Over 1 Horsepower Motors of ratings larger than specified in 430.53(A) shall be permitted to be connected to general-purpose branch circuits only where each motor is protected by overload protection selected to protect the motor as specified in 430.32. Both the controller and the motor overload device shall be approved for group installation with the short-circuit and ground-fault protective device selected in accordance with 430.53.

(C) Cord-and-Plug Connected Where a motor is connected to a branch circuit by means of an attachment plug and receptacle and individual overload protection is omitted as provided in 430.42(A), the rating of the attachment plug and receptacle shall not exceed 15 amperes at 125 volts or 250 volts. Where individual overload protection is required as provided in 430.42(B) for a motor or motor-operated appliance that is attached to the branch circuit through an attachment plug and receptacle, the overload device shall be an integral part of the motor or of the appliance. The rating of the attachment plug and receptacle shall determine the rating of the circuit to which the motor may be connected, as provided in Article 210.

(D) Time Delay The branch-circuit short-circuit and ground-fault protective device protecting a circuit to which a motor or motor-operated appliance is connected shall have sufficient time delay to permit the motor to start and accelerate its load.

430.43 Automatic Restarting

A motor overload device that can restart a motor automatically after overload tripping shall not be installed unless approved for use with the motor it protects. A motor overload device that can restart a motor automatically after overload tripping shall not be installed if automatic restarting of the motor can result in injury to persons.

An integral motor overload protective device may be of the type that, after tripping and sufficiently cooling, automatically restarts the motor, or it may be of the type that, after tripping, is closed by use of a manually operated reset button. See the commentary following 430.32(B)(2).

430.44 Orderly Shutdown

If immediate automatic shutdown of a motor by a motor overload protective device(s) would introduce additional or increased hazard(s) to a person(s) and continued motor operation is necessary for safe shutdown of equipment or process, a motor overload sensing device(s) conforming with the provisions of Part III of this article shall be permitted to be connected to a supervised alarm instead of causing immediate interruption of the motor circuit, so that corrective action or an orderly shutdown can be initiated.

IV. Motor Branch-Circuit Short-Circuit and Ground-Fault Protection

430.51 General

Part IV specifies devices intended to protect the motor branch-circuit conductors, the motor control apparatus, and the motors against overcurrent due to short circuits or grounds. These rules add to or amend the provisions of Article 240. The devices specified in Part IV do not include the types of devices required by 210.8, 230.95, and 527.6.

The provisions of Part IV shall not apply to motor circuits rated over 600 volts, nominal.

> FPN No. 1: For over 600 volts, nominal, see Part XI.
> FPN No. 2: See Annex D, Example D8.

430.52 Rating or Setting for Individual Motor Circuit

(A) General The motor branch-circuit short-circuit and ground-fault protective device shall comply with 430.52(B) and either 430.52(C) or 430.52(D), as applicable.

Section 430.52(A) establishes the maximum allowable ratings or settings of devices (fuses or circuit breakers) acceptable for motor branch-circuit short-circuit and ground-fault protection and states that these devices are expected to carry the starting current of the motor and provide short-circuit and ground-fault protection. For certain exceptions to the maximum rating or setting of these motor branch-circuit protective devices, as specified in Table 430.52, see 430.52, 430.53, and 430.54.

Section 430.6 requires that if the current rating of a motor is used to determine the ampacity of conductors or ampere ratings of switches, branch-circuit overcurrent devices, and so on, the values given in Table 430.248 through Table 430.250 (including notes) must be used instead of the actual motor nameplate current rating. Separate motor overload protection must be based on the motor nameplate current rating.

Exhibit 430.5 illustrates a typical motor circuit in which the branch-circuit short-circuit and ground-fault protective fuse or circuit breaker rating must carry the starting current and may be sized 150 to 300 percent of the motor full-load current (depending on the type of motor).

The rules for short-circuit and ground-fault protection are specific for particular situations. A *short circuit* is a fault between two conductors or between phases. A *ground fault* is a fault between an ungrounded conductor and ground. During a short-circuit or phase-to-ground condition, the extremely high current causes the protective fuses or circuit breakers to open the circuit. Excess current flow caused by an overload condition passes through the overload protective device at the motor controller, thereby causing the device to open the control circuit or motor circuit conductors.

Branch-circuit conductors with an ampacity of 125 percent (not 150 to 300 percent) of the motor full-load current are reasonably protected by motor-protective devices set to operate at nearly the same current as the ampacity of the conductors. Branch-circuit short-circuit and ground-fault protective devices will open the circuit under short-circuit conditions and thereby provide short-circuit and ground-fault protection for both the motor and overload protective device; however, the overload protective device is not intended to open short circuits or ground faults.

The selected rating or setting of the branch-circuit short-circuit and ground-fault protective device should be as low as possible for maximum protection. However, if the rating or setting specified in Table 430.52 or permitted by 430.52(C)(1), Exception No. 1, is not sufficient for the starting current of the motor, a higher rating or setting is allowed per 430.52(C)(1), Exception No. 2. For example, a higher rating would be allowed for a motor under severe starting conditions in which the motor and its driven machinery require an extended period of time to reach the desired speed.

(B) All Motors The motor branch-circuit short-circuit and ground-fault protective device shall be capable of carrying the starting current of the motor.

(C) Rating or Setting

(1) In Accordance with Table 430.52 A protective device that has a rating or setting not exceeding the value calculated according to the values given in Table 430.52 shall be used.

Exception No. 1: Where the values for branch-circuit short-circuit and ground-fault protective devices determined by Table 430.52 do not correspond to the standard sizes or ratings of fuses, nonadjustable circuit breakers, thermal protective devices, or possible settings of adjustable circuit breakers, a higher size, rating, or possible setting that does not exceed the next higher standard ampere rating shall be permitted.

Table 430.52 Maximum Rating or Setting of Motor Branch-Circuit Short-Circuit and Ground-Fault Protective Devices

Type of Motor	Percentage of Full-Load Current			
	Nontime Delay Fuse[1]	Dual Element (Time-Delay) Fuse[1]	Instantaneous Trip Breaker	Inverse Time Breaker[2]
Single-phase motors	300	175	800	250
AC polyphase motors other than wound-rotor				
Squirrel cage — other than Design B energy-efficient	300	175	800	250
Design B energy-efficient	300	175	1100	250
Synchronous[3]	300	175	800	250
Wound rotor	150	150	800	150
Direct current (constant voltage)	150	150	250	150

Note: For certain exceptions to the values specified, see 430.54.
[1]The values in the Nontime Delay Fuse column apply to Time-Delay Class CC fuses.
[2]The values given in the last column also cover the ratings of nonadjustable inverse time types of circuit breakers that may be modified as in 430.52(C), Exception No. 1 and No. 2.
[3]Synchronous motors of the low-torque, low-speed type (usually 450 rpm or lower), such as are used to drive reciprocating compressors, pumps, and so forth, that start unloaded, do not require a fuse rating or circuit-breaker setting in excess of 200 percent of full-load current.

Although Class CC fuses are rated as time delay, they are permitted to be sized according to the requirements of non-time-delay-rated fuses because they are so fast acting. Examples of Class CC fuses are shown in Exhibit 430.11.

The Design E motor standard was rescinded by NEMA in February 2000. All references to Design E motors have been removed from the *NEC* and from NEMA MG 1-2003, *Motors and Generators.*

Exception No. 2: Where the rating specified in Table 430.52, as modified by Exception No. 1, is not sufficient for the starting current of the motor:

(a) The rating of a nontime-delay fuse not exceeding 600 amperes or a time-delay Class CC fuse shall be permitted to be increased but shall in no case exceed 400 percent of the full-load current.

(b) The rating of a time-delay (dual-element) fuse shall be permitted to be increased but shall in no case exceed 225 percent of the full-load current.

Exhibit 430.11 Class CC fuses. (Courtesy of Bussmann Division, Cooper Industries)

(c) The rating of an inverse time circuit breaker shall be permitted to be increased but shall in no case exceed 400 percent for full-load currents of 100 amperes or less or 300 percent for full-load currents greater than 100 amperes.

(d) The rating of a fuse of 601–6000 ampere classification shall be permitted to be increased but shall in no case exceed 300 percent of the full-load current.

FPN: See Annex D, Example D8, and Figure 430.1.

(2) Overload Relay Table Where maximum branch-circuit short-circuit and ground-fault protective device ratings are shown in the manufacturer's overload relay table for use with a motor controller or are otherwise marked on the equipment, they shall not be exceeded even if higher values are allowed as shown above.

(3) Instantaneous Trip Circuit Breaker An instantaneous trip circuit breaker shall be used only if adjustable and if part of a listed combination motor controller having coordinated motor overload and short-circuit and ground-fault protection in each conductor, and the setting is adjusted to no more than the value specified in Table 430.52.

FPN: For the purpose of this article, instantaneous trip circuit breakers may include a damping means to accommodate a transient motor inrush current without nuisance tripping of the circuit breaker.

Exception No. 1: Where the setting specified in Table 430.52 is not sufficient for the starting current of the motor, the setting of an instantaneous trip circuit breaker shall be permitted to be increased but shall in no case exceed 1300

percent of the motor full-load current for other than Design B energy-efficient motors and no more than 1700 percent of full-load motor current for Design B energy-efficient motors. Trip settings above 800 percent for other than Design B energy-efficient motors and above 1100 percent for Design B energy-efficient motors shall be permitted where the need has been demonstrated by engineering evaluation. In such cases, it shall not be necessary to first apply an instantaneous-trip circuit breaker at 800 percent or 1100 percent.

FPN: For additional information on the requirements for a motor to be classified "energy efficient," see NEMA Standards Publication No. MG1-1993, Revision, *Motors and Generators,* Part 12.59.

Exception No. 2: Where the motor full-load current is 8 amperes or less, the setting of the instantaneous-trip circuit breaker with a continuous current rating of 15 amperes or less in a listed combination motor controller that provides coordinated motor branch-circuit overload and short-circuit and ground-fault protection shall be permitted to be increased to the value marked on the controller.

(4) Multispeed Motor For a multispeed motor, a single short-circuit and ground-fault protective device shall be permitted for two or more windings of the motor, provided the rating of the protective device does not exceed the above applicable percentage of the nameplate rating of the smallest winding protected.

Exception: For a multispeed motor, a single short-circuit and ground-fault protective device shall be permitted to be used and sized according to the full-load current of the highest current winding, where all of the following conditions are met:

(a) Each winding is equipped with individual overload protection sized according to its full-load current.

(b) The branch-circuit conductors supplying each winding are sized according to the full-load current of the highest full-load current winding.

(c) The controller for each winding has a horsepower rating not less than that required for the winding having the highest horsepower rating.

(5) Power Electronic Devices Suitable fuses shall be permitted in lieu of devices listed in Table 430.52 for power electronic devices in a solid state motor controller system, provided that the marking for replacement fuses is provided adjacent to the fuses.

(6) Self-Protected Combination Controller A listed self-protected combination controller shall be permitted in lieu of the devices specified in Table 430.52. Adjustable instantaneous-trip settings shall not exceed 1300 percent of full-load motor current for other than Design B energy-efficient motors and not more than 1700 percent of full-load motor current for Design B energy-efficient motors.

A self-protected combination controller combines the functions of short-circuit protection, disconnect, controller, and overload protection into a single unit. Only listed units are permitted; therefore, they must be applied within their ratings.

FPN: Proper application of self-protected combination controllers on 3-phase systems, other than solidly grounded wye, particularly on corner grounded delta systems, considers the self-protected combination controllers' individual pole-interrupting capability.

(7) Motor Short-Circuit Protector A motor short-circuit protector shall be permitted in lieu of devices listed in Table 430.52 if the motor short-circuit protector is part of a listed combination motor controller having coordinated motor overload protection and short-circuit and ground-fault protection in each conductor and it will open the circuit at currents exceeding 1300 percent of motor full-load current for other than Design B energy-efficient motors and 1700 percent of motor full-load motor current for Design B energy-efficient motors.

(D) Torque Motors Torque motor branch circuits shall be protected at the motor nameplate current rating in accordance with 240.4(B).

430.53 Several Motors or Loads on One Branch Circuit

Two or more motors or one or more motors and other loads shall be permitted to be connected to the same branch circuit under conditions specified in 430.53(D) and in 430.53(A), (B), or (C).

(A) Not Over 1 Horsepower Several motors, each not exceeding 1 hp in rating, shall be permitted on a nominal 120-volt branch circuit protected at not over 20 amperes or a branch circuit of 600 volts, nominal, or less, protected at not over 15 amperes, if all of the following conditions are met:

(1) The full-load rating of each motor does not exceed 6 amperes.
(2) The rating of the branch-circuit short-circuit and ground-fault protective device marked on any of the controllers is not exceeded.
(3) Individual overload protection conforms to 430.32.

Two or more motors or one or more motors and other loads may be connected to the same 120-volt, 15- or 20-ampere, single-phase lighting circuit as long as each motor is rated

not more than 1 hp, the full-load rating of each motor does not exceed 6 amperes, and the rating of the branch-circuit protective device is not exceeded.

The requirements for overload protection, as provided in 430.32, must be applied in all cases, regardless of the number (one or more) of motors or the type of branch circuit.

(B) If Smallest Rated Motor Protected If the branch-circuit short-circuit and ground-fault protective device is selected not to exceed that allowed by 430.52 for the smallest rated motor, two or more motors or one or more motors and other load(s), with each motor having individual overload protection, shall be permitted to be connected to a branch circuit where it can be determined that the branch-circuit short-circuit and ground-fault protective device will not open under the most severe normal conditions of service that might be encountered.

(C) Other Group Installations Two or more motors of any rating or one or more motors and other load(s), with each motor having individual overload protection, shall be permitted to be connected to one branch circuit where the motor controller(s) and overload device(s) are (1) installed as a listed factory assembly and the motor branch-circuit short-circuit and ground-fault protective device either is provided as part of the assembly or is specified by a marking on the assembly, or (2) the motor branch-circuit short-circuit and ground-fault protective device, the motor controller(s), and overload device(s) are field-installed as separate assemblies listed for such use and provided with manufacturers' instructions for use with each other, and (3) all of the following conditions are complied with:

(1) Each motor overload device is listed for group installation with a specified maximum rating of fuse, inverse time circuit breaker, or both.
(2) Each motor controller is listed for group installation with a specified maximum rating of fuse, circuit breaker, or both.
(3) Each circuit breaker is listed and is of the inverse time type.
(4) The branch circuit shall be protected by fuses or inverse time circuit breakers having a rating not exceeding that specified in 430.52 for the highest rated motor connected to the branch circuit plus an amount equal to the sum of the full-load current ratings of all other motors and the ratings of other loads connected to the circuit. Where this calculation results in a rating less than the ampacity of the supply conductors, it shall be permitted to increase the maximum rating of the fuses or circuit breaker to a value not exceeding that permitted by 240.4(B).
(5) The branch-circuit fuses or inverse time circuit breakers are not larger than allowed by 430.40 for the overload relay protecting the smallest rated motor of the group.

(6) Overcurrent protection for loads other than motor loads shall be in accordance with Parts I through VII of Article 240.

> FPN: See 110.10 for circuit impedance and other characteristics.

Section 110.10 addresses characteristics of components such as impedance and short-circuit current ratings. Devices with the same ampere rating may have significantly different short-circuit current ratings. Proper selection of components includes consideration of the characteristics of all components so that a fault will not cause unacceptable damage.

A new paragraph (6) was added to 430.53(C) in the 2005 *Code* to clarify that in group installations, where there are other loads that are not motor loads, the other loads must be provided with overcurrent protection in accordance with Part I and Part VII of Article 240. The ground-fault short-circuit protection for motors might be greater than is permitted for other loads in accordance with Article 240.

(D) Single Motor Taps For group installations described above, the conductors of any tap supplying a single motor shall not be required to have an individual branch-circuit short-circuit and ground-fault protective device, provided they comply with one of the following:

(1) No conductor to the motor shall have an ampacity less than that of the branch-circuit conductors.

(2) No conductor to the motor shall have an ampacity less than one-third that of the branch-circuit conductors, with a minimum in accordance with 430.22, the conductors to the motor overload device being not more than 7.5 m (25 ft) long and being protected from physical damage.

(3) Conductors from the branch-circuit short-circuit and ground-fault protective device to a listed manual motor controller additionally marked "Suitable for Tap Conductor Protection in Group Installations" shall be permitted to have an ampacity not less than ¹⁄₁₀ the rating or setting of the branch-circuit short-circuit and ground-fault protective device. The conductors from the controller to the motor shall have an ampacity in accordance with 430.22. The conductors from the branch-circuit short-circuit and ground-fault protective device to the controller shall (1) be suitably protected from physical damage and enclosed either by an enclosed controller or by a raceway and shall be not more than 3 m (10 ft) long or (2) shall have an ampacity not less than that of the branch circuit conductors.

For group motor applications covered in 430.53(C), the provisions of 430.53(D)(3) add a third alternative for tapping a branch circuit to supply a single motor. The conditions for applying this tap rule are similar to those in 430.28 covering motor supply conductors tapped to a feeder. The short-circuit ground-fault device on the line side of the tap conductors is protecting more than one set of conductors that supply individual motors, thus eliminating the need for an individual short-circuit ground-fault device for each set of conductors that supply a motor.

This approach requires that the tap conductors meet certain size, physical protection, length, and termination conditions. The tap conductors always have to meet the conductor size requirements of 430.22 and must have an ampacity not less than one-tenth the rating of the upstream short-circuit ground-fault protective device. Where the conductors are sized according to this provision, their length cannot exceed 10 ft, and they have to be enclosed in a raceway or by the motor controller. If the conductor ampacity is not less than the rating of the upstream short-circuit ground-fault protective device, the length of the conductor is not limited.

The tap conductors are permitted to terminate in a listed manual motor controller that is marked "Suitable for Tap Conductor Protection in Group Installations." This controller provides an instantaneous trip mechanism, motor overload protection, and provisions for disconnecting the motor.

Exhibit 430.12 illustrates main branch-circuit conductors supplying a motor that is part of a group installation. The tap conductors have an ampacity equal to the ampacity of the main branch-circuit conductors; therefore, branch-circuit short-circuit and ground-fault protective devices,

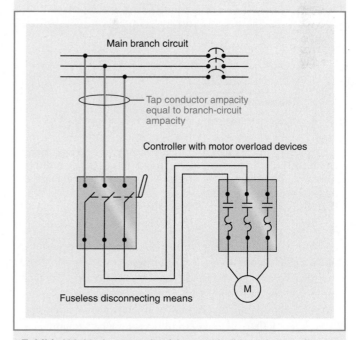

Exhibit 430.12 An example of the permissible omission of motor branch-circuit protective devices for tap conductors that have the same ampacity as the main conductors.

fuses, or circuit breakers for the conductors in the tap are not required at the point of connection of the tap conductors to the main conductors, provided that the motor controller and motor overload protective device are listed for group installation with the size of the main branch-circuit short-circuit and ground-fault protective device used.

Exhibit 430.13 also illustrates main branch-circuit conductors supplying a motor that is part of a group installation. Here, the tap conductors have an ampacity at least one-third the ampacity of the main branch-circuit conductors, are not more than 25 ft in length, and are suitably protected from physical damage. The motor controller and motor overload protective device must be listed for group installation with the size of the main branch-circuit short-circuit and ground-fault protective device used.

Single Motor Tap for Group Installation

Main branch circuit

Tap conductor ampacity at least 1/3 branch-circuit ampacity

Controller with motor overload devices

Fuseless disconnecting means

M

25 ft max.

Exhibit 430.13 An example of the permissible omission of motor branch-circuit protective devices for tap conductors that have at least one-third the ampacity of the main conductors, are not over 25 ft long, and are protected from physical damage.

In both examples, the main branch-circuit fuses or circuit breakers would operate in the event of a short circuit, and the overload protective device would operate to protect the motor and tap conductors under overload conditions.

The tap conductors should never be of a smaller size and ampacity than the branch-circuit conductors required by 430.22. That is, a tap conductor (25 ft or less) may be one-third the ampacity of the main branch-circuit conductor to which it is connected; however, this ampacity must be equal

to or larger than 125 percent of the motor's full-load current rating (see 430.22).

Example

A branch circuit sized at 2/0 AWG THW copper typically has an ampacity of 175 amperes (see Table 310.16). A tap conductor (25 ft or less) would normally be permitted to be sized at 6 AWG THW copper (65 amperes). But one-third of 175 amperes is 58 amperes, and the motor circuit conductors must have an ampacity of at least 85 amperes.

If a 25-hp, 230-volt, 3-phase squirrel-cage motor is to be supplied from this branch circuit, a 6 AWG tap conductor would not meet the requirements of 430.22. That is, 125 percent of the full-load current of the motor (68 amperes from Table 430.250) is 85 amperes (1.25 × 68 amperes = 85 amperes). Therefore, the branch-circuit tap conductors are not permitted to be smaller than 4 AWG THW copper, with a normal ampacity of 85 amperes (see Table 310.16).

Note that the ampacities in Table 310.16 are reduced for ambient temperatures above 30°C and for more than three conductors in the raceway or cable.

430.54 Multimotor and Combination-Load Equipment

The rating of the branch-circuit short-circuit and ground-fault protective device for multimotor and combination-load equipment shall not exceed the rating marked on the equipment in accordance with 430.7(D).

430.55 Combined Overcurrent Protection

Motor branch-circuit short-circuit and ground-fault protection and motor overload protection shall be permitted to be combined in a single protective device where the rating or setting of the device provides the overload protection specified in 430.32.

Fuses and circuit breakers are not permitted to be sized as overload protection according to the values of 430.32(C). Rather, fuses are only permitted to be sized as overload protection according to the values found in 430.32(A)(1), 430.32(B)(1), and 430.32(D)(1).

Either a circuit breaker with inverse time characteristics or a dual-element (time-delay) fuse may serve both as motor overload protection and as the branch-circuit short-circuit and ground-fault protection.

One-time, time-delay dual-element and Type S dual-element fuses and adapters are available with up to a 30-ampere rating. Type S fuses are designed to prevent oversize fusing. See 240.50 through 240.54 for more information about these fuses and adapters.

Exhibits 430.14 and 430.15 are examples of time-delay,

Exhibit 430.14 Fusetron cartridge-type fuses. (Redrawn courtesy of Bussmann Division, Cooper Industries)

Exhibit 430.15 A Class R dual-element fuse with physical rejection feature to prevent interchangeability. (Redrawn courtesy of Bussmann Division, Cooper Industries)

cartridge-type dual-element fuses that are able to withstand the normal motor starting current if sized at or near the motor full-load rating but that open when subjected to prolonged overload or blow quickly during a short circuit or ground fault. The dual-element characteristics are the thermal cutout element, which permits harmless high-inrush currents to flow for short periods (but which would open the circuit during a prolonged period), and the fuse link element, which has current-limiting ability for short-circuit currents (and which would blow quickly). Dual-element fuses may be

used in larger sizes to provide only short-circuit and ground-fault protection.

430.56 Branch-Circuit Protective Devices — In Which Conductor

Branch-circuit protective devices shall comply with the provisions of 240.20.

430.57 Size of Fuseholder

Where fuses are used for motor branch-circuit short-circuit and ground-fault protection, the fuseholders shall not be of a smaller size than required to accommodate the fuses specified by Table 430.52.

Exception: Where fuses having time delay appropriate for the starting characteristics of the motor are used, it shall be permitted to use fuseholders sized to fit the fuses that are used.

The use of dual-element (time-delay) fuses makes it possible to use smaller fuses, thereby providing better protection because of the smaller fuses' lower ratings. Dual-element fuses also save in installation cost by allowing smaller-size switches and panels, and they allow for easier arrangement of equipment where space is at a premium at motor control centers.

430.58 Rating of Circuit Breaker

A circuit breaker for motor branch-circuit short-circuit and ground-fault protection shall have a current rating in accordance with 430.52 and 430.110.

V. Motor Feeder Short-Circuit and Ground-Fault Protection

430.61 General

Part V specifies protective devices intended to protect feeder conductors supplying motors against overcurrents due to short circuits or grounds.

FPN: See Annex D, Example D8.

430.62 Rating or Setting — Motor Load

(A) Specific Load A feeder supplying a specific fixed motor load(s) and consisting of conductor sizes based on 430.24 shall be provided with a protective device having a rating or setting not greater than the largest rating or setting of the branch-circuit short-circuit and ground-fault protective device for any motor supplied by the feeder [based on the maximum permitted value for the specific type of a protective device in accordance with 430.52, or 440.22(A) for hermetic refrigerant motor-compressors], plus the sum of the full-load currents of the other motors of the group.

Where the same rating or setting of the branch-circuit short-circuit and ground-fault protective device is used on two or more of the branch circuits supplied by the feeder, one of the protective devices shall be considered the largest for the above calculations.

The rating of a motor feeder short-circuit ground-fault protective device is determined by adding the rating of the largest branch-circuit short-circuit ground-fault protective device for any motor supplied by the feeder to the sum of the full-load currents of all of the other motors supplied by that feeder. The largest branch-circuit short-circuit ground-fault protective device is based on 430.52 and Table 430.52. The largest rating can be based on either of the exceptions to 430.52(C)(1). For the purposes of sizing the feeder protective device, it is assumed that the same type of protective device is being used for the feeder and the branch circuits. This is necessary if the feeder protective device and the largest branch-circuit protective device are different types, for example, one is a fuse and the other is a circuit breaker.

Section 430.62(A) recognizes the lower setting for motor overload devices that is required for hermetic refrigerant motor-compressors.

Exception No. 1: Where one or more instantaneous trip circuit breakers or motor short-circuit protectors are used for motor branch-circuit short-circuit and ground-fault protection as permitted in 430.52(C), the procedure provided above for determining the maximum rating of the feeder protective device shall apply with the following provision: For the purpose of the calculation, each instantaneous trip circuit breaker or motor short-circuit protector shall be assumed to have a rating not exceeding the maximum percentage of motor full-load current permitted by Table 430.52 for the type of feeder protective device employed.

Exception No. 2: Where the feeder overcurrent protective device also provides overcurrent protection for a motor control center, the provisions of 430.94 shall apply.

FPN: See Annex D, Example D8.

(B) Other Installations Where feeder conductors have an ampacity greater than required by 430.24, the rating or setting of the feeder overcurrent protective device shall be permitted to be based on the ampacity of the feeder conductors.

Section 430.62(B) explains how to size a motor feeder that is larger than the minimum size required by the *Code*. If the motor feeder conductors are sized larger than the minimum

required, then the size of the overcurrent device for the feeder is based on the size of the feeder conductors.

Exception No. 2 to 430.62(A) correlates the requirements of 430.62(B) for determining feeder short-circuit ground-fault protection with the requirements of 430.94 covering overcurrent protection for motor-control centers. Where the motor feeder short-circuit ground-fault protective device is also the overcurrent protective device for a motor control center, its rating cannot exceed that allowed for protecting the common power bus of the motor control center.

430.63 Rating or Setting — Power and Lighting Loads

Where a feeder supplies a motor load and, in addition, a lighting or a lighting and appliance load, the feeder protective device shall have a rating sufficient to carry the lighting or lighting and appliance load, plus the following:

(1) For a single motor, the rating permitted by 430.52
(2) For a single hermetic refrigerant motor-compressor, the rating permitted by 440.22
(3) For two or more motors, the rating permitted by 430.62

Exception: Where the feeder overcurrent device provides the overcurrent protection for a motor control center, the provisions of 430.94 shall apply.

See commentary following 430.62(A) Exception No. 2.

VI. Motor Control Circuits

430.71 General

Part VI contains modifications of the general requirements and applies to the particular conditions of motor control circuits.

FPN: See 430.9(B) for equipment device terminal requirements.

430.72 Overcurrent Protection

(A) General A motor control circuit tapped from the load side of a motor branch-circuit short-circuit and ground-fault protective device(s) and functioning to control the motor(s) connected to that branch circuit shall be protected against overcurrent in accordance with 430.72. Such a tapped control circuit shall not be considered to be a branch circuit and shall be permitted to be protected by either a supplementary or branch-circuit overcurrent protective device(s). A motor control circuit other than such a tapped control circuit shall be protected against overcurrent in accordance with 725.23 or the notes to Table 11(A) and Table 11(B) in Chapter 9, as applicable.

(B) Conductor Protection The overcurrent protection for conductors shall be provided as specified in 430.72(B)(1) or (B)(2).

Exception No. 1: Where the opening of the control circuit would create a hazard as, for example, the control circuit of a fire pump motor, and the like, conductors of control circuits shall require only short-circuit and ground-fault protection and shall be permitted to be protected by the motor branch-circuit short-circuit and ground-fault protective device(s).

Exception No. 2: Conductors supplied by the secondary side of a single-phase transformer having only a two-wire (single-voltage) secondary shall be permitted to be protected by overcurrent protection provided on the primary (supply) side of the transformer, provided this protection does not exceed the value determined by multiplying the appropriate maximum rating of the overcurrent device for the secondary conductor from Table 430.72(B) by the secondary-to-primary voltage ratio. Transformer secondary conductors (other than two-wire) shall not be considered to be protected by the primary overcurrent protection.

(1) Separate Overcurrent Protection Where the motor branch-circuit short-circuit and ground-fault protective device does not provide protection in accordance with 430.72(B)(2), separate overcurrent protection shall be provided. The overcurrent protection shall not exceed the values specified in Column A of Table 430.72(B).

(2) Branch-Circuit Overcurrent Protective Device Conductors shall be permitted to be protected by the motor branch-circuit short-circuit and ground-fault protective device and shall require only short-circuit and ground-fault

protection. Where the conductors do not extend beyond the motor control equipment enclosure, the rating of the protective device(s) shall not exceed the value specified in Column B of Table 430.72(B). Where the conductors extend beyond the motor control equipment enclosure, the rating of the protective device(s) shall not exceed the value specified in Column C of Table 430.72(B).

(C) Control Circuit Transformer Where a motor control circuit transformer is provided, the transformer shall be protected in accordance with 430.72(C)(1), (C)(2), (C)(3), (C)(4), or (C)(5).

Exception: Overcurrent protection shall be omitted where the opening of the control circuit would create a hazard as, for example, the control circuit of a fire pump motor and the like.

(1) Compliance with Article 725 Where the transformer supplies a Class 1 power-limited circuit, Class 2, or Class 3 remote-control circuit conforming with the requirements of Article 725, protection shall comply with Article 725.

(2) Compliance with Article 450 Protection shall be permitted to be provided in accordance with 450.3.

(3) Less Than 50 Volt-Amperes Control circuit transformers rated less than 50 volt-amperes (VA) and that are an integral part of the motor controller and located within the motor controller enclosure shall be permitted to be protected by primary overcurrent devices, impedance limiting means, or other inherent protective means.

(4) Primary Less Than 2 Amperes Where the control circuit transformer rated primary current is less than 2 amperes,

Table 430.72(B) Maximum Rating of Overcurrent Protective Device in Amperes

	Protection Provided by Motor Branch-Circuit Protective Device(s)					
	Column A Separate Protection Provided		Column B Conductors Within Enclosure		Column C Conductors Extend Beyond Enclosure	
Control Circuit Conductor Size (AWG)	Copper	Aluminum or Copper-Clad Aluminum	Copper	Aluminum or Copper-Clad Aluminum	Copper	Aluminum or Copper-Clad Aluminum
18	7	—	25	—	7	—
16	10	—	40	—	10	—
14	(Note 1)	—	100	—	45	—
12	(Note 1)	(Note 1)	120	100	60	45
10	(Note 1)	(Note 1)	160	140	90	75
Larger than 10	(Note 1)	(Note 1)	(Note 2)	(Note 2)	(Note 3)	(Note 3)

Notes:
1. Value specified in 310.15 as applicable.
2. 400 percent of value specified in Table 310.17 for 60°C conductors.
3. 300 percent of value specified in Table 310.16 for 60°C conductors.

an overcurrent device rated or set at not more than 500 percent of the rated primary current shall be permitted in the primary circuit.

(5) Other Means Protection shall be permitted to be provided by other approved means.

Motor control circuits are allowed to receive their power either from the load side of the motor short-circuit and ground-fault protective device or from a separate source, such as a panelboard.

Motor control circuits that receive their power from a separate source are protected against overcurrent in accordance with 725.23 for Class 1 circuits. Conductor sizes 14 AWG and larger are protected according to their ampacity listed in Table 310.16 through Table 310.20. Conductor sizes 16 and 18 AWG must be protected at not more than 10 and 7 amperes, respectively, as specified in Table 430.72(B).

If a motor control circuit is tapped from the load side of the motor branch-circuit short-circuit and ground-fault protective device, the size of the tapped conductor and the rating of the overcurrent device are based on whether the conductor stays within the motor control enclosure or leaves it. The load on a motor control circuit is similar to a motor branch-circuit load in that there is a predetermined connected load. There is also an initial high inrush of current, until the armature of the relay is seated and the current decreases to a steady state. Therefore, the overcurrent protection is similar to the short-circuit and ground-fault protection provided for a motor and is allowed to be greater than the ampacity of the control circuit conductor.

430.73 Mechanical Protection of Conductor

Where damage to a motor control circuit would constitute a hazard, all conductors of such a remote motor control circuit that are outside the control device itself shall be installed in a raceway or be otherwise suitably protected from physical damage.

Where one side of the motor control circuit is grounded, the motor control circuit shall be arranged so that an accidental ground in the control circuit remote from the motor controller will (1) not start the motor and (2) not bypass manually operated shutdown devices or automatic safety shutdown devices.

If damage to the motor control circuit conductors would constitute a fire or accident hazard, then physical protection of the motor control circuit conductors is extremely important. If damage to the control circuit conductors could result in an accidental ground fault or short circuit, causing the device to operate or rendering the device inoperative (either condition constituting a hazard to persons or prop-

erty), conductors must be installed in a raceway. Where boilers or furnaces are equipped with an automatic safety control device, damage to the conductors of the low-voltage control circuit (for example, a thermostat) does not constitute a hazard (see Article 725, Part III).

The second paragraph of 430.73 requires that if one side of the motor control circuit is grounded, the circuit must be arranged so that an accidental ground in the remote control device will not start the motor. For example, see the control wiring illustrated in Exhibit 430.16. If the control circuit is a 120-volt, single-phase circuit derived from a 208-volt, 3-phase wye system supplying the motor, one side of the control circuit will be the grounded neutral. If the start button of the motor control circuit is in the grounded neutral, a ground fault on the coil side of the start button can short-circuit the start circuit and start the motor. The same condition exists if the ground fault is in the wiring rather than in the control device itself. This hazardous condition can be alleviated by locating the start button in the ungrounded side for the control circuit as shown in Exhibit 430.17.

Combinations of ground faults in motor and motor control circuits can also result in inadvertent motor starting. If the circuit is ungrounded, the first fault may go undetected. One solution is to use double-pole control devices, with one pole in each of the two control lines.

430.74 Disconnection

(A) General Motor control circuits shall be arranged so that they will be disconnected from all sources of supply when the disconnecting means is in the open position. The disconnecting means shall be permitted to consist of two or more separate devices, one of which disconnects the motor and the controller from the source(s) of power supply for the motor, and the other(s), the motor control circuit(s) from its power supply. Where separate devices are used, they shall be located immediately adjacent to each other.

Exception No. 1: Where more than 12 motor control circuit conductors are required to be disconnected, the disconnecting means shall be permitted to be located other than immediately adjacent to each other where all of the following conditions are complied with:

(a) Access to energized parts is limited to qualified persons in accordance with Part XI of this article.

(b) A warning sign is permanently located on the outside of each equipment enclosure door or cover permitting access to the live parts in the motor control circuit(s), warning that motor control circuit disconnecting means are remotely located and specifying the location and identification of each disconnect. Where energized parts are not in an equipment enclosure as permitted by 430.132 and 430.133, an additional warning sign(s) shall be located where visible

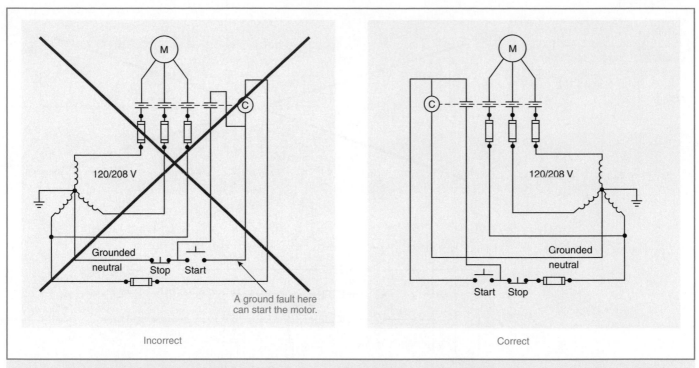

Incorrect   Correct

Exhibit 430.16 An example of control wiring in violation of 430.73, paragraph 2 (incorrect, left), and in compliance with 430.73, paragraph 2 (correct, right). (For simplification, motor overload elements and disconnecting means are not shown.)

to persons who may be working in the area of the energized parts.

Exception No. 2: The motor control circuit disconnecting means shall be permitted to be remote from the motor controller power supply disconnecting means where the opening of one or more motor control circuit disconnect means may result in potentially unsafe conditions for personnel or property and the conditions of items (a) and (b) of Exception No. 1 are complied with.

(B) Control Transformer in Controller Enclosure Where a transformer or other device is used to obtain a reduced voltage for the motor control circuit and is located in the controller enclosure, such transformer or other device shall be connected to the load side of the disconnecting means for the motor control circuit.

VII. Motor Controllers

430.81 General

Part VII is intended to require suitable controllers for all motors.

(A) Stationary Motor of ⅛ Horsepower or Less For a stationary motor rated at ⅛ hp or less that is normally left running and is constructed so that it cannot be damaged by overload or failure to start, such as clock motors and the

like, the branch-circuit protective device shall be permitted to serve as the controller.

(B) Portable Motor of ⅓ Horsepower or Less For a portable motor rated at ⅓ hp or less, the controller shall be permitted to be an attachment plug and receptacle.

430.82 Controller Design

(A) Starting and Stopping Each controller shall be capable of starting and stopping the motor it controls and shall be capable of interrupting the locked-rotor current of the motor.

(B) Autotransformer An autotransformer starter shall provide an "off" position, a running position, and at least one starting position. It shall be designed so that it cannot rest in the starting position or in any position that will render the overload device in the circuit inoperative.

(C) Rheostats Rheostats shall be in compliance with the following:

(1) Motor-starting rheostats shall be designed so that the contact arm cannot be left on intermediate segments. The point or plate on which the arm rests when in the starting position shall have no electrical connection with the resistor.

Exhibit 430.17 An example of control wiring using a 480/120-volt control power transformer. (The upper control circuit is not in compliance with 430.73, paragraph 2.)

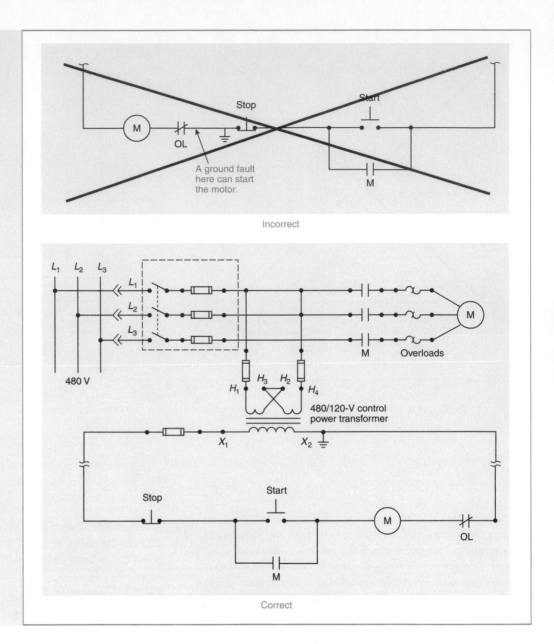

(2) Motor-starting rheostats for dc motors operated from a constant voltage supply shall be equipped with automatic devices that will interrupt the supply before the speed of the motor has fallen to less than one-third its normal rate.

430.83 Ratings

The controller shall have a rating as specified in 430.83(A), unless otherwise permitted in 430.83(B) or 430.83(C), or as specified in 430.83(D), under the conditions specified.

(A) General

(1) Horsepower Ratings Controllers, other than inverse time circuit breakers and molded case switches, shall have horsepower ratings at the application voltage not lower than the horsepower rating of the motor.

(2) Circuit Breaker A branch-circuit inverse time circuit breaker rated in amperes shall be permitted as a controller for all motors. Where this circuit breaker is also used for overload protection, it shall conform to the appropriate provisions of this article governing overload protection.

(3) Molded Case Switch A molded case switch rated in amperes shall be permitted as a controller for all motors.

A molded case switch has the same frame appearance as a molded case circuit breaker and is designed to fit in circuit breaker enclosures. However, where the device is marked with only a short-circuit current withstand rating, this rating indicates that the switch does not provide overcurrent protection. Those fused molded case switches that do provide overcurrent protection are marked with a short-circuit current interrupting rating. Both fused and unfused molded case switches can be used in motor circuits.

Molded case switches are permitted as motor disconnecting means per 430.109. In general, molded case switches are rated only in amperes and, where used in a motor circuit, must be sized at 115 percent of the motor full-load current rating. There are disconnecting means assemblies employing molded case switches that are marked with horsepower ratings that can be used instead of the ampere rating of the molded case switch.

(B) Small Motors Devices as specified in 430.81(A) and 430.81(B) shall be permitted as a controller.

(C) Stationary Motors of 2 Horsepower or Less For stationary motors rated at 2 hp or less and 300 volts or less, the controller shall be permitted to be either of the following:

(1) A general-use switch having an ampere rating not less than twice the full-load current rating of the motor
(2) On ac circuits, a general-use snap switch suitable only for use on ac (not general-use ac–dc snap switches) where the motor full-load current rating is not more than 80 percent of the ampere rating of the switch

(D) Torque Motors For torque motors, the controller shall have a continuous-duty, full-load current rating not less than the nameplate current rating of the motor. For a motor controller rated in horsepower but not marked with the foregoing current rating, the equivalent current rating shall be determined from the horsepower rating by using Tables 430.147, 430.148, 430.149, or 430.150.

(E) Voltage Rating A controller with a straight voltage rating, for example, 240 volts or 480 volts, shall be permitted to be applied in a circuit in which the nominal voltage between any two conductors does not exceed the controller's voltage rating. A controller with a slash rating, for example, 240/120 volts or 480Y/277 volts, shall only be applied in a solidly grounded circuit in which the nominal voltage to ground from any conductor does not exceed the lower of the two values of the controller's voltage rating and the nominal voltage between any two conductors does not exceed the higher value of the controller's voltage rating.

Section 430.83(E) requires that controllers identified with slash voltage ratings, such as 120/240-volt and 480/277-volt grounded systems, are allowed to be used only on electrical systems in which the nominal voltage to ground does not exceed the lower voltage rating of the controller, and the nominal voltage between any two phases of the electrical system is not greater than the higher value of the controller voltage rating.

430.84 Need Not Open All Conductors

The controller shall not be required to open all conductors to the motor.

Exception: Where the controller serves also as a disconnecting means, it shall open all ungrounded conductors to the motor as provided in 430.111.

A controller that does not also serve as a disconnecting means must open only as many motor circuit conductors as may be necessary to stop the motor, that is, one conductor for a dc or single-phase motor circuit, two conductors for a 3-phase motor circuit, and three conductors for a 2-phase motor circuit.

430.85 In Grounded Conductors

One pole of the controller shall be permitted to be placed in a permanently grounded conductor, provided the controller is designed so that the pole in the grounded conductor cannot be opened without simultaneously opening all conductors of the circuit.

Generally, one conductor of a 120-volt circuit is grounded, and a single-pole device must be connected in the ungrounded conductor to serve as a controller. A 2-pole controller is permitted for such a circuit, where both conductors (grounded and ungrounded) are opened simultaneously. The same requirement can be applied to other circuits, such as 240-volt, 3-wire circuits with one conductor grounded.

430.87 Number of Motors Served by Each Controller

Each motor shall be provided with an individual controller.

Exception: For motors rated 600 volts or less, a single controller rated at not less than the equivalent horsepower, as determined in accordance with 430.110(C)(1), of all the motors in the group shall be permitted to serve the group under any of the following conditions:

(a) Where a number of motors drive several parts of a single machine or piece of apparatus, such as metal and woodworking machines, cranes, hoists, and similar apparatus

(b) Where a group of motors is under the protection of one overcurrent device as permitted in 430.53(A)

(c) Where a group of motors is located in a single room within sight from the controller location

The conditions stated in the exception to 430.87 are similar to those specified in the exception to 430.112, which permit the use of a single disconnecting means for a group of motors.

430.88 Adjustable-Speed Motors

Adjustable-speed motors that are controlled by means of field regulation shall be equipped and connected so that they cannot be started under a weakened field.

Exception: Starting under a weakened field shall be permitted where the motor is designed for such starting.

The torque and speed of a motor depend on the amount of current passing through the armature. This current is a function of shunt field strength and rpm of the armature. A reduction of the shunt field magnetic flux causes a reduction of the counterelectromotive force in the armature, resulting in an increase in armature current, thereby increasing torque and thus increasing speed.

Because of excessive armature starting currents, field-regulated, adjustable-speed motors are not permitted to be started under a weakened field condition unless some means is provided to limit the speed to within safe limits.

430.89 Speed Limitation

Machines of the following types shall be provided with speed-limiting devices or other speed-limiting means:

(1) Separately excited dc motors
(2) Series motors
(3) Motor-generators and converters that can be driven at excessive speed from the dc end, as by a reversal of current or decrease in load

Exception: Separate speed-limiting devices or means shall not be required under either of the following conditions:

(1) Where the inherent characteristics of the machines, the system, or the load and the mechanical connection thereto are such as to safely limit the speed
(2) Where the machine is always under the manual control of a qualified operator

It is still fairly common for dc motors to be used where speed control is essential, such as in the case of electric railways and elevators, where a smooth start, controlled acceleration, and a smooth stop are necessary.

If the load is removed from a series motor when it is running, the speed of the motor will increase until it is dangerously high. To produce the necessary counterelectromotive force with a weakened field, the armature must turn correspondingly faster. Series motors are commonly used as gear-drive traction motors of electric locomotives and, thus, are continuously loaded.

Separately excited dc motors, series motors, and motor (compound-wound dc) generators and (synchronous) converters must be provided with speed-limiting devices (note exceptions), such as a centrifugal device on the shaft of the machine or a remotely located overspeed device, which may be set at a predetermined speed to operate a set of contacts and thereby trip a circuit breaker and de-energize the machine.

430.90 Combination Fuseholder and Switch as Controller

The rating of a combination fuseholder and switch used as a motor controller shall be such that the fuseholder will accommodate the size of the fuse specified in Part III of this article for motor overload protection.

Exception: Where fuses having time delay appropriate for the starting characteristics of the motor are used, fuseholders of smaller size than specified in Part III of this article shall be permitted.

Time-delay (dual-element) fuses can be used for both motor overload and branch-circuit short-circuit and ground-fault protection and can be sized in accordance with 430.32. See also 430.36, 430.55, and 430.57 for other requirements regarding fuses and fuseholders.

430.91 Motor Controller Enclosure Types

Table 430.91 provides the basis for selecting enclosures for use in specific locations other than hazardous (classified) locations. The enclosures are not intended to protect against conditions such as condensation, icing, corrosion, or contamination that may occur within the enclosure or enter via the conduit or unsealed openings. These internal conditions shall require special consideration by the installer and user.

See the commentary following the definition of *enclosure* in Article 100. Enclosure type numbers are described in greater detail in industry standards such as ANSI/NEMA ICS6, *Industrial Control and Systems: Enclosures*; NEMA 250-2003, *Enclosures for Electrical Equipment (1000 Volts Maximum)*; UL 508, *Standard for Industrial Control Equipment* and in controller manufacturers' literature. For other than general-use Type 1 enclosures, the type number must be marked on the motor-controller enclosure.

VIII. Motor Control Centers

430.92 General

Part VIII covers motor control centers installed for the control of motors, lighting, and power circuits.

Part VIII of Article 430 provides requirements for the installation of motor control centers. These requirements cover subjects that include overcurrent protection use as service

Table 430.91 Motor Controller Enclosure Selection

	For Outdoor Use									
Provides a Degree of Protection Against the Following Environmental Conditions	**Enclosure Type Number[1]**									
	3	**3R**	**3S**	**3X**	**3RX**	**3SX**	**4**	**4X**	**6**	**6P**
Incidental contact with the enclosed equipment	X	X	X	X	X	X	X	X	X	X
Rain, snow, and sleet	X	X	X	X	X	X	X	X	X	X
Sleet[2]	—	—	X	—	—	X	—	—	—	—
Windblown dust	X	—	X	X	—	X	X	X	X	X
Hosedown	—	—	—	—	—	—	X	X	X	X
Corrosive agents	—	—	—	X	X	X	—	X	—	X
Temporary submersion	—	—	—	—	—	—	—	—	X	X
Prolonged submersion	—	—	—	—	—	—	—	—	—	X

	For Indoor Use									
Provides a Degree of Protection Against the Following Environmental Conditions	**Enclosure Type Number[1]**									
	1	**2**	**4**	**4X**	**5**	**6**	**6P**	**12**	**12K**	**13**
Incidental contact with the enclosed equipment	X	X	X	X	X	X	X	X	X	X
Falling dirt	X	X	X	X	X	X	X	X	X	X
Falling liquids and light splashing	—	X	X	X	X	X	X	X	X	X
Circulating dust, lint, fibers, and flyings	—	—	X	X	—	X	X	X	X	X
Settling airborne dust, lint, fibers, and flyings	—	—	X	X	X	X	X	X	X	X
Hosedown and splashing water	—	—	X	X	—	X	X	—	—	—
Oil and coolant seepage	—	—	—	—	—	—	—	X	X	X
Oil or coolant spraying and splashing	—	—	—	—	—	—	—	—	—	X
Corrosive agents	—	—	—	X	—	—	X	—	—	—
Temporary submersion	—	—	—	—	—	X	X	—	—	—
Prolonged submersion	—	—	—	—	—	—	X	—	—	—

[1]Enclosure type number shall be marked on the motor controller enclosure.
[2]Mechanism shall be operable when ice covered.

FPN: The term *raintight* is typically used in conjunction with Enclosure Types 3, 3S, 3SX, 3X, 4, 4X, 6, 6P. The term *rainproof* is typically used in conjunction with Enclosure Type 3R, 3RX. The term *watertight* is typically used in conjunction with Enclosure Types 4, 4X, 6, 6P.

The term *driptight* is typically used in conjunction with Enclosure Types 2, 5, 12, 12K, 13. The term *dusttight* is typically used in conjunction with Enclosure Types 3, 3S, 3SX, 3X, 5, 12, 12K, 13.

equipment, grounding, and construction. In addition to Part VIII, 430.1, FPN No. 1, refers to installation requirements for motor control centers contained in 110.26(F). The requirements of 110.26(F) specify dedicated space for a motor control center and physical protection from mechanical systems that might leak or otherwise adversely affect a motor control center.

Motor control centers are made up of a number of motor starters, controls, and disconnect switches assembled in one large enclosure. Motor control centers are allowed to be used as service equipment if provided with a single main disconnecting means. A second service disconnecting means, however, is permitted in the motor control center if it is provided to serve other loads.

Access and working space clearances as well as dedicated space requirements of 110.26 are applicable to motor control centers.

430.94 Overcurrent Protection

Motor control centers shall be provided with overcurrent protection in accordance with Parts I, II, and IX of Article 240. The ampere rating or setting of the overcurrent protective device shall not exceed the rating of the common power bus. This protection shall be provided by (1) an overcurrent protective device located ahead of the motor control center or (2) a main overcurrent protective device located within the motor control center.

The overcurrent protection cannot exceed the rating of the common power bus of a motor control center. It is permitted to use an overcurrent protective device with a rating less than the common power bus, provided it is of sufficient size to carry the load determined in accordance with Part II of Article 430.

430.95 Service-Entrance Equipment

Where used as service equipment, each motor control center shall be provided with a single main disconnecting means to disconnect all ungrounded service conductors.

Exception: A second service disconnect shall be permitted to supply additional equipment.

Where a grounded conductor is provided, the motor control center shall be provided with a main bonding jumper, sized in accordance with 250.28(D), within one of the sections for connecting the grounded conductor, on its supply side, to the motor control center equipment ground bus.

Exception: High-impedance grounded neutral systems shall be permitted to be connected as provided in 250.36.

430.96 Grounding

Multisection motor control centers shall be bonded together with an equipment grounding conductor or an equivalent grounding bus sized in accordance with Table 250.122.

Equipment grounding conductors shall terminate on this grounding bus or to a grounding termination point provided in a single-section motor control center.

430.97 Busbars and Conductors

(A) Support and Arrangement Busbars shall be protected from physical damage and be held firmly in place. Other than for required interconnections and control wiring, only those conductors that are intended for termination in a vertical section shall be located in that section.

Exception: Conductors shall be permitted to travel horizontally through vertical sections where such conductors are isolated from the busbars by a barrier.

(B) Phase Arrangement The phase arrangement on 3-phase horizontal common power and vertical buses shall be A, B, C from front to back, top to bottom, or left to right, as viewed from the front of the motor control center. The B phase shall be that phase having the higher voltage to ground on 3-phase, 4-wire, delta-connected systems. Other busbar arrangements shall be permitted for additions to existing installations and shall be marked.

Exception: Rear-mounted units connected to a vertical bus that is common to front-mounted units shall be permitted to have a C, B, A phase arrangement where properly identified.

(C) Minimum Wire-Bending Space The minimum wire-bending space at the motor control center terminals and minimum gutter space shall be as required in Article 312.

(D) Spacings Spacings between motor control center bus terminals and other bare metal parts shall be not less than specified in Table 430.97.

(E) Barriers Barriers shall be placed in all service-entrance motor control centers to isolate service busbars and terminals from the remainder of the motor control center.

430.98 Marking

(A) Motor Control Centers Motor control centers shall be marked according to 110.21, and such marking shall be plainly visible after installation. Marking shall also include common power bus current rating and motor control center short-circuit rating.

Table 430.97 Minimum Spacing Between Bare Metal Parts

	Opposite Polarity Where Mounted on the Same Surface		Opposite Polarity Where Held Free in Air		Live Parts to Ground	
Nominal Voltage	mm	in.	mm	in.	mm	in.
Not over 125 volts, nominal	19.1	¾	12.7	½	12.7	½
Not over 250 volts, nominal	31.8	1¼	19.1	¾	12.7	½
Not over 600 volts, nominal	50.8	2	25.4	1	25.4	1

(B) Motor Control Units Motor control units in a motor control center shall comply with 430.8.

IX. Disconnecting Means

430.101 General

Part IX is intended to require disconnecting means capable of disconnecting motors and controllers from the circuit.

> FPN No. 1: See Figure 430.1.
>
> FPN No. 2: See 110.22 for identification of disconnecting means.

430.102 Location

(A) Controller An individual disconnecting means shall be provided for each controller and shall disconnect the controller. The disconnecting means shall be located in sight from the controller location.

The installation shown in Exhibit 430.18 is an example of compliance with the main requirement of 430.102(A).

Exhibit 430.18 The disconnecting means for each controller, which must be within sight of the controller location. (Courtesy of International Association of Electrical Inspectors)

Exception No. 1: For motor circuits over 600 volts, nominal, a controller disconnecting means capable of being locked in the open position shall be permitted to be out of sight of the controller, provided the controller is marked with a warning label giving the location of the disconnecting means.

Exception No. 2: A single disconnecting means shall be permitted for a group of coordinated controllers that drive several parts of a single machine or piece of apparatus. The disconnecting means shall be located in sight from the controllers, and both the disconnecting means and the controllers shall be located in sight from the machine or apparatus.

(B) Motor A disconnecting means shall be located in sight from the motor location and the driven machinery location.

Exception: The disconnecting means shall not be required to be in sight from the motor and the driven machinery location under either condition (a) or (b), provided the disconnecting means required in accordance with 430.102(A) is individually capable of being locked in the open position. The provision for locking or adding a lock to the disconnecting means shall be installed on or at the switch or circuit breaker used as the disconnecting means and shall remain in place with or without the lock installed.

(a) Where such a location of the disconnecting means is impracticable or introduces additional or increased hazards to persons or property

(b) In industrial installations, with written safety procedures, where conditions of maintenance and supervision ensure that only qualified persons service the equipment

> FPN No. 1: Some examples of increased or additional hazards include, but are not limited to, motors rated in excess of 100 hp, multimotor equipment, submersible motors, motors associated with adjustable speed drives, and motors located in hazardous (classified) locations.
>
> FPN No. 2: For information on lockout/tagout procedures, see NFPA 70E-2004, *Standard for Electrical Safety in the Workplace.*

The main rules of 430.102(A) and 430.102(B) require that the disconnecting means be in sight of the controller, the motor location, and the driven-machinery location. For motors over 600 volts, the controller disconnecting means may be out of sight of the controller, as illustrated in Exhibit 430.19, provided that the controller has a warning label

Exhibit 430.19 A motor installation over 600 volts, where the motor controller is not located within sight of its disconnecting means.

indicating the location and identification of the disconnecting means, which must be capable of being locked in the open position.

A single disconnecting means may be located adjacent to a group of coordinated controllers, as illustrated in Exhibit 430.20, where the controllers are mounted on a multimotor continuous process machine.

The exception to 430.102(B) was revised for the 2002 *Code*. The disconnecting means may only be out of sight of the motor, as illustrated in Exhibit 430.21, if the disconnecting means complying with 430.102(A) is individually capable of being locked in the open position and meets the criteria of either (a) or (b) in the exception. If locating the disconnecting means close to the motor location and driven machinery is impracticable due to the type of machinery, the type of facility, the lack of space for locating large equipment such as disconnecting means rated over 600 volts, or any increased hazard to persons or property, then the disconnecting means is permitted to be located remotely. Industrial facilities that comply with OSHA, CFR 1910.147, *The Control of Hazardous Energy (Lockout/Tagout)*, are permitted to have the disconnecting means located remotely.

Section 430.102 clearly requires that individual disconnect switches or circuit breakers must be capable of being locked in the open position. Disconnect switches or circuit breakers that are only located behind the locked door of a panelboard or located within locked rooms do not comply with the requirements of 430.102. The provision for locking or attaching a lock to the disconnecting means must be part

Exhibit 430.21 A controller disconnecting means that is out of sight of the motor.

of the disconnect and a permanent component of the switch or circuit breaker.

Fine Print Note No. 2 points out an important consideration and reference standard for employee safety in the workplace. NFPA 70E-2004, *Standard for Electrical Safety in the Workplace*, 120.2(A), requires in part that "All electrical circuit conductors and circuit parts shall not be considered to be in an electrically safe condition until all sources of energy are removed, the disconnecting means is under lockout/tagout, [and] the absence of voltage is verified by an approved voltage testing device." Further, it states, "Lockout/tagout requirements shall apply to fixed permanently installed equipment, to temporarily installed equipment, and to portable equipment." The principles and procedures set forth in NFPA 70E establish strict work rules requiring locking off and tagging out of disconnect switches.

The disconnecting means required in accordance with 430.102(A) shall be permitted to serve as the disconnecting means for the motor if it is located in sight from the motor location and the driven machinery location.

430.103 Operation

The disconnecting means shall open all ungrounded supply conductors and shall be designed so that no pole can be operated independently. The disconnecting means shall be permitted in the same enclosure with the controller.

> FPN: See 430.113 for equipment receiving energy from more than one source.

Exhibit 430.20 A single disconnecting means located adjacent to a group of coordinated controllers mounted on a multimotor continuous process machine.

The *Code* requires that a switch, circuit breaker, or other device serve as a disconnecting means for both the controller and the motor, thereby providing safety during maintenance and inspection shutdown periods. The disconnecting means also disconnects the controller; therefore, it cannot be a part of the controller. However, separate disconnects and controllers may be mounted on the same panel or contained in the same enclosure, such as combination fused-switch, magnetic-starter units.

Depending on the size of the motor and other conditions, the type of disconnecting means required may be a motor circuit switch, a circuit breaker, a general-use switch, an isolating switch, an attachment plug and receptacle, or a branch-circuit short-circuit and ground-fault protective device, as specified in 430.109.

If a motor is stalled or under heavy overload and the motor controller fails to properly open the circuit, the disconnecting means, which must be rated to interrupt locked-rotor current, can be used to open the circuit. For motors larger than 100 hp ac or 40 hp dc, the disconnecting means is, in accordance with 430.109(E), permitted to be a general-use or an isolating switch where plainly marked "Do not operate under load."

Exhibit 430.22 Heavy-duty safety switches, UL-listed for use on systems up to 200,000 amperes, fault current rms symmetrical with Class J or Class R fuses installed. (Courtesy of Square D Co.)

430.104 To Be Indicating

The disconnecting means shall plainly indicate whether it is in the open (off) or closed (on) position.

430.105 Grounded Conductors

One pole of the disconnecting means shall be permitted to disconnect a permanently grounded conductor, provided the disconnecting means is designed so that the pole in the grounded conductor cannot be opened without simultaneously disconnecting all conductors of the circuit.

430.107 Readily Accessible

At least one of the disconnecting means shall be readily accessible.

430.108 Every Disconnecting Means

Every disconnecting means in the motor circuit between the point of attachment to the feeder and the point of connection to the motor shall comply with the requirements of 430.109 and 430.110.

Exhibit 430.22 represents an example of a listed disconnect switch rated in horsepower meeting the requirements of 430.109 and 430.110.

430.109 Type

The disconnecting means shall be a type specified in 430.109(A), unless otherwise permitted in 430.109(B) through 430.109(G), under the conditions specified.

(A) General

(1) Motor Circuit Switch A listed motor-circuit switch rated in horsepower.

(2) Molded Case Circuit Breaker A listed molded case circuit breaker.

(3) Molded Case Switch A listed molded case switch.

(4) Instantaneous Trip Circuit Breaker An instantaneous trip circuit breaker that is part of a listed combination motor controller.

(5) Self-Protected Combination Controller Listed self-protected combination controller.

(6) Manual Motor Controller Listed manual motor controllers additionally marked "Suitable as Motor Disconnect" shall be permitted as a disconnecting means where installed between the final motor branch-circuit short-circuit protective device and the motor. Listed manual motor controllers additionally marked "Suitable as Motor Disconnect" shall be permitted as disconnecting means on the line side of the fuses permitted in 430.52(C)(5). In this case, the fuses permitted in 430.52(C)(5) shall be considered supplementary fuses, and suitable branch-circuit short-circuit and ground-fault protective devices shall be installed on the line side of the manual motor controller additionally marked "Suitable as Motor Disconnect."

(7) System Isolation Equipment System isolation equipment shall be listed for disconnection purposes. System isolation equipment shall be installed on the load side of the overcurrent protection and its disconnecting means. The disconnecting means shall be one of the types permitted by 430.109(A)(1) through (A)(3).

The impetus for development of *system isolation equipment (SIE)* comes from today's large and often complex machines where repeated operation of disconnecting means for maintenance or servicing is inherent to the process and the risk of injury to personnel is increased due to moving parts and multiple points of entry. The safety for personnel servicing this equipment includes detailed lock-out, tag-out protocols for all sources of mechanical and electrical energy.

System isolation equipment helps simplify electrical lock-out, tag-out procedures and can also be used as a means to isolate other energy sources such as pneumatic energy. In accordance with its definition in 430.2, system isolation equipment is "a redundantly monitored, remotely operated, contactor-isolating system, packaged to provide the disconnection/isolation function." This type of equipment was first introduced in 5.5.4 in the 2002 edition of NFPA 79, *Electrical Standard for Industrial Machinery,* as a means to disconnect and isolate separately operable parts of a large industrial machine. With its inclusion in 430.109 as a permitted type of disconnecting means, there is now recognition in the *Code* for use of this disconnection/isolation system in applications that do not fall within the scope of the industrial machinery standard.

Unlike other disconnecting means recognized by 430.109(A) where the operation of the disconnecting means directly opens the supply circuit at that specific location, system isolation equipment employs a lockable control circuit switch(es) (lockout switch) and a verification indication at the disconnecting means location (lockout station), and operation of the lockout switch causes power components such as a monitored magnetic contactor to open and isolate the electrical equipment associated with the machine from its power supply circuit. The system isolation equipment is classified according to its intended application with parameters that include the load characteristics, the method used to monitor the controlled load side power circuit, the number and maximum distance to the farthest lockout station, and the available control interface functions.

In a typical configuration, the contactor is located in the system power and control panel and may control power to the entire machine or to portions of a large machine. The control equipment may be provided in several options for configuration that can used to distribute lockout stations in single or multiplexed radial schemes according to the application.

Once an electrically safe condition is achieved (including discharge of any residual energy), verification of such condition is provided at the remote lock-out station through the use of an indicator light. In equipment that uses lockable guarding, the same verification signal could also be used as part of the guard access system.

In contrast to a simple start/stop station and control circuit operating a magnetic contactor, the control panel for this system provides a sophisticated level of monitoring upon actuation of the remote lockout switch. If any portion of the safety system cannot be verified for proper operation, the safe condition indicator light will not illuminate at the remote lockout station. As part of the standard operating procedure, the failure to receive the safe condition signal has to be considered by personnel as an indication of an unsafe condition.

Among the critical safety elements that are provided by the control panel for the isolation system is the diversity and redundancy that is integrated into the safe condition verification logic. Another element is reducing the possibility of externally induced failure modes through the electrical isolation of the internal safety related control circuits and the physical isolation of the equipment's internal components. The control panel modules are sealed as are the circuits between the SIE component enclosures in order to discourage tampering that could compromise the safe operation of the equipment and endanger personnel. Where the system includes multiple lockout stations, the controlled equipment cannot be re-energized until all of the lockout switches are returned to the on position. Nominal configurations of the SIE include provisions to prevent power from unexpectedly reaching the machine upon the restoration of power from the utility source; all lockout switches must be in the closed or on position while at least one lockout switch must have been in the off or open position or placed in the off or open position (and then moved to the on or closed position after the utility power had been restored) in order to re-energize the machine.

The system isolation equipment is required to be listed for disconnection purposes and must be installed on the load side of overcurrent protection and either a motor circuit switch, a molded case circuit breaker, or a molded case switch. UL 60947-4-20, *Standard for Low-voltage Switchgear and Controlgear — Part 4-20: Contactors and Motor Starters-Equipment used for System Isolation and Rated as a Single Unit,* is the standard used for evaluation of this equipment. This standard includes requirements and tests for the SIE power circuits to confirm the normal load and overload making and breaking capacities, rated conditional short-circuit current for coordination with short-circuit devices and dielectric with stand and leakage current as well as mechanical durability tests.

(B) Stationary Motors of ⅛ Horsepower or Less For stationary motors of ⅛ hp or less, the branch-circuit overcurrent device shall be permitted to serve as the disconnecting means.

(C) Stationary Motors of 2 Horsepower or Less For stationary motors rated at 2 hp or less and 300 volts or less, the disconnecting means shall be permitted to be one of the devices specified in (1), (2), or (3):

(1) A general-use switch having an ampere rating not less than twice the full-load current rating of the motor

(2) On ac circuits, a general-use snap switch suitable only for use on ac (not general-use ac–dc snap switches) where the motor full-load current rating is not more than 80 percent of the ampere rating of the switch

(3) A listed manual motor controller having a horsepower rating not less than the rating of the motor and marked "Suitable as Motor Disconnect"

(D) Autotransformer-Type Controlled Motors For motors of over 2 hp to and including 100 hp, the separate disconnecting means required for a motor with an autotransformer-type controller shall be permitted to be a general-use switch where all of the following provisions are met:

(1) The motor drives a generator that is provided with overload protection.

(2) The controller is capable of interrupting the locked-rotor current of the motors, is provided with a no voltage release, and is provided with running overload protection not exceeding 125 percent of the motor full-load current rating.

(3) Separate fuses or an inverse time circuit breaker rated or set at not more than 150 percent of the motor full-load current are provided in the motor branch circuit.

(E) Isolating Switches For stationary motors rated at more than 40 hp dc or 100 hp ac, the disconnecting means shall be permitted to be a general-use or isolating switch where plainly marked "Do not operate under load."

(F) Cord-and-Plug-Connected Motors For a cord-and-plug-connected motor, a horsepower-rated attachment plug and receptacle having ratings no less than the motor ratings shall be permitted to serve as the disconnecting means. A horsepower-rated attachment plug and receptacle shall not be required for a cord-and-plug-connected appliance in accordance with 422.32, a room air conditioner in accordance with 440.63, or a portable motor rated ⅓ hp or less.

Section 430.109 requires the disconnecting means to be a motor circuit switch, a circuit breaker, or a molded-case switch (nonautomatic circuit interrupter). A motor circuit switch is a horsepower-rated switch capable of interrupting the maximum overload current of a motor (see the definition of *switch, motor-circuit* in Article 100). A molded-case switch (nonautomatic circuit interrupter) is a circuit-breaker-like device without the overcurrent element and automatic-trip mechanism. It is rated in amperes and is suitable for use as a motor circuit disconnect based on its ampere rating, as is a circuit breaker. The disconnecting means must be listed.

Exhibits 430.23, 430.24, 430.25, and 430.26 illustrate various methods of providing motor disconnecting means as permitted by 430.109(B), 430.109(C), 430.109(E), and 430.109(F), respectively.

Where horsepower-rated fused switches are required, it

Exhibit 430.23 A branch-circuit overcurrent device serving as the disconnecting means for a stationary motor of ⅛ hp or less according to 430.109(B).

Stationary motor 1/8 hp or less

Disconnecting means: branch-circuit fuse or circuit breaker

Stationary motor 2 hp or less and 300 V or less

Controller, including motor overload protective device

Disconnecting means: general-use snap switch

Start Stop

Exhibit 430.24 A general-use snap switch having an ampere rating not less than twice the motor full-load current serving as the disconnecting means for a stationary motor rated at 2 hp or less and at 300 volts or less according to 430.109(C).

should be noted that marking within the enclosure usually permits a dual horsepower rating. The standard horsepower rating is based on the largest non-time-delay (non-dual-element) fuse rating that can be used in the switch and that will permit the motor to start. The maximum horsepower rating is based on the largest rated time-delay (dual-element) fuse that can be used in the switch and that will permit the motor to start. Thus, where time-delay fuses are used,

Exhibit 430.25 A general-use or an isolating switch serving as the disconnecting means for a stationary motor rated at more than 40 hp dc or 100 hp ac according to 430.109(E).

Exhibit 430.26 An attachment plug and receptacle of the proper rating serving as the disconnecting means for a certain cord-and-plug-connected motor according to 430.109(F).

smaller-size switches and fuseholders can be used (see 430.57, Exception).

(G) Torque Motors For torque motors, the disconnecting means shall be permitted to be a general-use switch.

430.110 Ampere Rating and Interrupting Capacity

(A) General The disconnecting means for motor circuits rated 600 volts, nominal, or less shall have an ampere rating not less than 115 percent of the full-load current rating of the motor.

Exception: A listed nonfused motor-circuit switch having a horsepower rating not less than the motor horsepower shall be permitted to have an ampere rating less than 115 percent of the full-load current rating of the motor.

(B) For Torque Motors Disconnecting means for a torque motor shall have an ampere rating of at least 115 percent of the motor nameplate current.

(C) For Combination Loads Where two or more motors are used together or where one or more motors are used in combination with other loads, such as resistance heaters, and where the combined load may be simultaneous on a single disconnecting means, the ampere and horsepower ratings of the combined load shall be determined as follows.

(1) Horsepower Rating The rating of the disconnecting means shall be determined from the sum of all currents, including resistance loads, at the full-load condition and also at the locked-rotor condition. The combined full-load current and the combined locked-rotor current so obtained shall be considered as a single motor for the purpose of this requirement as follows.

The full-load current equivalent to the horsepower rating of each motor shall be selected from Tables 430.247, 430.248, 430.249, or 430.250. These full-load currents shall be added to the rating in amperes of other loads to obtain an equivalent full-load current for the combined load.

The locked-rotor current equivalent to the horsepower rating of each motor shall be selected from Tables 430.251(A) or 430.251(B). The locked-rotor currents shall be added to the rating in amperes of other loads to obtain an equivalent locked-rotor current for the combined load. Where two or more motors or other loads cannot be started simultaneously, the largest sum of locked rotor currents of a motor or group of motors that can be started simultaneously and the full load currents of other concurrent loads shall be permitted to be used to determine the equivalent locked-rotor current for the simultaneous combined loads.

Exception: Where part of the concurrent load is resistance load, and where the disconnecting means is a switch rated in horsepower and amperes, the switch used shall be permitted to have a horsepower rating that is not less than the combined load of the motor(s), if the ampere rating of the switch is not less than the locked-rotor current of the motor(s) plus the resistance load.

(2) Ampere Rating The ampere rating of the disconnecting means shall not be less than 115 percent of the sum of all currents at the full-load condition determined in accordance with 430.110(C)(1).

Exception: A listed nonfused motor-circuit switch having a horsepower rating equal to or greater than the equivalent horsepower of the combined loads, determined in accordance with 430.110(C)(1), shall be permitted to have an ampere rating less than 115 percent of the sum of all currents at the full-load condition.

(3) Small Motors For small motors not covered by Tables 430.247, 430.248, 430.249, or 430.250, the locked-rotor current shall be assumed to be six times the full-load current.

A general-use switch, circuit breaker, molded-case switch (nonautomatic circuit interrupter), or attachment plug and receptacle used as a disconnecting means must have an ampere rating of not less than 115 percent of the motor full-load current.

Following is an example of how to determine the size of a disconnect for a combination load.

Example

Assume the load consists of one 5-hp, one 3-hp, and two ½-hp motors, plus a 10-kilowatt heater, all rated 240 volts, 3-phase. All motors are Design B motors. The full-load and locked-rotor current equivalents are calculated, using the appropriate tables, as follows:

Calculation of Equivalent Full-Load Current Rating

Motor or Other Load	Table	Amperes
5-hp motor	430.250	15.2
3-hp motor	430.250	9.6
½-hp motor	430.250	2.2
½-hp motor	430.250	2.2
10-kW heater		24.1
$\frac{10 \times 1000}{(240 \times 1.732)}$		
	Total	53.3

Calculation of Equivalent Locked-Rotor Current Rating

Motor or Other Load	Table	Amperes
5-hp motor	430.251(B)	92
3-hp motor	430.251(B)	64
½-hp motor	430.251(B)	20
½-hp motor	430.251(B)	20
10-kW heater		24.1
$\frac{10 \times 1000}{(240 \times 1.732)}$		
	Total	220.1

Solution

The disconnecting means for a motor must have an ampere rating that is not less than 115 percent and also have a horsepower rating that is not less than the combined load. Therefore, the minimum ampere rating of the disconnecting means is 61.3 amperes (1.15×53.3 amperes = 61.3 amperes). Using Table 430.251(B) to obtain an equivalent horsepower from a locked-rotor current rating, the closest value equal to or greater than 220.1 amperes under the 230-volt column is 232 amperes, which equates to 15 hp.

A 15-hp general-purpose or heavy-duty switch satisfies the minimum horsepower requirement but fails to satisfy the minimum current requirement of 61.3 amperes. Therefore, the next larger size disconnect switch must be used, which in horsepower rating is 20 hp. This is because the locked-rotor current for a 20-hp, 3-phase, 230-volt motor is 290 amperes (over the needed 220.1 amperes) and the full-load current is 54 amperes (over the full-load rating of 53.3 amperes).

Listed circuit breakers and molded-case switches are tested under overload conditions at six times their rating, to cover motor circuit applications, and are suitable for use as a motor disconnecting means.

430.111 Switch or Circuit Breaker as Both Controller and Disconnecting Means

A switch or circuit breaker shall be permitted to be used as both the controller and disconnecting means if it complies with 430.111(A) and is one of the types specified in 430.111(B).

(A) General The switch or circuit breaker complies with the requirements for controllers specified in 430.83, opens all ungrounded conductors to the motor, and is protected by an overcurrent device in each ungrounded conductor (which shall be permitted to be the branch-circuit fuses). The overcurrent device protecting the controller shall be permitted to be part of the controller assembly or shall be permitted to be separate. An autotransformer-type controller shall be provided with a separate disconnecting means.

(B) Type The device shall be one of the types specified in 430.111(B)(1), (B)(2), or (B)(3).

(1) Air-Break Switch An air-break switch, operable directly by applying the hand to a lever or handle.

(2) Inverse Time Circuit Breaker An inverse time circuit breaker operable directly by applying the hand to a lever or handle. The circuit breaker shall be permitted to be both power and manually operable.

(3) Oil Switch An oil switch used on a circuit whose rating does not exceed 600 volts or 100 amperes, or by special permission on a circuit exceeding this capacity where under expert supervision. The oil switch shall be permitted to be both power and manually operable.

If used as a controller, a switch or circuit breaker must meet all the requirements for controllers and be protected by branch-circuit short-circuit and ground-fault protective devices (fuses or a circuit breaker), which ensure that all ungrounded conductors will be opened.

If the controller consists of a manually operable air-break switch, an inverse time circuit breaker, or a 100-ampere maximum oil switch (higher rating by special permission), the controller is considered a satisfactory disconnecting means. It is the intent of 430.111 to permit omission of an additional device to serve as a disconnecting means.

Note that a separate disconnecting means must be provided if the controller is of the autotransformer or compensator type. (This switch may be combined in the same enclosure with a motor overload protective device.)

430.112 Motors Served by Single Disconnecting Means

Each motor shall be provided with an individual disconnecting means.

Exception: A single disconnecting means shall be permitted to serve a group of motors under any one of the conditions of (a), (b), and (c). The single disconnecting means shall be rated in accordance with 430.110(C) .

(a) Where a number of motors drive several parts of a single machine or piece of apparatus, such as metal- and woodworking machines, cranes, and hoists.

(b) Where a group of motors is under the protection of one set of branch-circuit protective devices as permitted by 430.53(A).

(c) Where a group of motors is in a single room within sight from the location of the disconnecting means.

The exception to 430.112 permits a single disconnecting means to serve a group of motors. The disconnecting means must have a rating equal to the sum of the horsepower or current of each motor in the group. If the sum is over 2 hp, a motor circuit switch (horsepower-rated) must be used; thus, for five 2-hp motors, the disconnecting means should be a motor circuit switch rated at not less than 10 hp.

Part (a) of the exception to 430.112 indicates that a single disconnecting means may be used where a number of motors drive several parts of a single machine, such as cranes (see 610.31 through 610.33), metal or woodworking machines, steel rolling mill machinery, and so on. The single disconnecting means for multimotor machinery provides a positive means of simultaneously de-energizing all motor branch circuits, including remote control circuits, interlocking circuits, limit-switch circuits, and operator control stations.

Part (b) of the exception to 430.112 refers to 430.53(A), which permits a group of motors under the protection of the same branch-circuit device, provided the device is rated not more than 20 amperes at 125 volts or 15 amperes at more

than 125 volts but not more than 600 volts. The motors must be rated 1 hp or less, and the full-load current for each motor is not permitted to exceed 6 amperes. A single disconnecting means is both practical and economical for a group of such small motors.

Part (c) of the exception to 430.112 covers the common situation in which a group of motors is located in one room, such as a pump room, compressor room, mixer room, and so on. It is therefore possible to design the layout of a single disconnecting means with an unobstructed view (not more than 50 ft) from each motor.

These conditions for an individual disconnecting means are similar to those specified in 430.87, which permits the use of a single controller for a group of motors.

430.113 Energy from More Than One Source

Motor and motor-operated equipment receiving electrical energy from more than one source shall be provided with disconnecting means from each source of electrical energy immediately adjacent to the equipment served. Each source shall be permitted to have a separate disconnecting means. Where multiple disconnecting means are provided, a permanent warning sign shall be provided on or adjacent to each disconnecting means.

Exception No. 1: Where a motor receives electrical energy from more than one source, the disconnecting means for the main power supply to the motor shall not be required to be immediately adjacent to the motor, provided the controller disconnecting means is capable of being locked in the open position.

Exception No. 2: A separate disconnecting means shall not be required for a Class 2 remote-control circuit conforming with Article 725, rated not more than 30 volts, and isolated and ungrounded.

Certain motors may require multiple separate sources of power to operate properly. For example, large synchronous motors commonly receive electrical energy from more than one source. Section 430.113 could apply as well to control circuits that supply power to sensors mounted within or otherwise attached to the motor or the driven machine. Exception No. 2 to 430.113 relieves the user from the disconnect requirement only for Class 2 circuits.

Where there are multiple disconnecting means for the individual sources, a permanent warning sign is required to warn the user that other power sources are present.

X. Adjustable-Speed Drive Systems

430.120 General

The installation provisions of Part I through Part IX are applicable unless modified or supplemented by Part X.

A new Part X was developed for the 2005 *Code* to address adjustable-speed drive systems. Existing Parts X, XI, XII, and XIII were renumbered XI, XII, XIII, and XIV respectively, and the sections within these parts were changed from the 100 series to the 200 series.

Adjustable-speed drives are used extensively in commercial, institutional, and industrial motor applications, and the new Part X was established to consolidate requirements that are unique to these drives. However, Part I through Part IX must be followed unless modified or supplemented in Part X. The new text provides rules regarding methods of overtemperature protection in motors. This is a critical area, because motors operating at reduced speed do not provide adequate air circulation over windings from a fan integral with the motor. An overload device that actuates on current in excess of full-load amperes will not operate because the operating current at slower speeds is reduced. A thermal-sensing device integral with the motor will sense a temperature rise in the motor windings.

FPN: Electrical resonance can result from the interaction of the nonsinusoidal currents from this type of load with power factor correction capacitors.

430.122 Conductors — Minimum Size and Ampacity

(A) Branch/Feeder Circuit Conductors Circuit conductors supplying power conversion equipment included as part of an adjustable-speed drive system shall have an ampacity not less than 125 percent of the rated input to the power conversion equipment.

(B) Bypass Device For an adjustable speed drive system that utilizes a bypass device, the conductor ampacity shall not be less than required by 430.6. The ampacity of circuit conductors supplying power conversion equipment included as part of an adjustable-speed drive system that utilizes a bypass device shall be the larger of either of the following:

(1) 125 percent of the rated input to the power conversion equipment
(2) 125 percent of the motor full-load current rating as determined by 430.6

430.124 Overload Protection

Overload protection of the motor shall be provided.

(A) Included in Power Conversion Equipment Where the power conversion equipment is marked to indicate that motor overload protection is included, additional overload protection shall not be required.

(B) Bypass Circuits For adjustable speed drive systems that utilize a bypass device to allow motor operation at rated full load speed, motor overload protection as described in Article 430, Part III, shall be provided in the bypass circuit.

(C) Multiple Motor Applications For multiple motor application, individual motor overload protection shall be provided in accordance with Article 430, Part III.

430.126 Motor Overtemperature Protection

(A) General Adjustable speed drive systems shall protect against motor overtemperature conditions. Overtemperature protection is in addition to the conductor protection required in 430.32. Protection shall be provided by one of the following means.

(1) Motor thermal protector in accordance with 430.32
(2) Adjustable speed drive controller with load and speed-sensitive overload protection and thermal memory retention upon shutdown or power loss
(3) Overtemperature protection relay utilizing thermal sensors embedded in the motor and meeting the requirements of 430.32(A)(2) or (B)(2)
(4) Thermal sensor embedded in the motor that is received and acted upon by an adjustable speed drive

(B) Motors with Cooling Systems Motors that utilize external forced air or liquid cooling systems shall be provided with protection that shall be continuously enabled or enabled automatically if the cooling system fails.

FPN: Protection against cooling system failure can take many forms. Some examples of protection against inoperative or failed cooling systems are direct sensing of the motor temperature as described in 430.32(A)(1), (A)(3), and (A)(4) or sensing of the presence or absence of the cooling media (flow or pressure sensing).

(C) Multiple Motor Applications For multiple motor application, individual motor overtemperature protection shall be provided.

FPN: The relationship between motor current and motor temperature changes when the motor is operated by an adjustable speed drive. When operated at reduced speed, overheating of motors may occur at current levels less than or equal to a motor's rated full load current. This is the result of reduced motor cooling when its shaft-mounted fan is operating less than rated nameplate RPM.

(D) Automatic Restarting and Orderly Shutdown The provisions of 430.43 and 430.44 shall apply to the motor overtemperature protection means.

430.128 Disconnecting Means

The disconnecting means shall be permitted to be in the incoming line to the conversion equipment and shall have a rating not less than 115 percent of the rated input current of the conversion unit.

XI. Over 600 Volts, Nominal

430.221 General

Part XI recognizes the additional hazard due to the use of higher voltages. It adds to or amends the other provisions of this article.

430.222 Marking on Controllers

In addition to the marking required by 430.8, a controller shall be marked with the control voltage.

430.223 Conductor Enclosures Adjacent to Motors

Flexible metal conduit or liquidtight flexible metal conduit not exceeding 1.8 m (6 ft) in length shall be permitted to be employed for raceway connection to a motor terminal enclosure.

430.224 Size of Conductors

Conductors supplying motors shall have an ampacity not less than the current at which the motor overload protective device(s) is selected to trip.

430.225 Motor-Circuit Overcurrent Protection

(A) General Each motor circuit shall include coordinated protection to automatically interrupt overload and fault currents in the motor, the motor-circuit conductors, and the motor control apparatus.

Exception: Where a motor is vital to operation of the plant and the motor should operate to failure if necessary to prevent a greater hazard to persons, the sensing device(s) shall be permitted to be connected to a supervised annunciator or alarm instead of interrupting the motor circuit.

(B) Overload Protection

(1) Type of Overload Device Each motor shall be protected against dangerous heating due to motor overloads and failure to start by a thermal protector integral with the motor or external current-sensing devices, or both.

(2) Wound-Rotor AC Motors The secondary circuits of wound-rotor ac motors, including conductors, controllers, and resistors rated for the application, shall be considered as protected against overcurrent by the motor overload protection means.

(3) Operation Operation of the overload interrupting device shall simultaneously disconnect all ungrounded conductors.

(4) Automatic Reset Overload sensing devices shall not automatically reset after trip unless resetting of the overload sensing device does not cause automatic restarting of the motor or there is no hazard to persons created by automatic restarting of the motor and its connected machinery.

(C) Fault-Current Protection

(1) Type of Protection Fault-current protection shall be provided in each motor circuit by one of the following means.

(a) A circuit breaker of suitable type and rating arranged so that it can be serviced without hazard. The circuit breaker shall simultaneously disconnect all ungrounded conductors. The circuit breaker shall be permitted to sense the fault current by means of integral or external sensing elements.

(b) Fuses of a suitable type and rating placed in each ungrounded conductor. Fuses shall be used with suitable disconnecting means, or they shall be of a type that can also serve as the disconnecting means. They shall be arranged so that they cannot be serviced while they are energized.

(2) Reclosing Fault-current interrupting devices shall not automatically reclose the circuit.

Exception: Automatic reclosing of a circuit shall be permitted where the circuit is exposed to transient faults and where such automatic reclosing does not create a hazard to persons.

(3) Combination Protection Overload protection and fault-current protection shall be permitted to be provided by the same device.

430.226 Rating of Motor Control Apparatus

The ultimate trip current of overcurrent (overload) relays or other motor-protective devices used shall not exceed 115 percent of the controller's continuous current rating. Where the motor branch-circuit disconnecting means is separate from the controller, the disconnecting means current rating shall not be less than the ultimate trip setting of the overcurrent relays in the circuit.

430.227 Disconnecting Means

The controller disconnecting means shall be capable of being locked in the open position.

XII. Protection of Live Parts — All Voltages

430.231 General

Part XII specifies that live parts shall be protected in a manner judged adequate for the hazard involved.

430.232 Where Required

Exposed live parts of motors and controllers operating at 50 volts or more between terminals shall be guarded against accidental contact by enclosure or by location as follows:

(1) By installation in a room or enclosure that is accessible only to qualified persons
(2) By installation on a suitable balcony, gallery, or platform, elevated and arranged so as to exclude unqualified persons
(3) By elevation 2.5 m (8 ft) or more above the floor

Exception: Live parts of motors operating at more than 50 volts between terminals shall not require additional guarding for stationary motors that have commutators, collectors, and brush rigging located inside of motor-end brackets and not conductively connected to supply circuits operating at more than 150 volts to ground.

430.233 Guards for Attendants

Where live parts of motors or controllers operating at over 150 volts to ground are guarded against accidental contact only by location as specified in 430.232, and where adjustment or other attendance may be necessary during the operation of the apparatus, suitable insulating mats or platforms shall be provided so that the attendant cannot readily touch live parts unless standing on the mats or platforms.

FPN: For working space, see 110.26 and 110.34.

XIII. Grounding — All Voltages

430.241 General

Part XIII specifies the grounding of exposed non–current-carrying metal parts, likely to become energized, of motor and controller frames to prevent a voltage above ground in the event of accidental contact between energized parts and frames. Insulation, isolation, or guarding are suitable alternatives to grounding of motors under certain conditions.

430.242 Stationary Motors

The frames of stationary motors shall be grounded under any of the following conditions:

Any motor in a wet location and subject to contact by personnel constitutes a serious hazard and, unless it is isolated, elevated, or guarded from reach, must be grounded.

Stationary motors are usually supplied by wiring that is enclosed in metal raceways, flexible metal conduit, or cables with metal sheaths. When effectively attached to the motor junction box or frame, the metal raceway or cable armor serves as the equipment grounding conductor. See 250.118 for more information on types of equipment grounding conductors.

(1) Where supplied by metal-enclosed wiring
(2) Where in a wet location and not isolated or guarded
(3) If in a hazardous (classified) location as covered in Articles 500 through 517
(4) If the motor operates with any terminal at over 150 volts to ground

Where the frame of the motor is not grounded, it shall be permanently and effectively insulated from the ground.

430.243 Portable Motors

The frames of portable motors that operate at over 150 volts to ground shall be guarded or grounded.

FPN No. 1: See 250.114(4) for grounding of portable appliances in other than residential occupancies.
FPN No. 2: See 250.119(C) for color of equipment grounding conductor.

430.244 Controllers

Controller enclosures shall be grounded regardless of voltage. Controller enclosures shall have means for attachment of an equipment grounding conductor termination in accordance with 250.8.

Exception: Enclosures attached to ungrounded portable equipment shall not be required to be grounded.

430.245 Method of Grounding

Where required, grounding shall be done in the manner specified in Part VI of Article 250.

(A) Grounding Through Terminal Housings Where the wiring to fixed motors is metal-enclosed cable or in metal raceways, junction boxes to house motor terminals shall be provided, and the armor of the cable or the metal raceways shall be connected to them in the manner specified in Article 250.

FPN: See 430.12(E) for equipment grounding connection means required at motor terminal housings.

(B) Separation of Junction Box from Motor The junction box required by 430.245(A) shall be permitted to be separated from the motor by not more than 1.8 m (6 ft), provided the leads to the motor are stranded conductors within Type AC cable, interlocked metal tape Type MC cable where listed and identified in accordance with 250.118(10)(a), or armored cord or are stranded leads enclosed in liquidtight flexible metal conduit, flexible metal conduit, intermediate metal conduit, rigid metal conduit, or electrical metallic tubing not smaller than metric designator 12 (trade size ⅜), the armor or raceway being connected both to the motor and to the box.

Liquidtight flexible nonmetallic conduit and rigid nonmetallic conduit shall be permitted to enclose the leads to the motor, provided the leads are stranded and the required equipment grounding conductor is connected to both the motor and to the box.

Where stranded leads are used, protected as specified above, each strand within the conductor shall be not larger than 10 AWG and shall comply with other requirements of this *Code* for conductors to be used in raceways.

(C) Grounding of Controller-Mounted Devices Instrument transformer secondaries and exposed non–current-carrying metal or other conductive parts or cases of instrument transformers, meters, instruments, and relays shall be grounded as specified in 250.170 through 250.178.

Most motors are subject to vibration, and good wiring practice requires that, in nearly all cases, the wiring to motors that are fixed be installed with a short section (not more than 6 ft) of liquidtight flexible metal or nonmetallic conduit or of flexible metal conduit to the motor terminal housing. Such use of flexible conduit requires an equipment grounding conductor.

XIV. Tables

Table 430.248 through Table 430.250 accurately reflect the typical and most used 4-pole and 2-pole induction motors in use.

Table 430.248 Full-Load Currents in Amperes, Single-Phase Alternating-Current Motors

The following values of full-load currents are for motors running at usual speeds and motors with normal torque characteristics. The voltages listed are rated motor voltages. The currents listed shall be permitted for system voltage ranges of 110 to 120 and 220 to 240 volts.

Horsepower	115 Volts	200 Volts	208 Volts	230 Volts
1/6	4.4	2.5	2.4	2.2
1/4	5.8	3.3	3.2	2.9
1/3	7.2	4.1	4.0	3.6
1/2	9.8	5.6	5.4	4.9
3/4	13.8	7.9	7.6	6.9
1	16	9.2	8.8	8.0
1 1/2	20	11.5	11.0	10
2	24	13.8	13.2	12
3	34	19.6	18.7	17
5	56	32.2	30.8	28
7 1/2	80	46.0	44.0	40
10	100	57.5	55.0	50

Table 430.247 Full-Load Current in Amperes, Direct-Current Motors

The following values of full-load currents* are for motors running at base speed.

Horsepower	Armature Voltage Rating*					
	90 Volts	120 Volts	180 Volts	240 Volts	500 Volts	550 Volts
1/4	4.0	3.1	2.0	1.6	—	—
1/3	5.2	4.1	2.6	2.0	—	—
1/2	6.8	5.4	3.4	2.7	—	—
3/4	9.6	7.6	4.8	3.8	—	—
1	12.2	9.5	6.1	4.7	—	—
1 1/2	—	13.2	8.3	6.6	—	—
2	—	17	10.8	8.5	—	—
3	—	25	16	12.2	—	—
5	—	40	27	20	—	—
7 1/2	—	58	—	29	13.6	12.2
10	—	76	—	38	18	16
15	—	—	—	55	27	24
20	—	—	—	72	34	31
25	—	—	—	89	43	38
30	—	—	—	106	51	46
40	—	—	—	140	67	61
50	—	—	—	173	83	75
60	—	—	—	206	99	90
75	—	—	—	25 5	123	111
100	—	—	—	341	164	148
125	—	—	—	425	205	185
150	—	—	—	506	246	222
200	—	—	—	675	330	294

*These are average dc quantities.

Table 430.249 Full-Load Current, Two-Phase Alternating-Current Motors (4-Wire)

The following values of full-load current are for motors running at speeds usual for belted motors and motors with normal torque characteristics. Current in the common conductor of a 2-phase, 3-wire system will be 1.41 times the value given. The voltages listed are rated motor voltages. The currents listed shall be permitted for system voltage ranges of 110 to 120, 220 to 240, 440 to 480, and 550 to 600 volts.

Horsepower	Induction-Type Squirrel Cage and Wound Rotor (Amperes)				
	115 Volts	230 Volts	460 Volts	575 Volts	2300 Volts
1/2	4.0	2.0	1.0	0.8	—
3/4	4.8	2.4	1.2	1.0	—
1	6.4	3.2	1.6	1.3	—
1 1/2	9.0	4.5	2.3	1.8	—
2	11.8	5.9	3.0	2.4	—
3	—	8.3	4.2	3.3	—
5	—	13.2	6.6	5.3	—
7 1/2	—	19	9.0	8.0	—
10	—	24	12	10	—
15	—	36	18	14	—
20	—	47	23	19	—
25	—	59	29	24	—
30	—	69	35	28	—
40	—	90	45	36	—
50	—	113	56	45	—
60	—	133	67	53	14
75	—	166	83	66	18
100	—	218	109	87	23
125	—	270	135	108	28
150	—	312	156	125	32
200	—	416	208	167	43

Table 430.250 Full-Load Current, Three-Phase Alternating-Current Motors
The following values of full-load currents are typical for motors running at speeds usual for belted motors and motors with normal torque characteristics.

The voltages listed are rated motor voltages. The currents listed shall be permitted for system voltage ranges of 110 to 120, 220 to 240, 440 to 480, and 550 to 600 volts.

Horsepower	Induction-Type Squirrel Cage and Wound Rotor (Amperes)							Synchronous-Type Unity Power Factor* (Amperes)			
	115 Volts	200 Volts	208 Volts	230 Volts	460 Volts	575 Volts	2300 Volts	230 Volts	460 Volts	575 Volts	2300 Volts
½	4.4	2.5	2.4	2.2	1.1	0.9	—	—	—	—	—
¾	6.4	3.7	3.5	3.2	1.6	1.3	—	—	—	—	—
1	8.4	4.8	4.6	4.2	2.1	1.7	—	—	—	—	—
1½	12.0	6.9	6.6	6.0	3.0	2.4	—	—	—	—	—
2	13.6	7.8	7.5	6.8	3.4	2.7	—	—	—	—	—
3	—	11.0	10.6	9.6	4.8	3.9	—	—	—	—	—
5	—	17.5	16.7	15.2	7.6	6.1	—	—	—	—	—
7½	—	25.3	24.2	22	11	9	—	—	—	—	—
10	—	32.2	30.8	28	14	11	—	—	—	—	—
15	—	48.3	46.2	42	21	17	—	—	—	—	—
20	—	62.1	59.4	54	27	22	—	—	—	—	—
25	—	78.2	74.8	68	34	27	—	53	26	21	—
30	—	92	88	80	40	32	—	63	32	26	—
40	—	120	114	104	52	41	—	83	41	33	—
50	—	150	143	130	65	52	—	104	52	42	—
60	—	177	169	154	77	62	16	123	61	49	12
75	—	221	211	192	96	77	20	155	78	62	15
100	—	285	273	248	124	99	26	202	101	81	20
125	—	359	343	312	156	125	31	253	126	101	25
150	—	414	396	360	180	144	37	302	151	121	30
200		552	528	480	240	192	49	400	201	161	40
250	—	—	—	—	302	242	60	—	—	—	—
300	—	—	—	—	361	289	72	—	—	—	—
350	—	—	—	—	414	336	83	—	—	—	—
400	—	—	—	—	477	382	95	—	—	—	—
450	—	—	—	—	515	412	103	—	—	—	—
500	—	—	—	—	590	472	118	—	—	—	—

*For 90 and 80 percent power factor, the figures shall be multiplied by 1.1 and 1.25, respectively.

Table 430.251(A) Conversion Table of Single-Phase Locked-Rotor Currents for Selection of Disconnecting Means and Controllers as Determined from Horsepower and Voltage Rating

For use only with 430.110, 440.12, 440.41, and 455.8(C).

Rated Horsepower	Maximum Locked-Rotor Current in Amperes, Single Phase		
	115 Volts	208 Volts	230 Volts
½	58.8	32.5	29.4
¾	82.8	45.8	41.4
1	96	53	48
1½	120	66	60
2	144	80	72
3	204	113	102
5	336	186	168
7½	480	265	240
10	600	332	300

Table 430.251(B) Conversion Table of Polyphase Design B, C, and D Maximum Locked-Rotor Currents for Selection of Disconnecting Means and Controllers as Determined from Horsepower and Voltage Rating and Design Letter

For use only with 430.110, 440.12, 440.41 and 455.8(C).

Rated Horsepower	Maximum Motor Locked-Rotor Current in Amperes, Two- and Three-Phase, Design B, C, and D*					
	115 Volts	200 Volts	208 Volts	230 Volts	460 Volts	575 Volts
	B, C, D	B, C, D	B, C, D	B, C, D	B, C, D	B, C, D
½	40	23	22.1	20	10	8
¾	50	28.8	27.6	25	12.5	10
1	60	34.5	33	30	15	12
1½	80	46	44	40	20	16
2	100	57.5	55	50	25	20
3	—	73.6	71	64	32	25.6
5	—	105.8	102	92	46	36.8
7½	—	146	140	127	63.5	50.8
10	—	186.3	179	162	81	64.8
15	—	267	257	232	116	93
20	—	334	321	290	145	116
25	—	420	404	365	183	146
30	—	500	481	435	218	174
40	—	667	641	580	290	232
50	—	834	802	725	363	290
60	—	1001	962	870	435	348
75	—	1248	1200	1085	543	434
100	—	1668	1603	1450	725	580
125	—	2087	2007	1815	908	726
150	—	2496	2400	2170	1085	868
200	—	3335	3207	2900	1450	1160
250	—	—	—	—	1825	1460
300	—	—	—	—	2200	1760
350	—	—	—	—	2550	2040
400	—	—	—	—	2900	2320
450	—	—	—	—	3250	2600
500	—	—	—	—	3625	2900

*Design A motors are not limited to a maximum starting current or locked rotor current.

ARTICLE 440
Air-Conditioning and Refrigerating Equipment

Summary of Changes

- **440.4(B):** Revised to require nameplate information to include the short-circuit rating of motor controller or industrial control panel associated with air-conditioning and refrigerating equipment.

- **440.14 Exception No. 1:** Revised to require written safety procedures where disconnecting means is not within sight of the ac equipment.

- **440.32:** Revised second paragraph to increase the percentage from 58 percent to 72 percent for sizing the circuit conductors of a wye-delta-connected ac motor-compressor.

Contents

440.65 Leakage Current Detection and Interruption
(LCDI) and Arc Fault Circuit Interrupter
(AFCI)

I. General

440.1 Scope

The provisions of this article apply to electric motor-driven
air-conditioning and refrigerating equipment and to the
branch circuits and controllers for such equipment. It provides for the special considerations necessary for circuits
supplying hermetic refrigerant motor-compressors and for
any air-conditioning or refrigerating equipment that is supplied from a branch circuit that supplies a hermetic refrigerant motor-compressor.

440.2 Definitions

Branch-Circuit Selection Current. The value in amperes
to be used instead of the rated-load current in determining
the ratings of motor branch-circuit conductors, disconnecting
means, controllers, and branch-circuit short-circuit and
ground-fault protective devices wherever the running overload protective device permits a sustained current greater
than the specified percentage of the rated-load current. The
value of branch-circuit selection current will always be equal
to or greater than the marked rated-load current.

Hermetic Refrigerant Motor-Compressor. A combination consisting of a compressor and motor, both of which
are enclosed in the same housing, with no external shaft or
shaft seals, the motor operating in the refrigerant.

**Leakage Current Detection and Interruption (LCDI)
Protection.** A device provided in a power supply cord or
cord set that senses leakage current flowing between or
from the cord conductors and interrupts the circuit at a
predetermined level of leakage current.

As indicated by the definition, power supply cords or cord
sets with this type of protection automatically interrupt the
circuit where leakage current between or from the conductors
exceeds a predetermined level. Opening of the circuit is
accomplished through the use of electronic switching or by
"air-break" contacts. The circuit remains open until the
cause of the leakage current is eliminated or the protection
device is manually reset. Leakage current detection and interrupter protection is one of the protection methods for the
power supply cord or cord set of a room air conditioner
specified in 440.65.

Rated-Load Current. The rated-load current for a hermetic
refrigerant motor-compressor is the current resulting when

the motor-compressor is operated at the rated load, rated
voltage, and rated frequency of the equipment it serves.

440.3 Other Articles

(A) Article 430 These provisions are in addition to, or
amendatory of, the provisions of Article 430 and other articles in this *Code*, which apply except as modified in this
article.

(B) Articles 422, 424, or 430 The rules of Articles 422,
424, or 430, as applicable, shall apply to air-conditioning and
refrigerating equipment that does not incorporate a hermetic
refrigerant motor-compressor. This equipment includes devices that employ refrigeration compressors driven by conventional motors, furnaces with air-conditioning evaporator
coils installed, fan-coil units, remote forced air-cooled condensers, remote commercial refrigerators, and so forth.

(C) Article 422 Equipment such as room air conditioners,
household refrigerators and freezers, drinking water coolers,
and beverage dispensers shall be considered appliances, and
the provisions of Article 422 shall also apply.

(D) Other Applicable Articles Hermetic refrigerant
motor-compressors, circuits, controllers, and equipment
shall also comply with the applicable provisions of Table
440.3(D).

Table 440.3(D) Other Articles

Equipment/Occupancy	Article	Section
Capacitors		460.9
Commercial garages, aircraft hangars,	511, 513,	
motor fuel dispensing facilities, bulk	514, 515,	
storage plants, spray application,	516, and 517	
dipping, and coating processes, and	Part IV	
inhalation anesthetizing locations		
Hazardous (classified) locations	500–503 and	
	505	
Motion picture and television studios	530	
and similar locations		
Resistors and reactors	470	

Article 440 provides special considerations necessary for
circuits supplying hermetic refrigerant motor-compressors
and is in addition to or amendatory of the provisions of
Article 430 and other applicable articles. However, many
requirements, such as disconnecting means, controllers, single or group installations, and sizing of conductors, are the
same as or similar to those applied in Article 430.

Article 440 does not apply unless a hermetic refrigerant
motor-compressor is supplied. Article 440 must be applied
in conjunction with Article 430.

Note the terms *rated-load current* and *branch-circuit*

selection current, defined in 440.2. When a branch-circuit selection current is marked on a nameplate, it must be used instead of the rated-load current to determine the size of the disconnecting means, the controller, the motor branch-circuit conductors, and the overcurrent protective devices for the branch-circuit conductors and the motor. The value of branch-circuit selection current is always greater than the marked rated-load current.

440.4 Marking on Hermetic Refrigerant Motor-Compressors and Equipment

(A) Hermetic Refrigerant Motor-Compressor Nameplate A hermetic refrigerant motor-compressor shall be provided with a nameplate that shall indicate the manufacturer's name, trademark, or symbol; identifying designation; phase; voltage; and frequency. The rated-load current in amperes of the motor-compressor shall be marked by the equipment manufacturer on either or both the motor-compressor nameplate and the nameplate of the equipment in which the motor-compressor is used. The locked-rotor current of each single-phase motor-compressor having a rated-load current of more than 9 amperes at 115 volts, or more than 4.5 amperes at 230 volts, and each polyphase motor-compressor shall be marked on the motor-compressor nameplate. Where a thermal protector complying with 440.52(A)(2) and (B)(2) is used, the motor-compressor nameplate or the equipment nameplate shall be marked with the words "thermally protected." Where a protective system complying with 440.52(A)(4) and (B)(4) is used and is furnished with the equipment, the equipment nameplate shall be marked with the words, "thermally protected system." Where a protective system complying with 440.52(A)(4) and (B)(4) is specified, the equipment nameplate shall be appropriately marked.

(B) Multimotor and Combination-Load Equipment Multimotor and combination-load equipment shall be provided with a visible nameplate marked with the maker's name, the rating in volts, frequency and number of phases, minimum supply circuit conductor ampacity, the maximum rating of the branch-circuit short-circuit and ground-fault protective device, and the short-circuit current rating of the motor controllers or industrial control panel. The ampacity shall be calculated by using Part IV and counting all the motors and other loads that will be operated at the same time. The branch-circuit short-circuit and ground-fault protective device rating shall not exceed the value calculated by using Part III. Multimotor or combination-load equipment for use on two or more circuits shall be marked with the above information for each circuit.

To verify compliance with 110.10, a motor controller(s) or a control panel associated with multimotor and combination-

load air-conditioning and refrigeration equipment is now required to be marked with its short-circuit current rating. At large commercial, institutional, and industrial complexes, the controllers and control panels of air-conditioning and refrigeration equipment are often supplied from a point on the electrical distribution system where significant short-circuit current is available.

As is the case with any electrical installation where high levels of short circuit current are available, the short circuit current rating marked on the air-conditioning equipment controllers and control panels enables those responsible for designing and approving the electrical installation with the necessary information to ensure compliance with the requirements of 110.10. Multimotor and combination load air-conditioning equipment used in one- and two-family dwellings, cord-and-attachment plug-connected AC equipment, and AC equipment supplied by branch circuits rated 60 amperes or less are not required to be marked with their short-circuit current rating. A similar requirement for marking the short circuit current rating on equipment has been added to 430.8 for motor controllers.

Exception No. 1: Multimotor and combination-load equipment that is suitable under the provisions of this article for connection to a single 15- or 20-ampere, 120-volt, or a 15-ampere, 208- or 240-volt, single-phase branch circuit shall be permitted to be marked as a single load.

Exception No. 2: The minimum supply circuit conductor ampacity and the maximum rating of the branch-circuit short-circuit and ground-fault protective device shall not be required to be marked on a room air conditioner conforming with 440.62(A).

Exception No. 3: Multimotor and combination-load equipment used in one- and two-family dwellings, cord-and-attachment-plug-connected equipment, or equipment supplied from a branch circuit protected at 60 A or less shall not be required to be marked with a short-circuit current rating.

(C) Branch-Circuit Selection Current A hermetic refrigerant motor-compressor, or equipment containing such a compressor, having a protection system that is approved for use with the motor-compressor that it protects and that permits continuous current in excess of the specified percentage of nameplate rated-load current given in 440.52(B)(2) or (B)(4) shall also be marked with a branch-circuit selection current that complies with 440.52(B)(2) or (B)(4). This marking shall be provided by the equipment manufacturer and shall be on the nameplate(s) where the rated-load current(s) appears.

440.5 Marking on Controllers

A controller shall be marked with the manufacturer's name, trademark, or symbol; identifying designation; voltage;

phase; full-load and locked-rotor current (or horsepower) rating; and such other data as may be needed to properly indicate the motor-compressor for which it is suitable.

440.6 Ampacity and Rating

The size of conductors for equipment covered by this article shall be selected from Tables 310.16 through 310.19 or calculated in accordance with 310.15 as applicable. The required ampacity of conductors and rating of equipment shall be determined according to 440.6(A) and 440.6(B).

(A) Hermetic Refrigerant Motor-Compressor For a hermetic refrigerant motor-compressor, the rated-load current marked on the nameplate of the equipment in which the motor-compressor is employed shall be used in determining the rating or ampacity of the disconnecting means, the branch-circuit conductors, the controller, the branch-circuit short-circuit and ground-fault protection, and the separate motor overload protection. Where no rated-load current is shown on the equipment nameplate, the rated-load current shown on the compressor nameplate shall be used.

Exception No. 1: Where so marked, the branch-circuit selection current shall be used instead of the rated-load current to determine the rating or ampacity of the disconnecting means, the branch-circuit conductors, the controller, and the branch-circuit short-circuit and ground-fault protection.

Exception No. 2: For cord-and-plug-connected equipment, the nameplate marking shall be used in accordance with 440.22(B), Exception No. 2.

> FPN: For disconnecting means and controllers, see 440.12 and 440.41.

(B) Multimotor Equipment For multimotor equipment employing a shaded-pole or permanent split-capacitor-type fan or blower motor, the full-load current for such motor marked on the nameplate of the equipment in which the fan or blower motor is employed shall be used instead of the horsepower rating to determine the ampacity or rating of the disconnecting means, the branch-circuit conductors, the controller, the branch-circuit short-circuit and ground-fault protection, and the separate overload protection. This marking on the equipment nameplate shall not be less than the current marked on the fan or blower motor nameplate.

440.7 Highest Rated (Largest) Motor

In determining compliance with this article and with 430.24, 430.53(B) and 430.53(C), and 430.62(A), the highest rated (largest) motor shall be considered to be the motor that has the highest rated-load current. Where two or more motors have the same highest rated-load current, only one of them shall be considered as the highest rated (largest) motor. For other than hermetic refrigerant motor-compressors, and fan or blower motors as covered in 440.6(B), the full-load cur-

rent used to determine the highest rated motor shall be the equivalent value corresponding to the motor horsepower rating selected from Tables 430.248, 430.249, or 430.250.

Exception: Where so marked, the branch-circuit selection current shall be used instead of the rated-load current in determining the highest rated (largest) motor-compressor.

440.8 Single Machine

An air-conditioning or refrigerating system shall be considered to be a single machine under the provisions of 430.87, Exception, and 430.112, Exception. The motors shall be permitted to be located remotely from each other.

II. Disconnecting Means

440.11 General

The provisions of Part II are intended to require disconnecting means capable of disconnecting air-conditioning and refrigerating equipment, including motor-compressors and controllers from the circuit conductors.

440.12 Rating and Interrupting Capacity

(A) Hermetic Refrigerant Motor-Compressor A disconnecting means serving a hermetic refrigerant motor-compressor shall be selected on the basis of the nameplate rated-load current or branch-circuit selection current, whichever is greater, and locked-rotor current, respectively, of the motor-compressor as follows.

(1) Ampere Rating The ampere rating shall be at least 115 percent of the nameplate rated-load current or branch-circuit selection current, whichever is greater.

Exception: A listed nonfused motor circuit switch having a horsepower rating not less than the equivalent horsepower determined in accordance with 440.12(A)(2) shall be permitted to have an ampere rating less than 115 percent of the specified current.

(2) Equivalent Horsepower To determine the equivalent horsepower in complying with the requirements of 430.109, the horsepower rating shall be selected from Tables 430.248, 430.249, or 430.250 corresponding to the rated-load current or branch-circuit selection current, whichever is greater, and also the horsepower rating from Tables 430.251(A) or 430.251(B) corresponding to the locked-rotor current. In case the nameplate rated-load current or branch-circuit selection current and locked-rotor current do not correspond to the currents shown in Tables 430.248, 430.249, 430.250, 430.251(A), or 430.251(B), the horsepower rating corresponding to the next higher value shall be selected. In case different horsepower ratings are obtained when applying these tables, a horsepower rating at least equal to the larger of the values obtained shall be selected.

(B) Combination Loads Where the combined load of two or more hermetic refrigerant motor-compressors or one or more hermetic refrigerant motor-compressor with other motors or loads may be simultaneous on a single disconnecting means, the rating for the disconnecting means shall be determined in accordance with 440.12(B)(1) and (B)(2).

(1) Horsepower Rating The horsepower rating of the disconnecting means shall be determined from the sum of all currents, including resistance loads, at the rated-load condition and also at the locked-rotor condition. The combined rated-load current and the combined locked-rotor current so obtained shall be considered as a single motor for the purpose of this requirement as follows.

(a) The full-load current equivalent to the horsepower rating of each motor, other than a hermetic refrigerant motor-compressor, and fan or blower motors as covered in 440.6(B) shall be selected from Tables 430.248, 430.249, or 430.250. These full-load currents shall be added to the motor-compressor rated-load current(s) or branch-circuit selection current(s), whichever is greater, and to the rating in amperes of other loads to obtain an equivalent full-load current for the combined load.

(b) The locked-rotor current equivalent to the horsepower rating of each motor, other than a hermetic refrigerant motor-compressor, shall be selected from Tables 430.251(A) or 430.251(B), and, for fan and blower motors of the shaded-pole or permanent split-capacitor type marked with the locked-rotor current, the marked value shall be used. The locked-rotor currents shall be added to the motor-compressor locked-rotor current(s) and to the rating in amperes of other loads to obtain an equivalent locked-rotor current for the combined load. Where two or more motors or other loads such as resistance heaters, or both, cannot be started simultaneously, appropriate combinations of locked-rotor and rated-load current or branch-circuit selection current, whichever is greater, shall be an acceptable means of determining the equivalent locked-rotor current for the simultaneous combined load.

Exception: Where part of the concurrent load is a resistance load and the disconnecting means is a switch rated in horsepower and amperes, the switch used shall be permitted to have a horsepower rating not less than the combined load to the motor-compressor(s) and other motor(s) at the locked-rotor condition, if the ampere rating of the switch is not less than this locked-rotor load plus the resistance load.

(2) Full-Load Current Equivalent The ampere rating of the disconnecting means shall be at least 115 percent of the sum of all currents at the rated-load condition determined in accordance with 440.12(B)(1).

Exception: A listed nonfused motor circuit switch having a horsepower rating not less than the equivalent horsepower

determined by 440.12(B)(1) shall be permitted to have an ampere rating less than 115 percent of the sum of all currents.

(C) Small Motor-Compressors For small motor-compressors not having the locked-rotor current marked on the nameplate, or for small motors not covered by Tables 430.247, 430.248, 430.249, or 430.250, the locked-rotor current shall be assumed to be six times the rated-load current.

(D) Disconnecting Means Every disconnecting means in the refrigerant motor-compressor circuit between the point of attachment to the feeder and the point of connection to the refrigerant motor-compressor shall comply with the requirements of 440.12.

(E) Disconnecting Means Rated in Excess of 100 Horsepower Where the rated-load or locked-rotor current as determined above would indicate a disconnecting means rated in excess of 100 hp, the provisions of 430.109(E) shall apply.

440.13 Cord-Connected Equipment

For cord-connected equipment such as room air conditioners, household refrigerators and freezers, drinking water coolers, and beverage dispensers, a separable connector or an attachment plug and receptacle shall be permitted to serve as the disconnecting means.

FPN: For room air conditioners, see 440.63.

440.14 Location

Disconnecting means shall be located within sight from and readily accessible from the air-conditioning or refrigerating equipment. The disconnecting means shall be permitted to be installed on or within the air-conditioning or refrigerating equipment.

The disconnecting means shall not be located on panels that are designed to allow access to the air-conditioning or refrigeration equipment.

Exception No. 1: Where the disconnecting means provided in accordance with 430.102(A) is capable of being locked in the open position, and the refrigerating or air-conditioning equipment is essential to an industrial process in a facility with written safety procedures, and where the conditions of maintenance and supervision ensure that only qualified persons service the equipment, a disconnecting means within sight from the equipment shall not be required. The provision for locking or adding a lock to the disconnecting means shall be permanently installed on or at the switch or circuit breaker used as the disconnecting means.

Exception No. 1 accommodates special conditions associated with process refrigeration equipment. Typically, this equipment is very large, so rated disconnects may not be available. Additionally, this equipment may be in hazardous locations, and locating disconnecting means within sight of the motor may introduce additional hazards. The provision for locking or attaching a lock to the disconnecting means must be part of the disconnect and a permanent component of the switch or circuit breaker. The term *permanent component* is used to preclude portable or transferrable-type lockout devices from being used as the method to provide the ability to lock the switch or circuit breaker in the open (off) position. An example of "permanent-type" locking hardware is shown in Exhibit 440.1.

Exhibit 440.1 Four circuit breakers equipped with locking hardware that is not readily removable or transferable. (Courtesy of International Association of Electrical Inspectors)

Exception No. 2: Where an attachment plug and receptacle serve as the disconnecting means in accordance with 440.13, their location shall be accessible but shall not be required to be readily accessible.

　　FPN:　See Parts VII and IX of Article 430 for additional requirements.

The references to Parts VII and IX of Article 430 in the fine print note are intended to call attention to the additional

disconnect location requirements in 430.102, 430.107, and 430.113. Because 440.3(A) makes the requirements in Article 440 in addition to or amendatory of the provisions of Article 430, the requirement of 440.14 mandates that the equipment disconnecting means be within sight from and readily accessible from the equipment, even if there is also a remote disconnect capable of being locked in the open position under the provision of 430.102(B), Exception.

This special requirement for air-conditioning and refrigeration equipment covered by Article 440 is more stringent than the provisions in Article 430 to provide protection for service personnel working on equipment located in attics, on roofs, or outside in a remote location where it is difficult to gain access to a remote lockable disconnect. See 440.14, Exception No. 1.

III. Branch-Circuit Short-Circuit and Ground-Fault Protection

440.21 General

The provisions of Part III specify devices intended to protect the branch-circuit conductors, control apparatus, and motors in circuits supplying hermetic refrigerant motor-compressors against overcurrent due to short circuits and grounds. They are in addition to or amendatory of the provisions of Article 240.

Where an air conditioner is listed by a qualified electrical testing laboratory with a nameplate that reads "maximum fuse size," the listing restricts the use of this unit to fuse protection only and does not cover its use with circuit breakers. If the air conditioner has been evaluated for both fuses and ordinary circuit breakers or both fuses and HACR-type circuit breakers, it may be so marked. UL-listed circuit breakers that have been found suitable for use with heating, air-conditioning, and refrigeration equipment comprising multimotor or combination loads are marked "Listed HACR Type." It is the intent of 110.3(B) to require that the manufacturer's installation specifications be closely followed and that any restriction of the listing be applied to the installation of the equipment, in order to comply with the *Code*.

The UL *Electrical Appliance and Utilization Equipment Directory* states the following under Air-Conditioners, Central Cooling (ACAV): "This marked protective device rating is the maximum for which the equipment has been investigated and found acceptable. Where the marking specifies fuses, or 'HACR Type' circuit breakers, the circuit is intended to be protected only by the type of protective device specified." Exhibit 440.2 illustrates three wiring configurations where the equipment is under fuse protection only.

The UL *Electrical Appliance and Utilization Equipment Directory* further indicates that "in units employing two or more motors or a motor(s) and other loads operating from a single supply circuit, the motor overload protective devices (including thermal protectors for motors) and other factory-installed motor circuit components and wiring are investigated on the basis of compliance with the motor branch-circuit short-circuit and ground-fault protection requirements of 430.53(C), Other Group Installations. Such multimotor and combination load equipment is to be connected only to a circuit protected by fuses or a circuit breaker with a rating that does not exceed the value marked on the data plate of the equipment.''

Current-limiting overcurrent devices, which may reduce the amount of fault current to which the equipment is subjected, can be installed in the branch circuit supplying the equipment. See 240.2 for the definition of *current-limiting overcurrent protective device* and commentary explaining short-circuit damage. This is important particularly for larger commercial and industrial installations. See also 110.3(B) and 110.10 and associated commentary regarding the installation and use of listed or labeled equipment and the selection of overcurrent protective devices (such as fuses and circuit breakers).

440.22 Application and Selection

(A) Rating or Setting for Individual Motor-Compressor

The motor-compressor branch-circuit short-circuit and ground-fault protective device shall be capable of carrying the starting current of the motor. A protective device having a rating or setting not exceeding 175 percent of the motor-compressor rated-load current or branch-circuit selection current, whichever is greater, shall be permitted, provided that, where the protection specified is not sufficient for the starting current of the motor, the rating or setting shall be permitted to be increased but shall not exceed 225 percent of the motor rated-load current or branch-circuit selection current, whichever is greater.

Exception: The rating of the branch-circuit short-circuit and ground-fault protective device shall not be required to be less than 15 amperes.

(B) Rating or Setting for Equipment

The equipment branch-circuit short-circuit and ground-fault protective device shall be capable of carrying the starting current of the equipment. Where the hermetic refrigerant motor-compressor is the only load on the circuit, the protection shall conform with 440.22(A). Where the equipment incorporates more than one hermetic refrigerant motor-compressor or a hermetic refrigerant motor-compressor and other motors or other loads, the equipment short-circuit and ground-fault protection shall conform with 430.53 and 440.22(B)(1) and (B)(2).

(1) Motor-Compressor Largest Load Where a hermetic refrigerant motor-compressor is the largest load connected to the circuit, the rating or setting of the branch-circuit short-circuit and ground-fault protective device shall not exceed the value specified in 440.22(A) for the largest motor-compressor plus the sum of the rated-load current or branch-circuit selection current, whichever is greater, of the other motor-compressor(s) and the ratings of the other loads supplied.

(2) Motor-Compressor Not Largest Load Where a hermetic refrigerant motor-compressor is not the largest load connected to the circuit, the rating or setting of the branch-circuit short-circuit and ground-fault protective device shall not exceed a value equal to the sum of the rated-load current

Exhibit 440.2 Three correct alternate wiring configurations satisfying a nameplate that specifies fuses, thus restricting the equipment to protection by fuses only. (Note that the fuse rating cannot exceed the maximum fuse size specified on the air-conditioner nameplate.)

or branch-circuit selection current, whichever is greater, rating(s) for the motor-compressor(s) plus the value specified in 430.53(C)(4) where other motor loads are supplied, or the value specified in 240.4 where only nonmotor loads are supplied in addition to the motor-compressor(s).

Exception No. 1: Equipment that starts and operates on a 15- or 20-ampere 120-volt, or 15-ampere 208- or 240-volt single-phase branch circuit, shall be permitted to be protected by the 15- or 20-ampere overcurrent device protecting the branch circuit, but if the maximum branch-circuit short-circuit and ground-fault protective device rating marked on the equipment is less than these values, the circuit protective device shall not exceed the value marked on the equipment nameplate.

Exception No. 2: The nameplate marking of cord-and-plug-connected equipment rated not greater than 250 volts, single-phase, such as household refrigerators and freezers, drinking water coolers, and beverage dispensers, shall be used in determining the branch-circuit requirements, and each unit shall be considered as a single motor unless the nameplate is marked otherwise.

(C) Protective Device Rating Not to Exceed the Manufacturer's Values Where maximum protective device ratings shown on a manufacturer's overload relay table for use with a motor controller are less than the rating or setting selected in accordance with 440.22(A) and 440.22(B), the protective device rating shall not exceed the manufacturer's values marked on the equipment.

IV. Branch-Circuit Conductors

440.31 General

The provisions of Part IV and Article 310 specify ampacities of conductors required to carry the motor current without overheating under the conditions specified, except as modified in 440.6(A), Exception No. 1.

The provisions of these articles shall not apply to integral conductors of motors, motor controllers and the like, or to conductors that form an integral part of approved equipment.

FPN: See 300.1(B) and 310.1 for similar requirements.

440.32 Single Motor-Compressor

Branch-circuit conductors supplying a single motor-compressor shall have an ampacity not less than 125 percent of either the motor-compressor rated-load current or the branch-circuit selection current, whichever is greater.

For a wye-start, delta-run connected motor-compressor, the selection of branch-circuit conductors between the controller and the motor-compressor shall be permitted to be based on 72 percent of either the motor-compressor rated-load current or the branch-circuit selection current, whichever is greater.

The use of wye-start, delta-run hermetic motor-compressors is a means of reducing starting current in large air-conditioning equipment. Starting the motor using a wye-configured winding results in a phase voltage that is 1/1.732, or 58 percent, of the line voltage. This reduction in phase voltage results in a directly proportional reduction in phase current. In the starting configuration, the phase current and the line current are equal.

The second paragraph of 440.32 permits the two sets of conductors between the controller and the compressor motor to be sized at 72 percent of the larger of either the motor-compressor rated-load current or the branch circuit selection current. The conductors on the line side of the controller are sized at not less than 125 percent of either the motor-compressor rated-load current or the branch circuit selection current, whichever is larger. For more information on wye-start, delta-run motors, see the commentary following 430.22(C).

FPN: The individual motor circuit conductors of wye-start, delta-run connected motor-compressors carry 58 percent of the rated load current. The multiplier of 72 percent is obtained by multiplying 58 percent by 1.25.

440.33 Motor-Compressor(s) With or Without Additional Motor Loads

Conductors supplying one or more motor-compressor(s) with or without an additional load(s) shall have an ampacity not less than the sum of the rated-load or branch-circuit selection current ratings, whichever is larger, of all the motor-compressors plus the full-load currents of the other motors, plus 25 percent of the highest motor or motor-compressor rating in the group.

Exception No. 1: Where the circuitry is interlocked so as to prevent the starting and running of a second motor-compressor or group of motor-compressors, the conductor size shall be determined from the largest motor-compressor or group of motor-compressors that is to be operated at a given time.

Exception No. 2: The branch circuit conductors for room air conditioners shall be in accordance with Part VII of Article 440.

Branch circuits for listed air-conditioning and refrigeration equipment that have a nameplate marked with the branch-circuit conductor size and branch-circuit short-circuit protective device size are not required to have the branch-circuit conductors sized in accordance with 440.33. The testing laboratory standard includes the 25 percent increase for the largest motor or compressor in the group plus the other nonmotor or noncompressor load; therefore, the actual name-

440.34 Combination Load

Conductors supplying a motor-compressor load in addition to a lighting or appliance load as calculated from Article 220 and other applicable articles shall have an ampacity sufficient for the lighting or appliance load plus the required ampacity for the motor-compressor load determined in accordance with 440.33 or, for a single motor-compressor, in accordance with 440.32.

Exception: Where the circuitry is interlocked so as to prevent simultaneous operation of the motor-compressor(s) and all other loads connected, the conductor size shall be determined from the largest size required for the motor-compressor(s) and other loads to be operated at a given time.

440.35 Multimotor and Combination-Load Equipment

The ampacity of the conductors supplying multimotor and combination-load equipment shall not be less than the minimum circuit ampacity marked on the equipment in accordance with 440.4(B).

V. Controllers for Motor-Compressors

440.41 Rating

(A) Motor-Compressor Controller A motor-compressor controller shall have both a continuous-duty full-load current rating and a locked-rotor current rating not less than the nameplate rated-load current or branch-circuit selection current, whichever is greater, and locked-rotor current, respectively, of the compressor. In case the motor controller is rated in horsepower but is without one or both of the foregoing current ratings, equivalent currents shall be determined from the ratings as follows. Tables 430.248, 430.249, and 430.250 shall be used to determine the equivalent full-load current rating. Tables 430.251(A) and 430.252(B) shall be used to determine the equivalent locked-rotor current ratings.

(B) Controller Serving More Than One Load A controller serving more than one motor-compressor or a motor-compressor and other loads shall have a continuous-duty full-load current rating and a locked-rotor current rating not less than the combined load as determined in accordance with 440.12(B).

VI. Motor-Compressor and Branch-Circuit Overload Protection

440.51 General

The provisions of Part VI specify devices intended to protect the motor-compressor, the motor-control apparatus, and the branch-circuit conductors against excessive heating due to motor overload and failure to start.

> FPN: See 240.4(G) for application of Parts III and VI of Article 440.

440.52 Application and Selection

(A) Protection of Motor-Compressor Each motor-compressor shall be protected against overload and failure to start by one of the following means:

(1) A separate overload relay that is responsive to motor-compressor current. This device shall be selected to trip at not more than 140 percent of the motor-compressor rated-load current.

(2) A thermal protector integral with the motor-compressor, approved for use with the motor-compressor that it protects on the basis that it will prevent dangerous overheating of the motor-compressor due to overload and failure to start. If the current-interrupting device is separate from the motor-compressor and its control circuit is operated by a protective device integral with the motor-compressor, it shall be arranged so that the opening of the control circuit will result in interruption of current to the motor-compressor.

(3) A fuse or inverse time circuit breaker responsive to motor current, which shall also be permitted to serve as the branch-circuit short-circuit and ground-fault protective device. This device shall be rated at not more than 125 percent of the motor-compressor rated-load current. It shall have sufficient time delay to permit the motor-compressor to start and accelerate its load. The equipment or the motor-compressor shall be marked with this maximum branch-circuit fuse or inverse time circuit breaker rating.

(4) A protective system, furnished or specified and approved for use with the motor-compressor that it protects on the basis that it will prevent dangerous overheating of the motor-compressor due to overload and failure to start. If the current-interrupting device is separate from the motor-compressor and its control circuit is operated by a protective device that is not integral with the current-interrupting device, it shall be arranged so that the opening of the control circuit will result in interruption of current to the motor-compressor.

(B) Protection of Motor-Compressor Control Apparatus and Branch-Circuit Conductors The motor-compressor controller(s), the disconnecting means, and the branch-circuit conductors shall be protected against overcurrent due to motor overload and failure to start by one of the following means, which shall be permitted to be the same device or system protecting the motor-compressor in accordance with 440.52(A):

Exception: Overload protection of motor-compressors and equipment on 15- and 20-ampere, single-phase, branch circuits shall be permitted to be in accordance with 440.54 and 440.55.

(1) An overload relay selected in accordance with 440.52(A)(1)

(2) A thermal protector applied in accordance with 440.52(A)(2), that will not permit a continuous current in excess of 156 percent of the marked rated-load current or branch-circuit selection current

(3) A fuse or inverse time circuit breaker selected in accordance with 440.52(A)(3)

(4) A protective system, in accordance with 440.52(A)(4), that will not permit a continuous current in excess of 156 percent of the marked rated-load current or branch-circuit selection current

440.53 Overload Relays

Overload relays and other devices for motor overload protection that are not capable of opening short circuits shall be protected by fuses or inverse time circuit breakers with ratings or settings in accordance with Part III unless approved for group installation or for part-winding motors and marked to indicate the maximum size of fuse or inverse time circuit breaker by which they shall be protected.

Exception: The fuse or inverse time circuit breaker size marking shall be permitted on the nameplate of approved equipment in which the overload relay or other overload device is used.

440.54 Motor-Compressors and Equipment on 15- or 20-Ampere Branch Circuits — Not Cord-and-Attachment-Plug-Connected

Overload protection for motor-compressors and equipment used on 15- or 20-ampere 120-volt, or 15-ampere 208- or 240-volt single-phase branch circuits as permitted in Article 210 shall be permitted as indicated in 440.54(A) and 440.54(B).

(A) Overload Protection The motor-compressor shall be provided with overload protection selected as specified in 440.52(A). Both the controller and motor overload protective device shall be approved for installation with the short-circuit and ground-fault protective device for the branch circuit to which the equipment is connected.

(B) Time Delay The short-circuit and ground-fault protective device protecting the branch circuit shall have sufficient time delay to permit the motor-compressor and other motors to start and accelerate their loads.

440.55 Cord-and-Attachment-Plug-Connected Motor-Compressors and Equipment on 15- or 20-Ampere Branch Circuits

Overload protection for motor-compressors and equipment that are cord-and-attachment-plug-connected and used on 15- or 20-ampere 120-volt, or 15-ampere 208- or 240-volt, single-phase branch circuits as permitted in Article 210 shall be permitted as indicated in 440.55(A), (B), and (C).

(A) Overload Protection The motor-compressor shall be provided with overload protection as specified in 440.52(A). Both the controller and the motor overload protective device shall be approved for installation with the short-circuit and ground-fault protective device for the branch circuit to which the equipment is connected.

(B) Attachment Plug and Receptacle Rating The rating of the attachment plug and receptacle shall not exceed 20 amperes at 125 volts or 15 amperes at 250 volts.

(C) Time Delay The short-circuit and ground-fault protective device protecting the branch circuit shall have sufficient time delay to permit the motor-compressor and other motors to start and accelerate their loads.

VII. Provisions for Room Air Conditioners

440.60 General

The provisions of Part VII shall apply to electrically energized room air conditioners that control temperature and humidity. For the purpose of Part VII, a room air conditioner (with or without provisions for heating) shall be considered as an ac appliance of the air-cooled window, console, or in-wall type that is installed in the conditioned room and that incorporates a hermetic refrigerant motor-compressor(s). The provisions of Part VII cover equipment rated not over 250 volts, single phase, and such equipment shall be permitted to be cord-and-attachment-plug-connected.

A room air conditioner that is rated three phase or rated over 250 volts shall be directly connected to a wiring method recognized in Chapter 3, and provisions of Part VII shall not apply.

440.61 Grounding

Room air conditioners shall be grounded in accordance with 250.110, 250.112, and 250.114.

440.62 Branch-Circuit Requirements

(A) Room Air Conditioner as a Single Motor Unit A room air conditioner shall be considered as a single motor unit in determining its branch-circuit requirements where all the following conditions are met:

(1) It is cord-and-attachment-plug-connected.

(2) Its rating is not more than 40 amperes and 250 volts, single phase.

(3) Total rated-load current is shown on the room air-conditioner nameplate rather than individual motor currents.

(4) The rating of the branch-circuit short-circuit and ground-fault protective device does not exceed the ampacity of the branch-circuit conductors or the rating of the receptacle, whichever is less.

(B) Where No Other Loads Are Supplied The total marked rating of a cord-and-attachment-plug-connected room air conditioner shall not exceed 80 percent of the rating of a branch circuit where no other loads are supplied.

(C) Where Lighting Units or Other Appliances Are Also Supplied The total marked rating of a cord-and-attachment-plug-connected room air conditioner shall not exceed 50 percent of the rating of a branch circuit where lighting outlets, other appliances, or general-use receptacles are also supplied. Where the circuitry is interlocked to prevent simultaneous operation of the room air conditioner and energization of other outlets on the same branch circuit, a cord-and-attachment-plug-connected room air conditioner shall not exceed 80 percent of the branch-circuit rating.

440.63 Disconnecting Means

An attachment plug and receptacle shall be permitted to serve as the disconnecting means for a single-phase room air conditioner rated 250 volts or less if (1) the manual controls on the room air conditioner are readily accessible and located within 1.8 m (6 ft) of the floor or (2) an approved manually operable disconnecting means is installed in a readily accessible location within sight from the room air conditioner.

440.64 Supply Cords

Where a flexible cord is used to supply a room air conditioner, the length of such cord shall not exceed 3.0 m (10 ft) for a nominal, 120-volt rating or 1.8 m (6 ft) for a nominal, 208- or 240-volt rating.

440.65 Leakage Current Detection and Interruption (LCDI) and Arc Fault Circuit Interrupter (AFCI)

Single-phase cord-and-plug-connected room air conditioners shall be provided with factory-installed LCDI or AFCI protection. The LCDI or AFCI protection shall be an integral part of the attachment plug or be located in the power supply cord within 300 mm (12 in.) of the attachment plug.

Generally, portable room air conditioners are used only on a seasonal basis and are removed and stored at the end of the cooling season. During the life of a room air conditioner, the installation and removal occurs many times, and there is an increased likelihood of a damaged cord as a result of the unit being set on the cord or pushed against it. To provide enhanced protection against fires initiated by damaged supply cords, all single-phase cord-and-plug-connected room air conditioners are required to be equipped with either leakage current detection and interruption protection or arc-fault circuit interrupter protection. The protective device is required to be installed in the appliance cord and must be within 12 in. of the attachment plug.

ARTICLE 445
Generators

Summary of Changes

- **445.11:** Revised to require that the generator marking include the subtransient and transient impedances.
- **445.18:** Revised to allow multiple disconnecting means as the method to isolate the generator from the loads supplied.

Contents

445.1 Scope

This article covers the installation of generators.

Generators are generally associated with fire pumps (see Article 695), emergency systems (see Article 700), legally required standby systems (see Article 701), optional standby

445.10 Location

Generators shall be of a type suitable for the locations in which they are installed. They shall also meet the requirements for motors in 430.14.

445.11 Marking

Each generator shall be provided with a nameplate giving the manufacturer's name, the rated frequency, power factor, number of phases if of alternating current, the subtransient and transient impedances, the rating in kilowatts or kilovolt amperes, the normal volts and amperes corresponding to the rating, rated revolutions per minute, insulation system class and rated ambient temperature or rated temperature rise, and time rating.

445.12 Overcurrent Protection

(A) Constant-Voltage Generators Constant-voltage generators, except ac generator exciters, shall be protected from overloads by inherent design, circuit breakers, fuses, or other acceptable overcurrent protective means suitable for the conditions of use.

(B) Two-Wire Generators Two-wire, dc generators shall be permitted to have overcurrent protection in one conductor only if the overcurrent device is actuated by the entire current generated other than the current in the shunt field. The overcurrent device shall not open the shunt field.

(C) 65 Volts or Less Generators operating at 65 volts or less and driven by individual motors shall be considered as protected by the overcurrent device protecting the motor if these devices will operate when the generators are delivering not more than 150 percent of their full-load rated current.

(D) Balancer Sets Two-wire, dc generators used in conjunction with balancer sets to obtain neutrals for 3-wire systems shall be equipped with overcurrent devices that disconnect the 3-wire system in case of excessive unbalancing of voltages or currents.

(E) Three-Wire, Direct-Current Generators Three-wire, dc generators, whether compound or shunt wound, shall be equipped with overcurrent devices, one in each armature lead, and connected so as to be actuated by the entire current from the armature. Such overcurrent devices shall consist either of a double-pole, double-coil circuit breaker or of a 4-pole circuit breaker connected in the main and equalizer leads and tripped by two overcurrent devices, one in each

armature lead. Such protective devices shall be interlocked so that no one pole can be opened without simultaneously disconnecting both leads of the armature from the system.

Exception to (A) through (E): Where deemed by the authority having jurisdiction, a generator is vital to the operation of an electrical system and the generator should operate to failure to prevent a greater hazard to persons. The overload sensing device(s) shall be permitted to be connected to an annunciator or alarm supervised by authorized personnel instead of interrupting the generator circuit.

Alternating-current generators can be designed so that during short periods of time, when the generator may carry an excessive overload, the voltage will fall off sufficiently to limit the current and power output to values that will not damage the generator.

The connection of a 2-wire generator, protected by a single-pole circuit breaker, is illustrated in Exhibit 445.1. Where two or more dc generators are operated in parallel or in multiples, an equalizer conductor lead is connected to the positive terminal of each generator and, in effect, connects the series fields in parallel so as to maintain equal output voltage for each generator. The current could divide at the positive terminal, some flowing through the series field and positive lead and some flowing through the equalizer lead. The entire current generated flows through the negative lead; therefore, the fuse or circuit breaker (or at least the operating coil of the circuit breaker) must be placed in the negative lead. Overcurrent devices must be connected so as to be actuated by the entire armature output current.

Exhibit 445.1 A 2-wire dc generator protected by a single-pole circuit breaker.

An overcurrent device should not be placed in the shunt field circuit because, if the circuit were to open when the field was at full strength, an extremely high voltage would be induced that could damage the field winding insulation and the generator.

Section 445.12(C) indicates that generators operating at 65 volts or less are to be thought of as protected by the overcurrent devices that also protect the drive motor, provided these devices operate when the generator delivers 150 percent of its full-load rated current.

Exhibit 445.2 illustrates a two-pole circuit breaker with one pole connected in each lead of the main generator and with the operating coil properly designed to be connected in the neutral lead from the balancer, and arranged so as to be operated by either of the A coils or by the B coil. Each of the two generators used as a balancer set carries approximately half the unbalanced load and, thus, is always smaller than the main generator. During an excessive imbalance of the load, the balancer set would be overloaded, with no overload on the main generator; hence, a double-pole circuit breaker is connected (as noted) to guard against this condition.

Exhibit 445.2 A two-pole circuit breaker (one pole connected in each lead of the main generator) with the opening coil connected in the neutral of the balancer set.

Note that the authority having jurisdiction may judge that having the generator operate to failure is preferable to providing automatic means to shut it down, which, in many cases, could present a greater hazard to personnel. An overload sensing device(s) would be permitted to be connected to an annunciator or an alarm (instead of interrupting the generator) and allow operating personnel to shut down load-side equipment in a safe and orderly fashion.

445.13 Ampacity of Conductors

The ampacity of the conductors from the generator terminals to the first distribution device(s) containing overcurrent protection shall not be less than 115 percent of the nameplate current rating of the generator. It shall be permitted to size the neutral conductors in accordance with 220.61. Conductors that must carry ground-fault currents shall not be smaller than required by 250.24(C). Neutral conductors of dc generators that must carry ground-fault currents shall not be smaller than the minimum required size of the largest conductor.

Exception: Where the design and operation of the generator prevent overloading, the ampacity of the conductors shall not be less than 100 percent of the nameplate current rating of the generator.

445.14 Protection of Live Parts

Live parts of generators operated at more than 50 volts to ground shall not be exposed to accidental contact where accessible to unqualified persons.

445.15 Guards for Attendants

Where necessary for the safety of attendants, the requirements of 430.233 shall apply.

445.16 Bushings

Where wires pass through an opening in an enclosure, conduit box, or barrier, a bushing shall be used to protect the conductors from the edges of an opening having sharp edges. The bushing shall have smooth, well-rounded surfaces where it may be in contact with the conductors. If used where oils, grease, or other contaminants may be present, the bushing shall be made of a material not deleteriously affected.

445.17 Generator Terminal Housings

Generator terminal housings shall comply with 430.12. Where a horsepower rating is required to determine the required minimum size of the generator terminal housing, the full-load current of the generator shall be compared with comparable motors in Tables 430.247 through 430.250. The higher horsepower rating of Tables 430.247 and 430.250 shall be used whenever the generator selection is between two ratings.

445.18 Disconnecting Means Required for Generators

Generators shall be equipped with disconnect(s) by means of which the generator and all protective devices and control apparatus are able to be disconnected entirely from the circuits supplied by the generator except where both of the following conditions apply:

(1) The driving means for the generator can be readily shut down.
(2) The generator is not arranged to operate in parallel with another generator or other source of voltage.

Added to the 1999 *Code*, 445.18 requires that generators be equipped with a disconnect switch or circuit breaker unless the prime mover can be readily shut down and the generator is not operating in parallel with another generator or source of power.

Revised for the 2005 *Code*, 445.18 now permits multiple means to be used as the method of isolating the generator from the loads that it supplies. Article 700 permits the use of a single generator to supply emergency, legally required standby (Article 701) and optional standby (Article 702) loads. In such arrangements, there may be more than one feeder run from the generator to the distribution and transfer equipment. To allow such arrangements, this revision now permits multiple devices as the method to provide the required disconnecting means.

ARTICLE 450
Transformers and Transformer Vaults (Including Secondary Ties)

Summary of Changes

- **450.5:** Revised to prohibit placing zig-zag connections where they would be compromised by other line-to-ground fault current sources on the premises.

- **450.6(C):** Added new grounding requirement for transformers that supply secondary tie systems.

Contents

(E) Dampers
(F) Ducts
450.46 Drainage
450.47 Water Pipes and Accessories
450.48 Storage in Vaults

450.1 Scope

This article covers the installation of all transformers.

Exception No. 1: Current transformers.

See 110.23 for the requirement on energized current transformers that are not in use.

Exception No. 2: Dry-type transformers that constitute a component part of other apparatus and comply with the requirements for such apparatus.

Exception No. 3: Transformers that are an integral part of an X-ray, high-frequency, or electrostatic-coating apparatus.

Exception No. 4: Transformers used with Class 2 and Class 3 circuits that comply with Article 725.

Exception No. 5: Transformers for sign and outline lighting that comply with Article 600.

Exception No. 6: Transformers for electric-discharge lighting that comply with Article 410.

Exception No. 7: Transformers used for power-limited fire alarm circuits that comply with Part III of Article 760.

Exception No. 8: Transformers used for research, development, or testing, where effective arrangements are provided to safeguard persons from contacting energized parts.

This article covers the installation of transformers dedicated to supplying power to a fire pump installation as modified by Article 695.

This article also covers the installation of transformers in hazardous (classified) locations as modified by Articles 501 through 504.

I. General Provisions

450.2 Definition

For the purpose of this article, the following definition shall apply.

Transformer. An individual transformer, single- or polyphase, identified by a single nameplate, unless otherwise indicated in this article.

450.3 Overcurrent Protection

Overcurrent protection of transformers shall comply with 450.3(A), (B), or (C). As used in this section, the word

transformer shall mean a transformer or polyphase bank of two or more single-phase transformers operating as a unit.

Section 450.3 was reorganized for the 1999 *Code* to present transformer protection requirements in a simpler and more user-friendly table format. The fundamental protection requirements were not changed; only the presentation of the requirements is different. Instead of 11 paragraphs, 10 exceptions, and 2 tables, 450.3(A) and 450.3(B) use two basic tables with notes and one exception to provide overcurrent protection requirements for transformers. Notes to a table are part of the requirements of the table, unlike the information contained in fine print notes (FPNs), which per 90.5(C) are explanatory in nature and not enforceable.

FPN No. 1: See 240.4, 240.21, 240.100, and 240.101 for overcurrent protection of conductors.

The requirements for protection of transformer secondaries in Article 450 apply only to the protection of transformers, not to the protection of conductors. Article 240 applies only to the protection of conductors, not to the protection of transformers. It is possible that the overcurrent protection required by Article 450 also satisfies the requirements in Article 240 for protection of the conductors, and vice versa, but it is also possible that they do not.

FPN No. 2: Nonlinear loads can increase heat in a transformer without operating its overcurrent protective device.

The increased heating effects of nonlinear load currents must be taken into account when determining the load on a transformer. There are several methods for dealing with the heating effects of nonlinear loads, including derating equipment, oversizing equipment, increasing insulation ratings, installing thermal protection systems, and using K-factor transformers. The optimum method for dealing with transformer overheating will vary, depending on several technical and economic factors, and should be considered during the design phase of the electrical system.

(A) Transformers Over 600 Volts, Nominal Overcurrent protection shall be provided in accordance with Table 450.3(A).

For Note 1 of Table 450.3(A), concerning standard ratings of circuit breakers and fuses, see 240.6.

For Note 2 of the table, overcurrent protection to protect the secondary of a transformer is allowed to consist of not more than six sets of fuses or six circuit breakers, under the following conditions:

Table 450.3(A) Maximum Rating or Setting of Overcurrent Protection for Transformers Over 600 Volts (as a Percentage of Transformer-Rated Current)

| Location Limitations | Transformer Rated Impedance | Primary Protection Over 600 Volts | | Secondary Protection (See Note 2.) | | |
| | | | | Over 600 Volts | | 600 Volts or Less |
		Circuit Breaker (See Note 4.)	Fuse Rating	Circuit Breaker (See Note 4.)	Fuse Rating	Circuit Breaker or Fuse Rating
Any Location	Not more than 6%	600% (See Note 1.)	300% (See Note 1.)	300% (See Note 1.)	250% (See Note 1.)	125% (See Note 1.)
	More than 6% and not more than 10%	400% (See Note 1.)	300% (See Note 1.)	250% (See Note 1.)	225% (See Note 1.)	125% (See Note 1.)
Supervised locations only (See Note 3.)	Any	300% (See Note 1.)	250% (See Note 1.)	Not required	Not required	Not required
	Not more than 6%	600%	300%	300% (See Note 5.)	250% (See Note 5.)	250% (See Note 5.)
	More than 6% and not more than 10%	400%	300%	250% (See Note 5.)	225% (See Note 5.)	250% (See Note 5.)

Notes:
1. Where the required fuse rating or circuit breaker setting does not correspond to a standard rating or setting, a higher rating or setting that does not exceed the next higher standard rating or setting shall be permitted.
2. Where secondary overcurrent protection is required, the secondary overcurrent device shall be permitted to consist of not more than six circuit breakers or six sets of fuses grouped in one location. Where multiple overcurrent devices are utilized, the total of all the device ratings shall not exceed the allowed value of a single overcurrent device. If both circuit breakers and fuses are used as the overcurrent device, the total of the device ratings shall not exceed that allowed for fuses.
3. A supervised location is a location where conditions of maintenance and supervision ensure that only qualified persons monitor and service the transformer installation.
4. Electronically actuated fuses that may be set to open at a specific current shall be set in accordance with settings for circuit breakers.
5. A transformer equipped with a coordinated thermal overload protection by the manufacturer shall be permitted to have separate secondary protection omitted.

1. They must be grouped in one location.
2. The sum of the overcurrent devices must not exceed the maximum value of a single device.
3. Where a combination of fuses and circuit breakers is used, the maximum value will be that of a single set of fuses.

See Exhibits 450.1 and 450.2 for an application of Note 2. This note also appears in Table 450.3(B) and applies to transformers rated 600 volts and less as well.

For Note 3 of Table 450.3(A), a supervised location is where maintenance of the equipment is performed by personnel that have received safety training and are familiar with proper operation of the equipment and aware of the hazards associated with it.

For Note 4 of the table, *electronically actuated fuse* is defined in Article 100, Part II, Over 600 Volts, Nominal.

The ratings or settings obtained from Table 450.3(A) are based on the type of protective device (fuse, electronic fuse, or circuit breaker), transformer rated current and impedance, and primary and secondary voltages. According to Table 450.3(A), the maximum ratings or settings of an overcurrent protective device for transformers rated over 600 volts is separated into two broad categories: *any location* (or unsupervised) and *supervised locations only.*

The first category for over 600-volt transformers is not limited by location and is referred to as *any location*. The maximum ratings or settings for overcurrent devices permitted in the *any location* row are applicable for all unsupervised locations. An *any location* transformer installation must be

Exhibit 450.1 A transformer rated over 600 volts with a secondary rated over 600 volts, with secondary protection consisting of six circuit breakers. The sum of the ratings of the circuit breakers is not permitted to exceed 300 percent of the rated secondary current.

Exhibit 450.2 A transformer rated over 600 volts with a secondary rated over 600 volts, with secondary protection consisting of fuses and circuit breakers. The sum of the ratings of all the overcurrent devices is not permitted to exceed the rating permitted for fuses.

provided with overcurrent protection in both the primary and secondary circuit. Of course, the user may select the *any location* row even if the location qualifies for the *supervised locations only* category, because the requirements for the *any location* row are the same as or exceed the *supervised*

Exhibit 450.3 A transformer with 6 percent impedance and rated over 600 volts using circuit-breaker protection for both the primary and the secondary. For the transformer in this exhibit, both the primary and the secondary voltages are over 600 volts.

locations only row. See Exhibit 450.3 for an example of an installation using circuit breakers on the primary and the secondary for an over 600-volt transformer with 6 percent impedance.

The second category for over 600-volt transformers is *supervised locations only.* The maximum ratings or settings for overcurrent devices permitted in the *supervised locations only* rows are strictly limited to the conditions explained in Note 3 of Table 450.3(A). If the location of a transformer does not qualify as a supervised location, then it is necessary to select the general rows of *any location* in Table 450.3(A). Notice that the installation shown in Exhibit 450.3 fulfills the requirements of both *any location* and *supervised locations only.*

Article 240, Overcurrent Protection, contains Part VIII, Supervised Industrial Installations, which contains many revised overcurrent protection requirements for feeders and feeder taps associated with transformers. Also, requirements for the overcurrent protection of transformer secondary conductors are found in 240.4(F).

(B) Transformers 600 Volts, Nominal, or Less. Overcurrent protection shall be provided in accordance with Table 450.3(B).

The ratings or settings of the overcurrent protective device obtained from Table 450.3(B) are based on the transformer-rated current and whether or not secondary protection is provided. According to Table 450.3(B), the maximum ratings or settings of overcurrent protective devices for transformers rated 600 volts and less are separated into two categories: *primary only protection* and *primary and secondary protection.*

According to Table 450.3(B), transformers with currents

Table 450.3(B) Maximum Rating or Setting of Overcurrent Protection for Transformers 600 Volts and Less (as a Percentage of Transformer-Rated Current).

Protection Method	Primary Protection			Secondary Protection (See Note 2.)	
	Currents of 9 Amperes or More	Currents Less Than 9 Amperes	Currents Less Than 2 Amperes	Currents of 9 Amperes or More	Currents Less Than 9 Amperes
Primary only protection	125% (See Note 1.)	167%	300%	Not required	Not required
Primary and secondary protection	250% (See Note 3.)	250% (See Note 3.)	250% (See Note 3.)	125% (See Note 1.)	167%

Notes:
1. Where 125 percent of this current does not correspond to a standard rating of a fuse or nonadjustable circuit breaker, a higher rating that does not exceed the next higher standard rating shall be permitted.
2. Where secondary overcurrent protection is required, the secondary overcurrent device shall be permitted to consist of not more than six circuit breakers or six sets of fuses grouped in one location. Where multiple overcurrent devices are utilized, the total of all the device ratings shall not exceed the allowed value of a single overcurrent device. If both breakers and fuses are utilized as the overcurrent device, the total of the device ratings shall not exceed that allowed for fuses.
3. A transformer equipped with coordinated thermal overload protection by the manufacturer and arranged to interrupt the primary current shall be permitted to have primary overcurrent protection rated or set at a current value that is not more than six times the rated current of the transformer for transformers having not more than 6 percent impedance and not more than four times the rated current of the transformer for transformers having more than 6 percent but not more than 10 percent impedance.

of 9 amperes or more must be protected by either of two methods. Method 1 requires primary protection only and is set at not more than 125 percent of the primary side current rating. Method 1 does not require secondary side overcurrent protection. Method 2 requires secondary side overcurrent protection to be set at not more than 125 percent, provided the primary side overcurrent protection is set at not more than 250 percent of the primary side current rating.

Although not required, following either protection method will free the user from any further protection requirements of Table 450.3(B). According to this table, smaller transformers have protection requirements that are less restrictive. For overcurrent protection of motor control circuit transformers, see 430.72(C).

An example of *primary only protection* is shown in Exhibit 450.4. An example of *primary and secondary protection* is shown in Exhibit 450.5.

Questions frequently arise as to whether the overcurrent protection required for transformers, as specified in 450.3, provides satisfactory protection for the primary and secondary conductors. Where polyphase transformers are involved, primary and secondary conductors will usually not be properly protected. The rules in 450.3 are intended to protect the transformer alone. The primary overcurrent device provides short-circuit protection for the primary conductors and a degree of overload protection for the transformer, and secondary overcurrent devices prevent the transformer and secondary conductors from being overloaded. The transformer is considered the point of supply, and the conductors it supplies must be protected in accordance with their ampacity.

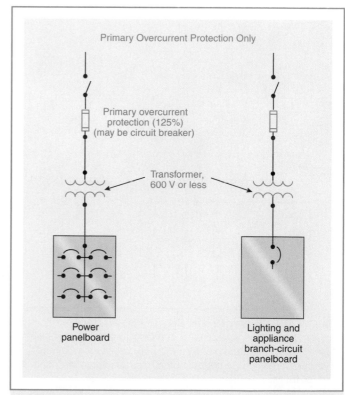

Exhibit 450.4 Two transformers (with currents of 9 amperes or more) rated 600 volts or less with only primary overcurrent protection, according to Table 450.3(B).

Section 240.4(F) permits the secondary circuit conductors from a transformer to be protected by overcurrent de-

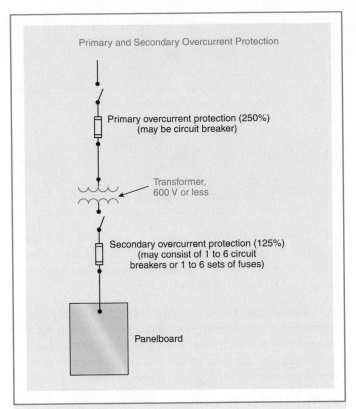

Exhibit 450.5 A transformer (9 amperes or more) rated 600 volts or less and protected by a combination of primary and secondary overcurrent protection, according to Table 450.3(B).

vices in the primary circuit conductors of the transformer only in two special cases. The first case is a transformer with a 2-wire primary and a 2-wire secondary, provided the transformer primary is protected in accordance with 450.3. The second case is a 3-phase, delta-delta-connected transformer having a 3-wire, single-voltage secondary, provided its primary is protected in accordance with 450.3. In cases where the primary feeder to the transformer incorporates overcurrent protective devices rated (or set) at a level not to exceed those prescribed herein, it is not necessary to duplicate them at the transformer.

Exception: Where the transformer is installed as a motor-control circuit transformer in accordance with 430.72(C)(1) through (C)(5).

(C) Voltage Transformers Voltage transformers installed indoors or enclosed shall be protected with primary fuses.

FPN: For protection of instrument circuits including voltage transformers, see 408.52.

450.4 Autotransformers 600 Volts, Nominal, or Less

(A) Overcurrent Protection Each autotransformer 600 volts, nominal, or less shall be protected by an individual

overcurrent device installed in series with each ungrounded input conductor. Such overcurrent device shall be rated or set at not more than 125 percent of the rated full-load input current of the autotransformer. Where this calculation does not correspond to a standard rating of a fuse or nonadjustable circuit breaker and the rated input current is 9 amperes or more, the next higher standard rating described in 240.6 shall be permitted. An overcurrent device shall not be installed in series with the shunt winding (the winding common to both the input and the output circuits) of the autotransformer between Points A and B as shown in Figure 450.4.

Figure 450.4 Autotransformer.

Because of the voltage feedback problem that may occur, an overcurrent device is not permitted between points A and B in *Code* Figure 450.4.

Exhibit 450.6 provides an example of overcurrent protection for an autotransformer. It shows a two-winding, single-phase transformer connected to boost a 208-volt supply to 240 volts. The autotransformer is provided with a two-pole disconnect switch with both overcurrent devices (OC-1a and OC-1b) located on the supply side of the autotransformer. If an overcurrent device were located between points A and B and this overcurrent device opened, the full 208-volt supply voltage would be applied across the 32-volt

Exhibit 450.6 A disconnect switch with overcurrent devices properly connected to protect an autotransformer and located to meet the requirements of 450.4(A), last sentence.

secondary winding in series with the load. Under those conditions, a higher-than-normal voltage would appear across the primary winding. If the load impedance were very low, this voltage could approach $208/32 \times 208 = 1352$ volts.

Exception: Where the rated input current of the autotransformer is less than 9 amperes, an overcurrent device rated or set at not more than 167 percent of the input current shall be permitted.

(B) Transformer Field-Connected as an Autotransformer A transformer field-connected as an autotransformer shall be identified for use at elevated voltage.

The requirement in 450.4(B) is necessary because of the dielectric voltage withstand test requirements applied to transformers. The test is conducted at 2500 volts for windings rated 250 volts or less and at 4000 volts for higher-rated windings. A transformer intended for buck or boost operation would require that the test for the low-voltage winding be based on the sum of the primary and secondary voltage ratings.

FPN: For information on permitted uses of autotransformers, see 210.9 and 215.11.

450.5 Grounding Autotransformers

Grounding autotransformers covered in this section are zig-zag or T-connected transformers connected to 3-phase, 3-wire ungrounded systems for the purpose of creating a 3-phase, 4-wire distribution system or providing a neutral reference for grounding purposes. Such transformers shall have a continuous per-phase current rating and a continuous neutral current rating. Zig-zag connected transformers shall not be installed on the load side of any system grounding connection, including those made in accordance with 250.24(B), 250.30(A)(1), or 250.32(B)(2).

A revision to 450.5 in the 2005 *Code* prohibits the installation of grounding autotransformers on the load side of a supply system grounding connection. This restriction applies to services, to separately derived systems, and to feeders and branch circuits that supply separate buildings or structures. Where a zig-zag transformer is used to create a neutral reference point on a circuit that is supplied from a grounded system, the current from a line-to-ground fault is shared through the supply system transformer and the zig-zag transformer.

Where the rating of the circuit in which the line-to-ground fault occurs exceeds the rating of the circuit in which the zig-zag transformer is used, the shared ground fault current through the zig-zag transformer has the potential to cause serious damage to the transformer. For instance, a zig-zag transformer is installed on an existing 50-ampere, three-phase, three-wire branch circuit to create a neutral. The branch circuit is derived from a grounded wye service, from which large capacity, 800-ampere and 1000-ampere feeders are also supplied. A line-to-ground fault in one of these feeder circuits can result in serious damage to the zig-zag transformer as a result of it sharing the fault current with the system supply transformer.

FPN: The phase current in a grounding autotransformer is one-third the neutral current.

(A) Three-Phase, 4-Wire System A grounding autotransformer used to create a 3-phase, 4-wire distribution system from a 3-phase, 3-wire ungrounded system shall conform to 450.5(A)(1) through (A)(4).

(1) Connections The transformer shall be directly connected to the ungrounded phase conductors and shall not be switched or provided with overcurrent protection that is independent of the main switch and common-trip overcurrent protection for the 3-phase, 4-wire system.

(2) Overcurrent Protection An overcurrent sensing device shall be provided that will cause the main switch or common-trip overcurrent protection referred to in 450.5(A)(1) to open if the load on the autotransformer reaches or exceeds 125 percent of its continuous current per-phase or neutral rating. Delayed tripping for temporary overcurrents sensed at the autotransformer overcurrent device shall be permitted for the purpose of allowing proper operation of branch or feeder protective devices on the 4-wire system.

(3) Transformer Fault Sensing A fault-sensing system that causes the opening of a main switch or common-trip overcurrent device for the 3-phase, 4-wire system shall be provided to guard against single-phasing or internal faults.

FPN: This can be accomplished by the use of two subtractive-connected donut-type current transformers installed to sense and signal when an unbalance occurs in the line current to the autotransformer of 50 percent or more of rated current.

(4) Rating The autotransformer shall have a continuous neutral-current rating that is sufficient to handle the maximum possible neutral unbalanced load current of the 4-wire system.

Exhibit 450.7 shows the proper method of protecting a grounding autotransformer used to provide a neutral for a 3-phase system where necessary to supply a group of single-phase, line-to-neutral loads. Separate overcurrent protection

Exhibit 450.7 A zigzag autotransformer used to create a 3-phase, 4-wire distribution system or to provide a neutral reference for grounding purposes.

Exhibit 450.8 A zigzag autotransformer used to establish a neutral connection for a 480Y/277-volt, 3-phase ungrounded system to supply single-phase line-to-neutral loads.

is not provided for the autotransformer, because there will be no control of the system line-to-neutral voltages if the autotransformer becomes disconnected. Consequently, simultaneous interruption of the power supply to all the line-to-neutral loads is necessary whenever the grounding autotransformer is switched off.

The donut-type current transformers CT-1, CT-2, and CT-3 shown in Exhibit 450.8 must be arranged to trip the circuit breaker located upstream of both the autotransformer and line-to-neutral connected loads, to satisfy the requirements of 450.5(A)(1). All three relays are intended to trip the main breaker if the current in any phase or the neutral conductor exceeds 125 percent of the rated current, as specified in 450.5(A)(2). The current transformers CT-2 and CT-3 are also differentially connected, to protect against an internal failure of the autotransformer, as required by 450.5(A)(3).

(B) Ground Reference for Fault Protection Devices A grounding autotransformer used to make available a specified magnitude of ground-fault current for operation of a ground-responsive protective device on a 3-phase, 3-wire ungrounded system shall conform to 450.5(B)(1) and (B)(2).

(1) Rating The autotransformer shall have a continuous neutral-current rating sufficient for the specified ground-fault current.

(2) Overcurrent Protection An overcurrent protective device of adequate short-circuit rating that will open simultaneously all ungrounded conductors when it operates shall be applied in the grounding autotransformer branch circuit and

shall be rated or set at a current not exceeding 125 percent of the autotransformer continuous per-phase current rating or 42 percent of the continuous-current rating of any series connected devices in the autotransformer neutral connection. Delayed tripping for temporary overcurrents to permit the proper operation of ground-responsive tripping devices on the main system shall be permitted but shall not exceed values that would be more than the short-time current rating of the grounding autotransformer or any series connected devices in the neutral connection thereto.

Exhibit 450.9 shows the proper method of protecting a grounding autotransformer where it is used as a ground reference for fault protection devices. The overcurrent protective device is to have a rating (or setting) not in excess of 125 percent of the rated phase current of the autotransformer (42 percent of the neutral current rating) and not more than 42 percent of the continuous current rating of the neutral grounding resistor or other current-carrying device in the neutral connection, as specified in 450.5(B)(2).

(C) Ground Reference for Damping Transitory Overvoltages A grounding autotransformer used to limit transitory overvoltages shall be of suitable rating and connected in accordance with 450.5(A)(1).

For installations involving a high-resistance grounding package, the functional performance of the installation parallels

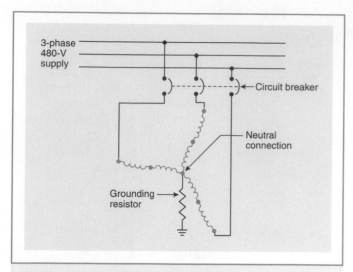

Exhibit 450.9 A zigzag autotransformer used to establish a reference ground-fault current for fault-protective-device operation or for damping-transitory-overvoltage surges.

that described in 450.5(B), differing only in that the magnitude of available ground-fault current would likely be a lower value. It would be appropriate to employ the connections displayed in Exhibit 450.9 and to conform with the overcurrent protection requirements prescribed in 450.5(B)(2).

With any of the grounding autotransformer applications covered by 450.5(A), 450.5(B), or 450.5(C), it is important to emphasize the use of a ganged 3-pole switching interrupter for connecting and disconnecting the autotransformer, in order to accomplish simultaneous connection (and disconnection) of the three line terminals. If, at any time, one or two of the line connections to the autotransformer should open, which could occur if the protective devices were single pole, the grounding autotransformer would cease to function in the desired fashion and would act as a high-inductive-reactance connection between the electrical system and ground. The latter connection is prone to create high-value transitory overvoltages, line-to-ground.

450.6 Secondary Ties

As used in this article, a secondary tie is a circuit operating at 600 volts, nominal, or less between phases that connects two power sources or power supply points, such as the secondaries of two transformers. The tie shall be permitted to consist of one or more conductors per phase or neutral. Conductors connecting the secondaries of transformers in accordance with 450.7 shall not be considered secondary ties.

As used in this section, the word *transformer* means a transformer or a bank of transformers operating as a unit.

(A) Tie Circuits Tie circuits shall be provided with overcurrent protection at each end as required in Parts I, II, and VIII of Article 240.

Under the conditions described in 450.6(A)(1) and 450.6(A)(2), the overcurrent protection shall be permitted to be in accordance with 450.6(A)(3).

(1) Loads at Transformer Supply Points Only Where all loads are connected at the transformer supply points at each end of the tie and overcurrent protection is not provided in accordance with Parts I, II, and VIII of Article 240, the rated ampacity of the tie shall not be less than 67 percent of the rated secondary current of the highest rated transformer supplying the secondary tie system.

(2) Loads Connected Between Transformer Supply Points Where load is connected to the tie at any point between transformer supply points and overcurrent protection is not provided in accordance with Parts I, II, and VIII of Article 240, the rated ampacity of the tie shall not be less than 100 percent of the rated secondary current of the highest rated transformer supplying the secondary tie system.

Exception: Tie circuits comprised of multiple conductors per phase shall be permitted to be sized and protected in accordance with 450.6(A)(4).

(3) Tie Circuit Protection Under the conditions described in 450.6(A)(1) and (A)(2), both supply ends of each ungrounded tie conductor shall be equipped with a protective device that opens at a predetermined temperature of the tie conductor under short-circuit conditions. This protection shall consist of one of the following: (1) a fusible link cable connector, terminal, or lug, commonly known as a limiter, each being of a size corresponding with that of the conductor and of construction and characteristics according to the operating voltage and the type of insulation on the tie conductors or (2) automatic circuit breakers actuated by devices having comparable time-current characteristics.

(4) Interconnection of Phase Conductors Between Transformer Supply Points Where the tie consists of more than one conductor per phase or neutral, the conductors of each phase or neutral shall comply with one of the following provisions.

(a) *Interconnected.* The conductors shall be interconnected in order to establish a load supply point, and the protective device specified in 450.6(A)(3) shall be provided in each ungrounded tie conductor at this point on both sides of the interconnection. The means of interconnection shall have an ampacity not less than the load to be served.

(b) *Not Interconnected.* The loads shall be connected to one or more individual conductors of a paralleled conductor tie without interconnecting the conductors of each phase or neutral and without the protection specified in 450.6(A)(3) at load connection points. Where this is done, the tie conductors of each phase or neutral shall have a combined capacity ampacity of not less than 133 percent of the rated secondary current of the highest rated transformer supplying the secondary tie system, the total load of such taps shall not exceed the rated secondary current of the highest rated transformer, and the loads shall be equally divided on each phase and on the individual conductors of each phase as far as practicable.

(5) Tie Circuit Control Where the operating voltage exceeds 150 volts to ground, secondary ties provided with limiters shall have a switch at each end that, when open, de-energizes the associated tie conductors and limiters. The current rating of the switch shall not be less than the rated current ampacity of the conductors connected to the switch. It shall be capable of interrupting its rated current, and it shall be constructed so that it will not open under the magnetic forces resulting from short-circuit current.

(B) Overcurrent Protection for Secondary Connections Where secondary ties are used, an overcurrent device rated or set at not more than 250 percent of the rated secondary current of the transformers shall be provided in the secondary connections of each transformer supplying the tie system. In addition, an automatic circuit breaker actuated by a reverse-current relay set to open the circuit at not more

than the rated secondary current of the transformer shall be provided in the secondary connection of each transformer.

The requirements of 450.6 apply specifically to network systems for power distribution commonly employed where the load density is high and reliability of service is important. Such a system is illustrated in Exhibit 450.10. This type of distribution system introduces a variety of problems not encountered in the more common radial-type distribution system and must be designed by experienced electrical engineers. Exhibit 450.10 shows a typical 3-phase network system for an industrial plant fed by two primary feeders, preferably from separate substations, energized at any standard voltage up to 34,500 volts. Each of the transformers is supplied by the two primary feeders, which are arranged by means of a double-throw switch at the transformer so that the transformer may be supplied by either feeder.

Each of the network transformers is rated in the range of 300 to 1000 kilovolt-amperes (kVA) and is required to be protected as illustrated in Exhibit 450.11. The primary

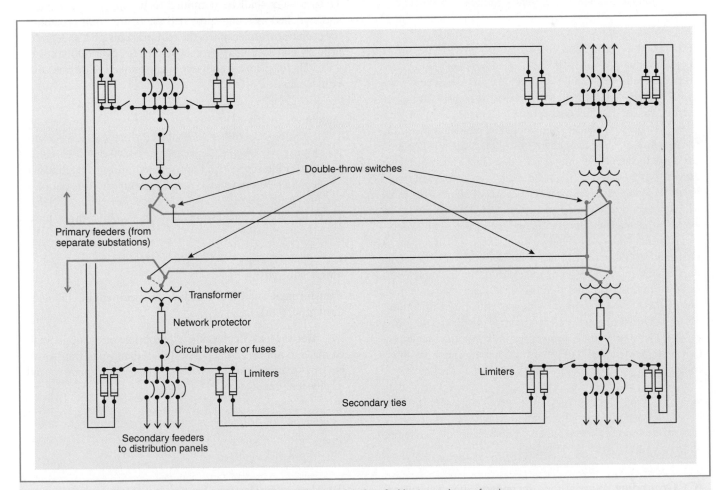

Exhibit 450.10 A typical 3-phase network system for an industrial plant fed by two primary feeders.

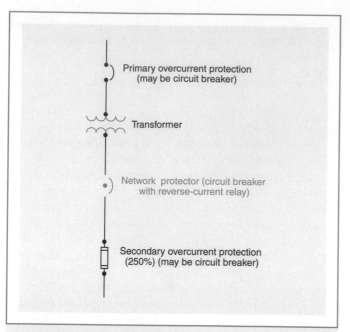

Exhibit 450.11 Primary and secondary overcurrent protection for a transformer in a network system, showing a network protector (an automatic circuit breaker actuated by a reverse-current relay).

Exhibit 450.12 A current limiter (a special type of high-interrupting-capacity fuse).

and secondary protection is in accordance with 450.3, but an additional protective device must be provided on the secondary side. This protective device is known as a *network protector* and consists of a circuit breaker and a reverse-current relay. The network protector operates on reverse current to prevent power from being fed back into the transformer through the secondary ties should a fault occur in the transformer or a primary feeder. The reverse-current relay is set to trip the circuit breaker at a current value not more than the rated secondary current of the transformer. The relay is not designed to trip the circuit breaker in the event of an overload on the secondary of the transformer.

The secondary ties shown in Exhibit 450.10 must be protected at each end with an overcurrent device, in accordance with 450.6(A)(3). The overcurrent device most commonly provided for this purpose is a special type of fuse known as a *current limiter*, which is illustrated in Exhibit 450.12. This high-interrupting-capacity device is designed to provide short-circuit protection only for the secondary ties, which will open safely before temperatures damaging to the cable insulation are reached. See 240.2 for the definition of *current-limiting overcurrent protective device* and its accompanying commentary. The secondary ties form a closed loop equipped with switching devices so that any part of the loop may be isolated when repairs are needed or a current limiter must be replaced.

(C) Grounding Where the secondary tie system is grounded, each transformer secondary supplying the tie sys-

tem shall be grounded in accordance with the requirements of 250.30 for separately derived systems.

450.7 Parallel Operation

Transformers shall be permitted to be operated in parallel and switched as a unit, provided the overcurrent protection for each transformer meets the requirements of 450.3(A) for primary and secondary protective devices over 600 volts, or 450.3(B) for primary and secondary protective devices 600 volts or less.

Parallel operation of transformers that are not switched as a unit can present dangerous backfeed situations for workers performing electrical maintenance. Appropriate lockout/tagout procedures must be implemented during maintenance of electrical equipment operated or connected in parallel. See NFPA 70E, *Standard for Electrical Safety in the Workplace*, for safety-related work practices and appropriate lockout/tagout procedures.

450.8 Guarding

Transformers shall be guarded as specified in 450.8(A) through 450.8(D).

(A) Mechanical Protection Appropriate provisions shall be made to minimize the possibility of damage to transformers from external causes where the transformers are exposed to physical damage.

One method of providing mechanical protection is to strategically place bollards around the transformer. This practice provides a degree of protection from vehicles.

(B) Case or Enclosure Dry-type transformers shall be provided with a noncombustible moisture-resistant case or enclosure that provides protection against the accidental insertion of foreign objects.

(C) Exposed Energized Parts Switches or other equipment operating at 600 volts, nominal, or less and serving only equipment within a transformer enclosure shall be permitted to be installed in the transformer enclosure if accessible to qualified persons only. All energized parts shall be guarded in accordance with 110.27 and 110.34.

(D) Voltage Warning The operating voltage of exposed live parts of transformer installations shall be indicated by signs or visible markings on the equipment or structures.

450.9 Ventilation

The ventilation shall be adequate to dispose of the transformer full-load losses without creating a temperature rise that is in excess of the transformer rating.

> FPN No. 1: See ANSI/IEEE C57.12.00-1993, *General Requirements for Liquid-Immersed Distribution, Power, and Regulating Transformers*, and ANSI/IEEE C57.12.01-1989, *General Requirements for Dry-Type Distribution and Power Transformers*.
>
> FPN No. 2: Additional losses may occur in some transformers where nonsinusoidal currents are present, resulting in increased heat in the transformer above its rating. See ANSI/IEEE C57.110-1993, *Recommended Practice for Establishing Transformer Capability When Supplying Nonsinusoidal Load Currents*, where transformers are utilized with nonlinear loads.

Transformers with ventilating openings shall be installed so that the ventilating openings are not blocked by walls or other obstructions. The required clearances shall be clearly marked on the transformer.

Section 450.9 is intended to clarify that transformers are not permitted to be installed directly against walls or other obstructions that block openings for ventilation and that the required clearances should be clearly marked on the transformer (see 450.11).

Fine Print Note No. 2 of 450.9 warns of increased heating of transformers. See the commentary following 450.3, FPN No. 2, and the commentary following 310.15(B)(4) for additional information concerning nonlinear loads.

450.10 Grounding

Exposed non–current-carrying metal parts of transformer installations, including fences, guards, and so forth, shall be grounded where required under the conditions and in the manner specified for electric equipment and other exposed metal parts in Article 250.

450.11 Marking

Each transformer shall be provided with a nameplate giving the name of the manufacturer, rated kilovolt-amperes, frequency, primary and secondary voltage, impedance of transformers 25 kVA and larger, required clearances for transformers with ventilating openings, and the amount and kind of insulating liquid where used. In addition, the nameplate of each dry-type transformer shall include the temperature class for the insulation system.

The information given on a transformer nameplate is necessary to determine whether special precautions must be used pertaining to clearances for ventilation, overcurrent protection, or liquid confinement.

450.12 Terminal Wiring Space

The minimum wire-bending space at fixed, 600-volt and below terminals of transformer line and load connections shall be as required in 312.6. Wiring space for pigtail connections shall conform to Table 314.16(B).

The requirement in 450.12 ensures adequate wire-bending space at fixed terminals of transformer line and load connections rated 600 volts or less, because this is a point of maximum mechanical and electrical stress on the conductor insulation.

450.13 Accessibility

All transformers and transformer vaults shall be readily accessible to qualified personnel for inspection and maintenance or shall meet the requirements of 450.13(A) or 450.13(B).

Transformers are not accessible if wiring methods or other equipment obstruct the access of a worker or prevent removal of the covers for inspection or maintenance. Practical clearance considerations required for removal and replacement of the transformer are also important.

(A) Open Installations Dry-type transformers 600 volts, nominal, or less, located in the open on walls, columns, or structures, shall not be required to be readily accessible.

(B) Hollow Space Installations Dry-type transformers 600 volts, nominal, or less and not exceeding 50 kVA shall be permitted in hollow spaces of buildings not permanently closed in by structure, provided they meet the ventilation requirements of 450.9 and separation from combustible materials requirements of 450.21(A). Transformers so installed shall not be required to be readily accessible.

Section 450.13(B) continues to permit the installation of dry-type transformers rated 600 volts or less and not exceeding 50 kVA in hollow spaces of hung ceiling areas, provided these spaces are fire resistant, ventilated, and accessible. According to 300.22(C)(2), transformers are permitted to be installed in hollow spaces where the space is used for environmental air, provided the transformer is in a metal enclosure (ventilated or nonventilated) and the transformer is suitable for the ambient air temperature within the hollow space. Of course, the requirement of 450.13(B) applies to transformer installations in "other space used for environmental air" per 300.22(C).

II. Specific Provisions Applicable to Different Types of Transformers

450.21 Dry-Type Transformers Installed Indoors

(A) Not Over 112½ kVA Dry-type transformers installed indoors and rated 112½ kVA or less shall have a separation of at least 305 mm (12 in.) from combustible material unless separated from the combustible material by a fire-resistant, heat-insulated barrier.

Exception: This rule shall not apply to transformers rated for 600 volts, nominal, or less that are completely enclosed, with or without ventilating openings.

(B) Over 112½ kVA Individual dry-type transformers of more than 112½ kVA rating shall be installed in a transformer room of fire-resistant construction. Unless specified otherwise in this article, the term *fire resistant* means a construction having a minimum fire rating of 1 hour.

Exception No. 1: Transformers with Class 155 or higher insulation systems and separated from combustible material by a fire-resistant, heat-insulating barrier or by not less than 1.83 m (6 ft) horizontally and 3.7 m (12 ft) vertically.

Exception No. 2: Transformers with Class 155 or higher insulation systems and completely enclosed except for ventilating openings.

Dry-type transformers with a Class 155 or higher insulation system rating are not required to be installed in transformer rooms or vaults if space separation or a fire-resistant heat-insulating barrier is provided. Although these units are designed for higher operating temperatures, the need for a transformer vault is mitigated by the fire-resistant characteristics of high-temperature insulations.

The two exceptions to 450.21(B) were revised for the 1999 *Code* by eliminating the reference to the specific temperature rating (80°C rise or higher rating) and substituting

the class insulation system (Class 155 or higher). The transformer class insulation system provides a more complete reference than simply using the permitted temperature rise. Further information on specific transformer class insulation systems may be found in UL 1561, *Dry-Type General Purpose and Power Transformers*.

> FPN: See ANSI/ASTM E119-1995, *Method for Fire Tests of Building Construction and Materials*, and NFPA 251-1999, *Standard Methods of Tests of Fire Endurance of Building Construction and Materials*.

(C) Over 35,000 Volts Dry-type transformers rated over 35,000 volts shall be installed in a vault complying with Part III of this article.

Dry-type transformers depend on the surrounding air for adequate ventilation and, where rated 112½ kVA or less, are not required to be installed in a fire-resistant transformer room but must comply with 450.9.

Dry-type transformers, or gas-filled or less-flammable liquid-insulated transformers (see 450.23), installed indoors with a primary voltage of not more than 35,000 volts are commonly used because a transformer vault is not required.

For the same reason, askarel-filled transformers have been extensively used indoors in the past. Askarel, which contains a polychlorinated biphenyl (PCB), is no longer being manufactured. Acceptable substitutes that comply with 450.23 are readily available.

Exhibit 450.13 shows a dry-type transformer with the outside casing in place and with the latest core and coil design for a typical dry-type power transformer rated at 1000 kVA, 13,800 volts to 480 volts, 3-phase, 60 Hz. This transformer has a high-voltage and low-voltage flange for connection to switchgear and a high-voltage, 2-position (double-throw), 3-pole-load air-break switch that may be attached to the case and arranged as a selector switch for the connection of the transformer primary to either of two feeder sources.

Dry-type transformers rated 112½ kVA or less require 12 in. of separation from combustible material or separation by fire-resistant barriers. Transformers rated less than 600 volts and completely enclosed, except for ventilating openings, are exempt from this requirement unless the manufacturer's installation instructions specify clearance distances. Noncombustible insulations used in transformers, such as mica, porcelain, and glass, which can withstand high temperatures, have permitted the application of larger dry-type transformers. Combustible materials, however, such as varnishes, may have been used with those insulations, and, under short-circuit conditions, flames can escape from the transformer enclosure. Transformers rated over 112½ kVA must be located in fire-resistant transformer rooms or vaults unless either of the exceptions to 450.21(B) applies.

Exhibit 450.13 A dry-type transformer with a core and coil design rated at 1000 kVA, 13,800 volts to 480 volts, 3-phase, 60 Hz. (Courtesy of Square D Co.)

450.22 Dry-Type Transformers Installed Outdoors

Dry-type transformers installed outdoors shall have a weatherproof enclosure.

Transformers exceeding 112½ kVA shall not be located within 305 mm (12 in.) of combustible materials of buildings unless the transformer has Class 155 insulation systems or higher and is completely enclosed except for ventilating openings.

See the commentary following 450.21(B), Exception No. 2, for an explanation of the change from specific temperature rating (80°C rise or higher rating) to the class insulation system (Class 155 or higher.)

450.23 Less-Flammable Liquid-Insulated Transformers

Transformers insulated with listed less-flammable liquids that have a fire point of not less than 300°C shall be permitted to be installed in accordance with 450.23(A) or 450.23(B).

(A) Indoor Installations Indoor installations shall be permitted in accordance with one of the following:

(1) In Type I or Type II buildings, in areas where all of the following requirements are met:

   a. The transformer is rated 35,000 volts or less.
   b. No combustible materials are stored.
   c. A liquid confinement area is provided.
   d. The installation complies with all restrictions provided for in the listing of the liquid.

(2) With an automatic fire extinguishing system and a liquid confinement area, provided the transformer is rated 35,000 volts or less

(3) In accordance with 450.26

(B) Outdoor Installations Less-flammable liquid-filled transformers shall be permitted to be installed outdoors, attached to, adjacent to, or on the roof of buildings, where installed in accordance with (1) or (2):

(1) For Type I and Type II buildings, the installation shall comply with all restrictions provided for in the listing of the liquid.

> FPN: Installations adjacent to combustible material, fire escapes, or door and window openings may require additional safeguards such as those listed in 450.27.

(2) In accordance with 450.27.

> FPN No. 1: As used in this section, *Type I and Type II buildings* refers to Type I and Type II building construction as defined in NFPA 220-1999, *Standard on Types of Building Construction*. *Combustible materials* refers to those materials not classified as noncombustible or limited-combustible as defined in NFPA 220-1999, *Standard on Types of Building Construction*.

NFPA 220, *Standard on Types of Building Construction*, defines Type I building construction as "that type in which the structural members, including walls, columns, beams, girders, trusses, arches, floors, and roofs, are of approved noncombustible or limited-combustible materials and have fire resistance ratings not less than those specified in Table 3-1." Table 3-1 is reprinted here as Commentary Table 450.1.

Type II building construction is defined in NFPA 220 as "that type not qualifying as Type I construction in which the structural members, including walls, columns, beams, girders, trusses, arches, floors, and roofs, are of approved noncombustible or limited-combustible materials and shall have fire resistance ratings not less than those specified in Table 3-1." See Commentary Table 450.1.

> FPN No. 2: See definition of *Listed* in Article 100.

Two listing agencies, Factory Mutual Research and Underwriters Laboratories Inc., list less-flammable liquids for transformers. These liquids have a fire point of at least

300°C. Exhibit 450.14 shows an example of a liquid-insulated transformer.

The Factory Mutual Research listing is based on the use of a Factory Mutual Research–approved less-flammable fluid in a transformer tank that meets certain criteria. Pressure-relief devices must be provided. Factory Mutual Research also recommends the use of enhanced electrical protection. Spacing from adjacent combustibles must be provided, based on the fluid capacity of the transformer tank, as illustrated in Exhibit 450.15. In the event of a leak, the liquid confinement area is intended to prevent transformer dielectric fluid from spreading beyond the vicinity of the transformer, as illustrated in Exhibit 450.16. Further information on applications may be found in the Factory Mutual Loss Prevention Data Sheet 5-4/14-8.

The Underwriters Laboratories Inc. listing is based on UL requirements that no tank rupture or noted fluid leakage occur during low- and high-current arcing fault tests. Further information may be obtained from the UL *Gas and Oil Equipment Directory*, under Transformer Fluids (EOVK), or from the manufacturer.

Exhibit 450.14 A liquid-insulated transformer filled with a listed less-flammable liquid having a fire point of at least 300°C. (Courtesy of Square D Co.)

450.24 Nonflammable Fluid-Insulated Transformers

Transformers insulated with a dielectric fluid identified as nonflammable shall be permitted to be installed indoors or outdoors. Such transformers installed indoors and rated over

Commentary Table 450.1 Fire Resistance Ratings (in Hours) for Type I and Type II Construction*

	Type I		Type II	
	443	332	222	111
Exterior Bearing Walls —				
Supporting more than one floor, columns, or other bearing walls	4	3	2	1
Supporting one floor only	4	3	2	1
Supporting a roof only	4	3	1	1
Interior Bearing Walls —				
Supporting more than one floor, columns, or other bearing walls	4	3	2	1
Supporting one floor only	3	2	2	1
Supporting roofs only	3	2	1	1
Columns —				
Supporting more than one floor, columns, or other bearing walls	4	3	2	1
Supporting one floor only	3	2	2	1
Supporting roofs only	3	2	1	1
Beams, Girders, Trusses & Arches —				
Supporting more than one floor, columns, or other bearing walls	4	3	2	1
Supporting one floor only	3	2	2	1
Supporting roofs only	3	2	1	1
Floor Construction	3	2	2	1
Roof Construction	2	1½	1	1
Exterior Nonbearing Walls	0	0	0	0

*For further information, see NFPA 220, *Standard on Types of Building Construction*.
Source: Reprinted from NFPA 220, *Standard on Types of Building Construction*, 1999 edition.

The arabic numbers at the top of the fire resistance rating columns of Commentary Table 450.1 reflect the fire resistance ratings of the following three building elements: exterior bearing walls; columns, beams, girders, trusses and arches, supporting bearing walls, columns, or loads from more than one floor; and the floor construction.

For example, a building of Type I, 443 construction has 4-hour fire-resistance-rated exterior bearing walls; 4-hour fire-resistance-rated columns, beams, girders, trusses, or arches; and 3-hour fire-resistance-rated floor construction. Whether a building is of Type I, Type II, or other type is determined by the requirements of the building construction code that is adopted by a jurisdiction.

35,000 volts shall be installed in a vault. Such transformers installed indoors shall be furnished with a liquid confinement area and a pressure-relief vent. The transformers shall be furnished with a means for absorbing any gases generated by arcing inside the tank, or the pressure-relief vent shall be connected to a chimney or flue that will carry such gases to an environmentally safe area.

FPN: Safety may be increased if fire hazard analyses are performed for such transformer installations.

For the purposes of this section, a nonflammable dielectric fluid is one that does not have a flash point or fire point and is not flammable in air.

Section 450.24 requires a liquid confinement area and a pressure-relief vent. The liquid confinement area is intended to limit the extent of a spill if the tank leaks or ruptures. If a means for absorbing gases generated by arcing within the transformer is not provided, the pressure-relief vent must be connected to a chimney or flue that vents to an environmentally safe area.

Exhibit 450.16 Side view of a transformer tank containing a Factory Mutual Research listed less-flammable fluid, where the volume of the confinement area is based on the capacity of the tank.

Exhibit 450.15 A transformer tank containing a Factory Mutual Research listed less-flammable fluid, where the spacing from adjacent combustibles to the liquid confinement area is based on the capacity of the tank.

The need for a gas absorption system or a chimney or flue that vents to an environmentally safe area is due to concerns about products generated during arcing. The high arc temperatures may cause the insulating medium to break down, resulting in the evolution of toxic or corrosive compounds.

450.25 Askarel-Insulated Transformers Installed Indoors

Askarel-insulated transformers installed indoors and rated over 25 kVA shall be furnished with a pressure-relief vent.

Where installed in a poorly ventilated place, they shall be furnished with a means for absorbing any gases generated by arcing inside the case, or the pressure-relief vent shall be connected to a chimney or flue that carries such gases outside the building. Askarel-insulated transformers rated over 35,000 volts shall be installed in a vault.

Askarel-insulated transformers are no longer manufactured. The information in the *Code* is for reference and for the modification of existing askarel-insulated installations.

450.26 Oil-Insulated Transformers Installed Indoors

Oil-insulated transformers installed indoors shall be installed in a vault constructed as specified in Part III of this article.

Exception No. 1: Where the total capacity does not exceed 112½ kVA, the vault specified in Part III of this article shall be permitted to be constructed of reinforced concrete that is not less than 100 mm (4 in.) thick.

Exception No. 2: Where the nominal voltage does not exceed 600, a vault shall not be required if suitable arrangements are made to prevent a transformer oil fire from igniting other materials and the total capacity in one location does not exceed 10 kVA in a section of the building classified as combustible or 75 kVA where the surrounding structure is classified as fire-resistant construction.

Exception No. 3: Electric furnace transformers that have a total rating not exceeding 75 kVA shall be permitted to be installed without a vault in a building or room of fire-resistant construction, provided suitable arrangements are made to prevent a transformer oil fire from spreading to other combustible material.

Exception No. 4: A transformer that has a total rating not exceeding 75 kVA and a supply voltage of 600 volts or less that is an integral part of charged-particle-accelerating equipment shall be permitted to be installed without a vault in a building or room of noncombustible or fire-resistant construction, provided suitable arrangements are made to prevent a transformer oil fire from spreading to other combustible material.

Exception No. 5: Transformers shall be permitted to be installed in a detached building that does not comply with Part III of this article if neither the building nor its contents present a fire hazard to any other building or property, and if the building is used only in supplying electric service and the interior is accessible only to qualified persons.

Exception No. 6: Oil-insulated transformers shall be permitted to be used without a vault in portable and mobile surface mining equipment (such as electric excavators) if each of the following conditions is met:

(a) Provision is made for draining leaking fluid to the ground.

(b) Safe egress is provided for personnel.

(c) A minimum 6-mm (¼-in.) steel barrier is provided for personnel protection.

450.27 Oil-Insulated Transformers Installed Outdoors

Combustible material, combustible buildings, and parts of buildings, fire escapes, and door and window openings shall be safeguarded from fires originating in oil-insulated transformers installed on roofs, attached to or adjacent to a building or combustible material.

In cases where the transformer installation presents a fire hazard, one or more of the following safeguards shall be applied according to the degree of hazard involved:

(1) Space separations
(2) Fire-resistant barriers
(3) Automatic fire suppression systems
(4) Enclosures that confine the oil of a ruptured transformer tank

Oil enclosures shall be permitted to consist of fire-resistant dikes, curbed areas or basins, or trenches filled with coarse, crushed stone. Oil enclosures shall be provided with trapped drains where the exposure and the quantity of oil involved are such that removal of oil is important.

> FPN: For additional information on transformers installed on poles or structures or under ground, see ANSI C2-2002, *National Electrical Safety Code*.

450.28 Modification of Transformers

When modifications are made to a transformer in an existing installation that change the type of the transformer with respect to Part II of this article, such transformer shall be marked to show the type of insulating liquid installed, and the modified transformer installation shall comply with the applicable requirements for that type of transformer.

Askarel-insulated transformers are permitted to be modified by replacing the askarel with either oil or a less-flammable liquid. Where such a modification takes place, the completed installation must have the same degree of safety as a new installation. For example, replacement of askarel with oil in an indoor installation without a vault may not be acceptable (see 450.26 and its exceptions). The same is true if the replacement liquid is a less-flammable liquid (see 450.23).

III. Transformer Vaults

450.41 Location

Vaults shall be located where they can be ventilated to the outside air without using flues or ducts wherever such an arrangement is practicable.

450.42 Walls, Roofs, and Floors

The walls and roofs of vaults shall be constructed of materials that have adequate structural strength for the conditions with a minimum fire resistance of 3 hours. The floors of vaults in contact with the earth shall be of concrete that is not less than 100 mm (4 in.) thick, but where the vault is constructed with a vacant space or other stories below it, the floor shall have adequate structural strength for the load imposed thereon and a minimum fire resistance of 3 hours. For the purposes of this section, studs and wallboard construction shall not be acceptable.

Exception: Where transformers are protected with automatic sprinkler, water spray, carbon dioxide, or halon, construction of 1-hour rating shall be permitted.

> FPN No. 1: For additional information, see ANSI/ASTM E119-1995, *Method for Fire Tests of Building Construction and Materials*, and NFPA 251-1999, *Standard Methods of Tests of Fire Endurance of Building Construction and Materials*.
> FPN No. 2: A typical 3-hour construction is 150 mm (6 in.) thick reinforced concrete.

Vaults are intended primarily as passive fire protection. The need for vaults is dictated by the combustibility of the dielectric media and the size of the transformer. Transformers insulated with mineral oil have the greatest need for passive protection, to prevent the spread of burning oil to other combustible materials.

Although it may be possible to construct a 3-hour-rated wall using studs and wallboard, this construction method is not permitted for transformer vaults. A reduction in fire-resistance rating from 3 hours to 1 hour is permitted for vaults equipped with an automatic fire suppression system.

There is less need for a vault around a dry-type transformer of less than 35,000 volts. If the transformer has adequate clearance from combustible construction and storage (see 450.21), a fire would be confined to the transformer.

Askarel is no longer manufactured as a transformer-insulating fluid. Askarel-insulated transformers of less than 35,000 volts do not require vaults, because askarel is considered a noncombustible fluid. Transformers with a listed less-flammable liquid insulation may be installed without a vault, as permitted in 450.23. See the commentary following 450.23(B)(2), which relates to Type I and Type II building construction.

450.43 Doorways

Vault doorways shall be protected in accordance with 450.43(A), (B), and (C).

(A) Type of Door Each doorway leading into a vault from the building interior shall be provided with a tight-fitting door that has a minimum fire rating of 3 hours. The authority having jurisdiction shall be permitted to require such a door for an exterior wall opening where conditions warrant.

Exception: Where transformers are protected with automatic sprinkler, water spray, carbon dioxide, or halon, construction of 1-hour rating shall be permitted.

> FPN: For additional information, see NFPA 80-1999, *Standard for Fire Doors and Fire Windows.*

(B) Sills A door sill or curb that is of sufficient height to confine the oil from the largest transformer within the vault shall be provided, and in no case shall the height be less than 100 mm (4 in.).

(C) Locks Doors shall be equipped with locks, and doors shall be kept locked, access being allowed only to qualified persons. Personnel doors shall swing out and be equipped with panic bars, pressure plates, or other devices that are normally latched but open under simple pressure.

Section 450.43 prohibits the use of conventional rotation-type door knobs on transformer vault doors. It is believed that an injured worker attempting to escape from a transformer vault may not be able to operate a rotating-type door knob but would be able to escape through a door equipped with panic-type hardware.

450.45 Ventilation Openings

Where required by 450.9, openings for ventilation shall be provided in accordance with 450.45(A) through 450.45(F).

(A) Location Ventilation openings shall be located as far as possible from doors, windows, fire escapes, and combustible material.

(B) Arrangement A vault ventilated by natural circulation of air shall be permitted to have roughly half of the total area of openings required for ventilation in one or more openings near the floor and the remainder in one or more openings in the roof or in the sidewalls near the roof, or all of the area required for ventilation shall be permitted in one or more openings in or near the roof.

(C) Size For a vault ventilated by natural circulation of air to an outdoor area, the combined net area of all ventilating openings, after deducting the area occupied by screens, gratings, or louvers, shall not be less than 1900 mm² (3 in.²) per kVA of transformer capacity in service, and in no case shall the net area be less than 0.1 m² (1 ft²) for any capacity under 50 kVA.

(D) Covering Ventilation openings shall be covered with durable gratings, screens, or louvers, according to the treatment required in order to avoid unsafe conditions.

(E) Dampers All ventilation openings to the indoors shall be provided with automatic closing fire dampers that operate in response to a vault fire. Such dampers shall possess a standard fire rating of not less than 1½ hours.

> FPN: See ANSI/UL 555-1995, *Standard for Fire Dampers.*

(F) Ducts Ventilating ducts shall be constructed of fire-resistant material.

450.46 Drainage

Where practicable, vaults containing more than 100 kVA transformer capacity shall be provided with a drain or other means that will carry off any accumulation of oil or water in the vault unless local conditions make this impracticable. The floor shall be pitched to the drain where provided.

450.47 Water Pipes and Accessories

Any pipe or duct system foreign to the electrical installation shall not enter or pass through a transformer vault. Piping or other facilities provided for vault fire protection, or for transformer cooling, shall not be considered foreign to the electrical installation.

Section 450.47 permits automatic sprinkler protection for transformer vaults. Piping or ductwork for cooling of the transformer is also permitted to be installed in a transformer vault. No other piping or ductwork is permitted to enter or pass through a transformer vault.

450.48 Storage in Vaults

Materials shall not be stored in transformer vaults.

ARTICLE 455
Phase Converters

Contents

I. General

455.1 Scope

This article covers the installation and use of phase converters.

A *phase converter* is an electrical device that converts single-phase electrical power to 3-phase, for the operation of equipment that normally operates from a 3-phase electrical supply. Phase converters are of two types: *static*, with no moving parts, and *rotary*, with an internal rotor that must be rotating before a load is applied (see 455.2 for definitions of *rotary-phase converter* and *static-phase converter*).

Phase converters are most commonly used to supply 3-phase motor loads in locations where only single-phase power is available from the local utility. Electrical installations on farms and in other remote or rural areas are examples of such locations. Although their most common loads are motors, phase converters are increasingly used to supply such loads as cellular telephone and other communication transmitter sites.

455.2 Definitions

Manufactured Phase. The manufactured or derived phase originates at the phase converter and is not solidly connected to either of the single-phase input conductors.

Phase Converter. An electrical device that converts single-phase power to 3-phase electrical power.

> FPN: Phase converters have characteristics that modify the starting torque and locked-rotor current of motors served, and consideration is required in selecting a phase converter for a specific load.

Rotary-Phase Converter. A device that consists of a rotary transformer and capacitor panel(s) that permits the operation of 3-phase loads from a single-phase supply.

Static-Phase Converter. A device without rotating parts, sized for a given 3-phase load to permit operation from a single-phase supply.

455.3 Other Articles

All applicable requirements of this *Code* shall apply to phase converters except as amended by this article.

455.4 Marking

Each phase converter shall be provided with a permanent nameplate indicating the following:

(1) Manufacturer's name
(2) Rated input and output voltages
(3) Frequency
(4) Rated single-phase input full-load amperes
(5) Rated minimum and maximum single load in kilovolt-amperes (kVA) or horsepower
(6) Maximum total load in kilovolt-amperes (kVA) or horsepower
(7) For a rotary-phase converter, 3-phase amperes at full load

455.5 Equipment Grounding Connection

A means for attachment of an equipment grounding conductor termination in accordance with 250.8 shall be provided.

455.6 Conductors

(A) Ampacity The ampacity of the single-phase supply conductors shall be determined by 455.6(A)(1) or (A)(2).

> FPN: Single-phase conductors sized to prevent a voltage drop not exceeding 3 percent from the source of supply

to the phase converter may help ensure proper starting and operation of motor loads.

(1) Variable Loads Where the loads to be supplied are variable, the conductor ampacity shall not be less than 125 percent of the phase converter nameplate single-phase input full-load amperes.

(2) Fixed Loads Where the phase converter supplies specific fixed loads, and the conductor ampacity is less than 125 percent of the phase converter nameplate single-phase input full-load amperes, the conductors shall have an ampacity not less than 250 percent of the sum of the full-load, 3-phase current rating of the motors and other loads served where the input and output voltages of the phase converter are identical. Where the input and output voltages of the phase converter are different, the current as determined by this section shall be multiplied by the ratio of output to input voltage.

(B) Manufactured Phase Marking The manufactured phase conductors shall be identified in all accessible locations with a distinctive marking. The marking shall be consistent throughout the system and premises.

455.7 Overcurrent Protection

The single-phase supply conductors and phase converter shall be protected from overcurrent by 455.7(A) or 455.7(B). Where the required fuse or nonadjustable circuit breaker rating or settings of adjustable circuit breakers do not correspond to a standard rating or setting, a higher rating or setting that does not exceed the next higher standard rating shall be permitted.

(A) Variable Loads Where the loads to be supplied are variable, overcurrent protection shall be set at not more than 125 percent of the phase converter nameplate single-phase input full-load amperes.

(B) Fixed Loads Where the phase converter supplies specific fixed loads and the conductors are sized in accordance with 455.6(A)(2), the conductors shall be protected in accordance with their ampacity. The overcurrent protection determined from this section shall not exceed 125 percent of the phase converter nameplate single-phase input amperes.

455.8 Disconnecting Means

Means shall be provided to disconnect simultaneously all ungrounded single-phase supply conductors to the phase converter.

(A) Location The disconnecting means shall be readily accessible and located in sight from the phase converter.

(B) Type The disconnecting means shall be a switch rated in horsepower, a circuit breaker, or a molded-case switch.

Where only nonmotor loads are served, an ampere-rated switch shall be permitted.

(C) Rating The ampere rating of the disconnecting means shall not be less than 115 percent of the rated maximum single-phase input full-load amperes or, for specific fixed loads, shall be permitted to be selected from 455.8(C)(1) or (C)(2).

(1) Current Rated Disconnect The disconnecting means shall be a circuit breaker or molded-case switch with an ampere rating not less than 250 percent of the sum of the following:

(1) Full-load, 3-phase current ratings of the motors
(2) Other loads served

(2) Horsepower Rated Disconnect The disconnecting means shall be a switch with a horsepower rating. The equivalent locked rotor current of the horsepower rating of the switch shall not be less than 200 percent of the sum of the following:

(1) Nonmotor loads
(2) The 3-phase, locked-rotor current of the largest motor as determined from 430.251(B)
(3) The full-load current of all other 3-phase motors operating at the same time

(D) Voltage Ratios The calculations in 455.8(C) shall apply directly where the input and output voltages of the phase converter are identical. Where the input and output voltages of the phase converter are different, the current shall be multiplied by the ratio of the output to input voltage.

455.9 Connection of Single-Phase Loads

Where single-phase loads are connected on the load side of a phase converter, they shall not be connected to the manufactured phase.

455.10 Terminal Housings

A terminal housing in accordance with the provisions of 430.12 shall be provided on a phase converter.

II. Specific Provisions Applicable to Different Types of Phase Converters

455.20 Disconnecting Means

The single-phase disconnecting means for the input of a static phase converter shall be permitted to serve as the disconnecting means for the phase converter and a single load if the load is within sight of the disconnecting means.

455.21 Start-Up

Power to the utilization equipment shall not be supplied until the rotary-phase converter has been started.

455.22 Power Interruption

Utilization equipment supplied by a rotary-phase converter shall be controlled in such a manner that power to the equipment will be disconnected in the event of a power interruption.

> FPN: Magnetic motor starters, magnetic contactors, and similar devices, with manual or time delay restarting for the load, provide restarting after power interruption.

455.23 Capacitors

Capacitors that are not an integral part of the rotary-phase conversion system but are installed for a motor load shall be connected to the line side of that motor overload protective device.

ARTICLE 460
Capacitors

Contents

460.1 Scope

This article covers the installation of capacitors on electric circuits.

Surge capacitors or capacitors included as a component part of other apparatus and conforming with the requirements of such apparatus are excluded from these requirements.

This article also covers the installation of capacitors in hazardous (classified) locations as modified by Articles 501 through 503.

460.2 Enclosing and Guarding

(A) Containing More Than 11 L (3 gal) of Flammable Liquid Capacitors containing more than 11 L (3 gal) of flammable liquid shall be enclosed in vaults or outdoor fenced enclosures complying with Article 110, Part III. This limit shall apply to any single unit in an installation of capacitors.

(B) Accidental Contact Where capacitors are accessible to unauthorized and unqualified persons, they shall be enclosed, located, or guarded so that persons cannot come into accidental contact or bring conducting materials into accidental contact with exposed energized parts, terminals, or buses associated with them. However, no additional guarding is required for enclosures accessible only to authorized and qualified persons.

> Means are required to drain off the stored charge in a capacitor after the supply circuit has been opened. Otherwise, a person servicing the equipment could receive a severe shock, or damage could occur to the equipment.
>
> Exhibit 460.1, diagram (a), shows a method in which capacitors are connected in a motor circuit so that they may be switched with the motor. In this arrangement, the stored charge drains off through the windings when the circuit is opened. Diagram (b) shows another arrangement in which the capacitor is connected to the line side of the motor starter contacts. An automatic discharge device and a separate disconnecting means are required.
>
> As shown in Exhibit 460.2, capacitors are often equipped with built-in resistors to drain off the stored charge, although this type of capacitor is not needed where connected as shown in Exhibit 460.1, diagram (a).

I. 600 Volts, Nominal, and Under

460.6 Discharge of Stored Energy

Capacitors shall be provided with a means of discharging stored energy.

(A) Time of Discharge The residual voltage of a capacitor shall be reduced to 50 volts, nominal, or less within 1 minute after the capacitor is disconnected from the source of supply.

Exhibit 460.1 Methods of connecting capacitors in induction motor circuit for power factor correction.

(B) Means of Discharge The discharge circuit shall be either permanently connected to the terminals of the capacitor or capacitor bank or provided with automatic means of connecting it to the terminals of the capacitor bank on removal of voltage from the line. Manual means of switching or connecting the discharge circuit shall not be used.

460.8 Conductors

(A) Ampacity The ampacity of capacitor circuit conductors shall not be less than 135 percent of the rated current of the capacitor. The ampacity of conductors that connect a capacitor to the terminals of a motor or to motor circuit conductors shall not be less than one-third the ampacity of the motor circuit conductors and in no case less than 135 percent of the rated current of the capacitor.

(B) Overcurrent Protection An overcurrent device shall be provided in each ungrounded conductor for each capacitor bank. The rating or setting of the overcurrent device shall be as low as practicable.

Except as permitted in the exception to 460.8(B), it is intended that the overcurrent device be separate from the overcurrent device protecting any other equipment or conductor. See Exhibit 460.1, diagrams (a) and (b).

Exception: A separate overcurrent device shall not be required for a capacitor connected on the load side of a motor overload protective device.

Exhibit 460.2 Power factor correction capacitors with discharge resistors.

(C) Disconnecting Means A disconnecting means shall be provided in each ungrounded conductor for each capacitor bank and shall meet the following requirements:

(1) The disconnecting means shall open all ungrounded conductors simultaneously.
(2) The disconnecting means shall be permitted to disconnect the capacitor from the line as a regular operating procedure.
(3) The rating of the disconnecting means shall not be less than 135 percent of the rated current of the capacitor.

Exception: A separate disconnecting means shall not be required where a capacitor is connected on the load side of a motor controller.

Capacitors are rated in reactive kilovolt-amperes, or kilovars, which is abbreviated "kVAr." Sometimes capacitors are rated in kilovolt amperes capacitive (kVAc). Both ratings are synonymous. The kVAr rating shows how many reactive kilovolt-amperes the capacitor will supply to cancel out the reactive kilovolt-amperes caused by inductance. For example, a 20-kVAr capacitor will cancel out 20 kVA of inductive reactive kilovolt-amperes.

The basic capacitor is 3-phase and delta-connected internally, but single-phase and 2-phase units are also available. The capacitors are constructed with built-in fuses for short-circuit protection and discharge resistors that reduce the voltage to a 50-volt crest or less when disconnected from the power supply. This occurs within 1 minute on 600-volt units and within 5 minutes on 2400- and 4160-volt units.

The capacitor circuit conductors and disconnecting means must have an ampacity not less than 135 percent of the rated current of the capacitor. The reason is that all capacitors are manufactured with a tolerance of zero percent to 15 percent, so a 100-kVAr capacitor may actually draw a current equivalent to that of a 115-kVAr capacitor. In addition, the current drawn by a capacitor varies directly with the line voltage, and any variation in the line voltage from a pure sine wave form causes the capacitor to draw an increased current. Considering these several factors, the increased current can amount to 135 percent of the rated current of the capacitor.

The current corresponding to the kVAr rating of a 3-phase capacitor, I_c, is computed from the following formula:

$$I_c = \frac{\text{kVAr} \times 1000}{1.73 \text{ volts}}$$

The ampacity of the conductors and the switching device is then determined by multiplying I_c by 1.35. The most effective power factor correction is obtained where the individual capacitors are connected directly to the terminals of the motors, transformers, and other inductive machinery.

Where capacitors are connected and operated as a unit, no complicated calculations are needed to determine the proper size capacitor to use. Capacitor manufacturers publish tables in which the required capacitor value is obtained by referring to the speed and the horsepower of the motor. These values improve the motor power factor to approximately 95 percent. To improve a plant power factor, capacitor manufacturers also publish tables to assist in calculating the total kVAr rating of capacitors required to improve the power factor to any desired value.

Care should be given to using capacitors where harmonic-producing loads are present. Adding capacitors to an electrical system can place the system in a harmonic resonance condition. The harmonic loads can excite the electrical system at the harmonic resonance frequency and cause over-current and overvoltage conditions. If capacitors are to be placed on electrical systems with harmonic loads, an engineering study that evaluates the size and placement of capacitors and the reactive impedance and load of the system should be conducted. Capacitors may need a reactor placed in series with them to help detune the electrical system from a harmonic resonance condition.

460.9 Rating or Setting of Motor Overload Device

Where a motor installation includes a capacitor connected on the load side of the motor overload device, the rating or setting of the motor overload device shall be based on the improved power factor of the motor circuit.

The effect of the capacitor shall be disregarded in determining the motor circuit conductor rating in accordance with 430.22.

Where a capacitor is connected on the load side of the overload relays, as shown in Exhibit 460.1, diagram (a), consideration must be given when selecting the rating or setting of the motor overload device because the line current will be reduced due to an improved power factor. A value lower than that indicated in 430.32 should be used for proper protection of the motor.

460.10 Grounding

Capacitor cases shall be grounded in accordance with Article 250.

Exception: Capacitor cases shall not be grounded where the capacitor units are supported on a structure designed to operate at other than ground potential.

460.12 Marking

Each capacitor shall be provided with a nameplate giving the name of the manufacturer, rated voltage, frequency, kilovar or amperes, number of phases, and, if filled with a combustible liquid, the volume of liquid. Where filled with a nonflammable liquid, the nameplate shall so state. The nameplate shall also indicate whether a capacitor has a discharge device inside the case.

II. Over 600 Volts, Nominal

460.24 Switching

(A) Load Current Group-operated switches shall be used for capacitor switching and shall be capable of the following:

(1) Carrying continuously not less than 135 percent of the rated current of the capacitor installation

(2) Interrupting the maximum continuous load current of each capacitor, capacitor bank, or capacitor installation that will be switched as a unit

(3) Withstanding the maximum inrush current, including contributions from adjacent capacitor installations

(4) Carrying currents due to faults on capacitor side of switch

(B) Isolation

(1) General A means shall be installed to isolate from all sources of voltage each capacitor, capacitor bank, or capacitor installation that will be removed from service as a unit. The isolating means shall provide a visible gap in the electrical circuit adequate for the operating voltage.

(2) Isolating or Disconnecting Switches with No Interrupting Rating Isolating or disconnecting switches (with no interrupting rating) shall be interlocked with the load-interrupting device or shall be provided with prominently displayed caution signs in accordance with 490.22 to prevent switching load current.

(C) Additional Requirements for Series Capacitors The proper switching sequence shall be ensured by use of one of the following:

(1) Mechanically sequenced isolating and bypass switches

(2) Interlocks

(3) Switching procedure prominently displayed at the switching location

460.25 Overcurrent Protection

(A) Provided to Detect and Interrupt Fault Current A means shall be provided to detect and interrupt fault current likely to cause dangerous pressure within an individual capacitor.

(B) Single Pole or Multipole Devices Single-pole or multipole devices shall be permitted for this purpose.

(C) Protected Individually or in Groups Capacitors shall be permitted to be protected individually or in groups.

(D) Protective Devices Rated or Adjusted Protective devices for capacitors or capacitor equipment shall be rated or adjusted to operate within the limits of the safe zone for individual capacitors. If the protective devices are rated or adjusted to operate within the limits for Zone 1 or Zone 2, the capacitors shall be enclosed or isolated.

In no event shall the rating or adjustment of the protective devices exceed the maximum limit of Zone 2.

> FPN: For definitions of *Safe Zone, Zone 1,* and *Zone 2,* see ANSI/IEEE 18-1992, *Shunt Power Capacitors.*

The reference to Zones 1 and 2 of ANSI/IEEE 18-1992 in the fine print note pertains to the performance of the capaci-

tors under fault conditions. If a fault current exceeds the limit established for Zone 2, the capacitor tank may burst.

460.26 Identification

Each capacitor shall be provided with a permanent nameplate giving the manufacturer's name, rated voltage, frequency, kilovar or amperes, number of phases, and the volume of liquid identified as flammable, if such is the case.

460.27 Grounding

Capacitor neutrals and cases, if grounded, shall be grounded in accordance with Article 250.

Exception: Where the capacitor units are supported on a structure that is designed to operate at other than ground potential.

460.28 Means for Discharge

(A) Means to Reduce the Residual Voltage A means shall be provided to reduce the residual voltage of a capacitor to 50 volts or less within 5 minutes after the capacitor is disconnected from the source of supply.

(B) Connection to Terminals A discharge circuit shall be either permanently connected to the terminals of the capacitor or provided with automatic means of connecting it to the terminals of the capacitor bank after disconnection of the capacitor from the source of supply. The windings of motors, transformers, or other equipment directly connected to capacitors without a switch or overcurrent device interposed shall meet the requirements of 460.28(A).

ARTICLE 470
Resistors and Reactors

Contents

I. 600 Volts, Nominal, and Under

470.1 Scope

This article covers the installation of separate resistors and reactors on electric circuits.

> Resistors are made in many sizes and shapes and for different purposes. They may be wire or ribbon wound, form wound, edgewise wound, cast grid, punched steel grid, or box resistors. They may be mounted in the open or in ventilated metal boxes or cabinets, depending on their use and location. Because they give off heat, resistors must be guarded and located at safe distances from combustible materials. Where mounted on switchboards or installed in control panels, they are not required to have additional guards.
>
> Reactors are installed in a circuit to introduce inductance for motor starting, combined with a capacitor to make a filter, controlling the current, and paralleling transformers. Current-limiting reactors are installed to limit the amount of current that can flow in a circuit when a short circuit occurs. Reactors can be divided into two classes: those with iron cores and those with no magnetic materials in the windings. Either type may be air cooled or oil immersed.
>
> Mechanical stresses exist between adjacent air-core reactors due to their external fields, and the manufacturer's recommendations should be followed in spacing and bracing units and fastening supporting insulators.
>
> Saturable reactors may be used for theater dimming [see 520.25(A) and commentary]. These reactors have, in addition to the ac winding, an auxiliary winding connected line-to-line or line-to-ground, to neutralize charging current and prevent a voltage rise. Those reactors used on high-voltage systems may be oil immersed.

Exception: Resistors and reactors that are component parts of other apparatus.

This article also covers the installation of resistors and reactors in hazardous (classified) locations as modified by Articles 501 through 504.

470.2 Location

Resistors and reactors shall not be placed where exposed to physical damage.

470.3 Space Separation

A thermal barrier shall be required if the space between the resistors and reactors and any combustible material is less than 305 mm (12 in.).

470.4 Conductor Insulation

Insulated conductors used for connections between resistance elements and controllers shall be suitable for an operating temperature of not less than 90°C (194°F).

Exception: Other conductor insulations shall be permitted for motor starting service.

II. Over 600 Volts, Nominal

470.18 General

(A) Protected Against Physical Damage Resistors and reactors shall be protected against physical damage.

(B) Isolated by Enclosure or Elevation Resistors and reactors shall be isolated by enclosure or elevation to protect personnel from accidental contact with energized parts.

(C) Combustible Materials Resistors and reactors shall not be installed in close enough proximity to combustible materials to constitute a fire hazard and shall have a clearance of not less than 305 mm (12 in.) from combustible materials.

(D) Clearances Clearances from resistors and reactors to grounded surfaces shall be adequate for the voltage involved.

> FPN: See Article 490.

(E) Temperature Rise from Induced Circulating Currents Metallic enclosures of reactors and adjacent metal parts shall be installed so that the temperature rise from induced circulating currents is not hazardous to personnel or does not constitute a fire hazard.

470.19 Grounding

Resistor and reactor cases or enclosures shall be grounded in accordance with Article 250.

Exception: Resistor or reactor cases or enclosures supported on a structure designed to operate at other than ground potential shall not be grounded.

470.20 Oil-Filled Reactors

Installation of oil-filled reactors, in addition to the above requirements, shall comply with applicable requirements of Article 450.

ARTICLE 480
Storage Batteries

Contents

480.1 Scope

The provisions of this article shall apply to all stationary installations of storage batteries.

There are two general types of storage cells: the *lead-acid* type and the *alkali* (nickel-cadmium) type. Basically, a lead-acid cell consists of a positive plate, usually lead peroxide (a semisolid compound) mounted on a framework or grid for support, and a negative plate, made of sponge lead mounted on a grid. Grids are generally made of a lead alloy, such as lead-calcium, lead-antimony, or lead-selenium. The electrolyte is sulfuric acid and distilled water.

Lead-acid cells may be of the vented or sealed (valve-regulated) type. Under normal charging conditions, the vented type will liberate gases, hydrogen at the negative plate and oxygen at the positive plate. The valve-regulated type provides a means to recombine this gas, thus minimizing emissions from the cell.

In the alkali, or nickel-cadmium, battery the principal active material in the positive plate is nickelous hydroxide; in the negative plate, it is cadmium hydroxide. The electrolyte is potassium hydroxide (an alkali).

In stationary installations, nickel-cadmium cells are generally of the vented type and liberate hydrogen and oxygen during normal charging. Hermetically sealed nickel-cadmium cells are sometimes used, but they require special charging equipment to prevent gas emissions.

480.2 Definitions

Nominal Battery Voltage. The voltage calculated on the basis of 2 volts per cell for the lead-acid type and 1.2 volts per cell for the alkali type.

Sealed Cell or Battery. A sealed cell or battery is one that has no provision for the addition of water or electrolyte or for external measurement of electrolyte specific gravity. The individual cells shall be permitted to contain a venting arrangement as described in 480.10(B).

Storage Battery. A battery comprised of one or more rechargeable cells of the lead-acid, nickel-cadmium, or other rechargeable electrochemical types.

480.3 Wiring and Equipment Supplied from Batteries

Wiring and equipment supplied from storage batteries shall be subject to the requirements of this *Code* applying to wiring and equipment operating at the same voltage, unless otherwise permitted by 480.4.

480.4 Overcurrent Protection for Prime Movers

Overcurrent protection shall not be required for conductors from a battery rated less than 50 volts if the battery provides power for starting, ignition, or control of prime movers. Section 300.3 shall not apply to these conductors.

The overcurrent protection requirements of Article 240 do not apply to field-installed conductors used for starting, ignition, or control of prime movers, provided that the supply source for these conductors is a battery rated less than 50 volts. In addition, the requirement to use single conductors only in conjunction with a Chapter 3 wiring method is not applicable to battery-powered conductors. For example, if it were necessary at a generator location to extend the conductors from the battery to the prime mover starting solenoid, these conductors would not be required to have overcurrent protection and could be run as open, single conductors.

480.6 Insulation of Batteries Not Over 250 Volts

This section shall apply to storage batteries having cells connected so as to operate at a nominal battery voltage of not over 250 volts.

(A) Vented Lead-Acid Batteries Cells and multicompartment batteries with covers sealed to containers of nonconductive, heat-resistant material shall not require additional insulating support.

(B) Vented Alkaline-Type Batteries Cells with covers sealed to jars of nonconductive, heat-resistant material shall require no additional insulation support. Cells in jars of conductive material shall be installed in trays of nonconductive material with not more than 20 cells (24 volts, nominal) in the series circuit in any one tray.

(C) Rubber Jars Cells in rubber or composition containers shall require no additional insulating support where the total nominal voltage of all cells in series does not exceed 150 volts. Where the total voltage exceeds 150 volts, batteries shall be sectionalized into groups of 150 volts or less, and each group shall have the individual cells installed in trays or on racks.

(D) Sealed Cells or Batteries Sealed cells and multicompartment sealed batteries constructed of nonconductive, heat-resistant material shall not require additional insulating support. Batteries constructed of a conducting container shall have insulating support if a voltage is present between the container and ground.

480.7 Insulation of Batteries of Over 250 Volts

The provisions of 480.6 shall apply to storage batteries having the cells connected so as to operate at a nominal voltage exceeding 250 volts, and, in addition, the provisions of this section shall also apply to such batteries. Cells shall be installed in groups having a total nominal voltage of not over 250 volts. Insulation, which can be air, shall be provided between groups and shall have a minimum separation between live battery parts of opposite polarity of 50 mm (2 in.) for battery voltages not exceeding 600 volts.

480.8 Racks and Trays

Racks and trays shall comply with 480.8(A) and 480.8(B).

(A) Racks Racks, as required in this article, are rigid frames designed to support cells or trays. They shall be substantial and be made of one of the following:

(1) Metal, treated so as to be resistant to deteriorating action by the electrolyte and provided with nonconducting members directly supporting the cells or with continuous insulating material other than paint on conducting members

(2) Other construction such as fiberglass or other suitable nonconductive materials

(B) Trays Trays are frames, such as crates or shallow boxes usually of wood or other nonconductive material, constructed or treated so as to be resistant to deteriorating action by the electrolyte.

480.9 Battery Locations

Battery locations shall conform to 480.9(A), (B), and (C).

(A) Ventilation Provisions shall be made for sufficient diffusion and ventilation of the gases from the battery to prevent the accumulation of an explosive mixture.

Compliance with 480.9(A) is necessary to prevent classification of a battery location as a hazardous (classified) location, in accordance with Article 500.

It is not the intent of 480.9(A) to mandate mechanical ventilation. Hydrogen disperses rapidly and requires little air movement to prevent accumulation. Unrestricted natural air movement in the vicinity of the battery, together with normal air changes for occupied spaces or heat removal, normally is sufficient. If the space is confined, mechanical ventilation may be required in the vicinity of the battery.

Hydrogen is lighter than air and tends to concentrate at ceiling level, so some form of ventilation should be provided at the upper portion of the structure. Ventilation can be a fan, roof ridge vent, or louvered area.

Although valve-regulated batteries are often referred to as "sealed," they actually emit very small quantities of hydrogen gas under normal operation and are capable of liberating large quantities of explosive gases if overcharged. These batteries therefore require the same amount of ventilation as their vented counterparts.

(B) Live Parts Guarding of live parts shall comply with 110.27.

Batteries should be located in clean, dry rooms. Batteries must be arranged to provide sufficient work space for inspection and maintenance. Provisions must also be made for adequate ventilation, to prevent an accumulation of an explosive mixture of the gases from the batteries.

The fumes given off by storage batteries are very corrosive; therefore, wiring and its insulation must be of a type that withstands corrosive action (see 310.9). Special precautions are necessary to ensure that all metalwork (metal raceways, metal racks, etc.) is designed or treated so as to be corrosion resistant. Manufacturers sometimes suggest that aluminum or plastic conduit be used to withstand the corrosive battery fumes, or, if steel conduit is used, that it be zinc coated and corrosion protected with a coating of an asphaltum-type paint (see 300.6).

Overcharging heats a battery and causes gassing and loss of water. A battery should not be allowed to reach temperatures over 110°F, because heat causes a shedding of active materials from the plates, which will eventually form a sediment buildup in the bottom of the case and short circuit the plates and the cell. Because mixtures of oxygen and hydrogen are highly explosive, flame or sparks should never be allowed near a cell, especially if the filler cap is removed.

(C) Working Space Working space about the battery systems shall comply with 110.26. Working clearance shall be measured from the edge of the battery rack.

480.10 Vents

(A) Vented Cells Each vented cell shall be equipped with a flame arrester that is designed to prevent destruction of the cell due to ignition of gases within the cell by an external spark or flame under normal operating conditions.

(B) Sealed Cells Sealed battery or cells shall be equipped with a pressure-release vent to prevent excessive accumulation of gas pressure, or the battery or cell shall be designed to prevent scatter of cell parts in event of a cell explosion.

ARTICLE 490
Equipment, Over 600 Volts, Nominal

Summary of Changes

- **490.46:** Revised to require a ground bus in the high-voltage service compartment of high-voltage switchgear for the attachment of service cable shields and the attachment of safety grounds for personnel protection.

Contents

I. General

490.1 Scope

This article covers the general requirements for equipment operating at more than 600 volts, nominal.

> FPN No. 1: See NFPA 70E-2004, *Standard for Electrical Safety in the Workplace*, for electrical safety requirements for employee workplaces.
>
> FPN No. 2: For further information on hazard signs and labels, see ANSI Z535-4, *Product Signs and Safety Labels*.

490.2 Definition

High Voltage. For the purposes of this article, more than 600 volts, nominal.

490.3 Oil-Filled Equipment

Installation of electrical equipment, other than transformers covered in Article 450, containing more than 38 L (10 gal) of flammable oil per unit shall meet the requirements of Parts II and III of Article 450.

II. Equipment — Specific Provisions

490.21 Circuit-Interrupting Devices

(A) Circuit Breakers

(1) Location

(a) Circuit breakers installed indoors shall be mounted either in metal-enclosed units or fire-resistant cell-mounted units, or they shall be permitted to be open-mounted in locations accessible to qualified persons only.

(b) Circuit breakers used to control oil-filled transformers shall either be located outside the transformer vault or be capable of operation from outside the vault.

(c) Oil circuit breakers shall be arranged or located so that adjacent readily combustible structures or materials are safeguarded in an approved manner.

(2) Operating Characteristics Circuit breakers shall have the following equipment or operating characteristics:

(1) An accessible mechanical or other approved means for manual tripping, independent of control power.
(2) Be release free (trip free).
(3) If capable of being opened or closed manually while energized, main contacts that operate independently of the speed of the manual operation.
(4) A mechanical position indicator at the circuit breaker to show the open or closed position of the main contacts.
(5) A means of indicating the open and closed position of the breaker at the point(s) from which they may be operated.

(3) Nameplate A circuit breaker shall have a permanent and legible nameplate showing manufacturer's name or trademark, manufacturer's type or identification number, continuous current rating, interrupting rating in megavolt-amperes (MVA) or amperes, and maximum voltage rating. Modification of a circuit breaker affecting its rating(s) shall be accompanied by an appropriate change of nameplate information.

(4) Rating Circuit breakers shall have the following ratings:

(1) The continuous current rating of a circuit breaker shall not be less than the maximum continuous current through the circuit breaker.
(2) The interrupting rating of a circuit breaker shall not be less than the maximum fault current the circuit breaker

will be required to interrupt, including contributions from all connected sources of energy.
(3) The closing rating of a circuit breaker shall not be less than the maximum asymmetrical fault current into which the circuit breaker can be closed.
(4) The momentary rating of a circuit breaker shall not be less than the maximum asymmetrical fault current at the point of installation.
(5) The rated maximum voltage of a circuit breaker shall not be less than the maximum circuit voltage.

(B) Power Fuses and Fuseholders

(1) Use Where fuses are used to protect conductors and equipment, a fuse shall be placed in each ungrounded conductor. Two power fuses shall be permitted to be used in parallel to protect the same load if both fuses have identical ratings and both fuses are installed in an identified common mounting with electrical connections that divide the current equally. Power fuses of the vented type shall not be used indoors, under ground, or in metal enclosures unless identified for the use.

(2) Interrupting Rating The interrupting rating of power fuses shall not be less than the maximum fault current the fuse is required to interrupt, including contributions from all connected sources of energy.

(3) Voltage Rating The maximum voltage rating of power fuses shall not be less than the maximum circuit voltage. Fuses having a minimum recommended operating voltage shall not be applied below this voltage.

(4) Identification of Fuse Mountings and Fuse Units Fuse mountings and fuse units shall have permanent and legible nameplates showing the manufacturer's type or designation, continuous current rating, interrupting current rating, and maximum voltage rating.

(5) Fuses Fuses that expel flame in opening the circuit shall be designed or arranged so that they function properly without hazard to persons or property.

(6) Fuseholders Fuseholders shall be designed or installed so that they are de-energized while a fuse is being replaced.

Exception: Fuses and fuseholders designed to permit fuse replacement by qualified persons using equipment designed for the purpose without de-energizing the fuseholder shall be permitted.

(7) High-Voltage Fuses Metal-enclosed switchgear and substations that utilize high-voltage fuses shall be provided with a gang-operated disconnecting switch. Isolation of the fuses from the circuit shall be provided by either connecting a switch between the source and the fuses or providing roll-out switch and fuse-type construction. The switch shall be of the load-interrupter type, unless mechanically or electrically interlocked with a load-interrupting device arranged to reduce the load to the interrupting capability of the switch.

Exception: More than one switch shall be permitted as the disconnecting means for one set of fuses where the switches are installed to provide connection to more than one set of supply conductors. The switches shall be mechanically or electrically interlocked to permit access to the fuses only when all switches are open. A conspicuous sign shall be placed at the fuses identifying the presence of more than one source.

(C) Distribution Cutouts and Fuse Links — Expulsion Type

(1) Installation Cutouts shall be located so that they may be readily and safely operated and re-fused, and so that the exhaust of the fuses does not endanger persons. Distribution cutouts shall not be used indoors, underground, or in metal enclosures.

(2) Operation Where fused cutouts are not suitable to interrupt the circuit manually while carrying full load, an approved means shall be installed to interrupt the entire load. Unless the fused cutouts are interlocked with the switch to prevent opening of the cutouts under load, a conspicuous sign shall be placed at such cutouts identifying that they shall not be operated under load.

(3) Interrupting Rating The interrupting rating of distribution cutouts shall not be less than the maximum fault current the cutout is required to interrupt, including contributions from all connected sources of energy.

(4) Voltage Rating The maximum voltage rating of cutouts shall not be less than the maximum circuit voltage.

(5) Identification Distribution cutouts shall have on their body, door, or fuse tube a permanent and legible nameplate or identification showing the manufacturer's type or designation, continuous current rating, maximum voltage rating, and interrupting rating.

(6) Fuse Links Fuse links shall have a permanent and legible identification showing continuous current rating and type.

(7) Structure Mounted Outdoors The height of cutouts mounted outdoors on structures shall provide safe clearance between lowest energized parts (open or closed position) and standing surfaces, in accordance with 110.34(E).

(D) Oil-Filled Cutouts

(1) Continuous Current Rating The continuous current rating of oil-filled cutouts shall not be less than the maximum continuous current through the cutout.

(2) Interrupting Rating The interrupting rating of oil-filled cutouts shall not be less than the maximum fault current the oil-filled cutout is required to interrupt, including contributions from all connected sources of energy.

(3) Voltage Rating The maximum voltage rating of oil-filled cutouts shall not be less than the maximum circuit voltage.

(4) Fault Closing Rating Oil-filled cutouts shall have a fault closing rating not less than the maximum asymmetrical fault current that can occur at the cutout location, unless suitable interlocks or operating procedures preclude the possibility of closing into a fault.

(5) Identification Oil-filled cutouts shall have a permanent and legible nameplate showing the rated continuous current, rated maximum voltage, and rated interrupting current.

(6) Fuse Links Fuse links shall have a permanent and legible identification showing the rated continuous current.

(7) Location Cutouts shall be located so that they are readily and safely accessible for re-fusing, with the top of the cutout not over 1.5 m (5 ft) above the floor or platform.

(8) Enclosure Suitable barriers or enclosures shall be provided to prevent contact with nonshielded cables or energized parts of oil-filled cutouts.

(E) Load Interrupters Load-interrupter switches shall be permitted if suitable fuses or circuit breakers are used in conjunction with these devices to interrupt fault currents. Where these devices are used in combination, they shall be coordinated electrically so that they will safely withstand the effects of closing, carrying, or interrupting all possible currents up to the assigned maximum short-circuit rating.

Where more than one switch is installed with interconnected load terminals to provide for alternate connection to different supply conductors, each switch shall be provided with a conspicuous sign identifying this hazard.

(1) Continuous Current Rating The continuous current rating of interrupter switches shall equal or exceed the maximum continuous current at the point of installation.

(2) Voltage Rating The maximum voltage rating of interrupter switches shall equal or exceed the maximum circuit voltage.

(3) Identification Interrupter switches shall have a permanent and legible nameplate including the following information: manufacturer's type or designation, continuous current rating, interrupting current rating, fault closing rating, maximum voltage rating.

(4) Switching of Conductors The switching mechanism shall be arranged to be operated from a location where the operator is not exposed to energized parts and shall be arranged to open all ungrounded conductors of the circuit simultaneously with one operation. Switches shall be arranged to be locked in the open position. Metal-enclosed switches shall be operable from outside the enclosure.

(5) Stored Energy for Opening The stored-energy operator shall be permitted to be left in the uncharged position after the switch has been closed if a single movement of the operating handle charges the operator and opens the switch.

(6) Supply Terminals The supply terminals of fused interrupter switches shall be installed at the top of the switch enclosure, or, if the terminals are located elsewhere, the equipment shall have barriers installed so as to prevent persons from accidentally contacting energized parts or dropping tools or fuses into energized parts.

See Exhibits 490.1 and 490.2 for an example of a fused interrupter switch and the fuseholder components.

Exhibit 490.1 Group-operated interrupter-switch and powerfuse combination rated at 13.8 kV, 600 amperes continuous and interrupting, 40,000 amperes momentary, 40,000 amperes fault closing. (Courtesy of S&C Electric Co.)

Exhibit 490.2 Components of the indoor solid-material (SM) power fuseholder (boric-acid arc-extinguishing type) with a 14.4 kV, 400E-ampere maximum, 40,000-ampere rms asymmetrical interrupting rating. Shown here are the spring and cable assembly, refill unit, holder, and snuffler. (Courtesy of S&C Electric Co.)

490.22 Isolating Means

Means shall be provided to completely isolate an item of equipment. The use of isolating switches shall not be required where there are other ways of de-energizing the equipment for inspection and repairs, such as draw-out-type metal-enclosed switchgear units and removable truck panels.

Isolating switches not interlocked with an approved circuit-interrupting device shall be provided with a sign warning against opening them under load.

A fuseholder and fuse, designed for the purpose, shall be permitted as an isolating switch.

490.23 Voltage Regulators

Proper switching sequence for regulators shall be ensured by use of one of the following:

(1) Mechanically sequenced regulator bypass switch(es)
(2) Mechanical interlocks
(3) Switching procedure prominently displayed at the switching location

490.24 Minimum Space Separation

In field-fabricated installations, the minimum air separation between bare live conductors and between such conductors and adjacent grounded surfaces shall not be less than the values given in Table 490.24. These values shall not apply to interior portions or exterior terminals of equipment designed, manufactured, and tested in accordance with accepted national standards.

III. Equipment — Metal-Enclosed Power Switchgear and Industrial Control Assemblies

490.30 General

This part covers assemblies of metal-enclosed power switchgear and industrial control, including but not limited to switches, interrupting devices and their control, metering, protection and regulating equipment, where an integral part of the assembly, with associated interconnections and supporting structures. This part also includes metal-enclosed power switchgear assemblies that form a part of unit substations, power centers, or similar equipment.

For the control and protection of feeders leaving a substation, Exhibit 490.3 illustrates a typical example of modern, metal-enclosed switchgear. This industrial unit substation includes a high-voltage disconnect switch, transformer, and low-voltage switchgear with a fully functioning ground-fault relay protection system.

Indicator instruments, such as voltmeters, ammeters, wattmeters, and protective relays, may be mounted on the panel doors as desired. This switchgear affords a high degree of safety because all live parts are metal-enclosed, and interlocks are provided for safe operation.

490.31 Arrangement of Devices in Assemblies

Arrangement of devices in assemblies shall be such that individual components can safely perform their intended function without adversely affecting the safe operation of other components in the assembly.

490.32 Guarding of High-Voltage Energized Parts Within a Compartment

Where access for other than visual inspection is required to a compartment that contains energized high-voltage parts,

Table 490.24 Minimum Clearance of Live Parts*

Nominal Voltage Rating (kV)	Impulse Withstand, B.I.L (kV)		Minimum Clearance of Live Parts							
			Phase-to-Phase				Phase-to-Ground			
			Indoors		Outdoors		Indoors		Outdoors	
	Indoors	Outdoors	mm	in.	mm	in.	mm	in.	mm	in.
2.4–4.16	60	95	115	4.5	180	7	80	3.0	155	6
7.2	75	95	140	5.5	180	7	105	4.0	155	6
13.8	95	110	195	7.5	305	12	130	5.0	180	7
14.4	110	110	230	9.0	305	12	170	6.5	180	7
23	125	150	270	10.5	385	15	190	7.5	255	10
34.5	150	150	320	12.5	385	15	245	9.5	255	10
	200	200	460	18.0	460	18	335	13.0	335	13
46	—	200	—	—	460	18	—	—	335	13
	—	250	—	—	535	21	—	—	435	17
69	—	250	—	—	535	21	—	—	435	17
	—	350	—	—	790	31	—	—	635	25
115	—	550	—	—	1350	53	—	—	1070	42
138	—	550	—	—	1350	53	—	—	1070	42
	—	650	—	—	1605	63	—	—	1270	50
161	—	650	—	—	1605	63	—	—	1270	50
	—	750	—	—	1830	72	—	—	1475	58
230	—	750	—	—	1830	72	—	—	1475	58
	—	900	—	—	2265	89	—	—	1805	71
	—	1050	—	—	2670	105	—	—	2110	83

*The values given are the minimum clearance for rigid parts and bare conductors under favorable service conditions. They shall be increased for conductor movement or under unfavorable service conditions or wherever space limitations permit. The selection of the associated impulse withstand voltage for a particular system voltage is determined by the characteristics of the surge protective equipment.

Exhibit 490.3 An assembly of metal-enclosed switchgear. (Courtesy of Square D Co.)

An example of a high-voltage pad-mounted transformer and enclosure that may contain primary and secondary switches or circuit breakers is shown in Exhibit 490.4.

Exhibit 490.4 A 300-kVA, 15-kV pad-mounted transformer integral unit containing a primary hookstick-operated switch with a limited number of secondary breakers or switches. (Courtesy of Square D Co.)

barriers shall be provided to prevent accidental contact by persons, tools, or other equipment with energized parts. Exposed live parts shall only be permitted in compartments accessible to qualified persons. Fuses and fuseholders designed to enable future replacement without de-energizing the fuse holder shall only be permitted for use by qualified persons.

490.33 Guarding of Low-Voltage Energized Parts Within a Compartment

Energized bare parts mounted on doors shall be guarded where the door must be opened for maintenance of equipment or removal of draw-out equipment.

490.34 Clearance for Cable Conductors Entering Enclosure

The unobstructed space opposite terminals or opposite raceways or cables entering a switchgear or control assembly shall be adequate for the type of conductor and method of termination.

490.35 Accessibility of Energized Parts

(A) High-Voltage Equipment Doors that would provide unqualified persons access to high-voltage energized parts shall be locked.

(B) Low-Voltage Control Equipment Low-voltage control equipment, relays, motors, and the like shall not be installed in compartments with exposed high-voltage energized parts or high-voltage wiring unless either of the following conditions is met:

(1) The access means is interlocked with the high-voltage switch or disconnecting means to prevent the access means from being opened or removed.
(2) The high-voltage switch or disconnecting means is in the isolating position.

(C) High-Voltage Instruments or Control Transformers and Space Heaters High-voltage instrument or control transformers and space heaters shall be permitted to be installed in the high-voltage compartment without access restrictions beyond those that apply to the high-voltage compartment generally.

490.36 Grounding

Frames of switchgear and control assemblies shall be grounded.

490.37 Grounding of Devices

Devices with metal cases or frames, or both, such as instruments, relays, meters, and instrument and control transformers, located in or on switchgear or control, shall have the frame or case grounded.

490.38 Door Stops and Cover Plates

External hinged doors or covers shall be provided with stops to hold them in the open position. Cover plates intended to be removed for inspection of energized parts or wiring shall be equipped with lifting handles and shall not exceed 1.1 m² (12 ft²) in area or 27 kg (60 lb) in weight, unless they are hinged and bolted or locked.

490.39 Gas Discharge from Interrupting Devices

Gas discharged during operating of interrupting devices shall be directed so as not to endanger personnel.

490.40 Inspection Windows

Windows intended for inspection of disconnecting switches or other devices shall be of suitable transparent material.

490.41 Location of Devices

(A) Control and Instrument Transfer Switch Handles or Pushbuttons Control and instrument transfer switch handles or pushbuttons other than those covered in 490.41(B) shall be in a readily accessible location at an elevation of not over 2.0 m (78 in.).

Exception: Operating handles requiring more than 23 kg (50 lb) of force shall be located no higher than 1.7 m (66 in.) in either the open or closed position.

(B) Infrequently Operated Devices Operating handles for infrequently operated devices, such as drawout fuses, fused potential or control transformers and their primary disconnects, and bus transfer switches, shall be permitted to be located where they are safely operable and serviceable from a portable platform.

490.42 Interlocks — Interrupter Switches

Interrupter switches equipped with stored energy mechanisms shall have mechanical interlocks to prevent access to the switch compartment unless the stored energy mechanism is in the discharged or blocked position.

490.43 Stored Energy for Opening

The stored energy operator shall be permitted to be left in the uncharged position after the switch has been closed if a single movement of the operating handle charges the operator and opens the switch.

490.44 Fused Interrupter Switches

(A) Supply Terminals The supply terminals of fused interrupter switches shall be installed at the top of the switch enclosure or, if the terminals are located elsewhere, the equipment shall have barriers installed so as to prevent persons from accidentally contacting energized parts or dropping tools or fuses into energized parts.

(B) Backfeed Where fuses can be energized by backfeed, a sign shall be placed on the enclosure door identifying this hazard.

(C) Switching Mechanism The switching mechanism shall be arranged to be operated from a location outside the enclosure where the operator is not exposed to energized parts and shall be arranged to open all ungrounded conductors of the circuit simultaneously with one operation. Switches shall be capable of being locked in the open position.

490.45 Circuit Breakers — Interlocks

(A) Circuit Breakers Circuit breakers equipped with stored energy mechanisms shall be designed to prevent the release of the stored energy unless the mechanism has been fully charged.

(B) Mechanical Interlocks Mechanical interlocks shall be provided in the housing to prevent the complete withdrawal of the circuit breaker from the housing when the stored energy mechanism is in the fully charged position, unless a suitable device is provided to block the closing function of the circuit breaker before complete withdrawal.

490.46 Metal Enclosed and Metal Clad Service Equipment

Metal enclosed and metal clad switchgear installed as high-voltage service equipment shall include a ground bus for the connection of service cable shields and to facilitate the attachment of safety grounds for personnel protection. This bus shall be extended into the compartment where the service conductors are terminated.

Section 490.46 was added to the 2005 *Code* to require metal-enclosed and metal-clad switchgear to include a ground bus for the service cable shields. It also provides a location for the connection of safety grounds for personnel protection during servicing of the equipment. The bus must extend to the compartment where the service conductors terminals are located.

IV. Mobile and Portable Equipment

490.51 General

(A) Covered The provisions of this part shall apply to installations and use of high-voltage power distribution and utilization equipment that is portable, mobile, or both, such as substations and switch houses mounted on skids, trailers, or cars; mobile shovels; draglines; cranes; hoists; drills; dredges; compressors; pumps; conveyors; underground excavators; and the like.

(B) Other Requirements The requirements of this part shall be additional to, or amendatory of, those prescribed in

Articles 100 through 725 of this *Code*. Special attention shall be paid to Article 250.

(C) Protection Adequate enclosures, guarding, or both, shall be provided to protect portable and mobile equipment from physical damage.

(D) Disconnecting Means Disconnecting means shall be installed for mobile and portable high-voltage equipment according to the requirements of Part VIII of Article 230 and shall disconnect all ungrounded conductors.

490.52 Overcurrent Protection

Motors driving single or multiple dc generators supplying a system operating on a cyclic load basis do not require overload protection, provided that the thermal rating of the ac drive motor cannot be exceeded under any operating condition. The branch-circuit protective device(s) shall provide short-circuit and locked-rotor protection and shall be permitted to be external to the equipment.

490.53 Enclosures

All energized switching and control parts shall be enclosed in effectively grounded metal cabinets or enclosures. These cabinets or enclosures shall be marked "DANGER — HIGH VOLTAGE — KEEP OUT" and shall be locked so that only authorized and qualified persons can enter. Circuit breakers and protective equipment shall have the operating means projecting through the metal cabinet or enclosure so these units can be reset without opening locked doors. With doors closed, reasonable safe access for normal operation of these units shall be provided.

490.54 Collector Rings

The collector ring assemblies on revolving-type machines (shovels, draglines, etc.) shall be guarded to prevent accidental contact with energized parts by personnel on or off the machine.

490.55 Power Cable Connections to Mobile Machines

A metallic enclosure shall be provided on the mobile machine for enclosing the terminals of the power cable. The enclosure shall include provisions for a solid connection for the ground wire(s) terminal to effectively ground the machine frame. Ungrounded conductors shall be attached to insulators or be terminated in approved high-voltage cable couplers (which include ground wire connectors) of proper voltage and ampere rating. The method of cable termination used shall prevent any strain or pull on the cable from stressing the electrical connections. The enclosure shall have provision for locking so only authorized and qualified persons may open it and shall be marked

DANGER — HIGH VOLTAGE — KEEP OUT.

490.56 High-Voltage Portable Cable for Main Power Supply

Flexible high-voltage cable supplying power to portable or mobile equipment shall comply with Article 250 and Article 400, Part III.

V. Electrode-Type Boilers

490.70 General

The provisions of this part shall apply to boilers operating over 600 volts, nominal, in which heat is generated by the passage of current between electrodes through the liquid being heated.

490.71 Electric Supply System

Electrode-type boilers shall be supplied only from a 3-phase, 4-wire solidly grounded wye system, or from isolating transformers arranged to provide such a system. Control circuit voltages shall not exceed 150 volts, shall be supplied from a grounded system, and shall have the controls in the ungrounded conductor.

490.72 Branch-Circuit Requirements

(A) Rating Each boiler shall be supplied from an individual branch circuit rated not less than 100 percent of the total load.

(B) Common-Trip Fault-Interrupting Device The circuit shall be protected by a 3-phase, common-trip fault-interrupting device, which shall be permitted to automatically reclose the circuit upon removal of an overload condition but shall not reclose after a fault condition.

(C) Phase-Fault Protection Phase-fault protection shall be provided in each phase, consisting of a separate phase-overcurrent relay connected to a separate current transformer in the phase.

(D) Ground Current Detection Means shall be provided for detection of the sum of the neutral and ground currents and shall trip the circuit-interrupting device if the sum of those currents exceeds the greater of 5 amperes or $7\frac{1}{2}$ percent of the boiler full-load current for 10 seconds or exceeds an instantaneous value of 25 percent of the boiler full-load current.

(E) Grounded Neutral Conductor The grounded neutral conductor shall be as follows:

(1) Connected to the pressure vessel containing the electrodes
(2) Insulated for not less than 600 volts

(3) Have not less than the ampacity of the largest ungrounded branch-circuit conductor

(4) Installed with the ungrounded conductors in the same raceway, cable, or cable tray, or, where installed as open conductors, in close proximity to the ungrounded conductors

(5) Not used for any other circuit

490.73 Pressure and Temperature Limit Control

Each boiler shall be equipped with a means to limit the maximum temperature, pressure, or both, by directly or indirectly interrupting all current flow through the electrodes.

Such means shall be in addition to the temperature, pressure, or both, regulating systems and pressure relief or safety valves.

490.74 Grounding

All exposed non–current-carrying metal parts of the boiler and associated exposed grounded structures or equipment shall be bonded to the pressure vessel or to the neutral conductor to which the vessel is connected in accordance with 250.102, except the ampacity of the bonding jumper shall not be less than the ampacity of the neutral conductor.

Special Occupancies

ARTICLE 500
Hazardous (Classified) Locations, Classes I, II, and III, Divisions 1 and 2

Summary of Changes

- **500.2:** Revised definition of *purged and pressurized* for correlation with NFPA 496 and also to clarify that pressurization is an acceptable protection technique for Class II and Class III locations.

- **500.7(K):** Added requirement that the type of detection equipment, its listing, installation location(s), alarm and shutdown criteria, and calibration frequency be documented when combustible gas detectors are used as a protection technique.

Contents

FPN: Rules that are followed by a reference in brackets contain text that has been extracted from NFPA 497, *Recommended Practice for the Classification of Flammable Liquids, Gases, or Vapors and of Hazardous (Classified) Locations for Electrical Installations in Chemical Process Areas,* 2004 edition, and NFPA 499, *Recommended Practice for the Classification of Combustible Dusts and of Hazardous (Classified) Locations for Electrical Installation in Chemical Process Areas,* 2004 edition. Only editorial changes were made to the extracted text to make it consistent with this *Code.*

500.1 Scope — Articles 500 Through 504

Articles 500 through 504 cover the requirements for electrical and electronic equipment and wiring for all voltages in Class I, Divisions 1 and 2; Class II, Divisions 1 and 2; and Class III, Divisions 1 and 2 locations where fire or explosion hazards may exist due to flammable gases or vapors, flammable liquids, combustible dust, or ignitible fibers or flyings.

Article 500 is limited to locations classified as Class I, II, or III, Division 1 or 2. Article 505 contains the requirements for using the Class I, Zone 0, Zone 1, and Zone 2 method of area classification in accordance with the International Electrotechnical Commission standard IEC 60079-10, *Electrical Apparatus for Explosive Gas Atmospheres.*

See the commentary following 505.1. New in the 2005 *Code,* Article 506 provides the requirements for using the Class II or Class III, Zone 20, Zone 21, and Zone 22 method of area classification in accordance with IEC 61241-10, *Electrical Apparatus for Use in the Presence of Combustible Dust — Part 10: Classification of Areas Where Combustible Dusts Are or May Be Present.*

The *Code* does not classify areas where the manufacture, transportation, storage, or use of explosive materials, such as ammunition, dynamite, and blasting powder, occurs. Areas where such materials are present are not considered hazardous (classified) locations in accordance with Article 500. However, many organizations responsible for the safety of such areas require equipment and wiring methods suitable for hazardous locations as part of many safety precautions, even though the equipment and wiring have not been investigated for such locations. The logic is that hazardous location equipment and wiring methods are safer than ordinary location equipment and wiring methods. Further information on these locations can be found in NFPA 495, *Explosive Materials Code.*

The *NEC* does not classify specific Class I, Class II, and Class III locations. NFPA technical committees with the experience and expertise of working with flammable liquids, gases, vapors, dusts, and flyings that are inherent to a process or that may be present under abnormal conditions of operation determine the parameters, distances, and degrees of hazard associated with classified locations. Some of this information has been extracted from other NFPA documents and is included in Articles 511 through 517 of the *Code*.

Articles 500 through 506 cover the requirements for electrical installations in locations that are classified as hazardous locations due to the materials handled, processed, or stored in those locations. Some of the most common materials encountered in hazardous (classified) locations are flammable liquids. A flammable liquid is one that has a flash point below 100°F. A flammable or combustible liquid must be at its flash point for an explosion to occur. No. 1-D diesel fuel oil and kerosene have flash points higher than 100°F and therefore do not emit flammable vapors unless heated above their flash points. They are considered combustible liquids.

For information on hazardous locations in general, including background on the classification of areas, equipment protection systems, ignition sources, static electricity and lightning, and requirements that apply outside the United States, see *Electrical Installations in Hazardous Locations*, by Peter J. Schram and Mark W. Earley. This book is available from NFPA.

FPN No. 1: The unique hazards associated with explosives, pyrotechnics, and blasting agents are not addressed in this article.

FPN No. 1 was added to the 2005 *Code* to alert users that explosives, pyrotechnics, and blasting agents are not covered by the *Code*. These materials are unique in that they can be set off by bumping, jarring, and static charges from various sources, in addition to electrical arcs, sparks, or heat.

FPN No. 2: For the requirements for electrical and electronic equipment and wiring for all voltages in Class I, Zone 0, Zone 1, and Zone 2 hazardous (classified) locations where fire or explosion hazards may exist due to flammable gases or vapors or flammable liquids, refer to Article 505.

FPN No. 3: For the requirements for electrical and electronic equipment and wiring for all voltages in Zone 20, Zone 21, and Zone 22 hazardous (classified) locations where fire or explosion hazards may exist due to combustible dusts or ignitible fibers or flyings, refer to Article 506.

FPN No. 3 was added to the 2005 *Code* to alert users that a new article, Article 506, was developed to cover the requirements for zone classification in locations where combustible dusts, fibers, and flyings may be present.

Combustible metallic dusts are not covered by Article 506. This is similar to Article 505, which covers zone classification as an alternative to the division classification system covered in Article 500.

500.2 Definitions

For purposes of Articles 500 through 504 and Articles 510 through 516, the following definitions apply.

Associated Nonincendive Field Wiring Apparatus. Apparatus in which the circuits are not necessarily nonincendive themselves but that affect the energy in nonincendive field wiring circuits and are relied upon to maintain nonincendive energy levels. Associated nonincendive field wiring apparatus may be either of the following:

(1) Electrical apparatus that has an alternative type of protection for use in the appropriate hazardous (classified) location
(2) Electrical apparatus not so protected that shall not be used in a hazardous (classified) location

FPN: Associated nonincendive field wiring apparatus has designated associated nonincendive field wiring apparatus connections for nonincendive field wiring apparatus and may also have connections for other electrical apparatus.

Combustible Gas Detection System. A protection technique utilizing stationary gas detectors in industrial establishments.

Control Drawing. A drawing or other document provided by the manufacturer of the intrinsically safe or associated apparatus, or of the nonincendive field wiring apparatus or associated nonincendive field wiring apparatus, that details the allowed interconnections between the intrinsically safe and associated apparatus or between the nonincendive field wiring apparatus or associated nonincendive field wiring apparatus.

Dust-Ignitionproof. Equipment enclosed in a manner that excludes dusts and does not permit arcs, sparks, or heat otherwise generated or liberated inside of the enclosure to cause ignition of exterior accumulations or atmospheric suspensions of a specified dust on or in the vicinity of the enclosure.

FPN: For further information on dust-ignitionproof enclosures, see Type 9 enclosure in ANSI/NEMA 250-1991, *Enclosures for Electrical Equipment*, and ANSI/UL 1203-1994, *Explosionproof and Dust-Ignitionproof Electrical Equipment for Hazardous (Classified) Locations*.

Dusttight. Enclosures constructed so that dust will not enter under specified test conditions.

FPN: See ANSI/ISA 12.12.01-2000, *Nonincendive Electrical Equipment for Use in Class I and II, Division 2, and Class III, Divisions 1 and 2 Hazardous (Classified) Locations*, and UL 1604-1994, *Electrical Equipment for Use in Class I and II, Division 2 and Class III Hazardous (Classified) Locations*.

Electrical and Electronic Equipment. Materials, fittings, devices, appliances, and the like that are part of, or in connection with, an electrical installation.

FPN: Portable or transportable equipment having self-contained power supplies, such as battery-operated equipment, could potentially become an ignition source in hazardous (classified) locations. See ISA-RP12.12.03-2002, *Portable Electronic Products Suitable for Use in Class I and II, Division 2, Class I Zone 2 and Class III, Division 1 and 2 Hazardous (Classified) Locations*.

Explosionproof Apparatus. Apparatus enclosed in a case that is capable of withstanding an explosion of a specified gas or vapor that may occur within it and of preventing the ignition of a specified gas or vapor surrounding the enclosure by sparks, flashes, or explosion of the gas or vapor within, and that operates at such an external temperature that a surrounding flammable atmosphere will not be ignited thereby.

FPN: For further information, see ANSI/UL 1203-1994, *Explosion-Proof and Dust-Ignition-Proof Electrical Equipment for Use in Hazardous (Classified) Locations*.

Hermetically Sealed. Equipment sealed against the entrance of an external atmosphere where the seal is made by fusion, for example, soldering, brazing, welding, or the fusion of glass to metal.

FPN: For further information, see ANSI/ISA 12.12.01-2000, *Nonincendive Electrical Equipment for Use in Class I and II, Division 2, and Class III, Division 1 and 2 Hazardous (Classified) Locations*.

Nonincendive Circuit. A circuit, other than field wiring, in which any arc or thermal effect produced under intended operating conditions of the equipment is not capable, under specified test conditions, of igniting the flammable gas–air, vapor–air, or dust–air mixture.

FPN: Conditions are described in ANSI/ISA 12.12.01-2000, *Nonincendive Electrical Equipment for Use in Class I and II, Division 2, and Class III, Divisions 1 and 2 Hazardous (Classified) Locations*.

Nonincendive Component. A component having contacts for making or breaking an incendive circuit and the contacting mechanism is constructed so that the component is incapable of igniting the specified flammable gas–air or vapor–air mixture. The housing of a nonincendive component is not intended to exclude the flammable atmosphere or contain an explosion.

FPN: For further information, see UL 1604-1994, *Electrical Equipment for Use in Class I and II, Division 2, and Class III Hazardous (Classified) Locations*.

Nonincendive Equipment. Equipment having electrical/electronic circuitry that is incapable, under normal operating conditions, of causing ignition of a specified flammable gas–air, vapor–air, or dust–air mixture due to arcing or thermal means.

FPN: For further information, see ANSI/ISA 12.12.01-2000, *Nonincendive Electrical Equipment for Use in Class I and II, Division 2, and Class III, Divisions 1 and 2 Hazardous (Classified) Locations*.

Nonincendive Field Wiring. Wiring that enters or leaves an equipment enclosure and, under normal operating conditions of the equipment, is not capable, due to arcing or thermal effects, of igniting the flammable gas–air, vapor–air, or dust–air mixture. Normal operation includes opening, shorting, or grounding the field wiring.

A nonincendive circuit employs a protection technique that prevents electrical circuits from causing a fire or explosion in a hazardous (classified) location. Section 500.7(F) points out that this protection technique is permitted only for Division 2 areas of Class I or Class II locations and for Division 1 or 2 areas of Class III locations. Nonincendive circuits are not permitted in Division 1 areas of Class I or Class II locations. Nonincendive circuits and equipment are tested or evaluated essentially in the same way that intrinsically safe circuits and equipment are tested and evaluated, except that abnormal conditions are not considered. Because of its definition, a nonincendive circuit is a low-energy circuit.

The definition of *nonincendive field wiring* was added to Article 100 of the 1999 *Code* to alert users that, although the circuits and equipment may have been evaluated and approved as nonincendive, field wiring is not generally approved by a testing laboratory. The definition was moved from Article 100 and now appears in 500.2. Field wiring meeting this definition would require limitations of energy on the wiring under conditions such as opening, shorting, or grounding. For example, stored energy in the form of mutual inductance or capacitance could be released during an opening, shorting, or grounding of nonincendive field wiring, thus defeating the purpose of this protection technique. Further information regarding nonincendive protection techniques is found in the 500.2 definitions of *nonincendive circuit*, *nonincendive component*, *nonincendive equipment*, and *nonincendive field wiring apparatus*.

Nonincendive Field Wiring Apparatus. Apparatus intended to be connected to nonincendive field wiring.

FPN: For further information see ANSI/ISA 12.12.01-2000, *Nonincendive Electrical Equipment for Use in Class I and II, Division 2, and Class III, Divisions 1 and 2 Hazardous (Classified) Locations*.

Oil Immersion. Electrical equipment immersed in a protective liquid in such a way that an explosive atmosphere that may be above the liquid or outside the enclosure cannot be ignited.

> FPN: For further information, see ANSI/UL 698-1995, *Industrial Control Equipment for Use in Hazardous (Classified) Locations.*

Purged and Pressurized. The process of (1) purging, supplying an enclosure with a protective gas at a sufficient flow and positive pressure to reduce the concentration of any flammable gas or vapor initially present to an acceptable level; and (2) pressurization, supplying an enclosure with a protective gas with or without continuous flow at sufficient pressure to prevent the entrance of a flammable gas or vapor, a combustible dust, or an ignitible fiber.

The definition of *purged and pressurized* has been revised to correlate with the definitions of these terms as specified in NFPA 496, *Standard for Purged and Pressurized Enclosures for Electrical Equipment.* The revised definition makes it clear that an enclosure can be purged continuously by a supply of clean air or inert gas or it can be pressurized with a supply of clean air or inert gas sufficient to prevent the entrance of flammable gas or vapors, combustible dust, or flyings into the enclosure.

> FPN: For further information, see ANSI/NFPA 496-2003, *Purged and Pressurized Enclosures for Electrical Equipment.*

Unclassified Locations. Locations determined to be neither Class I, Division 1; Class I, Division 2; Class I, Zone 0; Class I, Zone 1; Class I, Zone 2; Class II, Division 1; Class II, Division 2; Class III, Division 1; Class III, Division 2; or any combination thereof.

500.3 Other Articles

Except as modified in Articles 500 through 504, all other applicable rules contained in this *Code* shall apply to electrical equipment and wiring installed in hazardous (classified) locations.

The first four chapters of the *Code* cover general installation requirements for all electrical equipment and wiring (see 90.3). This means that materials and equipment must be suitable for environmental conditions such as rain, snow, ice, altitude, and heat; deteriorating effects on the conductors and equipment; and interrupting rating sufficient for the nominal circuit voltage and available fault current, just to name a few. The requirements in Articles 500 through 506 amend or modify the general rules to ensure the integrity of the electrical installation and to minimize the possibility

of the electrical equipment being an ignition source in the volatile or potentially volatile environments covered in these articles. An example of how the hazardous (classified) location articles of Chapter 5 modify the general requirements is found in the wiring method requirements of 501.10, 502.10, and 503.10. These sections of the *Code* limit the wiring methods that can be used in Class I, Class II, and Class III locations in order to provide the highest degree of protection against physical damage. The installation of any wiring methods allowed by Articles 500 through 506 must be in accordance with the Chapter 3 article that governs that particular wiring method and with any modifications in Articles 500 through 517.

500.4 General

(A) Documentation All areas designated as hazardous (classified) locations shall be properly documented. This documentation shall be available to those authorized to design, install, inspect, maintain, or operate electrical equipment at the location.

One type of documentation consists of area classification drawings. This type of documentation provides the necessary information for installers, service personnel, and authorities having jurisdiction to ensure that electrical equipment installed or maintained in classified areas is of the proper type. See the fine print note to 505.4(A).

(B) Reference Standards Important information relating to topics covered in Chapter 5 may be found in other publications.

The NFPA and ANSI standards referenced in Articles 500 through 517 are essential for proper application of those articles. The following NFPA codes, standards, and recommended practices include information on hazardous (classified) locations and the extent of hazardous (classified) locations in specific occupancies or industries.

> NFPA 30, *Flammable and Combustible Liquids Code*
>
> NFPA 30A, *Code for Motor Fuel Dispensing Facilities and Repair Garages*
>
> NFPA 32, *Standard for Drycleaning Plants*
>
> NFPA 33, *Standard for Spray Application Using Flammable or Combustible Materials*
>
> NFPA 34, *Standard for Dipping and Coating Processes Using Flammable or Combustible Liquids*
>
> NFPA 35, *Standard for the Manufacture of Organic Coatings*
>
> NFPA 36, *Standard for Solvent Extraction Plants*

NFPA 45, *Standard on Fire Protection for Laboratories Using Chemicals*

NFPA 50A, *Standard for Gaseous Hydrogen Systems at Consumer Sites*

NFPA 50B, *Standard for Liquefied Hydrogen Systems at Consumer Sites*

NFPA 51, *Standard for the Design and Installation of Oxygen–Fuel Gas Systems for Welding, Cutting, and Allied Processes*

NFPA 51A, *Standard for Acetylene Cylinder Charging Plants*

NFPA 52, *Compressed Natural Gas (CNG) Vehicular Fuel Systems Code*

NFPA 54, *Natural Fuel Gas Code*

NFPA 58, *Liquefied Petroleum Gas Code*

NFPA 59, *Utility LP-Gas Plant Code*

NFPA 59A, *Standard for the Production, Storage, and Handling of Liquefied Natural Gas (LNG)*

NFPA 61, *Standard for the Prevention of Fires and Dust Explosions in Agricultural and Food Products Facilities*

NFPA 85, *Boiler and Combustion Systems Hazards Code*

NFPA 88A, *Standard for Parking Structures*

NFPA 99, *Standard for Health Care Facilities*

NFPA 407, *Standard for Aircraft Fuel Servicing*

NFPA 409, *Standard on Aircraft Hangars*

NFPA 495, *Explosive Materials Code*

NFPA 496, *Standard for Purged and Pressurized Enclosures for Electrical Equipment*

NFPA 497, *Recommended Practice for the Classification of Flammable Liquids, Gases, or Vapors and of Hazardous (Classified) Locations for Electrical Installations in Chemical Process Areas*

NFPA 499, *Recommended Practice for the Classification of Combustible Dusts and of Hazardous (Classified) Locations for Electrical Installations in Chemical Process Areas*

NFPA 654, *Standard for the Prevention of Fire and Dust Explosions from the Manufacturing, Processing, and Handling of Combustible Particulate Solids*

NFPA 655, *Standard for Prevention of Sulfur Fires and Explosions*

FPN No. 1: It is important that the authority having jurisdiction be familiar with recorded industrial experience as well as with the standards of the National Fire Protection Association (NFPA), the American Petroleum

Institute (API), and the Instrumentation, Systems, and Automation Society (ISA) that may be of use in the classification of various locations, the determination of adequate ventilation, and the protection against static electricity and lightning hazards.

FPN No. 2: For further information on the classification of locations, see NFPA 30-2003, *Flammable and Combustible Liquids Code*; NFPA 32-2004, *Standard for Drycleaning Plants*; NFPA 33-2003, *Standard for Spray Application Using Flammable or Combustible Materials*; NFPA 34-2003, *Standard for Dipping and Coating Processes Using Flammable or Combustible Liquids*; NFPA 35-1999, *Standard for the Manufacture of Organic Coatings*; NFPA 36-2004, *Standard for Solvent Extraction Plants*; NFPA 45-2004, *Standard on Fire Protection for Laboratories Using Chemicals*; NFPA 50A-1999, *Standard for Gaseous Hydrogen Systems at Consumer Sites*; NFPA 50B-1999, *Standard for Liquefied Hydrogen Systems at Consumer Sites*; NFPA 58-2004, *Liquefied Petroleum Gas Code*; NFPA 59-2004, *Utility LP-Gas Plant Code*; NFPA 497-2004, *Recommended Practice for the Classification of Flammable Liquids, Gases, or Vapors and of Hazardous (Classified) Locations for Electrical Installations in Chemical Process Areas*; NFPA 499-2004, *Recommended Practice for the Classification of Combustible Dusts and of Hazardous (Classified) Locations for Electrical Installations in Chemical Process Areas*; NFPA 820-2003, *Standard for Fire Protection in Wastewater Treatment and Collection Facilities*; ANSI/API RP 500-1997, *Recommended Practice for Classification of Locations of Electrical Installations at Petroleum Facilities Classified as Class I, Division 1 and Division 2*; ISA 12.10-1988, *Area Classification in Hazardous (Classified) Dust Locations*.

FPN No. 3: For further information on protection against static electricity and lightning hazards in hazardous (classified) locations, see NFPA 77-2000, *Recommended Practice on Static Electricity*; NFPA 780-2004, *Standard for the Installation of Lightning Protection Systems*; and API RP 2003-1998, *Protection Against Ignitions Arising Out of Static Lightning and Stray Currents*.

FPN No. 4: For further information on ventilation, see NFPA 30-2003, *Flammable and Combustible Liquids Code*; and API RP 500-1997, *Recommended Practice for Classification of Locations for Electrical Installations at Petroleum Facilities Classified as Class I, Division 1 and Division 2*.

FPN No. 5: For further information on electrical systems for hazardous (classified) locations on offshore oil- and gas-producing platforms, see ANSI/API RP 14F-1999, *Recommended Practice for Design and Installation of Electrical Systems for Fixed and Floating Offshore Petroleum Facilities for Unclassified and Class I, Division 1 and Division 2 Locations*.

500.5 Classifications of Locations

(A) Classifications of Locations Locations shall be classified depending on the properties of the flammable vapors, liquids, or gases, or combustible dusts or fibers that may be present, and the likelihood that a flammable or combustible concentration or quantity is present. Where pyrophoric materials are the only materials used or handled, these locations shall not be classified. Each room, section, or area shall be considered individually in determining its classification.

Pyrophoric materials ignite spontaneously upon contact with air. The use of electrical equipment that is suitable for a hazardous (classified) location will not prevent ignition of pyrophoric materials. The process containment system should be designed to prevent contact between pyrophoric material and air.

FPN: Through the exercise of ingenuity in the layout of electrical installations for hazardous (classified) locations, it is frequently possible to locate much of the equipment in a reduced level of classification or in an unclassified location and, thus, to reduce the amount of special equipment required.

Rooms and areas containing ammonia refrigeration systems that are equipped with adequate mechanical ventilation may be classified as "unclassified" locations.

FPN: For further information regarding classification and ventilation of areas involving ammonia, see ANSI/ASHRAE 15-1994, *Safety Code for Mechanical Refrigeration*, and ANSI/CGA G2.1-1989, *Safety Requirements for the Storage and Handling of Anhydrous Ammonia*.

(B) Class I Locations Class I locations are those in which flammable gases or vapors are or may be present in the air in quantities sufficient to produce explosive or ignitible mixtures. Class I locations shall include those specified in 500.5(B)(1) and (B)(2).

(1) Class I, Division 1 A Class I, Division 1 location is a location

(1) In which ignitible concentrations of flammable gases or vapors can exist under normal operating conditions, or

(2) In which ignitible concentrations of such gases or vapors may exist frequently because of repair or maintenance operations or because of leakage, or

(3) In which breakdown or faulty operation of equipment or processes might release ignitible concentrations of flammable gases or vapors and might also cause simultaneous failure of electrical equipment in such a way as to directly cause the electrical equipment to become a source of ignition.

FPN No. 1: This classification usually includes the following locations:

(1) Where volatile flammable liquids or liquefied flammable gases are transferred from one container to another

(2) Interiors of spray booths and areas in the vicinity of spraying and painting operations where volatile flammable solvents are used

(3) Locations containing open tanks or vats of volatile flammable liquids

(4) Drying rooms or compartments for the evaporation of flammable solvents

(5) Locations containing fat- and oil-extraction equipment using volatile flammable solvents

(6) Portions of cleaning and dyeing plants where flammable liquids are used

(7) Gas generator rooms and other portions of gas manufacturing plants where flammable gas may escape

(8) Inadequately ventilated pump rooms for flammable gas or for volatile flammable liquids

(9) The interiors of refrigerators and freezers in which volatile flammable materials are stored in open, lightly stoppered, or easily ruptured containers

(10) All other locations where ignitible concentrations of flammable vapors or gases are likely to occur in the course of normal operations

FPN No. 2: In some Division 1 locations, ignitible concentrations of flammable gases or vapors may be present continuously or for long periods of time. Examples include the following:

(1) The inside of inadequately vented enclosures containing instruments normally venting flammable gases or vapors to the interior of the enclosure

(2) The inside of vented tanks containing volatile flammable liquids

(3) The area between the inner and outer roof sections of a floating roof tank containing volatile flammable fluids

(4) Inadequately ventilated areas within spraying or coating operations using volatile flammable fluids

(5) The interior of an exhaust duct that is used to vent ignitible concentrations of gases or vapors

Experience has demonstrated the prudence of avoiding the installation of instrumentation or other electric equipment in these particular areas altogether or where it cannot be avoided because it is essential to the process and other locations are not feasible [see 500.5(A), FPN] using electric equipment or instrumentation approved for the specific application or consisting of intrinsically safe systems as described in Article 504.

Fine print note No. 2 describes locations that are defined in 500.5(B)(1) as Class I, Division 1 locations. These locations are Class I, Zone 0 locations in accordance with 505.5(B)(1). Where a location is classified as a Zone 0 location, only intrinsically safe equipment and wiring suitable for Zone 0 locations are permitted to be used.

(2) Class I, Division 2 A Class I, Division 2 location is a location

(1) In which volatile flammable liquids or flammable gases are handled, processed, or used, but in which the liquids,

vapors, or gases will normally be confined within closed containers or closed systems from which they can escape only in case of accidental rupture or breakdown of such containers or systems or in case of abnormal operation of equipment, or

(2) In which ignitible concentrations of gases or vapors are normally prevented by positive mechanical ventilation and which might become hazardous through failure or abnormal operation of the ventilating equipment, or

(3) That is adjacent to a Class I, Division 1 location, and to which ignitible concentrations of gases or vapors might occasionally be communicated unless such communication is prevented by adequate positive-pressure ventilation from a source of clean air and effective safeguards against ventilation failure are provided.

FPN No. 1: This classification usually includes locations where volatile flammable liquids or flammable gases or vapors are used but that, in the judgment of the authority having jurisdiction, would become hazardous only in case of an accident or of some unusual operating condition. The quantity of flammable material that might escape in case of accident, the adequacy of ventilating equipment, the total area involved, and the record of the industry or business with respect to explosions or fires are all factors that merit consideration in determining the classification and extent of each location.

FPN No. 2: Piping without valves, checks, meters, and similar devices would not ordinarily introduce a hazardous condition even though used for flammable liquids or gases. Depending on factors such as the quantity and size of the containers and ventilation, locations used for the storage of flammable liquids or liquefied or compressed gases in sealed containers may be considered either hazardous (classified) or unclassified locations. See NFPA 30-2003, *Flammable and Combustible Liquids Code*, and NFPA 58-2004, *Liquefied Petroleum Gas Code*.

(C) Class II Locations Class II locations are those that are hazardous because of the presence of combustible dust. Class II locations shall include those specified in 500.5(C)(1) and (C)(2).

(1) Class II, Division 1 A Class II, Division 1 location is a location

(1) In which combustible dust is in the air under normal operating conditions in quantities sufficient to produce explosive or ignitible mixtures, or

(2) Where mechanical failure or abnormal operation of machinery or equipment might cause such explosive or ignitible mixtures to be produced, and might also provide a source of ignition through simultaneous failure of electric equipment, through operation of protection devices, or from other causes, or

(3) In which Group E combustible dusts may be present in quantities sufficient to be hazardous.

FPN: Dusts containing magnesium or aluminum are particularly hazardous, and the use of extreme precaution is necessary to avoid ignition and explosion.

Section 500.5(C)(1) was revised for the 2005 *Code* to clarify what conditions would require a space to be classified as Class II, Division 1. One condition is dust in suspension in the air under normal operating conditions in sufficient quantities to produce an explosive or ignitible mixture. Another condition is the malfunction of processing equipment that causes combustible dust to be released and, at the same time, an ignition source is created by the malfunction or some other ignition source occurs.

(2) Class II, Division 2 A Class II, Division 2 location is a location

(1) In which combustible dust due to abnormal operations may be present in the air in quantities sufficient to produce explosive or ignitible mixtures; or

(2) Where combustible dust accumulations are present but are normally insufficient to interfere with the normal operation of electrical equipment or other apparatus, but could as a result of infrequent malfunctioning of handling or processing equipment become suspended in the air; or

(3) In which combustible dust accumulations on, in, or in the vicinity of the electrical equipment could be sufficient to interfere with the safe dissipation of heat from electrical equipment, or could be ignitible by abnormal operation or failure of electrical equipment.

FPN No. 1: The quantity of combustible dust that may be present and the adequacy of dust removal systems are factors that merit consideration in determining the classification and may result in an unclassified area.

FPN No. 2: Where products such as seed are handled in a manner that produces low quantities of dust, the amount of dust deposited may not warrant classification.

(D) Class III Locations Class III locations are those that are hazardous because of the presence of easily ignitible fibers or flyings, but in which such fibers or flyings are not likely to be in suspension in the air in quantities sufficient to produce ignitible mixtures. Class III locations shall include those specified in 500.5(D)(1) and (D)(2).

(1) Class III, Division 1 A Class III, Division 1 location is a location in which easily ignitible fibers or materials producing combustible flyings are handled, manufactured, or used.

FPN No. 1: Such locations usually include some parts of rayon, cotton, and other textile mills; combustible fiber manufacturing and processing plants; cotton gins and cotton-seed mills; flax-processing plants; clothing manufacturing plants; woodworking plants; and establishments and industries involving similar hazardous processes or conditions.

FPN No. 2: Easily ignitible fibers and flyings include rayon, cotton (including cotton linters and cotton waste), sisal or henequen, istle, jute, hemp, tow, cocoa fiber, oakum, baled waste kapok, Spanish moss, excelsior, and other materials of similar nature.

(2) Class III, Division 2 A Class III, Division 2 location is a location in which easily ignitible fibers are stored or handled other than in the process of manufacture.

Sections 500.5(B), 500.5(C), and 500.5(D) describe three classes of hazardous (classified) locations based on the type of material involved. Within each class there are varying degrees of hazard, so each class is subdivided into two divisions. The classification by division is based on the likelihood the material will be present. The requirements for Division 1 of each class are more stringent than those for Division 2.

The materials in the three classes are defined as follows: Class I, flammable gases or vapors; Class II, combustible dust; and Class III, combustible fibers or flyings.

When a given location is classified as hazardous, it should be easy to determine to which of the three classes it belongs (it may belong in more than one class). Common sense and good judgment must prevail in classifying an area that is likely to become hazardous and in determining those portions of the premises to be classified Division 1 or Division 2. However, if different types of material exist in a process, such as flammable liquids, gases, or vapors and combustible dust, the area must be classified as both a Class I and a Class II location. If a location is classified due to two different hazards, such as a dust hazard and a flammable liquids hazard, protection must be provided for both hazards. Equipment that is approved for a Class I location may not be suitable for a Class II location and vice versa.

500.6 Material Groups

For purposes of testing, approval, and area classification, various air mixtures (not oxygen-enriched) shall be grouped in accordance with 500.6(A) and 500.6(B).

Oxygen enrichment can drastically change the explosion characteristics of materials. It lowers the minimum ignition energies, increases explosion pressures, and can reduce the maximum experimental safe gap, rendering both intrinsically safe and explosionproof equipment unsafe unless the equipment has been tested for the conditions involved.

Exception: Equipment identified for a specific gas, vapor, or dust.

FPN: This grouping is based on the characteristics of the materials. Facilities are available for testing and iden-

tifying equipment for use in the various atmospheric groups.

Determining the proper group classification for flammable gases and vapors involves evaluating explosion pressures and maximum safe clearances between parts of a clamped joint under several conditions and comparison of the values obtained with those obtained for classified materials under the same test conditions. Although some work has been done on the classification of flammable materials on the basis of chemical structure, this method is not sufficiently refined or accurate enough to ensure proper classification of all flammable materials.

For additional information on the rationale for classification, reference may be made to the following:

1. *Rationale for Classification of Combustible Gases, Vapors, and Dusts with Reference to the National Electrical Code*, Publication NMAB 353-6, 1982, a report of the Committee on Evaluation of Industrial Hazards, The National Materials Advisory Board, Commission on Engineering and Technical Systems, National Research Council. This publication is available from the National Technical Information Service (NTIS), Springfield, VA 22151.
2. *An Investigation of Fifteen Flammable Gases or Vapors with Respect to Explosion-Proof Electrical Equipment*, Bulletin of Research No. 58 by Underwriters Laboratories Inc., August 1969 (also Bulletin of Research Nos. 58A and 58B, which supplement No. 58). These publications are available from Underwriters Laboratories Inc., Publications Stock, Northbrook, IL 60062.
3. *Electrical Installations in Hazardous Locations*, Chapter 2, Peter J. Schram and Mark W. Earley. This book is available from NFPA.

(A) Class I Group Classifications Class I groups shall be according to 500.6(A)(1) through (A)(4).

FPN No. 1: FPN Nos. 2 and 3 apply to 500.6(A).

FPN No. 2: The explosion characteristics of air mixtures of gases or vapors vary with the specific material involved. For Class I locations, Groups A, B, C, and D, the classification involves determinations of maximum explosion pressure and maximum safe clearance between parts of a clamped joint in an enclosure. It is necessary, therefore, that equipment be identified not only for class but also for the specific group of the gas or vapor that will be present.

FPN No. 3: Certain chemical atmospheres may have characteristics that require safeguards beyond those required for any of the Class I groups. Carbon disulfide is one of these chemicals because of its low ignition temperature [100°C (212°F)] and the small joint clearance permitted to arrest its flame.

(1) Group A Acetylene. [NFPA 497:3.3]

(2) Group B Flammable gas, flammable liquid–produced vapor, or combustible liquid–produced vapor mixed with air that may burn or explode, having either a maximum experimental safe gap (MESG) value less than or equal to 0.45 mm or a minimum igniting current ratio (MIC ratio) less than or equal to 0.40. [NFPA 497:3.3]

> FPN: A typical Class I, Group B material is hydrogen.

Exception No. 1: Group D equipment shall be permitted to be used for atmospheres containing butadiene, provided all conduit runs into explosionproof equipment are provided with explosionproof seals installed within 450 mm (18 in.) of the enclosure.

Exception No. 2: Group C equipment shall be permitted to be used for atmospheres containing allyl glycidyl ether, n-butyl glycidyl ether, ethylene oxide, propylene oxide, and acrolein, provided all conduit runs into explosionproof equipment are provided with explosionproof seals installed within 450 mm (18 in.) of the enclosure.

The specific materials identified in Exception No. 1 and Exception No. 2 to 500.6(A)(2) produce high pressures because of pressure piling in unsealed conduits. If all conduits are sealed, the volume of air and gas between the sealing fittings and the arcing contacts is limited, thereby minimizing the pressure that can build within the enclosure and the raceway.

(3) Group C Flammable gas, flammable liquid–produced vapor, or combustible liquid–produced vapor mixed with air that may burn or explode, having either a maximum experimental safe gap (MESG) value greater than 0.45 mm and less than or equal to 0.75 mm, or a minimum igniting current ratio (MIC ratio) greater than 0.40 and less than or equal to 0.80. [NFPA 497:3.3]

> FPN: A typical Class I, Group C material is ethylene.

(4) Group D Flammable gas, flammable liquid–produced vapor, or combustible liquid–produced vapor mixed with air that may burn or explode, having either a maximum experimental safe gap (MESG) value greater than 0.75 mm or a minimum igniting current ratio (MIC ratio) greater than 0.80. [NFPA 497:3.3]

> FPN No. 1: A typical Class I, Group D material is propane.
>
> FPN No. 2: For classification of areas involving ammonia atmospheres, see ANSI/ASHRAE 15-1994, *Safety Code for Mechanical Refrigeration*, and ANSI/CGA G2.1-1989, *Safety Requirements for the Storage and Handling of Anhydrous Ammonia*.

Class I flammable gases or vapors are separated into four different groups—A, B, C, and D. The *Code* requirements for Class I locations do not vary for different kinds of gas or vapor contained in the group, except in those cases where seals may be used in all conduits to change the group classification. See Exception No. 1 and Exception No. 2 to 500.6(A)(2) for materials such as butadiene and ethylene oxide. It is necessary to select equipment designed for use in the particular group involved. The reason for designating the groups this way is that explosive mixtures have different igniting current ratios and maximum safe clearances between parts of a joint in an enclosure.

Underwriters Laboratories Inc. and Factory Mutual Research Corp. list or approve electrical equipment suitable for use in all groups of Class I locations. Further information is available from the UL *Hazardous Locations Equipment Directory* and the FM *Approval Guide*. It should be noted (from the UL Directory) that "only those products bearing the appropriate listing mark and the company's name, trade name, trademark, or other recognized identification should be considered as covered by UL's Listing and Follow-Up Service." Other testing laboratories may also list equipment for hazardous locations for use in the United States. The acceptance of the listing agency is the responsibility of the authority having jurisdiction.

Several testing laboratories outside the United States also provide listing, approval, or certification of equipment for use in hazardous locations. However, they may not be testing and investigating the equipment for use in hazardous locations, as defined in Article 500. Some international laboratories certify equipment for installation where the classification is according to the IEC classification scheme covered in Articles 505 and 506. Commentary Table 5.1 is an alphabetical listing of selected combustible materials with their group classification and relevant physical properties. All the materials included in this table have been evaluated for the purpose of designating the appropriate gas group.

This information is used to properly select electrical equipment for use in Class I locations. The combustible materials in Commentary Table 5.1 are classified into four Class I, division groups—A, B, C, and D—or three Class I, zone groups—IIC, IIB, and IIA—depending on their properties. Commentary Table 5.1 is extracted from NFPA 497, *Recommended Practice for the Classification of Flammable Liquids, Gases, or Vapors and of Hazardous (Classified) Locations for Electrical Installations in Chemical Process Areas*. For the definitions of terms used in this table, refer to NFPA 497.

Publication NMAB 353-4, *Classifications of Dusts Relative to Electrical Equipment in Class II Hazardous Locations*, published in 1982 by the Committee on Evaluation of Industrial Hazards, National Materials Advisory Board, Commission on Engineering and Technical Systems, National Research Council, and National Academy of Sciences,

Commentary Table 5.1 Selected Chemicals

Chemical	CAS No.	Class I Division Group	Type[a]	Flash Point (°C)	AIT (°C)	%LFL	%UFL	Vapor Density (Air = 1)	Vapor Pressure[b] (mm Hg)	Class I Zone Group[c]	MIE (mJ)	MIC Ratio	MESG (mm)
Acetaldehyde	75-07-0	C[d]	I	−38	175	4.0	60.0	1.5	874.9	IIA	0.37	0.98	0.92
Acetic Acid	64-19-7	D[d]	II	43	464	4.0	19.9	2.1	15.6	IIA		2.67	1.76
Acetic Acid-tert-Butyl Ester	540-88-5	D	II			1.7	9.8	4.0	40.6				
Acetic Anhydride	108-24-7	D	II	54	316	2.7	10.3	3.5	4.9				
Acetone	67-64-1	D[d]	I		465	2.5	12.8	2.0	230.7	IIA	1.15	1.00	1.02
Acetone Cyanohydrin	75-86-5	D	IIIA	74	688	2.2	12.0	2.9	0.3				
Acetonitrile	75-05-8	D	I	6	524	3.0	16.0	1.4	91.1	IIA			1.50
Acetylene	74-86-2	A[d]	GAS		305	2.5	99.9	0.9	36,600.0	IIC	0.017	0.28	0.25
Acrolein (Inhibited)	107-02-8	B(C)[d]	I		235	2.8	31.0	1.9	274.1	IIB	0.13		
Acrylic Acid	79-10-7	D	II	54	438	2.4	8.0	2.5	4.3				
Acrylonitrile	107-13-1	D[d]	I	−26	481	3.0	17.0	1.8	108.5	IIB	0.16	0.78	0.87
Adiponitrile	111-69-3	D	IIIA	93	550			1.0	0.002				
Allyl Alcohol	107-18-6	C[d]	I	22	378	2.5	18.0	2.0	25.4				0.84
Allyl Chloride	107-05-1	D	I	−32	485	2.9	11.1	2.6	366.0			1.33	1.17
Allyl Glycidyl Ether	106-92-3	B(C)e	II		57			3.9					
Alpha-Methyl Styrene	98-83-9	D	II		574	0.8	11.0	4.1	2.7				
n-Amyl Acetate	628-63-7	D	I	25	360	1.1	7.5	4.5	4.2				1.02
sec-Amyl Acetate	626-38-0	D	I		23		1.1	7.5	4.5	IIA			
Ammonia	7664-41-7	D[d,f]	I		498	15.0	28.0	0.6	7498.0	IIA	680.0	6.85	3.17
Aniline	62-53-3	D	IIIA	70	615	1.3	11.0	3.2	0.7	IIA			
Benzene	71-43-2	D[d]	I	−11	498	1.2	7.8	2.8	94.8	IIA	0.20	1.00	0.99
Benzyl Chloride	98-87-3	D	IIIA		585	1.1		4.4	0.5				
Bromopropyne	106-96-7	D	I	10	324	3.0							
n-Butane	3583-47-9	D[d,g]	GAS		288	1.9	8.5	2.0			0.25	0.94	1.07
1,3-Butadiene	106-99-0	B(D)[d,e]	GAS	−76	420	2.0	12.0	1.9		IIB	0.13	0.76	0.79
1-Butanol	71-36-3	D[d]	I	36	343	1.4	11.2	2.6	7.0	IIA			0.91
2-Butanol	78-92-2	D[d]	I	36	405	1.7	9.8	2.6		IIA			
Butylamine	109-73-9	D	GAS	−12	312	1.7	9.8	2.5	92.9			1.13	
Butylene	25167-67-3	D	I		385	1.6	10.0	1.9	2214.6				
n-Butyraldehyde	123-72-8	C[d]	I	−12	218	1.9	12.5	2.5	112.2				0.92
n-Butyl Acetate	123-86-4	D[d]	I	22	421	1.7	7.6	4.0	11.5	IIA		1.08	1.04
sec-Butyl Acetate	105-46-4	D	II	−8		1.7	9.8	4.0	22.2				
tert-Butyl Acetate	540-88-5	D	II			1.7	9.8	4.0	40.6				
n-Butyl Acrylate (Inhibited)	141-32-2	D	II	49	293	1.7	9.9	4.4	5.5				
n-Butyl Glycidyl Ether	2426-08-6	B(C)e	II										
n-Butyl Formal	110-62-3	C	IIIA						34.3				
Butyl Mercaptan	109-79-5	C	I	2				3.1	46.4				
Butyl-2-Propenoate	141-32-2	D	II	49		1.7	9.9	4.4	5.5				
para tert-Butyl Toluene	98-51-1	D	IIIA										
n-Butyric Acid	107-92-6	D[d]	IIIA	72	443	2.0	10.0	3.0	0.8				
Carbon Disulfide	75-15-0	d,h	I	−30	90	1.3	50.0	2.6	358.8	IIC	0.009	0.39	0.20
Carbon Monoxide	630-08-0	C[d]	GAS		609	12.5	74.0	0.97		IIA			0.84
Chloroacetaldehyde	107-20-0	C	IIIA	88					63.1				
Chlorobenzene	108-90-7	D	I	29	593	1.3	9.6	3.9	11.9				
1-Chloro-1-Nitropropane	2425-66-3	C	IIIA										
Chloroprene	126-99-8	D	GAS	−20		4.0	20.0	3.0					
Cresol	1319-77-3	D	IIIA	81	559	1.1		3.7					
Crotonaldehyde	4170-30-3	C[d]	I	13	232	2.1	15.5	2.4	33.1	IIB			0.81
Cumene	98-82-8	D	I	36	424	0.9	6.5	4.1	4.6	IIA			
Cyclohexane	110-82-7	D	I	−17	245	1.3	8.0	2.9	98.8	IIA	0.22	1.0	0.94

(continues)

Commentary Table 5.1 *Continued*

Chemical	CAS No.	Class I Division Group	Type[a]	Flash Point (°C)	AIT (°C)	%LFL	%UFL	Vapor Density (Air = 1)	Vapor Pressure[b] (mm Hg)	Class I Zone Group[c]	MIE (mJ)	MIC Ratio	MESG (mm)
Cyclohexanol	108-93-0	D	IIIA	68	300			3.5	0.7	IIA			
Cyclohexanone	108-94-1	D	II	44	245	1.1	9.4	3.4	4.3	IIA			0.98
Cyclohexene	110-83-8	D	I	−6	244	1.2		2.8	89.4			0.97	
Cyclopropane	75-19-4	D[d]	I		503	2.4	10.4	1.5	5430.0	IIB	0.17	0.84	0.91
p-Cymene	99-87-6	D	II	47	436	0.7	5.6	4.6	1.5	IIA			
Decene	872-05-9	D	II		235			4.8	1.7				
n-Decaldehyde	112-31-2	C	IIIA						0.09				
n-Decanol	112-30-1	D	IIIA	82	288			5.3	0.008				
Decyl Alcohol	112-30-1	D	IIIA	82	288			5.3	0.008				
Diacetone Alcohol	123-42-2	D	IIIA	64	603	1.8	6.9	4.0	1.4				
Di-Isobutylene	25167-70-8	D[d]	I	2	391	0.8	4.8	3.8			0.96		
Di-Isobutyl Ketone	108-83-8	D	II	60	396	0.8	7.1	4.9	1.7				
o-Dichlorobenzene	955-50-1	D	IIIA	66	647	2.2	9.2	5.1		IIA			
1,4-Dichloro-2,3 Epoxybutane	3583-47-9	D[d]	I			1.9	8.5	2.0			0.25	0.98	1.07
1,1-Dichloroethane	1300-21-6	D	I		438	6.2	16.0	3.4	227.0				1.82
1,2-Dichloroethylene	156-59-2	D	I	97	460	5.6	12.8	3.4	204.0	IIA			
1,1-Dichloro-1-Nitroethane	594-72-9	C	IIIA	76				5.0					
1,3-Dichloropropene	10061-02-6	D	I	35		5.3	14.5	3.8					
Dicyclopentadiene	77-73-6	C	I	32	503				2.8				0.91
Diethylamine	109-87-9	C[d]	I	−28	312	1.8	10.1	2.5		IIA			1.15
Diethylaminoethanol	100-37-8	C	IIIA	60	320			4.0	1.6	IIA			
Diethyl Benzene	25340-17-4	D	II	57	395			4.6					
Diethyl Ether (Ethyl Ether)	60-29-7	C[d]	I	−45	160	1.7	48.0	2.6	538.0	IIB	0.19	0.88	0.83
Diethylene Glycol Monobutyl Ether	112-34-5	C	IIIA	78	228	0.9	24.6	5.6	0.02				
Diethylene Glycol Monomethyl Ether	111-77-3	C	IIIA	93	241				0.2				
n-n-Dimethyl Aniline	121-69-7	C	IIIA	63	371	1.0		4.2	0.7				
Dimethyl Formamide	68-12-2	D	II	58	455	2.2	15.2	2.5	4.1				1.08
Dimethyl Sulfate	77-78-1	D	IIIA	83	188			4.4	0.7				
Dimethylamine	124-40-3	C	GAS		400	2.8	14.4	1.6		IIA			
2,2-Dimethylbutane	75-83-2	D[g]	I	−48	405				319.3				
2,3-Dimethylbutane	78-29-8	D[g]	I		396								
3,3-Dimethylheptane	1071-26-7	D[g]	I		325				10.8				
2,3-Dimethylhexane	31394-54-4	D[g]	I		438								
2,3-Dimethylpentane	107-83-5	D[g]	I		335				211.7				
Di-N-Propylamine	142-84-7	C	I	17	299				27.1				
1,4-Dioxane	123-91-1	C[d]	I	12	180	2.0	22.0	3.0	38.2	IIB	0.19		0.70
Dipentene	138-86-3	D	II	45	237	0.7	6.1	4.7					1.18
Dipropylene Glycol Methyl Ether	34590-94-8	C	IIIA	85		1.1	3.0	5.1	0.5				
Diisopropylamine	108-18-9	C	GAS	−6	316	1.1	7.1	3.5					
Dodecene	6842-15-5	D	IIIA	100	255								
Epichlorohydrin	3132-64-7	C[d]	I	33	411	3.8	21.0	3.2	13.0				
Ethane	74-84-0	D[d]	GAS	−29	472	3.0	12.5	1.0		IIA	0.24	0.82	0.91
Ethanol	64-17-5	D[d]	I	13	363	3.3	19.0	1.6	59.5	IIA		0.88	0.89
Ethylamine	75-04-7	D[d]	I	−18	385	3.5	14.0	1.6	1048.0		2.4		
Ethylene	74-85-1	C[d]	GAS	0	450	2.7	36.0	1.0		IIB	0.070	0.53	0.65
Ethylenediamine	107-15-3	D[d]	I	33	385	2.5	12.0	2.1	12.5				

Commentary Table 5.1 *Continued*

Chemical	CAS No.	Class I Division Group	Type[a]	Flash Point (°C)	AIT (°C)	%LFL	%UFL	Vapor Density (Air = 1)	Vapor Pressure[b] (mm Hg)	Class I Zone Group[c]	MIE (mJ)	MIC Ratio	MESG (mm)
Ethylenimine	151-56-4	C[d]	I	−11	320	3.3	54.8	1.5	211.0		0.48		
Ethylene Chlorohydrin	107-07-3	D	IIIA	59	425	4.9	15.9	2.8	7.2				
Ethylene Dichloride	107-06-2	D[d]	I	13	413	6.2	16.0	3.4	79.7				
Ethylene Glycol Monoethyl Ether Acetate	111-15-9	C	II	47	379	1.7		4.7	2.3			0.53	0.97
Ethylene Glycol Monobutyl Ether Acetate	112-07-2	C	IIIA		340	0.9	8.5		0.9				
Ethylene Glycol Monobutyl Ether	111-76-2	C	IIIA		238	1.1	12.7	4.1	1.0				
Ethylene Glycol Monoethyl Ether	110-80-5	C	II		235	1.7	15.6	3.0	5.4	0.84			
Ethylene Glycol Monomethyl Ether	109-86-4	D	II		285	1.8	14.0	2.6	9.2				0.85
Ethylene Oxide	75-21-8	B(C)[d,e]	I	−20	429	3.0	99.9	1.5	1314.0	IIB	0.065	0.47	0.59
2-Ethylhexaldehyde	123-05-7	C	II	52	191	0.8	7.2	4.4	1.9				
2-Ethylhexanol	104-76-7	D	IIIA	81		0.9	9.7	4.5	0.2				
2-Ethylhexyl Acrylate	103-09-3	D	IIIA	88	252				0.3				
Ethyl Acetate	141-78-6	D[d]	I	−4	427	2.0	11.5	3.0	93.2		0.46		0.99
Ethyl Acrylate (Inhibited)	140-88-5	D[d]	I	9	372	1.4	14.0	3.5	37.5	IIA			0.86
Ethyl Alcohol	64-17-5	D[d]	I	13	363	3.3	19.0	1.6	59.5				0.89
Ethyl Sec-Amyl Ketone	541-85-5	D	II	59									
Ethyl Benzene	100-41-4	D	I	21	432	0.8	6.7	3.7	9.6				
Ethyl Butanol	97-95-0	D	II	57		1.2	7.7	3.5	1.5				
Ethyl Butyl Ketone	106-35-4	D	II	46				4.0	3.6				
Ethyl Chloride	75-00-3	D	GAS	−50	519	3.8	15.4	2.2					
Ethyl Formate	109-94-4	D	GAS	−20	455	2.8	16.0	2.6		IIA			0.94
Ethyl Mercaptan	75-08-1	C[d]	I	−18	300	2.8	18.0	2.1	527.4			0.90	0.90
n-Ethyl Morpholine	100-74-3	C	I	32				4.0					
2-Ethyl-3-Propyl Acrolein	645-62-5	C	IIIA	68				4.4					
Ethyl Silicate	78-10-4	D	II					7.2					
Formaldehyde (Gas)	50-00-0	B	GAS	60	429	7.0	73.0	1.0	0.57				
Formic Acid	64-18-6	D	II	50	434	18.0	57.0	1.6	42.7				1.86
Fuel Oil 1	8008-20-6	D	II	72	210	0.7	5.0						
Furfural	98-01-1	C	IIIA	60	316	2.1	19.3	3.3	2.3				0.94
Furfuryl Alcohol	98-00-0	C	IIIA	75	490	1.8	16.3	3.4	0.6				
Gasoline	8006-61-9	D[d]	I	−46	280	1.4	7.6	3.0					
n-Heptane	142-82-5	D[d]	I	−4	204	1.0	6.7	3.5	45.5	IIA	0.24	0.88	0.91
n-Heptene	81624-04-6	D[g]	I	−1	204			3.4					0.97
n-Hexane	110-54-3	D[d,g]	I	−23	225	1.1	7.5	3.0	152.0	IIA	0.24	0.88	0.93
Hexanol	111-27-3	D	IIIA	63				3.5	0.8	IIA			0.98
2-Hexanone	591-78-6	D	I	35	424	1.2	8.0	3.5	10.6				
Hexene	592-41-6	D	I	−26	245	1.2	6.9		186.0				
sec-Hexyl Acetate	108-84-9	D	II	45				5.0					
Hydrazine	302-01-2	C	II	38	23		98.0	1.1	14.4				
Hydrogen	1333-74-0	B[d]	GAS		520	4.0	75.0	0.1		IIC	0.019	0.25	0.28
Hydrogen Cyanide	74-90-8	C[d]	GAS	−18	538	5.6	40.0	0.9		IIB			0.80

(continues)

Commentary Table 5.1 *Continued*

Chemical	CAS No.	Class I Division Group	Type[a]	Flash Point (°C)	AIT (°C)	%LFL	%UFL	Vapor Density (Air = 1)	Vapor Pressure[b] (mm Hg)	Class I Zone Group[c]	MIE (mJ)	MIC Ratio	MESG (mm)
Hydrogen Selenide	7783-07-5	C	I						7793.0				
Hydrogen Sulfide	7783-06-4	C[d]	GAS		260	4.0	44.0	1.2			0.068		0.90
Isoamyl Acetate	123-92-2	D	I	25	360	1.0	7.5	4.5	6.1				
Isoamyl Alcohol	123-51-3	D	II	43	350	1.2	9.0	3.0	3.2				1.02
Isobutane	75-28-5	D[g]	GAS		460	1.8	8.4	2.0					
Isobutyl Acetate	110-19-0	D[d]	I	18	421	2.4	10.5	4.0	17.8				
Isobutyl Acrylate	106-63-8	D	I		427			4.4	7.1				
Isobutyl Alcohol	78-83-1	D[d]	I	−40	416	1.2	10.9	2.5	10.5			0.92	0.98
Isobutyraldehyde	78-84-2	C	GAS	−40	196	1.6	10.6	2.5					
Isodecaldehyde	112-31-2	C	IIIA					5.4	0.09				
Isohexane	107-83-5	D[g]			264				211.7			1.00	
Isopentane	78-78-4	D[g]			420				688.6				
Isooctyl Aldehyde	123-05-7	C	II		197				1.9				
Isophorone	78-59-1	D		84	460	0.8	3.8	4.8	0.4				
Isoprene	78-79-5	D[d]	I	−54	220	1.5	8.9	2.4	550.6				
Isopropyl Acetate	108-21-4	D	I		460	1.8	8.0	3.5	60.4				
Isopropyl Ether	108-20-3	D[d]	I	−28	443	1.4	7.9	3.5	148.7		1.14		0.94
Isopropyl Glycidyl Ether	4016-14-2	C	I										
Isopropylamine	75-31-0	D	GAS	−26	402	2.3	10.4	2.0			2.0		
Kerosene	8008-20-6	D	II	72	210	0.7	5.0			IIA			
Liquefied Petroleum Gas	68476-85-7	D	I		405								
Mesityl Oxide	141-97-9	D[d]	I	31	344	1.4	7.2	3.4	47.6				
Methane	74-82-8	D[d]	GAS	−223	630	5.0	15.0	0.6		IIA	0.28	1.00	1.12
Methanol	67-56-1	D[d]	I	12	385	6.0	36.0	1.1	126.3	IIA	0.14	0.82	0.92
Methyl Acetate	79-20-9	D	GAS	−10	454	3.1	16.0	2.6		IIB		1.08	0.99
Methyl Acrylate	96-33-3	D	GAS	−3	468	2.8	25.0	3.0				0.98	0.85
Methyl Alcohol	67-56-1	D[d]	I		385	6.0	36.0	1.1	126.3				0.91
Methyl Amyl Alcohol	108-11-2	D	II	41		1.0	5.5	3.5	5.3				1.01
Methyl Chloride	74-87-3	D	GAS	−46	632	8.1	17.4	1.7					1.00
Methyl Ether	115-10-6	C[d]	GAS	−41	350	3.4	27.0	1.6				0.85	0.84
Methyl Ethyl Ketone	78-93-3	D[d]	I	−6	404	1.4	11.4	2.5	92.4		0.53	0.92	0.84
Methyl Formal	534-15-6	C[d]	I	1	238			3.1					
Methyl Formate	107-31-3	D	GAS	−19	449	4.5	23.0	2.1					0.94
2-Methylhexane	31394-54-4	D[g]	I		280								
Methyl Isobutyl Ketone	141-79-7	D[d]	I	31	440	1.2	8.0	3.5	11.0				
Methyl Isocyanate	624-83-9	D	GAS	−15	534	5.3	26.0	2.0					
Methyl Mercaptan	74-93-1	C	GAS	−18		3.9	21.8	1.7					
Methyl Methacrylate	80-62-6	D	I	10	422	1.7	8.2	3.6	37.2	IIA			0.95
Methyl N-Amyl Ketone	110-43-0	D	II	49	393	1.1	7.9	3.9	3.8				
Methyl Tertiary Butyl Ether	1634-04-4	D	I	−80	435	1.6	8.4	0.2	250.1				
2-Methyloctane	3221-61-2				220				6.3				
2-Methylpropane	75-28-5	D[g]	I		460				2639.0				
Methyl-1-Propanol	78-83-1	D[d]	I	−40	416	1.2	10.9	2.5	10.1				0.98
Methyl-2-Propanol	75-65-0	D[d]	I	10	360	2.4	8.0	2.6	42.2				
2-Methyl-5-Ethyl Pyridine	104-90-5	D		74		1.1	6.6	4.2					
Methylacetylene	74-99-7	C[d]	I			1.7		1.4	4306.0		0.11		
Methylacetylene-Propadiene	27846-30-6	C	I										0.74
Methylal	109-87-5	C	I	−18	237	1.6	17.6	2.6	398.0				

Commentary Table 5.1 *Continued*

Chemical	CAS No.	Class I Division Group	Type[a]	Flash Point (°C)	AIT (°C)	%LFL	%UFL	Vapor Density (Air = 1)	Vapor Pressure[b] (mm Hg)	Class I Zone Group[c]	MIE (mJ)	MIC Ratio	MESG (mm)
Methylamine	74-89-5	D	GAS		430	4.9	20.7	1.0		IIA			1.10
2-Methylbutane	78-78-4	D[g]		−56	420	1.4	8.3	2.6	688.6				
Methylcyclohexane	208-87-2	D	I	−4	250	1.2	6.7	3.4			0.27		
Methylcyclohexanol	25630-42-3	D		68	296			3.9					
2-Methycyclohexanone	583-60-8	D	II					3.9					
2-Methylheptane		D[g]			420								
3-Methylhexane	589-34-4	D[g]			280				61.5				
3-Methylpentane	94-14-0	D[g]			278								
2-Methylpropane	75-28-5	D[g]	I		460				2639.0				
2-Methyl-1-Propanol	78-83-1	D[d]	I	−40	223	1.2	10.9	2.5	10.5				
2-Methyl-2-Propanol	75-65-0	D[d]	I		478	2.4	8.0	2.6	42.2				
2-Methyloctane	2216-32-2	D[g]			220								
3-Methyloctane	2216-33-3	D[g]			220				6.3				
4-Methyloctane	2216-34-4	D[g]			225				6.8				
Monoethanolamine	141-43-5	D		85	410			2.1	0.4	IIA			
Monoisopropanolamine	78-96-6	D		77	374			2.6	1.1				
Monomethyl Aniline	100-61-8	C			482				0.5				
Monomethyl Hydrazine	60-34-4	C	I	23	194	2.5	92.0	1.6					
Morpholine	110-91-8	C[d]	II	35	310	1.4	11.2	3.0	10.1				0.95
Naphtha (Coal Tar)	8030-30-6	D	II	42	277					IIA			
Naphtha (Petroleum)	8030-30-6	D[d,i]	I	42	288	1.1	5.9	2.5		IIA			
Neopentane	463-82-1	D[g]		−65	450	1.4	8.3	2.6	1286.0				
Nitrobenzene	98-95-3	D		88	482	1.8		4.3	0.3				0.94
Nitroethane	79-24-3	C	I	28	414	3.4		2.6	20.7	IIA			0.87
Nitromethane	75-52-5	C	I	35	418	7.3		2.1	36.1	IIA		0.92	1.17
1-Nitropropane	108-03-2	C	I	34	421	2.2		3.1	10.1				0.84
2-Nitropropane	79-46-9	C[d]	I	28	428	2.6	11.0	3.1	17.1				
n-Nonane	111-84-2	D[g]	I	31	205	0.8	2.9	4.4	4.4	IIA			
Nonene	27214-95-8	D	I			0.8		4.4					
Nonyl Alcohol	143-08-8	D				0.8	6.1	5.0	0.02	IIA			
n-Octane	111-65-9	D[d,g]	I	13	206	1.0	6.5	3.9	14.0	IIA			0.94
Octene	25377-83-7	D	I	8	230	0.9		3.9					
n-Octyl Alcohol	111-87-5	D						4.5	0.08	IIA			1.05
n-Pentane	109-66-0	D[d,g]	I	−40	243	1.5	7.8	2.5	513.0		0.28	0.97	0.93
1-Pentanol	71-41-0	D[d]	I	33	300	1.2	10.0	3.0	2.5	IIA			
2-Pentanone	107-87-9	D	I	7	452	1.5	8.2	3.0	35.6				0.99
1-Pentene	109-67-1	D	I	−18	275	1.5	8.7	2.4	639.7				
2-Pentene	109-68-2	D	I	−18				2.4					
2-Pentyl Acetate	626-38-0	D	I	23		1.1	7.5	4.5					
Phenylhydrazine	100-63-0	D		89				3.7	0.03				
Process Gas > 30% H2	1333-74-0	B[j]	GAS		520	4.0	75.0	0.1			0.019	0.45	
Propane	74-98-6	D[d]	GAS	−104	450	2.1	9.5	1.6		IIA	0.25	0.82	0.97
1-Propanol	71-23-8	D[d]	I	15	413	2.2	13.7	2.1	20.7	IIA			0.89
2-Propanol	67-63-0	D[d]	I	12	399	2.0	12.7	2.1	45.4		0.65		1.00
Propiolactone	57-57-8	D				2.9		2.5	2.2				
Propionaldehyde	123-38-6	C	I	−9	207	2.6	17.0	2.0	318.5				
Propionic Acid	79-09-4	D	II	54	466	2.9	12.1	2.5	3.7				
Propionic Anhydride	123-62-6	D		74	285	1.3	9.5	4.5	1.4				
n-Propyl Acetate	109-60-4	D	I	14	450	1.7	8.0	3.5	33.4				1.05
n-Propyl Ether	111-43-3	C[d]	I	21	215	1.3	7.0	3.5	62.3				

(continues)

Commentary Table 5.1 *Continued*

Chemical	CAS No.	Class I Division Group	Type[a]	Flash Point (°C)	AIT (°C)	%LFL	%UFL	Vapor Density (Air = 1)	Vapor Pressure[b] (mm Hg)	Class I Zone Group[c]	MIE (mJ)	MIC Ratio	MESG (mm)
Propyl Nitrate	627-13-4	B[d]	I	20	175	2.0	100.0						
Propylene	115-07-1	D[d]	GAS	−108	455	2.0	11.1	1.5			0.28		0.91
Propylene Dichloride	78-87-5	D	I	16	557	3.4	14.5	3.9	51.7				1.32
Propylene Oxide	75-56-9	B(C)[d,e]	I	−37	449	2.3	36.0	2.0	534.4		0.13		0.70
Pyridine	110-86-1	D[d]	I	20	482	1.8	12.4	2.7	20.8	IIA			
Styrene	100-42-5	D[d]	I	31	490	0.9	6.8	3.6	6.1	IIA		1.21	
Tetrahydrofuran	109-99-9	C[d]	I	−14	321	2.0	11.8	2.5	161.6	IIB	0.54		0.87
Tetrahydronaphthalene	119-64-2	D	IIIA		385	0.8	5.0	4.6	0.4				
Tetramethyl Lead	75-74-1	C	II	38				9.2					
Toluene	108-88-3	D[d]	I	4	480	1.1	7.1	3.1	28.53	IIA	0.24		
n-Tridecene	2437-56-1	D	IIIA			0.6		6.4	593.4				
Triethylamine	121-44-8	C[d]	I	−9	249	1.2	8.0	3.5	68.5	IIA	0.75		
Triethylbenzene	25340-18-5	D		83			56.0	5.6					
2,2,3-Trimethylbutane		D[g]			442								
2,2,4-Trimethylbutane		D[g]			407								
2,2,3-Trimethylpentane		D[g]			396								
2,2,4-Trimethylpentane		D[g]			415								
2,3,3-Trimethylpentane		D[g]			425								
Tripropylamine	102-69-2	D	II	41				4.9	1.5				1.13
Turpentine	8006-64-2	D	I	35	253	0.8			4.8				
n-Undecene	28761-27-5	D	IIIA			0.7		5.5					
Unsymmetrical Dimethyl Hydrazine	57-14-7	C[d]	I	−15	249	2.0	95.0	1.9					0.85
Valeraldehyde	110-62-3	C	I	280	222			3.0	34.3				
Vinyl Acetate	108-05-4	D[d]	I	−6	402	2.6	13.4	3.0	113.4	IIA	0.70		0.94
Vinyl Chloride	75-01-4	D[d]	GAS	−78	472	3.6	33.0	2.2					0.96
Vinyl Toluene	25013-15-4	D		52	494	0.8	11.0	4.1					
Vinylidene Chloride	75-35-4	D	I		570	6.5	15.5	3.4	599.4				3.91
Xylene	1330-20-7	D[d]	I	25	464	0.9	7.0	3.7		IIA	0.2		
Xylidine	121-69-7	C	IIIA	63	371	1.0		4.2	0.7				

Notes:

[a]Type is used to designate if the material is a gas, flammable liquid, or combustible liquid. *(See 4.2.6 and 4.2.7.)*

[b]Vapor pressure reflected in units of mm Hg at 25°C (77°F) unless stated otherwise.

[c]Class I, Zone Groups are based on 1996 IEC TR3 60079-20, *Electrical apparatus for explosive gas atmospheres — Part 20: Data for flammable gases and vapors, relating to the use of electrical apparatus*, which contains additional data on MESG and group classifications.

[d]Material has been classified by test.

[e]Where all conduit runs into explosionproof equipment are provided with explosionproof seals installed within 450 mm (18 in.) of the enclosure, equipment for the group classification shown in parentheses is permitted.

[f]For classification of areas involving ammonia, see ASHRAE 15, *Safety Code for Mechanical Refrigeration*, and ANSI/CGA G2.1, *Safety Requirements for the Storage and Handling of Anhydrous Ammonia.*

[g]Commercial grades of aliphatic hydrocarbon solvents are mixtures of several isomers of the same chemical formula (or molecular weight). The autoignition temperatures of the individual isomers are significantly different. The electrical equipment should be suitable for the AIT of the solvent mixture. *(See A.4.4.2.)*

[h]Certain chemicals have characteristics that require safeguards beyond those required for any of the above groups. Carbon disulfide is one of these chemicals because of its low autoignition temperature and the small joint clearance necessary to arrest its flame propagation.

[i]Petroleum naphtha is a saturated hydrocarbon mixture whose boiling range is 20°C to 135°C (68°F to 275°F). It is also known as benzine, ligroin, petroleum ether, and naphtha.

[j]Fuel and process gas mixtures found by test not to present hazards similar to those of hydrogen may be grouped based on the test results.

Source: Table 4.4.2 in NFPA 497, *Recommended Practice for the Classification of Flammable Liquids, Gases, or Vapors and of Hazardous (Classified) Locations for Electrical Installations in Chemical Process Areas*, 2004 edition.

includes a description of the test methods used to determine the ignition temperatures and electrical resistivity of combustible dusts. The publication is available from the National Technical Information Service (NTIS), Springfield, VA 22151.

(B) Class II Group Classifications Class II groups shall be in accordance with 500.6(B)(1) through (B)(3).

(1) Group E Atmospheres containing combustible metal dusts, including aluminum, magnesium, and their commercial alloys, or other combustible dusts whose particle size, abrasiveness, and conductivity present similar hazards in the use of electrical equipment. [NFPA 499:3.3]

> FPN: Certain metal dusts may have characteristics that require safeguards beyond those required for atmospheres containing the dusts of aluminum, magnesium, and their commercial alloys. For example, zirconium, thorium, and uranium dusts have extremely low ignition temperatures [as low as 20°C (68°F)] and minimum ignition energies lower than any material classified in any of the Class I or Class II groups.

(2) Group F Atmospheres containing combustible carbonaceous dusts that have more than 8 percent total entrapped volatiles (see ASTM D 3175-89, *Standard Test Method for Volatile Material in the Analysis Sample for Coal and Coke*, for coal and coke dusts) or that have been sensitized by other materials so that they present an explosion hazard. Coal, carbon black, charcoal, and coke dusts are examples of carbonaceous dusts. [NFPA 499:3.3]

(3) Group G Atmospheres containing combustible dusts not included in Group E or F, including flour, grain, wood, plastic, and chemicals.

> FPN No. 1: For additional information on group classification of Class II materials, see NFPA 499-2004, *Recommended Practice for the Classification of Combustible Dusts and of Hazardous (Classified) Locations for Electrical Installations in Chemical Process Areas*.
>
> FPN No. 2: The explosion characteristics of air mixtures of dust vary with the materials involved. For Class II locations, Groups E, F, and G, the classification involves the tightness of the joints of assembly and shaft openings to prevent the entrance of dust in the dust-ignitionproof enclosure, the blanketing effect of layers of dust on the equipment that may cause overheating, and the ignition temperature of the dust. It is necessary, therefore, that equipment be identified not only for the class, but also for the specific group of dust that will be present.
>
> FPN No. 3: Certain dusts may require additional precautions due to chemical phenomena that can result in the generation of ignitible gases. See ANSI C2-2002, *National Electrical Safety Code*, Section 127A, Coal Handling Areas.

Commentary Table 5.2 is an alphabetical listing of selected combustible materials with their group classification and relevant physical properties. All the materials included in this table have been evaluated for the purpose of designating the appropriate dust group. This information is used to properly select electrical equipment for use in Class II locations. Combustible dusts are classified into three Class II, division groups — E, F, and G — depending on their properties. Commentary Table 5.2 is extracted from NFPA 499, *Recommended Practice for the Classification of Combustible Dusts and of Hazardous (Classified) Locations for Electrical Installations in Chemical Process Areas*. For the definitions of terms used in the table, refer to NFPA 499.

As in Class I locations, equipment must be approved not only for the class but also for the specific group. It is important that, in addition to the proper selection of equipment, high standards of installation be maintained for subsequent additions or alterations.

500.7 Protection Techniques

Section 500.7(A) through 500.7(L) shall be acceptable protection techniques for electrical and electronic equipment in hazardous (classified) locations.

(A) Explosionproof Apparatus This protection technique shall be permitted for equipment in Class I, Division 1 or 2 locations.

(B) Dust Ignitionproof This protection technique shall be permitted for equipment in Class II, Division 1 or 2 locations.

(C) Dusttight This protection technique shall be permitted for equipment in Class II, Division 2 or Class III, Division 1 or 2 locations.

(D) Purged and Pressurized This protection technique shall be permitted for equipment in any hazardous (classified) location for which it is identified.

NFPA 496, *Standard for Purged and Pressurized Enclosures for Electrical Equipment*, covers purged and pressurized enclosures for electrical equipment in Class I and Class II hazardous (classified) locations.

In Class I locations, purged and pressurized enclosures are used to eliminate or reduce, within the enclosure, a Class I hazardous (classified) location classification, as defined in Article 500 of the *Code*. Purged and pressurized enclosures make it possible for equipment that is not otherwise acceptable for hazardous (classified) locations to be used in these locations, in accordance with the *Code*.

Purging is the process of supplying an enclosure with a protective gas at a sufficient flow and positive pressure to reduce the concentration of any flammable gas or vapor

Commentary Table 5.2 Selected Combustible Materials

Chemical Name	CAS No.	NEC Group	Code	Layer or Cloud Ignition Temp. (°C)
Acetal, Linear		G	NL	440
Acetoacet-p-phenetidide	122-82-7	G	NL	560
Acetoacetanilide	102-01-2	G	M	440
Acetylamino-t-nitrothiazole		G		450
Acrylamide Polymer		G		240
Acrylonitrile Polymer		G		460
Acrylonitrile-Vinyl Chloride-Vinylidenechloride copolymer (70-20-10)		G		210
Acrylonitrile-Vinyl Pyridine Copolymer		G		240
Adipic Acid	124-04-9	G	M	550
Alfalfa Meal		G		200
Alkyl Ketone Dimer Sizing Compound		G		160
Allyl Alcohol Derivative (CR-39)		G	NL	500
Almond Shell		G		200
Aluminum, A422 Flake	7429-90-5	E		320
Aluminum, Atomized Collector Fines		E	CL	550
Aluminum—cobalt alloy (60-40)		E		570
Aluminum—copper alloy (50-50)		E		830
Aluminum—lithium alloy (15% Li)		E		400
Aluminum—magnesium alloy (Dowmetal)		E	CL	430
Aluminum—nickel alloy (58-42)		E		540
Aluminum—silicon alloy (12% Si)		E	NL	670
Amino-5-nitrothiazole	121-66-4	G		460
Anthranilic Acid	118-92-3	G	M	580
Apricot Pit		G		230
Aryl-nitrosomethylamide		G	NL	490
Asphalt	8052-42-4	F		510
Aspirin [acetol (2)]	50-78-2	G	M	660
Azelaic Acid	109-31-9	G	M	610
Azo-bis-butyronitrile	78-67-1	G		350
Benzethonium Chloride		G	CL	380
Benzoic Acid	65-85-0	G	M	440
Benzotriazole	95-14-7	G	M	440
Beta-naphthalene-axo-dimethylaniline		G		175
Bis(2-hydroxy-5-chlorophenyl) Methane	97-23-4	G	NL	570
Bisphenol-A	80-05-7	G	M	570
Boron, Commercial Amorphous (85% B)	7440-42-8	E		400
Calcium Silicide		E		540
Carbon Black (More Than 8% Total Entrapped Volatiles)		F		
Carboxymethyl Cellulose	9000-11-7	G		290
Carboxypolymethylene		G	NL	520
Cashew Oil, Phenolic, Hard		G		180
Cellulose		G		260
Cellulose Acetate		G		340
Cellulose Acetate Butyrate		G	NL	370
Cellulose Triacetate		G	NL	430
Charcoal (Activated)	64365-11-3	F		180
Charcoal (More Than 8% Total Entrapped Volatiles)		F		
Cherry Pit		G		220
Chlorinated Phenol		G	NL	570
Chlorinated Polyether Alcohol		G		460
Chloroacetoacetanilide	101-92-8	G	M	640

Commentary Table 5.2 *Continued*

Chemical Name	CAS No.	*NEC* Group	Code	Layer or Cloud Ignition Temp. (°C)
Chromium (97%) Electrolytic, Milled	7440-47-3	E		400
Cinnamon		G		230
Citrus Peel		G		270
Coal, Kentucky Bituminous		F		180
Coal, Pittsburgh Experimental		F		170
Coal, Wyoming		F		
Cocoa Bean Shell		G		370
Cocoa, Natural, 19% Fat		G		240
Coconut Shell		G		220
Coke (More Than 8% Total Entrapped Volatiles)		F		
Cork		G		210
Corn		G		250
Corn Dextrine		G		370
Corncob Grit		G		240
Cornstarch, Commercial		G		330
Cornstarch, Modified		G		200
Cottonseed Meal		G		200
Coumarone-Indene, Hard		G	NL	520
Crag No. 974	533-74-4	G	CL	310
Cube Root, South America	83-79-4	G		230
Di-alphacumyl Peroxide, 40-60 on CA	80-43-3	G		180
Diallyl Phthalate	131-17-9	G	M	480
Dicyclopentadiene Dioxide		G	NL	420
Dieldrin (20%)	60-57-1	G	NL	550
Dihydroacetic Acid		G	NL	430
Dimethyl Isophthalate	1459-93-4	G	M	580
Dimethyl Terephthalate	120-61-6	G	M	570
Dinitro-o-toluamide	148-01-6	G	NL	500
Dinitrobenzoic Acid		G	NL	460
Diphenyl	92-52-4	G	M	630
Ditertiary-butyl-paracresol	128-37-0	G	NL	420
Dithane m-45	8018-01-7	G		180
Epoxy		G	NL	540
Epoxy-bisphenol A		G	NL	510
Ethyl Cellulose		G	CL	320
Ethyl Hydroxyethyl Cellulose		G	NL	390
Ethylene Oxide Polymer		G	NL	350
Ethylene-maleic Anhydride Copolymer		G	NL	540
Ferbam™	14484-64-1	G		150
Ferromanganese, Medium Carbon	12604-53-4	E		290
Ferrosilicon (88% Si, 9% Fe)	8049-17-0	E		800
Ferrotitanium (19% Ti, 74.1% Fe, 0.06% C)		E	CL	380
Flax Shive		G		230
Fumaric Acid	110-17-8	G	M	520
Garlic, Dehydrated		G	NL	360
Gilsonite	12002-43-6	F		500
Green Base Harmon Dye		G		175
Guar Seed		G	NL	500
Gulasonic Acid, Diacetone		G	NL	420
Gum, Arabic		G		260

(continues)

Commentary Table 5.2 *Continued*

Chemical Name	CAS No.	*NEC* Group	Code	Layer or Cloud Ignition Temp. (°C)
Gum, Karaya		G		240
Gum, Manila		G	CL	360
Gum, Tragacanth	9000-65-1	G		260
Hemp Hurd		G		220
Hexamethylene Tetramine	100-97-0	G	S	410
Hydroxyethyl Cellulose		G	NL	410
Iron, 98% H2 Reduced		E		290
Iron, 99% Carbonyl	13463-40-6	E		310
Isotoic Anhydride		G	NL	700
L-sorbose		G	M	370
Lignin, Hydrolized, Wood-type, Fine		G	NL	450
Lignite, California		F		180
Lycopodium		G		190
Malt Barley		G		250
Manganese	7439-96-5	E		240
Magnesium, Grade B, Milled		E		430
Manganese Vancide		G		120
Mannitol	69-65-8	G	M	460
Methacrylic Acid Polymer		G		290
Methionine (l-methionine)	63-68-3	G		360
Methyl Cellulose		G		340
Methyl Methacrylate Polymer	9011-14-7	G	NL	440
Methyl Methacrylate-ethyl Acrylate		G	NL	440
Methyl Methacrylate-styrene-butadiene		G	NL	480
Milk, Skimmed		G		200
N,N-Dimethylthio-formamide		G		230
Nitropyridone	100703-82-0	G	M	430
Nitrosamine		G	NL	270
Nylon Polymer	63428-84-2	G		430
Para-oxy-benzaldehyde	123-08-0	G	CL	380
Paraphenylene Diamine	106-50-3	G	M	620
Paratertiary Butyl Benzoic Acid	98-73-7	G	M	560
Pea Flour		G		260
Peach Pit Shell		G		210
Peanut Hull		G		210
Peat, Sphagnum	94114-14-4	G		240
Pecan Nut Shell	8002-03-7	G		210
Pectin	5328-37-0	G		200
Pentaerythritol	115-77-5	G	M	400
Petrin Acrylate Monomer	7659-34-9	G	NL	220
Petroleum Coke (More Than 8% Total Entrapped Volatiles)		F		
Petroleum Resin	64742-16-1	G		500
Phenol Formaldehyde	9003-35-4	G	NL	580
Phenol Formaldehyde, Polyalkylene-p	9003-35-4	G		290
Phenol Furfural	26338-61-4	G		310
Phenylbetanaphthylamine	135-88-6	G	NL	680
Phthalic Anhydride	85-44-9	G	M	650
Phthalimide	85-41-6	G	M	630
Pitch, Coal Tar	65996-93-2	F	NL	710
Pitch, Petroleum	68187-58-6	F	NL	630
Polycarbonate		G	NL	710
Polyethylene, High Pressure Process	9002-88-4	G		380

Commentary Table 5.2 *Continued*

Chemical Name	CAS No.	*NEC* Group	Code	Layer or Cloud Ignition Temp. (°C)
Polyethylene, Low Pressure Process	9002-88-4	G	NL	420
Polyethylene Terephthalate	25038-59-9	G	NL	500
Polyethylene Wax	68441-04-8	G	NL	400
Polypropylene (no antioxidant)	9003-07-0	G	NL	420
Polystyrene Latex	9003-53-6	G		500
Polystyrene Molding Compound	9003-53-6	G	NL	560
Polyurethane Foam, Fire Retardant	9009-54-5	G		390
Polyurethane Foam, No Fire Retardant	9009-54-5	G		440
Polyvinyl Acetate	9003-20-7	G	NL	550
Polyvinyl Acetate/Alcohol	9002-89-5	G		440
Polyvinyl Butyral	63148-65-2	G		390
Polyvinyl Chloride-dioctyl Phthalate		G	NL	320
Potato Starch, Dextrinated	9005-25-8	G	NL	440
Pyrethrum	8003-34-7	G		210
Rayon (Viscose) Flock	61788-77-0	G		250
Red Dye Intermediate		G		175
Rice		G		220
Rice Bran		G	NL	490
Rice Hull		G		220
Rosin, DK	8050-09-7	G	NL	390
Rubber, Crude, Hard	9006-04-6	G	NL	350
Rubber, Synthetic, Hard (33% S)	64706-29-2	G	NL	320
Safflower Meal		G		210
Salicylanilide	87-17-2	G	M	610
Sevin	63-25-2	G		140
Shale, Oil	68308-34-9	F		
Shellac	9000-59-3	G	NL	400
Sodium Resinate	61790-51-0	G		220
Sorbic Acid (Copper Sorbate or Potash)	110-44-1	G		460
Soy Flour	68513-95-1	G		190
Soy Protein	9010-10-0	G		260
Stearic Acid, Aluminum Salt	637-12-7	G		300
Stearic Acid, Zinc Salt	557-05-1	G	M	510
Styrene Modified Polyester-Glass Fiber	100-42-5	G		360
Styrene-acrylonitrile (70-30)	9003-54-7	G	NL	500
Styrene-butadiene Latex (>75% styrene)	903-55-8	G	NL	440
Styrene-maleic Anhydride Copolymer	9011-13-6	G	CL	470
Sucrose	57-50-1	G	CL	350
Sugar, Powdered	57-50-1	G	CL	370
Sulfur	7704-34-9	G		220
Tantalum	7440-25-7	E		300
Terephthalic Acid	100-21-0	G	NL	680
Thorium, 1.2% O2	7440-29-1	E	CL	280
Tin, 96%, Atomized (2% Pb)	7440-31-5	E		430
Titanium, 99% Ti	7440-32-6	E	CL	330
Titanium Hydride (95% Ti, 3.8% H2)	7704-98-5	E	CL	480
Trithiobisdimethylthio-formamide		G		230
Tung, Kernels, Oil-free	8001-20-5	G		240
Urea Formaldehyde Molding Compound	9011-05-6	G	NL	460
Urea Formaldehyde-phenol Formaldehyde	25104-55-6	G		240

(continues)

Commentary Table 5.2 *Continued*

Chemical Name	CAS No.	*NEC* Group	Code	Layer or Cloud Ignition Temp. (°C)
Vanadium, 86.4%	7440-62-2	E		490
Vinyl Chloride-acrylonitrile Copolymer	9003-00-3	G		470
Vinyl Toluene-acrylonitrile Butadiene	76404-69-8	G	NL	530
Violet 200 Dye		G		175
Vitamin B1, Mononitrate	59-43-8	G	NL	360
Vitamin C	50-81-7	G		280
Walnut Shell, Black		G		220
Wheat		G		220
Wheat Flour	130498-22-5	G		360
Wheat Gluten, Gum	100684-25-1	G	NL	520
Wheat Starch		G	NL	380
Wheat Straw		G		220
Wood Flour		G		260
Woodbark, Ground		G		250
Yeast, Torula	68602-94-8	G		260
Zirconium Hydride	7704-99-6	E		270
Zirconium		E	CL	330

Notes:
1. Normally, the minimum ignition temperature of a layer of a specific dust is lower than the minimum ignition temperature of a cloud of that dust. Since this is not universally true, the lower of the two minimum ignition temperatures is listed. If no symbol appears between the two temperature columns, then the layer ignition temperature is shown. "CL" means the cloud ignition temperature is shown. "NL" means that no layer ignition temperature is available, and the cloud ignition temperature is shown. "M" signifies that the dust layer melts before it ignites; the cloud ignition temperature is shown. "S" signifies that the dust layer sublimes before it ignites; the cloud ignition temperature is shown.
2. Certain metal dusts may have characteristics that require safeguards beyond those required for atmospheres containing the dusts of aluminum, magnesium, and their commercial alloys. For example, zirconium, thorium, and uranium dusts have extremely low ignition temperatures [as low as 20°C (68°F)] and minimum ignition energies lower than any material classified in any of the Class I or Class II groups.
Source: Table 4.5.2 in NFPA 499, *Recommended Practice for the Classification of Combustible Dusts and of Hazardous (Classified) Locations for Electrical Installations in Chemical Process Areas*, 2004 edition.

initially present to an acceptable level. The types of pressurizing are as follows:

1. Type X pressurizing reduces the classification within a protected enclosure from Division 1 or Zone 1 to unclassified.
2. Type Y pressurizing reduces the classification within a protected enclosure from Division 1 to Division 2 or from Zone 1 to Zone 2.
3. Type Z pressurizing reduces the classification within a protected enclosure from Division 2 or Zone 2 to unclassified.

In Class II hazardous (classified) locations, pressurized enclosures prevent the entrance of dusts into an enclosure. Pressurized enclosures make it possible for equipment that is not otherwise acceptable for hazardous (classified) locations to be used in these locations, in accordance with the

Code. Pressurization, for the purposes of NFPA 496, is the process of supplying an enclosure with a protective gas, with or without continuous flow, at sufficient pressure to prevent the entrance of a flammable gas or vapor, a combustible dust, or an ignitible fiber.

It should be noted that an atmosphere that is made hazardous by combustible dust inside an enclosure cannot be reduced to a safe level by supplying a flow of protective gas in the same manner as with gases or vapors. Supplying a flow of air into the enclosure could stir up the dust that has accumulated at the bottom of the enclosure and create a dust cloud within the enclosure that could explode if an ignition source occurs. The enclosure must be opened, and the dust must be removed. Visual inspection can determine if the dust has been removed. Positive pressure then prevents dust from entering a clean enclosure.

Field-installed devices and equipment, such as pushbutton controls and pilot lights, are permitted in purged and pressurized enclosures, provided the conductor terminals are within the purged or pressurized atmosphere. Section 7.3.4 of NFPA 30, *Flammable and Combustible Liquids Code*, and Section 4.6 of ANSI/API RP 500, *Recommended Practice for Classification of Locations for Electrical Installations at Petroleum Facilities Classified as Class I, Division 1 and Division 2*, provide guidelines on what is considered adequate ventilation.

(E) Intrinsic Safety This protection technique shall be permitted for equipment in Class I, Division 1 or 2; or Class II, Division 1 or 2; or Class III, Division 1 or 2 locations. The provisions of Articles 501 through 503 and Articles 510 through 516 shall not be considered applicable to such installations, except as required by Article 504, and installation of intrinsically safe apparatus and wiring shall be in accordance with the requirements of Article 504.

(F) Nonincendive Circuit This protection technique shall be permitted for equipment in Class I, Division 2; Class II, Division 2; or Class III, Division 1 or 2 locations.

(G) Nonincendive Equipment This protection technique shall be permitted for equipment in Class I, Division 2; Class II, Division 2; or Class III, Division 1 or 2 locations.

(H) Nonincendive Component This protection technique shall be permitted for equipment in Class I, Division 2; Class II, Division 2; or Class III, Division 1 or 2 locations.

(I) Oil Immersion This protection technique shall be permitted for current-interrupting contacts in Class I, Division 2 locations as described in 501.115(B)(1)(2).

(J) Hermetically Sealed This protection technique shall be permitted for equipment in Class I, Division 2; Class II, Division 2; or Class III, Division 1 or 2 locations.

(K) Combustible Gas Detection System A combustible gas detection system shall be permitted as a means of protection in industrial establishments with restricted public access and where the conditions of maintenance and supervision ensure that only qualified persons service the installation. Gas detection equipment shall be listed for detection of the specific gas or vapor to be encountered. Where such a system is installed, equipment specified in 500.7(K)(1), (K)(2), or (K)(3) shall be permitted.

The type of detection equipment, its listing, installation location(s), alarm and shutdown criteria, and calibration frequency shall be documented when combustible gas detectors are used as a protection technique.

Where combustible gas detection systems are utilized for protection, documentation must be provided indicating the type of equipment, its listing, the location where it is installed, the type of alarm or signal, and the shutdown procedure. A schedule for calibration of the system is also required to be documented.

FPN No. 1: For further information, see ANSI/ISA 12.13.01, *Performance Requirements, Combustible Gas Detectors*.
FPN No. 2: For further information, see ANSI/API RP 500, *Recommended Practice for Classification of Locations for Electrical Installations at Petroleum Facilities Classified as Class I, Division I or Division 2*.
FPN No. 3: For further information, see ISA-RP12.13.02, *Installation, Operation, and Maintenance of Combustible Gas Detection Instruments*.

(1) Inadequate Ventilation In a Class I, Division 1 location that is so classified due to inadequate ventilation, electrical equipment suitable for Class I, Division 2 locations shall be permitted.

(2) Interior of a Building In a building located in, or with an opening into, a Class I, Division 2 location where the interior does not contain a source of flammable gas or vapor, electrical equipment for unclassified locations shall be permitted.

(3) Interior of a Control Panel In the interior of a control panel containing instrumentation utilizing or measuring flammable liquids, gases, or vapors, electrical equipment suitable for Class I, Division 2 locations shall be permitted.

(L) Other Protection Techniques Other protection techniques used in equipment identified for use in hazardous (classified) locations.

Some listed equipment employs unique protection techniques or a combination of protection techniques, such as listed attachment plugs for use in hazardous locations and listed battery-operated two-way radios, cell phones, flashlights, and lanterns.

500.8 Equipment

Articles 500 through 504 require equipment construction and installation that ensure safe performance under conditions of proper use and maintenance.

FPN No. 1: It is important that inspection authorities and users exercise more than ordinary care with regard to installation and maintenance.
FPN No. 2: Since there is no consistent relationship between explosion properties and ignition temperature, the two are independent requirements.
FPN No. 3: Low ambient conditions require special consideration. Explosionproof or dust-ignitionproof equipment may not be suitable for use at temperatures lower

than $-25°C$ ($-13°F$) unless they are identified for low-temperature service. However, at low ambient temperatures, flammable concentrations of vapors may not exist in a location classified as Class I, Division 1 at normal ambient temperature.

At low ambient temperatures, such as those encountered in the Arctic, explosion pressures increase at very low temperatures. The strengths of materials change, and the explosion pressure in explosionproof equipment may increase beyond the safe operating strength of the material. In addition, some sealing materials for sealing fittings may become brittle. However, the extent of the hazardous (classified) location may also change under low ambient conditions. The material may be used in a location where the temperature range is so low that no vapors are produced based on the flash point of the material involved.

(A) Approval for Class and Properties

(1) Equipment shall be identified not only for the class of location but also for the explosive, combustible, or ignitible properties of the specific gas, vapor, dust, fiber, or flyings that will be present. In addition, Class I equipment shall not have any exposed surface that operates at a temperature in excess of the ignition temperature of the specific gas or vapor. Class II equipment shall not have an external temperature higher than that specified in 500.8(C)(2). Class III equipment shall not exceed the maximum surface temperatures specified in 503.5.

> FPN: Luminaires (lighting fixtures) and other heat-producing apparatus, switches, circuit breakers, and plugs and receptacles are potential sources of ignition and are investigated for suitability in classified locations. Such types of equipment, as well as cable terminations for entry into explosionproof enclosures, are available as listed for Class I, Division 2 locations. Fixed wiring, however, may utilize wiring methods that are not evaluated with respect to classified locations. Wiring products such as cable, raceways, boxes, and fittings, therefore, are not marked as being suitable for Class I, Division 2 locations. Also see 500.8(B)(6)(a).

Suitability of identified equipment shall be determined by any of the following:

(1) Equipment listing or labeling
(2) Evidence of equipment evaluation from a qualified testing laboratory or inspection agency concerned with product evaluation
(3) Evidence acceptable to the authority having jurisdiction such as a manufacturer's self-evaluation or an owner's engineering judgment

(2) Equipment that has been identified for a Division 1 location shall be permitted in a Division 2 location of the same class, group, and temperature class and shall comply with (a) or (b) as applicable.

(a) Intrinsically safe apparatus having a control drawing requiring the installation of associated apparatus for a Division 1 installation shall be permitted to be installed in a Division 2 location if the same associated apparatus is used for the Division 2 installation.

(b) Equipment that is required to be explosionproof shall incorporate seals per 501.15(A) or 501.15(D) when the wiring methods of 501.10(B) are employed.

(3) Where specifically permitted in Articles 501 through 503, general-purpose equipment or equipment in general-purpose enclosures shall be permitted to be installed in Division 2 locations if the equipment does not constitute a source of ignition under normal operating conditions.

(4) Equipment that depends on a single compression seal, diaphragm, or tube to prevent flammable or combustible fluids from entering the equipment shall be identified for a Class I, Division 2 location even if installed in an unclassified location. Equipment installed in a Class I, Division 1 location shall be identified for the Class I, Division 1 location.

> FPN: Equipment used for flow measurement is an example of equipment having a single compression seal, diaphragm, or tube.

(5) Unless otherwise specified, normal operating conditions for motors shall be assumed to be rated full-load steady conditions.

It is not intended that locked-rotor or other motor overload conditions, such as single phasing, be considered when evaluating motor-operating temperatures (internal and external) in Class I, Division 2 locations. However, such abnormal load conditions must be considered when evaluating the external temperatures of explosionproof motors for Class I, Division 1 locations and motors such as dust-ignitionproof motors for Class II, Division 1 locations. It is important to be aware of the increase in temperature in some variable-speed motors when they are operated at the lower speed and are dependent on the fan for cooling.

(6) Where flammable gases or combustible dusts are or may be present at the same time, the simultaneous presence of both shall be considered when determining the safe operating temperature of the electrical equipment.

Examples of where flammable liquid and dust can be present at the same time are at a coal-handling facility, where there is methane gas and coal dust, and in an automotive paint spray shop, where flammable paint and powdered metal flecks are sprayed.

FPN: The characteristics of various atmospheric mixtures of gases, vapors, and dusts depend on the specific material involved.

(B) Marking Equipment shall be marked to show the environment for which it has been evaluated. Unless otherwise specified or allowed in (B)(6), the marking shall include the information specified in (B)(1) through (B)(5).

The marked operating temperature or temperature range is normally referenced to a 104°F ambient. Unless the equipment is provided with thermally actuated sensors that limit the temperature to that marked on the equipment, operation in ambient temperatures higher than 104°F increases the operating temperature of the equipment. Many explosionproof and dust-ignitionproof motors are equipped with thermal protectors. In like manner, operation in ambient temperatures lower than 104°F usually reduces the operating temperature.

(1) Class The marking shall specify the class(es) for which the equipment is suitable.

(2) Division The marking shall specify the division if the equipment is suitable for Division 2 only. Equipment suitable for Division 1 shall be permitted to omit the division marking.

FPN: Equipment not marked to indicate a division, or marked "Division 1" or "Div. 1," is suitable for both Division 1 and 2 locations; see 500.8(A)(2). Equipment marked "Division 2" or "Div. 2" is suitable for Division 2 locations only.

(3) Material Classification Group The marking shall specify the applicable material classification group(s) in accordance with 500.6.

Exception: Fixed luminaires (lighting fixtures) marked for use only in Class I, Division 2 or Class II, Division 2 locations shall not be required to indicate the group.

(4) Equipment Temperature The marking shall specify the temperature class or operating temperature at a 40°C ambient temperature, or at the higher ambient temperature if the equipment is rated and marked for an ambient temperature of greater than 40°C. The temperature class, if provided, shall be indicated using the temperature class (T Codes) shown in Table 500.8(B). Equipment for Class I and Class II shall be marked with the maximum safe operating temperature, as determined by simultaneous exposure to the combinations of Class I and Class II conditions.

Exception: Equipment of the non–heat-producing type, such as junction boxes, conduit, and fittings, and equipment of the heat-producing type having a maximum temperature not

Table 500.8(B) Classification of Maximum Surface Temperature

Maximum Temperature		Temperature Class (T Code)
°C	°F	
450	842	T1
300	572	T2
280	536	T2A
260	500	T2B
230	446	T2C
215	419	T2D
200	392	T3
180	356	T3A
165	329	T3B
160	320	T3C
135	275	T4
120	248	T4A
100	212	T5
85	185	T6

more than 100°C shall not be required to have a marked operating temperature or temperature class.

FPN: More than one marked temperature class or operating temperature, for gases and vapors, dusts, and different ambient temperatures, may appear.

(5) Ambient Temperature Range For equipment rated for a temperature range other than –25°C to +40°C, the marking shall specify the special range of ambient temperatures. The marking shall include either the symbol "Ta" or "Tamb."

FPN: As an example, such a marking might be "–30°C ≤ Ta ≤ +40°C."

(6) Special Allowances

Exception No. 2 and Exception No. 3 to 500.8(B) in the 2002 *Code* were changed to mandatory *Code* language and relocated in a new section, 500.8(B)(6) Special Allowances, in the 2005 *Code*. A squirrel-cage induction motor without brushes, switching mechanisms, or similar arc-producing devices is an example of fixed general-purpose equipment. See 501.125(B) and its associated commentary for more information on motors in Class I, Division 2 locations.

(a) General Purpose Equipment. Fixed general-purpose equipment in Class I locations, other than fixed luminaires (lighting fixtures), that is acceptable for use in Class I, Division 2 locations shall not be required to be marked with the class, division, group, temperature class, or ambient temperature range.

(b) Dusttight Equipment. Fixed dusttight equipment, other than fixed luminaires (lighting fixtures), that is acceptable for use in Class II, Division 2 and Class III locations shall not be required to be marked with the class, division, group, temperature class, or ambient temperature range.

(c) Associated Apparatus. Associated intrinsically safe apparatus and associated nonincendive field wiring apparatus that are not protected by an alternative type of protection shall not be marked with the class, division, group, or temperature class. Associated intrinsically safe apparatus and associated nonincendive field wiring apparatus shall be marked with the class, division, and group of the apparatus to which it is to be connected.

(d) Simple Apparatus. "Simple apparatus" as defined in Article 504, shall not be required to be marked with class, division, group, temperature class, or ambient temperature range.

(C) Temperature

(1) Class I Temperature The temperature marking specified in 500.8(B) shall not exceed the ignition temperature of the specific gas or vapor to be encountered.

The ignition temperature of a solid, liquid, or gaseous substance is the minimum temperature required to initiate or cause self-sustained combustion independent of the heating or heated element. The flash point is the temperature at which the material gives off vapors that will ignite when the temperature reaches the ignition temperature, provided the air-to-fuel ratio is within the proper range. The ignition temperature and the flash point are unrelated properties, except that the flash point is always lower than the ignition temperature.

FPN: For information regarding ignition temperatures of gases and vapors, see NFPA 497-2004, *Recommended Practice for the Classification of Flammable Liquids, Gases, or Vapors, and of Hazardous (Classified) Locations for Electrical Installations in Chemical Process Areas.*

(2) Class II Temperature The temperature marking specified in 500.8(B) shall be less than the ignition temperature of the specific dust to be encountered. For organic dusts that may dehydrate or carbonize, the temperature marking shall not exceed the lower of either the ignition temperature or 165°C (329°F).

FPN: See NFPA 499-2004, *Recommended Practice for the Classification of Combustible Dusts and of Hazardous (Classified) Locations for Electrical Installations in Chemical Process Areas*, for minimum ignition temperatures of specific dusts.

The ignition temperature for which equipment was approved prior to this requirement shall be assumed to be as shown in Table 500.8(C)(2).

Table 500.8(C)(2) Class II Temperatures

Class II Group	Equipment Not Subject to Overloading		Equipment (Such as Motors or Power Transformers) That May Be Overloaded			
			Normal Operation		Abnormal Operation	
	°C	°F	°C	°F	°C	°F
E	200	392	200	392	200	392
F	200	392	150	302	200	392
G	165	329	120	248	165	329

(D) Threading All NPT threaded conduit and fittings referred to herein shall be threaded with a National (American) Standard Pipe Taper (NPT) thread that provides a taper of 1 in 16 (¾-in. taper per foot). Conduit and fittings shall be made wrenchtight to prevent sparking when fault current flows through the conduit system, and to ensure the explosionproof integrity of the conduit system where applicable. Equipment provided with threaded entries for field wiring connections shall be installed in accordance with 500.8(D)(1) or (D)(2). Threaded entries into explosionproof equipment shall be made up with at least five threads fully engaged.

Exception: For listed explosionproof equipment, factory threaded NPT entries shall be made up with at least 4½ threads fully engaged.

(1) Equipment Provided with Threaded Entries for NPT Threaded Conduit or Fittings For equipment provided with threaded entries for NPT threaded conduit or fittings, listed conduit, conduit fittings, or cable fittings shall be used.

FPN: Thread form specifications for NPT threads are located in ANSI/ASME B1.20.1-1983, *Pipe Threads, General Purpose (Inch).*

(2) Equipment Provided with Threaded Entries for Metric Threaded Conduit or Fittings For equipment with metric threaded entries, such entries shall be identified as being metric, or listed adapters to permit connection to conduit or NPT-threaded fittings shall be provided with the equipment. Adapters shall be used for connection to conduit or NPT-threaded fittings. Listed cable fittings that have metric threads shall be permitted to be used.

FPN: Threading specifications for metric threaded entries are located in ISO 965/1-1980, *Metric Screw Threads*, and ISO 965/3-1980, *Metric Screw Threads.*

All conduit joints must be made up wrenchtight to prevent arcing between the conduit and the coupling, fitting, or enclosure of the conduit under ground-fault conditions. The use of a bonding jumper in lieu of a wrenchtight connection is not permitted. The integrity of the ground-fault current path is critical in hazardous locations in order to prevent ignition-capable arcing or sparking.

The information on metric threads in 500.8(D) has been included to allow for safe electrical and mechanical connections where the enclosure has metric threads and the raceway or cable has NPT threads. Equipment with metric threaded entries must be identified or provided with suitable adapters that permit the connection of conduit and fittings that are NPT threaded.

(E) Fiber Optic Cable Assembly Where a fiber optic cable assembly contains conductors that are capable of carrying current, the fiber optic cable assembly shall be installed in accordance with the requirements of Articles 500, 501, 502, or 503, as applicable.

The requirements of Articles 500, 501, 502, or 503 apply, even if the conductor is grounded.

500.9 Specific Occupancies

Articles 510 through 517 cover garages, aircraft hangars, motor fuel dispensing facilities, bulk storage plants, spray application, dipping and coating processes, and health care facilities.

ARTICLE 501
Class I Locations

Summary of Changes

- **General:** Restructured and renumbered to provide a scope section and parallel numbering systems for Articles 501, 502, and 503.

- **501.10(B)(6):** Revised to require single conductor Type MV cables to be shielded or metallic-armored.

- **501.15(B)(2):** Revised to permit seals identified for the purpose of minimizing the passage of gases and vapors (does not have to be of the explosionproof type) within the Division 2 portion of the conduit.

- **501.25:** Revised to provide guidance concerning protection against explosion and against electric shock in order to limit the voltage of exposed live parts.

- **501.35:** Revised to include TVSS devices.

Contents

I. General
    501.1 Scope
    501.5 General
II. Wiring
    501.10 Wiring Methods
        (A) Class I, Division 1
        (B) Class I, Division 2
    501.15 Sealing and Drainage
        (A) Conduit Seals, Class I, Division 1
        (B) Conduit Seals, Class I, Division 2
        (C) Class I, Divisions 1 and 2
        (D) Cable Seals, Class I, Division 1
        (E) Cable Seals, Class I, Division 2
        (F) Drainage
    501.20 Conductor Insulation, Class I, Divisions 1 and 2
    501.25 Uninsulated Exposed Parts, Class I, Divisions 1 and 2
    501.30 Grounding and Bonding, Class I, Divisions 1 and 2
        (A) Bonding
        (B) Types of Equipment Grounding Conductors
    501.35 Surge Protection
        (A) Class I, Division 1
        (B) Class I, Division 2
    501.40 Multiwire Branch Circuits
III. Equipment
    501.100 Transformers and Capacitors
        (A) Class I, Division 1
        (B) Class I, Division 2
    501.105 Meters, Instruments, and Relays
        (A) Class I, Division 1
        (B) Class I, Division 2
    501.115 Switches, Circuit Breakers, Motor Controllers, and Fuses
        (A) Class I, Division 1
        (B) Class I, Division 2
    501.120 Control Transformers and Resistors
        (A) Class I, Division 1
        (B) Class I, Division 2
    501.125 Motors and Generators
        (A) Class I, Division 1
        (B) Class I, Division 2
    501.130 Luminaires (Lighting Fixtures)
        (A) Class I, Division 1
        (B) Class I, Division 2
    501.135 Utilization Equipment
        (A) Class I, Division 1
        (B) Class I, Division 2
    501.140 Flexible Cords, Class I, Divisions 1 and 2
        (A) Permitted Uses
        (B) Installation

501.145 Receptacles and Attachment Plugs, Class I,
         Divisions 1 and 2
501.150 Signaling, Alarm, Remote-Control, and
         Communications Systems
    (A) Class I, Division 1
    (B) Class I, Division 2

I. General

501.1 Scope

Article 501 covers the requirements for electrical and electronic equipment and wiring for all voltages in Class I, Division 1 and 2 locations where fire or explosion hazards may exist due to flammable gases or vapors or flammable liquids.

> FPN: For the requirements for electrical and electronic equipment and wiring for all voltages in Class I, Zone 0, Zone 1, or Zone 2 hazardous (classified) locations where fire or explosion hazards may exist due to flammable gases or vapors or flammable liquids, refer to Article 505.

For the 2005 *Code*, Article 501 was divided into three parts: I General, II Wiring, and III Equipment. This division provides a chronological order to the requirements used in an electrical installation. Sections 501.2 and 501.3 were moved to Part III, Equipment, and renumbered 501.100 and 501.105, respectively. Section 501.4, Wiring Methods, was relocated from 501.4 to Part II, 501.10.

501.5 General

The general rules of this *Code* shall apply to the electric wiring and equipment in locations classified as Class I in 500.5.

Exception: As modified by this article.

The most common Class I locations are those areas involved in the handling or processing of volatile flammable liquids such as gasoline, naphtha, benzene, diethyl ether, and acetone, or flammable gases such as hydrogen, methane, and propane.

Where ignitible concentrations (concentrations within the flammable or explosive limits) of flammable gases or vapors are present, atmospheres exist that are explosive when ignited by an arc, a spark, or high temperature. NFPA 497, *Recommended Practice for the Classification of Flammable Liquids, Gases, or Vapors and of Hazardous (Classified) Locations for Electrical Installations in Chemical Process Areas*, includes information on the explosive limits of flammable liquids and gases.

All electrical equipment that may cause ignition-capable arcs or sparks should be kept out of Class I locations where practicable. If this is not practicable, such apparatus must be approved for the purpose and installed properly. The arc produced at the contacts of listed or labeled intrinsically safe equipment is not ignition-capable because the energy available is insufficient to cause ignition.

Hermetic sealing of all electrical equipment is impractical, because equipment such as motors, conventional switches, and circuit breakers has movable parts that must be operated through the enclosing case; that is, the lever of a switch or the shaft of a motor must have sufficient clearance to operate freely. In addition, in many cases, it is necessary to have access to the inside of enclosures for installation, servicing, or alterations.

It is practically impossible to make threaded conduit joints gastight. The conduit system and apparatus enclosure "breathe" due to temperature changes, and any flammable gases or vapors in the room may slowly enter the conduit or enclosure, creating an explosive mixture. Should an arc occur, an explosion could take place.

When an explosion occurs within the enclosure or conduit system, the burning mixture or hot gases must be sufficiently confined within the system to prevent ignition of any explosive mixture that might be present in the area outside the enclosures or conduit system. An apparatus enclosure must be designed with sufficient strength to withstand the maximum pressure generated by an internal explosion in order to prevent rupture and the release of burning or hot gases. Enclosures have been designed to withstand such internal explosions. The ability to withstand an internal explosion is one criterion by which explosionproof enclosures are evaluated.

During an explosion within an enclosure, gases escape through any paths or openings that exist, but the gases are sufficiently cooled if they are carried out through an opening that is long in proportion to its width; that is, the spiral path of at least five fully engaged threads of a screw-on type junction box cover, as illustrated in Exhibit 501.1. This principle is also applied in the design of explosionproof enclosures for apparatus in which a wide machined flange on the body of the enclosure and a similar machined flange on the cover are provided, as illustrated in Exhibit 501.2. These machined flanges are ground so that when the cover is seated in place, the clearance between the two surfaces at no point exceeds, for example, 0.0015 in. If an explosion occurs within the enclosure, escaping gas travels a considerable distance through a very small opening. The gas therefore is cooled sufficiently when it enters and mixes with the surrounding atmosphere, thus preventing ignition of the external explosive mixture.

The clearance between flat surfaces may increase somewhat under explosion conditions because the internal pressures created by the explosion tend to force the surfaces

Exhibit 501.1 Cooling of hot gases as they pass through the threads of a screw-type cover of an explosionproof junction box.

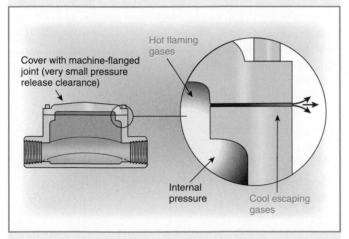

Exhibit 501.2 Cooling of hot gases as they pass across a machine-flanged joint. The clearance between the machined surfaces is kept very small.

apart, as shown in Exhibit 501.3. The amount of increase in the joint clearance depends on the stiffness of the enclosure parts; the size, strength, and spacing of the bolts; and the explosion pressure. Simply measuring the joint width and clearance when there are no internal pressures does not indicate the actual clearances under the dynamic conditions of an explosion. Explosion tests are usually needed to demonstrate the acceptability of the design.

Equipment listed and marked in accordance with 505.9(C)(2) for use in Class I, Zone 0, 1, or 2 locations shall be permitted in Class I, Division 2 locations for the same gas and with a suitable temperature class. Equipment listed and marked in accordance with 505.9(C)(2) for use in Class I, Zone 0 locations shall be permitted in Class I, Division 1 or Division 2 locations for the same gas and with a suitable temperature class.

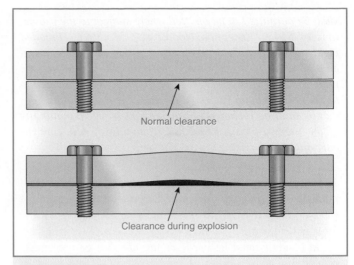

Exhibit 501.3 Effect of internal explosion (bottom) on cover-to-body joint clearance in an explosionproof enclosure. (Redrawn courtesy of Underwriters Laboratories Inc.)

II. Wiring

501.10 Wiring Methods

Wiring methods shall comply with 501.10(A) or 501.10(B).

(A) Class I, Division 1

(1) General In Class I, Division 1 locations, the wiring methods in (a) through (d) shall be permitted.

(a) Threaded rigid metal conduit or threaded steel intermediate metal conduit.

Rigid metal conduit and intermediate metal conduit must be threaded with an NPT standard conduit cutting die that provides a ¾-in. taper per foot. For enclosures that are field-threaded, five full threads must be engaged, while for listed enclosures with factory-threaded NPT entries, 4½ threads must be fully engaged. See 500.8(D).

The *Code* recognizes electrical equipment with metric threaded entries [see 500.8(D)(2)]. Equipment with metric threaded entries must be identified to indicate that metric threads are provided or be provided with listed adapters to allow the connection of NPT-threaded conduit or fittings to the equipment. Each joint must be made up wrenchtight at couplings and unions, threaded hubs of junction boxes, device boxes, conduit bodies, and so on.

Exception: Rigid nonmetallic conduit complying with Article 352 shall be permitted where encased in a concrete envelope a minimum of 50 mm (2 in.) thick and provided with not less than 600 mm (24 in.) of cover measured from the top of the conduit to grade. The concrete encasement shall be permitted to be omitted where subject to the provisions of 514.8, Exception No. 2; and 515.8(A). Threaded rigid metal conduit or threaded steel intermediate metal conduit shall be used for the last 600 mm (24 in.) of the underground run to emergence or to the point of connection to the aboveground raceway. An equipment grounding conductor shall be included to provide for electrical continuity of the raceway system and for grounding of non–current-carrying metal parts.

The exception to 501.10(A)(1)(a) permits the use of rigid nonmetallic conduit in some underground installations. If rigid nonmetallic conduit is used for underground wiring, threaded rigid metal conduit or threaded steel intermediate metal conduit must be used for the last 2 ft of the underground run to the point of emergence or to the point of connection to the aboveground raceway. The rigid nonmetallic conduit, including rigid nonmetallic conduit elbows and fittings, must be located not less than 2 ft below grade. The conduit must also be encased in not less than 2 in. of concrete.

The requirements covering the use of rigid nonmetallic conduit in underground locations per 514.8, Exception No. 2, and 515.8 do not require concrete encasement. These provisions are for specific occupancies where there has been considerable experience with underground nonmetallic conduit. The exception to 501.10(A)(1)(a) applies to other occupancies where rigid nonmetallic conduit is installed in an underground location that has been classified as a Class I, Division 1 location.

If rigid nonmetallic conduit is used, an equipment grounding conductor must be included and must be bonded to the metal raceways that extend from the underground rigid nonmetallic conduit.

(b) Type MI cable with termination fittings listed for the location. Type MI cable shall be installed and supported in a manner to avoid tensile stress at the termination fittings.

The requirement in 501.10(A)(1)(b) specifies that termination fittings used with Type MI cable must be listed for use in Class I, Division 1 hazardous (classified) locations. In editions of the *Code* before 2002, termination fittings used with Type MI cable were required to be approved. (See the definition of *approved* in Article 100.) This change means that MI cable fittings must be evaluated in accordance with an appropriate product standard or be tested for the specific use. A fitting that is approved must be acceptable to the authority having jurisdiction, which, in many cases, requires the use of listed equipment as a basis for approval. However, the term *approved* does not mandate product evaluation or testing. The requirement that fittings used with MI cables be specifically listed for use in the particular hazardous (classified) location class and group involved provides a more objective basis for selecting the proper fitting. Type MI cable fittings, as shown in Exhibit 501.4, have a clamp-type joint that must be investigated to determine that it is explosionproof. Type MI cable fittings not investigated for use in hazardous locations may not be explosionproof. Type MI cable fittings that are suitable for nonhazardous locations may not be suitable for Class I, Division 1 hazardous (classified) locations.

Exhibit 501.5 shows an explosionproof junction box with two hubs and a threaded opening for the screw-type cover. Unused openings must be effectively closed by inserting threaded metal plugs that engage at least five full threads [4½ permitted in accordance with 500.8(D), Exception] and afford protection equivalent to that of the wall of the box.

(c) In industrial establishments with restricted public access, where the conditions of maintenance and supervision ensure that only qualified persons service the installation, Type MC-HL cable, listed for use in Class I, Division 1 locations, with a gas/vaportight continuous corrugated metallic sheath, an overall jacket of suitable polymeric material,

Exhibit 501.4 Type MI cable and fitting listed for use in hazardous locations. The screw-on pot contains field-installed sealing compound to seal the end of the cable. The threaded gland has threads for connection to explosionproof enclosures. (Courtesy of Pyrotenax Cables, Ltd.)

Exhibit 501.5 An explosionproof junction box with a screw-type cover. (Courtesy of O-Z/Gedney, a division of EGS Electrical Group)

separate grounding conductors in accordance with 250.122, and provided with termination fittings listed for the application.

> FPN: See 330.12 for restrictions on use of Type MC cable.

Due to the potential for physical damage to Type MC cable, its application is limited to cable that is listed specifically for use in Class I, Division 1 locations and installed at facilities that have full-time, qualified maintenance personnel. This special-use cable is identified as Type MC-HL. Qualified maintenance personnel are those who, in the course of regular maintenance procedures, would notice whether cables were damaged, understand the associated hazards, and are able to de-energize the circuit to repair the installation.

(d) In industrial establishments with restricted public access, where the conditions of maintenance and supervision ensure that only qualified persons service the installation, Type ITC-HL cable, listed for use in Class I, Division 1 locations, with a gas/vaportight continuous corrugated metallic sheath, an overall jacket of suitable polymeric material and provided with termination fittings listed for the application.

Type ITC-HL cable has a gastight/vaportight continuous corrugated metallic sheath and a polymeric jacket and is listed for use in Class I, Division 1 locations. Type ITC-HL cable is permitted to be used in industrial establishments that have specific conditions of operation. The conditions

under which Type ITC cable can be used in Class I, Division 1 locations parallel those in 501.10(A)(1)(c) for Type MC-HL cable.

(2) Flexible Connections Where necessary to employ flexible connections, as at motor terminals, flexible fittings listed for Class I, Division 1 locations or flexible cord in accordance with the provisions of 501.140 shall be permitted.

Flexible connection fittings are available in lengths up to 3 ft for use in Class I, Division 1 locations. A flexible connection fitting consists of a deeply corrugated bronze tube with an internal nonmetallic tubular protective liner and an outer cover of braided fine bronze wires. A threaded fitting is securely attached to each end of the flexible tube. This type of flexible fitting is commonly used at motor connections, can withstand continuous vibration for long periods, is explosionproof, and affords maximum protection to any enclosed conductors. Limited use of flexible cord in accordance with 501.140 is also permitted for specific applications where flexibility of the wiring method is made necessary by the type of equipment being supplied.

(3) Boxes and Fittings All boxes and fittings shall be approved for Class I, Division 1.

(B) Class I, Division 2

(1) General In Class I, Division 2 locations, the following wiring methods shall be permitted:

(1) All wiring methods permitted in Article 501.10(A).
(2) Threaded rigid metal conduit, threaded steel intermediate metal conduit.
(3) Enclosed gasketed busways, enclosed gasketed wireways.
(4) Type PLTC cable in accordance with the provisions of Article 725, or in cable tray systems. PLTC shall be installed in a manner to avoid tensile stress at the termination fittings.
(5) Type ITC cable as permitted in 727.4.
(6) Type MI, MC, MV, or TC cable with termination fittings, or in cable tray systems and installed in a manner to avoid tensile stress at the termination fittings. Single conductor Type MV cables shall be shielded or metallic armored.

(2) Flexible Connections Where provision must be made for limited flexibility, one or more of the following shall also be permitted:

(1) Flexible metal fittings
(2) Flexible metal conduit with listed fittings
(3) Liquidtight flexible metal conduit with listed fittings

(4) Liquidtight flexible nonmetallic conduit with listed fittings

(5) Flexible cord listed for extra-hard usage and provided with listed bushed fittings. An additional conductor for grounding shall be included in the flexible cord.

FPN: See 501.30(B) for grounding requirements where flexible conduit is used.

(3) Nonincendive Field Wiring Nonincendive field wiring shall be permitted using any of the wiring methods permitted for unclassified locations. Nonincendive field wiring systems shall be installed in accordance with the control drawing(s). Simple apparatus, not shown on the control drawing, shall be permitted in a nonincendive field wiring circuit, provided the simple apparatus does not interconnect the nonincendive field wiring circuit to any other circuit.

The installation of nonincendive field wiring is covered in 501.10(B)(3). See the commentary following 501.105(B)(1), Exception, and the definitions of *nonincendive circuit* and *nonincendive field wiring* in 500.2. Many low-voltage, low-energy circuits are of the nonincendive type. However, a Class 2 circuit, as defined in Article 725, is not necessarily nonincendive. Testing laboratories, such as Factory Mutual Research Corp. and Underwriters Laboratories Inc., list many types of equipment that have nonincendive circuits intended for connection of nonincendive field wiring. This equipment is evaluated for use in one or more of the Class I gas or vapor groups and is permitted for use only in Division 2 locations. Some common telephone circuits and thermocouple circuits are also nonincendive.

FPN: Simple apparatus is defined in 504.2.

Separate nonincendive field wiring circuits shall be installed in accordance with one of the following:

(1) In separate cables

(2) In multiconductor cables where the conductors of each circuit are within a grounded metal shield

(3) In multiconductor cables, where the conductors of each circuit have insulation with a minimum thickness of 0.25 mm (0.01 in.)

(4) Boxes and Fittings Boxes and fittings shall not be required to be explosionproof except as required by 501.105(B)(1), 501.115(B)(1), and 501.150(B)(1).

In Class I, Division 2 locations, boxes, fittings, and joints are not required to be explosionproof at lighting outlets or at enclosures containing no arcing devices, such as solenoids and control transformers, if the maximum operating temperature of any exposed surface does not exceed 80 percent of

the ignition temperature in degrees Celsius. Where general-purpose enclosures are permitted by 501.10(B)(4), rigid or intermediate metal conduit may be used with locknuts and bushings. However, a bonding jumper with proper fittings or bonding-type locknuts is required to be used between the enclosure and the raceway to ensure adequate bonding from the hazardous area to the point of grounding at the service equipment or separately derived system. See 501.30 for grounding and bonding requirements.

Where limited flexibility is necessary and approved fittings are required for use with flexible metal conduit, liquidtight flexible conduit, and extra-hard-usage flexible cord, the fittings are not required to be specifically approved for Class I locations. Also, where flexible conduit or liquidtight flexible conduit is used, internal or external bonding jumpers with proper fittings must be provided, in accordance with 501.30(B), unless liquidtight flexible conduit is installed under the conditions described in 501.30(B), Exception.

Section 501.10(B)(1) permits a variety of cable types, cable tray systems in accordance with 392.3(D), enclosed gasketed wireways, and enclosed gasketed busways. The cable and cable fittings, cable trays, wireways, and busways are not required to be specifically listed or labeled for Class I, Division 2 locations. For example, if Type ITC or MC cable is used, neither the cable nor the fittings need to be listed for use in hazardous (classified) locations. Type AC cable is not a permitted wiring method in 501.10(B) because of concern about arcing between convolutions during ground-fault conditions.

Any wiring method suitable for ordinary locations may be used for nonincendive field wiring. See 501.10(B)(3).

501.15 Sealing and Drainage

Seals in conduit and cable systems shall comply with 501.15(A) through 501.15(F). Sealing compound shall be used in Type MI cable termination fittings to exclude moisture and other fluids from the cable insulation.

FPN No. 1: Seals are provided in conduit and cable systems to minimize the passage of gases and vapors and prevent the passage of flames from one portion of the electrical installation to another through the conduit. Such communication through Type MI cable is inherently prevented by construction of the cable. Unless specifically designed and tested for the purpose, conduit and cable seals are not intended to prevent the passage of liquids, gases, or vapors at a continuous pressure differential across the seal. Even at differences in pressure across the seal equivalent to a few inches of water, there may be a slow passage of gas or vapor through a seal and through conductors passing through the seal. See 501.15(E)(2). Temperature extremes and highly corrosive liquids and vapors can affect the ability of seals to perform their intended function. See 501.15(C)(2).

FPN No. 2: Gas or vapor leakage and propagation of flames may occur through the interstices between the strands of standard stranded conductors larger than 2 AWG. Special conductor constructions, for example, compacted strands or sealing of the individual strands, are means of reducing leakage and preventing the propagation of flames.

The sealing compound used in conduit seal fittings is somewhat porous, so that gases, particularly those under slight pressure and those with small molecules such as hydrogen, can pass slowly through the sealing compound. Also, the seal is around the insulation on the conductor, and gases can be transmitted slowly through the air spaces (the interstices) between strands of stranded conductors. The cable core does not include the interstices of the conductor strands.

Experience has shown, however, that under normal conditions for smaller conductors, and with only normal atmospheric pressure differentials across the seal, the passage of gas through a seal is not sufficient to result in a hazard. For larger conductors, however, gas or vapor leakage and flame propagation may occur through the interstices between the strands, of stranded conductors. Special conductor constructions, such as compacted strands or sealing individual strands, may reduce leakage and prevent flame propagation.

Sealing fittings should be used only with the sealing compound or compounds recommended by the fitting manufacturer. Different sealing compounds have different rates of expansion and contraction that may affect their performance within a given fitting. Sealing compound must be used as soon as possible on Type MI cable terminations to exclude moisture from cable insulation.

The use of Teflon tapes or joint compounds on conduit threads may weaken the seal fitting and interrupt the equipment grounding path. Cracks have developed in fittings during hydrostatic testing in which these materials were used.

(A) Conduit Seals, Class I, Division 1 In Class I, Division 1 locations, conduit seals shall be located in accordance with 501.15(A)(1) through (A)(4).

(1) Entering Enclosures In each conduit entry into an explosionproof enclosure where either of the following apply:

(1) The enclosure contains apparatus, such as switches, circuit breakers, fuses, relays, or resistors, that may produce arcs, sparks, or high temperatures that are considered to be an ignition source in normal operation.
(2) The entry is metric designator 53 (trade size 2) or larger and the enclosure contains terminals, splices, or taps.

In each 2-in. or larger conduit, a sealing fitting must be placed within 18 in. of the conduit entrance to any explosionproof enclosure, regardless of whether the enclosure contains arcing or sparking equipment or only splices, taps, or terminals.

For the purposes of this section, high temperatures shall be considered to be any temperatures exceeding 80 percent of the autoignition temperature in degrees Celsius of the gas or vapor involved.

Exception to 501.15(A)(1)(1): Seals shall not be required for conduit entering an enclosure where such switches, circuit breakers, fuses, relays, or resistors comply with one of the following:

(1) Are enclosed within a chamber hermetically sealed against the entrance of gases or vapors
(2) Are immersed in oil in accordance with 501.115(B)(1)(2)
(3) Are enclosed within a factory-sealed explosionproof chamber located within the enclosure, identified for the location, and marked "factory sealed" or equivalent, unless the enclosure entry is metric designator 53 (trade size 2) or larger
(4) Are in nonincendive circuits

Factory-sealed enclosures shall not be considered to serve as a seal for another adjacent explosionproof enclosure that is required to have a conduit seal.

Conduit seals shall be installed within 450 mm (18 in.) from the enclosure. Only explosionproof unions, couplings, reducers, elbows, capped elbows, and conduit bodies similar to L, T, and Cross types that are not larger than the trade size of the conduit shall be permitted between the sealing fitting and the explosionproof enclosure.

(2) Pressurized Enclosures In each conduit entry into a pressurized enclosure where the conduit is not pressurized as part of the protection system. Conduit seals shall be installed within 450 mm (18 in.) from the pressurized enclosure.

> FPN No. 1: Installing the seal as close as possible to the enclosure will reduce problems with purging the dead airspace in the pressurized conduit.
>
> FPN No. 2: For further information, see NFPA 496-2003, *Standard for Purged and Pressurized Enclosures for Electrical Equipment.*

(3) Two or More Explosionproof Enclosures Where two or more explosionproof enclosures for which conduit seals are required under 501.15(A)(1) are connected by nipples or by runs of conduit not more than 900 mm (36 in.) long, a single conduit seal in each such nipple connection or run of conduit shall be considered sufficient if located not more than 450 mm (18 in.) from either enclosure.

An example of 501.15(A) requirements for the location of conduit seals in Class I, Division 1 locations is illustrated in Exhibit 501.6. In the example shown in Exhibit 501.6,

Exhibit 501.6 Two seals required so that the run of conduit between Enclosure No. 1 and Enclosure No. 2 is sealed. Even if Enclosure No. 3 were not required to be sealed, the vertical seal in the vertical run of conduit to Enclosure No. 3 would be required to be sealed within 18 in. of Enclosure No. 1, because the vertical conduit run to the "T" fitting is a conduit run to Enclosure No. 1.

two seals are required so that the run of conduit between Enclosure No. 1 and Enclosure No. 2 is sealed.

A seal in a conduit prevents an explosion from traveling through the conduit to another enclosure and minimizes the passage of gases or vapors from a hazardous (classified) location to a nonhazardous location. If the conduit enters an enclosure that contains arcing or high-temperature equipment, a sealing fitting must be placed within 18 in. of the enclosure it isolates; conduit bodies ("L," "T," etc.), couplings, unions, and elbows are the only enclosures or fittings permitted between the seal and the enclosure. Exhibit 501.6 illustrates the placement of conduit seals. See Exhibit 501.7 for an approved type of union. If two enclosures are spaced not more than 36 in. apart as is the case with enclosures No. 1 and No. 2, a single seal may be placed between two connecting nipples if the seal is located not more than 18 in. from either enclosure.

(4) Class I, Division 1 Boundary In each conduit run leaving a Class I, Division 1 location. The sealing fitting shall be permitted on either side of the boundary of such location

Exhibit 501.7 An explosionproof union. (Courtesy of Thomas & Betts Corp.)

within 3.05 m (10 ft) of the boundary and shall be designed and installed so as to minimize the amount of gas or vapor within the Division 1 portion of the conduit from being communicated to the conduit beyond the seal. Except for listed explosionproof reducers at the conduit seal, there shall be no union, coupling, box, or fitting between the conduit seal and the point at which the conduit leaves the Division 1 location.

Exception No. 1: Metal conduit that contains no unions, couplings, boxes, or fittings, and passes completely through a Class I, Division 1 location with no fittings less than 300 mm (12 in.) beyond each boundary, shall not require a conduit seal if the termination points of the unbroken conduit are in unclassified locations.

Exception No. 2: For underground conduit installed in accordance with 300.5 where the boundary is beneath the ground, the sealing fitting shall be permitted to be installed after the conduit leaves the ground, but there shall be no union, coupling, box, or fitting, other than listed explosionproof reducers at the sealing fitting, in the conduit between the sealing fitting and the point at which the conduit leaves the ground.

A sealing fitting is also required at the point where the conduit leaves a Division 1 location or passes from a Division 2 location to an unclassified location. The sealing fitting is permitted on either side of the boundary, and no union, coupling, box, or similar fitting is permitted between the seal and the boundary. However, approved explosionproof reducers are permitted to be installed in conduit seals.

It is preferable to locate the sealing fitting on the nonhazardous side of the boundary. A sealing fitting located as such serves two purposes: It completes the explosionproof wiring method, and it completes the explosionproof enclosure system. Note that a ½-in. conduit connected to an explosionproof box that contains only splices, even in a Division 1 location, is not required to be sealed within 18 in. of the box. The sealing fitting at the boundary of the Division 1 location serves to complete the explosionproof system. The sealing fitting at the boundary also prevents the conduit system from serving as a pipe to transmit flammable mixtures from a Division 1 or Division 2 location to an unclassified location.

Exhibit 501.8 illustrates a Class I, Division 1 location using threaded rigid metal conduit or threaded intermediate metal conduit and explosionproof fittings and equipment, including motors, motor controllers, pushbutton stations, lighting outlets, and junction boxes. The enclosures for the disconnecting means and motor controller for the motor (right portion of the drawing) are placed on the other side of the wall in a nonhazardous location and are thus not required to be explosionproof.

In Exhibit 501.8, each of the three conduits is sealed on the nonhazardous side before passing into the hazardous (classified) location. The pigtail leads of both motors are factory sealed at the motor-terminal housing, and, unless the size of the flexible fitting entering the motor-terminal housing is trade size 2 or larger, no other seals are needed at this point. Because the push-button control station and the motor controller and disconnect (left portion of the drawing) are considered arc-producing devices, conduits are sealed within 18 in. of the entrance to these enclosures. Seals are required even though the contacts may be immersed in oil.

Additionally Exhibit 501.8 shows a seal provided within 18 in. of the switch controlling the lighting. The design of the luminaire, as required by ANSI/UL 844, *Electric Lighting Fixtures for Use in Hazardous (Classified) Locations*, is such that the explosionproof chamber for the wiring must be separated or sealed from the lamp compartment; hence, a separate seal is not required adjacent to luminaires that comply with ANSI/UL 844. The luminaire is suspended on a conduit stem threaded into the cover of an explosionproof ceiling box. See 501.130 for luminaire requirements.

(B) Conduit Seals, Class I, Division 2 In Class I, Division 2 locations, conduit seals shall be located in accordance with 501.15(B)(1) and (B)(2).

(1) Entering Enclosures For connections to enclosures that are required to be explosionproof, a conduit seal shall be provided in accordance with 501.15(A)(1) and (A)(3). All portions of the conduit run or nipple between the seal and such enclosure shall comply with 501.10(A).

In Class I, Division 2 locations, a seal is required in each conduit entering an enclosure that is required to be explosionproof, in order to complete the explosionproof enclosure.

(2) Class I, Division 2 Boundary In each conduit run passing from a Class I, Division 2 location into an unclassified location. The sealing fitting shall be permitted on either side

Exhibit 501.8 A Class I, Division 1 location where threaded metal conduits, sealing fittings, explosionproof fittings, and equipment for power and lights are used.

of the boundary of such location within 3.05 m (10 ft) of the boundary. Rigid metal conduit or threaded steel intermediate metal conduit shall be used between the sealing fitting and the point at which the conduit leaves the Division 2 location, and a threaded connection shall be used at the sealing fitting. Except for listed reducers at the conduit seal, there shall be no union, coupling, box, or fitting between the conduit seal and the point at which the conduit leaves the Division 2 location. Conduits shall be sealed to minimize the amount of gas or vapor within the Division 2 portion of the conduit from being communicated to the conduit beyond the seal. Such seals shall not be required to be explosionproof but shall be identified for the purpose of minimizing passage of gases under normal operating conditions and shall be accessible.

Exception No. 1: Metal conduit that contains no unions, couplings, boxes, or fittings, and passes completely through a Class I, Division 2 location with no fittings less than 300 mm (12 in.) beyond each boundary, shall not be required to be sealed if the termination points of the unbroken conduit are in unclassified locations.

Exception No. 2: Conduit systems terminating at an unclassified location where a wiring method transition is made to cable tray, cablebus, ventilated busway, Type MI cable, or cable not installed in any cable tray or raceway system, shall not be required to be sealed where passing from the Class I, Division 2 location into the unclassified location. The unclassified location shall be outdoors or, if the conduit system is all in one room, it shall be permitted to be indoors. The conduits shall not terminate at an enclosure containing an ignition source in normal operation.

Exception No. 3: Conduit systems passing from an enclosure or room that is unclassified as a result of pressurization into a Class I, Division 2 location shall not require a seal at the boundary.

> FPN: For further information, refer to NFPA 496-2003, *Standard for Purged and Pressurized Enclosures for Electrical Equipment.*

Exception No. 4: Segments of aboveground conduit systems shall not be required to be sealed where passing from a Class I, Division 2 location into an unclassified location if all of the following conditions are met:

(1) No part of the conduit system segment passes through a Class I, Division 1 location where the conduit contains unions, couplings, boxes, or fittings within 300 mm (12 in.) of the Class I, Division 1 location.

(2) The conduit system segment is located entirely in outdoor locations.

(3) The conduit system segment is not directly connected to canned pumps, process or service connections for flow, pressure, or analysis measurement, and so forth, that depend on a single compression seal, diaphragm,

or tube to prevent flammable or combustible fluids from entering the conduit system.

(4) The conduit system segment contains only threaded metal conduit, unions, couplings, conduit bodies, and fittings in the unclassified location.

(5) The conduit system segment is sealed at its entry to each enclosure or fitting housing terminals, splices, or taps in Class I, Division 2 locations.

(C) Class I, Divisions 1 and 2 Seals installed in Class I, Division 1 and Division 2 locations shall comply with 501.15(C)(1) through (C)(6).

Exception: Seals not required to be explosionproof by 501.15(B)(2) or 504.70.

(1) Fittings Enclosures for connections or equipment shall be provided with an integral means for sealing, or sealing fittings listed for the location shall be used. Sealing fittings shall be listed for use with one or more specific compounds and shall be accessible.

(2) Compound The compound shall provide a seal against passage of gas or vapors through the seal fitting, shall not be affected by the surrounding atmosphere or liquids, and shall not have a melting point of less than 93°C (200°F).

(3) Thickness of Compounds Except for listed cable sealing fittings, the thickness of the sealing compound in a completed seal shall not be less than the metric designator (trade size) of the sealing fitting expressed in the units of measurement employed, and in no case less than 16 mm (⅝ in.).

(4) Splices and Taps Splices and taps shall not be made in fittings intended only for sealing with compound, nor shall other fittings in which splices or taps are made be filled with compound.

(5) Assemblies In an assembly where equipment that may produce arcs, sparks, or high temperatures is located in a compartment separate from the compartment containing splices or taps, and an integral seal is provided where conductors pass from one compartment to the other, the entire assembly shall be identified for the location. Seals in conduit connections to the compartment containing splices or taps shall be provided in Class I, Division 1 locations where required by 501.15(A)(1)(2).

(6) Conductor Fill The cross-sectional area of the conductors permitted in a seal shall not exceed 25 percent of the cross-sectional area of a rigid metal conduit of the same trade size unless it is specifically identified for a higher percentage of fill.

The maximum permitted fill for the conduit is 40 percent; the maximum permitted fill for most conduit seals is 25 percent. If the conduit fill exceeds 25 percent of the cross-sectional area of the sealing fitting, a larger trade size seal may be required. Reducers are allowed for connection of a larger trade size sealing fitting to the conduit. Exhibit 501.9 illustrates the proper sealing of a fitting. A dam must be provided to prevent the sealing material, while still in the liquid state, from running out of the fitting. All conductors must be separated to permit the sealing material to run between them. The sealing compound must have a minimum thickness of not less than the trade size of the conduit and in no case less than ⅝ in. Conduit fittings for sealing are to be used only with sealing compound that is supplied with the fitting and specified by the manufacturer in instructions furnished with the fitting.

Exhibit 501.9 A sealing fitting placed in a run of conduit to minimize the passage of gases from one portion of the electrical installation to another. (Courtesy of Appleton Electric Co., EGS Electrical Group)

(D) Cable Seals, Class I, Division 1 In Class I, Division 1 locations, cable seals shall be located according to 501.15(D)(1) through (D)(3).

(1) At Terminations Cable shall be sealed at all terminations. The sealing fitting shall comply with 501.15(C). Multiconductor Type MC-HL cables with a gas/vaportight continuous corrugated metallic sheath and an overall jacket of suitable polymeric material shall be sealed with a listed fitting after removing the jacket and any other covering so that the sealing compound surrounds each individual insulated conductor in such a manner as to minimize the passage of gases and vapors.

In accordance with 501.10(A)(1)(c), Type MC-HL cable is permitted as a wiring method in Class I, Division 1 areas.

Type MC-HL cable has a continuous corrugated metallic sheath that is gastight/vaportight and an outer nonmetallic material that enables it to be installed in wet locations. Type MC-HL cables are available with ratings up to 35,000 volts. The cable is specifically listed for use in Class I, Division 1 locations.

The provisions of 501.15(D) contain the sealing requirements for cables installed in Class I, Division 1 locations, which differ from the requirements for sealing conduits. Conduits entering explosionproof enclosures must be sealed if the enclosure contains equipment that produces arcs, sparks, or high temperatures or if the conduit entering the enclosure is trade size 2 or larger. In accordance with 501.15(D)(1), cables must be sealed at all terminations, irrespective of the type of equipment contained in the enclosure or the diameter of the cable. Exhibit 501.10 shows an example of a cable sealing fitting for Type MC-HL cable.

Exhibit 501.10 An explosionproof sealing fitting for Type MC-HL cable. (Courtesy of Cooper Crouse-Hinds)

Exception: Shielded cables and twisted pair cables shall not require the removal of the shielding material or separation of the twisted pairs, provided the termination is by an approved means to minimize the entrance of gases or vapors and prevent propagation of flame into the cable core.

Shielded cables and twisted pair cables, commonly used for signaling and instrumentation circuits, are permitted by the exception to 501.15(D) to be sealed without removal of the outer sheathing or the separation of the twisted conductors. The need to provide a suitable seal while not adversely affecting the operational performance of these cables is accomplished through this exception.

(2) Cables Capable of Transmitting Gases or Vapors Cables in conduit with a gas/vaportight continuous sheath capable of transmitting gases or vapors through the cable core shall be sealed in the Division 1 location after removing the jacket and any other coverings so that the sealing compound will surround each individual insulated conductor and the outer jacket.

Exception: Multiconductor cables with a gas/vaportight continuous sheath capable of transmitting gases or vapors through the cable core shall be permitted to be considered as a single conductor by sealing the cable in the conduit within 450 mm (18 in.) of the enclosure and the cable end within the enclosure by an approved means to minimize the entrance of gases or vapors and prevent the propagation of flame into the cable core, or by other approved methods. For shielded cables and twisted pair cables, it shall not be required to remove the shielding material or separate the twisted pair.

The intent of the exception to 501.15(D)(2) is to permit flat computer cables, coaxial cables, and twisted pair cables to be treated as single conductors if installed in conduit, because separating the individual conductors or removing the outer jacket (of a coaxial cable or a twisted pair, for example) is impractical and can destroy the electrical properties of the cable.

In addition to the cable seal, the end of the cable within the enclosure must also be sealed.

(3) Cables Incapable of Transmitting Gases or Vapors Each multiconductor cable in conduit shall be considered as a single conductor if the cable is incapable of transmitting gases or vapors through the cable core. These cables shall be sealed in accordance with 501.15(A).

(E) Cable Seals, Class I, Division 2 In Class I, Division 2 locations, cable seals shall be located in accordance with 501.15(E)(1) through (E)(4).

(1) Terminations Cables entering enclosures that are required to be explosionproof shall be sealed at the point of entrance. The sealing fitting shall comply with 501.15(B)(1). Multiconductor cables with a gas/vaportight continuous sheath capable of transmitting gases or vapors through the cable core shall be sealed in a listed fitting in the Division 2 location after removing the jacket and any other coverings so that the sealing compound surrounds each individual insulated conductor in such a manner as to minimize the passage of gases and vapors. Multiconductor cables in conduit shall be sealed as described in 501.15(D).

Exception No. 1: Cables passing from an enclosure or room that is unclassified as a result of Type Z pressurization into a Class I, Division 2 location shall not require a seal at the boundary.

If cables are run from a Type Z pressurized room or enclosure into a Class I, Division 2 location, Exception No. 1 to 501.15(E)(1) allows the cables to be installed without a sealing fitting at the boundary. This exception correlates with a similar allowance for conduit systems found in 501.15(B)(2), Exception No. 3.

Exception No. 2: Shielded cables and twisted pair cables shall not require the removal of the shielding material or separation of the twisted pairs, provided the termination is by an approved means to minimize the entrance of gases or vapors and prevent propagation of flame into the cable core.

(2) Cables That Do Not Transmit Gases or Vapors Cables that have a gas/vaportight continuous sheath and do not transmit gases or vapors through the cable core in excess of the quantity permitted for seal fittings shall not be required to be sealed except as required in 501.15(E)(1). The minimum length of such cable run shall not be less than that length that limits gas or vapor flow through the cable core to the rate permitted for seal fittings [200 cm^3/hr (0.007 ft^3/hr) of air at a pressure of 1500 pascals (6 in. of water)].

The ability of a cable to transmit gases or vapors through the core (primarily between insulated conductors) depends not only on how tightly packed the conductors are within the outer sheaths and the location and composition of fillers but also on how the cable has been handled and the geometry of the cable run. If there is any question as to whether or not the cable run is capable of transmitting gases or vapors through the core, a sealing fitting should be installed. See the commentary following 501.15, FPN No. 1.

FPN No. 1: See ANSI/UL 886-1994, *Outlet Boxes and Fittings for Use in Hazardous (Classified) Locations.*

In a leak rate test, the ends of each individual conductor in the cable are sealed to prevent migration of gases or vapors between the individual strands of wire. Sealing can be achieved by dipping the cable end in hot wax. The rate of flow through the filler between the insulated conductors can now be accurately measured, excluding any leakage through the conductor strands. If this is done, the wax should be removed before the connections are made and the system is placed in service.

FPN No. 2: The cable core does not include the interstices of the conductor strands.

(3) Cables Capable of Transmitting Gases or Vapors Cables with a gas/vaportight continuous sheath capable of transmitting gases or vapors through the cable core shall not be required to be sealed except as required in 501.15(E)(1), unless the cable is attached to process equipment or devices that may cause a pressure in excess of 1500 pascals (6 in. of water) to be exerted at a cable end, in which case a seal, barrier, or other means shall be provided to prevent migration of flammables into an unclassified location.

Exception: Cables with an unbroken gas/vaportight continuous sheath shall be permitted to pass through a Class I, Division 2 location without seals.

(4) Cables Without Gas/Vaportight Sheath Cables that do not have gas/vaportight continuous sheath shall be sealed at the boundary of the Division 2 and unclassified location in such a manner as to minimize the passage of gases or vapors into an unclassified location.

(F) Drainage

Unless the additional seal or barrier, described in 501.15(F)(3), and interconnecting enclosures meet the performance requirements of the primary seal, the application of pressure or exposure to extreme temperatures must be prevented at the additional seal or barrier so that the process fluid will not enter the conduit system if the primary seal fails. If the process fluid is a gas or can become a gas under ordinary atmospheric conditions (liquefied natural gas, for example), the drain mentioned in 501.15(F)(3) should be a vent. See the commentary following 501.15(F)(3). The necessary sealing may be accomplished by a sealing fitting and compound.

To eliminate the time-consuming task of field-poured seals, a factory-sealed device with the seal designed into the device is permissible. A wide selection of factory-sealed devices are available for a variety of installations in hazardous (classified) locations. For example, explosionproof motors are normally factory sealed and therefore require no additional seal. Factory-sealed devices are usually marked as such. If a conduit terminates in a motor, however, and if the conduit is 2 in. or larger, a seal must be placed within 18 in. of the motor terminal housing.

Exhibit 501.11 shows a sealing fitting designed for use in a vertical run of conduit to provide drainage for any condensation of moisture trapped by the seal above the enclosure. Any accumulation of water runs down over the surface of the sealing compound, flowing through an explosionproof drain.

Exhibit 501.12 shows a combination drain and breather fitting. This type of fitting is specifically designed to serve as a water drain and air vent while providing positive explosionproof protection. The fitting permits the escape of accumulated water through its drain, and the breather allows the continuous circulation of air, preventing condensation of any moisture that may be present. Individual drain or breather

Exhibit 501.11 A sealing fitting with an automatic drain plug. (Courtesy of Appleton Electric Co., EGS Electrical Group)

Exhibit 501.12 A combination breather-drainage fitting. (Courtesy of Appleton Electric Co., EGS Electrical Group)

fittings are also available. It is good practice to consider the installation of drain, breather, or combination fittings to guard against water accumulation, which can cause future insulation failures, even though prevalent conditions may not indicate a need.

(1) Control Equipment Where there is a probability that liquid or other condensed vapor may be trapped within enclosures for control equipment or at any point in the raceway system, approved means shall be provided to prevent accumulation or to permit periodic draining of such liquid or condensed vapor.

(2) Motors and Generators Where the authority having jurisdiction judges that there is a probability that liquid or condensed vapor may accumulate within motors or generators, joints and conduit systems shall be arranged to minimize the entrance of liquid. If means to prevent accumulation or to permit periodic draining are judged necessary, such means shall be provided at the time of manufacture and shall be considered an integral part of the machine.

(3) Canned Pumps, Process, or Service Connections, etc.
For canned pumps, process, or service connections for flow, pressure, or analysis measurement, and so forth, that depend on a single compression seal, diaphragm, or tube to prevent flammable or combustible fluids from entering the electrical raceway or cable system capable of transmitting fluids, an additional approved seal, barrier, or other means shall be provided to prevent the flammable or combustible fluid from entering the raceway or cable system capable of transmitting fluids beyond the additional devices or means, if the primary seal fails. The additional approved seal or barrier and the interconnecting enclosure shall meet the temperature and pressure conditions to which they will be subjected upon failure of the primary seal, unless other approved means are provided to accomplish this purpose. Drains, vents, or other devices shall be provided so that primary seal leakage will be obvious.

FPN: See also the fine print notes to 501.15.

Canned pumps and other process equipment that operate above atmospheric pressure are provided with a primary seal where the electrical conductors enter the pump or equipment containing flammable liquids or gases under pressure. Seal-

ant for sealing fittings is not designed to withstand high pressure or extremely low temperatures, which might be encountered in canned pump installations. Therefore, a second seal or barrier is required to prevent fluid from entering the electrical conduit or cable system. In addition to this seal or barrier, a drain, vent, or other similar device that indicates failure of the primary seal must be provided.

This redundant protection system may be achieved by installing a vented junction (box) enclosure within the classified area where the conductors terminate on busbars. Terminating the conductors in this manner allows any fluid that escapes through the primary seal and that has traveled through the stranding of the conductors to vent at the terminations. The circuit continues through the vented enclosure at normal atmospheric pressure to another set of conductors that also must be sealed with a sealing fitting if they travel into a different classified area. A gas detector can be installed in the vicinity of the vented termination box to signal that the primary seal has failed and allow an orderly shutdown of the process system either automatically or manually.

Commentary Table 5.3 summarizes the sealing requirements of 501.15.

Process-connected equipment that is listed and marked "Dual Seal" shall not require additional process sealing when used within the manufacturer's ratings.

FPN: For construction and testing requirements for dual seal process connected equipment, refer to ISA 12.27.01,

Commentary Table 5.3 Conduit and Cable Sealing Requirements

Classification	Application	Location of Seal
Conduit Seals		In conduit run within 18 in. of enclosure
Class I, Division 1	Switch enclosure Circuit breaker enclosure Fuse enclosure Relay enclosure Resistor enclosure Arcing or sparking apparatus High-temperature apparatus	
	Explosionproof enclosure containing arcing or sparking contacts that are hermetically sealed against gas or vapor entry Explosionproof enclosure containing arcing or sparking contacts that are immersed in oil, in accordance with 501.115(B)(1)(2)	In conduit runs of 1½ in. and smaller, no seal is required. If conduit is larger than 1½ in., in conduit run within 18 in. of enclosure
	Enclosure containing terminals, splices, or taps fitting containing terminals, splices, or taps	In conduit runs smaller than trade size 2, no seal is required. If conduit is 2 in. or larger, in conduit run within 18 in. of enclosure

Commentary Table 5.3 *Continued*

Classification	Application	Location of Seal
	Two explosionproof enclosures with a conduit run between them of 36 in. or less	In conduit run within 18 in. of each enclosure. Permitted to use a single seal in each run as long as the seal is within 18 in. of each enclosure
	Two explosionproof enclosures with a conduit run between them greater than 36 in.	In conduit run within 18 in. of each enclosure
	Conduit run leaving Division 1 location	On either side of boundary. No unions, couplings, boxes, or fittings (other than explosionproof reducers) permitted between the seal fitting and the point where the conduit leaves the Division 1 location.
	Metal conduit containing no unions, couplings, boxes, or fittings that passes completely through a Class I, Division 1 location, with no fittings less than 12 in. beyond each boundary	Not required to be sealed if the termination points of the unbroken conduit are in unclassified locations.
Class I, Division 2	Enclosure required to be explosionproof	Seal as required for similar equipment in Division 1 location
	Conduit run leaving Division 2 location	On either side of boundary. No unions, couplings, boxes, or fittings (other than explosionproof reducers) permitted between the seal fitting and the point where the conduit leaves the Division 2 location. Not required to be explosionproof seal.
	Metal conduit containing no unions, couplings, boxes, or fittings that passes completely through a Division 2 location with no fittings less than 12 in. beyond each boundary	Not required to be sealed if the termination points of the unbroken conduit are in unclassified locations.
	Conduit systems terminating at an unclassified location where a wiring method transition is made to cable tray, cablebus, ventilated busway, Type MI cable, or open wiring	Not required to be sealed if passing from the Class I, Division 2 location into an outdoor unclassified location or an indoor location if the conduit system is all in one room. The conduits do not terminate at an enclosure containing an ignition source in normal operation.
Cable Seals Class I, Division 1	Enclosure with integral seal Multiconductor Type MC-HL cables with a gastight/vaportight continuous corrugated metallic sheath and an overall jacket of suitable polymeric material	Conduit seal fitting not required Seal at all terminations with an approved fitting after removing the jacket and any other covering, so that the sealing compound surrounds each individual insulated conductor.
	Cables in conduit with a gas/vaportight continuous sheath capable of transmitting gases or vapors through the cable core	Seal in the Division 1 location after removing the jacket and any other coverings, so that the sealing compound surrounds each individual insulated conductor and the outer jacket.
	Multiconductor cables with a gastight/vaportight continuous sheath capable of transmitting gases or vapors through the cable core	Permitted to be considered a single conductor by sealing the cable in the conduit within 18 in. of the enclosure and the cable end within the enclosure by an approved means, to minimize the entrance of gases or vapors and prevent the propagation of flame into the cable core, or by other approved methods.

(continues)

Commentary Table 5.3 *Continued*

Classification	Application	Location of Seal
	For shielded cables and twisted pair cables	Removal of the shielding material or separation of the twisted pair is not required. Sealing the cable in the conduit within 18 in. of the enclosure and the cable end within the enclosure by an approved means, to minimize the entrance of gases or vapors and prevent the propagation of flame into the cable core, or by other approved methods.
	Each multiconductor cable in conduit if the cable is incapable of transmitting gases or vapors through the cable core	Considered a single conductor. These cables are sealed in accordance with 501.15(A)
Class I, Division 2	Cables entering enclosures that are required to be approved for Class I locations	Sealed at the point of entrance
	Multiconductor cables in conduit	Sealed in accordance with the requirements for Division 1 locations
	Multiconductor cables with a gastight/vaportight continuous sheath capable of transmitting gases or vapors through the cable core	Sealed in an approved fitting in the Division 2 location after removing the jacket and any other coverings, so that the sealing compound surrounds each individual insulated conductor.
	Cables with a gas/vaportight continuous sheath that will not transmit gases or vapors through the cable core in excess of the quantity permitted for seal fittings. The minimum length of such cable run is not less than that length that limits gas or vapor flow through the cable core to the rate permitted for seal fittings (0.007 cu ft per hour of air at a pressure of 6 in. of water).	Not required to be sealed unless entering an enclosure that is required to be approved for Class I locations.
	Cables with a gastight/vaportight continuous sheath capable of transmitting gases or vapors through the cable core	Not required to be sealed unless entering an enclosure that is required to be approved for Class I locations or unless the cable is attached to process equipment or devices that may cause a pressure in excess of 6 in. of water to be exerted at a cable end, in which case a seal, barrier, or other means is provided to prevent migration of flammables into an unclassified area.
	Cables with an unbroken gastight/vaportight continuous sheath that pass through a Class I, Division 2 location	No seal required
	Cables that do not have a gastight/vaportight continuous sheath	Sealed at the boundary of the Division 2 and unclassified location in such a manner as to minimize the passage of gases or vapors into an unclassified location.

Requirements for Process Sealing Between Electrical Systems and Potentially Flammable or Combustible Process Fluids.

501.20 Conductor Insulation, Class I, Divisions 1 and 2

Where condensed vapors or liquids may collect on, or come in contact with, the insulation on conductors, such insulation shall be of a type identified for use under such conditions; or the insulation shall be protected by a sheath of lead or by other approved means.

Nylon-jacketed conductors, such as Type THWN, that are suitable for use where exposed to gasoline have gained widespread acceptance because of their ease in handling and application as well as for economic reasons.

The UL *Electrical Construction Materials Directory* states in part:

Wires, Thermoplastic-Insulated:

THWN — wire that is suitable for exposure to mineral oil, and to liquid gasoline and gasoline vapors at ordinary ambient temperature, is marked "Gasoline and Oil Resistant I" if suitable for exposure to mineral oil at 60°C or "Gasoline and Oil Resistant II" if the compound is suitable for exposure to mineral oil at 75°C. Gasoline-resistant wire has been tested at 23°C when immersed in gasoline. It is considered inherently resistant to gasoline vapors within the limits of the temperature rating of the wire type.

501.25 Uninsulated Exposed Parts, Class I, Divisions 1 and 2

There shall be no uninsulated exposed parts, such as electric conductors, buses, terminals, or components, that operate at more than 30 volts (15 volts in wet locations). These parts shall additionally be protected by a protection technique according to 500.7(E), 500.7(F), or 500.7(G) that is suitable for the location.

Exposed live parts are allowed in Class I, Division 1 and 2 locations provided that the voltage does not exceed 30 volts in dry locations or 15 volts in wet locations. However, caution must be observed in working around uninsulated exposed parts because tool or other conductive items that come in contact with the circuit could produce sparks and cause an explosion in a hazardous (classified) location.

501.30 Grounding and Bonding, Class I, Divisions 1 and 2

Wiring and equipment in Class I, Division 1 and 2 locations shall be grounded as specified in Article 250 and with the requirements in 501.30(A) and 501.30(B).

(A) Bonding The locknut-bushing and double-locknut types of contacts shall not be depended on for bonding purposes, but bonding jumpers with proper fittings or other approved means of bonding shall be used. Such means of bonding shall apply to all intervening raceways, fittings, boxes, enclosures, and so forth between Class I locations and the point of grounding for service equipment or point of grounding of a separately derived system.

Exception: The specific bonding means shall be required only to the nearest point where the grounded circuit conductor and the grounding electrode are connected together on the line side of the building or structure disconnecting means as specified in 250.32(A), (B), and (C), provided the branch-circuit overcurrent protection is located on the load side of the disconnecting means.

Bonding requirements intended to address the unique hazards inherent to electrical equipment installed in a Class I, Division 1 or Division 2 location are covered by 501.30(A) and also by 250.100. These specific bonding methods, intended to provide a mechanical/electrical connection that is low impedance and free from accidental arcing due to loose connections, apply to raceways and raceway-to-enclosure connections both inside and outside the classified location. Section 250.100 specifies this enhanced level of bonding for all raceways and enclosures, and the requirement is not contingent on the circuit voltage. Therefore metal raceways and enclosures containing signaling, communications, or other power-limited circuits are covered.

A revision to 250.100 for the 2005 *Code* clarifies that the installation of a wire-type equipment grounding conductor in a metal raceway does not negate the special raceway and enclosure bonding requirements. Unless it is an isolated equipment grounding conductor as permitted by 250.96(B) or 250.146(D), the wire-type equipment grounding conductor and the metal raceway will be electrically in parallel and will share ground-fault current based on the impedance of the wire and of the metal raceway. For that reason, the electrical continuity of raceways and raceway-to-enclosure connections must always be ensured through compliance with 250.100 and 501.30(A), regardless of whether a supplementary equipment grounding conductor has been installed in the raceway.

The exception to 501.30(A) covers the grounding and bonding requirements that are specific to hazardous (classified) locations where the installation occurs at a multibuilding or multistructure setting. If the service equipment and the electrical equipment operating in the hazardous (classified) location are not located in the same building or structure, it is not necessary to apply the bonding requirement of 501.30(A) from the hazardous location back to the service equipment. It is necessary only to apply the bonding requirement from the hazardous location back to the grounding electrode on the line side of the building or structure disconnecting means. This connection must be ahead of the branch circuits that are on the load side of the disconnecting means for the building or structure.

FPN: See 250.100 for additional bonding requirements in hazardous (classified) locations.

(B) Types of Equipment Grounding Conductors Where flexible metal conduit or liquidtight flexible metal conduit is used as permitted in 501.10(B) and is to be relied on to complete a sole equipment grounding path, it shall be installed with internal or external bonding jumpers in parallel with each conduit and complying with 250.102.

Special consideration is necessary in the grounding and bonding of exposed non–current-carrying metal parts of equipment, such as the frames or metal exteriors of motors, fixed or portable lamps, luminaires, enclosures, and raceways, to ensure permanent and effective mechanical and electrical connections in order to prevent the possibility of arcs or sparks caused by ineffective or poor grounding methods. One example of an external bonding jumper is shown in Exhibit 501.13.

Exhibit 501.13 A fitting for the connection of an external bonding jumper used with liquidtight flexible metal conduit.

To be effective, proper grounding and bonding apply to all interconnected raceways, fittings, enclosures, and so on, between hazardous (classified) locations and the point of grounding for service equipment or the point of grounding of the building disconnecting means that supplies the branch-circuit overcurrent protection. If conduit is used in hazardous (classified) locations, it is preferable that threaded connections also be employed in the nonhazardous location.

Exception: In Class I, Division 2 locations, the bonding jumper shall be permitted to be deleted where all of the following conditions are met:

(1) Listed liquidtight flexible metal conduit 1.8 m (6 ft) or less in length, with fittings listed for grounding, is used.

(2) Overcurrent protection in the circuit is limited to 10 amperes or less.

(3) The load is not a power utilization load.

501.35 Surge Protection

(A) Class I, Division 1 Surge arresters, transient voltage surge suppressors (TVSS), and capacitors shall be installed in enclosures identified for Class I, Division 1 locations. Surge-protective capacitors shall be of a type designed for specific duty.

(B) Class I, Division 2 Surge arresters and TVSS shall be nonarcing, such as metal-oxide varistor (MOV) sealed type,

and surge-protective capacitors shall be of a type designed for specific duty. Enclosures shall be permitted to be of the general-purpose type. Surge protection of types other than described in this paragraph shall be installed in enclosures identified for Class I, Division 1 locations.

Article 285, Transient Voltage Surge Suppressors: TVSSs, was added to the 2002 *Code*. Nonarcing, sealed type TVSSs are included as a device permitted in Class I, Division 2 locations where installed in a general purpose–type enclosure. Some surge arresters, such as older style lightning arresters, are spark-producing devices. Others, such as solid-state types, are not. Nonsparking-type surge arresters need no special enclosure in a Class I, Division 2 location. Surge arresters should be connected to the service conductors outside the building and should be bonded to the service-entrance raceway system. For services less than 1000 volts, the arrester grounding conductor is connected as provided in Article 280, Part II. Where the service voltage is less than 600 volts, the supply system is a secondary system. Therefore, the grounded service conductor should always be bonded to the equipment grounding conductor, as required by Article 250.

In Class I, Division 1 locations, all surge arresters must be installed in explosionproof or purged enclosures. In Class I, Division 2 locations, only the spark-producing types of surge arresters require such protection. Surge arresters can also be installed in oil-filled enclosures or have the arcing or sparking contacts enclosed in hermetically sealed chambers.

501.40 Multiwire Branch Circuits

In a Class I, Division 1 location, a multiwire branch circuit shall not be permitted.

Exception: Where the disconnect device(s) for the circuit opens all ungrounded conductors of the multiwire circuit simultaneously.

The requirement in 501.40 does not permit multiwire branch circuits in Class I hazardous locations unless the disconnecting device opens all the ungrounded conductors of the multiwire branch circuit simultaneously. See the definition of *branch circuit, multiwire* in Article 100. The requirement in 501.40 addresses the inherent hazard of working on a multiwire branch circuit in which all the ungrounded conductors are not opened by the disconnecting means. Although the ungrounded conductor that is supplying equipment being serviced may be disconnected, the common grounded conductor in this circuit configuration is still a current-carrying conductor in a circuit with an ungrounded conductor(s) that is not disconnected. In a Division 1 or 2 location, the opening of the grounded conductor in this scenario could result in an ignition-capable arc.

III. Equipment

501.100 Transformers and Capacitors

(A) Class I, Division 1 In Class I, Division 1 locations, transformers and capacitors shall comply with 501.100(A)(1) and (A)(2).

(1) Containing Liquid That Will Burn Transformers and capacitors containing a liquid that will burn shall be installed only in vaults that comply with 450.41 through 450.48 and with (1) through (4) as follows:

(1) There shall be no door or other communicating opening between the vault and the Division 1 location.
(2) Ample ventilation shall be provided for the continuous removal of flammable gases or vapors.
(3) Vent openings or ducts shall lead to a safe location outside of buildings.
(4) Vent ducts and openings shall be of sufficient area to relieve explosion pressures within the vault, and all portions of vent ducts within the buildings shall be of reinforced concrete construction.

(2) Not Containing Liquid That Will Burn Transformers and capacitors that do not contain a liquid that will burn shall be installed in vaults complying with 501.100(A)(1) or be approved for Class I locations.

(B) Class I, Division 2 In Class I, Division 2 locations, transformers and capacitors shall comply with 450.21 through 450.27.

501.105 Meters, Instruments, and Relays

(A) Class I, Division 1 In Class I, Division 1 locations, meters, instruments, and relays, including kilowatt-hour meters, instrument transformers, resistors, rectifiers, and thermionic tubes, shall be provided with enclosures identified for Class I, Division 1 locations. Enclosures for Class I, Division 1 locations include explosionproof enclosures and purged and pressurized enclosures.

FPN: See NFPA 496-2003, *Standard for Purged and Pressurized Enclosures for Electrical Equipment.*

See the commentary on purged and pressurized enclosures for electrical equipment in hazardous (classified) locations following 500.7(D) and the commentary on explosionproof enclosures following the exception to 501.5.

(B) Class I, Division 2 In Class I, Division 2 locations, meters, instruments, and relays shall comply with 501.105(B)(1) through (B)(6).

(1) Contacts Switches, circuit breakers, and make-and-break contacts of pushbuttons, relays, alarm bells, and horns shall have enclosures identified for Class I, Division 1 locations in accordance with 501.105(A).

Exception: General-purpose enclosures shall be permitted if current-interrupting contacts comply with one of the following:

(1) Are immersed in oil
(2) Are enclosed within a chamber that is hermetically sealed against the entrance of gases or vapors

Generally speaking, there are several types of hermetic seals, including fusion seals such as the glass-to-metal seals in mercury-tube switches and some reed switches, welded seals, soldered seals, and seals made with gaskets. Seals of the glass-to-metal-fusion type are usually the most reliable. Soft soldered seals can be relatively porous, and their effectiveness is highly dependent on workmanship. Although gasketed seals can be very effective, depending on the gasket material used, gasket materials can be damaged and deteriorate rapidly if exposed to atmospheres that contain solvent vapors. Gasketed enclosures may be considered hermetically sealed under some conditions; however, in accordance with the 500.2 definition of *hermetically sealed*, such enclosures cannot be used to satisfy those requirements in which hermetic sealing is recognized as a protection technique.

(3) Are in nonincendive circuits
(4) Are listed for Division 2

See the definitions of the terms *nonincendive circuit, nonincendive component,* and *nonincendive equipment* in 500.2. *Nonincendive* is similar to *intrinsically safe,* which is defined in 504.2 and ANSI/UL 913, *Intrinsically Safe Apparatus and Associated Apparatus for Use in Class I, II, and III, Division 1, Hazardous (Classified) Locations,* but *nonincendive* does not include consideration of faults and all the abnormal conditions inherent in the definition of *intrinsically safe.* A circuit may be nonincendive in a Group D atmosphere but not in a Group C atmosphere, due to the differences in minimum ignition energies for the various flammable materials.

(2) Resistors and Similar Equipment Resistors, resistance devices, thermionic tubes, rectifiers, and similar equipment that are used in or in connection with meters, instruments, and relays shall comply with 501.105(A).

Exception: General-purpose-type enclosures shall be permitted if such equipment is without make-and-break or sliding contacts [other than as provided in 501.105(B)(1)] and if the maximum operating temperature of any exposed surface will not exceed 80 percent of the ignition temperature in degrees Celsius of the gas or vapor involved or has been tested and found incapable of igniting the gas or vapor. This exception shall not apply to thermionic tubes.

(3) Without Make-or-Break Contacts Transformer windings, impedance coils, solenoids, and other windings that do not incorporate sliding or make-or-break contacts shall be provided with enclosures. General-purpose-type enclosures shall be permitted.

(4) General-Purpose Assemblies Where an assembly is made up of components for which general-purpose enclosures are acceptable as provided in 501.105(B)(1), (B)(2), and (B)(3), a single general-purpose enclosure shall be acceptable for the assembly. Where such an assembly includes any of the equipment described in 501.105(B)(2), the maximum obtainable surface temperature of any component of the assembly shall be clearly and permanently indicated on the outside of the enclosure. Alternatively, equipment shall be permitted to be marked to indicate the temperature class for which it is suitable, using the temperature class (T Code) of Table 500.8(B).

(5) Fuses Where general-purpose enclosures are permitted in 501.105(B)(1) through (B)(4), fuses for overcurrent protection of instrument circuits not subject to overloading in normal use shall be permitted to be mounted in general-purpose enclosures if each such fuse is preceded by a switch complying with 501.105(B)(1).

(6) Connections To facilitate replacements, process control instruments shall be permitted to be connected through flexible cord, attachment plug, and receptacle, provided all of the following conditions apply:

(1) A switch complying with 501.105(B)(1) is provided so that the attachment plug is not depended on to interrupt current.
(2) The current does not exceed 3 amperes at 120 volts, nominal.
(3) The power-supply cord does not exceed 900 mm (3 ft), is of a type listed for extra-hard usage or for hard usage if protected by location, and is supplied through an attachment plug and receptacle of the locking and grounding type.
(4) Only necessary receptacles are provided.
(5) The receptacle carries a label warning against unplugging under load.

501.115 Switches, Circuit Breakers, Motor Controllers, and Fuses

(A) Class I, Division 1 In Class I, Division 1 locations, switches, circuit breakers, motor controllers, and fuses, including pushbuttons, relays, and similar devices, shall be

provided with enclosures, and the enclosure in each case, together with the enclosed apparatus, shall be identified as a complete assembly for use in Class I locations.

Exhibit 501.14 shows an explosionproof panelboard that consists of an assembly of branch-circuit devices enclosed in a cast metal explosionproof housing. Explosionproof panelboards are provided with bolted access covers and threaded conduit-entry hubs designed to withstand the force of an internal explosion.

Exhibit 501.15 shows a cylindrical-type (spin-top) combination motor controller, motor control starter, and circuit breaker in an explosionproof enclosure. The top and bottom covers are threaded on for quick removal for installation and servicing. Exhibit 501.16 shows the same type of equipment in a rectangular enclosure with a hinged, bolted-on cover. These types of housings are designed to accommodate a wide range of manually or magnetically operated across-the-line types of motor starters in a variety of ratings.

Exhibit 501.17 illustrates a standard toggle switch in an explosionproof enclosure.

Exhibit 501.14 An explosionproof panelboard. (Courtesy of Appleton Electric Co., EGS Electrical Group)

Exhibit 501.16 A magnetic motor starter for use in a Class I, Group D location. Note the number of securing bolts and the width of the flange. (Courtesy of O-Z/Gedney, a division of EGS Electrical Group)

Exhibit 501.15 An explosionproof enclosure for a motor control starter and circuit breaker. (Courtesy of Appleton Electric Co., EGS Electrical Group)

Fuses used for supplementary ballast protection are permitted to be installed in high-intensity-discharge and fluorescent fixtures in accordance with 501.115(B)(4).

(B) Class I, Division 2 Switches, circuit breakers, motor controllers, and fuses in Class I, Division 2 locations shall comply with 501.115(B)(1) through (B)(4).

(1) Type Required Circuit breakers, motor controllers, and switches intended to interrupt current in the normal performance of the function for which they are installed shall be provided with enclosures identified for Class I, Division 1 locations in accordance with 501.105(A), unless general-purpose enclosures are provided and any of the following apply:

(1) The interruption of current occurs within a chamber hermetically sealed against the entrance of gases and vapors.
(2) The current make-and-break contacts are oil-immersed and of the general-purpose type having a 50-mm (2-in.) minimum immersion for power contacts and a 25-mm (1-in.) minimum immersion for control contacts.

Exhibit 501.17 A standard toggle switch in an explosionproof enclosure. (Courtesy of Appleton Electric Co., EGS Electrical Group)

(3) The interruption of current occurs within a factory-sealed explosionproof chamber.
(4) The device is a solid state, switching control without contacts, where the surface temperature does not exceed 80 percent of the ignition temperature in degrees Celsius of the gas or vapor involved.

(2) Isolating Switches Fused or unfused disconnect and isolating switches for transformers or capacitor banks that are not intended to interrupt current in the normal performance of the function for which they are installed shall be permitted to be installed in general-purpose enclosures.

(3) Fuses For the protection of motors, appliances, and lamps, other than as provided in 501.115(B)(4), standard plug or cartridge fuses shall be permitted, provided they are placed within enclosures identified for the location; or fuses shall be permitted if they are within general-purpose enclosures, and if they are of a type in which the operating element is immersed in oil or other approved liquid, or the operating element is enclosed within a chamber hermetically sealed against the entrance of gases and vapors, or the fuse is a nonindicating, filled, current-limiting type.

(4) Fuses Internal to Luminaires (Lighting Fixtures) Listed cartridge fuses shall be permitted as supplementary protection within luminaires (lighting fixtures).

501.120 Control Transformers and Resistors

Transformers, impedance coils, and resistors used as, or in conjunction with, control equipment for motors, generators, and appliances shall comply with 501.120(A) and 501.120(B).

(A) Class I, Division 1 In Class I, Division 1 locations, transformers, impedance coils, and resistors, together with any switching mechanism associated with them, shall be provided with enclosures identified for Class I, Division 1 locations in accordance with 501.105(A).

(B) Class I, Division 2 In Class I, Division 2 locations, control transformers and resistors shall comply with 501.120(B)(1) through (B)(3).

(1) Switching Mechanisms Switching mechanisms used in conjunction with transformers, impedance coils, and resistors shall comply with 501.115(B).

(2) Coils and Windings Enclosures for windings of transformers, solenoids, or impedance coils shall be permitted to be of the general-purpose type.

(3) Resistors Resistors shall be provided with enclosures; and the assembly shall be identified for Class I locations, unless resistance is nonvariable and maximum operating temperature, in degrees Celsius, will not exceed 80 percent of the ignition temperature of the gas or vapor involved or has been tested and found incapable of igniting the gas or vapor.

501.125 Motors and Generators

(A) Class I, Division 1 In Class I, Division 1 locations, motors, generators, and other rotating electric machinery shall be one of the following:

(1) Identified for Class I, Division 1 locations
(2) Of the totally enclosed type supplied with positive-pressure ventilation from a source of clean air with discharge to a safe area, so arranged to prevent energizing of the machine until ventilation has been established and the enclosure has been purged with at least 10 volumes of air, and also arranged to automatically de-energize the equipment when the air supply fails
(3) Of the totally enclosed inert gas-filled type supplied with a suitable reliable source of inert gas for pressurizing the enclosure, with devices provided to ensure a positive pressure in the enclosure and arranged to automatically de-energize the equipment when the gas supply fails
(4) Of a type designed to be submerged in a liquid that is flammable only when vaporized and mixed with air, or in a gas or vapor at a pressure greater than atmospheric and that is flammable only when mixed with air; and the machine is arranged so to prevent energizing it until it has been purged with the liquid or gas to exclude air, and also arranged to automatically de-energize the equipment when the supply of liquid or gas or vapor fails or the pressure is reduced to atmospheric

The intent of 501.125(A)(4) is to permit nonexplosionproof motors to be submerged in liquefied natural gas (LNG), liquefied petroleum gas (LP-Gas), gasoline, and other flammable liquids. The provisions of 501.125(A)(4) do not permit nonexplosionproof motors under water, such as in wet pits, unless the motors are provided with some other system of explosion protection, for example, if they are purged and pressurized per NFPA 496, *Standard for Purged and Pressurized Enclosures for Electrical Equipment*.

The ASTM test procedure is used to determine the ignition temperature of some flammable and combustible liquids.

Totally enclosed motors of the types specified in 501.125(A)(2) or 501.125(A)(3) shall have no external surface with an operating temperature in degrees Celsius in excess of 80 percent of the ignition temperature of the gas or vapor involved. Appropriate devices shall be provided to detect and automatically de-energize the motor or provide an adequate alarm if there is any increase in temperature of the motor beyond designed limits. Auxiliary equipment shall be of a type identified for the location in which it is installed.

FPN: See D 2155-69, *ASTM Test Procedure*.

(B) Class I, Division 2 In Class I, Division 2 locations, motors, generators, and other rotating electric machinery in which are employed sliding contacts, centrifugal or other types of switching mechanism (including motor overcurrent, overloading, and overtemperature devices), or integral resistance devices, either while starting or while running, shall be identified for Class I, Division 1 locations, unless such sliding contacts, switching mechanisms, and resistance devices are provided with enclosures identified for Class I, Division 2 locations in accordance with 501.105(B). The exposed surface of space heaters used to prevent condensation of moisture during shutdown periods shall not exceed 80 percent of the ignition temperature in degrees Celsius of the gas or vapor involved when operated at rated voltage, and the maximum surface temperature [based on a 40°C (104°F) ambient] shall be permanently marked on a visible nameplate mounted on the motor. Otherwise, space heaters shall be identified for Class I, Division 2 locations. In Class I, Division 2 locations, the installation of open or nonexplosionproof enclosed motors, such as squirrel-cage induction motors without brushes, switching mechanisms, or similar arc-producing devices that are not identified for use in a Class I, Division 2 location, shall be permitted.

It is intended that the phrase "other rotating electric machinery" include electric brakes. Listed and labeled electric brakes are available for Class I, Division 1, Group C and D locations.

Many motor heaters are de-energized automatically when the motor is running. However, the heater ratings are usually low when compared with the normal heat generated during motor operation. Unless otherwise indicated on the motor wiring diagram or in instructions provided with the motor, there is no need to de-energize the heater except to save energy. Note the requirement that the heater temperature be marked on the motor or that the heaters be identified for the location.

FPN No. 1: It is important to consider the temperature of internal and external surfaces that may be exposed to the flammable atmosphere.

FPN No. 2: It is important to consider the risk of ignition due to currents arcing across discontinuities and overheating of parts in multisection enclosures of large motors and generators. Such motors and generators may need equipotential bonding jumpers across joints in the enclosure and from enclosure to ground. Where the presence of ignitible gases or vapors is suspected, clean-air purging may be needed immediately prior to and during start-up periods.

Exhibit 501.18 shows a totally enclosed fan-cooled motor listed for use in explosive atmospheres. The main frame and end-bells are designed with sufficient strength to withstand an internal explosion. Flames or hot gases are cooled while escaping because of the wide metal-to-metal joints between the frame and the end-bells and the long, close-tolerance clearance provided for the free turn of the shaft. Air circulation outside the motor is maintained by a nonsparking (aluminum, bronze, or non-static-generating-type plastic) fan on

the end opposite the shaft end of the motor. A sheet metal housing surrounds the fan to reduce the likelihood of an individual or object coming into contact with the moving blades and to direct the flow of air. An internal fan on the shaft, as shown in Exhibit 501.19, circulates air around the windings.

Motors that have arc- or spark-producing devices, such as commutators, internal switches, or other control devices,

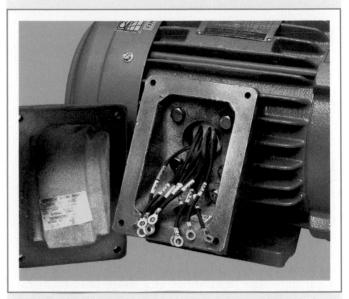

Exhibit 501.18 Terminal housing of a motor listed for use in specific hazardous locations. Note integral sealing of the motor. (Courtesy of General Electric Co.)

Exhibit 501.19 View showing internal fan of motor in Exhibit 501.18. (Courtesy of General Electric Co.)

must be explosionproof. General-purpose squirrel-cage induction motors without arc- or spark-producing devices may be used in Division 2 locations.

Some open-type motors are permitted in Class I, Division 2 locations. Motor types used where flammable gases or vapors with very low ignition temperatures may be present should be selected with great care. Modern motors with high-temperature insulation systems, such as Class H [180°C (356°F)], may operate close to or above the ignition temperature of the flammable mixture.

FPN No. 3: For further information on the application of electric motors in Class I, Division 2 hazardous (classified) locations, see IEEE Std. 1349-2001, *IEEE Guide for the Application of Electric Motors in Class I, Division 2 Hazardous (Classified) Locations.*

501.130 Luminaires (Lighting Fixtures)

Luminaires (lighting fixtures) shall comply with 501.130(A) or (B).

(A) Class I, Division 1 In Class I, Division 1 locations, luminaires (lighting fixtures) shall comply with 501.130(A)(1) through (A)(4).

(1) Luminaires (Lighting Fixtures) Each luminaire (lighting fixture) shall be identified as a complete assembly for the Class I, Division 1 location and shall be clearly marked to indicate the maximum wattage of lamps for which it is identified. Luminaires (lighting fixtures) intended for portable use shall be specifically listed as a complete assembly for that use.

(2) Physical Damage Each luminaire (lighting fixture) shall be protected against physical damage by a suitable guard or by location.

(3) Pendant Luminaires (Lighting Fixtures) Pendant luminaires (lighting fixtures) shall be suspended by and supplied through threaded rigid metal conduit stems or threaded steel intermediate conduit stems, and threaded joints shall be provided with set-screws or other effective means to prevent loosening. For stems longer than 300 mm (12 in.), permanent and effective bracing against lateral displacement shall be provided at a level not more than 300 mm (12 in.) above the lower end of the stem, or flexibility in the form of a fitting or flexible connector identified for the Class I, Division 1 location shall be provided not more than 300 mm (12 in.) from the point of attachment to the supporting box or fitting.

(4) Supports Boxes, box assemblies, or fittings used for the support of luminaires (lighting fixtures) shall be identified for Class I locations.

(B) Class I, Division 2 In Class I, Division 2 locations, luminaires (lighting fixtures) shall comply with 501.130(B)(1) through 501.130(B)(6).

(1) Luminaires (Lighting Fixtures) Where lamps are of a size or type that may, under normal operating conditions, reach surface temperatures exceeding 80 percent of the ignition temperature in degrees Celsius of the gas or vapor involved, fixtures shall comply with 501.130(A)(1) or shall be of a type that has been tested in order to determine the marked operating temperature or temperature class (T Code).

(2) Physical Damage Luminaires (lighting fixtures) shall be protected from physical damage by suitable guards or by location. Where there is danger that falling sparks or hot metal from lamps or fixtures might ignite localized concentrations of flammable vapors or gases, suitable enclosures or other effective protective means shall be provided.

(3) Pendant Luminaires (Fixtures) Pendant luminaires (lighting fixtures) shall be suspended by threaded rigid metal conduit stems, threaded steel intermediate metal conduit stems, or other approved means. For rigid stems longer than 300 mm (12 in.), permanent and effective bracing against lateral displacement shall be provided at a level not more than 300 mm (12 in.) above the lower end of the stem, or flexibility in the form of an identified fitting or flexible connector shall be provided not more than 300 mm (12 in.) from the point of attachment to the supporting box or fitting.

(4) Portable Lighting Equipment Portable lighting equipment shall comply with 501.130(A)(1).

Exception: Where portable lighting equipment is mounted on movable stands and is connected by flexible cords, as covered in 501.140, it shall be permitted, where mounted in any position, if it conforms to 501.130(B)(2).

(5) Switches Switches that are a part of an assembled fixture or of an individual lampholder shall comply with 501.115(B)(1).

(6) Starting Equipment Starting and control equipment for electric-discharge lamps shall comply with 501.120(B).

Exhibit 501.20 shows a typical luminaire for Class I, Group C and D locations. The outlet boxes have an internally threaded opening designed to receive the cover. A pendant fixture is attached to the cover by threaded rigid metal conduit or threaded intermediate metal conduit. To prevent loosening from vibration or lamp changing, threaded joints must be provided with set-screws. The set-screws should not interrupt the explosionproof joint. Rigid metal conduit or intermediate metal conduit stems longer than 12 in. require effective

Exhibit 501.20 A typical lighting fixture for use in Class I, Group C and D locations. (Courtesy of Appleton Electric Co., EGS Electrical Group)

Exhibit 501.21 An explosionproof hand lamp for use in Class I locations. (Courtesy of Appleton Electric Co., EGS Electrical Group)

bracing or a flexible fitting approved for the purpose, placed not more than 12 in. from the point of attachment to the supporting box, cover, or fitting.

A globe holder is threaded onto the body of the fixture housing and supports a heavy glass globe, guard, and reflector. It is available in sizes suitable for lamps from 40 watts through 500 watts. In designing any hazardous (classified) location lighting system, operating temperatures must be considered. Therefore, if the area is Class I, Division 1, fixtures approved for this location that are properly marked must be used. Generally, enclosed and gasketed fixtures (previously called vaportight fixtures) without guards, if breakage is unlikely, or fixtures approved for Class I, Division 2 locations are required in Division 2 locations.

Fixtures listed by Underwriters Laboratories Inc. for use in any of the groups under Class I, either Division 1 or 2 locations, or both, are designed to operate without igniting surrounding flammable gas or vapor atmospheres and are marked with the operating temperature or temperature class, as shown in Table 500.8(B).

Exhibit 501.21 shows an explosionproof hand lamp. Lamp compartments must be sealed from the terminal compartment. Provisions must be made for the connection of 3-conductor (one must be a grounding conductor) flexible, extra-hard-usage cord. See 501.140(A)(1).

Exception: A thermal protector potted into a thermally protected fluorescent lamp ballast if the luminaire (lighting fixture) is identified for the location.

501.135 Utilization Equipment

(A) Class I, Division 1 In Class I, Division 1 locations, all utilization equipment shall be identified for Class I, Division 1 locations.

(B) Class I, Division 2 In Class I, Division 2 locations, all utilization equipment shall comply with 501.135(B)(1) through (B)(3).

(1) Heaters Electrically heated utilization equipment shall conform with either item (1) or (2):

(1) The heater shall not exceed 80 percent of the ignition temperature in degrees Celsius of the gas or vapor involved on any surface that is exposed to the gas or vapor when continuously energized at the maximum rated ambient temperature. If a temperature controller is not provided, these conditions shall apply when the heater is operated at 120 percent of rated voltage.

Exception No. 1: For motor-mounted anticondensation space heaters, see 501.125.

Exception No. 2: Where a current-limiting device is applied to the circuit serving the heater to limit the current in the heater to a value less than that required to raise the heater surface temperature to 80 percent of the ignition temperature.

(2) The heater shall be identified for Class I, Division 1 locations.

Exception: Electrical resistance heat tracing identified for Class I, Division 2 locations.

(2) Motors Motors of motor-driven utilization equipment shall comply with 501.125(B).

(3) Switches, Circuit Breakers, and Fuses Switches, circuit breakers, and fuses shall comply with 501.115(B).

The requirements for utilization equipment in Class I locations are virtually identical for Division 1 and 2 locations, except for heaters. Electric pipe heat-tracing systems listed for Class I, Division 2 locations and complying with 501.135(B)(1)(2), Exception, are available.

501.140 Flexible Cords, Class I, Divisions 1 and 2

(A) Permitted Uses Flexible cord shall be permitted:

(1) For connection between portable lighting equipment or other portable utilization equipment and the fixed portion of their supply circuit.
(2) For that portion of the circuit where the fixed wiring methods of 501.10(A) cannot provide the necessary degree of movement for fixed and mobile electrical utilization equipment, and the flexible cord is protected by location or by a suitable guard from damage and only in an industrial establishment where conditions of maintenance and engineering supervision ensure that only qualified persons install and service the installation.
(3) For electric submersible pumps with means for removal without entering the wet-pit. The extension of the flexible cord within a suitable raceway between the wet-pit and the power source shall be permitted.

Section 501.140(A)(3) recognizes a wet-pit type of installation that is finding increasing acceptance for use in wastewater systems. Section 501.125(A)(4) permits a flexible cord to be run in raceway to a junction box outside the wet pit. See the commentary following 501.125(A)(4) for more information on motors installed in wet pits.

Due to the operating conditions associated with electric mixers used in mixing tanks, the mixers are considered portable utilization equipment and may be wired with flexible cord.

(4) For electric mixers intended for travel into and out of open-type mixing tanks or vats.

(B) Installation Where flexible cords are used, the cords shall comply with all of the following:

(1) Be of a type listed for extra-hard usage
(2) Contain, in addition to the conductors of the circuit, a grounding conductor complying with 400.23
(3) Be connected to terminals or to supply conductors in an approved manner
(4) Be supported by clamps or by other suitable means in such a manner that there is no tension on the terminal connections
(5) Be provided with suitable seals where the flexible cord enters boxes, fittings, or enclosures of the explosionproof type

Exception to (5): Seals shall not be required as provided in 501.10(B) and 501.105(B)(6).

(6) Be of continuous length

FPN: See 501.20 for flexible cords exposed to liquids having a deleterious effect on the conductor insulation.

501.145 Receptacles and Attachment Plugs, Class I, Divisions 1 and 2

Receptacles and attachment plugs shall be of the type providing for connection to the grounding conductor of a flexible cord and shall be identified for the location.

Exception: As provided in 501.105(B)(6).

Exhibit 501.22 shows an explosionproof receptacle and attachment plug with an interlocking switch. The design of this device is such that when the switch is in the on position, the plug cannot be removed. Also, the switch cannot be placed in the on position when the plug has been removed; that is, the plug cannot be inserted or removed unless the switch is in the off position. The receptacle is factory sealed, with a provision for threaded-conduit entry to the switch

Exhibit 501.22 A receptacle and attachment plug of the explosionproof type with an interlocking switch. The switch must be in the off position before the attachment plug can be removed. (Courtesy of Appleton Electric Co., EGS Electrical Group)

compartment. The plug is to be used with Type S or equivalent extra-hard-service flexible cord having an equipment grounding conductor.

Exhibit 501.23 shows a 30-ampere, 4-pole receptacle and attachment plug assembly that is suitable for use without a switch. The design is such that the mating parts of the receptacle and plug are enclosed in a chamber that seals the arc and, by delayed-action construction, prevents complete removal of the plug until the arc or hot metal has cooled. The receptacle is factory sealed, and the attachment plug is designed for use with a 4-conductor cord (3-conductor, 3-phase circuit with one grounding conductor) or a 3-conductor cord (two circuit conductors and one grounding conductor).

Exhibit 501.23 A 4-pole (delayed action) explosionproof receptacle and attachment plug suitable for use without a switch. (Courtesy of Appleton Electric Co., EGS Electrical Group)

501.150 Signaling, Alarm, Remote-Control, and Communications Systems

(A) Class I, Division 1 In Class I, Division 1 locations, all apparatus and equipment of signaling, alarm, remote-control, and communications systems, regardless of voltage, shall be identified for Class I, Division 1 locations, and all wiring shall comply with 501.10(A), 501.15(A), and 501.15(C).

(B) Class I, Division 2 In Class I, Division 2 locations, signaling, alarm, remote-control, and communications systems shall comply with 501.150(B)(1) through (B)(4).

(1) Contacts Switches, circuit breakers, and make-and-break contacts of pushbuttons, relays, alarm bells, and horns shall have enclosures identified for Class I, Division 1 locations in accordance with 501.105(A).

Exception: General-purpose enclosures shall be permitted if current-interrupting contacts are one of the following:

(1) Immersed in oil

(2) Enclosed within a chamber hermetically sealed against the entrance of gases or vapors

(3) In nonincendive circuits

See the commentary following 501.105(B)(1), Exception (4), for information on nonincendive circuits.

(4) Part of a listed nonincendive component

(2) Resistors and Similar Equipment Resistors, resistance devices, thermionic tubes, rectifiers, and similar equipment shall comply with 501.105(B)(2).

(3) Protectors Enclosures shall be provided for lightning protective devices and for fuses. Such enclosures shall be permitted to be of the general-purpose type.

(4) Wiring and Sealing All wiring shall comply with 501.10(B), 501.15(B), and 501.15(C).

Audible signaling devices, such as bells, sirens, and horns, other than electronic types, usually involve make-and-break contacts that are capable of producing a spark of sufficient energy to cause ignition of a hazardous atmospheric mixture. If used in Class I locations, therefore, this type of equipment, as shown in Exhibit 501.24, must be contained in explo-

Exhibit 501.24 An audible signaling device for use in hazardous (classified) locations. (Courtesy of Cooper Crouse-Hinds)

sionproof or purged and pressurized enclosures, wiring methods must comply with 501.10, and sealing fittings must be provided in accordance with 501.15.

Explosionproof devices or explosionproof enclosures may prove more practical than oil-immersed contacts because maintaining the condition and level of the oil can be a problem. Hermetically sealed enclosures, such as float-operated mercury-tube switches, are available for some applications. Electronic signal devices without make-and-break contacts usually do not require explosionproof enclosures in Division 2 locations.

ARTICLE 502
Class II Locations

Summary of Changes

- **General:** Restructured and renumbered to provide a scope section and parallel numbering systems for Articles 501, 502, and 503.

- **502.10(A)(2):** Revised to add interlocked armor Type MC cable as an additional method to provide a flexible connection in Class II, Division 1 locations.

- **502.15(4):** Added item (4) permitting a raceway to extend horizontally and downward as an equivalent to a horizontally installed raceway or a vertically installed raceway as a method of sealing dust-ignition proof enclosures.

- **502.25:** Revised to provide guidance concerning protection against explosion and against electric shock in order to limit the voltage of exposed live parts.

Contents

I. General

502.1 Scope

Article 502 covers the requirements for electrical and electronic equipment and wiring for all voltages in Class II, Division 1 and 2 locations where fire or explosion hazards may exist due to combustible dust.

Two different types of dust environments typically warrant a Class II, Division 1 area classification. The first is where a cloud of combustible dust is likely to be present continuously or intermittently under normal operating conditions as a result of repair or maintenance operations or leakage. The other environment is one in which a dust layer is likely to accumulate to a depth greater than 1/8 in. on major horizontal surfaces over a defined period of time, generally 24 hours. The size of the dust particle is the primary factor in determining whether the area should be classified as Class II or Class III. Combustible dust, as defined in NFPA 499, *Recom-*

mended Practice for the Classification of Combustible Dusts and of Hazardous (Classified) Locations for Electrical Installations in Chemical Process Areas, is any finely divided solid material 420 microns (0.0165 in.) or less in diameter (i.e., material passing through a U.S. No. 40 standard sieve) that presents a fire or explosion hazard when dispersed.

502.5 General

The general rules of this *Code* shall apply to the electric wiring and equipment in locations classified as Class II locations in 500.5(C).

Exception: As modified by this article.

Equipment installed in Class II locations shall be able to function at full rating without developing surface temperatures high enough to cause excessive dehydration or gradual carbonization of any organic dust deposits that may occur.

> FPN: Dust that is carbonized or excessively dry is highly susceptible to spontaneous ignition.

Explosionproof equipment and wiring shall not be required and shall not be acceptable in Class II locations unless identified for such locations.

Class II, Division 1 and 2 locations are defined in 500.5(C) as hazardous due to the presence of combustible dust. These locations are separated into three groups: Group E, Group F, and Group G [see 500.6(B)].

It should be noted that equipment suitable for one class and group is not necessarily suitable for any other class and group. Class I equipment is not necessarily better, or even suitable, for a Class II location, because the hazard contemplated in the equipment design is different. Class II equipment is designed to prevent the ignition of layers of dust, which may increase the operating temperature of the equipment. Class I equipment is not designed for dust layering unless it is also designed and approved for Class II locations. To protect against explosions in hazardous (classified) locations, all electrical equipment exposed to the hazardous atmosphere must be suitable for such locations. Grain dust, for example, ignites at a temperature lower than that of most flammable vapors. Motors listed for use in Class I locations may not have dust shields on the bearings to prevent entrance of dust into the bearing race, thereby causing overheating of the bearing and resulting in ignition of dust on the motor.

One or more of the following four hazards may be present in a Class II location:

1. An explosive mixture of air and dust in suspension
2. Accumulation of dust that acts as a thermal blanket and interferes with the safe dissipation of heat from electrical equipment
3. Accumulation of electrically conductive dust lodged between terminals that have a difference of potential, thereby causing tracking and glowing hot particles, short-circuits, or ground faults that may ignite dust accumulated in the vicinity
4. Deposits of dust that could be ignited by arcs or sparks

In the layout of electrical installations for hazardous (classified) locations, it is preferable to locate service equipment, switchboards, panelboards, and much of the electrical equipment in less hazardous areas, usually in a separate room. The use of pressurized rooms, as described in NFPA 496, *Standard for Purged and Pressurized Enclosures for Electrical Equipment*, is a common method of protecting panelboards and switchboards in grain elevators and similar locations.

II. Wiring

502.10 Wiring Methods

Wiring methods shall comply with 502.10(A) or 502.10(B).

(A) Class II, Division 1

(1) General In Class II, Division 1 locations, the wiring methods in (1) through (4) shall be permitted.

(1) Threaded rigid metal conduit, or threaded steel intermediate metal conduit.
(2) Type MI cable with termination fittings listed for the location. Type MI cable shall be installed and supported in a manner to avoid tensile stress at the termination fittings.
(3) In industrial establishments with limited public access, where the conditions of maintenance and supervision ensure that only qualified persons service the installation, Type MC cable, listed for use in Class II, Division 1 locations, with a gas/vaportight continuous corrugated metallic sheath, an overall jacket of suitable polymeric material, separate grounding conductors in accordance with 250.122, and provided with termination fittings listed for the application, shall be permitted.
(4) Fittings and boxes shall be provided with threaded bosses for connection to conduit or cable terminations and shall be dusttight. Fittings and boxes in which taps, joints, or terminal connections are made, or that are used in Group E locations, shall be identified for Class II locations.

In Class II, Division 1 locations, boxes or fittings, such as conduit "L," "T," or "C" fittings, that contain splices, taps, or terminations must be of the threaded type with close-fitting covers. There can be no holes in the box or fitting, such as mounting holes, that would allow dust to enter or sparks or burning material to be ejected from the enclosure. Where used as stated above, the equipment must be approved for Class II locations, usually of the dust-ignitionproof or pressurized type.

Boxes or fittings of the type described in the preceding paragraph that do not enclose splices, taps, or terminations must be dusttight, with threaded bosses for connection of conduit or cable connectors.

(2) Flexible Connections Where necessary to employ flexible connections, one or more of the following shall also be permitted:

(1) Dusttight flexible connectors
(2) Liquidtight flexible metal conduit with listed fittings
(3) Liquidtight flexible nonmetallic conduit with listed fittings
(4) Interlocked armor Type MC cable having an overall jacket of suitable polymeric material and provided with termination fittings listed for Class II, Division 1 locations.
(5) Flexible cord listed for extra-hard usage and provided with bushed fittings. Where flexible cords are used, they shall comply with 502.140.

FPN: See 502.30(B) for grounding requirements where flexible conduit is used.

Due to the potential for physical damage to Type MC cable, its application in Class II, Division 1 locations is limited to cable that is listed specifically for use in Class II, Division 1 locations and installed at facilities that have full-time, qualified maintenance personnel. Qualified maintenance personnel are those who, in the course of regular maintenance procedures, would notice if cables were damaged, understand the associated hazards, and be able to de-energize the circuit to repair the installation.

(B) Class II, Division 2

(1) General In Class II, Division 2 locations, the following wiring methods shall be permitted:

(1) All wiring methods permitted in 502.10(A).
(2) Rigid metal conduit, intermediate metal conduit, electrical metallic tubing, dusttight wireways.
(3) Type MC or MI cable with listed termination fittings.
(4) Type PLTC in cable trays.
(5) Type ITC in cable trays.
(6) Type MC, MI, or TC cable installed in ladder, ventilated trough, or ventilated channel cable trays in a single layer, with a space not less than the larger cable diameter between the two adjacent cables, shall be the wiring method employed.

Exception to (6): Type MC cable listed for use in Class II, Division 1 locations shall be permitted to be installed without the spacings required by (6).

(2) Flexible Connections Where provision must be made for flexibility, 502.10(A)(2) shall apply.

Where it is necessary to use flexible connections, liquidtight flexible conduit or extra-hard-usage flexible cord is permitted. Another method would be to use a flexible fitting, as described in the commentary following 501.10(A)(2). Where liquidtight flexible conduit is used, a bonding jumper (internal or external) must be provided around such conduit. See 502.30. An additional conductor for grounding must be provided where flexible cord is used.

(3) Nonincendive Field Wiring Nonincendive field wiring shall be permitted using any of the wiring methods permitted for unclassified locations. Nonincendive field wiring systems shall be installed in accordance with the control drawing(s). Simple apparatus, not shown on the control drawing, shall be permitted in a nonincendive field wiring circuit, provided the simple apparatus does not interconnect the nonincendive field wiring circuit to any other circuit.

FPN: Simple apparatus is defined in 504.2.

Separate nonincendive field wiring circuits shall be installed in accordance with one of the following:

(1) In separate cables
(2) In multiconductor cables where the conductors of each circuit are within a grounded metal shield
(3) In multiconductor cables where the conductors of each circuit have insulation with a minimum thickness of 0.25 mm (0.01 in.)

(4) Boxes and Fittings All boxes and fittings shall be dusttight.

In Division 1 locations, boxes containing taps, joints, or terminal connections must be dusttight and must be provided with threaded hubs, as shown in Exhibit 502.1, and must be identified for use in Class II locations. Threaded hubs also provide adequate bonding in Division 2 locations (see Exhibit 502.1.)

Exhibit 502.1 also shows a close-fitting cover, which is required for Class II locations. Standard pressed-steel boxes that are not identified for Class II locations are permitted as long as they do not contain taps, joints, or terminal connections, are dusttight, and are provided with a bonding jumper around the box in order to compensate for the absence of threaded hubs.

Exhibit 502.1 Junction box with threaded hubs, suitable for use in Class II, Group E hazardous atmospheres. (Courtesy of Appleton Electric Co., EGS Electrical Group)

502.15 Sealing, Class II, Divisions 1 and 2

Where a raceway provides communication between an enclosure that is required to be dust-ignitionproof and one that is not, suitable means shall be provided to prevent the entrance of dust into the dust-ignitionproof enclosure through the raceway. One of the following means shall be permitted:

(1) A permanent and effective seal
(2) A horizontal raceway not less than 3.05 m (10 ft) long
(3) A vertical raceway not less than 1.5 m (5 ft) long and extending downward from the dust-ignitionproof enclosure
(4) A raceway installed in a manner equivalent to (2) or (3) that extends only horizontally and downward from the dust-ignition proof enclosures.

Where a raceway provides communication between an enclosure that is required to be dust-ignitionproof and an enclosure in an unclassified location, seals shall not be required.

Sealing fittings shall be accessible.

Seals shall not be required to be explosionproof.

FPN: Electrical sealing putty is a method of sealing.

The requirements of 502.15 provide three suitable ways to prevent dust from entering dust-ignitionproof enclosures through the raceway. A seal in the raceway, a horizontal separation between enclosures of not less than 10 ft, or a

vertical separation between enclosures of not less than 5 ft in a raceway extending downward from a dust-ignitionproof enclosure are methods of sealing in Class II locations and are shown in Exhibit 502.2.

The requirement to provide a seal applies if a raceway connects an enclosure that is required to be dust-ignitionproof to one that is not required to be dust-ignitionproof. If a raceway extends from a dust-ignitionproof enclosure to an unclassified location, it is not necessary to provide a seal in that raceway. Sealing fittings designed for use in Class I locations are acceptable. However, because the Class I location pressure-piling considerations are not inherent to Class II locations, conduit seals are not required to be explosionproof. Conduit seals are expected only to prevent the migration of dust into dust-ignitionproof enclosures. No sealing method is needed in the special, but not unusual, situation in which dust cannot enter the raceway in the hazardous (classified) location.

Exhibit 502.2 Three methods for preventing dust from entering a dust-ignitionproof enclosure through the raceway.

502.25 Uninsulated Exposed Parts, Class II, Divisions 1 and 2

There shall be no uninsulated exposed parts, such as electric conductors, buses, terminals, or components, that operate at more than 30 volts (15 volts in wet locations). These parts shall additionally be protected by a protection technique according to 500.7(E), 500.7(F), or 500.7(G) that is suitable for the location.

502.30 Grounding and Bonding, Class II, Divisions 1 and 2

Wiring and equipment in Class II, Division 1 and 2 locations shall be grounded as specified in Article 250 and with the requirements in 502.30(A) and 502.30(B).

(A) Bonding The locknut-bushing and double-locknut types of contact shall not be depended on for bonding purposes, but bonding jumpers with proper fittings or other approved means of bonding shall be used. Such means of bonding shall apply to all intervening raceways, fittings, boxes, enclosures, and so forth, between Class II locations and the point of grounding for service equipment or point of grounding of a separately derived system.

For information regarding bonding in hazardous (classified) locations, see the commentary on 501.30(A). Although that section addresses bonding in Class I locations, the same requirement for enhanced bonding in Class II locations is found in 502.30(A), and the requirements of 250.100 apply to Class I, Class II, and Class III hazardous locations.

Exception: The specific bonding means shall only be required to the nearest point where the grounded circuit conductor and the grounding electrode conductor are connected together on the line side of the building or structure disconnecting means as specified in 250.32(A), (B), and (C), if the branch-circuit overcurrent protection is located on the load side of the disconnecting means.

> FPN: See 250.100 for additional bonding requirements in hazardous (classified) locations.

(B) Types of Equipment Grounding Conductors Where flexible conduit is used as permitted in 502.10, it shall be installed with internal or external bonding jumpers in parallel with each conduit and complying with 250.102.

Exception: In Class II, Division 2 locations, the bonding jumper shall be permitted to be deleted where all of the following conditions are met:

(1) Listed liquidtight flexible metal conduit 1.8 m (6 ft) or less in length, with fittings listed for grounding, is used.
(2) Overcurrent protection in the circuit is limited to 10 amperes or less.
(3) The load is not a power utilization load.

In Class II locations, a raceway connected to an enclosure via a double locknut connection or a single locknut and bushing connection is not an acceptable method of bonding. Bonding jumpers or other approved means with proper fittings are required for the interconnection of all raceways, junction boxes, fittings, enclosures, and so on, installed between the hazardous area and the grounding electrode conductor connection point at the service equipment, the grounding electrode connection at a feeder or branch circuit disconnecting means at separate buildings, or the grounding electrode connection to the source of a separately derived system. If installed outside the raceway or enclosure, the

grounding conductor must not exceed 6 ft and must be routed with the raceway or enclosure. See 250.102(E) for equipment bonding jumper installation requirements.

502.35 Surge Protection — Class II, Divisions 1 and 2

Surge arresters and transient voltage surge suppressors (TVSS) installed in a Class II, Division 1 location shall be in suitable enclosures. Surge-protective capacitors shall be of a type designed for specific duty.

502.40 Multiwire Branch Circuits

In a Class II, Division 1 location, a multiwire branch circuit shall not be permitted.

Exception: Where the disconnect device(s) for the circuit opens all ungrounded conductors of the multiwire circuit simultaneously.

III. Equipment

502.100 Transformers and Capacitors

(A) Class II, Division 1 In Class II, Division 1 locations, transformers and capacitors shall comply with 502.100(A)(1) through (A)(3).

(1) Containing Liquid That Will Burn Transformers and capacitors containing a liquid that will burn shall be installed only in vaults complying with 450.41 through 450.48, and, in addition, (1), (2), and (3) shall apply.

(1) Doors or other openings communicating with the Division 1 location shall have self-closing fire doors on both sides of the wall, and the doors shall be carefully fitted and provided with suitable seals (such as weather stripping) to minimize the entrance of dust into the vault.
(2) Vent openings and ducts shall communicate only with the outside air.
(3) Suitable pressure-relief openings communicating with the outside air shall be provided.

(2) Not Containing Liquid That Will Burn Transformers and capacitors that do not contain a liquid that will burn shall be installed in vaults complying with 450.41 through 450.48 or be identified as a complete assembly, including terminal connections for Class II locations.

(3) Metal Dusts No transformer or capacitor shall be installed in a location where dust from magnesium, aluminum, aluminum bronze powders, or other metals of similarly hazardous characteristics may be present.

(B) Class II, Division 2 In Class II, Division 2 locations, transformers and capacitors shall comply with 502.100(B)(1) through (B)(3).

(1) Containing Liquid That Will Burn Transformers and capacitors containing a liquid that will burn shall be installed in vaults that comply with 450.41 through 450.48.

(2) Containing Askarel Transformers containing askarel and rated in excess of 25 kVA shall be as follows:

(1) Provided with pressure-relief vents
(2) Provided with a means for absorbing any gases generated by arcing inside the case, or the pressure-relief vents shall be connected to a chimney or flue that will carry such gases outside the building
(3) Have an airspace of not less than 150 mm (6 in.) between the transformer cases and any adjacent combustible material

(3) Dry-Type Transformers Dry-type transformers shall be installed in vaults or shall have their windings and terminal connections enclosed in tight metal housings without ventilating or other openings and shall operate at not over 600 volts, nominal.

Where it is necessary to install a transformer, it may be possible to use a small, low-voltage, dusttight (without ventilating openings) dry-type transformer, but transformers that have a primary voltage rating of over 600 volts must be either less flammable liquid-insulated or installed in a vault. In almost all cases, transformers can be remotely located from dust atmospheres.

Capacitors used for power-factor correction of individual motors are of sealed construction, but if installed in Class II, Division 1 locations, they must also be identified as a complete assembly, including dusttight terminal enclosures. The only special requirement for capacitors in Division 2 locations is that they are not permitted to contain oil or any other liquid that burns; otherwise, they are to be installed in vaults.

502.115 Switches, Circuit Breakers, Motor Controllers, and Fuses

(A) Class II, Division 1 In Class II, Division 1 locations, switches, circuit breakers, motor controllers, and fuses shall comply with 502.115(A)(1) through (A)(3).

(1) Type Required Switches, circuit breakers, motor controllers, and fuses, including pushbuttons, relays, and similar devices that are intended to interrupt current during normal operation or that are installed where combustible dusts of an electrically conductive nature may be present, shall be provided with identified dust-ignitionproof enclosures.

(2) Isolating Switches Disconnecting and isolating switches containing no fuses and not intended to interrupt current and not installed where dusts may be of an electrically conductive nature shall be provided with tight metal enclosures that shall be designed to minimize the entrance of dust and that shall (1) be equipped with telescoping or close-fitting covers or with other effective means to prevent the escape of sparks or burning material and (2) have no openings (such as holes for attachment screws) through which, after installation, sparks or burning material might escape or through which exterior accumulations of dust or adjacent combustible material might be ignited.

(3) Metal Dusts In locations where dust from magnesium, aluminum, aluminum bronze powders, or other metals of similarly hazardous characteristics may be present, fuses, switches, motor controllers, and circuit breakers shall have enclosures identified for such locations.

(B) Class II, Division 2 In Class II, Division 2 locations, enclosures for fuses, switches, circuit breakers, and motor controllers, including pushbuttons, relays, and similar devices, shall be dusttight.

The electrical equipment required in Class II locations is different from that required for Class I locations. Dust-ignitionproof enclosures for Class II locations are not required to be explosionproof. However, explosionproof equipment is allowed to be used in Class II locations if the equipment is dual rated and identified as suitable for the Class II division and group. Explosionproof enclosures are not necessarily dust-ignitionproof. They may have a different shape to minimize the accumulation of dust on top of the enclosure or on any protrusions or ledges.

Explosionproof enclosures, where used in an environment that is only a Class II location, are not required to be sealed within 18 in. of the enclosure to complete the explosionproof assembly, as required in a Class I environment. However, explosionproof enclosures must be provided with seals to prevent the entrance of dust into the dust-ignitionproof enclosure where a raceway provides a means for dust to enter the system, such as in a conduit run between a dust-ignitionproof enclosure and a general-purpose junction box.

Where equipment is located in a Class II, Division 1 location and is likely to produce arcs or sparks during normal operation, it must be installed in a dust-ignitionproof or pressurized enclosure. In addition, heat-generating equipment, such as control transformers, solenoids, impedance coils, resistors, and any associated overcurrent devices or switching mechanisms, must have dust-ignitionproof enclosures or pressurized enclosures installed in accordance with NFPA 496, *Standard for Purged and Pressurized Enclosures for Electrical Equipment*. Caution should be used where metal dusts, such as magnesium, aluminum, or other metals with similar hazardous characteristics, may be present. Enclosures must be approved specifically for that environment (suitable for Class II, Division 1, Group E).

Exhibit 502.3 shows a dust-ignitionproof pushbutton station with pilot light that is suitable for use in Class II, Division 1 hazardous (classified) locations. Dust-ignitionproof equipment enclosures for switching devices can be used in Class II, Division 2 locations, but because of the reduced level of hazard associated with Division 2, dusttight equipment enclosures are also permitted. In addition to being suitable for the specific class and division, this type of equipment must also be suitable for the dust group(s) (i.e., Groups E, F, and G) that will be present in a specific hazardous (classified) location.

Exhibit 502.4 shows a panelboard that is suitable for use in Class II locations only. Many, but not all, of the switches or circuit breakers and their associated enclosures that are approved for Class I, Division 1 locations are also approved for Class II locations. It is always important to look for the listing and identification of the hazardous (classified) locations in which the equipment is listed for use.

Exhibit 502.4 A dust-ignitionproof panelboard for use in Class II, Group E, F, and G locations. (Courtesy of Cooper Crouse-Hinds)

Exhibit 502.3 A dust-ignitionproof pushbutton control station suitable for use in Class II, Group E, F, and G locations. (Courtesy of Appleton Electric Co., EGS Electrical Group)

502.120 Control Transformers and Resistors

(A) Class II, Division 1 In Class II, Division 1 locations, control transformers, solenoids, impedance coils, resistors, and any overcurrent devices or switching mechanisms associated with them shall have dust-ignitionproof enclosures identified for Class II locations. No control transformer, impedance coil, or resistor shall be installed in a location where dust from magnesium, aluminum, aluminum bronze powders, or other metals of similarly hazardous characteristics may be present unless provided with an enclosure identified for the specific location.

(B) Class II, Division 2 In Class II, Division 2 locations, transformers and resistors shall comply with 502.120(B)(1) through (B)(3).

(1) Switching Mechanisms Switching mechanisms (including overcurrent devices) associated with control transformers, solenoids, impedance coils, and resistors shall be provided with dusttight enclosures.

(2) Coils and Windings Where not located in the same enclosure with switching mechanisms, control transformers, solenoids, and impedance coils shall be provided with tight metal housings without ventilating openings.

(3) Resistors Resistors and resistance devices shall have dust-ignitionproof enclosures identified for Class II locations.

Exception: Where the maximum normal operating temperature of the resistor will not exceed 120°C (248°F), nonadjust-

able resistors or resistors that are part of an automatically timed starting sequence shall be permitted to have enclosures complying with 502.120(B)(2).

502.125 Motors and Generators

(A) Class II, Division 1 In Class II, Division 1 locations, motors, generators, and other rotating electrical machinery shall be in conformance with either of the following:

(1) Identified for Class II, Division 1 locations
(2) Totally enclosed pipe-ventilated, meeting temperature limitations in 502.5

It is intended that the phrase "other rotating electrical machinery" include electric brakes. Listed and labeled electric brakes are available for Class II, Group E, F, and G locations.

Although some explosionproof motors for Class I, Division 1 locations are dust-ignitionproof and approved for both Class I and II locations, this is not true of all motors. The marking should always be checked to be sure the motor is designed and tested for the Class II location involved. If control wiring to the motor is necessary, according to the motor installation instructions, the control circuit must be installed and connected properly. Most motors for Class II locations require internal thermal protection to comply with the temperature limitations in 500.8(C)(2). Integral horsepower Class II motors may require both power and control circuit wiring from the motor controller to the motor.

(B) Class II, Division 2 In Class II, Division 2 locations, motors, generators, and other rotating electrical equipment shall be totally enclosed nonventilated, totally enclosed pipe-ventilated, totally enclosed water-air-cooled, totally enclosed fan-cooled or dust-ignitionproof for which maximum full-load external temperature shall be in accordance with 500.8(C)(2) for normal operation when operating in free air (not dust blanketed) and shall have no external openings.

Exception: If the authority having jurisdiction believes accumulations of nonconductive, nonabrasive dust will be moderate and if machines can be easily reached for routine cleaning and maintenance, the following shall be permitted to be installed:

(1) Standard open-type machines without sliding contacts, centrifugal or other types of switching mechanism (including motor overcurrent, overloading, and overtemperature devices), or integral resistance devices
(2) Standard open-type machines with such contacts, switching mechanisms, or resistance devices enclosed within dusttight housings without ventilating or other openings
(3) Self-cleaning textile motors of the squirrel-cage type

The requirements in 502.125(B) permit all types of totally enclosed motors in Class II, Division 2 locations if the external surface temperatures, without a dust blanket, do not exceed the temperatures indicated under the maximum full-load (normal operation) conditions in 500.8(C)(2). Totally enclosed fan-cooled (TEFC) motors are specifically mentioned. The motor should be examined carefully to be sure there are no external openings, even though the motor may be marked TEFC.

Totally enclosed motors with no special provision for cooling may be used in Class II, Division 2 locations, but to deliver the same horsepower, they must be considerably larger than an open-type, fan-cooled, or pipe-ventilated motor.

502.128 Ventilating Piping

Ventilating pipes for motors, generators, or other rotating electric machinery, or for enclosures for electric equipment, shall be of metal not less than 0.53 mm (0.021 in.) in thickness or of equally substantial noncombustible material and shall comply with all of the following:

(1) Lead directly to a source of clean air outside of buildings
(2) Be screened at the outer ends to prevent the entrance of small animals or birds
(3) Be protected against physical damage and against rusting or other corrosive influences

Ventilating pipes shall also comply with 502.128(A) and 502.128(B).

(A) Class II, Division 1 In Class II, Division 1 locations, ventilating pipes, including their connections to motors or to the dust-ignitionproof enclosures for other equipment, shall be dusttight throughout their length. For metal pipes, seams and joints shall comply with one of the following:

(1) Be riveted and soldered
(2) Be bolted and soldered
(3) Be welded
(4) Be rendered dusttight by some other equally effective means

(B) Class II, Division 2 In Class II, Division 2 locations, ventilating pipes and their connections shall be sufficiently tight to prevent the entrance of appreciable quantities of dust into the ventilated equipment or enclosure and to prevent the escape of sparks, flame, or burning material that might ignite dust accumulations or combustible material in the vicinity. For metal pipes, lock seams and riveted or welded joints shall be permitted; and tight-fitting slip joints shall be permitted where some flexibility is necessary, as at connections to motors.

502.130 Luminaires (Lighting Fixtures)

Luminaires (lighting fixtures) shall comply with 502.130(A) and 502.130(B).

(A) Class II, Division 1 In Class II, Division 1 locations, luminaires (lighting fixtures) for fixed and portable lighting shall comply with 502.130(A)(1) through (A)(4).

(1) Fixtures Each luminaire (fixture) shall be identified for Class II locations and shall be clearly marked to indicate the maximum wattage of the lamp for which it is designed. In locations where dust from magnesium, aluminum, aluminum bronze powders, or other metals of similarly hazardous characteristics may be present, luminaires (fixtures) for fixed or portable lighting and all auxiliary equipment shall be identified for the specific location.

(2) Physical Damage Each luminaire (fixture) shall be protected against physical damage by a suitable guard or by location.

(3) Pendant Luminaires (Fixtures) Pendant luminaires (fixtures) shall be suspended by threaded rigid metal conduit stems, threaded steel intermediate metal conduit stems, by chains with approved fittings, or by other approved means. For rigid stems longer than 300 mm (12 in.), permanent and effective bracing against lateral displacement shall be provided at a level not more than 300 mm (12 in.) above the lower end of the stem, or flexibility in the form of a fitting or a flexible connector listed for the location shall be provided not more than 300 mm (12 in.) from the point of attachment to the supporting box or fitting. Threaded joints shall be provided with set-screws or other effective means to prevent loosening. Where wiring between an outlet box or fitting and a pendant luminaire (fixture) is not enclosed in conduit, flexible cord listed for hard usage shall be used, and suitable seals shall be provided where the cord enters the luminaire (fixture) and the outlet box or fitting. Flexible cord shall not serve as the supporting means for a fixture.

(4) Supports Boxes, box assemblies, or fittings used for the support of luminaires (lighting fixtures) shall be identified for Class II locations.

(B) Class II, Division 2 In Class II, Division 2 locations, luminaires (lighting fixtures) shall comply with 502.130(B)(1) through (B)(5).

(1) Portable Lighting Equipment Portable lighting equipment shall be identified for Class II locations. They shall be clearly marked to indicate the maximum wattage of lamps for which they are designed.

(2) Fixed Lighting Luminaires (lighting fixtures) for fixed lighting, where not of a type identified for Class II locations, shall provide enclosures for lamps and lampholders that shall be designed to minimize the deposit of dust on lamps and to prevent the escape of sparks, burning material, or hot metal. Each fixture shall be clearly marked to indicate the maximum wattage of the lamp that shall be permitted without exceeding an exposed surface temperature in accordance with 500.8(C)(2) under normal conditions of use.

(3) Physical Damage Luminaires (lighting fixtures) for fixed lighting shall be protected from physical damage by suitable guards or by location.

(4) Pendant Luminaires (Fixtures) Pendant luminaires (fixtures) shall be suspended by threaded rigid metal conduit stems, threaded steel intermediate metal conduit stems, by chains with approved fittings, or by other approved means. For rigid stems longer than 300 mm (12 in.), permanent and effective bracing against lateral displacement shall be provided at a level not more than 300 mm (12 in.) above the lower end of the stem, or flexibility in the form of an identified fitting or a flexible connector shall be provided not more than 300 mm (12 in.) from the point of attachment to the supporting box or fitting. Where wiring between an outlet box or fitting and a pendant luminaire (fixture) is not enclosed in conduit, flexible cord listed for hard usage shall be used. Flexible cord shall not serve as the supporting means for a fixture.

(5) Electric-Discharge Lamps Starting and control equipment for electric-discharge lamps shall comply with the requirements of 502.120(B).

Exhibit 502.5 shows a listed fixture suitable for use in Class II, Group E, F, and G locations. Luminaires, fixed or portable, and auxiliary equipment (such as ballasts) must be approved for use in Group E atmospheres if metal dusts are present.

Other than the requirement that the fixture be marked to indicate maximum lamp wattage, the only requirement for fixtures in Division 2 locations is that lamps be enclosed in suitable globes to minimize dust deposits on the lamps and prevent the escape of sparks or burning material. Metal guards must be provided, unless globe breakage is unlikely.

Flexible cord of the hard-usage type is permitted with approved sealed connections for the wiring of chain-suspended or hook-and-eye-suspended fixtures. Flexible cords are not intended to be used as cord pendants or drop cords.

The portable hand lamp shown in Exhibit 501.21 is approved as a complete assembly for use in Class I locations and also in any Class II, Group F or G location.

502.135 Utilization Equipment

(A) Class II, Division 1 In Class II, Division 1 locations, all utilization equipment shall be identified for Class II locations. Where dust from magnesium, aluminum, aluminum bronze powders, or other metals of similarly hazardous char-

Exhibit 502.5 A typical luminaire for use in Class II, Division 1 locations. (Courtesy of Cooper Crouse-Hinds)

actcristics may be present, such equipment shall be identified for the specific location.

(B) Class II, Division 2 In Class II, Division 2 locations, all utilization equipment shall comply with 502.135(B)(1) through (B)(4).

(1) Heaters Electrically heated utilization equipment shall be identified for Class II locations.

Exception: Metal-enclosed radiant heating panel equipment shall be dusttight and marked in accordance with 500.8(B).

(2) Motors Motors of motor-driven utilization equipment shall comply with 502.125(B).

(3) Switches, Circuit Breakers, and Fuses Enclosures for switches, circuit breakers, and fuses shall be dusttight.

(4) Transformers, Solenoids, Impedance Coils, and Resistors Transformers, solenoids, impedance coils, and resistors shall comply with 502.120(B).

502.140 Flexible Cords — Class II, Divisions 1 and 2

Flexible cords used in Class II locations shall comply with all of the following:

(1) Be of a type listed for extra-hard usage

Exception: Flexible cord listed for hard usage as permitted by 502.130(A)(3) and (B)(4).

(2) Contain, in addition to the conductors of the circuit, a grounding conductor complying with 400.23
(3) Be connected to terminals or to supply conductors in an approved manner
(4) Be supported by clamps or by other suitable means in such a manner that there will be no tension on the terminal connections
(5) Be provided with suitable seals to prevent the entrance of dust where the flexible cord enters boxes or fittings that are required to be dust-ignitionproof

502.145 Receptacles and Attachment Plugs

(A) Class II, Division 1 In Class II, Division 1 locations, receptacles and attachment plugs shall be of the type providing for connection to the grounding conductor of the flexible cord and shall be identified for Class II locations.

(B) Class II, Division 2 In Class II, Division 2 locations, receptacles and attachment plugs shall be of the type that provides for connection to the grounding conductor of the flexible cord and shall be designed so that connection to the supply circuit cannot be made or broken while live parts are exposed.

502.150 Signaling, Alarm, Remote-Control, and Communications Systems; and Meters, Instruments, and Relays

> FPN: See Article 800 for rules governing the installation of communications circuits.

(A) Class II, Division 1 In Class II, Division 1 locations, signaling, alarm, rcmotc-control, and communications systems; and meters, instruments, and relays shall comply with 502.150(A)(1) through (A)(6).

(1) Wiring Methods The wiring method shall comply with 502.10(A).

(2) Contacts Switches, circuit breakers, relays, contactors, fuses and current-breaking contacts for bells, horns, howlers, sirens, and other devices in which sparks or arcs may be produced shall be provided with enclosures identified for a Class II location.

Exception: Where current-breaking contacts are immersed in oil or where the interruption of current occurs within a chamber sealed against the entrance of dust, enclosures shall be permitted to be of the general-purpose type.

(3) Resistors and Similar Equipment Resistors, transformers, choke coils, rectifiers, thermionic tubes, and other heat-generating equipment shall be provided with enclosures identified for Class II locations.

Exception: Where resistors or similar equipment are immersed in oil or enclosed in a chamber sealed against the entrance of dust, enclosures shall be permitted to be of the general-purpose type.

(4) Rotating Machinery Motors, generators, and other rotating electric machinery shall comply with 502.125(A).

(5) Combustible, Electrically Conductive Dusts Where dusts are of a combustible, electrically conductive nature, all wiring and equipment shall be identified for Class II locations.

(6) Metal Dusts Where dust from magnesium, aluminum, aluminum bronze powders, or other metals of similarly hazardous characteristics may be present, all apparatus and equipment shall be identified for the specific conditions.

(B) Class II, Division 2 In Class II, Division 2 locations, signaling, alarm, remote-control, and communications systems; and meters, instruments, and relays shall comply with 502.150(B)(1) through (B)(5).

(1) Contacts Enclosures shall comply with 502.150(A)(2), or contacts shall have tight metal enclosures designed to minimize the entrance of dust and shall have telescoping or tight-fitting covers and no openings through which, after installation, sparks or burning material might escape.

Exception: In nonincendive circuits, enclosures shall be permitted to be of the general-purpose type.

(2) Transformers and Similar Equipment The windings and terminal connections of transformers, choke coils, and similar equipment shall be provided with tight metal enclosures without ventilating openings.

(3) Resistors and Similar Equipment Resistors, resistance devices, thermionic tubes, rectifiers, and similar equipment shall comply with 502.130(A)(3).

Exception: Enclosures for thermionic tubes, nonadjustable resistors, or rectifiers for which maximum operating temperature will not exceed 120°C (248°F) shall be permitted to be of the general-purpose type.

(4) Rotating Machinery Motors, generators, and other rotating electric machinery shall comply with 502.125(B).

(5) Wiring Methods The wiring method shall comply with 502.10(B).

ARTICLE 503
Class III Locations

Summary of Changes

- **General:** Restructured and renumbered to provide a scope section and parallel numbering systems for Articles 501, 502, and 503.

- **503.25:** Revised to provide guidance concerning protection against explosion and against electric shock in order to limit the voltage of exposed live parts.

Contents

I. General

503.1 Scope

Article 503 covers the requirements for electrical and electronic equipment and wiring for all voltages in Class III, Division 1 and 2 locations where fire or explosion hazards may exist due to ignitible fibers or flyings.

503.5 General

The general rules of this *Code* shall apply to electric wiring and equipment in locations classified as Class III locations in 500.5(D).

Exception: As modified by this article.

Equipment installed in Class III locations shall be able to function at full rating without developing surface temperatures high enough to cause excessive dehydration or gradual carbonization of accumulated fibers or flyings. Organic material that is carbonized or excessively dry is highly susceptible to spontaneous ignition. The maximum surface temperatures under operating conditions shall not exceed 165°C (329°F) for equipment that is not subject to overloading, and 120°C (248°F) for equipment (such as motors or power transformers) that may be overloaded.

FPN: For electric trucks, see NFPA 505-2002, *Fire Safety Standard for Powered Industrial Trucks Including Type Designations, Areas of Use, Conversions, Maintenance, and Operation.*

Class III locations usually include textile mills that process cotton, rayon, and so on, where easily ignitible fibers or combustible flyings are present in the manufacturing process. Sawmills and other woodworking plants, where sawdust, wood shavings, and combustible fibers or flyings are present, may also become hazardous. If wood flour (dust) is also present, the location is a Class II, Group G location, not a Class III location.

Fibers or flyings are hazardous not only because they are easily ignited, but also because flames spread through them quickly. Such fires travel with a rapidity approaching an explosion and are commonly called flash fires.

Class III, Division 1 applies to locations where material is handled, manufactured, or used. Division 2 applies to locations where material is stored or handled but where no manufacturing processes are performed. Unlike Class I locations (Groups A, B, C, and D) and Class II locations (Groups E, F, and G), there are no material group designations in Class III locations.

II. Wiring

503.10 Wiring Methods

Wiring methods shall comply with 503.10(A) or 503.10(B).

(A) Class III, Division 1 In Class III, Division 1 locations, the wiring method shall be rigid metal conduit, rigid nonmetallic conduit, intermediate metal conduit, electrical metallic tubing, dusttight wireways, or Type MC or MI cable with listed termination fittings.

(1) Boxes and Fittings All boxes and fittings shall be dusttight.

(2) Flexible Connections Where necessary to employ flexible connections, dusttight flexible connectors, liquidtight flexible metal conduit with listed fittings, liquidtight flexible nonmetallic conduit with listed fittings, or flexible cord in conformance with 503.140 shall be used.

FPN: See 503.30(B) for grounding requirements where flexible conduit is used.

(3) Nonincendive Field Wiring Nonincendive field wiring shall be permitted using any of the wiring methods permitted for unclassified locations. Nonincendive field wiring systems shall be installed in accordance with the control drawing(s). Simple apparatus, not shown on the control drawing, shall be permitted in a nonincendive field wiring circuit, provided the simple apparatus does not interconnect the nonincendive field wiring circuit to any other circuit.

This section was added to the 2005 *Code* regarding nonincendive field wiring. The requirements for nonincendive field wiring for Class I and Class II were modified and expanded for the 2002 *Code*, but they were not included for Class III locations. Section 503.10(A)(3) ensures that nonincendive field wiring is permitted in Class III locations.

FPN: Simple apparatus is defined in 504.2.

Separate nonincendive field wiring circuits shall be installed in accordance with one of the following:

(1) In separate cables

(2) In multiconductor cables where the conductors of each circuit are within a grounded metal shield

(3) In multiconductor cables where the conductors of each circuit have insulation with a minimum thickness of 0.25 mm (0.01 in.)

(B) Class III, Division 2 In Class III, Division 2 locations, the wiring method shall comply with 503.10(A).

Exception: In sections, compartments, or areas used solely for storage and containing no machinery, open wiring on insulators shall be permitted where installed in accordance with Article 398, but only on condition that protection as required by 398.15(C) be provided where conductors are not run in roof spaces and are well out of reach of sources of physical damage.

503.25 Uninsulated Exposed Parts, Class III, Divisions 1 and 2

There shall be no uninsulated exposed parts, such as electric conductors, buses, terminals, or components, that operate at more than 30 volts (15 volts in wet locations). These parts shall additionally be protected by a protection technique according to 500.7(E), 500.7(F), or 500.7(G) that is suitable for the location.

Exception: As provided in 503.155.

Exposed live parts are allowed in Class III, Division 1 and 2 locations provided that the voltage does not exceed 30 volts in dry locations or 15 volts in wet locations. However, caution must be observed in working around uninsulated exposed parts because tool or other conductive items that come in contact with the circuit could produce sparks and be a source of ignition in a hazardous (classified) location.

503.30 Grounding and Bonding — Class III, Divisions 1 and 2

Wiring and equipment in Class III, Division 1 and 2 locations shall be grounded as specified in Article 250 and with the following additional requirements in 503.30(A) and 503.30(B).

(A) Bonding The locknut-bushing and double-locknut types of contacts shall not be depended on for bonding purposes, but bonding jumpers with proper fittings or other approved means of bonding shall be used. Such means of bonding shall apply to all intervening raceways, fittings, boxes, enclosures, and so forth, between Class III locations and the point of grounding for service equipment or point of grounding of a separately derived system.

For information regarding bonding in hazardous (classified) locations, see the commentary on 501.30(A). Although that section addresses bonding in Class I locations, the same requirement for enhanced bonding in Class III locations is found in 502.30(A), and the requirements of 250.100 apply to Class I, Class II, and Class III locations.

Exception: The specific bonding means shall only be required to the nearest point where the grounded circuit conductor and the grounding electrode conductor are connected together on the line side of the building or structure disconnecting means as specified in 250.32(A), (B), and (C), if the branch-circuit overcurrent protection is located on the load side of the disconnecting means.

> FPN: See 250.100 for additional bonding requirements in hazardous (classified) locations.

(B) Types of Equipment Grounding Conductors Where flexible conduit is used as permitted in 503.10, it shall be installed with internal or external bonding jumpers in parallel with each conduit and complying with 250.102.

Exception: In Class III, Division 1 and 2 locations, the bonding jumper shall be permitted to be deleted where all of the following conditions are met:

(1) *Listed liquidtight flexible metal 1.8 m (6 ft) or less in length, with fittings listed for grounding, is used.*

(2) *Overcurrent protection in the circuit is limited to 10 amperes or less.*

(3) *The load is not a power utilization load.*

III. Equipment

503.100 Transformers and Capacitors — Class III, Divisions 1 and 2

Transformers and capacitors shall comply with 502.100(B).

503.115 Switches, Circuit Breakers, Motor Controllers, and Fuses — Class III, Divisions 1 and 2

Switches, circuit breakers, motor controllers, and fuses, including pushbuttons, relays, and similar devices, shall be provided with dusttight enclosures.

See the definition of *dusttight* in Article 100.

503.120 Control Transformers and Resistors — Class III, Divisions 1 and 2

Transformers, impedance coils, and resistors used as or in conjunction with control equipment for motors, generators, and appliances shall be provided with dusttight enclosures complying with the temperature limitations in 503.5.

See the definition of *dusttight* in Article 100.

503.125 Motors and Generators — Class III, Divisions 1 and 2

In Class III, Divisions 1 and 2 locations, motors, generators, and other rotating machinery shall be totally enclosed nonventilated, totally enclosed pipe ventilated, or totally enclosed fan cooled.

Exception: In locations where, in the judgment of the authority having jurisdiction, only moderate accumulations of lint or flyings are likely to collect on, in, or in the vicinity of a rotating electric machine and where such machine is readily accessible for routine cleaning and maintenance, one of the following shall be permitted:

(1) Self-cleaning textile motors of the squirrel-cage type

(2) Standard open-type machines without sliding contacts, centrifugal or other types of switching mechanisms, including motor overload devices

(3) Standard open-type machines having such contacts, switching mechanisms, or resistance devices enclosed within tight housings without ventilating or other openings

It is intended that the phrase "other rotating machinery" in 503.125 include electric brakes. Listed and labeled electric brakes are available for Class II, Group G locations, and, according to the UL *Hazardous Location Equipment Directory*, these brakes are suitable for Class III locations.

503.128 Ventilating Piping — Class III, Divisions 1 and 2

Ventilating pipes for motors, generators, or other rotating electric machinery, or for enclosures for electric equipment, shall be of metal not less than 0.53 mm (0.021 in.) in thickness, or of equally substantial noncombustible material, and shall comply with the following:

(1) Lead directly to a source of clean air outside of buildings

(2) Be screened at the outer ends to prevent the entrance of small animals or birds

(3) Be protected against physical damage and against rusting or other corrosive influences

Ventilating pipes shall be sufficiently tight, including their connections, to prevent the entrance of appreciable quantities of fibers or flyings into the ventilated equipment or enclosure and to prevent the escape of sparks, flame, or burning material that might ignite accumulations of fibers or flyings or combustible material in the vicinity. For metal pipes, lock seams and riveted or welded joints shall be permitted; and tight-fitting slip joints shall be permitted where some flexibility is necessary, as at connections to motors.

503.130 Luminaires (Lighting Fixtures) — Class III, Divisions 1 and 2

(A) Fixed Lighting Luminaires (lighting fixtures) for fixed lighting shall provide enclosures for lamps and lampholders that are designed to minimize entrance of fibers and flyings and to prevent the escape of sparks, burning material, or hot metal. Each luminaire (fixture) shall be clearly marked to show the maximum wattage of the lamps that shall be permitted without exceeding an exposed surface temperature of 165°C (329°F) under normal conditions of use.

(B) Physical Damage A luminaire (fixture) that may be exposed to physical damage shall be protected by a suitable guard.

(C) Pendant Luminaires (Fixtures) Pendant luminaires (fixtures) shall be suspended by stems of threaded rigid metal conduit, threaded intermediate metal conduit, threaded metal tubing of equivalent thickness, or by chains with approved fittings. For stems longer than 300 mm (12 in.), permanent and effective bracing against lateral displacement shall be provided at a level not more than 300 mm (12 in.) above the lower end of the stem, or flexibility in the form of an identified fitting or a flexible connector shall be provided not more than 300 mm (12 in.) from the point of attachment to the supporting box or fitting.

(D) Portable Lighting Equipment Portable lighting equipment shall be equipped with handles and protected with substantial guards. Lampholders shall be of the unswitched type with no provision for receiving attachment plugs. There shall be no exposed current-carrying metal parts, and all exposed non–current-carrying metal parts shall be grounded. In all other respects, portable lighting equipment shall comply with 503.130(A).

503.135 Utilization Equipment — Class III, Divisions 1 and 2

(A) Heaters Electrically heated utilization equipment shall be identified for Class III locations.

(B) Motors Motors of motor-driven utilization equipment shall comply with 503.125.

(C) Switches, Circuit Breakers, Motor Controllers, and Fuses Switches, circuit breakers, motor controllers, and fuses shall comply with 503.115.

503.140 Flexible Cords — Class III, Divisions 1 and 2

Flexible cords shall comply with the following:

(1) Be of a type listed for extra-hard usage
(2) Contain, in addition to the conductors of the circuit, a grounding conductor complying with 400.23
(3) Be connected to terminals or to supply conductors in an approved manner
(4) Be supported by clamps or other suitable means in such a manner that there will be no tension on the terminal connections
(5) Be provided with suitable means to prevent the entrance of fibers or flyings where the cord enters boxes or fittings

503.145 Receptacles and Attachment Plugs — Class III, Divisions 1 and 2

Receptacles and attachment plugs shall be of the grounding type and shall be designed so as to minimize the accumulation or the entry of fibers or flyings, and shall prevent the escape of sparks or molten particles.

Exception: In locations where, in the judgment of the authority having jurisdiction, only moderate accumulations of lint or flyings will be likely to collect in the vicinity of a receptacle, and where such receptacle is readily accessible for routine cleaning, general-purpose grounding-type receptacles mounted so as to minimize the entry of fibers or flyings shall be permitted.

503.150 Signaling, Alarm, Remote-Control, and Local Loudspeaker Intercommunications Systems — Class III, Divisions 1 and 2

Signaling, alarm, remote-control, and local loudspeaker intercommunications systems shall comply with the requirements of Article 503 regarding wiring methods, switches, transformers, resistors, motors, luminaires (lighting fixtures), and related components.

503.155 Electric Cranes, Hoists, and Similar Equipment — Class III, Divisions 1 and 2

Where installed for operation over combustible fibers or accumulations of flyings, traveling cranes and hoists for material handling, traveling cleaners for textile machinery, and similar equipment shall comply with 503.155(A) through (D).

(A) Power Supply Power supply to contact conductors shall be electrically isolated from all other systems, ungrounded, and shall be equipped with an acceptable ground detector that gives an alarm and automatically de-energizes the contact conductors in case of a fault to ground or gives a visual and audible alarm as long as power is supplied to the contact conductors and the ground fault remains.

(B) Contact Conductors Contact conductors shall be located or guarded so as to be inaccessible to other than authorized persons and shall be protected against accidental contact with foreign objects.

(C) Current Collectors Current collectors shall be arranged or guarded so as to confine normal sparking and prevent escape of sparks or hot particles. To reduce sparking, two or more separate surfaces of contact shall be provided for each contact conductor. Reliable means shall be provided to keep contact conductors and current collectors free of accumulations of lint or flyings.

(D) Control Equipment Control equipment shall comply with 503.115 and 503.120.

In a Class III location, cranes that are installed over accumulations of fibers or flyings and equipped with rolling or sliding collectors that make contact with bare conductors introduce two hazards.

The first hazard results from any arcing between a conductor and a collector rail igniting combustible fibers or lint that has accumulated on or near the bare conductor. This hazard may be prevented by maintaining the proper alignment of the bare conductor, by using a collector designed so that proper contact is always maintained, and by using guards or shields to confine hot metal particles that result from arcing.

The second hazard occurs if enough moisture is present and fibers and flyings accumulating on the insulating supports of the bare conductors form a conductive path between the conductors or from one conductor to ground, permitting enough current to flow to ignite the fibers. If the system is ungrounded, a current flow to ground is unlikely to start a fire.

A suitable recording ground detector sounds an alarm and automatically de-energizes contact conductors when the insulation resistance is lowered by an accumulation of fibers on the insulators or in case of a fault to ground. A ground-fault indicator is permitted that maintains an alarm until the system is de-energized or the ground fault is cleared.

503.160 Storage Battery Charging Equipment — Class III, Divisions 1 and 2

Storage battery charging equipment shall be located in separate rooms built or lined with substantial noncombustible materials. The rooms shall be constructed to prevent the entrance of ignitible amounts of flyings or lint and shall be well ventilated.

ARTICLE 504
Intrinsically Safe Systems

Summary of Changes

- **504.10(B):** Revised to add requirements for the installation of simple apparatus in a hazardous (classified) location.
- **504.30(B)(3):** Added requirement for spacing between terminals as a means to provide separation between different intrinsically safe circuits.

Contents

504.1 Scope

This article covers the installation of intrinsically safe (I.S.) apparatus, wiring, and systems for Class I, II, and III locations.

> FPN: For further information, see ANSI/ISA RP 12.06.01-2002, *Wiring Methods for Hazardous (Classified) Locations Instrumentation — Part 1: Intrinsic Safety.*

Article 504 was first included in the 1990 *NEC*. Previously, the installation requirements for intrinsically safe systems were in ANSI/ISA RP 12.6-03, *Recommended Practice for Wiring Methods for Hazardous (Classified) Locations Instrumentation, Part 1: Intrinsic Safety.*

504.2 Definitions

The standard used in the United States for construction and performance requirements for intrinsically safe systems is ANSI/UL 913, *Intrinsically Safe Apparatus and Associated Apparatus for Use in Class I, II, and III, Division 1, Hazardous (Classified) Locations.* ANSI/UL 913 is similar to standards for intrinsically safe equipment in other countries; all standards are based on the IEC (International Electrotechnical Commission) standard. ANSI/UL 60079-11, *Electrical Apparatus for Explosive Gas Atmospheres — Part II: Intrinsic Safety "i"* is based on the IEC 60079-11 standard and also contains the U.S. deviations that allow it to be compatible for installations in the United States. The *NEC* offers the choice of designating hazardous (classified) locations as two divisions (1 and 2) or three zones (0, 1, and 2). ANSI/UL 913 requirements, however, are based on the IEC Zone 0 requirements, which are the most stringent. Equipment certified by a testing laboratory for Zone 1 would not necessarily meet ANSI/UL 913 requirements for Division 1.

Associated Apparatus. Apparatus in which the circuits are not necessarily intrinsically safe themselves but that affect the energy in the intrinsically safe circuits and are relied on to maintain intrinsic safety. Associated apparatus may be either of the following:

(1) Electrical apparatus that has an alternative-type protection for use in the appropriate hazardous (classified) location
(2) Electrical apparatus not so protected that shall not be used within a hazardous (classified) location

> FPN No. 1: Associated apparatus has identified intrinsically safe connections for intrinsically safe apparatus and also may have connections for nonintrinsically safe apparatus.

> FPN No. 2: An example of associated apparatus is an intrinsic safety barrier, which is a network designed to limit the energy (voltage and current) available to the protected circuit in the hazardous (classified) location, under specified fault conditions.

For an illustration of an intrinsic safety barrier, see Exhibit 504.1.

Control Drawing. See definition in 500.2.

Different Intrinsically Safe Circuits. Intrinsically safe circuits in which the possible interconnections have not been evaluated and identified as intrinsically safe.

Exhibit 504.1 A typical intrinsic safety barrier that limits the energy available to the hazardous location. (Courtesy of Cooper Crouse-Hinds)

Intrinsically Safe Apparatus. Apparatus in which all the circuits are intrinsically safe.

Intrinsically Safe Circuit. A circuit in which any spark or thermal effect is incapable of causing ignition of a mixture of flammable or combustible material in air under prescribed test conditions.

> FPN: Test conditions are described in ANSI/UL 913-1997, *Standard for Safety, Intrinsically Safe Apparatus and Associated Apparatus for Use in Class I, II, and III, Division 1, Hazardous (Classified) Locations.*

Due to its physical and electrical characteristics, an intrinsically safe circuit does not develop sufficient electrical energy (millijoules) in an arc or spark to cause ignition or sufficient thermal energy resulting from an overload condition to cause the temperature of the installed circuit to exceed the ignition temperature of a specified gas or vapor under normal or abnormal operating conditions.

An abnormal condition may be due to accidental damage, failure of electrical components, excessive voltage, or improper adjustment or maintenance of the equipment.

Intrinsically Safe System. An assembly of interconnected intrinsically safe apparatus, associated apparatus, and interconnecting cables in that those parts of the system that may be used in hazardous (classified) locations are intrinsically safe circuits.

Although low-energy devices, such as thermocouples, crystal stain transducers, or pressure transducers, generate milli-

volts and currents in the microampere range, they are not necessarily intrinsically safe. Low-energy devices are normally connected to amplifiers and power supplies that are connected to 120-volt or higher circuits. Should a fault occur within the amplifier or power supply, or a voltage surge occur in the electrical supply system, high-energy arcing, sparking, or overheating of the low-energy portion of the circuit could occur.

> FPN: An intrinsically safe system may include more than one intrinsically safe circuit.

Simple Apparatus. An electrical component or combination of components of simple construction with well-defined electrical parameters that does not generate more than 1.5 volts, 100 milliamps, and 25 milliwatts, or a passive component that does not dissipate more than 1.3 watts and is compatible with the intrinsic safety of the circuit in which it is used.

> FPN: The following apparatus are examples of simple apparatus:
>
> (a) Passive components, for example, switches, junction boxes, resistance temperature devices, and simple semiconductor devices such as LEDs
>
> (b) Sources of generated energy, for example, thermocouples and photocells, which do not generate more than 1.5 V, 100 mA, and 25 mW

The definition of *simple apparatus* clarifies the use of the term in 504.4, Exception, and 504.10, Exception. The intent is to permit the use of apparatus that stores little or no energy without requiring the apparatus to be listed or to comply with the control drawing. See the fine print note following the definition of *simple apparatus* for examples of simple apparatus.

504.3 Application of Other Articles
Except as modified by this article, all applicable articles of this *Code* shall apply.

Because intrinsically safe wiring must be low-energy wiring to be intrinsically safe, the wiring itself is most likely to be Class 2, in accordance with 725.41, or, in a fire-protective signaling system, power-limited in accordance with 760.41. See Article 725 or 760, as appropriate, for the requirements for such wiring. The installation may also fall under the scope of Article 800. The intrinsically safe apparatus and associated apparatus, on the other hand, may be supplied by ordinary power circuits, in which case other *Code* rules may apply. It is common for intrinsically safe apparatus or associated apparatus supplied by a power circuit to be located in a hazardous (classified) location, with the apparatus protected by one of the protection systems required by Articles 500 through 503, 505, or 506, that is, explosionproof, dust-ignitionproof, purged, and pressurized. In this case, Articles 500 through 503, 505, and 506 also apply. Also, intrinsically safe systems are not exempt from the grounding and bonding requirements of 501.30, 502.30, 503.30, and 505.25.

504.4 Equipment

All intrinsically safe apparatus and associated apparatus shall be listed.

Exception: Simple apparatus, as described on the control drawing, shall not be required to be listed.

504.10 Equipment Installation

(A) Control Drawing Intrinsically safe apparatus, associated apparatus, and other equipment shall be installed in accordance with the control drawing(s).

Exception: A simple apparatus that does not interconnect intrinsically safe circuits.

> FPN: The control drawing identification is marked on the apparatus.

The control drawing may put limitations on cables and on the separation of circuits in an intrinsically safe system. The control drawing also illustrates what is permitted to be connected in the system. Compliance with all the provisions in the control drawing is essential if intrinsic safety is to be maintained. The investigation of the equipment by third-party testing laboratories is based on installation in accordance with the control drawing.

(B) Location Intrinsically safe apparatus shall be permitted to be installed in any hazardous (classified) location for which it has been identified. General-purpose enclosures shall be permitted for intrinsically safe apparatus.

Associated apparatus shall be permitted to be installed in any hazardous (classified) location for which it has been identified or, if protected by other means, permitted by Articles 501 through 503 and Article 505.

Simple apparatus shall be permitted to be installed in any hazardous (classified) location in which the maximum surface temperature of the simple apparatus does not exceed the ignition temperature of the flammable gases or vapors, flammable liquids, combustible dusts, or ignitible fibers or flyings present.

For simple apparatus, the maximum surface temperature can be determined from the values of the output power from the associated apparatus or apparatus to which it is connected to obtain the temperature class. The temperature class can be determined by:

Section 504.10(B) in the 2005 *Code* has been expanded to provide guidance for designers, installers, and inspectors pertaining to simple apparatus installed in hazardous (classified) locations. See the definition of *simple apparatus* and the accompanying fine print note in 504.2. Table 504.10(B) provides the surface temperature allowed for T4 classification based on the total surface area of the equipment.

(1) Reference to Table 504.10(B)
(2) Calculation using the formula:

$$T = P_o R_{th} + T_{amb}$$

where:

T is the surface temperature

P_o is the output power marked on the associated apparatus or intrinsically safe apparatus

R_{th} is the thermal resistance of the simple apparatus

T_{amb} is the ambient temperature (normally 40°C) and reference Table 500.8(B)

Table 504.10(B) Assessment for T4 Classification According to Component Size and Temperature

Total Surface Area Excluding Lead Wires	Requirement for T4 Classification (Based on 40°C Ambient Temperature)
<20 mm²	Surface temperature ≤275°C
≥20 mm² ≤10 cm²	Surface temperature ≤200°C
≥20 mm²	Power not exceeding 1.3 W*

*Reduce to 1.2 W with an ambient of 60°C or 1.0 W with 80°C ambient temperature.

In addition, components with a surface area smaller than 10 cm² (excluding lead wires) may be classified as T5 if their surface temperature does not exceed 150°C.

> FPN: The following apparatus are examples of simple apparatus:
> (1) Passive components, for example, switches, junction boxes, resistance temperature devices, and simple semiconductor devices such as LEDs
> (2) Sources of generated energy, for example, thermocouples and photocells, which do not generate more than 1.5 V, 100 mA, and 25 mW

504.20 Wiring Methods

Intrinsically safe apparatus and wiring shall be permitted to be installed using any of the wiring methods suitable for unclassified locations, including Chapter 7 and Chapter 8. Sealing shall be as provided in 504.70, and separation shall be as provided in 504.30.

See the commentary following 504.3 for more information on the types of circuits typically used as part of an intrinsically safe system.

504.30 Separation of Intrinsically Safe Conductors

(A) From Nonintrinsically Safe Circuit Conductors

(1) In Raceways, Cable Trays, and Cables Conductors of intrinsically safe circuits shall not be placed in any raceway, cable tray, or cable with conductors of any nonintrinsically safe circuit.

Exception No. 1: Where conductors of intrinsically safe circuits are separated from conductors of nonintrinsically safe circuits by a distance of at least 50 mm (2 in.) and secured, or by a grounded metal partition or an approved insulating partition.

> FPN: No. 20 gauge sheet metal partitions 0.91 mm (0.0359 in.) or thicker are generally considered acceptable.

Exception No. 2: Where either (1) all of the intrinsically safe circuit conductors or (2) all of the nonintrinsically safe circuit conductors are in grounded metal-sheathed or metal-clad cables where the sheathing or cladding is capable of carrying fault current to ground.

> FPN: Cables meeting the requirements of Articles 330 and 332 are typical of those considered acceptable.

Type MI cable with a copper sheath and Type MC cable of smooth metallic sheath or corrugated metallic sheath construction meet the conditions prescribed in 504.30(A)(1), Exception No. 2. The metallic sheath of interlocked-tape Type MC cable (as opposed to the smooth or corrugated continuous outer sheath) is generally not investigated for suitability as an equipment grounding conductor and, therefore, may not be capable of carrying a fault current to ground as required by Exception No. 2.

(2) Within Enclosures

(1) Conductors of intrinsically safe circuits shall be separated at least 50 mm (2 in.) from conductors of any nonintrinsically safe circuits, or as specified in 504.30(A)(2).

(2) All conductors shall be secured so that any conductor that might come loose from a terminal cannot come in contact with another terminal.

> FPN No. 1: The use of separate wiring compartments for the intrinsically safe and nonintrinsically safe terminals is the preferred method of complying with this requirement.
> FPN No. 2: Physical barriers such as grounded metal partitions or approved insulating partitions or approved restricted access wiring ducts separated from other such ducts by at least 19 mm (¾ in.) can be used to help ensure the required separation of the wiring.

The intent of 504.30(A) is to prevent the intrusion of unsafe energy into the intrinsically safe system as a result of a wiring fault. Because low-voltage, low-energy but nonintrinsically safe circuits are permitted by Articles 725 and 800 to be minimally insulated and may not be protected by being installed in raceways or cables, particularly in nonhazardous locations, it is essential that nonintrinsically safe circuits and intrinsically safe circuits be physically and electrically separated.

(3) Other (Not in Raceway or Cable Tray Systems) Conductors and cables of intrinsically safe circuits run in other than raceway or cable tray systems shall be separated by at least 50 mm (2 in.) and secured from conductors and cables of any nonintrinsically safe circuits.

The provisions of 504.30(A)(3) provide separation requirements for intrinsically safe circuit conductors and cables that are installed using a wiring method other than a raceway or a support system other than cable tray.

Exception: Where either (1) all of the intrinsically safe circuit conductors are in Type MI or MC cables or (2) all of the nonintrinsically safe circuit conductors are in raceways or Type MI or MC cables where the sheathing or cladding is capable of carrying fault current to ground.

(B) From Different Intrinsically Safe Circuit Conductors Different intrinsically safe circuits shall be in separate cables or shall be separated from each other by one of the following means:

(1) The conductors of each circuit are within a grounded metal shield.
(2) The conductors of each circuit have insulation with a minimum thickness of 0.25 mm (0.01 in.).

Exception: Unless otherwise identified.

(3) The clearance between two terminals for connection of field wiring of different intrinsically safe circuits shall be at least 6 mm (0.25 in.) unless this clearance is permitted to be reduced by the control drawing.

New in the 2005 *NEC*, terminals for different intrinsically safe circuits are to have at least a ¼- (0.25-) inch separation. This separation can be reduced if shown in the control drawing for the intrinsically safe circuits that are connected to the adjacent terminals. The minimum clearance requirement provides a safeguard against an inadvertent connection between adjacent terminals (with different IS circuits) that could occur during maintenance or connection of a new circuit to an existing terminal block for intrinsically safe circuits.

504.50 Grounding

(A) Intrinsically Safe Apparatus, Associated Apparatus, and Raceways Intrinsically safe apparatus, associated ap-

paratus, cable shields, enclosures, and raceways, if of metal, shall be grounded.

FPN: Supplementary bonding to the grounding electrode may be needed for some associated apparatus, for example, zener diode barriers, if specified in the control drawing. See ANSI/ISA RP 12.06.01-2002, *Wiring Methods for Hazardous (Classified) Locations Instrumentation Part 1: Intrinsic Safety.*

It is important to maintain a low-impedance path to ground for zener diode barrier systems because such systems shunt fault currents to ground.

Exhibit 504.2 illustrates a common type of zener diode barrier system. F_1 limits the duration of the current in Z_1 and Z_2 to the power rating of the zener diodes under fault conditions, so that the diodes need not be excessively large. Although the diodes themselves can be considered subject to open-circuit fault, they are redundant components, and either one alone provides the necessary protection in the event of a first fault, that is, high voltage across input terminals 1 and 2. Resistor R_2 is the current-limiting resistor and has been investigated as a protective component not subject to short-circuit fault. Resistor R_1 is used primarily to permit testing to determine that the diodes are intact. Terminals 2 and 4 and the ends of the two diodes are connected to a ground bus that, in turn, is connected to a ground system to which all grounds in the intrinsically safe system are connected. A very low impedance (1 ohm or less is usually recommended) is necessary so that the voltage level on the ground bus will not be raised to an unsafe level under high-current fault conditions.

A shunt diode barrier system is commonly tested with 250 volts ac across terminals 1 and 2, representing a fault in associated apparatus on the nonhazardous location side of the barrier, which imposes a voltage up to 250 volts ac on these terminals. Diode Z_1 conducts at its rated voltage, typically 14 or 15 volts for a barrier designed for a 12-volt (normal operation) system, thus limiting the output voltage of the circuit to 14 or 15 volts. The voltage at which this diode conducts is designed to be higher than the rated input voltage (under no-fault conditions) at terminals 1 and 2. The fuse is selected so that it will open before the power rating of Z_1 is exceeded. Diode Z_2 usually conducts within 1 or 2 volts of Z_1 and serves as a backup to Z_1 in the event Z_1 fails in an open-circuit condition for any reason.

Exhibit 504.2 Fused zener diode barrier.

Resistor R_2 limits the current in the intrinsically safe circuit to the required value at the output voltage of Z_1 and Z_2. Therefore, even though up to 250 volts ac may be applied to the input of the circuit (terminals 1 and 2) as a result of a fault in the equipment or circuit connected to terminals 1 and 2, the output at terminals 3 and 4 cannot exceed the voltage and current permitted by the diodes and resistor.

Shunt diode barriers normally have maximum inductance and capacitance ratings on the intrinsically safe side because, even though both the voltage and the current are limited at terminals 3 and 4, too much inductance in the circuit connected to terminals 3 and 4 could result in the release of an ignition-capable spark when the circuit is opened. In a like manner, too much capacitance between terminals 3 and 4 could result in the release of an ignition-capable spark if there were a short-circuit between wiring connected to terminals 3 and 4 or between wiring connected to terminal 3 and ground.

Wiring always has inductance and capacitance associated with it, depending on the spacing between conductors, size of conductors, and length of conductors. It is necessary, therefore, to limit the length of conductors connected to terminals 3 and 4, just as it is necessary to limit the inductance and capacitance of connected equipment. The length limitation is usually on the order of thousands of feet. The manufacturer's instructions provide the information on installation limitations.

By adjusting the values of its components, the barrier can be designed for a variety of uses, including different types of instrument systems and thermocouples. Barriers such as described are available commercially from a number of manufacturers.

(B) Connection to Grounding Electrodes Where connection to a grounding electrode is required, the grounding electrode shall be as specified in 250.52(A)(1), (A)(2), (A)(3), and (A)(4) and shall comply with 250.30(A)(7). Section 250.52(A)(5), (A)(6), and (A)(7) shall not be used if electrodes specified in 250.52(A)(1), (A)(2), (A)(3), or (A)(4) are available.

The electrodes specified in 250.52(A)(1) through 250.52(A)(4) include metal underground water pipes, the metal frames of building, concrete-encased electrodes, and ground rings. These electrodes usually provide lower resistance grounds than electrodes, such as ground rods and plate electrodes, covered in 250.52(A)(5) and 250.52(A)(6).

(C) Shields Where shielded conductors or cables are used, shields shall be grounded.

Exception: Where a shield is part of an intrinsically safe circuit.

504.60 Bonding

(A) Hazardous Locations In hazardous (classified) locations, intrinsically safe apparatus shall be bonded in the hazardous (classified) location in accordance with 250.100.

(B) Unclassified In unclassified or nonhazardous locations, where metal raceways are used for intrinsically safe system wiring in hazardous (classified) locations, associated apparatus shall be bonded in accordance with 501.30(A), 502.30(A), 503.30(A), or 505.25, as applicable.

504.70 Sealing

Conduits and cables that are required to be sealed by 501.15, 502.15, and 505.16 shall be sealed to minimize the passage of gases, vapors, or dusts. Such seals shall not be required to be explosionproof or flameproof.

Exception: Seals shall not be required for enclosures that contain only intrinsically safe apparatus, except as required by 501.15(F)(3).

504.80 Identification

Labels required by this section shall be suitable for the environment where they are installed with consideration given to exposure to chemicals and sunlight.

(A) Terminals Intrinsically safe circuits shall be identified at terminal and junction locations in a manner that will prevent unintentional interference with the circuits during testing and servicing.

(B) Wiring Raceways, cable trays, and other wiring methods for intrinsically safe system wiring shall be identified with permanently affixed labels with the wording "Intrinsic Safety Wiring" or equivalent. The labels shall be located so as to be visible after installation and placed so that they may be readily traced through the entire length of the installation. Intrinsic safety circuit labels shall appear in every section of the wiring system that is separated by enclosures, walls, partitions, or floors. Spacing between labels shall not be more than 7.5 m (25 ft).

Exception: Circuits run underground shall be permitted to be identified where they become accessible after emergence from the ground.

> FPN No. 1: Wiring methods permitted in unclassified locations may be used for intrinsically safe systems in hazardous (classified) locations. Without labels to identify the application of the wiring, enforcement authorities cannot determine that an installation is in compliance with this *Code*.
>
> FPN No. 2: In unclassified locations, identification is necessary to ensure that nonintrinsically safe wire will not be inadvertently added to existing raceways at a later date.

(C) Color Coding Color coding shall be permitted to identify intrinsically safe conductors where they are colored light blue and where no other conductors colored light blue are used. Likewise, color coding shall be permitted to identify raceways, cable trays, and junction boxes where they are colored light blue and contain only intrinsically safe wiring.

ARTICLE 505
Class I, Zone 0, 1, and 2 Locations

Summary of Changes

- **505.8(I):** Added requirement that the type of detection equipment, its listing, installation location(s), alarm and shutdown criteria, and calibration frequency be documented when combustible gas detectors are used as a protection technique.

- **505.9(C)(2):** Added two exceptions for associated apparatus not suitable for hazardous locations and for marking of simple apparatus.

- **505.9(E):** Added exception for listed explosionproof and flameproof equipment regarding the minimum number of threads required to be engaged.

- **505.15(C)(1)(b):** Revised to require single conductor Type MV cables to be shielded or metallic-armored.

- **505.16(B)(1):** Revised to require conduit seals within 2 in. of protection "d" or "e" enclosures, with two exceptions (Nos. 2 and 3) for protection "e."

- **505.18:** Revised to require every conductor (including spares) entering type "e" equipment to be terminated at a type "e" terminal.

- **505.19:** Revised to provide guidance concerning protection against explosion and against electric shock in order to limit the voltage of exposed live parts.

Contents

FPN: Rules that are followed by a reference in brackets contain text that has been extracted from NFPA 497-2004, *Recommended Practice for the Classification of Flammable Liquids, Gases, or Vapors and of Hazardous (Classified) Locations for Electrical Installations in Chemical Process Areas*. Only editorial changes were made to the extracted text to make it consistent with this *Code*.

505.1 Scope

This article covers the requirements for the zone classification system as an alternative to the division classification system covered in Article 500 for electrical and electronic equipment and wiring for all voltages in Class I, Zone 0, Zone 1, and Zone 2 hazardous (classified) locations where fire or explosion hazards may exist due to flammable gases, vapors, or liquids.

> FPN: For the requirements for electrical and electronic equipment and wiring for all voltages in Class I, Division 1 or Division 2; Class II, Division 1 or Division 2; and Class III, Division 1 or Division 2 hazardous (classified) locations where fire or explosion hazards may exist due to flammable gases or vapors, flammable liquids, or combustible dusts or fibers, refer to Articles 500 through 504.

The requirements covering wiring methods, sealing, and flexible cord use were added to Article 505 for the 2002 *Code* so that the Class I, Zone 0, Zone 1, and Zone 2 area classification concept can be used as a separate set of requirements that parallel those in Articles 500 and 501 covering installations in which the Class I, Division 1 and Division 2 area classification concept is used. The zone classification concept offers an alternative method of classifying Class I hazardous locations. The zone classification method is based on the standards for area classification used by the International Electrotechnical Commission (IEC).

The IEC classification scheme includes underground mines. In the United States, mines are under the jurisdiction of the Mine Safety and Health Administration (MSHA) and are outside of the scope of the *NEC*.

505.2 Definitions

For purposes of this article, the following definitions apply.

Combustible Gas Detection System. A protection technique utilizing stationary gas detectors in industrial establishments.

Electrical and Electronic Equipment. Materials, fittings, devices, appliances, and the like that are part of, or in connection with, an electrical installation.

> FPN: Portable or transportable equipment having self-contained power supplies, such as battery-operated equipment, could potentially become an ignition source in hazardous (classified) locations.

Encapsulation "m." Type of protection where electrical parts that could ignite an explosive atmosphere by either sparking or heating are enclosed in a compound in such a way that this explosive atmosphere cannot be ignited.

> FPN: See ANSI/ISA 12.23.01-2002, *Electrical Apparatus for Use in Class I, Zone 1 Hazardous (Classified)*

Locations, Type of Protection — Encapsulation "m"; and ANSI/UL 60079–18, Electrical apparatus for explosive gas atmospheres — Part 18: Encapsulation "m."

Flameproof "d." Type of protection where the enclosure will withstand an internal explosion of a flammable mixture that has penetrated into the interior, without suffering damage and without causing ignition, through any joints or structural openings in the enclosure, of an external explosive gas atmosphere consisting of one or more of the gases or vapors for which it is designed.

> FPN: See ANSI/ISA 12.22.01-2002, Electrical Apparatus for Use in Class I, Zone 1 and 2 Hazardous (Classified) Locations, Type of Protection — Flameproof "d"; and ANSI/UL 60079–1, Electrical apparatus for explosive gas atmospheres — Part 1: Flameproof enclosures "d."

Increased Safety "e." Type of protection applied to electrical equipment that does not produce arcs or sparks in normal service and under specified abnormal conditions, in which additional measures are applied so as to give increased security against the possibility of excessive temperatures and of the occurrence of arcs and sparks.

> FPN: See ANSI/ISA — 12.16.01-2002, Electrical Apparatus for Use in Class I, Zone 1 Hazardous (Classified) Locations, Type of Protection — Increased Safety "e"; and ANSI/UL 60079-7, Electrical apparatus for explosive gas atmospheres — Part 7: Increased Safety "e."

Intrinsic Safety "i." Type of protection where any spark or thermal effect is incapable of causing ignition of a mixture of flammable or combustible material in air under prescribed test conditions.

> FPN No. 1: See ANSI/UL 913-1997, Intrinsically Safe Apparatus and Associated Apparatus for Use in Class I, II, and III, Hazardous Locations; ISA 12.02.01-1999, Electrical Apparatus for Use in Class I, Zones 0, 1 and 2 Hazardous (Classified) Locations — Intrinsic Safety "i"; and ANSI/UL 60079-11, Electrical apparatus for explosive gas atmospheres — Part II: Intrinsic safety "i."
>
> FPN No. 2: Intrinsic safety is designated type of protection "ia" for use in Zone 0 locations. Intrinsic safety is designated type of protection "ib" for use in Zone 1 locations.
>
> FPN No. 3: Intrinsically safe associated apparatus, designated by [ia] or [ib], is connected to intrinsically safe apparatus ("ia" or "ib," respectively) but is located outside the hazardous (classified) location unless also protected by another type of protection (such as flameproof).

Oil Immersion "o." Type of protection where electrical equipment is immersed in a protective liquid in such a way that an explosive atmosphere that may be above the liquid or outside the enclosure cannot be ignited.

> FPN: See ISA 12.26.01-1998, Electrical Apparatus for Use in Class I, Zone 1 Hazardous (Classified) Locations,

Type of Protection — Oil-Immersion "o"; and ANSI/UL 60079-6, Electrical apparatus for explosive gas atmospheres — Part 6: Oil-immersion "o."

Powder Filling "q." Type of protection where electrical parts capable of igniting an explosive atmosphere are fixed in position and completely surrounded by filling material (glass or quartz powder) to prevent the ignition of an external explosive atmosphere.

> FPN: See ANSI/ISA-12.25.01-2002, Electrical Apparatus for Use in Class I, Zone 1 Hazardous (Classified) Locations Type of Protection — Powder Filling "q"; and ANSI/UL 60079-5, Electrical apparatus for explosive gas atmospheres — Part 5: Powder filling "q."

Purged and Pressurized. Type of protection for electrical equipment that uses the technique of guarding against the ingress of the external atmosphere, which may be explosive, into an enclosure by maintaining a protective gas therein at a pressure above that of the external atmosphere.

> FPN No. 1: See NFPA 496-2003, Standard for Purged and Pressurized Enclosures for Electrical Equipment.
>
> FPN No. 2: See IEC 60079-2-2000, Electrical Apparatus for Explosive Gas Atmospheres — Part 2: Electrical Apparatus, Type of Protection "p"; and IEC 60079-13-1982, Electrical Apparatus for Explosive Gas Atmospheres — Part 13: Construction and Use of Rooms or Buildings Protected by Pressurization.

Type of Protection "n." Type of protection where electrical equipment, in normal operation, is not capable of igniting a surrounding explosive gas atmosphere and a fault capable of causing ignition is not likely to occur.

> FPN: See ANSI/UL 60079-15-2002, Electrical apparatus for explosive gas atmospheres — Part 15: Type of protection "n"; and ANSI/ISA 12.12.02-2003, Electrical apparatus for use in Class I, Zone 2 Hazardous (Classified) Locations: Type of protection "n."

Unclassified Locations. Locations determined to be neither Class I, Division 1; Class I, Division 2; Class I, Zone 0; Class I, Zone 1; Class I, Zone 2; Class II, Division 1; Class II, Division 2; Class III, Division 1; Class III, Division 2; or any combination thereof.

505.3 Other Articles

All other applicable rules contained in this *Code* shall apply to electrical equipment and wiring installed in hazardous (classified) locations.

> See the commentary following 500.3 for more information on other applicable *Code* requirements that apply to electrical equipment and wiring.

Exception: As modified by Article 504 and this article.

505.4 General

(A) Documentation for Industrial Occupancies All areas in industrial occupancies designated as hazardous (classified) locations shall be properly documented. This documentation shall be available to those authorized to design, install, inspect, maintain, or operate electrical equipment at the location.

> FPN: For examples of area classification drawings, see ANSI/API RP 505-1997, *Recommended Practice for Classification of Locations for Electrical Installations at Petroleum Facilities Classified as Class I, Zone 0, Zone 1, or Zone 2*; ISA RP12.24.01-1998, *Recommended Practice for Classification of Locations for Electrical Installations Classified as Class I, Zone 0, Zone 1, or Zone 2*; IEC 60079-10-1995, *Electrical Apparatus for Explosive Gas Atmospheres, Classification of Hazardous Areas*; and *Model Code of Safe Practice in the Petroleum Industry, Part 15: Area Classification Code for Petroleum Installations*, IP 15, The Institute of Petroleum, London.

(B) Reference Standards Important information relating to topics covered in Chapter 5 may be found in other publications.

> FPN No. 1: It is important that the authority having jurisdiction be familiar with recorded industrial experience as well as with standards of the National Fire Protection Association (NFPA), the American Petroleum Institute (API), the Instrumentation, Systems, and Automation Society (ISA), and the International Electrotechnical Commission (IEC) that may be of use in the classification of various locations, the determination of adequate ventilation, and the protection against static electricity and lightning hazards.
>
> FPN No. 2: For further information on the classification of locations, see ANSI/API RP 505-1997, *Recommended Practice for Classification of Locations for Electrical Installations at Petroleum Facilities Classified as Class I, Zone 0, Zone 1, or Zone 2*; ISA RP 12.24.01-1998, *Recommended Practice for Classification of Locations for Electrical Installations Classified as Class I, Zone 0, Zone 1, or Zone 2*; IEC 60079-10-1995, *Electrical Apparatus for Explosive Gas Atmospheres, Classification of Hazardous Areas*; and *Model Code of Safe Practice in the Petroleum Industry, Part 15: Area Classification Code for Petroleum Installations*, IP 15, The Institute of Petroleum, London.
>
> FPN No. 3: For further information on protection against static electricity and lightning hazards in hazardous (classified) locations, see NFPA 77-2000, *Recommended Practice on Static Electricity*; NFPA 780-2004, *Standard for the Installation of Lightning Protection Systems*; and API RP 2003-1998, *Protection Against Ignitions Arising Out of Static Lightning and Stray Currents*.
>
> FPN No. 4: For further information on ventilation, see NFPA 30-2003, *Flammable and Combustible Liquids Code*, and ANSI/API RP 505-1997, *Recommended Practice for Classification of Locations for Electrical Installations at Petroleum Facilities Classified as Class I, Zone 0, Zone 1, or Zone 2*.
>
> FPN No. 5: For further information on electrical systems for hazardous (classified) locations on offshore oil and gas producing platforms, see ANSI/API RP 14FZ-2000, *Recommended Practice for Design and Installation of Electrical Systems for Fixed and Floating Offshore Petro-*

leum Facilities for Unclassified and Class I, Zone 0, Zone 1, and Zone 2 Locations.

> FPN No. 6: For further information on the installation of electrical equipment in hazardous (classified) locations in general, see IEC 60079-14-1996, *Electrical Apparatus for Explosive Gas Atmospheres — Part 14: Electrical Installations in Explosive Gas Atmospheres (Other Than Mines)*, and IEC 60079-16-1990, *Electrical Apparatus for Explosive Gas Atmospheres — Part 16: Artificial Ventilation for the Protection of Analyzer(s) Houses*.
>
> FPN No. 7: For further information on application of electrical equipment in hazardous (classified) locations in general, see ANSI/ISA 12.00.01-2002, *Electrical Apparatus for Use in Class I, Zones 0 and 1, Hazardous (Classified) Locations: General Requirements*; ISA 12.01.01-1999, *Definitions and Information Pertaining to Electrical Apparatus in Hazardous (Classified) Locations*; and ANSI/UL 60079-0, *Electrical apparatus for explosive gas atmospheres — Part 0: General requirements*.

505.5 Classifications of Locations

(A) Classification of Locations Locations shall be classified depending on the properties of the flammable vapors, liquids, or gases that may be present and the likelihood that a flammable or combustible concentration or quantity is present. Where pyrophoric materials are the only materials used or handled, these locations shall not be classified. Each room, section, or area shall be considered individually in determining its classification.

> FPN No. 1: See 505.7 for restrictions on area classification.
>
> FPN No. 2: Through the exercise of ingenuity in the layout of electrical installations for hazardous (classified) locations, it is frequently possible to locate much of the equipment in reduced level of classification or in an unclassified location and, thus, to reduce the amount of special equipment required.

Rooms and areas containing ammonia refrigeration systems that are equipped with adequate mechanical ventilation may be classified as "unclassified" locations.

> FPN: For further information regarding classification and ventilation of areas involving ammonia, see ANSI/ASHRAE 15-1994, *Safety Code for Mechanical Refrigeration*; and ANSI/CGA G2.1-1989 (14-39), *Safety Requirements for the Storage and Handling of Anhydrous Ammonia*.

(B) Class I, Zone 0, 1, and 2 Locations Class I, Zone 0, 1, and 2 locations are those in which flammable gases or vapors are or may be present in the air in quantities sufficient to produce explosive or ignitible mixtures. Class I, Zone 0, 1, and 2 locations shall include those specified in 505(B)(1), (B)(2), and (B)(3).

(1) Class I, Zone 0 A Class I, Zone 0 location is a location in which

(1) Ignitible concentrations of flammable gases or vapors are present continuously, or

(2) Ignitible concentrations of flammable gases or vapors are present for long periods of time.

FPN No. 1: As a guide in determining when flammable gases or vapors are present continuously or for long periods of time, refer to ANSI/API RP 505-1997, *Recommended Practice for Classification of Locations for Electrical Installations of Petroleum Facilities Classified as Class I, Zone 0, Zone 1 or Zone 2*; ISA 12.24.01-1998, *Recommended Practice for Classification of Locations for Electrical Installations Classified as Class I, Zone 0, Zone 1, or Zone 2*; IEC 60079-10-1995, *Electrical Apparatus for Explosive Gas Atmospheres, Classifications of Hazardous Areas*; and *Area Classification Code for Petroleum Installations, Model Code, Part 15*, Institute of Petroleum.

FPN No. 2: This classification includes locations inside vented tanks or vessels that contain volatile flammable liquids; inside inadequately vented spraying or coating enclosures, where volatile flammable solvents are used; between the inner and outer roof sections of a floating roof tank containing volatile flammable liquids; inside open vessels, tanks and pits containing volatile flammable liquids; the interior of an exhaust duct that is used to vent ignitible concentrations of gases or vapors; and inside inadequately ventilated enclosures that contain normally venting instruments utilizing or analyzing flammable fluids and venting to the inside of the enclosures.

FPN No. 3: It is not good practice to install electrical equipment in Zone 0 locations except when the equipment is essential to the process or when other locations are not feasible. [See 505.5(A) FPN No. 2.] If it is necessary to install electrical systems in a Zone 0 location, it is good practice to install intrinsically safe systems as described by Article 504.

(2) Class I, Zone 1 A Class I, Zone 1 location is a location

(1) In which ignitible concentrations of flammable gases or vapors are likely to exist under normal operating conditions; or

(2) In which ignitible concentrations of flammable gases or vapors may exist frequently because of repair or maintenance operations or because of leakage; or

(3) In which equipment is operated or processes are carried on, of such a nature that equipment breakdown or faulty operations could result in the release of ignitible concentrations of flammable gases or vapors and also cause simultaneous failure of electrical equipment in a mode to cause the electrical equipment to become a source of ignition; or

(4) That is adjacent to a Class I, Zone 0 location from which ignitible concentrations of vapors could be communicated, unless communication is prevented by adequate positive pressure ventilation from a source of clean air and effective safeguards against ventilation failure are provided.

FPN No. 1: Normal operation is considered the situation when plant equipment is operating within its design parameters. Minor releases of flammable material may be part of normal operations. Minor releases include the releases from mechanical packings on pumps. Failures that involve repair or shutdown (such as the breakdown of pump seals and flange gaskets, and spillage caused by accidents) are not considered normal operation.

FPN No. 2: This classification usually includes locations where volatile flammable liquids or liquefied flammable gases are transferred from one container to another. In areas in the vicinity of spraying and painting operations where flammable solvents are used; adequately ventilated drying rooms or compartments for evaporation of flammable solvents; adequately ventilated locations containing fat and oil extraction equipment using volatile flammable solvents; portions of cleaning and dyeing plants where volatile flammable liquids are used; adequately ventilated gas generator rooms and other portions of gas manufacturing plants where flammable gas may escape; inadequately ventilated pump rooms for flammable gas or for volatile flammable liquids; the interiors of refrigerators and freezers in which volatile flammable materials are stored in the open, lightly stoppered, or in easily ruptured containers; and other locations where ignitible concentrations of flammable vapors or gases are likely to occur in the course of normal operation but not classified Zone 0.

(3) Class I, Zone 2 A Class I, Zone 2 location is a location

(1) In which ignitible concentrations of flammable gases or vapors are not likely to occur in normal operation and, if they do occur, will exist only for a short period; or

(2) In which volatile flammable liquids, flammable gases, or flammable vapors are handled, processed, or used but in which the liquids, gases, or vapors normally are confined within closed containers of closed systems from which they can escape, only as a result of accidental rupture or breakdown of the containers or system, or as a result of the abnormal operation of the equipment with which the liquids or gases are handled, processed, or used; or

(3) In which ignitible concentrations of flammable gases or vapors normally are prevented by positive mechanical ventilation but which may become hazardous as a result of failure or abnormal operation of the ventilation equipment; or

(4) That is adjacent to a Class I, Zone 1 location, from which ignitible concentrations of flammable gases or vapors could be communicated, unless such communication is prevented by adequate positive-pressure ventilation from a source of clean air and effective safeguards against ventilation failure are provided.

FPN: The Zone 2 classification usually includes locations where volatile flammable liquids or flammable gases or vapors are used but which would become hazardous only in case of an accident or of some unusual operating condition.

505.6 Material Groups

For purposes of testing, approval, and area classification, various air mixtures (not oxygen enriched) shall be grouped as required in 505.6(A), (B), and (C).

> FPN: Group I is intended for use in describing atmospheres that contain firedamp (a mixture of gases, composed mostly of methane, found underground, usually in mines). This *Code* does not apply to installations underground in mines. See 90.2(B).

Group II shall be subdivided into IIC, IIB, and IIA, as noted in 505.6(A), (B), and (C), according to the nature of the gas or vapor, for protection techniques "d," "ia," "ib," "[ia]," and "[ib]," and, where applicable, "n" and "o."

> FPN No. 1: The gas and vapor subdivision as described above is based on the maximum experimental safe gap (MESG), minimum igniting current (MIC), or both. Test equipment for determining the MESG is described in IEC 60079-1A-1975, Amendment No. 1 (1993), *Construction and Verification Tests of Flameproof Enclosures of Electrical Apparatus*; and *UL Technical Report No. 58* (1993). The test equipment for determining MIC is described in IEC 60079-11-1999, *Electrical Apparatus for Explosive Gas Atmospheres — Part 11: Intrinsic Safety "i."* The classification of gases or vapors according to their maximum experimental safe gaps and minimum igniting currents is described in IEC 60079-12-1978, *Classification of Mixtures of Gases or Vapours with Air According to Their Maximum Experimental Safe Gaps and Minimum Igniting Currents*.
>
> FPN No. 2: Verification of electrical equipment utilizing protection techniques "e," "m," "p," and "q," due to design technique, does not require tests involving MESG or MIC. Therefore, Group II is not required to be subdivided for these protection techniques.
>
> FPN No. 3: It is necessary that the meanings of the different equipment markings and Group II classifications be carefully observed to avoid confusion with Class I, Divisions 1 and 2, Groups A, B, C, and D.

In the zone classification system, the gas or vapor group order is the approximate inverse of the gas or vapor groups specified in Article 500. For example, Group IIC includes Article 500, Groups A and B. Determination of a gas or vapor for the purposes of grouping includes the evaluation of the maximum safe experimental gap ratio as well as minimum igniting current ratio. Although the maximum safe experimental gap for Group A is less than that for Group B in some circumstances, the minimum igniting current ratio is less for hydrogen (Group B) than it is for acetylene (Group A). This difference has been accounted for in ANSI/UL 913, *Intrinsically Safe Apparatus and Associated Apparatus for*

Use in Class I, II, III, Division 1, Hazardous (Classified) Locations, because it is a factor that must be considered in the evaluation of intrinsically safe apparatus.

Class I, Zone 0, 1, and 2, groups shall be as follows:

(A) Group IIC Atmospheres containing acetylene, hydrogen, or flammable gas, flammable liquid-produced vapor, or combustible liquid-produced vapor mixed with air that may burn or explode, having either a maximum experimental safe gap (MESG) value less than or equal to 0.50 mm or minimum igniting current ratio (MIC ratio) less than or equal to 0.45. [NFPA 497:3.3]

> FPN: Group IIC is equivalent to a combination of Class I, Group A, and Class I, Group B, as described in 500.6(A)(1) and 500.6(A)(2).

(B) Group IIB Atmospheres containing acetaldehyde, ethylene, or flammable gas, flammable liquid-produced vapor, or combustible liquid-produced vapor mixed with air that may burn or explode, having either maximum experimental safe gap (MESG) values greater than 0.50 mm and less than or equal to 0.90 mm or minimum igniting current ratio (MIC ratio) greater than 0.45 and less than or equal to 0.80. [NFPA 497:3.3]

> FPN: Group IIB is equivalent to Class I, Group C, as described in 500.6(A)(3).

(C) Group IIA Atmospheres containing acetone, ammonia, ethyl alcohol, gasoline, methane, propane, or flammable gas, flammable liquid-produced vapor, or combustible liquid-produced vapor mixed with air that may burn or explode, having either a maximum experiment safe gap (MESG) value greater than 0.90 mm or minimum igniting current ratio (MIC ratio) greater than 0.80. [NFPA 497:3.3]

> FPN: Group IIA is equivalent to Class I, Group D as described in 500.6(A)(4).

505.7 Special Precaution

Article 505 requires equipment construction and installation that ensures safe performance under conditions of proper use and maintenance.

> FPN No. 1: It is important that inspection authorities and users exercise more than ordinary care with regard to the installation and maintenance of electrical equipment in hazardous (classified) locations.
>
> FPN No. 2: Low ambient conditions require special consideration. Electrical equipment depending on the protection techniques described by 505.8(A) may not be suitable for use at temperatures lower than $-20°C$ ($-4°F$) unless they are identified for use at lower temperatures. However, at low ambient temperatures, flammable

concentrations of vapors may not exist in a location classified Class I, Zones 0, 1, or 2 at normal ambient temperature.

(A) Supervision of Work Classification of areas and selection of equipment and wiring methods shall be under the supervision of a qualified Registered Professional Engineer.

Note that 505.7(A) requires area classification, wiring, and equipment selection to be under the supervision of a qualified Registered Professional Engineer for installations in Class I, Zone 0, 1, and 2 locations.

(B) Dual Classification In instances of areas within the same facility classified separately, Class I, Zone 2 locations shall be permitted to abut, but not overlap, Class I, Division 2 locations. Class I, Zone 0 or Zone 1 locations shall not abut Class I, Division 1 or Division 2 locations.

An installation is permitted to be designed using either the classification scheme of Article 500 or the classification scheme of Article 505. Both schemes cannot be used for classifying the same area. In areas within the same facility, Class I, Zone 2 locations are allowed to be adjacent to and share the same border, but not overlap Class I, Division 2 locations. However, Class I, Zone 0 or Zone 1 locations are not allowed to be adjacent to and share the same border with Class I, Division 1 or Division 2 locations.

(C) Reclassification Permitted A Class I, Division 1 or Division 2 location shall be permitted to be reclassified as a Class I, Zone 0, Zone 1, or Zone 2 location, provided all of the space that is classified because of a single flammable gas or vapor source is reclassified under the requirements of this article.

(D) Solid Obstacles Flameproof equipment with flanged joints shall not be installed such that the flange openings are closer than the distances shown in Table 505.7(D) to any solid obstacle that is not a part of the equipment (such as steelworks, walls, weather guards, mounting brackets,

Table 505.7(D) Minimum Distance of Obstructions from Flameproof "d" Flange Openings

Gas Group	Minimum Distance	
	mm	in.
IIC	40	$1^{37}/_{64}$
IIB	30	$1^{3}/_{16}$
IIA	10	$^{25}/_{64}$

pipes, or other electrical equipment) unless the equipment is listed for a smaller distance of separation.

505.8 Protection Techniques

Acceptable protection techniques for electrical and electronic equipment in hazardous (classified) locations shall be as described in 505.8(A) through 505.8(I).

> FPN: For additional information, see ANSI/ISA 12.00.01-2002, *Electrical Apparatus for Use in Class I, Zones 0 and 1 Hazardous (Classified) Locations, General Requirements*; ANSI/ISA 12.01.01-2002, *Definitions and Information Pertaining to Electrical Apparatus in Hazardous (Classified) Locations*; and ANSI/UL 60079–0, *Electrical apparatus for explosive gas atmospheres — Part 0: General requirements*.

Where the area is classified in accordance with the zone method, electrical and electronic equipment may be protected by the following methods:

1. In Class I, Zone 1, equipment approved as flameproof "d," which is very similar to explosionproof equipment in the United States
2. Purged and pressurized equipment for Class I, Zone 1 or Zone 2 locations for which it is approved
3. Intrinsic safety techniques for the Class I, Zone 0, Zone 1, or Zone 2 for which the technique is listed
4. In Class I, Zone 2, equipment approved as type "n," which under normal operation is not capable of igniting a surrounding explosive gas atmosphere and is not likely to create a fault that is capable of causing ignition
5. Oil immersion "o" technique for Class I, Zone 1 or Zone 2, where the equipment or part of the equipment is immersed in a protective liquid so that explosive gases above or outside the enclosure cannot be ignited
6. A type of protection technique of increased safety "e" approved for Class I, Zone 1 or Zone 2, in which the electrical equipment involved does not produce arcs or sparks or excessive temperature under normal operation, as well as under abnormal conditions, under specified conditions of operation
7. Encapsulation "m" technique in which the arcing, sparking, or hot parts are completely surrounded in a compound in such a way that an explosive gas or vapor cannot be ignited in a Class I, Zone 1 or Zone 2 area
8. A technique using powder filling "q" in which the arcing, sparking, or hot parts of the equipment are surrounded by a material such as glass or quartz powder to prevent ignition of an external explosive atmosphere in a Class I, Zone 1 or Zone 2 area

The letter in quotation marks following the protection method is the letter used in the marking on the equipment. See 505.9(C) for marking requirements.

Of the many protection techniques, intrinsic safety, flameproof, and increased safety are the most common for Zone 1 locations.

(A) Flameproof "d" This protection technique shall be permitted for equipment in Class I, Zone 1 or Zone 2 locations.

Flameproof protection is commonly combined with increased safety protection. For example, motor control and other switching contacts are commonly protected by flameproof enclosures with the field wiring terminals protected in a separate but attached enclosure by increased safety. The conductors between the enclosures are protected by flameproof feed-through insulators. The equipment shown in Exhibit 505.1 is an example of where this combination of protection techniques is employed.

Exhibit 505.1 Typical control stations with the combination of flameproof and increased safety types of protection suitable for use in Class I, Zone 1 areas. (Courtesy of Cooper Crouse-Hinds)

(B) Purged and Pressurized This protection technique shall be permitted for equipment in those Class I, Zone 1 or Zone 2 locations for which it is identified.

(C) Intrinsic Safety This protection technique shall be permitted for apparatus and associated apparatus in Class I, Zone 0, Zone 1, or Zone 2 locations for which it is listed.

The identifying letter for intrinsic safety is "i" followed by either "a" or "b," identifying whether the equipment is suitable for Zone 0 (ia) or Zone 1 (ib). The associated apparatus is identified by the same letters in brackets, that is, [ia] or [ib].

(D) Type of Protection "n" This protection technique shall be permitted for equipment in Class I, Zone 2 locations. Type of protection "n" is further subdivided into nA, nC, and nR.

FPN: See Table 505.9(C)(2)(4) for the descriptions of subdivisions for type of protection "n".

(E) Oil Immersion "o" This protection technique shall be permitted for equipment in Class I, Zone 1 or Zone 2 locations.

(F) Increased Safety "e" This protection technique shall be permitted for equipment in Class I, Zone 1 or Zone 2 locations.

The increased safety protection technique is commonly used for motors and generators (see 505.22) and fluorescent luminaires. It is also used for terminal boxes.

(G) Encapsulation "m" This protection technique shall be permitted for equipment in Class I, Zone 1 or Zone 2 locations.

(H) Powder Filling "q" This protection technique shall be permitted for equipment in Class I, Zone 1 or Zone 2 locations.

(I) Combustible Gas Detection System A combustible gas detection system shall be permitted as a means of protection in industrial establishments with restricted public access and where the conditions of maintenance and supervision ensure that only qualified persons service the installation. Gas detection equipment shall be listed for detection of the specific gas or vapor to be encountered. Where such a system is installed, equipment specified in 505.8(I)(1), I(2), or I(3) shall be permitted. The type of detection equipment, its listing, installation location(s), alarm and shutdown criteria, and calibration frequency shall be documented when combustible gas detectors are used as a protection technique.

FPN No. 1: For further information, see ANSI/ISA-12.13.01, *Performance Requirements, Combustible Gas Detectors*.
FPN No. 2: For further information, see ANSI/API RP 505, *Recommended Practice for Classification of Locations for Electrical Installations at Petroleum Facilities Classified as Class I, Zone 0, Zone 1, and Zone 2*.
FPN No. 3: For further information, see ISA-RP12.13.02, *Installation, Operation, and Maintenance of Combustible Gas Detection Instruments*.

(1) Inadequate Ventilation In a Class I, Zone 1 location that is so classified due to inadequate ventilation, electrical equipment suitable for Class I, Zone 2 locations shall be permitted.

(2) Interior of a Building In a building located in, or with an opening into, a Class I, Zone 2 location where the interior does not contain a source of flammable gas or vapor, electrical equipment for unclassified locations shall be permitted.

(3) Interior of a Control Panel In the interior of a control panel containing instrumentation utilizing or measuring flammable liquids, gases, or vapors, electrical equipment suitable for Class I, Zone 2 locations shall be permitted.

505.9 Equipment

(A) Suitability Suitability of identified equipment shall be determined by one of the following:

(1) Equipment listing or labeling
(2) Evidence of equipment evaluation from a qualified testing laboratory or inspection agency concerned with product evaluation
(3) Evidence acceptable to the authority having jurisdiction such as a manufacturer's self-evaluation or an owner's engineering judgment

(B) Listing

(1) Equipment that is listed for a Zone 0 location shall be permitted in a Zone 1 or Zone 2 location of the same gas or vapor, provided that it is installed in accordance with the requirements for the marked type of protection. Equipment that is listed for a Zone 1 location shall be permitted in a Zone 2 location of the same gas or vapor, provided that it is installed in accordance with the requirements for the marked type of protection.
(2) Equipment shall be permitted to be listed for a specific gas or vapor, specific mixtures of gases or vapors, or any specific combination of gases or vapors.

> FPN: One common example is equipment marked for "IIB. + H2."

(C) Marking Equipment shall be marked in accordance with 505.9(C)(1) or (C)(2).

(1) Division Equipment Equipment identified for Class I, Division 1 or Class I, Division 2 shall, in addition to being marked in accordance with 500.8(B), be permitted to be marked with all of the following:

(1) Class I, Zone 1 or Class I, Zone 2 (as applicable)
(2) Applicable gas classification group(s) in accordance with Table 505.9(C)

Table 505.9(C) Gas Classification Groups

Gas Group	Comment
IIC	See 505.6(A)(1)
IIB	See 505.6(A)(2)
IIA	See 505.6(A)(3)

(3) Temperature classification in accordance with 505.9(D)(1)

(2) Zone Equipment Equipment meeting one or more of the protection techniques described in 505.8 shall be marked with all of the following in the order shown:

(1) Class
(2) Zone
(3) Symbol "AEx"
(4) Protection technique(s) in accordance with Table 505.9(C)(2)(4)
(5) Applicable gas classification group(s) in accordance with Table 505.9(C)
(6) Temperature classification in accordance with 505.9(D)(1)

> The symbol AEx identifies the equipment as meeting American standards. In European Common Market countries, the symbol is EEx. In the IEC standards, on which American and European standards are based, the symbol is Ex.

Exception No. 1: Associated apparatus NOT suitable for installation in a hazardous (classified) locations shall be required to be marked only with (3), (4), and (5), but BOTH the symbol AEx (3) and the symbol for the type of protection (4) shall be enclosed within the same square brackets, for example, [AEx ia] IIC.

Exception No. 2: Simple apparatus as defined in 504.2 shall not be required to have a marked operating temperature or temperature class.

Electrical equipment of types of protection "e," "m," "p," or "q" shall be marked Group II. Electrical equipment

Table 505.9(C)(2)(4) Types of Protection Designation

Designation	Technique	Zone*
d	Flameproof enclosure	1
e	Increased safety	1
ia	Intrinsic safety	0
ib	Intrinsic safety	1
[ia]	Associated apparatus	Unclassified
[ib]	Associated apparatus	Unclassified
m	Encapsulation	1
nA	Nonsparking equipment	2
nC	Sparking equipment in which the contacts are suitably protected other than by restricted breathing enclosure	2
nR	Restricted breathing enclosure	2
o	Oil immersion	1
p	Purged and pressurized	1 or 2
q	Powder filled	1

*Does not address use where a combination of techniques is used.

of types of protection "d," "ia," "ib," "[ia]," or "[ib]" shall be marked Group IIA, IIB, or IIC, or for a specific gas or vapor. Electrical equipment of types of protection "n" shall be marked Group II unless it contains enclosed-break devices, nonincendive components, or energy-limited equipment or circuits, in which case it shall be marked Group IIA, IIB, or IIC, or a specific gas or vapor. Electrical equipment of other types of protection shall be marked Group II unless the type of protection utilized by the equipment requires that it be marked Group IIA, IIB, or IIC, or a specific gas or vapor.

FPN No. 1: An example of the required marking for intrinsically safe apparatus for installation in Class I, Zone 0 is "Class I, Zone 0, AEx ia IIC T6." An explanation of the marking that is required is shown in FPN Figure 505.9(C)(2).

FPN Figure 505.9(C)(2) Zone Equipment Marking

FPN No. 2: An example of the required marking for intrinsically safe associated apparatus mounted in a flameproof enclosure for installation in Class I, Zone 1 is "Class I, Zone 1 AEx d[ia] IIC T4."

FPN No. 3: An example of the required marking for intrinsically safe associated apparatus NOT for installation in a hazardous (classified) location is "[AEx ia] IIC."

(D) Class I Temperature The temperature marking specified below shall not exceed the ignition temperature of the specific gas or vapor to be encountered.

FPN: For information regarding ignition temperatures of gases and vapors, see NFPA 497-2004, *Recommended Practice for the Classification of Flammable Liquids, Gases, or Vapors and of Hazardous (Classified) Locations for Electrical Installations in Chemical Process Areas*; and IEC 60079-20-1996, *Electrical Apparatus for Explosive Gas Atmospheres, Data for Flammable Gases and Vapours, Relating to the Use of Electrical Apparatus.*

(1) Temperature Classifications Equipment shall be marked to show the operating temperature or temperature class referenced to a 40°C (104°F) ambient. The temperature class, if provided, shall be indicated using the temperature class (T Code) shown in Table 505.9(D)(1).

Table 505.9(D)(1) Classification of Maximum Surface Temperature for Group II Electrical Equipment

Temperature Class (T Code)	Maximum Surface Temperature (°C)
T1	≤450
T2	≤300
T3	≤200
T4	≤135
T5	≤100
T6	≤85

Electrical equipment designed for use in the ambient temperature range between −20°C and +40°C shall require no additional ambient temperature marking.

Electrical equipment that is designed for use in a range of ambient temperatures other than −20°C and +40°C is considered to be special; and the ambient temperature range shall then be marked on the equipment, including either the symbol "Ta" or "Tamb" together with the special range of ambient temperatures. As an example, such a marking might be " −30°C ≤ Ta ≤ +40°C."

Electrical equipment suitable for ambient temperatures exceeding 40°C (104°F) shall be marked with both the maximum ambient temperature and the operating temperature or temperature class at that ambient temperature.

Exception No. 1: Equipment of the non–heat-producing type, such as conduit fittings, and equipment of the heat-producing type having a maximum temperature of not more than 100°C (212°F) shall not be required to have a marked operating temperature or temperature class.

Exception No. 2: Equipment identified for Class I, Division 1 or Division 2 locations as permitted by 505.20(B) and 505.20(D) shall be permitted to be marked in accordance with 500.8(B) and Table 500.8(B).

(E) Threading All NPT threaded conduit and fittings referred to herein shall be threaded with a National (American) Standard Pipe Taper (NPT) thread that provides a taper of 1 in 16 (¾-in. taper per foot). Conduit and fittings shall be made wrenchtight to prevent sparking when fault current flows through the conduit system, and to ensure the explosionproof or flameproof integrity of the conduit system where applicable. Equipment provided with threaded entries for field wiring connections shall be installed in accordance with 505.9(E)(1) or 505.9(E)(2). Threaded entries into explosionproof or flameproof equipment shall be made up with at least five threads fully engaged.

Exception: For listed explosionproof or flameproof equipment, factory threaded NPT entries shall be made up with at least 4½ threads fully engaged.

(1) Equipment Provided with Threaded Entries for NPT Threaded Conduit or Fittings For equipment provided with threaded entries for NPT threaded conduit or fittings, listed conduit fittings or cable fittings shall be used.

> FPN: Thread form specifications for NPT threads are located in ANSI/ASME B1.20.1-1983, *Pipe Threads, General Purpose (Inch).*

(2) Equipment Provided with Threaded Entries for Metric Threaded Conduit or Fittings For equipment with metric threaded entries, such entries shall be identified as being metric, or listed adapters to permit connection to conduit or NPT-threaded fittings shall be provided with the equipment. Adapters shall be used for connection to conduit or NPT-threaded fittings. Listed cable fittings that have metric threads shall be permitted to be used.

See the commentary on 500.8(D)(2) FPN for information on metric threads. Exhibit 505.2 is an example of an adapter that provides a means of connecting conduit or fitting with NPT threads to an "increased safety" enclosure that has metric threads.

Exhibit 505.2 A typical hub providing an NPT threaded entry for conduit or cable into an increased safety enclosure. (Courtesy of Cooper Crouse-Hinds)

> FPN: Threading specifications for metric threaded entries are located in ISO 965/1-1980, *Metric Screw Threads*; and ISO 965/3-1980, *Metric Screw Threads.*

505.15 Wiring Methods

Wiring methods shall maintain the integrity of protection techniques and shall comply with 505.15(A) through 505.15 (C).

(A) Class I, Zone 0 In Class I, Zone 0 locations, only intrinsically safe wiring methods in accordance with Article 504 shall be permitted.

> FPN: Article 504 only includes protection technique "ia."

The provision in 505.15(A) is one of the most significant differences between the zone and division area classification requirements. The degree of hazard within a Zone 0 area is considered so severe that all wiring is this area must be intrinsically safe. In general, only instrumentation and signaling circuits installed in accordance with Article 504 can be used in a Zone 0 area. Explosionproof power utilization equipment, such as motors and luminaires, is not permitted in Class I, Zone 0 locations.

(B) Class I, Zone 1

(1) General In Class I, Zone 1 locations, the wiring methods in (B)(1)(a) through (B)(1)(f) shall be permitted.

(a) All wiring methods permitted by 505.15(A).

(b) In industrial establishments with restricted public access, where the conditions of maintenance and supervision ensure that only qualified persons service the installation, and where the cable is not subject to physical damage, Type MC-HL cable listed for use in Class I, Zone 1 or Division 1 locations, with a gas/vaportight continuous corrugated metallic sheath, an overall jacket of suitable polymeric material, separate grounding conductors in accordance with 250.122, and provided with termination fittings listed for the application.

> FPN: See 330.12 for restrictions on use of Type MC cable.

(c) In industrial establishments with restricted public access, where the conditions of maintenance and supervision ensure that only qualified persons service the installation, and where the cable is not subject to physical damage, Type ITC-HL cable, listed for use in Class I, Zone 1 or Division 1 locations, with a gas/vaportight continuous corrugated metallic sheath, an overall jacket of suitable polymeric material and provided with termination fittings listed for the application.

(d) Type MI cable with termination fittings listed for Class I, Zone 1 or Division 1 locations. Type MI cable shall be installed and supported in a manner to avoid tensile stress at the termination fittings.

(e) Threaded rigid metal conduit, or threaded steel intermediate metal conduit.

(f) Rigid nonmetallic conduit complying with Article 352 shall be permitted where encased in a concrete envelope a minimum of 50 mm (2 in.) thick and provided with not less than 600 mm (24 in.) of cover measured from the top of the conduit to grade. Threaded rigid metal conduit or threaded steel intermediate metal conduit shall be used for the last 600 mm (24 in.) of the underground run to emergence or to the point of connection to the aboveground raceway. An equipment grounding conductor shall be included to provide for electrical continuity of the raceway system and for grounding of non–current-carrying metal parts.

(2) Flexible Connections Where necessary to employ flexible connections, flexible fittings listed for Class I, Zone 1 or Division 1 locations or flexible cord in accordance with the provisions of 505.17 shall be permitted.

(C) Class I, Zone 2

(1) General In Class I, Zone 2 locations, the wiring methods in (C)(1)(a) through (C)(1)(g) shall be permitted.

(a) All wiring methods permitted by 505.15(B).

(b) Types MI, MC, MV, or TC cable with termination fittings, or in cable tray systems and installed in a manner to avoid tensile stress at the termination fittings. Single conductor Type MV cables shall be shielded or metallic-armored.

(c) Type ITC cable as permitted in 727.4.

(d) Type PLTC cable in accordance with the provisions of Article 725, or in cable tray systems. PLTC shall be installed in a manner to avoid tensile stress at the termination fittings.

(e) Enclosed gasketed busways, enclosed gasketed wireways.

(f) Threaded rigid metal conduit, threaded steel intermediate metal conduit.

(g) Nonincendive field wiring shall be permitted using any of the wiring methods permitted for unclassified locations. Nonincendive field wiring systems shall be installed in accordance with the control drawing(s). Simple apparatus, not shown on the control drawing, shall be permitted in a nonincendive field wiring circuit, provided the simple apparatus does not interconnect the nonincendive field wiring circuit to any other circuit.

FPN: Simple apparatus is defined in 504.2.

Separate nonincendive field wiring circuits shall be installed in accordance with one of the following:

(1) In separate cables
(2) In multiconductor cables where the conductors of each circuit are within a grounded metal shield
(3) In multiconductor cables where the conductors of each circuit have insulation with a minimum thickness of 0.25 mm (0.01 in.)

(2) Flexible Connections Where provision must be made for limited flexibility, flexible metal fittings, flexible metal conduit with listed fittings, liquidtight flexible metal

with listed fittings, liquidtight flexible nonmetallic conduit with listed fittings, or flexible cord in accordance with the provisions of 505.17 shall be permitted.

FPN: See 505.25(B) for grounding requirements where flexible conduit is used.

505.16 Sealing and Drainage

Seals in conduit and cable systems shall comply with 505.16(A) through 505.16(E). Sealing compound shall be used in Type MI cable termination fittings to exclude moisture and other fluids from the cable insulation.

FPN No. 1: Seals are provided in conduit and cable systems to minimize the passage of gases and vapors and prevent the passage of flames from one portion of the electrical installation to another through the conduit. Such communication through Type MI cable is inherently prevented by construction of the cable. Unless specifically designed and tested for the purpose, conduit and cable seals are not intended to prevent the passage of liquids, gases, or vapors at a continuous pressure differential across the seal. Even at differences in pressure across the seal equivalent to a few inches of water, there may be a slow passage of gas or vapor through a seal and through conductors passing through the seal. See 505.16(C)(2)(b). Temperature extremes and highly corrosive liquids and vapors can affect the ability of seals to perform their intended function. See 505.16(D)(2).

FPN No. 2: Gas or vapor leakage and propagation of flames may occur through the interstices between the strands of standard stranded conductors larger than 2 AWG. Special conductor constructions, for example, compacted strands or sealing of the individual strands, are means of reducing leakage and preventing the propagation of flames.

(A) Zone 0 In Class I, Zone 0 locations, seals shall be located according to 505.16(A)(1), (A)(2), and (A)(3).

(1) Conduit Seals Seals shall be provided within 3.05 m (10 ft) of where a conduit leaves a Zone 0 location. There shall be no unions, couplings, boxes, or fittings, except listed reducers at the seal, in the conduit run between the seal and the point at which the conduit leaves the location.

Exception: A rigid unbroken conduit that passes completely through the Zone 0 location with no fittings less than 300 mm (12 in.) beyond each boundary shall not be required to be sealed if the termination points of the unbroken conduit are in unclassified locations.

(2) Cable Seals Seals shall be provided on cables at the first point of termination after entry into the Zone 0 location.

(3) Not Required to Be Explosionproof or Flameproof Seals shall not be required to be explosionproof or flameproof.

(B) Zone 1 In Class I, Zone 1 locations, seals shall be located in accordance with 505.16(B)(1) through (B)(8).

(1) Type of Protection "d" or "e" Enclosures Conduit seals shall be provided within 50 mm (2 in.) for each conduit entering enclosures having type of protection "d" or "e."

Exception No. 1: Where the enclosure having type of protection "d" is marked to indicate that a seal is not required.

Exception No. 2: For type of protection "e," conduit and fittings employing only NPT to NPT raceway joints or fittings listed for type of protection "e" shall be permitted between the enclosure and the seal, and the seal shall not be required to be within 50 mm (2 in.) of the entry.

> FPN: Examples of fittings employing other than NPT threads include conduit couplings, capped elbows, unions, and breather drains.

Exception No. 3: For conduit installed between type of protection "e" enclosures employing only NPT to NPT raceway joints or conduit fittings listed for type of protection "e," a seal shall not be required.

(2) Explosionproof Equipment Conduit seals shall be provided for each conduit entering explosionproof equipment according to (B)(2)(a), (B)(2)(b), and (B)(2)(c).

(a) In each conduit entry into an explosionproof enclosure where either (1) the enclosure contains apparatus, such as switches, circuit breakers, fuses, relays, or resistors, that may produce arcs, sparks, or high temperatures that are considered to be an ignition source in normal operation, or (2) the entry is metric designator 53 (trade size 2) or larger and the enclosure contains terminals, splices, or taps. For the purposes of this section, high temperatures shall be considered to be any temperatures exceeding 80 percent of the autoignition temperature in degrees Celsius of the gas or vapor involved.

Exception: Conduit entering an enclosure where such switches, circuit breakers, fuses, relays, or resistors comply with one of the following:

(1) Are enclosed within a chamber hermetically sealed against the entrance of gases or vapors.
(2) Are immersed in oil.
(3) Are enclosed within a factory-sealed explosionproof chamber located within the enclosure, identified for the location, and marked "factory sealed" or equivalent, unless the entry is metric designator 53 (trade size 2) or larger. Factory-sealed enclosures shall not be considered to serve as a seal for another adjacent explosionproof enclosure that is required to have a conduit seal.

(b) Conduit seals shall be installed within 450 mm (18 in.) from the enclosure. Only explosionproof unions, couplings, reducers, elbows, capped elbows, and conduit bodies similar to L, T, and cross types that are not larger than the

trade size of the conduit shall be permitted between the sealing fitting and the explosionproof enclosure.

(c) Where two or more explosionproof enclosures for which conduit seals are required under 505.16(B)(2) are connected by nipples or by runs of conduit not more than 900 mm (36 in.) long, a single conduit seal in each such nipple connection or run of conduit shall be considered sufficient if located not more than 450 mm (18 in.) from either enclosure.

(3) Pressurized Enclosures Conduit seals shall be provided in each conduit entry into a pressurized enclosure where the conduit is not pressurized as part of the protection system. Conduit seals shall be installed within 450 mm (18 in.) from the pressurized enclosure.

> FPN No. 1: Installing the seal as close as possible to the enclosure reduces problems with purging the dead airspace in the pressurized conduit.
>
> FPN No. 2: For further information, see NFPA 496-2003, *Standard for Purged and Pressurized Enclosures for Electrical Equipment.*

(4) Class I, Zone 1 Boundary Conduit seals shall be provided in each conduit run leaving a Class I, Zone 1 location. The sealing fitting shall be permitted on either side of the boundary of such location within 3.05 m (10 ft) of the boundary and shall be designed and installed so as to minimize the amount of gas or vapor within the Zone 1 portion of the conduit from being communicated to the conduit beyond the seal. Except for listed explosionproof reducers at the conduit seal, there shall be no union, coupling, box, or fitting between the conduit seal and the point at which the conduit leaves the Zone 1 location.

Exception: Metal conduit containing no unions, couplings, boxes, or fittings and passing completely through a Class I, Zone 1 location with no fittings less than 300 mm (12 in.) beyond each boundary shall not require a conduit seal if the termination points of the unbroken conduit are in unclassified locations.

(5) Cables Capable of Transmitting Gases or Vapors Conduits containing cables with a gas/vaportight continuous sheath capable of transmitting gases or vapors through the cable core shall be sealed in the Zone 1 location after removing the jacket and any other coverings so that the sealing compound surrounds each individual insulated conductor and the outer jacket.

Exception: Multiconductor cables with a gas/vaportight continuous sheath capable of transmitting gases or vapors through the cable core shall be permitted to be considered as a single conductor by sealing the cable in the conduit within 450 mm (18 in.) of the enclosure and the cable end within the enclosure by an approved means to minimize the entrance of gases or vapors and prevent the propagation of flame into the cable core, or by other approved methods. For shielded cables and twisted pair cables, it shall not be required to remove the shielding material or separate the twisted pair.

(6) Cables Incapable of Transmitting Gases or Vapors Each multiconductor cable in conduit shall be considered as a single conductor if the cable is incapable of transmitting gases or vapors through the cable core. These cables shall be sealed in accordance with 505.16(D).

(7) Cables Entering Enclosures Cable seals shall be provided for each cable entering flameproof or explosionproof enclosures. The seal shall comply with 505.16(D).

(8) Class I, Zone 1 Boundary Cables shall be sealed at the point at which they leave the Zone 1 location.

Exception: Where cable is sealed at the termination point.

(C) Zone 2 In Class I, Zone 2 locations, seals shall be located in accordance with 505.16(C)(1) and (C)(2).

(1) Conduit Seals Conduit seals shall be located in accordance with (C)(1)(a) and (C)(1)(b).

(a) For connections to enclosures that are required to be flameproof or explosionproof, a conduit seal shall be provided in accordance with 505.16(B)(1) and 505.16(B)(2). All portions of the conduit run or nipple between the seal and such enclosure shall comply with 505.16(B).

(b) In each conduit run passing from a Class I, Zone 2 location into an unclassified location. The sealing fitting shall be permitted on either side of the boundary of such location within 3.05 m (10 ft) of the boundary and shall be designed and installed so as to minimize the amount of gas or vapor within the Zone 2 portion of the conduit from being communicated to the conduit beyond the seal. Rigid metal conduit or threaded steel intermediate metal conduit shall be used between the sealing fitting and the point at which the conduit leaves the Zone 2 location, and a threaded connection shall be used at the sealing fitting. Except for listed explosionproof reducers at the conduit seal, there shall be no union, coupling, box, or fitting between the conduit seal and the point at which the conduit leaves the Zone 2 location.

Exception No. 1: Metal conduit containing no unions, couplings, boxes, or fittings and passing completely through a Class I, Zone 2 location with no fittings less than 300 mm (12 in.) beyond each boundary shall not be required to be sealed if the termination points of the unbroken conduit are in unclassified locations.

Exception No. 2: Conduit systems terminating at an unclassified location where a wiring method transition is made to cable tray, cablebus, ventilated busway, Type MI cable, or cable that is not installed in a raceway or cable tray system

shall not be required to be sealed where passing from the Class I, Zone 2 location into the unclassified location. The unclassified location shall be outdoors or, if the conduit system is all in one room, it shall be permitted to be indoors. The conduits shall not terminate at an enclosure containing an ignition source in normal operation.

Exception No. 3: Conduit systems passing from an enclosure or room that is unclassified as a result of pressurization into a Class I, Zone 2 location shall not require a seal at the boundary.

> FPN: For further information, refer to NFPA 496-2003, *Standard for Purged and Pressurized Enclosures for Electrical Equipment.*

Exception No. 4: Segments of aboveground conduit systems shall not be required to be sealed where passing from a Class I, Zone 2 location into an unclassified location if all the following conditions are met:

(1) No part of the conduit system segment passes through a Class I, Zone 0 or Class I, Zone 1 location where the conduit contains unions, couplings, boxes, or fittings within 300 mm (12 in.) of the Class I, Zone 0 or Class I, Zone 1 location.

(2) The conduit system segment is located entirely in outdoor locations.

(3) The conduit system segment is not directly connected to canned pumps, process or service connections for flow, pressure, or analysis measurement, and so forth, that depend on a single compression seal, diaphragm, or tube to prevent flammable or combustible fluids from entering the conduit system.

(4) The conduit system segment contains only threaded metal conduit, unions, couplings, conduit bodies, and fittings in the unclassified location.

(5) The conduit system segment is sealed at its entry to each enclosure or fitting housing terminals, splices, or taps in Class I, Zone 2 locations.

(2) Cable Seals Cable seals shall be located in accordance with (C)(2)(a), (C)(2)(b), and (C)(2)(c).

(a) Explosionproof and Flameproof Enclosures. Cables entering enclosures required to be flameproof or explosionproof shall be sealed at the point of entrance. The seal shall comply with 505.16(D). Multiconductor cables with a gas/vaportight continuous sheath capable of transmitting gases or vapors through the cable core shall be sealed in the Zone 2 location after removing the jacket and any other coverings so that the sealing compound surrounds each individual insulated conductor in such a manner as to minimize the passage of gases and vapors. Multiconductor cables in conduit shall be sealed as described in 505.16(B)(4).

Exception No. 1: Cables passing from an enclosure or room that is unclassified as a result of Type Z pressurization into a Class I, Zone 2 location shall not require a seal at the boundary.

Exception No. 2: Shielded cables and twisted pair cables shall not require the removal of the shielding material or separation of the twisted pairs, provided the termination is by an approved means to minimize the entrance of gases or vapors and prevent propagation of flame into the cable core.

(b) **Cables That Will Not Transmit Gases or Vapors.** Cables with a gas/vaportight continuous sheath and that will not transmit gases or vapors through the cable core in excess of the quantity permitted for seal fittings shall not be required to be sealed except as required in 505.16(C)(2)(a). The minimum length of such cable run shall not be less than the length that limits gas or vapor flow through the cable core to the rate permitted for seal fittings [200 cm^3/hr (0.007 ft^3/hr) of air at a pressure of 1500 pascals (6 in. of water)].

> FPN No. 1: See ANSI/UL 886-1994, *Outlet Boxes and Fittings for Use in Hazardous (Classified) Locations.*
> FPN No. 2: The cable core does not include the interstices of the conductor strands.

(c) **Cables Capable of Transmitting Gases or Vapors.** Cables with a gas/vaportight continuous sheath capable of transmitting gases or vapors through the cable core shall not be required to be sealed except as required in 505.16(C)(2)(a), unless the cable is attached to process equipment or devices that may cause a pressure in excess of 1500 pascals (6 in. of water) to be exerted at a cable end, in which case a seal, barrier, or other means shall be provided to prevent migration of flammables into an unclassified area.

Exception: Cables with an unbroken gas/vaportight continuous sheath shall be permitted to pass through a Class I, Zone 2 location without seals.

(d) **Cables Without Gas/Vaportight Continuous Sheath.** Cables that do not have gas/vaportight continuous sheath shall be sealed at the boundary of the Zone 2 and unclassified location in such a manner as to minimize the passage of gases or vapors into an unclassified location.

> FPN: The cable sheath may be either metal or a nonmetallic material.

(D) Class I, Zones 0, 1, and 2 Where required, seals in Class I, Zones 0, 1, and 2 locations shall comply with 505.16(D)(1) through (D)(5).

(1) Fittings Enclosures for connections or equipment shall be provided with an integral means for sealing, or sealing fittings listed for the location shall be used. Sealing fittings shall be listed for use with one or more specific compounds and shall be accessible.

(2) Compound The compound shall provide a seal against passage of gas or vapors through the seal fitting, shall not

be affected by the surrounding atmosphere or liquids, and shall not have a melting point less than 93°C (200°F).

(3) Thickness of Compounds In a completed seal, the minimum thickness of the sealing compound shall not be less than the trade size of the sealing fitting and, in no case, less than 16 mm (⅝ in.).

Exception: Listed cable sealing fittings shall not be required to have a minimum thickness equal to the trade size of the fitting.

(4) Splices and Taps Splices and taps shall not be made in fittings intended only for sealing with compound, nor shall other fittings in which splices or taps are made be filled with compound.

(5) Conductor Fill The cross-sectional area of the conductors permitted in a seal shall not exceed 25 percent of the cross-sectional area of a rigid metal conduit of the same trade size unless it is specifically listed for a higher percentage of fill.

(E) Drainage

(1) Control Equipment Where there is a probability that liquid or other condensed vapor may be trapped within enclosures for control equipment or at any point in the raceway system, approved means shall be provided to prevent accumulation or to permit periodic draining of such liquid or condensed vapor.

(2) Motors and Generators Where the authority having jurisdiction judges that there is a probability that liquid or condensed vapor may accumulate within motors or generators, joints and conduit systems shall be arranged to minimize entrance of liquid. If means to prevent accumulation or to permit periodic draining are judged necessary, such means shall be provided at the time of manufacture and shall be considered an integral part of the machine.

(3) Canned Pumps, Process, or Service Connections, and So Forth For canned pumps, process, or service connections for flow, pressure, or analysis measurement, and so forth, that depend on a single compression seal, diaphragm, or tube to prevent flammable or combustible fluids from entering the electrical conduit system, an additional approved seal, barrier, or other means shall be provided to prevent the flammable or combustible fluid from entering the conduit system beyond the additional devices or means if the primary seal fails.

The additional approved seal or barrier and the interconnecting enclosure shall meet the temperature and pressure conditions to which they will be subjected upon failure of the primary seal, unless other approved means are provided to accomplish the purpose in the preceding paragraph.

Drains, vents, or other devices shall be provided so that primary seal leakage is obvious.

FPN: See also the fine print notes to 505.16.

Process-connected equipment that is listed and marked "Dual Seal" shall not require additional process sealing when used within the manufacturer's ratings.

FPN: For construction and testing requirements for dual seal process, connected equipment, refer to ISA 12.27.01, *Requirements for Process Sealing Between Electrical Systems and Potentially Flammable or Combustible Process Fluids.*

505.17 Flexible Cords, Class I, Zones 1 and 2

A flexible cord shall be permitted for connection between portable lighting equipment or other portable utilization equipment and the fixed portion of their supply circuit. Flexible cord shall also be permitted for that portion of the circuit where the fixed wiring methods of 505.15(B) cannot provide the necessary degree of movement for fixed and mobile electrical utilization equipment, in an industrial establishment where conditions of maintenance and engineering supervision ensure that only qualified persons install and service the installation, and the flexible cord is protected by location or by a suitable guard from damage. The length of the flexible cord shall be continuous. Where flexible cords are used, the cords shall comply with all of the following:

(1) Be of a type listed for extra-hard usage
(2) Contain, in addition to the conductors of the circuit, a grounding conductor complying with 400.23
(3) Be connected to terminals or to supply conductors in an approved manner
(4) Be supported by clamps or by other suitable means in such a manner that there will be no tension on the terminal connections
(5) Be provided with listed seals where the flexible cord enters boxes, fittings, or enclosures that are required to be explosionproof or flameproof

Exception: As provided in 505.16.

Electric submersible pumps with means for removal without entering the wet-pit shall be considered portable utilization equipment. The extension of the flexible cord within a suitable raceway between the wet-pit and the power source shall be permitted.

Electric mixers intended for travel into and out of open-type mixing tanks or vats shall be considered portable utilization equipment.

FPN: See 505.18 for flexible cords exposed to liquids having a deleterious effect on the conductor insulation.

505.18 Conductors and Conductor Insulation

(A) **Conductors** For type of protection "e," field wiring conductors shall be copper. Every conductor (including

spares) that enters Type "e" equipment shall be terminated at a Type "e" terminal.

(B) **Conductor Insulation** Where condensed vapors or liquids may collect on, or come in contact with, the insulation on conductors, such insulation shall be of a type identified for use under such conditions, or the insulation shall be protected by a sheath of lead or by other approved means.

505.19 Uninsulated Exposed Parts

There shall be no uninsulated exposed parts, such as electric conductors, buses, terminals, or components that operate at more than 30 volts (15 volts in wet locations). These parts shall additionally be protected by type of protection ia, ib, or nA that is suitable for the location.

505.20 Equipment Requirements

(A) **Zone 0** In Class I, Zone 0 locations, only equipment specifically listed and marked as suitable for the location shall be permitted.

Exception: Intrinsically safe apparatus listed for use in Class I, Division 1 locations for the same gas, or as permitted by 505.9(B)(2), and with a suitable temperature class shall be permitted.

The exception to 505.20(A) results from the fact that ANSI/UL 913, the standard used to evaluate intrinsically safe systems for Class I, Division 1 locations, is based on the IEC requirements for intrinsically safe equipment for Class I, Zone 0 locations.

(B) **Zone 1** In Class I, Zone 1 locations, only equipment specifically listed and marked as suitable for the location shall be permitted.

Exception No. 1: Equipment identified for use in Class I, Division 1 or listed for use in Class I, Zone 0 locations for the same gas, or as permitted by 505.9(B)(2), and with a suitable temperature class shall be permitted.

Exception No. 2: Equipment identified for Class I, Zone 1, or Zone 2 type of protection "p" shall be permitted.

(C) **Zone 2** In Class I, Zone 2 locations, only equipment specifically listed and marked as suitable for the location shall be permitted.

Exception No. 1: Equipment listed for use in Class I, Zone 0 or Zone 1 locations for the same gas, or as permitted by 505.9(B)(2), and with a suitable temperature class, shall be permitted.

Exception No. 2: Equipment identified for Class I, Zone 1 or Zone 2 type of protection "p" shall be permitted.

Exception No. 3: Equipment identified for use in Class I, Division 1 or Division 2 locations for the same gas, or as permitted by 505.9(B)(2), and with a suitable temperature class shall be permitted.

Exception No. 4: In Class I, Zone 2 locations, the installation of open or nonexplosionproof or nonflameproof enclosed motors, such as squirrel-cage induction motors without brushes, switching mechanisms, or similar arc-producing devices that are not identified for use in a Class I, Zone 2 location shall be permitted.

> FPN No. 1: It is important to consider the temperature of internal and external surfaces that may be exposed to the flammable atmosphere.
>
> FPN No. 2: It is important to consider the risk of ignition due to currents arcing across discontinuities and overheating of parts in multisection enclosures of large motors and generators. Such motors and generators may need equipotential bonding jumpers across joints in the enclosure and from enclosure to ground. Where the presence of ignitible gases or vapors is suspected, clean air purging may be needed immediately prior to and during start-up periods.

See the commentary following 501.125(B), FPN No. 2, for more information on electric motors installed in hazardous (classified) locations.

(D) Manufacturer's Instructions Electrical equipment installed in hazardous (classified) locations shall be installed in accordance with the instructions (if any) provided by the manufacturer.

505.21 Multiwire Branch Circuits

In a Class I, Zone 1 location, a multiwire branch circuit shall not be permitted.

Exception: Where the disconnect device(s) for the circuit opens all ungrounded conductors of the multiwire circuit simultaneously.

505.22 Increased Safety "e" Motors and Generators

In Class I, Zone 1 locations, Increased Safety "e" motors and generators of all voltage ratings shall be listed for Class I, Zone 1 locations, and shall comply with all of the following:

(1) Motors shall be marked with the current ratio, I_A/I_N, and time, t_E.
(2) Motors shall have controllers marked with the model or identification number, output rating (horsepower or kilowatt), full-load amperes, starting current ratio (I_A/I_N), and time (t_E) of the motors that they are intended to protect; the controller marking shall also include the specific overload protection type (and setting, if applicable) that is listed with the motor or generator.
(3) Connections shall be made with the specific terminals listed with the motor or generator.
(4) Terminal housings shall be permitted to be of substantial, nonmetallic, nonburning material, provided an in-

ternal grounding means between the motor frame and the equipment grounding connection is incorporated within the housing.

(5) The provisions of Part III of Article 430 shall apply regardless of the voltage rating of the motor.
(6) The motors shall be protected against overload by a separate overload device that is responsive to motor current. This device shall be selected to trip or shall be rated in accordance with the listing of the motor and its overload protection.
(7) Sections 430.32(C) and 430.44 shall not apply to such motors.
(8) The motor overload protection shall not be shunted or cut out during the starting period.

505.25 Grounding and Bonding

Grounding and bonding shall comply with Article 250 and the requirements in 505.25(A) and 505.25(B).

(A) Bonding The locknut-bushing and double-locknut types of contacts shall not be depended on for bonding purposes, but bonding jumpers with proper fittings or other approved means of bonding shall be used. Such means of bonding shall apply to all intervening raceways, fittings, boxes, enclosures, and so forth, between Class I locations and the point of grounding for service equipment or point of grounding of a separately derived system.

Exception: The specific bonding means shall be required only to the nearest point where the grounded circuit conductor and the grounding electrode are connected together on the line side of the building or structure disconnecting means as specified in 250.32(A), (B), and (C), provided the branch-circuit overcurrent protection is located on the load side of the disconnecting means.

> FPN: See 250.100 for additional bonding requirements in hazardous (classified) locations.

(B) Types of Equipment Grounding Conductors Where flexible metal conduit or liquidtight flexible metal conduit is used as permitted in 505.15(C) and is to be relied on to complete a sole equipment grounding path, it shall be installed with internal or external bonding jumpers in parallel with each conduit and complying with 250.102.

Exception: In Class I, Zone 2 locations, the bonding jumper shall be permitted to be deleted where all of the following conditions are met:

(a) Listed liquidtight flexible metal conduit 1.8 m (6 ft) or less in length, with fittings listed for grounding, is used.

(b) Overcurrent protection in the circuit is limited to 10 amperes or less.

(c) The load is not a power utilization load.

See the commentary for 501.30 for information on grounding and bonding requirements in Class I locations.

ARTICLE 506
Zone 20, 21, and 22 Locations for Combustible Dusts, Fibers, and Flyings

Summary of Changes

- Added article providing an alternative zone (IEC-based) system for locations that include combustible concentrations of combustible dust, or of ignitible fibers and flyings.

Contents

506.1 Scope

This article covers the requirements for the zone classification system as an alternative to the division classification system covered in Article 500, Article 502, and Article 503 for electrical and electronic equipment and wiring for all voltages in Zone 20, Zone 21, and Zone 22 hazardous (classified) locations where fire and explosion hazards may exist due to combustible dusts, or ignitible fibers or flyings. Combustible metallic dusts are not covered by the requirements of this article.

> FPN No. 1: For the requirements for electrical and electronic equipment and wiring for all voltages in Class I, Division 1 or Division 2; Class II, Division 1 or Division 2; Class III, Division 1 or Division 2; and Class I, Zone 0 or Zone 1 or Zone 2 hazardous (classified) locations where fire or explosion hazards may exist due to flammable gases or vapors, flammable liquids, or combustible dusts or fibers, refer to Articles 500 through 505.
>
> FPN No. 2: Zone 20, Zone 21, and Zone 22 area classifications are based on the modified IEC area classification system as defined in ISA 12.10.05, *Electrical Apparatus for Use in Zone 20, Zone 21, and Zone 22 Hazardous (Classified) Locations — Classification of Zone 20, Zone 21, and Zone 22 Hazardous (Classified) Locations (IEC61241-10 Mod)*.
>
> FPN No. 3: The unique hazards associated with explosives, pyrotechnics, and blasting agents are not addressed in this article.

506.2 Definitions

For purposes of this article, the following definitions apply.

Associated Nonincendive Field Wiring Apparatus. Apparatus in which the circuits are not necessarily nonincendive themselves but that affect the energy in nonincendive field wiring circuits and are relied upon to maintain nonincendive energy levels. Associated nonincendive field wiring apparatus may be either of the following:

(1) Electrical apparatus that has an alternative type of protection for use in the appropriate hazardous (classified) location

(2) Electrical apparatus not so protected that shall not be used in a hazardous (classified) location

FPN: Associated nonincendive field wiring apparatus has designated associated nonincendive field wiring apparatus connections for nonincendive field wiring apparatus and may also have connections for other electrical apparatus.

Dust-Ignitionproof. Equipment enclosed in a manner that excludes dusts and does not permit arcs, sparks, or heat otherwise generated or liberated inside of the enclosure to cause ignition of exterior accumulations or atmospheric suspensions of a specified dust on or in the vicinity of the enclosure.

FPN: For further information on dust-ignitionproof enclosures, see Type 9 enclosure in ANSI/NEMA 250-1991, *Enclosures for Electrical Equipment*, and ANSI/ UL 1203-1994, *Explosionproof and Dust-Ignitionproof Electrical Equipment for Hazardous (Classified) Locations*.

Dusttight. Enclosures constructed so that dust will not enter under specified test conditions.

FPN No. 1: See ANSI/ISA 12.12.01-2000, *Nonincendive Electrical Equipment for Use in Class I and II, Division 2, and Class III, Divisions 1 and 2 Hazardous (Classified) Locations*, and UL 1604-1994, *Electrical Equipment for Use in Class I and II, Division 2 and Class III Hazardous (Classified) Locations*.

Nonincendive Circuit. A circuit, other than field wiring, in which any arc or thermal effect produced under intended operating conditions of the equipment is not capable, under specified test conditions, of igniting the flammable gas–air, vapor–air, or dust–air mixture.

FPN: Conditions are described in ANSI/ISA 12.12.01-2000, *Nonincendive Electrical Equipment for Use in Class I and II, Division 2, and Class III, Divisions 1 and 2 Hazardous (Classified) Locations*.

Nonincendive Equipment. Equipment having electrical/ electronic circuitry that is incapable, under normal operating conditions, of causing ignition of a specified flammable gas–air, vapor–air, or dust–air mixture due to arcing or thermal means.

FPN: Conditions are described in ANSI/ISA 12.12.01-2000, *Nonincendive Electrical Equipment for Use in Class I and II, Division 2, and Class III, Divisions 1 and 2 Hazardous (Classified) Locations*.

Nonincendive Field Wiring. Wiring that enters or leaves an equipment enclosure and, under normal operating conditions of the equipment, is not capable, due to arcing or thermal effects, of igniting the flammable gas–air, vapor–air, or dust–air mixture. Normal operation includes opening, shorting, or grounding the field wiring.

Nonincendive Field Wiring Apparatus. Apparatus intended to be connected to nonincendive field wiring.

FPN: Conditions are described in ANSI/ISA 12.12.01-2000, *Nonincendive Electrical Equipment for Use in Class I and II, Division 2, and Class III, Divisions 1 and 2 Hazardous (Classified) Locations*.

Pressurized. The process of supplying an enclosure with a protective gas with or without continuous flow at sufficient pressure to prevent the entrance of combustible dust, or an ignitible fiber or flying.

FPN: For further information, see ANSI/ NFPA 496-2003, *Purged and Pressurized Enclosures for Electrical Equipment*.

Zone 20 Hazardous (Classified) Location. An area where combustible dust or ignitible fibers and flyings are present continuously or for long periods of time in quantities sufficient to be hazardous, as classified by 506.5(B)(1).

Zone 21 Hazardous (Classified) Location. An area where combustible dust or ignitible fibers and flyings are likely to exist occasionally under normal operation in quantities sufficient to be hazardous, as classified by 506.5(B)(2).

Zone 22 Hazardous (Classified) Location. An area where combustible dust or ignitible fibers and flyings are not likely to occur under normal operation in quantities sufficient to be hazardous, as classified by 506.5(B)(3).

506.4 General

(A) Documentation for Industrial Occupancies Areas designated as hazardous (classified) locations shall be properly documented. This documentation shall be available to those authorized to design, install, inspect, maintain or operate electrical equipment.

(B) Reference Standards Important information relating to topics covered in Chapter 5 are found in other publications.

FPN: It is important that the authority having jurisdiction be familiar with the recorded industrial experience as well as with standards of the National Fire Protection Association (NFPA), the ISA, International Society for Measurement and Control, and the International Electrotechnical Commission (IEC) that may be of use in the classification of various locations, the determination of adequate ventilation, and the protection against static electricity and lightning hazards.

506.5 Classification of Locations

(A) Classifications of Locations Locations shall be classified on the basis of the properties of the combustible dust, ignitible fibers or flyings that may be present, and the likelihood that a combustible or combustible concentration or quantity is present. Each room, section, or area shall be considered individually in determining its classification. Where pyrophoric materials are the only materials used or handled, these locations are outside of the scope of this article.

(B) Zone 20, Zone 21, and Zone 22 Locations Zone 20, Zone 21, and Zone 22 locations are those in which combustible dust, ignitible fibers, or flyings are or may be present in the air or in layers, in quantities sufficient to produce explosive or ignitible mixtures. Zone 20, Zone 21, and Zone 22 locations shall include those specified in 506.5(B)(1), (B)(2), and (B)(3).

> FPN: Through the exercise of ingenuity in the layout of electrical installations for hazardous (classified) locations, it is frequently possible to locate much of the equipment in a reduced level of classification, and, thus, to reduce the amount of special equipment required.

(1) Zone 20 A Zone 20 location is a location in which

(a) Ignitible concentrations of combustible dust or ignitible fibers or flyings are present continuously.

(b) Ignitible concentrations of combustible dust or ignitible fibers or flyings are present for long periods of time.

> FPN No. 1: As a guide to classification of Zone 20 locations, refer to ISA 12.10.05, *Electrical Apparatus for Use in Zone 20, Zone 21, and Zone 22 Hazardous (Classified) Locations — Classification of Zone 20, Zone 21, and Zone 22 Hazardous (Classified) Locations (IEC61241-10 Mod)*.
>
> FPN No. 2: Zone 20 classification includes locations inside dust containment systems; hoppers, silos, etc., cyclones and filters, dust transport systems, except some parts of belt and chain conveyors, etc.; blenders, mills, dryers, bagging equipment, etc.

(2) Zone 21 A Zone 21 location is a location

(a) In which ignitible concentrations of combustible dust or ignitible fibers or flyings are likely to exist occasionally under normal operating conditions; or

(b) In which ignitible concentrations of combustible dust or ignitible fibers or flyings may exist frequently because of repair or maintenance operations or because of leakage; or

(c) In which equipment is operated or processes are carried on, of such a nature that equipment breakdown or faulty operations could result in the release of ignitible concentrations of combustible dust, or ignitible fibers or flyings and also cause simultaneous failure of electrical equipment in a mode to cause the electrical equipment to become a source of ignition; or

(d) That is adjacent to a Zone 20 location from which ignitible concentrations of dust or ignitible fibers or flyings could be communicated, unless communication is prevented by adequate positive pressure ventilation from a source of clean air and effective safeguards against ventilation failure are provided.

> FPN No. 1: As a guide to classification of Zone 21 locations, refer to ISA 12.10.05, *Electrical Apparatus for Use In Zone 20, Zone 21, and Zone 22 Hazardous (Classified) Locations — Classification of Zone 20, Zone 21, and Zone 22 Hazardous (Classified) Locations (IEC61241-10 Mod)*.
>
> FPN No. 2: This classification usually includes locations outside dust containment and in the immediate vicinity of access doors subject to frequent removal or opening for operation purposes when internal combustible mixtures are present; locations outside dust containment in the proximity of filling and emptying points, feed belts, sampling points, truck dump stations, belt dump over points, etc. where no measures are employed to prevent the formation of combustible mixtures; locations outside dust containment where dust accumulates and where due to process operations the dust layer is likely to be disturbed and form combustible mixtures; locations inside dust containment where explosive dust clouds are likely to occur (but neither continuously, nor for long periods, nor frequently) as, for example, silos (if filled and/or emptied only occasionally) and the dirty side of filters if large self-cleaning intervals are occurring.

(3) Zone 22 A Zone 22 location is a location

(a) In which ignitible concentrations of combustible dust or ignitible fibers or flyings are not likely to occur in normal operation, and if they do occur, will only persist for a short period; or

(b) In which combustible dust, or fibers, or flyings are handled, processed, or used but in which the dust, fibers, or flyings are normally confined within closed containers of closed systems from which they can escape only as a result of the abnormal operation of the equipment with which the dust, or fibers, or flyings are handled, processed, or used; or

(c) That is adjacent to a Zone 21 location, from which ignitible concentrations of dust or fibers or flyings could be communicated, unless such communication is prevented by adequate positive pressure ventilation from a source of clean air and effective safeguards against ventilation failure are provided.

> FPN No. 1: As a guide to classification of Zone 22 locations, refer to ISA 12.10.05, *Electrical Apparatus for Use in Zone 20, Zone 21, and Zone 22 Hazardous (Classified) Locations — Classification of Zone 20, Zone 21, and Zone 22 Hazardous (Classified) Locations (IEC61241-10 Mod)*.
>
> FPN No. 2: Zone 22 locations usually include outlets from bag filter vents, because in the event of a malfunction there can be emission of combustible mixtures; locations near equipment that has to be opened at infrequent intervals or equipment that from experience can easily form leaks where, due to pressure above atmospheric, dust will blow out; pneumatic equipment, flexible connections that can become damaged, etc.; storage locations for bags containing dusty product, since failure of bags can occur during handling, causing dust leakage; and locations where controllable dust layers are formed that are likely to be raised into explosive dust/air mixtures. Only if the layer is removed by cleaning before hazardous

dust–air mixtures can be formed is the area designated non-hazardous.

FPN No. 3: Locations that normally are classified as Zone 21 can fall into Zone 22 when measures are employed to prevent the formation of explosive dust–air mixtures. Such measures include exhaust ventilation. The measures should be used in the vicinity of (bag) filling and emptying points, feed belts, sampling points, truck dump stations, belt dump over points, etc.

506.6 Special Precaution

Article 506 requires equipment construction and installation that ensures safe performance under conditions of proper use and maintenance.

> FPN: It is important that inspection authorities and users exercise more than ordinary care with regard to the installation and maintenance of electrical equipment in hazardous (classified) locations.

(A) Implementation of Zone Classification System Classification of areas, engineering and design, selection of equipment and wiring methods, installation, and inspection shall be performed by qualified persons.

(B) Dual Classification In instances of areas within the same facility classified separately, Zone 22 locations shall be permitted to abut, but not overlap, Class II or Class III, Division 2 locations. Zone 20 or Zone 21 locations shall not abut Class II or Class III, Division 1 or Division 2 locations.

(C) Reclassification Permitted A Class II or Class III, Division 1 or Division 2 location shall be permitted to be reclassified as a Zone 20, Zone 21, or Zone 22 location, provided that all of the space that is classified because of a single combustible dust or ignitible fiber or flying source is reclassified under the requirements of this article.

(D) Simultaneous Presence of Flammable Gases and Combustible Dusts, Fibers, or Flyings Where flammable gases or combustible dusts, fibers, or flyings are or may be present at the same time, the simultaneous presence shall be considered during the selection and installation of the electrical equipment and the wiring methods, including the determination of the safe operating temperature of the electrical equipment.

506.8 Protection Techniques

Acceptable protection techniques for electrical and electronic equipment in hazardous (classified) locations shall be as described in 506.8(A) through 506.8(F).

(A) Dust Ignitionproof This protection technique shall be permitted for equipment in Zone 20, Zone 21, and Zone 22 locations for which it is identified.

(B) Pressurized This protection technique shall be permitted for equipment in Zone 21, and Zone 22 locations for which it is identified.

(C) Intrinsic Safety This protection technique shall be permitted for equipment in Zone 20, Zone 21, and Zone 22 locations for which it is identified. Installation of intrinsically safe apparatus and wiring shall be in accordance with the requirements of Article 504.

(D) Dusttight This protection technique shall be permitted for equipment in Zone 22 locations for which it is identified.

(E) Nonincendive Circuit This protection technique shall be permitted for equipment in Zone 22 locations for which it is identified.

(F) Nonincendive Equipment This protection technique shall be permitted for equipment in Zone 22 locations for which it is identified.

506.9 Equipment Requirements

(A) Suitability Suitability of identified equipment shall be determined by one of the following:

(1) Equipment listing or labeling
(2) Evidence of equipment evaluation from a qualified testing laboratory or inspection agency concerned with product evaluation
(3) Evidence acceptable to the authority having jurisdiction such as a manufacturer's self-evaluation or an owner's engineering judgment

(B) Listing

(1) Equipment that is listed for Zone 20 shall be permitted in a Zone 21 or Zone 22 location of the same dust, or ignitible fiber, or flying. Equipment that is listed for Zone 21 may be used in a Zone 22 location of the same dust, fiber, or flying.
(2) Equipment shall be permitted to be listed for a specific dust, or ignitible fiber or flying, or any specific combination of dusts, fibers, or flyings.

(C) Marking Equipment identified for Class II, Division 1 or Class II, Division 2 shall, in addition to being marked in accordance with 500.8(B), be permitted to be marked with both of the following:

(1) Zone 20, 21, or 22 (as applicable)
(2) Temperature classification in accordance with 506.9(D)

(D) Temperature Classifications Equipment shall be marked to show the operating temperature referenced to a 40°C (104°F) ambient. Electrical equipment designed for use in the ambient temperature range between −20°C and +40°C shall require no additional ambient temperature marking. Electrical equipment that is designed for use in a range of ambient temperatures other than −20°C and +40°C is considered to be special; and the ambient temperature range shall then be marked on the equipment, including either the symbol "Ta" or "Tamb" together with the special range of ambient temperatures. As an example, such a marking might be "−30°C ≤ Ta ≤ +40°C." Electrical equipment suitable for ambient temperatures exceeding 40°C (104°F) shall be marked with both the maximum ambient temperature and the operating temperature at that ambient temperature.

Exception No. 1: Equipment of the non–heat-producing type, such as conduit fittings, shall not be required to have a marked operating temperature.

Exception No. 2: Equipment identified for Class II, Division 1 or Class II, Division 2 locations as permitted by 506.20(B) and 506.20(C) shall be permitted to be marked in accordance with 500.8(B) and Table 500.8(B).

(E) Threading All NPT threads referred to herein shall be threaded with a National (American) Standard Pipe Taper (NPT) thread that provides a taper of 1 in 16 (¾-in. taper per foot). Conduit and fittings shall be made wrenchtight to prevent sparking when the fault current flows through the conduit system, and to ensure the integrity of the conduit system. Equipment provided with threaded entries for field wiring connections shall be installed in accordance with 506.9(E)(1) or (E)(2).

(1) Equipment Provided with Threaded Entries for NPT Threaded Conduit or Fittings For equipment provided with threaded entries for NPT threaded conduit or fittings, listed conduit fittings, or cable fittings shall be used.

(2) Equipment Provided with Threaded Entries for Metric Threaded Conduit or Fittings For equipment with metric threaded entries, such entries shall be identified as being metric, or listed adapters to permit connection to conduit or NPT-threaded fittings shall be provided with the equipment. Adapters shall be used for connection to conduit or NPT-threaded fittings. Listed cable fittings that have metric threads shall be permitted to be used.

506.15 Wiring Methods

Wiring methods shall maintain the integrity of the protection techniques and shall comply with 506.15(A), (B), or (C).

(A) Zone 20 In Zone 20 locations, the wiring methods in (1) through (5) shall be permitted.

(1) Threaded rigid metal conduit, or threaded steel intermediate metal conduit.
(2) Type MI cable with termination fittings listed for the location. Type MI cable shall be installed and supported

in a manner to avoid tensile stress at the termination fittings.

Exception: MI cable and fittings listed for Class II, Division 1 locations are permitted to be used.

(3) In industrial establishments with limited public access, where the conditions of maintenance and supervision ensure that only qualified persons service the installation, Type MC cable, listed for continuous use in Zone 20 locations, with a gas/vaportight continuous corrugated metallic sheath, and overall jacket of suitable polymeric material, separate grounding conductors in accordance with 250.122, and provided with termination fittings listed for the application, shall be permitted.

Exception: MC cable and fittings listed for Class II, Division 1 locations are permitted to be used.

(4) Fittings and boxes shall be identified for use in Zone 20 locations.

Exception: Boxes and fittings listed for Class II, Division 1 locations are permitted to be used.

(5) Where necessary to employ flexible connections, liquidtight flexible metal conduit with listed fittings, liquidtight flexible nonmetallic conduit with listed fittings, or flexible cord listed for extra-hard usage and provided with listed fittings shall be used. Where flexible cords are used, they shall also comply with 506.17. Where flexible connections are subject to oil or other corrosive conditions, the insulation of the conductors shall be of a type listed for the condition or shall be protected by means of a suitable sheath.

Exception: Flexible conduit and flexible conduit and cord fittings listed for Class II, Division 1 locations are permitted to be used.

FPN: See 506.25 for grounding requirements where flexible conduit is used.

(B) Zone 21 In Zone 21 locations, the wiring methods in (B)(1) and (B)(2) shall be permitted.

(1) All wiring methods permitted in 506.15(A)
(2) Fittings and boxes that are dusttight, provided with threaded bosses for connection to conduit, in which taps, joints, or terminal connections are not made, and are not used in locations where metal dust is present, may be used

(C) Zone 22 In Zone 22 locations, the wiring methods in (1) through (8) shall be permitted.

(1) All wiring methods permitted in 506.15(B).
(2) Rigid metal conduit, intermediate metal conduit, electrical metallic tubing, dusttight wireways.

(3) Type MC or MI cable with listed termination fittings.

(4) Type PLTC in cable trays.

(5) Type ITC in cable trays.

(6) Type MC, MI, MV, or TC cable installed in ladder, ventilated trough, or ventilated channel cable trays in a single layer, with a space not less than the larger cable diameter between two adjacent cables, shall be the wiring method employed. Single conductor Type MV cables shall be shielded or metallic armored.

(7) Nonincendive field wiring shall be permitted using any of the wiring methods permitted for unclassified locations. Nonincendive field wiring systems shall be installed in accordance with the control drawing(s). Simple apparatus, not shown on the control drawing, shall be permitted in a nonincendive field wiring circuit, provided the simple apparatus does not interconnect the nonincendive field wiring circuit to any other circuit.

FPN: *Simple apparatus* is defined in 504.2.

Separation of nonincendive field wiring circuits shall be in accordance with one of the following:

a. Be in separate cables

b. Be in multiconductor cables where the conductors of each circuit are within a grounded metal shield

c. Be in multiconductor cables where the conductors have insulation with a minimum thickness of 0.25 mm (0.01 in.)

(8) Boxes and fittings shall be dusttight.

506.16 Sealing

Where necessary to protect the ingress of combustible dust, or ignitible fibers, or flyings, or to maintain the type of protection, seals shall be provided. The seal shall be identified as capable of preventing the ingress of combustible dust or ignitible fibers or flyings and maintaining the type of protection but need not be explosionproof or flameproof.

506.17 Flexible Cords

Flexible cords used in Zone 20, Zone 21, and Zone 22 locations shall comply with all of the following:

(1) Be of a type listed for extra-hard usage

(2) Contain, in addition to the conductors of the circuit, a grounding conductor in complying with 400.23

(3) Be connected to terminals or to supply conductors in an approved manner

(4) Be supported by clamps or by other suitable means in such a manner to minimize tension on the terminal connections

(5) Be provided with suitable seals to prevent the entrance of combustible dust, or ignitible fibers, or flyings where the flexible cord enters boxes or fittings

506.20 Equipment Installation

(A) Zone 20 In Zone 20 locations, only equipment listed and marked as suitable for the location shall be permitted.

Exception: Intrinsically safe apparatus listed for use in Class II, Division 1 locations with a suitable temperature class shall be permitted.

(B) Zone 21 In Zone 21 locations, only equipment listed and marked as suitable for the location shall be permitted.

Exception No. 1: Apparatus listed for use in Class II, Division 1 locations with a suitable temperature class shall be permitted.

Exception No. 2: Pressurized equipment identified for Class II, Division 1 shall be permitted.

(C) Zone 22 In Zone 22 locations, only equipment listed and marked as suitable for the location shall be permitted.

Exception No. 1: Apparatus listed for use in Class II, Division 1 or Class II, Division 2 locations with a suitable temperature class shall be permitted.

Exception No. 2: Pressurized equipment identified for Class II, Division 1 or Division 2 shall be permitted.

(D) Manufacturer's Instructions Electrical equipment installed in hazardous (classified) locations shall be installed in accordance with the instructions (if any) provided by the manufacturer.

(E) Temperature The temperature marking specified in 506.9(C)(2)(5) shall comply with (E)(1) or (E)(2).

(1) For combustible dusts, less than the lower of either the layer or cloud ignition temperature of the specific combustible dust. For organic dusts that may dehydrate or carbonize, the temperature marking shall not exceed the lower of either the ignition temperature or 165°C (329°F).

(2) For ignitible fibers or flyings, less than 165°C (329°F) for equipment that is not subject to overloading, or 120°C (248°F) for equipment (such as motors or power transformers) that may be overloaded.

FPN: See NFPA 499-2004, *Recommended Practice for the Classification of Combustible Dusts and of Hazardous (Classified) Locations for Electrical Installations in Chemical Processing Areas*, for minimum ignition temperatures of specific dusts.

506.21 Multiwire Branch Circuits

In Zone 20 and Zone 21 locations, a multiwire branch circuit shall not be permitted.

Exception: Where the disconnect device(s) for the circuit opens all ungrounded conductors of the multiwire circuit simultaneously.

506.25 Grounding and Bonding

Grounding and bonding shall comply with Article 250 and the requirements in 506.25(A) and 506.25(B).

(A) Bonding The locknut-bushing and double-locknut types of contacts shall not be depended on for bonding purposes, but bonding jumpers with proper fittings or other approved means of bonding shall be used. Such means of bonding shall apply to all intervening raceways, fittings, boxes, enclosures, and so forth, between Zone 20, Zone 21, and Zone 22 locations and the point of grounding for service equipment or point of grounding of a separately derived system.

Exception: The specific bonding means shall be required only to the nearest point where the grounded circuit conductor and the grounding electrode conductor are connected together on the line side of the building or structure disconnecting means as specified in 250.32(A), (B), and (C), if the branch side overcurrent protection is located on the load side of the disconnecting means.

> FPN: See 250.100 for additional bonding requirements in hazardous (classified) locations.

(B) Types of Equipment Grounding Conductors Where flexible conduit is used as permitted in 506.15, it shall be installed with internal or external bonding jumpers in parallel with each conduit and complying with 250.102.

Exception: In Zone 22 locations, the bonding jumper shall be permitted to be deleted where all of the following conditions are met:

(1) *Listed liquidtight flexible metal conduit 1.8 m (6 ft) or less in length, with fittings listed for grounding, is used.*
(2) *Overcurrent protection in the circuit is limited to 10 amperes or less.*
(3) *The load is not a power utilization load.*

ARTICLE 510
Hazardous (Classified)
Locations — Specific

Contents

510.1 Scope
510.2 General

510.1 Scope

Articles 511 through 517 cover occupancies or parts of occupancies that are or may be hazardous because of atmospheric concentrations of flammable liquids, gases, or vapors, or because of deposits or accumulations of materials that may be readily ignitible.

510.2 General

The general rules of this *Code* and the provisions of Articles 500 through 504 shall apply to electric wiring and equipment in occupancies within the scope of Articles 511 through 517, except as such rules are modified in Articles 511 through 517. Where unusual conditions exist in a specific occupancy, the authority having jurisdiction shall judge with respect to the application of specific rules.

Some of the requirements contained in Articles 511 through 517 have been extracted from other NFPA codes and standards. For example, Table 514.3(B)(1), Table 514.3(B)(2), and Table 515.3 are extracted from NFPA 30A, *Code for Motor Fuel Dispensing Facilities and Repair Garages*, and NFPA 30, *Flammable and Combustible Liquids Code*. The tables were developed by the NFPA Technical Committees responsible for those documents. The documents were developed through the same process as the *NEC*; however, the National Electrical Code Committee is not directly responsible for the technical content of extracted material.

NFPA publishes a number of standards and recommended practices that provide requirements or guidance on the classification of hazardous locations in specific occupancies. Information and copies of standards may be obtained from NFPA, 1 Batterymarch Park, Quincy, MA 02169-7471.

ARTICLE 511
Commercial Garages, Repair
and Storage

Summary of Changes

- **511.3:** Revised requirements on which areas are classified as Class I, Division 2 and the areas that are unclassified.
- **511.4(A):** Deleted requirement for raceways embedded in a masonry wall or beneath the floor of a classified area in a commercial garage to be considered as being within the Class I location above the floor.

Contents

511.1 Scope
511.3 Classifications of Locations
  (A) Unclassified Locations
  (B) Classified Locations

FPN: Rules that are followed by a reference in brackets contain text that has been extracted from NFPA 30A-2003, *Code for Motor Fuel Dispensing Facilities and Repair Garages*. Only editorial changes were made to the extracted text to make it consistent with this *Code*.

511.1 Scope

These occupancies shall include locations used for service and repair operations in connection with self-propelled vehicles (including, but not limited to, passenger automobiles, buses, trucks, and tractors) in which volatile flammable liquids or flammable gases are used for fuel or power.

Article 100 defines *garage* as "a building or portion of a building in which one or more self-propelled vehicles can be kept for use, sale, storage, rental, repair, exhibition, or demonstration purposes." Article 511 applies to commercial garages in which the primary operation is the service and repair of self-propelled vehicles that use flammable gases or liquids for fuel. The commercial garages covered by Article 511 include automotive service centers; repair garages for commercial vehicles, such as trucks and tractors; and service garages for fleet vehicles, such as buses, cars, and trucks.

The requirements of Article 511 are intended to mitigate the potential for an ignition-capable arc or spark from electrical wiring or equipment used in or above hazardous (classified) locations. Additionally, there are requirements for personnel protection in occupancies that are frequently wet or damp in which service personnel are subject to contact with large grounded surfaces, such as concrete slabs in direct contact with the earth. The increasing number of service operations in which minor repairs, such as oil changes, occur

is covered under the requirements of this article plus the belowgrade work area classification requirements of Article 514. See 511.3(A)(3) and its associated commentary.

Parking, storage, and similar occupancies are not required to be classified, provided that any repair that occurs is minor and does not involve the use of electrical equipment. In accordance with NFPA 88A, *Standard for Parking Structures*, a mechanical ventilating system that is capable of continuously providing a ventilation rate of 1 ft^3 per minute for each square foot of floor area is required for all enclosed, basement, and underground parking garages.

Operations that involve open flames or electric arcs, including fusion gas welding and electric welding, previously were covered by NFPA 88B, *Standard for Repair Garages*. The NFPA 88B requirements were incorporated into the 2003 edition of NFPA 30A, *Code for Motor Fuel Dispensing Facilities and Repair Garages*, and NFPA 88B was withdrawn as a standard at the 2002 NFPA Fall Meeting. Repair work that involves an open flame or electric arcs must be restricted to areas specifically provided for such purposes.

Approved suspended unit heaters may be used in commercial garages, provided they are located not less than 8 ft above the floor and are installed in accordance with the conditions of their approval. This requirement also came from NFPA 88B.

For the 2005 *Code*, new Section 555.22 requires that the electrical repair facilities for boats and other marine craft comply with the requirements of Article 511.

511.3 Classifications of Locations

(A) Unclassified Locations

Section 511.3 was reorganized in the 2005 *Code* to make the classification of areas clearer. Unclassified locations are described in 511.3(A), and classified locations are found in 511.3(B). The Technical Committee on Automotive and Marine Service Stations; NFPA 30A, *Code for Motor Fuel Dispensing Facilities and Repair Garages*; and NEC Code-Making Panel 14 have correlated and updated all the electrical requirements between the 2002 *Code* and the 2005 *Code* to eliminate any potential conflicts. Section 511.3 uses the term *transferred* in place of the old terminology *dispensing* in determining the requirements for classifying locations when a significant quantity of flammable or gaseous liquids will be exposed to the atmosphere by a motor vehicle repair operation (e.g., major engine overhauls or repairs that require draining of the motor vehicle fuel tank). Minor repair garages, by definition under NFPA 30A, would not be permitted to conduct such types of repair operations involving the transfer of flammable or gaseous liquids and thus would not require a classified location designation.

Section 511.3 correlates the type of repair garage with the NFPA 30A definitions for *major* and *minor* repair garages. The definitions from NFPA 30A are as follows:

Major Repair Garage. A building or portions of a building where major repairs, such as engine overhauls, painting, body and fender work, and repairs that require draining of the motor vehicle fuel tank are performed on motor vehicles, including associated floor space used for offices, parking, or showrooms. [NFPA 30A-2003, 3.3.12.1]

Minor Repair Garage. A building or portions of a building used for lubrication, inspection, and minor automotive maintenance work, such as engine tune-ups, replacement of parts, fluid changes (e.g., oil, antifreeze, transmission fluid, brake fluid, air conditioning refrigerants, etc.), brake system repairs, tire rotation, and similar routine maintenance work, including associated floor space used for offices, parking, or showrooms. [NFPA 30A-2003, 3.3.12.2]

The term *Class I liquid* used in 511.3(A) and 511.3(B) refers to flammable liquids as defined in NFPA 30–2003, *Flammable and Combustible Liquids Code*. Gasoline is a common Class I liquid, whereas diesel fuel, because of its flashpoint being above 100°F, is classified as a Class II combustible liquid. The use of Class I, Class II, and Class III in NFPA 30 for the classification of liquids has no direct correlation to the use of Class I, Class II, and Class III in the *NEC* to designate hazardous (classified) locations.

Flammable Liquid. Any liquid that has a closed-cup flash point below 100°F (37.8°C), as determined by the test procedures and apparatus set forth in 1.7.4. Flammable liquids are classified as Class I as follows: Class I Liquid — any liquid that has a closed-cup flash point below 100°F (37.8°C) and a Reid vapor pressure not exceeding 40 psia (2068.6 mm Hg) at 100°F (37.8°C), as determined by ASTM D 323, *Standard Method of Test for Vapor Pressure of Petroleum Products (Reid Method)*. Class I liquids are further classified as follows:

1. Class IA liquids. Those liquids that have flash points below 73°F (22.8°C) and boiling points below 100°F (37.8°C)
2. Class IB liquids. Those liquids that have flash points below 73°F (22.8°C) and boiling points at or above 100°F (37.8°C);
3. Class IC liquids. Those liquids that have flash points at or above 73°F (22.8°C), but below 100°F (37.8°C). [NFPA 30-2003, 3.3.25.2]

(1) Parking and Repair Garages. Parking garages used for parking or storage shall be permitted to be unclassi-

fied. Repair garages shall be permitted to be unclassified when designed in accordance with 511.3(A)(2) through 511.3(A)(7).

> FPN: For further information, see NFPA 88A-2002, *Standard for Parking Structures*, and NFPA 30A-2003, *Code for Motor Fuel Dispensing Facilities and Repair Garages*.

(2) Alcohol-Based Windshield Washer Fluid The storage, handling, or dispensing into motor vehicles of alcohol-based windshield washer fluid in areas used for the service and repair operations of the vehicles shall not cause such areas to be classified as hazardous (classified) locations.

> FPN: For further information, see 8.3.5, Exception, of NFPA 30A-2003, *Code for Motor Fuel Dispensing Facilities and Repair Garages*.

(3) Specific Areas Adjacent to Classified Locations Areas adjacent to classified locations in which flammable vapors are not likely to be released, such as stock rooms, switchboard rooms, and other similar locations, shall not be classified where mechanically ventilated at a rate of four or more air changes per hour, or designed with positive air pressure, or where effectively cut off by walls or partitions.

Section 511.3(A)(3) provides guidance for determining the classification of facilities that primarily offer oil and filter change and lubrication-type service, not the transfer of fuel. If the lower-level work area of a lubritorium is provided with exhaust ventilation at a rate specified in Table 514.3(B)(1) under "Lubrication or Service Room — Without Dispensing," the lower level is not a hazardous (classified) location. Table 514.3(B)(1) is extracted from NFPA 30A, *Code for Motor Fuel Dispensing Facilities and Repair Garages*.

(4) Pits in Lubrication or Service Room Where Class I Liquids Are Not Transferred Any pit, belowgrade work area, or subfloor work area that is provided with exhaust ventilation at a rate of not less than 0.3 m³/min/m² (1 cfm/ft²) of floor area at all times that the building is occupied or when vehicles are parked in or over this area and where exhaust air is taken from a point within 300 mm (12 in.) of the floor of the pit, belowgrade work area, or subfloor work area is unclassified. [NFPA 30A:7.4.5.4 and Table 8.3.1]

(5) Up to a Level of 450 mm (18 in.) Above the Floor in Lubrication or Service Rooms Where Class I Liquids Are Transferred For each floor, the entire area up to a level of 450 mm (18 in.) above the floor shall be considered unclassified where there is mechanical ventilation providing a minimum of four air changes per hour or one cubic foot per minute of exchanged air for each square foot of floor area. Ventilation shall provide for air exchange across the entire floor area, and exhaust air shall be taken at a point within 0.3 m (12 in.) of the floor.

(6) Flammable Liquids Having Flash Points Below 38°C (100°F) Where flammable liquids having a flash point below 38°C (100°F) (such as gasoline) or gaseous fuels (such as natural gas, hydrogen, or LPG) will not be transferred, such location shall be considered to be unclassified. unless the location is required to be classified in accordance with 511.3(B)(2) or (B)(4).

(7) Within 450 mm (18 in.) of the Ceiling In major repair garages, where lighter-than-air gaseous fuels (such as natural gas or hydrogen) vehicles are repaired or stored, the area within 450 mm (18 in.) of the ceiling shall be considered unclassified where ventilation of at least 1 cfm/sq ft of ceiling area taken from a point within 450 mm (18 in.) of the highest point in the ceiling is provided.

> FPN: For further information on the definition of *major repair garage*, see 3.3.12.1 of NFPA 30A-2003, *Code for Motor Fuel Dispensing Facilities and Repair Garages.*

(B) Classified Locations

(1) Flammable Fuel Dispensing Areas Areas in which flammable fuel is dispensed into vehicle fuel tanks shall conform to Article 514.

(2) Lubrication or Service Room Where Class I Liquids or Gaseous Fuels (Such as Natural Gas, Hydrogen, or LPG) Are Not Transferred The following spaces that are not designed in accordance with 511.3(A)(4) shall be classified as Class I, Division 2:

(1) Entire area within any unventilated pit, belowgrade work area, or subfloor area.
(2) Area up to 450 mm (18 in.) above any such unventilated pit, belowgrade work area, or subfloor work area and extending a distance of 900 mm (3 ft) horizontally from the edge of any such pit, belowgrade work area, or subfloor work area.

(3) Lubrication or Service Room Where Class I Liquids or Gaseous Fuels (Such as Natural Gas, Hydrogen, or LPG) Are Transferred The following spaces that are not designed in accordance with 511.3(A)(5) shall be classified as follows:

(1) Up to a Level of 450 mm (18 in.) Above the Floor. For each floor, the entire area up to a level of 450 mm (18 in.) above the floor shall be a Class I, Division 2 location.
(2) Any Unventilated Pit or Depression Below Floor Level. Any unventilated pit or depression below floor level shall be a Class I, Division 1 location and shall extend up to said floor level.

(3) Any Ventilated Pit or Depression Below Floor Level. Any ventilated pit or depression in which six air changes per hour are exhausted from a point within 300 mm (12 in.) of the floor level of the pit shall be a Class I, Division 2 location.
(4) Space Above an Unventilated Pit or Depression Below Floor Level. Above a pit, or depression below floor level, the space up to 450 mm (18 in.) above the floor or grade level and 900 mm (3 ft) horizontally from a lubrication pit shall be a Class I, Division 2 location.
(5) Dispenser for Class I Liquids, Other Than Fuels. Within 900 mm (3 ft) of any fill or dispensing point, extending in all directions shall be a Class I, Division 2 location. See also 511.3(B)(1).

Section 511.3 was updated in the 2005 *Code* to address repair areas where gaseous-fueled vehicles, such as those fueled by natural gas, hydrogen, or LPG, are serviced. It is not the intent of the *Code* to assume that all repair garages service gaseous-fueled vehicles. Only such repair garages that intend to service such types of gaseous-fueled vehicles need to comply with the requirements for gaseous fuels. Note that the classified location requirements for gaseous-fueled vehicles are based on LPG being a heavier-than-air fuel, whereas natural gas and hydrogen are lighter-than-air fuels.

(4) Within 450 mm (18 in.) of the Ceiling In major repair garages where lighter-than-air gaseous fuel (such as natural gas or hydrogen) vehicles are repaired or stored, ceiling spaces that are not designed in accordance with 511.3(A)(7) shall be classified as Class I, Division 2.

> FPN: For further information on the definition of *major repair garage*, see 3.3.12.1 of NFPA 30A, 2003, *Code for Motor Fuel Dispensing Facilities and Repair Garages.*

511.4 Wiring and Equipment in Class I Locations

(A) Wiring Located in Class I Locations Within Class I locations as classified in 511.3, wiring shall conform to applicable provisions of Article 501.

In accordance with 511.3(B), specific area classification is provided for lubrication and service rooms within a commercial garage. The classified locations covered in 511.3(B)(2) typically apply to *minor repair garages,* while the classified locations covered in 511.3(B)(3) typically apply to *major repair garages.* [See the commentary on 511.3(A) for definitions of major and minor repair garages.]

The benchmark for distinguishing which classification to use and the extent of the classified area is whether there will be a transfer of Class I liquids or gaseous fuels. Transfer of liquids or gaseous fuels includes the draining of a fuel tank that is inherent to a tank repair or replacement. In accordance with 511.3(A)(4), (5), and (7), the need to establish classified locations can be mitigated through the use of mechanical ventilation that meets the specified air exchange parameters. Where it is established that there will be classified locations within a commercial garage, all applicable requirements of Article 501 for installing wiring and equipment in Class I, Division 1 and 2 locations of a commercial garage must be followed.

In the 2002 and previous editions of the *Code,* raceways installed under the Class I, Division 2 areas of a commercial garage were considered to be in that classified location if the raceway penetrated through the slab or floor and extended into the classified location, even if the extension passed unbroken (without any fittings less than 12 in. from the boundary) through the classified location and into an unclassified location. This below-floor classification is no longer required based on the substantiation that there is typically a very low volume of air in the underground locations and the air–fuel mixture is below the lower flammable limit (LFL), in other words, too lean to burn or ignite.

It should be noted that the general rules of 501.15(A)(4) and (B)(2) on providing seals at classified location boundaries do apply where raceway installations in a commercial garage pass from classified to unclassified locations and conduit fittings, outlet boxes, or both are installed less than 12 in. from either side of the boundary. In accordance with 501.15(A)(4) Exception No. 4, where a raceway runs from a Class I, Division 1 location into an underground unclassified location and then emerges from below ground into an unclassified location, the boundary seal is permitted to be located more than 10 ft from the actual boundary, provided the seal is located at the point the conduit emerges from below grade into the unclassified location. The seal and, if necessary, an associated explosionproof union are required to be the first fitting(s) at the point the conduit emerges from below ground into the unclassified location.

Exhibit 511.1 depicts two receptacle outlet enclosures that are located at least 12 in. above an area that has been classified as Class I. The rigid metal conduit passes unbroken from the outlet boxes through the Class I location into the unclassified underground location beneath the floor. The conduit coupling is located 12 in. or more from the penetration into the classified location. No seals are required for this installation in accordance with 501.15(A)(4) Exception No. 1.

(B) Equipment Located in Class I Locations Within Class I locations as defined in 511.3, equipment shall conform to applicable provisions of Article 501.

(1) Fuel-Dispensing Units Where fuel-dispensing units (other than liquid petroleum gas, which is prohibited) are located within buildings, the requirements of Article 514 shall govern.

Exhibit 511.1 Seals not required for conduits that pass unbroken through the Class I location.

Where mechanical ventilation is provided in the dispensing area, the control shall be interlocked so that the dispenser cannot operate without ventilation, as prescribed in 500.5(B)(2).

See Figure 514.3 in the *Code* and Exhibit 514.1 for information on classified areas in the vicinity of dispensing units.

(2) Portable Lighting Equipment Portable lighting equipment shall be equipped with handle, lampholder, hook, and substantial guard attached to the lampholder or handle. All exterior surfaces that might come in contact with battery terminals, wiring terminals, or other objects shall be of nonconducting material or shall be effectively protected with insulation. Lampholders shall be of an unswitched type and shall not provide means for plug-in of attachment plugs. The outer shell shall be of molded composition or other suitable material. Unless the lamp and its cord are supported or arranged in such a manner that they cannot be used in the locations classified in 511.3, they shall be of a type identified for Class I, Division 1 locations.

Often, a reel-type portable handlamp is used for supplemental lighting during vehicle servicing. See Exhibit 511.2 for an illustration of a cord reel assembly.

Where there is no transfer of flammable liquids to vehicle fuel tanks but there is use of electrical diagnostic equipment, electrically operated tools or machinery, or open flames such as used for welding or cutting, the area must be classified in accordance with 511.3. Fuel dispensing units located within the garage are governed by the requirements

Exhibit 511.2 Cord reel. This cord, which is part of a portable lamp assembly, must be arranged so that the lamp cannot be used in a Class I location. Otherwise, the lamp must be an explosionproof type approved for Class I, Division 1 hazardous locations. (Courtesy of Appleton Electric Co., EGS Electrical Group)

of Article 514.

In accordance with 511.3(A) and (B), the Class I, Division 2 location above grade within a commercial garage in which Class I liquids or gaseous fuels are transferred extends 18 in. above floor level, unless the authority having jurisdiction determines otherwise because mechanical ventilation provides at least four air changes per hour.

The Class I, Division 1 location below grade extends from the floor of the pit or depression to floor level, unless the authority having jurisdiction permits the pit or depression to be classified as Class I, Division 2 because ventilation providing at least six air changes per hour exhausts air at the floor level of the pit or depression. Areas suitably cut off and areas adjacent to unclassified, ventilated garages are not classified as hazardous. See Exhibit 511.3 for an illustration of classified and unclassified locations in a commercial garage.

511.7 Wiring and Equipment Installed Above Class I Locations

For the installation of electrical wiring in areas above those designated by 511.3(B) as Class I locations, 511.7 specifies the types of raceway and cable systems that are allowed. The integrity of the wiring system above the classified location

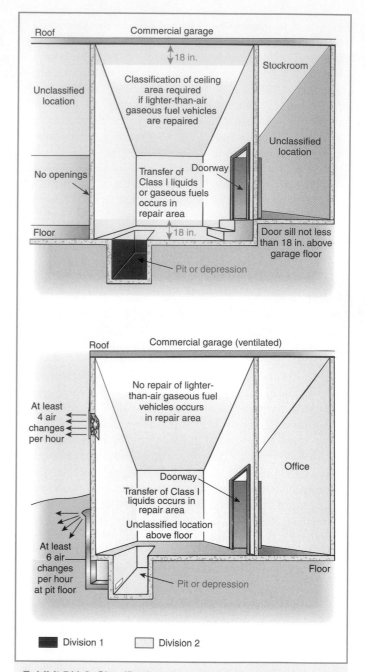

Exhibit 511.3 Classification of locations in commercial garages.

is necessary to ensure that an ignition-capable arc or spark does not migrate into the classified location. Wiring installed above unclassified areas can be selected from the methods covered in Chapter 3, provided the article covering that wiring method does not contain any restrictions that would limit its use in commercial garages.

(A) Wiring in Spaces Above Class I Locations

(1) Fixed Wiring Above Class I Locations All fixed wiring above Class I locations shall be in metal raceways, rigid nonmetallic conduit, electrical nonmetallic tubing, flexible metal conduit, liquidtight flexible metal conduit, or liquidtight flexible nonmetallic conduit, or shall be Type MC, AC, MI, manufactured wiring systems, or PLTC cable in accordance with Article 725, or Type TC cable or Type ITC cable in accordance with Article 727. Cellular metal floor raceways or cellular concrete floor raceways shall be permitted to be used only for supplying ceiling outlets or extensions to the area below the floor, but such raceways shall have no connections leading into or through any Class I location above the floor.

(2) Pendant For pendants, flexible cord suitable for the type of service and listed for hard usage shall be used.

(B) Electrical Equipment Installed Above Class I Locations

(1) Fixed Electrical Equipment Electrical equipment in a fixed position shall be located above the level of any defined Class I location or shall be identified for the location.

(a) Arcing Equipment. Equipment that is less than 3.7 m (12 ft) above the floor level and that may produce arcs, sparks, or particles of hot metal, such as cutouts, switches, charging panels, generators, motors, or other equipment (excluding receptacles, lamps, and lampholders) having make-and-break or sliding contacts, shall be of the totally enclosed type or constructed so as to prevent the escape of sparks or hot metal particles.

(b) Fixed Lighting. Lamps and lampholders for fixed lighting that is located over lanes through which vehicles are commonly driven or that may otherwise be exposed to physical damage shall be located not less than 3.7 m (12 ft) above floor level, unless of the totally enclosed type or constructed so as to prevent escape of sparks or hot metal particles.

511.9 Sealing

Seals conforming to the requirements of 501.15 and 501.15(B)(2) shall be provided and shall apply to horizontal as well as vertical boundaries of the defined Class I locations.

Seals are required if any part of the raceway is in or passes through a Class I, Division 2 location. See the commentary on seals following 501.15(F)(3).

511.10 Special Equipment

(A) Battery Charging Equipment Battery chargers and their control equipment, and batteries being charged, shall not be located within locations classified in 511.3.

(B) Electric Vehicle Charging Equipment

(1) General All electrical equipment and wiring shall be installed in accordance with Article 625, except as noted in 511.10(B)(2) and (B)(3). Flexible cords shall be of a type identified for extra-hard usage.

(2) Connector Location No connector shall be located within a Class I location as defined in 511.3.

(3) Plug Connections to Vehicles Where the cord is suspended from overhead, it shall be arranged so that the lowest point of sag is at least 150 mm (6 in.) above the floor. Where an automatic arrangement is provided to pull both cord and plug beyond the range of physical damage, no additional connector shall be required in the cable or at the outlet.

511.12 Ground-Fault Circuit-Interrupter Protection for Personnel

All 125-volt, single-phase, 15- and 20-ampere receptacles installed in areas where electrical diagnostic equipment, electrical hand tools, or portable lighting equipment are to be used shall have ground-fault circuit-interrupter protection for personnel.

Ground-fault circuit interrupters (GFCIs) intended to protect personnel from shock hazards are designed to trip when a ground-fault current of 5 milliamperes (plus or minus 1 mA) or greater is detected. The GFCI necessary to comply with the requirement in 511.12 may be either a receptacle type or a circuit-breaker type. This requirement applies to receptacles supplying specific types of utilization equipment that will be in use by repair personnel in environments where the floor surface (typically, concrete slabs with direct or indirect earth contact) and the possibility of dampness or even standing water increase the potential for electric shock.

See the definition of *ground-fault circuit interrupter* in Article 100; also see Exhibit 210.11 and Exhibit 210.12.

511.16 Grounded and Grounding Requirements

(A) General Grounding Requirements All metal raceways, the metal armor or metallic sheath on cables, and all non–current-carrying metal parts of fixed or portable electrical equipment, regardless of voltage, shall be grounded as provided in Article 250.

(B) Supplying Circuits with Grounded and Grounding Conductors in Class I Locations Grounding in Class I locations shall comply with 501.30.

(1) Circuits Supplying Portable Equipment or Pendants Where a circuit supplies portables or pendants and includes a grounded conductor as provided in Article 200, receptacles, attachment plugs, connectors, and similar devices shall be of the grounding type, and the grounded conductor of the flexible cord shall be connected to the screw shell of any lampholder or to the grounded terminal of any utilization equipment supplied.

(2) Approved Means Approved means shall be provided for maintaining continuity of the grounding conductor between the fixed wiring system and the non–current-carrying metal portions of pendant luminaires (fixtures), portable lamps, and portable utilization equipment.

ARTICLE 513
Aircraft Hangars

Summary of Changes

- **513.8(B):** Added requirement covering area classification for raceways embedded in or buried beneath a hangar floor.

- **513.12:** Added requirement for GFCI protection of 125-volt, 15- and 20-ampere receptacles that supply diagnostic equipment, electric hand tools, or portable lighting equipment. Applies only to those receptacles that are a part of the 50/60 Hz distribution system.

Contents

513.1 Scope

This article shall apply to buildings or structures in any part of which aircraft containing Class I (flammable) liquids or Class II (combustible) liquids whose temperatures are above their flash points are housed or stored and in which aircraft might undergo service, repairs, or alterations. It shall not apply to locations used exclusively for aircraft that have never contained fuel or unfueled aircraft.

The scope of Article 513 clarifies that it is not necessary to classify areas in which the only fuel contained in the aircraft is a Class II combustible liquid, unless the fuel is going to be used or stored above its flash point. A Class II liquid has a closed-cup flash point at or above 100°F. See the definition of *combustible liquid* in NFPA 30, *Flammable and Combustible Liquids Code*. Some aviation fuel, such as that used in jet engines, is a Class II combustible liquid. An aircraft manufacturing plant in which the aircraft under construction have never contained fuel is an example of a facility that is not covered by the requirements of Article 513.

For further information, see NFPA 409, *Standard on Aircraft Hangars*.

> FPN No. 1: For definitions of aircraft hangar and unfueled aircraft, see NFPA 409-2004, *Standard on Aircraft Hangars*.
> FPN No. 2: For further information on fuel classification see NFPA 30-2003, *Flammable and Combustible Liquids Code*.

513.2 Definitions

For the purpose of this article, the following definitions shall apply.

The definitions in 513.2 apply only to the portable and mobile equipment covered in Article 513.

Mobile Equipment. Equipment with electric components suitable to be moved only with mechanical aids or is provided with wheels for movement by person(s) or powered devices.

Portable Equipment. Equipment with electric components suitable to be moved by a single person without mechanical aids.

513.3 Classification of Locations

(A) Below Floor Level Any pit or depression below the level of the hangar floor shall be classified as a Class I, Division 1 or Zone 1 location that shall extend up to said floor level.

(B) Areas Not Cut Off or Ventilated The entire area of the hangar, including any adjacent and communicating areas not suitably cut off from the hangar, shall be classified as a Class I, Division 2 or Zone 2 location up to a level 450 mm (18 in.) above the floor.

(C) Vicinity of Aircraft The area within 1.5 m (5 ft) horizontally from aircraft power plants or aircraft fuel tanks hall be classified as a Class I, Division 2 or Zone 2 locations that shall extend upward from the floor to a level 1.5 m

(5 ft) above the upper surface of wings and of engine enclosures.

To properly classify the area in accordance with 513.3(C), it is necessary to obtain information on the aircraft parking patterns, the types of aircraft, and the operations to be performed in the hangar. See Exhibit 513.1 for area classification in aircraft hangars.

Consideration of future changes in aircraft types and locations is appropriate to avoid the need for costly wiring and equipment alterations as a result of changes in the area classification.

(D) Areas Suitably Cut Off and Ventilated Adjacent areas in which flammable liquids or vapors are not likely to be released, such as stock rooms, electrical control rooms, and other similar locations, shall not be classified where adequately ventilated and where effectively cut off from the hangar itself by walls or partitions.

Exhibit 513.1 Area classification in aircraft hangars.

513.4 Wiring and Equipment in Class I Locations

(A) General All wiring and equipment that is or may be installed or operated within any of the Class I locations defined in 513.3 shall comply with the applicable provisions of Article 501 or Article 505 for the division or zone in which they are used.

Attachment plugs and receptacles in Class I locations shall be identified for Class I locations or shall be designed such that they cannot be energized while the connections are being made or broken.

(B) Stanchions, Rostrums, and Docks Electric wiring, outlets, and equipment (including lamps) on or attached to stanchions, rostrums, or docks that are located or likely to be located in a Class I location, as defined in 513.3(C), shall comply with the applicable provisions of Article 501 or Article 505 for the division or zone in which they are used.

513.7 Wiring and Equipment Not Installed in Class I Locations

(A) Fixed Wiring All fixed wiring in a hangar but not installed in a Class I location as classified in 513.3 shall be installed in metal raceways or shall be Type MI, TC, or MC cable.

Exception: Wiring in unclassified locations, as described in 513.3(D), shall be permitted to be any suitable type wiring method recognized in Chapter 3.

(B) Pendants For pendants, flexible cord suitable for the type of service and identified for hard usage or extra-hard usage shall be used. Each such cord shall include a separate equipment grounding conductor.

(C) Arcing Equipment In locations above those described in 513.3, equipment that is less than 3.0 m (10 ft) above wings and engine enclosures of aircraft and that may produce arcs, sparks, or particles of hot metal, such as lamps and lampholders for fixed lighting, cutouts, switches, receptacles, charging panels, generators, motors, or other equipment having make-and-break or sliding contacts, shall be of the totally enclosed type or constructed so as to prevent the escape of sparks or hot metal particles.

Exception: Equipment in areas described in 513.3(D) shall be permitted to be of the general-purpose type.

(D) Lampholders Lampholders of metal-shell, fiber-lined types shall not be used for fixed incandescent lighting.

(E) Stanchions, Rostrums, or Docks Where stanchions, rostrums, or docks are not located or likely to be located in a Class I location, as defined in 513.3(C), wiring and equipment shall comply with 513.7, except that such wiring and equipment not more than 457 mm (18 in.) above the floor

in any position shall comply with 513.4(B). Receptacles and attachment plugs shall be of a locking type that will not readily disconnect.

(F) Mobile Stanchions Mobile stanchions with electric equipment complying with 513.7(E) shall carry at least one permanently affixed warning sign with the following words or equivalent:

> WARNING
> KEEP 5 FT CLEAR OF AIRCRAFT
> ENGINES AND FUEL TANK AREAS

or

> WARNING
> KEEP 1.5 METERS CLEAR OF AIRCRAFT
> ENGINES AND FUEL TANK AREAS

513.8 Underground Wiring

(A) Wiring and Equipment Embedded, Under Slab, or Under Ground All wiring installed in or under the hangar floor shall comply with the requirements for Class I, Division 1 locations. Where such wiring is located in vaults, pits, or ducts, adequate drainage shall be provided.

(B) Uninterrupted Raceways, Embedded, Under Slab, or Under Ground Uninterrupted raceways that are embedded in a hangar floor or buried beneath the hangar floor shall be considered to be within the Class I location above the floor, regardless of the point at which the raceway descends below or rises above the floor.

Section 513.8(B) was added to the 2005 *Code* to clarify that raceways embedded in the hangar floor or buried below the floor are considered to be in a hazardous (classified) location above the floor. Raceways that rise out of the floor in unclassified locations must be provided with boundary seals in accordance with 501.15.

513.9 Sealing

Seals shall be provided in accordance with 501.15 or 505.16, as applicable. Sealing requirements specified shall apply to horizontal as well as to vertical boundaries of the defined Class I locations.

513.10 Special Equipment

(A) Aircraft Electrical Systems

(1) De-energizing Aircraft Electrical Systems Aircraft electrical systems shall be de-energized when the aircraft is stored in a hangar and, whenever possible, while the aircraft is undergoing maintenance.

(2) Aircraft Batteries Aircraft batteries shall not be charged where installed in an aircraft located inside or partially inside a hangar.

(B) Aircraft Battery Charging and Equipment Battery chargers and their control equipment shall not be located or operated within any of the Class I locations defined in 513.3 and shall preferably be located in a separate building or in an area such as defined in 513.3(D). Mobile chargers shall carry at least one permanently affixed warning sign with the following words or equivalent:

WARNING
KEEP 5 FT CLEAR OF AIRCRAFT ENGINES
AND FUEL TANK AREAS

or

WARNING
KEEP 1.5 METERS CLEAR OF AIRCRAFT
ENGINES AND FUEL TANK AREAS

Tables, racks, trays, and wiring shall not be located within a Class I location and, in addition, shall comply with Article 480.

(C) External Power Sources for Energizing Aircraft

(1) Not Less Than 450 mm (18 in.) Above Floor Aircraft energizers shall be designed and mounted such that all electric equipment and fixed wiring will be at least 450 mm (18 in.) above floor level and shall not be operated in a Class I location as defined in 513.3(C).

(2) Marking for Mobile Units Mobile energizers shall carry at least one permanently affixed warning sign with the following words or equivalent:

WARNING
KEEP 5 FT CLEAR OF AIRCRAFT
ENGINES AND FUEL TANK AREAS

or

WARNING
KEEP 1.5 METERS CLEAR OF AIRCRAFT
ENGINES AND FUEL TANK AREAS

(3) Cords Flexible cords for aircraft energizers and ground support equipment shall be identified for the type of service and extra-hard usage and shall include an equipment grounding conductor.

(D) Mobile Servicing Equipment with Electric Components

(1) General Mobile servicing equipment (such as vacuum cleaners, air compressors, air movers) having electric wiring and equipment not suitable for Class I, Division 2 or Zone 2 locations shall be so designed and mounted that all such fixed wiring and equipment will be at least 450 mm (18 in.) above the floor. Such mobile equipment shall not be operated within the Class I location defined in 513.3(C) and shall carry at least one permanently affixed warning sign with the following words or equivalent:

WARNING
KEEP 5 FT CLEAR OF AIRCRAFT ENGINES
AND FUEL TANK AREAS

or

WARNING
KEEP 1.5 METERS CLEAR OF AIRCRAFT ENGINES
AND FUEL TANK AREAS

(2) Cords and Connectors Flexible cords for mobile equipment shall be suitable for the type of service and identified for extra-hard usage and shall include an equipment grounding conductor. Attachment plugs and receptacles shall be identified for the location in which they are installed and shall provide for connection of the equipment grounding conductor.

(3) Restricted Use Equipment that is not identified as suitable for Class I, Division 2 locations shall not be operated in locations where maintenance operations likely to release flammable liquids or vapors are in progress.

(E) Portable Equipment

(1) Portable Lighting Equipment Portable lighting equipment that is used within a hangar shall be identified for the location in which they are used. For portable lamps, flexible cord suitable for the type of service and identified for extra-hard usage shall be used. Each such cord shall include a separate equipment grounding conductor.

(2) Portable Utilization Equipment Portable utilization equipment that is or may be used within a hangar shall be of a type suitable for use in Class I, Division 2 or Zone 2 locations. For portable utilization equipment, flexible cord suitable for the type of service and approved for extra-hard usage shall be used. Each such cord shall include a separate equipment grounding conductor.

513.12 Ground-Fault Circuit-Interrupter Protection for Personnel

All 125-volt, 50/60 Hz, single phase, 15- and 20-ampere receptacles installed in areas where electrical diagnostic equipment, electrical hand tools, or portable lighting equipment are to be used shall have ground-fault circuit-interrupter protection for personnel.

Section 513.12 was added to the 2005 *Code* to require all 125-volt, 50/60-Hz, single-phase, 15- to 20-ampere receptacles to be provided with ground-fault protection for personnel in locations where electrical tools and diagnostic equipment are used. GFCI protection is not required on circuits supplied by 400 Hz and higher.

513.16 Grounded and Grounding Requirements

(A) General Grounding Requirements All metal raceways, the metal armor or metallic sheath on cables, and all non–current-carrying metal parts of fixed or portable electrical equipment, regardless of voltage, shall be grounded as provided in Article 250. Grounding in Class I locations shall comply with 501.30 for Class I, Division 1 and 2 locations and 505.25 for Class I, Zone 0, 1, and 2 locations.

(B) Supplying Circuits with Grounded and Grounding Conductors in Class I Locations

(1) Circuits Supplying Portable Equipment or Pendants Where a circuit supplies portables or pendants and includes a grounded conductor as provided in Article 200, receptacles, attachment plugs, connectors, and similar devices shall be of the grounding type, and the grounded conductor of the flexible cord shall be connected to the screw shell of any lampholder or to the grounded terminal of any utilization equipment supplied.

(2) Approved Means Approved means shall be provided for maintaining continuity of the grounding conductor between the fixed wiring system and the non–current-carrying metal portions of pendant luminaires (fixtures), portable lamps, and portable utilization equipment.

ARTICLE 514
Motor Fuel Dispensing Facilities

Summary of Changes

- **514.1:** Revised scope to include marine motor fuel dispensing facilities.
- **514.2:** Revised definition of *motor fuel dispensing facility* to include facilities that serve marine craft.
- **514.8:** Revised to specify where sealing is required for underground installations below classified locations instead of defining the underground area as a hazardous (classified) location.
- **514.13:** Revised to require that the dispensing device disconnecting means be capable of being locked in the open position.

Contents

FPN: Rules that are followed by a reference in brackets contain text that has been extracted from NFPA 30A-2003, *Code for Motor Fuel Dispensing Facilities and Repair Garages*. Only editorial changes were made to the extracted text to make it consistent with this *Code*.

514.1 Scope

This article shall apply to motor fuel dispensing facilities, marine/motor fuel dispensing facilities, motor fuel dispensing facilities located inside buildings, and fleet vehicle motor fuel dispensing facilities.

The title of Article 514 was revised for the 2002 *Code* to encompass all locations where volatile flammable liquids or gases are dispensed into the fuel tanks of self-propelled vehicles or other approved fuel tanks. For the 2005 *Code*, the scope has been revised to include dispensing at marine facilities such as marinas and boatyards. The phrase "approved containers" covers portable gasoline containers and is also intended to apply to dispensing locations for liquefied petroleum gas (LPG), including those locations that do not serve self-propelled vehicles. The popularity of outdoor cooking appliances using LPG has resulted in a marked increase in portable-container dispensing sites. Electrical area classification for these sites is specified in Table 514.3(B)(2).

FPN: For further information regarding safeguards for motor fuel dispensing facilities, see NFPA 30A-2003, *Code for Motor Fuel Dispensing Facilities and Repair Garages*.

514.2 Definition

Motor Fuel Dispensing Facility. That portion of a property where motor fuels are stored and dispensed from fixed equipment into the fuel tanks of motor vehicles or marine craft or into approved containers, including all equipment used in connection therewith. [NFPA 30A:3.3.11]

FPN: Refer to Articles 510 and 511 with respect to electric wiring and equipment for other areas used as lubritoriums, service rooms, repair rooms, offices, salesrooms, compressor rooms, and similar locations.

514.3 Classification of Locations

(A) Unclassified Locations Where the authority having jurisdiction can satisfactorily determine that flammable liquids having a flash point below 38°C (100°F), such as gasoline, will not be handled, such location shall not be required to be classified.

(B) Classified Locations

(1) Class I Locations Table 514.3(B)(1) shall be applied where Class I liquids are stored, handled, or dispensed and shall be used to delineate and classify motor fuel dispensing facilities and commercial garages as defined in Article 511. Table 515.3 shall be used for the purpose of delineating and classifying aboveground tanks. A Class I location shall not extend beyond an unpierced wall, roof, or other solid partition. [NFPA 30A:8.1, 8.3]

(2) Compressed Natural Gas, Liquefied Natural Gas, and Liquefied Petroleum Gas Areas Table 514.3(B)(2) shall be used to delineate and classify areas where compressed natural gas (CNG), liquefied natural gas (LNG), or liquefied petroleum gas (LPG) are stored, handled, or dispensed. Where CNG or LNG dispensers are installed beneath a canopy or enclosure, either the canopy or the enclosure shall be designed to prevent accumulation or entrapment of ignitible vapors, or all electrical equipment installed beneath the canopy or enclosure shall be suitable for Class I, Division 2 hazardous (classified) locations. Dispensing devices for liquefied petroleum gas shall be located not less than 1.5 m (5 ft) from any dispensing device for Class I liquids. [NFPA 30A:12.1, 12.4, 12.5]

Table 514.3(B)(1) Class I Locations — Motor Fuel Dispensing Facilities

Location	Class I, Group D Division	Extent of Classified Location[1]
Underground Tank		
Fill opening	1	Any pit or box below grade level, any part of which is within the Division 1 or Division 2, Zone 1 or Zone 2 classified location
	2	Up to 450 mm (18 in.) above grade level within a horizontal radius of 3.0 m (10 ft) from a loose fill connection and within a horizontal radius of 1.5 m (5 ft) from a tight fill connection
Vent — discharging upward	1	Within 900 mm (3 ft) of open end of vent, extending in all directions
	2	Space between 900 mm (3 ft) and 1.5 m (5 ft) of open end of vent, extending in all directions
Dispensing Device[2,5] **(except overhead type)**[3]		
Pits	1	Any pit or box below grade level, any part of which is within the Division 1 or Division 2, Zone 1 or Zone 2 classified location
Dispenser		FPN: Space classification inside the dispenser enclosure is covered in ANSI/UL 87-1995, *Power Operated Dispensing Devices for Petroleum Products.*
	2	Within 450 mm (18 in.) horizontally in all directions extending to grade from the dispenser enclosure or that portion of the dispenser enclosure containing liquid-handling components
		FPN: Space classification inside the dispenser enclosure is covered in ANSI/UL 87-1995, *Power Operated Dispensing Devices for Petroleum Products.*
Outdoor	2	Up to 450 mm (18 in.) above grade level within 6.0 m (20 ft) horizontally of any edge of enclosure.
Indoor with mechanical ventilation	2	Up to 450 mm (18 in.) above grade or floor level within 6.0 m (20 ft) horizontally of any edge of enclosure
with gravity ventilation	2	Up to 450 mm (18 in.) above grade or floor level within 7.5 m (25 ft) horizontally of any edge of enclosure
Dispensing Device[5]		
Overhead type[3]	1	The space within the dispenser enclosure, and all electrical equipment integral with the dispensing hose or nozzle
	2	A space extending 450 mm (18 in.) horizontally in all directions beyond the enclosure and extending to grade
	2	Up to 450 mm (18 in.) above grade level within 6.0 m (20 ft) horizontally measured from a point vertically below the edge of any dispenser enclosure

(continues)

Table 514.3(B)(1) *Continued*

Location	Class I, Group D Division	Extent of Classified Location[1]
Remote Pump — Outdoor	1	Any pit or box below grade level if any part is within a horizontal distance of 3.0 m (10 ft) from any edge of pump
	2	Within 900 mm (3 ft) of any edge of pump, extending in all directions. Also up to 450 mm (18 in.) above grade level within 3.0 m (10 ft) horizontally from any edge of pump
Remote Pump — Indoor	1	Entire space within any pit
	2	Within 1.5 m (5 ft) of any edge of pump, extending in all directions. Also up to 900 mm (3 ft) above grade level within 7.5 m (25 ft) horizontally from any edge of pump
Lubrication or Service Room — Without Dispensing	2	Entire area within any pit used for lubrication or similar services where Class I liquids may be released
	2	Area up to 450 mm (18 in.) above any such pit and extending a distance of 900 mm (3 ft) horizontally from any edge of the pit
	2	Entire unventilated area within any pit, belowgrade area, or subfloor area
	2	Area up to 450 mm (18 in.) above any such unventilated pit, belowgrade work area, or subfloor work area and extending a distance of 900 mm (3 ft) horizontally from the edge of any such pit, belowgrade work area, or subfloor work area
	Unclassified	Any pit, belowgrade work area, or subfloor work area that is provided with exhaust ventilation at a rate of not less than 0.3 m³/min/m² (1 cfm/ft²) of floor area at all times that the building is occupied or when vehicles are parked in or over this area and where exhaust air is taken from a point within 300 mm (12 in.) of the floor of the pit, belowgrade work area, or subfloor work area
Special Enclosure Inside Building[4]	1	Entire enclosure
Sales, Storage, and Rest Rooms	Unclassified	If there is any opening to these rooms within the extent of a Division 1 location, the entire room shall be classified as Division 1
Vapor Processing Systems Pits	1	Any pit or box below grade level, any part of which is within a Division 1 or Division 2 classified location or that houses any equipment used to transport or process vapors
Vapor Processing Equipment Located Within Protective Enclosures FPN: See 10.1.7 of NFPA 30A-2003, *Code for Motor Fuel Dispensing Facilities and Repair Garages.*	2	Within any protective enclosure housing vapor processing equipment
Vapor Processing Equipment Not Within Protective Enclosures (excluding piping and combustion devices)	2	The space within 450 mm (18 in.) in all directions of equipment containing flammable vapor or liquid extending to grade level. Up to 450 mm (18 in.) above grade level within 3.0 m (10 ft) horizontally of the vapor processing equipment
Equipment Enclosures	1	Any space within the enclosure where vapor or liquid is present under normal operating conditions
Vacuum-Assist Blowers	2	The space within 450 mm (18 in.) in all directions extending to grade level. Up to 450 mm (18 in.) above grade level within 3.0 m (10 ft) horizontally

[1]For marine application, *grade level* means the surface of a pier extending down to water level.
[2]Refer to Figure 514.3 for an illustration of classified location around dispensing devices.
[3]Ceiling mounted hose reel.
[4]FPN: See 4.3.9 of NFPA 30A-2003, *Code for Motor Fuel Dispensing Facilities and Repair Garages.*
[5]FPN: Area classification inside the dispenser enclosure is covered in ANSI/UL 87-1995, *Power-Operated Dispensing Devices for Petroleum Products.* [NFPA 30A:Table 8.3.1]

Table 514.3(B)(2) Electrical Equipment Classified Areas for Dispensing Devices

| Dispensing Device | Extent of Classified Area | |
	Class I, Division 1	Class I, Division 2
Compressed natural gas	Entire space within the dispenser enclosure	1.5 m (5 ft) in all directions from dispenser enclosure
Liquefied natural gas	Entire space within the dispenser enclosure and 1.5 m (5 ft) in all directions from the dispenser enclosure	From 1.5 m to 3.0 m (5 ft to 10 ft) in all directions from the dispenser enclosure
Liquefied petroleum gas	Entire space within the dispenser enclosure; 450 mm (18 in.) from the exterior surface of the dispenser enclosure to an elevation of 1.2 m (4 ft) above the base of the dispenser; the entire pit or open space beneath the dispenser and within 6.0 m (20 ft) horizontally from any edge of the dispenser when the pit or trench is not mechanically ventilated.	Up to 450 mm (18 in.) aboveground and within 6.0 m (20 ft) horizontally from any edge of the dispenser enclosure, including pits or trenches within this area when provided with adequate mechanical ventilation

[NFPA 30A:Table 12.6.2]

Figure 514.3 Classified Areas Adjacent to Dispensers as Detailed in Table 514.3(B)(1) [NFPA 30A:Figure 8.3.1]

FPN No. 1: For information on area classification where liquefied petroleum gases are dispensed, see NFPA 58-2004, *Liquefied Petroleum Gas Code.*

FPN No. 2: For information on classified areas pertaining to LP-Gas systems other than residential or commercial, see NFPA 58-2004, *Liquefied Petroleum Gas Code,* and NFPA 59-2004, *Utility LP-Gas Plant Code.*

FPN No. 3: See 555.21 for motor fuel dispensing stations in marinas and boatyards.

Table 514.3(B)(1) and Table 514.3(B)(2) are extracted from NFPA 30A, *Code for Motor Fuel Dispensing Facilities and Repair Garages.* See the commentary following 511.3(A)(3) for information regarding the classification of facilities that offer primarily lubrication-type service. See Exhibit 514.1 for an illustration of the Class I location around overhead motor fuel dispensing units.

The use of aboveground tanks with dispensing equipment is an acceptable alternative when underground tanks are impractical. Aboveground tanks with dispensing equipment are frequently used at fleet motor fuel–dispensing facilities. Hazardous area classification for aboveground tank installations is performed in accordance with Table 515.3.

514.4 Wiring and Equipment Installed in Class I Locations

All electrical equipment and wiring installed in Class I locations as classified in 514.3 shall comply with the applicable provisions of Article 501.

Exception: As permitted in 514.8.

FPN: For special requirements for conductor insulation, see 501.20.

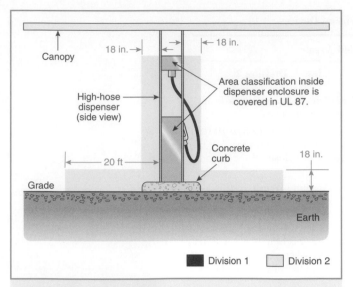

Exhibit 514.1 Extent of Class I location around overhead motor fuel dispensing units, in accordance with Table 514.3(B)(1).

The wiring methods and equipment required in Article 501 must be used within the Class I areas at a motor fuel dispensing facility. An explosionproof outlet box of the type frequently used in gasoline dispensing units is shown in Exhibit 514.2. The branch-circuit conductors for dispenser power,

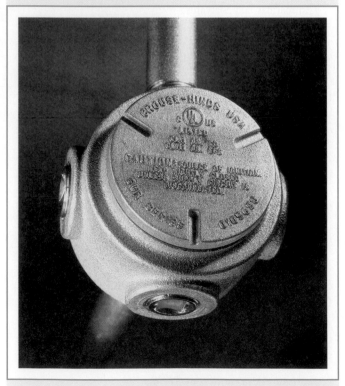

Exhibit 514.2 A typical explosionproof outlet box used in dispenser applications. (Courtesy of Cooper Crouse-Hinds)

lighting, or both connect to the internal wiring of the dispenser in the explosionproof outlet box. For gasoline- and oil-resistant insulated conductors, see the commentary following 501.120.

514.7 Wiring and Equipment Above Class I Locations

Wiring and equipment above the Class I locations as classified in 514.3 shall comply with 511.7.

514.8 Underground Wiring

Underground wiring shall be installed in threaded rigid metal conduit or threaded steel intermediate metal conduit. Any portion of electrical wiring that is below the surface of a Class I, Division 1, or a Class I, Division 2, location [as classified in Tables 514.3(B)(1) and 514.3(B)(2)] shall be sealed within 3.05 m (10 ft) of the point of emergence above grade. Except for listed explosionproof reducers at the conduit seal, there shall be no union, coupling, box, or fitting between the conduit seal and the point of emergence above grade. Refer to Table 300.5.

Experience has shown that if fuel spilled in the vicinity of gasoline dispensers seeps into the ground, it may migrate into underground electrical conduits and accumulate in voids. Therefore, 514.8 requires that all conduits installed below the classified locations of a motor fuel dispensing facility be sealed within 10 ft of the point of emergence from below grade. This boundary seal minimizes the passage of gasoline or other fuel vapors into unclassified locations where the electrical equipment is not explosionproof or otherwise protected. Table 514.3(B)(1) and Table 514.3(B)(2) define the extent of the aboveground Class I, Divisions 1 and 2 locations.

Exception No. 1: Type MI cable shall be permitted where it is installed in accordance with Article 332.

Exception No. 2: Rigid nonmetallic conduit shall be permitted where buried under not less than 600 mm (2 ft) of cover. Where rigid nonmetallic conduit is used, threaded rigid metal conduit or threaded steel intermediate metal conduit shall be used for the last 600 mm (2 ft) of the underground run to emergence or to the point of connection to the aboveground raceway, and an equipment grounding conductor shall be included to provide electrical continuity of the raceway system and for grounding of non–current-carrying metal parts.

Exception No. 2 to 514.8 makes it clear that if rigid nonmetallic conduit is used for underground wiring, threaded rigid metal conduit or threaded steel intermediate metal conduit must be used for the last 2 ft of the underground run to the point of emergence or to the point of connection to the aboveground raceway. The rigid nonmetallic conduit, including rigid nonmetallic conduit elbows and fittings, must be located not less than 2 ft below grade, as shown in Exhibit 514.3.

Exhibit 514.3 Use of rigid nonmetallic conduit in accordance with 514.8, Exception No. 2.

If rigid nonmetallic conduit is used, an equipment grounding conductor must be included and must be bonded to the explosionproof raceway system inside the dispenser. This is accomplished by terminating the equipment grounding conductor on the ground screw (or other means) provided in the dispenser junction box.

514.9 Sealing

(A) At Dispenser A listed seal shall be provided in each conduit run entering or leaving a dispenser or any cavities or enclosures in direct communication therewith. The sealing fitting shall be the first fitting after the conduit emerges from the earth or concrete.

(B) At Boundary Additional seals shall be provided in accordance with 501.15. Sections 501.15(A)(4) and (B)(2) shall apply to horizontal as well as to vertical boundaries of the defined Class I locations.

Sealing fittings are required in all conduits leaving a Class I location. All conduits passing under the boundaries of the hazardous (classified) locations (20-ft radius from dispenser) or the tank fill-pipe (10-ft radius from a loose-fill connection and 5-ft radius from a tight-fill connection) are considered

in a Class I, Division 1 location, and the seal is to be the first fitting at the point of emergence. A seal must be provided in each conduit run entering or leaving a dispenser. Therefore, even though a conduit runs from dispenser to dispenser and does not leave the hazardous (classified) location, a seal is necessary where the conduit leaves, and again where it enters, the dispenser.

Panelboards are generally located in a room classified as a nonhazardous location; however, any conduit coming from the dispenser or passing under the hazardous (classified) location boundaries from the dispenser or tank fill-pipe would require a seal at the panelboard location to minimize the likelihood of gas migration into the remote location. If the panelboard is located in the lube or repair room, all conduits emerging into the 18-in. hazardous (classified) location would require seals. See Exhibits 514.4 and 514.5.

514.11 Circuit Disconnects

(A) General Each circuit leading to or through dispensing equipment, including equipment for remote pumping systems, shall be provided with a clearly identified and readily accessible switch or other acceptable means, located remote from the dispensing devices, to disconnect simultaneously from the source of supply, all conductors of the circuits, including the grounded conductor, if any.

Single-pole breakers utilizing handle ties shall not be permitted.

The disconnecting means required by 514.11(A) must be clearly marked and readily accessible. The disconnecting means also must be remote from the dispensing device, so that if an emergency occurs that requires rapid shutdown of the dispensing equipment, the person operating the disconnecting means is not exposed to the hazard.

It is important to note that all conductors of a circuit, including the grounded conductor, that may be present within a dispensing device must be provided with a switch or special-type circuit breaker that simultaneously disconnects all conductors. Handle ties on single-pole circuit breakers are not permitted. The intent is that no energized conductors are in the dispenser vicinity during maintenance or alteration. Considering possible accidental reversal of the polarities of conductors at panelboards, the grounded conductor must be able to be switched to the open or off position. Grounded conductors may be present in old-style pump motors, or they may pass through a dispenser as part of a circuit for the dispensing island lighting.

(B) Attended Self-Service Motor Fuel Dispensing Facilities Emergency controls as specified in 514.11(A) shall be installed at a location acceptable to the authority having

Exhibit 514.4 A gasoline dispenser installation indicating locations for sealing fittings. Emergency controls are required for self-service stations.

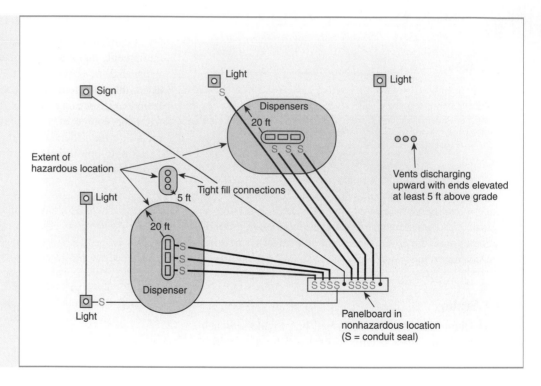

Exhibit 514.5 Required seals at points marked "S." Seals are not required at the sign and two of the lights because conduit runs do not pass through a hazardous location.

jurisdiction, but controls shall not be more than 30 m (100 ft) from dispensers. [NFPA 30A:6.7.1]

(C) Unattended Self-Service Motor Fuel Dispensing Facilities Emergency controls as specified in 514.11(A) shall be installed at a location acceptable to the authority having jurisdiction, but the control shall be more than 6 m (20 ft) but less than 30 m (100 ft) from the dispensers. Additional emergency controls shall be installed on each group of dispensers or the outdoor equipment used to control the dispensers. Emergency controls shall shut off all power to all dispensing equipment at the station. Controls shall be

manually reset only in a manner approved by the authority having jurisdiction. [NFPA 30A:6.7.2]

FPN: For additional information, see 6.7.1 and 6.7.2 of NFPA 30A-2003, *Code for Motor Fuel Dispensing Facilities and Repair Garages.*

Because a fire or large gasoline spill at the dispensing island may make it impossible for a person to approach and shut off the flow of gasoline by operating a disconnecting means located at the dispensing island, Section 6.7 of NFPA 30A,

Code for Motor Fuel Dispensing Facilities and Repair Garages, requires an easily accessible and clearly identified emergency power shutoff to be provided at a location remote from the dispensing device. The requirements in 514.11(B) and 514.11(C) are extracted from NFPA 30A and provide the maximum and minimum distances for the location of the emergency control for attended and unattended self-service dispensing facilities.

The term *clearly identified* means that a sign must be posted indicating where the shutoff switch is located. This emergency power shutoff must be readily accessible and not blocked by the storage of such things as tires, cases of lubricating oil, or merchandise on display. All dispensing facility operators as well as responding fire fighters should know the location of the emergency power shutoff.

514.13 Provisions for Maintenance and Service of Dispensing Equipment

Each dispensing device shall be provided with a means to remove all external voltage sources, including feedback, during periods of maintenance and service of the dispensing equipment. The location of this means shall be permitted to be other than inside or adjacent to the dispensing device. The means shall be capable of being locked in the open position.

This requirement is intended to enhance the level of safety for personnel servicing dispensing equipment. As more sophisticated control circuitry is integrated into dispensing equipment, simply shutting off the main power source to the dispenser or remote pump does not necessarily ensure that the equipment being worked on has been isolated from all sources of voltage. To ensure that the equipment is completely isolated from all voltage sources, a means must be provided to remove all external voltage sources from each dispensing device, including sources that may backfeed into the dispenser.

The disconnecting means required by this section must be capable of being locked in the open position. In order to ensure that the ability to lock the disconnecting means in the open position is always available, the method employed should not be portable or easily removed from the disconnecting means.

514.16 Grounding

All metal raceways, the metal armor or metallic sheath on cables, and all non–current-carrying metal parts of fixed portable electrical equipment, regardless of voltage, shall be grounded as provided in Article 250. Grounding in Class I locations shall comply with 501.130.

ARTICLE 515
Bulk Storage Plants

Contents

FPN: Rules that are followed by a reference in brackets contain text that has been extracted from NFPA 30-2003, *Flammable and Combustible Liquids Code*. Only editorial changes were made to the extracted text to make it consistent with this *Code*.

515.1 Scope

This article covers a property or portion of a property where flammable liquids are received by tank vessel, pipelines, tank car, or tank vehicle and are stored or blended in bulk for the purpose of distributing such liquids by tank vessel, pipeline, tank car, tank vehicle, portable tank, or container.

515.2 Definition

Bulk Plant or Terminal. That portion of a property where liquids are received by tank vessel, pipelines, tank car, or tank vehicle and are stored or blended in bulk for the purpose of distributing such liquids by tank vessel, pipeline, tank car, tank vehicle, portable tank, or container. [NFPA 30:3.3.32.1]

FPN: For further information, see NFPA 30-2003, *Flammable and Combustible Liquids Code*.

515.3 Class I Locations

Table 515.3 shall be applied where Class I liquids are stored, handled, or dispensed and shall be used to delineate and classify bulk storage plants. The class location shall not extend beyond a floor, wall, roof, or other solid partition that has no communicating openings. [NFPA 30:8.1, 8.2.2]

Table 515.3 is extracted from Table 8.2.2 of NFPA 30, *Flammable and Combustible Liquids Code*. Section 7.3.4.5 of

Table 515.3 Electrical Area Classifications

Location	NEC Class I Division	Zone	Extent of Classified Area
Indoor equipment installed in accordance with Section 5.3 of NFPA 30 where flammable vapor–air mixtures can exist under normal operation	1	0	The entire area associated with such equipment where flammable gases or vapors are present continuously or for long periods of time
	1	1	Area within 1.5 m (5 ft) of any edge of such equipment, extending in all directions
	2	2	Area between 1.5 m and 2.5 m (5 ft and 8 ft) of any edge of such equipment, extending in all directions; also, space up to 900 mm (3 ft) above floor or grade level within 1.5 m to 7.5 m (5 ft to 25 ft) horizontally from any edge of such equipment[1]
Outdoor equipment of the type covered in Section 5.3 of NFPA 30 where flammable vapor–air mixtures may exist under normal operation	1	0	The entire area associated with such equipment where flammable gases or vapors are present continuously or for long periods of time
	1	1	Area within 900 mm (3 ft) of any edge of such equipment, extending in all directions
	2	2	Area between 900 mm (3 ft) and 2.5 m (8 ft) of any edge of such equipment, extending in all directions; also, space up to 900 mm (3 ft) above floor or grade level within 900 mm to 3.0 m (3 ft to 10 ft) horizontally from any edge of such equipment
Tank storage installations inside buildings	1	1	All equipment located below grade level
	2	2	Any equipment located at or above grade level
Tank – aboveground	1	0	Inside fixed roof tank
	1	1	Area inside dike where dike height is greater than the distance from the tank to the dike for more than 50 percent of the tank circumference
Shell, ends, or roof and dike area	2	2	Within 3.0 m (10 ft) from shell, ends, or roof of tank; also, area inside dike to level of top of tank
Vent	1	0	Area inside of vent piping or opening
	1	1	Within 1.5 m (5 ft) of open end of vent, extending in all directions
	2	2	Area between 1.5 m and 3.0 m (5 ft and 10 ft) from open end of vent, extending in all directions
Floating roof with fixed outer roof	1	0	Area between the floating and fixed roof sections and within the shell
Floating roof with no fixed outer roof	1	1	Area above the floating roof and within the shell
Underground tank fill opening	1	1	Any pit, or space below grade level, if any part is within a Division 1 or 2, or Zone 1 or 2, classified location
	2	2	Up to 450 mm (18 in.) above grade level within a horizontal radius of 3.0 m (10 ft) from a loose fill connection, and within a horizontal radius of 1.5 m (5 ft) from a tight fill connection
Vent – discharging upward	1	0	Area inside of vent piping or opening
	1	1	Within 900 mm (3 ft) of open end of vent, extending in all directions
	2	2	Area between 900 mm and 1.5 m (3 ft and 5 ft) of open end of vent, extending in all directions

Table 515.3 *Continued*

Location	NEC Class I Division	Zone	Extent of Classified Area
Drum and container filling – outdoors or indoors	1	0	Area inside the drum or container
	1	1	Within 900 mm (3 ft) of vent and fill openings, extending in all directions
	2	2	Area between 900 mm and 1.5 m (3 ft and 5 ft) from vent or fill opening, extending in all directions; also, up to 450 mm (18 in.) above floor or grade level within a horizontal radius of 3.0 m (10 ft) from vent or fill opening
Pumps, bleeders, withdrawal fittings, Indoors	2	2	Within 1.5 m (5 ft) of any edge of such devices, extending in all directions; also, up to 900 mm (3 ft) above floor or grade level within 7.5 m (25 ft) horizontally from any edge of such devices
Outdoors	2	2	Within 900 mm (3 ft) of any edge of such devices, extending in all directions. Also, up to 450 mm (18 in.) above grade level within 3.0 m (10 ft) horizontally from any edge of such devices
Pits and sumps Without mechanical ventilation	1	1	Entire area within a pit or sump if any part is within a Division 1 or 2, or Zone 1 or 2, classified location
With adequate mechanical ventilation	2	2	Entire area within a pit or sump if any part is within a Division 1 or 2, or Zone 1 or 2, classified location
Containing valves, fittings, or piping, and not within a Division 1 or 2, or Zone 1 or 2, classified location	2	2	Entire pit or sump
Drainage ditches, separators, impounding basins Outdoors	2	2	Area up to 450 mm (18 in.) above ditch, separator, or basin; also, area up to 450 mm (18 in.) above grade within 4.5 m (15 ft) horizontally from any edge
Indoors			Same classified area as pits
Tank vehicle and tank car[2] loading through open dome	1	0	Area inside of the tank
	1	1	Within 900 mm (3 ft) of edge of dome, extending in all directions
	2	2	Area between 900 mm and 4.5 m (3 ft and 15 ft) from edge of dome, extending in all directions
Loading through bottom connections with atmospheric venting	1	0	Area inside of the tank
	1	1	Within 900 mm (3 ft) of point of venting to atmosphere, extending in all directions
	2	2	Area between 900 mm and 4.5 m (3 ft and 15 ft) from point of venting to atmosphere, extending in all directions; also, up to 450 mm (18 in.) above grade within a horizontal radius of 3.0 m (10 ft) from point of loading connection

(continues)

Table 515.3 *Continued*

Location	NEC Class I Division	Zone	Extent of Classified Area
Office and rest rooms	Ordinary		If there is any opening to these rooms within the extent of an indoor classified location, the room shall be classified the same as if the wall, curb, or partition did not exist.
Loading through closed dome with atmospheric venting	1	1	Within 900 mm (3 ft) of open end of vent, extending in all directions
	2	2	Area between 900 mm and 4.5 m (3 ft and 15 ft) from open end of vent, extending in all directions; also, within 900 mm (3 ft) of edge of dome, extending in all directions
Loading through closed dome with vapor control	2	2	Within 900 mm (3 ft) of point of connection of both fill and vapor lines extending in all directions
Bottom loading with vapor control or any bottom unloading	2	2	Within 900 mm (3 ft) of point of connections, extending in all directions; also up to 450 mm (18 in.) above grade within a horizontal radius of 3.0 m (10 ft) from point of connections
Storage and repair garage for tank vehicles	1	1	All pits or spaces below floor level
	2	2	Area up to 450 mm (18 in.) above floor or grade level for entire storage or repair garage
Garages for other than tank vehicles	Ordinary		If there is any opening to these rooms within the extent of an outdoor classified location, the entire room shall be classified the same as the area classification at the point of the opening.
Outdoor drum storage	Ordinary		
Inside rooms or storage lockers used for the storage of Class I liquids	2	2	Entire room
Indoor warehousing where there is no flammable liquid transfer	Ordinary		If there is any opening to these rooms within the extent of an indoor classified location, the room shall be classified the same as if the wall, curb, or partition did not exist.
Piers and wharves			See Figure 515.3.

[1]The release of Class I liquids may generate vapors to the extent that the entire building, and possibly an area surrounding it, should be considered a Class I, Division 2 or Zone 2 location.
[2]When classifying extent of area, consideration shall be given to fact that tank cars or tank vehicles may be spotted at varying points. Therefore, the extremities of the loading or unloading positions shall be used.
[NFPA 30:Table 8.2.2]

15 m (50 ft) 15 m (50 ft)

7.5 m (25 ft) 7.5 m (25 ft)

Deck

7.5 m (25 ft)

Operating envelope and stored position of loading arms or hoses

Open sump in deck for draining lines and hoses

15 m (50 ft)

600 mm (2 ft)

7.5 m (25 ft)

Approach

Pier

Shore

Water level

▢XX Division 1   ▢// Division 2   ▢ Unclassified

Notes:
(1) The "source of vapor" shall be the operating envelope and stored position of the outboard flange connection of the loading arm (or hose).
(2) The berth area adjacent to tanker and barge cargo tanks is to be Division 2 to the following extent:
    a. 7.6 m (25 ft) horizontally in all directions on the pier side from that portion of the hull containing cargo tanks
    b. From the water level to 7.6 m (25 ft) above the cargo tanks at their highest position
(3) Additional locations may have to be classified as required by the presence of other sources of flammable liquids on the berth, by Coast Guard, or other regulations.

Figure 515.3 Marine Terminal Handling Flammable Liquids [NFPA 30:Figure 7.7.16]

NFPA 30, as referenced in the first two items in column 1 of Table 515.3, states:

7.3.4.5 Where equipment such as dispensing stations, open centrifuges, plate and frame filters, and open vacuum filters is used in a building, the equipment and ventilation of the building shall be designed to limit flammable vapor–air mixtures under normal operating conditions to the interior of equipment and to not more than 1.5 m (5 ft) from equipment that exposes Class I liquids to the air. [NFPA 30:7.3.4.5]

Exhibits 515.1 through 515.7 illustrate the hazardous (classified) locations associated with several types of flammable liquid containers and operations.

FPN No. 1: The area classifications listed in Table 515.3 are based on the premise that the installation meets the applicable requirements of NFPA 30-2003, *Flammable and Combustible Liquids Code*, Chapter 5, in all respects.

Should this not be the case, the authority having jurisdiction has the authority to classify the extent of the classified space.
FPN No. 2: See 555.21 for gasoline dispensing stations in marinas and boatyards.

515.4 Wiring and Equipment Located in Class I Locations

All electrical wiring and equipment within the Class I locations defined in 515.3 shall comply with the applicable provisions of Article 501 or Article 505 for the division or zone in which they are used.

Exception: As permitted in 515.8.

515.7 Wiring and Equipment Above Class I Locations

(A) Fixed Wiring All fixed wiring above Class I locations shall be in metal raceways or PVC Schedule 80 rigid nonmetallic conduit, or equivalent, or be Type MI, TC, or MC cable.

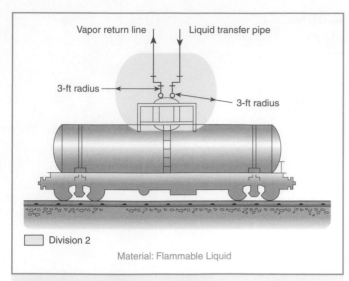

Exhibit 515.1 Tank car/tank truck loading and unloading via closed system. Transfer through dome only.

Exhibit 515.2 Tank car/tank truck loading and unloading via closed system. Bottom product transfer only.

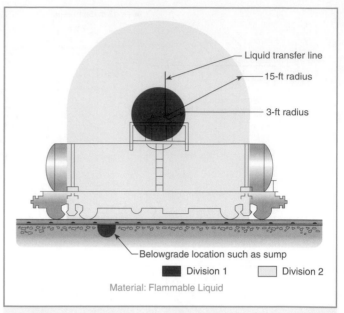

Exhibit 515.3 Tank car/tank truck loading and unloading via open system. Top or bottom product transfer.

Exhibit 515.4 Tank car/tank truck loading and unloading via closed system. Transfer through dome only.

(B) Fixed Equipment Fixed equipment that may produce arcs, sparks, or particles of hot metal, such as lamps and lampholders for fixed lighting, cutouts, switches, receptacles, motors, or other equipment having make-and-break or sliding contacts, shall be of the totally enclosed type or be constructed so as to prevent the escape of sparks or hot metal particles.

(C) Portable Lamps or Other Utilization Equipment Portable lamps or other utilization equipment and their flexible cords shall comply with the provisions of Article 501 or Article 505 for the class of location above which they are connected or used.

515.8 Underground Wiring

(A) Wiring Method Underground wiring shall be installed in threaded rigid metal conduit or threaded steel intermediate metal conduit or, where buried under not less than 600 mm (2 ft) of cover, shall be permitted in rigid nonmetallic conduit or a listed cable. Where rigid nonmetallic conduit is used, threaded rigid metal conduit or threaded steel intermediate

Exhibit 515.5 Drum filling station, outdoors or indoors, with adequate ventilation.

Exhibit 515.6 Storage tanks for cryogenic liquids. [From NFPA 59A, *Standard for the Production, Storage, and Handling of Liquefied Natural Gas (LNG)*]

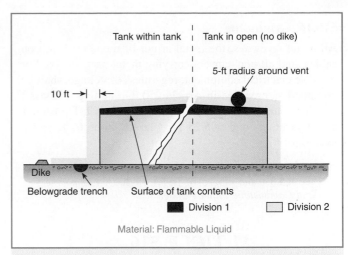

Exhibit 515.7 Storage tanks, outdoors at grade. (From API RP 500, *Recommended Practice for Classification of Locations for Electrical Installations at Petroleum Facilities Classified as Class I, Division 1 and Division 2*)

metal conduit shall be used for the last 600 mm (2 ft) of the conduit run to emergence or to the point of connection to the aboveground raceway. Where cable is used, it shall be enclosed in threaded rigid metal conduit or threaded steel intermediate metal conduit from the point of lowest buried cable level to the point of connection to the aboveground raceway.

See the commentary following 514.8 for more information on underground wiring installed below hazardous (classified) locations.

(B) Insulation Conductor insulation shall comply with 501.20.

(C) Nonmetallic Wiring Where rigid nonmetallic conduit or cable with a nonmetallic sheath is used, an equipment grounding conductor shall be included to provide for electrical continuity of the raceway system and for grounding of non–current-carrying metal parts.

515.9 Sealing

Sealing requirements shall apply to horizontal as well as to vertical boundaries of the defined Class I locations. Buried raceways and cables under defined Class I locations shall be considered to be within a Class I, Division 1 or Zone 1 location.

515.10 Special Equipment — Gasoline Dispensers

Where gasoline or other volatile flammable liquids or liquefied flammable gases are dispensed at bulk stations, the applicable provisions of Article 514 shall apply.

515.16 Grounding

All metal raceways, the metal armor or metallic sheath on cables, and all non–current-carrying metal parts of fixed or portable electrical equipment, regardless of voltage, shall be grounded as provided in Article 250. Grounding in Class I locations shall comply with 501.30 for Class I, Division 1 and 2 locations and 505.25 for Class I, Zone 0, 1, and 2 locations.

> FPN: For information on grounding for static protection, see 4.5.3.4 and 4.5.3.5 of NFPA 30-2003, *Flammable and Combustible Liquids Code.*

ARTICLE 516
Spray Application, Dipping, and Coating Processes

Summary of Changes

- **516.3(A):** Added and revised text to allow the use of the zone area classification concept to be applied to spray application, dipping, and coating processes.
- **516.3(B)(6):** Revised text to classify specific locations surrounding Class I, Division 1 or Zone 0 areas as Class I, Division 1 or Zone 1 locations.
- **516.3(C)(7):** Added requirement for classification of space near open containers as a Class I, Division 2 or Zone 2 location.

Contents

> FPN: Rules that are followed by a reference in brackets contain text that has been extracted from NFPA 33-2003, *Standard for Spray Application Using Flammable and Combustible Materials*, or NFPA 34-2003, *Standard for Dipping and Coating Processes Using Flammable or Combustible Liquids*. Only editorial changes were made to the extracted text to make it consistent with this *Code.*

516.1 Scope

This article covers the regular or frequent application of flammable liquids, combustible liquids, and combustible powders by spray operations and the application of flammable liquids, or combustible liquids at temperatures above their flashpoint, by dipping, coating, or other means.

> FPN: For further information regarding safeguards for these processes, such as fire protection, posting of warning signs, and maintenance, see NFPA 33-2003, *Standard for Spray Application Using Flammable and Combustible Materials*, and NFPA 34-2003, *Standard for Dipping and Coating Processes Using Flammable or Combustible Liquids*. For additional information regarding ventilation, see NFPA 91-2004, *Standard for Exhaust Systems for Air Conveying of Vapors, Gases, Mists, and Noncombustible Particulate Solids.*

516.2 Definitions

For the purpose of this article, the following definitions shall apply.

Spray Area. Normally, locations outside of buildings or localized operations within a larger room or space. Such are normally provided with some local vapor extraction/ventilation system. In automated operations, the area limits shall be the maximum area in the direct path of spray operations. In manual operations, the area limits shall be the maximum area of spray when aimed at 180 degrees to the application surface.

Spray Booth. An enclosure or insert within a larger room used for spray/coating/dipping applications. A spray booth may be fully enclosed or have open front or face and may include separate conveyor entrance and exit. The spray booth is provided with a dedicated ventilation exhaust but may draw supply air from the larger room or have a dedicated air supply.

Spray Room. A purposefully enclosed room built for spray/coating/dipping applications provided with dedicated ventilation supply and exhaust. Normally the room is configured to house the item to be painted, providing reasonable access around the item/process. Depending on the size of the item being painted, such rooms may actually be the entire building or the major portion thereof.

516.3 Classification of Locations

Classification is based on dangerous quantities of flammable vapors, combustible mists, residues, dusts, or deposits.

(A) Class I, Division 1 or Class I, Zone 0 Locations The following spaces shall be considered Class I, Division 1, or Class I, Zone 0, as applicable:

(1) The interior of any open or closed container of a flammable liquid
(2) The interior of any dip tank or coating tank

> FPN: For additional guidance and explanatory diagrams, see 4.3.5 of NFPA 33-2003, *Standard for Spray Application Using Flammable or Combustible Materials*, and Sections 4.2, 4.3, and 4.4 of NFPA 34-2003, *Standard for Dipping and Coating Processes Using Flammable or Combustible Liquids.*

Article 516 was revised for the 2005 *Code* to correlate with the other hazardous (classified) locations covered by the *Code*. The Zone 0, Zone 1, and Zone 2 classification for Class I material used in spray, dipping, and coating processes is designated in the requirements.

(B) Class I or Class II, Division 1 Locations The following spaces shall be considered Class I, Division 1, or Class I, Zone 1, or Class II, Division 1 locations, as applicable:

(1) The interior of spray booths and rooms except as specifically provided in 516.3(D).
(2) The interior of exhaust ducts.
(3) Any area in the direct path of spray operations.
(4) For open dipping and coating operations, all space within a 1.5-m (5-ft) radial distance from the vapor sources extending from these surfaces to the floor. The vapor source shall be the liquid exposed in the process and the drainboard, and any dipped or coated object from which it is possible to measure vapor concentrations exceeding 25 percent of the lower flammable limit at a distance of 300 mm (1 ft), in any direction, from the object.
(5) Sumps, pits, or belowgrade channels within 7.5 m (25 ft) horizontally of a vapor source. If the sump, pit, or channel extends beyond 7.5 m (25 ft) from the vapor source, it shall be provided with a vapor stop or it shall be classified as Class I, Division 1 for its entire length.
(6) All space in all directions outside of but within 900 mm (3 ft) of open containers, supply containers, spray gun cleaners, and solvent distillation units containing flammable liquids.

(C) Class I or Class II, Division 2 Locations The following spaces shall be considered Class I, Division 2 or Class I, Zone 2, or Class II, Division 2 as applicable.

(1) Open Spraying For open spraying, all space outside of but within 6 m (20 ft) horizontally and 3 m (10 ft) vertically of the Class I, Division 1 or Class I, Zone 1 location as defined in 516.3(A), and not separated from it by partitions. See Figure 516.3(B)(1). [NFPA 33:6.5.1]

Figure 516.3(B)(1) Electrical Area Classification for Open Spray Areas [NFPA 33:Figure 6.5.1]

(2) Closed-Top, Open-Face, and Open-Front Spraying If spray application operations are conducted within a closed-top, open-face, or open-front booth or room, any electrical wiring or utilization equipment located outside of the booth or room but within the boundaries designated as Division 2 or Zone 2 in Figure 516.3(B)(2) shall be suitable for Class I, Division 2, Class I, Zone 2, or Class II, Division 2 locations, whichever is applicable. The Class I, Division 2, Class I, Zone 2, or Class II, Division 2 locations shown in Figure 516.3(B)(2) shall extend from the edges of the open face or open front of the booth or room in accordance with the following:

(a) If the exhaust ventilation system is interlocked with the spray application equipment, the Division 2 or Zone 2

Figure 516.3(B)(2) Class I, Division 2, Class I, Zone 2, or Class II, Division 2 Locations Adjacent to a Closed Top, Open Face, or Open Front Spray Booth or Room [NFPA 33:Figures 6.5.2(a) and 6.5.2(b)]

location shall extend 1.5 m (5 ft) horizontally and 900 mm (3 ft) vertically from the open face or open front of the booth or room, as shown in Figure 516.3(B)(2), top.

(b) If the exhaust ventilation system is not interlocked with the spray application equipment, the Division 2 or Zone 2 location shall extend 3 m (10 ft) horizontally and 900 mm (3 ft) vertically from the open face or open front of the booth or room, as shown in Figure 516.3(B)(2), bottom.

For the purposes of this subsection, *interlocked* shall mean that the spray application equipment cannot be operated unless the exhaust ventilation system is operating and functioning properly and spray application is automatically stopped if the exhaust ventilation system fails. [NFPA 33:6.5.2]

(3) Open-Top Spraying For spraying operations conducted within an open top spray booth, the space 900 mm (3 ft) vertically above the booth and within 900 mm (3 ft)

of other booth openings shall be considered Class I, Division 2, Class I, Zone 2, or Class II, Division 2. [NFPA 33:6.5.3]

(4) Enclosed Booths and Rooms For spraying operations confined to an enclosed spray booth or room, the space within 900 mm (3 ft) in all directions from any openings shall be considered Class I, Division 2, Class I, or Zone 2, or Class II, Division 2 as shown in Figure 516.3(B)(4). [NFPA 33:6.5.4]

Figure 516.3(B)(4) Class I, Division 2, Class I, Zone 2, or Class II, Division 2 Locations Adjacent to an Enclosed Spray Booth or Spray Room [NFPA 33:Figure 6.5.4]

(5) Dip Tanks and Drain Boards — Surrounding Space For dip tanks and drain boards, the 914-mm (3-ft) space surrounding the Class I, Division 1 or Class I, Zone 1 location as defined in 516.3(A)(4) and as shown in Figure 516.3(B)(5). [NFPA 34:6.4.3]

(6) Dip Tanks and Drain Boards — Space Above Floor For dip tanks and drain boards, the space 900 mm (3 ft) above the floor and extending 6 m (20 ft) horizontally in all directions from the Class I, Division 1 or Class I, Zone 1 location.

Exception: This space shall not be required to be considered a hazardous (classified) location where the vapor source area is 0.46 m^2 (5 ft^2) or less and where the contents of the open tank trough or container do not exceed 19 L (5 gal).

Figure 516.3(B)(5) Electrical Area Classification for Open Processes Without Vapor Containment or Ventilation [NFPA 34:Figure 6.4(a)]

In addition, the vapor concentration during operation and shutdown periods shall not exceed 25 percent of the lower flammable limit outside the Class I location specified in 516.3(A)(4). [NFPA 34:6.4.4]

(7) Open Containers All space in all directions within 600 mm (2 ft) of the Division 1 or Zone 1 area surrounding open containers, supply containers, spray gun cleaners, and solvent distillation units containing flammable liquids, as well as the area extending 1.5 m (5 ft) beyond the Division 1 or Zone 1 area up to a height of 460 mm (18 in.) above the floor or grade level. [NFPA 33:6.5.5.1(2)]

> The requirement in 516.3(C)(7) was added to the 2005 *Code* to define the parameters of the hazardous (classified) location around open containers containing flammable liquids. This area classification requirement is extracted from 6.5.5.1(2) of NFPA 33-2003, *Standard for Spray Application Using Flammable and Combustible Materials.*

(D) Enclosed Coating and Dipping Operations The space adjacent to an enclosed dipping or coating process or apparatus shall be considered unclassified. [NFPA 34:6.5.2]

Exception: The space within 900 mm (3 ft) in all directions from any opening in the enclosures shall be classified as Class I, Division 2 or Class I, Zone 2, as applicable. [NFPA 34:6.5.3]

(E) Adjacent Locations Adjacent locations that are cut off from the defined Class I or Class II locations by tight partitions without communicating openings, and within which

flammable vapors or combustible powders are not likely to be released, shall be unclassified.

(F) Unclassified Locations Locations using drying, curing, or fusion apparatus and provided with positive mechanical ventilation adequate to prevent accumulation of flammable concentrations of vapors, and provided with effective interlocks to de-energize all electrical equipment (other than equipment identified for Class I locations) in case the ventilating equipment is inoperative, shall be permitted to be unclassified where the authority having jurisdiction so judges.

> FPN: For further information regarding safeguards, see NFPA 86-2003, *Standard for Ovens and Furnaces.*

516.4 Wiring and Equipment in Class I Locations

(A) Wiring and Equipment — Vapors All electric wiring and equipment within the Class I location (containing vapor only — not residues) defined in 516.3 shall comply with the applicable provisions of Article 501 or Article 505, as applicable.

(B) Wiring and Equipment — Vapors and Residues Unless specifically listed for locations containing deposits of dangerous quantities of flammable or combustible vapors, mists, residues, dusts, or deposits (as applicable), there shall be no electrical equipment in any spray area as herein defined whereon deposits of combustible residue may readily accumulate, except wiring in rigid metal conduit, intermediate metal conduit, Type MI cable, or in metal boxes or fittings containing no taps, splices, or terminal connections. [NFPA 33:6.4.2]

Only electrical equipment that is specifically listed for the location is permitted where deposits of combustible residue may accumulate, as stated in 516.4(B). The latter part of the requirement in 516.4(B), however, permits wiring in rigid metal conduit, intermediate metal conduit, Type MI cable, or metal boxes or fittings without splices, taps, or terminal connections.

(C) Illumination Illumination of readily ignitible areas through panels of glass or other transparent or translucent material shall be permitted only if it complies with the following:

(1) Fixed lighting units are used as the source of illumination.
(2) The panel effectively isolates the Class I location from the area in which the lighting unit is located.
(3) The lighting unit is identified for its specific location.
(4) The panel is of a material or is protected so that breakage is unlikely.
(5) The arrangement is such that normal accumulations of hazardous residue on the surface of the panel will not be raised to a dangerous temperature by radiation or conduction from the source of illumination.

The UL *Hazardous Locations Equipment Directory* indicates that listed hazardous location fixtures suitable for use in paint spray booths are evaluated for deposits of readily combustible paint residues on the side of the fixture that forms part of the interior ceiling or wall surface of the spray booth.

(D) Portable Equipment Portable electric lamps or other utilization equipment shall not be used in a spray area during spray operations.

Exception No. 1: Where portable electric lamps are required for operations in spaces not readily illuminated by fixed lighting within the spraying area, they shall be of the type identified for Class I, Division 1 or Class 1, Zone 1 locations where readily ignitible residues may be present. [NFPA 33:6.9]

Exception No. 2: Where portable electric drying apparatus is used in automobile refinishing spray booths and the following requirements are met.

(a) The apparatus and its electrical connections are not located within the spray enclosure during spray operations.

(b) Electrical equipment within 450 mm (18 in.) of the floor is identified for Class I, Division 2 or Class I, Zone 2 locations.

(c) All metallic parts of the drying apparatus are electrically bonded and grounded.

(d) Interlocks are provided to prevent the operation of spray equipment while drying apparatus is within the spray enclosure, to allow for a 3-minute purge of the enclosure before energizing the drying apparatus and to shut off drying apparatus on failure of ventilation system.

(E) Electrostatic Equipment Electrostatic spraying or detearing equipment shall be installed and used only as provided in 516.10.

FPN: For further information, see NFPA 33-2003, *Standard for Spray Application Using Flammable or Combustible Materials.*

516.7 Wiring and Equipment Not Within Class I and II Locations

(A) Wiring All fixed wiring above the Class I and II locations shall be in metal raceways, rigid nonmetallic conduit, or electrical nonmetallic tubing, or shall be Type MI, TC, or MC cable. Cellular metal floor raceways shall be permitted only for supplying ceiling outlets or extensions to the area below the floor of a Class I or II location, but such raceways shall have no connections leading into or through the Class I or II location above the floor unless suitable seals are provided.

(B) Equipment Equipment that may produce arcs, sparks, or particles of hot metal, such as lamps and lampholders for fixed lighting, cutouts, switches, receptacles, motors, or other equipment having make-and-break or sliding contacts, where installed above a Class I or II location or above a location where freshly finished goods are handled, shall be of the totally enclosed type or be constructed so as to prevent the escape of sparks or hot metal particles.

516.10 Special Equipment

(A) Fixed Electrostatic Equipment This section shall apply to any equipment using electrostatically charged elements for the atomization, charging, and/or precipitation of hazardous materials for coatings on articles or for other similar purposes in which the charging or atomizing device is attached to a mechanical support or manipulator. This shall include robotic devices. This section shall not apply to devices that are held or manipulated by hand. Where robot or programming procedures involve manual manipulation of the robot arm while spraying with the high voltage on, the provisions of 516.10(B) shall apply. The installation of electrostatic spraying equipment shall comply with 516.10(A)(1) through (A)(10). Spray equipment shall be listed. All automatic electrostatic equipment systems shall comply with 516.4(A)(1) through (A)(9).

(1) Power and Control Equipment Transformers, high-voltage supplies, control apparatus, and all other electric portions of the equipment shall be installed outside of the Class I location as defined in 516.3 or be of a type identified for the location.

Exception: High-voltage grids, electrodes, electrostatic atomizing heads, and their connections shall be permitted within the Class I location.

(2) Electrostatic Equipment Electrodes and electrostatic atomizing heads shall be adequately supported in permanent locations and shall be effectively insulated from ground. Electrodes and electrostatic atomizing heads that are permanently attached to their bases, supports, reciprocators, or robots shall be deemed to comply with this section.

(3) High-Voltage Leads High-voltage leads shall be properly insulated and protected from mechanical damage or exposure to destructive chemicals. Any exposed element at high voltage shall be effectively and permanently supported on suitable insulators and shall be effectively guarded against accidental contact or grounding.

(4) Support of Goods Goods being coated using this process shall be supported on conveyors or hangers. The conveyors or hangers shall be arranged (1) to ensure that the parts being coated are electrically connected to ground with a resistance of 1 megohm or less and (2) to prevent parts from swinging.

(5) Automatic Controls Electrostatic apparatus shall be equipped with automatic means that will rapidly de-energize the high-voltage elements under any of the following conditions:

(1) Stoppage of ventilating fans or failure of ventilating equipment from any cause
(2) Stoppage of the conveyor carrying goods through the high-voltage field unless stoppage is required by the spray process
(3) Occurrence of excessive current leakage at any point in the high-voltage system
(4) De-energizing the primary voltage input to the power supply

(6) Grounding All electrically conductive objects in the spray area, except those objects required by the process to be at high voltage, shall be adequately grounded. This requirement shall apply to paint containers, wash cans, guards, hose connectors, brackets, and any other electrically conductive objects or devices in the area.

(7) Isolation Safeguards such as adequate booths, fencing, railings, interlocks, or other means shall be placed about the equipment or incorporated therein so that they, either by their location, character, or both, ensure that a safe separation of the process is maintained.

(8) Signs Signs shall be conspicuously posted to convey the following:

(1) Designate the process zone as dangerous with regard to fire and accident
(2) Identify the grounding requirements for all electrically conductive objects in the spray area
(3) Restrict access to qualified personnel only

(9) Insulators All insulators shall be kept clean and dry.

(10) Other Than Nonincendive Equipment Spray equipment that cannot be classified as nonincendive shall comply with (A)(10)(a) and (A)(10)(b).

(a) Conveyors or hangers shall be arranged so as to maintain a safe distance of at least twice the sparking distance between goods being painted and electrodes, electrostatic atomizing heads, or charged conductors. Warnings defining this safe distance shall be posted.

(b) The equipment shall provide an automatic means of rapidly de-energizing the high-voltage elements in the event the distance between the goods being painted and the electrodes or electrostatic atomizing heads falls below that specified in (a). [NFPA 33:Chapter 11]

(B) Electrostatic Hand-Spraying Equipment This section shall apply to any equipment using electrostatically charged elements for the atomization, charging, and/or precipitation of materials for coatings on articles, or for other similar purposes in which the atomizing device is hand-held or manipulated during the spraying operation. Electrostatic hand-spraying equipment and devices used in connection with paint-spraying operations shall be of listed types and shall comply with 516.10(B)(1) through (B)(5).

(1) General The high-voltage circuits shall be designed so as not to produce a spark of sufficient intensity to ignite the most readily ignitible of those vapor–air mixtures likely to be encountered, or result in appreciable shock hazard upon coming in contact with a grounded object under all normal operating conditions. The electrostatically charged exposed elements of the handgun shall be capable of being energized only by an actuator that also controls the coating material supply.

(2) Power Equipment Transformers, power packs, control apparatus, and all other electric portions of the equipment shall be located outside of the Class I location or be identified for the location.

Exception: The handgun itself and its connections to the power supply shall be permitted within the Class I location.

(3) Handle The handle of the spraying gun shall be electrically connected to ground by a metallic connection and be constructed so that the operator in normal operating position is in intimate electrical contact with the grounded handle to prevent buildup of a static charge on the operator's body. Signs indicating the necessity for grounding other persons entering the spray area shall be conspicuously posted.

(4) Electrostatic Equipment All electrically conductive objects in the spraying area shall be adequately grounded. This requirement shall apply to paint containers, wash cans, and any other electrical conductive objects or devices in the area. The equipment shall carry a prominent, permanently installed warning regarding the necessity for this grounding feature.

(5) Support of Objects Objects being painted shall be maintained in metallic contact with the conveyor or other grounded support. Hooks shall be regularly cleaned to ensure adequate grounding of 1 megohm or less. Areas of contact shall be sharp points or knife edges where possible. Points of support of the object shall be concealed from random spray where feasible; and, where the objects being sprayed are supported from a conveyor, the point of attachment to the conveyor shall be located so as to not collect spray material during normal operation. [NFPA 33:Chapter 12]

(C) Powder Coating This section shall apply to processes in which combustible dry powders are applied. The hazards associated with combustible dusts are present in such a process to a degree, depending on the chemical composition of the material, particle size, shape, and distribution.

(1) Electric Equipment and Sources of Ignition Electric equipment and other sources of ignition shall comply with the requirements of Article 502. Portable electric lamps and other utilization equipment shall not be used within a Class II location during operation of the finishing processes. Where such lamps or utilization equipment are used during cleaning or repairing operations, they shall be of a type identified for Class II, Division 1 locations, and all exposed metal parts shall be effectively grounded.

Exception: Where portable electric lamps are required for operations in spaces not readily illuminated by fixed lighting within the spraying area, they shall be of the type listed for Class II, Division 1 locations where readily ignitible residues may be present.

(2) Fixed Electrostatic Spraying Equipment The provisions of 516.10(A) and 516.10(C)(1) shall apply to fixed electrostatic spraying equipment.

(3) Electrostatic Hand-Spraying Equipment The provisions of 516.10(B) and 516.10(C)(1) shall apply to electrostatic hand-spraying equipment.

(4) Electrostatic Fluidized Beds Electrostatic fluidized beds and associated equipment shall be of identified types.

The high-voltage circuits shall be designed so that any discharge produced when the charging electrodes of the bed are approached or contacted by a grounded object shall not be of sufficient intensity to ignite any powder–air mixture likely to be encountered or to result in an appreciable shock hazard.

(a) Transformers, power packs, control apparatus, and all other electric portions of the equipment shall be located outside the powder-coating area or shall otherwise comply with the requirements of 516.10(C)(1).

Exception: The charging electrodes and their connections to the power supply shall be permitted within the powder-coating area.

(b) All electrically conductive objects within the powder-coating area shall be adequately grounded. The powder-coating equipment shall carry a prominent, permanently installed warning regarding the necessity for grounding these objects.

(c) Objects being coated shall be maintained in electrical contact (less than 1 megohm) with the conveyor or other support in order to ensure proper grounding. Hangers shall be regularly cleaned to ensure effective electrical contact. Areas of electrical contact shall be sharp points or knife edges where possible.

(d) The electric equipment and compressed air supplies shall be interlocked with a ventilation system so that the equipment cannot be operated unless the ventilating fans are in operation. [NFPA 33:Chapter 15]

516.16 Grounding

All metal raceways, the metal armors or metallic sheath on cables, and all non–current-carrying metal parts of fixed or portable electrical equipment, regardless of voltage, shall be grounded as provided in Article 250. Grounding shall comply with 501.30, 502.30, or 505.25, as applicable.

NFPA 33, *Standard for Spray Application Using Flammable or Combustible Materials*, covers the spray application of flammable or combustible materials by means of compressed air atomization, airless or hydraulic atomization, electrostatic application methods, or any other means in continuous or intermittent processes. NFPA 33 also covers the application of combustible powders applied by powder spray guns, electrostatic powder spray guns, and fluidized bed application method or electrostatic fluidized bed application method. NFPA 33 also contains requirements for the maintenance of safe conditions as well as personal safety.

The proper maintenance and operation of processes and process areas where flammable and combustible materials are handled and applied are critical with respect to the protection of life and property from fire or explosion. An analysis of actual experience in industry demonstrates that the largest fire losses and frequency of fires have occurred where the proper codes and standards have not been used or applied properly.

Notes on Electrical Installations

The safety of life and property from fire or explosion as a result of spray applications of flammable and combustible materials, such as paints, finishes, and adhesives, depends on the arrangement and operation of a particular installation. The principal hazards of spray application operations originate from flammable or combustible liquids or powders and their vapors or mists, as well as from highly combustible residues or powders.

Properly constructed spray booths, with adequate mechanical ventilation, may be used to discharge vapors or powder to a safe location and reduce the possibility of an explosion. In like manner, the accumulation of overspray residues, many of which are not only highly combustible but also subject to spontaneous ignition, can be controlled.

The elimination of all sources of ignition in those areas where flammable or combustible liquids, vapors, mists, or combustible residues are present, together with constant supervision and maintenance, is essential to the safe operation of spraying.

The human element necessitates careful consideration of the location of the operation and the installation of extinguishing equipment to reduce the possibility of fire spreading to other property and to minimize the probability of damage to other property.

No open flames or spark-producing equipment should be in any area where, because of inadequate ventilation, explosive vapor–air mixtures or mists are present. Equally important, no open flames or spark-producing equipment should be located where highly combustible spray residues will be deposited on them. Because some residues may be ignited at very low temperatures, additional consideration must be given to operating temperatures of equipment subject to residue deposits.

Many deposits may be ignited at temperatures produced by low-pressure steam pipes or incandescent light globes, even fixtures of the explosionproof type.

It should be noted that electrical equipment is generally not permitted inside any spray booth, in the exhaust duct from a spray booth, in the entrained air of an exhaust system from a spraying operation, or in the direct path of spray, unless such equipment is specifically listed for both readily ignitible deposits and flammable vapor.

The determination of the extent of hazardous areas involved in spray application requires an understanding of the multiple hazards of flammable vapors, mists, powders, and highly combustible deposits applied at each individual location.

Where electrical equipment is installed in locations not subject to deposits of combustible residues but, due to inadequate ventilation, is subject to explosive concentrations of flammable vapors or mists, only approved explosionproof or other types of equipment approved for Class I, Division 1 locations (e.g., purged, pressurized, or intrinsically safe equipment or systems) are permitted.

Where spray areas contain dangerous quantities of flammable or combustible vapors, mists, residues, dusts, or deposits under normal operation, the adjacent unpartitioned areas, which are safe under normal operating conditions but which may become dangerous due to accident or careless operation, should be given consideration. Equipment known to produce sparks or flames under normal operating conditions should not be installed in these adjacent unpartitioned areas.

Where spraying operations are confined to adequately ventilated spray booths or rooms, there should be no deposits of combustible residues or dangerous concentrations of flammable vapors, mists, or dusts outside the spray booth under normal operating conditions. In the interest of safety, however, unless separated by partitions, an area within a certain distance [see 516.3(C)] of the Class I (or Class II), Division 1 spraying area, depending on the arrangement, is classified as Division 2; that is, the area should contain no equipment that produces ignition-capable sparks under normal operation. Furthermore, within this distance, electric lamps must be enclosed to prevent hot particles from falling on freshly painted stock or other readily ignitible material and, if subject to physical damage, must be properly guarded. See 516.7(B).

Even though it is assumed that areas adjacent to spray booths (particularly where coating-material stocks are located) will be provided with ventilation sufficient to prevent the presence of flammable vapors or deposits, it is nevertheless advisable that electric lamps be totally enclosed to prevent hot particles from falling in any area where there may be freshly painted stock, accidentally spilled flammable or combustible materials, readily ignitible refuse, or flammable or combustible liquid containers that have been left open accidentally. See 516.7(B).

Where electric lamps are in areas subject to atmospheres of flammable vapor, lamps should be replaced while electricity is off; otherwise, there may be a spark from this source.

Sufficient lighting for coating operations, booth cleaning, and booth repair work should be provided at the time the equipment is installed, to avoid the use of temporary or emergency lamps connected to ordinary extension cords in this area. See 516.4(D). A satisfactory and practical method of lighting is the use of ¼-in.-thick wired or tempered glass panels in the top or sides of the spray booth, with electrical luminaires located outside the booth, avoiding the direct path of the spray. See 516.4(C).

To prevent sparks from the accumulation of static electricity, all electrically conductive objects, including metal parts of spray booths, exhaust ducts, piping systems conveying flammable or combustible liquids or paint, solvent tanks, and canisters, should be properly grounded. See Section 6.5 of NFPA 33 and Section 9.3 of NFPA 77, *Recommended Practice on Static Electricity.*

ARTICLE 517
Health Care Facilities

Summary of Changes

- **517.13(A) & (B):** Deleted reference to specific cable types. Any metal wiring method recognized by 250.118 as an equipment grounding conductor is suitable for use in patient care areas. Paragraph (B) has been revised to require that the insulated equipment grounding conductor be installed in a metal raceway or as part of a listed cable with a metallic armor or sheath assembly.

- **517.17(A):** Added requirement detailing the applicability of 517.17 to multiple-occupancy buildings containing health care facilities with critical care areas or life-support equipment.

- **517.26:** Added requirement that the essential electrical system meet the requirements of Article 700 unless amended by Article 517.

- **517.30(C)(3):** Revised to permit the use of listed Types AC and MC cables as a wiring method for the emergency systems of health care facilities where the cable is fished into existing walls or ceilings.

- **517.34(A)(6):** Added requirement that certain ventilating system equipment be arranged for delayed automatic connection to the alternate power source.

- **517.34(C):** Added requirement that ac generator accessories be arranged for non-delayed automatic connection to the alternate power source.

Contents

FPN: Rules that are followed by a reference in brackets contain text that has been extracted from NFPA 99-2002, *Standard for Health Care Facilities.* Only editorial changes were made to the extracted text to make it consistent with this *Code.*

I. General

517.1 Scope

The provisions of this article shall apply to electrical construction and installation criteria in health care facilities that provide services to human beings.

The requirements in Parts II and III not only apply to single-function buildings but are also intended to be individually applied to their respective forms of occupancy within a multifunction building (e.g., a doctor's examining room located within a limited care facility would be required to meet the provisions of 517.10).

> FPN: For information concerning performance, maintenance, and testing criteria, refer to the appropriate health care facilities documents.

The requirements of Article 517 are intended to apply to all types of health care facilities. The requirements for each type of health care facility are nevertheless intended to be applied in a very specific manner. For example, in a suite of doctors' offices within an office building, a doctor's business office would be treated as an ordinary occupancy and would be required to meet only the applicable portions of other parts of this *Code.* However, the examining rooms attached to the doctor's business office would be required to meet the provisions of Part II and 517.45 in Article 517.

The scope of Article 517 also includes health care facilities that may be mobile or supply very limited outpatient services. However, the scope does not include animal hospitals or veterinary offices.

Other standards referenced in Article 517 are NFPA 99, *Standard for Health Care Facilities;* NFPA 101®, *Life Safety Code®;* and NFPA 20, *Standard for the Installation of Stationary Pumps for Fire Protection.*

517.2 Definitions

Alternate Power Source. One or more generator sets, or battery systems where permitted, intended to provide power during the interruption of the normal electrical services or the public utility electrical service intended to provide power during interruption of service normally provided by the generating facilities on the premises.

As stated in 517.30(B)(6), alternate power sources are permitted to serve essential electrical systems of contiguous or multiple building facilities.

Ambulatory Health Care Facility. A building or part thereof used to provide services or treatment to four or more patients at the same time and meeting either (1) or (2).

(1) Those facilities that provide, on an outpatient basis, treatment for patients that would render them incapable of taking action for self-preservation under emergency conditions without assistance from others, such as hemodialysis units or freestanding emergency medical units.

(2) Those facilities that provide, on an outpatient basis, surgical treatment requiring general anesthesia.

Ambulatory health care facilities, such as outpatient surgery centers, freestanding emergency medical centers, and hemodialysis units, are subject to the requirements of Part II and 517.45. The definition of *ambulatory health care facility* in Article 517 correlates with the definition of the same term in NFPA 99, *Standard for Health Care Facilities.*

Anesthetizing Location. Any area of a facility that has been designated to be used for the administration of any flammable or nonflammable inhalation anesthetic agent in the course of examination or treatment, including the use of such agents for relative analgesia.

The definition of *anesthetizing location* recognizes that in an emergency it may be necessary to administer an anesthetic almost anywhere in a health care facility. However, only those areas in a health care facility that are set aside specifically for the induction of anesthetics are required to meet the provisions of Part IV of Article 517. This definition and the provisions of Part IV are not intended to apply to the administering of analgesic or local anesthetics, such as might be used in minor medical or dental procedures.

The definition of *anesthetizing location* applies to health care facilities where inhalation anesthetics are used for *relative analgesia.* The term *relative analgesia* (see definition of this term in 517.2) is sometimes referred to as "conscious sedation" and is a state of sedation in which the perception of pain is partially blocked and the patient does not lose consciousness. Oral surgeons often use this form of anesthesia. For guidance on flammable anesthetizing locations, see Annex E of NFPA 99, *Standard for Health Care Facilities.*

Critical Branch. A subsystem of the emergency system consisting of feeders and branch circuits supplying energy to task illumination, special power circuits, and selected receptacles serving areas and functions related to patient care, and which are connected to alternate power sources by one or more transfer switches during interruption of the normal power source.

Electrical Life-Support Equipment. Electrically powered equipment whose continuous operation is necessary to maintain a patient's life.

Emergency System. A system of circuits and equipment intended to supply alternate power to a limited number of prescribed functions vital to the protection of life and safety.

The requirements for sizing a generator and a feeder capacity for the essential electrical system in a health care facility are found in 517.30(D). Emergency systems in occupancies other than health care are installed primarily for life safety and building evacuation. The emergency systems in hospitals are for life safety systems as well as maintenance of power for critical patient care systems.

Equipment System. A system of circuits and equipment arranged for delayed, automatic, or manual connection to the alternate power source and that serves primarily 3-phase power equipment.

Essential Electrical System. A system comprised of alternate sources of power and all connected distribution systems and ancillary equipment, designed to ensure continuity of electrical power to designated areas and functions of a health care facility during disruption of normal power sources, and also designed to minimize disruption within the internal wiring system.

Exposed Conductive Surfaces. Those surfaces that are capable of carrying electric current and that are unprotected, unenclosed, or unguarded, permitting personal contact. Paint, anodizing, and similar coatings are not considered suitable insulation, unless they are listed for such use.

Fault Hazard Current. See *Hazard Current*.

Flammable Anesthetics. Gases or vapors, such as fluroxene, cyclopropane, divinyl ether, ethyl chloride, ethyl ether, and ethylene, which may form flammable or explosive mixtures with air, oxygen, or reducing gases such as nitrous oxide.

Flammable Anesthetizing Location. Any area of the facility that has been designated to be used for the administration of any flammable inhalation anesthetic agents in the normal course of examination or treatment.

Hazard Current. For a given set of connections in an isolated power system, the total current that would flow through a low impedance if it were connected between either isolated conductor and ground.

Fault Hazard Current. The hazard current of a given isolated system with all devices connected except the line isolation monitor.

Monitor Hazard Current. The hazard current of the line isolation monitor alone.

Total Hazard Current. The hazard current of a given isolated system with all devices, including the line isolation monitor, connected.

Health Care Facilities. Buildings or portions of buildings in which medical, dental, psychiatric, nursing, obstetrical, or surgical care are provided. Health care facilities include, but are not limited to, hospitals, nursing homes, limited care facilities, clinics, medical and dental offices, and ambulatory care centers, whether permanent or movable.

Definitions of *hospital, nursing home, ambulatory health care facility,* and *limited care facility* correlate with the corresponding definitions found in NFPA *101, Life Safety Code.*

Hospital. A building or part thereof used for the medical, psychiatric, obstetrical, or surgical care, on a 24-hour basis, of four or more inpatients. *Hospital,* wherever used in this *Code,* shall include general hospitals, mental hospitals, tuberculosis hospitals, children's hospitals, and any such facilities providing inpatient care.

Isolated Power System. A system comprising an isolating transformer or its equivalent, a line isolation monitor, and its ungrounded circuit conductors.

Isolation Transformer. A transformer of the multiple-winding type, with the primary and secondary windings physically separated, which inductively couples its secondary winding to the grounded feeder systems that energize its primary winding.

Life Safety Branch. A subsystem of the emergency system consisting of feeders and branch circuits, meeting the requirements of Article 700 and intended to provide adequate power needs to ensure safety to patients and personnel, and which are automatically connected to alternate power sources during interruption of the normal power source.

Limited Care Facility. A building or part thereof used on a 24-hour basis for the housing of four or more persons who are incapable of self-preservation because of age, physical limitation due to accident or illness, or mental limitations, such as mental retardation/developmental disability, mental illness, or chemical dependency.

Line Isolation Monitor. A test instrument designed to continually check the balanced and unbalanced impedance from each line of an isolated circuit to ground and equipped with a built-in test circuit to exercise the alarm without adding to the leakage current hazard.

Monitor Hazard Current. See *Hazard Current*.

Nurses' Stations. Areas intended to provide a center of nursing activity for a group of nurses serving bed patients, where the patient calls are received, nurses are dispatched, nurses' notes written, inpatient charts prepared, and medications prepared for distribution to patients. Where such activities are carried on in more than one location within a nursing unit, all such separate areas are considered a part of the nurses' station.

Nursing Home. A building or part thereof used for the lodging, boarding, and nursing care, on a 24-hour basis, of four or more persons who, because of mental or physical incapacity, may be unable to provide for their own needs and safety without the assistance of another person. *Nursing home*, wherever used in this *Code*, shall include nursing and convalescent homes, skilled nursing facilities, intermediate care facilities, and infirmaries of homes for the aged.

Patient Bed Location. The location of an inpatient sleeping bed; or the bed or procedure table used in a critical patient care area.

Patient Care Area. Any portion of a health care facility wherein patients are intended to be examined or treated. Areas of a health care facility in which patient care is administered are classified as general care areas or critical care areas, either of which may be classified as a wet location. The governing body of the facility designates these areas in accordance with the type of patient care anticipated and with the following definitions of the area classification.

> FPN: Business offices, corridors, lounges, day rooms, dining rooms, or similar areas typically are not classified as patient care areas.

General Care Areas. Patient bedrooms, examining rooms, treatment rooms, clinics, and similar areas in which it is intended that the patient will come in contact with ordinary appliances such as a nurse call system, electrical beds, examining lamps, telephone, and entertainment devices. In such areas, it may also be intended that patients be connected to electromedical devices (such as heating pads, electrocardiographs, drainage pumps, monitors, otoscopes, ophthalmoscopes, intravenous lines, etc.).

Critical Care Areas. Those special care units, intensive care units, coronary care units, angiography laboratories, cardiac catheterization laboratories, delivery rooms, operating rooms, and similar areas in which patients are intended to be subjected to invasive procedures and connected to line-operated, electromedical devices.

Wet Locations. Those patient care areas that are normally subject to wet conditions while patients are present. These include standing fluids on the floor or drenching of the work area, either of which condition is intimate to the patient or staff. Routine housekeeping procedures and incidental spillage of liquids do not define a wet location.

The definition of *patient care area* applies to hospitals as well as patient care areas in outpatient facilities. A patient bed location in a nursing home can be considered a patient care area if a person is examined or treated in that location. However, it excludes such areas as laundry rooms, boiler rooms, and utility areas, which, although routinely wet, are not patient care areas. The governing body of the health care facility may elect to include such areas as hydrotherapy areas, dialysis laboratories, and certain wet laboratories under this definition. Lavatories or bathrooms within a health care facility are not intended to be classified as wet locations. For infection control purposes, many patient and treatment areas have a sink for hand washing, which also is not intended to be a wet location.

Patient Equipment Grounding Point. A jack or terminal bus that serves as the collection point for redundant grounding of electric appliances serving a patient vicinity or for grounding other items in order to eliminate electromagnetic interference problems.

Patient Vicinity. In an area in which patients are normally cared for, the *patient vicinity* is the space with surfaces likely to be contacted by the patient or an attendant who can touch the patient. Typically in a patient room, this encloses a space within the room not less than 1.8 m (6 ft) beyond the perimeter of the bed in its nominal location, and extending vertically not less than 2.3 m (7½ ft) above the floor.

The patient vicinity is limited to patient beds in their normal location, that is, the location of the bed as called for in the architect's plans, rather than the temporary location of the bed subject to movement by housekeeping staff or for the convenience of the medical staff.

Psychiatric Hospital. A building used exclusively for the psychiatric care, on a 24-hour basis, of four or more inpatients.

Reference Grounding Point. The ground bus of the panelboard or isolated power system panel supplying the patient care area.

Relative Analgesia. A state of sedation and partial block of pain perception produced in a patient by the inhalation of concentrations of nitrous oxide insufficient to produce loss of consciousness (conscious sedation).

Selected Receptacles. A minimum number of electric receptacles to accommodate appliances ordinarily required for local tasks or likely to be used in patient care emergencies.

Task Illumination. Provision for the minimum lighting required to carry out necessary tasks in the described areas, including safe access to supplies and equipment, and access to exits.

Therapeutic High Frequency Diathermy Equipment.
Therapeutic high-frequency diathermy equipment is therapeutic induction and dielectric heating equipment.

Total Hazard Current. See *Hazard Current.*

X-Ray Installations, Long-Time Rating. A rating based on an operating interval of 5 minutes or longer.

X-Ray Installations, Mobile. X-ray equipment mounted on a permanent base with wheels, casters, or a combination of both to facilitate moving the equipment while completely assembled.

X-Ray Installations, Momentary Rating. A rating based on an operating interval that does not exceed 5 seconds.

X-Ray Installations, Portable. X-ray equipment designed to be hand carried.

X-Ray Installations, Transportable. X-ray equipment to be installed in a vehicle or that may be readily disassembled for transport in a vehicle.

II. Wiring and Protection

Formal Interpretation 99-1
Reference: Article 517, Part II

Question: Does Part II of Article 517 of the *NEC* apply to patient sleeping rooms of nursing homes or limited care facilities where patient care activities do not involve the use of electrical or electronic life support systems; or invasive procedures where patients are electrically connected to line connected electromedical devices?

Answer: No

Issue Edition: 1999

Reference: Article 517

Issue Date: August 1, 2000

Effective Date: August 21, 2000

517.10 Applicability

(A) Applicability Part II shall apply to patient care areas of all health care facilities.

(B) Not Covered Part II shall not apply to the following:

(1) Business offices, corridors, waiting rooms, and the like in clinics, medical and dental offices, and outpatient facilities

(2) Areas of nursing homes and limited care facilities wired in accordance with Chapters 1 through 4 of this *Code* where these areas are used exclusively as patient sleeping rooms

FPN: See NFPA *101*®-2003, *Life Safety Code*®.

In accordance with the conditions specified in 517.10(B)(2), areas in nursing homes that are designated as patient sleeping rooms are not considered to be patient care areas. This is often the case in nursing homes where some residents require assistance to attend to their personal needs and safety but do not require special medical care. However, areas of nursing homes in which it is intended that patients be examined or treated can be considered patient bedrooms or examining rooms and are classified in the broader category as patient care areas. See the definition of *patient care area* in 517.2.

See the Formal Interpretation on Article 517, Part II applicability.

517.11 General Installation — Construction Criteria

The purpose of this article is to specify the installation criteria and wiring methods that minimize electrical hazards by the maintenance of adequately low potential differences only between exposed conductive surfaces that are likely to become energized and could be contacted by a patient.

FPN: In a health care facility, it is difficult to prevent the occurrence of a conductive or capacitive path from the patient's body to some grounded object, because that path may be established accidentally or through instrumentation directly connected to the patient. Other electrically conductive surfaces that may make an additional contact with the patient, or instruments that may be connected to the patient, then become possible sources of electric currents that can traverse the patient's body. The hazard is increased as more apparatus is associated with the patient, and, therefore, more intensive precautions are needed. Control of electric shock hazard requires the limitation of electric current that might flow in an electric circuit involving the patient's body by raising the resistance of the conductive circuit that includes the patient, or by insulating exposed surfaces that might become energized, in addition to reducing the potential difference that can appear between exposed conductive surfaces in the patient vicinity, or by combinations of these methods. A special problem is presented by the patient with an externalized direct conductive path to the heart muscle. The patient may be electrocuted at current levels so low that additional protection in the design of appliances, insulation of the catheter, and control of medical practice is required.

This fine print note recognizes the possibility of increased sensitivity to electric shock by patients whose body resistance may be compromised either accidentally or by a necessary medical procedure. For example, incontinence or the insertion of a catheter may render a patient much more vulnerable to the effects of an electric current. Therefore, it is essential that those responsible for the design, installation, and maintenance of the electrical system in patient care areas be well acquainted with at least the rudiments of the hazard as explained in this note.

Since the original recognition of this hazard in the 1971 *Code*, continued clinical evaluation of the problem has provided a better understanding of the extent of the hazard, bringing about the changes in wiring methods found in the *Code*.

The *Code* assigns responsibility for the designation of the types of patient care areas to the governing body of the health care facility. Both the design and the inspection of a patient care area, therefore, must be based on the governing body's designation rather than the superficial appearance of the area.

517.12 Wiring Methods

Except as modified in this article, wiring methods shall comply with the applicable requirements of Chapters 1 through 4 of this *Code*.

517.13 Grounding of Receptacles and Fixed Electric Equipment in Patient Care Areas

Wiring in patient care areas shall comply with 517.13(A) and 517.13(B).

(A) Wiring Methods All branch circuits serving patient care areas shall be provided with a ground path for fault current by installation in a metal raceway system, or a cable having a metallic armor or sheath assembly. The metal raceway system, or metallic cable armor, or sheath assembly shall itself qualify as an equipment grounding return path in accordance with 250.118.

The wiring method provisions of 517.13(A) apply to the branch circuits in areas used for patient care and are not limited to patient rooms. Additional areas, such as examining rooms, therapy areas, recreational areas, solaria, and certain patient corridors, are also included. The branch circuit wiring method used in these areas is one component of a two-part redundant grounding scheme unique to patient care areas. The second component in this approach is the separate insulated copper conductor required by 517.13(B). Therefore, the metal raceway or metal cable armor or sheath must qualify as an equipment grounding conductor in accordance with 250.118, independent of any separate wire-type equipment grounding conductor. Metal-sheathed cable assemblies are not permitted as a general wiring method for emergency circuits in the patient vicinity because 517.30(C)(3) requires such wiring to be protected by installation in metal raceways. However, flexible raceway and metal-sheathed cable assemblies that are listed are allowed for limited application in a health care facility emergency system where used in listed prefabricated headwalls, in office furnishings, or where they are fished into existing walls or ceilings and are not subject to physical damage.

(B) Insulated Equipment Grounding Conductor The grounding terminals of all receptacles and all non–current-carrying conductive surfaces of fixed electric equipment likely to become energized that are subject to personal contact, operating at over 100 volts, shall be grounded by an insulated copper conductor. The equipment grounding conductor shall be sized in accordance with Table 250.122 and installed in metal raceways or as a part of listed cables having a metallic armor or sheath assembly with the branch-circuit conductors supplying these receptacles or fixed equipment.

The requirements in 517.13(B) cover the second component of the redundant grounding approach. An insulated, copper equipment grounding conductor sized in accordance with 250.122 must be installed with the branch-circuit conductors in the wiring method that meets the provisions of 517.13(A). The conductor can be either solid or stranded. It is not required to run a separate, insulated equipment grounding conductor to the branch-circuit panelboard where the feeder wiring method is recognized as an equipment grounding conductor per 250.118 or where the feeder wiring method contains an equipment grounding conductor.

The grounding requirements of 517.13 for patient care areas are not limited to hospitals. They are also required for patient care areas in other health care facilities, such as nursing homes, clinics, and medical and dental offices.

Exception No. 1 to 517.13(B) permits metal faceplates to be grounded by means of the metal mounting screws rather than by having a separate equipment grounding conductor or bonding jumper run to the metal plate. See 404.9(B), which requires switches and their metal faceplates to be effectively grounded.

Exception No. 2 to 517.13(B) exempts luminaires mounted 7½ ft above the floor and switches located outside the patient vicinity from redundant grounding requirements because it is unlikely that a patient would contact these items or that an attendant would contact these items and a patient at the same time. The patient vicinity space consists of a volume 6 ft horizontally in all directions from the bed and vertically to 7½ ft above the floor.

Exception No. 1: Metal faceplates shall be permitted to be grounded by means of a metal mounting screw(s) securing the faceplate to a grounded outlet box or grounded wiring device.

Exception No. 2: Luminaires (light fixtures) more than 2.3 m (7½ ft) above the floor and switches located outside of the patient vicinity shall not be required to be grounded by an insulated equipment grounding conductor.

517.14 Panelboard Bonding

The equipment grounding terminal buses of the normal and essential branch-circuit panelboards serving the same individual patient vicinity shall be bonded together with an insulated continuous copper conductor not smaller than 10 AWG. Where two or more panelboards serving the same individual patient vicinity are served from separate transfer switches on the emergency system, the equipment grounding terminal buses of those panelboards shall be bonded together with an insulated continuous copper conductor not smaller than 10 AWG. This conductor shall be permitted to be broken in order to terminate on the equipment grounding terminal bus in each panelboard.

Section 517.14 was expanded for the 2005 *Code* to require that where two or more panelboards supply the same patient vicinity and are supplied from different emergency transfer switches, the equipment grounding terminal bars must be bonded together with an insulated conductor not smaller than 10 AWG.

517.16 Receptacles with Insulated Grounding Terminals

Receptacles with insulated grounding terminals, as permitted in 250.146(D), shall be identified; such identification shall be visible after installation.

> FPN: Caution is important in specifying such a system with receptacles having insulated grounding terminals, since the grounding impedance is controlled only by the equipment grounding conductors and does not benefit functionally from any parallel grounding paths. This type of installation is typically used where a reduction of electrical noise (electromagnetic interference) is necessary and parallel grounding paths are to be avoided.

The requirement in 517.16 prevents indiscriminate use of isolated ground receptacles. Isolated ground receptacles are required to be identified by an orange triangle located on the face of the receptacle. Older isolated ground-type receptacles may be identified differently. The fine print note to 517.16 was revised in the 2005 *Code* to clarify that the isolated ground is to minimize electrical noise, which can cause interference in sensitive electronic equipment.

517.17 Ground-Fault Protection

Wherever ground-fault protection of equipment is applied to the service providing power to a health care facility, whether by design or by reason of the requirements of 215.10 or 230.95, an additional level of ground-fault protection is required downstream. Under this rule, ground-fault protection must be applied to every feeder, and additional ground-fault protective devices may be applied farther downstream at the option of the governing body of the health care facility. Unlike the requirements of 215.10 and 230.95 where mandatory GFPE protection is predicated on the rating of the

disconnecting means (1000 amperes or more), the second level of GFPE is not based on the rating of the feeder disconnecting means, because the function of the second level of GFPE is to provide the desired selectivity between the feeder protective devices and the service or building supply protective devices.

With proper coordination, this additional ground-fault protection is intended to limit a ground fault to a single feeder and thereby prevent a total power outage of the entire health facility. Coordination includes consideration of the trip setting, the time setting, and the time required for operation (opening time) of each level of the ground-fault protection system.

For the 2005 *Code*, the requirement for providing two levels of ground-fault protection has been extended to include health care facilities that are located in multiple-occupancy buildings and whose building service disconnecting means or the disconnecting means for a feeder supplying the building is provided with ground-fault protection of equipment (GFPE) in accordance with 215.10 or 230.95 or GFPE has been provided at the discretion of the designer. These multiple-occupancy buildings may be multiple medical office or clinic-type occupancies or may be multiple occupancies of mixed use in which one or more of the occupancies are health care facilities. As discussed, the selectivity required by 517.17(C) is accomplished through the installation of GFPE for all feeder disconnecting means in the first level of distribution downstream of the GFPE-protected service equipment or building disconnecting means. Therefore, any feeder supplied from this level of distribution will be required to have GFPE protection regardless of the occupancy type or use group.

It should be noted that the requirement for the second level of GFPE does not apply to all health care facilities located in multiple-occupancy buildings. Application of this requirement applies only where the health care facility governing authority has classified certain portions of the facility as critical care areas or where the types of procedures for which the facility is approved or licensed requires the use of life support equipment. If there are no designated critical care areas or life support equipment is not used, second level GFPE is not required. Based on these criteria, most general medical and dental practices located in multiple-occupancy buildings will not be impacted by this requirement.

In addition, if the service is not provided with GFPE (either because it is not required by the *Code* or, if optional, has not been incorporated as part of the design), then, of course, the second level of GFPE becomes moot. In accordance with 517.45, health care facilities, such as clinics or ambulatory care facilities, that use life support equipment or have areas designated as critical care are required to be provided with an essential electrical system that includes an alternate power source.

In the case of existing multiple-occupancy buildings

that have GFPE for the service equipment, a tenant fit-up or renovation for a new health care occupancy may result in the need to also provide second-level GFPE for the feeders supplying all other occupancy types in order to provide the selectivity of GFPE operation required by this section. Careful analysis of the impact of this requirement on the existing service equipment may warrant an alternative approach such as installation of another service if permitted under the provisions of 230.2(A) through (D).

It is not intended that ground-fault protection be installed between the on-site generator and the transfer switch or on the load side of the essential electrical system transfer switch. FPN No. 3 to 230.95(C) calls attention to problems that may arise when ground-fault-protected systems are transferred to another supply system.

(A) Applicability The requirements of 517.17 shall apply to hospitals and other buildings (including multiple occupancy buildings) with critical care areas or utilizing electrical life support equipment, and buildings that provide the required essential utilities or services for the operation of critical care areas or electrical life support equipment.

(B) Feeders Where ground-fault protection is provided for operation of the service disconnecting means or feeder disconnecting means as specified by 230.95 or 215.10, an additional step of ground-fault protection shall be provided in all next level feeder disconnecting means downstream toward the load. Such protection shall consist of overcurrent devices and current transformers or other equivalent protective equipment that shall cause the feeder disconnecting means to open.

The additional levels of ground-fault protection shall not be installed as follows:

(1) On the load side of an essential electrical system transfer switch
(2) Between the on-site generating unit(s) described in 517.35(B) and the essential electrical system transfer switch(es)
(3) On electrical systems that are not solidly grounded wye systems with greater than 150 volts to ground but not exceeding 600 volts phase-to-phase

(C) Selectivity Ground-fault protection for operation of the service and feeder disconnecting means shall be fully selective such that the feeder device, but not the service device, shall open on ground faults on the load side of the feeder device. A six-cycle minimum separation between the service and feeder ground-fault tripping bands shall be provided. Operating time of the disconnecting devices shall be considered in selecting the time spread between these two bands to achieve 100 percent selectivity.

See 230.95(C) and its commentary for more information on performance testing.

> FPN: See 230.95, fine print note, for transfer of alternate source where ground-fault protection is applied.

(D) Testing When equipment ground-fault protection is first installed, each level shall be performance tested to ensure compliance with 517.17(C).

517.18 General Care Areas

(A) Patient Bed Location Each patient bed location shall be supplied by at least two branch circuits, one from the emergency system and one from the normal system. All branch circuits from the normal system shall originate in the same panelboard.

Exception No. 1: Branch circuits serving only special-purpose outlets or receptacles, such as portable X-ray outlets, shall not be required to be served from the same distribution panel or panels.

Exception No. 2: Requirements of 517.18(A) shall not apply to patient bed locations in clinics, medical and dental offices, and outpatient facilities; psychiatric, substance abuse, and rehabilitation hospitals; sleeping rooms of nursing homes and limited care facilities meeting the requirements of 517.10(B)(2).

Exception No. 3: A general care patient bed location served from two separate transfer switches on the emergency system shall not be required to have circuits from the normal system.

Patient bed locations in general care areas are prohibited from deriving all their branch circuits from the emergency system. At least one branch circuit for each patient bed location must originate in a normal system panelboard. This is a reflection of the requirements in 517.33.

Exception No. 3 to 517.18 allows both of the two required branch circuits for the general care area to be supplied by the emergency system, provided they are supplied by two separate transfer switches. A normal branch circuit is not required in this case. Two emergency branch circuits have a higher reliability than one normal and one emergency branch circuit.

(B) Patient Bed Location Receptacles Each patient bed location shall be provided with a minimum of four receptacles. They shall be permitted to be of the single or duplex types or a combination of both. All receptacles, whether four or more, shall be listed "hospital grade" and so identified. Each receptacle shall be grounded by means of an insulated copper conductor sized in accordance with Table 250.122.

Exception No. 1: Requirements of 517.18(B) shall not apply to psychiatric, substance abuse, and rehabilitation hospitals meeting the requirements of 517.10(B)(2).

Exception No. 2: Psychiatric security rooms shall not be required to have receptacle outlets installed in the room.

> FPN: It is not intended that there be a total, immediate replacement of existing non–hospital grade receptacles. It is intended, however, that non–hospital grade receptacles be replaced with hospital grade receptacles upon modification of use, renovation, or as existing receptacles need replacement.

Since the 1990 *Code*, the provisions of 517.18(B) have required hospital-grade receptacles in general care patient bed locations. See the commentary following 517.19(B)(2) for more information.

(C) Pediatric Locations Receptacles located within the rooms, bathrooms, playrooms, activity rooms, and patient care areas of pediatric wards shall be listed tamper resistant or shall employ a listed tamper resistant cover.

The receptacle safeguarding requirement of 517.18(C) was revised to cover all receptacles installed in the patient care areas of pediatric locations. This safeguarding can be achieved through the use of either listed tamper-resistant receptacles or listed tamper-resistant covers. The use of locking covers over ordinary receptacles does not meet this requirement. The 2005 *Code* has been revised to be more explicit in describing the types of pediatric locations that are subject to this requirement. Exhibit 517.1 shows a listed tamper-resistant receptacle that can be used to comply with 517.18(C).

517.19 Critical Care Areas

(A) Patient Bed Location Branch Circuits Each patient bed location shall be supplied by at least two branch circuits, one or more from the emergency system and one or more circuits from the normal system. At least one branch circuit from the emergency system shall supply an outlet(s) only at that bed location. All branch circuits from the normal system shall be from a single panelboard. Emergency system receptacles shall be identified and shall also indicate the panelboard and circuit number supplying them.

Exception No. 1: Branch circuits serving only special-purpose receptacles or equipment in critical care areas shall be permitted to be served by other panelboards.

Exception No. 2: Critical care locations served from two separate transfer switches on the emergency system shall not be required to have circuits from the normal system.

Exhibit 517.1 A tamper-resistant hospital-grade receptacle, identified by a green dot on its face, that fulfills the requirements of 517.18(B) and 517.18(C). (Courtesy of Pass & Seymour/Legrand®)

Exception No. 2 to 517.19(A) covers the special case in which two separate transfer switches supply a single patient care area. Branch circuits supplied from two separate transfer switches provide the same level of redundancy required by the main requirement.

The requirements in 517.19(A) are similar to those in 517.18(A). See the commentary following 517.18(A), Exception No. 3.

(B) Patient Bed Location Receptacles

(1) Minimum Number and Supply Each patient bed location shall be provided with a minimum of six receptacles, at least one of which shall be connected to either of the following:

(1) The normal system branch circuit required in 517.19(A)
(2) An emergency system branch circuit supplied by a different transfer switch than the other receptacles at the same location

The number and type of branch circuits required for patient bed locations in critical care areas are covered in 517.19(A). The number of receptacles and the type of circuit supplying them are covered in 517.19(B). Each patient bed location must be provided with at least six receptacles. A single receptacle counts as one receptacle, and a duplex receptacle counts as two receptacles; therefore, three duplex receptacles meet the requirement.

Each patient bed location must be supplied by at least two branch circuits, one from the normal panel and one from the emergency panel, as shown in Exhibit 517.2. The normal circuits must be supplied from the same panel (L-1). The emergency circuits are permitted to be supplied from different panels (EML-1 and EML-2). However, the emergency branch circuit to patient bed location A cannot supply emergency receptacles for patient bed location B. The patient bed location receptacles can also be supplied by two different emergency circuits, instead of one emergency and one normal, provided the emergency circuits are supplied from two different emergency transfer switches.

Receptacles may be of the single or the duplex type, provided they are listed hospital-grade type and are identified as such. A typical method of marking hospital-grade receptacles is a green dot on the face of the receptacle, as shown in Exhibit 517.1. The emergency system receptacles must be marked to indicate the panelboard and circuit number supplying them. The requirements for the number and type of branch circuits in critical care areas are intended to ensure that critical care patients will not be without electrical power regardless of whether the equipment, the branch circuits, or the normal system itself is at fault.

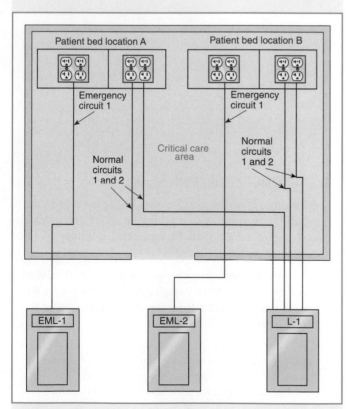

Exhibit 517.2 Examples of normal and emergency circuits supplying patient bed locations in a critical care area.

(2) Receptacle Requirements The receptacles required in 517.19(B)(1) shall be permitted to be of the single or duplex types or a combination of both. All receptacles, whether six or more, shall be listed "hospital grade" and so identified. Each receptacle shall be grounded to the reference grounding point by means of an insulated copper equipment grounding conductor.

Installing the conductor inside the raceway with the conductors delivering the fault current reduces the impedance of the grounding path.

(C) Patient Vicinity Grounding and Bonding (Optional) A patient vicinity shall be permitted to have a patient equipment grounding point. The patient equipment grounding point, where supplied, shall be permitted to contain one or more listed grounding and bonding jacks. An equipment bonding jumper not smaller than 10 AWG shall be used to connect the grounding terminal of all grounding-type receptacles to the patient equipment grounding point. The bonding conductor shall be permitted to be arranged centrically or looped as convenient.

> FPN: Where there is no patient equipment grounding point, it is important that the distance between the reference grounding point and the patient vicinity be as short as possible to minimize any potential differences.

(D) Panelboard Grounding Where a grounded electrical distribution system is used and metal feeder raceway or Type MC or MI cable is installed, grounding of a panelboard or switchboard shall be ensured by one of the following means at each termination or junction point of the raceway or Type MC or MI cable:

(1) A grounding bushing and a continuous copper bonding jumper, sized in accordance with 250.122, with the bonding jumper connected to the junction enclosure or the ground bus of the panel

(2) Connection of feeder raceways or Type MC or MI cable to threaded hubs or bosses on terminating enclosures

(3) Other approved devices such as bonding-type locknuts or bushings

(E) Additional Protective Techniques in Critical Care Areas (Optional) Isolated power systems shall be permitted to be used for critical care areas, and, if used, the isolated power system equipment shall be listed as isolated power equipment. The isolated power system shall be designed and installed in accordance with 517.160.

Exception: The audible and visual indicators of the line isolation monitor shall be permitted to be located at the nursing station for the area being served.

(F) Isolated Power System Grounding Where an isolated ungrounded power source is used and limits the first-fault current to a low magnitude, the grounding conductor associated with the secondary circuit shall be permitted to be run outside of the enclosure of the power conductors in the same circuit.

> FPN: Although it is permitted to run the grounding conductor outside of the conduit, it is safer to run it with the power conductors to provide better protection in case of a second ground fault.

(G) Special-Purpose Receptacle Grounding The equipment grounding conductor for special-purpose receptacles, such as the operation of mobile X-ray equipment, shall be extended to the reference grounding points of branch circuits for all locations likely to be served from such receptacles. Where such a circuit is served from an isolated ungrounded system, the grounding conductor shall not be required to be run with the power conductors; however, the equipment grounding terminal of the special-purpose receptacle shall be connected to the reference grounding point.

517.20 Wet Locations

In areas that are designated patient care wet locations by the governing body of the facility, GFCI protection is required for the protection of receptacles and fixed equipment if a circuit interruption can be tolerated. Otherwise, an isolated power system is required. See the commentary following the definition of *patient care area* in 517.2.

(A) Receptacles and Fixed Equipment All receptacles and fixed equipment within the area of the wet location shall have ground-fault circuit-interrupter protection for personnel if interruption of power under fault conditions can be tolerated, or be served by an isolated power system if such interruption cannot be tolerated.

Exception: Branch circuits supplying only listed, fixed, therapeutic and diagnostic equipment shall be permitted to be supplied from a normal grounded service, single- or 3-phase system, provided that

(a) Wiring for grounded and isolated circuits does not occupy the same raceway, and

(b) All conductive surfaces of the equipment are grounded.

(B) Isolated Power Systems Where an isolated power system is utilized, the isolated power equipment shall be listed as isolated power equipment, and the isolated power system shall be designed and installed in accordance with 517.160.

> FPN: For requirements for installation of therapeutic pools and tubs, see Part VI of Article 680.

517.21 Ground-Fault Circuit-Interrupter Protection for Personnel

Ground-fault circuit-interrupter protection for personnel shall not be required for receptacles installed in those critical care areas where the toilet and basin are installed within the patient room.

In critical care areas, patients are bedridden. Bathroom accommodations in the critical care area for patients are not the same as those in other patient areas. Because of the unique use of the bathroom facility, only those receptacles in the specific location are exempt from the GFCI requirement. The provisions of 517.21 do not exempt the requirements for GFCI-protected receptacles in other bathrooms for patients, staff, or the public, as required by 210.8(B)(1).

III. Essential Electrical System

517.25 Scope

The essential electrical system for these facilities shall comprise a system capable of supplying a limited amount of lighting and power service, which is considered essential for life safety and orderly cessation of procedures during the time normal electrical service is interrupted for any reason. This includes clinics, medical and dental offices, outpatient facilities, nursing homes, limited care facilities, hospitals, and other health care facilities serving patients.

> FPN: For information on the need for an essential electrical system, see NFPA 99-2002, *Standard for Health Care Facilities*.

517.26 Application of Other Articles

The essential electrical system shall meet the requirements of Article 700, except as amended by Article 517.

With the addition of 517.26 to the 2005 *Code*, there is now a correlation between 700.6(D) and 517.30(B)(4), which permits a single transfer switch to supply the emergency system and also the equipment system in a small hospital with a maximum demand on the essential electrical system of 150 kVa or less.

Similar to the physical separation requirements for emergency systems specified in 700.9(B), the physical separation requirements for circuits supplied by essential electrical systems (including the emergency system of a hospital) are covered in 517.30(C)(1) for hospitals and 517.41(D) for nursing homes and limited care facilities.

517.30 Essential Electrical Systems for Hospitals

(A) Applicability The requirements of Part III, 517.30 through 517.35, shall apply to hospitals where an essential electrical system is required.

> FPN No. 1: For performance, maintenance, and testing requirements of essential electrical systems in hospitals, see NFPA 99-2002, *Standard for Health Care Facilities*. For installation of centrifugal fire pumps, see NFPA 20-2002, *Standard for the Installation of Stationary Fire Pumps for Fire Protection*.

FPN No. 2: For additional information, see NFPA 99-2002, *Standard for Health Care Facilities.*

(B) General

(1) Separate Systems Essential electrical systems for hospitals shall be comprised of two separate systems capable of supplying a limited amount of lighting and power service, which is considered essential for life safety and effective hospital operation during the time the normal electrical service is interrupted for any reason. These two systems shall be the emergency system and the equipment system.

(2) Emergency Systems The emergency system shall be limited to circuits essential to life safety and critical patient care. These are designated the life safety branch and the critical branch.

(3) Equipment System The equipment system shall supply major electrical equipment necessary for patient care and basic hospital operation.

(4) Transfer Switches The number of transfer switches to be used shall be based on reliability, design, and load considerations. Each branch of the emergency system and each equipment system shall have one or more transfer switches. One transfer switch shall be permitted to serve one or more branches or systems in a facility with a maximum demand on the essential electrical system of 150 kVA.

> FPN No. 1: See NFPA 99-2002, *Standard for Health Care Facilities*: 4.4.3.2, Transfer Switch Operation Type I; 4.4.2.1.4, Automatic Transfer Switch Features; and 4.4.2.1.6, Nonautomatic Transfer Device Features.
> FPN No. 2: See FPN Figure 517.30, No. 1.
> FPN No. 3: See FPN Figure 517.30, No. 2.

FPN Figure 517.30, No. 1 Hospital — Minimum Requirement for Transfer Switch Arrangement

FPN Figure 517.30, No. 2 Hospital — Minimum Requirement (150 kVA or less) for Transfer Switch Arrangement

FPN Figure 517.30, No. 1 and FPN Figure 517.30, No. 2 illustrate possible electrical system connections for hospitals. For a small electrical system having a maximum demand on the essential electrical system of 150 kVA, see FPN Figure 517.30, No. 2. A small load can be served by a single transfer switch that can handle the loads associated with both the emergency system and the equipment system. This, of course, is based on the assumption that the transfer switch has sufficient capacity to handle the combined loads and that the alternate source of power is sufficiently large to withstand the impact of the simultaneous transfer of both systems in the event of a normal power loss. For further explanation of loads permitted on an emergency system, see NFPA 99, *Standard for Health Care Facilities*, 4.4.2.2.

The requirement in 517.30(B)(6) correlates with 12.3.3.2 of NFPA 99.

(5) Other Loads Loads served by the generating equipment not specifically named in Article 517 shall be served by their own transfer switches such that the following conditions apply:

(1) These loads shall not be transferred if the transfer will overload the generating equipment.
(2) These loads shall be automatically shed upon generating equipment overloading.

(6) Contiguous Facilities Hospital power sources and alternate power sources shall be permitted to serve the essential electrical systems of contiguous or same site facilities. [NFPA 99:13.3.4.3]

(C) Wiring Requirements

(1) Separation from Other Circuits The life safety branch and critical branch of the emergency system shall be kept entirely independent of all other wiring and equipment and shall not enter the same raceways, boxes, or cabinets with each other or other wiring.

Wiring of the life safety branch and the critical branch shall be permitted to occupy the same raceways, boxes, or cabinets of other circuits not part of the branch where such wiring complies with one of the following:

(1) Is in transfer equipment enclosures
(2) Is in exit or emergency luminaires (lighting fixtures) supplied from two sources
(3) Is in a common junction box attached to exit or emergency luminaires (lighting fixtures) supplied from two sources
(4) Is for two or more emergency circuits supplied from the same branch

The wiring of the equipment system shall be permitted to occupy the same raceways, boxes, or cabinets of other circuits that are not part of the emergency system.

Multiwire branch circuits supplied by the same panelboard are permitted by 517.30(C)(1)(4).

(2) Isolated Power Systems Where isolated power systems are installed in any of the areas in 517.33(A)(1) and (A)(2), each system shall be supplied by an individual circuit serving no other load.

(3) Mechanical Protection of the Emergency System The wiring of the emergency system in hospitals shall be mechanically protected. Where installed as branch circuits in patient care areas, the installation shall comply with the requirements of 517.13(A) and 517.13(B). The following wiring methods shall be permitted:

The wiring of emergency systems in hospitals requires additional protection not normally required in other occupancies. Only metal raceways of the nonflexible type and Type MI cable were permitted as a wiring method for hospital emergency systems in the 2002 *Code*. (Emergency system circuits can include services, feeders, and branch circuits.) As described in 517.31, the emergency system in a hospital consists of the life safety branch and the critical branch. The five exceptions to the requirement in 517.30(C)(3) became permissive rules in the 2005 *Code* to provide other means of wiring branch circuits and Class 2 and Class 3 systems.

(1) Nonflexible metal raceways, Type MI cable, or Schedule 80 rigid nonmetallic conduit. Nonmetallic raceways

shall not be used for branch circuits that supply patient care areas.
(2) Where encased in not less than 50 mm (2 in.) of concrete, Schedule 40 rigid nonmetallic conduit, flexible nonmetallic or jacketed metallic raceways, or jacketed metallic cable assemblies listed for installation in concrete. Nonmetallic raceways shall not be used for branch circuits that supply patient care areas.
(3) Listed flexible metal raceways and listed metal sheathed cable assemblies in any of the following:
   a. Where used in listed prefabricated medical headwalls
   b. In listed office furnishings
   c. Where fished into existing walls or ceilings, not otherwise accessible and not subject to physical damage
   d. Where necessary for flexible connection to equipment
(4) Flexible power cords of appliances or other utilization equipment connected to the emergency system.
(5) Secondary circuits of Class 2 or Class 3 communication or signaling systems

Section 517.30(C)(3), Exception No. 1 through Exception No. 5 in the 2002 *Code* have been revised as five conditions that do not require the installation of conductors in nonflexible metal raceways.

New in the 2005 *Code* is permission to fish flexible metal raceways and metal sheathed cables in existing installations. This provision facilitates installations in renovated areas where the existing walls or ceilings remain intact.

FPN: See 517.13 for additional grounding requirements in patient care areas.

The secondary conductors of limited energy systems, such as nurse call, telephone, and alarm circuits, are exempt from being run in metal raceways, provided they comply with their applicable articles elsewhere in the *Code*. Although this requirement allows substantial latitude in the wiring method, it should be noted that the restrictions of 300.22 (ducts, plenums, and other air-handling spaces) apply, unless cables specifically listed for use in these environments are used. See the requirements for these cables in Articles 725, 760, 770, 800, and 820.

(D) Capacity of Systems The essential electrical system shall have adequate capacity to meet the demand for the operation of all functions and equipment to be served by each system and branch.

Feeders shall be sized in accordance with Articles 215 and 220. The generator set(s) shall have sufficient capacity and proper rating to meet the demand produced by the load of the essential electrical system(s) at any given time.

Demand calculations for sizing of the generator set(s) shall be based on any of the following:

(1) Prudent demand factors and historical data
(2) Connected load
(3) Feeder calculation procedures described in Article 220
(4) Any combination of the above

The sizing requirements in 700.5 and 701.6 shall not apply to hospital generator set(s).

The intent of 517.30(D) is to permit the sizing of generators based on actual demand likely to be produced by the connected load of the system at any one time. This method of calculation facilitates practical sizing of generators in health care facilities and helps eliminate prime mover operational problems associated with lightly loaded generators.

(E) Receptacle Identification The cover plates for the electrical receptacles or the electrical receptacles themselves supplied from the emergency system shall have a distinctive color or marking so as to be readily identifiable. [NFPA 99:4.4.2.2.4.2(B)]

The 1999 *Code* required identification of cover plates or the receptacles themselves where supplied by the critical branch of the emergency system. This identification requirement now applies to all receptacles supplied by the emergency system and thus now also includes those receptacles supplied by the life safety branch.

517.31 Emergency System

Those functions of patient care depending on lighting or appliances that are connected to the emergency system shall be divided into two mandatory branches: the life safety branch and the critical branch, described in 517.32 and 517.33.

The branches of the emergency system shall be installed and connected to the alternate power source so that all functions specified herein for the emergency system shall be automatically restored to operation within 10 seconds after interruption of the normal source. [NFPA 99:4.4.2.2.2.1, 4.4.3.1]

517.32 Life Safety Branch

No function other than those listed in 517.32(A) through 517.32(G) shall be connected to the life safety branch. The life safety branch of the emergency system shall supply power for the following lighting, receptacles, and equipment.

(A) Illumination of Means of Egress Illumination of means of egress, such as lighting required for corridors, passageways, stairways, and landings at exit doors, and all necessary ways of approach to exits. Switching arrangements to transfer patient corridor lighting in hospitals from general illumination circuits to night illumination circuits shall be permitted, provided only one of two circuits can be selected and both circuits cannot be extinguished at the same time.

FPN: See NFPA *101*-2003, *Life Safety Code*, Sections 7.8 and 7.9.

(B) Exit Signs Exit signs and exit directional signs.

FPN: See NFPA *101*-2003, *Life Safety Code*, Section 7.10.

(C) Alarm and Alerting Systems Alarm and alerting systems including the following:

(1) Fire alarms

FPN: See NFPA *101*-2003, *Life Safety Code*, Section 9.6 and 18.3.4.

(2) Alarms required for systems used for the piping of nonflammable medical gases

FPN: See NFPA 99-2002, *Standard for Health Care Facilities*, 4.4.2.2.2.2(3).

(D) Communications Systems Hospital communications systems, where used for issuing instructions during emergency conditions.

(E) Generator Set Location Task illumination battery charger for emergency battery-powered lighting unit(s) and selected receptacles at the generator set location.

(F) Elevators Elevator cab lighting, control, communications, and signal systems.

(G) Automatic Doors Automatically operated doors used for building egress. [NFPA 99:4.4.2.2.2.2]

517.33 Critical Branch

(A) Task Illumination and Selected Receptacles The critical branch of the emergency system shall supply power for task illumination, fixed equipment, selected receptacles, and special power circuits serving the following areas and functions related to patient care:

(1) Critical care areas that utilize anesthetizing gases — task illumination, selected receptacles, and fixed equipment
(2) The isolated power systems in special environments
(3) Patient care areas — task illumination and selected receptacles in the following:

    a. Infant nurseries
    b. Medication preparation areas
    c. Pharmacy dispensing areas
    d. Selected acute nursing areas

e. Psychiatric bed areas (omit receptacles)

f. Ward treatment rooms

g. Nurses' stations (unless adequately lighted by corridor luminaires)

(4) Additional specialized patient care task illumination and receptacles, where needed

(5) Nurse call systems

(6) Blood, bone, and tissue banks

(7) Telephone equipment rooms and closets

(8) Task illumination, selected receptacles, and selected power circuits for the following:

a. General care beds (at least one duplex receptacle per patient bedroom)

b. Angiographic labs

c. Cardiac catheterization labs

d. Coronary care units

e. Hemodialysis rooms or areas

f. Emergency room treatment areas (selected)

g. Human physiology labs

h. Intensive care units

i. Postoperative recovery rooms (selected)

(9) Additional task illumination, receptacles, and selected power circuits needed for effective hospital operation. Single-phase fractional horsepower motors shall be permitted to be connected to the critical branch. [NFPA 99:4.4.2.2.2.3(9)]

> The critical branch is intended to serve a limited number of receptacles and locations, to reduce the load, and to minimize the chances of a fault condition. Receptacles in general patient care area corridors are permitted on the critical branch, but they must be identified in some manner (color-coded or labeled) as part of the emergency system, in accordance with 517.30(E).

(B) Subdivision of the Critical Branch It shall be permitted to subdivide the critical branch into two or more branches.

> FPN: It is important to analyze the consequences of supplying an area with only critical care branch power when failure occurs between the area and the transfer switch. Some proportion of normal and critical power or critical power from separate transfer switches may be appropriate.

517.34 Equipment System Connection to Alternate Power Source

The equipment system shall be installed and connected to the alternate power source such that the equipment described in 517.34(A) is automatically restored to operation at appropriate time-lag intervals following the energizing of the emergency system. Its arrangement shall also provide for the

subsequent connection of equipment described in 517.34(B). [NFPA 99:4.4.2.2.3.2]

Exception: For essential electrical systems under 150 kVA, deletion of the time-lag intervals feature for delayed automatic connection to the equipment system shall be permitted.

(A) Equipment for Delayed Automatic Connection The following equipment shall be arranged for delayed automatic connection to the alternate power source:

(1) Central suction systems serving medical and surgical functions, including controls. Such suction systems shall be permitted on the critical branch.

(2) Sump pumps and other equipment required to operate for the safety of major apparatus, including associated control systems and alarms.

(3) Compressed air systems serving medical and surgical functions, including controls. Such air systems shall be permitted on the critical branch.

(4) Smoke control and stair pressurization systems, or both.

(5) Kitchen hood supply or exhaust systems, or both, if required to operate during a fire in or under the hood. [NFPA 99:4.4.2.2.3.4(5)]

(6) Supply, return, and exhaust ventilating systems for airborne infectious/isolation rooms, protective environment rooms, exhaust fans for laboratory fume hoods, nuclear medicine areas where radioactive material is used, ethylene oxide evacuation and anesthesia evacuation. Where delayed automatic connection is not appropriate, such ventilation systems shall be permitted to be placed on the critical branch. [NFPA 99:4.4.2.2.3.4(6)]

> Section 517.34(A)(6) was added in the 2005 *Code* for the connection of equipment to the alternate power source. It requires a delay in connecting supply, return, and exhaust ventilating systems for the areas identified in this section.

Exception: Sequential delayed automatic connection to the alternate power source to prevent overloading the generator shall be permitted where engineering studies indicate it is necessary.

(B) Equipment for Delayed Automatic or Manual Connection The following equipment shall be arranged for either delayed automatic or manual connection to the alternate power source:

(1) Heating equipment to provide heating for operating, delivery, labor, recovery, intensive care, coronary care, nurseries, infection/isolation rooms, emergency treatment spaces, and general patient rooms and pressure maintenance (jockey or make-up) pump(s) for water-based fire protection systems.

Exception: Heating of general patient rooms and infection/ isolation rooms during disruption of the normal source shall not be required under any of the following conditions:

(1) The outside design temperature is higher than −6.7°C (20°F).

(2) The outside design temperature is lower than −6.7°C (20°F), and where a selected room(s) is provided for the needs of all confined patients, only such room(s) need be heated.

(3) The facility is served by a dual source of normal power.

> FPN No. 1: The design temperature is based on the 97½ percent design value as shown in Chapter 24 of the ASHRAE *Handbook of Fundamentals* (1997).

In some areas, it is common practice to install individual room heating/air conditioning rather than to have a central heating/air-conditioning plant. If these individual units are electrically powered, it may not be practical to apply their high-demand load to the generator. If the governing body of the nursing home has full-time skilled attendants who can move people to one or more room(s) that will be heated when this smaller load is picked up by the generator, the intent of the *Code* is satisfied. The provisions for limited heating during emergency conditions are based on consideration of outside design temperature.

> FPN No. 2: For a description of a dual source of normal power, see 517.35(C), FPN.

(2) An elevator(s) selected to provide service to patient, surgical, obstetrical, and ground floors during interruption of normal power. In instances where interruption of normal power would result in other elevators stopping between floors, throw-over facilities shall be provided to allow the temporary operation of any elevator for the release of patients or other persons who may be confined between floors.

(3) Hyperbaric facilities.

(4) Hypobaric facilities.

(5) Automatically operated doors.

(6) Minimal electrically heated autoclaving equipment shall be permitted to be arranged for either automatic or manual connection to the alternate source.

(7) Controls for equipment listed in 517.34.

(8) Other selected equipment shall be permitted to be served by the equipment system. [NFPA 99:4.4.2.2.3.5(9)]

(C) AC Equipment for Nondelayed Automatic Connection Generator accessories, including but not limited to, the transfer fuel pump, electrically operated louvers, and other generator accessories essential for generator operation, shall be arranged for automatic connection to the alternate power source. [NFPA 99:4.4.2.2.3.3]

Section 517.34(C) was added to the 2005 *Code* to correlate with NFPA 99, *Standard for Health Care Facilities*, 4.4.2.2.3.3, which pertains to associated support equipment for proper operation of the emergency generator and requires this equipment to be connected to the alternate power source. Some of the accessories essential to the operation of the generator are the fuel transfer pump, electrically operated ventilation louvers, and illumination at the generator.

517.35 Sources of Power

(A) Two Independent Sources of Power Essential electrical systems shall have a minimum of two independent sources of power: a normal source generally supplying the entire electrical system and one or more alternate sources for use when the normal source is interrupted. [NFPA 99:4.4.1.1.4]

(B) Alternate Source of Power The alternate source of power shall be one of the following:

(1) Generator(s) driven by some form of prime mover(s) and located on the premises

(2) Another generating unit(s) where the normal source consists of a generating unit(s) located on the premises

(3) An external utility service when the normal source consists of a generating unit(s) located on the premises

(4) A battery system located on the premises [NFPA 99:4.4.1.2]

(C) Location of Essential Electrical System Components Careful consideration shall be given to the location of the spaces housing the components of the essential electrical system to minimize interruptions caused by natural forces common to the area (e.g., storms, floods, earthquakes, or hazards created by adjoining structures or activities). Consideration shall also be given to the possible interruption of normal electrical services resulting from similar causes as well as possible disruption of normal electrical service due to internal wiring and equipment failures.

> FPN: Facilities in which the normal source of power is supplied by two or more separate central station-fed services experience greater than normal electrical service reliability than those with only a single feed. Such a dual source of normal power consists of two or more electrical services fed from separate generator sets or a utility distribution network that has multiple power input sources and is arranged to provide mechanical and electrical separation so that a fault between the facility and the generating sources is not likely to cause an interruption of more than one of the facility service feeders.

517.40 Essential Electrical Systems for Nursing Homes and Limited Care Facilities

(A) Applicability The requirements of Part III, 517.40(C) through 517.44, shall apply to nursing homes and limited care facilities.

Exception: The requirements of Part III, 517.40(C) through 517.44, shall not apply to freestanding buildings used as nursing homes and limited care facilities, provided that the following apply:

(a) Admitting and discharge policies are maintained that preclude the provision of care for any patient or resident who may need to be sustained by electrical life-support equipment.

(b) No surgical treatment requiring general anesthesia is offered.

(c) An automatic battery-operated system(s) or equipment is provided that shall be effective for at least 1½ hours and is otherwise in accordance with 700.12 and that shall be capable of supplying lighting for exit lights, exit corridors, stairways, nursing stations, medical preparation areas, boiler rooms, and communications areas. This system shall also supply power to operate all alarm systems. [NFPA 99:17.3.4.1.2(3), 18.3.4.1.2(3)]

NFPA 99, *Standard for Health Care Facilities*, recognizes two classes of nursing homes or limited care facilities. For the smaller, less complex facility, only a minimum alternate lighting and alarm service needs to be furnished.

At nursing homes or limited care facilities where patients are sustained by electrical life-support equipment or inpatient hospital care is provided, the requirements of 517.41 through 517.44 apply. The branches of the emergency system for this class of occupancy bear identical titles to their counterparts for hospital-type occupancies.

FPN: See NFPA *101*-2003, *Life Safety Code*.

(B) Inpatient Hospital Care Facilities Nursing homes and limited care facilities that provide inpatient hospital care shall comply with the requirements of Part III, 517.30 through 517.35.

Regardless of the name given to the facility, the type of electrical system depends on the type of patient care provided. If such care is clearly inpatient hospital care, a hospital-type electrical system must be installed. The type of care that can be provided at a nursing home or limited care facility is generally regulated through the administrative agency that licenses the facility.

(C) Facilities Contiguous or Located on the Same Site with Hospitals Nursing homes and limited care facilities that are contiguous or located on the same site with a hospital shall be permitted to have their essential electrical systems supplied by that of the hospital.

If a nursing home or limited care facility shares essentially the same building with a hospital, the nursing home is not required to have its own essential electrical system as long as it derives its power supply from the hospital. It should be noted, however, that this rule applies only to the electrical supply and does not permit the sharing of transfer devices and the like.

FPN: For performance, maintenance, and testing requirements of essential electrical systems in nursing homes and limited care facilities, see NFPA 99-2002, *Standard for Health Care Facilities*.

517.41 Essential Electrical Systems

(A) General Essential electrical systems for nursing homes and limited care facilities shall be comprised of two separate branches capable of supplying a limited amount of lighting and power service, which is considered essential for the protection of life safety and effective operation of the institution during the time normal electrical service is interrupted for any reason. These two separate branches shall be the life safety branch and the critical branch. [NFPA 99:Annex A, 4.5.2.2.1]

(B) Transfer Switches The number of transfer switches to be used shall be based on reliability, design, and load considerations. Each branch of the essential electrical system shall be served by one or more transfer switches. One transfer switch shall be permitted to serve one or more branches or systems in a facility with a maximum demand on the essential electrical system of 150 kVA. [NFPA 99:4.5.2.2.1]

FPN No. 1: See NFPA 99-2002, *Standard for Health Care Facilities*, 4.5.3.2, Transfer Switch Operation Type II; 4.4.2.1.4, Automatic Transfer Switch Features; and 4.4.2.1.6, Nonautomatic Transfer Device Features.
FPN No. 2: See FPN Figure 517.41, No. 1.
FPN No. 3: See FPN Figure 517.41, No. 2.

(C) Capacity of System The essential electrical system shall have adequate capacity to meet the demand for the operation of all functions and equipment to be served by each branch at one time.

(D) Separation from Other Circuits The life safety branch shall be kept entirely independent of all other wiring and equipment and shall not enter the same raceways, boxes, or cabinets with other wiring except as follows:

(1) In transfer switches
(2) In exit or emergency luminaires (lighting fixtures) supplied from two sources
(3) In a common junction box attached to exit or emergency luminaires (lighting fixtures) supplied from two sources

FPN Figure 517.41, No. 1 Nursing Home and Limited Health Care Facilities — Minimum Requirement for Transfer Switch Arrangement

FPN Figure 517.41, No. 2 Nursing Home and Limited Health Care Facilities — Minimum Requirement (150 kVA or less) for Transfer Switch Arrangement

The wiring of the critical branch shall be permitted to occupy the same raceways, boxes, or cabinets of other circuits that are not part of the life safety branch.

(E) Receptacle Identification The cover plates for the electrical receptacles or the electrical receptacles themselves supplied from the emergency system shall have a distinctive color or marking so as to be readily identifiable. [NFPA 99: 4.5.2.2.4.2]

517.42 Automatic Connection to Life Safety Branch

The life safety branch shall be installed and connected to the alternate source of power so that all functions specified

herein shall be automatically restored to operation within 10 seconds after the interruption of the normal source. No functions other than those listed in 517.42(A) through 517.42(G) shall be connected to the life safety branch. The life safety branch shall supply power for the following lighting, receptacles, and equipment.

> FPN: The life safety branch is called the emergency system in NFPA 99-2002, *Standard for Health Care Facilities.*

(A) Illumination of Means of Egress Illumination of means of egress as is necessary for corridors, passageways, stairways, landings, and exit doors and all ways of approach to exits. Switching arrangement to transfer patient corridor lighting from general illumination circuits shall be permitted, providing only one of two circuits can be selected and both circuits cannot be extinguished at the same time.

> FPN: See NFPA *101*-2003, *Life Safety Code*, Sections 7.8 and 7.9.

(B) Exit Signs Exit signs and exit directional signs.

> FPN: See NFPA *101*-2003, *Life Safety Code*, Section 7.10.

(C) Alarm and Alerting Systems Alarm and alerting systems, including the following:

(1) Fire alarms

> FPN: See NFPA *101*-2003, *Life Safety Code*, Sections 9.6 and 18.3.4.

(2) Alarms required for systems used for the piping of nonflammable medical gases

> FPN: See NFPA 99-2002, *Standard for Health Care Facilities*, 4.4.2.2.2.2(3).

(D) Communications Systems Communications systems, where used for issuing instructions during emergency conditions.

(E) Dining and Recreation Areas Sufficient lighting in dining and recreation areas to provide illumination to exit ways.

(F) Generator Set Location Task illumination and selected receptacles in the generator set location.

(G) Elevators Elevator cab lighting, control, communications, and signal systems. [NFPA 99:4.4.2.2.2.2(6), 4.5.2.2.2(7)]

517.43 Connection to Critical Branch

The critical branch shall be installed and connected to the alternate power source so that the equipment listed in 517.43(A) shall be automatically restored to operation at appropriate time-lag intervals following the restoration of the life safety branch to operation. Its arrangement shall also provide for the additional connection of equipment listed in 517.43(B) by either delayed automatic or manual operation.

Exception: For essential electrical systems under 150 kVA, deletion of the time-lag intervals feature for delayed automatic connection to the equipment system shall be permitted.

(A) Delayed Automatic Connection The following equipment shall be connected to the critical branch and shall be arranged for delayed automatic connection to the alternate power source:

(1) Patient care areas — task illumination and selected receptacles in the following:

    a. Medication preparation areas

    b. Pharmacy dispensing areas

    c. Nurses' stations (unless adequately lighted by corridor luminaires)

(2) Sump pumps and other equipment required to operate for the safety of major apparatus and associated control systems and alarms

(3) Smoke control and stair pressurization systems

(4) Kitchen hood supply and/or exhaust systems, if required to operate during a fire in or under the hood

(5) Supply, return, and exhaust ventilating systems for airborne infectious isolation rooms

(B) Delayed Automatic or Manual Connection The following equipment shall be connected to the critical branch and shall be arranged for either delayed automatic or manual connection to the alternate power source.

(1) Heating equipment to provide heating for patient rooms.

Exception: Heating of general patient rooms during disruption of the normal source shall not be required under any of the following conditions:

(1) The outside design temperature is higher than −6.7°C (20°F).

(2) The outside design temperature is lower than −6.7°C (20°F) and where a selected room(s) is provided for the needs of all confined patients, only such room(s) need be heated.

(3) The facility is served by a dual source of normal power as described in 517.44(C), FPN.

> FPN: The outside design temperature is based on the 97½ percent design values as shown in Chapter 24 of the ASHRAE *Handbook of Fundamentals* (1997).

(2) Elevator service — in instances where disruption of power would result in elevators stopping between floors, throw-over facilities shall be provided to allow the temporary operation of any elevator for the release of passengers. For elevator cab lighting, control, and signal system requirements, see 517.42(G).

(3) Additional illumination, receptacles, and equipment shall be permitted to be connected only to the critical branch. [NFPA 99:4.5.2.2.3.3(C)]

517.44 Sources of Power

(A) Two Independent Sources of Power Essential electrical systems shall have a minimum of two independent sources of power: a normal source generally supplying the entire electrical system and one or more alternate sources for use when the normal source is interrupted. [NFPA 99:4.4.1.1.4]

(B) Alternate Source of Power The alternate source of power shall be a generator(s) driven by some form of prime mover(s) and located on the premises.

Exception No. 1: Where the normal source consists of generating units on the premises, the alternate source shall be either another generator set or an external utility service.

Exception No. 2: Nursing homes or limited care facilities meeting the requirements of 517.40(A), Exception, shall be permitted to use a battery system or self-contained battery integral with the equipment. [NFPA 99:17.3.4.1.3, 18.3.4.1.1]

(C) Location of Essential Electrical System Components Careful consideration shall be given to the location of the spaces housing the components of the essential electrical system to minimize interruptions caused by natural forces common to the area (e.g., storms, floods, earthquakes, or hazards created by adjoining structures or activities). Consideration shall also be given to the possible interruption of normal electrical services resulting from similar causes as well as possible disruption of normal electrical service due to internal wiring and equipment failures.

> FPN: Facilities in which the normal source of power is supplied by two or more separate central station-fed services experience greater than normal electrical service reliability than those with only a single feed. Such a dual source of normal power consists of two or more electrical services fed from separate generator sets or a utility distribution network that has multiple power input sources and is arranged to provide mechanical and electrical separation so that a fault between the facility and the generating sources will not likely cause an interruption of more than one of the facility service feeders.

517.45 Essential Electrical Systems for Other Health Care Facilities

(A) Essential Electrical Distribution The essential electrical distribution system shall be a battery or generator system.

> FPN: See NFPA 99–2002, *Standard for Health Care Facilities.*

(B) Electrical Life Support Equipment Where electrical life support equipment is required, the essential electrical distribution system shall be as described in 517.30 through 517.35. [NFPA 99:14.3.4.2.1]

(C) Critical Care Areas Where critical care areas are present, the essential electrical distribution system shall be as described in 517.30 through 517.35. [NFPA 99:14.3.4.2.2]

(D) Power Systems Battery systems shall be installed in accordance with the requirements of Article 700, and generator systems shall be as described in 517.30 through 517.35.

The provisions of 517.45 and NFPA 99, *Standard for Health Care Facilities*, provide the essential electrical system requirements for health care facilities other than hospitals, nursing homes, and limited care facilities. These facilities include medical and dental offices and ambulatory health care facilities. Depending on the type and level of patient care, these facilities may require an alternate source of power since the patients are treated in essentially the same manner as in hospitals, even though the facility may not be called a hospital.

IV. Inhalation Anesthetizing Locations

FPN: For further information regarding safeguards for anesthetizing locations, see NFPA 99-2002, *Standard for Health Care Facilities*.

517.60 Anesthetizing Location Classification

FPN: If either of the anesthetizing locations in 517.60(A) or 517.60(B) is designated a wet location, refer to 517.20.

(A) Hazardous (Classified) Location

(1) Use Location In a location where flammable anesthetics are employed, the entire area shall be considered to be a Class I, Division 1 location that extends upward to a level 1.52 m (5 ft) above the floor. The remaining volume up to the structural ceiling is considered to be above a hazardous (classified) location. [NFPA 99:Annex E, E.1, and E.2]

(2) Storage Location Any room or location in which flammable anesthetics or volatile flammable disinfecting agents are stored shall be considered to be a Class I, Division 1 location from floor to ceiling.

Past editions of NFPA 99, *Standard for Health Care Facilities*, contained requirements for flammable anesthetizing locations throughout the standard. Now, all of these requirements are located in Annex E of NFPA 99. The reason for this change is explained in the note to Annex E in NFPA 99, which reads as follows:

The text of this annex is a compilation of requirements included in previous editions of NFPA 99 on safety practices for facilities that used flammable inhalation anesthetics. This material is being retained in this annex by the Technical Committee on Anesthesia Ser-

vices for the following reasons: (1) the Committee is aware that some countries outside the United States still use this type of anesthetics and rely on the safety measures herein; and (2) while the Committee is unaware of any medical schools in the U.S. still teaching the proper use of flammable anesthetics or any health care facilities in the U.S. using flammable anesthetics, retaining this material will serve as a reminder of the precautions that would be necessary should the use of this type of anesthetics be re-instituted.

In 517.60, anesthetizing locations are designated either as hazardous (classified) locations, where flammable or nonflammable anesthetics may be interchangeably employed [517.60(A)], or as other-than-hazardous (classified) locations, where only nonflammable anesthetics are used [517.60(B)]. In the case of the flammable anesthetizing location, the entire volume of the room, extending upward from a level 5 ft above the floor to the surface of the structural ceiling of the room and including the space between a drop ceiling and the structural ceiling, is considered to be above a hazardous (classified) location.

(B) Other-Than-Hazardous (Classified) Location Any inhalation anesthetizing location designated for the exclusive use of nonflammable anesthetizing agents shall be considered to be an other-than-hazardous (classified) location.

517.61 Wiring and Equipment

(A) Within Hazardous (Classified) Anesthetizing Locations

(1) Isolation Except as permitted in 517.160, each power circuit within, or partially within, a flammable anesthetizing location as referred to in 517.60 shall be isolated from any distribution system by the use of an isolated power system. [NFPA 99:Annex E, E.6.6.2]

(2) Design and Installation Where an isolated power system is utilized, the isolated power equipment shall be listed as isolated power equipment, and the isolated power system shall be designed and installed in accordance with 517.160.

(3) Equipment Operating at More Than 10 Volts In hazardous (classified) locations referred to in 517.60, all fixed wiring and equipment and all portable equipment, including lamps and other utilization equipment, operating at more than 10 volts between conductors shall comply with the requirements of 501.1 through 501.25, and 501.100 through 501.150, and 501.30(A) and 501.30(B) for Class I, Division 1 locations. All such equipment shall be specifically approved for the hazardous atmospheres involved. [NFPA 99:Annex E, E.2.1, E.4.5, E.4.6, and E.4.7]

(4) Extent of Location Where a box, fitting, or enclosure is partially, but not entirely, within a hazardous (classified) location(s), the hazardous (classified) location(s) shall be considered to be extended to include the entire box, fitting, or enclosure.

(5) Receptacles and Attachment Plugs Receptacles and attachment plugs in a hazardous (classified) location(s) shall be listed for use in Class I, Group C hazardous (classified) locations and shall have provision for the connection of a grounding conductor.

(6) Flexible Cord Type Flexible cords used in hazardous (classified) locations for connection to portable utilization equipment, including lamps operating at more than 8 volts between conductors, shall be of a type approved for extra-hard usage in accordance with Table 400.4 and shall include an additional conductor for grounding.

(7) Flexible Cord Storage A storage device for the flexible cord shall be provided and shall not subject the cord to bending at a radius of less than 75 mm (3 in.).

(B) Above Hazardous (Classified) Anesthetizing Locations

(1) Wiring Methods Wiring above a hazardous (classified) location referred to in 517.60 shall be installed in rigid metal conduit, electrical metallic tubing, intermediate metal conduit, Type MI cable, or Type MC cable that employs a continuous, gas/vaportight metal sheath.

(2) Equipment Enclosure Installed equipment that may produce arcs, sparks, or particles of hot metal, such as lamps and lampholders for fixed lighting, cutouts, switches, generators, motors, or other equipment having make-and-break or sliding contacts, shall be of the totally enclosed type or be constructed so as to prevent escape of sparks or hot metal particles.

Exception: Wall-mounted receptacles installed above the hazardous (classified) location in flammable anesthetizing locations shall not be required to be totally enclosed or have openings guarded or screened to prevent dispersion of particles.

(3) Luminaires (Lighting Fixtures) Surgical and other luminaires (lighting fixtures) shall conform to 501.130(B).

Exception No. 1: The surface temperature limitations set forth in 501.130(B)(1) shall not apply.

Exception No. 2: Integral or pendant switches that are located above and cannot be lowered into the hazardous (classified) location(s) shall not be required to be explosionproof.

(4) Seals Approved seals shall be provided in conformance with 501.15, and 501.15(A)(4) shall apply to horizontal as well as to vertical boundaries of the defined hazardous (classified) locations.

(5) Receptacles and Attachment Plugs Receptacles and attachment plugs located above hazardous (classified) anesthetizing locations shall be listed for hospital use for services of prescribed voltage, frequency, rating, and number of conductors with provision for the connection of the grounding conductor. This requirement shall apply to attachment plugs and receptacles of the 2-pole, 3-wire grounding type for single-phase, 120-volt, nominal, ac service.

See the commentary following 517.19(B)(2) regarding receptacles listed for hospital use.

(6) 250-Volt Receptacles and Attachment Plugs Rated 50 and 60 Amperes Receptacles and attachment plugs rated 250 volts, for connection of 50-ampere and 60-ampere ac medical equipment for use above hazardous (classified) locations, shall be arranged so that the 60-ampere receptacle will accept either the 50-ampere or the 60-ampere plug. Fifty-ampere receptacles shall be designed so as not to accept the 60-ampere attachment plug. The attachment plugs shall be of the 2-pole, 3-wire design with a third contact connecting to the insulated (green or green with yellow stripe) equipment grounding conductor of the electrical system.

(C) Other-Than-Hazardous (Classified) Anesthetizing Locations

(1) Wiring Methods Wiring serving other-than-hazardous (classified) locations, as defined in 517.60, shall be installed in a metal raceway system or cable assembly. The metal raceway system or cable armor or sheath assembly shall qualify as an equipment grounding return path in accordance with 250.118. Type MC and Type MI cable shall have an outer metal armor or sheath that is identified as an acceptable grounding return path.

Exception: Pendant receptacle constructions that employ at least Type SJO or equivalent flexible cords suspended not less than 1.8 m (6 ft) from the floor shall not be required to be installed in a metal raceway or cable assembly.

(2) Receptacles and Attachment Plugs Receptacles and attachment plugs installed and used in other-than-hazardous (classified) locations shall be listed for hospital use for services of prescribed voltage, frequency, rating, and number of conductors with provision for connection of the grounding conductor. This requirement shall apply to 2-pole, 3-wire grounding type for single-phase, 120-, 208-, or 240-volt, nominal, ac service.

See the commentary following 517.19(B)(2) regarding receptacles listed for hospital use.

(3) 250-Volt Receptacles and Attachment Plugs Rated 50 Amperes and 60 Amperes Receptacles and attachment plugs rated 250 volts, for connection of 50-ampere and 60-ampere ac medical equipment for use in other-than-hazardous (classified) locations, shall be arranged so that the 60-ampere receptacle will accept either the 50-ampere or the 60-ampere plug. Fifty-ampere receptacles shall be designed so as not to accept the 60-ampere attachment plug. The attachment plugs shall be of the 2-pole, 3-wire design with a third contact connecting to the insulated (green or green with yellow stripe) equipment grounding conductor of the electrical system.

517.62 Grounding

In any anesthetizing area, all metal raceways and metal-sheathed cables and all non–current-carrying conductive portions of fixed electric equipment shall be grounded. Grounding in Class I locations shall comply with 501.30.

Exception: Equipment operating at not more than 10 volts between conductors shall not be required to be grounded.

The grounding requirements for anesthetizing locations apply only to metal raceways, metal-sheathed cables, and electrical equipment. Carts, tables, and other nonelectrical items are not required to be grounded. In flammable anesthetizing locations, however, portable carts and tables usually have a resistance to ground of not over 1,000,000 ohms, through the use of conductive tires and wheels and conductive flooring, to avoid the buildup of static electrical charges. See NFPA 99, *Standard for Health Care Facilities*, Annex E, E.6.6.8, for electrostatic safeguards.

517.63 Grounded Power Systems in Anesthetizing Locations

(A) Battery-Powered Emergency Lighting Units One or more battery-powered emergency lighting units shall be provided in accordance with 700.12(F).

The failure of the emergency circuit feeder that supplies the operating room will ordinarily plunge the room into darkness. The requirement to install at least one battery-operated emergency lighting unit results in immediate illumination upon loss of power, which helps mitigate the potentially dangerous impact of total power interruption.

(B) Branch-Circuit Wiring Branch circuits supplying only listed, fixed, therapeutic and diagnostic equipment, permanently installed above the hazardous (classified) location and in other-than-hazardous (classified) locations, shall be permitted to be supplied from a normal grounded service, single- or three-phase system, provided the following apply:

(1) Wiring for grounded and isolated circuits does not occupy the same raceway or cable.

(2) All conductive surfaces of the equipment are grounded.
(3) Equipment (except enclosed X-ray tubes and the leads to the tubes) are located at least 2.5 m (8 ft) above the floor or outside the anesthetizing location.
(4) Switches for the grounded branch circuit are located outside the hazardous (classified) location.

Exception: Sections 517.63(B)(3) and (B)(4) shall not apply in other-than-hazardous (classified) locations.

(C) Fixed Lighting Branch Circuits Branch circuits supplying only fixed lighting shall be permitted to be supplied by a normal grounded service, provided the following apply:

(1) Such luminaires (fixtures) are located at least 2.5 m (8 ft) above the floor.
(2) All conductive surfaces of luminaires (fixtures) are grounded.
(3) Wiring for circuits supplying power to luminaires (fixtures) does not occupy the same raceway or cable for circuits supplying isolated power.
(4) Switches are wall-mounted and located above hazardous (classified) locations.

Exception: Sections 517.63(C)(1) and (C)(4) shall not apply in other-than-hazardous (classified) locations.

(D) Remote-Control Stations Wall-mounted remote-control stations for remote-control switches operating at 24 volts or less shall be permitted to be installed in any anesthetizing location.

(E) Location of Isolated Power Systems Where an isolated power system is utilized, the isolated power equipment shall be listed as isolated power equipment. Isolated power equipment and its grounded primary feeder shall be permitted to be located in an anesthetizing location, provided it is installed above a hazardous (classified) location or in an other-than-hazardous (classified) location.

(F) Circuits in Anesthetizing Locations Except as permitted above, each power circuit within, or partially within, a flammable anesthetizing location as referred to in 517.60 shall be isolated from any distribution system supplying other-than-anesthetizing locations.

517.64 Low-Voltage Equipment and Instruments

(A) Equipment Requirements Low-voltage equipment that is frequently in contact with the bodies of persons or has exposed current-carrying elements shall comply with one of the following:

(1) Operate on an electrical potential of 10 volts or less
(2) Approved as intrinsically safe or double-insulated equipment
(3) Be moisture resistant

(B) Power Supplies Power shall be supplied to low-voltage equipment from one of the following:

(1) An individual portable isolating transformer (autotransformers shall not be used) connected to an isolated power circuit receptacle by means of an appropriate cord and attachment plug

(2) A common low-voltage isolating transformer installed in an other-than-hazardous (classified) location

(3) Individual dry-cell batteries

(4) Common batteries made up of storage cells located in an other-than-hazardous (classified) location

(C) Isolated Circuits Isolating-type transformers for supplying low-voltage circuits shall have both of the following:

(1) Approved means for insulating the secondary circuit from the primary circuit

(2) The core and case grounded

(D) Controls Resistance or impedance devices shall be permitted to control low-voltage equipment but shall not be used to limit the maximum available voltage to the equipment.

(E) Battery-Powered Appliances Battery-powered appliances shall not be capable of being charged while in operation unless their charging circuitry incorporates an integral isolating-type transformer.

(F) Receptacles or Attachment Plugs Any receptacle or attachment plug used on low-voltage circuits shall be of a type that does not permit interchangeable connection with circuits of higher voltage.

> FPN: Any interruption of the circuit, even circuits as low as 10 volts, either by any switch or loose or defective connections anywhere in the circuit, may produce a spark that is sufficient to ignite flammable anesthetic agents.

V. X-Ray Installations

Nothing in this part shall be construed as specifying safeguards against the useful beam or stray X-ray radiation.

> FPN No. 1: Radiation safety and performance requirements of several classes of X-ray equipment are regulated under Public Law 90-602 and are enforced by the Department of Health and Human Services.
>
> FPN No. 2: In addition, information on radiation protection by the National Council on Radiation Protection and Measurements is published as *Reports of the National Council on Radiation Protection and Measurement*. These reports are obtainable from NCRP Publications, P.O. Box 30175, Washington, DC 20014.

517.71 Connection to Supply Circuit

(A) Fixed and Stationary Equipment Fixed and stationary X-ray equipment shall be connected to the power supply by means of a wiring method that meets the general requirements of this *Code*.

Exception: Equipment properly supplied by a branch circuit rated at not over 30 amperes shall be permitted to be supplied through a suitable attachment plug and hard-service cable or cord.

(B) Portable, Mobile, and Transportable Equipment Individual branch circuits shall not be required for portable, mobile, and transportable medical X-ray equipment requiring a capacity of not over 60 amperes.

(C) Over 600-Volt Supply Circuits and equipment operated on a supply circuit of over 600 volts shall comply with Article 490.

517.72 Disconnecting Means

(A) Capacity A disconnecting means of adequate capacity for at least 50 percent of the input required for the momentary rating or 100 percent of the input required for the long-time rating of the X-ray equipment, whichever is greater, shall be provided in the supply circuit.

(B) Location The disconnecting means shall be operable from a location readily accessible from the X-ray control.

(C) Portable Equipment For equipment connected to a 120-volt branch circuit of 30 amperes or less, a grounding-type attachment plug and receptacle of proper rating shall be permitted to serve as a disconnecting means.

517.73 Rating of Supply Conductors and Overcurrent Protection

(A) Diagnostic Equipment

(1) Branch Circuits The ampacity of supply branch-circuit conductors and the current rating of overcurrent protective devices shall not be less than 50 percent of the momentary rating or 100 percent of the long-time rating, whichever is greater.

(2) Feeders The ampacity of supply feeders and the current rating of overcurrent protective devices supplying two or more branch circuits supplying X-ray units shall not be less than 50 percent of the momentary demand rating of the largest unit plus 25 percent of the momentary demand rating of the next largest unit plus 10 percent of the momentary demand rating of each additional unit. Where simultaneous biplane examinations are undertaken with the X-ray units, the supply conductors and overcurrent protective devices shall be 100 percent of the momentary demand rating of each X-ray unit.

> FPN: The minimum conductor size for branch and feeder circuits is also governed by voltage regulation requirements. For a specific installation, the manufacturer usually specifies minimum distribution transformer and

conductor sizes, rating of disconnecting means, and over-current protection.

(B) Therapeutic Equipment The ampacity of conductors and rating of overcurrent protective devices shall not be less than 100 percent of the current rating of medical X-ray therapy equipment.

> FPN: The ampacity of the branch-circuit conductors and the ratings of disconnecting means and overcurrent protection for X-ray equipment are usually designated by the manufacturer for the specific installation.

517.74 Control Circuit Conductors

(A) Number of Conductors in Raceway The number of control circuit conductors installed in a raceway shall be determined in accordance with 300.17.

(B) Minimum Size of Conductors Size 18 AWG or 16 AWG fixture wires as specified in 725.27 and flexible cords shall be permitted for the control and operating circuits of X-ray and auxiliary equipment where protected by not larger than 20-ampere overcurrent devices.

517.75 Equipment Installations

All equipment for new X-ray installations and all used or reconditioned X-ray equipment moved to and reinstalled at a new location shall be of an approved type.

517.76 Transformers and Capacitors

Transformers and capacitors that are part of X-ray equipment shall not be required to comply with Articles 450 and 460.

Capacitors shall be mounted within enclosures of insulating material or grounded metal.

517.77 Installation of High-Tension X-Ray Cables

Cables with grounded shields connecting X-ray tubes and image intensifiers shall be permitted to be installed in cable trays or cable troughs along with X-ray equipment control and power supply conductors without the need for barriers to separate the wiring.

517.78 Guarding and Grounding

(A) High-Voltage Parts All high-voltage parts, including X-ray tubes, shall be mounted within grounded enclosures. Air, oil, gas, or other suitable insulating media shall be used to insulate the high-voltage from the grounded enclosure. The connection from the high-voltage equipment to X-ray tubes and other high-voltage components shall be made with high-voltage shielded cables.

(B) Low-Voltage Cables Low-voltage cables connecting to oil-filled units that are not completely sealed, such as

transformers, condensers, oil coolers, and high-voltage switches, shall have insulation of the oil-resistant type.

(C) Noncurrent–Carrying Metal Parts Noncurrent-carrying metal parts of X-ray and associated equipment (controls, tables, X-ray tube supports, transformer tanks, shielded cables, X-ray tube heads, etc.) shall be grounded in the manner specified in Article 250, as modified by 517.13(A) and 517.13(B).

VI. Communications, Signaling Systems, Data Systems, Fire Alarm Systems, and Systems Less Than 120 Volts, Nominal

517.80 Patient Care Areas

Equivalent insulation and isolation to that required for the electrical distribution systems in patient care areas shall be provided for communications, signaling systems, data system circuits, fire alarm systems, and systems less than 120 volts, nominal.

> FPN: An acceptable alternate means of providing isolation for patient/nurse call systems is by the use of nonelectrified signaling, communications, or control devices held by the patient or within reach of the patient.

517.81 Other-Than-Patient-Care Areas

In other-than-patient-care areas, installations shall be in accordance with the appropriate provisions of Articles 640, 725, 760, and 800.

517.82 Signal Transmission Between Appliances

(A) General Permanently installed signal cabling from an appliance in a patient location to remote appliances shall employ a signal transmission system that prevents hazardous grounding interconnection of the appliances.

> FPN: See 517.13(A) for additional grounding requirements in patient care areas.

(B) Common Signal Grounding Wire Common signal grounding wires (i.e., the chassis ground for single-ended transmission) shall be permitted to be used between appliances all located within the patient vicinity, provided the appliances are served from the same reference grounding point.

VII. Isolated Power Systems

517.160 Isolated Power Systems

(A) Installations

(1) Isolated Power Circuits Each isolated power circuit shall be controlled by a switch that has a disconnecting pole in each isolated circuit conductor to simultaneously disconnect all power. Such isolation shall be accomplished by means of one or more transformers having no electrical connection between primary and secondary windings, by means of motor generator sets, or by means of suitably isolated batteries.

(2) Circuit Characteristics Circuits supplying primaries of isolating transformers shall operate at not more than 600 volts between conductors and shall be provided with proper overcurrent protection. The secondary voltage of such transformers shall not exceed 600 volts between conductors of each circuit. All circuits supplied from such secondaries shall be ungrounded and shall have an approved overcurrent device of proper ratings in each conductor. Circuits supplied directly from batteries or from motor generator sets shall be ungrounded and shall be protected against overcurrent in the same manner as transformer-fed secondary circuits. If an electrostatic shield is present, it shall be connected to the reference grounding point. [NFPA 99:4.3.2.6.1]

(3) Equipment Location The isolating transformers, motor generator sets, batteries and battery chargers, and associated primary or secondary overcurrent devices shall not be installed in hazardous (classified) locations. The isolated secondary circuit wiring extending into a hazardous anesthetizing location shall be installed in accordance with 501.10.

(4) Isolation Transformers An isolation transformer shall not serve more than one operating room except as covered in (A)(4)(a) and (A)(4)(b).

For purposes of this section, anesthetic induction rooms are considered part of the operating room or rooms served by the induction rooms.

(a) Induction Rooms. Where an induction room serves more than one operating room, the isolated circuits of the induction room shall be permitted to be supplied from the isolation transformer of any one of the operating rooms served by that induction room.

(b) Higher Voltages. Isolation transformers shall be permitted to serve single receptacles in several patient areas where the following apply:

(1) The receptacles are reserved for supplying power to equipment requiring 150 volts or higher, such as portable X-ray units.
(2) The receptacles and mating plugs are not interchangeable with the receptacles on the local isolated power system. [NFPA 99:13.4.1.2.6.6]

(5) Conductor Identification The isolated circuit conductors shall be identified as follows:

(1) Isolated Conductor No. 1 — Orange
(2) Isolated Conductor No. 2 — Brown

For 3-phase systems, the third conductor shall be identified as yellow. Where isolated circuit conductors supply 125-volt, single-phase, 15- and 20-ampere receptacles, the orange conductor(s) shall be connected to the terminal(s) on the receptacles that are identified in accordance with 200.10(B) for connection to the grounded circuit conductor.

(6) Wire-Pulling Compounds Wire-pulling compounds that increase the dielectric constant shall not be used on the secondary conductors of the isolated power supply.

> FPN No. 1: It is desirable to limit the size of the isolation transformer to 10 kVA or less and to use conductor insulation with low leakage to meet impedance requirements.
>
> FPN No. 2: Minimizing the length of branch-circuit conductors and using conductor insulations with a dielectric constant less than 3.5 and insulation resistance constant greater than 6100 megohm-meters (20,000 megohm-feet) at 16°C (60°F) reduces leakage from line to ground, reducing the hazard current.

See Exhibit 517.3 for an example of a hospital isolated power system panel.

Exhibit 517.3 An example of a hospital isolated power system panel with built-in isolation transformer, line isolation monitor, load center, and grounded busbar. (Courtesy of Square D Co.)

(B) Line Isolation Monitor

(1) Characteristics In addition to the usual control and overcurrent protective devices, each isolated power system shall be provided with a continually operating line isolation monitor that indicates total hazard current. The monitor shall be designed such that a green signal lamp, conspicuously visible to persons in each area served by the isolated power system, remains lighted when the system is adequately isolated from ground. An adjacent red signal lamp and an audible warning signal (remote if desired) shall be energized when the total hazard current (consisting of possible resistive and capacitive leakage currents) from either isolated conductor to ground reaches a threshold value of 5 mA under nominal line voltage conditions. The line monitor shall not alarm for a fault hazard of less than 3.7 mA or for a total hazard current of less than 5 mA.

Exception: A system shall be permitted to be designed to operate at a lower threshold value of total hazard current. A line isolation monitor for such a system shall be permitted to be approved with the provision that the fault hazard current shall be permitted to be reduced but not to less than 35 percent of the corresponding threshold value of the total hazard current, and the monitor hazard current is to be correspondingly reduced to not more than 50 percent of the alarm threshold value of the total hazard current.

(2) Impedance The line isolation monitor shall be designed to have sufficient internal impedance such that, when properly connected to the isolated system, the maximum internal current that can flow through the line isolation monitor, when any point of the isolated system is grounded, shall be 1 mA.

Exception: The line isolation monitor shall be permitted to be of the low-impedance type such that the current through the line isolation monitor, when any point of the isolated system is grounded, will not exceed twice the alarm threshold value for a period not exceeding 5 milliseconds.

> FPN: Reduction of the monitor hazard current, provided this reduction results in an increased "not alarm" threshold value for the fault hazard current, will increase circuit capacity.

(3) Ammeter An ammeter calibrated in the total hazard current of the system (contribution of the fault hazard current plus monitor hazard current) shall be mounted in a plainly visible place on the line isolation monitor with the "alarm on" zone at approximately the center of the scale.

Exception: The line isolation monitor shall be permitted to be a composite unit, with a sensing section cabled to a separate display panel section on which the alarm or test functions are located.

> FPN: It is desirable to locate the ammeter so that it is conspicuously visible to persons in the anesthetizing location.

ARTICLE 518
Assembly Occupancies

Summary of Changes

- **Article Title, 518.1, 518.2(A), 518.2(B):** Aligned *NEC* with NFPA 1, NFPA *101®*, and *NFPA 5000®* by changing "Places of Assembly" to "Assembly Occupancies."

- **518.2(A):** Revised to include all auditoriums, to replace church chapels with places of religious worship, and to include drinking facilities as examples of assembly occupancies.

- **518.2(B):** Revised to make it clear that only a part of a building needs to be classified as an assembly occupancy in buildings containing an assembly occupancy plus other occupancy types.

- **518.4(C):** Revised to delete college and university classrooms, drinking establishments, and passenger stations as assembly occupancies in which nonmetallic wiring methods can be installed within spaces with a 15-minute finish rating.

Contents

518.1 Scope

Except for the assembly occupancies explicitly covered by 520.1, this article covers all buildings or portions of buildings or structures designed or intended for the gathering together of 100 or more persons for such purposes as deliberation, worship, entertainment, eating, drinking, amusement, awaiting transportation, or similar purposes.

Article 518 applies to assembly occupancies designed or intended for 100 or more persons. Article 518 would apply, for example, to a church chapel or auditorium for occupancy of 100 or more persons, its capacity determined by the methods for occupancy population capacity in accordance with NFPA *101*, *Life Safety Code*. Article 518 does not apply to supermarkets; even though a supermarket may contain 100 or more persons, it is not specifically designed or intended for the assembly of persons, nor is it considered to be an auditorium. Article 518 does not apply to office buildings or schools, even though such buildings, as a rule, are designed for occupancy by 100 or more persons. Article 518 does, however, apply to assembly halls, restaurants, and so on, within an office or school building if these parts of the building are designed or intended for the assembly of 100 or more persons.

The following information for determining new assembly occupancy capacity is extracted from NFPA *101*, *Life Safety Code*:

12.1.7 Occupant Load.

12.1.7.1 General. The occupant load, in number of persons for whom means of egress and other provisions are required, shall be determined on the basis of the occupant load factors of [Commentary Table 5.4] that are characteristic of the use of the space or shall be determined as the maximum probable population

of the space under consideration, whichever is greater. [NFPA *101*:12.1.7]

⋮

7.3.1 Occupant Load.

7.3.1.1 Sufficient Capacity for Occupant Load. The total capacity of the means of egress for any story, balcony, tier, or other occupied space shall be sufficient for the occupant load thereof.

7.3.1.2 Occupant Load Factor. The occupant load in any building or portion thereof shall be not less than the number of persons determined by dividing the floor area assigned to that use by the occupant load factor for that use as specified in Table 7.3.1.2 [shown here as Commentary Table 5.4]. Where both gross and net area figures are given for the same occupancy, calculations shall be made by applying the gross area figure to the gross area of the portion of the building devoted to the use for which the gross area figure is specified and by applying the net area figure to the net area of the portion of the building devoted to the use for which the net area figure is specified. [NFPA *101*:12.1.7]

Commentary Table 5.4 Occupant Load Factor

Use	m² (per person)[1]	ft² (per person)[1]
Assembly Use		
Concentrated use, without fixed seating	0.65 net	7 net
Less concentrated use, without fixed seating	1.4 net	15 net
Bench-type seating	1 person/455 linear mm	1 person/18 linear in.
Fixed seating	Number of fixed seats	Number of fixed seats
Waiting spaces	See 12.1.7.2 and 13.1.7.2	See 12.1.7.2 and 13.1.7.2
Kitchens	9.3	100
Library stack areas	9.3	100
Library reading rooms	4.6 net	50 net
Swimming pools	4.6 (water surface)	50 (water surface)
Swimming pool decks	2.8	30
Exercise rooms with equipment	4.6	50
Exercise rooms without equipment	1.4	15
Stages	1.4 net	15 net
Lighting and access catwalks, galleries, gridirons	9.3 net	100 net
Casinos and similar gaming areas	1	11
Skating rinks	4.6	50

(continues)

Commentary Table 5.4 *Continued*

Use	m² (per person)[1]	ft² (per person)[1]
Educational Use		
Classrooms	1.9 net	20 net
Shops, laboratories, vocational rooms	4.6 net	50 net
Day-Care Use	3.3 net	35 net
Health Care Use		
Inpatient treatment departments	22.3	240
Sleeping departments	11.1	120
Detention and Correctional Use	11.1	120
Residential Use		
Hotels and dormitories	18.6	200
Apartment buildings	18.6	200
Board and care, large	18.6	200
Industrial Use		
General and high hazard industrial	9.3	100
Special purpose industrial	NA	NA
Business Use	9.3	100
Storage Use		
(other than mercantile storerooms)	NA	NA
Mercantile Use		
Sales area on street floor[2,3]	2.8	30
Sales area on two or more street floors[3]	3.7	40
Sales area on floor below street floor[3]	2.8	30
Sales area on floors above street floor[3]	5.6	60
Floors or portions of floors used only for offices	See business use	See business use
Floors or portions of floors used only for storage, receiving, and shipping, and not open to general public	27.9	300
Mall buildings[4]	Per factors applicable to use of space[5]	Per factors applicable to use of space[5]

Note: NA = not applicable. The occupant load is the maximum probable number of occupants present at any time.

[1]All factors are expressed in gross area unless marked "net."

[2]For the purpose of determining occupant load in mercantile occupancies where, due to differences in grade of streets on different sides, two or more floors directly accessible from streets (not including alleys or similar back streets) exist, each such floor is permitted to be considered a street floor. The occupant load factor is one person for each 3.7 m²(40 ft²) of gross floor area of sales space.

[3]For the purpose of determining occupant load in mercantile occupancies with no street floor, as defined in 3.3.216, but with access directly from the street by stairs or escalators, the floor at the point of entrance to the mercantile occupancy is considered the street floor.

[4]For any food court or other assembly use areas located in the mall that are not included as a portion of the gross leasable area of the mall building, the occupant load is calculated based on the occupant load factor for that use as specified in Table 7.3.1.2. The remaining mall area is not required to be assigned an occupant load.

[5]The portions of the mall that are considered a pedestrian way and not used as gross leasable area are not required to be assessed an occupant load based on Table 7.3.1.2. However, means of egress from a mall pedestrian way are required to be provided for an occupant load determined by dividing the gross leasable area of the mall building (not including anchor stores) by the appropriate lowest whole number occupant load factor from Figure 7.3.1.2(a) or Figure 7.3.1.2(b).

Each individual tenant space is required to have means of egress to the outside or to the mall based on occupant loads calculated by using the appropriate occupant load factor from Table 7.3.1.2.

Each individual anchor store is required to have means of egress independent of the mall.

518.2 General Classification

(A) Examples Assembly occupancies shall include, but not be limited to, the following:

A task group consisting of members of Code-Making Panel 15 and the NFPA *101* and *NFPA 5000* committees was formed to discuss the definition of *assembly occupancy* and the classification of various assembly occupancies. The classification examples in 518.2(A) were revised for the 2005 *Code*, and the terminology was changed to specifically identify those portions of a building that are assembly occupancies. This terminology change helps to correlate the *Code* with other NFPA standards.

Armories	Exhibition halls
Assembly halls	Gymnasiums
Auditoriums	Mortuary chapels
Bowling lanes	Multipurpose rooms
Club rooms	Museums
Conference rooms	Places of awaiting transportation
Courtrooms	Places of religious worship
Dance halls	Pool rooms
Dining and drinking	Restaurants
facilities	Skating rinks

(B) Multiple Occupancies Where an assembly occupancy forms a portion of a building containing other occupancies, Article 518 applies only to that portion of the building considered an assembly occupancy. Occupancy of any room or space for assembly purposes by less than 100 persons in a building of other occupancy, and incidental to such other occupancy, shall be classified as part of the other occupancy and subject to the provisions applicable thereto.

(C) Theatrical Areas Where any such building structure, or portion thereof, contains a projection booth or stage platform or area for the presentation of theatrical or musical productions, either fixed or portable, the wiring for that area, including associated audience seating areas, and all equipment that is used in the referenced area, and portable equipment and wiring for use in the production that will not be connected to permanently installed wiring, shall comply with Article 520.

See the commentary following 520.1.

> FPN: For methods of determining population capacity, see local building code or, in its absence, NFPA *101*-2003, *Life Safety Code*.

518.3 Other Articles

(A) Hazardous (Classified) Areas Electrical installations in hazardous (classified) areas located in assembly occupancies shall comply with Article 500.

(B) Temporary Wiring In exhibition halls used for display booths, as in trade shows, the temporary wiring shall be installed in accordance with Article 590. Flexible cables and cords approved for hard or extra-hard usage shall be permitted to be laid on floors where protected from contact by the general public. The ground-fault circuit-interrupter requirements of 590.6 shall not apply.

A treadle, such as the one shown in Exhibit 518.1, is an example of a protection technique used to protect cords from abuse in areas where the cords are laid across pedestrian ways. For example, treadles may be used to protect temporary wiring used in exhibition halls.

Exhibit 518.1 A treadle used to protect cords. (Courtesy of Daniel Woodhead Co.)

Exception: Where conditions of supervision and maintenance ensure that only qualified persons will service the installation, flexible cords or cables identified in Table 400.4 for hard usage or extra-hard usage shall be permitted in cable trays used only for temporary wiring. All cords or cables shall be installed in a single layer. A permanent sign shall be attached to the cable tray at intervals not to exceed 7.5 m (25 ft). The sign shall read

CABLE TRAY FOR TEMPORARY WIRING ONLY

(C) Emergency Systems Control of emergency systems shall comply with Article 700.

518.4 Wiring Methods

(A) General The fixed wiring methods shall be metal raceways, flexible metal raceways, nonmetallic raceways encased in not less than 50 mm (2 in.) of concrete, Type MI, MC, or AC cable containing an insulated equipment grounding conductor sized in accordance with Table 250.122.

Exception: Fixed wiring methods shall be as provided in

(a) Audio signal processing, amplification, and reproduction equipment — Article 640

(b) Communications circuits — Article 800

(c) Class 2 and Class 3 remote-control and signaling circuits — Article 725

(d) Fire alarm circuits — Article 760

(B) Nonrated Construction In addition to the wiring methods of 518.4(A), nonmetallic-sheathed cable, Type AC cable, electrical nonmetallic tubing, and rigid nonmetallic conduit shall be permitted to be installed in those buildings or portions thereof that are not required to be of fire-rated construction by the applicable building code.

> FPN: Fire-rated construction is the fire-resistive classification used in building codes.

(C) Spaces with Finish Rating Electrical nonmetallic tubing and rigid nonmetallic conduit shall be permitted to be installed in club rooms, conference and meeting rooms in hotels or motels, courtrooms, dining facilities, restaurants, mortuary chapels, museums, libraries, and places of religious worship where the following apply:

(1) The electrical nonmetallic tubing or rigid nonmetallic conduit is installed concealed within walls, floors, and ceilings where the walls, floors, and ceilings provide a thermal barrier of material that has at least a 15-minute finish rating as identified in listings of fire-rated assemblies.

(2) The electrical nonmetallic tubing or rigid nonmetallic conduit is installed above suspended ceilings where the suspended ceilings provide a thermal barrier of material that has at least a 15-minute finish rating as identified in listings of fire-rated assemblies.

Electrical nonmetallic tubing and rigid nonmetallic conduit are not recognized for use in other space used for environmental air in accordance with 300.22(C).

> FPN: A finish rating is established for assemblies containing combustible (wood) supports. The finish rating is defined as the time at which the wood stud or wood joist reaches an average temperature rise of 121°C (250°F) or an individual temperature rise of 163°C (325°F) as measured on the plane of the wood nearest the fire. A finish rating is not intended to represent a rating for a membrane ceiling.

In Exhibit 518.2, the wash rooms and office area of the single-story facility shown are not assembly occupancies, as defined in 518.2, and therefore require no special wiring methods. Ordinary wiring methods can be used on the inside surface of the storage area walls and on or in the partitions between storage areas, because those also are not assembly

occupancies. Inside any hollow spaces of the fire-rated storage area walls, however, the main requirements of 518.4 apply, because the serving corridors are part of the assembly occupancies as a result of this particular building design. If the hollow spaces of fire-rated walls or ceiling also provide a 15-minute finish rating and are not other spaces for environmental air, as described in 300.22(C), then electrical nonmetallic tubing as well as rigid nonmetallic conduit would be permitted for specifically described occupancies. See 362.10.

Exhibit 518.2 The walls (represented by wide gold lines) required by the local building code to be of fire-rated construction. The thin black lines represent walls not required by the local building code to be of fire-rated construction.

In like manner, wiring in ceilings or floors that is required to be of fire-rated construction in an assembly occupancy as defined in 518.2 must also comply with 518.4, except as noted in 518.4(A), Exception.

The intent is that, within an assembly occupancy, as defined in 518.2, 518.4(A) and the exception to 518.4(A) apply to any wall, floor, or ceiling. The provisions of 518.4(B) apply to those portions of the building and those assembly occupancies not required to be fire rated. The permission to use electrical nonmetallic tubing and rigid nonmetallic conduit as specified in 518.4(C) applies only to the specific occupancies described, provided these wiring methods are installed concealed behind a surface that has a 15-minute finish rating.

Assembly occupancies frequently require emergency wiring, particularly for emergency illumination and exit lighting. The requirements of 700.9(D) contain special fire protection requirements for emergency circuits in assembly occupancies with an occupant capacity of 1000 or more.

518.5 Supply

Portable switchboards and portable power distribution equipment shall be supplied only from listed power outlets of sufficient voltage and ampere rating. Such power outlets shall be protected by overcurrent devices. Such overcurrent devices and power outlets shall not be accessible to the general public. Provisions for connection of an equipment grounding conductor shall be provided. The neutral of feeders supplying solid-state, 3-phase, 4-wire dimmer systems shall be considered a current-carrying conductor.

In accordance with 518.5, portable switchboards and portable power distribution equipment must be supplied only from listed power outlets rated for the voltage and current they are used for. See Exhibit 518.3. The overcurrent devices and power outlets used to supply such equipment must be located so as not to be accessible to the general public.

ARTICLE 520
Theaters, Audience Areas of Motion Picture and Television Studios, Performance Areas, and Similar Locations

Contents

Exhibit 518.3 A listed power outlet for connection of portable switchboards in an assembly occupancy. (Courtesy of Union Connector Co., Inc.)

I. General

520.1 Scope

This article covers all buildings or that part of a building or structure, indoor or outdoor, designed or used for presentation, dramatic, musical, motion picture projection, or similar purposes and to specific audience seating areas within motion picture or television studios.

The special requirements of Article 520 apply only to that part of a building used as a theater or for a similar purpose and do not necessarily apply to the entire building. For example, the requirements of Article 520 would apply to an auditorium in a school building used for dramatic or other performances. The special requirements of this article apply to the stage, auditorium, dressing rooms, and main corridors leading to the auditorium, but not to other parts of the building that are not involved in the use of the auditorium for performances or entertainment. The theater space may be a traditional theater, where the audience sits in the auditorium (house) facing the proscenium arch and views the performance on the stage on the other side of the arch, or other spaces, such as a simple stage platform, either indoors or outdoors, with seats on three or four sides facing the platform.

The audience areas of motion picture and television studios (as defined and covered in Article 530) are also covered by the requirements of Article 520.

520.2 Definitions

Border Light. A permanently installed overhead strip light.

Breakout Assembly. An adapter used to connect a multipole connector containing two or more branch circuits to multiple individual branch-circuit connectors.

Bundled. Cables or conductors that are physically tied, wrapped, taped or otherwise periodically bound together.

Connector Strip. A metal wireway containing pendant or flush receptacles.

Drop Box. A box containing pendant- or flush-mounted receptacles attached to a multiconductor cable via strain relief or a multipole connector.

Footlight. A border light installed on or in the stage.

Grouped. Cables or conductors positioned adjacent to one another but not in continuous contact with each other.

Performance Area. The stage and audience seating area associated with a temporary stage structure, whether indoors or outdoors, constructed of scaffolding, truss, platforms, or similar devices, that is used for the presentation of theatrical or musical productions or for public presentations.

Portable Equipment. Equipment fed with portable cords or cables intended to be moved from one place to another.

Portable Power Distribution Unit. A power distribution box containing receptacles and overcurrent devices.

Proscenium. The wall and arch that separates the stage from the auditorium (house).

Stand Lamp (Work Light). A portable stand that contains a general-purpose luminaire (lighting fixture) or lampholder with guard for the purpose of providing general illumination on the stage or in the auditorium.

Strip Light. A luminaire (lighting fixture) with multiple lamps arranged in a row.

Two-Fer. An adapter cable containing one male plug and two female cord connectors used to connect two loads to one branch circuit.

A two-fer, as shown in Exhibit 520.1, consists of two cord connectors on separate cords connected to a single supply cord.

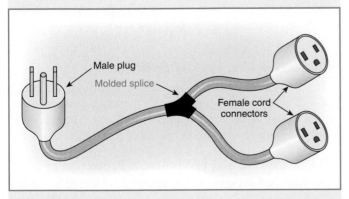

Exhibit 520.1 A two-fer.

520.3 Motion Picture Projectors

Motion picture equipment and its installation and use shall comply with Article 540.

520.4 Audio Signal Processing, Amplification, and Reproduction Equipment

Audio signal processing, amplification, and reproduction equipment and its installation shall comply with Article 640.

520.5 Wiring Methods

(A) General The fixed wiring method shall be metal raceways, nonmetallic raceways encased in at least 50 mm (2 in.) of concrete, Type MI cable, MC cable, or AC cable containing an insulated equipment grounding conductor sized in accordance with Table 250.122.

Exception: Fixed wiring methods shall be as provided in Article 640 for audio signal processing, amplification, and reproduction equipment, in Article 800 for communication circuits, in Article 725 for Class 2 and Class 3 remote-control and signaling circuits, and in Article 760 for fire alarm circuits.

(B) Portable Equipment The wiring for portable switchboards, stage set lighting, stage effects, and other wiring not fixed as to location shall be permitted with approved flexible cords and cables as provided elsewhere in Article 520. Fastening such cables and cords by uninsulated staples or nailing shall not be permitted.

(C) Nonrated Construction Nonmetallic-sheathed cable, Type AC cable, electrical nonmetallic tubing, and rigid nonmetallic conduit shall be permitted to be installed in those buildings or portions thereof that are not required to be of fire-rated construction by the applicable building code.

Theaters and similar buildings are usually required to be of fire-rated construction, as determined by applicable building codes; therefore, the fixed wiring methods are limited. See 518.4 for the requirements on wiring methods.

The exception to the requirements for metal-enclosed or concrete-encased fixed wiring permits the installation of communications circuits, Class 2 and Class 3 remote-control and signaling circuits, sound-reproduction wiring, and fire alarm circuits using wiring methods from the respective articles covering these systems in Chapters 7 and 8. Where portability, flexibility, and adjustments are necessary for portable switchboards, stage lighting, and special effects, suitable cords and cables are permitted. In accordance with 520.5(C), Type NM cable, Type AC cable, ENT, and RNC are permitted as the wiring method in buildings or portions of buildings that are not required to be of fire-rated construction. In this application, Type AC cable is not required to contain an insulated equipment grounding conductor.

520.6 Number of Conductors in Raceway

The number of conductors permitted in any metal conduit, rigid nonmetallic conduit as permitted in this article, or electrical metallic tubing for border or stage pocket circuits or for remote-control conductors shall not exceed the percentage fill shown in Table 1 of Chapter 9. Where contained within an auxiliary gutter or a wireway, the sum of the cross-sectional areas of all contained conductors at any cross section shall not exceed 20 percent of the interior cross-sectional area of the auxiliary gutter or wireway. The 30-conductor limitation of 366.22 and 376.22 shall not apply.

520.7 Enclosing and Guarding Live Parts

Live parts shall be enclosed or guarded to prevent accidental contact by persons and objects. All switches shall be of the externally operable type. Dimmers, including rheostats, shall be placed in cases or cabinets that enclose all live parts.

520.8 Emergency Systems

Control of emergency systems shall comply with Article 700.

520.9 Branch Circuits

A branch circuit of any size supplying one or more receptacles shall be permitted to supply stage set lighting. The voltage rating of the receptacles shall not be less than the circuit voltage. Receptacle ampere ratings and branch-circuit conductor ampacity shall not be less than the branch-circuit overcurrent device ampere rating. Table 210.21(B)(2) shall not apply.

The stage set lighting and associated equipment, such as stage effects, both fixed and portable, must be as flexible as possible. Connectors are often used for different purposes and are therefore marked on a show-by-show basis as to the voltage, current, and type of current actually employed. The provisions of 520.9 require only that connectors be rated sufficiently for the parameters involved, thus permitting connectors with voltage and current ratings higher than the branch-circuit rating to be used.

The intent of 520.9 is to exclude the occupancies referenced in Article 520 from all the general requirements relating to connector rating and branch-circuit loading found elsewhere in the *Code*, such as in Table 210.21(B)(2). The requirements of 520.9 modify several other sections, such as 210.23(C) and 210.23(D), which would disallow 40-ampere and larger branch circuits from serving 5000-watt and larger portable stage lighting equipment found in theaters.

Stage set lighting is usually planned in advance, and the loads on each receptacle are known. Loads are not casually connected, as they might be at a typical general-use wall receptacle. Care is taken to ensure that circuits are not overloaded, thereby avoiding nuisance tripping during a performance.

520.10 Portable Equipment

Portable stage and studio lighting equipment and portable power distribution equipment shall be permitted for temporary use outdoors, provided the equipment is supervised by qualified personnel while energized and barriered from the general public.

In accordance with 520.10, portable indoor stage or studio equipment that is not marked suitable for wet or damp locations is permitted to be used temporarily in outdoor locations. If rain occurs, this equipment is typically de-energized, and a protective cover is installed before it is re-energized. At the end of the day, this equipment is either de-energized and protected or dismantled and stored.

II. Fixed Stage Switchboards

520.21 Dead Front

Stage switchboards shall be of the dead-front type and shall comply with Part IV of Article 408 unless approved based on suitability as a stage switchboard as determined by a qualified testing laboratory and recognized test standards and principles.

Early stage switchboards were vertical marble or slate slabs mounted on the stage near the proscenium wall, with exposed knife switches and fuseholders mounted on them and with exposed resistance-type dimmer plates across the top. The "dead front," "guarded back," and "metal hood" requirements of the *Code* are intended to provide the operator with some sort of protection from shock and the heat-producing equipment with some sort of protection from flammable curtains and scenery likely to be above and around the equipment. For these reasons, modern switchboards are totally enclosed.

520.22 Guarding Back of Switchboard

Stage switchboards having exposed live parts on the back of such boards shall be enclosed by the building walls, wire mesh grilles, or by other approved methods. The entrance to this enclosure shall be by means of a self-closing door.

520.23 Control and Overcurrent Protection of Receptacle Circuits

Means shall be provided at a stage-lighting switchboard to which load circuits are connected for overcurrent protection of stage-lighting branch circuits, including branch circuits supplying stage and auditorium receptacles used for cord- and plug-connected stage equipment. Where the stage switchboard contains dimmers to control nonstage lighting, the locating of the overcurrent protective devices for these branch circuits at the stage switchboard shall be permitted.

The purpose of 520.23 is to ensure that the overcurrent protection devices are readily accessible to stage personnel during the presentation.

520.24 Metal Hood

A stage switchboard that is not completely enclosed dead-front and dead-rear or recessed into a wall shall be provided with a metal hood extending the full length of the board to protect all equipment on the board from falling objects.

Because stages are usually crowded and a great deal of flammable material is often present, a stage switchboard is not permitted to have exposed live parts on its front. Moreover, the space at the rear of a stage switchboard must be guarded to prevent entrance or contact by unqualified and unauthorized persons. One accepted method of accomplishing this is by enclosing the space between the rear of the switchboard and the wall in a sheet-steel housing with a door at one end.

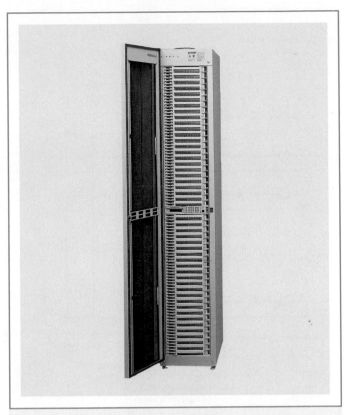

Exhibit 520.2 A typical high-density digital SCR dimmer switchboard. (Courtesy of Electronic Theatre Controls, Inc.)

520.25 Dimmers

Dimmers shall comply with 520.25(A) through (D).

(A) Disconnection and Overcurrent Protection Where dimmers are installed in ungrounded conductors, each dimmer shall have overcurrent protection not greater than 125 percent of the dimmer rating and shall be disconnected from all ungrounded conductors when the master or individual switch or circuit breaker supplying such dimmer is in the open position.

A modern, high-density digital dimmer rack typically contains one dimmer (usually of 20-, 50-, or 100-ampere capacity) for each branch circuit connected to it. The rack is usually serviced by a 3-phase, 4-wire-plus-ground feeder, which is distributed via buses to all dimmers in the rack. Typical dimmer racks contain between 12 and 96 dimmers and may have total power capacities of up to 288 kW. In large theatrical systems, many racks may be bused together. A central control electronics module drives multiple dimmers in the rack. A digital data link may connect the dimmer rack to the remotely located computer control console. Exhibit 520.2 shows a high-density digital SCR dimmer switchboard, and Exhibit 520.3 shows its schematic diagram.

(B) Resistance- or Reactor-Type Dimmers Resistance- or series reactor-type dimmers shall be permitted to be placed in either the grounded or the ungrounded conductor of the circuit. Where designed to open either the supply circuit to the dimmer or the circuit controlled by it, the dimmer shall then comply with 404.2(B). Resistance- or reactor-type dimmers placed in the grounded neutral conductor of the circuit shall not open the circuit.

(C) Autotransformer-Type Dimmers The circuit supplying an autotransformer-type dimmer shall not exceed 150 volts between conductors. The grounded conductor shall be common to the input and output circuits.

> FPN: See 210.9 for circuits derived from autotransformers.

Circuits supplying autotransformer-type dimmers are not permitted to exceed 150 volts between conductors. Any desired voltage may be applied to the lamps, from full-line voltage to voltage so low that the lamps provide no illumination, by means of a movable contact tap. Typical connections for an autotransformer-type dimmer are shown in Exhibit 520.4 This type of dimmer produces very little heat and operates at high efficiency. Its dimming effect,

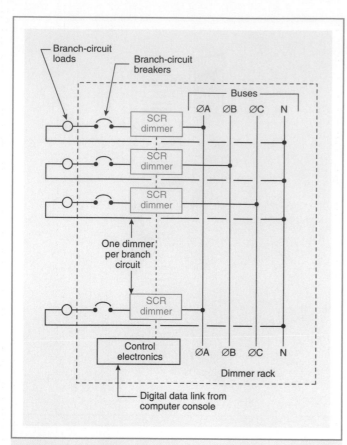

Exhibit 520.3 Schematic diagram of a typical high-density SCR dimmer switchboard. (Redrawn courtesy of Production Arts Lighting, Inc. and Production Resource Group, L.L.C.)

within its maximum rating, is independent of the wattage of the load. Autotransformer-type dimmers are currently seldom used. See the commentary that follows 470.1, which discusses saturable reactors that are sometimes used for stage dimmers.

Exhibit 520.4 Typical connections for an autotransformer-type dimmer. (Courtesy of Kliegl Bros.)

(D) Solid-State-Type Dimmers The circuit supplying a solid-state dimmer shall not exceed 150 volts between conductors unless the dimmer is listed specifically for higher voltage operation. Where a grounded conductor supplies a dimmer, it shall be common to the input and output circuits. Dimmer chassis shall be connected to the equipment grounding conductor.

Modern stage switchboards are usually of the remote-control type. The switchboard is operated from a remote console, typically a computer system such as the one shown in Exhibit 520.5. The switchboard or dimmer rack is normally located offstage in a dimmer room, where proper climate control can be furnished and noise from the rack cooling fans will not interfere with the performance onstage. Branch circuits are usually connected to the dimmer rack on a dimmer-per-circuit basis. A digital control cable connects the computer and the dimmer rack, allowing the operator to be positioned on stage or in the auditorium for easy viewing of the performance.

Exhibit 520.5 An electronic computer lighting control console for remotely controlling solid-state-type dimmers. (Courtesy of Electronic Theatre Controls, Inc.)

A front view of a typical high-density digital SCR dimmer rack is shown in Exhibit 520.2. A schematic for this type of dimmer rack is shown in Exhibit 520.3. Dimmers for individual circuits are contained in dual plug-in dimmer modules. These modules also contain circuit breakers for overcurrent protection and filter chokes to eliminate acoustic noise from the lamp filaments. The digital control electronics are contained in a plug-in module with front-panel controls for configuration and testing.

520.26 Type of Switchboard

A stage switchboard shall be either one or a combination of the types specified in 520.26(A), (B), and (C).

(A) Manual Dimmers and switches are operated by handles mechanically linked to the control devices.

Manual-type switchboards usually contain resistance-type or autotransformer-type dimmers. Exhibit 520.4 is a schematic of a manual autotransformer-type dimmer.

(B) Remotely Controlled Devices are operated electrically from a pilot-type control console or panel. Pilot control panels either shall be part of the switchboard or shall be permitted to be at another location.

(C) Intermediate A stage switchboard with circuit interconnections is a secondary switchboard (patch panel) or panelboard remote to the primary stage switchboard. It shall contain overcurrent protection. Where the required branch-circuit overcurrent protection is provided in the dimmer panel, it shall be permitted to be omitted from the intermediate switchboard.

An intermediate stage switchboard, usually called a patch panel, is located between the dimmer switchboard and the branch circuits. Its purpose is to break down larger dimmer circuits to smaller branch circuits, to select the branch circuits to be controlled by a dimmer, or both.

520.27 Stage Switchboard Feeders

(A) Type of Feeder Feeders supplying stage switchboards shall be one of the types in 520.27(A)(1) through (A)(3).

(1) Single Feeder A single feeder disconnected by a single disconnect device.

(2) Multiple Feeders to Intermediate Stage Switchboard (Patch Panel) Multiple feeders of unlimited quantity shall be permitted, provided that all multiple feeders are part of a single system. Where combined, neutral conductors in a given raceway shall be of sufficient ampacity to carry the maximum unbalanced current supplied by multiple feeder conductors in the same raceway, but they need not be greater than the ampacity of the neutral supplying the primary stage switchboard. Parallel neutral conductors shall comply with 310.4.

The feeders to patch panels are often many dimmer-controlled circuits at 100 amperes or less, single phase, so they can be distributed to different combinations of the same size or smaller branch circuits. This type of installation usually requires a common neutral, and because of the quantity of circuits, many installations require several parallel neutrals running in several raceways. Generally, these parallel neutrals are sized as follows:

1. Size the common neutral to the feeder of the primary switchboard.
2. Split this neutral into multiple parallel conductors, one per raceway.

3. Equally divide, per phase, and size each ungrounded conductor of the many single-phase circuits among the raceways.

In no case is it acceptable to install the ungrounded conductors in one raceway and the common neutral in another.

(3) Separate Feeders to Single Primary Stage Switchboard (Dimmer Bank) Installations with separate feeders to a single primary stage switchboard shall have a disconnecting means for each feeder. The primary stage switchboard shall have a permanent and obvious label stating the number and location of disconnecting means. If the disconnecting means are located in more than one distribution switchboard, the primary stage switchboard shall be provided with barriers to correspond with these multiple locations.

Larger primary stage switchboards usually consist of several sections, often called dimmer racks, which form a dimmer bank. See Exhibit 520.2. These dimmer racks may be fed separately or may be bused together to accept one or more feeder circuits. If an intermediate stage switchboard is connected to a primary stage switchboard, a single large feeder usually supplies the primary stage switchboard, because the intermediate stage switchboard patches only the ungrounded conductors and requires a common neutral. Modern theaters do not use intermediate stage switchboards, and dimmer banks may have one or several feeders.

(B) Neutral The neutral of feeders supplying solid-state, 3-phase, 4-wire dimming systems shall be considered a current-carrying conductor.

(C) Supply Capacity For the purposes of calculating supply capacity to switchboards, it shall be permissible to consider the maximum load that the switchboard is intended to control in a given installation, provided that the following apply:

(1) All feeders supplying the switchboard shall be protected by an overcurrent device with a rating not greater than the ampacity of the feeder.
(2) The opening of the overcurrent device shall not affect the proper operation of the egress or emergency lighting systems.

FPN: For calculation of stage switchboard feeder loads, see 220.40.

The feeder for single, primary stage switchboards is sized in accordance with the maximum load the switchboard is intended to control for a specific location. The feeder(s) must be protected by an overcurrent device that has a rating not greater than the feeder ampacity. Operation of the overcurrent device is not allowed to have any effect on egress or emergency lighting systems. The neutral of feeders supplying solid-state, 3-phase, 4-wire dimming systems carries third-harmonic currents that are present even under balanced load conditions.

III. Fixed Stage Equipment Other Than Switchboards

520.41 Circuit Loads

(A) **Circuits Rated 20 Amperes or Less** Footlights, border lights, and proscenium sidelights shall be arranged so that no branch circuit supplying such equipment carries a load exceeding 20 amperes.

(B) **Circuits Rated Greater Than 20 Amperes** Where only heavy-duty lampholders are used, such circuits shall be permitted to comply with Article 210 for circuits supplying heavy-duty lampholders.

In accordance with 210.23(B) and 210.23(C), 30-, 40-, or 50-ampere branch circuits are permitted if heavy-duty lampholders, such as medium- or mogul-base Edison screw shell types, are used for fixed lighting.

520.42 Conductor Insulation

Foot, border, proscenium, or portable strip lights and connector strips shall be wired with conductors that have insulation suitable for the temperature at which the conductors are operated, but not less than 125°C (257°F). The ampacity of the 125°C (257°F) conductors shall be that of 60°C (140°F) conductors. All drops from connector strips shall be 90°C (194°F) wire sized to the ampacity of 60°C (140°F) cords and cables with no more than 150 mm (6 in.) of conductor extending into the connector strip. Section 310.15(B)(2)(a) shall not apply.

   FPN: See Table 310.13 for conductor types.

The 257°F minimum temperature rating is based on the heat from the lamps raising the ambient temperature in which the wiring is located. Drops from connector strips are usually flexible cord. Although the 90°C-rated cord is also in the higher ambient, it is not in sufficient contact with other circuits that might also heat it. The derating factors of 310.15(B)(2)(a) are judged unnecessary because the conductors are not all energized at one time, are not often energized at full intensity (dimmed), and are not energized continuously.

520.43 Footlights

(A) **Metal Trough Construction** Where metal trough construction is employed for footlights, the trough containing the circuit conductors shall be made of sheet metal not lighter than 0.81 mm (0.032 in.) and treated to prevent oxidation.

Lampholder terminals shall be kept at least 13 mm (½ in.) from the metal of the trough. The circuit conductors shall be soldered to the lampholder terminals.

(B) **Other-Than-Metal Trough Construction** Where the metal trough construction specified in 520.43(A) is not used, footlights shall consist of individual outlets with lampholders wired with rigid metal conduit, intermediate metal conduit, or flexible metal conduit, Type MC cable, or mineral-insulated, metal-sheathed cable. The circuit conductors shall be soldered to the lampholder terminals.

(C) **Disappearing Footlights** Disappearing footlights shall be arranged so that the current supply is automatically disconnected when the footlights are replaced in the storage recesses designed for them.

The footlights described in 520.43(A) and (B) are generally obsolete units that were built in the field. Modern footlights are compartmentalized, factory-wired assemblies for field installation, as shown in Exhibit 520.6. Footlight assemblies may be permanently exposed or be of the disappearing type. Disappearing footlights are arranged to automatically disconnect the current supply when the footlights are in the closed position, thereby preventing heat entrapment that could cause a fire. Disconnection is accomplished by mercury switches in the terminal compartment.

Exhibit 520.6 Disappearing footlights.

520.44 Borders and Proscenium Sidelights

(A) **General** Borders and proscenium sidelights shall be as follows:

(1) Constructed as specified in 520.43
(2) Suitably stayed and supported
(3) Designed so that the flanges of the reflectors or other adequate guards protect the lamps from mechanical damage and from accidental contact with scenery or other combustible material

Exhibit 520.7 shows a modern border light installed over a stage. Exhibit 520.8 is a cross-sectional view that illustrates construction details. This particular border light is designed

Exhibit 520.7 A suspended border light assembly for installation over a stage. (Courtesy of Kliegl Bros.)

Lamp (200 watts)

Reflector

Glass roundel

9 in.

8 in.

Lampholder Reflector Assembly

7 in.

6 in. 6 in. 6 in. 6 in. 6 in. 6 in.

Spacing of Lights

Exhibit 520.8 A cross-sectional view of a typical light in the border light assembly shown in Exhibit 520.7. (Redrawn courtesy of Kliegl Bros.)

for 200-watt lamps. To obtain the highest illumination efficiency, each lamp is provided with its own reflector. Fitted to each reflector is a glass roundel available in any color. Commonly, lampholders are wired alternately on three or four circuits. A splice box is provided on top of the housing for enclosing connections between the cable supplying the border light and the border light's internal wiring, which consists of wiring from the splice box to the lamp sockets in a trough extending the length of the border.

(B) Cords and Cables for Border Lights

(1) General Cords and cables for supply to border lights shall be listed for extra-hard usage. The cords and cables shall be suitably supported. Such cords and cables shall be employed only where flexible conductors are necessary. Ampacity of the conductors shall be as provided in 400.5.

To facilitate height adjustment for cleaning and lamp replacement, border lights are usually supported by steel cables, as shown in Exhibit 520.9. Therefore, the circuit conductors supplying the border lights must be carried to the border light in a flexible cable. Each of these flexible cables usually contains many circuits; however, its overall size is limited by its ability to travel up and down without getting tangled.

(2) Cords and Cables Not in Contact with Heat-Producing Equipment Listed multiconductor extra-hard-usage-type cords and cables not in direct contact with equipment containing heat-producing elements shall be permitted to have their ampacity determined by Table 520.44. Maximum load current in any conductor with an ampacity determined by Table 520.44 shall not exceed the values in Table 520.44.

The provisions of 520.44(B)(2) permit extra-hard-usage cords not in direct contact with heat-producing equipment to have their ampacity determined by Table 520.44 instead of Table 400.5.

Table 520.44 is based on a minimum 50 percent diversity factor. It includes the fact that not all circuits are on at the same time, not all circuits are at full intensity (dimmed), and not all circuits are on for a long period of time. If the load diversity does not follow this pattern, such as border lights that are all left on at full intensity to light the stage for rehearsal, lecture, or classroom purposes, Table 520.44 must not be used.

520.45 Receptacles

Receptacles for electrical equipment on stages shall be rated in amperes. Conductors supplying receptacles shall be in accordance with Articles 310 and 400.

Table 520.44 Ampacity of Listed Extra-Hard-Usage Cords and Cables with Temperature Ratings of 75°C (167°F) and 90°C (194°F)* [Based on Ambient Temperature of 30°C (86°F)]

Size (AWG)	Temperature Rating of Cords and Cables		Maximum Rating of Overcurrent Device
	75°C (167°F)	90°C (194°F)	
14	24	28	15
12	32	35	20
10	41	47	25
8	57	65	35
6	77	87	45
4	101	114	60
2	133	152	80

* Ampacity shown is the ampacity for multiconductor cords and cables where only three copper conductors are current-carrying as described in 400.5. If the number of current-carrying conductors in a cord or cable exceeds three and the load diversity factor is a minimum of 50 percent, the ampacity of each conductor shall be reduced as shown in the following table.

Number of Conductors	Percent of Ampacity
4–6	80
7–24	70
25–42	60
43 and above	50

Note: Ultimate insulation temperature. In no case shall conductors be associated together in such a way with respect to the kind of circuit, the wiring method used, or the number of conductors such that the temperature limit of the conductors is exceeded.

A neutral conductor that carries only the unbalanced current from other conductors of the same circuit need not be considered as a current-carrying conductor.

In a 3-wire circuit consisting of two phase wires and the neutral of a 4-wire, 3-phase, wye-connected system, a common conductor carries approximately the same current as the line-to-neutral currents of the other conductors and shall be considered to be a current-carrying conductor.

On a 4-wire, 3-phase, wye circuit where the major portion of the load consists of nonlinear loads such as electric-discharge lighting, electronic computer/data processing, or similar equipment, there are harmonic currents present in the neutral conductor, and the neutral shall be considered to be a current-carrying conductor.

520.46 Connector Strips, Drop Boxes, Floor Pockets, and Other Outlet Enclosures

Receptacles for the connection of portable stage-lighting equipment shall be pendant or mounted in suitable pockets or enclosures and shall comply with 520.45. Supply cables for connector strips and drop boxes shall be as specified in 520.44(B).

Exhibit 520.9 shows a hanging connector strip with its associated border light cable. Border lights are hung and supplied

in a similar manner. Exhibits 520.9, 520.10, and 520.11 illustrate different types of connections for portable stage lighting equipment.

520.47 Backstage Lamps (Bare Bulbs)

Lamps (bare bulbs) installed in backstage and ancillary areas where they can come in contact with scenery shall be located and guarded so as to be free from physical damage and shall provide an air space of not less than 50 mm (2 in.) between such lamps and any combustible material.

Exception: Decorative lamps installed in scenery shall not be considered to be backstage lamps for the purpose of this section.

520.48 Curtain Machines

Curtain machines shall be listed.

520.49 Smoke Ventilator Control

Where stage smoke ventilators are released by an electrical device, the circuit operating the device shall be normally closed and shall be controlled by at least two externally operable switches, one switch being placed at a readily accessible location on stage and the other where designated by the authority having jurisdiction. The device shall be designed for the full voltage of the circuit to which it is connected, no resistance being inserted. The device shall be located in the loft above the scenery and shall be enclosed in a suitable metal box having a tight, self-closing door.

In addition to the smoke ventilators being controlled from two externally operable switches at different locations, the design of a normally closed circuit ensures that the smoke ventilators will operate when the circuit opens for any reason, such as a circuit breaker tripping or a fuse blowing.

IV. Portable Switchboards on Stage

520.50 Road Show Connection Panel (A Type of Patch Panel)

A panel designed to allow for road show connection of portable stage switchboards to fixed lighting outlets by means of permanently installed supplementary circuits. The panel, supplementary circuits, and outlets shall comply with 520.50(A) through (D).

Also known as a road show interconnect or intercept panel, a road show connection panel is designed to connect the load side of a portable switchboard to the fixed building

Exhibit 520.9 A suspended connector strip with border light cable attached. (Redrawn courtesy of Colortran, Inc.)

branch circuits and associated outlets. It may also provide for the fixed branch circuits to be connected to a fixed switchboard when the portable switchboard is not installed.

(A) Load Circuits Circuits shall terminate in grounding-type polarized inlets of current and voltage rating that match the fixed-load receptacle.

The grounding-type polarized inlets may be flush or pendant. The fixed-load receptacle is on the other end of the branch circuit that emanates from the panel.

(B) Circuit Transfer Circuits that are transferred between fixed and portable switchboards shall have all circuit conductors transferred simultaneously.

In accordance with 520.50(B), simultaneous transfer of all conductors of the circuit, including any grounded conductors, is required.

(C) Overcurrent Protection The supply devices of these supplementary circuits shall be protected by branch-circuit overcurrent protective devices. The individual supplementary circuit, within the road show connection panel and theater, shall be protected by branch-circuit overcurrent protective devices of suitable ampacity installed within the road show connection panel.

The branch-circuit overcurrent protection normally should be in the switchboard, but because some older units do

Exhibit 520.10 A 4-gang, 4-receptacle pin-plug outlet box designed for flush mounting. (Courtesy of Electronic Theatre Controls, Inc.)

Exhibit 520.11 A typical three-circuit connector strip designed for wall or pipe mounting. (Courtesy of Electronic Theatre Controls, Inc.)

not have this protection, backup overcurrent protection is provided by 520.50(C).

(D) Enclosure Panel construction shall be in accordance with Article 408.

520.51 Supply

Portable switchboards shall be supplied only from power outlets of sufficient voltage and ampere rating. Such power outlets shall include only externally operable, enclosed fused switches or circuit breakers mounted on stage or at the permanent switchboard in locations readily accessible from the stage floor. Provisions for connection of an equipment grounding conductor shall be provided. The neutral of feed-

ers supplying solid-state, 3-phase, 4-wire dimmer systems shall be considered a current-carrying conductor.

Power outlets, known in the entertainment industry as company switches or bull switches, are the point in the wiring system where portable feeder cables connect to the fixed building wiring. They may be as simple as an overcurrent-protected multipole receptacle designed to accept the supply cable described in 520.53(P), Exception, or they may be multiple sets of parallel single-conductor feeder cables. These single-conductor feeder cables, as described in 520.53(H), may be terminated via single-pole separable connectors, as described in 520.53(K), or directly to busbars, fused disconnect switches, or circuit breakers with wire connectors (lugs).

520.52 Overcurrent Protection

Circuits from portable switchboards directly supplying equipment containing incandescent lamps of not over 300 watts shall be protected by overcurrent protective devices having a rating or setting of not over 20 amperes. Circuits for lampholders over 300 watts shall be permitted where overcurrent protection complies with Article 210.

520.53 Construction and Feeders

Portable switchboards and feeders for use on stages shall comply with 520.53(A) through (P).

See Exhibit 520.12 for an example of a portable switchboard.

(A) Enclosure Portable switchboards shall be placed within an enclosure of substantial construction, which shall be permitted to be arranged so that the enclosure is open during operation. Enclosures of wood shall be completely lined with sheet metal of not less than 0.51 mm (0.020 in.) and shall be well galvanized, enameled, or otherwise properly coated to prevent corrosion or be of a corrosion-resistant material.

(B) Energized Parts There shall not be exposed energized parts within the enclosure.

(C) Switches and Circuit Breakers All switches and circuit breakers shall be of the externally operable, enclosed type.

(D) Circuit Protection Overcurrent devices shall be provided in each ungrounded conductor of every circuit supplied through the switchboard. Enclosures shall be provided for all overcurrent devices in addition to the switchboard enclosure.

(E) Dimmers The terminals of dimmers shall be provided with enclosures, and dimmer faceplates shall be arranged

Exhibit 520.12 A large, portable SCR dimmer switchboard (rolling rack). (Courtesy of Electronic Theatre Controls, Inc.)

so that accidental contact cannot be readily made with the faceplate contacts.

(F) Interior Conductors

(1) Type All conductors other than busbars within the switchboard enclosure shall be stranded. Conductors shall be approved for an operating temperature at least equal to the approved operating temperature of the dimming devices used in the switchboard and in no case less than the following:

(1) Resistance-type dimmers — 200°C (392°F); or
(2) Reactor-type, autotransformer, and solid-state dimmers — 125°C (257°F)

All control wiring shall comply with Article 725.

(2) Protection Each conductor shall have an ampacity not less than the rating of the circuit breaker, switch, or fuse that it supplies. Circuit interrupting and bus bracing shall

be in accordance with 110.9 and 110.10. The short-circuit current rating shall be marked on the switchboard.

Conductors shall be enclosed in metal wireways or shall be securely fastened in position and shall be bushed where they pass through metal.

(G) Pilot Light A pilot light shall be provided within the enclosure and shall be connected to the circuit supplying the board so that the opening of the master switch does not cut off the supply to the lamp. This lamp shall be on an individual branch circuit having overcurrent protection rated or set at not over 15 amperes.

The requirement of 520.53(G) applies only to switchboards with a main disconnect, if provided, on the switchboard. The pilot light serves as a warning at the switchboard that indicates the presence of power before the main disconnect is activated.

(H) Supply Conductors

(1) General The supply to a portable switchboard shall be by means of listed extra-hard usage cords or cables. The supply cords or cable shall terminate within the switchboard enclosure, in an externally operable fused master switch or circuit breaker or in a connector assembly identified for the purpose. The supply cords or cable (and connector assembly) shall have sufficient ampacity to carry the total load connected to the switchboard and shall be protected by overcurrent devices.

As with the supply end described in 520.51, the connection described in 520.53(H)(1) may be as simple as permanently terminated multiconductor supply cord or multipole connector assembly (inlet) or as complex as a set of parallel single-conductor feeder cables. These cables may be field-connected to an assembly of single-pole connectors (inlet) or directly connected, with wire connectors, to busbars or a fused switch or breaker.

The requirement of 520.53(H)(1) permits road shows with fixed lighting plans to size the feeder to the actual connected load.

(2) Single-Conductor Cables Single-conductor portable supply cable sets shall not be smaller than 2 AWG conductors. The equipment grounding conductor shall not be smaller than 6 AWG conductor. Single-conductor grounded neutral cables for a supply shall be sized in accordance with 520.53(O)(2). Where single conductors are paralleled for increased ampacity, the paralleled conductors shall be of the same length and size. Single-conductor supply cables shall be grouped together but not bundled. The equipment grounding conductor shall be permitted to be of a different type, provided it meets the other requirements of this section, and it shall be permitted to be reduced in size as permitted by 250.122. Grounded (neutral) and equipment grounding conductors shall be identified in accordance with 200.6, 250.119, and 310.12. Grounded conductors shall be permitted to be identified by marking at least the first 150 mm (6 in.) from both ends of each length of conductor with white or gray. Equipment grounding conductors shall be permitted to be identified by marking at least the first 150 mm (6 in.) from both ends of each length of conductor with green or green with yellow stripes. Where more than one nominal voltage exists within the same premises, each ungrounded conductor shall be identified by system.

(3) Supply Conductors Not Over 3.0 m (10 ft) Long
Where supply conductors do not exceed 3.0 m (10 ft) in length between supply and switchboard or supply and a subsequent overcurrent device, the supply conductors shall be permitted to be reduced in size where all of the following conditions are met:

(1) The ampacity of the supply conductors shall be at least one-quarter of the ampacity of the supply overcurrent protection device.

(2) The supply conductors shall terminate in a single overcurrent protection device that will limit the load to the ampacity of the supply conductors. This single overcurrent device shall be permitted to supply additional overcurrent devices on its load side.

(3) The supply conductors shall not penetrate walls, floors, or ceilings or be run through doors or traffic areas. The supply conductors shall be adequately protected from physical damage.

(4) The supply conductors shall be suitably terminated in an approved manner.

(5) Conductors shall be continuous without splices or connectors.

(6) Conductors shall not be bundled.

(7) Conductors shall be supported above the floor in an approved manner.

(4) Supply Conductors Not Over 6.0 m (20 ft) Long
Where supply conductors do not exceed 6.0 m (20 ft) in length between supply and switchboard or supply and a subsequent overcurrent device, the supply conductors shall be permitted to be reduced in size where all of the following conditions are met:

(1) The ampacity of the supply conductors shall be at least one-half of the ampacity of the supply overcurrent protection device.

(2) The supply conductors shall terminate in a single overcurrent protection device that limits the load to the ampacity of the supply conductors. This single overcurrent device shall be permitted to supply additional overcurrent devices on its load side.

(3) The supply conductors shall not penetrate walls, floors, or ceilings or be run through doors or traffic areas. The supply conductors shall be adequately protected from physical damage.

(4) The supply conductors shall be suitably terminated in an approved manner.

(5) The supply conductors shall be supported in an approved manner at least 2.1 m (7 ft) above the floor except at terminations.

(6) The supply conductors shall not be bundled.

(7) Tap conductors shall be in unbroken lengths.

Loads of 144 kVA and greater are not uncommon, even on portable switchboard equipment. Installations in the field include lighting for theatrical-type productions with large numbers of stage lighting fixtures. However, only a fraction of the many fixtures installed are used at any one time. The intent of 520.53(H)(3) and 520.53(H)(4) is that the supply conductors must be sized according to their overcurrent protection and not by the total connected load. These requirements are similar to the requirements for taps found in 240.21.

The tap rules of 520.53(H)(3) and 520.53(H)(4) are designed to allow one or more switchboards with smaller feeders to be connected to larger supplies (company switches). If these "rules" are not complied with, proper overcurrent protection devices, either fixed or portable, must be provided for each of the smaller switchboards.

The requirement that the conductors not be bundled is so that column D of Table 400.5(B) can be employed. If the conductors were bundled, column F and all applicable derating factors would apply. Most devices used in the theater to terminate single-conductor cables are rated for use at 90°C ampacity. However, if single-conductor cables are terminated directly to a circuit breaker or fused switch, a 75°C ampacity or lower most likely would apply.

(5) Supply Conductors Not Reduced in Size Supply conductors not reduced in size under provisions of 520.53(H)(3) or 520.53(H)(4) shall be permitted to pass through holes in walls specifically designed for the purpose. If penetration is through the fire-resistant–rated wall, it shall be in accordance with 300.21.

(I) Cable Arrangement Cables shall be protected by bushings where they pass through enclosures and shall be arranged so that tension on the cable is not transmitted to the connections. Where power conductors pass through metal, the requirements of 300.20 shall apply.

Tension on the connections is removed by using conventional strain relief devices or, often, by lashing the cable to the enclosure with rope.

(J) Number of Supply Interconnections Where connectors are used in a supply conductor, there shall be a maximum number of three interconnections (mated connector pairs) where the total length from supply to switchboard does not exceed 30 m (100 ft). In cases where the total length from supply to switchboard exceeds 30 m (100 ft), one additional interconnection shall be permitted for each additional 30 m (100 ft) of supply conductor.

The addition of excessive numbers of interconnections could jeopardize the mechanical and electrical integrity of the supply conductors.

(K) Single-Pole Separable Connectors Where single-pole portable cable connectors are used, they shall be listed and of the locking type. Sections 400.10, 406.6, and 406.7 shall not apply to listed single-pole separable connectors and single-conductor cable assemblies utilizing listed single-pole separable connectors. Where paralleled sets of current-carrying, single-pole separable connectors are provided as input devices, they shall be prominently labeled with a warning indicating the presence of internal parallel connections. The use of single-pole separable connectors shall comply with at least one of the following conditions:

(1) Connection and disconnection of connectors are only possible where the supply connectors are interlocked to the source, and it is not possible to connect or disconnect connectors when the supply is energized.

(2) Line connectors are of the listed sequential-interlocking type so that load connectors shall be connected in the following sequence:

   a. Equipment grounding conductor connection
   b. Grounded circuit conductor connection, if provided
   c. Ungrounded conductor connection, and that disconnection shall be in the reverse order

(3) A caution notice shall be provided adjacent to the line connectors indicating that plug connection shall be in the following order:

   a. Equipment grounding conductor connectors
   b. Grounded circuit conductor connectors, if provided
   c. Ungrounded conductor connectors, and that disconnection shall be in the reverse order

The requirements in 520.53(K) provide for a special type of connection device suitable for connecting single-conductor feeder cables. The connection device must be listed and of the locking type, reducing the likelihood of its separating while under load. The connectors must be used in sets, because they are only single-pole types. It is important that the grounding conductor be connected first and disconnected last and that the grounded conductor be connected next-to-

first and disconnected next-to-last. The connector sets must be arranged so as to reduce the likelihood that the connections will be made in the incorrect order, in accordance with one of the following methods:

1. A scheme whereby the main disconnect cannot be energized until all conductors are connected
2. A scheme whereby the connectors are precluded from being connected in any order other than the proper one
3. A scheme whereby the individual connectors, free of any special electromechanical intervention, are marked with instructions to the user regarding proper connection

Single-pole separable connectors are quick-connect feeder splicing and terminating devices, not attachment plugs or receptacles. They are designed to be sized, terminated, and inspected by a qualified person before being energized and are to be guarded from accidental disconnection before being de-energized.

(L) Protection of Supply Conductors and Connectors All supply conductors and connectors shall be protected against physical damage by an approved means. This protection shall not be required to be raceways.

Rubber mats and commercially available rubber bridges often are used for the protection of supply conductors and connectors.

(M) Flanged Surface Inlets Flanged surface inlets (recessed plugs) that are used to accept the power shall be rated in amperes.

(N) Terminals Terminals to which stage cables are connected shall be located so as to permit convenient access to the terminals.

The requirement in 520.53(N) facilitates the field connection and disconnection of the large feeder cables as a show travels from place to place.

(O) Neutral

(1) Neutral Terminal In portable switchboard equipment designed for use with 3-phase, 4-wire with ground supply, the supply neutral terminal, its associated busbar, or equivalent wiring, or both, shall have an ampacity equal to at least twice the ampacity of the largest ungrounded supply terminal.

Exception: Where portable switchboard equipment is specifically constructed and identified to be internally converted in the field, in an approved manner, from use with a balanced 3-phase, 4-wire with ground supply to a balanced single-phase, 3-wire with ground supply, the supply neutral terminal and its associated busbar, equivalent wiring, or both, shall have an ampacity equal to at least that of the largest ungrounded single-phase supply terminal.

(2) Supply Neutral The power supply conductors for portable switchboards shall be sized considering the neutral as a current-carrying conductor. Where single-conductor feeder cables, not installed in raceways, are used on multiphase circuits, the grounded neutral conductor shall have an ampacity of at least 130 percent of the ungrounded circuit conductors feeding the portable switchboard.

The requirement in 520.53(O)(1) requires careful study because overlapping concepts are involved. If a 3-phase, 4-wire switchboard of any kind is brought into a space that has only single-phase, 3-wire service, the switchboard most likely will be connected with two phases to one leg and one phase to the other. This connection could double the current flowing through the neutral, so the neutral must be double size to allow for that possibility. The exception to 520.53(O)(1) provides for a smaller neutral sized for the single-phase feed where a switchboard contains switching devices that can divide the B-phase load equally between the A-phase and C-phase buses for single-phase operation.

Additionally, 3-phase, 4-wire switchboards that contain solid-state dimming devices must, when connected to a 3-phase, 4-wire supply, be connected to that supply with a multiconductor cable sized by counting the neutral as a current-carrying conductor or with a set of single-conductor cables where the neutral is sized 130 percent greater than the phases.

For example, a 3-phase, 4-wire switchboard containing six 50-ampere SCR dimmers (100 amperes per phase) without a reassignment switching system would have to have a 200-ampere neutral. (A single-phase, 3-wire-only switchboard would not have to meet this special requirement.) This 200-percent rule would cover all the components making up the neutral conductor system inside or permanently attached to the switchboard, to allow for a full-size, single-phase, 3-wire feed when two of the 3-phase, 4-wire phase conductors are terminated to one single-phase, 3-wire leg. Note that the 200-percent neutral already covers the derating requirements (125 percent for a multiconductor feeder system and 130 percent for a single-conductor feeder system) when used in the 3-phase mode. If a reassignment system were added, the neutral would be required to be only 150 amperes. Again, when used in the 3-phase mode, the derating factors would be covered.

Note that the double-neutral requirement covers the terminal and associated busbar or wiring. This requirement begins at the main input terminals or busing, main input inlet connector, or attached main input cord-and-plug set and includes all wiring on the load side of that point.

Power supply feeders easily detached at the terminals or inlet connector need not adhere to the 200-percent neutral rule because they can easily be sized on a show-by-show basis for the type of supply encountered. These cables must, however, adhere to the requirements of the neutral as a current-carrying conductor or to the requirements of the 130-percent single-conductor-cable neutral.

(P) Qualified Personnel The routing of portable supply conductors, the making and breaking of supply connectors and other supply connections, and the energization and de-energization of supply services shall be performed by qualified personnel, and portable switchboards shall be so marked, indicating this requirement in a permanent and conspicuous manner.

Exception: A portable switchboard shall be permitted to be connected to a permanently installed supply receptacle by other than qualified personnel, provided that the supply receptacle is protected for its rated ampacity by an overcurrent device of not greater than 150 amperes, and where the receptacle, interconnection, and switchboard comply with all of the following:

(a) Employ listed multipole connectors suitable for the purpose for every supply interconnection

(b) Prevent access to all supply connections by the general public

(c) Employ listed extra-hard usage multiconductor cords or cables with an ampacity suitable for the type of load and not less than the ampere rating of the connectors.

The intent of 520.53(P) is to divide the acceptable practices in what are most likely to be professional and professional-grade educational venues from those in amateur or amateur-grade educational venues. The basic requirements allow for such things as single-conductor feeder systems, feeders sized for the current-connected load, tap rules, and so on, and require the services of a qualified person. The exception to 520.53(P) provides for a conventional feeder system suitable for use by an untrained person.

V. Portable Stage Equipment Other Than Switchboards

520.61 Arc Lamps

Arc lamps, including enclosed arc lamps and associated ballasts, shall be listed. Interconnecting cord sets and interconnecting cords and cables shall be extra-hard usage type and listed.

520.62 Portable Power Distribution Units

Portable power distribution units shall comply with 520.62(A) through (E).

(A) Enclosure The construction shall be such that no current-carrying part will be exposed.

(B) Receptacles and Overcurrent Protection Receptacles shall comply with 520.45 and shall have branch-circuit overcurrent protection in the box. Fuses and circuit breakers shall be protected against physical damage. Cords or cables supplying pendant receptacles shall be listed for extra-hard usage.

(C) Busbars and Terminals Busbars shall have an ampacity equal to the sum of the ampere ratings of all the circuits connected to the busbar. Lugs shall be provided for the connection of the master cable.

(D) Flanged Surface Inlets Flanged surface inlets (recessed plugs) that are used to accept the power shall be rated in amperes.

(E) Cable Arrangement Cables shall be adequately protected where they pass through enclosures and be arranged so that tension on the cable is not transmitted to the terminations.

520.63 Bracket Fixture Wiring

(A) Bracket Wiring Brackets for use on scenery shall be wired internally, and the fixture stem shall be carried through to the back of the scenery where a bushing shall be placed on the end of the stem. Externally wired brackets or other fixtures shall be permitted where wired with cords designed for hard usage that extend through scenery and without joint or splice in canopy of fixture back and terminate in an approved-type stage connector located, where practical, within 450 mm (18 in.) of the fixture.

(B) Mounting Fixtures shall be securely fastened in place.

520.64 Portable Strips

Portable strips shall be constructed in accordance with the requirements for border lights and proscenium sidelights in 520.44(A). The supply cable shall be protected by bushings where it passes through metal and shall be arranged so that tension on the cable will not be transmitted to the connections.

FPN No. 1: See 520.42 for wiring of portable strips.
FPN No. 2: See 520.68(A)(3) for insulation types required on single conductors.

520.65 Festoons

Joints in festoon wiring shall be staggered. Lamps enclosed in lanterns or similar devices of combustible material shall be equipped with guards.

Festoon lighting is defined in Article 100. Joints in festoon wiring must be staggered and properly insulated. This ar-

rangement ensures that connections will not be opposite one another, which could cause sparking due to improper insulation or unraveling of insulation, which, in turn, could ignite lanterns or other combustible material enclosing lamps. Where lampholders have terminals of a type that puncture the conductor insulation and make contact with the conductors, stranded conductors should be used.

520.66 Special Effects

Electrical devices used for simulating lightning, waterfalls, and the like shall be constructed and located so that flames, sparks, or hot particles cannot come in contact with combustible material.

520.67 Multipole Branch-Circuit Cable Connectors

Multipole branch-circuit cable connectors, male and female, for flexible conductors shall be constructed so that tension on the cord or cable is not transmitted to the connections. The female half shall be attached to the load end of the power supply cord or cable. The connector shall be rated in amperes and designed so that differently rated devices cannot be connected together; however, a 20-ampere T-slot receptacle shall be permitted to accept a 15-ampere attachment plug of the same voltage rating. Alternating-current multipole connectors shall be polarized and comply with 406.6 and 406.9.

FPN: See 400.10 for pull at terminals.

520.68 Conductors for Portables

(A) Conductor Type

(1) General Flexible conductors, including cable extensions, used to supply portable stage equipment shall be listed extra-hard usage cords or cables.

(2) Stand Lamps Listed, hard usage cord shall be permitted to supply stand lamps where the cord is not subject to physical damage and is protected by an overcurrent device rated at not over 20 amperes.

Section 520.68 was revised for the 2005 *Code* to permit listed hard-usage cord to supply stand lamps, where the cord is not subject to physical damage and is protected with an overcurrent device rated at not more than 20 amperes. The term *reinforced cord* was deleted because it is obsolete.
See 520.2 for the definition of *stand lamp (work light)*.

(3) High-Temperature Applications A special assembly of conductors in sleeving not longer than 1.0 m (3.3 ft) shall be permitted to be employed in lieu of flexible cord if the individual wires are stranded and rated not less than 125°C (257°F) and the outer sleeve is glass fiber with a wall thickness of at least 0.635 mm (0.025 in.).

Portable stage equipment requiring flexible supply conductors with a higher temperature rating where one end is permanently attached to the equipment shall be permitted to employ alternate, suitable conductors as determined by a qualified testing laboratory and recognized test standards.

The requirements of 520.68(A)(3) cover the connection of high-temperature equipment, including stage lighting fixtures, which often operate at elevated temperatures. High-temperature (150°C to 250°C) extra-hard-usage cords are, in general, not available. Less-than-extra-hard-usage cords are limited to 3.3 ft in length to reduce the likelihood that they would be placed on the floor or other area where they might be damaged by traffic or moving scenery.

(4) Breakouts Listed, hard usage (junior hard service) cords shall be permitted in breakout assemblies where all of the following conditions are met:

(1) The cords are utilized to connect between a single multipole connector containing two or more branch circuits and multiple 2-pole, 3-wire connectors.
(2) The longest cord in the breakout assembly does not exceed 6.0 m (20 ft).
(3) The breakout assembly is protected from physical damage by attachment over its entire length to a pipe, truss, tower, scaffold, or other substantial support structure.
(4) All branch circuits feeding the breakout assembly are protected by overcurrent devices rated at not over 20 amperes.

The provisions of 520.68(A)(4) apply to multiconductor cable assemblies with multipole connectors that contain more than one branch circuit. The breakout assembly is a multipole connector with several pendant receptacles connected to it, separating the multiple branch circuits into individual branch circuits. It is also possible to use a similar arrangement of pendant plugs to form a breaking assembly on the other end of the multiconductor cable.

(B) Conductor Ampacity The ampacity of conductors shall be as given in 400.5, except multiconductor, listed, extra-hard usage portable cords that are not in direct contact with equipment containing heat-producing elements shall be permitted to have their ampacity determined by Table 520.44. Maximum load current in any conductor with an ampacity determined by Table 520.44 shall not exceed the values in Table 520.44.

In accordance with 520.68(B), portable, multicircuit, multiconductor cable is permitted to be sized in accordance with Table 520.44, similar to the method used for border light

cable. If portable, multicircuit, multiconductor cable is located horizontally directly above heat-producing equipment, in lieu of a connector strip, it should be spaced sufficiently above that equipment to avoid the elevated temperatures or sized in accordance with 400.5.

Exception: Where alternate conductors are allowed in 520.68(A)(3), their ampacity shall be as given in the appropriate table in this Code for the types of conductors employed.

520.69 Adapters

Adapters, two-fers, and other single- and multiple-circuit outlet devices shall comply with 520.69(A), (B), and (C).

(A) No Reduction in Current Rating Each receptacle and its corresponding cable shall have the same current and voltage rating as the plug supplying it. It shall not be utilized in a stage circuit with a greater current rating.

(B) Connectors All connectors shall be wired in accordance with 520.67.

Adapters are available where cords and connector bodies of one ampacity are connected to a plug of a larger rating. For example, a 12 AWG conductor with an ampacity of 20 amperes could be connected to a 100-ampere circuit. An overload could result in a fire because the circuit breaker or fuse would not provide adequate protection. The plug and receptacle must be of the same rating, in accordance with 520.69(B).

(C) Conductor Type Conductors for adapters and two-fers shall be listed, extra-hard usage or listed, hard usage (junior hard service) cord. Hard usage (junior hard service) cord shall be restricted in overall length to 1.0 m (3.3 ft).

VI. Dressing Rooms

520.71 Pendant Lampholders

Pendant lampholders shall not be installed in dressing rooms.

520.72 Lamp Guards

All exposed incandescent lamps in dressing rooms, where less than 2.5 m (8 ft) from the floor, shall be equipped with open-end guards riveted to the outlet box cover or otherwise sealed or locked in place.

Because of the many types of flammable materials present in dressing rooms, such as costumes and wigs, pendant lampholders are not permitted. Lamps must be provided with suitable open-end guards that permit relamping and that are not easily removed. This makes it difficult to circumvent the guard's intended purpose of preventing contact between the lamps and flammable material.

520.73 Switches Required

All lights and any receptacles adjacent to the mirror(s) and above the dressing table counter(s) installed in dressing rooms shall be controlled by wall switches installed in the dressing room(s). Each switch controlling receptacles adjacent to the mirror(s) and above the dressing table counter(s) shall be provided with a pilot light located outside the dressing room, adjacent to the door to indicate when the receptacles are energized. Other outlets installed in the dressing room shall not be required to be switched.

The requirement in 520.73 addresses only receptacles located adjacent to the mirror and on the countertop. The receptacles located elsewhere in the room are not subject to the disconnect and pilot light requirements of 520.73. The purpose of the switching requirement is to make sure that all coffee pots, curling irons, hair dryers, and other similar countertop appliances can be readily disconnected at the end of a performance.

VII. Grounding

520.81 Grounding

All metal raceways and metal-sheathed cables shall be grounded. The metal frames and enclosures of all equipment, including border lights and portable luminaires (lighting fixtures), shall be grounded. Grounding, where used, shall be in accordance with Article 250.

ARTICLE 525
Carnivals, Circuses, Fairs, and Similar Events

Summary of Changes

- **525.10:** Revised to remove redundant reference to requirements in Chapters 1–4 of *NEC*. Requirements in Chapters 1–4 must be followed in accordance with 90.3.
- **525.11:** Added to require that rides, attractions, and structures less than 12 ft apart, and powered by multiple services, multiple separately derived systems, or both, be bonded to the same grounding electrode system.

Contents

I. General Requirements

525.1 Scope

This article covers the installation of portable wiring and equipment for carnivals, circuses, fairs, and similar functions, including wiring in or on all structures.

Article 525 addresses the installation of portable wiring and equipment for temporary attractions, such as carnivals, circuses, and fairs. Article 525 is intended to apply to all wiring in or on portable structures, whereas Articles 518 and 520 apply to permanent structures. Prior to the 2002 edition of the *Code*, installations for portable equipment used at carnivals, circuses, fairs, and the like, were covered under Article 590, Temporary Installations. Article 525 was developed because the requirements for temporary installations contained in Article 590 apply more to construction sites than to events that are open to the general public, such as fairs and carnivals. Additionally, Article 525 significantly expands the requirements and scope for installing electrical equipment at these locations.

525.3 Other Articles

(A) Portable Wiring and Equipment Wherever the requirements of other articles of this *Code* and Article 525 differ, the requirements of Article 525 shall apply to the portable wiring and equipment.

(B) Permanent Structures Articles 518 and 520 shall apply to wiring in permanent structures.

(C) Audio Signal Processing, Amplification, and Reproduction Equipment Article 640 shall apply to the wiring and installation of audio signal processing, amplification, and reproduction equipment.

(D) Attractions Utilizing Pools, Fountains, and Similar Installations with Contained Volumes of Water This equipment shall be installed to comply with the applicable requirements of Article 680.

525.5 Overhead Conductor Clearances

(A) Vertical Clearances Conductors shall have a vertical clearance to ground in accordance with 225.18. These clearances shall apply only to wiring installed outside of tents and concessions.

(B) Clearance to Rides and Attractions Amusement rides and amusement attractions shall be maintained not less than 4.5 m (15 ft) in any direction from overhead conductors operating at 600 volts or less, except for the conductors supplying the amusement ride or attraction. Amusement rides or attractions shall not be located under or within

4.5 m (15 ft) horizontally of conductors operating in excess of 600 volts.

525.6 Protection of Electrical Equipment

Electrical equipment and wiring methods in or on rides, concessions, or other units shall be provided with mechanical protection where such equipment or wiring methods are subject to physical damage.

II. Power Sources

525.10 Services

Services shall comply with 525.10(A) and 525.10(B).

(A) Guarding Service equipment shall not be installed in a location that is accessible to unqualified persons, unless the equipment is lockable.

(B) Mounting and Location Service equipment shall be mounted on solid backing and be installed so as to be protected from the weather, unless of weatherproof construction.

Part II, Power Sources, was reorganized for the 2005 *Code* and provides some of the requirements for services and multiple sources of supply, such as generators and transformers, that are separately derived systems. In addition to the requirements in 525.10(A) and 525.10(B), the requirements for services in Article 230 also are applicable.

Service equipment must be installed in accordance with Article 230 and must be lockable where accessible to unqualified persons. At fairgrounds, carnivals, and similar events, there is significant pedestrian traffic throughout the site, including through those areas where electrical equipment is located. This requirement helps safeguard the general public from accidentally coming in contact with energized service equipment.

525.11 Multiple Sources of Supply

Where multiple services or separately derived systems or both supply rides, attractions, and other structures, all sources of supply that serve rides, attractions, or other structures separated by less than 3.7 m (12 ft) shall be bonded to the same grounding electrode system.

In order to maintain an equal potential between exposed, noncurrent-carrying metal parts of rides, amusements, or other structures that have a physical separation less than 12 ft, this section requires the use of a common grounding electrode system where there are multiple power sources supplying equipment at the carnival or fair.

III. Wiring Methods

525.20 Wiring Methods

(A) Type Where flexible cords or cables are used, they shall be listed for extra hard usage. Where flexible cords or cables are used and are not subject to physical damage, they shall be permitted to be listed for hard usage. Where used outdoors, flexible cords and cables shall also be listed for wet locations and shall be sunlight resistant. Extra-hard usage flexible cords or cables shall be permitted for use as permanent wiring on portable amusement rides and attractions where not subject to physical damage.

(B) Single-Conductor Single-conductor cable shall be permitted only in sizes 2 AWG or larger.

(C) Open Conductors Open conductors are prohibited except as part of a listed assembly or festoon lighting installed in accordance with Article 225.

(D) Splices Flexible cords or cables shall be continuous without splice or tap between boxes or fittings.

(E) Cord Connectors Cord connectors shall not be laid on the ground unless listed for wet locations. Connectors and cable connections shall not be placed in audience traffic paths or within areas accessible to the public unless guarded.

(F) Support Wiring for an amusement ride, attraction, tent, or similar structure shall not be supported by any other ride or structure unless specifically designed for the purpose.

(G) Protection Flexible cords or cables accessible to the public shall be arranged to minimize the tripping hazard and shall be permitted to be covered with nonconductive matting, provided that the matting does not constitute a greater tripping hazard than the uncovered cables. It shall be permitted to bury cables. The requirements of 300.5 shall not apply.

(H) Boxes and Fittings A box or fitting shall be installed at each connection point, outlet, switchpoint, or junction point.

525.21 Rides, Tents, and Concessions

(A) Disconnecting Means Each ride and concession shall be provided with a fused disconnect switch or circuit breaker located within sight and within 1.8 m (6 ft) of the operator's station. The disconnecting means shall be readily accessible to the operator, including when the ride is in operation. Where accessible to unqualified persons, the enclosure for the switch or circuit breaker shall be of the lockable type. A shunt trip device that opens the fused disconnect or circuit breaker when a switch located in the ride operator's console is closed shall be a permissible method of opening the circuit.

(B) Portable Wiring Inside Tents and Concessions Electrical wiring for lighting, where installed inside of tents and concessions, shall be securely installed and, where subject to physical damage, shall be provided with mechanical protection. All lamps for general illumination shall be protected from accidental breakage by a suitable fixture or lampholder with a guard.

525.22 Portable Distribution or Termination Boxes

Portable distribution or termination boxes shall comply with 525.22(A) through 525.22(D).

Portable distribution or termination equipment must be mounted so that the bottom of the enclosure is at least 6 in. above the ground. This requirement prevents excessive moisture from entering the equipment and allows for proper radius of bend on conductors entering and exiting the equipment from below.

(A) Construction Boxes shall be designed so that no live parts are exposed to accidental contact. Where installed outdoors, the box shall be of weatherproof construction and mounted so that the bottom of the enclosure is not less than 150 mm (6 in.) above the ground.

(B) Busbars and Terminals Busbars shall have an ampere rating not less than the overcurrent device supplying the feeder supplying the box. Where conductors terminate directly on busbars, busbar connectors shall be provided.

(C) Receptacles and Overcurrent Protection Receptacles shall have overcurrent protection installed within the box. The overcurrent protection shall not exceed the ampere rating of the receptacle, except as permitted in Article 430 for motor loads.

(D) Single-Pole Connectors Where single-pole connectors are used, they shall comply with 530.22.

525.23 Ground-Fault Circuit-Interrupter (GFCI) Protection

(A) Where GFCI Protection Is Required The ground-fault circuit interrupter shall be permitted to be an integral part of the attachment plug or located in the power-supply cord, within 300 mm (12 in.) of the attachment plug. Listed cord sets incorporating ground-fault circuit-interrupter for personnel shall be permitted.

(1) 125-volt, single-phase, 15- and 20-ampere non-locking type receptacles used for disassembly and reassembly or readily accessible to the general public.

(2) Equipment that is readily accessible to the general public and supplied from a 125-volt, single-phase, 15- or 20-ampere branch circuit.

(B) Where GFCI Protection Is Not Required Receptacles that only facilitate quick disconnecting and reconnecting of electrical equipment shall not be required to be provided with GFCI protection. These receptacles shall be of the locking type.

(C) Where GFCI Protection Is Not Permitted Egress lighting shall not be protected by a GFCI.

Section 525.23 has been reorganized in the 2005 *Code* and provides three categories: where GFCIs are required, where GFCIs are not required, and where GFCIs are not permitted to be installed. GFCI protection is not allowed on circuits that supply means-of-egress illumination.

IV. Grounding and Bonding

525.30 Equipment Bonding

The following equipment connected to the same source shall be bonded:

(1) Metal raceways and metal-sheathed cable
(2) Metal enclosures of electric equipment
(3) Metal frames and metal parts of rides, concessions, tents, trailers, trucks, or other equipment that contain or support electrical equipment

525.31 Equipment Grounding

All equipment requiring grounding shall be grounded by an equipment grounding conductor of a type and size recognized by 250.118 and installed in accordance with Article 250. The equipment grounding conductor shall be bonded to the system grounded conductor at the service disconnecting means or, in the case of a separately derived system such as a generator, at the generator or first disconnecting means supplied by the generator. The grounded circuit conductor shall not be connected to the equipment grounding conductor on the load side of the service disconnecting means or on the load side of a separately derived system disconnecting means.

525.32 Grounding Conductor Continuity Assurance

The continuity of the grounding conductor system used to reduce electrical shock hazards as required by 250.114, 250.138, 406.3(C), and 590.4(D) shall be verified each time that portable electrical equipment is connected.

The transient nature of amusements and, in some cases, the entire electrical distribution system associated with fairs, carnivals, and circuses increase the possibility that continuity of the equipment grounding conductor system could be interrupted. The verification of the grounding system continuity helps ensure the safety of workers and members of the general public who may come in contact with exposed non–current-carrying surfaces of electrical equipment or equipment that is electrically powered. The verification of the grounding system continuity is required each time portable equipment is reconnected.

ARTICLE 530
Motion Picture and Television Studios and Similar Locations

Contents

I. General

530.1 Scope

The requirements of this article shall apply to television studios and motion picture studios using either film or electronic cameras, except as provided in 520.1, and exchanges, factories, laboratories, stages, or a portion of the building in which film or tape more than 22 mm (⅞ in.) in width is exposed, developed, printed, cut, edited, rewound, repaired, or stored.

> FPN: For methods of protecting against cellulose nitrate film hazards, see NFPA 40-2001, *Standard for the Storage and Handling of Cellulose Nitrate Motion Picture Film.*

The requirements for motion picture studios and those for television studios are virtually the same and are intended to apply only to those locations presenting special hazards, that is, temporary structures constructed of wood or other combustible material. Otherwise, the conditions are similar to those for theater stages. Therefore, the applicable provisions of Article 520, such as those for stages and dressing rooms, should be observed.

530.2 Definitions

Alternating-Current Power Distribution Box (Alternating-Current Plugging Box, Scatter Box). An ac distribution center or box that contains one or more grounding-type polarized receptacles that may contain overcurrent protection devices.

Bull Switch. An externally operated wall-mounted safety switch that may or may not contain overcurrent protection and is designed for the connection of portable cables and cords.

Location (Shooting Location). A place outside a motion picture studio where a production or part of it is filmed or recorded.

Location Board (Deuce Board). Portable equipment containing a lighting contactor or contactors and overcurrent protection designed for remote control of stage lighting.

Motion Picture Studio (Lot). A building or group of buildings and other structures designed, constructed, or permanently altered for use by the entertainment industry for the purpose of motion picture or television production.

Plugging Box. A dc device consisting of one or more 2-pole, 2-wire, nonpolarized, nongrounding-type receptacles intended to be used on dc circuits only.

Portable Equipment. Equipment intended to be moved from one place to another.

Single-Pole Separable Connector. A device that is installed at the ends of portable, flexible, single-conductor cable that is used to establish connection or disconnection between two cables or one cable and a single-pole, panel-mounted separable connector.

Spider (Cable Splicing Block). A device that contains busbars that are insulated from each other for the purpose of splicing or distributing power to portable cables and cords that are terminated with single-pole busbar connectors.

Stage Effect (Special Effect). An electrical or electromechanical piece of equipment used to simulate a distinctive visual or audible effect such as wind machines, lightning simulators, sunset projectors, and the like.

Stage Property. An article or object used as a visual element in a motion picture or television production, except painted backgrounds (scenery) and costumes.

Stage Set. A specific area set up with temporary scenery and properties designed and arranged for a particular scene in a motion picture or television production.

Stand Lamp (Work Light). A portable stand that contains a general-purpose luminaire (lighting fixture) or lampholder with guard for the purpose of providing general illumination in the studio or stage.

Television Studio or Motion Picture Stage (Sound Stage). A building or portion of a building usually insulated from the outside noise and natural light for use by the entertainment industry for the purpose of motion picture, television, or commercial production.

530.6 Portable Equipment

Portable stage and studio lighting equipment and portable power distribution equipment shall be permitted for temporary use outdoors if the equipment is supervised by qualified personnel while energized and barriered from the general public.

See the commentary following 520.10 for more information on portable equipment.

II. Stage or Set

530.11 Permanent Wiring

The permanent wiring shall be Type MC cable, Type AC cable containing an insulated equipment grounding conductor sized in accordance with Table 250.122, Type MI cable, or in approved raceways.

Exception: Communications circuits; audio signal processing, amplification, and reproduction circuits; Class 1, Class 2, and Class 3 remote-control or signaling circuits and power-limited fire alarm circuits shall be permitted to be wired in accordance with Articles 640, 725, 760, and 800.

530.12 Portable Wiring

(A) Stage Set Wiring The wiring for stage set lighting and other supply wiring not fixed as to location shall be done with listed hard usage flexible cords and cables. Where subject to physical damage, such wiring shall be listed extra-hard usage flexible cords and cables. Splices or taps in cables shall be permitted if the total connected load does not exceed the maximum ampacity of the cable.

(B) Stage Effects and Electrical Equipment Used as Stage Properties The wiring for stage effects and electrical equipment used as stage properties shall be permitted to be wired with single- or multiconductor listed flexible cords or cables if the conductors are protected from physical damage and secured to the scenery by approved cable ties or by insulated staples. Splices or taps shall be permitted where such are made with listed devices and the circuit is protected at not more than 20 amperes.

(C) Other Electrical Equipment Cords and cables other than extra-hard usage, where supplied as a part of a listed assembly, shall be permitted.

530.13 Stage Lighting and Effects Control

Switches used for studio stage set lighting and effects (on the stages and lots and on location) shall be of the externally operable type. Where contactors are used as the disconnecting means for fuses, an individual externally operable switch, such as a tumbler switch, for the control of each contactor shall be located at a distance of not more than 1.8 m (6 ft) from the contactor, in addition to remote-control switches. A single externally operable switch shall be permitted to simultaneously disconnect all the contactors on any one location board, where located at a distance of not more than 1.8 m (6 ft) from the location board.

530.14 Plugging Boxes

Each receptacle of dc plugging boxes shall be rated at not less than 30 amperes.

530.15 Enclosing and Guarding Live Parts

(A) Live Parts Live parts shall be enclosed or guarded to prevent accidental contact by persons and objects.

(B) Switches All switches shall be of the externally operable type.

(C) Rheostats Rheostats shall be placed in approved cases or cabinets that enclose all live parts, having only the operating handles exposed.

(D) Current-Carrying Parts Current-carrying parts of bull switches, location boards, spiders, and plugging boxes shall be enclosed, guarded, or located so that persons cannot accidentally come into contact with them or bring conductive material into contact with them.

530.16 Portable Lamps

Portable lamps and work lights shall be equipped with flexible cords, composition or metal-sheathed porcelain sockets, and substantial guards.

Exception: Portable lamps used as properties in a motion picture set or television stage set, on a studio stage or lot, or on location shall not be considered to be portable lamps for the purpose of this section.

530.17 Portable Arc Lamps

(A) Portable Carbon Arc Lamps Portable carbon arc lamps shall be substantially constructed. The arc shall be provided with an enclosure designed to retain sparks and carbons and to prevent persons or materials from coming into contact with the arc or bare live parts. The enclosures shall be ventilated. All switches shall be of the externally operable type.

(B) Portable Noncarbon Arc Electric-Discharge Lamps Portable noncarbon arc lamps, including enclosed arc lamps, and associated ballasts shall be listed. Interconnecting cord sets and interconnecting cords and cables shall be extra-hard usage type and listed.

530.18 Overcurrent Protection — General

Automatic overcurrent protective devices (circuit breakers or fuses) for motion picture studio stage set lighting and the stage cables for such stage set lighting shall be as given in 530.18(A) through (G). The maximum ampacity allowed on a given conductor, cable, or cord size shall be as given in the applicable tables of Articles 310 and 400.

(A) Stage Cables Stage cables for stage set lighting shall be protected by means of overcurrent devices set at not more than 400 percent of the ampacity given in the applicable tables of Articles 310 and 400.

(B) Feeders In buildings used primarily for motion picture production, the feeders from the substations to the stages shall be protected by means of overcurrent devices (generally located in the substation) having a suitable ampere rating. The overcurrent devices shall be permitted to be multipole or single-pole gang operated. No pole shall be required in the neutral conductor. The overcurrent device setting for each feeder shall not exceed 400 percent of the ampacity of the feeder, as given in the applicable tables of Article 310.

An overcurrent device setting of up to 400 percent is permitted if the loads are of short duration. In accordance with 530.18(B), the use of short-term ratings where the equipment operates for 20 minutes or less is permitted. A longer period of operation may pose a fire hazard.

(C) Cable Protection Cables shall be protected by bushings where they pass through enclosures and shall be arranged so that tension on the cable is not transmitted to the connections. Where power conductors pass through metal, the requirements of 300.20 shall apply.

Portable feeder cables shall be permitted to temporarily penetrate fire-rated walls, floors, or ceilings provided that all of the following apply:

(1) The opening is of noncombustible material.
(2) When in use, the penetration is sealed with a temporary seal of a listed firestop material.
(3) When not in use, the opening shall be capped with a material of equivalent fire rating.

(D) Location Boards Overcurrent protection (fuses or circuit breakers) shall be provided at the location boards. Fuses in the location boards shall have an ampere rating of not over 400 percent of the ampacity of the cables between the location boards and the plugging boxes.

(E) Plugging Boxes Cables and cords supplied through plugging boxes shall be of copper. Cables and cords smaller than 8 AWG shall be attached to the plugging box by means of a plug containing two cartridge fuses or a 2-pole circuit

breaker. The rating of the fuses or the setting of the circuit breaker shall not be over 400 percent of the rated ampacity of the cables or cords as given in the applicable tables of Articles 310 and 400. Plugging boxes shall not be permitted on ac systems.

(F) Alternating-Current Power Distribution Boxes Alternating-current power distribution boxes used on sound stages and shooting locations shall contain connection receptacles of a polarized, grounding type.

(G) Lighting Work lights, stand lamps, and luminaires (fixtures) rated 1000 watts or less and connected to dc plugging boxes shall be by means of plugs containing two cartridge fuses not larger than 20 amperes, or they shall be permitted to be connected to special outlets on circuits protected by fuses or circuit breakers rated at not over 20 amperes. Plug fuses shall not be used unless they are on the load side of the fuse or circuit breakers on the location boards.

530.19 Sizing of Feeder Conductors for Television Studio Sets

(A) General It shall be permissible to apply the demand factors listed in Table 530.19(A) to that portion of the maximum possible connected load for studio or stage set lighting for all permanently installed feeders between substations and stages and to all permanently installed feeders between the main stage switchboard and stage distribution centers or location boards.

Table 530.19(A) Demand Factors for Stage Set Lighting

Portion of Stage Set Lighting Load to Which Demand Factor Applied (volt-amperes)	Feeder Demand Factor
First 50,000 or less at	100%
From 50,001 to 100,000 at	75%
From 100,001 to 200,000 at	60%
Remaining over 200,000 at	50%

(B) Portable Feeders A demand factor of 50 percent of maximum possible connected load shall be permitted for all portable feeders.

530.20 Grounding

Type MC cable, Type MI cable, metal raceways, and all non–current-carrying metal parts of appliances, devices, and equipment shall be grounded as specified in Article 250. This shall not apply to pendant and portable lamps, to stage lighting and stage sound equipment, or to other portable and special stage equipment operating at not over 150 volts dc to ground.

530.21 Plugs and Receptacles

(A) Rating Plugs and receptacles shall be rated in amperes. The voltage rating of the plugs and receptacles shall not be less than the circuit voltage. Plug and receptacle ampere ratings for ac circuits shall not be less than the feeder or branch-circuit overcurrent device ampere rating. Table 210.21(B)(2) shall not apply.

(B) Interchangeability Plugs and receptacles used in portable professional motion picture and television equipment shall be permitted to be interchangeable for ac or dc use on the same premises, provided they are listed for ac/dc use and marked in a suitable manner to identify the system to which they are connected.

530.22 Single-Pole Separable Connectors

(A) General Where ac single-pole portable cable connectors are used, they shall be listed and of the locking type. Sections 400.10, 406.6, and 406.7 shall not apply to listed single-pole separable connections and single-conductor cable assemblies utilizing listed single-pole separable connectors. Where paralleled sets of current-carrying single-pole separable connectors are provided as input devices, they shall be prominently labeled with a warning indicating the presence of internal parallel connections. The use of single-pole separable connectors shall comply with at least one of the following conditions:

(1) Connection and disconnection of connectors are only possible where the supply connectors are interlocked to the source and it is not possible to connect or disconnect connectors when the supply is energized.

(2) Line connectors are of the listed sequential-interlocking type so that load connectors shall be connected in the following sequence:
   a. Equipment grounding conductor connection
   b. Grounded circuit conductor connection, if provided
   c. Ungrounded conductor connection, and that disconnection shall be in the reverse order

(3) A caution notice shall be provided adjacent to the line connectors, indicating that plug connection shall be in the following order:
   a. Equipment grounding conductor connectors
   b. Grounded circuit-conductor connectors, if provided
   c. Ungrounded conductor connectors, and that disconnection shall be in the reverse order

(B) Interchangeability Single-pole separable connectors used in portable professional motion picture and television equipment shall be permitted to be interchangeable for ac or dc use or for different current ratings on the same premises, provided they are listed for ac/dc use and marked in a suitable manner to identify the system to which they are connected.

530.23 Branch Circuits

A branch circuit of any size supplying one or more receptacles shall be permitted to supply stage set lighting loads.

III. Dressing Rooms

530.31 Dressing Rooms

Fixed wiring in dressing rooms shall be installed in accordance with the wiring methods covered in Chapter 3. Wiring for portable dressing rooms shall be approved.

IV. Viewing, Cutting, and Patching Tables

530.41 Lamps at Tables

Only composition or metal-sheathed, porcelain, keyless lampholders equipped with suitable means to guard lamps from physical damage and from film and film scrap shall be used at patching, viewing, and cutting tables.

V. Cellulose Nitrate Film Storage Vaults

530.51 Lamps in Cellulose Nitrate Film Storage Vaults

Lamps in cellulose nitrate film storage vaults shall be installed in rigid fixtures of the glass-enclosed and gasketed type. Lamps shall be controlled by a switch having a pole in each ungrounded conductor. This switch shall be located outside of the vault and provided with a pilot light to indicate whether the switch is on or off. This switch shall disconnect from all sources of supply all ungrounded conductors terminating in any outlet in the vault.

530.52 Electrical Equipment in Cellulose Nitrate Film Storage Vaults

Except as permitted in 530.51, no receptacles, outlets, heaters, portable lights, or other portable electric equipment shall be located in cellulose nitrate film storage vaults. Electric motors shall be permitted, provided they are listed for the application and comply with Article 500, Class I, Division 2.

VI. Substations

530.61 Substations

Wiring and equipment of over 600 volts, nominal, shall comply with Article 490.

530.62 Portable Substations

Wiring and equipment in portable substations shall conform to the sections applying to installations in permanently fixed substations, but, due to the limited space available, the working spaces shall be permitted to be reduced, provided that the equipment shall be arranged so that the operator can work safely and so that other persons in the vicinity cannot accidentally come into contact with current-carrying parts or bring conducting objects into contact with them while they are energized.

530.63 Overcurrent Protection of Direct-Current Generators

Three-wire generators shall have overcurrent protection in accordance with 445.12(E).

530.64 Direct-Current Switchboards

(A) General Switchboards of not over 250 volts dc between conductors, where located in substations or switchboard rooms accessible to qualified persons only, shall not be required to be dead-front.

(B) Circuit Breaker Frames Frames of dc circuit breakers installed on switchboards shall not be required to be grounded.

ARTICLE 540
Motion Picture Projection Rooms

Contents

I. General

540.1 Scope

The provisions of this article apply to motion picture projection rooms, motion picture projectors, and associated equipment of the professional and nonprofessional types using incandescent, carbon arc, xenon, or other light source equipment that develops hazardous gases, dust, or radiation.

> FPN: For further information, see NFPA 40-2001, *Standard for the Storage and Handling of Cellulose Nitrate Motion Picture Film.*

Hazardous (classified) locations, as defined in Article 500, do not include motion picture projection rooms, even though some of the older types of film, such as cellulose nitrate film (rarely used now), are highly flammable. Cellulose acetate film, called safety film, is in wide use today. Because film is not volatile at ordinary temperatures and no flammable gases are present, the wiring installation is not required to be suitable for hazardous (classified) locations, as defined in Article 500, but should be installed with special care to protect against the hazards of fire.

540.2 Definitions

Nonprofessional Projector. Nonprofessional projectors are those types other than as described in 540.2.

Professional Projector. A type of projector using 35- or 70-mm film that has a minimum width of 35 mm (1⅜ in.) and has on each edge 212 perforations per meter (5.4 perforations per inch), or a type using carbon arc, xenon, or other light source equipment that develops hazardous gases, dust, or radiation.

II. Equipment and Projectors of the Professional Type

540.10 Motion Picture Projection Room Required

Every professional-type projector shall be located within a projection room. Every projection room shall be of permanent construction, approved for the type of building in which the projection room is located. All projection ports, spotlight ports, viewing ports, and similar openings shall be provided with glass or other approved material so as to completely close the opening. Such rooms shall not be considered as hazardous (classified) locations as defined in Article 500.

> FPN: For further information on protecting openings in projection rooms handling cellulose nitrate motion picture film, see NFPA *101*-2003, *Life Safety Code.*

540.11 Location of Associated Electrical Equipment

(A) Motor Generator Sets, Transformers, Rectifiers, Rheostats, and Similar Equipment Motor generator sets, transformers, rectifiers, rheostats, and similar equipment for the supply or control of current to projection or spotlight equipment shall, where nitrate film is used, be located in a separate room. Where placed in the projection room, they shall be located or guarded so that arcs or sparks cannot come in contact with film, and the commutator end or ends of motor generator sets shall comply with one of the conditions in 540.11(A)(1) through (A)(6).

(1) Types Be of the totally enclosed, enclosed fan-cooled, or enclosed pipe-ventilated type.

(2) Separate Rooms or Housings Be enclosed in separate rooms or housings built of noncombustible material constructed so as to exclude flyings or lint, and properly ventilated from a source of clean air.

(3) Solid Metal Covers Have the brush or sliding-contact end of motor-generator enclosed by solid metal covers.

(4) Tight Metal Housings Have brushes or sliding contacts enclosed in substantial, tight metal housings.

(5) Upper and Lower Half Enclosures Have the upper half of the brush or sliding-contact end of the motor-generator enclosed by a wire screen or perforated metal and the lower half enclosed by solid metal covers.

(6) Wire Screens or Perforated Metal Have wire screens or perforated metal placed at the commutator of brush ends. No dimension of any opening in the wire screen or perforated metal shall exceed 1.27 mm (0.05 in.), regardless of the shape of the opening and of the material used.

(B) Switches, Overcurrent Devices, or Other Equipment
Switches, overcurrent devices, or other equipment not normally required or used for projectors, sound reproduction, flood or other special effect lamps, or other equipment shall not be installed in projection rooms.

Exception No. 1: In projection rooms approved for use only with cellulose acetate (safety) film, the installation of appurtenant electrical equipment used in conjunction with the operation of the projection equipment and the control of lights, curtains, and audio equipment, and so forth, shall be permitted. In such projection rooms, a sign reading "Safety Film Only Permitted in This Room" shall be posted on the outside of each projection room door and within the projection room itself in a conspicuous location.

Exception No. 2: Remote-control switches for the control of auditorium lights or switches for the control of motors operating curtains and masking of the motion picture screen shall be permitted to be installed in projection rooms.

(C) Emergency Systems Control of emergency systems shall comply with Article 700.

The plan of a projection room of a motion picture theater is illustrated in Exhibit 540.1. This plan shows one stereopticon, or effect machine (L); two spot machines (S); and three motion picture projectors (P), which are supplied from the dc panelboard.

A dc arc lamp is the light source in each of the six machines shown in Exhibit 540.1. The dc supply is furnished by two motor-generator sets, which are usually installed in soundproof areas to avoid interfering with the sound-reproducing equipment and are controlled from the general control panel in the projection room. Two 500-kcmil feeder cables are run from each generator to the dc panelboard.

A branch circuit consisting of two 2/0 AWG cables runs from the dc panelboard to each projector (P) and to each spot machine (S). One of the two branch-circuit conductors runs directly to a projector (P); the other passes through an auxiliary gutter to the bank of resistors in the rheostat room and then to the projector. The resistors are equipped with short-circuiting switches, so that the total resistance in series with each arc may be preset to a desired value.

Because the stereopticon, or effect machine, (L) contains two arc lamps, two circuits with 1 AWG conductors are routed to this machine.

The provisions of 540.13 require that the conductors supplying outlets for arc and xenon lamps of the professional type not be smaller than 8 AWG and be of sufficient size for the lamps employed. In each case, therefore, the maximum current drawn by the lamps should be determined. In this example, with the arc lamps sized for a large picture, the arc in each projector draws nearly 150 amperes.

Four outlets, in addition to the main outlet for supplying the arc, are located at each projector for the following auxiliary circuits:

1. Outlet C supplies a small incandescent lamp inside each lamphouse and/or projector.
2. Outlet G is for the 8 AWG equipment grounding conductor, which is connected to each projector frame and to a metal water pipe.
3. Outlet F supplies a foot switch that controls a solenoid-operated shutter behind each lens, for changing from one projector to another.
4. Outlet M supplies the motor used to operate each projector.

Two exhaust fans and two duct systems, one exhausting from the ceiling of the projection room and one connected to the arc lamp housing of each machine, provide ventilation.

540.12 Work Space

Each motion picture projector, floodlight, spotlight, or similar equipment shall have clear working space not less than 750 mm (30 in.) wide on each side and at the rear thereof.

Exhibit 540.1 A typical layout of a projection room, including associated generator-supplied equipment. Modern projectors contain rectifiers as an integral part of their equipment, thereby eliminating generators and other associated equipment.

Exception: One such space shall be permitted between adjacent pieces of equipment.

540.13 Conductor Size

Conductors supplying outlets for arc and xenon projectors of the professional type shall not be smaller than 8 AWG and shall be of sufficient size for the projector employed. Conductors for incandescent-type projectors shall conform to normal wiring standards as provided in 210.24.

540.14 Conductors on Lamps and Hot Equipment

Insulated conductors having a rated operating temperature of not less than 200°C (392°F) shall be used on all lamps or other equipment where the ambient temperature at the conductors as installed will exceed 50°C (122°F).

540.15 Flexible Cords

Cords approved for hard usage, as provided in Table 400.4, shall be used on portable equipment.

540.20 Approval

Projectors and enclosures for arc, xenon and incandescent lamps and rectifiers, transformers, rheostats, and similar equipment shall be listed.

540.21 Marking

Projectors and other equipment shall be marked with the manufacturer's name or trademark and with the voltage and current for which they are designed in accordance with 110.21.

III. Nonprofessional Projectors

540.31 Motion Picture Projection Room Not Required

Projectors of the nonprofessional or miniature type, where employing cellulose acetate (safety) film, shall be permitted to be operated without a projection room.

540.32 Approval

Projection equipment shall be listed.

IV. Audio Signal Processing, Amplification, and Reproduction Equipment

540.50 Audio Signal Processing, Amplification, and Reproduction Equipment

Audio signal processing, amplification, and reproduction equipment shall be installed as provided in Article 640.

ARTICLE 545
Manufactured Buildings

Contents

545.1 Scope

This article covers requirements for a manufactured building and building components as herein defined.

The term *manufactured building* is defined in 545.2. With respect to this construction method being used for dwelling units, it is important to make the distinction between manufactured buildings covered in Article 545 and manufactured homes as covered and defined in Article 550. The most distinguishing feature between the two types of structures is how they are placed on the building site. Manufactured homes are built on a chassis and installed on site with or without a permanent foundation. Manufactured buildings are generally constructed within a factory or assembly plant and then transported to the building site. They are not built on a chassis and are designed to be installed on a permanent foundation.

In addition, the organizations responsible for construction standards for these units differ. In the case of manufactured homes, the U.S. Department of Housing and Urban Development, 24 CFR 3280, *Manufactured Home Construction and Safety Standards*, contains construction requirements for manufactured homes. Manufactured homes bear a nameplate documenting that the unit was constructed in accordance with the federal standard. In accordance with federal law, this identifying mark is universally recognized throughout the United States.

Manufactured building construction standards generally are promulgated through state or local units of government

and typically have an information sheet (often inside the cabinet below the kitchen sink) indicating the applicable building, electrical, plumbing, and mechanical codes to which the building was constructed. Manufactured building construction can be directly affected by differences in building construction regulations among those jurisdictions responsible for regulating this type of construction.

545.2 Definitions

Building Component. Any subsystem, subassembly, or other system designed for use in or integral with or as part of a structure, which can include structural, electrical, mechanical, plumbing, and fire protection systems, and other systems affecting health and safety.

Building System. Plans, specifications, and documentation for a system of manufactured building or for a type or a system of building components, which can include structural, electrical, mechanical, plumbing, and fire protection systems, and other systems affecting health and safety, and including such variations thereof as are specifically permitted by regulation, and which variations are submitted as part of the building system or amendment thereto.

Closed Construction. Any building, building component, assembly, or system manufactured in such a manner that all concealed parts of processes of manufacture cannot be inspected before installation at the building site without disassembly, damage, or destruction.

Manufactured Building. Any building that is of closed construction and is made or assembled in manufacturing facilities on or off the building site for installation, or for assembly and installation on the building site, other than manufactured homes, mobile homes, park trailers, or recreational vehicles.

545.4 Wiring Methods

(A) Methods Permitted All raceway and cable wiring methods included in this *Code* and such other wiring systems specifically intended and listed for use in manufactured buildings shall be permitted with listed fittings and with fittings listed and identified for manufactured buildings.

(B) Securing Cables In closed construction, cables shall be permitted to be secured only at cabinets, boxes, or fittings where 10 AWG or smaller conductors are used and protection against physical damage is provided.

545.5 Supply Conductors

Provisions shall be made to route the service-entrance, service-lateral, feeder, or branch-circuit supply to the service or building disconnecting means conductors.

545.6 Installation of Service-Entrance Conductors

Service-entrance conductors shall be installed after erection at the building site.

Exception: Where point of attachment is known prior to manufacture.

545.7 Service Equipment

Service equipment shall be installed in accordance with 230.70.

545.8 Protection of Conductors and Equipment

Protection shall be provided for exposed conductors and equipment during processes of manufacturing, packaging, in transit, and erection at the building site.

545.9 Boxes

(A) Other Dimensions Boxes of dimensions other than those required in Table 314.16(A) shall be permitted to be installed where tested, identified, and listed to applicable standards.

(B) Not Over 1650 cm³ (100 in.³) Any box not over 1650 cm³ (100 in.³) in size, intended for mounting in closed construction, shall be affixed with anchors or clamps so as to provide a rigid and secure installation.

545.10 Receptacle or Switch with Integral Enclosure

A receptacle or switch with integral enclosure and mounting means, where tested, identified, and listed to applicable standards, shall be permitted to be installed.

See the commentary following 300.15(E) for additional discussion about wiring devices with integral enclosures.

545.11 Bonding and Grounding

Prewired panels and building components shall provide for the bonding, or bonding and grounding, of all exposed metals likely to become energized, in accordance with Article 250, Parts V, VI, and VII.

545.12 Grounding Electrode Conductor

Provisions shall be made to route a grounding electrode conductor from the service, feeder, or branch-circuit supply to the point of attachment to the grounding electrode.

545.13 Component Interconnections

Fittings and connectors that are intended to be concealed at the time of on-site assembly, where tested, identified, and listed to applicable standards, shall be permitted for on-site

interconnection of modules or other building components. Such fittings and connectors shall be equal to the wiring method employed in insulation, temperature rise, and fault-current withstand and shall be capable of enduring the vibration and minor relative motions occurring in the components of manufactured building.

Structural components or modules are usually constructed in manufacturing facilities and then transported over the road to a building site for complete assembly of a structure, such as a dwelling unit, motel, or office building. At the on-site location, approved wiring methods are used to interconnect two or more modules. Exhibit 545.1 shows a type of nonmetallic-sheathed cable connector permitted for such interconnections.

Exhibit 545.1 A type of nonmetallic-sheathed cable connector used for interconnecting modules in a manufacturing building. (Courtesy of Pass & Seymour/Legrand®)

ARTICLE 547
Agricultural Buildings

Summary of Changes

- **547.5(A):** Revised to delete open wiring on insulators as an acceptable wiring method.

- **547.5(G):** Revised to require ground-fault protection in dirt livestock confinement areas for protection of personnel.

- **547.9:** Revised to require that site-isolating devices be pole-mounted. Where a service disconnecting means is located at the distribution point, the device shall be readily accessible at grade level and shall meet all the provisions for service equipment, including overcurrent protection and short-circuit current ratings.

• **547.10:** Revised requirements specifying where equipotential planes are required. Equipotential planes are required to be installed in a livestock building only where there is a concrete floor. For outdoor confinement areas consisting of either concrete or dirt surfaces, an equipotential plane is required to be installed only where the area contains metallic equipment that may become energized and the equipment is accessible to the livestock. In this case, the equipotential plane only has to encompass the area where the livestock stands while accessing the metallic equipment that may become energized.

Contents

547.1 Scope

The provisions of this article shall apply to the following agricultural buildings or that part of a building or adjacent areas of similar or like nature as specified in 547.1(A) and 547.1(B).

(A) Excessive Dust and Dust with Water Agricultural buildings where excessive dust and dust with water may accumulate, including all areas of poultry, livestock, and fish confinement systems, where litter dust or feed dust, including mineral feed particles, may accumulate.

(B) Corrosive Atmosphere Agricultural buildings where a corrosive atmosphere exists. Such buildings include areas where the following conditions exist:

(1) Poultry and animal excrement may cause corrosive vapors.
(2) Corrosive particles may combine with water.
(3) The area is damp and wet by reason of periodic washing for cleaning and sanitizing with water and cleansing agents.
(4) Similar conditions exist.

Article 547 applies not only to buildings but also to adjacent areas of similar or like nature. The requirements in Article 547 address the severe environmental conditions that regularly exist on agricultural premises. Damp and wet conditions, dust from feed and litter, and corrosive agents from livestock excrement are all present in these settings as part of normal operating conditions. Grounding and bonding requirements unique to agricultural settings are necessary due to the sensitivity of livestock to differences in potential between surfaces with which they are in direct contact. The wet or damp concrete common to animal confinement areas enhances this sensitivity.

547.2 Definitions

Distribution Point. An electrical supply point from which service drops, service laterals, feeders, or branch circuits to agricultural buildings, associated farm dwelling(s), and associated buildings under single management are supplied.

A definition for the term *site-isolating device* has been added in the 2005 *Code*. The concept of a site-isolating device was introduced in the 2002 edition of the *Code*. The purpose of the site-isolating device is to provide a means to disconnect and isolate the agricultural premises wiring system from the serving utility under emergency conditions, for maintenance of the load side wiring system or to allow for the connection of an alternate power source when there is a power outage. In accordance with 547.9(A)(1), the site-isolating device must be pole-mounted, and, as its name implies, it is an isolating switch and is not considered to be the service disconnecting means for the agricultural premises.

FPN No. 1: Distribution points are also known as the center yard pole, meterpole, or the common distribution point.

FPN No. 2: The service point as defined in Article 100 is typically at the distribution point.

Equipotential Plane. An area where wire mesh or other conductive elements are embedded in or placed under concrete, bonded to all metal structures and fixed nonelectrical equipment that may become energized, and connected to the electrical grounding system to prevent a difference in voltage from developing within the plane.

Site-Isolating Device. A disconnecting means installed at the distribution point for the purposes of isolation, system maintenance, emergency disconnection, or connection of optional standby systems.

547.3 Other Articles

For agricultural buildings not having conditions as specified in 547.1, the electrical installations shall be made in accordance with the applicable articles in this *Code*.

547.4 Surface Temperatures

Electrical equipment or devices installed in accordance with the provisions of this article shall be installed in a manner such that they will function at full rating without developing surface temperatures in excess of the specified normal safe operating range of the equipment or device.

547.5 Wiring Methods

(A) Wiring Systems Types UF, NMC, copper SE cables, jacketed Type MC cable, rigid nonmetallic conduit, liquidtight flexible nonmetallic conduit, or other cables or raceways suitable for the location, with approved termination fittings, shall be the wiring methods employed. The wiring methods of Article 502, Part II, shall be permitted for areas described in 547.1(A).

FPN: See 300.7 and 352.44 for installation of raceway systems exposed to widely different temperatures.

(B) Mounting All cables shall be secured within 200 mm (8 in.) of each cabinet, box, or fitting. The 6-mm (¼-in.) airspace required for nonmetallic boxes, fittings, conduit, and cables in 300.6(C) shall not be required in buildings covered by this article.

Cables must be secured within 8 in. of cabinets, boxes, or fittings installed in agricultural buildings. This distance is less than that required for cables in other types of occupancies. The requirement for a ¼-in. airspace in 300.6(D) is judged unnecessary in agricultural buildings, provided nonmetallic wiring methods are used. Decreasing the support

spacing requirements coupled with eliminating the ¼-in. airspace requirement reduces the potential for physical damage to cable-type wiring methods. Locating the wiring methods directly on the interior surface of the building allows a sealant to be placed along the wiring method to facilitate cleaning. See also 300.6(D), Exception.

(C) Equipment Enclosures, Boxes, Conduit Bodies, and Fittings

(1) Excessive Dust Equipment enclosures, boxes, conduit bodies, and fittings installed in areas of buildings where excessive dust may be present shall be designed to minimize the entrance of dust and shall have no openings (such as holes for attachment screws) through which dust could enter the enclosure.

(2) Damp or Wet Locations In damp or wet locations, equipment enclosures, boxes, conduit bodies, and fittings shall be placed or equipped so as to prevent moisture from entering or accumulating within the enclosure, box, conduit body, or fitting. In wet locations, including normally dry or damp locations where surfaces are periodically washed or sprayed with water, boxes, conduit bodies, and fittings shall be listed for use in wet locations and equipment enclosures shall be weatherproof.

(3) Corrosive Atmosphere Where wet dust, excessive moisture, corrosive gases or vapors, or other corrosive conditions may be present, equipment enclosures, boxes, conduit bodies, and fittings shall have corrosion resistance properties suitable for the conditions.

FPN No. 1: See Table 430.91 for appropriate enclosure type designations.

FPN No. 2: Aluminum and magnetic ferrous materials may corrode in agricultural environments.

Gasketed weatherproof enclosures may not provide adequate ventilation for certain types of sensitive equipment, such as electronic equipment. In some cases, it may be necessary to provide ventilation by using other types of enclosures. In accordance with 547.5(C), appropriate enclosures for the conditions encountered are required. This requirement includes consideration of both the type of enclosure and the proper materials used in the construction of the enclosure.

(D) Flexible Connections Where necessary to employ flexible connections, dusttight flexible connectors, liquidtight flexible conduit, or flexible cord listed and identified for hard usage shall be used. All connectors and fittings used shall be listed and identified for the purpose.

(E) Physical Protection All electrical wiring and equipment subject to physical damage shall be protected.

(F) Separate Equipment Grounding Conductor Non–current-carrying metal parts of equipment, raceways, and other enclosures, where required to be grounded, shall be grounded by a copper equipment grounding conductor installed between the equipment and the building disconnecting means. If installed underground, the equipment grounding conductor shall be insulated or covered.

The requirements in 547.5(F) improve the longevity of equipment grounding conductors installed aboveground and underground in the highly corrosive locations that are typical of many farm buildings.

(G) Receptacles All 125-volt, single-phase, 15- and 20-ampere general-purpose receptacles installed in the following locations shall have ground-fault circuit-interrupter protection for personnel:

(1) In areas having an equipotential plane
(2) Outdoors
(3) Damp or wet locations
(4) Dirt confinement areas for livestock

547.6 Switches, Receptacles, Circuit Breakers, Controllers, and Fuses

Switches, including pushbuttons, relays, and similar devices, receptacles, circuit breakers, controllers, and fuses, shall be provided with enclosures as specified in 547.5(C).

547.7 Motors

Motors and other rotating electrical machinery shall be totally enclosed or designed so as to minimize the entrance of dust, moisture, or corrosive particles.

547.8 Luminaires (Lighting Fixtures)

Luminaires (lighting fixtures) shall comply with 547.8(A) through 547.8(C).

(A) Minimize the Entrance of Dust Luminaires (lighting fixtures) shall be installed to minimize the entrance of dust, foreign matter, moisture, and corrosive material.

(B) Exposed to Physical Damage Any luminaire (lighting fixture) that may be exposed to physical damage shall be protected by a suitable guard.

(C) Exposed to Water A luminaire (fixture) that may be exposed to water from condensation, building cleansing water, or solution shall be watertight.

547.9 Electrical Supply to Building(s) or Structure(s) from a Distribution Point

Overhead electrical supply shall comply with 547.9(A) and 547.9(B), or with 547.9(C). Underground electrical supply shall comply with 547.9(C) and 547.9(D).

(A) Site-Isolating Device Site-isolating devices shall comply with 547.9(A)(1) through (A)(9).

(1) Where Required A site-isolating device shall be installed at the distribution point where two or more agricultural buildings, structures, associated farm dwelling(s), or other buildings are supplied from the distribution point.

(2) Location The site-isolating device shall be pole-mounted and shall meet the clearance requirements of 230.24.

(3) Operation The site-isolating device shall simultaneously disconnect all ungrounded service conductors from the premises wiring.

(4) Bonding Provisions The site-isolating device enclosure shall be bonded to the grounded circuit conductor and the grounding electrode system.

(5) Grounding At the site-isolating device, the system grounded conductor shall be connected to a grounding electrode system via a grounding electrode conductor.

(6) Rating The site-isolating device shall be rated for the calculated load as determined by Part V of Article 220.

(7) Overcurrent Protection The site-isolating device shall not be required to provide overcurrent protection.

(8) Accessibility Where the site-isolating device is not readily accessible, it shall be capable of being remotely operated by an operating handle installed at a readily accessible location. The operating handle of the site-isolating device, when in its highest position, shall not be more than 2.0 m (6 ft 7 in.) above grade or a working platform.

(9) Series Devices An additional site-isolating device for the premises wiring system shall not be required where a site-isolating device meeting all applicable requirements of this section is provided by the serving utility as part of their service requirements.

(B) Service Disconnecting Means and Overcurrent Protection at the Building(s) or Structure(s) Where the service disconnecting means and overcurrent protection are located at the building(s) or structure(s), the requirements of 547.9(B)(1) through (B)(3) shall apply.

(1) Conductor Sizing The supply conductors shall be sized in accordance with Part V of Article 220.

(2) Conductor Installation The supply conductors shall be installed in accordance with the requirements of Part II of Article 225.

(3) Grounding and Bonding For each building or structure, the conditions in either (B)(3)(a) or (B)(3)(b) shall be permitted.

(a) System with grounded neutral conductor. The grounded circuit conductor shall be connected to the building disconnecting means and to the grounding electrode system of that building or structure where all the requirements of 250.32(B)(2) are met.

FPN: A system with a grounded neutral conductor is commonly referred to as a "3-wire system" in single-phase applications.

(b) System with separate equipment grounding conductor. A separate equipment-grounding conductor shall be run with the supply conductors to the building(s) or structure(s), and the following conditions shall be met:

FPN: A system with a separate equipment grounding conductor is commonly referred to as a "4-wire system" in single-phase applications.

(1) The equipment grounding conductor shall be the same size as the largest supply conductor if of the same material, or adjusted in size in accordance with the equivalent size columns of Table 250.122 if of different materials.

(2) The equipment grounding conductor is bonded to the grounded circuit conductor and the site-isolating device at the distribution point.

(3) A grounding electrode system is provided in accordance with Part III of Article 250 and connected to the equipment-grounding conductor at the building(s) or structure(s) disconnecting means.

(4) The grounded circuit conductor is not connected to a grounding electrode or to any equipment-grounding conductor on the load side of the distribution point.

(C) Service Disconnecting Means and Overcurrent Protection at the Distribution Point Where the service disconnecting means and overcurrent protection for each set of feeder conductors are located at the distribution point, feeders to building(s) or structure(s) shall meet the requirements of 250.32 and Article 225, Parts I and II.

FPN: Methods to reduce neutral-to-earth voltages in livestock facilities include supplying buildings or structures with 4-wire single-phase services, sizing 3-wire single-phase service and feeder conductors to limit voltage drop to 2 percent, and connecting loads line-to-line.

(D) Direct-Buried Equipment Grounding Conductors Where livestock is housed, any portion of a direct-buried equipment grounding conductor run to the building or structure shall be insulated or covered copper.

Section 547.9 was revised and reorganized for the 2005 *Code*. The requirements in 547.9 cover the installation of conductors that originate from an electrical distribution point

and supply agricultural buildings. The term *distribution point* is defined in 547.2 as a supply point for service conductors, feeders, or branch circuits supplying agricultural premises.

Many agricultural sites consist of multiple buildings that are directly related to the agricultural operation or that support the operation, as is the case with dwelling units. A distribution point, sometimes referred to as the center yard pole, is often used as a means of centrally locating the origin of the electrical distribution system that supplies multiple buildings at an agricultural site via an overhead distribution system. A means to disconnect all ungrounded conductors run to the buildings and structures that are supplied from the distribution point is required. This site-isolating device provides a means to disconnect all power to the buildings at the agricultural site from a single location. This is useful in the event of an emergency, for the purposes of maintenance, or for connection to a stand-by power source.

The site-isolating device is required to be pole-mounted and therefore in most cases will not be readily accessible. The remote operating handle for the site-isolating device must be capable of being reached by personnel and must be located not more than 6 ft 7 in. above finished grade. A site-isolating device provided by an electric utility is permitted, provided that it meets the requirements of 547.9. An additional site-isolating device is not required if the device is provided by the serving utility. If the supply system includes a grounded conductor, it must be connected to a grounding electrode system at the site-isolating device. The intent of referring to this disconnecting means as the site-isolating device is so that it is not considered the service disconnecting means.

The site-isolating device is not required to have provisions for overcurrent protection, nor is overcurrent protection for the load-side conductors required to be located immediately adjacent to this device. Based on the requirements in 547.9(A) and (B), the location of the service disconnecting means and overcurrent protection is on the load side of the site-isolating switch and is located either at the distribution point or at the building or structure supplied.

A site-isolating device is not required where the service equipment is located at the distribution point in accordance with 547.9(C).

In Exhibit 547.1, the pole-mounted site-isolating device is located at the distribution point. A set of overload service conductors is run to each of the three structures on the premises, and at each of the buildings the service disconnecting means and overcurrent protection are installed. The supply conductors in Exhibit 547.1 are service conductors because there is no overcurrent protection at the site-isolating device. A grounding electrode system is required at the distribution point, and a grounding electrode conductor connection to the supply system grounded conductor must be made at the site-isolating device. A grounding electrode system is also required at each of the buildings. At these

Exhibit 547.1 Site-isolating device located at the distribution point with service conductors run to each building. A service disconnecting means and overcurrent protection is installed at each building.

Exhibit 547.2 Service disconnecting means and overcurrent protection located at the distribution point with feeders run to each building.

locations, a grounding electrode conductor connection to the system grounded conductor is permitted, provided such a connection does not create a parallel path for neutral current between the distribution point and the structure disconnecting means. If a parallel path for neutral current is created through multiple grounding connections, 547.9(B)(3)(b) requires the installation of a separate equipment grounding conductor between the distribution point and the disconnecting means at each building.

The equipment grounding conductor must be the same size as the largest supply conductor, or if it is of different conductor material than the supply conductors, it must be adjusted in size based on Table 250.122.

Any portion of the equipment grounding conductor installed under ground in raceways, in cable assemblies, or direct-buried must be insulated or covered copper in accordance with 547.5(F) and 547.9(D). This provision applies to the equipment grounding conductor of all underground branch circuits and feeders covered by the requirements of Article 547. The intent of this requirement is to protect the equipment grounding conductor from the corrosive influences inherent to agricultural premises and to reduce leakage current in those areas where livestock are kept, because prevention of stray voltage at agricultural premises is of the utmost importance.

In Exhibit 547.2, the service equipment and overcurrent protection are located at the distribution point. This distribution arrangement does not require a site-isolating device per

547.9. A set of feeder conductors is run to each of the three structures on the premises. At the distribution point, a grounding electrode system is required, and in accordance with 250.32 a grounding electrode system is required at each of the buildings. In addition, a disconnecting means at each of the structures is required per 225.31 and 225.32. This distribution arrangement for agricultural sites is covered in 547.9(C).

547.10 Equipotential Planes and Bonding of Equipotential Planes

The installation and bonding of equipotential planes shall comply with 547.10(A) and 547.10(B). For the purposes of this section, the term *livestock* shall not include poultry.

(A) Where Required Equipotential planes shall be installed in all concrete floor confinement areas in livestock buildings, and in all outdoor confinement areas such as feedlots, containing metallic equipment that may become energized and is accessible to livestock. The equipotential plane shall encompass the area where the livestock stands while accessing metallic equipment that may become energized.

(B) Bonding Equipotential planes shall be bonded to the electrical grounding system. The bonding conductor shall be copper, insulated, covered or bare, and not smaller than 8 AWG. The means of bonding to wire mesh or conductive elements shall be by pressure connectors or clamps of brass, copper, copper alloy, or an equally substantial approved means. Slatted floors that are supported by structures that are a part of an equipotential plane shall not require bonding.

FPN No. 1: Methods to establish equipotential planes are described in American Society of Agricultural Engineers (ASAE) EP473-2001, *Equipotential Planes in Animal Containment Areas.*

FPN No. 2: Low grounding electrode system resistances may reduce potential differences in livestock facilities.

ARTICLE 550
Mobile Homes, Manufactured Homes, and Mobile Home Parks

Summary of Changes

- **550.12(C):** Revised to prohibit the laundry circuit from supplying other than laundry receptacle outlets.

- **550.13(F)(1):** Revised requirement to prohibit receptacle outlets within a bathtub or shower space; deleted the phrase "within reach of such space."

- **550.13(G):** Revised to not require, rather than prohibit, receptacle outlets in the identified spaces.

- **550.15(H), Exception:** Revised to include MI cable as an under-chassis wiring method.

- **550.32(A):** Revised to require that the rating of the permitted additional site disconnecting means be not less than the rating required for site service equipment in accordance 550.32(C).

Contents

I. General

550.1 Scope

The provisions of this article cover the electrical conductors and equipment installed within or on mobile and manufactured homes, the conductors that connect mobile and manufactured homes to a supply of electricity, and the installation of electrical wiring, luminaires (fixtures), equipment, and appurtenances related to electrical installations within a mobile home park up to the mobile home service-entrance conductors or, if none, the mobile home service equipment.

> FPN: For additional information on manufactured housing see NFPA 501-2003, *Standard on Manufactured Housing*, and Part 3280, *Manufactured Home Construction and Safety Standards*, of the Federal Department of Housing and Urban Development.

The *Federal Mobile Home Construction and Safety Standard*, issued by the Federal Housing and Urban Development Administration (HUD), incorporates many of the provisions of Article 550 of the *NEC*. The federal standard contains the requirements for electrical systems, conductors, and equipment installed within or on mobile homes and the conductors that connect mobile homes to a supply of electric-

ity. Mobile homes are defined as manufactured homes in the HUD regulations. For the purposes of this *Code*, and unless otherwise indicated, the term *mobile home* includes manufactured homes.

The regulations pertaining to electrical systems are located in 24 CFR 3280.801–3280.816. They require that new manufactured homes comply with the federal standard. In some cases, HUD has delegated the enforcement of this standard to state and private inspection agencies and qualified testing laboratories. The service equipment and feeders installed at the mobile or manufactured home site is covered by the requirements in Part III of this article.

See Article 545 for the requirements covering electrical systems in manufactured buildings.

550.2 Definitions

Appliance, Fixed. An appliance that is fastened or otherwise secured at a specific location.

Appliance, Portable. An appliance that is actually moved or can easily be moved from one place to another in normal use.

> FPN: For the purpose of this article, the following major appliances, other than built-in, are considered portable if cord connected: refrigerators, range equipment, clothes washers, dishwashers without booster heaters, or other similar appliances.

Appliance, Stationary. An appliance that is not easily moved from one place to another in normal use.

Distribution Panelboard. See definition of panelboard in Article 100.

Feeder Assembly. The overhead or under-chassis feeder conductors, including the grounding conductor, together with the necessary fittings and equipment or a power-supply cord listed for mobile home use, designed for the purpose of delivering energy from the source of electrical supply to the distribution panelboard within the mobile home.

Laundry Area. An area containing or designed to contain a laundry tray, clothes washer, or a clothes dryer.

Manufactured Home. A structure, transportable in one or more sections, that is 2.5 m (8 body ft) or more in width or 12 m (40 body ft) or more in length in the traveling mode or, when erected on site, is 30 m² (320 ft²) or more; which is built on a chassis and designed to be used as a dwelling, with or without a permanent foundation, when connected to the required utilities, including the plumbing, heating, air conditioning, and electrical systems contained therein. Calculations used to determine the number of square meters (square feet) in a structure will be based on the structure's exterior dimensions, measured at the largest horizontal projections when erected on site. These dimensions include all expandable rooms, cabinets, and other projections containing interior space but do not include inside bay windows.

For the purpose of this *Code* and unless otherwise indicated, the term *mobile home* includes manufactured homes.

FPN No. 1: See the applicable building code for definition of the term *permanent foundation*.
FPN No. 2: See Part 3280, *Manufactured Home Construction and Safety Standards*, of the Federal Department of Housing and Urban Development, for additional information on the definition.

Mobile Home. A factory-assembled structure or structures transportable in one or more sections that is built on a permanent chassis and designed to be used as a dwelling without a permanent foundation where connected to the required utilities and that includes the plumbing, heating, air-conditioning, and electric systems contained therein.

For the purpose of this *Code* and unless otherwise indicated, the term *mobile home* includes manufactured homes.

Mobile Home Accessory Building or Structure. Any awning, cabana, ramada, storage cabinet, carport, fence, windbreak, or porch established for the use of the occupant of the mobile home on a mobile home lot.

Mobile home is the original term covering a structure that is built on a chassis, designed to be transportable and intended for installation on a site with or without a permanent foundation. The *Code* has covered these units and their associated site equipment since the 1965 edition. Manufactured homes (not to be confused with manufactured buildings, covered in Article 545) are also covered by Article 550 and, for the purposes of this article, are considered mobile homes. The term *manufactured home* is used in the federal standard that covers their construction. The requirements in Article 550 treat mobile and manufactured homes the same unless specifically stated otherwise. An example of a distinction between the two is found in 550.32(B), which contains provisions exclusive to manufactured homes that cover mounting of service equipment on the structure.

The requirements contained in Article 550, Part III cover the installation of service equipment and feeders at mobile and manufactured home sites. It is intended that mobile and manufactured homes constructed in accordance with the requirements of Article 550, Parts I and II (or under the HUD 24 CFR 3280 regulations) be installed at their sites in accordance with the requirements of Part III. For more information on the distinction between manufactured homes and manufactured buildings, see the commentary following 545.1.

Mobile Home Lot. A designated portion of a mobile home park designed for the accommodation of one mobile home

and its accessory buildings or structures for the exclusive use of its occupants.

Mobile Home Park. A contiguous parcel of land that is used for the accommodation of occupied mobile homes.

Mobile Home Service Equipment. The equipment containing the disconnecting means, overcurrent protective devices, and receptacles or other means for connecting a mobile home feeder assembly.

Park Electrical Wiring Systems. All of the electrical wiring, luminaires (fixtures), equipment, and appurtenances related to electrical installations within a mobile home park, including the mobile home service equipment.

550.4 General Requirements

(A) Mobile Home Not Intended as a Dwelling Unit A mobile home not intended as a dwelling unit — for example, those equipped for sleeping purposes only, contractor's on-site offices, construction job dormitories, mobile studio dressing rooms, banks, clinics, mobile stores, or intended for the display or demonstration of merchandise or machinery — shall not be required to meet the provisions of this article pertaining to the number or capacity of circuits required. It shall, however, meet all other applicable requirements of this article if provided with an electrical installation intended to be energized from a 120-volt or 120/240-volt ac power supply system. Where different voltage is required by either design or available power supply system, adjustment shall be made in accordance with other articles and sections for the voltage used.

(B) In Other Than Mobile Home Parks Mobile homes installed in other than mobile home parks shall comply with the provisions of this article.

(C) Connection to Wiring System The provisions of this article shall apply to mobile homes intended for connection to a wiring system rated 120/240 volts, nominal, 3-wire ac, with grounded neutral.

(D) Listed or Labeled All electrical materials, devices, appliances, fittings, and other equipment shall be listed or labeled by a qualified testing agency and shall be connected in an approved manner when installed.

II. Mobile and Manufactured Homes

550.10 Power Supply

(A) Feeder The power supply to the mobile home shall be a feeder assembly consisting of not more than one listed 50-ampere mobile home power-supply cord with an integrally molded or securely attached plug cap or a permanently installed feeder.

Exception No. 1: A mobile home that is factory equipped with gas or oil-fired central heating equipment and cooking appliances shall be permitted to be provided with a listed mobile home power-supply cord rated 40 amperes.

Exception No. 2: A feeder assembly shall not be required for manufactured homes constructed in accordance with 550.32(B).

Exception No. 2 modifies the requirement of 550.10(A) only for manufactured homes. The installation of service equipment is permitted in or on a manufactured home per 550.32(B).

(B) Power-Supply Cord If the mobile home has a power-supply cord, it shall be permanently attached to the distribution panelboard or to a junction box permanently connected to the distribution panelboard, with the free end terminating in an attachment plug cap.

Cords with adapters and pigtail ends, extension cords, and similar items shall not be attached to, or shipped with, a mobile home.

A suitable clamp or the equivalent shall be provided at the distribution panelboard knockout to afford strain relief for the cord to prevent strain from being transmitted to the terminals when the power-supply cord is handled in its intended manner.

The cord shall be a listed type with four conductors, one of which shall be identified by a continuous green color or a continuous green color with one or more yellow stripes for use as the grounding conductor.

(C) Attachment Plug Cap The attachment plug cap shall be a 3-pole, 4-wire, grounding type, rated 50 amperes, 125/250 volts with a configuration as shown in Figure 550.10(C) and intended for use with the 50-ampere, 125/250-volt receptacle configuration shown in Figure 550.10(C). It shall be listed, by itself or as part of a power-supply cord assembly, for the purpose and shall be molded to or installed on the flexible cord so that it is secured tightly to the cord at the point where the cord enters the attachment plug cap. If a

Receptacle   Cap

125/250-V, 50-A, 3-pole, 4-wire, grounding type

Figure 550.10(C) 50-Ampere, 125/250-Volt Receptacle and Attachment Plug Cap Configurations, 3-Pole, 4-Wire, Grounding-Types, Used for Mobile Home Supply Cords and Mobile Home Parks

right-angle cap is used, the configuration shall be oriented so that the grounding member is farthest from the cord.

> FPN: Complete details of the 50-ampere plug and receptacle configuration can be found in the National Electrical Manufacturers Association *Standard for Dimensions of Attachment Plugs and Receptacles*, ANSI/NEMA WD 6-1989, Figure 14-50.

(D) Overall Length of a Power-Supply Cord The overall length of a power-supply cord, measured from the end of the cord, including bared leads, to the face of the attachment plug cap shall not be less than 6.4 m (21 ft) and shall not exceed 11 m (36½ ft). The length of the cord from the face of the attachment plug cap to the point where the cord enters the mobile home shall not be less than 6.0 m (20 ft).

(E) Marking The power-supply cord shall bear the following marking:

FOR USE WITH MOBILE HOMES — 40 AMPERES.

or

FOR USE WITH MOBILE HOMES — 50 AMPERES.

(F) Point of Entrance The point of entrance of the feeder assembly to the mobile home shall be in the exterior wall, floor, or roof.

(G) Protected Where the cord passes through walls or floors, it shall be protected by means of conduits and bushings or equivalent. The cord shall be permitted to be installed within the mobile home walls, provided a continuous raceway having a maximum size of 32 mm (1¼ in.) is installed from the branch-circuit panelboard to the underside of the mobile home floor.

(H) Protection Against Corrosion and Mechanical Damage Permanent provisions shall be made for the protection of the attachment plug cap of the power-supply cord and any connector cord assembly or receptacle against corrosion and mechanical damage if such devices are in an exterior location while the mobile home is in transit.

(I) Mast Weatherhead or Raceway Where the calculated load exceeds 50 amperes or where a permanent feeder is used, the supply shall be by means of either of the following:

(1) One mast weatherhead installation, installed in accordance with Article 230, containing four continuous, insulated, color-coded feeder conductors, one of which shall be an equipment grounding conductor

(2) A metal raceway or rigid nonmetallic conduit from the disconnecting means in the mobile home to the underside of the mobile home, with provisions for the attachment to a suitable junction box or fitting to the raceway on the underside of the mobile home [with or without

conductors as in 550.10(I)(1)]. The manufacturer shall provide written installation instructions stating the proper feeder conductor sizes for the raceway and the size of the junction box to be used.

Today's mobile and manufactured homes are generally larger and have more electrically powered equipment than earlier units, which in most cases results in a feeder calculation that exceeds 50 amperes. Cord-and-plug connection is permitted for units in which the calculated load does not exceed 50 amperes. If the calculated load exceeds 50 amperes, the *Code* requires a permanently connected feeder, as covered in 550.10(I) and 550.33. Local requirements must be checked for the approved method of installing overhead and underground feeder assemblies.

A raceway is required from the distribution panelboard in the mobile home to the underside of the mobile home. Typically, the feeder conductors in this raceway are installed when the mobile home is located at its site. The raceway provides a means to install the feeder conductors to the mobile home panelboard without having to damage the interior finish. The feeder assembly must comprise four continuous, insulated, color-coded conductors, as indicated in 550.10(I)(1) and 550.33(A).

550.11 Disconnecting Means and Branch-Circuit Protective Equipment

The branch-circuit equipment shall be permitted to be combined with the disconnecting means as a single assembly. Such a combination shall be permitted to be designated as a distribution panelboard. If a fused distribution panelboard is used, the maximum fuse size for the mains shall be plainly marked with lettering at least 6 mm (¼ in.) high and visible when fuses are changed.

Where plug fuses and fuseholders are used, they shall be tamper-resistant Type S, enclosed in dead-front fuse panelboards. Electrical distribution panelboards containing circuit breakers shall also be dead-front type.

FPN: See 110.22 concerning identification of each disconnecting means and each service, feeder, or branch circuit at the point where it originated and the type marking needed.

(A) Disconnecting Means A single disconnecting means shall be provided in each mobile home consisting of a circuit breaker, or a switch and fuses and its accessories installed in a readily accessible location near the point of entrance of the supply cord or conductors into the mobile home. The main circuit breakers or fuses shall be plainly marked "Main." This equipment shall contain a solderless type of grounding connector or bar for the purposes of grounding, with sufficient terminals for all grounding conductors. The neutral bar termination of the grounded circuit conductors

shall be insulated in accordance with 550.16(A). The disconnecting equipment shall have a rating not less than the calculated load. The distribution equipment, either circuit breaker or fused type, shall be located a minimum of 600 mm (24 in.) from the bottom of such equipment to the floor level of the mobile home.

FPN: See 550.20(B) for information on disconnecting means for branch circuits designed to energize heating or air-conditioning equipment, or both, located outside the mobile home, other than room air conditioners.

A distribution panelboard shall be rated not less than 50 amperes and employ a 2-pole circuit breaker rated 40 amperes for a 40-ampere supply cord, or 50 amperes for a 50-ampere supply cord. A distribution panelboard employing a disconnect switch and fuses shall be rated 60 amperes and shall employ a single 2-pole, 60-ampere fuseholder with 40- or 50-ampere main fuses for 40- or 50-ampere supply cords, respectively. The outside of the distribution panelboard shall be plainly marked with the fuse size.

The distribution panelboard shall be located in an accessible location but shall not be located in a bathroom or a clothes closet. A clear working space at least 750 mm (30 in.) wide and 750 mm (30 in.) in front of the distribution panelboard shall be provided. This space shall extend from the floor to the top of the distribution panelboard.

(B) Branch-Circuit Protective Equipment Branch-circuit distribution equipment shall be installed in each mobile home and shall include overcurrent protection for each branch circuit consisting of either circuit breakers or fuses.

The branch-circuit overcurrent devices shall be rated as follows:

(1) Not more than the circuit conductors; and
(2) Not more than 150 percent of the rating of a single appliance rated 13.3 amperes or more that is supplied by an individual branch circuit; but
(3) Not more than the overcurrent protection size and of the type marked on the air conditioner or other motor-operated appliance.

(C) Two-Pole Circuit Breakers Where circuit breakers are provided for branch-circuit protection, 240-volt circuits shall be protected by a 2-pole common or companion trip, or handle-tied paired circuit breakers.

(D) Electrical Nameplates A metal nameplate on the outside adjacent to the feeder assembly entrance shall read:

THIS CONNECTION FOR 120/240-VOLT,
3-POLE, 4-WIRE, 60-HERTZ,
_____ AMPERE SUPPLY

The correct ampere rating shall be marked in the blank space.

Exception: For manufactured homes, the manufacturer shall provide in its written installation instructions or in the data plate the minimum ampere rating of the feeder assembly or, where provided, the service entrance conductors intended for connection to the manufactured home. The rating provided shall not be less than the minimum load calculated in accordance with 550.18.

550.12 Branch Circuits

The number of branch circuits required shall be determined in accordance with 550.12(A) through (E).

(A) Lighting The number of branch circuits shall be based on 33 volt-amperes/m^2 (3 VA/ft^2) times outside dimensions of the mobile home (coupler excluded) divided by 120 volts to determine the number of 15- or 20-ampere lighting area circuits, for example,

$$\frac{3 \times \text{length} \times \text{width}}{120 \times 15 \text{ (or 20)}} = \frac{\text{No. of 15- (or 20-)}}{\text{ampere circuits}}$$

(B) Small Appliances In kitchens, pantries, dining rooms, and breakfast rooms, two or more 20-ampere small-appliance circuits, in addition to the number of circuits required elsewhere in this section, shall be provided for all receptacle outlets required by 550.13(D) in these rooms. Such circuits shall have no other outlets.

Exception No. 1: Receptacle outlets installed solely for the electrical supply and support of an electric clock in any the rooms specified in 550.12(B) shall be permitted.

Exception No. 2: Receptacle outlets installed to provide power for supplemental equipment and lighting on gas-fired ranges, ovens, or counter-mounted cooking units shall be permitted.

Exception No. 3: A single receptacle for refrigeration equipment shall be permitted to be supplied from an individual branch circuit rated 15 amperes or greater.

Countertop receptacle outlets installed in the kitchen shall be supplied by not less than two small-appliance circuit branch circuits, either or both of which shall be permitted to supply receptacle outlets in the kitchen and other locations specified in 550.12(B).

(C) Laundry Area Where a laundry area is provided, a 20-ampere branch circuit shall be provided to supply the laundry receptacle outlet(s). This circuit shall have no other outlets.

Section 550.12(C) was expanded in the 2005 *Code* to clarify that a 20-ampere laundry circuit is not allowed to supply other outlets beyond the laundry area.

(D) General Appliances (Including furnace, water heater, range, and central or room air conditioner, etc.) There shall be one or more circuits of adequate rating in accordance with the following:

FPN: For central air conditioning, see Article 440.

(1) The ampere rating of fixed appliances shall not be over 50 percent of the circuit rating if lighting outlets (receptacles, other than kitchen, dining area, and laundry, considered as lighting outlets) are on the same circuit.

(2) For fixed appliances on a circuit without lighting outlets, the sum of rated amperes shall not exceed the branch-circuit rating. Motor loads or continuous loads shall not exceed 80 percent of the branch-circuit rating.

(3) The rating of a single cord-and-plug-connected appliance on a circuit having no other outlets shall not exceed 80 percent of the circuit rating.

(4) The rating of a range branch circuit shall be based on the range demand as specified for ranges in 550.18(B)(5).

(E) Bathrooms Bathroom receptacle outlets shall be supplied by at least one 20-ampere branch circuit. Such circuits shall have no other outlets other than as provided for in 550.13(E)(2).

550.13 Receptacle Outlets

(A) Grounding-Type Receptacle Outlets All receptacle outlets shall comply with the following:

(1) Be of grounding type
(2) Be installed according to 406.3
(3) Except where supplying specific appliances, be 15- or 20-ampere, 125-volt, either single or multiple type, and accept parallel-blade attachment plugs

(B) Ground-Fault Circuit Interrupters (GFCI) All 125-volt, single-phase, 15- and 20-ampere receptacle outlets installed outdoors, in compartments accessible from outside the unit, or in bathrooms, including receptacles in luminaires (light fixtures), shall have GFCI protection. GFCI protection shall be provided for receptacle outlets serving countertops in kitchens, and receptacle outlets located within 1.8 m (6 ft) of a wet bar sink.

Exception: Receptacles installed for appliances in dedicated spaces, such as for dishwashers, disposals, refrigerators, freezers, and laundry equipment.

Feeders supplying branch circuits shall be permitted to be protected by a ground-fault circuit-interrupter in lieu of the provision for such interrupters specified herein.

(C) Cord-Connected Fixed Appliance A grounding-type receptacle outlet shall be provided for each cord-connected fixed appliance installed.

(D) Receptacle Outlets Required Except in the bath, closet, and hall areas, receptacle outlets shall be installed at wall spaces 600 mm (2 ft) wide or more so that no point along the floor line is more than 1.8 m (6 ft) measured horizontally from an outlet in that space. In addition, a receptacle outlet shall be installed in the following locations:

(1) Over or adjacent to countertops in the kitchen [at least one on each side of the sink if countertops are on each side and are 300 mm (12 in.) or over in width].
(2) Adjacent to the refrigerator and freestanding gas-range space. A duplex receptacle shall be permitted to serve as the outlet for a countertop and a refrigerator.
(3) At countertop spaces for built-in vanities.
(4) At countertop spaces under wall-mounted cabinets.
(5) In the wall at the nearest point to where a bar-type counter attaches to the wall.
(6) In the wall at the nearest point to where a fixed room divider attaches to the wall.
(7) In laundry areas within 1.8 m (6 ft) of the intended location of the laundry appliance(s).
(8) At least one receptacle outlet located outdoors and accessible at grade level and not more than 2.0 m (6½ ft) above grade. A receptacle outlet located in a compartment accessible from the outside of the unit shall be considered an outdoor receptacle.
(9) At least one receptacle outlet shall be installed in bathrooms within 900 mm (36 in.) of the outside edge of each basin. The receptacle outlet shall be located above or adjacent to the basin location. This receptacle shall be in addition to any receptacle that is a part of a luminaire (fixture) or appliance. The receptacle shall not be enclosed within a bathroom cabinet or vanity.

(E) Pipe Heating Cable(s) Outlet For the connection of pipe heating cable(s), a receptacle outlet shall be located on the underside of the unit as follows:

(1) Within 600 mm (2 ft) of the cold water inlet.
(2) Connected to an interior branch circuit, other than a small appliance branch circuit. It shall be permitted to use a bathroom receptacle circuit for this purpose.
(3) On a circuit where all of the outlets are on the load side of the ground-fault circuit-interrupter.
(4) This outlet shall not be considered as the receptacle required by 550.13(D)(8).

The provisions of 550.13(E) require a receptacle outlet on the underside of mobile homes to supply cord-and-plug-connected pipe-heating cables (sometimes referred to as heat tape outlets). The receptacle must be GFCI protected and connected to a branch circuit that serves the interior of the mobile home. All the outlets supplied by this branch circuit must be on the load (downstream) side of a GFCI. The

purpose of arranging the supply circuit to the pipe heating cable outlet in this manner is to allow supervision of the power supply to, and GFCI protection of, this outlet from the interior of the mobile home. If the overcurrent protective device or GFCI device opens, the occupants of the mobile home are more likely to notice it than if the outlet were supplied by a dedicated circuit and the GFCI device were located at the outlet.

(F) Receptacle Outlets Not Permitted Receptacle outlets shall not be permitted in the following locations:

(1) Receptacle outlets shall not be installed within a bathtub or shower space.
(2) A receptacle shall not be installed in a face-up position in any countertop.
(3) Receptacle outlets shall not be installed above electric baseboard heaters, unless provided for in the listing or manufacturer's instructions.

(G) Receptacle Outlets Not Required Receptacle outlets shall not be required in the following locations:

(1) In the wall space occupied by built-in kitchen or wardrobe cabinets
(2) In the wall space behind doors that can be opened fully against a wall surface
(3) In room dividers of the lattice type that are less than 2.5 m (8 ft) long, not solid, and within 150 mm (6 in.) of the floor
(4) In the wall space afforded by bar-type counters

550.14 Luminaires (Fixtures) and Appliances

(A) Fasten Appliances in Transit Means shall be provided to securely fasten appliances when the mobile home is in transit. (See 550.16 for provisions on grounding.)

(B) Accessibility Every appliance shall be accessible for inspection, service, repair, or replacement without removal of permanent construction.

(C) Pendants Listed pendant-type luminaires (fixtures) or pendant cords shall be permitted.

(D) Bathtub and Shower Luminaires (Fixtures) Where a luminaire (lighting fixture) is installed over a bathtub or in a shower stall, it shall be of the enclosed and gasketed type listed for wet locations.

550.15 Wiring Methods and Materials

Except as specifically limited in this section, the wiring methods and materials included in this *Code* shall be used in mobile homes. Aluminum conductors, aluminum alloy conductors, and aluminum core conductors such as copper-clad aluminum shall not be acceptable for use as branch-circuit wiring.

(A) Nonmetallic Boxes Nonmetallic boxes shall be permitted only with nonmetallic cable or nonmetallic raceways.

(B) Nonmetallic Cable Protection Nonmetallic cable located 380 mm (15 in.) or less above the floor, if exposed, shall be protected from physical damage by covering boards, guard strips, or raceways. Cable likely to be damaged by stowage shall be so protected in all cases.

(C) Metal-Covered and Nonmetallic Cable Protection Metal-covered and nonmetallic cables shall be permitted to pass through the centers of the wide side of 2 by 4 studs. However, they shall be protected where they pass through 2 by 2 studs or at other studs or frames where the cable or armor would be less than 32 mm (1¼ in.) from the inside or outside surface of the studs where the wall covering materials are in contact with the studs. Steel plates on each side of the cable, or a tube, with not less than 1.35 mm (0.053 in.) wall thickness shall be required to protect the cable. These plates or tubes shall be securely held in place.

(D) Metal Faceplates Where metal faceplates are used, they shall be effectively grounded.

(E) Installation Requirements If a range, clothes dryer, or similar appliance is connected by metal-covered cable or flexible metal conduit, a length of not less than 900 mm (3 ft) of free cable or conduit shall be provided to permit moving the appliance. The cable or flexible metal conduit shall be secured to the wall. Type NM or Type SE cable shall not be used to connect a range or dryer. This shall not prohibit the use of Type NM or Type SE cable between the branch-circuit overcurrent-protective device and a junction box or range or dryer receptacle.

(F) Raceways Where rigid metal conduit or intermediate metal conduit is terminated at an enclosure with a locknut and bushing connection, two locknuts shall be provided, one inside and one outside of the enclosure. Rigid nonmetallic conduit, electrical nonmetallic tubing, or surface raceway shall be permitted. All cut ends of conduit and tubing shall be reamed or otherwise finished to remove rough edges.

(G) Switches Switches shall be rated as follows:

(1) For lighting circuits, switches shall be rated not less than 10 amperes, 120 to 125 volts, and in no case less than the connected load.
(2) Switches for motor or other loads shall comply with the provisions of 404.14.

(H) Under-Chassis Wiring (Exposed to Weather) Where outdoor or under-chassis line-voltage (120 volts, nominal, or higher) wiring is exposed to moisture or physical damage,

it shall be protected by rigid metal conduit or intermediate metal conduit. The conductors shall be suitable for wet locations.

Exception: Type MI cable, electrical metallic tubing, or rigid nonmetallic conduit shall be permitted where closely routed against frames and equipment enclosures.

(I) Boxes, Fittings, and Cabinets Boxes, fittings, and cabinets shall be securely fastened in place and shall be supported from a structural member of the home, either directly or by using a substantial brace.

Exception: Snap-in-type boxes. Boxes provided with special wall or ceiling brackets and wiring devices with integral enclosures that securely fasten to walls or ceilings and are identified for the use shall be permitted without support from a structural member or brace. The testing and approval shall include the wall and ceiling construction systems for which the boxes and devices are intended to be used.

(J) Appliance Terminal Connections Appliances having branch-circuit terminal connections that operate at temperatures higher than 60°C (140°F) shall have circuit conductors as described in the following:

(1) Branch-circuit conductors having an insulation suitable for the temperature encountered shall be permitted to be run directly to the appliance.
(2) Conductors having an insulation suitable for the temperature encountered shall be run from the appliance terminal connection to a readily accessible outlet box placed at least 300 mm (1 ft) from the appliance. These conductors shall be in a suitable raceway or Type AC or MC cable of at least 450 mm (18 in.) but not more than 1.8 m (6 ft) in length.

(K) Component Interconnections Fittings and connectors that are intended to be concealed at the time of assembly shall be listed and identified for the interconnection of building components. Such fittings and connectors shall be equal to the wiring method employed in insulation, temperature rise, and fault-current withstanding and shall be capable of enduring the vibration and shock occurring in mobile home transportation.

FPN: See 550.19 for interconnection of multiple section units.

550.16 Grounding

Grounding of both electrical and nonelectrical metal parts in a mobile home shall be through connection to a grounding bus in the mobile home distribution panelboard. The grounding bus shall be grounded through the green-colored insulated conductor in the supply cord or the feeder wiring to the service ground in the service-entrance equipment located adjacent to the mobile home location. Neither the frame of the mobile home nor the frame of any appliance shall be connected to the grounded circuit conductor (neutral) in the mobile home. Where the distribution panelboard is the service equipment as permitted by 550.32(B), the neutral conductors and the equipment grounding bus shall be connected.

(A) Grounded (Neutral) Conductor

(1) Insulated The grounded circuit conductor (neutral) shall be insulated from the grounding conductors and from equipment enclosures and other grounded parts. The grounded (neutral) circuit terminals in the distribution panelboard and in ranges, clothes dryers, counter-mounted cooking units, and wall-mounted ovens shall be insulated from the equipment enclosure. Bonding screws, straps, or buses in the distribution panelboard or in appliances shall be removed and discarded. Where the distribution panelboard is the service equipment as permitted by 550.32(B), the neutral conductors and the equipment grounding bus shall be connected.

(2) Connections of Ranges and Clothes Dryers Connections of ranges and clothes dryers with 120/240-volt, 3-wire ratings shall be made with 4-conductor cord and 3-pole, 4-wire, grounding-type plugs or by Type AC cable, Type MC cable, or conductors enclosed in flexible metal conduit.

The provisions of 550.33(A) require that the feeder assembly for a mobile home consist of a listed cord or four color-coded insulated conductors, one of which is the grounded conductor (white) and one of which is used for grounding purposes (green). Thus, the grounded and grounding conductors are kept independent of each other and are connected only at the service equipment (at the point of connection of the grounding electrode conductor). Grounding of both electrical and nonelectrical metal parts, including the frame of the mobile home or the frame of any appliance, is accomplished by connection to the equipment grounding bus [never to the grounded conductor (neutral bus)]. The purpose of this requirement is to prevent incidental contact between the grounded conductor and non–current-carrying metal parts of electrical equipment. Without the separation of the grounded and grounding conductors, this contact could result in the metal structure or metal sheathing of the mobile home becoming a parallel path for neutral current.

Bonding screws, straps, or buses, which bond the grounded (neutral) circuit conductors to the non–current-carrying metal parts in the mobile home panelboard or to the metal frame of an appliance (ranges, clothes dryers), must not be installed or, in the case of ranges and clothes dryers, must be removed. In general, new ranges and clothes dryers have a factory-installed bonding jumper. Removal of the factory-installed bonding jumper does not compromise or void the listing of the product, because isolation of the metal appliance frame from the grounded circuit conductor is required by the *Code*. There is limited application in existing branch circuits of conventional or "site-built" construction where the factory-installed bonding jumper can remain intact.

(B) Equipment Grounding Means

(1) Supply Cord or Permanent Feeder The green-colored insulated grounding wire in the supply cord or permanent feeder wiring shall be connected to the grounding bus in the distribution panelboard or disconnecting means.

(2) Electrical System In the electrical system, all exposed metal parts, enclosures, frames, lamp fixture canopies, and so forth shall be effectively bonded to the grounding terminal or enclosure of the distribution panelboard.

(3) Cord-Connected Appliances Cord-connected appliances, such as washing machines, clothes dryers, and refrigerators, and the electrical system of gas ranges and so forth, shall be grounded by means of a cord with grounding conductor and grounding-type attachment plug.

(C) Bonding of Non–Current-Carrying Metal Parts

(1) Exposed Non–Current-Carrying Metal Parts All exposed non–current-carrying metal parts that may become energized shall be effectively bonded to the grounding terminal or enclosure of the distribution panelboard. A bonding conductor shall be connected between the distribution panelboard and accessible terminal on the chassis.

(2) Grounding Terminals Grounding terminals shall be of the solderless type and listed as pressure-terminal connectors recognized for the wire size used. The bonding conductor shall be solid or stranded, insulated or bare, and shall be 8 AWG copper minimum, or equivalent. The bonding conductor shall be routed so as not to be exposed to physical damage.

(3) Metallic Piping and Ducts Metallic gas, water, and waste pipes and metallic air-circulating ducts shall be considered bonded if they are connected to the terminal on the chassis [see 550.16(C)(1)] by clamps, solderless connectors, or by suitable grounding-type straps.

(4) Metallic Roof and Exterior Coverings Any metallic roof and exterior covering shall be considered bonded if the following conditions are met:

(1) The metal panels overlap one another and are securely attached to the wood or metal frame parts by metallic fasteners.

(2) The lower panel of the metallic exterior covering is secured by metallic fasteners at a cross member of the chassis by two metal straps per mobile home unit or section at opposite ends.

The bonding strap material shall be a minimum of 100 mm (4 in.) in width of material equivalent to the skin or a material of equal or better electrical conductivity. The straps shall be fastened with paint-penetrating fittings such as screws and starwashers or equivalent.

550.17 Testing

(A) Dielectric Strength Test The wiring of each mobile home shall be subjected to a 1-minute, 900-volt, dielectric strength test (with all switches closed) between live parts (including neutral) and the mobile home ground. Alternatively, the test shall be permitted to be performed at 1080 volts for 1 second. This test shall be performed after branch circuits are complete and after luminaires (fixtures) or appliances are installed.

Exception: Listed luminaires (fixtures) or appliances shall not be required to withstand the dielectric strength test.

(B) Continuity and Operational Tests and Polarity Checks Each mobile home shall be subjected to all of the following:

(1) An electrical continuity test to ensure that all exposed electrically conductive parts are properly bonded
(2) An electrical operational test to demonstrate that all equipment, except water heaters and electric furnaces, is connected and in working order
(3) Electrical polarity checks of permanently wired equipment and receptacle outlets to determine that connections have been properly made

550.18 Calculations

The following method shall be employed in calculating the supply-cord and distribution-panelboard load for each feeder assembly for each mobile home in lieu of the procedure shown in Article 220 and shall be based on a 3-wire, 120/240-volt supply with 120-volt loads balanced between the two legs of the 3-wire system.

(A) Lighting, Small Appliance, and Laundry Load

(1) Lighting Volt-Amperes Length times width of mobile home floor (outside dimensions) times 33 volt-amperes/m² (3 VA/ft²), for example, length × width × 3 = lighting volt-amperes.

(2) Small Appliance Volt-Amperes Number of circuits times 1500 volt-amperes for each 20-ampere appliance receptacle circuit (see definition of Appliance, Portable, with

note in 550.2), for example, number of circuits × 1500 = small appliance volt-amperes.

(3) Laundry Area Circuit Volt-Amperes 1500 volt-amperes.

(4) Total Volt-Amperes Lighting volt-amperes plus small appliance volt-amperes plus laundry area volt-amperes equals total volt-amperes.

(5) Net Volt-Amperes First 3000 total volt-amperes at 100 percent plus remainder at 35 percent equals volt-amperes to be divided by 240 volts to obtain current (amperes) per leg.

(B) Total Load for Determining Power Supply Total load for determining power supply is the sum of the following:

(1) Lighting and small appliance load as calculated in 550.18(A)(5).
(2) Nameplate amperes for motors and heater loads (exhaust fans, air conditioners, electric, gas, or oil heating). Omit smaller of the heating and cooling loads, except include blower motor if used as air-conditioner evaporator motor. Where an air conditioner is not installed and a 40-ampere power-supply cord is provided, allow 15 amperes per leg for air conditioning.
(3) Twenty-five percent of current of largest motor in (2).
(4) Total of nameplate amperes for waste disposer, dishwasher, water heater, clothes dryer, wall-mounted oven, cooking units. Where the number of these appliances exceeds three, use 75 percent of total.
(5) Derive amperes for freestanding range (as distinguished from separate ovens and cooking units) by dividing the following values by 240 volts:

Nameplate Rating (watts)	Use (volt-amperes)
0 – 10,000	80 percent of rating
Over 10,000 – 12,500	8,000
Over 12,500 – 13,500	8,400
Over 13,500 – 14,500	8,800
Over 14,500 – 15,500	9,200
Over 15,500 – 16,500	9,600
Over 16,500 – 17,500	10,000

(6) If outlets or circuits are provided for other than factory-installed appliances, include the anticipated load.

FPN: Refer to Annex D, Example D11, for an illustration of the application of this calculation.

(C) Optional Method of Calculation for Lighting and Appliance Load The optional method for calculating lighting and appliance load shown in 220.82 shall be permitted.

550.19 Interconnection of Multiple-Section Mobile or Manufactured Home Units

(A) Wiring Methods Approved and listed fixed-type wiring methods shall be used to join portions of a circuit that must be electrically joined and are located in adjacent sections after the home is installed on its support foundation. The circuit's junction shall be accessible for disassembly when the home is prepared for relocation.

FPN: See 550.15(K) for component interconnections.

(B) Disconnecting Means Expandable or multi-unit manufactured homes not having permanently installed feeders that are to be moved from one location to another shall be permitted to have disconnecting means with branch-circuit protective equipment in each unit when so located that after assembly or joining together of units, the requirements of 550.10 will be met.

550.20 Outdoor Outlets, Luminaires (Fixtures), Air-Cooling Equipment, and So Forth

(A) Listed for Outdoor Use Outdoor luminaires (fixtures) and equipment shall be listed for outdoor use. Outdoor receptacle or convenience outlets shall be of a gasketed-cover type for use in wet locations. Where located on the underside of the home or located under roof extensions or similarly protected locations, outdoor luminaires (fixtures) and equipment shall be listed for use in damp locations.

(B) Outside Heating Equipment, Air-Conditioning Equipment, or Both A mobile home provided with a branch circuit designed to energize outside heating equipment, air-conditioning equipment, or both, located outside the mobile home, other than room air conditioners, shall have such branch-circuit conductors terminate in a listed outlet box, or disconnecting means, located on the outside of the mobile home. A label shall be permanently affixed adjacent to the outlet box and shall contain the following information:

> THIS CONNECTION IS FOR HEATING
> AND/OR AIR-CONDITIONING EQUIPMENT.
> THE BRANCH CIRCUIT IS RATED AT NOT
> MORE THAN _____ AMPERES, AT _____ VOLTS, 60
> HERTZ, _____ CONDUCTOR AMPACITY.
> A DISCONNECTING MEANS SHALL BE LOCATED
> WITHIN SIGHT OF THE EQUIPMENT.

The correct voltage and ampere rating shall be given. The tag shall be not less than 0.51 mm (0.020 in.) thick etched brass, stainless steel, anodized or alclad aluminum, or equivalent. The tag shall not be less than 75 mm by 45 mm (3 in. by 1¾ in.) minimum size.

550.25 Arc-Fault Circuit-Interrupter Protection

(A) Definition Arc-fault circuit interrupters are defined in Article 210.12(A).

(B) Bedrooms of Mobile Homes and Manufactured Homes All branch circuits that supply 125-volt, single-phase, 15- and 20-ampere outlets installed in bedrooms of mobile homes and manufactured homes shall be protected by arc-fault circuit interrupter(s).

The requirement for arc-fault circuit-interrupter protection in mobile and manufactured homes is similar to that found in 210.12 for site-built dwelling units. Arc-fault circuit-interrupter protection must be provided for the branch circuits supplying 125-volt, 15- and 20-ampere outlets in bedrooms. The branch circuits covered by this requirement are all those that fall within the voltage and current ratings specified and supply bedroom lighting outlets, receptacle outlets, and other power outlets to which devices or utilization equipment are connected. It should be noted that this requirement does not supersede the current HUD 3280 requirements for factory-installed wiring in manufactured homes.

III. Services and Feeders

550.30 Distribution System

The mobile home park secondary electrical distribution system to mobile home lots shall be single-phase, 120/240 volts, nominal. For the purpose of Part III, where the park service exceeds 240 volts, nominal, transformers and secondary distribution panelboards shall be treated as services.

Mobile homes are intended for connection to a wiring system nominally rated 120/240 volts, 3-wire ac, with a grounded neutral; therefore, distribution systems at mobile home parks must supply 120/240 volts to the mobile home lot. Because appliances and other equipment are usually installed during the manufacturing process of mobile homes and are rated 120/240 volts, a 120/208-volt supply derived from a 4-wire, 120/208-volt wye system is unsuitable.

550.31 Allowable Demand Factors

Park electrical wiring systems shall be calculated (at 120/240 volts) on the larger of the following:

(1) 16,000 volt-amperes for each mobile home lot
(2) The load calculated in accordance with 550.18 for the largest typical mobile home that each lot will accept

It shall be permissible to calculate the feeder or service load in accordance with Table 550.31. No demand factor shall be allowed for any other load, except as provided in this *Code*.

Service and feeder conductors to a mobile home in compliance with 310.15(B)(6) shall be permitted.

In accordance with 550.31, park electrical wiring systems must be calculated on the basis of the larger of (1) not less

Table 550.31 Demand Factors for Services and Feeders

Number of Mobile Homes	Demand Factor (percent)
1	100
2	55
3	44
4	39
5	33
6	29
7–9	28
10–12	27
13–15	26
16–21	25
22–40	24
41–60	23
61 and over	22

than 16,000 volt-amperes (at 120/240 volts) for each mobile home lot or (2) the calculated load of the largest typical mobile home the lot will accommodate. However, the ampacity of the feeder-circuit conductors to each mobile home lot cannot be less than 100 amperes (at 120/240 volts), per 550.33(B).

550.32 Service Equipment

(A) Mobile Home Service Equipment The mobile home service equipment shall be located adjacent to the mobile home and not mounted in or on the mobile home. The service equipment shall be located in sight from and not more than 9.0 m (30 ft) from the exterior wall of the mobile home it serves. The service equipment shall be permitted to be located elsewhere on the premises, provided that a disconnecting means suitable for use as service equipment is located within sight from and not more than 9.0 m (30 ft) from the exterior wall of the mobile home it serves and is rated not less than that required for service equipment per 550.32(C). Grounding at the disconnecting means shall be in accordance with 250.32.

Mobile home service equipment must be located in sight of the mobile home, but the equipment can be up to 30 ft from any point on the exterior wall of the mobile home. This requirement recognizes the use of feeder raceways that are external to the mobile home. Service equipment may be located more than 30 ft from the mobile home if an additional disconnecting means is located within 30 ft of the mobile home and grounding and bonding of this additional disconnecting means are performed in accordance with the provisions of 250.32. In a mobile home park, this arrangement facilitates locating service equipment at one or more centralized locations that are not within the required 30-ft proximity

to the mobile home. Feeders are installed from this service equipment to the properly located mobile home site disconnecting means.

(B) Manufactured Home Service Equipment The manufactured home service equipment shall be permitted to be installed in or on a manufactured home, provided that all of the following conditions are met:

(1) The manufacturer shall include in its written installation instructions information indicating that the home shall be secured in place by an anchoring system or installed on and secured to a permanent foundation.
(2) The installation of the service equipment shall comply with Article 230.
(3) Means shall be provided for the connection of a grounding electrode conductor to the service equipment and routing it outside the structure.
(4) Bonding and grounding of the service shall be in accordance with Article 250.
(5) The manufacturer shall include in its written installation instructions one method of grounding the service equipment at the installation site. The instructions shall clearly state that other methods of grounding are found in Article 250.
(6) The minimum size grounding electrode conductor shall be specified in the instructions.
(7) A red warning label shall be mounted on or adjacent to the service equipment. The label shall state the following:

WARNING
DO NOT PROVIDE ELECTRICAL POWER
UNTIL THE GROUNDING ELECTRODE(S)
IS INSTALLED AND CONNECTED
(SEE INSTALLATION INSTRUCTIONS).

Where the service equipment is not installed in or on the unit, the installation shall comply with the other provisions of this section.

The provisions of 550.32(B) specify the conditions required in order to install the service equipment in or on a manufactured home. The concern over the unit being moved off site (intentionally or unintentionally) without the ability to disconnect the electrical supply is addressed in condition (1). A manufactured home with a service in or on the unit must be anchored in place or secured to a permanent foundation.

The other conditions specified by 550.32(B) cover the need to provide proper grounding and bonding conductors, systems, and connections and to install the service equipment in accordance with the applicable requirements in Article

230. The provisions of 550.32(B) apply only to manufactured homes as defined in 550.2.

(C) Rating Mobile home service equipment shall be rated at not less than 100 amperes at 120/240 volts, and provisions shall be made for connecting a mobile home feeder assembly by a permanent wiring method. Power outlets used as mobile home service equipment shall also be permitted to contain receptacles rated up to 50 amperes with appropriate overcurrent protection. Fifty-ampere receptacles shall conform to the configuration shown in Figure 550.10(C).

> FPN: Complete details of the 50-ampere plug and receptacle configuration can be found in ANSI/NEMA WD 6-1989, National Electrical Manufacturers Association *Standard for Wiring Devices — Dimensional Requirements*, Figure 14-50.

(D) Additional Outside Electrical Equipment Means for connecting a mobile home accessory building or structure or additional electrical equipment located outside a mobile home by a fixed wiring method shall be provided in either the mobile home service equipment or the local external disconnecting means permitted in 550.32(A).

(E) Additional Receptacles Additional receptacles shall be permitted for connection of electrical equipment located outside the mobile home, and all such 125-volt, single-phase, 15- and 20-ampere receptacles shall be protected by a listed ground-fault circuit interrupter.

(F) Mounting Height Outdoor mobile home disconnecting means shall be installed so the bottom of the enclosure containing the disconnecting means is not less than 600 mm (2 ft) above finished grade or working platform. The disconnecting means shall be installed so that the center of the grip of the operating handle, when in the highest position, is not more than 2.0 m (6 ft 7 in.) above the finished grade or working platform.

(G) Marking Where a 125/250-volt receptacle is used in mobile home service equipment, the service equipment shall be marked as follows:

TURN DISCONNECTING SWITCH OR
CIRCUIT BREAKER OFF BEFORE INSERTING
OR REMOVING PLUG. PLUG MUST BE FULLY
INSERTED OR REMOVED.

The marking shall be located on the service equipment adjacent to the receptacle outlet.

550.33 Feeder

(A) Feeder Conductors Feeder conductors shall consist of either a listed cord, factory installed in accordance with 550.10(B), or a permanently installed feeder consisting of four insulated, color-coded conductors that shall be identified by the factory or field marking of the conductors in compliance with 310.12. Equipment grounding conductors shall not be identified by stripping the insulation.

Exception: Where a feeder is installed between service equipment and a disconnecting means as covered in 550.32(A), it shall be permitted to omit the equipment grounding conductor where the grounded circuit conductor is grounded at the disconnecting means as required in 250.32(B).

(B) Feeder Capacity Mobile home and manufactured home lot feeder circuit conductors shall have a capacity not less than the loads supplied and shall be rated at not less than 100 amperes at 120/240 volts.

ARTICLE 551
Recreational Vehicles and
Recreational Vehicle Parks

Summary of Changes

- **551.1:** Revised to reflect deletion of Part II covering the low-voltage electrical systems of recreational vehicles.

- **Part II, Low-Voltage Systems:** Deleted Part II, comprising 551.10(A) through (H) and covering the low voltage wiring system of recreational vehicles.

- **551.32:** Revised to include engine generators.

- **551.46(E):** Added Exception No. 3 covering electrical point of entrance for RVs designed for transporting livestock.

- **551.47(N):** Added Type MI cable as wiring method for outdoor and under-chassis locations.

- **551.47(P)(2):** Added requirement covering direct wiring connection of expandable units to the main body of the vehicle.

- **551.51(B):** Added requirement prohibiting switches within tub or shower space unless part of a listed tub or shower assembly.

- **551.71:** Revised to increase from a 5% to a 20% minimum for the number of sites equipped with 50-ampere receptacles. New FPN warns that more than 20 percent may be needed.

- **551.73(A):** Added second paragraph pertaining to sites supplying two RVs.

- **551.76(D):** Revised to reference Article 250 for requirements connecting the neutral conductor to a grounding electrode system at separately derived systems and at separate buildings and structures.

Contents

I. General

551.1 Scope

The provisions of this article cover the electrical conductors and equipment other than low-voltage and automotive vehicle circuits or extensions thereof, installed within or on recreational vehicles, the conductors that connect recreational vehicles to a supply of electricity, and the installation of equipment and devices related to electrical installations within a recreational vehicle park.

The scope of Article 551 was revised in the 2005 *Code* to eliminate the requirements covering 12-volt low-voltage electrical systems. Laws in many states require a factory inspection by either a governmental or a private inspection agency. NFPA 1192-2002, *Standard on Recreational Vehicles*, is widely accepted by the recreational vehicle (RV) industry and those authorities having jurisdiction responsible for ensuring that RVs are built to a recognized safety standard. Section 4.4 of NFPA 1192 requires compliance with Parts I, III, IV, V, and VI of Article 551 and requires compliance with ANSI/RVIA 12V, *Low Voltage Systems in Conversion and Recreational Vehicles*, for the RV electrical systems rated 24 volts, nominal, or less.

FPN: For information on low-voltage systems, see NFPA 1192-2002, *Standard for Recreational Vehicles,* and ANSI/RVIA 12V, *Low Voltage Systems in Conversion and Recreational Vehicles*, 2002 edition.

551.2 Definitions

(See Article 100 for additional definitions.)

Air-Conditioning or Comfort-Cooling Equipment. All of that equipment intended or installed for the purpose of processing the treatment of air so as to control simultaneously its temperature, humidity, cleanliness, and distribution to meet the requirements of the conditioned space.

Appliance, Fixed. An appliance that is fastened or otherwise secured at a specific location.

Appliance, Portable. An appliance that is actually moved or can easily be moved from one place to another in normal use.

FPN: For the purpose of this article, the following major appliances, other than built-in, are considered portable if cord connected: refrigerators, range equipment, clothes washers, dishwashers without booster heaters, or other similar appliances.

Appliance, Stationary. An appliance that is not easily moved from one place to another in normal use.

Camping Trailer. A vehicular portable unit mounted on wheels and constructed with collapsible partial side walls that fold for towing by another vehicle and unfold at the campsite to provide temporary living quarters for recreational, camping, or travel use. *(See Recreational Vehicle.)*

Converter. A device that changes electrical energy from one form to another, as from alternating current to direct current.

Dead Front (as applied to switches, circuit breakers, switchboards, and distribution panelboards). Designed, constructed, and installed so that no current-carrying parts are normally exposed on the front.

Disconnecting Means. The necessary equipment usually consisting of a circuit breaker or switch and fuses, and their accessories, located near the point of entrance of supply conductors in a recreational vehicle and intended to constitute the means of cutoff for the supply to that recreational vehicle.

Distribution Panelboard. A single panel or group of panel units designed for assembly in the form of a single panel, including buses, and with or without switches and/or automatic overcurrent-protective devices for the control of light, heat, or power circuits of small individual as well as aggregate capacity; designed to be placed in a cabinet or cutout box placed in or against a wall or partition and accessible only from the front.

Frame. Chassis rail and any welded addition thereto of metal thickness of 1.35 mm (0.053 in.) or greater.

Low Voltage. An electromotive force rated 24 volts, nominal, or less, supplied from a transformer, converter, or battery.

Motor Home. A vehicular unit designed to provide temporary living quarters for recreational, camping, or travel use built on or permanently attached to a self-propelled motor vehicle chassis or on a chassis cab or van that is an integral part of the completed vehicle. *(See Recreational Vehicle.)*

Power-Supply Assembly. The conductors, including ungrounded, grounded, and equipment grounding conductors, the connectors, attachment plug caps, and all other fittings, grommets, or devices installed for the purpose of delivering energy from the source of electrical supply to the distribution panel within the recreational vehicle.

Recreational Vehicle. A vehicular-type unit primarily designed as temporary living quarters for recreational, camping, or travel use, which either has its own motive power or is mounted on or drawn by another vehicle. The basic entities are travel trailer, camping trailer, truck camper, and motor home.

Recreational Vehicle Park. A plot of land upon which two or more recreational vehicle sites are located, established, or maintained for occupancy by recreational vehicles of the

general public as temporary living quarters for recreation or vacation purposes.

Recreational Vehicle Site. A plot of ground within a recreational vehicle park set aside for the accommodation of a recreational vehicle on a temporary basis. It can be used as either a recreational vehicle site or as a camping unit site.

Recreational Vehicle Site Feeder Circuit Conductors. The conductors from the park service equipment to the recreational vehicle site supply equipment.

Recreational Vehicle Site Supply Equipment. The necessary equipment, usually a power outlet, consisting of a circuit breaker or switch and fuse and their accessories, located near the point of entrance of supply conductors to a recreational vehicle site and intended to constitute the disconnecting means for the supply to that site.

Recreational Vehicle Stand. That area of a recreational vehicle site intended for the placement of a recreational vehicle.

Transformer. A device that, when used, raises or lowers the voltage of alternating current of the original source.

Travel Trailer. A vehicular unit, mounted on wheels, designed to provide temporary living quarters for recreational, camping, or travel use, of such size or weight as not to require special highway movement permits when towed by a motorized vehicle, and of gross trailer area less than 30 m^2 (320 ft^2). *(See Recreational Vehicle.)*

Truck Camper. A portable unit constructed to provide temporary living quarters for recreational, travel, or camping use, consisting of a roof, floor, and sides, designed to be loaded onto and unloaded from the bed of a pick-up truck. *(See Recreational Vehicle.)*

551.4 General Requirements

(A) Not Covered A recreational vehicle not used for the purposes as defined in 551.2 shall not be required to meet the provisions of Part I pertaining to the number or capacity of circuits required. It shall, however, meet all other applicable requirements of this article if the recreational vehicle is provided with an electrical installation intended to be energized from a 120- or 120/240-volt, nominal, ac power-supply system.

(B) Systems This article covers combination electrical systems, generator installations, and 120- or 120/240-volt, nominal, systems.

The wiring requirements for the 12-volt low-voltage system that were located in 551.10(A) and other sections of the 2002 *Code* have been deleted in the 2005 *Code*.

FPN: For information on low-voltage systems, refer to NFPA 1192-2002, *Standard on Recreational Vehicles*, and ANSI/RVIA 12V-2002, *Standard for Low-Voltage Systems in Conversion and Recreational Vehicles*.

II. Combination Electrical Systems

551.20 Combination Electrical Systems

(A) General Vehicle wiring suitable for connection to a battery or dc supply source shall be permitted to be connected to a 120-volt source, provided the entire wiring system and equipment are rated and installed in full conformity with Parts I, II, III, IV, and V requirements of this article covering 120-volt electrical systems. Circuits fed from ac transformers shall not supply dc appliances.

(B) Voltage Converters (120-Volt Alternating Current to Low-Voltage Direct Current) The 120-volt ac side of the voltage converter shall be wired in full conformity with Parts I, II, III, IV, and V requirements of this article for 120-volt electrical systems.

Exception: Converters supplied as an integral part of a listed appliance shall not be subject to 551.20(B).

All converters and transformers shall be listed for use in recreation vehicles and designed or equipped to provide over-temperature protection. To determine the converter rating, the following formula shall be applied to the total connected load, including average battery charging rate, of all 12-volt equipment:

The first 20 amperes of load at 100 percent, plus

The second 20 amperes of load at 50 percent, plus

All load above 40 amperes at 25 percent

Exception: A low-voltage appliance that is controlled by a momentary switch (normally open) that has no means for holding in the closed position or refrigerators with a 120-volt function shall not be considered as a connected load when determining the required converter rating. Momentarily energized appliances shall be limited to those used to prepare the vehicle for occupancy or travel.

(C) Bonding Voltage Converter Enclosures The non–current-carrying metal enclosure of the voltage converter shall be bonded to the frame of the vehicle with a minimum 8 AWG copper conductor. The voltage converter shall be provided with a separate chassis bonding conductor that shall not be used as a current-carrying conductor.

The intent of 551.20(C) is to reduce the possibility of damage to the power supply cord by large dc fault currents that may find their way back to the vehicle frame or battery through the ac grounding conductor of the converter. Metal enclosures of listed converters are provided with an external pressure terminal connector for this purpose.

(D) Dual-Voltage Fixtures, Including Luminaires or Appliances Fixtures, including luminaires, or appliances having both 120-volt and low-voltage connections shall be listed for dual voltage.

In the dual-voltage fixtures described in 551.20(D), barriers are used to separate the 120-volt and the 12-volt wiring connections.

(E) Autotransformers Autotransformers shall not be used.

(F) Receptacles and Plug Caps Where a recreational vehicle is equipped with a 120-volt or 120/240-volt ac system, a low-voltage system, or both, receptacles and plug caps of the low-voltage system shall differ in configuration from those of the 120- or 120/240-volt system. Where a vehicle equipped with a battery or other low-voltage system has an external connection for low-voltage power, the connector shall have a configuration that will not accept 120-volt power.

III. Other Power Sources

551.30 Generator Installations

(A) Mounting Generators shall be mounted in such a manner as to be effectively bonded to the recreational vehicle chassis.

(B) Generator Protection Equipment shall be installed to ensure that the current-carrying conductors from the engine generator and from an outside source are not connected to a vehicle circuit at the same time.

Receptacles used as disconnecting means shall be accessible (as applied to wiring methods) and capable of interrupting their rated current without hazard to the operator.

(C) Installation of Storage Batteries and Generators Storage batteries and internal-combustion-driven generator units (subject to the provisions of this *Code*) shall be secured in place to avoid displacement from vibration and road shock.

(D) Ventilation of Generator Compartments Compartments accommodating internal-combustion-driven generator units shall be provided with ventilation in accordance with instructions provided by the manufacturer of the generator unit.

FPN: For generator compartment construction requirements, see NFPA 1192-2002, *Standard on Recreational Vehicles*.

(E) Supply Conductors The supply conductors from the engine generator to the first termination on the vehicle shall be of the stranded type and be installed in listed flexible conduit or listed liquidtight flexible conduit. The point of first termination shall be in one of the following:

(1) Panelboard
(2) Junction box with a blank cover
(3) Junction box with a receptacle
(4) Enclosed transfer switch
(5) Receptacle assembly listed in conjunction with the generator

The panelboard or junction box with a receptacle shall be installed within the vehicle's interior and within 450 mm (18 in.) of the compartment wall but not inside the compartment. If the generator is below the floor level and not in a compartment, the panelboard or junction box with receptacle shall be installed within the vehicle interior within 450 mm (18 in.) of the point of entry into the vehicle. A junction box with a blank cover shall be mounted on the compartment wall and shall be permitted inside or outside the compartment. A receptacle assembly listed in conjunction with the generator shall be mounted in accordance with its listing. If the generator is below floor level and not in a compartment, the junction box with blank cover shall be mounted either to any part of the generator supporting structure (but not to the generator) or to the vehicle floor within 450 mm (18 in.) of any point directly above the generator on either the inside or outside of the floor surface. Overcurrent protection in accordance with 240.4 shall be provided for supply conductors as an integral part of a listed generator or shall be located within 450 mm (18 in.) of their point of entry into the vehicle.

551.31 Multiple Supply Source

(A) Multiple Supply Sources Where a multiple supply system consisting of an alternate power source and a power-supply cord is installed, the feeder from the alternate power source shall be protected by an overcurrent-protective device. Installation shall be in accordance with 551.30(A), 551.30(B), and 551.40.

(B) Multiple Supply Sources Capacity The multiple supply sources shall not be required to be of the same capacity.

(C) Alternate Power Sources Exceeding 30 Amperes If an alternate power source exceeds 30 amperes, 120 volts, nominal, it shall be permissible to wire it as a 120-volt, nominal, system or a 120/240-volt, nominal, system, provided an overcurrent-protective device of the proper rating is installed in the feeder.

(D) Power-Supply Assembly Not Less Than 30 Amperes The external power-supply assembly shall be permitted to be less than the calculated load but not less than 30 amperes and shall have overcurrent protection not greater than the capacity of the external power-supply assembly.

551.32 Other Sources

Other sources of ac power, such as inverters, motor generators, or engine generators, shall be listed for use in recreational vehicles and shall be installed in accordance with the terms of the listing. Other sources of ac power shall be wired in full conformity with the requirements in Parts I, II, III, IV, and V of this article covering 120-volt electrical systems.

551.33 Alternate Source Restriction

Transfer equipment, if not integral with the listed power source, shall be installed to ensure that the current-carrying conductors from other sources of ac power and from an outside source are not connected to the vehicle circuit at the same time.

IV. Nominal 120-Volt or 120/240-Volt Systems

551.40 120-Volt or 120/240-Volt, Nominal, Systems

(A) General Requirements The electrical equipment and material of recreational vehicles indicated for connection to a wiring system rated 120 volts, nominal, 2-wire with ground, or a wiring system rated 120/240 volts, nominal, 3-wire with ground, shall be listed and installed in accordance with the requirements of Parts I, II, III, IV, and V of this article. Electrical equipment connected line-to-line shall have a voltage rating of 208–230 volts.

Section 551.40(A) was revised in the 2005 *Code* to require electrical equipment to be rated 208–230 volts when it is to be connected line to line. This rating allows for compatibility with recreational vehicle parks that have 208Y/120-volt, 3-phase, 4-wire electrical distribution systems.

(B) Materials and Equipment Electrical materials, devices, appliances, fittings, and other equipment installed in, intended for use in, or attached to the recreational vehicle shall be listed. All products shall be used only in the manner in which they have been tested and found suitable for the intended use.

(C) Ground-Fault Circuit-Interrupter Protection The internal wiring of a recreational vehicle having only one 15- or 20-ampere branch circuit as permitted in 551.42(A) and 551.42(B) shall have ground-fault circuit-interrupter protection for personnel. The ground-fault circuit interrupter shall be installed at the point where the power supply assembly terminates within the recreational vehicle. Where a separable cord set is not employed, the ground-fault circuit interrupter shall be permitted to be an integral part of the attachment plug of the power supply assembly. The ground-fault circuit interrupter shall provide protection also under the conditions of an open grounded circuit conductor, interchanged circuit conductors, or both.

551.41 Receptacle Outlets Required

(A) Spacing Receptacle outlets shall be installed at wall spaces 600 mm (2 ft) wide or more so that no point along the floor line is more than 1.8 m (6 ft), measured horizontally, from an outlet in that space.

Exception No. 1: Bath and hall areas.

Exception No. 2: Wall spaces occupied by kitchen cabinets, wardrobe cabinets, built-in furniture, behind doors that may open fully against a wall surface, or similar facilities.

(B) Location Receptacle outlets shall be installed as follows:

(1) Adjacent to countertops in the kitchen [at least one on each side of the sink if countertops are on each side and are 300 mm (12 in.) or over in width]
(2) Adjacent to the refrigerator and gas range space, except where a gas-fired refrigerator or cooking appliance, requiring no external electrical connection, is factory installed
(3) Adjacent to countertop spaces of 300 mm (12 in.) or more in width that cannot be reached from a receptacle required in 551.41(B)(1) by a cord of 1.8 m (6 ft) without crossing a traffic area, cooking appliance, or sink

(C) Ground-Fault Circuit-Interrupter Protection Where provided, each 125-volt, single-phase, 15- or 20-ampere receptacle outlet shall have ground-fault circuit-interrupter protection for personnel in the following locations:

(1) Adjacent to a bathroom lavatory
(2) Where the receptacles are installed to serve the countertop surfaces and are within 1.8 m (6 ft) of any lavatory or sink

Exception No. 1: Receptacles installed for appliances in dedicated spaces, such as for dishwashers, disposals, refrigerators, freezers, and laundry equipment.

Exception No. 2: Single receptacles for interior connections of expandable room sections.

Exception No. 3: De-energized receptacles that are within 1.8 m (6 ft) of any sink or lavatory due to the retraction of the expandable room section.

(3) In the area occupied by a toilet, shower, tub, or any combination thereof

(4) On the exterior of the vehicle

Exception: Receptacles that are located inside of an access panel that is installed on the exterior of the vehicle to supply power for an installed appliance shall not be required to have ground-fault circuit-interrupter protection.

The receptacle outlet shall be permitted in a listed luminaire (lighting fixture). A receptacle outlet shall not be installed in a tub or combination tub–shower compartment.

In accordance with 551.41(C)(4), a bathroom receptacle is permitted to be mounted in the side of a lavatory cabinet where installation of a receptacle outlet is not possible in a wall that does not provide the necessary depth. The receptacle must be GFCI protected in accordance with 551.41(C)(1). Receptacles of any type are not permitted in a tub or tub-shower compartment.

(D) Face-Up Position A receptacle shall not be installed in a face-up position in any countertop or similar horizontal surfaces within the living area.

551.42 Branch Circuits Required

Each recreational vehicle containing a 120-volt electrical system shall contain one of the following.

(A) One 15-Ampere Circuit One 15-ampere circuit to supply lights, receptacle outlets, and fixed appliances. Such recreational vehicles shall be equipped with one 15-ampere switch and fuse or one 15-ampere circuit breaker.

(B) One 20-Ampere Circuit One 20-ampere circuit to supply lights, receptacle outlets, and fixed appliances. Such recreational vehicles shall be equipped with one 20-ampere switch and fuse or one 20-ampere circuit breaker.

(C) Two to Five 15- or 20-Ampere Circuits A maximum of five 15- or 20-ampere circuits to supply lights, receptacle outlets, and fixed appliances shall be permitted. Such recreational vehicles shall be equipped with a distribution panelboard rated at 120 volts maximum with a 30-ampere rated main power supply assembly. Not more than two 120-volt thermostatically controlled appliances (i.e., air conditioner and water heater) shall be installed in such systems unless appliance isolation switching, energy management systems, or similar methods are used.

Exception: Additional 15- or 20-ampere circuits shall be permitted where a listed energy management system rated at 30-ampere maximum is employed within the system.

FPN: See 210.23(A) for permissible loads. See 551.45(C) for main disconnect and overcurrent protection requirements.

(D) More Than Five Circuits Without a Listed Energy Management System A 50-ampere, 120/240-volt power-supply assembly shall be used where six or more circuits are employed. The load distribution shall ensure a reasonable current balance between phases.

Experience and field data indicate that supply calculations are not necessary; in some cases, such calculations have resulted in oversized power supplies. In recreational vehicles with six or more circuits, the power supply assembly shall have a minimum rating of 50 amperes. Reasonable balancing of the electrical load between phases is required.

551.43 Branch-Circuit Protection

(A) Rating The branch-circuit overcurrent devices shall be rated as follows:

(1) Not more than the circuit conductors, and
(2) Not more than 150 percent of the rating of a single appliance rated 13.3 amperes or more and supplied by an individual branch circuit, but
(3) Not more than the overcurrent protection size marked on an air conditioner or other motor-operated appliances

(B) Protection for Smaller Conductors A 20-ampere fuse or circuit breaker shall be permitted for protection for fixtures, including luminaires, leads, cords, or small appliances, and 14 AWG tap conductors, not over 1.8 m (6 ft) long for recessed luminaires (lighting fixtures).

(C) Fifteen-Ampere Receptacle Considered Protected by 20 Amperes If more than one receptacle or load is on a branch circuit, a 15-ampere receptacle shall be permitted to be protected by a 20-ampere fuse or circuit breaker.

551.44 Power-Supply Assembly

Each recreational vehicle shall have only one of the following main power-supply assemblies.

(A) Fifteen-Ampere Main Power-Supply Assembly Recreational vehicles wired in accordance with 551.42(A) shall use a listed 15-ampere or larger main power-supply assembly.

(B) Twenty-Ampere Main Power-Supply Assembly Recreational vehicles wired in accordance with 551.42(B) shall use a listed 20-ampere or larger main power-supply assembly.

(C) Thirty-Ampere Main Power-Supply Assembly Recreational vehicles wired in accordance with 551.42(C) shall use a listed 30-ampere or larger main power-supply assembly.

(D) Fifty-Ampere Power-Supply Assembly Recreational vehicles wired in accordance with 551.42(D) shall use a listed 50-ampere, 120/240-volt main power-supply assembly.

551.45 Distribution Panelboard

(A) Listed and Appropriately Rated A listed and appropriately rated distribution panelboard or other equipment specifically listed for this purpose shall be used. The grounded conductor termination bar shall be insulated from the enclosure as provided in 551.54(C). An equipment grounding terminal bar shall be attached inside the metal enclosure of the panelboard.

(B) Location The distribution panelboard shall be installed in a readily accessible location. Working clearance for the panelboard shall be not less than 600 mm (24 in.) wide and 750 mm (30 in.) deep.

Exception No. 1: Where the panelboard cover is exposed to the inside aisle space, one of the working clearance dimensions shall be permitted to be reduced to a minimum of 550 mm (22 in.). A panelboard is considered exposed where the panelboard cover is within 50 mm (2 in.) of the aisle's finished surface.

Exception No. 2: Compartment doors used for access to a generator shall be permitted to be equipped with a locking system.

(C) Dead-Front Type The distribution panelboard shall be of the dead-front type and shall consist of one or more circuit breakers or Type S fuseholders. A main disconnecting means shall be provided where fuses are used or where more than two circuit breakers are employed. A main overcurrent protective device not exceeding the power-supply assembly rating shall be provided where more than two branch circuits are employed.

551.46 Means for Connecting to Power Supply

(A) Assembly The power-supply assembly or assemblies shall be factory supplied or factory installed and be of one of the types specified herein.

(1) Separable Where a separable power-supply assembly consisting of a cord with a female connector and molded attachment plug cap is provided, the vehicle shall be equipped with a permanently mounted, flanged surface inlet (male, recessed-type motor-base attachment plug) wired directly to the distribution panelboard by an approved wiring method. The attachment plug cap shall be of a listed type.

(2) Permanently Connected Each power-supply assembly shall be connected directly to the terminals of the distribution panelboard or conductors within a junction box and provided with means to prevent strain from being transmitted to the terminals. The ampacity of the conductors between each junction box and the terminals of each distribution panelboard shall be at least equal to the ampacity of the power-supply cord. The supply end of the assembly shall be equipped with an attachment plug of the type described in 551.46(C). Where the cord passes through the walls or floors, it shall be protected by means of conduit and bushings or equivalent. The cord assembly shall have permanent provisions for protection against corrosion and mechanical damage while the vehicle is in transit.

(B) Cord The cord exposed usable length shall be measured from the point of entrance to the recreational vehicle or the face of the flanged surface inlet (motor-base attachment plug) to the face of the attachment plug at the supply end.

The cord exposed usable length, measured to the point of entry on the vehicle exterior, shall be a minimum of 7.5 m (25 ft) where the point of entrance is at the side of the vehicle or shall be a minimum 9.0 m (30 ft) where the point of entrance is at the rear of the vehicle.

Where the cord entrance into the vehicle is more than 900 mm (3 ft) above the ground, the minimum cord lengths above shall be increased by the vertical distance of the cord entrance heights above 900 mm (3 ft).

FPN: See 551.46(E).

(C) Attachment Plugs

(1) Units with One 15-Ampere Branch Circuit Recreational vehicles having only one 15-ampere branch circuit as permitted by 551.42(A) shall have an attachment plug that shall be 2-pole, 3-wire grounding type, rated 15 amperes, 125 volts, conforming to the configuration shown in Figure 551.46(C).

FPN: Complete details of this configuration can be found in National Electrical Manufacturers Association's ANSI/NEMA WD 6-1989, *Standard for Dimensions of Attachment Plugs and Receptacle*, Figure 5.15.

(2) Units with One 20-Ampere Branch Circuit Recreational vehicles having only one 20-ampere branch circuit as permitted in 551.42(B) shall have an attachment plug that shall be 2-pole, 3-wire grounding type, rated 20 amperes, 125 volts, conforming to the configuration shown in Figure 551.46(C).

FPN: Complete details of this configuration can be found in ANSI/NEMA WD 6-1989, National Electrical Manufacturers Association's *Standard for Dimensions of Attachment Plugs and Receptacles*, Figure 5.20.

(3) Units with Two to Five 15- or 20-Ampere Branch Circuits Recreational vehicles wired in accordance with 551.42(C) shall have an attachment plug that shall be 2-pole, 3-wire grounding type, rated 30 amperes, 125 volts, conforming to the configuration shown in Figure 551.46(C), intended for use with units rated at 30 amperes, 125 volts.

Receptacles   Caps

20-A, 125-V, 2-pole, 3-wire, grounding type

125-V, 20-A, 2-pole, 3-wire, grounding type

125-V, 15-A, 2-pole, 3-wire, grounding type

30-A, 125-V, 2-pole, 3-wire, grounding type

50-A, 125/250-V, 3-pole, 4-wire, grounding type

Figure 551.46(C) Configurations for Grounding-Type Receptacles and Attachment Plug Caps Used for Recreational Vehicle Supply Cords and Recreational Vehicle Lots

FPN: Complete details of this configuration can be found in ANSI/NEMA WD 6-1989, National Electrical Manufacturers Association's *Standard for Dimensions of Attachment Plugs and Receptacles*, Figure TT.

(4) Units with 50-Ampere Power Supply Assembly Recreational vehicles having a power-supply assembly rated 50 amperes as permitted by 551.42(D) shall have a 3-pole, 4-wire grounding-type attachment plug rated 50 amperes, 125/250 volts, conforming to the configuration shown in Figure 551.46(C).

FPN: Complete details of this configuration can be found in ANSI/NEMA WD 6-1989, National Electrical Manufacturers Association's *Standard for Dimensions of Attachment Plugs and Receptacles*, Figure 14.50.

The 20- and 50-ampere receptacle and plug configurations used for recreational vehicles appear in Exhibit 406.3 (NEMA configuration chart); however, the 30-ampere plug and receptacle configurations are unique to recreational vehicles and are described in UL 498, *Standard for Attachment Plugs and Receptacles.*

(D) Labeling at Electrical Entrance Each recreational vehicle shall have permanently affixed to the exterior skin, at or near the point of entrance of the power-supply cord(s), a label 75 mm × 45 mm (3 in. × 1¾ in.) minimum size, made of etched, metal-stamped, or embossed brass, stainless steel, or anodized or alclad aluminum not less than 0.51 mm (0.020 in.) thick, or other suitable material [e.g., 0.13 mm

(0.005 in.) thick plastic laminate] that reads, as appropriate, either

THIS CONNECTION IS FOR 110–125-VOLT AC,
60 HZ, _____ AMPERE SUPPLY.

or

THIS CONNECTION IS FOR 120/240-VOLT AC, 3-POLE,
4-WIRE, 60 HZ, _____ AMPERE SUPPLY.

The correct ampere rating shall be marked in the blank space.

(E) Location The point of entrance of a power-supply assembly shall be located within 4.5 m (15 ft) of the rear, on the left (road) side or at the rear, left of the longitudinal center of the vehicle, within 450 mm (18 in.) of the outside wall.

Exception No. 1: A recreational vehicle equipped with only a listed flexible drain system or a side-vent drain system shall be permitted to have the electrical point of entrance located on either side, provided the drain(s) for the plumbing system is (are) located on the same side.

Exception No. 2: A recreational vehicle shall be permitted to have the electrical point of entrance located more than 4.5 m (15 ft) from the rear. Where this occurs, the distance beyond the 4.5-m (15-ft) dimension shall be added to the cord's minimum length as specified in 551.46(B).

Exception No. 3: Recreational vehicles designed for transporting livestock shall be permitted to have the electrical point of entrance located on either side or the front.

Exception No. 3 to 551.46(E), which is new in the 2005 *Code*, gives permission to locate the power supply entrance on either side or the front of a recreational vehicle that also has provisions for transporting livestock.

551.47 Wiring Methods

(A) Wiring Systems Cables and raceways installed in accordance with Articles 320, 322, 330 through 340, 342 through 362, 386, and 388 shall be permitted in accordance with their applicable article, except as otherwise specified in this article. An equipment grounding means shall be provided in accordance with 250.118.

See the commentary on 348.60 for information regarding the use of flexible metal conduit as an equipment grounding conductor.

(B) Conduit and Tubing Where rigid metal conduit or intermediate metal conduit is terminated at an enclosure with a locknut and bushing connection, two locknuts shall be provided, one inside and one outside of the enclosure. All cut ends of conduit and tubing shall be reamed or otherwise finished to remove rough edges.

See the commentary following 358.28(A) and 300.4(F) for information on protecting conductor insulation against abrasion at conduit and tubing terminations.

(C) Nonmetallic Boxes Nonmetallic boxes shall be acceptable only with nonmetallic-sheathed cable or nonmetallic raceways.

(D) Boxes In walls and ceilings constructed of wood or other combustible material, boxes and fittings shall be flush with the finished surface or project therefrom.

(E) Mounting Wall and ceiling boxes shall be mounted in accordance with Article 314.

Exception No. 1: Snap-in-type boxes or boxes provided with special wall or ceiling brackets that securely fasten boxes in walls or ceilings shall be permitted.

Exception No. 2: A wooden plate providing a 38-mm (1½-in.) minimum width backing around the box and of a thickness of 13 mm (½ in.) or greater (actual) attached directly to the wall panel shall be considered as approved means for mounting outlet boxes.

Exception No. 2 to 551.47(E) permits the mounting of outlet boxes by screws to a wooden plate that is secured directly to the back of the wall panel. The wooden plate must be not less than ½ in. thick and must extend at least 1½ in. around the box. This requirement recognizes the special construction of recreational vehicle walls, which often makes it difficult or impossible to attach an outlet box to a structural member, as required by 314.23(B).

(F) Raceway and Cable Continuity Raceways and cable sheaths shall be continuous between boxes and other enclosures.

(G) Protected Metal-clad, Type AC, or nonmetallic-sheathed cables and electrical nonmetallic tubing shall be permitted to pass through the centers of the wide side of 2 by 4 wood studs. However, they shall be protected where they pass through 2 by 2 wood studs or at other wood studs or frames where the cable or tubing would be less than 32 mm (1¼ in.) from the inside or outside surface. Steel plates on each side of the cable or tubing or a steel tube, with not less than 1.35 mm (0.053 in.) wall thickness, shall be installed to protect the cable or tubing. These plates or tubes shall be securely held in place. Where nonmetallic-sheathed cables pass through punched, cut, or drilled slots or holes in metal members, the cable shall be protected by bushings or grommets securely fastened in the opening prior to installation of the cable.

(H) Bends No bend shall have a radius of less than five times the cable diameter.

(I) Cable Supports Where connected with cable connectors or clamps, cables shall be supported within 300 mm (12 in.) of outlet boxes, distribution panelboards, and splice boxes on appliances. Supports shall be provided every 1.4 m (4½ ft) at other places.

(J) Nonmetallic Box Without Cable Clamps Nonmetallic-sheathed cables shall be supported within 200 mm (8 in.) of a nonmetallic outlet box without cable clamps. Where wiring devices with integral enclosures are employed with a loop of extra cable to permit future replacement of the device, the cable loop shall be considered as an integral portion of the device.

(K) Physical Damage Where subject to physical damage, exposed nonmetallic cable shall be protected by covering boards, guard strips, raceways, or other means.

(L) Metal Faceplates Metal faceplates shall be of ferrous metal not less than 0.76 mm (0.030 in.) in thickness or of nonferrous metal not less than 1.0 mm (0.040 in.) in thickness. Nonmetallic faceplates shall be listed.

(M) Metal Faceplates Effectively Grounded Where metal faceplates are used, they shall be effectively grounded.

(N) Moisture or Physical Damage Where outdoor or under-chassis wiring is 120 volts, nominal, or over and is exposed to moisture or physical damage, the wiring shall be protected by rigid metal conduit, by intermediate metal conduit, or by electrical metallic tubing, rigid nonmetallic conduit, or Type MI cable, that is closely routed against frames and equipment enclosures or other raceway or cable identified for the application.

(O) Component Interconnections Fittings and connectors that are intended to be concealed at the time of assembly shall be listed and identified for the interconnection of building components. Such fittings and connectors shall be equal to the wiring method employed in insulation, temperature rise, and fault-current withstanding and shall be capable of enduring the vibration and shock occurring in recreational vehicles.

(P) Method of Connecting Expandable Units The method of connecting expandable units to the main body of the vehicle shall comply with 551.47(P)(1) or (P)(2):

(1) Cord-and-Plug Connected. Cord and plug connections shall comply with (a) through (d).

   a. That portion of a branch circuit that is installed in an expandable unit shall be permitted to be connected to the portion of the branch circuit in the main body

of the vehicle by means of an attachment plug and cord listed for hard usage. The cord and its connections shall conform to all provisions of Article 400 and shall be considered as a permitted use under 400.7. Where the attachment plug and cord are located within the vehicle's interior, use of plastic thermoset or elastomer parallel cord Type SPT-3, SP-3, or SPE shall be permitted.

   b. Where the receptacle provided for connection of the cord to the main circuit is located on the outside of the vehicle, it shall be protected with a ground-fault circuit interrupter for personnel and be listed for wet locations. A cord located on the outside of a vehicle shall be identified for outdoor use.

   c. Unless removable or stored within the vehicle interior, the cord assembly shall have permanent provisions for protection against corrosion and mechanical damage while the vehicle is in transit.

   d. The attachment plug and cord shall be installed so as not to permit exposed live attachment plug pins.

(2) Direct Wired. That portion of a branch circuit that is installed in an expandable unit shall be permitted to be connected to the portion of the branch circuit in the main body of the vehicle by means of flexible cord in accordance with 551.47(P)(2)(a) through (P)(2)(d).

Section 551.47(P) was revised for the 2005 *Code* to expand the provisions covering the interconnection between the main body of the vehicle and an expandable unit. Two methods of interconnection are permitted: one by means of cord-and-plug connections with the cord listed for hard usage, the other by means of a direct-wired connection using flexible cord with the outer jacket intact, installed in nonflexible conduit or tubing.

   a. The flexible cord shall be listed for hard usage and for use in wet locations.

   b. The flexible cord shall be permitted to pass through the interior of a wall or through a floor in lengths not to exceed 600 mm (24 in.) before terminating at an outlet.

   c. The flexible cord shall be installed in a nonflexible conduit or tubing that runs continuously from the outlet box inside the recreational vehicle to a strain relief connector listed for use in wet locations that is located on the underside of the recreational vehicle.

   d. The outer jacket of the flexible cord shall not be removed for that portion that is installed in the conduit or tubing.

(Q) Prewiring for Air-Conditioning Installation Prewiring installed for the purpose of facilitating future air-conditioning installation shall conform to the applicable portions of this article and the following:

(1) An overcurrent protective device with a rating compatible with the circuit conductors shall be installed in the distribution panelboard and wiring connections completed.

(2) The load end of the circuit shall terminate in a junction box with a blank cover or other listed enclosure. Where a junction box with a blank cover is used, the free ends of the conductors shall be adequately capped or taped.

(3) A label conforming to 551.46(D) shall be placed on or adjacent to the junction box and shall read

> AIR-CONDITIONING CIRCUIT.
> THIS CONNECTION IS FOR AIR CONDITIONERS
> RATED 110–125-VOLT AC, 60 HZ,
> _____ AMPERES MAXIMUM. DO NOT EXCEED
> CIRCUIT RATING.

An ampere rating, not to exceed 80 percent of the circuit rating, shall be legibly marked in the blank space.

(4) The circuit shall serve no other purpose.

(R) Prewiring for Generator Installation Prewiring installed for the purpose of facilitating future generator installation shall conform to the other applicable portions of this article and the following:

(1) Circuit conductors shall be appropriately sized in relation to the anticipated load and shall be protected by an overcurrent device in accordance with their ampacities.

(2) Where junction boxes are utilized at either of the circuit originating or terminus points, free ends of the conductors shall be adequately capped or taped.

(3) Where devices such as receptacle outlet, transfer switch, and so forth, are installed, the installation shall be complete, including circuit conductor connections. All devices shall be listed and appropriately rated.

(4) A label conforming to 551.46(D) shall be placed on the cover of each junction box containing incomplete circuitry and shall read, as appropriate, either

> GENERATOR CIRCUIT. THIS CONNECTION
> IS FOR GENERATORS RATED 110–125-VOLT AC,
> 60 HZ, _____ AMPERES MAXIMUM.

or

> GENERATOR CIRCUIT. THIS CONNECTION
> IS FOR GENERATORS RATED 120/240-VOLT AC,
> 60 HZ, _____ AMPERES MAXIMUM.

The correct ampere rating shall be legibly marked in the blank space.

551.48 Conductors and Boxes

The maximum number of conductors permitted in boxes shall be in accordance with 314.16.

551.49 Grounded Conductors

The identification of grounded conductors shall be in accordance with 200.6.

551.50 Connection of Terminals and Splices

Conductor splices and connections at terminals shall be in accordance with 110.14.

551.51 Switches

(A) Rating Switches shall be rated in accordance with 551.51(A)(1) and (A)(2).

(1) Lighting Circuits For lighting circuits, switches shall be rated not less than 10 amperes, 120–125 volts and in no case less than the connected load.

(2) Motors or Other Loads Switches for motor or other loads shall comply with the provisions of 404.14.

(B) Location Switches shall not be installed within wet locations in tub or shower spaces unless installed as part of a listed tub or shower assembly.

551.52 Receptacles

All receptacle outlets shall be of the grounding type and installed in accordance with 406.3 and 210.21.

551.53 Luminaires (Lighting Fixtures)

(A) General Any combustible wall or ceiling finish exposed between the edge of a luminaire (fixture) canopy, or pan and the outlet box, shall be covered with noncombustible material or a material identified for the purpose.

(B) Shower Luminaires (Fixtures) If a luminaire (lighting fixture) is provided over a bathtub or in a shower stall, it shall be of the enclosed and gasketed type and listed for the type of installation, and it shall be ground-fault circuit-interrupter protected.

Many shower luminaires have a metal base and, due to the low ceilings in recreational vehicles, may be easily reached by most persons under the shower or standing in the bathtub. In accordance with 551.53(B), the installation of luminaires that are listed for wet locations and have GFCI protection is permitted above a tub or shower enclosure.

(C) Outdoor Outlets, Luminaires (Fixtures), Air-Cooling Equipment, and So On Outdoor luminaires (fixtures) and other equipment shall be listed for outdoor use.

551.54 Grounding

(See also 551.56 on bonding of non–current-carrying metal parts.)

(A) Power-Supply Grounding The grounding conductor in the supply cord or feeder shall be connected to the grounding bus or other approved grounding means in the distribution panelboard.

(B) Distribution Panelboard The distribution panelboard shall have a grounding bus with sufficient terminals for all grounding conductors or other approved grounding means.

(C) Insulated Neutral The grounded circuit conductor (neutral) shall be insulated from the equipment grounding conductors and from equipment enclosures and other grounded parts. The grounded (neutral) circuit terminals in the distribution panelboard and in ranges, clothes dryers, counter-mounted cooking units, and wall-mounted ovens shall be insulated from the equipment enclosure. Bonding screws, straps, or buses in the distribution panelboard or in appliances shall be removed and discarded. Connection of electric ranges and electric clothes dryers utilizing a grounded (neutral) conductor, if cord-connected, shall be made with 4-conductor cord and 3-pole, 4-wire grounding-type plug caps and receptacles.

551.55 Interior Equipment Grounding

(A) Exposed Metal Parts In the electrical system, all exposed metal parts, enclosures, frames, luminaire (lighting fixture) canopies, and so forth, shall be effectively bonded to the grounding terminals or enclosure of the distribution panelboard.

(B) Equipment Grounding and Bonding Conductors Bare wires, insulated wire with an outer finish that is green or green with one or more yellow stripes, shall be used for equipment grounding or bonding conductors only.

(C) Grounding of Electrical Equipment Grounding of electrical equipment shall be accomplished by one or more of the following methods:

(1) Connection of metal raceway, the sheath of Type MC and Type MI cable where the sheath is identified for grounding, or the armor of Type AC cable to metal enclosures.
(2) A connection between the one or more equipment grounding conductors and a metal enclosure by means of a grounding screw, which shall be used for no other purpose, or a listed grounding device.
(3) The equipment grounding conductor in nonmetallic-sheathed cable shall be permitted to be secured under a screw threaded into the luminaire (fixture) canopy other than a mounting screw or cover screw, or attached to a listed grounding means (plate) in a nonmetallic outlet box for luminaire (fixture) mounting. [Grounding means shall also be permitted for luminaire (fixture) attachment screws.]

(D) Grounding Connection in Nonmetallic Box A connection between the one or more grounding conductors brought into a nonmetallic outlet box shall be so arranged that a connection can be made to any fitting or device in that box that requires grounding.

(E) Grounding Continuity Where more than one equipment grounding or bonding conductor of a branch circuit enters a box, all such conductors shall be in good electrical contact with each other, and the arrangement shall be such that the disconnection or removal of a receptacle, luminaire (fixture), or other device fed from the box will not interfere with or interrupt the grounding continuity.

(F) Cord-Connected Appliances Cord-connected appliances, such as washing machines, clothes dryers, refrigerators, and the electrical system of gas ranges, and so forth, shall be grounded by means of an approved cord with equipment grounding conductor and grounding-type attachment plug.

551.56 Bonding of Non–Current-Carrying Metal Parts

(A) Required Bonding All exposed non–current-carrying metal parts that may become energized shall be effectively bonded to the grounding terminal or enclosure of the distribution panelboard.

(B) Bonding Chassis A bonding conductor shall be connected between any distribution panelboard and an accessible terminal on the chassis. Aluminum or copper-clad aluminum conductors shall not be used for bonding if such conductors or their terminals are exposed to corrosive elements.

Exception: Any recreational vehicle that employs a unitized metal chassis-frame construction to which the distribution panelboard is securely fastened with a bolt(s) and nut(s) or by welding or riveting shall be considered to be bonded.

(C) Bonding Conductor Requirements Grounding terminals shall be of the solderless type and listed as pressure terminal connectors recognized for the wire size used. The bonding conductor shall be solid or stranded, insulated or bare, and shall be 8 AWG copper minimum, or equal.

(D) Metallic Roof and Exterior Bonding The metal roof and exterior covering shall be considered bonded where both of the following conditions apply:

(1) The metal panels overlap one another and are securely attached to the wood or metal frame parts by metal fasteners.
(2) The lower panel of the metal exterior covering is secured by metal fasteners at each cross member of the chassis, or the lower panel is bonded to the chassis by a metal strap.

(E) Gas, Water, and Waste Pipe Bonding The gas, water, and waste pipes shall be considered grounded if they are bonded to the chassis.

(F) Furnace and Metal Air Duct Bonding Furnace and metal circulating air ducts shall be bonded.

551.57 Appliance Accessibility and Fastening

Every appliance shall be accessible for inspection, service, repair, and replacement without removal of permanent construction. Means shall be provided to securely fasten appliances in place when the recreational vehicle is in transit.

V. Factory Tests

551.60 Factory Tests (Electrical)

Each recreational vehicle designed with a 120-volt or a 120/240-volt electrical system shall withstand the applied potential without electrical breakdown of a 1-minute, 900-volt dielectric strength test, or a 1-second, 1080-volt dielectric strength test, with all switches closed, between ungrounded and grounded conductors and the recreational vehicle ground. During the test, all switches and other controls shall be in the "on" position. Fixtures, including luminaires and permanently installed appliances, shall not be required to withstand this test. The test shall be performed after branch circuits are complete prior to energizing the system and again after all outer coverings and cabinetry have been secured.

Each recreational vehicle shall be subjected to all of the following:

(1) A continuity test to ensure that all metal parts are properly bonded
(2) Operational tests to demonstrate that all equipment is properly connected and in working order
(3) Polarity checks to determine that connections have been properly made

VI. Recreational Vehicle Parks

551.71 Type Receptacles Provided

Every recreational vehicle site with electrical supply shall be equipped with at least one 20-ampere, 125-volt receptacle. A minimum of 20 percent of all recreational vehicle sites, with electrical supply, shall each be equipped with a 50-ampere, 125/250-volt receptacle conforming to the configuration as identified in Figure 551.46(C). These electrical supplies shall be permitted to include additional receptacles that have configurations in accordance with 551.81. A minimum of 70 percent of all recreational vehicle sites with electrical supply shall each be equipped with a 30-ampere, 125-volt receptacle conforming to Figure 551.46(C). This supply shall be permitted to include additional receptacle configurations conforming to 551.81. The remainder of all recreational vehicle sites with electrical supply shall be equipped with one or more of the receptacle configurations conforming to 551.81. Dedicated tent sites with a 15- or 20-ampere electrical supply shall be permitted to be excluded when determining the percentage of recreational vehicle sites with 30- or 50-ampere receptacles.

Additional receptacles shall be permitted for the connec-

tion of electrical equipment outside the recreational vehicle within the recreational vehicle park.

In accordance with 551.71, at least one 20-ampere, 125-volt receptacle must be installed at each recreational vehicle campsite. Existing recreational vehicle campgrounds may have some sites that are equipped with 30-ampere receptacles only. Adapter plugs or "cheater" cords are often used to connect a recreational vehicle with a 20-ampere supply cord to a 30-ampere receptacle outlet. This practice does not provide adequate overload protection for the cord or the connected load. The requirement in 551.71 ensures the availability of the properly rated receptacle.

Some newer recreational vehicles have a 50-ampere, 120/240-volt supply installed, and 551.71 contains provisions that require a limited number of RV sites with 50-ampere, 125/250-volt receptacles. Section 551.71 was revised for the 2005 *Code* to increase the number of required 50-ampere receptacles from 5 percent to 20 percent to accommodate the larger electrical systems that are installed in today's recreational vehicles. This revision will also result in increased load capacity for RV park services and feeders. Receptacle configurations are shown in Figure 551.46(C), and receptacle ratings are described in 551.81.

All 125-volt, single-phase, 15- and 20-ampere receptacles shall have listed ground-fault circuit-interrupter protection for personnel.

> FPN: The percentage of 50 ampere sites required by 551.71 may be inadequate for seasonal recreational vehicle sites serving a higher percentage of recreational vehicles with 50 ampere electrical systems. In that type of recreational vehicle park, the percentage of 50 ampere sites could approach 100 percent.

551.72 Distribution System

Receptacles rated at 50 amperes shall be supplied from a branch circuit of the voltage class and rating of the receptacle. Other recreational vehicle sites with 125-volt, 20- and 30-ampere receptacles shall be permitted to be derived from any grounded distribution system that supplies 120-volt single-phase power. The neutral conductors shall not be reduced in size below the size of the ungrounded conductors for the site distribution. The neutral conductors shall be permitted to be reduced in size below the minimum required size of the ungrounded conductors for 240-volt, line-to-line, permanently connected loads only.

The distribution system of a recreational vehicle park has to be configured such that the sites are supplied by a 120-volt, 2-wire circuit; a 120/240 volt, 3-wire circuit; or a 3-wire 120/208-volt circuit derived from a 208Y/120-volt, 3-

phase, 4-wire system. See the commentary on 551.40(A) for information regarding the voltage rating of line-to-line connected appliances in recreational vehicles.

551.73 Calculated Load

(A) **Basis of Calculations** Electrical service and feeders shall be calculated on the basis of not less than 9600 volt-amperes per site equipped with 50-ampere, 120/240-volt supply facilities; 3600 volt-amperes per site equipped with both 20-ampere and 30-ampere supply facilities; 2400 volt-amperes per site equipped with only 20-ampere supply facilities; and 600 volt-amperes per site equipped with only 20-ampere supply facilities that are dedicated to tent sites. The demand factors set forth in Table 551.73 shall be the minimum allowable demand factors that shall be permitted in calculating load for service and feeders. Where the electrical supply for a recreational vehicle site has more than one receptacle, the calculated load shall only be calculated for the highest rated receptacle.

A second paragraph was added to 551.73(A) in the 2005 *Code* to require that, where receptacle outlets are installed at a location that can serve two recreational vehicles, the load is to be calculated for the two receptacles with the highest rating.

Where the electrical supply is in a location that serves two recreational vehicles, the equipment for both sites must

Table 551.73 Demand Factors for Site Feeders and Service-Entrance Conductors for Park Sites

Number of Recreational Vehicle Sites	Demand Factor (percent)
1	100
2	90
3	80
4	75
5	65
6	60
7–9	55
10–12	50
13–15	48
16–18	47
19–21	45
22–24	43
25–35	42
36 plus	41

Dedicated tent sites supplied with electricity are not intended to accommodate recreational vehicles; therefore, the calculated load for these sites can be smaller.

comply with 551.77 and the calculated load shall only be computed for the two receptacles with the highest rating.

(B) **Transformers and Secondary Distribution Panelboards** For the purpose of this *Code*, where the park service exceeds 240 volts, transformers and secondary distribution panelboards shall be treated as services.

(C) **Demand Factors** The demand factor for a given number of sites shall apply to all sites indicated. For example, 20 sites calculated at 45 percent of 3600 volt-amperes results in a permissible demand of 1620 volt-amperes per site or a total of 32,400 volt-amperes for 20 sites.

> FPN: These demand factors may be inadequate in areas of extreme hot or cold temperature with loaded circuits for heating or air conditioning.

(D) **Feeder-Circuit Capacity** Recreational vehicle site feeder-circuit conductors shall have adequate ampacity for the loads supplied and shall be rated at not less than 30 amperes. The grounded conductors shall have the same ampacity as the ungrounded conductors.

> FPN: Due to the long circuit lengths typical in most recreational vehicle parks, feeder conductor sizes found in the ampacity tables of Article 310 may be inadequate to maintain the voltage regulation suggested in the fine print note to 210.19. Total circuit voltage drop is a sum of the voltage drops of each serial circuit segment, where the load for each segment is calculated using the load that segment sees and the demand factors of 551.73(A).

Loads for other amenities such as, but not limited to, service buildings, recreational buildings, and swimming pools shall be sized separately and then be added to the value calculated for the recreational vehicle sites where they are all supplied by one service.

551.74 Overcurrent Protection

Overcurrent protection shall be provided in accordance with Article 240.

551.75 Grounding

All electrical equipment and installations in recreational vehicle parks shall be grounded as required by Article 250.

551.76 Grounding — Recreational Vehicle Site Supply Equipment

(A) **Exposed Non–Current-Carrying Metal Parts** Exposed non–current-carrying metal parts of fixed equipment, metal boxes, cabinets, and fittings that are not electrically connected to grounded equipment shall be grounded by a continuous equipment grounding conductor run with the circuit conductors from the service equipment or from the transformer of a secondary distribution system. Equipment grounding conductors shall be sized in accordance with 250.122 and shall be permitted to be spliced by listed means.

The arrangement of equipment grounding connections shall be such that the disconnection or removal of a receptacle or other device will not interfere with, or interrupt, the grounding continuity.

(B) Secondary Distribution System Each secondary distribution system shall be grounded at the transformer.

(C) Neutral Conductor Not to Be Used as an Equipment Ground The neutral conductor shall not be used as an equipment ground for recreational vehicles or equipment within the recreational vehicle park.

(D) No Connection on the Load Side No connection to a grounding electrode shall be made to the neutral conductor on the load side of the service disconnecting means except as covered in 250.30(A) for separately derived systems and 250.32(B)(2) for separate buildings.

551.77 Recreational Vehicle Site Supply Equipment

(A) Location Where provided on back-in sites, the recreational vehicle site electrical supply equipment shall be located on the left (road) side of the parked vehicle, on a line that is 1.5 m to 2.1 m (5 ft to 7 ft) from the left edge (driver's side of the parked RV) of the stand and shall be located at any point on this line from the rear of the stand to 4.5 m (15 ft) forward of the rear of the stand.

For pull-through sites, the electrical supply equipment shall be permitted to be located at any point along the line that is 1.5 m to 2.1 m (5 ft to 7 ft) from the left edge (driver's side of the parked RV) from 4.9 m (16 ft) forward of the rear of the stand to the center point between the two roads that gives access to and egress from the pull-through sites.

The left edge (driver's side of the parked RV) of the stand shall be marked.

The requirements of 551.77(A) are intended to accommodate vehicles towing boats or other trailers. The location of the site supply equipment permitted for pull-through sites reduces the use of extension cords.

(B) Disconnecting Means A disconnecting switch or circuit breaker shall be provided in the site supply equipment for disconnecting the power supply to the recreational vehicle.

(C) Access All site supply equipment shall be accessible by an unobstructed entrance or passageway not less than 600 mm (2 ft) wide and 2.0 m (6 ft 6 in.) high.

(D) Mounting Height Site supply equipment shall be located not less than 600 mm (2 ft) or more than 2.0 m (6 ft 6 in.) above the ground.

(E) Working Space Sufficient space shall be provided and maintained about all electrical equipment to permit ready and safe operation, in accordance with 110.26.

(F) Marking Where the site supply equipment contains a 125/250-volt receptacle, the equipment shall be marked as follows: "Turn disconnecting switch or circuit breaker off before inserting or removing plug. Plug must be fully inserted or removed." The marking shall be located on the equipment adjacent to the receptacle outlet.

Partially engaged attachment plugs may result in intermittent neutral (grounded conductor) contact. Loss of the neutral could momentarily apply the line-to-line voltage (240 volts) across 125-volt equipment, causing malfunction or damage.

551.78 Protection of Outdoor Equipment

(A) Wet Locations All switches, circuit breakers, receptacles, control equipment, and metering devices located in wet locations or outside of a building shall be rainproof equipment.

(B) Meters If secondary meters are installed, meter sockets without meters installed shall be blanked off with an approved blanking plate.

551.79 Clearance for Overhead Conductors

Open conductors of not over 600 volts, nominal, shall have a vertical clearance of not less than 5.5 m (18 ft) and a horizontal clearance of not less than 900 mm (3 ft) in all areas subject to recreational vehicle movement. In all other areas, clearances shall conform to 225.18 and 225.19.

FPN: For clearances of conductors over 600 volts, nominal, see 225.60 and 225.61.

551.80 Underground Service, Feeder, Branch-Circuit, and Recreational Vehicle Site Feeder-Circuit Conductors

(A) General All direct-burial conductors, including the equipment grounding conductor if of aluminum, shall be insulated and identified for the use. All conductors shall be continuous from equipment to equipment. All splices and taps shall be made in approved junction boxes or by use of material listed and identified for the purpose.

(B) Protection Against Physical Damage Direct-buried conductors and cables entering or leaving a trench shall be protected by rigid metal conduit, intermediate metal conduit, electrical metallic tubing with supplementary corrosion protection, rigid nonmetallic conduit, liquidtight flexible non-metallic conduit, liquidtight flexible metal conduit, or other approved raceways or enclosures. Where subject to physical damage, the conductors or cables shall be protected by rigid metal conduit, intermediate metal conduit, or Schedule 80 rigid nonmetallic conduit. All such protection shall extend at least 450 mm (18 in.) into the trench from finished grade.

> FPN: See 300.5 and Article 340 for conductors or Type UF cable used underground or in direct burial in earth.

551.81 Receptacles

A receptacle to supply electric power to a recreational vehicle shall be one of the configurations shown in Figure 551.46(C) in the following ratings.

(1) 50-ampere — 125/250-volt, 50-ampere, 3-pole, 4-wire grounding type for 120/240-volt systems
(2) 30-ampere — 125-volt, 30-ampere, 2-pole, 3-wire grounding type for 120-volt systems
(3) 20-ampere — 125-volt, 20-ampere, 2-pole, 3-wire grounding type for 120-volt systems

> FPN: Complete details of these configurations can be found in ANSI/NEMA WD 6-1989, National Electrical Manufacturers Association's *Standard for Dimensions of Attachment Plugs and Receptacles*, Figures 14-50, TT, and 5-20.

ARTICLE 552
Park Trailers

Summary of Changes

- **552.41(C)(1):** Deleted redundant requirement for GFCI protection of receptacles adjacent to bathrooms and added requirement for GFCI protection of receptacles installed to serve kitchen countertop surfaces.

- **552.45(A):** Deleted reference to "other equipment specifically listed for the purpose."

Contents

I. General

552.1 Scope

The provisions of this article cover the electrical conductors and equipment installed within or on park trailers not covered fully under Articles 550 and 551.

The scope of Article 552 covers park trailers that have a single chassis and wheels, that do not exceed 400 ft² (set up), and that are not used as permanent residences. Additionally, Article 552 does not apply to units that meet the definition of *park trailer* (see 552.2) but are used for commercial purposes.

It is not uncommon for park trailers to be equipped with electrical loads similar to those used in mobile homes. It is also not uncommon for the park trailer to be located in the same park trailer community for several years without relocation.

Park trailers are somewhat similar to mobile homes and recreational vehicles, and many requirements in Article 552 are the same or similar to those contained in Articles 550 and 551. Article 552, therefore, is similar in structure to Articles 550 and 551.

552.2 Definition

(See Articles 100, 550, and 551 for additional definitions.)

Park Trailer. A unit that is built on a single chassis mounted on wheels and has a gross trailer area not exceeding 37 m² (400 ft²) in the set-up mode.

552.3 Other Articles

Wherever the provisions of other articles and Article 552 differ, the provisions of Article 552 shall apply.

552.4 General Requirements

A park trailer as specified in 552.2 is intended for seasonal use. It is not intended as a permanent dwelling unit or for commercial uses such as banks, clinics, offices, or similar.

II. Low-Voltage Systems

552.10 Low-Voltage Systems

(A) Low-Voltage Circuits Low-voltage circuits furnished and installed by the park trailer manufacturer, other than those related to braking, are subject to this *Code*. Circuits supplying lights subject to federal or state regulations shall comply with applicable government regulations and this *Code*.

In accordance with 552.10(A), the requirements of Article 552, Part II apply to the low-voltage wiring within the park trailer that would be used in place of 120-volt ac supplies. These requirements do not apply to the trailer braking circuits.

(B) Low-Voltage Wiring

(1) Material Copper conductors shall be used for low-voltage circuits.

Exception: A metal chassis or frame shall be permitted as the return path to the source of supply.

It is not the intent of 552.10(B)(1) to permit the sidewalls or the roof of a park trailer to serve as the ground return path. See the definition of *frame* in 551.2.

(2) Conductor Types Conductors shall conform to the requirements for Type GXL, HDT, SGT, SGR, or Type SXL or shall have insulation in accordance with Table 310.13 or the equivalent. Conductor sizes 6 AWG through 18 AWG or SAE shall be listed. Single-wire, low-voltage conductors shall be of the stranded type.

> FPN: See SAE Standard J1128-1995 for Types GXL, HDT, and SXL and SAE Standard J1127-1995 for Types SGT and SGR.

(3) Marking All insulated low-voltage conductors shall be surface marked at intervals not greater than 1.2 m (4 ft) as follows:

(1) Listed conductors shall be marked as required by the listing agency.
(2) SAE conductors shall be marked with the name or logo of the manufacturer, specification designation, and wire gauge.
(3) Other conductors shall be marked with the name or logo of the manufacturer, temperature rating, wire gauge, conductor material, and insulation thickness.

(C) Low-Voltage Wiring Methods

(1) Physical Protection Conductors shall be protected against physical damage and shall be secured. Where insulated conductors are clamped to the structure, the conductor insulation shall be supplemented by an additional wrap or layer of equivalent material, except that jacketed cables shall not be required to be so protected. Wiring shall be routed away from sharp edges, moving parts, or heat sources.

(2) Splices Conductors shall be spliced or joined with splicing devices that provide a secure connection or by brazing, welding, or soldering with a fusible metal or alloy. Soldered splices shall first be spliced or joined to be mechanically and electrically secure without solder, and then soldered. All splices, joints, and free ends of conductors shall be

covered with an insulation equivalent to that on the conductors.

(3) Separation Battery and other low-voltage circuits shall be physically separated by at least a 13-mm (½-in.) gap or other approved means from circuits of a different power source. Acceptable methods shall be by clamping, routing, or equivalent means that ensure permanent total separation. Where circuits of different power sources cross, the external jacket of the nonmetallic-sheathed cables shall be deemed adequate separation.

(4) Ground Connections Ground connections to the chassis or frame shall be made in an accessible location and shall be mechanically secure. Ground connections shall be by means of copper conductors and copper or copper-alloy terminals of the solderless type identified for the size of wire used. The surface on which ground terminals make contact shall be cleaned and be free from oxide or paint or shall be electrically connected through the use of a cadmium, tin, or zinc-plated internal/external-toothed lockwasher or locking terminals. Ground terminal attaching screws, rivets or bolts, nuts, and lockwashers shall be cadmium, tin, or zinc-plated except rivets shall be permitted to be unanodized aluminum where attaching to aluminum structures.

The chassis-grounding terminal of the battery shall be bonded to the unit chassis with a minimum 8 AWG copper conductor. In the event the power lead from the battery exceeds 8 AWG, the bonding conductor shall be of an equal size.

The provisions of 552.10(C)(4) require that the chassis-grounding terminal of the battery be bonded to the vehicle chassis in a mechanically secure manner and be placed in an accessible location using a minimum 8 AWG copper conductor. This minimizes the possibility of low-voltage circuit-fault currents passing through the ac panelboard bonding conductor and the equipment grounding conductor of the combination ac/dc appliance and subsequently passing through the negative dc conductor feeding the appliance that also may be bonded to the external metal cover of the appliance. The ac equipment grounding conductor of the appliance may not have sufficient ampacity to safely conduct the dc fault current, which would necessitate installation of the battery bonding conductor. Some recreational vehicles already have one side of the battery circuit bonded to the frame by an 8 AWG or larger copper conductor.

(D) Battery Installations Storage batteries subject to the provisions of this *Code* shall be securely attached to the unit and installed in an area vaportight to the interior and ventilated directly to the exterior of the unit. Where batteries are installed in a compartment, the compartment shall be ventilated with openings having a minimum area of 1100 mm² (1.7 in.²) at both the top and at the bottom. Where compartment doors are equipped for ventilation, the openings shall be within 50 mm (2 in.) of the top and bottom. Batteries shall not be installed in a compartment containing spark- or flame-producing equipment.

(E) Overcurrent Protection

(1) Rating Low-voltage circuit wiring shall be protected by overcurrent protective devices rated not in excess of the ampacity of copper conductors, in accordance with Table 552.10(E)(1).

Table 552.10(E)(1) Low-Voltage Overcurrent Protection

Wire Size (AWG)	Ampacity	Wire Type
18	6	Stranded only
16	8	Stranded only
14	15	Stranded or solid
12	20	Stranded or solid
10	30	Stranded or solid

(2) Type Circuit breakers or fuses shall be of an approved type, including automotive types. Fuseholders shall be clearly marked with maximum fuse size and shall be protected against shorting and physical damage by a cover or equivalent means.

> FPN: For further information, see ANSI/SAE J554-1987, *Standard for Electric Fuses (Cartridge Type)*; SAE J1284-1988, *Standard for Blade Type Electric Fuses*; and UL 275-1993, *Standard for Automotive Glass Tube Fuses*.

The requirement for protection of fuseholders by a cover or equivalent means is intended to reduce the possibility of the low-voltage system shorting to ground.

(3) Appliances Appliances such as pumps, compressors, heater blowers, and similar motor-driven appliances shall be installed in accordance with the manufacturer's instructions.

Motors that are controlled by automatic switching or by latching-type manual switches shall be protected in accordance with 430.32(B).

(4) Location The overcurrent protective device shall be installed in an accessible location on the unit within 450 mm (18 in.) of the point where the power supply connects to the unit circuits. If located outside the park trailer, the device shall be protected against weather and physical damage.

Exception: External low-voltage supply shall be permitted to have the overcurrent protective device within 450 mm (18 in.) after entering the unit or after leaving a metal raceway.

(F) Switches Switches shall have a dc rating not less than the connected load.

(G) Luminaires (Lighting Fixtures) All low-voltage interior luminaires (lighting fixtures) rated more than 4 watts, employing lamps rated more than 1.2 watts, shall be listed.

Twelve-volt systems for running and signal lights, similar to those used in conventional automobiles, are covered in 552.10 and 552.20. In some park trailers, 12-volt systems are also used for interior lighting and other small loads. The 12-volt system is often supplied from an on-board battery or through a transfer switch from a 120/12-volt transformer in conjunction with a full-wave rectifier.

III. Combination Electrical Systems

552.20 Combination Electrical Systems

(A) General Unit wiring suitable for connection to a battery or other low-voltage supply source shall be permitted to be connected to a 120-volt source, provided that the entire wiring system and equipment are rated and installed in full conformity with Parts I, III, IV, and V requirements of this article covering 120-volt electrical systems. Circuits fed from ac transformers shall not supply dc appliances.

(B) Voltage Converters (120-Volt Alternating Current to Low-Voltage Direct Current) The 120-volt ac side of the voltage converter shall be wired in full conformity with Parts I, III, IV, and V requirements of this article for 120-volt electrical systems.

Exception: Converters supplied as an integral part of a listed appliance shall not be subject to 552.20(B).

All converters and transformers shall be listed for use in recreation units and designed or equipped to provide over-temperature protection. To determine the converter rating, the following formula shall be applied to the total connected load, including average battery charging rate, of all 12-volt equipment:

The first 20 amperes of load at 100 percent; plus

The second 20 amperes of load at 50 percent; plus

All load above 40 amperes at 25 percent

Exception: A low-voltage appliance that is controlled by a momentary switch (normally open) that has no means for holding in the closed position shall not be considered as a connected load when determining the required converter rating. Momentarily energized appliances shall be limited to those used to prepare the unit for occupancy or travel.

(C) Bonding Voltage Converter Enclosures The non–current-carrying metal enclosure of the voltage converter shall be bonded to the frame of the unit with an 8 AWG copper conductor minimum. The grounding conductor for the battery and the metal enclosure shall be permitted to be the same conductor.

(D) Dual-Voltage Fixtures Including Luminaires or Appliances Fixtures, including luminaires, or appliances having both 120-volt and low-voltage connections shall be listed for dual voltage.

In the dual-voltage fixtures described in 552.20(D), barriers are used to separate the 120-volt and the 12-volt wiring connections.

(E) Autotransformers Autotransformers shall not be used.

(F) Receptacles and Plug Caps Where a park trailer is equipped with a 120-volt or 120/240-volt ac system, a low-voltage system, or both, receptacles and plug caps of the low-voltage system shall differ in configuration from those of the 120-volt or 120/240-volt system. Where a unit equipped with a battery or dc system has an external connection for low-voltage power, the connector shall have a configuration that will not accept 120-volt power.

IV. Nominal 120-Volt or 120/240-Volt Systems

552.40 120-Volt or 120/240-Volt, Nominal, Systems

(A) General Requirements The electrical equipment and material of park trailers indicated for connection to a wiring system rated 120 volts, nominal, 2-wire with ground, or a wiring system rated 120/240 volts, nominal, 3-wire with ground, shall be listed and installed in accordance with the requirements of Parts I, III, IV, and V of this article.

(B) Materials and Equipment Electrical materials, devices, appliances, fittings, and other equipment installed, intended for use in, or attached to the park trailer shall be listed. All products shall be used only in the manner in which they have been tested and found suitable for the intended use.

552.41 Receptacle Outlets Required

(A) Spacing Receptacle outlets shall be installed at wall spaces 600 mm (2 ft) wide or more so that no point along the floor line is more than 1.8 m (6 ft), measured horizontally, from an outlet in that space.

Exception No. 1: Bath and hall areas.

Exception No. 2: Wall spaces occupied by kitchen cabinets, wardrobe cabinets, built-in furniture; behind doors that may open fully against a wall surface; or similar facilities.

(B) Location Receptacle outlets shall be installed as follows:

(1) Adjacent to countertops in the kitchen [at least one on each side of the sink if countertops are on each side and are 300 mm (12 in.) or over in width]
(2) Adjacent to the refrigerator and gas range space, except where a gas-fired refrigerator or cooking appliance, requiring no external electrical connection, is factory-installed
(3) Adjacent to countertop spaces of 300 mm (12 in.) or more in width that cannot be reached from a receptacle required in 552.41(B)(1) by a cord of 1.8 m (6 ft) without crossing a traffic area, cooking appliance, or sink

(C) Ground-Fault Circuit-Interrupter Protection Each 125-volt, single-phase, 15- or 20-ampere receptacle shall have ground-fault circuit-interrupter protection for personnel in the following locations:

(1) Where the receptacles are installed to serve kitchen countertop surfaces
(2) Within 1.8 m (6 ft) of any lavatory or sink

Exception: Receptacles installed for appliances in dedicated spaces, such as for dishwashers, disposals, refrigerators, freezers, and laundry equipment.

(3) In the area occupied by a toilet, shower, tub, or any combination thereof
(4) On the exterior of the unit

Exception: Receptacles that are located inside of an access panel that is installed on the exterior of the unit to supply power for an installed appliance shall not be required to have ground-fault circuit-interrupter protection.

The receptacle outlet shall be permitted in a listed luminaire (lighting fixture). A receptacle outlet shall not be installed in a tub or combination tub–shower compartment.

In accordance with 552.41(C)(4), a bathroom receptacle is permitted to be mounted in the side of a lavatory cabinet where installation of a receptacle outlet is not possible in a wall that does not provide the necessary depth. The receptacle must be GFCI protected in accordance with 551.41(C). Receptacles of any type are not permitted in a tub or tub-shower compartment.

(D) Pipe Heating Cable Outlet Where a pipe heating cable outlet is installed, the outlet shall be as follows:

(1) Located within 600 mm (2 ft) of the cold water inlet
(2) Connected to an interior branch circuit, other than a small appliance branch circuit

(3) On a circuit where all of the outlets are on the load side of the ground-fault circuit-interrupter protection for personnel

(4) Mounted on the underside of the park trailer and shall not be considered to be the outdoor receptacle outlet required in 552.41(E)

(E) Outdoor Receptacle Outlets At least one receptacle outlet shall be installed outdoors. A receptacle outlet located in a compartment accessible from the outside of the park trailer shall be considered an outdoor receptacle. Outdoor receptacle outlets shall be protected as required in 552.41(C)(4).

(F) Receptacle Outlets Not Permitted

(1) Shower or Bathtub Space Receptacle outlets shall not be installed in or within reach [750 mm (30 in.)] of a shower or bathtub space.

(2) Face-Up Position A receptacle shall not be installed in a face-up position in any countertop.

552.43 Power Supply

(A) Feeder The power supply to the park trailer shall be a feeder assembly consisting of not more than one listed 30-ampere or 50-ampere park trailer power-supply cord with an integrally molded or securely attached cap, or a permanently installed feeder.

(B) Power-Supply Cord If the park trailer has a power-supply cord, it shall be permanently attached to the distribution panelboard or to a junction box permanently connected to the distribution panelboard, with the free end terminating in a molded-on attachment plug cap.

Cords with adapters and pigtail ends, extension cords, and similar items shall not be attached to, or shipped with, a park trailer.

A suitable clamp or the equivalent shall be provided at the distribution panelboard knockout to afford strain relief for the cord to prevent strain from being transmitted to the terminals when the power-supply cord is handled in its intended manner.

The cord shall be a listed type with 3-wire, 120-volt or 4-wire, 120/240-volt conductors, one of which shall be identified by a continuous green color or a continuous green color with one or more yellow stripes for use as the grounding conductor.

(C) Mast Weatherhead or Raceway Where the calculated load exceeds 50 amperes or where a permanent feeder is used, the supply shall be by means of one of the following:

(1) One mast weatherhead installation, installed in accordance with Article 230, containing four continuous, insulated, color-coded feeder conductors, one of which shall be an equipment grounding conductor

(2) A metal raceway, rigid nonmetallic conduit, or liquidtight flexible nonmetallic conduit from the disconnecting means in the park trailer to the underside of the park trailer, with provisions for the attachment to a suitable junction box or fitting to the raceway on the underside of the park trailer [with or without conductors as in 550.10(I)(1)]

552.44 Cord

(A) Permanently Connected Each power-supply assembly shall be factory supplied or factory installed and connected directly to the terminals of the distribution panelboard or conductors within a junction box and provided with means to prevent strain from being transmitted to the terminals. The ampacity of the conductors between each junction box and the terminals of each distribution panelboard shall be at least equal to the ampacity of the power-supply cord. The supply end of the assembly shall be equipped with an attachment plug of the type described in 552.44(C). Where the cord passes through the walls or floors, it shall be protected by means of conduit and bushings or equivalent. The cord assembly shall have permanent provisions for protection against corrosion and mechanical damage while the unit is in transit.

(B) Cord Length The cord-exposed usable length shall be measured from the point of entrance to the park trailer or the face of the flanged surface inlet (motor-base attachment plug) to the face of the attachment plug at the supply end.

The cord-exposed usable length, measured to the point of entry on the unit exterior, shall be a minimum of 7.0 m (23 ft) where the point of entrance is at the side of the unit, or shall be a minimum 8.5 m (28 ft) where the point of entrance is at the rear of the unit. The maximum length shall not exceed 11 m (36½ ft).

Where the cord entrance into the unit is more than 900 mm (3 ft) above the ground, the minimum cord lengths above shall be increased by the vertical distance of the cord entrance heights above 900 mm (3 ft).

(C) Attachment Plugs

(1) Units with Two to Five 15- or 20-Ampere Branch Circuits Park trailers wired in accordance with 552.46(A) shall have an attachment plug that shall be 2-pole, 3-wire grounding-type, rated 30 amperes, 125 volts, conforming to the configuration shown in Figure 552.44(C) intended for use with units rated at 30 amperes, 125 volts.

> FPN: Complete details of this configuration can be found in ANSI/NEMA WD 6-1989, National Electrical Manufacturers Association's *Standard for Dimensions of Attachment Plugs and Receptacles*, Figure TT.

(2) Units with 50-Ampere Power Supply Assembly Park trailers having a power-supply assembly rated 50 amperes

Receptacles   Caps

30-A,125-V, 2-pole, 3-wire, grounding type

50-A,125/250-V, 3-pole, 4-wire, grounding type

Figure 552.44(C) Attachment Cap and Receptacle Configurations

as permitted by 552.43(B) shall have a 3-pole, 4-wire grounding-type attachment plug rated 50 amperes, 125/250 volts, conforming to the configuration shown in Figure 552.44(C).

> FPN: Complete details of this configuration can be found in ANSI/NEMA WD 6-1989, National Electrical Manufacturers Association *Standard for Dimensions of Attachment Plugs and Receptacles*, Figure 14-50.

The 50-ampere receptacle and plug configurations used for park trailers appear in Exhibit 406.3 (NEMA configuration chart); however, the 30-ampere plug and receptacle configurations are unique to recreational vehicles and are described in UL 498, *Standard for Attachment Plugs and Receptacles*.

(D) Labeling at Electrical Entrance Each park trailer shall have permanently affixed to the exterior skin, at or near the point of entrance of the power-supply assembly, a label 75 mm × 45 mm (3 in. × 1¾ in.) minimum size, made of etched, metal-stamped, or embossed brass, stainless steel, or anodized or alclad aluminum not less than 0.51 mm (0.020 in.) thick, or other suitable material [e.g., 0.13 mm (0.005 in.) thick plastic laminate], that reads, as appropriate, either

> THIS CONNECTION IS FOR 110–125-VOLT AC,
> 60 HZ, 30 AMPERE SUPPLY

or

> THIS CONNECTION IS FOR 120/240-VOLT AC,
> 3-POLE, 4-WIRE, 60 HZ, _____ AMPERE SUPPLY.

The correct ampere rating shall be marked in the blank space.

(E) Location The point of entrance of a power-supply assembly shall be located within 4.5 m (15 ft) of the rear, on the left (road) side or at the rear, left of the longitudinal center of the unit, within 450 mm (18 in.) of the outside wall.

Exception: A park trailer shall be permitted to have the electrical point of entrance located more than 4.5 m (15 ft) from the rear. Where this occurs, the distance beyond the 4.5-m (15-ft) dimension shall be added to the cord's minimum length as specified in 551.46(B).

552.45 Distribution Panelboard

(A) Listed and Appropriately Rated A listed and appropriately rated distribution panelboard shall be used. The grounded conductor termination bar shall be insulated from the enclosure as provided in 552.55(C). An equipment grounding terminal bar shall be attached inside the metal enclosure of the panelboard.

(B) Location The distribution panelboard shall be installed in a readily accessible location. Working clearance for the panelboard shall be not less than 600 mm (24 in.) wide and 750 mm (30 in.) deep.

Exception: Where the panelboard cover is exposed to the inside aisle space, one of the working clearance dimensions shall be permitted to be reduced to a minimum of 550 mm (22 in.). A panelboard shall be considered exposed where the panelboard cover is within 50 mm (2 in.) of the aisle's finished surface.

(C) Dead-Front Type The distribution panelboard shall be of the dead-front type. A main disconnecting means shall be provided where fuses are used or where more than two circuit breakers are employed. A main overcurrent protective device not exceeding the power-supply assembly rating shall be provided where more than two branch circuits are employed.

552.46 Branch Circuits

Branch circuits shall be determined in accordance with 552.46(A) and 552.46(B).

(A) Two to Five 15- or 20-Ampere Circuits Two to five 15- or 20-ampere circuits to supply lights, receptacle outlets, and fixed appliances shall be permitted. Such park trailers shall be equipped with a distribution panelboard rated at 120 volts maximum with a 30-ampere rated main power supply assembly. Not more than two 120-volt thermostatically controlled appliances (e.g., air conditioner and water heater) shall be installed in such systems unless appliance isolation switching, energy management systems, or similar methods are used.

Exception: Additional 15- or 20-ampere circuits shall be permitted where a listed energy management system rated at 30 amperes maximum is employed within the system.

(B) More Than Five Circuits Where more than five circuits are needed, they shall be determined in accordance with 552.46(B)(1), (B)(2), and (B)(3).

(1) Lighting Based on 33 volt-amperes/m^2 (3 VA/ft^2) multiplied by the outside dimensions of the park trailer (coupler excluded) divided by 120 volts to determine the number of 15- or 20-ampere lighting area circuits, for example,

$$\frac{3 \times length \times width}{120 \times 15 \ (or \ 20)} = \frac{\text{No. of 15- (or 20-)}}{\text{ampere circuits}}$$

The lighting circuits shall be permitted to serve built-in gas ovens with electric service only for lights, clocks or timers, or listed cord-connected garbage disposal units.

(2) Small Appliances Small appliance branch circuits shall be installed in accordance with 210.11(C)(1).

(3) General Appliances (including furnace, water heater, space heater, range, and central or room air conditioner, etc.) An individual branch circuit shall be permitted to supply any load for which it is rated. There shall be one or more circuits of adequate rating in accordance with (a) through (d).

> FPN No. 1: For the laundry branch circuit, see 210.11(C)(2).
> FPN No. 2: For central air conditioning, see Article 440.

(a) The total rating of fixed appliances shall not exceed 50 percent of the circuit rating if lighting outlets, general-use receptacles, or both, are also supplied.

(b) For fixed appliances with a motor(s) larger than ⅛ horsepower, the total calculated load shall be based on 125 percent of the largest motor plus the sum of the other loads. Where a branch circuit supplies continuous load(s) or any combination of continuous and noncontinuous loads, the branch-circuit conductor size shall be in accordance with 210.19(A).

(c) The rating of a single cord- and plug-connected appliance supplied by other than an individual branch circuit shall not exceed 80 percent of the circuit rating.

(d) The rating of a range branch circuit shall be based on the range demand as specified for ranges in 552.47(B)(5).

552.47 Calculations

The following method shall be employed in computing the supply-cord and distribution-panelboard load for each feeder assembly for each park trailer in lieu of the procedure shown in Article 220 and shall be based on a 3-wire, 120/240-volt supply with 120-volt loads balanced between the two phases of the 3-wire system.

(A) Lighting and Small Appliance Load Lighting Volt-Amperes: Length times width of park trailer floor (outside dimensions) times 33 volt-amperes/m^2 (3 VA/ft^2). For example,

$$Length \times width \times 3 = lighting \ volt\text{-}amperes$$

Small Appliance Volt-Amperes: Number of circuits times 1500 volt-amperes for each 20-ampere appliance receptacle circuit (see definition of *Appliance, Portable* with note) including 1500 volt-amperes for laundry circuit. For example,

$$No. \ of \ circuits \times 1500 = small \ appliance \ volt\text{-}amperes$$

Total: Lighting volt-amperes plus small appliance volt-amperes = total volt-amperes

First 3000 total volt-amperes at 100 percent plus remainder at 35 percent = volt-amperes to be divided by 240 volts to obtain current (amperes) per leg.

(B) Total Load for Determining Power Supply Total load for determining power supply is the sum of the following:

(1) Lighting and small appliance load as calculated in 552.47(A).
(2) Nameplate amperes for motors and heater loads (exhaust fans, air conditioners, electric, gas, or oil heating). Omit smaller of the heating and cooling loads, except include blower motor if used as air-conditioner evaporator motor. Where an air conditioner is not installed and a 50-ampere power-supply cord is provided, allow 15 amperes per phase for air conditioning.
(3) Twenty-five percent of current of largest motor in (B)(2).
(4) Total of nameplate amperes for disposal, dishwasher, water heater, clothes dryer, wall-mounted oven, cooking units. Where the number of these appliances exceeds three, use 75 percent of total.
(5) Derive amperes for freestanding range (as distinguished from separate ovens and cooking units) by dividing the following values by 240 volts:

Nameplate Rating (watts)	Use (volt-amperes)
0–10,000	80 percent of rating
Over 10,000–12,500	8,000
Over 12,500–13,500	8,400
Over 13,500–14,500	8,800
Over 14,500–15,500	9,200
Over 15,500–16,500	9,600
Over 16,500–17,500	10,000

(6) If outlets or circuits are provided for other than factory-installed appliances, include the anticipated load.

> FPN: Refer to Annex D, Example D12, for an illustration of the application of this calculation.

(C) Optional Method of Calculation for Lighting and Appliance Load For park trailers, the optional method for calculating lighting and appliance load shown in 220.82 shall be permitted.

552.48 Wiring Methods

(A) Wiring Systems Cables and raceways installed in accordance with Articles 320, 322, 330 through 340, 342 through 362, 386, and 388 shall be permitted in accordance with their applicable article, except as otherwise specified in this article. An equipment grounding means shall be provided in accordance with 250.118.

See the commentary on 348.60 for information regarding the use of flexible metal conduit as an equipment grounding conductor.

(B) Conduit and Tubing Where rigid metal conduit or intermediate metal conduit is terminated at an enclosure with a locknut and bushing connection, two locknuts shall be provided, one inside and one outside of the enclosure. All cut ends of conduit and tubing shall be reamed or otherwise finished to remove rough edges.

See the commentary following 358.28(A) and 300.4(F) for information on protecting conductor insulation against abrasion at conduit and tubing terminations.

(C) Nonmetallic Boxes Nonmetallic boxes shall be acceptable only with nonmetallic-sheathed cable or nonmetallic raceways.

(D) Boxes In walls and ceilings constructed of wood or other combustible material, boxes and fittings shall be flush with the finished surface or project therefrom.

(E) Mounting Wall and ceiling boxes shall be mounted in accordance with Article 314.

Exception No. 1: Snap-in-type boxes or boxes provided with special wall or ceiling brackets that securely fasten boxes in walls or ceilings shall be permitted.

Exception No. 2: A wooden plate providing a 38-mm (1½-in.) minimum width backing around the box and of a thickness of 13 mm (½ in.) or greater (actual) attached directly to the wall panel shall be considered as approved means for mounting outlet boxes.

Exception No. 2 to 552.48(E) permits the mounting of outlet boxes by screws to a wooden plate that is secured directly to the back of the wall panel. The wooden plate must be not less than ½ in. thick and must extend at least 1½ in. around the box. This requirement recognizes the special construction of recreational vehicle walls, which often makes it difficult or impossible to attach an outlet box to a structural member, as required by 314.23(B).

(F) Sheath Armor The sheath of nonmetallic-sheathed cable, metal-clad cable, and Type AC cable shall be continuous between outlet boxes and other enclosures.

(G) Protected Metal-clad, Type AC, or nonmetallic-sheathed cables and electrical nonmetallic tubing shall be permitted to pass through the centers of the wide side of 2 by 4 wood studs. However, they shall be protected where they pass through 2 by 2 wood studs or at other wood studs or frames where the cable or tubing would be less than 32 mm (1¼ in.) from the inside or outside surface. Steel plates on each side of the cable or tubing, or a steel tube, with not less than 1.35 mm (0.053 in.) wall thickness, shall be installed to protect the cable or tubing. These plates or tubes shall be securely held in place. Where nonmetallic-sheathed cables pass through punched, cut, or drilled slots or holes in metal members, the cable shall be protected by bushings or grommets securely fastened in the opening prior to installation of the cable.

(H) Cable Supports Where connected with cable connectors or clamps, cables shall be supported within 300 mm (12 in.) of outlet boxes, distribution panelboards, and splice boxes on appliances. Supports shall be provided every 1.4 m (4½ ft) at other places.

(I) Nonmetallic Box Without Cable Clamps Nonmetallic-sheathed cables shall be supported within 200 mm (8 in.) of a nonmetallic outlet box without cable clamps.

Exception: Where wiring devices with integral enclosures are employed with a loop of extra cable to permit future replacement of the device, the cable loop shall be considered as an integral portion of the device.

(J) Physical Damage Where subject to physical damage, exposed nonmetallic cable shall be protected by covering boards, guard strips, raceways, or other means.

(K) Metal Faceplates Metal faceplates shall be of ferrous metal not less than 0.76 mm (0.030 in.) in thickness or of nonferrous metal not less than 1.0 mm (0.040 in.) in thickness. Nonmetallic faceplates shall be listed.

(L) Metal Faceplates Effectively Grounded Where metal faceplates are used, they shall be effectively grounded.

(M) Moisture or Physical Damage Where outdoor or under-chassis wiring is 120 volts, nominal, or over and is exposed to moisture or physical damage, the wiring shall be protected by rigid metal conduit, by intermediate metal conduit, or by electrical metallic tubing or rigid nonmetallic conduit that is closely routed against frames and equipment enclosures or other raceway or cable identified for the application.

(N) Component Interconnections Fittings and connectors that are intended to be concealed at the time of assembly shall be listed and identified for the interconnection of building components. Such fittings and connectors shall be equal to the wiring method employed in insulation, temperature rise, and fault-current withstanding, and shall be capable of enduring the vibration and shock occurring in park trailers.

(O) Method of Connecting Expandable Units The method of connecting expandable units to the main body of the vehicle shall comply with the following as applicable:

(1) That portion of a branch circuit that is installed in an expandable unit shall be permitted to be connected to the branch circuit in the main body of the vehicle by means of a flexible cord or attachment plug and cord listed for hard usage. The cord and its connections shall conform to all provisions of Article 400 and shall be considered as a permitted use under 400.7.

(2) If the receptacle provided for connection of the cord to the main circuit is located on the outside of the unit, it shall be protected with a ground-fault circuit interrupter for personnel and be listed for wet locations. A cord located on the outside of a unit shall be identified for outdoor use.

(3) Unless removable or stored within the unit interior, the cord assembly shall have permanent provisions for protection against corrosion and mechanical damage while the unit is in transit.

(4) If an attachment plug and cord is used, it shall be installed so as not to permit exposed live attachment plug pins.

(P) Prewiring for Air-Conditioning Installation Prewiring installed for the purpose of facilitating future air-conditioning installation shall conform to the applicable portions of this article and the following:

(1) An overcurrent protective device with a rating compatible with the circuit conductors shall be installed in the distribution panelboard and wiring connections completed.

(2) The load end of the circuit shall terminate in a junction box with a blank cover or other listed enclosure. Where a junction box with a blank cover is used, the free ends of the conductors shall be adequately capped or taped.

(3) A label conforming to 552.44(D) shall be placed on or adjacent to the junction box and shall read

AIR-CONDITIONING CIRCUIT.
THIS CONNECTION IS FOR AIR CONDITIONERS
RATED 110–125-VOLT AC, 60 HZ,
_____ AMPERES MAXIMUM.
DO NOT EXCEED CIRCUIT RATING.

An ampere rating, not to exceed 80 percent of the circuit rating, shall be legibly marked in the blank space.
(4) The circuit shall serve no other purpose.

552.49 Maximum Number of Conductors in Boxes

The maximum number of conductors permitted in boxes shall be in accordance with 314.16.

552.50 Grounded Conductors

The identification of grounded conductors shall be in accordance with 200.6.

552.51 Connection of Terminals and Splices

Conductor splices and connections at terminals shall be in accordance with 110.14.

552.52 Switches

Switches shall be rated as required by 552.52(A) and 552.52(B).

(A) Lighting Circuits For lighting circuits, switches shall be rated not less than 10 amperes, 120/125 volts, and in no case less than the connected load.

(B) Motors or Other Loads For motors or other loads, switches shall have ampere or horsepower ratings, or both, adequate for loads controlled. (An ac general-use snap switch shall be permitted to control a motor 2 hp or less with full-load current not over 80 percent of the switch ampere rating.)

552.53 Receptacles

All receptacle outlets shall be of the grounding type and installed in accordance with 210.21 and 406.3.

552.54 Luminaires (Lighting Fixtures)

(A) General Any combustible wall or ceiling finish exposed between the edge of a luminaire (fixture) canopy or pan and the outlet box shall be covered with noncombustible material or a material identified for the purpose.

(B) Shower Luminaires (Fixtures) If a luminaire (lighting fixture) is provided over a bathtub or in a shower stall, it shall be of the enclosed and gasketed type and listed for the type of installation, and it shall be ground-fault circuit-interrupter protected.

The switch for shower luminaires (lighting fixtures) and exhaust fans, located over a tub or in a shower stall, shall be located outside the tub or shower space.

Many shower luminaires have a metal base and, due to the low ceilings in recreational vehicles, may be easily reached by most persons under the shower or standing in the bathtub.

In accordance with 552.54(B), the installation of luminaires that are listed for wet locations and have GFCI protection is permitted above a tub or shower enclosure.

(C) Outdoor Outlets, Luminaires (Fixtures), Air-Cooling Equipment, and So On Outdoor luminaires (fixtures) and other equipment shall be listed for outdoor use.

552.55 Grounding

(See also 552.57 on bonding of non–current-carrying metal parts.)

(A) Power-Supply Grounding The grounding conductor in the supply cord or feeder shall be connected to the grounding bus or other approved grounding means in the distribution panelboard.

(B) Distribution Panelboard The distribution panelboard shall have a grounding bus with sufficient terminals for all grounding conductors or other approved grounding means.

(C) Insulated Neutral The grounded circuit conductor (neutral) shall be insulated from the equipment grounding conductors and from equipment enclosures and other grounded parts. The grounded (neutral) circuit terminals in the distribution panelboard and in ranges, clothes dryers, counter-mounted cooking units, and wall-mounted ovens shall be insulated from the equipment enclosure. Bonding screws, straps, or buses in the distribution panelboard or in appliances shall be removed and discarded. Connection of electric ranges and electric clothes dryers utilizing a grounded (neutral) conductor, if cord-connected, shall be made with 4-conductor cord and 3-pole, 4-wire, grounding-type plug caps and receptacles.

552.56 Interior Equipment Grounding

(A) Exposed Metal Parts In the electrical system, all exposed metal parts, enclosures, frames, luminaire (lighting fixture) canopies, and so forth, shall be effectively bonded to the grounding terminals or enclosure of the distribution panelboard.

(B) Equipment Grounding Conductors Bare wires, green-colored wires, or green wires with a yellow stripe(s) shall be used for equipment grounding conductors only.

(C) Grounding of Electrical Equipment Where grounding of electrical equipment is specified, it shall be permitted as follows:

(1) Connection of metal raceway (conduit or electrical metallic tubing), the sheath of Type MC and Type MI cable where the sheath is identified for grounding, or the armor of Type AC cable to metal enclosures.

(2) A connection between the one or more equipment grounding conductors and a metal box by means of a grounding screw, which shall be used for no other purpose, or a listed grounding device.

(3) The equipment grounding conductor in nonmetallic-sheathed cable shall be permitted to be secured under a screw threaded into the luminaire (fixture) canopy other than a mounting screw or cover screw or attached to a listed grounding means (plate) in a nonmetallic outlet box for luminaire (fixture) mounting [grounding means shall also be permitted for luminaire (fixture) attachment screws].

(D) Grounding Connection in Nonmetallic Box A connection between the one or more grounding conductors brought into a nonmetallic outlet box shall be arranged so that a connection can be made to any fitting or device in that box that requires grounding.

(E) Grounding Continuity Where more than one equipment grounding conductor of a branch circuit enters a box, all such conductors shall be in good electrical contact with each other, and the arrangement shall be such that the disconnection or removal of a receptacle, fixture, including a luminaire, or other device fed from the box will not interfere with or interrupt the grounding continuity.

(F) Cord-Connected Appliances Cord-connected appliances, such as washing machines, clothes dryers, refrigerators, and the electrical system of gas ranges, and so on, shall be grounded by means of an approved cord with equipment grounding conductor and grounding-type attachment plug.

552.57 Bonding of Non–Current-Carrying Metal Parts

(A) Required Bonding All exposed non–current-carrying metal parts that may become energized shall be effectively bonded to the grounding terminal or enclosure of the distribution panelboard.

(B) Bonding Chassis A bonding conductor shall be connected between any distribution panelboard and an accessible terminal on the chassis. Aluminum or copper-clad aluminum conductors shall not be used for bonding if such conductors or their terminals are exposed to corrosive elements.

Exception: Any park trailer that employs a unitized metal chassis-frame construction to which the distribution panelboard is securely fastened with a bolt(s) and nut(s) or by welding or riveting shall be considered to be bonded.

(C) Bonding Conductor Requirements Grounding terminals shall be of the solderless type and listed as pressure terminal connectors recognized for the wire size used. The bonding conductor shall be solid or stranded, insulated or bare, and shall be 8 AWG copper minimum or equivalent.

(D) Metallic Roof and Exterior Bonding The metal roof and exterior covering shall be considered bonded where both of the following conditions apply:

(1) The metal panels overlap one another and are securely attached to the wood or metal frame parts by metal fasteners.
(2) The lower panel of the metal exterior covering is secured by metal fasteners at each cross member of the chassis, or the lower panel is bonded to the chassis by a metal strap.

(E) Gas, Water, and Waste Pipe Bonding The gas, water, and waste pipes shall be considered grounded if they are bonded to the chassis.

(F) Furnace and Metal Air Duct Bonding Furnace and metal circulating air ducts shall be bonded.

552.58 Appliance Accessibility and Fastening

Every appliance shall be accessible for inspection, service, repair, and replacement without removal of permanent construction. Means shall be provided to securely fasten appliances in place when the park trailer is in transit.

552.59 Outdoor Outlets, Fixtures, Including Luminaires, Air-Cooling Equipment, and So On

(A) Listed for Outdoor Use Outdoor fixtures, including luminaires, and equipment shall be listed for outdoor use. Outdoor receptacle or convenience outlets shall be of a gasketed-cover type for use in wet locations.

(B) Outside Heating Equipment, Air-Conditioning Equipment, or Both A park trailer provided with a branch circuit designed to energize outside heating equipment or air-conditioning equipment, or both, located outside the park trailer, other than room air conditioners, shall have such branch-circuit conductors terminate in a listed outlet box or disconnecting means located on the outside of the park trailer. A label shall be permanently affixed within 150 mm (6 in.) from the listed box or disconnecting means, and shall contain the following information:

> THIS CONNECTION IS FOR HEATING
> AND/OR AIR-CONDITIONING EQUIPMENT.
> THE BRANCH CIRCUIT IS RATED AT NOT MORE
> THAN _____ AMPERES, AT _____ VOLTS,
> 60 HZ, _____ CONDUCTOR AMPACITY.
> A DISCONNECTING MEANS SHALL BE
> LOCATED WITHIN SIGHT OF THE EQUIPMENT.

The correct voltage and ampere rating shall be given. The tag shall not be less than 0.51 mm (0.020 in.) thick etched brass, stainless steel, anodized or alclad aluminum, or equivalent. The tag shall not be less than 75 mm × 45 mm (3 in. × 1¾ in.) minimum size.

V. Factory Tests

552.60 Factory Tests (Electrical)

Each park trailer shall be subjected to the tests required by 552.60(A) and 552.60(B).

(A) Circuits of 120 Volts or 120/240 Volts Each park trailer designed with a 120-volt or a 120/240-volt electrical system shall withstand the applied potential without electrical breakdown of a 1-minute, 900-volt dielectric strength test, or a 1-second, 1080-volt dielectric strength test, with all switches closed, between ungrounded and grounded conductors and the park trailer ground. During the test, all switches and other controls shall be in the on position. Fixtures, including luminaires, and permanently installed appliances shall not be required to withstand this test.

Each park trailer shall be subjected to the following:

(1) A continuity test to ensure that all metal parts are properly bonded
(2) Operational tests to demonstrate that all equipment is properly connected and in working order
(3) Polarity checks to determine that connections have been properly made
(4) Receptacles requiring GFCI protection shall be tested for correct function by the use of a GFCI testing device

(B) Low-Voltage Circuits Low-voltage circuit conductors in each park trailer shall withstand the applied potential without electrical breakdown of a 1-minute, 500-volt or a 1-second, 600-volt dielectric strength test. The potential shall be applied between ungrounded and grounded conductors.

The test shall be permitted on running light circuits before the lights are installed, provided the unit's outer covering and interior cabinetry have been secured. The braking circuit shall be permitted to be tested before being connected to the brakes, provided the wiring has been completely secured.

ARTICLE 553
Floating Buildings

Summary of Changes

- **553.4:** Revised to prohibit installation of service equipment on any floating building or floating structure.
- **553.8(C):** Revised to permit re-identification of insulated conductors larger than 6 AWG or conductors in multiconductor cables as equipment grounding conductors.

Contents

I. General

553.1 Scope

This article covers wiring, services, feeders, and grounding for floating buildings.

553.2 Definition

Floating Building. A building unit as defined in Article 100 that floats on water, is moored in a permanent location, and has a premises wiring system served through connection by permanent wiring to an electricity supply system not located on the premises.

553.3 Application of Other Articles

Wiring for floating buildings shall comply with the applicable provisions of other articles of this *Code*, except as modified by this article.

II. Services and Feeders

553.4 Location of Service Equipment

The service equipment for a floating building shall be located adjacent to, but not in or on, the building or any floating structure.

The intent of 553.4 is to ensure that supply conductors to a floating building can be disconnected in an emergency, such as during a storm, when the floating building has to be moved quickly. A revision to this section for the 2005 *Code* prohibits service equipment from being installed on the floating building and any other floating structure such as a wharf or pier.

Overcurrent protection for supply conductors is provided, since these conductors may develop excessive leakage where located underwater or where a short circuit or ground fault occurs.

553.5 Service Conductors

One set of service conductors shall be permitted to serve more than one set of service equipment.

553.6 Feeder Conductors

Each floating building shall be supplied by a single set of feeder conductors from its service equipment.

Exception: Where the floating building has multiple occupancy, each occupant shall be permitted to be supplied by a single set of feeder conductors extended from the occupant's service equipment to the occupant's panelboard.

553.7 Installation of Services and Feeders

(A) Flexibility Flexibility of the wiring system shall be maintained between floating buildings and the supply conductors. All wiring shall be installed so that motion of the water surface and changes in the water level will not result in unsafe conditions.

(B) Wiring Methods Liquidtight flexible metal conduit or liquidtight flexible nonmetallic conduit with approved fittings shall be permitted for feeders and where flexible connections are required for services. Extra-hard usage portable power cable listed for both wet locations and sunlight resistance shall be permitted for a feeder to a floating building where flexibility is required. Other raceways suitable for the location shall be permitted to be installed where flexibility is not required.

    FPN: See 555.1 and 555.13.

Where Type W cables from Table 400.4 are used, it is necessary to verify that the cable is listed for use in wet locations. Not all Type W cables are listed for wet location application. Liquidtight flexible nonmetallic conduit with approved fittings is permitted where flexible connections are required.

III. Grounding

553.8 General Requirements

Grounding at floating buildings shall comply with 553.8(A) through (D).

(A) Grounding of Electrical and Nonelectrical Parts
Grounding of both electrical and nonelectrical parts in a floating building shall be through connection to a grounding bus in the building panelboard.

(B) Installation and Connection of Equipment Grounding Conductor The equipment grounding conductor shall be installed with the feeder conductors and connected to a grounding terminal in the service equipment.

(C) Identification of Equipment Grounding Conductor The equipment grounding conductor shall be an insulated copper conductor with a continuous outer finish that is either green or green with one or more yellow stripes. For conductors larger than 6 AWG, or where multiconductor cables are used, re-identification of conductors as allowed in 250.119(A)(2)(b) and (A)(2)(c) or 250.119(B)(2) and (B)(3) shall be permitted.

(D) Grounding Electrode Conductor Connection The grounding terminal in the service equipment shall be grounded by connection through an insulated grounding electrode conductor to a grounding electrode on shore.

Section 553.8 was revised for the 2005 *Code* and subdivided into four paragraphs in order to provide grounding requirements in a better organized format. Grounding of electrical and nonelectrical parts must be connected to a grounding bus at the building panelboard. An equipment grounding conductor must be included in the feeder supplying the building and connected to the grounding terminal at the service equipment. For conductor sizes 6 AWG and smaller, the equipment grounding conductor must be provided with green insulation or green insulation with a yellow tracer. An insulated grounding electrode conductor must be installed between the service equipment and the grounding electrode on shore.

553.9 Insulated Neutral

The grounded circuit conductor (neutral) shall be an insulated conductor identified in conformance with 200.6. The neutral conductor shall be connected to the equipment grounding terminal in the service equipment, and, except for that connection, it shall be insulated from the equipment grounding conductors, equipment enclosures, and all other grounded parts. The neutral circuit terminals in the panelboard and in ranges, clothes dryers, counter-mounted cooking units, and the like shall be insulated from the enclosures.

553.10 Equipment Grounding

(A) Electrical Systems All enclosures and exposed metal parts of electrical systems shall be bonded to the grounding bus.

(B) Cord-Connected Appliances Where required to be grounded, cord-connected appliances shall be grounded by means of an equipment grounding conductor in the cord and a grounding-type attachment plug.

553.11 Bonding of Non–Current-Carrying Metal Parts

All metal parts in contact with the water, all metal piping, and all non–current-carrying metal parts that may become energized shall be bonded to the grounding bus in the panelboard.

ARTICLE 555
Marinas and Boatyards

Summary of Changes

- **555.21:** Revised to change "gasoline" to "motor fuel," and to require all electrical wiring for power and lighting to be installed on the side of the wharf, pier, or dock opposite from the fuel piping system. New FPN added referencing NFPA 30 and NFPA 303.

- **555.22:** Added section to require repair facilities to comply with Article 511 if marine craft located at the repair facility contain flammable or combustible liquids or gases.

Contents

555.1 Scope

This article covers the installation of wiring and equipment in the areas comprising fixed or floating piers, wharves, docks, and other areas in marinas, boatyards, boat basins, boathouses, yacht clubs, boat condominiums, docking facilities associated with residential condominiums, any multiple docking facility, or similar occupancies, and facilities that are used, or intended for use, for the purpose of repair, berthing, launching, storage, or fueling of small craft and the moorage of floating buildings.

Private, noncommercial docking facilities constructed or occupied for the use of the owner or residents of the associated single-family dwelling are not covered by this article.

FPN: See NFPA 303-2000, *Fire Protection Standard for Marinas and Boatyards*, for additional information.

The requirements of Article 555 apply to public and private docking, storage, repair, and fueling facilities for small craft. The term *small craft* is not defined in the *Code*; however, based on the scope of NFPA 303, *Fire Protection Standard for Marinas and Boatyards*, the small craft include recreational and commercial boats, yachts, and other craft that do not exceed 300 gross tons. For facilities that serve larger craft and ships, see NFPA 307, *Standard for the Construction and Fire Protection of Marine Terminals, Piers, and Wharves*. See Article 553 for requirements for floating buildings, including floating dwelling units.

The requirements of Article 555 apply to stand-alone boathouses except those constructed and used in association with a single-family dwelling. See 210.8(A)(8) for the GFCI protection requirement of 125-volt, single-phase, 15- and 20-ampere receptacles installed in or on boathouses located at a single-family dwelling. Electrical installations on docks and piers located at a single-family dwelling are not subject to the requirements of Article 555; however, all applicable requirements in Chapters 1 through 4 for these outdoor, wet

locations, including the GFCI requirements of 210.8(A)(3) for outdoor receptacles, are applicable.

555.2 Definitions

Electrical Datum Plane. The electrical datum plane is defined as follows:

Throughout Article 555, the physical location of electrical equipment is referenced to the electrical datum plane, which is used as a horizontal benchmark on land and on floating piers. The definition of *electrical datum plane* encompasses areas subject to tidal movement and areas in which the water level is affected only by conditions such as climate (rain or snow fall) or by human intervention (the opening or closing of dams and floodgates). In either case, the term covers the normal highest water level, such as astronomical high tides. The term does not cover extremes due to natural or manmade disasters.

(1) In land areas subject to tidal fluctuation, the electrical datum plane is a horizontal plane 606 mm (2 ft) above the highest tide level for the area occurring under normal circumstances, that is, highest high tide.
(2) In land areas not subject to tidal fluctuation, the electrical datum plane is a horizontal plane 606 mm (2 ft) above the highest water level for the area occurring under normal circumstances.
(3) The electrical datum plane for floating piers and landing stages that are (a) installed to permit rise and fall response to water level, without lateral movement, and (b) that are so equipped that they can rise to the datum plane established for (1) or (2), is a horizontal plane 762 mm (30 in.) above the water level at the floating pier or landing stage and a minimum of 305 mm (12 in.) above the level of the deck.

Marine Power Outlet. An enclosed assembly that can include receptacles, circuit breakers, fused switches, fuses, watt-hour meter(s), and monitoring means approved for marine use.

555.4 Distribution System

Yard and pier distribution systems shall not exceed 600 volts phase to phase.

555.5 Transformers

Transformers and enclosures shall be specifically approved for the intended location. The bottom of enclosures for transformers shall not be located below the electrical datum plane.

555.7 Location of Service Equipment

The service equipment for floating docks or marinas shall be located adjacent to, but not on or in, the floating structure.

The requirement covering service equipment location in 555.7 is similar to that in 553.4 for service equipment supplying floating buildings.

555.9 Electrical Connections

All electrical connections shall be located at least 305 mm (12 in.) above the deck of a floating pier. All electrical connections shall be located at least 305 mm (12 in.) above the deck of a fixed pier but not below the electrical datum plane.

555.10 Electrical Equipment Enclosures

(A) Securing and Supporting Electrical equipment enclosures installed on piers above deck level shall be securely and substantially supported by structural members, independent of any conduit connected to them. If enclosures are not attached to mounting surfaces by means of external ears or lugs, the internal screw heads shall be sealed to prevent seepage of water through mounting holes.

(B) Location Electrical equipment enclosures on piers shall be located so as not to interfere with mooring lines.

555.11 Circuit Breakers, Switches, Panelboards, and Marine Power Outlets

Circuit breakers and switches installed in gasketed enclosures shall be arranged to permit required manual operation without exposing the interior of the enclosure. All such enclosures shall be arranged with a weep hole to discharge condensation.

555.12 Load Calculations for Service and Feeder Conductors

General lighting and other loads shall be calculated in accordance with Article 220, and, in addition, the demand factors set forth in Table 555.12 shall be permitted for each service and/or feeder circuit supplying receptacles that provide shore power for boats. These calculations shall be permitted to be modified as indicated in notes (1) and (2) to Table 555.12.

A revision to this section for the 2005 *Code* clarifies that the use of the demand factors from Table 555.12 is permitted but not mandatory. It should be noted, however, that for receptacles that supply shore power for boats, the demand factors of Table 555.12 are the only ones permitted by the *Code*.

FPN: These demand factors may be inadequate in areas of extreme hot or cold temperatures with loaded circuits for heating, air-conditioning, or refrigerating equipment.

Table 555.12 Demand Factors

Number of Receptacles	Sum of the Rating of the Receptacles (percent)
1–4	100
5–8	90
9–14	80
15–30	70
31–40	60
41–50	50
51–70	40
71-plus	30

Notes:
1. Where shore power accommodations provide two receptacles specifically for an individual boat slip and these receptacles have different voltages (for example, one 30 ampere, 125 volt and one 50 ampere, 125/250 volt), only the receptacle with the larger kilowatt demand shall be required to be calculated.
2. If the facility being installed includes individual kilowatt-hour submeters for each slip and is being calculated using the criteria listed in Table 555.12, the total demand amperes may be multiplied by 0.9 to achieve the final demand amperes.

555.13 Wiring Methods and Installation

(A) Wiring Methods

(1) General Wiring methods of Chapter 3 shall be permitted where identified for use in wet locations.

(2) Portable Power Cables Extra-hard usage portable power cables rated not less than 167°F (75°C), 600 volts; listed for both wet locations and sunlight resistance; and having an outer jacket rated to be resistant to temperature extremes, oil, gasoline, ozone, abrasion, acids, and chemicals shall be permitted as follows:

(1) As permanent wiring on the underside of piers (floating or fixed)
(2) Where flexibility is necessary as on piers composed of floating sections

Table 400.4 identifies Types G, PPE, and W as portable power cables suitable for extra-hard usage. The requirements in 555.13(A)(2) for cable construction are necessary due to the cable's exposure to extremes in weather conditions and to operational hazards such as oil and gasoline spills. The use of these cables on floating piers and docks provides the necessary degree of flexibility to compensate for tidal and wave action.

(3) Temporary Wiring Temporary wiring, except as permitted by Article 590, shall not be used to supply power to boats.

(B) Installation

(1) Overhead Wiring Overhead wiring shall be installed to avoid possible contact with masts and other parts of boats being moved in the yard.

Conductors and cables shall be routed to avoid wiring closer than 6.0 m (20 ft) from the outer edge or any portion of the yard that can be used for moving vessels or stepping or unstepping masts.

(2) Outside Branch Circuits and Feeders Outside branch circuits and feeders shall comply with Article 225 except that clearances for overhead wiring in portions of the yard other than those described in 555.13(B)(1) shall not be less than 5.49 m (18 ft) above grade.

(3) Wiring Over and Under Navigable Water Wiring over and under navigable water shall be subject to approval by the authority having jurisdiction.

Some federal and local agencies, such as the Army Corps of Engineers, the Coast Guard, or local harbormasters, have specific authority over navigable waterways. Therefore, approval of any proposed installation over or under such a waterway should be obtained from the appropriate authority.

FPN: See NFPA 303-2000, *Fire Protection Standard for Marinas and Boatyards*, for warning sign requirements.

(4) Portable Power Cables

(a) Where portable power cables are permitted by 555.13(A)(2), the installation shall comply with the following:

(1) Cables shall be properly supported.
(2) Cables shall be located on the underside of the pier.
(3) Cables shall be securely fastened by nonmetallic clips to structural members other than the deck planking.
(4) Cables shall not be installed where subject to physical damage.
(5) Where cables pass through structural members, they shall be protected against chafing by a permanently installed oversized sleeve of nonmetallic material.

(b) Where portable power cables are used as permitted in 555.13(A)(2)(2), there shall be an approved junction box of corrosion-resistant construction with permanently installed terminal blocks on each pier section to which the feeder and feeder extensions are to be connected. Metal junction boxes and their covers, and metal screws and parts that are exposed externally to the boxes, shall be of corrosion-resistant materials or protected by material resistant to corrosion.

(5) Protection Rigid metal or nonmetallic conduit suitable for the location shall be installed to protect wiring above decks of piers and landing stages and below the enclosure that it serves. The conduit shall be connected to the enclosure by full standard threads. The use of special fittings of nonmetallic material to provide a threaded connection into enclosures on rigid nonmetallic conduit, employing joint design as recommended by the conduit manufacturer, for attachment of the fitting to the conduit shall be acceptable, provided the equipment and method of attachment are approved and the assembly meets the requirements of installation in damp or wet locations as applicable.

555.15 Grounding

Wiring and equipment within the scope of this article shall be grounded as specified in Article 250 and as required by 555.15(A) through 555.15(E).

(A) Equipment to Be Grounded The following items shall be connected to an equipment grounding conductor run with the circuit conductors in the same raceway, cable, or trench:

(1) Metal boxes, metal cabinets, and all other metal enclosures
(2) Metal frames of utilization equipment
(3) Grounding terminals of grounding-type receptacles

(B) Type of Equipment Grounding Conductor The equipment grounding conductor shall be an insulated copper conductor with a continuous outer finish that is either green or green with one or more yellow stripes. The equipment grounding conductor of Type MI cable shall be permitted to be identified at terminations. For conductors larger than 6 AWG, or where multiconductor cables are used, re-identification of conductors as allowed in 250.119(A)(2)(b) and (A)(2)(c) or 250.119(B)(2) and (B)(3) shall be permitted.

The provisions of 555.15(B) require an insulated equipment grounding conductor that ensures a high-integrity path for ground-fault current. Because of the corrosive conditions in marinas and boatyards, metal raceways are not permitted to serve as the sole equipment grounding conductor.

(C) Size of Equipment Grounding Conductor The insulated copper equipment grounding conductor shall be sized in accordance with 250.122 but not smaller than 12 AWG.

(D) Branch-Circuit Equipment Grounding Conductor The insulated equipment grounding conductor for branch circuits shall terminate at a grounding terminal in a remote panelboard or the grounding terminal in the main service equipment.

(E) Feeder Equipment Grounding Conductors Where a feeder supplies a remote panelboard, an insulated equipment grounding conductor shall extend from a grounding terminal in the service equipment to a grounding terminal in the remote panelboard.

555.17 Disconnecting Means for Shore Power Connection(s)

Disconnecting means shall be provided to isolate each boat from its supply connection(s).

(A) Type The disconnecting means shall be permitted to consist of a circuit breaker, switch, or both, and shall be properly identified as to which receptacle it controls.

(B) Location The disconnecting means shall be readily accessible, located not more than 762 mm (30 in.) from the receptacle it controls, and shall be located in the supply circuit ahead of the receptacle. Circuit breakers or switches located in marine power outlets complying with this section shall be permitted as the disconnecting means.

555.19 Receptacles

Receptacles shall be mounted not less than 305 mm (12 in.) above the deck surface of the pier and not below the electrical datum plane on a fixed pier.

In accordance with 555.19, the location of enclosures for receptacles on fixed and floating piers is based on the electrical datum plane, as defined in 555.2. For floating piers, the datum plane is 12 in. above the deck of the pier. The purpose of this requirement is to prevent submersion of receptacle enclosures.

The requirements for enclosures in 555.19(A)(1) address their exposure to the severe weather (wind-driven rain) and environmental conditions (splashing from breaking waves or wakes) frequently encountered at marine locations.

(A) Shore Power Receptacles

(1) Enclosures Receptacles intended to supply shore power to boats shall be housed in marine power outlets listed as marina power outlets or listed for set locations, or shall be installed in listed enclosures protected from the weather or in listed weatherproof enclosures. The integrity of the assembly shall not be affected when the receptacles are in use with any type of booted or nonbooted attachment plug/cap inserted.

(2) Strain Relief Means shall be provided where necessary to reduce the strain on the plug and receptacle caused by the weight and catenary angle of the shore power cord.

(3) Branch Circuits Each single receptacle that supplies shore power to boats shall be supplied from a marine power outlet or panelboard by an individual branch circuit of the voltage class and rating corresponding to the rating of the receptacle.

> FPN: Supplying receptacles at voltages other than the voltages marked on the receptacle may cause overheating or malfunctioning of connected equipment, for example, supplying single-phase, 120/240-volt, 3-wire loads from a 208Y/120-volt, 3-wire source.

Each single receptacle that supplies shore power to boats must be supplied from an individual branch circuit. The requirement for shore power receptacles to be supplied by individual branch circuits can be met through the use of multiwire branch circuits derived from single-phase, 3-wire systems or from 3-phase, 4-wire systems. Although the ungrounded conductors of a multiwire branch circuit share the same grounded (neutral) conductor, this configuration can be considered multiple branch circuits in accordance with 210.4(A). See the commentary following 300.13(B) regarding device removal for multiwire branch circuits.

Locking- and grounding-type receptacles and attachment caps must ensure proper connections to prevent unintentional disconnection of on-board equipment, such as bilge pumps, refrigerators, and so on.

(4) Ratings Shore power for boats shall be provided by single receptacles rated not less than 30 amperes.

> FPN: For locking- and grounding-type receptacles for auxiliary power to boats, see NFPA 303-2000, *Fire Protection Standard for Marinas and Boatyards.*

(a) Receptacles rated not less than 30 amperes or more than 50 amperes shall be of the locking and grounding type.

> FPN: For various configurations and ratings of locking and grounding-type receptacles and caps, see ANSI/NEMA 18WD 6-1989, National Electrical Manufacturers Association's *Standard for Dimensions of Attachment Plugs and Receptacles.*

(b) Receptacles rated for 60 amperes or 100 amperes shall be of the pin and sleeve type.

> FPN: For various configurations and ratings of pin and sleeve receptacles, see ANSI/UL 1686, *UL Standard for Safety Pin and Sleeve Configurations.*

Single locking- and grounding-type receptacles are required for providing shore power to boats. Exhibit 555.1 and Exhibit 406.3 together illustrate a complete chart of grounding-type locking plug and receptacle configurations. Exhibit 555.2 shows pin-and-sleeve-type receptacle configurations.

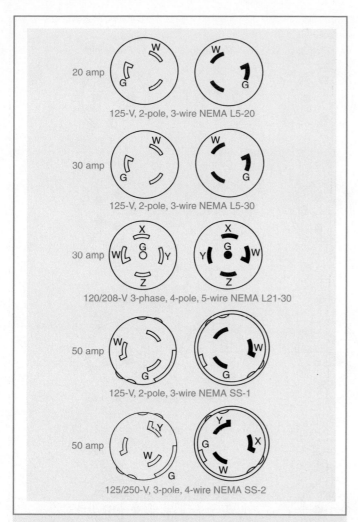

20 amp
125-V, 2-pole, 3-wire NEMA L5-20

30 amp
125-V, 2-pole, 3-wire NEMA L5-30

30 amp
120/208-V 3-phase, 4-pole, 5-wire NEMA L21-30

50 amp
125-V, 2-pole, 3-wire NEMA SS-1

50 amp
125/250-V, 3-pole, 4-wire NEMA SS-2

Exhibit 555.1 Typical configurations for single locking- and grounding-type receptacles and attachment plug caps used to provide shore power for boats in marinas and boatyards. These configurations are 30 amperes to 50 amperes.

(B) Other Than Shore Power

(1) Ground-Fault Circuit-Interrupter (GFCI) Protection for Personnel Fifteen- and 20-ampere, single-phase, 125-volt receptacles installed outdoors, in boathouses, in buildings used for storage, maintenance, or repair where portable electrical hand tools, electrical diagnostic equipment, or portable lighting equipment are to be used shall be provided with GFCI protection for personnel. Receptacles in other locations shall be protected in accordance with 210.8(B).

Fifteen- and 20-ampere, single-phase, 125-volt receptacles, other than those supplying shore power to boats and used for maintenance or other purposes at piers, wharves, and so on, may be of the general-purpose, nonlocking type and must be protected by GFCIs. See Exhibit 210.18.

(2) Marking Receptacles other than those supplying shore power to boats shall be permitted to be housed in marine power outlets with the receptacles that provide shore power to boats, provided they are marked to clearly indicate that they are not to be used to supply power to boats.

555.21 Motor Fuel Dispensing Stations — Hazardous (Classified) Locations

Electrical wiring and equipment located at or serving motor fuel dispensing stations shall comply with Article 514 in addition to the requirements of this article. All electrical wiring for power and lighting shall be installed on the side of the wharf, pier, or dock opposite from the liquid piping system.

Section 555.21 was expanded for the 2005 *Code* to require all electrical wiring for power and lighting to be installed on the opposite side of the wharf or pier from motor fuel dispensers and fuel piping. In addition to the requirements in Article 514, Section 6.3 in NFPA 303, *Fire Protection Standard for Marinas and Boatyards*, and NFPA 30A, *Code for Motor Fuel Dispensing Facilities and Repair Garages*, contain requirements pertaining to gasoline dispensing facilities and operations.

FPN: For additional information, see NFPA 303-2000, *Fire Protection Standard for Marinas and Boatyards*, and NFPA 30A-2003, *Motor Fuel Dispensing Facilities and Repair Garages*.

555.22 Repair Facilities — Hazardous (Classified) Locations

Electrical wiring and equipment located at facilities for the repair of marine craft containing flammable or combustible liquids or gases shall comply with Article 511 in addition to the requirements of this article.

Section 555.22 was added to the 2005 *Code* to require that facilities for the repair of marine craft that contain flammable or combustible liquids or gases comply with the requirements of Article 511.

555.23 Marine Hoists, Railways, Cranes, and Monorails

Motors and controls for marine hoists, railways, cranes, and monorails shall not be located below the electrical datum plane. Where it is necessary to provide electric power to a mobile crane or hoist in the yard and a trailing cable is utilized, it shall be a listed portable power cable rated for the conditions of use and be provided with an outer jacket of distinctive color for safety.

Exhibit 555.2 Typical configurations for safety pin-and-sleeve-type receptacles, plugs, connectors, and power inlets used to provide shore power for boats in marinas and boatyards. These configurations are 60 amperes or 100 amperes.

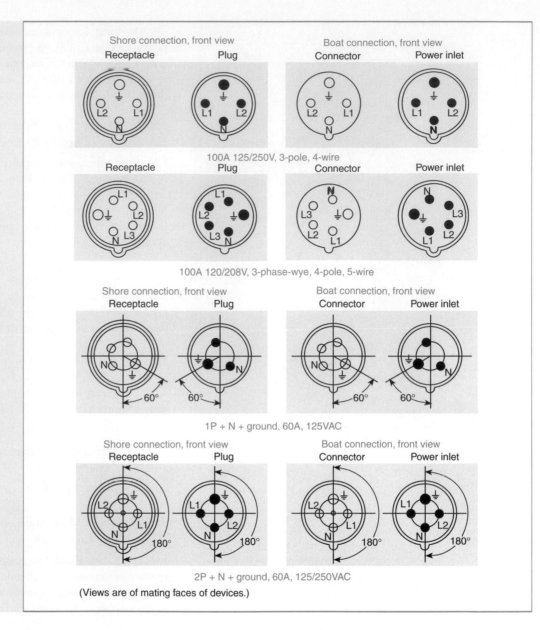

Shore connection, front view · Boat connection, front view

100A 125/250V, 3-pole, 4-wire

100A 120/208V, 3-phase-wye, 4-pole, 5-wire

1P + N + ground, 60A, 125VAC

2P + N + ground, 60A, 125/250VAC

(Views are of mating faces of devices.)

ARTICLE 590
Temporary Installations

Summary of Changes

- Relocated Article 527 as Article 590.
- **590.4(B) and (C):** Revised to include an exemption for building construction type where Type NM or NMC cable is used for temporary installations, and to permit Type NM and NMC cables to be used without concealment.
- **590.4(J):** Added exception to permit trees as a support method for overhead spans of branch circuits supplying holiday lighting.

- **590.5:** Added requirement for listing of decorative lighting used for holiday or similar displays.

Contents

590.1 Scope

The provisions of this article apply to temporary electrical power and lighting installations.

590.2 All Wiring Installations

(A) Other Articles Except as specifically modified in this article, all other requirements of this *Code* for permanent wiring shall apply to temporary wiring installations.

Temporary installations of electrical equipment must be installed in accordance with all applicable permanent installation requirements except as modified by the rules in this article. For example, the requirements of 300.15 specify that a box or other enclosure must be used where splices are made. This rule is amended by 590.4(G, which, for construction sites, permits splices to be made in multiconductor cords and cables without the use of a box.

(B) Approval Temporary wiring methods shall be acceptable only if approved based on the conditions of use and any special requirements of the temporary installation.

The provisions of 590.2(B) require that all temporary wiring methods be approved based on criteria such as length of time in service, severity of physical abuse, exposure to weather, and other special requirements. Special requirements may range from tunnel construction projects and tent cities constructed after a natural disaster to flammable hazardous material reclamation projects.

590.3 Time Constraints

(A) During the Period of Construction Temporary electrical power and lighting installations shall be permitted during the period of construction, remodeling, maintenance, repair, or demolition of buildings, structures, equipment, or similar activities.

(B) 90 Days Temporary electrical power and lighting installations shall be permitted for a period not to exceed 90 days for holiday decorative lighting and similar purposes.

Note that the 90-day time limit in 590.3(B) applies only to temporary electrical installations associated with holiday displays. Construction and emergency and test temporary wiring installations are not bound by this time limit.

(C) Emergencies and Tests Temporary electrical power and lighting installations shall be permitted during emergencies and for tests, experiments, and developmental work.

(D) Removal Temporary wiring shall be removed immediately upon completion of construction or purpose for which the wiring was installed.

Due to the modifications permitted by Article 590, temporary wiring installations may not meet all of the requirements for a permanent installation. Therefore, all temporary wiring not only must be disconnected but also must be removed from the building, structure, or other location of installation.

590.4 General

(A) Services Services shall be installed in conformance with Article 230.

(B) Feeders Overcurrent protection shall be provided in accordance with 240.4, 240.5, 240.100, and 240.101. They shall originate in an approved distribution center. Conductors shall be permitted within cable assemblies or within multiconductor cords or cables of a type identified in Table 400.4 for hard usage or extra-hard usage. For the purpose of this section, Type NM and Type NMC cables shall be permitted to be used in any dwelling, building, or structure without any height limitation or limitation by building construction type and without concealment within walls, floors, or ceilings.

Section 590.4(B) was revised for the 2005 *Code* by allowing Type NM and Type NMC cable to be used in any building or structure regardless of building height and construction type in which the cable is used. This change appeared as a Tentative Interim Amendment, dated October 23, 2002, for the 2002 *Code*.

Temporary feeders are permitted to be cable assemblies, multiconductor cords, or single-conductor cords. Cords used as feeders must be identified for hard or extra-hard usage according to Table 400.4. Individual conductors, as described in Table 310.13, are not permitted as open conductors but, rather, must be part of a cable assembly or used in a raceway system. Open or individual conductor feeders are permitted only during emergencies or tests.

All temporary wiring methods must be approved by the authority having jurisdiction. [See 590.2(B).]

Exception: Single insulated conductors shall be permitted where installed for the purpose(s) specified in 590.3(C), where accessible only to qualified persons.

(C) Branch Circuits All branch circuits shall originate in an approved power outlet or panelboard. Conductors shall be permitted within cable assemblies or within multiconductor cord or cable of a type identified in Table 400.4 for hard usage or extra-hard usage. Conductors shall be protected from overcurrent as provided in 240.4, 240.5, and 240.100. For the purposes of this section, Type NM and Type NMC cables shall be permitted to be used in any dwelling, building, or structure without any height limitation or limitation by building construction type and without concealment within walls, floors, or ceilings.

Section 590.4(C) was revised for the 2005 *Code* by allowing Type NM and Type NMC cable to be used in any building or structure regardless of the height or construction type of the building in which the cable is used. This change appeared as a Tentative Interim Amendment, dated October 23, 2002, for the 2002 *Code*.

The basic requirement for safety in 590.4(C) is that temporary wiring be located and installed so that it will not be physically damaged. In accordance with 590.2(A), temporary wiring must be installed in accordance with the appropriate Chapter 3 article for the wiring method employed (unless modified in Article 590).

Note that hard-usage or extra-hard-usage extension cords are permitted to be laid on the floor.

Exception: Branch circuits installed for the purposes specified in 590.3(B) or 590.3(C) shall be permitted to be run as single insulated conductors. Where the wiring is installed in accordance with 590.3(B), the voltage to ground shall not exceed 150 volts, the wiring shall not be subject to physical damage, and the conductors shall be supported on insulators at intervals of not more than 3.0 m (10 ft); or, for festoon lighting, the conductors shall be so arranged that excessive strain is not transmitted to the lampholders.

(D) Receptacles All receptacles shall be of the grounding type. Unless installed in a continuous grounded metal raceway or metal-covered cable, all branch circuits shall contain a separate equipment grounding conductor, and all receptacles shall be electrically connected to the equipment grounding conductors. Receptacles on construction sites shall not be installed on branch circuits that supply temporary lighting. Receptacles shall not be connected to the same ungrounded conductor of multiwire circuits that supply temporary lighting.

The intent of the branch-circuit provisions in 590.4(D) is to require separate ungrounded conductors for lighting and receptacle loads so that the activation of a fuse, circuit breaker, or GFCI, due to a fault or equipment overload, does not de-energize the lighting circuit.

(E) Disconnecting Means Suitable disconnecting switches or plug connectors shall be installed to permit the disconnection of all ungrounded conductors of each temporary circuit. Multiwire branch circuits shall be provided with a means to disconnect simultaneously all ungrounded conductors at the power outlet or panelboard where the branch circuit originated. Approved handle ties shall be permitted.

(F) Lamp Protection All lamps for general illumination shall be protected from accidental contact or breakage by a suitable fixture or lampholder with a guard.

Brass shell, paper-lined sockets, or other metal-cased sockets shall not be used unless the shell is grounded.

(G) Splices On construction sites, a box shall not be required for splices or junction connections where the circuit conductors are multiconductor cord or cable assemblies, provided that the equipment grounding continuity is maintained with or without the box. See 110.14(B) and 400.9. A box, conduit body, or terminal fitting having a separately bushed hole for each conductor shall be used wherever a change is made to a conduit or tubing system or a metal-sheathed cable system.

(H) Protection from Accidental Damage Flexible cords and cables shall be protected from accidental damage. Sharp corners and projections shall be avoided. Where passing through doorways or other pinch points, protection shall be provided to avoid damage.

Unlike the requirement in 400.8, flexible cords and cables, because of the nature of their use, are permitted to pass through doorways, in accordance with 590.4(H).

(I) Termination(s) at Devices Flexible cords and cables entering enclosures containing devices requiring termination shall be secured to the box with fittings designed for the purpose.

(J) Support Cable assemblies and flexible cords and cables shall be supported in place at intervals that ensure that they will be protected from physical damage. Support shall be in the form of staples, cable ties, straps, or similar type fittings installed so as not to cause damage. Vegetation shall not be used for support of overhead spans of branch circuits or feeders.

Section 590.4(J), Exception was added to the 2005 *Code* to allow holiday lighting to be installed and supported by trees for a period of not more than 90 days, provided the wiring is arranged with proper strain relief devices, tension take-up devices, or other means to prevent damage to the conductors from the tree swaying.

According to 590.4(J), temporary wiring methods do not have to be supported in accordance with the permanent installation requirements (from Chapter 3) for the particular wiring method. It should be noted that the temporary wiring must be removed upon completion of construction and adequate support is needed only to minimize the possibility of damage to the wiring method during its temporary period of use. It is not permitted to use vegetation as a support structure for overhead spans of branch-circuit and feeder conductors.

Exception: For holiday lighting in accordance with 590.3(B), where the conductors or cables are arranged with proper strain relief devices, tension take-up devices, or other approved means to avoid damage from the movement of the live vegetation, trees shall be permitted to be used for support of overhead spans of branch circuit conductors or cables.

590.5 Listing of Decorative Lighting

Decorative lighting used for holiday lighting and similar purposes, in accordance with 590.3(B), shall be listed.

590.6 Ground-Fault Protection for Personnel

Ground-fault protection for personnel for all temporary wiring installations shall be provided to comply with 590.6(A) and 590.6(B). This section shall apply only to temporary wiring installations used to supply temporary power to equipment used by personnel during construction, remodeling, maintenance, repair, or demolition of buildings, structures, equipment, or similar activities.

(A) Receptacle Outlets All 125-volt, single-phase, 15-, 20-, and 30-ampere receptacle outlets that are not a part of the permanent wiring of the building or structure and that are in use by personnel shall have ground-fault circuit interrupter protection for personnel. If a receptacle(s) is installed or exists as part of the permanent wiring of the building or

structure and is used for temporary electric power, ground-fault circuit-interrupter protection for personnel shall be provided. For the purposes of this section, cord sets or devices incorporating listed ground-fault circuit interrupter protection for personnel identified for portable use shall be permitted.

Exception: In industrial establishments only, where conditions of maintenance and supervision ensure that only qualified personnel are involved, an assured equipment grounding conductor program as specified in 590.6(B)(2) shall be permitted for only those receptacle outlets used to supply equipment that would create a greater hazard if power was interrupted or having a design that is not compatible with GFCI protection.

(B) Use of Other Outlets Receptacles other than 125-volt, single-phase, 15-, 20-, and 30-ampere receptacles shall have protection in accordance with (B)(1) or the assured equipment grounding conductor program in accordance with (B)(2).

(1) GFCI Protection Ground-fault circuit interrupter protection for personnel.

(2) Assured Equipment Grounding Conductor Program A written assured equipment grounding conductor program continuously enforced at the site by one or more designated persons to ensure that equipment grounding conductors for all cord sets, receptacles that are not a part of the permanent wiring of the building or structure, and equipment connected by cord and plug are installed and maintained in accordance with the applicable requirements of 250.114, 250.138, 406.3(C), and 590.4(D).

(a) The following tests shall be performed on all cord sets, receptacles that are not part of the permanent wiring of the building or structure, and cord-and-plug-connected equipment required to be grounded:

(1) All equipment grounding conductors shall be tested for continuity and shall be electrically continuous.
(2) Each receptacle and attachment plug shall be tested for correct attachment of the equipment grounding conductor. The equipment grounding conductor shall be connected to its proper terminal.
(3) All required tests shall be performed as follows:
   a. Before first use on site
   b. When there is evidence of damage
   c. Before equipment is returned to service following any repairs
   d. At intervals not exceeding 3 months

(b) The tests required in item (2)(a) shall be recorded and made available to the authority having jurisdiction.

Due to the more severe environmental conditions often encountered by personnel using temporary wiring while performing activities such as construction, remodeling, maintenance, repair, and demolition, there is generally an elevated exposure to electrical shock or electrocution hazards. The requirement of 590.6(A) for GFCI protection of all temporarily installed, 125-volt, single-phase, 15-, 20-, and 30-ampere receptacles is intended to protect personnel using these receptacles from shock hazards that may be encountered during construction and maintenance activities.

The exception to 590.6(A) is limited in scope and application. The exception applies only to those industrial occupancies in which qualified persons will be using 125-volt, single-phase, 15-, 20-, and 30-ampere receptacles. Additionally, either the nature of the equipment being supplied by these receptacles has to be of such importance that the hazard of power interruption outweighs the benefits of GFCI protection or the equipment has been demonstrated to be incompatible with the proper operation of GFCI protective devices. In those instances where the conditions specified by the exception are present, the use of the assured equipment grounding conductor program specified in 590.6(B)(2) is permitted. An electrically operated air supply for personnel working in toxic environments is an example of where the loss of power is the greater hazard. Some electrically operated testing equipment has proved to be incompatible with GFCI protection.

Receptacle configurations, other than the 125-volt, single-phase, 15-, 20-, and 30-ampere types, must be GFCI protected or installed and maintained in accordance with the assured equipment grounding conductor program of 590.6(B)(2).

According to OSHA 29 CFR 1926.404(b)(1)(iii):

> The employer shall establish and implement an assured equipment grounding conductor program on construction sites covering all cord sets, receptacles which are not a part of the building or structure, and equipment connected by cord and plug which are available for use or used by employees. This program shall comply with the following minimum requirements:
>
> (A) A written description of the program, including the specific procedures adopted by the employer, shall be available at the jobsite for inspection and copying by the Assistant Secretary and any affected employee.
>
> (B) The employer shall designate one or more competent persons.

These OSHA requirements are very similar to the present *NEC* requirements for an assured grounding program.

GFCI protection for construction or maintenance personnel using receptacles that are part of the permanent wiring

Exhibit 590.1 A raintight GFCI with open neutral protection that is designed for use on the line end of a flexible cord. (Courtesy of Pass & Seymour/Legrand®)

Exhibit 590.2 A temporary power outlet unit commonly used on construction sites with a variety of configurations, including GFCI protection. (Courtesy of Hubbell, Inc.)

Exhibit 590.3 A watertight plug and connector used to prevent tripping of GFCI protective devices in wet or damp weather. (Courtesy of Hubbell, Inc.)

Exhibit 590.4 A 15-ampere duplex receptacle with integral GFCI that also protects downstream loads. (Courtesy of Pass & Seymour/Legrand®)

and that are not GFCI protected may be provided by using cord sets or listed portable GFCIs identified for portable use. An example of a GFCI cord set that is identified for portable use is shown in Exhibit 590.1.

Exhibits 590.1 through 590.4 show some examples of ways to implement the temporary wiring requirements of 590.6.

590.7 Guarding

For wiring over 600 volts, nominal, suitable fencing, barriers, or other effective means shall be provided to limit access only to authorized and qualified personnel.

Special Equipment

ARTICLE 600
Electric Signs and Outline Lighting

Summary of Changes

- **600.1:** Revised to include all installations and equipment using neon tubing, such as signs, decorative elements, skeleton tubing, or art forms within the scope.

- **600.2:** Added new definition for *section sign*.

- **600.12:** Added requirement for field-installed secondary circuit wiring of section signs.

- **600.24:** Added requirement that signs and outline lighting systems supplied by Class 2 transformers, etc., also comply with Article 725.

Contents

I. General

600.1 Scope

This article covers the installation of conductors and equipment for electric signs and outline lighting. All installations and equipment using neon tubing, such as signs, decorative elements, skeleton tubing, or art forms, are covered by this article.

Article 600 provides construction and installation requirements for electric signs and outline lighting. Covered under these requirements are signs of the fixed, stationary, and portable self-contained type. Electric signs and outline lighting frequently include sources of illumination identical to those of luminaires; however, the structure and electrical operation of many of today's electric signs are far more complex than simply a set of fluorescent lamps within an enclosure. The terms *electric sign* and *outline lighting* as defined in Article 100 clearly distinguish the function and use of equipment covered by the requirements of Article 600 from the equipment covered by the requirements of Article 410. It should be noted, however, that 600.3(B) does defer to the requirements of Chapters 3 and 4 where individual listed luminaires are used for outline lighting.

Section 600.1 has been revised to clearly describe the types of equipment covered by the requirements of this article and now includes text that was formerly expressed as a nonmandatory fine print note to 600.1. Neon tubing is used extensively in today's sign industry, and its uses go far beyond the typical electric sign or outline lighting applications. Today, decorative and artistic use of neon tubing is extremely popular, and it is being installed indoors and outdoors to enhance the appearance of buildings and structures. These neon art forms are mounted on enclosures, sign bodies, and other support structures, or they may be field-installed skeleton tubing. Depending on how these neon art forms are constructed and installed, they are subject to either the requirements of Part I or the requirements of Parts I and II of Article 600. The revised scope statement makes it clear that all neon installations and applications are covered by the requirements of Article 600.

600.2 Definitions

Electric-Discharge Lighting. Systems of illumination utilizing fluorescent lamps, high-intensity discharge (HID) lamps, or neon tubing.

Neon Tubing. Electric-discharge tubing manufactured into shapes that form letters, parts of letters, skeleton tubing, outline lighting, other decorative elements, or art forms, and filled with various inert gases.

Section Sign. A sign or outline lighting system, shipped as subassemblies, that requires field-installed wiring between the subassemblies to complete the overall sign.

In many cases, the size or other logistical factors affecting the installation of an electric sign dictate that it be constructed in multiple factory-wired subassemblies that can be assembled at the sign-installation location. The subassemblies may be physically joined to form a single sign unit, or they may be installed as separate parts of an overall sign. The new definition of *section sign* clarifies that the multiple parts of a section sign are referred to as subassemblies and the only field wiring involved are the connections between subassemblies and connection of the subassemblies to the power source.

The power source may be a line voltage branch circuit or the secondary wiring from a sign power supply. In accordance with 600.3, section signs are required to be listed. In accordance with UL 48, *Electric Signs*, each subassembly is provided with installation instructions containing detailed information on the mechanical and electrical connections that will be performed when the subassemblies are installed to form the completed section sign. A section sign, transported in multiple subassemblies, is shown in Exhibit 600.1. The size of this sign requires that it be constructed in multiple sections and assembled at the installation site.

Exhibit 600.1 Example of a section sign. (Courtesy of Kieffer & Co. Inc.)

Sign Body. A portion of a sign that may provide protection from the weather but is not an electrical enclosure.

Skeleton Tubing. Neon tubing that is itself the sign or outline lighting and not attached to an enclosure or sign body.

600.3 Listing

Electric signs, section signs, and outline lighting — fixed, mobile, or portable — shall be listed and installed in conformance with that listing, unless otherwise approved by special permission.

(A) Field-Installed Skeleton Tubing Field-installed skeleton tubing shall not be required to be listed where installed in conformance with this *Code*.

(B) Outline Lighting Outline lighting shall not be required to be listed as a system when it consists of listed luminaires (lighting fixtures) wired in accordance with Chapter 3.

Section 600.3 requires listing of electric signs and outline lighting unless the sign or lighting is approved through special permission (i.e., written consent) from the authority having jurisdiction. Electric signs come in many different types and configurations and, depending on their construction and operation, contain electrical components such as transformers, ballasts, fluorescent lamps, channel letters, plastic faces, neon tubing, glass tube supports, raceways, glass cups or insulating boots (if provided), and in some cases factory-installed disconnecting means. Like other requirements in the *NEC* specifying listed products, construction and evaluation of electric signs in accordance with a recognized product standard and field installation in accordance with instructions provided as part of the product listing help ensure that electrical equipment does not pose a shock or fire hazard.

Section signs, which may be large signs constructed in multiple sections or remote wired letters, are often transported in several parts from the factory to the location where they are to be field-assembled into a single sign. According to the definition of *section sign* in 600.2, the individual parts of a section sign are *subassemblies*. The subassemblies are wired together in the field to form the completed section sign. The applicable product standard (UL 48; see Annex A) specifies that all necessary materials and detailed installation instructions, often including wiring and assembly diagrams, be included with each subassembly. At its installation destination, the installer and the approving authority must verify that the section sign installation complies with all applicable requirements of the *Code* and that the assembly of the section sign is in conformity with the installation instructions.

Skeleton tubing (neon tubing mounted directly on a

building) that has not been listed is permitted to be field-installed by 600.3(A), provided the installation is done in accordance with the applicable requirements of the *Code*. In 600.3(B), installation of unlisted outline lighting systems constructed of listed lighting fixtures is permitted, provided the installation is performed in accordance with Chapters 3 and 4 of the *Code*.

600.4 Markings

(A) Signs and Outline Lighting Systems Signs and outline lighting systems shall be marked with the manufacturer's name, trademark, or other means of identification; and input voltage and current rating.

(B) With Incandescent Lamp Holders Signs and outline lighting systems with incandescent lamp holders shall be marked to indicate the maximum allowable wattage of lamps. The markings shall be permanently installed, in letters at least 6 mm (¼ in.) high, and shall be located where visible during relamping.

600.5 Branch Circuits

(A) Required Branch Circuit Each commercial building and each commercial occupancy accessible to pedestrians shall be provided with at least one outlet in an accessible location at each entrance to each tenant space for sign or outline lighting system use. The outlet(s) shall be supplied by a branch circuit rated at least 20 amperes that supplies no other load. Service hallways or corridors shall not be considered accessible to pedestrians.

Electric signs and outline lighting to attract the attention of consumers are used extensively by commercial enterprises. The need for an accessible provision to facilitate the connection of an electric sign at a commercial building is covered by 600.5(A). A 20-ampere outlet on a circuit dedicated to the purpose of supplying an electric sign(s) is required to be installed at the entrance to single-occupant commercial buildings and at the entrance to each occupancy of multiple-occupant commercial buildings (e.g., shopping malls). This requirement is not contingent on whether an electric sign will be installed at the time an occupant moves into a commercial space, since it is not uncommon for an electric sign to be installed after the space is occupied or when a new occupant moves into an existing space. For branch circuit, feeder and service calculations, a minimum load of 1200 VA per 20-ampere sign branch circuit is required to be added. Due to the nature of their use, electric signs are considered to be continuous loads, and as such the conductors and overcurrent devices supplying sign loads have to be adjusted in size accordingly.

(B) Rating Branch circuits that supply signs shall be rated in accordance with 600.5(B)(1) or (B)(2).

(1) Incandescent and Fluorescent Branch circuits that supply signs and outline lighting systems containing incandescent and fluorescent forms of illumination shall be rated not to exceed 20 amperes.

(2) Neon Branch circuits that supply neon tubing installations shall not be rated in excess of 30 amperes.

Branch circuits that supply electric signs are limited by 600.5(B)(1) and 600.5(B)(2) to ratings of 20 and 30 amperes, respectively. Large signs often have load requirements that exceed the ratings permitted by 600.5(B). It is typical for signs having large electrical loads to be supplied by a feeder that in turn supplies branch circuits operating within the parameters specified by this requirement. In some cases, particularly for signs installed along highways, a utility service dedicated to the sign is provided. The rating of the feeder or service is not limited by this requirement.

(C) Wiring Methods Wiring methods used to supply signs shall comply with 600.5(C)(1), (C)(2), and (C)(3).

(1) Supply The wiring method used to supply signs and outline lighting systems shall terminate within a sign, an outline lighting system enclosure, a suitable box, or a conduit body.

(2) Enclosures as Pull Boxes Signs and transformer enclosures shall be permitted to be used as pull or junction boxes for conductors supplying other adjacent signs, outline lighting systems, or floodlights that are part of a sign and shall be permitted to contain both branch and secondary circuit conductors.

(3) Metal Poles Metal poles used to support signs shall be permitted to enclose supply conductors, provided the poles and conductors are installed in accordance with 410.15(B).

600.6 Disconnects

Each sign and outline lighting system, or feeder circuit or branch circuit supplying a sign or outline lighting system, shall be controlled by an externally operable switch or circuit breaker that will open all ungrounded conductors. Signs and outline lighting systems located within fountains shall have the disconnect located in accordance with 680.12.

Exception No. 1: A disconnecting means shall not be required for an exit directional sign located within a building.

Exception No. 2: A disconnecting means shall not be required for cord-connected signs with an attachment plug.

(A) Location

(1) Within Sight of the Sign The disconnecting means shall be within sight of the sign or outline lighting system that it controls. Where the disconnecting means is out of the line of sight from any section that is able to be energized, the disconnecting means shall be capable of being locked in the open position.

The requirement of 600.6(A)(1) covers sign installations where the branch circuit or feeder is run directly to the sign. Each branch circuit or feeder supplying a sign must have an externally operable switch or circuit breaker to open the ungrounded conductors. Two options are permitted for locating the sign disconnecting means. The disconnecting means is required either to be located within sight of the sign or to be equipped with the provision to lock it in the open position. Exhibit 600.2 depicts a sign with two supply circuits, which could be feeders or branch circuits. Each circuit is provided with an externally operable switch located within sight of the sign.

Exhibit 600.3 illustrates three compliant alternatives. The supply circuit disconnecting means shown in Example 1 is externally operable and located at and within sight of the sign. The disconnecting means in Example 2 is externally operable, and its location, though not at or on the sign, is acceptable because it meets the Article 100 definition of *within sight*. Where the disconnecting means is not located within sight of the sign, as shown in Example 3, it is required to be located within sight of the controller and must be capable of being locked in the open position.

Exhibit 600.2 Supply circuit disconnecting means located at or on an electric sign.

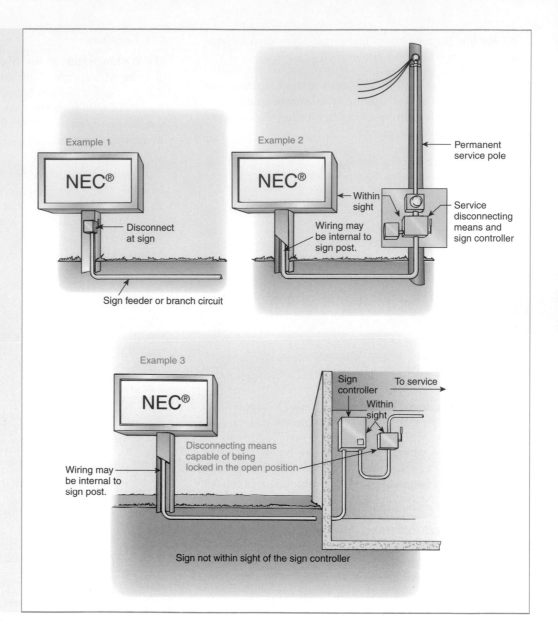

Exhibit 600.3 Three acceptable methods of providing the disconnecting means for an electric sign.

Example 1

NEC®

Disconnect at sign

Sign feeder or branch circuit

Example 2

NEC®

Within sight

Wiring may be internal to sign post.

Permanent service pole

Service disconnecting means and sign controller

Example 3

NEC®

Sign controller

To service

Within sight

Disconnecting means capable of being locked in the open position

Wiring may be internal to sign post.

Sign not within sight of the sign controller

(2) Within Sight of the Controller The following shall apply for signs or outline lighting systems operated by electronic or electromechanical controllers located external to the sign or outline lighting system:

(1) The disconnecting means shall be permitted to be located within sight of the controller or in the same enclosure with the controller.
(2) The disconnecting means shall disconnect the sign or outline lighting system and the controller from all ungrounded supply conductors.
(3) The disconnecting means shall be designed such that no pole can be operated independently and shall be capable of being locked in the open position.

For signs or outline lighting systems operated by mechanical or electromechanical controllers located external to the sign, the disconnecting means is required to be located within sight of or in the same enclosure as the controller and must be capable of being locked in the open position. This requirement enhances safe working conditions for persons servicing the controller or the sign.

(B) Control Switch Rating Switches, flashers, and similar devices controlling transformers and electronic power supplies shall be rated for controlling inductive loads or have a current rating not less than twice the current rating of the transformer.

FPN: See 404.14 for rating of snap switches.

A switching device that controls the primary circuit of a transformer supplying a luminous gas tube is subject to a highly inductive load that causes severe arcing of its contacts. Therefore, the switch or flasher is required to be rated for the inductive load (a general-use ac snap switch used in accordance with 404.14 is permitted) or must have a current rating that is at least twice the current rating of the transformer it controls.

600.7 Grounding

Signs and metal equipment of outline lighting systems shall be grounded.

The requirements of 600.7 cover grounding and bonding of signs and metal equipment of outline lighting. All metal parts larger than 2 in. are to be bonded. It is common in the sign industry to use flexible metal raceways to enclose the conductors supplied from the secondary circuit of a transformer or electronic power supply. In addition to providing protection from physical damage, enclosing the secondary conductors in flexible metal conduit or liquidtight flexible metal conduit is permitted as the bonding means for non–current-carrying metal parts of electric signs. Where used for bonding in the secondary circuit, the total length of flexible metal conduit in the secondary circuit is not permitted to exceed 100 ft.

Where rigid or flexible nonmetallic conduit is used on secondary wiring of transformers and power supplies, copper bonding conductors must be located external to the nonmetallic raceway. Bonding conductor spacing is required to be at least 1½ in. for circuit frequencies of 100 Hz or less and 1¾ in. for circuit frequencies over 100 Hz. These raceways normally contain only one conductor, which is connected to one side of the neon tube. Where rigid nonmetallic conduit or liquidtight flexible nonmetallic conduit is used and any sign parts are required to be bonded, the bonding conductor(s) must be run outside of and be separated from the nonmetallic conduit. Installing bonding conductors inside the nonmetallic conduit with secondary power supply conductors is not permitted, since it could increase the chance of failure of the conductor or nonmetallic tubing.

(A) Flexible Metal Conduit Length Listed flexible metal conduit or listed liquidtight flexible metal conduit that encloses the secondary circuit conductor from a transformer or power supply for use with electric discharge tubing shall be permitted as a bonding means if the total accumulative length of the conduit in the secondary circuit does not exceed 30 m (100 ft).

(B) Small Metal Parts Small metal parts not exceeding 50 mm (2 in.) in any dimension, not likely to be energized, and spaced at least 19 mm (¾ in.) from neon tubing shall not require bonding.

(C) Nonmetallic Conduit Where listed nonmetallic conduit is used to enclose the secondary circuit conductor from a transformer or power supply and a bonding conductor is required, the bonding conductor shall be installed separate and remote from the nonmetallic conduit and be spaced at least 38 mm (1½ in.) from the conduit when the circuit is operated at 100 Hz or less or 45 mm (1¾ in.) when the circuit is operated at over 100 Hz.

(D) Bonding Conductors Bonding conductors shall be copper and not smaller than 14 AWG.

(E) Metal Building Parts Metal parts of a building shall not be permitted as a secondary return conductor or an equipment grounding conductor.

(F) Signs in Fountains Signs or outline lighting installed inside a fountain shall have all metal parts and equipment grounding conductors bonded to the equipment grounding conductor for the fountain recirculating system. The bonding connection shall be as near as practicable to the fountain and shall be permitted to be made to metal piping systems that are bonded in accordance with 680.53.

FPN: Refer to 600.32(J) for restrictions on length of high-voltage secondary conductors.

600.8 Enclosures

Live parts, other than lamps, and neon tubing shall be enclosed. Transformers and power supplies provided with an integral enclosure, including a primary and secondary circuit splice enclosure, shall not require an additional enclosure.

(A) Strength Enclosures shall have ample structural strength and rigidity.

(B) Material Sign and outline lighting system enclosures shall be constructed of metal or shall be listed.

(C) Minimum Thickness of Enclosure Metal Sheet copper or aluminum shall be at least 0.51 mm (0.020 in.) thick. Sheet steel shall be at least 0.41 mm (0.016 in.) thick.

(D) Protection of Metal Metal parts of equipment shall be protected from corrosion.

600.9 Location

(A) Vehicles Sign or outline lighting system equipment shall be at least 4.3 m (14 ft) above areas accessible to vehicles unless protected from physical damage.

(B) Pedestrians Neon tubing, other than dry-location portable signs, accessible to pedestrians shall be protected from physical damage.

(C) Adjacent to Combustible Materials Signs and outline lighting systems shall be installed so that adjacent combustible materials are not subjected to temperatures in excess of 90°C (194°F).

The spacing between wood or other combustible materials and an incandescent or HID lamp or lampholder shall not be less than 50 mm (2 in.).

(D) Wet Location Signs and outline lighting system equipment for wet location use, other than listed watertight type, shall be weatherproof and have drain holes, as necessary, in accordance with the following:

(1) Drain holes shall not be larger than 13 mm (½ in.) or smaller than 6 mm (¼ in.).
(2) Every low point or isolated section of the equipment shall have at least one drain hole.
(3) Drain holes shall be positioned such that there will be no external obstructions.

600.10 Portable or Mobile Signs

Portable or mobile electric signs, such as those mounted on trailers, are subject to the requirements of 600.10. These requirements address the safety concerns associated with equipment that is frequently moved and that may be used in damp or wet environments.

(A) Support Portable or mobile signs shall be adequately supported and readily movable without the use of tools.

(B) Attachment Plug An attachment plug shall be provided for each portable or mobile sign.

(C) Wet or Damp Location Portable or mobile signs in wet or damp locations shall comply with 600.10(C)(1) and (C)(2).

(1) Cords All cords shall be junior hard service or hard service types as designated in Table 400.4 and have an equipment grounding conductor.

(2) Ground-Fault Circuit Interrupter Portable or mobile signs shall be provided with factory-installed ground-fault circuit-interrupter protection for personnel. The ground-fault circuit interrupter shall be an integral part of the attachment plug or shall be located in the power-supply cord within 300 mm (12 in.) of the attachment plug.

The ground-fault circuit-interrupters required for portable electric signs must have integral *open-neutral* protection in accordance with UL 48, *Standard for Electric Signs*. The

internal circuitry of a GFCI requires the presence of the 120-volt supply circuit for proper operation. An interruption of the neutral conductor on the supply side of the GFCI disables the protection circuitry. Open-neutral protection ensures that if there is damage to the supply cord that causes a break in the grounded conductor, both conductors on the load side circuit to the portable sign will be opened and there will be no voltage present at the sign. These protective devices are required to be original equipment installed by the manufacturer and can be located in-line with the supply cord as shown in Exhibit 600.4 or an attachment plug with an integrated GFCI device can be used.

Exhibit 600.4 A factory-installed GFCI device located in the power supply cord within 12 in. of the attachment plug.

(D) Dry Location Portable or mobile signs in dry locations shall meet the following:

(1) Cords shall be SP-2, SPE-2, SPT-2, or heavier, as designated in Table 400.4.
(2) The cord shall not exceed 4.5 m (15 ft) in length.

600.12 Field-Installed Secondary Wiring

The field-installed secondary circuit wiring of section signs shall comply with 600.31 if 1000 volts or less, or with 600.32 if over 1000 volts.

This new provision requires field-installed secondary circuit wiring of section signs to be installed in accordance with either 600.31 or 600.32, depending on the secondary circuit voltage. The two referenced sections are located in Part II of Article 600. In general, the requirements of Part II are intended for application only to field-installed skeleton tubing. However, since 600.31 and 600.32 provide detailed requirements for the safe installation of neon secondary conductors, they are referenced in 600.12 to ensure that the secondary circuits associated with section signs are installed correctly.

600.21 Ballasts, Transformers, and Electronic Power Supplies

(A) Accessibility Ballasts, transformers, and electronic power supplies shall be located where accessible and shall be securely fastened in place.

(B) Location Ballasts, transformers, and electronic power supplies shall be installed as near to the lamps or neon tubing as practicable to keep the secondary conductors as short as possible.

(C) Wet Location Ballasts, transformers, and electronic power supplies used in wet locations shall be of the weatherproof type or be of the outdoor type and protected from the weather by placement in a sign body or separate enclosure.

(D) Working Space A working space at least 900 mm (3 ft) high, 900 mm (3 ft) wide, by 900 mm (3 ft) deep shall be provided at each ballast, transformer, and electronic power supply or at its enclosure where not installed in a sign.

(E) Attic and Soffit Locations Ballasts, transformers, and electronic power supplies shall be permitted to be located in attics and soffits, provided there is an access door at least 900 mm by 600 mm (3 ft by 2 ft) and a passageway of at least 900 mm (3 ft) high by 600 mm (2 ft) wide with a suitable permanent walkway at least 300 mm (12 in.) wide extending from the point of entry to each component.

(F) Suspended Ceilings Ballasts, transformers, and electronic power supplies shall be permitted to be located above suspended ceilings, provided their enclosures are securely fastened in place and not dependent on the suspended ceiling grid for support. Ballasts, transformers, and electronic power supplies installed in suspended ceilings shall not be connected to the branch circuit by flexible cord.

600.22 Ballasts

(A) Type Ballasts shall be identified for the use and shall be listed.

(B) Thermal Protection Ballasts shall be thermally protected.

600.23 Transformers and Electronic Power Supplies

(A) Type Transformers and electronic power supplies shall be identified for the use and shall be listed.

(B) Secondary-Circuit Ground-Fault Protection Transformers and electronic power supplies other than the following shall have secondary-circuit ground-fault protection:

(1) Transformers with isolated ungrounded secondaries and with a maximum open circuit voltage of 7500 volts or less
(2) Transformers with integral porcelain or glass secondary housing for the neon tubing and requiring no field wiring of the secondary circuit

(C) Voltage Secondary-circuit voltage shall not exceed 15,000 volts, nominal, under any load condition. The voltage to ground of any output terminals of the secondary circuit shall not exceed 7500 volts, under any load condition.

(D) Rating Transformers and electronic power supplies shall have a secondary-circuit current rating of not more than 300 mA.

(E) Secondary Connections Secondary circuit outputs shall not be connected in parallel or in series.

(F) Marking A transformer or power supply shall be marked to indicate that it has secondary-circuit ground-fault protection.

600.24 Class 2 Power Sources

In addition to the requirements of Article 600, signs and outline lighting systems supplied by Class 2 transformers, power supplies, and power sources shall comply with 725.41.

II. Field-Installed Skeleton Tubing

600.30 Applicability

Part II of this article shall apply to field-installed skeleton tubing. These requirements are in addition to the requirements of Part I.

Field-installed skeleton tubing often involves the use of electrically isolated metal components, such as tube supports, fasteners, or decorative channel over the tubing. Such metallic components should be located so as to minimize the possibility of their becoming energized.

600.31 Neon Secondary-Circuit Conductors, 1000 Volts or Less, Nominal

(A) Wiring Method Conductors shall be installed using any wiring method included in Chapter 3 suitable for the conditions.

(B) Insulation and Size Conductors shall be listed, insulated, and not smaller than 18 AWG.

(C) Number of Conductors in Raceway The number of conductors in a raceway shall be in accordance with Table 1 of Chapter 9.

(D) Installation Conductors shall be installed so they are not subject to physical damage.

(E) Protection of Leads Bushings shall be used to protect wires passing through an opening in metal.

600.32 Neon Secondary Circuit Conductors, Over 1000 Volts, Nominal

(A) Wiring Methods

(1) Installation Conductors shall be installed on insulators, in rigid metal conduit, intermediate metal conduit, rigid nonmetallic conduit, liquidtight flexible nonmetallic conduit, flexible metal conduit, liquidtight flexible metal conduit, electrical metallic tubing, metal enclosures, or other equipment listed for the purpose and shall be installed in accordance with the requirements of Chapter 3.

(2) Number of Conductors Conduit or tubing shall contain only one conductor.

(3) Size Conduit or tubing shall be a minimum of metric designator 16 (trade size ½).

(4) Spacing from Ground Other than at the location of connection to a metal enclosure or sign body, nonmetallic conduit or flexible nonmetallic conduit shall be spaced no less than 38 mm (1½ in.) from grounded or bonded parts when the conduit contains a conductor operating at 100 Hz or less, and shall be spaced no less than 45 mm (1¾ in.) from grounded or bonded parts when the conduit contains a conductor operating at more than 100 Hz.

Locating GTO cable in close proximity to a grounded surface will result in damaging stress to the cable insulation. The requirements in 600.32(A)(4) provide for minimum separation between nonmetallic raceways containing neon secondary circuits and grounded or bonded metal parts.

(5) Metal Building Parts Metal parts of a building shall not be permitted as a secondary return conductor or an equipment grounding conductor.

See the commentary following 600.7.

(B) Insulation and Size Conductors shall be insulated, listed as gas tube sign and ignition cable type GTO, rated for 5, 10, or 15 kV, not smaller than 18 AWG, and have a minimum temperature rating of 105°C (221°F).

The 2002 *Code* revised the requirement for the type of conductors permitted in neon secondary circuits. The requirement previously specified that the use of conductors listed for the purpose be installed in the secondary circuit. The revised text requires the use of cable listed as gas tube sign and ignition cable, Type GTO, rated at 5, 10, or 15 kV. Annex A identifies the product standard for this cable as UL 814, *Gas-Tube-Sign Cable*.

(C) Installation Conductors shall be so installed that they are not subject to physical damage.

(D) Bends in Conductors Sharp bends in insulated conductors shall be avoided.

(E) Spacing Secondary conductors shall be separated from each other and from all objects other than insulators or neon tubing by a spacing of not less than 38 mm (1½ in.). GTO cable installed in metal conduit or tubing requires no spacing between the cable insulation and the conduit or tubing.

(F) Insulators and Bushings Insulators and bushings for conductors shall be listed for the purpose.

(G) Conductors in Raceways

(1) Damp or Wet Locations In damp or wet locations, the insulation on all conductors shall extend not less than 100 mm (4 in.) beyond the metal conduit or tubing.

(2) Dry Locations In dry locations, the insulation on all conductors shall extend not less than 65 mm (2½ in.) beyond the metal conduit or tubing.

(H) Between Neon Tubing and Midpoint Return Conductors shall be permitted to run between the ends of neon tubing or to the secondary circuit midpoint return of listed transformers or listed electronic power supplies and provided with terminals or leads at the midpoint.

(I) Dwelling Occupancies Equipment having an open circuit voltage exceeding 1000 volts shall not be installed in or on dwelling occupancies.

(J) Length of Secondary Circuit Conductors

(1) Secondary Conductor to the First Electrode The length of secondary circuit conductors from a high-voltage terminal or lead of a transformer or electronic power supply to the first neon tube electrode shall not exceed the following:

(1) 6 m (20 ft) where installed in metal conduit or tubing
(2) 15 m (50 ft) where installed in nonmetallic conduit

(2) Other Secondary Circuit Conductors All other sections of secondary circuit conductor in a neon tube circuit shall be as short as practicable.

600.41 Neon Tubing

(A) Design The length and design of the tubing shall not cause a continuous overcurrent beyond the design loading of the transformer or electronic power supply.

(B) Support Tubing shall be supported by listed tube supports.

(C) Spacing A spacing of not less than 6 mm (¼ in.) shall be maintained between the tubing and the nearest surface, other than its support.

Electric discharge tubing is required to be of such length and design that it will not cause a continuous overvoltage on the transformer. A tube that is too long or too small in diameter increases the impedance of the load and thus stresses the transformer insulation. Gas tube sign transformers are designed to operate at or near short-circuit current. Generally, the primary voltage of the transformers is 120 volts, and proper installation and maintenance of transformers and high-voltage secondary conductors minimize the possibility of injury or fire. Precautions should be taken to ensure that secondary conductors are properly terminated to the tube electrodes and that the connections are protected from contact by unauthorized persons or contact with any flammable or combustible material. Broken tubes should be replaced or de-energized.

600.42 Electrode Connections

(A) Accessibility Terminals of the electrode shall not be accessible to unqualified persons.

(B) Electrode Connections Connections shall be made by use of a connection device, twisting of the wires together, or use of an electrode receptacle. Connections shall be electrically and mechanically secure and shall be in an enclosure listed for the purpose.

(C) Support The neon tubing and conductor shall be supported not more than 150 mm (6 in.) from the electrode connection.

(D) Receptacles Electrode receptacles shall be listed.

(E) Bushings Where electrodes penetrate an enclosure, bushings listed for the purpose shall be used unless receptacles are provided.

(F) Wet Locations A listed cap shall be used to close the opening between neon tubing and a receptacle where the receptacle penetrates a building. Where a bushing or neon tubing penetrates a building, the opening between neon tubing and the bushing shall be sealed.

(G) Electrode Enclosures Electrode enclosures shall be listed.

ARTICLE 604
Manufactured Wiring Systems

Summary of Changes

- **604.6(A)(1) and (A)(2):** Revised maximum conductor size for manufactured wiring systems from 10 AWG to 8 AWG.

- **604.6(A)(3):** Revised to clarify that the provisions of this section limiting the size and length of flexible cord are not applicable to luminaires. See new 604.6(F).

- **604.6(E):** Revised to require cables and conduits used in the manufactured wiring system to be secured and supported in accordance with the requirements for the particular cable or conduit type used.

- **604.6(F):** Added section permitting listed electric-discharge luminaires to be supplied from listed manufactured wiring system cord assemblies. Installation of cord assembly must comply with 410.30(C).

Contents

604.1 Scope

The provisions of this article apply to field-installed wiring using off-site manufactured subassemblies for branch circuits, remote-control circuits, signaling circuits, and communications circuits in accessible areas.

604.2 Definition

Manufactured Wiring System. A system containing component parts that are assembled in the process of manufacture and cannot be inspected at the building site without damage or destruction to the assembly.

604.3 Other Articles

Except as modified by the requirements of this article, all other applicable articles of this *Code* shall apply.

604.4 Uses Permitted

Manufactured wiring systems shall be permitted in accessible and dry locations and in ducts, plenums, and other air-handling spaces where listed for this application and installed in accordance with 300.22.

Article 604 covers manufactured wiring systems, typically constructed of Types AC or MC cables that are provided with factory connectors and receptacles. The connection devices used with these systems facilitate ease of initial installation and future relocation of equipment such as luminaires. These systems are used extensively for the installation of branch-circuit and tap conductors supplying luminaires in accessible locations, including open and suspended-ceiling construction.

For further information, see Article 100 for the definition of *accessible (as applied to wiring methods)*.

Exception No. 1: In concealed spaces, one end of tapped cable shall be permitted to extend into hollow walls for direct termination at switch and outlet points.

Exception No. 2: Manufactured wiring system assemblies installed outdoors shall be listed for use in outdoor locations.

The marking "Outdoor" on a listed manufactured wiring system assembly indicates it has been evaluated and is suitable for use in an outdoor wet or damp location.

604.5 Uses Not Permitted

Manufactured wiring system types shall not be permitted where limited by the applicable article in Chapter 3 for the wiring method used in its construction.

604.6 Construction

(A) Cable or Conduit Types

(1) Cables Cable shall be listed Type AC cable or listed Type MC cable containing nominal 600-volt, 8 to 12 AWG insulated copper conductors with a bare or insulated copper equipment grounding conductor equivalent in size to the ungrounded conductor.

Other cables as listed in 725.61, 800.113, 820.113, and 830.179 shall be permitted in manufactured wiring systems for wiring of equipment within the scope of their respective articles.

> The text of 604.6(A)(1) has been revised by a tentative interim amendment (TIA). See page 1.

(2) Conduits Conduit shall be listed flexible metal conduit or listed liquidtight flexible conduit containing nominal 600-

volt, 8 to 12 AWG insulated copper conductors with a bare or insulated copper equipment grounding conductor equivalent in size to the ungrounded conductor.

Revisions to 604.6(A)(1) and 600.4(A)(2) permit 8 AWG copper circuit conductors to be installed in cable- and raceway-type manufactured wiring systems. The increase from 10 AWG to 8 AWG for the maximum conductor size permitted is specifically aimed at improving the ability of the manufactured wiring system to overcome voltage drop.

Exception No. 1 to (1) and (2): A luminaire (fixture) tap, no longer than 1.8 m (6 ft) and intended for connection to a single luminaire (fixture), shall be permitted to contain conductors smaller than 12 AWG but not smaller than 18 AWG.

Exception No. 1 permits 6-ft lengths of the wiring methods covered in 604.6(A)(1) and 604.6(A)(2) to contain conductors smaller than 12 AWG but not smaller than 18 AWG for use as tap conductors to supply a single luminaire.

Exception No. 2 to (1) and (2): Listed manufactured wiring assemblies containing conductors smaller than 12 AWG shall be permitted for remote-control, signaling, or communication circuits.

(3) Flexible Cord Flexible cord suitable for hard usage, with minimum 12 AWG conductors, shall be permitted as part of a listed factory-made assembly not exceeding 1.8 m (6 ft) in length when making a transition between components of a manufactured wiring system and utilization equipment, other than luminaires (fixtures), not permanently secured to the building structure. The cord shall be visible for its entire length and shall not be subject to strain or physical damage.

This provision facilitates a transition between manufactured wiring systems and utilization equipment found in display cases, merchandise racks, temporary work stations, and the like. This transition is limited, however, to hard-usage cord not over 6 ft in length, to minimize damage, as illustrated in Exhibit 604.1.

(B) Marking Each section shall be marked to identify the type of cable, flexible cord, or conduit.

(C) Receptacles and Connectors Receptacles and connectors shall be of the locking type, uniquely polarized and identified for the purpose, and shall be part of a listed assembly for the appropriate system.

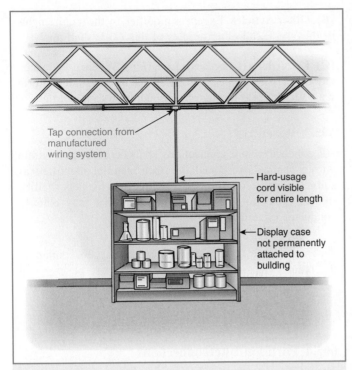

Exhibit 604.1 Transition wiring between a manufactured wiring system and utilization equipment.

Examples of polarized receptacles and connectors are shown in Exhibit 604.2.

(D) Other Component Parts Other component parts shall be listed for the appropriate system.

(E) Securing and Supporting Manufactured wiring systems shall be secured and supported in accordance with the applicable cable or conduit article for the cable or conduit type employed.

Manufactured wiring systems are permitted to be constructed using any of the wiring methods described in 604.6(A). The installation requirements for these wiring methods are contained in Chapter 3. A revision to this section for the 2005 *Code* adds the requirement to secure the wiring methods of manufactured wiring systems. The securing and supporting requirements are taken from the Chapter 3 article that covers the wiring method employed in construction of the manufactured wiring system.

(F) Luminaires (Fixtures) Installation of listed electric-discharge luminaires (fixtures) complying with 410.30(C) shall be permitted.

This new provision and a corresponding revision to 410.30(C)(2) permit the use of flexible cord equipped with a manufactured wiring system connector as a means to supply listed electric-discharge luminaires such as fluorescent or high-intensity discharge types. In this application, the cord-equipped luminaires are supplied from branch-circuit conductors installed using a manufactured wiring system. To minimize the length of flexible cord used in this application, the cord-connected luminaire must be located directly below where it connects to the manufactured wiring system. This method of supplying electric-discharge luminaires is permitted only where the cord is visible for its entire length, from its attachment to the luminaire to its interface with the branch-circuit conductors of the manufactured wiring system.

604.7 Unused Outlets

All unused outlets shall be capped to effectively close the connector openings.

Exhibit 604.2 Polarized receptacles and connectors of a manufactured wiring system. (Courtesy of Lithonia Lighting, Reloc Wiring Systems)

ARTICLE 605
Office Furnishings (Consisting of Lighting Accessories and Wired Partitions)

Summary of Changes

- **605.6 and 605.7:** Revised to require that all ungrounded conductors of multiwire branch circuits supplying power to fixed-type and freestanding-type office partitions be simultaneously disconnected at the panelboard where the circuit originates.

Contents

605.1 Scope

This article covers electrical equipment, lighting accessories, and wiring systems used to connect, or contained within, or installed on relocatable wired partitions.

605.2 General

Wiring systems shall be identified as suitable for providing power for lighting accessories and appliances in wired partitions. These partitions shall not extend from floor to ceiling.

Exception: Where permitted by the authority having jurisdiction, these relocatable wired partitions shall be permitted to extend to, but shall not penetrate, the ceiling.

(A) Use These assemblies shall be installed and used only as provided for by this article.

(B) Other Articles Except as modified by the requirements of this article, all other articles of this *Code* shall apply.

(C) Hazardous (Classified) Locations Where used in hazardous (classified) locations, these assemblies shall conform with Articles 500 through 517 in addition to this article.

605.3 Wireways

All conductors and connections shall be contained within wiring channels of metal or other material identified as suitable for the conditions of use. Wiring channels shall be free of projections or other conditions that may damage conductor insulation.

A wiring channel that is separate from the channel containing the branch circuits for light and power may be provided within the system components for the routing of communications, signaling, and fiber optic cables.

605.4 Partition Interconnections

The electrical connection between partitions shall be a flexible assembly identified for use with wired partitions or shall be permitted to be installed using flexible cord, provided all the following conditions are met:

(1) The cord is extra-hard usage type with 12 AWG or larger conductors, with an insulated grounding conductor.
(2) The partitions are mechanically contiguous.
(3) The cord is not longer than necessary for maximum positioning of the partitions but is in no case to exceed 600 mm (2 ft).
(4) The cord is terminated at an attachment plug and cord connector with strain relief.

605.5 Lighting Accessories

Lighting equipment listed and identified for use with wired partitions shall comply with 605.5(A), (B), and (C).

(A) Support A means for secure attachment or support shall be provided.

(B) Connection Where cord and plug connection is provided, the cord length shall be suitable for the intended application but shall not exceed 2.7 m (9 ft) in length. The cord shall not be smaller than 18 AWG, shall contain an equipment grounding conductor, and shall be of the hard usage type. Connection by other means shall be identified as suitable for the condition of use.

(C) Receptacle Outlet Convenience receptacles shall not be permitted in lighting accessories.

605.6 Fixed-Type Partitions

Wired partitions that are fixed (secured to building surfaces) shall be permanently connected to the building electrical system by one of the wiring methods of Chapter 3. Multiwire branch circuits supplying power to the partition shall be provided with a means to disconnect simultaneously all ungrounded conductors at the panelboard where the branch circuit originates.

605.7 Freestanding-Type Partitions

Partitions of the freestanding type (not fixed) shall be permitted to be connected to the building electrical system by one of the wiring methods of Chapter 3. Multiwire branch circuits supplying power to permanently connected freestanding partitions shall be provided with a means to disconnect simultaneously all ungrounded conductors at the panelboard where the branch circuit originates.

Intended to enhance the safety of those servicing permanently connected office partitions, a new requirement for the 2005 *Code* specifies simultaneous disconnection of all ungrounded conductors of multiwire branch circuits supplying fixed and freestanding-type office partitions.

605.8 Freestanding-Type Partitions, Cord-and-Plug-Connected

Individual partitions of the freestanding type, or groups of individual partitions that are electrically connected, are mechanically contiguous, and do not exceed 9.0 m (30 ft) when assembled, shall be permitted to be connected to the building electrical system by a single flexible cord and plug, provided all of the conditions of 605.8(A) through 605.8(D) are met.

(A) Flexible Power-Supply Cord The flexible power-supply cord shall be extra-hard usage type with 12 AWG or larger conductors with an insulated equipment grounding conductor and shall not exceed 600 mm (2 ft) in length.

(B) Receptacle Supplying Power The receptacle(s) supplying power shall be on a separate circuit serving only panels and no other loads and shall be located not more than 300 mm (12 in.) from the partition that is connected to it.

(C) Receptacle Outlets, Maximum Individual partitions or groups of interconnected individual partitions shall not contain more than thirteen 15-ampere, 125-volt receptacle outlets.

(D) Multiwire Circuits, Not Permitted Individual partitions or groups of interconnected individual partitions shall not contain multiwire circuits.

FPN: See 210.4 for circuits supplying partitions in 605.6 and 605.7.

Cord-and-plug-connected freestanding-type office partitions are not permitted to contain multiwire branch circuits.

ARTICLE 610
Cranes and Hoists

Summary of Changes

- **610.61:** Revised to no longer permit the trolley frame and bridge frame to be grounded through the bridge and trolley wheels and their respective tracks. A separate wire-type bonding conductor is required.

Contents

I. General

610.1 Scope

This article covers the installation of electrical equipment and wiring used in connection with cranes, monorail hoists, hoists, and all runways.

> FPN: For further information, see ANSI B30, *Safety Code for Cranes, Derricks, Hoists, Jacks, and Slings.*

610.3 Special Requirements for Particular Locations

(A) Hazardous (Classified) Locations All equipment that operates in a hazardous (classified) location shall conform to Article 500.

(1) Class I Locations Equipment used in locations that are hazardous because of the presence of flammable gases or vapors shall conform to Article 501.

(2) Class II Locations Equipment used in locations that are hazardous because of combustible dust shall conform to Article 502.

(3) Class III Locations Equipment used in locations that are hazardous because of the presence of easily ignitible fibers or flyings shall conform to Article 503.

See the commentary following 503.155(D).

(B) Combustible Materials Where a crane, hoist, or monorail hoist operates over readily combustible material, the resistors shall be located as permitted in the following:

(1) A well-ventilated cabinet composed of noncombustible material constructed so that it does not emit flames or molten metal

(2) A cage or cab constructed of noncombustible material that encloses the sides of the cage or cab from the floor to a point at least 150 mm (6 in.) above the top of the resistors

(C) Electrolytic Cell Lines See 668.32.

Special precautions are necessary on electrolytic cell lines to prevent the introduction of exposed grounded parts, as described in 668.32.

II. Wiring

610.11 Wiring Method

Conductors shall be enclosed in raceways or be Type AC cable with insulated grounding conductor, Type MC cable, or Type MI cable unless otherwise permitted or required in 610.11(A) through 610.11(E).

Type AC cable with an insulated grounding conductor is permitted as an acceptable wiring method for cranes and hoists. The insulated wire-type equipment grounding conductor, terminated on the grounding terminals of crane- and hoist-associated equipment, is required to ensure the continuity of the grounding and bonding connection to equipment that is often subject to vibration.

(A) Contact Conductor Contact conductors shall not be required to be enclosed in raceways.

(B) Exposed Conductors Short lengths of exposed conductors at resistors, collectors, and other equipment shall not be required to be enclosed in raceways.

(C) Flexible Connections to Motors and Similar Equipment Where flexible connections are necessary, flexible stranded conductors shall be used. Conductors shall be in flexible metal conduit, liquidtight flexible metal conduit, liquidtight flexible nonmetallic conduit, multiconductor cable, or an approved nonmetallic flexible raceway.

The use of short lengths of open wiring on cranes and hoists is permitted by 610.11(B). Short runs of open conductors facilitate connection to resistors, collectors, and similar equipment. Each conductor is required to be provided with separately bushed holes in boxes, as well as in cable and raceway fittings used where the transition to open wiring is made. In addition to other types of raceways and cables, 610.11(C) permits flexible metal conduit, liquidtight flexible metal conduit, liquidtight flexible nonmetallic conduit, multiconductor cable, or an approved nonmetallic enclosure (tubing) where flexibility is necessary.

(D) Pushbutton Stations Multiconductor Cable Where multiconductor cable is used with a suspended pushbutton station, the station shall be supported in some satisfactory manner that protects the electric conductors against strain.

Exhibit 610.1 shows an example of suitable strain relief for a cord that supports a control pushbutton station.

Exhibit 610.1 A suitable strain relief grip for a cord-suspended pushbutton station. (Courtesy of Hubbell Inc., Kellems Division)

(E) Flexibility to Moving Parts Where flexibility is required for power or control to moving parts, a cord suitable for the purpose shall be permitted, provided the following apply:

(1) Suitable strain relief and protection from physical damage is provided.
(2) In Class I, Division 2 locations, the cord is approved for extra-hard usage.

610.12 Raceway or Cable Terminal Fittings

Conductors leaving raceways or cables shall comply with either 610.12(A) or 610.12(B).

(A) Separately Bushed Hole A box or terminal fitting that has a separately bushed hole for each conductor shall be used wherever a change is made from a raceway or cable to exposed wiring. A fitting used for this purpose shall not contain taps or splices and shall not be used at luminaire (fixture) outlets.

(B) Bushing in Lieu of a Box A bushing shall be permitted to be used in lieu of a box at the end of a rigid metal conduit, intermediate metal conduit, or electrical metallic tubing where the raceway terminates at unenclosed controls or similar equipment, including contact conductors, collectors, resistors, brakes, power-circuit limit switches, and dc split-frame motors.

610.13 Types of Conductors

Conductors shall comply with Table 310.13 unless otherwise permitted in 610.13(A) through 610.13(D).

(A) Exposed to External Heat or Connected to Resistors A conductor(s) exposed to external heat or connected to resistors shall have a flame-resistant outer covering or be covered with flame-resistant tape individually or as a group.

(B) Contact Conductors Contact conductors along runways, crane bridges, and monorails shall be permitted to be bare and shall be copper, aluminum, steel, or other alloys or combinations thereof in the form of hard-drawn wire, tees, angles, tee rails, or other stiff shapes.

(C) Flexibility Where flexibility is required, flexible cord or cable shall be permitted to be used and, where necessary, cable reels or take-up devices shall be used.

(D) Class 1, Class 2, and Class 3 Circuits Conductors for Class 1, Class 2, and Class 3 remote-control, signaling, and power-limited circuits, installed in accordance with Article 725, shall be permitted.

Section 610.13(D) references Article 725 for Class 1, Class 2, and Class 3 control circuit conductors. This cross-reference is necessary because 90.3 does not allow Article 725 to modify Article 610. The change allows the methods of wiring control and signaling circuits permitted in Article 725 to be used for crane- and hoist-control circuits.

610.14 Rating and Size of Conductors

(A) Ampacity The allowable ampacities of conductors shall be as shown in Table 610.14(A).

> FPN: For the ampacities of conductors between controllers and resistors, see 430.23.

(B) Secondary Resistor Conductors Where the secondary resistor is separate from the controller, the minimum size of the conductors between controller and resistor shall be calculated by multiplying the motor secondary current by the appropriate factor from Table 610.14(B) and selecting a wire from Table 610.14(A).

(C) Minimum Size Conductors external to motors and controls shall not be smaller than 16 AWG unless otherwise permitted in (1) or (2):

(1) 18 AWG wire in multiconductor cord shall be permitted for control circuits not exceeding 7 amperes.
(2) Wires not smaller than 20 AWG shall be permitted for electronic circuits.

(D) Contact Conductors Contact wires shall have an ampacity not less than that required by Table 610.14(A) for 75°C (167°F) wire, and in no case shall be smaller than as shown in Table 610.14(D).

(E) Calculation of Motor Load

(1) Single Motor For one motor, 100 percent of motor nameplate full-load ampere rating shall be used.

(2) Multiple Motors on Single Crane or Hoist For multiple motors on a single crane or hoist, the minimum ampacity of the power supply conductors shall be the nameplate full-load ampere rating of the largest motor or group of motors for any single crane motion, plus 50 percent of the nameplate full-load ampere rating of the next largest motor or group of motors, using that column of Table 610.14(A) that applies to the longest time-rated motor.

(3) Multiple Cranes or Hoists on a Common Conductor System For multiple cranes, hoists, or both, supplied by a common conductor system, calculate the motor minimum ampacity for each crane as defined in 610.14(E), add them together, and multiply the sum by the appropriate demand factor from Table 610.14(E).

(F) Other Loads Additional loads, such as heating, lighting, and air conditioning, shall be provided for by application of the appropriate sections of this *Code*.

(G) Nameplate Each crane, monorail, or hoist shall be provided with a visible nameplate marked with the manufacturer's name, rating in volts, frequency, number of phases, and circuit amperes as calculated in 610.14(E) and 610.14(F).

610.15 Common Return

Where a crane or hoist is operated by more than one motor, a common-return conductor of proper ampacity shall be permitted.

III. Contact Conductors

610.21 Installation of Contact Conductors

Contact conductors shall comply with 610.21(A) through 610.21(H).

(A) Locating or Guarding Contact Conductors Runway contact conductors shall be guarded, and bridge contact conductors shall be located or guarded in such a manner that persons cannot inadvertently touch energized current-carrying parts.

(B) Contact Wires Wires that are used as contact conductors shall be secured at the ends by means of approved strain insulators and shall be mounted on approved insulators so that the extreme limit of displacement of the wire does not bring the latter within less than 38 mm (1½ in.) from the surface wired over.

(C) Supports Along Runways Main contact conductors carried along runways shall be supported on insulating supports placed at intervals not exceeding 6.0 m (20 ft) unless otherwise permitted in 610.21(F).

Such conductors shall be separated at not less than 150 mm (6 in.), other than for monorail hoists where a spacing of not less than 75 mm (3 in.) shall be permitted. Where necessary, intervals between insulating supports shall be permitted to be increased up to 12 m (40 ft), the separation between conductors being increased proportionately.

(D) Supports on Bridges Bridge wire contact conductors shall be kept at least 65 mm (2½ in.) apart, and, where the span exceeds 25 m (80 ft), insulating saddles shall be placed at intervals not exceeding 15 m (50 ft).

(E) Supports for Rigid Conductors Conductors along runways and crane bridges, that are of the rigid type specified in 610.13(B) and not contained within an approved enclosed assembly, shall be carried on insulating supports spaced at intervals of not more than 80 times the vertical dimension of the conductor, but in no case greater than 4.5 m (15 ft), and spaced apart sufficiently to give a clear electrical separation of conductors or adjacent collectors of not less than 25 mm (1 in.).

(F) Track as Circuit Conductor Monorail, tram rail, or crane runway tracks shall be permitted as a conductor of

Table 610.14(A) Ampacities of Insulated Copper Conductors Used with Short-Time Rated Crane and Hoist Motors. Based on Ambient Temperature of 30°C (86°F)

Maximum Operating Temperature	Up to Four Conductors in Raceway or Cable[1]				Up to Three ac[2] or Four dc[1] Conductors in Raceway or Cable		Maximum Operating Temperature
	75°C (167°F)		90°C (194°F)		125°C (257°F)		
	Types MTW, RHW, THW, THWN, XHHW, USE, ZW		Types TA, TBS, SA, SIS, PFA, FEP, FEPB, RHH, THHN, XHHW, Z, ZW		Types FEP, FEPB, PFA, PFAH, SA, TFE, Z, ZW		
Size (AWG or kcmil)	60 Min	30 Min	60 Min	30 Min	60 Min	30 Min	Size (AWG or kcmil)
16	10	12	—	—	—	—	16
14	25	26	31	32	38	40	14
12	30	33	36	40	45	50	12
10	40	43	49	52	60	65	10
8	55	60	63	69	73	80	8
6	76	86	83	94	101	119	6
5	85	95	95	106	115	134	5
4	100	117	111	130	133	157	4
3	120	141	131	153	153	183	3
2	137	160	148	173	178	214	2
1	143	175	158	192	210	253	1
1/0	190	233	211	259	253	304	1/0
2/0	222	267	245	294	303	369	2/0
3/0	280	341	305	372	370	452	3/0
4/0	300	369	319	399	451	555	4/0
250	364	420	400	461	510	635	250
300	455	582	497	636	587	737	300
350	486	646	542	716	663	837	350
400	538	688	593	760	742	941	400
450	600	765	660	836	818	1042	450
500	660	847	726	914	896	1143	500

AMPACITY CORRECTION FACTORS

Ambient Temperature (°C)	For ambient temperatures other than 30°C (86°F), multiply the ampacities shown above by the appropriate factor shown below.						Ambient Temperature (°F)
21–25	1.05	1.05	1.04	1.04	1.02	1.02	70–77
26–30	1.00	1.00	1.00	1.00	1.00	1.00	79–86
31–35	0.94	0.94	0.96	0.96	0.97	0.97	88–95
36–40	0.88	0.88	0.91	0.91	0.95	0.95	97–104
41–45	0.82	0.82	0.87	0.87	0.92	0.92	106–113
46–50	0.75	0.75	0.82	0.82	0.89	0.89	115–122
51–55	0.67	0.67	0.76	0.76	0.86	0.86	124–131
56–60	0.58	0.58	0.71	0.71	0.83	0.83	133–140
61–70	0.33	0.33	0.58	0.58	0.76	0.76	142–158
71–80	—	—	0.41	0.41	0.69	0.69	160–176
81–90	—	—	—	—	0.61	0.61	177–194
91–100	—	—	—	—	0.51	0.51	195–212
101–120	—	—	—	—	0.40	0.40	213–248

Note: Other insulations shown in Table 310.13 and approved for the temperature and location shall be permitted to be substituted for those shown in Table 610.14(A). The allowable ampacities of conductors used with 15-minute motors shall be the 30-minute ratings increased by 12 percent.

[1] For 5 to 8 simultaneously energized power conductors in raceway or cable, the ampacity of each power conductor shall be reduced to a value of 80 percent of that shown in this table.

[2] For 4 to 6 simultaneously energized 125°C (257°F) ac power conductors in raceway or cable, the ampacity of each power conductor shall be reduced to a value of 80 percent of that shown in this table.

Table 610.14(B) Secondary Conductor Rating Factors

Time in Seconds		Ampacity of Wire in Percent of Full-Load Secondary Current
On	**Off**	
5	75	35
10	70	45
15	75	55
15	45	65
15	30	75
15	15	85
Continuous Duty		110

Table 610.14(D) Contact Conductor Supports

Distance Between End Strain Insulators or Clamp-Type Intermediate Supports	Size of Wire (AWG)
Less than 9.0 m (30 ft)	6
9.0 m–18 m (30 ft–60 ft)	4
Over 18 m (60 ft)	2

Table 610.14(E) Demand Factors

Number of Cranes or Hoists	Demand Factor
2	0.95
3	0.91
4	0.87
5	0.84
6	0.81
7	0.78

current for one phase of a 3-phase, ac system furnishing power to the carrier, crane, or trolley, provided all of the following conditions are met:

(1) The conductors supplying the other two phases of the power supply are insulated.
(2) The power for all phases is obtained from an insulating transformer.
(3) The voltage does not exceed 300 volts.
(4) The rail serving as a conductor is effectively grounded at the transformer and also shall be permitted to be grounded by the fittings used for the suspension or attachment of the rail to a building or structure.

Crane runway tracks are permitted as a current-carrying conductor where part of a 3-phase system that is furnishing power to the crane. Exhibit 610.2 illustrates a 3-phase, iso-lated-delta secondary with one phase grounded (a corner-grounded delta secondary system) at the transformer. The

track is also permitted to be grounded through the metal supporting means attached to the metal frame of a building.

Exhibit 610.2 Three-phase, delta isolating transformer.

(G) Electrical Continuity of Contact Conductors All sections of contact conductors shall be mechanically joined to provide a continuous electrical connection.

(H) Not to Supply Other Equipment Contact conductors shall not be used as feeders for any equipment other than the crane(s) or hoist(s) that they are primarily designed to serve.

610.22 Collectors

Collectors shall be designed so as to reduce to a minimum sparking between them and the contact conductor; and, where operated in rooms used for the storage of easily ignitible combustible fibers and materials, they shall comply with 503.155.

IV. Disconnecting Means

610.31 Runway Conductor Disconnecting Means

A disconnecting means that has a continuous ampere rating not less than that calculated in 610.14(E) and 610.14(F) shall be provided between the runway contact conductors and the power supply. Such disconnecting means shall consist of a motor-circuit switch, circuit breaker, or molded case switch. This disconnecting means shall be as follows:

(1) Readily accessible and operable from the ground or floor level
(2) Capable of being locked in the open position
(3) Open all ungrounded conductors simultaneously
(4) Placed within view of the runway contact conductors

610.32 Disconnecting Means for Cranes and Monorail Hoists

A motor-circuit switch, molded-case switch, or circuit breaker shall be provided in the leads from the runway contact conductors or other power supply on all cranes and monorail hoists. The disconnecting means shall be capable of being locked in the open position.

Where a monorail hoist or hand-propelled crane bridge installation meets all of the following, the disconnecting means shall be permitted to be omitted:

(1) The unit is controlled from the ground or floor level.
(2) The unit is within view of the power supply disconnecting means.
(3) No fixed work platform has been provided for servicing the unit.

Where the disconnecting means is not readily accessible from the crane or monorail hoist operating station, means shall be provided at the operating station to open the power circuit to all motors of the crane or monorail hoist.

Many crane installations are not arranged so that the unit is within view of the power supply disconnecting means. Therefore, a disconnecting means (lock-open type) must be provided in the contact conductors. However, personnel should be aware that when one crane is being serviced, another unit on the same system could remain energized and be run into the person performing maintenance on the locked-out unit.

610.33 Rating of Disconnecting Means

The continuous ampere rating of the switch or circuit breaker required by 610.32 shall not be less than 50 percent of the combined short-time ampere rating of the motors or less than 75 percent of the sum of the short-time ampere rating of the motors required for any single motion.

V. Overcurrent Protection

610.41 Feeders, Runway Conductors

(A) **Single Feeder** The runway supply conductors and main contact conductors of a crane or monorail shall be protected by an overcurrent device(s) that shall not be greater than the largest rating or setting of any branch-circuit protective device plus the sum of the nameplate ratings of all the other loads with application of the demand factors from Table 610.14(E).

(B) **More Than One Feeder Circuit** Where more than one feeder circuit is installed to supply runway conductors, each feeder circuit shall be sized and protected in compliance with 610.41(A).

Multiple feeders are sometimes used to supply long runway conductors to minimize voltage drops on the runway conductors.

610.42 Branch-Circuit Short-Circuit and Ground-Fault Protection

Branch circuits shall be protected in accordance with 610.42(A). Branch-circuit taps, where made, shall comply with 610.42(B).

(A) **Fuse or Circuit Breaker Rating** Crane, hoist, and monorail hoist motor branch circuits shall be protected by fuses or inverse-time circuit breakers that have a rating in accordance with Table 430.52. Where two or more motors operate a single motion, the sum of their nameplate current ratings shall be considered as that of a single motor.

(B) **Taps**

(1) **Multiple Motors** Where two or more motors are connected to the same branch circuit, each tap conductor to an individual motor shall have an ampacity not less than one-third that of the branch circuit. Each motor shall be protected from overload according to 610.43.

(2) **Control Circuits** Where taps to control circuits originate on the load side of a branch-circuit protective device, each tap and piece of equipment shall be protected in accordance with 430.72.

(3) **Brake Coils** Taps without separate overcurrent protection shall be permitted to brake coils.

610.43 Overload Protection

(A) **Motor and Branch-Circuit Overload Protection** Each motor, motor controller, and branch-circuit conductor shall be protected from overload by one of the following means:

(1) A single motor shall be considered as protected where the branch-circuit overcurrent device meets the rating requirements of 610.42.
(2) Overload relay elements in each ungrounded circuit conductor, with all relay elements protected from short circuit by the branch-circuit protection.
(3) Thermal sensing devices, sensitive to motor temperature or to temperature and current, that are thermally in contact with the motor winding(s). A hoist or trolley shall be considered to be protected if the sensing device is connected in the hoist's upper limit switch circuit so as to prevent further hoisting during an overload condition of either motor.

(B) **Manually Controlled Motor** If the motor is manually controlled, with spring return controls, the overload protective device shall not be required to protect the motor against stalled rotor conditions.

(C) **Multimotor** Where two or more motors drive a single trolley, truck, or bridge and are controlled as a unit and protected by a single set of overload devices with a rating equal to the sum of their rated full-load currents, a hoist or trolley shall be considered to be protected if the sensing device is connected in the hoist's upper limit switch circuit so as to prevent further hoisting during an overtemperature condition of either motor.

(D) Hoists and Monorail Hoists Hoists and monorail hoists and their trolleys that are not used as part of an overhead traveling crane shall not require individual motor overload protection, provided the largest motor does not exceed 7½ hp and all motors are under manual control of the operator.

VI. Control

610.51 Separate Controllers

Each motor shall be provided with an individual controller unless otherwise permitted in 610.51(A) or 610.51(B).

(A) Motions with More Than One Motor Where two or more motors drive a single hoist, carriage, truck, or bridge, they shall be permitted to be controlled by a single controller.

(B) Multiple Motion Controller One controller shall be permitted to be switched between motors, under the following conditions:

(1) The controller has a horsepower rating that is not lower than the horsepower rating of the largest motor.
(2) Only one motor is operated at one time.

610.53 Overcurrent Protection

Conductors of control circuits shall be protected against overcurrent. Control circuits shall be considered as protected by overcurrent devices that are rated or set at not more than 300 percent of the ampacity of the control conductors, unless otherwise permitted in 610.53(A) or 610.53(B).

(A) Taps to Control Transformers Taps to control transformers shall be considered as protected where the secondary circuit is protected by a device rated or set at not more than 200 percent of the rated secondary current of the transformer and not more than 200 percent of the ampacity of the control circuit conductors.

(B) Continuity of Power Where the opening of the control circuit would create a hazard, as for example, the control circuit of a hot metal crane, the control circuit conductors shall be considered as being properly protected by the branch-circuit overcurrent devices.

610.55 Limit Switch

A limit switch or other device shall be provided to prevent the load block from passing the safe upper limit of travel of all hoisting mechanisms.

610.57 Clearance

The dimension of the working space in the direction of access to live parts that are likely to require examination, adjustment, servicing, or maintenance while energized shall be a minimum of 750 mm (2½ ft). Where controls are enclosed in cabinets, the door(s) shall either open at least 90 degrees or be removable.

VII. Grounding

610.61 Grounding

All exposed non–current-carrying metal parts of cranes, monorail hoists, hoists, and accessories, including pendant controls, shall be metallically joined together into a continuous electrical conductor so that the entire crane or hoist will be grounded in accordance with Article 250. Moving parts, other than removable accessories, or attachments that have metal-to-metal bearing surfaces, shall be considered to be electrically connected to each other through bearing surfaces for grounding purposes. The trolley frame and bridge frame shall not be considered as electrically grounded through the bridge and trolley wheels and its respective tracks. A separate bonding conductor shall be provided.

It is not intended that the trolley frame or bridge frame serve as the equipment grounding conductor for electrical equipment (such as motors, motor controllers, lighting fixtures, and transformers) on a crane. The equipment grounding conductors that are run with the circuit conductors are required to be one of the types described in 250.118. Metal-to-metal bearing surfaces of moving parts are considered to be a suitable grounding and bonding connection. However, a revision to the 2005 *Code* no longer permits the bridge and trolley wheel contact with their tracks as a reliable grounding and bonding connection. Due to dirt or other foreign surfaces impeding the effectiveness of the wheel-to-track contact as a reliable grounding and bonding connection, the bridge and trolley frames of an electric crane are now required to be bonded through the use of a separate conductor.

ARTICLE 620
Elevators, Dumbwaiters, Escalators, Moving Walks, Wheelchair Lifts, and Stairway Chair Lifts

Summary of Changes

• **620.2:** Added definitions for *control room*, *control space*, *machine room*, and *machinery space* as applied to elevators and dumbwaiters.

• **620.22(A):** Added requirement prohibiting required lighting in elevator car to be supplied from circuit conductors connected to the load side of a GFCI.

- **620.61(B):** Revised to require motor and branch-circuit overload protection to conform to Article 430, Part III, as well as 620.61(B).
- **620.61(B)(1) and (B)(4):** Revised to permit, rather than require, protection per 430.33.

Contents

I. General

620.1 Scope

This article covers the installation of electrical equipment and wiring used in connection with elevators, dumbwaiters, escalators, moving walks, wheelchair lifts, and stairway chair lifts.

Since the 1996 edition of the *Code*, there has been a continuing effort to revise Article 620 to harmonize elevator requirements throughout North America and to keep pace with modern advances in elevator technology. These changes continue to reflect time-honored safety concerns and more accurately address the use of modern equipment.

> FPN No. 1: For further information, see ASME/ANSI A17.1-2000, *Safety Code for Elevators and Escalators*.
>
> FPN No. 2: For further information, see ASME/ANSI A17.5-1996 (CSA B44.1-1996), *Elevator and Escalator Electrical Equipment Certification Standard*.

FPN No. 1 to 620.1 points out the existence of the widely recognized elevator code, ASME A17.1-2004, *Safety Code for Elevators and Escalators*. ASME A17.1 contains many construction and maintenance requirements, from required machine room lighting to seismic considerations.

620.2 Definitions

New definitions for elevator and dumbwaiter control rooms, control spaces, machine rooms, and machinery spaces have been included in 620.2. These four new definitions describe the types of elevator or dumbwaiter equipment (controllers, driving machines, other associated electrical equipment, and mechanical and hydraulic apparatus) that is and is not located in each of those respective areas. These terms are used in various requirements of Article 620 and correlate with their use in ASME A17.1-2004.

Control Room (for Elevator, Dumbwaiter). An enclosed control space outside the hoistway, intended for full bodily entry, that contains the elevator motor controller. The room could also contain electrical and/or mechanical equipment used directly in connection with the elevator or dumbwaiter but not the electric driving machine or the hydraulic machine.

Control Space (for Elevator, Dumbwaiter). A space inside or outside the hoistway, intended to be accessed with or without full bodily entry, that contains the elevator motor controller. This space could also contain electrical and/or mechanical equipment used directly in connection with the elevator or dumbwaiter but not the electric driving machine or the hydraulic machine.

Control System. The overall system governing the starting, stopping, direction of motion, acceleration, speed, and retardation of the moving member.

Controller, Motion. The electric device(s) for that part of the control system that governs the acceleration, speed, retardation, and stopping of the moving member.

Controller, Motor. The operative units of the control system comprised of the starter device(s) and power conversion equipment used to drive an electric motor, or the pumping unit used to power hydraulic control equipment.

Controller, Operation. The electric device(s) for that part of the control system that initiates the starting, stopping, and direction of motion in response to a signal from an operating device.

Machine Room (for Elevator, Dumbwaiter). An enclosed machinery space outside the hoistway, intended for full bodily entry, that contains the electrical driving machine or the hydraulic machine. The room could also contain electrical and/or mechanical equipment used directly in connection with the elevator or dumbwaiter.

Machinery Space (for Elevator, Dumbwaiter). A space inside or outside the hoistway, intended to be accessed with or without full bodily entry, that contains elevator or dumbwaiter mechanical equipment, and could also contain electrical equipment used directly in connection with the elevator or dumbwaiter. This space could also contain the electric driving machine or the hydraulic machine.

Operating Device. The car switch, pushbuttons, key or toggle switch(s), or other devices used to activate the operation controller.

Signal Equipment. Includes audible and visual equipment such as chimes, gongs, lights, and displays that convey information to the user.

> FPN No. 1: The motor controller, motion controller, and operation controller may be located in a single enclosure or a combination of enclosures.
>
> FPN No. 2: FPN Figure 620.2 is for information only.

FPN Figure 620.2 illustrates a typical elevator control system. The figure does not suggest that every elevator control system should have these exact components.

The definitions in 620.2 and FPN Figure 620.2 separate the control system into its functional parts. Article 620 uses the definitions to address and specify the proper safety concerns of motor controllers, motion controllers, and operation controllers.

620.3 Voltage Limitations

The supply voltage shall not exceed 300 volts between conductors unless otherwise permitted in 620.3(A) through 620.3(C).

(A) Power Circuits Branch circuits to door operator controllers and door motors and branch circuits and feeders to motor controllers, driving machine motors, machine brakes, and motor-generator sets shall not have a circuit voltage in excess of 600 volts. Internal voltages of power conversion and functionally associated equipment, including the interconnecting wiring, shall be permitted to have higher voltages, provided that all such equipment and wiring shall be listed for the higher voltages. Where the voltage exceeds 600 volts, warning labels or signs that read "DANGER — HIGH VOLTAGE" shall be attached to the equipment and shall be plainly visible.

(B) Lighting Circuits Lighting circuits shall comply with the requirements of Article 410.

(C) Heating and Air-Conditioning Circuits Branch circuits for heating and air-conditioning equipment located on the elevator car shall not have a circuit voltage in excess of 600 volts.

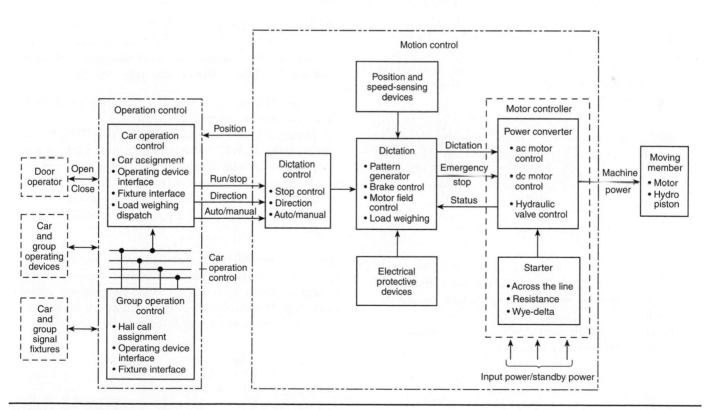

FPN Figure 620.2 Control System.

The voltage limitations for power, lighting, and heating and air-conditioning circuits are described in 620.3. Voltage limitations for power conversion units and functionally associated equipment have been replaced with a warning sign requirement for voltages above 600 volts.

620.4 Live Parts Enclosed

All live parts of electrical apparatus in the hoistways, at the landings, in or on the cars of elevators and dumbwaiters, in the wellways or the landings of escalators or moving walks, or in the runways and machinery spaces of wheelchair lifts and stairway chairlifts shall be enclosed to protect against accidental contact.

> FPN: See 110.27 for guarding of live parts (600 volts, nominal, or less).

620.5 Working Clearances

Working space shall be provided about controllers, disconnecting means, and other electrical equipment. The minimum working space shall not be less than that specified in 110.26(A).

Where conditions of maintenance and supervision ensure that only qualified persons examine, adjust, service, and maintain the equipment, the clearance requirements of 110.26(A) shall be waived as permitted in 620.5(A) through 620.5(D).

(A) Flexible Connections to Equipment Electrical equipment in (A)(1) through (A)(4) shall be permitted to be provided with flexible leads to all external connections so that it can be repositioned to meet the clear working space requirements of 110.26(A):

Due to the physical constraints of the locations where this equipment is typically installed and the necessity of performing diagnostic work on it while it is energized, 620.5(A) permits flexible leads on equipment so it can be moved to a location that meets the working clearance requirements of 110.26(A).

(1) Controllers and disconnecting means for dumbwaiters, escalators, moving walks, wheelchair lifts, and stairway chairlifts installed in the same space with the driving machine
(2) Controllers and disconnecting means for elevators installed in the hoistway or on the car
(3) Controllers for door operators
(4) Other electrical equipment installed in the hoistway or on the car

(B) Guards Live parts of the electrical equipment are suitably guarded, isolated, or insulated, and the equipment can be examined, adjusted, serviced, or maintained while energized without removal of this protection.

> FPN: See definition of *Exposed* in Article 100.

(C) Examination, Adjusting, and Servicing Electrical equipment is not required to be examined, adjusted, serviced, or maintained while energized.

(D) Low Voltage Uninsulated parts are at a voltage not greater than 30 volts rms, 42 volts peak, or 60 volts dc.

II. Conductors

620.11 Insulation of Conductors

The insulation of conductors shall comply with 620.11(A) through 620.11(D).

> FPN: One method of determining that conductors are flame retardant is by testing the conductors to the VW-1 (Vertical-Wire) Flame Test in ANSI/UL 1581-1991, *Reference Standard for Electrical Wires, Cables, and Flexible Cords*.

(A) Hoistway Door Interlock Wiring The conductors to the hoistway door interlocks from the hoistway riser shall be flame retardant and suitable for a temperature of not less than 200°C (392°F). Conductors shall be Type SF or equivalent.

(B) Traveling Cables Traveling cables used as flexible connections between the elevator or dumbwaiter car or counterweight and the raceway shall be of the types of elevator cable listed in Table 400.4 or other approved types.

(C) Other Wiring All conductors in raceways shall have flame-retardant insulation.

Conductors shall be Type MTW, TF, TFF, TFN, TFFN, THHN, THW, THWN, TW, XHHW, hoistway cable, or any other conductor with insulation designated as flame retardant. Shielded conductors shall be permitted if such conductors are insulated for the maximum nominal circuit voltage applied to any conductor within the cable or raceway system.

(D) Insulation All conductors shall have an insulation voltage rating equal to at least the maximum nominal circuit voltage applied to any conductor within the enclosure, cable, or raceway. Insulations and outer coverings that are marked for limited smoke and are so listed shall be permitted.

Hoistway door interlock wiring is required to be suitable for 200°C (392°F). See Table 310.13, Conductor Application and Insulations. See also Table 310.18.

See Table 400.4 for approved types of elevator cables for use in hazardous (classified) and nonhazardous locations.

See also Note 5 to Table 400.4. A characteristic equally important with respect to safety is the need to prevent twisting of cables as they travel up and down with the elevator or dumbwaiter.

620.12 Minimum Size of Conductors

The minimum size of conductors, other than conductors that form an integral part of control equipment, shall be in accordance with 620.12(A) and 620.12(B).

(A) Traveling Cables

(1) Lighting Circuits For lighting circuits, 14 AWG copper. 20 AWG copper or larger conductors shall be permitted in parallel, provided the ampacity is equivalent to at least that of 14 AWG copper.

(2) Other Circuits For other circuits, 20 AWG copper.

(B) Other Wiring 24 AWG copper. Smaller size listed conductors shall be permitted.

In general, the requirements of 310.4 provide the conditions under which conductors can be installed in parallel for power and lighting circuits. One of those conditions stipulates that the minimum size for parallel conductors is 1/0 AWG. In high-rise structures, the length of the elevator traveling cables is problematic with respect to maintaining an acceptable level of voltage drop for equipment on or within the car. To require compliance with the general rules for parallel conductors would result in exceedingly large traveling cables.

Section 620.12(B) amends the general requirements for parallel conductors and permits 20 AWG and larger conductors to be installed in parallel for lighting circuits, provided that the combined ampacity of the paralleled conductors is not less than that of a 14 AWG conductor. This provision is unique to Article 620 and is an example of the structure of the *Code* as set forth in 90.3, in that a requirement in Chapter 6 modifies a general requirement from Chapter 3.

The extensive use of electronics with correspondingly lower currents permits the use of smaller wire sizes. The use of conductors smaller than 24 AWG is permitted by 620.12(B), where they are listed for the purpose. One application may be the shielded cables interconnecting various microprocessors in an elevator distributed control system. All conductors should, of course, have the necessary strength and durability for the conditions to which they will be exposed.

620.13 Feeder and Branch-Circuit Conductors

Conductors shall have an ampacity in accordance with 620.13(A) through 620.13(D). With generator field control,

the conductor ampacity shall be based on the nameplate current rating of the driving motor of the motor-generator set that supplies power to the elevator motor.

> FPN No. 1: The heating of conductors depends on root-mean-square current values, which, with generator field control, are reflected by the nameplate current rating of the motor-generator driving motor rather than by the rating of the elevator motor, which represents actual but short-time and intermittent full-load current values.

Figure 620.13 depicts the appropriate reference for each part of an elevator circuit.

FPN No. 2: See Figure 620.13.

Figure 620.13 Single-Line Diagram.

(A) Conductors Supplying Single Motor Conductors supplying a single motor shall have an ampacity not less than the percentage of motor nameplate current determined from 430.22(A) and 430.22(E).

> FPN: Elevator motor currents, or those of similar functions, may exceed the nameplate value, but because they are inherently intermittent duty and the heating of the

motor and conductors is dependent on the root-mean-square (rms) current value, conductors are sized for duty cycle service as shown in Table 430.22(E).

(B) Conductors Supplying a Single Motor Controller Conductors supplying a single motor controller shall have an ampacity not less than the motor controller nameplate current rating, plus all other connected loads.

> FPN: Motor controller nameplate current rating may be derived based on the rms value of the motor current using an intermittent duty cycle and other control system loads, if applicable.

(C) Conductors Supplying a Single Power Transformer Conductors supplying a single power transformer shall have an ampacity not less than the nameplate current rating of the power transformer plus all other connected loads.

> FPN No. 1: The nameplate current rating of a power transformer supplying a motor controller reflects the nameplate current rating of the motor controller at line voltage (transformer primary).
>
> FPN No. 2: See Annex D, Example No. D10.

(D) Conductors Supplying More Than One Motor, Motor Controller, or Power Transformer Conductors supplying more than one motor, motor controller, or power transformer shall have an ampacity not less than the sum of the nameplate current ratings of the equipment plus all other connected loads. The ampere ratings of motors to be used in the summation shall be determined from Table 430.22(E), and 430.24 and 430.24, Exception No. 1.

> FPN: See Annex D, Example Nos. D9 and D10.

620.14 Feeder Demand Factor

Feeder conductors of less ampacity than required by 620.13 shall be permitted, subject to the requirements of Table 620.14.

> FPN: Demand factors are based on 50 percent duty cycle (i.e., half time on and half time off).

Table 620.14 Feeder Demand Factors for Elevators

Number of Elevators on a Single Feeder	Demand Factor
1	1.00
2	0.95
3	0.90
4	0.85
5	0.82
6	0.79
7	0.77
8	0.75
9	0.73
10 or more	0.72

620.15 Motor Controller Rating

The motor controller rating shall comply with 430.83. The rating shall be permitted to be less than the nominal rating of the elevator motor, when the controller inherently limits the available power to the motor and is marked as power limited.

> FPN: For controller markings, see 430.8.

The inherent power-limiting ability of certain adjustable-speed drive controllers is the basis for permitting the controller to have a lower current or horsepower rating than that of the motor. To use a controller in this manner, the manufacturer's marking must indicate that it is power limiting.

III. Wiring

620.21 Wiring Methods

Conductors and optical fibers located in hoistways, in escalator and moving walk wellways, in wheelchair lifts, stairway chairlift runways, machinery spaces, control spaces, in or on cars, in machine rooms and control rooms, not including the traveling cables connecting the car or counterweight and hoistway wiring, shall be installed in rigid metal conduit, intermediate metal conduit, electrical metallic tubing, rigid nonmetallic conduit, or wireways, or shall be Type MC, MI, or AC cable unless otherwise permitted in 620.21(A) through 620.21(C).

(A) Elevators

(1) Hoistways

(a) Flexible metal conduit, liquidtight flexible metal conduit, or liquidtight flexible nonmetallic conduit shall be permitted in hoistways between risers and limit switches, interlocks, operating buttons, and similar devices.

(b) Cables used in Class 2 power-limited circuits shall be permitted to be installed between risers and signal equipment and operating devices, provided the cables are supported and protected from physical damage and are of a jacketed and flame-retardant type.

(c) Flexible cords and cables that are components of listed equipment and used in circuits operating at 30 volts rms or less or 42 volts dc or less shall be permitted in lengths not to exceed 1.8 m (6 ft), provided the cords and cables are supported and protected from physical damage and are of a jacketed and flame-retardant type.

Limited (6 ft) lengths of flexible cord and cable are permitted to be used on elevator cars, where the cord or cable is part of listed equipment such as transducers (position, velocity, direction) and the circuit is limited to 30 volts rms or 42 volts dc.

(d) Flexible metal conduit, liquidtight flexible metal conduit, liquidtight flexible nonmetallic conduit or flexible cords and cables, or conductors grouped together and taped or corded that are part of listed equipment, a driving machine, or a driving machine brake shall be permitted in the hoistway, in lengths not to exceed 1.8 m (6 ft), without being installed in a raceway and where located to be protected from physical damage and are of a flame-retardant type.

(2) Cars

(a) Flexible metal conduit, liquidtight flexible metal conduit, or liquidtight flexible nonmetallic conduit of metric designator 12 (trade size ⅜), or larger, not exceeding 1.8 m (6 ft) in length, shall be permitted on cars where so located as to be free from oil and if securely fastened in place.

Exception: Liquidtight flexible nonmetallic conduit of metric designator 12 (trade size ⅜), or larger, as defined by 356.2(2), shall be permitted in lengths in excess of 1.8 m (6 ft).

(b) Hard-service cords and junior hard-service cords that conform to the requirements of Article 400 (Table 400.4) shall be permitted as flexible connections between the fixed wiring on the car and devices on the car doors or gates. Hard-service cords only shall be permitted as flexible connections for the top-of-car operating device or the car-top work light. Devices or luminaires (fixtures) shall be grounded by means of an equipment grounding conductor run with the circuit conductors. Cables with smaller conductors and other types and thicknesses of insulation and jackets shall be permitted as flexible connections between the fixed wiring on the car and devices on the car doors or gates, if listed for this use.

(c) Flexible cords and cables that are components of listed equipment and used in circuits operating at 30 volts rms or less or 42 volts dc or less shall be permitted in lengths not to exceed 1.8 m (6 ft), provided the cords and cables are supported and protected from physical damage and are of a jacketed and flame-retardant type.

(d) Flexible metal conduit, liquidtight flexible metal conduit, liquidtight flexible nonmetallic conduit or flexible cords and cables, or conductors grouped together and taped or corded that are part of listed equipment, a driving machine, or a driving machine brake shall be permitted on the car assembly, in lengths not to exceed 1.8 m (6 ft) without being installed in a raceway and where located to be protected from physical damage and are of a flame-retardant type.

The requirements of 620.21(A)(2)(d) describe the permitted wiring methods where a driving machine or driving machine brake is located on the car. In addition to flexible metal and nonmetallic conduits, the use of single conductors that are taped or corded together is permitted. This use includes rack-

and-pinion or screw column drives located on cars. See also 620.71(B).

(3) Within Machine Rooms, Control Rooms, and Machinery Spaces and Control Spaces

(a) Flexible metal conduit, liquidtight flexible metal conduit, or liquidtight flexible nonmetallic conduit of metric designator 12 (trade size ⅜), or larger, not exceeding 1.8 m (6 ft) in length, shall be permitted between control panels and machine motors, machine brakes, motor-generator sets, disconnecting means, and pumping unit motors and valves.

Exception: Liquidtight flexible nonmetallic conduit metric designator 12 (trade size ⅜) or larger, as defined in 356.2(2), shall be permitted to be installed in lengths in excess of 1.8 m (6 ft).

(b) Where motor-generators, machine motors, or pumping unit motors and valves are located adjacent to or underneath control equipment and are provided with extra-length terminal leads not exceeding 1.8 m (6 ft) in length, such leads shall be permitted to be extended to connect directly to controller terminal studs without regard to the carrying-capacity requirements of Articles 430 and 445. Auxiliary gutters shall be permitted in machine and control rooms between controllers, starters, and similar apparatus.

(c) Flexible cords and cables that are components of listed equipment and used in circuits operating at 30 volts rms or less or 42 volts dc or less shall be permitted in lengths not to exceed 1.8 m (6 ft), provided the cords and cables are supported and protected from physical damage and are of a jacketed and flame-retardant type.

(d) On existing or listed equipment, conductors shall also be permitted to be grouped together and taped or corded without being installed in a raceway. Such cable groups shall be supported at intervals not over 900 mm (3 ft) and located so as to be protected from physical damage.

(4) Counterweight Flexible metal conduit, liquidtight flexible metal conduit, liquidtight flexible nonmetallic conduit or flexible cords and cables, or conductors grouped together and taped or corded that are part of listed equipment, a driving machine, or a driving machine brake shall be permitted on the counterweight assembly, in lengths not to exceed 1.8 m (6 ft) without being installed in a raceway and where located to be protected from physical damage and are of a flame-retardant type.

(B) Escalators

(1) Wiring Methods Flexible metal conduit, liquidtight flexible metal conduit, or liquidtight flexible nonmetallic conduit shall be permitted in escalator and moving walk wellways. Flexible metal conduit or liquidtight flexible conduit of metric designator 12 (trade size ⅜) shall be permitted in lengths not in excess of 1.8 m (6 ft).

Exception: Metric designator 12 (trade size ⅜), nominal, or larger liquidtight flexible nonmetallic conduit, as defined in 356.2(2), shall be permitted to be installed in lengths in excess of 1.8 m (6 ft).

(2) Class 2 Circuit Cables Cables used in Class 2 power-limited circuits shall be permitted to be installed within escalators and moving walkways, provided the cables are supported and protected from physical damage and are of a jacketed and flame-retardant type.

(3) Flexible Cords Hard-service cords that conform to the requirements of Article 400 (Table 400.4) shall be permitted as flexible connections on escalators and moving walk control panels and disconnecting means where the entire control panel and disconnecting means are arranged for removal from machine spaces as permitted in 620.5.

(C) Wheelchair Lifts and Stairway Chairlift Raceways

(1) Wiring Methods Flexible metal conduit or liquidtight flexible metal conduit shall be permitted in wheelchair lifts and stairway chairlift runways and machinery spaces. Flexible metal conduit or liquidtight flexible conduit of metric designator 12 (trade size ⅜) shall be permitted in lengths not in excess of 1.8 m (6 ft).

Exception: Metric designator 12 (trade size ⅜) or larger liquidtight flexible nonmetallic conduit, as defined in 356.2(2), shall be permitted to be installed in lengths in excess of 1.8 m (6 ft).

(2) Class 2 Circuit Cables Cables used in Class 2 power-limited circuits shall be permitted to be installed within wheelchair lifts and stairway chairlift runways and machinery spaces, provided the cables are supported and protected from physical damage and are of a jacketed and flame-retardant type.

620.22 Branch Circuits for Car Lighting, Receptacle(s), Ventilation, Heating, and Air Conditioning

(A) Car Light Source A separate branch circuit shall supply the car lights, receptacle(s), auxiliary lighting power source, and ventilation on each elevator car. The overcurrent device protecting the branch circuit shall be located in the elevator machine room or control room/machinery space or control space.

Required lighting shall not be connected to the load side of a ground-fault circuit interrupter.

Section 620.22 has been revised to prohibit elevator car lighting from being supplied from wiring connected to the

load side of a ground-fault circuit interrupter. A service receptacle installed on an elevator car top is required to be a GFCI-type device, and since the car lights and receptacle are supplied from the same branch circuit, this requirement ensures that GFCI operation will not extinguish the car lighting. The same requirement is found in 620.23(A) and 620.24(A) for machine room lighting and hoistway pit lighting.

(B) Air-Conditioning and Heating Source A dedicated branch circuit shall supply the air-conditioning and heating units on each elevator car. The overcurrent device protecting the branch circuit shall be located in the elevator machine room or control room/machinery space or control space.

The requirements in 620.22(A) and 620.22(B) specify that the overcurrent devices protecting the branch circuits for elevator car lighting and heating and air conditioning are to be located in the elevator machine room or control room or in the elevator machinery or control space. This requirement facilitates ease of maintenance and troubleshooting and correlates with similar requirements in other North American standards. See Exhibit 620.1.

Exhibit 620.1 Disconnecting means and overcurrent devices for elevator power, car lighting, and car heating/air-conditioning branch circuits in the elevator machine room.

620.23 Branch Circuits for Machine Room or Control Room/Machinery Space or Control Space Lighting and Receptacle(s)

(A) Separate Branch Circuit A separate branch circuit shall supply the machine room or control room/machinery space or control space lighting and receptacle(s).

Required lighting shall not be connected to the load side of a ground-fault circuit interrupter.

Luminaires are not permitted to be connected to the load side of a GFCI device of any kind, because the machine room lighting could be de-energized during certain fault conditions.

(B) Lighting Switch The machine room or control room/ machinery space or control space lighting switch shall be located at the point of entry.

(C) Duplex Receptacle At least one 125-volt, single-phase, duplex receptacle shall be provided in each machine room or control room and machinery space or control space.

> FPN: See ANSI/ASME A17.1-2000, *Safety Code for Elevators and Escalators*, for illumination levels.

The receptacles required by 620.23 and 620.24 are also required to be provided with GFCI protection for personnel (see 620.85). Luminaires are not permitted to be connected to the load side of GFCI devices. See also the commentary following 620.23(A).

ASME A17.1-2004, *Safety Code for Elevators and Escalators*, requires a minimum of 5 ft-candles (54 lux) at the pit floor and requires luminaires to be externally guarded to prevent accidental breakage.

Luminaires in pits should be mounted so that the car or counterweight does not strike them when on fully compressed buffers. It is recommended that luminaires be mounted in the corners of the pit.

620.24 Branch Circuit for Hoistway Pit Lighting and Receptacle(s)

(A) Separate Branch Circuit A separate branch circuit shall supply the hoistway pit lighting and receptacle(s).

Required lighting shall not be connected to the load side of a ground-fault circuit interrupter.

(B) Lighting Switch The lighting switch shall be so located as to be readily accessible from the pit access door.

(C) Duplex Receptacle At least one 125-volt, single-phase, duplex receptacle shall be provided in the hoistway pit.

> FPN: See ANSI/ASME A17.1-2000, *Safety Code for Elevators and Escalators*, for illumination levels.

620.25 Branch Circuits for Other Utilization Equipment

(A) Additional Branch Circuits Additional branch circuit(s) shall supply utilization equipment not identified in

620.22, 620.23, and 620.24. Other utilization equipment shall be restricted to that equipment identified in 620.1.

(B) Overcurrent Devices The overcurrent devices protecting the branch circuit(s) shall be located in the elevator machinery room or control room/machinery space or control space.

IV. Installation of Conductors

620.32 Metal Wireways and Nonmetallic Wireways

The sum of the cross-sectional area of the individual conductors in a wireway shall not be more than 50 percent of the interior cross-sectional area of the wireway.

Vertical runs of wireways shall be securely supported at intervals not exceeding 4.5 m (15 ft) and shall have not more than one joint between supports. Adjoining wireway sections shall be securely fastened together to provide a rigid joint.

620.33 Number of Conductors in Raceways

The sum of the cross-sectional area of the individual conductors in raceways shall not exceed 40 percent of the interior cross-sectional area of the raceway, except as permitted in 620.32 for wireways.

620.34 Supports

Supports for cables or raceways in a hoistway or in an escalator or moving walk wellway or wheelchair lift and stairway chairlift runway shall be securely fastened to the guide rail; escalator or moving walk truss; or to the hoistway, wellway, or runway construction.

620.35 Auxiliary Gutters

Auxiliary gutters shall not be subject to the restrictions of 366.12(2) covering length or of 366.22 covering number of conductors.

620.36 Different Systems in One Raceway or Traveling Cable

Optical fiber cables and conductors for operating devices, operation and motion control, power, signaling, fire alarm, lighting, heating, and air-conditioning circuits of 600 volts or less shall be permitted to be run in the same traveling cable or raceway system if all conductors are insulated for the maximum voltage applied to any conductor within the cables or raceway system and if all live parts of the equipment are insulated from ground for this maximum voltage. Such a traveling cable or raceway shall also be permitted to include shielded conductors and/or one or more coaxial cables, if such conductors are insulated for the maximum voltage applied to any conductor within the cable or raceway system. Conductors shall be permitted to be covered with suitable shielding for telephone, audio, video, or higher frequency communications circuits.

With the use of greater numbers of individual cables and the use of much longer cables in tall buildings, there is a possibility of intertwisting cable loops. To eliminate the practice of tying a cable to the traveling cable, one elevator cable or raceway is permitted to enclose optical fiber cables and all the conductors for power, control, lighting, video, fire alarm, and communications circuits if all conductors are insulated for the maximum voltage applied to any conductor within the cable or raceway and all live parts are also insulated from ground for the maximum voltage present.

The revision of 620.36 in the 1999 *Code* clarified that power-limited and non–power-limited fire alarm conductors are permitted in the same raceway or traveling cable as power and other types of signaling conductors. All power, signaling, and fire alarm conductors are to be insulated for the maximum voltage applied to any conductor in the raceway or cable.

620.37 Wiring in Hoistways, Machine Rooms, Control Rooms, Machinery Spaces, and Control Spaces

(A) Uses Permitted Only such electric wiring, raceways, and cables used directly in connection with the elevator or dumbwaiter, including wiring for signals, for communication with the car, for lighting, heating, air conditioning, and ventilating the elevator car, for fire detecting systems, for pit sump pumps, and for heating, lighting, and ventilating the hoistway, shall be permitted inside the hoistway, machine rooms, control rooms, machinery spaces, and control spaces.

(B) Lightning Protection Bonding of elevator rails (car and/or counterweight) to a lightning protection system grounding down conductor(s) shall be permitted. The lightning protection system grounding down conductor(s) shall not be located within the hoistway. Elevator rails or other hoistway equipment shall not be used as the grounding down conductor for lightning protection systems.

Elevator hoistways are often convenient locations to run wiring from the basement to the roof. However, 620.37(B) prohibits equipment and wiring not associated with the elevator from being installed in elevator machine rooms and hoistways. Only electrical equipment and wiring used directly in connection with the elevator may be installed inside the hoistway and the machine room.

Where a lightning protection system is provided, and where a lightning protection system grounding "down" conductor(s) located outside the hoistway is within a critical

horizontal distance of the elevator rails, bonding of the rails to the lightning protection system grounding "down" conductor(s) is required by NFPA 780, *Standard for the Installation of Lightning Protection Systems.* Bonding prevents a dangerous side flash between the lightning protection system grounding "down" conductor(s) and the elevator rails. A lightning strike on the building air terminal will be conducted through the lightning protection system grounding "down" conductor(s), and, if the elevator rails are not at the same potential as the lightning protection system grounding "down" conductor(s), a side flash may occur. Generally, "down" conductors are installed vertically near the perimeter of the structure, so this requirement may have application to "outside" elevators.

FPN: See 250.106 for bonding requirements. For further information, see NFPA 780-2004, *Standard for the Installation of Lightning Protection Systems.*

(C) Main Feeders Main feeders for supplying power to elevators and dumbwaiters shall be installed outside the hoistway unless as follows:

(1) By special permission, feeders for elevators shall be permitted within an existing hoistway if no conductors are spliced within the hoistway.
(2) Feeders shall be permitted inside the hoistway for elevators with driving machine motors located in the hoistway or on the car or counterweight.

620.38 Electrical Equipment in Garages and Similar Occupancies

Electrical equipment and wiring used for elevators, dumbwaiters, escalators, moving walks, and wheelchair lifts and stairway chairlifts in garages shall comply with the requirements of Article 511.

FPN: Garages used for parking or storage and where no repair work is done in accordance with 511.3 are not classified.

V. Traveling Cables

620.41 Suspension of Traveling Cables

Traveling cables shall be suspended at the car and hoistways' ends, or counterweight end where applicable, so as to reduce the strain on the individual copper conductors to a minimum.

Traveling cables shall be supported by one of the following means:

(1) By their steel supporting member(s)
(2) By looping the cables around supports for unsupported lengths less than 30 m (100 ft)
(3) By suspending from the supports by a means that automatically tightens around the cable when tension is increased for unsupported lengths up to 60 m (200 ft)

FPN: Unsupported length for the hoistway suspension means is that length of cable as measured from the point of suspension in the hoistway to the bottom of the loop, with the elevator car located at the bottom landing. Unsupported length for the car suspension means is that length of cable as measured from the point of suspension on the car to the bottom of the loop, with the elevator car located at the top landing.

Traveling cables between fixed suspension points are not required to be installed in a raceway. If the fixed suspension point is on top of the car, the cables on the side of the car might be exposed. Suitable guards are necessary to protect these cables from damage. If the suspension point is under the car, the cables might be run up the side of the car to the car's top junction box. If the runs are over 6 ft, the cables are required to be run in a raceway. Refer to 620.44 and Exhibit 620.2.

620.42 Hazardous (Classified) Locations

In hazardous (classified) locations, traveling cables shall be of a type approved for hazardous (classified) locations and shall comply with 501.140, 502.140, or 503.140, as applicable.

620.43 Location of and Protection for Cables

Traveling cable supports shall be located so as to reduce to a minimum the possibility of damage due to the cables coming in contact with the hoistway construction or equipment in the hoistway. Where necessary, suitable guards shall be provided to protect the cables against damage.

620.44 Installation of Traveling Cables

Traveling cable shall be permitted to be run without the use of a raceway for a distance not exceeding 1.8 m (6 ft) in length as measured from the first point of support on the elevator car or hoistway wall, or counterweight where applicable, provided the conductors are grouped together and taped or corded, or in the original sheath.

Traveling cables shall be permitted to be continued as fixed wiring to elevator controller enclosures and to elevator car and machine room, control room, machinery space, and control space connections, provided they are suitably supported and protected from physical damage.

VI. Disconnecting Means and Control

620.51 Disconnecting Means

A single means for disconnecting all ungrounded main power supply conductors for each unit shall be provided and be designed so that no pole can be operated independently. Where multiple driving machines are connected to a single elevator, escalator, moving walk, or pumping unit, there shall be one disconnecting means to disconnect the motor(s) and control valve operating magnets.

The disconnecting means for the main power supply conductors shall not disconnect the branch circuit required in 620.22, 620.23, and 620.24.

Exhibit 620.2 Unsupported lengths of traveling cable.

The branch circuits that supply elevator car lighting, receptacles, ventilation, air conditioning, and heating are required to be independent of the control portion of the elevator. In addition, the branch circuits supplying hoistway pit lighting and receptacles and machine room or control room lights and receptacles are not permitted to be disconnected by the main elevator power disconnect. This requirement provides for passenger safety and comfort and for the safety of elevator maintenance personnel during an inadvertent or emergency shutdown of the main power circuit to the elevator.

(A) Type The disconnecting means shall be an enclosed externally operable fused motor circuit switch or circuit breaker capable of being locked in the open position. The disconnecting means shall be a listed device.

> FPN: For additional information, see ASME/ANSI A17.1-2000, *Safety Code for Elevators and Escalators.*

Exception: Where an individual branch circuit supplies a wheelchair lift, the disconnecting means required by 620.51(C)(4) shall be permitted to comply with 430.109(C). This disconnecting means shall be listed and shall be capable of being locked in the open position.

(B) Operation No provision shall be made to open or close this disconnecting means from any other part of the premises. If sprinklers are installed in hoistways, machine rooms, control rooms, machinery spaces, or control spaces, the disconnecting means shall be permitted to automatically open the power supply to the affected elevator(s) prior to the application of water. No provision shall be made to automatically close this disconnecting means. Power shall only be restored by manual means.

> FPN: To reduce hazards associated with water on live elevator electrical equipment.

ASME A17.1-2004, *Safety Code for Elevators and Escalators,* Rule 2.8.2.3 requires that where sprinklers are installed in hoistways, machine rooms, or machinery spaces, a means must be provided to automatically disconnect the main line power supply to the affected elevator(s) upon or prior to the application of water. Water on elevator electrical equipment can result in hazards such as uncontrolled car movement (wet machine brakes), movement of elevator with open doors (water on safety circuits bypassing car and/or hoistway door interlocks), and shock hazards.

Automatic disconnection of the main line power supply is not required by ASME A17.1-2004, *Safety Code for Elevators and Escalators,* where hoistways and machine rooms are not sprinklered. NFPA 13, *Standard for the Installation of Sprinkler Systems,* provides requirements for the installation of sprinklers in machine rooms, hoistways, and pits.

Elevator shutdown is generally accomplished through the use of heat detectors located near sprinkler heads. The heat detectors are designed to actuate and generate an alarm signal prior to water discharge from the sprinkler heads. An output control relay powered by the fire alarm system then provides a monitored output to the main line disconnecting means control circuit, which activates the shunt trip. This practice ensures that all components have secondary power and are monitored for integrity, as required by *NFPA 72®, National Fire Alarm Code®.* Stand-alone heat detectors connected directly to the elevator disconnecting means control circuit are not monitored for integrity, have no secondary power supply, and are not permitted by *NFPA 72.*

Elevator shutdown can occur even if the car is not at a landing. However, to avoid trapping occupants in the car(s), it is highly desirable to recall the car(s) to the designated landing prior to disconnecting the main line power. Most fires produce detectable quantities of smoke before there is sufficient heat to activate a sprinkler head. Therefore, ASME A17.1-2004, *Safety Code for Elevators and Escalators,* Rule 102.2(c), requires smoke detectors to be installed in hoistways that are sprinklered for the purposes of recalling the elevator car(s) before the main line power is disconnected. See 6.15.4 of *NFPA 72, National Fire Alarm Code,* for additional requirements relating to the fire alarm system and elevator shutdown.

Exhibit 620.3 illustrates a typical method of supervising control power using a fire alarm system. Loss of control power produces a supervisory signal at the fire alarm control unit that then would be investigated.

(C) Location The disconnecting means shall be located where it is readily accessible to qualified persons.

(1) On Elevators Without Generator Field Control On elevators without generator field control, the disconnecting means shall be located within sight of the motor controller. Driving machines or motion and operation controllers not within sight of the disconnecting means shall be provided with a manually operated switch installed in the control circuit to prevent starting. The manually operated switch(es) shall be installed adjacent to this equipment.

Where the driving machine of an electric elevator or the hydraulic machine of a hydraulic elevator is located in a remote machine room or remote machinery space, a single means for disconnecting all ungrounded main power supply conductors shall be provided and be capable of being locked in the open position.

See Exhibit 620.4 for disconnecting means for driving machines or motion and operation controllers not within sight of the main line disconnecting means.

Exhibit 620.3 Typical method of control power supervision using a fire alarm control unit.

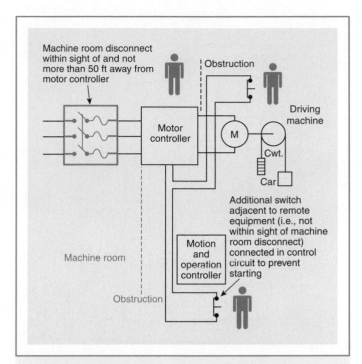

Exhibit 620.4 Disconnecting means for driving machines or motion and operation controllers not within sight of the main line disconnecting means. (Redrawn courtesy of ASME)

(2) On Elevators with Generator Field Control On elevators with generator field control, the disconnecting means shall be located within sight of the motor controller for the driving motor of the motor-generator set. Driving machines, motor-generator sets, or motion and operation controllers not within sight of the disconnecting means shall be provided with a manually operated switch installed in the control circuit to prevent starting. The manually operated switch(es) shall be installed adjacent to this equipment.

Where the driving machine or the motor-generator set is located in a remote machine room or remote machinery space, a single means for disconnecting all ungrounded main power supply conductors shall be provided and be capable of being locked in the open position.

See Exhibit 620.5 and Exhibit 620.6 for examples of disconnecting means for a motor-generator set and for driving machines in remote locations.

(3) On Escalators and Moving Walks On escalators and moving walks, the disconnecting means shall be installed in the space where the controller is located.

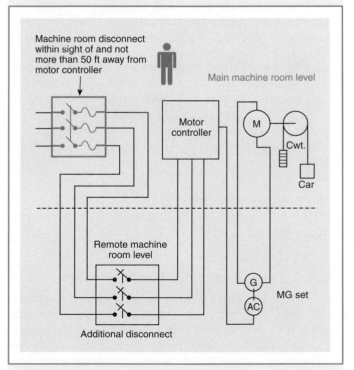

Exhibit 620.5 Disconnecting means for a motor-generator (MG) set in a remote location. (Redrawn courtesy of ASME)

Machine room disconnect
within sight of and not
more than 50 ft away from
motor controller

Main machine room level

Motor
controller

AC
MG set
G

Remote machine
room level

M

Additional disconnect

Cwt.
Car

Exhibit 620.6 Disconnecting means for driving machines in a remote location. (Redrawn courtesy of ASME)

(4) On Wheelchair Lifts and Stairway Chairlifts On wheelchair lifts and stairway chairlifts, the disconnecting means shall be located within sight of the motor controller.

(D) Identification and Signs Where there is more than one driving machine in a machine room, the disconnecting means shall be numbered to correspond to the identifying number of the driving machine that they control.

The disconnecting means shall be provided with a sign to identify the location of the supply side overcurrent protective device.

Sign requirements for the location of supply-side overcurrent devices assist the elevator mechanic in troubleshooting during a power loss.

620.52 Power from More Than One Source

(A) Single-Car and Multicar Installations On single-car and multicar installations, equipment receiving electrical power from more than one source shall be provided with a disconnecting means for each source of electrical power. The disconnecting means shall be within sight of the equipment served.

(B) Warning Sign for Multiple Disconnecting Means Where multiple disconnecting means are used and parts of the controllers remain energized from a source other than

the one disconnected, a warning sign shall be mounted on or next to the disconnecting means. The sign shall be clearly legible and shall read as follows:

WARNING
PARTS OF THE CONTROLLER ARE NOT
DE-ENERGIZED BY THIS SWITCH.

(C) Interconnection Multicar Controllers Where interconnections between controllers are necessary for the operation of the system on multicar installations that remain energized from a source other than the one disconnected, a warning sign in accordance with 620.52(B) shall be mounted on or next to the disconnecting means.

620.53 Car Light, Receptacle(s), and Ventilation Disconnecting Means

Elevators shall have a single means for disconnecting all ungrounded car light, receptacle(s), and ventilation power-supply conductors for that elevator car.

The disconnecting means shall be an enclosed externally operable fused motor circuit switch or circuit breaker capable of being locked in the open position and shall be located in the machine room or control room for that elevator car. Where there is no machine room or control room, the disconnecting means shall be located in the same space as the disconnecting means required by 620.51.

This requirement specifies the location of the disconnecting means for lighting, receptacle, and ventilation branch circuits associated with elevators that do not have a machine room. This type of installation includes those designs using drive systems located on the car, on the counterweight, or in the hoistway. Such designs include screw drive or linear induction motor drives. See ASME A17.1-2004, *Safety Code for Elevators and Escalators*, for more information on this type of arrangement.

Disconnecting means shall be numbered to correspond to the identifying number of the elevator car whose light source they control.

The disconnecting means shall be provided with a sign to identify the location of the supply side overcurrent protective device.

620.54 Heating and Air-Conditioning Disconnecting Means

Elevators shall have a single means for disconnecting all ungrounded car heating and air-conditioning power-supply conductors for that elevator car.

The disconnecting means shall be an enclosed externally operable fused motor circuit switch or circuit breaker capable of being locked in the open position and shall be located in the machine room or control room for that elevator car.

Where there is no machine room or control room, the disconnecting means shall be located in the same space as the disconnecting means required by 620.51.

Where there is equipment for more than one elevator car in the machine room, the disconnecting means shall be numbered to correspond to the identifying number of the elevator car whose heating and air-conditioning source they control.

The disconnecting means shall be provided with a sign to identify the location of the supply side overcurrent protective device.

620.55 Utilization Equipment Disconnecting Means

Each branch circuit for other utilization equipment shall have a single means for disconnecting all ungrounded conductors. The disconnecting means shall be capable of being locked in the open position and shall be located in the machine room or control room/machine space or control space. Where there is more than one branch circuit for other utilization equipment, the disconnecting means shall be numbered to correspond to the identifying number of the equipment served. The disconnecting means shall be provided with a sign to identify the location of the supply side overcurrent protective device.

VII. Overcurrent Protection

620.61 Overcurrent Protection

Overcurrent protection shall be provided in accordance with 620.61(A) through 620.61(D).

(A) Operating Devices and Control and Signaling Circuits Operating devices and control and signaling circuits shall be protected against overcurrent in accordance with the requirements of 725.23 and 725.24.

Class 2 power-limited circuits shall be protected against overcurrent in accordance with the requirements of Chapter 9, Notes to Tables 11(A) and 11(B).

(B) Overload Protection for Motors Motor and branch-circuit overload protection shall conform to Article 430, Part III, and the following:

This new text clarifies that overload protection as specified in Part III of Article 430 applies to motors associated with elevators, dumbwaiters, escalators, and wheelchair and stairway lifts. Specific permission to use the provisions of 430.33 for intermittent duty motors is given in 620.61(B)(1) and 620.61(B)(4).

(1) Duty Rating on Elevator, Dumbwaiter, and Motor-Generator Sets Driving Motors Duty on elevator and dumbwaiter driving machine motors and driving motors of motor-generators used with generator field control shall be rated as intermittent. Such motors shall be permitted to be protected against overload in accordance with 430.33.

(2) Duty Rating on Escalator Motors Duty on escalator and moving walk driving machine motors shall be rated as continuous. Such motors shall be protected against overload in accordance with 430.32.

(3) Overload Protection Escalator and moving walk driving machine motors and driving motors of motor-generator sets shall be protected against running overload as provided in Table 430.37.

(4) Duty Rating and Overload Protection on Wheelchair and Stairway Chairlift Motors Duty on wheelchair lift and stairway chairlift driving machine motors shall be rated as intermittent. Such motors shall be permitted to be protected against overload in accordance with 430.33.

FPN: For further information, see 430.44 for orderly shutdown.

(C) Motor Feeder Short-Circuit and Ground-Fault Protection Motor feeder short-circuit and ground-fault protection shall be as required in Article 430, Part V.

(D) Motor Branch-Circuit Short-Circuit and Ground-Fault Protection Motor branch-circuit short-circuit and ground-fault protection shall be as required in Article 430, Part IV.

620.62 Selective Coordination

Where more than one driving machine disconnecting means is supplied by a single feeder, the overcurrent protective devices in each disconnecting means shall be selectively coordinated with any other supply side overcurrent protective devices.

Coordination of the overcurrent protective devices is important. For example, if a building contains three elevators and a fault occurs in the circuit conductors to one of the elevators, only the overcurrent device ahead of that faulted circuit should open. Coordination leaves the remaining two elevators in operation. This is especially important because elevators are commonly used to carry fire fighters and equipment closer to the fire during fire-fighting operations.

Where the overcurrent devices in the elevator room do not have proper coordination with the upstream feeder overcurrent device, there is increased potential for interruption of power to all three elevators.

For selective coordination of overcurrent protective devices, the manufacturer's time-current curves, let-through and withstand capacity data, and unlatching times data must be used for sizing or setting overcurrent devices. See Exhibit 620.7.

Exhibit 620.7 Example of selective coordination, in which only overcurrent devices in switch A should open for a fault at X_1. Feeder overcurrent devices in the main distribution panel should not open, so two elevators can remain in use. See 620.51(A) for the requirement regarding power supply disconnecting means.

VIII. Machine Rooms, Control Rooms, Machinery Spaces, and Control Spaces

620.71 Guarding Equipment

Elevator, dumbwaiter, escalator, and moving walk driving machines; motor-generator sets; motor controllers; and disconnecting means shall be installed in a room or space set aside for that purpose unless otherwise permitted in 620.71(A) or 620.71(B). The room or space shall be secured against unauthorized access.

(A) Motor Controllers Motor controllers shall be permitted outside the spaces herein specified, provided they are in enclosures with doors or removable panels that are capable of being locked in the closed position and the disconnecting means is located adjacent to or is an integral part of the motor controller. Motor controller enclosures for escalator or moving walks shall be permitted in the balustrade on the side located away from the moving steps or moving treadway. If the disconnecting means is an integral part of the motor controller, it shall be operable without opening the enclosure.

(B) Driving Machines Elevators with driving machines located on the car, on the counterweight, or in the hoistway, and driving machines for dumbwaiters, wheelchair lifts, and stairway lifts shall be permitted outside the spaces herein specified.

IX. Grounding

620.81 Metal Raceways Attached to Cars

Metal raceways, Type MC cable, Type MI cable, or Type AC cable attached to elevator cars shall be bonded to grounded metal parts of the car that they contact.

620.82 Electric Elevators

For electric elevators, the frames of all motors, elevator machines, controllers, and the metal enclosures for all electrical equipment in or on the car or in the hoistway shall be grounded in accordance with Article 250.

620.83 Nonelectric Elevators

For elevators other than electric having any electric conductors attached to the car, the metal frame of the car, where normally accessible to persons, shall be grounded in accordance with Article 250.

620.84 Escalators, Moving Walks, Wheelchair Lifts, and Stairway Chairlifts

Escalators, moving walks, wheelchair lifts, and stairway chairlifts shall comply with Article 250.

620.85 Ground-Fault Circuit-Interrupter Protection for Personnel

Each 125-volt, single-phase, 15- and 20-ampere receptacle installed in pits, in hoistways, on elevator car tops, and in escalator and moving walk wellways shall be of the ground-fault circuit-interrupter type.

All 125-volt, single-phase, 15- and 20-ampere receptacles installed in machine rooms and machinery spaces shall have ground-fault circuit-interrupter protection for personnel.

A single receptacle supplying a permanently installed sump pump shall not require ground-fault circuit-interrupter protection.

The GFCI requirements of 620.85 are intended to reduce the shock hazard to maintenance personnel who service elevator equipment using portable hand tools and temporary lighting.

The first paragraph of 620.85 requires a GFCI-type receptacle for each 15- and 20-ampere receptacle installed in pits, on elevator car tops, and in escalator and moving-walk wellways. This requirement is based on the premise that the reset pushbutton for a tripped GFCI receptacle should be within easy reach of an elevator mechanic working in confined spaces.

The second paragraph of 620.85 requires that all 15- and 20-ampere receptacles installed in machine rooms and machinery spaces have GFCI protection for personnel. This protection can be afforded by either a GFCI-type circuit breaker or a GFCI-type receptacle because machine spaces usually do not cause access hazards for service personnel.

X. Emergency and Standby Power Systems

620.91 Emergency and Standby Power Systems

An elevator(s) shall be permitted to be powered by an emergency or standby power system.

FPN: See ASME/ANSI A17.1-2000, *Rule 2.27.2*, and CAN/CSA-B44-1994, *Clause 3.12.13*, for additional information.

(A) Regenerative Power For elevator systems that regenerate power back into the power source that is unable to absorb the regenerative power under overhauling elevator load conditions, a means shall be provided to absorb this power.

(B) Other Building Loads Other building loads, such as power and lighting, shall be permitted as the energy absorption means required in 620.91(A), provided that such loads are automatically connected to the emergency or standby power system operating the elevators and are large enough to absorb the elevator regenerative power.

(C) Disconnecting Means The disconnecting means required by 620.51 shall disconnect the elevator from both the emergency or standby power system and the normal power system.

Where an additional power source is connected to the load side of the disconnecting means, the disconnecting means required in 620.51 shall be provided with an auxiliary contact that is positively opened mechanically, and the opening shall not be solely dependent on springs. This contact shall cause the additional power source to be disconnected from its load when the disconnecting means is in the open position.

ARTICLE 625
Electric Vehicle Charging System

Summary of Changes

- **625.26:** Added new requirement allowing an electric vehicle to serve as an optional standby system or an electric power production source or provide for bi-directional power feed if listed as suitable for that purpose and it complies with Article 702 and Article 705.

Contents

I. General

A variety of street- and highway-worthy electric and combination electric/fossil fuel vehicles are becoming available to consumers. New and proposed legislation in several regions around the United States calls for increasing deployment of electric vehicles as a way to reduce air pollution. Other states have adopted similar requirements. In addition, the Clean Air Act Amendments of 1990 and the National Energy Policy Act of 1992 have requirements for public and private purchases of clean-fuel vehicles and alternatively fueled vehicles, respectively. Electric vehicles fulfill both those requirements. It is apparent that electric vehicle charging will be occurring in all occupancies, including residential, commercial, retail, and public sites.

Article 625 sets forth installation safety requirements for typical hard-wired conductive connections of battery charging equipment, as well as the safety concerns of the new "smart" inductive coupling connections of battery charging equipment. In particular, this article covers the wiring methods, equipment construction, control and protection, and equipment locations for automotive-type vehicle charging equipment. Throughout Article 625, the intent is to prevent the users of electric equipment associated with the vehicle charging system from being exposed to energized live parts and to provide for a safe vehicle charging environment.

625.1 Scope

The provisions of this article cover the electrical conductors and equipment external to an electric vehicle that connect an electric vehicle to a supply of electricity by conductive or inductive means, and the installation of equipment and devices related to electric vehicle charging.

The scope of Article 625 is intended to cover all electrical wiring and equipment installed between the service point and the skin of the automotive-type electric vehicle. Automotive-type electric vehicles are emphasized because they are much different from other electric vehicles commonly used today. Most existing electric vehicles are off-road types, such as industrial forklifts, hoists, lifts, transports, golf carts, and airport personnel trams. The charging requirements and other exterior electrical connections are usually serviced and maintained by trained mechanics or technicians. The *NEC* has adequate provisions to allow the authority having jurisdiction to make interpretations that provide the safety levels needed for these installations.

Article 625 specifically excludes off-road vehicles, to avoid conflict with existing articles. Motorcycles are not covered by Article 625 because motorcycles typically have smaller propulsion systems that operate at lower voltages, 12 to 24 volts dc versus 100 to 350 volts dc for electric automotive vehicles. Typically, motorcycles are charged from standard 120-volt, 15-ampere receptacles due to lower battery capacity. GFCI protection is not mandatory for charging electric motorcycles. However, 210.8(A)(2) and 210.8(A)(3) require GFCI protection of receptacles in the locations where an electric motorcycle would typically be charged.

FPN: For industrial trucks, see NFPA 505-2002, *Fire Safety Standard for Powered Industrial Trucks Including Type Designations, Areas of Use, Conversions, Maintenance, and Operation.*

625.2 Definitions

Several of the definitions in 625.2 correlate with industry standards such as those from the Society of Automotive Engineers, SAE J1772, *SAE Electric Vehicle Conductive Charge Coupler*, and SAE J1773, *SAE Electric Vehicle In-*

ductively Coupled Charging, and from Underwriters Laboratories, UL 2231-1, *Standard for Personnel Protection Systems for Electric Vehicle Supply Circuits: General Requirements*, and UL 2231-2, *Standard for Personnel Protection Systems for Electric Vehicle Supply Circuits: Particular Requirements for Protection Devices for Use in Charging Systems.*

Electric Vehicle. An automotive-type vehicle for highway use, such as passenger automobiles, buses, trucks, vans, neighborhood electric vehicles, and the like, primarily powered by an electric motor that draws current from a rechargeable storage battery, fuel cell, photovoltaic array, or other source of electric current. For the purpose of this article, electric motorcycles and similar type vehicles and off-road self-propelled electric vehicles, such as industrial trucks, hoists, lifts, transports, golf carts, airline ground support equipment, tractors, boats, and the like, are not included.

The primary difference between electric vehicles as defined in Article 625 and electric vehicles covered by other sections in the *NEC* is in their road and highway worthiness. The automotive electric vehicles under consideration are comparable in performance and function to the conventional automobiles and light trucks in use today. The automotive electric vehicles under development must be capable of complying with the Federal Motor Vehicle Safety Standards and other Department of Transportation, National Highway Traffic Safety Administration, and U.S. Environmental Protection Agency requirements. The definition of *electric vehicle* has been revised for the 2005 *Code* to include neighborhood electric vehicles, which are low-speed, limited-use electric vehicles similar to golf carts but provided with automotive-grade headlights, seat belts, windshields, brakes, and other safety equipment. Neighborhood electric vehicles are increasing in popularity as low-cost, energy-efficient, zero-polluting alternatives to traditional automobiles. Under National Highway Traffic Safety Administration guidelines, the intended use for these vehicles is shopping and recreation in inner-city areas and planned and retirement communities where the street speed limit is 35 mph or less. Electric vehicles, such as lift trucks and golf carts, are not covered by Article 625.

Electric Vehicle Connector. A device that, by insertion into an electric vehicle inlet, establishes an electrical connection to the electric vehicle for the purpose of charging and information exchange. This device is part of the electric vehicle coupler.

Electric Vehicle Coupler. A mating electric vehicle inlet and electric vehicle connector set.

Electric Vehicle Inlet. The device on the electric vehicle into which the electric vehicle connector is inserted for charging and information exchange. This device is part of the electric vehicle coupler. For the purposes of this *Code*, the electric vehicle inlet is considered to be part of the electric vehicle and not part of the electric vehicle supply equipment.

Electric Vehicle Nonvented Storage Battery. A hermetically sealed battery comprised of one or more rechargeable electrochemical cells that has no provision for the release of excessive gas pressure, or for the addition of water or electrolyte, or for external measurements of electrolyte specific gravity.

Electric Vehicle Supply Equipment. The conductors, including the ungrounded, grounded, and equipment grounding conductors and the electric vehicle connectors, attachment plugs, and all other fittings, devices, power outlets, or apparatus installed specifically for the purpose of delivering energy from the premises wiring to the electric vehicle.

Electric vehicle supply equipment, as illustrated in Exhibit 625.1, comprises the components between the skin of the electric vehicle and the premises wiring, including any flexible cable, disconnecting means, enclosures, power outlet,

Exhibit 625.1 Example of electric vehicle supply equipment. (Courtesy of Ford Motor Co.)

and electric vehicle connector. The definition of *electric vehicle* includes all off-vehicle charging equipment and does not include charging equipment installed on the vehicle.

Personnel Protection System. A system of personnel protection devices and constructional features that when used together provide protection against electric shock of personnel.

625.3 Other Articles

Wherever the requirements of other articles of this *Code* and Article 625 differ, the requirements of Article 625 shall apply.

625.4 Voltages

Unless other voltages are specified, the nominal ac system voltages of 120, 120/240, 208Y/120, 240, 480Y/277, 480, 600Y/347, and 600 volts shall be used to supply equipment covered by this article.

625.5 Listed or Labeled

All electrical materials, devices, fittings, and associated equipment shall be listed or labeled.

II. Wiring Methods

625.9 Electric Vehicle Coupler

The electric vehicle coupler shall comply with 625.9(A) through 625.9(F).

The electric vehicle connector is the device that inserts into the electric vehicle inlet (charge port) of the vehicle. The electric vehicle inlet is not a premises wiring receptacle or an attachment cap. An electric vehicle coupler is the mating set of the electric vehicle connector and electric vehicle inlet. A requirement was added to the 1999 *Code* to require the coupler to be noninterchangeable, to prevent equipment damage or personal injury. Couplers for the electric vehicle charging equipment are not permitted to be standard NEMA-configuration wiring devices.

(A) Polarization The electric vehicle coupler shall be polarized unless part of a system identified and listed as suitable for the purpose.

(B) Noninterchangeability The electric vehicle coupler shall have a configuration that is noninterchangeable with wiring devices in other electrical systems. Nongrounding-type electric vehicle couplers shall not be interchangeable with grounding-type electric vehicle couplers.

(C) Construction and Installation The electric vehicle coupler shall be constructed and installed so as to guard against inadvertent contact by persons with parts made live from the electric vehicle supply equipment or the electric vehicle battery.

The requirements for coupler construction in 625.9(C) provide a safe interface component for persons connecting the vehicle to or disconnecting the vehicle from the charging system. This type of activity generally is performed daily by persons who typically do not have any knowledge of the equipment operation and its associated hazards.

(D) Unintentional Disconnection The electric vehicle coupler shall be provided with a positive means to prevent unintentional disconnection.

(E) Grounding Pole The electric vehicle coupler shall be provided with a grounding pole, unless part of a system identified and listed as suitable for the purpose in accordance with Article 250.

(F) Grounding Pole Requirements If a grounding pole is provided, the electric vehicle coupler shall be so designed that the grounding pole connection is the first to make and the last to break contact.

III. Equipment Construction

625.13 Electric Vehicle Supply Equipment

Electric vehicle supply equipment rated at 125 volts, single phase, 15 or 20 amperes or part of a system identified and listed as suitable for the purpose and meeting the requirements of 625.18, 625.19, and 625.29 shall be permitted to be cord-and-plug connected. All other electric vehicle supply equipment shall be permanently connected and fastened in place. This equipment shall have no exposed live parts.

Some manufacturers produce 125-volt, single-phase, 15- or 20-ampere portable charging units for convenience charging. These charging units may be stored in the vehicle. However, 625.13 makes it clear that nonportable equipment must be mounted and permanently wired. This equipment may be physically attached to the wall, floor, or ceiling. The provision for no exposed live parts is a safety concern for the general public.

625.14 Rating

Electric vehicle supply equipment shall have sufficient rating to supply the load served. For the purposes of this article, electric vehicle charging loads shall be considered to be continuous loads.

Considering both near-term and long-term requirements for electric vehicle (EV) charging, three methods have been identified for recommended development. Referred to as Level 1, Level 2, and Level 3 EV charging, they cover the range of power levels anticipated for charging EVs.

Level 1. This method, which allows broad access to charge an EV, permits plugging into a common, grounded 120-volt electrical receptacle (NEMA 5-15R or 5-20R). The maximum load on this receptacle is 12 amperes or 1.4 kVA. The minimum circuit and overcurrent rating for this connection is 15 amperes for a 15-ampere receptacle and 20 amperes for a 20-ampere receptacle.

Level 2. This is the primary and preferred method of EV charging at both private and public facilities. It requires special equipment and connection to an electric power supply dedicated to EV charging. The voltage of this connection is either 240 volts or 208 volts. The maximum load is 32 amperes (7.7 kVA at 240 volts or 6.7 kVA at 208 volts). The minimum circuit and overcurrent rating for this connection is 40 amperes ($32 \times 1.25 = 40$ amperes). Electric vehicles are treated as continuous loads. See 625.21 for sizing overcurrent protection devices.

Level 3. The EV equivalent of a commercial gasoline dispensing station, this high-speed, high-power method charges an EV in about the same time it takes to refuel a conventional vehicle. Because of individual supply requirements and available source voltages, exact voltage and load specifications for Level 3 charging have not been defined as in Level 1 and Level 2. These power requirements are specified by the equipment manufacturer.

625.15 Markings

The electric vehicle supply equipment shall comply with 625.15(A) through 625.15(C).

(A) General All electric vehicle supply equipment shall be marked by the manufacturer as follows:

FOR USE WITH ELECTRIC VEHICLES

(B) Ventilation Not Required Where marking is required by 625.29(C), the electric vehicle supply equipment shall be clearly marked by the manufacturer as follows:

VENTILATION NOT REQUIRED

The marking shall be located so as to be clearly visible after installation.

(C) Ventilation Required Where marking is required by 625.29(D), the electric vehicle supply equipment shall be clearly marked by the manufacturer "Ventilation Required." The marking shall be located so as to be clearly visible after installation.

625.16 Means of Coupling

The means of coupling to the electric vehicle shall be either conductive or inductive. Attachment plugs, electric vehicle connectors, and electric vehicle inlets shall be listed or labeled for the purpose.

625.17 Cable

The electric vehicle supply equipment cable shall be Type EV, EVJ, EVE, EVJE, EVT, or EVJT flexible cable as specified in Article 400 and Table 400.4. Ampacities shall be as specified in Table 400.5(A) for 10 AWG and smaller, and in Table 400.5(B) for 8 AWG and larger. The overall length of the cable shall not exceed 7.5 m (25 ft) unless equipped with a cable management system that is listed as suitable for the purpose. Other cable types and assemblies listed as being suitable for the purpose, including optional hybrid communications, signal, and optical fiber cables, shall be permitted.

The 25-ft cable length is established by adding the 15-ft car length to the 7-ft car width, plus 3 ft to the power outlet securement point. This requirement limits excessive cable lengths, which may be exposed to damage. To use a single electric vehicle charging system for multiple electric vehicles, the 2002 *Code* permitted cable lengths in excess of 25 ft where a listed cable management system is installed. For commercial parking areas, this change allows flexibility in site planning and meeting any legislated requirements that may be in place on the number of charging spaces that must be provided.

625.18 Interlock

Electric vehicle supply equipment shall be provided with an interlock that de-energizes the electric vehicle connector and its cable whenever the electric connector is uncoupled from the electric vehicle. An interlock shall not be required for portable cord-and-plug-connected electric vehicle supply equipment intended for connection to receptacle outlets rated at 125 volts, single phase, 15 and 20 amperes.

To reduce shock hazard, a pilot or communications interlock establishes power through the electric vehicle supply equipment. Loss of the pilot or communications circuit locks out power, isolating possible hazardous situations in the electric vehicle supply equipment. See 625.29(D) for mechanical ventilation interlock requirements.

For ventilation interlock, see 625.29(D)(3) for 125-volt receptacles intended to charge electric vehicles.

625.19 Automatic De-Energization of Cable

The electric vehicle supply equipment or the cable-connector combination of the equipment shall be provided with an automatic means to de-energize the cable conductors and electric vehicle connector upon exposure to strain that could result in either cable rupture or separation of the cable from the electric connector and exposure of live parts. Automatic means to de-energize the cable conductors and electric vehicle connector shall not be required for portable cord-and-plug-connected electric vehicle supply equipment intended for connection to receptacle outlets rated at 125 volts, single phase, 15 and 20 amperes.

IV. Control and Protection

625.21 Overcurrent Protection

Overcurrent protection for feeders and branch circuits supplying electric vehicle supply equipment shall be sized for continuous duty and shall have a rating of not less than 125 percent of the maximum load of the electric vehicle supply equipment. Where noncontinuous loads are supplied from the same feeder or branch circuit, the overcurrent device shall have a rating of not less than the sum of the noncontinuous loads plus 125 percent of the continuous loads.

625.22 Personnel Protection System

The electric vehicle supply equipment shall have a listed system of protection against electric shock of personnel. The personnel protection system shall be composed of listed personnel protection devices and constructional features. Where cord-and-plug-connected electric vehicle supply equipment is used, the interrupting device of a listed personnel protection system shall be provided and shall be an integral part of the attachment plug or shall be located in the power supply cable not more than 300 mm (12 in.) from the attachment plug.

The personnel protection system may consist of one or more components that provide protection against electric shock for different portions of the electric vehicle supply equipment circuitry, which may be operating at frequencies other than 50/60 Hz, at direct current potentials, and/or voltages above 150 volts to ground.

Standard GFCI devices do not provide the range of protection needed for the various types of charging systems being developed. Devices or methods that may be used include basic insulation, double insulation, grounding monitors, insulation monitors with interrupters, and leakage current monitors. Many combinations and variations of these devices can be used to provide the personnel protection required. For systems operating above 150 volts to ground, the protective system may include monitoring systems to ensure that proper grounding is provided and maintained during charging.

625.23 Disconnecting Means

For electric vehicle supply equipment rated more than 60 amperes or more than 150 volts to ground, the disconnecting means shall be provided and installed in a readily accessible location. The disconnecting means shall be capable of being locked in the open position.

625.25 Loss of Primary Source

Means shall be provided such that, upon loss of voltage from the utility or other electric system(s), energy cannot be back fed through the electric vehicle and the supply equipment to the premises wiring system unless permitted by 625.26.

625.26 Interactive Systems

Electric vehicle supply equipment and other parts of a system, either on-board or off-board the vehicle, that are identified for and intended to be interconnected to a vehicle and also serve as an optional standby system or an electric power production source or provide for bi-directional power feed shall be listed as suitable for that purpose. When used as an optional standby system, the requirements of Article 702 shall apply, and when used as an electric power production source, the requirements of Article 705 shall apply.

The on-board power production system of some electric vehicles is capable of operating as a stand-alone or interactive power supply for premises wiring systems. Such systems are required to be listed for this type of use, and 625.26, which is new in the 2005 *Code*, requires compliance with the provisions of Article 702 or Article 705, depending on how the system connects to premises wiring system and/or the primary source of electricity.

V. Electric Vehicle Supply Equipment Locations

625.28 Hazardous (Classified) Locations

Where electric vehicle supply equipment or wiring is installed in a hazardous (classified) location, the requirements of Articles 500 through 516 shall apply.

The installation of EV charging equipment is permitted in hazardous locations where the installation is made in accordance with the requirements of Chapter 5. The increased use of electric vehicles makes this provision necessary to cover installations at commercial repair garages and combination gasoline/EV charging stations (see 511.10).

625.29 Indoor Sites

Indoor sites shall include, but not be limited to, integral, attached, and detached residential garages; enclosed and underground parking structures; repair and nonrepair commercial garages; and agricultural buildings.

(A) Location The electric vehicle supply equipment shall be located to permit direct connection to the electric vehicle.

(B) Height Unless specifically listed for the purpose and location, the coupling means of the electric vehicle supply equipment shall be stored or located at a height of not less than 450 mm (18 in.) and not more than 1.2 m (4 ft) above the floor level.

(C) Ventilation Not Required Where electric vehicle nonvented storage batteries are used or where the electric vehicle supply equipment is listed or labeled as suitable for charging electric vehicles indoors without ventilation and marked in accordance with 625.15(B), mechanical ventilation shall not be required.

Major auto manufacturers are taking the necessary steps to make electric vehicle systems safe. Most batteries used in electric vehicles manufactured by major automakers do not emit hydrogen gas in quantities that could cause an explosion. Preventive measures such as mechanical or passive ventilation are not required, because the electric vehicle batteries and charging systems are designed to prevent or limit the emission of hydrogen during charging. The Society of Automotive Engineers' recommended practice SAE J-1718, *Measurement of Hydrogen Gas Emission from Battery-Powered Passenger Cars and Light Trucks During Battery Charging*, can be used to assess suitability for indoor charging. This standard includes provisions for tests during normal charging operations and potential equipment failure modes.

See Exhibit 625.2 for an illustration of a garage without ventilation. In this application, when the electric vehicle is connected to the charging equipment, a signal is received at the electric vehicle charging equipment. The signal indicates that the electric vehicle either is equipped with a nonvented storage battery or is listed or labeled as suitable to be charged indoors. Failure to receive a verification signal from the vehicle prevents initiation of the charging operation. The electric vehicle supply equipment is required to be marked in accordance with 625.15(B).

(D) Ventilation Required Where the electric vehicle supply equipment is listed or labeled as suitable for charging electric vehicles that require ventilation for indoor charging, and is marked in accordance with 625.15(C), mechanical ventilation, such as a fan, shall be provided. The ventilation shall include both supply and exhaust equipment and shall be permanently installed and located to intake from, and vent directly to, the outdoors. Positive pressure ventilation systems shall be permitted only in buildings or areas that

Exhibit 625.2 Garage interior with electric vehicle supply equipment where ventilation is not required.

ac power supply from premises wiring system

Electric vehicle supply equipment

Electric vehicle supply cable

Electric vehicle connector

have been specifically designed and approved for that application. Mechanical ventilation requirements shall be determined by one of the methods specified in 625.29(D)(1) through (D)(4).

The interlock described in 625.29(D) prevents a vehicle requiring ventilation from being charged unless ventilation is provided. This interlock feature is included in the pilot connection of the standard electric vehicle supply equipment connection. See Exhibit 625.3 for an illustration of electric vehicle supply equipment with interlocked ventilation.

(1) Table Values For supply voltages and currents specified in Table 625.29(D)(1) or (D)(2), the minimum ventilation requirements shall be as specified in Table 625.29(D)(1) or (D)(2) for each of the total number of electric vehicles that can be charged at one time.

(2) Other Values For supply voltages and currents other than specified in Table 625.29(D)(1) or (D)(2), the minimum ventilation requirements shall be calculated by means of the following general formulas as applicable:

(1) Single phase: Ventilation$_{single phase}$ in cubic meters per minute (m^3/min) =

$$\frac{(volts)(amperes)}{1718}$$

Ventilation $_{single phase}$ in cubic feet per minute (cfm) =

$$\frac{(volts)(amperes)}{48.7}$$

(2) Three phase: Ventilation $_{three phase}$ in cubic meters per minute (m^3/min) =

$$\frac{1.732(volts)(amperes)}{1718}$$

Ventilation$_{three phase}$ in cubic feet per minute (cfm) =

$$\frac{1.732(volts)(amperes)}{48.7}$$

(3) Engineered Systems For an electric vehicle supply equipment ventilation system designed by a person qualified to perform such calculations as an integral part of a building's total ventilation system, the minimum ventilation requirements shall be permitted to be determined per calculations specified in the engineering study.

(4) Supply Circuits The supply circuit to the mechanical ventilation equipment shall be electrically interlocked with the electric vehicle supply equipment and shall remain energized during the entire electric vehicle charging cycle. Electric vehicle supply equipment shall be marked in accordance with 625.15. Electric vehicle supply equipment receptacles rated at 125 volts, single phase, 15 and 20 amperes shall be marked in accordance with 625.15(C) and shall be switched, and the mechanical ventilation system shall be electrically interlocked through the switch supply power to the receptacle.

The intent of 625.29(D) is to ensure sufficient diffusion and dilution of hydrogen gas from gas-emitting batteries to prevent a hazardous condition. Certain batteries used in some

Exhibit 625.3 Garage interior with electric vehicle supply equipment and interlocked ventilation.

Exhaust fan to outside

ac power supply from premises wiring system

ac power line to fan

Electric vehicle supply equipment

Electric vehicle supply cable

Electric vehicle connector

Outside air intake

Table 625.29(D)(1) Minimum Ventilation Required in Cubic Meters per Minute (m^3/min) for Each of the Total Number of Electric Vehicles That Can Be Charged at One Time

Branch-Circuit Ampere Rating	Branch-Circuit Voltage						
	Single Phase			3 Phase			
	120 V	208 V	240 V or 120/240 V	208 V or 208Y/120 V	240 V	480 V or 480Y/277 V	600 V or 600Y/347 V
15	1.1	1.8	2.1	—	—	—	—
20	1.4	2.4	2.8	4.2	4.8	9.7	12
30	2.1	3.6	4.2	6.3	7.2	15	18
40	2.8	4.8	5.6	8.4	9.7	19	24
50	3.5	6.1	7.0	10	12	24	30
60	4.2	7.3	8.4	13	15	29	36
100	7.0	12	14	21	24	48	60
150	—	—	—	31	36	73	91
200	—	—	—	42	48	97	120
250	—	—	—	52	60	120	150
300	—	—	—	63	73	145	180
350	—	—	—	73	85	170	210
400	—	—	—	84	97	195	240

electric vehicles emit hydrogen gas during the charging process.

Hydrogen is a colorless, odorless, tasteless, nontoxic flammable gas. At atmospheric pressure, the flammable range for hydrogen is 4 to 75 percent by volume in air.

NFPA 69, *Standard on Explosion Prevention Systems*, establishes requirements to ensure safety with flammable mixtures. The provisions of 6.3, Design and Operating Requirements, of NFPA 69 specify that combustible gas concentrations be restricted to 25 percent of the lower flammable limit. This design criterion provides a safety margin for personnel working with atmospheres containing hydrogen. Safety is accomplished by keeping the concentration of hydrogen below 25 percent of the lower flammability limit, or 1 percent (25 percent $\times$ 4 percent = 1 percent) hydrogen by volume in air, that is, below 10,000 ppm hydrogen.

A ventilation system for a typical residential-type garage includes both supply and mechanical exhaust equipment and is permanently installed. The equipment is located in the space such that it takes in air from outdoors to the space,

Table 625.29(D)(2) Minimum Ventilation Required in Cubic Feet per Minute (cfm) for Each of the Total Number of Electric Vehicles That Can Be Charged at One Time

Branch-Circuit Ampere Rating	Branch-Circuit Voltage						
	Single Phase			3 Phase			
	120 V	208 V	240 V or 120/240 V	208 V or 208Y/120 V	240 V	480 V or 480Y/277 V	600 V or 600Y/347 V
15	37	64	74	—	—	—	—
20	49	85	99	148	171	342	427
30	74	128	148	222	256	512	641
40	99	171	197	296	342	683	854
50	123	214	246	370	427	854	1066
60	148	256	296	444	512	1025	1281
100	246	427	493	740	854	1708	2135
150	—	—	—	1110	1281	2562	3203
200	—	—	—	1480	1708	3416	4270
250	—	—	—	1850	2135	4270	5338
300	—	—	—	2221	2562	5125	6406
350	—	—	—	2591	2989	5979	7473
400	—	—	—	2961	3416	6832	8541

circulates the air through the space, and exhausts the air directly to the outdoors. Typically, the equipment includes a passive vent for intake on one side of the enclosed space and an exhaust fan vented to the outside on the other side of the space.

In enclosed commercial garages and other structures, additional ventilation is not required if the exhaust, as required by the building code for carbon monoxide or other purposes, is greater than the quantity listed in the table. Other engineered electric vehicle ventilation systems are allowed, provided they are designed properly. The electric vehicle charging area is permitted to be ventilated by the building ventilation system. The ventilation system and the charging system must be interlocked to prevent charging if the ventilation is not operating, as shown in Exhibit 625.4. This charging arrangement can be used with electric vehicles

Exhibit 625.4 An example of ventilation equipment electrically interlocked with the electric vehicle charging equipment receptacle.

Exhaust fan to outside

ac power line to fan

ac power supply from premises wiring system

125-V, 15- or 20-A single-pole switch

125-V, 15- or 20-A single receptacle (duplex okay)

Outside air intake

equipped with a self-contained charging system in which activation of the charging system does not depend on a signal from the electric vehicle. A manually operated switch controls the receptacle used to supply the vehicle charging system, and it is also interlocked with the power supply to the ventilation fan. This arrangement ensures that the ventilation fan is operating whenever the vehicle charging receptacle is energized. A qualified person must perform the calculation of the ventilation requirements.

625.30 Outdoor Sites

Outdoor sites shall include but not be limited to residential carports and driveways, curbside, open parking structures, parking lots, and commercial charging facilities.

Where the operation is conducted in outdoor or open locations, the off-gassing of hydrogen resulting from battery charging does not pose the same risk of creating an ignitible environment compared to indoor locations. The lighter-than-air hydrogen readily diffuses into the atmosphere. In addition to driveways and parking lots, structures with adequate natural ventilation, such as carports and open parking structures, do not require mechanical ventilation. NFPA 88A, *Standard for Parking Structures*, provides a quantifiable definition of the term *open parking structure*.

(A) Location The electric vehicle supply equipment shall be located to permit direct connection to the electric vehicle.

(B) Height Unless specifically listed for the purpose and location, the coupling means of electric vehicle supply equipment shall be stored or located at a height of not less than 600 mm (24 in.) and not more than 1.2 m (4 ft) above the parking surface.

ARTICLE 630
Electric Welders

Summary of Changes

- **630.1:** Revised to include plasma cutting.
- **630.11(B):** Revised to clarify that the minimum ampacity is based on the required calculations and that larger size conductors are permitted.

Contents

I. General

630.1 Scope

This article covers apparatus for electric arc welding, resistance welding, plasma cutting, and other similar welding and cutting process equipment that is connected to an electric supply system.

The two general types of electric welding are resistance welding and arc welding. Resistance welding, or "spot" welding, is the process of joining, or electrically fusing, two or more metal sheets or parts without any preparation of stock. The metal parts are placed between two electrodes or welding points, and a heavy current at a low voltage is passed through the electrodes. The metal parts offer resistance to the flow of current such that they heat to a molten state, and a weld is made.

Arc welding is the butting of two metal parts to be welded and then the striking of an arc at this joint with a metal electrode (a flux-coated wire rod). The electrode itself is melted and supplies the extra metal necessary for joining the metal parts.

A transformer supplies current for one ac arc welder, and a generator or rectifier supplies current for one or more dc arc welders.

A revision to 630.1 for the 2005 *Code* expands the

scope of Article 630 to cover electrically supplied equipment associated with plasma cutting operations. This electrically powered equipment controls the flammable gas or gases used for cutting.

II. Arc Welders

630.11 Ampacity of Supply Conductors

The ampacity of conductors for arc welders shall be in accordance with 630.11(A) and 630.11(B).

(A) Individual Welders The ampacity of the supply conductors shall not be less than the $I_{1\text{eff}}$ value on the rating plate. Alternatively, if the $I_{1\text{eff}}$ is not given, the ampacity of the supply conductors shall not be less than the current value determined by multiplying the rated primary current in amperes given on the welder rating plate by the factor shown in Table 630.11(A) based on the duty cycle of the welder.

Table 630.11(A) Duty Cycle Multiplication Factors for Arc Welders

Duty Cycle	Multiplier for Arc Welders	
	Nonmotor Generator	Motor Generator
100	1.00	1.00
90	0.95	0.96
80	0.89	0.91
70	0.84	0.86
60	0.78	0.81
50	0.71	0.75
40	0.63	0.69
30	0.55	0.62
20 or less	0.45	0.55

(B) Group of Welders Minimum conductor ampacity shall be based on the individual currents determined in 630.11(A) as the sum of 100 percent of the two largest welders, plus 85 percent of the third largest welder, plus 70 percent of the fourth largest welder, plus 60 percent of all remaining welders.

Exception: Percentage values lower than those given in 630.11(B) shall be permitted in cases where the work is such that a high-operating duty cycle for individual welders is impossible.

> FPN: Duty cycle considers welder loading based on the use to be made of each welder and the number of welders supplied by the conductors that will be in use at the same time. The load value used for each welder considers both the magnitude and the duration of the load while the welder is in use.

Even under high-production conditions, the loads on transformer arc welders are considered intermittent; therefore, it

is permitted to determine the minimum ampacity of feeder conductors supplying several transformers (three or more) by applying the percentage values specified in 630.11(B). It is obvious that intermittent transformer arc welder loads would be considerably less than a continuous load equal to the sum of the full-load current ratings of all the transformers. See also 630.31(B). The ampacity of conductors supplying welders is based on the $I_{1\text{eff}}$ rating on the welder rating plate. If the $I_{1\text{eff}}$ rating is not available, the size of supply conductors to welders may be calculated. The calculation is done by selecting the appropriate factor from 630.11(A) based on the type of welder and the duty cycle of the welder. The selected factor is then multiplied by the primary current rating from the welder rating plate to determine the minimum ampacity of the supply conductors.

Obsolete terms and ratings such as *nameplate* and *1-hour duty cycle* were removed from 630.11 in the 1999 *NEC*. The terms $I_{1\text{eff}}$ and $I_{1\text{max}}$ were added to the marking requirements. A new FPN includes a formula for $I_{1\text{eff}}$, which is preferred over the value derived from using the factors in 630.11. The calculated value is still allowed. $I_{1\text{max}}$ is basically the same as the rated primary current.

630.12 Overcurrent Protection

Overcurrent protection for arc welders shall be as provided in 630.12(A) and 630.12(B). Where the values as determined by this section do not correspond with the standard ampere ratings provided in 240.6 or the rating or setting specified results in unnecessary opening of the overcurrent device, the next higher standard rating or setting shall be permitted.

(A) For Welders Each welder shall have overcurrent protection rated or set at not more than 200 percent of $I_{1\text{max}}$. Alternatively, if the $I_{1\text{max}}$ is not given, the overcurrent protection shall be rated or set at not more than 200 percent of the rated primary current of the welder.

An overcurrent device shall not be required for a welder that has supply conductors protected by an overcurrent device rated or set at not more than 200 percent of $I_{1\text{max}}$ or at the rated primary current of the welder.

If the supply conductors for a welder are protected by an overcurrent device rated or set at not more than 200 percent of $I_{1\text{max}}$ or at the rated primary current of the welder, a separate overcurrent device shall not be required.

(B) For Conductors Conductors that supply one or more welders shall be protected by an overcurrent device rated or set at not more than 200 percent of the conductor ampacity.

Some arc welding machines have a welding range involving an excess secondary-current output capacity beyond that indicated by the secondary rating marked on the machines. This excess capacity (generally not more than 150 percent of the marked output capacity) is usually supplied by means of one or more secondary taps in addition to the tap(s) intended for normal output current; the higher currents thus available are intended to provide for heavier welding work, including the use of larger-sized electrodes. This excess capacity is somewhat analogous to the inherent overload capacity of motors and transformers. However, the use of this excess current capacity and the overloading of welding machines, except for relatively short periods of time, could be hazardous and should be undertaken with caution.

FPN: $I_{1\text{max}}$ is the maximum value of the rated supply current at maximum rated output. $I_{1\text{eff}}$ is the maximum value of the effective supply current, calculated from the rated supply current (I_1), the corresponding duty cycle (duty factor) (X), and the supply current at no-load (I_0) by the following formula:

$$I_{1\text{eff}} = \sqrt{I_1^2 + I_0^2(1 - X)}$$

630.13 Disconnecting Means

A disconnecting means shall be provided in the supply circuit for each arc welder that is not equipped with a disconnect mounted as an integral part of the welder.

The disconnecting means shall be a switch or circuit breaker, and its rating shall not be less than that necessary to accommodate overcurrent protection as specified under 630.12.

630.14 Marking

A rating plate shall be provided for arc welders giving the following information:

(1) Name of manufacturer
(2) Frequency
(3) Number of phases
(4) Primary voltage
(5) $I_{1\text{max}}$ and $I_{1\text{eff}}$, or rated primary current
(6) Maximum open-circuit voltage
(7) Rated secondary current and
(8) Basis of rating, such as the duty cycle

630.15 Grounding of Welder Secondary Circuit

The secondary circuit conductors of an arc welder, consisting of the electrode conductor and the work conductor, shall not be considered as premises wiring for the purpose of applying Article 250.

FPN: Connecting welder secondary circuits to grounded objects can create parallel paths and can cause objectionable current over equipment grounding conductors.

In theory and in accordance with the *NEC* definition of *separately derived system* in Article 100, the secondary cir-

cuit of an arc welder could be viewed as such a system. However, the intended operation of a welder is to create a high-current circuit between the electrode and the work surface. In the normal operation of an ac power distribution system, such an event would be considered a fault, and the operation of an overcurrent device to open the circuit and clear the fault is a fundamental concept of Articles 240 and 250. In the case of an arc welder, the opening of an overcurrent device is not intended unless the welding operation significantly exceeds the operating parameters of the welder. Grounding of a welder secondary terminal has the potential to cause excessive and potentially degrading parallel currents on power system equipment grounding conductors.

This requirement clarifies that for the purposes of Article 250, specifically the requirements covering grounding of separately derived systems, the secondary circuit of a welder is not treated as premises wiring and is not required to be grounded as such. The new wording modified Article 250 for the purposes of electric welder secondary circuits and thereby removed any potential conflict where grounding in the welder secondary circuit occurs at the work object.

III. Resistance Welders

630.31 Ampacity of Supply Conductors

The ampacity of the supply conductors for resistance welders necessary to limit the voltage drop to a value permissible for the satisfactory performance of the welder is usually greater than that required to prevent overheating as covered in 630.31(A) and 630.31(B).

(A) Individual Welders The rated ampacity for conductors for individual welders shall comply with the following:

(1) The ampacity of the supply conductors for a welder that may be operated at different times at different values of primary current or duty cycle shall not be less than 70 percent of the rated primary current for seam and automatically fed welders, and 50 percent of the rated primary current for manually operated nonautomatic welders.

(2) The ampacity of the supply conductors for a welder wired for a specific operation for which the actual primary current and duty cycle are known and remain unchanged shall not be less than the product of the actual primary current and the multiplier specified in Table 630.31(A)(2) for the duty cycle at which the welder will be operated.

(B) Groups of Welders The ampacity of conductors that supply two or more welders shall not be less than the sum of the value obtained in accordance with 630.31(A) for the largest welder supplied and 60 percent of the values obtained for all the other welders supplied.

Table 630.31(A)(2) Duty Cycle Multiplication Factors for Resistance Welders

Duty Cycle (percent)	Multiplier
50	0.71
40	0.63
30	0.55
25	0.50
20	0.45
15	0.39
10	0.32
7.5	0.27
5 or less	0.22

FPN: **Explanation of Terms**

(1) The *rated primary current* is the rated kilovolt-amperes (kVA) multiplied by 1000 and divided by the rated primary voltage, using values given on the nameplate.

(2) The *actual primary current* is the current drawn from the supply circuit during each welder operation at the particular heat tap and control setting used.

(3) The *duty cycle* is the percentage of the time during which the welder is loaded. For instance, a spot welder supplied by a 60-Hz system (216,000 cycles per hour) and making four hundred 15-cycle welds per hour would have a duty cycle of 2.8 percent (400 multiplied by 15, divided by 216,000, multiplied by 100). A seam welder operating 2 cycles "on" and 2 cycles "off" would have a duty cycle of 50 percent.

The ampacity of supply conductors for a welder that is not wired for a specific function (i.e., one operated at varying intervals for different applications, such as dissimilar metals or thicknesses) is permitted to be 70 percent of the rated primary current for automatically fed welders and 50 percent of the rated primary current for manually operated welders. The rated primary current can be determined using the following equation with the values given on the welder nameplate:

$$\text{Rated primary current} = \frac{\text{welder kVA} \times 1000}{\text{rated primary voltage}}$$

Where the actual primary current and the duty cycle are known, such as for a welder wired for a specific operation, the ampacity of the supply conductors is not permitted to be less than the product of the actual primary current (current drawn during weld operation) and the multiplier, as provided in 630.31(A)(2), for the duty cycle at which the welder will be operated. For example, a spot welder is specifically set to perform 300 welds per hour on a 60-Hz system. Each weld draws current for 16 cycles. During the 1-hour period, the welder draws current for 4800 cycles (300

× 16). There are 216,000 cycles per hour (60 × 60 × 60). The duty cycle is calculated as follows:

$$\frac{4800}{216,000} \times 100\% = 2.2\% \text{ (duty cycle)}$$

For a seam welder that draws current for 3 cycles and is off for 4 cycles during every 7-cycle period, the duty cycle is calculated as follows:

$$\frac{3}{7} \times 100\% = 42.9\% \text{ (duty cycle)}$$

An instrument capable of measuring current impulses for 3 cycles (1/20 second), as shown in the preceding example, is required to measure the actual primary current. The duty cycle is set for a specific operation by adjusting the controller for the welder. For the sizing of supply conductors, voltage drop should be limited to a value permissible for the satisfactory performance of the welder.

630.32 Overcurrent Protection

Overcurrent protection for resistance welders shall be as provided in 630.32(A) and 630.32(B). Where the values as determined by this section do not correspond with the standard ampere ratings provided in 240.6 or where the rating or setting specified results in unnecessary opening of the overcurrent device, a higher rating or setting that does not exceed the next higher standard ampere rating shall be permitted.

(A) For Welders Each welder shall have an overcurrent device rated or set at not more than 300 percent of the rated primary current of the welder. If the supply conductors for a welder are protected by an overcurrent device rated or set at not more than 200 percent of the rated primary current of the welder, a separate overcurrent device shall not be required.

(B) For Conductors Conductors that supply one or more welders shall be protected by an overcurrent device rated or set at not more than 300 percent of the conductor rating.

630.33 Disconnecting Means

A switch or circuit breaker shall be provided by which each resistance welder and its control equipment can be disconnected from the supply circuit. The ampere rating of this disconnecting means shall not be less than the supply conductor ampacity determined in accordance with 630.31. The supply circuit switch shall be permitted as the welder disconnecting means where the circuit supplies only one welder.

630.34 Marking

A nameplate shall be provided for each resistance welder, giving the following information:

(1) Name of manufacturer
(2) Frequency
(3) Primary voltage
(4) Rated kilovolt-amperes (kVA) at 50 percent duty cycle
(5) Maximum and minimum open-circuit secondary voltage
(6) Short-circuit secondary current at maximum secondary voltage
(7) Specified throat and gap setting

IV. Welding Cable

630.41 Conductors

Insulation of conductors intended for use in the secondary circuit of electric welders shall be flame retardant.

630.42 Installation

Cables shall be permitted to be installed in a dedicated cable tray as provided in 630.42(A), (B), and (C).

(A) Cable Support The cable tray shall provide support at not greater than 150-mm (6-in.) intervals.

(B) Spread of Fire and Products of Combustion The installation shall comply with 300.21.

(C) Signs A permanent sign shall be attached to the cable tray at intervals not greater than 6.0 m (20 ft). The sign shall read as follows:

CABLE TRAY FOR WELDING CABLES ONLY

ARTICLE 640
Audio Signal Processing, Amplification, and Reproduction Equipment

Summary of Changes

• **640.6:** Revised to clarify the support requirements for exposed cables and new reference requiring cable installations to comply with 300.11.

Contents

640.42 Use of Flexible Cords and Cables

   (A) Between Equipment and Branch-Circuit Power

   (B) Between Loudspeakers and Amplifiers, or Between Loudspeakers

   (C) Between Equipment and/or Between Equipment Racks

   (D) Between Equipment, Equipment Racks, and Power Supplies Other Than Branch-Circuit Power

   (E) Between Equipment Racks and Branch-Circuit Power

640.43 Wiring of Equipment Racks

640.44 Environmental Protection of Equipment

640.45 Protection of Wiring

640.46 Equipment Access

I. General

640.1 Scope

This article covers equipment and wiring for audio signal generation, recording, processing, amplification and reproduction; distribution of sound; public address; speech input systems; temporary audio system installations; and electronic organs or other electronic musical instruments. This also includes audio systems subject to Article 517, Part VI, and Articles 518, 520, 525, and 530.

Equipment covered by Article 640 includes amplifiers; public address (PA) systems and centralized sound systems used in schools, factories, businesses, stadiums, and similar locations; intercommunications devices and systems; and devices used for recording or reproducing voice or music. The scope remains limited to equipment whose main function is the processing, distribution, amplification, and reproduction of audio frequency bandwidth signals. This does not preclude equipment that uses radio frequency or other forms of transmission between equipment components, such as wireless microphone systems.

Electronic organs are synthesizers, and synthesizers also generate audio signals. For the sake of clarity, electronic organs are still uniquely cited in the scope, and electronic musical instruments have been added to cover all other forms of electronic tone generation. Electronic musical instruments create an electronic signal as their sole or primary output and require amplification and reproduction equipment to be audible.

FPN No. 1: Examples of permanently installed distributed audio system locations include, but are not limited to, restaurant, hotel, business office, commercial and retail sales environments, churches, and schools. Both portable and permanently installed equipment locations include, but are not limited to, residences, auditoriums, theaters, stadiums, and movie and television studios. Temporary installations include, but are not limited to, auditoriums, theaters, stadiums (which use both temporary and permanently installed systems), and outdoor events such as fairs, festivals, circuses, public events, and concerts.

FPN No. 2: Fire and burglary alarm signaling devices are specifically not encompassed by this article.

640.2 Definitions

For purposes of this article, the following definitions apply.

Abandoned Audio Distribution Cable. Installed audio distribution cable that is not terminated at equipment and not identified for future use with a tag.

Audio Amplifier or Pre-Amplifier. Electronic equipment that increases the current or voltage, or both, potential of an audio signal intended for use by another piece of audio equipment. *Amplifier* is the term used to denote an audio amplifier within this article.

Audio Autotransformer. A transformer with a single winding and multiple taps intended for use with an amplifier loudspeaker signal output.

Audio Signal Processing Equipment. Electrically operated equipment that produces, processes, or both, electronic signals that, when appropriately amplified and reproduced by a loudspeaker, produce an acoustic signal within the range of normal human hearing (typically 20–20 kHz). Within this article, the terms *equipment* and *audio equipment* are assumed to be equivalent to audio signal processing equipment.

FPN: This equipment includes, but is not limited to, loudspeakers; headphones; pre-amplifiers; microphones and their power supplies; mixers; MIDI (musical instrument digital interface) equipment or other digital control systems; equalizers, compressors, and other audio signal processing equipment; and audio media recording and playback equipment, including turntables, tape decks and disk players (audio and multimedia), synthesizers, tone generators, and electronic organs. Electronic organs and synthesizers may have integral or separate amplification and loudspeakers. With the exception of amplifier outputs, virtually all such equipment is used to process signals (utilizing analog or digital techniques) that have nonhazardous levels of voltage or current potential.

The definition of *audio signal processing equipment* clarifies the limits of signal processing (frequency bandwidth), which falls under Article 640. The fine print note enumerates the breadth of equipment considered to fall within the defined scope of Article 640. It is intended to provide a sufficiently broad list of current technology equipment to assist in determining the applicability of Article 640 to the equipment under review.

"MIDI (musical instrument digital interface) equipment or other digital control systems" is mentioned specifically because, while MIDI or similar digital control signals may issue from an electronic musical instrument, such signals also can be obtained from a computer that is appropriately configured to perform similar controlling functions.

Audio System. Within this article, the totality of all equipment and interconnecting wiring used to fabricate a fully functional audio signal processing, amplification, and reproduction system.

Audio Transformer. A transformer with two or more electrically isolated windings and multiple taps intended for use with an amplifier loudspeaker signal output.

The definition of *audio transformer* is included to clearly state that such transformers are intended only for use with audio signals, not light and power.

Equipment Rack. A framework for the support, enclosure, or both, of equipment; may be portable or stationary. See ANSI/EIA/310-D-1992, *Cabinets, Racks, Panels and Associated Equipment*.

ANSI/EIA 310-D-1992, *Cabinets, Racks, Panels and Associated Equipment*, defines *commercial equipment racks*. However, custom-fabricated racks that utilize equipment mounting hole patterns that generally comply with this standard might not be constructed of steel. Within Article 640, both *equipment rack* and *rack* are used to refer to equipment enclosures that are conceptually similar in intended use to those defined by the ANSI/EIA standard.

Loudspeaker. Equipment that converts an ac electric signal into an acoustic signal. The term *speaker* is commonly used to mean *loudspeaker*.

Exhibit 640.1 is an example of a surface-mounted loudspeaker.

Maximum Output Power. The maximum output power delivered by an amplifier into its rated load as determined under specified test conditions. This may exceed the manufacturer's rated output power for the same amplifier.

Mixer. Equipment used to combine and level match a multiplicity of electronic signals, such as from microphones, electronic instruments, and recorded audio.

Exhibit 640.1 Surface-mounted loudspeaker for indoor or outdoor use. (Courtesy of Bose Corp.)

Typical peak signal operating voltages for such equipment vary from a few millivolts for microphones to 2 to 4 volts for disk players. A mixer's purpose is to balance these inputs to provide (typically) a 1-volt peak output signal to an amplifier.

Mixer–Amplifier. Equipment that combines the functions of a mixer and amplifier within a single enclosure.

Portable Equipment. Equipment fed with portable cords or cables intended to be moved from one place to another.

Powered Loudspeaker. Equipment that consists of a loudspeaker and amplifier within the same enclosure. Other signal processing may also be included.

Rated Load Impedance. The amplifier manufacturer's stated or marked speaker impedance into which an amplifier will deliver its rated output power. 2Ω, 4Ω, and 8Ω are typical ratings.

Rated Output Power. The amplifier manufacturer's stated or marked output power capability into its rated load.

Rated Output Voltage. For audio amplifiers of the constant-voltage type, the nominal output voltage when the amplifier is delivering full rated power. Rated output voltage is used for determining approximate acoustic output in distributed speaker systems that typically employ impedance matching transformers. Typical ratings are 25 volts, 70.7 volts, and 100 volts.

Technical Power System. An electrical distribution system with grounding in accordance with 250.146(D), where the equipment grounding conductor is isolated from the premises grounded conductor except at a single grounded termination point within a branch-circuit panelboard, at the originating (main breaker) branch-circuit panelboard, or at the premises grounding electrode.

The terms *technical power* and *technical ground* are commonly used by audio/video technicians and electricians to designate a wiring system that is in compliance with 250.146(D). Including the definition of *technical power system* in Article 640 is intended to broaden the scope of this term to include the commonly employed distribution systems fabricated in compliance with 250.146(D).

Temporary Equipment. Portable wiring and equipment intended for use with events of a transient or temporary nature where all equipment is presumed to be removed at the conclusion of the event.

Temporary equipment may be used in facilities of a permanent or temporary nature or with no facilities other than a source of electrical power. Locations would include indoor and outdoor areas such as athletic facilities, halls, auditoriums, concert shells, athletic fields, beaches, and other places designated for public assembly.

640.3 Locations and Other Articles

Circuits and equipment shall comply with 640.3(A) through 640.3(L), as applicable.

(A) Spread of Fire or Products of Combustion The accessible portion of abandoned audio distribution cables shall be removed. See 300.21.

(B) Ducts, Plenums, and Other Air-Handling Spaces See 300.22 for circuits and equipment installed in ducts or plenums or other space used for environmental air.

> FPN: NFPA 90A-2002, *Standard for the Installation of Air Conditioning and Ventilation Systems*, Section 4.3.10.2.6.5, permits loudspeakers, loudspeaker assemblies, and their accessories listed in accordance with UL 2043-1996, *Fire Test for Heat and Visible Smoke Release for Discrete Products and Their Accessories Installed in Air-Handling Spaces*, to be installed in other spaces used for environmental air (ceiling cavity plenums).

(C) Cable Trays Cable trays shall be used in accordance with Article 392.

> FPN: See 725.61(C) for the use of Class 2, Class 3, and Type PLTC cable in cable trays.

(D) Hazardous (Classified) Locations Equipment used in hazardous (classified) locations shall comply with the applicable requirements of Chapter 5.

(E) Assembly Occupancies Equipment used in assembly occupancies shall comply with Article 518.

The examples of assembly occupancies described in 518.2 are some of the most common locations for the installation of both distributed audio systems (e.g., background music) and centralized systems (permanently installed sound reinforcement systems for meeting rooms, auditoriums, gymnasiums, and the like).

(F) Theaters, Audience Areas of Motion Picture and Television Studios, and Similar Locations Equipment used in theaters, audience areas of motion picture and television studios, and similar locations shall comply with Article 520.

(G) Carnivals, Circuses, Fairs, and Similar Events Equipment used in carnivals, circuses, fairs, and similar events shall comply with Article 525.

(H) Motion Picture and Television Studios Equipment used in motion picture and television studios shall comply with Article 530.

(I) Swimming Pools, Fountains, and Similar Locations Audio equipment used in or near swimming pools, fountains, and similar locations shall comply with Article 680.

The underwater installation of audio equipment in swimming pools is found in 680.27. The acceptable placement, wiring, and use of equipment used near (rather than immersed in) bodies of water, both natural and artificial, are covered in 640.10.

(J) Combination Systems Where the authority having jurisdiction permits audio systems for paging or music, or both, to be combined with fire alarm systems, the wiring shall comply with Article 760.

In addition to alarm tones, fire alarm systems frequently use loudspeakers for verbal announcements. All such systems must comply with Article 760 and *NFPA 72, National Fire Alarm Code*. Audio systems that use a paging or background music system are permitted to be used as part of a fire alarm warning system, but they require compliance with Article 760. The installation of fire alarm systems is governed by *NFPA 72*. Refer to NFPA *101*®, *Life Safety Code*®, for multipurpose systems.

> FPN: For installation requirements for such combination systems, refer to *NFPA 72*®-2002, *National Fire Alarm Code*®, and NFPA *101*®-2003, *Life Safety Code*®.

(K) Antennas Equipment used in audio systems that contain an audio or video tuner and an antenna input shall comply with Article 810. Wiring other than antenna wiring that connects such equipment to other audio equipment shall comply with this article.

The term *receiver* is commonly used in the consumer market to mean an amplifier combined with a radio tuner (typically AM/FM) and other signal processing and/or switching functions. Except for the tuner function and the antenna input, the signal processing functions are the same as those provided by equipment in Article 640. Article 810, Part II, covers the antenna installation for such equipment, and 810.3 references Article 640 as appropriate for wiring requirements (other than for the antenna).

(L) Generators Generators shall be installed in accordance with 445.10 through 445.12, 445.14 through 445.16, and 445.18. Grounding of portable and vehicle-mounted generators shall be in accordance with 250.34.

640.4 Protection of Electrical Equipment

Amplifiers, loudspeakers, and other equipment shall be so located or protected as to guard against environmental exposure or physical damage, such as might result in fire, shock, or personal hazard.

640.5 Access to Electrical Equipment Behind Panels Designed to Allow Access

Access to equipment shall not be denied by an accumulation of wires and cables that prevents removal of panels, including suspended ceiling panels.

640.6 Mechanical Execution of Work

Equipment and cables shall be installed in a neat workmanlike manner. Cables installed exposed on the surface of ceilings and sidewalls shall be supported in such a manner that the cables will not be damaged by normal building use. Such cables shall be supported by straps, staples, hangers, or similar fittings designed and installed so as not to damage the cable. The installation shall conform to 300.4(D) and 300.11.

> FPN: Accepted industry practices are described in ANSI/NECA/BICSI 568-2001, *Standard for Installing Commercial Building Telecommunications Cabling*, and other ANSI-approved installation standards.

Where installed in the hollow space above a suspended, dropped, or similar ceiling, cables of audio systems covered in Article 640 are required to be supported in accordance with 300.11. Without documentation from the manufacturer of the ceiling system providing specific instruction on the use of the ceiling system support wires as a means to support wiring methods, an independent support system for the cables shall be installed. Additional ceiling wires installed specifically for the purpose of supporting the audio system wiring are required to be secured in place. Attachment of the additional support wires to the ceiling system and to the building structure above the ceiling provides the secure support required by 300.11(A)(1) and 300.11(A)(2).

640.7 Grounding

(A) General Wireways and auxiliary gutters shall be grounded and bonded in accordance with the requirements of Article 250. Where the wireway or auxiliary gutter does not contain power-supply wires, the equipment grounding conductor shall not be required to be larger than 14 AWG copper or its equivalent. Where the wireway or auxiliary gutter contains power-supply wires, the equipment grounding conductor shall not be smaller than specified in 250.122.

(B) Separately Derived Systems with 60 Volts to Ground Grounding of separately derived systems with 60 volts to ground shall be in accordance with 647.6.

(C) Isolated Ground Receptacles Isolated grounding-type receptacles shall be permitted as described in 250.146(D), and for the implementation of other technical power systems in compliance with Article 250. For separately derived systems with 60 volts to ground, the branch-circuit equipment grounding conductor shall be terminated as required in 647.6(B).

The reference to 647.6 in 640.7(B) provides guidance for grounding of separately derived systems operating at 60 volts to ground. The requirements for these systems, formerly located in Part G of Article 530 in the 1999 *Code*, were relocated to Article 647, Sensitive Electronic Equipment, as a revision to the 2002 *Code*. One of the most significant impacts of this relocation is that these systems can now be used in other than motion picture and television studios. This separately derived system is used as a technique for the reduction of electromagnetic noise in audio systems and in video systems.

The intent of 640.7(C) is to clarify the proper use of isolated ground receptacles when used with technical power systems of the type that are separately derived systems with 60 volts to ground.

> FPN: See 406.2(D) for grounding-type receptacles and required identification.

640.8 Grouping of Conductors

Insulated conductors of different systems grouped or bundled so as to be in close physical contact with each other in the same raceway or other enclosure, or in portable cords or cables, shall comply with 300.3(C)(1).

640.9 Wiring Methods

(A) Wiring to and Between Audio Equipment

(1) Power Wiring Wiring and equipment from source of power to and between devices connected to the premises wiring systems shall comply with the requirements of Chapters 1 through 4, except as modified by this article.

(2) Separately Derived Power Systems Separately derived systems shall comply with the applicable articles of this *Code*, except as modified by this article. Separately derived systems with 60 volts to ground shall be permitted for use in audio system installations as specified in Article 647.

(3) Other Wiring All wiring not connected to the premises wiring system or to a wiring system separately derived from the premises wiring system shall comply with Article 725.

(B) Auxiliary Power Supply Wiring Equipment that has a separate input for an auxiliary power supply shall be wired in compliance with Article 725. Battery installation shall be in accordance with Article 480.

> FPN No. 1: This section does not apply to the use of uninterruptible power supply (UPS) equipment, or other sources of supply, that are intended to act as a direct replacement for the primary circuit power source and are connected to the primary circuit input.

Audio equipment with a separate input for an auxiliary power supply is typically used for emergency paging or fire alarm systems. These auxiliary power supply inputs typically range from 12 to 48 volts dc. Article 480 adequately covers installation and overcurrent protection of battery circuits of this type. The term *auxiliary* is used to indicate that the equipment is also capable of being powered by the premises wiring system through an independent input connector, cord, or cable.

FPN No. 1 to 640.9(B) clarifies that, while the equipment can be powered by a replacement source for the premises wiring system, such a source (a UPS or a standby generator) is not considered to be supplying the equipment auxiliary power supply unless it is directly connected to the auxiliary power supply input and supplying a dc voltage.

> FPN No. 2: Refer to *NFPA 72-2002, National Fire Alarm Code*, where equipment is used for a fire alarm system.

(C) Output Wiring and Listing of Amplifiers Amplifiers with output circuits carrying audio program signals shall be permitted to employ Class 1, Class 2, or Class 3 wiring where the amplifier is listed and marked for use with the specific class of wiring method. Such listing shall ensure

the energy output is equivalent to the shock and fire risk of the same class as stated in Article 725. Overcurrent protection shall be provided and shall be permitted to be inherent in the amplifier.

Audio amplifier output circuits wired using Class 1 wiring methods shall be considered equivalent to Class 1 circuits and shall be installed in accordance with 725.25, where applicable.

Audio amplifier output circuits wired using Class 2 or Class 3 wiring methods shall be considered equivalent to Class 2 or Class 3 circuits, respectively. They shall use conductors insulated at not less than the requirements of 725.82 and shall be installed in accordance with 725.54 and 725.61.

> FPN No. 1: ANSI/UL 1711-1994, *Amplifiers for Fire Protective Signaling Systems*, contains requirements for the listing of amplifiers used for fire alarm systems in compliance with *NFPA 72-2002, National Fire Alarm Code*.
>
> FPN No. 2: Examples of requirements for listing amplifiers used in residential, commercial, and professional use are found in ANSI/UL 813-1996, *Commercial Audio Equipment*; ANSI/UL 1419-1997, *Professional Video and Audio Equipment*; ANSI/UL 1492-1996, *Audio-Video Products and Accessories*; and ANSI/UL 6500-1996, *Audio/Video and Musical Instrument Apparatus for Household, Commercial, and Similar Use*.

(D) Use of Audio Transformers and Autotransformers Audio transformers and autotransformers shall be used only for audio signals in a manner so as not to exceed the manufacturer's stated input or output voltage, impedance, or power limitations. The input or output wires of an audio transformer or autotransformer shall be allowed to connect directly to the amplifier or loudspeaker terminals. No electrical terminal or lead shall be required to be grounded or bonded.

Audio transformers and autotransformers are commonly used between the amplifier output and the loudspeaker input for the following reasons:

1. At the output of the amplifier to change the amplifier's operating voltage to match the design impedance of the loudspeaker
2. At the loudspeaker, where the inherently low voice coil impedance is raised, to match the output voltage of the amplifier (or autotransformer)
3. Between the amplifier output and loudspeaker input as an attenuating device (volume control)

Audio autotransformers are similar in concept to autotransformers used for light and power, but they are not designed for such use. Audio transformers are commonly used to provide electrical isolation of the speakers from the signal source. Either type of audio transformer (two winding or autotransformer) is frequently referred to as an "impedance matching transformer." The last sentence of 640.9(D) specifically addresses the fact that electrical terminals are not to be treated in the same manner as transformers used for light and power might be (e.g., grounding the common terminal of an autotransformer). Some amplifier outputs are deliberately isolated from equipment ground, in which case such a connection could damage the amplifier and violate the manufacturer's recommended use. The frame of the transformer may or may not require bonding, depending on the manufacturer's installation instructions.

640.10 Audio Systems Near Bodies of Water

Audio systems near bodies of water, either natural or artificial, shall be subject to the restrictions specified in 640.10(A) and 640.10(B).

Exception: This section does not include audio systems intended for use on boats, yachts, or other forms of land or water transportation used near bodies of water, whether or not supplied by branch-circuit power.

> FPN: See 680.27(A) for installation of underwater audio equipment.

(A) Equipment Supplied by Branch-Circuit Power Audio system equipment supplied by branch-circuit power shall not be placed laterally within 1.5 m (5 ft) of the inside wall of a pool, spa, hot tub, or fountain, or within 1.5 m (5 ft) of the prevailing or tidal high water mark. The equipment shall be provided with branch-circuit power protected by a ground-fault circuit interrupter where required by other articles.

Article 680 is limited by its scope to the use of underwater loudspeakers. This particular application for audio equipment is unique in construction and wiring to pools and is appropriate for Article 680.

Other locations where audio equipment might be used "near bodies of water" are not addressed in Article 680. The purpose of 640.10(A), in conjunction with the other requirements of Article 640, is to address locations where audio equipment is used near bodies of water.

The exception to 640.10(A) excludes sound systems on vehicles used in or near water, such as amphibious vehicles and boats of all sizes.

The term *prevailing or tidal high water mark* recognizes that the edges of natural bodies of water can advance or recede. This phrase makes clear that such changes can be anticipated.

The requirement for use of a GFCI is placed here to ensure that personnel are properly protected. Where the equipment (specifically an amplifier or a receiver) is installed in an electrical cabinet or a room not near a body of water, the requirement for GFCI protection does not apply, unless required by other sections of the *Code*.

(B) Equipment Not Supplied by Branch-Circuit Power Audio system equipment powered by a listed Class 2 power supply or by the output of an amplifier listed as permitting the use of Class 2 wiring shall be restricted in placement only by the manufacturer's recommendations.

> FPN: Placement of the power supply or amplifier, if supplied by branch-circuit power, is still subject to 640.10(A).

II. Permanent Audio System Installations

Permanent audio systems are characterized by fixed locations for the wiring, signal processing equipment, and reproduction equipment. Wiring is attached to the building structure and is frequently concealed. Speakers in commercial, hospital, school, and restaurant areas are commonly recessed into ceiling or wall surfaces, or they are mounted to structure surfaces using brackets, usually beyond the reach of a standing person.

640.21 Use of Flexible Cords and Cables

(A) Between Equipment and Branch-Circuit Power Power supply cords for audio equipment shall be suitable for the use and shall be permitted to be used where the interchange, maintenance, or repair of such equipment is facilitated through the use of a power supply cord.

(B) Between Loudspeakers and Amplifiers or Between Loudspeakers Cables used to connect loudspeakers to each other or to an amplifier shall comply with Article 725. Other listed cable types and assemblies, including optional hybrid communications, signal, and optical fiber cables, shall be permitted.

Some loudspeakers are specifically identified as being for outdoor use. The requirements of 110.11 specify that electrical equipment and conductors be identified for use in the operating environment. This general requirement applies to equipment and conductors used in conjunction with the installation of audio equipment. Exhibit 640.2 shows a loudspeaker identified for outdoor use and located partially inground. The conductors supplying this outdoor speaker must also be identified for the environment.

Exhibit 640.2 Loudspeaker for outdoor use above ground or partially in ground. (Courtesy of Bose Corp.)

(C) Between Equipment Cables used for the distribution of audio signals between equipment shall comply with Article 725. Other listed cable types and assemblies, including optional hybrid communications, signal, and optical fiber cables, shall be permitted. Other cable types and assemblies specified by the equipment manufacturer as acceptable for the use shall be permitted in accordance with 110.3(B).

(D) Between Equipment and Power Supplies Other Than Branch-Circuit Power The following power supplies, other than branch-circuit power supplies, shall be installed and wired between equipment in accordance with the requirements of this *Code* for the voltage and power delivered:

(1) Storage batteries
(2) Transformers
(3) Transformer rectifiers
(4) Other ac or dc power supplies

> FPN: For some equipment, these sources such as in items (1) and (2) serve as the only source of power. These could, in turn, be supplied with intermittent or continuous branch-circuit power.

(E) Between Equipment Racks and Premises Wiring System Flexible cords and cables shall be permitted for the electrical connection of permanently installed equipment racks to the premises wiring system to facilitate access to equipment or for the purpose of isolating the technical power system of the rack from the premises ground. Connection shall be made either by using approved plugs and receptacles or by direct connection within an approved enclosure. Flexible cords and cables shall not be subjected to physical manipulation or abuse while the rack is in use.

640.22 Wiring of Equipment Racks and Enclosures

Metal equipment racks and enclosures shall be grounded. Bonding shall not be required if the rack is connected to a technical power ground.

Equipment racks shall be wired in a neat and workmanlike manner. Wires, cables, structural components, or other equipment shall not be placed in such a manner as to prevent reasonable access to equipment power switches and resettable or replaceable circuit overcurrent protection devices.

Supply cords or cables, if used, shall terminate within the equipment rack enclosure in an identified connector assembly. The supply cords or cable (and connector assembly, if used) shall have sufficient ampacity to carry the total load connected to the equipment rack and shall be protected by overcurrent devices.

640.23 Conduit or Tubing

(A) Number of Conductors The number of conductors permitted in a single conduit or tubing shall not exceed the percentage fill specified in Table 1, Chapter 9.

(B) Nonmetallic Conduit or Tubing and Insulating Bushings The use of nonmetallic conduit or tubing and insulating bushings shall be permitted where a technical power system is employed and shall comply with applicable articles.

640.24 Wireways, Gutters, and Auxiliary Gutters

The use of metallic and nonmetallic wireways, gutters, and auxiliary gutters shall be permitted for use with audio signal conductors and shall comply with applicable articles with respect to permitted locations, construction, and fill.

640.25 Loudspeaker Installation in Fire Resistance-Rated Partitions, Walls, and Ceilings

Loudspeakers installed in a fire resistance-rated partition, wall, or ceiling shall be listed for that purpose or installed in an enclosure or recess that maintains the fire resistance rating.

> FPN: Fire-rated construction is the fire-resistive classification used in building codes. One method of determining fire rating is testing in accordance with NFPA 251-1999, *Standard Methods of Tests of Fire Endurance of Building Construction and Materials.*

Revised in the 1999 *Code*, 640.25 requires an enclosure that maintains the requisite fire resistance rating of the wall or ceiling in which a flush-mount loudspeaker is installed. Listed enclosures are available for this purpose; however, prior to their availability, site-built enclosures were installed with the approval of the authority having jurisdiction and have been used as a method to maintain the fire-resistance rating of the wall or ceiling.

The reference to NFPA 251, *Standard Methods of Tests of Fire Endurance of Building Construction and Materials*, is intended to clarify a typical classification method for a component of the ceiling or for components installed therein.

III. Portable and Temporary Audio System Installations

Portable and temporary audio system installations are characterized by the portable nature of the signal processing equipment and the reproduction equipment. While the equipment may not differ fundamentally from that used in permanent installations, the enclosures that serve as portable equipment racks provide both transit protection and mechanical protection while the equipment is in use. Such enclosures may accommodate one or multiple pieces of equipment and may be of metal, wood, plastic, or reinforced plastic construction. The nonmetal construction enclosures frequently do not comply with ANSI/EIA 310-D-1992, *Cabinets, Racks, Panels and Associated Equipment*, except for the mounting attachment points for the equipment.

Three conditions distinguish a portable installation from a temporary installation. The first is the nature of use and possession. For example, a theater might employ portable equipment that is moved during the course of a performance but is never used for a nontheater event. The second condition is the transitory nature of the event. A performer could use his or her personal portable equipment for a performance in a theater, but because the presence of the equipment within the theater terminates with the end of the performance, the equipment's use qualifies as temporary. Finally, an event that is considered temporary will obviously use temporary audio systems. Examples include fairs, circuses, outdoor concerts, and folk festivals. Part III of Article 640 covers portable and temporary installations, the differences being primarily ones of intent and, for outdoor temporary use, adequate protection of the equipment from environmental hazards.

640.41 Multipole Branch-Circuit Cable Connectors

Multipole branch-circuit cable connectors, male and female, for power supply cords and cables shall be so constructed that tension on the cord or cable is not transmitted to the connections. The female half shall be attached to the load end of the power supply cord or cable. The connector shall be rated in amperes and designed so that differently rated devices cannot be connected together. Alternating-current

multipole connectors shall be polarized and comply with 406.6(A) and 406.6(B) and 406.9. Alternating-current or direct-current multipole connectors utilized for connection between loudspeakers and amplifiers, or between loudspeakers, shall not be compatible with nonlocking 15- or 20-ampere rated connectors intended for branch-circuit power or with connectors rated 250 volts or greater and of either the locking or nonlocking type. Signal cabling not intended for such loudspeaker and amplifier interconnection shall not be permitted to be compatible with multipole branch-circuit cable connectors of any accepted configuration.

FPN: See 400.10 for pull at terminals.

640.42 Use of Flexible Cords and Cables

(A) Between Equipment and Branch-Circuit Power Power supply cords for audio equipment shall be listed and shall be permitted to be used where the interchange, maintenance, or repair of such equipment is facilitated through the use of a power supply cord.

(B) Between Loudspeakers and Amplifiers, or Between Loudspeakers Flexible cords and cables used to connect loudspeakers to each other or to an amplifier shall comply with Article 400 and Article 725, respectively. Cords and cables listed for portable use, either hard or extra-hard usage as defined by Article 400, shall also be permitted. Other listed cable types and assemblies, including optional hybrid communications, signal, and optical fiber cables, shall be permitted.

(C) Between Equipment and/or Between Equipment Racks Flexible cords and cables used for the distribution of audio signals between equipment shall comply with Article 400 and Article 725, respectively. Cords and cables listed for portable use, either hard or extra-hard service as defined by Article 400, shall also be permitted. Other listed cable types and assemblies, including optional hybrid communications, signal, and optical fiber cables, shall be permitted.

(D) Between Equipment, Equipment Racks, and Power Supplies Other Than Branch-Circuit Power Wiring between the following power supplies, other than branch-circuit power supplies, shall be installed, connected, or wired in accordance with the requirements of this *Code* for the voltage and power required:

(1) Storage batteries
(2) Transformers
(3) Transformer rectifiers
(4) Other ac or dc power supplies

(E) Between Equipment Racks and Branch-Circuit Power The supply to a portable equipment rack shall be by means of listed extra-hard usage cords or cables, as defined in Table 400.4. For outdoor portable or temporary use, the cords or cables shall be further listed as being suitable for wet locations and sunlight resistant. Sections 520.5, 520.10, and 525.3 shall apply as appropriate when the following conditions exist:

(1) Where equipment racks include audio and lighting and/or power equipment
(2) When using or constructing cable extensions, adapters, and breakout assemblies

640.43 Wiring of Equipment Racks

Equipment racks fabricated of metal shall be grounded. Non-metallic racks with covers (if provided) removed shall not allow access to Class 1, Class 3, or primary circuit power without the removal of covers over terminals or the use of tools.

Equipment racks shall be wired in a neat and workmanlike manner. Wires, cables, structural components, or other equipment shall not be placed in such a manner as to prevent reasonable access to equipment power switches and resettable or replaceable circuit overcurrent protection devices.

Wiring that exits the equipment rack for connection to other equipment or to a power supply shall be relieved of strain or otherwise suitably terminated such that a pull on the flexible cord or cable shall not increase the risk of damage to the cable or connected equipment such as to cause an unreasonable risk of fire or electric shock.

640.44 Environmental Protection of Equipment

Portable equipment not listed for outdoor use shall be permitted only where appropriate protection of such equipment from adverse weather conditions is provided to prevent risk of fire or electrical shock. Where the system is intended to remain operable during adverse weather, arrangements shall be made for maintaining operation and ventilation of heat-dissipating equipment.

Most portable audio equipment used in temporary audio systems is not listed for use in an outdoor environment. The use of such equipment with an appropriate enclosure or other means to protect the equipment from anticipated adverse weather conditions is permitted by 640.44.

640.45 Protection of Wiring

Where accessible to the public, flexible cords and cables laid or run on the ground or on the floor shall be covered with approved nonconductive mats. Cables and mats shall be arranged so as not to present a tripping hazard.

640.46 Equipment Access

Equipment likely to present a risk of fire, electrical shock, or physical injury to the public shall be protected by barriers

or supervised by qualified personnel so as to prevent public access.

ARTICLE 645
Information Technology Equipment

Summary of Changes

- **645.5(D)(3):** Revised to correlate with 645.2(2) regarding other areas permitted to be served by ITE room HVAC where provided with proper smoke/fire dampers that isolate ITE room upon detection of products of combustion.
- **645.5(D)(5)(c):** Deleted Type MP cable.
- **645.17:** Added new section covering number of overcurrent devices permitted to be installed in listed power distribution units.

Contents

645.1 Scope

This article covers equipment, power-supply wiring, equipment interconnecting wiring, and grounding of information technology equipment and systems, including terminal units, in an information technology equipment room.

The term *information technology equipment* (ITE) replaced other terms that describe computer-based business, personal, and industrial equipment. This terminology is also used by UL 60950-1, *Safety of Information Technology, Part 1: General Requirements*, as well as international standards, as a more inclusive term for the equipment addressed by Article 645.

> FPN: For further information, see NFPA 75-2003, *Standard for the Protection of Information Technology Equipment.*

645.4 Special Requirements for Information Technology Equipment Room

This article shall apply, provided all of the following conditions are met:

(1) Disconnecting means complying with 645.10 are provided.
(2) A separate heating/ventilating/air-conditioning (HVAC) system is provided that is dedicated for information technology equipment use and is separated from other areas of occupancy. Any HVAC system that serves other occupancies shall be permitted to also serve the information technology equipment room if fire/smoke dampers are provided at the point of penetration of the room boundary. Such dampers shall operate on activation of smoke detectors and also by operation of the disconnecting means required by 645.10.

> FPN: For further information, see NFPA 75-2003, *Standard for the Protection of Information Technology Equipment*, Chapter 10, 10.1, 10.1.1, 10.1.2, and 10.1.3.

(3) Listed information technology equipment is installed.
(4) The room is occupied only by those personnel needed for the maintenance and functional operation of the installed information technology equipment.
(5) The room is separated from other occupancies by fire-resistant-rated walls, floors, and ceilings with protected openings.

> FPN: For further information on room construction requirements, see NFPA 75-2003, *Standard for the Protection of Information Technology Equipment*, Chapter 5.

Use of the requirements in Article 645 is based on the assumption that construction of the information technology equipment (ITE) room complies with NFPA 75, *Standard for the Protection of Information Technology Equipment.* For those ITE rooms constructed in accordance with NFPA 75, Article 645 contains electrical installation requirements that are less stringent than the requirements in the first four chapters of the *Code.* Application of these provisions is contingent on the ITE room construction and equipment meeting all five provisions specified in 645.4. For example, the provisions for wiring installations in the space beneath the raised floor of an ITE room where that space is also used for environmental air are less stringent than those in Chapters 1 through 4 for that same type of space. The less

restrictive provisions in Article 645 cannot be taken advantage of if any one of the conditions specified in 645.4 is absent.

Article 645 applies only to equipment and wiring located within the ITE room. An ITE room is an enclosed area, with one or more means of entry, that contains computer-based business and industrial equipment. It is designed to comply with the special construction and fire protection provisions of NFPA 75, as well as 645.4 of the *Code.*

Small terminals, such as remote telephone terminal units, remote data terminals, personal computers, and cash registers in stores and supermarkets, are not covered by Article 645.

645.5 Supply Circuits and Interconnecting Cables

(A) Branch-Circuit Conductors The branch-circuit conductors supplying one or more units of a data processing system shall have an ampacity not less than 125 percent of the total connected load.

(B) Cord-and-Plug Connections The data processing system shall be permitted to be connected to a branch circuit by any of the following listed means:

(1) Flexible cord and attachment plug cap not to exceed 4.5 m (15 ft).
(2) Cord set assembly. Where run on the surface of the floor, they shall be protected against physical damage.

(C) Interconnecting Cables Separate data processing units shall be permitted to be interconnected by means of listed cables and cable assemblies. Where exposed to physical damage, the installation shall be protected by approved means.

(D) Under Raised Floors Power cables, communications cables, connecting cables, interconnecting cables, and receptacles associated with the information technology equipment shall be permitted under a raised floor, provided the following conditions are met:

(1) The raised floor is of suitable construction, and the area under the floor is accessible.
(2) The branch-circuit supply conductors to receptacles or field-wired equipment are in rigid metal conduit, rigid nonmetallic conduit, intermediate metal conduit, electrical metallic tubing, electrical nonmetallic tubing, metal wireway, nonmetallic wireway, surface metal raceway with metal cover, nonmetallic surface raceway, flexible metal conduit, liquidtight flexible metal conduit, or liquidtight flexible nonmetallic conduit, Type MI cable, Type MC cable, or Type AC cable. These supply conductors shall be installed in accordance with the requirements of 300.11.

Branch-circuit conductors installed under the raised floor of an ITE room using any of the wiring methods listed in 645.5(D)(2) are required to conform to the specific article for the wiring method used. In addition, Article 300 applies, except where modified by Article 645. For example, 300.11 requires raceways, cables, and boxes to be securely fastened in place, even though they are installed below a raised floor.

(3) Ventilation in the underfloor area is used for the information technology equipment room only, except as provided in 645.4(2). The ventilation system shall be so arranged, with approved smoke detection devices, that upon the detection of fire or products of combustion in the underfloor space the circulation of air will cease.

The underfloor area of the ITE room is required to be provided with smoke detection device(s). Upon detection of smoke, the circulation of air in the underfloor area must be interrupted. The most common method of interrupting air circulation is to open the circuit that supplies power to the air circulation fan. In addition to causing cessation of air circulation in the underfloor area, the smoke detectors may provide other fire protection functions as part of a complete building fire alarm system.

A revision to 645.5(D)(3) for the 2005 *Code* recognizes that a ventilation system can serve the underfloor areas of an ITE room as well as other areas of a building, if the ventilation system is equipped with the requisite smoke and fire dampers at the ITE room boundaries. These fire protection features isolate the underfloor area from other areas served by the ventilation system upon detection of smoke or activation of the ITE room disconnecting means. This revision correlates 645.5(D)(3) with 645.4(2).

(4) Openings in raised floors for cables protect cables against abrasions and minimize the entrance of debris beneath the floor.
(5) Cables, other than those covered in (D)(2) and those complying with (D)(5)(a), (D)(5)(b), or (D)(5)(c), shall be listed as Type DP cable having adequate fire-resistant characteristics suitable for use under raised floors of an information technology equipment room.
   a. Interconnecting cables enclosed in a raceway.
   b. Interconnecting cables listed with equipment manufactured prior to July 1, 1994, being installed with that equipment.
   c. Cable type designations Type TC (Article 336); Types CL2, CL3, and PLTC (Article 725); Type ITC (Article 727); Types NPLF and FPL (Article 760); Types OFC and OFN (Article 770); Type CM (Article 800); and Type CATV (Article 820). These designations shall be permitted to have an additional letter

P or R or G. Green, or green with one or more yellow stripes, insulated single conductor cables, 4 AWG and larger, marked "for use in cable trays" or "for CT use" shall be permitted for equipment grounding.

FPN: One method of defining fire resistance is by establishing that the cables do not spread fire to the top of the tray in the "Vertical Tray Flame Test" referenced in ANSI/UL 1581-2001, *Standard for Electrical Wires, Cables, and Flexible Cords*. Another method of defining fire resistance is for the damage (char length) not to exceed 1.5 m (4 ft 11 in.) when performing the CSA "Vertical Flame Test — Cables in Cable Trays," as described in CSA C22.2 No. 0.3-M-2001, *Test Methods for Electrical Wires and Cables*.

Interconnecting cables used under raised floors (other than branch-circuit conductors) are required by 645.5(D)(5) to be listed as Type DP cables. Cables listed as part of equipment manufactured before the effective date of July 1, 1994, were not required to be listed. Cables in raceways are also exempt. Cables that pass the Vertical Tray Flame Test referenced in ANSI/UL 1581-2001, *Standard for Electrical Wires, Cables, and Flexible Cords*, or the Vertical Flame Test — Cables in Cable Trays as described in CSA C22.2, No. 0.3-M-2001, *Test Method for Electrical Wires and Cables* (where not more than 4 ft 11 in. of cable is damaged during the CSA test), are permitted to be installed under raised floors of computer rooms. Type DP cables that satisfy these tests are also permitted under raised floors.

(6) Abandoned cables shall be removed unless contained in metal raceways.

(E) Securing in Place Power cables; communications cables; connecting cables; interconnecting cables; and associated boxes, connectors, plugs, and receptacles that are listed as part of, or for, information technology equipment shall not be required to be secured in place.

645.6 Cables Not in Information Technology Equipment Room

Cables extending beyond the information technology equipment room shall be subject to the applicable requirements of this *Code*.

FPN: For signaling circuits, refer to Article 725; for fiber optic circuits, refer to Article 770; and for communications circuits, refer to Article 800. For fire alarm systems, refer to Article 760.

645.7 Penetrations

Penetrations of the fire-resistant room boundary shall be in accordance with 300.21.

645.10 Disconnecting Means

A means shall be provided to disconnect power to all electronic equipment in the information technology equipment room. There shall also be a similar means to disconnect the power to all dedicated HVAC systems serving the room and cause all required fire/smoke dampers to close. The control for these disconnecting means shall be grouped and identified and shall be readily accessible at the principal exit doors. A single means to control both the electronic equipment and HVAC systems shall be permitted. Where a pushbutton is used as a means to disconnect power, pushing the button in shall disconnect the power.

In 645.10, two separate disconnecting means are required, but a single control, such as one pushbutton, is permitted to electrically operate both disconnecting means. The disconnecting means is required to disconnect the conductors of each circuit from their supply source and close all required fire/smoke dampers. (See the definition of *disconnecting means* in Article 100.) The disconnecting means is permitted to be remote-controlled switching devices, such as relays, with pushbutton stations at the principal exit doors. The 2002 *Code* specified that the actuation of the emergency pushbutton(s) be accomplished by pushing the button in, rather than pulling it out. The requirement recognizes that in an emergency situation the intuitive reaction to operating the control is to push, not pull, the button.

The requirements of 645.10 and those of 645.7 for sealing penetrations are intended to minimize the passage of smoke or fire to other parts of the building.

Exception: Installations qualifying under the provisions of Article 685.

645.11 Uninterruptible Power Supplies (UPS)

Except for installations and constructions covered in 645.11(1) or 645.11(2), UPS systems installed within the information technology equipment room, and their supply and output circuits, shall comply with 645.10. The disconnecting means shall also disconnect the battery from its load.

(1) Installations qualifying under the provisions of Article 685
(2) Power sources limited to 750 volt-amperes or less derived either from UPS equipment or from battery circuits integral to electronic equipment

645.15 Grounding

All exposed non–current-carrying metal parts of an information technology system shall be grounded in accordance with Article 250 or shall be double insulated. Power systems derived within listed information technology equipment that supply information technology systems through receptacles or cable assemblies supplied as part of this equipment shall not be considered separately derived for the purpose of applying 250.20(D). Where signal reference structures are installed, they shall be bonded to the equipment grounding system provided for the information technology equipment.

The last sentence of 645.15 recognizes that properly bonded high-frequency signal reference structures provide additional safety measures.

> FPN No. 1: The bonding and grounding requirements in the product standards governing this listed equipment ensure that it complies with Article 250.
> FPN No. 2: Where isolated grounding-type receptacles are used, see 250.146(D) and 406.2(D).

645.16 Marking

Each unit of an information technology system supplied by a branch circuit shall be provided with a manufacturer's nameplate, which shall also include the input power requirements for voltage, frequency, and maximum rated load in amperes.

645.17 Power Distribution Units

Power distribution units that are used for information technology equipment shall be permitted to have multiple panelboards within a single cabinet, provided that each panelboard has no more than 42 overcurrent devices and the power distribution unit is utilization equipment listed for information technology application.

Power distribution units (PDUs) are specialized electrical distribution equipment used to supply multiple bays of rack-mounted modules installed in an ITE room. Due to the large number of overcurrent protective devices used in this type of application, PDUs are built with multiple panelboards installed in a single cabinet. In a single cabinet or enclosure, the general requirement of 408.35 limits the number of overcurrent devices protecting lighting and appliance branch circuits to no more than 42. This provision amends that requirement for this specialized equipment provided the PDU is specifically listed for use with information technology equipment and there are no more than 42 overcurrent protective devices installed in any one panelboard located in the overall PDU enclosure.

ARTICLE 647
Sensitive Electronic Equipment

Contents

647.1 Scope

This article covers the installation and wiring of separately derived systems operating at 120 volts line-to-line and 60 volts to ground for sensitive electronic equipment.

The use of this type of supply system as a means to reduce objectionable noise and its adverse effect on the performance of electronic audio and video equipment has been recognized since the 1996 edition of the *Code*. Until the 2002 *Code*, the use of separately derived, 3-wire, 120-volt line-to-line, 60-volt-to-ground technical power systems was limited to motion picture and television studios. Article 647 permits the use of this type of supply system for all commercial and industrial applications where sensitive audio/video or similar electronic equipment is used. Such systems can be used only under the close supervision of qualified individuals.

Unlike electrical distribution systems that supply lighting and appliance branch circuits, the supply systems covered by Article 647 are subject to mandatory voltage-drop requirements. The voltage-drop requirements are needed to ensure the operation of overcurrent devices to protect conductors and equipment supplied by these systems. Because the use of standard overcurrent devices and distribution equipment with higher voltage ratings is permitted, the impedance in circuits supplied by these systems under fault conditions is a primary concern, hence the mandatory voltage-drop requirement.

647.3 General

Use of a separately derived 120-volt single-phase 3-wire system with 60 volts on each of two ungrounded conductors to a grounded neutral conductor shall be permitted for the purpose of reducing objectionable noise in sensitive electronic equipment locations, provided that the following conditions apply:

(1) The system is installed only in commercial or industrial occupancies.
(2) The system's use is restricted to areas under close supervision by qualified personnel.
(3) All of the requirements in 647.4 through 647.8 are met.

647.4 Wiring Methods

(A) Panelboards and Overcurrent Protection Use of standard single-phase panelboards and distribution equipment with a higher voltage rating shall be permitted. The system shall be clearly marked on the face of the panel or on the inside of the panel doors. Common trip two-pole circuit breakers or a combination two-pole fused disconnecting means that are identified for use at the system voltage shall be provided for both ungrounded conductors in all feeders and branch circuits. Branch circuits and feeders shall be provided with a means to simultaneously disconnect all ungrounded conductors.

This section was revised for the 2005 *Code* to recognize both circuit breakers and fuses as acceptable means of providing overcurrent protection for technical power circuits. Additionally, all technical power feeder and branch circuits are now required to be provided with a disconnecting means that simultaneously opens all ungrounded conductors of the circuit.

(B) Junction Boxes All junction box covers shall be clearly marked to indicate the distribution panel and the system voltage.

(C) Color Coding All feeders and branch-circuit conductors installed under this section shall be identified as to system at all splices and terminations by color, marking, tagging, or equally effective means. The means of identification shall be posted at each branch-circuit panelboard and at the disconnecting means for the building.

(D) Voltage Drop The voltage drop on any branch circuit shall not exceed 1.5 percent. The combined voltage drop of feeder and branch-circuit conductors shall not exceed 2.5 percent.

(1) Fixed Equipment The voltage drop on branch circuits supplying equipment connected using wiring methods in Chapter 3 shall not exceed 1.5 percent. The combined voltage drop of feeder and branch-circuit conductors shall not exceed 2.5 percent.

(2) Cord-Connected Equipment The voltage drop on branch circuits supplying receptacles shall not exceed 1 percent. For the purposes of making this calculation, the load connected to the receptacle outlet shall be considered to be 50 percent of the branch-circuit rating. The combined voltage drop of feeder and branch-circuit conductors shall not exceed 2.0 percent.

> FPN: The purpose of this provision is to limit voltage drop to 1.5 percent where portable cords may be used as a means of connecting equipment.

647.5 Three-Phase Systems

Where 3-phase power is supplied, a separately derived 6-phase "wye" system with 60 volts to ground installed under this article shall be configured as three separately derived 120-volt single-phase systems having a combined total of no more than six main disconnects.

647.6 Grounding

(A) General The system shall be grounded as provided in 250.30 as a separately derived single-phase 3-wire system.

(B) Grounding Conductors Required Permanently wired utilization equipment and receptacles shall be grounded by means of an equipment grounding conductor run with the circuit conductors to an equipment grounding bus prominently marked "Technical Equipment Ground" in the originating branch-circuit panelboard. The grounding bus shall be connected to the grounded conductor on the line side of the separately derived system's disconnecting means. The grounding conductor shall not be smaller than that specified in Table 250.122 and run with the feeder conductors. The technical equipment grounding bus need not be bonded to the panelboard enclosure. Other grounding methods authorized elsewhere in this *Code* shall be permitted where the impedance of the grounding return path does not exceed the impedance of equipment grounding conductors sized and installed in accordance with this article.

> FPN No. 1: See 250.122 for equipment grounding conductor sizing requirements where circuit conductors are adjusted in size to compensate for voltage drop.

> FPN No. 2: These requirements limit the impedance of the ground fault path where only 60 volts apply to a fault condition instead of the usual 120 volts.

647.7 Receptacles

(A) General Where receptacles are used as a means of connecting equipment, the following conditions shall be met:

(1) All 15- and 20-ampere receptacles shall be GFCI protected.

(2) All outlet strips, adapters, receptacle covers, and faceplates shall be marked with the following words or equivalent:

> WARNING — TECHNICAL POWER
> Do not connect to lighting equipment.
> For electronic equipment use only.
> 60/120 V. 1φac
> GFCI protected

(3) A 125-volt, single-phase, 15- or 20-ampere-rated receptacle outlet having one of its current-carrying poles connected to a grounded circuit conductor shall be located within 1.8 m (6 ft) of all permanently installed 15- or 20-ampere-rated 60/120-volt technical power-system receptacles.

(4) All 125-volt receptacles used for 60/120-volt technical power shall have a unique configuration and be identified for use with this class of system. All 125-volt, single-phase, 15- or 20-ampere-rated receptacle outlets and attachment plugs that are identified for use with grounded circuit conductors shall be permitted in machine rooms, control rooms, equipment rooms, equipment racks, and other similar locations that are restricted to use by qualified personnel.

(B) Isolated Ground Receptacles Isolated ground receptacles shall be permitted as described in 250.146(D); however, the branch circuit equipment grounding conductor shall be terminated as required in 647.6(B).

647.8 Lighting Equipment

Lighting equipment installed under this article for the purpose of reducing electrical noise originating from lighting equipment shall meet the conditions of 647.8(A) through 647.8(C).

(A) Disconnecting Means All luminaires (lighting fixtures) connected to separately derived systems operating at 60 volts to ground, and associated control equipment if provided, shall have a disconnecting means that simultaneously opens all ungrounded conductors. The disconnecting means shall be located within sight of the luminaire (lighting fixture) or be capable of being locked in the open position.

(B) Luminaires (Lighting Fixtures) All luminaires (lighting fixtures) shall be permanently installed and listed for connection to a separately derived system at 120 volts line-to-line and 60 volts to ground.

(C) Screw-Shell Luminaires installed under this section shall not have an exposed lamp screw-shell.

ARTICLE 650
Pipe Organs

Contents

650.1 Scope

This article covers those electrical circuits and parts of electrically operated pipe organs that are employed for the control of the sounding apparatus and keyboards.

650.3 Other Articles

Electronic organs shall comply with the appropriate provisions of Article 640.

650.4 Source of Energy

The source of power shall be a transformer-type rectifier, the dc potential of which shall not exceed 30 volts dc.

650.5 Grounding

The rectifier shall be grounded according to the provisions in 250.112(B).

650.6 Conductors

Conductors shall comply with 650.6(A) through 650.6(D).

Note that 14 AWG wire is required only for a main common return conductor in the electromagnetic supply.

(A) Size Conductors shall be not less than 28 AWG for electronic signal circuits and not less than 26 AWG for electromagnetic valve supply and the like. A main common-return conductor in the electromagnetic supply shall not be less than 14 AWG.

(B) Insulation Conductors shall have thermoplastic or thermosetting insulation.

(C) Conductors to Be Cabled Except for the common-return conductor and conductors inside the organ proper, the organ sections and the organ console conductors shall be cabled. The common-return conductors shall be permitted under an additional covering enclosing both cable and return conductor, or they shall be permitted as a separate conductor and shall be permitted to be in contact with the cable.

(D) Cable Covering Each cable shall be provided with an outer covering, either overall or on each of any subassemblies of grouped conductors. Tape shall be permitted in place of a covering. Where not installed in metal raceway, the covering shall be resistant to flame spread, or the cable or each cable subassembly shall be covered with a closely wound listed fireproof tape.

> FPN: One method of determining that cable is resistant to flame spread is by testing the cable to the VW-1 (vertical-wire) flame test in the ANSI/UL 1581-1991, *Reference Standard for Electrical Wires, Cables and Flexible Cords.*

650.7 Installation of Conductors

Cables shall be securely fastened in place and shall be permitted to be attached directly to the organ structure without insulating supports. Cables shall not be placed in contact with other conductors.

Insulating supports are not required; however, measures should be taken to prevent contact between the pipe organ cables and conductors of other systems.

650.8 Overcurrent Protection

Circuits shall be so arranged that 26 AWG and 28 AWG conductors shall be protected by an overcurrent device rated at not more than 6 amperes. Other conductor sizes shall be protected in accordance with their ampacity. A common return conductor shall not require overcurrent protection.

ARTICLE 660
X-Ray Equipment

Contents

I. General

660.1 Scope

This article covers all X-ray equipment operating at any frequency or voltage for industrial or other nonmedical or nondental use.

> FPN: See Article 517, Part V, for X-ray installations in health care facilities.

Nothing in this article shall be construed as specifying safeguards against the useful beam or stray X-ray radiation.

> FPN No. 1: Radiation safety and performance requirements of several classes of X-ray equipment are regulated under Public Law 90-602 and are enforced by the Department of Health and Human Services.
>
> FPN No. 2: In addition, information on radiation protection by the National Council on Radiation Protection and Measurements is published as *Reports of the National Council on Radiation Protection and Measurement*. These reports can be obtained from NCRP Publications, 7910 Woodmont Ave., Suite 1016, Bethesda, MD 20814.

660.2 Definitions

Long-Time Rating. A rating based on an operating interval of 5 minutes or longer.

Mobile. X-ray equipment mounted on a permanent base with wheels and/or casters for moving while completely assembled.

Momentary Rating. A rating based on an operating interval that does not exceed 5 seconds.

Portable. X-ray equipment designed to be hand-carried.

Transportable. X-ray equipment that is to be installed in a vehicle or that may be readily disassembled for transport in a vehicle.

660.3 Hazardous (Classified) Locations

Unless approved for the location, X-ray and related equipment shall not be installed or operated in hazardous (classified) locations.

X-ray equipment in industrial establishments or similar locations is commonly used for inspecting a process or product. This method permits nondestructive testing without dismantling or applying stress to detect cracks, flaws, or structural defects. Welded joints are frequently inspected with X-ray equipment to detect hidden defects that can cause failure under stress.

Among the industrial applications of X-rays, the most common is radiography, in which shadow pictures of the object are produced on photographic film. The type and thickness of the material involved govern the voltage to be employed, which can range from a few thousand volts (kilovolts) to millions of volts (megavolts). It is possible to X-ray metal objects that are 20 in. thick.

Fluoroscopy is another X-ray technique used for industrial and commercial applications. Fluoroscopy is similar to radiography, but it operates at a much lower voltage range (less than 250 kilovolts). Instead of producing a film, it projects a shadow picture on a screen, similar to those used for security checks of luggage at airport terminals. Fluoroscopy is capable of detecting minute flaws or defects.

> FPN: See Article 517, Part IV.

660.4 Connection to Supply Circuit

(A) Fixed and Stationary Equipment Fixed and stationary X-ray equipment shall be connected to the power supply by means of a wiring method meeting the general requirements of this *Code*. Equipment properly supplied by a branch circuit rated at not over 30 amperes shall be permitted to be supplied through a suitable attachment plug cap and hard-service cable or cord.

(B) Portable, Mobile, and Transportable Equipment Individual branch circuits shall not be required for portable, mobile, and transportable X-ray equipment requiring a capacity of not over 60 amperes. Portable and mobile types of X-ray equipment of any capacity shall be supplied through a suitable hard-service cable or cord. Transportable X-ray equipment of any capacity shall be permitted to be connected to its power supply by suitable connections and hard-service cable or cord.

(C) Over 600 Volts, Nominal Circuits and equipment operated at more than 600 volts, nominal, shall comply with Article 490.

660.5 Disconnecting Means

A disconnecting means of adequate capacity for at least 50 percent of the input required for the momentary rating or 100 percent of the input required for the long-time rating of the X-ray equipment, whichever is greater, shall be provided in the supply circuit. The disconnecting means shall be operable from a location readily accessible from the X-ray control. For equipment connected to a 120-volt, nominal, branch circuit of 30 amperes or less, a grounding-type attachment plug cap and receptacle of proper rating shall be permitted to serve as a disconnecting means.

660.6 Rating of Supply Conductors and Overcurrent Protection

(A) Branch-Circuit Conductors The ampacity of supply branch-circuit conductors and the overcurrent protective devices shall not be less than 50 percent of the momentary rating or 100 percent of the long-time rating, whichever is greater.

(B) Feeder Conductors The rated ampacity of conductors and overcurrent devices of a feeder for two or more branch circuits supplying X-ray units shall not be less than 100 percent of the momentary demand rating [as determined by 660.6(A)] of the two largest X-ray apparatus plus 20 percent of the momentary ratings of other X-ray apparatus.

> FPN: The minimum conductor size for branch and feeder circuits is also governed by voltage regulation requirements. For a specific installation, the manufacturer usually specifies minimum distribution transformer and conductor sizes, rating of disconnect means, and overcurrent protection.

660.7 Wiring Terminals

X-ray equipment not provided with a permanently attached cord or cord set shall be provided with suitable wiring terminals or leads for the connection of power-supply conductors of the size required by the rating of the branch circuit for the equipment.

660.8 Number of Conductors in Raceway

The number of control circuit conductors installed in a raceway shall be determined in accordance with 300.17.

660.9 Minimum Size of Conductors

Size 18 AWG or 16 AWG fixture wires, as specified in 725.27, and flexible cords shall be permitted for the control and operating circuits of X-ray and auxiliary equipment where protected by not larger than 20-ampere overcurrent devices.

660.10 Equipment Installations

All equipment for new X-ray installations and all used or reconditioned X-ray equipment moved to and reinstalled at a new location shall be of an approved type.

II. Control

660.20 Fixed and Stationary Equipment

(A) Separate Control Device A separate control device, in addition to the disconnecting means, shall be incorporated in the X-ray control supply or in the primary circuit to the high-voltage transformer. This device shall be a part of the X-ray equipment but shall be permitted in a separate enclosure immediately adjacent to the X-ray control unit.

(B) Protective Device A protective device, which shall be permitted to be incorporated into the separate control device, shall be provided to control the load resulting from failures in the high-voltage circuit.

660.21 Portable and Mobile Equipment

Portable and mobile equipment shall comply with 660.20, but the manually controlled device shall be located in or on the equipment.

660.23 Industrial and Commercial Laboratory Equipment

(A) Radiographic and Fluoroscopic Types All radiographic- and fluoroscopic-type equipment shall be effectively enclosed or shall have interlocks that de-energize the equipment automatically to prevent ready access to live current-carrying parts.

(B) Diffraction and Irradiation Types Diffraction- and irradiation-type equipment or installations not effectively enclosed or not provided with interlocks to prevent access to live current-carrying parts during operation shall be provided with a positive means to indicate when they are energized. The indicator shall be a pilot light, readable meter deflection, or equivalent means.

660.24 Independent Control

Where more than one piece of equipment is operated from the same high-voltage circuit, each piece or each group of equipment as a unit shall be provided with a high-voltage switch or equivalent disconnecting means. This disconnecting means shall be constructed, enclosed, or located so as to avoid contact by persons with its live parts.

A control device provides means for initiating and terminating X-ray exposures and automatically times their duration.

III. Transformers and Capacitors

660.35 General

Transformers and capacitors that are part of an X-ray equipment shall not be required to comply with Articles 450 and 460.

High-ratio step-up transformers that are an integral part of an X-ray are not required to comply with Article 450 and are generally used to provide the high voltage necessary for X-ray tubes. Because there is a lesser degree of fire hazard due to the low primary voltage, X-ray transformers are not required to be installed in fire-resistant vaults.

660.36 Capacitors

Capacitors shall be mounted within enclosures of insulating material or grounded metal.

IV. Guarding and Grounding

660.47 General

(A) High-Voltage Parts All high-voltage parts, including X-ray tubes, shall be mounted within grounded enclosures. Air, oil, gas, or other suitable insulating media shall be used to insulate the high voltage from the grounded enclosure. The connection from the high-voltage equipment to X-ray tubes and other high-voltage components shall be made with high-voltage shielded cables.

(B) Low-Voltage Cables Low-voltage cables connecting to oil-filled units that are not completely sealed, such as transformers, condensers, oil coolers, and high-voltage switches, shall have insulation of the oil-resistant type.

Grounded enclosures are required to be provided for all high-voltage X-ray equipment, including X-ray tubes. High-voltage shielded cables are required to be used to connect high-voltage equipment to X-ray tubes, and the shield is required to be grounded, as specified in 660.48.

660.48 Grounding

Non–current-carrying metal parts of X-ray and associated equipment (controls, tables, X-ray tube supports, transformer tanks, shielded cables, X-ray tube heads, and so forth) shall

be grounded in the manner specified in Article 250. Portable and mobile equipment shall be provided with an approved grounding-type attachment plug cap.

Exception: Battery-operated equipment.

ARTICLE 665
Induction and Dielectric Heating Equipment

Contents

I. General

665.1 Scope

This article covers the construction and installation of dielectric heating, induction heating, induction melting, and induction welding equipment and accessories for industrial and scientific applications. Medical or dental applications, appliances, or line frequency pipeline and vessel heating are not covered in this article.

> FPN: See Article 427, Part V, for line frequency induction heating of pipelines and vessels.

To prevent spurious radiation caused by induction and dielectric heating equipment and to ensure that the frequency spectrum is utilized equitably, the Federal Communications

Commission (FCC) has established rules that govern the use of this type of industrial heating equipment operating above 10 kHz (FCC, 47 CFR 18).

665.2 Definitions

Induction and dielectric heating are used for ovens, furnaces, and industrial equipment where pieces of material are heated by a rapidly alternating magnetic or electric field. For further information on electric heating systems using an induction heater or a dielectric heater in ovens and furnaces, see NFPA 86, *Standard for Ovens and Furnaces*.

Theory of Operation — Solid-State Converter Power Circuit

A solid-state power converter consists of three sections: the rectifier section, the inverter section, and the output section, which includes the load coil and is usually located outside the power supply. Exhibit 665.1 is an example of an enclosed power supply for an induction heating process. There are two basic types of inverters: voltage fed (Exhibit 665.2) and current fed (Exhibit 665.3).

Exhibit 665.1 Enclosed power supply for an induction heating process. (Courtesy of Ajax Tocco Magnethermic, Park Ohio Industries)

The rectifier section converts 3-phase, line frequency voltage to direct current. The input voltage can be any desired voltage—480 volts, 575 volts, and 1150 volts are some typical voltages. The dc output is filtered with either a large choke (current-fed inverter) or a filter capacitor (voltage-fed inverter). The output of the rectifier section supplies energy for the inverter section. The inverter section converts the energy stored in the magnetic field of the choke (current-fed inverter) or in the electric field of the filter capacitor (voltage-fed inverter) to a variable frequency for the output circuit. The variable output frequency controls the power delivered to the load.

The output section consists of a capacitor in parallel (current fed) or in series (voltage fed) with the coil. Because the output capacitance and coil inductance have a resonant frequency, as the output frequency approaches this resonant frequency, the power to the load approaches the maximum output. At minimum frequency, the output power is very low.

Induction Heating

Induction heating occurs when an electrically conductive material (the load) is placed in a varying magnetic field. The magnetic field is generated by a coil (inductor) around or adjacent to the workpiece to be heated. The varying magnetic field induces current in the electrically conductive load. Heat in the load is generated by the resulting I2R losses in the load. Induction heating can be further subdivided into heating, melting, and welding.

Induction heating raises the temperature of the load to some temperature below its melting point, usually for the purposes of hardening, tempering, annealing, forging, extruding, or rolling. Frequencies used for induction heating range from about 50 Hz to 500 kHz. Power levels range from 5 kW to 42,000 kW.

Induction melting raises the temperature of the load to some temperature above its melting point, so the molten material can be alloyed, homogenized, and/or poured. Frequencies used for induction melting range from about 50 Hz to 10 kHz. Power levels range from 5 kW to 16,500 kW.

Induction welding is primarily used in the manufacture of welded pipe and tubing. In this process, a high-frequency current is passed through an induction coil in the proximity of the conducting metal surfaces to be joined. By proper arrangement of the coil and by the addition of ferrite elements to control the currents induced in the surfaces, selected portions are heated nearly instantaneously to the forging temperature. The surfaces are then joined under pressure to produce a forge weld. Frequencies used for induction welding range from about 100 kHz to 800 kHz. Power levels range from 20 kW to 1000 kW. Induction heating (or melting or welding) of magnetic loads, such as iron or carbon steel, must contend with the change in magnetic permeability when

Exhibit 665.2 Schematic of a half-bridge, voltage-fed inverter. (Redrawn courtesy of Inductotherm Corp.)

Exhibit 665.3 Schematic of a full-bridge, current-fed inverter. (Redrawn courtesy of Radyne Corp.)

the temperature of the load material passes its Curie temperature (at which point the load changes from the magnetic to the nonmagnetic state). The solid-state converter varies its output frequency to maintain constant output power through the entire range of temperatures of the load. See Exhibit 665.4.

Dielectric Heating

Dielectric heating equipment is similar to induction heating equipment, except that it is used to heat nonmetallic materials as opposed to metals. Typical applications include the drying of textiles after dyeing, drying of water-based coatings on paper, preheating of wood fibers for the medium-density fiberboard (MDF) industry, welding of plastic materials, food processing, and many other diverse applications.

At radio frequencies, the material to be heated forms a lossy dielectric when placed between metal capacitor plates connected across the output of the generator. A high-frequency alternating electric field is created between the electrode plates. The molecules in the dielectric field are made to vibrate, causing dissipation of energy through the material and frictional heating of the dielectric material.

At higher (microwave) frequencies, the process is similar, but the generator is coupled to a resonant cavity into which the dielectric material is placed.

Exhibit 665.4 Simplified diagram of a solid-state inverter used for induction welding. (Redrawn courtesy of Thermatool Corp.)

The frequency of operation of dielectric heating equipment is considerably higher than for induction heating. By international agreement, these machines operate at the assigned radio frequencies of 13.56 MHz, 27.12 MHz, and 40.68 MHz or at microwave frequencies of 915 MHz and 2450 MHz.

All industrial dielectric heating equipment should meet appropriate U.S. federal regulations for radio frequency (RF) emissions (FCC, 47 CFR 18) and occupational safety (OSHA, 29 CFR 1910.97).

The majority of installed machines use vacuum tube generators, and powers range from 0.5 kW to 1 MW. Solid-state generators also have been installed, although powers have been limited to 5 kW or less. Exhibit 665.5 illustrates the components of a vacuum tube generator in a dielectric heating process.

Converting Device. That part of the heating equipment that converts input mechanical or electrical energy to the voltage, current, and frequency suitable for the heating applicator. A converting device shall consist of equipment using mains frequency, all static multipliers, oscillator-type units using vacuum tubes, inverters using solid state devices, or motor generator equipment.

Dielectric Heating. Heating of a nominally insulating material due to its own dielectric losses when the material is placed in a varying electric field.

Heating Equipment. As used in this article, any equipment that is used for heating purposes and whose heat is generated by induction or dielectric methods.

Heating Equipment Applicator. The device used to transfer energy between the output circuit and the object or mass to be heated.

Induction Heating, Melting, and Welding. The heating, melting, or welding of a nominally conductive material due to its own I^2R losses when the material is placed in a varying electromagnetic field.

665.3 Other Articles

Unless specifically amended by this article, wiring from the source of power to the heating equipment shall comply with Chapters 1 through 4.

Exhibit 665.5 Simplified diagram of the components of a vacuum tube generator used for dielectric heating. (Redrawn courtesy of Strayfield Ltd.)

665.4 Hazardous (Classified) Locations

Heating equipment shall not be installed in hazardous (classified) locations as defined in Article 500 unless the equipment and wiring are designed and approved for the hazardous (classified) locations.

665.5 Output Circuit

The output circuit shall include all output components external to the converting device, including contactors, switches, bus bars, and other conductors. The current flow from the output circuit to ground under operating and ground-fault conditions shall be limited to a value that does not cause 50 volts or more to ground to appear on any accessible part of the heating equipment and its load. The output circuit shall be permitted to be isolated from ground.

If the load (the object or mass being heated) accidentally comes in contact with the output coil, a voltage to ground will appear on the load, depending on the various impedances to ground of the coil and the load. If the voltage on the load is limited to less than 50 volts, guarding per 110.27(A) is not required. If the coil is isolated from ground and the load is grounded through an impedance that is low (less than 1 percent) relative to the coil impedance to ground, the voltage of the load to ground will be low no matter where the load contacts the coil.

In induction melting furnaces, an additional reason for isolating the coil from ground is to limit the fault current when a coil does go to ground. Limiting the fault current is

intended to prevent severe damage to the water-cooled coil, resulting in a water leak and the potential for a water–molten metal explosion. If water is trapped under molten metal, the rapid transfer of heat to the water causes the water to turn almost instantly into steam. The resulting 1600-to-1 expansion of the steam results in the ejection of molten metal from the furnace.

665.7 Remote Control

(A) Multiple Control Points Where multiple control points are used for applicator energization, a means shall be provided and interlocked so that the applicator can be energized from only one control point at a time. A means for de-energizing the applicator shall be provided at each control point.

(B) Foot Switches Switches operated by foot pressure shall be provided with a shield over the contact button to avoid accidental closing of a foot switch.

665.10 Ampacity of Supply Conductors

The ampacity of supply conductors shall be determined by 665.10(A) or 665.10(B).

(A) Nameplate Rating The ampacity of conductors supplying one or more pieces of equipment shall not be less than the sum of the nameplate ratings for the largest group of machines capable of simultaneous operation, plus 100 percent of the standby currents of the remaining machines. Where standby currents are not given on the nameplate, the nameplate rating shall be used as the standby current.

(B) Motor-Generator Equipment The ampacity of supply conductors for motor generator equipment shall be determined in accordance with Article 430, Part II.

665.11 Overcurrent Protection

Overcurrent protection for the heating equipment shall be provided as specified in Article 240. This overcurrent protection shall be permitted to be provided separately or as a part of the equipment.

665.12 Disconnecting Means

A readily accessible disconnecting means shall be provided to disconnect each heating equipment from its supply circuit. The disconnecting means shall be located within sight from the controller or be capable of being locked in the open position. The rating of this disconnecting means shall not be less than the nameplate rating of the heating equipment. Motor-generator equipment shall comply with Article 430, Part IX. The supply circuit disconnecting means shall be permitted to serve as the heating equipment disconnecting means where only one heating equipment is supplied.

II. Guarding, Grounding, and Labeling

665.19 Component Interconnection

The interconnection components required for a complete heating equipment installation shall be guarded.

665.20 Enclosures

The converting device (excluding the component interconnections) shall be completely contained within an enclosure(s) of noncombustible material.

665.21 Control Panels

All control panels shall be of dead-front construction.

665.22 Access to Internal Equipment

Access doors or detachable access panels shall be employed for internal access to heating equipment. Access doors to internal compartments containing equipment employing voltages from 150 volts to 1000 volts ac or dc shall be capable of being locked closed or shall be interlocked to prevent the supply circuit from being energized while the door(s) is open. Access doors to internal compartments containing equipment employing voltages exceeding 1000 volts ac or dc shall be provided with a disconnecting means equipped with mechanical lockouts to prevent access while the heating equipment is energized, or the access doors shall be capable of being locked closed and interlocked to prevent the supply circuit from being energized while the door(s) is open. Detachable panels not normally used for access to such parts shall be fastened in a manner that makes them inconvenient to remove.

665.23 Warning Labels or Signs

Warning labels or signs that read "DANGER — HIGH VOLTAGE — KEEP OUT" shall be attached to the equipment and shall be plainly visible where persons might come in contact with energized parts when doors are open or closed or when panels are removed from compartments containing over 150 volts ac or dc.

665.24 Capacitors

The time and means of discharge shall be in accordance with 460.6 for capacitors rated 600 volts, nominal, and under. The time and means of discharge shall be in accordance with 460.28 for capacitors rated over 600 volts, nominal. Capacitor internal pressure switches connected to a circuit-interrupter device shall be permitted for capacitor overcurrent protection.

It should be noted that it is necessary to provide enhanced protection against rupture of capacitor cases when capacitors are operated at the higher frequencies used for induction and dielectric heating. A high-resistance fault condition can cause case pressure to build up inside the capacitor over a very short time. Capacitor internal pressure switches are the preferred method to detect this type of failure.

Consider a 5000-kVAR, 2500-V, 300-Hz capacitor. Nominal current is 2000 amperes. A "high-resistance" fault of 10 ohms results in 250 amperes of resistive current, or a total capacitor current of 2016 amperes rms. This small increase in rms current will not result in the opening of an overcurrent device even though 625 kW of thermal energy is being generated inside the capacitor, which is designed to dissipate about 1.5 kW of losses.

665.25 Dielectric Heating Applicator Shielding

Protective cages or adequate shielding shall be used to guard dielectric heating applicators. Interlock switches shall be used on all hinged access doors, sliding panels, or other easy means of access to the applicator. All interlock switches shall be connected in such a manner as to remove all power from the applicator when any one of the access doors or panels is open.

665.26 Grounding and Bonding

Grounding or inter-unit bonding, or both, shall be used wherever required for circuit operation, for limiting to a safe value radio frequency voltages between all exposed non–current-carrying parts of the equipment and earth ground, between all equipment parts and surrounding objects, and between such objects and earth ground. Such grounding and bonding shall be installed in accordance with Article 250, Parts II and V.

Because of stray currents flowing between units of the equipment or to the ground, bonding presents special problems at radio frequencies. Special bonding requirements are particularly needed at dielectric heating frequencies (100 to 200 MHz) because of the differences in radio frequency potential that can exist between the equipment and surrounding metal units or other units of the installation. Satisfactory bonding can be accomplished by placing all units of the equipment on a flooring or base consisting of a copper or aluminum sheet and then thoroughly bonding by soldering, welding, or bolting.

Such special bonding holds the radio frequency resistance and reactance between units to a minimum, and any stray circulating currents flowing through the bonding will not cause a dangerous voltage drop.

It is necessary to protect the operator from high radio frequency potentials by shielding at dielectric heating frequencies. Interference with radio communications systems at such high frequencies can be eliminated by totally enclosing all components in a shielding of copper or aluminum.

FPN: Under certain conditions, contact between the object being heated and the applicator results in an unsafe condition, such as eruption of heated materials. This unsafe condition may be prevented by grounding of the object being heated and ground detection.

665.27 Marking

Each heating equipment shall be provided with a nameplate giving the manufacturer's name and model identification and the following input data: line volts, frequency, number of phases, maximum current, full-load kilovolt-amperes (kVA), and full-load power factor. Additional data shall be permitted.

ARTICLE 668
Electrolytic Cells

Contents

668.1 Scope

The provisions of this article apply to the installation of the electrical components and accessory equipment of electrolytic cells, electrolytic cell lines, and process power supply for the production of aluminum, cadmium, chlorine, copper, fluorine, hydrogen peroxide, magnesium, sodium, sodium chlorate, and zinc.

Not covered by this article are cells used as a source of electric energy and for electroplating processes and cells used for the production of hydrogen.

FPN No. 1: In general, any cell line or group of cell lines operated as a unit for the production of a particular metal, gas, or chemical compound may differ from any other cell lines producing the same product because of variations in the particular raw materials used, output capacity, use of proprietary methods or process practices, or other modifying factors to the extent that detailed *Code* requirements become overly restrictive and do not accomplish the stated purpose of this *Code.*

FPN No. 2: For further information, see IEEE 463-1993, *Standard for Electrical Safety Practices in Electrolytic Cell Line Working Zones.*

Within a cell line working zone, both an electrolytic cell line and its dc process power supply circuit are treated as an individual machine supplied from a single source, even though they might cover acres of space, have a load current in excess of 400,000 amperes dc, or have a circuit voltage in excess of 1000 volts dc. The cell line process current passes through each cell in a series connection, and the load current cannot be subdivided, as it can in the heating circuit of a resistance-type electric furnace. Because a cell line is supplied by its individual dc rectifier system, the rectifier or the entire cell line circuit is de-energized by removing its source of primary power.

In some electrolytic cell systems, the terminal voltage of the process supply can be appreciable. The voltage to ground of exposed live parts from one end of a cell line to the other is variable between the limits of the terminal voltage. Hence, operating and maintenance personnel and their tools are required to be insulated from ground. See Exhibit 668.1 for an example of a pot room in an aluminum reduction plant.

668.2 Definitions

Cell Line. An assembly of electrically interconnected electrolytic cells supplied by a source of direct-current power.

Cell Line Attachments and Auxiliary Equipment. As applied to this article, a term that includes, but is not limited to, auxiliary tanks; process piping; ductwork; structural supports; exposed cell line conductors; conduits and other race-ways; pumps, positioning equipment, and cell cutout or bypass electrical devices. Auxiliary equipment includes tools, welding machines, crucibles, and other portable equipment used for operation and maintenance within the electrolytic cell line working zone.

In the cell line working zone, auxiliary equipment includes the exposed conductive surfaces of ungrounded cranes and crane-mounted cell-servicing equipment.

Electrically Connected. A connection capable of carrying current as distinguished from connection through electromagnetic induction.

Electrolytic Cell. A tank or vat in which electrochemical reactions are caused by applying electrical energy for the purpose of refining or producing usable materials.

Electrolytic Cell Line Working Zone. The space envelope wherein operation or maintenance is normally performed on or in the vicinity of exposed energized surfaces of electrolytic cell lines or their attachments.

668.3 Other Articles

(A) Lighting, Ventilating, Material Handling Chapters 1 through 4 shall apply to services, feeders, branch circuits, and apparatus for supplying lighting, ventilating, material handling, and the like that are outside the electrolytic cell line working zone.

(B) Systems Not Electrically Connected Those elements of a cell line power-supply system that are not electrically

Exhibit 668.1 A typical pot room in an aluminum reduction plant. (Courtesy of Alcoa Inc.)

connected to the cell supply system, such as the primary winding of a two-winding transformer, the motor of a motor-generator set, feeders, branch circuits, disconnecting means, motor controllers, and overload protective equipment, shall be required to comply with all applicable provisions of this *Code*.

(C) Electrolytic Cell Lines Electrolytic cell lines shall comply with the provisions of Chapters 1, 2, 3, and 4 except as amended in 668.3(C)(1), (C)(2), (C)(3), or (C)(4).

(1) Conductors The electrolytic cell line conductors shall not be required to comply with the provisions of Articles 110, 210, 215, 220, and 225. See 668.11.

(2) Overcurrent Protection Overcurrent protection of electrolytic cell dc process power circuits shall not be required to comply with the requirements of Article 240.

(3) Grounding Equipment located or used within the electrolytic cell line working zone or associated with the cell line dc power circuits shall not be required to comply with the provisions of Article 250.

(4) Working Zone The electrolytic cells, cell line attachments, and the wiring of auxiliary equipment and devices within the cell line working zone shall not be required to comply with the provisions of Articles 110, 210, 215, 220, and 225. See 668.30.

> FPN: See 668.15 for equipment, apparatus, and structural component grounding.

668.10 Cell Line Working Zone

(A) Area Covered The space envelope of the cell line working zone shall encompass spaces that meet any of the following conditions:

(1) Is within 2.5 m (96 in.) above energized surfaces of electrolytic cell lines or their energized attachments
(2) Is below energized surfaces of electrolytic cell lines or their energized attachments, provided the headroom in the space beneath is less than 2.5 m (96 in.)
(3) Is within 1.0 m (42 in.) horizontally from energized surfaces of electrolytic cell lines or their energized attachments or from the space envelope described in 668.10(A)(1) or (A)(2)

(B) Area Not Covered The cell line working zone shall not be required to extend through or beyond walls, floors, roofs, partitions, barriers, or the like.

668.11 Direct-Current Cell Line Process Power Supply

(A) Not Grounded The dc cell line process power-supply conductors shall not be required to be grounded.

(B) Metal Enclosures Grounded All metal enclosures of dc cell line process power-supply apparatus operating at a power-supply potential between terminals of over 50 volts shall be grounded by either of the following means:

(1) Through protective relaying equipment
(2) By a minimum 2/0 AWG copper grounding conductor or a conductor of equal or greater conductance

(C) Grounding Requirements The grounding connections required by 668.11(B) shall be installed in accordance with 250.8, 250.10, 250.12, 250.68, and 250.70.

668.12 Cell Line Conductors

(A) Insulation and Material Cell line conductors shall be either bare, covered, or insulated and of copper, aluminum, copper-clad aluminum, steel, or other suitable material.

(B) Size Cell line conductors shall be of such cross-sectional area that the temperature rise under maximum load conditions and at maximum ambient shall not exceed the safe operating temperature of the conductor insulation or the material of the conductor supports.

(C) Connections Cell line conductors shall be joined by bolted, welded, clamped, or compression connectors.

668.13 Disconnecting Means

(A) More Than One Process Power Supply Where more than one dc cell line process power supply serves the same cell line, a disconnecting means shall be provided on the cell line circuit side of each power supply to disconnect it from the cell line circuit.

(B) Removable Links or Conductors Removable links or removable conductors shall be permitted to be used as the disconnecting means.

668.14 Shunting Means

(A) Partial or Total Shunting Partial or total shunting of cell line circuit current around one or more cells shall be permitted.

(B) Shunting One or More Cells The conductors, switches, or combination of conductors and switches used for shunting one or more cells shall comply with the applicable requirements of 668.12.

668.15 Grounding

For equipment, apparatus, and structural components that are required to be grounded by provisions of Article 668, the provisions of Article 250 shall apply, except a water pipe electrode shall not be required to be used. Any electrode or combination of electrodes described in 250.52 shall be permitted.

668.20 Portable Electrical Equipment

(A) Portable Electrical Equipment Not to Be Grounded The frames and enclosures of portable electrical equipment used within the cell line working zone shall not be grounded.

Exception No. 1: Where the cell line voltage does not exceed 200 volts dc, these frames and enclosures shall be permitted to be grounded.

Exception No. 2: These frames and enclosures shall be permitted to be grounded where guarded.

(B) Isolating Transformers Electrically powered, hand-held, cord-connected portable equipment with ungrounded frames or enclosures used within the cell line working zone shall be connected to receptacle circuits that have only ungrounded conductors such as a branch circuit supplied by an isolating transformer with an ungrounded secondary.

(C) Marking Ungrounded portable electrical equipment shall be distinctively marked and shall employ plugs and receptacles of a configuration that prevents connection of this equipment to grounding receptacles and that prevents inadvertent interchange of ungrounded and grounded portable electrical equipments.

668.21 Power Supply Circuits and Receptacles for Portable Electrical Equipment

(A) Isolated Circuits Circuits supplying power to ungrounded receptacles for hand-held, cord-connected equipments shall be electrically isolated from any distribution system supplying areas other than the cell line working zone and shall be ungrounded. Power for these circuits shall be supplied through isolating transformers. Primaries of such transformers shall operate at not more than 600 volts between conductors and shall be provided with proper overcurrent protection. The secondary voltage of such transformers shall not exceed 300 volts between conductors, and all circuits supplied from such secondaries shall be ungrounded and shall have an approved overcurrent device of proper rating in each conductor.

(B) Noninterchangeability Receptacles and their mating plugs for ungrounded equipment shall not have provision for a grounding conductor and shall be of a configuration that prevents their use for equipment required to be grounded.

(C) Marking Receptacles on circuits supplied by an isolating transformer with an ungrounded secondary shall be a distinctive configuration, distinctively marked, and shall not be used in any other location in the plant.

668.30 Fixed and Portable Electrical Equipment

(A) Electrical Equipment Not Required to Be Grounded Alternating-current systems supplying fixed and portable electrical equipment within the cell line working zone shall not be required to be grounded.

(B) Exposed Conductive Surfaces Not Required to Be Grounded Exposed conductive surfaces, such as electrical equipment housings, cabinets, boxes, motors, raceways, and the like, that are within the cell line working zone shall not be required to be grounded.

(C) Wiring Methods Auxiliary electrical equipment such as motors, transducers, sensors, control devices, and alarms, mounted on an electrolytic cell or other energized surface, shall be connected to premises wiring systems by any of the following means:

(1) Multiconductor hard usage cord.
(2) Wire or cable in suitable raceways or metal or nonmetallic cable trays. If metal conduit, cable tray, armored cable, or similar metallic systems are used, they shall be installed with insulating breaks such that they do not cause a potentially hazardous electrical condition.

(D) Circuit Protection Circuit protection shall not be required for control and instrumentation that are totally within the cell line working zone.

(E) Bonding Bonding of fixed electrical equipment to the energized conductive surfaces of the cell line, its attachments, or auxiliaries shall be permitted. Where fixed electrical equipment is mounted on an energized conductive surface, it shall be bonded to that surface.

668.31 Auxiliary Nonelectric Connections

Auxiliary nonelectric connections, such as air hoses, water hoses, and the like, to an electrolytic cell, its attachments, or auxiliary equipments shall not have continuous conductive reinforcing wire, armor, braids, and the like. Hoses shall be of a nonconductive material.

668.32 Cranes and Hoists

(A) Conductive Surfaces to Be Insulated from Ground The conductive surfaces of cranes and hoists that enter the cell line working zone shall not be required to be grounded. The portion of an overhead crane or hoist that contacts an energized electrolytic cell or energized attachments shall be insulated from ground.

(B) Hazardous Electrical Conditions Remote crane or hoist controls that could introduce hazardous electrical conditions into the cell line working zone shall employ one or more of the following systems:

(1) Isolated and ungrounded control circuit in accordance with 668.21(A)
(2) Nonconductive rope operator

Article 670 — Industrial Machinery

(3) Pendant pushbutton with nonconductive supporting means and having nonconductive surfaces or ungrounded exposed conductive surfaces
(4) Radio

668.40 Enclosures

General-purpose electrical equipment enclosures shall be permitted where a natural draft ventilation system prevents the accumulation of gases.

ARTICLE 669
Electroplating

669.1 Scope

The provisions of this article apply to the installation of the electrical components and accessory equipment that supply the power and controls for electroplating, anodizing, electropolishing, and electrostripping. For purposes of this article, the term *electroplating* shall be used to identify any or all of these processes.

Because of the extremely high currents and low voltages normally involved, conventional wiring methods cannot be used in electroplating, anodizing, electropolishing, and electrostripping processes. Note the permission in 669.6(A) and 669.6(B) to use bare conductors, even in systems exceeding 50 volts dc. Some systems in the aluminum anodizing process have potentials up to 240 volts. Warning signs are required to be posted to indicate the presence of bare conductors.

669.3 General

Equipment for use in electroplating processes shall be identified for such service.

669.5 Branch-Circuit Conductors

Branch-circuit conductors supplying one or more units of equipment shall have an ampacity of not less than 125 per-

cent of the total connected load. The ampacities for busbars shall be in accordance with 366.10.

669.6 Wiring Methods

Conductors connecting the electrolyte tank equipment to the conversion equipment shall be in accordance with 669.6(A) and 669.6(B).

(A) Systems Not Exceeding 50 Volts Direct Current Insulated conductors shall be permitted to be run without insulated support, provided they are protected from physical damage. Bare copper or aluminum conductors shall be permitted where supported on insulators.

(B) Systems Exceeding 50 Volts Direct Current Insulated conductors shall be permitted to be run on insulated supports, provided they are protected from physical damage. Bare copper or aluminum conductors shall be permitted where supported on insulators and guarded against accidental contact up to the point of termination in accordance with 110.27.

669.7 Warning Signs

Warning signs shall be posted to indicate the presence of bare conductors.

669.8 Disconnecting Means

(A) More Than One Power Supply Where more than one power supply serves the same dc system, a disconnecting means shall be provided on the dc side of each power supply.

(B) Removable Links or Conductors Removable links or removable conductors shall be permitted to be used as the disconnecting means.

669.9 Overcurrent Protection

Direct-current conductors shall be protected from overcurrent by one or more of the following:

(1) Fuses or circuit breakers
(2) A current-sensing device that operates a disconnecting means
(3) Other approved means

ARTICLE 670
Industrial Machinery

Summary of Changes

• **670.3(A):** Reorganized text into a numbered list and added requirement to provide the short-circuit current rating for the control panel or control assembly on the machine nameplate.

- **670.4:** Placed requirements for machine disconnecting means in separate paragraph (B) and re-identified machine supply circuit overcurrent protection requirements as paragraph (C).

Contents

670.1 Scope

This article covers the definition of, the nameplate data for, and the size and overcurrent protection of supply conductors to industrial machinery.

> FPN: For information on the workspace requirements for equipment containing supply conductor terminals, see 110.26. For information on the workspace requirements for machine power and control equipment, see NFPA 79–2002, *Electrical Standard for Industrial Machinery.*

An important point concerning the revised fine print note is to recognize that 670.5, which covered working clearances unique to industrial machinery in previous editions of the *Code*, does not appear in the 2005 *Code*. The requirements of 670.5 amended those of 110.26 and under specific conditions permitted a reduction in the minimum amount of working clearance needed about energized parts of industrial machinery. This rule is no longer necessary in the *NEC* because

12.5.1.1 in the 2002 edition of NFPA 79, *Electrical Standard for Industrial Machinery*, contains working clearance requirements for the equipment covered in the scope of that standard, and working space about equipment covered in the scope of the *NEC* is specified in 110.26.

670.2 Definitions

Exhibit 670.1 shows an example of an industrial machine. The scope of Article 670, described in 670.1, and the definitions contained in 670.2 permit the inclusion of other types of industrial machines without the need to continuously modify the scope of Article 670 and NFPA 79, *Electrical Standard for Industrial Machinery*. Also, the scope and definitions are more in harmony with NFPA 79 and IEC 60204-1, *Safety of Machinery — Electrical Equipment of Machines — Part 1: General Requirements.*

Industrial Machinery (Machine). A power-driven machine (or a group of machines working together in a coordinated manner), not portable by hand while working, that is used to process material by cutting; forming; pressure; electrical, thermal, or optical techniques; lamination; or a combination of these processes. It can include associated equipment used to transfer material or tooling, including fixtures, to assemble/disassemble, to inspect or test, or to package. [The associated electrical equipment, including the logic controller(s) and associated software or logic together with the machine actuators and sensors, are considered as part of the industrial machine.]

Industrial Manufacturing System. A systematic array of one or more industrial machines that is not portable by hand and includes any associated material handling, manipulating, gauging, measuring, or inspection equipment.

670.3 Machine Nameplate Data

(A) Permanent Nameplate A permanent nameplate shall be attached to the control equipment enclosure or machine

Exhibit 670.1 All-electric injection molding machine with Programmable Controller–based control. (Photograph courtesy of Milacron)

and shall be plainly visible after installation. The nameplate shall include the following information:

(1) Supply voltage, phase, frequency, and full-load current
(2) Maximum ampere rating of the short-circuit and ground-fault protective device
(3) Ampere rating of largest motor, from the motor nameplate, or load
(4) Short circuit current rating of the machine industrial control panel based on one of the following:
   a. Short circuit current rating of a listed and labeled machine control enclosure or assembly
   b. Short circuit current rating established utilizing an approved method

   FPN: UL 508A-2001, Supplement SB, is an example of an approved method.

(5) Electrical diagram number(s) or the number of the index to the electrical drawings

The full-load current shown on the nameplate shall not be less than the sum of the full-load currents required for all motors and other equipment that may be in operation at the same time under normal conditions of use. Where unusual type loads, duty cycles, and so forth require oversized conductors or permit reduced-size conductors, the required capacity shall be included in the marked "full-load current." Where more than one incoming supply circuit is to be provided, the nameplate shall state the preceding information for each circuit.

To better convey the types of information required on the nameplate of an industrial machine, the first paragraph of 670.3(A) has been reorganized into a list describing the required information that must be included on the nameplate of an industrial machine. In addition, the required short-circuit current nameplate information has been revised to clarify what part of the industrial machine the marking applies to and how the short-circuit current rating is to be determined. The nameplate information must now provide the short-circuit current rating of the machine's industrial control panel. That rating is established either by the rating determined as part of the listing of the control enclosure or assembly or, for assemblies that are not listed, by an approved method of determining the short-circuit current rating.

Based on the *NEC* definition of *approved*, such a method is one that is acceptable to the authority having jurisdiction. In the absence of product listing, Supplement SB to UL 508A, *Standard for Industrial Control Panels*, is referred to as one example of a method for determining the short-circuit current rating of a control panel or assembly that could be used as a basis for equipment approval.

The second paragraph of 670.3(A) recognizes that the

operating characteristics of an industrial machine may permit the use of a feeder demand factor, as covered in 430.26. An industrial machine containing motors sized for high torque but in normal operation run at close to no-load current values is an example of where it may be appropriate to reduce the full-load current marking on the machine nameplate.

   FPN: See 430.22(E) and 430.26 for duty cycle requirements.

(B) Overcurrent Protection Where overcurrent protection is provided in accordance with 670.4(B), the machine shall be marked "overcurrent protection provided at machine supply terminals."

670.4 Supply Conductors and Overcurrent Protection

(A) Size The size of the supply conductor shall be such as to have an ampacity not less than 125 percent of the full-load current rating of all resistance heating loads plus 125 percent of the full-load current rating of the highest rated motor plus the sum of the full-load current ratings of all other connected motors and apparatus based on their duty cycle that may be in operation at the same time.

In conjunction with the revision to 670.3, the requirements for determining the minimum ampacity of a supply circuit conductor for an industrial machine now specifically reference the duty cycle of motors and apparatus as a consideration. Depending on the operating characteristics of the motor, the duty cycle of the apparatus might not always result in reduction of the supply conductor ampacity. Where motors are used in other than a continuous-duty mode of operation, Table 430.22(E) provides percentages by which the full-load current (FLC) of a given motor is increased or decreased for the purpose of sizing motor circuit conductors. A motor that is loaded continuously under any conditions of use is regarded to be a continuous-duty application.

   FPN No. 1: See the 0–2000-volt ampacity tables of Article 310 for ampacity of conductors rated 600 volts and below.
   FPN No. 2: See 430.22(E) and 430.26 for duty cycle requirements.

(B) Disconnecting Means A machine shall be considered as an individual unit and therefore shall be provided with disconnecting means. The disconnecting means shall be permitted to be supplied by branch circuits protected by either fuses or circuit breakers. The disconnecting means shall not be required to incorporate overcurrent protection.

NFPA 79, *Electrical Standard for Industrial Machinery*, states, in part:

An operating handle of the disconnecting means shall . . . be readily accessible . . . The center of the grip of the operating handle of the disconnecting means, when in its highest position, shall be not more than 2.0 m (6 ft 7 in.) above the floor. A permanent operating platform, readily accessible by means of a permanent stair or ladder, shall be considered as the floor for the purpose of this requirement. . . . The operating handle shall be capable of being locked only in the open (off) position. . . . [NFPA 79-2002, 5.3.3 and 5.3.4]

(C) Overcurrent Protection Where furnished as part of the machine, overcurrent protection for each supply circuit shall consist of a single circuit breaker or set of fuses, the machine shall bear the marking required in 670.3 and the supply conductors shall be considered either as feeders or taps as covered by 240.21.

A commonly asked question is, "When does the *NEC* apply to wiring for machinery defined in Article 670?" The equipment and wiring of industrial machinery, for which different component parts may be purchased and assembled at the location of use, must be installed in accordance with the applicable articles in the *NEC*.

Machinery assembled by the manufacturer, in accordance with NFPA 79, *Electrical Standard for Industrial Machinery*, then disassembled for shipping and reassembled at its place of use, comes only under Article 670 and any *NEC* sections referenced therein. In this case, the machinery is treated as a package unit. The nameplate provides the necessary information to size the branch-circuit conductors, disconnecting means, and overcurrent protection. The computation of motor and nonmotor loads is reflected on the nameplate as full-load amperes, and no further calculation is necessary.

The rating or setting of the overcurrent protective device for the circuit supplying the machine shall not be greater than the sum of the largest rating or setting of the branch-circuit short-circuit and ground-fault protective device provided with the machine, plus 125 percent of the full-load current rating of all resistance heating loads, plus the sum of the full-load currents of all other motors and apparatus that could be in operation at the same time.

Exception: Where one or more instantaneous trip circuit breakers or motor short-circuit protectors are used for motor branch-circuit short-circuit and ground-fault protection as permitted by 430.52(C), the procedure specified above for determining the maximum rating of the protective device for the circuit supplying the machine shall apply with the following provision: For the purpose of the calculation, each instantaneous trip circuit breaker or motor short-circuit

protector shall be assumed to have a rating not exceeding the maximum percentage of motor full-load current permitted by Table 430.52 for the type of machine supply circuit protective device employed.

Where no branch-circuit short-circuit and ground-fault protective device is provided with the machine, the rating or setting of the overcurrent protective device shall be based on 430.52 and 430.53, as applicable.

ARTICLE 675
Electrically Driven or Controlled Irrigation Machines

Summary of Changes
- **675.8(A)(2) and (B):** Added exceptions permitting use of listed molded case switches as main controller and as main disconnecting means.

Contents

I. General

675.1 Scope

The provisions of this article apply to electrically driven or controlled irrigation machines, and to the branch circuits and controllers for such equipment.

Electric pump motors used to supply water to irrigation machines are governed by the general requirements of the *NEC*, not by Article 675.

675.2 Definitions

Center Pivot Irrigation Machine. A multimotored irrigation machine that revolves around a central pivot and employs alignment switches or similar devices to control individual motors.

Collector Rings. An assembly of slip rings for transferring electrical energy from a stationary to a rotating member.

Irrigation Machine. An electrically driven or controlled machine, with one or more motors, not hand portable, and used primarily to transport and distribute water for agricultural purposes.

675.3 Other Articles

These provisions are in addition to, or amendatory of, the provisions of Article 430 and other articles in this *Code* that apply except as modified in this article.

The requirements of Article 675 apply to special equipment for a particular condition; they supplement or modify the general rules. See 90.3, Code Arrangement.

675.4 Irrigation Cable

(A) Construction The cable used to interconnect enclosures on the structure of an irrigation machine shall be an assembly of stranded, insulated conductors with nonhygroscopic and nonwicking filler in a core of moisture- and

flame-resistant nonmetallic material overlaid with a metallic covering and jacketed with a moisture-, corrosion-, and sunlight-resistant nonmetallic material.

The conductor insulation shall be of a type listed in Table 310.13 for an operating temperature of 75°C (167°F) and for use in wet locations. The core insulating material thickness shall not be less than 0.76 mm (30 mils), and the metallic overlay thickness shall not be less than 0.20 mm (8 mils). The jacketing material thickness shall not be less than 1.27 mm (50 mils).

A composite of power, control, and grounding conductors in the cable shall be permitted.

(B) Alternate Wiring Methods Installation of other listed cables complying with the construction requirements of 675.4(A) shall be permitted.

(C) Supports Irrigation cable shall be secured by straps, hangers, or similar fittings identified for the purpose and so installed as not to damage the cable. Cable shall be supported at intervals not exceeding 1.2 m (4 ft).

(D) Fittings Fittings shall be used at all points where irrigation cable terminates. The fittings shall be designed for use with the cable and shall be suitable for the conditions of service.

675.5 More Than Three Conductors in a Raceway or Cable

The signal and control conductors of a raceway or cable shall not be counted for the purpose of derating the conductors as required in 310.15(B)(2)(a).

675.6 Marking on Main Control Panel

The main control panel shall be provided with a nameplate that shall give the following information:
(1) The manufacturer's name, the rated voltage, the phase, and the frequency
(2) The current rating of the machine
(3) The rating of the main disconnecting means and size of overcurrent protection required

675.7 Equivalent Current Ratings

Where intermittent duty is not involved, the provisions of Article 430 shall be used for determining ratings for controllers, disconnecting means, conductors, and the like. Where irrigation machines have inherent intermittent duty, the determinations of equivalent current ratings in 675.7(A) and 675.7(B) shall be used.

(A) Continuous-Current Rating The equivalent continuous-current rating for the selection of branch-circuit conductors and overcurrent protection shall be equal to 125 percent of the motor nameplate full-load current rating of the largest motor, plus a quantity equal to the sum of each of the motor nameplate full-load current ratings of all remaining motors on the circuit, multiplied by the maximum percent duty cycle at which they can continuously operate.

(B) Locked-Rotor Current The equivalent locked-rotor current rating shall be equal to the numerical sum of the locked-rotor current of the two largest motors plus 100 percent of the sum of the motor nameplate full-load current ratings of all the remaining motors on the circuit.

675.8 Disconnecting Means

(A) Main Controller A controller that is used to start and stop the complete machine shall meet all of the following requirements:

(1) An equivalent continuous current rating not less than specified in 675.7(A) or 675.22(A)
(2) A horsepower rating not less than the value from Tables 430.251(A) and 430.251(B), based on the equivalent locked-rotor current specified in 675.7(B) or 675.22(B)

Exception: A listed molded case switch shall not require a horsepower rating.

A molded case switch used as a motor controller is not required to have a horsepower rating, but it is required to have a continuous current (ampere) rating not less than that specified by 675.7(A) or 675.22(A).

(B) Main Disconnecting Means The main disconnecting means for the machine shall provide overcurrent protection, shall be at the point of connection of electrical power to the machine or shall be visible and not more than 15 m (50 ft) from the machine, and shall be readily accessible and capable of being locked in the open position. This disconnecting means shall have a horsepower and current rating not less than required for the main controller.

In accordance with 675.8(B), the main disconnecting means is permitted to be up to 50 ft from the machine but must be readily accessible and capable of being locked in the open position. This eliminates one set of overcurrent protective devices and one disconnecting means where the circuit originates at the motor control panel for the irrigation pump and the panel is located within 50 ft of the center pivot machine. It also alleviates some potential problems with machines designed to be towed to a second site.

Exception No. 1: Circuit breakers without marked horsepower ratings shall be permitted in accordance with 430.109.

Exception No. 2: A listed fusible molded case switch without marked horsepower ratings shall be permitted.

(C) Disconnecting Means for Individual Motors and Controllers A disconnecting means shall be provided to simultaneously disconnect all ungrounded conductors for each motor and controller and shall be located as required by Article 430, Part IX. The disconnecting means shall not be required to be readily accessible.

Article 430, Part IX, provides for safety during maintenance and inspection shutdown periods. See the commentary following 430.103.

675.9 Branch-Circuit Conductors

The branch-circuit conductors shall have an ampacity not less than specified in 675.7(A) or 675.22(A).

675.10 Several Motors on One Branch Circuit

The requirements of 430.53 provide for motor branch-circuit short-circuit and ground-fault protection where several motors are supplied by one branch circuit. A combination of these requirements, which are modified for this special equipment application, is found in 675.10.

(A) Protection Required Several motors, each not exceeding 2 hp rating, shall be permitted to be used on an irrigation machine circuit protected at not more than 30 amperes at 600 volts, nominal, or less, provided all of the following conditions are met:

(1) The full-load rating of any motor in the circuit shall not exceed 6 amperes.
(2) Each motor in the circuit shall have individual overload protection in accordance with 430.32.
(3) Taps to individual motors shall not be smaller than 14 AWG copper and not more than 7.5 m (25 ft) in length.

(B) Individual Protection Not Required Individual branch-circuit short-circuit protection for motors and motor controllers shall not be required where the requirements of 675.10(A) are met.

675.11 Collector Rings

(A) Transmitting Current for Power Purposes Collector rings shall have a current rating not less than 125 percent of the full-load current of the largest device served plus the full-load current of all other devices served, or as determined from 675.7(A) or 675.22(A).

(B) Control and Signal Purposes Collector rings for control and signal purposes shall have a current rating not less than 125 percent of the full-load current of the largest device served plus the full-load current of all other devices served.

(C) Grounding The collector ring used for grounding shall have a current rating not less than that sized in accordance with 675.11(A).

(D) Protection Collector rings shall be protected from the expected environment and from accidental contact by means of a suitable enclosure.

675.12 Grounding

The following equipment shall be grounded:

(1) All electrical equipment on the irrigation machine
(2) All electrical equipment associated with the irrigation machine
(3) Metal junction boxes and enclosures
(4) Control panels or control equipment that supplies or controls electrical equipment to the irrigation machine

Exception: Grounding shall not be required on machines where all of the following provisions are met:

(a) The machine is electrically controlled but not electrically driven.

(b) The control voltage is 30 volts or less.

(c) The control or signal circuits are current limited as specified in Chapter 9, Tables 11(A) and 11(B).

675.13 Methods of Grounding

Machines that require grounding shall have a non–current-carrying equipment grounding conductor provided as an integral part of each cord, cable, or raceway. This grounding conductor shall be sized not less than the largest supply conductor in each cord, cable, or raceway. Feeder circuits supplying power to irrigation machines shall have an equipment grounding conductor sized according to Table 250.122.

675.14 Bonding

Where electrical grounding is required on an irrigation machine, the metallic structure of the machine, metallic conduit, or metallic sheath of cable shall be bonded to the grounding conductor. Metal-to-metal contact with a part that is bonded to the grounding conductor and the non–current-carrying parts of the machine shall be considered as an acceptable bonding path.

675.15 Lightning Protection

If an irrigation machine has a stationary point, a grounding electrode system in accordance with Article 250, Part III, shall be connected to the machine at the stationary point for lightning protection.

Where the electrical power supply to irrigation machine equipment is a service, the requirements of Article 250 for grounding the system and equipment are applicable. Due to the physical location of irrigation equipment, the most likely grounding electrode of the types covered in 250.52 is a driven ground rod or ground plate. Consideration should be given to the requirements of 250.60 and NFPA 780, *Standard for the Installation of Lightning Protection Systems*, in areas where lightning protection is critical. A common electrode system is not permitted to be used for the dual function of grounding the electric service and grounding the lightning protection system. The separate electrode systems are required to be bonded together.

675.16 Energy from More Than One Source

Equipment within an enclosure receiving electrical energy from more than one source shall not be required to have a disconnecting means for the additional source, provided that its voltage is 30 volts or less and it meets the requirements of Part III of Article 725.

675.17 Connectors

External plugs and connectors on the equipment shall be of the weatherproof type.

Unless provided solely for the connection of circuits meeting the requirements of Part III of Article 725, external plugs and connectors shall be constructed as specified in 250.124(A).

II. Center Pivot Irrigation Machines

675.21 General

The provisions of Part II are intended to cover additional special requirements that are peculiar to center pivot irrigation machines. See 675.2 for the definition of *Center Pivot Irrigation Machine*.

675.22 Equivalent Current Ratings

To establish ratings of controllers, disconnecting means, conductors, and the like, for the inherent intermittent duty of center pivot irrigation machines, the determinations in 675.22(A) and 675.22(B) shall be used.

The ratings of electrical components of any circuit should be selected so as to avoid extensive damage to the equipment during a short circuit or ground fault. Requirements for establishing ratings of components of special equipment for inherent intermittent duty are covered in 675.22. Also see the commentary following 110.10 and 430.52.

(A) Continuous-Current Rating The equivalent continuous-current rating for the selection of branch-circuit conductors and branch-circuit devices shall be equal to 125 percent of the motor nameplate full-load current rating of the largest motor plus 60 percent of the sum of the motor nameplate full-load current ratings of all remaining motors on the circuit.

(B) Locked-Rotor Current The equivalent locked-rotor current rating shall be equal to the numerical sum of two times the locked-rotor current of the largest motor plus 80 percent of the sum of the motor nameplate full-load current ratings of all the remaining motors on the circuit.

ARTICLE 680
Swimming Pools, Fountains, and Similar Installations

Summary of Changes

- **680.8:** Added new requirement that specified clearances be measured from the maximum water level.
- **680.12:** Revised to require that the maintenance disconnecting means be "readily" accessible.
- **680.21(A)(1):** Revised to require that the branch circuit equipment grounding conductor for swimming pool pump motors be an insulated conductor.
- **680.21(A)(4):** Revised to correlate with the change in 680.21(A) requiring insulated equipment grounding conductor for all wiring methods other than cable assemblies installed in the interior of a one-family dwelling.
- **680.23(B)(6):** Added new requirement that luminaires be removable from the water for relamping or normal maintenance and that they be installed so that relamping, maintenance, or inspection can be accomplished from the deck or equivalently dry location.
- **680.23(E):** Revised to clarify the purpose of the nonmetallic hub.
- **680.23(F)(1):** Revised to permit Type MC cable where installed within building as a branch circuit wiring method for underwater luminaires.
- **680.24(A)(1)(1) and (B)(1)(1):** Revised to delete the text "listed for the purpose" associated with nonmetallic hubs.
- **680.25(B)(2):** Revised to require that swimming pool feeder circuits comply with 250.32(B)(1) where run to a separate building or structure, and prohibits option of connecting the grounded conductor to a grounding electrode system at the second building or structure.
- **680.26(B)(1):** Revised to clarify the conditions under which an alternative means for eliminating voltage gradient is required.
- **680.26(C):** Changed title to "Equipotential" Bonding Grid. Conduit made of brass or other corrosion-resistant

metal is permitted as a method for connecting metal parts of equipment to the equipotential bonding grid. The bonding grid is now required to extend at least 3 ft horizontally under paved walking surfaces. Subparagraph (3) revised to permit alternate means of creating equipotential bonding grid.

- **680.32:** Added new text to provide GFCI protection requirement for all 125-volt receptacles located within 20 ft of a storable swimming pool and provides method by which to determine the proximity of receptacle outlets to a storable pool.
- **680.34:** Added new requirement prohibiting all receptacles from within 10 ft of the storable pool, and provides method by which to determine the proximity of receptacle outlets to a storable pool.
- **680.51(A):** Revised to specify types of equipment required to be GFCI protected.
- **680.57(A):** Revised to clarify that section applies only to signs in a fountain or within 10 ft of a fountain edge.
- **680.58:** Added new requirement for GFCI protection of 125 and 250 volt, 15- and 20-ampere receptacles located within 20 ft of a fountain.
- **680.74:** Revised to delete portion of bonding requirement addressing double-insulated equipment.

Contents

I. General

680.1 Scope

The provisions of this article apply to the construction and installation of electrical wiring for and equipment in or adjacent to all swimming, wading, therapeutic, and decorative pools; fountains; hot tubs; spas; and hydromassage bathtubs, whether permanently installed or storable, and to metallic auxiliary equipment, such as pumps, filters, and similar equipment. The term *body of water* used throughout Part I applies to all bodies of water covered in this scope unless otherwise amended.

Article 680 applies to decorative pools and fountains; swimming, wading, and wave pools; therapeutic tubs and tanks; hot tubs; spas; hydromassage bathtubs; and similar installations. The installations covered by this article can be indoors or outdoors, permanent or storable, and may or may not be directly supplied by electrical circuits of any nature. Requirements for natural and artificially made bodies of water not covered by Article 680 are contained in Article 682, new for the *Code*.

Studies conducted by Underwriters Laboratories, various manufacturers, and others indicate that a person in a swimming pool can receive a severe electric shock by reaching out and touching the energized casing of a faulty appliance — such as a radio or a hair dryer — as the person's body, which due to immersion has a lower resistance to electric current, establishes a conductive path through the water to earth. Also, a person not in contact with a faulty appliance or any grounded object can receive an electric shock and be rendered immobile by a potential gradient in the water itself. This level of electrical current necessary to cause immobilization may not be fatal from an electrical shock perspective but is dangerous from an accidental drowning standpoint. Shock hazards in and around a swimming pool can result from faulty electrical equipment directly associated with the pool or from faulty electrical equipment not associated with but in close proximity to the pool.

Accordingly, the requirements of Article 680 covering effective bonding and grounding, installation of receptacles and luminaires, use of GFCIs, modified wiring methods, and so on, apply not only to the installation of the pool but also to installations and equipment adjacent to or associated with the pool.

Enhanced electric shock protection in this wet environment, where it is intended that people are immersed in bodies of water that also contain electrical equipment, is provided through one or more of the following means:

- GFCI protection and low-voltage equipment
- Double-insulated equipment
- Insulation and isolation
- Equipotenial bonding
- Physical separation and restricted locations
- Robust physical protection requirements for circuit conductors

680.2 Definitions

Cord-and-Plug-Connected Lighting Assembly. A lighting assembly consisting of a luminaire (lighting fixture) intended for installation in the wall of a spa, hot tub, or storable pool, and a cord-and-plug-connected transformer.

Dry-Niche Luminaire (Lighting Fixture). A luminaire (lighting fixture) intended for installation in the wall of a pool or fountain in a niche that is sealed against the entry of pool water.

Equipment, Fixed. Equipment that is fastened or otherwise secured at a specific location.

Equipment, Portable. Equipment that is actually moved or can easily be moved from one place to another in normal use.

Equipment, Stationary. Equipment that is not easily moved from one place to another in normal use.

Forming Shell. A structure designed to support a wet-niche luminaire (lighting fixture) assembly and intended for mounting in a pool or fountain structure.

Fountain. Fountains, ornamental pools, display pools, and reflection pools. The definition does not include drinking fountains.

Hydromassage Bathtub. A permanently installed bathtub equipped with a recirculating piping system, pump, and associated equipment. It is designed so it can accept, circulate, and discharge water upon each use.

See the commentary following 680.71, 680.73, and 680.74.

Maximum Water Level. The highest level that water can reach before it spills out.

No-Niche Luminaire (Lighting Fixture). A luminaire (lighting fixture) intended for installation above or below the water without a niche.

Packaged Spa or Hot Tub Equipment Assembly. A factory-fabricated unit consisting of water-circulating, heating, and control equipment mounted on a common base, intended to operate a spa or hot tub. Equipment can include pumps, air blowers, heaters, lights, controls, sanitizer generators, and so forth.

The definition of *packaged spa or hot tub equipment assembly* clarifies which assemblies are subject to the requirements of 680.44.

Packaged Therapeutic Tub or Hydrotherapeutic Tank Equipment Assembly. A factory-fabricated unit consisting of water-circulating, heating, and control equipment mounted on a common base, intended to operate a therapeutic tub or hydrotherapeutic tank. Equipment can include pumps, air blowers, heaters, lights, controls, sanitizer generators, and so forth.

Permanently Installed Decorative Fountains and Reflection Pools. Those that are constructed in the ground, on the ground, or in a building in such a manner that the fountain cannot be readily disassembled for storage, whether or not served by electrical circuits of any nature. These units are primarily constructed for their aesthetic value and are not intended for swimming or wading.

Permanently Installed Swimming, Wading, and Therapeutic Pools. Those that are constructed in the ground or partially in the ground, and all others capable of holding water in a depth greater than 1.0 m (42 in.), and all pools installed inside of a building, regardless of water depth, whether or not served by electrical circuits of any nature.

See the commentary following Part VI, Pools and Tubs for Therapeutic Use.

Pool. Manufactured or field-constructed equipment designed to contain water on a permanent or semipermanent basis and used for swimming, wading, or other purposes.

Pool Cover, Electrically Operated. Motor-driven equipment designed to cover and uncover the water surface of a pool by means of a flexible sheet or rigid frame.

The requirements for electrically operated pool covers are found in 680.27(B).

Self-Contained Spa or Hot Tub. Factory-fabricated unit consisting of a spa or hot tub vessel with all water-circulating, heating, and control equipment integral to the unit. Equipment can include pumps, air blowers, heaters, lights, controls, sanitizer generators, and so forth.

The definition of *self-contained spa or hot tub* clarifies which assemblies are subject to the requirements of 680.44.

Self-Contained Therapeutic Tubs or Hydrotherapeutic Tanks. A factory-fabricated unit consisting of a therapeutic tub or hydrotherapeutic tank with all water-circulating, heating, and control equipment integral to the unit. Equipment may include pumps, air blowers, heaters, light controls, sanitizer generators, and so forth.

Spa or Hot Tub. A hydromassage pool, or tub for recreational or therapeutic use, not located in health care facilities, designed for immersion of users, and usually having a filter, heater, and motor-driven blower. It may be installed indoors or outdoors, on the ground or supporting structure, or in the ground or supporting structure. Generally, a spa or hot tub is not designed or intended to have its contents drained or discharged after each use.

See the commentary following 680.41, 680.42, 680.43, and 680.44.

Storable Swimming or Wading Pool. Those that are constructed on or above the ground and are capable of holding water to a maximum depth of 1.0 m (42 in.), or a pool with nonmetallic, molded polymeric walls or inflatable fabric walls regardless of dimension.

See the commentary following 680.30 and Exhibit 680.16.

Originally, storable pools were not specifically addressed in the *NEC*. Article 680 was written to provide guidance relative to permanent, in-ground pools and their unique construction requirements because of the unusual earth-water-electricity-human body environment created in the finished product. The conductivity of moist concrete or metal walls buried in the ground, the incorporation of large masses of reinforcing steel, and the inclusion of stainless-steel handrails and diving-board stands, as well as 120-volt lights in the pool structure, all called for the strict wiring, bonding, and grounding requirements of Article 680.

Storable pools, on the other hand, are intended to be temporary structures, without the need for special wiring or modification to the pool site. They are usually sold as a complete package, consisting of the pool walls, vinyl liner, plumbing kit, and pump/filter device. A storable pool is often disassembled and stored during the winter months. Regional preferences, weather patterns, economic considerations, and design characteristics of the pool are all factors influencing this action. The original Article 680 definition of a storable pool was "One that is so constructed that it may be readily disassembled for storage and reassembled to its original integrity."

Part III of Article 680 was created to address the special equipment specifications of storable pools, and Underwriters Laboratories developed testing and labeling criteria for listing the pump/filter units designed especially for these pools. This equipment has the following characteristics:

1. It must have an approved system of double insulation or the equivalent.
2. It is permitted to have a flexible cord equipped with a parallel-blade, grounding-type attachment plug for electrical connection.
3. It must have a grounding conductor included in the flexible cord.
4. The flexible cord is not limited to 3 ft, as required in 680.7, and is specified by UL to be not less than 25 ft long. This length was chosen to discourage the use of extension cords.

The UL labeling requirement for these listed units includes the wording "Do Not Use with Permanently Installed Pools." In some cases, consumers and swimming pool installers, however, have found it desirable to use these pump/ filter units on any aboveground or on-ground pool, regardless of the pool's dimensions or "storability."

Storable pools are supplied as two distinct types. One type is intended to be disassembled at the end of each swimming season. The second type, by the nature of its construction, can be disassembled, but manufacturers recommend leaving it assembled. The pools in the latter category frequently require special modification to and preparation of the pool site, making them impractical to disassemble. Draining these pools, especially the larger ones, increases the likelihood of costly damage caused by shrinkage of the vinyl liner material.

The main factor differentiating the two types of pools is wall height. Generally, pools, other than the inflatable type, intended to be disassembled at season's end have wall heights of 42 in. or less, while those not intended for disassembly have wall heights of 48 in. or more. The surface area of the pools is not a factor. Inflatable pools are treated as storable pools regardless of their wall height.

Through-Wall Lighting Assembly. A lighting assembly intended for installation above grade, on or through the

wall of a pool, consisting of two interconnected groups of components separated by the pool wall.

Wet-Niche Luminaire (Lighting Fixture). A luminaire (lighting fixture) intended for installation in a forming shell mounted in a pool or fountain structure where the luminaire (fixture) will be completely surrounded by water.

680.3 Other Articles

Except as modified by this article, wiring and equipment in or adjacent to pools and fountains shall comply with other applicable provisions of this *Code*, including those provisions identified in Table 680.3.

Table 680.3 Other Articles

Topic	Section or Article
Wiring	Chapters 1–4
Junction box support	314.23
Rigid nonmetallic conduit	352.12
Audio Equipment	Article 640, Parts I and II
Adjacent to pools and fountains	640.10
Underwater speakers*	

*Underwater loudspeakers shall be installed in accordance with 680.27(A).

Note that 314.23 (E) specifies the requirements for the support of threaded boxes that do not contain devices and that 352.12 (B) does not permit luminaires or most other electrical equipment to be supported by rigid nonmetallic conduit. Exhibit 680.1 shows a properly supported junction box for

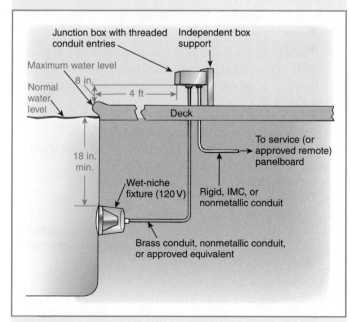

Exhibit 680.1 Wet-niche fixture installation with junction box supported above pool deck.

a wet-niche fixture. Also see the commentary following 314.23(E).

680.4 Approval of Equipment

All electrical equipment installed in the water, walls, or decks of pools, fountains, and similar installations shall comply with the provisions of this article.

680.5 Ground-Fault Circuit Interrupters

Ground-fault circuit interrupters (GFCIs) shall be self-contained units, circuit-breaker or receptacle types, or other listed types.

See the definition of *ground-fault circuit interrupter (GFCI)* in Article 100.

A GFCI is intended to be used only in a circuit that has a solidly grounded conductor; however, an equipment grounding conductor is not necessary for the GFCI to function. A Class A GFCI trips where the current to ground has a value in the range of 4 through 6 mA; it is suitable for use in swimming pool circuits. It should be noted, however, that circuits supplying pool equipment that were installed before local adoption of the 1965 edition of the *Code* may have sufficient leakage current to cause a Class A GFCI to trip. A Class B GFCI trips if the current to ground exceeds 20 mA; it is suitable for use only with underwater swimming pool lighting fixtures installed before the local adoption of the 1965 *Code*.

680.6 Grounding

Electrical equipment shall be grounded in accordance with Parts V, VI, and VII of Article 250 and connected by wiring methods of Chapter 3, except as modified by this article. The following equipment shall be grounded:

(1) Through-wall lighting assemblies and underwater luminaires (lighting fixtures), other than those low-voltage systems listed for the application without a grounding conductor
(2) All electrical equipment located within 1.5 m (5 ft) of the inside wall of the specified body of water
(3) All electrical equipment associated with the recirculating system of the specified body of water
(4) Junction boxes
(5) Transformer enclosures
(6) Ground-fault circuit interrupters
(7) Panelboards that are not part of the service equipment and that supply any electrical equipment associated with the specified body of water

Electrical equipment other than underwater lighting fixtures and pool-associated motors is required to be connected by the wiring methods of Chapter 3 and grounded in accordance

with Article 250. For example, an outdoor receptacle installed to meet the requirements of 680.22(A)(3) is permitted to be wired with Type UF cable containing an insulated or bare conductor for equipment grounding purposes. Circuits for pools may be derived from an existing remote panelboard supplied by an approved cable assembly, as specified in 680.25(A), Exception. The requirements of 680.6 permit Type UF cable to be used for the receptacle required by 680.22(A)(3) and for some pool-related equipment, but circuit conductors for underwater lighting fixtures are required to be run in raceways. Circuit conductors for pool-associated motors other than flexible cord, as permitted by 680.7, are required to be installed in raceways except in the interior of one-family dwelling units, where any raceway or cable assembly permitted by Chapter 3 is acceptable if the equipment grounding conductor is at least 12 AWG copper and is enclosed by the wiring method.

Equipment grounding requirements are contained in 680.6, 680.21(A)(1), 680.23(F)(2), and 680.25(B). These requirements specify that equipment grounding conductors be connected to non–current-carrying metal parts of the specified equipment. These equipment grounding conductors are required to be run with the circuit conductors in rigid metal conduit, intermediate conduit, listed MC cable (for motors only), or rigid nonmetallic conduit (electrical metallic tubing is permitted in or on buildings, and electrical nonmetallic tubing is permitted inside buildings), and they must be terminated at the grounding terminal bus of the service panelboard, the source of the separately derived system, or the subpanel. This equipment grounding conductor provides a path of low impedance that limits the voltage to ground and facilitates operation of the circuit overcurrent protective device(s). The equipment grounding conductor is required to be an insulated copper conductor not smaller than 12 AWG.

The requirements of 680.6, 680.21(A)(1), 680.23(F)(2), and 680.25 (B) are in addition to the bonding requirements in 680.26. The intent of the bonding requirements is to establish an equipotential plane to limit the voltage between all non–current-carrying parts of electrical and nonelectrical equipment in the pool area.

Bonding conductors may be insulated, covered, or bare and are required to be 8 AWG solid copper or larger. They may be direct buried, and, if connected to metal parts of the pool structure or metal parts of electrical equipment, they may be externally clamped or attached and are not required to be accessible. All these parts form a common bonding grid that establishes an equipotential grounding system, and they do not have to be run to the equipment grounding terminals of panelboards or service equipment.

680.7 Cord-and-Plug-Connected Equipment

Fixed or stationary equipment other than an underwater luminaire (lighting fixture) for a permanently installed pool shall be permitted to be connected with a flexible cord to facilitate the removal or disconnection for maintenance or repair.

(A) Length For other than storable pools, the flexible cord shall not exceed 900 mm (3 ft) in length.

(B) Equipment Grounding The flexible cord shall have a copper equipment grounding conductor sized in accordance with 250.122 but not smaller than 12 AWG. The cord shall terminate in a grounding-type attachment plug.

(C) Construction The equipment grounding conductors shall be connected to a fixed metal part of the assembly. The removable part shall be mounted on or bonded to the fixed metal part.

In some climates, it is preferable to disconnect and remove a permanent pool's filter pump during cold-weather months. A 3-ft cord is permitted, to facilitate the removal of fixed or stationary equipment for maintenance and storage. The 3-ft cord limitation does not apply to cord-and-plug-connected filter pumps used with storable-type pools (covered in Part III of Article 680), since these pumps are neither fixed nor stationary. Listed filter pumps for use with storable pools are considered portable and are permitted to be equipped with cords longer than 3 ft.

680.8 Overhead Conductor Clearances

Overhead conductors shall meet the clearance requirements in this section. Where a minimum clearance from the water level is given, the measurement shall be taken from the maximum water level of the specified body of water.

(A) Power With respect to service drop conductors and open overhead wiring, swimming pool and similar installations shall comply with the minimum clearances given in Table 680.8 and illustrated in Figure 680.8.

Figure 680.8 Clearances from Pool Structures.

FPN: Open overhead wiring as used in this article typically refers to conductor(s) not in an enclosed raceway.

(B) Communications Systems Communication, radio, and television coaxial cables within the scope of Articles 800 through 820 shall be permitted at a height of not less than 3.0 m (10 ft) above swimming and wading pools, diving structures, and observation stands, towers, or platforms.

(C) Network-Powered Broadband Communications Systems The minimum clearances for overhead network-

Table 680.8 Overhead Conductor Clearances

	Clearance Parameters	Insulated Cables, 0–750 Volts to Ground, Supported on and Cabled Together with an Effectively Grounded Bare Messenger or Effectively Grounded Neutral Conductor		All Other Conductors Voltage to Ground			
				0 through 15 kV		Over 15 through 50 kV	
		m	ft	m	ft	m	ft
A.	Clearance in any direction to the water level, edge of water surface, base of diving platform, or permanently anchored raft	6.9	22.5	7.5	25	8.0	27
B.	Clearance in any direction to the observation stand, tower, or diving platform	4.4	14.5	5.2	17	5.5	18
C.	Horizontal limit of clearance measured from inside wall of the pool	This limit shall extend to the outer edge of the structures listed in A and B of this table but not to less than 3 m (10 ft).					

powered broadband communications systems conductors from pools or fountains shall comply with the provisions in Table 680.8 for conductors operating at 0 to 750 volts to ground.

Service drop conductors, conductors of network-powered broadband communications systems, and aerial feeders and branch circuits are permitted to be located above a swimming pool and associated pool structures where provided with the clearances specified in Table 680.8. Overhead conductors of communications systems are required to comply with 680.8(B). These clearances consider such factors as the use of skimmers with aluminum handles and provide sufficient separation between the conductors and the pool. In some instances, locating a swimming pool below electric conductors is unavoidable, for example, on a building lot with limited area or an existing lot where the electric supply lines are already in place. The clearances for conductors from pools and pool structures were increased in the 1999 *Code* to harmonize the *NEC* with ANSI C2, *National Electrical Safety Code (NESC)*. The maximum water level of the body of water (pool, spa, hot tub, or other) is used to determine compliance with 680.8. For the definition of maximum water level, see 680.2.

680.9   Electric Pool Water Heaters

All electric pool water heaters shall have the heating elements subdivided into loads not exceeding 48 amperes and protected at not over 60 amperes. The ampacity of the branch-circuit conductors and the rating or setting of overcurrent protective devices shall not be less than 125 percent of the total nameplate-rated load.

680.10   Underground Wiring Location

Underground wiring shall not be permitted under the pool or within the area extending 1.5 m (5 ft) horizontally from the inside wall of the pool unless this wiring is necessary to supply pool equipment permitted by this article. Where space limitations prevent wiring from being routed a distance 1.5 m (5 ft) or more from the pool, such wiring shall be permitted where installed in rigid metal conduit, intermediate metal conduit, or a nonmetallic raceway system. All metal conduit shall be corrosion resistant and suitable for the location. The minimum burial depth shall be as given in Table 680.10.

This section allows wiring within 5 ft of the inside walls of the swimming pool under two conditions. The first condition permits wiring to pool-associated equipment such as an underwater luminaire. The second condition permits wiring not associated with the pool within this area where spatial

Table 680.10 Minimum Burial Depths

Wiring Method	Minimum Burial	
	mm	in.
Rigid metal conduit	150	6
Intermediate metal conduit	150	6
Nonmetallic raceways listed for direct burial without concrete encasement	450	18
Other approved raceways*	450	18

*Raceways approved for burial only where concrete encased shall require a concrete envelope not less than 50 mm (2 in.) thick.

constraints such as property lines preclude the 5-ft minimum separation. Under the second condition, underground wiring located within the 5-ft zone is required to be installed in rigid metal conduit, intermediate metal conduit, or rigid nonmetallic conduit and must be buried to a depth not less than that required by Table 680.10 for these permitted wiring methods. Beyond the 5-ft zone, the minimum cover requirements of Table 300.5 apply to the underground wiring methods used for circuits rated 600 volts and less.

As indicated by the title of this section ("Underground Wiring Location"), the focus of 680.10 is to mitigate shock hazards that may occur as a result of a faulty or damaged underground installation that is in close proximity to the swimming pool. Due to water splashing out of the pool and water dripping off those who have been in the pool, the area within 5 ft of the inside walls is generally the wettest location; as a result, electrical leakage from underground installations presents a greater shock hazard in this continuously wet environment.

680.11   Equipment Rooms and Pits

Electric equipment shall not be installed in rooms or pits that do not have drainage that adequately prevents water accumulation during normal operation or filter maintenance.

680.12   Maintenance Disconnecting Means

One or more means to disconnect all ungrounded conductors shall be provided for all utilization equipment other than lighting. Each means shall be readily accessible and within sight from its equipment.

A readily accessible disconnecting means is required to be located within sight of pool, spa, and hot tub equipment in order to provide service personnel with the ability to safely disconnect power while servicing equipment such as motors, heaters, and control panels. Underwater luminaires are not subject to this requirement. See Exhibit 680.2.

Exhibit 680.2 Required pool equipment disconnect. The disconnect for pool equipment must be located within sight of the pool equipment and at least 5 ft from the pool.

II. Permanently Installed Pools

680.20 General

Electrical installations at permanently installed pools shall comply with the provisions of Part I and Part II of this article.

680.21 Motors

(A) Wiring Methods

(1) General The branch circuits for pool-associated motors shall be installed in rigid metal conduit, intermediate metal conduit, rigid nonmetallic conduit, or Type MC cable listed for the location. Other wiring methods and materials shall be permitted in specific locations or applications as covered in this section. Any wiring method employed shall contain an insulated copper equipment grounding conductor sized in accordance with 250.122 but not smaller than 12 AWG.

Type MC cable listed for the application is permitted as a wiring method for swimming pool motor circuits. This listing requirement ensures that the MC cable is suitable for the environmental and installation conditions typically encountered with swimming pool pump motors. Type MC cables listed for installation in direct sunlight or direct burial are marked to indicate suitability for such applications.

Other than cable assemblies installed on the interior of a one-family dwelling per 680.21(A)(4), wiring methods used for the supply circuit to a swimming pool pump motor must include an insulated, copper equipment grounding conductor not less than 12 AWG.

(2) On or Within Buildings Where installed on or within buildings, electrical metallic tubing shall be permitted.

(3) Flexible Connections Where necessary to employ flexible connections at or adjacent to the motor, liquidtight flexible metal or nonmetallic conduit with approved fittings shall be permitted.

(4) One-Family Dwellings In the interior of one-family dwellings, or in the interior of accessory buildings associated with a one-family dwelling, any of the wiring methods recognized in Chapter 3 of this *Code* that comply with the provisions of this paragraph shall be permitted. Where run in a cable assembly, the equipment grounding conductor shall be permitted to be uninsulated, but it shall be enclosed within the outer sheath of the cable assembly.

(5) Cord-and-Plug Connections Pool-associated motors shall be permitted to employ cord-and-plug connections. The flexible cord shall not exceed 900 mm (3 ft) in length. The flexible cord shall include an equipment grounding conductor sized in accordance with 250.122 and shall terminate in a grounding-type attachment plug.

(B) Double Insulated Pool Pumps
A listed cord-and-plug-connected pool pump incorporating an approved system of double insulation that provides a means for grounding only the internal and nonaccessible, non–current-carrying metal parts of the pump shall be connected to any wiring method recognized in Chapter 3 that is suitable for the location. Where the bonding grid is connected to the equipment grounding conductor of the motor circuit in accordance with the second paragraph of 680.26(B)(4), the branch circuit wiring shall comply with 680.21(A).

Cord-and-plug-connected double-insulated swimming pool filter pumps have been used with permanently installed aboveground pools and some storable pools, regardless of the pool's size, for many years without any known field-related problems. The internal metal parts of a swimming pool filter pump incorporating a system of double insulation are grounded; however, they are not required to be incorporated into the bonding system required by 680.26(B), since the act of bonding compromises the double-insulation system.

680.22 Area Lighting, Receptacles, and Equipment

(A) Receptacles

(1) Circulation and Sanitation System, Location Receptacles that provide power for water-pump motors or for other loads directly related to the circulation and sanitation system shall be located at least 3.0 m (10 ft) from the inside walls of the pool, or not less than 1.5 m (5 ft) from the inside walls of the pool if they meet all of the following conditions:

(1) Consist of single receptacles

(2) Employ a locking configuration

(3) Are of the grounding type

(4) Have GFCI protection

(2) Other Receptacles, Location Other receptacles shall be not less than 3.0 m (10 ft) from the inside walls of a pool.

(3) Dwelling Unit(s) Where a permanently installed pool is installed at a dwelling unit(s), no fewer than one 125-volt 15- or 20-ampere receptacle on a general-purpose branch circuit shall be located not less than 3.0 m (10 ft) from, and not more than 6.0 m (20 ft) from, the inside wall of the pool. This receptacle shall be located not more than 2.0 m (6 ft 6 in.) above the floor, platform, or grade level serving the pool.

(4) Restricted Space Where a pool is within 3.0 m (10 ft) of a dwelling and the dimensions of the lot preclude meeting the required clearances, not more than one receptacle outlet shall be permitted if not less than 1.5 m (5 ft) measured horizontally from the inside wall of the pool.

(5) GFCI Protection All 15- and 20-ampere, single-phase, 125-volt receptacles located within 6.0 m (20 ft) of the inside walls of a pool shall be protected by a ground-fault circuit interrupter. Receptacles that supply pool pump motors and that are rated 15 or 20 amperes, 125 volts through 250 volts, single phase, shall be provided with GFCI protection.

All single-phase, 15- and 20-ampere, 125-volt through 250-volt receptacles that supply swimming pool pump motors are required to have GFCI protection. While this requirement applied only to installations at other than dwellings in the 1999 *Code*, the 2002 *Code* was revised to require GFCI protection of these receptacles for all swimming pool installations. It should be noted that 680.22(A)(5) applies to these receptacles regardless of their proximity to the swimming pool and that it applies only to cord-and-plug-connected pump motors.

(6) Measurements In determining the dimensions in this section addressing receptacle spacings, the distance to be measured shall be the shortest path the supply cord of an appliance connected to the receptacle would follow without piercing a floor, wall, ceiling, doorway with hinged or sliding door, window opening, or other effective permanent barrier.

The requirements of 680.22(A) apply to receptacles located near a permanently installed pool or fountain. They do not apply to direct-connected equipment. Permission is given in 680.22(A)(1) to allow a single locking- and grounding-type receptacle to supply a recirculation pump motor where the receptacle is located not less than 5 ft from the inside walls of the pool or fountain and is protected by a GFCI.

As required by 680.22(A)(3), each permanently installed pool in a residential setting is required to have at least one receptacle, which must be located at least 10 ft from the pool and not more than 20 ft from the pool. The intent of this requirement is to permit ordinary appliances to be safely plugged in and used near the pool but to avoid the need for extension cords in the vicinity of the pool. The 10-ft minimum dimension was chosen so that an appliance with a 6-ft cord could not be accidentally knocked into the pool.

The provision of 680.22(A)(5) covers receptacle outlet installation at dwelling units where the spatial constraints prevent locating the required receptacle 10 ft or more from the inside walls of the pool. Where this condition exists, one GFCI-protected receptacle is permitted to be located closer than 10 ft but not less than 5 ft from the inside walls of the pool.

GFCI protection of all 125-volt receptacles located within 20 ft of a pool or fountain is required by 680.22(A)(5). This rule applies to pools located outdoors or indoors, permanently installed or storable, and for residential or commercial use. Since people within 20 ft of a pool are normally subjected to dampness and moisture, the GFCI requirement within the 20-ft space is warranted.

Examples of receptacles meeting the requirements of 680.22 (A) are shown in Exhibits 680.3 and 680.4. Exhibit 680.5 illustrates that the determination of the minimum distance for receptacles from a pool does not include receptacles within a structure. The receptacles within the structure are permitted to be less than 10 ft from the pool. Where this installation is at a dwelling unit, it is necessary to provide at least one receptacle between 10 ft and 20 ft from the inside walls of the pool. This location precludes having to run the cord of an appliance used on the pool deck through a doorway.

(B) Luminaires (Lighting Fixtures), Lighting Outlets, and Ceiling-Suspended (Paddle) Fans

(1) New Outdoor Installation Clearances In outdoor pool areas, luminaires (lighting fixtures), lighting outlets, and ceiling-suspended (paddle) fans installed above the pool or the area extending 1.5 m (5 ft) horizontally from the inside walls of the pool shall be installed at a height not less than 3.7 m (12 ft) above the maximum water level of the pool.

(2) Indoor Clearances For installations in indoor pool areas, the clearances shall be the same as for outdoor areas unless modified as provided in this paragraph. If the branch circuit supplying the equipment is protected by a ground-fault circuit interrupter, the following equipment shall be permitted at a height not less than 2.3 m (7 ft 6 in.) above the maximum pool water level:

Exhibit 680.3 An example of a receptacle installed according to 680.22(A). For permanently installed pools at a dwelling unit(s), it is mandatory to install a 125-volt receptacle between 10 ft and 20 ft from the inside wall of the pool.

Locking- and grounding-type single receptacle to supply pool pump motor

Required 125 V, 15 A or 20 A for pools installed at dwellings

GFCI required

5 ft

10 ft

20 ft

(1) Totally enclosed luminaires (fixtures)
(2) Ceiling-suspended (paddle) fans identified for use beneath ceiling structures such as provided on porches or patios

(3) Existing Installations Existing luminaires (lighting fixtures) and lighting outlets located less than 1.5 m (5 ft) measured horizontally from the inside walls of a pool shall be not less than 1.5 m (5 ft) above the surface of the maximum water level, shall be rigidly attached to the existing structure, and shall be protected by a ground-fault circuit interrupter.

(4) GFCI Protection in Adjacent Areas Luminaires (lighting fixtures), lighting outlets, and ceiling-suspended (paddle) fans installed in the area extending between 1.5 m (5 ft) and 3.0 m (10 ft) horizontally from the inside walls of a pool shall be protected by a ground-fault circuit interrupter unless installed not less than 1.5 m (5 ft) above the maximum water level and rigidly attached to the structure adjacent to or enclosing the pool.

(5) Cord-and-Plug-Connected Luminaires (Lighting Fixtures) Cord-and-plug-connected luminaires (lighting fixtures) shall comply with the requirements of 680.7 where installed within 4.9 m (16 ft) of any point on the water surface, measured radially.

See Exhibit 680.6 for diagrams that clarify the limitations applicable to certain zones surrounding outdoor and indoor pools.

(C) Switching Devices Switching devices shall be located at least 1.5 m (5 ft) horizontally from the inside walls of a

pool unless separated from the pool by a solid fence, wall, or other permanent barrier. Alternatively, a switch that is listed as being acceptable for use within 1.5 m (5 ft) shall be permitted.

Panelboards containing circuit breakers, time clocks, pool light switches, and similar switching devices, where located not less than 5 ft horizontally from the inside walls of a pool without a solid fence, wall, or other permanent barrier, must be out of reach of persons who are in the pool, thereby preventing contact and possible shock hazards.

680.23 Underwater Luminaires (Lighting Fixtures)

This section covers all luminaires (lighting fixtures) installed below the normal water level of the pool.

(A) General

(1) Luminaire (Fixture) Design, Normal Operation The design of an underwater luminaire (lighting fixture) supplied from a branch circuit either directly or by way of a transformer meeting the requirements of this section shall be such that, where the luminaire (fixture) is properly installed without a ground-fault circuit interrupter, there is no shock hazard with any likely combination of fault conditions during normal use (not relamping).

Dry-niche, no-niche, or wet-niche underwater luminaires operating at more than 15 volts require GFCI protection. See the commentary following 680.5.
Branch-circuit conductors for dry-niche fixtures are re-

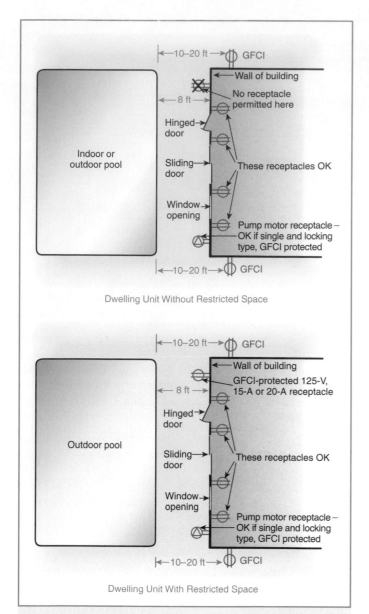

Exhibit 680.4 Acceptable receptacle locations within 20 ft of a permanently installed swimming pool.

Exhibit 680.5 Permitted receptacle location less than 10 ft from the inside wall of a permanently installed pool. The minimum distance required by 680.22(A) does not apply to a receptacle located in a structure.

quired to be installed in approved rigid metal conduit, intermediate metal conduit, or rigid nonmetallic conduit from the fixture to a panelboard or the service equipment. Branch-circuit conductors for wet-niche fixtures leaving the pool junction box are required to be enclosed in rigid metal conduit, intermediate metal conduit, liquidtight flexible nonmetallic conduit, or rigid nonmetallic conduit, except where located in or on buildings, where the conductors are permitted to be installed in electrical metallic tubing or electrical nonmetallic tubing. Unlike wet-niche fixtures, a junction box is not required for dry-niche fixtures. If one is used, it is not required to be elevated or located as specified in 680.24(A)(2). (See Exhibit 680.1 and Exhibit 680.7.)

(2) **Transformers** Transformers used for the supply of underwater luminaires (fixtures), together with the transformer enclosure, shall be listed as a swimming pool and spa transformer. The transformer shall be an isolated winding type with an ungrounded secondary that has a grounded metal barrier between the primary and secondary windings.

Unless marked otherwise, UL-listed swimming pool and spa transformers are not suitable for connection to a conduit that extends directly to an underwater pool light forming shell. Swimming pool and spa transformers are not permitted to be used outdoors unless marked "For Outdoor Use" or in an equivalent manner that signifies that they have been found acceptable for both outdoor and indoor use. See 110.3(B).

(3) **GFCI Protection, Relamping** A ground-fault circuit interrupter shall be installed in the branch circuit supplying luminaires (fixtures) operating at more than 15 volts such that there is no shock hazard during relamping. The installation of the ground-fault circuit interrupter shall be such that there is no shock hazard with any likely fault-condition combination that involves a person in a conductive path from any ungrounded part of the branch circuit or the luminaire (fixture) to ground.

(4) **Voltage Limitation** No luminaires (lighting fixtures) shall be installed for operation on supply circuits over 150 volts between conductors.

(5) **Location, Wall-Mounted Luminaires (Fixtures)** Luminaires (lighting fixtures) mounted in walls shall be

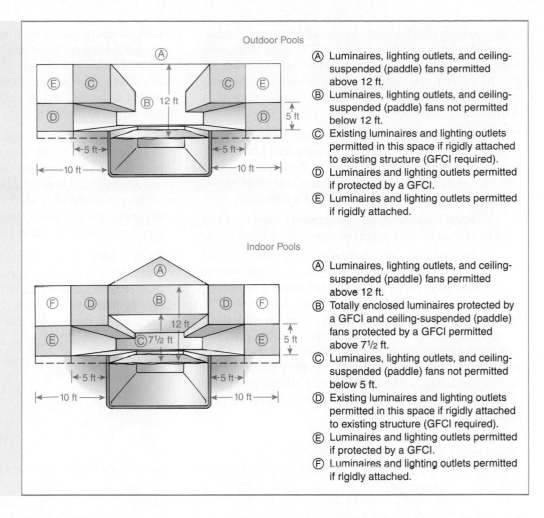

Exhibit 680.6 Limitations that apply to the placement of luminaires, lighting outlets, and ceiling-suspended fans in the area surrounding outdoor and indoor pools.

Outdoor Pools

Ⓐ 12 ft

5 ft
←5 ft→   ←5 ft→
|←— 10 ft —→| |←— 10 ft —→|

Ⓐ Luminaires, lighting outlets, and ceiling-suspended (paddle) fans permitted above 12 ft.
Ⓑ Luminaires, lighting outlets, and ceiling-suspended (paddle) fans not permitted below 12 ft.
Ⓒ Existing luminaires and lighting outlets permitted in this space if rigidly attached to existing structure (GFCI required).
Ⓓ Luminaires and lighting outlets permitted if protected by a GFCI.
Ⓔ Luminaires and lighting outlets permitted if rigidly attached.

Indoor Pools

Ⓐ

12 ft
Ⓒ 7½ ft

5 ft
|←5 ft→| |←5 ft→|
|←— 10 ft —→| |←— 10 ft —→|

Ⓐ Luminaires, lighting outlets, and ceiling-suspended (paddle) fans permitted above 12 ft.
Ⓑ Totally enclosed luminaires protected by a GFCI and ceiling-suspended (paddle) fans protected by a GFCI permitted above 7½ ft.
Ⓒ Luminaires, lighting outlets, and ceiling-suspended (paddle) fans not permitted below 5 ft.
Ⓓ Existing luminaires and lighting outlets permitted in this space if rigidly attached to existing structure (GFCI required).
Ⓔ Luminaires and lighting outlets permitted if protected by a GFCI.
Ⓕ Luminaires and lighting outlets permitted if rigidly attached.

installed with the top of the luminaire (fixture) lens not less than 450 mm (18 in.) below the normal water level of the pool, unless the luminaire (lighting fixture) is listed and identified for use at lesser depths. No luminaire (fixture) shall be installed less than 100 mm (4 in.) below the normal water level of the pool.

The reason for the 18-in. minimum submergence requirement is to reduce the likelihood that a person in the water and hanging onto the side of the pool directly over the fixture will have his or her chest in line with the fixture. This section covers fixtures that have been investigated and found acceptable for use where a person's chest may be directly in front of the fixture. The highest level of leakage current in the pool coming from a wet-niche fixture with a broken lens and bulb is found directly in front of the fixture.

(6) Bottom-Mounted Luminaires (Fixtures) A luminaire (lighting fixture) facing upward shall have the lens adequately guarded to prevent contact by any person.

(7) Dependence on Submersion Luminaires (fixtures) that depend on submersion for safe operation shall be inherently

protected against the hazards of overheating when not submerged.

Fixtures that depend on submersion for safe operation are required to be inherently protected against the hazards of overheating when not submerged, for example, during a relamping process. Protection against overheating is required to be built into a fixture or to be a part of it. A remotely located low-water cutoff switch does not provide the intended protection.

(8) Compliance Compliance with these requirements shall be obtained by the use of a listed underwater luminaire (lighting fixture) and by installation of a listed ground-fault circuit interrupter in the branch circuit or a listed transformer for luminaires (fixtures) operating at not more than 15 volts.

(B) Wet-Niche Luminaires (Fixtures)

(1) Forming Shells Forming shells shall be installed for the mounting of all wet-niche underwater luminaires (fixtures) and shall be equipped with provisions for conduit entries. Metal parts of the luminaire (fixture) and forming shell in contact with the pool water shall be of brass or other approved corrosion-resistant metal. All forming shells used with nonmetallic conduit systems, other than those that are part of a listed low-voltage lighting system not requiring grounding, shall include provisions for terminating an 8 AWG copper conductor.

(2) Wiring Extending Directly to the Forming Shell Conduit shall be installed from the forming shell to a junction box or other enclosure conforming to the requirements in 680.24. Conduit shall be rigid metal, intermediate metal, liquidtight flexible nonmetallic, or rigid nonmetallic.

(a) Metal Conduit. Metal conduit shall be approved and shall be of brass or other approved corrosion-resistant metal.

(b) Nonmetallic Conduit. Where a nonmetallic conduit is used, an 8 AWG insulated solid or stranded copper bonding jumper shall be installed in this conduit unless a listed low-voltage lighting system not requiring grounding is used. The bonding jumper shall be terminated in the forming shell, junction box or transformer enclosure, or ground-fault circuit-interrupter enclosure. The termination of the 8 AWG bonding jumper in the forming shell shall be covered with, or encapsulated in, a listed potting compound to protect the connection from the possible deteriorating effect of pool water.

Where rigid nonmetallic conduit or liquidtight flexible nonmetallic conduit is used between a forming shell for a wet-niche fixture and a junction box or other enclosure, an 8 AWG insulated copper bonding jumper is required to be installed in the conduit to provide electrical continuity between the forming shell and the junction box or other enclosure. The conduit must be sized large enough to enclose both the 8 AWG insulated copper bonding jumper and the approved flexible cord that supplies the wet-niche fixture, to facilitate easy withdrawal and insertion of the bonding conductor and the cord. Low-voltage lighting systems are exempt from this equipment grounding conductor requirement.

(3) Equipment Grounding Provisions for Cords Wet-niche luminaires (lighting fixtures) that are supplied by a flexible cord or cable shall have all exposed non–current-carrying metal parts grounded by an insulated copper equipment grounding conductor that is an integral part of the cord or cable. This grounding conductor shall be connected to a grounding terminal in the supply junction box, transformer enclosure, or other enclosure. The grounding conductor shall not be smaller than the supply conductors and not smaller than 16 AWG.

(4) Luminaire (Fixture) Grounding Terminations The end of the flexible-cord jacket and the flexible-cord conductor terminations within a luminaire (fixture) shall be covered with, or encapsulated in, a suitable potting compound to prevent the entry of water into the luminaire (fixture) through the cord or its conductors. In addition, the grounding connection within a luminaire (fixture) shall be similarly treated to protect such connection from the deteriorating effect of pool water in the event of water entry into the luminaire (fixture).

(5) Luminaire (Fixture) Bonding The luminaire (fixture) shall be bonded to and secured to the forming shell by a positive locking device that ensures a low-resistance contact and requires a tool to remove the luminaire (fixture) from the forming shell. Bonding shall not be required for luminaires (fixtures) that are listed for the application and have no non–current-carrying metal parts.

(6) Servicing All luminaires shall be removable from the water for relamping or normal maintenance. Luminaires shall be installed in such a manner that personnel can reach the luminaire for relamping, maintenance, or inspection while on the deck or equivalently dry location.

Custom swimming pool installations where the pool is incorporated as an architectural feature of a building or structure can present access problems for those who have to change the lamps of an underwater luminaire. In some cases, the length of the flexible cord connected to a wet-niche luminaire does not permit the luminaire to be removed from the pool for relamping or servicing. To address the concern over a person having to be in the pool in order to change lamps, this new requirement specifies that the underwater luminaire installation has to be made so that changing of the lamp can occur on the pool deck or other location outside of the pool.

(C) Dry-Niche Luminaires (Fixtures)

(1) Construction A dry-niche luminaire (lighting fixture) shall be provided with a provision for drainage of water and a means for accommodating one equipment grounding conductor for each conduit entry.

(2) Junction Box A junction box shall not be required but, if used, shall not be required to be elevated or located as specified in 680.24(A)(2) if the luminaire (fixture) is specifically identified for the purpose.

(D) No-Niche Luminaires (Fixtures) A no-niche luminaire (fixture) shall meet the construction requirements of 680.23(B)(3) and be installed in accordance with the requirements of 680.23(B). Where connection to a forming shell is specified, the connection shall be to the mounting bracket.

(E) Through-Wall Lighting Assembly A through-wall lighting assembly shall be equipped with a threaded entry or hub, or a nonmetallic hub, for the purpose of accommodating the termination of the supply conduit. A through-wall lighting assembly shall meet the construction requirements and not smaller than 12 AWG. The grounding terminals within the deck (junction) box are used to terminate and bond together all of conductors.

Exhibit 680.7 illustrates an installation of a forming shell for a wet-niche luminaire and a flush junction (deck) box. (See Exhibit 680.1 for surface deck boxes.)

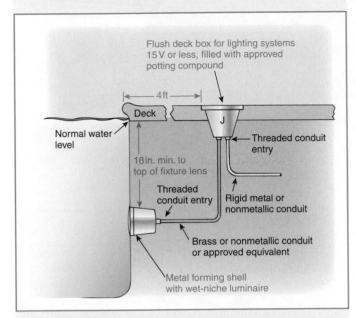

Flush deck box for lighting systems 15 V or less, filled with approved potting compound

4 ft

Deck

J

Normal water level

18 in. min. to top of fixture lens

Threaded conduit entry

Threaded conduit entry

Rigid metal or nonmetallic conduit

Brass or nonmetallic conduit or approved equivalent

Metal forming shell with wet-niche luminaire

Exhibit 680.7 A flush junction (deck) box and a forming shell for a wet-niche luminaire installed according to 680.24(A)(2).

680.25 Feeders

These provisions shall apply to any feeder on the supply side of panelboards supplying branch circuits for pool equipment covered in Part II of this article and on the load side of the service equipment or the source of a separately derived system.

(A) Wiring Methods Feeders shall be installed in rigid metal conduit, intermediate metal conduit, liquidtight flexible nonmetallic conduit, or rigid nonmetallic conduit. Electrical metallic tubing shall be permitted where installed on or within a building, and electrical nonmetallic tubing shall be permitted where installed within a building.

Exception: An existing feeder between an existing remote panelboard and service equipment shall be permitted to run in flexible metal conduit or an approved cable assembly that includes an equipment grounding conductor within its outer sheath. The equipment grounding conductor shall comply with 250.24(A)(5).

(B) Grounding An equipment grounding conductor shall be installed with the feeder conductors between the grounding terminal of the pool equipment panelboard and the

grounding terminal of the applicable service equipment or source of a separately derived system. For other than (1) existing feeders covered in 680.25(A), Exception, or (2) feeders to separate buildings that do not utilize an insulated equipment grounding conductor in accordance with 680.25(B)(2), this equipment grounding conductor shall be insulated.

(1) Size This conductor shall be sized in accordance with 250.122 but not smaller than 12 AWG. On separately derived systems, this conductor shall be sized in accordance with Table 250.66 but not smaller than 8 AWG.

(2) Separate Buildings A feeder to a separate building or structure shall be permitted to supply swimming pool equipment branch circuits, or feeders supplying swimming pool equipment branch circuits, if the grounding arrangements in the separate building meet the requirements in 250.32(B)(1). Where installed in other than existing feeders covered in 680.25(A), Exception, a separate equipment grounding conductor shall be an insulated conductor.

The insulated equipment grounding conductor can be aluminum or copper and is required to be installed in a raceway. It should be understood that for an existing remote panelboard, the 680.25(A) Exception permits an approved cable assembly with an insulated or covered aluminum or copper equipment grounding conductor. See Exhibit 680.8.

Swimming pool equipment supplied by a separately derived system is covered in 680.25(B). Where a remote panelboard supplying a pool is supplied by a separately derived system, the rules covering the grounding conductor apply only to the feeder between the separately derived system and the panelboard, not all the way back to the service, which might be high voltage.

The general rule in 680.25(B) requires an equipment grounding conductor to be installed between a panelboard serving swimming pool equipment and the service or the source of a separately derived system. Added in the 1999 *Code*, 680.25(B)(2) allows pool equipment to be supplied from a remote panelboard in a separate building where an insulated equipment grounding conductor is installed with the feeder circuit conductors run from the service (or derived system) to the panelboard and the installation complies with 250.32(B)(1). See Exhibit 680.9.

680.26 Equipotential Bonding

(A) Performance The equipotential bonding required by this section shall be installed to eliminate voltage gradients in the pool area as prescribed.

FPN: The 8 AWG or larger solid copper bonding conductor shall not be required to be extended or attached to any remote panelboard, service equipment, or any electrode.

Exhibit 680.8 An existing remote panelboard supplying new pool equipment. A raceway is not required for this application if the existing feeder wiring method contains an insulated or covered equipment grounding conductor.

Existing remote panelboard

Service

Type UF cable permitted for general purpose receptacle

Existing cable assembly or flexible metal conduit

Panelboard in separate structure

Insulated equipment grounding conductor run with circuit conductors

Grounding electrode conductor

Feeder from service

Exhibit 680.9 Grounding requirements per 680.25(B)(2) for remote panelboard and swimming pool equipment located in a structure remote from the service equipment.

It is important to understand the difference between the terms *bonding* and *grounding* as they apply to Article 680. As defined in Article 100, bonding is "the permanent joining of metallic parts to form an electrically conductive path that ensures electrical continuity and the capacity to conduct

safely any current likely to be imposed." As described in 680.26(A), the function of equipotential bonding differs from the function of bonding to meet the requirements of Article 250 in that providing a path for ground fault current is not the function of the equipotential bonding grid and associated bonding conductors.

Creating an electrically safe environment in and around permanently installed swimming pools requires the installation of a bonding system with the sole function of establishing equal electrical potential (voltage) in the vicinity of the swimming pool. A person who is immersed in a pool or who is dripping wet, has a large amount of exposed skin, and is lying or walking on a concrete deck is extremely susceptible to any differences in electrical potential that may be present in the pool area.

The primary purpose of bonding in and around swimming pools is to ensure that voltage gradients in the pool area are not present. The fine print note explains that the 8 AWG conductor's only function is equipotential bonding to eliminate the voltage gradient in the pool area.

The reason for connecting metal parts (ladders, handrails, water-circulating equipment, forming shells, diving boards, etc.) to a common bonding grid [pool reinforcing steel, pool metal wall, or an alternative bonding grid as described in 680.26(C)(3)] is to ensure that all such metal parts are at the same electrical potential. The grid reduces possible injurious or disabling shock hazards created by stray currents in the ground or piping connected to the swimming pool. Stray currents can also exist in nonmetallic piping because of the low resistivity of chlorinated water. See Exhibit 680.10.

Exhibit 680.10 Bonding in a swimming pool.

conductive compound or another conductive material is not available, provisions shall be made for an alternative means to eliminate voltage gradients that would otherwise be provided by unencapsulated, bonded reinforcing steel.

Encapsulated reinforcing steel might not provide the conductivity necessary to establish the required common bonding grid. A common bonding grid will not be formed if the steel is effectively encapsulated by a listed compound during installation and manufacturing. Therefore, a bonding connection to the encapsulated reinforcing steel is not required for this type of application. See Exhibit 680.12.

In Exhibit 680.13, the structural reinforcing steel serves as the common bonding grid to which all metal appurtenances associated with the pool are connected. Safety-rope hooks are not required to be bonded, as specified in 680.26(B)(3). The flush deck box meets the provisions of 680.24(A).

(B) Bonded Parts The parts specified in 680.26(B)(1) through (B)(5) shall be bonded together.

The metal parts required to be bonded per 680.26(B) include all metal parts of electrical equipment associated with the water-circulating system of the pool, all metal parts of the pool structure, and all fixed metal parts, which include conduit and piping, metal door frames, and metal window frames, within 5 ft of the inside walls of the pool and not separated by a permanent barrier. The bonding of these parts does not mean they are required to be connected to each other; rather, it means they are required to be connected to a common bonding grid by using an insulated, covered, or bare solid copper conductor not smaller than 8 AWG or by using brass or other corrosion-resistant metal conduit. See Exhibit 680.11. Connections are required to be made by exothermic welds or by listed pressure connectors, clamps, or other listed means, in accordance with 250.8.

(1) Metallic Structural Components All metallic parts of the pool structure, including the reinforcing metal of the pool shell, coping stones, and deck, shall be bonded. The usual steel tie wires shall be considered suitable for bonding the reinforcing steel together, and welding or special clamping shall not be required. These tie wires shall be made tight. If reinforcing steel is effectively insulated by an encapsulating nonconductive compound at the time of manufacture and installation, it shall not be required to be bonded. Where reinforcing steel of the pool shell or the reinforcing steel of coping stones and deck is encapsulated with a non-

(2) Underwater Lighting All metal forming shells and mounting brackets of no-niche luminaires (fixtures) shall be bonded unless a listed low-voltage lighting system with nonmetallic forming shells not requiring bonding is used.

(3) Metal Fittings All metal fittings within or attached to the pool structure shall be bonded. Isolated parts that are not over 100 mm (4 in.) in any dimension and do not penetrate into the pool structure more than 25 mm (1 in.) shall not require bonding.

(4) Electrical Equipment Metal parts of electrical equipment associated with the pool water circulating system, including pump motors and metal parts of equipment associated with pool covers, including electric motors, shall be bonded. Accessible metal parts of listed equipment incorporating an approved system of double insulation and providing a means for grounding internal nonaccessible, non–current-carrying metal parts shall not be bonded by a direct connection to the equipotential bonding grid. The means for grounding internal nonaccessible, non–current carrying metal parts shall be an equipment grounding conductor run with the power-supply conductors in the case of motors supplied with a flexible cord, or a grounding terminal in the case of motors intended for permanent connection.

Where a double-insulated water-pump motor is installed under the provisions of this rule, a solid 8 AWG copper conductor that is of sufficient length to make a bonding connection to a replacement motor shall be extended from the bonding grid to an accessible point in the motor vicinity. Where there is no connection between the swimming pool bonding grid and the equipment grounding system for the premises, this bonding conductor shall be connected to the equipment grounding conductor of the motor circuit.

Exhibit 680.11 A metal-perimeter (e.g., steel or aluminum) pool with bolted or welded sections. The metal perimeter serves as the common bonding grid to which the metal ladder, metal diving board, and pump motor are connected.

Pump with filter

Service (or approved remote) panelboard

Raceway with equipment grounding conductor

8 AWG bonding conductor

8 AWG bonding conductor

Bolted metal perimeter (vinyl liner)

Epoxy coating applied by reinforcing steel manufacturer

Exhibit 680.12 Epoxy-coated rebar, which does not require bonding.

(5) Metal Wiring Methods and Equipment Metal-sheathed cables and raceways, metal piping, and all fixed metal parts that are within the following distances of the pool, except those separated from the pool by a permanent barrier, shall be bonded.

Examples of fixed metal parts bound by this requirement include metal fences, metal awnings, metal door frames, and metal window frames.

(1) Within 1.5 m (5 ft) horizontally of the inside walls of the pool

(2) Within 3.7 m (12 ft) measured vertically above the maximum water level of the pool, or any observation stands, towers, or platforms, or any diving structures

(C) Equipotential Bonding Grid The parts specified in 680.26(B) shall be connected to an equipotential bonding grid with a solid copper conductor, insulated, covered, or bare, not smaller than 8 AWG or rigid metal conduit of brass or other identified corrosion-resistant metal conduit. Connection shall be made by exothermic welding or by listed pressure connectors or clamps that are labeled as being suitable for the purpose and are of stainless steel, brass, copper, or copper alloy. The equipotential common bonding grid shall extend under paved walking surfaces for 1 m (3 ft) horizontally beyond the inside walls of the pool and shall be permitted to be any of the following:

The requirements for creating the equipotential bonding grid are substantially revised in the 2005 *Code*. First, the use of an 8 AWG, solid copper conductor or brass rigid metal conduit as the equipotential bonding grid is no longer recognized. This change precludes, for example, the use of an 8 AWG solid copper conductor encircling the pool perimeter as an equipotential bonding grid. The 8 AWG conductor and/or the metal conduit can be used as the method for connecting electrical and nonelectrical equipment to the bonding grid. Exhibit 680.14 illustrates the use of brass rigid metal conduit or other corrosion-resistant metal conduit as

Exhibit 680.13 A poured-concrete pool with structural reinforcing steel that serves as the common bonding grid.

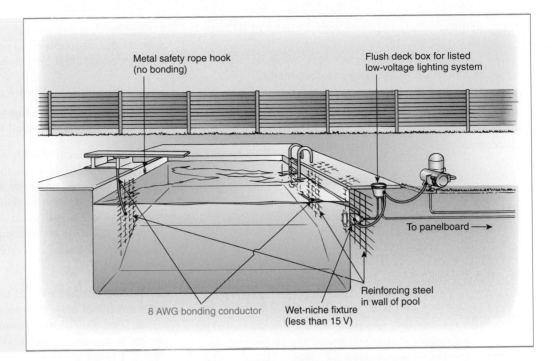

Metal safety rope hook (no bonding)

Flush deck box for listed low-voltage lighting system

To panelboard →

Reinforcing steel in wall of pool

8 AWG bonding conductor

Wet-niche fixture (less than 15 V)

a means to connect electrical equipment, such as the forming shell of a wet niche luminaire, to a common bonding grid comprised of the pool reinforcing steel.

The second change regarding the types of permitted equipotential bonding grids is the recognition of a field-fabricated bonding structure that can be employed in the absence of structural reinforcing steel (which could be a result of nonconductive encapsulation) or bolted or welded metal pool walls. This "alternate means" as described in 680.26(C)(3) is required to "cover the contour" of the pool. What that means is that the field-fabricated bonding grid has to cover the entire outside outline of the pool structure, as would be the case with reinforcing steel or with bolted or welded metal walls. In addition to covering the pool contour, the bonding grid is required to extend horizontally into the deck area for not less than 3 ft. This 3-ft horizontal extension for the pool deck is required for all pool installations, including those with exposed reinforcing steel and those with bolted or metal walls (see Exhibit 680.14).

(1) Structural Reinforcing Steel. The structural reinforcing steel of a concrete pool where the reinforcing rods are bonded together by the usual steel tie wires or the equivalent

(2) Bolted or Welded Metal Pools. The wall of a bolted or welded metal pool

(3) Alternate Means. This system shall be permitted to be constructed as specified in (a) through (c):

    a. Materials and Connections. The grid shall be constructed of minimum 8 AWG bare solid copper conductors. Conductors shall be bonded to each other

Deck box

Wet-niche fixture

Brass conduit bonded to rebar

Reinforcing steel

Exhibit 680.14 Brass rigid metal conduit used as a means to connect electrical equipment to the common bonding grid.

at all points of crossing. Connections shall be made as required by 680.26(D).

    b. Grid Structure. The equipotential bonding grid shall cover the contour of the pool and the pool deck extending 1 m (3 ft) horizontally from the inside walls of the pool. The equipotential bonding grid shall be arranged in a 300 mm (12 in.) by 300 mm (12 in.) network of conductors in a uniformly spaced perpendicular grid pattern with tolerance of 100 mm (4 in.).

c. Securing. The below-grade grid shall be secured within or under the pool and deck media.

(D) Connections Where structural reinforcing steel or the walls of bolted or welded metal pool structures are used as an equipotential bonding grid for nonelectrical parts, the connections shall be made in accordance with 250.8.

As specified in 250.8 for the grounding and bonding connections required by Article 250, exothermic welding, pressure connectors and clamps specifically listed for the purpose, and other listed means are permitted as the method of connecting swimming pool bonding conductors to a common bonding grid. Connections in pool areas must be suitable for wet conditions and high levels of chlorine. High concentrations of chlorine in swimming pool water make the wet locations in the vicinity of swimming pool areas (including many pool pump rooms) a corrosive environment. The integrity of the bonding connections should be periodically inspected, particularly those bonding connections between the 8 AWG copper conductor and, for instance, an aluminum (or other dissimilar metal) ladder.

See Exhibit 680.15 for an illustration of two acceptable methods of making swimming pool bonding connections.

Exhibit 680.15 Bonding connections in a swimming pool.

(E) Pool Water Heaters For pool water heaters rated at more than 50 amperes and having specific instructions re-

garding bonding and grounding, only those parts designated to be bonded shall be bonded and only those parts designated to be grounded shall be grounded.

680.27 Specialized Pool Equipment

(A) Underwater Audio Equipment All underwater audio equipment shall be identified for the purpose.

(1) Speakers Each speaker shall be mounted in an approved metal forming shell, the front of which is enclosed by a captive metal screen, or equivalent, that is bonded to, and secured to, the forming shell by a positive locking device that ensures a low-resistance contact and requires a tool to open for installation or servicing of the speaker. The forming shell shall be installed in a recess in the wall or floor of the pool.

(2) Wiring Methods Rigid metal conduit or intermediate metal conduit of brass or other identified corrosion-resistant metal, liquidtight flexible nonmetallic conduit (LFNC-B), or rigid nonmetallic conduit shall extend from the forming shell to a listed junction box or other enclosure as provided in 680.24. Where rigid nonmetallic conduit or liquidtight flexible nonmetallic conduit is used, an 8 AWG insulated solid or stranded copper bonding jumper shall be installed in this conduit. The bonding jumper shall be terminated in the forming shell and the junction box. The termination of the 8 AWG bonding jumper in the forming shell shall be covered with, or encapsulated in, a listed potting compound to protect such connection from the possible deteriorating effect of pool water.

(3) Forming Shell and Metal Screen The forming shell and metal screen shall be of brass or other approved corrosion-resistant metal. All forming shells shall include provisions for terminating an 8 AWG copper conductor.

(B) Electrically Operated Pool Covers

(1) Motors and Controllers The electric motors, controllers, and wiring shall be located not less than 1.5 m (5 ft) from the inside wall of the pool unless separated from the pool by a wall, cover, or other permanent barrier. Electric motors installed below grade level shall be of the totally enclosed type. The device that controls the operation of the motor for an electrically operated pool cover shall be located such that the operator has full view of the pool.

FPN No. 1: For cabinets installed in damp and wet locations, see 312.2(A).

FPN No. 2: For switches or circuit breakers installed in wet locations, see 404.4.

FPN No. 3: For protection against liquids, see 430.11.

(2) Protection The electric motor and controller shall be connected to a circuit protected by a ground-fault circuit interrupter.

(C) Deck Area Heating These provisions of this section shall apply to all pool deck areas, including a covered pool, where electrically operated comfort heating units are installed within 6.0 m (20 ft) of the inside wall of the pool.

(1) Unit Heaters Unit heaters shall be rigidly mounted to the structure and shall be of the totally enclosed or guarded types. Unit heaters shall not be mounted over the pool or within the area extending 1.5 m (5 ft) horizontally from the inside walls of a pool.

(2) Permanently Wired Radiant Heaters Radiant electric heaters shall be suitably guarded and securely fastened to their mounting device(s). Heaters shall not be installed over a pool or within the area extending 1.5 m (5 ft) horizontally from the inside walls of the pool and shall be mounted at least 3.7 m (12 ft) vertically above the pool deck unless otherwise approved.

(3) Radiant Heating Cables Not Permitted Radiant heating cables embedded in or below the deck shall not be permitted.

Only unit heaters and permanently connected radiant heaters are permitted in the area that extends 5 ft to 20 ft horizontally from the inside walls of a pool. Radiant heat cables embedded in the deck are not permitted.

III. Storable Pools

680.30 General

Electrical installations at storable pools shall comply with the provisions of Part I and Part III of this article.

Storable pools can be readily disassembled and are limited (other than inflatable type) to a maximum water depth of 42 in. Pools of any dimension with inflatable walls are considered storable. See the definition of *storable swimming or wading pool* in 680.2. This type of pool and its associated equipment do not require bonding conductors. However, the filter pump must be double insulated, and the provision of grounding means consisting of an equipment grounding conductor that is an integral part of the flexible cord also is required. There are portable filter pumps for use with storable pools listed by Underwriters Laboratories. All electrical equipment used with a storable pool is required to have GFCI protection for personnel. Exhibit 680.16 illustrates the requirements for a storable-type pool.

680.31 Pumps

A cord-connected pool filter pump shall incorporate an approved system of double insulation or its equivalent and shall be provided with means for grounding only the internal

Exhibit 680.16 The requirements for a storable-type pool. Metal appurtenances are not required to be bonded. The 3-wire cord may be longer than 3 ft (listed filter pumps are equipped with cords 25 ft long). The receptacle shown can be a GFCI-type receptacle, a receptacle supplied through a GFCI-type receptacle, or a receptacle protected by a GFCI-type circuit breaker. The water depth restriction of 42 in. does not apply to inflatable swimming pools.

and nonaccessible non–current-carrying metal parts of the appliance.

The means for grounding shall be an equipment grounding conductor run with the power-supply conductors in the flexible cord that is properly terminated in a grounding-type attachment plug having a fixed grounding contact member.

680.32 Ground-Fault Circuit Interrupters Required

All electrical equipment, including power-supply cords, used with storable pools shall be protected by ground-fault circuit interrupters.

All 125-volt receptacles located within 6.0 m (20 ft) of the inside walls of a storable pool shall be protected by a ground-fault circuit interrupter. In determining these dimensions, the distance to be measured shall be the shortest path the supply cord of an appliance connected to the receptacle would follow without piercing a floor, wall, ceiling, doorway with hinged or sliding door, window opening, or other effective permanent barrier.

FPN: For flexible cord usage, see 400.4.

680.33 Luminaires (Lighting Fixtures)

An underwater luminaire (lighting fixture), if installed, shall be installed in or on the wall of the storable pool. It shall comply with either 680.33(A) or 680.33(B).

(A) 15 Volts or Less A luminaire (lighting fixture) shall be part of a cord-and-plug-connected lighting assembly. This assembly shall be listed as an assembly for the purpose and have the following construction features:

(1) No exposed metal parts
(2) A luminaire (fixture) lamp that operates at 15 volts or less
(3) An impact-resistant polymeric lens, luminaire (fixture) body, and transformer enclosure
(4) A transformer meeting the requirements of 680.23(A)(2) with a primary rating not over 150 volts

(B) Over 15 Volts But Not Over 150 Volts A lighting assembly without a transformer and with the luminaire (fixture) lamp(s) operating at not over 150 volts shall be permitted to be cord-and-plug connected where the assembly is listed as an assembly for the purpose. The installation shall comply with 680.23(A)(5), and the assembly shall have the following construction features:

(1) No exposed metal parts
(2) An impact-resistant polymeric lens and luminaire (fixture) body
(3) A ground-fault circuit interrupter with open neutral protection as an integral part of the assembly
(4) The luminaire (fixture) lamp permanently connected to the ground-fault circuit interrupter with open-neutral protection
(5) Compliance with the requirements of 680.23(A)

This requirement permits lighting fixtures to be installed in or on storable pools. These cord-and-plug-connected fixtures are required to be listed as an assembly.

680.34 Receptacle Locations

Receptacles shall not be less than 3.0 m (10 ft) from the inside walls of a pool. In determining these dimensions, the distance to be measured shall be the shortest path the supply cord of an appliance connected to the receptacle would follow without piercing a floor, wall, ceiling, doorway with hinged or sliding door, window opening, or other effective permanent barrier.

IV. Spas and Hot Tubs

680.40 General

Electrical installations at spas and hot tubs shall comply with the provisions of Part I and Part IV of this article.

680.41 Emergency Switch for Spas and Hot Tubs

A clearly labeled emergency shutoff or control switch for the purpose of stopping the motor(s) that provide power to the recirculation system and jet system shall be installed at a point readily accessible to the users and not less than 1.5 m (5 ft) away, adjacent to, and within sight of the spa or

hot tub. This requirement shall not apply to single-family dwellings.

The provisions of 680.41 require a local disconnecting device for spas and hot tubs that is capable of being used in an emergency. This requirement was added to address entrapment hazards associated with spas and hot tubs. The definitive publication on this issue, *Guideline for Entrapment Hazards: Making Pools and Spas Safer* (Pub. No. 363), is available from the U.S. Consumer Product Safety Commission, Washington, DC 20207, or on-line at www.cpsc.gov.

The emergency shutoff switch must be installed within sight of and at least 5 ft from the spa or hot tub and must be clearly labeled "Emergency Shutoff." See Exhibit 680.17 for an illustration of the switch location. The shutoff switch can be either a line-operated device or a remote-control circuit that causes the pump circuit to open. This requirement does not apply to one-family dwellings.

680.42 Outdoor Installations

A spa or hot tub installed outdoors shall comply with the provisions of Parts I and II of this article, except as permitted in 680.42(A) and 680.42(B), that would otherwise apply to pools installed outdoors.

(A) Flexible Connections Listed packaged spa or hot tub equipment assemblies or self-contained spas or hot tubs utilizing a factory-installed or assembled control panel or panelboard shall be permitted to use flexible connections as covered in 680.42(A)(1) and (A)(2).

(1) Flexible Conduit Liquidtight flexible metal conduit or liquidtight flexible nonmetallic conduit shall be permitted in lengths of not more than 1.8 m (6 ft).

The use of liquidtight flexible metal or nonmetallic conduit is permitted by 680.41(A)(1). This modifies the requirements for wiring methods in 680.25(A).

(2) Cord-and-Plug Connections Cord-and-plug connections with a cord not longer than 4.6 m (15 ft) shall be permitted where protected by a ground-fault circuit interrupter.

(B) Bonding Bonding by metal-to-metal mounting on a common frame or base shall be permitted. The metal bands or hoops used to secure wooden staves shall not be required to be bonded as required in 680.26.

(C) Interior Wiring to Outdoor Installations In the interior of a one-family dwelling or in the interior of another building or structure associated with a one-family dwelling,

Exhibit 680.17 Location of
the emergency shutoff device
required by 680.41.

Switch labeled as
emergency shutoff

5 ft min. – 50 ft max.
and within sight

Spa or hot tub

any of the wiring methods recognized in Chapter 3 of this *Code* that contain a copper equipment grounding conductor that is insulated or enclosed within the outer sheath of the wiring method and not smaller than 12 AWG shall be permitted to be used for the connection to motor, heating, and control loads that are part of a self-contained spa or hot tub or a packaged spa or hot tub equipment assembly. Wiring to an underwater light shall comply with 680.23 or 680.33.

680.43 Indoor Installations

A spa or hot tub installed indoors shall comply with the provisions of Parts I and II of this article except as modified by this section and shall be connected by the wiring methods of Chapter 3.

Exception: Listed spa and hot tub packaged units rated 20 amperes or less shall be permitted to be cord-and-plug connected to facilitate the removal or disconnection of the unit for maintenance and repair.

(A) Receptacles At least one 125-volt, 15- or 20-ampere receptacle on a general-purpose branch circuit shall be located not less than 1.5 m (5 ft) from, and not exceeding 3.0 m (10 ft) from, the inside wall of the spa or hot tub.

(1) Location Receptacles shall be located at least 1.5 m (5 ft) measured horizontally from the inside walls of the spa or hot tub.

(2) Protection, General Receptacles rated 125 volts and 30 amperes or less and located within 3.0 m (10 ft) of the inside walls of a spa or hot tub shall be protected by a ground-fault circuit interrupter.

(3) Protection, Spa or Hot Tub Supply Receptacle Receptacles that provide power for a spa or hot tub shall be ground-fault circuit-interrupter protected.

(4) Measurements In determining the dimensions in this section addressing receptacle spacings, the distance to be measured shall be the shortest path the supply cord of an appliance connected to the receptacle would follow without piercing a floor, wall, ceiling, doorway with hinged or sliding door, window opening, or other effective permanent barrier.

(B) Installation of Luminaires (Lighting Fixtures), Lighting Outlets, and Ceiling-Suspended (Paddle) Fans

(1) Elevation Luminaires (lighting fixtures), except as covered in 680.43(B)(2), lighting outlets, and ceiling-suspended (paddle) fans located over the spa or hot tub or within 1.5 m (5 ft) from the inside walls of the spa or hot tub shall comply with the clearances specified in (B)(1)(a), (B)(1)(b), and (B)(1)(c) above the maximum water level.

(a) Without GFCI. Where no GFCI protection is provided, the mounting height shall be not less than 3.7 m (12 ft).

(b) With GFCI. Where GFCI protection is provided, the mounting height shall be permitted to be not less than 2.3 m (7 ft 6 in.).

(c) Below 2.3 m (7 ft 6 in.). Luminaires (lighting fixtures) meeting the requirements of item (1) or (2) and protected by a ground-fault circuit interrupter shall be permitted to be installed less than 2.3 m (7 ft 6 in.) over a spa or hot tub:

(1) Recessed luminaires (fixtures) with a glass or plastic lens, nonmetallic or electrically isolated metal trim, and suitable for use in damp locations

(2) Surface-mounted luminaires (fixtures) with a glass or plastic globe, a nonmetallic body, or a metallic body isolated from contact, and suitable for use in damp locations

(2) Underwater Applications Underwater luminaires (lighting fixtures) shall comply with the provisions of 680.23 or 680.33.

(C) Wall Switches Switches shall be located at least 1.5 m (5 ft), measured horizontally, from the inside walls of the spa or hot tub.

Receptacles, wall switches, and electrical devices and controls not associated with a spa or hot tub are required to be located at least 5 ft from the inside wall of the spa or hot tub. Receptacles within 10 ft are required to be protected by a GFCI. Receptacles supplying power to a spa or hot tub are also required to be protected by a GFCI unless the unit is a listed package unit with integral GFCI protection.

Lighting fixtures, lighting outlets, and ceiling-suspended (paddle) fans located less than 12 ft over a spa or hot tub and within 5 ft horizontally from the inside walls of the spa or hot tub are required to be protected by a GFCI.

(D) Bonding The following parts shall be bonded together:

(1) All metal fittings within or attached to the spa or hot tub structure
(2) Metal parts of electrical equipment associated with the spa or hot tub water circulating system, including pump motors
(3) Metal conduit and metal piping that are within 1.5 m (5 ft) of the inside walls of the spa or hot tub and that are not separated from the spa or hot tub by a permanent barrier
(4) All metal surfaces that are within 1.5 m (5 ft) of the inside walls of the spa or hot tub and that are not separated from the spa or hot tub area by a permanent barrier

Exception: Small conductive surfaces not likely to become energized, such as air and water jets and drain fittings, where not connected to metallic piping, towel bars, mirror frames, and similar nonelectrical equipment, shall not be required to be bonded.

(5) Electrical devices and controls that are not associated with the spas or hot tubs and that are located not less than 1.5 m (5 ft) from such units; otherwise they shall be bonded to the spa or hot tub system

Bonding and grounding requirements are similar to those in Parts I and II of Article 680, except that metal-to-metal mounting on a common frame or base is an acceptable bonding method.

Small conductive surfaces such as air and water jets, drain fittings, and towel bars are not required to be bonded. See 680.43(D)(4), Exception.

Listed packaged units are permitted to be cord connected.

(E) Methods of Bonding All metal parts associated with the spa or hot tub shall be bonded by any of the following methods:

(1) The interconnection of threaded metal piping and fittings
(2) Metal-to-metal mounting on a common frame or base
(3) The provisions of a copper bonding jumper, insulated, covered, or bare, not smaller than 8 AWG solid.

(F) Grounding The following equipment shall be grounded:

(1) All electric equipment located within 1.5 m (5 ft) of the inside wall of the spa or hot tub
(2) All electric equipment associated with the circulating system of the spa or hot tub

(G) Underwater Audio Equipment Underwater audio equipment shall comply with the provisions of Part II of this article.

680.44 Protection

Except as otherwise provided in this section, the outlet(s) that supplies a self-contained spa or hot tub, a packaged spa or hot tub equipment assembly, or a field-assembled spa or hot tub shall be protected by a ground-fault circuit interrupter.

The requirements of 680.44 specify that field-assembled spas and hot tubs with heater loads of 50 amperes or less are to be GFCI protected. Spas and hot tubs utilizing voltages over 250 volts or 3-phase power are not required to have GFCI protection because GFCI devices are not available in all voltage, amperage, and phasing arrangements. Combination spa-pool or hot tub–pool arrangements are not required to have GFCI protection if they share a common bonding grid.

(A) Listed Units If so marked, a listed self-contained unit or listed packaged equipment assembly that includes integral ground-fault circuit-interrupter protection for all electrical parts within the unit or assembly (pumps, air blowers, heaters, lights, controls, sanitizer generators, wiring, and so forth) shall be permitted without additional GFCI protection.

(B) Other Units A field assembled spa or hot tub rated 3 phase or rated over 250 volts or with a heater load of more than 50 amperes shall not require the supply to be protected by a ground-fault circuit interrupter.

(C) Combination Pool and Spa or Hot Tub A combination pool/hot tub or spa assembly commonly bonded need not be protected by a ground-fault circuit interrupter.

FPN: See 680.2 for definitions of *self-contained spa or hot tub* and for *packaged spa or hot tub equipment assembly.*

V. Fountains

Part V applies to permanently installed decorative fountains and reflecting pools in the ground, partially in the ground, or in a building. These units are primarily for aesthetic value and are not intended for swimming or wading.

Part V does not cover installations in natural lakes, rivers, or ponds. Such installations are covered by the requirements of new Article 682, Natural and Artificially Made Bodies of Water.

680.50 General

The provisions of Part I and Part V of this article shall apply to all permanently installed fountains as defined in 680.2. Fountains that have water common to a pool shall additionally comply with the requirements in Part II of this article. Part V does not cover self-contained, portable fountains not larger than 1.5 m (5 ft) in any dimension. Portable fountains shall comply with Parts II and III of Article 422.

680.51 Luminaires (Lighting Fixtures), Submersible Pumps, and Other Submersible Equipment

(A) Ground-Fault Circuit Interrupter Luminaires (lighting fixtures), submersible pumps, and other submersible equipment, unless listed for operation at 15 volts or less and supplied by a transformer that complies with 680.23(A)(2), shall be protected by a ground-fault circuit interrupter.

(B) Operating Voltage No luminaires (lighting fixtures) shall be installed for operation on supply circuits over 150 volts between conductors. Submersible pumps and other submersible equipment shall operate at 300 volts or less between conductors.

(C) Luminaire (Lighting Fixture) Lenses Luminaires (lighting fixtures) shall be installed with the top of the luminaire (fixture) lens below the normal water level of the fountain unless listed for above-water locations. A luminaire (lighting fixture) facing upward shall have the lens adequately guarded to prevent contact by any person.

(D) Overheating Protection Electrical equipment that depends on submersion for safe operation shall be protected against overheating by a low-water cutoff or other approved means when not submerged.

(E) Wiring Equipment shall be equipped with provisions for threaded conduit entries or be provided with a suitable flexible cord. The maximum length of exposed cord in the fountain shall be limited to 3.0 m (10 ft). Cords extending beyond the fountain perimeter shall be enclosed in approved wiring enclosures. Metal parts of equipment in contact with water shall be of brass or other approved corrosion-resistant metal.

(F) Servicing All equipment shall be removable from the water for relamping or normal maintenance. Luminaires (fixtures) shall not be permanently embedded into the fountain structure such that the water level must be reduced or the fountain drained for relamping, maintenance, or inspection.

(G) Stability Equipment shall be inherently stable or be securely fastened in place.

680.52 Junction Boxes and Other Enclosures

(A) General Junction boxes and other enclosures used for other than underwater installation shall comply with 680.24.

(B) Underwater Junction Boxes and Other Underwater Enclosures Junction boxes and other underwater enclosures shall meet the requirements of 680.52(B)(1) and (B)(2).

(1) Construction

(a) Underwater enclosures shall be equipped with provisions for threaded conduit entries or compression glands or seals for cord entry.

(b) Underwater enclosures shall be submersible and made of copper, brass, or other approved corrosion-resistant material.

(2) Installation Underwater enclosure installations shall comply with (a) and (b).

(a) Underwater enclosures shall be filled with an approved potting compound to prevent the entry of moisture.

(b) Underwater enclosures shall be firmly attached to the supports or directly to the fountain surface and bonded as required. Where the junction box is supported only by the conduit, the conduit shall be of copper, brass, stainless steel, or other approved corrosion-resistant metal. Where the box is fed by nonmetallic conduit, it shall have additional supports and fasteners of copper, brass, or other approved corrosion-resistant material.

FPN: See 314.23 for support of enclosures.

680.53 Bonding

All metal piping systems associated with the fountain shall be bonded to the equipment grounding conductor of the branch circuit supplying the fountain.

FPN: See 250.122 for sizing of these conductors.

680.54 Grounding

The following equipment shall be grounded:

(1) All electrical equipment located within the fountain or within 1.5 m (5 ft) of the inside wall of the fountain
(2) All electrical equipment associated with the recirculating system of the fountain
(3) Panelboards that are not part of the service equipment and that supply any electrical equipment associated with the fountain

680.55 Methods of Grounding

(A) Applied Provisions The provisions of 680.21(A), 680.23(B)(3), 680.23(F)(1) and (F)(2), 680.24(F), and 680.25 shall apply.

(B) Supplied by a Flexible Cord Electrical equipment that is supplied by a flexible cord shall have all exposed non–current-carrying metal parts grounded by an insulated copper equipment grounding conductor that is an integral part of this cord. The grounding conductor shall be connected to a grounding terminal in the supply junction box, transformer enclosure, or other enclosure.

680.56 Cord-and-Plug-Connected Equipment

(A) Ground-Fault Circuit Interrupter All electrical equipment, including power-supply cords, shall be protected by ground-fault circuit interrupters.

(B) Cord Type Flexible cord immersed in or exposed to water shall be of a type for extra-hard usage, as designated in Table 400.4, and shall be a listed type with a "W" suffix.

(C) Sealing The end of the flexible cord jacket and the flexible cord conductor termination within equipment shall be covered with, or encapsulated in, a suitable potting compound to prevent the entry of water into the equipment through the cord or its conductors. In addition, the ground connection within equipment shall be similarly treated to protect such connections from the deteriorating effect of water that may enter into the equipment.

(D) Terminations Connections with flexible cord shall be permanent, except that grounding-type attachment plugs and receptacles shall be permitted to facilitate removal or disconnection for maintenance, repair, or storage of fixed or stationary equipment not located in any water-containing part of a fountain.

680.57 Signs

(A) General This section covers electric signs installed within a fountain or within 3.0 m (10 ft) of the fountain edge.

(B) Ground-Fault Circuit-Interrupter Protection for Personnel All circuits supplying the sign shall have ground-fault circuit-interrupter protection for personnel.

(C) Location

(1) Fixed or Stationary A fixed or stationary electric sign installed within a fountain shall be not less than 1.5 m (5 ft) inside the fountain measured from the outside edges of the fountain.

(2) Portable A portable electric sign shall not be placed within a pool or fountain or within 1.5 m (5 ft) measured horizontally from the inside walls of the fountain.

(D) Disconnect A sign shall have a local disconnecting means in accordance with 600.6 and 680.12.

(E) Bonding and Grounding A sign shall be grounded and bonded in accordance with 600.7.

The use of electric signs in fountains has become increasingly popular. The requirements for electric signs in fountains were added for the 1999 *Code*. Signs in fountains were previously addressed by Article 600, and fountains were addressed by Article 680. Now 680.57 combines the requirements into a one location. Electric signs in fountains are required to have GFCI protection for personnel. This protection may be provided in the feeder or branch circuit. To prevent contact by persons around the fountain, the sign must be at least 5 ft from the edge of the fountain (see Exhibit 680.18). Disconnecting and bonding requirements in Article 600 apply, and grounding must be provided in accordance with Article 250.

680.58 GFCI Protection for Adjacent Receptacle Outlets

All 15- or 20-ampere, single-phase 125-volt through 250-volt receptacles located within 6.0 m (20 ft) of a fountain edge shall be provided with GFCI protection.

VI. Pools and Tubs for Therapeutic Use

Part VI recognizes therapeutic equipment in other locations, such as athletic training rooms and health care facilities, where conditions of use are the same or similar. Portable therapeutic appliances, which are covered by the provisions of Article 422, are required to provide protection from elec-

Exhibit 680.18 Electric sign located in a fountain as described in 680.57.

trocution while in the on or off position. The device used is an immersion-detection circuit interrupter (IDCI).

Permanently installed therapeutic pools that cannot be readily disassembled are required to comply with Parts I and II of Article 680. The limitations regarding lighting fixtures over and around a swimming pool do not apply to therapeutic pools and tubs. The lighting fixtures in the tub area are required to be totally enclosed. Therapeutic tubs that cannot easily be moved are subject to the same basic requirements.

Bonding and grounding requirements are similar to those in Parts I and II of Article 680, except that metal-to-metal mounting on a common frame or base is permitted. Where equipment is connected by a flexible cord, the equipment grounding conductor is required to be connected to a fixed metal part of the assembly.

680.60 General

The provisions of Part I and Part VI of this article shall apply to pools and tubs for therapeutic use in health care facilities, gymnasiums, athletic training rooms, and similar areas. Portable therapeutic appliances shall comply with Parts II and III of Article 422.

FPN: See 517.2 for definition of health care facilities.

680.61 Permanently Installed Therapeutic Pools

Therapeutic pools that are constructed in the ground, on the ground, or in a building in such a manner that the pool cannot be readily disassembled shall comply with Parts I and II of this article.

Exception: The limitations of 680.22(B)(1) through (B)(4) shall not apply where all luminaires (lighting fixtures) are of the totally enclosed type.

680.62 Therapeutic Tubs (Hydrotherapeutic Tanks)

Therapeutic tubs, used for the submersion and treatment of patients, that are not easily moved from one place to another in normal use or that are fastened or otherwise secured at a specific location, including associated piping systems, shall conform to this part.

(A) Protection Except as otherwise provided in this section, the outlet(s) that supplies a self-contained therapeutic tub or hydrotherapeutic tank, a packaged therapeutic tub or hydrotherapeutic tank, or a field-assembled therapeutic tub or hydrotherapeutic tank shall be protected by a ground-fault circuit interrupter.

(1) Listed Units If so marked, a listed self-contained unit or listed packaged equipment assembly that includes integral ground-fault circuit-interrupter protection for all electrical parts within the unit or assembly (pumps, air blowers, heaters, lights, controls, sanitizer generators, wiring, and so forth) shall be permitted without additional GFCI protection.

(2) Other Units A therapeutic tub or hydrotherapeutic tank rated 3 phase or rated over 250 volts or with a heater load of more than 50 amperes shall not require the supply to be protected by a ground-fault circuit interrupter.

The requirements in 680.62(A)(2) for large therapeutic tanks and therapeutic tubs are similar to the requirements for large hot tubs and spas. Large field-assembled therapeutic tubs are not required to have GFCI protection in their electrical supply when the heater load is over 50 amperes.

(B) Bonding The following parts shall be bonded together:
(1) All metal fittings within or attached to the tub structure
(2) Metal parts of electrical equipment associated with the tub water circulating system, including pump motors
(3) Metal-sheathed cables and raceways and metal piping that are within 1.5 m (5 ft) of the inside walls of the tub and not separated from the tub by a permanent barrier
(4) All metal surfaces that are within 1.5 m (5 ft) of the inside walls of the tub and not separated from the tub area by a permanent barrier
(5) Electrical devices and controls that are not associated with the therapeutic tubs and located within 1.5 m (5 ft) from such units.

(C) Methods of Bonding All metal parts required to be bonded by this section shall be bonded by any of the following methods:
(1) The interconnection of threaded metal piping and fittings
(2) Metal-to-metal mounting on a common frame or base

(3) Connections by suitable metal clamps

(4) By the provisions of a solid copper bonding jumper, insulated, covered, or bare, not smaller than 8 AWG

(D) Grounding

(1) Fixed or Stationary Equipment The equipment specified in (D)(1)(a) and (D)(1)(b) shall be grounded.

(a) Location. All electrical equipment located within 1.5 m (5 ft) of the inside wall of the tub shall be grounded.

(b) Circulation System. All electrical equipment associated with the circulating system of the tub shall be grounded.

(2) Portable Equipment Portable therapeutic appliances shall meet the grounding requirements in 250.114.

(E) Receptacles All receptacles within 1.5 m (5 ft) of a therapeutic tub shall be protected by a ground-fault circuit interrupter.

(F) Luminaires (Lighting Fixtures) All luminaires (lighting fixtures) used in therapeutic tub areas shall be of the totally enclosed type.

VII. Hydromassage Bathtubs

680.70 General

Hydromassage bathtubs as defined in 680.2 shall comply with Part VII of this article. They shall not be required to comply with other parts of this article.

680.71 Protection

Hydromassage bathtubs and their associated electrical components shall be protected by a ground-fault circuit interrupter. All 125-volt, single-phase receptacles not exceeding 30 amperes and located within 1.5 m (5 ft) measured horizontally of the inside walls of a hydromassage tub shall be protected by a ground-fault circuit interrupter(s).

Hydromassage bathtubs (see definition in 680.2) are required to be protected by a GFCI. In addition, all 125-volt, single-phase, 15-, 20-, and 30-ampere receptacles within 5 ft of the inside wall of the hydromassage bathtub are required to be GFCI protected. Hydromassage bathtubs are treated the same as ordinary bathtubs in regard to the installation of luminaires, switches, and other electrical equipment. See 410.4(D) for special requirements relating to cord-connected fixtures, hanging fixtures, and pendants near bathtubs. Also see 210.8(A)(1) and 210.8(B)(1) for requirements for GFCI protection of bathroom receptacles.

680.72 Other Electrical Equipment

Luminaires (lighting fixtures), switches, receptacles, and other electrical equipment located in the same room, and not directly associated with a hydromassage bathtub, shall be installed in accordance with the requirements of Chapters 1 through 4 in this *Code* covering the installation of that equipment in bathrooms.

680.73 Accessibility

Hydromassage bathtub electrical equipment shall be accessible without damaging the building structure or building finish.

This section requires access to electrical equipment associated with the hydromassage tub. Building codes and plumbing codes might not require access to this equipment. This requirement is intended to ensure that the electrical equipment associated with hydromassage bathtubs can be accessed for maintenance and repair without damaging the finish or structure of the building. The access may be either an integral part of the tub or one that is provided in the finish that encloses the tub. See Exhibit 680.19.

Access plate or cover

Exhibit 680.19 Access plate for hydromassage tub electrical equipment, located as described in 680.73.

680.74 Bonding

All metal piping systems and all grounded metal parts in contact with the circulating water shall be bonded together using a copper bonding jumper, insulated, covered, or bare, not smaller than 8 AWG solid.

The bonding requirement for hydromassage bathtubs requires interconnection between metal piping systems and metal parts associated with the water recirculation system only at the hydromassage bathtub location. As is the case with swimming pool bonding, this section does not require the installation of a bonding conductor from the hydromassage bathtub pump motor to the service equipment or panelboard from which the hydromassage bathtub branch circuit originates even if there is no metal piping or metal parts in the vicinity of the hydromassage bathtub. The bonding required by 680.74 is intended to create a local equipotential plane, and the equipment grounding conductor of the branch circuit supplying the hydromassage tub provides the path for ground fault current.

ARTICLE 682
Natural and Artificially Made Bodies of Water

Summary of Changes

- Added new article covering electrical wiring and equipment in or adjacent to bodies of water not covered by other *Code* articles, particularly Article 680.

Contents

I. General

682.1 Scope

This article applies to the installation of electrical wiring for, and equipment in and adjacent to, natural or artificially made bodies of water not covered by other articles in this *Code*, such as but not limited to aeration ponds, fish farm ponds, storm retention basins, treatment ponds, irrigation (channels) facilities.

The scope of Article 680 does not include bodies of water other than those described in 680.1 and defined in 680.2. Therefore, in previous editions of the *NEC*, only the general requirements in Chapters 1 through 4 applied to electrical installations in and around bodies of water that were not covered by Article 680. A task group assigned to study the need to include specific requirements covering other bodies of water recommended that, instead of expanding the scope of Article 680, a new article, Natural and Artificially Made Bodies of Water, be developed for the 2005 *Code*.

Electrical equipment such as pumps, luminaires and their associated supply wiring are frequently installed in lakes, ponds, aeration and treatment basins, and similar bodies of water, and the new requirements are designed to minimize the shock hazards inherent in those wet and damp locations. Safety concepts contained in the requirements of Articles 553, 555, 590, and 680 provided the foundation for the requirements in Article 682.

682.2 Definitions

Artificially Made Bodies of Water. Bodies of water that have been constructed or modified to fit some decorative or commercial purpose such as, but not limited to, aeration ponds, fish farm ponds, storm retention basins, treatment ponds, irrigation (channel) facilities. Water depths may vary seasonally or be controlled.

The term *artificially made bodies of water* includes all bodies of water that are not naturally created and that are not covered by the requirements of Article 680. The uses of artificially made bodies of water include decorative, agricultural, municipal infrastructure, and industrial. The decorative pond shown in Exhibit 682.1 is an example of an artificially made body of water because it was constructed and filled with water and did not occur naturally.

Electrical Datum Plane. The electrical datum plane as used in this article is defined as follows:

See the commentary on the electrical datum plane in 555.2.

(1) In land areas subject to tidal fluctuation, the electrical datum plane is a horizontal plane 600 mm (2 ft) above

Exhibit 682.1 An artificially made body of water. The electrical equipment associated with pumps used to circulate water in this artificial pond is subject to the requirements of Article 682.

the highest tide level for the area occurring under normal circumstances, that is, highest high tide.

(2) In land areas not subject to tidal fluctuation, the electrical datum plane is a horizontal plane 600 mm (2 ft) above the highest water level for the area occurring under normal circumstances.

(3) In land areas subject to flooding, the electrical datum plane based on (1) or (2) above is a horizontal plane 600 mm (2 ft) above the point identified as the prevailing high water mark or an equivalent benchmark based on seasonal or storm-driven flooding from the authority having jurisdiction.

(4) The electrical datum plane for floating structures and landing stages that are (1) installed to permit rise and fall response to water level, without lateral movement, and (2) that are so equipped that they can rise to the datum plane established for (1) or (2) above, is a horizontal plane 750 mm (30 in.) above the water level at the floating structure or landing stage and a minimum of 300 mm (12 in.) above the level of the deck.

Equipotential Plane. An area where wire mesh or other conductive elements are on, embedded in, or placed under the walk surface within 75 mm (3 in.), bonded to all metal structures and fixed nonelectrical equipment that may become energized, and connected to the electrical grounding system to prevent a difference in voltage from developing within the plane.

Natural Bodies of Water. Bodies of water such as lakes, streams, ponds, rivers, and other naturally occurring bodies of water, which may vary in depth throughout the year.

Shoreline. The farthest extent of standing water under the applicable conditions that determine the electrical datum plane for the specified body of water.

682.3 Other Articles

Wiring and equipment in or adjacent to natural or artificially made bodies of water shall comply with the applicable provisions of other articles of this *Code*, except as modified by this article. If the water is subject to boat traffic, the wiring shall comply with 555.13(B).

II. Installation

682.10 Electrical Equipment and Transformers

Electrical equipment and transformers, including their enclosures, shall be specifically approved for the intended location. No portion of an enclosure for electrical equipment not identified for operation while submerged shall be located below the electrical datum plane.

See 640.10 for requirements covering the installation of audio system equipment near a body of water.

682.11 Location of Service Equipment

On land, the service equipment for floating structures and submersible electrical equipment shall be located no closer than 1.5 m (5 ft) horizontally from the shoreline and live parts elevated a minimum of 300 mm (12 in.) above the electrical datum plane. Service equipment shall disconnect when the water level reaches the height of the established electrical datum plane.

682.12 Electrical Connections

All electrical connections not intended for operation while submerged shall be located at least 300 mm (12 in.) above the deck of a floating or fixed structure, but not below the electrical datum plane.

682.13 Wiring Methods and Installation

Wiring methods and installations of Chapter 3 and Articles 553, 555, and 590 shall be permitted where identified for use in wet locations.

682.14 Disconnecting Means for Floating Structures or Submersible Electrical Equipment

(A) Type The disconnecting means shall be permitted to consist of a circuit breaker, switch, or both and shall be properly identified as to which structure or equipment it controls.

(B) Location The disconnecting means shall be readily accessible on land and shall be located in the supply circuit ahead of the structure or the equipment connection. The disconnecting means shall be located not more than 750 mm (30 in.) from the structure or equipment connection. The disconnecting means shall be within sight of, but not closer than, 1.5 m (5 ft) horizontally from the edge of the shoreline and live parts elevated a minimum of 300 mm (12 in.) above the electrical datum plane.

682.15 Ground Fault Circuit Interrupter (GFCI) Protection

Fifteen- and 20-ampere single-phase, 125-volt through 250-volt receptacles installed outdoors and in or on floating buildings or structures within the electrical datum plane area that are used for storage, maintenance, or repair where portable electric hand tools, electrical diagnostic equipment, or portable lighting equipment are to be used shall be provided with GFCI protection. The GFCI protection device shall be located not less than 300 mm (12 in.) above the established electrical datum plane.

III. Grounding and Bonding

682.30 Grounding

Wiring and equipment within the scope of this article shall be grounded as specified in Articles 250, 553, and 555 and with the requirements in Part III.

682.31 Equipment Grounding Conductors

(A) **Type** Equipment grounding conductors shall be insulated copper conductors sized in accordance with 250.122 but not smaller than 12 AWG.

(B) **Feeders** Where a feeder supplies a remote panelboard, an insulated equipment grounded conductor shall extend from a grounding terminal in the service to a grounding terminal and busbar in the remote panelboard.

(C) **Branch Circuits** The insulated equipment grounding conductor for branch circuits shall terminate at a grounding terminal in a remote panelboard or the grounding terminal in the main service equipment.

(D) **Cord-and-Plug-Connected Appliances** Where required to be grounded, cord-and-plug-connected appliances shall be grounded by means of an equipment grounding conductor in the cord and a grounding-type attachment plug.

682.32 Bonding of Non–Current-Carrying Metal Parts

All metal parts in contact with the water, all metal piping, tanks, and all non–current-carrying metal parts that may become energized shall be bonded to the grounding bus in the panelboard.

682.33 Equipotential Planes and Bonding of Equipment Planes

An equipotential plane shall be installed where required in this section to mitigate step and touch voltages at electrical equipment.

(A) **Areas Requiring Equipotential Planes** Equipotential planes shall be installed adjacent to all outdoor service equipment or disconnecting means that control equipment in or on water, that have a metallic enclosure and controls accessible to personnel, and that are likely to become energized. The equipotential plane shall encompass the area around the equipment and shall extend from the area directly below the equipment out not less than 900 mm (36 in.) in all directions from which a person would be able to stand and come in contact with the equipment.

(B) **Areas Not Requiring Equipotential Planes** Equipotential planes shall not be required for the controlled equipment supplied by the service equipment or disconnecting means. All circuits rated not more than 60 amperes at 120 through 250 volts, single phase, shall have GFCI protection.

(C) **Bonding** Equipotential planes shall be bonded to the electrical grounding system. The bonding conductor shall be solid copper, insulated, covered or bare, and not smaller than 8 AWG. Connections shall be made by exothermic welding or by listed pressure connectors or clamps that are labeled as being suitable for the purpose and are of stainless steel, brass, copper, or copper alloy.

ARTICLE 685
Integrated Electrical Systems

Summary of Changes
- **685.1(2):** Revised to include prescriptive requirements intended to define the conditions under which a person is considered to be a "qualified person" and how this information is to be documented.

Contents

I. General

685.1 Scope

This article covers integrated electrical systems, other than unit equipment, in which orderly shutdown is necessary to ensure safe operation. An *integrated electrical system* as used in this article is a unitized segment of an industrial wiring system where all of the following conditions are met:

(1) An orderly shutdown is required to minimize personnel hazard and equipment damage.
(2) The conditions of maintenance and supervision ensure that qualified persons service the system. The name(s) of the qualified person(s) shall be kept in a permanent record at the office of the establishment in charge of the completed installation.

    A person designated as a qualified person shall possess the skills and knowledge related to the construction and operation of the electrical equipment and installation and shall have received documented safety training on the hazards involved. Documentation of their qualifications shall be on file with the office of the establishment in charge of the completed installation.
(3) Effective safeguards acceptable to the authority having jurisdiction are established and maintained.

The integrated electrical systems commonly used in large and complex industrial processes are designed, installed, and operated under stringent on-site engineering supervision. The control equipment, including overcurrent devices, is located so as to be accessible to qualified personnel, but that location might not meet and is not required to meet the conditions described in the Article 100 definition of *readily accessible*. Locating overcurrent devices and their associated disconnecting means so as not to be readily accessible to unqualified personnel is one of the preventative measures used to help maintain continuity of operation.

    For some industrial processes, the sudden loss of electrical power to vital equipment is an unacceptable level of risk, and an orderly shutdown procedure is necessary to prevent severe equipment damage, injury to personnel, or in some extreme cases catastrophic failure. Orderly shutdown is commonly employed in nuclear power–generating facilities, paper mills, and other areas with hazardous processes.

685.3   Application of Other Articles

The articles/sections in Table 685.3 apply to particular cases of installation of conductors and equipment, where there are orderly shutdown requirements that are in addition to those of this article or are modifications of them.

II.   Orderly Shutdown

685.10   Location of Overcurrent Devices in or on Premises

Location of overcurrent devices that are critical to integrated electrical systems shall be permitted to be accessible, with

Table 685.3 Application of Other Articles

Conductor/Equipment	Section
More than one building or other structure	225, Part II
Ground-fault protection of equipment	230.95, Exception No. 1
Protection of conductors	240.4
Electrical system coordination	240.12
Ground-fault protection of equipment	240.13(1)
Grounding ac systems of 50 volts to 1000 volts	250.21
Equipment protection	427.22
Orderly shutdown	430.44
Disconnection	430.74, Exception Nos. 1 and 2
Disconnecting means in sight from controller	430.102(A), Exception No. 2
Energy from more than one source	430.113, Exception Nos. 1 and 2
Disconnecting means	645.10, Exception
Uninterruptible power supplies (UPS)	645.11(1)
Point of connection	705.12(A)

mounting heights permitted to ensure security from operation by unqualified personnel.

685.12   Direct-Current System Grounding

Two-wire dc circuits shall be permitted to be ungrounded.

685.14   Ungrounded Control Circuits

Where operational continuity is required, control circuits of 150 volts or less from separately derived systems shall be permitted to be ungrounded.

ARTICLE 690
Solar Photovoltaic Systems

Summary of Changes

- **690.2:** Added new definition for *building integrated photovoltaics*.
- **690.3:** Added new exception for installations in hazardous locations requiring compliance with applicable requirements of Articles 500, 505, and 510.
- **690.13:** Revised to permit an automatic switch or circuit breaker in grounded conductor if it is part of a ground-fault detection system.
- **690.14(D):** Added new requirement specifying conditions under which utility-interactive inverters can be installed in locations that are not readily accessible.

- **690.17:** Deleted Exception No.1 permitting dc side disconnecting means to have lower interrupting rating under specified conditions.

- **690.31(E):** Added new requirement for mechanical protection of dc photovoltaic source and output circuits inside building on line side of source of output disconnecting means.

- **690.35:** Added new requirements for operation of ungrounded photovoltaic power systems.

- **690.41:** Added Exception permitting ungrounded systems complying with new 690.35.

- **690.47(C):** Added new requirement covering PV systems where both ac and dc circuits are required to be grounded.

- **690.48:** Added new requirement for maintaining continuity of equipment grounding systems.

- **690.49:** Added new requirement for maintaining continuity of photovoltaic source and output current grounded conductor.

- **690.53:** Revised to specify that the field-marking is required for dc power sources.

- **690.64(B)(5):** Added new requirements for mechanically securing backfed circuit breakers supplied from listed utility-interactive inverters.

- **690.71(B)(1):** Revised to clarify the number of series-connected, lead-acid battery 2-volt cells permitted to be installed in dwellings.

- **690.72(B)(2)(1):** Revised to limit current rating of the diversion load to less than or equal to the rating of the charge controller.

- **Section 690.72(B)(3):** Added new requirements covering PV systems using utility-interactive inverters to divert excess power into a utility system.

Contents

I. General

690.1 Scope

The provisions of this article apply to solar photovoltaic electrical energy systems, including the array circuit(s), inverter(s), and controller(s) for such systems. [See Figures 690.1(A) and 690.1(B).] Solar photovoltaic systems covered by this article may be interactive with other electrical power production sources or stand-alone, with or without electrical energy storage such as batteries. These systems may have ac or dc output for utilization.

Notes:
1. These diagrams are intended to be a means of identification for photovoltaic system components, circuits, and connections.
2. Disconnecting means required by Article 690, Part III, are not shown.
3. System grounding and equipment grounding are not shown. See Article 690, Part V.

Figure 690.1(A) Identification of Solar Photovoltaic System Components.

The use of photovoltaic (PV) systems as utility-interactive or as stand-alone power supply systems has steadily increased as the technology and availability of the PV equipment have evolved. The requirements of Article 690 cover the use of stand-alone and utility-interactive PV systems. In

Interactive system

Hybrid system

Stand-alone system

Notes:
1. These diagrams are intended to be a means of identification for photovoltaic system components, circuits, and connections.
2. Disconnecting means and overcurrent protection required by Article 690 are not shown.
3. System grounding and equipment grounding are not shown. See Article 690, Part V.
4. Custom designs occur in each configuration, and some components are optional.

Figure 690.1(B) Identification of Solar Photovoltaic System Components in Common System Configurations.

accordance with Exception No. 1 to 705.3, the requirements for utility-interactive PV systems are covered in Article 690, and these systems are not subject to the requirements of Article 705.

Exhibit 690.1 shows a custom-designed home with a PV electrical system.

Typical solar PV systems are illustrated in Exhibits 690.2 through 690.5. Other circuit arrangements are permissible.

690.2 Definitions

Alternating-Current (ac) Module (Alternating-Current Photovoltaic Module). A complete, environmentally protected unit consisting of solar cells, optics, inverter, and other components, exclusive of tracker, designed to generate ac power when exposed to sunlight.

Exhibit 690.1 The south roof of a solar house. The integrated array of glass modules generates electricity directly from the sun using PV modules. (Courtesy of Solar Design Associates, Inc.)

An ac PV module consists of a single integrated mechanical unit. Because there is no accessible, field-installed dc wiring in this single unit, the dc PV source-circuit requirements in this *Code* are not applicable to the dc wiring in an ac PV module.

Array. A mechanically integrated assembly of modules or panels with a support structure and foundation, tracker, and other components, as required, to form a direct-current power-producing unit.

The building blocks of an array are illustrated in Exhibit 690.6.

Bipolar Photovoltaic Array. A photovoltaic array that has two outputs, each having opposite polarity to a common reference point or center tap.

Blocking Diode. A diode used to block reverse flow of current into a photovoltaic source circuit.

Blocking diodes are not required by this *Code*, although the instructions or labels supplied with the PV module may require them. Blocking diodes are not overcurrent devices and may not be substituted for any overcurrent device required by this *Code*.

Building Integrated Photovoltaics. Photovoltaic cells, devices, modules, or modular materials that are integrated into the outer surface or structure of a building and serve as the outer protective surface of that building.

Charge Controller. Equipment that controls dc voltage or dc current, or both, used to charge a battery.

Diversion Charge Controller. Equipment that regulates the charging process of a battery by diverting power from energy

Exhibit 690.2 Simplified circuit diagram of a small residential stand-alone system.

storage to direct-current or alternating-current loads or to an interconnected utility service.

Electrical Production and Distribution Network. A power production, distribution, and utilization system, such as a utility system and connected loads, that is external to and not controlled by the photovoltaic power system.

Hybrid System. A system comprised of multiple power sources. These power sources may include photovoltaic, wind, micro-hydro generators, engine-driven generators, and others, but do not include electrical production and distribution network systems. Energy storage systems, such as batteries, do not constitute a power source for the purpose of this definition.

Interactive System. A solar photovoltaic system that operates in parallel with and may deliver power to an electrical production and distribution network. For the purpose of this definition, an energy storage subsystem of a solar photovoltaic system, such as a battery, is not another electrical production source.

Inverter. Equipment that is used to change voltage level or waveform, or both, of electrical energy. Commonly, an inverter [also known as a power conditioning unit (PCU) or power conversion system (PCS)] is a device that changes dc input to an ac output. Inverters may also function as battery chargers that use alternating current from another

source and convert it into direct current for charging batteries.

Inverter Input Circuit. Conductors between the inverter and the battery in stand-alone systems or the conductors between the inverter and the photovoltaic output circuits for electrical production and distribution network.

Inverter Output Circuit. Conductors between the inverter and an ac panelboard for stand-alone systems or the conductors between the inverter and the service equipment or another electric power production source, such as a utility, for electrical production and distribution network.

Module. A complete, environmentally protected unit consisting of solar cells, optics, and other components, exclusive of tracker, designed to generate dc power when exposed to sunlight.

Panel. A collection of modules mechanically fastened together, wired, and designed to provide a field-installable unit.

Photovoltaic Output Circuit. Circuit conductors between the photovoltaic source circuit(s) and the inverter or dc utilization equipment.

Photovoltaic Power Source. An array or aggregate of arrays that generates dc power at system voltage and current.

Exhibit 690.3 Simplified circuit diagram of a medium-sized residential hybrid system.

Roof-mounted
PV modules

PV power center

dc loads

Charge controller relay

Roof junction box

Ground-fault system

Typical subarray 1 of 5

Battery bank

Inverter with battery charger

H N

120 Vac out

Generator 240 V

240/120

Wall

ac panelboard

H

N

- - - - Equipment grounding   * Main disconnects

Surge arrester   Current-limiting fuse

Photovoltaic Source Circuit. Circuits between modules and from modules to the common connection point(s) of the dc system.

Photovoltaic System Voltage. The direct current (dc) voltage of any photovoltaic source or photovoltaic output circuit. For multiwire installations, the photovoltaic system voltage is the highest voltage between any two dc conductors.

Solar Cell. The basic photovoltaic device that generates electricity when exposed to light.

Solar Photovoltaic System. The total components and subsystems that, in combination, convert solar energy into electrical energy suitable for connection to a utilization load.

Stand-Alone System. A solar photovoltaic system that supplies power independently of an electrical production and distribution network.

The simplified circuit diagrams in Exhibits 690.2 through 690.5 demonstrate the use of various components in a PV system. Specific requirements for overcurrent protection, disconnecting means, and grounding are covered in other sections of Article 690 and should not be assumed based on these diagrams. Instructions for or labels on the PV module might require additional overcurrent devices that may not be shown.

690.3 Other Articles

Wherever the requirements of other articles of this *Code* and Article 690 differ, the requirements of Article 690 shall apply and, if the system is operated in parallel with a primary source(s) of electricity, the requirements in 705.14, 705.16, 705.32, and 705.43 shall apply.

Exception: Solar photovoltaic systems, equipment, or wiring installed in a hazardous (classified) location shall also comply with 500.1, 505.1, and 510.1.

690.4 Installation

(A) Solar Photovoltaic System A solar photovoltaic system shall be permitted to supply a building or other structure in addition to any service(s) of another electricity supply system(s).

Exhibit 690.4 Simplified circuit diagram of a remote-cabin dc-only system.

Junction box

Control panel

Grounding electrode

Charge controller

dc loads

Fused disconnect with current-limiting fuses

Batteries

PV modules

----- Equipment grounding

🔲 Surge arrester

⊰ Current-limiting fuse

** Overcurrent device may be required by UL 1703

(B) Conductors of Different Systems Photovoltaic source circuits and photovoltaic output circuits shall not be contained in the same raceway, cable tray, cable, outlet box, junction box, or similar fitting as feeders or branch circuits of other systems, unless the conductors of the different systems are separated by a partition or are connected together.

For example, 690.4(B) does not permit the conductors supplying an exterior lighting fixture located in close proximity of a roof-mounted PV array to be installed in the same raceway or cable with the conductors of PV source circuits or PV output circuits.

Conductors directly related to a specific PV system, such as those in dc and ac output power circuits, may be contained in the same raceway as PV source and output conductors providing they meet the requirements of 300.3(C).

(C) Module Connection Arrangement The connections to a module or panel shall be arranged so that removal of

a module or panel from a photovoltaic source circuit does not interrupt a grounded conductor to another photovoltaic source circuit. Sets of modules interconnected as systems rated at 50 volts or less, with or without blocking diodes, and having a single overcurrent device shall be considered as a single-source circuit. Supplementary overcurrent devices used for the exclusive protection of the photovoltaic modules are not considered as overcurrent devices for the purpose of this section.

In general, 690.4(C) requires that a jumper be installed between a module terminal or lead and the connection point to the grounded PV source circuit conductor. That way, a module can be removed without interrupting the grounded conductor to other PV source circuits. If interrupted, such conductors, although identified as grounded, would be operating at the system potential with respect to ground, and a shock hazard could result. The reverse current protection on nearly all PV modules (as indicated by the fuse requirement labeled on the back of each module) generally dictates that

Exhibit 690.5 Simplified circuit diagram of a rooftop grid-connected system.

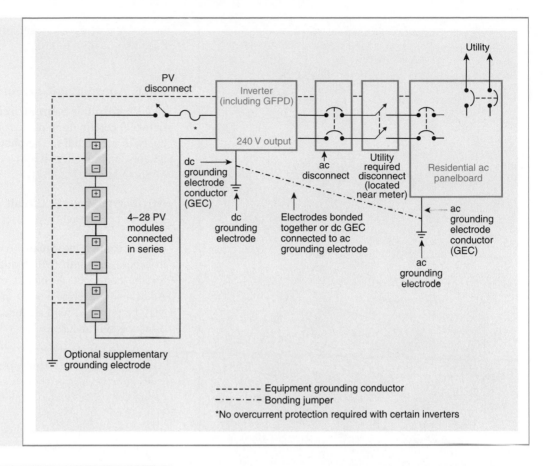

each module or string of modules has a series overcurrent device and becomes a source circuit.

(D) Equipment Inverters or motor generators shall be identified for use in solar photovoltaic systems.

Equipment listed for marine, mobile, telecommunications, or other applications may not be suitable for installation in permanent PV power systems complying with this *Code*.

690.5 Ground-Fault Protection

Roof-mounted dc photovoltaic arrays located on dwellings shall be provided with dc ground-fault protection to reduce fire hazards.

Ground-fault detection and interruption for PV systems should not be confused with the requirements for GFCI protection, as defined in Article 100. A GFCI is intended for the protection of personnel in single-phase ac systems. It functions to open the ungrounded conductor when a 5-mA fault current is detected. In contrast, devices meeting 690.5 are intended to prevent fires in dc PV circuits due to ground faults.

(A) Ground-Fault Detection and Interruption The ground-fault protection device or system shall be capable of detecting a ground fault, interrupting the flow of fault current, and providing an indication of the fault.

Typical ground-fault protection devices meeting the requirements of 690.5(A) operate by opening the main dc bonding jumper. They sense dc ground faults anywhere on the dc system and may be mounted anywhere in that system. They are usually installed inside the utility-interactive inverters or in the dc power center in stand-alone PV systems.

(B) Disconnection of Conductors The ungrounded conductors of the faulted source circuit shall be automatically disconnected. If the grounded conductors of the faulted source circuit are disconnected to comply with the requirements of 690.5(A), all conductors of the faulted source circuit shall be opened automatically and simultaneously. Opening the grounded conductor of the array or opening the faulted sections of the array shall be permitted to interrupt the ground-fault current path.

Typical ground-fault protection devices operating in PV systems (48-volt nominal and below) automatically disconnect

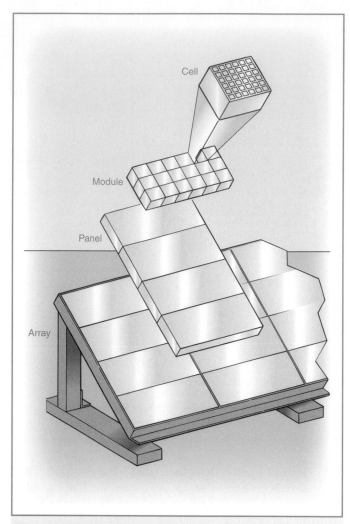

Exhibit 690.6 Components of an array.

the ungrounded conductor with a circuit breaker mechanically linked to a ground-fault-sensing circuit breaker. Equipment operating above 48 volts nominal usually automatically shuts off the connected equipment to meet this requirement when a ground fault is sensed. Both methods provide the intended second indication that something needs to be corrected.

(C) Labels and Markings Labels and markings shall be applied near the ground-fault indicator at a visible location, stating that, if a ground fault is indicated, the normally grounded conductors may be energized and ungrounded.

Many types of ground-fault detection and interruption equipment break the negative-to-ground bond to interrupt the fault currents, and the now ungrounded PV negative conductor

generally is at open-circuit voltage below the ground reference (e.g., –400 volts).

690.6 Alternating-Current (ac) Modules

(A) Photovoltaic Source Circuits The requirements of Article 690 pertaining to photovoltaic source circuits shall not apply to ac modules. The photovoltaic source circuit, conductors, and inverters shall be considered as internal wiring of an ac module.

(B) Inverter Output Circuit The output of an ac module shall be considered an inverter output circuit.

(C) Disconnecting Means A single disconnecting means, in accordance with 690.15 and 690.17, shall be permitted for the combined ac output of one or more ac modules. Additionally, each ac module in a multiple ac-module system shall be provided with a connector, bolted, or terminal-type disconnecting means.

Alternating-current PV modules, as utility-interactive devices, are designed to produce ac power only when they are connected to an external source of ac power at the correct voltage and frequency. A single disconnecting means removes the external source and turns off all ac PV modules connected to that disconnecting device.

(D) Ground-Fault Detection Alternating-current-module systems shall be permitted to use a single detection device to detect only ac ground faults and to disable the array by removing ac power to the ac module(s).

The permissive language of 690.6 (D) for ac PV modules replaces the requirements of 690.5 that apply only to conventional dc PV modules. As in 690.5, this is a fire prevention device and is not intended to be a shock prevention device. Existing GFCI and equipment ground-fault protection devices are generally not listed for back feeding and are not suitable for meeting this requirement.

(E) Overcurrent Protection The output circuits of ac modules shall be permitted to have overcurrent protection and conductor sizing in accordance with 240.5(B)(2).

II. Circuit Requirements

690.7 Maximum Voltage

(A) Maximum Photovoltaic System Voltage In a dc photovoltaic source circuit or output circuit, the maximum photovoltaic system voltage for that circuit shall be calculated as the sum of the rated open-circuit voltage of the series-connected photovoltaic modules corrected for the lowest expected ambient temperature. For crystalline and multicrystalline silicon modules, the rated open-circuit voltage shall be multiplied by the correction factor provided in Table 690.7. This voltage shall be used to determine the voltage rating of cables, disconnects, overcurrent devices, and other equipment. Where the lowest expected ambient temperature is below −40°C (−40°F), or where other than crystalline or multicrystalline silicon photovoltaic modules are used, the system voltage adjustment shall be made in accordance with the manufacturer's instructions.

Table 690.7 Voltage Correction Factors for Crystalline and Multicrystalline Silicon Modules

Ambient Temperature (°C)	Correction Factors for Ambient Temperatures Below 25°C (77°F) (Multiply the rated open-circuit voltage by the appropriate correction factor shown below.)	Ambient Temperature (°F)
25 to 10	1.06	77 to 50
9 to 0	1.10	49 to 32
−1 to −10	1.13	31 to 14
−11 to −20	1.17	13 to −4
−21 to −40	1.25	−5 to −40

A PV source is not a constant-voltage source, and the difference between the rated operating voltage determined under controlled laboratory conditions and the open-circuit voltage under field-installed conditions can be significant. Consequently, the higher-rated open-circuit voltage must be used to select circuit components with proper voltage ratings.

The voltage potential (both open circuit and operating) of a PV power source increases with decreasing temperature. The installer should note the temperature conditions under which the PV device was rated. If the anticipated lowest temperature at the installation site is lower than the rating condition (25°C), Table 690.7 should be used to adjust the maximum open-circuit voltage of the crystalline system before conductors, overcurrent devices, and switchgear are selected.

The temperature adjustment requirement is also included in the instructions provided with listed PV modules and is established at a fixed 125 percent, regardless of temperature. Table 690.7 supersedes the fixed 125 percent where the lowest temperature is above –40°C.

Bipolar PV systems (with positive and negative voltages) must use the sum of the absolute values of the open-circuit voltages to determine the rated open-circuit system voltage. For example, a system with open-circuit voltages

of +480 volts and -480 volts with respect to ground would have a system open-circuit voltage of 960 volts. This voltage should be multiplied by a temperature-dependent factor from Table 690.7, yielding a system design voltage of up to 1200 volts. This system design voltage should be used in the selection of cables and other equipment. Also, see the definition of *photovoltaic system voltage* in 690.2. Certain methods of connecting bipolar PV arrays meeting the requirements of 690.7(E) may have different requirements for calculating the maximum system voltage.

(B) Direct-Current Utilization Circuits The voltage of dc utilization circuits shall conform with 210.6.

The requirements of 690.7(B) cover installations where the PV output is connected to dc utilization circuits.

(C) Photovoltaic Source and Output Circuits In one- and two-family dwellings, photovoltaic source circuits and photovoltaic output circuits that do not include lampholders, fixtures, or receptacles shall be permitted to have a maximum photovoltaic system voltage up to 600 volts. Other installations with a maximum photovoltaic system voltage over 600 volts shall comply with Article 690, Part I.

PV dc circuits in buildings are permanently connected using wiring systems recognized by this *Code*. Requirements for protecting unqualified persons from contact with these circuits are included in 690.7(B) and 690.7(D). Unqualified persons are not likely to service equipment in these circuits due to its complexity. There is a significant difference between the rated open-circuit voltage and the operating voltage in PV dc circuits. For the PV system to perform its intended function, rated dc open-circuit voltages of up to 600 volts may be present.

(D) Circuits Over 150 Volts to Ground In one- and two-family dwellings, live parts in photovoltaic source circuits and photovoltaic output circuits over 150 volts to ground shall not be accessible to other than qualified persons while energized.

Where dc circuitry over 150 volts to ground is present in one- and two-family dwellings, additional protection for unqualified persons may be needed. Protection may be in the form of conduit, a closed cabinet, or an enclosure that requires the use of tools to open it and that permits entry only by qualified persons.

FPN: See 110.27 for guarding of live parts, and 210.6 for voltage to ground and between conductors.

(E) Bipolar Source and Output Circuits For 2-wire circuits connected to bipolar systems, the maximum system voltage shall be the highest voltage between the conductors of the 2-wire circuit if all of the following conditions apply:

(1) One conductor of each circuit is solidly grounded.
(2) Each circuit is connected to a separate subarray.
(3) The equipment is clearly marked with a label as follows:

> WARNING
> BIPOLAR PHOTOVOLTAIC ARRAY.
> DISCONNECTION OF NEUTRAL OR
> GROUNDED CONDUCTORS MAY RESULT IN
> OVERVOLTAGE ON ARRAY OR INVERTER.

690.8 Circuit Sizing and Current

(A) Calculation of Maximum Circuit Current The maximum current for the specific circuit shall be calculated in accordance with 690.8(A)(1) through (A)(4).

(1) Photovoltaic Source Circuit Currents The maximum current shall be the sum of parallel module rated short-circuit currents multiplied by 125 percent.

The use of the array short-circuit current allows for proper sizing of conductors to handle the current generated during extended periods of operation under a short-circuit current operating point.

The 125 percent factor mentioned in 690.8(A)(1) is required because PV modules, PV source circuits, and PV output circuits can deliver output currents higher than the rated short-circuit currents for more than 3 hours near solar noon. Prior to the 1999 edition of the *Code*, this requirement was specified in the instructions provided with each listed module. This 125 percent requirement is in addition to the 125 percent factor required by 690.8(B).

PV modules in hot climates operate at temperatures of 60°C to 80°C due to solar heating. Conductors with insulation types rated at least 90°C should be used, and these conductors should have the ampacity corrected in accordance with Table 310.16 or Table 310.17.

(2) Photovoltaic Output Circuit Currents The maximum current shall be the sum of parallel source circuit maximum currents as calculated in 690.8(A)(1).

(3) Inverter Output Circuit Current The maximum current shall be the inverter continuous output current rating.

Both stand-alone and utility-interactive inverters are power-limited devices. Output circuits connected to these devices are sized on the continuous rated outputs of these devices

and are not based on load calculations or reduced-size PV arrays or battery banks, if any. Some inverters may have specifications listing sustained maximum output currents, and the higher of this number or the rated output should be used.

(4) Stand-Alone Inverter Input Circuit Current The maximum current shall be the stand-alone continuous inverter input current rating when the inverter is producing rated power at the lowest input voltage.

Stand-alone inverters are nearly constant-output-voltage devices. As the input battery voltage decreases, the input battery current increases to maintain a constant ac output power. The input current for such inverters is calculated by taking the rated full-power output of the inverter in watts and dividing it by the lowest operating battery voltage and then by the rated efficiency of the inverter under those operating conditions. For example, the input current for a 4000-watt, 24-volt inverter that is 85 percent efficient at 22 volts can be calculated as follows:

$$\text{Ampere input} = \frac{\text{watt output}}{\text{voltage} \times \text{efficiency}}$$
$$= \frac{4000 \text{ W}}{22 \text{ V} \times 0.85}$$
$$= 214 \text{ A}$$

Ripple currents might be present in the dc-input circuits of single-phase, stand-alone inverters. These ripple currents might cause nuisance operation of overcurrent devices at continuous high inverter outputs. In such cases, the measured maximum true rms (root mean square) value of the total (ac + dc) input current, which will be greater than the average current calculated here, should be used to determine conductor sizes and overcurrent device ratings.

(B) Ampacity and Overcurrent Device Ratings Photovoltaic system currents shall be considered to be continuous.

(1) Sizing of Conductors and Overcurrent Devices The circuit conductors and overcurrent devices shall be sized to carry not less than 125 percent of the maximum currents as calculated in 690.8(A). The rating or setting of overcurrent devices shall be permitted in accordance with 240.4(B) and (C).

Exception: Circuits containing an assembly, together with its overcurrent device(s), that is listed for continuous operation at 100 percent of its rating shall be permitted to be utilized at 100 percent of its rating.

The exception to 690.8(B)(1) permits use at the full rating of assemblies, such as panelboards, incorporating overcurrent devices listed for continuous operation at 100 percent of the rating.

(2) Internal Current Limitation Overcurrent protection for photovoltaic output circuits with devices that internally limit the current from the photovoltaic output circuit shall be permitted to be rated at less than the value calculated in 690.8(B)(1). This reduced rating shall be at least 125 percent of the limited current value. Photovoltaic output circuit conductors shall be sized in accordance with 690.8(B)(1).

Exception: An overcurrent device in an assembly listed for continuous operation at 100 percent of its rating shall be permitted to be utilized at 100 percent of its rating.

(C) Systems with Multiple Direct-Current Voltages For a photovoltaic power source that has multiple output circuit voltages and employs a common-return conductor, the ampacity of the common-return conductor shall not be less than the sum of the ampere ratings of the overcurrent devices of the individual output circuits.

(D) Sizing of Module Interconnection Conductors Where a single overcurrent device is used to protect a set of two or more parallel-connected module circuits, the ampacity of each of the module interconnection conductors shall not be less than the sum of the rating of the single fuse plus 125 percent of the short-circuit current from the other parallel-connected modules.

Normally, labeling or module instructions require reverse overcurrent protection for each module or string of modules. In some cases, modules with low rated short-circuit currents and high values of the required series protective fuse may allow the use of one overcurrent device to provide reverse current protection for two modules or strings of modules and overcurrent protection for the conductors. The PV module manufacturer should be contacted for specific information.

690.9 Overcurrent Protection

(A) Circuits and Equipment Photovoltaic source circuit, photovoltaic output circuit, inverter output circuit, and storage battery circuit conductors and equipment shall be protected in accordance with the requirements of Article 240. Circuits connected to more than one electrical source shall have overcurrent devices located so as to provide overcurrent protection from all sources.

Exception: An overcurrent device shall not be required for circuit conductors sized in accordance with 690.8(B) and located where one of the following apply:

(a) There are no external sources such as parallel-connected source circuits, batteries, or backfeed from inverters.

(b) The short-circuit currents from all sources do not exceed the ampacity of the conductors.

FPN: Possible backfeed of current from any source of supply, including a supply through an inverter into the photovoltaic output circuit and photovoltaic source circuits, is a consideration in determining whether adequate overcurrent protection from all sources is provided for conductors and modules.

In the circuits illustrated in Exhibits 690.2, 690.3, 690.4, 690.5, and 690.7, the PV source-circuit overcurrent devices are required to be rated so that the source-circuit conductors are protected in accordance with Article 240 and the overcurrent device ratings do not exceed the maximum overcurrent device rating marked on the modules. Possible backfeed currents from the other PV source circuits, other supply sources through the inverter, and storage-battery circuits, if any, have to be considered.

Blocking diodes (possibly required by the module manufacturer for specific applications) can lose their blocking ability because of overtemperature conditions or internal breakdown. Therefore, overcurrent protection has to be considered with a condition of shorted blocking diodes if they are used in the circuit.

At the inverter or battery/charge controller end of the PV output circuit, the need for overcurrent protection has to be considered with respect to the maximum backfeed fault current available from other sources.

Exhibit 690.7 Application of blocking diodes and source-circuit overcurrent devices.

(B) Power Transformers Overcurrent protection for a transformer with a source(s) on each side shall be provided in accordance with 450.3 by considering first one side of the transformer, then the other side of the transformer, as the primary.

Exception: A power transformer with a current rating on the side connected toward the photovoltaic power source, not less than the short-circuit output current rating of the inverter, shall be permitted without overcurrent protection from that source.

(C) Photovoltaic Source Circuits Branch-circuit or supplementary-type overcurrent devices shall be permitted to provide overcurrent protection in photovoltaic source circuits. The overcurrent devices shall be accessible but shall not be required to be readily accessible.

Standard values of supplementary overcurrent devices allowed by this section shall be in one ampere size increments, starting at one ampere up to and including 15 amperes. Higher standard values above 15 amperes for supplementary overcurrent devices shall be based on the standard sizes provided in 240.6(A).

If the overcurrent protection of PV source circuits is considered supplementary overcurrent protection, use of overcurrent devices with ratings other than those suitable for branch-circuit protection is permitted. The use of such devices permits module protection closer to the specified ratings required on the labels attached to listed modules. It is anticipated that only qualified service personnel will replace or reset overcurrent devices in PV source circuits. Consequently, ready access to the user need not be provided. These supplementary overcurrent devices must be listed for dc operation and have appropriate voltage and current ratings.

(D) Direct-Current Rating Overcurrent devices, either fuses or circuit breakers, used in any dc portion of a photovoltaic power system shall be listed for use in dc circuits and shall have the appropriate voltage, current, and interrupt ratings.

Direct-current fault currents are considerably harder to interrupt than ac faults. Overcurrent devices marked or listed only for ac use should not be used in dc circuits. Automotive- and marine-type fuses, although used in these dc systems, are not suitable for use in permanently wired residential or commercial electrical power systems meeting the requirements of the *Code*.

(E) Series Overcurrent Protection In series-connected strings of two or more modules, a single overcurrent protection device shall be permitted.

The single overcurrent device (when required) may provide both the reverse-current protection required for the series-connected PV modules and the overcurrent protection required for the interconnecting conductors.

690.10 Stand-Alone Systems

The premises wiring system shall be adequate to meet the requirements of this *Code* for a similar installation connected to a service. The wiring on the supply side of the building or structure disconnecting means shall comply with this *Code* except as modified by 690.10(A), (B), and (C).

(A) Inverter Output The ac inverter output from a stand-alone system shall be permitted to supply ac power to the building or structure disconnecting means at current levels below the rating of that disconnecting means.

A stand-alone residential or commercial PV installation may have an ac output and be connected to a building wired in full compliance with all articles of this *Code*. Even though such an installation may have service-entrance equipment rated at 100 or 200 amperes at 120/240 volts, there is no requirement that the PV source provide either the rated full current or the dual voltages of the service equipment. While safety requirements dictate full compliance with the ac wiring sections of the *Code*, a PV installation is usually designed so that the actual ac demands on the system are sized to the output rating of the PV system.

(B) Sizing and Protection The circuit conductors between the inverter output and the building or structure disconnecting means shall be sized based on the output rating of the inverter. These conductors shall be protected from overcurrents in accordance with Article 240. The overcurrent protection shall be located at the output of the inverter.

(C) Single 120-Volt Supply The inverter output of a stand-alone solar photovoltaic system shall be permitted to supply 120 volts to single-phase, 3-wire, 120/240-volt service equipment or distribution panels where there are no 240-volt outlets and where there are no multiwire branch circuits. In all installations, the rating of the overcurrent device connected to the output of the inverter shall be less than the rating of the neutral bus in the service equipment. This equipment shall be marked with the following words or equivalent:

WARNING
SINGLE 120-VOLT SUPPLY. DO NOT CONNECT
MULTIWIRE BRANCH CIRCUITS!

Multiwire branch circuits are common in one- and two-family dwelling units. When connected to a normal 120/240-volt ac service, the currents in the neutral conductors of these multiwire branch circuits (typically 14-3 AWG) subtract or are, at most, no larger than the rating of the branch-circuit overcurrent device. When these electrical systems are connected to a single 120-volt PV power system inverter by paralleling the two ungrounded conductors in the service entrance load center, the currents in the neutral conductor for each multiwire branch circuit add rather than subtract. The currents in the neutral conductor may be as high as twice the rating of the branch-circuit overcurrent device. With this configuration, neutral conductor overloading is possible.

III. Disconnecting Means

690.13 All Conductors

Means shall be provided to disconnect all current-carrying conductors of a photovoltaic power source from all other conductors in a building or other structure. A switch or circuit breaker shall not be installed in a grounded conductor unless that switch or circuit breaker is part of a ground-fault detection system required by 690.5 and that switch or circuit breaker is automatically opened and indicated as a normal function of the device in responding to ground faults.

> FPN: The grounded conductor may have a bolted or terminal disconnecting means to allow maintenance or troubleshooting by qualified personnel.

690.14 Additional Provisions

Photovoltaic disconnecting means shall comply with 690.14(A) through 690.14(D).

(A) Disconnecting Means The disconnecting means shall not be required to be suitable as service equipment and shall be rated in accordance with 690.17.

A disconnecting means rated in accordance with 690.17 may be used, but it is not required to be marked suitable for use as service equipment.

(B) Equipment Equipment such as photovoltaic source circuit isolating switches, overcurrent devices, and blocking diodes shall be permitted on the photovoltaic side of the photovoltaic disconnecting means.

In general, equipment that needs servicing must be disconnected from sources of supply. In a PV system, however, some equipment, as indicated, is permitted to be located on the PV power source side of the disconnecting means. See Exhibit 690.8. Servicing the exempted equipment might require disabling all or portions of the array, as explained in the commentary following 690.18.

There is no intent or requirement to have a disconnecting means located in each PV source circuit or located physically at each PV module location. Unlike load circuits (e.g., rooftop air conditioners), PV source-circuit conductors may be

energized at any time from the PV modules. A centrally located disconnect meeting the requirements of 690.14(C)(1) near the inverter or batteries serves to disconnect the PV source circuits from the other portions of the electrical power system.

Exhibit 690.8 Equipment permitted on the PV power source side of the PV power source disconnecting means.

(C) Requirements for Disconnecting Means Means shall be provided to disconnect all conductors in a building or other structure from the photovoltaic system conductors.

(1) Location The photovoltaic disconnecting means shall be installed at a readily accessible location either on the outside of a building or structure or inside nearest the point of entrance of the system conductors.

The photovoltaic system disconnecting means shall not be installed in bathrooms.

These requirements generally prohibit long runs of PV source and output circuits inside a building before reaching the required PV disconnect. A short conductor run through a wall at the point of first penetration to reach a disconnect mounted inside the building is allowed. Section 690.31(E) permits these circuits to be run inside a building when installed in metal conduit from the point of entrance to the system disconnecting means.

Exception: Installations that comply with 690.31(E) shall be permitted to have the disconnecting means located remote from the point of entry of the system conductors.

(2) Marking Each photovoltaic system disconnecting means shall be permanently marked to identify it as a photovoltaic system disconnect.

(3) Suitable for Use Each photovoltaic system disconnecting means shall be suitable for the prevailing conditions. Equipment installed in hazardous (classified) locations shall comply with the requirements of Articles 500 through 517.

(4) Maximum Number of Disconnects The photovoltaic system disconnecting means shall consist of not more than six switches or six circuit breakers mounted in a single enclosure, in a group of separate enclosures, or in or on a switchboard.

(5) Grouping The photovoltaic system disconnecting means shall be grouped with other disconnecting means for the system to comply with 690.14(C)(4). A photovoltaic disconnecting means shall not be required at the photovoltaic module or array location.

PV systems may be one of multiple sources of power for a building or structure, including the utility, the PV array, a backup generator, and a wind system. No more than six disconnects for each source of power to the building are allowed, and the disconnects for each source should be grouped together. A PV system may be considered a source of supply separate from a utility source, and each source may have up to six grouped disconnects. See 230.2 and 230.72.

(D) Utility-Interactive Inverters Mounted in Not-Readily-Accessible Locations Utility-interactive inverters shall be permitted to be mounted on roofs or other exterior areas that are not readily accessible. These installations shall comply with (1) through (4):

(1) A direct-current photovoltaic disconnecting means shall be mounted within sight of or in the inverter.

(2) An alternating-current disconnecting means shall be mounted within sight of or in the inverter.

The requirements in 690.14(D)(1) and 690.14(D)(2) provide for servicing disconnects at the inverter.

(3) The alternating-current output conductors from the inverter and an additional alternating-current disconnecting means for the inverter shall comply with 690.14(C)(1).

The disconnect required by 690.14(C)(1) allows the inverter and the circuit to it to be de-energized from a readily accessible location.

(4) A plaque shall be installed in accordance with 705.10.

690.15 Disconnection of Photovoltaic Equipment

Means shall be provided to disconnect equipment, such as inverters, batteries, charge controllers, and the like, from all ungrounded conductors of all sources. If the equipment is energized from more than one source, the disconnecting means shall be grouped and identified.

A single disconnecting means in accordance with 690.17 shall be permitted for the combined ac output of one or more inverters or ac modules in an interactive system.

690.16 Fuses

Disconnecting means shall be provided to disconnect a fuse from all sources of supply if the fuse is energized from both directions and is accessible to other than qualified persons. Such a fuse in a photovoltaic source circuit shall be capable of being disconnected independently of fuses in other photovoltaic source circuits.

Switches, pullouts, or similar devices that have suitable ratings may serve as means to disconnect fuses from all sources of supply.

690.17 Switch or Circuit Breaker

The disconnecting means for ungrounded conductors shall consist of a manually operable switch(es) or circuit breaker(s) complying with all of the following requirements:

(1) Located where readily accessible
(2) Externally operable without exposing the operator to contact with live parts
(3) Plainly indicating whether in the open or closed position
(4) Having an interrupting rating sufficient for the nominal circuit voltage and the current that is available at the line terminals of the equipment

Where all terminals of the disconnecting means may be energized in the open position, a warning sign shall be mounted on or adjacent to the disconnecting means. The sign shall be clearly legible and have the following words or equivalent:

WARNING
ELECTRIC SHOCK HAZARD.
DO NOT TOUCH TERMINALS. TERMINALS
ON BOTH THE LINE AND
LOAD SIDES MAY BE ENERGIZED
IN THE OPEN POSITION.

Exception: A connector shall be permitted to be used as an ac or a dc disconnecting means, provided that it complies with the requirements of 690.33 and is listed and identified for the use.

690.18 Installation and Service of an Array

Open circuiting, short circuiting, or opaque covering shall be used to disable an array or portions of an array for installation and service.

> FPN: Photovoltaic modules are energized while exposed to light. Installation, replacement, or servicing of array components while a module(s) is irradiated may expose persons to electric shock.

To prevent contact by personnel with energized parts during installation, servicing, or other procedures, a number of methods can be used to disable an array or portions of an array. One method, used infrequently because of the expense in time and materials, is to cover all of the array or portions of it with an opaque material. Care must be taken that all of the area to be covered is shielded from light.

Another method divides the array into nonhazardous segments, which can be accomplished by switches or connectors. Also see 690.33.

Short-circuiting all or portions of an array by means of switches or plug-in connectors, in conjunction with bypass diodes, can also provide the necessary disablement. (Bypass diodes are incorporated in PV modules for performance purposes.)

IV. Wiring Methods

690.31 Methods Permitted

(A) **Wiring Systems** All raceway and cable wiring methods included in this *Code* and other wiring systems and fittings specifically intended and identified for use on photovoltaic arrays shall be permitted. Where wiring devices with integral enclosures are used, sufficient length of cable shall be provided to facilitate replacement.

> FPN: Photovoltaic modules operate at elevated temperatures when exposed to high ambient temperatures and to bright sunlight. These temperatures may routinely exceed 70°C (158°F) in many locations. Module interconnection conductors are available with insulation rated for wet locations and a temperature rating of 90°C (194°F) or greater.

(B) **Single-Conductor Cable** Types SE, UF, USE, and USE-2 single-conductor cable shall be permitted in photovoltaic source circuits where installed in the same manner as a Type UF multiconductor cable in accordance with Part II of Article 340. Where exposed to sunlight, Type UF cable identified as sunlight-resistant shall be used.

Some PV modules are designed for a direct series connection by having terminations located at both ends. To accommo-

date such a direct series connection without the waste of one or more conductors in a multiconductor cable, use of a single-conductor Type USE, SE, UF, or USE-2 cable is permitted in PV source circuits. The Article 340 reference to installation as a multiconductor cable permits the single-conductor cable to be routed separately. Long runs of separated conductors (with loop inductance and distributed capacitance) and the resulting long-time constants in dc circuits may result in improper operation of overcurrent devices. It is suggested that, wherever possible, both positive and negative conductors of each circuit and the equipment grounding conductor be routed as close together as possible to minimize the circuit time constant. The smaller loop resulting from the close routing also decreases induced currents from nearby lightning strikes. Because PV modules may operate at high temperatures and are installed in outdoor, exposed locations, the use of high-temperature, wet-rated conductors such as USE-2 is suggested. UF conductors might not have the necessary temperature ratings required.

(C) **Flexible Cords and Cables** Flexible cords and cables, where used to connect the moving parts of tracking PV modules, shall comply with Article 400 and shall be of a type identified as a hard-service cord or portable power cable; they shall be suitable for extra-hard usage, listed for outdoor use, water resistant, and sunlight resistant. Allowable ampacities shall be in accordance with 400.5. For ambient temperatures exceeding 30°C (86°F), the ampacities shall be derated by the appropriate factors given in Table 690.31(C).

(D) **Small-Conductor Cables** Single-conductor cables listed for outdoor use that are sunlight resistant and moisture resistant in sizes 16 AWG and 18 AWG shall be permitted for module interconnections where such cables meet the ampacity requirements of 690.8. Section 310.15 shall be used to determine the cable ampacity and temperature derating factors.

Because these smaller cables are not normally marked with standard *Code*-recognized markings (e.g., USE-2, UF, SE), the PV module manufacturer or installer should provide information establishing that these cables have the necessary sunlight and moisture resistance and are suitable for exposed, outdoor use.

In accordance with 200.6(A), grounded conductors smaller than 6 AWG that are used in PV source circuits are permitted to be marked with a white marking.

(E) **Direct-Current Photovoltaic Source and Output Circuits Inside a Building** Where direct current photovol-

Table 690.31(C) Correction Factors

Ambient Temperature (°C)	Temperature Rating of Conductor				Ambient Temperature (°F)
	60°C (140°F)	75°C (167°F)	90°C (194°F)	105°C (221°F)	
30	1.00	1.00	1.00	1.00	86
31–35	0.91	0.94	0.96	0.97	87–95
36–40	0.82	0.88	0.91	0.93	96–104
41–45	0.71	0.82	0.87	0.89	105–113
46–50	0.58	0.75	0.82	0.86	114–122
51–55	0.41	0.67	0.76	0.82	123–131
56–60	—	0.58	0.71	0.77	132–140
61–70	—	0.33	0.58	0.68	141–158
71–80	—	—	0.41	0.58	159–176

taic source or output circuits of a utility-interactive inverter from a building-integrated or other photovoltaic system are run inside a building or structure, they shall be contained in metallic raceways or enclosures from the point of penetration of the surface of the building or structure to the first readily accessible disconnecting means. The disconnecting means shall comply with 690.14(A) through 690.14(D).

The use of metallic raceways inside a building provides additional physical protection for these circuits. Metallic raceways also provide additional fire resistance should faults develop in the cable, and they provide an additional ground fault detection path for the ground-fault protection device required by 690.5.

690.32 Component Interconnections

Fittings and connectors that are intended to be concealed at the time of on-site assembly, where listed for such use, shall be permitted for on-site interconnection of modules or other array components. Such fittings and connectors shall be equal to the wiring method employed in insulation, temperature rise, and fault-current withstand, and shall be capable of resisting the effects of the environment in which they are used.

690.33 Connectors

The connectors permitted by Article 690 shall comply with 690.33(A) through 690.33(E).

(A) Configuration The connectors shall be polarized and shall have a configuration that is noninterchangeable with receptacles in other electrical systems on the premises.

(B) Guarding The connectors shall be constructed and installed so as to guard against inadvertent contact with live parts by persons.

(C) Type The connectors shall be of the latching or locking type.

(D) Grounding Member The grounding member shall be the first to make and the last to break contact with the mating connector.

(E) Interruption of Circuit The connectors shall be capable of interrupting the circuit current without hazard to the operator.

Connectors marked "Do not open under load" are generally acceptable.

690.34 Access to Boxes

Junction, pull, and outlet boxes located behind modules or panels shall be so installed that the wiring contained in them can be rendered accessible directly or by displacement of a module(s) or panel(s) secured by removable fasteners and connected by a flexible wiring system.

690.35 Ungrounded Photovoltaic Power Systems

Photovoltaic power systems shall be permitted to operate with ungrounded photovoltaic source and output circuits where the system complies with 690.35(A) through 690.35(G).

(A) Disconnects All photovoltaic source and output circuit conductors shall have disconnects complying with 690, Part III.

(B) Overcurrent Protection All photovoltaic source and output circuit conductors shall have overcurrent protection complying with 690.9.

(C) Ground-Fault Protection All photovoltaic source and output circuits shall be provided with a ground-fault protection device or system that complies with (1) through (3):

(1) Detects a ground fault.
(2) Indicates that a ground fault has occurred

(3) Automatically disconnects the conductors and/or shuts off the utility-interactive inverter or charge controller for that portion of the faulted array

(D) The photovoltaic source and output conductors shall consist of sheathed (jacketed) multi-conductor cables or shall be installed in a raceway.

Since there are no double-insulated, single-conductor cables available or listed in the United States, a multiconductor sheathed (jacketed) cable was included as an option to provide the double layers of insulation without using conduit. Single-conductor, double-insulated cables, when listed to an appropriate UL standard, meet the intent of this requirement.

(E) The photovoltaic power system direct-current circuits shall be permitted to be used with ungrounded battery systems complying with 690.71(G).

(F) The photovoltaic power source shall be labeled with the following warning at each junction box, combiner box, disconnect, and device where the ungrounded circuits may be exposed during service:

WARNING
ELECTRIC SHOCK HAZARD. THE DIRECT
CURRENT CIRCUIT CONDUCTORS OF THIS
PHOTOVOLTAIC POWER SYSTEM ARE
UNGROUNDED BUT MAY BE ENERGIZED
WITH RESPECT TO GROUND DUE TO
LEAKAGE PATHS AND/OR GROUND FAULTS.

PV dc circuits operate in outdoor environments and are expected to be energized for 40 years or more. Aging of the conductors, dust and dirt infiltration, and moisture and water intrusion create leakage paths from the conductors to ground. These high-resistance leakage paths can result in leakage current values less than those detected by the required ground-fault detection device, but they can cause any ungrounded conductor to become a potential shock hazard with respect to ground.

(G) The inverters or charge controllers used in systems with ungrounded photovoltaic source and output circuits shall be listed for the purpose.

Many types of PV equipment are designed to operate only on grounded systems. Equipment to be used on ungrounded systems must be tested and evaluated for such use.

V. Grounding

690.41 System Grounding

For a photovoltaic power source, one conductor of a two-wire system with a photovoltaic system voltage over 50

volts and the reference (center tap) conductor of a bipolar system shall be solidly grounded or shall use other methods that accomplish equivalent system protection in accordance with 250.4(A) and that utilize equipment listed and identified for the use.

Low-voltage systems that are not grounded must have overcurrent protection in each of the ungrounded conductors, as required by 240.21.

Other methods that employ available equipment may be used to achieve objectives contained in 250.4(A), thereby providing protection for the PV power source circuits equivalent to solid grounding.

Exception: Systems complying with 690.35.

690.42 Point of System Grounding Connection

The dc circuit grounding connection shall be made at any single point on the photovoltaic output circuit.

FPN: Locating the grounding connection point as close as practicable to the photovoltaic source better protects the system from voltage surges due to lightning.

If other than solid grounding is utilized, as permitted by 690.41, the connections should be made in accordance with the markings found on the equipment or in the installation instructions.

Stand-alone PV systems might require the grounding connection point to be located close to the high-current conductors associated with the battery and the inverter.

PV systems on the roofs of dwellings that require ground-fault protection devices (see 690.5) might require that the single-point grounding connection be made inside the ground-fault protection equipment or inside the utility-interactive inverter. Connections should be made in accordance with markings on the equipment or in the installation instructions.

690.43 Equipment Grounding

Exposed non–current-carrying metal parts of module frames, equipment, and conductor enclosures shall be grounded in accordance with 250.134 or 250.136(A) regardless of voltage.

Equipment grounding is required even in low-voltage (12- and 24-volt) systems not otherwise required to have a system ground. A grounding electrode must be added to an ungrounded system to accommodate the equipment grounds.

To maintain the shortest electrical time constant in each

dc circuit, the equipment grounding conductor should be routed as close as possible to the circuit conductors. This facilitates the operation of overcurrent devices.

690.45 Size of Equipment Grounding Conductor

Where not protected by the ground-fault protection equipment required by 690.5, the equipment-grounding conductor for photovoltaic source and photovoltaic output circuits shall be sized for 125 percent of the photovoltaic-originated short-circuit currents in that circuit. Where protected by the ground-fault protection equipment required by 690.5, the equipment-grounding conductors for photovoltaic source and photovoltaic output circuits shall be sized in accordance with 250.122.

In systems where a ground-fault protection device according to 690.5 is not used, and the circuit conductors are oversized for voltage drop, the requirements of 250.122(B) must be followed.

Some PV systems (primarily utility-interactive) with ground-fault protection devices do not require overcurrent devices in the PV source or output circuits, and 250.122 cannot be used to size the equipment grounding conductors. In these cases, the equipment grounding conductor for a PV source or output circuit should be sized at least 125 percent of the PV-originated short-circuit current in that circuit.

690.47 Grounding Electrode System

(A) Alternating-Current Systems If installing an ac system, a grounding electrode system shall be provided in accordance with 250.50 through 250.60. The grounding electrode conductor shall be installed in accordance with 250.64.

(B) Direct-Current Systems If installing a dc system, a grounding electrode system shall be provided in accordance with 250.166 for grounded systems or 250.169 for ungrounded systems. The grounding electrode conductor shall be installed in accordance with 250.64.

(C) Systems with Alternating-Current and Direct-Current Grounding Requirements Photovoltaic power systems with both alternating-current and direct-current (dc) grounding requirements shall be permitted to be grounded as described in (1) or (2):

Inverters used in PV power systems usually contain a transformer that isolates the dc grounded circuit conductor from the ac grounded circuit conductor. This isolation necessitates that both a dc and an ac grounding system be installed. The two grounding systems are to be bonded together or have a common grounding electrode so that all ac and dc grounded

circuit conductors and equipment grounding conductors have the same near-zero potential to earth.

(1) A grounding-electrode conductor shall be connected between the identified dc grounding point to a separate dc grounding electrode. The dc grounding-electrode conductor shall be sized according to 250.166. The dc grounding electrode shall be bonded to the ac grounding electrode to make a grounding electrode system according to 250.52 and 250.53. The bonding conductor shall be no smaller than the largest grounding electrode conductor, either ac or dc.

(2) The dc grounding electrode conductor and ac grounding electrode conductor shall be connected to a single grounding electrode. The separate grounding electrode conductors shall be sized as required by 250.66 (ac) and 250.166 (dc).

690.48 Continuity of Equipment Grounding Systems

Where the removal of equipment disconnects the bonding connection between the grounding electrode conductor and exposed conducting surfaces in the photovoltaic source or output circuit equipment, a bonding jumper shall be installed while the equipment is removed.

PV source and output circuits are energized anytime the PV modules are exposed to light. The equipment grounding system is a primary line of defense against electric shocks and fires. In many PV systems, the main bonding jumper and the equipment grounding bus bar are located in the inverter or a dc power center that may require removal for service. The continuity of the equipment grounding conductors should be maintained even when the equipment is removed.

690.49 Continuity of Photovoltaic Source and Output Circuit Grounded Conductors

Where the removal of the utility-interactive inverter or other equipment disconnects the bonding connection between the grounding electrode conductor and the photovoltaic source and/or photovoltaic output circuit grounded conductor, a bonding jumper shall be installed to maintain the system grounding while the inverter or other equipment is removed.

PV source and output circuits are energized anytime the PV modules are exposed to light. The marked, grounded circuit conductors should always remain grounded because they may be energized daily and cannot be easily disconnected from the source. In many PV systems, the main bonding jumper is located in the inverter or a dc power center that may require removal for service. The continuity to ground of the grounded circuit conductors should be maintained even when the equipment is removed.

VI. Marking

690.51 Modules

Modules shall be marked with identification of terminals or leads as to polarity, maximum overcurrent device rating for module protection, and with the following ratings:

(1) Open-circuit voltage
(2) Operating voltage
(3) Maximum permissible system voltage
(4) Operating current
(5) Short-circuit current
(6) Maximum power

690.52 Alternating-Current Photovoltaic Modules

Alternating-current modules shall be marked with identification of terminals or leads and with identification of the following ratings:

(1) Nominal operating ac voltage
(2) Nominal operating ac frequency
(3) Maximum ac power
(4) Maximum ac current
(5) Maximum overcurrent device rating for ac module protection

690.53 Direct-Current Photovoltaic Power Source

A marking for the direct-current photovoltaic power source indicating items (1) through (4) shall be provided by the installer at an accessible location at the disconnecting means for this power source:

(1) Operating current
(2) Operating voltage
(3) Maximum system voltage
(4) Short-circuit current

> FPN: Reflecting systems used for irradiance enhancement may result in increased levels of output current and power.

After installation of PV arrays, it may be difficult to determine the system's rated voltage and current. These ratings, along with the open-circuit voltage and short-circuit current, are necessary to size the remainder of the system components, as specified elsewhere in Article 690.

Generally, the marking described in 690.53 is required to be provided by the installer. The rated values for the PV power source can be calculated by adding voltage ratings of series-connected modules and adding current ratings of parallel-connected modules or PV source circuits.

With respect to the fine print note, a deliberate increase in the level of irradiance by reflectors or the like can cause the power source to operate at levels above those recommended by the manufacturer. See 110.3.

690.54 Interactive System Point of Interconnection

All interactive system(s) points of interconnection with other sources shall be marked at an accessible location at the disconnecting means as a power source with the maximum ac output operating current and the operating ac voltage.

690.55 Photovoltaic Power Systems Employing Energy Storage

Photovoltaic power systems employing energy storage shall also be marked with the maximum operating voltage, including any equalization voltage and the polarity of the grounded circuit conductor.

690.56 Identification of Power Sources

(A) Facilities with Stand-Alone Systems Any structure or building with a photovoltaic power system that is not connected to a utility service source and is a stand-alone system shall have a permanent plaque or directory installed on the exterior of the building or structure at a readily visible location acceptable to the authority having jurisdiction. The plaque or directory shall indicate the location of system disconnecting means and that the structure contains a stand-alone electrical power system.

(B) Facilities with Utility Services and PV Systems Buildings or structures with both utility service and a photovoltaic system shall have a permanent plaque or directory providing the location of the service disconnecting means and the photovoltaic system disconnecting means, if not located at the same location.

VII. Connection to Other Sources

690.60 Identified Interactive Equipment

Only inverters and ac modules listed and identified as interactive shall be permitted in interactive systems.

690.61 Loss of Interactive System Power

An inverter or an ac module in an interactive solar photovoltaic system shall automatically de-energize its output to the connected electrical production and distribution network upon loss of voltage in that system and shall remain in that state until the electrical production and distribution network voltage has been restored.

A normally interactive solar photovoltaic system shall be permitted to operate as a stand-alone system to supply loads that have been disconnected from electrical production and distribution network sources.

The requirement of 690.61 prevents energizing of otherwise de-energized system conductors or output conductors of other off-site sources (e.g., an electrical utility) and is intended to prevent electric shock. This feature normally is provided as part of the utility-interactive inverter.

690.62 Ampacity of Neutral Conductor

If a single-phase, 2-wire inverter output is connected to the neutral and one ungrounded conductor (only) of a 3-wire system or of a 3-phase, 4-wire wye-connected system, the maximum load connected between the neutral and any one ungrounded conductor plus the inverter output rating shall not exceed the ampacity of the neutral conductor.

690.63 Unbalanced Interconnections

(A) Single Phase Single-phase inverters for photovoltaic systems and ac modules in interactive solar photovoltaic systems shall not be connected to 3-phase power systems unless the interconnected system is designed so that significant unbalanced voltages cannot result.

(B) Three Phase Three-phase inverters and 3-phase ac modules in interactive systems shall have all phases automatically de-energized upon loss of, or unbalanced, voltage in one or more phases unless the interconnected system is designed so that significant unbalanced voltages will not result.

690.64 Point of Connection

The output of a photovoltaic power source shall be connected as specified in 690.64(A) or 690.64(B).

(A) Supply Side A photovoltaic power source shall be permitted to be connected to the supply side of the service disconnecting means as permitted in 230.82(6).

(B) Load Side A photovoltaic power source shall be permitted to be connected to the load side of the service disconnecting means of the other source(s) at any distribution equipment on the premises, provided that all of the following conditions are met:

(1) Each source interconnection shall be made at a dedicated circuit breaker or fusible disconnecting means.
(2) The sum of the ampere ratings of overcurrent devices in circuits supplying power to a busbar or conductor shall not exceed the rating of the busbar or conductor.

Exception: For a dwelling unit, the sum of the ampere ratings of the overcurrent devices shall not exceed 120 percent of the rating of the busbar or conductor.

(3) The interconnection point shall be on the line side of all ground-fault protection equipment.

Exception: Connection shall be permitted to be made to the load side of ground-fault protection, provided that there is ground-fault protection for equipment from all ground-fault current sources.

Load-side connection of energy sources to commonly available ac GFCI and ac equipment ground-fault protection circuit breakers may result in backfeed currents from the PV system output. Tests have shown that backfeed currents through these devices may damage them and prevent proper operation. Load-side connections to these devices should not be made.

(4) Equipment containing overcurrent devices in circuits supplying power to a busbar or conductor shall be marked to indicate the presence of all sources.

Exception: Equipment with power supplied from a single point of connection.

(5) Circuit breakers, if backfed, shall be identified for such operation. Dedicated circuit breakers backfed from listed utility-interactive inverters complying with 690.60 shall not be required to be individually clamped to the panelboard busbars. A front panel shall clamp all circuit breakers to the panelboard busbars. Main circuit breakers connected directly to energized feeders shall also be individually clamped.

Circuit breakers *not* marked "Line" and "Load" are considered to be identified as suitable for backfeeding.

VIII. Storage Batteries

690.71 Installation

(A) General Storage batteries in a solar photovoltaic system shall be installed in accordance with the provisions of Article 480. The interconnected battery cells shall be considered grounded where the photovoltaic power source is installed in accordance with 690.41.

Batteries in PV power systems are usually grounded when the PV power system is grounded in accordance with Article 690, Part VI.

(B) Dwellings

(1) Operating Voltage Storage batteries for dwellings shall have the cells connected so as to operate at less than 50 volts nominal. Lead-acid storage batteries for dwellings shall have no more than twenty-four 2-volt cells connected in series (48 volts nominal).

Exception: Where live parts are not accessible during routine battery maintenance, a battery system voltage in accordance with 690.7 shall be permitted.

(2) Guarding of Live Parts Live parts of battery systems for dwellings shall be guarded to prevent accidental contact by persons or objects, regardless of voltage or battery type.

> FPN: Batteries in solar photovoltaic systems are subject to extensive charge–discharge cycles and typically require frequent maintenance, such as checking electrolyte and cleaning connections.

At any voltage, a primary safety concern in battery systems is that a fault (e.g., a metal tool dropped onto a terminal) might cause a fire or an explosion. *Guarded*, as defined in Article 100, describes the best method to reduce this hazard.

(C) Current Limiting A listed, current-limiting, overcurrent device shall be installed in each circuit adjacent to the batteries where the available short-circuit current from a battery or battery bank exceeds the interrupting or withstand ratings of other equipment in that circuit. The installation of current-limiting fuses shall comply with 690.16.

Large banks of storage batteries can deliver significant amounts of short-circuit current. Current-limiting overcurrent devices should be used if necessary.

(D) Battery Nonconductive Cases and Conductive Racks Flooded, vented, lead-acid batteries with more than twenty-four 2-volt cells connected in series (48 volts, nominal) shall not use conductive cases or shall not be installed in conductive cases. Conductive racks used to support the nonconductive cases shall be permitted where no rack material is located within 150 mm (6 in.) of the tops of the nonconductive cases.

This requirement shall not apply to any type of valve-regulated lead-acid (VRLA) battery or any other types of sealed batteries that may require steel cases for proper operation.

Grounded metal trays and cases or containers (as normally required by 250.110) in flooded, lead-acid battery systems operating over 48 volts, nominal, have been shown to be a contributing factor in ground faults. Nonconductive racks, trays, and cases minimize this problem.

(E) Disconnection of Series Battery Circuits Battery circuits subject to field servicing, where more than twenty-

four 2-volt cells are connected in series (48 volts, nominal), shall have provisions to disconnect the series-connected strings into segments of 24 cells or less for maintenance by qualified persons. Non–load-break bolted or plug-in disconnects shall be permitted.

(F) Battery Maintenance Disconnecting Means Battery installations, where there are more than twenty-four 2-volt cells connected in series (48 volts, nominal), shall have a disconnecting means, accessible only to qualified persons, that disconnects the grounded circuit conductor(s) in the battery electrical system for maintenance. This disconnecting means shall not disconnect the grounded circuit conductor(s) for the remainder of the photovoltaic electrical system. A non–load-break-rated switch shall be permitted to be used as the disconnecting means.

(G) Battery Systems of More Than 48 Volts On photovoltaic systems where the battery system consists of more than twenty-four 2-volt cells connected in series (more than 48 volts, nominal), the battery system shall be permitted to operate with ungrounded conductors, provided the following conditions are met:

(1) The photovoltaic array source and output circuits shall comply with 690.41.
(2) The dc and ac load circuits shall be solidly grounded.
(3) All main ungrounded battery input/output circuit conductors shall be provided with switched disconnects and overcurrent protection.
(4) A ground-fault detector and indicator shall be installed to monitor for ground faults in the battery bank.

690.72 Charge Control

(A) General Equipment shall be provided to control the charging process of the battery. Charge control shall not be required where the design of the photovoltaic source circuit is matched to the voltage rating and charge current requirements of the interconnected battery cells and the maximum charging current multiplied by 1 hour is less than 3 percent of the rated battery capacity expressed in ampere-hours or as recommended by the battery manufacturer.

All adjusting means for control of the charging process shall be accessible only to qualified persons.

> FPN: Certain battery types such as valve-regulated lead acid or nickel cadmium can experience thermal failure when overcharged.

(B) Diversion Charge Controller

(1) Sole Means of Regulating Charging A photovoltaic power system employing a diversion charge controller as the sole means of regulating the charging of a battery shall be equipped with a second independent means to prevent overcharging of the battery.

(2) Circuits with Direct-Current Diversion Charge Controller and Diversion Load Circuits containing a dc diversion charge controller and a dc diversion load shall comply with the following:

(1) The current rating of the diversion load shall be less than or equal to the current rating of the diversion load charge controller. The voltage rating of the diversion load shall be greater than the maximum battery voltage. The power rating of the diversion load shall be at least 150 percent of the power rating of the photovoltaic array.
(2) The conductor ampacity and the rating of the overcurrent device for this circuit shall be at least 150 percent of the maximum current rating of the diversion charge controller.

If any portion of a diversion charge control system fails, the batteries may be overcharged and can create a potentially hazardous condition. Requiring a second, independent charge control method (usually a series regulator) and robust diversion controller circuits minimize the potential problems.

(3) PV Systems Using Utility-Interactive Inverters Photovoltaic power systems using utility-interactive inverters to control battery state-of-charge by diverting excess power into the utility system shall comply with (1) and (2):

(1) These systems shall not be required to comply with 690.72(B)(2). The charge regulation circuits used shall comply with the requirements of 690.8.
(2) These systems shall have a second, independent means of controlling the battery charging process for use when the utility is not present or when the primary charge controller fails or is disabled.

690.74 Battery Interconnections

Flexible cables, as identified in Article 400, in sizes 2/0 AWG and larger shall be permitted within the battery enclosure from battery terminals to a nearby junction box where they shall be connected to an approved wiring method. Flexible battery cables shall also be permitted between batteries and cells within the battery enclosure. Such cables shall be listed for hard-service use and identified as moisture resistant.

Battery plates and terminals are frequently constructed of relatively soft lead and lead alloys encased in plastics that are sealed with asphalt. Large-size, low-stranding stiff copper conductors attached to these components can cause them to be distorted. The use of flexible cables (see Article 400) reduces such distortions. Listed cables with the appropriate

physical and chemical-resistant properties should be used. Welding and "battery" cables are not allowed or described in the *NEC* for this use. Flexible "building wire"–type cables (Chapter 3) are also available and suitable for this use.

IX. Systems Over 600 Volts

690.80 General

Solar photovoltaic systems with a maximum system voltage over 600 volts dc shall comply with Article 490 and other requirements applicable to installations rated over 600 volts.

690.85 Definitions

For the purposes of Part IX of this article, the voltages used to determine cable and equipment ratings are as follows.

Battery Circuits. In battery circuits, the highest voltage experienced under charging or equalizing conditions.

Photovoltaic Circuits. In dc photovoltaic source circuits and photovoltaic output circuits, the maximum system voltage.

ARTICLE 692
Fuel Cell Systems

Summary of Changes

- **692.1:** Revised to indicate the system may have ac or dc output.
- **692.54:** Revised to clarify that marking of the fuel shutoff valve location, but not the shutoff valve, is required at the primary disconnect location.

Contents

I. General

692.1 Scope

This article identifies the requirements for the installation of fuel cell power systems, which may be stand-alone or interactive with other electrical power production sources and may be with or without electrical energy storage such as batteries. These systems may have ac or dc output for utilization.

The rising demand for electrical power has led to the development of power sources that are viable alternatives to or can be interconnected with electric utility distribution systems. Article 692, which was new in the 2002 *Code*, covers the installation of on-premises electrical supply systems where the power is derived from an emerging technology—fuel cells.

The principle of operation is that dc is generated through a chemical reaction in which fuel such as natural gas or LP-Gas is consumed. As opposed to internal combustion prime movers, the consumption of the fuel gas is via an electrochemical process rather than a combustion process. A power inverter converts the dc to ac. The installation requirements of Article 692 allow power derived from fuel cells to be safely delivered into residential and light commercial occupancies as the sole source of electrical power or as an integrated source with a utility or other power source.

692.2 Definitions

Fuel Cell. An electrochemical system that consumes fuel to produce an electric current. The main chemical reaction used in a fuel cell for producing electric power is not combustion. However, there may be sources of combustion used within the overall fuel cell system such as reformers/fuel processors.

Fuel Cell System. The complete aggregate of equipment used to convert chemical fuel into usable electricity. A fuel cell system typically consists of a reformer, stack, power inverter, and auxiliary equipment.

Interactive System. A fuel cell system that operates in parallel with and may deliver power to an electrical production and distribution network. For the purpose of this definition, an energy storage subsystem of a fuel cell system, such as a battery, is not another electrical production source.

Maximum System Voltage. The highest fuel cell inverter output voltage between any ungrounded conductors present at accessible output terminals.

Output Circuit. The conductors used to connect the fuel cell system to its electrical point of delivery. In the case of sites that have series- or parallel-connected multiple units, the term *output circuit* also refers to the conductors used to electrically interconnect the fuel cell system(s).

Point of Common Coupling. The point at which the power production and distribution network and the customer interface occurs in an interactive system. Typically, this is the load side of the power network meter.

Stand-Alone System. A fuel cell system that supplies power independently of an electrical production and distribution network.

692.3 Other Articles

Wherever the requirements of other articles of this *Code* and Article 692 differ, the requirements of Article 692 shall apply, and, if the system is operated in parallel with a primary source(s) of electricity, the requirements in 705.14, 705.16, 705.32, and 705.43 shall apply.

692.4 Installation

(A) Fuel Cell System A fuel cell system shall be permitted to supply a building or other structure in addition to any service(s) of another electricity supply system(s).

(B) Identification A permanent plaque or directory, denoting all electrical power sources on or in the premises, shall be installed at each service equipment location.

692.6 Listing Requirement

The fuel cell system shall be evaluated and listed for its intended application prior to installation.

Because the fuel cell system typically is a component of the premises wiring system, to facilitate its implementation into commercial use, 692.6 requires that the system be evaluated and listed for its intended use.

II. Circuit Requirements

692.8 Circuit Sizing and Current

(A) Nameplate Rated Circuit Current The nameplate(s) rated circuit current shall be the rated current indicated on the fuel cell nameplate(s).

(B) Conductor Ampacity and Overcurrent Device Ratings The ampacity of the feeder circuit conductors from the fuel cell system(s) to the premises wiring system shall not be less than the greater of (1) nameplate(s) rated circuit current or (2) the rating of the fuel cell system(s) overcurrent protective device(s).

(C) Ampacity of Grounded or Neutral Conductor If an interactive single-phase, 2-wire fuel cell output(s) is connected to the grounded or neutral conductor and a single ungrounded conductor of a 3-wire system or of a 3-phase, 4-wire wye-connected system, the maximum unbalanced neutral load current plus the fuel cell system(s) output rating shall not exceed the ampacity of the grounded or neutral conductor.

692.9 Overcurrent Protection

(A) Circuits and Equipment If the fuel cell system is provided with overcurrent protection sufficient to protect the circuit conductors that supply the load, additional circuit overcurrent devices shall not be required. Equipment and conductors connected to more than one electrical source shall be protected.

(B) Accessibility Overcurrent devices shall be readily accessible.

692.10 Stand-Alone Systems

The premises wiring system shall meet the requirements of this *Code* except as modified by 692.10(A), (B), and (C).

(A) Fuel Cell System Output The fuel cell system output from a stand-alone system shall be permitted to supply ac power to the building or structure disconnecting means at current levels below the rating of that disconnecting means.

(B) Sizing and Protection The circuit conductors between the fuel cell system(s) output and the building or structure disconnecting means shall be sized based on the output rating of the fuel cell system(s). These conductors shall be protected from overcurrents in accordance with 240.4. The overcurrent protection shall be located at the output of the fuel cell system(s).

(C) Single 120-Volt Nominal Supply The inverter output of a stand-alone fuel cell system shall be permitted to supply 120 volts, nominal, to single-phase, 3-wire 120/240-volt service equipment or distribution panels where there are no 240-volt loads and where there are no multiwire branch circuits. In all installations, the rating of the overcurrent device connected to the output of the fuel cell system(s) shall be less than the rating of the service equipment. This equipment shall be marked as follows:

> WARNING
> SINGLE 120-VOLT SUPPLY.
> DO NOT CONNECT MULTIWIRE
> BRANCH CIRCUITS!

III. Disconnecting Means

692.13 All Conductors

Means shall be provided to disconnect all current-carrying conductors of a fuel cell system power source from all other conductors in a building or other structure.

692.14 Provisions

The provisions of 225.31 and 225.33 through 225.40 shall apply to the fuel cell source disconnecting means. The disconnecting means shall not be required to be suitable as service equipment and shall be rated in accordance with 692.17.

692.17 Switch or Circuit Breaker

The disconnecting means for ungrounded conductors shall consist of readily accessible, manually operable switch(es) or circuit breaker(s).

Where all terminals of the disconnecting means may be energized in the open position, a warning sign shall be mounted on or adjacent to the disconnecting means. The sign shall be clearly legible and shall have the following words or equivalent:

> DANGER
> ELECTRIC SHOCK HAZARD.
> DO NOT TOUCH TERMINALS.
> TERMINALS ON BOTH THE LINE AND
> LOAD SIDES MAY BE ENERGIZED
> IN THE OPEN POSITION.

IV. Wiring Methods

692.31 Wiring Systems

All raceway and cable wiring methods included in Chapter 3 of this *Code* and other wiring systems and fittings specifically intended and identified for use with fuel cell systems shall be permitted. Where wiring devices with integral enclosures are used, sufficient length of cable shall be provided to facilitate replacement.

V. Grounding

692.41 System Grounding

For a fuel cell system output circuit, one conductor of a 2-wire system rated over 50 volts and a neutral conductor of a 3-wire system shall be solidly grounded by either 692.41(A) or 692.41(B).

(A) Stand-Alone Systems Grounding and bonding shall be in accordance with 250.30.

(B) Other Than Stand-Alone Systems

(1) Two-Wire Systems One conductor shall be terminated at the grounded circuit conductor terminal of the premises wiring system.

(2) Three-Wire Systems The neutral conductor shall be terminated at the grounded circuit conductor terminal of the premises wiring system.

692.44 Equipment Grounding Conductor

A separate equipment grounding conductor shall be installed.

692.45 Size of Equipment Grounding Conductor

The equipment grounding conductor shall be sized in accordance with 250.122.

692.47 Grounding Electrode System

Any supplementary grounding electrode(s) required by the manufacturer shall be connected to the equipment grounding conductor specified in 250.118.

VI. Marking

692.53 Fuel Cell Power Sources

A marking specifying the fuel cell system, output voltage, output power rating, and continuous output current rating shall be provided at the disconnecting means for the fuel cell power source at an accessible location on the site.

692.54 Fuel Shut-Off

The location of the manual fuel shut-off valve shall be marked at the location of the primary disconnecting means of the building or circuits supplied.

692.56 Stored Energy

A fuel cell system that stores electrical energy shall require the following warning sign, or equivalent, at the location of the service disconnecting means of the premises:

WARNING
FUEL CELL POWER SYSTEM CONTAINS
ELECTRICAL ENERGY STORAGE DEVICES.

VII. Connection to Other Circuits

692.59 Transfer Switch

A transfer switch shall be required in non–grid-interactive systems that use utility grid backup. The transfer switch shall maintain isolation between the electrical production and distribution network and the fuel cell system. The transfer switch shall be permitted to be located externally or internally to the fuel cell system unit. When the utility service conductors of the structure are connected to the transfer switch, the switch shall comply with Article 230, Part V.

692.60 Identified Interactive Equipment

Only fuel cell systems listed and identified as interactive shall be permitted in interactive systems.

692.61 Output Characteristics

The output of a fuel cell system operating in parallel with an electric supply system shall be compatible with the voltage, wave shape, and frequency of the system to which it is connected.

> FPN: The term *compatible* does not necessarily mean matching the primary source wave shape.

692.62 Loss of Interactive System Power

The fuel cell system shall be provided with a means of detecting when the electrical production and distribution network has become de-energized and shall not feed the electrical production and distribution network side of the point of common coupling during this condition. The fuel cell system shall remain in that state until the electrical production and distribution network voltage has been restored.

A normally interactive fuel cell system shall be permitted to operate as a stand-alone system to supply loads that have been disconnected from electrical production and distribution network sources.

692.64 Unbalanced Interconnections

(A) Single Phase Single-phase interactive fuel cell systems shall not be connected to a 3-phase power system unless the interactive system is so designed that significant unbalanced voltages cannot result.

(B) Three Phase Three-phase interactive fuel cell systems shall have all phases automatically de-energized upon loss of voltage, or upon unbalance of voltage in one or more phases, unless the interactive system is designed so that significant unbalanced voltages will not result.

692.65 Point of Connection

The output of a fuel cell system power source shall be connected as specified in 692.65(A) or 692.65(B).

(A) Supply Side A fuel cell system power source shall be permitted to be connected to the supply side of the service disconnecting means as permitted in 230.82(6).

(B) Load Side A fuel cell system power source shall be permitted to be connected to the load side of the service disconnecting means of the other source(s) at any distribution equipment on the premises, provided that all of the following conditions are met:

(1) Each source interconnection shall be made at a dedicated circuit breaker or fusible disconnecting means.
(2) The sum of the ampere ratings of overcurrent devices in circuits supplying power to a busbar or conductor shall not exceed the rating of the busbar or conductor.

Exception: For a dwelling unit, the sum of the ampere ratings of the overcurrent devices shall not exceed 120 percent of the rating of the busbar or conductor.

(3) The interconnection point shall be on the line side of all ground-fault protection equipment.
(4) Equipment containing overcurrent devices in circuits supplying power to a busbar or conductor shall be marked to indicate the presence of all sources.
(5) Equipment such as circuit breakers, if backfed, shall be identified for such operation.
(6) The circuit breaker on the dedicated output of a utility-interactive inverter shall be positioned in the distribution equipment at the opposite (load) end from the input feeder connection or main circuit location. A permanent warning label shall be applied to the distribution equipment with the following, or equivalent:

WARNING
FUEL CELL POWER SYSTEM OUTPUT.
DO NOT RELOCATE THIS CIRCUIT BREAKER.

VIII. Outputs Over 600 Volts

692.80 General

Fuel cell systems with a maximum output voltage over 600 volts ac shall comply with the requirements of other articles applicable to such installations.

ARTICLE 695
Fire Pumps

Summary of Changes

- **695.2:** Revised to make it clear that "fault tolerant" applies to the entire circuit, not just the conductors.
- **695.4(B)(1), 695.5(B), and 695.5(C)(2):** Revised to specify that the requirement to carry locked-rotor currents indefinitely applies only to overcurrent devices in the fire pump motor supply circuit.
- **695.6(D):** Revised to clarify that fire pump feeder and branch circuit conductors are required to have short-circuit protection only.
- **695.6(E):** Revised to permit listed Type MC cable with an impervious covering as a fire pump motor wiring method on the load side of the controller.
- **695.6(H):** Added new section prohibits ground-fault protection of equipment for fire pump supply circuits.
- **695.14(E):** Revised to permit listed Type MC cable with an impervious covering as a wiring method for fire pump control wiring.

Contents

Contents header

695.7 Voltage Drop
695.10 Listed Equipment
695.12 Equipment Location
  (A) Controllers and Transfer Switches
  (B) Engine-Drive Controllers
  (C) Storage Batteries
  (D) Energized Equipment
  (E) Protection Against Pump Water
  (F) Mounting
695.14 Control Wiring
  (A) Control Circuit Failures
  (B) Sensor Functioning
  (C) Remote Device(s)
  (D) Engine-Drive Control Wiring
  (E) Electric Fire Pump Control Wiring Methods
  (F) Generator Control Wiring Methods

FPN: Rules that are followed by a reference in brackets contain text that has been extracted from NFPA 20-2003, *Standard for the Installation of Stationary Pumps for Fire Protection*. Only editorial changes were made to the extracted text to make it consistent with this *Code*.

695.1 Scope

(A) Covered This article covers the installation of the following:

(1) Electric power sources and interconnecting circuits
(2) Switching and control equipment dedicated to fire pump drivers

(B) Not Covered This article does not cover the following:

(1) The performance, maintenance, and acceptance testing of the fire pump system, and the internal wiring of the components of the system
(2) Pressure maintenance (jockey or makeup) pumps

FPN: See NFPA 20-2003, *Standard for the Installation of Stationary Pumps for Fire Protection*, for further information.

Since it was first included in the 1996 *NEC*, Article 695 has been the subject of a number of important revisions, particularly in the requirements covering reliable power supplies for electric fire pump motors. Many of those revisions correlate the requirements in the *NEC* with those in NFPA 20, *Standard for the Installation of Stationary Pumps for Fire Protection*. One of the most significant changes that has occurred in the evolution of Article 695 is the requirement of 695.3(B)(2) covering reliable power for electric fire pumps installed in multibuilding campus-style arrangements such as those found in many institutional and industrial settings.

It should be noted that there is a distinct division of responsibility for fire pump requirements between the *NEC*

and NFPA 20. Performance issues, including the determination of power supply reliability, are under the jurisdiction of the NFPA Technical Committee on Fire Pumps, while electrical installation requirements are within the purview of the National Electrical Code Committee. Many of the requirements in Article 695 are extracted from the 2003 edition of NFPA 20, as indicated by a bracketed reference at the end of the *NEC* requirement. Proposals for changes to the extracted requirements in Article 695 have to be processed by the NFPA 20 committee because they have the primary responsibility for that technical content. The scope of Article 695, stated in 695.1, speaks to this division of responsibility.

Article 695 does not apply to pumps used to supply sprinkler systems in one- and two-family dwellings. NFPA 13D, *Standard for the Installation of Sprinkler Systems in One- and Two-Family Dwellings and Manufactured Homes*, does not require the use of a listed pump; thus, neither NFPA 20 nor Article 695 is applicable. Although jockey pumps are not covered by Article 695, they are permitted to be connected to the fire pump feeder.

The requirements of Article 695 are independent of those in Article 700 unless otherwise mandated by the authority having jurisdiction. Installation of the supply wiring to an electric motor-driven fire pump such as the one shown in Exhibit 695.1 is covered by the requirements of Article 695.

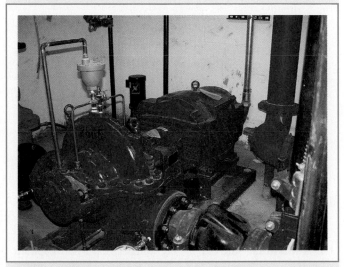

Exhibit 695.1 An electric motor specifically listed for fire pump service. (Courtesy of International Association of Electrical Inspectors)

695.2 Definitions

Fault Tolerant External Control Circuits. Those control circuits either entering or leaving the fire pump controller enclosure, which if broken, disconnected, or shorted will not prevent the controller from starting the fire pump from all other internal or external means and may cause the controller to start the pump under these conditions.

On-Site Power Production Facility. The normal supply of electric power for the site that is expected to be constantly producing power.

On-Site Standby Generator. A facility producing electric power on site as the alternate supply of electric power. It differs from an on-site power production facility, in that it is not constantly producing power.

695.3 Power Source(s) for Electric Motor-Driven Fire Pumps

Electric motor-driven fire pumps shall have a reliable source of power.

The power source for an electric motor-driven fire pump must be reliable and have adequate capacity to carry the locked-rotor currents of the fire pump motor and accessory equipment. These two main requirements ensure that the fire pump will operate in the event of a fire without being accidentally disconnected and that the fire pump will continue to operate until the fire is extinguished, the fire pump is purposely shut down, or the pump itself is destroyed.

One or more of the following three basic types of power sources are permitted. If only one source is used, that source must be capable of supplying locked-rotor current to the fire pump and accessory equipment. Power may be supplied by the following:

1. A separate utility service or connection ahead of the main disconnecting means
2. An on-site power production system
3. Multiple sources (two or more feeders) on a multibuilding campus where utility or power produced on site is not reasonably available and this method of power supply is approved by the authority having jurisdiction

If none of these options is individually capable of providing reliable power with adequate capacity, a combination (two or more) of these sources or a combination of one or more of these sources with an on-site standby generator must be used.

(A) Individual Sources Where reliable, and where capable of carrying indefinitely the sum of the locked-rotor current of the fire pump motor(s) and the pressure maintenance pump motor(s) and the full-load current of the associated fire pump accessory equipment when connected to this power supply, the power source for an electric motor-driven fire pump shall be one or more of the following.

(1) Electric Utility Service Connection A fire pump shall be permitted to be supplied by a separate service, or from a connection located ahead of and not within the same cabinet,

enclosure, or vertical switchboard section as the service disconnecting means. The connection shall be located and arranged so as to minimize the possibility of damage by fire from within the premises and from exposing hazards. A tap ahead of the service disconnecting means shall comply with 230.82(5). The service equipment shall comply with the labeling requirements in 230.2 and the location requirements in 230.72(B). [NFPA 20:9.2.2]

(2) On-Site Power Production Facility A fire pump shall be permitted to be supplied by an on-site power production facility. The source facility shall be located and protected to minimize the possibility of damage by fire. [NFPA 20:9.2.3]

Examples of power sources are illustrated in Exhibits 695.2 and 695.3.

The determination of whether the serving electric utility is a reliable source of power is an issue for the authority having jurisdiction. Annex A of NFPA 20, *Standard for the Installation of Stationary Pumps for Fire Protection*, provides guidance for evaluating the reliability of the power supply.

Performance requirements for an alternative source of electric power can be found in NFPA 110, *Standard for Emergency and Standby Power Systems*.

Most important in 695.3(A)(2) is the requirement that a fire in one source of power not affect the reliability of another source. For example, if an emergency feeder were physically routed above the normal switchboard and a fire occurred in the normal switchboard, the reliability of the emergency feeder during the fire would be questionable. There are many other examples for such interruption scenarios. The requirements for fire protection of the fire pump supply conductors are located in 695.6(A) and 695.6(B).

(B) Multiple Sources Where reliable power cannot be obtained from a source described in 695.3(A), power shall be supplied from an approved combination of two or more of either of such sources, or from an approved combination of feeders constituting two or more power sources as covered in 695.3(B)(2), or from an approved combination of one or more of such power sources in combination with an on-site standby generator complying with 695.3(B)(1) and (B)(3).

(1) Generator Capacity An on-site generator(s) used to comply with this section shall be of sufficient capacity to allow normal starting and running of the motor(s) driving the fire pump(s) while supplying all other simultaneously operated load. Automatic shedding of one or more optional standby loads in order to comply with this capacity requirement shall be permitted. A tap ahead of the on-site generator disconnecting means shall not be required. The requirements of 430.113 shall not apply. [NFPA 20:9.6.1]

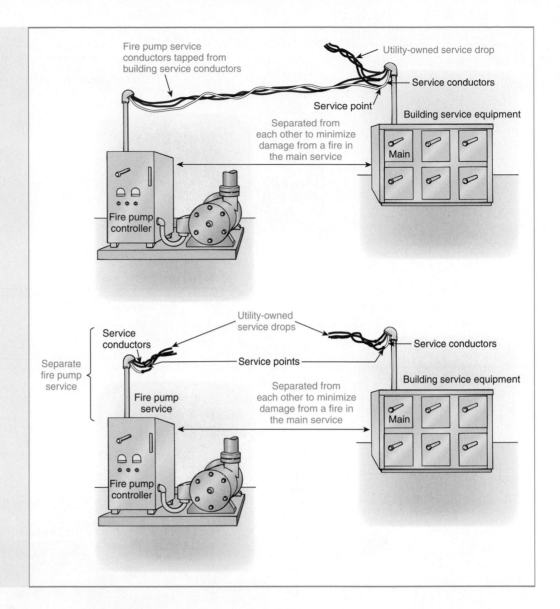

Exhibit 695.2 Two permitted configurations for connecting to electric utility–owned service drops.

Fire pump service conductors tapped from building service conductors

Utility-owned service drop

Service conductors

Service point

Building service equipment

Separated from each other to minimize damage from a fire in the main service

Main

Fire pump controller

Utility-owned service drops

Service conductors

Service conductors

Service points

Separate fire pump service

Building service equipment

Separated from each other to minimize damage from a fire in the main service

Fire pump service

Main

Fire pump controller

Where the alternative source of power is an on-site generator, the alternative source disconnecting means and the alternative source overcurrent protective device(s) for the electric-drive fire pump are not required to be sized for locked-rotor current of the fire pump motor(s). Rather, the circuit components of the alternative source are permitted to be sized according to Article 430, provided they "allow instantaneous pickup of the full [fire] pump room load," as specified by 9.6.5 in NFPA 20. See 445.12(A).

(2) Feeder Sources This section applies to multibuilding campus-style complexes with fire pumps at one or more buildings. Where sources in 695.3(A) are not practicable, and with the approval of the authority having jurisdiction,

two or more feeder sources shall be permitted as one power source or as more than one power source where such feeders are connected to or derived from separate utility services. The connection(s), overcurrent protective device(s), and disconnecting means for such feeders shall meet the requirements of 695.4(B). [NFPA 20:9.2.5.3]

The requirements of 695.3(B)(2) were an important revision to the 1999 *Code*, permitting the use of feeder sources for campus-style applications. Chapter 9 of NFPA 20, *Standard for the Installation of Stationary Pumps for Fire Protection*, permits the use of a reliable feeder to supply a fire pump if it is acceptable to the authority having jurisdiction. See Exhibit 695.4. In NFPA 20, the use of a reliable feeder as

On-site power production facility located where it is protected from fire, such as outdoors or in a room of fire-rated construction

Generator

Fire pump controller

Fire pump supply conductors

13.8-kV utility service No. 1

13.8-kV utility service No. 2

Multibuilding campus-style complex

13.8 kV to 480Y/277 volt

13.8 kV to 480Y/277 volt

Feeder 1

Feeder 2

To fire pump controller

a power supply for a fire pump is based on performance criteria, such as the following:

1. Redundant power supply features
2. Special requirements for generators
3. Special requirements for transfer switches
4. Generator overcurrent protection

(3) Arrangement The power sources shall be arranged so that a fire at one source will not cause an interruption at the other source. [NFPA 20:9.2.5.1]

695.4 Continuity of Power

Circuits that supply electric motor-driven fire pumps shall be supervised from inadvertent disconnection as covered in 695.4(A) or 695.4(B).

(A) Direct Connection The supply conductors shall directly connect the power source to either a listed fire pump controller or listed combination fire pump controller and power transfer switch. [NFPA 20:9.3.2.2.2]

(B) Supervised Connection A single disconnecting means and associated overcurrent protective device(s) shall be permitted to be installed between a remote power source and one of the following:

(1) A listed fire pump controller

The requirement for sizing the overcurrent protection in a supervised fire pump disconnecting means to be able to carry locked-rotor current indefinitely is a key factor in the reliable power source equation. Due to the critical life safety and property protection function of a fire pump, opening of the circuit by an overcurrent device installed between the point of connection to the power source and the fire pump controller cannot be tolerated and has to perform as if there were a direct connection to the power source. A change in the 2005 *Code* clarifies that sizing for locked-rotor current applies only to overcurrent protective devices and does not extend to conductors or other devices in the fire pump motor circuit. Similar revisions to clarify this point are made in 695.5(B) and 695.5(C)(2).

(2) A listed fire pump power transfer switch
(3) A listed combination fire pump controller and power transfer switch

For systems installed under the provisions of 695.3(B)(2) only, such additional disconnecting means and associated overcurrent protective device(s) shall be permitted as required to comply with other provisions of this *Code*. Overcurrent protective devices between an on-site standby generator and a fire pump controller shall be selected and sized according to 430.62 to provide short-circuit protection only. All disconnecting devices and overcurrent protective devices that are unique to the fire pump loads shall comply with 695.4(B)(1) through (B)(5).

(1) Overcurrent Device Selection The overcurrent protective device(s) shall be selected or set to carry indefinitely the sum of the locked-rotor current of the fire pump motor(s) and the pressure maintenance pump motor(s) and the full-load current of the associated fire pump accessory equipment when connected to this power supply. The requirement to carry the locked-rotor currents indefinitely shall not apply to conductors or devices other than overcurrent devices in the fire pump motor circuit(s).

(2) Disconnecting Means The disconnecting means shall comply with all of the following:

(1) Be identified as suitable for use as service equipment
(2) Be lockable in the closed position
(3) Not be located within equipment that feeds loads other than the fire pump

A revision in the 2005 *Code* prohibits the supervised disconnecting means from being installed in distribution equipment that supplies other than fire pump loads. This change provides clear direction on how compliance with 695.4(B)(2)(4) can be achieved. "Sufficiently remote from other building or other fire pump source disconnecting means" can no longer be interpreted as permitting a supervised fire pump disconnecting means to be located in a switchboard or panelboard that supplies other than fire pump loads.

(4) Be located sufficiently remote from other building or other fire pump source disconnecting means such that inadvertent contemporaneous operation would be unlikely

(3) Disconnect Marking The disconnecting means shall be marked "Fire Pump Disconnecting Means." The letters shall be at least 25 mm (1 in.) in height, and they shall be visible without opening enclosure doors or covers.

(4) Controller Marking A placard shall be placed adjacent to the fire pump controller, stating the location of this disconnecting means and the location of the key (if the disconnecting means is locked).

(5) Supervision The disconnecting means shall be supervised in the closed position by one of the following methods:

Minimizing the locations where the power supply to the fire pump can be automatically or manually interrupted is the objective of 695.4. Ideally, the supply conductors from the power source are run directly to the listed fire pump control and/or transfer equipment, without the need for an additional service disconnecting means and overcurrent protection. However, this arrangement is not always possible; therefore, a single supervised disconnecting means is permitted, provided all the conditions in 695.4(B) are met.

Where installed in the fire pump circuit, the supervised disconnecting means with overcurrent protection must meet the following conditions:

1. The overcurrent device is sized to carry the locked-rotor currents of all the fire pump motors.
2. The fire pump disconnecting means is located away from the other service disconnecting means. This disconnecting means is rated as service equipment and is lockable in the on position.
3. The equipment containing the disconnecting means supplies only the fire pump and associated loads.
4. The disconnect and controller are marked and placarded as described.
5. The circuit is supervised in the closed position, in accordance with 695.4(B)(5).

Specifically, the size of the overcurrent protective device is the sum of the locked-rotor currents of all the permitted motors plus the sum of any other fire pump auxiliary loads.

Example

A fusible service disconnect switch supplies power to a 100-hp, 460-volt, 3-phase fire pump and to a 1½-hp, 460-volt, 3-phase jockey pump. The fire pump feeder circuit will be installed in a raceway between the disconnecting means and fire pump controller. The raceway is considered outside the building per 230.6. Using the requirements of Article 695, determine the sizes of the disconnecting means and overcurrent protection device. Also determine the minimum ampacity of the feeder conductors.

Solution

STEP 1. Determine the minimum ratings of the disconnecting means and the overcurrent protective device. According to the motor nameplates, the locked-rotor current (LRC) is 725 amperes for the 100-hp motor and 20 amperes for the 1½-hp motor. If the locked-rotor amperes are not on the nameplates, the locked-rotor currents found in Table 430.251(B) must be used. Calculate the size by summing the locked-rotor currents of both motors and then going to the next larger standard-size overcurrent device, as follows:

$$100\text{-hp, 3-phase LRC} = 725 \text{ A}$$
$$1\tfrac{1}{2}\text{-hp, 3-phase LRC} = \underline{20 \text{ A}}$$
$$\text{Total LRC} = 745 \text{ A}$$

The next larger standard-size disconnect switch and overcurrent device is 800 amperes. An adjustable-trip circuit breaker

of 750 amperes is also permitted, because it, too, will carry the locked-rotor current indefinitely.

STEP 2. Determine the minimum ampacity for the fire pump feeder conductor. Even though the disconnect switch, fuse, and circuit breakers are sized according to locked-rotor currents, the feeder conductors to the fire pump and associated equipment are required to have an ampacity not less than 125 percent of the full-load current (FLC) rating of the fire pump motor(s) and pressure maintenance pump motor(s), plus 100 percent of associated accessory equipment. Using the same motors as above, calculate the size of the feeder to the fire pump controller as follows, using 430.6(A)(1) and Table 430.250 for the full-load currents of the motors:

100-hp, 3-phase FLC
$$124 \text{ A} \times 1.25 = 155.00 \text{ A}$$

1½-hp, 3-phase FLC
$$3 \text{ A} \times 1.25 = \underline{3.75 \text{ A}}$$
$$\text{Total FLC} = 158.75 \text{ A} \quad \text{or} \quad 159 \text{ A}$$

Thus, the minimum ampacity for the feeder conductors is 159 amperes. Using the 75°C column from Table 310.16, a 2/0 copper conductor is the minimum size required, per 110.14(C)(1)(b).

Supervision of the disconnecting means by a local (protected premises) fire alarm system, central station, proprietary supervising station, or remote supervising station requires a connection to the premises fire alarm system. The connection is generally made through the use of dry contacts in the fire pump controller that are connected to a fire alarm system initiating device circuit. This circuit is programmed to generate a supervisory signal at the fire alarm control unit on loss of voltage to the fire pump controller. A supervisory signal indicates that the suppression system is "off-normal." Connection of this device must initiate a supervisory signal and may also initiate a trouble signal. For more information, see *NFPA 72, National Fire Alarm Code.*

(1) Central station, proprietary, or remote station signal device
(2) Local signaling service that causes the sounding of an audible signal at a constantly attended point
(3) Locking the disconnecting means in the closed position
(4) Sealing of disconnecting means and approved weekly recorded inspections when the disconnecting means are located within fenced enclosures or in buildings under the control of the owner [NFPA 20:9.3.2.2.3]

695.5 Transformers

Where the service or system voltage is different from the utilization voltage of the fire pump motor, transformer(s) protected by disconnecting means and overcurrent protective devices shall be permitted to be installed between the system supply and the fire pump controller in accordance with 695.5(A) and (B), or (C). Only transformers covered in 695.5(C) shall be permitted to supply loads not directly associated with the fire pump system.

(A) Size Where a transformer supplies an electric motor-driven fire pump, it shall be rated at a minimum of 125 percent of the sum of the fire pump motor(s) and pressure maintenance pump(s) motor loads, and 100 percent of the associated fire pump accessory equipment supplied by the transformer.

(B) Overcurrent Protection The primary overcurrent protective device(s) shall be selected or set to carry indefinitely the sum of the locked-rotor current of the fire pump motor(s) and the pressure maintenance pump motor(s) and the full-load current of the associated fire pump accessory equipment when connected to this power supply. Secondary overcurrent protection shall not be permitted. The requirement to carry the locked-rotor currents indefinitely shall not apply to conductors or devices other than overcurrent devices in the fire pump motor circuit(s).

See Exhibit 695.5.

Dedicated transformer and overcurrent protection sizing can be broken down into three basic requirements. Generally stated, they are as follows:

1. The transformer must be sized to at least 125 percent of the sum of the loads.
2. The transformer primary overcurrent device must be at least a specified minimum size.
3. The transformer secondary must not contain any overcurrent devices whatsoever.

Example

A 4160/480-volt, 3-phase, dedicated transformer supplies power to a 100-hp, 460-volt, 3-phase, code letter G fire pump and to a 1½-hp, 460-volt, 3-phase, code letter H jockey pump. Using the requirements of Article 695, determine the sizes of the dedicated transformer and its primary overcurrent protection.

Solution

STEP 1. Determine the minimum standard-size transformer. According to 695.5(A), to determine the minimum current value to use in the 3-phase power calculation, add the full-load currents of the fire pump motor(s) and jockey pump motor(s), then increase the total to 125 percent. The result of this calculation provides the minimum kVA rating of a transformer dedicated to a fire pump installation. Any transformer with a kVA rating equal to or greater than the

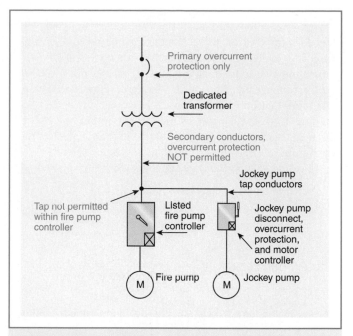

Exhibit 695.5 The overcurrent device in the primary of a transformer supplying a fire pump installation. The device is required to be sized to carry the locked-rotor current motor(s) and associated fire pump accessory equipment indefinitely.

value determined by this calculation is acceptable. First, add the full-load currents of the two motors, using the full-load current (FLC) values from Table 430.250, as follows:

$$
\begin{aligned}
\text{100-hp, 3-phase FLC} &= 124\ \text{A} \\
\text{1½-hp, 3-phase FLC} &= \underline{\ \ \ 3\ \text{A}} \\
\text{Total FLC} &= 127\ \text{A}
\end{aligned}
$$

Now, increase the sum of the fire pump motor and the jockey pump motor to 125 percent, as follows:

$$127\ \text{A} \times 1.25 = 158.75\ \text{A}$$

Then, size the transformer as follows:

$$\text{Transformer kVA} = \frac{\text{volts} \times \text{amperes} \times \sqrt{3}}{1000}$$

$$\text{Transformer kVA} = \frac{480 \times 158.75 \times \sqrt{3}}{1000}$$

$$= 131.98\ \text{kVA}$$

Thus, the minimum-size transformer permitted by 695.5(A) is 131.98 kVA. The next larger standard-size transformer available is 150 kVA, but any larger size is permitted.

STEP 2. Calculate the minimum-size primary overcurrent protection device permitted for this transformer. According to 695.5(B), the minimum primary overcurrent protection device must allow the transformer secondary to supply the locked-rotor current (LRC) to the fire pump and, in this case, the jockey pump. The locked-rotor current of each motor must be individually calculated if it is not available on the motor nameplate. In this example, however, we are assuming that only the kVA code letters are available. According to 430.7(B) and using the maximum values for the individual code letters per Table 430.7(B), calculate the maximum locked-rotor currents, as follows.

For the 100-hp motor, code letter G:

$$LRC = motor\ hp \times max.\ code\ letter\ value$$
$$\times \frac{1000}{motor\ voltage \times 3\text{-phase factor}}$$

$$LRC = 100\ hp \times 6.29\ \frac{kVA}{hp} \times \frac{1000}{460 \times \sqrt{3}} = 789.49A$$

For the 1½-hp motor, code letter H (using the same formula):

$$LRC = 1\frac{1}{2}\ hp \times 7.09\ \frac{kVA}{hp} \times \frac{1000}{460 \times \sqrt{3}} = 13.35\ A$$

For the total LRC:

$$100\text{-hp LRC} = 789.49\ A$$
$$1\frac{1}{2}\text{-hp LRC} = \underline{13.35\ A}$$
$$Total\ LRC = 802.84\ A \quad or \quad 803\ A$$

Now, calculate the equivalent LRC on the primary side of the transformer, based on the calculated LRC of the secondary of the transformer, as follows:

$$LRC_{primary} = \frac{secondary\ voltage}{primary\ voltage} \times LRC_{secondary}$$
$$= \frac{480\ V}{4160\ V} \times 803\ A$$
$$= 92.65\ A \quad or \quad 93A$$

This value of 93 amperes represents the secondary LRC reflected to the primary side of the transformer. Because this value is the absolute smallest overcurrent protective device permitted, the next larger standard size, according to 240.6, is 100 amperes. Thus, the minimum standard-size overcurrent protective device is 100 amperes.

Conclusion

The calculation for a 4160/480-volt, 3-phase transformer supplying a 100-hp fire pump and a 1½-hp jockey pump, both at 460 volts, 3 phase, can be summarized as follows:

1. The smallest standard-size transformer that is permitted is 150 kVA.

2. The smallest standard-size overcurrent protective device permitted on the primary of the transformer is 100 amperes.
3. A secondary overcurrent protective device is not permitted.

(C) Feeder Source Where a feeder source is provided in accordance with 695.3(B)(2), transformers supplying the fire pump system shall be permitted to supply other loads. All other loads shall be calculated in accordance with Article 220, including demand factors as applicable.

(1) Size Transformers shall be rated at a minimum of 125 percent of the sum of the fire pump motor(s) and pressure maintenance pump(s) motor loads, and 100 percent of the remaining load supplied by the transformer.

(2) Overcurrent Protection The transformer size, the feeder size, and the overcurrent protective device(s) shall be coordinated such that overcurrent protection is provided for the transformer in accordance with 450.3 and for the feeder in accordance with 215.3, and such that the overcurrent protective device(s) is selected or set to carry indefinitely the sum of the locked-rotor current of the fire pump motor(s), the pressure maintenance pump motor(s), the full-load current of the associated fire pump accessory equipment, and 100 percent of the remaining loads supplied by the transformer. The requirement to carry the locked-rotor currents indefinitely shall not apply to conductors or devices other than overcurrent devices in the fire pump motor circuit(s).

695.6 Power Wiring

Power circuits and wiring methods shall comply with the requirements in 695.6(A) through (H), and as permitted in 230.90(A), Exception No. 4; 230.94, Exception No. 4; 230.95, Exception No. 2; 240.13; 230.208; 240.4(A); and 430.31.

(A) Service Conductors Supply conductors shall be physically routed outside a building(s) and shall be installed as service entrance conductors in accordance with Article 230. Where supply conductors cannot be physically routed outside buildings, they shall be permitted to be routed through buildings where installed in accordance with 230.6(1) or 230.6(2). Where a fire pump is wired under the provisions of 695.3(B)(2), this requirement shall apply to all supply conductors on the load side of the service disconnecting means that constitute the normal source of supply to that fire pump.

Exception: Where there are multiple sources of supply with means for automatic connection from one source to the other, the requirement shall apply only to those conductors on the load side of that point of automatic connection between sources.

(B) Circuit Conductors Fire pump supply conductors on the load side of the final disconnecting means and overcurrent device(s) permitted by 695.4(B) shall be kept entirely independent of all other wiring. They shall supply only loads that are directly associated with the fire pump system, and they shall be protected to resist potential damage by fire, structural failure, or operational accident. They shall be permitted to be routed through a building(s) using one of the following methods:

The circuit conductors covered by 695.6(B) are those that are installed from the load side of the supervised disconnecting means permitted by 695.4(B) to the fire pump controller or combination controller/transfer switch. Unlike service conductors, these supply conductors are installed on the load side of a short-circuit protective device and are not required to be installed on the outside of a building or structure. However, if these supply conductors are run through a building from the disconnecting means to the fire pump equipment room, they are required to be protected against the effects of a fire to ensure that power to the fire pump is not compromised.

Encasement in 2 in. of concrete, as recognized in 230.6 for service conductors, is one method permitted to isolate and protect conductors. In addition, enclosure of a wiring method within 1-hour fire-resistive building construction and the use of a wiring method listed as an electrical circuit protective system with a minimum 1-hour fire rating are methods permitted for these circuit conductors that do have short-circuit protection. These latter two wiring methods are recognized for protecting conductors against the effects of fire, but they are not recognized as means by which conductors can be considered as being outside a building or other structure. Therefore, if the installation involves service conductors that have to be run through a building, the only available options, short of installing a supervised disconnecting means, are to install the conductors beneath not less than 2 in. of concrete or to encase the conductors in 2 in. of concrete.

It is important to understand the difference between a 1-hour fire rating of an electrical circuit, such as a conduit with wires, and a 1-hour fire-resistance rating of a structural member, such as a wall. Simply stated, at the end of a 1-hour fire test on an electrical conduit with wires, the circuit must function electrically (no short circuits, grounds, or opens are permitted). The circuit and its insulation must be intact and electrically functioning. A wall subjected to a 1-hour fire-resistance test must only prevent a fire from passing through or past the wall, without regard to damage to the wall. All fire ratings and fire-resistance ratings are

based on the assumption that the structural supports for the assembly are not impaired by the effects of the fire.

The UL 2004 *Fire Resistance Directory,* Volume 2, describes three categories of products that can be used in the fire protection of electrical circuits for fire pumps: various electrical circuit protective systems (FHIT), electrical circuit protective materials (FHIY), and fire resistive cables (FHJR). (The four-letter codes in parentheses are the UL product category guide designations.) For information on electrical circuit protective systems, see UL Subject 1724, *Fire Tests for Electrical Circuit Protective Systems.*

(1) Be encased in a minimum 50 mm (2 in.) of concrete
(2) Be within an enclosed construction dedicated to the fire pump circuit(s) and having a minimum of a 1-hour fire resistive rating
(3) Be a listed electrical circuit protective system with a minimum 1-hour fire rating

Exception: The supply conductors located in the electrical equipment room where they originate and in the fire pump room shall not be required to have the minimum 1-hour fire separation or fire resistance rating, unless otherwise required by 700.9(D) of this Code.

(C) Conductor Size

(1) Fire Pump Motors and Other Equipment Conductors supplying a fire pump motor(s), pressure maintenance pumps, and associated fire pump accessory equipment shall have a rating not less than 125 percent of the sum of the fire pump motor(s) and pressure maintenance motor(s) full-load current(s), and 100 percent of the associated fire pump accessory equipment.

(2) Fire Pump Motors Only Conductors supplying only a fire pump motor shall have a minimum ampacity in accordance with 430.22 and shall comply with the voltage drop requirements in 695.7.

Listed fire pump controller and pump combinations are available in a wye-start, delta-run configuration. From the controller to the motor there are six circuit conductors, and when the motor is in the run mode, the conductors that supply each winding are connected in parallel. Prior to the 2005 edition of the *Code,* the general rule in 695.6(C)(2) for sizing the conductors supplying a single fire pump motor required a minimum conductor ampacity not less than 125 percent of the pump motor full load current rating. A revision to this requirement for the 2005 *Code* references 430.22 for sizing the conductors supplying a single motor. For a wye-start, delta-run fire pump motor, the minimum conductor ampacity for each of the six leads between the controller and the motor is calculated as shown in the following example.

Background Information

- Fire pump motor nameplate information: 30 hp, 3-phase, 460 volt
- Table 430.250 full load current for 50 hp motor: 65 amperes
- Section 430.22(A) specifies motor circuit conductor minimum ampacity based on 125 percent of motor full load current
- Section 430.22(C) specifies full load current for conductors on the load side of controller based on 58 percent of motor full load current

Calculation

- 65 amperes $\times$ 0.58 $\times$ 1.25 = 47 amperes

Minimum Conductor Size

- Assuming 75°C terminations in the controller and conductor insulation type THWN, ampacity selection from Table 310.16 = 8 AWG copper circuit conductors (50 amperes)

Analysis

- There will be six 8 AWG copper conductors between the controller and the motor.
- The combined ampacity of the two circuit conductors connected in parallel to each winding in the run mode is 100 amperes (combined ampacity of two 8 AWG conductors).
- Comparison of controller load side circuit conductors with minimum ampacity for circuit conductors from the power source to the line side of fire pump controller:

  65 amperes $\times$ 1.25 = 81 amperes
  4 AWG THWN copper circuit conductors

- The minimum size for the conductors as determined here may have to be increased to comply with the mandatory voltage drop performance requirements specified in 695.7.

(D) Overload Protection Power circuits shall not have automatic protection against overloads. Branch-circuit and feeder conductors shall be protected against short circuit only. Where a tap is made to supply a fire pump, the wiring shall be treated as service conductors in accordance with 230.6. The applicable distance and size restrictions in 240.21 shall not apply.

Exception No. 1: Conductors between storage batteries and the engine shall not require overcurrent protection or disconnecting means.

Exception No. 2: For on-site standby generator(s) rated to produce continuous current in excess of 225 percent of the full-load amperes of the fire pump motor, the conductors between the on-site generator(s) and the combination fire pump transfer switch controller or separately mounted trans-

fer switch shall be installed in accordance with 695.6(B) or protected in accordance with 430.52.

The protection provided shall be in accordance with the short-circuit current rating of the combination fire pump transfer switch controller or separately mounted transfer switch.

(E) Pump Wiring All wiring from the controllers to the pump motors shall be in rigid metal conduit, intermediate metal conduit, liquidtight flexible metal conduit, or liquidtight flexible nonmetallic conduit Type LFNC-B, listed Type MC cable with an impervious covering, or Type MI cable.

The requirement of 695.6(E) does not apply to light switches, convenience receptacles, telephone outlets, fire detectors, and similar equipment located in the fire pump room.

(F) Junction Points Where wire connectors are used in the fire pump circuit, the connectors shall be listed. A fire pump controller or fire pump power transfer switch, where provided, shall not be used as a junction box to supply other equipment, including a pressure maintenance (jockey) pump(s). A fire pump controller and fire pump power transfer switch, where provided, shall not serve any load other than the fire pump for which it is intended.

(G) Mechanical Protection All wiring from engine controllers and batteries shall be protected against physical damage and shall be installed in accordance with the controller and engine manufacturer's instructions.

(H) Ground Fault Protection of Equipment Ground fault protection of equipment shall not be permitted for fire pumps.

Ground-fault protection of equipment as specified in several sections of the *NEC* is not permitted to be used to protect components of a fire pump installation. (The function of ground-fault protection of equipment protection should not be confused with the function of GFCI protection for personnel.) Although the protection afforded to equipment by this type of ground-fault protection is a mandatory safety requirement in certain circumstances, the need for an uninterrupted source of power takes precedence for fire pump installations. See 230.95, Exception No. 2, and 240.13(3).

695.7 Voltage Drop

The voltage at the controller line terminals shall not drop more than 15 percent below normal (controller-rated voltage) under motor starting conditions. The voltage at the motor terminals shall not drop more than 5 percent below the voltage rating of the motor when the motor is operating at 115 percent of the full-load current rating of the motor.

Exception: This limitation shall not apply for emergency run mechanical starting. [NFPA 20:9.4]

695.10 Listed Equipment

Diesel engine fire pump controllers, electric fire pump controllers, electric motors, fire pump power transfer switches, foam pump controllers, and limited service controllers shall be listed for fire pump service. [NFPA 20:9.5.1.1, 10.1.2.1, 12.1.3.1]

This requirement parallels those provisions in NFPA 20, *Standard for the Installation of Stationary Pumps for Fire Protection*, that require components used in fire pump systems to be listed. The fire pump controller shown in Exhibit 695.6 is an example of the listed equipment covered under 695.10.

Exhibit 695.6 Listed fire pump controller and power transfer switch.

Prior to being shipped to the installation site, listed fire pump controllers are matched with the listed electric motor(s) they will control to ensure compatibility of the individually listed components.

695.12 Equipment Location

(A) Controllers and Transfer Switches Electric motor-driven fire pump controllers and power transfer switches shall be located as close as practicable to, and within sight of, the motors that they control.

(B) Engine-Drive Controllers Engine-drive fire pump controllers shall be located as close as is practical to, and within sight of, the engines that they control.

(C) Storage Batteries Storage batteries for fire pump engine drives shall be supported above the floor, secured against displacement, and located where they are not subject to physical damage, flooding with water, excessive temperature, or excessive vibration.

(D) Energized Equipment All energized equipment parts shall be located at least 300 mm (12 in.) above the floor level.

(E) Protection Against Pump Water Fire pump controllers and power transfer switches shall be located or protected so that they are not damaged by water escaping from pumps or pump connections.

(F) Mounting All fire pump control equipment shall be mounted in a substantial manner on noncombustible supporting structures.

Neither NFPA 20, *Standard for the Installation of Stationary Pumps for Fire Protection*, nor this *Code* mandates a dedicated room for the fire pump. However, NFPA 20 does specify a suitable space for this equipment.

Compliance with the 695.12(A) requirement that fire pump controllers and transfer switches be located "as close as practicable" to the fire pump motor that they control may require that additional space be available to achieve the minimum maintenance working space set forth in 110.26.

Generally, fire pump controllers are housed in substantial enclosures suitable to protect the contents against limited amounts of falling water and dirt. In addition, all energized parts in the enclosure must be mounted at least 12 in. above the floor. Typically, the floor space for this area is equipped with a floor drain.

The requirement of 695.12(F) does not permit fire pump control equipment to be mounted on combustible backboards (such as plywood).

695.14 Control Wiring

(A) Control Circuit Failures External control circuits that extend outside the fire pump room shall be arranged so that failure of any external circuit (open or short circuit) shall not prevent the operation of a pump(s) from all other internal or external means. Breakage, disconnecting, shorting of the wires, or loss of power to these circuits could cause continuous running of the fire pump but shall not prevent the controller(s) from starting the fire pump(s) due to causes other than these external control circuits. All control conductors within the fire pump room that are not fault tolerant shall be protected against physical damage. [NFPA 20:10.5.2.6, 12.5.2.5]

(B) Sensor Functioning No undervoltage, phase-loss, frequency-sensitive, or other sensor(s) shall be installed that automatically or manually prohibit actuation of the motor contactor. [NFPA 20:10.4.5.6]

Exception: A phase loss sensor(s) shall be permitted only as a part of a listed fire pump controller.

(C) Remote Device(s) No remote device(s) shall be installed that will prevent automatic operation of the transfer switch. [NFPA 20:10.8.1.3]

(D) Engine-Drive Control Wiring All wiring between the controller and the diesel engine shall be stranded and sized to continuously carry the charging or control currents as required by the controller manufacturer. Such wiring shall be protected against physical damage. Controller manufacturer's specifications for distance and wire size shall be followed. [NFPA 20:12.3.5.1]

(E) Electric Fire Pump Control Wiring Methods All electric motor-driven fire pump control wiring shall be in rigid metal conduit, intermediate metal conduit, liquidtight flexible metal conduit, liquidtight flexible nonmetallic conduit Type B (LFNC-B), listed Type MC cable with an impervious covering, or Type MI cable.

The wiring methods described in 695.14(E) apply only to the control wiring for electric motor–driven fire pumps. They do not apply to the control wiring for engine-driven fire pumps, because there are no similar requirements in NFPA 20, *Standard for the Installation of Stationary Pumps for Fire Protection.*

(F) Generator Control Wiring Methods Control conductors installed between the fire pump power transfer switch and the standby generator supplying the fire pump during normal power loss shall be kept entirely independent of all other wiring. They shall be protected to resist potential damage by fire or structural failure. They shall be permitted to be routed through a building(s) encased in 50 mm (2 in.) of concrete or within enclosed construction dedicated to the fire pump circuits and having a minimum 1-hour fire resistance rating, or circuit protective systems with a minimum of 1-hour fire resistance. The installation shall comply with any restrictions provided in the listing of the electrical circuit protective system used.

Special Conditions

ARTICLE 700
Emergency Systems

Summary of Changes

- **700.5(B):** Revised to permit alternate power source to supply legally required standby and optional standby system loads without the use of selective load pickup and load shedding, where the source has adequate capacity.

- **700.9(D)(1):** Revised to permit installations in spaces or areas of a building protected by automatic fire protection, not just where the entire building is so protected.

- **700.12(B)(6):** Revised to cover ungrounded conductors that pass through the building or structure.

- **700.12(E):** Added new paragraph permitting fuel cell systems as a source of supply for emergency systems.

- **700.21:** Revised to limit requirement for location of the emergency switch in a lobby or location conveniently accessible to the lobby to locations covered by Articles 518 and 520.

- **700.27:** Added new section requiring that emergency system overcurrent devices be selectively coordinated with all supply side overcurrent devices.

Contents

I. General

700.1 Scope

The provisions of this article apply to the electrical safety of the installation, operation, and maintenance of emergency systems consisting of circuits and equipment intended to supply, distribute, and control electricity for illumination, power, or both, to required facilities when the normal electrical supply or system is interrupted.

Emergency systems are those systems legally required and classed as emergency by municipal, state, federal, or other codes, or by any governmental agency having jurisdiction. These systems are intended to automatically supply illumination, power, or both, to designated areas and equipment in the event of failure of the normal supply or in the event of accident to elements of a system intended to supply, distribute, and control power and illumination essential for safety to human life.

Emergency systems are designed and installed to maintain a specific degree of illumination or provide power for essential equipment if the normal power supply fails. Examples of essential equipment include fire pumps and operating room and life-support equipment in hospitals.

Article 700 applies to the installation of emergency systems that are essential for safety to human life and are legally required by municipal, state, federal, or other codes or by a governmental agency having jurisdiction.

Article 700 does not dictate whether emergency systems are required or where emergency or exit lights should be located. These determinations may be made by using NFPA 101®, *Life Safety Code*®.

If authorities determine that emergency lighting, including the proper placement of exit signs, is required for safe egress from various classes of buildings or parts of buildings, then corridors, passageways, stairways, lobbies, and so on must also be sufficiently illuminated.

FPN No. 1: For further information regarding wiring and installation of emergency systems in health care facilities, see Article 517.

FPN No. 2: For further information regarding performance and maintenance of emergency systems in health care facilities, see NFPA 99-2002, *Standard for Health Care Facilities*.

FPN No. 3: Emergency systems are generally installed in places of assembly where artificial illumination is required for safe exiting and for panic control in buildings subject to occupancy by large numbers of persons, such as hotels, theaters, sports arenas, health care facilities, and similar institutions. Emergency systems may also provide power for such functions as ventilation where essential to maintain life, fire detection and alarm systems, elevators, fire pumps, public safety communications systems, industrial processes where current interruption would produce serious life safety or health hazards, and similar functions.

FPN No. 4: For specification of locations where emergency lighting is considered essential to life safety, see NFPA 101®-2003, *Life Safety Code*®.

FPN No. 5: For further information regarding performance of emergency and standby power systems, see NFPA 110-2002, *Standard for Emergency and Standby Power Systems*.

700.2 Application of Other Articles

Except as modified by this article, all applicable articles of this *Code* shall apply.

700.3 Equipment Approval

All equipment shall be approved for use on emergency systems.

700.4 Tests and Maintenance

(A) Conduct or Witness Test The authority having jurisdiction shall conduct or witness a test of the complete system upon installation and periodically afterward.

(B) Tested Periodically Systems shall be tested periodically on a schedule acceptable to the authority having jurisdiction to ensure the systems are maintained in proper operating condition.

(C) Battery Systems Maintenance Where battery systems or unit equipments are involved, including batteries used for starting, control, or ignition in auxiliary engines, the authority having jurisdiction shall require periodic maintenance.

(D) Written Record A written record shall be kept of such tests and maintenance.

(E) Testing Under Load Means for testing all emergency lighting and power systems during maximum anticipated load conditions shall be provided.

FPN: For testing and maintenance procedures of emergency power supply systems (EPSSs), see NFPA 110-2002, *Standard for Emergency and Standby Power Systems*.

Emergency system testing may be broken down into two general categories — acceptance testing and operational testing. Section 700.4 requires both types of testing as well as written records of both types of testing and of maintenance.

Acceptance testing is performed after the emergency system has been installed but before the system is used. Acceptance testing ensures that the emergency system meets or exceeds the original installation specification. Portable load banks may be used for the acceptance testing of the system to maximum design load.

Operational testing, which is performed during the life of the system, ensures that the emergency system remains functional and that maintenance is performed adequately. One method of operational testing is running the generating system to power the load of the facility. Generally, actual emergency system loads are smaller than the design capacity of the emergency generator system. Actual peak loads of the emergency system should be kept as part of the written record.

Further information on tests and maintenance may be found in *NFPA 72*®, *National Fire Alarm Code*®; NFPA 99, *Standard for Health Care Facilities*; NFPA 101®, *Life Safety Code*®; NFPA 110, *Standard for Emergency and Standby Power Systems*; and NFPA 111, *Standard on Stored Electrical Energy Emergency and Standby Power Systems*.

700.5 Capacity

(A) Capacity and Rating An emergency system shall have adequate capacity and rating for all loads to be operated simultaneously. The emergency system equipment shall be suitable for the maximum available fault current at its terminals.

The emergency system must be designed with adequate capacity and rating to safely carry the entire load connected to the emergency system at one time. It must be capable of restarting emergency loads, such as motors, that may have stopped, and it must be suitable for the available fault current. Using devices that limit the available fault current is one method of achieving suitability.

(B) Selective Load Pickup, Load Shedding, and Peak Load Shaving The alternate power source shall be permitted to supply emergency, legally required standby, and optional standby system loads where the source has adequate capacity or where automatic selective load pickup and load shedding is provided as needed to ensure adequate power to (1) the emergency circuits, (2) the legally required standby circuits, and (3) the optional standby circuits, in that order of priority. The alternate power source shall be permitted to be used for peak load shaving, provided these conditions are met.

Peak load-shaving operation shall be permitted for satisfying the test requirement of 700.4(B), provided all other conditions of 700.4 are met.

A portable or temporary alternate source shall be available whenever the emergency generator is out of service for major maintenance or repair.

Section 700.5(B) permits a generator to serve more than one level of emergency, legally required standby, or other operational standby system loads. Section 700.5(B) also permits the use of a generator for peak load shaving, supplying backup power, and other uses. However, assurance is required that priority loads are properly and reliably served. This section was revised for the 2005 *Code*. It no longer requires selective load pickup and load shedding where the generator has the capacity to supply all loads served.

If a generator is used for peak load shaving or in a cogeneration system, major downtime for maintenance of the generator must be anticipated. On the other hand, using the emergency generator on a regular basis for nonemergency loads provides assurance that the emergency generator will supply emergency power when it is needed. The requirement for a portable or temporary alternate source is intended to provide emergency power when the generator set is out of service for a long time. A major maintenance or repair procedure is one that keeps the generator set out of service for more than a few hours.

700.6 Transfer Equipment

(A) General Transfer equipment, including automatic transfer switches, shall be automatic, identified for emergency use, and approved by the authority having jurisdiction. Transfer equipment shall be designed and installed to prevent the inadvertent interconnection of normal and emergency sources of supply in any operation of the transfer equipment. Transfer equipment and electric power production systems installed to permit operation in parallel with the normal source shall meet the requirements of Article 705.

The transfer equipment described in 700.6(A) is permitted to allow parallel operation of the generation equipment with the normal source as long as the requirements of Article 705 are met. Traditional automatic transfer switches are not designed to permit parallel operation of generation equipment and the normal source. Therefore, traditional automatic transfer switches need not comply with Article 705. However, certain automatic transfer switch configurations are intentionally designed to briefly (a few cycles) parallel the generation equipment with the normal source upon load transfer from generator to normal source. This load transfer can occur with minimal disturbance or effect on the load. Transfer switches that employ this type of paralleling must comply with Article 705.

If continuous parallel operation of generation equipment and the source is desired, paralleling switchgear or paralleling equipment with appropriate protection is required. (See Article 705.)

(B) Bypass Isolation Switches Means shall be permitted to bypass and isolate the transfer equipment. Where bypass isolation switches are used, inadvertent parallel operation shall be avoided.

(C) Automatic Transfer Switches Automatic transfer switches shall be electrically operated and mechanically held.

Section 700.6(C) is intended to ensure that relay contacts are mechanically held in the event of coil failure. This change correlates the *NEC* with the requirements found in NFPA 110, *Standard for Emergency and Standby Power Systems*.

(D) Use Transfer equipment shall supply only emergency loads.

Although the alternate power source is permitted to supply emergency loads as well as other loads, the transfer switch used for the emergency system is strictly limited to emergency loads only, that is, loads classified as emergency in accordance with 700.1. Other loads, such as legally required standby loads or optional standby loads (covered by Articles 701 and 702), are not permitted to be supplied from the emergency system transfer switch. If a single generator is used to supply both emergency and nonemergency loads, then multiple transfer switches are required.

700.7 Signals

Audible and visual signal devices shall be provided, where practicable, for the purpose described in 700.7(A) through 700.7(D).

(A) Derangement To indicate derangement of the emergency source.

(B) Carrying Load To indicate that the battery is carrying load.

(C) Not Functioning To indicate that the battery charger is not functioning.

The major causes of emergency equipment failure are improper testing or lack of testing, inadequate maintenance, and the failure of attendants to see the signals that indicate malfunctioning battery or battery-charging equipment. One way to minimize equipment failure is to install signal devices to annunciate trouble where attendants or other personnel familiar with the operation of the emergency equipment can see or hear them. In theaters, assembly halls, or similar locations, audible signal bells or horns that annunciate the functions specified in 700.7 should be located where their sounding will not cause panic.

Battery-operated unit equipment generally has a test switch to simulate a failure of the normal system and an indicating light that glows brightly while charging and dims when ready. Transparent cases for lead-acid batteries allow easy viewing of electrolyte levels.

A storage battery system is normally capable of delivering 12 volts, 24 volts, 32 volts, or 120 volts and consists of monitoring and distribution cabinets and a console with battery and charger. A storage battery system generally includes audio, visual, and remote signal devices and a test switch, and it may include a trouble bell and silence switch.

(D) Ground Fault To indicate a ground fault in solidly grounded wye emergency systems of more than 150 volts to ground and circuit-protective devices rated 1000 amperes or more. The sensor for the ground-fault signal devices shall be located at, or ahead of, the main system disconnecting means for the emergency source, and the maximum setting of the signal devices shall be for a ground-fault current of 1200 amperes. Instructions on the course of action to be taken in event of indicated ground fault shall be located at or near the sensor location.

FPN: For signals for generator sets, see NFPA 110-2002, *Standard for Emergency and Standby Power Systems.*

Although 700.26 indicates that ground-fault protection of equipment is not required on the alternate source for emergency systems, ground faults can occur on such systems,

and they can result in equipment burndown. Because of the emergency nature of such systems, automatic disconnect in the event of a ground fault is inappropriate. Detection of such a fault, however, is desirable so that the condition can be corrected.

700.8 Signs

(A) Emergency Sources A sign shall be placed at the service entrance equipment, indicating type and location of on-site emergency power sources.

Exception: A sign shall not be required for individual unit equipment as specified in 700.12(F).

(B) Grounding Where the grounded circuit conductor connected to the emergency source is connected to a grounding electrode conductor at a location remote from the emergency source, there shall be a sign at the grounding location that shall identify all emergency and normal sources connected at that location.

Section 700.8(B) requires a sign at the grounding location if the emergency source is a separately derived system and is connected to a grounding electrode conductor at a location that is remote from the emergency source.

II. Circuit Wiring

700.9 Wiring, Emergency System

(A) Identification All boxes and enclosures (including transfer switches, generators, and power panels) for emergency circuits shall be permanently marked so they will be readily identified as a component of an emergency circuit or system.

The marking may be by color code, the words "emergency system," or any other type of identification that identifies the box or enclosure as a component of the emergency system.

(B) Wiring Wiring of two or more emergency circuits supplied from the same source shall be permitted in the same raceway, cable, box, or cabinet. Wiring from an emergency source or emergency source distribution overcurrent protection to emergency loads shall be kept entirely independent of all other wiring and equipment, unless otherwise permitted in (1) through (4):

(1) Wiring from the normal power source located in transfer equipment enclosures
(2) Wiring supplied from two sources in exit or emergency luminaires (lighting fixtures)

(3) Wiring from two sources in a common junction box, attached to exit or emergency luminaires (lighting fixtures)

(4) Wiring within a common junction box attached to unit equipment, containing only the branch circuit supplying the unit equipment and the emergency circuit supplied by the unit equipment

Where an alternate power source supplies a switchboard from a single feeder or feeders in parallel, that switchboard may further distribute and provide power for the emergency, legally required, and optional standby systems.

Switchboards provide the physical separation requirements of both system and wiring from a common power source. This physical separation cannot occur within a panelboard enclosure because of its open design. Individual overcurrent sections can supply emergency, legally required, and optional standby transfer switches.

Where generators operate in parallel, frequency and voltage are synchronized and supply a common bus array. From this bus, emergency, legally required, and optional standby transfer switches can be supplied.

Emergency circuit wiring is not permitted to enter the same raceway, cable, box, or cabinet as the regular or normal wiring of the building. Wiring for the emergency circuits (see Exhibit 700.1) must be completely independent of all other wiring and equipment. This practice ensures that any fault on the normal wiring circuits will not affect the performance of the emergency wiring or equipment.

To effect an immediate transfer from one system to the other, both the normal source and the emergency source must be present within a transfer switch enclosure per 700.9(B)(1).

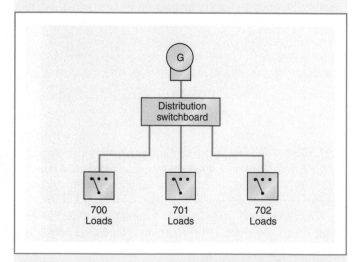

Exhibit 700.1 Illustration of a single generator supplying distribution switchboard from which emergency (Article 700), legally required standby (Article 701), and optional standby (Article 702) loads are supplied.

Sections 700.9(B)(2) and 700.9(B)(3) permit the use of two-lamp exit or two-lamp emergency fixtures, where one lamp is connected to the normal supply and one lamp is connected to the alternate supply. Both lamps may be illuminated as part of the regular lighting operation.

Wiring on the load side of a transfer switch serves as both the emergency circuit wiring and the normal circuit wiring. It is not intended that two sets of wiring supply emergency loads from the transfer switch to the emergency load distribution panel, as shown in Exhibit 700.3 and Exhibit 700.4, or from these emergency distribution panels to the emergency loads.

(C) Wiring Design and Location Emergency wiring circuits shall be designed and located so as to minimize the hazards that might cause failure due to flooding, fire, icing, vandalism, and other adverse conditions.

The purpose of 700.9(C) is to minimize the likelihood of impairment of the emergency system due to flood, fire, icing, vandalism, or other adverse conditions. The same requirement applies to sources of power covered in 700.12.

(D) Fire Protection Emergency systems shall meet the additional requirements in 700.9(D)(1) and (D)(2) assembly occupancies for not less than 1000 persons or in buildings above 23 m (75 ft) in height with any of the following occupancy classes: assembly, educational, residential, detention and correctional, business, and mercantile.

(1) Feeder-Circuit Wiring Feeder-circuit wiring shall meet one of the following conditions:

(1) Be installed in spaces or areas that are fully protected by an approved automatic fire suppression system

This section was revised for the 2005 *Code*. Previously, it required wiring to be in buildings that were fully protected by an approved automatic fire suppression system. However, such buildings may have spaces, such as the space above a suspended ceiling, that are not fully protected by the suppression system. The revised wording clarifies that the wiring must be in a space that is protected by the automatic suppression system.

(2) Be a listed electrical circuit protective system with a minimum 1-hour fire rating

(3) Be protected by a listed thermal barrier system for electrical system components

(4) Be protected by a fire-rated assembly listed to achieve a minimum fire rating of 1 hour

(5) Be embedded in not less than 50 mm (2 in.) of concrete

(6) Be a cable listed to maintain circuit integrity for not less than 1 hour when installed in accordance with the listing requirements

(2) Feeder-Circuit Equipment Equipment for feeder circuits (including transfer switches, transformers, and panelboards) shall be located either in spaces fully protected by approved automatic fire suppression systems (including sprinklers, carbon dioxide systems) or in spaces with a 1-hour fire resistance rating.

> FPN: For the definition of *occupancy classification*, see Section 6.1 of NFPA *101-2003, Life Safety Code.*

The proper operation of emergency electrical systems is critical for densely populated occupancies and for high-rise occupancies. Therefore, fire protection requirements for both emergency system feeder circuits and equipment ensure the integrity as well as the performance of the emergency electrical system. If feeders and equipment are located in buildings that are fully protected by an approved fire suppression system, then no further fire protection techniques are generally required.

Sprinkler systems are the most common fire suppression systems, and they are covered in NFPA 13, *Standard for the Installation of Sprinkler Systems.* Buildings that are fully protected by automatic sprinkler systems meet the requirements of 700.9(D). Additional fire suppression systems are included in the following standards:

1. NFPA 11, *Standard for Low-, Medium-, and High-Expansion Foam*
2. NFPA 12, *Standard on Carbon Dioxide Extinguishing Systems*
3. NFPA 12A, *Standard on Halon 1301 Fire Extinguishing Systems*
4. NFPA 15, *Standard for Water Spray Fixed Systems for Fire Protection*
5. NFPA 17, *Standard for Dry Chemical Extinguishing Systems*
6. NFPA 2001, *Standard on Clean Agent Fire Extinguishing Systems*

If feeders and equipment are not located in buildings that are fully protected by an approved fire suppression system, other methods and protection techniques are available to comply with the fire protection requirements of 700.9(D)(2). These additional fire protection methods and techniques include the following.

(1) Listed Electrical Circuit Protective Systems. These systems are described in the UL *Building Materials Directory.* The four-letter code (shown in parentheses) following each category heading in the directory is the UL product category guide designation. Examples of these systems include electrical circuit protective systems (FHIT), electrical circuit protective materials (FHIY), and fire-resistive cables (FHJR).

(2) Listed Thermal Barrier Systems. These systems are described in the UL *Building Materials Directory* as thermal barrier systems (XCLF). Examples of the thermal barrier protection technique include batts and blankets (XCLR), packing material (XCMD), and preformed mineral and fiber units (XCMK) wrapped or otherwise formed over the conduit to achieve a predetermined fire rating.

(3) Fire-Rated Assembly. These systems are described in the UL *Fire Resistance Directory,* Volumes 1 and 2. The assemblies found in Volume 1 include hourly ratings for beams, floors, roofs, columns, and walls and partitions. Volume 2 of the directory includes hourly ratings for joint systems and through-penetration firestop systems. All fire ratings and fire resistance ratings are based on the assumption that the structural supports for the assembly are not impaired by the fire.

(4) Embedded in Concrete. Embedding a conduit in concrete is most effective when implemented during original construction. This method has been successful for many years in protecting premises from service conductors. According to 230.6, conductors embedded in not less than 2 in. of concrete are considered to be outside the building.

(5) Cables Listed to Maintain Circuit Integrity. Circuit integrity cables are classified by the UL *Building Materials Directory* under the existing product category of fire-resistive cables (FHJR).

It is important to understand the difference between a 1-hour fire rating of an electrical cable and a 1-hour fire resistance rating of a structural member, such as a wall. Simply stated, at the end of a 1-hour fire rating test on an electrical cable, the circuit and its insulation must be intact and electrically functioning. (No short circuits, grounds, or opens are permitted.) However, a wall subjected to a 1-hour fire resistance test must only prevent a fire from passing through or past the wall, without regard to damage to the wall.

III. Sources of Power

700.12 General Requirements

Current supply shall be such that, in the event of failure of the normal supply to, or within, the building or group of buildings concerned, emergency lighting, emergency power, or both shall be available within the time required for the application but not to exceed 10 seconds. The supply system for emergency purposes, in addition to the normal services to the building and meeting the general requirements of this section, shall be one or more of the types of systems described in 700.12(A) through 700.12(E). Unit equipment in accordance with 700.12(F) shall satisfy the applicable requirements of this article.

In selecting an emergency source of power, consideration shall be given to the occupancy and the type of service to be rendered, whether of minimum duration, as for evacuation of a theater, or longer duration, as for supplying emergency power and lighting due to an indefinite period of current failure from trouble either inside or outside the building.

Equipment shall be designed and located so as to minimize the hazards that might cause complete failure due to flooding, fires, icing, and vandalism.

An alternate source of power must be designed and located such that it is not exposed to hazards that could impair its reliability. Such hazards include fire, flooding, icing, and vandalism. See Exhibit 700.2.

Exhibit 700.2 Alternate source of power in a heated and secure building that is located in an area not subject to flooding.

Equipment for sources of power as described in 700.12(A) through 700.12(E) where located within assembly occupancies for greater than 1000 persons or in buildings above 23 m (75 ft) in height with any of the following occupancy classes — assembly, educational, residential, detention and correctional, business, and mercantile — shall be installed either in spaces fully protected by approved automatic fire suppression systems (sprinklers, carbon dioxide systems, and so forth) or in spaces with a 1-hour fire rating.

> FPN No. 1: For the definition of occupancy classification, see Section 6.1 of NFPA *101*-2003, *Life Safety Code*.
> FPN No. 2: Assignment of degree of reliability of the recognized emergency supply system depends on the

careful evaluation of the variables at each particular installation.

(A) Storage Battery Storage batteries used as a source of power for emergency systems shall be of suitable rating and capacity to supply and maintain the total load for a minimum period of 1½ hours, without the voltage applied to the load falling below 87½ percent of normal.

Batteries, whether of the acid or alkali type, shall be designed and constructed to meet the requirements of emergency service and shall be compatible with the charger for that particular installation.

For a sealed battery, the container shall not be required to be transparent. However, for the lead acid battery that requires water additions, transparent or translucent jars shall be furnished. Automotive-type batteries shall not be used.

An automatic battery charging means shall be provided.

(B) Generator Set

(1) Prime Mover-Driven For a generator set driven by a prime mover acceptable to the authority having jurisdiction and sized in accordance with 700.5, means shall be provided for automatically starting the prime mover on failure of the normal service and for automatic transfer and operation of all required electrical circuits. A time-delay feature permitting a 15-minute setting shall be provided to avoid retransfer in case of short-time reestablishment of the normal source.

(2) Internal Combustion as Prime Movers Where internal combustion engines are used as the prime mover, an on-site fuel supply shall be provided with an on-premise fuel supply sufficient for not less than 2 hours' full-demand operation of the system. Where power is needed for the operation of the fuel transfer pumps to deliver fuel to a generator set day tank, this pump shall be connected to the emergency power system.

Engine-driven generators that require a fuel pump may not start or continue operating if the fuel pump is not operating. The last sentence of 700.12(B)(2) ensures that fuel transfer pumps have power when needed.

(3) Dual Supplies Prime movers shall not be solely dependent on a public utility gas system for their fuel supply or municipal water supply for their cooling systems. Means shall be provided for automatically transferring from one fuel supply to another where dual fuel supplies are used.

Exception: Where acceptable to the authority having jurisdiction, the use of other than on-site fuels shall be permitted where there is a low probability of a simultaneous failure of both the off-site fuel delivery system and power from the outside electrical utility company.

(4) Battery Power and Dampers Where a storage battery is used for control or signal power or as the means of starting the prime mover, it shall be suitable for the purpose and shall be equipped with an automatic charging means independent of the generator set. Where the battery charger is required for the operation of the generator set, it shall be connected to the emergency system. Where power is required for the operation of dampers used to ventilate the generator set, the dampers shall be connected to the emergency system.

Engine-driven generators that are not equipped with an alternator to provide battery charging require a separate battery-charging system. Failure of the power supply to a battery-charging system or failure of the battery-charging system itself could render the engine-driven generator inoperable. The second sentence of 700.12(B)(4) ensures that the charging system always has power.

Ventilation dampers may require electrical power in order to operate. If this is the case, 700.12(B)(4) also requires them to be powered by the emergency power source to ensure operation when the engine-driven generator operates. These additional loads require the system designer to ensure that the engine-driven generator is of adequate capacity in accordance with 700.5(A).

(5) Auxiliary Power Supply Generator sets that require more than 10 seconds to develop power shall be permitted if an auxiliary power supply energizes the emergency system until the generator can pick up the load.

Exhibit 700.3 illustrates a typical generator installation supplying standby power.

(6) Outdoor Generator Sets Where an outdoor housed generator set is equipped with a readily accessible disconnecting means located within sight of the building or structure supplied, an additional disconnecting means shall not be required where ungrounded conductors serve or pass through the building or structure.

Typical Generator Installation

Exhibit 700.3 A typical generator installation supplying standby power in ratings from 55 kW to 930 kW, 60 Hz. (Courtesy of Caterpillar)

Section 700.12(B)(6) clarifies the requirements for the disconnecting means for an outdoor generator set. The disconnecting means on the generator can be used as the disconnecting means required in 225.31, provided the disconnecting means is readily accessible and is within sight of the building. See the definitions of the terms *accessible*, *readily*, and *in sight from* in Article 100.

(C) Uninterruptible Power Supplies Uninterruptible power supplies used to provide power for emergency systems shall comply with the applicable provisions of 700.12(A) and 700.12(B).

(D) Separate Service Where acceptable to the authority having jurisdiction as suitable for use as an emergency source of power, an additional service shall be permitted. This service shall be in accordance with the applicable provisions of Article 230 and the following additional requirements:

(1) Separate service drop or service lateral
(2) Service conductors sufficiently remote electrically and physically from any other service conductors to minimize the possibility of simultaneous interruption of supply

(E) Fuel Cell System Fuel cell systems used as a source of power for emergency systems shall be of suitable rating and capacity to supply and maintain the total load for not less than 2 hours of full-demand operation.

Installation of a fuel cell system shall meet the requirements of Parts II through VIII of Article 692.

Where a single fuel cell system serves as the normal supply for the building or group of buildings concerned, it shall not serve as the sole source of power for the emergency standby system.

When designing emergency systems, whether for lighting or power, or both, the type of source that is needed must be considered.

Supply systems for emergency systems can be designed as one or more of the following:

1. One storage battery or a group of storage batteries provided with an automatic battery-charging means. (See also Article 480.)
2. A generator set driven by a prime mover, acceptable to the authority having jurisdiction, and with adequate capacity to carry the maximum load connected. Prime movers may be internal-combustion engines, steam or gas turbines, or other approved types of mechanical drivers. A storage battery used to start the prime mover must be provided with an automatic battery-charging means. An on-site fuel supply that is sufficient to operate internal-combustion engines at full load for 2 hours must also be available.

Off-site fuel supplies such as natural gas or piped steam may be used where experience has demonstrated their reliability. Off-site fuel supplies may also be used where they will provide greater reliability than gasoline or diesel engines or in isolated areas where maintenance or refueling could be a problem.

Some types of drivers, particularly large ones, may take longer than 10 seconds to accelerate and develop generator voltage. Gas and steam turbines and large internal-combustion engines may have prolonged starting times. Depending on the specific loads, short-time supply could be provided by an uninterruptible power supply; a generator shared with other loads; or a generator with limited emergency supply, such as an expander, a steam turbine, or a waste heat system.

3. Uninterruptible power supplies (UPS), which generally include a rectifier, a storage battery, and an inverter to ac. Uninterruptible power supplies may be very complex systems with redundant components and high-speed solid-state switching. It is common practice to include an automatic bypass for UPS malfunction to permit maintenance.
4. The use of a separate service, which requires a judgment by the authority having jurisdiction. Such judgment should be based on the nature of the emergency loads and the expected reliability of the other available sources.
5. Unit equipment wired with a flexible cord (not longer than 3 ft) and attachment plug cap. Unit equipment must be permanently fixed in place, usually by mounting screws that are accessible only from within the unit. One or more lamps may be mounted on or remote from the unit. The unit should be located where it can be readily checked or tested for proper performance. See Exhibit 700.4.

Exhibit 700.4 Self-contained, fully automatic unit equipment for operating emergency lighting located on the unit or for remotely located exit signs or lighting heads. (Courtesy of Dual-Lite, Inc.)

Unit equipment is intended to provide illumination for the area where it is installed. For instance, if a unit is located in a corridor, it must be connected to the branch circuit supplying the normal corridor lights (on the line side of any switching arrangements). If normal power fails, the unit automatically energizes the unit lamps, restoring illumination to the corridor. A separate circuit is not permitted for unit equipment [except as noted in the exception to 700.12(F)] because, if applied to the preceding example, failure of the normal corridor circuit would not affect the unit equipment, and the corridor would remain dark. The branch circuit feeding the unit must be identified at the panelboard.

Notes on General Requirements for Emergency Lighting Systems

At least two sources of power must be provided — one normal supply and one or more emergency systems described in 700.12. The sources may be (1) two services — one normal supply and one emergency supply (preferably from separate utility stations), (2) one normal service and a storage battery (or unit equipment) system, or (3) one normal service and a generator set. (See Exhibits 700.5 and 700.6.)

A transfer means (or throw-over switch) must be

Exhibit 700.6 Emergency load arranged to be supplied from two widely separated services, as permitted by 700.12(D). When one service fails, the emergency load is transferred to the other service.

provided to energize the emergency equipment from the alternate supply when the normal source of supply is interrupted.

If a separate service is used, both may operate normally, but equipment for emergency lighting and power must be arranged to be energized from either service.

If the alternate or emergency source of supply is a storage battery or generator set, the single emergency system is usually operated on the normal service, and the battery (or batteries) or generator operates only if the normal service fails. A generator may be used, however, for peak load shaving and the like in accordance with 700.5.

Two or more separate and complete systems may provide power for emergency lighting, but means must be provided for energizing one system if the other one fails.

It should be noted that provisions for disconnecting means and overcurrent protection (see Exhibits 700.5 and 700.6) are to be provided for emergency systems as required by Article 230.

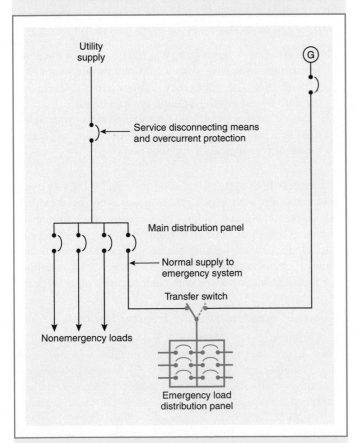

Exhibit 700.5 Emergency load arranged to be supplied from a generator, as permitted by 700.12(B).

(F) Unit Equipment Individual unit equipment for emergency illumination shall consist of the following:

(1) A rechargeable battery
(2) A battery charging means
(3) Provisions for one or more lamps mounted on the equipment, or shall be permitted to have terminals for remote lamps, or both
(4) A relaying device arranged to energize the lamps automatically upon failure of the supply to the unit equipment

The batteries shall be of suitable rating and capacity to supply and maintain at not less than 87½ percent of the nominal battery voltage for the total lamp load associated with the unit for a period of at least 1½ hours, or the unit equipment shall supply and maintain not less than 60 percent of the initial emergency illumination for a period of at least 1½ hours. Storage batteries, whether of the acid or alkali type, shall be designed and constructed to meet the requirements of emergency service.

Unit equipment shall be permanently fixed in place (i.e., not portable) and shall have all wiring to each unit installed in accordance with the requirements of any of the wiring methods in Chapter 3. Flexible cord-and-plug connection shall be permitted, provided that the cord does not exceed 900 mm (3 ft) in length. The branch circuit feeding the unit equipment shall be the same branch circuit as that serving the normal lighting in the area and connected ahead of any local switches. The branch circuit that feeds unit equipment shall be clearly identified at the distribution panel. Emergency luminaires (illumination fixtures) that obtain power from a unit equipment and are not part of the unit equipment shall be wired to the unit equipment as required by 700.9 and by one of the wiring methods of Chapter 3.

Exception: In a separate and uninterrupted area supplied by a minimum of three normal lighting circuits, a separate branch circuit for unit equipment shall be permitted if it originates from the same panelboard as that of the normal lighting circuits and is provided with a lock-on feature.

IV. Emergency System Circuits for Lighting and Power

700.15 Loads on Emergency Branch Circuits

No appliances and no lamps, other than those specified as required for emergency use, shall be supplied by emergency lighting circuits.

700.16 Emergency Illumination

Emergency illumination shall include all required means of egress lighting, illuminated exit signs, and all other lights specified as necessary to provide required illumination.

Emergency lighting systems shall be designed and installed so that the failure of any individual lighting element,

such as the burning out of a light bulb, cannot leave in total darkness any space that requires emergency illumination.

Where high-intensity discharge lighting such as high- and low-pressure sodium, mercury vapor, and metal halide is used as the sole source of normal illumination, the emergency lighting system shall be required to operate until normal illumination has been restored.

Exception: Alternative means that ensure emergency lighting illumination level is maintained shall be permitted.

High-intensity discharge (HID) fixtures take some time to start once they are energized. Therefore, if HID fixtures are the sole source of normal illumination in an area, the *Code* requires that the emergency lighting system operate not only until the normal system is returned to service but also until the HID fixtures provide illumination. This may require a timing circuit, photoelectric monitoring system, or the equivalent.

700.17 Circuits for Emergency Lighting

Branch circuits that supply emergency lighting shall be installed to provide service from a source complying with 700.12 when the normal supply for lighting is interrupted. Such installations shall provide either of the following:

(1) An emergency lighting supply, independent of the general lighting supply, with provisions for automatically transferring the emergency lights upon the event of failure of the general lighting system supply
(2) Two or more separate and complete systems with independent power supply, each system providing sufficient current for emergency lighting purposes

Unless both systems are used for regular lighting purposes and are both kept lighted, means shall be provided for automatically energizing either system upon failure of the other. Either or both systems shall be permitted to be a part of the general lighting system of the protected occupancy if circuits supplying lights for emergency illumination are installed in accordance with other sections of this article.

General Considerations for Transfer Switches

Automatic transfer switches of double-throw construction are used primarily for emergency and standby power generation systems rated 600 volts or less. These transfer switches do not normally incorporate overcurrent protection and are designed and applied in accordance with the *Code*, particularly Articles 700 and 701. Automatic transfer switches are available in ratings from 30 to 3000 amperes. For reliability, most automatic transfer switches rated above 100 amperes are mechanically held and electrically operated from the power source to which the load is to be transferred.

An automatic transfer switch is usually located in the main or secondary distribution bus that feeds the branch circuits. Because of its location in the system, the capabilities that must be designed into the transfer switch are unique and extensive compared with the design requirements for other branch-circuit and feeder devices. For example, special consideration should be given to the following characteristics of an automatic transfer:

1. Its ability to close against high inrush currents
2. Its ability to carry full-rated current continuously from normal and emergency sources
3. Its ability to withstand fault currents
4. Its ability to interrupt six times the full-load current

In addition to considering each of the preceding characteristics individually, it is also necessary to consider the effect each has on the others.

In arrangements that provide protection against failure of the utility service, consideration should also be given to the following:

1. An open circuit within the building area on the load side of the incoming service
2. Overload or fault condition
3. Electrical or mechanical failure of the electric power distribution system within the building

It is, therefore, desirable to locate transfer switches close to the load and to keep the operation of the transfer switches independent of overcurrent protection. It is often desirable to use multiple transfer switches of lower current rating located near the load rather than one large transfer switch at the point of incoming service.

Location of Overcurrent Devices

The location of overcurrent devices for both normal and emergency power is covered by 240.21 and is not affected by the installation of an automatic transfer switch. Transfer switches should be rated for continuous duty and have low contact temperature rise.

Solid Neutral on Alternating-Current and Direct-Current Systems

If automatic ac-to-ac transfer switches are used, then solid neutrals can be used with the grounding connections, as required in 250.24. If multiple grounding creates objectionable ground current, then corrective action, as specified in 250.6(B), must be made.

Section 230.95 requires ground-fault protection of equipment. Because the normal source and the emergency

source are typically grounded at their locations, the multiple neutral-to-ground connections usually require some additional means or devices to ensure proper ground-fault sensing by the ground-fault protection device. Additional means or devices are generally required because the usual alterations used to stop objectionable current, per 250.6(B), do not apply if the objectionable current is a ground-fault current. [See 250.6(C).] Rather, solutions often include adding an overlapping neutral transfer pole or conventional fourth pole to the transfer switch. Other solutions include using isolation transformers and special ground-fault circuits or using the service ground to also ground the generator neutral with 3-pole transfer switches.

On ac-to-dc automatic transfer switches, a solid neutral tie between the ac and dc neutrals is not permitted where both sources of supply are exterior distribution systems. Section 250.164, which addresses the location of grounds for dc exterior systems, clearly specifies that the dc system can be grounded only at the supply station. If the dc system is an interior isolated system, such as a storage battery, then solid neutral connection between the ac system neutral and the dc source is acceptable.

On an ac-to-dc automatic transfer switch where the neutral must be switched, the size of the neutral switching pole must be considered. A 4-pole, double-throw switch must be used where a 3-phase, 4-wire normal source and a 2-wire dc emergency source are transferred. A 4-pole, double-throw transfer switch is required because the neutral is switched. In this instance, one pole of the dc emergency source carries three times the current of the other poles.

Close Differential Voltage Supervision of Normal Source

Most often, the normal source is an electric utility company whose power is transmitted many miles to the point of utilization. The automatic transfer switch control panel at the utility company continuously monitors the voltage of all phases. (Because utility frequency is, for all practical purposes, constant, only the voltage needs to be monitored.) For single-phase power systems, the line-to-line voltage is monitored. For 3-phase power systems, all three line-to-line voltages should be monitored to provide full-phase protection.

Monitoring protects against operation at reduced voltage, such as during brownouts, which can damage loads such as motors. Because the voltage sensitivity of loads varies, the pickup (acceptable) voltage setting and dropout (unacceptable) voltage setting of the monitors should be adjustable. The typical range of adjustment for the pickup is 85 to 100 percent of nominal, while the dropout setting, which is a function of the pickup setting, is 75 to 98 percent of the pickup selected. Typical settings for most loads are

95 percent of nominal for pickup and 85 percent of nominal for dropout (90 percent of pickup). Consideration must be given to voltage supervision at closer differential for many installations where the load circuits are critical to voltage.

Electronic equipment load is frequently voltage critical. Installations that use such equipment include patient care equipment in health care facilities, X-ray equipment, television stations, cable TV centers, microwave communications, telephone communications, computers, computer-operated equipment, computer centers, and similar applications.

Polyphase motors operating at low load have a tendency to single phase, despite the loss of voltage in one phase, leading to burnout of the motor. A close differential of voltage supervision should be applied to automatic transfer switches for motor installations of the polyphase type. Differential voltage relays with a close adjustment of 2 percent for transfer and retransfer values aid in the detection of phase outages and provide protection from single phasing.

Automatic Transfer Switches with Emergency Source on Automatically Started Power Plant

In installations that require an emergency power source, the normal source is usually a utility power line, and the emergency source is an automatically started engine generator set that starts when the normal source fails. To ensure maximum reliability, a minimum installation should be arranged to do the following:

1. Initiate engine starting of the power plant from a contact on the automatic transfer switch control panel.
2. Sustain connection of load circuits to the normal source during the starting period to provide utilization of any existing service from the normal source.
3. Measure output voltage and frequency of emergency source through the use of a voltage-frequency-sensitive monitor and effect transfer of the load circuits to the power plant only when both voltage and frequency of the power plant are approximately normal. Sensing of the emergency source need only be single phase, because most applications involve an on-site engine generator with a relatively short line run to the automatic transfer switch. In addition to monitoring voltage, the emergency source's frequency should be monitored. Unlike the utility power, the engine generator frequency can vary during start-up. Frequency monitoring avoids overloading the engine generator while it is starting and can thus prevent stalling the engine. Combined frequency and voltage monitoring prevents loads from being transferred to an engine generator set with an unacceptable output.
4. Provide visual signal and auxiliary contact for remote indication when the power plant is feeding the load.

Time-Delay Devices on Automatic Transfer Switches

Time-delay controls are essential to the operation of the automatic transfer switch. To avoid unnecessary starting and transfer to the alternate supply, a nominal 1-second time delay, adjustable up to 6 seconds, can override momentary interruptions and temporary reductions in normal source voltage but still allow starting and transfer if the reduction or outage is sustained. Local electric utilities can provide circuit protection schemes and timing information. Because momentary outages may last 2 seconds, nominal, a correctly set time delay should prevent the transfer switch from operating (and prevent generator starting) during utility automatic circuit protection operation. However, the time delay should be set fast enough to effectively operate the transfer switch and provide backup power for long-term outages.

The advantages of a time delay are realized in all types of automatic transfer installations. In standby plant installations, the reduced number of false starts is especially important to minimize wear on the starting gear, battery, and associated equipment. This delay is generally set at 1 second but may be set higher if reclosers or circuit breakers on the utility power lines take longer to operate or if momentary power dips exceed 1 second. If longer delay settings are used, care must be taken to ensure that sufficient time remains to meet 10-second power restoration requirements. The authority having jurisdiction may determine that a longer-term power failure does not occur until the utility automatic protective devices fail to restore power to the facility. For example, the 10-second power restoration requirements would become effective after the 2-second, nominal, recloser or circuit breaker cycle.

Once the load is transferred to the alternate source, another timer delays retransfer to the normal source until that source has time to stabilize. See the commentary following the exception to 700.16. This timer is required by 700.12(B)(1) and is controlled by the preferred source voltage monitors. The timer is adjustable from 0 to 30 minutes and is normally set at 30 minutes. Another important function of this retransfer timer is to allow an engine generator to operate under load for a preselected minimum time to ensure continued good performance of the set and its starting system. This delay should be automatically nullified if the alternate source fails and the normal source is available, as determined by the voltage monitors.

Engine generator manufacturers often recommend a cool-down period for their sets that allows them to run unloaded after the load is retransferred to the normal source. A third time delay, usually 5 minutes, is provided for this purpose. Running an unloaded engine for more than 5 minutes is neither necessary nor recommended because it can cause deterioration in engine performance.

If more than one automatic transfer switch is connected

to the same engine generator, it is sometimes recommended that transfer of the loads be purposely sequenced to the alternate source. Using such a sequencing scheme can reduce starting kVA capacity requirements of the generator. A fourth timer, adjustable from 0 to 5 minutes, will delay transfer to the emergency supply source for this and other, similar requirements.

700.18 Circuits for Emergency Power

For branch circuits that supply equipment classed as emergency, there shall be an emergency supply source to which the load will be transferred automatically upon the failure of the normal supply.

V. Control — Emergency Lighting Circuits

700.20 Switch Requirements

The switch or switches installed in emergency lighting circuits shall be arranged so that only authorized persons have control of emergency lighting.

Exception No. 1: Where two or more single-throw switches are connected in parallel to control a single circuit, at least one of these switches shall be accessible only to authorized persons.

Exception No. 2: Additional switches that act only to put emergency lights into operation but not disconnect them shall be permissible.

Switches connected in series or 3- and 4-way switches shall not be used.

700.21 Switch Location

All manual switches for controlling emergency circuits shall be in locations convenient to authorized persons responsible for their actuation. In facilities covered by Articles 518 and 520, a switch for controlling emergency lighting systems shall be located in the lobby or at a place conveniently accessible thereto.

In no case shall a control switch for emergency lighting be placed in a motion-picture projection booth or on a stage or platform.

Exception: Where multiple switches are provided, one such switch shall be permitted in such locations where arranged so that it can only energize the circuit, but cannot de-energize the circuit.

700.22 Exterior Lights

Those lights on the exterior of a building that are not required for illumination when there is sufficient daylight shall be permitted to be controlled by an automatic light-actuated device.

VI. Overcurrent Protection

700.25 Accessibility

The branch-circuit overcurrent devices in emergency circuits shall be accessible to authorized persons only.

700.26 Ground-Fault Protection of Equipment

The alternate source for emergency systems shall not be required to have ground-fault protection of equipment with automatic disconnecting means. Ground-fault indication of the emergency source shall be provided per 700.7(D).

700.27 Coordination

Emergency system(s) overcurrent devices shall be selectively coordinated with all supply side overcurrent protective devices.

This requirement, new in the 2005 *NEC*, requires that all emergency system overcurrent devices be selectively coordinated with the overcurrent devices installed on their supply side. The term *coordination (selective)*, as defined in Article 100, indicates that a selectively coordinated system is one where the operation of the overcurrent protective scheme localizes an overcurrent condition to the circuit conductors or equipment in which an overload or fault (short-circuit or ground fault) has occurred. Because the purpose of an emergency system is to provide power to essential life safety systems in a building or facility, a selectively coordinated overcurrent protection scheme that localizes and minimizes the extent of an interruption of power due to the opening of a protective device is a critical safety element. Continuity of operation of illumination for occupant evacuation or maintaining continuity of operation of essential safety equipment such as smoke evacuation systems is necessary for occupant safety during a fire or other emergency. Simply put, an overcurrent event (overload, short-circuit, or ground-fault) in a 20 ampere branch circuit cannot cause the feeder protective device supplying the branch circuit panelboard to open. This coordination must be carried through each level of distribution that supplies power to the emergency system.

Design and subsequent verification of electrical system coordination can be achieved only through a coordination study that entails detailed knowledge of electrical supply system fault current characteristics and a design that integrates overcurrent protective devices that react to overcurrent and interact with each other in such a manner that the objective of minimizing outages by localizing the overcurrent problem and isolating that part of the emergency system can be achieved. It is important to note that modifications to the electrical system subsequent to the initial design and installation can significantly impact the original implementation of the coordinated system. For additional discussion on selective coordination, see the commentary for 620.62 and Exhibit 620.7.

ARTICLE 701
Legally Required Standby Systems

Summary of Changes

- **701.11(B)(5):** Revised to cover ungrounded conductors that pass through, the building or structure.
- **701.11(F):** Added new paragraph permitting fuel cell systems as a source of supply.
- **701.18:** Added new section requiring that standby system(s) overcurrent devices be selectively coordinated with all supply side overcurrent protective devices.

Contents

I. General

701.1 Scope

The provisions of this article apply to the electrical safety of the installation, operation, and maintenance of legally required standby systems consisting of circuits and equipment intended to supply, distribute, and control electricity to required facilities for illumination or power, or both, when the normal electrical supply or system is interrupted.

The systems covered by this article consist only of those that are permanently installed in their entirety, including the power source.

> FPN No. 1: For additional information, see NFPA 99-2002, *Standard for Health Care Facilities.*
>
> FPN No. 2: For further information regarding performance of emergency and standby power systems, see NFPA 110-2002, *Standard for Emergency and Standby Power Systems.*
>
> FPN No. 3: For further information, see ANSI/IEEE 446-1995, *Recommended Practice for Emergency and Standby Power Systems for Industrial and Commercial Applications.*

Legally required standby systems are intended to provide electric power to aid in fire fighting, rescue operations, control of health hazards, and similar operations. In comparison, emergency systems (see Article 700) are those systems essential for safety to life. Optional standby systems (see Article 702) are those in which failure can cause physical discomfort, serious interruption of an industrial process, damage to process equipment, or disruption of business, for example.

The requirements for legally required standby systems are much the same as for emergency systems. There are, however, some differences. When normal power is lost, legally required systems must be able to supply standby power in 60 seconds or less, instead of the 10 seconds or less required of emergency systems. Wiring for legally required standby systems may occupy the same raceways, cables, boxes, and cabinets as other general wiring. Wiring for emergency systems must be kept entirely independent of other wiring. Legally required standby systems take second priority to emergency systems if they are involved in sharing an alternate supply and/or load shedding or peak shaving schemes.

701.2 Definition

Legally Required Standby Systems. Those systems required and so classed as legally required standby by municipal, state, federal, or other codes or by any governmental agency having jurisdiction. These systems are intended to automatically supply power to selected loads (other than those classed as emergency systems) in the event of failure of the normal source.

> FPN: Legally required standby systems are typically installed to serve loads, such as heating and refrigeration systems, communications systems, ventilation and smoke removal systems, sewage disposal, lighting systems, and industrial processes, that, when stopped during any interruption of the normal electrical supply, could create hazards or hamper rescue or fire-fighting operations.

701.3 Application of Other Articles

Except as modified by this article, all applicable articles of this *Code* shall apply.

701.4 Equipment Approval

All equipment shall be approved for the intended use.

701.5 Tests and Maintenance for Legally Required Standby Systems

(A) Conduct or Witness Test The authority having jurisdiction shall conduct or witness a test of the complete system upon installation.

(B) Tested Periodically Systems shall be tested periodically on a schedule and in a manner acceptable to the authority having jurisdiction to ensure the systems are maintained in proper operating condition.

(C) Battery Systems Maintenance Where batteries are used for control, starting, or ignition of prime movers, the authority having jurisdiction shall require periodic maintenance.

(D) Written Record A written record shall be kept on such tests and maintenance.

(E) Testing Under Load Means for testing legally required standby systems under load shall be provided.

> FPN: For testing and maintenance procedures of emergency power supply systems (EPSSs), see NFPA 110-2002, *Standard for Emergency and Standby Power Systems*.

701.6 Capacity and Rating

A legally required standby system shall have adequate capacity and rating for the supply of all equipment intended to be operated at one time. Legally required standby system equipment shall be suitable for the maximum available fault current at its terminals.

The legally required standby alternate power source shall be permitted to supply both legally required standby and optional standby system loads under either of the following conditions:

(1) Where the alternate source has adequate capacity to handle all connected loads
(2) Where automatic selective load pickup and load shedding is provided that will ensure adequate power to the legally required standby circuits

This section was revised for the 2005 *Code*. Previously, 701.6(2) required selective load pickup and load shedding where a legally required standby system also supplied optional standby loads. Selective load pickup and load shedding are no longer required if the generator has sufficient capacity to supply all connected loads.

701.7 Transfer Equipment

(A) General Transfer equipment, including automatic transfer switches, shall be automatic and identified for standby use and approved by the authority having jurisdiction. Transfer equipment shall be designed and installed to prevent the inadvertent interconnection of normal and alternate sources of supply in any operation of the transfer equipment. Transfer equipment and electric power production systems installed to permit operation in parallel with the normal source shall meet the requirements of Article 705.

Section 701.7(A) permits transfer equipment to allow parallel operation of the generation equipment with the normal source as long as the requirements of Article 705 are met. Traditional automatic transfer switches are not designed to permit parallel operation of generation equipment and the normal source. Therefore, traditional automatic transfer switches need not comply with Article 705. However, certain automatic transfer switch configurations are intentionally designed to briefly (for a few cycles) parallel the generation equipment with the normal source upon load transfer from generator to normal source. This load transfer can occur with minimal disturbance or effect on the load. Transfer switches that employ this type of paralleling must comply with Article 705.

(B) Bypass Isolation Switches Means to bypass and isolate the transfer switch equipment shall be permitted. Where bypass isolation switches are used, inadvertent parallel operation shall be avoided.

(C) Automatic Transfer Switches Automatic transfer switches shall be electrically operated and mechanically held.

The intent of 701.7(C) is to ensure that relay contacts are mechanically held in the event of coil failure. This requirement also correlates the *NEC* with NFPA 110, *Standard for Emergency and Standby Power Systems.*

701.8 Signals

Audible and visual signal devices shall be provided, where practicable, for the purposes described in 701.8(A), (B), and (C).

(A) Derangement To indicate derangement of the standby source.

(B) Carrying Load To indicate that the standby source is carrying load.

(C) Not Functioning To indicate that the battery charger is not functioning.

> FPN: For signals for generator sets, see NFPA 110-2002, *Standard for Emergency and Standby Power Systems.*

701.9 Signs

(A) Mandated Standby A sign shall be placed at the service entrance indicating type and location of on-site legally required standby power sources.

Exception: A sign shall not be required for individual unit equipment as specified in 701.11(G).

(B) Grounding Where the grounded circuit conductor connected to the legally required standby power source is connected to a grounding electrode conductor at a location remote from the legally required standby power source, there shall be a sign at the grounding location that shall identify all legally required standby power and normal sources connected at that location.

II. Circuit Wiring

701.10 Wiring Legally Required Standby Systems

The legally required standby system wiring shall be permitted to occupy the same raceways, cables, boxes, and cabinets with other general wiring.

III. Sources of Power

701.11 Legally Required Standby Systems

Current supply shall be such that, in the event of failure of the normal supply to, or within, the building or group of buildings concerned, legally required standby power will be available within the time required for the application but not to exceed 60 seconds. The supply system for legally required standby purposes, in addition to the normal services to the building, shall be permitted to comprise one or more of the types of systems described in 701.11(A) through 701.11(F). Unit equipment in accordance with 701.11(G) shall satisfy the applicable requirements of this article.

In selecting a legally required standby source of power, consideration shall be given to the type of service to be rendered, whether of short-time duration or long duration.

Consideration shall be given to the location or design, or both, of all equipment to minimize the hazards that might cause complete failure due to floods, fires, icing, and vandalism.

> FPN: Assignment of degree of reliability of the recognized legally required standby supply system depends on the careful evaluation of the variables at each particular installation.

(A) Storage Battery A storage battery shall be of suitable rating and capacity to supply and maintain at not less than 87½ percent of system voltage the total load of the circuits supplying legally required standby power for a period of at least 1½ hours.

Batteries, whether of the acid or alkali type, shall be designed and constructed to meet the service requirements of emergency service and shall be compatible with the charger for that particular installation.

For a sealed battery, the container shall not be required to be transparent. However, for the lead acid battery that requires water additions, transparent or translucent jars shall be furnished. Automotive-type batteries shall not be used.

An automatic battery charging means shall be provided.

(B) Generator Set

(1) Prime Mover-Driven For a generator set driven by a prime mover acceptable to the authority having jurisdiction and sized in accordance with 701.6, means shall be provided for automatically starting the prime mover upon failure of the normal service and for automatic transfer and operation of all required electrical circuits. A time-delay feature permitting a 15-minute setting shall be provided to avoid retransfer in case of short-time re-establishment of the normal source.

(2) Internal Combustion Engines as Prime Mover Where internal combustion engines are used as the prime mover, an on-site fuel supply shall be provided with an on-premise fuel supply sufficient for not less than 2 hours' full-demand operation of the system.

(3) Dual Fuel Supplies Prime movers shall not be solely dependent on a public utility gas system for their fuel supply or municipal water supply for their cooling systems. Means shall be provided for automatically transferring one fuel supply to another where dual fuel supplies are used.

Exception: Where acceptable to the authority having jurisdiction, the use of other than on-site fuels shall be permitted where there is a low probability of a simultaneous failure of both the off-site fuel delivery system and power from the outside electrical utility company.

(4) Battery Power Where a storage battery is used for control or signal power or as the means of starting the prime mover, it shall be suitable for the purpose and shall be equipped with an automatic charging means independent of the generator set.

(5) Outdoor Generator Sets Where an outdoor housed generator set is equipped with a readily accessible disconnecting means located within sight of the building or structure supplied, an additional disconnecting means shall not be required where ungrounded conductors serve or pass through the building or structure.

The disconnecting means on the outdoor generator can be used as the disconnecting means required in 225.31, provided the disconnecting means is readily accessible and is within sight of the building. See the definition of the terms *accessible*, *readily*, and *in sight from* in Article 100.

(C) Uninterruptible Power Supplies Uninterruptible power supplies used to provide power for legally required standby systems shall comply with the applicable provisions of 701.11(A) and 701.11(B).

(D) Separate Service Where acceptable to the authority having jurisdiction as a source of power, an additional service shall be permitted. This service shall be in accordance with the applicable provisions of Article 230, with separate service drop or lateral sufficiently remote electrically and physically from any other service to minimize the possibility of simultaneous interruption of supply from an occurrence in another service.

(E) Connection Ahead of Service Disconnecting Means Where acceptable to the authority having jurisdiction, connections located ahead of and not within the same cabinet, enclosure, or vertical switchboard section as the service disconnecting means shall be permitted. The legally required standby service shall be sufficiently separated from the normal main service disconnecting means to prevent simultaneous interruption of supply through an occurrence within the building or groups of buildings served.

FPN: See 230.82 for equipment permitted on the supply side of a service disconnecting means.

Where a legally required standby system is supplied by conductors tapped to the normal service conductors (connection required to be on line side of the normal service disconnecting means), 230.82 requires that the tapped conductors be installed in accordance with all of the requirements for service-entrance conductors and the conductors terminate in equipment suitable for use as service equipment. These requirements help ensure that the legally required standby system disconnecting means can safely interrupt the fault current available from the utility and that the tapped conductors, which do not have short-circuit and ground-fault protection, are not run through the interior of a building.

(F) Fuel Cell System Fuel cell systems used as a source of power for legally required standby systems shall be of suitable rating and capacity to supply and maintain the total load for not less than 2 hours of full-demand operation.

Installation of a fuel cell system shall meet the requirements of Parts II through VIII of Article 692.

Where a single fuel cell system serves as the normal supply for the building or group of buildings concerned, it shall not serve as the sole source of power for the legally required standby system.

(G) Unit Equipment Individual unit equipment for legally required standby illumination shall consist of the following:

(1) A rechargeable battery
(2) A battery charging means
(3) Provisions for one or more lamps mounted on the equipment and shall be permitted to have terminals for remote lamps
(4) A relaying device arranged to energize the lamps automatically upon failure of the supply to the unit equipment

The batteries shall be of suitable rating and capacity to supply and maintain at not less than 87½ percent of the nominal battery voltage for the total lamp load associated with the unit for a period of at least 1½ hours, or the unit equipment shall supply and maintain not less than 60 percent of the initial legally required standby illumination for a period of at least 1½ hours. Storage batteries, whether of the acid or alkali type, shall be designed and constructed to meet the requirements of emergency service.

Unit equipment shall be permanently fixed in place (i.e., not portable) and shall have all wiring to each unit installed in accordance with the requirements of any of the wiring methods in Chapter 3. Flexible cord-and-plug connection shall be permitted, provided that the cord does not exceed 900 mm (3 ft) in length. The branch circuit feeding the unit equipment shall be the same branch circuit as that serving the normal lighting in the area and connected ahead of any local switches. Legally required standby luminaires (illumination fixtures) that obtain power from a unit equipment and are not part of the unit equipment shall be wired to the unit equipment by one of the wiring methods of Chapter 3.

Exception: In a separate and uninterrupted area supplied by a minimum of three normal lighting circuits, a separate branch circuit for unit equipment shall be permitted if it originates from the same panelboard as that of the normal lighting circuits and is provided with a lock-on feature.

IV. Overcurrent Protection

701.15 Accessibility

The branch-circuit overcurrent devices in legally required standby circuits shall be accessible to authorized persons only.

701.17 Ground-Fault Protection of Equipment

The alternate source for legally required standby systems shall not be required to have ground-fault protection of equipment.

701.18 Coordination

Legally required standby system(s) overcurrent devices shall be selectively coordinated with all supply side overcurrent protective devices.

See the commentary regarding selective coordination for emergency systems following 700.27.

ARTICLE 702
Optional Standby Systems

Summary of Changes

- **702.6:** Added exception to permit the connection of a portable generator without transfer equipment under specified conditions.
- **702.7:** Added exception to indicate that signals are not required for portable standby sources.
- **702.11:** Added requirement for disconnecting means associated with outdoor generator sets.

Contents

I. General

702.1 Scope

The provisions of this article apply to the installation and operation of optional standby systems.

    The systems covered by this article consist of those that are permanently installed in their entirety, including prime movers, and those that are arranged for a connection to a premises wiring system from a portable alternate power supply.

Article 702 applies not only to permanently installed generators and prime movers but also to portable alternate power supplies that can be connected to an optional standby system. For example, upon failure of an optional standby generator at a frozen food processing plant, a vehicle-mounted generator can be brought in and connected to the plant's optional standby system, which has provisions for such a connection.

Optional standby systems are those in which failure can cause physical discomfort, serious interruption of an industrial process, damage to process equipment, or disruption of business.

702.2 Definition

Optional Standby Systems. Those systems intended to supply power to public or private facilities or property where life safety does not depend on the performance of the system. Optional standby systems are intended to supply on-site generated power to selected loads either automatically or manually.

    FPN: Optional standby systems are typically installed to provide an alternate source of electric power for such facilities as industrial and commercial buildings, farms, and residences and to serve loads such as heating and refrigeration systems, data processing and communications systems, and industrial processes that, when stopped during any power outage, could cause discomfort, serious interruption of the process, damage to the product or process, or the like.

702.3 Application of Other Articles

Except as modified by this article, all applicable articles of this *Code* shall apply.

702.4 Equipment Approval

All equipment shall be approved for the intended use.

702.5 Capacity and Rating

An optional standby system shall have adequate capacity and rating for the supply of all equipment intended to be operated at one time. Optional standby system equipment shall be suitable for the maximum available fault current at its terminals. The user of the optional standby system shall be permitted to select the load connected to the system.

702.6 Transfer Equipment

Transfer equipment shall be suitable for the intended use and designed and installed so as to prevent the inadvertent interconnection of normal and alternate sources of supply in any operation of the transfer equipment. Transfer equipment and electric power production systems installed to permit operation in parallel with the normal source shall meet the requirements of Article 705.

Transfer equipment, located on the load side of branch circuit protection, shall be permitted to contain supplementary overcurrent protection having an interrupting rating sufficient for the available fault current that the generator can deliver. The supplementary overcurrent protection devices shall be part of a listed transfer equipment.

Transfer equipment shall be required for all standby systems subject to the provisions of this article and for which an electric-utility supply is either the normal or standby source.

Revised for the 2002 *NEC*, 702.6 now permits transfer equipment to allow parallel operation of the generation equipment with the normal source as long as the requirements of Article 705 are met. Traditional automatic transfer switches are not designed to permit parallel operation of generation equipment and the normal source. Therefore, traditional automatic transfer switches need not comply with Article 705. However, certain automatic transfer switch configurations are intentionally designed to briefly (for a few cycles) parallel the generation equipment with the normal source upon load transfer from generator to normal source. This load transfer can occur with minimal disturbance or effect on the load. Transfer switches that employ this type of paralleling must comply with Article 705.

Exception: Temporary connection of a portable generator without transfer equipment shall be permitted where conditions of maintenance and supervision ensure that only qualified persons service the installation and where the normal supply is physically isolated by a lockable disconnect means or by disconnection of the normal supply conductors.

This exception was added in the 2005 *Code*. This exception provides requirements for the connection of loads to a generator without the use of a transfer switch, where the installation is under the supervision of qualified service personnel. Such applications often occur when necessary for equipment maintenance or breakdown or when there is an extended power outage. In such instances, a portable generator can be brought to a facility and connected to the existing distribution system. The supervision by qualified personnel is critical to ensuring that a dangerous backfeed condition is not created by connecting the generator to the system without the benefit of transfer equipment.

702.7 Signals

Audible and visual signal devices shall be provided, where practicable, for the following purposes.

(1) Derangement To indicate derangement of the optional standby source.

(2) Carrying Load To indicate that the optional standby source is carrying load.

Exception: Signals shall not be required for portable standby power sources.

702.8 Signs

(A) Standby A sign shall be placed at the service-entrance equipment that indicates the type and location of on-site optional standby power sources. A sign shall not be required for individual unit equipment for standby illumination.

(B) Grounding Where the grounded circuit conductor connected to the optional standby power source is connected to a grounding electrode conductor at a location remote from the optional standby power source, there shall be a sign at the grounding location that shall identify all optional standby power and normal sources connected at that location.

II. Circuit Wiring

702.9 Wiring Optional Standby Systems

The optional standby system wiring shall be permitted to occupy the same raceways, cables, boxes, and cabinets with other general wiring.

III. Grounding

702.10 Portable Generator Grounding

(A) Separately Derived System Where a portable optional standby source is used as a separately derived system, it shall be grounded to a grounding electrode in accordance with 250.30.

(B) Nonseparately Derived System Where a portable optional standby source is used as a nonseparately derived system, the equipment grounding conductor shall be bonded to the system grounding electrode.

IV. Sources of Power

702.11 Outdoor Generator Sets

Where an outdoor housed generator set is equipped with a readily accessible disconnecting means located within sight of the building or structure supplied, an additional disconnecting means shall not be required where ungrounded conductors serve or pass through the building or structure.

ARTICLE 705
Interconnected Electric Power Production Sources

Contents

705.1 Scope

This article covers installation of one or more electric power production sources operating in parallel with a primary source(s) of electricity.

> FPN: Examples of the types of primary sources are a utility supply, on-site electric power source(s), or other sources.

Article 705 sets forth basic safety requirements for the installation of generators and other types of power production sources that are interconnected and operate in parallel as distributed generation. Power sources include any systems that produce electric power, including not only electric utility sources but also on-premises sources ranging from rotating generators (see Article 445) to solar photovoltaic systems (see Article 690) to fuel cells (see Article 692).

Article 705 addresses the basic safety requirements specifically related to parallel operation for the generators and other power sources, the power system that interconnects the power sources, and the equipment that is connected to these systems. The proper application of these systems requires a thorough review of the entire power system.

705.2 Definition

Interactive System. An electric power production system that is operating in parallel with and capable of delivering energy to an electric primary source supply system.

705.3 Other Articles

Interconnected electric power production sources shall comply with this article and also with the applicable requirements of the articles in Table 705.3.

Exception No. 1: Installation of solar photovoltaic systems operated as interconnected power sources shall be in accordance with Article 690.

Exception No. 2: Installation of fuel cell systems operated as interconnected power sources shall be in accordance with Article 692.

705.10 Directory

A permanent plaque or directory, denoting all electrical power sources on or in the premises, shall be installed at

Table 705.3 Other Articles

Equipment/System	Article
Generators	445
Emergency systems	700
Legally required standby systems	701
Optional standby systems	702

each service equipment location and at locations of all electric power production sources capable of being interconnected.

Exception: Installations with large numbers of power production sources shall be permitted to be designated by groups.

705.12 Point of Connection

The outputs of electric power production systems shall be interconnected at the premises service disconnecting means.

(A) Integrated Electric System The outputs shall be permitted to be interconnected at a point or points elsewhere on the premises where the system qualifies as an integrated electric system and incorporates protective equipment in accordance with all applicable sections of Article 685.

(B) General The outputs shall be permitted to be interconnected at a point or points elsewhere on the premises where all of the following conditions are met:

(1) The aggregate of nonutility sources of electricity has a capacity in excess of 100 kW, or the service is above 1000 volts.
(2) The conditions of maintenance and supervision ensure that qualified persons service and operate the system.
(3) Safeguards and protective equipment are established and maintained.

In general, the point of interconnection must be at the premises service disconnecting means, which may be difficult to accomplish for systems larger than 100 kW. This requirement is intended to prevent the indiscriminate interconnection of small generators or other sources of power without proper protection against fire and electric shock. (See Exhibit 705.1.) It is important to use disconnect devices (switches, etc.) that are suitable for the purpose.

The requirement specifying "at the premises service disconnecting means" permits connection ahead of the disconnect or on the load side. This practice accommodates the safe work practices of many utilities, which provide a readily accessible disconnect for distributed generation.

Sections 705.12(A) and 705.12(B) recognize that generators and other power sources can be safely connected elsewhere on the premises system. These locations include where the premises has an integrated electrical system as set forth in Article 685, where the total generator capacity on premises is greater than 100 kW, and where the service is greater than 1000 volts.

705.14 Output Characteristics

The output of a generator or other electric power production source operating in parallel with an electric supply system

Exhibit 705.1 The point of interconnection located at the premises service disconnecting means, as required by 705.12.

shall be compatible with the voltage, wave shape, and frequency of the system to which it is connected.

> FPN: The term *compatible* does not necessarily mean matching the primary source wave shape.

The level and quality of output power of interconnected sources must be controlled to facilitate proper operation of the interconnected generator or other power production source and the electric system. Control of the generator or power production source should include real power, reactive power, and harmonic content of the output. Section 705.14 states that the interconnected equipment must be compatible with the electric supply system in voltage, wave shape, and frequency.

The output characteristics of a rotating generator are significantly different from those of a solid-state power source. Their compatibility with other sources and with different types of loads is limited in different ways. Control of the driver speed causes real power (kW) to flow from an induction generator. Control of the prime mover torque causes real power (kW) to flow from a synchronous generator. Control of voltage causes reactive power (kVAr) to flow to or from a synchronous generator. Induction generators have no means to control reactive power (kVAr) flow and continuously draw reactive power. The parallel operation of generators is a complex balance of several variables that are design parameters and therefore beyond the scope of the *Code*.

Where either the power source or the loads have solid-state equipment, such as inverters, uninterruptible power supplies (UPS), or solid-state variable-speed drives, harmonic currents may flow in the system. (See Exhibits 705.2, 705.3, and 705.4.) These multiples of the basic supply frequency (usually 60 Hz) cause additional heating, which may require derating of generators, transformers, cables, and motors. Special generator voltage control systems are required to avoid erratic operation or destruction of control devices. Circuit breakers may require derating if the higher harmonics become significant.

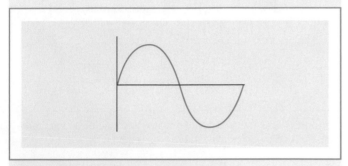

Exhibit 705.2 Typical output wave shape of rotating generator and system wave shape normally encountered with motor, lighting, and heating loads.

Exhibit 705.3 Typical output wave shape with inverter source. Motors and transformers will be driven by harmonic-rich voltage and may require derating.

Exhibit 705.4 Wave shape typical of system with variable speed drive, rectifier, elevator, and uninterruptible power to supply loads. Source generator may require derating, and special voltage control may be needed.

Significant magnitudes of harmonics may be inadvertently matched to system resonance and result in the opening of capacitor fuses, overheating of circuits, and erratic operation of controls. The usual solution is to detune the system by rearrangement or installation of reactors or both.

705.16 Interrupting and Short-Circuit Current Rating

Consideration shall be given to the contribution of fault currents from all interconnected power sources for the interrupting and short-circuit current ratings of equipment on interactive systems.

705.20 Disconnecting Means, Sources

Means shall be provided to disconnect all ungrounded conductors of an electric power production source(s) from all other conductors.

705.21 Disconnecting Means, Equipment

Means shall be provided to disconnect equipment, such as inverters or transformers associated with a power production source, from all ungrounded conductors of all sources of supply. Equipment intended to be operated and maintained as an integral part of a power production source exceeding 1000 volts shall not be required to have a disconnecting means.

An example of equipment that is covered under the second sentence of 705.21 is a generator designed for 4160 volts and connected to a 13,800-volt system through a transformer. In this example, the transformer and generator are to be operated as a unit; therefore, a disconnecting means is not required between the generator and the transformer. A disconnecting means, however, is required between the transformer and the point of connection to the power system. Section 445.18 addresses the general requirements for disconnecting means associated with generators.

705.22 Disconnect Device

The disconnecting means for ungrounded conductors shall consist of a manually or power operable switch(es) or circuit breaker(s) with the following features:

(1) Located where accessible
(2) Externally operable without exposing the operator to contact with live parts and, if power operable, of a type that can be opened by hand in the event of a power supply failure
(3) Plainly indicating whether in the open or closed position
(4) Having ratings not less than the load to be carried and the fault current to be interrupted

For disconnect equipment energized from both sides, a marking shall be provided to indicate that all contacts of the disconnect equipment may be energized.

FPN No. 1: In parallel generation systems, some equipment, including knife blade switches and fuses, is likely to be energized from both directions. See 240.40.

FPN No. 2: Interconnection to an off-premises primary source could require a visibly verifiable disconnecting device.

The requirements for disconnects in 705.22 are important. A disconnecting means must serve each generating source. This disconnecting means will be the service-entrance disconnect. Still another disconnecting means may be applied to separate the generating systems.

The basic requirement in 705.22 recognizes the success of applying switches as well as circuit breakers as the disconnecting means for ungrounded conductors. Most safe work practices on the premises use these disconnect devices.

The disconnect at the service entrance is required for disconnecting the premises wiring system from the utility. The utility safe work practices may also use this disconnect device. Utility safe work practices may require a visibly verifiable disconnect device. For this reason, some utility contracts require that a visible break be provided. The second fine print note to 705.22 brings attention to this common utility requirement.

705.30 Overcurrent Protection

Conductors shall be protected in accordance with Article 240. Equipment and conductors connected to more than one electrical source shall have a sufficient number of overcurrent devices located so as to provide protection from all sources.

(A) Generators Generators shall be protected in accordance with 445.12.

(B) Solar Photovoltaic Systems Solar photovoltaic systems shall be protected in accordance with Article 690.

(C) Transformers Overcurrent protection for a transformer with a source(s) on each side shall be provided in accordance with 450.3 by considering first one side of the transformer, then the other side of the transformer, as the primary.

(D) Fuel Cell Systems Fuel cell systems shall be protected in accordance with Article 692.

705.32 Ground-Fault Protection

Where ground-fault protection is used, the output of an interactive system shall be connected to the supply side of the ground-fault protection.

Exception: Connection shall be permitted to be made to the load side of ground-fault protection, provided that there is ground-fault protection for equipment from all ground-fault current sources.

Larger optional standby systems have become more sophisticated and may involve many special relays. The more common switchgear-type relays include under- and overvoltage, under- and overfrequency, voltage-restrained overcurrent, anti-motoring, loss of excitation, overtemperature, and shutdown for derangement of the mechanical driver.

Small generator installations cannot justify the cost of switchgear-type relays. Therefore, small generator protection comprises more common devices with fewer features. Application guides for relay protection are available in manufacturers' technical literature.

The requirements in Article 705 are only for protection against conditions that may occur because generating sources are being operated in parallel. The installation must also meet the requirements of the referenced articles to provide the basic protection and safeguards for all equipment, whether or not the equipment is involved with more than one source.

705.40 Loss of Primary Source

Upon loss of primary source, an electric power production source shall be automatically disconnected from all ungrounded conductors of the primary source and shall not be reconnected until the primary source is restored.

FPN No. 1: Risks to personnel and equipment associated with the primary source could occur if an interactive electric power production source can operate as an island. Special detection methods can be required to determine that a primary source supply system outage has occurred and whether there should be automatic disconnection. When the primary source supply system is restored, special detection methods can be required to limit exposure of power production sources to out-of-phase reconnection.

FPN No. 2: Induction-generating equipment on systems with significant capacitance can become self-excited upon loss of primary source and experience severe overvoltage as a result.

When two interconnected power systems separate, they drift out of synchronism. When the utility and the interconnected power system separate, there is a risk of damage to the system if restoration of the utility occurs out of phase. If the timing of the reconnection is random, violent electromechanical stresses can destroy mechanical components such as gears, couplings, and shafts and can displace coils. Therefore, the premises wiring system or generator must be disconnected from the primary source.

Many technical guides are available from which to select appropriate protective systems and equipment for large systems. A limited choice of low-cost devices is still available for application to small systems. Induction generators are commonly used today. Induction generators have characteristics quite different from those of synchronous machines. They are more rugged because of the construction of the rotor, and they are less expensive because of their basic design, availability, and type of starting and control equipment. Theoretically, an induction machine can continue to run on an isolated system if a large capacitor bank provides excitation. In reality, an induction machine will probably lose stability and be shut down quickly by one of the protective devices.

705.42 Unbalanced Interconnections

A 3-phase electric power production source shall be automatically disconnected from all ungrounded conductors of the interconnected systems when one of the phases of that source opens. This requirement shall not be applicable to an electric power production source providing power for an emergency or legally required standby system.

705.43 Synchronous Generators

Synchronous generators in a parallel system shall be provided with the necessary equipment to establish and maintain a synchronous condition.

705.50 Grounding

Interconnected electric power production sources shall be grounded in accordance with Article 250.

Exception: For direct-current systems connected through an inverter directly to a grounded service, other methods that accomplish equivalent system protection and that utilize equipment listed and identified for the use shall be permitted.

ARTICLE 720
Circuits and Equipment Operating at Less Than 50 Volts

Contents

720.1 Scope

This article covers installations operating at less than 50 volts, direct current or alternating current.

720.2 Other Articles

Installations operating at less than 50 volts, direct current or alternating current, as covered in Articles 411, 517, 550, 551, 552, 650, 669, 690, 725, and 760 shall not be required to comply with this article.

Article 411 covers lighting systems operating at 30 volts or less.

720.3 Hazardous (Classified) Locations

Installations coming within the scope of this article and installed in hazardous (classified) locations shall also comply with the appropriate provisions of Articles 500 through 517.

Low voltage alone does not render a circuit incapable of igniting flammable atmospheres. Ordinary flashlights using two 1½-volt D-cell batteries can become a source of ignition in some hazardous (classified) locations.

720.4 Conductors

Conductors shall not be smaller than 12 AWG copper or equivalent. Conductors for appliance branch circuits supplying more than one appliance or appliance receptacle shall not be smaller than 10 AWG copper or equivalent.

720.5 Lampholders

Standard lampholders that have a rating of not less than 660 watts shall be used.

720.6 Receptacle Rating

Receptacles shall have a rating of not less than 15 amperes.

720.7 Receptacles Required

Receptacles of not less than 20-ampere rating shall be provided in kitchens, laundries, and other locations where portable appliances are likely to be used.

720.8 Overcurrent Protection

Overcurrent protection shall comply with Article 240.

720.9 Batteries

Installations of storage batteries shall comply with Article 480.

720.10 Grounding

Grounding shall be as provided in Article 250.

720.11 Mechanical Execution of Work

Circuits operating at less than 50 volts shall be installed in a neat and workmanlike manner. Cables shall be supported by the building structure in such a manner that the cable will not be damaged by normal building use.

Section 720.11 requires cables to be installed in a manner that is consistent with standard industry practice.

ARTICLE 725
Class 1, Class 2, and Class 3 Remote-Control, Signaling, and Power-Limited Circuits

Summary of Changes

- **725.2:** Added definition for circuit integrity (CI) cable.
- **725.3(C):** Revised to permit listed plenum signaling raceways containing CL2P and CL3P cables for Class 2 and Class 3 circuits in other spaces used for environmental air.
- **725.8:** Revised requirement covering mechanical execution of work.
- **725.41:** Added new Exception No. 2 covering limited power circuits of listed equipment as power sources for Class 2 and Class 3 circuits.
- **725.56(F):** Added new section prohibiting audio system circuits described in 640.90(C), and using Class 2 and Class 3 wiring methods, in same cable or raceway as Class 2 or Class 3 circuits.
- **725.61(A):** Revised to permit listed plenum signaling raceways for CL2P and CL3P cables in other space used for environmental air.
- **725.61(B)(1):** Revised to include listed riser signaling raceways under specified conditions.
- **725.61(B)(3):** Revised to permit listed general-purpose signaling raceways for CL2, CL3, CL2X, and CL3X cables.
- **725.61(C):** Revised to permit listed signaling raceways in cable trays.
- **725.61(C)(2):** Revised to include intrinsically safe circuits.
- **725.61(H):** Added new requirement permitting circuit integrity (CI) cable.

- **725.82 and 725.82(I), (J), (K):** Added new text covering nonmetallic, plenum, riser, and general-purpose signaling raceways.
- **725.82(F):** Added new requirement permitting circuit integrity (CI) cable.

Contents

I. General

725.1 Scope

This article covers remote-control, signaling, and power-limited circuits that are not an integral part of a device or appliance.

> FPN: The circuits described herein are characterized by usage and electrical power limitations that differentiate them from electric light and power circuits; therefore, alternative requirements to those of Chapters 1 through 4 are given with regard to minimum wire sizes, derating factors, overcurrent protection, insulation requirements, and wiring methods and materials.

The scope of Article 725 may include such systems as burglar alarm circuits (see Exhibit 725.1), access control circuits, sound circuits, nurse call circuits, intercom circuits, some computer network systems, some control circuits for lighting dimmer systems, and some low-voltage control circuits that originate from listed appliances or from listed computer equipment. Article 760 covers fire alarm circuits and systems. Article 800 covers communications circuits. Article 830 covers network-powered broadband communications circuits.

The installation requirements for the low-voltage wiring of information technology equipment (electronic data processing and computer equipment) located within the confines of a room that is constructed according to the requirements of NFPA 75, *Standard for the Protection of Information*

Exhibit 725.1 Typical burglar alarm system keypad. (Courtesy of NAPCO Security Systems, Inc.)

Technology Equipment, are not covered by Article 725. Low-voltage wiring within these specially constructed rooms is covered in Article 645.

Also, if listed computer equipment is interconnected and all the interconnected equipment is in close proximity, the wiring is considered an integral part of the equipment and therefore not subject to the requirements of Article 725. If the wiring leaves the group of equipment to connect to other devices in the same room or elsewhere in the building, the wiring is considered "wiring within buildings" and is subject to the requirements of Article 725.

General Discussion of Remote-Control, Signaling, and Power-Limited Circuits

The wiring methods required by Chapters 1 through 4 of the *Code* apply to remote-control, signaling, and power-limited circuits, except as amended by Article 725 for specified conditions.

A remote-control, signaling, or power-limited circuit is the portion of the wiring system between the load side of the overcurrent device or the power-limited supply and all connected equipment. The circuit is categorized as Class 1, Class 2, or Class 3.

Class 1 circuits are not permitted to exceed 600 volts. In many cases, Class 1 circuits are extensions of power systems and are subject to the requirements of the power systems, except under the following conditions:

1. Conductors sized 16 AWG and 18 AWG may be used if properly protected against overcurrent (see 725.23).
2. Where damage to the circuit would introduce a hazard, the circuit must be mechanically protected by a suitable means. [See 725.11(B).]
3. The adjustment factors of 310.15(B)(2) apply only if such conductors carry a continuous load. (See 725.28

for the exact requirements for adjustment factors affecting ampacity.)

Class 1 remote-control circuits are commonly used to operate motor controllers in conjunction with moving equipment or mechanical processes, elevators, conveyors, and other such equipment. Class 1 remote-control circuits may also be used as shunt trip circuits for circuit breakers. Class 1 signaling circuits often operate at 120 volts but are not limited to this value.

Conductors and equipment on the supply side of overcurrent protection, transformers, or current-limiting devices of Class 2 and Class 3 circuits must be installed according to the applicable requirements of Chapter 3. Load-side conductors and equipment must comply with Article 725. Class 2 and Class 3 conductors are required to be separated from and not occupy the same raceways, cable trays, cables, or enclosures as electric light, power, and Class 1 conductors, except as noted in 725.55.

Dry-cell batteries are considered Class 2 power supplies, provided the voltage is 30 volts or less and the capacity is equal to or less than that available from series-connected No. 6 carbon-zinc cells. (A No. 6 dry-cell battery is cylindrically shaped with nominal dimensions of 2½ in. in diameter by 6 in. tall and weighing just over 2 lb. A No. 6 dry-cell battery is about 10 times the volume of the standard D-cell battery commonly used in flashlights.)

Circuits originating from thermocouples are classified Class 2 circuits. Neither dry-cell batteries nor thermocouples are required to be listed.

725.2 Definitions

Abandoned Class 2, Class 3, and PLTC Cable. Installed Class 2, Class 3, and PLTC cable that is not terminated at equipment and not identified for future use with a tag.

This definition is for use with 725.3(B), which requires removal of accessible abandoned Class 2, Class 3, and PLTC cables. Abandoned cable increases fire loading unnecessarily, and, where installed in plenums, it can affect airflow. Similar requirements can be found in Articles 640, 645, 725, 760, 770, 820, and 830.

Circuit Integrity (CI) Cable. Cable(s) used for remote control, signaling, or power-limited systems that supply critical circuits to ensure survivability for continued circuit operation for a specified time under fire conditions.

This definition is for use with 725.61(H) and 725.82(F), which permits circuit integrity (CI) cables to be used for remote control, signaling, or power-limited systems that supply critical circuits where survivability of the critical system functions under fire conditions is needed. Such cables are likely to be used in systems that are critical to life safety, property protection, or emergency management.

Class 1 Circuit. The portion of the wiring system between the load side of the overcurrent device or power-limited supply and the connected equipment. The voltage and power limitations of the source are in accordance with 725.21.

Class 2 Circuit. The portion of the wiring system between the load side of a Class 2 power source and the connected equipment. Due to its power limitations, a Class 2 circuit considers safety from a fire initiation standpoint and provides acceptable protection from electric shock.

Class 3 Circuit. The portion of the wiring system between the load side of a Class 3 power source and the connected equipment. Due to its power limitations, a Class 3 circuit considers safety from a fire initiation standpoint. Since higher levels of voltage and current than for Class 2 are permitted, additional safeguards are specified to provide protection from an electric shock hazard that could be encountered.

725.3 Other Articles

Circuits and equipment shall comply with the articles or sections listed in 725.3(A) through 725.3(G). Only those sections of Article 300 referenced in this article shall apply to Class 1, Class 2, and Class 3 circuits.

(A) Number and Size of Conductors in Raceway Section 300.17.

(B) Spread of Fire or Products of Combustion Section 300.21. The accessible portion of abandoned Class 2, Class 3, and PLTC cables shall be removed.

(C) Ducts, Plenums, and Other Air-Handling Spaces Class 1, Class 2, and Class 3 circuits installed in ducts, plenums, or other space used for environmental air shall comply with 300.22. Type CL2P or CL3P cables and plenum signaling raceways shall be permitted for Class 2 and Class 3 circuits installed in other spaces used for environmental air.

See the commentary following 300.22(B), 300.22(C), and 725.82(A), FPN for more information on wiring in ducts, plenums, and other air-handling spaces.

(D) Hazardous (Classified) Locations Articles 500 through 516 and Article 517, Part IV, where installed in hazardous (classified) locations.

(E) Cable Trays Article 392, where installed in cable tray.

(F) Motor Control Circuits Article 430, Part VI, where tapped from the load side of the motor branch-circuit protective device(s) as specified in 430.72(A).

(G) Instrumentation Tray Cable See Article 727.

725.7 Access to Electrical Equipment Behind Panels Designed to Allow Access

Access to electrical equipment shall not be denied by an accumulation of wires and cables that prevents removal of panels, including suspended ceiling panels.

The excess accumulation of wires and cables can limit access to electrical equipment by preventing the removal of access panels. To service, rearrange, or install electrical equipment, the worker must have an accessible work space for working safely on energized equipment. See 300.11(A), which permits the use of support wires and approved fittings that are independent of the suspended ceiling support wires. (See Exhibit 725.2; also see 725.8.)

Exhibit 725.2 Incorrect installation of wires and cables (upper diagram), which can prevent access to equipment or cables. Correct method is shown in the lower diagram.

725.8 Mechanical Execution of Work

Class 1, Class 2, and Class 3 circuits shall be installed in a neat and workmanlike manner. Cables and conductors installed exposed on the surface of ceilings and sidewalls shall be supported by the building structure in such a manner that the cable will not be damaged by normal building use. Such cables shall be supported by straps, staples, hangers, or similar fittings designed and installed so as not to damage the cable. The installation shall also conform with 300.4(D).

> The requirement in 725.8 makes it clear that the *Code* requires cables to be installed in a neat and workmanlike manner. This section provides definitive requirements for workmanship. Cable must be attached to or supported by the structure by straps, clamps, hangers, and the like. The installation method must not damage the cable. In addition, the location of the cable should be carefully evaluated to ensure that activities and processes within the building do not cause damage to the cable. (See Exhibit 725.2 and 725.58.) The reference to 300.4(D) calls attention to the hazards to which cables are exposed where they are installed on framing members. Such cables are required to be installed in a manner that protects them from nail or screw penetration. In the 2005 *Code*, this section was changed to permit attachment to baseboards and non–load bearing walls, which are not structural components.

725.10 Class 1, Class 2, and Class 3 Circuit Identification

Class 1, Class 2, and Class 3 circuits shall be identified at terminal and junction locations, in a manner that prevents unintentional interference with other circuits during testing and servicing.

725.11 Safety-Control Equipment

(A) Remote-Control Circuits Remote-control circuits for safety-control equipment shall be classified as Class 1 if the failure of the equipment to operate introduces a direct fire or life hazard. Room thermostats, water temperature regulating devices, and similar controls used in conjunction with electrically controlled household heating and air conditioning shall not be considered safety-control equipment.

(B) Physical Protection Where damage to remote-control circuits of safety control equipment would introduce a hazard, as covered in 725.11(A), all conductors of such remote-control circuits shall be installed in rigid metal conduit, intermediate metal conduit, rigid nonmetallic conduit, electrical metallic tubing, Type MI cable, Type MC cable, or be otherwise suitably protected from physical damage.

> The remote-control circuits to safety-control devices may be required to be classified as Class 1 if failure of the safety-control circuit could cause a direct fire or life hazard. One example of a direct fire hazard could be a boiler explosion caused by failure of the low-water cutoff circuit. (See Exhibit

725.3.) This is an example of a direct link between failure and the initiation of a hazard, that is, where the failure of the equipment itself causes the hazard.

Generally, fire alarm systems and nurse call systems do not fit this category. These systems do not have a direct link to the initiation of fire or the initiation of a life hazard but, rather, serve as the reporting or warning link of a hazard initiated by some other (indirect) cause.

Exhibit 725.3 Typical installation of an automatic oil burner unit for a boiler employing a safety shutdown circuit, according to the requirements of 725.11.

725.15 Class 1, Class 2, and Class 3 Circuit Requirements

A remote-control, signaling, or power-limited circuit shall comply with the following parts of this article:

(1) Class 1 Circuits: Parts I and II
(2) Class 2 and Class 3 Circuits: Parts I and III

II. Class 1 Circuits

725.21 Class 1 Circuit Classifications and Power Source Requirements

Class 1 circuits shall be classified as either Class 1 power-limited circuits where they comply with the power limitations of 725.21(A) or as Class 1 remote-control and signaling circuits where they are used for remote-control or signaling purposes and comply with the power limitations of 725.21(B).

(A) Class 1 Power-Limited Circuits These circuits shall be supplied from a source that has a rated output of not more than 30 volts and 1000 volt-amperes.

(1) Class 1 Transformers Transformers used to supply power-limited Class 1 circuits shall comply with the applicable sections within Parts I and II of Article 450.

(2) Other Class 1 Power Sources Power sources other than transformers shall be protected by overcurrent devices rated at not more than 167 percent of the volt-ampere rating of the source divided by the rated voltage. The overcurrent devices shall not be interchangeable with overcurrent devices of higher ratings. The overcurrent device shall be permitted to be an integral part of the power supply.

To comply with the 1000 volt-ampere limitation of 725.21(A), the maximum output (VA_{max}) of power sources other than transformers shall be limited to 2500 volt-amperes, and the product of the maximum current (I_{max}) and maximum voltage (V_{max}) shall not exceed 10,000 volt-amperes. These ratings shall be determined with any overcurrent-protective device bypassed.

VA_{max} is the maximum volt-ampere output after one minute of operation regardless of load and with overcurrent protection bypassed, if used. Current-limiting impedance shall not be bypassed when determining VA_{max}.

I_{max} is the maximum output current under any noncapacitive load, including short circuit, and with overcurrent protection bypassed, if used. Current-limiting impedance should not be bypassed when determining I_{max}. Where a current-limiting impedance, listed for the purpose or as part of a listed product, is used in combination with a stored energy source, for example, storage battery, to limit the output current, I_{max} limits apply after 5 seconds.

V_{max} is the maximum output voltage regardless of load with rated input applied.

(B) Class 1 Remote-Control and Signaling Circuits These circuits shall not exceed 600 volts. The power output of the source shall not be required to be limited.

Often, remote-control and signaling circuits that do not meet the requirements of Class 2 or Class 3 circuits are classified Class 1 circuits by default. For example, a listed nurse call system may contain a power supply with an output of 500 watts at 24 volts. Because this power supply obviously exceeds the maximum permitted output of a Class 2 power supply and the output terminals are not marked to indicate the equipment as suitable for a Class 2 power supply, the output circuit wiring is classified as Class 1 and subject to all the Class 1 circuit requirements.

725.23 Class 1 Circuit Overcurrent Protection

Overcurrent protection for conductors 14 AWG and larger shall be provided in accordance with the conductor ampacity, without applying the derating factors of 310.15 to the ampacity calculation. Overcurrent protection shall not exceed 7

amperes for 18 AWG conductors and 10 amperes for 16 AWG.

Exception: Where other articles of this Code permit or require other overcurrent protection.

> FPN: For example, see 430.72 for motors, 610.53 for cranes and hoists, and 517.74(B) and 660.9 for X-ray equipment.

725.24 Class 1 Circuit Overcurrent Device Location

Overcurrent devices shall be located as specified in 725.24(A), (B), (C), (D), or (E).

(A) Point of Supply Overcurrent devices shall be located at the point where the conductor to be protected receives its supply.

(B) Feeder Taps Class 1 circuit conductors shall be permitted to be tapped, without overcurrent protection at the tap, where the overcurrent device protecting the circuit conductor is sized to protect the tap conductor.

(C) Branch Circuit Taps Class 1 circuit conductors 14 AWG and larger that are tapped from the load side of the overcurrent-protective device(s) of a controlled light and power circuit shall require only short-circuit and ground-fault protection and shall be permitted to be protected by the branch-circuit overcurrent protective device(s) where the rating of the protective device(s) is not more than 300 percent of the ampacity of the Class 1 circuit conductor.

Section 725.24(C) correlates with 240.4(G).

(D) Primary Side of Transformer Class 1 circuit conductors supplied by the secondary of a single-phase transformer having only a 2-wire (single-voltage) secondary shall be permitted to be protected by overcurrent protection provided on the primary side of the transformer, provided this protection is in accordance with 450.3 and does not exceed the value determined by multiplying the secondary conductor ampacity by the secondary-to-primary transformer voltage ratio. Transformer secondary conductors other than 2 wire shall not be considered to be protected by the primary overcurrent protection.

Section 725.24(D) correlates with 240.4(F) and 430.72(B), Exception No. 2.

(E) Input Side of Electronic Power Source Class 1 circuit conductors supplied by the output of a single-phase, listed electronic power source, other than a transformer, having only a 2-wire (single voltage) output for connection to Class

1 circuits shall be permitted to be protected by overcurrent protection provided on the input side of the electronic power source, provided this protection does not exceed the value determined by multiplying the Class 1 circuit conductor ampacity by the output-to-input voltage ratio. Electronic power source outputs, other than 2 wire (single voltage), shall not be considered to be protected by the primary overcurrent protection.

Electronic power supplies that do not supply energy directly through the use of transformers are now covered by 725.24(E), which permits overcurrent protection for the Class 1 circuit conductors to be installed on the input side of the power source rather than the output side.

725.25 Class 1 Circuit Wiring Methods

Class 1 circuits shall be installed in accordance with Part I of Article 300 and with the wiring methods from the appropriate articles in Chapter 3.

Exception No. 1: The provisions of 725.26 through 725.28 shall be permitted to apply in installations of Class 1 circuits.

Exception No. 2: Methods permitted or required by other articles of this Code shall apply to installations of Class 1 circuits.

725.26 Conductors of Different Circuits in the Same Cable, Cable Tray, Enclosure, or Raceway

Class 1 circuits shall be permitted to be installed with other circuits as specified in 725.26(A) and 725.26(B).

(A) Two or More Class 1 Circuits Class 1 circuits shall be permitted to occupy the same cable, cable tray, enclosure, or raceway without regard to whether the individual circuits are alternating current or direct current, provided all conductors are insulated for the maximum voltage of any conductor in the cable, cable tray, enclosure, or raceway.

(B) Class 1 Circuits with Power Supply Circuits Class 1 circuits shall be permitted to be installed with power supply conductors as specified in 725.26(B)(1) through (B)(4).

(1) In a Cable, Enclosure, or Raceway Class 1 circuits and power supply circuits shall be permitted to occupy the same cable, enclosure, or raceway only where the equipment powered is functionally associated.

(2) In Factory- or Field-Assembled Control Centers Class 1 circuits and power supply circuits shall be permitted to be installed in factory- or field-assembled control centers.

(3) In a Manhole Class 1 circuits and power supply circuits shall be permitted to be installed as underground conductors in a manhole in accordance with one of the following:

(1) The power-supply or Class 1 circuit conductors are in a metal-enclosed cable or Type UF cable.
(2) The conductors are permanently separated from the power-supply conductors by a continuous firmly fixed nonconductor, such as flexible tubing, in addition to the insulation on the wire.
(3) The conductors are permanently and effectively separated from the power supply conductors and securely fastened to racks, insulators, or other approved supports.

Section 725.26(B)(3) permits Class 1 power-limited circuit conductors to be installed in manholes with wiring of non–power-limited systems where suitable separation provides the same degree of protection as required for the following:

1. Class 2 and Class 3 power-limited circuits in 725.55
2. Communications circuits in 800.133(A)
3. Radio/television antennas and lead-in conductors in 810.18
4. CATV conductors in 820.133(A)

(4) In Cable Trays In cable trays, where the Class 1 circuit conductors and power-supply conductors not functionally associated with them are separated by a solid fixed barrier of a material compatible with the cable tray, or where the power-supply or Class 1 circuit conductors are in a metal-enclosed cable.

725.27 Class 1 Circuit Conductors

(A) Sizes and Use Conductors of sizes 18 AWG and 16 AWG shall be permitted to be used, provided they supply loads that do not exceed the ampacities given in 402.5 and are installed in a raceway, an approved enclosure, or a listed cable. Conductors larger than 16 AWG shall not supply loads greater than the ampacities given in 310.15. Flexible cords shall comply with Article 400.

(B) Insulation Insulation on conductors shall be suitable for 600 volts. Conductors larger than 16 AWG shall comply with Article 310. Conductors in sizes 18 AWG and 16 AWG shall be Type FFH-2, KF-2, KFF-2, PAF, PAFF, PF, PFF, PGF, PGFF, PTF, PTFF, RFH-2, RFHH-2, RFHH-3, SF-2, SFF-2, TF, TFF, TFFN, TFN, ZF, or ZFF. Conductors with other types and thicknesses of insulation shall be permitted if listed for Class 1 circuit use.

Section 725.27 requires all Class 1 circuit conductors to be rated at 600 volts. This effectively requires Class 1 circuits to be wired using the wiring methods found in Chapter 3 or the use of conductors specifically listed for Class 1 circuit use.

725.28 Number of Conductors in Cable Trays and Raceway, and Derating

(A) Class 1 Circuit Conductors Where only Class 1 circuit conductors are in a raceway, the number of conductors shall be determined in accordance with 300.17. The derating factors given in 310.15(B)(2)(a) shall apply only if such conductors carry continuous loads in excess of 10 percent of the ampacity of each conductor.

(B) Power-Supply Conductors and Class 1 Circuit Conductors Where power-supply conductors and Class 1 circuit conductors are permitted in a raceway in accordance with 725.26, the number of conductors shall be determined in accordance with 300.17. The derating factors given in 310.15(B)(2)(a) shall apply as follows:

(1) To all conductors where the Class 1 circuit conductors carry continuous loads in excess of 10 percent of the ampacity of each conductor and where the total number of conductors is more than three

(2) To the power-supply conductors only, where the Class 1 circuit conductors do not carry continuous loads in excess of 10 percent of the ampacity of each conductor and where the number of power-supply conductors is more than three

(C) Class 1 Circuit Conductors in Cable Trays Where Class 1 circuit conductors are installed in cable trays, they shall comply with the provisions of 392.9 through 392.11.

725.29 Circuits Extending Beyond One Building

Class 1 circuits that extend aerially beyond one building shall also meet the requirements of Article 225.

III. Class 2 and Class 3 Circuits

725.41 Power Sources for Class 2 and Class 3 Circuits

(A) Power Source The power source for a Class 2 or a Class 3 circuit shall be as specified in 725.41(A)(1), (A)(2), (A)(3), (A)(4), or (A)(5):

> FPN No. 1: Figure 725.41 illustrates the relationships between Class 2 or Class 3 power sources, their supply, and the Class 2 or Class 3 circuits.
>
> FPN No. 2: Tables 11(A) and 11(B) in Chapter 9 provide the requirements for listed Class 2 and Class 3 power sources.

(1) A listed Class 2 or Class 3 transformer

(2) A listed Class 2 or Class 3 power supply

(3) Other listed equipment marked to identify the Class 2 or Class 3 power source

Exception No. 1: Thermocouples shall not require listing as a Class 2 power source.

Figure 725.41 Class 2 and Class 3 Circuits.

Exception No. 2: Limited power circuits of listed equipment where these circuits have energy levels rated at or below the limits established in Chapter 9, Tables 11(A) and 11(B).

> FPN: Examples of other listed equipment are as follows:
>
> (1) A circuit card listed for use as a Class 2 or Class 3 power source where used as part of a listed assembly
> (2) A current-limiting impedance, listed for the purpose, or part of a listed product, used in conjunction with a non–power-limited transformer or a stored energy source, for example, storage battery, to limit the output current
> (3) A thermocouple
> (4) Limited voltage/current or limited impedance secondary communications circuits of listed industrial control equipment

(4) Listed information technology (computer) equipment limited power circuits.

> FPN: One way to determine applicable requirements for listing of information technology (computer) equipment is to refer to UL 1950-1995, *Standard for Safety of Information Technology Equipment, Including Electrical Business Equipment.* Typically such circuits are used to interconnect information technology equipment for the purpose of exchanging information (data).

(5) A dry cell battery shall be considered an inherently limited Class 2 power source, provided the voltage is 30 volts or less and the capacity is equal to or less than that available from series connected No. 6 carbon zinc cells.

(B) Interconnection of Power Sources Class 2 or Class 3 power sources shall not have the output connections paralleled or otherwise interconnected unless listed for such interconnection.

Except for dry-cell batteries and thermocouples, 725.41(B) requires listed power sources for Class 2 and Class 3 circuits.

725.42 Circuit Marking

The equipment supplying the circuits shall be durably marked where plainly visible to indicate each circuit that is a Class 2 or Class 3 circuit.

Section 725.42 requires equipment to be marked to clearly identify Class 2 and Class 3 circuits. This section was revised for the 2005 *Code* to clarify that the marking is on the equipment where the circuits originate, rather than on the individual raceways, cables, and fittings.

725.51 Wiring Methods on Supply Side of the Class 2 or Class 3 Power Source

Conductors and equipment on the supply side of the power source shall be installed in accordance with the appropriate requirements of Chapters 1 through 4. Transformers or other devices supplied from electric light or power circuits shall be protected by an overcurrent device rated not over 20 amperes.

Listed Class 2 and Class 3 transformers must be protected by an overcurrent device not exceeding 20 amperes, unless the transformers are fed from circuits other than power or lighting.

Exception: The input leads of a transformer or other power source supplying Class 2 and Class 3 circuits shall be permitted to be smaller than 14 AWG, but not smaller than 18 AWG if they are not over 12 in. (305 mm) long and if they have insulation that complies with 725.27(B).

725.52 Wiring Methods and Materials on Load Side of the Class 2 or Class 3 Power Source

Class 2 and Class 3 circuits on the load side of the power source shall be permitted to be installed using wiring methods and materials in accordance with either 725.52(A) or 725.52(B).

(A) Class 1 Wiring Methods and Materials Installation shall be in accordance with 725.25.

Exception No. 1: The derating factors given in 310.15(B)(2)(a) shall not apply.

Exception No. 2: Class 2 and Class 3 circuits shall be permitted to be reclassified and installed as Class 1 circuits if the Class 2 and Class 3 markings required in 725.42 are eliminated and the entire circuit is installed using the wiring methods and materials in accordance with Part II, Class 1 circuits.

FPN: Class 2 and Class 3 circuits reclassified and installed as Class 1 circuits are no longer Class 2 or Class

3 circuits, regardless of the continued connection to a Class 2 or Class 3 power source.

(B) Class 2 and Class 3 Wiring Methods Conductors on the load side of the power source shall be insulated at not less than the requirements of 725.82 and shall be installed in accordance with 725.54 and 725.61.

Exception No. 1: As provided for in 620.21 for elevators and similar equipment.

Exception No. 2: Other wiring methods and materials installed in accordance with the requirements of 725.3 shall be permitted to extend or replace the conductors and cables described in 725.82 and permitted by 725.52(B).

Section 725.52 permits the following two distinct wiring methods for circuits connected to Class 2 or Class 3 power sources:

• Class 1 wiring methods
• Class 2 and Class 3 wiring methods

Where it is necessary to locate Class 2 or Class 3 circuits inside the same cable or raceway as a Class 1 circuit, 725.52(A), Exception No. 2, permits a Class 2 or Class 3 circuit to be reclassified and installed as Class 1, providing that the Class 2 or Class 3 marking is removed, that overcurrent protection is provided in accordance with 725.23, and that the reclassified circuit maintains separation from other Class 2 and Class 3 circuits in accordance with 725.55.

725.54 Installation of Conductors and Equipment in Cables, Compartments, Cable Trays, Enclosures, Manholes, Outlet Boxes, Device Boxes, and Raceways for Class 2 and Class 3 Circuits

Conductors and equipment for Class 2 and Class 3 circuits shall be installed in accordance with 725.55 through 725.58.

725.55 Separation from Electric Light, Power, Class 1, Non–Power-Limited Fire Alarm Circuit Conductors, and Medium Power Network-Powered Broadband Communications Cables

(A) General Cables and conductors of Class 2 and Class 3 circuits shall not be placed in any cable, cable tray, compartment, enclosure, manhole, outlet box, device box, raceway, or similar fitting with conductors of electric light, power, Class 1, non–power-limited fire alarm circuits, and medium power network-powered broadband communications circuits unless permitted by 725.55(B) through 725.55(J).

Section 725.55(A) specifically includes cables of Class 2 and Class 3 circuits. Jackets of listed Class 2 and Class 3 cables do not have sufficient construction specifications to permit them to be installed with electric light, power, Class 1, non–power-limited fire alarm circuits, and medium power network-powered broadband communications cables. Failure of the cable insulation due to a fault could lead to hazardous voltages being imposed on the Class 2 or Class 3 circuit conductors.

(B) Separated by Barriers Class 2 and Class 3 circuits shall be permitted to be installed together with the conductors of electric light, power, Class 1, non–power-limited fire alarm and medium power network-powered broadband communications circuits where they are separated by a barrier.

(C) Raceways Within Enclosures In enclosures, Class 2 and Class 3 circuits shall be permitted to be installed in a raceway to separate them from Class 1, non–power-limited fire alarm and medium power network-powered broadband communications circuits.

(D) Associated Systems Within Enclosures Class 2 and Class 3 circuit conductors in compartments, enclosures, device boxes, outlet boxes, or similar fittings shall be permitted to be installed with electric light, power, Class 1, non–power-limited fire alarm, and medium power network-powered broadband communications circuits where they are introduced solely to connect the equipment connected to Class 2 and Class 3 circuits, and where (1) or (2) applies:

(1) The electric light, power, Class 1, non–power-limited fire alarm, and medium power network-powered broadband communications circuit conductors are routed to maintain a minimum of 6 mm (0.25 in.) separation from the conductors and cables of Class 2 and Class 3 circuits.
(2) The circuit conductors operate at 150 volts or less to ground and also comply with one of the following:

  a. The Class 2 and Class 3 circuits are installed using Type CL3, CL3R, or CL3P or permitted substitute cables, provided these Class 3 cable conductors extending beyond the jacket are separated by a minimum of 6 mm (0.25 in.) or by a nonconductive sleeve or nonconductive barrier from all other conductors.
  b. The Class 2 and Class 3 circuit conductors are installed as a Class 1 circuit in accordance with 725.21.

(E) Enclosures with Single Opening Class 2 and Class 3 circuit conductors entering compartments, enclosures, device boxes, outlet boxes, or similar fittings shall be permitted to be installed with Class 1, non–power-limited fire alarm and medium power network-powered broadband communications circuits where they are introduced solely to connect the equipment connected to Class 2 and Class 3 circuits.

Where Class 2 and Class 3 circuit conductors must enter an enclosure that is provided with a single opening, they shall be permitted to enter through a single fitting (such as a tee), provided the conductors are separated from the conductors of the other circuits by a continuous and firmly fixed nonconductor, such as flexible tubing.

Power circuit and Class 2 circuit conductors can be permitted in the same motor-starter enclosure, where the Class 2 circuit source is the secondary of a control transformer in the same motor-starter enclosure. In such an installation, the Class 2 conductor insulation is not required to have the same voltage rating as the insulation on the power conductors in the same enclosure.

(F) Manholes Underground Class 2 and Class 3 circuit conductors in a manhole shall be permitted to be installed with Class 1, non–power-limited fire alarm and medium power network-powered broadband communications circuits where one of the following conditions is met:

(1) The electric light, power, Class 1, non–power-limited fire alarm and medium power network-powered broadband communications circuit conductors are in a metal-enclosed cable or Type UF cable.
(2) The Class 2 and Class 3 circuit conductors are permanently and effectively separated from the conductors of other circuits by a continuous and firmly fixed nonconductor, such as flexible tubing, in addition to the insulation or covering on the wire.
(3) The Class 2 and Class 3 circuit conductors are permanently and effectively separated from conductors of the other circuits and securely fastened to racks, insulators, or other approved supports.

Section 725.55(F) permits the installation of Class 2 and Class 3 power-limited circuit conductors in manholes that have wiring for electric light, power, Class 1, and non–power-limited fire alarm circuit systems, and from medium power network-powered broadband communications cables where suitable separation provides the same degree of protection as required in 725.55(A). [See the commentary following 725.26(B)(3).]

(G) Closed-Loop and Programmed Power Distribution Class 2 and Class 3 conductors shall be permitted to be installed in accordance with 780.6.

(H) Cable Trays Class 2 and Class 3 circuit conductors shall be permitted to be installed in cable trays, where the conductors of the electric light, Class 1, and non–powerlimited fire alarm circuits are separated by a solid fixed barrier of a material compatible with the cable tray or where the Class 2 or Class 3 circuits are installed in Type MC cable.

Section 725.55(H) permits the mixing of circuits in some cable tray applications where a physical barrier within the tray provides the necessary separation, without the need for a separate tray. Type MC cable also provides suitable physical protection so that jacket damage that results in faults between circuits is unlikely.

(I) In Hoistways In hoistways, Class 2 or Class 3 circuit conductors shall be installed in rigid metal conduit, rigid nonmetallic conduit, intermediate metal conduit, liquidtight flexible nonmetallic conduit, or electrical metallic tubing. For elevators or similar equipment, these conductors shall be permitted to be installed as provided in 620.21.

(J) Other Applications For other applications, conductors of Class 2 and Class 3 circuits shall be separated by at least 50 mm (2 in.) from conductors of any electric light, power, Class 1 non–power-limited fire alarm or medium power network-powered broadband communications circuits unless one of the following conditions is met:

(1) Either (a) all of the electric light, power, Class 1, non–power-limited fire alarm and medium power network-powered broadband communications circuit conductors or (b) all of the Class 2 and Class 3 circuit conductors are in a raceway or in metal-sheathed, metal-clad, non–metallic-sheathed, or Type UF cables.

(2) All of the electric light, power, Class 1 non–power-limited fire alarm, and medium power network-powered broadband communications circuit conductors are permanently separated from all of the Class 2 and Class 3 circuit conductors by a continuous and firmly fixed nonconductor, such as porcelain tubes or flexible tubing, in addition to the insulation on the conductors.

725.56 Installation of Conductors of Different Circuits in the Same Cable, Enclosure, or Raceway

(A) Two or More Class 2 Circuits Conductors of two or more Class 2 circuits shall be permitted within the same cable, enclosure, or raceway.

(B) Two or More Class 3 Circuits Conductors of two or more Class 3 circuits shall be permitted within the same cable, enclosure, or raceway.

(C) Class 2 Circuits with Class 3 Circuits Conductors of one or more Class 2 circuits shall be permitted within the same cable, enclosure, or raceway with conductors of Class

3 circuits, provided that the insulation of the Class 2 circuit conductors in the cable, enclosure, or raceway is at least that required for Class 3 circuits.

(D) Class 2 and Class 3 Circuits with Communications Circuits

(1) Classified as Communications Circuits Class 2 and Class 3 circuit conductors shall be permitted in the same cable with communications circuits, in which case the Class 2 and Class 3 circuits shall be classified as communications circuits and shall be installed in accordance with the requirements of Article 800. The cables shall be listed as communications cables or multipurpose cables.

(2) Composite Cables Cables constructed of individually listed Class 2, Class 3, and communications cables under a common jacket shall be permitted to be classified as communications cables. The fire resistance rating of the composite cable shall be determined by the performance of the composite cable.

In a typical office environment consisting of a group of computers in a local area network, data wiring is as prevalent as telephone wiring. A common way to minimize the amount of cabling is to run the telephone and data circuits in the same cable. Exhibit 725.4 illustrates such an arrangement. Section 725.56(D) requires that either a communications cable or a multipurpose cable be used for this purpose.

Exhibit 725.4 Telephone and data circuits in the same cable.

(E) Class 2 or Class 3 Cables with Other Circuit Cables Jacketed cables of Class 2 or Class 3 circuits shall be permitted in the same enclosure or raceway with jacketed cables of any of the following:

(1) Power-limited fire alarm systems in compliance with Article 760
(2) Nonconductive and conductive optical fiber cables in compliance with Article 770
(3) Communications circuits in compliance with Article 800
(4) Community antenna television and radio distribution systems in compliance with Article 820
(5) Low-power, network-powered broadband communications in compliance with Article 830

(F) Class 2 or Class 3 Conductors or Cables and Audio System Circuits Audio system circuits described in 640.9(C), and installed using Class 2 or Class 3 wiring methods in compliance with Sections 725.54 and 725.61, shall not be permitted to be installed in the same cable or raceway with Class 2 or Class 3 conductors or cables.

725.57 Installation of Circuit Conductors Extending Beyond One Building

Where Class 2 or Class 3 circuit conductors extend beyond one building and are run so as to be subject to accidental contact with electric light or power conductors operating over 300 volts to ground, or are exposed to lightning on interbuilding circuits on the same premises, the requirements of the following shall also apply:

(1) Sections 800.44, 800.50, 800.53, 800.93, 800.100, 800.170(A), and 800.170(B) for other than coaxial conductors
(2) Sections 820.44, 820.93, and 820.100 for coaxial conductors

725.58 Support of Conductors

Class 2 or Class 3 circuit conductors shall not be strapped, taped, or attached by any means to the exterior of any conduit or other raceway as a means of support. These conductors shall be permitted to be installed as permitted by 300.11(B)(2).

See the commentary following 725.8 for more information on the support of conductors.

725.61 Applications of Listed Class 2, Class 3, and PLTC Cables

Class 2, Class 3, and PLTC cables shall comply with any of the requirements described in 725.61(A) through 725.61(H).

It should be noted that Section 725.3(B) requires the removal of accessible abandoned cable. Abandoned cable increases fire loading unnecessarily, and, where installed in plenums, it can affect airflow. Similar requirements can be found in Articles 640, 645, 760, 770, 800, 820, and 830. See the definition of *abandoned Class 2, Class 3, and PLTC cable* in 725.2.

(A) Plenums Cables installed in ducts, plenums, and other spaces used for environmental air shall be Type CL2P or CL3P. Listed wires and cables installed in compliance with 300.22 shall be permitted. Listed plenum signaling raceways shall be permitted to be installed in other spaces used for environmental air as described in 300.22(C). Only Type CL2P or CL3P cable shall be permitted to be installed in these raceways.

(B) Riser Cables installed in risers shall be as described in any of (B)(1), (B)(2), or (B)(3):

(1) Cables installed in vertical runs and penetrating more than one floor, or cables installed in vertical runs in a shaft, shall be Type CL2R or CL3R. Floor penetrations requiring Type CL2R or CL3R shall contain only cables suitable for riser or plenum use. Listed riser signaling raceways shall be permitted to be installed in vertical riser runs in a shaft from floor to floor. Only Type CL2R, CL3R, CL2P, or CL3P cables shall be permitted to be installed in these raceways.
(2) Other cables as covered in Table 725.61 and other listed wiring methods as covered in Chapter 3 shall be installed in metal raceways, or located in a fireproof shaft having firestops at each floor.

Table 725.61 Cable Substitutions

Cable Type	Permitted Substitutions
CL3P	CMP
CL2P	CMP, CL3P
CL3R	CMP, CL3P, CMR
CL2R	CMP, CL3P, CL2P, CMR, CL3R
PLTC	
CL3	CMP, CL3P, CMR, CL3R, CMG, CM, PLTC
CL2	CMP, CL3P, CL2P, CMR, CL3R, CL2R, CMG, CM, PLTC, CL3
CL3X	CMP, CL3P, CMR, CL3R, CMG, CM, PLTC, CL3, CMX
CL2X	CMP, CL3P, CL2P, CMR, CL3R, CL2R, CMG, CM, PLTC, CL3, CL2, CMX, CL3X

(3) Type CL2, CL3, CL2X, and CL3X cables shall be permitted in one- and two-family dwellings. Listed general purpose signaling raceways shall be permitted for use with Type CL2, CL3, CL2X, and CL3X cables.

FPN: See 300.21 for firestop requirements for floor penetrations.

(C) Cable Trays Cables installed in cable trays outdoors shall be Type PLTC. Cables installed in cable trays indoors shall be Types PLTC, CL3P, CL3R, CL3, CL2P, CL2R, and CL2.

Listed signaling raceways shall be permitted for use with cable trays.

FPN: See 800.133(B) for cables permitted in cable trays.

(D) Hazardous (Classified) Locations Cables installed in hazardous locations shall be as described in 725.61(D)(1) through (D)(4).

(1) Type PLTC Cables installed in hazardous (classified) locations shall be Type PLTC. Where the use of Type PLTC cable is permitted by 501.10(B), 502.10(B), and 504.20, the cable shall be installed in cable trays, in raceways, supported by messenger wire, or otherwise adequately supported and mechanically protected by angles, struts, channels, or other mechanical means. The cable shall be permitted to be directly buried where the cable is listed for this use.

(2) Intrinsically Safe Circuits and Nonincendive Field Wiring Wiring for nonincendive circuits as permitted by 501.10(B)(3), and wiring for intrinsically safe circuits as permitted by 504.20, shall be permitted for circuits derived from Class 2 sources.

This section was revised for the 2005 *Code* to clarify that Class 2 wiring methods, including but not limited to Type PLTC cable, may be used for Class 2 circuits that are also intrinsically safe or nonincendive. The requirements for Class 2 sources are covered in 725.41, and the definitions for nonincendive and intrinsically safe circuits are covered in 500.2 and 504.2, respectively. Type PLTC cable is permitted and required for Class 2 circuits in Division 2 areas if the circuits are nonincendive according to 501.10(B) and 725.61(D). Other methods are required in Division 1 areas if the circuits are not intrinsically safe according to 501.10(A).

(3) Thermocouple Circuits Conductors in Type PLTC cables used for Class 2 thermocouple circuits shall be permitted to be any of the materials used for thermocouple extension wire.

(4) In Industrial Establishments In industrial establishments where the conditions of maintenance and supervision ensure that only qualified persons service the installation, and where the cable is not subject to physical damage, Type PLTC cable that complies with the crush and impact require-

ments of Type MC cable and is identified for such use shall be permitted to be exposed between the cable tray and utilization equipment or device. The cable shall be continuously supported and protected against physical damage using mechanical protection such as dedicated struts, angles, or channels. The cable shall be secured at intervals not exceeding 1.8 m (6 ft).

Section 725.61(D)(4) was revised for the 2005 *Code*. It now allows limited use of a special Type PLTC cable as exposed wiring between a cable tray and utilization equipment in industrial establishments. This allowance offers the installer an alternative to Type MC cable. If used in the manner described in 725.61(D)(4), Type PLTC cable must meet the crush and impact requirements of Type MC cable. Type PLTC cable must also be continuously supported and mechanically protected from damage. The cable must be secured at intervals not exceeding 6 ft.

(E) Other Wiring Within Buildings Cables installed in building locations other than those covered in 725.61(A) through 725.61(D) shall be as described in any of (1) through (7).

(1) Type CL2 or CL3 shall be permitted.
(2) Type CL2X or CL3X shall be permitted to be installed in a raceway or in accordance with other wiring methods covered in Chapter 3.
(3) Cables shall be permitted to be installed in nonconcealed spaces where the exposed length of cable does not exceed 3 m (10 ft).
(4) Listed Type CL2X cables less than 6 mm (0.25 in.) in diameter and listed Type CL3X cables less than 6 mm (0.25 in.) in diameter shall be permitted to be installed in one- and two-family dwellings.
(5) Listed Type CL2X cables less than 6 mm (0.25 in.) in diameter and listed Type CL3X cables less than 6 mm (0.25 in.) in diameter shall be permitted to be installed in nonconcealed spaces in multifamily dwellings.
(6) Type CMUC undercarpet communications wires and cables shall be permitted to be installed under carpet.
(7) In industrial establishments where the conditions of maintenance and supervision ensure that only qualified persons service the installation, and where the cable is not subject to physical damage, Type PLTC cable that complies with the crush and impact requirements of Type MC cable and is identified for such use shall be permitted to be exposed between the cable tray and the utilization equipment or device. The cable shall be continuously supported and protected against physical damage using mechanical protection such as dedicated struts, angles, or channels. The cable shall be supported and secured at intervals not exceeding 1.8 m (6 ft).

(F) Cross-Connect Arrays Type CL2 or CL3 conductors or cables shall be used for cross-connect arrays.

Cross-connect arrays and patch panels located within building spaces must be cross wired, jumpered, or interconnected using listed cables.

(G) Class 2 and Class 3 Cable Substitutions The substitutions for Class 2 and Class 3 cables listed in Table 725.61 shall be permitted. Where substitute cables are installed, the wiring requirements of Article 725, Parts I and III, shall apply.

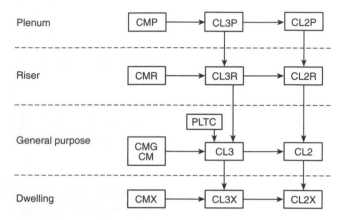

Type CM—Communications wires and cables
Type CL2 and CL3—Class 2 and Class 3 remote-control, signaling, and power-limited cables
Type PLTC—Power-limited tray cable

[A]——▶[B]   Cable A shall be permitted to be used in place of Cable B.

Figure 725.61 Cable Substitution Hierarchy.

FPN: For information on Types CMP, CMR, CM, and CMX cables, see 800.179.

(H) Class 2, Class 3, PLTC Circuit Integrity (CI) Cable or Electrical Circuit Protective System Circuit integrity (CI) cable or a listed electrical circuit protective system shall be permitted for use in remote control, signaling, or power limited systems that supply critical circuits to ensure survivability for continued circuit operation for a specified time under fire conditions.

Section 725.61(H) was added to the 2005 *Code* to permit the use of circuit integrity cable for applications where continuity of the operations of critical circuits is needed during a fire. Such circuits could be essential to fire fighting operations or could be circuits whose interruption could cause a more dangerous condition to occur. An example is a smoke removal system.

IV. Listing Requirements

725.82 Listing and Marking of Class 2, Class 3, and Type PLTC Cables

Class 2, Class 3, and Type PLTC cables and nonmetallic signaling raceways installed as wiring methods within buildings shall be listed as being resistant to the spread of fire and other criteria in accordance with 725.82(A) through 725.82(K) and shall be marked in accordance with 725.82(L).

(A) Types CL2P and CL3P Types CL2P and CL3P plenum cable shall be listed as being suitable for use in ducts, plenums, and other space for environmental air and shall also be listed as having adequate fire-resistant and low smoke-producing characteristics.

FPN: One method of defining low smoke-producing cable is by establishing an acceptable value of the smoke produced when tested in accordance with NFPA 262-2002, *Standard Method of Test for Flame Travel and Smoke of Wires and Cables for Use in Air-Handling Spaces*, to a maximum peak optical density of 0.5 and a maximum average optical density of 0.15. Similarly, one method of defining fire-resistant cables is by establishing a maximum allowable flame travel distance of 1.52 m (5 ft) when tested in accordance with the same test.

Section 725.82, previously numbered 725.71, has been revised. NFPA 262, *Standard Method of Test for Flame Travel and Smoke of Wires and Cables for Use in Air-Handling Spaces*, is a test method for electrical wires and cables that are to be installed without raceways in plenums and other spaces used for environmental air. NFPA 262 was originally developed as UL 910, *Test for Flame Propagation and Smoke-Density Values for Electrical and Optical-Fiber Cables Used in Spaces Transporting Environmental Air*, which is an adaptation of the Steiner Tunnel test (NFPA 255/ASTM E84/UL 723, *Standard Method of Test of Surface Burning Characteristics of Building Materials*).

The test is conducted in a 25-ft-long horizontal duct lined with insulating masonry faced with a row of refractory fire brick. See Exhibit 725.5. One side of the duct is provided with a row of double-paned observation windows that permit monitoring of the progress of the test. The end of the chamber designated the fire end is provided with two gas burners that deliver flames upward to engulf the test specimen. The burners are positioned on each side of the centerline of the furnace so that flame is evenly distributed across the sample. Test specimens are installed in a cable tray that is 23.9 ft long.

Smoke output is measured in a metal vent pipe at the vent end of the furnace, which is opposite the fire end. Where smoke passes through the vent pipe, it passes between a light source and a photocell. The output signal of the photocell is directly proportional to the amount of light received.

Exhibit 725.5 Steiner Tunnel test chamber used in NFPA 262, *Standard Method of Test for Flame Travel and Smoke of Wires and Cables for Use in Air-Handling Spaces.*

The cable specimens are exposed to a fire source with an output of 300,000 Btu/hr for a period of 20 minutes. A graph is plotted that illustrates the flame distance beyond 4½ ft (measured from the end of the flame delivered by the burners) versus the time taken for the duration of the test. This graph is plotted against a flame spread curve representa- tive of a red oak specimen. A graph is also plotted for optical density versus time for the sample. Optical density is calculated as follows:

$$\text{Optical density} = \log_{10} \frac{T_0}{T}$$

where:

T_0 = *initial light transmission*

T = *light transmission during test*

NFPA 262 does not list pass/fail criteria. The criteria for acceptance of a given application are given in the appropriate sections of the *Code*. See the fine print notes that follow 725.82(A), 760.82(D), 770.182(A), 800.182(A), and 820.179(A).

A Class 2 or Class 3 cable that has passed the requirements of this test may be used in ducts, plenums, or other air-handling spaces. In addition, such cable may be used anywhere in a building where Class 2 or Class 3 cable is permitted. (See Table 725.61.)

(B) Types CL2R and CL3R Types CL2R and CL3R riser cables shall be marked as Type CL2R or CL3R, respectively, and be listed as suitable for use in a vertical run in a shaft or from floor to floor and shall also be listed as having fire-resistant characteristics capable of preventing the carrying of fire from floor to floor.

> FPN: One method of defining fire-resistant characteristics capable of preventing the carrying of fire from floor to floor is that the cables pass the requirements of ANSI/UL 1666-2002, *Test for Flame Propagation Height of Electrical and Optical-Fiber Cable Installed Vertically in Shafts.*

In the fire test covered in UL 1666, *Test for Flame Propagation Height of Electrical and Optical-Fiber Cables Installed Vertically in Shafts,* cables are arranged in a simulated vertical shaft and subjected to an ignition source. The shaft is a 19-ft-high concrete shaft divided into two compartments at the 12-ft level. There is a 1-ft by 2-ft opening between the two compartments. The ignition source is a burner with a heat output rate of 495,000 Btu/hr. See Exhibit 725.6.

The individual cable lengths are suspended from a support system on the second floor and held in place just below the support system and at the first-floor slot.

The support frame consists of a steel bar located on the second floor above the slot opening. The individual cable lengths are suspended from the support frame by one of the following methods:

1. The cables are draped over the support frame.
2. The cables are arranged in a clamping device that is suspended from the support frame by two hooks.
3. In the case of large-diameter cables, each individual cable is placed in a wire-mesh grip. Each grip is attached to a hook that is suspended from the support frame.

To pass, cables must not propagate flame to the top of the 12-ft-high compartment during the 30-minute test.

Details of Fire Test Chamber

Details of Fire Test Chamber, View A-A

Exhibit 725.6 Fire test chamber used in ANSI/UL 1666, *Test for Flame Propagation Height of Electrical and Optical-Fiber Cables Installed Vertically in Shafts.*

(C) Types CL2 and CL3 Types CL2 and CL3 cables shall be marked as Type CL2 or CL3, respectively, and be listed as suitable for general-purpose use, with the exception of risers, ducts, plenums, and other space used for environmental air, and shall also be listed as being resistant to the spread of fire.

> FPN: One method of defining *resistant to the spread of fire* is that the cables do not spread fire to the top of the tray in the vertical tray flame test in ANSI/UL 1581-1991, *Reference Standard for Electrical Wires, Cables and Flexible Cords.*
>
> Another method of defining *resistant to the spread of fire* is for the damage (char length) not to exceed 1.5 m (4 ft 11 in.) when performing the CSA vertical flame test for cables in cable trays, as described in CSA C22.2 No. 0.3-M-1985, *Test Methods for Electrical Wires and Cables.*

The UL vertical tray flame test determines whether cables installed in a ladder-type cable tray will propagate fire from a given exposure. The test flame is supplied by a strip or ribbon-type propane gas burner.

The cables are installed in a steel ladder-type cable tray that is 12 in. wide, 3 in. deep, and 8 ft long. Cables are installed in the tray spaced half a diameter apart, filling 75 percent of the tray width. A burner supplying 70,000 Btu/hr (20.32 kW/hr) is positioned 18 in. above the bottom of the tray, midway between two rungs. The burner flame is applied to the samples for a period of 20 minutes. See Exhibit 725.7. The samples are considered to have passed the test if flame has not propagated to the top of the cable tray by the test's conclusion. See also the commentary following 725.82(D), FPN.

A cable that passes this testing may be listed as a Type CL2 or Type CL3 cable.

Exhibit 725.7 UL vertical tray flame test.

(D) Types CL2X and CL3X Types CL2X and CL3X limited-use cables shall be marked as Type CL2X or CL3X, respectively, and be listed as being suitable for use in dwell-

ings and for use in raceway and shall also be listed as being resistant to flame spread.

> FPN: One method of determining that cable is resistant to flame spread is by testing the cable to the VW-1 (vertical wire) flame test in ANSI/UL 1581-1991, *Reference Standard for Electrical Wires, Cables and Flexible Cords*.

UL 1581, *Reference Standard for Electrical Wires, Cables and Flexible Cords*, contains basic requirements for conductors, insulation, jackets, and other coverings and the methods of sample preparation, specimen selection and conditioning, and measurements and calculations required in UL 44, *Thermoset-Insulated Wires and Cables*; UL 83, *Thermoplastic-Insulated Wires and Cables*; and UL 62, *Flexible Cord and Fixture Wire*. The flame test methods of these standards include the vertical wire flame test (VW-1) and the vertical tray flame test. [See the commentary following 725.82(C), FPN.]

The VW-1 test is a small-scale test of the flammability of insulating material. It uses a single specimen, secured in a vertical position in a three-sided metal enclosure that has a layer of untreated surgical cotton on the bottom. See Exhibit 725.8. For thermoplastic- or rubber-insulated wire and cable, the specimen used is an 18-in. insulated 14 AWG copper or

Exhibit 725.8 UL vertical wire flame test.

12 AWG aluminum conductor. A small kraft-paper indicator flag is attached to the specimen near its top. A flame is impinged 10 in. below the bottom of the flag for a period of 15 seconds. The flame is removed for 15 seconds and then reapplied for 15 seconds. This procedure is repeated for five 15-second flame applications. The specimen is considered to be capable of propagating fire if one of the following occurs:

1. More than 25 percent of the indicator flag is burned away or charred.
2. Flaming or glowing particles are emitted that ignite the surgical cotton.
3. The specimen continues to flame longer than 60 seconds after removal of the burner.

A sample that passes this test may be listed as Type CL2X or Type CL3X, suitable for use as a limited-use, power-limited cable. This cable may be used in one-family, two-family, and multifamily dwellings.

(E) Type PLTC Type PLTC nonmetallic-sheathed, power-limited tray cable shall be listed as being suitable for cable trays and shall consist of a factory assembly of two or more insulated conductors under a nonmetallic jacket. The insulated conductors shall be 22 AWG through 12 AWG. The conductor material shall be copper (solid or stranded). Insulation on conductors shall be suitable for 300 volts. The cable core shall be either (1) two or more parallel conductors, (2) one or more group assemblies of twisted or parallel conductors, or (3) a combination thereof. A metallic shield or a metallized foil shield with drain wire(s) shall be permitted to be applied either over the cable core, over groups of conductors, or both. The cable shall be listed as being resistant to the spread of fire. The outer jacket shall be a sunlight- and moisture-resistant nonmetallic material.

Exception No. 1: Where a smooth metallic sheath, continuous corrugated metallic sheath, or interlocking tape armor is applied over the nonmetallic jacket, an overall nonmetallic jacket shall not be required. On metallic-sheathed cable without an overall nonmetallic jacket, the information required in 310.11 shall be located on the nonmetallic jacket under the sheath.

Exception No. 2: Conductors in PLTC cables used for Class 2 thermocouple circuits shall be permitted to be any of the materials used for thermocouple extension wire.

FPN: One method of defining *resistant to the spread of fire* is that the cables do not spread fire to the top of the tray in the vertical tray flame test in ANSI/UL 1581-1991, *Reference Standard for Electrical Wires, Cables and Flexible Cords.*

Another method of defining *resistant to the spread of fire* is for the damage (char length) not to exceed 1.5

m (4 ft 11 in.) when performing the CSA vertical flame test for cables in cable trays, as described in CSA C22.2 No. 0.3-M-1985, *Test Methods for Electrical Wires and Cables.*

(F) Circuit Integrity (CI) Cable or Electrical Circuit Protective System Cables used for survivability of critical circuits shall be listed as circuit integrity (CI) cable. Cables specified in 725.61(A), (B), (D)(1), and (E), and used for circuit integrity, shall have the additional classification using the suffix "-CI." Cables that are part of a listed electrical circuit protective system shall be considered to meet the requirements of survivability.

FPN: One method of defining circuit integrity is by establishing a minimum 2-hour fire resistance rating when tested in accordance with UL 2196-2002, *Standard for Tests of Fire Resistive Cables.*

Section 725.82(F) was added to the 2005 *Code* to permit the use of circuit integrity (CI) cable for applications where continuity of the operations of critical circuits is needed during a fire. Such circuits could be essential to fire fighting operations or could be circuits whose interruption could cause a more dangerous condition to occur. A smoke removal system is an example of where it may be necessary to use circuit integrity cables for control circuits to ensure that the dampers will operate during a fire.

(G) Class 2 and Class 3 Cable Voltage Ratings Class 2 cables shall have a voltage rating of not less than 150 volts. Class 3 cables shall have a voltage rating of not less than 300 volts.

(H) Class 3 Single Conductors Class 3 single conductors used as other wiring within buildings shall not be smaller than 18 AWG and shall be Type CL3. Conductor types described in 725.27(B) that are also listed as Type CL3 shall be permitted.

FPN: One method of defining *resistant to the spread of fire* is that the cables do not spread fire to the top of the tray in the vertical tray flame test in ANSI/UL 1581-1991, *Reference Standard for Electrical Wires, Cables and Flexible Cords.*

Another method of defining *resistant to the spread of fire* is for the damage (char length) not to exceed 1.5 m (4 ft 11 in.) when performing the CSA vertical flame test for cables in cable trays as described in CSA C22.2 No. 0.3-M-1985, *Test Methods for Electrical Wires and Cables.*

Section 725.82(H) clarifies that Class 3 single conductors must also be resistant to the spread of fire.

(I) Plenum Signaling Raceways Plenum signaling raceways shall be listed as having adequate fire-resistant and low smoke-producing characteristics.

Section 725.82(I) was added to the 2005 *Code* to recognize listed nonmetallic signaling raceways for use in plenums. Similar raceways are currently recognized for optical fiber and communications cables in Articles 770 and 800, respectively. Raceways listed for plenums must pass fire tests similar to those required of cables listed for plenums.

(J) Riser Signaling Raceways Riser signaling raceways shall be listed as having adequate fire-resistant characteristics capable of preventing the carrying of fire from floor to floor.

> FPN: One method of defining fire-resistant characteristics capable of preventing the carrying of fire from floor to floor is that the raceways pass the requirements of the Test for Flame Propagation (Riser) in UL 2024, *Standard for Optical Fiber Cable Raceway.*

Section 725.82(J) was added to the 2005 *Code* to recognize listed nonmetallic signaling raceways for use in risers. Similar raceways are currently recognized for optical fiber and communications cables in Articles 770 and 800, respectively. Raceways listed for risers must pass fire tests similar to those required of cables listed for risers.

(K) General-Purpose Signaling Raceways General-purpose signaling raceways shall be listed as being resistant to the spread of fire.

> FPN: One method of defining resistance to the spread of fire is that the raceways pass the requirements of the Vertical-Tray Flame Test (General use) in UL 2024, *Standard for Optical Fiber Cable Raceway.*

Section 725.82(K) was added to the 2005 *Code* to recognize listed nonmetallic signaling raceways for use in general-purpose applications. Similar raceways are currently recognized for optical fiber and communications cables in Articles 770 and 800, respectively. Raceways listed for general-purpose applications must pass fire tests similar to those required of cables listed for general-purpose applications.

(L) Marking Cables shall be marked in accordance with 310.11(A)(2), (A)(3), (A)(4), and (A)(5) and Table 725.82. Voltage ratings shall not be marked on the cables.

> FPN: Voltage markings on cables may be misinterpreted to suggest that the cables may be suitable for Class 1 electric light and power applications.

Exception: Voltage markings shall be permitted where the cable has multiple listings and a voltage marking is required for one or more of the listings.

Table 725.82 Cable Marking

Cable Marking	Type
CL3P	Class 3 plenum cable
CL2P	Class 2 plenum cable
CL3R	Class 3 riser cable
CL2R	Class 2 riser cable
PLTC	Power-limited tray cable
CL3	Class 3 cable
CL2	Class 2 cable
CL3X	Class 3 cable, limited use
CL2X	Class 2 cable, limited use

FPN: Class 2 and Class 3 cable types are listed in descending order of fire resistance rating, and Class 3 cables are listed above Class 2 cables because Class 3 cables can substitute for Class 2 cables.

ARTICLE 727
Instrumentation Tray Cable: Type ITC

Contents
727.1 Scope
727.2 Definition
727.3 Other Articles
727.4 Uses Permitted
727.5 Uses Not Permitted
727.6 Construction
727.7 Marking
727.8 Allowable Ampacity
727.9 Overcurrent Protection
727.10 Bends

727.1 Scope

This article covers the use, installation, and construction specifications of instrumentation tray cable for application to instrumentation and control circuits operating at 150 volts or less and 5 amperes or less.

Article 727 permits an alternate wiring method for circuits that do not exceed 5 amperes and 150 volts. Instrument tray cable is particularly suited for instrumentation circuits in industrial establishments where qualified persons perform service and maintenance.

727.2 Definition

Type ITC Instrumentation Tray Cable. A factory assembly of two or more insulated conductors, with or without a grounding conductor(s), enclosed in a nonmetallic sheath.

727.3 Other Articles

In addition to the provisions of this article, installation of Type ITC cable shall comply with other applicable articles of this *Code*, such as Articles 240, 250, 300, and 392.

727.4 Uses Permitted

Type ITC cable shall be permitted to be used as follows in industrial establishments where the conditions of maintenance and supervision ensure that only qualified persons service the installation:

(1) In cable trays.
(2) In raceways.
(3) In hazardous locations as permitted in 501.10, 502.10, 503.10, 504.20, 504.30, 504.80, and 505.15.
(4) Enclosed in a smooth metallic sheath, continuous corrugated metallic sheath, or interlocking tape armor applied over the nonmetallic sheath in accordance with 727.6. The cable shall be supported and secured at intervals not exceeding 1.8 m (6 ft).
(5) Without a metallic sheath or armor between cable tray and equipment in lengths not to exceed 15 m (50 ft), where the cable is supported and protected against physical damage using mechanical protection, such as struts, angles, or channels. The cable shall be supported and secured at intervals not exceeding 1.8 m (6 ft).
(6) Between cable tray and equipment in lengths not to exceed 15 m (50 ft), where the cable complies with the crush and impact requirements of Type MC cable and is identified for such use. The cable shall be supported and secured at intervals not exceeding 1.8 m (6 ft).
(7) As aerial cable on a messenger.
(8) Direct buried where identified for the use.
(9) Under raised floors in rooms containing industrial process control equipment and rack rooms where arranged to prevent damage to the cable.
(10) Under raised floors in information technology equipment rooms in accordance with 645.5(D)(5)(c).

727.5 Uses Not Permitted

Type ITC cable shall not be installed on circuits operating at more than 150 volts or more than 5 amperes.

Installation of Type ITC cable with other cables shall be subject to the stated provisions of the specific articles for the other cables. Where the governing articles do not contain stated provisions for installation with Type ITC cable, the installation of Type ITC cable with the other cables shall not be permitted.

Type ITC cable shall not be installed with power, lighting, Class 1, or non–power-limited circuits.

Exception No. 1: Where terminated within equipment or junction boxes and separations are maintained by insulating barriers or other means.

Exception No. 2: Where a metallic sheath or armor is applied over the nonmetallic sheath of the Type ITC cable.

727.6 Construction

The insulated conductors of Type ITC cable shall be in sizes 22 AWG through 12 AWG. The conductor material shall be copper or thermocouple alloy. Insulation on the conductors shall be rated for 300 volts. Shielding shall be permitted.

The cable shall be listed as being resistant to the spread of fire. The outer jacket shall be sunlight and moisture resistant.

Where a smooth metallic sheath, continuous corrugated metallic sheath, or interlocking tape armor is applied over the nonmetallic sheath, an overall nonmetallic jacket shall not be required.

727.7 Marking

The cable shall be marked in accordance with 310.11(A)(2), (A)(3), (A)(4), and (A)(5). Voltage ratings shall not be marked on the cable.

727.8 Allowable Ampacity

The allowable ampacity of the conductors shall be 5 amperes, except for 22 AWG conductors, which shall have an allowable ampacity of 3 amperes.

727.9 Overcurrent Protection

Overcurrent protection shall not exceed 5 amperes for 20 AWG and larger conductors, and 3 amperes for 22 AWG conductors.

727.10 Bends

Bends in Type ITC cables shall be made so as not to damage the cable.

ARTICLE 760
Fire Alarm Systems

Summary of Changes
- **760.8:** Revised requirement covering mechanical execution of work.
- **760.21:** Revised to prohibit arc-fault circuit interrupters (AFCIs) in addition to GFCIs in the supply circuit for the power source of non–power-limited fire alarm circuits.
- **760.30(B)(2), (B)(3), and (B)(4):** Added new Exception No. 3 permitting circuit integrity (CI) cable.
- **760.41:** Revised to prohibit arc-fault circuit interrupters (AFCIs) in addition to GFCIs in the supply circuit for the power source of power-limited fire alarm circuits.

- **760.52:** Revised to permit combinations of 760.52(A) and (B) wiring methods on load side of PLFA power source.
- **760.56(D):** Added new requirement prohibiting audio system circuits using Class 2 or Class 3 conductors and PLFA circuits in the same raceway.
- **760.61(A), (B)(1), and (C):** Revised to permit circuit integrity (CI) cable.
- **760.81(F) and 760.82(G):** Added new requirements for NPLFA and PLFA circuit integrity (CI) cable.

Contents

    (B) Insulated Conductors
    (C) Type NPLFP
    (D) Type NPLFR
    (E) Type NPLF
    (F) Fire Alarm Circuit Integrity (CI) Cable or Electrical Circuit Protective System
    (G) NPLFA Cable Markings
760.82 Listing and Marking of PLFA Cables and Insulated Continuous Line-Type Fire Detectors
    (A) Conductor Materials
    (B) Conductor Size
    (C) Ratings
    (D) Type FPLP
    (E) Type FPLR
    (F) Type FPL
    (G) Fire Alarm Circuit Integrity (CI) Cable or Electrical Circuit Protective System
    (H) Coaxial Cables
    (I) Cable Marking
    (J) Insulated Continuous Line-Type Fire Detectors

I. General

760.1 Scope

This article covers the installation of wiring and equipment of fire alarm systems including all circuits controlled and powered by the fire alarm system.

> FPN No. 1: Fire alarm systems include fire detection and alarm notification, guard's tour, sprinkler waterflow, and sprinkler supervisory systems. Circuits controlled and powered by the fire alarm system include circuits for the control of building systems safety functions, elevator capture, elevator shutdown, door release, smoke doors and damper control, fire doors and damper control and fan shutdown, but only where these circuits are powered by and controlled by the fire alarm system. For further information on the installation and monitoring for integrity requirements for fire alarm systems, refer to the *NFPA 72®-2002, National Fire Alarm Code®*.
>
> FPN No. 2: Class 1, 2, and 3 circuits are defined in Article 725.

Article 760 covers only those circuits that are powered and controlled by the fire alarm system, including fire safety features such as smoke door control, damper control, fan shutdown, and elevator recall. Circuits powered and controlled by other building systems such as heating, ventilating, and air conditioning (HVAC); security; lighting controls; and time recording are covered by Article 725.

Article 760 covers the wiring between the devices and equipment required by *NFPA 72®, National Fire Alarm Code®. NFPA 72* provides the requirements for the selection, installation, performance, use, and testing and maintenance of fire alarm system components. The provision for whether an occupancy is required to have a fire alarm system is found in NFPA *101®, Life Safety Code®,* or other local codes.

Examples of fire alarm devices and equipment are shown in Exhibits 760.1 and 760.2. The installation of these system components is covered by *NFPA 72*, but the circuit wiring associated with these components must be installed in accordance with the requirements of Article 760. Single- and multiple-station smoke alarms, such as those commonly installed in dwelling units, are supplied through 120-volt branch circuits rather than through a non–power-limited or power-limited fire alarm signaling circuit that is powered and controlled by a fire alarm control panel. Branch circuits supplying power to single- and multiple-station smoke alarms are not subject to the requirements of Article 760.

NFPA 72 requires that all wiring, cable, and equipment be in accordance with NFPA 70, *National Electrical Code®*, and specifically with Article 760. Additionally, *NFPA 72* requires all equipment to be listed for its intended purpose.

NFPA 1221, *Standard for the Installation, Maintenance, and Use of Emergency Services Communications Systems*, covers the installation, maintenance, and use of all public fire service communications facilities. These facilities in-

Exhibit 760.1 Typical fire alarm control unit.

Exhibit 760.2 Typical spot-type smoke detector.

clude public reporting, dispatching, telephone, and two-way and microwave radio systems, all of which fulfill two principal functions: the receipt of fire alarms or other emergency calls from the public and the retransmission of these alarms and emergency calls to fire companies and other appropriate agencies.

760.2 Definitions

Abandoned Fire Alarm Cable. Installed fire alarm cable that is not terminated at equipment other than a connector and not identified for future use with a tag.

This definition was added to the 2002 *Code* for use with 760.3(A), which requires removal of accessible abandoned fire alarm cable. Abandoned cable increases fire loading unnecessarily, and, where installed in plenums, it can affect airflow. Similar requirements can be found in Articles 640, 645, 725, 770, 800, 820, and 830.

Fire Alarm Circuit. The portion of the wiring system between the load side of the overcurrent device or the power-limited supply and the connected equipment of all circuits powered and controlled by the fire alarm system. Fire alarm circuits are classified as either non–power-limited or power-limited.

Fire Alarm Circuit Integrity (CI) Cable. Cable used in fire alarm systems to ensure continued operation of critical circuits during a specified time under fire conditions.

Non–Power-Limited Fire Alarm Circuit (NPLFA). A fire alarm circuit powered by a source that complies with 760.21 and 760.23.

Power-Limited Fire Alarm Circuit (PLFA). A fire alarm circuit powered by a source that complies with 760.41.

760.3 Other Articles

Circuits and equipment shall comply with 760.3(A) through 760.3(F). Only those sections of Article 300 referenced in this article shall apply to fire alarm systems.

(A) Spread of Fire or Products of Combustion Section 300.21. The accessible portion of abandoned fire alarm cables shall be removed.

(B) Ducts, Plenums, and Other Air-Handling Spaces Section 300.22, where installed in ducts or plenums or other spaces used for environmental air.

Exception: As permitted in 760.30(B)(1) and (B)(2) and 760.61(A).

See the commentary following 300.22(B) and 300.22(C) for more information on wiring installed in ducts, plenums, or other spaces used for environmental air.

(C) Hazardous (Classified) Locations Articles 500 through 516 and Article 517, Part IV, where installed in hazardous (classified) locations.

(D) Corrosive, Damp, or Wet Locations Sections 110.11, 300.6, and 310.9 where installed in corrosive, damp, or wet locations.

Section 760.3(D) requires cables and equipment that are used in wet or damp locations, high ambient temperature areas, or corrosive locations to be identified as suitable for the particular use. Underground installations are considered wet locations.

(E) Building Control Circuits Article 725 where building control circuits (e.g., elevator capture, fan shutdown) are associated with the fire alarm system.

(F) Optical Fiber Cables Where optical fiber cables are utilized for fire alarm circuits, the cables shall be installed in accordance with Article 770.

760.7 Access to Electrical Equipment Behind Panels Designed to Allow Access

Access to electrical equipment shall not be denied by an accumulation of conductors and cables that prevents removal of panels, including suspended ceiling panels.

An excess accumulation of wires and cables can limit access to equipment by preventing the removal of access panels. See Exhibit 760.3.

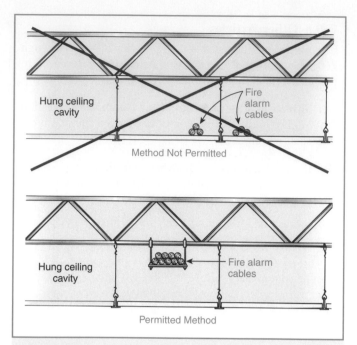

Hung ceiling cavity

Fire alarm cables

Method Not Permitted

Hung ceiling cavity

Fire alarm cables

Permitted Method

Exhibit 760.3 Incorrect installation of conductors and cables (upper diagram), which can prevent access to equipment or cables. Correct method is shown in lower diagram.

760.8 Mechanical Execution of Work

Fire alarm circuits shall be installed in a neat workmanlike manner. Cables and conductors installed exposed on the surface of ceilings and sidewalls shall be supported by the building structure in such a manner that the cable will not be damaged by normal building use. Such cables shall be supported by straps, staples, hangers, or similar fittings designed and installed so as not to damage the cable. The installation shall also conform with 300.4(D).

> FPN: One source of information describing industry practices is ANCI/NECA 305-2001, *Standard for Fire Alarm System Job Practice.*

This section provides definitive requirements for workmanship. Cable must be attached to or supported by the structure by straps, clamps, hangers, and the like. The installation method must not damage the cable. In addition, the location of the cable should be carefully evaluated to ensure that activities and processes within the building do not cause damage to the cable.

The reference to 300.4(D) calls attention to the hazards to which cables are exposed where they are installed on framing members. Such cables are required to be installed in a manner that protects them from nail or screw penetration. In the 2005 *Code,* there was a change to this section to permit attachment to baseboards and non–load bearing walls, which are not structural components.

760.9 Fire Alarm Circuit and Equipment Grounding

Fire alarm circuits and equipment shall be grounded in accordance with Article 250.

760.10 Fire Alarm Circuit Identification

Fire alarm circuits shall be identified at terminal and junction locations in a manner that will prevent unintentional interference with the signaling circuit during testing and servicing.

760.11 Fire Alarm Circuits Extending Beyond One Building

Power-limited fire alarm circuits that extend beyond one building and run outdoors either shall meet the installation requirements of Parts II, III, and IV of Article 800 or shall meet the installation requirements of Part I of Article 300. Non–power-limited fire alarm circuits that extend beyond one building and run outdoors shall meet the installation requirements of Part I of Article 300 and the applicable sections of Part I of Article 225.

760.15 Fire Alarm Circuit Requirements

Fire alarm circuits shall comply with 760.15(A) and 760.15(B).

(A) Non–Power-Limited Fire Alarm (NPLFA) Circuits See Parts I and II.

(B) Power-Limited Fire Alarm (PLFA) Circuits See Parts I and III.

Exact power source limitations for power-limited fire alarm circuits used by testing laboratories are found in Chapter 9, Table 12(A) and Table 12(B). Table 12(A) covers alternating-current source limitations, and Table 12(B) covers direct-current source limitations.

II. Non–Power-Limited Fire Alarm (NPLFA) Circuits

760.21 NPLFA Circuit Power Source Requirements

The power source of non–power-limited fire alarm circuits shall comply with Chapters 1 through 4, and the output voltage shall not be more than 600 volts, nominal. These circuits shall not be supplied through ground-fault circuit interrupters or arc-fault circuit interrupters.

> FPN: See 210.8(A)(5), Exception No. 3, for receptacles in dwelling-unit unfinished basements that supply power for fire alarm systems.

Section 760.21 was revised for the 2005 *Code.* It prohibits supplying a non–power-limited fire alarm circuit through a ground-fault circuit interrupter or an arc-fault circuit interrupter. The intent is to supply a fire alarm system through a source that is not subject to interruption. This requirement does not apply to single- or multiple-station smoke alarms. Single- and multiple-station smoke alarms are not supplied through a non–power-limited fire alarm circuit. Smoke alarms in new construction are required by *NFPA 72®, National Fire Alarm Code®*, to have a backup battery that will supply power in the event that the branch circuit power is interrupted due to the operation of a GFCI or AFCI device.

760.23 NPLFA Circuit Overcurrent Protection

Overcurrent protection for conductors 14 AWG and larger shall be provided in accordance with the conductor ampacity without applying the derating factors of 310.15 to the ampacity calculation. Overcurrent protection shall not exceed 7 amperes for 18 AWG conductors and 10 amperes for 16 AWG conductors.

Exception: Where other articles of this Code permit or require other overcurrent protection.

760.24 NPLFA Circuit Overcurrent Device Location

Overcurrent devices shall be located at the point where the conductor to be protected receives its supply.

Exception No. 1: Where the overcurrent device protecting the larger conductor also protects the smaller conductor.

Exception No. 2: Transformer secondary conductors. Non–power-limited fire alarm circuit conductors supplied by the secondary of a single-phase transformer that has only a 2-wire (single-voltage) secondary shall be permitted to be protected by overcurrent protection provided by the primary (supply) side of the transformer, provided the protection is in accordance with 450.3 and does not exceed the value determined by multiplying the secondary conductor ampacity by the secondary-to-primary transformer voltage ratio. Transformer secondary conductors other than 2-wire shall not be considered to be protected by the primary overcurrent protection.

Exception No. 3: Electronic power source output conductors. Non–power-limited circuit conductors supplied by the output of a single-phase, listed electronic power source, other than a transformer, having only a 2-wire (single-voltage) output for connection to non–power-limited circuits shall be permitted to be protected by overcurrent protection provided on the input side of the electronic power source, provided this protection does not exceed the value determined by multiplying the non–power-limited circuit conductor ampacity by the output-to-input voltage ratio. Electronic power source outputs, other than 2-wire (single voltage), connected to non–power-limited circuits shall not be consid-

ered to be protected by overcurrent protection on the input of the electronic power source.

> FPN: A single-phase, listed electronic power supply whose output supplies a 2-wire (single-voltage) circuit is an example of a non–power-limited power source that meets the requirements of 760.21.

Non–power-limited electronic power supplies that do not supply energy directly through the use of transformers are covered by Exception No. 3 to 760.24, which permits overcurrent protection for the non–power-limited circuit conductors to be installed on the input side of the electronic power source rather than on the output side for 2-wire circuits only.

760.25 NPLFA Circuit Wiring Methods

Installation of non–power-limited fire alarm circuits shall be in accordance with 110.3(B), 300.11, 300.15, 300.17, and other appropriate articles of Chapter 3.

Exception No. 1: As provided in 760.26 through 760.30.

Exception No. 2: Where other articles of this Code require other methods.

Section 760.25 requires that the appropriate wiring methods in Chapter 3 be used for non–power-limited circuits. However, 760.25, Exception No. 1, permits special non–power-limited cable types to be used in place of Chapter 3 wiring methods.

Section 760.25 requires that devices be mounted in accordance with Chapter 3. Section 300.11(A) requires devices and equipment to be securely mounted. Section 300.15(B) is referenced to require non–power-limited circuit terminations to be made in a box or conduit body. However, 300.15(E) permits devices with integral terminal enclosures and mounting brackets to be used without a box. Devices must be mounted on a box or conduit body where the instructions or listing require the use of a box. Fire alarm system components such as manual fire alarm boxes are frequently tested. Therefore, secure mounting of the back box is necessary to ensure that the manual fire alarm device will remain in place. (See Exhibit 760.4.)

760.26 Conductors of Different Circuits in Same Cable, Enclosure, or Raceway

(A) Class 1 with NPLFA Circuits Class 1 and non–power-limited fire alarm circuits shall be permitted to occupy the same cable, enclosure, or raceway without regard to whether the individual circuits are alternating current or direct current, provided all conductors are insulated for the maximum voltage of any conductor in the enclosure or raceway.

(B) Fire Alarm with Power-Supply Circuits Power-supply and fire alarm circuit conductors shall be permitted

Exhibit 760.4 Typical manual fire alarm box.

in the same cable, enclosure, or raceway only where connected to the same equipment.

760.27 NPLFA Circuit Conductors

(A) Sizes and Use Only copper conductors shall be permitted to be used for fire alarm systems. Size 18 AWG and 16 AWG conductors shall be permitted to be used, provided they supply loads that do not exceed the ampacities given in Table 402.5 and are installed in a raceway, an approved enclosure, or a listed cable. Conductors larger than 16 AWG shall not supply loads greater than the ampacities given in 310.15, as applicable.

The minimum size of conductors permitted to be used on non–power-limited fire protective signaling circuits is 18 AWG. The load must not exceed the conductor ampacities specified in Table 402.5.

NFPA 72, National Fire Alarm Code, requires fire alarm device and appliance voltages to be between 85 and 110 percent of nominal rated voltage. Calculations should be made to ensure that all devices or appliances will be operating within these limits at full circuit load. Where future circuit extensions are anticipated, larger conductors should be considered. Some manufacturers specify maximum circuit loop resistances. The equipment specifications should be consulted to ensure that maximum allowable loop resistances are not exceeded.

(B) Insulation Insulation on conductors shall be suitable for 600 volts. Conductors larger than 16 AWG shall comply with Article 310. Conductors 18 AWG and 16 AWG shall

be Type KF-2, KFF-2, PAFF, PTFF, PF, PFF, PGF, PGFF, RFH-2, RFHH-2, RFHH-3, SF-2, SFF-2, TF, TFF, TFN, TFFN, ZF, or ZFF. Conductors with other types and thickness of insulation shall be permitted if listed for non–power-limited fire alarm circuit use.

> FPN: For application provisions, see Table 402.3.

(C) Conductor Materials Conductors shall be solid or stranded copper.

Exception to (B) and (C): Wire Types PAF and PTF shall be permitted only for high-temperature applications between 90°C (194°F) and 250°C (482°F).

760.28 Number of Conductors in Cable Trays and Raceways, and Derating

(A) NPLFA Circuits and Class 1 Circuits Where only non–power-limited fire alarm circuit and Class 1 circuit conductors are in a raceway, the number of conductors shall be determined in accordance with 300.17. The derating factors given in 310.15(B)(2)(a) shall apply if such conductors carry continuous load in excess of 10 percent of the ampacity of each conductor.

(B) Power-Supply Conductors and Fire Alarm Circuit Conductors Where power-supply conductors and fire alarm circuit conductors are permitted in a raceway in accordance with 760.26, the number of conductors shall be determined in accordance with 300.17. The derating factors given in 310.15(B)(2)(a) shall apply as follows:

(1) To all conductors where the fire alarm circuit conductors carry continuous loads in excess of 10 percent of the ampacity of each conductor and where the total number of conductors is more than three

(2) To the power-supply conductors only, where the fire alarm circuit conductors do not carry continuous loads in excess of 10 percent of the ampacity of each conductor and where the number of power-supply conductors is more than three

(C) Cable Trays Where fire alarm circuit conductors are installed in cable trays, they shall comply with 392.9 through 392.11.

760.30 Multiconductor NPLFA Cables

Multiconductor non–power-limited fire alarm cables that meet the requirements of 760.81 shall be permitted to be used on fire alarm circuits operating at 150 volts or less and shall be installed in accordance with 760.30(A) and 760.30(B).

(A) NPLFA Wiring Method Multiconductor non–power-limited fire alarm circuit cables shall be installed in accordance with 760.30(A)(1), (A)(2), and (A)(3).

(1) Exposed or Fished in Concealed Spaces In raceway or exposed on surface of ceiling and sidewalls or fished in concealed spaces. Cable splices or terminations shall be made in listed fittings, boxes, enclosures, fire alarm devices, or utilization equipment. Where installed exposed, cables shall be adequately supported and installed in such a way that maximum protection against physical damage is afforded by building construction such as baseboards, door frames, ledges, and so forth. Where located within 2.1 m (7 ft) of the floor, cables shall be securely fastened in an approved manner at intervals of not more than 450 mm (18 in.).

(2) Passing Through a Floor or Wall In metal raceway or rigid nonmetallic conduit where passing through a floor or wall to a height of 2.1 m (7 ft) above the floor unless adequate protection can be afforded by building construction such as detailed in 760.30(A)(1) or unless an equivalent solid guard is provided.

(3) In Hoistways In rigid metal conduit, rigid nonmetallic conduit, intermediate metal conduit, liquidtight flexible non-metallic conduit, or electrical metallic tubing where installed in hoistways.

Exception: As provided for in 620.21 for elevators and similar equipment.

(B) Applications of Listed NPLFA Cables The use of non–power-limited fire alarm circuit cables shall comply with 760.30(B)(1) through (B)(4).

(1) Ducts and Plenums Multiconductor non–power-limited fire alarm circuit cables, Types NPLFP, NPLFR, and NPLF, shall not be installed exposed in ducts or plenums.

Wiring methods for non–power-limited circuits in ducts and plenums must be in accordance with the Chapter 3 wiring methods covered by 300.22(B). It is important to note that cables marked NPLFP may not be installed in plenums. While the "P" designation was used for consistency, the higher possible voltages and currents of non–power-limited fire alarm circuits preclude the use of the listed cables inside plenums.

FPN: See 300.22(B).

(2) Other Spaces Used for Environmental Air Cables installed in other spaces used for environmental air shall be Type NPLFP.

Exception No. 1: Types NPLFR and NPLF cables installed in compliance with 300.22(C).

Exception No. 2: Other wiring methods in accordance with 300.22(C) and conductors in compliance with 760.27(C).

Other spaces used for environmental air are covered by 300.22(C) and the related fine print note. Spaces over sus-

pended ceilings used as an environmental air-handling return are considered by the *Code* as "other spaces used for environmental air." Non–power-limited cables used in other spaces used for environmental air must, however, be marked NPLFP. [See 760.81(C).]

Exception No. 3: Type NPLFP-CI cable shall be permitted to be installed to provide a 2-hour circuit integrity rated cable.

(3) Riser Cables installed in vertical runs and penetrating more than one floor or cables installed in vertical runs in a shaft shall be Type NPLFR. Floor penetrations requiring Type NPLFR shall contain only cables suitable for riser or plenum use.

Exception No. 1: Type NPLF or other cables that are specified in Chapter 3 and are in compliance with 760.27(C) and encased in metal raceway.

Exception No. 2: Type NPLF cables located in a fireproof shaft having firestops at each floor.

FPN: See 300.21 for firestop requirements for floor penetrations.

Exception No. 3: Type NPLFR-CI cable shall be permitted to be installed to provide a 2-hour circuit integrity rated cable.

(4) Other Wiring Within Buildings Cables installed in building locations other than the locations covered in 760.30(B)(1), (B)(2), and (B)(3) shall be Type NPLF.

Exception No. 1: Chapter 3 wiring methods with conductors in compliance with 760.27(C).

Exception No. 2: Type NPLFP or Type NPLFR cables shall be permitted.

Exception No. 3: Type NPLF-CI cable shall be permitted to be installed to provide a 2-hour circuit integrity rated cable.

III. Power-Limited Fire Alarm (PLFA) Circuits

760.41 Power Sources for PLFA Circuits

The power source for a power-limited fire alarm circuit shall be as specified in 760.41(A), (B), or (C). These circuits shall not be supplied through ground-fault circuit interrupters or arc-fault circuit interrupters.

FPN No. 1: Tables 12(A) and 12(B) in Chapter 9 provide the listing requirements for power-limited fire alarm circuit sources.

FPN No. 2: See 210.8(A)(5), Exception No. 3, for receptacles in dwelling-unit unfinished basements that supply power for fire alarm systems.

(A) Transformers A listed PLFA or Class 3 transformer.

(B) Power Supplies A listed PLFA or Class 3 power supply.

(C) Listed Equipment Listed equipment marked to identify the PLFA power source.

> FPN: Examples of listed equipment are a fire alarm control panel with integral power source; a circuit card listed for use as a PLFA source, where used as part of a listed assembly; a current-limiting impedance, listed for the purpose or part of a listed product, used in conjunction with a non–power-limited transformer or a stored energy source, for example, storage battery, to limit the output current.

760.42 Circuit Marking

The equipment supplying PLFA circuits shall be durably marked where plainly visible to indicate each circuit that is a power-limited fire alarm circuit.

> FPN: See 760.52(A), Exception No. 3, where a power-limited circuit is to be reclassified as a non–power-limited circuit.

760.51 Wiring Methods on Supply Side of the PLFA Power Source

Conductors and equipment on the supply side of the power source shall be installed in accordance with the appropriate requirements of Part II and Chapters 1 through 4. Transformers or other devices supplied from power-supply conductors shall be protected by an overcurrent device rated not over 20 amperes.

Exception: The input leads of a transformer or other power source supplying power-limited fire alarm circuits shall be permitted to be smaller than 14 AWG, but not smaller than 18 AWG, if they are not over 300 mm (12 in.) long and if they have insulation that complies with 760.27(B).

760.52 Wiring Methods and Materials on Load Side of the PLFA Power Source

Fire alarm circuits on the load side of the power source shall be permitted to be installed using wiring methods and materials in accordance with 760.52(A), 760.52(B), or a combination of (A) and (B).

Section 760.52 has been revised for the 2005 *Code*. It now permits individual power-limited circuits to be installed using Chapter 3 wiring methods, non–power-limited fire alarm circuit wiring methods, power-limited circuit wiring methods, or a combination. If it is desirable to run power-limited circuits in the same cable or raceway with non–power-limited circuits, the power-limited circuits may be reclassified as permitted by 760.52(A), Exception No. 3.

Also note the information contained in the fine print note that follows 760.52(A), Exception No. 3, regarding circuit classification.

(A) NPLFA Wiring Methods and Materials Installation shall be in accordance with 760.25, and conductors shall be solid or stranded copper.

Section 760.52(A) requires power-limited fire alarm circuits using non–power-limited fire alarm wiring methods and materials to be installed in accordance with 760.25. Section 760.25 requires that non–power-limited devices be mounted in accordance with Chapter 3. Section 300.11(A) requires these devices and equipment to be securely mounted. Section 300.15(B) is also referenced to require non–power-limited circuit terminations to be made in a box or conduit body. However, 300.15(E) permits devices with integral terminal enclosures and mounting brackets to be used without a box. Devices are required to be mounted on a box or conduit body where the instructions or listing indicates the use of a box. Fire alarm system components such as manual fire alarm boxes are frequently tested. Secure mounting is necessary to ensure that they will remain in place.

Exception No. 1: The derating factors given in 310.15(B)(2)(a) shall not apply.

Exception No. 2: Conductors and multiconductor cables described in and installed in accordance with 760.27 and 760.30 shall be permitted.

Exception No. 3: Power-limited circuits shall be permitted to be reclassified and installed as non–power-limited circuits if the power-limited fire alarm circuit markings required by 760.42 are eliminated and the entire circuit is installed using the wiring methods and materials in accordance with Part II, Non–Power-Limited Fire Alarm Circuits.

> FPN: Power-limited circuits reclassified and installed as non–power-limited circuits are no longer power-limited circuits, regardless of the continued connection to a power-limited source.

Section 760.52(A) permits any of the wiring methods in Chapter 3 to be used for power-limited circuits. In addition, 760.52(A), Exception No. 3, allows power-limited circuits to be reclassified and installed in accordance with the requirements for non–power-limited circuits. Where installed as non–power-limited circuits, the power-limited marking must be removed from equipment, overcurrent protection must be provided in accordance with 760.23, and reclassified circuits must maintain separation from power-limited circuits, in accordance with 760.26 and 760.54.

(B) PLFA Wiring Methods and Materials Power-limited fire alarm conductors and cables described in 760.82 shall be installed as detailed in 760.52(B)(1), (B)(2), or (B)(3) of this section. Devices shall be installed in accordance with 110.3(B), 300.11(A), and 300.15.

Section 760.52(B) requires mechanical protection at splices and termination points. Because failure of a circuit often occurs at splices or termination points, this requirement offers more protection and strain relief for these cable connections.

(1) Exposed or Fished in Concealed Spaces In raceway or exposed on the surface of ceiling and sidewalls or fished in concealed spaces. Cable splices or terminations shall be made in listed fittings, boxes, enclosures, fire alarm devices, or utilization equipment. Where installed exposed, cables shall be adequately supported and installed in such a way that maximum protection against physical damage is afforded by building construction such as baseboards, door frames, ledges, and so forth. Where located within 2.1 m (7 ft) of the floor, cables shall be securely fastened in an approved manner at intervals of not more than 450 mm (18 in.).

(2) Passing Through a Floor or Wall In metal raceways or rigid nonmetallic conduit where passing through a floor or wall to a height of 2.1 m (7 ft) above the floor, unless adequate protection can be afforded by building construction such as detailed in 760.52(B)(1) or unless an equivalent solid guard is provided.

(3) In Hoistways In rigid metal conduit, rigid nonmetallic conduit, intermediate metal conduit, or electrical metallic tubing where installed in hoistways.

Exception: As provided for in 620.21 for elevators and similar equipment.

760.54 Installation of Conductors and Equipment in Cables, Compartments, Cable Trays, Enclosures, Manholes, Outlet Boxes, Device Boxes, and Raceways for Power-Limited Circuits

Conductors and equipment for power-limited fire alarm circuits shall be installed in accordance with 760.55 through 760.58.

760.55 Separation from Electric Light, Power, Class 1, NPLFA, and Medium Power Network-Powered Broadband Communications Circuit Conductors

(A) General Power-limited fire alarm circuit cables and conductors shall not be placed in any cable, cable tray, compartment, enclosure, manhole, outlet box, device box, raceway, or similar fitting with conductors of electric light, power, Class 1, non–power-limited fire alarm circuits, and medium power network-powered broadband communications circuits unless permitted by 760.55(B) through 760.55(G).

Jackets of listed power-limited fire alarm cables do not have sufficient construction specifications to permit them to be installed with electric light, power, Class 1, non–power-limited fire alarm circuits, and medium power network-powered broadband communications cables. Failure of the cable insulation due to a fault could lead to hazardous voltages being imposed on the power-limited fire alarm circuit conductors.

(B) Separated by Barriers Power-limited fire alarm circuit cables shall be permitted to be installed together with Class 1, non–power-limited fire alarm, and medium power network-powered broadband communications circuits where they are separated by a barrier.

(C) Raceways Within Enclosures In enclosures, power-limited fire alarm circuits shall be permitted to be installed in a raceway within the enclosure to separate them from Class 1, non–power-limited fire alarm, and medium power network-powered broadband communications circuits.

(D) Associated Systems Within Enclosures Power-limited fire alarm conductors in compartments, enclosures, device boxes, outlet boxes, or similar fittings shall be permitted to be installed with electric light, power, Class 1, non–power-limited fire alarm, and medium power network-powered broadband communications circuits where they are introduced solely to connect the equipment connected to power-limited fire alarm circuits, and comply with either of the following conditions:

(1) The electric light, power, Class 1, non–power-limited fire alarm, and medium power network-powered broadband communications circuit conductors are routed to maintain a minimum of 6 mm (0.25 in.) separation from the conductors and cables of power-limited fire alarm circuits.

(2) The circuit conductors operate at 150 volts or less to ground and also comply with one of the following:

a. The fire alarm power-limited circuits are installed using Type FPL, FPLR, FPLP, or permitted substitute cables, provided these power-limited cable conductors extending beyond the jacket are separated by a minimum of 6 mm (0.25 in.) or by a nonconductive sleeve or nonconductive barrier from all other conductors.

b. The power-limited fire alarm circuit conductors are installed as non–power-limited circuits in accordance with 760.25.

(E) Enclosures with Single Opening Power-limited fire alarm circuit conductors entering compartments, enclosures, device boxes, outlet boxes, or similar fittings shall be permitted to be installed with electric light, power, Class 1, non–power-limited fire alarm, and medium power network-powered broadband communications circuits where they are introduced solely to connect the equipment connected to power-limited fire alarm circuits or to other circuits controlled by the fire alarm system to which the other conductors in the enclosure are connected. Where power-limited fire alarm circuit conductors must enter an enclosure that is provided with a single opening, they shall be permitted to enter through a single fitting (such as a tee), provided the conductors are separated from the conductors of the other circuits by a continuous and firmly fixed nonconductor, such as flexible tubing.

(F) In Hoistways In hoistways, power-limited fire alarm circuit conductors shall be installed in rigid metal conduit, rigid nonmetallic conduit, intermediate metal conduit, liquidtight flexible nonmetallic conduit, or electrical metallic tubing. For elevators or similar equipment, these conductors shall be permitted to be installed as provided in 620.21.

(G) Other Applications For other applications, power-limited fire alarm circuit conductors shall be separated by at least 50 mm (2 in.) from conductors of any electric light, power, Class 1, non–power-limited fire alarm, or medium power network-powered broadband communications circuits unless one of the following conditions is met:

(1) Either (a) all of the electric light, power, Class 1, non–power-limited fire alarm, and medium power network-powered broadband communications circuit conductors or (b) all of the power-limited fire alarm circuit conductors are in a raceway or in metal-sheathed, metal-clad, nonmetallic-sheathed, or Type UF cables.

(2) All of the electric light, power, Class 1, non–power-limited fire alarm, and medium power network-powered broadband communications circuit conductors are permanently separated from all of the power-limited fire alarm circuit conductors by a continuous and firmly fixed nonconductor, such as porcelain tubes or flexible tubing, in addition to the insulation on the conductors.

760.56 Installation of Conductors of Different PLFA Circuits, Class 2, Class 3, and Communications Circuits in the Same Cable, Enclosure, or Raceway

(A) Two or More PLFA Circuits Cable and conductors of two or more power-limited fire alarm circuits, communi-

cations circuits, or Class 3 circuits shall be permitted within the same cable, enclosure, or raceway.

(B) Class 2 Circuits with PLFA Circuits Conductors of one or more Class 2 circuits shall be permitted within the same cable, enclosure, or raceway with conductors of power-limited fire alarm circuits, provided that the insulation of the Class 2 circuit conductors in the cable, enclosure, or raceway is at least that required by the power-limited fire alarm circuits.

(C) Low-Power Network-Powered Broadband Communications Cables and PLFA Cables Low-power network-powered broadband communications circuits shall be permitted in the same enclosure or raceway with PLFA cables.

(D) Audio System Circuits and PLFA Circuits Audio system circuits described in 640.9(C) and installed using Class 2 or Class 3 wiring methods in compliance with 725.54 and 725.61 shall not be permitted to be installed in the same cable or raceway with power-limited conductors or cables.

Section 760.56(D) is new in the 2005 *Code*. This section prohibits audio circuits that are installed as Class 2 or Class 3 circuits from being installed in the same cable or raceway with power-limited fire alarm wiring. A fault between audio amplifier circuits and power-limited fire alarm circuits has the potential to impair the fire alarm system.

760.57 Conductor Size

Conductors of 26 AWG shall be permitted only where spliced with a connector listed as suitable for 26 AWG to 24 AWG or larger conductors that are terminated on equipment or where the 26 AWG conductors are terminated on equipment listed as suitable for 26 AWG conductors. Single conductors shall not be smaller than 18 AWG.

See the commentary following 760.8.

760.58 Support of Conductors

Power-limited fire alarm circuit conductors shall not be strapped, taped, or attached by any means to the exterior of any conduit or other raceway as a means of support.

Due to a signaling method called *multiplexing* used with digitally addressable fire alarm systems, power-limited fire alarm cable may contain circuit conductors as small as 26 AWG. In the past, these small conductors were typically reserved for communications circuits, but due to recent technological advances, they now have application within the fire alarm industry. Of course, these small circuit conductors are permitted to be used only as specified in 760.58 and as permitted by the listing or installation instructions of specific fire alarm equipment.

760.59 Current-Carrying Continuous Line-Type Fire Detectors

(A) Application Listed continuous line-type fire detectors, including insulated copper tubing of pneumatically operated detectors, employed for both detection and carrying signaling currents shall be permitted to be used in power-limited circuits.

(B) Installation Continuous line-type fire detectors shall be installed in accordance with 760.42 through 760.52 and 760.54.

760.61 Applications of Listed PLFA Cables

PLFA cables shall comply with the requirements described in either 760.61(A), (B), or (C) or where cable substitutions are made as shown in 760.61(D).

It should be noted that 760.3(A) requires the removal of accessible abandoned fire alarm cable. Abandoned cable increases fire loading unnecessarily, and, where installed in plenums, it can affect airflow. Similar requirements can be found in Articles 640, 645, 725, 770, 800, 820, and 830. See the definition of *abandoned fire alarm cable* in 760.2.

(A) Plenum Cables installed in ducts, plenums, and other spaces used for environmental air shall be Type FPLP. Types FPLP, FPLR, and FPL cables installed in compliance with 300.22 shall be permitted. Type FPLP-CI cable shall be permitted to be installed to provide a 2-hour circuit integrity rated cable.

(B) Riser Cables installed in risers shall be as described in either (1), (2), or (3):

(1) Cables installed in vertical runs and penetrating more than one floor, or cables installed in vertical runs in a shaft, shall be Type FPLR. Floor penetrations requiring Type FPLR shall contain only cables suitable for riser or plenum use. Type FPLR-CI cable shall be permitted to be installed to provide a 2-hour circuit integrity rated cable.
(2) Other cables shall be installed in metal raceways or located in a fireproof shaft having firestops at each floor.
(3) Type FPL cable shall be permitted in one- and two-family dwellings.

   FPN: See 300.21 for firestop requirements for floor penetrations.

(C) Other Wiring Within Buildings Cables installed in building locations other than those covered in 760.61(A) or

760.61(B) shall be as described in either (1), (2), (3), or (4). Type FPL-CI cable shall be permitted to be installed as described in either (1), (2), (3), or (4) to provide a 2-hour circuit integrity rated cable.

Section 760.61(C) was added to the 2005 *Code* to permit the use of circuit integrity cable for applications where survivability of fire alarm circuits is needed during a fire. Such circuits could be essential to communicating evacuation or relocation instructions to building occupants under fire or other emergency conditions.

(1) Type FPL shall be permitted.
(2) Cables shall be permitted to be installed in raceways.
(3) Cables specified in Chapter 3 and meeting the requirements of 760.82(A) and 760.82(B) shall be permitted to be installed in nonconcealed spaces where the exposed length of cable does not exceed 3 m (10 ft).
(4) A portable fire alarm system provided to protect a stage or set when not in use shall be permitted to use wiring methods in accordance with 530.12.

(D) Fire Alarm Cable Substitutions The substitutions for fire alarm cables listed in Table 760.61 shall be permitted. Where substitute cables are installed, the wiring requirements of Article 760, Parts I and III shall apply.

Table 760.61 Cable Substitutions

Cable Type	References	Permitted Substitutions
FPLP	760.61(A)	CMP
FPLR	760.61(B)	CMP, FPLP, CMR
FPL	760.61(C)	CMP, FPLP, CMR, FPLR, CMG, CM

Type CM—Communications wires and cables
Type FPL—Power-limited fire alarm cables

[A]→[B] Cable A shall be permitted to be used in place of Cable B.
26 AWG minimum

Figure 760.61 Cable Substitution Hierarchy.

FPN: For information on communications cables (CMP, CMR, CMG, CM), see 800.179.

IV. Listing Requirements

760.81 Listing and Marking of NPLFA Cables

Non–power-limited fire alarm cables installed as wiring within buildings shall be listed in accordance with 760.81(A) and 760.81(B) and as being resistant to the spread of fire in accordance with 760.81(C) through 760.81(F), and shall be marked in accordance with 760.81(G).

(A) NPLFA Conductor Materials Conductors shall be 18 AWG or larger solid or stranded copper.

(B) Insulated Conductors Insulated conductors shall be suitable for 600 volts. Insulated conductors 14 AWG and larger shall be one of the types listed in Table 310.13 or one that is identified for this use. Insulated conductors 18 AWG and 16 AWG shall be in accordance with 760.27.

(C) Type NPLFP Type NPLFP non–power-limited fire alarm cable for use in other space used for environmental air shall be listed as being suitable for use in other space used for environmental air as described in 300.22(C) and shall also be listed as having adequate fire-resistant and low smoke-producing characteristics.

FPN: One method of defining low smoke-producing cable is by establishing an acceptable value of the smoke produced when tested in accordance with NFPA 262-2002, *Standard Method of Test for Flame Travel and Smoke of Wires and Cables for Use in Air-Handling Spaces*, to a maximum peak optical density of 0.5 and a maximum average optical density of 0.15. Similarly, one method of defining fire-resistant cables is by establishing a maximum allowable flame travel distance of 1.52 m (5 ft) when tested in accordance with the same test.

For further information on the fire test method for Type NPLFP cables, see the commentary following 725.82(A), FPN. Also see the commentary following 760.30(B)(2), Exception No. 2, which discusses spaces used for environmental air.

(D) Type NPLFR Type NPLFR non–power-limited fire alarm riser cable shall be listed as being suitable for use in a vertical run in a shaft or from floor to floor and shall also be listed as having fire-resistant characteristics capable of preventing the carrying of fire from floor to floor.

FPN: One method of defining fire-resistant characteristics capable of preventing the carrying of fire from floor to floor is that the cables pass ANSI/UL 1666-2002, *Test for Flame Propagation Height of Electrical and Optical-Fiber Cables Installed Vertically in Shafts*.

For further information on the fire test method for Type NPLFR cables, see the commentary following 725.82(B), FPN.

(E) Type NPLF Type NPLF non–power-limited fire alarm cable shall be listed as being suitable for general-purpose fire alarm use, with the exception of risers, ducts, plenums, and other space used for environmental air, and shall also be listed as being resistant to the spread of fire.

FPN No. 1: One method of defining *resistant to the spread of fire* is that the cables do not spread fire to the top of the tray in the vertical-tray flame test in ANSI/UL 1581-1991, *Reference Standard for Electrical Wires, Cables and Flexible Cords*.

FPN No. 2: Another method of defining *resistant to the spread of fire* is for the damage (char length) not to exceed 1.5 m (4 ft 11 in.) when performing the CSA vertical flame test for cables in cable trays, as described in CSA C22.2 No. 0.3-M-1985, *Test Methods for Electrical Wires and Cables*.

For further information on the fire test method for Type NPLF cables used as other wiring within buildings, see the commentary following 725.82(C), FPN.

(F) Fire Alarm Circuit Integrity (CI) Cable or Electrical Circuit Protective System Cables used for survivability of critical circuits shall be listed as circuit integrity (CI) cable. Cables specified in 760.81(C), (D), and (E), and used for circuit integrity shall have the additional classification using the suffix "-CI." Cables that are part of a listed electrical circuit protective system shall be considered to meet the requirements of survivability.

FPN No. 1: Fire alarm circuit integrity (CI) cable and electrical circuit protective systems may be used for fire alarm circuits to comply with the survivability requirements of *NFPA 72®-2002, National Fire Alarm Code®*, 6.9.4.3 and 6.9.4.6, that the circuit maintain its electrical function during fire conditions for a defined period of time.

FPN No. 2: One method of defining circuit integrity (CI) cable is by establishing a minimum 2-hour fire resistance rating for the cable when tested in accordance with UL 2196-1995, *Standard for Tests of Fire Resistive Cables*.

Circuit integrity (CI) cable is intended to meet the performance requirements for survivability required by *NFPA 72, National Fire Alarm Code*. This type of cable is designed to retain vital electrical performance during and immediately after fire exposure. Circuit integrity cable, which carries the CI suffix, is considered a 2-hour-rated cable assembly and is an alternative to fire-rated mineral-insulated cable (Type MI).

(G) NPLFA Cable Markings Multiconductor non–power-limited fire alarm cables shall be marked in accordance with Table 760.81(G). Non–power-limited fire alarm circuit cables shall be permitted to be marked with a maximum usage voltage rating of 150 volts. Cables that are listed for circuit integrity shall be identified with the suffix "CI" as defined in 760.81(F).

Table 760.81(G) NPLFA Cable Markings

Cable Marking	Type	Reference
NPLFP	Non–power-limited fire alarm circuit cable for use in "other space used for environmental air"	760.81(C) and (G)
NPLFR	Non–power-limited fire alarm circuit riser cable	760.81(D) and (G)
NPLF	Non–power-limited fire alarm circuit cable	760.81(E) and (G)

Note: Cables identified in 760.81(C), (D), and (E) and meeting the requirements for circuit integrity shall have the additional classification using the suffix "CI" (for example, NPLFP-CI, NPLFR-CI, and NPLF-CI).

FPN: Cable types are listed in descending order of fire resistance rating.

760.82 Listing and Marking of PLFA Cables and Insulated Continuous Line-Type Fire Detectors

Type FPL cables installed as wiring within buildings shall be listed as being resistant to the spread of fire and other criteria in accordance with 760.82(A) through 760.82(H) and shall be marked in accordance with 760.82(I). Insulated continuous line-type fire detectors shall be listed in accordance with 760.82(J).

(A) Conductor Materials Conductors shall be solid or stranded copper.

Some line-type fire detectors may not be made exclusively of copper but are listed for the application nevertheless.

(B) Conductor Size The size of conductors in a multiconductor cable shall not be smaller than 26 AWG. Single conductors shall not be smaller than 18 AWG.

(C) Ratings The cable shall have a voltage rating of not less than 300 volts.

(D) Type FPLP Type FPLP power-limited fire alarm plenum cable shall be listed as being suitable for use in ducts, plenums, and other space used for environmental air and

shall also be listed as having adequate fire-resistant and low smoke-producing characteristics.

FPN: One method of defining low smoke-producing cable is by establishing an acceptable value of the smoke produced when tested in accordance with NFPA 262-2002, *Standard Method of Test for Flame Travel and Smoke of Wires and Cables for Use in Air-Handling Spaces*, to a maximum peak optical density of 0.5 and a maximum average optical density of 0.15. Similarly, one method of defining fire-resistant cables is by establishing a maximum allowable flame travel distance of 1.52 m (5 ft) when tested in accordance with the same test.

For further information on the fire test method for Type FPLP cables, see the commentary following 725.82(A), FPN.

(E) Type FPLR Type FPLR power-limited fire alarm riser cable shall be listed as being suitable for use in a vertical run in a shaft or from floor to floor and shall also be listed as having fire-resistant characteristics capable of preventing the carrying of fire from floor to floor.

FPN: One method of defining fire-resistant characteristics capable of preventing the carrying of fire from floor to floor is that the cables pass the requirements of ANSI/UL 1666-2002, *Standard Test for Flame Propagation Height of Electrical and Optical-Fiber Cable Installed Vertically in Shafts*.

For further information on the fire test method for Type FPLR cables, see the commentary following 725.82(B), FPN.

(F) Type FPL Type FPL power-limited fire alarm cable shall be listed as being suitable for general-purpose fire alarm use, with the exception of risers, ducts, plenums, and other spaces used for environmental air, and shall also be listed as being resistant to the spread of fire.

FPN: One method of defining *resistant to the spread of fire* is that the cables do not spread fire to the top of the tray in the vertical-tray flame test in ANSI/UL 1581-1991, *Reference Standard for Electrical Wires, Cables and Flexible Cords*.
Another method of defining *resistant to the spread of fire* is for the damage (char length) not to exceed 1.5 m (4 ft 11 in.) when performing the CSA vertical flame test for cables in cable trays, as described in CSA C22.2 No. 0.3-M-1985, *Test Methods for Electrical Wires and Cables*.

For further information on the fire test method for Type FPL cables used as other wiring within buildings, see the commentary following 725.82(C), FPN.

(G) Fire Alarm Circuit Integrity (CI) Cable or Electrical Circuit Protective System Cables used for survivability of critical circuits shall be listed as circuit integrity (CI) cable. Cables specified in 760.82(D), (E), (F), and (H) and used for circuit integrity shall have the additional classification using the suffix "-CI." Cables that are part of a listed electrical circuit protective system shall be considered to meet the requirements of survivability.

> FPN No. 1: Fire alarm circuit integrity (CI) cable and electrical circuit protective systems may be used for fire alarm circuits to comply with the survivability requirements of *NFPA 72®-2002, National Fire Alarm Code®*, 6.9.4.3 and 6.9.4.6, that the circuit maintain its electrical function during fire conditions for a defined period of time.
>
> FPN No. 2: One method of defining circuit integrity (CI) cable is by establishing a minimum 2-hour fire resistance rating for the cable when tested in accordance with UL 2196-1995, *Standard for Tests of Fire Resistive Cables*.

There are provisions in *NFPA 72, National Fire Alarm Code*, that require continued operation of the fire alarm system, including circuit wiring, under severe conditions such as attack by fire. To provide this integrity, *NFPA 72* recognizes the use of 2-hour fire-rated cable assemblies. FPN No. 2 to 760.82(G) refers to cables tested in accordance with UL 2196, *Standard for Tests for Fire Resistive Cables*, as an example of the type of wiring method that would qualify as circuit integrity (CI) cable. For one such example of CI cable, see Exhibit 760.5.

Low-smoke flame-retardant jacket

Electrical-grade ceramifiable silicone

Foil shield   Fire barrier tape   Drain wire

Exhibit 760.5 Circuit integrity cable. (Redrawn courtesy of Rockbestos-Suprenant Cable Corp.)

(H) Coaxial Cables Coaxial cables shall be permitted to use 30 percent conductivity copper-covered steel center conductor wire and shall be listed as Type FPLP, FPLR, or FPL cable.

(I) Cable Marking The cable shall be marked in accordance with Table 760.82(I). The voltage rating shall not be marked on the cable. Cables that are listed for circuit integrity shall be identified with the suffix CI as defined in 760.82(G).

> FPN: Voltage ratings on cables may be misinterpreted to suggest that the cables may be suitable for Class 1, electric light, and power applications.

Exception: Voltage markings shall be permitted where the cable has multiple listings and voltage marking is required for one or more of the listings.

Table 760.82(I) Cable Markings

Cable Marking	Type
FPLP	Power-limited fire alarm plenum cable
FPLR	Power-limited fire alarm riser cable
FPL	Power-limited fire alarm cable

Note: Cables identified in 760.82(D), (E), and (F) as meeting the requirements for circuit integrity shall have the additional classification using the suffix "CI" (for example, FPLP-CI, FPLR-CI, and FPL-CI).

> FPN: Cable types are listed in descending order of fire-resistance rating.

(J) Insulated Continuous Line-Type Fire Detectors Insulated continuous line-type fire detectors shall be rated in accordance with 760.82(C), listed as being resistant to the spread of fire in accordance with 760.82(D) through 760.82(F), marked in accordance with 760.82(I), and the jacket compound shall have a high degree of abrasion resistance.

ARTICLE 770
Optical Fiber Cables and Raceways

Summary of Changes

- General restructuring of Article 770 to make it consistent with Articles 800, 820, and 830.
- **770.12:** Completely revised wording of 770.6 of 2002 *NEC* to become 770.12. Section 770.12(C) accepts listed optical fiber raceways as innerduct.
- **770.12(D):** Revised to require firestopping of unlisted outside plant innerduct at the point of entrance to a building or structure.
- **770.113, Exception No. 2:** Revised to specifically identify the types of Chapter 3 raceways permitted to be used with unlisted nonconductive optical fiber cables entering a building from the outside.

• **770.133(A):** Revised to permit Type ITC cable to be mixed with nonconductive optical fiber cables in the same raceway or cable tray.

Contents

I. General

770.1 Scope

The provisions of this article apply to the installation of optical fiber cables and raceways. This article does not cover the construction of optical fiber cables and raceways.

Article 770 permits the orderly development and use of optical fiber technology in conjunction with electrical conductors for communications, signaling, and control circuits in lieu of metallic conductors. The most common optical fiber cable used in buildings is nonconductive. (See Exhibit 770.1.)

Because they are not affected by electrical noise, optical fiber cables may be desirable in some circumstances to transmit data or other communications where electrical noise is a problem. Optical fiber cables may be nonconductive, or they may contain electrical conductors. See Exhibits 770.1 and 770.2.

Exhibit 770.1 An example of a nonconductive optical fiber cable.

770.2 Definitions

Abandoned Optical Fiber Cable. Installed optical fiber cable that is not terminated at equipment other than a connector and not identified for future use with a tag.

Exhibit 770.2 An example of a composite optical fiber cable. This cable also meets the requirements of Article 330 and is referred to as Type MC cable. (Courtesy of AFC Cable Systems, Inc.)

The definition of *abandoned optical fiber cable* is for use with 770.3(A), which requires removal of accessible abandoned optical fiber cable. Abandoned cable increases fire loading unnecessarily, and, where installed in plenums, it can affect airflow. Similar requirements can be found in Articles 640, 645, 725, 760, 800, 820, and 830.

Exposed. The circuit is in such a position that, in case of failure of supports and insulation, contact with another circuit may result.

> FPN: See Article 100 for two other definitions of *Exposed.*

Optical Fiber Raceway. A raceway designed for enclosing and routing listed optical fiber cables.

Point of Entrance. The point at which the cable emerges from an external wall, from a concrete floor slab, or from a rigid metal conduit or an intermediate metal conduit grounded to an electrode in accordance with 800.100(B).

770.3 Other Articles

Circuits and equipment shall comply with 770.3(A) and 770.3(B). Only those sections of Article 300 referenced in this article shall apply to optical fiber cables and raceways.

(A) Spread of Fire or Products of Combustion The requirements of 300.21 for electrical installations shall also apply to installations of optical fiber cables and raceways. The accessible portion of abandoned optical fiber cables shall be removed.

(B) Ducts, Plenums, and Other Air-Handling Spaces The requirements of 300.22 for electric wiring shall also apply to installations of optical fiber cables and raceways where they are installed in ducts or plenums or other space used for environmental air.

Exception: As permitted in 770.154(A).

See the commentary following 300.22(B) and 300.22(C) for more information on wiring systems installed in ducts, plenums, or other spaces used for environmental air.

770.6 Optical Fiber Cables

Optical fiber cables transmit light for control, signaling, and communications through an optical fiber.

770.9 Types

Optical fiber cables can be grouped into three types.

(A) Nonconductive These cables contain no metallic members and no other electrically conductive materials.

(B) Conductive These cables contain non–current-carrying conductive members such as metallic strength members, metallic vapor barriers, and metallic armor or sheath.

(C) Composite These cables contain optical fibers and current-carrying electrical conductors, and shall be permitted to contain non–current-carrying conductive members such as metallic strength members and metallic vapor barriers. Composite optical fiber cables shall be classified as electrical cables in accordance with the type of electrical conductors.

770.12 Raceways for Optical Fiber Cables

Installations of raceways shall comply with 770.12(A) through 770.12(D).

(A) Listed Chapter 3 Raceways Listed optical fiber cable shall be permitted to be installed in any type of listed raceway permitted in Chapter 3 where that listed raceway is installed in accordance with Chapter 3. Where optical fiber cables are installed within raceway without current-carrying conductors, the raceway fill tables of Chapter 3 and Chapter 9 shall not apply. Where nonconductive optical fiber cables are installed with electric conductors in a raceway, the raceway fill tables of Chapter 3 and Chapter 9 shall apply.

Conduit fill requirements apply where the optical fiber is installed in a raceway with electrical conductors.

(B) Optical Fiber Raceways Listed optical fiber cable shall be permitted to be installed in listed plenum optical fiber raceway, listed riser optical fiber raceway, or listed general-purpose optical fiber raceway installed in accordance with 770.154 and 362.24 through 362.56, where the requirements applicable to electrical nonmetallic tubing shall apply.

(C) Innerduct Listed plenum optical fiber raceway, listed riser optical fiber raceway, or listed general-purpose optical fiber raceway installed in accordance with 770.154 shall be permitted to be installed as innerduct in any type of listed raceway permitted in Chapter 3.

(D) Entering Buildings Unlisted underground or outside plant construction plastic innerduct entering the building from the outside shall be terminated and firestopped at the point of entrance.

770.21 Access to Electrical Equipment Behind Panels Designed to Allow Access

Access to electrical equipment shall not be denied by an accumulation of cables that prevents removal of panels, including suspended ceiling panels.

See the commentary following 725.7 for information on safe access to electrical equipment behind panels.

770.24 Mechanical Execution of Work

Optical fiber cables shall be installed in a neat and workmanlike manner. Cables installed exposed on the surface of ceilings and sidewalls shall be supported by the building structure in such a manner that the cable will not be damaged by normal building use. Such cables shall be secured by straps, staples, hangers, or similar fittings designed and installed so as not to damage the cable. The installation shall also conform with 300.4(D) and 300.11.

> FPN: Accepted industry practices are described in ANSI/NECA/BICSI 568-2001, *Standard for Installing Commercial Building Telecommunications Cabling*, and other ANSI-approved installation standards.

Section 770.24 makes it clear that the *Code* requires optical fiber cables to be installed in a neat and workmanlike manner. Revised for the 2002 *Code*, this section provides definitive requirements for workmanship. Cable must be attached to or supported by the structure by straps, clamps, hangers, and the like. The installation method must not damage the cable. In addition, the location of the cable should be carefully evaluated to ensure that activities and processes within the building do not cause damage to the cable. The reference to 300.4(D) calls attention to the hazard to which cables are exposed where they are installed on framing members. Such cables are required to be installed in a manner that protects them from nail or screw penetration.

II. Protection

770.93 Grounding of Entrance Cables

Where exposed to contact with electric light or power conductors, the non–current-carrying metallic members of optical fiber cables entering buildings shall be grounded as close to the point of entrance as practicable or shall be interrupted as close to the point of entrance as practicable by an insulating joint or equivalent device.

III. Cables Within Buildings

770.113 Installation and Marking of Listed Optical Fiber Cables

Listed optical fiber cables shall be installed as wiring within buildings. Optical fiber cables shall be marked in accordance with Table 770.113.

Table 770.113 Cable Markings

Cable Marking	Type	Reference
OFNP	Nonconductive optical fiber plenum cable	770.179(A) and 770.154
OFCP	Conductive optical fiber plenum cable	770.179(A) and 770.154(A)
OFNR	Nonconductive optical fiber riser cable	770.179(B) and 770.154(B)
OFCR	Conductive optical fiber riser cable	770.179(B) and 770.154(B)
OFNG	Nonconductive optical fiber general-purpose cable	770.179(C) and 770.154(C)
OFCG	Conductive optical fiber general-purpose cable	770.179(C) and 770.154(C)
OFN	Nonconductive optical fiber general-purpose cable	770.179(D) and 770.154(C)
OFC	Conductive optical fiber general-purpose cable	770.179(D) and 770.154(C)

FPN No. 1: Cables types are listed in descending order of fire resistance rating. Within each fire resistance rating, nonconductive cable is listed first because it may substitute for the conductive cable.

FPN No. 2: See the referenced sections for requirements and permitted uses.

Exception No. 1: Optical fiber cables shall not be required to be listed and marked where the length of the cable within the building, measured from its point of entrance, does not exceed 15 m (50 ft) and the cable enters the building from the outside and is terminated in an enclosure.

> FPN: Splice cases or terminal boxes, both metallic and plastic types, typically are used as enclosures for splicing or terminating optical fiber cables.

Exception No. 2: Nonconductive optical fiber cables shall not be required to be listed and marked where the cable enters the building from the outside and is run in raceway systems installed in compliance with any of the following articles in Chapter 3: Article 342, Intermediate Metal Conduit: Type IMC; Article 344, Rigid Metal Conduit: Type RMC; Article 352, Rigid Nonmetallic Conduit: Type RNC; and Article 358, Electrical Metallic Tubing: Type EMT.

770.133 Installation of Optical Fibers and Electrical Conductors

(A) With Conductors for Electric Light, Power, Class 1, Non–Power-Limited Fire Alarm, or Medium Power Network-Powered Broadband Communications Circuits Optical fibers shall be permitted within the same composite cable for electric light, power, Class 1, non–power-limited fire alarm, or medium power network-powered broadband communications circuits operating at 600 volts or less only where the functions of the optical fibers and the electrical conductors are associated.

Nonconductive optical fiber cables shall be permitted to occupy the same cable tray or raceway with conductors for electric light, power, Class 1, non–power-limited fire alarm, Type ITC, or medium power network-powered broadband communications circuits, operating at 600 volts or less. Conductive optical fiber cables shall not be permitted to occupy the same cable tray or raceway with conductors for electric light, power, Class 1, non–power-limited fire alarm, Type ITC, or medium power network-powered broadband communications circuits.

Composite optical fiber cables containing only current-carrying conductors for electric light, power, Class 1 circuits rated 600 volts or less shall be permitted to occupy the same cabinet, cable tray, outlet box, panel, raceway, or other termination enclosure with conductors for electric light, power, or Class 1 circuits operating at 600 volts or less.

Nonconductive optical fiber cables shall not be permitted to occupy the same cabinet, outlet box, panel, or similar enclosure housing the electrical terminations of an electric light, power, Class 1, non–power-limited fire alarm, or medium power network-powered broadband communications circuit.

Exception No. 1: Occupancy of the same cabinet, outlet box, panel, or similar enclosure shall be permitted where nonconductive optical fiber cable is functionally associated with the electric light, power, Class 1, non–power-limited fire alarm, or medium power network-powered broadband communications circuit.

Exception No. 2: Occupancy of the same cabinet, outlet box, panel, or similar enclosure shall be permitted where nonconductive optical fiber cables are installed in factory- or field-assembled control centers.

Exception No. 3: In industrial establishments only, where conditions of maintenance and supervision ensure that only

qualified persons service the installation, nonconductive optical fiber cables shall be permitted with circuits exceeding 600 volts.

Exception No. 4: In industrial establishments only, where conditions of maintenance and supervision ensure that only qualified persons service the installation, composite optical fiber cables shall be permitted to contain current-carrying conductors operating over 600 volts.

(B) With Other Conductors Optical fibers shall be permitted in the same cable, and conductive and nonconductive optical fiber cables shall be permitted in the same cable tray, enclosure, or raceway with conductors of any of the following:

(1) Class 2 and Class 3 remote-control, signaling, and power-limited circuits in compliance with Article 725
(2) Power-limited fire alarm systems in compliance with Article 760
(3) Communications circuits in compliance with Article 800
(4) Community antenna television and radio distribution systems in compliance with Article 820
(5) Low-power network-powered broadband communications circuits in compliance with Article 830

(C) Grounding Non–current-carrying conductive members of optical fiber cables shall be grounded in accordance with Article 250.

770.154 Applications of Listed Optical Fiber Cables and Raceways

Nonconductive and conductive optical fiber cables shall comply with any of the requirements given in 770.154(A) through 770.154(E) or where cable substitutions are made as shown in 770.154(F).

It should be noted that 770.3(A) requires the removal of accessible abandoned optical fiber cable. Abandoned cable increases fire loading unnecessarily, and, where installed in plenums, it can affect airflow. Similar requirements can be found in Articles 640, 645, 725, 760, 800, 820, and 830. See the definition of *abandoned communications cable* in 800.2.

(A) Plenums Cables installed in ducts, plenums, and other spaces used for environmental air shall be Type OFNP or OFCP. Abandoned cables shall not be permitted to remain. Types OFNR, OFCR, OFNG, OFN, OFCG, and OFC cables installed in compliance with 300.22 shall be permitted. Listed plenum optical fiber raceways shall be permitted to be installed in ducts and plenums as described in 300.22(B) and in other spaces used for environmental air as described in 300.22(C). Only type OFNP and OFCP cables shall be permitted to be installed in these raceways.

> FPN: See 8.14.1 of NFPA 13 (2002), *Installation of Sprinkler Systems*, for requirements for sprinklers in concealed spaces containing exposed combustibles.

(B) Riser Cables installed in risers shall be as described in any of the following:

(1) Cables installed in vertical runs and penetrating more than one floor, or cables installed in vertical runs in a shaft, shall be Type OFNR or OFCR. Floor penetrations requiring Type OFNR or OFCR shall contain only cables suitable for riser or plenum use. Abandoned cables shall not be permitted to remain. Listed riser optical fiber raceways shall be permitted to be installed in vertical riser runs in a shaft from floor to floor. Only Type OFNP, OFCP, OFNR, and OFCR cables shall be permitted to be installed in these raceways.

(2) Type OFNG, OFN, OFCG, and OFC cables shall be permitted to be encased in a metal raceway or located in a fireproof shaft having firestops at each floor.

(3) Type OFNG, OFN, OFCG, and OFC cables shall be permitted in one- and two-family dwellings.

> FPN: See 300.21 for firestop requirements for floor penetrations.

(C) Other Wiring Within Buildings Cables installed in building locations other than the locations covered in 770.154(A) and 770.154(B) shall be Type OFNG, OFN, OFCG, or OFC. Such cables shall be permitted to be installed in listed general-purpose optical fiber raceways.

(D) Hazardous (Classified) Locations Cables installed in hazardous (classified) locations shall be any type indicated in Table 770.154.

Table 770.154 Cable Substitutions

Cable Type	Permitted Substitutions
OFNP	None
OFCP	OFNP
OFNR	OFNP
OFCR	OFNP, OFCP, OFNR
OFNG, OFN	OFNP, OFNR
OFCG, OFC	OFNP, OFCP, OFNR, OFCR, OFNG, OFN

(E) Cable Trays Optical fiber cables of the types listed in Table 770.113 shall be permitted to be installed in cable trays.

> FPN: It is not the intent to require that these optical fiber cables be listed specifically for use in cable trays.

(F) Cable Substitutions The substitutions for optical fiber cables listed in Table 770.154 shall be permitted.

Figure 770.154 Cable Substitution Hierarchy.

IV. Listing Requirements

770.179 Optical Fiber Cables

Optical fiber cables shall be listed in accordance with 770.179(A) through 770.179(D).

(A) Types OFNP and OFCP Types OFNP and OFCP nonconductive and conductive optical fiber plenum cables shall be listed as being suitable for use in ducts, plenums, and other space used for environmental air and shall also be listed as having adequate fire resistant and low smoke producing characteristics.

> FPN: One method of defining a cable that is low smoke producing cable and fire-resistant cable is that the cable exhibits a maximum peak optical density of 0.5 or less, an average optical density of 0.15 or less, and a maximum flame spread distance of 1.52 m (5 ft) or less when tested in accordance with NFPA 262–2002, *Standard Method of Test for Flame Travel and Smoke of Wires and Cables for Use in Air-Handling Spaces*.

For further information on the fire test method for optical fiber plenum cables, see the commentary following 725.82(A), FPN.

(B) Types OFNR and OFCR Types OFNR and OFCR nonconductive and conductive optical fiber riser cables shall be listed as being suitable for use in a vertical run in a shaft or from floor to floor and shall also be listed as having the fire-resistant characteristics capable of preventing the carrying of fire from floor to floor.

> FPN: One method of defining fire-resistant characteristics capable of preventing the carrying of fire from floor to floor is that the cables pass the requirements of ANSI/UL 1666-2002, *Standard Test for Flame Propagation*

Height of Electrical and Optical-Fiber Cable Installed Vertically in Shafts.

For further information on the fire test method for optical fiber riser cables, see the commentary following 725.82(B), FPN.

(C) Types OFNG and OFCG Types OFNG and OFCG nonconductive and conductive general-purpose optical fiber cables shall be listed as being suitable for general-purpose use, with the exception of risers and plenums, and shall also be listed as being resistant to the spread of fire.

> FPN: One method of defining *resistance to the spread of fire* is for the damage (char length) not to exceed 1.5 m (4 ft 11 in.) when performing the vertical flame test for cables in cable trays, as described in CSA C22.2 No. 0.3-M-1985, *Test Methods for Electrical Wires and Cables.*

For further information on the fire test method for optical fiber cables used as other wiring within buildings, see the commentary following 725.82(C), FPN.

(D) Types OFN and OFC Types OFN and OFC nonconductive and conductive optical fiber cables shall be listed as being suitable for general-purpose use, with the exception of risers, plenums, and other spaces used for environmental air, and shall also be listed as being resistant to the spread of fire.

> FPN: One method of defining *resistant to the spread of fire* is that the cables do not spread fire to the top of the tray in the vertical-tray flame test in ANSI/UL 1581-1991, *Reference Standard for Electrical Wires, Cables, and Flexible Cords.*
>
> Another method of defining *resistant to the spread of fire* is for the damage (char length) not to exceed 1.5 m (4 ft 11 in.) when performing the vertical flame test for cables in cable trays, as described in CSA C22.2 No. 0.3-M-1985, *Test Methods for Electrical Wires and Cables.*

770.182 Optical Fiber Raceways

Optical fiber raceways shall be listed in accordance with 770.182(A) through 770.182(C).

The optical fiber raceways covered in 770.182 are listed raceways used in plenum, riser, or general-purpose applications. This listing includes raceways and fittings for installations of nonconductive optical fiber cables in accordance with Article 770. These raceways are not suitable for installation of current-carrying conductors, cords, or cables. Nor are these raceways suitable for installations of hybrid cables that contain optical fiber members and current-carrying conductors.

Plenum Raceway

A raceway marked "plenum" is suitable for use in ducts, plenums, or other spaces used for environmental air in accordance with 770.154(A) where used to enclose optical fiber cables marked OFNP. This plenum raceway exhibits a maximum peak optical density of 0.5, a maximum average optical density of 0.15, and a maximum flame-spread distance of 5 ft when tested in accordance with the plenum test in UL 2024, *Standard for Optical Fiber Cable Raceway.* This raceway is identified by a marking on the surface of the raceway or on a marker tape indicating "plenum." A raceway marked "plenum" is also suitable for installation in risers where used to enclose optical fiber cables marked OFNP or OFNR and for general-purpose use where used to enclose optical fiber cables marked OFNP, OFNR, OFNG, or OFN.

Riser Raceway

A raceway marked "riser" is suitable for installation in risers in accordance with 770.154(B) where used to enclose optical fiber cable marked OFNP or OFNR. This raceway has fire-resistant characteristics capable of preventing the carrying of fire from floor to floor. Riser raceway meets the flame propagation test requirements in UL 2024, *Standard for Optical Fiber Cable Raceway.* This raceway is identified by a marking on the surface of the raceway or on a marker tape indicating "riser." A raceway marked "riser" is also suitable for general-purpose use when used to enclose optical fiber cable marked OFNP, OFNR, OFNG, or OFN.

General-Purpose Raceway

A raceway marked "general purpose" is suitable for installation in general-purpose areas in accordance with 770.154(C) where used to enclose optical fiber cable marked OFNP, OFNR, OFNG, or OFN. General-purpose raceway has fire-resistant characteristics that are capable of preventing the spread of fire as determined by the Vertical-Tray Flame Test in UL 2024, *Standard for Optical Fiber Raceway.*

Pliable raceway is raceway that can be bent by hand without the use of tools. The smallest radius of the curve of the inner edge of any bend to which the raceway may be bent without cracking either on the outer surface or internally is not less than 2½ times the outside diameter of the raceway.

(A) Plenum Optical Fiber Raceway Plenum optical fiber raceways shall be listed as having adequate fire-resistant and low smoke-producing characteristics.

> FPN: One method of defining that an optical fiber raceway is a low smoke producing raceway and a fire-resistant raceway is that the raceway exhibits a maximum peak optical density of 0.5 or less, an average optical density of 0.15 or less, and a maximum flame spread distance of 1.52 m (5 ft) or less when tested in accordance

with the plenum test in UL 2024, *Standard for Optical Fiber Cable Raceway.*

(B) Riser Optical Fiber Raceway Riser optical fiber raceways shall be listed as having fire-resistant characteristics capable of preventing the carrying of fire from floor to floor.

FPN: One method of defining fire-resistant characteristics capable of preventing the carrying of fire from floor to floor is that the raceways pass the requirements of the test for Flame Propagation (riser) in UL 2024, *Standard for Optical Fiber Cable Raceway.*

(C) General-Purpose Optical Fiber Cable Raceway General-purpose optical fiber cable raceway shall be listed as being resistant to the spread of fire.

FPN: One method of defining resistance to the spread of fire is that the raceways pass the requirements of the Vertical-Tray Flame Test (General Use) in UL 2024, *Standard for Optical Fiber Cable Raceway.*

ARTICLE 780
Closed-Loop and Programmed Power Distribution

Contents

780.1 Scope

The provisions of this article apply to premise power distribution systems jointly controlled by a signaling between the energy controlling equipment and utilization equipment.

Article 780 provides requirements for the "smart house" concept, which involves universal cable terminating in universal outlets.

Buildings wired by conventional methods require separate sets of conductors for different systems. In a smart house, however, multiple conductors for 120-volt ac power, 24-volt dc UPS, telephone, remote-control, and signaling, as well as coaxial cable, are combined in a single construction known as *hybrid cabling.*

Hybrid cabling serves multipurpose receptacle outlets known as *convenience centers*, which are capable of supplying different types of energy and signals to specific appliances or equipment.

A smart house uses an energy safety technique called *closed-loop control* to reduce shock hazard. In conventional wiring, receptacles are energized at all times under normal operating conditions. In the closed-loop configuration, receptacles are not energized until the insertion of an attachment plug generates a characteristic electrical identification.

Exhibit 780.1 illustrates a typical smart house installation. Present smart house technology uses 120/240-volts ac, with 24-volts dc UPS, to maintain system electronics in the event of a transient or utility power outage.

780.2 General

(A) Other Articles Except as modified by the requirements of this article, all other applicable articles of this *Code* shall apply.

(B) Component Parts All equipment and conductors shall be listed and identified.

780.3 Control

The control equipment and all power switching devices operated by the control equipment shall be listed and identified. The system shall operate in accordance with 780.3(A) through 780.3(D).

(A) Characteristic Electrical Identification Required Outlets shall not be energized unless the utilization equipment first exhibits a characteristic electrical identification.

Receptacles are energized with 120-volt ac power only when electronic circuitry in the convenience center receives this characteristic identification.

(B) Conditions for De-Energization Outlets shall be de-energized when any of the following conditions occur:

(1) A nominal-operation acknowledgment signal is not being received from the utilization equipment connected to the outlet.
(2) A ground-fault condition exists.

Convenience center receptacles are de-energized when the characteristic electrical identification ceases (when the attachment plug is withdrawn). In addition, appliances with

Exhibit 780.1 A typical smart house installation. (Courtesy of Smart House)

built-in smart house communications chips transmit a continuous nominal-operation signal to the convenience center electronics. If this signal is interrupted, indicating a possible malfunction or safety problem, the receptacle is automatically de-energized.

(3) An overcurrent condition exists.

(C) Additional Conditions for De-Energization When an Alternate Source of Power Is Used In addition to the requirements in 780.3(B), outlets shall be de-energized when any of the following conditions occur:

(1) The grounded conductor is not properly grounded.
(2) Any ungrounded conductor is not at nominal voltage.

(D) Controller Malfunction In the event of a controller malfunction, all associated outlets shall be de-energized.

780.5 Power Limitation in Signaling Circuits

For signaling circuits not exceeding 24 volts, the current required shall not exceed 1 ampere where protected by an overcurrent device or an inherently limited power source.

780.6 Cables and Conductors

(A) Hybrid Cable Listed hybrid cable consisting of power, communications, and signaling conductors shall be permitted under a common jacket. The jacket shall be applied so as to separate the power conductors from the communications and signaling conductors. An optional outer jacket shall be permitted to be applied. The individual conductors of a hybrid cable shall conform to the *Code* provisions applicable to their current, voltage, and insulation rating. The signaling conductors shall not be smaller than 24 AWG copper.

(B) Cables and Conductors in the Same Cabinet, Panel, or Box The power, communications, and signaling conductors of listed hybrid cable are permitted to occupy the same cabinet, panel, or outlet box (or similar enclosure housing the electrical terminations of electric light or power circuits) only if connectors specifically listed for hybrid cable are employed.

780.7 Noninterchangeability

Receptacles, cord connectors, and attachment plugs used on closed-loop power distribution systems shall be constructed so that they are not interchangeable with other receptacles, cord connectors, and attachment plugs.

Convenience center receptacles are constructed so that they do not accept an attachment plug with a different voltage or current rating than that for which the device is intended. Attachment plugs for use with closed-loop power distribution systems do not fit into conventional receptacles, which ensures that "smart" appliances are not used on other power distribution systems that lack closed-loop control features.

Communications Systems

ARTICLE 800
Communications Circuits

Summary of Changes

- Reorganized and renumbered as part of an effort to achieve parallel structure among Articles 770, 800, 820, and 830.
- **800.2:** Definitions for *air duct, communications circuit integrity (CI) cable,* and *communications equipment* added.
- **800.24:** Removed requirement that the cable must be supported by a structural component. Clarification added that this section applies to ceilings regardless of which side of the ceiling is supporting the cable. Revised to apply the requirement of 300.11 to Article 800 cables.
- **800.100A(4):** FPN added to clarify the requirement of a 20 ft maximum grounding conductor in one- and two-family dwellings.
- **800.100(D):** Revised to remove permission to bond together all separate electrodes.
- **Table 800.113:** Deleted multipurpose cable Types MPP, MPR, MPG, and MP.
- **Figure 800.154:** Revised to delete multipurpose cable Types MPP, MPR, MPG, and MP.
- **800.179:** Revised to delete multipurpose cable Types MPP, MPR, MPG, and MP.
- **800.179(H):** Revised to permit a 2-hour fire-rated communications circuit integrity cable.

Contents

(E) Type CMX
(F) Type CMUC Undercarpet Wire and Cable
(G) Multipurpose (MP) Cables
(H) Communications Circuit Integrity (CI) Cable
(I) Wires
(J) Hybrid Power and Communications Cable
800.182 Communications Raceways
(A) Plenum Communications Raceways
(B) Riser Communications Raceways
(C) General-Purpose Communications Raceways

FPN: Rules that are followed by a reference in brackets contain text that has been extracted from NFPA 97–2003, *Standard Glossary of Terms Relating to Chimneys, Vents, and Heat-Producing Appliances*. Only editorial changes were made to the extracted text to make it consistent with this *Code*.

I. General

800.1 Scope

This article covers telephone, telegraph (except radio), outside wiring for fire alarm and burglar alarm, and similar central station systems; and telephone systems not connected to a central station system but using similar types of equipment, methods of installation, and maintenance.

Section 90.3, Code Arrangement, states that Chapter 8, which comprises Articles 800, 810, 820, and 830, covers communications systems and is not subject to the requirements of Chapters 1–7 except where a requirement from these chapters is specifically referenced in Chapter 8. For instance, 800.44(A)(3) references 225.14(D), 800.90(C) references Article 500, and 800.3(D) references 300.22(C).

Although information technology equipment systems are often used for or with communications systems, Article 800 does not cover wiring of this equipment. Instead, Article 645 provides requirements for wiring contained solely within an information technology equipment (computer) room. (See 645.4 for a description of the type of information technology equipment room to which Article 645 applies.) Article 725 provides requirements for wiring that extends beyond a computer room and also covers wiring of local area networks within buildings. Article 760 covers wiring requirements for fire alarm systems.

In some cases, telephone system wiring is also used for data transmission; this use is covered by Article 800. Telephone company central offices are exempt from the requirements of Article 800 by 90.2(B)(4). The format of Article 800 is similar to that of Articles 725, 760, 770, and 820.

Article 830 covers network-powered broadband communications systems.

FPN No. 1: For further information for fire alarm, sprinkler waterflow, and sprinkler supervisory systems, see Article 760.
FPN No. 2: For installation requirements of optical fiber cables, see Article 770.
FPN No. 3: For installation requirements for network-powered broadband communications circuits, see Article 830.

800.2 Definitions

See Article 100. For purposes of this article, the following additional definitions apply.

Abandoned Communications Cable. Installed communications cable that is not terminated at both ends at a connector or other equipment and not identified for future use with a tag.

The term *abandoned communications cable* applies to 800.154, which requires removal of accessible abandoned communications cable. Abandoned cable increases fire loading unnecessarily, and, where installed in plenums, it can affect airflow. Similar requirements can be found in Articles 640, 645, 725, 760, 770, 820, and 830.

Air Duct. A conduit or passageway for conveying air to or from heating, cooling, air conditioning, or ventilating equipment, but not including the plenum. [NFPA 97:1.2.6]

The definition of *air duct* was added to the 2005 *Code* to provide a term to distinguish between electrical ducts and ducts that form part of an environmental air distribution system.

Block. A square or portion of a city, town, or village enclosed by streets and including the alleys so enclosed, but not any street.

Cable. A factory assembly of two or more conductors having an overall covering.

Cable Sheath. A covering over the conductor assembly that may include one or more metallic members, strength members, or jackets.

Communications Circuit Integrity (CI) Cable. Cable used in communications systems to ensure continued operation of critical circuits during a specified time under fire conditions.

The definition of *communications circuit integrity (CI) cable* was added to the 2005 *Code* to define a term used in 800.179(H). CI cables are used to maintain communications throughout the entire time of an emergency. Such cable is intended to ensure the survivability of certain critical communications circuits during a fire in a building.

Communications Equipment. The electronic equipment that performs the telecommunications operations for the transmission of audio, video, and data, and including power equipment (e.g., dc converters, inverters and batteries) and technical support equipment (e.g., computers).

The definition of *communications equipment* was added to the 2005 *Code* to clearly define what associated equipment is considered part of the communications equipment. The definition clearly indicates that the power supplies and computers are considered part of the communications equipment and thus are subject to requirements that apply to communications equipment. The telephone switch shown in Exhibit 800.1 also is considered part of telecommunications equipment and so is subject to the same requirements. The definition correlates with NFPA 76, *Recommended Practice for the Fire Protection of Telecommunications Facilities*.

Exhibit 800.1 A private automatic branch exchange, one part of telecommunications equipment.

Exposed. A circuit that is in such a position that, in case of failure of supports and insulation, contact with another circuit may result.

> FPN: See Article 100 for two other definitions of *Exposed*.

Point of Entrance. Within a building, the point at which the wire or cable emerges from an external wall, from a concrete floor slab, or from a rigid metal conduit or an intermediate metal conduit grounded to an electrode in accordance with 800.100(B).

Premises. The land and buildings of a user located on the user side of the utility-user network point of demarcation.

Wire. A factory assembly of one or more insulated conductors without an overall covering.

See Article 100 for the definitions of *conductor*, *equipment*, and *raceway*.

800.3 Other Articles

(A) Hybrid Power and Communications Cables The provisions of 780.6 shall apply for listed hybrid power and communications cables in closed-loop and programmed power distribution.

See 800.179(J) for listing requirements and applications of hybrid power and communications cable in one- and two-family residences for other than closed-loop and programmed power distribution.

> FPN: See 800.179(J) for hybrid power and communications cable in other applications.

(B) Hazardous (Classified) Locations Communications circuits and equipment installed in a location that is classified in accordance with Article 500 shall comply with the applicable requirements of Chapter 5.

Paragraph 800.3(B) alerts users that communications circuits installed in locations classified in accordance with Article 500 must conform to the applicable requirements of Chapter 5.

(C) Spread of Fire or Products of Combustion Section 300.21 shall apply. The accessible portion of abandoned communications cables shall not be permitted to remain.

Section 800.3(C) was revised for the 2005 *Code* for use with the definition of *abandoned communications cable* in

800.2. Section 800.3(C) requires the removal of accessible abandoned communications cable. Abandoned cable increases fire loading unnecessarily, and, where installed in plenums, it can affect airflow. Similar requirements can be found in Articles 640, 645, 725, 760, 770, 820, and 830. (See the definition of *abandoned communications cable* in 800.2.)

(D) Equipment in Other Space Used for Environmental Air Section 300.22(C) shall apply.

800.18 Installation of Equipment

Equipment electrically connected to a telecommunications network shall be listed in accordance with 800.170. Installation of equipment shall also comply with 110.3(B).

UL 1863, *Communication Circuit Accessories*, and UL 60950, *Safety of Information Technology Equipment, Part 1: General Requirements*, are two safety standards that contain requirements for determining whether equipment connected to a telecommunications network is suitable for the intended purpose. Listed equipment that is connected to the telecommunications network and evaluated according to other U.S. safety standards is also subject to telecommunications requirements appropriate for the equipment. Examples of this equipment include information technology equipment, audio-video equipment, and signaling equipment connected to a central station. The appropriate requirements contained within the applicable safety standard are extracted from UL 1863, UL 60950, or both.

Except for test equipment, all permanently installed electrical components of the communications network are subject to the listing requirements of 800.170.

Exception: This listing requirement shall not apply to test equipment that is intended for temporary connection to a telecommunications network by qualified persons during the course of installation, maintenance, or repair of telecommunications equipment or systems.

800.21 Access to Electrical Equipment Behind Panels Designed to Allow Access

Access to electrical equipment shall not be denied by an accumulation of wires and cables that prevents removal of panels, including suspended ceiling panels.

An excess accumulation of wires and cables can limit access to equipment by preventing the removal of access panels. (See Exhibit 800.2.)

Exhibit 800.2 Installations of conductors and cables, which can prevent access to equipment or cables. Correct and incorrect methods are shown.

800.24 Mechanical Execution of Work

Communications circuits and equipment shall be installed in a neat and workmanlike manner. Cables installed exposed on the surface of ceilings and sidewalls shall be supported by the building structure in such a manner that the cable will not be damaged by normal building use. Such cables shall be secured by straps, staples, hangers, or similar fittings designed and installed so as not to damage the cable. The installation shall also conform with 300.4(D) and 300.11.

Section 800.24 provides definitive requirements for workmanship. Cable must be attached to or supported by the structure by straps, clamps, hangers, and the like. The installation method must not damage the cable. In addition, the location of the cable must be carefully evaluated to ensure that activities and processes within the building do not damage the cable. In the 2005 *Code*, there was a change to this section to permit attachment to baseboards and non-load bearing walls, which are not structural components. The equipment illustrated in Exhibit 800.3 is used by installers of telecommunications systems to organize cables and make connections in a neat and workmanlike manner.

FPN: Accepted industry practices are described in ANSI/NECA/BICSI 568-2001, *Standard for Installing Commercial Building Telecommunications Cabling,* and other ANSI-approved installation standards.

Exhibit 800.3 An example of punch blocks used on a private automatic branch exchange.

II. Wires and Cables Outside and Entering Buildings

800.44 Overhead Communications Wires and Cables

Overhead communications wires and cables entering buildings shall comply with 800.44(A) and 800.44(B).

(A) On Poles and In-Span Where communications wires and cables and electric light or power conductors are supported by the same pole or run parallel to each other in-span, the conditions described in 800.44(A)(1) through (A)(4) shall be met.

(1) Relative Location Where practicable, the communications wires and cables shall be located below the electric light or power conductors.

(2) Attachment to Crossarms Communications wires and cables shall not be attached to a cross-arm that carries electric light or power conductors.

(3) Climbing Space The climbing space through communications wires and cables shall comply with the requirements of 225.14(D).

(4) Clearance Supply service drops of 0–750 volts running above and parallel to communications service drops shall have a minimum separation of 300 mm (12 in.) at any point in the span, including the point of and at their attachment to the building, provided the nongrounded conductors are insulated and that a clearance of not less than 1.0 m (40 in.) is maintained between the two services at the pole.

(B) Above Roofs Communications wires and cables shall have a vertical clearance of not less than 2.5 m (8 ft) from all points of roofs above which they pass.

Exception No. 1: Auxiliary buildings, such as garages and the like.

Exception No. 2: A reduction in clearance above only the overhanging portion of the roof to not less than 450 mm (18 in.) shall be permitted if (a) not more than 1.2 m (4 ft) of communications service-drop conductors pass above the roof overhang and (b) they are terminated at a through- or above-the-roof raceway or approved support.

Exception No. 3: Where the roof has a slope of not less than 100 mm in 300 mm (4 in. in 12 in.), a reduction in clearance to not less than 900 mm (3 ft) shall be permitted.

> FPN: For additional information regarding overhead wires and cables, see ANSI C2-2002, *National Electric Safety Code*, Part 2, Safety Rules for Overhead Lines.

800.47 Underground Circuits Entering Buildings

Underground communications wires and cables entering buildings shall comply with 800.47(A) and 800.47(B).

(A) With Electric Light or Power Conductors Underground communications wires and cables in a raceway, handhole enclosure, or manhole containing electric light, power, Class 1, or non–power-limited fire alarm circuit conductors shall be in a section separated from such conductors by means of brick, concrete, or tile partitions or by means of a suitable barrier.

(B) Underground Block Distribution Where the entire street circuit is run underground and the circuit within the block is placed so as to be free from the likelihood of accidental contact with electric light or power circuits of over 300 volts to ground, the insulation requirements of 800.50(A) and 800.50(C) shall not apply, insulating supports shall not be required for the conductors, and bushings shall not be required where the conductors enter the building.

800.50 Circuits Requiring Primary Protectors

Circuits that require primary protectors as provided in 800.90 shall comply with 800.50(A), (B), and (C).

(A) Insulation, Wires, and Cables Communications wires and cables without a metallic shield, running from the last outdoor support to the primary protector, shall be listed.

(B) On Buildings Communications wires and cables in accordance with 800.50(A) shall be separated at least 100 mm (4 in.) from electric light or power conductors not in a raceway or cable or be permanently separated from conductors of the other system by a continuous and firmly fixed nonconductor in addition to the insulation on the wires, such as porcelain tubes or flexible tubing. Communications wires and cables in accordance with 800.50(A) exposed to accidental contact with electric light and power conductors operating at over 300 volts to ground and attached to buildings shall be separated from woodwork by being supported on glass, porcelain, or other insulating material.

Exception: Separation from woodwork shall not be required where fuses are omitted as provided for in 800.90(A)(1), or where conductors are used to extend circuits to a building from a cable having a grounded metal sheath.

(C) Entering Buildings Where a primary protector is installed inside the building, the communications wires and cables shall enter the building either through a noncombustible, nonabsorbent insulating bushing or through a metal raceway. The insulating bushing shall not be required where the entering communications wires and cables (1) are in metal-sheathed cable, (2) pass through masonry, (3) meet the requirements of 800.50(A) and fuses are omitted as provided in 800.90(A)(1), or (A)(4) meet the requirements of 800.50(A) and are used to extend circuits to a building from a cable having a grounded metallic sheath. Raceways or bushings shall slope upward from the outside or, where this cannot be done, drip loops shall be formed in the communications wires and cables immediately before they enter the building.

Raceways shall be equipped with an approved service head. More than one communications wire and cable shall be permitted to enter through a single raceway or bushing. Conduits or other metal raceways located ahead of the primary protector shall be grounded.

800.53 Lightning Conductors

Where practicable, a separation of at least 1.8 m (6 ft) shall be maintained between communications wires and cables on buildings and lightning conductors.

III. Protection

800.90 Protective Devices

(A) Application A listed primary protector shall be provided on each circuit run partly or entirely in aerial wire or aerial cable not confined within a block. Also, a listed primary protector shall be provided on each circuit, aerial or underground, located within the block containing the building served so as to be exposed to accidental contact with electric light or power conductors operating at over 300 volts to ground. In addition, where there exists a lightning exposure, each interbuilding circuit on a premises shall be protected by a listed primary protector at each end of the interbuilding circuit. Installation of primary protectors shall also comply with 110.3(B).

Telephone utility companies ordinarily provide primary protectors where telephone lines are exposed to lightning. Installers of private networks that include interbuilding cable should also install primary protectors where cables are exposed to lightning. Generally, cable is considered to be exposed to lightning unless one or more of the conditions in FPN No. 2 exist. A primary protector is required at each end of an interbuilding communications circuit where lightning exposure exists.

FPN No. 1: On a circuit not exposed to accidental contact with power conductors, providing a listed primary protector in accordance with this article helps protect against other hazards, such as lightning and above-normal voltages induced by fault currents on power circuits in proximity to the communications circuit.

FPN No. 2: Interbuilding circuits are considered to have a lightning exposure unless one or more of the following conditions exist:

(1) Circuits in large metropolitan areas where buildings are close together and sufficiently high to intercept lightning.
(2) Interbuilding cable runs of 42 m (140 ft) or less, directly buried or in underground conduit, where a continuous metallic cable shield or a continuous metallic conduit containing the cable is bonded to each building grounding electrode system.
(3) Areas having an average of five or fewer thunderstorm days per year and earth resistivity of less than 100 ohm-meters. Such areas are found along the Pacific coast.

(1) Fuseless Primary Protectors Fuseless-type primary protectors shall be permitted under any of the conditions given in (A)(1)(a) through (A)(1)(e).

(a) Where conductors enter a building through a cable with grounded metallic sheath member(s) and where the conductors in the cable safely fuse on all currents greater than the current-carrying capacity of the primary protector and of the primary protector grounding conductor

(b) Where insulated conductors in accordance with 800.50(A) are used to extend circuits to a building from a cable with an effectively grounded metallic sheath member(s) and where the conductors in the cable or cable stub, or the connections between the insulated conductors and the exposed plant, safely fuse on all currents greater than the current-carrying capacity of the primary protector, or the associated insulated conductors and of the primary protector grounding conductor

(c) Where insulated conductors in accordance with 800.50(A) or 800.50(B) are used to extend circuits to a building from other than a cable with metallic sheath member(s), where (1) the primary protector is listed as being suitable for this purpose for application with circuits extending from other than a cable with metallic sheath members, and (2) the connections of the insulated conductors to the exposed plant or the conductors of the exposed plant safely fuse on all currents greater than the current-carrying capacity of the primary protector, or associated insulated conductors and of the primary protector grounding conductor

(d) Where insulated conductors in accordance with 800.50(A) are used to extend circuits aerially to a building from an unexposed buried or underground circuit

(e) Where insulated conductors in accordance with 800.50(A) are used to extend circuits to a building from cable with an effectively grounded metallic sheath member(s), and where (1) the combination of the primary protector and insulated conductors is listed as being suitable for this purpose for application with circuits extending from a cable with an effectively grounded metallic sheath member(s), and (2) the insulated conductors safely fuse on all currents greater than the current-carrying capacity of the primary protector and of the primary protector grounding conductor

> The term *effectively grounded* (listed as *Grounded, Effectively*) is defined in Article 100.

(2) Fused Primary Protectors Where the requirements listed under 800.90(A)(1)(a) through (A)(1)(e) are not met, fused-type primary protectors shall be used. Fused-type primary protectors shall consist of an arrester connected between each line conductor and ground, a fuse in series with each line conductor, and an appropriate mounting arrangement. Primary protector terminals shall be marked to indicate line, instrument, and ground, as applicable.

(B) Location The primary protector shall be located in, on, or immediately adjacent to the structure or building served and as close as practicable to the point of entrance.

FPN: See 800.2 for the definition of *point of entrance*.

> Exhibit 800.4 shows an example of a primary protector unit typically installed in commercial buildings. Exhibit 800.5 shows an example of applications of listed communications and multipurpose cable.

For purposes of this section, primary protectors located at mobile home service equipment located in sight from and not more than 9.0 m (30 ft) from the exterior wall of the mobile home it serves, or at a mobile home disconnecting means grounded in accordance with 250.32 and located in sight from and not more than 9.0 m (30 ft) from the exterior wall of the mobile home it serves, shall be considered to meet the requirements of this section.

FPN: Selecting a primary protector location to achieve the shortest practicable primary protector grounding conductor helps limit potential differences between communications circuits and other metallic systems.

(C) Hazardous (Classified) Locations The primary protector shall not be located in any hazardous (classified) loca-

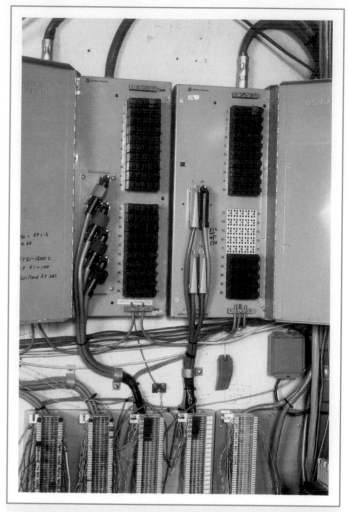

Exhibit 800.4 A primary protector unit typically installed in commercial buildings. This unit is the interface to the outside plant cable.

tion as defined in Article 500 or in the vicinity of easily ignitible material.

Exception: As permitted in 501.150, 502.150, and 503.150.

(D) Secondary Protectors Where a secondary protector is installed in series with the indoor communications wire and cable between the primary protector and the equipment, it shall be listed for the purpose in accordance with 800.170(B).

FPN: Secondary protectors on exposed circuits are not intended for use without primary protectors.

800.93 Cable Grounding

The metallic sheath of communications cables entering buildings shall be grounded as close as practicable to the point of entrance or shall be interrupted as close to the point of entrance as practicable by an insulating joint or equivalent device.

FPN: See 800.2 for the definition of *point of entrance*.

Exhibit 800.5 An example of applications of listed communications cables.

① CMP
(not in conduit)

② CMR
(not in conduit)

③ CMR
(in conduit only in plenum)

④ CM, CMG
(not in conduit and not in same penetration as riser cables)

Plenum or other space used for environmental air

Basement

Outside plant cable (not to exceed 50 ft within building)

Primary protector (as close as practicable to cable entrance)

IV. Grounding Methods

800.100 Cable and Primary Protector Grounding

The metallic member(s) of the cable sheath, where required to be grounded by 800.93, and primary protectors shall be grounded as specified in 800.100(A) through 800.100(D).

(A) Grounding Conductor.

(1) Insulation The grounding conductor shall be insulated and shall be listed as suitable for the purpose.

(2) Material The grounding conductor shall be copper or other corrosion-resistant conductive material, stranded or solid.

(3) Size The grounding conductor shall not be smaller than 14 AWG.

(4) Length The primary protector grounding conductor shall be as short as practicable. In one- and two-family dwellings, the primary protector grounding conductor shall be as short as practicable, not to exceed 6.0 m (20 ft) in length.

For one- and two-family dwellings, 800.100(A)(4) restricts the length of the primary protector grounding conductor to 20 ft. This restricted conductor length reduces the impedance

of the grounding conductor, resulting in a lower potential difference between the communications system conductors and equipment and the electrical conductors and equipment in the building. The low impedance bonding connection will reduce the fire hazard and shock hazard to persons in the event that electric utility power lines come in contact with communications conductors. Section 800.100(D) requires bonding of communications and power grounding electrodes at the same building or structure.

See the commentary following 250.52(A)(1) for information on water pipes as grounding electrodes.

FPN: Similar grounding conductor length limitations applied at apartment buildings and commercial buildings help to reduce voltages that may be developed between the building's power and communications systems during lightning events.

When the 20-ft limitation was instituted in the 2002 *Code*, the predominant application was in one- and two-family dwellings; apartment and commercial buildings were specifically not addressed. In the 2005 *Code*, some guidance is provided for apartment and commercial buildings, without being overly restrictive because of intersystem bonding situations that may exist at these facilities. The FPN to 800.100(A)(4) provides guidance for the treatment of the cable and primary protector grounding conductor length at apartment and commercial buildings that is consistent with the 20-ft rule for one- and two-family dwellings. However, a specific length is not specified in the *Code* because such a length limitation may not be practical in some installations.

Exception: In one- and two-family dwellings where it is not practicable to achieve an overall maximum primary protector grounding conductor length of 6.0 m (20 ft), a separate communications ground rod meeting the minimum dimensional criteria of 800.100(B)(2)(2) shall be driven, the primary protector shall be grounded to the communications ground rod in accordance with 800.100(C), and the communications ground rod shall be bonded to the power grounding electrode system in accordance with 800.100(D).

(5) Run in Straight Line The grounding conductor shall be run to the grounding electrode in as straight a line as practicable.

(6) Physical Damage Where necessary, the grounding conductor shall be guarded from physical damage. Where the grounding conductor is run in a metal raceway, both ends of the raceway shall be bonded to the grounding conductor or the same terminal or electrode to which the grounding conductor is connected.

(B) Electrode The grounding conductor shall be connected in accordance with 800.100(B)(1) and (B)(2).

(1) In Buildings or Structures with Grounding Means To the nearest accessible location on the following:

(1) The building or structure grounding electrode system as covered in 250.50
(2) The grounded interior metal water piping system, within 1.5 m (5 ft) from its point of entrance to the building, as covered in 250.52
(3) The power service accessible means external to enclosures as covered in 250.94
(4) The metallic power service raceway
(5) The service equipment enclosure
(6) The grounding electrode conductor or the grounding electrode conductor metal enclosure
(7) The grounding conductor or the grounding electrode of a building or structure disconnecting means that is grounded to an electrode as covered in 250.32

For purposes of this section, the mobile home service equipment or the mobile home disconnecting means, as described in 800.90(B), shall be considered accessible.

(2) In Buildings or Structures Without Grounding Means If the building or structure served has no grounding means, as described in 800.100(B)(1), the grounding conductor shall be connected to either of the following:

(1) To any one of the individual electrodes described in 250.52(A)(1), (A)(2), (A)(3), or (A)(4)
(2) If the building or structure served has no grounding means, as described in 800.100(B)(1) or (B)(2)(1), to an effectively grounded metal structure or to a ground rod or pipe not less than 1.5 m (5 ft) in length and 12.7 mm (½ in.) in diameter, driven, where practicable, into permanently damp earth and separated from lightning conductors as covered in 800.53 and at least 1.8 m (6 ft) from electrodes of other systems. Steam or hot water pipes or air terminal conductors (lightning-rod conductors) shall not be employed as electrodes for protectors.

(C) Electrode Connection Connections to grounding electrodes shall comply with 250.70.

(D) Bonding of Electrodes A bonding jumper not smaller than 6 AWG copper or equivalent shall be connected between the communications grounding electrode and power grounding electrode system at the building or structure served where separate electrodes are used.

Exception: At mobile homes as covered in 800.106.

> FPN No. 1: See 250.60 for use of air terminals (lightning rods).
>
> FPN No. 2: Bonding together of all separate electrodes limits potential differences between them and between their associated wiring systems.

800.106 Primary Protector Grounding and Bonding at Mobile Homes

(A) Grounding Where there is no mobile home service equipment located in sight from, and not more than 9.0 m (30 ft) from, the exterior wall of the mobile home it serves, or there is no mobile home disconnecting means grounded in accordance with 250.32 and located within sight from, and not more than 9.0 m (30 ft) from, the exterior wall of the mobile home it serves, the primary protector ground shall be in accordance with 800.100(B)(2).

(B) Bonding The primary protector grounding terminal or grounding electrode shall be bonded to the metal frame or available grounding terminal of the mobile home with a copper grounding conductor not smaller than 12 AWG under either of the following conditions:

(1) Where there is no mobile home service equipment or disconnecting means as in 800.106(A)
(2) Where the mobile home is supplied by cord and plug

V. Communications Wires and Cables Within Buildings

Data circuits between computers are classified as Class 2 circuits. In a typical office environment consisting of a group of computers connected to a local area network, data wiring is as prevalent as telephone wiring. One common way to minimize the amount of cabling is to run the telephone and data circuits in the same cable, as illustrated in Exhibit 800.6. Section 725.56(D) requires that either a communications cable or a multipurpose cable be used for this purpose.

Exhibit 800.6 An example of telephone and data circuits in the same cable.

800.110 Raceways for Communications Wires and Cables

Where communications wires and cables are installed in a raceway, the raceway shall be either of a type permitted in Chapter 3 and installed in accordance with Chapter 3 or a listed nonmetallic raceway complying with 800.182, and installed in accordance with 362.24 through 362.56, where the requirements applicable to electrical nonmetallic tubing apply.

Exception: Conduit fill restrictions shall not apply.

800.113 Installation and Marking of Communications Wires and Cables

Listed communications wires and cables and listed multipurpose cables shall be installed as wiring within buildings. Communications cables and undercarpet communications wires shall be marked in accordance with Table 800.113. The cable voltage rating shall not be marked on the cable or on the undercarpet communications wire.

> FPN: Voltage markings on cables may be misinterpreted to suggest that the cables may be suitable for Class 1, electric light, and power applications.

Table 800.113 Cable Markings

Cable Marking	Type	Reference
CMP	Communications plenum cable	800.179(A) and 800.154(A)
CMR	Communications riser cable	800.179(B) and 800.154(B)
CMG	Communications general-purpose cable	800.179(C) and 800.154(D) and (E)(1)
CM	Communications general-purpose cable	800.179(D) and 800.154(D) and (E)(1)
CMX	Communications cable, limited use	800.179(E) and 800.154(E)(2), (3), (4), and (5)
CMUC	Undercarpet communications wire and cable	800.179(F) and 800.154(E)(6)

FPN No. 1: Cable types are listed in descending order of fire resistance rating.

FPN No. 2: See the referenced sections for permitted uses.

Exception No. 1: Voltage markings shall be permitted where the cable has multiple listings and voltage marking is required for one or more of the listings.

Exception No. 2: Listing and marking shall not be required where the length of the cable within the building, measured from its point of entrance, does not exceed 15 m (50 ft) and the cable enters the building from the outside and is terminated in an enclosure or on a listed primary protector.

> FPN No. 1 to Exception No. 2: Splice cases or terminal boxes, both metallic and plastic types, are typically used as enclosures for splicing or terminating telephone cables.
>
> FPN No. 2 to Exception No. 2: This exception limits the length of unlisted outside plant cable to 15 m (50 ft), while 800.90(B) requires that the primary protector be located as close as practicable to the point at which the cable enters the building. Therefore, in installations requiring a primary protector, the outside plant cable may not be permitted to extend 15 m (50 ft) into the building if it is practicable to place the primary protector closer than 15 m (50 ft) to the entrance point.

800.133 Installation of Communications Wires, Cables, and Equipment

Communications wires and cables from the protector to the equipment or, where no protector is required, communications wires and cables attached to the outside or inside of the building shall comply with 800.133(A) through 800.133(D).

Section 800.133 includes non–power-limited fire alarm circuits covered by Article 760 and network-powered broadband communications circuits covered by Article 830.

(A) Separation from Other Conductors

(1) In Raceways, Boxes, and Cables

(a) Other Power-Limited Circuits. Communications cables shall be permitted in the same raceway or enclosure with cables of any of the following:

(1) Class 2 and Class 3 remote-control, signaling, and power-limited circuits in compliance with Article 725
(2) Power-limited fire alarm systems in compliance with Article 760
(3) Nonconductive and conductive optical fiber cables in compliance with Article 770
(4) Community antenna television and radio distribution systems in compliance with Article 820
(5) Low-power network-powered broadband communications circuits in compliance with Article 830

(b) Class 2 and Class 3 Circuits. Class 1 circuits shall not be run in the same cable with communications circuits. Class 2 and Class 3 circuit conductors shall be permitted in the same cable with communications circuits, in which case the Class 2 and Class 3 circuits shall be classified as communications circuits and shall meet the requirements of this article. The cables shall be listed as communications cables or multipurpose cables.

Exception: Cables constructed of individually listed Class 2, Class 3, and communications cables under a common jacket shall not be required to be classified as communications cable. The fire-resistance rating of the composite cable shall be determined by the performance of the composite cable.

(c) Electric Light, Power, Class 1, Non–Power-Limited Fire Alarm, and Medium Power Network-Powered Broadband Communications Circuits in Raceways, Compartments, and Boxes. Communications conductors shall not be placed in any raceway, compartment, outlet box, junction box, or similar fitting with conductors of electric light, power, Class 1, non–power-limited fire alarm, or medium power network-powered broadband communications circuits.

Exception No. 1: Where all of the conductors of electric light, power, Class 1, non–power-limited fire alarm, and medium power network-powered broadband communications circuits are separated from all of the conductors of communications circuits by a permanent barrier or listed divider.

This exception recognizes the use of a listed field-installed divider to separate the communications circuits from the power circuits.

Exception No. 2: Power conductors in outlet boxes, junction boxes, or similar fittings or compartments where such con-

ductors are introduced solely for power supply to communications equipment. The power circuit conductors shall be routed within the enclosure to maintain a minimum of 6 mm (0.25 in.) separation from the communications circuit conductors.

Exception No. 3: As permitted by 620.36.

(2) Other Applications

Communications wires and cables shall be separated at least 50 mm (2 in.) from conductors of any electric light, power, Class 1, non–power-limited fire alarm, or medium power network-powered broadband communications circuits.

Exception No. 1: Where either (1) all of the conductors of the electric light, power, Class 1, non–power-limited fire alarm, and medium power network-powered broadband communications circuits are in a raceway or in metal-sheathed, metal-clad, nonmetallic-sheathed, Type AC, or Type UF cables, or (2) all of the conductors of communications circuits are encased in raceway.

Exception No. 2: Where the communications wires and cables are permanently separated from the conductors of electric light, power, Class 1, non–power-limited fire alarm, and medium power network-powered broadband communications circuits by a continuous and firmly fixed nonconductor, such as porcelain tubes or flexible tubing, in addition to the insulation on the wire.

(B) Cable Trays Types CMP, CMR, CMG, and CM communications cables shall be permitted to be installed in cable trays. Communications raceways, as described in 800.179, shall be permitted to be installed in cable trays.

Exhibit 800.7 shows overhead ladder–type cable tray that contains communications cables.

(C) Support of Conductors Raceways shall be used for their intended purpose. Communications cables or wires shall not be strapped, taped, or attached by any means to the exterior of any conduit or raceway as a means of support.

See 800.21 and 800.24, which require that communications cable be supported by the building structure in such a manner that it will not be damaged by ordinary building use.

Exception: Overhead (aerial) spans of communications cables or wires shall be permitted to be attached to the exterior of a raceway-type mast intended for the attachment and support of such conductors.

In some instances, the only way to achieve the proper clearance above roadways, driveways, or structures is by use of a mast. The exception to 800.133(C) permits overhead spans

Exhibit 800.7 Overhead ladder–type cable tray containing communications cables.

of communications cable to be attached to the exterior of a raceway-type mast only if the mast is installed to support communications cable. Section 230.28 prohibits the attachment of communications cable to a service mast.

(D) Wiring in Ducts for Dust, Loose Stock, or Vapor Removal Section 300.22(A) shall apply.

800.154 Applications of Listed Communications Wires and Cables and Communications Raceways

Communications wires and cables shall comply with the requirements of 800.154(A) through 800.154(F) or where cable substitutions are made in accordance with 800.154(G).

Note that the length of unlisted outside-plant cable permitted in a building depends on the location of the primary protector, in accordance with 800.90(B) and 800.113 Exception No. 2.

Section 800.154(A) covers listed plenum communications raceways. These raceways provide limited mechanical protection and ease of installation, but they are limited to Type CMP plenum-rated cable if installed in ducts and plenums.

Section 800.154(B) covers riser raceways. Riser raceways provide limited mechanical protection and ease of installation, but they are limited to Type CMP plenum-rated cable or Type CMR riser-rated cable if installed in risers. Table 800.154 lists the permitted uses of field applications for various cable types.

(A) Plenum Cables installed in ducts, plenums, and other spaces used for environmental air shall be Type CMP. Aban-

doned cables shall not be permitted to remain. Types CMP, CMR, CMG, CM, and CMX and communications wire installed in compliance with 300.22 shall be permitted. Listed plenum communications raceways shall be permitted to be installed in ducts and plenums as described in 300.22(B) and in other spaces used for environmental air as described in 300.22(C). Only Type CMP cable shall be permitted to be installed in raceways.

> FPN: See 8.14.1 of NFPA 13-2002, *Installation of Sprinkler Systems*, for requirements for sprinklers in concealed spaces containing exposed combustibles.

(B) Riser Cables installed in risers shall comply with 800.154(B)(1), (B)(2), or (B)(3).

(1) Cables in Vertical Runs Cables installed in vertical runs and penetrating more than one floor, or cables installed in vertical runs in a shaft, shall be Type CMR. Floor penetrations requiring Type CMR shall contain only cables suitable for riser or plenum use. Abandoned cables shall not be permitted to remain. Listed riser communications raceways shall be permitted to be installed in vertical riser runs in a shaft from floor to floor. Only Type CMR and CMP cables shall be permitted to be installed in these raceways.

(2) Metal Raceways or Fireproof Shafts Listed communications cables shall be encased in a metal raceway or located in a fireproof shaft having firestops at each floor.

(3) One- and Two-Family Dwellings Type CM and CMX cable shall be permitted in one- and two-family dwellings.

> FPN: See 800.3(C) for firestop requirements for floor penetrations.

(C) Distributing Frames and Cross-Connect Arrays Listed communications wire and Types CMP, CMR, CMG, and CM communications cables shall be used in distributing frames and cross-connect arrays.

(D) Cable Trays Types CMP, CMR, CMG, and CM communications cables shall be permitted to be installed in cable trays.

(E) Other Wiring Within Buildings Cables installed in building locations other than the locations covered in 800.154(A) through 800.154(D) shall be in accordance with 800.154(E)(1) through (E)(6).

(1) General Cables shall be Type CMG or Type CM. Listed communications general-purpose raceways shall be permitted. Only Types CMG, CM, CMR, or CMP cables shall be permitted to be installed in general-purpose communications raceways.

(2) In Raceways Listed communications wires that are enclosed in a raceway of a type included in Chapter 3 shall be permitted.

(3) Nonconcealed Spaces Type CMX communications cable shall be permitted to be installed in nonconcealed spaces where the exposed length of cable does not exceed 3 m (10 ft).

(4) One- and Two-Family Dwellings Type CMX communications cable less than 6 mm (0.25 in.) in diameter shall be permitted to be installed in one- and two-family dwellings.

(5) Multi-Family Dwellings Type CMX communications cable less than 6 mm (0.25 in.) in diameter shall be permitted to be installed in nonconcealed spaces in multi-family dwellings.

(6) Under Carpets Type CMUC undercarpet communications wires and cables shall be permitted to be installed under carpet.

(F) Hybrid Power and Communications Cable Hybrid power and communications cable listed in accordance with 800.179(I) shall be permitted to be installed in one- and two-family dwellings.

(G) Cable Substitutions The uses and permitted substitutions for communications cables listed in Table 800.154 shall be considered suitable for the purpose and shall be permitted.

Table 800.154 Cable Substitutions

Cable Type	Use	References	Permitted Substitutions
CMR	Communications riser cable	800.154(B)	CMP
CMG, CM	Communications general-purpose cable	800.154(E)(1)	CMP, CMR
CMX	Communications cable, limited use	800.154(E)	CMP, CMR, CMG, CM

FPN: See Figure 800.154, Cable Substitution Hierarchy.

FPN: For information on Types CMP, CMR, CMG, CM, and CMX cables, see 800.179.

VI. Listing Requirements

800.170 Equipment

Communications equipment shall be listed as being suitable for electrical connection to a telecommunications network.

FPN: One way to determine applicable requirements is to refer to UL 1950-1993, *Standard for Safety of Information Technology Equipment, Including Electrical Business Equipment*, third edition; UL 1459-1995, *Standard for Safety, Telephone Equipment*, third edition; or UL 1863-1995, *Standard for Safety, Communications Circuit Accessories*, second edition. For information on listing requirements for communications raceways, see UL 2024-1995, *Standard for Optical Fiber Raceways*.

Type CM—Communications cables

Cable A shall be permitted to be used in place of cable B.

Figure 800.154 Cable Substitution Hierarchy.

(A) Primary Protectors The primary protector shall consist of an arrester connected between each line conductor and ground in an appropriate mounting. Primary protector terminals shall be marked to indicate line and ground as applicable.

FPN: One way to determine applicable requirements for a listed primary protector is to refer to ANSI/UL 497-1995, *Standard for Protectors for Paired Conductor Communications Circuits*.

(B) Secondary Protectors The secondary protector shall be listed as suitable to provide means to safely limit currents to less than the current-carrying capacity of listed indoor communications wire and cable, listed telephone set line cords, and listed communications terminal equipment having ports for external wire line communications circuits. Any overvoltage protection, arresters, or grounding connection shall be connected on the equipment terminals side of the secondary protector current-limiting means.

FPN: One way to determine applicable requirements for a listed secondary protector is to refer to UL 497A-1996, *Standard for Secondary Protectors for Communications Circuits*.

800.173 Drop Wire and Cable

Communications wires and cables without a metallic shield, running from the last outdoor support to the primary protector, shall be listed as being suitable for the purpose and shall have current-carrying capacity as specified in 800.90(A)(1)(b) or (A)(1)(c).

800.179 Communications Wires and Cables

Communications wires and cables shall have a voltage rating of not less than 300 volts and shall be listed in accordance with 800.179(A) through 800.179(J). Conductors in communications cables, other than in a coaxial cable, shall be copper.

Section 800.179 requires a rating of 300 volts for the following reasons:

1. To coordinate with protector installation requirements (i.e., protectors are not required within a block unless the cable is exposed to over 300 volts)
2. To recognize the fact that primary protectors are designed to allow voltages below 300 to pass
3. To accommodate the voltages ordinarily found on a telephone line (48 volts dc plus ringing voltage up to 130 volts rms)
4. To permit communications cable to substitute for 300-volt power-limited fire-protective signaling cable

FPN: See 800.170 for listing requirement for equipment.

(A) Type CMP Type CMP communications plenum cable shall be listed as being suitable for use in ducts, plenums, and other spaces used for environmental air and shall also be listed as having adequate fire-resistant and low smoke-producing characteristics.

See the commentary following 725.82(A), FPN, for information on a test method for wires and cables to be installed without raceways in plenums and other spaces used for environmental air.

FPN: One method of defining a cable that is low smoke-producing cable and fire-resistant cable is that the cable exhibits a maximum peak optical density of 0.5 or less, an average optical density of 0.15 or less, and a maximum flame spread distance of 1.52 m (5 ft) or less when tested in accordance with NFPA 262-2002, *Standard Method of Test for Flame Travel and Smoke of Wires and Cables for Use in Air-Handling Spaces.*

(B) Type CMR Type CMR communications riser cable shall be listed as being suitable for use in a vertical run in a shaft or from floor to floor and shall also be listed as having fire-resistant characteristics capable of preventing the carrying of fire from floor to floor.

See the commentary following 725.82(B), FPN, for information on a test for defining fire-resistant characteristics capable of preventing fire spread from floor to floor.

FPN: One method of defining fire-resistant characteristics capable of preventing the carrying of fire from floor to floor is that the cables pass the requirements of ANSI/UL 1666-2002, *Standard Test for Flame Propagation Height of Electrical and Optical-Fiber Cable Installed Vertically in Shafts.*

(C) Type CMG Type CMG general-purpose communications cable shall be listed as being suitable for general-purpose communications use, with the exception of risers and plenums, and shall also be listed as being resistant to the spread of fire.

See the commentary following 725.82(C), FPN, for information on the UL vertical tray flame test.

FPN: One method of defining *resistant to the spread of fire* is for the damage (char length) not to exceed 1.5 m (4 ft 11 in.) when performing the vertical flame test for cables in cable trays, as described in CSA C22.2 No. 0.3-M-1985, *Test Methods for Electrical Wires and Cables.*

(D) Type CM Type CM communications cable shall be listed as being suitable for general-purpose communications use, with the exception of risers and plenums, and shall also be listed as being resistant to the spread of fire.

See the commentary following 725.82(D), FPN, for information on test methods for determining whether cable is resistant to the spread of fire.

FPN: One method of defining *resistant to the spread of fire* is that the cables do not spread fire to the top of the tray in the vertical-tray flame test in ANSI/UL 1581-1991, *Reference Standard for Electrical Wires, Cables and Flexible Cords.* Another method of defining *resistant to the spread of fire* is for the damage (char length) not to exceed 1.5 m (4 ft 11 in.) when performing the vertical flame test for cables in cable trays, as described in CSA C22.2 No. 0.3-M-1985, *Test Method for Electrical Wires and Cables.*

(E) Type CMX Type CMX limited-use communications cable shall be listed as being suitable for use in dwellings and for use in raceway and shall also be listed as being resistant to flame spread.

FPN: One method of determining that cable is resistant to flame spread is by testing the cable to the VW-1 (vertical-wire) flame test in ANSI/UL 1581-1991, *Reference Standard for Electrical Wires, Cables and Flexible Cords.*

(F) Type CMUC Undercarpet Wire and Cable Type CMUC undercarpet communications wire and cable shall be listed as being suitable for undercarpet use and shall also be listed as being resistant to flame spread.

FPN: One method of determining that cable is resistant to flame spread is by testing the cable to the VW-1 (vertical-wire) flame test in ANSI/UL 1581-1991, *Reference Standard for Electrical Wires, Cables and Flexible Cords.*

(G) Multipurpose (MP) Cables Until July 1, 2003, cables that meet the requirements for Types CMP, CMR, CMG, and CM and also satisfy the requirements of 760.82(B) for multiconductor cables and 760.82(H) for coaxial cables shall be permitted to be listed and marked as multipurpose cable Types MPP, MPR, MPG, and MP, respectively.

The deletion of stranding requirements for fire alarm cable resulted in an increased number of copper communications cables, such as Types MPP, MPR, MPG, and MP, that qualify for listing as multipurpose cable.

(H) Communications Circuit Integrity (CI) Cable Cables suitable for use in communications systems to ensure survivability of critical circuits during a specified time under fire conditions shall be listed as circuit integrity (CI) cable. Cables identified in 800.179 (A), (B), (C), (D), and (E) that meet the requirements for circuit integrity shall have the additional classification using the suffix "CI."

> FPN: One method of defining circuit integrity (CI) cable is by establishing a minimum 2-hour fire resistance rating for the cable when tested in accordance with UL 2196-1995, *Standard for Tests of Fire Resistive Cables.*

(I) Communications Wires Communications wires, such as distributing frame wire and jumper wire, shall be listed as being resistant to the spread of fire.

> FPN: One method of defining *resistant to the spread of fire* is that the cables do not spread fire to the top of the tray in the vertical-tray flame test in ANSI/UL 1581-1991, *Reference Standard for Electrical Wires, Cables and Flexible Cords.* Another method of defining *resistant to the spread of fire* is for the damage (char length) not to exceed 1.5 m (4 ft 11 in.) when performing the vertical flame test for cables in cable trays, as described in CSA C22.2 No. 0.3-M-1985, *Test Methods for Electrical Wires and Cables.*

(J) Hybrid Power and Communications Cable Listed hybrid power and communications cable shall be permitted where the power cable is a listed Type NM or NM-B conforming to the provisions of Article 334, and the communications cable is a listed Type CM, the jackets on the listed NM or NM-B and listed CM cables are rated for 600 volts minimum, and the hybrid cable is listed as being resistant to the spread of fire.

> FPN: One method of defining *resistant to the spread of fire* is that the cables do not spread fire to the top of the tray in the vertical-tray flame test in ANSI/UL 1581-1991, *Reference Standard for Electrical Wires, Cables and Flexible Cords.* Another method of defining *resistant to the spread of fire* is for the damage (char length) not to exceed 1.5 m (4 ft 11 in.) when performing the vertical flame test for cables in cable trays, as described in CSA C22.2 No. 0.3-M-1985, *Test Methods for Electrical Wires and Cables.*

800.182 Communications Raceways

Communications raceways shall be listed in accordance with 800.182(A) through 800.182(C).

(A) Plenum Communications Raceways Plenum communications raceways listed as plenum optical fiber raceways shall be permitted for use in ducts, plenums, and other spaces used for environmental air and shall also be listed as having adequate fire-resistant and low smoke-producing characteristics.

> FPN: One method of defining that an optical fiber raceway is a low smoke producing raceway and a fire-resistant raceway is that the raceway exhibits a maximum peak optical density of 0.5 or less, an average optical density of 0.15 or less, and a maximum flame spread distance of 1.52 m (5 ft) or less when tested in accordance with the plenum test in UL 2024, *Standard for Optical Fiber Cable Raceway.*

(B) Riser Communications Raceways Riser communications raceways shall be listed as having adequate fire-resistant characteristics capable of preventing the carrying of fire from floor to floor.

> FPN: One method of defining fire-resistant characteristics capable of preventing the carrying of fire from floor to floor is that the raceways pass the requirements of the test for Flame Propagation (riser) in UL 2024, *Standard for Optical Fiber Cable Raceway.*

(C) General-Purpose Communications Raceways General-purpose communications raceways shall be listed as being resistant to the spread of fire.

The communications raceways covered in 800.182(A) through (C) are listed raceways used in plenum, riser, or general-purpose applications. This listing includes raceways and fittings for the installation of communications cable in accordance with Article 800. These raceways are not suitable for installation of wires, cords, or cabling with or without communications members.

A raceway marked "plenum" is suitable for use in ducts, plenums, or other spaces used for environmental air in accordance with 800.154(A) when used to enclose communications cable marked CMP. This raceway exhibits a maximum peak optical density of 0.5, a maximum average optical density of 0.15, and a maximum flame-spread distance of 5 ft when tested in accordance with UL 2024, *Standard for Optical-Fiber Cable Raceway.* This raceway is identified by a marking on its surface or on a marker tape indicating "plenum." A raceway marked "plenum" is also suitable for installation in risers when used to enclose communications cable marked CMP or CMR; for general-purpose use when used to enclose communications cable marked CMP, CMR, CMG, or CM; and for dwellings when used to enclose communications cable marked CMP, CMR, CMG, CM, or CMX.

A raceway marked "riser" is suitable for installation in risers in accordance with 800.154(B) when used to enclose communications cable marked CMP or CMR. This raceway has fire-resistant characteristics capable of preventing the carrying of fire from floor to floor, and it meets the test requirements of UL 2024, *Standard for Optical-Fiber Cable Raceway*. This raceway is identified by a marking on its surface or on a marker tape indicating "riser." A raceway marked "riser" is also suitable for general-purpose use when used to enclose communications cable marked CMP, CMR, CMG, or CM, and for dwellings when used to enclose communications cable marked CMP, CMR, CMG, CM, or CMX.

A raceway marked "general purpose" is suitable for installation in general-purpose areas in accordance with 800.154(D) when used to enclose communications cable marked CMP, CMR, CMG, or CM, and for dwellings when used to enclose communications cable marked CMP, CMR, CMG, CM, or CMX.

Pliable raceway is raceway that can be bent by hand without the use of tools. The smallest radius of the curve of the inner edge of any bend to which the raceway can be bent without cracking either on the outer surface or internally is not less than 2½ times the outside diameter of the raceway.

FPN: One method of defining *resistance to the spread of fire* is that the raceways pass the requirements of the Vertical-Tray Flame Test (General Use) in UL 2024, *Standard for Optical Fiber Cable Raceway*.

ARTICLE 810
Radio and Television Equipment

Contents

I. General

810.1 Scope

This article covers antenna systems for radio and television receiving equipment, amateur radio transmitting and receiving equipment, and certain features of transmitter safety. This article covers antennas such as multi-element, vertical rod, and dish, and also covers the wiring and cabling that connects them to equipment. This article does not cover equipment and antennas used for coupling carrier current to power line conductors.

Article 810 covers wiring requirements for television and radio receiving equipment, specifically including digital satellite receiving equipment for television signals, and wiring for amateur radio equipment. Chapters 1 through 4 cover wiring for the power supply. Article 640 contains requirements for sound distribution systems. The interior wiring of coaxial cable is covered by Article 820.

810.2 Definitions

For definitions applicable to this article, see Article 100.

In accordance with 90.3, the articles in Chapter 8 are not subject to the requirements contained in the other articles of the *Code* unless specifically referenced. Section 810.2 clearly states that the definitions in Article 100 also apply to Article 810.

810.3 Other Articles

Wiring from the source of power to and between devices connected to the interior wiring system shall comply with Chapters 1 through 4 other than as modified by Parts I and II of Article 640. Wiring for audio signal processing, amplification, and reproduction equipment shall comply with Article 640. Coaxial cables that connect antennas to equipment shall comply with Article 820.

810.4 Community Television Antenna

The antenna shall comply with this article. The distribution system shall comply with Article 820.

810.5 Radio Noise Suppressors

Radio interference eliminators, interference capacitors, or noise suppressors connected to power-supply leads shall be of a listed type. They shall not be exposed to physical damage.

II. Receiving Equipment — Antenna Systems

810.11 Material

Antennas and lead-in conductors shall be of hard-drawn copper, bronze, aluminum alloy, copper-clad steel, or other high-strength, corrosion-resistant material.

Exception: Soft-drawn or medium-drawn copper shall be permitted for lead-in conductors where the maximum span between points of support is less than 11 m (35 ft).

810.12 Supports

Outdoor antennas and lead-in conductors shall be securely supported. The antennas or lead-in conductors shall not be attached to the electric service mast. They shall not be attached to poles or similar structures carrying open electric light or power wires or trolley wires of over 250 volts between conductors. Insulators supporting the antenna conductors shall have sufficient mechanical strength to safely support the conductors. Lead-in conductors shall be securely attached to the antennas.

810.13 Avoidance of Contacts with Conductors of Other Systems

Outdoor antennas and lead-in conductors from an antenna to a building shall not cross over open conductors of electric light or power circuits and shall be kept well away from all such circuits so as to avoid the possibility of accidental contact. Where proximity to open electric light or power service conductors of less than 250 volts between conductors cannot be avoided, the installation shall be such as to provide a clearance of at least 600 mm (2 ft).

Where practicable, antenna conductors shall be installed so as not to cross under open electric light or power conductors.

One of the leading causes of electric shock and electrocution, according to statistical reports, is the accidental contact of radio, television, and amateur radio transmitting and receiving antennas and equipment with light or power conductors. Extreme caution should be exercised during this type of installation, and periodic visual inspections should be conducted thereafter.

810.14 Splices

Splices and joints in antenna spans shall be made mechanically secure with approved splicing devices or by such other means as will not appreciably weaken the conductors.

Section 810.14 requires that splices and joints be made with approved connectors or other means that provide sufficient mechanical strength so that conductors are not weakened appreciably, a condition that could cause them to break and come into contact with higher-voltage conductors. Conductor spans from antennas must be of sufficient size and strength to maintain clearances and avoid possible contact with light or power conductors.

810.15 Grounding

Masts and metal structures supporting antennas shall be grounded in accordance with 810.21.

810.16 Size of Wire-Strung Antenna — Receiving Station

(A) Size of Antenna Conductors Outdoor antenna conductors for receiving stations shall be of a size not less than given in Table 810.16(A).

Table 810.16(A) Size of Receiving Station Outdoor Antenna Conductors

Material	Minimum Size of Conductors (AWG) Where Maximum Open Span Length Is		
	Less Than 11 m (35 ft)	11 m to 45 m (35 ft to 150 ft)	Over 45 m (150 ft)
Aluminum alloy, hard-drawn copper	19	14	12
Copper-clad steel, bronze, or other high-strength material	20	17	14

(B) Self-Supporting Antennas Outdoor antennas, such as vertical rods, dishes, or dipole structures, shall be of corrosion-resistant materials and of strength suitable to withstand ice and wind loading conditions and shall be located well away from overhead conductors of electric light and power circuits of over 150 volts to ground, so as to avoid the possibility of the antenna or structure falling into or making accidental contact with such circuits.

Section 810.16(B) includes dish-type (parabolic) antennas.

810.17 Size of Lead-in — Receiving Station

Lead-in conductors from outside antennas for receiving stations shall, for various maximum open span lengths, be of such size as to have a tensile strength at least as great as that of the conductors for antennas as specified in 810.16. Where the lead-in consists of two or more conductors that are twisted together, are enclosed in the same covering, or are concentric, the conductor size shall, for various maximum open span lengths, be such that the tensile strength of the combination is at least as great as that of the conductors for antennas as specified in 810.16.

810.18 Clearances — Receiving Stations

(A) Outside of Buildings Lead-in conductors attached to buildings shall be installed so that they cannot swing closer than 600 mm (2 ft) to the conductors of circuits of 250 volts or less between conductors, or 3.0 m (10 ft) to the conductors of circuits of over 250 volts between conductors, except that

in the case of circuits not over 150 volts between conductors, where all conductors involved are supported so as to ensure permanent separation, the clearance shall be permitted to be reduced but shall not be less than 100 mm (4 in.). The clearance between lead-in conductors and any conductor forming a part of a lightning rod system shall not be less than 1.8 m (6 ft) unless the bonding referred to in 250.60 is accomplished. Underground conductors shall be separated at least 300 mm (12 in.) from conductors of any light or power circuits or Class 1 circuits.

Exception: Where the electric light or power conductors, Class 1 conductors, or lead-in conductors are installed in raceways or metal cable armor.

(B) Antennas and Lead-ins — Indoors Indoor antennas and indoor lead-ins shall not be run nearer than 50 mm (2 in.) to conductors of other wiring systems in the premises.

Exception No. 1: Where such other conductors are in metal raceways or cable armor.

Exception No. 2: Where permanently separated from such other conductors by a continuous and firmly fixed nonconductor, such as porcelain tubes or flexible tubing.

(C) In Boxes or Other Enclosures Indoor antennas and indoor lead-ins shall be permitted to occupy the same box or enclosure with conductors of other wiring systems where separated from such other conductors by an effective permanently installed barrier.

810.19 Electric Supply Circuits Used in Lieu of Antenna — Receiving Stations

Where an electric supply circuit is used in lieu of an antenna, the device by which the radio receiving set is connected to the supply circuit shall be listed.

The approved connecting device is usually a small, fixed capacitor connecting the antenna terminal of the receiver and one wire of the supply circuit. As is the case with most receivers, the capacitor should be designed for operation at not less than 300 volts. This rating ensures a high degree of safety and minimizes the possibility of a breakdown in the capacitor, thereby avoiding a short circuit to ground through the antenna coil of the set.

810.20 Antenna Discharge Units — Receiving Stations

(A) Where Required Each conductor of a lead-in from an outdoor antenna shall be provided with a listed antenna discharge unit.

Exception: Where the lead-in conductors are enclosed in a continuous metallic shield that either is permanently and effectively grounded or is protected by an antenna discharge unit.

(B) Location Antenna discharge units shall be located outside the building or inside the building between the point of entrance of the lead-in and the radio set or transformers and as near as practicable to the entrance of the conductors to the building. The antenna discharge unit shall not be located near combustible material or in a hazardous (classified) location as defined in Article 500.

An antenna discharge unit (lightning arrester) is not required if the lead-in conductors are enclosed in a continuous metal shield, such as rigid or intermediate metal conduit, electrical metallic tubing, or any metal raceway or metal-shielded cable that is effectively grounded. A lightning discharge will take the path of lower impedance and jump from the lead-in conductors to the metal raceway or shield rather than take the path through the antenna coil of the receiver.

(C) Grounding The antenna discharge unit shall be grounded in accordance with 810.21.

810.21 Grounding Conductors — Receiving Stations

Grounding conductors shall comply with 810.21(A) through 810.21(K).

(A) Material The grounding conductor shall be of copper, aluminum, copper-clad steel, bronze, or similar corrosion-resistant material. Aluminum or copper-clad aluminum grounding conductors shall not be used where in direct contact with masonry or the earth or where subject to corrosive conditions. Where used outside, aluminum or copper-clad aluminum shall not be installed within 450 mm (18 in.) of the earth.

(B) Insulation Insulation on grounding conductors shall not be required.

(C) Supports The grounding conductors shall be securely fastened in place and shall be permitted to be directly attached to the surface wired over without the use of insulating supports.

Exception: Where proper support cannot be provided, the size of the grounding conductors shall be increased proportionately.

(D) Mechanical Protection The grounding conductor shall be protected where exposed to physical damage, or the size of the grounding conductors shall be increased proportionately to compensate for the lack of protection. Where the grounding conductor is run in a metal raceway, both ends

of the raceway shall be bonded to the grounding conductor or to the same terminal or electrode to which the grounding conductor is connected.

If metal enclosures such as steel conduit are used to enclose the grounding conductor, bonding must be provided at both ends to ensure an adequate low-impedance current path.

(E) Run in Straight Line The grounding conductor for an antenna mast or antenna discharge unit shall be run in as straight a line as practicable from the mast or discharge unit to the grounding electrode.

(F) Electrode The grounding conductor shall be connected as follows:

See the commentary following 250.52(A)(1).

(1) To the nearest accessible location on the following:
   a. The building or structure grounding electrode system as covered in 250.50
   b. The grounded interior metal water piping systems, within 1.52 m (5 ft) from its point of entrance to the building, as covered in 250.52
   c. The power service accessible means external to the building, as covered in 250.94
   d. The metallic power service raceway
   e. The service equipment enclosure, or
   f. The grounding electrode conductor or the grounding electrode conductor metal enclosures; or
(2) If the building or structure served has no grounding means, as described in 810.21(F)(1), to any one of the individual electrodes described in 250.52; or
(3) If the building or structure served has no grounding means, as described in 810.21(F)(1) or (F)(2), to an effectively grounded metal structure or to any of the individual electrodes described in 250.52

(G) Inside or Outside Building The grounding conductor shall be permitted to be run either inside or outside the building.

(H) Size The grounding conductor shall not be smaller than 10 AWG copper, 8 AWG aluminum, or 17 AWG copper-clad steel or bronze.

(I) Common Ground A single grounding conductor shall be permitted for both protective and operating purposes.

(J) Bonding of Electrodes A bonding jumper not smaller than 6 AWG copper or equivalent shall be connected between the radio and television equipment grounding electrode and the power grounding electrode system at the building or structure served where separate electrodes are used.

The requirements for grounding are in accordance with Article 250. Antenna masts must be grounded to the same grounding electrode used for the building's electrical system, to ensure that all exposed, non–current-carrying metal parts are at the same potential. In many cases, masts are connected incorrectly to conveniently located vent pipes, metal gutters, or downspouts. Such a connection could create potential differences between lead-in conductors and various metal parts located in or on buildings, resulting in possible shock and fire hazards. An underground gas piping system is not permitted to be used as a grounding electrode.

Section 810.21(J) clarifies that the bonding requirement applies only to electrodes at the same building or structure. The use of separate radio/television grounding electrodes is not required.

(K) Electrode Connection Connections to grounding electrodes shall comply with 250.70.

This section is new in the 2005 *Code*. This requirement is similar to the requirements for grounding electrode conductors in Articles 800, 820, and 830.

III. Amateur Transmitting and Receiving Stations — Antenna Systems

810.51 Other Sections

In addition to complying with Part III, antenna systems for amateur transmitting and receiving stations shall also comply with 810.11 through 810.15.

810.52 Size of Antenna

Antenna conductors for transmitting and receiving stations shall be of a size not less than given in Table 810.52.

Table 810.52 Size of Amateur Station Outdoor Antenna Conductors

Material	Minimum Size of Conductors (AWG) Where Maximum Open Span Length Is	
	Less Than 45 m (150 ft)	Over 45 m (150 ft)
Hard-drawn copper	14	10
Copper-clad steel, bronze, or other high-strength material	14	12

810.53 Size of Lead-in Conductors

Lead-in conductors for transmitting stations shall, for various maximum span lengths, be of a size at least as great as that of conductors for antennas as specified in 810.52.

810.54 Clearance on Building

Antenna conductors for transmitting stations, attached to buildings, shall be firmly mounted at least 75 mm (3 in.) clear of the surface of the building on nonabsorbent insulating supports, such as treated pins or brackets equipped with insulators having not less than 75-mm (3-in.) creepage and airgap distances. Lead-in conductors attached to buildings shall also comply with these requirements.

Exception: Where the lead-in conductors are enclosed in a continuous metallic shield that is permanently and effectively grounded, they shall not be required to comply with these requirements. Where grounded, the metallic shield shall also be permitted to be used as a conductor.

Creepage distance is measured from the conductor across the face of the supporting insulator to the building surface. Airgap distance is measured from the conductor (at its closest point) across the air space (not necessarily in a straight line) to the surface of the building. This exception covers coaxial cable with the shield permanently and effectively grounded.

810.55 Entrance to Building

Except where protected with a continuous metallic shield that is permanently and effectively grounded, lead-in conductors for transmitting stations shall enter buildings by one of the following methods:

(1) Through a rigid, noncombustible, nonabsorbent insulating tube or bushing
(2) Through an opening provided for the purpose in which the entrance conductors are firmly secured so as to provide a clearance of at least 50 mm (2 in.)
(3) Through a drilled window pane

810.56 Protection Against Accidental Contact

Lead-in conductors to radio transmitters shall be located or installed so as to make accidental contact with them difficult.

810.57 Antenna Discharge Units — Transmitting Stations

Each conductor of a lead-in for outdoor antennas shall be provided with an antenna discharge unit or other suitable means that drain static charges from the antenna system.

If an antenna discharge unit is not installed at a transmitting station, protection against lightning may be provided by a switch that connects the lead-in conductors to ground during the times the station is not in operation.

Exception No. 1: Where protected by a continuous metallic shield that is permanently and effectively grounded.

Exception No. 2: Where the antenna is permanently and effectively grounded.

810.58 Grounding Conductors — Amateur Transmitting and Receiving Stations

Grounding conductors shall comply with 810.58(A) through 810.58(C).

(A) Other Sections All grounding conductors for amateur transmitting and receiving stations shall comply with 810.21(A) through 810.21(K).

(B) Size of Protective Grounding Conductor The protective grounding conductor for transmitting stations shall be as large as the lead-in but not smaller than 10 AWG copper, bronze, or copper-clad steel.

(C) Size of Operating Grounding Conductor The operating grounding conductor for transmitting stations shall not be less than 14 AWG copper or its equivalent.

IV. Interior Installation — Transmitting Stations

810.70 Clearance from Other Conductors

All conductors inside the building shall be separated at least 100 mm (4 in.) from the conductors of any electric light, power, or signaling circuit.

Exception No. 1: As provided in Article 640.

Exception No. 2: Where separated from other conductors by raceway or some firmly fixed nonconductor, such as porcelain tubes or flexible tubing.

810.71 General

Transmitters shall comply with 810.71(A) through 810.71(C).

(A) Enclosing The transmitter shall be enclosed in a metal frame or grille, or separated from the operating space by a barrier or other equivalent means, all metallic parts of which are effectively connected to ground.

(B) Grounding of Controls All external metal handles and controls accessible to the operating personnel shall be effectively grounded.

(C) Interlocks on Doors. All access doors shall be provided with interlocks that disconnect all voltages of over 350 volts between conductors when any access door is opened.

ARTICLE 820
Community Antenna Television and Radio Distribution Systems

Summary of Changes

- Reorganized and renumbered as part of an effort to achieve parallel structure among Articles 770, 800, 820, and 830.
- **820.24:** Removed requirement that the cable must be supported by a structural component. Clarification added that this section applies to ceilings regardless of which side of the ceiling is supporting the cable. Revised to apply the requirement of 300.11 to Article 820.
- **820.100(A)(4):** FPN added to clarify the requirement of a 20-ft maximum grounding conductor in one- and two-family dwellings.
- **820.154(A) and (B):** Revised to prohibit unlisted cables in risers, air ducts, plenums, and other space used for environmental air, even though the length does not exceed 50 ft.
- **820.182:** Plenum CATV raceways accepted for use in other spaces used for environmental air, riser CATV raceways accepted for use in vertical risers, and general-purpose CATV raceways accepted for use in other areas of a building.

Contents

FPN: Rules that are followed by a reference in brackets contain text that has been extracted from NFPA 97-2003, *Standard Glossary of Terms Relating to Chimneys, Vents, and Heat-Producing Appliances.* Only editorial changes were made to the extracted text to make it consistent with this *Code.*

I. General

820.1 Scope

This article covers coaxial cable distribution of radio frequency signals typically employed in community antenna television (CATV) systems.

Article 820 covers the installation of coaxial cable for the distribution of radio frequency (RF) signals associated with closed-circuit television, cable television, and security television cameras. This article also covers interior coaxial cable for radio and television receiving equipment. Article 830 was added to the 1999 *Code* to cover network-powered broadband system installations.

820.2 Definitions

See Article 100. For the purposes of this article, the following additional definitions apply.

Abandoned Coaxial Cable. Installed coaxial cable that is not terminated at equipment other than a coaxial connector and not identified for future use with a tag.

The definition of *abandoned coaxial cable* is used with 820.3(A) and 820.154, which require removal of accessible abandoned communications cable. Abandoned cable increases fire loading unnecessarily, and, where installed in plenums, it can affect airflow. Similar requirements can be found in Articles 640, 645, 725, 760, 770, 800, and 830.

Air Duct. A conduit or passageway for conveying air to or from heating, cooling, air conditioning, or ventilating equipment, but not including the plenum. [NFPA 97:1.2.6]

The definition of *air duct* was added to the 2005 *Code* to provide a term to distinguish between electrical ducts and ducts that form part of an environmental air distribution system.

Exposed. An exposed cable is one that is in such a position that, in case of failure of supports and insulation, contact with another circuit could result.

    FPN: See Article 100 for two other definitions of *exposed.*

Point of Entrance. The point within a building at which the cable emerges from an external wall, from a concrete floor slab, or from a rigid metal conduit or an intermediate metal conduit grounded to an electrode in accordance with 820.100(B).

Premises. The land and buildings of a user located on the user side of utility-user network point of demarcation.

820.3 Other Articles

Circuits and equipment shall comply with 820.3(A) through 820.3(G).

Paragraph 820.3(G) permits Article 830 wiring methods to substitute for those covered in Article 820. The substitution of these wiring methods facilitates an upgrade of Article 820 installations to network-powered broadband applications. It is important to note that 820.3(A) requires that accessible abandoned CATV cable be removed.

(A) Spread of Fire or Products of Combustion Section 300.21 shall apply. The accessible portion of abandoned coaxial cables shall be removed.

(B) Ducts, Plenums, and Other Air-Handling Spaces Section 300.22, where installed in ducts, plenums, or other spaces used for environmental air, shall apply.

Exception: As permitted in 820.154(A).

(C) Installation and Use Section 110.3 shall apply.

(D) Installations of Conductive and Nonconductive Optical Fiber Cables Article 770 shall apply.

(E) Communications Circuits Article 800 shall apply.

(F) Network-Powered Broadband Communications Systems Article 830 shall apply.

(G) Alternate Wiring Methods The wiring methods of Article 830 shall be permitted to substitute for the wiring methods of Article 820.

> FPN: Use of Article 830 wiring methods will facilitate the upgrading of Article 820 installations to network-powered broadband applications.

820.15 Energy Limitations

Coaxial cable shall be permitted to deliver low-energy power to equipment that is directly associated with the radio frequency distribution system if the voltage is not over 60 volts and if the current supply is from a transformer or other device that has energy-limiting characteristics.

820.21 Access to Electrical Equipment Behind Panels Designed to Allow Access

Access to electrical equipment shall not be denied by an accumulation of wires and cables that prevents removal of panels, including suspended ceiling panels.

An excess accumulation of wires and cables can limit access to equipment by preventing the removal of access panels. (See Exhibit 820.1.)

Exhibit 820.1 Installations of conductors and cables, which can prevent access to equipment or cables. Correct and incorrect methods are shown.

820.24 Mechanical Execution of Work

Community antenna television and radio distribution systems shall be installed in a neat and workmanlike manner. Cables installed exposed on the surface of ceiling and sidewalls shall be supported by the building structure in such a manner that the cable will not be damaged by normal building use. Such cables shall be secured by straps, staples, hangers, or similar fittings designed and installed so as not to damage the cable. The installation shall also conform with 300.4(D) and 300.11.

Section 820.24 provides clear requirements for workmanship. Cables are required to be attached to or supported by the structure by straps, clamps, hangers, and the like. The installation method must not damage the cable. In addition, the location of the cable must be carefully evaluated to ensure that activities and processes within the building do not damage the cable. In the 2005 *Code*, there was a change to this section to permit attachment to baseboards and non–load bearing walls, which are not structural components.

> FPN: Accepted industry practices are described in ANSI/NECA/BICSI 568–2001, *Standard for Installing Com-*

mercial Building Telecommunications Cabling, and other ANSI-approved installation standards.

II. Cables Outside and Entering Buildings

820.44 Overhead Cables

Coaxial cables, prior to the point of grounding, as defined in 820.93, shall comply with 820.44(A) through 820.44(F).

(A) On Poles Where practicable, conductors on poles shall be located below the electric light, power, Class 1, or non–power-limited fire alarm circuit conductors and shall not be attached to a crossarm that carries electric light or power conductors.

(B) Lead-in Clearance Lead-in or aerial-drop cables from a pole or other support, including the point of initial attachment to a building or structure, shall be kept away from electric light, power, Class 1, or non–power-limited fire alarm circuit conductors so as to avoid the possibility of accidental contact.

Exception: Where proximity to electric light, power, Class 1, or non–power-limited fire alarm circuit service conductors cannot be avoided, the installation shall be such as to provide clearances of not less than 300 mm (12 in.) from light, power, Class 1, or non–power-limited fire alarm circuit service drops. The clearance requirement shall apply at all points along the drop, and it shall increase to 1.02 m (40 in.) at the pole.

(C) On Masts Aerial cable shall be permitted to be attached to an above-the-roof raceway mast that does not enclose or support conductors of electric light or power circuits.

(D) Above Roofs Cables shall have a vertical clearance of not less than 2.5 m (8 ft) from all points of roofs above which they pass.

Exception No. 1: Auxiliary buildings such as garages and the like.

Exception No. 2: A reduction in clearance above only the overhanging portion of the roof to not less than 450 mm (18 in.) shall be permitted if (1) not more than 1.2 m (4 ft) of communications service drop conductors pass above the roof overhang, and (2) they are terminated at a raceway mast or other approved support.

Exception No. 3: Where the roof has a slope of not less than 100 mm in 300 mm (4 in. in 12 in.), a reduction in clearance to not less than 900 mm (3 ft) shall be permitted.

(E) Between Buildings Cables extending between buildings and also the supports or attachment fixtures shall be acceptable for the purpose and shall have sufficient strength to withstand the loads to which they may be subjected.

Wind and ice loads, which can be excessive, should be considered.

Exception: Where a cable does not have sufficient strength to be self-supporting, it shall be attached to a supporting messenger cable that, together with the attachment fixtures or supports, shall be acceptable for the purpose and shall have sufficient strength to withstand the loads to which they may be subjected.

(F) On Buildings Where attached to buildings, cables shall be securely fastened in such a manner that they will be separated from other conductors in accordance with 820.44(F)(1), (F)(2), and (F)(3).

(1) Electric Light or Power The coaxial cable shall have a separation of at least 100 mm (4 in.) from electric light, power, Class 1, or non–power-limited fire alarm circuit conductors not in raceway or cable, or shall be permanently separated from conductors of the other system by a continuous and firmly fixed nonconductor in addition to the insulation on the wires.

(2) Other Communications Systems Coaxial cable shall be installed so that there will be no unnecessary interference in the maintenance of the separate systems. In no case shall the conductors, cables, messenger strand, or equipment of one system cause abrasion to the conductors, cable, messenger strand, or equipment of any other system.

(3) Lightning Conductors Where practicable, a separation of at least 1.8 m (6 ft) shall be maintained between any coaxial cable and lightning conductors.

> FPN: For additional information regarding overhead wires and cables, see ANSI C2-2002, *National Electric Safety Code*, Part 2, Safety Rules for Overhead Lines.

820.47 Underground Circuits Entering Buildings

(A) Underground Systems Underground coaxial cables in a duct, pedestal, handhole enclosure, or manhole that contains electric light or power conductors or Class 1 circuits shall be in a section permanently separated from such conductors by means of a suitable barrier.

(B) Direct-Buried Cables and Raceways Direct-buried coaxial cable shall be separated at least 300 mm (12 in.) from conductors of any light or power or Class 1 circuit.

Exception No. 1: Where electric service conductors or coaxial cables are installed in raceways or have metal cable armor.

Exception No. 2: Where electric light or power branch-circuit or feeder conductors or Class 1 circuit conductors are installed in a raceway or in metal-sheathed, metal-clad, or Type UF or Type USE cables; or the coaxial cables have metal cable armor or are installed in a raceway.

III. Protection

820.93 Grounding of Outer Conductive Shield of a Coaxial Cable

The outer conductive shield of the coaxial cable shall be grounded at the building premises as close to the point of cable entrance or attachment as practicable.

For purposes of this section, grounding located at mobile home service equipment located in sight from, and not more than 9.0 m (30 ft) from, the exterior wall of the mobile home it serves, or at a mobile home disconnecting means grounded in accordance with 250.32 and located in sight from and not more than 9.0 m (30 ft) from the exterior wall of the mobile home it serves, shall be considered to meet the requirements of this section.

FPN: Selecting a grounding location to achieve the shortest practicable grounding conductor helps limit potential differences between CATV and other metallic systems.

(A) Shield Grounding Where the outer conductive shield of a coaxial cable is grounded, no other protective devices shall be required.

(B) Shield Protection Devices Grounding of a coaxial drop cable shield by means of a protective device that does not interrupt the grounding system within the premises shall be permitted.

Section 820.93(B) permits the use of a shield protection device that does not interrupt the grounding system within the premises. The electric utility supply, the CATV system, and the premises wiring are all grounded. When a ground fault occurs at the premises, the current tries to return to its source and follows the multiple paths through the different premises metal water piping systems or earth. Such ground faults can cause current on the CATV shield, whose primary function is to prevent RF leakage out of the cable. The fault current in the cable shield can cause the shield to burn open and also damage the cable insulation. A device that can safely conduct current at 60 Hz and block current at the higher frequencies can be connected between the cable shield and ground, thereby maintaining grounding integrity. An ordinary fuse, for example, would not be suitable.

IV. Grounding Methods

820.100 Cable Grounding

Where required by 820.93, the shield of the coaxial cable shall be grounded as specified in 820.100(A) through 820.100(D).

(A) Grounding Conductor

(1) Insulation The grounding conductor shall be insulated and shall be listed as suitable for the purpose.

(2) Material The grounding conductor shall be copper or other corrosion-resistant conductive material, stranded or solid.

(3) Size The grounding conductor shall not be smaller than 14 AWG. It shall have a current-carrying capacity approximately equal to that of the outer conductor of the coaxial cable. The grounding conductor shall not be required to exceed 6 AWG.

(4) Length The grounding conductor shall be as short as practicable. In one- and two-family dwellings, the grounding conductor shall be as short as practicable, not to exceed 6.0 m (20 ft) in length.

The limitation on length will result in a lower impedance, which will limit the potential difference between communications systems and other systems during a lightning strike. Large potential differences between grounding conductors can result in increased damage if a lightning strike were to occur.

FPN: Similar grounding conductor length limitations applied at apartment buildings and commercial buildings will help to reduce voltages that may be developed between the building's power and communications systems during lightning events.

When the 20-ft limitation was instituted in the 2002 *Code*, the predominant application was in one- and two-family dwellings; apartment and commercial buildings were specifically not addressed. In the 2005 *Code*, some guidance is provided for apartment and commercial buildings, without being overly restrictive because of intersystem bonding situations that may exist at these facilities. This FPN provides guidance for the treatment of the cable and primary protector grounding conductor length at apartment and commercial buildings that is consistent with the 20-ft rule for one- and two-family dwellings. However, a specific length is not specified in the *Code* because such a length limitation may not be practical in some installations.

Exception: In one- and two-family dwellings where it is not practicable to achieve an overall maximum grounding conductor length of 6.0 m (20 ft), a separate ground as specified in 250.52(A)(5), (A)(6), or (A)(7) shall be used, the grounding conductor shall be grounded to the separate ground in accordance with 250.70, and the separate ground bonded to the power grounding electrode system in accordance with 820.100(D).

(5) Run in Straight Line The grounding conductor shall be run to the grounding electrode in as straight a line as practicable.

(6) Physical Protection Where subject to physical damage, the grounding conductor shall be adequately protected. Where the grounding conductor is run in a metal raceway, both ends of the raceway shall be bonded to the grounding conductor or the same terminal or electrode to which the grounding conductor is connected.

(B) Electrode The grounding conductor shall be connected in accordance with 820.100(B)(1) and (B)(2).

(1) In Buildings or Structures with Grounding Means To the nearest accessible location on the following:

(1) The building or structure grounding electrode system as covered in 250.50
(2) The grounded interior metal water piping system, within 1.52 m (5 ft) from its point of entrance to the building, as covered in 250.52
(3) The power service accessible means external to enclosures as covered in 250.94
(4) The metallic power service raceway
(5) The service equipment enclosure
(6) The grounding electrode conductor or the grounding electrode conductor metal enclosure, or
(7) The grounding conductor or the grounding electrode of a building or structure disconnecting means that is grounded to an electrode as covered in 250.32

See commentary following 250.52(A)(1).

(2) In Buildings or Structures Without Grounding Means If the building or structure served has no grounding means, as described in 820.100(B)(1):

(1) To any one of the individual electrodes described in 250.52(A)(1), (A)(2), (A)(3), (A)(4); or,
(2) If the building or structure served has no grounding means, as described in 820.100(B)(1) or (B)(2)(1), to an effectively grounded metal structure or to any one of the individual electrodes described in 250.52(A)(5), (A)(6), and (A)(7).

(C) Electrode Connection Connections to grounding electrodes shall comply with 250.70.

(D) Bonding of Electrodes A bonding jumper not smaller than 6 AWG copper or equivalent shall be connected between the community antenna television system's grounding electrode and the power grounding electrode system at the building or structure served where separate electrodes are used.

Paragraph 820.100(D) requires bonding of CATV and power grounding electrodes at the same building or structure.

Exception: At mobile homes as covered in 820.106.

FPN No. 1: See 250.60 for use of air terminals (lightning rods).
FPN No. 2: Bonding together of all separate electrodes limits potential differences between them and between their associated wiring systems.

The most common error made in grounding CATV systems is connecting the coaxial cable sheath to a ground rod driven by the CATV installer at a convenient location near the point of cable entry to the building instead of bonding it to the electrical service grounding electrode system, service raceway, or other components that make up the grounding electrode system. A separate grounding electrode is permitted by the 2005 *Code* only if the building or structure has none of the grounding means described in 820.100(B)(1) or 820.100(B)(2), which is rare. The *Code* requires that some means that is accessible and external to the service equipment be provided for making the bonding and grounding connection for other systems. One of the following means must be provided:

1. Exposed service raceways
2. Exposed grounding electrode conductor
3. An approved means for the external connection of a conductor (A 6 AWG copper conductor with one end bonded to the service raceway or equipment with about 6 in. exposed is acceptable.)

Proper bonding of the CATV system coaxial cable sheath to the electrical power ground is needed to prevent potential fire and shock hazards. The earth cannot be used as an equipment grounding conductor or bonding conductor because it does not have the low-impedance path required. (See 250.54.)

Both CATV systems and power systems are subject to current surges as a result of, for example, induced voltages from lightning in the vicinity of the usually extensive outside distribution systems. Surges also result from switching operations on power systems. If the grounded conductors and parts of the two systems are not bonded by a low-impedance path, such line surges can raise the potential difference between the two systems to many thousands of volts. This can result in arcing between the two systems, for example, wherever the coaxial cable jacket contacts a grounded part, such as a metal water pipe or metal structural member, inside the building.

If a person is the interface between the two systems and the bonding has not been done in accordance with the *Code*, the high-voltage surge could result in electric shock.

More common, however, is burnout of a television tuner, a part that is almost always an interface between the two systems. The tuner is connected to the power system ground through the grounded neutral of the power supply, even if the television set itself is not provided with an equipment grounding conductor.

Also see the commentary following 250.92(B) and 820.93(B).

820.103 Equipment Grounding

Unpowered equipment and enclosures or equipment powered by the coaxial cable shall be considered grounded where connected to the metallic cable shield.

820.106 Bonding and Grounding at Mobile Homes

(A) Grounding Where there is no mobile home service equipment located in sight from, and not more than 9.0 m (30 ft) from, the exterior wall of the mobile home it serves or there is no mobile home disconnecting means grounded in accordance with 250.32 and located within sight from, and not more than 9.0 m (30 ft) from, the exterior wall of the mobile home it serves, the coaxial cable shield ground, or surge arrester ground, shall be in accordance with 820.100(B)(2).

(B) Bonding The coaxial cable shield grounding terminal, surge arrester grounding terminal, or grounding electrode shall be bonded to the metal frame or available grounding terminal of the mobile home with a copper grounding conductor not smaller than 12 AWG under any of the following conditions:

(1) Where there is no mobile home service equipment or disconnecting means as in 820.106(A)
(2) Where the mobile home is supplied by cord and plug

V. Cables Within Buildings

820.110 Raceways for Coaxial Cables

Where coaxial cables are installed in a raceway, the raceway shall be either of a type permitted in Chapter 3 and installed in accordance with Chapter 3 or a listed nonmetallic raceway complying with 820.182(A), (B), or (C), as applicable, and installed in accordance with 362.24 through 362.56, where the requirements applicable to electrical nonmetallic tubing apply.

Exception: Conduit fill restrictions shall not apply.

820.113 Installation and Marking of Coaxial Cables

Listed coaxial cables shall be installed as wiring within buildings. Coaxial cables shall be marked in accordance

with Table 820.113. The cable voltage rating shall not be marked on the cable.

FPN: Voltage markings on cables could be misinterpreted to suggest that the cables may be suitable for Class 1, electric light, and power applications.

Table 820.113 Cable Markings

Cable Marking	Type	Reference
CATVP	CATV plenum cable	820.179(A) and 820.154(A)
CATVR	CATV riser cable	820.179(B) and 820.154(B)
CATV	CATV cable	820.179(C) and 820.154(D)
CATVX	CATV cable, limited use	820.179(D) and 820.154(D)

Exception No. 1: Voltage markings shall be permitted where the cable has multiple listings and voltage marking is required for one or more of the listings.

Exception No. 2: Listing and marking shall not be required where the length of the cable within the building, measured from its point of entrance, does not exceed 15 m (50 ft) and the cable enters the building from the outside and is terminated at a grounding block.

FPN No. 1: Cable types are listed in descending order of fire-resistance rating.
FPN No. 2: See the referenced sections for listing requirements and permitted uses.

820.133 Installation of Cables and Equipment

Beyond the point of grounding, as defined in 820.93, the cable installation shall comply with 820.133(A) through 820.133(C).

Section 820.133 specifically includes separation from network-powered broadband communications circuits. Jackets of coaxial cable do not have sufficient construction specifications to permit them to be installed with electric light, power, Class 1, non–power-limited fire alarm circuits, and medium- and high-power network-powered broadband communications cable. Failure of the cable insulation due to a fault could lead to hazardous voltages being imposed on the Class 2 or Class 3 circuit conductors.

(A) Separation from Other Conductors

(1) In Raceways and Boxes

(1) Other Circuits. Coaxial cables shall be permitted in the same raceway or enclosure with jacketed cables of any of the following:

a. Class 2 and Class 3 remote-control, signaling, and power-limited circuits in compliance with Article 725

b. Power-limited fire alarm systems in compliance with Article 760

c. Nonconductive and conductive optical fiber cables in compliance with Article 770

d. Communications circuits in compliance with Article 800

e. Low-power network-powered broadband communications circuits in compliance with Article 830

(2) Electric Light, Power, Class 1, Non–Power-Limited Fire Alarm, and Medium Power Network-Powered Broadband Communications Circuits. Coaxial cable shall not be placed in any raceway, compartment, outlet box, junction box, or other enclosures with conductors of electric light, power, Class 1, non–power-limited fire alarm, or medium power network-powered broadband communications circuits.

Exception No. 1: Where all of the conductors of electric light, power, Class 1, non–power-limited fire alarm, and medium power network-powered broadband communications circuits are separated from all of the coaxial cables by a permanent barrier or listed divider.

This exception has been revised to recognize the use of a listed field-installed divider to separate the communications circuits from the power circuits.

Exception No. 2: Power circuit conductors in outlet boxes, junction boxes, or similar fittings or compartments where such conductors are introduced solely for power supply to the coaxial cable system distribution equipment. The power circuit conductors shall be routed within the enclosure to maintain a minimum 6-mm (0.25-in.) separation from coaxial cables.

(2) Other Applications Coaxial cable shall be separated at least 50 mm (2 in.) from conductors of any electric light, power, Class 1, non–power-limited fire alarm, or medium power network-powered broadband communications circuits.

Exception No. 1: Where either (1) all of the conductors of electric light, power, Class 1, non–power-limited fire alarm, and medium power network-powered broadband communications circuits are in a raceway, or in metal-sheathed, metal-clad, nonmetallic-sheathed Type AC or Type UF cables, or (2) all of the coaxial cables are encased in raceway.

Exception No. 2: Where the coaxial cables are permanently separated from the conductors of electric light, power, Class 1, non–power-limited fire alarm, and medium power net-

work-powered broadband communications circuits by a continuous and firmly fixed nonconductor, such as porcelain tubes or flexible tubing, in addition to the insulation on the wire.

(B) Hybrid Power and Coaxial Cabling The provisions of 780.6 shall apply for listed hybrid power and coaxial cabling in closed-loop and programmed power distribution.

(C) Support of Cables Raceways shall be used for their intended purpose. Coaxial cables shall not be strapped, taped, or attached by any means to the exterior of any conduit or raceway as a means of support.

Exception: Overhead (aerial) spans of coaxial cables shall be permitted to be attached to the exterior of a raceway-type mast intended for the attachment and support of such cables.

820.154 Applications of Listed CATV Cables and CATV Raceways

CATV cables shall comply with the requirements of 820.154(A) through 820.154(D) or where cable substitutions are made as shown in Table 820.154.

Table 820.154 Coaxial Cable Uses and Permitted Substitutions

Cable Type	Permitted Substitutions
CATVP	CMP, BLP
CATVR	CATVP, CMP, CMR, BMR, BLP, BLR
CATV	CATVP, CMP, CATVR, CMR, CMG, CM, BMR, BM, BLP, BLR, BL
CATVX	CATVP, CMP, CATVR, CMR, CATV, CMG, CM, BMR, BM, BLP, BLR, BL, BLX

(A) Plenums Cables installed in ducts, plenums, and other spaces used for environmental air shall be Type CATVP. Abandoned cables shall not be permitted to remain. Types CATVP, CATVR, CATV, and CATVX cables installed in compliance with 300.22 shall be permitted. Listed plenum CATV raceways shall be permitted to be installed in ducts and plenums as described in 300.22(B) and in other spaces used for environmental air as described in 300.22(C). Only Type CATVP cable shall be permitted to be installed in these raceways.

> FPN: See 8.14.1 of NFPA 13-2002, *Installation of Sprinkler Systems*, for requirements for sprinklers in concealed spaces containing exposed combustibles.

(B) Riser Cables installed in risers shall comply with any of the requirements of 820.154(B)(1) through (B)(3).

(1) Cables in Vertical Runs Cables installed in vertical runs and penetrating more than one floor, or cables installed in vertical runs in a shaft, shall be Type CATVR. Floor penetrations requiring Type CATVR shall contain only cables suitable for riser or plenum use. Abandoned cables shall not be permitted to remain. Listed riser CATV raceways shall be permitted to be installed in vertical riser runs in a shaft from floor to floor. Only Type CATVR and CATVP cables shall be permitted to be installed in these raceways.

(2) Metal Raceways or Fireproof Shafts Types CATV and CATVX cables shall be permitted to be encased in a metal raceway or located in a fireproof shaft having firestops at each floor.

(3) One- and Two-Family Dwellings Types CATV and CATVX cables shall be permitted in one- and two-family dwellings.

> FPN: See 820.3(A) for the firestop requirements for floor penetrations.

(C) Cable Trays Cables installed in cable trays shall be Types CATVP, CATVR, and CATV.

(D) Other Wiring Within Buildings Cables installed in building locations other than the locations covered in 820.154(A) and 820.154(B) shall be with any of the requirements in 820.154(D)(1) through (D)(5). Abandoned cables in hollow spaces shall not be permitted to remain.

Used in conjunction with the definition of *abandoned coaxial cable* in 820.2 and the general requirement in 820.3(A), this section requires removal of accessible abandoned CATV cable. Abandoned cable increases fire loading unnecessarily, and, where installed in plenums, it can affect airflow. Similar requirements can be found in Articles 640, 645, 725, 760, 770, 800, and 830.

(1) General Type CATV shall be permitted. Listed CATV general-purpose raceways shall be permitted. Only Types CATV, CATVX, CATVR, or CATVP cables shall be permitted to be installed in general-purpose communications raceways.

(2) In Raceways Type CATVX shall be permitted to be installed in a raceway.

(3) Nonconcealed Spaces Type CATVX shall be permitted to be installed in nonconcealed spaces where the exposed length of cable does not exceed 3 m (10 ft).

(4) One- and Two-Family Dwellings Type CATVX cables less than 10 mm (0.375 in.) in diameter shall be permitted to be installed in one- and two-family dwellings.

(5) Multifamily Dwellings Type CATVX cables less than 10 mm (0.375 in.) in diameter shall be permitted to be installed in multifamily dwellings.

> FPN No. 1: See Figure 820.154, Cable Substitution Hierarchy.
>
> FPN No. 2: The substitute cables in Table 820.154 are only coaxial-type cables.

Type CATV—Community antenna television cables
Type CM—Communications cables

[A]→[B]   Coaxial cable A shall be permitted to be used in place of coaxial cable B.

Figure 820.154 Cable Substitution Hierarchy.

VI. Listing Requirements

820.179 Coaxial Cables

Cables shall be listed in accordance with 820.179(A) through 820.179(D).

(A) Type CATVP Type CATVP community antenna television plenum cable shall be listed as being suitable for use in ducts, plenums, and other spaces used for environmental air and shall also be listed as having adequate fire-resistant and low smoke-producing characteristics.

See the commentary following 725.82(A), FPN, for more information on a test for wires and cables to be installed in plenums and other spaces used for environmental air.

> FPN: One method of defining a cable that is low smoke-producing cable and fire-resistant cable is that the cable exhibits a maximum peak optical density of 0.5 or less, an average optical density of 0.15 or less, and a maximum flame spread distance of 1.52 m (5 ft) or less when tested in accordance with NFPA 262, *Standard Method of Test for Flame Travel and Smoke of Wires and Cables for Use in Air Handling Spaces.*

(B) Type CATVR Type CATVR community antenna television riser cable shall be listed as being suitable for use in a vertical run in a shaft or from floor to floor and shall also be listed as having fire-resistant characteristics capable of preventing the carrying of fire from floor to floor.

> See the commentary following 725.82(B), FPN, for information on a test method for defining fire-resistant characteristics capable of preventing fire spread from floor to floor.

> FPN: One method of defining fire-resistant characteristics capable of preventing the carrying of fire from floor to floor is that the cables pass the requirements of ANSI/UL 1666-2002, *Standard Test for Flame Propagation Height of Electrical and Optical-Fiber Cable Installed Vertically in Shafts.*

(C) Type CATV Type CATV community antenna television cable shall be listed as being suitable for general-purpose CATV use, with the exception of risers and plenums, and shall also be listed as being resistant to the spread of fire.

> See the commentary following 725.82(C), FPN, for information on the UL vertical flame test.

> FPN: One method of defining *resistant to the spread of fire* is that the cables do not spread fire to the top of the tray in the vertical-tray flame test in ANSI/UL 1581-1991, *Reference Standard for Electrical Wires, Cables and Flexible Cords.*
>   Another method of defining *resistant to the spread of fire* is for the damage (char length) not to exceed 1.5 m (4 ft 11 in.) when performing the vertical flame test for cables in cable trays, as described in CSA C22.2 No. 0.3-M-1985, *Test Methods for Electrical Wires and Cables.*

(D) Type CATVX Type CATVX limited-use community antenna television cable shall be listed as being suitable for use in dwellings and for use in raceway and shall also be listed as being resistant to flame spread.

> See the commentary following 725.82(D), FPN, for information on test methods for determining whether cable is resistant to fire spread.

> FPN: One method of determining that cable is resistant to flame spread is by testing the cable to the VW-1 (vertical-wire) flame test in ANSI/UL 1581-1991, *Reference Standard for Electrical Wires, Cables and Flexible Cords.*

820.182 CATV Raceways

CATV raceways shall be listed in accordance with 820.182(A), through 820.182(C).

(A) Plenum CATV Raceways Plenum CATV raceways shall be listed for use in other spaces used for environmental air and shall also be listed as having adequate fire-resistant and low smoke-producing characteristics.

> FPN: One method of defining that an optical fiber raceway is a low smoke producing raceway and a fire-resistant raceway is that the raceway exhibits a maximum peak optical density of 0.5 or less, an average optical density of 0.15 or less, and a maximum flame spread distance of 1.52 m (5 ft) or less when tested in accordance with the plenum test in UL 2024, *Standard for Optical-Fiber Cable Raceway.*

(B) Riser CATV Raceways Riser CATV raceways shall be listed for use in risers and shall also be listed as having adequate fire-resistant characteristics capable of preventing the carrying of fire from floor to floor.

> FPN: One method of defining fire-resistant characteristics capable of preventing the carrying of fire from floor to floor is that the raceways pass the requirements of the Test for Flame Propagation (Riser) in UL 2024, *Standard for Optical-Fiber Cable Raceway.*

(C) General-Purpose CATV Raceways General purpose CATV raceways shall be listed suitable for general-purpose use and shall also be listed as being resistant to the spread of fire.

> FPN: One method of defining resistance to the spread of fire is that the raceway pass the requirements of the Vertical-Tray Flame Test (General Use) in UL 2024, *Standard for Optical-Fiber Cable Raceway.*

ARTICLE 830
Network-Powered Broadband Communications Systems

Summary of Changes

- Reorganized and renumbered as part of an effort to achieve parallel structure among Articles 770, 800, and 820.

- **830.24:** Requirement that the cable must be supported by a structural component removed. Clarification added that this section applies to ceilings regardless of which side of the ceiling is supporting the cable. Revised to apply the requirement of 300.11 to Article 830.

- **830.100(A)(4):** FPN added to clarify the requirement of a 20-ft maximum grounding conductor in one- and two-family dwellings.

- **830.133(C) and Table 830.133:** Permitted broadband cable substitutions expanded.

Contents

I. General

830.1 Scope

This article covers network-powered broadband communications systems that provide any combination of voice, audio, video, data, and interactive services through a network interface unit.

Article 830 covers network-powered broadband communications circuits, which provide a wide array of subscriber services, including voice, data (such as internet access), interactive services, and television signals.

FPN No. 1: A typical basic system configuration includes a cable supplying power and broadband signal to a network interface unit that converts the broadband signal to the component signals. Typical cables are coaxial cable with both broadband signal and power on the center conductor, composite metallic cable with a coaxial member for the broadband signal and a twisted pair for power, and composite optical fiber cable with a pair of conductors for power. Larger systems may also include network components such as amplifiers that require network power.

FPN No. 2: See 90.2(B)(4) for installations of broadband communications systems that are not covered.

830.2 Definitions

See Article 100. For purposes of this article, the following additional definitions apply.

Abandoned Network-Powered Broadband Communications Cable. Installed network-powered broadband communications cable that is not terminated at equipment other than a connector and not identified for future use with a tag.

The definition of *abandoned network-powered broadband communications cable* is used with 830.3(A) and 830.154, which require removal of accessible abandoned broadband communications cable. Abandoned cable increases fire loading unnecessarily, and, where installed in plenums, it can affect airflow. Similar requirements can be found in Articles 640, 645, 725, 760, 770, 800, and 820.

Block. A square or portion of a city, town, or village enclosed by streets, including the alleys so enclosed but not any street.

Exposed to Accidental Contact with Electrical Light or Power Conductors. A circuit in such a position that, in case of failure of supports or insulation, contact with another circuit may result.

Fault Protection Device. An electronic device that is intended for the protection of personnel and functions under fault conditions, such as network-powered broadband communications cable short or open circuit, to limit the current or voltage, or both, for a low-power network-powered broadband communications circuit and provide acceptable protection from electric shock.

Network Interface Unit (NIU). A device that converts a broadband signal into component voice, audio, video, data, and interactive services signals. The NIU provides isolation between the network power and the premises signal circuits. The NIU may also contain primary and secondary protectors.

Network-Powered Broadband Communications Circuit. The circuit extending from the communications utility's serving terminal or tap up to and including the NIU.

> FPN: A typical single-family network-powered communications circuit consists of a communications drop or communications service cable and an NIU and includes the communications utility's serving terminal or tap where it is not under the exclusive control of the communications utility.

Point of Entrance. The point within a building at which the cable emerges from an external wall, from a concrete floor slab, or from a rigid metal conduit or an intermediate metal conduit grounded to an electrode in accordance with 830.100(B).

Premises Wiring. The circuits located on the user side of the network interface unit.

830.3 Other Articles

Circuits and equipment shall comply with 830.3(A) through 830.3(E).

Article 830 contains provisions for wiring both the inside and the outside of buildings. Other articles cover the wiring derived from the network interface unit (NIU) into the premises. For example, Article 725 covers wiring of Class 2 and Class 3 circuits, Article 760 covers wiring of fire alarm systems, Article 770 covers the installation of optical fiber cable, Article 800 covers telephone wiring, and Article 820 covers coaxial cable installations for television signals. The major difference between Article 820 and Article 830 is the voltage present on the circuit conductors. Article 820 systems are limited to 60 volts, but Article 830 systems are permitted to have ratings as high as 150 volts. Higher voltages allow systems to power more sophisticated electronics and to provide a wider variety of services. (See Exhibit 830.1.)

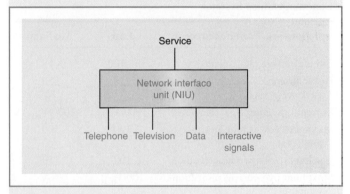

Exhibit 830.1 A network interface unit (NIU) diagram showing derived circuits.

(A) Spread of Fire or Products of Combustion Section 300.21 shall apply. The accessible portion of abandoned network-powered broadband communications cables shall be removed.

(B) Ducts, Plenums, and Other Air-Handling Spaces Section 300.22 shall apply where installed in ducts, plenums, or other spaces used for environmental air.

Exception: As permitted in 830.154(B).

(C) Equipment in Other Space Used for Environmental Air Section 300.22(C) shall apply.

(D) Output Circuits As appropriate for the services provided, the output circuits derived from the network interface unit shall comply with the requirements of the following:

(1) Installations of communications circuits — Article 800
(2) Installations of community antenna television and radio distribution circuits — Article 820

Exception: 830.90(B)(3) shall apply where protection is provided in the output of the NIU.

(3) Installations of optical fiber cables — Article 770

(4) Installations of Class 2 and Class 3 circuits — Article 725

(5) Installations of power-limited fire alarm circuits — Article 760

(E) Hazardous (Classified) Locations Network-powered broadband communications circuits and equipment installed in a location that is classified in accordance with Article 500 shall comply with the applicable requirements of Chapter 5.

830.15 Power Limitations

Network-powered broadband communications systems shall be classified as having low or medium power sources as defined in Table 830.15.

Table 830.15 Limitations for Network-Powered Broadband Communications Systems

Network Power Source	Low	Medium
Circuit voltage, V_{max} (volts)[1]	0–100	0–150
Power limitation, VA_{max} (volt-amperes)[1]	250	250
Current limitation, I_{max} (amperes)[1]	$1000/V_{max}$	$1000/V_{max}$
Maximum power rating (volt-amperes)	100	100
Maximum voltage rating (volts)	100	150
Maximum overcurrent protection (amperes)[2]	$100/V_{max}$	NA

[1] V_{max}, I_{max}, and VA_{max} are determined with the current-limiting impedance in the circuit (not bypassed) as follows:

V_{max}—Maximum system voltage regardless of load with rated input applied.

I_{max}—Maximum system current under any noncapacitive load, including short circuit, and with overcurrent protection bypassed if used. I_{max} limits apply after 1 minute of operation.

VA_{max}—Maximum volt-ampere output after 1 minute of operation regardless of load and overcurrent protection bypassed if used.

[2] Overcurrent protection is not required where the current-limiting device provides equivalent current limitation and the current-limiting device does not reset until power or the load is removed.

High-power network-powered broadband communications circuits are not covered by the 2005 *Code*.

830.21 Access to Electrical Equipment Behind Panels Designed to Allow Access

Access to electrical equipment shall not be denied by an accumulation of wires and cables that prevents removal of panels, including suspended ceiling panels.

An excess accumulation of wires and cables can limit access to equipment by preventing the removal of access panels. (See Exhibit 830.2.)

Exhibit 830.2 Installations of conductors and cables, which can prevent access to equipment or cables. Correct and incorrect methods are shown.

830.24 Mechanical Execution of Work

Network-powered broadband communications circuits and equipment shall be installed in a neat and workmanlike manner. Cables installed exposed on the surface of ceilings and sidewalls shall be supported by the building structure in such a manner that the cable will not be damaged by normal building use. Such cables shall be secured by straps, staples, hangers, or similar fittings designed and installed so as not to damage the cable. The installation shall also conform with 300.4(D) and 300.11.

Section 830.24 of the 2005 *Code* now provides definitive requirements for workmanship. Cable must be attached to or supported on the structure by straps, clamps, hangers, and the like. The installation method must not damage the cable. In addition, the location of the cable must be carefully evaluated to ensure that activities and processes within the building do not damage the cable. In the 2005 *Code*, there was a change to this section to permit attachment to baseboards and non–load bearing walls, which are not "structural components."

FPN: Accepted industry practices are described in ANSI/NECA/BICSI 568–2001, *Standard for Installing Com-*

mercial Building Telecommunications Cabling, and other ANSI-approved installation standards.

II. Cables Outside and Entering Buildings

830.40 Entrance Cables

Network-powered broadband communications cables located outside and entering buildings shall comply with 830.40(A) and 830.40(B).

(A) Medium Power Circuits Medium power network-powered broadband communications circuits located outside and entering buildings shall be installed using Type BMU, Type BM, or Type BMR network-powered broadband communications medium power cables.

(B) Low-Power Circuits Low-power network-powered broadband communications circuits located outside and entering buildings shall be installed using Type BLU or Type BLX low-power network-powered broadband communications cables. Cables shown in Table 830.133 shall be permitted to substitute.

Exception: Outdoor community antenna television and radio distribution system coaxial cables installed prior to January 1, 2000, and installed in accordance with Article 820, shall be permitted for low-power-type, network-powered broadband communications circuits.

Network-powered broadband communications systems may contain sufficient energy to pose injurious or fatal electrical shock hazards. For that reason, they are subject to requirements similar to those for other high-powered circuits.

Section 830.40 requires that conductor spans be of sufficient size and strength to maintain clearances and avoid possible contact with light or power conductors. Splices and joints must be made with approved connectors or other means that provide sufficient mechanical strength so that conductors are not weakened appreciably, which could cause them to break and come into contact with higher-voltage conductors. Wind and ice loads, which can be excessive, also must be considered.

830.44 Aerial Cables

Aerial network-powered broadband communications cables shall comply with 830.44(A) through 830.44(I).

FPN: For additional information regarding overhead wires and cables, see ANSI C2-2002, *National Electric Safety Code*, Part 2, Safety Rules For Overhead Lines.

(A) On Poles Where practicable, network-powered broadband communications cables on poles shall be located below the electric light, power, Class 1, or non–power-limited fire alarm circuit conductors and shall not be attached to a cross-arm that carries electric light or power conductors.

(B) Climbing Space The climbing space through network-powered broadband communications cables shall comply with the requirements of 225.14(D).

(C) Lead-in Clearance Lead-in or aerial-drop network-powered broadband communications cables from a pole or other support, including the point of initial attachment to a building or structure, shall be kept away from electric light, power, Class 1, or non–power-limited fire alarm circuit conductors so as to avoid the possibility of accidental contact.

Exception: Where proximity to electric light, power, Class 1, or non–power-limited fire alarm circuit service conductors cannot be avoided, the installation shall be such as to provide clearances of not less than 300 mm (12 in.) from light, power, Class 1, or non–power-limited fire alarm circuit service drops. The clearance requirement shall apply to all points along the drop, and it shall increase to 1.02 m (40 in.) at the pole.

(D) Clearance from Ground Overhead spans of network-powered broadband communication cables shall conform to not less than the following:

(1) 2.9 m (9.5 ft) — above finished grade, sidewalks, or from any platform or projection from which they might be reached and accessible to pedestrians only
(2) 3.5 m (11.5 ft) — over residential property and driveways, and those commercial areas not subject to truck traffic
(3) 4.7 m (15.5 ft) — over public streets, alleys, roads, parking areas subject to truck traffic, driveways on other than residential property, and other land traversed by vehicles such as cultivated, grazing, forest, and orchard

FPN: These clearances have been specifically chosen to correlate with ANSI C2-2002, *National Electrical Safety Code*, Table 232-1, which provides for clearances of wires, conductors, and cables above ground and roadways, rather than using the clearances referenced in 225.18. Because Article 800 and Article 820 have had no required clearances, the communications industry has used the clearances from the NESC for their installed cable plant.

(E) Over Pools Clearance of network-powered broadband communications cable in any direction from the water level, edge of pool, base of diving platform, or anchored raft shall comply with those clearances in 680.8.

(F) Above Roofs Network-powered broadband communications cables shall have a vertical clearance of not less than 2.5 m (8 ft) from all points of roofs above which they pass.

Exception No. 1: Auxiliary buildings such as garages and the like.

Exception No. 2: A reduction in clearance above only the overhanging portion of the roof to not less than 450 mm (18 in.) shall be permitted if (1) not more than 1.2 m (4 ft) of the broadband communications drop cables pass above the roof overhang, and (2) they are terminated at a through-the-roof raceway or support.

Exception No. 3: Where the roof has a slope of not less than 100 mm in 300 mm (4 in. in 12 in.), a reduction in clearance to not less than 900 mm (3 ft) shall be permitted.

(G) Final Spans Final spans of network-powered broadband communications cables without an outer jacket shall be permitted to be attached to the building, but they shall be kept not less than 900 mm (3 ft) from windows that are designed to be opened, doors, porches, balconies, ladders, stairs, fire escapes, or similar locations.

Exception: Conductors run above the top level of a window shall be permitted to be less than the 900-mm (3-ft) requirement above.

Overhead network-powered broadband communications cables shall not be installed beneath openings through which materials may be moved, such as openings in farm and commercial buildings, and shall not be installed where they will obstruct entrance to these building openings.

(H) Between Buildings Network-powered broadband communications cables extending between buildings and also the supports or attachment fixtures shall be acceptable for the purpose and shall have sufficient strength to withstand the loads to which they may be subjected.

Wind and ice loads, which can be excessive, must be considered.

Exception: Where a network-powered broadband communications cable does not have sufficient strength to be self-supporting, it shall be attached to a supporting messenger cable that, together with the attachment fixtures or supports, shall be acceptable for the purpose and shall have sufficient strength to withstand the loads to which they may be subjected.

(I) On Buildings Where attached to buildings, network-powered broadband communications cables shall be securely fastened in such a manner that they are separated from other conductors in accordance with 830.44(I)(1) through (I)(4).

(1) Electric Light or Power The network-powered broadband communications cable shall have a separation of at least 100 mm (4 in.) from electric light, power, Class 1, or non–power-limited fire alarm circuit conductors not in raceway or cable, or be permanently separated from conduc-

tors of the other system by a continuous and firmly fixed nonconductor in addition to the insulation on the wires.

(2) Other Communications Systems Network-powered broadband communications cables shall be installed so that there will be no unnecessary interference in the maintenance of the separate systems. In no case shall the conductors, cables, messenger strand, or equipment of one system cause abrasion to the conductors, cables, messenger strand, or equipment of any other system.

(3) Lightning Conductors Where practicable, a separation of at least 1.8 m (6 ft) shall be maintained between any network-powered broadband communications cable and lightning conductors.

(4) Protection from Damage Network-powered broadband communications cables attached to buildings and located within 2.5 m (8 ft) of finished grade shall be protected by enclosures, raceways, or other approved means.

Exception: A low-power network-powered broadband communications circuit that is equipped with a listed fault protection device, appropriate to the network-powered broadband communications cable used, and located on the network side of the network-powered broadband communications cable being protected.

830.47 Underground Circuits Entering Buildings

(A) Underground Systems Underground network-powered broadband communications cables in a duct, pedestal, handhole enclosure, or manhole that contains electric light, power conductors, non–power-limited fire alarm circuit conductors, or Class 1 circuits shall be in a section permanently separated from such conductors by means of a suitable barrier.

(B) Direct-Buried Cables and Raceways Direct-buried network-powered broadband communications cables shall be separated at least 300 mm (12 in.) from conductors of any light, power, non–power-limited fire alarm circuit conductors or Class 1 circuit.

Exception No. 1: Where electric service conductors or network-powered broadband communications cables are installed in raceways or have metal cable armor.

Exception No. 2: Where electric light or power branch-circuit or feeder conductors, non–power-limited fire alarm circuit conductors, or Class 1 circuit conductors are installed in a raceway or in metal-sheathed, metal-clad, or Type UF or Type USE cables; or the network-powered broadband communications cables have metal cable armor or are installed in a raceway.

(C) Mechanical Protection Direct-buried cable, conduit, or other raceways shall be installed to meet the minimum cover requirements of Table 830.47. In addition, direct-buried cables emerging from the ground shall be protected by enclosures, raceways, or other approved means extending from the minimum cover distance required by Table 830.47 below grade to a point at least 2.5 m (8 ft) above finished grade. In no case shall the protection be required to exceed 450 mm (18 in.) below finished grade. Type BMU and BLU direct-buried cables emerging from the ground shall be installed in rigid metal conduit, intermediate metal conduit, rigid nonmetallic conduit, or other approved means extending from the minimum cover distance required by Table 830.47 below grade to the point of entrance.

Exception: A low-power network-powered broadband communications circuit that is equipped with a listed fault protection device, appropriate to the network-powered broadband communications cable used, and located on the network side of the network-powered broadband communications cable being protected.

(D) Pools Cables located under the pool or within the area extending 1.5 m (5 ft) horizontally from the inside wall of the pool shall meet those clearances and requirements specified in 680.10.

III. Protection

830.90 Primary Electrical Protection

(A) Application Primary electrical protection shall be provided on all network-powered broadband communications conductors that are neither grounded nor interrupted and are run partly or entirely in aerial cable not confined within a block. Also, primary electrical protection shall be provided on all aerial or underground network-powered broadband communications conductors that are neither grounded nor interrupted and are located within the block containing the building served so as to be exposed to lightning or accidental contact with electric light or power conductors operating at over 300 volts to ground.

Exception: Where electrical protection is provided on the derived circuit(s) (output side of the NIU) in accordance with 830.90(B)(3).

FPN No. 1: On network-powered broadband communications conductors not exposed to lightning or accidental contact with power conductors, providing primary electrical protection in accordance with this article helps protect against other hazards, such as ground potential rise caused by power fault currents, and above-normal

Table 830.47 Network-Powered Broadband Communications Systems Minimum Cover Requirements
(*Cover* is the shortest distance measured between a point on the top surface of any direct-buried cable, conduit, or other raceway and the top surface of finished grade, concrete, or similar cover.)

Location of Wiring Method or Circuit	Direct Burial Cables		Rigid Metal Conduit or Intermediate Metal Conduit		Nonmetallic Raceways Listed for Direct Burial; Without Concrete Encasement or Other Approved Raceways	
	mm	in.	mm	in.	mm	in.
All locations not specified below	450	18	150	6	300	12
In trench below 50-mm (2-in.) thick concrete or equivalent	300	12	150	6	150	6
Under a building (in raceway only)	0	0	0	0	0	0
Under minimum of 100-mm (4-in.) thick concrete exterior slab with no vehicular traffic and the slab extending not less than 150 mm (6 in.) beyond the underground installation	300	12	100	4	100	4
One- and two-family dwelling driveways and outdoor parking areas and used only for dwelling-related purposes	300	12	300	12	300	12

Notes:
1. Raceways approved for burial only where concrete encased shall require a concrete envelope not less than 50 mm (2 in.) thick.
2. Lesser depths shall be permitted where cables rise for terminations or splices or where access is otherwise required.
3. Where solid rock is encountered, all wiring shall be installed in metal or nonmetallic raceway permitted for direct burial. The raceways shall be covered by a minimum of 50 mm (2 in.) of concrete extending down to rock.
4. Low-power network-powered broadband communications circuits using directly buried community antenna television and radio distribution system coaxial cables that were installed outside and entering buildings prior to January 1, 2000, in accordance with Article 820 shall be permitted where buried to a minimum depth of 300 mm (12 in.).

voltages induced by fault currents on power circuits in proximity to the network-powered broadband communications conductors.

FPN No. 2: Network-powered broadband communications circuits are considered to have a lightning exposure unless one or more of the following conditions exist:

(1) Circuits in large metropolitan areas where buildings are close together and sufficiently high to intercept lightning.
(2) Areas having an average of five or fewer thunderstorm days per year and earth resistivity of less than 100 ohm-meters. Such areas are found along the Pacific coast.

Utility companies may provide primary protectors if conductors are exposed to lightning. Generally, cable is not considered to be exposed to lightning unless one or both of the conditions in FPN No. 2 exist. A primary protector is required at each end of a communications circuit if lightning exposure exists, unless protection is provided on the output side of the network interface unit (NIU).

(1) Fuseless Primary Protectors Fuseless-type primary protectors shall be permitted where power fault currents on all protected conductors in the cable are safely limited to a value no greater than the current-carrying capacity of the primary protector and of the primary protector grounding conductor.

(2) Fused Primary Protectors Where the requirements listed in 830.90(A)(1) are not met, fused-type primary protectors shall be used. Fused-type primary protectors shall consist of an arrester connected between each conductor to be protected and ground, a fuse in series with each conductor to be protected, and an appropriate mounting arrangement. Fused primary protector terminals shall be marked to indicate line, instrument, and ground, as applicable.

(B) Location The location of the primary protector, where required, shall comply with (B)(1), (B)(2), or (B)(3):

(1) A listed primary protector shall be applied on each network-powered broadband communications cable external to and on the network side of the network interface unit.
(2) The primary protector function shall be an integral part of and contained in the network interface unit. The network interface unit shall be listed as being suitable for application with network-powered broadband communications systems and shall have an external marking indicating that it contains primary electrical protection.
(3) The primary protector(s) shall be provided on the derived circuit(s) (output side of the NIU), and the combination of the NIU and the protector(s) shall be listed as

being suitable for application with network-powered broadband communications systems.

A primary protector, whether provided integrally or external to the network interface unit, shall be located as close as practicable to the point of entrance.

For purposes of this section, a network interface unit and any externally provided primary protectors located at mobile home service equipment located in sight from and not more than 9.0 m (30 ft) from the exterior wall of the mobile home it serves, or at a mobile home disconnecting means grounded in accordance with 250.32 and located in sight from and not more than 9.0 m (30 ft) from the exterior wall of the mobile home it serves, shall be considered to meet the requirements of this section.

FPN: Selecting a network interface unit and primary protector location to achieve the shortest practicable primary protector grounding conductor helps limit potential differences between communications circuits and other metallic systems.

(C) Hazardous (Classified) Locations The primary protector or equipment providing the primary protection function shall not be located in any hazardous (classified) location as defined in Article 500 or in the vicinity of easily ignitible material.

Exception: As permitted in 501.150, 502.150, and 503.150.

830.93 Grounding or Interruption of Metallic Members of Network-Powered Broadband Communications Cables

The shields of network-powered broadband communications cables used for communications or powering shall be grounded at the building as close as practicable to the point of entrance or attachment of the NIU. Metallic cable members not used for communications or powering shall be grounded or interrupted by an insulating joint or equivalent device as close as practicable to the point of entrance or attachment of the NIU.

For purposes of this section, grounding or interruption of network-powered broadband communications cable metallic members installed at mobile home service equipment located in sight from and no more than 9.0 m (30 ft) from the exterior wall of the mobile home it serves, or at a mobile home disconnecting means grounded in accordance with 250.32 and located in sight from and not more than 9.0 m (30 ft) from the exterior wall of the mobile home it serves, shall be considered to meet the requirements of this section.

FPN: Selecting a grounding location to achieve the shortest practicable grounding conductor helps limit potential differences between the network-powered broadband communications circuits and other metallic systems.

IV. Grounding Methods

830.100 Cable, Network Interface Unit, and Primary Protector Grounding

Network interface units containing protectors, NIUs with metallic enclosures, primary protectors, and the metallic members of the network-powered broadband communications cable that are intended to be grounded shall be grounded as specified in 830.100(A) through 830.100(D).

(A) Grounding Conductor

(1) Insulation The grounding conductor shall be insulated and shall be listed as suitable for the purpose.

(2) Material The grounding conductor shall be copper or other corrosion-resistant conductive material, stranded or solid.

(3) Size The grounding conductor shall not be smaller than 14 AWG and shall have a current-carrying capacity approximately equal to that of the grounded metallic member(s) and protected conductor(s) of the network-powered broadband communications cable. The grounding conductor shall not be required to exceed 6 AWG.

(4) Length The grounding conductor shall be as short as practicable. In one-family and multifamily dwellings, the grounding conductor shall be as short as permissible, not to exceed 6.0 m (20 ft) in length.

The limitation on length will result in a lower impedance, which will limit the potential difference between broadband communications systems and other systems during a lightning strike. Large potential differences between grounding conductors can result in increased damage if a lightning strike were to occur.

FPN: Similar grounding conductor length limitations applied at apartment buildings and commercial buildings will help to reduce voltages that may be developed between the building's power and communications systems during lightning events.

Exception: In one- and two-family dwellings where it is not practicable to achieve an overall maximum grounding conductor length of 6.0 m (20 ft), a separate communications ground rod meeting the minimum dimensional criteria of 830.100(B)(2)(2) shall be driven, and the grounding conductor shall be connected to the communications ground rod in accordance with 830.100(C). The communications ground rod shall be bonded to the power grounding electrode system in accordance with 830.100(D).

(5) Run in Straight Line The grounding conductor shall be run to the grounding electrode in as straight a line as practicable.

(6) Physical Protection Where subject to physical damage, the grounding conductor shall be adequately protected. Where the grounding conductor is run in a metal raceway, both ends of the raceway shall be bonded to the grounding conductor or the same terminal or electrode to which the grounding conductor is connected.

(B) Electrode The grounding conductor shall be connected as follows.

(1) In Buildings or Structures with Grounding Means To the nearest accessible location on the following:

(1) The building or structure grounding electrode system as covered in 250.50
(2) The grounded interior metal water piping system, within 1.5 m (5 ft) from its point of entrance to the building, as covered in 250.52

See the commentary following 250.52(A)(1).

(3) The power service accessible means external to enclosures as covered in 250.94
(4) The metallic power service raceway
(5) The service equipment enclosure
(6) The grounding electrode conductor or the grounding electrode metal enclosure, or
(7) The grounding conductor or the grounding electrode of a building or structure disconnecting means that is grounded to an electrode as covered in 250.32

For purposes of this section, the mobile home service equipment or the mobile home disconnecting means, as described in 830.93, shall be considered accessible.

(2) In Buildings or Structures Without Grounding Means If the building or structure served has no grounding means, as described in (B)(1), the grounding conductor shall be connected to either of the following:

(1) To any one of the individual electrodes described in 250.52(A)(1), (A)(2), (A)(3), or (A)(4)
(2) If the building or structure served has no grounding means, as described in 830.100(B)(1) or (B)(2)(1), to an effectively grounded metal structure or to a ground rod or pipe not less than 1.5 m (5 ft) in length and 12.7 mm (½ in.) in diameter, driven, where practicable, into permanently damp earth and separated from lightning conductors as covered in 800.53 and at least 1.8 m (6 ft) from electrodes of other systems. Steam or hot water pipes or lightning-rod conductors shall not be employed as electrodes for protectors, NIUs with integral protection, grounded metallic members, NIUs with metallic enclosures, and other equipment.

(C) Electrode Connection Connections to grounding electrodes shall comply with 250.70.

(D) Bonding of Electrodes A bonding jumper not smaller than 6 AWG copper or equivalent shall be connected between the network-powered broadband communications system grounding electrode and the power grounding electrode system at the building or structure served where separate electrodes are used.

Exception: At mobile homes as covered in 830.106.

> FPN No. 1: See 250.60 for use of lightning rods.
> FPN No. 2: Bonding together of all separate electrodes limits potential differences between them and between their associated wiring systems.

The most common error made in grounding network-powered broadband communications systems is connecting the cable sheath to a ground rod driven by the utility installer at a convenient location near the point of cable entry to the building instead of bonding it to the electrical service grounding electrode system, service raceway, or other components that make up the grounding electrode system. A separate grounding electrode is permitted by the 2005 *Code* only if the building or structure has none of the grounding means described in 830.100(B). The *Code* requires that some means that is accessible and external to the service equipment be provided for making the bonding and grounding connection for other systems. One of the following means must be provided:

1. Exposed service raceways
2. An exposed grounding electrode conductor
3. An approved means for the external connection of a conductor (A 6 AWG copper conductor with one end bonded to the service raceway or equipment with about 6 in. exposed is acceptable.)

Proper bonding of the network-powered broadband communications system cable sheath to the electrical power ground is needed to prevent potential fire and shock hazards. The earth cannot be used as an equipment grounding conductor or bonding conductor because it does not have the required low-impedance path. (See 250.54.)

Both network-powered broadband communications systems and power systems are subject to current surges as a result of, for example, induced voltages from lightning in the vicinity of the usually extensive outside distribution systems. Surges also result from switching operations on power systems. If the grounded conductors and parts of the two systems are not bonded by a low-impedance path, such line surges can raise the potential difference between the two systems to many thousands of volts. This can result in arcing between the two systems, for example, wherever the coaxial cable jacket contacts a grounded part, such as a metal water pipe or metal structural member, inside the building.

If a person is the interface between the two systems and the bonding has not been done in accordance with the *Code*, the high-voltage surge could result in electric shock. More common, however, is burnout of a television tuner, a part that is almost always an interface between the two systems. The tuner is connected to the power system ground through the grounded neutral of the power supply, even if the television itself is not provided with an equipment grounding conductor.

Also see the commentary following 250.92(B) and 820.93(B).

830.106 Bonding and Grounding at Mobile Homes

(A) Grounding Where there is no mobile home service equipment located in sight from and not more than 9.0 m (30 ft) from the exterior wall of the mobile home it serves, or there is no mobile home disconnecting means grounded in accordance with 250.32 and located within sight from and not more than 9.0 m (30 ft) from the exterior wall of the mobile home it serves, the network-powered broadband communications cable, network interface unit, and primary protector ground shall be installed in accordance with 830.106(B)(2).

(B) Bonding The network-powered broadband communications cable grounding terminal, network interface unit grounding terminal, if present, and primary protector grounding terminal shall be bonded together with a copper bonding conductor not smaller than 12 AWG. The network-powered broadband communications cable grounding terminal, network interface unit grounding terminal, primary protector grounding terminal, or the grounding electrode shall be bonded to the metal frame or available grounding terminal of the mobile home with a copper bonding conductor not smaller than 12 AWG under any of the following conditions:

(1) Where there is no mobile home service equipment or disconnecting means as in 830.106(A)
(2) Where the mobile home is supplied by cord and plug

V. Wiring Methods Within Buildings

830.133 Installation of Network-Powered Broadband Communications Cables and Equipment

Cable and equipment installations within buildings shall comply with 830.133(A) through 830.133(D), as applicable.

(A) Separation of Conductors

(1) In Raceways and Enclosures

    (a) Low and Medium Power Network-Powered Broadband Communications Circuit Cables. Low and medium power network-powered broadband communications cables shall be permitted in the same raceway or enclosure.

(b) Low-Power Network-Powered Broadband Communications Circuit Cables. Low-power network-powered broadband communications cables shall be permitted in the same raceway or enclosure with jacketed cables of any of the following circuits:

(1) Class 2 and Class 3 remote-control, signaling, and power-limited circuits in compliance with Article 725
(2) Power-limited fire alarm systems in compliance with Article 760
(3) Communications circuits in compliance with Article 800
(4) Nonconductive and conductive optical fiber cables in compliance with Article 770
(5) Community antenna television and radio distribution systems in compliance with Article 820

(c) Medium Power Network-Powered Broadband Communications Circuit Cables. Medium power network-powered broadband communications cables shall not be permitted in the same raceway or enclosure with conductors of any of the following circuits:

(1) Class 2 and Class 3 remote-control, signaling, and power-limited circuits in compliance with Article 725
(2) Power-limited fire alarm systems in compliance with Article 760
(3) Communications circuits in compliance with Article 800
(4) Conductive optical fiber cables in compliance with Article 770
(5) Community antenna television and radio distribution systems in compliance with Article 820

(d) Electric Light, Power, Class 1, Non–Powered Broadband Communications Circuit Cables. Network-powered broadband communications cable shall not be placed in any raceway, compartment, outlet box, junction box, or similar fittings with conductors of electric light, power, Class 1, or non–power-limited fire alarm circuit cables.

Exception No. 1: Where all of the conductors of electric light, power, Class 1, non–power-limited fire alarm circuits are separated from all of the network-powered broadband communications cables by a permanent barrier or listed divider.

This exception has been revised to recognize the use of a listed field-installed divider to separate the communication circuits from the power circuits.

Exception No. 2: Power circuit conductors in outlet boxes, junction boxes, or similar fittings or compartments where such conductors are introduced solely for power supply to the network-powered broadband communications system distribution equipment. The power circuit conductors shall be routed within the enclosure to maintain a minimum 6-mm (0.25-in.) separation from network-powered broadband communications cables.

(2) Other Applications Network-powered broadband communications cable shall be separated at least 50 mm (2 in.) from conductors of any electric light, power, Class 1, and non–power-limited fire alarm circuits.

Exception No. 1: Where either (1) all of the conductors of electric light, power, Class 1, and non–power-limited fire alarm circuits are in a raceway, or in metal-sheathed, metal-clad, nonmetallic-sheathed, Type AC, or Type UF cables, or (2) all of the network-powered broadband communications cables are encased in raceway.

Exception No. 2: Where the network-powered broadband communications cables are permanently separated from the conductors of electric light, power, Class 1, and non-power-limited fire alarm circuits by a continuous and firmly fixed nonconductor, such as porcelain tubes or flexible tubing, in addition to the insulation on the wire.

(B) Support of Conductors Raceways shall be used for their intended purpose. Network-powered broadband communications cables shall not be strapped, taped, or attached by any means to the exterior of any conduit or raceway as a means of support.

(C) Cable Substitutions The substitutions for network-powered broadband cables listed in Table 830.133 shall be permitted. All cables in Table 830.133, other than network-powered broadband cables, shall be coaxial cables.

(D) Installation and Use Section 110.3(B) shall apply.

See the commentary following 300.11(B)(3).

Table 830.133 Cable Substitutions

Cable Type	Permitted Cable Substitutions
BM	BMR
BLP	CMP, CL3P
BLR	CMP, CL3P, CMR, CL3R, BLP, BMR
BL	CMP, CMR, CM, CMG, CL3P, CL3R, CL3, BMR, BM, BLP, BLR
BLX	CMP, CMR, CM, CMG, CMX, CL3P, CL3R, CL3, CL3X, BMR, BM, BLP, BRP, BL

830.151 Medium Power Network-Powered Broadband Communications System Wiring Methods

Medium power network-powered broadband communications systems shall be installed within buildings using listed Type BM or Type BMR, network-powered broadband communications medium power cables.

Section 300.22 requires that the appropriate methods given in Chapter 3 be used for the installation of medium-power network-powered broadband communications circuits in plenums and other spaces used for environmental air within buildings. There is no plenum-rated cable for medium-power network-powered broadband communications circuits. However, if the wiring downstream of the NIU is covered by other articles, then listed cable may be available.

(A) Ducts, Plenums, and Other Air-Handling Spaces Section 300.22 shall apply.

(B) Riser Cables installed in vertical runs and penetrating more than one floor, or cables installed in vertical runs in a shaft, shall be Type BMR. Floor penetrations requiring Type BMR shall contain only cables suitable for riser or plenum use.

Exception No. 1: Type BM cables encased in metal raceway or located in a fireproof shaft that has firestops at each floor.

Exception No. 2: Type BM cables in one- and two-family dwellings.

(C) Other Wiring Cables installed in locations other than the locations covered in 830.151(A) and 830.151(B) shall be Type BM.

Exception: Type BMU cable where the cable enters the building from the outside and is run in rigid metal conduit or intermediate metal conduit, and such conduits are grounded to an electrode in accordance with 830.100(B).

830.154 Low-Power Network-Powered Broadband Communications System Wiring Methods

Low-power network-powered broadband communications systems shall comply with any of the requirements of 830.154(A) through 830.154(D).

(A) In Buildings Low-power network-powered broadband communications systems shall be installed within buildings using listed Type BLX, Type BL, Type BLR, or Type BLP network-powered broadband communications low-power cables.

Section 830.154(A) permits the use of listed Type BLP cable for low-power network-powered broadband communications circuits within plenums and other spaces used for environmental air within buildings.

(B) Ducts, Plenums, and Other Air-Handling Spaces Cables installed in ducts, plenums, and other spaces used for environmental air shall be Type BLP. Type BLX cable installed in compliance with 300.22 shall be permitted.

Used in conjunction with the definition of *abandoned network-powered broadband communications cable* in 830.2 and the general requirement in 830.3(A), 830.154(B) and 830.154(C) require removal of accessible abandoned broadband communications cable. Abandoned cable increases fire loading unnecessarily, and, where installed in plenums, it can affect airflow. Similar requirements can be found in Articles 640, 645, 725, 760, 770, 800, and 820.

(C) Riser Cables installed in risers shall comply with any of the requirements in 830.154(C)(1), (C)(2), or (C)(3).

(1) Cables in Vertical Runs Cables installed in vertical runs and penetrating more than one floor, or cables installed in vertical runs in a shaft, shall be Type BLP, BLR, or BMR. Floor penetrations requiring Type BMR or BLR shall contain only cables suitable for riser or plenum use.

(2) Metal Raceways or Fireproof Shafts Type BLX cables shall be permitted to be encased in a metal raceway or located in a fireproof shaft having firestops at each floor.

(3) One- and Two-Family Dwellings Type BLX or BL cables less than 10 mm (0.375 in.) in diameter shall be permitted in one- and two-family dwellings.

(D) Other Wiring Cables installed in locations other than those covered in 830.154(A), (B), and (C) shall comply with the requirements of 830.154(D)(1) through (D)(5).

(1) General Type BLP, BL, or BM shall be permitted.

(2) In Raceways Type BLX shall be permitted to be installed in a raceway.

(3) Type BLU Cable Type BLU cable entering the building from outside shall be permitted to be run in rigid metal conduit or intermediate metal conduit. Such conduits shall be grounded to an electrode in accordance with 830.100(B).

(4) One- and Two-Family Dwellings Type BLX or BL cables less than 10 mm (0.375 in.) in diameter shall be permitted to be installed in one- and two-family dwellings.

(5) Type BLX Cable Type BLX cable entering the building from outside and terminated at a grounding block or a primary protection location shall be permitted to be installed, provided that the length of cable within the building does not exceed 15 m (50 ft).

> FPN: This provision limits the length of Type BLX cable to 15 m (50 ft), while 830.90(B) requires that the primary protector, or NIU with integral protection, be located as close as practicable to the point at which the cable enters the building. Therefore, in installations requiring a primary protector, or NIU with integral protection, Type BLX cable may not be permitted to extend 15 m (50 ft) into the building if it is practicable to place the primary protector closer than 15 m (50 ft) to the entrance point.

830.157 Protection Against Physical Damage

Section 300.4 shall apply.

830.160 Bends

Bends in network broadband cable shall be made so as not to damage the cable.

VI. Listing Requirements

830.179 Network-Powered Broadband Communications Equipment and Cables

Network-powered broadband communications equipment and cables shall be listed as suitable for the purpose.

Exception No. 1: This listing requirement shall not apply to community antenna television and radio distribution system coaxial cables that were installed prior to January 1, 2000, in accordance with Article 820 and are used for low-power network-powered broadband communications circuits.

Exception No. 2: Substitute cables for network-powered broadband communications cables shall be permitted as shown in Table 830.133.

(A) Listing and Marking Listing and marking of network-powered broadband communications cables shall comply with 830.179(A)(1) or (A)(2).

Article 830 permits three types of listed cable: coaxial, coaxial with a twisted pair of conductors, and fiber optic with a twisted pair of conductors. In the coaxial configuration, power and signal are carried on the coaxial center conductor. With coaxial and a twisted pair of conductors, the signal is carried on the coaxial center conductor, and the power is carried on the twisted pair. The configuration of fiber optic and twisted-pair conductors carries the signal on the fiber and the power on the twisted pair.

(1) Type BMU, Type BM, and Type BMR Cables Network-powered broadband communications medium power underground cable, Type BMU; network-powered broadband communications medium power cable, Type BM; and network-powered broadband communications medium

power riser cable, Type BMR, shall be factory-assembled cables consisting of a jacketed coaxial cable, a jacketed combination of coaxial cable and multiple individual conductors, or a jacketed combination of an optical fiber cable and multiple individual conductors. The insulation for the individual conductors shall be rated for 300 volts minimum. Cables intended for outdoor use shall be listed as suitable for the application. Cables shall be marked in accordance with 310.11. Type BMU cables shall be jacketed and listed as being suitable for outdoor underground use. Type BM cables shall be listed as being suitable for general-purpose use, with the exception of risers and plenums, and shall also be listed as being resistant to the spread of fire. Type BMR cables shall be listed as being suitable for use in a vertical run in a shaft or from floor to floor and shall also be listed as having fire-resistant characteristics capable of preventing the carrying of fire from floor to floor.

A rating of 300 volts is necessary for the following reasons:

1. To coordinate with protector installation requirements (i.e., protectors are not required within a block unless the cable is exposed to over 300 volts)
2. To recognize the fact that primary protectors are designed to allow voltages below 300 to pass
3. To accommodate the voltages ordinarily found on a network-powered broadband communications circuit (voltage up to 150 volts rms)

> FPN No. 1: One method of defining *resistant to spread of fire* is that the cables do not spread fire to the top of the tray in the vertical tray flame test in ANSI/UL 1581-1991, *Reference Standard for Electrical Wires, Cables and Flexible Cords.* Another method of defining *resistant to the spread of fire* is for the damage (char length) not to exceed 1.5 m (4 ft 11 in.) when performing the CSA vertical flame test for cables in cable trays, as described in CSA C22.2 No. 0.3-M-1985, *Test Methods for Electrical Wires and Cables.*

See the commentary following 725.82(C), FPN, for information on the UL vertical tray flame test.

> FPN No. 2: One method of defining fire-resistant characteristics capable of preventing the carrying of fire from floor to floor is that the cables pass the requirements of ANSI/UL 1666-2002, *Standard Test for Flame Propagation Height of Electrical and Optical-Fiber Cable Installed Vertically in Shafts.*

See the commentary following 725.82(B), FPN, for information on a test for defining fire-resistant characteristics capable of preventing fire spread from floor to floor.

(2) Type BLU, Type BLX, Type BL, and Type BLP Cables Network-powered broadband communications low-power underground cable, Type BLU; limited use network-powered broadband communications low-power cable, Type BLX; network-powered broadband communications low-power cable, Type BL; network-powered broadband communications low-power riser cable, Type BLR; and network-powered broadband communications low-power plenum cable, Type BLP, shall be factory-assembled cables consisting of a jacketed coaxial cable, a jacketed combination of coaxial cable and multiple individual conductors, or a jacketed combination of an optical fiber cable and multiple individual conductors. The insulation for the individual conductors shall be rated for 300 volts minimum. Cables intended for outdoor use shall be listed as suitable for the application. Cables shall be marked in accordance with 310.11. Type BLU cables shall be jacketed and listed as being suitable for outdoor underground use. Type BLX limited-use cables shall be listed as being suitable for use outside, for use in dwellings, and for use in raceways and shall also be listed as being resistant to flame spread. Type BL cables shall be listed as being suitable for general-purpose use, with the exception of risers and plenums, and shall also be listed as being resistant to the spread of fire. Type BLR cables shall be listed as being suitable for use in a vertical run in a shaft or from floor to floor and shall also be listed as having fire-resistant characteristics capable of preventing the carrying of fire from floor to floor. Type BLP cables shall be listed as being suitable for use in ducts, plenums, and other spaces used for environmental air and shall also be listed as having adequate fire-resistant and low smoke-producing characteristics.

FPN No. 1: One method of determining that cable is resistant to flame spread is by testing the cable to VW-1 (vertical-wire) flame test in ANSI/UL 1581-1991, *Reference Standard for Electrical Wires, Cables and Flexible Cords*.

See the commentary following 725.82(D), FPN, for information on test methods for determining the fire resistance of cable.

FPN No. 2: One method of defining *resistant to spread of fire* is that the cables do not spread fire to the top of the tray in the vertical tray flame test in ANSI/UL 1584-1991, *Reference Standard for Electrical Wires, Cables and Flexible Cords*.

See the commentary following 725.82(A), FPN, for information on a test method for wires and cables to be installed without raceways in plenums and other spaces used for environmental air.

FPN No. 3: One method of defining fire-resistant characteristics capable of preventing the carrying of fire from floor to floor is that the cables pass the requirements of ANSI/UL 1666-1997, *Standard Test for Flame Propagation Height of Electrical and Optical-Fiber Cable Installed Vertically in Shafts*.

FPN No. 4: One method of defining a cable that is low smoke-producing cable and fire-resistant cable is that the cable exhibits a maximum peak optical density of 0.5 or less, an average optical density of 0.15 or less, and a maximum flame spread distance of 1.52 m (5 ft) or less when tested in accordance with NFPA 262-1999, *Standard Method of Test for Flame Travel and Smoke of Wires and Cables for Use in Air Handling Spaces*.

Tables

Contents

As the adoption and use of the *NEC* increase in areas of the world where the metric system is the standard, it becomes necessary to provide means to allow for the use of electrical products with metric measurements or designations. The revision to how conduit and tubing sizes are given in the *NEC* is a prime example of a change designed to allow for assimilation of *NEC* requirements in the metric world. The 1996 and 1999 editions of the *NEC* provided guidance in the form of fine print notes on metric conduit and tubing designations. Beginning with the 2002 *NEC*, for every provision where the size of a conduit or tubing with a cylindrical cross-section size was specified, two size designations were given. The two sizing indicators, referred to as metric designator and trade size, per Table 300.1(C), replaced the traditional *NEC* method of using of an actual dimension for the size designation of conduit or tubing.

This change affects all metal, nonmetallic, rigid, and flexible conduit and tubing types that have a cylindrical cross-section. For example, ½ in. conduit and tubing are now referred to as metric designator 16 or trade size ½; ¾ in. is now metric designator 21 or trade size ¾; 4 in. is now metric designator 103 or trade size 4, and so forth.

Throughout the NEC, including Table 4 in Chapter 9 and Annex C, both the metric designator and the trade size of conduit and tubing are provided.

For further information on metrication and the revision to how conduit and tubing sizes are treated in the *NEC*, see the commentary following 90.9(D) FPN No. 2 and Table 300.1(C).

Because conduits and tubing from different manufacturers have different internal diameters for the same trade size, Table 4 provides the diameter and the actual area of different conduit and tubing types at 100, 60, 53 (one wire), 31 (two wires), and 40 (more than two wires) percent fill. The 60 percent fill is provided in Table 4 to correlate with Note 4 to the conduit and tubing fill tables, which permits conduit or tubing nipples 24 in. or less in length to have a conductor fill of up to 60 percent. The order of the fill percentages for the different raceway types covered in Table 4 have been revised into a more logical order. There are separate sections in Table 4, to cover all metal, nonmetallic, rigid, and flexible conduit and tubing types. Examples of how to use the conduit and tubing conductor fill tables are included in the commentary both here and in Annex C.

Annex C contains conductor fill tables for 12 different types of conduit and tubing. The Annex C tables, which are based on the dimensions given in Tables 1 and 4 of Chapter 9 for conduit and tubing fill and on the dimensions for conductors in Table 5 of Chapter 9, provide conductor fill information based on the specific conduit or tubing and on the conductor insulation type, size, and stranding characteristics. There are separate tables for metallic- and nonmetallic-type conduit and tubing raceways and for flexible raceways. Examples of how to use these tables are included in the commentary both here and in Annex C.

Table 1 Percent of Cross Section of Conduit and Tubing for Conductors

Number of Conductors	All Conductor Types
1	53
2	31
Over 2	40

FPN No. 1: Table 1 is based on common conditions of proper cabling and alignment of conductors where the length of the pull and the number of bends are within reasonable limits. It should be recognized that, for certain conditions, a larger size conduit or a lesser conduit fill should be considered.

Table 1 establishes the maximum fill permitted for the circular conduit and tubing types. It is the basis for Table 4 and for the information on conduit and tubing fill provided in the Annex C tables. FPN No. 1 advises that factors such as the length of the run or the number and total radius of bends can increase the difficulty of pulling conductors into the raceway and in extreme cases could result in damage to conductor insulation. To mitigate such adverse effects and to facilitate the ease of installing the conductors in the conduit or tubing, it is recommended that where a difficult installation is anticipated the maximum number of conductors permitted not be installed nor the size of the conduit or tubing be increased. Experienced personnel can attest to the wisdom of this advice.

> FPN No. 2: When pulling three conductors or cables into a raceway, if the ratio of the raceway (inside diameter) to the conductor or cable (outside diameter) is between 2.8 and 3.2, jamming can occur. While jamming can occur when pulling four or more conductors or cables into a raceway, the probability is very low.

FPN No. 2 warns of another potential pitfall associated with pulling conductors into conduit or tubing. Conductor jamming may occur during the installation (pulling) of conductors into a conduit even if fill allowances of 40 percent are observed. During the installation of three conductors or cables into the raceway, it is possible that one conductor may slip between the other two conductors. This is more likely to take place at bends, where the raceway may be slightly oval.

As an example, Table C.1 in Annex C permits three 8 AWG conductors in trade size ½ electrical metallic tubing (EMT). Using this example, ½ in. EMT has an internal diameter (ID) of 0.622 in. (from Table 4), and an 8 AWG conductor has an outside diameter (OD) of 0.216 in. (from Table 5).

The EMT in a straight run has an internal diameter of 0.622 in., but because it may not be round at a bend, one conductor may slip between the other two and cause a jam as the conductors exit the bend. In a straight run, assuming no variation in the EMT's internal diameter or in a conductor's outside diameter, one conductor usually cannot slip between the other two, because the total of the outside diameters of the conductors (3×0.216 in. $= 0.648$ in.) is greater than the EMT's internal diameter of 0.622 in. At a bend, however, the major internal diameter of the raceway may increase due to bending, particularly in tubing, to a diameter slightly larger than 0.648 in., permitting the middle conductor to be pulled between the outer two conductors. As the conductors exit the bend and the raceway returns to its normal shape with an internal diameter of 0.622 in., the conductors may jam. This can also occur in straight runs where the ratio of the raceway's internal diameter to the conductor's outside diameter approaches 3. The jam ratio is calculated as follows:

$$\text{Jam ratio} = \frac{\text{ID of raceway}}{\text{OD of conductor}} = \frac{0.622}{0.216} = 2.88$$

To avoid difficult conductor installations and potential conductor insulation damage due to jamming within the conduit or tubing, a jam ratio between 2.8 and 3.2 is recommended.

Notes to Tables

(1) See Annex C for the maximum number of conductors and fixture wires, all of the same size (total cross-sectional area including insulation) permitted in trade sizes of the applicable conduit or tubing.

Annex C is a useful tool for determining the correct size of conduit or tubing based on the size and number of conductors to be installed in the raceway. See the commentary in Annex C for examples of how to use the tables.

(2) Table 1 applies only to complete conduit or tubing systems and is not intended to apply to sections of conduit or tubing used to protect exposed wiring from physical damage.

The maximum fill requirements do not apply to short sections of conduit or tubing used for the physical protection of conductors and cables. Cables are commonly protected from physical damage by conduit or tubing sleeves sized to enable the cable to be passed through with relative ease without injuring or abrading the protective jacket of the cable. The requirement of 300.5(D)(1) regarding physical protection of direct buried cables and conductors as they emerge from below grade is an example of conduit or tubing being used as a protective sleeve and not as a continuous raceway system per 300.12. However, a fitting is required on the end(s) of the conduit or tubing to protect the conductors or cables from abrasion. [See 300.15(C).]

(3) Equipment grounding or bonding conductors, where installed, shall be included when calculating conduit or tubing fill. The actual dimensions of the equipment grounding or bonding conductor (insulated or bare) shall be used in the calculation.

All insulated, covered, and bare conductors occupy space within a raceway. Therefore, all installed conductors have to be included in the raceway fill calculation, including non-current-carrying conductors such as equipment grounding conductors, bonding conductors, and bonding jumpers. The only exception to this rule is the addition of an uninsulated grounding conductor permitted in trade size ⅜ flexible metal conduit (see the footnote to Table 348.22). The dimensions of bare conductors are given in Table 8.

(4) Where conduit or tubing nipples having a maximum length not to exceed 600 mm (24 in.) are installed between boxes, cabinets, and similar enclosures, the nipples shall be permitted to be filled to 60 percent of their total cross-sectional area, and 310.15(B)(2)(a) adjustment factors need not apply to this condition.

(5) For conductors not included in Chapter 9, such as multiconductor cables, the actual dimensions shall be used.

For conductors not included in Chapter 9, such as high-voltage types, the cross-sectional area can be calculated in the following manner, using the actual dimensions of each conductor:

$$\text{cross-sectional area} = d^2 \text{ cmil}$$

where:

d = outside diameter of a conductor (including insulation) [1 in. = 1000 mil (1 mil = 0.001 in.)]

cmil = circular mil, a unit measure of area equal to $\pi/4$ (3.1416/4 = 0.7854) square mil. In other words, 1 cmil = 0.7854 square mil.

Example

Three 15 kV single conductors are to be installed in rigid metal conduit (RMC). The outside diameter of each conductor measures 1⅝ in., or 1.625 in. What size RMC will accommodate the three conductors?

Solution

Step 1. Find the area within the cross-section of the conduit that will be displaced by the three conductors:

$$1.625 \text{ in.} \times 1.625 \text{ in.} \times 0.7854 \times 3$$
$$= 6.2218 \text{ in.}^2, \quad \text{or } 6.222 \text{ in.}^2$$

Step 2. Determine the size conduit that will accommodate the three conductors. Table 1 allows 40 percent conduit fill for three or more conductors, and Table 4 indicates that 40 percent of trade size 5 RMC is 8.085 in.² Thus, trade size 5 RMC will accommodate three 15 kV single conductors.

(6) For combinations of conductors of different sizes, use Tables 5 and 5A for dimensions of conductors and Table 4 for the applicable conduit or tubing dimensions.

The following two examples demonstrate how to calculate the minimum trade size conduit or tubing required where there are conductors of different sizes.

Example

A 200-ampere feeder is routed in various wiring methods [electrical metallic tubing (EMT); rigid nonmetallic conduit (RNC), Schedule 40; and rigid metal conduit (RMC)] from the main switchboard in one building to a distribution panelboard in another building. The circuit consists of four 4/0 AWG XHHW copper conductors and one 6 AWG XHHW copper conductor. Select the proper trade size for the various types of conduit and tubing to be used for the feeder.

Solution

All the raceways for this example require conduit fill to be calculated according to Table 1 in Chapter 9. (See 344.22 for RMC, 352.22 for RNC, and 358.22 for EMT.) Table 1 of Chapter 9 permits conduit fill to a maximum of 40 percent where more than two conductors are installed in the conduit or tubing. Note 6 refers to Table 5 for the area required for each insulated conductor. For each respective conductor size, multiply the number of conductors of that size by their Table 5 area and then add all the conductor areas to determine the total conductor-occupied space within the conduit. Note 6 refers to Table 4 for selection of the appropriate trade size conduit or tubing. Table 4 contains the allowable cross-sectional area for conduit and tubing based on conductor-occupied space (40 percent maximum in this example).

Step 1. Calculate the total area occupied by the conductors, using the approximate areas listed in Table 5.

Four 4/0 AWG XHHW:

$$4 \times 0.3197 \text{ in.}^2 = 1.2788 \text{ in.}^2$$

One 6 AWG XHHW:

$$1 \times 0.0590 \text{ in.}^2 = 0.0590 \text{ in.}^2$$

Total area = 1.3378 in.², or 1.338 in.²

Step 2. Determine the proper trade size EMT, RMC, and RNC (Schedule 40) from Table 4. The portion of this feeder installed in EMT requires a minimum trade size 2. Trade size 2 EMT has 1.342 in.² of available space for over two conductors, and the minimum required space is 1.338 in.², which is less than the trade size 2 EMT 40 percent fill. RMC also requires a minimum trade size 2, because trade size 2 RMC has 1.363 in.² of available space for over two conductors. RNC (Schedule 40), however, requires a minimum trade size 2½. Trade size 2 RNC has 1.316 in.² allowable space for over two conductors and is less than the 1.338 in.² required for this combination of conductors. Therefore, it is necessary to increase the RNC size to 2½ trade size, the next standard size increment.

Example

Determine the minimum size rigid metal conduit (RMC) allowed for the 10 mixed conductor sizes and types described in Commentary Table 9.1.

Commentary Table 9.1 Conductor Sizes and Types for Example

Quantity	Wire Size and Type	Cross-Sectional Area of Each Wire (from Table 5)	Cross-Sectional Area
4	12 AWG THWN	0.0133	0.0532
3	8 AWG TW	0.0437	0.1311
3	6 AWG THW	0.0726	0.2178
		Total	0.4021

Solution

The "Over 2 Wires" column in Table 4 indicates that 40 percent of a trade size 1¼ RMC is 0.610 in.² Therefore, trade size 1¼ is the minimum size RMC allowed for this combination of 10 conductors.

(7) When calculating the maximum number of conductors permitted in a conduit or tubing, all of the same size (total cross-sectional area including insulation), the next higher whole number shall be used to determine the maximum number of conductors permitted when the calculation results in a decimal of 0.8 or larger.

Example

Determine how many 10 AWG THHN conductors are permitted in a trade size 1¼ rigid metal conduit (RMC).

Solution

Table 1 permits 40 percent fill for over two conductors. From Table 4, 40 percent fill for trade size 1¼ RMC is 0.610 in., and from Table 5, the cross-sectional area of a 10 AWG THHN conductor is 0.0211 in.² The number of conductors permitted is calculated as follows:

$$\frac{0.610 \text{ in.}^2}{0.0211 \text{ in.}^2 \text{ per conductor}} = 28.910 \text{ conductors}$$

Based on the maximum allowable fill of 40 percent from Table 1, the number of 10 AWG THHN conductors in trade size 1¼ RMC cannot exceed 28. However, in accordance with Note 7, an increase to the next whole number of 29 conductors is permitted in this case. That is because 0.910 is greater than 0.8, which is the benchmark for determining whether an increase to the next whole number for the maximum number of conductors is permitted. Although increasing the total to 29 conductors results in the raceway fill exceeding 40 percent, the amount by which it is exceeded is a fraction of 1 percent and will not adversely affect the installation of the conductors. This number of conductors does have to be addressed from a mutual heating effect in accordance with 310.15(B)(2)(a). It is important to bear in mind that this is the maximum number of conductors permitted, and in accordance with FPN No.1 to Table 1 in Chapter 9, an installation with fewer than the maximum number of conductors allowed may be prudent.

Verification of this solution can be found in Annex C, Table C.8, which lists twenty-nine 10 AWG THWN conductors as the maximum number of conductors permitted in trade size 1¼ RMC. Several examples of the application of Note 7 are found in the Annex C tables. Where the calculation results in a decimal value less than 0.8, the maximum number of conductors permitted is based on the next lower whole number.

(8) Where bare conductors are permitted by other sections of this *Code*, the dimensions for bare conductors in Table 8 shall be permitted.

(9) A multiconductor cable of two or more conductors shall be treated as a single conductor for calculating percentage conduit fill area. For cables that have elliptical cross sections, the cross-sectional area calculation shall be based on using the major diameter of the ellipse as a circle diameter.

Table 2 was formerly included in Article 344 as Table 344.24. Although its location in past editions of the *NEC* was in Article 344, Rigid Metal Conduit, the requirements on minimum bending radius (depending on the type of bending equipment employed) contained in that table applied to all the rigid, flexible, metallic, and nonmetallic conduit and tubing types via references to Table 344.24 found in the xxx.24 sections of the respective conduit and tubing articles. Because of its widespread application for all the circular conduit and tubing types, this table is more appropriately located in Chapter 9 with the other tables that apply to these types of raceways.

Table 2 Radius of Conduit and Tubing Bends

Conduit or Tubing Size		One Shot and Full Shoe Benders		Other Bends	
Metric Designator	Trade Size	mm	in.	mm	in.
16	½	101.6	4	101.6	4
21	¾	114.3	4½	127	5
27	1	146.05	5¾	152.4	6
35	1¼	184.15	7¼	203.2	8
41	1½	209.55	8¼	254	10
53	2	241.3	9½	304.8	12
63	2½	266.7	10½	381	15
78	3	330.2	13	457.2	18
91	3½	381	15	533.4	21
103	4	406.4	16	609.6	24
129	5	609.6	24	762	30
155	6	762	30	914.4	36

Table 4 Dimensions and Percent Area of Conduit and Tubing (Areas of Conduit or Tubing for the Combinations of Wires Permitted in Table 1, Chapter 9)

Article 358 — Electrical Metallic Tubing (EMT)

Metric Designator	Trade Size	Nominal Internal Diameter		Total Area 100%		60%		1 Wire 53%		2 Wires 31%		Over 2 Wires 40%	
		mm	in.	mm²	in.²	mm²	in.²	mm²	in.²	mm²	in.²	mm²	in.²
16	½	15.8	0.622	196	0.304	118	0.182	104	0.161	61	0.094	78	0.122
21	¾	20.9	0.824	343	0.533	206	0.320	182	0.283	106	0.165	137	0.213
27	1	26.6	1.049	556	0.864	333	0.519	295	0.458	172	0.268	222	0.346
35	1¼	35.1	1.380	968	1.496	581	0.897	513	0.793	300	0.464	387	0.598
41	1½	40.9	1.610	1314	2.036	788	1.221	696	1.079	407	0.631	526	0.814
53	2	52.5	2.067	2165	3.356	1299	2.013	1147	1.778	671	1.040	866	1.342
63	2½	69.4	2.731	3783	5.858	2270	3.515	2005	3.105	1173	1.816	1513	2.343
78	3	85.2	3.356	5701	8.846	3421	5.307	3022	4.688	1767	2.742	2280	3.538
91	3½	97.4	3.834	7451	11.545	4471	6.927	3949	6.119	2310	3.579	2980	4.618
103	4	110.1	4.334	9521	14.753	5712	8.852	5046	7.819	2951	4.573	3808	5.901

Article 362 — Electrical Nonmetallic Tubing (ENT)

Metric Designator	Trade Size	Nominal Internal Diameter		Total Area 100%		60%		1 Wire 53%		2 Wires 31%		Over 2 Wires 40%	
		mm	in.	mm²	in.²	mm²	in.²	mm²	in.²	mm²	in.²	mm²	in.²
16	½	14.2	0.560	158	0.246	95	0.148	84	0.131	49	0.076	63	0.099
21	¾	19.3	0.760	293	0.454	176	0.272	155	0.240	91	0.141	117	0.181
27	1	25.4	1.000	507	0.785	304	0.471	269	0.416	157	0.243	203	0.314
35	1¼	34.0	1.340	908	1.410	545	0.846	481	0.747	281	0.437	363	0.564
41	1½	39.9	1.570	1250	1.936	750	1.162	663	1.026	388	0.600	500	0.774
53	2	51.3	2.020	2067	3.205	1240	1.923	1095	1.699	641	0.993	827	1.282
63	2½	—	—	—	—	—	—	—	—	—	—	—	—
78	3	—	—	—	—	—	—	—	—	—	—	—	—
91	3½	—	—	—	—	—	—	—	—	—	—	—	—

Article 348 — Flexible Metal Conduit (FMC)

Metric Designator	Trade Size	Nominal Internal Diameter		Total Area 100%		60%		1 Wire 53%		2 Wires 31%		Over 2 Wires 40%	
		mm	in.	mm²	in.²	mm²	in.²	mm²	in.²	mm²	in.²	mm²	in.²
12	⅜	9.7	0.384	74	0.116	44	0.069	39	0.061	23	0.036	30	0.046
16	½	16.1	0.635	204	0.317	122	0.190	108	0.168	63	0.098	81	0.127
21	¾	20.9	0.824	343	0.533	206	0.320	182	0.283	106	0.165	137	0.213
27	1	25.9	1.020	527	0.817	316	0.490	279	0.433	163	0.253	211	0.327
35	1¼	32.4	1.275	824	1.277	495	0.766	437	0.677	256	0.396	330	0.511
41	1½	39.1	1.538	1201	1.858	720	1.115	636	0.985	372	0.576	480	0.743
53	2	51.8	2.040	2107	3.269	1264	1.961	1117	1.732	653	1.013	843	1.307
63	2½	63.5	2.500	3167	4.909	1900	2.945	1678	2.602	982	1.522	1267	1.963
78	3	76.2	3.000	4560	7.069	2736	4.241	2417	3.746	1414	2.191	1824	2.827
91	3½	88.9	3.500	6207	9.621	3724	5.773	3290	5.099	1924	2.983	2483	3.848
103	4	101.6	4.000	8107	12.566	4864	7.540	4297	6.660	2513	3.896	3243	5.027

Table 4 *Continued*

Article 342 — Intermediate Metal Conduit (IMC)

Metric Designator	Trade Size	Nominal Internal Diameter mm	Nominal Internal Diameter in.	Total Area 100% mm²	Total Area 100% in.²	60% mm²	60% in.²	1 Wire 53% mm²	1 Wire 53% in.²	2 Wires 31% mm²	2 Wires 31% in.²	Over 2 Wires 40% mm²	Over 2 Wires 40% in.²
12	⅜	—	—	—	—	—	—	—	—	—	—	—	—
16	½	16.8	0.660	222	0.342	133	0.205	117	0.181	69	0.106	89	0.137
21	¾	21.9	0.864	377	0.586	226	0.352	200	0.311	117	0.182	151	0.235
27	1	28.1	1.105	620	0.959	372	0.575	329	0.508	192	0.297	248	0.384
35	1¼	36.8	1.448	1064	1.647	638	0.988	564	0.873	330	0.510	425	0.659
41	1½	42.7	1.683	1432	2.225	859	1.335	759	1.179	444	0.690	573	0.890
53	2	54.6	2.150	2341	3.630	1405	2.178	1241	1.924	726	1.125	937	1.452
63	2½	64.9	2.557	3308	5.135	1985	3.081	1753	2.722	1026	1.592	1323	2.054
78	3	80.7	3.176	5115	7.922	3069	4.753	2711	4.199	1586	2.456	2046	3.169
91	3½	93.2	3.671	6822	10.584	4093	6.351	3616	5.610	2115	3.281	2729	4.234
103	4	105.4	4.166	8725	13.631	5235	8.179	4624	7.224	2705	4.226	3490	5.452

Article 356— Liquidtight Flexible Nonmetallic Conduit (LFNC-B*)

Metric Designator	Trade Size	Nominal Internal Diameter mm	Nominal Internal Diameter in.	Total Area 100% mm²	Total Area 100% in.²	60% mm²	60% in.²	1 Wire 53% mm²	1 Wire 53% in.²	2 Wires 31% mm²	2 Wires 31% in.²	Over 2 Wires 40% mm²	Over 2 Wires 40% in.²
12	⅜	12.5	0.494	123	0.192	74	0.115	65	0.102	38	0.059	49	0.077
16	½	16.1	0.632	204	0.314	122	0.188	108	0.166	63	0.097	81	0.125
21	¾	21.1	0.830	350	0.541	210	0.325	185	0.287	108	0.168	140	0.216
27	1	26.8	1.054	564	0.873	338	0.524	299	0.462	175	0.270	226	0.349
35	1¼	35.4	1.395	984	1.528	591	0.917	522	0.810	305	0.474	394	0.611
41	1½	40.3	1.588	1276	1.981	765	1.188	676	1.050	395	0.614	510	0.792
53	2	51.6	2.033	2091	3.246	1255	1.948	1108	1.720	648	1.006	836	1.298

*Corresponds to 356.2(2)

Article 356 — Liquidtight Flexible Nonmetallic Conduit (LFNC-A*)

Metric Designator	Trade Size	Nominal Internal Diameter mm	Nominal Internal Diameter in.	Total Area 100% mm²	Total Area 100% in.²	60% mm²	60% in.²	1 Wire 53% mm²	1 Wire 53% in.²	2 Wires 31% mm²	2 Wires 31% in.²	Over 2 Wires 40% mm²	Over 2 Wires 40% in.²
12	⅜	12.6	0.495	125	0.192	75	0.115	66	0.102	39	0.060	50	0.077
16	½	16.0	0.630	201	0.312	121	0.187	107	0.165	62	0.097	80	0.125
21	¾	21.0	0.825	346	0.535	208	0.321	184	0.283	107	0.166	139	0.214
27	1	26.5	1.043	552	0.854	331	0.513	292	0.453	171	0.265	221	0.342
35	1¼	35.1	1.383	968	1.502	581	0.901	513	0.796	300	0.466	387	0.601
41	1½	40.7	1.603	1301	2.018	781	1.211	690	1.070	403	0.626	520	0.807
53	2	52.4	2.063	2157	3.343	1294	2.006	1143	1.772	669	1.036	863	1.337

*Corresponds to 356.2(1)

(continues)

Table 4 *Continued*

Article 350 — Liquidtight Flexible Metal Conduit (LFMC)

Metric Designator	Trade Size	Nominal Internal Diameter		Total Area 100%		60%		1 Wire 53%		2 Wires 31%		Over 2 Wires 40%	
		mm	in.	mm²	in.²	mm²	in.²	mm²	in.²	mm²	in.²	mm²	in.²
12	⅜	12.5	0.494	123	0.192	74	0.115	65	0.102	38	0.059	49	0.077
16	½	16.1	0.632	204	0.314	122	0.188	108	0.166	63	0.097	81	0.125
21	¾	21.1	0.830	350	0.541	210	0.325	185	0.287	108	0.168	140	0.216
27	1	26.8	1.054	564	0.873	338	0.524	299	0.462	175	0.270	226	0.349
35	1¼	35.4	1.395	984	1.528	591	0.917	522	0.810	305	0.474	394	0.611
41	1½	40.3	1.588	1276	1.981	765	1.188	676	1.050	395	0.614	510	0.792
53	2	51.6	2.033	2091	3.246	1255	1.948	1108	1.720	648	1.006	836	1.298
63	2½	63.3	2.493	3147	4.881	1888	2.929	1668	2.587	976	1.513	1259	1.953
78	3	78.4	3.085	4827	7.475	2896	4.485	2559	3.962	1497	2.317	1931	2.990
91	3½	89.4	3.520	6277	9.731	3766	5.839	3327	5.158	1946	3.017	2511	3.893
103	4	102.1	4.020	8187	12.692	4912	7.615	4339	6.727	2538	3.935	3275	5.077
129	5	—	—	—	—	—	—	—	—	—	—	—	—
155	6	—	—	—	—	—	—	—	—	—	—	—	—

Article 344 — Rigid Metal Conduit (RMC)

Metric Designator	Trade Size	Nominal Internal Diameter		Total Area 100%		60%		1 Wire 53%		2 Wires 31%		Over 2 Wires 40%	
		mm	in.	mm²	in.²	mm²	in.²	mm²	in.²	mm²	in.²	mm²	in.²
12	⅜	—	—	—	—	—	—	—	—	—	—	—	—
16	½	16.1	0.632	204	0.314	122	0.188	108	0.166	63	0.097	81	0.125
21	¾	21.2	0.836	353	0.549	212	0.329	187	0.291	109	0.170	141	0.220
27	1	27.0	1.063	573	0.887	344	0.532	303	0.470	177	0.275	229	0.355
35	1¼	35.4	1.394	984	1.526	591	0.916	522	0.809	305	0.473	394	0.610
41	1½	41.2	1.624	1333	2.071	800	1.243	707	1.098	413	0.642	533	0.829
53	2	52.9	2.083	2198	3.408	1319	2.045	1165	1.806	681	1.056	879	1.363
63	2½	63.2	2.489	3137	4.866	1882	2.919	1663	2.579	972	1.508	1255	1.946
78	3	78.5	3.090	4840	7.499	2904	4.499	2565	3.974	1500	2.325	1936	3.000
91	3½	90.7	3.570	6461	10.010	3877	6.006	3424	5.305	2003	3.103	2584	4.004
103	4	102.9	4.050	8316	12.882	4990	7.729	4408	6.828	2578	3.994	3326	5.153
129	5	128.9	5.073	13050	20.212	7830	12.127	6916	10.713	4045	6.266	5220	8.085
155	6	154.8	6.093	18821	29.158	11292	17.495	9975	15.454	5834	9.039	7528	11.663

Article 352 — Rigid PVC Conduit (RNC), Schedule 80

Metric Designator	Trade Size	Nominal Internal Diameter		Total Area 100%		60%		1 Wire 53%		2 Wires 31%		Over 2 Wires 40%	
		mm	in.	mm²	in.²	mm²	in.²	mm²	in.²	mm²	in.²	mm²	in.²
12	⅜	—	—	—	—	—	—	—	—	—	—	—	—
16	½	13.4	0.526	141	0.217	85	0.130	75	0.115	44	0.067	56	0.087
21	¾	18.3	0.722	263	0.409	158	0.246	139	0.217	82	0.127	105	0.164
27	1	23.8	0.936	445	0.688	267	0.413	236	0.365	138	0.213	178	0.275
35	1¼	31.9	1.255	799	1.237	480	0.742	424	0.656	248	0.383	320	0.495
41	1½	37.5	1.476	1104	1.711	663	1.027	585	0.907	342	0.530	442	0.684
53	2	48.6	1.913	1855	2.874	1113	1.725	983	1.523	575	0.891	742	1.150
63	2½	58.2	2.290	2660	4.119	1596	2.471	1410	2.183	825	1.277	1064	1.647
78	3	72.7	2.864	4151	6.442	2491	3.865	2200	3.414	1287	1.997	1660	2.577
91	3½	84.5	3.326	5608	8.688	3365	5.213	2972	4.605	1738	2.693	2243	3.475
103	4	96.2	3.786	7268	11.258	4361	6.755	3852	5.967	2253	3.490	2907	4.503
129	5	121.1	4.768	11518	17.855	6911	10.713	6105	9.463	3571	5.535	4607	7.142
155	6	145.0	5.709	16513	25.598	9908	15.359	8752	13.567	5119	7.935	6605	10.239

Table 4 *Continued*

Articles 352 and 353 — Rigid PVC Conduit (RNC), Schedule 40, and HDPE Conduit

Metric Designator	Trade Size	Nominal Internal Diameter mm	in.	Total Area 100% mm²	in.²	60% mm²	in.²	1 Wire 53% mm²	in.²	2 Wires 31% mm²	in.²	Over 2 Wires 40% mm²	in.²
12	⅜	—	—	—	—	—	—	—	—	—	—	—	—
16	½	15.3	0.602	184	0.285	110	0.171	97	0.151	57	0.088	74	0.114
21	¾	20.4	0.804	327	0.508	196	0.305	173	0.269	101	0.157	131	0.203
27	1	26.1	1.029	535	0.832	321	0.499	284	0.441	166	0.258	214	0.333
35	1¼	34.5	1.360	935	1.453	561	0.872	495	0.770	290	0.450	374	0.581
41	1½	40.4	1.590	1282	1.986	769	1.191	679	1.052	397	0.616	513	0.794
53	2	52.0	2.047	2124	3.291	1274	1.975	1126	1.744	658	1.020	849	1.316
63	2½	62.1	2.445	3029	4.695	1817	2.817	1605	2.488	939	1.455	1212	1.878
78	3	77.3	3.042	4693	7.268	2816	4.361	2487	3.852	1455	2.253	1877	2.907
91	3½	89.4	3.521	6277	9.737	3766	5.842	3327	5.161	1946	3.018	2511	3.895
103	4	101.5	3.998	8091	12.554	4855	7.532	4288	6.654	2508	3.892	3237	5.022
129	5	127.4	5.016	12748	19.761	7649	11.856	6756	10.473	3952	6.126	5099	7.904
155	6	153.2	6.031	18433	28.567	11060	17.140	9770	15.141	5714	8.856	7373	11.427

Article 352 — Type A, Rigid PVC Conduit (RNC)

Metric Designator	Trade Size	Nominal Internal Diameter mm	in.	Total Area 100% mm²	in.²	60% mm²	in.²	1 Wire 53% mm²	in.²	2 Wires 31% mm²	in.²	Over 2 Wires 40% mm²	in.²
16	½	17.8	0.700	249	0.385	149	0.231	132	0.204	77	0.119	100	0.154
21	¾	23.1	0.910	419	0.650	251	0.390	222	0.345	130	0.202	168	0.260
27	1	29.8	1.175	697	1.084	418	0.651	370	0.575	216	0.336	279	0.434
35	1¼	38.1	1.500	1140	1.767	684	1.060	604	0.937	353	0.548	456	0.707
41	1½	43.7	1.720	1500	2.324	900	1.394	795	1.231	465	0.720	600	0.929
53	2	54.7	2.155	2350	3.647	1410	2.188	1245	1.933	728	1.131	940	1.459
63	2½	66.9	2.635	3515	5.453	2109	3.272	1863	2.890	1090	1.690	1406	2.181
78	3	82.0	3.230	5281	8.194	3169	4.916	2799	4.343	1637	2.540	2112	3.278
91	3½	93.7	3.690	6896	10.694	4137	6.416	3655	5.668	2138	3.315	2758	4.278
103	4	106.2	4.180	8858	13.723	5315	8.234	4695	7.273	2746	4.254	3543	5.489
129	5	—	—	—	—	—	—	—	—	—	—	—	—
155	6	—	—	—	—	—	—	—	—	—	—	—	—

Article 352 — Type EB, PVC Conduit (RNC)

Metric Designator	Trade Size	Nominal Internal Diameter mm	in.	Total Area 100% mm²	in.²	60% mm²	in.²	1 Wire 53% mm²	in.²	2 Wires 31% mm²	in.²	Over 2 Wires 40% mm²	in.²
16	½	—	—	—	—	—	—	—	—	—	—	—	—
21	¾	—	—	—	—	—	—	—	—	—	—	—	—
27	1	—	—	—	—	—	—	—	—	—	—	—	—
35	1¼	—	—	—	—	—	—	—	—	—	—	—	—
41	1½	—	—	—	—	—	—	—	—	—	—	—	—
53	2	56.4	2.221	2498	3.874	1499	2.325	1324	2.053	774	1.201	999	1.550
63	2½	—	—	—	—	—	—	—	—	—	—	—	—
78	3	84.6	3.330	5621	8.709	3373	5.226	2979	4.616	1743	2.700	2248	3.484
91	3½	96.6	3.804	7329	11.365	4397	6.819	3884	6.023	2272	3.523	2932	4.546
103	4	108.9	4.289	9314	14.448	5589	8.669	4937	7.657	2887	4.479	3726	5.779
129	5	135.0	5.316	14314	22.195	8588	13.317	7586	11.763	4437	6.881	5726	8.878
155	6	160.9	6.336	20333	31.530	12200	18.918	10776	16.711	6303	9.774	8133	12.612

Table 5 Dimensions of Insulated Conductors and Fixture Wires

Type	Size (AWG or kcmil)	Approximate Diameter		Approximate Area	
		mm	in.	mm²	in.²
Type: FFH-2, RFH-1, RFH-2, RHH*, RHW*, RHW-2*, RHH, RHW, RHW-2, SF-1, SF-2, SFF-1, SFF-2, TF, TFF, THHW, THW, THW-2, TW, XF, XFF					
RFH-2, FFH-2	18	3.454	0.136	9.355	0.0145
	16	3.759	0.148	11.10	0.0172
RHH, RHW, RHW-2	14	4.902	0.193	18.90	0.0293
	12	5.385	0.212	22.77	0.0353
	10	5.994	0.236	28.19	0.0437
	8	8.280	0.326	53.87	0.0835
	6	9.246	0.364	67.16	0.1041
	4	10.46	0.412	86.00	0.1333
	3	11.18	0.440	98.13	0.1521
	2	11.99	0.472	112.9	0.1750
	1	14.78	0.582	171.6	0.2660
	1/0	15.80	0.622	196.1	0.3039
	2/0	16.97	0.668	226.1	0.3505
	3/0	18.29	0.720	262.7	0.4072
	4/0	19.76	0.778	306.7	0.4754
	250	22.73	0.895	405.9	0.6291
	300	24.13	0.950	457.3	0.7088
	350	25.43	1.001	507.7	0.7870
	400	26.62	1.048	556.5	0.8626
	500	28.78	1.133	650.5	1.0082
	600	31.57	1.243	782.9	1.2135
	700	33.38	1.314	874.9	1.3561
	750	34.24	1.348	920.8	1.4272
	800	35.05	1.380	965.0	1.4957
	900	36.68	1.444	1057	1.6377
	1000	38.15	1.502	1143	1.7719
	1250	43.92	1.729	1515	2.3479
	1500	47.04	1.852	1738	2.6938
	1750	49.94	1.966	1959	3.0357
	2000	52.63	2.072	2175	3.3719
SF-2, SFF-2	18	3.073	0.121	7.419	0.0115
	16	3.378	0.133	8.968	0.0139
	14	3.759	0.148	11.10	0.0172
SF-1, SFF-1	18	2.311	0.091	4.194	0.0065
RFH-1, XF, XFF	18	2.692	0.106	5.161	0.0080
TF, TFF, XF, XFF	16	2.997	0.118	7.032	0.0109
TW, XF, XFF, THHW, THW, THW-2	14	3.378	0.133	8.968	0.0139
TW, THHW, THW, THW-2	12	3.861	0.152	11.68	0.0181
	10	4.470	0.176	15.68	0.0243
	8	5.994	0.236	28.19	0.0437
RHH*, RHW*, RHW-2*	14	4.140	0.163	13.48	0.0209
RHH*, RHW*, RHW-2*, XF, XFF	12	4.623	0.182	16.77	0.0260

Table 5 *Continued*

Type	Size (AWG or kcmil)	Approximate Diameter		Approximate Area	
		mm	in.	mm²	in.²
Type: RHH*, RHW*, RHW-2*, THHN, THHW, THW, THW-2, TFN, TFFN, THWN, THWN-2, XF, XFF					
RHH,* RHW,* RHW-2,* XF, XFF	10	5.232	0.206	21.48	0.0333
RHH*, RHW*, RHW-2*	8	6.756	0.266	35.87	0.0556
TW, THW, THHW, THW-2, RHH*, RHW*, RHW-2*	6	7.722	0.304	46.84	0.0726
	4	8.941	0.352	62.77	0.0973
	3	9.652	0.380	73.16	0.1134
	2	10.46	0.412	86.00	0.1333
	1	12.50	0.492	122.6	0.1901
	1/0	13.51	0.532	143.4	0.2223
	2/0	14.68	0.578	169.3	0.2624
	3/0	16.00	0.630	201.1	0.3117
	4/0	17.48	0.688	239.9	0.3718
	250	19.43	0.765	296.5	0.4596
	300	20.83	0.820	340.7	0.5281
	350	22.12	0.871	384.4	0.5958
	400	23.32	0.918	427.0	0.6619
	500	25.48	1.003	509.7	0.7901
	600	28.27	1.113	627.7	0.9729
	700	30.07	1.184	710.3	1.1010
	750	30.94	1.218	751.7	1.1652
	800	31.75	1.250	791.7	1.2272
	900	33.38	1.314	874.9	1.3561
	1000	34.85	1.372	953.8	1.4784
	1250	39.09	1.539	1200	1.8602
	1500	42.21	1.662	1400	2.1695
	1750	45.11	1.776	1598	2.4773
	2000	47.80	1.882	1795	2.7818
TFN, TFFN	18	2.134	0.084	3.548	0.0055
	16	2.438	0.096	4.645	0.0072
THHN, THWN, THWN-2	14	2.819	0.111	6.258	0.0097
	12	3.302	0.130	8.581	0.0133
	10	4.166	0.164	13.61	0.0211
	8	5.486	0.216	23.61	0.0366
	6	6.452	0.254	32.71	0.0507
	4	8.230	0.324	53.16	0.0824
	3	8.941	0.352	62.77	0.0973
	2	9.754	0.384	74.71	0.1158
	1	11.33	0.446	100.8	0.1562
	1/0	12.34	0.486	119.7	0.1855
	2/0	13.51	0.532	143.4	0.2223
	3/0	14.83	0.584	172.8	0.2679
	4/0	16.31	0.642	208.8	0.3237
	250	18.06	0.711	256.1	0.3970
	300	19.46	0.766	297.3	0.4608
Type: FEP, FEPB, PAF, PAFF, PF, PFA, PFAH, PFF, PGF, PGFF, PTF, PTFF, TFE, THHN, THWN, THWN-2, Z, ZF, ZFF					
THHN, THWN, THWN-2	350	20.75	0.817	338.2	0.5242
	400	21.95	0.864	378.3	0.5863
	500	24.10	0.949	456.3	0.7073

(continues)

Table 5 *Continued*

Type	Size (AWG or kcmil)	Approximate Diameter		Approximate Area	
		mm	in.	mm²	in.²
	600	26.70	1.051	559.7	0.8676
	700	28.50	1.122	637.9	0.9887
	750	29.36	1.156	677.2	1.0496
	800	30.18	1.188	715.2	1.1085
	900	31.80	1.252	794.3	1.2311
	1000	33.27	1.310	869.5	1.3478
PF, PGFF, PGF, PFF, PTF, PAF, PTFF, PAFF	18	2.184	0.086	3.742	0.0058
	16	2.489	0.098	4.839	0.0075
PF, PGFF, PGF, PFF, PTF, PAF, PTFF, PAFF, TFE, FEP, PFA, FEPB, PFAH	14	2.870	0.113	6.452	0.0100
TFE, FEP, PFA, FEPB, PFAH	12	3.353	0.132	8.839	0.0137
	10	3.962	0.156	12.32	0.0191
	8	5.232	0.206	21.48	0.0333
	6	6.198	0.244	30.19	0.0468
	4	7.417	0.292	43.23	0.0670
	3	8.128	0.320	51.87	0.0804
	2	8.941	0.352	62.77	0.0973
TFE, PFAH	1	10.72	0.422	90.26	0.1399
TFE, PFA, PFAH, Z	1/0	11.73	0.462	108.1	0.1676
	2/0	12.90	0.508	130.8	0.2027
	3/0	14.22	0.560	158.9	0.2463
	4/0	15.70	0.618	193.5	0.3000
ZF, ZFF	18	1.930	0.076	2.903	0.0045
	16	2.235	0.088	3.935	0.0061
Z, ZF, ZFF	14	2.616	0.103	5.355	0.0083
Z	12	3.099	0.122	7.548	0.0117
	10	3.962	0.156	12.32	0.0191
	8	4.978	0.196	19.48	0.0302
	6	5.944	0.234	27.74	0.0430
	4	7.163	0.282	40.32	0.0625
	3	8.382	0.330	55.16	0.0855
	2	9.195	0.362	66.39	0.1029
	1	10.21	0.402	81.87	0.1269
Type: KF-1, KF-2, KFF-1, KFF-2, XHH, XHHW, XHHW-2, ZW					
XHHW, ZW, XHHW-2, XHH	14	3.378	0.133	8.968	0.0139
	12	3.861	0.152	11.68	0.0181
	10	4.470	0.176	15.68	0.0243
	8	5.994	0.236	28.19	0.0437
	6	6.960	0.274	38.06	0.0590
	4	8.179	0.322	52.52	0.0814
	3	8.890	0.350	62.06	0.0962
	2	9.703	0.382	73.94	0.1146
XHHW, XHHW-2, XHH	1	11.23	0.442	98.97	0.1534
	1/0	12.24	0.482	117.7	0.1825
	2/0	13.41	0.528	141.3	0.2190
	3/0	14.73	0.58	170.5	0.2642
	4/0	16.21	0.638	206.3	0.3197

Table 5 *Continued*

Type	Size (AWG or kcmil)	Approximate Diameter		Approximate Area	
		mm	in.	mm²	in.²
	250	17.91	0.705	251.9	0.3904
	300	19.30	0.76	292.6	0.4536
	350	20.60	0.811	333.3	0.5166
	400	21.79	0.858	373.0	0.5782
	500	23.95	0.943	450.6	0.6984
	600	26.75	1.053	561.9	0.8709
	700	28.55	1.124	640.2	0.9923
	750	29.41	1.158	679.5	1.0532
	800	30.23	1.190	717.5	1.1122
	900	31.85	1.254	796.8	1.2351
	1000	33.32	1.312	872.2	1.3519
	1250	37.57	1.479	1108	1.7180
	1500	40.69	1.602	1300	2.0157
	1750	43.59	1.716	1492	2.3127
	2000	46.28	1.822	1682	2.6073
KF-2, KFF-2	18	1.600	0.063	2.000	0.0031
	16	1.905	0.075	2.839	0.0044
	14	2.286	0.090	4.129	0.0064
	12	2.769	0.109	6.000	0.0093
	10	3.378	0.133	8.968	0.0139
KF-1, KFF-1	18	1.448	0.057	1.677	0.0026
	16	1.753	0.069	2.387	0.0037
	14	2.134	0.084	3.548	0.0055
	12	2.616	0.103	5.355	0.0083
	10	3.226	0.127	8.194	0.0127

*Types RHH, RHW, and RHW-2 without outer covering.

Table 5A Compact Aluminum Building Wire Nominal Dimensions* and Areas

Size (AWG or kcmil)	Bare Conductor Diameter		Types THW and THHW				Type THHN				Type XHHW				Size (AWG or kcmil)
			Approximate Diameter		Approximate Area		Approximate Diameter		Approximate Area		Approximate Diameter		Approximate Area		
	mm	in.	mm	in.	mm²	in.²	mm	in.	mm²	in.²	mm	in.	mm²	in.²	
8	3.404	0.134	6.477	0.255	32.90	0.0510	—	—	—	—	5.690	0.224	25.42	0.0394	8
6	4.293	0.169	7.366	0.290	42.58	0.0660	6.096	0.240	29.16	0.0452	6.604	0.260	34.19	0.0530	6
4	5.410	0.213	8.509	0.335	56.84	0.0881	7.747	0.305	47.10	0.0730	7.747	0.305	47.10	0.0730	4
2	6.807	0.268	9.906	0.390	77.03	0.1194	9.144	0.360	65.61	0.1017	9.144	0.360	65.61	0.1017	2
1	7.595	0.299	11.81	0.465	109.5	0.1698	10.54	0.415	87.23	0.1352	10.54	0.415	87.23	0.1352	1
1/0	8.534	0.336	12.70	0.500	126.6	0.1963	11.43	0.450	102.6	0.1590	11.43	0.450	102.6	0.1590	1/0
2/0	9.550	0.376	13.84	0.545	150.5	0.2332	12.57	0.495	124.1	0.1924	12.45	0.490	121.6	0.1885	2/0
3/0	10.74	0.423	14.99	0.590	176.3	0.2733	13.72	0.540	147.7	0.2290	13.72	0.540	147.7	0.2290	3/0
4/0	12.07	0.475	16.38	0.645	210.8	0.3267	15.11	0.595	179.4	0.2780	14.99	0.590	176.3	0.2733	4/0
250	13.21	0.520	18.42	0.725	266.3	0.4128	17.02	0.670	227.4	0.3525	16.76	0.660	220.7	0.3421	250
300	14.48	0.570	19.69	0.775	304.3	0.4717	18.29	0.720	262.6	0.4071	18.16	0.715	259.0	0.4015	300
350	15.65	0.616	20.83	0.820	340.7	0.5281	19.56	0.770	300.4	0.4656	19.30	0.760	292.6	0.4536	350
400	16.74	0.659	21.97	0.865	379.1	0.5876	20.70	0.815	336.5	0.5216	20.32	0.800	324.3	0.5026	400
500	18.69	0.736	23.88	0.940	447.7	0.6939	22.48	0.885	396.8	0.6151	22.35	0.880	392.4	0.6082	500
600	20.65	0.813	26.67	1.050	558.6	0.8659	25.02	0.985	491.6	0.7620	24.89	0.980	486.6	0.7542	600
700	22.28	0.877	28.19	1.110	624.3	0.9676	26.67	1.050	558.6	0.8659	26.67	1.050	558.6	0.8659	700
750	23.06	0.908	29.21	1.150	670.1	1.0386	27.31	1.075	585.5	0.9076	27.69	1.090	602.0	0.9331	750
900	25.37	0.999	31.09	1.224	759.1	1.1766	30.33	1.194	722.5	1.1196	29.69	1.169	692.3	1.0733	900
1000	26.92	1.060	32.64	1.285	836.6	1.2968	31.88	1.255	798.1	1.2370	31.24	1.230	766.6	1.1882	1000

*Dimensions are from industry sources.

Most aluminum building wire in Types THW, THHW, THWN/THHN, and XHHW conductors is compact stranded. Table 5A provides appropriate dimensions for these types of wire.

Table 8 Conductor Properties

Size (AWG or kcmil)	Area		Conductors							Direct-Current Resistance at 75°C (167°F)					
			Stranding			Overall				Copper				Aluminum	
				Diameter		Diameter		Area		Uncoated		Coated			
	mm²	Circular mils	Quantity	mm	in.	mm	in.	mm²	in.²	ohm/km	ohm/kFT	ohm/km	ohm/kFT	ohm/km	ohm/kFT
18	0.823	1620	1	—	—	1.02	0.040	0.823	0.001	25.5	7.77	26.5	8.08	42.0	12.8
18	0.823	1620	7	0.39	0.015	1.16	0.046	1.06	0.002	26.1	7.95	27.7	8.45	42.8	13.1
16	1.31	2580	1	—	—	1.29	0.051	1.31	0.002	16.0	4.89	16.7	5.08	26.4	8.05
16	1.31	2580	7	0.49	0.019	1.46	0.058	1.68	0.003	16.4	4.99	17.3	5.29	26.9	8.21
14	2.08	4110	1	—	—	1.63	0.064	2.08	0.003	10.1	3.07	10.4	3.19	16.6	5.06
14	2.08	4110	7	0.62	0.024	1.85	0.073	2.68	0.004	10.3	3.14	10.7	3.26	16.9	5.17
12	3.31	6530	1	—	—	2.05	0.081	3.31	0.005	6.34	1.93	6.57	2.01	10.45	3.18
12	3.31	6530	7	0.78	0.030	2.32	0.092	4.25	0.006	6.50	1.98	6.73	2.05	10.69	3.25
10	5.261	10380	1	—	—	2.588	0.102	5.26	0.008	3.984	1.21	4.148	1.26	6.561	2.00
10	5.261	10380	7	0.98	0.038	2.95	0.116	6.76	0.011	4.070	1.24	4.226	1.29	6.679	2.04
8	8.367	16510	1	—	—	3.264	0.128	8.37	0.013	2.506	0.764	2.579	0.786	4.125	1.26
8	8.367	16510	7	1.23	0.049	3.71	0.146	10.76	0.017	2.551	0.778	2.653	0.809	4.204	1.28
6	13.30	26240	7	1.56	0.061	4.67	0.184	17.09	0.027	1.608	0.491	1.671	0.510	2.652	0.808
4	21.15	41740	7	1.96	0.077	5.89	0.232	27.19	0.042	1.010	0.308	1.053	0.321	1.666	0.508
3	26.67	52620	7	2.20	0.087	6.60	0.260	34.28	0.053	0.802	0.245	0.833	0.254	1.320	0.403
2	33.62	66360	7	2.47	0.097	7.42	0.292	43.23	0.067	0.634	0.194	0.661	0.201	1.045	0.319
1	42.41	83690	19	1.69	0.066	8.43	0.332	55.80	0.087	0.505	0.154	0.524	0.160	0.829	0.253
1/0	53.49	105600	19	1.89	0.074	9.45	0.372	70.41	0.109	0.399	0.122	0.415	0.127	0.660	0.201
2/0	67.43	133100	19	2.13	0.084	10.62	0.418	88.74	0.137	0.3170	0.0967	0.329	0.101	0.523	0.159
3/0	85.01	167800	19	2.39	0.094	11.94	0.470	111.9	0.173	0.2512	0.0766	0.2610	0.0797	0.413	0.126
4/0	107.2	211600	19	2.68	0.106	13.41	0.528	141.1	0.219	0.1996	0.0608	0.2050	0.0626	0.328	0.100
250	127	—	37	2.09	0.082	14.61	0.575	168	0.260	0.1687	0.0515	0.1753	0.0535	0.2778	0.0847
300	152	—	37	2.29	0.090	16.00	0.630	201	0.312	0.1409	0.0429	0.1463	0.0446	0.2318	0.0707
350	177	—	37	2.47	0.097	17.30	0.681	235	0.364	0.1205	0.0367	0.1252	0.0382	0.1984	0.0605
400	203	—	37	2.64	0.104	18.49	0.728	268	0.416	0.1053	0.0321	0.1084	0.0331	0.1737	0.0529
500	253	—	37	2.95	0.116	20.65	0.813	336	0.519	0.0845	0.0258	0.0869	0.0265	0.1391	0.0424
600	304	—	61	2.52	0.099	22.68	0.893	404	0.626	0.0704	0.0214	0.0732	0.0223	0.1159	0.0353
700	355	—	61	2.72	0.107	24.49	0.964	471	0.730	0.0603	0.0184	0.0622	0.0189	0.0994	0.0303
750	380	—	61	2.82	0.111	25.35	0.998	505	0.782	0.0563	0.0171	0.0579	0.0176	0.0927	0.0282
800	405	—	61	2.91	0.114	26.16	1.030	538	0.834	0.0528	0.0161	0.0544	0.0166	0.0868	0.0265
900	456	—	61	3.09	0.122	27.79	1.094	606	0.940	0.0470	0.0143	0.0481	0.0147	0.0770	0.0235
1000	507	—	61	3.25	0.128	29.26	1.152	673	1.042	0.0423	0.0129	0.0434	0.0132	0.0695	0.0212
1250	633	—	91	2.98	0.117	32.74	1.289	842	1.305	0.0338	0.0103	0.0347	0.0106	0.0554	0.0169
1500	760	—	91	3.26	0.128	35.86	1.412	1011	1.566	0.02814	0.00858	0.02814	0.00883	0.0464	0.0141
1750	887	—	127	2.98	0.117	38.76	1.526	1180	1.829	0.02410	0.00735	0.02410	0.00756	0.0397	0.0121
2000	1013	—	127	3.19	0.126	41.45	1.632	1349	2.092	0.02109	0.00643	0.02109	0.00662	0.0348	0.0106

Notes:

1. These resistance values are valid **only** for the parameters as given. Using conductors having coated strands, different stranding type, and, especially, other temperatures changes the resistance.

2. Formula for temperature change: $R_2 = R_1 [1 + \alpha (T_2 - 75)]$ where $\alpha_{cu} = 0.00323$, $\alpha_{AL} = 0.00330$ at 75°C.

3. Conductors with compact and compressed stranding have about 9 percent and 3 percent, respectively, smaller bare conductor diameters than those shown. See Table 5A for actual compact cable dimensions.

4. The IACS conductivities used: bare copper = 100%, aluminum = 61%.

5. Class B stranding is listed as well as solid for some sizes. Its overall diameter and area is that of its circumscribing circle.

FPN: The construction information is per NEMA WC8-1992 or ANSI/UL 1581-1998. The resistance is calculated per National Bureau of Standards Handbook 100, dated 1966, and Handbook 109, dated 1972.

Traditionally, wire sizes have been expressed as American Wire Gage (AWG), circular mil (cmil) area, or thousands of circular mil (kcmil) area. Today, wire is available with its cross-sectional area expressed in square millimeters (mm^2) as well.

The 2005 *Code* specifically requires that insulated conductors be marked with their sizes and that the sizes be expressed in either AWG or circular mil area. [See 110.6 and 310.11(A)(4).] There are no exceptions to either of these two requirements. Because Article 310 does not specifically prohibit optional marking on insulated conductors, the *Code* permits square millimeter (mm^2) markings on conductors, but only if they are in addition to the required traditional markings of AWG or circular mil area.

According to IEEE/ASTM SI 10-2002, *Standard for Use of the International System of Units (SI): The Modern Metric System*, conversion from circular mils to square meters is done by multiplying circular mils by 5.067075×10^{-10}. However, because square millimeters, rather than square meters, is the standard marking for wire size and because the reciprocal is more appropriate for this conversion, a simpler conversion factor to convert from square millimeters to circular mils (approximately) follows:

$$k = 1973.53 \frac{\text{circular mils}}{mm^2}$$

The following example provides a comparison of the square millimeter wire gauge to traditional wire sizes.

Example

What traditional wire size does the size 125 mm^2 represent (approximately)?

Solution

$$
\begin{aligned}
\text{Circular mil area} &= \text{wire size } (mm^2) \times \text{conversion factor} \\
&= 125 \ mm^2 \times 1973.53 \frac{\text{circular mils}}{mm^2} \\
&= 246{,}691 \text{ circular mills} \\
&\quad \text{or } 246.691 \text{ kcmil}
\end{aligned}
$$

Therefore, the 125 mm^2 wire is larger than 4/0 AWG but smaller than a 250 kcmil conductor.

Conclusion: If a 125 mm^2 wire is determined to be the minimum or recommended size conductor, it is important to understand that size 250 kcmil would be the only Table 8 conductor with equivalent cross-sectional area. Because 4/0 AWG is simply not enough metal, 250 kcmil would be the choice for minimum equivalency. It is important, however, to note that the 250 kcmil conductor ampacity could not be used for a 125 mm^2 conductor, because the metric conductor size is smaller. The 4/0 AWG ampacity can be used, or the ampacity can be calculated under engineering supervision.

Table 9 Alternating-Current Resistance and Reactance for 600-Volt Cables, 3-Phase, 60 Hz, 75°C (167°F) — Three Single Conductors in Conduit

	Ohms to Neutral per Kilometer / Ohms to Neutral per 1000 Feet															
	X_L (Reactance) for All Wires		Alternating-Current Resistance for Uncoated Copper Wires			Alternating-Current Resistance for Aluminum Wires			Effective Z at 0.85 PF for Uncoated Copper Wires			Effective Z at 0.85 PF for Aluminum Wires			
Size (AWG or kcmil)	PVC, Aluminum Conduits	Steel Conduit	PVC Conduit	Aluminum Conduit	Steel Conduit	PVC Conduit	Aluminum Conduit	Steel Conduit	PVC Conduit	Aluminum Conduit	Steel Conduit	PVC Conduit	Aluminum Conduit	Steel Conduit	Size (AWG or kcmil)
14	0.190 0.058	0.240 0.073	10.2 3.1	10.2 3.1	10.2 3.1	— —	— —	— —	8.9 2.7	8.9 2.7	8.9 2.7	— —	— —	— —	14
12	0.177 0.054	0.223 0.068	6.6 2.0	6.6 2.0	6.6 2.0	10.5 3.2	10.5 3.2	10.5 3.2	5.6 1.7	5.6 1.7	5.6 1.7	9.2 2.8	9.2 2.8	9.2 2.8	12
10	0.164 0.050	0.207 0.063	3.9 1.2	3.9 1.2	3.9 1.2	6.6 2.0	6.6 2.0	6.6 2.0	3.6 1.1	3.6 1.1	3.6 1.1	5.9 1.8	5.9 1.8	5.9 1.8	10
8	0.171 0.052	0.213 0.065	2.56 0.78	2.56 0.78	2.56 0.78	4.3 1.3	4.3 1.3	4.3 1.3	2.26 0.69	2.26 0.69	2.30 0.70	3.6 1.1	3.6 1.1	3.6 1.1	8
6	0.167 0.051	0.210 0.064	1.61 0.49	1.61 0.49	1.61 0.49	2.66 0.81	2.66 0.81	2.66 0.81	1.44 0.44	1.48 0.45	1.48 0.45	2.33 0.71	2.36 0.72	2.36 0.72	6
4	0.157 0.048	0.197 0.060	1.02 0.31	1.02 0.31	1.02 0.31	1.67 0.51	1.67 0.51	1.67 0.51	0.95 0.29	0.95 0.29	0.98 0.30	1.51 0.46	1.51 0.46	1.51 0.46	4
3	0.154 0.047	0.194 0.059	0.82 0.25	0.82 0.25	0.82 0.25	1.31 0.40	1.35 0.41	1.31 0.40	0.75 0.23	0.79 0.24	0.79 0.24	1.21 0.37	1.21 0.37	1.21 0.37	3
2	0.148 0.045	0.187 0.057	0.62 0.19	0.66 0.20	0.66 0.20	1.05 0.32	1.05 0.32	1.05 0.32	0.62 0.19	0.62 0.19	0.66 0.20	0.98 0.30	0.98 0.30	0.98 0.30	2
1	0.151 0.046	0.187 0.057	0.49 0.15	0.52 0.16	0.52 0.16	0.82 0.25	0.85 0.26	0.82 0.25	0.52 0.16	0.52 0.16	0.52 0.16	0.79 0.24	0.79 0.24	0.82 0.25	1
1/0	0.144 0.044	0.180 0.055	0.39 0.12	0.43 0.13	0.39 0.12	0.66 0.20	0.69 0.21	0.66 0.20	0.43 0.13	0.43 0.13	0.43 0.13	0.62 0.19	0.66 0.20	0.66 0.20	1/0
2/0	0.141 0.043	0.177 0.054	0.33 0.10	0.33 0.10	0.33 0.10	0.52 0.16	0.52 0.16	0.52 0.16	0.36 0.11	0.36 0.11	0.36 0.11	0.52 0.16	0.52 0.16	0.52 0.16	2/0
3/0	0.138 0.042	0.171 0.052	0.253 0.077	0.269 0.082	0.259 0.079	0.43 0.13	0.43 0.13	0.43 0.13	0.289 0.088	0.302 0.092	0.308 0.094	0.43 0.13	0.43 0.13	0.46 0.14	3/0
4/0	0.135 0.041	0.167 0.051	0.203 0.062	0.220 0.067	0.207 0.063	0.33 0.10	0.36 0.11	0.33 0.10	0.243 0.074	0.256 0.078	0.262 0.080	0.36 0.11	0.36 0.11	0.36 0.11	4/0
250	0.135 0.041	0.171 0.052	0.171 0.052	0.187 0.057	0.177 0.054	0.279 0.085	0.295 0.090	0.282 0.086	0.217 0.066	0.230 0.070	0.240 0.073	0.308 0.094	0.322 0.098	0.33 0.10	250
300	0.135 0.041	0.167 0.051	0.144 0.044	0.161 0.049	0.148 0.045	0.233 0.071	0.249 0.076	0.236 0.072	0.194 0.059	0.207 0.063	0.213 0.065	0.269 0.082	0.282 0.086	0.289 0.088	300
350	0.131 0.040	0.164 0.050	0.125 0.038	0.141 0.043	0.128 0.039	0.200 0.061	0.217 0.066	0.207 0.063	0.174 0.053	0.190 0.058	0.197 0.060	0.240 0.073	0.253 0.077	0.262 0.080	350
400	0.131 0.040	0.161 0.049	0.108 0.033	0.125 0.038	0.115 0.035	0.177 0.054	0.194 0.059	0.180 0.055	0.161 0.049	0.174 0.053	0.184 0.056	0.217 0.066	0.233 0.071	0.240 0.073	400
500	0.128 0.039	0.157 0.048	0.089 0.027	0.105 0.032	0.095 0.029	0.141 0.043	0.157 0.048	0.148 0.045	0.141 0.043	0.157 0.048	0.164 0.050	0.187 0.057	0.200 0.061	0.210 0.064	500
600	0.128 0.039	0.157 0.048	0.075 0.023	0.092 0.028	0.082 0.025	0.118 0.036	0.135 0.041	0.125 0.038	0.131 0.040	0.144 0.044	0.154 0.047	0.167 0.051	0.180 0.055	0.190 0.058	600
750	0.125 0.038	0.157 0.048	0.062 0.019	0.079 0.024	0.069 0.021	0.095 0.029	0.112 0.034	0.102 0.031	0.118 0.036	0.131 0.040	0.141 0.043	0.148 0.045	0.161 0.049	0.171 0.052	750
1000	0.121 0.037	0.151 0.046	0.049 0.015	0.062 0.019	0.059 0.018	0.075 0.023	0.089 0.027	0.082 0.025	0.105 0.032	0.118 0.036	0.131 0.040	0.128 0.039	0.138 0.042	0.151 0.046	1000

Notes:

1. These values are based on the following constants: UL-Type RHH wires with Class B stranding, in cradled configuration. Wire conductivities are 100 percent IACS copper and 61 percent IACS aluminum, and aluminum conduit is 45 percent IACS. Capacitive reactance is ignored, since it is negligible at these voltages. These resistance values are valid only at 75°C (167°F) and for the parameters as given, but are representative for 600-volt wire types operating at 60 Hz.

2. *Effective Z* is defined as $R \cos(\theta) + X \sin(\theta)$, where θ is the power factor angle of the circuit. Multiplying current by effective impedance gives a good approximation for line-to-neutral voltage drop. Effective impedance values shown in this table are valid only at 0.85 power factor. For another circuit power factor (PF), effective impedance (Ze) can be calculated from R and X_L values given in this table as follows: $Ze = R \times PF + X_L \sin[\arccos(PF)]$.

Voltage-drop calculations using the dc-resistance formula are not always accurate for ac circuits, especially for those with a less-than-unity power factor or for those that use conductors larger than 2 AWG. Table 9 allows *Code* users to perform simple ac voltage-drop calculations. Table 9 was compiled using the Neher-McGrath ac-resistance calculation method, and the values presented are both reliable and conservative. This table contains completed calculations of effective impedance (Z) for the average ac circuit with an 85 percent power factor (see Example 1). If calculations with a different power factor are necessary, Table 9 also contains the appropriate values of inductive reactance and ac resistance (see Example 2).

The basic assumptions and the limitations of Table 9 are as follows:

1. Capacitive reactance is ignored.
2. There are three conductors in a raceway.
3. The calculated voltage-drop values are approximate.
4. For circuits with other parameters, the Neher-McGrath ac-resistance calculation method is used.

Example 1

A feeder has a 100-ampere continuous load. The system source is 240 volts, 3 phase, and the supplying circuit breaker is 125 amperes. The feeder is in a trade size 1¼ aluminum conduit with three 1 AWG THHN copper conductors operating at their maximum temperature rating of 75°C. The circuit length is 150 ft, and the power factor is 85 percent. Using Table 9, determine the approximate voltage drop of this circuit.

Solution

Step 1. Find the approximate line-to-neutral voltage drop. Using the Table 9 column "Effective Z at 0.85 PF for Uncoated Copper Wires," select aluminum conduit and size 1 AWG copper wire. Use the given value of 0.16 ohm per 1000 ft in the following formula:

$$\text{Voltage drop}_{(\text{line-to-neutral})} = \text{table value} \times \frac{\text{circuit length}}{1000 \text{ ft}}$$
$$\times \text{ circuit load}$$
$$= 0.16 \text{ ohm} \times \frac{150 \text{ ft}}{1000 \text{ ft}}$$
$$\times 100 \text{ A}$$
$$= 2.40 \text{ V}$$

Step 2. Find the line-to-line voltage drop.

$$\text{Voltage drop}_{(\text{line-to-line})} = \text{voltage drop}_{(\text{line-to-neutral})} \times \sqrt{3}$$
$$= 2.40 \text{ V} \times 1.732$$
$$= 4.157 \text{ V}$$

Step 3. Find the voltage drop expressed as a percentage of the circuit voltage.

$$\text{Percentage voltage drop}_{(\text{line-to-line})} = \frac{4.157\text{V}}{240\text{V}} \times 100$$
$$= 1.73\% \text{ VD}$$

Step 4. Find the voltage present at the load end of the circuit.

$$240 \text{ V} - 4.157 \text{ V} = 235.84 \text{ V}$$

Example 2

A 270-ampere continuous load is present on a feeder. The circuit consists of a single 4 in. PVC conduit with three 600 kcmil XHHW/USE aluminum conductors fed from a 480-volt, 3-phase, 3-wire source. The conductors are operating at their maximum rated temperature of 75°C. If the power factor is 0.7 and the circuit length is 250 ft, is the voltage drop excessive?

Solution

Step 1. Using the Table 9 column "X_L (Reactance) for All Wires," select PVC conduit and the row for size 600 kcmil. A value of 0.039 ohm per 1000 ft is given as this X_L. Next, using the column "Alternating-Current Resistance for Aluminum Wires," select PVC conduit and the row for size 600 kcmil. A value of 0.036 ohm per 1000 ft is given as this R.

Step 2. Find the angle representing a power factor of 0.7. Using a calculator with trigonometric functions or a trigonometric function table, find the arccosine ($\cos^{-1}$) θ of 0.7, which is 45.57 degrees. For this example, call this angle θ. Use the table or a calculator to find the sine of 45.57 degrees, which is 0.7141.

Step 3. Find the impedance (Z) corrected to 0.7 power factor (Z_c).

$$Z_c = (R \times \cos \theta) + (X_L \times \sin \theta)$$
$$= (0.036 \times 0.7) + (0.039 \times 0.7141)$$
$$= 0.0252 + 0.0279$$
$$= 0.0531 \text{ ohm to neutral}$$

Step 4. As in Example 1, find the approximate line-to-neutral voltage drop.

$$\text{Voltage drop}_{(\text{line-to-neutral})} = Z_c \times \frac{\text{circuit length}}{1000 \text{ ft}}$$
$$\times \text{ circuit load}$$
$$= 0.0530 \times \frac{250 \text{ ft}}{1000 \text{ ft}} \times 270 \text{ A}$$
$$= 3.577 \text{ V}$$

Step 5. Find the approximate line-to-line voltage drop.

$$\text{Voltage drop}_{(\text{line-to-line})} = \text{voltage drop}_{(\text{line-to-neutral})}$$
$$\times \sqrt{3}$$
$$= 3.577 \text{ V} \times 1.732$$
$$= 6.196 \text{ V}$$

Step 6. Find the approximate voltage drop expressed as a percentage of the circuit voltage.

$$\text{Percentage voltage drop}_{(\text{line-to-line})} = \frac{6.196 \text{ V}}{480 \text{ V}} \times 100$$
$$= 1.25\% \text{ VD}$$

Step 7. Find the voltage present at the load end of the circuit.

$$480 \text{ V} - 6.196 \text{ V} = 473.8 \text{ V}$$

Conclusion: According to 210.19(A)(1), FPN No. 4, this voltage drop does not appear to be excessive.

Tables 11(A) and 11(B)

For listing purposes, Table 11(A) and Table 11(B) provide the required power source limitations for Class 2 and Class 3 power sources. Table 11(A) applies for alternating-current sources, and Table 11(B) applies for direct-current sources.

The power for Class 2 and Class 3 circuits shall be either (1) inherently limited, requiring no overcurrent protection, or (2) not inherently limited, requiring a combination of power source and overcurrent protection. Power sources designed for interconnection shall be listed for the purpose.

As part of the listing, the Class 2 or Class 3 power source shall be durably marked where plainly visible to indicate the class of supply and its electrical rating. A Class 2 power source not suitable for wet location use shall be so marked.

Exception: Limited power circuits used by listed information technology equipment.

Overcurrent devices, where required, shall be located at the point where the conductor to be protected receives its supply and shall not be interchangeable with devices of higher ratings. The overcurrent device shall be permitted as an integral part of the power source.

Until the 1996 edition of the *Code,* the information in Table 11(A) and Table 11(B) was included in Article 725. Because of the listing requirements for Class 2 and Class 3 power supplies per 725.41, this information has little or no utility for the typical installer. It has been retained in Chapter 9 as Table 11(A) and Table 11(B) to provide direction for organizations properly equipped and qualified to evaluate and list these products.

Table 11(A) Class 2 and Class 3 Alternating-Current Power Source Limitations

Power Source		Inherently Limited Power Source (Overcurrent Protection Not Required)				Not Inherently Limited Power Source (Overcurrent Protection Required)			
		Class 2			Class 3	Class 2		Class 3	
Source voltage V_{max} (volts) (see Note 1)		0 through 20*	Over 20 and through 30*	Over 30 and through 150	Over 30 and through 100	0 through 20*	Over 20 and through 30*	Over 30 and through 100	Over 100 and through 150
Power limitations VA_{max} (volt-amperes) (see Note 1)		—	—	—	—	250 (see Note 3)	250	250	N.A.
Current limitations I_{max} (amperes) (see Note 1)		8.0	8.0	0.005	$150/V_{max}$	$1000/V_{max}$	$1000/V_{max}$	$1000/V_{max}$	1.0
Maximum overcurrent protection (amperes)		—	—	—	—	5.0	$100/V_{max}$	$100/V_{max}$	1.0
Power source maximum nameplate rating	VA (volt-amperes)	$5.0 \times V_{max}$	100	$0.005 \times V_{max}$	100	$5.0 \times V_{max}$	100	100	100
	Current (amperes)	5.0	$100/V_{max}$	0.005	$100/V_{max}$	5.0	$100/V_{max}$	$100/V_{max}$	$100/V_{max}$

*Voltage ranges shown are for sinusoidal ac in indoor locations or where wet contact is not likely to occur.
For nonsinusoidal or wet contact conditions, see Note 2.

Table 11(B) Class 2 and Class 3 Direct-Current Power Source Limitations

Power Source		Inherently Limited Power Source (Overcurrent Protection Not Required)					Not Inherently Limited Power Source (Overcurrent Protection Required)			
		Class 2				Class 3	Class 2		Class 3	
Source voltage V_{max} (volts) (see Note 1)		0 through 20*	Over 20 and through 30*	Over 30 and through 60*	Over 60 and through 150	Over 60 and through 100	0 through 20*	Over 20 and through 60*	Over 60 and through 100	Over 100 and through 150
Power limitations VA_{max} (volt-amperes) (see Note 1)		—	—	—	—	—	250 (see Note 3)	250	250	N.A.
Current limitations I_{max} (amperes) (see Note 1)		8.0	8.0	$150/V_{max}$	0.005	$150/V_{max}$	$1000/V_{max}$	$1000/V_{max}$	$1000/V_{max}$	1.0
Maximum overcurrent protection (amperes)		—	—	—	—	—	5.0	$100/V_{max}$	$100/V_{max}$	1.0
Power source maximum nameplate rating	VA (volt-amperes)	$5.0 \times V_{max}$	100	100	$0.005 \times V_{max}$	100	$5.0 \times V_{max}$	100	100	100
	Current (amperes)	5.0	$100/V_{max}$	$100/V_{max}$	0.005	$100/V_{max}$	5.0	$100/V_{max}$	$100/V_{max}$	$100/V_{max}$

*Voltage ranges shown are for continuous dc in indoor locations or where wet contact is not likely to occur.
For interrupted dc or wet contact conditions, see Note 4.

Notes for Tables 11(A) and 11(B)

1. V_{max}, I_{max}, and VA_{max} are determined with the current-limiting impedance in the circuit (not bypassed) as follows:
V_{max}: Maximum output voltage regardless of load with rated input applied.
I_{max}: Maximum output current under any noncapacitive load, including short circuit, and with overcurrent protection bypassed if used. Where a transformer limits the output current, I_{max} limits apply after 1 minute of operation. Where a current-limiting impedance, listed for the purpose, or as part of a listed product, is used in combination with a nonpower-limited transformer or a stored energy source, e.g., storage battery, to limit the output current, I_{max} limits apply after 5 seconds.
VA_{max}: Maximum volt-ampere output after 1 minute of operation regardless of load and overcurrent protection bypassed if used.
2. For nonsinusoidal ac, V_{max} shall not be greater than 42.4 volts peak. Where wet contact (immersion not included) is likely to occur, Class 3 wiring methods shall be used or V_{max} shall not be greater than 15 volts for sinusoidal ac and 21.2 volts peak for nonsinusoidal ac.
3. If the power source is a transformer, VA_{max} is 350 or less when V_{max} is 15 or less.
4. For dc interrupted at a rate of 10 to 200 Hz, V_{max} shall not be greater than 24.8 volts peak. Where wet contact (immersion not included) is likely to occur, Class 3 wiring methods shall be used, or V_{max} shall not be greater than 30 volts for continuous dc; 12.4 volts peak for dc that is interrupted at a rate of 10 to 200 Hz.

Tables 12(A) and 12(B)

For listing purposes, Tables 12(A) and 12(B) provide the required power source limitations for power-limited fire alarm sources. Table 12(A) applies for alternating-current sources, and Table 12(B) applies for direct-current sources.

The power for power-limited fire alarm circuits shall be either (1) inherently limited, requiring no overcurrent protection, or (2) not inherently limited, requiring the power to be limited by a combination of power source and overcurrent protection.

As part of the listing, the PLFA power source shall be durably marked where plainly visible to indicate that it is a power-limited fire alarm power source. The overcurrent device, where required, shall be located at the point where the conductor to be protected receives its supply and shall not be interchangeable with devices of higher ratings. The overcurrent device shall be permitted as an integral part of the power source.

Until the 1996 edition of the *Code*, the information in Table 12(A) and Table 12(B) was included in Article 760. Because of the listing requirements for power-limited fire alarm (PLFA) power supplies, this information has little or no utility for the typical installer. The tables were moved to Chapter 9 as Table 12(A) and Table 12(B) to provide direction for organizations properly equipped and qualified to evaluate and list these products.

Table 12(A) PLFA Alternating-Current Power Source Limitations

Power Source		Inherently Limited Power Source (Overcurrent Protection Not Required)			Not Inherently Limited Power Source (Overcurrent Protection Required)		
Circuit voltage V_{max}(volts) (see Note 1)		0 through 20	Over 20 and through 30	Over 30 and through 100	0 through 20	Over 20 and through 100	Over 100 and through 150
Power limitations VA_{max}(volt-amperes) (see Note 1)		—	—	—	250 (see Note 2)	250	N.A.
Current limitations I_{max} (amperes) (see Note 1)		8.0	8.0	$150/V_{max}$	$1000/V_{max}$	$1000/V_{max}$	1.0
Maximum overcurrent protection (amperes)		—	—	—	5.0	$100/V_{max}$	1.0
Power source maximum nameplate ratings	VA (volt-amperes)	$5.0 \times V_{max}$	100	100	$5.0 \times V_{max}$	100	100
	Current (amperes)	5.0	$100/V_{max}$	$100/V_{max}$	5.0	$100/V_{max}$	$100/V_{max}$

Table 12(B) PLFA Direct-Current Power Source Limitations

Power Source		Inherently Limited Power Source (Overcurrent Protection Not Required)			Not Inherently Limited Power Source (Overcurrent Protection Required)		
Circuit voltage V_{max} (volts) (see Note 1)		0 through 20	Over 20 and through 30	Over 30 and through 100	0 through 20	Over 20 and through 100	Over 100 and through 150
Power limitations VA_{max} (volt-amperes) (see Note 1)		—	—	—	250 (see Note 2)	250	N.A.
Current limitations I_{max} (amperes) (see Note 1)		8.0	8.0	$150/V_{max}$	$1000/V_{max}$	$1000/V_{max}$	1.0
Maximum overcurrent protection (amperes)		—	—	—	5.0	$100/V_{max}$	1.0
Power source maximum nameplate ratings	VA (volt-amperes)	$5.0 \times V_{max}$	100	100	$5.0 \times V_{max}$	100	100
	Current (amperes)	5.0	$100/V_{max}$	$100/V_{max}$	5.0	$100/V_{max}$	$100/V_{max}$

Notes for Tables 12(A) and 12(B)

1. V_{max}, I_{max}, and VA_{max} are determined as follows:

V_{max}: Maximum output voltage regardless of load with rated input applied.

I_{max}: Maximum output current under any noncapacitive load, including short circuit, and with overcurrent protection bypassed if used. Where a transformer limits the output current, I_{max} limits apply after 1 minute of operation. Where a current-limiting impedance, listed for the purpose, is used in combination with a nonpower-limited transformer or a stored energy source, e.g., storage battery, to limit the output current, I_{max} limits apply after 5 seconds.

VA_{max}: Maximum volt-ampere output after 1 minute of operation regardless of load and overcurrent protection bypassed if used. Current limiting impedance shall not be bypassed when determining I_{max} and VA_{max}.

2. If the power source is a transformer, VA_{max} is 350 or less when V_{max} is 15 or less.

Annex A

Product Safety Standards

Annex A is not a part of the requirements of this NFPA document but is included for informational purposes only.

This informational annex provides a list of product safety standards used for product listing where that listing is required by this *Code*. It is recognized that this list is current at the time of publication but that new standards or modifications to existing standards can occur at any time while this edition of the *Code* is in effect.

This annex does not form a mandatory part of the requirements of this *Code* but is intended only to provide *Code* users with informational guidance about the product characteristics about which *Code* requirements have been based.

A key element to a safe and *Code*-compliant electrical installation is adherence to product installation requirements imposed by product-testing organizations as part of their evaluation of an electrical product. Section 110.3(B) requires compliance with the installation and use instructions that are included with listed and labeled products. Numerous requirements in the *Code* specify the use of listed products. For those *Code* requirements where product listing is mandatory, Annex A is a compilation of applicable product safety standards. It is important to understand that the product safety standards included in Annex A are only those for which there is a mandatory listing requirement in the *Code*. There are many other safety standards associated with prod-

ucts that do not have a mandatory listing or labeling requirement in the *Code*. For more information on product standards, consult the product directories available from testing organizations.

Product safety standards, installation codes such as the *NEC*, and qualified electrical inspection are separate but not mutually exclusive components of the North American electrical safety system. The effectiveness of this system strongly depends on a close working relationship among the organizations responsible for the development of product standards and installation codes and the electrical inspection community. All three components must be in place for the electrical safety system to be effective.

Annex A first appeared in the 2002 *Code* and was updated for the 2005 edition to include references to product standards for which new requirements for listed products were added in the 2005 *Code*. The function of Annex A is to provide users of the *Code* with the name, number, and developing organization for all product standards related to mandatory *Code* requirements that require the use of listed products. For more information on listing and labeling, see the commentary following 110.3(B).

Product Standard Name	Product Standard Number
Antenna-Discharge Units	UL 452
Arc-Fault Circuit-Interrupters	UL 1699
Armored Cable	UL 4
Attachment Plugs and Receptacles	UL 498
Audio/Video and Musical Instrument Apparatus for Household, Commercial, and Similar General Use	UL 60065
Audio-Video Products and Accessories	UL 1492
Busways	UL 857
Cables—Thermoplastic-Insulated Underground Feeder and Branch-Circuit Cables	UL 493
Cables—Thermoplastic-Insulated Wires and Cables	UL 83
Cables—Thermoset-Insulated Wires and Cables	UL 44
Cables for Non–Power-Limited Fire-Alarm Circuits	UL 1425
Cables for Power-Limited Fire-Alarm Circuits	UL 1424
Capacitors	UL 810
Cellular Metal Floor Raceways and Fittings	UL 209
Class 2 and Class 3 Transformers	UL 1585
Class 2 Power Units	UL 1310
Commercial Audio Equipment	UL 813
Communication Circuit Accessories	UL 1863
Communications Cables	UL 444
Community-Antenna Television Cables	UL 1655
Conduit—Type EB and A Rigid PVC Conduit and HDPE Conduit	UL 651A
Continuous Length HDPE Conduit	UL 651B HDPE
Control Centers for Changing Message Type Electric Signs	UL 1433
Cord Sets and Power-Supply Cords	UL 817
Data-Processing Cable	UL 1690
Dead-Front Switchboards	UL 891
Electric Motors	UL 1004
Electric Signs	UL 48
Electric Spas, Equipment Assemblies, and Associated Equipment	UL 1563
Electric Vehicle (EV) Charging System Equipment	UL 2202
Electric Water Heaters for Pools and Tubs	UL 1261
Electrical Apparatus for Explosive Gas Atmospheres	(UL 60079 Series)
Electrical Apparatus for Use in Class I, Zone 1 Hazardous (Classified) Locations Type of Protection— Encapsulation "m"	ISA S12.23.01
Electrical Apparatus for Use in Class I, Zones 0 & 1 Hazardous (Classified) Locations: General Requirements	ISA 12.0.01
Electrical Apparatus for Use in Class I, Zone 1 Hazardous (Classified) Locations: Type of Protection— Increased Safety "e"	ISA S12.16.01
Electrical Apparatus for Use in Class I, Zone 1 Hazardous (Classified) Locations: Type of Protection— Flameproof "d"	ISA S12.22.01
Electrical Apparatus for Use in Class I, Zone 1 Hazardous (Classified) Locations: Type of Protection— Powder Filling "q"	ISA S12.25.01
Electrical Apparatus for Use in Class I, Zone 1 Hazardous (Classified) Locations: Type of Protection— Oil-Immersion "O"	ISA S12.26.01
Electrical Heating Appliances	UL 499
Electrical Metallic Tubing—Aluminum	UL 797A
Electrical Metallic Tubing—Steel	UL 797
Electrical Nonmetallic Tubing	UL 1653
Electric-Battery-Powered Industrial Trucks	UL 583
Electrode Receptacles for Gas-Tube Signs	UL 879
Emergency Lighting and Power Equipment	UL 924
Enclosed and Dead-Front Switches	UL 98
Enclosures for Electrical Equipment	UL 50
Energy Management Equipment	UL 916
Fire Pump Controllers	UL 218

Product Standard Name	Product Standard Number
Fittings for Cable and Conduit	UL 514B
Flat-Plate Photovoltaic Modules and Panels	UL 1703
Flexible Cord and Fixture Wire	UL 62
Flexible Lighting Products	UL 2388
Flexible Metal Conduit	UL 1
Fluorescent-Lamp Ballasts	UL 935
Gas-Burning Heating Appliances for Manufactured Homes and Recreational Vehicles	UL 307B
Gas-Fired Cooking Appliances for Recreational Vehicles	UL 1075
Gas-Tube-Sign and Ignition Cable	UL 814
General-Use Snap Switches	UL 20
Ground-Fault Circuit-Interrupters	UL 943
Ground-Fault Sensing and Relaying Equipment	UL 1053
Grounding and Bonding Equipment	UL 467
Heating and Cooling Equipment	UL 1995
High-Intensity-Discharge Lamp Ballasts	UL 1029
High Voltage Industrial Control Equipment	UL 347
Household Refrigerators and Freezers	UL 250
Industrial Battery Chargers	UL 1564
Industrial Control Equipment	UL 508
Industrial Control Panels	UL 508A
Instrumentation Tray Cable	UL 2250
Insulated Wire Connector Systems for Underground Use or in Damp or Wet Locations	UL 486D
Intermediate Metal Conduit—Steel	UL 1242
Inverters, Converters, and Controllers for Use in Independent Power Systems	UL 1741
Isolated Power Systems Equipment	UL 1047
Junction Boxes for Swimming Pool Luminaires	UL 1241
Liquid Fuel-Burning Heating Appliances for Manufactured Homes and Recreational Vehicles	UL 307A
Liquid-Tight Flexible Nonmetallic Conduit	UL 1660
Liquid-Tight Flexible Steel Conduit	UL 360
Low Voltage Landscape Lighting Systems	UL 1838
Low-Voltage Fuses—Part 1: General Requirements	UL 248-1
Low-Voltage Fuses—Part 2: Class C Fuses	UL 248-2
Low-Voltage Fuses—Part 3: Class CA and CB Fuses	UL 248-3
Low-Voltage Fuses—Part 4: Class CC Fuses	UL 248-4
Low-Voltage Fuses—Part 5: Class G Fuses	UL 248-5
Low-Voltage Fuses—Part 6: Class H Non-Renewable Fuses	UL 248-6
Low-Voltage Fuses—Part 7: Class H Renewable Fuses	UL 248-7
Low-Voltage Fuses—Part 8: Class J Fuses	UL 248-8
Low-Voltage Fuses—Part 9: Class K Fuses	UL 248-9
Low-Voltage Fuses—Part 10: Class L Fuses	UL 249-10
Low-Voltage Fuses—Part 11: Plug Fuses	UL 248-11
Low-Voltage Fuses—Part 12: Class R Fuses	UL 248-12
Low-Voltage Fuses—Part 13: Semiconductor Fuses	UL 248–13
Low-Voltage Fuses—Part 14: Supplemental Fuses	UL 248–14
Low-Voltage Fuses—Part 15: Class T Fuses	UL 248-15
Low-Voltage Fuses—Part 16: Test Limiters	UL 248-16
Low-Voltage Lighting Fixtures for Use in Recreational Vehicles	UL 234
Luminaire Reflector Kits for Installation on Previously Installed Fluorescent Luminaires, Supplemental Requirements	UL 1598
Machine-Tool Wires and Cables	UL 1063
Manufactured Wiring Systems	UL 183
Medical Electrical Equipment—Part 1: General Requirements	UL 60601–1
Medium-Voltage Power Cables	UL 1072
Metal-Clad Cables	UL 1569
Metal-Clad Cables and Cable-Sealing Fittings for Use in Hazardous (Classified) Locations	UL 2225
Metallic Outlet Boxes	UL 514A
Mobile Home Pipe Heating Cable	UL 1462

Product Standard Name	Product Standard Number
Molded-Case Circuit Breakers, Molded-Case Switches, and Circuit-Breaker Enclosures	UL 489
Motor Control Centers	UL 845
Motor-Operated Appliances	UL 73
Neon Transformers and Power Supplies	UL 2161
Nonincendive Electrical Equipment for Use in Class I and II, Division 2 and Class III, Divisions 1 and 2 Hazardous (Classified) Locations	ISA S12.12
Nonmetallic Outlet Boxes, Flush-Device Boxes, and Covers	UL 514C
Nonmetallic Surface Raceways and Fittings	UL 5A
Nonmetallic Underground Conduit with Conductors	UL 1990
Office Furnishings	UL 1286
Optical Fiber Cable	UL 1651
Optical Fiber Cable Raceway	UL 2024
Panelboards	UL 67
Personal Protection Systems for Electric Vehicle Supply Circuits: General Requirements	UL 2231–1
Personal Protection Systems for Electric Vehicle Supply Circuits: Particular Requirements for Protection Devices for Use in Charging Systems	UL 2231–2
Plugs, Receptacles and Couplers for Electrical Vehicles	UL 2251
Portable Electric Luminaires	UL 153
Potting Compounds for Swimming Pool, Fountain, and Spa Equipment	UL 676A
Power Conversion Equipment	UL 508C
Power Outlets	UL 231
Power Units Other Than Class 2	UL 1012
Power-Limited Circuit Cables	UL 13
Professional Video and Audio Equipment	UL 1419
Protectors for Coaxial Communications Circuits	UL 497C
Protectors for Data Communication and Fire Alarm Circuits	UL 497B
Protectors for Paired Conductor Communications Circuits	UL 497
Reference Standard for Electrical Wires, Cables, and Flexible Cords	UL 1581
Reinforced Thermosetting Resin Conduit (RTRC) and Fittings	UL 1684
Residential Pipe Heating Cable	UL 2049
Rigid Metal Conduit—Steel	UL 6
Roof and Gutter De-Icing Cable Units	UL 1588
Room Air Conditioners	UL 484
Safety of Information Technology Equipment, Part 1: General Requirements	UL 60950–1
Schedule 40 and 80 Rigid PVC Conduit	UL 651
Seasonal and Holiday Decorative Products	UL 588
Secondary Protectors for Communications Circuits	UL 497A
Service-Entrance Cables	UL 854
Smoke Detectors for Fire Alarm Signaling Systems	UL 268
Specialty Transformers	UL 506
Splicing Wire Connectors	UL 486C
Stationary Engine Generator Assemblies	UL 2200
Strut-Type Channel Raceways and Fittings	UL 5B
Surface Metal Raceways and Fittings	UL 5
Surface Raceways and Fittings for Use with Data, Signal and Control Circuits	UL 5C
Surge Arresters—Gapped Silicon-Carbide Surge Arresters for AC Power Circuits	IEEE C62.1
Surge Arresters—Metal-Oxide Surge Arresters for AC Power Circuits	IEEE C62.11
Swimming Pool Pumps, Filters, and Chlorinators	UL 1081
Transfer Switch Equipment	UL 1008
Transient Voltage Surge Suppressors	UL 1449
Underfloor Raceways and Fittings	UL 884
Underwater Luminaires and Submersible Junction Boxes	UL 676
Uninterruptible Power Systems	UL 1778
Vacuum Cleaners, Blower Cleaners, and Household Floor Finishing Machines	UL 1017
Waste Disposers	UL 430
Wire Connectors and Soldering Lugs for Use with Copper Conductors	UL 486A
Wire Connectors for Use with Aluminum Conductors	UL 486B
Wireways, Auxiliary Gutters, and Associated Fittings	UL 870

Annex B

Application Information for Ampacity Calculation

This annex is not a part of the requirements of this NFPA document but is included for informational purposes only.

B.310.15(B)(1) Formula Application Information. This annex provides application information for ampacities calculated under engineering supervision.

The data in Annex B are based on calculations using the Neher-McGrath conductor ampacity formula. The use of this method to calculate conductor ampacity is permitted under engineering supervision per 310.15(A)(1) and 310.15(C). The formula is found in 310.15(C).

Although conductor ampacities calculated using this formula may exceed those found in a table of allowable ampacities, such as Table 310.16, the limitations for connecting to equipment terminals specified in 110.14(C) have to be followed. For equipment 600 volts and under, the ampacities for use with equipment terminals are based on Table 310.16. See the commentary following 110.14(C) and 310.15(C).

B.310.15(B)(2) Typical Applications Covered by Tables. Typical ampacities for conductors rated 0 through 2000 volts are shown in Tables B.310.1 through B.310.10. Underground electrical duct bank configurations, as detailed in Figures B.310.3, B.310.4, and B.310.5, are utilized for conductors rated 0 through 5000 volts. In Figures B.310.2 through B.310.5, where adjacent duct banks are used, a separation of 1.5 m (5 ft) between the centerlines of the closest ducts in each bank or 1.2 m (4 ft) between the extremities of the concrete envelopes is sufficient to prevent derating of the conductors due to mutual heating. These ampacities were calculated as detailed in the basic ampacity paper, AIEE Paper 57-660, *The Calculation of the Temperature Rise and Load Capability of Cable Systems,* by J. H. Neher and M. H. McGrath. For additional informa-

tion concerning the application of these ampacities, see IEEE/ICEA Standard S-135/P-46-426, *Power Cable Ampacities*, and IEEE Standard 835-1994, *Standard Power Cable Ampacity Tables*.

If other factors remain the same, a soil resistivity higher than 90 reduces ampacities to values below those listed in Table B.310.5 through Table B.310.10 for underground ampacity. Conversely, a load factor less than 100 percent will increase ampacities if other factors remain the same. See B.310.15(B)(7) for allowable adjustments if the load factor is less than 100 percent. Reduced load factors are used in Figure B.310.3 through Figure B.310.5.

Typical values of thermal resistivity (Rho) are as follows:

Average soil (90 percent of USA) = 90

Concrete = 55

Damp soil (coastal areas, high water table) = 60

Paper insulation = 550

Polyethylene (PE) = 450

Polyvinyl chloride (PVC) = 650

Rubber and rubber-like = 500

Very dry soil (rocky or sandy) = 120

Thermal resistivity, as used in this annex, refers to the heat transfer capability through a substance by conduction. It is the reciprocal of thermal conductivity and is normally expressed in the units °C-cm/watt. For additional information on determining soil thermal resistivity (Rho), see ANSI/IEEE Standard 442-1996, *Guide for Soil Thermal Resistivity Measurements.*

B.310.15(B)(3) Criteria Modifications. Where values of load factor and Rho are known for a particular electrical duct bank installation and they are different from those shown in a specific table or figure, the ampacities shown in the table or figure can be modified by the application of factors derived from the use of Figure B.310.1.

Where two different ampacities apply to adjacent portions of a circuit, the higher ampacity can be used beyond the point of transition, a distance equal to 3 m (10 ft) or 10 percent of the circuit length figured at the higher ampacity, whichever is less.

The information provided in B.310.15(B)(3) on how to address different conductor ampacities that apply to adjacent portions of a circuit is also specified in the exception to 310.15(A)(2). See the commentary following that exception.

Exhibit B.1 illustrates an installation where a change in the burial depth results in an analysis of whether a modification to the conductor ampacity is necessary. If that portion deeper than 30 in. does not exceed 25 percent of the total run length, no decrease in ampacity is required, even if part of the run is more than 30 in. deep, which is the maximum depth assumed in Figure B.310.2, Note 1, to maintain the accuracy of the tables.

Where the burial depth of direct burial or electrical duct bank circuits are modified from the values shown in a figure or table, ampacities can be modified as shown in (a) and (b) as follows.

(a) Where burial depths are increased in part(s) of an electrical duct run to avoid underground obstructions, no decrease in ampacity of the conductors is needed, provided the total length of parts of the duct run increased in depth to avoid obstructions is less than 25 percent of the total run length.

(b) Where burial depths are deeper than shown in a specific underground ampacity table or figure, an ampacity derating factor of 6 percent per increased 300 mm (foot) of depth for all values of Rho can be utilized. No rating change is needed where the burial depth is decreased.

For example, in accordance with Table B.310.7 and Figure B.310.2, the ampacity of six parallel runs of 500 kcmil, Type XHHW copper conductors (as shown in Detail 3 in Figure B.310.2), where Rho = 90, is 6 × 273 amperes = 1638 amperes at a depth measuring no more than 30 in. to the top duct in the bank. If the burial depth is 6 ft, the ampacity is calculated as follows: 1638 amperes − (3.5 × 0.06 × 1638 amperes) = 1294 amperes.

B.310.15(B)(4) Electrical Ducts. The term *electrical duct(s)* is defined in 310.60.

B.310.15(B)(5) Tables B.310.6 and B.310.7.

(a) To obtain the ampacity of cables installed in two electrical ducts in one horizontal row with 190-mm (7.5-in.) center-to-center

Exhibit B.1 The presence of a drain pipe results in a change of burial depth for that portion of the electrical installation under the drain pipe.

spacing between electrical ducts, similar to Figure B.310.2, Detail 1, multiply the ampacity shown for one duct in Tables B.310.6 and B.310.7 by 0.88.

(b) To obtain the ampacity of cables installed in four electrical ducts in one horizontal row with 190-mm (7.5-in.) center-to-center spacing between electrical ducts, similar to Figure B.310.2, Detail 2, multiply the ampacity shown for three electrical ducts in Tables B.310.6 and B.310.7 by 0.94.

The underground ampacity tables (Table B.310.5 through Table B.310.10) are based on the 7.5 in. center-to-center spacing illustrated in Figure B.310.2. Although moving directly buried cables or electrical ducts farther apart will increase ampacities, the effect is surprisingly small. One calculation indicates that two side-by-side electrical ducts buried 30 in. below grade would have to be spaced about 5 ft apart before they could be considered single electrical ducts as shown in Detail 1, Figure B.310.2. Decreasing the burial depth, decreasing the thermal resistivity of the earth or other surrounding medium, and decreasing the load factor each has a much greater effect in increasing ampacity than does increasing the horizontal spacing.

B.310.15(B)(6) Electrical Ducts Used in Figure B.310.2. If spacing between electrical ducts, as shown in Figure B.310.2, is less than specified in Figure B.310.2, where electrical ducts enter equipment enclosures from underground, the ampacity of conductors contained within such electrical ducts need not be reduced.

B.310.15(B)(7) Examples Showing Use of Figure B.310.1 for Electrical Duct Bank Ampacity Modifications. Figure B.310.1 is used for interpolation or extrapolation for values of Rho and load factor for cables installed in electrical ducts. The upper family of curves shows the variation in ampacity and Rho at unity load factor in terms of I_1, the ampacity for Rho = 60, and 50 percent load factor. Each curve is designated for a particular ratio I_2/I_1, where I_2 is the ampacity at Rho = 120 and 100 percent load factor.

The lower family of curves shows the relationship between Rho and load factor that will give substantially the same ampacity as the indicated value of Rho at 100 percent load factor.

As an example, to find the ampacity of a 500 kcmil copper cable circuit for six electrical ducts as shown in Table B.310.5: At the Rho = 60, LF = 50, I_1 = 583; for Rho = 120 and LF = 100, I_2 = 400. The ratio I_2/I_1 = 0.686. Locate Rho = 90 at the bottom of the chart and follow the 90 Rho line to the intersection with

100 percent load factor where the equivalent Rho = 90. Then follow the 90 Rho line to I_2/I_1 ratio of 0.686 where F = 0.74. The desired ampacity = 0.74 × 583 = 431, which agrees with the table for Rho = 90, LF = 100.

To determine the ampacity for the same circuit where Rho = 80 and LF = 75, using Figure B.310.1, the equivalent Rho = 43, F = 0.855, and the desired ampacity = 0.855 × 583 = 498 amperes. Values for using Figure B.310.1 are found in the electrical duct bank ampacity tables of this annex.

Where the load factor is less than 100 percent and can be verified by measurement or calculation, the ampacity of electrical duct bank installations can be modified as shown. Different values of Rho can be accommodated in the same manner.

Table B.310.1 Ampacities of Two or Three Insulated Conductors, Rated 0 Through 2000 Volts, Within an Overall Covering (Multiconductor Cable), in Raceway in Free Air Based on Ambient Air Temperature of 30°C (86°F)

Size (AWG or kcmil)	Temperature Rating of Conductor. (See Table 310.13.)						Size (AWG or kcmil)
	60°C (140°F)	75°C (167°F)	90°C (194°F)	60°C (140°F)	75°C (167°F)	90°C (194°F)	
	Types TW, UF	Types RHW, THHW, THW, THWN, XHHW, ZW	Types THHN, THHW, THW-2, THWN-2, RHH, RWH-2, USE-2, XHHW, XHHW-2, ZW-2	Type TW	Types RHW, THHW, THW, THWN, XHHW	Types THHN, THHW, THW-2, THWN-2, RHH, RWH-2,USE-2, XHHW, XHHW-2, ZW-2	
	COPPER			ALUMINUM OR COPPER-CLAD ALUMINUM			
14	16*	18*	21*	—	—	—	14
12	20*	24*	27*	16*	18*	21*	12
10	27*	33*	36*	21*	25*	28*	10
8	36	43	48	28	33	37	8
6	48	58	65	38	45	51	6
4	66	79	89	51	61	69	4
3	76	90	102	59	70	79	3
2	88	105	119	69	83	93	2
1	102	121	137	80	95	106	1
1/0	121	145	163	94	113	127	1/0
2/0	138	166	186	108	129	146	2/0
3/0	158	189	214	124	147	167	3/0
4/0	187	223	253	147	176	197	4/0
250	205	245	276	160	192	217	250
300	234	281	317	185	221	250	300
350	255	305	345	202	242	273	350
400	274	328	371	218	261	295	400
500	315	378	427	254	303	342	500
600	343	413	468	279	335	378	600
700	376	452	514	310	371	420	700
750	387	466	529	321	384	435	750
800	397	479	543	331	397	450	800
900	415	500	570	350	421	477	900
1000	448	542	617	382	460	521	1000

Correction Factors

Ambient Temp. (°C)	For ambient temperatures other than 30°C (86°F), multiply the ampacities shown above by the appropriate factor shown below.						Ambient Temp. (°F)
21–25	1.08	1.05	1.04	1.08	1.05	1.04	70–77
26–30	1.00	1.00	1.00	1.00	1.00	1.00	79–86
31–35	0.91	0.94	0.96	0.91	0.94	0.96	88–95
36–40	0.82	0.88	0.91	0.82	0.88	0.91	97–104
41–45	0.71	0.82	0.87	0.71	0.82	0.87	106–113
46–50	0.58	0.75	0.82	0.58	0.75	0.82	115–122
51–55	0.41	0.67	0.76	0.41	0.67	0.76	124–131
56–60	—	0.58	0.71	—	0.58	0.71	133–140
61–70	—	0.33	0.58	—	0.33	0.58	142–158
71–80	—	—	0.41	—	—	0.41	160–176

*Unless otherwise specifically permitted elsewhere in this *Code*, the overcurrent protection for these conductor types shall not exceed 15 amperes for 14 AWG, 20 amperes for 12 AWG, and 30 amperes for 10 AWG copper; or 15 amperes for 12 AWG and 25 AWG amperes for 10 AWG aluminum and copper-clad aluminum.

Table B.310.3 Ampacities of Multiconductor Cables with Not More Than Three Insulated Conductors, Rated 0 Through 2000 Volts, in Free Air Based on Ambient Air Temperature of 40°C (104°F) (For Types TC, MC, MI, UF, and USE Cables)

| | Temperature Rating of Conductor. (See Table 310.13.) | | | | | | | | |
| | 60°C (140°F) | 75°C (167°F) | 85°C (185°F) | 90°C (194°F) | 60°C (140°F) | 75°C (167°F) | 85°C (185°F) | 90°C (194°F) | |
Size (AWG or kcmil)	COPPER				ALUMINUM OR COPPER-CLAD ALUMINUM				Size (AWG or kcmil)
18	—	—	—	11*	—	—	—	—	18
16	—	—	—	16*	—	—	—	—	16
14	18*	21*	24*	25*	—	—	—	—	14
12	21*	28*	30*	32*	18*	21*	24*	25*	12
10	28*	36*	41*	43*	21*	28*	30*	32*	10
8	39	50	56	59	30	39	44	46	8
6	52	68	75	79	41	53	59	61	6
4	69	89	100	104	54	70	78	81	4
3	81	104	116	121	63	81	91	95	3
2	92	118	132	138	72	92	103	108	2
1	107	138	154	161	84	108	120	126	1
1/0	124	160	178	186	97	125	139	145	1/0
2/0	143	184	206	215	111	144	160	168	2/0
3/0	165	213	238	249	129	166	185	194	3/0
4/0	190	245	274	287	149	192	214	224	4/0
250	212	274	305	320	166	214	239	250	250
300	237	306	341	357	186	240	268	280	300
350	261	337	377	394	205	265	296	309	350
400	281	363	406	425	222	287	317	334	400
500	321	416	465	487	255	330	368	385	500
600	354	459	513	538	284	368	410	429	600
700	387	502	562	589	306	405	462	473	700
750	404	523	586	615	328	424	473	495	750
800	415	539	604	633	339	439	490	513	800
900	438	570	639	670	362	469	514	548	900
1000	461	601	674	707	385	499	558	584	1000

Correction Factors

Ambient Temp. (°C)	For ambient temperatures other than 40°C (104°F), multiply the ampacities shown above by the appropriate factor shown below.								Ambient Temp. (°F)
21–25	1.32	1.20	1.15	1.14	1.32	1.20	1.15	1.14	70–77
26–30	1.22	1.13	1.11	1.10	1.22	1.13	1.11	1.10	79–86
31–35	1.12	1.07	1.05	1.05	1.12	1.07	1.05	1.05	88–95
36–40	1.00	1.00	1.00	1.00	1.00	1.00	1.00	1.00	97–104
41–45	0.87	0.93	0.94	0.95	0.87	0.93	0.94	0.95	106–113
46–50	0.71	0.85	0.88	0.89	0.71	0.85	0.88	0.89	115–122
51–55	0.50	0.76	0.82	0.84	0.50	0.76	0.82	0.84	124–131
56–60	—	0.65	0.75	0.77	—	0.65	0.75	0.77	133–140
61–70	—	0.38	0.58	0.63	—	0.38	0.58	0.63	142–158
71–80	—	—	0.33	0.44	—	—	0.33	0.44	160–176

*Unless otherwise specifically permitted elsewhere in this *Code*, the overcurrent protection for these conductor types shall not exceed 15 amperes for 14 AWG, 20 amperes for 12 AWG, and 30 amperes for 10 AWG copper; or 15 amperes for 12 AWG and 25 amperes for 10 AWG aluminum and copper-clad aluminum.

Table B.310.5 Ampacities of Single Insulated Conductors, Rated 0 through 2000 Volts, in Nonmagnetic Underground Electrical Ducts (One Conductor per Electrical Duct), Based on Ambient Earth Temperature of 20°C (68°F), Electrical Duct Arrangement per Figure B.310.2, Conductor Temperature 75°C (167°F)

Size (kcmil)	3 Electrical Ducts (Fig. B.310.2, Detail 2) Types RHW, THHW, THW, THWN, XHHW, USE			6 Electrical Ducts (Fig. B.310.2, Detail 3) Types RHW, THHW, THW, THWN, XHHW, USE			9 Electrical Ducts (Fig. B.310.2, Detail 4) Types RHW, THHW, THW, THWN, XHHW, USE			3 Electrical Ducts (Fig. B.310.2, Detail 2) Types RHW, THHW, THW, THWN, XHHW, USE			6 Electrical Ducts (Fig. B.310.2, Detail 3) Types RHW, THHW, THW, THWN, XHHW, USE			9 Electrical Ducts (Fig. B.310.2, Detail 4) Types RHW, THHW, THW, THWN, XHHW, USE			Size (kcmil)
	COPPER									ALUMINUM OR COPPER-CLAD ALUMINUM									
	RHO 60 LF 50	RHO 90 LF 100	RHO 120 LF 100	RHO 60 LF 50	RHO 90 LF 100	RHO 120 LF 100	RHO 60 LF 50	RHO 90 LF 100	RHO 120 LF 100	RHO 60 LF 50	RHO 90 LF 100	RHO 120 LF 100	RHO 60 LF 50	RHO 90 LF 100	RHO 120 LF 100	RHO 60 LF 50	RHO 90 LF 100	RHO 120 LF 100	
250	410	344	327	386	295	275	369	270	252	320	269	256	302	230	214	288	211	197	250
350	503	418	396	472	355	330	446	322	299	393	327	310	369	277	258	350	252	235	350
500	624	511	484	583	431	400	545	387	360	489	401	379	457	337	313	430	305	284	500
750	794	640	603	736	534	494	674	469	434	626	505	475	581	421	389	538	375	347	750
1000	936	745	700	864	617	570	776	533	493	744	593	557	687	491	453	629	432	399	1000
1250	1055	832	781	970	686	632	854	581	536	848	668	627	779	551	508	703	478	441	1250
1500	1160	907	849	1063	744	685	918	619	571	941	736	689	863	604	556	767	517	477	1500
1750	1250	970	907	1142	793	729	975	651	599	1026	796	745	937	651	598	823	550	507	1750
2000	1332	1027	959	1213	836	768	1030	683	628	1103	850	794	1005	693	636	877	581	535	2000

Ambient Temp. (°C)	Correction Factors						Ambient Temp. (°F)
6–10	1.09	1.09	1.09	1.09	1.09	1.09	43–50
11–15	1.04	1.04	1.04	1.04	1.04	1.04	52–59
16–20	1.00	1.00	1.00	1.00	1.00	1.00	61–68
21–25	0.95	0.95	0.95	0.95	0.95	0.95	70–77
26–30	0.90	0.90	0.90	0.90	0.90	0.90	79–86

Table B.310.6 Ampacities of Three Insulated Conductors, Rated 0 through 2000 Volts, Within an Overall Covering (Three-Conductor Cable) in Underground Electrical Ducts (One Cable per Electrical Duct) Based on Ambient Earth Temperature of 20°C (68°F), Electrical Duct Arrangement per Figure B.310.2, Conductor Temperature 75°C (167°F)

Size (AWG or kcmil)	1 Electrical Duct (Fig. B.310.2, Detail 1) Types RHW, THHW, THW, THWN, XHHW, USE			3 Electrical Ducts (Fig. B.310.2, Detail 2) Types RHW, THHW, THW, THWN, XHHW, USE			6 Electrical Ducts (Fig. B.310.2, Detail 3) Types RHW, THHW, THW, THWN, XHHW, USE			1 Electrical Duct (Fig. B.310.2, Detail 1) Types RHW, THHW, THW, THWN, XHHW, USE			3 Electrical Ducts (Fig. B.310.2, Detail 2) Types RHW, THHW, THW, THWN, XHHW, USE			6 Electrical Ducts (Fig. B.310.2, Detail 3) Types RHW, THHW, THW, THWN, XHHW, USE			Size (AWG or kcmil)
	COPPER									ALUMINUM OR COPPER-CLAD ALUMINUM									
	RHO 60 LF 50	RHO 90 LF 100	RHO 120 LF 100	RHO 60 LF 50	RHO 90 LF 100	RHO 120 LF 100	RHO 60 LF 50	RHO 90 LF 100	RHO 120 LF 100	RHO 60 LF 50	RHO 90 LF 100	RHO 120 LF 100	RHO 60 LF 50	RHO 90 LF 100	RHO 120 LF 100	RHO 60 LF 50	RHO 90 LF 100	RHO 120 LF 100	
8	58	54	53	56	48	46	53	42	39	45	42	41	43	37	36	41	32	30	8
6	77	71	69	74	63	60	70	54	51	60	55	54	57	49	47	54	42	39	6
4	101	93	91	96	81	77	91	69	65	78	72	71	75	63	60	71	54	51	4
2	132	121	118	126	105	100	119	89	83	103	94	92	98	82	78	92	70	65	2
1	154	140	136	146	121	114	137	102	95	120	109	106	114	94	89	107	79	74	1
1/0	177	160	156	168	137	130	157	116	107	138	125	122	131	107	101	122	90	84	1/0
2/0	203	183	178	192	156	147	179	131	121	158	143	139	150	122	115	140	102	95	2/0
3/0	233	210	204	221	178	158	205	148	137	182	164	159	172	139	131	160	116	107	3/0
4/0	268	240	232	253	202	190	234	168	155	209	187	182	198	158	149	183	131	121	4/0
250	297	265	256	280	222	209	258	184	169	233	207	201	219	174	163	202	144	132	250
350	363	321	310	340	267	250	312	219	202	285	252	244	267	209	196	245	172	158	350
500	444	389	375	414	320	299	377	261	240	352	308	297	328	254	237	299	207	190	500
750	552	478	459	511	388	362	462	314	288	446	386	372	413	314	293	374	254	233	750
1000	628	539	518	579	435	405	522	351	321	521	447	430	480	361	336	433	291	266	1000

Ambient Temp. (°C)	Correction Factors						Ambient Temp. (°F)
6–10	1.09	1.09	1.09	1.09	1.09	1.09	43–50
11–15	1.04	1.04	1.04	1.04	1.04	1.04	52–59
16–20	1.00	1.00	1.00	1.00	1.00	1.00	61–68
21–25	0.95	0.95	0.95	0.95	0.95	0.95	70–77
26–30	0.90	0.90	0.90	0.90	0.90	0.90	79–86

Table B.310.7 Ampacities of Three Single Insulated Conductors, Rated 0 Through 2000 Volts, in Underground Electrical Ducts (Three Conductors per Electrical Duct) Based on Ambient Earth Temperature of 20°C (68°F), Electrical Duct Arrangement per Figure B.310.2, Conductor Temperature 75°C (167°F)

Size (AWG or kcmil)	1 Electrical Duct (Fig. B.310.2, Detail 1) Types RHW, THHW, THW, THWN, XHHW, USE			3 Electrical Ducts (Fig. B.310.2, Detail 2) Types RHW, THHW, THW, THWN, XHHW, USE			6 Electrical Ducts (Fig. B.310.2, Detail 3) Types RHW, THHW, THW, THWN, XHHW, USE			1 Electrical Duct (Fig. B.310.2, Detail 1) Types RHW, THHW, THW, THWN, XHHW, USE			3 Electrical Ducts (Fig. B.310.2, Detail 2) Types RHW, THHW, THW, THWN, XHHW, USE			6 Electrical Ducts (Fig. B.310.2, Detail 3) Types RHW, THHW, THW, THWN, XHHW, USE			Size (AWG or kcmil)
	COPPER									ALUMINUM OR COPPER-CLAD ALUMINUM									
	RHO 60 LF 50	RHO 90 LF 100	RHO 120 LF 100	RHO 60 LF 50	RHO 90 LF 100	RHO 120 LF 100	RHO 60 LF 50	RHO 90 LF 100	RHO 120 LF 100	RHO 60 LF 50	RHO 90 LF 100	RHO 120 LF 100	RHO 60 LF 50	RHO 90 LF 100	RHO 120 LF 100	RHO 60 LF 50	RHO 90 LF 100	RHO 120 LF 100	
8	63	58	57	61	51	49	57	44	41	49	45	44	47	40	38	45	34	32	8
6	84	77	75	80	67	63	75	56	53	66	60	58	63	52	49	59	44	41	6
4	111	100	98	105	86	81	98	73	67	86	78	76	79	67	63	77	57	52	4
3	129	116	113	122	99	94	113	83	77	101	91	89	83	77	73	84	65	60	3
2	147	132	128	139	112	106	129	93	86	115	103	100	108	87	82	101	73	67	2
1	171	153	148	161	128	121	149	106	98	133	119	115	126	100	94	116	83	77	1
1/0	197	175	169	185	146	137	170	121	111	153	136	132	144	114	107	133	94	87	1/0
2/0	226	200	193	212	166	156	194	136	126	176	156	151	165	130	121	151	106	98	2/0
3/0	260	228	220	243	189	177	222	154	142	203	178	172	189	147	138	173	121	111	3/0
4/0	301	263	253	280	215	201	255	175	161	235	205	198	219	168	157	199	137	126	4/0
250	334	290	279	310	236	220	281	192	176	261	227	218	242	185	172	220	150	137	250
300	373	321	308	344	260	242	310	210	192	293	252	242	272	204	190	245	165	151	300
350	409	351	337	377	283	264	340	228	209	321	276	265	296	222	207	266	179	164	350
400	442	376	361	394	302	280	368	243	223	349	297	284	321	238	220	288	191	174	400
500	503	427	409	460	341	316	412	273	249	397	338	323	364	270	250	326	216	197	500
600	552	468	447	511	371	343	457	296	270	446	373	356	408	296	274	365	236	215	600
700	602	509	486	553	402	371	492	319	291	488	408	389	443	321	297	394	255	232	700
750	632	529	505	574	417	385	509	330	301	508	425	405	461	334	309	409	265	241	750
800	654	544	520	597	428	395	527	338	308	530	439	418	481	344	318	427	273	247	800
900	692	575	549	628	450	415	554	355	323	563	466	444	510	365	337	450	288	261	900
1000	730	605	576	659	472	435	581	372	338	597	494	471	538	385	355	475	304	276	1000

Ambient Temp. (°C)	Correction Factors						Ambient Temp. (°F)
6–10	1.09	1.09	1.09	1.09	1.09	1.09	43–50
11–15	1.04	1.04	1.04	1.04	1.04	1.04	52–59
16–20	1.00	1.00	1.00	1.00	1.00	1.00	61–68
21–25	0.95	0.95	0.95	0.95	0.95	0.95	70–77
26–30	0.90	0.90	0.90	0.90	0.90	0.90	79–86

Table B.310.8 Ampacities of Two or Three Insulated Conductors, Rated 0 Through 2000 Volts, Cabled Within an Overall (Two- or Three-Conductor) Covering, Directly Buried in Earth, Based on Ambient Earth Temperature of 20°C (68°F), Arrangement per Figure B.310.2, 100 Percent Load Factor, Thermal Resistance (Rho) of 90

	1 Cable (Fig. B.310.2, Detail 5)		2 Cables (Fig. B.310.2, Detail 6)		1 Cable (Fig. B.310.2, Detail 5)		2 Cables (Fig. B.310.2, Detail 6)		
	60°C (140°F)	75°C (167°F)	60°C (140°F)	75°C (167°F)	60°C (140°F)	75°C (167°F)	60°C (140°F)	75°C (167°F)	
	TYPES				TYPES				
	UF	RHW, THHW, THW, THWN, XHHW, USE	UF	RHW, THHW, THW, THWN, XHHW, USE	UF	RHW, THHW, THW, THWN, XHHW, USE	UF	RHW, THHW, THW, THWN, XHHW, USE	
Size (AWG or kcmil)	COPPER				ALUMINUM OR COPPER-CLAD ALUMINUM				Size (AWG or kcmil)
8	64	75	60	70	51	59	47	55	8
6	85	100	81	95	68	75	60	70	6
4	107	125	100	117	83	97	78	91	4
2	137	161	128	150	107	126	110	117	2
1	155	182	145	170	121	142	113	132	1
1/0	177	208	165	193	138	162	129	151	1/0
2/0	201	236	188	220	157	184	146	171	2/0
3/0	229	269	213	250	179	210	166	195	3/0
4/0	259	304	241	282	203	238	188	220	4/0
250	—	333	—	308	—	261	—	241	250
350	—	401	—	370	—	315	—	290	350
500	—	481	—	442	—	381	—	350	500
750	—	585	—	535	—	473	—	433	750
1000	—	657	—	600	—	545	—	497	1000
Ambient Temp. (°C)	Correction Factors								Ambient Temp. (°F)
6–10	1.12	1.09	1.12	1.09	1.12	1.09	1.12	1.09	43–50
11–15	1.06	1.04	1.06	1.04	1.06	1.04	1.06	1.04	52–59
16–20	1.00	1.00	1.00	1.00	1.00	1.00	1.00	1.00	61–68
21–25	0.94	0.95	0.94	0.95	0.94	0.95	0.94	0.95	70–77
26–30	0.87	0.90	0.87	0.90	0.87	0.90	0.87	0.90	79–86

Note: For ampacities of Type UF cable in underground electrical ducts, multiply the ampacities shown in the table by 0.74.

Table B.310.9 Ampacities of Three Triplexed Single Insulated Conductors, Rated 0 Through 2000 Volts, Directly Buried in Earth Based on Ambient Earth Temperature of 20°C (68°F), Arrangement per Figure B.310.2, 100 Percent Load Factor, Thermal Resistance (Rho) of 90

Size (AWG or kcmil)	See Fig. B.310.2, Detail 7		See Fig. B.310.2, Detail 8		See Fig. B.310.2, Detail 7		See Fig. B.310.2, Detail 8		Size (AWG or kcmil)
	60°C (140°F)	75°C (167°F)	60°C (140°F)	75°C (167°F)	60°C (140°F)	75°C (167°F)	60°C (140°F)	75°C (167°F)	
	TYPES				TYPES				
	UF	USE	UF	USE	UF	USE	UF	USE	
	COPPER				ALUMINUM OR COPPER-CLAD ALUMINUM				
8	72	84	66	77	55	65	51	60	8
6	91	107	84	99	72	84	66	77	6
4	119	139	109	128	92	108	85	100	4
2	153	179	140	164	119	139	109	128	2
1	173	203	159	186	135	158	124	145	1
1/0	197	231	181	212	154	180	141	165	1/0
2/0	223	262	205	240	175	205	159	187	2/0
3/0	254	298	232	272	199	233	181	212	3/0
4/0	289	339	263	308	226	265	206	241	4/0
250	—	370	—	336	—	289	—	263	250
350	—	445	—	403	—	349	—	316	350
500	—	536	—	483	—	424	—	382	500
750	—	654	—	587	—	525	—	471	750
1000	—	744	—	665	—	608	—	544	1000
Ambient Temp. (°C)	Correction Factors								**Ambient Temp. (°F)**
6–10	1.12	1.09	1.12	1.09	1.12	1.09	1.12	1.09	43–50
11–15	1.06	1.04	1.06	1.04	1.06	1.04	1.06	1.04	52–59
16–20	1.00	1.00	1.00	1.00	1.00	1.00	1.00	1.00	61–68
21–25	0.94	0.95	0.94	0.95	0.94	0.95	0.94	0.95	70–77
26–30	0.87	0.90	0.87	0.90	0.87	0.90	0.87	0.90	79–86

Table B.310.10 Ampacities of Three Single Insulated Conductors, Rated 0 Through 2000 Volts, Directly Buried in Earth Based on Ambient Earth Temperature of 20°C (68°F), Arrangement per Figure B.310.2, 100 Percent Load Factor, Thermal Resistance (Rho) of 90

Size (AWG or kcmil)	See Fig. B.310.2, Detail 9		See Fig. B.310.2, Detail 10		See Fig. B.310.2, Detail 9		See Fig. B.310.2, Detail 10		Size (AWG or kcmil)
	60°C (140°F)	75°C (167°F)	60°C (140°F)	75°C (167°F)	60°C (140°F)	75°C (167°F)	60°C (140°F)	75°C (167°F)	
	TYPES				TYPES				
	UF	USE	UF	USE	UF	USE	UF	USE	
	COPPER				ALUMINUM OR COPPER-CLAD ALUMINUM				
8	84	98	78	92	66	77	61	72	8
6	107	126	101	118	84	98	78	92	6
4	139	163	130	152	108	127	101	118	4
2	178	209	165	194	139	163	129	151	2
1	201	236	187	219	157	184	146	171	1
1/0	230	270	212	249	179	210	165	194	1/0
2/0	261	306	241	283	204	239	188	220	2/0
3/0	297	348	274	321	232	272	213	250	3/0
4/0	336	394	309	362	262	307	241	283	4/0
250	—	429	—	394	—	335	—	308	250
350	—	516	—	474	—	403	—	370	350
500	—	626	—	572	—	490	—	448	500
750	—	767	—	700	—	605	—	552	750
1000	—	887	—	808	—	706	—	642	1000
1250	—	979	—	891	—	787	—	716	1250
1500	—	1063	—	965	—	862	—	783	1500
1750	—	1133	—	1027	—	930	—	843	1750
2000	—	1195	—	1082	—	990	—	897	2000
Ambient Temp. (°C)	Correction Factors								Ambient Temp. (°F)
6–10	1.12	1.09	1.12	1.09	1.12	1.09	1.12	1.09	43–50
11–15	1.06	1.04	1.06	1.04	1.06	1.04	1.06	1.04	52–59
16–20	1.00	1.00	1.00	1.00	1.00	1.00	1.00	1.00	61–68
21–25	0.94	0.95	0.94	0.95	0.94	0.95	0.94	0.95	70–77
26–30	0.87	0.90	0.87	0.90	0.87	0.90	0.87	0.90	79–86

Table B.310.11 Adjustment Factors for More Than Three Current-Carrying Conductors in a Raceway or Cable with Load Diversity

Number of Current-Carrying Conductors	Percent of Values in Tables as Adjusted for Ambient Temperature if Necessary
4 – 6	80
7 – 9	70
10 – 24	70*
25 – 42	60*
43 – 85	50*

*These factors include the effects of a load diversity of 50 percent.

FPN: The ampacity limit for the number of current-carrying conductors in 10 through 85 is based on the following formula. For greater than 85 conductors, special calculations are required that are beyond the scope of this table.

$$A_2 = \sqrt{\frac{0.5N}{E}} \times (A_1) \text{ or } A_1, \text{ whichever is less}$$

where:

A_1 = ampacity from Tables 310.16; 310.18; B.310.1; B.310.6; and B.310.7 multiplied by the appropriate factor from Table B.310.11.

N = total number of conductors used to obtain multiplying factor from Table B.310.11

E = desired number of current-carrying conductors in the raceway or cable

A_2 = ampacity limit for the current-carrying conductors in the raceway or cable

Example 1

Calculate the ampacity limit for twelve 14 AWG THWN current-carrying conductors (75°C) in a raceway that contains 24 conductors.

$$A_2 = \sqrt{\frac{(0.5)(24)}{12}} \times 20(0.7)$$
$$= 14 \text{ amperes (i.e., 50 percent diversity)}$$

Example 2

Calculate the ampacity limit for eighteen 14 AWG THWN current-carrying conductors (75°C) in a raceway that contains 24 conductors.

$$A_2 = \sqrt{\frac{(0.5)(24)}{18}} \times 20(0.7) = 11.5 \text{ amperes}$$

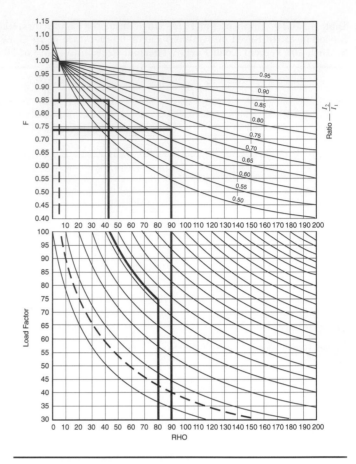

Figure B.310.1 Interpolation Chart for Cables in a Duct Bank
I_1 = ampacity for Rho = 60, 50 LF; I_2 = ampacity for Rho = 120, 100 LF (load factor); desired ampacity = F × I_1.

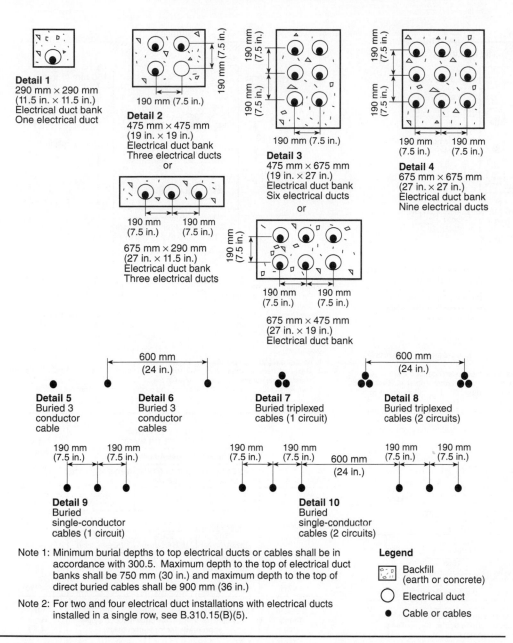

Detail 1
290 mm × 290 mm
(11.5 in. × 11.5 in.)
Electrical duct bank
One electrical duct

Detail 2
475 mm × 475 mm
(19 in. × 19 in.)
Electrical duct bank
Three electrical ducts
or

675 mm × 290 mm
(27 in. × 11.5 in.)
Electrical duct bank
Three electrical ducts

Detail 3
475 mm × 675 mm
(19 in. × 27 in.)
Electrical duct bank
Six electrical ducts
or

675 mm × 475 mm
(27 in. × 19 in.)
Electrical duct bank

Detail 4
675 mm × 675 mm
(27 in. × 27 in.)
Electrical duct bank
Nine electrical ducts

Detail 5
Buried 3
conductor
cable

Detail 6
Buried 3
conductor
cables

Detail 7
Buried triplexed
cables (1 circuit)

Detail 8
Buried triplexed
cables (2 circuits)

Detail 9
Buried
single-conductor
cables (1 circuit)

Detail 10
Buried
single-conductor
cables (2 circuits)

Note 1: Minimum burial depths to top electrical ducts or cables shall be in accordance with 300.5. Maximum depth to the top of electrical duct banks shall be 750 mm (30 in.) and maximum depth to the top of direct buried cables shall be 900 mm (36 in.).

Note 2: For two and four electrical duct installations with electrical ducts installed in a single row, see B.310.15(B)(5).

Legend

⬚ Backfill
(earth or concrete)

◯ Electrical duct

● Cable or cables

Figure B.310.2 Cable Installation Dimensions for Use with Table B.310.5 through Table B.310.10.

Design Criteria
Neutral and Equipment
   Grounding conductor (EGC)
   Duct = 150 mm (6 in.)
Phase Ducts = 75 to 125 mm (3 to 5 in.)
Conductor Material = Copper
Number of Cables per Duct = 3

Number of Cables per Phase = 9
Rho concrete = Rho Earth − 5

Rho PVC Duct = 650
Rho Cable Insulation = 500
Rho Cable Jacket = 650

Notes:
1. Neutral configuration per 300.5(I), Exception No. 2, for isolated phase installations in nonmagnetic ducts.
2. Phasing is A, B, C in rows or columns. Where magnetic electrical ducts are used, conductors are installed A, B, C per electrical duct with the neutral and all equipment grounding conductors in the same electrical duct. In this case, the 6-in. trade size neutral duct is eliminated.
3. Maximum harmonic loading on the neutral conductor cannot exceed 50 percent of the phase current for the ampacities shown in the table.
4. Metallic shields of Type MV-90 cable shall be grounded at one point only where using A, B, C phasing in rows or columns.

Size kcmil	TYPES RHW, THHW, THW, THWN, XHHW, USE, OR MV-90*			Size kcmil
	Total per Phase Ampere Rating			
	RHO EARTH 60 LF 50	RHO EARTH 90 LF 100	RHO EARTH 120 LF 100	
250	2340 (260A/Cable)	1530 (170A/Cable)	1395 (155A/Cable)	250
350	2790 (310A/Cable)	1800 (200A/Cable)	1665 (185A/Cable)	350
500	3375 (375A/Cable)	2160 (240A/Cable)	1980 (220A/Cable)	500

Ambient Temp. (°C)	For ambient temperatures other than 20°C (68°F), multiply the ampacities shown above by the appropriate factor shown below.					Ambient Temp. (°F)
6–10	1.09	1.09	1.09	1.09	1.09	43–50
11–15	1.04	1.04	1.04	1.04	1.04	52–59
16–20	1.00	1.00	1.00	1.00	1.00	61–68
21–25	0.95	0.95	0.95	0.95	0.95	70–77
26–30	0.90	0.90	0.90	0.90	0.90	79–86

*Limited to 75°C conductor temperature.

Figure B.310.3 Ampacities of Single Insulated Conductors Rated 0 through 5000 Volts in Underground Electrical Ducts (Three Conductors per Electrical Duct), Nine Single-Conductor Cables per Phase Based on Ambient Earth Temperature of 20°C (68°F), Conductor Temperature 75°C (167°F).

Design Criteria
Neutral and Equipment
   Grounding Conductor (EGC)
   Duct = 150 mm (6 in.)
Phase Ducts = 75 mm (3 in.)
Conductor Material = Copper
Number of Cables per Duct = 1

Number of Cables per Phase = 4
Rho Concrete = Rho Earth − 5
Rho PVC Duct = 650

Rho Cable Insulation = 500
Rho Cable Jacket = 650

Notes:
1. Neutral configuration per 300.5(I), Exception No. 2.
2. Maximum harmonic loading on the neutral conductor cannot exceed 50 percent of the phase current for the ampacities shown in the table.
3. Metallic shields of Type MV-90 cable shall be grounded at one point only.

Size kcmil	TYPES RHW, THHW, THW, THWN, XHHW, USE, OR MV-90*			Size kcmil
	Total per Phase Ampere Rating			
	RHO EARTH 60 LF 50	RHO EARTH 90 LF 100	RHO EARTH 120 LF 100	
750	2820 (705A/Cable)	1860 (465A/Cable)	1680 (420A/Cable)	750
1000	3300 (825A/Cable)	2140 (535A/Cable)	1920 (480A/Cable)	1000
1250	3700 (925A/Cable)	2380 (595A/Cable)	2120 (530A/Cable)	1250
1500	4060 (1015A/Cable)	2580 (645A/Cable)	2300 (575A/Cable)	1500
1750	4360 (1090A/Cable)	2740 (685A/Cable)	2460 (615A/Cable)	1750

Ambient Temp. (°C)	For ambient temperatures other than 20°C (68°F), multiply the ampacities shown above by the appropriate factor shown below.					Ambient Temp. (°F)
6–10	1.09	1.09	1.09	1.09	1.09	43–50
11–15	1.04	1.04	1.04	1.04	1.04	52–59
16–20	1.00	1.00	1.00	1.00	1.00	61–68
21–25	0.95	0.95	0.95	0.95	0.95	70–77
26–30	0.90	0.90	0.90	0.90	0.90	79–86

*Limited to 75°C conductor temperature.

Figure B.310.4 Ampacities of Single Insulated Conductors Rated 0 through 5000 Volts in Nonmagnetic Underground Electrical Ducts (One Conductor per Electrical Duct), Four Single-Conductor Cables per Phase Based on Ambient Earth Temperature of 20°C (68°F), Conductor Temperature 75°C (167°F).

Surface

150 mm (6 in.)

N and EGC

AØ
250 mm (10 in.)
BØ
250 mm (10 in.)
CØ

150 mm (6 in.)

600 mm (24 in.)

950 mm (38 in.)

250 mm (10 in.)   250 mm (10 in.)   250 mm (10 in.)   250 mm (10 in.)

1.1 m (44 in.)

Design Criteria
Neutral and Equipment
   Grounding Conductor (EGC)
   Duct = 150 mm (6 in.)
Phase Ducts = 75 mm (3 in.)
Conductor Material = Copper
Number of Cables per Duct = 1

Number of Cables per Phase = 5
Rho Concrete = Rho Earth – 5
Rho PVC Duct = 650

Rho Cable Insulation = 500
Rho Cable Jacket = 650

Notes:
1. Neutral configuration per 300.5(I), Exception No. 2.
2. Maximum harmonic loading on the neutral conductor cannot exceed 50 percent of the phase current for the ampacities shown in the table.
3. Metallic shields of Type MV-90 cable shall be grounded at one point only.

Size kcmil	TYPES RHW,THHW,THW,THWN, XHHW, USE, OR MV-90*			Size kcmil
	Total per Phase Ampere Rating			
	RHO EARTH 60 LF 50	RHO EARTH 90 LF 100	RHO EARTH 120 LF 100	
2000	5575 (1115A/Cable)	3375 (675A/Cable)	3000 (600A/Cable)	2000

Ambient Temp. (°C)	For ambient temperatures other than 20°C (68°F), multiply the ampacities shown above by the appropriate factor shown below.					Ambient Temp. (°F)
6–10	1.09	1.09	1.09	1.09	1.09	43–50
11–15	1.04	1.04	1.04	1.04	1.04	52–59
16–20	1.00	1.00	1.00	1.00	1.00	61–68
21–25	0.95	0.95	0.95	0.95	0.95	70–77
26–30	0.90	0.90	0.90	0.90	0.90	79–86

*Limited to 75°C conductor temperature.

Figure B.310.5 Ampacities of Single Insulated Conductors Rated 0 through 5000 Volts in Nonmagnetic Underground Electrical Ducts (One Conductor per Electrical Duct), Five Single-Conductor Cables per Phase Based on Ambient Earth Temperature of 20°C (68°F), Conductor Temperature 75°C (167°F).

Table B.310.11 Adjustment Factors for More Than Three Current-Carrying Conductors in a Raceway or Cable with Load Diversity

Number of Current-Carrying Conductors	Percent of Values in Tables as Adjusted for Ambient Temperature if Necessary
4 – 6	80
7 – 9	70
10 – 24	70*
25 – 42	60*
43 – 85	50*

*These factors include the effects of a load diversity of 50 percent.

Annex C

Conduit and Tubing Fill Tables for Conductors and Fixture Wires of the Same Size

This annex is not a part of the requirements of this NFPA document but is included for informational purposes only.

Contents

Table C.9   Maximum Number of Conductors or Fixture Wires in Rigid PVC Conduit, Schedule 80 (Based on Table 1, Chapter 9)

Table C.9(A)   Maximum Number of Compact Conductors in Rigid PVC Conduit, Schedule 80 (Based on Table 1, Chapter 9)

Table C.10   Maximum Number of Conductors or Fixture Wires in Rigid PVC Conduit, Schedule 40 and HDPE Conduit (Based on Table 1, Chapter 9)

Table C.10(A)   Maximum Number of Compact Conductors in Rigid PVC Conduit, Schedule 40 and HDPE Conduit (Based on Table 1, Chapter 9)

Table C.11   Maximum Number of Conductors or Fixture Wires in Type A, Rigid PVC Conduit (Based on Table 1, Chapter 9)

Table C.11(A)   Maximum Number of Compact Conductors in Type A, Rigid PVC Conduit (Based on Table 1, Chapter 9)

Table C.12   Maximum Number of Conductors in Type EB, PVC Conduit (Based on Table 1, Chapter 9)

Table C.12(A)   Maximum Number of Compact Conductors in Type EB, PVC Conduit (Based on Table 1, Chapter 9)

*Where this table is used in conjunction with Tables C.1 through C.12, the conductors installed must be of the compact type.

The Annex C conduit and tubing conductor fill tables are provided only as an informational tool and are not part of the mandatory requirements of the *Code*. The conductor fill values specified in the Annex C tables have been calculated based on the conductor fill requirements, conduit and tubing dimensions, and conductor dimensions from Chapter 9, Tables 1, 4, 5, and 5A. As such, where the conductors in conduit or tubing are all of the same physical size and insulation characteristic, the use of the Annex C tables to determine the maximum number of conductors ensures compliance with the *Code* requirements for raceway fill found in 300.17 and the respective conduit and tubing articles. For that reason, *NEC* users rely on the widely used and referred-to tables in Annex C as though they were part of the mandatory requirements of the *Code*. Although the values for maximum conduit or tubing fill do not exceed that permitted by Chapter 9, Table 1, the advice provided in FPN No.1 to Table 1 on considering a lesser conductor fill or a larger size conduit to that table should always be taken into consideration where the wire-pulling conditions are not optimum.

The Annex C tables came about because of the different internal dimensions of the many types of conduit and tubing. Table 3A, Table 3B, and Table 3C in previous editions of the *Code* were generic in nature and did not distinguish among the different conduit and tubing types. Although more comprehensive than the approach taken in past editions of the *Code*, the Annex C tables provide a far more precise approach to conductor fill tables.

As in past editions of the *Code*, Chapter 9, Table 1, sets forth the percentage fill required, and Table 4, Table 5, and Table 5A list the accurate conduit, tubing, and wire dimensions. Users can calculate the percent fill, as permitted in past editions of the *Code* or use the tables in Annex C, all of which were generated using Chapter 9, Table 1, Table 4, Table 5, and Table 5A.

The 12 sets of tables in Annex C correspond to the 12 conduit and tubing wiring methods in Chapter 3 of the *Code*. Each set of tables is subdivided into three conductor categories: (1) conductors for general wiring (Article 310), which appear in an order similar to Chapter 9, Table 3A, Table 3B, and Table 3C of the *NEC* prior to the 1996 edition; (2) fixture wires (Article 402), in an order similar to the pre-1996 edition Chapter 9, Table 2; and (3) compact stranded conductors (Article 310), in an order similar to the pre-1996 edition Chapter 9, Table 5A. In Annex C, tables that use compact stranding are listed as "A" tables.

To select the correct metric designator or trade size conduit or tubing, proceed as follows:

STEP 1. Select the appropriate wiring method from Tables C.1 through C.12 using the lists of metallic and nonmetallic wiring methods given in Commentary Table C.1 and Commentary Table C.2.

Commentary Table C.1 Metallic Wiring Methods

Type of Wiring	Appropriate Table
Electrical metallic tubing	Table C.1
Flexible metal conduit	Table C.3
Intermediate metal conduit	Table C.4
Liquidtight flexible metal conduit	Table C.7
Rigid metal conduit	Table C.8

Commentary Table C.2 Nonmetallic Wiring Methods

Type of Wiring	Appropriate Table
Electrical nonmetallic tubing	Table C.2
Liquidtight flexible nonmetallic conduit (LFNC-B)	Table C.5
Liquidtight flexible nonmetallic conduit (LFNC-A)	Table C.6
Rigid PVC conduit, Schedule 80	Table C.9
Rigid PVC conduit, Schedule 40 and HDPE conduit	Table C.10
Type A, rigid PVC conduit	Table C.11
Type EB, PVC conduit	Table C.12

STEP 2. Choose the appropriate conductors (general wiring conductors, fixture wires, or compact conductors).

STEP 3. Choose the appropriate insulation.

STEP 4. Select the correct trade size conduit or tubing for the given quantity and size of conductors required.

The following examples show how to determine the correct trade size conduit or tubing.

Example 1

An installation requires ten 10 AWG THWN-2 copper conductors in an underground conduit across a parking lot for exterior lighting. What size PVC conduit will be required?

Solution

Annex C lists the types of PVC conduit as given in Commentary Table C.3.

Example 2

An underground service lateral requires four 600 kcmil XHHW compact-stranded aluminum conductors. What trade

Commentary Table C.3 Types of PVC Conduit

Wiring Method	Table	Minimum Trade Size Conduit or Tubing
PVC Schedule 40 and HDPE	Table C.10	1
PVC Type A	Table C.11	¾
PVC Schedule 80	Table C.9	1
PVC Type EB	Table C.12	Available only in trade sizes 2 and larger

size conduit will be required for RMC, IMC, PVC Schedule 40, PVC Schedule 80, and PVC Type EB?

Solution

Annex C lists the minimum trade size for the respective conduit types needed for the four 600 kcmil aluminum conductors.

Commentary Table C.4

Wiring Method	Table	Minimum Trade Size Conduit or Tubing
RMC	Table C.8A	3
IMC	Table C.4A	3
PVC Schedule 40	Table C.10A	3
PVC Schedule 80	Table C.9A	3½
PVC Type EB	Table C.12A	3

Most aluminum building wire in Types THW, THWN/THHN, and XHHW is compact stranded.

Example 3

A three-phase, 480 volt, 40 hp motor will be supplied by three 4 AWG THW conductors. What size metal conduit or tubing will be required? What size metal flex will be required at the motor termination?

Solution

Annex C lists various types of metal conduit and metal flex as shown in Commentary Table C.5.

Commentary Table C.5 Types of Metal Conduit and Metal Flex

Wiring Method	Table	Trade Size Conduit or Tubing
EMT	Table C.1	1
IMC	Table C.4	1
RMC	Table C.8	1
FMC	Table C.3	1
Liquidtight FMC	Table C.7	1

Where a wire-type equipment grounding conductor (likely to be smaller than 4 AWG based on rating or size of motor circuit short-circuit, ground-fault protective device, and Table 250.122) is installed in any one of these raceway types, there is now a mixture of conductor sizes. For that reason, the minimum size conduit or tubing has to be calculated based on Chapter 9, Table 4, Table 5, and Table 8 if a bare equipment grounding conductor is installed. Of course, based on the Annex C tables, the conduit size could be increased to trade size 1¼. The rationale is that because six 4 AWG THW conductors can be installed in that size conduit or tubing, trade size 1¼ would be of sufficient size for the three circuit conductors and the wire-type equipment grounding conductor. According to 250.122(A) a wire-type equip-

ment grounding conductor is not required to be larger than the circuit conductors. This approach may not yield the minimum conduit or tubing size but will result in a compliant installation and may facilitate an easier installation if the conduit run is particularly long or has a substantial amount of bends.

Example 4

A fire alarm system installation requires the riser to contain twenty-one 16 AWG TFF conductors. If the riser conductors are installed in electrical metallic tubing, what is the minimum size tubing required?

Solution

According to Table C.1, trade size 1 EMT is required.

Table C.1 Maximum Number of Conductors or Fixture Wires in Electrical Metallic Tubing (EMT) (*Based on Table 1, Chapter 9*)

		CONDUCTORS									
		Metric Designator (Trade Size)									
Type	Conductor Size (AWG kcmil)	16 (½)	21 (¾)	27 (1)	35 (1¼)	41 (1½)	53 (2)	63 (2½)	78 (3)	91 (3½)	103 (4)
RHH, RHW, RHW-2	14	4	7	11	20	27	46	80	120	157	201
	12	3	6	9	17	23	38	66	100	131	167
	10	2	5	8	13	18	30	53	81	105	135
	8	1	2	4	7	9	16	28	42	55	70
	6	1	1	3	5	8	13	22	34	44	56
	4	1	1	2	4	6	10	17	26	34	44
	3	1	1	1	4	5	9	15	23	30	38
	2	1	1	1	3	4	7	13	20	26	33
	1	0	1	1	1	3	5	9	13	17	22
	1/0	0	1	1	1	2	4	7	11	15	19
	2/0	0	1	1	1	2	4	6	10	13	17
	3/0	0	0	1	1	1	3	5	8	11	14
	4/0	0	0	1	1	1	3	5	7	9	12
	250	0	0	0	1	1	1	3	5	7	9
	300	0	0	0	1	1	1	3	5	6	8
	350	0	0	0	1	1	1	3	4	6	7
	400	0	0	0	1	1	1	2	4	5	7
	500	0	0	0	0	1	1	2	3	4	6
	600	0	0	0	0	1	1	1	3	4	5
	700	0	0	0	0	0	1	1	2	3	4
	750	0	0	0	0	0	1	1	2	3	4
	800	0	0	0	0	0	1	1	2	3	4
	900	0	0	0	0	0	1	1	1	3	3
	1000	0	0	0	0	0	1	1	1	2	3
	1250	0	0	0	0	0	0	1	1	1	2
	1500	0	0	0	0	0	0	1	1	1	1
	1750	0	0	0	0	0	0	1	1	1	1
	2000	0	0	0	0	0	0	1	1	1	1
TW	14	8	15	25	43	58	96	168	254	332	424
	12	6	11	19	33	45	74	129	195	255	326
	10	5	8	14	24	33	55	96	145	190	243
	8	2	5	8	13	18	30	53	81	105	135
RHH*, RHW*, RHW-2*, THHW, THW, THW-2	14	6	10	16	28	39	64	112	169	221	282
RHH*, RHW*, RHW-2*, THHW, THW	12	4	8	13	23	31	51	90	136	177	227
	10	3	6	10	18	24	40	70	106	138	177
RHH*, RHW*, RHW-2*, THHW, THW, THW-2	8	1	4	6	10	14	24	42	63	83	106

(Continues)

Table C.1 *Continued*

CONDUCTORS

Type	Conductor Size (AWG kcmil)	Metric Designator (Trade Size)									
		16 (½)	21 (¾)	27 (1)	35 (1¼)	41 (1½)	53 (2)	63 (2½)	78 (3)	91 (3½)	103 (4)
RHH*, RHW*, RHW-2*, TW, THW, THHW, THW-2	6	1	3	4	8	11	18	32	48	63	81
	4	1	1	3	6	8	13	24	36	47	60
	3	1	1	3	5	7	12	20	31	40	52
	2	1	1	2	4	6	10	17	26	34	44
	1	1	1	1	3	4	7	12	18	24	31
	1/0	0	1	1	2	3	6	10	16	20	26
	2/0	0	1	1	1	3	5	9	13	17	22
	3/0	0	1	1	1	2	4	7	11	15	19
	4/0	0	0	1	1	1	3	6	9	12	16
	250	0	0	1	1	1	3	5	7	10	13
	300	0	0	1	1	1	2	4	6	8	11
	350	0	0	0	1	1	1	4	6	7	10
	400	0	0	0	1	1	1	3	5	7	9
	500	0	0	0	1	1	1	3	4	6	7
	600	0	0	0	1	1	1	2	3	4	6
	700	0	0	0	0	1	1	1	3	4	5
	750	0	0	0	0	1	1	1	3	4	5
	800	0	0	0	0	1	1	1	3	3	5
	900	0	0	0	0	0	1	1	2	3	4
	1000	0	0	0	0	0	1	1	2	3	4
	1250	0	0	0	0	0	1	1	1	2	3
	1500	0	0	0	0	0	1	1	1	1	2
	1750	0	0	0	0	0	0	1	1	1	2
	2000	0	0	0	0	0	0	1	1	1	1
THHN, THWN, THWN-2	14	12	22	35	61	84	138	241	364	476	608
	12	9	16	26	45	61	101	176	266	347	443
	10	5	10	16	28	38	63	111	167	219	279
	8	3	6	9	16	22	36	64	96	126	161
	6	2	4	7	12	16	26	46	69	91	116
	4	1	2	4	7	10	16	28	43	56	71
	3	1	1	3	6	8	13	24	36	47	60
	2	1	1	3	5	7	11	20	30	40	51
	1	1	1	1	4	5	8	15	22	29	37
	1/0	1	1	1	3	4	7	12	19	25	32
	2/0	0	1	1	2	3	6	10	16	20	26
	3/0	0	1	1	1	3	5	8	13	17	22
	4/0	0	1	1	1	2	4	7	11	14	18
	250	0	0	1	1	1	3	6	9	11	15
	300	0	0	1	1	1	3	5	7	10	13
	350	0	0	1	1	1	2	4	6	9	11
	400	0	0	0	1	1	1	4	6	8	10
	500	0	0	0	1	1	1	3	5	6	8
	600	0	0	0	1	1	1	2	4	5	7
	700	0	0	0	1	1	1	2	3	4	6
	750	0	0	0	0	1	1	1	3	4	5
	800	0	0	0	0	1	1	1	3	4	5
	900	0	0	0	0	1	1	1	3	3	4
	1000	0	0	0	0	1	1	1	2	3	4

Table C.1 *Continued*

		\multicolumn{10}{c}{CONDUCTORS}

| | Conductor Size (AWG kcmil) | \multicolumn{10}{c}{Metric Designator (Trade Size)} |

Type	Conductor Size (AWG kcmil)	16 (½)	21 (¾)	27 (1)	35 (1¼)	41 (1½)	53 (2)	63 (2½)	78 (3)	91 (3½)	103 (4)
FEP, FEPB, PFA, PFAH, TFE	14	12	21	34	60	81	134	234	354	462	590
	12	9	15	25	43	59	98	171	258	337	430
	10	6	11	18	31	42	70	122	185	241	309
	8	3	6	10	18	24	40	70	106	138	177
	6	2	4	7	12	17	28	50	75	98	126
	4	1	3	5	9	12	20	35	53	69	88
	3	1	2	4	7	10	16	29	44	57	73
	2	1	1	3	6	8	13	24	36	47	60
PFA, PFAH, TFE	1	1	1	2	4	6	9	16	25	33	42
PFAH, TFE PFA, PFAH, TFE, Z	1/0	1	1	1	3	5	8	14	21	27	35
	2/0	0	1	1	3	4	6	11	17	22	29
	3/0	0	1	1	2	3	5	9	14	18	24
	4/0	0	1	1	1	2	4	8	11	15	19
Z	14	14	25	41	72	98	161	282	426	556	711
	12	10	18	29	51	69	114	200	302	394	504
	10	6	11	18	31	42	70	122	185	241	309
	8	4	7	11	20	27	44	77	117	153	195
	6	3	5	8	14	19	31	54	82	107	137
	4	1	3	5	9	13	21	37	56	74	94
	3	1	2	4	7	9	15	27	41	54	69
	2	1	1	3	6	8	13	22	34	45	57
	1	1	1	2	4	6	10	18	28	36	46
XHH, XHHW, XHHW-2, ZW	14	8	15	25	43	58	96	168	254	332	424
	12	6	11	19	33	45	74	129	195	255	326
	10	5	8	14	24	33	55	96	145	190	243
	8	2	5	8	13	18	30	53	81	105	135
	6	1	3	6	10	14	22	39	60	78	100
	4	1	2	4	7	10	16	28	43	56	72
	3	1	1	3	6	8	14	24	36	48	61
	2	1	1	3	5	7	11	20	31	40	51
XHH, XHHW, XHHW-2	1	1	1	1	4	5	8	15	23	30	38
	1/0	1	1	1	3	4	7	13	19	25	32
	2/0	0	1	1	2	3	6	10	16	21	27
	3/0	0	1	1	1	3	5	9	13	17	22
	4/0	0	1	1	1	2	4	7	11	14	18
	250	0	0	1	1	1	3	6	9	12	15
	300	0	0	1	1	1	3	5	8	10	13
	350	0	0	1	1	1	2	4	7	9	11
	400	0	0	0	1	1	1	4	6	8	10
	500	0	0	0	1	1	1	3	5	6	8
	600	0	0	0	1	1	1	2	4	5	6
	700	0	0	0	0	1	1	2	3	4	6
	750	0	0	0	0	1	1	1	3	4	5
	800	0	0	0	0	1	1	1	3	4	5
	900	0	0	0	0	1	1	1	3	3	4
	1000	0	0	0	0	0	1	1	2	3	4
	1250	0	0	0	0	0	1	1	1	2	3
	1500	0	0	0	0	0	1	1	1	1	3
	1750	0	0	0	0	0	0	1	1	1	2
	2000	0	0	0	0	0	0	1	1	1	1

(Continues)

Table C.1 *Continued*

		FIXTURE WIRES					
		Metric Designator (Trade Size)					
Type	**Conductor Size (AWG/ kcmil)**	**16** (½)	**21** (¾)	**27** (1)	**35** (1¼)	**41** (1½)	**53** (2)
FFH-2, RFH-2, RFHH-3	18	8	14	24	41	56	92
	16	7	12	20	34	47	78
SF-2, SFF-2	18	10	18	30	52	71	116
	16	8	15	25	43	58	96
	14	7	12	20	34	47	78
SF-1, SFF-1	18	18	33	53	92	125	206
RFH-1, RFHH-2, TF, TFF, XF, XFF	18	14	24	39	68	92	152
RFHH-2, TF, TFF, XF, XFF	16	11	19	31	55	74	123
XF, XFF	14	8	15	25	43	58	96
TFN, TFFN	18	22	38	63	108	148	244
	16	17	29	48	83	113	186
PF, PFF, PGF, PGFF, PAF, PTF, PTFF, PAFF	18	21	36	59	103	140	231
	16	16	28	46	79	108	179
	14	12	21	34	60	81	134
ZF, ZFF, ZHF, HF, HFF	18	27	47	77	133	181	298
	16	20	35	56	98	133	220
	14	14	25	41	72	98	161
KF-2, KFF-2	18	39	69	111	193	262	433
	16	27	48	78	136	185	305
	14	19	33	54	93	127	209
	12	13	23	37	64	87	144
	10	8	15	25	43	58	96
KF-1, KFF-1	18	46	82	133	230	313	516
	16	33	57	93	161	220	362
	14	22	38	63	108	148	244
	12	14	25	41	72	98	161
	10	9	16	27	47	64	105
XF, XFF	12	4	8	13	23	31	51
	10	3	6	10	18	24	40

Notes:
1. This table is for concentric stranded conductors only. For compact stranded conductors, Table C.1(A) should be used.
2. Two-hour fire-rated RHH cable has ceramifiable insulation which has much larger diameters than other RHH wires. Consult manufacturer's conduit fill tables.
*Types RHH, RHW, and RHW-2 without outer covering.

**Table C.1(A) Maximum Number of Compact Conductors in Electrical Metallic Tubing
(EMT)** (*Based on Table 1, Chapter 9*)

		COMPACT CONDUCTORS									
		Metric Designator (Trade Size)									
Type	Conductor Size (AWG/ kcmil)	16 (½)	21 (¾)	27 (1)	35 (1¼)	41 (1½)	53 (2)	63 (2½)	78 (3)	91 (3½)	103 (4)
THW,	8	2	4	6	11	16	26	46	69	90	115
THW-2,	6	1	3	5	9	12	20	35	53	70	89
THHW	4	1	2	4	6	9	15	26	40	52	67
	2	1	1	3	5	7	11	19	29	38	49
	1	1	1	1	3	4	8	13	21	27	34
	1/0	1	1	1	3	4	7	12	18	23	30
	2/0	0	1	1	2	3	5	10	15	20	25
	3/0	0	1	1	1	3	5	8	13	17	21
	4/0	0	1	1	1	2	4	7	11	14	18
	250	0	0	1	1	1	3	5	8	11	14
	300	0	0	1	1	1	3	5	7	9	12
	350	0	0	1	1	1	2	4	6	8	11
	400	0	0	0	1	1	1	4	6	8	10
	500	0	0	0	1	1	1	3	5	6	8
	600	0	0	0	1	1	1	2	4	5	7
	700	0	0	0	1	1	1	2	3	4	6
	750	0	0	0	0	1	1	1	3	4	5
	900	0	0	0	0	1	1	2	3	4	5
	1000	0	0	0	0	1	1	1	2	3	4
THHN,	8	—	—	—	—	—	—	—	—	—	—
THWN,	6	2	4	7	13	18	29	52	78	102	130
THWN-2	4	1	3	4	8	11	18	32	48	63	81
	2	1	1	3	6	8	13	23	34	45	58
	1	1	1	2	4	6	10	17	26	34	43
	1/0	1	1	1	3	5	8	14	22	29	37
	2/0	1	1	1	3	4	7	12	18	24	30
	3/0	0	1	1	2	3	6	10	15	20	25
	4/0	0	1	1	1	3	5	8	12	16	21
	250	0	1	1	1	1	4	6	10	13	16
	300	0	0	1	1	1	3	5	8	11	14
	350	0	0	1	1	1	3	5	7	10	12
	400	0	0	1	1	1	2	4	6	9	11
	500	0	0	0	1	1	1	4	5	7	9
	600	0	0	0	1	1	1	3	4	6	7
	700	0	0	0	1	1	1	2	4	5	7
	750	0	0	0	1	1	1	2	4	5	6
	900	0	0	0	0	1	1	2	3	4	5
	1000	0	0	0	0	1	1	1	3	3	4
XHHW,	8	3	5	8	15	20	34	59	90	117	149
XHHW-2	6	1	4	6	11	15	25	44	66	87	111
	4	1	3	4	8	11	18	32	48	63	81
	2	1	1	3	6	8	13	23	34	45	58
	1	1	1	2	4	6	10	17	26	34	43
	1/0	1	1	1	3	5	8	14	22	29	37
	2/0	1	1	1	3	4	7	12	18	24	31
	3/0	0	1	1	2	3	6	10	15	20	25
	4/0	0	1	1	1	3	5	8	13	17	21
	250	0	1	1	1	2	4	7	10	13	17
	300	0	0	1	1	1	3	6	9	11	14
	350	0	0	1	1	1	3	5	8	10	13
	400	0	0	1	1	1	2	4	7	9	11
	500	0	0	0	1	1	1	4	6	7	9
	600	0	0	0	1	1	1	3	4	6	8
	700	0	0	0	1	1	1	2	4	5	7
	750	0	0	0	1	1	1	2	3	5	6
	900	0	0	0	0	1	1	2	3	4	5
	1000	0	0	0	0	1	1	1	3	4	5

Definition: *Compact stranding* is the result of a manufacturing process where the standard conductor
is compressed to the extent that the interstices (voids between strand wires) are virtually eliminated.

Table C.2 Maximum Number of Conductors or Fixture Wires in Electrical Nonmetallic Tubing (ENT) *(Based on Table 1, Chapter 9)*

		CONDUCTORS					
		Metric Designator (Trade Size)					
Type	Conductor Size (AWG/ kcmil)	16 (½)	21 (¾)	27 (1)	35 (1¼)	41 (1½)	53 (2)
RHH, RHW, RHW-2	14	3	6	10	19	26	43
	12	2	5	9	16	22	36
	10	1	4	7	13	17	29
	8	1	1	3	6	9	15
	6	1	1	3	5	7	12
	4	1	1	2	4	6	9
	3	1	1	1	3	5	8
	2	0	1	1	3	4	7
	1	0	1	1	1	3	5
	1/0	0	0	1	1	2	4
	2/0	0	0	1	1	1	3
	3/0	0	0	1	1	1	3
	4/0	0	0	1	1	1	2
	250	0	0	0	1	1	1
	300	0	0	0	1	1	1
	350	0	0	0	1	1	1
	400	0	0	0	1	1	1
	500	0	0	0	0	1	1
	600	0	0	0	0	1	1
	700	0	0	0	0	0	1
	750	0	0	0	0	0	1
	800	0	0	0	0	0	1
	900	0	0	0	0	0	1
	1000	0	0	0	0	0	1
	1250	0	0	0	0	0	0
	1500	0	0	0	0	0	0
	1750	0	0	0	0	0	0
	2000	0	0	0	0	0	0
TW	14	7	13	22	40	55	92
	12	5	10	17	31	42	71
	10	4	7	13	23	32	52
	8	1	4	7	13	17	29
RHH*, RHW*, RHW-2*, THHW, THW, THW-2	14	4	8	15	27	37	61
RHH*, RHW*, RHW-2*, THHW, THW	12	3	7	12	21	29	49
	10	3	5	9	17	23	38
RHH*, RHW*, RHW-2*, THHW, THW, THW-2	8	1	3	5	10	14	23

Table C.2 *Continued*

		CONDUCTORS					
		Metric Designator (Trade Size)					
Type	Conductor Size (AWG/ kcmil)	16 (½)	21 (¾)	27 (1)	35 (1¼)	41 (1½)	53 (2)
RHH*, RHW*, RHW-2*, TW, THW, THHW, THW-2	6	1	2	4	7	10	17
	4	1	1	3	5	8	13
	3	1	1	2	5	7	11
	2	1	1	2	4	6	9
	1	0	1	1	3	4	6
	1/0	0	1	1	2	3	5
	2/0	0	1	1	1	3	5
	3/0	0	0	1	1	2	4
	4/0	0	0	1	1	1	3
	250	0	0	1	1	1	2
	300	0	0	0	1	1	2
	350	0	0	0	1	1	1
	400	0	0	0	1	1	1
	500	0	0	0	1	1	1
	600	0	0	0	0	1	1
	700	0	0	0	0	1	1
	750	0	0	0	0	1	1
	800	0	0	0	0	1	1
	900	0	0	0	0	0	1
	1000	0	0	0	0	0	1
	1250	0	0	0	0	0	1
	1500	0	0	0	0	0	0
	1750	0	0	0	0	0	0
	2000	0	0	0	0	0	0
THHN, THWN, THWN-2	14	10	18	32	58	80	132
	12	7	13	23	42	58	96
	10	4	8	15	26	36	60
	8	2	5	8	15	21	35
	6	1	3	6	11	15	25
	4	1	1	4	7	9	15
	3	1	1	3	5	8	13
	2	1	1	2	5	6	11
	1	1	1	1	3	5	8
	1/0	0	1	1	3	4	7
	2/0	0	1	1	2	3	5
	3/0	0	1	1	1	3	4
	4/0	0	0	1	1	2	4
	250	0	0	1	1	1	3
	300	0	0	1	1	1	2
	350	0	0	0	1	1	2
	400	0	0	0	1	1	1
	500	0	0	0	1	1	1
	600	0	0	0	1	1	1
	700	0	0	0	0	1	1
	750	0	0	0	0	1	1
	800	0	0	0	0	1	1
	900	0	0	0	0	1	1
	1000	0	0	0	0	0	1

(Continues)

Table C.2 *Continued*

CONDUCTORS

Type	Conductor Size (AWG/kcmil)	Metric Designator (Trade Size)					
		16 (½)	21 (¾)	27 (1)	35 (1¼)	41 (1½)	53 (2)
FEP, FEPB, PFA, PFAH, TFE	14	10	18	31	56	77	128
	12	7	13	23	41	56	93
	10	5	9	16	29	40	67
	8	3	5	9	17	23	38
	6	1	4	6	12	16	27
	4	1	2	4	8	11	19
	3	1	1	4	7	9	16
	2	1	1	3	5	8	13
PFA, PFAH, TFE	1	1	1	1	4	5	9
PFA, PFAH, TFE, Z	1/0	0	1	1	3	4	7
	2/0	0	1	1	2	4	6
	3/0	0	1	1	1	3	5
	4/0	0	1	1	1	2	4
Z	14	12	22	38	68	93	154
	12	8	15	27	48	66	109
	10	5	9	16	29	40	67
	8	3	6	10	18	25	42
	6	1	4	7	13	18	30
	4	1	3	5	9	12	20
	3	1	1	3	6	9	15
	2	1	1	3	5	7	12
	1	1	1	2	4	6	10
XHH, XHHW, XHHW-2, ZW	14	7	13	22	40	55	92
	12	5	10	17	31	42	71
	10	4	7	13	23	32	52
	8	1	4	7	13	17	29
	6	1	3	5	9	13	21
	4	1	1	4	7	9	15
	3	1	1	3	6	8	13
	2	1	1	2	5	6	11
XHH, XHHW, XHHW-2	1	1	1	1	3	5	8
	1/0	0	1	1	3	4	7
	2/0	0	1	1	2	3	6
	3/0	0	1	1	1	3	5
	4/0	0	0	1	1	2	4
	250	0	0	1	1	1	3
	300	0	0	1	1	1	3
	350	0	0	1	1	1	2
	400	0	0	0	1	1	1
	500	0	0	0	1	1	1
	600	0	0	0	1	1	1
	700	0	0	0	0	1	1
	750	0	0	0	0	1	1
	800	0	0	0	0	1	1
	900	0	0	0	0	1	1
	1000	0	0	0	0	0	1
	1250	0	0	0	0	0	1
	1500	0	0	0	0	0	1
	1750	0	0	0	0	0	0
	2000	0	0	0	0	0	0

Table C.2 *Continued*

FIXTURE WIRES

Type	Conductor Size (AWG/ kcmil)	Metric Designator (Trade Size)					
		16 (½)	21 (¾)	27 (1)	35 (1¼)	41 (1½)	53 (2)
FFH-2, RFH-2, RFHH-3 SF-2, SFF-2	18	6	12	21	39	53	88
	16	5	10	18	32	45	74
	18	8	15	27	49	67	111
	16	7	13	22	40	55	92
	14	5	10	18	32	45	74
SF-1, SFF-1	18	15	28	48	86	119	197
RFH-1, RFHH-2, TF, TFF, XF, XFF	18	11	20	35	64	88	145
RFHH-2, TF, TFF, XF, XFF	16	9	16	29	51	71	117
XF, XFF	14	7	13	22	40	55	92
TFN, TFFN	18	18	33	57	102	141	233
	16	13	25	43	78	107	178
PF, PFF, PGF, PGFF, PAF, PTF, PTFF, PAFF	18	17	31	54	97	133	221
	16	13	24	42	75	103	171
	14	10	18	31	56	77	128
ZF, ZFF, ZHF, HF, HFF	18	22	40	70	125	172	285
	16	16	29	51	92	127	210
	14	12	22	38	68	93	154
KF-2, KFF-2	18	31	58	101	182	250	413
	16	22	41	71	128	176	291
	14	15	28	49	88	121	200
	12	10	19	33	60	83	138
	10	7	13	22	40	55	92
KF-1, KFF-1	18	38	69	121	217	298	493
	16	26	49	85	152	209	346
	14	18	33	57	102	141	233
	12	12	22	38	68	93	154
	10	7	14	24	44	61	101
XF, XFF	12	3	7	12	21	29	49
	10	3	5	9	17	23	38

Notes:

1. This table is for concentric stranded conductors only. For compact stranded conductors, Table C.2(A) should be used.

2. Two-hour fire-rated RHH cable has ceramifiable insulation which has much larger diameters than other RHH wires.

Consult manufacturer's conduit fill tables.

*Types RHH, RHW, and RHW-2 without outer covering.

Table C.2(A) Maximum Number of Compact Conductors in Electrical Nonmetallic Tubing (ENT) (*Based on Table 1, Chapter 9*)

COMPACT CONDUCTORS

Type	Conductor Size (AWG/ kcmil)	Metric Designator (Trade Size)					
		16 (½)	21 (¾)	27 (1)	35 (1¼)	41 (1½)	53 (2)
THW, THW-2, THHW	8	1	3	6	11	15	25
	6	1	2	4	8	11	19
	4	1	1	3	6	8	14
	2	1	1	2	4	6	10
	1	0	1	1	3	4	7
	1/0	0	1	1	3	4	6
	2/0	0	1	1	2	3	5
	3/0	0	1	1	1	3	4
	4/0	0	0	1	1	2	4
	250	0	0	1	1	1	3
	300	0	0	1	1	1	2
	350	0	0	0	1	1	2
	400	0	0	0	1	1	1
	500	0	0	0	1	1	1
	600	0	0	0	1	1	1
	700	0	0	0	0	1	1
	750	0	0	0	0	1	1
	900	0	0	0	0	1	1
	1000	0	0	0	0	0	1
THHN, THWN, THWN-2	8	—	—	—	—	—	—
	6	1	4	7	12	17	28
	4	1	2	4	7	10	17
	2	1	1	3	5	7	12
	1	1	1	2	4	5	9
	1/0	1	1	1	3	5	8
	2/0	0	1	1	3	4	6
	3/0	0	1	1	2	3	5
	4/0	0	1	1	1	2	4
	250	0	0	1	1	1	3
	300	0	0	1	1	1	3
	350	0	0	1	1	1	2
	400	0	0	0	1	1	2
	500	0	0	0	1	1	1
	600	0	0	0	1	1	1
	700	0	0	0	1	1	1
	750	0	0	0	1	1	1
	900	0	0	0	0	1	1
	1000	0	0	0	0	1	1
XHHW, XHHW-2	8	2	4	8	14	19	32
	6	1	3	6	10	14	24
	4	1	2	4	7	10	17
	2	1	1	3	5	7	12
	1	1	1	2	4	5	9
	1/0	1	1	1	3	5	8
	2/0	0	1	1	3	4	7
	3/0	0	1	1	2	3	5
	4/0	0	1	1	1	3	4
	250	0	0	1	1	1	3
	300	0	0	1	1	1	3
	350	0	0	1	1	1	3
	400	0	0	1	1	1	2
	500	0	0	0	1	1	1
	600	0	0	0	1	1	1
	700	0	0	0	1	1	1
	750	0	0	0	1	1	1
	900	0	0	0	0	1	1
	1000	0	0	0	0	1	1

Definition: *Compact stranding* is the result of a manufacturing process where the standard conductor is compressed to the extent that the interstices (voids between strand wires) are virtually eliminated.

Table C.3 Maximum Number of Conductors or Fixture Wires in Flexible Metal Conduit (FMC) (*Based on Table 1, Chapter 9*)

		CONDUCTORS									
		Metric Designator (Trade Size)									
Type	Conductor Size (AWG/ kcmil)	16 (½)	21 (¾)	27 (1)	35 (1¼)	41 (1½)	53 (2)	63 (2½)	78 (3)	91 (3½)	103 (4)
RHH, RHW, RHW-2	14	4	7	11	17	25	44	67	96	131	171
	12	3	6	9	14	21	37	55	80	109	142
	10	3	5	7	11	17	30	45	64	88	115
	8	1	2	4	6	9	15	23	34	46	60
	6	1	1	3	5	7	12	19	27	37	48
	4	1	1	2	4	5	10	14	21	29	37
	3	1	1	1	3	5	8	13	18	25	33
	2	1	1	1	3	4	7	11	16	22	28
	1	0	1	1	1	2	5	7	10	14	19
	1/0	0	1	1	1	2	4	6	9	12	16
	2/0	0	1	1	1	1	3	5	8	11	14
	3/0	0	0	1	1	1	3	5	7	9	12
	4/0	0	0	1	1	1	2	4	6	8	10
	250	0	0	0	1	1	1	3	4	6	8
	300	0	0	0	1	1	1	2	4	5	7
	350	0	0	0	1	1	1	2	3	5	6
	400	0	0	0	0	1	1	1	3	4	6
	500	0	0	0	0	1	1	1	3	4	5
	600	0	0	0	0	1	1	1	2	3	4
	700	0	0	0	0	0	1	1	1	3	3
	750	0	0	0	0	0	1	1	1	2	3
	800	0	0	0	0	0	1	1	1	2	3
	900	0	0	0	0	0	1	1	1	2	3
	1000	0	0	0	0	0	1	1	1	1	3
	1250	0	0	0	0	0	0	1	1	1	1
	1500	0	0	0	0	0	0	1	1	1	1
	1750	0	0	0	0	0	0	1	1	1	1
	2000	0	0	0	0	0	0	0	1	1	1
TW	14	9	15	23	36	53	94	141	203	277	361
	12	7	11	18	28	41	72	108	156	212	277
	10	5	8	13	21	30	54	81	116	158	207
	8	3	5	7	11	17	30	45	64	88	115
RHH*, RHW*, RHW-2*, THHW, THW, THW-2	14	6	10	15	24	35	62	94	135	184	240
RHH*, RHW*, RHW-2*, THHW, THW	12	5	8	12	19	28	50	75	108	148	193
	10	4	6	10	15	22	39	59	85	115	151
RHH*, RHW*, RHW-2*, THHW, THW, THW-2	8	1	4	6	9	13	23	35	51	69	90

(Continues)

Table C.3 *Continued*

CONDUCTORS

Type	Conductor Size (AWG/ kcmil)	Metric Designator (Trade Size)									
		16 (½)	21 (¾)	27 (1)	35 (1¼)	41 (1½)	53 (2)	63 (2½)	78 (3)	91 (3½)	103 (4)
RHH*, RHW*, RHW-2*, TW, THW, THHW, THW-2	6	1	3	4	7	10	18	27	39	53	69
	4	1	1	3	5	7	13	20	29	39	51
	3	1	1	3	4	6	11	17	25	34	44
	2	1	1	2	4	5	10	14	21	29	37
	1	1	1	1	2	4	7	10	15	20	26
	1/0	0	1	1	1	3	6	9	12	17	22
	2/0	0	1	1	1	3	5	7	10	14	19
	3/0	0	1	1	1	2	4	6	9	12	16
	4/0	0	0	1	1	1	3	5	7	10	13
	250	0	0	1	1	1	3	4	6	8	11
	300	0	0	1	1	1	2	3	5	7	9
	350	0	0	0	1	1	1	3	4	6	8
	400	0	0	0	1	1	1	3	4	6	7
	500	0	0	0	1	1	1	2	3	5	6
	600	0	0	0	0	1	1	1	3	4	5
	700	0	0	0	0	1	1	1	2	3	4
	750	0	0	0	0	1	1	1	2	3	4
	800	0	0	0	0	1	1	1	1	3	4
	900	0	0	0	0	0	1	1	1	3	3
	1000	0	0	0	0	0	1	1	1	2	3
	1250	0	0	0	0	0	1	1	1	1	2
	1500	0	0	0	0	0	0	1	1	1	1
	1750	0	0	0	0	0	0	1	1	1	1
	2000	0	0	0	0	0	0	1	1	1	1
THHN, THWN, THWN-2	14	13	22	33	52	76	134	202	291	396	518
	12	9	16	24	38	56	98	147	212	289	378
	10	6	10	15	24	35	62	93	134	182	238
	8	3	6	9	14	20	35	53	77	105	137
	6	2	4	6	10	14	25	38	55	76	99
	4	1	2	4	6	9	16	24	34	46	61
	3	1	1	3	5	7	13	20	29	39	51
	2	1	1	3	4	6	11	17	24	33	43
	1	1	1	1	3	4	8	12	18	24	32
	1/0	1	1	1	2	4	7	10	15	20	27
	2/0	0	1	1	1	3	6	9	12	17	22
	3/0	0	1	1	1	2	5	7	10	14	18
	4/0	0	1	1	1	1	4	6	8	12	15
	250	0	0	1	1	1	3	5	7	9	12
	300	0	0	1	1	1	3	4	6	8	11
	350	0	0	1	1	1	2	3	5	7	9
	400	0	0	0	1	1	1	3	5	6	8
	500	0	0	0	1	1	1	2	4	5	7
	600	0	0	0	0	1	1	1	3	4	5
	700	0	0	0	0	1	1	1	3	4	5
	750	0	0	0	0	1	1	1	2	3	4
	800	0	0	0	0	1	1	1	2	3	4
	900	0	0	0	0	0	1	1	1	3	4
	1000	0	0	0	0	0	1	1	1	3	3

Table C.3 *Continued*

CONDUCTORS

Type	Conductor Size (AWG/ kcmil)	Metric Designator (Trade Size)									
		16 (½)	21 (¾)	27 (1)	35 (1¼)	41 (1½)	53 (2)	63 (2½)	78 (3)	91 (3½)	103 (4)
FEP, FEPB, PFA, PFAH, TFE	14	12	21	32	51	74	130	196	282	385	502
	12	9	15	24	37	54	95	143	206	281	367
	10	6	11	17	26	39	68	103	148	201	263
	8	4	6	10	15	22	39	59	85	115	151
	6	2	4	7	11	16	28	42	60	82	107
	4	1	3	5	7	11	19	29	42	57	75
	3	1	2	4	6	9	16	24	35	48	62
	2	1	1	3	5	7	13	20	29	39	51
PFA, PFAH, TFE	1	1	1	2	3	5	9	14	20	27	36
PFA, PFAH, TFE, Z	1/0	1	1	1	3	4	8	11	17	23	30
	2/0	1	1	1	2	3	6	9	14	19	24
	3/0	0	1	1	1	3	5	8	11	15	20
	4/0	0	1	1	1	2	4	6	9	13	16
Z	14	15	25	39	61	89	157	236	340	463	605
	12	11	18	28	43	63	111	168	241	329	429
	10	6	11	17	26	39	68	103	148	201	263
	8	4	7	11	17	24	43	65	93	127	166
	6	3	5	7	12	17	30	45	65	89	117
	4	1	3	5	8	12	21	31	45	61	80
	3	1	2	4	6	8	15	23	33	45	58
	2	1	1	3	5	7	12	19	27	37	49
	1	1	1	2	4	6	10	15	22	30	39
XHH, XHHW, XHHW-2, ZW	14	9	15	23	36	53	94	141	203	277	361
	12	7	11	18	28	41	72	108	156	212	277
	10	5	8	13	21	30	54	81	116	158	207
	8	3	5	7	11	17	30	45	64	88	115
	6	1	3	5	8	12	22	33	48	65	85
	4	1	2	4	6	9	16	24	34	47	61
	3	1	1	3	5	7	13	20	29	40	52
	2	1	1	3	4	6	11	17	24	33	44
XHH, XHHW, XHHW-2	1	1	1	1	3	5	8	13	18	25	32
	1/0	1	1	1	2	4	7	10	15	21	27
	2/0	0	1	1	2	3	6	9	13	17	23
	3/0	0	1	1	1	3	5	7	10	14	19
	4/0	0	1	1	1	2	4	6	9	12	15
	250	0	0	1	1	1	3	5	7	10	13
	300	0	0	1	1	1	3	4	6	8	11
	350	0	0	1	1	1	2	4	5	7	9
	400	0	0	0	1	1	1	3	5	6	8
	500	0	0	0	1	1	1	3	4	5	7
	600	0	0	0	0	1	1	1	3	4	5
	700	0	0	0	0	1	1	1	3	4	5
	750	0	0	0	0	1	1	1	2	3	4
	800	0	0	0	0	1	1	1	2	3	4
	900	0	0	0	0	0	1	1	1	3	4
	1000	0	0	0	0	0	1	1	1	3	3
	1250	0	0	0	0	0	1	1	1	1	3
	1500	0	0	0	0	0	1	1	1	1	2
	1750	0	0	0	0	0	0	1	1	1	1
	2000	0	0	0	0	0	0	1	1	1	1

(Continues)

Table C.3 *Continued*

FIXTURE WIRES

Type	Conductor Size (AWG/ kcmil)	Metric Designator (Trade Size)					
		16 (½)	21 (¾)	27 (1)	35 (1¼)	41 (1½)	53 (2)
FFH-2, RFH-2, RFHH-3	18	8	14	22	35	51	90
	16	7	12	19	29	43	76
SF-2, SFF-2	18	11	18	28	44	64	113
	16	9	15	23	36	53	94
	14	7	12	19	29	43	76
SF-1, SFF-1	18	19	32	50	78	114	201
RFH-1, RFHH-2, TF, TFF, XF, XFF	18	14	24	37	58	84	148
RFHH-2, TF, TFF, XF, XFF	16	11	19	30	47	68	120
XF, XFF	14	9	15	23	36	53	94
TFN, TFFN	18	23	38	59	93	135	237
	16	17	29	45	71	103	181
PF, PFF, PGF, PGFF, PAF, PTF, PTFF, PAFF	18	22	36	56	88	128	225
	16	17	28	43	68	99	174
	14	12	21	32	51	74	130
ZF, ZFF, ZHF, HF, HFF	18	28	47	72	113	165	290
	16	20	35	53	83	121	214
	14	15	25	39	61	89	157
KF-2, KFF-2	18	41	68	105	164	239	421
	16	28	48	74	116	168	297
	14	19	33	51	80	116	204
	12	13	23	35	55	80	140
	10	9	15	23	36	53	94
KF-1, KFF-1	18	48	82	125	196	285	503
	16	34	57	88	138	200	353
	14	23	38	59	93	135	237
	12	15	25	39	61	89	157
	10	10	16	25	40	58	103
XF, XFF	12	5	8	12	19	28	50
	10	4	6	10	15	22	39

Notes:
1. This table is for concentric stranded conductors only. For compact stranded conductors, Table C.3(A) should be used.
2. Two-hour fire-rated RHH cable has ceramifiable insulation which has much larger diameters than other RHH wires.
Consult manufacturer's conduit fill tables.
*Types RHH, RHW, and RHW-2 without outer covering.

Table C.3(A) Maximum Number of Compact Conductors in Flexible Metal Conduit (FMC)
(Based on Table 1, Chapter 9)

COMPACT CONDUCTORS

Type	Conductor Size (AWG/ kcmil)	Metric Designator (Trade Size)									
		16 (½)	21 (¾)	27 (1)	35 (1¼)	41 (1½)	53 (2)	63 (2½)	78 (3)	91 (3½)	103 (4)
THW, THHW, THW-2	8	2	4	6	10	14	25	38	55	75	98
	6	1	3	5	7	11	20	29	43	58	76
	4	1	2	3	5	8	15	22	32	43	57
	2	1	1	2	4	6	11	16	23	32	42
	1	1	1	1	3	4	7	11	16	22	29
	1/0	1	1	1	2	3	6	10	14	19	25
	2/0	0	1	1	1	3	5	8	12	16	21
	3/0	0	1	1	1	2	4	7	10	14	18
	4/0	0	1	1	1	1	4	6	8	11	15
	250	0	0	1	1	1	3	4	7	9	12
	300	0	0	1	1	1	2	4	6	8	10
	350	0	0	1	1	1	2	3	5	7	9
	400	0	0	0	1	1	1	3	5	6	8
	500	0	0	0	1	1	1	3	4	5	7
	600	0	0	0	0	1	1	1	3	4	6
	700	0	0	0	0	1	1	1	3	4	5
	750	0	0	0	0	1	1	1	2	3	5
	900	0	0	0	0	1	1	1	2	3	4
	1000	0	0	0	0	0	1	1	1	3	4
THHN, THWN, THWN-2	8	—	—	—	—	—	—	—	—	—	—
	6	3	4	7	11	16	29	43	62	85	111
	4	1	3	4	7	10	18	27	38	52	69
	2	1	1	3	5	7	13	19	28	38	49
	1	1	1	2	3	5	9	14	21	28	37
	1/0	1	1	1	3	4	8	12	17	24	31
	2/0	1	1	1	2	4	6	10	14	20	26
	3/0	0	1	1	1	3	5	8	12	17	22
	4/0	0	1	1	1	2	4	7	10	14	18
	250	0	1	1	1	1	3	5	8	11	14
	300	0	0	1	1	1	3	5	7	9	12
	350	0	0	1	1	1	3	4	6	8	10
	400	0	0	1	1	1	2	3	5	7	9
	500	0	0	0	1	1	1	3	4	6	8
	600	0	0	0	1	1	1	2	3	5	6
	700	0	0	0	0	1	1	1	3	4	6
	750	0	0	0	0	1	1	1	3	4	5
	900	0	0	0	0	1	1	1	2	3	4
	1000	0	0	0	0	0	1	1	1	3	4
XHHW, XHHW-2	8	3	5	8	13	19	33	50	71	97	127
	6	2	4	6	9	14	24	37	53	72	95
	4	1	3	4	7	10	18	27	38	52	69
	2	1	1	3	5	7	13	19	28	38	49
	1	1	1	2	3	5	9	14	21	28	37
	1/0	1	1	1	3	4	8	12	17	24	31
	2/0	1	1	1	2	4	7	10	15	20	26
	3/0	0	1	1	1	3	5	8	12	17	22
	4/0	0	1	1	1	2	4	7	10	14	18
	250	0	1	1	1	1	4	5	8	11	14
	300	0	0	1	1	1	3	5	7	9	12
	350	0	0	1	1	1	3	4	6	8	11
	400	0	0	1	1	1	2	4	5	7	10
	500	0	0	0	1	1	1	3	4	6	8
	600	0	0	0	1	1	1	2	3	5	6
	700	0	0	0	0	1	1	1	3	4	6
	750	0	0	0	0	1	1	1	3	4	5
	900	0	0	0	0	1	1	2	2	3	4
	1000	0	0	0	0	1	1	1	2	3	4

Definition: *Compact stranding* is the result of a manufacturing process where the standard conductor is compressed to the extent that the interstices (voids between strand wires) are virtually eliminated.

Table C.4 Maximum Number of Conductors or Fixture Wires in Intermediate Metal Conduit (IMC) (*Based on Table 1, Chapter 9*)

CONDUCTORS

Type	Conductor Size (AWG/ kcmil)	Metric Designator (Trade Size)									
		16 (½)	21 (¾)	27 (1)	35 (1¼)	41 (1½)	53 (2)	63 (2½)	78 (3)	91 (3½)	103 (4)
RHH, RHW, RHW-2	14	4	8	13	22	30	49	70	108	144	186
	12	4	6	11	18	25	41	58	89	120	154
RHH, RHW, RHW-2	10	3	5	8	15	20	33	47	72	97	124
	8	1	3	4	8	10	17	24	38	50	65
	6	1	1	3	6	8	14	19	30	40	52
	4	1	1	3	5	6	11	15	23	31	41
	3	1	1	2	4	6	9	13	21	28	36
	2	1	1	1	3	5	8	11	18	24	31
	1	0	1	1	2	3	5	7	12	16	20
	1/0	0	1	1	1	3	4	6	10	14	18
	2/0	0	1	1	1	2	4	6	9	12	15
	3/0	0	0	1	1	1	3	5	7	10	13
	4/0	0	0	1	1	1	3	4	6	9	11
	250	0	0	1	1	1	1	3	5	6	8
	300	0	0	0	1	1	1	3	4	6	7
	350	0	0	0	1	1	1	2	4	5	7
	400	0	0	0	1	1	1	2	3	5	6
	500	0	0	0	1	1	1	1	3	4	5
	600	0	0	0	0	1	1	1	2	3	4
	700	0	0	0	0	1	1	1	2	3	4
	750	0	0	0	0	1	1	1	1	3	4
	800	0	0	0	0	0	1	1	1	3	3
	900	0	0	0	0	0	1	1	1	2	3
	1000	0	0	0	0	0	1	1	1	2	3
	1250	0	0	0	0	0	1	1	1	1	2
	1500	0	0	0	0	0	0	1	1	1	1
	1750	0	0	0	0	0	0	1	1	1	1
	2000	0	0	0	0	0	0	1	1	1	1
TW	14	10	17	27	47	64	104	147	228	304	392
	12	7	13	21	36	49	80	113	175	234	301
	10	5	9	15	27	36	59	84	130	174	224
	8	3	5	8	15	20	33	47	72	97	124
RHH*, RHW*, RHW-2*, THHW, THW, THW-2	14	6	11	18	31	42	69	98	151	202	261
RHH*, RHW*, RHW-2*, THHW, THW	12	5	9	14	25	34	56	79	122	163	209
	10	4	7	11	19	26	43	61	95	127	163
RHH*, RHW*, RHW-2*, THHW, THW, THW-2	8	2	4	7	12	16	26	37	57	76	98
RHH*, RHW*, RHW-2*, TW, THHW, THW, THW-2	6	1	3	5	9	12	20	28	43	58	75
	4	1	2	4	6	9	15	21	32	43	56

Table C.4 *Continued*

CONDUCTORS

Type	Conductor Size (AWG/ kcmil)	Metric Designator (Trade Size)									
		16 (½)	21 (¾)	27 (1)	35 (1¼)	41 (1½)	53 (2)	63 (2½)	78 (3)	91 (3½)	103 (4)
RHH*, RHW*, RHW-2*, TW, THW, THHW, THW-2	3	1	1	3	6	8	13	18	28	37	48
	2	1	1	3	5	6	11	15	23	31	41
	1	1	1	1	3	4	7	11	16	22	28
	1/0	1	1	1	3	4	6	9	14	19	24
	2/0	0	1	1	2	3	5	8	12	16	20
	3/0	0	1	1	1	3	4	6	10	13	17
	4/0	0	1	1	1	2	4	5	8	11	14
	250	0	0	1	1	1	3	4	7	9	12
	300	0	0	1	1	1	2	4	6	8	10
	350	0	0	1	1	1	2	3	5	7	9
	400	0	0	0	1	1	1	3	4	6	8
	500	0	0	0	1	1	1	2	4	5	7
	600	0	0	0	1	1	1	1	3	4	5
	700	0	0	0	0	1	1	1	3	4	5
	750	0	0	0	0	1	1	1	2	3	4
	800	0	0	0	0	1	1	1	2	3	4
	900	0	0	0	0	1	1	1	2	3	4
	1000	0	0	0	0	0	1	1	1	3	3
	1250	0	0	0	0	0	1	1	1	1	3
	1500	0	0	0	0	0	1	1	1	1	2
	1750	0	0	0	0	0	0	1	1	1	1
	2000	0	0	0	0	0	0	1	1	1	1
THHN, THWN, THWN-2	14	14	24	39	68	91	149	211	326	436	562
	12	10	17	29	49	67	109	154	238	318	410
	10	6	11	18	31	42	68	97	150	200	258
	8	3	6	10	18	24	39	56	86	115	149
	6	2	4	7	13	17	28	40	62	83	107
	4	1	3	4	8	10	17	25	38	51	66
	3	1	2	4	6	9	15	21	32	43	56
	2	1	1	3	5	7	12	17	27	36	47
	1	1	1	2	4	5	9	13	20	27	35
	1/0	1	1	1	3	4	8	11	17	23	29
	2/0	1	1	1	3	4	6	9	14	19	24
	3/0	0	1	1	2	3	5	7	12	16	20
	4/0	0	1	1	1	2	4	6	9	13	17
	250	0	0	1	1	1	3	5	8	10	13
	300	0	0	1	1	1	3	4	7	9	12
	350	0	0	1	1	1	2	4	6	8	10
	400	0	0	1	1	1	2	3	5	7	9
	500	0	0	0	1	1	1	3	4	6	7
	600	0	0	0	1	1	1	2	3	5	6
	700	0	0	0	1	1	1	1	3	4	5
	750	0	0	0	1	1	1	1	3	4	5
	800	0	0	0	0	1	1	1	3	4	5
	900	0	0	0	0	1	1	1	2	3	4
	1000	0	0	0	0	1	1	1	2	3	4

(Continues)

Table C.4 *Continued*

CONDUCTORS

Type	Conductor Size (AWG/kcmil)	Metric Designator (Trade Size)									
		16 (½)	21 (¾)	27 (1)	35 (1¼)	41 (1½)	53 (2)	63 (2½)	78 (3)	91 (3½)	103 (4)
FEP, FEPB, PFA, PFAH, TFE	14	13	23	38	66	89	145	205	317	423	545
	12	10	17	28	48	65	106	150	231	309	398
	10	7	12	20	34	46	76	107	166	221	285
	8	4	7	11	19	26	43	61	95	127	163
	6	3	5	8	14	19	31	44	67	90	116
	4	1	3	5	10	13	21	30	47	63	81
	3	1	3	4	8	11	18	25	39	52	68
	2	1	2	4	6	9	15	21	32	43	56
PFA, PFAH, TFE	1	1	1	2	4	6	10	14	22	30	39
PFA, PFAH, TFE, Z	1/0	1	1	1	4	5	8	12	19	25	32
	2/0	1	1	1	3	4	7	10	15	21	27
	3/0	0	1	1	2	3	6	8	13	17	22
	4/0	0	1	1	1	3	5	7	10	14	18
Z	14	16	28	46	79	107	175	247	381	510	657
	12	11	20	32	56	76	124	175	271	362	466
	10	7	12	20	34	46	76	107	166	221	285
	8	4	7	12	21	29	48	68	105	140	180
	6	3	5	9	15	20	33	47	73	98	127
	4	1	3	6	10	14	23	33	50	67	87
	3	1	2	4	7	10	17	24	37	49	63
	2	1	1	3	6	8	14	20	30	41	53
	1	1	1	3	5	7	11	16	25	33	43
XHH, XHHW, XHHW-2, ZW	14	10	17	27	47	64	104	147	228	304	392
	12	7	13	21	36	49	80	113	175	234	301
	10	5	9	15	27	36	59	84	130	174	224
	8	3	5	8	15	20	33	47	72	97	124
	6	1	4	6	11	15	24	35	53	71	92
	4	1	3	4	8	11	18	25	39	52	67
	3	1	2	4	7	9	15	21	33	44	56
	2	1	1	3	5	7	12	18	27	37	47
XHH, XHHW, XHHW-2	1	1	1	2	4	5	9	13	20	27	35
	1/0	1	1	1	3	5	8	11	17	23	30
	2/0	1	1	1	3	4	6	9	14	19	25
	3/0	0	1	1	2	3	5	7	12	16	20
	4/0	0	1	1	1	2	4	6	10	13	17
	250	0	0	1	1	1	3	5	8	11	14
	300	0	0	1	1	1	3	4	7	9	12
	350	0	0	1	1	1	3	4	6	8	10
	400	0	0	1	1	1	2	3	5	7	9
	500	0	0	0	1	1	1	3	4	6	8
	600	0	0	0	1	1	1	2	3	5	6
	700	0	0	0	1	1	1	1	3	4	5
	750	0	0	0	1	1	1	1	3	4	5
	800	0	0	0	0	1	1	1	3	4	5
	900	0	0	0	0	1	1	1	2	3	4
	1000	0	0	0	0	1	1	1	2	3	4
	1250	0	0	0	0	0	1	1	1	2	3
	1500	0	0	0	0	0	1	1	1	1	2
	1750	0	0	0	0	0	1	1	1	1	2
	2000	0	0	0	0	0	0	1	1	1	1

Table C.4 *Continued*

FIXTURE WIRES

Type	Conductor Size (AWG/ kcmil)	Metric Designator (Trade Size)					
		16 (½)	21 (¾)	27 (1)	35 (1¼)	41 (1½)	53 (2)
FHH-2, RFH-2, RFHH-3	18	9	16	26	45	61	100
	16	8	13	22	38	51	84
SF-2, SFF-2	18	12	20	33	57	77	126
	16	10	17	27	47	64	104
	14	8	13	22	38	51	84
SF-1, SFF-1	18	21	36	59	101	137	223
RFH-1, RFHH-2, TF, TFF, XF, XFF	18	15	26	43	75	101	165
RFH-2, TF, TFF, XF, XFF	16	12	21	35	60	81	133
XF, XFF	14	10	17	27	47	64	104
TFN, TFFN	18	25	42	69	119	161	264
	16	19	32	53	91	123	201
PF, PFF, PGF, PGFF, PAF, PTF, PTFF, PAFF	18	23	40	66	113	153	250
	16	18	31	51	87	118	193
	14	13	23	38	66	89	145
ZF, ZFF, ZHF, HF, HFF	18	30	52	85	146	197	322
	16	22	38	63	108	145	238
	14	16	28	46	79	107	175
KF-2, KFF-2	18	44	75	123	212	287	468
	16	31	53	87	149	202	330
	14	21	36	60	103	139	227
	12	14	25	41	70	95	156
	10	10	17	27	47	64	104
KF-1, KFF-1	18	52	90	147	253	342	558
	16	37	63	103	178	240	392
	14	25	42	69	119	161	264
	12	16	28	46	79	107	175
	10	10	18	30	52	70	114
XF, XFF	12	5	9	14	25	34	56
	10	4	7	11	19	26	43

Notes:
1. This table is for concentric stranded conductors only. For compact stranded conductors, Table C.4(A) should be used.
2. Two-hour fire-rated RHH cable has ceramifiable insulation which has much larger diameters than other RHH wires.
Consult manufacturer's conduit fill tables.
*Types RHH, RHW, and RHW-2 without outer covering.

Table C.4(A) Maximum Number of Compact Conductors in Intermediate Metal Conduit (IMC) (*Based on Table 1, Chapter 9*)

COMPACT CONDUCTORS

Type	Conductor Size (AWG/ kcmil)	16 (½)	21 (¾)	27 (1)	35 (1¼)	41 (1½)	53 (2)	63 (2½)	78 (3)	91 (3½)	103 (4)
THW, THW-2, THHW	8	2	4	7	13	17	28	40	62	83	107
	6	1	3	6	10	13	22	31	48	64	82
	4	1	2	4	7	10	16	23	36	48	62
	2	1	1	3	5	7	12	17	26	35	45
	1	1	1	1	4	5	8	12	18	25	32
	1/0	1	1	1	3	4	7	10	16	21	27
	2/0	0	1	1	3	4	6	9	13	18	23
	3/0	0	1	1	2	3	5	7	11	15	20
	4/0	0	1	1	1	2	4	6	9	13	16
	250	0	0	1	1	1	3	5	7	10	13
	300	0	0	1	1	1	3	4	6	9	11
	350	0	0	1	1	1	2	4	6	8	10
	400	0	0	1	1	1	2	3	5	7	9
	500	0	0	0	1	1	1	3	4	6	8
	600	0	0	0	1	1	1	2	3	5	6
	700	0	0	0	1	1	1	1	3	4	5
	750	0	0	0	1	1	1	1	3	4	5
	900	0	0	0	0	1	1	1	2	3	4
	1000	0	0	0	0	1	1	1	2	3	4
THHN, THWN, THWN-2	8	—	—	—	—	—	—	—	—	—	—
	6	3	5	8	14	19	32	45	70	93	120
	4	1	3	5	9	12	20	28	43	58	74
	2	1	1	3	6	8	14	20	31	41	53
	1	1	1	3	5	6	10	15	23	31	40
	1/0	1	1	2	4	5	9	13	20	26	34
	2/0	1	1	1	3	4	7	10	16	22	28
	3/0	0	1	1	3	4	6	9	14	18	24
	4/0	0	1	1	2	3	5	7	11	15	19
	250	0	1	1	1	2	4	6	9	12	15
	300	0	0	1	1	1	3	5	7	10	13
	350	0	0	1	1	1	3	4	7	9	11
	400	0	0	1	1	1	2	4	6	8	10
	500	0	0	1	1	1	2	3	5	7	9
	600	0	0	0	1	1	1	2	4	5	7
	700	0	0	0	1	1	1	2	3	5	6
	750	0	0	0	1	1	1	1	3	4	6
	900	0	0	0	0	1	1	2	3	3	5
	1000	0	0	0	0	1	1	1	2	3	4
XHHW, XHHW-2	8	3	6	9	16	22	37	52	80	107	138
	6	2	4	7	12	16	27	38	59	80	103
	4	1	3	5	9	12	20	28	43	58	74
	2	1	1	3	6	8	14	20	31	41	53
	1	1	1	3	5	6	10	15	23	31	40
	1/0	1	1	2	4	5	9	13	20	26	34
	2/0	1	1	1	3	4	7	11	17	22	29
	3/0	0	1	1	3	4	6	9	14	18	24
	4/0	0	1	1	2	3	5	7	11	15	20
	250	0	1	1	1	2	4	6	9	12	16
	300	0	0	1	1	1	3	5	8	10	13
	350	0	0	1	1	1	3	4	7	9	12
	400	0	0	1	1	1	3	4	6	8	11
	500	0	0	1	1	1	2	3	5	7	9
	600	0	0	0	1	1	1	2	4	5	7
	700	0	0	0	1	1	1	2	3	5	6
	750	0	0	0	1	1	1	1	3	4	6
	900	0	0	0	0	1	1	2	3	4	5
	1000	0	0	0	0	1	1	1	2	3	4

Definition: *Compact stranding* is the result of a manufacturing process where the standard conductor is compressed to the extent that interstices (voids between strand wires) are virtually eliminated.

Table C.5 Maximum Number of Conductors or Fixture Wires in Liquidtight Flexible Nonmetallic Conduit (Type LFNC-B*) (*Based on Table 1, Chapter 9*)

		CONDUCTORS						
		Metric Designator (Trade Size)						
Type	Conductor Size (AWG/ kcmil)	12 (⅜)	16 (½)	21 (¾)	27 (1)	35 (1¼)	41 (1½)	53 (2)
RHH, RHW, RHW-2	14	2	4	7	12	21	27	44
	12	1	3	6	10	17	22	36
	10	1	3	5	8	14	18	29
	8	1	1	2	4	7	9	15
	6	1	1	1	3	6	7	12
	4	0	1	1	2	4	6	9
	3	0	1	1	1	4	5	8
	2	0	1	1	1	3	4	7
	1	0	0	1	1	1	3	5
	1/0	0	0	1	1	1	2	4
	2/0	0	0	1	1	1	1	3
	3/0	0	0	0	1	1	1	3
	4/0	0	0	0	1	1	1	2
	250	0	0	0	0	1	1	1
	300	0	0	0	0	1	1	1
	350	0	0	0	0	1	1	1
	400	0	0	0	0	1	1	1
	500	0	0	0	0	1	1	1
	600	0	0	0	0	0	1	1
	700	0	0	0	0	0	0	1
	750	0	0	0	0	0	0	1
	800	0	0	0	0	0	0	1
	900	0	0	0	0	0	0	1
	1000	0	0	0	0	0	0	1
	1250	0	0	0	0	0	0	0
	1500	0	0	0	0	0	0	0
	1750	0	0	0	0	0	0	0
	2000	0	0	0	0	0	0	0
TW	14	5	9	15	25	44	57	93
	12	4	7	12	19	33	43	71
	10	3	5	9	14	25	32	53
	8	1	3	5	8	14	18	29
RHH†, RHW†, RHW-2†, THHW, THW, THW-2	14	3	6	10	16	29	38	62
RHH†, RHW†, RHW-2†, THHW, THW	12	3	5	8	13	23	30	50
	10	1	3	6	10	18	23	39
RHH†, RHW†, RHW-2†, THHW, THW, THW-2	8	1	1	4	6	11	14	23
RHH†, RHW†, RHW-2†, TW, THW, THHW, THW-2	6	1	1	3	5	8	11	18
	4	1	1	1	3	6	8	13
	3	1	1	1	3	5	7	11

(Continues)

Table C.5 *Continued*

		CONDUCTORS						
		Metric Designator (Trade Size)						
Type	Conductor Size (AWG/ kcmil)	12 (⅜)	16 (½)	21 (¾)	27 (1)	35 (1¼)	41 (1½)	53 (2)
RHH†, RHW†, RHW-2†, TW, THW, THHW, THW-2	2	0	1	1	2	4	6	9
	1	0	1	1	1	3	4	7
	1/0	0	0	1	1	2	3	6
	2/0	0	0	1	1	2	3	5
	3/0	0	0	1	1	1	2	4
	4/0	0	0	0	1	1	1	3
	250	0	0	0	1	1	1	3
	300	0	0	0	1	1	1	2
	350	0	0	0	0	1	1	1
	400	0	0	0	0	1	1	1
	500	0	0	0	0	1	1	1
	600	0	0	0	0	1	1	1
	700	0	0	0	0	0	1	1
	750	0	0	0	0	0	1	1
	800	0	0	0	0	0	1	1
	900	0	0	0	0	0	0	1
	1000	0	0	0	0	0	0	1
	1250	0	0	0	0	0	0	1
	1500	0	0	0	0	0	0	0
	1750	0	0	0	0	0	0	0
	2000	0	0	0	0	0	0	0
THHN, THWN, THWN-2	14	8	13	22	36	63	81	133
	12	5	9	16	26	46	59	97
	10	3	6	10	16	29	37	61
	8	1	3	6	9	16	21	35
	6	1	2	4	7	12	15	25
	4	1	1	2	4	7	9	15
	3	1	1	1	3	6	8	13
	2	1	1	1	3	5	7	11
	1	0	1	1	1	4	5	8
	1/0	0	1	1	1	3	4	7
	2/0	0	0	1	1	2	3	6
	3/0	0	0	1	1	1	3	5
	4/0	0	0	1	1	1	2	4
	250	0	0	0	1	1	1	3
	300	0	0	0	1	1	1	3
	350	0	0	0	1	1	1	2
	400	0	0	0	0	1	1	1
	500	0	0	0	0	1	1	1
	600	0	0	0	0	1	1	1
	700	0	0	0	0	1	1	1
	750	0	0	0	0	0	1	1
	800	0	0	0	0	0	1	1
	900	0	0	0	0	0	1	1
	1000	0	0	0	0	0	0	1
FEP, FEPB, PFA, PFAH, TFE	14	7	12	21	35	61	79	129
	12	5	9	15	25	44	57	94
	10	4	6	11	18	32	41	68
	8	1	3	6	10	18	23	39
	6	1	2	4	7	13	17	27
	4	1	1	3	5	9	12	19
	3	1	1	2	4	7	10	16
	2	1	1	1	3	6	8	13

Table C.5 *Continued*

CONDUCTORS

Type	Conductor Size (AWG/ kcmil)	Metric Designator (Trade Size)						
		12 (⅜)	16 (½)	21 (¾)	27 (1)	35 (1¼)	41 (1½)	53 (2)
PFA, PFAH, TFE	1	0	1	1	2	4	5	9
PFA, PFAH	1/0	0	1	1	1	3	4	7
TFE, Z	2/0	0	1	1	1	3	4	6
	3/0	0	0	1	1	2	3	5
	4/0	0	0	1	1	1	2	4
Z	14	9	15	26	42	73	95	156
	12	6	10	18	30	52	67	111
	10	4	6	11	18	32	41	68
	8	2	4	7	11	20	26	43
	6	1	3	5	8	14	18	30
	4	1	1	3	5	9	12	20
	3	1	1	2	4	7	9	15
	2	0	1	1	3	6	7	12
	1	0	1	1	2	5	6	10
XHH, XHHW, XHHW-2, ZW	14	5	9	15	25	44	57	93
	12	4	7	12	19	33	43	71
	10	3	5	9	14	25	32	53
	8	1	3	5	8	14	18	29
	6	1	1	3	6	10	13	22
	4	1	1	2	4	7	9	16
	3	1	1	1	3	6	8	13
	2	1	1	1	3	5	7	11
XHH, XHHW, XHHW-2	1	0	1	1	1	4	5	8
	1/0	0	1	1	1	3	4	7
	2/0	0	0	1	1	2	3	6
	3/0	0	0	1	1	1	3	5
	4/0	0	0	1	1	1	2	4
	250	0	0	0	1	1	1	3
	300	0	0	0	1	1	1	3
	350	0	0	0	1	1	1	2
	400	0	0	0	0	1	1	1
	500	0	0	0	0	1	1	1
	600	0	0	0	0	1	1	1
	700	0	0	0	0	1	1	1
	750	0	0	0	0	0	1	1
	800	0	0	0	0	0	1	1
	900	0	0	0	0	0	1	1
	1000	0	0	0	0	0	0	1
	1250	0	0	0	0	0	0	1
	1500	0	0	0	0	0	0	1
	1750	0	0	0	0	0	0	0
	2000	0	0	0	0	0	0	0

(Continues)

Table C.5 *Continued*

FIXTURE WIRES

Type	Conductor Size (AWG/ kcmil)	Metric Designator (Trade Size)						
		12 (⅜)	16 (½)	21 (¾)	27 (1)	35 (1¼)	41 (1½)	53 (2)
FFH-2, RFH-2	18	5	8	15	24	42	54	89
	16	4	7	12	20	35	46	75
SF-2, SFF-2	18	6	11	19	30	53	69	113
	16	5	9	15	25	44	57	93
	14	4	7	12	20	35	46	75
SF-1, SFF-1	18	11	19	33	53	94	122	199
RFH-1, RFHH-2, TF, TFF, XF, XFF	18	8	14	24	39	69	90	147
RFHH-2, TF, TFF, XF, XFF	16	7	11	20	32	56	72	119
XF, XFF	14	5	9	15	25	44	57	93
TFN, TFFN	18	14	23	39	63	111	144	236
	16	10	17	30	48	85	110	180
PF, PFF, PGF, PGFF, PAF, PTF, PTFF, PAFF	18	13	21	37	60	105	136	223
	16	10	16	29	46	81	105	173
	14	7	12	21	35	61	79	129
HF, HFF, ZF, ZFF, ZHF	18	17	28	48	77	136	176	288
	16	12	20	35	57	100	129	212
	14	9	15	26	42	73	95	156
KF-2, KFF-2	18	24	40	70	112	197	255	418
	16	17	28	49	79	139	180	295
	14	12	19	34	54	95	123	202
	12	8	13	23	37	65	85	139
	10	5	9	15	25	44	57	93
KF-1, KFF-1	18	29	48	83	134	235	304	499
	16	20	34	58	94	165	214	350
	14	14	23	39	63	111	144	236
	12	9	15	26	42	73	95	156
	10	6	10	17	27	48	62	102
XF, XFF	12	3	5	8	13	23	30	50
	10	1	3	6	10	18	23	39

Notes:
1. This table is for concentric stranded conductors only. For compact stranded conductors, Table C.5(A) should be used.
2. Two-hour fire-rated RHH cable has ceramifiable insulation which has much larger diameters than other RHH wires.
Consult manufacturer's conduit fill tables.
*Corresponds to 356.2(2).
†Types RHH, RHW, and RHW-2 without outer covering.

Table C.5(A) Maximum Number of Compact Conductors in Liquidtight Flexible Nonmetallic Conduit (Type LFNC-B*) (*Based on Table 1, Chapter 9*)

COMPACT CONDUCTORS

Type	Conductor Size (AWG/ kcmil)	Metric Designator (Trade Size)						
		12 (⅜)	16 (½)	21 (¾)	27 (1)	35 (1¼)	41 (1½)	53 (2)
THW, THW-2, THHW	8	1	2	4	7	12	15	25
	6	1	1	3	5	9	12	19
	4	1	1	2	4	7	9	14
	2	1	1	1	3	5	6	11
	1	0	1	1	1	3	4	7
	1/0	0	1	1	1	3	4	6
	2/0	0	0	1	1	2	3	5
	3/0	0	0	1	1	1	3	4
	4/0	0	0	1	1	1	2	4
	250	0	0	0	1	1	1	3
	300	0	0	0	1	1	1	2
	350	0	0	0	1	1	1	2
	400	0	0	0	0	1	1	1
	500	0	0	0	0	1	1	1
	600	0	0	0	0	1	1	1
	700	0	0	0	0	1	1	1
	750	0	0	0	0	0	1	1
	900	—	0	0	0	0	1	1
	1000	0	0	0	0	0	1	1
THHN, THWN, THWN-2	8	—	—	—	—	—	—	—
	6	1	2	4	7	13	17	28
	4	1	1	3	4	8	11	17
	2	1	1	1	3	6	7	12
	1	0	1	1	2	4	6	9
	1/0	0	1	1	1	4	5	8
	2/0	0	1	1	1	3	4	6
	3/0	0	0	1	1	2	3	5
	4/0	0	0	1	1	1	3	4
	250	0	0	1	1	1	1	3
	300	0	0	0	1	1	1	3
	350	0	0	0	1	1	1	2
	400	0	0	0	1	1	1	2
	500	0	0	0	0	1	1	1
	600	0	0	0	0	1	1	1
	700	0	0	0	0	1	1	1
	750	0	0	0	0	1	1	1
	900	—	0	0	0	0	1	1
	1000	0	0	0	0	0	1	1
XHHW, XHHW-2	8	1	3	5	9	15	20	33
	6	1	2	4	6	11	15	24
	4	1	1	3	4	8	11	17
	2	1	1	1	3	6	7	12
	1	0	1	1	2	4	6	9
	1/0	0	1	1	1	4	5	8
	2/0	0	1	1	1	3	4	7
	3/0	0	0	1	1	2	3	5
	4/0	0	0	1	1	1	3	4
	250	0	0	1	1	1	1	3
	300	0	0	0	1	1	1	3
	350	0	0	0	1	1	1	3
	400	0	0	0	1	1	1	2
	500	0	0	0	0	1	1	1
	600	0	0	0	0	1	1	1
	700	0	0	0	0	1	1	1
	750	0	0	0	0	1	1	1
	900	—	0	0	0	0	1	1
	1000	0	0	0	0	0	1	1

*Corresponds to 356.2(2).
Definition: *Compact stranding* is the result of a manufacturing process where the standard conductor is compressed to the extent that the interstices (voids between strand wires) are virtually eliminated.

Table C.6 Maximum Number of Conductors or Fixture Wires in Liquidtight Flexible Nonmetallic Conduit (Type LFNC-A*) (*Based on Table 1, Chapter 9*)

CONDUCTORS

Type	Conductor Size (AWG/ kcmil)	Metric Designator (Trade Size)						
		12 (⅜)	16 (½)	21 (¾)	27 (1)	35 (1¼)	41 (1½)	53 (2)
RHH, RHW, RHW-2	14	2	4	7	11	20	27	45
	12	1	3	6	9	17	23	38
	10	1	3	5	8	13	18	30
	8	1	1	2	4	7	9	16
	6	1	1	1	3	5	7	13
	4	0	1	1	2	4	6	10
	3	0	1	1	1	4	5	8
	2	0	1	1	1	3	4	7
	1	0	0	1	1	1	3	5
	1/0	0	0	1	1	1	2	4
	2/0	0	0	1	1	1	1	4
	3/0	0	0	0	1	1	1	3
	4/0	0	0	0	1	1	1	3
	250	0	0	0	0	1	1	1
	300	0	0	0	0	1	1	1
	350	0	0	0	0	1	1	1
	400	0	0	0	0	1	1	1
	500	0	0	0	0	0	1	1
	600	0	0	0	0	0	1	1
	700	0	0	0	0	0	0	1
	750	0	0	0	0	0	0	1
	800	0	0	0	0	0	0	1
	900	0	0	0	0	0	0	1
	1000	0	0	0	0	0	0	1
	1250	0	0	0	0	0	0	0
	1500	0	0	0	0	0	0	0
	1750	0	0	0	0	0	0	0
	2000	0	0	0	0	0	0	0
TW	14	5	9	15	24	43	58	96
	12	4	7	12	19	33	44	74
	10	3	5	9	14	24	33	55
	8	1	3	5	8	13	18	30
RHH[†], RHW[†], RHW-2[†], THHW, THW, THW-2	14	3	6	10	16	28	38	64
RHH[†], RHW[†], RHW-2[†], THHW, THW	12	3	4	8	13	23	31	51
	10	1	3	6	10	18	24	40
RHH[†], RHW[†], RHW-2[†], THHW, THW, THW-2	8	1	1	4	6	10	14	24

Table C.6 *Continued*

		CONDUCTORS						
		Metric Designator (Trade Size)						
Type	Conductor Size (AWG/ kcmil)	12 (⅜)	16 (½)	21 (¾)	27 (1)	35 (1¼)	41 (1½)	53 (2)
RHH†, RHW†, RHW-2†, TW, THW, THHW, THW-2	6	1	1	3	4	8	11	18
	4	1	1	1	3	6	8	13
	3	1	1	1	3	5	7	11
	2	0	1	1	2	4	6	10
	1	0	1	1	1	3	4	7
	1/0	0	0	1	1	2	3	6
	2/0	0	0	1	1	1	3	5
	3/0	0	0	1	1	1	2	4
	4/0	0	0	0	1	1	1	3
	250	0	0	0	1	1	1	3
	300	0	0	0	1	1	1	2
	350	0	0	0	0	1	1	1
	400	0	0	0	0	1	1	1
	500	0	0	0	0	1	1	1
	600	0	0	0	0	1	1	1
	700	0	0	0	0	0	1	1
	750	0	0	0	0	0	1	1
	800	0	0	0	0	0	1	1
	900	0	0	0	0	0	0	1
	1000	0	0	0	0	0	0	1
	1250	0	0	0	0	0	0	1
	1500	0	0	0	0	0	0	1
	1750	0	0	0	0	0	0	0
	2000	0	0	0	0	0	0	0
THHN, THWN, THWN-2	14	8	13	22	35	62	83	137
	12	5	9	16	25	45	60	100
	10	3	6	10	16	28	38	63
	8	1	3	6	9	16	22	36
	6	1	2	4	6	12	16	26
	4	1	1	2	4	7	9	16
	3	1	1	1	3	6	8	13
	2	1	1	1	3	5	7	11
	1	0	1	1	1	4	5	8
	1/0	0	1	1	1	3	4	7
	2/0	0	0	1	1	2	3	6
	3/0	0	0	1	1	1	3	5
	4/0	0	0	1	1	1	2	4
	250	0	0	0	1	1	1	3
	300	0	0	0	1	1	1	3
	350	0	0	0	1	1	1	2
	400	0	0	0	0	1	1	1
	500	0	0	0	0	1	1	1
	600	0	0	0	0	1	1	1
	700	0	0	0	0	1	1	1
	750	0	0	0	0	0	1	1
	800	0	0	0	0	0	1	1
	900	0	0	0	0	0	1	1
	1000	0	0	0	0	0	0	1

(Continues)

Table C.6 *Continued*

CONDUCTORS

Type	Conductor Size (AWG/ kcmil)	Metric Designator (Trade Size)						
		12 (⅜)	16 (½)	21 (¾)	27 (1)	35 (1¼)	41 (1½)	53 (2)
FEP, FEPB, PFA, PFAH, TFE	14	7	12	21	34	60	80	133
	12	5	9	15	25	44	59	97
	10	4	6	11	18	31	42	70
	8	1	3	6	10	18	24	40
	6	1	2	4	7	13	17	28
	4	1	1	3	5	9	12	20
	3	1	1	2	4	7	10	16
	2	1	1	1	3	6	8	13
PFA, PFAH, TFE	1	0	1	1	2	4	5	9
PFA, PFAH, TFE, Z	1/0	0	1	1	1	3	5	8
	2/0	0	1	1	1	3	4	6
	3/0	0	0	1	1	2	3	5
	4/0	0	0	1	1	1	2	4
Z	14	9	15	25	41	72	97	161
	12	6	10	18	29	51	69	114
	10	4	6	11	18	31	42	70
	8	2	4	7	11	20	26	44
	6	1	3	5	8	14	18	31
	4	1	1	3	5	9	13	21
	3	1	1	2	4	7	9	15
	2	1	1	1	3	6	8	13
	1	1	1	1	2	4	6	10
XHH, XHHW, XHHW-2, ZW	14	5	9	15	24	43	58	96
	12	4	7	12	19	33	44	74
	10	3	5	9	14	24	33	55
	8	1	3	5	8	13	18	30
	6	1	1	3	5	10	13	22
	4	1	1	2	4	7	10	16
	3	1	1	1	3	6	8	14
	2	1	1	1	3	5	7	11
XHH, XHHW, XHHW-2	1	0	1	1	1	4	5	8
	1/0	0	1	1	1	3	4	7
	2/0	0	0	1	1	2	3	6
	3/0	0	0	1	1	1	3	5
	4/0	0	0	1	1	1	2	4
	250	0	0	0	1	1	1	3
	300	0	0	0	1	1	1	3
	350	0	0	0	1	1	1	2
	400	0	0	0	0	1	1	1
	500	0	0	0	0	1	1	1
	600	0	0	0	0	1	1	1
	700	0	0	0	0	1	1	1
	750	0	0	0	0	0	1	1
	800	0	0	0	0	0	1	1
	900	0	0	0	0	0	1	1
	1000	0	0	0	0	0	0	1
	1250	0	0	0	0	0	0	1
	1500	0	0	0	0	0	0	1
	1750	0	0	0	0	0	0	0
	2000	0	0	0	0	0	0	0

Table C.6 *Continued*

FIXTURE WIRES

Type	Conductor Size (AWG/ kcmil)	Metric Designator (Trade Size)						
		12 (⅜)	16 (½)	21 (¾)	27 (1)	35 (1¼)	41 (1½)	53 (2)
FFH-2, RFH-2, RFHH-3	18	5	8	14	23	41	55	92
	16	4	7	12	20	35	47	77
SF-2, SFF-2	18	6	11	18	29	52	70	116
	16	5	9	15	24	43	58	96
	14	4	7	12	20	35	47	77
SF-1, SFF-1	18	12	19	33	52	92	124	205
RFH-1, RFHH-2, TF, TFF, XF, XFF	18	8	14	24	39	68	91	152
RFHH-2, TF, TFF, XF, XFF	16	7	11	19	31	55	74	122
XF, XFF	14	5	9	15	24	43	58	96
TFN, TFFN	18	14	22	39	62	109	146	243
	16	10	17	29	47	83	112	185
PF, PFF, PGF, PGFF, PAF, PTF, PTFF, PAFF	18	13	21	37	59	103	139	230
	16	10	16	28	45	80	107	178
	14	7	12	21	34	60	80	133
HF, HFF, ZF, ZFF, ZHF	18	17	27	47	76	133	179	297
	16	12	20	35	56	98	132	219
	14	9	15	25	41	72	97	161
KF-2, KFF-2	18	25	40	69	110	193	260	431
	16	17	28	48	77	136	183	303
	14	12	19	33	53	94	126	209
	12	8	13	23	36	64	86	143
	10	5	9	15	24	43	58	96
KF-1, KFF-1	18	29	48	82	131	231	310	514
	16	21	33	57	92	162	218	361
	14	14	22	39	62	109	146	243
	12	9	15	25	41	72	97	161
	10	6	10	17	27	47	63	105
XF, XFF	12	3	4	8	13	23	31	51
	10	1	3	6	10	18	24	40

Notes:
1. This table is for concentric stranded conductors only. For compact stranded conductors, Table C.6(A) should be used.
2. Two-hour fire-rated RHH cable has ceramifiable insulation which has much larger diameters than other RHH wires.
Consult manufacturer's conduit fill tables.
*Corresponds to 356.2(1).
†Types RHH, RHW, and RHW-2 without outer covering.

Table C.6(A) Maximum Number of Compact Conductors in Liquidtight Flexible Nonmetallic Conduit (Type LFNC-A*) (*Based on Table 1, Chapter 9*)

COMPACT CONDUCTORS

Type	Conductor Size (AWG/ kcmil)	Metric Designator (Trade Size)						
		12 (⅜)	16 (½)	21 (¾)	27 (1)	35 (1¼)	41 (1½)	53 (2)
THW, THW-2, THHW	8	1	2	4	6	11	16	26
	6	1	1	3	5	9	12	20
	4	1	1	2	4	7	9	15
	2	1	1	1	3	5	6	11
	1	0	1	1	1	3	4	8
	1/0	0	1	1	1	3	4	7
	2/0	0	0	1	1	2	3	5
	3/0	0	0	1	1	1	3	5
	4/0	0	0	1	1	1	2	4
	250	0	0	0	1	1	1	3
	300	0	0	0	1	1	1	3
	350	0	0	0	1	1	1	2
	400	0	0	0	0	1	1	1
	500	0	0	0	0	1	1	1
	600	0	0	0	0	1	1	1
	700	0	0	0	0	1	1	1
	750	0	0	0	0	0	1	1
	900	—	0	0	0	0	1	1
	1000	0	0	0	0	0	1	1
THHN, THWN, THWN-2	8	—	—	—	—	—	—	—
	6	1	2	4	7	13	18	29
	4	1	1	3	4	8	11	18
	2	1	1	1	3	6	8	13
	1	0	1	1	2	4	6	10
	1/0	0	1	1	1	3	5	8
	2/0	0	1	1	1	3	4	7
	3/0	0	0	1	1	2	3	6
	4/0	0	0	1	1	1	3	5
	250	0	0	1	1	1	1	3
	300	0	0	0	1	1	1	3
	350	0	0	0	1	1	1	3
	400	0	0	0	1	1	1	2
	500	0	0	0	0	1	1	1
	600	0	0	0	0	1	1	1
	700	0	0	0	0	1	1	1
	750	0	0	0	0	1	1	1
	900	—	0	0	0	0	1	1
	1000	0	0	0	0	0	1	1
XHHW, XHHW-2	8	1	3	5	8	15	20	34
	6	1	2	4	6	11	15	25
	4	1	1	3	4	8	11	18
	2	1	1	1	3	6	8	13
	1	0	1	1	2	4	6	10
	1/0	0	1	1	1	3	5	8
	2/0	0	1	1	1	3	4	7
	3/0	0	0	1	1	2	3	6
	4/0	0	0	1	1	1	3	5
	250	0	0	1	1	1	2	4
	300	0	0	0	1	1	1	3
	350	0	0	0	1	1	1	3
	400	0	0	0	1	1	1	2
	500	0	0	0	0	1	1	1
	600	0	0	0	0	1	1	1
	700	0	0	0	0	1	1	1
	750	0	0	0	0	1	1	1
	900	—	0	0	0	0	1	1
	1000	0	0	0	0	0	1	1

*Corresponds to 356.2(1).

Definition: *Compact stranding* is the result of a manufacturing process where the standard conductor is compressed to the extent that the interstices (voids between strand wires) are virtually eliminated.

Table C.7 Maximum Number of Conductors or Fixture Wires in Liquidtight Flexible Metal Conduit (LFMC) (*Based on Table 1, Chapter 9*)

		CONDUCTORS									
		Metric Designator (Trade Size)									
Type	**Conductor Size (AWG/ kcmil)**	**16** (½)	**21** (¾)	**27** (1)	**35** (1¼)	**41** (1½)	**53** (2)	**63** (2½)	**78** (3)	**91** (3½)	**103** (4)
RHH, RHW, RHW-2	14	4	7	12	21	27	44	66	102	133	173
	12	3	6	10	17	22	36	55	84	110	144
	10	3	5	8	14	18	29	44	68	89	116
	8	1	2	4	7	9	15	23	36	46	61
	6	1	1	3	6	7	12	18	28	37	48
	4	1	1	2	4	6	9	14	22	29	38
	3	1	1	1	4	5	8	13	19	25	33
	2	1	1	1	3	4	7	11	17	22	29
	1	0	1	1	1	3	5	7	11	14	19
	1/0	0	1	1	1	2	4	6	10	13	16
	2/0	0	1	1	1	1	3	5	8	11	14
	3/0	0	0	1	1	1	3	4	7	9	12
	4/0	0	0	1	1	1	2	4	6	8	10
	250	0	0	0	1	1	1	3	4	6	8
	300	0	0	0	1	1	1	2	4	5	7
	350	0	0	0	1	1	1	2	3	5	6
	400	0	0	0	1	1	1	1	3	4	6
	500	0	0	0	1	1	1	1	3	4	5
	600	0	0	0	0	1	1	1	2	3	4
	700	0	0	0	0	0	1	1	1	3	3
	750	0	0	0	0	0	1	1	1	2	3
	800	0	0	0	0	0	1	1	1	2	3
	900	0	0	0	0	0	1	1	1	2	3
	1000	0	0	0	0	0	1	1	1	1	3
	1250	0	0	0	0	0	0	1	1	1	1
	1500	0	0	0	0	0	0	1	1	1	1
	1750	0	0	0	0	0	0	1	1	1	1
	2000	0	0	0	0	0	0	0	1	1	1
TW	14	9	15	25	44	57	93	140	215	280	365
	12	7	12	19	33	43	71	108	165	215	280
	10	5	9	14	25	32	53	80	123	160	209
	8	3	5	8	14	18	29	44	68	89	116
RHH*, RHW*, RHW-2*, THHW, THW, THW-2	14	6	10	16	29	38	62	93	143	186	243
RHH*, RHW*, RHW-2*, THHW, THW	12	5	8	13	23	30	50	75	115	149	195
	10	3	6	10	18	23	39	58	89	117	152
RHH*, RHW*, RHW-2*, THHW, THW, THW-2	8	1	4	6	11	14	23	35	53	70	91

(Continues)

Table C.7 *Continued*

CONDUCTORS

Type	Conductor Size (AWG/ kcmil)	Metric Designator (Trade Size)									
		16 (½)	21 (¾)	27 (1)	35 (¼)	41 (1½)	53 (2)	63 (2½)	78 (3)	91 (3½)	103 (4)
RHH*, RHW*, RHW-2*, TW, THW, THHW, THW-2	6	1	3	5	8	11	18	27	41	53	70
	4	1	1	3	6	8	13	20	30	40	52
	3	1	1	3	5	7	11	17	26	34	44
	2	1	1	2	4	6	9	14	22	29	38
	1	1	1	1	3	4	7	10	15	20	26
	1/0	0	1	1	2	3	6	8	13	17	23
	2/0	0	1	1	2	3	5	7	11	15	19
	3/0	0	1	1	1	2	4	6	9	12	16
	4/0	0	0	1	1	1	3	5	8	10	13
	250	0	0	1	1	1	3	4	6	8	11
	300	0	0	1	1	1	2	3	5	7	9
	350	0	0	0	1	1	1	3	5	6	8
	400	0	0	0	1	1	1	3	4	6	7
	500	0	0	0	1	1	1	2	3	5	6
	600	0	0	0	1	1	1	1	3	4	5
	700	0	0	0	0	1	1	1	2	3	4
	750	0	0	0	0	1	1	1	2	3	4
	800	0	0	0	0	1	1	1	2	3	4
	900	0	0	0	0	0	1	1	1	3	3
	1000	0	0	0	0	0	1	1	1	2	3
	1250	0	0	0	0	0	1	1	1	1	2
	1500	0	0	0	0	0	0	1	1	1	2
	1750	0	0	0	0	0	0	1	1	1	1
	2000	0	0	0	0	0	0	1	1	1	1
THHN, THWN, THWN-2	14	13	22	36	63	81	133	201	308	401	523
	12	9	16	26	46	59	97	146	225	292	381
	10	6	10	16	29	37	61	92	141	184	240
	8	3	6	9	16	21	35	53	81	106	138
	6	2	4	7	12	15	25	38	59	76	100
	4	1	2	4	7	9	15	23	36	47	61
	3	1	1	3	6	8	13	20	30	40	52
	2	1	1	3	5	7	11	17	26	33	44
	1	1	1	1	4	5	8	12	19	25	32
	1/0	1	1	1	3	4	7	10	16	21	27
	2/0	0	1	1	2	3	6	8	13	17	23
	3/0	0	1	1	1	3	5	7	11	14	19
	4/0	0	1	1	1	2	4	6	9	12	15
	250	0	0	1	1	1	3	5	7	10	12
	300	0	0	1	1	1	3	4	6	8	11
	350	0	0	1	1	1	2	3	5	7	9
	400	0	0	0	1	1	1	3	5	6	8
	500	0	0	0	1	1	1	2	4	5	7
	600	0	0	0	1	1	1	1	3	4	6
	700	0	0	0	1	1	1	1	3	4	5
	750	0	0	0	0	1	1	1	3	3	5
	800	0	0	0	0	1	1	1	2	3	4
	900	0	0	0	0	1	1	1	2	3	4
	1000	0	0	0	0	0	1	1	1	3	3

Table C.7 *Continued*

CONDUCTORS

Type	Conductor Size (AWG/ kcmil)	Metric Designator (Trade Size)									
		16 (½)	21 (¾)	27 (1)	35 (¼)	41 (1½)	53 (2)	63 (2½)	78 (3)	91 (3½)	103 (4)
FEP, FEPB, PFA, PFAH, TFE	14	12	21	35	61	79	129	195	299	389	507
	12	9	15	25	44	57	94	142	218	284	370
	10	6	11	18	32	41	68	102	156	203	266
	8	3	6	10	18	23	39	58	89	117	152
	6	2	4	7	13	17	27	41	64	83	108
	4	1	3	5	9	12	19	29	44	58	75
	3	1	2	4	7	10	16	24	37	48	63
	2	1	1	3	6	8	13	20	30	40	52
PFA, PFAH, TFE	1	1	1	2	4	5	9	14	21	28	36
PFA, PFAH, TFE, Z	1/0	1	1	1	3	4	7	11	18	23	30
	2/0	1	1	1	3	4	6	9	14	19	25
	3/0	0	1	1	2	3	5	8	12	16	20
	4/0	0	1	1	1	2	4	6	10	13	17
Z	14	20	26	42	73	95	156	235	360	469	611
	12	14	18	30	52	67	111	167	255	332	434
	10	8	11	18	32	41	68	102	156	203	266
	8	5	7	11	20	26	43	64	99	129	168
	6	4	5	8	14	18	30	45	69	90	118
	4	2	3	5	9	12	20	31	48	62	81
	3	2	2	4	7	9	15	23	35	45	59
	2	1	1	3	6	7	12	19	29	38	49
	1	1	1	2	5	6	10	15	23	30	40
XHH, XHHW, XHHW-2, ZW	14	9	15	25	44	57	93	140	215	280	365
	12	7	12	19	33	43	71	108	165	215	280
	10	5	9	14	25	32	53	80	123	160	209
	8	3	5	8	14	18	29	44	68	89	116
	6	1	3	6	10	13	22	33	50	66	86
	4	1	2	4	7	9	16	24	36	48	62
	3	1	1	3	6	8	13	20	31	40	52
	2	1	1	3	5	7	11	17	26	34	44
	1	1	1	1	4	5	8	12	19	25	33
XHH, XHHW, XHHW-2	1/0	1	1	1	3	4	7	10	16	21	28
	2/0	0	1	1	2	3	6	9	13	17	23
	3/0	0	1	1	1	3	5	7	11	14	19
	4/0	0	1	1	1	2	4	6	9	12	16
	250	0	0	1	1	1	3	5	7	10	13
	300	0	0	1	1	1	3	4	6	8	11
	350	0	0	1	1	1	2	3	5	7	10
	400	0	0	0	1	1	1	3	5	6	8
	500	0	0	0	1	1	1	2	4	5	7
	600	0	0	0	1	1	1	1	3	4	6
	700	0	0	0	1	1	1	1	3	4	5
	750	0	0	0	0	1	1	1	3	3	5
	800	0	0	0	0	1	1	1	2	3	4
	900	0	0	0	0	1	1	1	2	3	4
	1000	0	0	0	0	0	1	1	1	3	3
	1250	0	0	0	0	0	1	1	1	1	3
	1500	0	0	0	0	0	1	1	1	1	2
	1750	0	0	0	0	0	0	1	1	1	2
	2000	0	0	0	0	0	0	1	1	1	2

(Continues)

Table C.7 *Continued*

FIXTURE WIRES

Type	Conductor Size (AWG/ kcmil)	Metric Designator (Trade Size)					
		16 (½)	21 (¾)	27 (1)	35 (1¼)	41 (1½)	53 (2)
FFH-2, RFH-2, RFHH-3	18	8	15	24	42	54	89
	16	7	12	20	35	46	75
SF-2, SFF-2	18	11	19	30	53	69	113
	16	9	15	25	44	57	93
	14	7	12	20	35	46	75
SF-1, SFF-1	18	19	33	53	94	122	199
RFH-1, RFHH-2, TF, TFF, XF, XFF	18	14	24	39	69	90	147
RFHH-2, TF, TFF, XF, XFF	16	11	20	32	56	72	119
XF, XFF	14	9	15	25	44	57	93
TFN, TFFN	18	23	39	63	111	144	236
	16	17	30	48	85	110	180
PF, PFF, PGF, PGFF, PAF, PTF, PTFF, PAFF	18	21	37	60	105	136	223
	16	16	29	46	81	105	173
	14	12	21	35	61	79	129
HF, HFF, ZF, ZFF, ZHF	18	28	48	77	136	176	288
	16	20	35	57	100	129	212
	14	15	26	42	73	95	156
KF-2, KFF-2	18	40	70	112	197	255	418
	16	28	49	79	139	180	295
	14	19	34	54	95	123	202
	12	13	23	37	65	85	139
	10	9	15	25	44	57	93
KF-1, KFF-1	18	48	83	134	235	304	499
	16	34	58	94	165	214	350
	14	23	39	63	111	144	236
	12	15	26	42	73	95	156
	10	10	17	27	48	62	102
XF, XFF	12	5	8	13	23	30	50
	10	3	6	10	18	23	39

Notes:

1. This table is for concentric stranded conductors only. For compact stranded conductors, Table C.7(A) should be used.

2. Two-hour fire-rated RHH cable has ceramifiable insulation which has much larger diameters than other RHH wires.

Consult manufacturer's conduit fill tables.

*Types RHH, RHW, and RHW-2 without outer covering.

Table C.7(A) Maximum Number of Compact Conductors in Liquidtight Flexible Metal Conduit (LFMC) (*Based on Table 1, Chapter 9*)

		COMPACT CONDUCTORS										
		Metric Designator (Trade Size)										
Type	Conductor Size (AWG/ kcmil)	12 (⅜)	16 (½)	21 (¾)	27 (1)	35 (1¼)	41 (1½)	53 (2)	63 (2½)	78 (3)	91 (3½)	103 (4)
THW, THW-2, THHW	8	1	2	4	7	12	15	25	38	58	76	99
	6	1	1	3	5	9	12	19	29	45	59	77
	4	1	1	2	4	7	9	14	22	34	44	57
	2	1	1	1	3	5	6	11	16	25	32	42
	1	0	1	1	1	3	4	7	11	17	23	30
	1/0	0	1	1	1	3	4	6	10	15	20	26
	2/0	0	0	1	1	2	3	5	8	13	16	21
	3/0	0	0	1	1	1	3	4	7	11	14	18
	4/0	0	0	1	1	1	2	4	6	9	12	15
	250	0	0	0	1	1	1	3	4	7	9	12
	300	0	0	0	1	1	1	2	4	6	8	10
	350	0	0	0	1	1	1	2	3	5	7	9
	400	0	0	0	0	1	1	1	3	5	6	8
	500	0	0	0	0	1	1	1	3	4	5	7
	600	0	0	0	0	1	1	1	1	3	4	6
	700	0	0	0	0	1	1	1	1	3	4	5
	750	0	0	0	0	0	1	1	1	3	3	5
	900	—	0	0	0	0	1	1	1	2	3	4
	1000	0	0	0	0	0	1	1	1	1	3	4
THHN, THWN, THWN-2	8	—	—	—	—	—	—	—	—	—	—	—
	6	1	2	4	7	13	17	28	43	66	86	112
	4	1	1	3	4	8	11	17	26	41	53	69
	2	1	1	1	3	6	7	12	19	29	38	50
	1	0	1	1	2	4	6	9	14	22	28	37
	1/0	0	1	1	1	4	5	8	12	19	24	32
	2/0	0	1	1	1	3	4	6	10	15	20	26
	3/0	0	0	1	1	2	3	5	8	13	17	22
	4/0	0	0	1	1	1	3	4	7	10	14	18
	250	0	0	1	1	1	1	3	5	8	11	14
	300	0	0	0	1	1	1	3	4	7	9	12
	350	0	0	0	1	1	1	2	4	6	8	11
	400	0	0	0	1	1	1	2	3	5	7	9
	500	0	0	0	0	1	1	1	3	5	6	8
	600	0	0	0	0	1	1	1	2	4	5	6
	700	0	0	0	0	1	1	1	1	3	4	6
	750	0	0	0	0	1	1	1	1	3	4	5
	900	—	0	0	0	0	1	1	1	2	3	4
	1000	0	0	0	0	0	1	1	1	2	3	4
XHHW, XHHW-2	8	1	3	5	9	15	20	33	49	76	98	129
	6	1	2	4	6	11	15	24	37	56	73	95
	4	1	1	3	4	8	11	17	26	41	53	69
	2	1	1	1	3	6	7	12	19	29	38	50
	1	0	1	1	2	4	6	9	14	22	28	37
	1/0	0	1	1	1	4	5	8	12	19	24	32
	2/0	0	1	1	1	3	4	7	10	16	20	27
	3/0	0	0	1	1	2	3	5	8	13	17	22
	4/0	0	0	1	1	1	3	4	7	11	14	18
	250	0	0	1	1	1	1	3	5	8	11	15
	300	0	0	0	1	1	1	3	5	7	9	12
	350	0	0	0	1	1	1	3	4	6	8	11
	400	0	0	0	1	1	1	2	4	6	7	10
	500	0	0	0	0	1	1	1	3	5	6	8
	600	0	0	0	0	1	1	1	2	4	5	6
	700	0	0	0	0	1	1	1	1	3	4	6
	750	0	0	0	0	1	1	1	1	3	4	5
	900	—	0	0	0	0	1	1	2	2	3	4
	1000	0	0	0	0	0	1	1	1	2	3	4

Definition: *Compact stranding* is the result of a manufacturing process where the standard conductor is compressed to the extent that the interstices (voids between strand wires) are virtually eliminated.

Table C.8 Maximum Number of Conductors or Fixture Wires in Rigid Metal Conduit (RMC) (*Based on Table 1, Chapter 9*)

		CONDUCTORS											
		Metric Designator (Trade Size)											
Type	Conductor Size (AWG/ kcmil)	16 (½)	21 (¾)	27 (1)	35 (1¼)	41 (1½)	53 (2)	63 (2½)	78 (3)	91 (3½)	103 (4)	129 (5)	155 (6)
RHH, RHW, RHW-2	14	4	7	12	21	28	46	66	102	136	176	276	398
	12	3	6	10	17	23	38	55	85	113	146	229	330
	10	3	5	8	14	19	31	44	68	91	118	185	267
	8	1	2	4	7	10	16	23	36	48	61	97	139
	6	1	1	3	6	8	13	18	29	38	49	77	112
	4	1	1	2	4	6	10	14	22	30	38	60	87
	3	1	1	2	4	5	9	12	19	26	34	53	76
	2	1	1	1	3	4	7	11	17	23	29	46	66
	1	0	1	1	1	3	5	7	11	15	19	30	44
	1/0	0	1	1	1	2	4	6	10	13	17	26	38
	2/0	0	1	1	1	2	4	5	8	11	14	23	33
	3/0	0	0	1	1	1	3	4	7	10	12	20	28
	4/0	0	0	1	1	1	3	4	6	8	11	17	24
	250	0	0	0	1	1	1	3	4	6	8	13	18
	300	0	0	0	1	1	1	2	4	5	7	11	16
	350	0	0	0	1	1	1	2	4	5	6	10	15
	400	0	0	0	1	1	1	1	3	4	6	9	13
	500	0	0	0	1	1	1	1	3	4	5	8	11
	600	0	0	0	0	1	1	1	2	3	4	6	9
	700	0	0	0	0	1	1	1	1	3	4	6	8
	750	0	0	0	0	0	1	1	1	3	3	5	8
	800	0	0	0	0	0	1	1	1	2	3	5	7
	900	0	0	0	0	0	1	1	1	2	3	5	7
	1000	0	0	0	0	0	1	1	1	1	3	4	6
	1250	0	0	0	0	0	0	1	1	1	1	3	5
	1500	0	0	0	0	0	0	1	1	1	1	3	4
	1750	0	0	0	0	0	0	1	1	1	1	2	4
	2000	0	0	0	0	0	0	0	1	1	1	2	3
TW	14	9	15	25	44	59	98	140	216	288	370	581	839
	12	7	12	19	33	45	75	107	165	221	284	446	644
	10	5	9	14	25	34	56	80	123	164	212	332	480
	8	3	5	8	14	19	31	44	68	91	118	185	267
RHH*, RHW*, RHW-2* THHW, THW, THW-2	14	6	10	17	29	39	65	93	143	191	246	387	558
RHH*, RHW*, RHW-2*, THHW, THW	12	5	8	13	23	32	52	75	115	154	198	311	448
	10	3	6	10	18	25	41	58	90	120	154	242	350
RHH*, RHW*, RHW-2*, THHW, THW, THW-2	8	1	4	6	11	15	24	35	54	72	92	145	209

Table C.8 *Continued*

CONDUCTORS

		Metric Designator (Trade Size)											
Type	Conductor Size (AWG/ kcmil)	16 (½)	21 (¾)	27 (1)	35 (1¼)	41 (1½)	53 (2)	63 (2½)	78 (3)	91 (3½)	103 (4)	129 (5)	155 (6)
RHH*,	6	1	3	5	8	11	18	27	41	55	71	111	160
RHW*,	4	1	1	3	6	8	14	20	31	41	53	83	120
RHW-2*, TW,	3	1	1	3	5	7	12	17	26	35	45	71	103
THW, THHW,	2	1	1	2	4	6	10	14	22	30	38	60	87
THW-2	1	1	1	1	3	4	7	10	15	21	27	42	61
	1/0	0	1	1	2	3	6	8	13	18	23	36	52
	2/0	0	1	1	2	3	5	7	11	15	19	31	44
	3/0	0	1	1	1	2	4	6	9	13	16	26	37
	4/0	0	0	1	1	1	3	5	8	10	14	21	31
	250	0	0	1	1	1	3	4	6	8	11	17	25
	300	0	0	1	1	1	2	3	5	7	9	15	22
	350	0	0	0	1	1	1	3	5	6	8	13	19
	400	0	0	0	1	1	1	3	4	6	7	12	17
	500	0	0	0	1	1	1	2	3	5	6	10	14
	600	0	0	0	1	1	1	1	3	4	5	8	12
	700	0	0	0	0	1	1	1	2	3	4	7	10
	750	0	0	0	0	1	1	1	2	3	4	7	10
	800	0	0	0	0	1	1	1	2	3	4	6	9
	900	0	0	0	0	1	1	1	1	3	4	6	8
	1000	0	0	0	0	0	1	1	1	2	3	5	8
	1250	0	0	0	0	0	1	1	1	1	2	4	6
	1500	0	0	0	0	0	1	1	1	1	2	3	5
	1750	0	0	0	0	0	0	1	1	1	1	3	4
	2000	0	0	0	0	0	0	1	1	1	1	3	4
THHN,	14	13	22	36	63	85	140	200	309	412	531	833	1202
THWN,	12	9	16	26	46	62	102	146	225	301	387	608	877
THWN-2	10	6	10	17	29	39	64	92	142	189	244	383	552
	8	3	6	9	16	22	37	53	82	109	140	221	318
	6	2	4	7	12	16	27	38	59	79	101	159	230
	4	1	2	4	7	10	16	23	36	48	62	98	141
	3	1	1	3	6	8	14	20	31	41	53	83	120
	2	1	1	3	5	7	11	17	26	34	44	70	100
	1	1	1	1	4	5	8	12	19	25	33	51	74
	1/0	1	1	1	3	4	7	10	16	21	27	43	63
	2/0	0	1	1	2	3	6	8	13	18	23	36	52
	3/0	0	1	1	1	3	5	7	11	15	19	30	43
	4/0	0	1	1	1	2	4	6	9	12	16	25	36
	250	0	0	1	1	1	3	5	7	10	13	20	29
	300	0	0	1	1	1	3	4	6	8	11	17	25
	350	0	0	1	1	1	2	3	5	7	10	15	22
	400	0	0	1	1	1	2	3	5	7	8	13	20
	500	0	0	0	1	1	1	2	4	5	7	11	16
	600	0	0	0	1	1	1	1	3	4	6	9	13
	700	0	0	0	1	1	1	1	3	4	5	8	11
	750	0	0	0	0	1	1	1	3	4	5	7	11
	800	0	0	0	0	1	1	1	2	3	4	7	10
	900	0	0	0	0	1	1	1	2	3	4	6	9
	1000	0	0	0	0	1	1	1	1	3	4	6	8

(Continues)

Table C.8 *Continued*

CONDUCTORS

Type	Conductor Size (AWG/ kcmil)	Metric Designator (Trade Size)											
		16 (½)	21 (¾)	27 (1)	35 (1¼)	41 (1½)	53 (2)	63 (2½)	78 (3)	91 (3½)	103 (4)	129 (5)	155 (6)
FEP, FEPB, PFA, PFAH, TFE	14	12	22	35	61	83	136	194	300	400	515	808	1166
	12	9	16	26	44	60	99	142	219	292	376	590	851
	10	6	11	18	32	43	71	102	157	209	269	423	610
	8	3	6	10	18	25	41	58	90	120	154	242	350
	6	2	4	7	13	17	29	41	64	85	110	172	249
	4	1	3	5	9	12	20	29	44	59	77	120	174
	3	1	2	4	7	10	17	24	37	50	64	100	145
	2	1	1	3	6	8	14	20	31	41	53	83	120
PFA, PFAH, TFE	1	1	1	2	4	6	9	14	21	28	37	57	83
PFA, PFAH, TFE, Z	1/0	1	1	1	3	5	8	11	18	24	30	48	69
	2/0	1	1	1	3	4	6	9	14	19	25	40	57
	3/0	0	1	1	2	3	5	8	12	16	21	33	47
	4/0	0	1	1	1	2	4	6	10	13	17	27	39
Z	14	15	26	42	73	100	164	234	361	482	621	974	1405
	12	10	18	30	52	71	116	166	256	342	440	691	997
	10	6	11	18	32	43	71	102	157	209	269	423	610
	8	4	7	11	20	27	45	64	99	132	170	267	386
	6	3	5	8	14	19	31	45	69	93	120	188	271
	4	1	3	5	9	13	22	31	48	64	82	129	186
	3	1	2	4	7	9	16	22	35	47	60	94	136
	2	1	1	3	6	8	13	19	29	39	50	78	113
	1	1	1	2	5	6	10	15	23	31	40	63	92
XHH, XHHW, XHHW-2, ZW	14	9	15	25	44	59	98	140	216	288	370	581	839
	12	7	12	19	33	45	75	107	165	221	284	446	644
	10	5	9	14	25	34	56	80	123	164	212	332	480
	8	3	5	8	14	19	31	44	68	91	118	185	267
	6	1	3	6	10	14	23	33	51	68	87	137	197
	4	1	2	4	7	10	16	24	37	49	63	99	143
	3	1	1	3	6	8	14	20	31	41	53	84	121
	2	1	1	3	5	7	12	17	26	35	45	70	101
	1	1	1	1	4	5	9	12	19	26	33	52	76
XHH, XHHW, XHHW-2	1/0	1	1	1	3	4	7	10	16	22	28	44	64
	2/0	0	1	1	2	3	6	9	13	18	23	37	53
	3/0	0	1	1	1	3	5	7	11	15	19	30	44
	4/0	0	1	1	1	2	4	6	9	12	16	25	36
	250	0	0	1	1	1	3	5	7	10	13	20	30
	300	0	0	1	1	1	3	4	6	9	11	18	25
	350	0	0	1	1	1	2	3	6	7	10	15	22
	400	0	0	1	1	1	2	3	5	7	9	14	20
	500	0	0	0	1	1	1	2	4	5	7	11	16
	600	0	0	0	1	1	1	1	3	4	6	9	13
	700	0	0	0	1	1	1	1	3	4	5	8	11
	750	0	0	0	0	1	1	1	3	4	5	7	11
	800	0	0	0	0	1	1	1	2	3	4	7	10
	900	0	0	0	0	1	1	1	2	3	4	6	9
	1000	0	0	0	0	1	1	1	1	3	4	6	8
	1250	0	0	0	0	0	1	1	1	2	3	4	6
	1500	0	0	0	0	0	1	1	1	1	2	4	5
	1750	0	0	0	0	0	0	1	1	1	1	3	5
	2000	0	0	0	0	0	0	1	1	1	1	3	4

Table C.8 *Continued*

FIXTURE WIRES

Type	Conductor Size (AWG/ kcmil)	Metric Designator (Trade Size)					
		16 (½)	21 (¾)	27 (1)	35 (1¼)	41 (1½)	53 (2)
FFH-2, RFH-2, RFHH-3	18	8	15	24	42	57	94
	16	7	12	20	35	48	79
SF-2, SFF-2	18	11	19	31	53	72	118
	16	9	15	25	44	59	98
	14	7	12	20	35	48	79
SF-1, SFF-1	18	19	33	54	94	127	209
RFH-1, RFHH-2, TF, TFF, XF, XFF	18	14	25	40	69	94	155
RFHH-2, TF, TFF, XF, XFF	16	11	20	32	56	76	125
XF, XFF	14	9	15	25	44	59	98
TFN, TFFN	18	23	40	64	111	150	248
	16	17	30	49	84	115	189
PF, PFF, PGF, PGFF, PAF, PTF, PTFF, PAFF	18	21	38	61	105	143	235
	16	16	29	47	81	110	181
	14	12	22	35	61	83	136
HF, HFF, ZF, ZFF, ZHF	18	28	48	79	135	184	303
	16	20	36	58	100	136	223
	14	15	26	42	73	100	164
KF-2, KFF-2	18	40	71	114	197	267	439
	16	28	50	80	138	188	310
	14	19	34	55	95	129	213
	12	13	23	38	65	89	146
	10	9	15	25	44	59	98
KF-1, KFF-1	18	48	84	136	235	318	524
	16	34	59	96	165	224	368
	14	23	40	64	111	150	248
	12	15	26	42	73	100	164
	10	10	17	28	48	65	107
XF, XFF	12	5	8	13	23	32	52
	10	3	6	10	18	25	41

Notes:

1. This table is for concentric stranded conductors only. For compact stranded conductors, Table C.8(A) should be used.

2. Two-hour fire-rated RHH cable has ceramifiable insulation which has much larger diameters than other RHH wires.

Consult manufacturer's conduit fill tables.

*Types RHH, RHW, and RHW-2 without outer covering.

Table C.8(A) Maximum Number of Compact Conductors in Rigid Metal Conduit (RMC)
(Based on Table 1, Chapter 9)

COMPACT CONDUCTORS

	Conductor Size (AWG/	Metric Designator (Trade Size)											
Type	kcmil)	16 (½)	21 (¾)	27 (1)	35 (1¼)	41 (1½)	53 (2)	63 (2½)	78 (3)	91 (3½)	103 (4)	129 (5)	155 (6)
THW,	8	2	4	7	12	16	26	38	59	78	101	158	228
THW-2,	6	1	3	5	9	12	20	29	45	60	78	122	176
THHW	4	1	2	4	7	9	15	22	34	45	58	91	132
	2	1	1	3	5	7	11	16	25	33	43	67	97
	1	1	1	1	3	5	8	11	17	23	30	47	68
	1/0	1	1	1	3	4	7	10	15	20	26	41	59
	2/0	0	1	1	2	3	6	8	13	17	22	34	50
	3/0	0	1	1	1	3	5	7	11	14	19	29	42
	4/0	0	1	1	1	2	4	6	9	12	15	24	35
	250	0	0	1	1	1	3	4	7	9	12	19	28
	300	0	0	1	1	1	3	4	6	8	11	17	24
	350	0	0	1	1	1	2	3	5	7	9	15	22
	400	0	0	1	1	1	1	3	5	7	8	13	20
	500	0	0	0	1	1	1	3	4	5	7	11	17
	600	0	0	0	1	1	1	1	3	4	6	9	13
	700	0	0	0	1	1	1	1	3	4	5	8	12
	750	0	0	0	0	1	1	1	3	4	5	7	11
	900	0	0	0	0	1	1	1	2	3	4	6	9
	1000	0	0	0	0	1	1	1	1	3	4	6	9
THHN,	8	—	—	—	—	—	—	—	—	—	—	—	—
THWN,	6	2	5	8	13	18	30	43	66	88	114	179	258
THWN-2	4	1	3	5	8	11	18	26	41	55	70	110	159
	2	1	1	3	6	8	13	19	29	39	50	79	114
	1	1	1	2	4	6	10	14	22	29	38	60	86
	1/0	1	1	1	4	5	8	12	19	25	32	51	73
	2/0	1	1	1	3	4	7	10	15	21	26	42	60
	3/0	0	1	1	2	3	6	8	13	17	22	35	51
	4/0	0	1	1	1	3	5	7	10	14	18	29	42
	250	0	1	1	1	2	4	5	8	11	14	23	33
	300	0	0	1	1	1	3	4	7	10	12	20	28
	350	0	0	1	1	1	3	4	6	8	11	17	25
	400	0	0	1	1	1	2	3	5	7	10	15	22
	500	0	0	0	1	1	1	3	5	6	8	13	19
	600	0	0	0	1	1	1	2	4	5	6	10	15
	700	0	0	0	1	1	1	1	3	4	6	9	13
	750	0	0	0	1	1	1	1	3	4	5	9	13
	900	0	0	0	0	1	1	1	2	3	4	6	9
	1000	0	0	0	0	1	1	1	2	3	4	6	9
XHHW,	8	3	5	9	15	21	34	49	76	101	130	205	296
XHHW-2	6	2	4	6	11	15	25	36	56	75	97	152	220
	4	1	3	5	8	11	18	26	41	55	70	110	159
	2	1	1	3	6	8	13	19	29	39	50	79	114
	1	1	1	2	4	6	10	14	22	29	38	60	86
	1/0	1	1	1	4	5	8	12	19	25	32	51	73
	2/0	1	1	1	3	4	7	10	16	21	27	43	62
	3/0	0	1	1	2	3	6	8	13	17	22	35	51
	4/0	0	1	1	1	3	5	7	11	14	19	29	42
	250	0	1	1	1	2	4	5	8	11	15	23	34
	300	0	0	1	1	1	3	5	7	10	13	20	29
	350	0	0	1	1	1	3	4	6	9	11	18	25
	400	0	0	1	1	1	2	4	6	8	10	16	23
	500	0	0	0	1	1	1	3	5	6	8	13	19
	600	0	0	0	1	1	1	2	4	5	7	10	15
	700	0	0	0	1	1	1	1	3	4	6	9	13
	750	0	0	0	1	1	1	1	3	4	5	8	12
	900	0	0	0	0	1	1	2	2	3	5	7	10
	1000	0	0	0	0	1	1	1	2	3	4	7	10

Definition: *Compact stranding* is the result of a manufacturing process where the standard conductor is compressed to the extent that the interstices (voids between strand wires) are virtually eliminated.

Table C.9 Maximum Number of Conductors or Fixture Wires in Rigid PVC Conduit, Schedule 80 (*Based on Table 1, Chapter 9*)

CONDUCTORS

Type	Conductor Size (AWG/ kcmil)	Metric Designator (Trade Size)											
		16 (½)	21 (¾)	27 (1)	35 (1¼)	41 (1½)	53 (2)	63 (2½)	78 (3)	91 (3½)	103 (4)	129 (5)	155 (6)
RHH, RHW, RHW-2	14	3	5	9	17	23	39	56	88	118	153	243	349
	12	2	4	7	14	19	32	46	73	98	127	202	290
	10	1	3	6	11	15	26	37	59	79	103	163	234
	8	1	1	3	6	8	13	19	31	41	54	85	122
	6	1	1	2	4	6	11	16	24	33	43	68	98
	4	1	1	1	3	5	8	12	19	26	33	53	77
	3	0	1	1	3	4	7	11	17	23	29	47	67
	2	0	1	1	3	4	6	9	14	20	25	41	58
	1	0	1	1	1	2	4	6	9	13	17	27	38
	1/0	0	0	1	1	1	3	5	8	11	15	23	33
	2/0	0	0	1	1	1	3	4	7	10	13	20	29
	3/0	0	0	1	1	1	3	4	6	8	11	17	25
	4/0	0	0	0	1	1	2	3	5	7	9	15	21
	250	0	0	0	1	1	1	2	4	5	7	11	16
	300	0	0	0	1	1	1	2	3	5	6	10	14
	350	0	0	0	1	1	1	1	3	4	5	9	13
	400	0	0	0	0	1	1	1	3	4	5	8	12
	500	0	0	0	0	1	1	1	2	3	4	7	10
	600	0	0	0	0	0	1	1	1	3	3	6	8
	700	0	0	0	0	0	1	1	1	2	3	5	7
	750	0	0	0	0	1	1	1	1	2	3	5	7
	800	0	0	0	0	0	1	1	1	2	3	4	7
	1000	0	0	0	0	0	1	1	1	1	2	4	5
	1250	0	0	0	0	0	0	1	1	1	1	3	4
	1500	0	0	0	0	0	0	1	1	1	1	2	4
	1750	0	0	0	0	0	0	0	1	1	1	2	3
	2000	0	0	0	0	0	0	0	1	1	1	1	3
TW	14	6	11	20	35	49	82	118	185	250	324	514	736
	12	5	9	15	27	38	63	91	142	192	248	394	565
	10	3	6	11	20	28	47	67	106	143	185	294	421
	8	1	3	6	11	15	26	37	59	79	103	163	234
RHH*, RHW*, RHW-2*, THHW, THW, THW-2	14	4	8	13	23	32	55	79	123	166	215	341	490
RHH*, RHW*, RHW-2*, THHW, THW	12	3	6	10	19	26	44	63	99	133	173	274	394
	10	2	5	8	15	20	34	49	77	104	135	214	307
RHH*, RHW*, RHW-2*, THHW, THW, THW-2	8	1	3	5	9	12	20	29	46	62	81	128	184

(Continues)

Table C.9 *Continued*

CONDUCTORS

Type	Conductor Size (AWG/ kcmil)	Metric Designator (Trade Size)											
		16 (½)	21 (¾)	27 (1)	35 (1¼)	41 (1½)	53 (2)	63 (2½)	78 (3)	91 (3½)	103 (4)	129 (5)	155 (6)
RHH*, RHW*, RHW-2*, TW, THW, THHW, THW-2	6	1	1	3	7	9	16	22	35	48	62	98	141
	4	1	1	3	5	7	12	17	26	35	46	73	105
	3	1	1	2	4	6	10	14	22	30	39	63	90
	2	1	1	1	3	5	8	12	19	26	33	53	77
	1	0	1	1	2	3	6	8	13	18	23	37	54
	1/0	0	1	1	1	3	5	7	11	15	20	32	46
	2/0	0	1	1	1	2	4	6	10	13	17	27	39
	3/0	0	0	1	1	1	3	5	8	11	14	23	33
	4/0	0	0	1	1	1	3	4	7	9	12	19	27
	250	0	0	0	1	1	2	3	5	7	9	15	22
	300	0	0	0	1	1	1	3	5	6	8	13	19
	350	0	0	0	1	1	1	2	4	6	7	12	17
	400	0	0	0	1	1	1	2	4	5	7	10	15
	500	0	0	0	1	1	1	1	3	4	5	9	13
	600	0	0	0	0	1	1	1	2	3	4	7	10
	700	0	0	0	0	1	1	1	2	3	4	6	9
	750	0	0	0	0	0	1	1	1	3	4	6	8
	800	0	0	0	0	0	1	1	1	3	3	6	8
	900	0	0	0	0	0	1	1	1	2	3	5	7
	1000	0	0	0	0	0	1	1	1	2	3	5	7
	1250	0	0	0	0	0	1	1	1	1	2	4	5
	1500	0	0	0	0	0	0	1	1	1	1	3	4
	1750	0	0	0	0	0	0	1	1	1	1	3	4
	2000	0	0	0	0	0	0	0	1	1	1	2	3
THHN, THWN, THWN-2	14	9	17	28	51	70	118	170	265	358	464	736	1055
	12	6	12	20	37	51	86	124	193	261	338	537	770
	10	4	7	13	23	32	54	78	122	164	213	338	485
	8	2	4	7	13	18	31	45	70	95	123	195	279
	6	1	3	5	9	13	22	32	51	68	89	141	202
	4	1	1	3	6	8	14	20	31	42	54	86	124
	3	1	1	3	5	7	12	17	26	35	46	73	105
	2	1	1	2	4	6	10	14	22	30	39	61	88
	1	0	1	1	3	4	7	10	16	22	29	45	65
	1/0	0	1	1	2	3	6	9	14	18	24	38	55
	2/0	0	1	1	1	3	5	7	11	15	20	32	46
	3/0	0	1	1	1	2	4	6	9	13	17	26	38
	4/0	0	0	1	1	1	3	5	8	10	14	22	31
	250	0	0	1	1	1	3	4	6	8	11	18	25
	300	0	0	0	1	1	2	3	5	7	9	15	22
	350	0	0	0	1	1	1	3	5	6	8	13	19
	400	0	0	0	1	1	1	3	4	6	7	12	17
	500	0	0	0	1	1	1	2	3	5	6	10	14
	600	0	0	0	0	1	1	1	3	4	5	8	12
	700	0	0	0	0	1	1	1	2	3	4	7	10
	750	0	0	0	0	1	1	1	2	3	4	7	9
	800	0	0	0	0	1	1	1	2	3	4	6	9
	900	0	0	0	0	0	1	1	1	3	3	6	8
	1000	0	0	0	0	0	1	1	1	2	3	5	7

Table C.9 *Continued*

		CONDUCTORS											
		Metric Designator (Trade Size)											
Type	**Conductor Size (AWG/ kcmil)**	**16** **(½)**	**21** **(¾)**	**27** **(1)**	**35** **(1¼)**	**41** **(1½)**	**53** **(2)**	**63** **(2½)**	**78** **(3)**	**91** **(3½)**	**103** **(4)**	**129** **(5)**	**155** **(6)**
FEP, FEPB,	14	8	16	27	49	68	115	164	257	347	450	714	1024
PFA, PFAH,	12	6	12	20	36	50	84	120	188	253	328	521	747
TFE	10	4	8	14	26	36	60	86	135	182	235	374	536
	8	2	5	8	15	20	34	49	77	104	135	214	307
	6	1	3	6	10	14	24	35	55	74	96	152	218
	4	1	2	4	7	10	17	24	38	52	67	106	153
	3	1	1	3	6	8	14	20	32	43	56	89	127
	2	1	1	3	5	7	12	17	26	35	46	73	105
PFA, PFAH, TFE	1	1	1	1	3	5	8	11	18	25	32	51	73
PFA, PFAH,	1/0	0	1	1	3	4	7	10	15	20	27	42	61
TFE, Z	2/0	0	1	1	2	3	5	8	12	17	22	35	50
	3/0	0	1	1	1	2	4	6	10	14	18	29	41
	4/0	0	0	1	1	1	4	5	8	11	15	24	34
Z	14	10	19	33	59	82	138	198	310	418	542	860	1233
	12	7	14	23	42	58	98	141	220	297	385	610	875
	10	4	8	14	26	36	60	86	135	182	235	374	536
	8	3	5	9	16	22	38	54	85	115	149	236	339
	6	2	4	6	11	16	26	38	60	81	104	166	238
	4	1	2	4	8	11	18	26	41	55	72	114	164
	3	1	2	3	5	8	13	19	30	40	52	83	119
	2	1	1	2	5	6	11	16	25	33	43	69	99
	1	0	1	2	4	5	9	13	20	27	35	56	80
XHH, XHHW,	14	6	11	20	35	49	82	118	185	250	324	514	736
XHHW-2, ZW	12	5	9	15	27	38	63	91	142	192	248	394	565
	10	3	6	11	20	28	47	67	106	143	185	294	421
	8	1	3	6	11	15	26	37	59	79	103	163	234
	6	1	2	4	8	11	19	28	43	59	76	121	173
	4	1	1	3	6	8	14	20	31	42	55	87	125
	3	1	1	3	5	7	12	17	26	36	47	74	106
	2	1	1	2	4	6	10	14	22	30	39	62	89
XHH, XHHW,	1	0	1	1	3	4	7	10	16	22	29	46	66
XHHW-2	1/0	0	1	1	2	3	6	9	14	19	24	39	56
	2/0	0	1	1	1	3	5	7	11	16	20	32	46
	3/0	0	1	1	1	2	4	6	9	13	17	27	38
	4/0	0	0	1	1	1	3	5	8	11	14	22	32
	250	0	0	1	1	1	3	4	6	9	11	18	26
	300	0	0	1	1	1	2	3	5	7	10	15	22
	350	0	0	0	1	1	1	3	5	6	8	14	20
	400	0	0	0	1	1	1	3	4	6	7	12	17
	500	0	0	0	1	1	1	2	3	5	6	10	14
	600	0	0	0	0	1	1	1	3	4	5	8	11
	700	0	0	0	0	1	1	1	2	3	4	7	10
	750	0	0	0	0	1	1	1	2	3	4	6	9
	800	0	0	0	0	1	1	1	1	3	4	6	9
	900	0	0	0	0	0	1	1	—	3	3	5	8
	1000	0	0	0	0	0	1	1	1	2	3	5	7
	1250	0	0	0	0	0	1	1	1	1	2	4	6
	1500	0	0	0	0	0	0	1	1	1	1	3	5
	1750	0	0	0	0	0	0	1	1	1	1	3	4
	2000	0	0	0	0	0	0	1	1	1	1	2	4

(Continues)

Table C.9 *Continued*

FIXTURE WIRES

Type	Conductor Size (AWG/ kcmil)	Metric Designator (Trade Size)					
		16 (½)	21 (¾)	27 (1)	35 (1¼)	41 (1½)	53 (2)
FFH-2, RFH-2, RFHH-3	18	6	11	19	34	47	79
	16	5	9	16	28	39	67
SF-2, SFF-2	18	7	14	24	43	59	100
	16	6	11	20	35	49	82
	14	5	9	16	28	39	67
SF-1, SFF-1	18	13	25	42	76	105	177
RFH-1, RFHH-2, TF, TFF, XF, XFF	18	10	18	31	56	77	130
RFHH-2, TF, TFF, XF, XFF	16	8	15	25	45	62	105
XF, XFF	14	6	11	20	35	49	82
TFN, TFFN	18	16	29	50	90	124	209
	16	12	22	38	68	95	159
PF, PFF, PGF, PGFF, PAF, PTF, PTFF, PAFF	18	15	28	47	85	118	198
	16	11	22	36	66	91	153
	14	8	16	27	49	68	115
HF, HFF, ZF, ZFF, ZHF	18	19	36	61	110	152	255
	16	14	27	45	81	112	188
	14	10	19	33	59	82	138
KF-2, KFF-2	18	28	53	88	159	220	371
	16	19	37	62	112	155	261
	14	13	25	43	77	107	179
	12	9	17	29	53	73	123
	10	6	11	20	35	49	82
KF-1, KFF-1	18	33	63	106	190	263	442
	16	23	44	74	133	185	310
	14	16	29	50	90	124	209
	12	10	19	33	59	82	138
	10	7	13	21	39	54	90
XF, XFF	12	3	6	10	19	26	44
	10	2	5	8	15	20	34

Notes:

1. This table is for concentric stranded conductors only. For compact stranded conductors, Table C.9(A) should be used.

2. Two-hour fire-rated RHH cable has ceramifiable insulation which has much larger diameters than other RHH wires.

Consult manufacturer's conduit fill tables.

*Types RHH, RHW, and RHW-2 without outer covering.

Table C.9(A) Maximum Number of Compact Conductors in Rigid PVC Conduit, Schedule 80 (*Based on Table 1, Chapter 9*)

		COMPACT CONDUCTORS											
		Metric Designator (Trade Size)											
Type	Conductor Size (AWG/ kcmil)	16 (½)	21 (¾)	27 (1)	35 (1¼)	41 (1½)	53 (2)	63 (2½)	78 (3)	91 (3½)	103 (4)	129 (5)	155 (6)
THW, THW-2, THHW	8	1	3	5	9	13	22	32	50	68	88	140	200
	6	1	2	4	7	10	17	25	39	52	68	108	155
	4	1	1	3	5	7	13	18	29	39	51	81	116
	2	1	1	1	4	5	9	13	21	29	37	60	85
	1	0	1	1	3	4	6	9	15	20	26	42	60
	1/0	0	1	1	2	3	6	8	13	17	23	36	52
	2/0	0	1	1	1	3	5	7	11	15	19	30	44
	3/0	0	0	1	1	2	4	6	9	12	16	26	37
	4/0	0	0	1	1	1	3	5	8	10	13	22	31
	250	0	0	1	1	1	2	4	6	8	11	17	25
	300	0	0	0	1	1	2	3	5	7	9	15	21
	350	0	0	0	1	1	1	3	5	6	8	13	19
	400	0	0	0	1	1	1	3	4	6	7	12	17
	500	0	0	0	1	1	1	2	3	5	6	10	14
	600	0	0	0	0	1	1	1	3	4	5	8	12
	700	0	0	0	0	1	1	1	2	3	4	7	10
	750	0	0	0	0	1	1	1	2	3	4	7	10
	900	0	0	0	0	0	1	1	2	3	4	6	8
	1000	0	0	0	0	0	1	1	1	2	3	5	8
THHN, THWN, THWN-2	8	—	—	—	—	—	—	—	—	—	—	—	—
	6	1	3	6	11	15	25	36	57	77	99	158	226
	4	1	1	3	6	9	15	22	35	47	61	98	140
	2	1	1	2	5	6	11	16	25	34	44	70	100
	1	1	1	1	3	5	8	12	19	25	33	53	75
	1/0	0	1	1	3	4	7	10	16	22	28	45	64
	2/0	0	1	1	2	3	6	8	13	18	23	37	53
	3/0	0	1	1	1	3	5	7	11	15	19	31	44
	4/0	0	0	1	1	2	4	6	9	12	16	25	37
	250	0	0	1	1	1	3	4	7	10	12	20	29
	300	0	0	1	1	1	3	4	6	8	11	17	25
	350	0	0	0	1	1	2	3	5	7	9	15	22
	400	0	0	0	1	1	1	3	5	6	8	13	19
	500	0	0	0	1	1	1	2	4	5	7	11	16
	600	0	0	0	1	1	1	1	3	4	6	9	13
	700	0	0	0	0	1	1	1	3	4	5	8	12
	750	0	0	0	0	1	1	1	3	4	5	8	11
	900	0	0	0	0	0	1	1	2	3	4	6	8
	1000	0	0	0	0	0	1	1	1	3	3	5	8
XHHW, XHHW-2	8	1	4	7	12	17	29	42	65	88	114	181	260
	6	1	3	5	9	13	21	31	48	65	85	134	193
	4	1	1	3	6	9	15	22	35	47	61	98	140
	2	1	1	2	5	6	11	16	25	34	44	70	100
	1	1	1	1	3	5	8	12	19	25	33	53	75
	1/0	0	1	1	3	4	7	10	16	22	28	45	64
	2/0	0	1	1	2	3	6	8	13	18	24	38	54
	3/0	0	1	1	1	3	5	7	11	15	19	31	44
	4/0	0	0	1	1	2	4	6	9	12	16	26	37
	250	0	0	1	1	1	3	5	7	10	13	21	30
	300	0	0	1	1	1	3	4	6	8	11	17	25
	350	0	0	1	1	1	2	3	5	7	10	15	22
	400	0	0	0	1	1	1	3	5	7	9	14	20
	500	0	0	0	1	1	1	2	4	5	7	11	17
	600	0	0	0	1	1	1	1	3	4	6	9	13
	700	0	0	0	0	1	1	1	3	4	5	8	12
	750	0	0	0	0	1	1	1	2	3	5	7	11
	900	0	0	0	0	1	1	1	2	3	4	6	8
	1000	0	0	0	0	0	1	1	1	3	3	6	8

Definition: *Compact stranding* is the result of a manufacturing process where the standard conductor is compressed to the extent that the interstices (voids between strand wires) are virtually eliminated.

Table C.10 Maximum Number of Conductors or Fixture Wires in Rigid PVC Conduit, Schedule 40 and HDPE Conduit (*Based on Table 1, Chapter 9*)

CONDUCTORS

Type	Conductor Size (AWG/ kcmil)	Metric Designator (Trade Size)											
		16 (½)	21 (¾)	27 (1)	35 (1¼)	41 (1½)	53 (2)	63 (2½)	78 (3)	91 (3½)	103 (4)	129 (5)	155 (6)
RHH, RHW, RHW-2	14	4	7	11	20	27	45	64	99	133	171	269	390
	12	3	5	9	16	22	37	53	82	110	142	224	323
	10	2	4	7	13	18	30	43	66	89	115	181	261
	8	1	2	4	7	9	15	22	35	46	60	94	137
	6	1	1	3	5	7	12	18	28	37	48	76	109
	4	1	1	2	4	6	10	14	22	29	37	59	85
	3	1	1	1	4	5	8	12	19	25	33	52	75
	2	1	1	1	3	4	7	10	16	22	28	45	65
	1	0	1	1	1	3	5	7	11	14	19	29	43
	1/0	0	1	1	1	2	4	6	9	13	16	26	37
	2/0	0	0	1	1	1	3	5	8	11	14	22	32
	3/0	0	0	1	1	1	3	4	7	9	12	19	28
	4/0	0	0	1	1	1	2	4	6	8	10	16	24
	250	0	0	0	1	1	1	3	4	6	8	12	18
	300	0	0	0	1	1	1	2	4	5	7	11	16
	350	0	0	0	1	1	1	2	3	5	6	10	14
	400	0	0	0	1	1	1	1	3	4	6	9	13
	500	0	0	0	0	1	1	1	3	4	5	8	11
	600	0	0	0	0	1	1	1	2	3	4	6	9
	700	0	0	0	0	0	1	1	1	3	3	6	8
	750	0	0	0	0	0	1	1	1	2	3	5	8
	800	0	0	0	0	0	1	1	1	2	3	5	7
	900	0	0	0	0	0	1	1	1	2	3	5	7
	1000	0	0	0	0	0	1	1	1	1	3	4	6
	1250	0	0	0	0	0	0	1	1	1	1	3	5
	1500	0	0	0	0	0	0	1	1	1	1	3	4
	1750	0	0	0	0	0	0	1	1	1	1	2	3
	2000	0	0	0	0	0	0	0	1	1	1	2	3
TW	14	8	14	24	42	57	94	135	209	280	361	568	822
	12	6	11	18	32	44	72	103	160	215	277	436	631
	10	4	8	13	24	32	54	77	119	160	206	325	470
	8	2	4	7	13	18	30	43	66	89	115	181	261
RHH*, RHW*, RHW-2*, THHW, THW, THW-2	14	5	9	16	28	38	63	90	139	186	240	378	546
RHH*, RHW*, RHW-2*, THHW, THW	12	4	8	12	22	30	50	72	112	150	193	304	439
	10	3	6	10	17	24	39	56	87	117	150	237	343
RHH*, RHW*, RHW-2*, THHW, THW, THW-2	8	1	3	6	10	14	23	33	52	70	90	142	205

Table C.10 *Continued*

CONDUCTORS

Type	Conductor Size (AWG/ kcmil)	Metric Designator (Trade Size)											
		16 (½)	21 (¾)	27 (1)	35 (1¼)	41 (1½)	53 (2)	63 (2½)	78 (3)	91 (3½)	103 (4)	129 (5)	155 (6)
RHH*,	6	1	2	4	8	11	18	26	40	53	69	109	157
RHW*,	4	1	1	3	6	8	13	19	30	40	51	81	117
RHW-2* TW,	3	1	1	3	5	7	11	16	25	34	44	69	100
THW, THHW,	2	1	1	2	4	6	10	14	22	29	37	59	85
THW-2	1	0	1	1	3	4	7	10	15	20	26	41	60
	1/0	0	1	1	2	3	6	8	13	17	22	35	51
	2/0	0	1	1	1	3	5	7	11	15	19	30	43
	3/0	0	1	1	1	2	4	6	9	12	16	25	36
	4/0	0	0	1	1	1	3	5	8	10	13	21	30
	250	0	0	1	1	1	3	4	6	8	11	17	25
	300	0	0	1	1	1	2	3	5	7	9	15	21
	350	0	0	0	1	1	1	3	5	6	8	13	19
	400	0	0	0	1	1	1	3	4	6	7	12	17
	500	0	0	0	1	1	1	2	3	5	6	10	14
	600	0	0	0	0	1	1	1	3	4	5	8	11
	700	0	0	0	0	1	1	1	2	3	4	7	10
	750	0	0	0	0	1	1	1	2	3	4	6	10
	800	0	0	0	0	1	1	1	2	3	4	6	9
	900	0	0	0	0	0	1	1	1	3	3	6	8
	1000	0	0	0	0	0	1	1	1	2	3	5	7
	1250	0	0	0	0	0	1	1	1	1	2	4	6
	1500	0	0	0	0	0	1	1	1	1	1	3	5
	1750	0	0	0	0	0	0	1	1	1	1	3	4
	2000	0	0	0	0	0	0	1	1	1	1	3	4
THHN,	14	11	21	34	60	82	135	193	299	401	517	815	1178
THWN,	12	8	15	25	43	59	99	141	218	293	377	594	859
THWN-2	10	5	9	15	27	37	62	89	137	184	238	374	541
	8	3	5	9	16	21	36	51	79	106	137	216	312
	6	1	4	6	11	15	26	37	57	77	99	156	225
	4	1	2	4	7	9	16	22	35	47	61	96	138
	3	1	1	3	6	8	13	19	30	40	51	81	117
	2	1	1	3	5	7	11	16	25	33	43	68	98
	1	1	1	1	3	5	8	12	18	25	32	50	73
	1/0	1	1	1	3	4	7	10	15	21	27	42	61
	2/0	0	1	1	2	3	6	8	13	17	22	35	51
	3/0	0	1	1	1	3	5	7	11	14	18	29	42
	4/0	0	1	1	1	2	4	6	9	12	15	24	35
	250	0	0	1	1	1	3	4	7	10	12	20	28
	300	0	0	1	1	1	3	4	6	8	11	17	24
	350	0	0	1	1	1	2	3	5	7	9	15	21
	400	0	0	0	1	1	1	3	5	6	8	13	19
	500	0	0	0	1	1	1	2	4	5	7	11	16
	600	0	0	0	1	1	1	1	3	4	5	9	13
	700	0	0	0	0	1	1	1	3	4	5	8	11
	750	0	0	0	0	1	1	1	2	3	4	7	11
	800	0	0	0	0	1	1	1	2	3	4	7	10
	900	0	0	0	0	1	1	1	2	3	4	6	9
	1000	0	0	0	0	0	1	1	1	3	3	6	8

(Continues)

Table C.10 *Continued*

CONDUCTORS

Type	Conductor Size (AWG/ kcmil)	Metric Designator (Trade Size)											
		16 (½)	21 (¾)	27 (1)	35 (1¼)	41 (1½)	53 (2)	63 (2½)	78 (3)	91 (3½)	103 (4)	129 (5)	155 (6)
FEP, FEPB, PFA, PFAH, TFE	14	11	20	33	58	79	131	188	290	389	502	790	1142
	12	8	15	24	42	58	96	137	212	284	366	577	834
	10	6	10	17	30	41	69	98	152	204	263	414	598
	8	3	6	10	17	24	39	56	87	117	150	237	343
	6	2	4	7	12	17	28	40	62	83	107	169	244
	4	1	3	5	8	12	19	28	43	58	75	118	170
	3	1	2	4	7	10	16	23	36	48	62	98	142
	2	1	1	3	6	8	13	19	30	40	51	81	117
PFA, PFAH, TFE	1	1	1	2	4	5	9	13	20	28	36	56	81
PFA, PFAH, TFE, Z	1/0	1	1	1	3	4	8	11	17	23	30	47	68
	2/0	0	1	1	3	4	6	9	14	19	24	39	56
	3/0	0	1	1	2	3	5	7	12	16	20	32	46
	4/0	0	1	1	1	2	4	6	9	13	16	26	38
Z	14	13	24	40	70	95	158	226	350	469	605	952	1376
	12	9	17	28	49	68	112	160	248	333	429	675	976
	10	6	10	17	30	41	69	98	152	204	263	414	598
	8	3	6	11	19	26	43	62	96	129	166	261	378
	6	2	4	7	13	18	30	43	67	90	116	184	265
	4	1	3	5	9	12	21	30	46	62	80	126	183
	3	1	2	4	6	9	15	22	34	45	58	92	133
	2	1	1	3	5	7	12	18	28	38	49	77	111
	1	1	1	2	4	6	10	14	23	30	39	62	90
XHH, XHHW, XHHW-2, ZW	14	8	14	24	42	57	94	135	209	280	361	568	822
	12	6	11	18	32	44	72	103	160	215	277	436	631
	10	4	8	13	24	32	54	77	119	160	206	325	470
	8	2	4	7	13	18	30	43	66	89	115	181	261
	6	1	3	5	10	13	22	32	49	66	85	134	193
	4	1	2	4	7	9	16	23	35	48	61	97	140
	3	1	1	3	6	8	13	19	30	40	52	82	118
	2	1	1	3	5	7	11	16	25	34	44	69	99
XHH, XHHW, XHHW-2	1	1	1	1	3	5	8	12	19	25	32	51	74
	1/0	1	1	1	3	4	7	10	16	21	27	43	62
	2/0	0	1	1	2	3	6	8	13	17	23	36	52
	3/0	0	1	1	1	3	5	7	11	14	19	30	43
	4/0	0	1	1	1	2	4	6	9	12	15	24	35
	250	0	0	1	1	1	3	5	7	10	13	20	29
	300	0	0	1	1	1	3	4	6	8	11	17	25
	350	0	0	1	1	1	2	3	5	7	9	15	22
	400	0	0	0	1	1	1	3	5	6	8	13	19
	500	0	0	0	1	1	1	2	4	5	7	11	16
	600	0	0	0	1	1	1	1	3	4	5	9	13
	700	0	0	0	0	1	1	1	3	4	5	8	11
	750	0	0	0	0	1	1	1	2	3	4	7	11
	800	0	0	0	0	1	1	1	2	3	4	7	10
	900	0	0	0	0	1	1	1	2	3	4	6	9
	1000	0	0	0	0	0	1	1	1	3	3	6	8
	1250	0	0	0	0	0	1	1	1	1	3	4	6
	1500	0	0	0	0	0	1	1	1	1	2	4	5
	1750	0	0	0	0	0	0	1	1	1	1	3	5
	2000	0	0	0	0	0	0	1	1	1	1	3	4

Table C.10 *Continued*

FIXTURE WIRES

Type	Conductor Size (AWG/ kcmil)	Metric Designator (Trade Size)					
		16 (½)	21 (¾)	27 (1)	35 (1¼)	41 (1½)	53 (2)
FFH-2, RFH-2, RFHH-3	18	8	14	23	40	54	90
	16	6	12	19	33	46	76
SF-2, SFF-2	18	10	17	29	50	69	114
	16	8	14	24	42	57	94
	14	6	12	19	33	46	76
SF-1, SFF-1	18	17	31	51	89	122	202
RFHH-2, TF, TFF, XF, XFF RFH-1,	18	13	23	38	66	90	149
RFHH-2, TF, TFF, XF, XFF	16	10	18	30	53	73	120
XF, XFF	14	8	14	24	42	57	94
TFN, TFFN	18	20	37	60	105	144	239
	16	16	28	46	80	110	183
PF, PFF, PGF, PGFF, PAF, PTF, PTFF, PAFF	18	19	35	57	100	137	227
	16	15	27	44	77	106	175
	14	11	20	33	58	79	131
HF, HFF, ZF, ZFF, ZHF	18	25	45	74	129	176	292
	16	18	33	54	95	130	216
	14	13	24	40	70	95	158
KF-2, KFF-2	18	36	65	107	187	256	424
	16	26	46	75	132	180	299
	14	17	31	52	90	124	205
	12	12	22	35	62	85	141
	10	8	14	24	42	57	94
KF-1, KFF-1	18	43	78	128	223	305	506
	16	30	55	90	157	214	355
	14	20	37	60	105	144	239
	12	13	24	40	70	95	158
	10	9	16	26	45	62	103
XF, XFF	12	4	8	12	22	30	50
	10	3	6	10	17	24	39

Notes:
1. This table is for concentric stranded conductors only. For compact stranded conductors, Table C.10(A) should be used.
2. Two-hour fire-rated RHH cable has ceramifiable insulation which has much larger diameters than other RHH wires.
Consult manufacturer's conduit fill tables.
*Types RHH, RHW, and RHW-2 without outer covering.

Table C.10(A) Maximum Number of Compact Conductors in Rigid PVC Conduit, Schedule 40 and HDPE Conduit (*Based on Table 1, Chapter 9*)

COMPACT CONDUCTORS

Type	Conductor Size (AWG/ kcmil)	Metric Designator (Trade Size)											
		16 (½)	21 (¾)	27 (1)	35 (1¼)	41 (1½)	53 (2)	63 (2½)	78 (3)	91 (3½)	103 (4)	129 (5)	155 (6)
THW, THW-2, THHW	8	1	4	6	11	15	26	37	57	76	98	155	224
	6	1	3	5	9	12	20	28	44	59	76	119	173
	4	1	1	3	6	9	15	21	33	44	57	89	129
	2	1	1	2	5	6	11	15	24	32	42	66	95
	1	1	1	1	3	4	7	11	17	23	29	46	67
	1/0	0	1	1	3	4	6	9	15	20	25	40	58
	2/0	0	1	1	2	3	5	8	12	16	21	34	49
	3/0	0	1	1	1	3	5	7	10	14	18	29	42
	4/0	0	1	1	1	2	4	5	9	12	15	24	35
	250	0	0	1	1	1	3	4	7	9	12	19	27
	300	0	0	1	1	1	2	4	6	8	10	16	24
	350	0	0	1	1	1	2	3	5	7	9	15	21
	400	0	0	0	1	1	1	3	5	6	8	13	19
	500	0	0	0	1	1	1	2	4	5	7	11	16
	600	0	0	0	1	1	1	1	3	4	5	9	13
	700	0	0	0	0	1	1	1	3	4	5	8	12
	750	0	0	0	0	1	1	1	2	3	5	7	11
	900	0	0	0	0	1	1	1	2	3	4	6	9
	1000	0	0	0	0	1	1	1	1	3	4	6	9
THHN, THWN, THWN-2	8	—	—	—	—	—	—	—	—	—	—	—	—
	6	2	4	7	13	17	29	41	64	86	111	175	253
	4	1	2	4	8	11	18	25	40	53	68	108	156
	2	1	1	3	5	8	13	18	28	38	49	77	112
	1	1	1	2	4	6	9	14	21	29	37	58	84
	1/0	1	1	1	3	5	8	12	18	24	31	49	72
	2/0	0	1	1	3	4	7	9	15	20	26	41	59
	3/0	0	1	1	2	3	5	8	12	17	22	34	50
	4/0	0	1	1	1	3	4	6	10	14	18	28	41
	250	0	0	1	1	1	3	5	8	11	14	22	32
	300	0	0	1	1	1	3	4	7	9	12	19	28
	350	0	0	1	1	1	3	4	6	8	10	17	24
	400	0	0	1	1	1	2	3	5	7	9	15	22
	500	0	0	0	1	1	1	3	4	6	8	13	18
	600	0	0	0	1	1	1	2	4	5	6	10	15
	700	0	0	0	1	1	1	1	3	4	5	9	13
	750	0	0	0	1	1	1	1	3	4	5	8	12
	900	0	0	0	0	1	1	1	2	3	4	6	9
	1000	0	0	0	0	1	1	1	2	3	4	6	9
XHHW, XHHW-2	8	3	5	8	14	20	33	47	73	99	127	200	290
	6	1	4	6	11	15	25	35	55	73	94	149	215
	4	1	2	4	8	11	18	25	40	53	68	108	156
	2	1	1	3	5	8	13	18	28	38	49	77	112
	1	1	1	2	4	6	9	14	21	29	37	58	84
	1/0	1	1	1	3	5	8	12	18	24	31	49	72
	2/0	1	1	1	3	4	7	10	15	20	26	42	60
	3/0	0	1	1	2	3	5	8	12	17	22	34	50
	4/0	0	1	1	1	3	5	7	10	14	18	29	42
	250	0	0	1	1	1	4	5	8	11	14	23	33
	300	0	0	1	1	1	3	4	7	9	12	19	28
	350	0	0	1	1	1	3	4	6	8	11	17	25
	400	0	0	1	1	1	2	3	5	7	10	15	22
	500	0	0	0	1	1	1	3	4	6	8	13	18
	600	0	0	0	1	1	1	2	4	5	6	10	15
	700	0	0	0	1	1	1	1	3	4	5	9	13
	750	0	0	0	1	1	1	1	3	4	5	8	12
	900	0	0	0	0	1	1	1	2	3	4	6	9
	1000	0	0	0	0	1	1	1	2	3	4	6	9

Definition: *Compact stranding* is the result of a manufacturing process where the standard conductor is compressed to the extent that the interstices (voids between strand wires) are virtually eliminated.

Table C.11 Maximum Number of Conductors or Fixture Wires in Type A, Rigid PVC Conduit (*Based on Table 1, Chapter 9*)

		CONDUCTORS									
		Metric Designator (Trade Size)									
Type	Conductor Size (AWG/ kcmil)	16 (½)	21 (¾)	27 (1)	35 (1¼)	41 (1½)	53 (2)	63 (2½)	78 (3)	91 (3½)	103 (4)
RHH, RHW, RHW-2	14	5	9	15	24	31	49	74	112	146	187
	12	4	7	12	20	26	41	61	93	121	155
	10	3	6	10	16	21	33	50	75	98	125
	8	1	3	5	8	11	17	26	39	51	65
	6	1	2	4	6	9	14	21	31	41	52
	4	1	1	3	5	7	11	16	24	32	41
	3	1	1	3	4	6	9	14	21	28	36
	2	1	1	2	4	5	8	12	18	24	31
	1	0	1	1	2	3	5	8	12	16	20
	1/0	0	1	1	2	3	5	7	10	14	18
	2/0	0	1	1	1	2	4	6	9	12	15
	3/0	0	1	1	1	1	3	5	8	10	13
	4/0	0	0	1	1	1	3	4	7	9	11
	250	0	0	1	1	1	1	3	5	7	8
	300	0	0	1	1	1	1	3	4	6	7
	350	0	0	0	1	1	1	2	4	5	7
	400	0	0	0	1	1	1	2	4	5	6
	500	0	0	0	1	1	1	1	3	4	5
	600	0	0	0	0	1	1	1	2	3	4
	700	0	0	0	0	1	1	1	2	3	4
	750	0	0	0	0	1	1	1	1	3	4
	800	0	0	0	0	1	1	1	1	3	3
	900	0	0	0	0	0	1	1	1	2	3
	1000	0	0	0	0	0	1	1	1	2	3
	1250	0	0	0	0	0	1	1	1	1	2
	1500	0	0	0	0	0	0	1	1	1	1
	1750	0	0	0	0	0	0	1	1	1	1
	2000	0	0	0	0	0	0	1	1	1	1
TW	14	11	18	31	51	67	105	157	235	307	395
	12	8	14	24	39	51	80	120	181	236	303
	10	6	10	18	29	38	60	89	135	176	226
	8	3	6	10	16	21	33	50	75	98	125
RHH*, RHW*, RHW-2*, THHW, THW, THW-2	14	7	12	20	34	44	70	104	157	204	262
RHH*, RHW*, RHW-2*, THHW, THW	12	6	10	16	27	35	56	84	126	164	211
	10	4	8	13	21	28	44	65	98	128	165
RHH*, RHW*, RHW-2*, THHW, THW, THW-2	8	2	4	8	12	16	26	39	59	77	98
RHH*, RHW*, RHW-2*, TW, THHW, THW, THW-2	6	1	3	6	9	13	20	30	45	59	75

(Continues)

Table C.11 *Continued*

CONDUCTORS

Type	Conductor Size (AWG/kcmil)	Metric Designator (Trade Size)									
		16 (½)	21 (¾)	27 (1)	35 (1¼)	41 (1½)	53 (2)	63 (2½)	78 (3)	91 (3½)	103 (4)
RHH*, RHW*, RHW-2*, TW, THW, THHW, THW-2	4	1	2	4	7	9	15	22	33	44	56
	3	1	1	4	6	8	13	19	29	37	48
	2	1	1	3	5	7	11	16	24	32	41
	1	1	1	1	3	5	7	11	17	22	29
	1/0	1	1	1	3	4	6	10	14	19	24
	2/0	0	1	1	2	3	5	8	12	16	21
	3/0	0	1	1	1	3	4	7	10	13	17
	4/0	0	1	1	1	2	4	6	9	11	14
	250	0	0	1	1	1	3	4	7	9	12
	300	0	0	1	1	1	2	4	6	8	10
	350	0	0	1	1	1	2	3	5	7	9
	400	0	0	1	1	1	1	3	5	6	8
	500	0	0	0	1	1	1	2	4	5	7
	600	0	0	0	1	1	1	1	3	4	5
	700	0	0	0	1	1	1	1	3	4	5
	750	0	0	0	1	1	1	1	3	3	4
	800	0	0	0	0	1	1	1	2	3	4
	900	0	0	0	0	1	1	1	2	3	4
	1000	0	0	0	0	1	1	1	1	3	3
	1250	0	0	0	0	0	1	1	1	1	3
	1500	0	0	0	0	0	1	1	1	1	2
	1750	0	0	0	0	0	0	1	1	1	1
	2000	0	0	0	0	0	0	1	1	1	1
THHN, THWN, THWN-2	14	16	27	44	73	96	150	225	338	441	566
	12	11	19	32	53	70	109	164	246	321	412
	10	7	12	20	33	44	69	103	155	202	260
	8	4	7	12	19	25	40	59	89	117	150
	6	3	5	8	14	18	28	43	64	84	108
	4	1	3	5	8	11	17	26	39	52	66
	3	1	2	4	7	9	15	22	33	44	56
	2	1	1	3	6	8	12	19	28	37	47
	1	1	1	2	4	6	9	14	21	27	35
	1/0	1	1	2	4	5	8	11	17	23	29
	2/0	1	1	1	3	4	6	10	14	19	24
	3/0	0	1	1	2	3	5	8	12	16	20
	4/0	0	1	1	1	3	4	6	10	13	17
	250	0	1	1	1	2	3	5	8	10	14
	300	0	0	1	1	1	3	4	7	9	12
	350	0	0	1	1	1	2	4	6	8	10
	400	0	0	1	1	1	2	3	5	7	9
	500	0	0	1	1	1	1	3	4	6	7
	600	0	0	0	1	1	1	2	3	5	6
	700	0	0	0	1	1	1	1	3	4	5
	750	0	0	0	1	1	1	1	3	4	5
	800	0	0	0	1	1	1	1	3	4	5
	900	0	0	0	0	1	1	1	2	3	4
	1000	0	0	0	0	1	1	1	2	3	4

Table C.11 *Continued*

CONDUCTORS

Type	Conductor Size (AWG/ kcmil)	Metric Designator (Trade Size)									
		16 (½)	21 (¾)	27 (1)	35 (1¼)	41 (1½)	53 (2)	63 (2½)	78 (3)	91 (3½)	103 (4)
FEP, FEPB, PFA, PFAH, TFE	14	15	26	43	70	93	146	218	327	427	549
	12	11	19	31	51	68	106	159	239	312	400
	10	8	13	22	37	48	76	114	171	224	287
	8	4	8	13	21	28	44	65	98	128	165
	6	3	5	9	15	20	31	46	70	91	117
	4	1	4	6	10	14	21	32	49	64	82
	3	1	3	5	8	11	18	27	40	53	68
	2	1	2	4	7	9	15	22	33	44	56
PFA, PFAH, TFE	1	1	1	3	5	6	10	15	23	30	39
PFA, PFAH, TFE, Z	1/0	1	1	2	4	5	8	13	19	25	32
	2/0	1	1	1	3	4	7	10	16	21	27
	3/0	1	1	1	3	3	6	9	13	17	22
	4/0	0	1	1	2	3	5	7	11	14	18
Z	14	18	31	52	85	112	175	263	395	515	661
	12	13	22	37	60	79	124	186	280	365	469
	10	8	13	22	37	48	76	114	171	224	287
	8	5	8	14	23	30	48	72	108	141	181
	6	3	6	10	16	21	34	50	76	99	127
	4	2	4	7	11	15	23	35	52	68	88
	3	1	3	5	8	11	17	25	38	50	64
	2	1	2	4	7	9	14	21	32	41	53
	1	1	1	3	5	7	11	17	26	33	43
XHH, XHHW, XHHW-2, ZW	14	11	18	31	51	67	105	157	235	307	395
	12	8	14	24	39	51	80	120	181	236	303
	10	6	10	18	29	38	60	89	135	176	226
	8	3	6	10	16	21	33	50	75	98	125
	6	2	4	7	12	15	24	37	55	72	93
	4	1	3	5	8	11	18	26	40	52	67
	3	1	2	4	7	9	15	22	34	44	57
	2	1	1	3	6	8	12	19	28	37	48
XHH, XHHW, XHHW-2	1	1	1	3	4	6	9	14	21	28	35
	1/0	1	1	2	4	5	8	12	18	23	30
	2/0	1	1	1	3	4	6	10	15	19	25
	3/0	0	1	1	2	3	5	8	12	16	20
	4/0	0	1	1	1	3	4	7	10	13	17
	250	0	1	1	1	2	3	5	8	11	14
	300	0	0	1	1	1	3	5	7	9	12
	350	0	0	1	1	1	3	4	6	8	10
	400	0	0	1	1	1	2	3	5	7	9
	500	0	0	1	1	1	1	3	4	6	8
	600	0	0	0	1	1	1	2	3	5	6
	700	0	0	0	1	1	1	1	3	4	5
	750	0	0	0	1	1	1	1	3	4	5
	800	0	0	0	1	1	1	1	3	4	5
	900	0	0	0	0	1	1	1	2	3	4
	1000	0	0	0	0	1	1	1	2	3	4
	1250	0	0	0	0	0	1	1	1	2	3
	1500	0	0	0	0	0	1	1	1	1	2
	1750	0	0	0	0	0	1	1	1	1	2
	2000	0	0	0	0	0	0	1	1	1	1

(Continues)

Table C.11 *Continued*

FIXTURE WIRES

Type	Conductor Size (AWG/ kcmil)	Metric Designator (Trade Size)					
		16 (½)	21 (¾)	27 (1)	35 (1¼)	41 (1½)	53 (2)
FFH-2, RFH-2, RFHH-3	18	10	18	30	48	64	100
	16	9	15	25	41	54	85
SF-2, SFF-2	18	13	22	37	61	81	127
	16	11	18	31	51	67	105
	14	9	15	25	41	54	85
SF-1, SFF-1	18	23	40	66	108	143	224
RFH-1, RFHH-2, TF, TFF, XF, XFF	18	17	29	49	80	105	165
RFHH-2, TF, TFF, XF, XFF	16	14	24	39	65	85	134
XF, XFF	14	11	18	31	51	67	105
TFN, TFFN	18	28	47	79	128	169	265
	16	21	36	60	98	129	202
PF, PFF, PGF, PGFF, PAF, PTF, PTFF, PAFF	18	26	45	74	122	160	251
	16	20	34	58	94	124	194
	14	15	26	43	70	93	146
HF, HFF, ZF, ZFF, ZHF	18	34	58	96	157	206	324
	16	25	42	71	116	152	239
	14	18	31	52	85	112	175
KF-2, KFF-2	18	49	84	140	228	300	470
	16	35	59	98	160	211	331
	14	24	40	67	110	145	228
	12	16	28	46	76	100	157
	10	11	18	31	51	67	105
KF-1, KFF-1	18	59	100	167	272	357	561
	16	41	70	117	191	251	394
	14	28	47	79	128	169	265
	12	18	31	52	85	112	175
	10	12	20	34	55	73	115
XF, XFF	12	6	10	16	27	35	56
	10	4	8	13	21	28	44

Notes:
1. This table is for concentric stranded conductors only. For compact stranded conductors, Table C.11(A) should be used.
2. Two-hour fire-rated RHH cable has ceramifiable insulation which has much larger diameters than other RHH wires.
Consult manufacturer's conduit fill tables.
*Types RHH, RHW, and RWH-2 without outer covering.

Table C.11(A) Maximum Number of Compact Conductors in Type A, Rigid PVC Conduit
(*Based on Table 1, Chapter 9*)

COMPACT CONDUCTORS

Type	Conductor Size (AWG/ kcmil)	Metric Designator (Trade Size)									
		16 (½)	21 (¾)	27 (1)	35 (1¼)	41 (1½)	53 (2)	63 (2½)	78 (3)	91 (3½)	103 (4)
THW, THW-2, THHW	8	3	5	8	14	18	28	42	64	84	107
	6	2	4	6	10	14	22	33	49	65	83
	4	1	3	5	8	10	16	24	37	48	62
	2	1	1	3	6	7	12	18	27	36	46
	1	1	1	2	4	5	8	13	19	25	32
	1/0	1	1	1	3	4	7	11	16	21	28
	2/0	1	1	1	3	4	6	9	14	18	23
	3/0	0	1	1	2	3	5	8	12	15	20
	4/0	0	1	1	1	3	4	6	10	13	17
	250	0	1	1	1	1	3	5	8	10	13
	300	0	0	1	1	1	3	4	7	9	11
	350	0	0	1	1	1	2	4	6	8	10
	400	0	0	1	1	1	2	3	5	7	9
	500	0	0	1	1	1	1	3	4	6	8
	600	0	0	0	1	1	1	2	3	5	6
	700	0	0	0	1	1	1	1	3	4	5
	750	0	0	0	1	1	1	1	3	4	5
	900	0	0	0	0	1	1	2	2	3	4
	1000	0	0	0	0	1	1	1	2	3	4
THHN, THWN, THWN-2	8	—	—	—	—	—	—	—	—	—	—
	6	3	5	9	15	20	32	48	72	94	121
	4	1	3	6	9	12	20	30	45	58	75
	2	1	2	4	7	9	14	21	32	42	54
	1	1	1	3	5	7	10	16	24	31	40
	1/0	1	1	2	4	6	9	13	20	27	34
	2/0	1	1	1	3	5	7	11	17	22	28
	3/0	1	1	1	3	4	6	9	14	18	24
	4/0	0	1	1	2	3	5	8	11	15	19
	250	0	1	1	1	2	4	6	9	12	15
	300	0	1	1	1	1	3	5	8	10	13
	350	0	0	1	1	1	3	4	7	9	11
	400	0	0	1	1	1	2	4	6	8	10
	500	0	0	1	1	1	2	3	5	7	9
	600	0	0	0	1	1	1	3	4	5	7
	700	0	0	0	1	1	1	2	3	5	6
	750	0	0	0	1	1	1	2	3	4	6
	900	0	0	0	1	1	1	2	3	4	5
	1000	0	0	0	0	1	1	1	2	3	4
XHHW, XHHW-2	8	4	6	11	18	23	37	55	83	108	139
	6	3	5	8	13	17	27	41	62	80	103
	4	1	3	6	9	12	20	30	45	58	75
	2	1	2	4	7	9	14	21	32	42	54
	1	1	1	3	5	7	10	16	24	31	40
	1/0	1	1	2	4	6	9	13	20	27	34
	2/0	1	1	1	3	5	7	11	17	22	29
	3/0	1	1	1	3	4	6	9	14	18	24
	4/0	0	1	1	2	3	5	8	12	15	20
	250	0	1	1	1	2	4	6	9	12	16
	300	0	1	1	1	1	3	5	8	10	13
	350	0	0	1	1	1	3	5	7	9	12
	400	0	0	1	1	1	3	4	6	8	11
	500	0	0	1	1	1	2	3	5	7	9
	600	0	0	0	1	1	1	3	4	5	7
	700	0	0	0	1	1	1	2	3	5	6
	750	0	0	0	1	1	1	2	3	4	6
	900	0	0	0	1	1	1	2	3	4	5
	1000	0	0	0	0	1	1	1	2	3	4

Definition: *Compact stranding* is the result of a manufacturing process where the standard conductor is compressed to the extent that the interstices (voids between strand wires) are virtually eliminated.

Table C.12 Maximum Number of Conductors in Type EB, PVC Conduit (*Based on Table 1, Chapter 9*)

		CONDUCTORS					
		Metric Designator (Trade Size)					
Type	Conductor Size (AWG/ kcmil)	53 (2)	78 (3)	91 (3½)	103 (4)	129 (5)	155 (6)
RHH, RHW, RHW-2	14	53	119	155	197	303	430
	12	44	98	128	163	251	357
	10	35	79	104	132	203	288
	8	18	41	54	69	106	151
	6	15	33	43	55	85	121
	4	11	26	34	43	66	94
	3	10	23	30	38	58	83
	2	9	20	26	33	50	72
	1	6	13	17	21	33	47
	1/0	5	11	15	19	29	41
	2/0	4	10	13	16	25	36
	3/0	4	8	11	14	22	31
	4/0	3	7	9	12	18	26
	250	2	5	7	9	14	20
	300	1	5	6	8	12	17
	350	1	4	5	7	11	16
	400	1	4	5	6	10	14
	500	1	3	4	5	9	12
	600	1	3	3	4	7	10
	700	1	2	3	4	6	9
	750	1	2	3	4	6	9
	800	1	2	3	4	6	8
	900	1	1	2	3	5	7
	1000	1	1	2	3	5	7
	1250	1	1	1	2	3	5
	1500	0	1	1	1	3	4
	1750	0	1	1	1	3	4
	2000	0	1	1	1	2	3
TW	14	111	250	327	415	638	907
	12	85	192	251	319	490	696
	10	63	143	187	238	365	519
	8	35	79	104	132	203	288
RHH*, RHW*, RHW-2*, THHW, THW, THW-2	14	74	166	217	276	424	603
RHH*, RHW*, RHW-2*, THHW, THW	12	59	134	175	222	341	485
	10	46	104	136	173	266	378
RHH*, RHW*, RHW-2*, THHW, THW, THW-2	8	28	62	81	104	159	227

Table C.12 *Continued*

		CONDUCTORS					
		Metric Designator (Trade Size)					
Type	**Conductor Size (AWG/ kcmil)**	**53 (2)**	**78 (3)**	**91 (3½)**	**103 (4)**	**129 (5)**	**155 (6)**
RHH*, RHW*, RHW-2*, TW, THW, THHW, THW-2	6	21	48	62	79	122	173
	4	16	36	46	59	91	129
	3	13	30	40	51	78	111
	2	11	26	34	43	66	94
	1	8	18	24	30	46	66
	1/0	7	15	20	26	40	56
	2/0	6	13	17	22	34	48
	3/0	5	11	14	18	28	40
	4/0	4	9	12	15	24	34
	250	3	7	10	12	19	27
	300	3	6	8	11	17	24
	350	2	6	7	9	15	21
	400	2	5	7	8	13	19
	500	1	4	5	7	11	16
	600	1	3	4	6	9	13
	700	1	3	4	5	8	11
	750	1	3	4	5	7	11
	800	1	3	3	4	7	10
	900	1	2	3	4	6	9
	1000	1	2	3	4	6	8
	1250	1	1	2	3	4	6
	1500	1	1	1	2	4	6
	1750	1	1	1	2	3	5
	2000	0	1	1	1	3	4
THHN, THWN, THWN-2	14	159	359	468	595	915	1300
	12	116	262	342	434	667	948
	10	73	165	215	274	420	597
	8	42	95	124	158	242	344
	6	30	68	89	114	175	248
	4	19	42	55	70	107	153
	3	16	36	46	59	91	129
	2	13	30	39	50	76	109
	1	10	22	29	37	57	80
	1/0	8	18	24	31	48	68
	2/0	7	15	20	26	40	56
	3/0	5	13	17	21	33	47
	4/0	4	10	14	18	27	39
	250	4	8	11	14	22	31
	300	3	7	10	12	19	27
	350	3	6	8	11	17	24
	400	2	6	7	10	15	21
	500	1	5	6	8	12	18
	600	1	4	5	6	10	14
	700	1	3	4	6	9	12
	750	1	3	4	5	8	12
	800	1	3	4	5	8	11
	900	1	3	3	4	7	10
	1000	1	2	3	4	6	9

(Continues)

Table C.12 *Continued*

CONDUCTORS

Type	Conductor Size (AWG/ kcmil)	Metric Designator (Trade Size)					
		53 (2)	78 (3)	91 (3½)	103 (4)	129 (5)	155 (6)
FEP, FEPB, PFA, PFAH, TFE	14	155	348	454	578	888	1261
	12	113	254	332	422	648	920
	10	81	182	238	302	465	660
	8	46	104	136	173	266	378
	6	33	74	97	123	189	269
	4	23	52	68	86	132	188
	3	19	43	56	72	110	157
	2	16	36	46	59	91	129
PFA, PFAH, TFE	1	11	25	32	41	63	90
PFA, PFAH, TFE, Z	1/0	9	20	27	34	53	75
	2/0	7	17	22	28	43	62
	3/0	6	14	18	23	36	51
	4/0	5	11	15	19	29	42
Z	14	186	419	547	696	1069	1519
	12	132	297	388	494	759	1078
	10	81	182	238	302	465	660
	8	51	115	150	191	294	417
	6	36	81	105	134	206	293
	4	24	55	72	92	142	201
	3	18	40	53	67	104	147
	2	15	34	44	56	86	122
	1	12	27	36	45	70	99
XHH, XHHW, XHHW-2, ZW	14	111	250	327	415	638	907
	12	85	192	251	319	490	696
	10	63	143	187	238	365	519
	8	35	79	104	132	203	288
	6	26	59	77	98	150	213
	4	19	42	56	71	109	155
	3	16	36	47	60	92	131
	2	13	30	39	50	77	110
XHH, XHHW, XHHW-2	1	10	22	29	37	58	82
	1/0	8	19	25	31	48	69
	2/0	7	16	20	26	40	57
	3/0	6	13	17	22	33	47
	4/0	5	11	14	18	27	39
	250	4	9	11	15	22	32
	300	3	7	10	12	19	28
	350	3	6	9	11	17	24
	400	2	6	8	10	15	22
	500	1	5	6	8	12	18
	600	1	4	5	6	10	14
	700	1	3	4	6	9	12
	750	1	3	4	5	8	12
	800	1	3	4	5	8	11
	900	1	3	3	4	7	10
	1000	1	2	3	4	6	9
	1250	1	1	2	3	5	7
	1500	1	1	1	3	4	6
	1750	1	1	1	2	4	5
	2000	0	1	1	1	3	5

Notes:
1. This table is for concentric stranded conductors only. For compact stranded conductors, Table C.12(A) should be used.
2. Two-hour fire-rated RHH cable has ceramifiable insulation which has much larger diameters than other RHH wires.
Consult manufacturer's conduit fill tables.
*Types RHH, RHW, and RHW-2 without outer covering.

Table C.12(A) Maximum Number of Compact Conductors in Type EB, PVC Conduit
(*Based on Table 1, Chapter 9*)

COMPACT CONDUCTORS

Type	Conductor Size (AWG/ kcmil)	Metric Designator (Trade Size)					
		53 (2)	78 (3)	91 (3½)	103 (4)	129 (5)	155 (6)
THW, THW-2, THHW	8	30	68	89	113	174	247
	6	23	52	69	87	134	191
	4	17	39	51	65	100	143
	2	13	29	38	48	74	105
	1	9	20	26	34	52	74
	1/0	8	17	23	29	45	64
	2/0	6	15	19	24	38	54
	3/0	5	12	16	21	32	46
	4/0	4	10	14	17	27	38
	250	3	8	11	14	21	30
	300	3	7	9	12	19	26
	350	3	6	8	11	17	24
	400	2	6	7	10	15	21
	500	1	5	6	8	12	18
	600	1	4	5	6	10	14
	700	1	3	4	6	9	13
	750	1	3	4	5	8	12
	900	1	3	4	5	7	10
	1000	1	2	3	4	7	9
THHN, THWN, THWN-2	8	—	—	—	—	—	—
	6	34	77	100	128	196	279
	4	21	47	62	79	121	172
	2	15	34	44	57	87	124
	1	11	25	33	42	65	93
	1/0	9	22	28	36	56	79
	2/0	8	18	23	30	46	65
	3/0	6	15	20	25	38	55
	4/0	5	12	16	20	32	45
	250	4	10	13	16	25	35
	300	4	8	11	14	22	31
	350	3	7	9	12	19	27
	400	3	6	8	11	17	24
	500	2	5	7	9	14	20
	600	1	4	6	7	11	16
	700	1	4	5	6	10	14
	750	1	4	5	6	9	14
	900	1	3	4	5	7	10
	1000	1	3	3	4	7	10
XHHW, XHHW-2	8	39	88	115	146	225	320
	6	29	65	85	109	167	238
	4	21	47	62	79	121	172
	2	15	34	44	57	87	124
	1	11	25	33	42	65	93
	1/0	9	22	28	36	56	79
	2/0	8	18	24	30	47	67
	3/0	6	15	20	25	38	55
	4/0	5	12	16	21	32	46
	250	4	10	13	17	26	37
	300	4	8	11	14	22	31
	350	3	7	10	12	19	28
	400	3	7	9	11	17	25
	500	2	5	7	9	14	20
	600	1	4	6	7	11	16
	700	1	4	5	6	10	14
	750	1	3	5	6	9	13
	900	1	3	4	5	7	10
	1000	1	3	4	5	7	10

Definition: *Compact stranding* is the result of a manufacturing process where the standard conductor is compressed to the extent that the interstices (voids between strand wires) are virtually eliminated.

Annex D

Examples

This annex is not a part of the requirements of this NFPA document but is included for informational purposes only.

Contents

Selection of Conductors. In the following examples, the results are generally expressed in amperes (A). To select conductor sizes, refer to the 0 through 2000 volt (V) ampacity tables of Article 310 and the rules of 310.15 that pertain to these tables.

Voltage. For uniform application of Articles 210, 215, and 220, a nominal voltage of 120, 120/240, 240, and 208Y/120 V is used in calculating the ampere load on the conductor.

Fractions of an Ampere. Except where the calculations result in a major fraction of an ampere (0.5 or larger), such fractions are permitted to be dropped.

Power Factor. Calculations in the following examples are based, for convenience, on the assumption that all loads have the same power factor (PF).

Ranges. For the calculation of the range loads in these examples, Column C of Table 220.55 has been used. For optional methods, see Columns A and B of Table 220.55. Except where the calculations result in a major fraction of a kilowatt (0.5 or larger), such fractions are permitted to be dropped.

SI Units. For metric conversions, 0.093 m² = 1 ft² and 0.3048 m = 1 ft.

In the examples in this annex, loads are assumed to be properly balanced on the system. If loads are not properly balanced, additional feeder capacity may be required. The calculations are based on the standard method.

Example D1(a) One-Family Dwelling

The dwelling has a floor area of 1500 ft², exclusive of an unfinished cellar not adaptable for future use, unfinished attic, and open porches. Appliances are a 12-kW range and a 5.5-kW, 240-V dryer. Assume range and dryer kW ratings equivalent to kVA ratings in accordance with 220.54 and 220.55.

The general lighting and general-use receptacle load is computed from the outside dimensions of the building, apartment, or other area involved. For a dwelling unit, the computed floor area is not to include open porches, garages, or, as stated in the opening paragraph, "an unfinished cellar not adaptable for future use." A point to consider regarding this statement is that many of today's homes with basements or cellars do have space that is suitable for conversion to family rooms, bedrooms, home offices, or other habitable areas. In such instances, the basement or cellar space suitable for conversion needs to be included in the general lighting load calculation. See 220.12 and 220.14(J) for requirements on how to calculate the lighting and general-use receptacle load for this occupancy.

The two-story dwelling measures 30 ft × 30 ft for the first floor and 30 ft × 20 ft for the second floor.

First-floor area: 30 ft × 30 ft = 900 ft^2

Second-floor area: 30 ft × 20 ft = 600 ft^2

Total area = 1500 ft^2

Calculated Load *[see 220.40]*

General Lighting Load: 1500 ft^2 at 3 VA per ft^2 = 4500 VA

Minimum Number of Branch Circuits Required *[see 210.11(A)]*

General Lighting Load: 4500 VA ÷ 120 V = 37.5 A
This requires three 15-A, 2-wire or two 20-A, 2-wire circuits.
Small Appliance Load: Two 2-wire, 20-A circuits *[see 210.11(C)(1)]*
Laundry Load: One 2-wire, 20-A circuit *[see 210.11(C)(2)]*
Bathroom Branch Circuit: One 2-wire, 20-A circuit (no additional load calculation is required for this circuit) *[see 210.11(C)(3)]*

Minimum Size Feeder Required *[see 220.40]*

General Lighting		4,500 VA
Small Appliance		3,000 VA
Laundry		1,500 VA
	Total	9,000 VA
3000 VA at 100%		3,000 VA
9000 VA − 3000 VA = 6000 VA at 35%		2,100 VA
	Net Load	5,100 VA
Range *(see Table 220.55)*		8,000 VA
Dryer Load *(see Table 220.54)*		5,500 VA
	Net Calculated Load	18,600 VA

Net Calculated Load for 120/240-V, 3-wire, single-phase service or feeder

18,600 VA ÷ 240 V = 77.5 A

Sections 230.42(B) and 230.79 require service conductors and disconnecting means rated not less than 100 amperes.

Calculation for Neutral for Feeder and Service

Lighting and Small Appliance Load		5,100 VA
Range: 8000 VA at 70% *(see 220.61)*		5,600 VA
Dryer: 5500 VA at 70% *(see 220.61)*		3,850 VA
	Total	14,550 VA

Calculated Load for Neutral

14,550 VA ÷ 240 V = 60.6 A

Example D1(b) One-Family Dwelling

Assume same conditions as Example No. D1(a), plus addition of one 6-A, 230-V, room air-conditioning unit and one 12-A, 115-V, room air-conditioning unit,* one 8-A, 115-V, rated waste disposer, and one 10-A, 120-V, rated dishwasher. See Article 430 for general motors and Article 440, Part VII, for air-conditioning equipment. Motors have nameplate ratings of 115 V and 230 V for use on 120-V and 240-V nominal voltage systems.

*(For feeder neutral, use larger of the two appliances for unbalance.)
From Example D1(a), feeder current is 78 A (3-wire, 240 V).

	Line A	Neutral	Line B
Amperes from Example D1(a)	78	61	78
One 230-V air conditioner	6	—	6
One 115-V air conditioner and 120-V dishwasher	12	12	10
One 115-V disposer	—	8	8
25% of largest motor *(see 430.24)*	3	3	2
Total amperes per line	99	84	104

Therefore, the service would be rated 110 A.

The air-conditioning load is calculated at 100 percent and is calculated separately to comply with the requirements of 220.82(C)(1).

Example D2(a) Optional Calculation for One-Family Dwelling, Heating Larger Than Air Conditioning
[see 220.82]

The dwelling has a floor area of 1500 ft^2, exclusive of an unfinished cellar not adaptable for future use, unfinished attic, and open porches. It has a 12-kW range, a 2.5-kW water heater, a 1.2-kW dishwasher, 9 kW of electric space heating installed in five rooms, a 5-kW clothes dryer, and a 6-A, 230-V, room air-conditioning unit. Assume range, water heater, dishwasher, space heating, and clothes dryer kW ratings equivalent to kVA.

Air Conditioner kVA Calculation

6 A × 230 V ÷ 1000 = 1.38 kVA

This 1.38 kVA [item 1 from 220.82(C)] is less than 40% of 9 kVA of separately controlled electric heat [item 6 from 220.82(C)], so the 1.38 kVA need not be included in the service calculation.

General Load

1500 ft^2 at 3 VA	4,500 VA
Two 20-A appliance outlet circuits at 1500 VA each	3,000 VA
Laundry circuit	1,500 VA
Range (at nameplate rating)	12,000 VA
Water heater	2,500 VA
Dishwasher	1,200 VA
Clothes dryer	5,000 VA
Total	29,700 VA

Application of Demand Factor *[see 220.82(B)]*

First 10 kVA of general load at 100%	10,000 VA
Remainder of general load at 40%	
(19.7 kVA × 0.4)	7,880 VA
Total of general load	17,880 VA
9 kVA of heat at 40% (9000 VA × 0.4) =	3,600 VA
Total	21,480 VA

Calculated Load for Service Size

21.48 kVA = 21,480 VA

21,480 VA ÷ 240 V = 89.5 A

Therefore, the minimum service rating would be 100 A in accordance with 230.42 and 230.79.

Feeder Neutral Load, per 220.61

1500 ft^2 at 3 VA		4,500 VA
Three 20-A circuits at 1500 VA		4,500 VA
	Total	9,000 VA
3000 VA at 100%		3,000 VA
9000 VA − 3000 VA = 6000 VA at 35%		2,100 VA
	Subtotal	5,100 VA

Range: 8 kVA at 70%   5,600 VA
Clothes dryer: 5 kVA at 70%   3,500 VA
Dishwasher   1,200 VA
Total   15,400 VA

Calculated Load for Neutral

15,400 VA ÷ 240 V = 64.2 A

Example D2(b) Optional Calculation for One-Family Dwelling, Air Conditioning Larger Than Heating *[see 220.82(A) and 220.82(C)]*

The dwelling has a floor area of 1500 ft², exclusive of an unfinished cellar not adaptable for future use, unfinished attic, and open porches. It has two 20-A small appliance circuits, one 20-A laundry circuit, two 4-kW wall-mounted ovens, one 5.1-kW counter-mounted cooking unit, a 4.5-kW water heater, a 1.2-kW dishwasher, a 5-kW combination clothes washer and dryer, six 7-A, 230-V room air-conditioning units, and a 1.5-kW permanently installed bathroom space heater. Assume wall-mounted ovens, counter-mounted cooking unit, water heater, dishwasher, and combination clothes washer and dryer kW ratings equivalent to kVA.

Air Conditioning kVA Calculation

Total amperes = 6 units × 7 A = 42 A

42 A × 240 V ÷ 1000 = 10.08 kVA (assume PF = 1.0)

Load Included at 100%

Air Conditioning: Included below *[see item 1 in 220.82(C)]*
Space Heater: Omit *[see item 5 in 220.82(C)]*

General Load

1500 ft² at 3 VA   4,500 VA
Two 20-A small appliance circuits at 1500 VA each   3,000 VA
Laundry circuit   1,500 VA
Two ovens   8,000 VA
One cooking unit   5,100 VA
Water heater   4,500 VA
Dishwasher   1,200 VA
Washer/dryer   5,000 VA
Total general load   32,800 VA
First 10 kVA at 100%   10,000 VA
Remainder at 40% (22.8 kVA × 0.4 × 1000)   9,120 VA
Subtotal general load   19,120 VA
Air conditioning   10,080 VA
Total   29,200 VA

Calculated Load for Service

29,200 VA ÷ 240 V = 122 A (service rating)

Feeder Neutral Load, per 220.61

Assume that the two 4-kVA wall-mounted ovens are supplied by one branch circuit, the 5.1-kVA counter-mounted cooking unit by a separate circuit.

1500 ft² at 3 VA   4,500 VA
Three 20-A circuits at 1500 VA   4,500 VA
Subtotal   9,000 VA
3000 VA at 100%   3,000 VA
9000 VA − 3000 VA = 6000 VA at 35%   2,100 VA
Subtotal   5,100 VA

Two 4-kVA ovens plus one 5.1-kVA cooking unit = 13.1 kVA. Table 220.55 permits 55% demand factor or 13.1 kVA × 0.55 = 7.2 kVA feeder capacity.

Subtotal from above   5,100 VA
Ovens and cooking unit: 7200 VA × 70% for neutral load   5,040 VA
Clothes washer/dryer: 5 kVA × 70% for neutral load   3,500 VA
Dishwasher   1,200 VA
Total   14,840 VA

Calculated Load for Neutral

14,840 VA ÷ 240 V = 61.83 A (use 62 A)

Example D2(c) Optional Calculation for One-Family Dwelling with Heat Pump (Single-Phase, 240/120-Volt Service) *(see 220.82)*

The dwelling has a floor area of 2000 ft², exclusive of an unfinished cellar not adaptable for future use, unfinished attic, and open porches. It has a 12-kW range, a 4.5-kW water heater, a 1.2-kW dishwasher, a 5-kW clothes dryer, and a 2½-ton (24-A) heat pump with 15 kW of backup heat.

Heat Pump kVA Calculation

24 A × 240 V ÷ 1000 = 5.76 kVA

This 5.76 kVA is less than 15 kVA of the backup heat; therefore, the heat pump load need not be included in the service calculation *[see 220.82(C)]*.

General Load

2000 ft² at 3 VA   6,000 VA
Two 20-A appliance outlet circuits at 1500 VA each   3,000 VA
Laundry circuit   1,500 VA
Range (at nameplate rating)   12,000 VA
Water heater   4,500 VA
Dishwasher   1,200 VA
Clothes dryer   5,000 VA
Subtotal general load   33,200 VA
First 10 kVA at 100%   10,000 VA
Remainder of general load at 40%
(23,200 VA × 0.4)   9,280 VA
Total net general load   19,280 VA

Heat Pump and Supplementary Heat*

240 V × 24 A = 5760 VA

15 kW Electric Heat:

5760 VA + (15,000 VA × 65%) = 5.76 kVA
+ 9.75 kVA = 15.51 kVA

*If supplementary heat is not on at same time as heat pump, heat pump kVA need not be added to total.

Totals

Net general load   19,280 VA
Heat pump and supplementary heat   15,510 VA
Total   34,790 VA

Calculated Load for Service

34.79 kVA × 1000 ÷ 240 V = 144.96 A

Therefore, this dwelling unit would be permitted to be served by a 150-A service.

Example D3 Store Building

A store 50 ft by 60 ft, or 3000 ft², has 30 ft of show window. There are a total of 80 duplex receptacles. The service is 120/240 V, single phase 3-wire service. Actual connected lighting load is 8500 VA.

Calculated Load *(see 220.40)*

Noncontinuous Loads
Receptacle Load *(see 220.44)*
80 receptacles at 180 VA   14,400 VA
10,000 VA at 100%   10,000 VA
14,400 VA − 10,000 VA = 4400 at 50%   2,200 VA
Subtotal   12,200 VA

Continuous Loads
General Lighting*
3000 ft² at 3 VA per ft²   9,000 VA
Show Window Lighting Load
30 ft at 200 VA per ft   6,000 VA

Outside Sign Circuit *[see 220.14(F)]*   1,200 VA

Subtotal	16,200 VA
Subtotal from noncontinuous	12,200 VA
Total noncontinuous loads + continuous loads =	28,400 VA

*In the example, 125% of the actual connected lighting load (8500 VA × 1.25 = 10,625 VA) is less than 125% of the load from Table 220.12, so the minimum lighting load from Table 220.12 is used in the calculation. Had the actual lighting load been greater than the value calculated from Table 220.12, 125% of the actual connected lighting load would have been used.

Minimum Number of Branch Circuits Required

General Lighting: Branch circuits need only be installed to supply the actual connected load *[see 210.11(B)]*.

$$8500 \text{ VA} \times 1.25 = 10{,}625 \text{ VA}$$

$$10{,}625 \text{ VA} \div 240 \text{ V} = 44 \text{ A for 3-wire, } 120/240 \text{ V}$$

The lighting load would be permitted to be served by 2-wire or 3-wire, 15- or 20-A circuits with combined capacity equal to 44 A or greater for 3-wire circuits or 88 A or greater for 2-wire circuits. The feeder capacity as well as the number of branch-circuit positions available for lighting circuits in the panelboard must reflect the full calculated load of 9000 VA × 1.25 = 11,250 VA.

Show Window

$$6000 \text{ VA} \times 1.25 = 7500 \text{ VA}$$

$$7500 \text{ VA} \div 240 \text{ V} = 31 \text{ A for 3-wire, } 120/240 \text{ V}$$

The show window lighting is permitted to be served by 2-wire or 3-wire circuits with a capacity equal to 31 A or greater for 3-wire circuits or 62 A or greater for 2-wire circuits.

Receptacles required by 210.62 are assumed to be included in the receptacle load above if these receptacles do not supply the show window lighting load.

Receptacles

Receptacle Load:14,400 VA ÷ 240 V = 60 A for 3-wire, 120/240 V

The receptacle load would be permitted to be served by 2-wire or 3-wire circuits with a capacity equal to 60 A or greater for 3-wire circuits or 120 A or greater for 2-wire circuits.

Minimum Size Feeder (or Service) Overcurrent Protection
[see 215.3 or 230.90]

Subtotal noncontinuous loads	12,200 VA
Subtotal continuous load at 125% (16,200 VA × 1.25)	20,250 VA
Total	32,450 VA

$$32{,}450 \text{ VA} \div 240 \text{ V} = 135 \text{ A}$$

The next higher standard size is 150 A *(see 240.6)*.

Minimum Size Feeders (or Service Conductors) Required
[see 215.2, 230.42(A)]

For 120/240 V, 3-wire system,

$$32{,}450 \text{ VA} \div 240 \text{ V} = 135 \text{ A}$$

Service or feeder conductor is 1/0 Cu per 215.3 and Table 310.16 (with 75°C terminations).

Example D3(a) Industrial Feeders in a Common Raceway

An industrial multi-building facility has its service at the rear of its main building, and then provides 480Y/277-volt feeders to additional buildings behind the main building in order to segregate certain processes. The facility supplies its remote buildings through a partially enclosed access corridor that extends from the main switchboard rearward along a path that provides convenient access to services within 15 m (50 ft) of each additional building supplied. Two building feeders share a common raceway for approximately 45 m (150 ft) and run in the access corridor along with process steam and control and communications cabling. The steam raises the ambient temperature around the power raceway to as much as 35°C. At a tee fitting, the individual building feeders then run to each of the two buildings involved. The feeder neutrals are not connected to the equipment grounding conductors in the remote buildings. All distribution equipment terminations are listed as being suitable for 75°C connections.

Each of the two buildings has the following loads:

Lighting, 11,600 VA, comprised of electric-discharge luminaires connected at 277 V

Receptacles, 22 125-volt, 20 ampere receptacles on general-purpose branch circuits, supplied by separately derived systems in each of the buildings

1-Air compressor, 460 volt, three phase, 7.5 hp

1-Grinder, 460 volt, three phase, 1.5 hp

3-Welders, AC transformer type (nameplate: 23 amperes, 480 volts, 60 percent duty cycle)

3-Industrial Process Dryers, 480 volt, three phase, 15 kW each (assume continuous use throughout certain shifts)

Determine the overcurrent protection and conductor size for the feeders in the common raceway, assuming the use of XHHW-2 insulation (90°C):

Calculated Load {Note: For reasonable precision, volt-ampere calculations are carried to three significant figures only; where loads are converted to amperes, the results are rounded to the nearest ampere *[see 220.5(B)]*}.

Noncontinuous Loads

Receptacle Load (see 220.44)		
22 receptacles at 180 VA		3,960 VA
Welder Load [see 630.11(A), Table 630.11(A)]		
Each welder: 480V × 23A × 0.78 = 8,610 VA		
All 3 welders: [see 630.11(B)] (demand factors 100%, 100%, 85% respectively)		
8,610 VA + 8,610 VA + 7,320 VA =		24,500 VA
	Subtotal, Noncontinuous Loads	28,500 VA

Motor Loads (see 430.24, Table 430.250)

Air compressor: 11 A × 480 V × √3 =		9,150 VA
Grinder: 3 A × 480 V × √3 =		2,490 VA
Largest motor, additional 25%:		2,290 VA
	Subtotal, Motor Loads	13,900 VA

By using 430.24, the motor loads and the noncontinuous loads can be combined for the remaining calculation.

Subtotal for load calculations, Noncontinuous Loads	42,400 VA

Continuous Loads

General Lighting		11,600 VA
3 Industrial Process Dryers	15 kW each	45,000 VA
	Subtotal, Continuous Loads	56,600 VA

Overcurrent protection *(see 215.3)*

The overcurrent protective device must accommodate 125% of the continuous load, plus the noncontinuous load:

Continuous load	56,600 VA
Noncontinuous load	42,400 VA
Subtotal, actual load [actual load in amperes]	99,000 VA

99,000 VA ÷ (480V × √3) = 119 A]

(25% of 56,600 VA) *(See 215.3)*	14,200 VA
Total VA	113,200 VA

Conversion to amperes using three significant figures:

$$113{,}200 \text{ VA} / (480\text{V} \times \sqrt{3}) = 136 \text{ A}$$

Minimum size overcurrent protective device: 136 A

Minimum standard size overcurrent protective device *(see 240.6)*: 150 amperes

Where the overcurrent protective device and its assembly are listed for operation at 100 percent of its rating, a 125 ampere overcurrent protective device would be permitted. However, overcurrent protective device assemblies listed for 100 percent of their rating are typically not available at the 125-ampere rating. *(See 215.3 Exception.)*

Ungrounded Feeder Conductors

The conductors must independently meet requirements for (1) terminations, and (2) conditions of use throughout the raceway run.

Minimum size conductor at the overcurrrent device termination *[see 110.14(C) and 215.2(A)(1), using 75°C ampacity column in Table 310.16]*: 1/0 AWG.

Minimum size conductors in the raceway based on actual load *[see Article 100, Ampacity, and 310.15(B)(2)(a) and correction factors to Table 310.16]*:

$$99,000 \text{ VA} / 0.7 / 0.96 = 147,000 \text{ VA}$$

(70% = 310.15(B)(2)(a)) & (0.96 = Correction factors to Table 310.16)

Conversion to amperes:

$$147,000 \text{ VA} / (480 \text{ V} \times \sqrt{3}) = 177 \text{ A}$$

Note that the neutral conductors are counted as current-carrying conductors *[see 310.15(B)(4)(c)]* in this example because the discharge lighting has substantial nonlinear content. This requires a 2/0 AWG conductor based on the 90°C column of Table 310.16. Therefore, the worst case is given by the raceway conditions, and 2/0 AWG conductors must be used. If the utility corridor was at normal temperatures [(30°C (86°F)], and if the lighting at each building were supplied from the local separately derived system (thus requiring no neutrals in the supply feeders) the raceway result (99,000 VA / 0.8 = 124,000 VA; 124,000 VA / (480V × $\sqrt{3}$) = 149 A, or a 1 AWG conductor @ 90°C) could not be used because the termination result (1/0 AWG based on the 75°C column of Table 310.16) would become the worst case, requiring the larger conductor.

In every case, the overcurrent protective device shall provide overcurrent protection for the feeder conductors in accordance with their ampacity as provided by this *Code (see 240.4)*. A 90°C 2/0 AWG conductor has a Table 310.16 ampacity of 195 amperes. Adjusting for the conditions of use (35°C ambient temperature, 8 current-carrying conductors in the common raceway),

$$195 \text{ amperes} \times 0.96 \times 0.7 = 131 \text{ A}$$

The 150-ampere circuit breaker protects the 2/0 AWG feeder conductors, because 240.4(B) permits the use of the next higher standard size overcurrent protective device. Note that the feeder layout precludes the application of 310.15(A)(2) Exception.

Feeder Neutral Conductor *(see 220.61)*

Because 210.11(B) does not apply to these buildings, the load cannot be assumed to be evenly distributed across phases. Therefore the maximum imbalance must be assumed to be the full lighting load in this case, or 11,600 VA. (11,600 VA / 277V = 42 amperes.) The ability of the neutral to return fault current *[see 250.32(B)(2)(2)]* is not a factor in this calculation.

Although the neutral runs between the main switchboard and the building panelboard, likely terminating on a busbar at both locations, the busbar connections are part of listed devices and are not "separately installed pressure devices." Therefore 110.14(C)(2) does not apply, and the normal termination temperature limits apply. In addition, the listing requirement to gain exemption from the additional sizing allowance under continuous loading *(see 215.3 Exception)* covers not just the overcurrent protective device, but its entire assembly as well. Therefore, since the lighting load is continuous, the minimum conductor size is based on 1.25 × (11,600 VA / 277V) = 52 amperes, to be evaluated under the 75°C column of Table 310.16. The minimum size of the neutral is 6 AWG. This size is also the minimum size required by 215.2(A)(1), because the minimum size equipment grounding conductor for a 150 ampere circuit, as covered in Table 250.122, is 6 AWG.

Example D4(a) Multifamily Dwelling

A multifamily dwelling has 40 dwelling units.

Meters are in two banks of 20 each with individual feeders to each dwelling unit.

One-half of the dwelling units are equipped with electric ranges not exceeding 12 kW each. Assume range kW rating equivalent to kVA rating in accordance with 220.55. Other half of ranges are gas ranges.

Area of each dwelling unit is 840 ft^2.

Laundry facilities on premises are available to all tenants. Add no circuit to individual dwelling unit.

Calculated Load for Each Dwelling Unit *(see Article 220)*

General Lighting: 840 ft^2 at 3 VA per ft^2 = 2520 VA
Special Appliance: Electric range *(see 220.55)* = 8000 VA

Minimum Number of Branch Circuits Required for Each Dwelling Unit *[see 210.11(A)]*

General Lighting Load: 2520 VA ÷ 120 V = 21 A or two 15-A, 2-wire circuits; or two 20-A, 2-wire circuits
Small Appliance Load: Two 2-wire circuits of 12 AWG wire *[see 210.11(C)(1)]*
Range Circuit: 8000 VA ÷ 240 V = 33 A or a circuit of two 8 AWG conductors and one 10 AWG conductor per 210.19(A)(3)

Minimum Size Feeder Required for Each Dwelling Unit *(see 215.2)*

Calculated Load *(see Article 220)*:

General Lighting	2,520 VA
Small Appliance (two 20-ampere circuits)	3,000 VA
Subtotal Calculated Load (without ranges)	5,520 VA

Application of Demand Factor *(see Table 220.42)*

First 3000 VA at 100%	3,000 VA
5520 VA − 3000 VA = 2520 VA at 35%	882 VA
Net Calculated Load (without ranges)	3,882 VA
Range Load	8,000 VA
Net Calculated Load (with ranges)	11,882 VA

Size of Each Feeder *(see 215.2)*

For 120/240-V, 3-wire system (without ranges)

Net calculated load of 3882 VA ÷ 240 V = 16.2 A

For 120/240-V, 3-wire system (with ranges)

Net calculated load, 11,882 VA ÷ 240 V = 49.5 A

Feeder Neutral

Lighting and Small Appliance Load	3,882 VA
Range Load: 8000 VA at 70% *(see 220.61)*	5,600 VA
(only for apartments with electric range)	5,600 VA
Net Calculated Load (neutral)	9,482 VA

Calculated Load for Neutral

9482 VA ÷ 240 V= 39.5 A

Minimum Size Feeders Required from Service Equipment to Meter Bank (For 20 Dwelling Units — 10 with Ranges)

Total Calculated Load:

Lighting and Small Appliance	
20 units × 5520 VA	110,400 VA
Application of Demand Factor	
First 3000 VA at 100%	3,000 VA
110,400 VA − 3000 VA = 107,400 VA at 35%	37,590 VA
Net Calculated Load	40,590 VA
Range Load: 10 ranges (less than 12 kVA) *(see Col. C, Table 220.55)*	25,000 VA
Net Calculated Load (with ranges)	65,590 VA

Net calculated load for 120/240-V, 3-wire system,

$$65,590 \text{ VA} \div 240 \text{ V} = 273 \text{ A}$$

Feeder Neutral

Lighting and Small Appliance Load	40,590 VA
Range Load: 25,000 VA at 70% *[see 220.61(B)]*	17,500 VA
Calculated Load (neutral)	58,090 VA

Calculated Load for Neutral

$$58,090 \text{ VA} \div 240 \text{ V} = 242 \text{ A}$$

Further Demand Factor *[220.61(B)]*

200 A at 100%	200	A
242 A − 200 A = 42 A at 70%	29	A
Net Calculated Load (neutral)	229	A

Minimum Size Main Feeders (or Service Conductors) Required (Less House Load) (For 40 Dwelling Units — 20 with Ranges)

Total Calculated Load:

Lighting and Small Appliance Load	
40 units × 5520 VA	220,800 VA

Application of Demand Factor *(from Table 220.42)*

First 3000 VA at 100%	3,000 VA
Next 120,000 VA − 3000 VA = 117,000 VA at 35%	40,950 VA
Remainder 220,800 VA − 120,000 VA =	
100,800 VA at 25%	25,200 VA
Net Calculated Load	69,150 VA
Range Load: 20 ranges (less than 12 kVA)	
(see Col. C, Table 220.55)	35,000 VA
Net Calculated Load	104,150 VA

For 120/240-V, 3-wire system

$$\text{Net calculated load of } 104,150 \text{ VA} \div 240 \text{ V} = 434 \text{ A}$$

Feeder Neutral

Lighting and Small Appliance Load	69,150 VA
Range: 35,000 VA at 70% *[see 220.61(B)]*	24,500 VA
Calculated Load (neutral)	93,650 VA
93,650 VA ÷ 240 V = 390 A	

Further Demand Factor *[see 220.61(B)]*

200 A at 100%	200	A
390 A − 200 A = 190 A at 70%	133	A
Net Calculated Load (neutral)	333	A

[See Tables 310.16 through 310.21, and 310.15(B)(2) and (B)(4).]

Example D4(b) Optional Calculation for Multifamily Dwelling

A multifamily dwelling equipped with electric cooking and space heating or air conditioning has 40 dwelling units.

Meters are in two banks of 20 each plus house metering and individual feeders to each dwelling unit.

Each dwelling unit is equipped with an electric range of 8-kW nameplate rating, four 1.5-kW separately controlled 240-V electric space heaters, and a 2.5-kW, 240-V electric water heater. Assume range, space heater, and water heater kW ratings equivalent to kVA.

A common laundry facility is available to all tenants *[see 210.52(F), Exception No. 1]*.

Area of each dwelling unit is 840 ft^2.

Calculated Load for Each Dwelling Unit *(see Article 220)*

General Lighting Load:	
840 ft^2 at 3 VA per ft^2	2,520 VA
Electric range	8,000 VA
Electric heat: 6 kVA (or air conditioning if larger)	6,000 VA
Electric water heater	2,500 VA

Minimum Number of Branch Circuits Required for Each Dwelling Unit

General Lighting Load: 2520 VA ÷ 120 V = 21 A or two 15-A, 2-wire circuits, or two 20-A, 2-wire circuits

Small Appliance Load: Two 2-wire circuits of 12 AWG *[see 210.11(C)(1)]*

Range Circuit *(See Table 220.55, Column B)*:

$$8000 \text{ VA} \times 80\% \div 240 \text{ V} = 27 \text{ A on a circuit of three}$$
$$10 \text{ AWG conductors per } 210.19(A)(3)$$

Space Heating: 6000 VA ÷ 240 V = 25 A

Number of circuits *(see 210.11)*

Minimum Size Feeder Required for Each Dwelling Unit *(see 215.2)*

Calculated Load *(see Article 220)*:

General Lighting	2,520 VA
Small Appliance (two 20-A circuits)	3,000 VA
Subtotal Calculated Load (without range and space heating)	5,520 VA

Application of Demand Factor

First 3000 VA at 100%	3,000 VA
5520 VA − 3000 VA = 2520 VA at 35%	882 VA
Net Calculated Load (without range and space heating)	3,882 VA
Range	6,400 VA
Space Heating *(see 220.51)*	6,000 VA
Water Heater	2,500 VA
Net Calculated Load (for individual dwelling unit)	18,782 VA

Size of Each Feeder

For 120/240-V, 3-wire system,

$$\text{Net calculated load of } 18,782 \text{ VA} \div 240 \text{ V} = 78 \text{ A}$$

Feeder Neutral *(see 220.61)*

Lighting and Small Appliance	3,882 VA
Range Load: 6400 VA at 70% *[see 220.61(B)]*	4,480 VA
Space and Water Heating (no neutral): 240 V	0 VA
Net Calculated Load (neutral)	8,362 VA

Calculated Load for Neutral

$$8362 \text{ VA} \div 240 \text{ V} = 35 \text{ A}$$

Minimum Size Feeder Required from Service Equipment to Meter Bank (For 20 Dwelling Units)

Total Calculated Load:

Lighting and Small Appliance Load	
20 units × 5520 VA	110,400 VA
Water and Space Heating Load	
20 units × 8500 VA	170,000 VA
Range Load: 20 × 8000 VA	160,000 VA
Net Calculated Load (20 dwelling units)	440,400 VA
Net Calculated Load Using Optional Calculation	
(see Table 220.84)	
440,400 VA × 0.38	167,352 VA

$$167,352 \text{ VA} \div 240 \text{ V} = 697 \text{ A}$$

Minimum Size Main Feeder Required (Less House Load) (For 40 Dwelling Units)

Calculated Load:

Lighting and Small Appliance Load	
40 units × 5520 VA	220,800 VA
Water and Space Heating Load	
40 units × 8500 VA	340,000 VA
Range: 40 ranges × 8000 VA	320,000 VA
Net Calculated Load (40 dwelling units)	880,800 VA

Net Calculated Load Using Optional Calculation *(see Table 220.84)*

$$880,800 \text{ VA} \times 0.28 = 246,624 \text{ VA}$$
$$246,624 \text{ VA} \div 240 \text{ V} = 1028 \text{ A}$$

Feeder Neutral Load for Feeder from Service Equipment to Meter Bank (For 20 Dwelling Units)

Lighting and Small Appliance Load

20 units × 5520 VA	110,400 VA
First 3000 VA at 100%	3,000 VA
110,400 VA − 3000 VA = 107,400 VA at 35%	37,590 VA
Net Calculated Load	40,590 VA
20 ranges: 35,000 VA at 70%	
[see Table 220.55 and Section 220.61(B)]	24,500 VA
Total	65,090 VA

$$65,090 \text{ VA} \div 240 \text{ V} = 271 \text{ A}$$

Further Demand Factor *[see 220.61(B)]*

First 200 A at 100%	200 A
Balance: 271 A − 200 A = 71 A at 70%	50 A
Total	250 amperes

Feeder Neutral Load of Main Feeder (Less House Load) (For 40 Dwelling Units)

Lighting and Small Appliance Load

40 units × 5520 VA	220,800 VA
First 3000 VA at 100%	3,000 VA
Next 120,000 VA − 3000 VA = 117,000 VA at 35%	40,950 VA
Remainder 220,800 VA − 120,000 VA = 100,800 VA at 25%	25,200 VA
Net Calculated Load	69,150 VA
40 ranges: 55,000 VA at 70% *[see Table 220.55 and Section 220.61(B)]*	38,500 VA
Total	107,650 VA

$$107,650 \text{ VA} \div 240 \text{ V} = 449 \text{ A}$$

Further Demand Factor *[see 220.61(B)]*

First 200 A at 100%	200 A
Balance: 449 − 200 A = 249 A at 70%	174 A
Total	374 A

Example D5(a) Multifamily Dwelling Served at 208Y/120 Volts, Three Phase

All conditions and calculations are the same as for the multifamily dwelling [Example D4(a)] served at 120/240 V, single phase except as follows:

Service to each dwelling unit would be two phase legs and neutral.

Minimum Number of Branch Circuits Required for Each Dwelling Unit *(see 210.11)*

Range Circuit: 8000 VA ÷ 208 V = 38 A or a circuit of two 8 AWG conductors and one 10 AWG conductor per 210.19(A)(3)

Minimum Size Feeder Required for Each Dwelling Unit *(see 215.2)*

For 120/208-V, 3-wire system (without ranges),

Net calculated load of 3882 VA ÷ 2 legs ÷ 120 V/leg = 16.2 A

For 120/208-V, 3-wire system (with ranges),

Net calculated load (range) of 8000 VA ÷ 208 V = 38.5 A
Total load (range + lighting) = 38.5 A + 16.2 A = 54.7 A
Feeder neutral: (range) of 8000 VA × 70% = 5600 VA ÷ 208 V = 26.9 A
Total load: (range + lighting) = 26.9 A + 16.2 A = 43.1 A

Minimum Size Feeders Required from Service Equipment to Meter Bank (For 20 Dwelling Units — 10 with Ranges)

For 208Y/120-V, 3-phase, 4-wire system,

Ranges: Maximum number between any two phase legs = 4
2 × 4 = 8.
Table 220.55 demand = 23,000 VA
Per phase demand = 23,000 VA ÷ 2 = 11,500 VA
Equivalent 3-phase load = 34,500 VA

Net Calculated Load (total):

$$40,590 \text{ VA} + 34,500 \text{ VA} = 75,090 \text{ VA}$$
$$75,090 \text{ VA} \div (208 \text{ V})(1.732) = 208.4 \text{ A}$$

Feeder Neutral Size

Net Calculated Lighting and Appliance Load & Equivalent Range Load:

$$40,590 \text{ VA} + (34,500 \text{ VA at } 70\%) = 64,700 \text{ VA}$$

Net Calculated Neutral Load:

$$64,700 \text{ VA} \div (208 \text{ V})(1.732) = 179.7 \text{ A}$$

Minimum Size Main Feeder (Less House Load) (For 40 Dwelling Units — 20 with Ranges)

For 208Y/120-V, 3-phase, 4-wire system,
Ranges:
Maximum number between any two phase legs = 7
2 × 7 = 14.
Table 220.55 demand = 29,000 VA
Per phase demand = 29,000 VA ÷ 2 = 14,500 VA
Equivalent 3-phase load = 43,500 VA
Net Calculated Load (total):

$$69,150 \text{ VA} + 43,500 \text{ VA} = 112,650 \text{ VA}$$
$$112,650 \text{ VA} \div (208 \text{ V})(1.732) = 312.7 \text{ A}$$

Main Feeder Neutral Size:

$$69,150 \text{ VA} + (43,500 \text{ VA at } 70\%) = 99,600 \text{ VA}$$
$$99,600 \text{ VA} \div (208 \text{ V})(1.732) = 276.5 \text{ A}$$

Further Demand Factor *(see 220.61)*

200 A at 100%	200.0 A
276.5 A − 200 A = 76.5 A at 70%	53.6 A
Net Calculated Load (neutral)	253.6 A

Example D5(b) Optional Calculation for Multifamily Dwelling Served at 208Y/120 Volts, Three Phase

All conditions and calculations are the same as for Optional Calculation for the Multifamily Dwelling [Example D4(b)] served at 120/240 V, single phase except as follows:

Service to each dwelling unit would be two phase legs and neutral.

Minimum Number of Branch Circuits Required for Each Dwelling Unit *(see 210.11)*

Range Circuit *(see Table 220.55 Column B):* 8000 VA at 80% ÷ 208 V = 30.7 A or a circuit of two 8 AWG conductors and one 10 AWG conductor per 210.19(A)(3)

Space Heating: 6000 VA ÷ 208 V = 28.8 A
Two 20-ampere, 2-pole circuits required, 12 AWG conductors

Minimum Size Feeder Required for Each Dwelling Unit

120/208-V, 3-wire circuit

Net calculated load of 18,782 VA ÷ 208 V = 90.3 A
Net calculated load (lighting line to neutral):

$$3882 \text{ VA} \div 2 \text{ legs} \div 120 \text{ V per leg} = 16.2 \text{ amperes}$$

Line to line = 14,900 VA ÷ 208 V = 71.6 A
Total load = 16.2 A + 71.6 A = 87.8 A

Minimum Size Feeder Required for Service Equipment to Meter Bank (For 20 Dwelling Units)

Net Calculated Load

$$167,352 \text{ VA} \div (208 \text{ V})(1.732) = 464.9 \text{ A}$$

Feeder Neutral Load

$$65,080 \text{ VA} \div (208 \text{ V})(1.732) = 180.65 \text{ A}$$

Minimum Size Main Feeder Required (Less House Load) (For 40 Dwelling Units)

Net Calculated Load

$$246,624 \text{ VA} \div (208 \text{ V})(1.732) = 684.6 \text{ A}$$

Main Feeder Neutral Load

$$107,650 \text{ VA} \div (208 \text{ V})(1.732) = 298.8 \text{ A}$$

Further Demand Factor *[see 220.61(B)]*

200 A at 100%	200.0 A
298.8 A − 200 A = 98.8 A at 70%	69.2 A
Net Calculated Load (neutral)	269.2 A

Example D6 Maximum Demand for Range Loads

Table 220.55, Column C applies to ranges not over 12 kW. The application of Note 1 to ranges over 12 kW (and not over 27 kW) and Note 2 to ranges over 8¾ kW (and not over 27 kW) is illustrated in the following two examples.

A. Ranges All the Same Rating *(see Table 220.55, Note 1)*
Assume 24 ranges, each rated 16 kW.
From Table 220.55, Column C, the maximum demand for 24 ranges of 12-kW rating is 39 kW. 16 kW exceeds 12 kW by 4.
5% × 4 = 20% (5% increase for each kW in excess of 12)
39 kW × 20% = 7.8 kW increase
39 + 7.8 = 46.8 kW (value to be used in selection of feeders)

B. Ranges of Unequal Rating *(see Table 220.55, Note 2)*
Assume 5 ranges, each rated 11 kW; 2 ranges, each rated 12 kW; 20 ranges, each rated 13.5 kW; 3 ranges, each rated 18 kW.

5 ranges × 12 kW =	60 kW	(use 12 kW for range rated less than 12)
2 ranges × 12 kW =	24 kW	
20 ranges × 13.5 kW =	270 kW	
3 ranges × 18 kW =	54 kW	
30 ranges, Total kW =	408 kW	

408 ÷ 30 ranges = 13.6 kW (average to be used for calculation)

From Table 220.55, Column C, the demand for 30 ranges of 12-kW rating is 15 kW + 30 (1 kW × 30 ranges) = 45 kW. 13.6 kW exceeds 12 kW by 1.6 kW (use 2 kW).

5% × 2 = 10% (5% increase for each kW in excess of 12 kW)
45 kW × 10% = 4.5 kW increase

45 kW + 4.5 kW = 49.5 kW (value to be used in selection of feeders)

Example D8 Motor Circuit Conductors, Overload Protection, and Short-Circuit and Ground-Fault Protection
(see 240.6, 430.6, 430.22, 430.23, 430.24, 430.32, 430.52, and 430.62, Tables 430.52 and 430.250)

Determine the minimum required conductor ampacity, the motor overload protection, the branch-circuit short-circuit and ground-fault protection, and the feeder protection, for three induction-type motors on a 480-V, 3-phase feeder, as follows:

(a) One 25-hp, 460-V, 3-phase, squirrel-cage motor, nameplate full-load current 32 A, Design B, Service Factor 1.15

(b) Two 30-hp, 460-V, 3-phase, wound-rotor motors, nameplate primary full-load current 38 A, nameplate secondary full-load current 65 A, 40°C rise.

Conductor Ampacity
The full-load current value used to determine the minimum required conductor ampacity is obtained from Table 430.150 *[see 430.6(A)]* for the squirrel-

cage motor and the primary of the wound-rotor motors. To obtain the minimum required conductor ampacity, the full-load current is multiplied by 1.25 *[see 430.22 and 430.23(A)]*.

For the 25-hp motor,

$$34 \text{ A} \times 1.25 = 42.5 \text{ A}$$

For the 30-horsepower motors,

$$40 \text{ A} \times 1.25 = 50 \text{ A}$$
$$65 \text{ A} \times 1.25 = 81.25 \text{ A}$$

Motor Overload Protection
Where protected by a separate overload device, the motors are required to have overload protection rated or set to trip at not more than 125% of the nameplate full-load current *[see 430.6(A) and 430.32(A)(1)]*.

For the 25-hp motor,

$$32 \text{ A} \times 1.25 = 40.0 \text{ A}$$

For the 30-hp motors,

$$38 \text{ A} \times 1.25 = 47.5 \text{ A}$$

Where the separate overload device is an overload relay (not a fuse or circuit breaker), and the overload device selected at 125% is not sufficient to start the motor or carry the load, the trip setting is permitted to be increased in accordance with 430.32(C).

Branch-Circuit Short-Circuit and Ground-Fault Protection
The selection of the rating of the protective device depends on the type of protective device selected, in accordance with 430.52 and Table 430.52. The following is for the 25-hp motor.

(a) Nontime-Delay Fuse: The fuse rating is 300% × 34 A = 102 A. The next larger standard fuse is 110 A *[see 240.6 and 430.52(C)(1), Exception No. 1]*. If the motor will not start with a 110-A nontime-delay fuse, the fuse rating is permitted to be increased to 125 A because this rating does not exceed 400% *[see 430.52(C)(1), Exception No. 2(a)]*.

(b) Time-Delay Fuse: The fuse rating is 175% × 34 A = 59.5 A. The next larger standard fuse is 60 A *[see 240.6 and 430.52(C)(1), Exception No. 1]*. If the motor will not start with a 60-A time-delay fuse, the fuse rating is permitted to be increased to 70 A because this rating does not exceed 225% *[see 430.52(C)(1), Exception No. 2(b)]*.

Feeder Short-Circuit and Ground-Fault Protection
The rating of the feeder protective device is based on the sum of the largest branch-circuit protective device (example is 110 A) plus the sum of the full-load currents of the other motors, or 110 A + 40 A + 40 A = 190 A. The nearest standard fuse that does not exceed this value is 175 A *[see 240.6 and 430.62(A)]*.

Example D9 Feeder Ampacity Determination for Generator Field Control *[see 215.2, 430.24, 430.24 Exception No. 1, 620.13, 620.14, 620.61, and Table 430.22(E)]*

Determine the conductor ampacity for a 460-V 3-phase, 60-Hz ac feeder supplying a group of six elevators. The 460-V ac drive motor nameplate rating of the largest MG set for one elevator is 40 hp and 52 A, and the remaining elevators each have a 30-hp, 40-A, ac drive motor rating for their MG sets. In addition to a motor controller, each elevator has a separate motion/operation controller rated 10 A continuous to operate microprocessors, relays, power supplies, and the elevator car door operator. The MG sets are rated continuous.

Conductor Ampacity. Conductor ampacity is determined as follows:

(a) Per 620.13(D) and 620.61(B)(1), use Table 430.22(E), for intermittent duty (elevators). For intermittent duty using a continuous rated motor, the percentage of nameplate current rating to be used is 140%.

(b) For the 30-hp ac drive motor,

$$140\% \times 40 \text{ A} = 56 \text{ A.}$$

(c) For the 40-hp ac drive motor,

$$140\% \times 52 \text{ A} = 73 \text{ A.}$$

(d) The total conductor ampacity is the sum of all the motor currents:

$$(1 \text{ motor} \times 73 \text{ A}) + (5 \text{ motors} \times 56 \text{ A}) = 353 \text{ A}$$

(e) Per 620.14 and Table 620.14, the conductor (feeder) ampacity would be permitted to be reduced by the use of a demand factor. Constant loads are not included *(see 620.14, FPN)*. For six elevators, the demand factor is 0.79. The feeder diverse ampacity is, therefore, 0.79×353 A = 279 A.

(f) Per 430.24 and 215.3, the controller continuous current is $125\% \times 10$ A = 12.5 A

(g) The total feeder ampacity is the sum of the diverse current and all the controller continuous current.

$$I_{\text{total}} = 279 \text{ A} + (6 \text{ elevators} \times 12.5 \text{ A}) = 354 \text{ A}$$

(h) This ampacity would be permitted to be used to select the wire size. See Figure D9.

Figure D9 Generator Field Control.

Example D10 Feeder Ampacity Determination for Adjustable Speed Drive Control *[see 215.2, 430.24, 620.13, 620.14, 620.61, and Table 430.22(E)]*

Determine the conductor ampacity for a 460-V, 3-phase, 60-Hz ac feeder supplying a group of six identical elevators. The system is adjustable-speed SCR dc drive. The power transformers are external to the drive (motor controller) cabinet. Each elevator has a separate motion/operation controller connected to the load side of the main line disconnect switch rated 10 A continuous to operate microprocessors, relays, power supplies, and the elevator car door operator. Each transformer is rated 95 kVA with an efficiency of 90%.

Conductor Ampacity

Conductor ampacity is determined as follows:

(a) Calculate the nameplate rating of the transformer:

$$I = \frac{95 \text{ kVa} \times 1000}{\sqrt{3} \times 460 \text{ V} \times 0.90_{\text{eff.}}}$$

(b) Per 620.13(D), for six elevators, the total conductor ampacity is the sum of all the currents.

$$6 \text{ elevators} \times 133 \text{ A} = 798 \text{ A}$$

(c) Per 620.14 and Table 620.14, the conductor (feeder) ampacity would be permitted to be reduced by the use of a demand factor. Constant loads are not included *(see 620.13, FPN No. 2)*. For six elevators, the demand factor is 0.79. The feeder diverse ampacity is, therefore, 0.79×798 A = 630 A.

(d) Per 430.24 and 215.3, the controller continuous current is $125\% \times 10$ A = 12.5 A.

(e) The total feeder ampacity is the sum of the diverse current and all the controller constant current.

$$I_{\text{total}} = 630 \text{ A} + (6 \text{ elevators} \times 12.5 \text{ A}) = 705 \text{ A}$$

(f) This ampacity would be permitted to be used to select the wire size. See Figure D10.

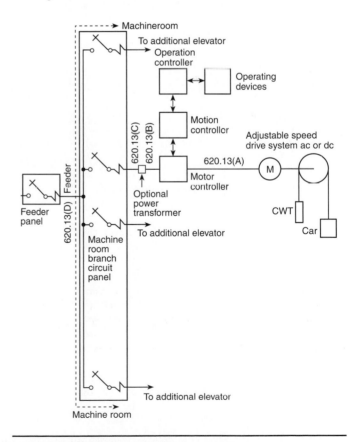

Figure D10 Adjustable Speed Drive Control.

Example D11 Mobile Home (see 550.18)

A mobile home floor is 70 ft by 10 ft and has two small appliance circuits; a 1000-VA, 240-V heater; a 200-VA, 120-V exhaust fan; a 400-VA, 120-V dishwasher; and a 7000-VA electric range.

Lighting and Small Appliance Load

Lighting (70 ft × 10 ft × 3 VA per ft²)		2,100 VA
Small appliance (1500 VA × 2 circuits)		3,000 VA
Laundry (1500 VA × 1 circuit)		1,500 VA
	Subtotal	6,600 VA
First 3000 VA at 100%		3,000 VA
Remainder (6600 VA – 3000 VA = 3600 VA) × 35%		1,260 VA
	Total	4,260 VA

4260 VA ÷ 240 V = 17.75 A per leg

Amperes per Leg	Leg A	Leg B
Lighting and appliances	17.75	17.75
Heater (1000 VA ÷ 240 V)	4.20	4.20
Fan (200 VA × 125% ÷ 120 V)	2.08	—
Dishwasher (400 VA ÷ 120 V)	—	3.30
Range (7000 VA × 0.8 ÷ 240 V)	23.30	23.30
Total amperes per leg	47.33	48.55

Based on the higher current calculated for either leg, a minimum 50-A supply cord would be required.

For SI units, 0.093 m² = 1 ft² and 0.3048 m = 1 ft.

Example D12 Park Trailer (see 552.47)

A park trailer floor is 40 ft by 10 ft and has two small appliance circuits, a 1000-VA, 240-V heater, a 200-VA, 120-V exhaust fan, a 400-VA, 120-V dishwasher, and a 7000-VA electric range.

Lighting and Small Appliance Load

Lighting (40 ft × 10 ft × 3 VA per ft²)		1,200 VA
Small appliance (1500 VA × 2 circuits)		3,000 VA
Laundry (1500 VA × 1 circuit)		1,500 VA
	Subtotal	5,700 VA
First 3000 VA at 100%		3,000 VA
Remainder (5700 VA – 3000 VA = 2700 VA) × 35%		945 VA
	Total	3,945 VA

3945 VA ÷ 240 V = 16.44 A per leg

Amperes per Leg	Leg A	Leg B
Lighting and appliances	16.44	16.44
Heater (1000 VA ÷ 240 V)	4.20	4.20
Fan (200 VA × 125% ÷ 120 V)	2.08	—
Dishwasher (400 VA ÷ 120 V)	—	3.3
Range (7000 VA × 0.8 ÷ 240 V)	23.30	23.30
Totals	46.02	47.24

Based on the higher current calculated for either leg, a minimum 50-A supply cord would be required.

For SI units, 0.093 m² = 1 ft² and 0.3048 m = 1 ft.

Annex E

Types of Construction

This annex is not a part of the requirements of this NFPA document but is included for informational purposes only.

The five different types of construction can be summarized briefly as follows (see also Table E.2):

Type I is a Fire-Resistive construction type. All structural elements and most interior elements are required to be noncombustible. Interior, nonbearing partitions are permitted to be 1 or 2 hour rated. For nearly all occupancy types, Type I construction can be of unlimited height.

Type II construction has 3 categories, Fire-Resistive,

Table E.1 Fire Resistance Ratings (in hours) for Type I through Type V Construction

	Type I		Type II			Type III		Type IV	Type V	
	443	332	222	111	000	211	200	2HH	111	000
Exterior Bearing Walls —										
Supporting more than one floor, columns, or other										
bearing walls	4	3	2	1	0^1	2	2	2	1	0^1
Supporting one floor only	4	3	2	1	0^1	2	2	2	1	0^1
Supporting a roof only	4	3	1	1	0^1	2	2	2	1	0^1
Interior Bearing Walls —										
Supporting more than one floor, columns, or other										
bearing walls	4	3	2	1	0	1	0	2	1	0
Supporting one floor only	3	2	2	1	0	1	0	2	1	0
Supporting a roof only	3	2	1	1	0	1	0	1	1	0
Columns —										
Supporting more than one floor, columns, or other										
bearing walls	4	3	2	1	0	1	0	H^2	1	0
Supporting one floor only	3	2	2	1	0	1	0	H^2	1	0
Supporting roofs only	3	2	1	1	0	1	0	H^2	1	0
Beams, Girders, Trusses & Arches —										
Supporting more than one floor, columns, or other										
bearing walls	4	3	2	1	0	1	0	H^2	1	0
Supporting one floor only	3	2	2	1	0	1	0	H^2	1	0
Supporting roofs only	3	2	1	1	0	1	0	H^2	1	0
Floor Construction	3	2	2	1	0	1	0	H^2	1	0
Roof Construction	2	1½	1	1	0	1	0	H^2	1	0
Exterior Nonbearing Walls³	0^1	0^1	0^1	0^1	0^1	0^1	0^1	0^1	0^1	0^1

     Those members that shall be permitted to be of approved combustible material.

Source: Table 3.1 from NFPA 220, *Standard on Types of Building Construction,* 1999.

[1]See A-3-1 in NFPA 220.

[2]"H" indicates heavy timber members; see text for requirements.

[3]Exterior nonbearing walls meeting the conditions of acceptance of NFPA 285, *Standard Method of Test for the Evaluation of Flammability Characteristics of Exterior Non-Load-Bearing Wall Assemblies Containing Combustible Components Using the Intermediate-Scale, Multistory Test Apparatus,* shall be permitted to be used.

Table E.2 Maximum Number of Stories for Types V, IV, and III Construction

Construction Type	Maximum Number of Stories Permitted
V Non-Rated	2
V Non-Rated, Sprinklered	3
V One-Hour Rated	3
V One-Hour Rated, Sprinklered	4
IV Heavy Timber	4
IV Heavy Timber, Sprinklered	5
III Non-Rated	2
III Non-Rated, Sprinklered	3
III One-Hour Rated	4
III One-Hour Rated, Sprinklered	5

One-Hour Rated and Non-Rated. The number of stories permitted for multifamily dwellings varies from 2 for Non-Rated and 4 for One-Hour Rated to 12 for Fire-Resistive construction.

Type III construction has two categories, One-Hour Rated and Non-Rated. Both categories require the structural framework and exterior walls to be of noncombustible material. One-Hour Rated construction requires all interior partitions to be one-hour rated. Non-Rated construction allows nonbearing interior partitions to be of non-rated construction. The maximum permitted number of stories for multifamily dwellings and other structures is 2 for Non-Rated and 4 for One-Hour Rated.

Type IV is a single construction category which provides for heavy timber construction. Both the structural framework and the exterior walls are required to be noncombustible except that wood members of certain minimum sizes are allowed. This construction type is seldom used for multifamily dwellings but, if used, would be permitted to be 4 stories high.

Type V construction has two categories, One-Hour Rated and Non-Rated. One-Hour Rated construction requires a minimum of one-hour rated construction throughout the building. Non-rated construction allows non-rated interior partitions with certain restrictions. The maximum permitted number of stories for multifamily dwellings and other structures is 2 for Non-Rated and 3 for One-Hour Rated.

Commentary Table E.1 is reproduced from the NFPA *101*®-2003, *Life Safety Code Handbook*®. This table cross-references the building construction types described in NFPA 220, *Standard on Types of Building Construction*, and *NFPA 5000*®-2003, *Building Construction and Safety Code*, to the construction types described in four other model building codes. For authorities having jurisdiction in a municipality where one of these model building codes is used, this table provides helpful information to assist in the proper application of 334.10(2) and (3) covering the permitted use of Type NM cable based on building construction type. The types of construction described in 334.10(2) and (3) are based on NFPA 220-1999, and this table facilitates assimilation of the 334.10 requirements with the construction types contained within the building code that is adopted by a jurisdiction.

The following is a description of the model building code acronyms contained in Commentary Table E.1:

UBC — *Uniform Building Code,* International Conference of Building Officials, Whittier, CA

BNBC — *BOCA National Building Code,* Building Officials and Code Administrators International Inc., Country Club Hills, IL

SBC — *Standard Building Code,* Southern Building Code Congress International, Inc., Birmingham, AL

IBC — *International Building Code,* International Code Council, Inc., Falls Church, VA

Commentary Table E.1 Cross-Reference of Building Construction Types

NFPA 220 & NFPA 5000	I(443)*	I(332)	II(222)	II(111)	II(000)	III(211)	III(200)	IV(2HH)	V(111)	V(000)
UBC	—	I FR	II FR	II 1-hr	II N	III 1-hr	III N	IV HT	V 1-hr	V-N
BNBC	1A	1B	2A	2B	2C	3A	3B	4	5A	5B
SBC	I	II	—	IV 1-hr	IV unprotected	V 1-hr	V unprotected	III	VI 1-hr	VI unprotected
IBC	I-A	I-B	—	II-A	II-B	III-A	III-B	IV	V-A	V-B

* I(442) per *NFPA 5000, Building Construction and Safety Code.*
Source: Life Safety Code® Handbook, 2003 edition, Ron Coté and Gregory Harrington, eds., National Fire Protection Association, Quincy, MA.

Annex F

Cross-Reference Tables

This annex is not a part of the requirements of this NFPA document but is included for informational purposes only.

Because Chapter 3 was extensively reorganized for the 2002 *Code*, Annex F is provided to make the *Code* more user friendly. This annex provides cross-references between the 2005, 2002, and 1999 editions of the *Code* as well as an alphabetical cross-index.

Table F.1 Chapter 3 Cross-Reference from the 2005 *NEC* to the 2002 *NEC* to the 1999 *NEC*

2005 *NEC*	2002 *NEC*	1999 *NEC*	Article Title
300	300	300	Wiring Methods
310	310	310	Conductors for General Wiring
312	312	373	Cabinets, Cutout Boxes, and Meter Socket Enclosures
314	314	370	Outlet, Device, Pull, and Junction Boxes; Conduit Bodies; Fittings; and Manholes
320	320	333	Armored Cable: Type AC
322	322	363	Flat Cable Assemblies: Type FC
324	324	328	Flat Conductor Cable: Type FCC
326	326	325	Integrated Gas Spacer Cable: Type IGS
328	328	326	Medium Voltage Cable: Type MV
330	330	334	Metal-Clad Cable: Type MC
332	332	330	Mineral-Insulated, Metal-Sheathed Cable: Type MI
334	334	336	Nonmetallic-Sheathed Cable: Types NM, NMC, and NMS
336	336	340	Power and Control Tray Cable: Type TC
338	338	338	Service-Entrance Cable: Types SE and USE
340	340	339	Underground Feeder and Branch-Circuit Cable: Type UF
342	342	345	Intermediate Metal Conduit: Type IMC
344	344	346	Rigid Metal Conduit: Type RMC
348	348	350	Flexible Metal Conduit: Type FMC
350	350	351 (Part A)	Liquidtight Flexible Metal Conduit: Type LFMC
352	352	347	Rigid Nonmetallic Conduit: Type RNC
353 (New)	—	—	High Density Polyethylene Conduit: Type HDPE Conduit
354	354	343	Nonmetallic Underground Conduit with Conductors: Type NUCC
356	356	351 (Part B)	Liquidtight Flexible Nonmetallic Conduit: Type LFNC
358	358	348	Electrical Metallic Tubing: Type EMT
360	360	349	Flexible Metallic Tubing: Type FMT
362	362	331	Electrical Nonmetallic Tubing: Type ENT
366	366	374	Auxiliary Gutters
368	368	364	Busways
370	370	365	Cablebus
372	372	358	Cellular Concrete Floor Raceways
374	374	356	Cellular Metal Floor Raceways
376	376	362 (Part A)	Metal Wireways

(continues)

Table F.1 *Continued*

2005 *NEC*	2002 *NEC*	1999 *NEC*	Article Title
378	378	362 (Part B)	Nonmetallic Wireways
380	380	353	Multioutlet Assembly
382	382	342	Nonmetallic Extensions
384	384	352 (Part C)	Strut-Type Channel Raceway
386	386	352 (Part A)	Surface Metal Raceways
388	388	352 (Part B)	Surface Nonmetallic Raceways
390	390	354	Underfloor Raceways
392	392	318	Cable Trays
394	394	324	Concealed Knob-and-Tube Wiring
396	396	321	Messenger Supported Wiring
398	398	320	Open Wiring on Insulators
404	404	380	Switches
408	408	384	Switchboards and Panelboards
590	527	305	Temporary Installations

Table F.2 Chapter 3 Cross-Reference, 1999 – 2002 – 2005 *NEC*

1999 *NEC*	2002 *NEC*	2005 *NEC*	Article Title
300	300	300	Wiring Methods
305	527	590	Temporary Installations
310	310	310	Conductors for General Wiring
318	392	392	Cable Trays
320	398	398	Open Wiring on Insulators
321	396	396	Messenger Supported Wiring
324	394	394	Concealed Knob-and-Tube Wiring
325	326	326	Integrated Gas Spacer Cable: Type IGS
326	328	328	Medium Voltage Cable: Type MV
328	324	324	Flat Conductor Cable: Type FCC
330	332	332	Mineral-Insulated, Metal-Sheathed Cable: Type MI
331	362	362	Electrical Nonmetallic Tubing: Type ENT
333	320	320	Armored Cable: Type AC
334	330	330	Metal-Clad Cable: Type MC
336	334	334	Nonmetallic-Sheathed Cable: Types NM, NMC, and NMS
338	338	338	Service-Entrance Cable: Types SE and USE
339	340	340	Underground Feeder and Branch-Circuit Cable: Type UF
340	336	336	Power and Control Tray Cable: Type TC
342	382	382	Nonmetallic Extensions
343	354	354	Nonmetallic Underground Conduit with Conductors: Type NUCC
345	342	342	Intermediate Metal Conduit: Type IMC
346	344	344	Rigid Metal Conduit: Type RMC
347	352	352	Rigid Nonmetallic Conduit: Type RNC
348	358	358	Electrical Metallic Tubing: Type EMT
349	360	360	Flexible Metallic Tubing: Type FMT
350	348	348	Flexible Metal Conduit: Type FMC
351 (Part A)	350	350	Liquidtight Flexible Metal Conduit: Type LFMC
351 (Part B)	356	356	Liquidtight Flexible Nonmetallic Conduit: Type LFNC
352 (Part C)	384	384	Strut-Type Channel Raceway
352 (Part A)	386	386	Surface Metal Raceways
352 (Part B)	388	388	Surface Nonmetallic Raceways
353	380	380	Multioutlet Assembly
—	—	353 (New)	High Density Polyethylene Conduit: Type HDPE Conduit
354	390	390	Underfloor Raceways
356	374	374	Cellular Metal Floor Raceways
358	372	372	Cellular Concrete Floor Raceways
362 (Part A)	376	376	Metal Wireways
362 (Part B)	378	378	Nonmetallic Wireways
363	322	322	Flat Cable Assemblies: Type FC

Table F.2 *Continued*

1999 *NEC*	2002 *NEC*	2005 *NEC*	Article Title
364	368	368	Busways
365	370	370	Cablebus
370	314	314	Outlet, Device, Pull, and Junction Boxes; Conduit Bodies; Fittings; and Manholes
373	312	312	Cabinets, Cutout Boxes, and Meter Socket Enclosures
374	366	366	Auxiliary Gutters
380	404	404	Switches
384	408	408	Switchboards and Panelboards

Table F.3 Chapter 3 Alphabetical Cross-Reference, 2005 – 2002 – 1999 *NEC*

Article Title	2005 *NEC*	2002 *NEC*	1999 *NEC*
Armored Cable: Type 320 AC	320	320	333
Auxiliary Gutters	366	366	374
Busways	368	368	364
Cabinets, Cutout Boxes, and Meter Socket Enclosures	312	312	373
Cable Trays	392	392	318
Cablebus	370	370	365
Cellular Concrete Floor Raceways	372	372	358
Cellular Metal Floor Raceways	374	374	356
Concealed Knob-and-Tube Wiring	394	394	324
Conductors for General Wiring	310	310	310
Electrical Metallic Tubing: Type EMT	358	358	348
Electrical Nonmetallic Tubing: Type ENT	362	362	331
Flat Cable Assemblies: Type FC	322	322	363
Flat Conductor Cable: Type FCC	324	324	328
Flexible Metal Conduit: Type FMC	348	348	350
Flexible Metallic Tubing: Type FMT	360	360	349
High Density Polyethylene Conduit: Type HDPE Conduit	353 (New)	—	—
Integrated Gas Spacer Cable: Type IGS	326	326	325
Intermediate Metal Conduit: Type IMC	342	342	345
Liquidtight Flexible Metal Conduit: Type LFMC	350	350	351 (Part A)
Liquidtight Flexible Nonmetallic Conduit: Type LFNC	356	356	351 (Part B)
Medium Voltage Cable: Type MV	328	328	326
Messenger Supported Wiring	396	396	321
Metal Wireways	376	376	362 (Part A)
Metal-Clad Cable: Type MC	330	330	334
Mineral-Insulated, Metal-Sheathed Cable: Type MI	332	332	330
Multioutlet Assembly	380	380	353
Nonmetallic Extensions	382	382	342
Nonmetallic Underground Conduit with Conductors: Type NUCC	354	354	343
Nonmetallic Wireways	378	378	362 (Part B)
Nonmetallic-Sheathed Cable: Types NM, NMC, and NMS	334	334	336
Open Wiring on Insulators	398	398	320
Outlet, Device, Pull, and Junction Boxes; Conduit Bodies; Fittings; and Manholes	314	314	370
Power and Control Tray Cable: Type TC	336	336	340
Rigid Metal Conduit: Type RMC	344	344	346
Rigid Nonmetallic Conduit: Type RNC	352	352	347
Service-Entrance Cable: Types SE and USE	338	338	338
Strut-Type Channel Raceway	384	384	352 (Part C)
Surface Metal Raceways	386	386	352 (Part A)
Surface Nonmetallic Raceways	388	388	352 (Part B)
Switchboards and Panelboards	408	408	384
Switches	404	404	380
Temporary Installations	590	527	305
Underfloor Raceways	390	390	354
Underground Feeder and Branch-Circuit Cable: Type UF	340	340	339
Wiring Methods	300	300	300

Article 250 underwent an extensive reorganization for the 1999 *NEC*. Commentary Table F.1 and Commentary Table F.2 provide a matrix of the Article 250 requirements for the 1996, 1999, 2002, and 2005 editions of the *Code*.

Commentary Table F.1 Article 250 Cross-References by 1996 Topic — 1996, 1999, 2002, and 2005 *NEC*

1996 Article 250 Topic	1996 *NEC*	1999 NEC	2002 NEC	2005 NEC
Part A. General				
Scope	-1	-1	.1	.1
Application of Other Articles	-2	-4	.3	.3
Part B. Circuit and System Grounding				
Direct-Current Systems	-3	-162	.162	.162
Alternating-Current Circuits and Systems to Be Grounded	-5	-20, -21	.20, .21	.20, .21
Portable and Vehicle-Mounted Generators	-6	-34	.34	.34
Circuits Not to Be Grounded	-7	-22	.22	.22
Part C. Location of System Grounding Connections				
Objectionable Current Over Grounding Conductors	-21	-6	.6	.6
Point of Connection for Direct-Current Systems	-22	-164	.164	.164
Grounding Service-Supplied Alternating-Current Systems	-23	-24	.24	.24
Two or More Buildings or Structures Supplied from a Common Service	-24	-32	.32	.32
Conductor to Be Grounded—Alternating-Current Systems	-25	-26	.26	.26
Grounding Separately Derived Alternating-Current Systems	-26	-30	.30	.30
High-Impedance Grounded Neutral System Connections	-27	-36	.36	.36
Part D. Enclosure and Raceway Grounding				
Service Raceways and Enclosures	-32	-80	.80	.80
Other Conductor Enclosures and Raceways	-33	-86	.86	.86
Part E. Equipment Grounding				
Equipment Fastened in Place or Connected by Permanent Wiring Methods (Fixed)	-42	-110	.110	.110
Fastened in Place or Connected by Permanent Wiring Methods (Fixed)—Specific	-43	-112	.112	.112
Nonelectric Equipment	-44	-116	.116	.116
Equipment Connected by Cord and Plug	-45	-114	.114	.114
Spacing from Lightning Rods	-46	-106	.106	.106
Part F. Methods of Grounding				
Equipment Grounding Conductor Connections	-50	-130	.130	.130
Effective Grounding Path	-51	-2	.4	.4
Grounding Path to Grounding Electrode at Services	-53	-24, -28	.24, .28.	24, .28
Common Grounding Electrode	-54	-58	.58	.58
Underground Service Cable	-55	-84	.84	.84
Short Sections of Raceway	-56	-132	.132	.132
Equipment Fastened in Place or Connected by Permanent Wiring Methods (Fixed)—Grounding	-57	-119, -134	.119, .134	.119, .134
Equipment Considered Effectively Grounded	-58	-136	.136	.136
Cord-and-Plug-Connected Equipment	-59	-138	.138	.138
Frames of Ranges and Clothes Dryers	-60	-140	.140	.140
Use of Grounded Circuit Conductor for Grounding Equipment	-61	-142	.142	.142
Multiple Circuit Connections	-62	-144	.144	.144
Part G. Bonding				
General	-70	-90	.90	.90

Commentary Table F.1 *Continued*

1996 Article 250 Topic	1996 NEC	1999 NEC	2002 NEC	2005 NEC
Service Equipment	-71	-92	.92	.92
Method of Bonding Service Equipment	-72	-94	.92(B)	.92(B)
Metal Armor or Tape of Service Cable	-73	(removed)	(removed)	(removed)
Connecting Receptacle Grounding Terminal to Box	-74	-146	.146	.146
Bonding Other Enclosures	-75	-96	.96	.96
Bonding for Over 250 Volts	-76	-97	.97	.97
Bonding Loosely Jointed Metal Raceways	-77	-98	.98	.98
Bonding in Hazardous (Classified) Locations	-78	-100	.100	.100
Main and Equipment Bonding Jumpers	-79	-28, -102, -168	.28, .102, .168	.28, .102, .168
Bonding of Piping Systems and Exposed Structural Steel	-80	-104	.104	.104
Part H. Grounding Electrode System				
Grounding Electrode System	-81	-50	.50	.50
Made and Other Electrodes	-83	-52	.50, .53	.50, .53
Resistance of Made Electrodes	-84	-56	.56	.56
Use of Lightning Rods	-86	-60	.60	.60
Part J. Grounding Conductors				
Material	-91	-54, -62, -64, -118	.54, .62, .64, .118	.54, .62, .64, .118
Installation	-92	-64, -120	.64, .120	.64, .120
Size of Direct-Current Grounding Electrode Conductor	-93	-166	.166	.166
Size of Alternating-Current Grounding Electrode Conductor	-94	-66	.66	.66
Size of Equipment Grounding Conductors	-95	-122	.122	.122
Outline Lighting	-97	(600-7)	(600.7)	(600.7)
Equipment Grounding Conductor Continuity	-99	-124	.124	.124
Part K. Grounding Conductor Connections				
To Grounding Electrode	-112	-68	.68	.68
To Conductors and Equipment	-113	-8	.8	.8
Continuity and Attachment of Equipment Grounding Conductors to Boxes	-114	-148	.148	.148
Connection to Electrodes	-115	-70	.70	.70
Protection of Attachment	-117	-10	.10	.10
Clean Surfaces	-118	-12	.12	.12
Identification of Wiring Device Terminals	-119	-126	.126	.126
Part L. Instrument Transformers, Relays, Etc.				
Instrument Transformer Circuits	-121	-170	.170	.170
Instrument Transformer Cases	-122	-172	.172	.172
Cases of Instruments, Meters, and Relays Operating at Less Than 1000 Volts	-123	-174	.174	.174
Cases of Instruments, Meters, and Relays Operating at Voltage 1 kV and Over	-124	-176	.176	.176
Instrument Grounding Conductor	-125	-178	.178	.178
Part M. Grounding of Systems and Circuits of 1 kV and Over (High Voltage)				
General	-150	-180	.180	.180
Derived Neutral Systems	-151	-182	.182	.182
Solidly Grounded Neutral Systems	-152	-184	.184	.184
Impedance Grounded Neutral Systems	-153	-186	.186	.186
Grounding of Systems Supplying Portable or Mobile Equipment	-154	-188	.188	.188
Grounding of Equipment	-155	-190	.190	.190

Commentary Table F.2 Article 250 Cross-References by 2005 Topic — 2005, 2002, 1999, and 1996 *NEC*

2005 Article 250 Topic	2005 *NEC*	2002 *NEC*	1999 *NEC*	1996 *NEC*
Part I. General				
Scope	.1	.1	-1	-1
Definitions	.2	.2	N/A	N/A
Application of Other Articles	.3	.3	-4	-2
General Requirements for Grounding and Bonding	.4	.4	-2	-51, -1 FPN 1 & 2
Objectionable Current Over Grounding Conductors	.6	.6	-6	-21
Connection of Grounding and Bonding Equipment	.8	.8	-8	-113
Protection of Ground Clamps and Fittings	.10	.10	-10	-117
Clean Surfaces	.12	.12	-12	-118
Part II. System Grounding				
Alternating-Current Circuits and Systems to Be Grounded	.20	.20	-20	-5
Alternating-Current Systems of 50 Volts to 1000 Volts Not Required to be Grounded	.21	.21	-21	-5
Circuits Not to Be Grounded	.22	.22	-22	-7
Grounding Service-Supplied Alternating-Current Systems	.24	.24	-24	-23, -53(a)
Conductor to Be Grounded—Alternating-Current Systems	.26	.26	-26	-25
Main Bonding Jumper and System Bonding Jumper (System Bonding Jumper new in 2005 *Code*)	.28	.28	-28	-79, -53(b)
Grounding Separately Derived Alternating-Current Systems	.30	.30	-30	-26
Buildings or Structures Supplied by Feeder(s) or Branch Circuit(s)	.32	.32	-32	-24
Portable and Vehicle-Mounted Generators	.34	.34	-34	-6
High-Impedance Grounded Neutral Systems	.36	.36	-36	-27
Part III. Grounding Electrode System and Grounding Electrode Conductor				
Grounding Electrode System	.50	.50	-50	-81
Grounding Electrodes	.52	.52	-50, -52	-81, -83
Grounding Electrode System Installation	.53	.53	-50, -52	-81, -83
Supplementary Grounding Electrodes	.54	.54	-54	-91(c)
Resistance of Rod, Pipe, and Plate Electrodes	.56	.56	-56	-84
Common Grounding Electrode	.58	.58	-58	-54
Use of Air Terminals	.60	.60	-60	-86
Grounding Electrode Conductor Material	.62	.62	-62	-91(a)
Grounding Electrode Conductor Installation	.64	.64	-64	-91(a), -92
Size of Alternating-Current Grounding Electrode Conductor	.66	.66	-66	-94
Grounding Electrode Conductor and Bonding Jumper Connection to Electrodes	.68	.68	-68	-112
Methods of Grounding and Bonding Conductor Connection to Electrodes	.70	.70	-70	-115
Part IV. Enclosure, Raceway, and Service Cable Grounding				
Service Raceways and Enclosures	.80	.80	-80	-32
Underground Service Cable or Raceway	.84	.84	-84	-55
Other Conductor Enclosures and Raceways	.86	.86	-86	-33
Part V. Bonding				
General	.90	.90	-90	-70
Services	.92	.92	-92	-71
Bonding for Other Systems	.94	.94	-92(b)	-72
Bonding Other Enclosures	.96	.96	-96	-75
Bonding for Over 250 Volts	.97	.97	-97	-76
Bonding Loosely Jointed Metal Raceways	.98	.98	-98	-77
Bonding in Hazardous (Classified) Locations	.100	.100	-100	-78
Equipment Bonding Jumpers	.102	.102	-102	-79

Commentary Table F.2 *Continued*

2005 Article 250 Topic	2005 *NEC*	2002 NEC	1999 NEC	1996 NEC
Bonding of Piping Systems and Exposed Structural Steel	.104	.104	-104	-80
Lightning Protection Systems	.106	.106	-106	-46
Part VI. Equipment Grounding and Equipment Grounding Conductors				
Equipment Fastened in Place or Connected by Permanent Wiring Methods (Fixed)	.110	.110	-110	-42
Fastened in Place or Connected by Permanent Wiring Methods (Fixed)—Specific	.112	.112	-112	-43
Equipment Connected by Cord and Plug	.114	.114	-114	-45
Nonelectric Equipment	.116	.116	-116	-44
Types of Equipment Grounding Conductors	.118	.118	-118	-91(b)
Identification of Equipment Grounding Conductors	.119	.119	-119	-57(b)
Equipment Grounding Conductor Installation	.120	.120	-120	-92(c)
Size of Equipment Grounding Conductors	.122	.122	-122	-95
Equipment Grounding Conductor Continuity	.124	.124	-124	-99
Identification of Wiring Device Terminals	.126	.126	-126	-119
Part VII. Methods of Equipment Grounding				
Equipment Grounding Conductor Connections	.130	.130	-130	-50
Short Sections of Raceway	.132	.132	-132	-56
Equipment Fastened in Place or Connected by Permanent Wiring Methods (Fixed)—Grounding	.134	.134	-134	-57
Equipment Considered Effectively Grounded	.136	.136	-136	-58
Cord-and-Plug-Connected Equipment	.138	.138	-138	-59
Frames of Ranges and Clothes Dryers	.140	.140	-140	-60
Use of Grounded Circuit Conductor for Grounding Equipment	.142	.142	-142	-61
Multiple Circuit Connections	.144	.144	-144	-62
Connecting Receptacle Grounding Terminal to Box	.146	.146	-146	-74
Continuity and Attachment of Equipment Grounding Conductors to Boxes	.148	.148	-148	-114
Part VIII. Direct-Current Systems				
General	.160	.160	-160	(new)
Direct-Current Circuits and Systems to Be Grounded	.162	.162	-162	-3
Point of Connection for Direct-Current Systems	.164	.164	-164	-22
Size of Direct-Current Grounding Electrode Conductor	.166	.166	-166	-93
Direct-Current Bonding Jumper	.168	.168	-168	-79(d)
Ungrounded Direct-Current Separately Derived Systems	.169	.169	-169	(new)
Part IX. Instruments, Meters, and Relays				
Instrument Transformer Circuits	.170	.170	-170	-121
Instrument Transformer Cases	.172	.172	-172	-122
Cases of Instruments, Meters, and Relays Operating at Less Than 1000 Volts	.174	.174	-174	-123
Cases of Instruments, Meters, and Relays — Operating Voltage 1 kV and Over	.176	.176	-176	-124
Instrument Grounding Conductor	.178	.178	-178	-125
Part X. Grounding of Systems and Circuits of 1 kV and Over (High Voltage)				
General	.180	.180	-180	-150
Derived Neutral Systems	.182	.182	-182	-151
Solidly Grounded Neutral Systems	.184	.184	-184	-152
Impedance Grounded Neutral Systems	.186	.186	-186	-153
Grounding of Systems Supplying Portable or Mobile Equipment	.188	.188	-188	-154
Grounding of Equipment	.190	.190	-190	-155

Annex G

Administration and Enforcement

Annex G is not a part of the requirements of this NFPA document and is included for informational purposes only. This annex is informative unless specifically adopted by the local jurisdiction adopting the National Electrical Code®.

Summary of Changes
- **Annex G:** Article 80 relocated as new Annex G.

Contents

Annex G is not new for the 2005 *Code*. It appeared as Article 80 and in Chapter 1 for the 2002 *Code*, but because the material is considered informative, it more appropriately belongs in an annex. The purpose of this administrative annex is to assist jurisdictions that do not have formalized electrical inspection procedures but want to amend their inspection laws. It is intended to serve as a guide to the adoption of the *Code*. Because Annex G, Article 80, is offered as a guide to the adoption process, its application is not mandatory unless it is specifically adopted by local law, as stated in the note after the title.

Annex G, Article 80, has an interesting history. Originally it was a model law that provided for inspection of electrical installations. It was prepared by the NFPA Electrical Field Service Advisory Committee to serve as a guide for those jurisdictions that either did not have formalized electrical inspection procedures or wanted to amend their electrical inspection laws, in addition to serving as a guide for adopting the *NEC*. The model law was intended for use by states as well as municipalities. The first edition was adopted by NFPA on May 15, 1973, and a second edition was approved on March 27, 1987.

Annex G, Article 80, continues to cover such issues as creation of an electrical board, plan review, and inspection. Professional qualifications of electrical inspectors and the investigation of fires attributed to electrical installations have been added.

Adoption of the *NEC* can occur in two ways. It can be incorporated in a law, or a law can be enacted authorizing a governmental agency or board to adopt it. To facilitate the drafting of such laws, as well as establishment of the accompanying inspection and enforcement procedures, NFPA offers Annex G, Article 80, Administration and Enforcement. Annex G, Article 80, may require modification to comply with the structure-writing rules of the adopting political jurisdiction.

Circumstances in a particular jurisdiction determine which alternative is more appropriate. Provisions that are less comprehensive may be adequate for smaller political subdivisions.

80.1 Scope. The following functions are covered:

(1) The inspection of electrical installations as covered by 90.2
(2) The investigation of fires caused by electrical installations
(3) The review of construction plans, drawings, and specifications for electrical systems
(4) The design, alteration, modification, construction, maintenance, and testing of electrical systems and equipment
(5) The regulation and control of electrical installations at special events including but not limited to exhibits, trade shows, amusement parks, and other similar special occupancies

80.2 Definitions.

Authority Having Jurisdiction. The organization, office, or individual responsible for approving equipment, materials, an installation, or a procedure.

Chief Electrical Inspector. An electrical inspector who either is the authority having jurisdiction or is designated by the authority having jurisdiction and is responsible for administering the requirements of this *Code*.

Electrical Inspector. An individual meeting the requirements of 80.27 and authorized to perform electrical inspections.

80.3 Purpose. The purpose of this article shall be to provide requirements for administration and enforcement of the *National Electrical Code*.

80.5 Adoption. Article 80 shall not apply unless specifically adopted by the local jurisdiction adopting the *National Electrical Code*.

80.7 Title. The title of this *Code* shall be NFPA 70, *National Electrical Code*®, of the National Fire Protection Association. The short title of this *Code* shall be the *NEC*®.

80.9 Application.

(A) New Installations. This *Code* applies to new installations. Buildings with construction permits dated after adoption of this *Code* shall comply with its requirements.

(B) Existing Installations. Existing electrical installations that do not comply with the provisions of this *Code* shall be permitted to be continued in use unless the authority having jurisdiction determines that the lack of conformity with this *Code* presents an imminent danger to occupants. Where changes are required for correction of hazards, a reasonable amount of time shall be given for compliance, depending on the degree of the hazard.

(C) Additions, Alterations, or Repairs. Additions, alterations, or repairs to any building, structure, or premises shall conform to that required of a new building without requiring the existing building to comply with all the requirements of this *Code*. Additions, alterations, installations, or repairs shall not cause an existing building to become unsafe or to adversely affect the performance of the building as determined by the authority having jurisdiction. Electrical wiring added to an existing service, feeder, or branch circuit shall not result in an installation that violates the provisions of the *Code* in force at the time the additions are made.

80.11 Occupancy of Building or Structure.

(A) New Construction. No newly constructed building shall be occupied in whole or in part in violation of the provisions of this *Code*.

(B) Existing Buildings. Existing buildings that are occupied at the time of adoption of this *Code* shall be permitted to remain in use provided the following conditions apply:

(1) The occupancy classification remains unchanged
(2) There exists no condition deemed hazardous to life or property that would constitute an imminent danger

80.13 Authority. Where used in this article, the term *authority having jurisdiction* shall include the chief electrical inspector or other individuals designated by the governing body. This *Code* shall be administered and enforced by the authority having jurisdiction designated by the governing authority as follows.

(1) The authority having jurisdiction shall be permitted to render interpretations of this *Code* in order to provide clarification to its requirements, as permitted by 90.4.

(2) When the use of any electrical equipment or its installations is found to be dangerous to human life or property, the authority having jurisdiction shall be empowered to have the premises disconnected from its source of electric supply, as established by the Board. When such equipment or installation has been so condemned or disconnected, a notice shall be placed thereon listing the causes for the condemnation, the disconnection, or both and the penalty under 80.23 for the unlawful use thereof. Written notice of such condemnation or disconnection and the causes therefor shall be given within 24 hours to the owners, the occupant, or both, of such building, structure, or premises. It shall be unlawful for any person to remove said notice, to reconnect the electric equipment to its source of electric supply, or to use or permit to be used electric power in any such electric equipment until such causes for the condemnation or disconnection have been remedied to the satisfaction of the inspection authorities.

(3) The authority having jurisdiction shall be permitted to delegate to other qualified individuals such powers as necessary for the proper administration and enforcement of this *Code*.

(4) Police, fire, and other enforcement agencies shall have authority to render necessary assistance in the enforcement of this *Code* when requested to do so by the authority having jurisdiction.

(5) The authority having jurisdiction shall be authorized to inspect, at all reasonable times, any building or premises for dangerous or hazardous conditions or equipment as set forth in this *Code*. The authority having jurisdiction shall be permitted to order any person(s) to remove or remedy such dangerous or hazardous condition or equipment. Any person(s) failing to comply with such order shall be in violation of this *Code*.

(6) Where the authority having jurisdiction deems that conditions hazardous to life and property exist, he or she shall be permitted to require that such hazardous conditions in violation of this *Code* be corrected.

(7) To the full extent permitted by law, any authority having jurisdiction engaged in inspection work shall be authorized at all reasonable times to enter and examine any building, structure, or premises for the purpose of making electrical inspections. Before entering a premises, the authority having jurisdiction shall obtain the consent of the occupant thereof or obtain a court warrant authorizing entry for the purpose of inspection except in those instances where an emergency exists. As used in this section, *emergency* means circumstances that the authority having jurisdiction knows, or has reason to believe, exist and that reasonably can constitute immediate danger to persons or property.

(8) Persons authorized to enter and inspect buildings, structures, and premises as herein set forth shall be identified by proper credentials issued by this governing authority.

(9) Persons shall not interfere with an authority having jurisdiction carrying out any duties or functions prescribed by this *Code*.

(10) Persons shall not use a badge, uniform, or other credentials to impersonate the authority having jurisdiction.

(11) The authority having jurisdiction shall be permitted to investigate the cause, origin, and circumstances of any fire, explosion, or other hazardous condition.

(12) The authority having jurisdiction shall be permitted to require plans and specifications to ensure compliance with this *Code*.

(13) Whenever any installation subject to inspection prior to use is covered or concealed without having first been inspected, the authority having jurisdiction shall be permitted to require that such work be exposed for inspection. The authority having jurisdiction shall be notified when the installation is ready for inspection and shall conduct the inspection within _____ days.

(14) The authority having jurisdiction shall be permitted to order the immediate evacuation of any occupied building deemed unsafe when such building has hazardous conditions that present imminent danger to building occupants.

(15) The authority having jurisdiction shall be permitted to waive specific requirements in this *Code* or permit alternative methods where it is assured that equivalent objectives can be achieved by establishing and maintaining effective safety. Technical documentation shall be submitted to the authority having jurisdiction to demonstrate equivalency and that the system, method, or device is approved for the intended purpose.

(16) Each application for a waiver of a specific electrical requirement shall be filed with the authority having jurisdiction and shall be accompanied by such evidence, letters, statements, results of tests, or other supporting information as required to justify the request. The authority having jurisdiction shall keep a record of actions on such applications, and a signed copy of the authority having jurisdiction's decision shall be provided for the applicant.

80.15 Electrical Board.

(A) Creation of the Electrical Board. There is hereby created the Electrical Board of the _____ of _____, hereinafter designated as the Board.

(B) Appointments. Board members shall be appointed by the Governor with the advice and consent of the Senate (or by the Mayor with the advice and consent of the Council, or the equivalent).

(1) Members of the Board shall be chosen in a manner to reflect a balanced representation of individuals or organizations. The Chair of the Board shall be elected by the Board membership.

(2) The Chief Electrical Inspector in the jurisdiction adopting this Article authorized in 80.15(B)(3)(a) shall be the nonvoting secretary of the Board. Where the Chief Electrical Inspector of a local municipality serves a Board at a state level, he or she shall be permitted to serve as a voting member of the Board.

(3) The board shall consist of not fewer than five voting members. Board members shall be selected from the following:

a. Chief Electrical Inspector from a local government (for State Board only)
b. An electrical contractor operating in the jurisdiction
c. A licensed professional engineer engaged primarily in the design or maintenance of electrical installations
d. A journeyman electrician

(4) Additional membership shall be selected from the following:

a. A master (supervising) electrician
b. The Fire Marshal (or Fire Chief)

c. A representative of the property/casualty insurance industry

d. A representative of an electric power utility operating in the jurisdiction

e. A representative of electrical manufacturers primarily and actively engaged in producing materials, fittings, devices, appliances, luminaires (fixtures), or apparatus used as part of or in connection with electrical installations

f. A member of the labor organization that represents the primary electrical workforce

g. A member from the public who is not affiliated with any other designated group

h. A representative of a telecommunications utility operating in the jurisdiction

(C) Terms. Of the members first appointed, _____ shall be appointed for a term of 1 year, _____ for a term of 2 years, _____ for a term of 3 years, and _____ for a term of 4 years, and thereafter each appointment shall be for a term of 4 years or until a successor is appointed. The Chair of the Board shall be appointed for a term not to exceed _____ years.

(D) Compensation. Each appointed member shall receive the sum of _____dollars ($_____) for each day during which the member attends a meeting of the Board and, in addition thereto, shall be reimbursed for direct lodging, travel, and meal expenses as covered by policies and procedures established by the jurisdiction.

(E) Quorum. A quorum as established by the Board operating procedures shall be required to conduct Board business. The Board shall hold such meetings as necessary to carry out the purposes of Article 80. The Chair or a majority of the members of the Board shall have the authority to call meetings of the Board.

(F) Duties. It shall be the duty of the Board to:

(1) Adopt the necessary rules and regulations to administer and enforce Article 80.

(2) Establish qualifications of electrical inspectors.

(3) Revoke or suspend the recognition of any inspector's certificate for the jurisdiction.

(4) After advance notice of the public hearings and the execution of such hearings, as established by law, the Board is authorized to establish and update the provisions for the safety of electrical installations to conform with the current edition of the *National Electrical Code* (NFPA 70) and other nationally recognized safety standards for electrical installations.

(5) Establish procedures for recognition of electrical safety standards and acceptance of equipment conforming to these standards.

(G) Appeals.

(1) Review of Decisions. Any person, firm, or corporation may register an appeal with the Board for a review of any decision of the Chief Electrical Inspector or of any Electrical Inspector, provided that such appeal is made in writing within fifteen (15) days after such person, firm, or corporation shall have been notified. Upon receipt of such appeal, said Board shall, if requested by the person making the appeal, hold a public hearing and proceed to determine whether the action of the

Board, or of the Chief Electrical Inspector, or of the Electrical Inspector complies with this law and, within fifteen (15) days after receipt of the appeal or after holding the hearing, shall make a decision in accordance with its findings.

(2) Conditions. Any person shall be permitted to appeal a decision of the authority having jurisdiction to the Board when it is claimed that any one or more of the following conditions exist:

a. The true intent of the codes or ordinances described in this *Code* has been incorrectly interpreted.

b. The provisions of the codes or ordinances do not fully apply.

c. A decision is unreasonable or arbitrary as it applies to alternatives or new materials.

(3) Submission of Appeals. A written appeal, outlining the *Code* provision from which relief is sought and the remedy proposed, shall be submitted to the authority having jurisdiction within 15 calendar days of notification of violation.

(H) Meetings and Records. Meetings and records of the Board shall conform to the following:

(1) Meetings of the Board shall be open to the public as required by law.

(2) Records of meetings of the Board shall be available for review during normal business hours, as required by law.

80.17 Records and Reports. The authority having jurisdiction shall retain records in accordance with 80.17(A) and (B).

(A) Retention. The authority having jurisdiction shall keep a record of all electrical inspections, including the date of such inspections and a summary of any violations found to exist, the date of the services of notices, and a record of the final disposition of all violations. All required records shall be maintained until their usefulness has been served or as otherwise required by law.

(B) Availability. A record of examinations, approvals, and variances granted shall be maintained by the authority having jurisdiction and shall be available for public review as prescribed by law during normal business hours.

80.19 Permits and Approvals. Permits and approvals shall conform to 80.19(A) through (H).

(A) Application.

(1) Activity authorized by a permit issued under this *Code* shall be conducted by the permittee or the permittee's agents or employees in compliance with all requirements of this *Code* applicable thereto and in accordance with the approved plans and specifications. No permit issued under this *Code* shall be interpreted to justify a violation of any provision of this *Code* or any other applicable law or regulation. Any addition or alteration of approved plans or specifications shall be approved in advance by the authority having jurisdiction, as evidenced by the issuance of a new or amended permit.

(2) A copy of the permit shall be posted or otherwise readily accessible at each work site or carried by the permit holder as specified by the authority having jurisdiction.

(B) Content. Permits shall be issued by the authority having jurisdiction and shall bear the name and signature of the authority having jurisdiction or that of the authority having jurisdiction's designated representative. In addition, the permit shall indicate the following:

(1) Operation or activities for which the permit is issued
(2) Address or location where the operation or activity is to be conducted
(3) Name and address of the permittee
(4) Permit number and date of issuance
(5) Period of validity of the permit
(6) Inspection requirements

(C) Issuance of Permits. The authority having jurisdiction shall be authorized to establish and issue permits, certificates, notices, and approvals, or orders pertaining to electrical safety hazards pursuant to 80.23, except that no permit shall be required to execute any of the classes of electrical work specified in the following:

(1) Installation or replacement of equipment such as lamps and of electric utilization equipment approved for connection to suitable permanently installed receptacles. Replacement of flush or snap switches, fuses, lamp sockets, and receptacles, and other minor maintenance and repair work, such as replacing worn cords and tightening connections on a wiring device
(2) The process of manufacturing, testing, servicing, or repairing electric equipment or apparatus

(D) Annual Permits. In lieu of an individual permit for each installation or alteration, an annual permit shall, upon application, be issued to any person, firm, or corporation regularly employing one or more employees for the installation, alteration, and maintenance of electric equipment in or on buildings or premises owned or occupied by the applicant for the permit. Upon application, an electrical contractor as agent for the owner or tenant shall be issued an annual permit. The applicant shall keep records of all work done, and such records shall be transmitted periodically to the Electrical Inspector.

(E) Fees. Any political subdivision that has been provided for electrical inspection in accordance with the provisions of Article 80 may establish fees that shall be paid by the applicant for a permit before the permit is issued.

(F) Inspection and Approvals.

(1) Upon the completion of any installation of electrical equipment that has been made under a permit other than an annual permit, it shall be the duty of the person, firm, or corporation making the installation to notify the Electrical Inspector having jurisdiction, who shall inspect the work within a reasonable time.
(2) Where the Inspector finds the installation to be in conformity with the statutes of all applicable local ordinances and all rules and regulations, the Inspector shall issue to the person, firm, or corporation making the installation a certificate of approval, with duplicate copy for delivery to the owner, authorizing the connection to the supply of electricity and shall send written notice of such authorization to the supplier of electric service. When a certificate of temporary approval is issued authorizing the connection of an installation, such certificates shall be issued to expire at a time to be stated therein and shall be revocable by the Electrical Inspector for cause.
(3) When any portion of the electrical installation within the jurisdiction of an Electrical Inspector is to be hidden from view by the permanent placement of parts of the building, the person, firm, or corporation installing the equipment shall notify the Electrical Inspector, and such equipment shall not be concealed until it has been approved by the Electrical Inspector or until _____ days have elapsed from the time of such notification, provided that on large installations, where the concealment of equipment proceeds continuously, the person, firm, or corporation installing the equipment shall give the Electrical Inspector due notice in advance, and inspections shall be made periodically during the progress of the work.
(4) At regular intervals, the Electrical Inspector having jurisdiction shall visit all buildings and premises where work may be done under annual permits and shall inspect all electric equipment installed under such permits since the date of the previous inspection. The Electrical Inspector shall issue a certificate of approval for such work as is found to be in conformity with the provisions of Article 80 and all applicable ordinances, orders, rules, and regulations, after payments of all required fees.
(5) If, upon inspection, any installation is found not to be fully in conformity with the provisions of Article 80, and all applicable ordinances, rules, and regulations, the Inspector making the inspection shall at once forward to the person, firm, or corporation making the installation a written notice stating the defects that have been found to exist.

(G) Revocation of Permits. Revocation of permits shall conform to the following:

(1) The authority having jurisdiction shall be permitted to revoke a permit or approval issued if any violation of this *Code* is found upon inspection or in case there have been any false statements or misrepresentations submitted in the application or plans on which the permit or approval was based.
(2) Any attempt to defraud or otherwise deliberately or knowingly design, install, service, maintain, operate, sell, represent for sale, falsify records, reports, or applications, or other related activity in violation of the requirements prescribed by this *Code* shall be a violation of this *Code*. Such violations shall be cause for immediate suspension or revocation of any related licenses, certificates, or permits issued by this jurisdiction. In addition, any such violation shall be subject to any other criminal or civil penalties as available by the laws of this jurisdiction.
(3) Revocation shall be constituted when the permittee is duly notified by the authority having jurisdiction.
(4) Any person who engages in any business, operation, or occupation, or uses any premises, after the permit issued therefor has been suspended or revoked pursuant to the provisions of this *Code*, and before such suspended permit has been reinstated or a new permit issued, shall be in violation of this *Code*.
(5) A permit shall be predicated upon compliance with the requirements of this *Code* and shall constitute written authority issued by the authority having jurisdiction to install electrical

equipment. Any permit issued under this *Code* shall not take the place of any other license or permit required by other regulations or laws of this jurisdiction.

(6) The authority having jurisdiction shall be permitted to require an inspection prior to the issuance of a permit.

(7) A permit issued under this *Code* shall continue until revoked or for the period of time designated on the permit. The permit shall be issued to one person or business only and for the location or purpose described in the permit. Any change that affects any of the conditions of the permit shall require a new or amended permit.

(H) Applications and Extensions. Applications and extensions of permits shall conform to the following:

(1) The authority having jurisdiction shall be permitted to grant an extension of the permit time period upon presentation by the permittee of a satisfactory reason for failure to start or complete the work or activity authorized by the permit.

(2) Applications for permits shall be made to the authority having jurisdiction on forms provided by the jurisdiction and shall include the applicant's answers in full to inquiries set forth on such forms. Applications for permits shall be accompanied by such data as required by the authority having jurisdiction, such as plans and specifications, location, and so forth. Fees shall be determined as required by local laws.

(3) The authority having jurisdiction shall review all applications submitted and issue permits as required. If an application for a permit is rejected by the authority having jurisdiction, the applicant shall be advised of the reasons for such rejection. Permits for activities requiring evidence of financial responsibility by the jurisdiction shall not be issued unless proof of required financial responsibility is furnished.

80.21 Plans Review. Review of plans and specifications shall conform to 80.21(A) through (C).

(A) Authority. For new construction, modification, or rehabilitation, the authority having jurisdiction shall be permitted to review construction documents and drawings.

(B) Responsibility of the Applicant. It shall be the responsibility of the applicant to ensure the following:

(1) The construction documents include all of the electrical requirements.

(2) The construction documents and drawings are correct and in compliance with the applicable codes and standards.

(C) Responsibility of the Authority Having Jurisdiction. It shall be the responsibility of the authority having jurisdiction to promulgate rules that cover the following:

(1) Review of construction documents and drawings within established time frames for the purpose of acceptance or to provide reasons for nonacceptance

(2) Review and approval by the authority having jurisdiction shall not relieve the applicant of the responsibility of compliance with this *Code*.

(3) Where field conditions necessitate any substantial change from the approved plan, the authority having jurisdiction shall be

permitted to require that the corrected plans be submitted for approval.

80.23 Notice of Violations, Penalties. Notice of violations and penalties shall conform to 80.23(A) and (B).

(A) Violations.

(1) Whenever the authority having jurisdiction determines that there are violations of this *Code*, a written notice shall be issued to confirm such findings.

(2) Any order or notice issued pursuant to this *Code* shall be served upon the owner, operator, occupant, or other person responsible for the condition or violation, either by personal service or mail or by delivering the same to, and leaving it with, some person of responsibility upon the premises. For unattended or abandoned locations, a copy of such order or notice shall be posted on the premises in a conspicuous place at or near the entrance to such premises and the order or notice shall be mailed by registered or certified mail, with return receipt requested, to the last known address of the owner, occupant, or both.

(B) Penalties.

(1) Any person who fails to comply with the provisions of this *Code* or who fails to carry out an order made pursuant to this *Code* or violates any condition attached to a permit, approval, or certificate shall be subject to the penalties established by this jurisdiction.

(2) Failure to comply with the time limits of an abatement notice or other corrective notice issued by the authority having jurisdiction shall result in each day that such violation continues being regarded as a new and separate offense.

(3) Any person, firm, or corporation who shall willfully violate any of the applicable provisions of this article shall be guilty of a misdemeanor and, upon conviction thereof, shall be punished by a fine of not less than _____ dollars ($_____) or more than _____ dollars ($_____) for each offense, together with the costs of prosecution, imprisonment, or both, for not less than _____ (_____) days or more than _____ (_____) days.

80.25 Connection to Electricity Supply. Connections to the electric supply shall conform to 80.25(A) through (E).

(A) Authorization. Except where work is done under an annual permit and except as otherwise provided in 80.25, it shall be unlawful for any person, firm, or corporation to make connection to a supply of electricity or to supply electricity to any electric equipment installation for which a permit is required or that has been disconnected or ordered to be disconnected.

(B) Special Consideration. By special permission of the authority having jurisdiction, temporary power shall be permitted to be supplied to the premises for specific needs of the construction project. The Board shall determine what needs are permitted under this provision.

(C) Notification. If, within _____ business days after the Electrical Inspector is notified of the completion of an installation of electric equipment, other than a temporary approval installation, the Electrical Inspector has neither authorized connection nor disapproved the installation, the supplier of electricity is authorized to make connections and supply electricity to such installation.

(D) Other Territories. If an installation or electric equipment is located in any territory where an Electrical Inspector has not been authorized or is not required to make inspections, the supplier of electricity is authorized to make connections and supply electricity to such installations.

(E) Disconnection. Where a connection is made to an installation that has not been inspected, as outlined in the preceding paragraphs of this section, the supplier of electricity shall immediately report such connection to the Chief Electrical Inspector. If, upon subsequent inspection, it is found that the installation is not in conformity with the provisions of Article 80, the Chief Electrical Inspector shall notify the person, firm, or corporation making the installation to rectify the defects and, if such work is not completed within fifteen (15) business days or a longer period as may be specified by the Board, the Board shall have the authority to cause the disconnection of that portion of the installation that is not in conformity.

80.27 Inspector's Qualifications.

(A) Certificate. All electrical inspectors shall be certified by a nationally recognized inspector certification program accepted by the Board. The certification program shall specifically qualify the inspector in electrical inspections. No person shall be employed as an Electrical Inspector unless that person is the holder of an Electrical Inspector's certificate of qualification issued by the Board, except that any person who on the date on which this law went into effect was serving as a legally appointed Electrical Inspector of _____ shall, upon application and payment of the prescribed fee and without examination, be issued a special certificate permitting him or her to continue to serve as an Electrical Inspector in the same territory.

(B) Experience. Electrical inspector applicants shall demonstrate the following:

(1) Have a demonstrated knowledge of the standard materials and methods used in the installation of electric equipment
(2) Be well versed in the approved methods of construction for safety to persons and property
(3) Be well versed in the statutes of _____ relating to electrical work and the *National Electrical Code*, as approved by the American National Standards Institute
(4) Have had at least _____ years' experience as an Electrical Inspector or _____ years in the installation of electrical equipment. In lieu of such experience, the applicant shall be a graduate in electrical engineering or of a similar curriculum of a college or university considered by the Board as having suitable requirements for graduation and shall have had two years' practical electrical experience.

(C) Recertification. Electrical inspectors shall be recertified as established by provisions of the applicable certification program.

(D) Revocation and Suspension of Authority. The Board shall have the authority to revoke an inspector's authority to conduct inspections within a jurisdiction.

80.29 Liability for Damages. Article 80 shall not be construed to affect the responsibility or liability of any party owning, designing, operating, controlling, or installing any electric equipment for damages to persons or property caused by a defect therein, nor shall the _____ or any of its employees be held as assuming any such liability by reason of the inspection, reinspection, or other examination authorized.

80.31 Validity. If any section, subsection, sentence, clause, or phrase of Article 80 is for any reason held to be unconstitutional, such decision shall not affect the validity of the remaining portions of Article 80.

80.33 Repeal of Conflicting Acts. All acts or parts of acts in conflict with the provisions of Article 80 are hereby repealed.

80.35 Effective Date. Article 80 shall take effect _____ (_____) days after its passage and publication.

Index

Electrical nonmetallic tubing, 362.22
Elevators, 620.32, 620.33
Fixture wire, 402.7
Flexible metal conduit, 348.22, Table 348.22
Flexible metallic tubing, 358.22
General installations, 300.17
Hazardous (classified) locations, sealing fittings, 501.15(C)(6), 505.16(D)(5)
High density polyethylene conduit, 353.22
Intermediate metal conduit, 342.22
Liquidtight flexible metal conduit, 350.22
Liquidtight flexible nonmetallic conduit, 356.22
Maximum permitted, 300.17
Nonmetallic underground conduit with conductors, 354.100(D)
Outlet boxes, etc., 314.16
Remote control circuits, 725.28
Rigid metal conduit, 344.22
Rigid nonmetallic conduit, 352.22
Signs, 600.31(C)
Surface raceways, 386.22, 388.22
Theaters, 520.6
Underfloor raceways, 390.5
Wireways, 376.22, 378.22
X-ray equipment, 660.8
Conductors. *see also* Cords, flexible; Fixture wires
Adjustment factors for more than three current-carrying conductors in raceway, cable, Table B.310.11
Aluminum, conductor material, 310.14
Aluminum, properties of, Chap. 9, Table 8
Ampacities of, 310.15, Tables 310.16 through 310.21, Tables 310.69 through 310.86, B.310.1 through B.310.10
Applications, 310.13
Armored cable, Type AC. *see* Armored cable (Type AC)
Bare. *see* Bare conductors
Bending radius, 300.34
Boxes and fittings, junction, 314.16, 314.17
Branch circuits, Art. 210. *see also* Branch circuits
Bundled, in theaters, etc. Definition, 520.2
Buried, 310.7
Busways. *see* Busways
Cabinets and cutout boxes, 312.5 through 312.7
Cablebus. *see* Cablebus
Calculations, examples, Annex D
Capacitors, 460.8
Cellular concrete floor raceways. *see* Cellular concrete floor raceways
Cellular metal floor raceways. *see* Cellular metal floor raceways
Circuit
Communications, Art. 800
Control, health care facilities, 517.74

Fire alarm, Art. 760
Less than 50 volts, Art. 720
Over 600 volts, Art. 490
Signal, remote-control, power-limited, Art. 725
Combinations, Chap. 9, Table 1
Communications circuits, 800–II, 800–V
Concealed knob-and-tube. *see* Knob-and-tube wiring
Conduit or tubing, number in, 342.22, 344.22, 348.22, Table 348.22, 350.22, 352.22, 353.22, 354.22, 356.22, 358.22, 360.22, 362.22, Chap. 9 Tables, Annex C
Construction, general, Table 310.13
Cooling of electric equipment, 110.13(B)
Copper, 110.5, 505.18(A), 760.27(C)
Ampacities, Tables 310.16 through 310.21, Tables 310.69 through 310.86, B.310.1 through B.310.10
Properties, Chap. 9, Table 8
Copper-clad aluminum Definition, Art. 100–I
Cords, flexible. *see* Cords, flexible
Corrosive conditions. *see* Corrosive conditions
Covered Definition, Art. 100–I
Crane and hoists, 610–II, 610–III, 610.41
Damp locations, 310.8(B), Table 310.13
Definitions, Art. 100–I
Deflection of. *see* Deflection of conductors
Different systems, 300.3(C), 725.26
Dimensions of, Chap. 9, Table 4
Dry locations, 310.8(B), Table 310.13
Electrical metallic tubing. *see* Electrical metallic tubing
Electrical nonmetallic tubing. *see* Electrical nonmetallic tubing
Elevators, dumbwaiters, escalators, moving walks, 620–II through 620–IV
Enclosure, grounding, 250–IV
Grounding electrode conductors, 250.64(E)
Equipment grounding. *see* Equipment grounding conductors
Feeder, Art. 215
Fire alarm systems, 760.26, 760.27, 760.52
Fixture wires, Art. 402
In free air, Tables 310.17, 310.19, 310.20
Gauges (AWG), general provisions, 110.6
General wiring, Art. 310
Generators, size, 445.13
Grounded
Alternating current systems, 250.24, 250.26
Identification, 200.6, 200.7, 210.5(A)
Change in size, 240.23

Continuity of, 300.13(B)
Definition, Art. 100–I
Different systems, 200.6(D)
Electrode-type boilers, over 600 volts, 490.72(E)
Fuses in, 430.36
Overcurrent protection, 230.90(B), 240.22
Separately derived systems, 250.30(A)
Services
Disconnection of, 230.70, 230.75
Overcurrent protection, 230.90(B)
Switches, disconnection, 404.2(B)
Use and identification, Art. 200
Grounded power systems in anesthetizing locations, 517.63
Grounding. *see* Grounding conductors
Grounding electrode. *see* Grounding electrode conductors
Grouped
Definition, 520.2
Identification, 310.12
Induction and dielectric heating equipment, ampacity of supply, 665.10
Industrial control panels, 409.20
Installation with other systems, 300.8
Insulated
Definition, Art. 100–I
Insulating materials, 310.13
Insulation, Art. 310
At bushings, 4 AWG and larger, 300.4(F), 312.6(C)
Construction and application, 310.13
Elevators, dumbwaiters, escalators, moving walks, wheelchair lifts, and stairway chair lifts, 620.11
Fire alarm systems, 760.27(B)
Fixtures, 410.24, 410.67
Hazardous (classified) locations, 501.13, 501.20, 505.18(B)
Identification, 310.12
Line-type fire detectors, 760.82(J)
Marking, 310.11
Motion picture projectors, 540.14
Service conductors, 230.22, 230.30, 230.41
Splices and joints, 110.14(B), 400.9, 530.12, 590.4(G)
Switchboards, 408.19
Theaters, 520.42
Intrinsically safe, 504.30, 504.50
Length in boxes, 300.14
Length in transient voltage surge suppressors, 285.12
Lightning rods, spacing from, 250.106, 800.53, 820.44(F)(3)
Liquidtight flexible metal conduit. *see* Liquidtight flexible metal conduit
Liquidtight flexible nonmetallic conduit. *see* Liquidtight flexible nonmetallic conduit
Metal-clad cable. *see* Metal-clad cable

Metal enclosures, spacing from lightning rods, 250.106 FPN No. 1 and No. 2
Mineral-insulated metal-sheathed cable. *see* Mineral-insulated metal-sheathed cable (Type MI)
Minimum size, 230.42, 310.5, 517.74(B), 620.12, 650.6(A), 660.9, 725.27(A), 760.27(A), 760.57, 810.17
Motion picture projectors, sizes, 540.13
Motor circuits, 430–II
Multioutlet assembly. *see* Multioutlet assembly
Multiple, 250.30(A)(8)(b), 250.102(C), 250.122(F), 300.20, 310.4, 392.8(D), 520.27(A)(2), 620.12(A)(1)
Nonmetallic extensions. *see* Nonmetallic extensions
Nonmetallic sheathed cable, Types NM, NMC, and NMS. *see* Nonmetallic-sheathed cable (Types NM, NMC, and NMS)
Number of, in. *see* Conductor fill
Organs, 650.6, 650.7
Outlet boxes, temperature limits, 410.11
Outside wiring, Art. 225
Overcurrent protection, 240.3, 240.4
Cablebus conductors, 370.5
Grounded conductors, 230.90(B), 240.22
Service-entrance conductors, 230.90, 230.91, 230.208
Overhead, 800.44
Over 600 volts, Art. 490
Paralleled, 250.30(A)(8)(b), 250.102(C), 250.122(F), 300.20, 310.4, 392.8(D), 520.27(A)(2), 620.12(A)(1)
Pendant fixtures, 410.27
Properties of, Chap. 9, Table 8
Raceways, number of conductors in. *see* Conductor fill
Radio and television
Antenna distribution systems, 810.4, Art. 820
Antenna systems, Art. 810
Equipment, Art. 810
Railway, power and light from, 110.19
Recreational vehicle site feeder circuit, 551.73, 551.80
Definition, 551.2
Remote-control circuits, Art. 725
Rigid metal conduit. *see* Rigid metal conduit
Same circuit, 300.5(I)
Selection of, examples, Annex D
Service. *see* Service-entrance cable (Types SE and USE); Service-entrance conductors
Definition, Art. 100–I
Service-entrance. *see* Service-entrance conductors
Signal circuits, Art. 725, Art. 760
Signs and outline lighting, 600.31, 600.32

2005 National Electrical Code Handbook

About The Editors

Mark W. Earley, P.E.

Mark W. Earley is Assistant Vice President and Chief Electrical Engineer at NFPA. He has served as Secretary of the *NEC* for the past 15 years and is the co-author of NFPA's reference book, *Electrical Installations in Hazardous Locations*. Prior to joining NFPA, he worked as an electrical engineer at Factory Mutual Research Corporation. Additionally, he has served on several of NFPA's electrical committees and NEC code-making panels. Earley is a registered professional engineer and a member of IAEI, IEEE, SFPE, the Standards Engineering Society, the Automatic Fire Alarm Association (AFAA), the UL Electrical Council, U.S. National Committee on the International Electrotechnical Commission, and the Canadian Electrical Code Committee. He serves as chair of Task Group 1—Electrical Installation Codes for the Pacific Area Standards Congress.

Jeffrey S. Sargent

Jeffrey S. Sargent is Senior Electrical Specialist at NFPA. He serves as staff liaison to several NFPA technical committees, including Electrical Equipment in Chemical Atmospheres and Standard on Manufactured Housing Electrical Committee. He is the executive secretary to the NFPA Electrical Section and has co-authored NFPA's *1999 NEC Changes* and the *NFPA Inspection Manual with Checklists* as well as being a regular contributor to *necdigest™*, NFPA's official magazine of the *National Electrical Code*. Prior to joining NFPA in 1997, he served as an electrical inspector in state and municipal positions and as an instructor in the electrical program at the New Hampshire Technical College. Sargent is a licensed master electrician, a certified electrical inspector, and a member of IAEI.

Joseph V. Sheehan, P.E.

Joseph V. Sheehan is Principal Electrical Engineer at NFPA and an instructor of NFPA's electrical code seminars. Prior to joining NFPA in 1993, he was employed for over 28 years in the electrical construction industry as an electrician, chief estimator, and project manager in the Boston area. He also taught electrical engineering for over 15 years in the evening division of the Franklin Institute of Boston. Sheehan now serves as staff liaison to the NFPA committee that deals with industrial machinery. He is a registered professional engineer and a licensed master electrician in three states and is a member of IAEI, IEEE, and IBEW.

John M. Caloggero

John M. Caloggero retired as Principal Electrical Specialist at NFPA in 2003. He served as staff liaison to the technical committee on lightning protection. He has contributed to seven editions of the *National Electrical Code Handbook,* developed several *NEC* workshops, and was technical adviser to the *NEC* videos. He served as an instructor for numerous *Code* classes for Spanish-speaking students in Latin America. Prior to joining NFPA in 1980, he was employed as a master electrician, electrical inspector, and electrical instructor. Caloggero is a member of IAEI, holds a journeyman electrician's license, and is a nationally certified electrical inspector.

Common Numbering System for Articles 770, 800, 820, and 830

ARTICLE 770 **Optical Fiber Cables and Raceways**	**ARTICLE 800** **Communications Circuits**	**ARTICLE 820** **Community Antenna Television and Radio Distribution Systems**	**ARTICLE 830** **Network-Powered Broadband Communications Systems**
I. General	I. General	I. General	I. General
770.1 Scope.	800.1 Scope.	820.1 Scope.	830.1 Scope.
770.2 Definitions.	800.2 Definitions.	820.2 Definitions.	830.2 Definitions.
770.3 Other Articles.	800.3 Other Articles. [800.8 & 800.52 (C) & (D)]	820.3 Other Articles.	830.3 Other Articles. (830.9)
770.6 Optical Fiber Cables. (770.4)	800.18 Installation of Equipment. (800.4)	820.15 Energy Limitations. (820.4)	830.15 Power Limitations. (830.4)
770.9 Types. (770.5)			
770.12 Raceways for Optical Fiber Cables. (770.6)			
770.21 Access to Electrical Equipment Behind Panels Designed to Allow Access. (770.7)	800.21 Access to Electrical Equipment Behind Panels Designed to Allow Access. (800.5)	820.21 Access to Electrical Equipment Behind Panels Designed to Allow Access. (820.5)	830.21 Access to Electrical Equipment Behind Panels Designed to Allow Access. (830.6)
770.24 Mechanical Execution of Work. (770.8)	800.24 Mechanical Execution of Work. (800.6)	820.24 Mechanical Execution of Work. (820.6)	830.24 Mechanical Execution of Work. (830.7)
	II. Wires and Cables Outside and Entering Buildings	II. Cables Outside and Entering Buildings	II. Cables Outside and Entering Buildings
			830.40 Entrance Cables. (830.10)
	800.44 Overhead Communications Wires and Cables. (800.10)	820.44 Overhead Cables. (820.10)	830.44 Aerial Cables. (830.11)
	800.47 Underground Circuits Entering Buildings. (800.11)	820.47 Underground Circuits Entering Buildings. (820.11)	830.47 Underground Circuits Entering Buildings. (830.12)
	800.50 Circuits Requiring Primary Protectors. (800.12)		
	800.53 Lightning Conductors. (800.13)		
II. Protection	III. Protection	III. Protection	III. Protection
	800.90 Protective Devices. (800.30 & 800.32)		830.90 Primary Electrical Protection. (800.30)
770.93 Grounding of Entrance Cables. (770.33)	800.93 Cable Grounding. (800.33)	820.93 Grounding of Outer Conductive Shield of a Coaxial Cable.	830.93 Grounding or Interruption of Metallic Members of Network-Powered Broadband Communications Cables. (830.33)

The 2002 edition section numbers are shown in parentheses.